BERGEY'S MANUAL OF
DETERMINATIVE BACTERIOLOGY

BERGEY'S MANUAL OF
DETERMINATIVE
BACTERIOLOGY

Eighth Edition

R. E. Buchanan† & N. E. Gibbons
CO-EDITORS

EDITORIAL BOARD
S. T. Cowan, J. G. Holt, J. Liston,
R. G. E. Murray, C. F. Niven,
A. W. Ravin & R. Y. Stanier

WITH CONTRIBUTIONS FROM
128 Colleagues

The Williams & Wilkins Company / Baltimore

First Edition, 1923
Second Edition, 1925
Third Edition, 1930
Fourth Edition, 1934
Preprint of pages ix + 79 of Fifth Edition, 1938
Fifth Edition, 1939
Sixth Edition, 1948
Seventh Edition, 1957
Eighth Edition, 1974
Reprinted, 1975

Made in United States of America

Library of Congress Cataloging in Publication Data
Main entry under title:

Bergey's manual of determinative bacteriology.

    First-7th ed. are entered under: American Society
for Microbiology.
    Bibliography: p.967
    1. Bacteriology—Classification. 2. Schizomycetes.
I. Bergey, David Hendricks, 1860–1937. II. Buchanan,
Robert Earle, 1883–1973 ed. III. Gibbons, Norman
Edwin, 1906–   ed. IV. American Society for
Microbiology. Bergey's manual of determinative
bacteriology. V. Title: Manual of determinative
bacteriology. [DNLM: 1. Bacteria—Classification.
2. Bacteriology—Terminology. QW4 B921b 1974]
QR81.B47 1974   589.9'001'2   73-20173
ISBN 0-683-01117-0

The illustration on the cover is one of the Myxobacteria, *Stigmatella aurantiaca* (*Stigmatella media*) ($\times$ 55), kindly supplied by Dr. Howard McCurdy, and reproduced by permission from the *Canadian Journal of Microbiology 15:* 1453–1461, Figure 20, 1969.

COMPOSED AND PRINTED AT THE
WAVERLY PRESS, INC.
Mt. Royal and Guilford Aves.
Baltimore, Md., U. S. A. 21202

LIST OF CONTRIBUTORS

Dr. O. N. Allen
4142 Hiawatha Drive
Madison, Wisconsin 53711 USA

Dr. T. V. Aristovskaya
Dokuchaev Centralny Musei Potchvovedenia
Birjevoi Proezd 6
Leningrad B-164, USSR

Dr. H.-D. Babenzien
Sektion Biologie
Ernst-Moritz-Arndt-Universität
Jahnstr. 15
DDR-22 Greifswald, Germany

Dr. A. C. Baird-Parker
Unilever Research Laboratory
Colworth House
Sharnbrook, Bedford
England, MK44 1LQ

Dr. Elio Baldacci
Istituto di Patologia Vegetale
Universita degli Studi di Mialno
Via Celoria 2
Milano, Italy

Dr. J. H. Becking
Institute for Atomic Sciences in Agriculture
6 Keyenbergseweg, Postbus 48
Wageningen, The Netherlands

Dr. E. L. Biberstein
School of Veterinary Medicine
Department of Veterinary Microbiology
University of California
Davis, California 95616 USA

Dr. C. E. Bland
Department of Biology
East Carolina University
Greenville, North Carolina 27834 USA

Dr. W. J. Brinley-Morgan
Central Veterinary Laboratory
New Haw, Weybridge
Surrey, England

Dr. Thomas D. Brock
Department of Bacteriology
University of Wisconsin
1550 Linden Drive
Madison, Wisconsin 53706 USA

Dr. Marion A. Brooks
Department of Entomology, Fisheries and Wildlife
University of Minnesota
St. Paul, Minnesota 55101 USA

Dr. M. P. Bryant
Department of Dairy Science
University of Illinois
Urbana, Illinois 61803 USA

Dr. R. E. Buchanan
(Deceased February 21, 1973)
Iowa State University
Ames, Iowa 50010 USA

Dr. Jeffrey C. Burnham
Department of Microbiology
Medical College of Ohio
P. O. Box 6190
Toledo, Ohio 43614 USA

Dr. L. Leon Campbell
104 Hullihen Hall
University of Delaware
Newark, Delaware 19711 USA

Dr. Ercole Canale-Parola
Department of Microbiology
University of Massachusetts
Amherst, Massachusetts 01002 USA

Dr. Patricia Carpenter
(Deceased July 4, 1971)
Central Public Health Laboratory
Colindale Avenue
London, England

Dr. J. Geoffrey Carr
University of Bristol
Research Station
Long Ashton, Bristol
BS18 9AF, England

Miss Sylvia G. Cary
Walter Reed Army Institute of Research
Washington, D.C. 20012 USA

Dr. J. N. Couch
Department of Botany
University of North Carolina
Chapel Hill, North Carolina 27514 USA

Dr. S. T. Cowan
Peacock Cottage, Queen Camel
Yeovil, Somerset BA22 7NQ
England

Dr. Koby T. Crabtree
University Center System
Department of Biology
University of Wisconsin
518 South 7th Avenue
Wausau, Wisconsin 54401 USA

Dr. Tom Cross
Postgraduate School of Studies
    in Biological Sciences
University of Bradford
Bradford, Yorkshire BD7 1DP, England

Dr. C. S. Cummins
Anaerobe Laboratory
College of Agriculture
Virginia Polytechnic Institute
    and State University
Blacksburg, Virginia 24061 USA

Dr. R. H. Deibel
Department of Bacteriology
University of Wisconsin
Madison, Wisconsin 53706 USA

Dr. J. De Ley
Laboratory of Microbiology and
    Microbial Genetics
State University
Ledeganckstraat 35
9000 Gent, Belgium

Dr. Robert B. Dienst
Department of Cell and Molecular Biology
Medical College of Georgia
Augusta, Georgia 30902 USA

Dr. M. Doudoroff
Department of Bacteriology
    and Immunology
University of California
Berkeley, California 94720 USA

Dr. Douglas W. Dye
Plant Diseases Division
Department of Scientific and
    Industrial Research
Private Bag
Auckland, New Zealand

Dr. James B. Evans
Department of Microbiology
North Carolina State University
Raleigh, North Carolina 47607 USA

Dr. Oscar Felsenfeld
Delta Regional Primate Research Center
Tulane University
Covington, Louisiana 70433 USA

Prof. émérite Joseph Frateur
5831 Bossière, Belgium

Prof. Dr. E. A. Freundt
International Reference Centre for
    Animal Mycoplasma
Institute of Medical Microbiology
University of Aarhus
DK 8000 Aarhus C, Denmark

Dr. Ellen I. Garvie
National Institute for Research in Dairying
Shinfield
Reading RG2 9AT, England

Dr. Lucille K. Georg
Research and Development Unit
Center for Disease Control
Atlanta, Georgia 30333 USA

Dr. N. E. Gibbons
64 Fuller Street
Ottawa, Ontario
Canada K1Y 3R8

Dr. Thomas Gibson
(Deceased November 9, 1973)
The Edinburgh School of Agriculture
West Mains Road
Edinburgh, EH9 3JG
Scotland

Dr. Marion Gilmour
Eastman Dental Center
800 E. Main Street
Rochester, New York 14603 USA

Dr. Morris A. Gordon
Division of Laboratories and Research
New York State Department of Health
New Scotland Avenue
Albany, New York 12201 USA

Dr. Ruth E. Gordon
Institute of Microbiology
Rutgers University
New Brunswick, New Jersey 08903 USA

Dr. David Gottlieb
Department of Plant Pathology
10 Horticulture Field Laboratory
University of Illinois
Urbana, Illinois 61801 USA

Mrs. Wilma Kane Hanton
Department of Zoology
202 Wilson Hall
University of North Carolina
Chapel Hill, North Carolina 27514 USA

Miss Margaret S. Hendrie
Torry Research Station
135 Abbey Road
Aberdeen AB9 8DG, Scotland

Prof. Dr. Aino Henssen
Herbarium der Universität
Biegenstr. 48
355 Marburg/Lahn, Germany

Dr. Peter Hirsch
Institut für Allgemeine Mikrobiologie
University of Kiel, Olhausenstr. 40–60
23 Kiel, West Germany

Dr. Geoffrey Hobbs
Department of Trade and Industry
Torry Research Station
Aberdeen AB9 8DG, Scotland

Dr. Lillian V. Holdeman
Anaerobe Laboratory
Virginia Polytechnic Institute
  and State University
Blacksburg, Virginia 24061 USA

Dr. A. J. Holding
University of Edinburgh
Department of Microbiology
School of Agriculture
West Mains Road
Edinburgh, EH9 3JG, Scotland

Dr. John G. Holt
Room 306, Science Bldg.
Iowa State University
Ames, Iowa 50010 USA

Dr. D. B. Johnstone
Department of Microbiology and Biochemistry
University of Vermont
Burlington, Vermont 05401 USA

Dr. D. C. Jordan
Microbiology Department
University of Guelph
Guelph, Ontario, Canada

Dr. R. M. Keddie
Department of Microbiology
University of Reading
London Road
Reading, RG1 5AQ, England

Dr. K. Kitahara
Tokyo University of Agriculture
1–1 Sakuragaoka, 1 Chome
Setagaya-Ku
Tokyo, Japan

Dr. Miroslav Kocur
Československá sbírka mikroorganismů
J. E. Purkyně University
tř. Obránců míru 10
Brno, Czechoslovakia

Dr. Julius P. Kreier
Department of Microbiology
Ohio State University
484 West 12th Avenue
Columbus, Ohio 43210 USA

Dr. Noel R. Krieg
Department of Biology
Virginia Polytechnic Institute
  and State University
Blacksburg, Virginia 24061 USA

Dr. George P. Kubica
Trudeau Institute, Inc.
P. O. Box 59
Saranac Lake, New York 12983 USA

Dr. Daisy A. Kuhn
Department of Biology
California State University
Northridge, California 91324 USA

Prof. Dr. E. Küster
Institut für landwirtschaftliche
  Mikrobiologie
Justus Liebig Universität
Landgraf Philipp Platz 4
D-63 Giessen, Germany

Dr. S. P. Lapage
National Collection of Type Cultures
Central Public Health Laboratory
Colindale Avenue
London NW9 5HT, England

Dr. J. W. M. la Rivière
Laboratorium voor Microbiologie
Technische Hoogeschool
Julianalaan 67A
Delft, The Netherlands

Dr. H. Lautrop
Statens Seruminstitut
Amager Boulevard 80
2300 Copenhagen S, Denmark

Dr. Edward R. Leadbetter
Amherst College
Department of Biology
Amherst, Massachusetts 01002 USA

Mr. R. A. Lelliott
Plant Pathology Laboratory
Hatching Green
Harpenden, Herts., England

Dr. L. Le Minor
Institut Pasteur
25 Rue du Docteur Roux
Paris F-75015, France

Dr. Ralph A. Lewin
Scripps Institution of Oceanography
University of California
La Jolla, California 92037 USA

Dr. John Liston
College of Fisheries
213 Fisheries Center
University of Washington
Seattle, Washington 98195 USA

Dr. R. Locci
Istituto di Patologia Vegetale
Universita degli Studi di Milano
Via Celoria 2
Milano, Italy

Dr. George M. Luedemann
Microbial Environmental Research, Inc.
46 Lincoln Street
Glen Ridge, New Jersey 07028 USA

Dr. S. Maier
Department of Zoology and Microbiology
Ohio University
Athens, Ohio 45701 USA

Dr. Karl Maramorosch
Boyce Thompson Institute
Yonkers, New York 10701 USA

Dr. Norvel M. McClung
Department of Biology
University of South Florida
Tampa, Florida 33620 USA

Dr. Elizabeth McCoy
Department of Bacteriology
University of Wisconsin
Madison, Wisconsin 53706 USA

Dr. N. B. McCullough
Department of Microbiology and
    Public Health
Michigan State University
East Lansing, Michigan 48823 USA

Dr. H. D. McCurdy
Department of Biology
University of Windsor
Windsor, Ontario, Canada

Dr. H. H. Mollaret
24 Rue Bertron
92-Sceaux, France

Dr. W. E. C. Moore
Anaerobe Laboratory
Virginia Polytechnic Institute and
    State University
Blacksburg, Virginia 24061 USA

Dr. J. W. Moulder
Department of Microbiology
The University of Chicago
Chicago, Illinois 60637 USA

Prof. E. G. Mulder
Laboratorium voor Microbiologie
Der Landbouwhogeschool
Hesselink van Suchtelenweg 4
Wageningen, Netherlands

Dr. R. G. E. Murray
Department of Bacteriology and Immunology
University of Western Ontario
London, Ontario, Canada N6A 3K7

Dr. C. F. Niven
Research Laboratories
Del Monte Corp.
205 N. Wiget Lane
Walnut Creek, California 94598 USA

Dr. Frits Ørskov
WHO International Escherichia Reference
    Center
Statens Seruminstitut
Amager Boulevard 80
2300 Copenhagen S, Denmark

Dr. Ida Ørskov
WHO International Escherichia Reference
    Center
Statens Seruminstitut
Amager Boulevard 80
2300 Copenhagen S, Denmark

Dr. Cora R. Owen
National Institute of Allergy
    and Infectious Diseases
Rocky Mountain Laboratory
Hamilton, Montana 59840 USA

Dr. Leslie A. Page
United States Department of Agriculture
Agricultural Research Service
National Animal Disease Laboratory
Ames, Iowa 50010 USA

Dr. N. J. Palleroni
Department of Bacteriology and Immunology
University of California
Berkeley, California 94720 USA

Dr. John E. Peterson
Dean School of Liberal Arts
    and Sciences
Kansas State Teachers College
Emporia, Kansas 66801 USA

Dr. Norbert Pfennig
Institut für Mikrobiologie der
    Universität und GSF
Grisebach Str. 8
34 Göttingen, Germany

Dr. C. B. Philip
California Academy of Sciences
Golden Gate Park
San Francisco, California 94118 USA

Dr. J. E. Phillips
Department of Veterinary Pathology
Royal (Dick) School of Veterinary
    Studies
Edinburgh EH9 1QH, Scotland

Dr. Leo S. Pine
Research and Development Unit
Center for Disease Control
Atlanta, Georgia 30333 USA

Dr. Margaret Pittman
Laboratory of Bacterial Products
Division of Biologics Standards
National Institutes of Health
Bethesda, Maryland 20014 USA

Dr. Jeanne S. Poindexter
Division of Natural Sciences
   and Mathematics
Medgar Evers College of
The City University of New York
1127 Carroll Street
Brooklyn, New York 11225 USA

Prof. J. R. Postgate
Unit of Nitrogen Fixation
University of Sussex
Brighton, Sussex
BN1 9QJ England

Dr. John R. Preer, Jr.
Department of Zoology
Jordan Hall 224
Bloomington, Indiana 47401 USA

Dr. T. G. Pridham
Fermentation Laboratory
Northern Regional Research Laboratory
United States Department of Agriculture
Peoria, Illinois 61604 USA

Dr. Alice Reyn
Statens Seruminstitut
Amager Boulevard 80
2300 Copenhagen S, Denmark

Dr. Miodrag Ristic
College of Veterinary Medicine
University of Illinois
Urbana, Illinois 61801 USA

Dr. John Robinson
Department of Bacteriology
   and Immunology
University of Western Ontario
London, Ontario
Canada

Mr. Morrison Rogosa
Laboratory of Microbiology and
   Immunology
National Institute of Dental Research
National Institutes of Health
Bethesda, Maryland 20014 USA

Dr. R. Rohde
Salmonella Centre
Institute of Hygiene
Gorch-Fock-Wall 15/17
2000 Hamburg 36, Germany

Dr. Ernest H. Runyon
Veterans Administration Hospital
Salt Lake City, Utah 84113 USA

Dr. Riichi Sakazaki
Department of Bacteriology
National Institute of Health
10–35 Kamiosaki, 2-Chome
Shinagawa-ku
Tokyo 141, Japan

Dr. R. H. W. Schubert
Abteilung für Hygiene
Hygiene-Institut der Universität Frankfurt
Paul Ehrlichstr. 40
6 Frankfurt a.M., Germany

Dr. Jiri Sedlák
Institute of Clinical and
   Experimental Medicine
Budějovická 800
Prague 4, Czechoslovakia

Dr. H. W. Seeley, Jr.
Microbiology Section
Cornell University
Stocking Hall
Ithaca, New York 14850 USA

Dr. H. P. R. Seeliger
Institut für Hygiene und
   Mikrobiologie
Josef Schneider Strasse 2
87 Würzburg, Germany

Dr. James Shewan
Department of Trade and Industry
Torry Research Station
135 Abbey Road
Aberdeen, AB9 8DG, Scotland

Prof. V. B. D. Skerman
Department of Microbiology
University of Queensland
St. Lucia, Brisbane 4067
Australia

Dr. H. L. Skuja (Deceased July 19, 1972)
Institute for Systematic Botany
Uppsala University
751 21 Uppsala 1, Sweden

Dr. J. M. Slack
Department of Microbiology
West Virginia Medical Center
West Virginia University
Morgantown, West Virginia 26506 USA

Dr. R. M. Smibert
Anaerobe Laboratory
Virginia Polytechnic Institute and
   State University
Blacksburg, Virginia 24061 USA

Dr. J. E. Smith
Department of Biological Sciences
University of Surrey
Guilford, Surrey
England

Dr. L. DS. Smith
Anaerobe Laboratory
Virginia Polytechnic Institute and
  State University
Blacksburg, Virginia 24061 USA

Prof. P. H. A. Sneath
MRC Microbial Systematics Unit
University of Leicester
Leicester LE1 7RH, England

Dr. James T. Staley
Department of Microbiology
University of Washington
Seattle, Washington 98105

Dr. R. Y. Stanier
Service da Physiologie Microbienne
Institut Pasteur
28 Rue du Dr. Roux
F-75015 Paris, France

Dr. Pamela D. Steed-Glaister
148 MacPherson Avenue
Toronto, Ontario
Canada M5R 1W8

Dr. E. Thal
Statens Veterinärmedicinska Anstalt
Fack, 10405
Stockholm 50, Sweden

Dr. Josef E. Thiemann
c/o Archifar
Via dei Colli 9
38068 Rovereto, Italy

Dr. H. D. Tresner
American Cyanamid Company
Research Division, Lederle Laboratories
Pearl River, New York 10865 USA

Prof. Dr. Hans G. Trüper
Institut für Mikrobiologie, Universität Bonn
Kurfürstenstr. 74
D-53 Bonn 1, Germany

Dr. L. H. Turner
London School of Hygiene and
  Tropical Medicine
Keppel Street
London WC1E 7HT, England

Dr. W. L. van Veen
Laboratorium voor Microbiologie
Hesselink van Suchtelenweg 4
Wageningen, The Netherlands

Dr. Méd. M. Véron
Laboratoire de Bactériologie
Faculté de Médecine Necker
156, rue de Vaugirard
75 730 Paris, Cedex 15, France

Dr. W. V. Vishniac
(Deceased December 10, 1973)
Department of Biology
University of Rochester
Rochester, New York 14627 USA

Dr. S. W. Watson
Woods Hole Oceanographic Institute
Woods Hole, Massachusetts 02543 USA

Dr. Lawrence G. Wayne
Tuberculosis Research Laboratory
Veterans Administration Hospital
5901 East Seventh Street
Long Beach, California 90801 USA

Dr. Owen B. Weeks
Arts and Sciences Research Center
New Mexico State University
Las Cruces, New Mexico 88003 USA

Dr. David Weinman
Department of Microbiology
Yale University
310 Cedar Street
New Haven, Connecticut 06510 USA

Dr. Emilio Weiss
Department of Microbiology
Naval Medical Research Institute
Bethesda, Maryland 20014 USA

Dr. H. J. Welshimer
Department of Microbiology
Health Sciences Division
Medical College of Virginia
Virginia Commonwealth University
Richmond, Virginia 23298 USA

Dr. Ruth G. Wittler
Walter Reed Army Institute
  of Research
Washington, D.C. 20012 USA

Dr. S. A. Zahler
Section of Microbiology
Stocking Hall
Cornell University
Ithaca, New York 14850 USA

Dr. G. A. Zavarzin
Institute of Microbiology
Academy of Sciences USSR
Profsojuznaya 7
Moscow B133, USSR

Dr. K. S. Zinnemann
45 Grove Lane
Leeds LS6 4EQ, England

LIST OF ADVISORY COMMITTEE MEMBERS

The Board of Trustees is grateful to all who served on the Advisory Committees and assisted materially in the preparation of this edition of THE MANUAL. Chairmen of committees are indicated by an asterisk.

Advisory Committee No. 1. Gliding Bacteria
Liaison Board Member: R. Y. Stanier

T. D. Brock	R. A. Lewin	J. E. Peterson
P. D. Glaister	H. D. McCurdy	H. Reichenbach
E. R. Leadbetter	E. J. Ordal*	

Advisory Committee No. 2. Gram-negative, Facultatively Anaerobic Rods
Liaison Board Member: S. T. Cowan

Subcommittee 2a. Enterobacters

L. S. Baron	H. Lautrop	R. Rohde
W. H. Ewing	L. LeMinor	P. H. A. Sneath
M. Kocur	H. H. Mollaret	Joan Taylor*

Subcommittee 2b. *Aeromonas et al.*

G. L. Bullock	R. W. Schubert	M. P. Starr
K. P. Carpenter	J. Shewan*	M. Véron
R. R. Colwell	R. M. Smibert	
J. C. Feeley	S. F. Snieszko	

Subcommittee 2c. *Erwinia*

D. W. Dye	D. C. Graham	M. P. Starr
W. Frederiksen	R. A. Lelliott*	

Advisory Committee No. 3. Aerobic Gram-negative Rods
Subcommittee 3a. Aerobic Pseudomonads
Liaison Board Member: J. Liston

O. N. Allen	A. J. Holding	M. P. Starr
E. Billing	R. Hugh	M. Thornley
J. De Ley	D. C. Jordan	O. B. Weeks*
M. Doudoroff	P. V. Liu	
W. C. Haynes	J. Shewan	

Subcommittee 3b. "Parvobacteria"
Liaison Board Member: S. T. Cowan
Actinobacillus: H. Haupt, J. E. Phillips, J. E. Smith

xii

Bordetella: M. Pittman

Brucella: W. J. Brinley-Morgan, L. M. Jones,* N. B. McCullough, M. E. Meyer

Haemophilus: E. L. Biberstein, S. K. Henriksen, D. C. White, K. S. Zinnemann*

Pasteurella: G. R. Carter, N. Mair, J. E. Smith

Francisella: C. R. Owen

Calymmatobacterium: R. B. Dienst

Subcommittee 3c. Azotobacters
Liaison Board Member: N. E. Gibbons

J. H. Becking	H. L. Jensen*	J. R. Norris
J. De Ley	D. B. Johnstone*	

Subcommittee 3d. Stalked, Budding and Sheathed Bacteria
Liaison Board Member: R. G. E. Murray

P. Hirsch*	J. S. Poindexter	G. A. Zavarzin
E. G. Mulder	H. L. Skuja	
E. J. Ordal	J. L. Stokes	

Subcommittee 3e. Spiral Bacteria
Liaison Board Member: R. G. E. Murray

D. Claus	N. R. Krieg	R. M. Smibert
H. W. Jannasch*	S. C. Rittenberg	

Advisory Committee No. 4, Actinomycetes
Liaison Board Member: R. E. Buchanan

J. N. Adams	R. E. Gordon	E. H. Runyon
E. Baldacci	D. Gottlieb*	E. B. Shirling
J. H. Becking	A. Hennsen	W. S. Silver
W. B. Bollen	G. P. Kubica	J. M. Slack
J. N. Couch	E. Kuster	W. H. Trejo
C. S. Cummins	H. Lechevalier	H. D. Tresner
L. Ettlinger	M. Lechevalier	M. Tsukamura
C. M. Gilmour	F. Mariat	L. G. Wayne
M. A. Gordon	T. G. Pridham	G. P. Youmans

Advisory Committee No. 5. Gram-positive Non-sporing Rods
Liaison Board Member: C. F. Niven

H. Beerens	O. Kandler	M. Rogosa*
C. S. Cummins	R. M. Keddie	H. P. R. Seeliger
F. Gasser	W. E. C. Moore	M. E. Sharpe
M. Gilmour	E. G. Mulder	

Advisory Committee No. 6. Gram-positive Cocci
Liaison Board Members: C. F. Niven & A. W. Ravin

A. C. Baird-Parker	J. B. Evans*	M. Kocur
H. Beerens	L. R. Hill	I. J. MacDonald
R. H. Deibel	K. Kitahara	H. W. Seeley

Advisory Committee No. 7. Sporeformers
Liaison Board Member: N. E. Gibbons

H. A. Barker	L. V. Holdeman	L. DS. Smith*
L. L. Campbell	L. S. McClung	A. T. Willis
V. Fredette	W. E. C. Moore	
G. Hobbs	C. L. Oakley	

Advisory Committee No. 8. Non-sporing, Gram-negative Anaerobes
Liaison Board Member: N. E. Gibbons

H. Beerens	R. J. Gibbons	J. R. Postgate
M. P. Bryant	R. E. Hungate*	M. Rogosa
V. Fredette	W. E. C. Moore	P. H. Smith

Advisory Committee No. 9. *Neisseria-Moraxella-Mimeae*
Liaison Board Members: R. G. E. Murray & R. Y. Stanier

B. W. Catlin	M. J. Pelczar*	A. Reyn
H. Lautrop	M. Piechaud	

Advisory Committee No. 10. Spirochaetes
Liaison Board Member: R. G. E. Murray

E. G. Hampp	T. A. Nevin	R. M. Smibert
P. H. Hardy	J. Pillot	S. S. Socransky*
M. A. Listgarten	T. Rosebury	L. H. Turner

Advisory Committee No. 11. Photosynthetic Bacteria
Liaison Board Member: R. Y. Stanier

E. N. Kondratcheva	H. G. Truper	
N. Pfennig*	C. B. van Niel	

Advisory Committee No. 12. Chemolithotrophs
Liaison Board Member: R. Y. Stanier

J. W. M. la Rivière	C. B. van Niel	G. A. Zavarzin
J. London	W. Vishniac	
S. C. Rittenberg*	S. W. Watson	

Advisory Committee No. 13. *Mycoplasma*
Liaison Board Member: A. W. Ravin

R. M. Chanock*	E. A. Freundt	S. Razin
D. G. ff. Edward	L. Hayflick	

Advisory Committee No. 14. *Rickettsia*
Liaison Board Member: C. F. Niven

A. C. Allison	Y. Mitsui	R. Schindler
L. H. Collier	J. W. Moulder*	D. N. Weinman
F. B. Gordon	L. A. Page	E. Weiss
H. Higashi	C. B. Philip	V. M. Zhdanov
E. Jawetz	M. Pollard	
J. P. Kreier	M. Ristic	

PREFACE TO EIGHTH EDITION

The eighth edition of BERGEY'S MANUAL OF DETERMINATIVE BACTERIOLOGY differs greatly from previous editions in format, in the presentation of information and in the approach to classification. To understand these changes some of "the thinking behind this edition" should be given.

The planning was done by the Board of Trustees acting as an Editorial Board. Before starting the Board considered various criticisms that had been made of earlier editions, some of which are listed below. (1) THE MANUAL did not meet the objective indicated by its title and often failed to help those trying to identify an isolate. (2) THE MANUAL fell between two stools: it was too detailed (and the information too scattered) for the junior student, and it did not contain sufficient detail for the research student. (3) THE MANUAL contained descriptions (often incomplete and based on old literature) of organisms no longer available for detailed study by modern methods. (4) Since the publication of INDEX BERGEYANA, much of the synonymy duplicated information readily available elsewhere, hence pre-1965 synonymy was required only where a name was regulated by an Opinion of the Judicial Commission, or by fiat of worldly-wise bodies such as the League of Nations Permanent Standards Commission of the Health Organization which decreed in 1931 the use of the name *Clostridium perfringens*, a name contrary to the botanical rules of nomenclature, applicable at that time to bacteria. (5) The keys to the higher categories were largely useless, and a user needed a working knowledge of bacteriology sufficient to place an organism into a genus before THE MANUAL became significantly helpful.

After prolonged and, let it be said in truth, sometimes acrimonious discussion, the Board decided to seek authoritative advice from those most familiar with the various groups of Bacteria. To establish advisory committees, the bacteria were divided on pragmatic grounds into 15 major groups, subsequently increased to 19. The 19 groups thus recognized for purposes of consultation later became the 19 Parts of the present edition. It was agreed that to avoid prejudice each part would be described by an appropriate vernacular rather than a latinized name.

The Advisory Committees were asked to advise on (1) the inclusion or exclusion of taxa within its purview, (2) the general arrangement within the group and (3) suitable authors for each genus.

Meanwhile, Board members discussed the problem of higher taxonomic categories. The majority held the view that for most groups of bacteria genera and species are the only categories that can now be recognized and defined with reasonable precision, but that families were sometimes useful, especially when, as in *Enterobacteriaceae*, the family is roughly equivalent to the genus in other groups of bacteria. It was felt that categories higher than family should be accepted only when pressed by the Advisory Committee or provided by the authors. However, the KINGDOM PROCARYOTAE is recognized with two Divisions having the vernacular names the Cyanobacteria (the blue-green bacteria or algae) and the Bacteria. Suggestions for a possible rearrangement are noted, perhaps for future editions on page 9.

In May 1968, the Board met with representatives of the Advisory Committees who presented their reports. That all experts and authorities do not think alike will be seen when the treatment of the enterobacteria by W. H. Ewing (in Edwards and Ewing, 3rd edition, 1972) is compared with that recommended to the Board (and followed in Part 8) by the Bergey's Manual Advisory Committee on enterobacteria, most of whose members were, like Ewing, members of the I.A.M.S. Subcommittee on the Taxonomy of the *Enterobacteriaceae*.

After consideration of the Advisory Committee reports, which sometimes conflicted on points of detail, the Board decided on the taxa to be included in each of the 19 Parts of THE MANUAL and selected authors for each taxon. At that time it was not considered necessary to send authors a detailed list of requirements or to suggest a definite arrangement of material in the descriptions. This error was soon pointed out by authors, and became more obvious when the first manuscripts arrived at the Editorial offices. Models were then prepared, perhaps too hastily, and with a set of recommendations, were sent to authors in December 1968.

Most of those invited accepted the invitation and the final descriptions of the taxa were prepared by 131 authors from 15 countries (61 authors are from countries other than the USA) (page vi). We regret to report the death of four contributors since work on THE MANUAL began; Dr. Carpenter in 1971, Dr. Skuja in 1972, and Dr. Gibson and Dr. Vishniac in 1973.

Names of authors are given at the beginning of their contributions, whether it be a genus or part thereof, a family, or a higher taxon. In this edition the introduction of new names of taxa has been discouraged but some new combinations are included; these should be attributed to the authors concerned and not to the editors of THE MANUAL. To all of these contributors the sincere thanks of the Board of Editors is due; the Editors are particularly grateful for the good grace with which the authors accepted comments and criticism of their manuscripts. Many authors have revised and up-dated their manuscripts voluntarily or have kept us informed of changes that should be made; their continued interest has been greatly appreciated.

In the descriptions, information is presented in a definite order as far as possible (see "On Using the Manual," page 1); much information is presented in tabular form and a number of figures have been added. Keys are used in many taxa and, since the presentation is by Parts rather than through a hierarchical arrangement, a "Key to the Parts" is given on page 18.

Dr. Skerman has revised his "A Key for the Determination of the Generic Position of Organisms Listed in The Manual" (page 1098) which provides another means of orientation.

Other new additions to THE MANUAL include a "Glossary" of terms used in the text, mention of type or reference strains, a list of the collections from which such strains may be obtained, and a consolidated list of references to authorities for names and to articles mentioned in the text.

The editing of the manuscripts was a joint effort; nomenclature, synonymy and etymology were undertaken by R. E. Buchanan, descriptions and general technical editing by N. E. Gibbons. John G. Holt was of great assistance in the Ames office, and S. T. Cowan's help and encouragement in the Ottawa office was invaluable during the final rush to get manuscripts to the printer.

The Editors and Editorial Board are grateful to many who have made their task easier during the six years THE MANUAL has been in preparation. These include in the Ames office, Mrs. Mildred McConnell, Mrs. Vlasta Krakowski, who prepared the consolidated list of references, and Mrs. Elza Zvirbulis, who checked most of the synonymy and the nomenclatural references; in the Ottawa office Mrs. Margaret Bergin, Miss Pat Moss and Miss Edna Finn who served as secretary at various times during this period. Invaluable assistance was received from Dr. E. F. Lessel of the American Type Culture Collection; Mr. R. H. Whitehead, photographer at the National Research Council, Ottawa, has given advice and help in preparing many of the plates; Mrs. N. E. Gibbons has prepared most of the index of scientific names.

Financial assistance is gratefully acknowledged from the National Science Foundation and The National Institutes of Health in the form of grants to Dr. Buchanan; from the American Society for Microbiology which made a grant for travel to meetings of the Board; from the Wellcome Trust of London which provided transportation for Dr. Cowan. The Trust is also grateful to The Upjohn Company of Kalamazoo, Michigan which made its Brook Lodge facility available for a meeting of the Board with its Advisory Committee Chairmen, and to the National Research Council of Canada which provided one of us (N. E. G.) with many facilities both before and after retirement.

Comments on this edition will be welcomed and should be directed to The Bergey Manual Trust, c/o The Williams and Wilkins Co., 428 E. Preston St., Baltimore, Md. 21202, USA.

PREFACE TO FIRST EDITION

The elaborate system of classification of the bacteria into families, tribes and genera by a Committee on Characterization and Classification of the Society of American Bacteriologists (1917, 1920) has made it very desirable to be able to place in the hands of students a more detailed key for the identification of species than any that is available at present. The valuable book on "Determinative Bacteriology" by Professor F. D. Chester, published in 1901, is now of very little assistance to the student, and all previous classifications are of still less value, especially as earlier systems of classification were based entirely on morphologic characters.

It is hoped that this manual will serve to stimulate efforts to perfect the classification of bacteria, especially by emphasizing the valuable features as well as the weaker points in the new system which the Committee of the Society of American Bacteriologists has promulgated. The Committee does not regard the classification of species offered here as in any sense final, but merely a progress report leading to more satisfactory classification in the future.

The Committee desires to express its appreciation and thanks to those members of the society who gave valuable aid in the compilation of material and the classification of certain species. . . .

The assistance of all bacteriologists is earnestly solicited in the correction of possible errors in the text; in the collection of descriptions of all bacteria that may have been omitted from the text; in supplying more detailed descriptions of such organisms as are described incompletely; and in furnishing complete descriptions of new organisms that may be discovered, or in directing the attention of the Committee to publications of such newly described bacteria.

DAVID H. BERGEY, *Chairman*
FRANCIS C. HARRISON
ROBERT S. BREED
BERNARD W. HAMMER
FRANK M. HUNTOON
Committee on Manual.

August, 1923.

ROBERT EARLE BUCHANAN
1883–1973

I SU!

Dr. Robert Earle Buchanan, who had served as Chairman of the Bergey's Manual Trust for 17 years, died February 21, 1973. He assumed the chairmanship following the death in 1956 of Dr. Robert S. Breed, at an almost identical point in the preparation of the seventh edition of THE MANUAL. Dr. Buchanan was thus responsible for seeing the seventh edition through the press, after which he embarked on the arduous task of compiling the INDEX BERGEYANA, which had been conceived and named by Dr. Breed, and which was eventually published in 1966. Despite his already advanced age, Dr. Buchanan then initiated the preparation of the present edition of BERGEY'S MANUAL, and took a very active part in the operation until his health began to fail in the course of his last year.

Throughout his long scientific career, Dr. Buchanan had always been interested in bacterial taxonomy and, more particularly, in the problems of nomenclature. Between 1916 and 1918, he published an extensive series of articles in the *Journal of Bacteriology* on bacterial nomenclature and classification, and from 1917 to 1920 he served as a member of the Winslow Committee of the Society of American Bacteriologists, which was charged with the characterization and classification of bacterial types. The proposals of this committee, published in 1917 and 1920, had a profound influence on the taxonomic plan of BERGEY'S MANUAL, which was the responsibility of a parallel committee chaired by Dr. D. H. Bergey. Buchanan's GENERAL SYSTEMATIC BACTERIOLOGY (1925) still remains a valuable historical source book on bacterial nomenclature and classification. The jacket carries a declaration of principle which epitomizes his approach to this field:

To present in Graphic Form the Phylogeny and the Relationships of Various Groups of Bacteria.

To Give Greater Stability to the Names for Groups of Organisms.

To Prevent Unnecessary Confusion in Nomenclature.

With the goal of implementing these principles, he was instrumental in creating, at the first Congress for Microbiology in 1930, what has now become the International Committee on Systematic Bacteriology; he was also largely responsible for preparing the International Code of Nomenclature of Bacteria.

The members of the Board wish to pay tribute to his role in the preparation of this edition; his experience, energy and devotion to the task have been invaluable.

DAVID HENDRICKS BERGEY
1860–1937
Bergey set up the Trust on January 2, 1936

ROBERT STANLEY BREED
1877–1956
Chairman, Bergey Manual Trust 1937–1956

CONTENTS

Introduction

On
Using the
Manual

R. E. BUCHANAN
N. E. GIBBONS

THE MANUAL is meant to assist in the identification of bacteria. No attempt has been made to provide a complete hierarchy, as in previous editions, because a complete and meaningful hierarchy is impossible. Instead THE MANUAL is presented in 19 parts based on a few readily determined criteria.

Each part bears a vernacular name and sometimes that of a taxon. All accepted genera have been placed in what seems the most appropriate part, although allocation of some genera presents difficulties; for example, Part 1 ("Phototrophic Bacteria") includes three genera of budding bacteria which are mostly in Part 4; the chemolithotrophic genus *Nitrobacter* (Part 12) also produces buds.

In Parts 1, 5, 11, 13, 14 and 18 all accepted genera have been placed in families (there may be an occasional exception covered in an addendum). In Parts 3 and 4 no attempt has been made to indicate families. In other parts the families are followed by "Genera of Uncertain Affiliation." While these genera belong in the part, as defined, they have not been accepted into any of the families and cannot themselves be grouped into families on the information now available. Also included in this section are the "lonely" or "solitary" genera that do not fit with any other genus or genera. The Editorial Board is well aware that the groupings used here will not be acceptable to all. Many will find the partially systematic approach unpalatable. It is hoped that these differences of opinion, as well as the uncertainties recognized in THE MANUAL, will stimulate further work in all phases of bacterial classification and taxonomy. It is not implied that the groupings accepted here are to be considered inviolable.

Entry into THE MANUAL may be achieved in several ways. Some will prefer the comprehensive key provided by Professor Skerman (Appendix). Others may prefer to study the titles of parts, which may be considered an elementary, but by no means perfect, key, or to use the short key provided on page 18. The index is the obvious resource in locating names of unfamiliar taxa or in discovering what has been done with a particular taxon.

Some recently used names have been reduced to synonyms. Many names have been relegated to appended lists of *"Species Incertae Sedis"*; the reasons for these actions are usually indicated. Many of the descriptions are based on recently completed or even ongoing work on all available strains of a group; authors of such groups have often placed named strains, which were unavailable to them for study, in appended lists. Problems of many kinds are discussed by authors under "Comments," which may apply to a species, a genus or a family. Thorny problems are at times mentioned under "Editorial Comments."

Accepted names in the descriptions and index are in **boldface**. The authority for the name follows with the year of the original description and the page on which the taxon is named and described. This is followed by the synonyms, both objective and subjective; the former are of interest to the nomenclaturist, the latter to the working bacteriologist. The synonymy may not always be complete as some authors felt that the earlier synonyms were covered adequately in *Index Bergeyana*. Errors in *Index Bergeyana* have been rectified where possible in the present text.

Etymologies are provided as in previous editions, although some feel this information is of little value. It is often difficult to determine why a name was chosen, or the nuance intended, if the details are not provided in the original publication. Those proposing new names are urged to give the exact Latin or Greek derivations rather than, as is so often done, a vague reference that the organism was so named because of the shape of the colony, the color or the source. Greek words may be given in the transliterated Greek form or as the corresponding Latin word.

The presentation of the descriptions is quite different from that of previous editions. A telegraphic, narrative style is used and the information presented insofar as it is available and wherever possible in a definite sequence: (1) structural characters, such as shape, size, special features, mode of movement, nature of resting stage, Gram reaction, appearance of macroscopic growth, if distinctive; (2) biochemical and nutritional characters—utilization of or action on compounds, usually in the order carbohydrate (presented alphabetically), protein and other compounds, with end products formed, followed by other biochemical information on components of the cell and on metabolites; (3) physiological characters such as relation to oxygen, temperature, pH and response to antibacterial substances; (4) ecological characters; (5) genetic data; (6) the type or reference strain and where it can be obtained. Very complete information is available on some taxa, very little on others.

Generic descriptions are intended to be full and to contain all characteristics common to all species of the genus. In all but a few of the generic descriptions, **most of the important distinguishing characteristics are presented in**

boldface. Common characters are NOT repeated in the descriptions of species. Also information given in tables is not necessarily repeated, so that, to obtain a complete picture of the species, both the generic and specific descriptions should be read, and the tables, including footnotes, consulted.

Some authors have used keys, some tables, some both. Keys, especially Skerman's comprehensive key, do not need symbols for their understanding, and some readers find them easy to use. The keys of this edition are, we believe, more practical than those of earlier editions.

Tables show either a few characteristics or a great many. The former may differentiate the next lower taxa, the latter become a part of the descriptions themselves, e.g. the large tables showing carbohydrate reactions of some groups; these are easier to use and take up less space than long lists of carbohydrates that are or are not attacked. In general, characters common to all units are excluded from the tables, but are indicated in footnotes.

Although a standard set of symbols for tables was suggested (p. 20), and generally used, some authors found it unsuitable for their material. The meaning of symbols, both standard and the variations therefrom, are indicated freely in footnotes throughout THE MANUAL.

Some authors have based their descriptions of species on the type strain. The concept of a species varies (see p. 17) but as the type strain is a nomenclatural entity, they are listed where possible. The addresses of culture collections mentioned are given in an Appendix.

A glossary has been added. Only terms which may cause difficulty (at least to one editor) have been included. Words found in the usual editions of *Webster's Dictionary* are not listed; more technical terms may be found in specialized dictionaries.

As well as the usual references to author citations in connection with the names of organisms, references are given to sources of specific information and to controversial areas. Nomenclatural references include a page number and are easily distinguished. Other references give only the year and multiple papers by the same author or authors in the same year are not differentiated; only rarely should this create any difficulty.

Suggestions on identification may be found under "The Mechanism of Identification."

Comments on the presentation will be welcomed and should be addressed to the Bergey Manual Board of Trustees, c/o The Williams and Wilkins Co., 428 East Preston St., Baltimore, Md. 21202, U.S.A.

A Place for
Bacteria in the
Living World

R. G. E. MURRAY

"What's in a name? that which we call a rose
By any other name would smell as sweet . . . ."
Romeo and Juliet, Act II. W. Shakespeare

One of the most difficult tasks for the Editor-Trustees of all previous editions of BERGEY'S MANUAL was the provision of an intelligible and recognizable description of the bacteria as a taxon and a scientifically acceptable placing among the realms of living things. Without doubt most thoughtful bacteriologist-taxonomists since the time of Ferdinand Cohn have been convinced on an intuitive base and slender morphological evidence, that there was something unique about bacteria and that a degree of morphological similarity existed between the bacteria (*Schizomycetes*) and the blue-green algae (*Cyanophyceae* or *Schizophyceae*) whatever the arrangements made at the highest hierarchical levels. This was recognized in the many and various versions of the familiar classifications of the living world by placing these two taxa together as equivalent classes in the Division *Protophyta* of the Plant Kingdom, as was done in the 7th edition. A consequence of the morphological alliance, together with the evident physiological similarity of photosynthesis by the blue-green algae to that of plants, was a continuing adherence to descriptions in a botanical context. The definition of the *Schizomycetes* in the 7th edition was framed in terms of this long established prejudice despite misgivings (see 7th edition, p. 9). The essential operative statement is that the bacteria are "typically unicellular plants," which was not then and certainly is not now defensible in terms of the description of a typical or idealized plant cell. In fact, it is surely true that the intuitive feeling of the bacteriologist rejected this view, but without solid evidence to buttress his faltering steps, as shown by the acceptance and continual refinement of the Code of Bacteriological Nomenclature independent of the Botanical Code. Resolution of the dilemma became possible as soon as the microbes could be described in terms of cellular organization.

The change in our view of the nature of bacteria was slow in coming to full expression. It derives from the application of three sets of experimental-observational approaches to the microbial world and bacteria in particular: (1) comparative cytology using the light microscope and classical staining methods to describe the form and behavior of the DNA-containing portions of nuclei; (2) the development of appropriate techniques of electron microscopy for the extension of comparative cytology to the ultrastructural level; and (3) the extension of biochemical and biophysical observations to the definition of unique features of cellular organization.

Cytologists who studied the protists became aware that the chromatin of bacteria, which they assumed correctly to be the equivalents of nuclei, was markedly different in appearance and behavior during segregation in comparison to the nuclei and component chromosomes of other protists, plant and animal cells. This was sufficient for some to make fresh attempts to circumscribe a taxon at the highest level to include the bacteria. Among these were the proposals of Copeland (1938), Stanier and van Niel (1941) and Whittaker (1959). In essence, these proposals (derived in part from attempts to rationalize Haeckel's concept of the Protists) placed the bacteria in a kingdom of anucleate organisms in contrast to those with true nuclei. Stanier and van Niel (1941) invoked two other negative features—the seeming absence of sexual reproduction and of plastids; however, it was soon to be established that photosynthetic bacteria possessed chromatophores derived from the plasma membrane and that the bacterial genome was capable of recombinative processes. Most significant of all, early electron microscopic observations of sections of cells showed that the genophore of the bacteria and the blue-green algae consisted of a nucleoplasm that was not separated from the cytoplasm by a nuclear membrane as found in fungi, protozoa, plants and animal cells. Further structural and biochemical studies established new constellations of features unique to bacterial cells in the form of heteropolymers and constituent molecules of the intimate make-up of cell walls (see Salton, 1964). In the face of this level of understanding, the kingdom designations of *Mychota* (Enderlein, 1925) or *Monera* (Copeland, 1938) were not appropriate in their definitions. However, Chatton (1937) had proposed a most appropriate conceptual basis for taxa at the highest level by recognizing two general patterns of *cellular organization*—the procaryotes and the eucaryotes. The truth of this prescient generalization was recognized by Stanier (1961) and is now amply supported by a wealth of data derived from comparative cytology involving microscopical, biochemical and physiological approaches. Stanier and van Niel (1962) put the matter in clear perspective, with an accompanying statement of the evidence: "The distinctive property of bacteria and blue-green algae is the procaryotic nature of their cells." Furthermore, it was possible to justify the use of the inclusive term, the cell, for the unit of structure of both procaryotic and eucaryotic organisms as an expression of equivalence of function; the differences between the broad groupings concern the detailed organization of the cellular machinery. The broad canvas was now sketched in and the bacteria could be confidently placed in perspective with other cellular organisms. The viruses, alone among organisms, remain without a clear position in the scheme of things.

There is no real need to repeat here all the arguments that have been presented to embroider and support the cytological recognition of procaryotic organization as the cornerstone for a coherent view of the nature of bacteria. The case has been made several times (Stanier, 1961; Stanier and van Niel, 1962; Murray, 1962; Allsopp, 1969; Stanier, 1970) and the student should consult these essays for details and references.

The essential features are:

I. The nature of the genophore (a term used by Ris (1961) to avoid the connotations of "chromosome" and "nucleus") constituting the morphologically distinct nucleoplasm of the procaryotic cell. This consists of a skein of double-stranded DNA fibrils that is not separated from cytoplasm by any membranous boundary. Structural and genetic evidence, based on bacterial examples, indicate that the genophore is in the form of a closed loop (often described as a "circular chromosome") and the constituent genes form a single linkage group. The fibrils are not associated in any regular way with a protein in contrast to the histones of the eucaryotic chromosome. For the moment it must be assumed, because of fragmentary genetic or cytogenetic evidence, that the morphologically similar nucleoplasm of blue-green algae has parallel properties.

II. The lack of unit membrane-bounded cytoplasmic organelles is the second most important distinction of the procaryotes. This belies their metabolic diversity and the simplest form is provided by some bacteria whose "membrane system" consists of the plasma membrane alone, smoothly enclosing the protoplasm. Simple intrusions of membrane are common and complex systems of unit membranes do exist (e.g. for photosynthesis, nitrification, etc.), but in all cases there is good support for the view that they are derived by invagination of the plasma membrane into the cytoplasm and in most cases the connection is maintained. There is a possible exception in the blue-green algae whose photosynthetic apparatus is located in an extensive system of thylakoids, which do not appear to be in direct continuity with the plasma membrane.

III. A further distinction attributable to all procaryotes is that the ribosomes are of the small 70 S type as opposed to the consistently larger 80 S ribosomes of the eucaryotes (Taylor and Storck, 1964). They are distributed in the cytoplasm and are not arrayed on membranes as in the endoplasmic reticulum of eucaryotes.

There are a number of positive characters that can be considered as features of many but not all procaryotes, which further strengthen the distinction. For example (1) *cell walls* are not confined to procaryotes but the components are unusual to a degree that is taxonomically significant. Without doubt the now well known peptidoglycan (murein or mucopeptide) component of the cell walls,

shared by the majority of bacteria and the blue-green algae, is a uniquely constituted heteropolymer with distinctive subunits in both the amino sugar backbone (muramic acid) and the peptide (D-amino acids and unusual diamino acids)
portions. It cannot be used to characterize the procaryotes because the wall-less
Mycoplasma and exceptional bacteria (e.g. *Halobacterium*) do not possess it.
Other constituents may yet prove to have taxonomic value but the peptidoglycan
is particularly important as a bridging character with the blue-green algae. (2)
Flagella are distinctive organs of swimming motility confined to certain genera
of bacteria. This complex tubular assembly of protein subunits with an elaborate
anchorage in both wall and plasma membrane cannot be confused with the
cilia of eucaryotes. (3) *Gas vacuoles* are common in blue-green algae and are
found in a few widely separated genera of bacteria. *Chlorobium* vesicles are
unique containers for the chlorophylls and carotenoids of the green bacteria.
What they have in common is that they are organellar inclusions that are not
bounded by unit membranes but rather by special single layers. They are without
counterpart among eucaryotes.

It is no longer possible to support two negative features that have been proposed as characterizing procaryotes: lack of cytoplasmic microtubules and an
inability to synthesize sterols. It is now apparent that some *Treponema* may
possess microtubules and sterols have been isolated from some *Mycoplasma* and
blue-green algae.

Bacteria are remarkable for an extraordinary variety of metabolic mechanisms
and, particularly, for a wide range of anaerobic energy-yielding reactions; these
are in marked contrast overall to the glycolysis utilized by the eucaryotic cell.
The result is that many bacteria are obligate anaerobes and share this condition
of existence with only a few protozoa among the eucaryotes. Apart from this
generalization and the synthesis of unique cell wall polymers, already mentioned,
it should be noted that the ability to fix nitrogen and to accumulate poly-β-
hydroxybutyrate as a reserve material are metabolic attributes widely distributed
among procaryotes but completely absent from eucaryotes.

The fossil record, although indicative of microbial life long ages before recognizable complex forms of life appeared, is not able to tell us anything of the order
of appearance and thus contribute to phylogeny. Those groupings that we can
observe today seem likely to represent a coherent segment of the terminal
branches of an evolutionary tree. Photosynthesis probably originated when the
procaryotic stem was already well developed. The precise forms of photosynthesis
represent the derivatives of, most likely, a single evolutionary event (Stanier,
1970) since many elements of the machinery are common to all existing phototrophs.

It has been pointed out by Stanier (1970) that the plasma membrane of
procaryotic organisms is not adapted to the transfer of particulates or large
molecules either in or out of the living cell; indeed, transforming DNA fragments
are about the only example. Furthermore, there is no evidence for endocytosis
(viz. phagocytosis and pinocytosis) and its directional counterpart, exocytosis,
as mechanisms of import or export of soluble material or particulates. Yet this is a
very generalized characteristic of eucaryotic cells. Many protozoa ingest pro-

caryotic organisms as food into food vacuoles. It is not surprising, then, that a good number of stable associations have been set up so that distinctly recognizable endosymbionts are characteristic of the cytoplasm of certain cells. These are also enclosed in a sac of host cell membrane. The most notable studies are on the endosymbionts of *Paramecium aurelia*. Although they are generally not cultivable, many of them are structurally recognizable as procaryotes and some even exhibit bacterial flagella and bacteriophages of the tailed type (Beale *et al.*, 1969; Preer *et al.*, 1972). Some endosymbionts have the characteristics of a eucaryotic green alga (as in *Paramecium bursaria*) and other protozoa have blue-green inclusions (cyanelles) that have been interpreted, with insufficient rigor, as blue-green algae. It is clear that endosymbiotic associations are widespread in a great variety of metazoa (Buchner, 1965), as well as in the vascular plants, such as the well known *Rhizobium* association in the root nodules of legumes. Identification, description and classification rests, in the absence of verifiable cultivation, on morphological and biochemical features. It appears that the capacity to take up external cells brings with it the possibility of endosymbiosis. No stable endosymbiosis within procaryotes has yet been identified. (Bacterial parasitism by *Bdellovibrio* is best described as a nonsymbiotic multiplication between cell wall and protoplast causing death of the host.)

One of the excitements of recent years is the realization that the mitochondria and the plastids of eucaryotes could represent the most extreme form of endosymbiotic parasitism by procaryotes. These unit membrane-bounded structures have physical, genetic and biochemical features that suggest the possibility of their procaryotic nature; furthermore, the DNA from the very small genophore can be isolated in circular form and has a very different base composition in comparison with that of the host cell. At this stage the observations provide the base for fascinating and possibly unverifiable speculation concerning an event (or set of events) of the greatest significance in eucaryotic evolution (see Stanier, 1970; Allsopp, 1969).

There is little doubt, at this time and in the face of the arguments that have been presented, that biologists can accept the division of cellular life (as opposed to virus "life") into two groupings at the highest level expressing the encompassing characters of procaryotic and eucaryotic cellular organization. Those who have written concerning the topic have generally avoided committing themselves to a formal nomenclature. So far as we know there are only two formal proposals, both based on recognizable descriptions. Murray (1968) proposed *Procaryotae* as a taxon "at the highest level" and described it as "a kingdom of microbes . . . characterized by the possession of nucleoplasm devoid of basic protein and not bounded from cytoplasm by a nuclear membrane." *Eucaryotae* was suggested as a possible taxon at the same level to include other protists, plants and animals. Allsopp (1969) declared that these groups merit the status of "kingdom or even superkingdom" and he proposed with an extended assessment of characters that the Kingdoms should be *Procaryota* and *Eucaryota* (Allsopp, 1969, 607). The Bergey's Manual Trust has discussed these names and the various alternatives on many occasions and agreed that *Procaryotae* was the most appropriate, as a plural feminine noun, for such a taxon.

The assumption of a new kingdom is both appropriate and helpful to the bacterial taxonomist, but a kingdom including all the eucaryotes would be disturbing to botanists and zoologists causing a realignment of their respective hierarchies. It is probably best to leave matters as they have been expressed above and only recognize, at the moment, the Kingdom *Procaryotae*.

How are the bacteria to be considered in relation to the blue-green algae within the Kingdom? There is no doubt of their relationship at that level and of the many parallels in the forms and arrangements of their cells. As proposed by Stanier *et al.* (1971), "the blue-green algae must now be recognized as a major group of bacteria, distinguished from other photosynthetic bacteria by the nature of their pigment system and by their performance of aerobic photosynthesis." It is a departure from tradition but it seems just, and it respects the fact that there are parallel groups of photosynthetic bacteria equally distinguished by the nature of their photopigments and their ability to perform anaerobic photosynthesis without oxygen production. For the present, the Bergey's Manual Trustees have no intention of attempting a reassessment of the blue-green algae. It is suggested, by the present designation (p. 22), that they are truly allied to bacteria and at a level of equal importance without hierarchical prejudice to the remainder of the Kingdom, the Bacteria.

The future will have to bring a regrouping of higher taxa to express a more coordinate view. This example is constructed without hierarchical prejudice (e.g. no attempt is made to indicate that non-photosynthetic procaryotes must have preceded the photosynthetic groups on the evolutionary scale) and using common names or tentative names in parentheses, as a suggestion of a possible arrangement.

Kingdom: *Procaryotae*

Division I: Phototrophic procaryotes ("Photobacteria")
      Class I: Blue-green photobacteria
      Class II: Red photobacteria ⎫
      Class III: Green photobacteria ⎬ Present Part 1

Division II: Procaryotes indifferent to light ("Scotobacteria")
      Class I: The bacteria—Present Parts 2 to 17
      Class II: Obligate intracellular Scotobacteria in eucaryotic cells—Rickettsias—Present Part 18
      Class III: Scotobacteria without cell walls—*Mollicutes*—Present Part 19

No doubt time will settle the problems involved in making a reasonable arrangement within the *Procaryotae*. Haste is unwise; all previous classifications seem to have suffered infinite rearrangement due to insufficient information. The groupings suggested above, even if the rank designations may need adjustment, might appear to be stable until a new level of understanding overtakes us. The new insights are likely to come from a clear understanding, on a comparative level, of the components of the genome of procaryotic cells. This may allow us to understand why and how vastly different physiological groupings of phototrophic, lithotrophic and organotrophic organisms have remarkably similar and complex structural components (as exemplified by the wide range of spirilla). Surprises may well be in store for us; hence, our adherence to simplicity.

The
Mechanism
of
Identification

S. T. COWAN
J. LISTON

Bacteria isolated directly from animals, plants or the general environment are rarely obtained in pure culture. The first step in identification is to obtain each organism in pure culture, since the reactions of mixed cultures in characterization tests cannot be used for identification. This step is so obvious that it is seldom mentioned, but in fact it cannot be stressed too much if we are to obtain clearcut characterization of our unknown organism. Techniques for obtaining pure cultures vary with the organism being studied; they range from repeated plating on non-inhibitory media to single cell isolations. The commonest single source of impure cultures is the picking of a colony from a selective or inhibitory medium; consequently, these media are to be avoided in the purification or checking process.

Characterizing tests range from the descriptive (morphology and tinctorial reactions) through simple biochemical tests for the detection of metabolic products or enzyme action, to highly specialized techniques required for the estimation of the GC percentage of the bacterial DNA; these advanced techniques are not likely to be within the competence of the worker responsible for making an identification, which must normally be based on characterizations made by simpler methods.

From a practical point of view, it is important that the number of tests applied to identify a bacterium be kept to a minimum. Hundreds of characterization tests have been proposed, and it is necessary to select a test pattern which will quickly and simply identify the organism under study. Knowledge of the source of the strain and the conditions under which it was originally isolated is often helpful in delimiting the genera to which the organism may belong. Indeed this information, together with details of morphology and Gram's staining reaction, which can easily be obtained during the purification procedure, will frequently enable the investigator to narrow his search to a small group of genera. Thus a Gram positive motile spore-forming rod isolated from soil by aerobic incubation of a non-selective medium is almost certainly a *Bacillus* and there is no need to apply tests to differentiate this organism from Gram negative rods, obligate anaerobes, Gram positive cocci, etc.

The desirability of careful morphological examination of strains as a first step in identification cannot be overemphasized. The phase contrast microscope is particularly useful, since it can be used to view living organisms without the distortion imposed by staining procedures or by drying and embedding, as in electron microscopy. It is simple to use and provides a rapid method for detecting motility, as well as shape and cell arrangements, in most cases. The Gram reaction is still the most generally useful of the staining methods and should be applied routinely to all fresh isolates. This may be done conveniently using the phase contrast preparations. Flagella staining and tests of acid-fastness are the only other two staining procedures which are likely to be useful with any great frequency, and they need only be applied when seen to be necessary.

Electron microscopy is not yet an everyday tool of the routine bacteriologist, but where facilities are available it can provide most useful information on the procaryotic nature of an unknown organism, wall structure, the exact mode or site of flagellar insertion and other taxonomically useful details.

Typically, the detailed classification of bacteria is based on results of physiological and biochemical tests, as well as the more specialized techniques often used for strain identification, such as serology, bacteriophage typing and genetic analysis, including measurement of percentage of GC and DNA-DNA homology analysis.

The major physiological divisions of bacteria into chemolithotrophic autotrophs, photosynthetic bacteria and chemoheterotrophic bacteria will normally present little difficulty. These are obvious separations, based on characteristics which should be apparent during the initial isolation procedures. Similarly, the striking morphological attributes of the larger and more complex organisms (formerly called "Higher Bacteria"), including holdfasts, slime-encased chains, tubules, etc., will distinguish these heterotrophic forms from the smaller, morphologically simpler types (formerly called "True Bacteria") routinely isolated in most bacteriological laboratories. It is these latter forms which consistently present most difficulty in identification. In practice the simple, or true, bacteria are classified primarily on a combination of morphology and physiology, with emphasis on oxygen requirement and the manner of utilization of carbohydrates (particularly glucose). Thus, we have the aerobic Gram negative rods, which are obligate aerobes utilizing sugars oxidatively, and the fermentative Gram negative rods which ferment sugars and are, in fact, facultative anaerobes. Each of these groups contains a number of genera further subdivided by other tests, and this is representative of the general pattern of classification within which the diagnostic bacteriologist must work. In practice, therefore, it is commonly useful to test whether an organism is obligately aerobic or anaerobic, or a facultative anaerobe, and to find out the mode of dissimilation of sugar (usually glucose).

The simple tests described lead, as noted, to groups of genera, and further classification depends on tests which vary with each group but usually involve more physiology and biochemistry. It is a good idea, wherever possible, to grow the pure strain in a relatively simple liquid medium such as peptone water to provide a non-interfering inoculum for these tests (this is also useful earlier for phase contrast examination). Again, however, it is important to apply common sense at this point. A polarly flagellate oxidative Gram negative rod which produces a green fluorescent pigment is obviously a *Pseudomonas* species and requires little further testing to identify to genus. A Gram positive facultatively anaerobic fermentative rod could belong to a number of genera, but a simple catalase test will establish with a high degree of probability whether or not it is a *Lactobacillus*. It is well to scan the general characteristics of the targeted group and select the tests necessary for identification. The tests which best distinguish between one group of organisms and another are shown in **boldface** in THE MANUAL.

In the case of bacteria suspected to be pathogenic, experimental inoculation of plants or animals may be a necessary step in identification. Here (as in most other tests) only a positive result is significant, since a negative result may merely indicate that the wrong animal or plant species was chosen or that conditions of inoculation were inappropriate.

Antigenic analysis is often complex and highly specialized, being frequently used for strain identification in epidemiological or clinical studies. However, the serological method can provide a rapid, unambiguous identification of bacterial isolates in the middle of an epidemic or where an early clue to the genus is available. A good example of this is in identification of *Salmonella* serotypes, for which there is a readily available supply of polyvalent and monospecific antisera.

Simpler biochemical and physiological tests which are frequently useful in identification of bacteria at the species level include oxidase, catalase, urease, nitrate reduction, H_2S production, acid or gas production from sugars, tests of amino acid metabolism, temperature range of growth, response to NaCl and antibiotic sensitivity. In some genera, the nutritional requirements, ability to utilize specific substrates and production of polyhydroxybutyrate are useful.

In summary, the rules of the game are:
1. Use all the information available to you.
2. Apply common sense at each step.
3. Use the minimum number of tests to make the identification.

The practical steps to follow are:
1. Make sure you have a pure culture.
2. On the basis of the isolation procedure, establish whether you have a chemolithotrophic autotroph, a photosynthetic organism or a chemoheterotrophic organism.

3. Examine living cells by phase contrast and Gram-stained cells by light microscopy. Apply other stains if this seems appropriate. If some outstanding morphological property, such as endospore production or possession of holdfasts, is obvious, confine your further efforts in identification to groups having these properties.
4. Examine gross growth appearances for pigments or other unique characteristics.
5. Test for oxygen requirements.
6. Test the dissimilation of glucose (or other simple sugar)—oxidative or fermentative.
7. Complete additional tests selected by scanning the characteristics for the cluster of genera to which you have assigned your organism on the basis of the tests listed above.

If, as may well happen, you cannot identify your isolate from the information contained in THE MANUAL, neither despair nor immediately assume that you have isolated a new species; many of the problems of microbial classification and nomenclature are the results of people jumping to this conclusion prematurely. Remember that descriptions in THE MANUAL are based on majorities, and give the reactions regarded, by those familiar with them, as characteristic of the species; the Editors have generally, but alas not always, deleted such words as generally, usually and sometimes, which so often befog descriptions of species. When you fail to identify your culture check (1) its purity, (2) that you have carried out the appropriate tests, (3) that your methods are satisfactory (in several places in THE MANUAL the need for specified methods is emphasized; one way to make sure is to include a known control culture in your tests) and (4) that you have used correctly the various keys and tables of THE MANUAL. It has been said that the most frequent cause of mistaken identity of bacteria is errors in the determination of shape, Gram reaction and motility. In most cases you should have little difficulty in placing your isolate into a genus; allocation to a species or subspecies may need the help of a specialized reference laboratory.

Reference Collections of Bacteria—The Need and Requirements for Type and Neotype Strains

N. E. GIBBONS

Background

As it became possible to grow bacteria in liquid and solid media, microbiologists began to exchange cultures with their colleagues for information and comparison. Each investigator kept his own isolates, added those received from others and in this way built up his own reference and working collection.

About the turn of the century, Professor František Král of Prague realized the value of a central collection and began to collect cultures which he made available for a fee to other workers. After Král's death in 1911 the collection was acquired by Professor Ernst Pribram and transferred to the University of Vienna in 1915. Pribram brought part of the collection to Loyola University in Chicago some years before the second World War. He was killed in a car accident in 1940, but the fate of his collection is not known. The cultures left in Vienna were destroyed during World War II.

The next oldest collection—Centraalbureau voor Schimmelcultures—was founded in 1906 by the Association internationale des Botanistes. Although the founding association did not survive the First World War, the collection is still in existence at Baarn under the auspices of The Royal Netherlands Academy of Sciences. This collection provides a holding and distribution center for fungi and an identification service. Originally it also kept yeast cultures but in 1922 these were moved to Delft where thorough study resulted in the well known manuals on yeast taxonomy and identification.

Since then many other collections have developed, some general, some specialized, some oriented to service. About 1946, Professor P. Hauduroy established a centralized information facility at Lausanne, the "Centre de Collections de Types Microbiens." This Centre did not maintain cultures (other than Professor Hauduroy's collection of mycobacteria) but provided an information service. When a request was received, the Lausanne Centre asked a collection, known to maintain the culture in question, to send it to the person making the request. The Centre also collected information on nomenclature, taxonomy and classification which was available to workers.

In 1947, the Lausanne Centre became associated with the International Association of Microbiological Societies (IAMS) and, in co-operation with it, an International Federation of Type Culture Collections was formed. The Federation planned to compile a world catalogue containing not only a list of strains but also corresponding descriptions. However, the curators of individual collections were apparently not as enthusiastic as Professor Hauduroy, and the Federation collapsed after a few years. The Lausanne Centre continued to operate.

In 1962, a Conference on Culture Collections, held after the VIIIth International Congress for Microbiology, asked IAMS to form a Section on Culture Collections. The Section was set up in 1963 and on the reorganization of IAMS in 1970 became the World Federation of Culture Collections. It has collected information on 329 collections in 52 countries and the first *World Directory of Collections of Cultures of Microorganisms* (Martin and Skerman, 1972) has been published.

The next stage is the collection of information on strains held by the collections, followed by detailed information on important strains and perhaps eventually on all strains.

Methods which allow the accumulation on IBM cards of information on individual strains have been tried (Quadling and Martin, 1968), and the methodology is expanding (Rogosa *et al.*, 1971). In the future THE MANUAL will undoubtedly be prepared by methods such as these, thus eliminating bias, confirming suspected and pointing out unsuspected relationships, as well as simplifying the editors' task. Even with computer techniques, the task of collecting information of this kind on a national and international basis is enormous and its success, of course, depends on the co-operation of all individuals interested in taxonomy and classification.

In a co-operative scheme, standard methods are of utmost importance. Each collection in the scheme should maintain cultures in an agreed manner and use an accepted system of taxonomy. In pursuit of these aims the World Federation will hold training courses for curators and workers in collections, and will work in close co-operation with the International Committee on Systematic Bacteriology (ICSB) and its Judicial Commission, the International Committee on Nomenclature of Viruses, and related bodies.

The Need for Culture Collections

It is essential for the orderly development of bacteriology that cultures of organisms described or mentioned in publications be available for independent study. Because microbiologists are mortal and their interests vary during their working life, collections are necessary to provide an element of stability and continuity.

While some microbiologists spend a lifetime on one or two groups of organisms and build up large specialized collections, others move from one organism to another, abandoning old favorites. Both approaches generate problems in the preservation of organisms. The specialized collection may become so large and so specialized that it is hard to find a willing successor to the original enthusiastic curator. The worker whose interests are more fickle seldom worries about the systematic aspects, which make the preservation of cultures so desirable to the taxonomist.

Until the 1920's, the main reason for the existence of collections was their value for taxonomic and epidemiological studies. In the 1930's, the burgeoning interest in microbial physiology and biochemistry gave rise to a need for preserving organisms that produced or gave better yields of specific compounds. This greatly increased the value of culture collections.

More recently, studies on bacterial genetics have resulted in the isolation of numerous mutants which have in turn necessitated specialized collections. Some of these mutants are concerned with genetic loci useful in studies of nutrition and of biochemical pathways.

The 1972 Stockholm Conference on the Environment recognized the importance of genetic pools and of collections of microorganisms.

Types and Neotypes

Many microbiologists are not particularly interested in nomenclature and are unaware of or indifferent to the problem they sometimes create. A few years ago I requested cultures of several new species of halophilic bacteria. The author replied that he had not kept cultures and would re-isolate them if he needed them, thus showing a disregard of taxonomic continuity and an unfortunate lack of responsibility. The organisms in question were not too well described and will remain as orphans to clutter up the taxonomic literature and provide headaches for authors of THE MANUAL and similar reference works.

The formal requirements governing types are laid down in the International Code of Nomenclature for Bacteria.

When an author describes a new organism and names it as a new species, the strain described becomes *the type of the species* and the name the nomenclatural type. If the description is based on several strains then the author should designate one of them as the type.

When a single species is assigned to a new genus, that species becomes the *type species of the genus*. If several new species are assigned to the new genus, the author should designate one of them as the type species. From then on the type species carries the specific epithet assigned to it. The organism may be reclassified into another genus but the specific epithet remains unchanged, except as required by grammar.

When only one species is described and named in the original description it automatically becomes the type whether the author so indicates or not. When several species are described in one genus and the author does not indicate the type, one of them can be designated as the type species by a later author.

If none of the original author's cultures is extant, workers who are familiar with the species or group may make a formal recommendation that a strain in existence be designated as a *neotype*. The recommendation becomes effective on publication in the *International Journal of Systematic Bacteriology*, unless objections are later received and sustained by the Judicial Commission. Needless to say, cultures of type and neotype strains should be deposited in one, or preferably several, of the main culture collections so that they are preserved and are available for study.

Type species and type strains are of such taxonomic and nomenclatural importance that an attempt has been made in this edition to identify as many of them as possible by their designation and catalogue number in the main collections; a list of collections mentioned is given in an Appendix. Other lists have been prepared (Sneath and Skerman, 1966; Hendrie *et al.*, 1966).

Because type cultures are designated in accordance with the rules of nomenclature, they are not necessarily the most useful for classification or identification purposes. In other words such formally designated cultures are not necessarily "typical" or characteristic although these aspects should be considered in choosing a neotype. Numerical taxonomists have proposed the concept of calculated hypothetical "mean" or "median" organisms based on the study of frequently occurring positive characteristics and of the similarity values between isolates (Liston *et al.*, 1963; Tsukamura and Mizuno, 1968). These are mathematical abstractions, not real organisms. Another possibility would be to designate as the physiological type an actual strain which is closely similar to the median organism (Liston *et al.*, 1963) or which is closest to the point centroid or "center of gravity" of a cluster of similar strains (Quadling, 1967).

Others prefer to consider the cluster and to designate several strains from the constellation to represent the variation acceptable within the species. Such a "population concept" was suggested by Gordon (1967). A modification may be feasible with the traditional nomenclatural type still being designated as the name bearer (nomenifer) accompanied by several reference strains that have a core of characteristics in common but illustrate the morphological, biochemical, physiological and/or antigenic variability acceptable within the species. This might be termed the type constellation.

Although the correct usage of "type" is given in the International Code, it has come to have a more general meaning. The first Type Culture Collection may have been named with the correct usage in mind; the term now has a more general meaning. While Culture Collections maintain type species, type strains and neotypes, they also keep typical and atypical strains, reference strains, biotypes, serotypes, phage types, etc. They are therefore of great value not only to systematists but to all bacteriologists, and are essential to the development of the subject.

Key
to the
19 Parts

---

aa. Cells not filamentous and ensheathed
  b. Products of binary fission not equivalent
     (have appendages other than flagella and
     pili or reproduce by budding)............. Part 4, p. 148
  bb. Not as above
    c. Cells not rigidly bound
       d. Cells spiral-shaped, have cell wall... Part 5, p. 167
       dd. Cells not spiral-shaped, no cell wall.. Part 19, p. 929
    cc. Cells rigidly bound
       d. Gram negative
          e. Obligate intracellular parasites.. Part 18, p. 882
          ee. Not as above
             f. Curved rods............... Part 6, p. 196
             ff. Not curved rods
                g. Rods
                   h. Aerobic.......... Part 7, p. 217
                   hh. Facultatively an-
                       aerobic.......... Part 8, p. 290
                   hhh. Anaerobic........ Part 9, p. 384
                gg. Cocci or coccobacilli
                   h. Aerobic.......... Part 10, p. 427
                                        Part 7, p. 217
                   hh. Anaerobic........ Part 11, p. 445
       dd. Gram positive
          e. Cocci
             f. Endospores produced........ Part 15, p. 529
             ff. Endospores not produced.... Part 14, p. 478
          ee. Rods or filaments
             f. Endospores produced........ Part 15, p. 529
             ff. Endospores not produced
                g. Straight rods.......... Part 16, p. 576
                                          Part 17, p. 599
                gg. Irregular rods (coryne-
                    form) or tend to form fila-
                    ments or filamentous..... Part 17, p. 599

Important Notes

for

Users of this Edition

1. Always read both generic and species descriptions because characters listed in the generic description are not usually listed in the species descriptions.

2. In tables, characters common to all taxa are not shown but may be listed in footnotes.

3. Generally in tables (exceptions are clearly indicated in footnotes, q.v.) the meanings of symbols are as follows:

+   more than 90 % strains positive
−   more than 90 % strains negative
d   11–89 % strains positive
( )   delayed reaction
w   weak reaction
D   Different reactions in different taxa (species of a genus or genera of a family)
v   strain instability (NOT differences between strains)

KINGDOM

PROCARYOTAE

Murray 1968, 252

R. G. E. MURRAY

Pro.car.y.o'tae. Gr. pref. *pro* before (primordial); Gr. n. *karyon* nut, kernel (nucleus); M.L. fem.pl.n. *Procaryotae* organism with primordial nucleus.

Single cells or simple associations of similar cells (0.2–10 μm in smallest dimension) forming a Kingdom defined by cellular, not organismal, properties. The nucleoplasm (genophore) is never separated from the cytoplasm by a unit-membrane system (nuclear membrane) and is not associated with a basic protein. Cell division is not accompanied by cyclical changes in the texture or staining properties of either nucleoplasm or cytoplasm; a microtubular (spindle) system is not formed. The plasma membrane is frequently complex in topology and forms vesicular, lamellar or tubular intrusions into the cytoplasm; vacuoles and replicating cytoplasmic organelles enclosed by unit membranes are absent. Cytoplasmic organelles independent of the plasma membrane system (chlorobium vesicles, gas vacuoles) are relatively rare, and are enclosed by non-unit membranes. Respiratory and photosynthetic functions are associated with the plasma membrane system in those members possessing these physiological attributes. Ribosomes of the 70 S type are dispersed in the cytoplasm; an endoplasmic reticulum with attached ribosomes is not present. The cytoplasm is immobile; cytoplasmic streaming, pseudo-podial movement, endocytosis and exocytosis are not observed. Nutrients are acquired in molecular form. Enclosure of the cell by a rigid wall is common but not universal.

In organismal terms, these ubiquitous inhabitants of moist environments are predominantly unicellular microorganisms, but filamentous, mycelial or colonial forms also occur. Differentiation is limited in scope (holdfast structures, resting forms and modifications in cell shape). Mechanisms of gene transfer and recombination occur but these processes never involve gametogenesis and zygote formation.

Editorial Note. Acceptance of the Kingdom *Procaryotae* logically calls for an integrated treatment of the blue-green algae and the bacteria. One possibility is outlined in "A Place for Bacteria in the Living World." However, in this edition of THE MANUAL, the Kingdom is divided into two divisions, one for the blue-green algae or cyanobacteria and one for "procaryotic organisms that are not blue-green algae." The Board of Editors is not prepared to subdivide the *Procaryotae* further at this time.

DIVISION I. THE CYANOBACTERIA

R. Y. STANIER

(*Myxophyceae* Wallroth 1833, 4; *Phycochromophyceae* Rabenhorst 1863; *Cyanophyceae* Sachs 1874, 249; *Schizophyceae* Cohn 1879.)

Phototrophic procaryotic organisms that use water as an electron donor and hence produce oxygen in the light.

Cells are always enclosed by a rigid, multilayered wall with an inner peptidoglycan layer. The wall may in turn be surrounded by a gelatinous or fibrous sheath. Most cyanobacteria are motile at some stage of development; motility is always of the gliding type, dependent on surface contact. The cytoplasmic region is traversed by an extensive system of paired photosynthetic lamellae (thylakoids), the outer surfaces bear characteristic granules (phycobilisomes) composed of aggregates of the phycobiliprotein pigments.

Some cyanobacteria are unicellular, others consist of chains of cells (filaments), either simple or branched. Reproduction of unicellular forms may occur by binary fission, by multiple fission or by the serial release of apical cells (exospores) from a sessile individual. Forms that consist of filaments grow by repeated intercalary cell divisions and reproduce either by random fragmentation of the filament or by terminal release of short motile chains of cells (hormogonia). Certain filament-formers can produce specialized cells, known as akinetes and heterocysts. Akinetes are larger than the vegetative cells in the filament and represent a resting stage, which germinates with release of a hormogonium. Heterocysts are non-reproductive cells, distinguishable from the adjoining vegetative cells by the presence of refractile polar granules and of a thick outer wall; they are believed to be physiologically specialized cells that serve as sites of nitrogen fixation.

The characteristic photopigments include chlorophyll *a* as the only chlorophyllous pigment and phycobiliproteins (allophycocyanin, phycocyanin and sometimes phycoerythrin). The cellular absorption spectrum has a peak at approximately 680 nm attributable to chlorophyll *a* and a broad absorption band with one or more peaks, attributable to phycobiliproteins, between 560 and 630 nm.

The Cyanobacteria are not considered further in this edition.

DIVISION II. THE BACTERIA

R. G. E. MURRAY

(*Bacteria* Haeckel 1894, 140.)

Unicellular procaryotic organisms or simple associations of similar cells based upon growth habit, planes of division and cell separation. Cell multiplication involves growth and division, usually binary but occasionally unequal and by budding. Cells may remain attached after division thus causing recognizable arrangements; true branching may occur in certain species. In fluid environments many exhibit swimming motility and possess flagella of bacterial type; others show gliding, twitching, snapping and darting movements on the surface of solid media. Endospores are formed by some species, arthrospores and cysts by some others; heterocysts are not formed.

With few exceptions (*Mollicutes*) the cells are contained within a rigid or semirigid cell wall conferring a constancy of form; with few exceptions the cell wall polymers include peptidoglycans.

Those that perform photosynthesis do so under anaerobic conditions, require an electron donor other than water, do not produce oxygen in the process, utilize one or more of the bacteriochlorophylls (*a, b, c* or *d*) and never contain chlorophyll *a* or phycobiliproteins.

Those that perform chemosynthesis may require either aerobic or anaerobic conditions or be facultative.

PART 1

THE PHOTOTROPHIC BACTERIA

NORBERT PFENNIG and HANS G. TRÜPER*

ORDER I. **RHODOSPIRILLALES** PFENNIG AND TRÜPER 1971, 17

Rho.do.spi.ril.lal'es. M.L. neut.n. *Rhodospirillaceae* type family of order; *-ales* ending to denote order; M.L. fem.pl.n. *Rhodospirillales* the *Rhodospirillaceae* order

The phototrophic (photosynthetic) bacteria represent a physiological community of different, predominantly aquatic bacteria. Cells spherical, rod-, vibrio- or spiral-shaped. Diameter of individual cells from 0.3 to over 6 μm. In most cases multiplication by binary fission, some species of green plants in that it occurs under anaerobic conditions and oxygen is not produced; it depends on the presence of oxidizable external electron donors, such as reduced sulfur compounds, molecular hydrogen or organic compounds. Carbon dioxide is photoassimilated through the reductive

TABLE 1.1

Nomenclature of bacterial chlorophylls

Designation of Jensen *et al.* (1964)	Former designations	Characteristic absorption maxima of living cells
		nm
Bacteriochlorophyll *a*	Bacteriochlorophyll	375, 590, 805, 830–890
Bacteriochlorophyll *b*	Bacteriochlorophyll *b*	400, 605, 835–850, 1020–1040
Bacteriochlorophyll *c*	Chlorobium chlorophyll 660	Long wave length abs. max. 745–755
Bacteriochlorophyll *d*	Chlorobium chlorophyll 650	Long wave length abs. max. 705–740

the *Rhodospirillaceae* have a polar type of cell growth and multiply by budding. Gram-negative. Colors of cell suspensions from purple-violet, purple, red, orange-brown to brown or green. May contain sulfur globules.

Common to all is the presence of bacteriochlorophylls (see Table 1.1) and of carotenoid pigments (see Table 1.2). The photosynthetic metabolism differs from that of the Cyanobacteria and pentose phosphate cycle and further carbon dioxide-incorporating reactions. As far as studied all strains contain cytochromes, ubiquinones and non-heme iron proteins of the ferredoxin type. Fixation of molecular nitrogen has been demonstrated in representatives of all groups.

The photopigments are located in internal membrane systems continuous with the cytoplasmic membrane or in vesicles characteristic of the genus

* The authors are greatly indebted to E. N. Kondrat'eva, C. B. van Niel and R. Y. Stanier for stimulating helpful discussions and correspondence.

TABLE 1.2

Carotenoid groups of phototrophic bacteria[a]

Group	Name	Major components
1	Normal spirilloxanthin series	Lycopene, rhodopin, spirilloxanthin
2	Alternative spirilloxanthin series and ketocarotenoids of spheroidenone type	Spheroidene, hydroxyspheroidene; spheroidenone, hydroxyspheroidenone, spirilloxanthin
3	Okenone series	Okenone
4	Rhodopinal series (former warmingone series)	Lycopenal, lycopenol, rhodopin, rhodopinal, rhodopinol
5	Chlorobactene series	Chlorobactene, hydroxychlorobactene, β-isorenieratene, isorenieratene

[a] Liaaen-Jensen, 1963, 1965; Schmidt *et al.*, 1965; Pfennig, 1967.

Chlorobium which underlie and are attached to the cytoplasmic membrane. DNA base ratios from 45–73 moles % guanine + cytosine (buoyant density). The type and neotype strains were proposed by Pfennig and Trüper (1971).

The present classification is—with the exception of a few species—based on pure culture studies. The phototrophic bacteria comprise the two essentially different groups: purple bacteria and green sulfur bacteria. Molisch's (1907) differentiation of two physiological groups among the purple bacteria, the *Chromatiaceae* (formerly *Thiorhodaceae*) and *Rhodospirillaceae* (formerly *Athiorhodaceae*) is maintained. For each of the three families an addendum has been added of species which are incompletely described, or are hardly recognizable, or are not available in enrichment or pure culture.

Key to the suborders and families of order **Rhodospirillales**

Cells capable of carrying out a photolithotrophic and/or a photoorganotrophic metabolism under anaerobic conditions.
I. Cells contain bacteriochlorophyll *a* or *b* and various carotenoids; photopigments located on internal membrane systems of different morphology, which are continuous with the cytoplasmic membrane.
   Purple bacteria: Suborder *Rhodospirillineae*
   A. Cells photoassimilate simple organic substances; most species unable to grow with sulfide as the sole photosynthetic electron donor. When sulfide or thiosulfate are used as an electron donor, elemental sulfur is not an intermediate oxidation product. When elemental sulfur is formed from sulfide, it is not further oxidized to sulfate. None of the species able to utilize elemental sulfur as an electron donor.
   Purple nonsulfur bacteria:
   Family I. *Rhodospirillaceae*, p. 26
   B. Cells able to grow with sulfide and sulfur as the sole photosynthetic electron donor. In the presence of sulfide, globules of elemental sulfur are formed inside or outside the cells and further oxidized to sulfate.
   Purple sulfur bacteria:
   Family II. *Chromatiaceae*, p. 34
II. Cells contain bacteriochlorophyll *c* or *d* as the major bacteriochlorophyll components and various carotenoids; photopigments located in "chlorobium vesicles" which underlie and are attached to the cytoplasmic membrane.
   A. Green sulfur bacteria: Suborder *Chlorobiineae*
   Family III. *Chlorobiaceae*, p. 51

FAMILY I. **RHODOSPIRILLACEAE** PFENNIG AND TRÜPER 1971, 17

Rho.do.spi.ril.la'ce.ae. M.L. neut.n. *Rhodospirillum* type genus of family; *-aceae* ending to denote family; M.L. fem.pl.n. *Rhodospirillaceae* the *Rhodospirillum* family.

Cells spherical, short or long rod-shaped, vibrio- or spiral-shaped; **multiplication by binary fission or budding.** Cells of two genera **motile** by means of polar flagella; in the third genus, *Rhodomicrobium*, cells peritrichously flagellated. **Internal photosynthetic membrane system continuous with the cytoplasmic membrane** and of vesicular, lamellar or tubular type (Plate 1.2). The known genera **do not contain gas vacuoles.**

Generally microaerophilic, although **many representatives may grow at full atmospheric oxygen tension in the light or dark.** In strains able to grow under microaerophilic or aerobic conditions, the photopigment content and the internal membrane system decrease as the concentration of dissolved oxygen increases; the formation of photopigments and internal membranes becomes derepressed below certain oxygen concentrations. Under strictly anaerobic conditions normal growth is possible only in illuminated cultures; however, very slow multiplication has been observed in the dark (Uffen and Wolfe, 1970). **Phototrophic development is dependent on simple organic substrates** which are photoassimilated or serve as electron donors for carbon dioxide assimilation. Molecular hydrogen can serve as an electron donor in many strains. When sulfide or thiosulfate is used as an electron donor, elemental sulfur is not an intermediate oxidation product. When elemental sulfur is formed from sulfide, it is not further oxidized to sulfate. **None of the species able to utilize elemental sulfur as an electron donor. Most strains require one or more vitamins** as growth factors; some do not require growth factors. The fixation of molecular nitrogen has been demonstrated in a number of strains.

Bacteriochlorophylls *a* or *b* and carotenoids of groups 1, 2 and 4, as well as some carotenoids not yet included in one of the existing series, **occur.** In general, cultures of strains with carotenoids of group 1 appear orange-brown to brownish red or pink to purple-red, of group 2 dirty greenish yellow under anaerobic conditions, brownish red to purple under aerobic conditions.

Storage materials: polysaccharides, poly-β-hydroxybutyrate and polyphosphate. Range of DNA base ratios: 61.2–72.4 moles % guanine + cytosine. Widely distributed in nature.

In the 7th edition of THE MANUAL the genus *Rhodomicrobium* Duchow and Douglas (1949) was included with the nonphototrophic Hyphomicrobiaceae. Following the definitions used here the genus *Rhodomicrobium* is now included with the phototrophic bacteria in *Rhodospirillaceae.*

The type genus: *Rhodospirillum* Molisch 1907, 24, designated by Pfennig and Trüper 1970, 33.

Key to the genera of family **Rhodospirillaceae**

I. Cells spiral-shaped

Genus I. *Rhodospirillum*

II. Cells rod-shaped, ovoid or spherical, not forming filaments

Genus II. *Rhodopseudomonas*, p. 29

III. Cells ovoid to elongated ovoid, forming filaments

Genus III. *Rhodomicrobium*, p. 33

Genus I. **Rhodospirillum** *Molisch 1907, 24**

Rho.do.spi.ril'lum. Gr. n. *rhodum* the rose; M.L. dim.neut.n. *Spirillum* a bacterial genus; M.L. neut.n. *Rhodospirillum* a red *Spirillum.*

Spiral-shaped bacteria, multiplying by binary fission, motile by means of polar flagella. Gram-negative. **Contain bacteriochlorophyll *a* and group 1 carotenoids,** both

---

* van Niel (1944, 86) in his description of the genus *Rhodospirillum* Molisch includes the following statement: "The type species is *Rhodospirillum rubrum* (Esmarch) Molisch. The genus includes *Phaeospirillum* Kluyver and van Niel." The reference is to Kluyver and van Niel (1936, 396). The generic name is proposed and described but no species is named; the statement reads, "The type species to be assigned in the near future." No record of a named species has been found. Under Rule 13 of the International Code the name apparently is not validly published.

pigments being located at internal membrane systems (vesicles, tubes, stacks of lamellae, thylakoids) which originate from and may be continuous with the cytoplasmic membrane (Plate 1.2, Figs. 1 and 2). Do not contain gas vacuoles.

Anaerobic phototrophs; in addition, **most species able to carry out an oxidative metabolism in the dark under microaerophilic to aerobic conditions.**

Capable of photosynthesis in the presence of simple organic carbon compounds which serve two functions: direct photoassimilation and source of reducing power for carbon dioxide fixation. Molecular hydrogen may be used as electron donor. Molecular oxygen is not evolved during photosynthesis. Cell suspensions appear in various shades of red to brown, due to the photosynthetic pigments.

Range of DNA base ratios: 61.7–65.8 moles % guanine + cytosine (buoyant density).

Type species of genus: *Rhodospirillum rubrum* (Esmarch) Molisch 1907, 25.

Key to the species of genus **Rhodospirillum**

I. Cultures deep red to purple red without brownish tinge; characteristic absorption band around 550 nm. Cells 0.7–1.0 μm wide, facultatively microaerophilic to aerobic:

1. *Rhodospirillum rubrum*

II. Cultures orange, reddish brown or purple-violet; characteristic absorption maximum around 530 not 550 nm.

A. Cells 0.3–0.5 μm wide, facultatively aerobic; cultures reddish brown or purple-violet.

2. *Rhodospirillum tenue*

B. Cells 0.5–0.7 μm wide, no growth under aerobic conditions.

3. *Rhodospirillum fulvum*

C. Cells more than 0.7 μm wide, no growth under aerobic conditions.

a. Cells 0.7–1.0 μm wide, 5–10 μm long.

4. *Rhodospirillum molischianum*

b. Cells 1.2–1.5 μm wide, 14–30 μm long.

5. *Rhodospirillum photometricum*

Description of the species of genus **Rhodospirillum**

1. Rhodospirillum rubrum (Esmarch) Molisch 1907, 25. (*Spirillum rubrum* Esmarch 1887, 230; *Rhodospirillum giganteum* Molisch 1907, 24; *Dicrospirillum rubrum* Enderlein 1925, 251; *Rhodospirillum longum* Hama 1933, 135; *Rhodospirillum gracile* Hama 1933, 159.)

rub′rum. L. neut.adj. *rubrum* red

The description is based in part on: van Niel, 1944; Schachman *et al.*, 1952; Liaaen-Jensen *et al.*; 1958; Hickman and Frenkel, 1959; Drews, 1960, Cohen-Bazire and Kunisawa, 1963.

Cells 0.8–1.0 μm wide; vibrio- to spiral-shaped, one complete turn of spiral 1.5–2.5 μm wide and 7–10 μm long. Under unfavorable conditions all kinds of abnormal cell forms and sizes may occur.

Internal photosynthetic membrane system of vesicular type (Plate 1.2, Fig. 1).

Anaerobic liquid cultures at first light pink, later deep purple-red without any brownish tinge; under aerobic conditions cells colorless to light pink.

Photoorganotrophs: facultatively aerobic, growing either anaerobically in the light or microaerophilically to aerobically in the dark.

Growth occurs in mineral media with simple organic substrates and bicarbonate, supplemented with biotin. For optimal development small amounts of yeast extract may be necessary in addition to the simple organic substrate. pH range: 6–8.5; optimum pH: 6.8–7.0. Optimum temperature: 30–35 C.

Organic substrates photoassimilated: fatty acids (above pH 7.0), most intermediates of the tricarboxylic acid cycle, ethanol, alanine, asparagine, aspartate, glutamate, fructose. Molecular hydrogen can serve as an electron donor for growth. No growth: citrate, gluconate, tartrate, sugars other than fructose and sugar-alcohols, thiosulfate, sulfide.

Nitrogen sources: ammonium salts, some amino acids.

Pigments: spirilloxanthin is quantitatively predominant and is responsible for the characteristic absorption maximum of intact cells at 550 nm.

Storage materials: polysaccharides, poly-β-hydroxybutyrate and polyphosphate.

Hydrogenase and catalase present.

DNA base ratios: 63.8–65.8 moles % guanine + cytosine (buoyant density).

Habitat: stagnant water and mud exposed to light.

Illustrations: Trüper, 1968, Fig. 1c.

Neotype: ATCC 11170; NCIB 8255.

2. Rhodospirillum tenue Pfennig 1969, 619.

te′nu.e. L. neut.adj. *tenue* thin, slender.

The description is based in part on: de Boer, 1969; Biebl and Drews, 1969.

Cells 0.3–0.5 μm by 3.0–6.0 μm, some longer. Cells curved in a spiral of one to two complete turns. One complete turn of spiral 0.8–1.0 μm wide and 3 μm long.

In addition to the cytoplasmic membrane only a few finger-like membranous intrusions present similar to those of *Rhodopseudomonas gelatinosa*.

Color of anaerobic liquid cultures brownish red or purple-violet; aerobically grown cells colorless to pale red.

Photolithotrophic (with molecular hydrogen), photoorganotrophic, facultatively aerobic, growing either anaerobically in the light or microaerophilically to aerobically in the dark. Optimum pH: 6.6–7.4. Grows well at 30 C. No growth factors required. Growth rate increased in the presence of complex organic nutrients or yeast extract. Cells tend to form clumps and a sticky sediment.

Organic substrates photoassimilated: acetate, arginine, butyrate, caproate, caprylate, casamino acids, fumarate, lactate, malate, pelargonate, propionate, pyruvate, succinate, valerate.

Not utilized: citrate, ethanol, glycerol, malonate, sugars and sugar-alcohols, thiosulfate.

Growth inhibited by: benzoate, cyclohexane carboxylate, formate, fructose, glycolate, methanol, aspartate and sulfide.

Pigments: absorption spectra of living cell suspensions show the maxima of bacteriochlorophyll *a*-containing organisms (375, 593, 805 and 855–877 nm; carotenoid maxima at 465, 495 and 530 nm); in addition to group 1 carotenoids, purple-violet strains contain group 4 carotenoids.

Hydrogenase and catalase present.

DNA base ratio: 64.8 moles % guanine + cytosine (buoyant density).

Habitat: muddy freshwater ponds and lakes.

Illustration: Pfennig, 1969.

Type strain: "Grünenplan"; SMG 109; ATCC 19137.

3. Rhodospirillum fulvum van Niel 1944, 108.

ful′vum. L. neut.adj. *fulvum* deep or reddish yellow, tawny.

The description is based in part on: Pfennig *et al.*, 1965; Cohen-Bazire and Sistrom, 1966.

Cells 0.5–0.7 μm wide (vibrio- to spiral-shaped). One complete turn of spiral 1.0–1.6 μm wide and 3.5 μm long, some strains 1.0–1.5 μm.

Internal photosynthetic membrane system con-

sisting of several stacks of short lamellae; the stacks are not parallel to the cytoplasmic membrane but form a sharp angle to it.

Anaerobic liquid cultures at first orange-brown, later deep brown.

Photoorganotrophs, growing anaerobically in the light. Unable to adapt to growth under aerobic conditions. In deep agar tube cultures in the dark growth may be microaerophilic, 5–20 mm below the surface (cells fully pigmented). Growth occurs in mineral media with simple organic substrates and bicarbonate, supplemented with *p*-aminobenzoate. For optimal development—particularly of small inocula—the addition of 0.05% ascorbate or thioglycolate as a reductant is necessary.

pH range: 6.0–8.5; optimum pH: 7.3. Growth temperature, 30 C.

Organic substrates photoassimilated: fatty acids up to pelargonate, benzoate, ethanol, fumarate, malate, succinate; glucose and aspartate used by some strains.

Not utilized: glutarate, sugars (except glucose), sugar-alcohols, thiosulfate and sulfide.

Nitrogen sources: ammonium salts.

Pigments: lycopene and rhodopin are major carotenoids, spirilloxanthin absent.

DNA base ratios: 64.3–65.3 moles % guanine + cytosine (type species 65.3) (buoyant density).

Habitat: bodies of stagnant water and mud exposed to light.

Illustrations: Pfennig *et al.*, 1965, Fig. 1.

Neotype strain: "Klein Kalden"; SMG 113; ATCC 15798.

4. Rhodospirillum molischianum Giesberger 1947, 142.

mo.li.schi.an′um. M.L. neut.adj. *molischianum* pertaining to Molisch; named for H. Molisch, an Austrian botanist.

The description is based in part on: Drews, 1960; Giesbrecht and Drews, 1962; Liaaen-Jensen, 1963; Pfennig *et al.*, 1965; Gibbs *et al.*, 1965.

Cells 0.7–1.0 μm wide; vibrio- to spiral-shaped, one complete turn 1.5–2.5 μm wide and 4–6 μm or 7–9 μm long (Plate 1.1, Fig. 1).

Internal photosynthetic membrane system consisting of several stacks of short lamellae; the stacks are not parallel to the cytoplasmic membrane but form a sharp angle to it (Plate 1.2, Fig. 2).

Anaerobic liquid cultures orange-brown to reddish brown.

Photoorganotrophs: growing anaerobically in the light. Unable to adapt to growth under aerobic conditions. In deep agar tube cultures in the dark growth may be microaerophilic, 5–10 mm below the surface (cells fully pigmented).

Growth occurs in mineral media with simple

organic substrates and bicarbonate; yeast extract or vitamin-free casamino acids stimulate growth considerably. For optimal development—particularly of small inocula—the addition of 0.05% ascorbate or thioglycolate as a reductant is necessary.

pH range: 6–8.5; optimum pH: 7.3. Growth temperature: 30 C.

Organic substrates photoassimilated: fatty acids up to pelargonate, ethanol, fumarate, malate, succinate, aspartate.

Not utilized: benzoate, sugars, sugar-alcohols, sulfide, thiosulfate.

Nitrogen sources: ammonium salts, some amino acids.

Pigments: lycopene and rhodopin are major components.

Hydrogenase and catalase present.

DNA base ratios: 61.7–64.8 moles % guanine + cytosine (type strain 61.7) (buoyant density).

Habitat: widely distributed in stagnant water and mud exposed to light.

Type strain: ATCC 14031.

5. Rhodospirillum photometricum Molisch 1907, 24.

pho.to.me'tri.cum. Gr. n. *phos* light; Gr. adj. *metricus* measuring; M.L. neut.adj. *photometricum* light-measuring.

The description is based in part on: Giesberger, 1947; Liaaen-Jensen, 1963.

Cells large 1.2–1.5 μm wide, spirals; one complete turn of spiral 4–6 μm wide and 7–10 μm long, cells 14–30 μm long not uncommon.

Internal photosynthetic membrane system consisting of several stacks of short lamellae; the stacks are not parallel to the cytoplasmic membrane but form a sharp angle to it.

Anaerobic liquid cultures orange-brown to reddish brown.

Photoorganotrophs growing anaerobically in the light. Unable to adapt to growth under aerobic conditions. In deep agar tube cultures in the dark may grow microaerophilically, 5–10 mm below the surface (cells fully pigmented). Capable of development under strictly anaerobic conditions in media containing citrate, ethanol, fatty acids, fructose, hydroxy acids and asparagine as substrates.

Not utilized: glucose, glycerol, sulfide, thiosulfate.

Lycopene and rhodopin are major carotenoids.

DNA base ratio: 65.8 moles % guanine + cytosine (neotype) (buoyant density).

Habitat: stagnant water and mud exposed to light.

Illustrations: Giesberger, 1947, Figs. 6 to 9.

Neotype: NTHC 132.

Genus II. Rhodopseudomonas Kluyver and van Niel in Czurda and Maresch 1937, 119 Nom. cons. proposed by Pfennig and Trüper 1969, 153.

(*Rhodobacillus* Molisch 1907, 14; *Rhodomonas* Kluyver and van Niel 1936, 397.)
Rho.do.pseu.do.mo'nas. Gr. n. *rhodum* the rose; M.L. fem.n. *Pseudomonas* a bacterial genus; M.L. fem.n. *Rhodopseudomonas* a red *Pseudomonas*.

Rod-shaped and ovoid to spherical bacteria, multiplying either by binary fission or asymmetrical division (budding without stalk formation). Motile by means of polar flagella. Gram-negative. **Contain bacteriochlorophyll *a* or *b* and carotenoids,** both types of pigments being located at internal membrane systems of either vesicular, tubular (in the species multiplying by binary fission) or lamellar type (in the species multiplying by budding) (Plate 1.2, Fig. 3). Do not contain gas vacuoles.

Anaerobic phototrophs; in addition, some species able to carry out an oxidative metabolism in the dark under microaerophilic or aerobic conditions.

Capable of photosynthesis in the presence of simple organic carbon compounds which serve two functions: direct photoassimilation and source of reducing power for carbon dioxide fixation. Molecular hydrogen may be used as electron donor. Not able to utilize elemental sulfur as a photosynthetic electron donor. Molecular oxygen is not evolved during photosynthesis. Cell suspensions appear in various shades of yellow-green to brown and red due to the photosynthetic pigments.

Range of DNA base ratios: 62.2–72.4 moles % guanine + cytosine (buoyant density).

Type species of genus: *Rhodopseudomonas palustris* (Molisch) van Niel 1944, 89.

Key to the species of genus **Rhodopseudomonas**

I. Cells clearly rod-shaped in all media; do not become spherical in media below pH 7.

  A. Multiplication by budding: daughter cells originate as buds at the pole opposite the flagellum-bearing end and are either sessile or grow at the tip of a slender opaque tube.

1. Cells rod-shaped to elongate-ovoid, somewhat curved; size of young and short cells 0.6–0.8 by 1.2–2.0 μm. Daughter cells (buds) grow at the tip of a slender tube, giving a dumbbell appearance before division.
   a. Cultures red to dark brown-red. Thiosulfate is used as an electron donor:
      1. *Rhodopseudomonas palustris*
   b. Cultures yellowish green, green or dirty green. Contain bacteriochlorophyll *b*; long wave length absorption maximum of living cells around 1020 nm.
      2. *Rhodopseudomonas viridis*
2. Cells rod-shaped to elongate-ovoid, somewhat curved, 1.0–1.3 μm wide and 2.0–6.0 μm long. Daughter cells (buds) sessile. Grows well in media of pH 5.0–5.5. Cultures orange-brown to purple-red:
      3. *Rhodopseudomonas acidophila*
   B. Multiplication by binary fission; cells slender rods, 0.5 by 1.2 μm, usually clumped together in extensive slime masses. Cultures pale brown to peach. Gelatin liquefied; propionate, mannitol and sorbitol not used:
      4. *Rhodopseudomonas gelatinosa*
II. Cells more or less spherical in media at pH below 7.
   A. In media above pH 7, cells clearly rod-shaped, 1.0 by 1–2.5 μm, frequently in chains with characteristic zigzag arrangement. 0.2% propionate and higher fatty acids (0.05%) used. Molecular hydrogen is usually a good electron donor:
      5. *Rhodopseudomonas capsulata*
   B. In media above pH 7, cells predominantly spherical, 0.7–4 μm in diameter. Mostly singles, little tendency to chain formation. Does not develop in media with 0.2% propionate; mannitol and sorbitol (0.2%) used:
      6. *Rhodopseudomonas sphaeroides*

Description of the species of genus **Rhodopseudomonas**

1. **Rhodopseudomonas palustris** (Molisch) · van Niel 1944, 89. (*Rhodobacillus palustris* Molisch 1907, 14; *Rhodobacterium capsulatum* Molisch 1907, 16; *Rhodovibrio parvus* Molisch 1907, 21; *Rhodomonas palustris* (Molisch) Kluyver and van Niel 1936, 397.)

pa.lus′tris. L. fem.adj. *palustris* marshy, swampy.

The description is based in part on: Whittenbury and McLee, 1967; Liaaen-Jensen, 1963; Tauschel and Drews, 1968; Qadri and Hoare, 1968.

Young individual cells (motile daughter cells) rod-shaped to ovoid, occasionally slightly curved, 0.6–0.9 μm wide and 1.2–2.0 μm long, motile by means of polar or subpolar flagella. Reproduction is by budding; the mother cell produces a slender tube, 1.5–2 times the length of the original cells, at the pole opposite that bearing the flagellum. The end of the tube swells (bud formation) and the daughter cell becomes as large and as opaque as the original cell, producing a dumbbell-shaped organism. Asymmetric division then takes place resulting in an ovoid daughter cell and the ovoid mother cell bearing the tube. The formation of rosettes and clusters in which the individual cells are attached to each other at their flagellated poles is characteristic for the organism in older cultures.

In certain complex organic media, individual cells may become up to 10 μm long, irregular in shape and may form branches.

Photosynthetic membrane system: parallel lamellae underlying and continuous with the cytoplasmic membrane; no lamellae present in the tube (Plate 1.2, Fig. 3).

Anaerobic liquid cultures at first light pink, later red to brownish red; old cultures dark reddish brown. Aerobic cultures colorless to pink.

Photoorganotrophic, facultatively aerobic, growing either anaerobically in the light or aerobically in the dark. Most strains are able to grow on agar plates or slants, although when first isolated, many strains appear sensitive to oxygen (air). Growth occurs in mineral media with simple organic substrates and bicarbonate, supplemented with *p*-aminobenzoate as a growth factor, some strains require biotin in addition. Yeast extract stimulates growth considerably. pH range: 5.5–8.5; fatty acids may prevent growth at a pH below 7.0. Optimum growth range: 30–37 C. Substrates utilized as carbon sources (photoassimilated) or photosynthetic electron donors: alcohols, fatty acids, C_4 dicarboxylic acids, amino acids, benzoate, cyclohexane carboxylate. Formate, molecular hydrogen and thiosulfate are utilized only in the presence of small amounts of yeast extract. Not utilized: monosaccharides and sugar-alcohols, sulfide.

Nitrogen source: ammonium salts.

Pigments: bacteriochlorophyll a and carotenoids of the normal spirilloxanthin series.

Storage materials: polysaccharides, poly-β-hydroxybutyrate.

Hydrogenase, catalase and formic hydrogen lyase present.

DNA base ratios: 64.8–66.3 moles % guanine + cytosine (buoyant density)

Habitat: widely distributed in mud and water exposed to light. Most common nonsulfur purple bacterium.

Illustrations: Whittenbury and McLee, 1967, 324.

Neotype strain: ATCC 17001.

2. Rhodopseudomonas viridis Drews and Giesbrecht 1966, 261.

vi'ri.dis. L. adj. *viridis* green.

The description is based in part on: Eimhjellen *et al.*, 1963; Drews and Giesbrecht, 1966; Giesbrecht and Drews, 1966; Pfennig, 1967; Whittenbury and McLee, 1967.

Morphology, size and fine structure as in *R. palustris*. Anaerobic liquid cultures at first yellowish green, later green to olive green. Aerobic cultures colorless to light yellowish green.

Photoorganotrophic, growing anaerobically in the light but also microaerophilically in the dark. Some strains able to grow aerobically in the dark. Growth occurs in mineral media with simple organic substrates and bicarbonate, supplemented with biotin and p-aminobenzoate; strains with more and with no growth factor requirements have been isolated. pH range: 6.3–8.0; optimum pH: 6.5–7.0. Growth temperature: 25–30 C.

Substrates utilized as carbon sources (photoassimilated) and/or photosynthetic electron donors: acetate, pyruvate, malate and succinate; poor growth in the presence of ethanol, glutamate, peptone, glucose and xylose; higher fatty acids inhibit growth. Not utilized: hydrogen, sulfide, thiosulfate.

Nitrogen sources: ammonium salts; nitrate not utilized.

Pigments: bacteriochlorophyll b; main carotenoids 1,2-dihydro derivatives of neurosporene and lycopene.

Storage materials: poly-β-hydroxybutyrate.

Hydrogenase and catalase present.

DNA base ratios: 66.3–71.4 moles % guanine + cytosine (type strain: 68.4) (buoyant density).

Habitat: mud and stagnant bodies of water exposed to light.

Illustrations: Drews and Giesbrecht, 1966, Fig. 1.

Type strain: "F" (Dreisam/Freiburg); ATCC 19567.

3. Rhodopseudomonas acidophila Pfennig 1969, 601.

a.ci.do'phi.la. L. adj. *acidus* sour; M.L. neut.n. *acidum* an acid; Gr. adj. *philus* loving, friendly; M.L. fem.adj. *acidophila* acid-loving.

Cells rod-shaped to elongate-ovoid slightly curved, 1.0–1.3 μm wide and 2.0–5.0 μm long (Plate 1.1, Fig. 2); daughter cells originate by polar growth as sessile buds at the pole opposite that bearing the flagellum; there is no tube or filament between mother and daughter cell. When the bud reaches the size of the mother cell, cell division is completed by constriction. Both mother and daughter cells form new buds at the poles of the former cell division. Under certain conditions the cells form rosettes and clusters reminiscent of *R. palustris*. In media lacking calcium ions the cells are immotile.

Photosynthetic membrane system: parallel lamellae underlying and possibly continuous with the cytoplasmic membrane. The lamellar membrane system of the daughter cell is newly formed in the growing bud.

Color of anaerobic liquid cultures purple-red to orange-brown. Aerobically grown cells colorless to light pink or orange.

Photolithotrophic with molecular hydrogen; photoorganotrophic, facultatively aerobic, growing either anaerobically in the light or aerobically in the dark. Some strains are microaerophilic. Growth occurs in mineral media with simple organic substrates and bicarbonate; no growth factors required; growth rate not increased in the presence of yeast extract or complex organic nutrients. pH range: 4.8–7.0; optimum pH: 5.8. Optimum temperature: 25–30 C.

Organic substrates photoassimilated: acetate, butyrate, ethanol, fumarate, lactate, malate, propionate, pyruvate, succinate, valerate; photolithotrophic growth with hydrogen and bicarbonate.

Not utilized: benzoate, caprylate, formate, glycerol, pelargonate, sugars and sugar-alcohols, g'utamate and other amino acids, sulfide, thiosulfate.

Nitrogen source: ammonium salts.

Pigments: absorption spectra of living cell suspensions show the maxima of bacteriochlorophyll a-containing organisms (375, 590, 855 nm and a shoulder at about 890 nm; carotenoid maxima at 460, 490 and 525 nm). Carotenoids of the normal spirilloxanthin series are present as well as glucosides of rhodopin and rhodopinal.

Storage material: poly-β-hydroxybutyrate.

Hydrogenase and catalase present.

DNA base ratios: 62.2–66.8 moles % guanine + cytosine (type strain: 65.3) (buoyant density).

Habitat: mud and water exposed to light, particularly acidic habitats and pools in peat bogs.

Type strain: "Crystal Lake"; SMG 137; ATCC 25092.

4. Rhodopseudomonas gelatinosa (Molisch) van Niel 1944, 98. (*Rhodocystis gelatinosa* Molisch 1907, 22.)

ge.la.ti.no′sa. L. part.adj. *gelatus* frozen, stiffened; M.L. n. *gelatinum* gelatin, that which stiffens; M.L. fem.adj. *gelatinosa* gelatinous.

The description is based in part on: Liaaen-Jensen, 1963; Klemme, 1968; Weckesser *et al.*, 1969; de Boer, 1969.

Cells rod-shaped, straight or slightly curved, 0.4–0.5 μm wide, 1–2 μm long, in older cultures up to 15 μm long and irregularly curved. Multiplication by binary fission. Many strains show abundant mucus production in all media which causes the cells to clump together and to appear immotile. In young cultures or when little slime is formed, cells motile by means of polar flagella.

Internal photosynthetic membrane system consisting of few tubular or finger-like intrusions of the cytoplasmic membrane.

Anaerobic liquid cultures pale peach to dirty yellowish brown. Strains which are able to grow aerobically appear colorless to light yellowish brown under aerobic conditions.

Photoorganotrophs, facultatively microaerophilic to aerobic, growth either anaerobic in the light or microaerophilic in the dark; some strains able to grow aerobically in the dark. Growth occurs in mineral media with simple organic substrates and bicarbonate, supplemented with biotin and thiamin; some strains require in addition pantothenate.

pH range: 6.0–8.5. Growth temperature: 30 C.

Organic substrates photoassimilated: most tricarboxylic acid cycle intermediates, sugars and a variety of amino acids, yeast extract and peptone; unique for this species is the liquefaction of gelatine. Molecular hydrogen is used as an electron donor for growth by some strains. Fatty acids utilized only at low concentrations.

Not utilized: benzoate, sulfide, thiosulfate.

Some strains are able to utilize butyrate, citrate, formate, glutarate, mannose and propionate, other strains utilize mannitol and sorbitol.

Nitrogen sources: ammonium salts, some amino acids.

Pigments: bacteriochlorophyll *a* and group 2 carotenoids.

Hydrogenase and catalase present.

DNA base ratios: 70.5–72.4 moles % guanine + cytosine (type strain: 72.4) (buoyant density).

Widely distributed in mud and water.

Illustrations: van Niel, 1944, Figs. 55 to 66.

Neotype: ATCC 17011.

5. Rhodopseudomonas capsulata (Molisch) van Niel 1944, 92. (*Rhodonostoc capsulatum* Molisch 1907, 23; *Rhodopseudomonas capsulatus* (*sic*) (Molisch) van Niel 1944, 92.)

cap.su.la′ta. L. dim.n. *capsula* a small chest, capsule; M.L. fem.adj. *capsulata* capsulated.

The description is based in part on: Liaaen-Jensen, 1963; Klemme, 1968.

Cells spherical, ovoid to rod-shaped, 0.5–1.2 μm wide and 2–2.5 μm long, sometimes up to 6 μm long. Spherical cells occur in media below pH 7.0 often regularly arranged in chains resembling streptococci. Ovoid and rod-shaped cells are characteristic in media above pH 7. Above pH 8.0 irregular filaments are formed and the media become mucoid. Outstandingly characteristic is the zigzag arrangement of the cells in chains. Multiplication by binary fission.

Internal photosynthetic membrane system of vesicular type.

Anaerobic liquid cultures light yellowish brown (some strains with a greenish tinge) later deep brown. When grown in the presence of air, the cultures are dark red to purple-red. Anaerobically grown cells change their color to distinct red when shaken with air for some hours; light enhances this color change.

Photoorganotrophs, facultatively aerobic growing either anaerobically in the light or aerobically in the dark.

Growth occurs in mineral media with simple organic substrates and bicarbonate, supplemented with thiamine; some strains require biotin and nicotinic acid in addition. pH range: 5.5–8.5; optimum pH: 7.0. Optimum temperature: 25–30 C.

Simple organic substrates photoassimilated: fatty acids up to pelargonate, a number of tricarboxylic acid cycle intermediates, fructose, glucose and a number of amino acids. Molecular hydrogen serves as an excellent electron donor for growth.

Not utilized: citrate, ethanol, gluconate, glycerol, mannitol, mannose, sorbitol, tartrate, leucine, sulfide, thiosulfate.

Pigments: bacteriochlorophyll *a* and group 2 carotenoids including spheroidene and hydroxyspheroidene; these two carotenoids are converted to the corresponding ketocarotenoids under aerobic conditions (color change!).

Storage materials: polysaccharides and poly-β-hydroxybutyrate.

Hydrogenase and catalase present.

DNA base ratios: 65.5–66.8 moles % guanine + cytosine (buoyant density).

Habitat: regularly found in mud and stagnant

bodies of water exposed to light, abundant in polluted waters.

Illustrations: van Niel, 1944, Figs. 4 to 6, 27 to 38.

Type strain: ATCC 11166.

6. **Rhodopseudomonas sphaeroides** van Niel 1944, 95. *Nom. cons.* proposed by Trüper and Pfennig 1969, 155. (*Rhodococcus capsulatus* Molisch 1907, 20; *Rhodococcus minor* Molisch 1907, 21; *Rhodosphaera capsulata* (Molisch) Buchanan 1918, 472; *Rhodosphaera minor* (Molisch) Bergey *et al.* 1923, 405; *Rhodorhagus capsulatus* (*sic*) (Molisch) Bergey *et al.* 1925, 415; *Rhodorhagus minor* (*sic*) (Molisch) Bergey *et al.* 1925, 415; *Rhodorrhagus capsulatus* Bergey *et al.* 1939, 905; *Rhodopseudomonas spheroides* (*sic*) van Niel 1944, 95; *Rhodorrhagus spheroides* (*sic*) (van Niel) Brisou 1955, 224.)

sphae.roi'des. Gr. adj. *sphaeroides* spherical.

The description is based in part on: Drews and Giesbrecht, 1963; Liaaen-Jensen, 1963.

Cells spherical from 0.7–4 μm in diameter; in sugar-containing media cells ovoid, 2–2.5 μm wide and 2.5–3.5 μm long. In young cultures motile by means of polar flagella; motility ceases in alkaline media in which copious slime production takes place. Multiplication by binary fission.

Internal photosynthetic membrane system, vesicular.

Anaerobic liquid cultures at first light, dirty greenish brown, later dark brown. Cultures grown in the presence of air are distinctly red. As in the case of *R. capsulata*, the brown color of anaerobic cultures can be changed to red by shaking with air, light stimulating the color change. Most cultures produce a water-soluble porphyrin type, bluish red pigment which diffuses into the culture medium.

Photoorganotrophs, facultatively aerobic, growing either anaerobically in the light or aerobically in the dark. Growth occurs in mineral media with simple organic substrates and bicarbonate and supplemented with thiamine, biotin and nicotinic acid. For optimal development yeast extract is required in addition to the simple organic substrate. pH range: 6.0–8.5; optimum pH 7.0. Optimum temperature: 25–30 C.

Organic substrates photoassimilated: ethanol, fructose, gluconate, glucose, glycerol, mannitol, mannose, sorbitol, tartrate, some intermediates of the tricarboxylic acid cycle and lower fatty acids in low concentrations; higher fatty acids are toxic.

Acids are produced in sugar-containing media under all conditions, but usually disappear later.

Molecular hydrogen can serve as an electron donor for growth.

Not utilized: benzoate, propionate, thiosulfate, sulfide.

Nitrogen sources: ammonium salts, some amino acids.

Pigments: bacteriochlorophyll *a* and group 2 carotenoids including spheroidene and hydroxyspheroidene, which are converted to the corresponding ketocarotenoids under aerobic conditions (color change!).

Hydrogenase and catalase present.

DNA base ratios: 68.4–69.9 moles % guanine + cytosine (buoyant density).

Habitat: mud and stagnant bodies of water exposed to light.

Illustrations: van Niel, 1944, Figs. 7 and 8, 39 to 54.

Type strain: "Ewart"; ATCC 17023.

Genus III. **Rhodomicrobium** *Duchow and Douglas 1949, 415*

Rho.do.mi.cro'bi.um. Gr. n. *rhodum* the rose; Gr. adj. *micrus* small; Gr. n. *bius* life; M.L. neut.n. *microbium* a microbe; M.L. neut.n. *Rodomicrobium* a red microbe.

Ovoid to elongate-ovoid **bacteria, multiplying by budding.** Daughter cells originate as spherical **buds at the end of filaments** from one to several times the length of the mother cell. Mature buds may separate from the filament; they are **motile** by means of **peritrichous flagella.** Gram-negative. **Contain bacteriochlorophyll a and carotenoids,** both pigments being located at **internal membranes of lamellar type.** Do not contain gas vacuoles.

Anaerobic phototrophs; in addition, most strains able to grow and to carry out an oxidative metabolism in the dark under microaerophilic to aerobic conditions.

Capable of photosynthesis in the presence of simple organic carbon compounds which serve two functions: direct photoassimilation and source of reducing power for carbon dioxide fixation. Molecular hydrogen may be used as electron donor. Molecular oxygen is not evolved during photosynthesis. Cell suspensions appear in various shades of pink to reddish brown due to the photosynthetic pigments.

Range of DNA base ratios: 61.8–63.8 moles % guanine + cytosine (buoyant density).

Type species of genus: *Rhodomicrobium vannielii* Duchow and Douglas 1949, 415.

Description of the species of genus **Rhodomicrobium**

1. Rhodomicrobium vannielii Duchow and Douglas 1949, 415.

van.niel'i.i. M.L. gen.n. *vannielii* of van Niel; named for C. B. van Niel, an American microbiologist.

The description is based in part on: Murray and Douglas, 1950; Volk and Pennington, 1950; Boatman and Douglas, 1961; Conti and Hirsch, 1965; Liaaen-Jensen, 1963; Gorlenko, 1969.

Mature cells ovoid to lemon-shaped, 1–1.2 μm wide and 2–2.8 μm long; in some strains cells up to 4 μm long. Young cells spherical, originating as buds at the end of filaments which are approximately 0.3 μm wide and one to several times as long as the mother cell (Plate 1.1, Fig. 3). Depending on the conditions, mature buds may remain attached to the filament and form another filament at the opposite pole, or separate from the filament as a motile, peritrichously flagellated swarm cell. A mature cell may produce as many as three daughter cells: one by formation of a primary filament from the pole of the cell and one or two more by lateral outgrowth of new filaments from the primary filament upon which the first daughter cell is borne. Because of the tendency of the cells to remain attached to the filament under certain conditions (e.g. high pH), the predominant growth habit is that of an aggregate containing many cells. In addition to the normal cells, several strains form characteristic polyhedral cells, 1–1.5 μm in diameter with 3–5 rounded facets; 1–4 of these cells occur as buds at a common branching point at the end of a filament. Their heat resistance is higher than that of the usual vegetative cells.

Photosynthetic membrane system: parallel lamellae underlying and continuous with the cytoplasmic membrane; no lamellae present in the filaments.

In agar shake tubes, colonies are dark orange-red to brown, irregular and have a rough, convoluted surface. Anaerobic liquid cultures at first turbid, becoming granular and flocculent; color from salmon pink to deep orange-red or orange-brown. Aerobically grown cells colorless to orange-brown.

Photoorganotrophs, facultatively microaerophilic to aerobic, some strains able to grow as anaerobes in the light or are microaerophilic to aerobic in the dark. Photolithotrophic growth with molecular hydrogen. Growth occurs in mineral media containing simple organic substrates and bicarbonate; organic growth factors not required. pH range: 5.2–7.5; optimum pH: 6.0. Optimum growth temperature: 30 C.

Organic substrates photoassimilated: acetate, butanol, butyrate, caproate, ethanol, fumarate, lactate, malate, propanol, propionate, succinate and valerate. Not utilized: citrate, formate, fructose, glucose, mannitol, mannose, tartrate, sulfide and thiosulfate.

Pigments: bacteriochlorophyll *a* and group 1 carotenoids plus β-carotene.

DNA base composition: 61.8–63.8 moles % guanine + cytosine (type strain: 62.8) (buoyant density).

Habitat: commonly found in mud, pond, lake and stream waters. Also isolated from brackish water and sea water habitats.

Type strain: ATCC 17100.

ADDENDUM TO THE FAMILY **RHODOSPIRILLACEAE**

Species that have been grown in pure culture but were lost before comparative taxonomic studies could be made.

a. *Vannielia aggregata* Pringsheim 1955, 288.

Rod-shaped pink phototrophic bacteria, 1.0 μm wide, 4.0–6.0 μm long, motile by means of polar flagella. Gram-negative. Form round compact radial colonies in natural media, but seldom form colonies in peptone yeast extract media.

Source: putrefying mud from Coe Fen pasture land near Cambridge, England.

Illustration: Pringsheim, 1955.

Comment: This is probably another species of *Rhodopseudomonas*.

FAMILY II. **CHROMATIACEAE** BAVENDAMM 1924, 125
Nom. cons. proposed by Pfennig and Trüper 1971, 16.

Chro.ma.ti.a'ce.ae. M.L. neut.n. *Chromatium* type genus of family; *-aceae* ending to denote a family; M.L. fem.pl.n. *Chromatiaceae* the *Chromatium* family.

Individual cells spherical, ovoid, rod-, vibrio- or spiral-shaped; motile or non-motile; with or without gas vacuoles.

Motile forms have polar flagella; in the largest forms, the tuft of polar flagella is visible under the light microscope. The very slow movements of groups of cells and jerky movements of individual cells seen in gas vacuole-containing forms and described by Winogradsky (1888) for *Amoebobacter* arise as artefacts of the microscopic observation technique and have no connection with the movements of gliding bacteria. **Cells contain bacteriochlorophylls a or b, carotenoids of groups 1, 3 and 4 or tetrahydrospirilloxanthin.** In general, cultures of strains with carotenoids of group 1 appear orange-brown to brownish red or pink, of group 3 purple-red and of group 4 purple-violet. **Internal photosynthetic membrane system continuous with the cytoplasmic membrane** and vesicular (in most species), tubular or lamellar (Plate 1.3, Figs. 1 and 2).

Most species are strictly anaerobic (only one species, *Thiocapsa roseopersicina*, is able to grow in the dark under microaerophilic conditions) and capable of **photolithotrophic carbon dioxide assimilation in the presence of elemental sulfur and sulfide. Elemental sulfur accumulates intermediarily as globules inside the cells or in one genus** (*Ectothiorhodospira*) **outside the cells.** Sulfate is the ultimate oxidation product of sulfur compounds. Many strains are able to use molecular hydrogen as an electron donor. **All strains photoassimilate** a number of **simple organic substrates** of which **acetate** and **pyruvate** are most widely used. Strains lacking assimilatory sulfate reduction utilize the organic substrates only in the presence of sulfide or other reduced sulfur compounds as a source of cell sulfur; the majority of strains is able to utilize organic substrates in the absence of hydrogen sulfide or sulfur; thus **all forms are potentially mixotrophic.** The fixation of molecular nitrogen has been demonstrated in a number of strains. Storage materials are polysaccharides, poly-β-hydroxybutyrate and polyphosphate.

Vitamin B_{12} is required by the larger forms.

Range of DNA base ratios: 45–70.4 moles % guanine + cytosine (buoyant density).

In nature, the *Chromatiaceae* occur in the anaerobic and sulfide-containing parts of all kinds of aquatic environments from moist and muddy soils to ditches, ponds, lakes, rivers, sulfur springs, salt lakes, estuaries and marine habitats.

Two physiological-ecological subgroups can be differentiated on the basis of the presence or absence of gas vacuoles and the associated selective advantage in enrichment cultures. Forms without gas vacuoles have a selective advantage in cultures with medium to high sulfide concentrations and light intensities. Forms containing gas vacuoles require very low sulfide concentrations, low light intensities and low incubation temperatures (10–20 C). The cells of the gas vacuole-containing forms rise to the top of the culture bottles at lower temperatures (4–10 C).

Depending on the culture conditions or environmental conditions in nature, all forms are able to develop either in the form of single cells or in non-motile cell aggregates or families of variable size and shape embedded in slime. The different modes of growth described by Winogradsky (1888) may be obtained experimentally by varying sulfide concentration, light intensity, pH, salinity, temperature and oxygen tension. At high sulfide concentration and high light intensity, all flagellated forms become non-motile as the result of slime formation embedding groups of cells. For a given light intensity the various genera and species differ from each other with respect to the sulfide concentration at which the cells develop functional flagella and become motile; for a given sulfide concentration they differ with respect to the light intensity and mode of illumination which allow the cells to become motile. The lower the light intensity the higher will be the sulfide concentration at which a given organism can persist in the motile stage. These characteristics must be considered in the classification of a particular organism.

For these reasons the classification of the *Chromaticeae* (*Thiorhodaceae*) used in previous editions of The Manual has not been followed. The present classification is based primarily on the characteristics of single cells (Pfennig, 1967).

The type genus: *Chromatium* Perty 1852, 174.

Key to the genera of family **Chromatiaceae**

I. Sulfur globules stored inside the cells when grown with sulfide as electron donor.
   A. Cells do not contain gas vacuoles
      1. Cells motile by means of polar flagella
         a. Cells ovoid to rod-shaped:
                    Genus I. *Chromatium*, p. 36
         b. Cells spherical; typically diplococcus-shaped before cell division:
                    Genus II. *Thiocystis*, p. 39
         c. Cells spherical to ovoid, grouped as regular sarcina packets:
                    Genus III. *Thiosarcina*, p. 40
         d. Cells clearly spiral-shaped:
                    Genus IV. *Thiospirillum*, p. 41

2. Cells non-motile
    a. Cells spherical; typically diplococcus-shaped before cell division; individual cells often surrounded by a more or less thick slime layer:
Genus V. Thiocapsa, p. 42

B. Cells contain gas vacuoles
  1. Cells motile by means of polar flagella
    a. Cells spherical; typically diplococcus-shaped before cell division:
Genus VI. Lamprocystis, p. 43

  2. Cells non-motile
    a. Cells rod-shaped:
Genus VII. Thiodictyon, p. 44

    b. Cells spherical to ovoid, characteristically arranged in regular platelets (flat sheets):
Genus VIII. Thiopedia, p. 45

    c. Cells spherical:
Genus IX. Amoebobacter, p. 46

II. Sulfur globules appearing outside the cells when grown with sulfide as electron donor.
Genus X. Ectothiorhodospira, p. 47

Genus I. Chromatium <i>Perty 1852, 174</i>

(In part, *Monas* Müller 1786, 1; includes *Rhabdomonas* Cohn 1875, 167; *Rhabdochromatium* Winogradsky 1888, 100.)
Chro.ma′ti.um. Gr. n. *chromatium* color, paint.

Ovoid, bean- or rod-shaped bacteria, multiplying by **binary fission, motile by means of polar flagella.** Under certain unfavorable conditions and in many natural environments cells **may be immotile and grow in** more or less regular **aggregates surrounded by slime.** Unfavorable mineral salt concentrations cause pronounced swellings, club- and spindle-shaped forms formerly recognized as a separate genus, *Rhabdomonas* Cohn 1875, 167 (synonym *Rhabdochromatium* Winogradsky 1888, 100). **Gram-negative. Contain bacteriochlorophyll a and carotenoids,** both pigments being located at **internal membranes of vesicular type. Do not contain gas vacuoles.**
  Anaerobic.

Capable of photosynthesis in the presence of hydrogen sulfide, during which they produce and **store,** as an intermediate oxidation product, **elemental sulfur in the form of globules inside the cells.** Molecular hydrogen may be used as electron donor. Molecular oxygen is not evolved during photosynthesis. Cell suspensions appear in various shades of orange-brown and pink to violet and purple-red, due to the photosynthetic pigments.
  Storage materials: polysaccharides, poly-β-hydroxybutyrate, polyphosphates.
  Range of DNA base ratios: 48.0–70.4 moles % guanine + cytosine (buoyant density).
  Type species of genus: *Chromatium okenii* (Ehrenberg) Perty 1852, 174.

Key to the species of genus **Chromatium**

I. Cells large, more than 2.5 μm wide; vitamin B_{12} usually required as a growth factor.
  A. Culture purple-red, major carotenoid okenone
    1. Cells 4.5–6 μm wide:
1. Chromatium okenii

    2. Cells 3.5–4.5 μm wide:
2. Chromatium weissei

  B. Culture purple-violet, major carotenoid rhodopinal
    1. Cells 3.5–4.0 μm wide; sulfur globules predominantly located at the two poles of the cell:
3. Chromatium warmingii

    2. Cells 3.0–4.5 μm wide; sulfur globules evenly distributed within the cell; 1–3% NaCl required, otherwise cells pleomorphic:
4. Chromatium buderi

II. Cells small, less than 2.5 μm wide. Vitamin B$_{12}$ usually not required as a growth factor.
  A. Culture purple-red, major carotenoid okenone
                              5. *Chromatium minus*
  B. Culture purple-violet, major carotenoid rhodopinal
                              6. *Chromatium violascens*
  C. Culture orange-brown to brownish red; carotenoids of the normal spirilloxanthin series
    1. Cells ovoid, about 2 μm wide; DNA base ratio 61–66 moles % guanine + cytosine:
                              7. *Chromatium vinosum*
    2. Cells slender, rod-shaped, 1.2–1.5 μm wide; DNA base ratio around 70 moles % guanine + cytosine:
                              8. *Chromatium gracile*
    3. Cells almost spherical to ovoid, about 1.2 μm wide; DNA base ratio around 64 moles % guanine + cytosine:
                              9. *Chromatium minutissimum*

Description of the species of genus **Chromatium**

1. Chromatium okenii (Ehrenberg) Perty 1852, 174. (*Monas okenii* Ehrenberg 1838, 15; *Rhabdomonas rosea* Cohn 1875, 167; *Spirillum violaceum* Warming 1875, 325; *Bacterium okenii* (Ehrenberg) Trevisan 1879, 140; *Beggiatoa roseopersicina* Zopf 1883, 79; *Rhabdochromatium roseum* (Cohn) Winogradsky 1888, 100; *Rhabdochromatium fusiforme* Winogradsky 1888, 102; *Bacillus okenii* (Ehrenberg) Trevisan 1889, 18; *Mantegazzea winogradskyi* Trevisan 1889, 12; *Mantegazzea rosea* (Cohn) Trevisan 1889, 12; *Pseudomonas okenii* (Ehrenberg) Migula 1895, 30; *Thiospirillum violaceum* (Warming) Migula 1900, 1050; *Chromatium densegranulatum* Skuja 1948, 18.)

o.ken′i.i. M.L. gen.n. *okenii* of Oken, named for L. Oken, a German naturalist.

The description is based in part on: Schlegel and Pfennig, 1961; Schlegel, 1962; Kran *et al.*, 1963; Cohen-Bazire, 1963; Schmidt *et al.*, 1963; Pfennig and Lippert, 1966; Thiele, 1968; Aasen and Liaaen-Jensen, 1967.

Cells 4.5–6.0 μm wide, 8–15 μm long, occasionally longer. Flagellar tuft usually 1.5–2 times the cell length, visible in the light microscope. Globules of elemental sulfur appear evenly distributed within the cell. Color of individual cells and of cell suspensions purple red.

Obligately phototrophic. Strictly anaerobic. Vitamin B$_{12}$ required. pH range: 6.5–7.6. Optimal growth temperature: 25–30 C.

Photosynthetic electron donors: sulfide, sulfur. In the presence of sulfide and bicarbonate, acetate and pyruvate are photoassimilated. Assimilatory sulfate reduction lacking. Not utilized: thiosulfate, thioglycolate, molecular hydrogen, sugars, sugar-alcohols, alcohols, higher fatty acids, amino acids, benzoate, formate and most intermediates of the tricarboxylic acid cycle.

Nitrogen sources: ammonium salts, urea.
Pigments: bacteriochlorophyll *a*, carotenoid: okenone.
Hydrogenase and catalase absent.
DNA base ratios: 48–50 moles % guanine + cytosine (type strain: 48.0) (buoyant density).
Habitat: stagnant fresh water containing hydrogen sulfide and exposed to light.
Illustrations: Pfennig and Lippert, 1966, Fig. 5.
Neotype strain: "Ostrau"; SMG 169.

2. Chromatium weissei (Perty) van Niel 1948, 857. (*Chromatium weissii* (*sic*) Perty 1852, 174; *Bacillus weissii* (*sic*) (Perty) Trevisan 1889, 18; *Bacterium weissii* (*sic*) (Perty) Trevisan in de Toni and Trevisan 1889, 1027; *Chromatium weisii* (*sic*) (Perty) Bergey *et al.* 1923, 401; *Chromatium obovatum* Skuja 1948, 17.)

weis′ se.i. M.L. gen.n. *weissei* of Weisse; named for J. F. Weisse, a German zoologist.

The description is based in part on: Schmidt *et al.*, 1965; Pfennig and Lippert, 1966; Thiele, 1968.

Description as for *C. okenii*, except: cells 3.5–4.5 μm wide, 7–9 μm long.

DNA base ratios: 48–50 moles % guanine + cytosine (type strain: 48.0) (buoyant density).
Illustrations: Winogradsky, 1888, Plate IV, Figs. 1 and 2;
Neotype strain: "Göttingen"; SMG 171.

3. Chromatium warmingii (Cohn) Migula 1900 1048. (*Monas warmingii* Cohn 1875, 167; *Chromatium cuculliferum* Gicklhorn 1920, 419; *Monas cucullifera* (Gicklhorn) de Rossi 1927, 887.)

war.min′gi.i. M.L. gen.n. *warmingii* of Warming; named for E. Warming, a Danish botanist.

The description is based in part on: Schlegel and Pfennig, 1961; Liaaen-Jensen and Schmidt,

1963; Schmidt *et al.*, 1965; Pfennig and Lippert, 1966; Thiele, 1968.

Cells 3.5–4.0 μm wide, 5–11 μm long, occasionally longer. Flagellar tuft usually 1.5–2 times the cell length, visible in the light microscope. Globules of elemental sulfur predominantly located at the two poles of the cell; dividing cells become apparent by additional sulfur globules located parallel to the central division plane. Color of individual cells grayish to very slightly pink, color of cell suspensions purple-violet.

Obligately phototrophic. Strictly anerobic. Vitamin B_{12} required. pH range: 6.5–7.6. Optimum growth temperature: 25–30 C.

Photosynthetic electron donors: sulfide, sulfur. In the presence of sulfide and bicarbonate, acetate and pyruvate are photoassimilated. Assimilatory sulfate reduction lacking. Not utilized: thiosulfate, thioglycolate, sugars, sugar-alcohols, alcohols, higher fatty acids, amino acids, benzoate, formate and most intermediates of the tricarboxylic acid cycle.

Nitrogen sources: ammonium salts, urea.

Pigments: bacteriochlorophyll *a*, carotenoids of the rhodopinal series.

Hydrogenase and catalase present.

DNA base ratios: 55.1–60.2 moles % guanine + cytosine (type strain: 55.1) (buoyant density).

Habitat: stagnant (predominantly fresh) water containing hydrogen sulfide and exposed to light.

Illustrations: Schlegel and Pfennig, 1961, Fig. 3.

Neotype strain: "Melbourne"; SMG 173; ATCC 14959.

4. Chromatium buderi Trüper and Jannasch 1968, 364.

bu'der.i. M.L. gen.n. *buderi* of Buder; named for J. Buder, a German plant physiologist.

Cells 3.5–4.5 μm wide, 4.5–9.0 μm long during exponential growth, about 3–4 μm during the stationary phase. Flagellar tuft usually 1.5–2 times the cell length, may be visible in the light microscope. Globules of elemental sulfur appear evenly distributed within the cell. Color of individual cells grayish yellow, color of cell suspensions purple-violet.

Obligately phototrophic, strictly anaerobic. Vitamin B_{12} required. Salinity of 1–3% w/v, NaCl required, otherwise extremely pleomorphic. pH range: 6.5–7.6. Optimum growth temperature: 25–30 C.

Photosynthetic electron donors: sulfide, sulfur. In the presence of sulfide and bicarbonate, acetate and pyruvate are photoassimilated. Incapable of assimilatory sulfate reduction. Not utilized: thiosulfate, thioglycolate, molecular hydrogen, sugars, sugar-alcohols, alcohols, higher fatty acids, amino acids, benzoate, formate and most intermediates of the tricarboxylic acid cycle.

Nitrogen sources: ammonium salts.

Pigments: bacteriochlorophyll *a* and group 4 carotenoids.

Hydrogenase and catalase absent.

DNA base ratios: 62.2–62.8 moles % guanine + cytosine (type strain: 62.2) (buoyant density).

Habitat: estuarine salt flats and salt marshes.

Illustration: Trüper and Jannasch, 1968, Figs. 1 and 3.

Type strain: "Santa Cruz/Galapagos"; SMG 176; ATCC 25588.

5. Chromatium minus Winogradsky 1888, 99. (*Bacillus minor* (Winogradsky) Trevisan 1889, 18; *Bacterium minus* Trevisan in de Toni and Trevisan 1889, 1027.)

mi'nus. L. comp.adj. *minor* (neut. *minus*) less, smaller.

The description is based in part on: Thiele, 1968.

Cells about 2 μm wide, 2.5–6 μm long. Globules of elemental sulfur appear evenly distributed within the cell. Single cells colorless to pinkish, cell suspensions purple-red.

Obligately phototrophic, strictly anaerobic. pH range: 6.5–7.6. Optimum growth temperature: 25–30 C.

Photosynthetic electron donors: sulfide, sulfur, thiosulfate. In the presence of sulfide and bicarbonate, acetate, pyruvate and glucose are photoassimilated. Incapable of assimilatory sulfate reduction. Not utilized: thioglycolate, sugar-alcohols, alcohols, higher fatty acids, amino acids, benzoate, formate and most intermediates of the tricarboxylic acid cycle.

Nitrogen sources: ammonium salts.

Pigments: bacteriochlorophyll *a* and carotenoid okenone.

Hydrogenase and catalase present.

DNA base ratio: 52.0–62.2 moles % guanine + cytosine (type strain: 52.0) (buoyant density).

Habitat: stagnant fresh water containing hydrogen sulfide and exposed to light.

Illustrations: Winogradsky, 1888, Plate IV, Fig. 5.

Neotype strain: "Reyershausen"; SMG 178.

6. Chromatium violascens Perty 1852, 174. (*Chromatium violaceum* Bergey *et al.* 1923, 401.)

vi.o.las'cens. L. part. *violascens* becoming violet.

The description is based in part on: Trüper and Genovese, 1968; Trüper and Jannasch, 1968.

Description same as for *C. vinosum* except:

Color of cell suspensions: purple-violet.

Pigments: bacteriochlorophyll *a* and group 4 carotenoids.

DNA base ratios: 61.8–64.3 moles % guanine + cytosine (type strain 62.2) (buoyant density).

Illustrations: Trüper and Genovese, 1968, Fig. 2B.

Neotype strain: "Carmel River"; SMG 198; ATCC 17096.

7. Chromatium vinosum (Ehrenberg) Winogradsky 1888, 99. (*Monas vinosa* Ehrenberg 1838, 11; *Bacterium sulfuratum* Warming 1875, 381; *Bacillus vinosus* (Ehrenberg) Trevisan 1889, 18; *Bacterium vinosum* (Ehrenberg) de Toni and Trevisan 1889, 1027; *Chromatium sulfuratum* (Warming) Isachenko 1927, 114.)

vi.no'sum. L. neut.adj. *vinosum* full of wine.

The description is based in part on: Hurlbert and Lascelles, 1963; Thiele, 1968.

Cells about 2 μm wide, 2.5–6 μm long, occasionally longer. Globules of elemental sulfur appear evenly distributed within the cell (Plate 1.1, Fig. 4). Single cells colorless. Color of cell suspensions at first yellowish to orange-brown, later brownish red.

Obligately phototrophic, anaerobic, some strains aerotolerant.

pH range: 6.5–7.6. Optimum growth temperature: 25–30 C.

Photosynthetic electron donors: sulfide, thiosulfate, sulfur, sulfite, molecular hydrogen. Acetate, pyruvate and one or more intermediates of the tricarboxylic acid cycle are photoassimilated. Some strains are able to utilize fructose, glucose, lactate, propionate, butyrate. Capable of assimilatory sulfate reduction.

Not utilized: thioglycolate, sugar-alcohols, alcohols, benzoate, citrate, amino acids.

Pigments: bacteriochlorophyll *a* and group 1 carotenoids.

Hydrogenase and catalase present.

DNA base ratios: 61.3–66.3 moles % guanine + cytosine (type strain 64.3) (buoyant density).

Habitat: stagnant fresh or sea water, sewage lagoons, estuaries, salt marshes—the most common species of the *Chromatiaceae*.

Neotype strain: "D" (Roelofsen, 1934, 660); ATCC 17899.

8. Chromatium gracile Strzeszewski 1913, 321.

gra'ci.le. L. neut.adj. *gracile* thin, slender.

Description same as for *C. vinosum* except:

Cells slender, 1–1.3 μm wide, 2–6 μm long.

DNA base ratios: 68.9–70.4 moles % guanine + cytosine (type strain 69.9) (buoyant density).

Illustrations: Strzeszewski, 1913, Plate XXXIX, Figs. 1 and 2.

Neotype strain: "Hadley Harbor"; SMG 203.

9. Chromatium minutissimum Winogradsky 1888, 100. (*Bacillus minutissimus* (Winogradsky) Trevisan 1889, 18; *Bacterium minutissimum* (Winogradsky) Trevisan in de Toni and Trevisan 1889, 1028.)

mi.nu.tis'si.mum. L. neut.sup.adj. *minutissimum* very small, smallest.

The description is based in part on: Moshentseva and Kondrat'eva, 1962; Uspenskaya and Kondrat'eva, 1962; Kondrat'eva and Malofeeva, 1964; Vanyushin *et al.*, 1966.

Description same as for *C. vinosum* except:

Cells 1–1.2 μm wide, 2 μm long.

pH optimum: 7–8.

DNA base ratio: 63.7 moles % guanine + cytosine (type strain) (chemical analysis).

Illustrations: Winogradsky, 1888, Plate IV, Fig. 8.

Neotype strain: "Glubokoe" (Microbiology Department, Moscow State University, Moscow, USSR).

Genus II. **Thiocystis** *Winogradsky 1888, 60*

(Includes *Thiothece* Winogradsky 1888, 82; *Thiosphaera* Miyoshi 1897, 170.)

Thi.o.cys'tis. Gr. n. *thium* sulfur; Gr. n. *cystis* the bladder, a bag; M.L. fem.n. *Thiocystis* sulfur bag.

Spherical to ovoid bacteria, before cell division typically diplococcus-shaped, multiplying by **binary fission, motile** by means of **polar flagella.** Under certain unfavorable conditions and in many natural environments, **cells may be immotile and grow in more or less regular aggregates surrounded by slime. Gram-negative. Contain bacteriochlorophyll a and carotenoids,** both pigments being located at **internal membranes of vesicular type. Do not contain gas vacuoles.**

Anaerobic.

Capable of photosynthesis in the presence of hydrogen sulfide, during which they produce and store, as an intermediate oxidation product, **elemental sulfur** in the form of **globules inside the cells.** Molecular hydrogen may be used as electron donor. Molecular oxygen is not evolved during photosynthesis. Cell suspensions appear in various shades of brown to purple-red and violet due to the photosynthetic pigments.

Storage materials: polysaccharides, poly-β-hydroxybutyrate, polyphosphates.

Catalase and hydrogenase present.

Range of DNA base ratios: 61.3–67.9 moles % guanine + cytosine (buoyant density).

Type species of genus: *Thiocystis violacea* Winogradsky 1888, 65.

Key to the species of genus **Thiocystis**

I. Sulfur globules appear evenly distributed throughout the cell; cells contain carotenoids of the rhodopinal series; culture purple-violet:

1. *Thiocystis violacea*

II. Sulfur globules occur only in the peripheral part of the cell; cells contain carotenoids of the okenone series; culture purple-red:

2. *Thiocystis gelatinosa*

Description of the species of genus **Thiocystis**

1. **Thiocystis violacea** Winogradsky 1888, 65. (*Planosarcina violacea* (Winogradsky) Migula 1895, 20.)

vi.o.la'ce.a. L. fem.adj. *violacea* violet-colored.

The description is based in part on: Thiele, 1968; Trüper and Jannasch, 1968.

Cells about 2.5–3 µm in diameter; larger individual cells may occur due to environmental influences. Under unfavorable conditions irregular aggregates of cells surrounded by slime capsules are formed. Sulfur globules appear evenly distributed throughout the cell.

Obligately phototrophic. Anaerobic.

pH-range: 4.3–8; optimum pH: 7–7.5. Growth temperature: 25–30 C.

Photosynthetic electron donors utilized: sulfide, sulfur, (sulfite, thiosulfate used by some strains). In the presence of sulfide and bicarbonate, acetate and pyruvate are photoassimilated; some strains utilize in addition certain intermediates of the tricarboxylic acid cycle, fatty acids and fructose or glucose. Some strains capable of assimilatory sulfate reduction.

Nitrogen sources: ammonium salts, urea, nitrate.

Pigments: bacteriochlorophyll *a* and group 4 carotenoids.

DNA base ratios: 62.8–67.9 moles % guanine + cytosine (type strain: 63.1) (buoyant density).

Habitat: mud and stagnant fresh and salt water containing hydrogen sulfide and exposed to light; sulfur springs.

Illustrations: Winogradsky, 1888, 65, Plate II, Figs. 1 to 7.

Neotype strain: "Grünenplan"; SMG 207.

2. **Thiocystis gelatinosa** (Winogradsky) Pfen-nig and Trüper 1971, 11. (*Thiothece gelatinosa* Winogradsky 1888, 82; *Thiosphaera gelatinosa* (Winogradsky) Miyoshi 1897, 170; *Lamprocystis gelatinosa* (Winogradsky) Migula 1900, 1044; *Chromatium sphaeroides* Hama 1933, 158.)

ge.la.ti.no'sa. L. part.adj. *gelatus* frozen, stiffened; M.L. n. *gelatinum* that which stiffens; M.L. fem.adj. *gelatinosa* gelatinous.

The description is based in part on: Pfennig *et al.*, 1968.

Cells spherical, about 3 µm in diameter; under unfavorable conditions, elongated ovoid cells may occur, due to delayed cell division. At high light intensity and sulfide concentration and unfavorable conditions cells immotile, growing in irregular aggregates surrounded by slime. Globules of elemental sulfur occur only in the peripheral part of the cytoplasm. Color of individual cells slightly pink, color of cell suspensions purple-red.

Obligately phototrophic, strictly anaerobic.

pH range: 6.5–7.6. Optimum growth temperature: 25–30 C. A salinity of 1% NaCl required by the neotype strain.

Photosynthetic electron donors: sulfide, sulfur. In the presence of sulfide and bicarbonate, acetate and pyruvate are photoassimilated.

Not utilized: thioglycolate, thiosulfate.

Nitrogen sources: ammonium salts.

Pigments: bacteriochlorophyll *a* and carotenoids of the okenone series.

DNA base ratio: 61.3 moles % guanine + cytosine (type strain) buoyant density.

Habitat: stagnant water containing hydrogen sulfide and exposed to light; hypolimnion of meromictic lakes.

Illustrations: Pfennig *et al.*, 1968, Fig. 1.

Neotype strain: "Langvik"; SMG 215.

Genus III. **Thiosarcina** *Winogradsky 1888, 104*

(*Rhodosarcina* Orla-Jensen 1909, 334; *Rhodothiosarcina* Ellis 1932, 163.)

Thi.o.sar.ci'na. Gr. n. *thium* sulfur; M.L. fem.n. *Sarcina* a bacterial genus; M.L. fem.n. *Thiosarcina* a sulfur *Sarcina*.

Individual bacteria spherical, forming regular cubical packets, resulting from consecutive divisions in three perpendicular planes. Packets commonly containing 8–64 cells. **Motile by means of polar flagella.** Not spore-forming. **Gram-negative. Contain bacteriochlorophyll**

a and carotenoids. **Do not contain gas vacuoles.**

Anaerobic.

Capable of photosynthesis in the presence of hydrogen sulfide, during which they produce and **store,** as an intermediate oxidation product, **elemental sulfur** in the form of **globules inside** the cells. Molecular hydrogen may be used as electron donor. Molecular oxygen is not evolved during photosynthesis. Cell suspensions appear in various shades of purplish to red due to the photosynthetic pigments.

Type species of genus: *Thiosarcina rosea* (Schroeter) Winogradsky 1888, 104.

Description of the species of genus **Thiosarcina**

1. Thiosarcina rosea (Schroeter) Winogradsky 1888, 104. (*Sarcina rosea* Schroeter 1886, 154; *Sarcina sulphurata* Winogradsky 1887, 576; *Sarcina rosacea* Migula 1900, 263; *Rhodothiosarcina rosea* (Schroeter) Ellis 1932, 163.)

ro'se.a. L. fem.adj. *rosea* rosy, rose-colored, pink.

Cells 2–3 μm in diameter. Color of cell suspensions purplish rose.

Obligately phototrophic.

Photosynthetic electron donors: sulfide, sulfur. Nitrogen sources: ammonium salts.

Habitat: occur less frequently than other purple sulfur bacteria in mud and stagnant bodies of water containing hydrogen sulfide and exposed to light, sulfur springs.

Illustrations: Isachenko, 1914, Plate II, Fig. 5.

Type strain: none; at present not in pure culture.

Genus IV. **Thiospirillum** *Winogradsky 1888, 104*

(*Thiorhodospirillum* Fuhrmann 1913, 323; not *Thiospirillum* Janke 1924, 68.)

Thi.o.spi.ril′lum. Gr. n. *thium* sulfur; M.L. dim.neut.n. *Spirillum* a bacterial genus; M.L. neut.n. *Thiospirillum* sulfur *Spirillum*.

Spiral-shaped bacteria, dividing by **binary fission, motile by means of polar flagella.** Under certain unfavorable conditions cells may be immotile and grow in more or less regular aggregates surrounded by slime. **Gram-negative. Contain bacteriochlorophyll *a* and carotenoids,** both pigments being located at **internal membranes of vesicular type. Do not contain gas vacuoles.**

Anaerobic.

Capable of photosynthesis in the presence of hydrogen sulfide, during which they produce and **store,** as an intermediate oxidation product, **elemental sulfur** in the form of globules **inside the cells.** Molecular hydrogen may be used as electron donor. Molecular oxygen is not evolved during photosynthesis. Cell suspensions appear in various shades of orange-brown to purple-red due to the photosynthetic pigments.

DNA base ratio: 45.5 (one species) moles % guanine + cytosine (buoyant density).

Type species of genus: *Thiospirillum sanguineum* (Ehrenberg) Winogradsky 1888, 104. (Designated by Enlows 1920, 94.)

Key to the species of genus **Thiospirillum**

I. Cells 2.5–4.0 μm wide, generally 30–40 μm long.

  A. Individual cells and cell masses purple-red:

                    1. *Thiospirillum sanguineum*

  B. Cultures orange-brown:

                    2. *Thiospirillum jenense*

II. Cells 1.5–2.5 μm wide, 4–12 μm long.

                    3. *Thiospirillum rosenbergii*

Description of the species of genus **Thiospirillum**

1. Thiospirillum sanguineum (Ehrenberg) Winogradsky 1888, 104. (*Ophidomonas sanguinea* Ehrenberg 1840, 201; *Spirillum sanguineum* (Ehrenberg) Cohn 1875, 169.)

san.gui′ne.um. L. neut.adj. *sanguineum* blood-colored, blood-red.

Cells cylindrical, sometimes attenuated at ends,

spirally coiled, 2.5–4.0 μm wide, commonly about 40 μm long with a range of from 10–100 μm. Size and shape of coils variable, a complete turn measures from 15–40 μm in length and from ½ to ⅒ of the length in width. Polarly flagellated, usually tufted at both ends. Individual cells definitely rose to purple-red, like cells of *Chromatium okenii*, and therefore probably contain carotenoids of the okenone series. Mass developments appear deep purple-red.

Habitat: mud and stagnant water, preferably sea water containing hydrogen sulfide and exposed to light. Rarely found in sulfur springs.

Illustrations: Buder, 1915, 537, Fig. 2.

Type: none; not in pure culture.

2. **Thiospirillum jenense** (Ehrenberg) Migula 1900, 1050. (*Ophidomonas jenensis* Ehrenberg 1838, 44; *Spirillum jenense* (Ehrenberg) Trevisan 1879, 149: *Rhodospirillum jenense* (Ehrenberg) Ellis 1932, 161; *Thiospirillum crassum* Hama 1933, 157.)

je.nen'se. M.L. neut.adj. *jenense* pertaining to Jena, Germany, the city where Ehrenberg discovered this organism.

The description is based in part on: Schlegel and Pfennig, 1961; Cohen-Bazire, 1963; Schmidt, 1963.

Cells 2.5–4.0 μm wide, cylindrical, sometimes pointed at the ends; coiled as spirals, generally 30–40 μm in length, but also up to 100 μm. Complete turns may measure from 15–40 μm with a coil depth of 3–7 μm. Polarly flagellated, sometimes tufted at both ends. Color of individual cells golden yellow; of cell suspensions orange-brown.

Obligately phototrophic, anaerobic. Require vitamin B_{12}. pH optimum: 7.0–7.5. Growth temperature: 20–25 C.

Photosynthetic electron donors: sulfide, sulfur. Acetate is photoassimilated in the presence of sulfide and bicarbonate.

Not utilized: thiosulfate.

Pigments: bacteriochlorophyll *a* and group 1 carotenoids (major components: rhodopin, lycopene).

Storage materials: polysaccharides, poly-β-hydroxybutyrate, polyphosphates.

Catalase and hydrogenase absent.

DNA base ratio: 45.5 moles % guanine + cytosine (buoyant density).

Habitat: mud and stagnant water (preferably fresh water) containing hydrogen sulfide and exposed to light; more rarely in sulfur springs.

Illustrations: Schlegel and Pfennig, 1961, 1, Figs. 2 and 9.

Neotype strain: "Ostrau"; SMG 216.

3. **Thiospirillum rosenbergii** (Warming) Migula 1900, 1050. (*Spirillum rosenbergii* Warming 1875, 346; *Thiospirillum coccineum* Hama 1933, 158.)

ro.sen.ber'gi.i. M.L. gen.n. *rosenbergii* of Rosenberg; named for Rosenberg, a Danish algologist.

Cells 1.5–2.5 μm wide, 4–12 μm long; coiled, with turns of about 6–7.5 μm in length and variable width up to 3–4 μm. Color of individual cells pale yellowish orange, at high levels of sulfur storage cells appear very dark. Accumulations of cells appear red.

Habitat: mud and stagnant sea and fresh water containing hydrogen sulfide and exposed to light.

Illustrations: Skuja, 1956, Plate III, Figs. 39 to 41.

Type strain: none; not in pure culture.

Genus V. **Thiocapsa** *Winogradsky 1888, 84*

Thi.o.cap'sa. Gr. n. *thium* sulfur; L. n. *capsa* box; M.L. fem.n. *Thiocapsa* sulfur box.

Spherical to slightly ovoid, multiplying by **binary fission, non-motile. Tetrads may be formed** as a result of consecutive divisions in two perpendicular planes. Under certain unfavorable conditions and in many natural environments cells **may grow in more or less regular aggregates surrounded by slime. Gram-negative. Contain bacteriochlorophyll *a* or *b* and carotenoids,** both types of pigments being located on **internal membranes of vesicular or tubular type. Do not contain gas vacuoles.**

Anaerobic phototrophs; in addition, one species, *T. roseopersicina*, able to carry out an oxidative metabolism in the dark under microaerophilic to aerobic conditions.

Capable of photosynthesis in the presence of hydrogen sulfide, during which they produce and **store,** as an intermediate oxidation product, **elemental sulfur** in the form of **globules inside the cells.** Molecular hydrogen may be used as electron donor. Molecular oxygen is not evolved during photosynthesis. Cell suspensions appear in various shades of orange-brown to pink and purple-red due to the photosynthetic pigments.

Range of DNA base ratios: 63.3–69.9 moles % guanine + cytosine (buoyant density).

Type species of genus: *Thiopcasa roseopersicina* Winogradsky 1888, 84.

Key to the species of genus **Thiocapsa**

I. Cells contain bacteriochlorophyll *a* and spirilloxanthin as the major carotenoid; culture definitely pink. Photosynthetic membrane system of vesicular type; all strains tested able to grow under microaerophilic to aerobic conditions in the dark:

1. *Thiocapsa roseopersicina*

II. Cells contain bacteriochlorophyll *b* and tetrahydrospirilloxanthin as the major carotenoid; culture light orange to orange-brown. Photosynthetic membrane system consisting of bundles of ribbon-like branched tubes; strictly anaerobic:

2. *Thiocapsa pfennigii*

Description of the species of genus **Thiocapsa**

1. **Thiocapsa roseopersicina** Winogradsky 1888, 84. (*Thiocapsa floridana* Uphof 1927, 84; *Thiocapsa minima* Isachenko 1929, 6.)

ro.se.o.per.si.ci′na. L. adj. *roseus* rosy; Gr. n. *persicus* the peach; M.L. fem.adj. *roseopersicina* rosy peach-colored.

The description is based in part on: Cohen-Bazire, 1963; Schmidt *et al.*, 1965; Trüper and Pfennig, 1966; Thiele, 1968.

Cells spherical, 1.2–3 µm, usually 1.5 µm in diameter (Plate 1.1, Fig. 5). Individual cells usually surrounded by a slime capsule. Diplococcus-shaped aggregates, tetrads and irregular clumps, usually surrounded by a slime layer, are common. Non-motile. Photosynthetic membrane system of vesicular type. Single cells under the microscope colorless. Cell suspensions and accumulations pink to rose-red.

Anaerobic phototrophs; all strains able to grow in the dark under microaerophilic to aerobic conditions in the presence of fructose, glycerol or organic acids. pH optimum: 7.0–7.5. Growth temperature: 25–30 C.

Photosynthetic electron donors: sulfide, thiosulfate, sulfur and molecular hydrogen. Acetate, fructose, fumarate, glycerol, malate, pyruvate and succinate are photoassimilated; most strains are not able to utilize sulfite, lactate, propionate or peptone. Primary alcohols, benzoate, butyrate, citrate, formate, α-ketoglutarate, tartrate and amino acids are not utilized. The majority of strains are capable of assimilatory sulfate reduction.

Pigments: bacteriochlorophyll *a*; group 1 carotenoids predominantly spirilloxanthin.

Storage materials: polysaccharides, poly-β-hydroxybutyrate, polyphosphates.

Catalase and hydrogenase present.

DNA base ratios: 63.3–66.3 moles % guanine + cytosine (type strain 65.3) (buoyant density).

Habitat: mud, stagnant water containing hydrogen sulfide and exposed to light. Rather common in estuarine environments and sewage lagoons.

Neotype strain: "Hardenberg"; SMG 217.

2. **Thiocapsa pfennigii** Eimhjellen 1970, 193. (*Thiococcus* sp. Eimhjellen *et al.* 1967, 84.)

pfen.nig′i.i. M.L. gen.n. *pfennigii* of Pfennig; named after N. Pfennig, a German microbiologist.

During the exponential growth phase, cells 1.2–1.5 µm in diameter, before division up to 2.5 µm long, occurring predominantly in pairs. Stationary phase cells 0.8–1.0 µm in diameter. Globules of elemental sulfur usually centrally located. Cells contain an extensive internal membrane system occupying the greater part of the cytoplasmic circumference of the cell, and consisting of bundles of ribbon-like branched tubes, continuous with the cytoplasmic membrane (Plate 1.3, Fig. 2). Individual cells colorless, color of cell suspensions orange-brown.

Obligately phototrophic, strictly anaerobic. pH range: 6.5–7.5. Optimum growth temperature: about 25 C.

Photosynthetic electron donors: sulfide, sulfur. In the presence of sulfide and bicarbonate, acetate and propionate are photoassimilated. Not utilized: thiosulfate, butyrate.

Nitrogen sources: ammonium salts.

Pigments: bacteriochlorophyll *b;* major carotenoid 3,4,3′,4′-tetrahydrospirilloxanthin.

Hydrogenase and catalase present.

DNA base ratios: 69.4–69.9 moles % guanine + cytosine (buoyant density).

Habitat: mud and water containing hydrogen sulfide and exposed to light.

Illustrations: (fine structure) Eimhjellen *et al.*, 1967, 82, Figs. 2 to 11.

Type strain: "Nidelven" (Department of Biochemistry, Technical University, Trondheim, Norway).

Genus VI. **Lamprocystis** *Schroeter 1886, 151*

Lam.pro.cys′tis. Gr. adj. *lamprus* bright, brilliant; Gr. n. *cystis* the bladder, a bag; M.L. fem.n. *Lamprocystis* brilliant bag.

Spherical to ovoid bacteria, before cell division typically diplococcus-shaped, multiplying by **binary fission, motile by** means of **polar flagella.** Depending on conditions, cells may remain as individuals or attached in tetrads which **aggregate into areas of considerable size, the whole being embedded in slime.** The elongated colonies may be branched or even somewhat netlike. The aggregates may break up into smaller clusters and **more or less spherical colonies which become motile** by the flagella of the composing individual cells. **Gram-negative. Contain bacteriochlorophyll *a* and carotenoids,** both pigments being located at **internal membranes of vesicular type. Contain gas vacuoles** in the central part of the cell.

Anaerobic.

Capable of photosynthesis in the presence of hydrogen sulfide, during which they produce and **store,** as an intermediate oxidation product, **elemental sulfur in the form of globules in the gas vacuole-free peripheral part of the cell.** Molecular hydrogen may be used as electron donor. Molecular oxygen is not evolved during photosynthesis. Cell suspensions appear in various shades of purple-red, due to the photosynthetic pigments.

DNA base ratio: 63.8 (one species) moles % guanine + cytosine (buoyant density).

Type species of genus: *Lamprocystis roseopersicina* (Kützing) Schroeter 1886, 151.

Description of the species of genus **Lamprocystis**

1. **Lamprocystis roseopersicina** (Kützing) Schroeter 1886, 151. (*Micraloa rosea* (*sic*) Kützing 1833, 371; *Cryptococcus roseus* (Kützing) Kützing 1845, 119; *Protococcus roseo-persicinus* Kützing 1849, 196; *Palmella persicina* Cohn 1864, 606; *Pleurococcus roseo-persicinus* (Kützing) Rabenhorst 1868, 28; *Bacterium rubescens* Lankester 1873, 410; *Clathrocystis roseo-persicina* (Kützing) Cohn 1875, 157; *Cohnia roseo-persicina* (Kützing) Winter 1884, 48; *Planosarcina roseo-persicina* (Kützing) Migula 1895, 20; *Clathrococcus roseo-persicinus* (Kützing) Schmidt and Weis 1902, 293; *Lankesteron rubescens* (Lankester) Ellis 1932, 135.)

ro.se.o.per.si.ci'na. L. adj. *roseus* rosy; Gr. n. *persicus* the peach; M.L. fem.adj. *roseopersicina* rosy peach-colored.

The description is based in part on: Pfennig *et al.*, 1968.

Cells 3.0–3.5 μm in diameter. Color of individual cells grayish, color of cell suspensions purple to purple-violet.

Obligately phototrophic, strictly anaerobic. pH range: 6.5–7.6. Optimum growth temperature; 20–25 C.

Photosynthetic electron donor: sulfide, sulfur. In the presence of sulfide and bicarbonate, acetate and pyruvate are photoassimilated. Incapable of assimilatory sulfate reduction.

Nitrogen sources: ammonium salts.

Pigments: bacteriochlorophyll *a;* carotenoids of the lycopenal series.

Storage product: poly-β-hydroxybutyrate.

DNA base ratio: 63.8 moles % guanine + cytosine (type strain: 63.8) (buoyant density).

Habitat: mud and stagnant water containing hydrogen sulfide and exposed to light; sulfur springs.

Illustrations: Pfennig *et al.*, 1968, Fig. 2.

Neotype strain: "Bergkamen"; SMG 229.

Genus VII. **Thiodictyon** *Winogradsky 1888, 80*

(Rhododictyon Orla-Jensen 1909, 334.)

Thi.o.dic'ty.on. Gr. n. *thium* sulfur; Gr. n. *dictyon* a net; M.L. neut.n. *Thiodictyon* sulfur net.

Individual bacteria **rod-shaped with rounded ends, sometimes** appearing **spindle-shaped,** multiplying by **binary fission, non-motile under all conditions. May form aggregates** in which the cells become arranged end to end in an **irregular netlike structure,** the shape of which is not constant; cells may also form more compact masses or break up into individual cells. **Gram-negative. Contain bacteriochlorophyll *a* and carotenoids,** both pigments being located at **internal membranes of vesicular type. Contain gas vacuoles in the central part of the cell.**

Anaerobic.

Capable of photosynthesis in the presence of hydrogen sulfide, during which they produce and **store,** as an intermediate oxidation product, **elemental sulfur** in the form of **globules in the gas vacuole-free peripheral part of the cells.** Molecular hydrogen may be used as electron donor. Molecular oxygen is not evolved during photosynthesis. Cell suspensions appear in various shades of purple-violet due to the photosynthetic pigments.

Range of DNA base ratios: 65.3–66.3 moles % guanine + cytosine (buoyant density).

Type species of genus: *Thiodictyon elegans* Winogradsky 1888, 82.

Key to the species of genus **Thiodictyon**

I. Cells able to form typical netlike aggregates:

1. *Thiodictyon elegans*

II. Cells not able to form netlike aggregates, usually present in the form of free individual cells:

2. *Thiodictyon bacillosum*

Description of the species of genus **Thiodictyon**

1. Thiodictyon elegans Winogradsky 1888, 82. (*Thiodictyon minus* Isachenko 1914, 251.)

e'le.gans. L. adj. *elegans* choice, elegant.

The description is based, in part, on: Schmidt *et al.*, 1965; Pfennig *et al.*, 1968.

Cells 1.5–2.0 μm wide, 3.0–8 μm long. Cells usually form aggregates in which they become arranged end to end in an irregular netlike structure somewhat reminiscent of the shape of the green alga *Hydrodictyon*. This shape of the aggregates is not constant; cells may also form more compact masses which may break up into individual cells. Color of individual cells grayish, color of cell suspensions purple-violet.

Obligately phototrophic, strictly anaerobic. pH range: 6.5–7.6. Optimum growth temperature: 20 C.

Photosynthetic electron donors: sulfide, sulfur. In the presence of sulfide and bicarbonate, acetate and pyruvate are photoassimilated. Incapable of assimilatory sulfate reduction.

Nitrogen sources: ammonium salts.

Pigments: bacteriochlorophyll *a*; carotenoids of the rhodopinal series.

Storage products: poly-β-hydroxybutyrate.

DNA base ratio: 65.3 moles % guanine + cytosine (type strain: 65.3) (buoyant density).

Habitat: mud and stagnant water containing hydrogen sulfide and exposed to light; sulfur springs.

Illustrations: Pfennig *et al.*, 1968, Fig. 3.

Neotype strain: "Bergkamen"; SMG 232.

2. Thiodictyon bacillosum (Winogradsky) Pfennig and Trüper 1971, 12. (*Amoebobacter bacillosus* Winogradsky 1888, 78; *Rhodocapsa suspensa* Molisch 1906, 223.)

ba.cil.lo'sum. L. dim.n. *bacillus* a small rod; M.L. neut.adj. *bacillosum* full of or made up of small rods.

The description is based in part on: Schmidt *et al.*, 1965.

Cells 1.5–2.0 μm wide, 3–6 μm long. Cells may form irregular aggregates surrounded by slime. Netlike aggregates are not formed. Color of individual cells grayish, color of cell suspensions purple-violet.

Obligately phototrophic, strictly anaerobic. pH range: 6.5–7.5. Growth temperature: 20–25 C.

Photosynthetic electron donors: sulfide, sulfur. In the presence of sulfide and bicarbonate, acetate and pyruvate are photoassimilated. Incapable of assimilatory sulfate reduction.

Nitrogen sources: ammonium salts.

Pigments: bacteriochlorophyll *a*; carotenoids of the rhodopinal series.

Storage product: poly-β-hydroxybutyrate

DNA base ratio: 66.3 moles % guanine + cytosine (type strain: 66.3) (buoyant density).

Habitat: mud and stagnant water containing hydrogen sulfide and exposed to light; sulfur springs.

Illustrations: Winogradsky, 1888, Plate III, Fig. 7.

Neotype strain: "Zeulenroda"; SMG 234.

Genus VIII. **Thiopedia** *Winogradsky 1888, 85*

Thi.o.pe'di.a. Gr. n. *thium* sulfur; Gr. n. *pedium* a plain, a flat area; M.L. fem.n. *Thiopedia* a sulfur plain.

The description is based in part on: Hirsch, 1969.

Individual bacteria **spherical to ovoid,** multiplying by **binary fission. Cells arranged in flat sheets with typical tetrads as the structural units;** due to consecutive divisions in two perpendicular planes, platelets with 4, 8, 16, 32 and 64 cells are formed. **Non-motile. Gram-negative. Contain bacteriochlorophyll *a* and carotenoids,** both pigments being located at **internal membranes of vesicular or stacklamellar type. Contain gas vacuoles in the central part of the cell. Anaerobic.**

Capable of photosynthesis in the presence of hydrogen sulfide, during which they produce and **store,** as an intermediate oxidation product, **elemental sulfur in the form of globules in the gas vacuole-free peripheral part of the cells.** Molecular hydrogen may be used as electron donor. Molecular oxygen is not evolved during photosynthesis. Cell suspensions appear in various shades of pink to purple due to the photosynthetic pigments.

Type species of genus: *Thiopedia rosea* Winogradsky 1888, 85.

Description of the species of genus **Thiopedia**

1. Thiopedia rosea Winogradsky 1888, 85. (*Erythroconis littoralis* (*sic*) Oersted 1841, 555; *Pediococcus roseus* (Winogradsky) Trevisan 1889, 28; *Lampropedia rosea* (Winogradsky) de Toni and Trevisan 1889, 1049; *Planococcus roseus* (Winogradsky) Migula 1895, 19.)

ro'se.a. L. adj. *rosea* rosy, rose-colored, pink.

Cells slightly elongated cocci or ovoid, 1–2 μm wide, 1.2–2.5 μm long, regularly arranged in platelets (Plate 1.1, Fig. 6). Color of individual cells grayish, color of cell suspensions pink to pinkish violet.

Obligately phototrophic, strictly anaerobic. pH range: 7.2–9.0. Optimum growth temperature: 20–25 C.

Photosynthetic electron donors: sulfide, sulfur. The amounts of hydrogen sulfide tolerated are very low.

Nitrogen sources: ammonium salts.

Pigments: *in vivo* spectra show typical maxima of bacteriochlorophyll *a*; carotenoids of the okenone series.

Storage product: poly-β-hydroxybutyrate.

Habitat: widely distributed in mud and stagnant bodies of fresh, brackish and salt water containing hydrogen sulfide and exposed to light; sulfur springs.

Type strain: none; not in pure culture.

Genus IX. **Amoebobacter** Winogradsky 1888, 71

A.moe.bo.bac'ter. Gr. n. *amoebe* change, transformation; M.L. n. *bacter* a rod; M.L. masc.n. *Amoebobacter* changeable rod.

The genus *Amoebobacter* Winogradsky (1888) has been redefined in accordance with the characteristics of the type species *A. roseus* as representing only the spherical, non-motile, gas vacuole-containing *Chromatiaceae*. Therefore, the rod-shaped species of the genus *Amoebobacter*, *A. bacillosus* Winogradsky (1888), was transferred to the genus *Thiodictyon* Winogradsky (1888) and became *Thiodictyon bacillosum*.

The genus *Rhodothece* Molisch (1907), comprising spherical, non-motile, gas vacuole-containing forms, is treated as a later synonym of the genus *Amoebobacter*.

Spherical to ovoid bacteria, multiplying by **binary fission, non-motile. Gram-negative. Contain bacteriochlorophyll *a* and group 1 carotenoids,** both pigments being located in **internal membranes of vesicular type. Con-**tain gas vacuoles in the central part of the **cell.**

Anaerobic.

Capable of photosynthesis in the presence of hydrogen sulfide, during which they produce and **store,** as an intermediate oxidation product, **elemental sulfur in the form of globules in the gas vacuole-free peripheral part of the cells.** Molecular hydrogen may be used as electron donor. Molecular oxygen is not evolved during photosynthesis. Cell suspensions appear in various shades of pink to purple due to the photosynthetic pigments.

Hydrogenase and catalase present.

Range of DNA base ratios: 64.3–65.3 moles % guanine + cytosine (buoyant density).

Type species of genus: *Amoebobacter roseus* Winogradsky 1888, 77.

Key to the species of genus **Amoebobacter**

I. Cells 2–3 μm in diameter:

1. *Amoebobacter roseus*

II. Cells 1.5–2.5 μm in diameter, pronounced slime formation under certain conditions resulting in a highly viscous culture medium:

2. *Amoebobacter pendens*

Description of the species of genus **Amoebobacter**

1. Amoebobacter roseus Winogradsky 1888, 77. (*Rhodothece conspicua* Skuja 1956, 32.)

ro'se.us. L. adj. *roseus* rosy, rose-colored, pink.

The description is based in part on: Schmidt *et al.*, 1965; Thiele, 1968.

Cells 2.0–3.0 μm in diameter, individual cells and irregular cell aggregates of different size may be surrounded by a slime capsule. Individual cells colorless, color of cell suspensions pink.

Obligately phototrophic, strictly anaerobic. pH range: 6.5–7.5. Growth temperature: 25–30 C.

Photosynthetic electron donors: sulfide, sulfur, thiosulfate, sulfite. In the presence of sulfide and bicarbonate, acetate, fructose, malate, pyruvate and certain amino acids are photoassimilated. Incapable of assimilatory sulfate reduction. Not utilized by the type strain: other sugars, sugar-alcohols, alcohols, most intermediates of the tricarboxylic acid cycle.

Nitrogen sources: ammonium salts, urea.

DNA base ratio: 64.3 moles % guanine + cytosine (type strain: 64.3) (buoyant density).

Habitat: mud and stagnant water containing hydrogen sulfide and exposed to light; sulfur springs.

Illustrations: Winogradsky, 1888, Plate III, Figs. 1 to 6; Skuja, 1956, Plate III, Fig. 46.

Neotype strain: "Davis"; SMG 235.

2. **Amoebobacter pendens** (Molisch) Pfennig and Trüper 1971, 13. (*Rhodothece pendens* Molisch 1906, 230; *Rhodothece nuda* Skuja 1956, 32.)

pen'dens. L. part.adj. *pendens* hanging.

The description is based in part on: Schmidt *et al.*, 1965; Thiele, 1968; Cherni *et al.*, 1969.

Cells 1.5–2.5 μm in diameter. Individual cells and irregular cell aggregates of different size surrounded by slime capsules. Depending on the culture conditions, slime formation may be very pronounced so that the culture becomes rather viscous. Individual cells colorless, color of cell suspensions pink.

Obligately phototrophic, strictly anaerobic. Vitamin B$_{12}$ required for growth. pH range: 7.0–7.5. Growth temperature: 25–30 C.

Photosynthetic electron donors: sulfide, sulfur, thiosulfate, sulfite. In the presence of sulfide and bicarbonate, pyruvate and glucose are photoassimilated. Incapable of assimilatory sulfate reduction. Not utilized (type strain): thioglycolate, acetate, other sugars, alcohols, sugar-alcohols, higher fatty acids, amino acids, most intermediates of the tricarboxylic acid cycle.

Nitrogen sources: ammonium salts.

DNA base ratio: 65.3 moles % guanine + cytosine (type strain: 65.3) (buoyant density).

Habitat: mud and stagnant water containing hydrogen sulfide and exposed to light.

Illustrations: Molisch, 1907, Plate II, Figs. 13 and 14; Skuja, 1956, Plate III, Fig. 45.

Neotype strain: "Klein-Kalden"; SMG 236.

Genus X. **Ectothiorhodospira** *Pelsh 1936, 63*

(Includes the "autotrophic" *Rhodopseudomonas* species of Kondrat'eva 1956, 393; Trüper 1968, 1910.)

Ec.to.thi.o.rho.do.spi'ra. Gr. prep. *ecto* outside; Gr. n. *thium* sulfur; Gr. n. *rhodum* the rose; Gr. n. *spira* the spiral; M.L. fem.n. *Ectothiorhodospira* the red spiral with sulfur outside.

Spiral to slightly bent rod-shaped bacteria, multiplying by **binary fission, motile** by means of **polar flagella.** Under certain unfavorable conditions and in many natural environments cells may be immotile and grow in more or less regular aggregates surrounded by slime. Gram-negative. **Contain bacteriochlorophyll *a* and carotenoids,** both pigments being located at **lamellar membrane stacks** which are continuous with the cytoplasmic membrane (Plate 1.3, Fig. 1). **Do not contain gas vacuoles.**

Anaerobic.

Capable of photosynthesis in the presence of hydrogen sulfide, during which they produce and deposit, as an intermediate oxidation product, **elemental sulfur in the form of globules outside the cells** in the medium. Molecular hydrogen may be used as electron donor. Molecular oxygen is not evolved during photosynthesis. Cell suspensions appear in various shades of brown to red due to the photosynthetic pigments.

Range of DNA base ratios: 62.2–69.9 moles % guanine + cytosine (buoyant density).

Type species of genus: *Ectothiorhodospira mobilis* Pelsh 1936, 63.

Key to the species of genus **Ectothiorhodospira**

I. Little or no NaCl required for growth. Polar tuft of flagella.
    A. Cells curved; 2–3% NaCl required for growth; DNA base ratio around 67 moles % guanine + cytosine:
1. *Ectothiorhodospira mobilis*
    B. Cells rather straight; no added NaCl required; DNA base ratio around 63 moles % guanine + cytosine:
2. *Ectothiorhodospira shaposhnikovii*
II. Extreme halophiles, 14–23% NaCl required for growth. One sheathed polar flagellum:
3. *Ectothiorhodospira halophila*

Description of the species of genus **Ectothiorhodospira**

1. Ectothiorhodospira mobilis Pelsh 1936, 63. (*Ectothiorhodospira mobile* (*sic*) Pelsh 1936, 63.)

mo′bi.lis. L. adj. *mobilis* motile.

The description is based in part on: Pelsh, 1937; Trüper, 1968; Remsen *et al.*, 1968; Holt *et al.*, 1968.

Cells weakly curved in a short spiral 0.7–1.0 μm wide, 2.0–2.6 μm long. Length of one full turn of spiral 3.6–4.8 μm, i.e. of two cells before division. Motile by means of a polar tuft of flagella. Individual cells colorless, color of cell suspensions at first yellowish to orange-brown, later brownish red to pink.

Obligately phototrophic, strictly anaerobic. Optimal pH range: 7.5–8.0. Optimum growth temperature: 25–30 C. Vitamin B_{12} and 2–3% NaCl are required.

Photosynthetic electron donors: sulfide, sulfur, thiosulfate, sulfite, molecular hydrogen, acetate, pyruvate, malate, fructose. Some strains also utilize butyrate, glucose, lactate, propionate. Capable of assimilatory sulfate reduction.

Not utilized: ethyl alcohol, benzoate, citrate.

Nitrogen sources: ammonium salts.

Pigments: bacteriochlorophyll a and group 1 carotenoids.

Storage products: polysaccharides, poly-β-hydroxybutyrate, polyphosphates.

Hydrogenase and catalase present.

DNA base ratios: 67.3–69.9 moles % guanine + cytosine (type strain: 67.3) (buoyant density).

Habitat: salt lakes, salt flats, estuaries containing hydrogen sulfide and exposed to light.

Illustrations: Trüper, 1968, Figs. 1 and 2.

Neotype strain: "Santa Cruz I"; SMG 237.

2. Ectothiorhodospira shaposhnikovii Cherni, Solov′eva, Fedorov and Kondrat′eva 1969 483. (Originally described as the "autotrophic Rhodopseudomonas spec." by Kondrat′eva 1956, 393.)

sha.posh.ni.kov′i.i. M.L. gen.n. *shaposhnikovii* of Shaposhnikov; named for D. I. Shaposhnikov, a Russian microbiologist.

The description is based in part on: Uspenskaya and Kondrat′eva, 1962; Moshentseva and Kondrat′eva, 1962; Kondrat′eva and Malofeeva, 1964; Vanyushin *et al.*, 1966; see also Yang Hui-Fang, 1962.

Cells rod-shaped, usually slightly bent, 0.8–0.9 μm wide, 1.5–2.5 μm long, in propionate media, cells vibrioid to spirilloid. Motile by means of a polar tuft of flagella. Individual cells colorless, color of cell suspensions at first yellowish to orange-brown, later brownish red to pink.

Obligately phototrophic, anaerobic, relatively aerotolerant.

pH range: 8.0–8.5. Optimum growth temperature: 30–35 C.

Photosynthetic electron donors: sulfide, sulfur, thiosulfate (most strains), molecular hydrogen, acetate, butyrate, propionate, malate, lactate. Capable of assimilatory sulfate reduction.

Nitrogen sources: ammonium salts, glutamate, aspartate.

Pigments: bacteriochlorophyll a and group 1 carotenoids.

Storage products: polysaccharides, poly-β-hydroxybutyrate, polyphosphates.

Hydrogenase and catalase present.

DNA base ratios: 62.3 (buoyant density); 64.0 (chemical analysis) moles % guanine + cytosine (type strain).

Habitat: fresh water ponds and lakes containing hydrogen sulfide and exposed to light.

Illustrations: Kondrat′eva, 1965, Fig. 12.

Type strain: "Moscow N 1" (Department of Microbiology, Moscow State University, Moscow, USSR; SMG 243.

3. Ectothiorhodospira halophila Raymond and Sistrom 1969, 125.

ha.lo′phi.la. Gr. n. *halus* salt; Gr. adj. *philus* loving; M.L. fem.adj. *halophila* salt-loving.

Cells curved in a spiral, 0.8 μm wide, about 5.0 μm long. Motile by means of a single, sheathed, polar flagellum at each end. Individual cells colorless, color of cell suspensions pink to brown-red.

Obligately phototrophic, strictly anaerobic. pH range: 7.6–8.0. Optimum temperature: 44 C; growth range: 25–47 C. Extreme halophile, optimal growth at 14–22% NaCl, no growth at 4% NaCl or less.

Photosynthetic electron donors: sulfide, sulfur, thiosulfate, acetate, succinate.

Nitrogen sources: ammonium salts.

Pigments: bacteriochlorophyll a and group 1 carotenoids.

Storage product: poly-β-hydroxybutyrate.

DNA base ratio: 68.4 moles % guanine + cytosine (type strain: 68.4).

Habitat: highly saline lakes, salt beds, containing sulfide and exposed to light.

Illustrations: Raymond and Sistrom, 1967, Figs. 5 to 7.

Type strain: "Summer Lake 1"; SL1, Biology Department, University of Oregon, Eugene, Ore. U.S.A.; SMG 244.

ADDENDA TO THE FAMILY **CHROMATIACEAE**

I. Species that have been in pure culture but were lost before comparative taxonomic studies could be made.

a. *Rhodopseudomonas vannielii* Scardovi 1950, 86. Rod-shaped phototrophic bacteria, 0.5 μm wide, 1.5 μm (up to 15.0 μm) long. Sometimes motile. Gram-negative. Although described as never curved or branching, rather pleomorphic in the illustration. During growth with sulfide, sulfur globules appear in the medium. Color in yeast autolysate media shell pink to peach-red, in sulfide medium greenish yellow to red-violet. Growth photolithotrophic (sulfide medium), facultatively photoorganotrophic (on high yeast extract concentrations), facultatively aerobic. Optimal pH: 6.5; temperature: 22–23 C. No growth factors required, no NaCl required. Photosynthetic electron donors: sulfide, elementary sulfur. Thiosulfate not utilized. The following organic carbon compounds were not utilized in the light: acetate, arabinose, citrate, fructose, glucose, lactate, maltose, mannitol, mannose, methionine, sorbitol, tartrate, α-alanine, β-alanine, aspartate, glutamate, glycine, histidine, leucine, methionine, tyrosine. Cysteine was slightly utilized. Gelatin: not liquefied. Nitrogen sources: ammonium salts, nitrates, urea. Asparagine and nitrite not utilized.

Source: from an enrichment for *Chromatium*.

Illustrations: Scardovi, 1950, 77.

Comment: If reisolated, this species must be renamed and placed with the *Chromatiaceae*.

b. *Rhodopseudomonas issatchenkoi* Osnitskaya 1954, 19. Rod-shaped pink phototrophic bacteria, 0.6 μm wide, 3.2 μm long, forming long filaments before division. Cells appear rather pleomorphic on the illustrations. Cells contain two to four inclusions. Contain bacteriochlorophyll *a* and carotenoids, when grown in the light; in the dark the amount of pigments was insignificant. Growth photolithotrophic (sulfide medium), facultatively photoorganotrophic. Slight but significant growth in the dark on heterotrophic media. No growth factors required, no NaCl required. Optimal growth temperature: 28–30 C; pH: 7.2–7.5. Photosynthetic electron donors: sulfide. In the presence of sulfide, malate, citrate, lactate, and, less effectively, succinate are also utilized. No growth in peptone media. Enzymes present: catalase, peroxidase.

Source: pink-colored stratal waters of a Russian oil field from a depth of 1367 meters.

Illustrations: Osnitskaya, 1954, 5.

Comment: If reisolated this species should be renamed and placed with the *Chromatiaceae*.

c. *Thiopedia sevani* Gambaryan 1962, 282. (*Thiopedia servani* (*sic*) Gambaryan 1962, 282.) Spherical to ovoid bacteria of an average diameter of 3.5 μm, occurring in pairs or tetrads encased in a slimy capsule. Multiplication by division in two mutually perpendicular directions. Motile colonies were observed. Anaerobic, phototrophic; sulfide and sulfite vigorously oxidized. Globules of elemental sulfur accumulated within the cells. Suspensions cherry red to dark brown.

Source: from deep hydrogen sulfide-containing mud of Lake Sevan, USSR.

Illustration: Gambaryan, 1962, 282.

Comment: If reisolated, more detailed characterization of the species is necessary to justify classification in the genus *Thiopedia*. Gambaryan's description does not mention gas vacuoles. It is not clear whether the observed movements of colonies are due to flagella or to the typical jerky movements observed in all gas vacuole-containing phototrophic bacteria.

II. Species never isolated and only incompletely described from samples of water and mud. Rarely found or mentioned again by other authors. Some hardly recognizable.

a. *Amoebobacter granula* Winogradsky 1888, 78. Spherical bacteria, small, about 0.5–1.0 μm in diameter. Faint pigmentation; the sulfur inclusions give the cell masses a black appearance under the microscope. Aggregates are apt to consist of closely knit masses which are difficult to separate. When sulfur is stored, a single globule usually fills most of the cell. Because of the high refractive index of this globule, it becomes difficult, if not impossible, to make accurate observations of the cell shape.

Habitat: mud and stagnant water containing hydrogen sulfide and exposed to light; sulfur springs.

Illustration: Winogradsky, 1888, Plate III, Fig. 8.

Comment: Position in the genus *Amoebobacter* questionable since no gas vacuoles are reported. If reisolated such an organism should be placed in *Thiocapsa*.

b. *Chromatiopsis elektron* Skuja 1948, 19. Cells ovoid, 3.5–5 μm wide, 4–6 μm long, motile by means of polar flagella of 1.5 times cell length. Motion trembling. Flagellum (probably flagellar tuft) described as apically inserted. Color of single cells yellowish brown. Intracellular sulfur globules concentrated centrally.

Habitat: water of Swedish lakes between rotting plant material.

Illustration: Skuja, 1948, Plate I, Fig. 14.

Comment: Similar forms have been observed in pure cultures of the large *Chromatium* species under conditions of vitamin B₁₂ deficiency (Pfennig and Lippert, 1966).

c. *Chromatiopsis cinerea* Skuja 1948, 20. Cells ovoid, 6–8 μm wide, 7–10 μm long, motile by means of polar flagella of ½ to ¾ cell length. Motion steady. Color of single cells slightly greyish yellow to slightly bluish. Otherwise like *Chromatiopsis elektron* Skuja and same comment applies.

Illustration: Skuja, 1948, Plate I, Fig. 13; Skuja, 1956, Plate III, Figs. 5 to 10.

d. *Chromatiopsis major* Skuja 1956, 28. Cells ovoid, 6.6–9 μm wide, 8–14 μm long, motile by means of polar flagella shorter than the cell. Cytoplasm appears opaque greyish orange-yellow. Sulfur granules (!) appear small and almost crystalline together with other granular inclusions positioned in the subperipheral layer of the cytoplasm. Otherwise like *Chromatiopsis elektron* Skuja.

Illustration: Skuja, 1956, Plate III, Figs. 11 to 16.

Comment: Similar forms have been observed in pure cultures of the large *Chromatium* species under conditions of vitamin B₁₂ deficiency, when, in addition to sulfur globules, the cells are filled with granular inclusions consisting of polysaccharides and poly-β-hydroxybutyrate (Pfennig and Lippert, 1966).

e. *Chromatium gobii* Isachenko 1914, 253. Cells 10 μm wide, 20–25 μm long. Found in sea water of the Arctic Ocean and presumably common in the colder portions of the ocean.

Illustration: Isachenko, 1914, Plate II, Fig. 12.

Comment: This is apparently a true *Chromatium*. If isolated the relation of this organism to *Chromatium okenii* must be examined; the cell size may be influenced by environmental factors.

f. *Chromatium linsbaueri* Gicklhorn 1921, 313. (*Rhabdochromatium linsbaueri* Gicklhorn 1921, 315; *Rhabdomonas linsbaueri* (Gicklhorn) van Niel 1948, 855.) Cells 6 μm wide, up to 15 μm long; also 6–8 μm in width (Ellis, 1932). Special characteristic is the occurrence of calcium carbonate inclusions. Otherwise resembles *Chromatium okenii*.

Found in a pool in the Stiftingtal, near Graz, Austria.

Illustrations: Gicklhorn, 1921, 314, Fig. 1; Ellis, 1932, 148, Fig. 31.

Comment: The similarity between *Chromatium okenii* (syn. *Rhabdomonas rosea*) and *Chromatium linsbaueri* (syn. *Rhabdomonas linsbaueri*) is so obvious from the descriptions that the latter may well be a variety of the former due to environmental influences.

g. *Lamprocystis rosea* (Miyoshi) Migula 1900, 1044. (*Thioderma roseum* Miyoshi 1897, 158.) Cells spherical, 1.5 μm wide, 2.5 μm long (!). Color slightly red. Cells form thin purple-red pellicles.

Habitat: grass and leaves.

Illustrations: Miyoshi, 1897, 170.

Comment: Very poorly described, hardly recognizable, pellicles may be formed by many other genera.

h. *Lamprocystis rubra* (Miyoshi) Migula 1900, 1044. (*Thioderma rubrum* Miyoshi 1897, 170.) Cells ovoid, 2–4 μm, form pellicles. Peachblossom red in color.

Habitat: mud.

Illustrations: Miyoshi, 1897, 170.

Comment: Very poorly described, hardly recognizable.

i. *Lamprocystis violacea* (Miyoshi) Migula 1900, 1044. (*Thiosphaerion violaceum* Miyoshi 1897, 170.) Cells spherical to ovoid, 1.8–2.5 μm diameter, form solid rounded colonies, violet color. Reported once from Yumoto Hot Springs near Nikko, Japan.

Illustrations: Miyoshi, 1897, 170.

Comment: Hardly recognizable, if motile, probably identical with *Thiocystis violacea*.

j. *Rhodopedia tetras* Skuja 1956, 33. Cells spherical to ovoid, 0.6–0.8 μm wide, up to 1.2 μm long. Non-motile. Color pink-violet. Do not contain gas vacuoles. Due to consecutive divisions in three perpendicular planes, first tetrads, then sarcinalike double tetrads are formed. The latter split up to tetrads again. Cells contain inclusions of elementary sulfur.

Source: hypolimnion of Lake Blankvatn, Norway.

Illustration: Skuja, 1956, Plate III, Fig. 47.

Comment: If isolated the organism must be placed with the genus *Thiocapsa*.

k. *Thiocystis rufa* Winogradsky 1888, 65. Cells spherical, less than 1.0 μm in diameter. Color red, usually darker than *Thiocystis violacea*. When the cells are stuffed with sulfur globules, the aggregates appear almost black under the microscope. A great number of closely packed individual colonies are contained in a common gelatinous capsule.

Habitat: mud and stagnant water containing hydrogen sulfide and exposed to light; sulfur springs.

Illustration: Winogradsky, 1888, Plate II, Fig. 8.

Comment: Description insufficient; hardly recognizable.

l. *Thiopolycoccus ruber* Winogradsky 1888, 79. Cells spherical, about 1.2 μm in diameter. No motility observed. Form dense aggregates of rather solid construction and irregular shape. Large clumps may fissure with the formation of irregular shreds and lobes, which continue to break up into smaller groups of cells. Elementary sulfur stored inside the cells. Color: distinct red.

Habitat: mud and stagnant water containing hydrogen sulfide and exposed to light; sulfur springs.

Illustrations: Winogradsky, 1888, Plate IV, Figs. 16 to 18; Isachenko, 1914, Plate II, Fig. 7.

Comment: The description fits the typical non-motile densely packed colonies of *Thiocystis* or *Thiocapsa* under natural conditions. If such an organism can be isolated, it should be placed in the genus *Thiocapsa*, if non-motile; in the genus *Thiocystis*, if motile.

m. *Thiospirillum rufum* (Perty) Migula 1900, 1050. (*Spirillum rufum* Perty 1852, 179.) Cells spiral-shaped, 1.0 μm wide, 8–18 μm long, coil width usually 4 μm and length of one turn 4 μm. General characteristics presumably those of the genus, although neither Perty, Migula, Bavendamm (1924, 132) nor Huber-Pestalozzi (1938, 304) mention that the cells contain sulfur globules. Only the red color is emphasized. The photograph given by Gietzen (1931, 183) and labeled *T. rufum* could probably be also identified with *T. rosenbergii*.

Habitat: found in red slime spots on the side of a well. Mud and stagnant bodies of water.

Illustration: Migula, 1900, Plate III, Fig. 7.

Comment: The given dimensions agree with those of *Rhodospirillum rubrum*, and it seems probable that the two organisms are identical.

III. Species whose photosynthetic character appears doubtful.

a. *Chromatium fallax* (Warming) Kolkwitz 1909, 161. (*Monas fallax* Warming 1875, 367.) Cells ovoid, 2 μm wide, several micrometers long, motile. Color of mass development: pure white, not reddish. Cells contain sulfur globules. Aerophilic.

Habitat: in ditch water on decaying leaves.

Illustration: Warming, 1875, Plate 10, Fig. 9.

Comment: Obviously nonphotosynthetic, may belong with the larger colorless sulfur bacteria.

b. *Chromatium gliscens* (Ehrenberg) Kolkwitz 1909, 161. (*Monas gliscens* Ehrenberg 1838, Plate I, Fig. 14.) Cells rod-shaped to ovoid, half the size of *Chromatium okenii*. Colorless.

Comment: Obviously nonphotosynthetic, may belong with the larger colorless sulfur bacteria.

c. *Chromatium molischii* (Bersa) van Niel 1948, 858. (*Pseudomonas molischii* Bersa 1926, 375.) Cells rod-shaped, 2–2.5 μm wide, 8 μm long. Contain calcium carbonate inclusions. Single cells colorless. Cell masses very slightly pink.

Source: ponds in Austria.

Illustration: Bersa, 1926, Fig. 3.

Comment: From Bersa's description the color of mass developments, pinkish white, could have been due to contaminating *Chromatium* cells. The photosynthetic character as well as the need for hydrogen sulfide were not clearly proven. Apparently similar colorless forms have been found by other authors.

FAMILY III. **CHLOROBIACEAE** COPELAND 1956, 31

(*Chlorobiacea* (*sic*) Copeland 1956, 31.)

Chlo.ro.bi.ac′e.ae. M.L. neut.n. *Chlorobium* type genus of the family; -aceae ending to denote family; M.L. fem.pl.n. *Chlorobiaceae* the *Chlorobium* family.

Cells spherical, ovoid- or rod-shaped; **multiplication by binary fission or binary plus ternary fission. In one genus (*Chloropseudomonas*) cells motile by means of polar flagella; all other genera non-motile.** Genera with or without gas vacuoles.

Depending on culture conditions or environmental conditions in nature, most forms are able to develop either in the form of single cells or in cell masses of various shapes and sizes with more or less slime. **In the presence of sulfide, globules of elemental sulfur are deposited outside the cells, never inside the cells. Single cells and cultures green or brown. The photopigments are characteristically located in the chlorobium vesicles which underlie and are** attached to the cytoplasmic membrane (Plate 1.4).

Bacteriochlorophylls *c* or *d* occur as major components in addition to small amounts of bacteriochlorophyll *a*; carotenoids of group 5 present. Major carotenoid of the green forms is chlorobactene; major components of the brown forms are the isorenieratenes. Many strains require vitamin B_{12} for growth.

All forms are **strictly anaerobic and obligately phototrophic;** capable of **photolithotrophic assimilation of carbon dioxide in the presence of sulfide or sulfur** which are photo-oxidized to sulfate. Many strains are able to use molecular hydrogen as an electron donor for growth provided hydrogen sulfide is present in addition as a source of cell sulfur. Most forms lack an assimilatory sulfate reduction. A number of simple organic substrates are photoassimilated in the presence of both sulfide and carbon dioxide; acetate is the most widely used compound; the *Chlorobiaceae* are therefore potentially mixotrophic. The fixation of molecular nitrogen has been demonstrated in a number of strains. Storage material: usually polyphosphates; poly-β-hydroxybutyrate has not been found in the strains studied. Polysaccharides were identified only in *Chloropseudomonas ethylica*. Range of DNA base ratios: 48.5–58.1 moles % guanine + cytosine (buoyant density).

Two physiological-ecological subgroups can be differentiated on the basis of the presence or absence of gas vacuoles and the resulting selective advantage in enrichment cultures. The first group comprises the forms without gas vacuoles which have a selective advantage in cultures with medium to high sulfide concentrations and light intensities. The second subgroup includes the gas vacuole-containing forms which require very low sulfide concentrations, low light intensities and growth temperatures (10–20 C). The cells of the gas vacuole-containing forms rise to the top of the culture bottles at lower temperatures (4–10 C).

In nature the green sulfur bacteria occur in the anaerobic and sulfide-containing parts of all kinds of aquatic environments, e.g. ditches, ponds, lakes, rivers, sulfur springs, estuaries and other marine habitats. The brown-colored forms occur only in deeper layers of ponds and lakes and in the hypolimnion of meromictic lakes.

The type genus: *Chlorobium* Nadson 1906, 190, designated by Trüper and Pfennig 1971, 9.

Further comments on the family

The former family name *Chlorobacteriaceae* Lauterborn 1913, 99 is illegitimate since it is based on the generic name *Chlorobacterium* which has been excluded from the phototrophic green sulfur bacteria (*cf.* Trüper and Pfennig, 1971) because of its uncertain position. Generic names applied to symbiotic associations (consortia) are not validly published. These associations are treated in the addenda. The genera *Chloropseudomonas* Czurda and Maresch 1937 and *Prosthecochloris* Gorlenko 1970 have been included in the *Chlorobiaceae*.

Key to the genera of family **Chlorobiaceae**

I. Cells free living not intimately associated with other microbes.
  A. Cells do not contain gas vacuoles.
    1. Cells non-motile.
      a. Cells ovoid to rod-shaped or vibrioid.
        Genus I. *Chlorobium*, p. 52
      b. Cells irregular spherical to starlike, covered with extrusions.
        Genus II. *Prosthecochloris*, p. 55
    2. Cells motile by means of polar flagella; rod-shaped.
        Genus III. *Chloropseudomonas*, p. 55
  B. Cells contain gas vacuoles; non-motile.
    1. Cells rod-shaped to ovoid; may unite into characteristic aggregates.
        Genus IV. *Pelodictyon*, p. 56
    2. Cells spherical to ovoid; united into characteristic aggregates.
        Genus V. *Clathrochloris*, p. 57
II. Cells live in symbiotic aggregates with other microorganisms (consortia).
        See addenda, p. 58

Genus I. **Chlorobium** *Nadson 1906, 190*

Chlo.ro'bi.um. Gr. adj. *chlorus* green, greenish yellow; Gr. n. *bios* life; M.L. neut.n. *Chlorobium* green life.

Rod-shaped, ovoid or vibrio-shaped bacteria, multiplying by **binary fission, non-motile.** Gram-negative. Contain either **bacteriochlorophyll c or d** as the major bacteriochlorophyll component as well as **carotenoids of group 5,** all pigments being located in elongated, ovoid vesicles ("chlorobium vesicles") underlying and attached to the cytoplasmic membrane. **Do not contain gas vacuoles.**

Anaerobic.

Capable of photosynthesis in the presence of hydrogen sulfide, during which they produce and **deposit** as an intermediate oxidation product, **elemental sulfur** in the form of globules **outside the cells in the medium.** Molecular hydrogen may be used as electron donor. Molecular oxygen is not evolved during photosynthesis. **Cell suspensions** appear in various shades of **yellowish green and chocolate brown** due to the photosynthetic pigments.

Storage material: polyphosphate, not polysaccharides or poly-β-hydroxybutyrate.

Range of DNA base ratios: 49.0–58.1 moles % guanine + cytosine (buoyant density).

Type species of genus: *Chlorobium limicola* Nadson 1906, 190.

Comment: The wide range of DNA base ratios indicate a marked genetic heterogeneity of the physiologically very homogeneous genus. However, no correlation can be detected when the strains, so far studied, are grouped according to taxonomically useful morphological (bacteroid or vibrioid cell form) and biochemical characteristics (occurrence of bacteriochlorophylls *c* or *d*, thiosulfate utilization) and compared with subgroups based on similar GC values. Further studies with more isolates are necessary to solve the inherent problems.

Key to the species of genus **Chlorobium**

I. Culture green; major carotenoid chlorobactene.
  A. Cells rod-shaped, 0.7–1.1 μm wide.
    1. Thiosulfate not utilized:
        1. *Chlorobium limicola*
    2. Thiosulfate utilized:
        1a. *Chlorobium limicola* forma sp. *thiosulfatophilum*
  B. Cells short vibrio-shaped, 0.5–0.7 μm wide.
    1. Thiosulfate not utilized:
        2. *Chlorobium vibrioforme*
    2. Thiosulfate utilized:
        2a. *Chlorobium vibrioforme* forma sp. *thiosulfatophilum*
II. Culture brown; major carotenoids isorenieratene and β-isorenieratene.
  A. Cells rod-shaped, 0.6–0.8 μm wide.
        3. *Chlorobium phaeobacteroides*
  B. Cells short vibrio-shaped, 0.3–0.4 μm wide.
        4. *Chlorobium phaeovibrioides*

Description of the species of genus **Chlorobium**

1. Chlorobium limicola Nadson 1906, 190.

li.mi'co.la. L. n. *limus* mud; L. suff., verbal n. *cola* dweller; M.L. masc.n. *limicola* the mud dweller.

The description is based in part on: Larsen, 1952, 1953; Liaaen-Jensen *et al.*, 1964; Cohen-Bazire *et al.*, 1964; Mandel *et al.*, 1965; Pfennig and Lippert, 1966; Pfennig, 1967.

Cells rod-shaped, sometimes slightly curved, 0.7–1.1 μm wide, 0.9–1.5 μm long, sometimes much longer (Plate 1.1, Fig. 7). Cells frequently united in chains resembling streptococci. Some strains produce slime dependent on the culture conditions. Color of individual cells light green, color of cell suspensions green.

Obligately phototrophic, strictly anaerobic. Vitamin B₁₂ may be required for growth. pH range: 6.0–7.0; optimum pH: 6.8. Optimum growth temperature: 25–30 C.

Photosynthetic electron donors: sulfide, sulfur.

In the presence of sulfide and bicarbonate, acetate and in most strains propionate are photoassimilated; few strains may utilize pyruvate, fructose, glutamate and peptone. Not utilized: thiosulfate, butyrate and higher fatty acids, ethanol, succinate. Incapable of assimilatory sulfate reduction.

Nitrogen sources: ammonium salts.

Predominant pigments: bacteriochlorophyll *c* (or *d*), carotenoid: chlorobactene (chlorobactene series).

Hydrogenase activity present in many strains.

DNA base ratios: 51.0–52.0 moles % guanine + cytosine (type strain: 51.0) (buoyant density).

Habitat: mud and stagnant water containing hydrogen sulfide and exposed to light.

Neotype strain: "Gilroy Hot Spring"; SMG 245.

1a. Chlorobium limicola forma sp. **thio-**

sulfatophilum (Larsen) Pfennig and Trüper 1971, 14. (*Chlorobium thiosulfatophilum* Larsen 1952, 189.)

thi.o.sul.fa.to'phi.lum. M.L. n. *thiosulfatum* thiosulfate; Gr. adj. *philus* loving; M.L. adj. *thiosulfatophilum* thiosulfate-loving.

The description is based in part on: Larsen, 1953; Kondrat'eva *et al.*, 1958; Shaposhnikov *et al.*, 1958; Moshentseva and Kondrat'eva, 1962; Liaaen-Jensen *et al.*, 1964; Cohen-Bazire *et al.*, 1964; Mandel *et al.*, 1965; Pfennig and Lippert, 1966; Tomina and Fedorov, 1967; Pfennig, 1967.

Description: as for *C. limicola*, except thiosulfate utilized as photosynthetic electron donor.

DNA base ratios: 52.5–58.1 moles % guanine + cytosine (type strain: 58.1) (buoyant density).

Neotype strain: "Tassajara"; SMG 249.

2. Chlorobium vibrioforme Pelsh 1936, 63. (*Chlorobium vibrioformis* (*sic*) Pelsh 1936, 63.)

vi.bri.o.for'me. L. v. *vibro* vibrate; M.L. n. *vibrio* that which vibrates; a generic name; L. adj. suffix *-formis* -like, of the shape of; M.L. neut.adj. *vibrioforme* of vibrio-shape.

Cells short vibrio-shaped, 0.5–0.7 μm wide, 1–2 μm long. In natural habitats and under unfavorable conditions in pure culture, spirally wound to tightly coiled involution forms occur. Color of individual cells light green, color of cell suspensions green.

Obligately phototrophic, strictly anaerobic. Most strains require vitamin B_{12} and at least 1% NaCl. pH range: 6.0–7.5. Optimum growth temperature: 25–30 C.

Photosynthetic electron donors: sulfide, sulfur. In the presence of sulfide and bicarbonate acetate and in most strains propionate are photoassimilated. Not utilized: thiosulfate, higher fatty acids, ethanol, succinate. Incapable of assimilatory sulfate reduction.

Nitrogen sources: ammonium salts.

Predominant pigments: bacteriochlorophyll *d* or *c*, chlorobactene (chlorobactene series)

Hydrogenase present in many strains.

DNA base ratios: 52.0–57.1 moles % guanine + cytosine (type strain: 53.5) (buoyant density).

Habitat: mud and stagnant water (predominantly salt water) containing hydrogen sulfide and exposed to light.

Illustrations: Pelsh, 1937, Fig. 2.

Neotype strain: "Moss Landing"; SMG 260.

2a. Chlorobium vibrioforme forma sp. **thiosulfatophilum** (Larsen) Pfennig and Trüper 1971, 14. (*Chlorobium thiosulfatophilum* Larsen 1952, 189.)

thi.o.sul.fa.to'phi.lum. M.L. n. *thiosulfatum*

thiosulfate; Gr. adj. *philus* loving; M.L. neut. adj. *thiosulfatophilum* thiosulfate-loving.

The description is based in part on: Pfennig, 1967.

Description: as for *C. vibrioforme*, except thiosulfate utilized as photosynthetic electron donor.

DNA base ratio: 53.5 moles % guanine + cytosine (type strain) (buoyant density).

Type strain: "Sehestedt"; SMG 265.

3. Chlorobium phaeobacteroides Pfennig 1968, 225.

phae.o.bac.te.ro.i'des. Gr. adj. *phaeus* brown; Gr. neut.n. *bakterion* rod; Gr. n. *eidus* form, shape; M.L. adj. *phaeobacteroides* brown, rod-shaped.

The description is based in part on: Trüper and Genovese, 1968; Pfennig, 1967; preliminary laboratory name "Phaeobium," Liaaen-Jensen, 1965.

Cells rod-shaped, occasionally slightly bent, 0.6–0.8 μm wide, 1.3–2.7 μm long. Color of cell suspensions yellowish to reddish brown to chocolate brown.

Obligately phototrophic, strictly anaerobic. Vitamin B_{12} required for growth, sea water strains may require at least 1% NaCl. pH range: 6.0–7.5. Optimum growth temperature: 25–30 C.

Photosynthetic electron donors: sulfide, sulfur. In the presence of sulfide and bicarbonate, acetate or fructose are photoassimilated. Not utilized: thiosulfate, higher fatty acids, propionate, pyruvate, alcohols, sugar-alcohols, hydrogen and methane. Incapable of assimilatory sulfate reduction.

Nitrogen sources: ammonium salts.

Predominant pigments: bacteriochlorophyll *d*, carotenoids of the chlorobactene series, predominantly isorenieratene and β-isorenieratene.

Hydrogenase may be present. Catalase negative.

DNA base ratios: 49.0–50.0 moles % guanine + cytosine (type strain: 49.0) (buoyant density).

Habitat: stagnant, hydrogen sulfide-containing waters, usually in the hypolimnion of meromictic lakes.

Illustrations: Trüper and Genovese, 1968, 225, Fig. 2a.

Type strain: "Blankvann"; SMG 266.

4. Chlorobium phaeovibrioides Pfennig 1968, 226.

phae.o.vi.bri.o'i.des. Gr. adj. *phaeus* brown; L. v. *vibro* vibrate; M.L. masc.n. *vibrio* that which vibrates, a generic name; Gr. n. *eidus* form, shape; M.L. adj. *phaeovibrioides* brown vibrioshaped.

The description is based in part on: Pfennig, 1967; preliminary laboratory name "Phaeobium" Liaaen-Jensen, 1965.

Description as for *C. phaeobacteroides*, except: Cells weakly curved to vibrio-shaped, 0.3–0.4 μm wide, 0.7–1.4 μm long (Plate 1.1, Fig. 8); under certain conditions cells remain attached and grow into more or less tightly wound coils and spirals.

In the presence of sulfide and bicarbonate, acetate and propionate are photoassimilated.

DNA base ratios: 52.0–53.0 moles % guanine + cytosine (type strain: 53.0) (buoyant density).

Type strain: "Langvikvann"; SMG 269.

Genus II. **Prosthecochloris** *Gorlenko 1970, 148*

Pros.the.co.chlo′ris. Gr. n. *prostheca* appendage; Gr. adj. *chloris* green; M.L. fem.m. *Prosthecochloris* green (organism) with appendages.

Spherical to ovoid bacteria, forming prosthecae, dividing by binary fissions in many directions. When separation is incomplete, cells form groups and branched chains, the configuration of which depends on the direction of fissions. **Non-motile. Gram-negative. Contain bacteriochlorophyll *c*** as the major bacteriochlorophyll component as well as **carotenoids**, all pigments being located in elongated, ovoid vesicles ("chlorobium vesicles") underlying and attached to the cytoplasmic membrane. **Do not contain gas vacuoles.**

Anaerobic.

Capable of photosynthesis in the presence of hydrogen sulfide, during which they produce and **deposit** as an intermediate oxidation product, elemental **sulfur** in the form of globules **outside the cells.** Molecular oxygen is not evolved during photosynthesis. **Cell suspensions** appear **green** due to the photosynthetic pigments.

DNA base ratio: 50.0–56.1 moles % guanine + cytosine (buoyant density).

Type species of genus: *Prosthecochloris aestuarii* Gorlenko 1970, 148.

Description of the species of genus **Prosthecochloris**

1. **Prosthecochloris aestuarii** Gorlenko 1970, 148.

ae.stu.a′ri. L. n. *aestuarium* estuary; M.L. gen.n. *aestuarii* of the estuary.

The description is based in part on: Gorlenko, 1968.

Cells spherical to ovoid, 0.5–0.7 μm wide, 1.0–1.2 μm long. Each cell produces about 20 prosthecae, 0.1–0.17 μm wide and 0.1–0.5 μm long. Cells usually embedded in slime capsules with slime filaments between them. When separation of cells after binary fission is incomplete, groups and branched chains may be formed, their configuration being determined by the direction of fission.

The "chlorobium vesicles" are located predominantly in the prosthecae. Color of individual cells light green, color of cell suspensions green.

Obligately phototrophic, strictly anaerobic. Vitamin B_{12} required for growth. Optimal pH

range: 6.7–7.0. Salinity required: 2–5% NaCl, a range of 1–8% is tolerated. Optimum growth temperature: 25–30 C.

Photosynthetic electron donors: sulfide, sulfur. In the presence of sulfide and carbon dioxide, acetate is photoassimilated. Thiosulfate is not utilized. Incapable of assimilatory sulfate reduction.

Nitrogen sources: ammonium salts.

Predominant pigments: bacteriochlorophyll *c*.

DNA base ratio: 50.0–56.1 moles % guanine + cytosine (type strain: 56.1).

Habitat: hydrogen sulfide-containing mud and stagnant water of brackish lakes and estuaries.

Illustrations: Gorlenko, 1968, Fig. 1.

Type strain: Sasyk-Sivash; SK-413, Department of Geol. Microbiol., Inst. of Microbiol., Academy of Sciences, Moscow, USSR; SMG 271.

Genus III. **Chloropseudomonas** *Czurda and Maresch 1937, 123*
Nom. cons. proposed by Pfennig and Trüper 1969, 153

Chlo.ro.pseu.do.mo′nas. Gr. adj. *chloris* green; M.L. fem.n. *Pseudomonas* a bacterial genus; M.L. fem.n. *Chloropseudomonas* a green *Pseudomonas*.

Rod-shaped bacteria, multiplying by **binary fission, motile by means of polar flagella. Gram-negative. Contain bacteriochlorophyll *c*** as the major bacteriochlorophyll component and **group 5 carotenoids,** all pigments being located in elongate-ovoid vesicles ("chlorobium vesicles")

underlying and attached to the cytoplasmic membrane. **Do not contain gas vacuoles.**

Anaerobic.

Capable of photosynthesis in the presence of hydrogen sulfide, during which they produce and **deposit** as an intermediate oxidation product,

elemental sulfur in the form of globules **outside the cells** in the medium. Molecular hydrogen may be used as electron donor. Molecular oxygen is not evolved during photosynthesis. **Cell suspensions** appear in various shades of **yellowish green**, due to the photosynthetic pigments.

Range of DNA base ratios: 57.2–57.6 (one species) moles % guanine + cytosine (chemical analysis), 55.1 (buoyant density).

Type species of genus: *Chloropseudomonas ethylica* Shaposhnikov *et al.* 1960, 168.

Description of the species of genus **Chloropseudomonas**

1. **Chloropseudomonas ethylica** Shaposhnikov, Kondrat'eva and Fedorov 1960, 168. (*Chlorobium ethylicum* Shaposhnikov *et al.* 1959, 1426; *Chloropseudomonas ethylicum* (*sic*) Shaposhnikov *et al.* 1960, 168.)

e.thyl'i.ca. M.L. n. *ethylicus* ethyl alcohol; M.L. fem.adj. *ethylica* pertaining to ethyl alcohol.

The description is based in part on: Moshentseva and Kondrat'eva, 1962; Liaaen-Jensen *et al.*, 1964; Holt *et al.*, 1966; Vanyushin *et al.*, 1966; Trotsenko, 1966; Kondrat'eva *et al.*, 1968.

Cells 0.7–0.9 μm wide, 1.0–1.5 μm long. Motile by means of one polar or subpolar flagellum. Individual cells colorless, color of cell suspensions green.

Obligately phototrophic, anaerobic. Optimum pH: 7.0–7.2. Optimum growth temperature: 25–30 C. Higher salt concentrations (NaCl, MgCl₂) may be required if present in the original habitat.

Photosynthetic electron donors: sulfide, sulfur, organic carbon compounds. In the presence of bicarbonate and traces of sulfide, ethanol, glucose, maltose and pyruvate are photoassimilated. Some strains utilize in addition thiosulfate, other sugars, acetate, formate, glycerol, lactate, mannitol, propanol.

Nitrogen sources: ammonium salts, some amino acids.

Predominant pigments: bacteriochlorophyll *c*, carotenoid: chlorobactene (chlorobactene series).

Storage products: polysaccharides, polyphosphates, not poly-β-hydroxybutyrate.

Hydrogenase may be present.

DNA base ratios: 55.1 (buoyant density), 57.3 (chemical analysis) moles % guanine + cytosine (type strain: 57.3)

Habitat: lakes, ponds, estuaries containing stagnant fresh or salt water with hydrogen sulfide and exposed to light.

Illustrations: Shaposhnikov *et al.*, 1960, 167; Skalinskii and Kondrat'eva, 1961, Fig. 8e–h.

Type strain: "1 M" (Microbiology Department, Moscow State University, Moscow, USSR).

Genus IV. **Pelodictyon** *Lauterborn 1913, 98*

Pe.lo.dic'ty.on. Gr. adj. *pelos* dark-colored; Gr. n. *dictyon* net; M.L. neut.n. *Pelodictyon* a dark-colored net.

Rod-shaped to ovoid bacteria, multiplying by **binary fission, non-motile. Branching may occur as a result of ternary fission and netlike three dimensional aggregates** or more or less **spherical colonies may be formed.** Gram-negative. Contain either **bacteriochlorophyll c or d** as the major bacteriochlorophyll components and **carotenoids of group 5,** both pigments being located in elongated-ovoid vesicles ("chlorobium vesicles") underlying and attached to the cell membrane (Plate 1.4). **Contain gas vacuoles.**

Anaerobic.

Capable of photosynthesis in the presence of hydrogen sulfide, during which they produce and **deposit** as an intermediate oxidation product, **elemental sulfur** in the form of globules **outside the cells** in the medium. Molecular hydrogen may be used as electron donor. Molecular oxygen is not evolved during photosynthesis. **Cell suspensions** appear in various shades of **yellowish green** due to the photosynthetic pigments.

Range of DNA base ratios: 48.5–58.1 moles % guanine + cytosine (buoyant density).

Type species of genus: *Pelodictyon clathratiforme* (Szafer) Lauterborn 1913, 98.

Key to the species of genus **Pelodictyon**

I. Cells rod-shaped, forming typical netlike aggregates due to their ability to undergo both binary and ternary fissions:

1. *Pelodictyon clathratiforme*

II. Cells ovoid to short vibrio-shaped, occurring either in the form of free individual cells or united into hollow spherical or irregular round colonies:

2. *Pelodictyon luteolum*

Description of the species of genus **Pelodictyon**

1. Pelodictyon clathratiforme (Szafer) Lauterborn 1913, 98. (*Aphanothece clathratiformis* Szafer 1911, 162; *Pelodictyon lauterbornii* Geitler in Geitler and Pascher 1925, 459.)

clath.ra.ti.for'me. L. part.adj. *clathratus* latticed; L. n. *forma* shape, form; M.L. neut.adj. *clathratiforme* lattice-like.

The description is based in part on: Pfennig and Cohen-Bazire, 1967.

Individual cells rod-shaped, 0.7–1.2 μm wide, 1.5–2.5 μm long, elongated cells up to 7 μm long may occur. Cells characteristically united in three dimensional nets which are formed in the following way. Successive binary fissions lead to the formation of chains of cells (Plate 1.1, Fig. 9). Occasionally, two adjacent cells in such a chain change their mode of growth. The contiguous poles start to branch simultaneously, leading to the formation, in the middle of the chain, of two Y-shaped cells, in apposition at the ends of both arms of the Y. If these two cells do not separate, the ring structure formed between them enlarges, by subsequent cell elongation and binary fission, into a typical, many celled mesh. The arrangement of cells at the branch points of fully formed nets implies that these cells eventually undergo ternary fission to yield three daughter cells, all in apposition at one pole. The colonial structure of *P. clathratiforme* is therefore caused by its ability to undergo both binary and ternary fissions. Color of cell suspensions green.

Obligately phototrophic, strictly anaerobic. pH range: 6.5–7.0. Growth temperature: 20–25 C.

Photosynthetic electron donors: sulfide, sulfur. In the presence of sulfide and bicarbonate, acetate is photoassimilated. Not utilized: thiosulfate, higher fatty acids, pyruvate, succinate, peptone. Incapable of assimilatory sulfate reduction.

Nitrogen sources: ammonium salts.

Predominant pigments: bacteriochlorophyll *c* or *d* and carotenoids.

Storage products: polyphosphates.

DNA base ratio: 48.5 moles % guanine + cytosine (buoyant density).

Habitat: mud and stagnant water containing hydrogen sulfide and exposed to light, sulfur springs.

Type strain: none; only partially purified cultures exist.

2. Pelodictyon luteolum (Schmidle) Pfennig and Trüper 1971, 13. (*Aphanothece luteola* Schmidle 1901, 179; *Schmidlea luteola* (Schmidle) Lauterborn 1913, 98; *Pelodictyon aggregatum* Perfil'ev 1914, 197.)

lu.te'o.lum. L. adj. *luteus* yellow; L. dim.adj. *luteolus* yellowish, somewhat yellow.

The description is based in part on: Pfennig, 1967.

Individual cells ovoid to short vibrio-shaped, 0.6–0.9 μm wide by 1.2–2.0 μm long, at times 1–1.5 μm wide by 2–4 μm long. In pure culture the organism may grow in the form of free individual cells; under certain conditions the cells produce slime and are united into hollow spherical or irregular round colonies with the cells in a single layer. The latter growth forms are occasionally observed in mud samples or enrichment cultures. Color of cell suspensions green.

Obligately phototrophic, strictly anaerobic. pH range: 6.5–7.0. Optimum growth temperature: 20–25 C.

Photosynthetic electron donors: sulfide, sulfur. In the presence of sulfide and bicarbonate, acetate and propionate are photoassimilated. Not utilized: thiosulfate, higher fatty acids, pyruvate, succinate, peptone. Incapable of assimilatory sulfate reduction.

Nitrogen sources: ammonium salts.

Predominant pigments: bacteriochlorophyll *c* or *d* and carotenoids.

Hydrogenase may be present.

DNA base ratios: 53.5–58.1 moles % guanine + cytosine (type strain: 58.1) (buoyant density).

Habitat: mud and stagnant water containing hydrogen sulfide and exposed to light; sulfur springs.

Illustrations: Lauterborn, 1916, Plate III, Figs. 29 to 31.

Neotype strain: "Polden"; SMG 273.

Genus V. **Clathrochloris** *Geitler in Geitler and Pascher 1925, 457*

Clath.ro.chlo'ris. L. pl.n. *clathri* lattice; Gr. adj. *chlorus* green; M.L. fem.n. *Clathrochloris* green lattice.

Spherical to ovoid bacteria, multiplying by **binary fission, non-motile. Cells in chains which are usually arranged in loose trellis-like aggregates.** Color is yellowish green. **Contain gas vacuoles** which have been mistaken for sulfur globules.

Anaerobic.

Capable of photosynthesis in the presence of hydrogen sulfide, during which they produce and **deposit** as an intermediate oxidation product, **elemental sulfur** in the form of globules **outside** the cells. **Cell suspensions** appear in various shades of **yellowish green.**

Type species of genus: *Clathrochloris sulfurica* (Szafer) Geitler in Geitler and Pascher 1925, 457.

Description of the species of genus **Clathrochloris**

1. **Clathrochloris sulfurica** (Szafer) Geitler in Geitler and Pascher 1925, 457. (*Aphanothece sulphurica* (sic) Szafer 1911, 162; *Clathrochloris sulphurica* (sic) Geitler in Geitler and Pascher 1925, 457; *Clathrochloris hypolimnica* Skuja 1948, 26.)

sul.fur′i.ca. L. n. *sulfur* sulfur; M.L. fem.adj. *sulfurica* sulfuric.

Cells about 0.5–1.5 μm in diameter. Color of cell masses and aggregates yellowish green.

Habitat: mud and stagnant water containing hydrogen sulfide and exposed to light. In sulfur springs and the hypolimnion of meromictic lakes.

Illustrations: Skuja, 1948, Plate I, Fig. 24.

Type strain: none; not in pure culture.

Comment: The forms described by Szafer and Skuja are most probably identical. As with other photosynthetic bacteria, the smaller size and abundant slime formation of Szafer's organism was probably due to the high sulfide content of the habitat.

ADDENDA TO THE FAMILY **CHLOROBIACEAE**

I. Green phototrophic bacteria living in consortia with other microorganisms.

The use of generic designations for these apparently stable complexes composed of two different organisms has been frequently questioned (Buder, 1914, 80; Perfil'ev, 1914, 223; van Niel, 1957, 66). According to the rules of bacterial nomenclature the generic designations are not validly published (Trüper and Pfennig, 1971). Therefore, the generic designations with the addition of the term consortium are used here as laboratory names without taxonomic significance. It is possible that these consortia represent fortuitous combinations whose success depends on environmental factors. If so, the green components should be placed in the appropriate genera. The isolation of *Chlorobium chlorochromatii* by Mechsner (1957) (*cf.* Addendum II, p. 59) was the first step in this direction.

a. *Chlorochromatium* Lauterborn 1906, 197.

Chlo.ro.chro.ma′ti.um. Gr. adj. *chlorus* green; Gr. n. *chromatium* color, paint; M.L. neut.n. *Chromatium* a bacterial genus; M.L. neut.n. *Chlorochromatium* a green *Chromatium.*

Green sulfur bacteria, ovoid- to rod-shaped with rounded ends occur as barrel-shaped aggregates consisting of a rather large, colorless, polar flagellated bacterium as the center which is surrounded by green bacteria, arranged in 4 to 6 rows, ordinarily from 2 to 4 cells high. The entire conglomerate behaves like a unit, is motile and multiplies by the more or less simultaneous fission of its components. The green constituents contain a chlorophyllous pigment which is not identical with the common green plant chlorophylls or with bacteriochlorophyll *a* or *b*. Capable of photosynthesis in the presence of hydrogen sulfide but do not store sulfur globules in the cells.

a1. *Chlorochromatium aggregatum* Lauterborn 1906, 197. (*Chloronium mirabile* Buder 1914, 80.)

ag.gre.ga′tum. L. part.adj. *aggregatum* flocked together, clumped.

Cells of the green component 0.5–1.0 by 1.0–2.5 μm, mostly from 8 to 16 individuals surrounding the central bacterium. Size of the total barrel-shaped unit variable, generally 2.5–5 by 7–12 μm. Occasionally a group of the complex colonies may remain attached in a chain.

Anaerobic.

Habitat: mud and stagnant water containing rather high concentrations of hydrogen sulfide and exposed to light.

There is at present no reason for distinguishing two varieties (forma *typica* and forma *minor*) on the basis of size differences of the consortium, as Geitler proposed (1925, 460). The reported and personally observed sizes of such units show that the extreme limits are linked by a complete series of transitions.

Illustrations: Buder, 1914, Plate XXIV, Figs. 1 to 5; Perfil'ev, 1914, Figs. 1 to 5, p. 213.

Comment: The green component was isolated by Mechsner and named *Chlorobium chlorochromatii* (see Addendum II, p. 59)

a2. *Chlorochromatium glebulum* (sic) Skuja 1956, 36.

gle′bu.lum. L. n. *glaeba,* also *gleba* a lump (of soil); L. dim.n. *glebula* small lump.

Cells of the green component 0.5–0.6 by 0.7–1.0 μm, non-motile, containing gas vacuoles; mostly from 7–40 individuals surround the curved central bacterium. Size of the barrel-shaped unit 3–4 μm by 4–8 μm. The central bacterium appears to be a polarly flagellated spirillum.

Source: hypolimnion of Swedish lakes.

Illustration: Skuja, 1956, Plate III, Figs. 52 and 53.

Comment: Skuja noted the similarity of the green component to *Pelodictyon luteolum* (see above).

a3. *Pelochromatium roseum* Lauterborn 1913,

99. (*Pelochromatium roseum* var. *minor* Skuja 1948, 29) (see also Pfennig, 1967; Utermöhl, 1924.)

ro.se'um. L. n. *rosa* the rose, L. adj. *roseum* rose-colored, rose, pink.

Cells of the pink-brown component slightly curved to vibrio-shaped 1 μm or less by 2 μm, mostly from 10–20 individuals surrounding the central bacterium. Size of the total barrel-shaped unit 2.5–4 by 4–8 μm.

Habitat: mud and hydrogen sulfide containing water of ponds and lakes.

Illustrations: Lauterborn, 1913, Plate III, Fig. 28, a–c.

Comment: The brown component contains bacteriochlorophyll *d* and appears similar to the brown-colored *Chlorobium* species (see above).

b. *Cylindrogloea* Perfil'ev 1914, 223.

cy.lin.dro.gloe'a. Gr. n. *cylindrus* cylinder; Gr. n. *gloea* gum; M.L. fem.n. *Cylindrogloea* cylindrical gum.

Green sulfur bacteria consisting of small ovoid to rod-shaped cells growing in association with a filamentous, colorless, central bacterium, thus forming a cylinder-shaped consortium. Non-motile. The green component contains a chlorophyllous pigment different from the common chlorophylls of green plants and from bacteriochlorophyll *a* or *b*. Capable of photosynthesis in the presence of hydrogen sulfide without depositing sulfur globules in the cells.

b1. *Cylindrogloea bacterifera* Perfil'ev, 1914, 223.

bac.te.ri'fe.ra. M.L. n. *bacter* rod (a combining form); L. verbal suf. *fer* bearing; M.L. fem.adj. *bacterifera* rod-bearing.

Individual green components ovoid- to rod-shaped, about 0.5–1 by 2–4 μm. The central filamentous bacterium is embedded in a slime capsule of considerable dimensions on the outside of which is a layer of green bacteria, usually one cell thick. The green organisms may form a very dense outer covering, or they may be more sparsely distributed over the slime capsule. This unit is again surrounded by a sizeable slime layer, the complete aggregate measuring 7–8 μm wide and up to 50 μm long; non-motile. Both components appear to be non-spore-forming.

Habitat: mud and stagnant water containing rather high concentrations of hydrogen sulfide and exposed to light.

Illustrations: Perfil'ev, 1914, Figs. 6 to 11, p. 213; Skuja, 1956, Plate IV, Figs. 19 to 23 (Skuja shows the green component with gas vacuoles).

Comment: The green component is probably a *Chlorobium* or *Pelodictyon* species. It is also possible that this is a variety of the *Chlorochromatium* consortia.

c. *Chlorobacterium* Lauterborn 1916, 429 (not *Chlorobacterium* Guillebeau 1890, 32).

chlo.ro.bac.te'ri.um. Gr. adj. *chlorus* green; L. n. *bacterium* a small rod; M.L. neut.n. *Chlorobacterium* a green rodlet.

Green sulfur bacteria (?) which grow symbiotically as an outside covering on cells of protozoa, such as amoeba and flagellates. Cells rod-shaped, often slightly curved, greenish. Non-motile.

c1. *Chlorobacterium symbioticum* Lauterborn 1916, 429.

sym.bi.o'ti.cum. Gr. adj. *symbioticum* of companionship, symbiotic.

Cells rod-shaped, about 0.5 by 2–5 μm, often slightly curved. Non-motile.

Occur as a peripheral covering of certain protozoa with which they may form a symbiotic unit.

It is not certain that this is a green sulfur bacterium. The descriptions of localities where it was found fail to mention the presence of hydrogen sulfide in the environment; this should be a prerequisite for a member of this group.

Source: reported from a number of pools in Germany.

Habitat: stagnant water.

Illustrations: Lauterborn, 1915, Plate III, Figs. 34 to 36; Pascher, in Geitler and Pascher, 1925, Fig. 149.

Comment: If the green component can be isolated and turns out to be a green sulfur bacterium it should be placed with the *Chlorobiaceae*.

II. Species that have been in pure culture but were lost before detailed taxonomic studies were made.

a. *Microchloris nadsonii* Pringsheim 1953, 362.

Rod-shaped bacteria, multiplying by binary fission, 0.4 μm wide, 2.3–4.5 (6.0) μm long, non-motile. Single cells colorless, in accumulations yellowish green. Cells surrounded by slime so that they are distinctly separated from each other.

Source: polluted ditches.

Illustration: Pringsheim, 1953, Fig. 4.

Comment: If reisolated this organism must be included in the genus *Chlorobium*, since a new genus, based on differences in size only, is not justified.

b. *Chlorobium chlorochromatii* Mechsner 1957, 46.

Spherical to rod-shaped bacteria, 0.7–0.9 μm wide, 0.9–1.1 μm long, non-motile. Usually arranged in chains. Gram-negative. Color: green to yellowish green. Pigments: bacteriochlorophyll *c* or *d* and carotenoids. Obligately anaerobic and phototrophic, facultatively heterotrophic. Hydrogen sulfide and thiosulfate are oxidized to sulfate via elemental sulfur. Organic substances (peptone, malate) are assimilated photosynthetically.

Habitat: mud and stagnant water containing

hydrogen sulfide. The organism lives in symbiosis with a colorless bacterium, together forming the *Chlorochromatium aggregatum*—consortium (cf. above).

Illustration: Mechsner, 1957, Figs. 1, 2, 4 and 5.
Comment: Unfortunately, this organism was lost before detailed metabolic and taxonomic studies were made.

III. Species never isolated and only incompletely described from samples of water and mud. Rarely found or mentioned again by other authors. Some hardly recognizable.

a. *Sorochloris aggregata* Pascher in Geitler and Pascher 1925, 455.

Cell spherical, 1.5 μm in diameter; enlarged cells may be found. Occur in spherical colonies or clusters of 10–25 μm diameter, embedded in thick slime layers. Color yellowish green to slightly brownish green.

Source: freshwater ponds with decaying plant material.

Illustration: Pascher in Geitler and Pascher, 1925, Fig. 1.
Comment: If such an organism is isolated it should be placed with the genus *Chlorobium* unless it contains gas vacuoles; in which case it would probably be identical with *Pelodictyon luteolum*, as already mentioned by Pascher in Geitler and Pascher, 1925, 455.

b. *Pelodictyon parallelum* (Szafer) Perfil'ev 1914, 198. (*Aphanothece parallela* Szafer 1911, 163; *Pediochloris parallela* (Szafer) Geitler in Geitler and Pascher 1925, 457.)

Ovoid- to rod-shaped bacteria; 0.6–0.7 μm wide, 1.0–1.2 μm long, non-motile. Contain gas vacuoles. Cells occur in chains which form two-dimensional plates. Color: yellowish green.

Source: sulfur springs near Lemberg.
Illustrations: Szafer, 1911, 161; Geitler and Pascher, 1925, 457, Fig. 4.

IV. Species whose classification as bacteria appears doubtful

a. *Tetrachloris inconstans* Pascher in Geitler and Pascher 1925, 456.

Cells spherical, 1.5 μm in diameter, sometimes up to 3 μm, non-motile. Usually arranged in tetrads, sometimes embedded in a thin slime layer. Color pale yellowish green.

Source: freshwater with decaying plant material.

Illustration: Pascher in Geitler and Pascher, 1925, Fig. 2.

b. *Tetrachloris merismopedioides* Skuja 1948, 27.

Cells ovoid, 0.6–0.7 μm wide, 0.7–0.9 μm long, non-motile. Older cells contain a centrally located gas vacuole. Usually arranged in flat platelets of 4, 8, 16 etc. cells. Color pale yellowish green.

Source: Swedish freshwater lakes.
Illustration: Skuja, 1948, Plate I, Fig. 23.
Comment: Pringsheim (1953, 353) placed the genus *Tetrachloris* in *Cyanophyceae* and described a new species, *Tetrachloris diplococcus*, as such.

c. *Pelogloea chlorina* Lauterborn 1913, 99.

Cells rod-shaped, 1.0 μm wide, 3–8 μm long, non-motile, usually forming chains embedded in a common slime capsule. Colonies up to 1 mm. Color yellowish green.

Source: mud from ponds with *Chara*.
Illustration: Lauterborn, 1916, Plate III, Fig. 32.

d. *Pelogloea bacillifera* Lauterborn 1916, 430.

Cells rod-shaped, 1–5 μm wide, 2–4 μm long. Otherwise like *P. chlorina*.

Illustration: Lauterborn, 1916, Plate III, Figs. 30 and 31.
Comment: Pringsheim (1953, 353) regards *Pelogloea* as a genus of *Cyanophyceae*.

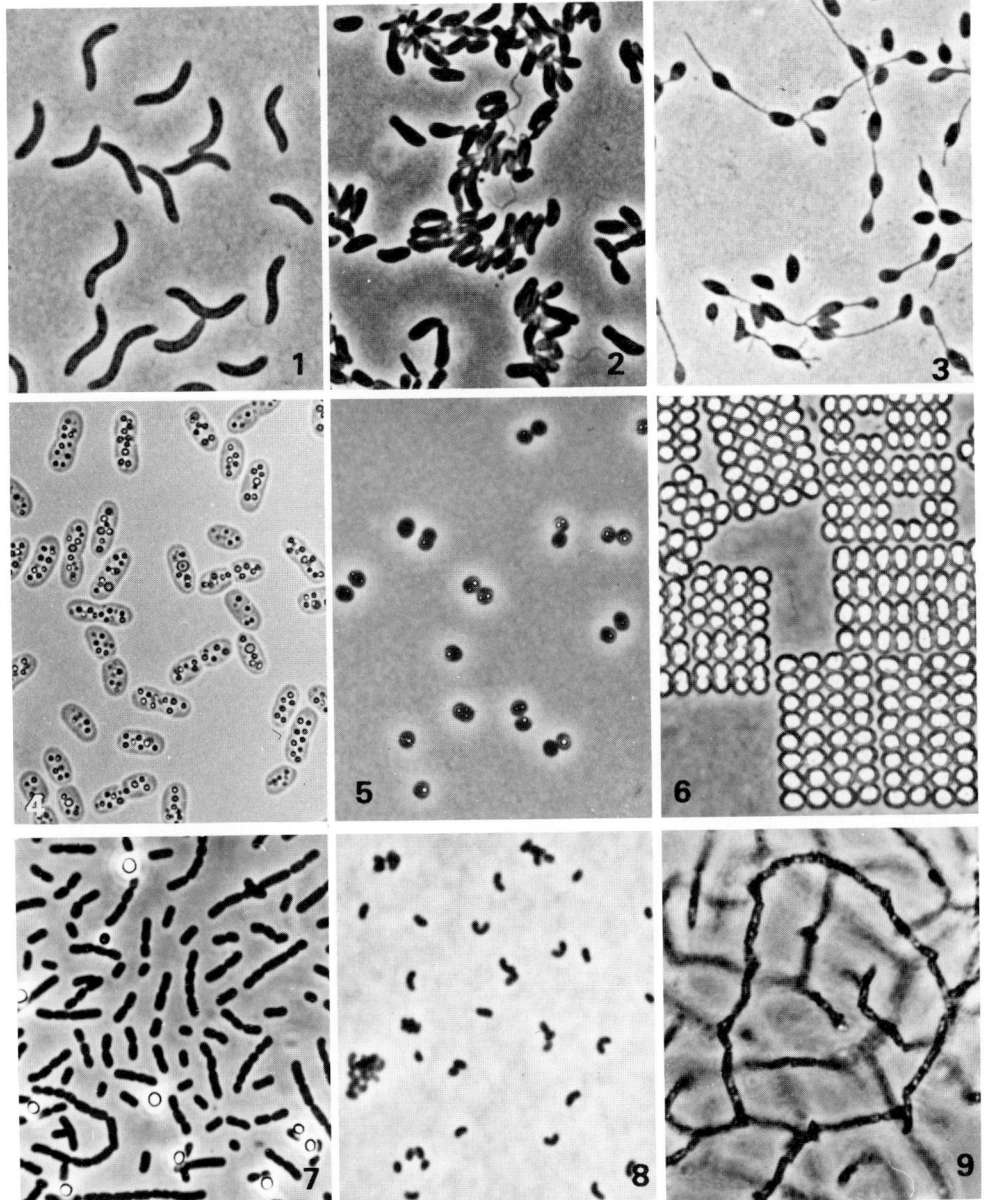

Plate 1.1. Morphology of typical representatives of the phototrophic bacteria during the exponential growth phase. (All organisms were grown anaerobically under optimum conditions.)

Fig. 1. *Rhodospirillum molischianum* strain NTHC 131. Phase contrast: × 2000.

Fig. 2. *Rhodopseudomonas acidophila* strain SMG 137. Phase contrast: × 2000.

Fig. 3. *Rhodomicrobium vannielii* strain SMG 163. Phase contrast: × 2000.

Fig. 4. *Chromatium vinosum* strain SMG 186. Light field: × 2000. Note the intracellular sulfur globules.

Fig. 5. *Thiocapsa roseopersicina* strain SMG 219. Phase contrast: × 2000. Note the intracellular sulfur globules.

Fig. 6. *Thiopedia rosea* from nature. Phase contrast: × 2000. Note the gas vacuoles.

Fig. 7. *Chlorobium limicola* strain SMG 249. Phase contrast: × 2000. Note the extracellular sulfur globules.

Fig. 8. *Chlorobium phaeovibrioides* strain SMG 269. Phase contrast: × 3000.

Fig. 9. *Pelodictyon clathratiforme* strain 1831. Phase contrast: × 2000. Note the gas vacuoles.

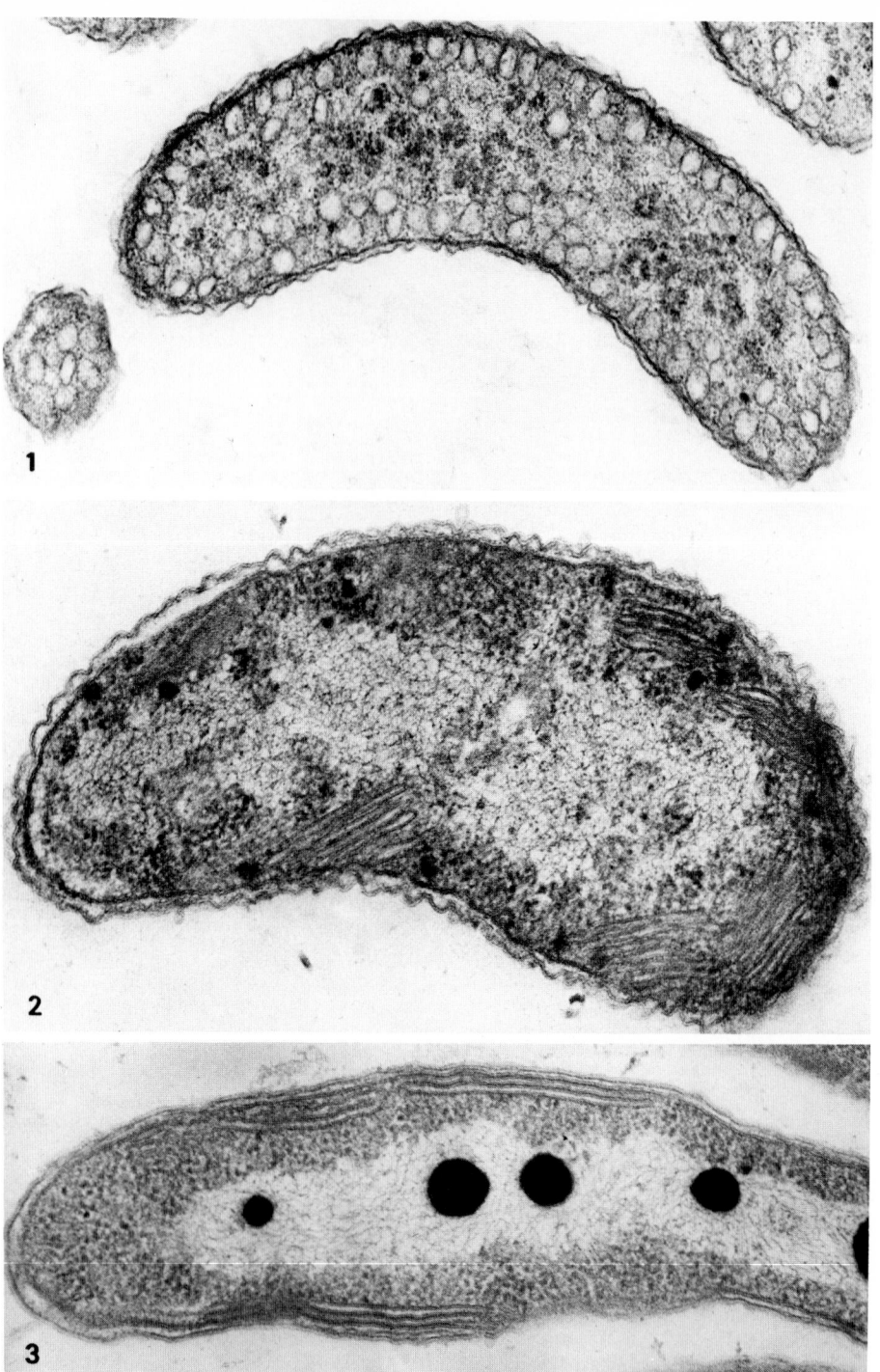

Plate 1.2. Fine structure of some Rhodospirillaceae

Fig. 1. *Rhodospirillum rubrum* strain FR1 grown anaerobically in the light. Note the vesicular structure of the intracytoplasmic membrane system. $\times$ 51,000. Courtesy of G. Drews and R. Ladwig.

Fig. 2. *Rhodospirillum molischianum* grown anaerobically in the light. Note the position and the lamellar stack type of the intracytoplasmic membrane system. $\times$ 90,000. Courtesy of G. Drews and R. Ladwig.

Fig. 3. *Rhodopseudomonas palustris* strain 11/1 grown semi-aerobically in the dark. Note position and type of intracytoplasmic membrane system. $\times$ 60,000. Courtesy of H. D. Tauschel and R. Ladwig.

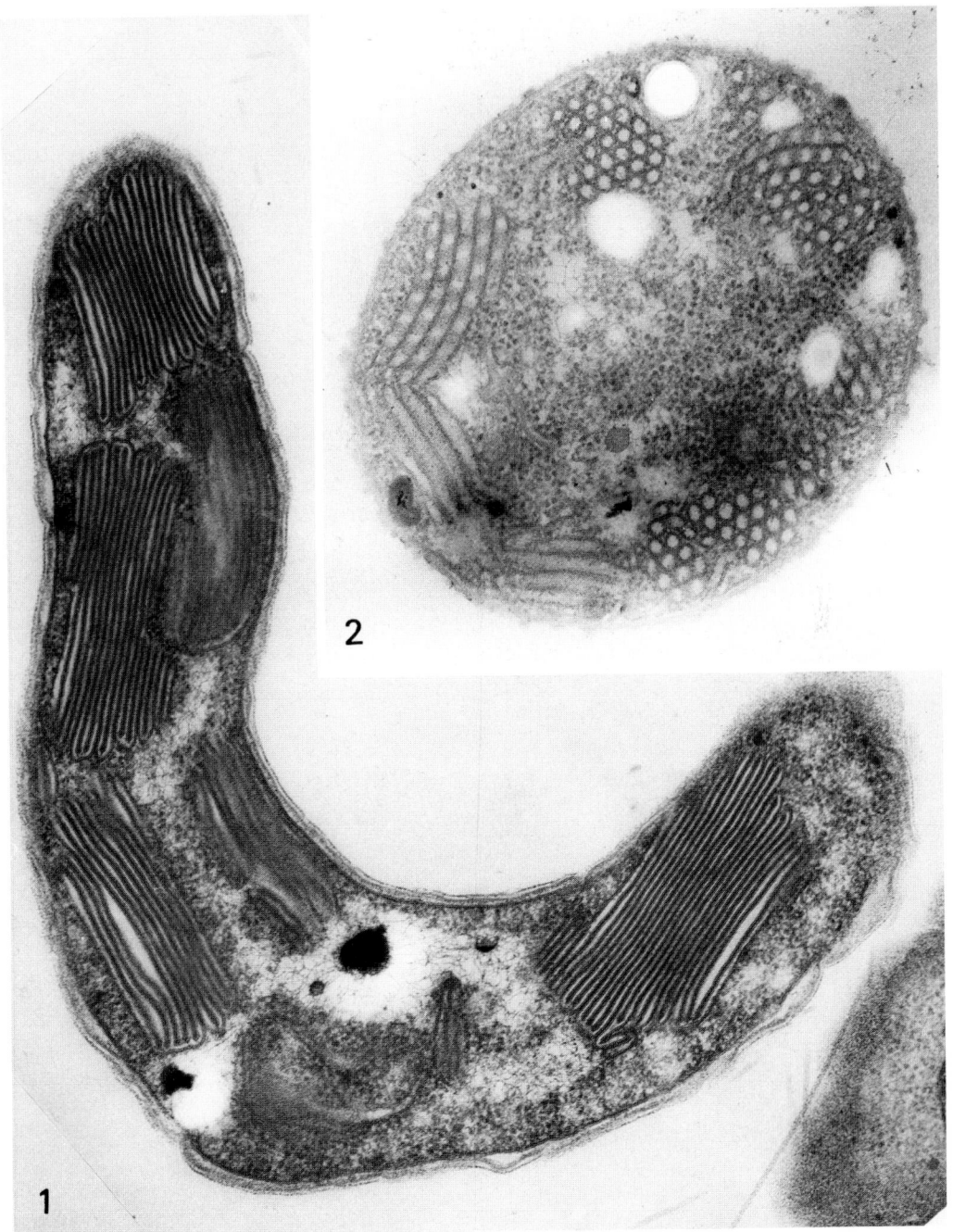

Plate 1.3. Fine structure of some Chromatiaceae

FIG. 1. *Ectothiorhodospira mobilis* strain SMG 237. Note the position and lamellar stack type of the intracytoplasmic membrane system. × 60,000. Courtesy of C. C. Remsen, S. W. Watson and J. B Waterbury.

Fig. 2. *Thiocapsa pfennigii* strain SMG 228. Note the bundled tube type of the intracytoplasmic membrane system. × 60,000. Courtesy of S. W. Watson, J. B. Waterbury and C. C. Remsen.

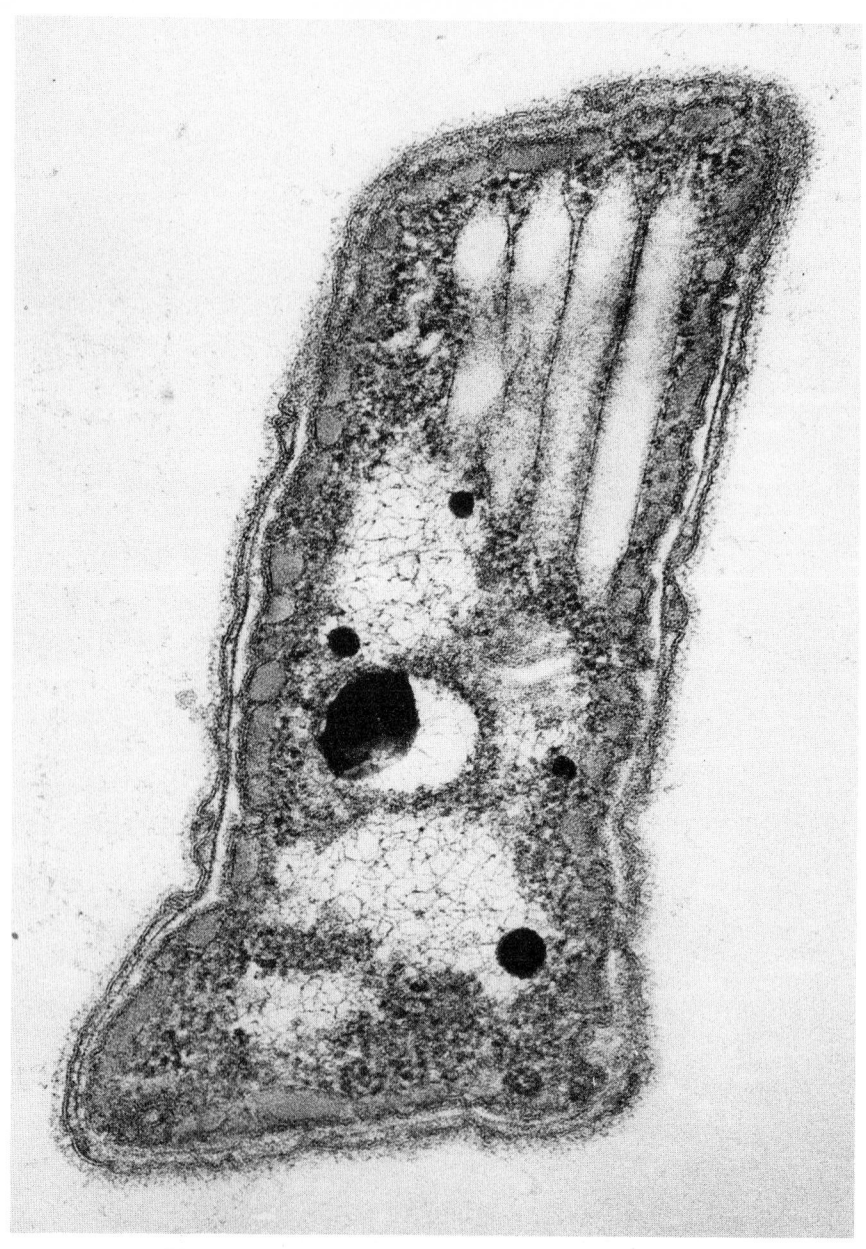

Plate 1.4. Fine structure of a green sulfur bacterium
Pelodictyon clathratiforme strain 1831. Note the Chlorobium vesicles (dark grey) underlying and attached to the cytoplasmic membrane, the bundle of gas vacuoles (light grey with pointed ends) in the upper part of the section, and the pointed ends of the cell indicating ternary fission. × 105,000. Courtesy of G. Cohen-Bazire.

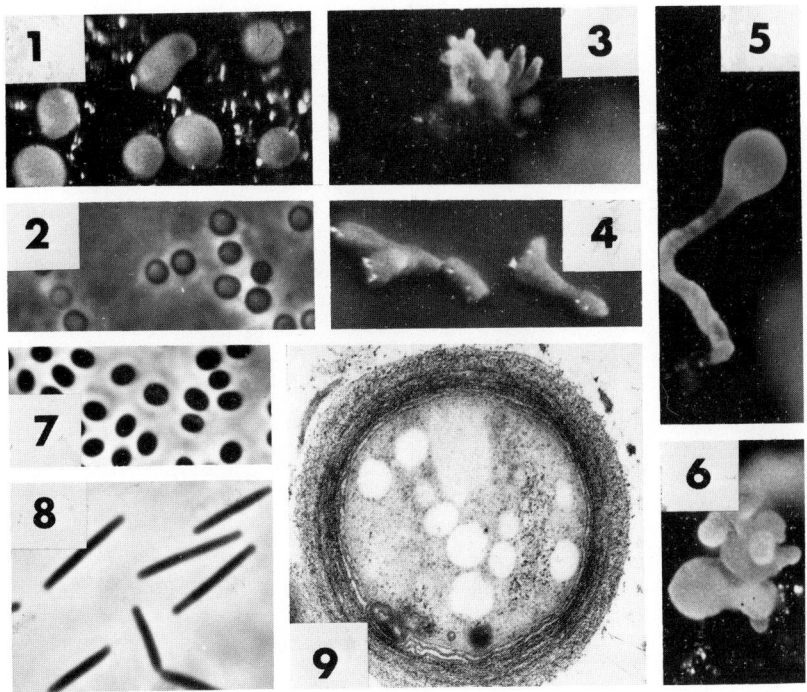

Plate 2.1. *Myxococcus*

Fig. 1. *Myxococcus fulvus* fruiting body on dung (× 30).
Fig. 2. *Myxococcus coralloides* microcysts (× 2000).
Fig. 3. *Myxococcus coralloides* fruiting body on bark (× 100).
Fig. 4. *Myxococcus coralloides* fruiting body on agar (× 100).
Fig. 5. *Myxococcus stipitatus* fruiting body on dung (× 100).
Fig. 6. *Myxococcus stipitatus* fruiting on dung (× 30).
Fig. 7. *Myxococcus stipitatus* microcysts (× 2000).
Fig. 8. *Myxococcus fulvus* vegetative cells (× 2000).
Fig. 9. *Myxococcus xanthus* electron micrograph of microcyst (ca. × 40,000).
Figs. 3, 4 and 5. Reproduced by permission of the National Research Council of Canada from McCurdy, *Canadian Journal of Microbiology 15:* 1453–1461, 1969.

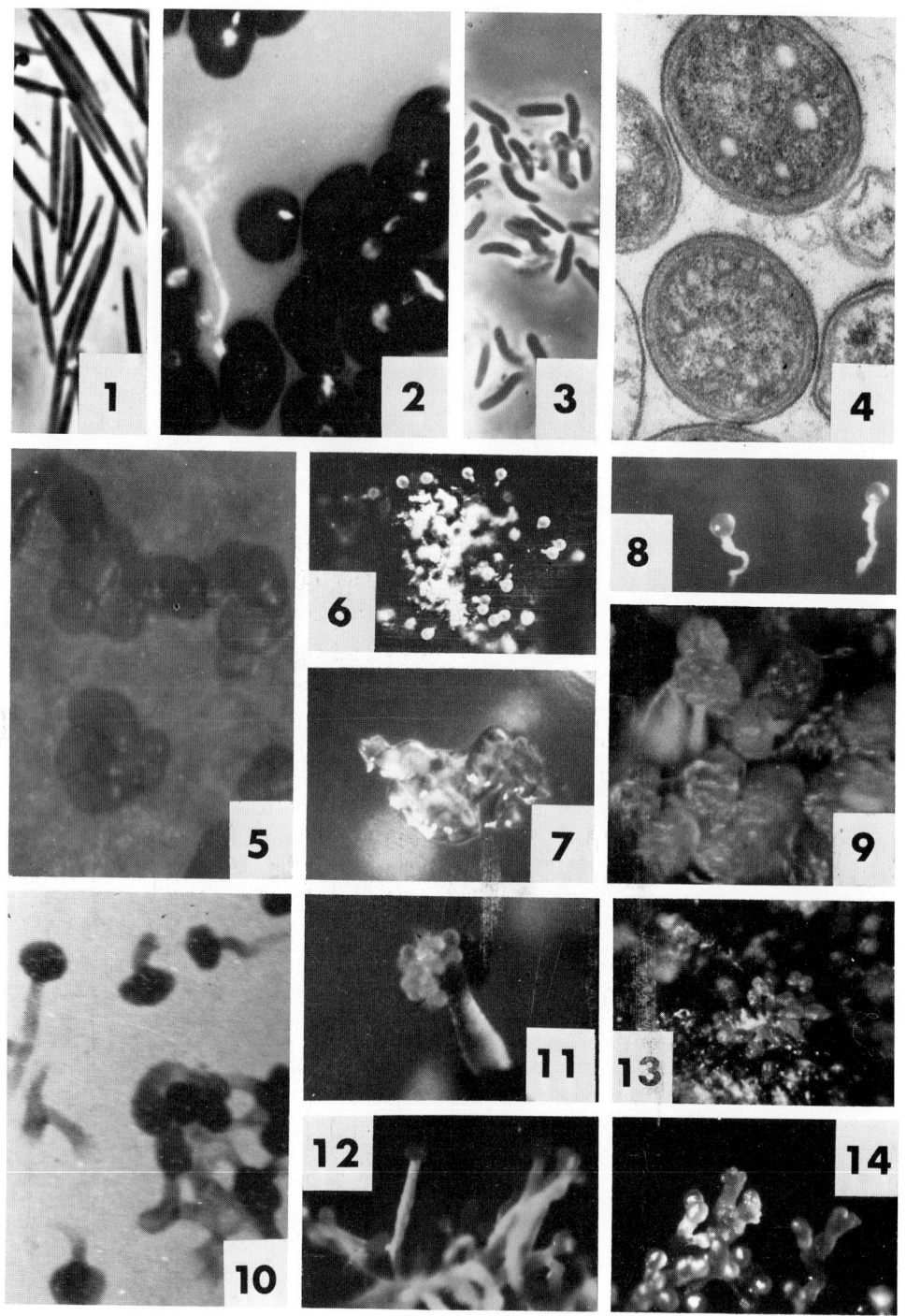

Plate 2.2.

Plate 2.2. *Cystobacter, Melittangium, Stigmatella*

Fig. 1. *Cystobacter fuscus* vegetative cells (× 2000).

Fig. 2. *Cystobacter fuscus* sporangia (× 100).

Fig. 3. *Cystobacter fuscus* microcysts (× 2000).

Fig. 4. *Cystobacter fuscus* microcyst, electron micrograph of cross-section (× 50,000).

Fig. 5. *Cystobacter ferrugineum* fruiting bodies (× 120).

Fig. 6. *Melittangium lichenicola* fruiting bodies on bark (× 100).

Fig. 7. *Melittangium lichenicola* archangium-like fruiting body on agar (× 25).

Fig. 8. *Melittangium alboraceum* fruiting bodies (ca. × 130).

Fig. 9. *Melittangium boletus* fruiting bodies (× 120).

Fig. 10. *Melittangium boletus* fruiting bodies on agar (ca. × 100).

Fig. 11. *Stigmatella aurantiaca* fruiting body on bark (× 90).

Fig. 12. *Stigmatella aurantiaca* fruiting bodies on agar (× 100).

Fig. 13. *Stigmatella erecta* fruiting body on bark (× 60).

Fig. 14. *Stigmatella erecta* fruiting body on agar (× 100).

Figs. 6 and 7. From McCurdy, *International Journal of Systematic Bacteriology 21:* 50–54, 1971.

Fig. 8. Courtesy of J. E. Peterson.

Figs. 9 and 10. From Krzemieniewska and Krzemieniewski, *Acta Soc Bot Polon 4:* 1–54, 1926.

Fig. 14. Reproduced by permission of the National Research Council of Canada from McCurdy and Khouw, *Canadian Journal of Microbiology 15:* 731–738, 1969.

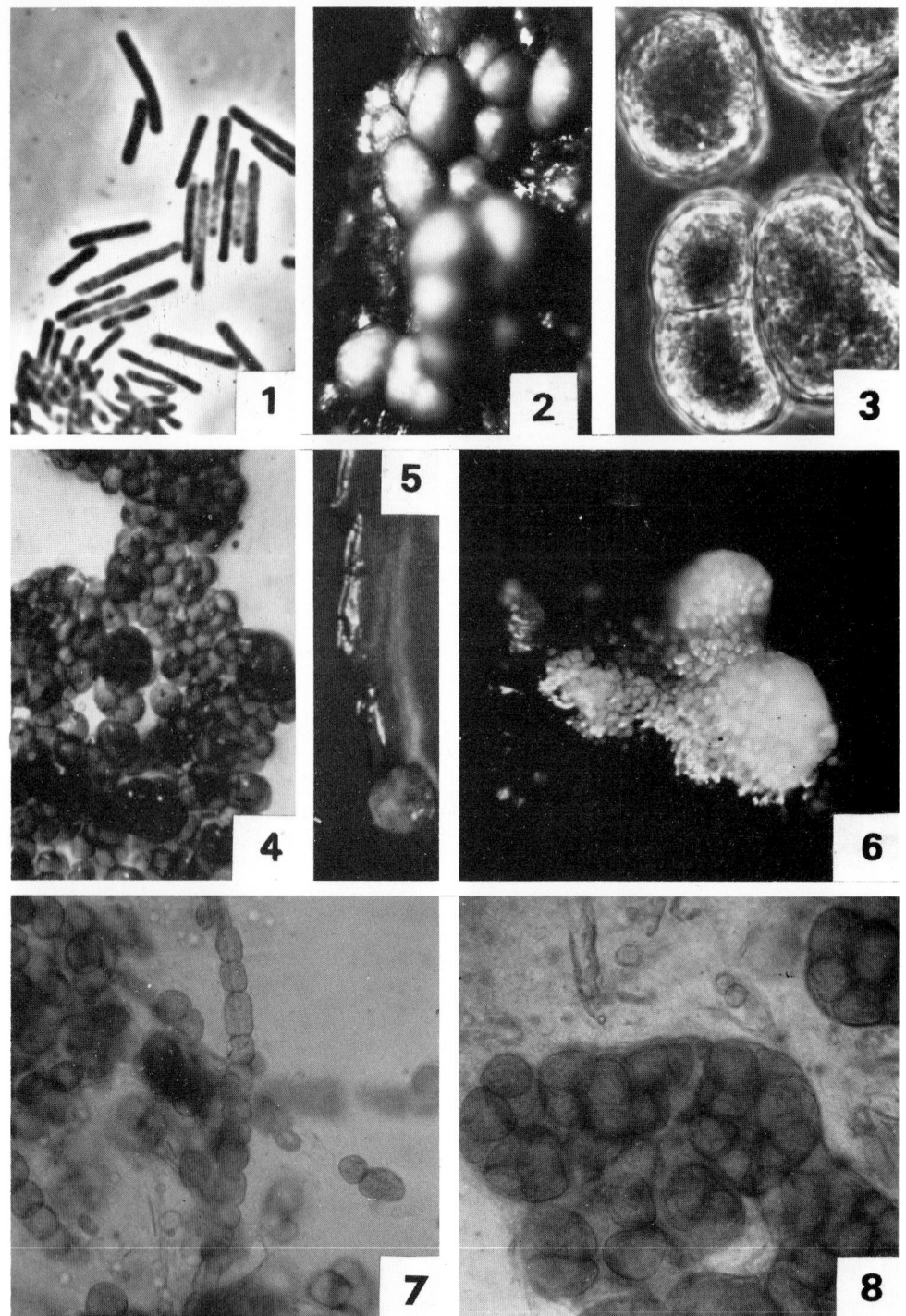

Plate 2.3. *Polyangium*

Fig. 1. *Polyangium vitellinum* vegetative cells (× 2000).

Fig. 2. *Polyangium vitellinum* sporangia (× 100).

Fig. 3. *Polyangium aureum* sporangia (× 1200).

Fig. 4. *Polyangium fumosum* sori (× 120).

Fig. 5. *Polyangium fumosum* migrating mass and grooved trail (× 50).

Fig. 6. *Polyangium sorediatum* fruiting body on dung (ca. × 80).

Fig. 7. *Polyangium cellulosum* (× 300).

Fig. 8. *Polyangium cellulosum* (× 300).

Figs. 2 and 3. From McCurdy, *International Journal of Systematic Bacteriology 20:* 283–296, 1970.

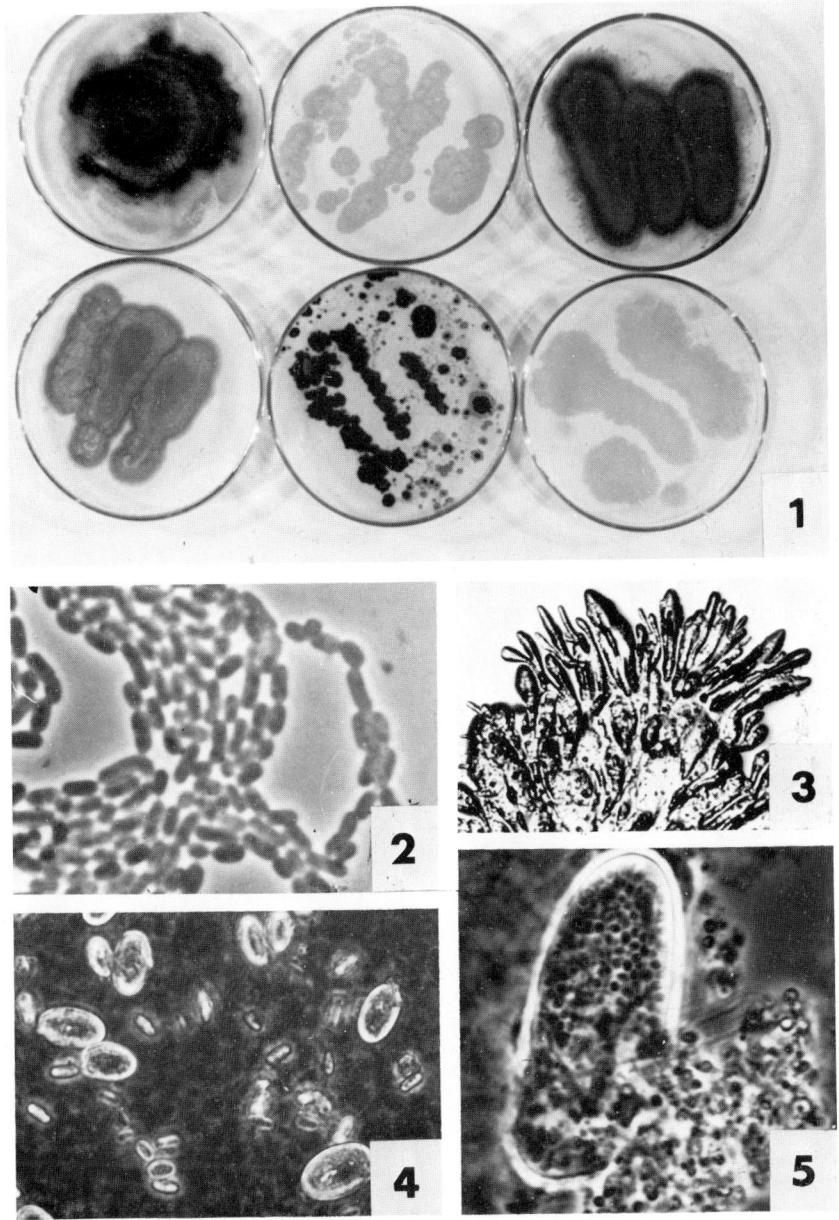

Plate 2.4. *Polyangium, Nannocystis*

Fig. 1. Color variants of *Polyangium cellulosum* on filter paper.
Fig. 2. *Nannocystis exedens* vegetative cells ($\times$ 2000).
Fig. 3. *Nannocystis exedens* swarm pattern etched into agar ($\times$ 4.7).
Fig. 4. *Nannocystis exedens* sporangia of various sizes embedded in agar ($\times$ 500).
Fig. 5. *Nannocystis exedens* squashed sporangium with myxospores ($\times$ 1500).
Fig. 1. Courtesy of J. E. Peterson.
Figs. 2 through 5. Courtesy of H. Reichenbach.

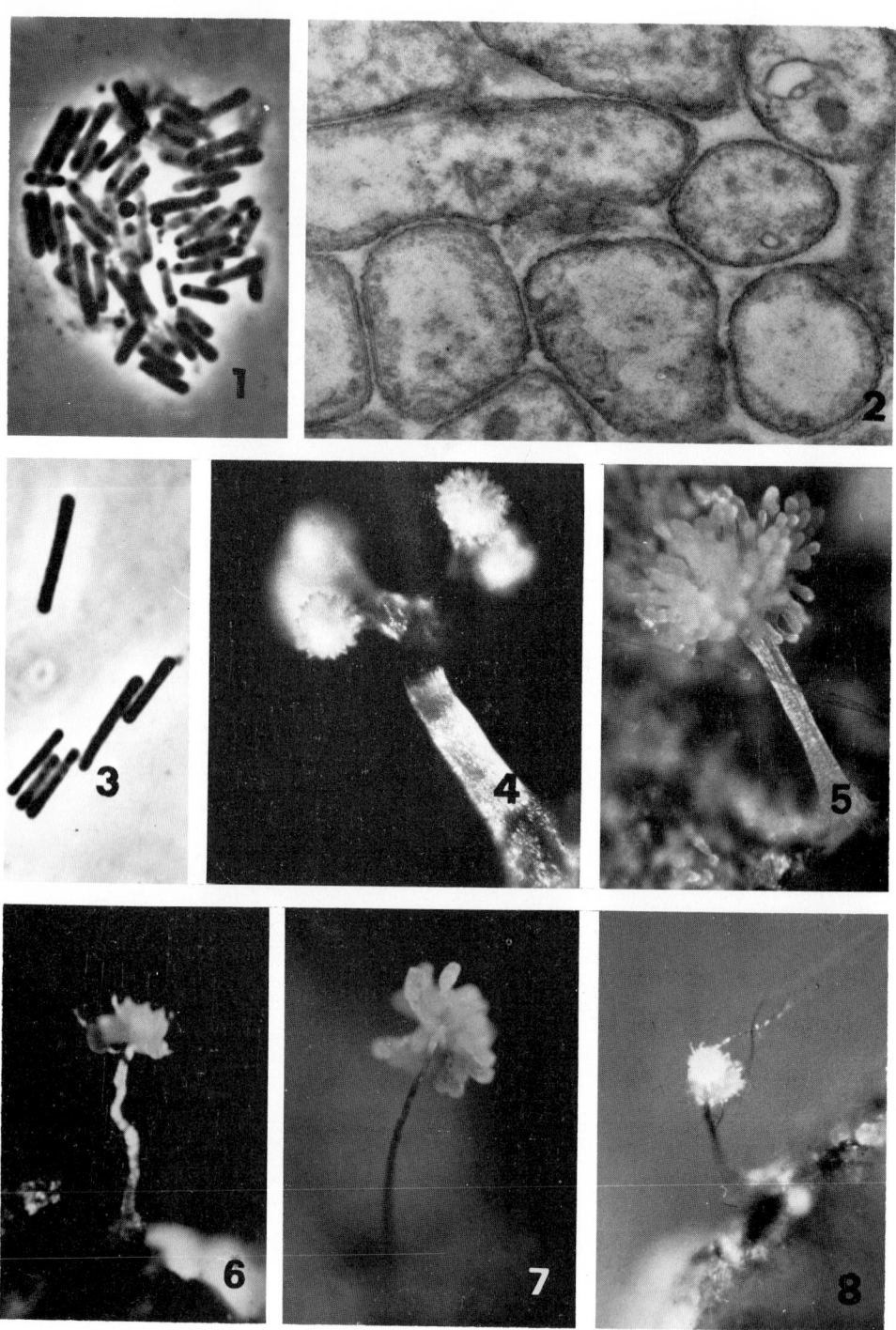

Plate 2.5.

70

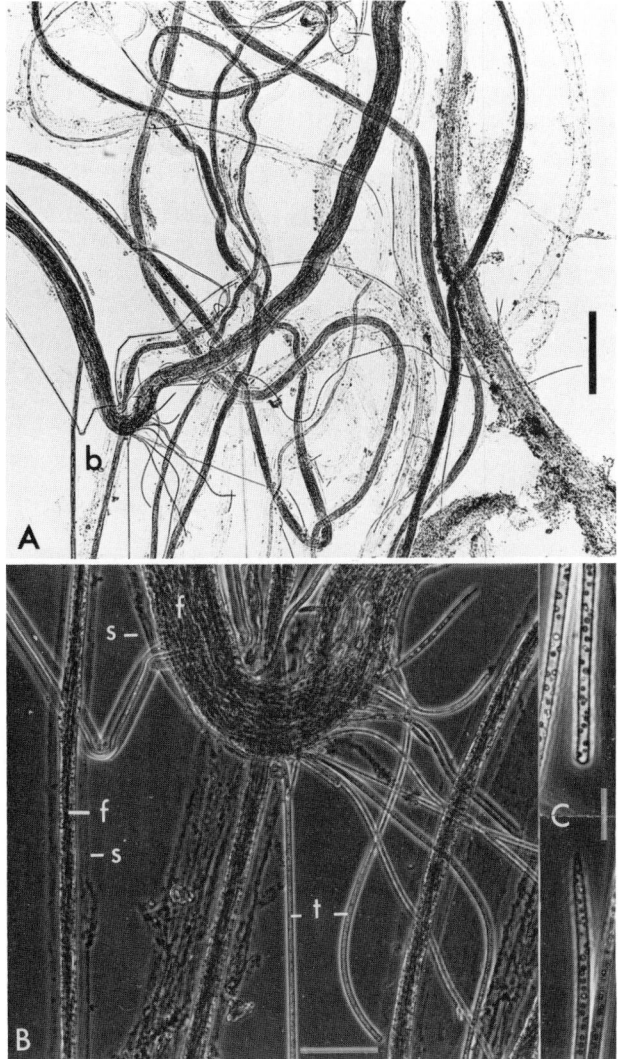

Plate 2.6. *Thioploca ingrica* from Lake Erie

Fig. A. Portion of mass of filaments washed free of mud. Scale 200 μm. *b*, area enlarged in Fig. B.

Fig. B. Fascicle of filaments (*f*) enclosed by the sheath (*s*); free filaments (*t*) emerged from a break in the sheath. Phase-contrast. Scale 50 μm.

Fig. C. Note sulfur inclusions and the obtuse (*top*) and tapered (*bottom*) terminal segments. Phase-contrast. Scale 10 μm.

Plate 2.5. *Chondromyces*

Fig. 1. *Chondromyces crocatus* myxospores (× 2000).
Fig. 2. *Chondromyces crocatus* myxospores in sporangium (electron micrograph, × 30,000).
Fig. 3. *Chondromyces crocatus* vegetative cells (× 2000).
Fig. 4. *Chondromyces crocatus* fruiting body (× 130).
Fig. 5. *Chondromyces catenulatus* fruiting body (× 120).
Fig. 6. *Chondromyces apiculatus* fruiting body (× 100).
Fig. 7. *Chondromyces pediculatus* fruiting body (× 100).
Fig. 8. *Chondromyces lanuginosus* fruiting body (× 90).

Figs. 1, 3, and 5. From McCurdy, *International Journal of Systematic Bacteriology 21:* 40–49, 1971.
Fig. 2. From McCurdy, *Arch Mikrobiol 65:* 380–390, 1968.
Figs. 6, 7, and 8. Reproduced by permission of the National Research Council of Canada from McCurdy, *Canadian Journal of Microbiology 15:* 1453–1461, 1969.

Plate 2.7. *Simonsiella, Alysiella*

Fig. 1. *Simonsiella crassa* short filament.

Fig. 2. Microcolony of *S. crassa* showing filaments coiled on edge and commencement of migration on flat side of filaments at left.

Fig. 3. *Alysiella filiformis* microcolony showing very short filaments.

Fig. 4. *A. filiformis* microcolony showing long filaments some on edge.

All young cultures on serum agar, phase-contrast microscopy. Lines = 10 μm. From *Journal of General Microbiology 29:* 624, 1962.

Plate 2.8. *Toxothrix trichogenes*

Figs. 1 and 2. Living trichomes during 1st min of laboratory observation; phase-contrast microscopy, wet mount.

Figs. 3 to 6. Fate of filaments during laboratory microscopic observation; Fig. 3 at beginning and Fig. 4 breakage after 5 min of observation; Fig. 5 after breakage and Fig. 6 disintegration 3 min after breakage.

Figs. 7 and 8. The excreted mucoid material partially encrusted with iron; Fig. 7 fan-shaped structure; Fig. 8 "twisted rope sheaths," fans and double tracks.

Figs. 9 and 10. Double tracks as produced by the ends of U-shaped filament moving over glass slide; dark-field. All μm lines = 10 μm, except as noted.

All figures from Krul *et al., Antonie van Leeuwenhoek Journal of Microbiology and Serology 36:* 409–420, 1970.

Plate 2.9. Cells of *Achromatium oxaliferum*

Fig. 1. Untreated.

Figs. 2 and 3. After immersion in 0.001 M acetic acid for 10 and 30 min, respectively. Scale line represents 20 μm.

Fig. 4. Slime and filaments, electron micrograph, osmic acid fixation, Pt shadowed, × 31,160. From de Boer, la Rivière and Schmidt, *Antonie van Leeuwenhoek Journal of Microbiology and Serology 37:* 553–563, 1970.

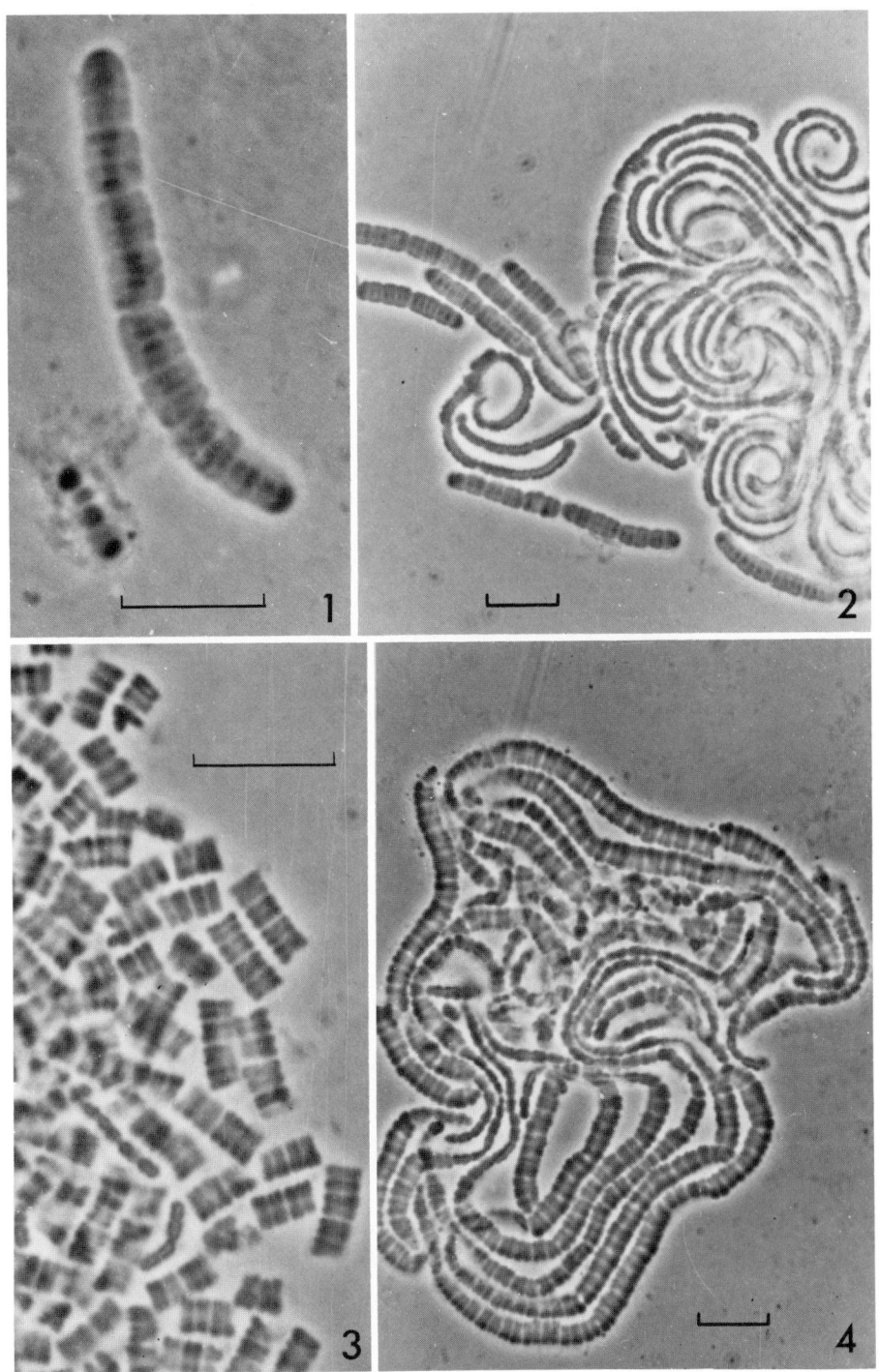

Plate 2.7.

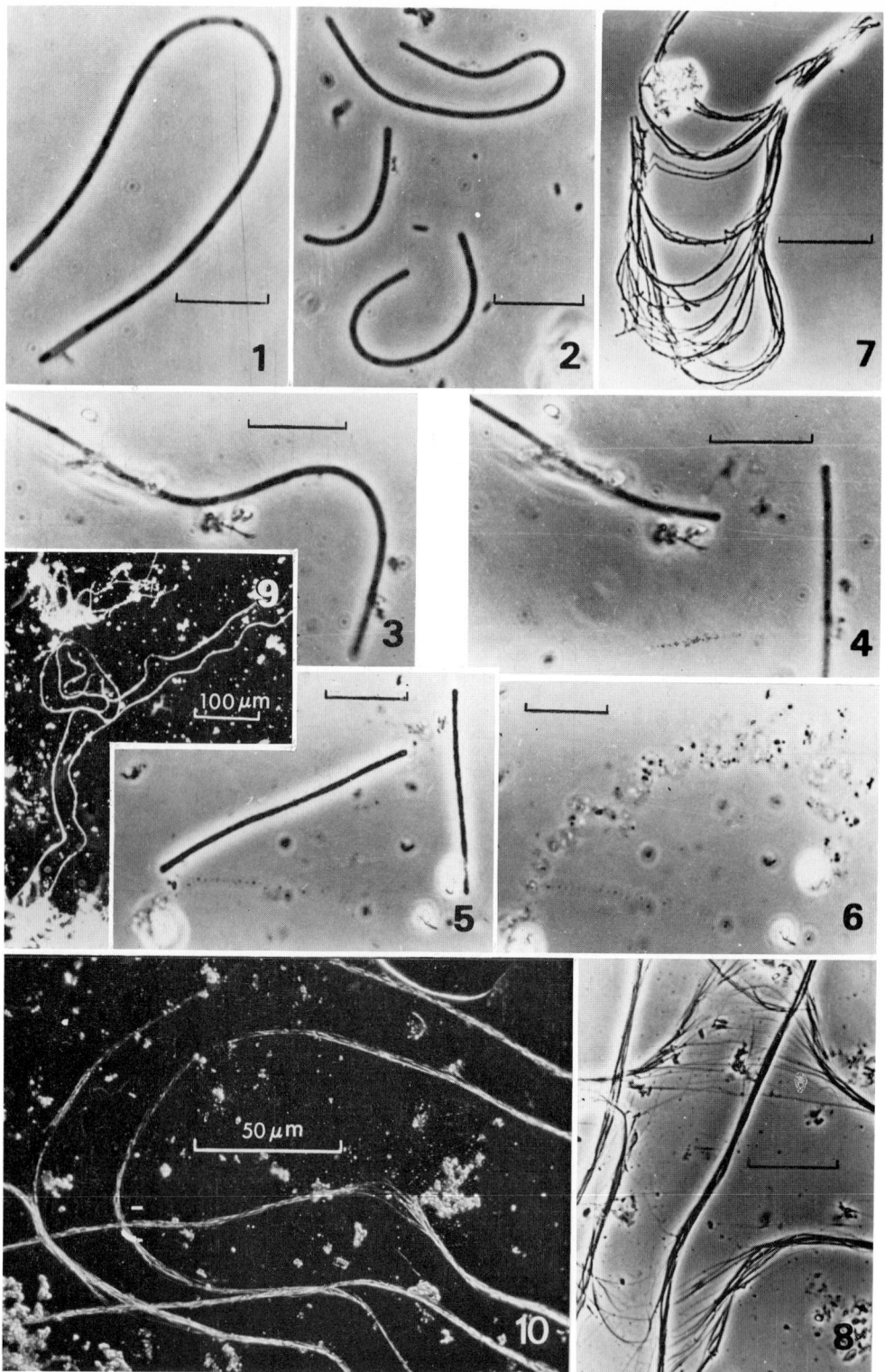

Plate 2.8.

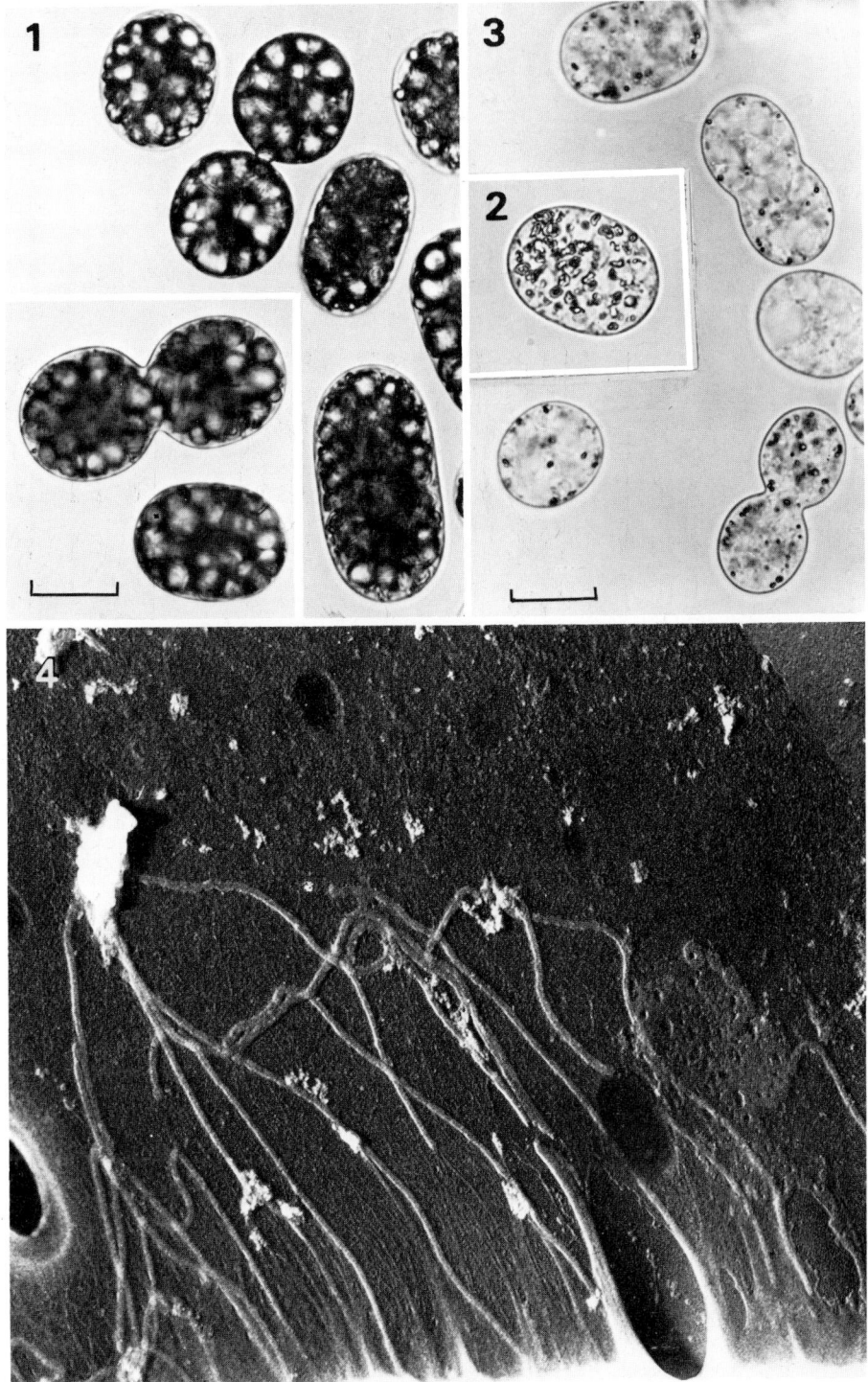

Plate 2.9.

PART 2

THE GLIDING BACTERIA

ORDER 1. MYXOBACTERALES
THAXTER *emend.* STANIER 1957, 854

The Fruiting Myxobacteria
HOWARD D. McCURDY

(*Myxobacteriaceae* (*sic*)* Thaxter 1892, 394; *Myxobactrales* (*sic*) Clements 1909, 8; *Myobacteriales* (*sic*) Jahn 1911, 187.)

Myx.o.bac.te.ra′les. M.L. masc.n. *Myxobacter* name of type genus; *-ales* ordinal ending; M.L. fem.pl.n. *Myxobacterales* the order based on the genus *Myxobacter*, the original type genus of the order.

Unicellular rods, <1.5 µm in diameter, typically embedded in a more or less tough layer of slime. The rods may be uniformly cylindrical with blunt rounded ends or somewhat tapered at the ends. Multiplication is by binary transverse fission. Capable of slow gliding movement in contact with a solid surface or air-water interface but lacking detectable locomotor organelles. Gram-negative.

Under appropriate environmental conditions the cells aggregate to form fruiting bodies constructed of slime and cells and which are often brightly colored and of macroscopic dimensions. The cells within the fruiting bodies become resting cells, termed myxospores. In some genera the myxospores are not readily distinguishable from vegetative cells; in others they are optically dense or refractile, encapsulated, shortened rods or spheres known as microcysts. The fruiting body may consist simply of a mass of slime and cells or the myxospores may be enclosed in sporangia of characteristic shape and dimensions which may be raised above the substrate on simple or branched stalks.

Chemoorganotrophs: strictly aerobic: energy-yielding metabolism respiratory, never fermentative. Typically produce enzymes capable of hy-drolyzing such macromolecules as proteins, nucleic acids, fatty acid esters and various polysaccharides including, in some species, cellulose. Many are capable of lysing other eucaryotic and procaryotic microorganisms. Photosynthetic pigments are lacking but carotenoid pigments characteristically, and melanin pigments often, are produced.

The G + C content of the DNA ranges from 67–71 moles % for all species examined.

Further Comments

Most known myxobacteria occur in soil and frequently develop on decomposing plant material, the bark of living trees or animal dung. Both in nature and in the laboratory their presence may be detected through the appearance of fruiting bodies. Although, one species has been described as aquatic (Geitler, 1924) and a number of myxobacteria have been isolated from fresh water environments (Jeffers, 1964; Shilo, 1970), further work will be required to determine whether these occurrences are incidental or reflect true aquatic habitats.

Most species of myxobacteria fail to grow or grow poorly on conventional bacteriological media

---

* Thaxter also referred to the family name as an order.

(e.g. nutrient broth, etc.). They may conveniently be divided into two physiological groups: the bacteriolytic group, which includes all *Myxococcaceae*, *Archangiaceae*, *Cystobacteraceae* and most *Polyangiaceae*; and the cellulolytic group. The bacteriolytic myxobacters may usually be cultivated on agar media containing living or killed bacterial or yeast cells, which they lyse. Their frequent observation on animal dungs probably reflects the abundance therein of bacteria as a source of food. Most will also grow well on agar media containing enzymatically hydrolyzed protein and salts. However cultivation in liquid media may present special problems and a solid medium which supported good growth, may fail to do so if agar is omitted.

The minimal nutritional requirements of the bacteriolytic myxobacters are still very poorly known. When cultivated on chemically defined media, the few strains studied (Dworkin, 1962; Mayer, 1967; Hemphill and Zahler, 1968; McCurdy and Khouw, 1969) required complex amino acid mixtures and relatively high concentrations of Mg^{++} or Ca^{++} for growth. No definite, obligate vitamin requirements have been demonstrated for most species but some of the *Polyangiaceae* seem to require vitamin B_{12} and perhaps other factors as well (McCurdy, 1964, 1969). None appears capable of utilizing carbohydrates, for carbon or energy. In spite of this the addition of a polysaccharide (as well as other soluble polymers) such as starch, glycogen, carboxymethylcellulose, etc. (Schurmann, 1967; Dworkin, 1962; McCurdy, 1969), will often stimulate growth in liquid media which would otherwise fail to support growth.

The cellulolytic myxobacters examined so far (Krzemieniewska and Krzemieniewski, 1937; Coucke, 1969) have simple requirements and can develop in a medium with an inorganic base supplemented with single carbohydrates (cellulose or a simple sugar).

Although distinguished from other bacteria by their gliding motility and the active bending and flexing which they frequently exhibit, the vegetative cells of myxobacters have not been found, either chemically or in ultrastructure, to differ, in any significant way, from other Gram-negative bacteria (Voelz and Dworkin, 1962; McCurdy, 1968; White *et al.*, 1968; Reichenbach and Voelz, 1969). There is a tendency to autolyse or to form spheroplasts under a variety of conditions (anaerobiosis, in old cultures, high temperature, in the presence of some cations). In general they stain poorly by conventional methods and are distorted by heat fixation; it is therefore best to study vegetative cells with phase-contrast microscopy. Motility is slight or absent in cells from liquid culture and is most conveniently observed in cells growing on the surfaces of (especially nutritionally dilute) agar media.

The vegetative colony of myxobacters, often termed a swarm or pseudoplasmodium, has a characteristic appearance as a result of slime production and gliding movement of the cells. It is usually flat and thin with many concentric folds and/or radiating lines and spreads extensively, sometimes rapidly, over the surface as a result of the outward movement of cells at the periphery. Frequently, the latter advance as groups forming tongue-like extensions or isolated clumps and streams. Especially in the *Polyangiaceae*, the underlying agar is etched, eroded and penetrated, thereby conferring an irregular topology to the colony.

Continued cultivation in nutrient-rich media frequently yields variants which produce mucoid colonies with little tendency to spread, which may no longer fruit and which therefore resemble non-gliding bacteria.

Media supporting good vegetative development are seldom suitable for fruiting body formation. In general fructification is favored by dilute media* (e.g. water agar, *E. coli* medium (ECM), dung pellet agar, etc.). In some cases, the suppression of fruiting body formation on nutrient-rich media furnished only with soluble nutrients, appears due to the inhibiting effect of certain amino acids (Dworkin, 1963; McCurdy, 1964; Hemphill and Zahler, 1968).

The fruiting bodies of the myxobacters are the resistant or resting stage. In those organisms producing microcysts, the hard slime capsule surrounding the cell confers considerably more resistance to heat, desiccation, physical disruption and ultraviolet radiation than is possessed by vegetative cells. Microcysts may in general be induced to form in liquid media without fruiting body formation by the addition of 0.5 M glycerol

---

* Media for Myxobacters: DPA, dung pellet agar. Three or four sterile antibiotic-free rabbit dung pellets are placed in a sterile Petri dish and partially covered with water agar or Ca^{++} agar (McCurdy, 1968).

ECM, *Escherichia coli* medium. Dried *E. coli* cells, 100 mg; agar, 1.5 g; water, 100 ml. Autoclave.

Cas.-Mg^{++}. Casitone (Difco), 2.5 g; soluble starch, 5.0 g; $MgSO_4 \cdot 7H_2O$, 0.5 g; K_2HPO_4, 0.25 g; agar, 15 g; water, 1000 ml. Autoclave.

W. A., water agar. K_2HPO_4, 0.25 g; $MgSO_4 \cdot 7H_2O$, 0.5 g; agar, 15 g; water 1000 ml. Autoclave. Living or killed bacterial cells (*E. coli*, *Micrococcus* (*Sarcina*) *luteus*, etc.) heavily streaked over the surface are the source of organic nutrients.

(Dworkin and Gibson, 1964; McCurdy, 1969; Reichenbach and Dworkin, 1970; McCurdy, unpublished results). In the *Polyangiaceae*, which do not produce microcysts, resistance appears to be a function of the modified slime which provides the wall material of the intact sporangium.

The taxonomy of the *Myxobacterales* is still largely based on the characteristics of the fruiting bodies, even though these structures are known to vary greatly in response to environmental influences and either may not be formed or are arrested in their development as a result of mutation or unfavorable conditions. Imperfectly formed fruiting bodies of certain species simulate the mature fruiting bodies of other members of the group. Accordingly it is essential, in identifying these organisms, to study fruiting body formation, preferably with pure cultures, under a wide range of defined nutritional and environmental conditions and over a considerable period of time. This principle has not always been followed in the taxonomy of the myxobacters; the biochemical data conventionally used in bacterial taxonomy are often lacking.

It should be noted that the *Myxobacterales* are unique among bacteria in that the type materials are, in most cases, represented by herbarium specimens and not by type cultures. Although often giving a good impression of fruiting body structure such specimens offer little information about vegetative cell morphology and other phenotypic properties. Since future work with these organisms will undoubtedly be based on the comparative study of pure cultures, it would be desirable, insofar as possible, to designate type or neotype strains of each species. However since the Code of Nomenclature of Bacteria does not cover this point and in the absence of a ruling by the Judicial Commission, we have noted, after the location of the type specimens, the location of representative strains for all those species for which cultures are now available.

Methods for the isolation, cultivation and characterization of myxobacters in pure culture may be found in Kühlwein and Reichenbach (1966), McCurdy (1969) and Peterson (1969). For a general review of the biology of the *Myxobacterales* see Dworkin (1966).

In the 7th edition of THE MANUAL the family *Cytophagaceae* and the genus *Sporocytophaga* (*Myxococcaceae*) were included in the order. These are transferred from the *Myxobacterales* in the present edition on the basis that they do not form fruiting bodies and have a significantly different G + C content (34–43 moles %) (Soriano and Lewin, 1965; Mandel and Leadbetter, 1965; McCurdy and Wolf, 1967; and McCurdy, 1969).

The family *Polyangiaceae* (McCurdy, 1970) is redefined to contain all myxobacters with cylindrical, blunt ended, vegetative cells and resting cells which differ little from them. It includes the genera *Chondromyces, Nannocystis* and *Polyangium*, incorporating within the latter all organisms previously in the family *Sorangiaceae*. A new family, the *Cystobacteraceae*, has been proposed for sporangial myxobacters with tapered vegetative cells which convert into microcysts. It includes the genera *Melittangium, Cystobacter* and *Stigmatella*; the last two include the microcyst-forming species formerly in the genera *Polyangium* and *Chondromyces*.

Many species listed in previous editions of THE MANUAL are not accepted in this because of inadequate descriptions or because the criteria used to distinguish them from accepted species are of doubtful validity. These are listed under *Species incertae sedis*; some of them may be found to deserve acceptance as a result of further study.

Key to the families of order **Myxobacterales**

I. Vegetative cells tapered, microcysts produced.
   A. Microcysts spherical or oval.
Family I. Myxococcaceae, p. 79

   B. Microcysts rod-shaped.
      1. Microcysts not in sporangia.
Family II. Archangiaceae, p. 83

      2. Microcysts in sporangia.
Family III. Cystobacteraceae, p. 86

II. Vegetative cells of uniform diameter with blunt, rounded ends. Myxospores resemble vegetative cells.
Family IV. Polyangiaceae, p. 92

FAMILY I. **MYXOCOCCACEAE** JAHN 1924, 84

S. A. ZAHLER AND H. D. McCURDY

Myx.o.coc.ca'ce.ae. M.L. masc.n. *Myxococcus* type genus of the family: *-aceae* ending to denote a family; M.L. fem.pl.n. *Myxococcaceae* the *Myxococcus* family.

Vegetative cells slender, straight to slightly tapered, flexible rods with rounded ends. **Refractile myxospores** (microcysts) **spherical or ellipsoidal.**

The type and only genus: *Myxococcus* Thaxter 1892, 403.

Further Comments

In the 7th edition of THE MANUAL, the Family included the genera *Sporocytophaga, Angiococcus, Chondrococcus* and *Myxococcus*.

Sporocytophaga is now placed in the order *Cytophagales*.

No specimens of *Angiococcus* are known to exist in stock collections, and Peterson and McDonald (1966) have cast serious doubts on the validity of the genus. It is not clear that the type species, *A. disciformis* Thaxter 1904, 412, has ever been isolated in pure culture. In a personal communication to J. E. Peterson, E. N. Mishustin has said that the organism he described as *A. cellulosum* (Mis-

hustin 1938, 437) is actually *Rhizophlyctis rosea* (de Bary and Woronin) Fischer, a member of the fungal order *Chytridiales*. The third and last species of *Angiococcus, A. moliroseus* Peterson (1959, 169), has been identified as *Streptosporangium roseum* Couch (1955, 148) by Peterson and McDonald (1966).

The name of the genus *Chondrococcus* Jahn 1924, 85, although illegitimate (*cf.* Jeffers and Holt, 1961), was retained in the 7th edition of THE MANUAL. The distinction based on consistency of the slime of the fruiting bodies, in *Chondrococcus* firm and not deliquescent, in *Myxococcus* deliquescent, no longer seems appropriate. The various species of the *Myxococcaceae* are sufficiently similar to be included within a single genus (*cf.* McCurdy and Wolf, 1967).

All species of *Myxococcaceae* can be isolated from soil (with or without dung as a source of nutrients) and are common on bark and decaying vegetation.

Genus I. **Myxococcus** Thaxter 1892, 403

Myx.o.coc'cus. Gr. fem.n. *myxa* mucus, slime: Gr. n. *coccus* berry; M.L. masc.n. *Myxococcus* slime coccus.

Vegetative cells slender rods with tapering or rounded ends, 0.4–0.7 by 2.0–10.0 μm. Motile by gliding motility. Fruiting bodies contain **refractile spherical or ellipsoidal microcysts** up to 2.3 μm in the largest diameter.

The **myxospores** are **not** enclosed **in a sporangium.**

Chemoorganotrophs: Metabolism respiratory. Those which have been studied require several amino acids, and are capable of hydrolyzing protein, starch, nucleic acids and various fatty acid esters. **Bacteriolytic; non-cellulolytic.**

Strict aerobes.

Fail to grow at 40 C or above.

Colonies adsorb congo red.

Resistant to 10-unit discs of penicillin. Sensitive to 10-μg discs of neomycin and tetracycline and to 5-μg discs of erythromycin.

Methods suitable for studying the physiology of members of the genus are described by McCurdy (1969).

The molar G + C content of the DNA of those

species examined ranges from 68–71% (Mandel and Leadbetter, 1965).

Type species: *Myxococcus fulvus* (Cohn) Jahn 1911, 198.

Further Comments

In the 7th edition of THE MANUAL, six species of *Myxococcus* plus three *species incertae sedis* and six of *Chondrococcus* were listed. The mis-identified *Chondrococcus columnaris* has been transferred to *Flexibacter*. The differences between species were sometimes based on morphological differences in fruiting bodies, but the variations in fruiting body structure within a single strain, when grown on different media, make this a characteristic of doubtful determinative value.

In the present descriptions only species which have been studied in pure culture are accepted. Some of the species listed as *incertae sedis* may deserve species rank after study has provided characteristics more substantial and easier to quantify than fruiting body morphology and color.

Description of the species of genus **Myxococcus**

1. Myxococcus fulvus (Cohn) Jahn 1911, 198. (*Micrococcus fulvus* Cohn 1875, 181.)

ful'vus. L. adj. *fulvus* reddish yellow.

Vegetative cells: Slender, only slightly tapering with rounded ends, 0.4–0.8 (average 0.6) by 5–9 (6) µm (Plate 2.1, Fig. 8).

Fruiting bodies (Plate 2.1, Fig. 1): Spherical, elongate or pear-shaped, constricted below, often with a definite slimy stalk which usually does not persist; at first coherent but later deliquescent if moist; flesh-red, reddish orange, pink or brownish red; when dry, deep red to brown; 150–400 µm.

Microcysts: Spherical to slightly oval, refractile, 1.1–1.7 (1.4) µm.

Colonies on Casitone-Mg^{++} agar thin with a filamentous border, at first translucent, gray-white to slightly pinkish, becoming more opaque, light flesh-colored to reddish orange to pink. Growth similar on *E. coli* medium, but less pigmented and with moderate lysis; concentric zones of fruiting bodies formed in central portion of colony.

Optimum pH 6.5–7.0.

Temperature range 18–37 C, optimum 26–30 C.

Resistant to 5-µg streptomycin discs. Sensitive to 10-µg chloramphenicol and kanamycin discs.

Very common on dung-soil plates and on bacterial streaks inoculated with soil. Also observed frequently on decaying plant material, the dung of various herbivores and on lichens.

Reference strain: University of Windsor M17; ATCC 25199. Other typical strains: University of Windsor M7, M16.

Illustrations: Cohn, 1875, Plate 6, Fig. 18; Baur, 1905, Figs. 1–3 and Plate 4, Figs. 1–13 and 16; Jahn, 1924, Figs. L–M, p. 43 and Fig. R, p. 47; Krzemieniewska and Krzemieniewski, 1927, Plate 1, Fig. 3, Plate 2, Fig. 14, Plate 3, Figs. 38 and 40; Kühlwein, 1950, Figs. 5 and 6; Oetker, 1953, Figs. 2–5, 7 and 8; Nolte, 1957, Figs. 1, 2, 4b, 5a, 6b; Reichenbach, 1966, Fig. 4; McCurdy, 1969, Fig. 3 (color).

Note: Jahn (1924, 85) noted two varieties, var. *albus* which is white and var. *miniatus* which is cinnabar red.

2. Myxococcus virescens Thaxter 1892, 404.

vi.res'cens. L. part.adj. *virescens* becoming green.

Vegetative cells: Slender, tapered rods, 0.4–0.8 (0.7) by 5–9 (7) µm.

Fruiting bodies: Spherical to elongate, yellow to greenish yellow, in culture on artificial media easily becoming white; deliquescent in continued moisture, 150–500 µm.

Microcysts refractile, spherical 1.7–2.3 µm diameter.

Vegetative colonies on Casitone-Mg^{++} agar thin, with many, fine intertwining radiating lines, edge indefinite, yellowish white to yellow, clearing in the central portion with formation of fruiting bodies. A diffusible green pigment develops on continued incubation. Colonies on *E. coli* medium (ECM), thinner, less pigmented with numerous fruiting bodies in concentric zones; lysis is moderate.

Optimum pH 6.5–7.0.

Temperature range 18–37 C, optimum 28 C.

Resistant to 5-µg streptomycin discs. Sensitive to 10-µg chloramphenicol and kanamycin discs.

TABLE 2.1

Differential characteristics of species of **Myxococcus**

Characteristics determined on Casitone-Mg^{++}. See McCurdy (1969) for methods.

	1. M. fulvus	2. M. virescens	3. M. xanthus	4. M. stipitatus	5. M. coralloides	6. M. macrosporus
Fruiting bodies deliquescent	+	+	+	+	−	−
Fruiting bodies raised on a well defined, persistent stalk	±	−	−	+	−	−
Myxospores 1.5 µm or more in diameter	−	+	+	−	−	+
Color of vegetative cell masses						
Yellow	−	+	+	+	−	−
Buff to tan	−	−	−	−	d	d
Flesh-colored to reddish orange	+	−	−	−	d	d
Greenish diffusible pigment on agar	−	+	−	−	−	−
Hydrolyzes starch in 3 days or less	−	−	−	−	+	+*
Hydrolyzes esculin	−	−	−	−	±	−
Oxidase	−	−	−	+	−	−

* Only one strain tested.

Originally obtained on decaying matter from soil and dung of various animals. One of the earliest and most frequently encountered myxobacters on dung in contact with soil.

Type material: Farlow Herbarium, Harvard University, Cambridge, Mass.

Reference strain: University of Windsor M22; ATCC 25203. Another typical strain: University of Windsor M100.

Illustrations: Krzemieniewska and Krzemieniewski, 1926, Plate 1, Figs. 6 and 9; Krzemieniewska and Krzemieniewski, 1927, Plate 1, Figs. 1 and 6, Plate 2, Fig. 26, Plate 3, Figs. 32–34; Badian, 1930, Plate 8; Oetker, 1953, Fig. 9; Reichenbach, 1966, Fig. 1; McCurdy, 1969, Fig. 1 (color).

3. **Myxococcus xanthus** Beebe 1941, 195.
xan'thus. Gr. adj. *xanthus* orange, golden.

Vegetative cells: Slender, flexible rods. 0.5–1.0 (0.75) by 4.0–10 (5) μm.

Fruiting bodies: Spherical to subspherical, constricted at the base, occasionally irregular from fusion of adjacent masses; up to 300–400 μm in diameter; color constant, light yellowish orange to bright orange, never greenish yellow.

Microcysts refractile, spherical 1.1–2.3 (2.0) μm in diameter (Plate 2.1, Fig. 9).

Vegetative colonies resemble those of *M. virescens* but no diffusible green pigment produced.

Can be cultivated on synthetic media (Dworkin, 1961; Hemphill and Zahler, 1968; Witkin and Rosenberg, 1970). Fruiting body formation occurs on synthetic media (Dworkin, 1963; Hemphill and Zahler, 1968).

Optimum pH 7.2–8.2 (Dworkin, 1962).

Optimum temperature 30 C.

Resistant to 5-μg streptomycin discs. Sensitive to 10-μg chloramphenicol and kanamycin discs.

Originally obtained on dried cow dung. Common on sterile dung in contact with soil or soil-inoculated bacterial streaks on water agar.

Reference strain: Dworkin (1962) strain FB; NCIB 9412; ATCC 19368.

Illustrations: Beebe, 1941, Figs. 1–28; Reichenbach, 1966, Figs. 5 and 8; McCurdy, 1969, Fig. 2 (color); fine structure: Voelz and Dworkin, 1962, Figs. 1–11; Voelz, 1966, Figs. 1–4.

Note: *M. xanthus* resembles *M. virescens* closely except for the absence of a green pigment.

4. **Myxococcus stipitatus** Thaxter 1897, 408. (*Myxococcus ovalisporus* Krzemieniewska and Krzemieniewski 1926, 15.)

sti.pi.ta'tus. L. masc.n. *stipes, stipitis* trunk, stalk; M.L. adj. *stipitatus* stalked.

Vegetative rods are slender, occasionally slightly tapering, flexible, 0.5–0.7 by 2.0–7.0 μm.

Fruiting bodies (Plate 2.1, Figs. 5 and 6): Spore mass nearly spherical, milky or yellowish white to slightly pink, up to 200 μm in diameter; raised on a stalk which may be as long as 200 μm and 30–50 μm in diameter. On artificial media fruiting bodies are variable; stalks may be poorly developed and adjacent fruiting bodies may fuse to form irregular rounded masses.

Microcysts are ellipsoidal, optically dense or slightly refractile, 1.1–1.4 by 1.3–1.8 μm (Plate 2.1, Fig. 7).

Vegetative colonies on Casitone-Mg^{++} (McCurdy, 1963) grayish or yellowish white, flat, mycelioid, almost mold-like. Fruiting body formation is seldom observed on ECM or Casitone-Mg^{++} agar.

In liquid media growth is luxuriant in the form of distinctive cottony, moldlike spheres.

Capable of growth on ordinary bacteriological media, e.g. nutrient broth or trypticase broth.

Temperature range 18–37 C.

Resistant to 5-μg streptomycin discs. Sensitive to 10-μg chloramphenicol and kanamycin discs.

Originally isolated on dung in laboratory cultures from Maine and Tennessee. Isolated from European soil also. Comparatively rare.

Type material: Farlow Herbarium, Harvard University, Cambridge, Mass.

Reference strain: University of Windsor M78.

Illustrations: Thaxter, 1897, Plate 31, Figs. 30–33; Krzemieniewska and Krzemieniewski, 1926, Plate 11, Figs. 13–14; McCurdy, 1969, Fig. 4 (color).

5. **Myxococcus coralloides** Thaxter 1892, 404. (*Chondrococcus coralloides* (Thaxter) Jahn 1924, 86; *Myxococcus clavatus* Quehl 1906, 18; *Myxococcus digitatus* Quehl 1906, 18; *Myxococcus polycystus* Kofler 1913, 865; *Myxococcus exiguus* Kofler 1913, 867; *Chondrococcus polycystus* (Kofler) Krzemieniewska and Kremieniewski 1926, 17.)

co.ral.lo.i'des or co.ral.loi'des. Gr. neut.n. *corallium* coral; Gr. n. *eidus* shape; M.L. adj. *coralloides* coral-like.

Vegetative cells: Slightly tapered rods 0.5–0.8 (0.6) by 4–8 (6) μm.

Fruiting bodies (Plate 2.1, Figs. 3 and 4): Firm, not deliquescent. Very variable in shape, size and color, ranging from simple, barely visible papillae (<25 μm) or straight or branched tubules recumbent on the substrate, to erect, simple or branched columnar structures (20–30 by 50–250 μm); or they may be flattened, cushion-like and constricted below to form a short stalk (25–40 μm); in other instances, they consist of irregular coral-like masses up to 300 μm in size with constricted lobes and finger-like outgrowths. They may be flesh-colored, pink, reddish orange, orange or buff in different strains. When formed on agar media the fruiting bodies often extend taproot-like into the underlying agar.

Microcysts spherical, optically dense or refractile, 1.0–1.5 μm in diameter (Plate 2.1, Fig. 2).

Colonies on Casitone-Mg⁺⁺ agar are flat, thin with densely radiating lines. The color corresponds to that of the fruiting bodies, which are densely arranged in concentric rings and various other patterns in the central portion of the colony. Growth on *E. coli* medium is similar but less pigmented; lysis is moderate and the underlying agar may be eroded. A brownish diffusible pigment is produced by some strains in old cultures.

Optimum pH 7.0–7.5.

Temperature range 18–37 C; optimum 25–30 C.

Sensitive to 5-μg streptomycin discs and to 10-μg chloramphenicol and kanamycin discs.

Originally reported on lichens but very common on sterile dung in contact with soil, on moist bark and on bacterial streaks over agar inoculated with soil.

Type material: Farlow Herbarium, Harvard University, Cambridge, Mass.

Reference strain: University of Windsor M2; ATCC 25202. Other typical strains: University of Windsor M1, M25.

Illustrations: Thaxter, 1892, Plate 24, Figs. 29–33; Quehl, 1906, Plate 1, Figs. 1 and 9; Jahn, 1924, Fig. Y, a–h; Reichenbach, 1962, Figs. 1–8; Reichenbach 1966, Figs. 6, 7 and 9; McCurdy, 1969, Figs. 5–7 (color).

6. **Myxococcus macrosporus** (Krzemieniewska and Krzemieniewski) comb. nov. (*Chondrococcus macrosporus* Krzemieniewska and Krzemieniewski 1926, 16; *not Myxococcus macrosporus* Zukal 1897, 551.)

ma.cro.spor'us. Gr. adj *macrus* long, large; Gr. n. *spora* seed; M.L. adj. *macrosporus* large spored.

Vegetative cells, slender rods, 7–13 by 0.5–0.8 μm.

Fruiting bodies: Like those of *Myxococcus coralloides* but with long branches, and yellow or light brown in color.

Microcysts: 1.6–2.0 μm diameter.

Vegetative colony resembles that of *M. coralloides*.

Resistant to 5-μg streptomycin discs and 10-μg chloramphenicol and kanamycin discs.

Originally obtained on leaves but also observed on sterile dung in contact with soil. Not common.

Reference strain: University of Windsor M271.

Illustrations: Krzemieniewska and Krzemieniewski, 1926, Plate II, Fig. 19.

Species incertae sedis

a. *Myxococcus cruentus* Thaxter 1897, 395.

cru.en'tus. L. adj. *cruentus* blood-red.

Vegetative cells: Rods 0.8 by 3.0–8.0 μm.

Fruiting bodies: Deliquescent, regularly spheri-

cal, 90–125 μm in diameter, blood-red. Slime on the surface of the fruiting body forms a more or less definite membrane within which the myxospores lie.

Myxospores ellipsoidal or irregularly oblong, 0.9–1.0 by 1.2–1.4 μm.

Comment: Possibly a variant form of *M. fulvus.*

b. *Myxococcus cirrhosus* Thaxter 1897, 409. (*Chondrococcus cirrhosus* Jahn 1924, 200.)

cir.rho'sus. Gr. *cirrhus* tawny; M.L. adj. *cirrhosus* tawny.

Vegetative cells: Rods 0.8 by 2.0–5.0 μm.

Fruiting bodies: Firm, not deliquescent; elongate, upright, thickened below, slender above, extended to a rounded point, 50–100 μm long, 20 μm in diameter at base, light red to flesh-colored.

Microcysts irregularly spherical, about 1 μm in diameter.

Comment: Possibly a small-fruited variant of *M. coralloides.* Forms corresponding to *M. cirrhosus* when first seen (on bark, for example), turn out to be *M. coralloides* after laboratory cultivation.

c. *Chondrococcus blasticus* Beebe 1941, 310.

blas'ti.cus. Gr. adj. *blasticus* budding.

Vegetative cells: Slender rods, 0.5–0.6 by 3.0–5.0 μm.

Fruiting bodies: Firm, not deliquescent, spherical to subspherical, usually sessile but occasionally with a short stalk; pale pink to bright salmon-pink, 300–600 μm in diameter. Smaller secondary fruiting bodies characteristically arise from the primary ones.

Myxospores 1.2–1.4 μm in diameter.

Comment: Possibly a large-fruited variant of *M. coralloides.*

d. *Mycococcus disciformis* Thaxter 1904, 412. (*Angiococcus disciformis* (Thaxter) Jahn 1924, 89.)

dis.ci.for'mis. Gr. n. *discus* a disc; L. n. *forma* form; M.L. adj. *disciformis* disc-shaped.

Vegetative cells: Rods 0.5–0.6 by 2.0–3.0 μm.

Fruiting bodies: Numerous disc-shaped cysts with thin cyst walls, containing myxospores. Cysts sessile, yellowish, aging to dark orange-yellow, about 10 by 35 μm.

Microcysts irregularly spherical, 0.9–1.0 μm.

Comment: Peterson and McDonald (1966) have cast grave doubts on the existence of this organism. They state that fruiting bodies which they have tentatively identified as *M. disciformis* have regularly turned out to be other *Myxococcus* species, unable to form cysts in culture. At present the question must be left open.

e. *Myxococcus albus* Finck 1950, 382.

al'bus. L. adj. *albus* white.

Vegetative cells: Rods about 1.2 (*sic*) by 6.5–9.0 μm.

Fruiting bodies: Deliquescent, roundish to

lengthened, dimensions not given, white to beige in color. Rapidly deliquescing at 30 C to slime masses with the appearance of oil drops.

Myxospores spherical, 2.0–2.5 μm in diameter.

Comment: Possibly a colorless variant of *M. xanthus* or *M. virescens.*

f. *Myxococcus viperus* Finck 1950, 383.

vi′pe.rus. L. n. *vipera* viper, snake; M.L. adj. *viperus* pertaining to a viper.

Vegetative cells: Rods 1.0–1.3 by 6.5–7.8 μm.

Fruiting bodies convoluted or club-shaped, mesenteric, yellow to copper-red in color; dimensions not given.

Microcysts: Spherical, 2 μm in diameter.

Vegetative colonies greenish yellow becoming red on exposure to light.

Comment: Possibly a variant of *Myxococcus virescens.*

g. *Myxococcus megalosporus* (Jahn) Kühlwein 1950, 681. (*Chondrococcus megalosporus* Jahn 1924, 86.)

me.ga.lo′spo.rus Gr. adj. *megas, megale, mega* big; Gr. n. *spora* seed; M.L. n. *spora* spore; M.L. adj. *megalosporus* large spored.

Vegetative cells not described.

Fruiting bodies: About 80–160 μm wide, rounded cushion-shaped, dark flesh-colored.

Microcysts: 2 μm.

Comment: Insufficient information for re-identification and not obtained in pure culture.

h. *Myxococcus lacteus* Peterson 1958, 631.

lac.te′us L. adj. *lacteus* milky.

Vegetative cells not described.

Fruiting bodies: Spore mass single, globose, pear-shaped milky white, 50–150 μm diameter, on a centrally attached crystalline or hyaline, striated stipe, 10–25 by 50–150 μm.

Microcysts: 1 μm or less in diameter.

Comment: The description is based upon a herbarium specimen in Farlow Herbarium, labeled *Chondromyces lacteus.* It has not been cultivated. Resembles *M. stipitatus* except for microcyst dimensions.

FAMILY II. **ARCHANGIACEAE** JAHN 1924, 66

Howard D. McCurdy

Ar.chan.gi.a′ce.ae. M.L. neut.n. *Archangium* type genus of the family; *-aceae* ending to denote a family; M.L. fem.pl.n. *Archangiaceae* the *Archangium* family.

The **vegetative rods** are slender **with tapered ends.** The **microcysts** are very **short rods, ellipsoids or spheres** which are **never enclosed in sporangia;** the **fruiting bodies** consist of **irregular masses** or projections of variable dimensions.

The type and only genus is *Archangium* Jahn 1924, 66.

Further Comments

Two genera, *Stelangium* with one species and *Archangium* with five, were recognized in the 7th edition of THE MANUAL. Only one species *Archangium gephyra,* is sufficiently well characterized and distinguished from other myxobacter species to warrant recognition here. It shows clear af-

finities to the *Myxococcaceae* and might well be included in that family. Four species, previously recognized as members of the genus *Archangium,* are here considered *species incertae sedis* because the lack of pure culture data, the generally inadequate descriptions and the absence of striking morphological features make difficult their differentiation from immature or variant fruiting bodies which are frequently observed in other, well characterized species (see for example McCurdy, 1969).

Similarly organisms included in the genus *Stelangium* could easily be commonly observed variants of *Stigmatella* or *Chondromyces,* especially if clear descriptions of vegetative cells and myxospores are lacking.

Genus I. **Archangium** *Jahn 1924, 66*

Ar.chan′gi.um. Gr. fem.n. *arche* beginning; primitive; Gr. neut.n. *angium* vessel, receptacle; M.L. neut.n. *Archangium,* primitive vessel.

Vegetative cells slender, **tapered,** flexible **rods.**

Sporangia lacking; fruiting bodies irregular in form, swollen or brainlike consisting internally of

intestine-like twisted or intertwined masses. There is **no definite slime wall.**

Microcysts are very **short rods, ellipsoids or spheres,** refractile or optically dense.

Vegetative colonies do not etch or erode agar media. Congo red is adsorbed.

Type species: *Archangium gephyra* Jahn 1924, 67.

Description of the species of genus **Archangium**

1. **Archangium gephyra** Jahn 1924, 67. (*Chondromyces serpens* Quehl 1906, 16; *Myxococcus cerebriformis* Kofler 1913, 886: *Chondrococcus cerebriformis* (Kofler) Jahn 1924, 86.)

ge'phy.ra. Gr. n. *gephyra* a bridge.

Vegetative cells have a characteristic shape. They are slightly tapered with slightly swollen rounded ends 0.4–0.7 by 6–15 μm.

The fruiting bodies are very variable in size and shape. Usually consisting of an irregular, brainlike or elongated mass with a padded or swollen surface; firm in consistency. The color is variable, light rose or reddish flesh-colored to orange by reflected light, later appearing bluish violet when observed on a dark background. By transmitted light the fruiting bodies are yellowish to light red. When dry, orange to red.

Internally, the fruiting body consists of a mesenteric mass of tubules which are periodically interrupted by cross-walls which may not entirely cut through the spore masses from one side to the other. When pressed between slide and coverslip, the tubules break up into a number of irregularly rounded masses 15–30 μm in diameter. Within these, the microcysts lie parallel in bundles.

Microcysts are spherical, oval or short, often bean-shaped, rods; optically dense or refractile, 1.0–2.0 by 1.5–2.8 μm.

Vegetative colonies on agar media are thin with many radiating ridges and concentric folds. The edge is thin with tongue-like extensions. The colonies are orange in areas of cell accumulations. Slime production is moderate; can be cut easily.

Growth on Casitone-Mg^{++} broth (McCurdy, 1969) is slight and in the form of tight orange balls in shake culture. Growth is substantially improved by the addition of 0.1–0.2% agar.

Easily cultivated on the usual complex media used for non-cellulolytic myxobacters. Nitrate is not reduced, catalase and oxidase are produced. Esculin, starch (3 days), casein, gelatin, Tween 80, indoxyl acetate, RNA and DNA are hydrolyzed. Non-cellulolytic.

Aerobic.

Optimum pH 7.5.

Temperature range 18–40 C; optimum 18–32 C.

Antibiotic sensitivity (discs): resistant to penicillin (10 units); sensitive to neomycin (10 μg), streptomycin (5 μg), tetracycline (10 μg), chloramphenicol (10 μg), kanamycin (10 μg) and erythromycin (5 μg).

Source and habitat: Frequently observed on sterile rabbit dung in contact with soil and on *E. coli* streaks over agar inoculated with soil.

G + C content 67.8–68.3 moles % by T_m.

Reference strain: Windsor M18; ATCC 25201.

Illustrations: Quehl, 1906, Plate 1, Fig. 7; Jahn, 1924, Plate 1, Fig. 5; and Krzemieniewska and Krzemieniewski, 1926, Plate III, Figs. 25–26.

Species incertae sedis

I. *Archangium*

a. *Archangium primigenium* (Quehl) Jahn 1924, 69; (*Polyangium primigenium* Quehl 1906, 16.)

pri.mi.ge'ni.um. L. adj. *primigenium* primitive.

Vegetative cells 4–8 μm in length. Shape not described.

The fruiting bodies are without definite shape or dimensions, up to 1 mm, with an irregularly swollen surface; when fresh they have a lively red color which is very prominent against a dark background. In transmitted light, flesh-colored to yellowish red. In transmitted light it becomes apparent that the fruiting body is made up of a complicated intestine-like mass of convolutions which are closely appressed but not always clearly delimited into tubules of 70–90 μm diameter. There is no wall-like slime layer. Quehl states that in culture the fruiting bodies are seldom formed on agar but tend to occur on the glass of the culture vessel.

Microcysts are 0.8 by 4 μm rods which tend to adhere together in fragments of various sizes when the fruiting body is crushed.

Vegetative colonies not described.

Cultivated only on dung agar.

Source and habitat: On rabbit dung, sometimes on roe dung. According to Jahn not particularly common.

Illustrations: Quehl, 1906, Plate 1, Fig. 5; Jahn, 1911, Plate 1, Fig. 5: Jahn, 1924, Plate 1, Fig. 4 and Fig. G, p. 37.

Note: Jahn described what he assumed to be a variety of this species, *A. primigenium* var. *assurgens*. On the basis of the descriptions available, *A. primigenium* is not distinguishable from *Archangium serpens* or variants of *Melittangium lichenicola* (see description) or species of *Stigmatella*.

b. *Archangium serpens* (Thaxter) Jahn 1924, 72; (*Chondromyces serpens* Thaxter 1892, 403.)

ser'pens. L. part.adj. *serpens* creeping.

Vegetative cells like those of *Melittangium lichenicola* i.e. tapering slightly, 5–7 by 0.6 μm.

Fruiting bodies up to 1 mm in diameter, sessile, consisting of a convoluted mass of intertwined 50 μm diameter tubules, confluent in an anastomosing coil, flesh-colored, when dry, dark red.

Microcysts not described.

Vegetative colonies not described. Apparently not cultivated in pure culture.

Note: Thaxter had at first considered this to be a variant of *Melittangium lichenicola* but decided in view of its constancy in culture, to recognize it as a species. However, see description of *Melittangium lichenicola*.

c. *Archangium flavum* (Kofler) Jahn 1924, 71. (*Polyangium flavum* Kofler 1913, 864.)

fla'vum. L. neut.adj. *flavum* yellow or golden.

Vegetative cells 2–4 μm long but shape not described.

Fruiting bodies are about 0.4 mm high and 0.6 mm in diameter, yellow, spherical or ellipsoidal with a humped or padded surface. Seen with the naked eye somewhat resembles *Myxococcus virescens* but the surface is not as smooth and shiny. The mass is quite homogeneous but upon pressure under cover glass tends to break into clumps. No sporangial wall is present although the rods are so tightly linked, especially at the periphery, that their arrangement as found in the fruiting body tends to be retained.

The original description does not clearly differentiate between vegetative cells and myxospores.

Vegetative colonies not described.

Not cultivated in pure culture.

Source and habitat: Originally described on hare dung in Danube meadows. Considered by Krzemieniewska and Krzemieniewski (1926) to be common in Polish soils but see below.

Illustrations: Kofler, Plate I, Fig. 5.

Comment: Krzemieniewska and Krzemieniewski (1926) describe isolates which they considered to be varieties (forms I and II) of *A. flavum* and which had cylindrical, blunt ended vegetative cells and similar myxospores (3.5–7 by 0.6–0.8 μm). If the Krzemieniewskis were correct then the description of *A. flavum* is strongly suggestive of an immature or malformed fruiting body of a *Polyangium*. It is not clear however that the Krzemieniewskis and Kofler were dealing with the same organism.

d. *Archangium thaxteri* Jahn 1924, 71.

thax'te.ri. M.L. gen.n. *thaxteri* of Thaxter; named for Dr. Roland Thaxter.

Vegetative cells not described.

Fruiting bodies usually 0.25–0.5 mm but occasionally 0.75 mm in diameter. Superficially sulfur yellow and of irregularly rounded shape. If crushed, numerous reddish convoluted tubules, without definite walls embedded in a yellowish slime, are observed. Diameter of tubules about 50 μm. Usually, the myxospores are located in the tubules but in smaller specimens they may be embedded in the surrounding slime. In well developed specimens the slime forms a stalk giving the whole the appearance of a morel but this form is rare.

Myxospores are very slender rods, 0.3(?) by 3 μm.

Cultivated only on dung pellets not on agar medium.

Source and habitat: Found only rarely on rabbit dung.

Illustrations: Jahn, 1924, Plate 1, Figs. 1–2.

Comment: The lack of pure culture data and the absence of precise descriptions of the vegetative cells and myxospores make re-identification difficult. Although Krzemieniewska and Krzemieniewski (1926, 1938) claim to have observed this organism and describe the cells as "*Sorangium*"-like, the dimensions given by them for the myxospores (0.9–1.3 by 2.3–4 μm) and the absence of a stalk in their material, cause considerable doubt. Both Jahn and the Krzemieniewski's considered *A. thaxteri* closely related, if not identical, to *A. flavum*.

e. *Archangium violaceum* (Kühlwein and Gallwitz) Kühlwein and Reichenbach 1964, 181. (*Polyangium violaceum* Kühlwein and Gallwitz 1958, 140.)

vi.o.la'ce.um. L. neut.adj. *violaceum* violet.

Vegetative cells are long, tapered flexible rods with pointed ends, 9–12 by 0.5–0.6 μm.

Fruiting bodies up to 1 mm in their largest dimension, in shape, similar to *A. gephyra*, occurring as irregular clumps, brainlike masses or swollen ridges. Varying in color similarly to *A. gephyra* but usually with a slight bluish violet cast, occasionally becoming dark violet.

Microcysts resemble those of *A. gephyra*, 1.8–4 by 0.75–1.8 μm.

Vegetative colonies on yeast potato agar uniformly dark violet due to melanin-related pigments. On peptone-rich media the colonies are red-brown.

Easily cultivated on the usual media employed for non-cellulolytic myxobacters. Growth has also been obtained on a medium containing 16 amino acids and salts (Mayer, 1967).

Nitrate is not reduced, catalase positive. Aesculin, starch (3 days), casein, gelatin, Tween 80, indoxyl acetate, RNA, DNA and carboxymethylcellulose but not cellulose are hydrolyzed.

Aerobic.

Optimum pH 7.5.

Temperature range 18–40 C.

Antibiotic sensitivity (discs): resistant to penicillin (10 units); sensitive to neomycin (10 μg), streptomycin (5 μg), tetracycline (10 μg), chloramphenicol (10 μg), kanamycin (10 μg) and erythromycin (5 μg).

Reported only once from hare dung from S. Africa.

Illustrations: Kühlwein and Gallwitz, 1958; Kühlwein and Reichenbach, 1964; McCurdy, 1969 (color).

Note: Very similar to, and perhaps only a color variant of, *A. gephyra*. When first reported however, it is said to have produced well developed sporangia suggesting perhaps that it is a *Cystobacter* which has lost the ability to produce typical fruiting bodies.

II. *Stelangium* Jahn 1924, 72.

Ste.lan'gi.um. Gr. fem.n. *stele* pillar, column; Gr. neut.n. *angium* vessel; M.L. neut.n. *Stelangium* columnar vessel.

Fruiting bodies are column- or finger-like, sometimes forked, without a definite stalk or enclosing wall, standing upright on the substrate.

a. *Stelangium muscorum* (Thaxter) Jahn 1924, 72. (*Chondromyces muscorum* Thaxter 1904, 411.)

mus.cor'.um. L. masc.n. *muscus* moss; L. gen.-pl.n. *muscorum* of mosses.

Vegetative cells not described.

Fruiting bodies are bright yellow-orange 90–300

μm high, 10–50 μm wide, simple or rarely furcate, sessile without a stalk, erect, elongate, stout or slender and tapering to a pointed apex.

Myxospores 4–6 by 1–1.3 μm, shape not given.

Source and habitat: Found on liverworts on living beech trees in Indiana.

Illustrations: Thaxter, 1904, Plate XXVII Figs. 16–18.

Comment: Description closely resembles the appearance of distorted fruiting bodies of a kind commonly observed in species of *Chondromyces* and *Stigmatella aurantiaca*.

b. *Stelangium vitreum* Peterson 1959, 163.

vit're.um. L. neut.adj. *vitreum* glassy.

Vegetative cells 4.0 by 0.7 μm shape not given.

Fruiting bodies crystalline, orange, globose or columnar, usually unbranched masses without a distinct wall, 60–70 μm in diameter and up to 150–200 μm in height. Within the fruiting body the myxospores are seen to be arranged in packs of 6–12 in parallel arrangement. Adjacent packs stand almost perpendicular to one another in basketweave fashion.

Myxospores are straight rods with squarish ends.

Not cultivated, observed only on original substrate.

Source and habitat: Found on the bark of various living trees in Missouri, U.S.A.

Illustrations: Peterson, 1959.

FAMILY III. **CYSTOBACTERACEAE** McCURDY 1970, 286

HOWARD D. McCURDY

Cys.to.bac.ter.a'ce.ae. M.L. n. *Cystobacter* type genus of the family; *-aceae* ending to denote a family; M.L. fem.pl.n. *Cystobacteraceae* the Cystobacter family.

Vegetative cells tapered flexible **rods** which are converted to refractile or phase-dense, **rod-shaped microcysts** enclosed **in sporangia** of definite shape. The sporangia may be sessile, occurring singly or in groups and enclosed in a slime membrane or envelope; or borne on stalks (sporangiophores) which may be simple or branched.

The sporangia may be solitary or in clusters at the tips of the stalks.

The vegetative colonies **do not etch or erode agar. Congo red is adsorbed** by the vegetative slime. All of the species examined grow well on media containing enzymatically hydrolyzed protein, starch and salts. **Bacteriolytic. Non-cellulolytic.**

Key to the genera of family **Cystobacteraceae**

I. Sporangia sessile.

II. Sporangia stalked.

A. Sporangia borne singly on an unbranched stalk.

B. Sporangia in clusters on branched or unbranched stalks.

Genus I. **Cystobacter** *Schroeter 1886, 170*

Cys.to.bac′ter. Gr. n. *cystis* bladder; M.L. n. *bacter* the masculine equivalent of the Gr. neut.n. *bakterion* a rod; M.L. masc.n. *Cystobacter* the bladder-forming rod.

Vegetative cells slender, **tapered,** flexible rods.

Sporangia sessile, occurring singly or in groups; rounded, elongate or coiled and surrounded by a definite slime envelope or wall; either free or embedded in a second slimy layer.

Microcysts rod-shaped, phase-dense or re-fractile, rigid.

Vegetative colonies do not etch or erode agar media, **congo red is adsorbed.**

The minimum nutritional requirements are not known but all species are easily cultivated on media containing enzymatically hydrolyzed protein, salts and starch or glycogen. The latter utilized. **Cellulose is not digested.**

Type species: *Cystobacter fuscus* Schroeter.

Key to the species of genus **Cystobacter**

I. Sporangia rounded to spherical.
 A. Mature sporangia dark brown, >60 μm diameter.
  1. *Cystobacter fuscus*
 B. Mature sporangia pink to light brown, <60 μm
  2. *Cystobacter minus*
II. Sporangia elongate, coiled.
  3. *Cystobacter ferrugineus*

Description of the species of genus **Cystobacter**

1. **Cystobacter fuscus** Schroeter 1886, 170. (*Polyangium fuscum* (Schroeter) Thaxter 1904, 414.)

fus′cus. L. adj. *fuscus* dark, tawny.

Vegetative cells distinctly tapered rods, 0.6–0.8 by 3–20 μm (exponential cells 0.65 by 8–10 μm) (Plate 2.2, Fig. 1).

Sporangia (Plate 2.2, Fig. 2) smooth, flesh-colored when young, when ripe light to dark chestnut brown, spherical, oval or elongate 50–150 μm in diameter with definite membrane, occurring 100 or more, in sori and embedded in a glossy gelatinous matrix. Sori flat or heaped, occasionally with chains of sporangia raised in finger-like projections. In some forms (*Cystobacter fuscus* subsp. *velatus* Krzemieniewska and Krzemieniewski 1926, 32) the matrix is replaced by a folded outer membrane-like layer.

Microcysts (Plate 2.2, Figs. 3 and 4), rod-shaped, often fusiform and bent, optically dense or definitely refractile, 0.4–1.5 by 2.5–5.0 μm.

Vegetative colonies on Casitone-Mg++ agar rapidly growing, thick, initially flat with a definite edge later exhibiting prominent veins, ridges and accumulations with pointed extensions at the edge. When young, yellow to yellowish orange becoming salmon or brownish orange. Older colonies are smooth, mucoid and translucent. A brownish pigment may be produced in the medium. The vegetative slime is very tenacious and difficult to cut; the colony is easily lifted intact from the agar surface. The colonies on *E. coli*

medium and dung pellet agar are similar but thinner and less pigmented. Clearing of *E. coli* medium is extensive. Fruiting bodies are produced on both media by freshly isolated strains.

Growth in Casitone-Mg++ broth is luxuriant, yellow to orange, in the form of loose spheres or as a dense turbidity; the medium becomes viscous and turns brown with age.

Easily cultivated on the usual complex media employed for the non-cellulolytic myxobacteria. Nitrate is not reduced, catalase positive, oxidase negative. Esculin, starch (3 days), casein, gelatin, Tween 80, indoxyl acetate, RNA, DNA are hydrolyzed. Urease is produced. Cellulose is not hydrolyzed.

Aerobic.

Optimum pH 6.9–8.2.

Temperature range 18–40 C, optimum 30 C.

Antibiotic sensitivity (discs); resistant to neomycin (10 μg), penicillin (10 units). Inhibited by tetracycline (10 μg), chloramphenicol (10 μg), kanamycin (10 μg), streptomycin (5 μg) and erythromycin (5 μg).

Source and habitat: Originally obtained on rabbit dung from California. Commonly obtained on moist bark and rabbit dung placed in contact with soil.

G + C content 68.3 moles % by T_m.

Reference strain: Windsor M-31; ATCC 25194.

Illustrations: Thaxter, 1897, Plate 31, Figs. 37–39; Baur, 1905, Plate 4, Figs. 14, 15 and 17; Quehl, 1906, Plate 1, Figs. 8 and 16; Jahn, 1924,

Plate 2, Fig. 12 and Fig. A, p. 9; Krzemieniewska and Krzemieniewski, 1926, Plate IV, Figs. 42–43; McCurdy, 1969, Fig. 26 (color); McCurdy, 1970, Figs. 4–6.

2. Cystobacter minus (Krzemieniewska and Krzemieniewski) McCurdy 1970, 288. (*Polyangium minus* Krzemieniewska and Krzemieniewski 1926, 33.)

mi'nus. L. comp.adj. *minor* less, smaller.

Vegetative cells 0.6–0.8 by 3–11 μm, slightly tapered with somewhat square ends as in *Archangium gephyra*.

Sporangia spherical or oval 20–70 by 20–50 μm, at first pale pink becoming brownish, walls definite, 0.5–1.0 μm thick, transparent revealing contents. Sporangia covered by a thin, transparent slime and occurring in flat accumulations of up to 0.5 μm². A secondary sporangium may be formed within the first as a result of contraction of the contents and the formation of a new wall.

Microcysts phase-dense or refractile, oval to short rod-shaped 0.8–1.2 by 1.3–2.7 μm.

Vegetative colonies on Casitone-Mg⁺⁺ agar are at first grayish translucent or slightly pink with a definite edge. Later the edge becomes thin and ill defined. The surface is marked by many loosely spiralling lines.

Cultivated on Casitone-Mg⁺⁺, *E. coli* medium and yeast-Ca⁺⁺ agars.

Nitrate not reduced. Catalase is produced. Oxidase negative, hydrolyzes starch, esculin, RNA and DNA. Urease negative. Does not digest cellulose.

Aerobic.

Temperature range 18–37 C, optimum 28–30 C.

Antibiotic sensitivity (discs); inhibited by tetracycline (10 μg) and chloramphenicol (10 μg). Resistant to neomycin (10 μg), kanamycin (10 μg), streptomycin (5 μg), erythromycin (5 μg).

Source and habitat: First isolated from sterilized rabbit dung placed in contact with soil (Poland). May be obtained from soil placed on *Micrococcus luteus* (*Sarcina lutea*) streaks on water agar. Relatively slow in appearance but not uncommon.

Reference strain: Windsor M-307.

Illustrations: Krzemieniewska and Krzemieniewski, 1926, Plate IV, Figs. 47–48, and Plate V, Fig. 49; McCurdy, 1969, Fig. 30; McCurdy, 1970, Figs. 9–11.

3. Cystobacter ferrugineus (Krzemieniewska and Krzemieniewski) McCurdy 1970, 288. (*Polyangium ferrugineum* Krzemieniewska and Krzemieniewski 1927, 89.)

fer.ru.gi'ne.us. L. adj. *ferrugineus* of the color of iron rust.

Vegetative cells tapered with rounded ends 0.6–0.8 by 4–15 μm.

Resting accumulations (Plate 2.2, Fig. 5) consist of irregular, branched and occasionally constricted coils. At first grayish or flesh-colored becoming bright orange-yellow, orange-red or reddish brown. The enclosing membrane may be absent or difficult to observe when present bearing the imprints of the enclosed microcysts. The external slime is colorless to yellow-orange.

Microcysts phase-dense or refractile, rigid, oval to short rod-shaped with rounded ends, 1.1–1.8 by 1.8–5 μm.

Vegetative colonies grayish white to slightly salmon-colored with many fine radiating ridges and concentric ripples, later clearing in the center. The edge is at first slightly heaped but becoming thin with tongue-like extensions.

Cultivated easily on Casitone-Mg⁺⁺, *E. coli* medium and dung pellet agar. Oxidase and urease negative. Esculin is not hydrolyzed.

Aerobic.

Maximum temperature 37 C, optimum 27–30 C.

Antibiotic sensitivity (discs) inhibited by tetracycline (10 μg), chloramphenicol (10 μg), erythromycin (5 μg). Resistant to neomycin (10 μg), streptomycin (5 μg), penicillin (10 units). Response to kanamycin (10 μg) variable.

Originally obtained from Polish soil on rabbit dung. Common on bacterial streaks over water agar inoculated with soil.

Reference strain: Windsor M-203.

Illustrations: Krzemieniewska and Krzemieniewski, 1927, Plate V, Fig. 21; McCurdy, 1969, Fig. 31 (color); McCurdy, 1970, Figs. 1–3.

Species incertae sedis

a. *Cystobacter stellatus* (Kofler) McCurdy 1970 289. (*Polyangium stellatum* Kofler 1913, 863.)

stel.la'tus. L. part. adj. *stellatus* stellate, set with stars.

Vegetative cells are not described.

Sporangia elongate 80–120 by 160–200 μm, flesh-colored when young, brownish red when old, in groups of two to nine forming a rosette or star-shaped cluster in which each sporangium is attached by its narrowed end to a common slime base or hypothallus.

Myxospores 4–6 μm long, but morphology not described.

Nutritional requirements not known. Not cultivated.

First found on hare dung from Vienna.

Illustrations: Kofler, 1913, Plate I, Fig. 6.

This species has been observed only once. Neither the vegetative nor resting cells are described. Kofler's description and the published illustration suggests a close resemblance to variants of *Stigmatella*.

b. *Cystobacter indivisus* (Krzemieniewska and Krzemieniewski) McCurdy 1970, 289. (*Polyangium indivisum* Krzemieniewska and Krzemieniewski 1927, 90.)

in.di.vi'sum. L. adj. *indivisus* undivided.

Vegetative rods not described.

Resting accumulations resemble those of *Cystobacter ferrugineus*, 50–65 by 190–400 μm, consisting of constricted widened coils, covered by slime which is at first colorless becoming bright yellow.

Membrane enclosing coils orange-red, 1.5 μm thick, bearing imprints of the microcysts which are said to be arranged perpendicularly to it as in *Melittangium boletus*.

Microcysts straight with rounded ends: 8–1.0 by 3–6 μm.

Vegetative colonies not described.

Cultivation not reported.

First found in Polish soil baited with rabbit dung.

Comment: Appears to be a variant of *Cystobacter ferrugineus*.

Genus II. **Melittangium** *Jahn 1924, 78*

Me.lit.tan'gi.um. Gr. n. *melitta* bee; Gr. n. *angium* vessel; M.L. neut.n. *Melittangium* resembling a bee hive.

Vegetative cells tapered rods.

Sporangia borne singly on a stalk.

Microcysts rod-shaped, optically dense or refractile.

Vegetative colonies do not etch or erode agar media; congo red is adsorbed.

Minimum nutritional requirements unknown but cultivatable on media containing enzymatically hydrolyzed protein.

Type species: *Melittangium boletus* Jahn 1924, 78.

Key to the species of genus **Melittangium**

I. Sporangia flattened like a mushroom cap.

                              1. *Melittangium boletus*

II. Sporangia spherical or ellipsoidal.

    A. Stalk short, <40 μm

                              2. *Melittangium lichenicola*

    B. Stalk long, >40 μm, bent.

                              3. *Melittangium alboraceum*

Description of the species of genus **Melittangium**

1. **Melittangium boletus** Jahn 1924, 78.

bo.le'tus. L. n. *boletus* a kind of mushroom.

Description based on: Jahn, 1924; Krzemieniewska and Krzemieniewski, 1928; Solntseva, 1941.

Vegetative rods slightly tapered with rounded ends (Krzemieniewska and Krzemieniewski, 1928) 0.7–4.5 by 10.5 μm (Solntseva, 1941).

Sporangia (Plate 2.2, Figs. 9 and 10) spherical or flattened and resembling a mushroom pileus, 45–50 by 50–100 μm. At first whitish becoming yellowish flesh-colored and finally yellowish brown to nut brown, when dried reddish brown. Sporangiophore white or yellowish 10–25 by up to 60 μm, occasionally poorly developed or absent. The resting cells in the sporangia stand at right angles to the enclosing wall in several layers, the wall has a honeycomb structure because of the impingement of the resting cells against it.

Myxospores rod-shaped 0.7–0.9 by 1.5–3.0 μm.

Vegetative colonies not described.

Cultivated with good growth on 10% dung decoction agar and potato agar (Solntseva, 1941).

Starch is hydrolyzed. Cellulose not digested (Solntseva, 1941).

Aerobic (Solntseva, 1941).

pH range 4.0–8.5 (Solntseva, 1941).

Temperature range 20–31 C (Solntseva, 1941).

Source and habitat: From dung of various herbivores, also from wet bark (Solntseva, 1941).

Illustrations: Jahn, 1924, Plate 2, Figs. 17 and 18 and Figs. B, C–F, O–Q, T–U; Krzemieniewska and Krzemieniewski, 1926, Plate V, Figs. 55 and 56; McCurdy, 1970, Figs. 9 and 10.

2. **Melittangium lichenicola** (Thaxter) McCurdy 1971, 53. (*Chondromyces lichenicolus* Thaxter 1892, 402; *Chondromyces gracilipes* Thaxter 1897, 406; *Podangium lichenicolum* (Thaxter) Jahn 1924, 81; *Podangium gracilipes* (Thaxter)

Jahn 1924; 82; *Chondromyces minor* Krzemieniew-ska and Krzemieniewski 1930, 267.)

li.che.ni′co.la. G. n. *lichen* lichen; L. n. *cola* dweller; M.L. n. *lichenicola* lichen-dweller.

Vegetative rods tapered, 0.6–0.9 by 7–12 μm (cells in logarithmic growth phase 0.6 by 6–8 μm). Fruiting bodies of two types. The first (Plate 2.2, Fig. 6) consists of a single orange-red to bright red, spherical, ellipsoidal or pear-shaped spo-rangium, 15–35 by 25–35 μm, on a white to orange, rigid stalk 5–10 by 10–40 μm. When crowded, the sporangia may be confluent. Sporangiophores occasionally lacking but when well developed tapering at the tip, persistent, the sporangia caducous. The second type of fruiting body (Plate 2.2, Fig. 7) often accompanies the first, is more usual on artificial media and consists of an orange to bright red irregular mass, up to 1 mm diameter with a swollen padded surface. Internally the microcysts and slime are oriented to form numer-ous, closely appressed, intestine-like convoluted tubes of about 50–90 μm in diameter. This form appears very similar to *Archangium primigenium* or *A. serpens*.

Microcysts irregularly rod-shaped, often bent and narrowing at the ends, rigid, phase-dense or refractile with a slime capsule. Usually in sheaves and difficult to separate.

Vegetative colonies on agar media very rapidly spreading; very thin, nearly transparent, with occasional reddish orange to bright red radial folds and accumulations. Characteristically ex-tends on the walls of the culture vessel and be-tween agar-glass interfaces. Lysis on *E. coli* medium is very extensive.

Growth in Casitone-Mg⁺⁺ broth is poor, in the form of compact red spheres, the medium remains clear.

Cultivated on ECM, dung pellet and Casitone-Mg⁺⁺ agars but with only limited growth.

Nitrate not reduced. Catalase is produced. Oxidase negative. Hydrolyzes starch, RNA, DNA and usually esculin. Urease negative. Does not digest cellulose.

Temperature range: 18–37 C, optimum 25–28 C.

Antibiotic sensitivity (discs): Inhibited by 10 μg of tetracycline, chloramphenicol, kanamycin and 5 μg of erythromycin. Response to neomycin (10 μg) and penicillin (10 units) variable.

Originally observed on rabbit dung in contact with soil and on lichens. Very common on moist tree bark.

Type: Acc. Nos. 4500 and 5170, Thaxter collec-tion, Farlow Herbarium, Harvard University.

Reference strain: Windsor M201; ATCC 25946.

Illustrations: Thaxter, 1892, Plate 23, Figs. 20–23; Thaxter, 1897, Plate 31, Figs. 20–24; Quehl, 1906, Plate 1, Figs. 6 and 9; Jahn, 1924, Plate II, Figs. 19 and 20; Krzemieniewska and Krzemie-niewski, 1926, Plate V, Fig. 54: McCurdy, 1969, Figs. 9–12, 42(color); McCurdy 1970, Figs. 1–8.

3. **Melittangium alboraceum** (Peterson) Mc-Curdy 1971, 54. (*Podangium alboraceum* Peterson 1959, 167.)

al.bo.ra′ce.um. L. adj. *albus* white; L. masc.n. *racemus* the stalk of a cluster; M.L. adj. *albora-ceum* implying a white stalk.

Vegetative cells 0.8–1.0 by 4.5–5.0 μm, no taper-ing apparent, square ends.

Fruiting body (Plate 2.2, Fig. 8) a single spo-rangium on a long, white, irregularly corkscrew-shaped sporangiophore 82–250 (average 125) by 20 μm. The sporangium is an irregular globe, pale orange, crystalline, 35 μm diameter, bounded by an elastic membrane which is difficult to see.

Myxospores, rod-shaped, 0.8 by 2.5 μm, slightly curved, difficult to separate from the sporangium and one another.

Not cultivated.

Source and habitat: Observed twice on elm bark.

Type: Three microscope slides, Peterson 72, University of Missouri herbarium.

Illustrations: Peterson, 1959, Figs. 4, 5 and 6.

Genus III. **Stigmatella** *Berkeley and Curtis in Berkeley 1875, 97*

Stig.ma.tel′la. L. neut.n. *stigma, stigmatis* brand or mark; M.L. fem.dim. ending *ella;* M.L. fem.n. *Stigmatella* a small brand or mark.

Vegetative cells are rods with tapered ends.

Sporangia borne singly or in clusters on stalked fruiting bodies (the stalks often occurring in groups arising from a common hypothallus).

Myxospores short, rigid, phase-dense or refrac-tile rods surrounded by a definite slime capsule.

Vegetative colonies do not etch, erode or pene-trate agar media. Congo red is adsorbed.

The minimum nutritional requirements are not known but easily cultivated on media containing enzymatically hydrolyzed protein.

Urea is usually hydrolyzed.

Aerobic.

Temperature range 18–37 C; optimum 30 C.

G + C content of species examined 68.5–68.7 by T_m (McCurdy and Wolf, 1967).

Type species: *Stigmatella aurantiaca* Berkeley and Curtis in Berkeley 1875, 97.

Description of the species of genus **Stigmatella**

1. **Stigmatella aurantiaca** Berkeley and Curtis in Berkeley 1875, 97. (*Chondromyces aurantiacus* (Berkeley and Curtis) Thaxter 1892, 401.)

au.ran.ti'a.ca. M.L. fem.adj. *aurantiaca* orange-colored.

Vegetative cells are rods with tapered ends, 0.6–1.0 by 4–10 μm.

On bark (Plate 2.2, Fig. 11) sporangia spherical, oval, pear-shaped or cylindrical; yellowish orange to bright orange-red or reddish brown, variable in size 16–70 by 25–102 μm. Pedicels may be absent at maturity, when present up to 40 μm in length. Stalks up to 400 μm long, usually unbranched sometimes arising from a common origin in a fascicled arrangement, granular, not striated, of hardened slime containing some cells, colorless to the color of the sporangia. Morphology of the fruiting bodies variable especially in culture; on plain agar in the presence of microbial contaminants, white to orange stalks are often observed up to 900 μm long, irregularly branched and either lacking sporangia or tipped by one or a few sporangia (Plate 2.2, Fig. 12); archangium-like forms are also common. When induced on Ca^{++}-water agar the fruiting bodies bear one or few sporangia.

Microcysts, 0.9–1.2 by 1.6–4.0 μm (average 1.0 by 2.8 μm). Vegetative colonies thin, flat with numerous radiating and concentric ridges, edge more or less definite or thin, filamentous and poorly delimited, light yellow or flesh-colored, occasionally producing a yellowish or brownish discoloration of agar media. Orange aggregates often form in old cultures on 0.1% Casitone or *E. coli* media but fruiting bodies fail to develop. Continued laboratory cultivation yields variants producing mucoid colonies unable to form fruiting bodies.

Easily cultivated on agar or in liquid media containing 0.1–0.2% hydrolyzed protein, starch and 0.01 M Mg^{++} or on agar media containing dead bacterial cells.

Nitrate is not reduced. Catalase positive, oxidase negative. Hydrolyzes starch (3 days), Tween 80, indoxyl acetate, RNA, DNA, gelatin, casein, urea and esculin.

Aerobic.

Temperature range 18–37 C, optimum 30 C. Optimum pH 7.0–7.2.

Antibiotic sensitivity (discs): Resistant to penicillin, (10 units). Sensitive to streptomycin (5 μg), tetracycline (10 μg), chloramphenicol (10 μg), kanamycin (10 μg) and erythromycin (5 μg). Response to neomycin variable.

Source and habitat: Originally observed on lichens. Most commonly observed on bark kept in a moisture chamber. Also from soil inoculated on streaks of living bacteria on filter paper over water agar.

G + C content: 68.5–68.7 moles % by T_m (McCurdy and Wolf, 1967).

Reference strain: Windsor M341.

Neotype: Acc. No. 4477, Thaxter collection, Farlow Herbarium, Harvard University.

Illustrations: Berkeley and Broome, 1873, Plate 4, Fig. 16; Kalchbrenner and Cooke, 1880, 23; Thaxter, 1892, Plates 23 and 24, Figs. 12–19, 25–28; Zukal, 1896, Plate 20; Quehl, 1906, Plate 1, Fig. 10; Jahn, 1924, Figs. V and W; Krzemieniewska and Krzemieniewski, 1926, Plate V, Figs. 57–60; Reichenbach, Voelz and Dworkin, 1969; Reichenbach and Dworkin, 1969, Plates 1 and 2, Figs. 1 and 4–9; McCurdy, 1969, Figs. 17–21 (color); McCurdy, 1971, Figs. 13–16.

2. **Stigmatella erecta** (Schroeter) McCurdy 1971, 48. (*Cystobacter erectus* Schroeter 1886, 170; *Chondromyces erectus* (Schroeter) Thaxter 1897, 407; *Podangium erectum* (Schroeter) Jahn 1924, 80; *Chondromyces aurantiacus* var. *frutescens* Krzemieniewska and Krzemieniewski 1927, 91; *Chondromyces brunneus* Krzemieniewska and Krzemieniewski 1946, 44; *Stigmatella brunnea* (Krzemieniewska and Krzemieniewski) McCurdy and Khouw 1969, 731.)

e.rec'ta. L. fem.adj. *erecta* erect.

Vegetative cells are slightly tapering, flexible rods 0.7–0.8 by 5–10 μm.

On bark (Plate 2.2, Fig. 13) the sporangia are spherical, oval or elongated, at first flesh-colored becoming orange-red and finally dark chestnut-brown or almost black at maturity, 30–90 by 35–140 μm. Sometimes (especially on rabbit dung) borne singly on opaque white, later yellowish, stalks arising in groups from a common hypothallus Plate 2.2, Fig. 14:). In other forms several sporangia are arranged in clusters on yellowish white to orange-red stalks. The sporangia may or may not be borne on pedicels. In both forms the stalks (and pedicels) commonly wither, depositing the cysts upon the substrate so that they appear sessile in masses of fifty to a hundred. On rabbit dung pellets the fruiting bodies may appear as archangium-like masses or coralloid accumulations of pinkish white or salmon-colored finger-like projections without clear-cut delimitation of sporangia. The tips eventually turn dark brown, the remainder becoming yellowish white. On Ca^{++}-water agar fruiting bodies with one to four sporangia on sparingly branched stalks (80–200 by 30–50 μm) are produced.

Microcysts are straight, curved or somewhat

fusiform, short rigid, phase-dense or refractile rods, 0.8–1.5 by 1.5–3.5 μm.

Vegetative colonies are at first thin transparent, later yellow or light flesh-colored with numerous radiating ridges. The edge is thin and indefinite. Continued laboratory cultivation selects variants with yellow or orange mucoid colonies. In old cultures the swarm turns dark brown and the surrounding medium also becomes darkened.

Easily cultivated on media containing enzymatically hydrolyzed protein, starch and 0.01 M Mg^{++} and salts. Growth has also been obtained on a medium containing 17 amino acids, thiamine, 0.01 M Mg^{++} and starch (McCurdy and Khouw, 1969). Strictly respiratory, utilizing complex amino acids as energy sources.

Nitrate is not reduced. Catalase positive. Oxidase negative.

Hydrolyzes starch (3 days), Tween 80, indoxyl acetate, DNA, RNA, gelatin, casein, urea and esculin.

Aerobic.

Temperature range 18–37 C, optimum 28–30 C. Optimum pH 7.0–7.2.

Antibiotic sensitivity (discs): Resistant to penicillin (10 units). Sensitive to streptomycin (5 μg), tetracycline (10 μg), chloramphenicol (10 μg), kanamycin (10 μg) and erythromycin (5 μg). Response to neomycin variable.

Source and habitat: Obtained on the dung of herbivores placed in contact with soil or bark incubated in a moist chamber.

G + C content 68.7 moles % by T_m (McCurdy and Wolf, 1967).

Neotype strain: Windsor M26; ATCC 25191.

Illustrations: Thaxter, 1897, Plate 31, Figs. 16–19; Quehl, 1906, Plate 4, Fig. 4; Jahn, 1924, Plate 1, Figs. 7–9; Krzemieniewska and Krzemieniewski, 1926, Plate V, Figs. 52 and 53; Krzemieniewska and Krzemieniewski, 1927, Plate VI, Figs. 27–35; Krzemieniewska and Krzemieniewski, 1946, Plate 1, Figs. 9 and 10; McCurdy and Khouw, 1969, Figs. 1–3; McCurdy, 1969, Figs. 13–16 (color). McCurdy, 1971, Figs. 17–19.

FAMILY IV. POLYANGIACEAE JAHN 1924, 75

HOWARD D. McCURDY*

Vegetative cells cylindrical, of uniform diameter **with blunt, rounded ends. Myxospores resemble vegetative cells,** not refractile or phase-dense, **lacking a slime capsule** in species examined (McCurdy, 1968; Peterson, 1969). **Sporangia sessile, solitary or in groups** (sori) bound by a common membrane or they may be borne singly or in clusters on branched or unbranched stalks.

The **colonies** of all species examined **etch, erode** and **penetrate agar media. Congo red** is **not adsorbed** by the vegetative slime of any of the species examined.

Cellulose may or may not be hydrolyzed.

The type genus is *Polyangium* Link 1809, 42.

Key to the genera of family **Polyangiaceae**

I. Sporangia sessile.
    A. Cells rod-shaped, never coccoid.

                Genus I. *Polyangium*

    B. Cells become coccoid.

                Genus II. *Nannocystis*

II. Sporangia stalked.

                Genus III. *Chondromyces*

Genus I. **Polyangium** *Link 1809, 42*

Po.ly.an'gi.um. Gr. adj. *poly* many; Gr. neut.n. *angium* vessel; M.L. neut.n. *Polyangium* many vessels.

Vegetative cells cylindrical, of uniform diameter **with blunt rounded ends.**

Sporangia sessile, solitary or in groups (sori) often bounded by a common envelope or layer of slime.

Myxospores resemble vegetative cells, not

---

* Species 9 and 10 of the genus *Polyangium* are described by J. E. Peterson.

refractile or phase-dense, **lacking a slime capsule** in species examined.

Colonies of all species examined etch, erode and penetrate agar. Congo red is not adsorbed.

Cellulose may or may not be digested.

Type species is *Polyangium vitellinum* Link 1809, 42.

Description of the species of genus **Polyangium**

1. **Polyangium vitellinum** Link 1809, 42. (*Myxobacter aureus* Thaxter 1892, 403; *Cystobacter aureus* (Thaxter) Thaxter 1897, 398.)

vi.tel.li′num. L. masc.n. *vitellus* egg yolk; M.L. neut.adj. *vitellinum* like an egg yolk.

Vegetative rods cylindrical with blunt rounded ends, 0.9–1.2 by 4–10 μm (Plate 2.3, Fig. 1).

Sporangia golden yellow, reddish yellow or brownish orange, oval, spherical or cushion-like, 75–400 μm, sporangial wall definite, showing imprints of enclosed rods (Plate 2.3, Fig. 2). The sporangia occur singly or in groups of up to 20 surrounded when fresh by a white slimy envelope. Contents at first flesh-colored later changing to an oily yellowish material.

Myxospores in young sporangia resemble vegetative cells 0.9 by 3–5 μm, often adhering together in sheaves. In older sporangia irregular shrunken rods are common and in some instances cells appear to be absent.

Colonies when rising to form fruiting bodies are white or slightly pinkish and creamy in consistency.

Cultivated only on original substrate. Not obtained in pure culture.

Most often observed on very wet wood or bark in swamps or moist ditches. Also obtained on bark kept in a moisture chamber and on rabbit dung in contact with soil.

Neotype: Acc. No. 4564, Thaxter collection, Farlow Herbarium, Harvard University.

Illustrations: Thaxter, 1892, Plate 25, Figs. 34–36; Zukal, 1897, Plate 27, Figs. 6–10; Jahn, 1911, Fig. 3; Jahn, 1924, p. 77 and Plate II, Fig. 13; McCurdy, 1969, Fig. 32 (color); McCurdy, 1970, Figs. 15 and 16.

2. **Polyangium luteum** Krzemieniewska and Krzemieniewski 1927, 90.

lu′te.um. L. adj. *luteum* saffron or golden yellow.

Vegetative cells 0.8 by 6.0–7.0 μm.

Sori golden yellow, containing spherical, oval or shapeless sporangia of variable size 37–115 by 39–90 μm with colorless thin walls. The common slime envelope enclosing the sporangia is bright yellow, double contoured, up to 10 μm thick.

Myxospores 0.7–0.8 by 3.8–5.8 μm.

Vegetative colonies colorless to light yellow, indefinite in outline and penetrating the agar.

Cultivated on bacterial cells. Not obtained in pure culture.

Not observed to digest cellulose.

Obtained from rabbit dung in contact with soil. Also on *Micrococcus luteus* (*Sarcina lutea*) streaks inoculated with soil.

Illustrations: Krzemieniewska and Krzemieniewski, 1927, Plate V, Figs. 22 and 23.

3. **Polyangium aureum** Krzemieniewska and Krzemieniewski 1930, 255.

au′re.um. L. neut.adj. *aureum* golden.

Vegetative rod, dimensions not given.

Sporangia (Plate 2.3, Fig. 3) spherical or oval, light brown to reddish brown, 20–60 by 15–50 μm (commonly 30 by 35 μm) each with an orange-yellow sporangial wall 3.5 μm thick, in sori covered by a transparent membrane but not appressed. Contents of older sporangia consist of colorless or light yellow oily liquid with few myxospores.

Myxospores resemble vegetative cells 0.7–0.9 by 2.8–5.3 μm.

Vegetative colonies indefinite, colorless, underlying agar etched and penetrated.

Growth is very slow on *Micrococcus luteus* (*Sarcina lutea*) cells. Pure cultures not obtained.

Originally obtained on rabbit dung in contact with soil. Also on *Micrococcus luteus* (*Sarcina lutea*) streaks inoculated with soil.

Illustrations: Krzemieniewska and Krzemieniewski, 1930, Plate XVII, Figs. 14 to 17; McCurdy, 1969, Fig. 31 (color); McCurdy, 1970, Fig. 7.

4. **Polyangium parasiticum** Geitler 1924, 67.

pa.ra.si′ti.cum. Gr. adj. *parasiticum* parasitic.

Vegetative rods 0.7 by 4–7 μm.

Sporangia spherical or elongate, 15–50 μm usually 25–50 μm, red-brown with distinct double contoured wall, usually occurring in groups of two to eight, appressed, enclosed by a colorless, distinct slime.

Myxospores resemble vegetative cells, slightly shorter.

Growth only on the alga *Cladophora* in water; at first externally as a saprophyte, later entering and destroying the algal cells. Not cultivated on artificial media.

Reported only once on *Cladophora* (*fracta?*) in pool in Vienna.

Illustrations: Geitler, 1924, Figs. A–K.

5. **Polyangium fumosum** Krzemieniewska and Krzemieniewski 1930, 253.

fu.mo'sum. L. neut.adj. *fumosum* smoky.

Vegetative cells 0.7–1.0 by 5.0–7.0 μm.

Sori (Plate 2.3, Fig. 4) usually flat, rounded or irregular in outline; consisting of from 2–50 or more sporangia, at first pale pink, later developing a brownish or smoky gray pigment outlining the sporangia. Sporangial wall definite often double contoured, slightly elongated to almost spherical, colorless, 17–72 by 13–60 μm (average 45 by 35 μm). The dark pigment is located in the common slime enclosure.

Myxospores 0.7–0.8 by 2.5–5.0 μm with conspicuous terminal granules. In old sporangia lysis may result in conversion of contents to an oily liquid devoid of cells.

Vegetative colonies light pink, delicate, becoming divided into migrating masses (200–500 μm) which leave radiating furrows and tunnels in the agar (Plate 2.3, Fig. 5).

Growth is slow in pure culture on yeast-Ca++ agar and *Micrococcus luteus* (*Sarcina lutea*) streaks over water agar.

Originally described on dung placed in contact with Polish soils. Commonly encountered on *M. luteus* streaks on water agar inoculated with soil.

Reference strain: Windsor M257.

Illustration: Krzemieniewska and Krzemieniewski, 1930, Plate XVI, Figs. 6–9 and 26; McCurdy, 1969, Figs. 28 and 29 (color): McCurdy, 1970, Figs. 12–14.

6. **Polyangium spumosum** (Krzemieniewska and Krzemieniewski) McCurdy 1970, 294. (*Sorangium spumosum* Krzemieniewska and Krzemieniewski 1927, 88.)

spu. mo'sum. L. adj. *spumosum* foamy or frothy.

Vegetative rods not described.

Fruiting bodies consist of numerous ellipsoidal or spherical sporangia, 8–26 by 7–20 μm, not surrounded by a common membrane but united into bodies by surrounding slime. Often in double or single rows. Sporangial walls colorless to slightly brownish, transparent.

Myxospores cylindrical with blunt ends, 2–3 by 0.4–0.5 μm.

Cultivated on filter paper which is digested, but not obtained in pure culture (Krzemieniewska and Krzemieniewski, 1937).

From Polish soil and decaying plant debris, most often observed on damp blotting paper inoculated with these materials.

Illustrations: Krzemieniewska and Krzemieniewski, 1927, Plate V, Fig. 19; Krzemieniewska and Krzemieniewski, 1937, Plate 3, Figs. 12–16.

7. **Polyangium minor** (Peterson) McCurdy 1970, 294. (*Haploangium minor* Peterson 1959, 3.)

mi'nor. L. comp.adj. *minor* less, smaller.

Vegetative cells with more or less square ends, sporangia sessile, solitary or in groups of 4–10 but usually without contact, globose, oval or bean-shaped, 60–140 μm, turgid, smooth, dull orange-brown when fresh, becoming collapsed and wrinkled when dry. Myxospores not visible through wall. Wall bright yellow-orange by transmitted light, about 2 μm thick.

Myxospores 0.7 by 2.5–3.5 μm, cylindrical with blunt ends becoming irregular in old sporangia, arranged in distinct spherical groups which are difficult to break up.

Colonies on rabbit dung pellet agar grayish yellow.

Not obtained in pure culture.

Source and habitat: Obtained on numerous pieces of Missouri bark.

Type specimen: Peterson 44, University of Missouri Herbarium.

Illustrations: Peterson, 1959.

8. **Polyangium rugiseptum** (Peterson) McCurdy 1970, 295. (*Haploangium rugiseptum* Peterson 1959, 5.)

ru.gi.sep'tum. L. fem.n. *ruga* wrinkle; L. neut.n. *septum* enclosure; M.L. neut.n. *rugiseptum* a wrinkled enclosure.

Vegetative cells not described.

Sporangia solitary, sessile, globose or oval, glistening orange-red, wrinkled when dry, up to 200 μm (average 85 μm). Sporangial wall of two distinct layers, the inner smooth and yellow, the outer irregular, flaky, dark orange.

Contents consist of fatty globules, amorphous material and few myxospores.

Myxospores cylindrical with blunt ends often becoming shrunken and irregular in shape (McCurdy, 1970).

Not cultivated.

Source and habitat: Bark of various Missouri trees.

Type specimen: Peterson 51, University of Missouri Herbarium.

Illustrations: Peterson, 1959; McCurdy, 1970, Fig. 8.

9. **Polyangium cellulosum** Imshenetski and Solntseva 1936, 1115.

BY J. E. PETERSON

(*Sorangium cellulosum* (Imshenetski and Solntseva) Imshenetski and Solntseva 1937, 7; *Sorangium nigrum* Krzemieniewska and Krzemieniewski 1938, 22; *Sorangium nigrescens* Krzemieniewska and Krzemieniewski 1938, 21.)

cel.lu.lo'sum. M.L. n. *cellulosum* cellulose.

Vegetative rods 0.8–1.2 by 3.0–10 μm.

Sporangia 20–30 μm in diameter, rounded to polygonal depending upon pressure from the mass

(Plate 2.3, Figs. 7 and 8). From four to several hundred sporangia (usually several dozen) clustered in sori which may or may not have discernible slime envelopes. The sori vary from slightly knobby rounded cushions, when formed free of constraints, to simple rows of sporangia when formed within cellulose fibers; color varies from pale yellow through shades of pink, orange, rusty red and brown to shades of gray and black.

Myxospores similar to vegetative cells, 1.0–3.0 μm in length.

Growth on filter paper overlaid on mineral salts medium is slow, appearing in from 3–14 days. The advancing periphery of the colony is lightly pigmented in shades of yellow, pink or orange; with the formation of sori (2–4 weeks) the central portion becomes more deeply pigmented (Plate 2.4, Fig. 1).

Strongly cellulolytic, marked decomposition of filter paper under the colony.

Most isolates may be grown on agar or in liquid media containing cellobiose, nitrate and salts. Ammonium salts, organic nitrogen and simpler carbon sources may or may not be utilized.

Temperature range 20–37 C, optimum 28–32 C. Optimum pH 6.8–7.2.

First isolated from Russian and Polish soils, has now been found elsewhere in Continental Europe and in North America. Closely associated with agricultural soils; found in large numbers in many arid and semi-arid soils both cultivated and uncultivated in the United States.

Reference strains: ATCC 15384; ATCC 25531; ATCC 25532; ATCC 25569.

Four subspecies have been described.

a. *Polyangium cellulosum* subsp. *ferrugineum* Mishustin 1938, 433; b. *Polyangium cellulosae* (*sic*) subsp. *fuscum* Mishustin 1938, 435; c. *Polyangium cellulosae* (*sic*) subsp. *fulvum* Mishustin 1938, 437; d. *Polyangium cellulosae* (*sic*) subsp. *luteum* Mishustin 1938, 438.

Illustrations: Imshenetski and Solntseva, 1936, Table II, Figs. 1–5; Krzemieniewska and Krzemieniewski, 1932, Plate IV, Figs. 22–26; McCurdy, 1969, Figs. 33–38 (color); Peterson, 1970, Figs. 7 and 8.

10. Polyangium sorediatum Thaxter 1904, 414.

BY J. E. PETERSON

(*Polyangium septatum* Thaxter 1904, 412; *Polyangium compositum* Thaxter 1904, 413; *Sorangium sorediatum* (Thaxter) Jahn 1924, 73; *Sorangium schroeteri* Jahn 1924, 73; *Sorangium compositum* (Thaxter) Jahn 1924, 74; *Sorangium septatum* (Thaxter) Jahn 1924, 75; *Polyangium schroeteri* (Jahn) McCurdy 1970, 294.)

so.re.di.a'tum. Gr. n. *sorus* a heap; M.L. dim.n. *soredium* a little heap; M.L. neut.adj. *sorediatum* having little heaps.

Vegetative rods 0.8–1.2 by 3.0–6.0 μm.

Fruiting bodies yellow-orange, up to 0.5 mm in diameter, flat or cushion-shaped, consisting of up to hundreds of sporangia (Plate 2.3, Fig. 6). The latter are often contained in slime-delimited groups of several sporangia each. Sporangia polygonal when appressed within the sorus but nearly spherical when free, 5–15 μm in diameter.

Myxospores shorter than vegetative cells but similar, 0.8 by 3–4 μm.

Vegetative growth not described.

Cultivation in pure culture not reported.

Noncellulolytic.

Obtained from animal dungs, decomposing plant debris, tree bark and soils rich in organic matter.

Illustrations: Thaxter, 1904, Plate 27, Figs. 22–30; Quehl, 1906, Plate 1, Fig. 2; Jahn, 1911, Figs. 1 and 2; Jahn, 1924, Plate I, Fig. 6; Krzemieniewska and Kzemieniewski, 1926, Plate III, Figs. 28–36, Plate IV, Figs. 37–41; Krzemieniewska and Krzemieniewski, 1927, Plates IV and V, Figs. 7–18; McCurdy, 1969, Fig. 38 (color).

Species incertae sedis

a. *Polyangium simplex* Thaxter 1904, 414. (*Myxobacter simplex* (Thaxter) Thaxter 1893, 29.)

sim'plex. L. adj. *simplex* simple.

Vegetative cells large cylindrical rods with rounded ends, 0.7–0.9 by 4.0–7.0 μm.

Sporangia solitary, large, 250–400 μm, bright reddish yellow, irregularly rounded.

Myxospores resemble vegetative cells, flesh-colored when in mass. Upon pressure adhere together in sheaves.

Not cultivated in pure culture.

Source: From very wet wood and bark in swamps.

Comment: This organism is probably a variant of *Polyangium vitellinum* Link 1809, 42.

b. *Polyangium morula* Jahn 1911, 202.

mo.ru'la. Gr. n. *mora* the black mulberry; L. dim.n. *morula* a small mulberry.

Vegetative rods not described.

Sporangia bright yellow, closely packed in a mulberry-shaped sorus, sporangia with thick membrane (3 μm), often polygonal by pressure, 20–35 μm, bound together by slime. The whole sorus is 100–200 μm broad.

Myxospores 3 μm in length, shape not given.

Source: Observed only once on rabbit dung.

Comment: This organism has been observed only once and was never cultivated. The description, which is based on old material and which does not give the morphology of either the vegetative rods or myxospores, is inadequate.

c. *Polyangium paraspumosum* (Krzemieniewska and Krzemieniewski) McCurdy 1970, 293. (*Polyangium spumosum* Krzemieniewska and Krzemieniewski 1930, 254.)

pa.ra.spu.mo'sum. Gr. pref. *para* near; M.L. neut.adj. *paraspumosum* near (*P.*) *spumosum*.

Vegetative rods 0.6–0.8 by 3.9–6.8 μm.

Sporangia almost spherical to slightly elongated, 20–50 by 18–38 μm (average 34 by 28 μm) each with a colorless sporangial wall, bound together by a colorless slime envelope in flat, irregular accumulations, 100–150 μm in diameter. Sporangia in addition to few developed rods, contain granular mass and colorless oily liquid.

Myxospores differ little from vegetative cells, 3.3–6.3 by 8 μm.

Has not been grown in pure culture.

Obtained from soil.

Illustrations: Krzemieniewska and Krzemieniewski, 1930, Plates XVI, XVII and XVIII.

Comment: The distinction between this species and *P. spumosum* is tenuous at best. They are said to differ chiefly in sporangium dimensions. In spite of the authors' written descriptions their published photographs show considerable heterogeneity and permit no distinction on the basis of size of the sporangia.

Genus II. **Nannocystis** *Reichenbach 1970, 136*

Nan.no.cys'tis. Gr. masc.n. *nannos* dwarf; Gr. fem.n. *cystis* bladder; M.L. fem.n. *Nannocystis* a dwarf (small) cyst.

Vegetative cells short, blunt ended rods or cocci. Sporangia solitary.

Type species is *Nannocystis exedens* Reichenbach 1970, 137.

Description of the species of genus **Nannocystis**

1. **Nannocystis exedens** Reichenbach 1970, 137.

ex.e'dens. L. v. *exedere* to eat away, L. part. adj. *exedens* corroding or eating away (the agar).

Vegetative cells (Plate 2.4, Fig. 2) short, fat rods with blunt, rounded ends, 1.4 by 2.5–3.5 μm, becoming oval, cube-shaped or coccoid.

Sporangia (Plate 2.4, Figs. 4 and 5) oval or spherical, highly variable in size, 3.5–6 by 40–110 μm, with the smallest predominating; embedded in the substrate. Myxospores differ little from vegetative cells, oval or spherical, optically dense.

Vegetative colonies light orange to red violet, eroding the agar in a characteristic fashion (Plate 2.4, Fig. 3). Cultivated on yeast-Ca^{++} agar supplemented with vitamin B$_{12}$. Cultivatable on Casitone media but inhibited by high concentrations.

Catalase produced. Gelatin and casein are hydrolyzed.

Lyses yeast cells and *Micrococcus luteus* (*Sarcina lutea*) almost completely.

Starch and cellulose are not hydrolyzed.

Optimum temperature 30–36 C.

Isolated from different soils from Samoa, The United States and Germany.

Type strain: KIMG Nacl.

Genus III. **Chondromyces** *Berkeley and Curtis in Berkeley 1874, 64*

(*Polycephalum* Kalchbrenner and Cooke 1880, 22; *Myxobotrys* Zukal 1896, 346.)

Chon.dro'my.ces or Chon.dro.my'ces. Gr. n. *chondrus* cartilage, gristle; Gr. n. *myces* fungus; M.L. masc.n. *Chondromyces* cartilaginous fungus.

Vegetative cells cylindrical, untapered rods, with blunt rounded ends.

Sporangia borne singly or in clusters **on** simple or branched **stalks.**

Myxospores lack capsules and resemble vegetative rods.

Vegetative swarms etch, erode and penetrate agar media. Vegetative slime does not adsorb congo red dye.

Aerobic.

Temperature range 18–37 C; optimum 28–30 C.

G + C content of species examined 69–70 moles % by T_m determinations.

Type species: *Chondromyces crocatus* Berkeley and Curtis in Berkeley 1874, 64.

Description of the species of genus **Chondromyces**

1. **Chondromyces crocatus** Berkeley and Curtis in Berkeley 1874, 64. (Putative synonym *Myxobotrys variabilis* Zukal 1896, 340.)

cro.ca'tus. L. adj. *crocatus* saffron yellow.

Vegetative cells (Plate 2.5, Fig. 3) 1.1–1.4 by 3–12 μm.

Sporangia broadly spindle-shaped, conical or nearly spherical 10–25 by 15–30 μm straw-colored initially, finally becoming golden yellow or orange. Sporangia borne in spherical clusters on usually branched stalks up to 700 μm or more in height (Plate 2.5, Fig. 4). Stalks orange to brown, striated, often spirally twisted, with several hollow ducts, containing few cells. Irregular forms with ramifying branches and few sporangia or with secondary fruiting structures arising from sporangia germinating *in situ*, may be observed in culture.

Myxospores lack capsules (Plate 2.5, Figs. 1 and 2), differ little from vegetative cells except for presence of conspicuous granules at one or both ends 1.0–1.3 by 3–6 μm.

Vegetative colonies on most media are initially translucent almost colorless, later becoming yellowish orange and heaped at the periphery to form a "front" which is particularly conspicuous in contact with masses of other living bacteria. The underlying agar is pitted, eroded and penetrated by columns of vegetative cells.

Growth on *E. coli* agar is poor with lysis generally limited to the immediate area of the colony. Lysis of living bacteria requires direct contact. Cultivated in pure culture on complex media containing enzymically hydrolyzed protein and Mg++. Growth is stimulated by, and initially requires, an extract from bacterial cells.

Nitrate is not reduced. Catalase and oxidase not produced. Hydrolyzes starch, Tween 80, indoxyl acetate, RNA, DNA, gelatin and casein. Does not hydrolyze urea, esculin or cellulose. Agar digestion not detected.

Temperature range 18–37 C, minimum 18 C, optimum 28–30 C.

Antibiotic sensitivity (discs): Resistant to neomycin (10 μg), kanamycin (10 μg), penicillin (10 units). Inhibited by streptomycin (5 μg), tetracycline (10 μg), chloramphenicol (10 μg) and erythromycin (5 μg).

Streptomycete-like odor produced different from that of myxobacters with tapered cells.

First observed on decayed melons from South Carolina. Later found by Thaxter (1892) on old straw from Ceylon. Commonly found on dung in contact with soil, bacterial streaks inoculated with soil or on bark.

Neotype: Acc. No. 601, Thaxter collection, Farlow Herbarium, Harvard University.

Reference strain: Windsor M38; ATCC 25193.

Illustrations: Berkeley, 1857, 313; Thaxter, 1892, Plates 22 and 23, Figs. 1–11; Quehl, 1906, Plate 1, Fig. 10; Jahn, 1924, Plate 2; McCurdy, 1969, Figs. 1–13; McCurdy, 1969, Fig. 22 (color); McCurdy, 1971, Figs. 1–3.

2. Chondromyces apiculatus Thaxter 1897, 405.

a.pi.cu.la′tus. L. n. *apex, apicis* point; M.L. adj. *apiculatus* having a small point.

Vegetative rods 1.1–1.4 by 3–14 μm.

Sporangia (Plate 2.5, Fig. 6) straw-colored to bright orange to brownish orange, variable in form and size, cylindrical to broadly turnip-shaped, 25–40 by 35–50 μm, with colorless, pointed, frequently branched, apical appendages up to 35 μm long. Pedicels absent or up to 30 μm long, colorless. Sporangia borne in spherical clusters of up to 60 or more, although usually fewer in number than in *C. crocatus*. Stalk seldom branched, up to 700 μm in height, diameter 15–40 μm, longitudinally striated, tunnelled and without cells internally. Forms without stalks, with basally fused sporangia, or with large, irregular, solitary sporangia are common. Secondary fruiting body formation as in *C. crocatus* or sporangium formation at tips of appendages is often observed.

Myxospores similar to vegetative cells, 1.0–1.3 by 3–6 μm.

In all other characteristics similar to *Chondromyces crocatus*.

Originally isolated from antelope dung from Liberia. Commonly observed on rabbit dung placed in contact with soil.

G + C content 69.3 moles % by T_m.

Type specimen: Acc. No. 4481, Thaxter Collection, Farlow Herbarium, Harvard University.

Reference strain: Windsor M-6.

Illustrations: Thaxter, 1897, Plate 30, Figs. 1–15; Quehl, 1906, Plate 1, Figs. 13 and 14; Jahn, 1924, Fig. 5; Kühlwein, 1952, 403; McCurdy, 1969, Fig. 23 (color); McCurdy, 1971, Figs. 6 and 7.

3. Chondromyces pediculatus Thaxter 1904, 410.

pe.di.cu.la′tus. L. dim.n. *pediculus* a small foot (stalk); M.L. adj. *pediculatus* having a small foot or stalk.

Vegetative cells 1.1–1.3 by 3–16 μm.

Sporangia (Plate 2.5, Fig. 7) pale yellow to orange, when dry orange-red, nearly spherical to long cylindrical, club-shaped or pyriform, usually broader and truncate at distal end, surface rough, 25–40 by 35–60 μm. Borne in groups of up to 60 in umbel-shaped heads on slender pedicels 20–40 μm long. Stalk unbranched, up to 750 μm in height, striated, sometimes twisted, colorless initially, becoming orange to brown when dry.

Myxospores 1.0–1.2 by 3–7 μm.

Vegetative colonies orange to reddish orange and resembling those of *C. crocatus*. Congo red is not adsorbed.

Cultivated on dung pellet agar and living bacteria. Growth is slow. Obtained once in pure culture on *E. coli* extract-enriched, Casitone-Mg⁺⁺ agar.

Urease is not produced.

Originally isolated on goose dung from South Carolina. Obtained on rabbit dung in contact with soil.

Type specimen: Acc. No. 4524, Thaxter collection, Farlow Herbarium, Harvard University.

Illustrations: Thaxter, 1904, Plate 26, Figs. 7–13; McCurdy, 1969, Fig. 24 (color); McCurdy, 1971, Figs. 8 and 9.

4. Chondromyces catenulatus Thaxter 1904, 410.

ca.te.nu.la′tus. L. n. *catena* chain; L. dim.n. *catenula* a small chain; M.L. adj. *catenulatus* having small chains.

Rods in "rising spore mass" 1.0–1.3 by 4–6 μm. Cells in vegetative colonies not described.

Sporangia light yellow to orange, fusiform, long elliptical or irregular in shape, dimensions 18 by 20–50 μm, united in chains up to 300 μm long which may be once or twice branched. Sporangia separated by shrivelled, membranous isthmuses. Stalk simple, orange to rust-colored, up to 400 μm in height, broad at base, tapering above and several times cleft into swollen, tapering parts each bearing one or several sporangial chains (Plate 2.5, Fig. 5).

Myxospores are nonrefractile cylindrical rods with blunt rounded ends 1.2–1.4 by 3–6 μm.

Cultivated only on decaying poplar wood from New Hampshire, not reported in pure culture.

Type specimen: Acc. No. 4517, Thaxter collection, Farlow Herbarium, Harvard University.

Illustrations: Thaxter, 1904, Plate 26, Figs. 1–5; McCurdy, 1969, Fig. 41 (color); McCurdy, 1971, Figs. 4 and 5.

5. Chondromyces lanuginosus Kofler 1913, 861. (*Chondromyces thaxteri* Faull 1916, 231; *Synangium lanuginosum* (Kofler) Jahn 1924, 79; *Synangium thaxteri* (Faull) Jahn 1924. 79.)

la.nu.gin.os′us. L. adj. *lanuginosus* downy, woolly.

Vegetative cells cylindrical with blunt rounded ends 0.9–1.0 by 3–8 μm.

Sporangia (Plate 2.5, Fig. 8) fused at their bases to form discoid or nearly spherical clusters containing up to 80 sporangia, each with an apical tuft of hairs. Diameter of clusters variable (40–250 μm) as is the length of the apical hairs (7–30 μm). Stalks simple or occasionally branched, bearing from 1–30 clusters. Sporangia initially white changing to yellow, light pink and eventually orange. Stalks are at first colorless but become

yellow. Sometimes the sporangial clusters give rise to secondary stalks which are thinner than the primary ones and which are tipped with smaller clusters.

Myxospores do not differ from vegetative cells, only slightly smaller 0.6–1.0 by 2.6 μm.

Cultivation: Grown in laboratory culture on hay (Krzemieniewska and Krzemieniewski, 1946, 27). Pure cultures not obtained.

Found on the dung of herbivores in Canada (Faull, 1916) and Austria (Kofler, 1913); from soil in Poland (Krzemieniewska and Krzemieniewski, 1946).

Neotype specimen: Acc. No. 4494 collected by J. H. Faull, Thaxter collection, Farlow Herbarium, Harvard University.

Illustrations: Kofler, 1913, Plate I, Figs. 1–3; Faull, 1916, Plates 5 and 6; Jahn, 1924, Fig. X; Krzemieniewska and Krzemieniewski, 1946, Plate 1, Figs. 1–3; McCurdy, 1969, Fig. 25 (color); McCurdy, 1971, Fig. 10.

Species incertae sedis

a. *Chondromyces sessilis* Thaxter 1904, 411.

Ses′si.lis. L. adj. *sessilis* sessile, stalkless.

Vegetative rods not described.

Sporangia yellow to reddish orange, forming a sessile rosette or tuft on the substrate without a clearly differentiated stalk although a poorly developed stalk is said occasionally to occur.

Sporangia very variable in shape, irregularly and broadly fusiform, often subapiculate, with a wrinkled surface, variable in size, coherent at base or more or less completely confluent in irregular masses. Dimensions of sporangia 18–55 by 25–75 μm, of rosettes 100–250 μm.

Myxospores are cylindrical with blunt rounded ends 0.8–1.0 by 3–5 μm.

Source: Rotten wood from Florida. Cultivation not reported. Observed only on natural substrate.

Illustrations: Thaxter, 1904, Plate 27, Figs. 14 and 15; McCurdy, 1969, Fig. 40 (color); McCurdy, 1971, Figs. 11a and 11b.

Comment: This organism was considered identical with *Chondromyces lanuginosus* in the 7th edition as suggested by Krzemieniewska and Krzemieniewski (1946). However comparison of the descriptions by Faull (1916), Kofler (1913) and Krzemieniewska and Krzemieniewski (1946) with that by Thaxter of *C. sessilis*, and of Faull's preserved specimens with Thaxter's type material, is not convincing. In fact Thaxter's organism, which was evidently observed only once, could well be a poorly developed variant of any of the described species of *Chondromyces*.

ORDER II. **CYTOPHAGALES** *Nomen novum*

E. R. Leadbetter

Cy.to.pha.ga'les. M.L. fem.n. *Cytophaga* type genus of the order; *-ales* ending to
denote an order; M.L. fem.pl.n. *Cytophagales* the *Cytophaga* order.

Cells rods, filaments. Hormogonia, gonidia or resting cells may be present. **Fruiting bodies are not produced. Motile by gliding in at least one morphological stage.** Gram-negative.

Chemolithotrophs, chemoorganotrophs or mixotrophs.

Further Comments

The organisms grouped in this order have in common one feature: gliding motility on solid surfaces. In some instances the gliding is rapid (ca. 10 or more μm/min), in other cases slow (ca. 2 μm/min) or jerky. The mechanism of the movement is not known, and it may well be that there are different bases for the trait. It may be noteworthy, however, that growth of many gliding bacteria (including members of the order *Myxobacterales*) is inhibited by low concentrations of the antibiotic actinomycin D (Dworkin, 1969), although one species each of *Vitreoscilla* and *Stigmatella* (order *Myxobacterales*) were not inhibited.

The degree of the natural relationships or affinities of organisms within this order is, at best, uncertain.

Key to the families of order **Cytophagales**

I. Cells motile with gliding motility; filaments not attached.
  A. Carotenoid pigments present; cells rods or filaments.
Family I. Cytophagaceae

  B. Carotenoid pigments absent; resting stages not known.
    1. Cells in cylindrical filaments; sulfur may or may not be present.
Family II. Beggiatoaceae, p. 112

    2. Flat filaments, thus far found only in oral cavities of vertebrates.
Family III. Simonsiellaceae, p. 116

II. Filaments generally attached at one end in natural habitats; only gonidia can glide.
Family IV. Leucotrichaceae, p. 118

Familiae incertae sedis

I. Cells motile with gliding motility. Carotenoid pigments absent; resting stages not known.
  A. Cells spherical, ovoid or cylindrical, 5–33 by 15–125 μm, may or may not contain sulfur and calcium carbonate. (May have flagella.)
Achromatiaceae, p. 120

II. Filaments, unbranched, occur singly or in bundles; some filaments are motile.
Pelonemataceae, p. 122

FAMILY I. **CYTOPHAGACEAE** STANIER *1940, 630, emend. mut. char.*

E. R. Leadbetter

Cy.to.pha.ga'ce.ae. M.L. fem.n. *Cytophaga* type genus of the family: *-aceae* ending
to denote a family: M.L. fem.pl.n. *Cytophagaceae* the *Cytophaga* family.

Cells rods, single or filaments which may be helical, branched or sheathed. Resting stages known to be formed in one genus. Unsheathed cells or filaments **motile by gliding.** Gram-negative.

Chemoorganotrophs: Metabolism is respiratory or fermentative. **Cells or cell masses are colored** some shade of yellow, orange or red.

Type genus: *Cytophaga* Winogradsky 1929, 577.

Further Comments

Essentially the only features common to the genera of this family are gliding motility and the presence of colored carotenoid pigments.

Soriano's excellent proposal (1945) that *Flexibacterales* and *Flexibacteraceae* be adopted for the classification of these organisms could not be utilized here because of the prior use of *Cytophaga* and *Cytophagaceae*.

The genus *Cytophaga* was originally described by Winogradsky, who did not work with pure cultures but stressed what he thought to be the obligate dependence on cellulose for growth. Stanier (1940, 1942) redefined the genus on the basis of morphological considerations and included organisms able to attack other complex polysaccharides. Lewin's (1969) further redefinition has sharpened the distinction between two apparently closely related groups, the cytophagas and the flexibacters. Organisms able to attack polysaccharides such as agar, cellulose, chitin and alginic acid are regarded as members of *Cytophaga*, while those which cannot are placed in *Flexibacter*.

As has already been pointed out by Soriano and Lewin (1965), *Flexibacter* Soriano 1945, 92, is practically indistinguishable from *Microscilla* Pringsheim 1951, 140. Lewin (1969) emended the definitions of *Flexibacter* and *Microscilla*, reserving for the latter genus organisms possessing, in addition to the properties of *Flexibacter*, greater length and (for some strains) the ability to attack alginate. This suggestion is *not* followed here for several reasons. Pringsheim indicated (1951, 124) that he had insufficient information about the characteristics of *Flexibacter* when he proposed the genus *Microscilla*. Only *Microscilla marina* was grown in pure culture, and one of Lewin's (1969) new isolates seems to possess the properties attributed to

the species by Pringsheim. Inasmuch as *M. marina* is pigmented (it is curious that Pringsheim used a somewhat pigmented organism as the type species of a genus in his colorless *Vitreoscillaceae*), glides and possesses other characteristics of *Flexibacter*, it seems reasonable to place the organism in this genus, which antedates *Microscilla*. The other *Microscilla* species were not studied in pure culture, are accordingly insufficiently characterized and may be regarded as potential members of *Flexibacter*.

The genera *Sphaerocytophaga* Gräf 1961, 457, and *Sphaeromyxa* Bauer 1962, 393, are not treated here because they are not sufficiently well characterized to be recognizable.

The distinction between round and tapered ends of cells does not seem sufficiently persuasive to justify the inclusion of some *Cytophaga* species in a separate genus *Promyxobacterium* Imshenetski and Solntseva 1945, 226.

Very little is known about these organisms and, until information of equivalent value is available, comparisons are difficult to make, to say nothing of speculations about relationships. Dimensions given in the literature may be for cells growing in well agitated liquid cultures, in stationary liquid cultures or on solid media. Unfortunately the cultural growth stage is rarely specified. Morphological descriptions are sometimes based on stained preparations, sometimes on living material. The terms utilization, digestion, hydrolysis and liquefaction are used in different senses by different authors. Often we can only infer whether a particular polysaccharide is simply being depolymerized or whether it is also being used as a source of carbon and energy. Truly comparative studies of the morphology, ecology, physiology and biochemistry of these organisms are badly needed.

Key to the genera of family **Cytophagaceae**

I. Resting stages (microcysts) not known.
  A. Cells, rods or filaments.
    1. Agar, cellulose or chitin attacked; cells not in sheaths.
                 Genus I. *Cytophaga*, p. 101
    2. Agar, cellulose or chitin not attacked; cells not in sheaths.
                 Genus .II. *Flexibacter*, p. 105
    3. Agar and chitin not attacked; cellulose may be attacked; filaments may be in sheaths.
                 Genus III. *Herpetosiphon*, p. 107
  B. Cells, filaments.
    1. Filaments may be sheathed, not helical.
                 Genus IV. *Flexithrix*, p. 109
    2. Filaments helical, not sheathed.
                 Genus V. *Saprospira*, p. 109
II. Microcysts formed.
                 Genus VI. *Sporocytophaga*, p. 111

Genus I. **Cytophaga** *Winogradsky 1929, 577; Lewin 1969, 191 emend. mut. char.*

E. R. LEADBETTER

(*Promyxobacterium* Imshenetski and Solntseva 1945, 220; in part, *Flexibacter* Soriano 1945, 92.)

Cy.toph'aga. Gr. n. *cytos* hollow vessel or cell; Gr. v. *phagein* to eat; M.L. fem.n. *Cytophaga* eater (=digester) of cell walls, and hence of cellulose.

Single, short or elongate **flexible rods or filaments, not branched, sheathed** or **helical.** Cells 0.3–0.7 by 5–50 μm with round or tapered ends. Resting stages not known. Motile by gliding. Gram-negative.

Chemoorganotrophs: **Metabolism is respiratory,** using molecular oxygen as electron acceptor and, in one known instance, nitrate as an alternate electron acceptor, *or* **respiratory and fermentative,** with acetate, propionate and succinate among the end products. Characteristics are shown in Table 2.2. One or more **polysaccharides such as agar, cellulose or chitin may be decomposed,** or **carboxymethylcellulose depolymerized.** Organic growth factors are required by some species.

Contain yellow, orange or red carotenoid pigment(s).

Strict aerobes *or* facultative anaerobes. Temperature optimum 20–30 C, where examined.

The G + C content of the DNA of the species examined ranges from 33–42 moles %.

Type species: *Cytophaga hutchinsonii* Winogradsky 1929, 578.

Further Comments

Members of the genus are probably common in soil and in both fresh water and marine environments, but are often overlooked in isolation procedures: (a) employing media rich in nutrients and (b) unless particular effort is made to detect the often thin, spreading growth of these organisms. The selective isolation techniques as described by Warke and Dhala (1968) may prove useful.

In addition to properties noted in the generic description, members of this genus, in general, are less refractile than are most bacteria and often form at the colony periphery finger-like projections on the agar surface. When given sufficient carbon, some produce extracellular polymers which render liquid media rather viscous and colonies on agar mucilaginous.

Unless otherwise indicated the species descriptions have been drawn from the studies of Stanier (1940, 1941, 1942, 1947), Lewin (1969), Lewin and Lounsbery (1969), Mandel and Leadbetter (1965) and Mandel and Lewin (1969).

TABLE 2.2

Characteristics of the species of genus **Cytophaga**

For the carbon compounds listed, + or − refers to utilization, or not, of a given substrate. Where the data permit more explicit information is conveyed by: D = decomposed; L = liquefied; C = serve as (−C will not serve as) source of carbon and energy; H = hydrolyzed; −H, not hydrolyzed; d = some strains positive, some strains negative.

		Agar	Cellulose	Chitin	Starch	Cellobiose	Glucose	Galactose	Xylose	Arabinose	Catalase	NaCl required
1.	*C. hutchinsonii*	−H	C	−H		+	+	−	−	−	+	−
2.	*C. rubra*	−H	C	−H		C	C	−	C	−	+	−
3.	*C. johnsonae*	−	−	C	C	C	C	C	C	C	−	−
4.	*C. krzemieniewskae*	L	D	−	D	+	+	+	+	−	+	+
5.	*C. diffluens*	L	dD	−	d		d	d			−	+
6.	*C. lytica*	H	−	−	H		+	+			+	+
7.	*C. salmonicolor*	dH	−	−	+	+		+		+	+	+
8a.	*C. fermentans* subsp. *fermentans*	H; −C	−C	−C	+	+	+	−	+		+	+
8b.	*C. fermentans* subsp. *agarovorans*	H	−	−H	+	+	+	+	+			+

Description of the species of genus **Cytophaga**

1. Cytophaga hutchinsonii Winogradsky 1929, 578.

hut.chin.so'ni.i. M.L. gen.n. *hutchinsonii* of Hutchinson; named for H. B. Hutchinson.

Single, often flexible rods 0.3–0.5 by 2.0–10.0 μm. Spheroplasts and abnormally long forms may occur in old cultures.

Growth on agar or silica gel media containing salts and cellulose (filter paper) or glucose is gummy, and liquid cultures become viscous as a result of extracellular slime production. Filter paper on the agar or silica gel surface is eventually dissolved around colonies so that translucent areas result. Colonies on glucose-salts media are raised.

Metabolism is respiratory, using molecular oxygen as terminal electron acceptor. Fructose, mannitol and mannose are not suitable sources of carbon and energy. Either ammonium or nitrate ions are suitable nitrogen sources, as are many amino acids, peptones and yeast extract. Amino acids, peptones, proteins, yeast extract or nutrient agar will not serve as sole carbon and energy sources. No organic growth factors are required.

Cells yellow, presumably due to carotenoid pigment(s).

Strict aerobe. Optimum temperature, 30 C. Optimum pH 7.0–7.5.

The G + C content of the DNA of the single strain examined is 39 moles % (buoyant density).

2. Cytophaga rubra Winogradsky 1929, 598.

ru'bra. L. fem.adj. *rubra* red.

Single, often flexible, rods 0.5–0.7 by 3.5–11 μm. Spheroplasts may occur in old cultures.

Little extracellular slime is produced on either cellulose agar or in liquid cultures. Filter paper is only partially decomposed. Colonies on glucose-salts agar remain small (2 mm diameter), are sunken into the agar and have hazily defined peripheries.

Metabolism is respiratory, using molecular oxygen as terminal electron acceptor. Mannose is a suitable source of carbon and energy, while fructose and mannitol are not. Ammonium and nitrate ions, various amino acids, peptones and yeast extract will serve as nitrogen sources. Amino acids, peptones, proteins, nutrient agar or yeast extract will not serve as sole carbon and energy sources. No organic growth factors are required.

Cells pink, presumably due to carotenoid pigment(s).

Strict aerobe. Optimum temperature 30 C.

3. Cytophaga johnsonae Stanier 1947, 306. (*Cytophaga johnsonii* (*sic*) Stanier 1957, 860; Vegetative myxobacteria, Johnson 1932, 340.)

john.so'nae. M.L. gen.n. *johnsonae* of Johnson; named for Miss Delia E. Johnson who first isolated the species.

Single, often flexible, rods 0.2–0.4 by 1.5–15 μm. Spheroplasts regularly formed in older cultures.

Colonial morphology markedly affected by nutrient and agar concentration(s). With low peptone concentrations, flat, thin, spreading colonies with uneven edges are formed. With higher peptone concentrations, the colony edge lacks the finger-like projections, and the colonies are raised, convex and confined. On chitin agar, colonies are similar to those on low peptone agar.

Metabolism is usually respiratory, using molecular oxygen as terminal electron acceptor; however, one strain can denitrify using nitrate as its alternate electron acceptor. Inulin, lactose, maltose, mannose, raffinose and sucrose are utilized as carbon and energy sources, although dulcitol and mannitol are not. Either ammonium or nitrate ions, or peptones are suitable nitrogen sources, and peptones and other complex nitrogenous materials can serve *both* as carbon and nitrogen sources. No organic growth factors are required.

Cells yellow, presumably due to carotenoid pigments.

Strict aerobe, except for one strain which is known to denitrify. Optimum temperature 25–30 C.

The G + C content of the DNA of the strains examined ranges from 33–35 moles % (buoyant density, T_m).

4. Cytophaga krzemieniewskae Stanier 1940, 623. (*Cytophaga krzemieniewskii* (*sic*) Stanier 1940, 623.)

krze.mi.en.i.ew'skae. M.L. gen.n. *krzemieniewskae* of Krzemieniewska; named for Helena Krzemieniewska.

Single, flexible rods with blunt ends 0.5–1.5 by 5–20 μm.

As a result of agar digestion and the gliding motility of the organism, clearly delimited colonies are not formed on agar. Growth begins as a pink, barely visible film which spreads rapidly. After a few days, cells in older portions of the swarm accumulate in masses resembling droplets, and cause the swarm to have an uneven topography. In about a week a diffusible brown-black pigment masks the pink color. Agar liquefaction is rapid and becomes nearly complete.

Metabolism is respiratory, utilizing molecular oxygen as terminal electron acceptor. Alginic acid is decomposed. Lactose and maltose are utilized, but sucrose is not. Yeast extract or peptone are the only known nitrogen source(s). Gelatin is liquefied. Indole is not formed. Nitrites are not produced from nitrates. Hydrogen sulfide is not produced. Minimal nutritional requirements unknown.

Catalase positive (weakly).

Cells pink, presumably due to carotenoid pigments.

Strict aerobe. Temperature optimum 22–25 C.

Sea water (or NaCl, from 1–5% w/v) required for growth.

5. Cytophaga diffluens Stanier 1940, 623.

dif'flu.ens. L. part.adj. *diffluens*, flowing away.

Single, often flexible, rods or filaments, 0.5 by 4–30 μm.

As a result of agar digestion and the gliding motility of the organism, clearly delimited colonies are not formed on agar. Growth begins as a pink, poorly visible film which may spread over the entire plate. Slightly sunken into the agar in early stages, the older growth becomes orange as it ages and sinks irregularly into the agar. Eventually agar liquefaction may be nearly complete.

Metabolism is respiratory, using molecular oxygen as terminal electron acceptor. Alginic acid is decomposed. Some strains not able to decompose cellulose may depolymerize sodium carboxymethylcellulose. The ability to utilize acetate, sucrose or other sugars, varies from strain to strain (Stanier, 1941, 546; Lewin and Lounsbery, 1969, 145). Glycerol and lactate are not sources of carbon and energy for the strains tested. Complex nitrogenous materials are used as sole carbon source(s) as well as nitrogen sources; nitrate rarely serves as sole nitrogen source. Gelatin is liquefied. Where examined, acid is produced from milk; the curd is not always digested; litmus is reduced. Crystalline suspension of tyrosine is not degraded; dihydroxyphenylalanine is degraded, with clearing of the medium. Indole is not formed. Nitrite is produced from nitrate. Ammonia is not produced. Hydrogen sulfide is produced by some strains. Organic growth factors are required by some strains.

Cells pink, becoming orange. Hexane extracts have absorption maximum at 471 nm. The carotenoid, saproxanthin, has been detected.

Strict aerobe, temperature optimum 20–25 C. Sea water or NaCl required for growth; one-half strength to double strength sea water (or 1.5–5% NaCl) is suitable for most strains.

The G + C content of the DNA ranges from 35–43 moles %; neotype strain 42.3% (buoyant density).

Suggested neotype strain: B-1 (Lewin and Lounsbery 1969, 158).

6. Cytophaga lytica Lewin 1969, 199.

ly'ti.ca. Gr. adj. *lytos* loosening; M.L. fem.adj. *lytica* loosening or dissolving.

Single, often flexible, rods or filaments 0.5 by 4–20 μm.

Metabolism is respiratory, using molecular oxygen as terminal electron acceptor. Alginic acid is hydrolyzed and carboxymethylcellulose is depolymerized. Glycerol and sucrose serve as carbon sources, while acetate and lactate do so only for some strains. Casamino acids, tryptone and glutamate can be used as sole nitrogen source; nitrate can also be used by some strains. Gelatin is liquefied. Acid is not produced from litmus milk; coagulation occurs, but the curd remains undigested by some species; litmus is reduced. Crystalline suspension of tyrosine is degraded, as is also dihydroxyphenylalanine. Indole is not formed. Nitrites are not produced from nitrates. Ammonia is not produced. Hydrogen sulfide is produced by some strains. Organic growth factors are not required.

Cells yellow; extracts, in hexane, have absorption maximum at 450 nm. The carotenoid, zeaxanthin, has been detected.

Strict aerobe. Temperature optimum 20–25 C. Sea water (or, presumably, 3% NaCl) required for growth.

The G + C content of the DNA varies from 33.2–34.2 moles % (buoyant density); that of the type strain 33.2%.

Type strain: LIM 21; ATCC 23178.

7. Cytophaga salmonicolor Veldkamp 1961, 339.

sal.mo.ni'co.lor. L. n. *salmo, salmonis* salmon; L. fem.n. *color* color; M.L. adj. *salmonicolor* salmon-colored.

Single, flexible rods with rounded ends 0.3–0.5 by 2–30 μm (average 6 μm). Spheroplasts occur in older cultures.

Metabolism is respiratory, using molecular oxygen as terminal electron acceptor, *or* fermentative, where glucose is fermented to form acetic, lactic, propionic and succinic acids, some ethanol and CO_2 and H_2. Under aerobic conditions agar is decomposed by some strains; chitin is not degraded. Anaerobically, inulin and many sugars including fructose, lactose, sucrose, trehalose and xylose are fermented. Mannitol, sorbitol, sorbose and rhamnose are not fermented. Ammonium or nitrate ions, yeast extract, nutrient broth and casamino acids will serve as nitrogen sources. Complex nitrogenous materials will also serve as sole carbon source(s) under aerobic conditions. No organic growth factors are required for fermentation in a vitamin-free glucose medium, although addition of a vitamin mixture stimulates growth. Under aerobic conditions no growth occurs in a chemically defined medium. The nature of the growth factors provided by complex extracts is undetermined. Carbon dioxide, supplied as bicarbonate, is an absolute requirement for fermentative growth.

Cells salmon-colored, presumably due to carotenoid pigment(s).

Facultatively anaerobic. Temperature range 28–37 C. NaCl, in the range of 1–3% (w/v), required.

Type strain: ATCC 19041.

8. Cytophaga fermentans Bachmann 1955, 549.

fer.men′tans. L. part.adj. *fermentans* fermenting.

Single, flexible rods with rounded ends, 0.5–0.7 by 2–30 μm.

Metabolism is respiratory, using molecular oxygen as terminal oxygen acceptor *or* fermentative with production of approximately equimolar quantities of acetic, propionic and succinic acids from glucose.

Cells grown aerobically on yeast extract agar are yellow, presumably due to the presence of carotenoid pigments, but the pigment is not produced under anaerobic conditions.

Facultatively anaerobic. Substrate amounts (ca. 1% NaHCO₃) of carbon dioxide are required for continued anaerobic growth in a chemically defined medium.

NaCl required for growth.

8a. *Cytophaga fermentans* subsp. *fermentans*.

Rods 8–15 μm long; forms up to 30 μm occur in older cultures.

Growth on yeast extract agar plates is a thin, transparent film, which spreads over the agar surface. Later, mucilaginous, opaque, bright yellow areas develop in central portions of the film. Colonies lie in shallow craters which extend 1–2 mm beyond the edges of the colony.

Alginic acid and agar are not utilized as sole source of carbon and energy. Fructose, lactose, sucrose and xylose serve as suitable carbon and energy sources for respiration *and* fermentation. Ammonium salts, glutamine and asparagine, but not amino acids, nitrates or urea, will serve as nitrogen sources. Nitrite is not produced from nitrate. Aerobically, yeast extract will serve as carbon source(s) in the absence of added sugar. For anaerobic growth, thiamine is the only organic growth factor required, but for aerobic growth, yeast extract is required; minimal nutritional requirements are unknown.

Catalase positive.

Temperature optimum ca. 30 C. pH optimum 7–8.

Optimum concentration of NaCl 2.5–3% (w/v).

The G + C content of the DNA is 39 moles % (T_m; buoyant density).

8b. *Cytophaga fermentans* subsp. *agarovorans* Veldkamp 1961, 340.

a.gar.o.vor′ans. Malayan n. *agar* agar, a jelly from sea weed; L. v. *voro* devour or consume; M.L. part.adj. *agarovorans* agar-digesting.

Single, flexible rods, 0.5 by 2–30 μm. Spheroplasts and distorted cells are formed in older cultures.

Growth on 1% (w/v) agar containing low concentrations of nutrients is gray, although the center of the colony may be pale yellow. Colonies develop below the surface, form deep craters in the agar and a wide extra-colonial area of agar is depressed, translucent and softened. Colony edges are usually sharp and complete, although swarming may occur. When the agar concentration is increased (2% w/v), flat, spreading, nearly colorless, translucent colonies develop.

Under anaerobic conditions and with 1% agar, gray to cream-colored spherical colonies up to 1 cm in diameter are formed, and the agar is softened.

Agar, but not chitin, hydrolyzed under aerobic conditions. Agar, cellobiose, fructose, galactose, glucose, lactose, maltose, mannitol, mannose, raffinose, sorbitol, sucrose and xylose, but not sorbose or trehalose, are fermented. Ammonium and nitrate ions, yeast extract, nutrient broth or Casamino acids are suitable nitrogen sources. Complex nitrogenous materials will also serve as sole carbon sources under aerobic conditions.

For anaerobic growth on glucose or other hexoses, no growth factors are required, but the addition of vitamins stimulates growth. For aerobic growth, complex extracts are required; minimal nutrient requirements are unknown.

Temperature range 28–37 C. Optimum pH around 7.

NaCl, in a range from 1–3% (w/v), is required.

Type strain: ATCC 19042.

Species incertae sedis

I. The following are probably species of *Cytophaga* but acceptance depends on their reisolation and further characterization.

  a. *C. albogilva* Fuller and Norman 1943, 566.
  b. *C. anularis* Stapp and Bortels 1934, 56.
  c. *C. aurantiaca* Winogradsky 1929, 597.
  d. *C. crocea* Stapp and Bortels 1934, 60.
  e. *C. deprimata* Fuller and Norman 1943, 566.
  f. *C. flavicula* Stapp and Bortels 1934, 62.
  g. *C. haloflava* Kadota 1953, 479.
  h. *C. lutea* Winogradsky 1929, 599.

i. *C. rosea* Kadota 1954, 126.

j. *C. sensitiva* Humm 1946, 64.

k. *C. sylvestris* Stapp and Bortels 1934, 55.

l. *C. tenuissima* Winogradsky 1929, 599.

m. *C. latercula* Lewin 1969, 200.
Type strain: S10 1; ATCC 23177.

n. *C. winogradskii* Verona 1934, 732.

o. *Microscilla arenaria* Lewin 1969, 197.
Type strain: HJ1; ATCC 23161.

p. *Microscilla sericea* Lewin 1969, 200.
Type strain: S10 7; ATCC 23182.

q. *Microscilla furvescens* Lewin 1969, 200.
Type strain: TV 2; ATCC 23129.

II. The following do not meet the description of *Cytophaga*, do not attack polysaccharides, do not glide.

r. *Cytophaga marinoflava* Colwell, Citarella and Chen 1966, 1102.

Genus II. **Flexibacter** *Soriano 1945, 92, Lewin 1969, 192 emend. mut. char.*

E. R. Leadbetter

Flex.i.bac'ter. L. part. *flexus*, flexible; M.L. masc.n. *bacter* masculine form of Gr. neut.n. *bactrum* rod; M.L. masc.n. *Flexibacter* flexible rod.

Flexible rods or **filaments** 0.5 by 5–100 μm. Resting stages not known. Motile by gliding. Gram-negative.

Cell masses are pink, red, orange or yellow due to carotenoid pigments.

Chemoorganotrophs: **Metabolism is usually respiratory,** utilizing molecular oxygen as terminal electron acceptor, but **one species is both respiratory and fermentative. Polymers such as agar, alginic acid, cellulose and chitin are not attacked.**

Characteristics of value in differentiating the species are given in Table 2.3.

Usually strict aerobes.

The G + C content of the DNA of the species examined ranges from 31.2–42.9 moles %.

Type species: *Flexibacter flexilis* Soriano 1945, 92.

Further Comments

Although in general the members described for this genus have been examined more uniformly than members of the genus *Cytophaga*, the same general comments about the paucity of critical information apply here and the plea for extensive, careful, comparative studies is equally valid.

Unless otherwise indicated the descriptive information has been drawn from the studies of Lewin (1969), Lewin and Lounsbery (1969) and Mandel and Lewin (1969).

It should be noted that for his characterization of *Cytophaga psychrophila*, an organism originally isolated by Borg (1960), Pacha obtained two of Borg's isolates from Ordal. Ordal also supplied a strain of the organism to Lewin and Lounsbery, who regarded it as *F. aurantiacus*. These fresh water isolates seem identical in all significant characteristics; a nearly identical soil isolate (DWO), supplied as *Cytophaga aurantiaca*, was also considered *F. aurantiacus* by Lewin and Lounsbery.

Attention is drawn also to the organism formerly regarded as the myxobacterium *Chondrococcus columnaris*. Ordal (personal communication) re-

---

TABLE 2.3

Differential characteristics of the species of genus **Flexibacter**

Reactions of *F. succinicans* and *F. columnaris* not known for tyrosine and dihydroxyphenylalanine degradation, indole or NH₃ production; negative for other species. Except for one strain of *F. succinicans*, all species fail to reduce nitrate to nitrite. + = 90% or more strains positive; − = 90% or more strains negative; d = some strains positive, some strains negative.

	Starch hydrolysis	Carboxymethylcellulose depolymerization	Carbon compounds utilized						Gelatin hydrolysis	H₂S production	Catalase presence
			Glucose	Galactose	Sucrose	Acetate	Lactate	Glycerol			
1. *F. flexilis*	−	−	d	d	+	d	−	−	+	+	−
2. *F. aggregans*	−	d	+	+	+	d	d	d	−	−	−
3. *F. giganteus*	+	−	+	d	−	−	−	−	+	d	−
4. *F. tractuosus*	d	−	+	d	d	d	−	d	+	d	−
5. *F. succinicans*	+		+	+	d		−	−			+
6. *F. columnaris*	−		+						+		

ports that microcysts are *not* formed. Accordingly, the columnar cell masses are no longer regarded as fruiting bodies, or the organism a member of *Myxobacterales*.

Description of the species of genus **Flexibacter**

1. Flexibacter flexilis Soriano 1945, 92.

flex'i.lis. L. adj. *flexilis* pliant, flexible.

Flexible rods or filaments 0.5 by 10–50 μm.

Casamino acids or tryptone, but not glutamate or nitrate can be used as sole nitrogen source. Acid is not produced and coagulation does not occur in litmus milk; litmus is reduced.

Arginine, isoleucine, leucine, methionine, thiamine, tryptophan and valine are essential growth factors.

Cell masses, orange; extracts, in hexane, have an absorption maximum of 471 nm. The carotenoid saproxanthin has been detected.

Strict aerobe. Temperature optimum ca. 20–25 C. pH optimum ca. 7.

The G + C content of the DNA ranges from 39.3–42.9 moles %; the suggested neotype 40.8 moles % (buoyant density).

Suggested neotype strain: CR63; ATCC 23079.

2. Flexibacter aggregans (Lewin) *comb. nov.* (*Microscilla aggregans* Lewin 1969, 197.)

ag'gre.gans. L. part. *aggregans* adding to, aggregating, forming clumps.

Flexible rods or filaments 0.5 μm wide by up to 100 μm long.

Casamino acids, tryptone, glutamate or (for some strains), nitrate, can be used as sole nitrogen source. Acid is not produced in litmus milk; for some strains coagulation does occur and some strains reduce litmus.

Growth factors are not required.

Cell masses, yellow. Extracts, in hexane, have an absorption maximum of 450 nm. The carotenoid zeaxanthin has been detected.

Strict aerobe. Temperature optimum ca. 20–25 C. pH optimum ca. 7. Sea water, ranging from one-half to double strength, is required for growth.

The G + C content of the DNA ranges from 35.7–42.3 moles %; type strain 36.7 % (buoyant density).

Type strain: NN 13; ATCC 23162.

3. Flexibacter giganteus Soriano 1945, 93.

gi.gan'teus. L. adj. *giganteus* of or belonging to giants, large.

Flexible rods or filaments 0.5 μm by up to 50 μm long.

Casamino acids or tryptone, but not glutamate or nitrate, can be used as sole nitrogen source. Acid is not produced, nor for most strains does coagulation occur in litmus milk; litmus is not reduced. Dihydroxyphenylalanine in a crystalline suspension inhibits growth.

Isoleucine, leucine, methionine, proline, thiamine and valine are essential growth factors for all strains and glutamine is required for some.

Cell masses, pink. Extracts, in hexane, have an absorption maximum of 478 nm. A carotenoid, "S. t. 483," has been detected.

Strict aerobe. Temperature optimum ca. 20–25 C. pH optimum ca. 7.

The G + C content of the DNA ranges from 31.2–33.2 moles %; suggested neotype strain 31.6% (buoyant density).

Suggested neotype: CR-104 (Lewin and Lounsbery, 1969, 159).

4. Flexibacter tractuosus (Lewin) *comb. nov.* (*Microscilla tractuosa* Lewin 1969, 199.)

trac.tu.o'sus. L. masc.adj. *tractuosus* that which draws to itself, clumping together

Flexible rods or filaments 0.5 μm by up to 40 μm long.

Casamino acids, tryptone or glutamate can be used as sole nitrogen source. Some strains produce acid and coagulation occurs in litmus milk; the curd is often digested; litmus is reduced. Organic growth factors are not required.

Cell masses, orange. Extracts, in hexane, have an absorption maximum of 471 nm. The carotenoid saproxanthin has been detected.

Strict aerobe. Temperature optimum ca. 20–25 C. pH optimum ca. 7. Organisms will grow either in fresh water or in double strength sea water.

The G + C content of the DNA ranges from 34.2–37.8 moles %; type strain 36.2% (buoyant density).

Type strain: H 43; ATCC 23168.

5. Flexibacter succinicans (Anderson and Ordal) *comb. nov.* (*Cytophaga succinicans* Anderson and Ordal 1961, 130.)

suc.cin'i.cans. L. n. *succinicum*, amber; M.L. n. *acidum succinicum* succinic acid (derived from amber); M.L. part. *succinicans* forming succinic acid.

Single rods with rounded ends 0.6 by 5 μm; longer forms occur in older cultures where spheroplasts also may occur.

Metabolism is respiratory, using molecular oxygen as terminal electron acceptor, *or* fermentative, glucose being fermented to acetic, formic and succinic acids. Nutrient broth and other complex nitrogenous materials will support aerobic growth, but a suitable carbohydrate must be present for anaerobic growth. Apparently, agar is not hydrolyzed; cellulose is not fermented. Galactose, glucose, lactose, maltose and starch are fermented by all strains examined; arabinose, cellobiose, sucrose by some; and fructose, glycerol, lactate, mannitol, raffinose, sorbose, xylose by none.

Organic growth factors are probably required.

Carbon dioxide, supplied as bicarbonate, is mandatory for fermentation (4 moles are used for the conversion of 5 moles of glucose to 6 moles of succinate, 4 moles of acetate and 2 moles of formate).

Cells yellow due to carotenoid pigment(s) when grown aerobically, but non-pigmented when grown anaerobically.

Facultatively anaerobic. Optimum temperature ca. 25 C; no growth at 37 C. pH optimum ca. 7–7.5.

The G + C content of the DNA of the strain examined is 38 moles % (T_m; Johnson and Ordal, 1968).

6. **Flexibacter columnaris** (Davis) *comb. nov.* (*Bacillus columnaris* Davis 1922, 263; *Chondrococcus columnaris* (Davis) Ordal and Rucker 1944, 18; *Cytophaga columnaris* (Davis) Garnjobst 1945, 127.)

co.lum.nar'is. L. adj. *columnaris* rising as a pillar.

Single, flexible rods, 0.5–0.7 by 4–8 μm.

In liquid media cells form columnar and sometimes branched masses where they are in contact with infected fish tissues or scales. Best growth reported (Anacker and Ordal, 1959) on a medium containing (% w/v): agar 0.9, tryptone 0.05, yeast extract 0.05, sodium acetate 0.02, beef extract 0.02 and adjusted to pH 7.2–7.4. Colonies flat, thin, spreading, yellow-green with uneven edges. Cell masses yellow-green.

Metabolism presumably respiratory, using molecular oxygen as the terminal electron acceptor. Sugars are not fermented, although glucose is oxidized. Minimal nutritional requirements not known.

Presumably a strict aerobe.

Pathogenic for some cold and warm water fishes.

Species incertae sedis

I. The following are probably species of *Flexibacter* but recognition depends on characterization of additional isolates.

a. *Flexibacter aurantiacus* Lewin 1969, 200.

b. *Flexibacter litoralis* Lewin 1969, 199. Type strain: S10 4; ATCC 23117.

c. *Flexibacter roseolus* Lewin 1969, 199. Type strain: CR155; ATCC 23088.

d. *Flexibacter ruber* Lewin 1969, 199. Type strain: GEY; ATCC 23103.

e. *Flexibacter sancti* Lewin 1969, 199. Type strain: BA3; ATCC 23092.

f. *Flexibacter albuminosus* Soriano 1945, 93.

g. *Flexibacter aureus* Soriano 1945, 93.

h. *Microscilla marina* Pringsheim 1951, 140; Lewin 1969, 198.

i. *Cytophaga psychrophila* Borg 1960, 66.

j. *Promyxobacterium flavum* Imshenetski and Solntseva 1945, 224.

k. *Promyxobacterium lanceolatus* (*sic*) Imshenetski and Solntseva 1945, 225.

II. The following organisms have not been studied in pure culture but may be members of the genus.

l. *Microscilla agilis* Pringsheim 1951, 142.

m. *Microscilla flagellum* (*sic*) Pringsheim 1951, 143.

Genus III. **Herpetosiphon** Holt and Lewin 1968, 2408; emend. mut. char.

R. A. LEWIN AND E. R. LEADBETTER

Her.pe.to.si'phon. Gr. n. *herpeton* gliding animal, reptile; Gr. masc.n. *siphon* tube or cylinder; M.L. masc.n. *Herpetosiphon* gliding cylinder.

Unbranched, flexible, sheathed rods or **filaments** 0.7–1.5 by 5–150 μm or more (to several millimeters). Resting stages not known. **Unsheathed segments motile** by gliding. Gramnegative.

Chemoorganotrophs: **Metabolism is respiratory, using molecular oxygen as terminal electron acceptor.** Agar, alginic acid and chitin are not known to be attacked. Cellulose may be degraded, carboxymethylcellulose depolymerized or starch hydrolyzed (Table 2.4). Gelatin is liquefied. In all species, except *H. geysericola* which has not been examined, indole, ammonia or H_2S

is not produced and nitrates are not reduced to nitrites. Cells possess yellow or orange carotenoid pigments.

Strict aerobes.

Temperature optima in the laboratory have not been determined. Where reported, pH optimum ca. 7.

Marine forms require sea water for growth.

Among the strains examined the G + C content of the DNA ranges from 44.9–53.1 moles % (Mandel and Lewin, 1969).

Type species: *Herpetosiphon aurantiacus* Holt and Lewin 1968, 2408.

TABLE 2.4
Differential characteristics of species of genus **Herpetosiphon**

	Color of cell mass	Cellulose hydro-lyzed	Starch hydro-lyzed	Tyrosine degraded	Catalase	Sea water required	Thermo-philic	Nitrate as sole nitrogen source	G + C
									moles %
1. *H. aurantiacus*	Orange	−	+	+	+	−			48.1
2. *H. geysericola*	Orange	+	+		+	−	+		48.5
3. *H. cohaerens*	Orange	−	−	−	−	+	−	−	44.9
4. *H. persicus*	Orange	−	−	−	−	+	−	+	52.6
5. *H. nigricans*	Yellow	−	−	+	−	+	−	+	53.1

Description of the species of genus **Herpetosiphon**

1. Herpetosiphon aurantiacus Holt and Lewin 1968, 2408.

au.ran.ti′a.cus. M.L. neut.n. *aurantium* specific name of the orange; M.L. adj. *aurantiacus* orange-colored.

Cells, 1.0–1.5 by 5–10 μm, in unbranched, flexible and usually sheathed filaments, which may exceed 500 μm in length.

Cellulose is not hydrolyzed. Starch is hydrolyzed. A crystalline suspension of tyrosine is degraded, with formation of a reddish brown pigment. Growth has been reported only in complex media; minimal nutritional requirements are unknown.

The G + C content of the DNA is 48.1 moles % (buoyant density).

Type strain: ATCC 23779.

2. Herpetosiphon geysericola (Copeland) Lewin 1970, 517. (*Phormidium geysericola* Copeland 1936, 186; *Herpetosiphon geyericolus* (*sic*) Lewin 1970, 517.)

gey.ser.i′cola. Icelandic n. *geysir* gusher, name of a hot spring; L. n. *cola* dweller; M.L. n. *geysericola* hot springs dweller.

Flexible, sheathed rods or filaments 0.5 by 10–150 μm or more.

Cellulose is digested. Carboxymethylcellulose is depolymerized. Starch is hydrolyzed.

High temperatures tolerated; recorded from hot springs at 60–80 C. pH tolerated (in nature) 8–9.

The G + C content of the DNA of the single strain (suggested neotype) examined is 48.5 moles % (buoyant density).

3. Herpetosiphon cohaerens Lewin 1970, 518.

co.hae′rens. L. part.adj. *cohaerens* cohering, uniting together.

Unbranched, flexible, sheathed rods or filaments 0.7 μm (1.0 μm if sheath is included) by 60–150 μm or longer.

Cellulose and starch are not attacked; carboxymethylcellulose is not depolymerized. Glucose and

sucrose promote growth, while acetate, galactose, glycerol and lactate do not. Tryptone or glutamate, but not nitrate, can serve as sole nitrogen source. No acid is produced in litmus milk, although coagulation occurs; the resultant curd is not digested but the litmus is reduced. A crystalline suspension of tyrosine or dihydroxyphenylalanine is not degraded. There are no known growth factor requirements.

Cell masses, orange. Extracts, in hexane, have an absorption maximum at 471 nm, attributable to the carotenoid saproxanthin.

Sea water, ranging from one-half to double strength, is required for growth.

The G + C content of the DNA of the type strain is 44.9 moles % (buoyant density).

Type strain: II-2; ATCC 23123.

4. Herpetosiphon persicus Lewin 1970, 518.

per′si.cus. L. adj. *persicus* Persian (of fruit = peach) *i.e.* peach-colored.

Unbranched, flexible, sheathed rods or filaments 0.7 μm (1.0 μm if sheath is included) by 30–150 μm, or longer.

Cellulose and starch are not attacked; carboxymethylcellulose is not depolymerized. Glucose is a suitable carbon source; galactose and sucrose also promote growth, but acetate, glycerol and lactate do not. Tryptone, glutamate or nitrate can serve as sole nitrogen source. Acid is not produced from litmus milk, although coagulation occurs; the curd is not digested but the litmus is reduced. A crystalline suspension of tyrosine or dihydroxyphenylalanine is not degraded. There are no known growth factor requirements.

Cell masses; orange; extracts, in hexane, have an absorption maximum at 471 nm, attributable to the carotenoid saproxanthin.

Sea water, ranging from one-half to double strength, is required for growth.

The G + C content of the DNA of the type strain is 52.6 moles %.

Type strain: T-3; ATCC 23167.

5. Herpetosiphon nigricans Lewin 1970, 518.

ni'gri.cans. L. part. adj. *nigricans* blackening.

Unbranched, flexible sheathed rods or filaments 0.5 μm (1.0 μm if sheath is included) by 5–50 μm.

Cellulose and starch are not attacked; carboxymethylcellulose is not depolymerized. Glucose is a suitable carbon source; galactose and sucrose also promote growth, but acetate, glycerol and lactate do not. Casamino acids, tryptone, glutamate or nitrate can serve as sole nitrogen source. Acid is not produced from litmus milk, although coagulation occurs; the curd is digested and the litmus is reduced. A crystalline suspension of tyrosine is degraded, with the formation of a pigmented product, but dihydroxyphenylalanine is not degraded. There are no known growth factor requirements.

Cell masses, yellow; extracts, in hexane, have an absorption maximum at 450 nm attributable to the carotenoid zeaxanthin.

Sea water, ranging from one-half to double strength, is required for growth.

The G + C content of the DNA of the type strain is 53.1 moles % (buoyant density).

Type strain: SS-2; ATCC 23147.

Genus IV. **Flexithrix** *Lewin 1970, 513*

R. A. LEWIN AND E. R. LEADBETTER

Flex'i.thrix. L. part.adj. *flexus* flexible; Gr. n. *thrix* hair; M.L. fem.n. *Flexithrix*, flexible hair.

Cells 0.3–0.5 by 5–15 μm, **usually as sheathed filaments** 0.5 μm in diameter and up to 500 μm long; may show false branching. Resting stages not known. Unsheathed cells 5–15 μm long, motile by gliding. Gram-negative.

Chemoorganotroph: Metabolism is respiratory, using molecular oxygen as terminal electron acceptor.

Strict aerobe.

The G + C content of the DNA of the single species examined is 37.2 moles % (Mandel and Lewin, 1969).

Type species: *Flexithrix dorotheae* Lewin 1970, 511.

Further Comments

Typical morphology is exhibited only in media in which the sugar concentration is low (e.g. 0.01%). In richer media, sheathed filaments are not produced, and the free, naked cells are then indistinguishable from those of species of *Cytophaga*. Under such conditions, *F. dorotheae* corresponds so closely to *Flexibacter aggregans*, in both cultural and nutritional features, that it might be regarded as a form variant of the latter species.

Description of the species of genus **Flexithrix**

1. Flexithrix dorotheae Lewin 1970, 511.

do.ro.the'ae. M.L. gen.n. *dorotheae* of Dorothy. Named after Mrs. Dorothy White, a technical assistant (deceased).

Morphology as for genus.

Agar, alginic acid, cellulose, chitin and starch are not attacked, but carboxymethyl cellulose is depolymerized. Gelatin is not liquefied. Galactose, glucose and sucrose are suitable sources of carbon and energy; acetate and glycerol are not. Tryptone, glutamate or nitrate can be used as sole nitrogen source. In litmus milk acid is not produced, coagulation does not occur and there is no reduction of the litmus. In a crystalline suspension neither tyrosine nor dihydroxyphenylalanine is degraded. Indole is not formed. Nitrites are not produced from nitrates. Ammonia is not produced. Hydrogen sulfide is not produced from cysteine-containing medium. No organic growth factor requirements are known.

Catalase negative.

Cell masses, yellow. Extracts, in hexane, have an absorption maximum of 450 nm attributable to the carotenoid zeaxanthin.

Temperature for growth, up to 35 C. pH range 7–9.

Sea water at concentrations from one-half to twice normal is required for growth.

The G + C content of the DNA of the type strain is 37.2 moles % (buoyant density).

Type strain: QQ-3; ATCC 23163.

Genus V. **Saprospira** *Gross 1911, 190; Lewin 1962, 560 emend. mut. char.*

R. A. LEWIN AND E. R. LEADBETTER

Sap.ro.spi'ra. Gr. adj. *sapros* rotten, putrid; Gr. n. *spira* a spiral; M.L. fem.n. *Saprospira* spiral associated with decaying matter.

Helical filaments 0.8–1.5 by 10–500 μm; unbranched, unsheathed. Motile by gliding. Resting stages not known. Gram-negative.

Chemoorganotrophs: Metabolism is respiratory, using molecular oxygen as terminal electron acceptor. Agar, alginic acid, cellulose and chitin are not known to be attacked.

S. albida has not been studied extensively. The other three species can use tryptone but not glutamate or nitrate as sole nitrogen source, liquefy gelatin, do not produce indole or ammonia, do not reduce nitrate to nitrite and are catalase negative. Characteristics of differential value are given in Table 2.5.

Cell masses are colored, presumably due to carotenoid pigments.

Strict aerobes.

Both fresh water and marine forms are known.

The G + C content of the DNA of the species examined ranges from 35.4–48.3 moles % (Mandel and Lewin, 1969).

Type species: *Saprospira grandis* Gross 1911, 190.

Description of the species of genus **Saprospira**

1. **Saprospira grandis** Gross 1911, 190.
gran'dis. L. adj. *grandis* large.

Cells 0.8 by 1–2.5 μm, in flexible helical filaments up to 500 μm in length. Helix 1.5 μm wide; pitch 4–9 μm; sense of helix dextral.

Acetate, galactose, glucose and/or sucrose promote growth of some strains; lactate and glycerol do not.

No acid is produced, although coagulation occurs, in litmus milk; the curd is usually not digested and litmus not reduced. A crystalline suspension of tyrosine is degraded with formation of a pigmented product; dihydroxyphenylalanine is inhibitory. Hydrogen sulfide is not produced from cysteine-containing media.

Apparently no vitamins are required. Arginine, asparagine, histidine, isoleucine, lysine, methionine, proline, tyrosine, tryptophan and valine are essential for growth, as are still undetermined compounds in an alkaline hydrolysate of yeast nucleic acid.

Cell masses are orange; extracts, in hexane, have an absorption maximum at 471 nm, attributable to an intracellular carotenoid, saproxanthin.

Temperature optimum ca. 30–37 C. pH optimum ca. 7.

Although some strains will grow in medium made with fresh water, most require sea water; tolerance of the latter varies from one-half to double strength.

The G + C content of the DNA of the strains examined varies from 45.9–48 moles % (buoyant density); that of the suggested neotype is 47.5%.

Suggested neotype: WH; ATCC 23119 (Lewin and Lounsbery, 1969, 170).

2. **Saprospira thermalis** Lewin 1965, 139.
ther.ma'lis. M.L. adj. *thermalis* pertaining to hot springs.

Cells 1.0 by 2–5 μm in flexible helical filaments up to 150 μm in length. Helix 1.5–2.5 μm wide; pitch 7–17 μm.

Arabinose, fructose, mannitol, rhamnose, ribose, trehalose, xylose and ethanol are not suitable sources of carbon and energy.

In a crystalline suspension, tyrosine or dihydroxyphenylalanine is not degraded. Hydrogen sulfide is produced from cysteine-containing medium.

Cobalamin, thiamine, isoleucine, leucine and valine are required for growth.

Cell masses are pink; extracts, in hexane, have an absorption maximum at 478 nm, due to a carotenoid provisionally designated "S. t. 483."

Temperature optimum ca. 35 C. pH optimum ca. 7.

The G + C content of the DNA of the strains

TABLE 2.5
Differential characteristics of three species of genus **Saprospira**

	Color of cell mass	Carboxymethylcellulose hydrolyzed	Starch hydrolyzed	Tyrosine degraded	Milk coagulated	Casamino acids as sole nitrogen source	Sources of carbon and energy					
							Acetate	Lactate	Galactose	Glucose	Sucrose	Glycerol
1. *S. grandis*	Orange	−	−	+	+	−	+	−	+	+	+	−
2. *S. thermalis*	Pink	−	+	−	−	+	±	−	−	+	−	+
3. *S. toviformis*	Orange-yellow	+	−	+	+	+	+	+	−	−	−	−
4. *S. albida*	Pale pink											

examined ranges from 32.7–36.7 moles % (buoyant density); that of the type strain is 36.7%.

Type strain: BEG (Lewin and Lounsbery, 1969, 199).

3. Saprospira toviformis Lewin and Mandel 1970, 510.

to.vi.for'mis. Tove a nonsense word from Lewis Carroll's "Jabberwocky" in "Through the Looking Glass," for a fictional organism which is "slithy" (i.e. slimy and lithe) and which can "gyre and gimble" (i.e. progress by rotating in a corkscrew fashion); L. n. *forma* form, shape; M.L. adj. *toviformis*.

Cells 0.8 by 1–2.5 μm in flexible helical filaments 10–500 μm in length. Helix 1.5 μm wide; pitch 4–9 μm; sense of helix usually dextral.

Acid is not produced in litmus milk although coagulation occurs; the curd is not digested and the litmus not reduced. Hydrogen sulfide is not produced from cysteine-containing medium.

The amino acids arginine, glutamine, histidine, isoleucine, leucine, lysine, methionine, proline, tryptophan and valine, but no other growth factors, are essential for growth.

Cell masses are orange-yellow; extracts, in hexane, have an absorption maximum at 471 nm. A carotenoid similar to saproxanthin is the predominant pigment.

Temperature optimum ca. 25 C. pH optimum ca. 9.

Sea water at normal to double strength is required for growth.

The G + C content of the DNA of the single strain examined is 38.4 moles % (buoyant density).

Type strain: A1 (Lewin 1969, 197, monotype).

4. Saprospira albida (Kolkwitz) Lewin 1962, 557. (*Spirulina albida* Kolkwitz 1909, 137; *Saprospira flexuosa* Dobell 1912, 161.)

al'bi.da. L. fem. adj. *albida* white.

Cells 0.8–1.0 by 2–3 μm in flexible helical filaments 10–50 μm or longer. Helix 1.5–2.0 μm wide; pitch 3–9 μm.

Growth reported only in complex media; minimal nutritional requirements are not known.

Cell masses are pale pink, due to carotenoid pigment(s).

Temperature optimum ca. 30 C. pH optimum ca. 7.

The G + C content of the DNA of various strains ranges from 39.8–42.9 moles % (buoyant density).

Species incertae sedis

a. *Saprospira flammula* Lewin 1965, 138.
   Probably a fresh water strain of *S. grandis*.
b. *Saprospira nana* Gross 1911, 195.
c. *Saprospira lepta* Dimitroff 1926, 144.
d. *Saprospira punctum* Dimitroff 1926, 146.
   S. lepta and *S. punctum* are probably spirochetes.

Genus VI. **Sporocytophaga** *Stanier 1940, 629*

E. R. LEADBETTER

Spo.ro.cy.toph'aga. M.L. n. *spora* a spore; M.L. fem.n. *Cytophaga* generic name; M.L. fem.n. *Sporocytophaga* the sporing *Cytophaga*.

Flexible rods with rounded ends, 0.3–0.5 by 5–8 μm, occurring singly. Spheroplasts and distorted cells occur in older cultures. A resting stage, the microcyst, is formed. Motile by gliding. Gram-negative.

Chemoorganotrophs: Metabolism is respiratory, using molecular oxygen as terminal electron acceptor. **Cellobiose, cellulose, glucose** and for some strains, **mannose, are the only known sources of carbon and energy.** Agar and chitin are not known to be attacked. Either ammonium or nitrate ions, or peptone, urea or yeast extract, can serve as sole nitrogen source. Amino acids, peptones, yeast extract or nutrient agar (Difco) cannot serve as sole carbon and energy sources. No organic growth factor requirements known.

Catalase positive.

Strict aerobe. Temperature optimum ca. 30 C.

The G + C content of the DNA of two strains examined was 36 moles % (buoyant density).

Type species: *Sporocytophaga myxococcoides* (Krzemieniewska) Stanier 1940, 629.

Further Comments

Only one species of the genus has been extensively examined. *S. myxococcoides* was shown by Stanier (1942) to grow on glucose sterilized by filtration, and by Kaars Sijpesteijn and Fahraeus (1949) to grow on glucose autoclaved separately from other components of the medium, thus, apparently, refuting the assertion that growth of the organism is obligately linked to cellulose utilization. Recent studies (Leadbetter, unpublished) indicate that, when isolated from nature, the organism is unable either to oxidize or to utilize glucose but that putative mutants, able to do so, arise in the population. These "mutants" are able to attack immediately either cellulose or glucose, irrespective of the substrate in which they are grown. These observations thus confirm and extend those of Kaars Sijpesteijn and Fahraeus (1949).

Recent studies of *S. myxococcoides* have demonstrated that the organism is able to form microcysts when either glucose or cellulose is the carbon and energy source (Leadbetter, 1963; Gallin and Leadbetter, 1966).

Description of the species of genus **Sporocytophaga**

1. Sporocytophaga myxococcoides (Krzemieniewska) Stanier 1940, 630. (*Cytophaga myxococcoides* Krzemieniewska 1933, 400; *Spirochaeta cytophaga* Hutchinson and Clayton 1919, 143.)

myx.o.coc.coi'des. M.L. masc.n. *Myxococcus* a generic name; Gr. n. *eidus* shape; M.L. adj. *myxococcoides* resembling *Myxococcus*.

Single, often flexible, rods with rounded ends 0.3–0.5 by 5–8 μm. Spheroplasts and abnormally long forms may occur in old cultures. The resting stage, the microcyst, is spherical and about 1.5 μm in diameter. Both the growing (vegetative) rod and the microcyst have noticeably smaller dimensions when grown on cellulose than when grown on glucose.

Electron microscopic studies indicate that the vegetative cell has a fine structure typical of Gram-negative bacteria while the microcyst has a thick, fibrillar capsule exterior to a highly convoluted cell wall.

Growth on cellulose (filter paper)-salts agar (or silica gel) or glucose-salts agar is gummy and liquid cultures become viscous as a result of extracellular slime production. Filter paper on the agar or silica gel surface is eventually dissolved around colonies so that translucent areas result. Colonies on glucose-salts agar medium are raised.

Other characteristics as for genus.

Species incertae sedis

The following may be species of *Sporocytophaga*. They are incompletely characterized.

a. *S. cauliformis* Knorr and Gräf in Gräf 1962, 124.

b. *S. congregata* subsp. *maroonicum* Akashi 1960, 899.

c. *S. ellipsospora* (Imshenetski and Solntseva) Stanier 1942, 190.

d. *S. ochracea* Ueda, Ishikawa, Itami and Asai 1952, 545.

FAMILY II. **BEGGIATOACEAE** MIGULA 1894, 238

E. R. LEADBETTER

Beg.gi.a.to.a'ce.ae. M.L. fem.n. *Beggiatoa* type genus of the family; -aceae ending to denote a family; M.L. fem.pl.n. *Beggiatoaceae* the *Beggiatoa* family.

Colorless filaments containing cells in chains. Filaments are flexible and motile by gliding. Resting stages not known. Gram-negative.

Mixotrophs or chemoorganotrophs: Metabolism is respiratory, using molecular oxygen as terminal electron acceptor. Poly-β-hydroxybutyrate or volutin granules may occur intracellularly.

Aerobes or microaerophiles.

Carotenoid pigments not formed.

Further Comments

The genera of the family are distinguished by their ability to deposit sulfur granules when grown in the presence of hydrogen sulfide or to form a well defined sheath around the filaments. Although the genera are apparently readily recognizable, this grouping may be precarious inasmuch as their affinities are uncertain, but this is a reflection of the paucity of *critical* information available about these organisms.

The genera *Bactoscilla* Pringsheim 1951, 144; *Flexoscilla* Pringsheim 1951, 145; and *Thiospirillopis* Uphof 1927, 81 which might also be placed in this family, are regarded as *incertae sedis*, pending reisolation and further characterization.

Key to the genera of family **Beggiatoaceae**

I. Filaments colorless, individual.

A. Cells contain sulfur when grown in presence of H₂S.

Genus I. *Beggiatoa*

B. Cells do not contain sulfur when grown in presence of H_2S.

Genus II. *Vitreoscilla*

II. One to many filaments in each sheath. Cells usually contain sulfur.

Genus III. *Thioploca*

Genus I. **Beggiatoa** Trevisan 1842, 56, *Nom. cons.* Opin. 13, Jud. Comm. 1954, 152

Beg.gi.a.to′a. M.L. fem.n. *Beggiatoa* named for F. S. Beggiatoa, a physician of Vicenza.

Cells colorless, in unattached filaments, 1–30 by 4–20 μm. Motile by gliding. Resting stages not known. Gram-negative.

Mixotrophs: Metabolism is respiratory, using molecular oxygen as terminal electron acceptor. Organic growth factors are required by some strains.

Cells contain granules of sulfur when grown in the presence of hydrogen sulfide. **Intracellular granules** of **poly-β-hydroxybutyric acid** or **volutin may be present.**

Aerobic to microaerophilic.

Marine and fresh water forms known.

The G + C content of DNA of a strain regarded as *B. leptomitiformis* is 37 moles % (Mandel and Lewin, 1969).

Type species: *Beggiatoa alba* (Vaucher) Trevisan 1845, 58.

Further Comments

Although Winogradsky's concept of chemolithotrophy grew out of his studies on *Beggiatoa*, only recently have uncertainties about their nutritional capabilities begun to be dispelled. Kowallik and Pringsheim (1966) and Pringsheim (1967) reported that several strains of *Beggiatoa* were able to grow autotrophically under carefully controlled conditions, but that all strains grew better in the presence of low levels of acetate. The

organisms studied, then, are not obligate chemolithotrophs, but rather mixotrophs in the concept of Rittenberg (1969). Inasmuch as Pringsheim's strains had been grown under chemoorganotrophic conditions, it is possible that the *Beggiatoa* strains studied by others (Faust and Wolfe, 1961; Scotten and Stokes, 1962; Morita and Stave, 1963; and Burton *et al.* 1966) may also display mixotrophy if examined under appropriate conditions. The ability to deposit and utilize intracellular sulfur is regarded here as a key feature for determining whether an organism is *Beggiatoa* or *Vitreoscilla*.

Pringsheim (1964) showed that filament width generally was a stable and, accordingly, useful characteristic (at least for the "thin" strains he examined). These and other comparative studies also indicate that there are strain differences in nutritional versatility, salt tolerance, growth rate, rate of motility, formation of poly-β-hydroxybutyrate, formation of gonidia (?), relationship to O_2, etc.

Many of the strains described in the 1800's were never cultured and it is difficult to compare recently studied organisms with them. Although one might then argue that all species are *incertae sedis*, it seems reasonable, if only for didactic purposes, to continue to recognize those species listed in the 7th edition of THE MANUAL, largely because they may represent the generic extremes of size and, perhaps, of physiology.

Description of the species of genus **Beggiatoa**

1. **Beggiatoa alba** (Vaucher) Trevisan 1845, 58. (*Oscillatoria alba* Vaucher 1803, 198; *Beggiatoa punctata* Trevisan 1842, 56.)

al′ba. L. fem.adj. *alba* white.

Filaments 2.5–5.0 μm wide; cell segments 3–9 μm long, except just after division when they are nearly square.

2. **Beggiatoa arachnoidea** (Agardh) Rabenhorst 1865, 94. (*Oscillatoria arachnoidea* Agardh 1827, 634; *Oscillaria versatilis* Kützing 1843, 184; *Beggiatoa versatilis* Trevisan 1845, 59.)

a.rach.noi′de.a. Gr. adj. *arachnoides* cobweb-like; M.L. fem.adj. *arachnoidea* cobweb-like.

Filaments 5–14 μm wide; cell segments 5–7 μm long.

3. **Beggiatoa gigantea** Klas 1937, 318.

gi.gan′te.a. Gr. fem.adj. *gigantea* gigantic.

Filaments 26–55 μm wide; cell segments 5–13 μm long.

4. **Beggiatoa leptomitiformis** (Meneghini) Trevisan 1842, 56. (*Oscillatoria leptomitiformis* Meneghini in Trevisan 1842, 56.)

lep.to.mi.ti.for′mis. M.L. n. *Leptomitus* a genus of water molds; L. n. *forma* shape; M.L. adj. *leptomitiformis Leptomitus*-like.

Filaments 1–2 μm wide; cell segments 4–8 μm long.

5. Beggiatoa minima Winogradsky 1888, 25. (*Beggiatoa minor* Uphof 1927, 79.)
mi′ni.ma. L. fem.sup.adj. *minima* least, smallest.
Filaments 1 μm or less in width; cell segments about 1 μm long.

6. Beggiatoa mirabilis Cohn 1865, 81.
mi.ra′bi.lis. L. adj. *mirabilis* marvelous.
Filaments 15–21 μm wide; cell segments 5–13 μm long.

Species incertae sedis

I. The following may be species and subspecies of *Beggiatoa*, but acceptance depends on better characterization.
  a. *B. alba* var. *marina* Cohn 1865, 82. (*B. cohnii* Trevisan 1889, 10.)
  b. *B. alba* var. *spiralis* Hansgirg 1892, 186.

  c. *B. arachnoidea* var. *marina* Hansgirg 1890, 21.
  d. *B. arachnoidea* var. *uncinata* Hansgirg 1888, 264.
  e. *B. dulcis* Meneghini acc. to Kützing 1849, 237.
  f. *B. iridescens* (Kützing) Meneghini in Rabenhorst 1865, 95.
  g. *B. leptomitiformis* var. *marina* Hansgirg 1893, 218.
  h. *B. major* Winogradsky 1888, 25.
  i. *B. marina* Molisch 1912, 59.
  j. *B. maxima* Uphof 1927, 80.
  k. *B. media* Winogradsky 1888, 25.
  l. *B. minima* Warming 1875, 356.
  m. *B. nodosa* van Tieghem 1880, 177.
  n. *B. pellucida* Cohn 1865, 82.
  o. *B. raineriana* Meneghini 1846, acc. to Trevisan in Saccardo 1889, 937.
  p. *B. ramosa* Gasperini 1899, 54.
  q. *B. tigrina* Rabenhorst 1865, 95.
  r. *B. uniguttata* Koppe 1924, 628.

Genus II. **Vitreoscilla** Pringsheim 1949, 70; emend. mut. char.

E. R. LEADBETTER

    Vit.re.os.cil′la. L. adj. *vitreus* glassy, clear; L. n. *oscillum* a swing; M.L. fem.n. *Vitreoscilla* transparent oscillator.

Colorless filaments, 1.2–2 by 3–70 μm, are composed of clearly delimited cylindrical or barrel-shaped cells. Reproduction is by fragmentation. Resting stages not known. Motile by gliding. Gram-negative.

Chemoorganotrophs: Metabolism is presumably solely respiratory. Proteins are not hydrolyzed. **Growth reported only on complex media;** minimal nutritional requirements unknown. **Intracellular sulfur granules not formed from hydrogen sulfide.**

The G + C content of two unnamed strains is 43.6 moles % (buoyant density) (Mandel and Lewin, 1969).

Type species: *Vitreoscilla beggiatoides* Pringsheim 1949, 70.

Further Comments

The species of the genus are differentiated essentially on morphological features although there are apparently some nutritional differences. Most strains may be isolated more easily if low concentrations of organic ingredients are used in the media (e.g. 0.05–0.1% peptone, w/v), although some strains grow luxuriantly in relatively rich media (e.g. 0.5%, w/v).

Description of the species of genus **Vitreoscilla**

1. Vitreoscilla beggiatoides Pringsheim 1949, 70.
beg.gi.a.toi′des. M.L. fem.n. *Beggiatoa* a generic name; Gr. n. *idos* shape, form; M.L. adj. *beggiatoides* *Beggiatoa*-like.
Filaments cylindrical, 1–2 by 10–150 μm.
Cells in filaments are cylindrical with blunt ends. Gliding motility, slow (ca. 2 μm/sec).

2. Vitreoscilla moniliformis Pringsheim 1951, 132.
mo.ni.li.for′mis. L. n. *monile* a necklace; L. n. *forma* shape; M.L. adj. *moniliformis* necklace-like.

Filaments in chains 2.2–3.0 by 30–150 μm. Sections of filaments may be sausage-shaped, up to 3.0 μm diameter. Gliding motility, slow (ca. 2 μm/min).

3. Vitreoscilla stercoraria Pringsheim 1951, 136.
ster.co.ra′ri.a. L. fem.adj. *stercoraria* pertaining to dung.
Filaments, 1.2–1.5 by up to 100 μm, often irregularly bent. Cells in filaments sausage-shaped, 1.2–1.5 by 1.2–12 μm.
Gliding motility, slow (ca. 2 μm/min).

4. **Vitreoscilla filiformis** Pringsheim 1951, 130.
fi.li.for'mis. L. n. *filum* a thread; L. n. *forma*
shape; M.L. adj. *filiformis* thread-shaped.

Filaments 1.2 by 5 μm or longer.

On agar may form rounded arches or loops, as
found in certain strains of *Oscillatoria*.

5. **Vitreoscilla catenula** Pringsheim 1951, 130.
ca.te'nu.la. M.L. n. *catenula* a small chain.

Filaments cylindrical and of uniform width
(1.5–2.0 μm) with constrictions between barrel-
shaped cells, which range from 3–6 μm in length.

Species incertae sedis

A. The following may be species of *Vitreoscilla*,
but acceptance depends on better characterization.

a. *V. stricta* Pringsheim 1949, 72.
b. *V. conica* Pringsheim 1951, 139.
c. *V. major* Pringsheim 1951, 138.

Genus III. **Thioploca** *Lauterborn 1907, 242*

Siegfried Maier

Thi.o.plo'ca. Gr. neut.n. *thion* sulfur; Gr. fem.n. *plokē* a braid, a twist; M.L. fem.n.
Thioploca a sulfur-braid.

Flexible filaments made up of **numerous
segments,** generally with **sulfur inclusions**
(Plate 2.6, Fig. C), occur in parallel or braided
fascicles, enclosed by a **common sheath** of
variable width (Plate 2.6, Figs. A and B). The
number of filaments embedded in one sheath is
variable. Within one sheath, filaments may be of
fairly uniform or of greatly differing diameters.
The sheath is frequently encrusted with detritus
(Plate 2.6, Fig. A). Individual filaments show
independent gliding movement; their terminal
segments are often tapered. Not isolated in pure
culture.

Type species: *Thioploca schmidlei* Lauterborn
1907, 242.

Further Comments

In gross morphology *Thioploca* resembles
Hydrocoleum and *Microcoleus* among the *Oscil-
latoriaceae*. Individual filaments of *Thioploca*
superficially resemble those of *Beggiatoa*, but *T.
ingrica* differs from *B. alba* in ultrastructure (Maier
and Murray, 1965) and in growth habit in extracted

hay medium (Maier, 1963). *Beggiatoa* possesses
neither the greenish blue color nor the tapered
terminal segments shown by *Thioploca*. The exact
taxonomic relationship of these two genera is
presently not clear. None of the species of *Thioploca* has been grown in pure culture. Consequently
neither the nature of the greenish blue coloration
nor their presumably chemolithotrophic nutrition
on hydrogen sulfide, nor the constancy of filament
diameter has been verified.

Members of the genus are found in the upper
layers of brackish and/or fresh water mud containing calcium carbonate and low concentrations
of hydrogen sulfide and are presumably microaerophilic. Specifically, *Thioploca* filaments occupy
both, the reducing and the oxidizing horizons of
the mud profile (Perfil'ev, 1965). Filaments show
a horizontal orientation in the reducing horizon,
changing to vertical in the oxidizing layer. The
motility of the filaments would allow transport of
hydrogen sulfide from the lower to the higher
horizon where oxygen is available for oxidation.
Perfil'ev (1965) suggested also that the detritus

TABLE 2.6

Differential characteristics of species of the genus **Thioploca**

	Diameter of		Length of	
	Individual filaments	Common sheath	Individual segments	Common sheath
	μm	μm	μm	cm
I. Filaments of fairly uni- form diameter				
1. *T. schmidlei*	5.0–9.0	50–160	5.0–8.0	Several
2. *T. ingrica*	2.0–4.5	–80	1.5–8.0	–1.0
3. *T. minima*	0.8–1.5	–30	1.0–2.0	–0.6
II. Filaments of greatly differing diameter				
4. *T. mixta*	1.0 and 6.0–8.0			

adhering to the sheaths in the oxidizing horizon contained iron and manganese oxides.

Species differentiation is based only on filament diameter (Table 2.6).

Description of the species of genus **Thioploca**

1. **Thioploca schmidlei** Lauterborn 1907, 242.
schmid'le.i. M.L. gen.n. *schmidlei* of Schmidle. Identified from various localities in Central Europe. Found in fresh water mud.

2. **Thioploca ingrica** Visloukh 1911, 2103.
in'gri.ca. M.L. adj. *ingrica* pertaining to Ingria, ancient district of Leningrad, Russia.
Identified from various localities in Central Europe and Lake Erie. Found in mud from fresh and brackish water.

3. **Thioploca minima** Koppe 1924, 630.

mi'ni.ma. L. sup.adj. *minima* least, smallest.
Identified from various localities in Central Europe and Lake Erie. Found in fresh water mud.

4. **Thioploca mixta** Koppe 1924, 630.
mix'ta. L. part.adj. *mixta* mixed.
Identified from Lake Constance. The existence in Lake Erie of a similar association of filaments of 1 μm and 2.0–4.5 μm in diameter, makes the species a doubtful taxonomic entity. Found in fresh water mud.

FAMILY III. **SIMONSIELLACEAE** STEED 1962, 615

PAMELA STEED-GLAISTER

Si.mon.si.el.la′ce.ae. M.L. fem.n. *Simonsiella* type genus of the family; *-aceae* ending to denote a family; M.L. fem.pl.n. *Simonsiellaceae* the *Simonsiella* family.

Cells arranged in apposition to form filaments which are flat, not cylindrical. Gliding motility when flat surface of filament is in contact with substrate, but not when filaments are coiled on edge as in colonies (Plate 2.7, Fig. 2).

Key to the genera of family **Simonsiellaceae**

I. Closely opposed cells form flat filaments which divide into hormogonia-like units with rounded terminal cells.

Genus I. *Simonsiella*

II. Cells arranged in pairs within flat filaments; terminal cells not rounded.

Genus II. *Alysiella*

Genus I. **Simonsiella** *Schmid 1922, 504*

Si.mon.si.el′la. M.L. dim. ending *-ella*; M.L. fem.n. *Simonsiella* named for Hellmuth Simons, who studied the species of this genus.

Cells 0.4–0.7 by 2–4 μm, **closely apposed to form filaments with free faces of terminal cells rounded. Filaments are flat, not cylindrical,** and divide into hormogonia-like units (Plate 2.7, Figs. 1 and 2). **Gliding motility** when flat surface in contact with substrate. No spores. No capsules. Gram-negative.

Chemoorganotrophs: **Metabolism fermentative;** growth best in presence of blood or serum; minimal nutritional requirements not known.

Strict aerobes.

Temperature optimum 30–37 C.

Occur among **oral flora** in man and animals.

Type species: *Simonsiella muelleri* Schmid 1922, 504.

Further Comments

Members of the genus *Simonsiella* inhabit oral cavities of man and animals, apparently as harmless saprophytes. By virtue of their characteristic morphology, they are readily identified and have been reported from horses, pigs, cattle, sheep, goats, guinea pigs, cats, dogs and fowls, as well as from man. Only recently, however, have they been studied in pure culture: *S. crassa* from sheep (Steed, 1962) and *S. muelleri* from guinea pigs (Berger, 1963).

In the 7th edition of THE MANUAL, the genus *Simonsiella* was inserted in the order *Caryopha-*
nales. However, the filamentous nature and gliding motility of these organisms warrant their present position within the order *Cytophagales*.

The genus contains two species, differentiated by width of the filament.

Description of the species of genus **Simonsiella**

1. **Simonsiella muelleri** Schmid 1922, 504. (*Caryophanon muelleri* (Schmid) Peshkoff 1948, 1004; Vernacular name: Scheibenbakterien Müller 1911, 2247.)

muel'le.ri. M.L. gen.n. *muelleri* of Müller; named for Reiner Müller, who first described these organisms.

Cells 0.4–0.7 μm long by 2–3 μm wide, closely apposed on the long dimension to form flat filaments 2–3 μm wide, which divide into hormogonia-like units with rounded terminal cells. Filaments often coil as a roll of tape, especially in colonies (*cf.* Plate 2.7, Fig. 2). Gliding motility. Gram-negative.

Blood agar colonies: Round, convex, silvery, sometimes slimy, 0.2–0.3 mm diameter. Narrow hemolyzed zone on sheep blood agar.

Serum broth: Poor but substantial growth in form of granular clusters adhering to test tube wall.

Oxidase and catalase positive.

On 5% sucrose agar, an iodine-positive polysaccharide is produced.

Habitat: Oral cavities of man and animals.

2. **Simonsiella crassa** Schmid 1922, 509. (*Caryophanon crassa* (Schmid) Peshkoff 1948, 1004.)

cras'sa. L. fem.adj. *crassa* thick.

Cells 0.6 μm long by 3–4 μm wide and 1–1.5 μm thick, closely apposed to form flat filaments 3–4 μm wide and 1–1.5 μm thick (Plate 2.7, Fig. 1). Filaments divide into hormogonia-like units with rounded terminal cells. Filaments may coil on their edges (Plate 2.7, Fig. 2). Gliding motility only when flat surface on substrate. Gram-negative.

Growth best in presence of blood or serum.

Blood agar colonies after 3–4 days: convex, 1–3 mm diameter, smooth, undulate, translucent; pigment and odor absent; β-hemolysis.

Serum broth: Granular and turbid, forming moderate sediment.

No growth on Simmon's citrate medium.

Acid but no gas produced from fructose, glucose, inulin, maltose, ribose, sucrose, trehalose and variable acid production from arabinose and mannitol, on agar or in shallow liquid media. Acid not produced from cellobiose, dulcitol, erythritol, galactose, glycerol, inositol, lactose, mannose, melezitose, melibiose, raffinose, rhamnose, salicin, sorbitol, sorbose or xylose.

Gelatin and inspissated serum liquefied, casein hydrolyzed, litmus milk peptonized. H_2S produced. Urease negative, indole not produced. Starch, sodium hippurate and esculin not hydrolyzed. Nitrate may be reduced to nitrite. Methylene blue not reduced. Methyl red and Voges-Proskauer tests negative, catalase positive.

Habitat: Oral cavities of animals.

Neotype strain: NCTC 10283.

Genus II. **Alysiella** Langeron 1923, 116

A.ly.si.el'la. Gr. fem.n. *alysion* small chain; M.L. dim. ending *-ella*; M.L. fem.n. *Alysiella* small chain.

Cells 0.6 by 2–3 μm **arranged side by side in pairs to form filaments which are flat, not cylindrical; terminal cells are not rounded. Gliding motility** when flat surface is presented to substrate. No spores. No capsules. **Gram-negative.**

Chemoorganotrophs: **Metabolism fermentative;** growth best in presence of blood or serum; minimal nutritional requirements unknown.

Strict aerobes.

Temperature optimum 37 C.

Habitat: **Oral cavities** of animals.

Type species: *Alysiella filiformis* (Schmid) Langeron 1923, 118.

Further Comments

The genus *Alysiella* contains a single species, *A. filiformis*, formerly classified in the genus *Simonsiella*. These organisms have been found in the oral cavities of horses, cows, pigs, goats, sheep, rabbits and guinea pigs, but only some strains from sheep (Steed, 1962) and from guinea pigs (Berger, 1963) have been studied in pure culture.

Description of the species of genus **Alysiella**

1. Alysiella filiformis (Schmid) Langeron 1923, 118. (*Simonsiella filiformis* Schmid 1922, 509; *Caryophanon filiformis* (Schmid) Peshkoff 1948, 1004.)

fi.li.for′mis. L. n. *filum* thread; L. n. *forma* shape; M.L. adj. *filiformis* filiform.

Cells 0.6 by 2–3 μm arranged side by side in pairs to form flat filaments (0.5–1 μm thick) and of variable length (Plate 2.7, Figs. 3 and 4). Terminal cells not morphologically differentiated from other cells within the filament. Gliding motility on solid substrate. Gram-negative.

Growth best in presence of blood or serum.

Agar colonies after 3–4 days, low convex, 1–1.5 mm diameter, smooth, undulate, translucent and often with a narrow fringe of gliding filaments; pigment and odor absent.

Blood agar: β-Hemolysis.

Serum broth: Granular and turbid, forming moderate sediment. No growth on Simmon's citrate medium.

Acid but no gas produced from fructose, glucose, maltose, sucrose, trehalose and variable acid production from inulin and ribose, on agar or in shallow liquid media.

Acid not produced from arabinose, cellobiose, dulcitol, erythritol, galactose, glycerol, inositol, lactose, mannitol, mannose, melezitose, melibiose, raffinose, rhamnose, salicin, sorbitol, sorbose or xylose.

Gelatin sometimes slightly liquefied, inspissated serum not liquefied, casein not hydrolyzed, litmus milk unchanged. H_2S may be produced, urease negative, indole not produced. Starch, sodium hippurate and esculin not hydrolyzed. Nitrate not reduced to nitrite. Methylene blue not reduced. Methyl red and Voges-Proskauer tests negative. Oxidase positive; catalase positive (Steed, 1962), catalase negative (Berger, 1963).

Habitat: Oral cavities of animals.

Neotype strain: NCTC 10282.

FAMILY IV. **LEUCOTRICHACEAE** BUCHANAN 1957, 850

THOMAS D. BROCK

Leu.co.trich.ac′e.ae. M.L. fem.n. *Leucothrix* type genus of the family; *-aceae* ending to denote a family; M.L. fem.pl.n. *Leucotrichaceae* the *Leucothrix* family.

Long filaments composed of short cylindrical or ovoid cells, cross-walls clearly in evidence, **colorless, unbranched;** filaments usually uniform in diameter throughout length although may taper from base to apex under some conditions. In nature filaments usually **attached to solid substrates by means of inconspicuous holdfasts, stalks absent. Filaments do not glide** but may wave sporadically from side to side. Dispersal by means of gonidia (single cells arising from cells of the filaments by rounding up, often released primarily from apices, but may also be intercalary); gonidia often but not always show jerky gliding motion on solid substrates. **Rosette formation is** a key diagnostic characteristic of the family, but is found rarely in nature, although frequently in culture. The rosettes may be formed of gonidia or, after conversion of gonidia to filaments, of several or more filaments attached at their bases. Filaments in culture often form true knots and knots also occur in nature although rarely. Resemble blue-green algae in many respects but differ from them in not forming photosynthetic pigments. **Strictly aerobic,** chemoorganotrophic or chemolithotrophic. **Aquatic.**

Genus I. **Leucothrix** *Oersted 1844, 44*

(*Pontothrix* Nadson and Krasil'nikov 1932. 246.)

Leu′co.thrix. Gr. adj. *leucus* clear, light; Gr. n. *thrix, trichis* hair; M.L. fem.n. *Leucothrix* colorless hair.

Morphological description as for family. **Sulfur granules are not formed,** even in habitats high in H_2S. **Marine,** requires NaCl for growth. Chemoorganotrophic.

Type species: *Leucothrix mucor* Oersted 1844, 44.

Description of the species of genus **Leucothrix**

1. Leucothrix mucor Oersted 1844, 44.

mu'cor. L. n. *mucor* mold; M.L. n. *Mucor* a genus of molds.

Filaments of variable lengths, often much greater than 100 μm, diameter 3–5 μm; colorless, unbranched, non-motile (although occasionally they wave back and forth), and lack a sheath although cells in regions of a filament may become emptied of their contents giving the appearance of a sheath. Filaments often grow intertwined or in dense tangles; true knots are usually formed by pure cultures when growing in organically rich media. Swollen cells often formed at random along filaments. Larger structures (bulbs) usually formed in knotty cultures, probably as a result of fusion of cells in the region of the knots. Filaments attached to solid substrates by means of an inconspicuous holdfast which can be seen by staining with primuline and viewing with blue light in a fluorescence microscope; the holdfast fluoresces red. Individual cells of filaments round up and form ovoid to spherical gonidia which when released acquire a jerky gliding motility. Gonidia frequently aggregate in cultures, probably chemotactically, to form rosettes.

A large variety of simple organic compounds may serve as sources of carbon and energy.

Most strains do not require growth factors, although occasional strains do.

Strictly aerobic.

Optimum temperature 25 C; maximum, 30–35 C, minimum, grows at 0 C to form visible colonies within 1–2 weeks. Strains from tropical waters are more stenothermal, not growing below 15 C.

All strains require NaCl for growth; optimal concentration for growth about 1.5% NaCl, grows at concentrations of 0.3–6.0% NaCl.

G + C content of the DNA for >30 strains ranges from 46–51%.

Usually grows epiphytically on sea weeds, especially in regions where good aeration exists due to wave action or current flow. Has not been found in fresh water habitats. Most common in temperate sea coasts, but has also been found in tropical waters.

Neotype strain: No. 1, ATCC 25107.

Comments: Pringsheim (1957) has described another species, *L. cohaerens*, which differs primarily in diameter of filaments. Cultures are no longer available and other isolates showing distinct differences in filament diameter have not been obtained. Filament diameter is somewhat variable in culture.

Genus II. **Thiothrix** *Winogradsky 1888, 39*

Thi'o.thrix. Gr. n. *thium* sulfur; Gr. n. *thrix* hair; M.L. fem.n. *Thiothrix* sulfur hair.

Morphologically very similar to *Leucothrix*, but **deposits sulfur granules inside the cells.** In H$_2$S-deficient habitats, *Thiothrix* may lose its sulfur granules and resemble very closely *Leucothrix*. Filaments usually attached to solid substrates. Dispersal by gonidia. Rosettes formed. Filaments non-motile. (May be confused with *Beggiatoa*, which also deposits sulfur, but the latter has motile gliding filaments.) *Thiothrix* is **found in both fresh water and marine habitats. It occurs commonly where H$_2$S concentrations are high.** In fresh water habitats it is found most frequently in sulfur springs and sewage plants and is especially adapted to flowing water. In marine environments, it is found where sulfur springs enter the sea or where H$_2$S is produced by sulfate reduction among rotting seaweeds, but usually where water flow occurs, and usually attached to marine algae. It apparently is an obligate autotroph but has not been consistently cultured.

Type species: *Thiothrix nivea* (Rabenhorst) Winogradsky.

Description of the species of genus **Thiothrix**

1. Thiothrix nivea (Rabenhorst) Winogradsky 1888, 39. (*Beggiatoa nivea* Rabenhorst 1865, 94.)

ni've.a. L. adj. *nivea* snow white.

Description as for genus.

Species incertae sedis

Other species have been proposed on the basis of morphological differences, but in the absence of pure culture data such species distinctions should be held in abeyance. The following are therefore considered *Species incertae sedis.*

    a. *Thiothrix annulata* Molisch 1912, 58.

    b. *Thiothrix longiarticulata* Klas 1936, 126.

    c. *Thiothrix marina* Molisch 1912, 58.

    d. *Thiothrix tenuis* Winogradsky 1888, 40.

    e. *Thiothrix tenuissima* Winogradsky 1888, 40.

    f. *Thiothrix voukii* Klas 1936, 123.

FAMILIES AND GENERA OF UNCERTAIN AFFILIATION

Genus **Toxothrix** *Molisch 1925, 144*

P. Hirsch and G. A. Zavarzin

Tox′o.thrix. Gr. n. *toxon* a bow; G. n. *thrix* a thread; M.L. fem.n. *Toxothrix* bent thread.

Cells cylindrical, colorless, 0.5–0.75 by 3–6 µm, in filaments up to 400 µm long. A dense body often located at either end of cell (Plate 2.8, Fig. 1). Gram reaction not recorded.

Filaments often U-shaped (Plate 2.8, Figs. 1 and 2) and rotate while slowly moving forward with the rounded part in the lead; a mucoid substance, excreted from several sites on the trailing ends, is deposited as a double track ("railroad track") of twisted strings each 0.2 µm wide (Plate 2.8, Figs. 8–10). Fan-shaped structures may be deposited laterally along the track as the arms of the U move from side to side, and between the tracks as a result of the middle section being lifted and then touched down again (Plate 2.8, Fig. 7) (Krul *et al.*, 1970).

Oxidized iron may be deposited onto the mucoid threads, rendering them yellowish brown and brittle and giving them a diameter of 2.5 µm.

Have not been obtained in pure culture but heterotrophic and psychrophilic; have been maintained for long periods at 5 and 10 C. But filaments are extremely fragile during laboratory examination and explosive disintegration has been observed after short periods (Plate 2.8, Figs. 3–6). Grow attached to surfaces and develop best at reduced oxygen tensions (Hasselbarth and Lüdemann, 1967) and around neutrality (pH 6.5-7.2).

Originally found in water reservoir near the Biological Station on the river Dneiper in the U.S.S.R. Widely distributed in cold springs, bogs, ponds and lakes containing ferrous iron.

Type species: *Toxothrix trichogenes* (Cholodny) Beger in Beger and Bringmann 1953, 332.

Editorial Note. The organism has not been cultivated and it is difficult to place it in a particular family. Krul *et al.* (1970) suggest it may belong near *Herpetosiphon.* Balashova (1968) has pointed out that in some respects it resembles *Gallionella.* As it does glide it is placed with the gliding bacteria as a genus of uncertain affiliation. Iron is apparently not needed for growth.

Description of the species of genus **Toxothrix**

1. **Toxothrix trichogenes** (Cholodny) Beger in Beger and Bringmann 1953, 332. (*Leptothrix trichogenes* Cholodny 1924, 296; *Toxothrix ferruginea* Molisch 1925, 144; *Chlamydothrix trichogenes* (Cholodny) Naumann 1929, 513; *Sphaerotilus trichogenes* (Cholodny) Pringsheim 1949, 234.)

tri.cho′ge.nes. Gr. n. *thrix* hair; Gr. v.suf.-*genes* producing M.L. adj. *trichogenes* hair-producing.

Description as for the genus.

Comments: Cholodny (1924) thought the organism had a thin, tubular sheath which split repeatedly longitudinally, thus giving rise to the "twisted thread rope." However, Krul *et al.* (1970) followed the formation of the double tracks and fan-shaped structures on living, undisturbed specimens. This species has been reported to have been cultivated by Teichmann (1935).

Toxothrix gelatinosa Beger 1953, 333 was described on the basis of smaller filaments (diameter with slime threads 1.5–1.7 µm) and the fan-shaped arrangements of the individual filaments in a gelatinous matrix. However, the individual cell size (0.5 by 3 µm) falls within the range given for *T. trichogenes.*

FAMILY **ACHROMATIACEAE** MASSART 1901, 256

E. R. Leadbetter

A.chro.ma.ti.a′ce.ae. M.L. neut.n. *Achromatium* type genus of the family; -*aceae* ending to denote a family; M.L. fem.pl.n. *Achromatiaceae* the *Achromatium* family.

Cells spherical to ovoid or cylindrical. Division by constriction. Motile by slow, jerky (gliding?) movements. Resting stages not known. Cells may contain sulfur and calcium carbonate inclusions.

Genus **Achromatium** *Schewiakoff 1893, 1*

J. W. M. LA RIVIÈRE

A.chro.ma'ti.um. Gr. pref. *a* not; Gr. n. *chromatium* color, paint; M.L. neut.n. *Achromatium* that which is not colored.

Cells **spherical** to **ovoid or cylindrical** with hemispherical ends (Plate 2.9, Fig. 1), 5–33 by 15–125 µm. Division by **constriction** in the middle. **Movements,** if any, are of a **slow rolling jerky type** dependent upon a substrate being present. **Photosynthetic pigments absent.** In their natural habitat cells may contain **sulfur droplets** and **large spherules** of **calcium carbonate.** Has not been cultivated in pure culture. **No resting stages known.**

Microaerophilic; apparently require sulfides.

Type species: *Achromatium oxaliferum* Schewiakoff 1893, 1.

Further Comments

Pending pure culture studies any classification can only be provisional. The present classification is based on the absence of pigment and the gliding type of movement seen on solid surfaces with no "apparent" means of locomotion. However, the application of careful fixation for electron microscopy has recently indicated the presence of peritrichous filaments within the slime of a high percentage of cells (de Boer, la Rivière and Schmidt, 1971) (Plate 2.9, Fig. 4). This finding would remove *Achromatium* from the gliding bacteria but pending further study they are left in this category, but as a *familia incertae sedis*.

There appears to be no justification for recognition of more than two species containing (a) the forms which contain the characteristic calcium carbonate inclusions as found in fresh and brackish waters and (b) those lacking these inclusions and usually found in marine environments.

Description of the species of genus **Achromatium**

1. Achromatium oxaliferum Schewiakoff 1893, 1. (*Hillhousia mirabilis* West and Griffiths 1909, 398; *Hillhousia palustris* West and Griffiths 1913, 84; *Achromatium gigas* Nadson 1913, 108.)

ox.al.if'er.um. M.L. n. *oxalatum* oxalate; Gr. n. *oxalis* sorrel, a sour plant; L. v. *fero* carry; M.L. adj. *oxaliferum* oxalate-containing.

Cells spherical, ovoid or cylindrical with a minimum width of short axis of 5 µm and a maximum length of long axis of 100 µm. Normally contain small sulfur globules and much larger, highly refractile calcium carbonate crystals. The latter may disappear under favorable environmental conditions. They may be removed by treating the cells with acetic acid (Plate 2.9, Figs. 2 and 3). Cells with calcium carbonate inclusions have a high specific gravity and therefore are found only on the bottom of pools and streams, usually in or on the mud; they can be easily concentrated by careful swirling of the mud with water in a beaker.

Gram-negative, catalase negative.

Described from fresh water and brackish mud environments but according to Nadson and Visloukh (1923, 33) also found in marine mud.

Further Comments

A slime layer was observed around the cells by Bersa (1920). Skuja (1948) found cells with one polar flagellum in natural populations which he thought could be swarmers of *A. oxaliferum*. Careful fixation has now shown peritrichous filaments within the slime (de Boer, la Rivière and Schmidt, 1971). The position of these organisms is thus in doubt but any change must await confirmation.

2. Achromatium volutans (Hinze) van Niel 1948, 999. (*Thiophysa volutans* Hinze 1903, 310; *Thiophysa macrophysa* Nadson 1913, 109.)

vol'u.tans. L. part.adj. *volutans* rolling.

Cells range from spheres about 5 µm in diameter to ovals up to 40 µm in length.

Gram reaction not known.

Normally contain sulfur globules but lack internal, calcium carbonate deposits.

Found in marine mud containing hydrogen sulfide and in decaying seaweeds.

FAMILY **PELONEMATACEAE** SKUJA 1956, 81

H. SKUJA†

Pel.o.ne.ma.ta′ce.ae. M.L. neut.n. *Pelonema* type genus of family; *-aceae* ending to denote family; M.L.fem.pl.n. *Pelonemataceae* the Pelonema family.

Cylindrical colorless cells, arranged in unbranched filaments, usually visible without staining. Filaments straight, flexuous or spirally coiled,

TABLE 2.7
Characteristics differentiating the genera of family **Pelonemataceae**

	Pelo-nema	Achroo-nema	Pelo-ploca	Desman-thos
Filaments				
Single	+	+	−	−
In bands or bundles	−	−	+	+
Gliding motility	∓	+	−	−
Holdfasts	−	−	−	+
Sheaths	−	Dᵃ	D	+
Gas vacuoles	+	D	D	−
Resting cells	−	D	−	−

ᵃ D, present in some species, not in others.

with or without a thin slime-sheath, occurring singly or aggregated in bundles or bands. Single filaments may show a gliding movement combined with rotation around the longitudinal axis. Cells may contain gas vacuoles but not pigments or sulfur globules.

Reproduction by fragmentation of the filaments, rarely by arthrospore-like resting cells.

Characteristics differentiating the genera are given in Table 2.7.

Aquatic. Have not been cultivated.

Comment: Possibly apochromatic cyanophytes related to those forming hormogonia.

Editorial Note: Since gliding motility occurs in only one genus and rarely in another, the position of these organisms in the present classification is in some doubt. Until they have been cultured and further studies made, they are grouped together as proposed by Professor Skuja.

Genus I. **Pelonema** *Lauterborn 1916, 408*

Pel.o.ne′ma. Gr. adj. *pelos* dark-colored, hence anaerobic mud; Gr. n. *nema*, thread; M.L. neut.n. *Pelonema* mud filament.

Colorless cells with a delicate smooth wall; usually contain gas vacuoles. In multicellular, unbranched straight or loosely flexuous filaments of uniform thickness, which may or may not be constricted at the cross-walls. The filaments may occasionally show a slow gliding motility. Propagation by fragmentation of filaments. Resting cells not known.

Type species: *Pelonema tenue* Lauterborn 1916, 408.

Key to the species of genus **Pelonema**

I. Filaments usually slightly constricted at cross walls and 300–600 μm long.
  A. Cells considerably longer than wide.
    1. Cells 1.9–2.2 by 3–19 μm. Filaments usually not more than 300 μm long.
      1. Pelonema tenue
    2. Cells 1.0–1.7 by 5–16 μm. Filaments up to 600 μm or more in length.
      2. Pelonema aphane
  B. Cells isodiametrical or only slightly longer than wide.
    1. Cells 1.5–2.0 by 2–4 μm. Filaments up to 500 μm or more in length.
      3. Pelonema pseudovacuolatum
II. Filaments not constricted at cross-walls, relatively short.
  A. Cells 0.6–0.7 by 1.5–6 μm. Filaments up to 120 μm long.
      4. Pelonema subtilissimum
  B. Cells 2.0 by 2–6 μm. Filaments rarely more than 200 μm long.
      5. Pelonema hyalinum

† Deceased.

Description of the species of genus **Pelonema**

1. **Pelonema tenue** Lauterborn 1916, 408.
ten'u.e. L. neut. adj. *tenue* thin, slender.
Filaments usually not motile, occasionally a very slow gliding motility, straight or slightly flexuous, only slightly constricted at cross walls. Protoplasts quite refractive. Gas vacuoles present.
Found in deeper fresh and brackish water, in hypolimnion and on bottom mud of lakes.

2. **Pelonema aphane** Skuja 1956, 92.
aph'an.e. Gr. adj. *aphanes* invisible, transparent; M.L. adj. *aphane* invisible, transparent.
Filaments not motile or only very slowly motile, straight to slightly flexuous, slight constrictions at cross-walls, uniform diameter. Cells colorless with a central gas vacuole.
Found in hypolimnion and on bottom mud of deep fresh water lakes; during the period of vernal circulation also in upper layers.

3. **Pelonema pseudovacuolatum** Lauterborn 1916, 408.
pseu.do.va.cu.o.la'tum. Gr. adj. *pseudes* false; M.L. dim.n. *vacuolum* a vacuole: M.L. adj. *pseudovacuolatum* false vacuolated.

Filaments not motile, uniform, straight or slightly and irregularly coiled. Constrictions at cross-walls minute. Cells cylindrical or slightly barrel-shaped, colorless with a central gas vacuole.
Found in deeper fresh water pools and on bottom mud of lakes, occasionally in plankton.

4. **Pelonema subtilissimum** Skuja 1956, 91.
sub.til.is'sim.um. L. sup. adj. *subtilissimum* finest, very slender.
Filaments not motile, uniform, straight, cross-walls not constricted. Cells colorless with a granular protoplasm and a central gas vacuole.
Found in hypolimnion of Blankvatn, a relic lake at Oslo, Norway.

5. **Pelonema hyalinum** Koppe 1924, 625.
hy.a.lin'um. Gr. adj. *hyalinos* glassy; M.L. adj. *hyalinum* glassy.
Filaments not motile or only very slowly motile, uniform, not constricted at cross-walls. Cells hyaline with a central gas vacuole.
Found on bottom mud of deep water lakes.

Genus II. **Achroonema** Skuja 1948, 30

A.chro.o.ne'ma. Gr. adj. *achroos* colorless; Gr. n. *nema* filament; M.L. neut.n. *Achroonema* colorless filament.

Cells have a delicate, smooth, sometimes distinctly porous cell wall; protoplasm homogeneous or granular, rarely with gas vacuoles. In multicellular, colorless, unbranched filaments, often constricted at cross-walls, straight, or nearly so or in very loose spirals.

Multiply by fragmentation of filaments, rarely by formation of arthrospore-like resting cells.
Aquatic. Have not been cultivated.
Type species: *Achroonema spiroideum* Skuja 1948, 31.

Key to the species of genus **Achroonema**

I. Filaments more or less regularly coiled but in loose spirals.
1. *Achroonema spiroideum*
II. Filaments straight or only slightly and irregularly coiled.
   A. Filaments not markedly constricted at cross-walls.
      1. Filaments 2.5 μm or less in diameter.
         a. Filaments 0.7–1.2 μm wide.
2. *Achroonema angustum*
         b. Filaments 1.3–1.6 μm wide.
3. *Achroonema proteiforme*
         c. Filaments ca. 2 μm wide.
4. *Achroonema profundum*
      2. Filaments more than 2.5 μm in diameter.
5. *Achroonema splendens*
   B. Filaments distinctly constricted at cross-walls.
      1. Produce resting cells.
         a. Wall of resting cells brown.
6. *Achroonema sporogenum*

b. Wall of resting cell colorless.
    bb. Resting cell long and cylindrical.
                  7. *Achroonema lentum*
    bbb. Resting cell short and more or less barrel-shaped.
                  8. *Achroonema inaequale*
2. Resting cells not known.
  a. Cells usually isodiametrical.
                  9. *Achroonema simplex*
  b. Cells usually longer than wide.
    bb. Protoplasts have a darker centroplast.
              10. *Achroonema articulatum*
    bbb. Protoplasts more or less homogeneous.
      i. Filaments 2.5–3 μm wide.
              11. *Achroonema gotlandicum*
      ii. Filaments 5–6.8 μm wide.
              12. *Achroonema macromeres*

Description of the species of genus **Achroonema**

1. Achroonema spiroideum Skuja 1948, 31.
spir.oi′de.um. L. n. *spira* spiral; L. adj.suff. *-oideus* form of; L. adj. *spiroideum* having a spiral form.

Cells 0.3–0.5 by 3–15 μm, with a homogeneous or finely granular protoplasm, in filaments, not constricted at the indistinct cross-walls and up to 250 μm long; wound in loose spirals. Filaments very motile.

Found in fresh water, particularly in hypolimnion of lakes, occasionally in the slime of planktonic organisms.

2. Achroonema angustum (Koppe) Skuja 1956, 84. (*Oscillatoria angusta* Koppe 1924, 641.)
an.gus′tum. L. adj. *angustum* slender.

Cells 0.7–1.2 by 2.5–8 μm, with a homogeneous protoplasm. In uniform, long, more or less straight filaments, not markedly constricted at the obscure cross-walls. Filaments motile.

Found in fresh water, particularly in hypolimnion of lakes, sometimes on bottom mud.

3. Achroonema proteiforme Skuja 1956, 84.
pro.te.i.for′me. Gr. comp.adj. *proteros* earlier, premature; L. n. *forma* shape, form; M.L. n. *proteiforme* earlier form.

Cells 1.3–1.6 by 3–13 μm, with a colorless protoplast which contains some large granules partly in a single row. In uniform filaments not constricted at the indistinct cross-walls, usually 200–300 μm long, rarely up to 600 μm, irregularly coiled. Filaments motile.

Found on bottom mud or occasionally floating in lakes.

4. Achroonema profundum (Kirchner) Skuja 1956, 85. (*Oscillatoria profunda* Kirchner 1896, 101.)
pro.fun′dum. L. adj. *profundum* of the depths.

Cells about 2 by 2–6 μm, with more or less homogeneous protoplasm. In uniform filaments, not constricted at the rather obscure cross-walls, nearly straight or slightly twisted. Filaments motile.

Found in the hypolimnion of lakes.

5. Achroonema splendens Skuja 1956, 88.
splen′dens. L. part.adj. *splendens* brilliant.

Cells 3.4–4.2 by 4–10 μm with a minutely, occasionally more coarsely, granulated protoplasm. In filaments not constricted at the distinct cross-walls, more or less straight or slightly coiled, up to 1 mm or more in length. Motile.

Found on bottom mud or heleoplanktonic in ponds containing hydrogen sulfide; probably a thiophilic organism.

Comment: A form of this species (forma *tenue* Skuja 1956, 88) has cells 5–10 μm long and filaments 2–2.5 μm wide.

6. Achroonema sporogenum Skuja 1956, 87.
spor.og′en.um. M.L. n. *spora* a spore; Gr. v. *gennao* produce; M.L. adj. *sporogenum* producing spores.

Cells about 2.5 by 2.5–4 μm; the protoplasm is usually differentiated into an ectoplasm and a coarser, lighter centroplasm. In uniform, straight or nearly straight filaments, distinctly constricted at the cross-walls so that the cells are almost separated. Filaments motile. Resting cell (hypnocysts or spores) are formed.

Spores short cylindrical or barrel-shaped with flattened or occasionally rounded ends, 2.5–2.8 by

2.7–5 μm, with a relatively thick, brown cell wall; occur singly or up to three in a row.

Originally found in a rivulet near Riga, Latvia, among decaying leaves of water plants; probably a heterotrophic organism.

7. Achroonema lentum Skuja 1956, 85.

len'tum. L. adj. *lentum* slow.

Cells 1.4–2 by 3–12 μm, in filaments slightly constricted at the cross-walls, uniform, straight or slightly flexuous. Filaments move slowly. Spores 2–2.5 by 5–10 μm, cylindrical with rounded ends and a thin colorless wall, usually occur singly.

Found in hypolimnion or on bottom mud of lakes and deep ponds.

8. Achroonema inaequale Skuja 1956, 86.

in.ae.qua'le. L. adj. *inaequale* unequal, referring to unequal thickness of sporiferous filaments.

Cells 2–2.7 by 1.5–3 μm, with a minutely but densely granular protoplasm. In uniform filaments, more or less constricted at the usually distinct cross-walls, nearly straight or slightly and irregularly flexuous. Filaments motile.

Spores short, barrel-shaped or discoid up to 3.5 or 4 μm wide by 2.5–4.5 μm long, with a relatively thin colorless and smooth wall.

Found in hypolimnion of eutrophic lakes.

9. Achroonema simplex Skuja 1956, 85.

sim'plex. L. adj. *simplex* simple.

Cells short, barrel-shaped, 1.8–2 by 1.5–3 μm. Filaments uniform and long, straight or slightly flexuous, definitely constricted at the distinct cross-walls. Filaments not or only slightly motile.

Found in hypolimnion of lakes, especially beneath bottom mud.

10. Achroonema articulatum Skuja 1956, 89.

ar.tic.u.la'tum. L. adj. *articulatum* divided into joints, articulate.

Cells cylindrical with slightly rounded ends, 2–3.5 by 2.8–13 μm, with a finely granular protoplasm, differentiated into a lighter ectoplasm and a darker axial centroplasm. In uniform filaments, usually 100–350 μm long, straight or nearly straight, definitely constricted at the thick and hyaline cross-walls. Filaments motile.

Found in deeper water, generally in the hypolimnion of lakes, sometimes, especially at the period of vernal circulation, also in upper layers of water.

11. Achroonema gotlandicum Skuja 1956, 88.

got.land'i.cum. L. adj. suff. *-icum* belonging to; *gotlandicum* belonging to Gotland, a Swedish island.

Cells cylindrical, 2.5–3 by 9–12 μm, with a uniformly and finely granular protoplasm, sometimes with a thin, lighter peripheral zone. In uniform filaments, elongate, not fragile, more or less constricted at the distinct cross-walls, straight or slightly flexuous. Motile.

Originally found in hypolimnic water of Lake Sigwaldeträsk on Gotland.

12. Achroonema macromeres Skuja 1956, 90.

mac.ro.mer'es. Gr. adj. *macros* large; Gr. n. *meros* part: M.L. n. *macromeres* large parts.

Cells cylindrical or slightly barrel-shaped, 5–6.8 by 4–19 μm, with a distinctly porous wall, protoplasm differentiated into a lighter ectoplasm and a darker, finely granular centroplasm. Filaments uniform but fragile and length therefore varies from a few cells to 5 mm long, more or less constricted at the evident cross-walls, straight or slightly flexuous. Gas vacuoles sometimes formed in older cells. Filaments motile.

Found in hypolimnion, but at periods of vernal circulation also in epilimnion of lakes; occasionally in pools and other stagnant waters.

Genus III. Peloploca Lauterborn 1913, 99

Pel.o.plo'ca. Gr. adj. *pelos* dark-colored, hence anaerobic mud; Gr. n. *plokē* braid, wickerwork; M.L. fem.n. *Peloploca* mud braid.

Cells of variable length with homogeneous or granular protoplasm, with or without gas vacuoles. Cells arranged in uniform filaments with or without a gelatinous sheath; no constrictions at cross-walls or slightly constricted; cell walls delicate. Filaments bound together laterally to form rigid bundles or flat ribbons which may be straight, undulate in one plane or spirally wound; the ends may appear more or less fibrous. Not motile.

Propagation by fragmentation of bundles or bands.

Aquatic, free floating. Has not been cultivated.

Type species: *Peloploca undulata* Lauterborn 1913, 99.

Key to the species of genus Peloploca

I. Separate filaments do not have a distinct gelatinous sheath; bind together in parallel to form a ribbon-like band.

A. Bands have wavy shape in one plane.
  1. Filaments firmly bonded and bands not fragile; cells with gas vacuoles.
                         1. *Peloploca undulata*
B. Bands more or less flat.
  1. Filaments not strongly bound, bands fragile usually breaking into several pieces that remain together; cells without gas vacuoles.
                         2. *Peloploca fibrata*
  2. Filaments firmly bonded and bands not fragile; cells with gas vacuoles.
                         3. *Peloploca taeniata*
II. Separate filaments have a distinct gelatinous sheath; bind together in more or less spirally wound bundles.
  A. Bundles small, only slightly twisted or sigmoid; cells and sheath without gas vacuoles.
                         4. *Peloploca ferruginea*
  B. Bundles longer, distinctly spirally wound; cells without gas vacuoles but with a vacuolated sheath.

                         5. *Peloploca pulchra*

Description of the species of genus **Peloploca**

1. Peloploca undulata Lauterborn 1913, 99.
un.du.la′ta. L. fem.adj. *undulata* undulated.

Cells 1 by 6–10 μm, in filaments, not constricted at cross-walls. Joined in bands up to 10 μm wide and 150 μm long, usually as a single ribbon but occasionally multiple bands occur. Colorless with homogeneous protoplasm and a central gas vacuole.

Found in deep fresh water pools and lakes on bottom mud.

2. Peloploca fibrata Skuja 1956, 95.
fi.bra′ta. L. adj. *fibrata* fibrous.

Cells, 0.4–0.6 by 0.4–3.0 μm, colorless with homogeneous protoplasm without gas vacuoles, in uniform, fragile filaments, not constricted or with minute constrictions at the indistinct cross-walls. Filaments form rigid bands up to 17 μm wide and up to 600 μm long, rarely 1 mm long. In the beginning in single ribbons, later multiple bands occasionally seen.

Found in hypolimnion of lakes and deeper pools.

3. Peloploca taeniata Lauterborn 1913, 99.
taen.i.a′ta. L. n. *taenia* a band; M.L. adj. *taeniata* stripped, banded.

Cells, 0.6–1.0 by 3–10 μm, colorless with homogeneous protoplasm and a central gas vacuole, in uniform filaments, minutely but distinctly constricted at the thickened cross-walls. Filaments form bands 3–15 μm wide, 1–3 μm thick and up to 1 mm long, flat, straight or slightly twisted, flexible and not fragile.

Found in deeper fresh and brackish water on bottom mud and in hypolimnion of lakes.

4. Peloploca ferruginea Skuja 1956, 94.
fer.ru.gin′e.a. L. adj. *ferruginea* dark red, rust-colored.

Cells 0.3–0.5 by 0.3–1.5 μm, colorless with a homogeneous or finely granulose protoplasm and no gas vacuole, in uniform filaments minutely constricted at the indistinct cross-walls. Filaments have a gelatinous, and often more or less ferruginous sheath 1–1.5 μm wide. Filaments in rigid, curved and twisted bundles of two to seven filaments and up to 300 μm long.

Found in hypolimnion of some eutrophic lakes in Sweden.

5. Peloploca pulchra Skuja 1956, 96.
pul′chra. L. adj. *pulcher, pulchra* beautiful.

Cells 0.4–0.6 by 0.4–2.4 μm, colorless with a homogeneous or finely granular protoplasm containing no gas vacuoles. In uniform filaments, often slightly constricted at the sites of the indistinct cross-walls, and with a gelatinous, hyaline or slightly brownish, vacuolated sheath 1.5–1.8 μm wide; 4–15 filaments in rigid, regularly and slightly curved bundles 3–20 μm wide and up to 400 μm in length; the curves are 30–57 μm in length with an amplitude of 3–20 μm. Propagation by fragmentation of individual bundles.

Found in deeper fresh water pools and in the hypolimnion and bottom mud of eutrophic lakes.

Genus IV. **Desmanthos** *Skuja 1958, 442*

Des.man'thos. Gr. n. *desmos* bundle; Gr. n. *anthos* flower; M.L. neut.n. *Desmanthos* flower (-like) bundle.

Colorless cells in unbranched, more or less straight filaments thicker at the base than at the apex; filaments in bundles, the basic part of which is enclosed in a common hyaline, gelatinous sheath of variable thickness; the filaments at the top are free and divergent. Bundles attached by a holdfast or partially buried in bottom mud. Propagation by fragmentation of filaments and probably by longitudinal separation of the bundle-filaments. Motility doubtful.

Type species: *Desmanthos thiocrenophilum* Skuja 1958, 442.

Description of the species of genus **Desmanthos**

1. **Desmanthos thiocrenophilum** Skuja 1958, 442.

thio.cren.oph'il.um. Gr. n. *thium* sulfur; Gr. n. *krene* a spring; Gr. adj. *philus* loving, fond of; M.L. adj. *thiocrenophilum* fond of sulfur springs.

Bundles composed of 7–10 filaments which are 50–160 μm long and measure at the top about 0.5 μm and at the base up to 1.5 μm in diameter with a rounded conical basal cell. Cells in middle of filament more or less isodiametrical but may be about 4 times longer than wide. Bundles 6–8 μm thick at sheathed base, at tips filaments are divergent. Cells have a colorless, clear protoplast, occasionally differentiated into a light centroplast and a somewhat darker ectoplasm. Gas vacuoles not present.

Has not been cultivated. Found in some sulfur springs at Kemeri, Latvia.

Important Notes

for

Users of this Edition

1. Always read both generic and species descriptions because characters listed in the generic description are not usually listed in the species descriptions.

2. In tables, characters common to all taxa are not shown but may be listed in footnotes.

3. Generally in tables (exceptions are clearly indicated in footnotes, q.v.) the meanings of symbols are as follows:

+ more than 90% strains positive
− more than 90% strains negative
d 11–89% strains positive
( ) delayed reaction
w weak reaction
D Different reactions in different taxa (species of a genus or genera of a family)
v strain instability (NOT differences between strains)

PART 3

THE SHEATHED BACTERIA

Key to the genera of sheathed bacteria

I. Single cells motile by means of a polar flagellum or subpolar flagella.
  A. Sheaths rarely encrusted with iron and not encrusted with manganese oxides.
      Sphaerotilus
  B. Sheaths encrusted with iron or manganese oxides.
      Leptothrix, p. 129.
II. Single cells not motile by flagella.
  A. Sheaths not attached.
    1. Sheaths not encrusted with metal oxides.
      Streptothrix, p. 133.
    2. Sheaths may be encrusted with metal oxides.
      Lieskeella, p. 134.
  B. Sheaths attached.
    1. Sheaths not encrusted with metal oxides.
      Phragmidiothrix, p. 134.
    2. Sheaths encrusted with metal oxides.
      a. Filaments may be swollen at tip.
        Crenothrix, p. 135.
      b. Filaments taper at tip.
        Clonothrix, p. 136.

Editorial Note. The above genera are grouped together for convenience in determination; whether they should be placed in a family is questionable. It has been suggested that, except for a sheath, *Leptothrix* and *Sphaerotilus* might belong in the family *Pseudomonadaceae*. Other problems in classification are mentioned in the *Comments* on the various genera.

Genus **Sphaerotilus** *Kützing 1833, 386*

E. G. Mulder and W. L. van Veen

(*Cladothrix* Cohn 1875, 185.)

Sphae.ro'ti.lus. Gr.n. *sphaera* a sphere; Gr.n. *tilus* anything shredded, flock, down; M.L. masc.n. *Sphaerotilus* spherical flock.

Straight rods, 0.7–2.4 by 3–10 μm, occurring in chains within a sheath of uniform width, which may be attached by means of a **holdfast** (Plate 3.1, Fig. 1) to submerged plants, stones, etc. **Sheaths usually thin without incrustation** by **ferric** and **manganic oxides** (Plate 3.1, Fig. 2). **Single cells motile** by means of a **bundle of subpolar flagella** (Plate 3.1, Fig. 3) often so intertwined as to give the appearance of a single large "unit flagellum." **Resting stages are not known. Gram-negative.**

Chemoorganotrophs: **metabolism respiratory,** never fermentative. Molecular oxygen is the universal electron acceptor. Alcohols, several organic acids and sugars are used as sources of carbon and energy. Inorganic nitrogen compounds (ammonium salts or nitrate) may serve as nitrogen sources in the presence of vitamin B_{12} or of methionine. Better growth is generally obtained with organic nitrogen sources: Casamino acids, peptone or a mixture of aspartic acid, glutamic acid and vitamin B_{12}. Gelatin is liquefied slowly.

Strict aerobes, but good growth can be obtained at low oxygen tensions. Temperature range: 15–37 C; optimum between 25 and 30 C.

The **G + C content** of the DNA of one strain of *S. natans* tested was **70.0 ± 0.5 moles** % (Mandel et al., 1966).

The normal habitat is slowly running fresh water contaminated with sewage or waste water from paper or dairy industries. Long tassels of strongly cohering sheaths largely filled with chains of cells are formed, attached to submerged plants, stones, etc. Poorly precipitating, so-called bulking, activated sludge may also contain many filaments of this organism.

Type species: *Sphaerotilus natans* Kützing 1833, 386.

Further Comments

The genus *Sphaerotilus* contains only one species: *Sphaerotilus natans*. *Sphaerotilus dichotomus* (Cohn) Migula 1900, 1035 (*Cladothrix dichotoma* Cohn), included in the 7th edition of THE MANUAL, is a variation of *S. natans* as a result of special nutritional conditions (Pringsheim, 1949).

Description of the species of genus **Sphaerotilus**

1. **Sphaerotilus natans** Kützing 1833, 386. (*Cladothrix dichotoma* Cohn 1875, 185; *Streptothrix fluitans* Migula 1895, 38; *Sphaerotilus fluitans* (Migula) Schikora 1899, 14; *Sphaerotilus dichotomus* (Cohn) Migula 1900, 1035.)

na′tans. L. part.adj. *natans* swimming.

Rods 0.7–2.4 by 3–10 μm, mostly in chains within sheaths, but sometimes free-swimming. The size of cells and the presence of a sheath depend on the nutrition of the organism (Mulder and van Veen, 1963). On a rich organic agar medium (basal medium with 0.5% each of glucose and peptone) the cells are large, have practically no sheaths (Plate 3.1, Fig. 6), and colonies are nearly circular with smooth edges (Plate 3.1, Fig. 5). On poor media (with 0.1% glucose and peptone) the cells are smaller, sheaths are present (Plate 3.1, Fig. 2) and colonies are rough and filamentous (Plate 3.1, Fig. 4). A mutational smooth-rough variation may also occur (Stokes, 1954).

In its normal habitat (slowly running polluted waters, bulking activated sludge) the sheaths are thin and colorless. In unpolluted, iron-containing waters, iron hydroxide may be deposited in or on the sheaths, which turn yellow-brown and sporadically become encrusted with ferric iron. *S. natans* may thus behave like an iron bacterium and according to some authors it is identical with *Leptothrix ochracea* (Pringsheim, 1949; Stokes, 1954). However, the present authors have clearly shown that the two organisms are distinct (compare description of *Leptothrix ochracea;* Mulder and van Veen, 1963 and 1965; Mulder, 1964). *S. natans* is unable to oxidize manganous compounds although the colonies of some strains may turn slightly brown when growing for a prolonged period on nutrient agar containing $MnCO_3$.

False branching of the filaments occurs in every strain of *S. natans*, but in some strains more than in others. It depends on cultural conditions (relatively poor media) rather than on strain specificity (Pringsheim, 1949).

Cells may contain large amounts of poly-β-hydroxybutyrate, either as numerous small globules or as a few large globules (Plate 3.1, Fig. 2). Polysaccharides may also accumulate. The synthesis of both compounds is stimulated by a high carbon-nitrogen ratio in the medium.

Acetate, alanine, asparagine, aspartic acid, butanol, butyrate, citrate, ethanol, fructose, fumarate, galactose, glucose, glutamine, glutamic acid, glycerol, β-hydroxybutyrate, lactate, malate, maltose, mannitol, pyruvate, sorbitol, succinate and sucrose are utilized as carbon and energy sources for growth (Stokes, 1954; Höhnl, 1955; Mulder and van Veen, 1963). However, different strains differ widely in their capacity to dissimilate the above-mentioned carbon compounds. In contrast with most *Leptothrix* strains, *Sphaerotilus natans* utilizes relatively high concentrations of assimilable substrates from which it synthesizes considerable amounts of cellular material.

Genus **Leptothrix** *Kützing 1843, 198*

E. G. MULDER

(*Detoniella* Trevisan in de Toni and Trevisan 1889, 929; *Chlamydothrix* Migula 1900, 1030.)

Lep′to.thrix. Gr. adj. *leptus* fine, small; Gr. n. *thrix* hair; M.L. fem.n. *Leptothrix* fine hair.

Straight rods, 0.6–1.5 by 3–12 μm, occurring in chains within a **sheath** or **free-swimming** as single cells, in pairs, and in some species as motile short chains containing up to eight cells. One species has well developed holdfasts. **Sheaths have a pronounced tendency to become impregnated or covered with hydrated ferric or manganic oxides. Free cells motile by means of one polar flagellum** (Plate 3.2, Fig. 1); one species has a subpolar tuft of several flagella (Table 3.1). Most species may contain globules of poly-β-hydroxybutyrate as intracellular reserve material (Plate 3.2, Fig. 3). Resting stages are not known. **Gram-negative.**

Chemoorganotrophs: **metabolism respiratory,** never fermentative. Molecular oxygen is the universal electron acceptor. A variety of organic acids and sugars can be utilized as sources of carbon and energy but the number of assimilable compounds is less and the response usually poorer than with organisms of the genus *Sphaerotilus*. Most strains require organic compounds as the nitrogen source (peptone or casamino acids). A mixture of aspartic and glutamic acids can also be utilized when vitamin B_{12} is present. Few strains assimilate inorganic nitrogen compounds. For some strains a requirement for biotin and thiamine has been reported (Rouf and Stokes, 1964). All isolated species oxidize manganous compounds to manganic oxide in contrast to members of the genus *Sphaerotilus*.

Strict aerobes, but growth and manganese oxidation may proceed at low oxygen tensions.

Temperature range: 10–35 C; optimum between 20 and 25 C.

The **G + C content** of the DNA of one species tested was **69.5 ± 0.5 moles** % (Mandel *et al.*, 1966).

Type species: *Leptothrix ochracea* (Roth) Kützing 1843, 198.

Further Comments

A number of species now placed in the genus *Leptothrix* should be transferred, according to some authors, to *Sphaerotilus*, eliminating the name *Leptothrix* (Pringsheim, 1949; Stokes, 1954; Höhnl, 1955). Pringsheim (1949) cultivated *Sphaerotilus natans* in ferrous-ammonium-citrate-containing soil extract. Under these conditions the organism resembled *L. ochracea* and Pringsheim concluded that *L. ochracea* was the iron-bacterium modification of *S. natans*. There is no proof of this hypothesis since all the efforts made by Pringsheim to isolate *S. natans* from iron hydroxide precipitates in which *L. ochracea* was seen microscopically failed. Stokes (1954) isolated *S. natans* from iron-containing ditch water by enrichment culture technique and Höhnl (1955) by plating. No efforts were made by these authors to demonstrate that the isolated *Sphaerotilus* strains were identical with *L. ochracea* by cultivating the former strains in flowing ferrous iron-containing ditch water.

The present authors have grown several strains of *Sphaerotilus natans* and of *Leptothrix* species in continuous culture in slowly running ferrous iron-containing soil extract and compared them with crude cultures of *L. ochracea* cultivated simultaneously in the same apparatus (Mulder and van Veen, 1963 and 1965; Mulder, 1964). Although under these conditions *S. natans* resembled *L. ochracea*, both organisms were considered to be not identical for the following reasons: (a) the sheaths of *S. natans* in the latter experiment were much longer than those of the crude culture of *L. ochracea* which apparently were more brittle; (b) the tendency of the *S. natans* cells to leave their sheaths was much less pronounced than that of cells of *L. ochracea* growing in crude culture; (c) *S. natans* was easily reisolated from its poorly developed iron-bacterium stage but not from the bulky enrichment culture of *L. ochracea*. The present authors are therefore of the opinion that *S. natans* and *L. ochracea* are not identical, as is suggested by Pringsheim (1949). The differences between *Sphaerotilus* and *Leptothrix* species are large enough for recognition of both genera.

Chemolithotrophy in sheath-forming iron bacteria: The concept of lithotrophy in sheath-forming iron bacteria has been a subject of controversy ever since it was put forward by Winogradsky in 1888. Winogradsky, Molisch (1910), Lieske (1919), Cholodny (1926), Präve (1957) and others have been unable to prove that organisms of the *Leptothrix* type are able to utilize the energy liberated during the oxidation of ferrous or manganous ions. The fact that these organisms only grow at pH values of 6 and higher, i.e. values at which ferrous ions are oxidized rapidly by purely chemical reactions, makes the study of the mechanism of the iron oxidation by these bacteria no easy task. In addition to the possible oxidation of the ferrous iron by the cells or by the slime layer surrounding the sheaths, it is highly probable that in many cases the so-called biological iron oxidation by these bacteria is confined to absorption of chemically oxidized iron by the sheaths or the slime layer surrounding the sheaths.

To study the principles of the oxidation of iron and manganese ions by sheath-forming iron bacteria, manganous compounds represent a more suitable substrate because they are oxidized chemically only at a strongly alkaline reaction (pH 8 and higher depending on the presence of catalytically active compounds). Organisms of the *Leptothrix* group, however, are able to convert manganous ions readily to manganic oxide at pH 6–7.5. Evidence is available that this conversion is due to the presence outside the sheaths of proteinous substances promoting manganese oxidation. The increased yield of cell nitrogen, observed in

Leptothrix cultures grown in media with organic nutrients, as a result of added manganous ions (Mulder, 1964; Johnson, 1966), has been shown by the present authors to be due to the formation of an insoluble MnO_2-protein complex which is not lost upon washing of the cells as contrasted with the proteinous material of cells grown without Mn^{++} which is lost upon washing.

Description of the species of genus **Leptothrix**

1. **Leptothrix ochracea** (Roth) Kützing 1843, 198. (*Conferva ochracea* Roth 1797, Table V, Fig. 2; *Detoniella ochracea* (Roth) Trevisan in de Toni and Trevisan 1889, 929; *Chlamydothrix ochracea* (Roth) Migula 1900, 1031; *Sphaerotilus natans* forma *ochracea* Pringsheim 1949, 203.)

o.chra′ce.a. Gr. n. *ochra* yellow ochre; M.L. adj. *ochracea* like ochre.

Since *L. ochracea* has never been isolated, only its morphological characteristics are known (Table 3.1).

When growing under natural conditions in iron-containing water it is characterized by the presence of flocculent masses of hydrated ferric hydroxide, containing many yellow-brown, smooth, relatively short, empty sheaths (Plate 3.1, Fig. 9). Colorless and slightly yellow sheaths containing chains of cells may be detected under the micro-scope. These cells may leave their sheaths and form new sheaths at the rate of about 1 μm/min (Plate 3.2, Fig. 2). In enrichment cultures in slowly running ferrous iron-containing soil extract, it is not exceptional to find more than 95% of the sheaths without cells. Thickening of sheaths with iron takes place after the cells have left the sheaths.

The organism is common in slowly running, iron-containing, uncontaminated, fresh water all over the world.

Comment: Molisch (1910), Lieske (1919) and Winogradsky (1922) isolated sheathed iron bacteria from crude cultures of *L. ochracea* and although they named them *L. ochracea* they were dealing with typical *L. discophora*. Cataldi (1939) isolated several sheathed iron bacteria, one of which was described as *L. ochracea*; the present authors be-

TABLE 3.1

Differential characteristics of species of **Leptothrix**[a]

	1. *L. ochracea*	2. *L. pseudo-ochracea*	3. *L. discophora*	4. *L. cholodnii*	5. *L. lopholea*
Cells					
Width, μm	0.8–1.0	0.8–1.3	0.6–0.8	0.9–1.3	1–1.4
Length, μm	2–3	5–12	2–4	2–5	3–6
Flagella					
Monotrichous, Polar		+	+	+	−
Polytrichous, Sub-polar		−	−	−	+
Sheaths in:					
Fe^{++}-containing soil extract	s	s–u	u	u	s
Mn^{++} medium		g	e	e	g
Mn^{++} oxidation		+	+	+	+(ret.)
Many holdfasts	−	−	−	−	+
Cohesion of sheaths in culture solution		+	−	−	−
Form and size (mm) of colonies on:					
Peptone-gluc. agar[b]		f–s(5)	s(0.1–0.3)	s(5)	f–s(1)
Mn-agar[c]		f(10)	s(0.5–2)	f–s(3–4)	s(2–3)
S-R dissociation by mutation		±	±	+	±
Response to nutrients		±	−	+	−

[a] Abbreviations: s, smooth; u, uneven, owing to precipitated ferric oxide; g, covered with small granules of MnO_2; e, encrusted with large masses of MnO_2; ret., retarded; f, filamentous.

[b] Basal medium containing 1 g of peptone and 1 g of glucose/liter.

[c] Nutrient medium containing 2 g of $MnCO_3$/liter.

lieve this organism was probably *L. pseudo-ochracea*.

2. Leptothrix pseudo-ochracea Mulder and van Veen 1963, 135.

pseu.do.o.chra′ce.a. Gr. adj. *pseudes* false; M.L. adj. *ochracea* specific epithet: M.L. adj. *pseudo-ochracea* not the true (*Leptothrix*) *ochracea*

Rods in chains within thin sheaths or free-swimming. On basal agar containing 0.1% peptone and 0.1% glucose, the sheathed cells grow in concentric rings (Plate 3.1, Fig. 8). Free cells are very motile by means of one polar flagellum; even free chains of as many as eight cells may show motility. In slowly running ferrous iron-containing soil extract (artificial ditch water) the sheaths become impregnated with hydrated ferric hydroxide and turn yellow-brown (Plate 3.1, Fig. 7). Their surface is smooth or slightly granular. Under these conditions many of the sheaths are empty. In media with manganous compounds the sheaths are covered with small granules of MnO_2. On Mn^{++}-containing agar the brown colonies are very filamentous and may exceed a width of 10 mm.

In basal media containing peptone or aspartic and glutamic acids as the nitrogen source and glucose as the carbon compound, growth is moderate.

The normal habitat is slowly running, unpolluted, fresh, iron-containing ditch, river or pond water. May also be found in weakly polluted water.

3. Leptothrix discophora (Schwers) Dorff 1934, 31. (*Megalothrix discophora* Schwers 1912, 273; *Leptothrix crassa* Cholodny 1924, 294.)

dis.coph′or.a. Gr. n. *discos* a disc; Gr. adj. *phoros* bearing; M.L. adj. *discophora* disc-bearing.

Rods, 0.5–1 by 1–6 μm, mostly 0.6–0.8 by 2–4 μm, in chains or as separate cells, in narrow sheaths of uniform width (approximately 1 μm). Free cells are motile by means of one thin polar flagellum (Plate 3.2, Fig. 1).

On glucose-peptone agar the sheaths are very thin; the colonies are small, often no more than 0.1–0.3 mm in diameter, with smooth edges. The manganese-oxidizing capacity of the organism is very pronounced. In the presence of manganous salts the sheaths are heavily encrusted with MnO_2 giving rise to sheaths of sometimes more than 10-μm thickness (Plate 3.2, Fig. 5). In running iron-containing soil extract the sheaths are surrounded with large amounts of flocculent ferric hydroxide (Plate 3.2, Fig. 6). On agar containing manganous ions the black-brown colonies are somewhat larger (0.5–2 mm) and sometimes filamentous.

Cells may contain granules of poly-β-hydroxybutyrate as reserve material.

Although requiring organic substrates for growth, cell yield is relatively poor with almost no response to increased concentration of nutrients.

The normal habitat is slowly running, unpolluted, iron-containing ditch, river or pond water.

4. Leptothrix cholodnii Mulder and van Veen 1963, 137.

cho.lod′ni.i. M.L. gen.n. *cholodnii* of Cholodny; named for N. Cholodny, a Russian bacteriologist.

Cells usually 1–1.3 by 2–5 μm (Plate 3.2, Fig. 3). Single cells motile by means of one thin polar flagellum. Chains of cells occur in sheaths which in the presence of manganous ions are heavily encrusted with manganic oxide (Plate 3.2, Fig. 4).

Colonies on peptone-glucose agar may be up to 5 mm in diameter. There is a strong tendency in many strains to dissociate spontaneously and to produce smooth colonies instead of the usual rough. The smooth strains oxidize manganous ions less vigorously than the rough strains (Mulder and van Veen, 1963; Rouf and Stokes, 1964; Stokes and Powers, 1965).

In contrast to other species, growth is stimulated by high concentration of organic nutrients.

May be isolated from running, unpolluted iron-containing fresh water as well as from polluted running water and from activated sludge.

5. Leptothrix lopholea Dorff 1934, 33.

loph.o.le′a. Gr. n. *lophos* a crest; M.L. dim.fem. adj. *lopholea* somewhat crested or tufted.

Single cells motile by means of a tuft of several flagella which are attached subpolarly. In this respect they resemble *Sphaerotilus natans*. Cells usually occur in short filaments radiating from a cluster of holdfasts (Plate 3.2, Fig. 7), giving rise to many tiny flocks when growing in liquid media. Under such conditions the accumulation of manganic and particularly ferric oxides is much more pronounced on the cluster of holdfasts than on the filaments (Plate 3.2, Fig. 8). In liquid cultures, false branching of filaments occurs regularly.

On agar media containing manganous ions, incrustation of sheaths with manganic oxide is retarded. As a result of this retardation, the colonies are first white, later becoming black-brown.

May be isolated from slowly running, polluted or unpolluted fresh water and from activated sludge.

Species incertae sedis

The majority of the 12 species of *Leptothrix* listed in the 7th edition of THE MANUAL have never been isolated and their descriptions have been based on crude cultures occurring under natural conditions. *Index Bergeyana* lists names of several other poorly defined organisms; some perhaps are synonyms of existing species of *Leptothrix*, others are unrelated organisms. In view of the variability of these organisms when grown under different conditions, it is highly probable that several of the names refer to the same organism.

A. Probably synonymous with accepted species:

 a. *Leptothrix skujae* Beger in Beger and Bringmann 1953, 331, and

 b. *L. pseudovacuolata* (Perfil'ev) Dorff 1934, 36, are presumably identical with *L. discophora* (Schwers) Dorff 1934, 31. The false vacuoles described as a typical characteristic of *L. pseudovacuolata* are apparently globules of poly-β-hydroxybutyrate, the reserve material of most *Leptothrix* strains.

 c. *Leptothrix thermalis* (Molisch) Dorff 1934, 38, might be a thermophilic variety of *L. discophora*.

 d. *Leptothrix sideropous* (Molisch) Cholodny 1926, 25.

 e. *L. echinata* Beger 1935, 401, may be closely related to or identical with *L. lopholea*.

B. Insufficiently described:

 f. *Leptothrix major* Dorff 1934, 35. Sheathed bacterium with large cells, 1.4 by 5–10 μm, showing false branching. Sheaths, partly irregularly encrusted with ferric oxide, may be attached to shells or stones.

 g. *Leptothrix winogradskii* Cataldi 1939, 64. Cells 0.9 μm in diameter. Motile, presumably polarly flagellated. Sheaths 1.5 μm thick; never attached. Colonies on iron-ammonium-citrate agar very filamentous.

C. Probably blue-green algae:

 h. *Leptothrix volubilis* Cholodny 1924, 297, and

 i. *Leptothrix epiphytica* (Migula) Schönichen and Kalberlah 1900, 46, may be blue-green algae belonging to the genus *Lyngbya*.

Genus **Streptothrix** Cohn 1875, 186, emend. mut. char. Migula 1895, 36

E. G. Mulder and W. L. van Veen

(Not *Streptothrix* Corda 1839, 27 (a fungus); not *Streptothrix* Cohn 1875, 186 (an actinomycete); *Chlamydothrix* Migula 1900, 1030.)

Strep′to.thrix. Gr. adj. *streptos* pliant; Gr. n. *thrix* hair; M.L. fem.n. *Streptothrix* pliant hair.

Thin rods, 0.35–0.45 by 3.2–4.6 μm, **occurring in chains in hardly visible hyaline sheaths** of 0.5–0.8 μm width (Plate 3.2, Figs. 9 and 10). **Free cells** outside the sheath occur only sporadically. Motility and flagella of these cells have not been observed. **No ferric** or **manganic oxides** present in or on the sheaths. **Branching** of the filaments may incidentally occur in stationary cultures. **Lateral branches** generally very **short** as compared with main filament. **No resting stages known. Gram-negative.**

Colonies on poor agar media <0.5 mm, highly filamentous; on a peptone, sucrose, vitamin medium 1–3 mm, faintly pink with slightly filamentous edge. Fast-growing cultures in liquid media turn pink owing to the presence of carotenoid pigments.

Chemoorganotrophs: **metabolism respiratory,** never fermentative. Molecular oxygen is the universal electron acceptor. Glucose, lactose, sucrose and to a less extent mannitol are used as sources of carbon and energy. Nitrates and ammonium compounds are moderately good nitrogen sources.

Better growth is obtained with organic nitrogen sources: glutamate or peptone. Thiamine and vitamin B_{12} are required for growth. Starch and gelatine are hydrolyzed.

Strict aerobes. Temperature range: 8–30 C; optimum: 25–27 C.

Widely distributed in fresh water and in activated sludge. Occurring in large masses in activated sludge loaded with residual water from meat industries or pig farms. Straight, sheathed chains of bacteria extending from the flocs interfere with settling of the sludge.

Type species: *Streptothrix hyalina* Migula 1895, 38.

Further Comments

Until recently strains have not been available. Four strains, which correspond morphologically with Migula's description of this species, were isolated recently by van Veen. The generic description is based on these strains. Further classification must await more study.

Description of the species of genus **Streptothrix**

1. **Streptothrix hyalina** Migula 1895, 38. (*Chlamydothrix hyalina* (Migula) Migula 1900, 1033; *Leptothrix hyalina* (Migula) Schönichen and Kalberlah 1900, 46.)

hy.a.li′na. Gr. adj. *hyalinos* glassy; M.L. fem.adj. *hyalina* glassy, hyaline.

Description as for genus.

Genus **Lieskeella** *Perfil'ev 1927, 335*

PETER HIRSCH

Lies.ke.el′la. M.L. dim.ending -*ella*; M.L. fem.n. *Lieskeella* named for Lieske, a German microbiologist.

Cells rod-shaped with rounded ends, 0.6 by 2–3 μm, in chains. Show bipolar staining when treated with methylene blue. Usually two chains of cells (filaments) are wound around one another to give a double spiral, which is surrounded by a yellowish, slimy capsule, often with heavy deposits of small granules of ferric hydroxide. When the deposits are dissolved with dilute HCl, the cell chains fragment and the individual rods appear as a double zigzag band. Cell chains may aggregate in distinct layers or even in more or less solid skeins. The filaments separate rapidly upon removal from the normal environment.

There is a slow but incessant motion similar to that of cyanophytes.

Have not been obtained in pure culture. Gram-reaction not recorded.

Originally found in the upper layers of mud in bodies of water around Alt-Peterhof, Russia; also observed by Perfil'ev and Gabe (1961) in several other locations in western Russia throughout the year, usually as thin layers of up to 15 chains. Observed at depths of 0.5 to 1.2 meters in the littoral zone of Kristatellevyi Pond, Russia; also in the mud water interface of Lake Windermere, England (Skerman, personal communication).

Type species: *Lieskeella bifida* Perfil'ev 1927, 335.

Description of the species of genus **Lieskeella**

1. **Lieskeella bifida** Perfil'ev 1927, 335.
bi′fi.da. L. adj. *bifida* cleft, divided.
Description as for genus.
Comment: Perfil'ev and Gabe (1961) point to an "extraordinary instability of the cells of this organism." The use of any of the usual fixatives,

even greatly diluted, causes the threads to separate explosively into cells which instantaneously shorten, swell and burst into granular remains. A similar instability has been reported for *Toxothrix trichogenes* (Krul *et al.*, 1970).

Genus **Phragmidiothrix** *Engler 1883, 192*

PETER HIRSCH

Phrag.mi.di′o.thrix. Gr. n. *phragma* fence; Gr. n. *eidus* form, shape; Gr. n. *thrix* hair; M.L. fem.n. *Phragmidiothrix* fence-like hair.

Filaments are articulate, **unbranched,** colorless and over 100 μm long. They are **attached,** forming grayish white tufts. The free end may be of larger diameter than the base, but there is no tapering in either direction: the diameter varies between 3 and 6 μm. Surrounded by a very thin, delicate, gelatinous, colorless sheath, which is not encrusted with iron or manganese compounds. The **cells are of variable size, usually small and disc-shaped,** the diameter being 4–6 times the thickness of the cell; cell walls are distinct and of even thickness throughout the filament. **Multiplication is by cross-septation and in certain regions of the filament by both cross- and longitudinal septation,** forming *Sarcina*-like aggregates of small, nearly cubical propagation

cells. Septating cells may be of greater diameter causing localized swelling of the filament; in areas septating in both planes the filaments may be up to 6 μm in diameter (Plate 3.3*A*).

Has not been cultivated on artificial media in pure culture.

Originally found attached to the surface of living *Gammarus locusta* collected from the anaerobic, H_2S-containing, polluted area called "Weisser Grund" or "Todten Grund" in the Kieler Förde, Germany. Apparently not rare; have been reported on seaweed from polluted water of the Northern Adriatic Sea (Beger, 1957).

Type species: *Phragmidiothrix multiseptata* Engler 1883, 192.

Description of the species of genus **Phragmidiothrix**

1. **Phragmidiothrix multiseptata** (Engler) Engler 1883, 192. (*Beggiatoa multiseptata* Engler 1883, 19.)

mul.ti.sep.ta′ta. L. masc.n. *multus* much; L. adj. *septatus* fenced; M.L. fem.adj. *multiseptata* much fenced, with many septa.

Morphology and characteristics as for genus. Comment: Engler reported outgrowths of slightly curved filaments of 4 to 10 cells, perpendicular to the main filament and seemingly a continuation of the adjacent cell. He suggested propagation cells in a particular row had grown out to form these thin, slightly curved twiglike outgrowths.

Multiple septation as a mode of reproduction is also found in *Crenothrix* and has been observed in the genera *Dermatophilus* and *Geodermatophilus*. Therefore, *Phragmidiothrix* may possibly be related to Gram-positive rather than Gram-negative bacteria. Further work is needed, especially on the cultivation of these organisms.

Genus **Crenothrix** *Cohn 1870, 108*

Peter Hirsch

Cre'no.thrix. Gr. n. *crenus* a fountain, spring; Gr. n. *thrix* a hair; M.L. fem.n. *Crenothrix* fountain hair.

Filaments up to 1 cm long, attached to a firm substrate and may be swollen at the free end. **Unbranched** but may show what appears to be false branching. The **very thin sheaths** surrounding the filaments may be colorless at the tip or encrusted with iron (or manganese) oxides at the base. **Cells cylindrical to disc-shaped, dividing by cross-septation in normal filaments and by cross- and longitudinal septation at the tips of enlarged filament ends.** Cells in enlarged filament ends are smaller and may round up; some-

times there is only one row of these present. Larger rod-shaped cells may also slip out of the sheath and form new filaments.

Gram-reaction has not been recorded. Not motile.

Has not been grown on artificial media in pure culture.

Found in stagnant and running waters containing organic matter and iron salts.

Type species: *Crenothrix polyspora* Cohn 1870, 131.

Description of the species of genus **Crenothrix**

1. **Crenothrix polyspora** Cohn 1870, 108.

po.ly.spo'ra. Gr. adj. *poly* many; Gr. n. *sporus* a seed; M.L. n. *spora* a spore; M.L. fem.adj. *polyspora* many-spored.

The diameter of individual filaments may vary from 1–6 μm at the base to 6–9 μm at the tip when swollen. Each filament is surrounded by a generally colorless sheath, which later may become rust-colored, especially at the base, due to iron deposition.

Cells within the filament appear more or less rectangular in shape (Plate 3.4, Fig. 1). Reproduction may occur either by transverse fission in the sheath or by both transverse and longitudinal fission at the swollen filament tip (Plate 3.4, Figs. 2–4). Reproductive cells have been termed "conidia," but their mode of formation appears to be different from that of true conidia in fungi. The smallest reproductive cells are usually rounded and 1–2 μm in diameter; occasionally they are only 0.6–0.8 μm wide. Larger reproductive cells may be up to 5 μm in diameter. After release, the reproductive cells may germinate upon the exterior of the terminal portion of the sheath, giving rise to new filaments that may simulate false branching. The reproductive cells may be held together by a slimy substance to form a zoogloeal mass.

Originally isolated from samples of spring water near Breslau, Germany. Widespread in water and

drainage pipes and in springs where the water contains iron and organic matter; may be found in stagnant and running waters and in city water supplies, where it grows as thick brownish masses.

Further Comments

According to Cohn (1870) reproduction may also occur by outgrowth of swollen, terminal cells of ellipsoidal shape as much as 7 times longer than wide; the short, colorless, *Oscillatoria*-like filaments growing out of such swollen cells had a characteristic slow gliding motion and lacked a clearly defined sheath. Subsequent authors usually fail to mention this type of reproduction.

Cohn also indicated the tip of the sheath was closed and retained the reproductive cells until rupture (see Figs. 9 and 10 in original description). This author has found no evidence for closed sheaths.

Other workers have described heavy iron and manganese deposition at the filament base of *C. polyspora*; Wolfe (1960) found only traces of ferric iron deposition and pointed out certain similarities of this organism with *Sphaerotilus* spp. which are known to occasionally deposit iron and/or manganese at their filament bases.

Cholodny (1926) believed this species to be identical with *Clonothrix fusca* Roze 1896, 325. However, the latter organism is characterized by tapering filaments, lack of typical reproductive cells and by frequent false branching. Kolk (1938) has described the differences clearly.

The observation of multiple division in several planes near the tips of *Crenothrix* filaments may point to a relationship with certain *Actinomycetales*

with similar reproductive processes, e.g. *Dermatophilus* and *Geodermatophilus*, and with *Blastococcus* spp. (Ahrens and Moll, 1970). Certain similarities with *Phragmidiothrix multiseptata* may well indicate the existence of a whole group of organisms capable of multiple septation.

Genus **Clonothrix** *Roze 1896, 329*

PETER HIRSCH

Clo'no.thrix. Gr. n. *clon* twig; Gr. n. *thrix* hair; M.L. fem.n. *Clonothrix* twig hair.

Filaments up to 1.5 cm long, attached or free, surrounded by a **more or less distinct sheath** which may be encrusted with iron or manganese compounds giving a yellowish brown color. Filaments tapering; they may be single or **with false branches.** Cells cylindrical, colorless or bluish; cell masses may appear in shades of brown. **Reproduction by separation of individual cells,** followed by breakage of the sheath and re-

lease to the outside; such cells frequently attach themselves parallel to the older cell and produce new filaments giving rise to "pseudo-branching." Most of these morphological features were noted in Roze's original drawings (Plate 3.3B).

Have not been grown in pure culture on artificial media.

Type species: *Clonothrix fusca* Roze 1896, 330.

Description of the species of genus **Clonothrix**

1. Clonothrix fusca Roze 1896, 330.

fus'ca. L. fem.adj. *fusca* dark, tawny.

Ensheathed filaments taper toward the tips; one or more may be attached to a common base. Older parts of the filaments appear yellowish brown (diameter 7 μm) and younger ones are faintly bluish (diameter 3 μm). Multiply by false branching, the "pseudo-branches" alternating. Cells 2–2.5 by 12–18 μm, being larger at the base and smaller at the tip of the filament. Rarely apparently unicellular ampoule-like bodies are produced, subterminally, by a budding process from cells of certain filaments. These may serve as a means of propagation but the mechanism is unknown.

Originally found attached to iron fittings in well water at 14 C, containing *Aplococcus natans.*

Comment: Originally described by Roze as a blue-green alga because of its faintly bluish coloration but subsequent authors have failed to find a pigment.

Schorler (1904) described an organism very similar to *Clonothrix fusca* Roze and independently gave it the same name.

Beger and Bringman (1953) compared *C. fusca* with *Glaucothrix putealis* and on the basis of similar size and the occurrence of false branching in both, concluded that *Clonothrix fusca* Roze 1896 was a later synonym of *Glaucothrix putealis* Kirchner 1878. However, a decision on this matter must be postponed until at least one organism, and prefer-

ably both, has been cultivated and studied in greater detail. Currently the relative importance (for taxonomic purposes) of the characteristics given is not known.

2. Clonothrix gracillima West and West 1898, 337.

gra.cil'li.ma. L. fem.superl.adj. *gracillima* most slender.

Filaments thin and never taper toward the tip. Sheath narrow, firm and colorless. Cells cylindrical, 1.5 by 4.5–6.0 μm with a bright bluish color. Pseudo-branches are elongated, flexible and alternating with alternating secondary pseudo-branches.

Originally isolated from a horse trough.

Comment: Beger and Bringmann (1953) considered this organism to be a species of *Leptothrix.* The lack of tapering filaments is not in disagreement with the description of the genus, although the type species does have tapering filaments. Discussion of its validity must await cultivation in pure culture.

Species incertae sedis

Several little understood organisms have been placed in this genus, but only one need be mentioned.

a. *Clonothrix tenuis* Kolkwitz 1909, 144. (*Crenothrix tenuis* (Kolkwitz) Dorff 1934, 42.) Observed in sewage.

Plate 3.1. *Sphaerotilus* and *Leptothrix*

Fig. 1. *Sphaerotilus natans* in nutrient medium containing ferric iron as the quinic acid complex; sheaths and holdfasts. × 812.

FIG. 2. *Sphaerotilus natans* in basal culture solution containing 2.5 g of glucose and 1 g of peptone/ liter. Cells within sheaths, as in rough colonies (Fig. 4). Globular inclusions are poly-β-hydroxybutyrate. × 812. From Poindexter, J. S. 1971. Microbiology, An Introduction to Protists. The Macmillan Company, New York.

Fig. 3. Flagella of *Sphaerotilus natans*. × 10,360. From Sykes, G. and Skinner, F. A. (editors). 1971. Microbiol aspects of pollution. The Society for Applied Bacteriology: Symposium Series 1. Academic Press, London.

Fig. 4. *Sphaerotilus natans*. Rough colony on basal agar containing 1 g of glucose and 1 g of peptone/ liter. × 14.5. From Antonie van Leeuwenhoek, 29, 121–153 (1963).

Fig. 5. *Sphaerotilus natans*. Smooth colony on basal agar containing 5 g of glucose and 5 g of peptone/ liter. × 14.5. From Antonie van Leeuwenhoek, 29, 121–153 (1963).

Fig. 6. *Sphaerotilus natans*. Cells without sheaths from a smooth colony (Fig. 5). × 812. From Antonie van Leeuwenhoek, 29, 121–153 (1963).

Fig. 7. *Leptothrix pseudo-ochracea* in slowly running sterilized iron-containing soil extract. × 812. From Antonie van Leeuwenhoek, 29, 121–153 (1963).

Fig. 8. *Leptothrix pseudo-ochracea* on basal agar medium containing 1 g of glucose and 1 g of peptone/ liter, showing circular growth. × 812. From Antonie van Leeuwenhoek, 29, 121–153 (1963).

Fig. 9. Empty sheaths of an enrichment culture of *Leptothrix ochracea* in slowly running iron-containing soil extract (artificial ditch water). × 812. From Journal of Applied Bacteriology 27, 151–173 (1964).

Plate 3.2. *Leptothrix and Streptothrix*

Fig. 1. Flagellum of *Leptothrix discophora*. × 27,100. From Revue d'écologie et de biologie du sol, 9, 1972.

Fig. 2. Formation of empty sheaths by an enrichment culture of *Leptothrix ochracea* in iron-containing soil extract. Cells leave old sheaths and form new sheaths with a rapidity of approximately 1 μm/min. × 812. From Anreicherungskultur und Mutantenauslese. 1965. Supplementheft 1 zum Zentralblatt für Bakteriologie, 1. Abt.

Fig. 3. *Leptothrix cholodnii* containing globules of poly-β-hydroxybutyrate. × 1620.

Fig. 4. Granular deposition of MnO_2 on sheaths of *Leptothrix cholodnii* grown in a basal medium containing 0.2 g of glucose, 0.2 g of peptone and 25 mg of $MnSO_4 \cdot H_2O$/liter. × 812. From Antonie van Leeuwenhoek, 29, 121–153 (1963).

Fig. 5. *Leptothrix discophora* in a basal medium containing 0.2 g of glucose, 0.2 g of peptone and 25 mg of $MnSO_4 \cdot H_2O$/liter. × 812. From Journal of Applied Bacteriology, 27, 151–173 (1964).

Fig. 6. *Leptothrix discophora* in slowly running sterilized iron-containing soil extract (artificial ditch water). × 812. From Antonie van Leeuwenhoek, 29, 121–153 (1963).

Fig. 7. *Leptothrix lopholea* in a basal culture solution supplied with 2.5 g of glucose and 2.5 g of peptone/liter. Chains of cells arising from a cluster of holdfasts. × 812. From Antonie van Leeuwenhoek, 29, 121–153 (1963).

Fig. 8. Iron-impregnated holdfasts of *Leptothrix lopholea* in a culture medium with ferric iron as the quinic acid complex. × 812.

Fig. 9. *Streptothrix hyalina* grown in soil extract enriched with trypticase soy broth, glucose and vitamins. × 325.

Fig. 10. *Streptothrix hyalina*, sheaths and cells. × 24,600.

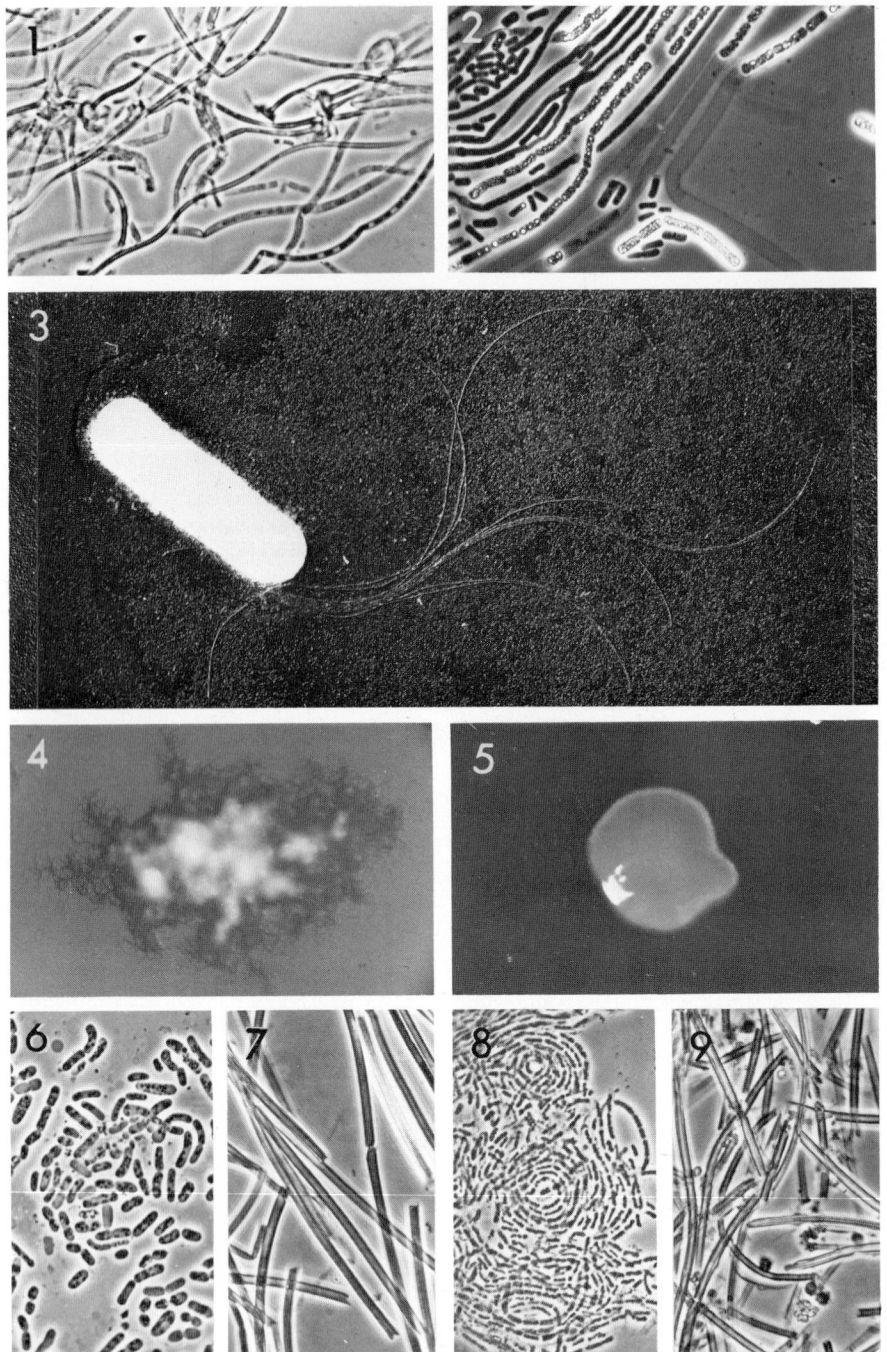

Plate 3.1

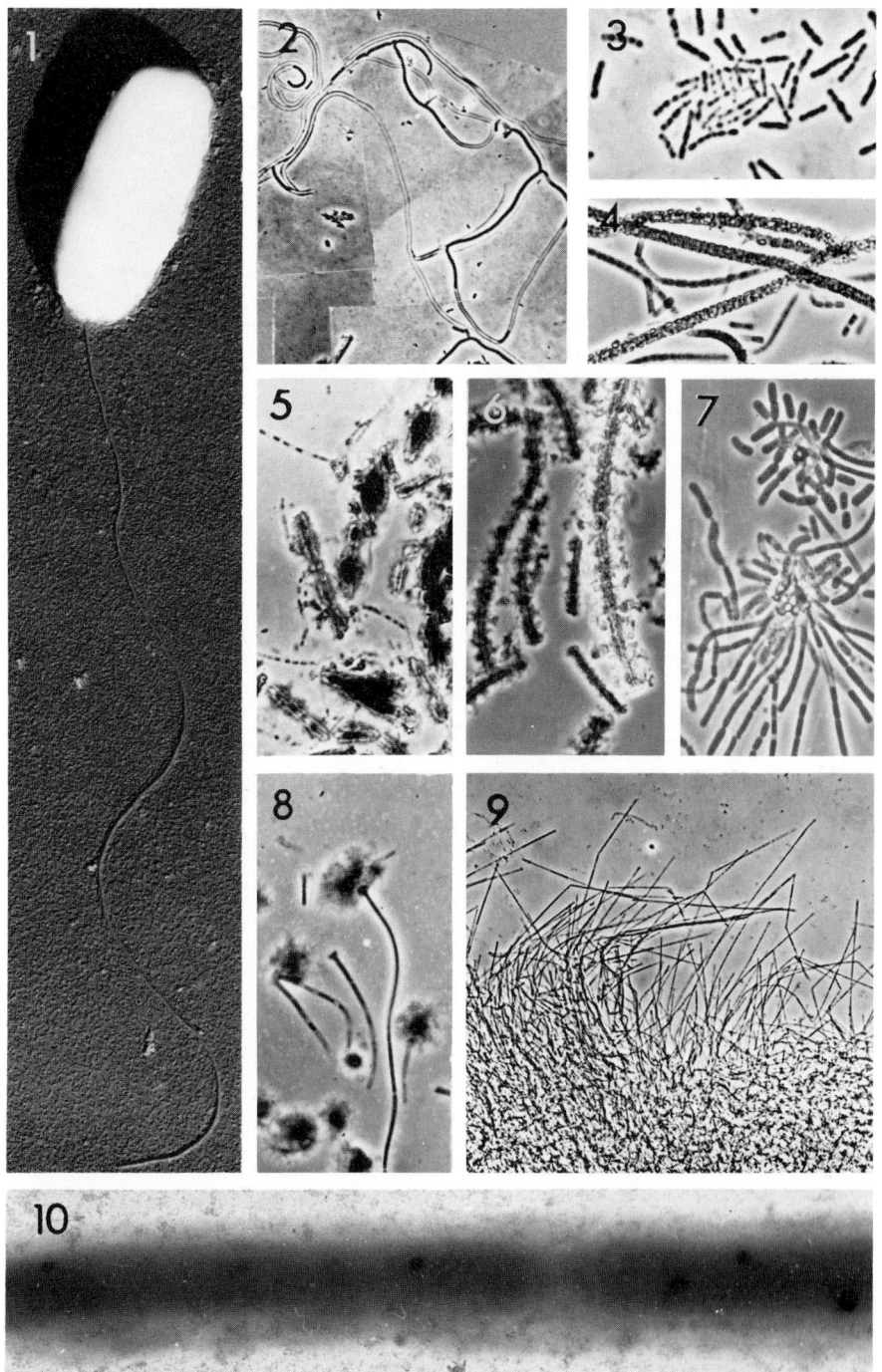

Plate 3.2

Plate 3.3*A*. *Phragmidiothrix multiseptata*. Original line drawings of Engler

Fig. 24. Filament showing arrangement and septation of disc-shaped cells.

Fig. 25. Part of filament showing enlarged end and individual disc-shaped cells of uneven diameter.

Fig. 26. Part of a filament with much cross- and longitudinal septation.

Fig. 26*a* shows a part at higher magnification.

Fig. 27. A bristle of *Gammarus locusta* with two attached filaments of *Phragmidiothrix* (wide filaments) and several narrow filaments of young "*Beggiatoa alba*"?.

Magnification: approximately × 330 (original magnification × 400). Reproduced from Engler (1883).

Plate 3.3*B*. *Clonothrix fusca*

Fig. 1. End of adult filament with false lateral branches.

Fig. 2. End of young filament with false lateral branches.

Fig. 3. Cell emerging from a broken filament.

Fig. 4. Young filament growing from a broken adult filament.

Fig. 5. Mode of insertion of two adult filaments, that on right forming a false lateral branch.

Fig. 6. A filament carrying an ascending series of ampoule-like swellings (buds).

Fig. 7. End of a filament terminating in an ampoule-like swelling.

Fig. 8. A filament terminating in swelling but also carrying a series of lateral swellings.

Figs. 9–11. Buds showing vacuoles, and granules.

Magnification: Figs. 1–4, 6–8 approximately × 365. Fig. 5, 9–11 approximately × 650. Reproduced from Roze, 1896; original magnification 450 and 800, respectively.

Plate 3.4. *Crenothrix polyspora*

Fig. 1. Filaments showing variation in size and shape of individual cells (Wolfe, 1960).

Figs. 2–4. Reproductive cells in terminal swellings. The thin sheath clearly visible in Fig. 4. All living filaments in water; phase contrast photomicrographs by R. S. Wolfe.

Reprinted from AWWA Journal 1960, 52, p. 917 by permission of the Association. Copyrighted 1960 by the American Water Works Association, Inc., 2 Park Avenue, New York, N. Y. 10016.

PHRAGMIDIOTHRIX   A

CLONOTHRIX   B

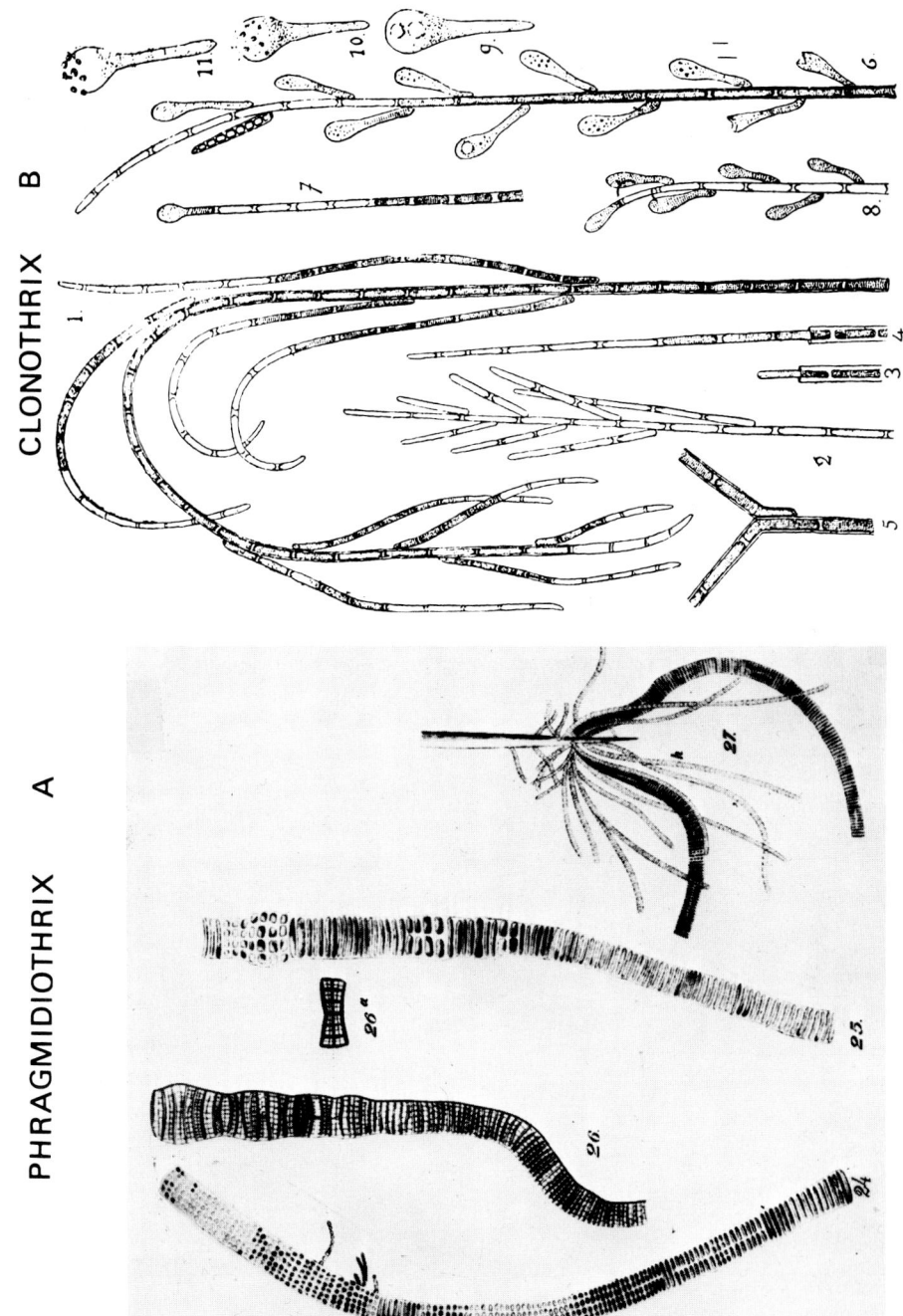

Plate 3.3

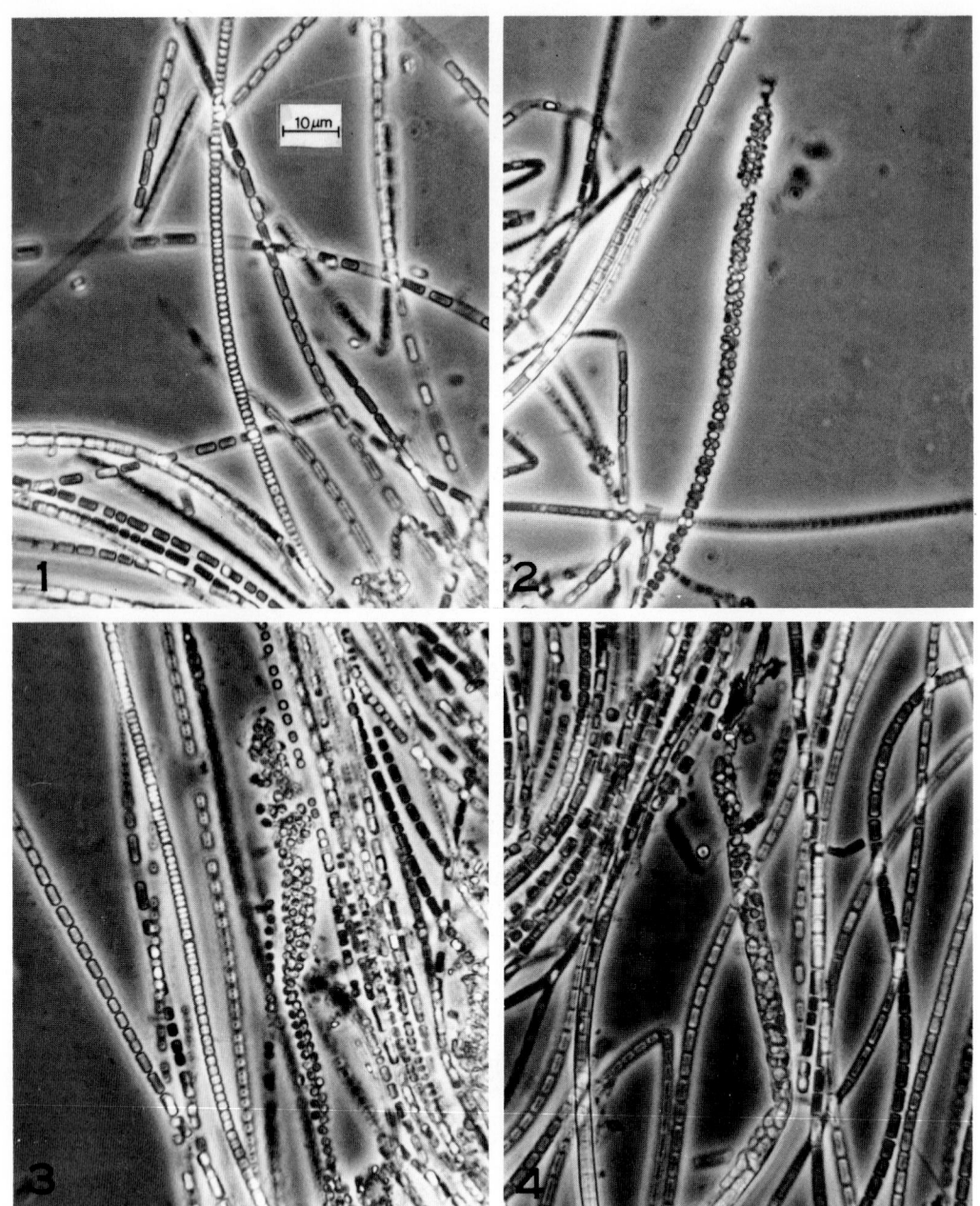

Plate 3.4

Plate 4.1. *Hyphomicrobium* and *Pasteuria*

Fig. 1. *Hyphomicrobium* sp. (strain P-546). Mature cell showing hyphae and terminal bud cells. Isolated from soil. Electron micrograph, negatively stained with phosphotungstate. $\times$ 20,500.

Figs. 2–4. *Pasteuria ramosa*. Spherical bodies multiplying by budding.

Fig. 4. On antennae of *Daphnia* sp. Wet mounts on agar slides. $\times$ 1250. From Hirsch, P. (1972) Int. J. Syst. Bacteriol. *22:* 112–116.

Plate 4.2. *Ancalomicrobium, Prosthecomicrobium, Thiodendron*

Fig. 1. *Ancalomicrobium adetum*. Electron micrograph, negatively stained. From Staley, J. T. (1968) J. Bacteriol. *95:* 1921–1924.

Fig. 2. *Prosthecomicrobium pneumaticum*. Electron transparent inclusions are vesicles of gas vacuoles. Electron micrograph, negatively stained. From Staley, J. T. (1968) J. Bacteriol. *95:* 1921–1924.

Figs. 3 and 4. *Thiodendron* sp. observed in mud and water samples from a pond. Stalks ca. 0.2 μm in diameter. Courtesy Dr. Hans Hippe.

Plate 4.3. *Seliberia*

Figs. 1 and 2. Star-shaped aggregates; stained with erythrosin. $\times$ 2000. Courtesy of T. V. Aristovskaya.

Fig. 3. Aggregates showing "buds" at ends and sides of rod-shaped cells. Phase contrast. $\times$ 2000. Courtesy of T. V. Aristovskaya.

Fig. 4. Electron micrograph showing twisted and aggregated cells. $\times$ 6000. Courtesy of T. V. Aristovskaya.

Fig. 5. *Seliberia* sp. from iron spring in Michigan showing spirally twisted cell and flagellum. $\times$ 50,000. From Hirsch, P. and Pankratz, H. (1970) Z. Allg. Mikrobiol. *10:* 589.

Plate 4.4. *Metallogenium*

Fig. 1. *Metallogenium symbioticum*. Free cells from 48-hr liquid culture. From Zavarzin, G. A. (1964) Z. Allg. Mikrobiol. *4:* 390–395.

Fig. 2. *M. symbioticum*. Part of young microcolony; old filaments are covered with manganese oxide; younger parts are less encrusted. An arai is visible at a rupture of the coating (*b*). From Zavarzin, G. A. (1964) Z. Allg. Mikrobiol. *4:* 390–395.

Figs. 3–5. *M. personatum*. Young colonies forming by germination of reproductive (round) cells. $\times$ 1000.

Fig. 6. *M. personatum*. Formation of reproductive cells on ends of young branches. $\times$ 2000.

Fig. 7. *M. personatum*. Young colonies. Show development of secondary microcolonies. $\times$ 1000.

Fig. 8. *M. personatum*. Young colonies. $\times$ 1000.

Fig. 9. *M. personatum*. Young colonies. Show development of secondary microcolonies. The beginning of mineralization. $\times$ 1000.

Figs. 3–9. From Perfil'ev, B. W., Gabe, D. R., Gal'perina, A. M., Rabinovich, V. A., Sapotnitskii, A. A., Sherman, É. É. and Troshanov, É. P. 1964 *Applied Capillary Microscopy*. Translated by F. L. Sinclair. 1965. Consultants Bureau, New York.

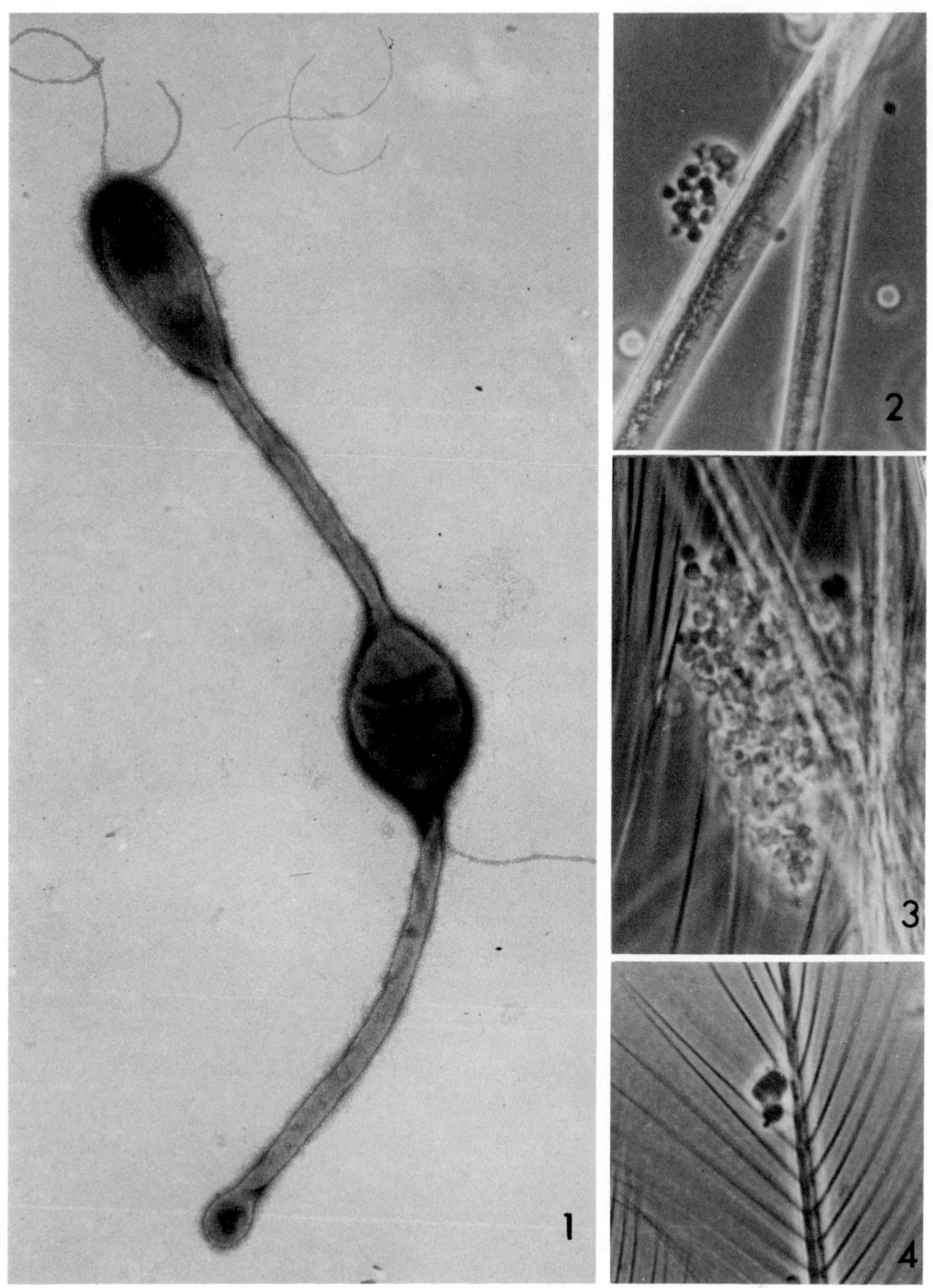

Plate 4.1

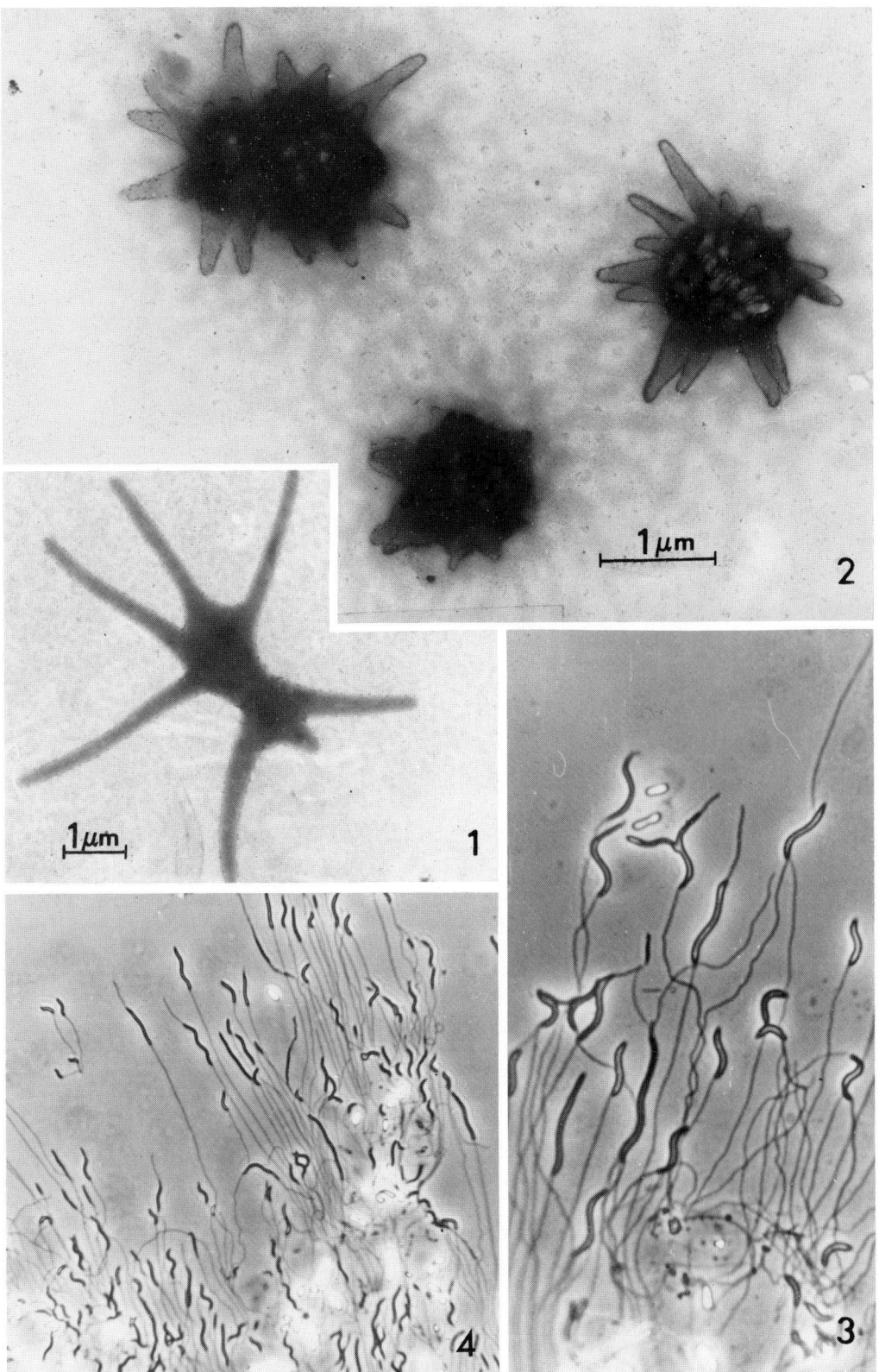

Plate 4.2

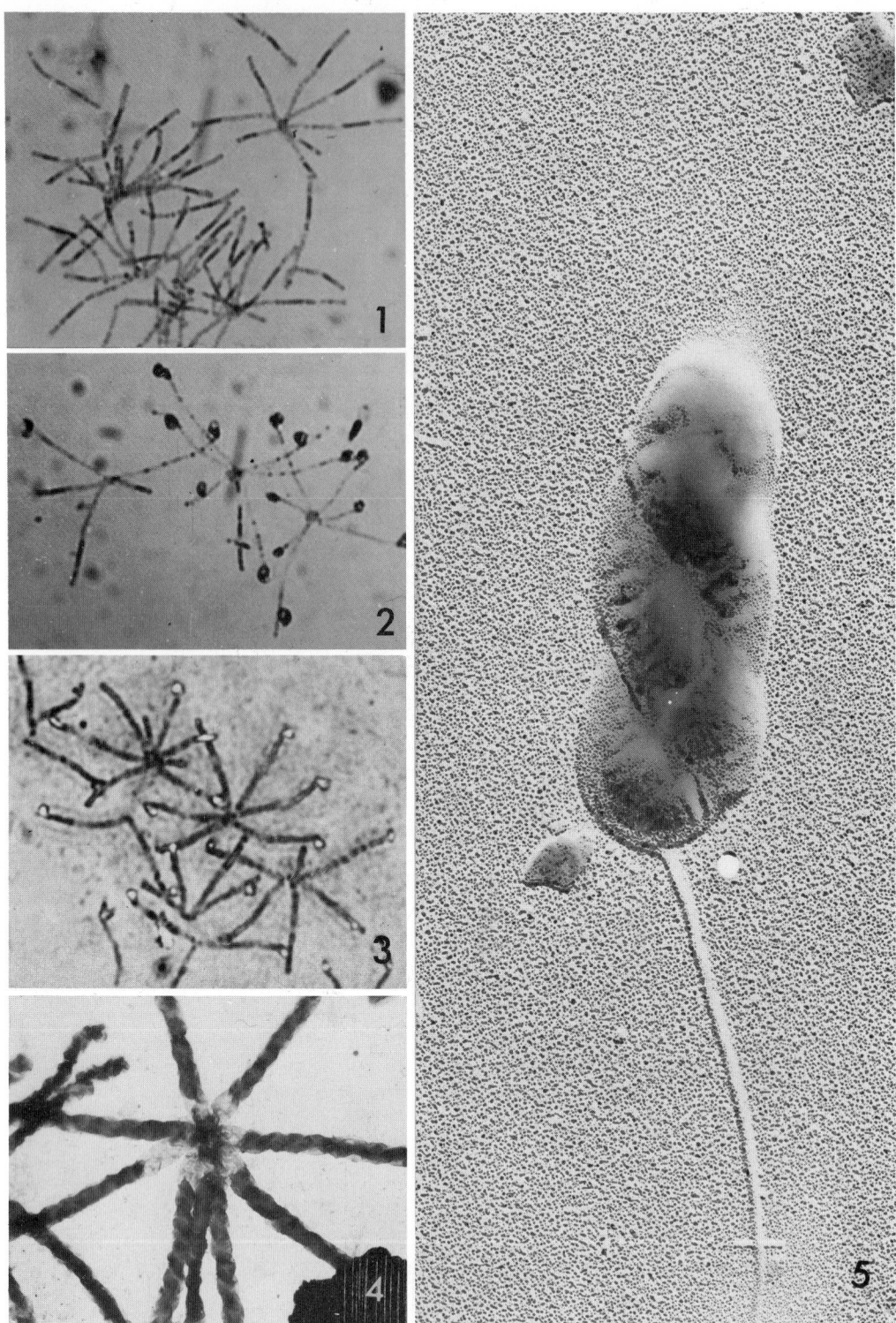

Plate 4.3

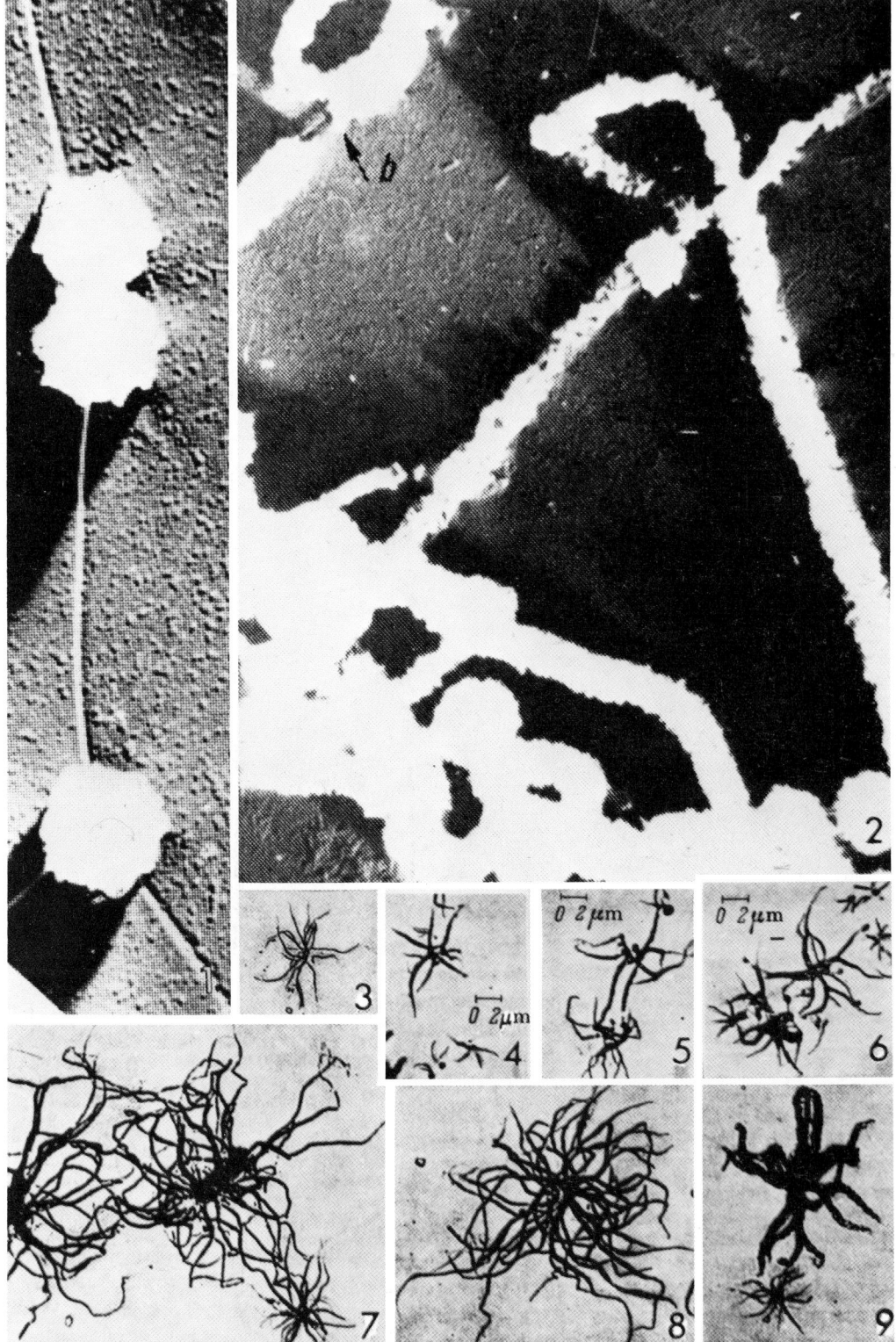

Plate 4.4

PART 4

BUDDING AND/OR APPENDAGED BACTERIA

Partial Key to Part 4

I. Bacteria in which products of binary fission are not equivalent.
  A. Prosthecate bacteria
    1. Prosthecae have a reproductive function; new cell formation by a budding process.
      Hyphomicrobium
      Hyphomonas
      Pedomicrobium
      (see also Part 1—*Rhodomicrobium*)
    2. Prosthecae have no reproductive function.
      Caulobacter
      Asticcacaulis
      Ancalomicrobium
      Prosthecomicrobium
    3. Prosthecae may have a reproductive function.
      Thiodendron
  B. Non-prosthecate bacteria
      Reproduction by a budding process.
      Pasteuria
      Blastobacter
      Seliberia
      (see also Part 1—*Rhodopseudomonas;* Part 7—*Methylomonadaceae;* Part 12—*Nitrobacter*)
II. Bacteria with excreted appendages and holdfasts; products of binary fission may not be equivalent.
  A. Reproduce by binary fission only.
      Gallionella
      Nevskia
  B. Reproduce by a budding process.
      Planctomyces
  Genera of uncertain affiliation
      Metallogenium
      Caulococcus
      Kusnezovia

Genus **Hyphomicrobium** *Stutzer and Hartleb 1898, 76*

PETER HIRSCH

Hy.pho.mi.cro′bi.um. Gr. *hyphe* thread; Gr. adj. *micrus* small; Gr. n. *bius* life; M.L. neut.n. *Hyphomicrobium* thread-producing microbe.

Cells 0.5–1.0 by 1–3 μm; rod-shaped with pointed ends, oval, egg- or bean-shaped forms; produce mono- or bipolar filamentous outgrowths (hyphae) of varying length 0.3–0.4 μm in diameter when stained. The hyphae are not septate but may show true branching. Cells stain well with carbolfuchsin but only feebly with aqueous aniline dyes. Gram reaction not recorded.

Multiplication by budding at tip(s) of hyphae (Plate 4.1, Fig. 1); mature buds become motile, break off and often attach themselves to surfaces or other cells to form clumps. Motility is lost soon after attachment.

Liquid media are never turbid; growth occurs as a surface pellicle or ring, which in older cultures falls to the bottom of the vessel.

On solid media colonies are small, even after long incubation, reflect light from the shiny but granular surface which may show folds or concentric rings; dirty white changing to brownish with age.

Chemoorganotrophic. Carbon dioxide is required for growth. Oligocarbophilic, i.e. growth can occur in a mineral salts medium without added C sources. Growth is stimulated by humus or soil extracts, if the pH is maintained near neutral. Good growth with 0.2% (w/v) formate, acetate, propionate, isobutyrate, valerate, lactate, succinate or mannitol. Slow growth with 0.2% oxalate, glycerol or 1% Liebig's meat extract. No growth with 0.5–1.0% beef extract plus peptone in nutrient or plain gelatin. Fructose and sucrose not utilized; fructose is inhibitory in the presence of KNO_3.

Good growth with NH_4^+, NO_3^- or NO_2^- as N source. Grows in dilute urine or a solution of asparagine but not on asparagine agar.

Does not oxidize ammonia to nitrate. Anaerobic growth occurs in the presence of nitrate but accumulation of nitrite cannot be detected.

Aerobic.

Temperature range 15–30 C; optimum 25–30 C. Neutral or slightly alkaline media preferred.

Osmotolerant; grows in 5% NaCl, 4% $NaNO_2$ or 9% $NaNO_3$.

Widely distributed in soils of all continents, particularly in soil with nitrification potential.

Type species: *Hyphomicrobium vulgare* Stutzer and Hartleb 1898, 76.

Further Comments

Since the original description by Stutzer and Hartleb (1898) several cultures have been isolated and labeled *H. vulgare*, primarily on the basis of morphology and the peculiar life cycle of the organism (Kingma-Boltjes, 1936; Mevius, 1953; Zavarzin, 1961; Hirsch and Conti, 1964). These new strains have not previously been compared closely with the original description of *H. vulgare*, but further study has shown great diversity among them, and an amended description of the generic characters, based on these more recent isolates, the description of new species and the designation of type cultures is in preparation. The description of the species in this edition of THE MANUAL is based on that of the original authors (Stutzer and Hartleb, 1898).

During stages of their life cycles or under special growth conditions, organisms of the genera *Pedomicrobium*, *Hyphomonas* and *Rhodomicrobium* may, by their morphology, resemble *Hyphomicrobium* species. *Rhodomicrobium* spp. can be distinguished by their red to orange photosynthetic and carotenoid pigments and by the presence of septa in the hyphae. *Hyphomonas* spp. need peptides and cannot use C-1 compounds as C source for growth, while *Pedomicrobium* spp. show an exclusive characteristic in the multiple formation of hyphae which arise from all parts of the cell surface; this can be seen especially well in older cultures.

Description of the species of genus **Hyphomicrobium**

1. **Hyphomicrobium vulgare** Stutzer and Hartleb 1898, 76.

vul.ga're. L. neut.adj. *vulgare* common.

Description as for the genus.

Species incertae sedis

The following species were described and named as species of *Hyphomicrobium* but recent work suggests that they were misplaced in that genus.

a. *Hyphomicrobium neptunium* Leifson 1964, 249. Cells round to oval, ca. 1 μm in diameter with one or more short hyphae up to three times the length of the mother cell. Buds are oval, with 1–2 polar or lateral flagella; buds motile until the hypha formed. An alternative multiplication process occurs by a swelling of short hyphae, which should not be confused with budding. Gram reaction not recorded.

On agar colonies are colorless, later brown, raised, semitranslucent or opaque and, after 3 days are up to 1.5 mm in diameter.

In liquid media growth forms a uniform turbidity without a pellicle or sediment.

Halotolerant; growth occurs in media containing sea water. Mg^{++} and Ca^{++} are essential; citrate inhibits growth. Does not grow on the mineral salts medium used for *H. vulgare*.

Good growth in vitamin-free casamino acids plus Mg^{++} or Ca^{++}, or sea water. Grows well in 0.2% casitone plus 0.1% yeast extract and 0.3% MgCl$_2$. Vitamins are not required.

Acid not formed from glucose, sucrose, lactose, xylose, maltose or mannitol. Gelatin, starch and cellulose not hydrolyzed. Lysine deaminated; phenylalanine may be deaminated. Nitrate reduced to nitrite. Catalase positive.

Aerobic. Heterotrophic. Temperature range: 20–40 C; optimum: 30–37 C. pH range: 6.0–9.5; optimum: ca. 8.

This organism was found to have a base ratio of 61.7% G + C (Mandel, Hirsch and Conti, 1972). In DNA base sequence homology studies *H. neptunium* was found to be more closely related to two strains of *Hyphomonas polymorpha* (Pongratz, 1957) than to 20 new *Hyphomicrobium* isolates (Moore and Hirsch, 1972). A detailed comparison of *H. neptunium* with 82 hyphomicrobia supported the proposed transfer of this organism to the genus *Hyphomonas*.

Originally isolated from sea water that had been stored for 18 months and frozen for 2 days.

Type strain: ATCC 15444.

b. *Hyphomicrobium indicum* Johnson and Weisrock 1969, 298. (*Hyphomicrobium indium* (*sic*) Weisrock and Johnson 1966, 22.) Cells rod-shaped, 0.7–1.0 by 2.0–6.0 μm. Cells up to 50 μm in length may appear after 6–9 hrs. of incubation in liquid media; although non-septate they apparently fragment later.

Broth and agar cultures may show coccoid cells (1–2.5 μm diameter) joined by slender filaments 0.5–2.0 μm long. This type of morphology seems to suggest a budding mechanism in reproduction. The cocci have a single polar flagellum; long cells are non-flagellate. Spores not formed. Cells stain well with crystal violet.

Gram-negative with Gram-positive granules in many cells. Not acid-fast.

Colonies up to 2 mm in 24 hrs., thin; white or cream in color, later become yellow. Growth in liquid media turbid without pellicle. In some media a ropy sediment may form.

No growth on MacConkey, Chapman-Stone or S-S agars. No growth in peptone yeast extract medium plus 3% NaCl, 0.1% CaCl$_2$ or 0.1% MgCl$_2$. No growth with glycine, L-cysteine, L-histidine, DL-valine, L-asparagine, DL-tyrosine, DL-serine or L-glutamic acid as single C source.

Acid formed from glucose (aerobically and anaerobically), maltose and sucrose. No growth on xylose.

Gelatin, starch and casein not hydrolyzed. Indole, nitrate and H$_2$S positive. Urease, catalase, and oxidase negative. Lysine not decarboxylated. Phenylalanine deaminated.

Marine; growth optimal with 50–100% sea water. Temperature range: 4–25 C. pH optimal: 4.5–9.5.

Growth inhibited by chloromycetin, neomycin, kanamycin, novobiocin and streptomycin; not inhibited by erythromycin, tetracycline, penicillin or pteridine 0/129.

The base ratio of 40% G + C distinguishes this organism from all other hyphomicrobia and related cultures which range from 59.2–66.8 moles % (Mandel, Hirsch and Conti, 1972). It can also be separated from the true-budding hyphal hyphomicrobia by the absence of buds and branching hyphae, and by its quite different metabolism. Future transfer of the organism to another genus is recommended.

Originally isolated from a sample of bottom mud, the Indian Ocean at 400 m depth.

Type strain: ATCC 19614.

Genus **Hyphomonas** *Pongratz 1957, 607*

PETER HIRSCH

Hy.pho.mo′nas. Gr. n. *hyphos* filament; M.L. n. *monas* small motile organism; M.L. fem.n. *Hyphomonas* hypha (bearing) monad.

Cells 0.6–1.0 μm in diameter and approximately 1.0 μm long, colorless, egg-shaped, oval- or pear-shaped, resembling human spermatozoa or cells of *Hyphomicrobium* spp. The narrow pole grows a tubelike and straight or wavy extension (prostheca) which may be branched like a fungal hypha and may be as long as 8 μm. Older cultures often contain abnormal cell types: giant cells, spindles or triangular forms, with the prosthecae (hyphae) highly branched.

Young cells and hyphae with dense, homogeneous cytoplasm; abnormal forms often vacuolated

or granular. Multiplication by a budding process and *not* by binary fission. Slide cultures on nutrient agar and at 20 C show the development of a distal terminal knob on the hypha. In 1–2 hrs. a bud (daughter cell) is completed and separates from the stalked mother cell. While the bud becomes motile shortly after separation ("swarmer cell"), the mother cell stalk elongates slightly before growing the next bud. Intercalary buds have been found occasionally, forming a rosary-like chain of cells interconnected by short hyphae. Direct budding without the formation of a hypha has also been observed. Motility is due to a single, subpolarly inserted flagellum which is often retained after the development of a stalk and new daughter cell. The mother cell then tows the daughter cell.

Endospores, conidia or chlamydospores have not been found.

Staining: Gram-negative, not acid-fast. Stain well with anilin dyes, neutral red, gentian violet, etc. The use of Lugol's iodine as a mordant facilitates the staining of the narrow hyphae. Best results were obtained with silver impregnation. Upon isolation on solid media, colonies appeared as smooth ("S") or rough ("R") types. The smooth colonies were round, convex, watery and translucent. They could be emulsified easily. The S organisms carried heavy capsules on the mother cells but usually not on hyphae or buds. Rough colonies were rare; they were smaller, foldy and dry, forming a central crater after several days. The organisms could not be readily emulsified and suspensions thus remained granular. They lacked capsules and were non-motile. While S forms were stable variants, R forms split into both S and R derivatives. The often voluminous capsules of the S form organisms were readily stainable with Ziehl's hot carbol fuchsin.

Growth on broth occurs in the form of a delicate veil.

Aerobic; catalase- and peroxidase-positive. Heterotrophic. Not proteolytic, hemolytic or saccharolytic. Gelatin is not digested, plasma is not coagulated. Pigments are not formed. Antibiotically not active against *Staphylococcus aureus*, *Streptococcus* sp., *Escherichia coli* or *Candida albicans*.

Optimum temperature 37 C. Growth slow at 18–20 C and better at 30 C. Optimum pH is 7.0–7.4, the range for growth is 6.5–8.5. Poor or no growth on acid media commonly used for the cultivation of fungi.

Growth is not influenced by light.

No growth on poor media such as tap water with KNO_3 and phosphate. Does not grow on mineral salts media (such as those of Korthof, van Niel or K. Boltjes) which contain $CaCO_3$, formate, acetate, propionate, lactate, methanol, ethanol, glycerol or glycine as the sole source of carbon.

Does not attack arabinose, rhamnose, xylose, glucose, fructose, galactose, lactose, saccharose, maltose, trehalose, melibiose, raffinose, amidon, inulin, dextrin, glycogen, esculin, glycerol, erythritol, mannitol, sorbitol or urea.

Good growth in coagulated serum, gelatin, blood agar, Loewenstein-Jensen medium, Dorset-agar and other rich media.

Reduce nitrate, nitrite, neutral red, methylene blue and janus green. Produce H_2S. Indol is formed from tryptophane. NH_3 is produced, the growth medium turns alkaline.

Resistant against therapeutic concentrations of penicillin, bacitracin or furadoine. Inhibited greatly by chloromycetin, streptomycin, neomycin and sulfonamides. The tetracyclins and erythromycin are slightly less inhibitory.

Not virulent against mice (5 ml), rats (1 ml), guinea pigs (1 ml) or rabbits (1 ml) if a suspension was injected subcutaneously or intraperitoneally which contained 5×10^9 cells/ml. The organisms could not be recovered from the treated animals.

Pure cultures survive at room temperature or in a refrigerator for several months if suspended in growth media.

Isolated only once from nasal mucus from a case of infectious sinusitis.

Type species: *Hyphomonas polymorpha* Pongratz 1957, 607.

Description of the species of genus **Hyphomonas**

1. **Hyphomonas polymorpha** Pongratz 1957, 607.

po.ly.mor′pha. Gr. adj. *poly* many; Gr. n. *morphe* shape, body; M.L. adj. *polymorpha* many shapes.

Description as for the genus.

Base ratios: 60.2 mole % G + C (S strain), and 61.2 mole % G + C (R strain), respectively (Mandel, Hirsch and Conti, 1972).

Further Comments

DNA-DNA base sequence homologies were investigated by Moore and Hirsch (1972) for several budding and prosthecate bacteria. While *Hyphomonas polymorpha* proved to be unrelated to a representative *Hyphomicrobium* (EA-617) despite great morphological similarities, a definite relationship appeared to exist between the two *Hyphomonas* cultures and an organism previously named *Hyphomicrobium neptunium* (ATCC 15444).

Genus **Pedomicrobium** *Aristovskaya 1961, 957*

T. V. ARISTOVSKAYA AND P. HIRSCH

Pe.do.mi.cro′bi.um. Gr. n. *pedon* soil, Gr. adj. *micrus* small; Gr. n. *bius* life; M.L. neut.n. *Pedomicrobium* soil microbe.

Cells spherical, oval, rod-shaped, pear- or bean-shaped, 0.4-2.0 μm wide, unevenly staining. **Multiplication is primarily by budding at the tips of cellular extensions (hyphae)** of constant diameter (0.15–0.3 μm). Daughter cells may remain attached to their mother hyphae or they may separate as uniflagellated "swarmers." After increasing in size, **mature swarmer cells grow from one to numerous hyphae from several sites of their cell surface.** Occasionally, division of single mother cells has been observed simultaneously with the budding process.

Gram-negative, microaerophilic to aerobic; **ferric and/or manganese depositions occur either on mother cells or on the hyphae.** Chemoheterotrophic, mesophilic; growth occurs on mineral salts media with simple organic compounds or with organomineral complexes of fulvenic acids and sesquioxides. Deposition of iron, manganese and slimes primarily on mother cells. Widely distributed in soils.

Type species: *Pedomicrobium ferrugineum* Aristovskaya 1961, 955 (translation p. 113).

Key to the species of genus **Pedomicrobium**

I. Buds form on hyphal tips only.
    A. Mother cell 0.6–2.0 μm wide.
       Ferric hydroxide deposited on mother cell.
                  1. *Pedomicrobium ferrugineum*
    B. Mother cell 0.4 μm in diameter.
       Manganese compounds deposited on mother cell.
                  2. *Pedomicrobium manganicum*
II. Buds form both on hyphal tips and directly on mother cells.
    Both iron and manganese deposited.
                  3. *Pedomicrobium podsolicum*

Description of the species of genus **Pedomicrobium**

1. Pedomicrobium ferrugineum Aristovskaya 1961, 957.

fer.ru.gin'e.um. M.L. neut.adj. *ferrugineum* resembling iron rust.

Mother cells spherical, oval- or rod-shaped, 0.6–2.0 μm in diameter, with one to four (or more) hyphae 0.2 μm or less in width. Cells stain unevenly. Single buds at the hyphal tips are initially spherical to oval and stain deeply They become motile by a single polar flagellum. Intercalary buds have been observed.

Surface colonies brownish, up to 0.5 mm in diameter after several months; develop deeper in the agar only when medium contains organomineral complexes of fulvenic acids with sesquioxides. Ferric hydroxide depositions occur primarily on the mother cells and only later on some hyphae. Manganese is not oxidized or deposited.

Originally isolated from soils of the Leningrad region, which contained ferromanganese deposits. Widely distributed.

2. Pedomicrobium manganicum Aristovskaya 1961, 954.

man.ga'ni.cum. M.L. neut.adj. *manganicum* of manganese.

Mother cells spherical to ovoid, 0.4 μm in diameter, with one to four or more hyphae and single spherical to ovoid buds at the hyphal tips. Hyphae 0.2 μm or less, frequently branched and of variable length; intercalary buds present.

On agar media containing organomineral complexes of fulvenic acids with sesquioxides, colonies are small and black.

Manganese is deposited primarily onto the mother cell; iron deposition has not been observed.

Chemoheterotrophic and microaerophilic.

Originally isolated from soils of the Leningrad region which contained manganese deposits. Widely distributed.

3. Pedomicrobium podsolicum Aristovskaya 1963, 30.

pod.so'li.cum. M.L. adj. *podsolicum* from podsol, a type of soil.

Mother cells spherical, rarely ovoid, 0.6–1.0 μm wide, with one to four hyphae and single, spherical buds at the hyphal tips. Budding of mother cells directly has also been observed. Hyphae 0.2 μm or less, very short and not always branched. Colonies on media with humus organomineral complexes are small and brownish black; iron and manganese hydroxides are deposited onto the mother cells.

Originally isolated from podsol soil; found only rarely.

Further Comments

Tyler and Marshall (1967) isolated a budding bacterium ("strain T₃₇") resembling *Pedomicro-*

bium sp. from manganese depositions in the Derwent hydroelectric pipelines, Lake Williams King, Tasmania. A study of this culture revealed pleomorphic response to growth conditions. On mineral salts media with 0.4% methylamine HCl or on media with 0.005% yeast extract and 0.002% $MnSO_4 \cdot H_2O$ cells resemble *Hyphomicrobium* spp., especially in younger cultures. Older cultures and when grown with Mn^{++} show multiple formation of hyphae from all over the cell surface, thus resembling *Pedomicrobium* spp. Similar organisms were also observed in cold springs, brook water, marl ponds and lakes of Lower Michigan, or in Lakes of North Germany.

Pedomicrobia have also been isolated from coastal surface water (2.7 g/liter NaCl) of Woodmont, Connecticut. Ferric iron was deposited primarily onto the hyphae (Hirsch, 1964).

Genus **Caulobacter** *Henrici and Johnson 1935, 83*

JEANNE S. POINDEXTER

Cau.lo.bac'ter. L. n. *caulis* stalk; M.L. *bacter* masc. form of Gr. neut.n. *bactrum* rod; M.L. masc.n. *Caulobacter* stalk rod.

Cells rod-shaped, fusiform or vibrioid, 0.4–0.5 μm by 1–2 μm. Typically with a **stalk,** ca. 0.15 μm diameter, varying in length among isolates and with environmental conditions, **extending from one pole** as a continuation of the long axis of the cell. The **stalk consists of the cell wall layers, the cell membrane and a core of twisted membranes. A small mass of adhesive material is present at the distal end of the stalk.** Occur singly; in dense populations, **cells may adhere to each other in rosettes,** the distal ends of their stalks embedded in a common mass of adhesive material.

Cell division by transverse, binary, **asymmetrical fission of stalked cells.** At the time of separation one cell possesses a stalk, the other a single polar flagellum. Each appendage occurs at the cell pole opposite the one formed during fission. The flagellated cell secretes adhesive material at the base of the flagellum, develops a stalk at this site, and enters the immotile vegetative phase.

Resting stages not known.

Gram-negative. The cell wall consists of an outer component similar in thickness and organization to the cell membrane, and an inner component whose thickness varies among isolates; the inner component is soluble in lysozyme-EDTA.

Colonies circular, convex, glistening; colorless, pink, red, orange or yellow.

Chemoorganotrophic. Strictly respiratory. The principal pathway of intermediary carbon metabolism is the Entner-Doudoroff pathway. Most strains can store carbon as poly-β-hydroxybutyric acid; a few also store carbon as polysaccharide.

Pigmentation common, but not universal; pigmented isolates contain carotenoids. A few isolates produce both nondiffusible red or orange pigments and diffusible brown or red-brown pigments.

Requirements for organic growth factors vary among species. All strains can grow on peptone-yeast extract media.

Strict aerobes.

Temperature range of most isolates is 15–35 C; 20–25 C is optimal. A few strains grow at 5–25 C; 15–20 C is optimal for these. Optimal pH for growth is around neutrality; pH range for all isolates is 6.0–9.0.

Fresh water and soil isolates typically grow well at low osmotic pressures, and multiplication can occur in distilled water. Growth is inhibited or cells become swollen and frequently lyse in media containing 1% (w/v) or more of organic material. Isolates from sea water require NaCl at 0.5 or 1% (w/v) for growth in organic media. Growth of most isolates is inhibited by streptomycin (0.1 mg/ml), penicillin G (1000 units/ml) and cycloserine (0.1 mg/ml).

Most species can be isolated from samples of fresh water containing low concentrations of organic materials and from soil. NaCl-requiring isolates have been obtained only from sea water.

The G + C content of the DNA is 64–67 mole % (buoyant density). Known bacteriophages, which include DNA- and RNA-containing viruses, are strain-specific among *Caulobacter* isolates. They do not lyse bacteria of other genera, nor have *Caulobacter* isolates been found susceptible to infection by phages lytic for other genera.

The type species is *Caulobacter vibrioides* Henrici and Johnson 1935, 84.

Further Comments

In their studies of fresh water bacteria which attached to slides submerged in a lake, Henrici and Johnson (1935) observed, in addition to typical *Caulobacter* cells, a stalked bacterium in which motility did not occur. Fission was preceded by the development of a stalk at the outer pole of the cell so that cell division was symmetric. Each of the products of fission bore a stalk, and neither was motile. In an extended study of the *Caulobacter* group, Poindexter (1964) was unable to isolate such an organism, but did observe it in enrichment cultures from which isolates of motile

Caulobacter and *Asticcacaulis* were obtained. She suggested that symmetrically dividing caulobacters should not be included in *Caulobacter*, but that a separate genus should be established when isolates were obtained and their immotility was determined in pure culture studies. Such an isolate is now available (de Bont *et al.*, 1970) and its properties are being investigated.

Key to the species of genus **Caulobacter**

I. Cells tapered.
   A. Long axis of the cell curved.
      1. Organic growth factors required.
         a. Vitamin B₂ necessary, but not sufficient.
              1. *C. vibrioides*
         b. Vitamin B₁₂ necessary and sufficient.
              2. *C. henricii*
         c. Biotin necessary, but not sufficient.
              3. *C. intermedius*
         d. Growth not stimulated by B vitamins.
              4. *C. subvibrioides*
      2. Organic growth factors not required
              5. *C. crescentus*
   B. Long axis of the cell not curved.
      1. Organic growth factors required.
              6. *C. fusiformis*
      2. Organic growth factors not required.
              7. *C. leidyi*
II. Cells not tapered.
   A. NaCl not required for growth in organic media.
              8. *C. bacteroides*
   B. NaCl required for growth in organic media.
      1. Growth inhibited by 4% (w/v) NaCl.
              9. *C. halobacteroides*
      2. Growth not inhibited by 4% (w/v) NaCl.
            10. *C. maris*

Description of the species of genus **Caulobacter**

1. Caulobacter vibrioides Henrici and Johnson 1935, 84.

vib.ri.oi′des. M.L. n. *vibrio* name of a genus; Gr. n. *eidus* form, shape; M.L. adj. *vibrioides* resembling a vibrio.

Cells vibrioid, slender or nearly ovoid. Colonies colorless or pale yellow. Vitamin B₂ essential for growth, but additional unidentified growth factors available in peptone-yeast extract also required by most strains.

Neotype strain: CB 51 (Poindexter and Lewis, 1966)

2. Caulobacter henricii Poindexter 1964, 288.

hen.ric′i.i. M.L. gen.n. *henricii* of Henrici; named for A. T. Henrici who observed stalked bacteria on slides which had been submerged in fresh water.

Cells vibrioid. Colonies of most strains bright yellow; others pale yellow or golden red. Vitamin B₁₂ the only organic growth factor required.

Type strain: CB4; ATCC 15253.

3. Caulobacter intermedius Poindexter 1964, 288.

in.ter.med.i′us. L. adj. *intermedius* in the middle degree, between extremes.

Cells vibrioid, often short. Colonies colorless. Biotin stimulates growth, but additional unidentified growth factors available in peptone-yeast extract also required.

Type strain: CB63; ATCC 15262.

4. Caulobacter subvibrioides Poindexter 1964, 289.

sub.vib.ri.oi′des. L. pref. *sub* almost, somewhat, near; M.L. n. *vibrio* name of a genus; Gr. n. *eidus* resembling; M.L. adj. *subvibrioides* less like a vibrio.

Cells tapered; within a clone, the long axis of some cells is curved, of others straight. Colonies orange or colorless. Organic growth factor requirements not satisfied by mixtures containing B vitamins, amino acids and purine and pyrimidine bases.

Type strain: CB81; ATCC 15264.

5. Caulobacter crescentus Poindexter 1964, 288.

cres.cen'tus. L. adj. *crescentus* of the moon in its first quarter, crescent.

Cells vibrioid, slender. Colonies colorless. Organic growth factors not required. Growth not inhibited by penicillin G (1000 units/ml).

Type strain: CB2; ATCC 15252.

6. Caulobacter fusiformis Poindexter 1964, 289.

fus.i.form'is. L. n. *fusus* spindle; L. n. *forma* shape, form; M.L. adj. *fusiformis* spindle-shaped.

Cells tapered, slender; long axis not curved. Colonies bright yellow. Known isolates do not utilize sugars as sources of carbon. Organic growth factor requirements not satisfied by mixtures containing B vitamins, amino acids and purine and pyrimidine bases.

Type strain: CB27; ATCC 15257.

7. Caulobacter leidyi Poindexter 1964, 289.

leid'y.i. M.L. gen.n. *leidyi* of Leidy; named for J. Leidy, who observed tufts of (bacterial) growth on fungi in insect guts in 1853.

Cells tapered, short; long axis not curved. Stalks very short (less than one-half cell length) in most environments. Colonies colorless. Organic growth factors not required. Growth not inhibited by

streptomycin (0.1 mg/ml) or penicillin G (1000 units/ml).

Type strain: CB37; ATCC 15260.

8. Caulobacter bacteroides Poindexter 1964, 289.

bac.ter.oi'des. M.L. n. *bacter* a rod; Gr. n. *eidus* shape, form; M.L. n. *bacteroides* rod-shaped.

Cells slender, rod-shaped. Colonies colorless, yellow or orange. Organic growth factor requirements not satisfied by mixtures containing B vitamins, amino acids and purine and pyrimidine bases.

Type strain: CB7; ATCC 15254.

9. Caulobacter halobacteroides Poindexter 1964, 289.

hal.o.bac.ter.oi'des. Gr. n. *hals* salt; M.L. n. *bacteroides* rod-shaped.

Cells slender, rod-shaped. Colonies colorless. Organic growth factor requirements not satisfied by mixtures containing B vitamins and amino acids. Growth requires NaCl (0.5–3%, w/v) in organic media. Starch hydrolyzed. Nitrate not reduced to nitrite anaerobically.

Type strain: CM13; ATCC 15269.

10. Caulobacter maris Poindexter 1964, 289.

mar'is. L. gen.n. *maris* of the sea.

Cells slender, rod-shaped. Colonies colorless. Known isolates do not utilize amino acids as sources of carbon. Organic growth factor requirements not satisfied by mixtures containing B vitamins and amino acids. Growth requires NaCl (1–4%, w/v) in organic media. Starch not hydrolyzed. Nitrate reduced to nitrite anaerobically.

Type strain: CM11; ATCC 15268.

Genus **Asticcacaulis** *Poindexter 1964, 282*

JEANNE S. POINDEXTER

A.stic.ca.cau'lis. Gr. α privative without; Anglo-Saxon n. *sticca* stick; L. n. *caulis* stalk; L. masc.n. *Asticcacaulis* stalk that does not stick.

Cells rod-shaped (0.5–0.7 μm by 1–3 μm). **Typically with an appendage** ca. 0.15 μm in diameter, varying in length among isolates and with environmental conditions, **arising excentrically from one pole, or laterally along the cell.** In the latter case, there may be two appendages per cell. Although structurally similar to stalks of *Caulobacter*, these appendages are called pseudostalks since they lack adhesive material and **cannot function in attachment.** Each cell typically possesses a small mass of adhesive material at a subpolar (rarely polar) position on the cell. Occur singly; in dense populations, **cells may adhere to each other in rosettes,** the poles of the cells associated with a common mass of adhesive material.

Cell division by transverse, binary, **disparate, asymmetrical fission of pseudostalked cells.** At the time of separation, one cell is longer and possesses one or two pseudostalks; the shorter sibling has a single subpolar flagellum. The flagellated cell develops a pseudostalk and enters the immotile vegetative phase.

Resting stages not known.

Gram-negative. The cell wall consists of an outer component similar in thickness and organization to the cell membrane, and an inner component whose thickness varies among isolates; the inner component is soluble in lysozyme-EDTA.

Colonies circular, convex, glistening; colorless.

Chemoorganotrophic. Strictly respiratory. The principal pathway of intermediary carbon metabolism is the Entner-Doudoroff pathway. Carbon is stored as poly-β-hydroxybutyric acid.

Strict aerobes.

Temperature range is 15–35 C; 25–30 C is optimal. Optimal pH for growth is around neutrality; pH range is 6.0–9.0.

All isolates typically grow well at low osmotic pressures, and multiplication can occur in distilled water. Growth is inhibited or cells become swollen and frequently lyse in media containing 1% (w/v) or more of organic material. Isolates have not been obtained from marine sources.

Growth is inhibited by streptomycin (0.1 mg/ml) and by penicillin G (1000 units/ml).

Observations of these bacteria in natural materials have not been reported. Presumably, their distribution is similar to that of *Caulobacter*, since all known *Asticcacaulis* isolates have been obtained inadvertently during attempts at isolation of *Caulobacter*.

The G + C content of the DNA is 55–60 moles % (buoyant density). Known bacteriophages, which are DNA viruses, are strain-specific among *Asticcacaulis* isolates. They do not lyse bacteria of other genera, nor have *Asticcacaulis* isolates been found susceptible to infection by phages lytic for other genera.

The type species is *Asticcacaulis excentricus* Poindexter 1964, 292.

Further Comments

The description of the genus presented here allows inclusion of organisms, such as that reported by Pate and Ordal (1965), which possess one or more lateral pseudostalks and can also be distinguished from the type species by physiological traits. A species epithet has not been proposed for this type.

Description of the species of genus **Asticcacaulis**

1. **Asticcacaulis excentricus** Poindexter 1964, 292.

ex.cen′tri.cus. L. pref. *ex* out, beyond; Gr. n. *centron* center of circle; M.L. adj. *excentricus* out from the center.

Each cell has a single, subpolar appendage; a flagellum in the motile stage and a pseudostalk in the nonmotile stage. Biotin is the only organic growth factor required.

Type strain: AC48; ATCC 15261.

Genus **Ancalomicrobium** *Staley 1968, 1940*

JAMES T. STALEY

An.ca′lo.mi.cro′bi.um. Gr. masc.n. *ancalos* arm; Gr. adj. *micros* small; Gr. masc.n. *bios* life; M.L. neut.n. *Ancalomicrobium* arm (-producing) microbe.

Unicellular, Gram-negative bacteria having from **two to eight prosthecae extending from the cell** (Plate 4.2, Fig. 1). **These prosthecae attain a length of about 3 μm at maturity,** about three times the diameter of the cell. Occasionally the prosthecae are bifurcated; the position of the bifurcation along the appendage is variable. **Cells reproduce by budding.** Buds are formed directly from one position on the mother cell, never from the prosthecae. Two to four prosthecae differentiate from the developing bud. Division occurs transversely when the bud has attained approximately the same size as the mother cell. Cells are **non-motile** and have **no holdfasts.** Gas vacuoles may be present in some cells. Found in natural waters.

Chemoorganotroph. Facultative anaerobe.

Type species: *Ancalomicrobium adetum* Staley 1968, 1940.

Description of the species of genus **Ancalomicrobium**

1. **Ancalomicrobium adetum** Staley 1968, 1940.

a.det′um. M.L. adj. *adetum* free, unattached.

Colonies are white, opaque, circular in form, convex in elevation, with an entire margin.

Ammonium can be used as the sole source of nitrogen. An organic energy and carbon source (Table 4.1), as well as vitamins (pantothenic acid absolutely required) are necessary for growth.

Ferments glucose. Nitrate not reduced. Catalase present.

Temperature range 9–39 C, More rapid growth and higher yields are obtained at temperatures above 30 C but growth is rapid at 30 C and it is suggested as closer to that of the natural habitat. pH range: <6.3–>7.5; optimum: 7.0.

The G + C content of the DNA of the type strain was 70.4 moles % (Staley and Mandel, 1973).

Type strain: Staley 4a; ATCC 23632, single isolate.

Genus **Prosthecomicrobium** Staley 1968, 1940

JAMES T. STALEY

Pros.the'co.mi.cro'bi.um. Gr. fem.n. *prosthece* appendage; Gr. adj. *micros* small; Gr. masc.n. *bios* life; M.L. neut.n. *Prosthecomicrobium* appendage (-producing) microbe.

Unicellular, Gram-negative bacteria that have **many prosthecae extending in all directions from the cell** (Plate 4.2, Fig. 2). These prosthecae are normally **less than 2 μm long** and are **conical in shape,** tapering distally from the cell toward a blunt tip. Occasionally longer, nontapering prosthecae are produced. **Cells divide by binary transverse fission.** Motile or non-motile; motile strains exhibit a tumbling, circular motility.

Chemoorganotroph, nonfermentative. Aerobic. Found in natural waters.

The G + C content of the DNA ranges from 65.8–69.9 moles % (buoyant density).

Type species: *Prosthecomicrobium pneumaticum* Staley 1968, 1940.

Description of the species of genus **Prosthecomicrobium**

1. Prosthecomicrobium pneumaticum Staley 1968, 1940.

pneu.mat'ic.um. M.L. adj. *pneumaticum* inflated.

The ratio of prosthecal length to cell diameter

TABLE 4.1
Carbon sources utilized by species of
Ancalomicrobium *and*
Prosthecomicrobium

Utilized by all species: L-arabinose, cellobiose, D-fructose, L-fucose, D-galactose, D-glucose, lactose, D-lyxose, maltose, D-mannose, L-rhamnose, D-ribose, D-xylose.

Not utilized by any species: raffinose, inulin, pectin, starch, adipate, benzoate, phthalate, arabitol, erythritol, methanol.

Utilized by some species	1. *Anca-lomicro-bium adetum*	1. *Pros-thecomi-crobium pneuma-ticum*	2. *Pros-thecomi-crobium enhy-drum*
D-Arabinose	+	+	?
Melibiose, glycogen	−	+	−
L-Sorbose, melezitose, arabin, dextrin	+	+	−
Xylan	?	−	−
Butyrate	−	−	+
Fumarate, succinate	+	−	+
Propionate	−	+	?
Adonitol	?	+	−
Dulcitol, ethanol	−	+	−
Glycerol, inositol, mannitol, sorbitol	+	+	−

? not tested or utilization uncertain.

(c a. 1.0 μm) is approximately 1.0 although occasionally much longer prosthecae are formed. Under some growth conditions some cells within a clone may not appear to have prosthecae. Cells have gas vacuoles; the vesicles comprising the gas vacuoles are approximately 100 × 300 nm. Colonies are not pigmented. Non-motile. Ammonium, but not nitrate, can be used as a sole source of nitrogen for growth. Organic carbon sources (Table 4.1) and vitamins (biotin, thiamine and B₁₂) are required for growth.

Temperature range 9–42 C. More rapid growth and higher yields are obtained at temperatures above 30 C but growth is rapid at 30 C and it is suggested as closer to that of natural habitats. pH range optimum 6.0–6.5; good growth at pH 7.0.

The G + C content of the DNA ranges from 69.4–69.9 moles %.

Type strain: Staley 3a; ATCC 23633.

2. Prosthecomicrobium enhydrum Staley 1968, 1941.

en.hyd'rum. M.L. adj. *enhydrum* living in water, aquatic.

The ratio of prosthecal length to cell diameter (ca. 1.0 μm) is less than 0.5. No gas vacuoles are produced. Motile by a single subpolar to polar flagellum with characteristic tumbling motility. Ammonium, but not nitrate, can be used as a sole source of nitrogen if appropriate vitamins and an organic carbon source (Table 4.1) are included in the growth medium. Thiamine is required. Colonies may be white (type strain), yellow or red.

Growth range: 9–37 C; more rapid growth and higher yields are obtained at temperatures above 30 C but growth is rapid at 30 C and it is suggested as closer to that of natural habitats. pH optimum: 7.0.

The G + C content of the DNA of one strain examined was 65.8 moles % (buoyant density) (see also Staley and Mandel, 1973).

Type strain: Staley 9b; ATCC 23634.

Genus **Thiodendron** *Perfil'ev and Gabe 1961, 162*

PETER HIRSCH

(NOT *Thiodendron* Lackey and Lackey 1961, 39.)

Thi.o.den′dron. Gr. n. *thium* sulfur; Gr. n. *dendron* a tree; M.L. n. *Thiodendron* a sulfur tree.

Vibrio-shaped cells, spirally twisted and with both ends somewhat tapered, ca. 10–15 times longer than wide, bear, thin threads ("stalks," 0.15–0.25 μm) on either one or both cell poles. The threads are straight or more or less flexuous, often of considerable length and occasionally appear to be branched; they are arranged radially to a common center. The vibrio-shaped cells may be motile with flagella.

Colonies concentrically layered, ± globular and grayish to bluish white with alternating lighter and darker zonation. The appearance may be that of thalli of the alga *Padina pavonia*. The layering is assumed to reflect rhythmical deposition of colloidal sulfur as a result of H_2S oxidation. In natural sulfur spring water, the colonies grow to a size of 4 cm in 3–4 days.

Gram reaction not recorded. Presumably aerobic to microaerophilic. **Sulfur** in form of H_2S is required for growth.

Originally isolated in pure culture from a natural sulfur spring near Chokrakskoye in Russia. Also observed in peat mud, sand or various fresh water samples.

Type species: *Thiodendron latens* Perfil'ev and Gabe 1961, 162.

Description of the species of genus **Thiodendron**

1. **Thiodendron latens** Perfil'ev and Gabe 1961, 162.

la′tens. L. part.adj. *latens* concealed, hidden.

Description as for the genus.

Further Comments

Colonies of organisms resembling the description of *T. latens* have been observed by Dr. P. Hippe, Göttingen (personal communication) in a mud and water sample from a pond, after storage in the laboratory for several months (Plate 4.2, Figs. 3 and 4). A morphological similarity to *Caulobacter* spp. is evident from these micrographs.

Genus **Pasteuria** *Metchnikoff 1888, 166, emend. mut. char. Hirsch 1971, 113*

PETER HIRSCH

Pas.teu′ri.a. M.L. gen.n. *Pasteuria* of Pasteur; named for Louis Pasteur, the French scientist.

Cells about 1–5 by 3–6 μm, **nearly spherical to pear-shaped** with one end broad and rounded, and the other pointed and attached. Older cells brighter near the rounded pole (DNA?). Occurs singly or in mass aggregations resembling grapes. **Multiplication is by budding from the top or side of the free rounded end** (Fig. 4A and Plate 4.1, Figs. 2–4) (Hirsch, 1971). Buds initially almost spherical. Not motile, not pigmented. Gram reaction not recorded.

Aerobic. Has not been cultured.

Originally isolated from *Daphnia pulex* and *Daphnia magna* and is probably pathogenic for them. Has recently been found in the body cavities and on antennae of *Daphnia* spp. from North America and England (Hirsch, 1971). Similar forms have been found attached to submerged surfaces in lake water (Henrici and Johnson, 1935; Hirsch, unpublished; Staley, personal communication).

Type species: *Pasteuria ramosa* Metchnikoff 1888, 166.

Further Comments

In his original description Metchnikoff (1888) attributed clusters of elongated, "branched" structures to the same organism. These forms supposedly multiplied by longitudinal fission and endospore formation. Hirsch (1971) re-examined *Daphnia* material containing structures which resembled those described by Metchnikoff and suggests that the elongated, flat and clustering forms in this material could be interpreted as aggregates of flat crystals and the "endospores" as buds formed at the distal unattached end of the spherical stage of the true *Pasteuria* organism.

Editorial Note. Staley (1973) has now isolated *Pasteuria ramosa* and confirmed the ovoid or pear-shaped cells, with buds formed at the large end only and an attachment at the narrow end. Newly formed buds are motile with a single flagellum at the large end. These cells attach and detach seemingly at random, but mature cells become irreversibly attached.

Chemoorganotrophic growing on a peptone-yeast extract medium with an appropriate carbon source; a variety of carbon sources are used. Glucose is metabolized oxidatively. Oxidase and catalase positive.

The G + C content of the DNA is 57.1 moles %.

Type strain: Michigan; ATCC 27377.

Staley discusses the relationship of *P. ramosa* and *Blastobacter henricii* and casts doubt on the need for the latter genus and species.

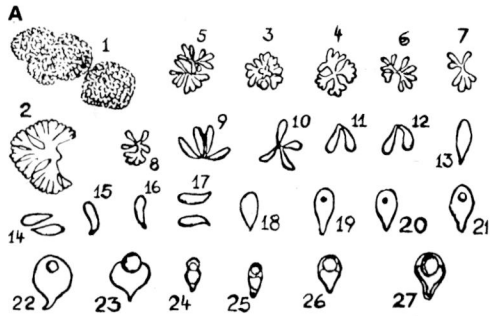

Fig. 4A. Stages in the life cycle of *Pasteuria ramosa* as redrawn from Plate I in the original description of Metchinkoff (1888). Stages 1–18 may be the crystals mentioned by Hirsch (1972). Stages 19–21 are also difficult to interpret. Stages 22–27 resemble the budding forms mentioned by Hirsch (1972). From Hirsch, P. 1972 Int. J. Syst. Bacteriol. *22:* 112–116.

Description of the species of genus **Pasteuria**

1. **Pasteuria ramosa** Metchnikoff 1888, 166 *emend. mut. char.* Hirsch 1971, 113.

ra.mo'sa. L. fem. adj. *ramosa* much branched. Morphology and characteristics as for genus.

Genus **Blastobacter** *Zavarzin 1961, 962*

P. Hirsch and G. A. Zavarzin

Blas.to.bac'ter. Gr. n. *blastos* bud shoot; M.L. n. *bacter*, masc. form of Gr. n. *bactrum* rod; M.L. masc.n. *blastobacter* budding rod.

Mature cells rod-shaped, wedge-shaped or club-shaped, often slightly curved, 0.7–1.0 (greatest width) by 2.0–4.5 μm (measurements from Fig. 13 of Zavarzin, 1961). Several cells attach by their narrow poles directly to a common base to form a rosette in the center of which there is usually a glistening corpuscle. Stalks absent. Multiplication by single buds appearing on the free cell pole. Buds spherical to oblong, 0.3 μm wide and non-motile.

Massive growth was observed in a cylinder containing iron water from a Northern Russian forest

brook to which shreds of filter paper had been added. Growth was best in a zone with reduced iron and a pH of 6.2.

Gram reaction not recorded.

Not obtained in pure culture.

Originally found in isolated samples of fresh water ponds. Probably widely distributed in fresh water as part of the attached microflora.

Type species: *Blastobacter henricii* Zavarzin 1961, 962.

Description of the species of genus **Blastobacter**

1. **Blastobacter henricii** Zavarzin 1961, 962.

hen.ri'ci.i. M.L. gen. n. *henricii* of Henrici, an American microbiologist.

Description as for the genus.

Further Comments

Zavarzin (1961) pointed out similarities to some

of the forms described and illustrated by Henrici and Johnson (1935). However, with the exception of their Figure 9, the organisms shown were stalked forms. Bacteria similar to *Blastobacter henricii* were also found by ZoBell and Upham (1944) and by Kriss (1959), in both cases in samples from deeper layers of the ocean.

Genus **Seliberia** *Aristovskaya and Parinkina 1963, 56*

T. V. ARISTOVSKAYA

Se.li.be'ria. M.L. fem.n. *Seliberia;* named for Prof. G. L. Seliber, a Russian microbiologist.

Rod-shaped, spirally twisted cells, 0.5–0.7 by 1–12 μm, forming **star-shaped figures** or rosettes (Plate 4.3, Figs. 1–4). Multiplies by transverse fission and by **budding.** When grown on soil media forms round to ovoid "generative" cells (Plate 4.3, Figs. 2 and 3). Resting stages not known. Gram reaction not determined.

Chemoorganotroph. Organic growth factors not required.

Facultatively anaerobic. Temperature optimum 24–25 C.

Type species: *Seliberia stellata* Aristovskaya and Parinkina 1963, 55.

Description of the species of genus **Seliberia**

1. **Seliberia stellata** Aristovskaya and Parinkina 1963, 55.

stel.la'ta. L. adj. *stellata* starred.

Morphology as for the genus. Size of cells depends on environmental conditions. Cells are very long on media containing organomineral complexes of fulvic acids and sesquioxides but shorter on media with ulmic acid complexes.

Motile cells are formed by transverse fission; these are motile with a subpolar flagellum (Plate 4.3, Fig. 5). The star-shaped aggregates arise from the budding process; buds originate from the end or sides of the rod-shaped cells and the rod-shaped buds remain attached to the mother cell.

Grows on media containing organomineral complexes of fulvic and ulmic acids with sesquioxides; on soil extract or on water agar, on very dilute broth and other organic media. Deposits iron hydroxide in iron-containing media and in soils.

Common inhabitants of soil where it transforms different forms of humus.

Type strain: N-9 (INMI).

Further Comment

Hirsch (unpublished) has found *Seliberia* spp. in laboratory-distilled water and in surface layers of a small forest pond. The twisted cells and flagellation are seen in Plate 4.3, Fig. 5.

Genus **Gallionella** *Ehrenberg 1838, 166 Nom. cons. Opin. 9, Jud. Comm. 1954, 147*

G. A. ZAVARZIN AND P. HIRSCH

Gal.li.o.nel'la. M.L. dim. ending -*ella*; M.L. fem.n. *Gallionella*; named for B. Gallion, a receiver of customs and zoologist (1782–1839) in Dieppe, France.

Cells kidney-shaped or rounded, on the ends of long "stalks" with the long axis of the cell transverse to that of the stalk. Stalks are formed of bundles of fibrils twisted around one another: they may be covered with iron hydroxide; manganese compounds are not deposited; fibrils may branch.

Multiplication is by fission of cells, the daughter cells remaining at first at the end of the stalk; later they may be liberated as swarmer cells, which are motile by one or two polar or subpolar flagella.

Gram-negative.

Probably chemolithotrophic since they oxidize ferrous to ferric iron while assimilating significant quantities of $C^{14}O_2$; grow in inorganic artificial media and in oligotrophic waters in nature. Iron hydroxide may make up 90% of dry weight of cell mass.

Microaerophilic, developing under oxygen concentrations of about 1 mg/liter. Optimum pH 6–7. Develop in cold water, often under snow, although a thermophilic strain has been reported.

Found in ferrous iron-containing waters and in soils; often associated with *Leptothrix ochracea* and with it is responsible for much iron oxide precipitation. Growth of these organisms may cause problems in water works.

Type species: *Gallionella ferruginea* Ehrenberg 1838, 166.

Further Comments

It is now generally accepted that the stalks are excreted by the cells as proposed by Cholodny (1924, 1953). However, as late as 1968, Hanert proposed that the stalk was the living component. Life cycles have been proposed by van Iterson (1958) and Balashova (1968, 1969).

The taxonomic position of the genus is uncertain. It may be related to *Metallogenium*.

Effective methods for elective enrichment cultures were suggested by Kucera and Wolfe (1957) and Perfil'ev and Gabe (1961).

Electron microscopic investigation revealed microfibrils in the stalk and extreme pleomorphism of the stalk. This makes uncertain the validity of several species and varieties which have been described without comparative studies under standard conditions. When grown in the medium of Kucera and Wolfe, only two species are recognized as being different.

Description of the species of genus **Gallionella**

1. Gallionella ferruginea Ehrenberg 1838, 166 *Nom. cons.* Opin 9, Jud. Comm 1954, 147. (*Spirillum ferrugineum* (Ehrenberg) de Toni and Trevisan 1889, 1007; *Chlamydothrix ferruginea* (Ehrenberg) Migula 1900, 1031; *Spirophyllum ferrugineum* (Ehrenberg) Ellis 1907, 507.)

fer.ru.gin'ea. L. fem.adj. *ferruginea* rust-colored.

Morphology as for genus. Stalk is a flat twisted band, 0.4–1.0 μm in width, formed by numerous (up to 90) fibrils. Stalk branches dichotomously; under suboptimal conditions branching is more, and twisting less, pronounced (*G. minor* Cholodny 1924, 42).

Widely distributed in iron-bearing waters.

2. Gallionella filamenta Balashova 1967, 650. (*Gloeotila ferruginea* (Ehrenberg) Kutzing 1849, 363; *Didymohelix ferruginea* (Ehrenberg) Griffith 1853, 438; *Gloeosphaera ferruginea* (Ehrenberg) Rabenhorst 1854, 43; *Spirulina ferruginea* (Ehrenberg) Kirchner 1878, 250; *Gallionella ferruginea* (Ehrenberg) *sensu* Ellis 1919, 17.)

fil.a.men'ta. L. n. *flamentum* a fine thread.

Morphology as for genus. Stalks are formed by usually less than 12 cylindrical fibrils having the appearance of spirally twisted hairpins.

Originally isolated from drainage water from swamps, Jachroma, near Moscow, USSR. Rather rare in iron-bearing waters.

Species incertae sedis

a. *Gallionella glomerata* Naumann 1921, 45.

b. *Gallionella reticulosa* Butkevich 1928, 58.

c. *Gallionella tortuosa* Butkevich 1928, 57.

d. *Gallionella planctonica* Razumov according to Krasil'nikov 1949, 671 (probably should be transferred to *Planctomyces*).

e. *Gallionella scheminzkyi* Vouk 1960, 95. A thermophilic organism, the relationship of which is unknown.

f. *Gallionella pyritica* Schopf, Ehlers, Stiles and Birla 1965, 304.

Genus **Nevskia** Famintzin 1892, 484

H.-D. BABENZIEN AND P. HIRSCH

Nev'ski.a. *Neva* a river in Leningrad; M.L. fem.n. *Nevskia* from the Neva.

Rod-shaped cells, 0.7–2.0 by 2.4–12 μm, often slightly bent. Under certain conditions, cells are surrounded with slime but more is produced on one side forming acellular, hyaline stalks perpendicular to the long axis of the organism (Figs. 4.B and 4.C). The stalks branch dichotomously as a result of division of mature cells and leads to the formation of colonies up to 80 μm in size, which float on the surface of the water. They may be spherical, branched or cup-shaped and often form more or less dense pellicles. There is no metal encrustation on the stalks.

The slime is soluble in 1% KOH. Very dilute methyl violet stains the slime but not the cells.

Cells are usually packed with strongly refractile colorless globules. The globules are not sulfur or poly-β-hydroxybutyrate; they stain with Burdon's fat stain and may be dissolved with 70% ethanol.

Young cultures motile by means of one to three polar flagella.

Gram-negative. Aerobic.

Originally isolated from aquarium water. Occur on the surface of fresh water bodies (neuston communities). Have not been grown in pure culture.

Type species: *Nevskia ramosa* Famintzin 1892, 484.

Description of the species of genus **Nevskia**

1. Nevskia ramosa Famintzin 1892, 484. (*Gallionella ramosa* (Famintzin) Krasil'nikov 1949, 672.)

ra.mo'sa. L. fem.adj. *ramosa* much branched.

Morphology and description as for genus.

Further Comments

Famintzin gave the average cell length as 12 μm and mature cells were 2–6 times longer than wide; according to Babenzien the cells are 0.7 by 2.4–2.7

μm. Famintzin has also seen a smaller form of *N. ramosa*.

Babenzien (1965, 1967) succeeded in stimulating the growth of *Nevskia* spp. from various fresh water habitats, especially the surface of swamp water, by adding 0.1% sodium lactate or acetate to the water, adjusting the pH to 5.9–6.5 and the temperature to 28 C. These organisms were strict aerobes. Slime formation depended on cell age and development; young cells lived submerged, were devoid of slime and were motile. Older cells developed the slime stalks and grew as part of the surface community of the neuston (Plate 4.4, Fig. 1).

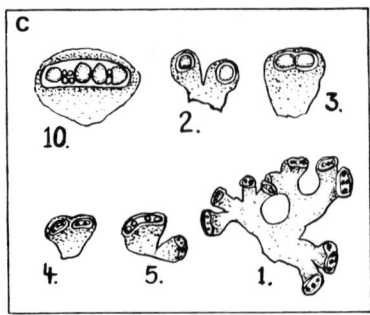

Fig. 4C. *Nevskia ramosa* as drawn by Famintzin (1892) (his Figures 1 thru 5 and 10).

Hirsch has confirmed these observations (unpublished). Smaller forms have also been observed but their relation to *N. ramosa* is not known.

Species incertae sedis

a. *Nevskia pediculata* (Koch and Hosaeus) Henrici and Johnson 1935, 83. (*Bacterium pediculatum* Koch and Hosaeus 1894, 225.) This may be a related form.

b. *Gallionella corneola* Dorff 1934, 26. (*Siderophacus corneolus* (Dorff) Beger 1949, 12.) May be a related form. Iron deposits were observed on its stalk.

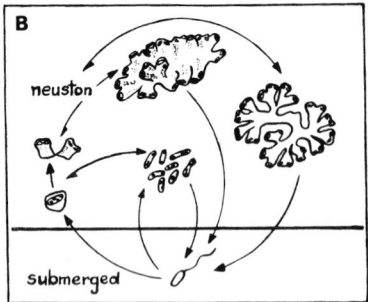

Fig. 4B. Life cycle of *Nevskia* sp. drawn schematically (Babenzien, 1967).

Genus **Planctomyces** Gimesi 1924, 4

P. Hirsch and H. Skuja

(*Blastocaulis* Henrici and Johnson 1935, 84; *Actinothrix* Novácek 1938, 66.)
Planc.to.my′ces. Gr. adj. *planktos* wandering floating; Gr. n. *mykes* fungus. M.L. masc.n. *Planctomyces* floating fungus.

Cells spherical to oblong or pear-shaped, 0.3–1.7 μm in diameter with long and slender stalks when in the mature state. Stalks are 0.3–0.9 μm wide and up to 11 μm long and probably excreted; they often attach to a common holdfast and thus form rosettes of up to 50 cells. Older parts of the stalks (in the center of the rosette) may have depositions of iron which thus render the stalks thicker at the base. Multiplication is by terminal or lateral budding of mature or occasionally even smaller, immature cells. Planktonic bacteria that appear primarily in the autumn in aerobic surface layers of lakes.

Have not been grown in artificial media.

Type species: *Planctomyces bekefii* Gimesi 1924, 4.

Further Comments

Originally this genus was placed among the fungi (Gimesi, 1924). A fresh water bacterium described as *Blastocaulis sphaerica* (Henrici and Johnson, 1935) appeared to be identical with *P. bekefii* and was regarded by Hirsch (1972) as a subjective synonym.

Key to the species of genus **Planctomyces**

I. Cells wider than 0.7 μm.
    A. Cells spherical; stalks up to 10 μm long and 0.4–0.9 μm wide, often encrusted with ferric hydroxide.

1. *Planctomyces bekefii*

    B. Cells globose, stalks very short or absent; cells attached to a central material containing iron; buds terminal.

2. *Planctomyces condensatus*

II. Cell diameter less than 0.7 μm.
    A. Cells spherical to oblong, the long axis perpendicular to the main stalk axis. Stalks may be slightly curved, attached at the base to form rosettes. Iron may be deposited.

                  3. *Planctomyces kljasmensis*

    B. Cells small and spherical; stalks thin and crsecent-shaped, up to 11 μm long; rosettes of few individuals. Iron deposition not reported:

                  4. *Planctomyces gracilis*

Description of the species of genus **Planctomyces**

1. **Planctomyces bekefii** Gimesi 1924, 4. (*Blastocaulis sphaerica* Henrici and Johnson 1935, 84; *Actinothrix globulifera* Nováçek 1938, 66; *Planctomyces subulatus* Wawrik 1952, 448; *Planctomyces crassus* Hortobágyi 1965, 111.)

be.ke'fi.i. M.L. gen.n *bekefii* of Bekefi, a Hungarian abbot.

Mature cells spherical, 1.4–1.7 μm in diameter, attached to long, slender stalks which radiate from a common center to form rosettes. Multiplication by terminal or lateral and often multiple budding. Cells colorless. Smaller individuals stain evenly and are Gram-positive. Larger and older cells are Gram-negative and stain deeply at the distal region. Stalks slightly flexible and of varying lengths; cells may fall off the stalks. Lengthening occurs from the end of origin (excretion?), older parts of the stalk may be heavily encrusted with iron.

Heterotrophic, psychrophilic, probably aerobic.

Originally observed in the September plankton of Lake Lágymányos (Budapest, Hungary). Widely distributed in aerobic zones of fresh water lakes.

Illustrations: Henrici and Johnson (1935) Plate 3, Fig. 2, and Plate 2, Fig. 14.

2. **Planctomyces condensatus** Skuja 1964, 16.

con.den.sa'tus. L. adj. *condensatus* (cells) close together.

Cells spherical, 0.7–1.7 μm in diameter, colorless, in groups attached to a common central material which may contain iron deposits. Pedicels very short and thin, gel-like, occasionally occur. Buds terminal. Cells appear to break up into two to four gonidia.

Originally found in Lake Vuolep Njakajaure (Northern Sweden) at a depth of 13.5 m. Frequently found in fresh water lakes of Michigan.

3. **Planctomyces kljasmensis** (Razumov) Hirsch 1972, 110. (*Galionella* (sic) *kljasmensis* Razumov 1949, 446; *Blastocaulis kljasmensis* Zavarzin 1961, 961; *Gallionella planctonica* Razumov acc. to Krasil'nikov 1949, 671.)

kljas.men'sis, M.L. adj.suff. *-ensis* belonging to; M.L. adj. *klajasmensis* pertaining to Klajasma, U.S.S.R.

Cells spherical to oblong, 0.3–0.5 μm in diameter, occasionally larger. Stalks slender and often encrusted with ferric hydroxide. Oblong cells attached with their long axis perpendicular to the main stalk axis. Rosettes of up to 50 cells have been observed. Multiplication presumably by budding.

Originally found in the plankton of Kljasma water reservoir near Moscow, Russia.

4. **Planctomyces gracilis** Hortobágyi 1965, 112.

gra'ci.lis. L. adj. *gracilis* slender.

Cells spherical 0.3–0.6 μm in diameter. Stalks slender (diameter of 0.3–0.4 μm), up to 11 μm long and crescent-shaped, joining at the base to form a small rosette.

Originally found in the autumn plankton of the Buzsák fish ponds (Hungary).

Species incertae sedis

A number of additional putative species have been described as *Planctomyces*.

    a. *Planctomyces ferrimorula* Wawrik 1956, 296, and

    b. *Planctomyces guttaeformis* Hortobágyi 1965, 111, do not have stalks.

    c. *Planctomyces stranskae* Wawrik 1952, 448, appears to be similar to the iron-depositing *Gallionella corneola* Dorff 1934, 26 (*Siderophacus corneolus* (Dorff) Beger 1949, 12) (see *Nevskia: Species incertae sedis*).

Genus **Metallogenium** Perfil'ev and Gabe 1961, 50

G. A. ZAVARZIN AND P. HIRSCH

Met.al.lo.ge'ni.um. Gr. n. *metallos* metal; Gr. v.suff. *-genium* producing; M.L. neut.n. *metallogenium* metal-producing.

Cells coccoid, 0.5–1.5 µm in diameter (Plate 4.4, Fig. 1) and do not have a rigid cell wall; attach to surfaces. They either germinate directly or give rise, by repeated budding, to a group of round "elementary bodies." Germination of these bodies results in the formation of one to several flexible filaments with tapering ends (Plate 4.4, Figs. 2–9) and, according to Perfil'ev, occasional branching. In fixed and stained preparations the filaments or arais (Gr. *araios* thin strand) have an approximate diameter of 0.02–0.25 µm (Plate 4.4, Fig. 1) and may give a beaded appearance, similar to a rosary. The cells appear highly refractile when growing in the presence of manganese.

Microcolonies or coenobia growing in microcapillaries show one to several central cocci and the thin tapering arais radiating outward (Plate 4.4, Figs. 3–9). Coenobia may grow to a diameter of 10 µm (Dubinina, personal communication) or up to 1 mm (Perfil'ev and Gabe, 1961).

Multiplication is probably by a budding process. Cocci may also be formed at the end of arais and may remain attached to the filament, thus giving rise to daughter microcolonies around the periphery of the mother coenobium, much like strawberry plants with their runners and daughter plants around them (Plate 4.4, Figs. 5, 7, 9).

Microcolonies assume a brownish to black color in the presence of oxidizable manganese, beginning in the center of the colony. Manganese and/or iron oxides are then deposited onto the arais, increasing the width of the filaments from the base to the tip; they may thicken to a point where they join in the center to form dense lobate masses.

Gram reaction not recorded. For light microscopy and staining the metal deposits must be removed by treatment with oxalic acid (0.2–1.0%) or EDTA, disodium salt (1–2%). Cells stain satisfactorily with carbol-gentian violet but poorly with carbol-erythrosine, methylene blue, aqueous or alcoholic gentian violet or crystal violet.

Heterotrophic. Growth in artificial medium is enhanced by the presence of living fungi. Growth of pure cultures in artificial culture media is slow; better growth is obtained in mixed populations. On ferrous ammonium citrate (1.5 g/liter) agar medium of pH 6.0–7.8 or on "Novogrudskii water agar" growth can be recognized after 4–6 weeks; manganese deposition (if present) is then meager. Colonies appear as irregularly shaped brownish spots. No growth on meat broth-peptone-KNO_3 agar at pH 7.0. Optimal pH: 6.8–7.0. Manganous compounds are not required but manganous carbonate and a low redox potential are stimulatory. Have been grown in pure culture in media containing serum such as is used for mycoplasmas.

Found in upper layers of bottom deposits in fresh water, planktonic in fresh water lakes and ponds and in microzones in soils of the northern hemisphere.

Type species: *Metallogenium personatum* Perfil'ev and Gabe 1961, 50.

Further Comments

The taxonomic position of these organisms is uncertain. Recent publications of Dubinina (1969, 1970) and Balashova (1969) point to similarities between *Metallogenium* and *Mycoplasmatales*, especially *Acholeplasma* spp. There is also some resemblance to *Caulococcus* and *Kusnezovia*, which may also be related to the mycoplasmas.

An acid-tolerant, iron-oxidizing organism, resembling *Metallogenium* spp., has recently been isolated in pure culture (Walsh and Mitchell, 1972).

Key to the species of genus **Metallogenium**

I. Tapering filaments contain minute cells which become visible after dissolution of the manganese oxides; grow in absence of living fungi although growth is stimulated by their presence.

1. Metallogenium personatum

II. Filaments never contain minute cells. Multiplication by formation of single buds terminally or laterally on the arais. Grow only in presence of living fungi, yeasts or bacteria.

2. Metallogenium symbioticum

Description of the species of genus **Metallogenium**

1. Metallogenium personatum Perfil'ev and Gabe 1961, 50. (*Metallogenium invisum* Perfil'ev 1952, 335; *Leptothrix echinata* Beger 1935, 401.)

per.so.na'tum. L. neut.adj. *personatum* wearing a shield, masked, referring to covering of manganese oxide.

Morphology and staining reactions as for the genus. Multiplication by formation of single coccal cells or rows of cells at the tips of filaments.

Dubinina (1970) reported good growth in a starch hydrolysate (1 g/liter), NH_4MgPO_4 (0.1 g/liter) containing serum and ox heart infusion. The elementary bodies were 0.02–0.3 µm in diameter and the arais 0.02–0.06 µm; growth was visible only with the electron microscope. Addition of manganese stimulated elongation of the arais to a length of 12 µm. These filaments showed lateral budding.

Originally found in mud from Lake Ukshezero in the southern part of the Karelian ASSR. Widely distributed in lake muds, in the plankton of fresh water lakes and in soils of the northern hemisphere.

Further Comment

Although Perfil'ev and Gabe (1961) indicated the organism was motile by flagella, flagella have not yet been demonstrated.

When first described the organism was not seen but its presence was deduced by the metal deposits, hence the original name *M. invisum*. After the "shield" was removed, a clear description was published and a more appropriate name given.

2. **Metallogenium symbioticum** Zavarzin 1961, 395.

sym.bi.ot'i.cum. M.L. neut.adj. *symbioticum* living together.

Arais do not contain minute elementary bodies. Multiplication by single bud formation at the tips or laterally on the arais (Zavarzin, 1963). The buds with a diameter of 0.5 μm already show manganese deposition.

Grow in the presence of fungi or bacteria. In the mixed cultures the fungi do not sporulate nor-mally. The coenobia appear as small spiders on the surface and between the hyphae of the fungus; they may also develop at a distance from the fungal hyphae. It is assumed that the organism can parasitize the fungus by intracellular growth.

Good growth on medium containing manganous acetate (0.1 g/liter) in water agar (2%) and no added source of nitrogen or phosphorus. Fungal colonies appear as concentric rings with the coenobia of *M. symbioticum* in between.

Slow but good growth occurs in liquid media containing 2% starch or 1–2% gum arabic plus freshly prepared manganous carbonate, pH 6.2. At first the fungal hyphae grow, then the medium suddenly becomes turbid from the coccoid bodies of *Metallogenium*. These bodies produce one or more arais, become encrusted with manganic oxide and settle as a precipitate within a few hours.

Oxidation of manganous salts was found to be proportional to the growth of *M. symbioticum* by direct colony count. Heating for 30 min at 60 C resulted in death of the fungi but a slow abiogenic oxidation of the manganous salts continued; the oxidation rate diminished upon transfer to fresh medium (Kossaya, 1967).

Mesophilic, aerobic; optimum pH 6–8.

Originally found in a swamp near Moscow, USSR. Widely distributed in lake mud and soil.

Genus **Caulococcus** *Perfil'ev 1964, 43*

G. A. ZAVARZIN AND P. HIRSCH

Cau.lo.coc'cus. Gr. n. *caulis* a stem; Gr. n. *coccus* a berry; M.L. masc.n. *Caulococcus* coccoid cells connected by filaments.

Coccoid cells ca. 0.5 μm diameter, which may be connected by fine threads of ca. 0.1 μm diameter. Multiply apparently by budding when in zoogloeal accumulations. Young cells motile, type of motility unknown. Microcolonies up to 0.3 mm with reticulate-ribbed surface structure; irregularly shaped and round, almost black (most common type), or radial-lobate (rare) or trichospherical (very rare). The margin of microcolonies often radially fringed by dichotomously branched filamentous cells.

The coccoid cells may be sparse (radial-lobate types of colonies) or frequent (round, heavily encrusted colonies).

Microaerophilic; deposits mainly Mn^{3+} (80–90%) and some Fe^{3+} (20–10%). Gram reaction not recorded.

Not cultured in the laboratory but can be transferred with mud samples (Perfil'ev and Gabe, 1961, 1964) and observed with the microcapillary technique.

Originally found abundantly in Lake Khepo-Yarvi (Karelian ASSR) especially in ore deposits; frequently observed in the upper layers of bottom mud deposits above the reducing horizon or in the bottom water over the mud surface. Also observed in several other lakes of that area.

Type species: *Caulococcus manganifer* Perfil'ev 1964, 43.

Description of the species of genus **Caulococcus**

1. **Caulococcus manganifer** Perfil'ev 1964, 43.

man.ga'ni.fer. M.L. n. *manganus* manganese; L. v. *fero* carry, bring, bear; M.L. n. *manganifer* the manganese-bearer.

Morphology and description as for genus.

Further Comments

Zavarzin considers it possible that *C. manganifer* represents another species of the genus *Metallogenium*. Hirsch (unpublished) relates this organism to *Acholeplasma* spp.

Genus **Kusnezovia** *Perfil'ev 1964, 40*

G. A. ZAVARZIN AND P. HIRSCH

Kus.ne.zo′vi.a. M.L. fem.n. *Kusnezovia;* named in honor of the Russian microbiologist, S. I. Kusnezov.

Cells coccoid, 0.5–1.5 μm in diameter, connected by filaments of ca. 0.1–0.2 μm diameter. Non-motile, attached, multiplication by budding. Stainable with carbol gentian violet after removal of the often heavy encrustations of manganese with oxalic acid. Gram reaction not recorded.

Microcolonies and the rows of coccoid cells can be observed with submerged peloscopes (Perfil'ev and Gabe, 1961, 1964); they are attached by a stalk to the previously exposed capillary wall and hang freely downward into the capillary canal. The shapes of microcolonies are extremely variable; the following types of manganese-encrusted micro-colonies have been observed: arrow-shaped shoots, open lily-of-the-valley leaves, goblets with a toothed margin, or compound, racemose structures resembling the lichen *Cetraria islandica*. In pelo-scopes, formations may also resemble clubs or candelabras, always pointing downward toward the reducing horizon with their wider parts. Layers of colonies in blackish brown horizons.

Budding can be observed in single-layered accumulations of the coccoid cells found in filamentous coenobia or leaflike formations.

Apparently microaerophilic. Development is stimulated by decomposing fungal hyphae. Manganese, and to much lesser extent iron, is oxidized and deposited.

Has not been cultured in the laboratory.

Originally found in some mud samples of Lake Ukshezero, Karelian ASSR; observed in secondary profiles in zones of manganese oxidation of mud samples stored in the laboratory. Rare.

Type species: *Kusnezovia polymorpha* Perfil'ev 1964, 40.

Description of the species of genus **Kusnezovia**

1. **Kusnezovia polymorpha** Perfil'ev 1964, 40.
po.ly.mor′pha. Gr. adj. *poly* many; Gr. n. *mor-phus* form, shape; M.L. fem.adj. *polymorpha* of many shapes.

Morphology and description as for genus.

Important Notes

for

Users of this Edition

1. Always read both generic and species descriptions because characters listed in the generic description are not usually listed in the species descriptions.

2. In tables, characters common to all taxa are not shown but may be listed in foot-notes.

3. Generally in tables (exceptions are clearly indicated in footnotes, q.v.) the meanings of symbols are as follows:

+   more than 90 % strains positive
−   more than 90 % strains negative
d   11–89 % strains positive
( )   delayed reaction
w   weak reaction
D   Different reactions in different taxa (species of a genus or genera of a family)
v   strain instability (NOT differences between strains)

PART 5

THE SPIROCHETES

ORDER I. **SPIROCHAETALES** *BUCHANAN 1917, 163*
ROBERT M. SMIBERT

Slender, flexuous, helically coiled, unicellular bacteria 3–500 μm in length (span of helical organism) with one or more complete turns in the helix (Plate 5.1, Figs. 1–5). May occur in chains held together by the outer envelope, and may assume shape of a planar wave. Multiplication is by transverse fission.

Electron microscopic studies revealed that the cells consist of a protoplasmic cylinder intertwined with one or more axial fibrils which originate in approximately equal numbers from subterminal attachment discs located at either end of the protoplasmic cylinder. Both the protoplasmic cylinder and axial fibrils are enclosed by the outer envelope. The axial fibrils extend toward the end of the cell opposite the points of insertion and overlap one another. The unattached ends of the axial fibrils may extend beyond the ends of the protoplasmic cylinder giving the impression of external polar flagella; however, they are enclosed within the outer envelope.

Motile. Motility is of three types; rapid rotation about the long axis of the helix, flexion of the cells, and locomotion along a helical (corkscrew) or serpentine path.

Do not form endospores.

Larger cells are Gram-negative. Some spirochetes possess inclusions and show cross-striations in stained preparations. Those spirochetes with a transverse cell diameter near the limit of resolution achievable in bright-field microscopy are best observed by phase-contrast or dark-field microscopy.

Tendency to develop involution forms. Distention of the outer envelope may result in terminal or central swellings or spirochetal spheres in which the protoplasmic cylinder is coiled within the outer envelope. Such forms are probably degenerate. Rupture of the outer envelope may release the axial fibrils simulating flagella. Neither genetic transfer nor a life cycle has been established.

May be aerobic, facultatively anaerobic or anaerobic.

Chemoheterotrophic.

Free-living, commensal or parasitic. Some species are pathogenic.

FAMILY I. **SPIROCHAETACEAE** SWELLENGREBEL 1907, 581

Description as for the order.

Key to the genera of family **Spirochaetaceae**

I. 5–500 μm in length, 0.2–0.75 μm wide, anaerobic or facultatively anaerobic. Free-living in H_2S-containing fresh water and marine environments, in sewage and polluted water.

Genus I. *Spirochaeta*

II. 30–150 μm in length; 0.5–3.0 μm wide with 3 to 10 complete turns. In living specimens ovoid inclusion bodies and large bundles of axial fibrils may be seen by phase-contrast microscopy. Commensal, usually in mollusks.

Genus II. *Cristispira*

III. 5–15 μm in length; 0.09–0.5 μm wide. Catalase and oxidase negative; anaerobic; commensal or parasitic. Some are pathogens.

Genus III. *Treponema*

IV. 3–15 μm in length; 0.2–0.5 μm wide; anaerobic; parasitic. Some are pathogens, transmitted by ticks and lice.

Genus IV. *Borrelia*

V. 6–20 μm in length; 0.1 μm wide; tightly coiled and may have bent or hooked ends; aerobic; free-living or parasitic. Some are pathogens.

Genus V. *Leptospira*

Genus I. **Spirochaeta** *Ehrenberg 1835, 313*

E. Canale-Parola

(*Spirochoeta* Dujardin 1841, 225, and *Spirochaete* Cohn 1872, 180, orthographic variants of *Spirochaeta*; *Ehrenbergia* Gieszczykiewicz 1939, 24.)

Spi.ro.chae'ta. Gr. n. *spira* a coil; Gr. n. *chaete* hair; M.L. fem.n. *Spirochaeta* coiled hair.

Helical cells, 0.20–0.75 by 5–500 μm (Plate 5.1, Fig. 1). The cells have **axial fibrils.** The protoplasmic cylinder is wound around or together with the axial fibrils. The cells do not possess either terminal hooks or cross-striations. Motile, probably by means of the axial fibrils. **Free-living** in H_2S-containing mud, in sewage and polluted water Non-parasitic.

Of the five species described, only the type species has not been grown in pure culture. The four species which have been cultured are **chemoorganotrophs, obligate and facultative anaerobes, ferment carbohydrates** and have a **guanine + cytosine content** in their DNA ranging **from 50–66 moles** % (buoyant density). Cells of these four species possess two axial fibrils, one inserted near one end of the protoplasmic cylinder and the other near the opposite end. The two fibrils overlap in the central portion of the cell (1-2-1 arrangement). The protoplasmic cylinder and axial fibrils of these species are enclosed in a sheath or outer cell envelope.

One species is pigmented.

Type species: *Spirochaeta plicatilis* Ehrenberg 1835, 313.

Further Comments

The genus includes two groups of organisms distinguishable on the basis of their relation to molecular oxygen. One of these groups comprises the strict anaerobes (e.g. *S. stenostrepta* and *S. zuelzerae*), the other the facultative anaerobes

(e.g. *S. aurantia* and similar spirochetes, see below). Each of these groups should constitute a separate genus (see Canale-Parola *et al.*, 1968). However, revision of the existing classification of organisms presently assigned to the genus *Spirochaeta* is hampered by the lack of information on *S. plicatilis*, the type species (e.g. the relation of this organism to O_2 is not known).

Among the free-living spirochetes capable of anaerobic growth there exists greater diversity than is reflected by the species recognized heretofore. A free-living, strictly anaerobic spirochete (strain Z4), resembling *S. zuelzerae* morphologically, but differing in certain physiological properties, has been isolated (Canale-Parola *et al.*, 1968). The spirochete ferments glucose to acetic, lactic, succinic and formic acids, ethanol, CO_2 and H_2. Unlike *S. zuelzerae*, it does not require an exogenous source of CO_2, and the G + C content of its DNA is 59.2 moles % (buoyant density). Two strains of free-living spirochetes physiologically similar to *S. aurantia*, but possessing fine, tight, regular coils (instead of the loose, irregular coils typical of the latter species) have been cultured. These tightly coiled spirochetes, presently named only by strain designation (J4T and J5), are facultatively anaerobic and form a yellow-orange pigment.

Many species which had originally been placed by various authors in the genus *Spirochaeta* were subsequently assigned to other genera (see Buchanan *et al.*, 1966, 1039). The descriptions of

Spirochaeta daxensis, Spirochaeta marina and *Spirochaeta eurystrepta*, which were included in the previous edition of Bergey's MANUAL, are omitted here. *S. daxensis* Cantacuzène 1910, 77 is not included because of the limited information available on this organism. *S. marina* Bergey *et al.* 1923, 420 and *S. eurystrepta*, also referred to as *S. plicatilis* subsp. *marina* and *S. plicatilis* subsp. *eurystrepta* by Zuelzer 1912, 17, are probably strains of *S. plicatilis*.

Key to the species of genus **Spirochaeta**

I. Large, helical cells, presumably obligate or facultative anaerobes.

          1. *Spirochaeta plicatilis*

II. Thin, helical cells.

  A. Obligate anaerobes.

      a. Ferment glucose to ethanol, acetic acid, lactic acid, CO_2 and H_2. Long cells (15–45 μm, up to 300 μm).

          2. *Spirochaeta stenostrepta*

      b. Ferment glucose to acetic acid, lactic acid, succinic acid, CO_2 and H_2.

          3. *Spirochaeta zuelzerae*

      c. Ferment glucose to ethanol, acetic acid, CO_2, H_2 and small amounts of lactic, formic and pyruvic acids. Require exogenous supplements of Na^+, Cl^-, biotin, niacin and coenzyme A for growth.

          4. *Spirochaeta litoralis*

  B. Facultative anaerobes.

          5. *Spirochaeta aurantia*

Description of the species of genus **Spirochaeta**

1. Spirochaeta plicatilis Ehrenberg 1835, 313.

pli.ca′ti.lis. L. adj. *plicatilis* flexible.

Helical cells, 0.5–0.75 by 100–200 μm, rarely as long as 500 μm. Cells have regular primary coils which are stable (they persist both in the presence and absence of movement). Cells in motion may exhibit broad secondary coils or waves on which the smaller primary coils are superimposed. Metachromatic granules are present in the cells.

Not cultivated in pure culture. Presumed to be an obligate or facultative anaerobe because of its occurrence in anaerobic environments. First seen by Ehrenberg (1835) in "überwinterten" water. Observed in H_2S-containing fresh and sea water mud (Zuelzer, 1912).

Photomicrographs and drawings of this species have been published (Ehrenberg, 1838; Zuelzer, 1912).

2. Spirochaeta stenostrepta Zuelzer 1912, 17.

ste.no.strep′ta. Gr. adj. *stenus* narrow; Gr. adj. *streptus*, pliant, easily bent; M.L. adj. *stenostrepta*, tightly coiled.

Helical cells, 0.2–0.3 by 15–45 μm. A small percentage of the cells present in cultures is shorter than 15 μm. In the late exponential and stationary phases the cells increase in length (up to 300 μm). Long organisms occasionally pair and become entwined, or a single organism becomes partially wrapped around itself. The cells have regular, stable primary coils. Cells in motion occasionally

exhibit broader secondary coils or waves, on which the smaller primary coils are superimposed. Spherical bodies generally 1–3 μm in diameter are occasionally observed in cultures. The spherical bodies occur either free or in association with helical cells. Two subterminally inserted axial fibrils are present in a 1-2-1 arrangement. Subsurface colonies in glucose-yeast extract-peptone-thioglycolate agar (see Canale-Parola *et al.*, 1968) are white, spherical, fluffy, approximately 2 mm in diameter when fully developed. Smaller spherical colonies, lacking the characteristic fluffiness, are also present.

Obligate anaerobe. Fermentative metabolism. Gram-negative. Catalase negative. L-Arabinose, D-ribose, D-xylose, D-fructose, D-galactose, D-glucose, D-mannose, cellobiose, lactose, maltose, sucrose are fermented. Products of glucose fermentation (strain Z1, growing cells, micromoles/100 μmoles of glucose): ethanol, 84; acetic acid, 93; lactic acid, 10; CO_2, 140; H_2, 180 (Canale-Parola *et al.*, 1967). Glucose is fermented to pyruvate via the Embden-Meyerhof pathway. Pyruvate is metabolized to acetyl-coenzyme A, CO_2 and H_2 by means of a phosphoroclastic system similar to that present in saccharolytic clostridia. Acetyl-coenzyme A is reduced to acetaldehyde and this intermediate to ethanol. Acetate is produced via the formation of acetyl-phosphate from acetyl-coenzyme A (Hespell and Canale-Parola, 1970).

Growth reported only on complex media. Minimal requirements unknown. Grows between 15–40 C; optimum temperature range 35–37 C. Optimum growth yields result when the initial pH of the medium is between 7.0–7.5.

The G + C content of the DNA is 60.2 moles % (strain Z1, buoyant density).

Originally isolated from H_2S-containing mud of a fresh water pond.

Phase-contrast photomicrographs and electron micrographs of this species have been published (Canale-Parola *et al.*, 1967, 1968; Holt and Canale-Parola, 1968).

Reference strain: Z1; ATCC 25083.

3. **Spirochaeta zuelzerae** (Veldkamp) Canale-Parola, Udris and Mandel 1968, 389. (*Treponema zuelzerae* Veldkamp 1960, 122.)

zuel′ze.rae. M.L. gen.n. *zuelzerae* of Zuelzer; named for Margarete Zuelzer who described the occurrence of morphologically diverse spirochetes in sulfide-containing environments.

Helical cells, 0.20–0.35 by 8–16 μm. Shorter cells (as short as 2–3 μm) are occasionally observed in cultures. Long organisms, up to 80 μm, are present in old cultures. Exponentially growing cells have fairly regular, stable primary coils. Secondary coils or waves are present infrequently. Spherical bodies, generally not exceeding 3–4 μm in diameter, are formed, usually at the end of cells in the stationary phase of growth. Two subterminally inserted axial fibrils are present in a 1-2-1 arrangement. Subsurface colonies in agar media (for medium composition see Veldkamp, 1960) are white, fluffy, spherical, with a tendency to diffuse in the agar medium. Disc-shaped colonies are present occasionally.

Obligate anaerobe. Fermentative metabolism. Gram-negative. Catalase negative. L-Arabinose, D-xylose, D-galactose, D-glucose, D-mannose, cellobiose, maltose, trehalose and starch are fermented. L-Rhamnose, D-fructose, L-sorbose, lactose, sucrose, raffinose, inulin, mannitol and sorbitol are not fermented. Products of glucose fermentation (Veldkamp's strain, cells growing in medium including 0.5% $NaHCO_3$, micromoles/100 μmoles of glucose): acetic acid, 82; lactic acid, 87; succinic acid, 13; CO_2, 68; H_2, 164 (Veldkamp, 1960).

Grows at 20 C, does not grow at 45 C; optimum temperature range 37–40 C. Optimum growth yields result when the initial pH of the medium is between 7 and 8. Inorganic ammonium salts or nitrates not utilized as sole nitrogen sources. Added CO_2 is an absolute requirement for growth. Growth reported only on complex media. Minimal requirements unknown.

Has a protein antigen that gives a positive complement fixation reaction with syphilitic serum.

The G + C content of the DNA is 56.1 moles % (ATCC strain 19044, buoyant density).

Originally isolated from an enrichment culture for green photosynthetic bacteria inoculated with sulfide-containing mud from a fresh water pond.

Phase-contrast photomicrographs and electron micrographs of this species have been published (Veldkamp, 1960; Canale-Parola *et al.*, 1968).

Type strain: ATCC 19044.

4. **Spirochaeta litoralis** Hespell and Canale-Parola 1970, 1.

li.to.ra′lis. L. adj. *litoralis* of the shore.

Helical cells, 0.4–0.5 by 5.5–7 μm. Cells are regularly and tightly coiled during the exponential phase of growth. Spherical bodies (2–3.5 μm in diameter) are present in the stationary growth phase and under unfavorable growth conditions (i.e. in the presence of O_2). Two subterminally inserted axial fibrils are present in a 1-2-1 arrangement. Subsurface colonies in agar media are spherical, fluffy, cream-colored, 1–5 mm in diameter. Surface colonies (anaerobic) are round, growing partially within the agar medium, cream-colored, 2–5 mm in diameter.

Obligate anaerobe. Fermentative metabolism. Gram-negative. Catalase negative. D-Arabinose, L-arabinose, L-fucose, L-rhamnose, D-fructose, D-galactose, D-glucose, D-mannose, cellobiose, lactose, maltose, sucrose, trehalose, inulin, raffinose are fermented. D-Lyxose, L-lyxose, D-ribose, L-xylose, ethanol, methanol, pectin, α-ketoglutarate, citrate, fumarate, malate, oxaloacetate, succinate, gluconate, tartrate, allantoin, uric acid, orotic acid, glucosamine, amino acids and sugar alcohols are not fermented. Products of glucose fermentation (strain R1, cell suspensions, micromoles/100 μmoles of glucose): ethanol, 140.5; acetate, 57; CO_2, 201.8; H_2, 74.4; and trace amounts of lactate, formate and pyruvate (Hespell and Canale-Parola, 1970). Nitrite is not accumulated in the medium by cells growing in the presence of nitrate.

Grows in media prepared with sea water but not with fresh water unless NaCl is added (minimum concentration, 0.05 M; optimum, 0.35 M). Has specific requirements for Na^+ and Cl^-. Exogenous supplements of biotin, niacin and coenzyme A are required for growth. Coenzyme A may be replaced by pantothenate, but the resulting cell yields are low. Added thiamine is stimulatory for growth. A reducing agent (sulfide or cysteine) is required for growth in laboratory media. Grows in chemically defined media containing glucose, NH_4Cl or amino acids, sulfide, NaCl, vitamins, coenzyme A and inorganic salts.

Temperature optimum: near 30 C. Slow growth at 15, no growth at 5 or 40 C. Optimum growth yields result when the initial pH of the medium is between 7 and 7.5.

The G + C content of the DNA is 50.5 moles % (strain R1, buoyant density).

Isolated from sulfide-containing marine mud. Phase-contrast photomicrographs and electron micrographs of this species have been published (Hespell and Canale-Parola, 1970).

Type strain: R1.

5. Spirochaeta aurantia Vinzent 1926, 1473.

au.ran'tia. M.L. n. *aurantium* the orange; M.L. adj. *aurantia* gold-colored, orange-colored.

Helical cells, 0.3 by 5–35 µm. Most cells measure 10–20 µm during exponential growth. Generally the cells are loosely and irregularly coiled. Regularly coiled cells are present in young cultures. Spherical bodies 0.5–2 µm in diameter are present, especially in the stationary phase of growth. The spherical bodies are either in association with cells or free. Two subterminally inserted axial fibrils are present in a 1-2-1 arrangement. Colonies on aerobic plates (maltose-peptone-yeast extract medium containing 1 g of agar/100 ml, see Canale-Parola *et al.*, 1968) are 2–4 mm in diameter, yellow-orange, round with slightly irregular edges, growing partially under the surface, and with a raised central portion. At low carbohydrate concentrations (see Breznak and Canale-Parola, 1969) the colonies are larger, diffuse in the agar medium in the shape of almost perfect circles, and their pigmentation is not readily apparent. Subsurface anaerobic colonies are white, spherical, fluffy, approximately 1 mm in diameter.

Facultative anaerobe. Metabolism fermentative and, probably, respiratory. Gram-negative. Catalase positive. L-Arabinose, D-xylose, L-rhamnose, D-fructose, D-galactose, D-glucose, D-mannose, cellobiose, lactose, maltose, sucrose, trehalose, dextrin, inulin, glycerol and mannitol are utilized for growth. D-Ribose, L-sorbose, raffinose, ethanol, dulcitol, sorbitol, allantoin, sorbitan monooleate polyoxyethylene (Tween 80) and acetic, α-ketoglutaric, fumaric, lactic, orotic, pyruvic, succinic, uric acids are not utilized. Amino acids are not used as energy sources. Amino acids serve as sole nitrogen sources; inorganic ammonium salts and nitrate do not. Exogenous biotin and thiamine are required for growth. Nitrate is reduced to nitrite. Products of glucose fermentation (strain J1 growing anaerobically, micromoles/100 µmoles of glucose): ethanol, 151; acetic acid, 69; formic acid, 5; lactic acid, 1; CO_2, 165; H_2, 107. Trace amounts of acetoin and diacetyl are formed (Breznak and Canale-Parola, 1969).

Slow growth at 15 C, no or poor growth at 37 C. Optimum temperature is 30 C. Optimum growth yields result when the initial pH of the medium is 7.0–7.3.

A yellow-orange, carotenoid pigment is produced aerobically.

The G + C content of the DNA is 66 moles % (strain J1, buoyant density).

Isolated from sewage and from mud of a duck pond.

Phase-contrast photomicrographs and electron micrographs of this species have been published (Canale-Parola *et al.*, 1968; Breznak and Canale-Parola, 1969).

Reference strain: J1 (Breznak and Canale-Parola, 1969); ATCC 25082.

Genus II. **Cristispira** *Gross 1910, 44*

Daisy A. Kuhn

Cris.ti.spi'ra. L. fem.n. *crista* a crest; Gr. fem.n. *spira* a coil; M.L. fem.n. *Cristispira* crested coil.

Cells single, flexuous, 0.5–3.0 µm in diameter and **helically coiled** or **planar wave-shaped** (Plate 5.1, Fig. 2). **Helix** spans **more than 30** and **up to 150 µm**, and has **2–10 complete turns.** Ends of cells either **rounded**, or **tapered**, or **pointed** with a rigid spicule. Divide by **transverse fission.**

The cells may contain 30–80 seriately aligned **ovoid inclusions** of unknown composition, which **displace** the **nuclear structure** so that it appears as **cross-striations** in stained preparations.

The normal movement is a **forward and backward locomotion** by means of an irrotational traveling helical wave. The helix may also rotate on its long axis or flex. There are **over 100 axial fibrils** intertwined with the protoplasmic cylinder; when grouped in a bundle, these may be visible by phase-contrast microscopy.

When removed from or disturbed in its habitat, **involutionary changes readily occur.** The axial fibrils may become detached from the protoplasmic cylinder within a distended spirochetal envelope and form the **so-called crista**; central and terminal bullae and spirochetal spheres ("cysts") may develop. Motility becomes irregular and cells may lyze.

Commensal: Widely distributed among **marine and fresh water** species of **bivalve and univalve mollusks,** usually found in the crystalline style (a rod-shaped structure composed of "mucus" secretions) or the fluid of the digestive tract. May also occur in other animals: large spirochetes have been observed in an echinoderm (Collier, 1921) and in tunicates (Hellmann, 1913). No species of *Cristispira* has thus far been grown in pure culture.

Proposed type species: *Cristispira pectinis* Gross 1910, 44 (Kuhn, 1970).

Further Comments

The genus *Cristispira* includes the large spirochetes living in the digestive tract of many mollusks. Certes (1882) discovered a cristispira in oysters, but named it *Trypanosoma balbianii*. In fixed and stained preparations he had observed a crest along the helical body, which he interpreted as an undulating membrane. Laveran and Mesnil (1901) considered the likelihood of the trypanosome of oysters being a bacterium related to spirilla and spirochetes. Gross (1910) proposed the genus *Cristispira* for these helically coiled, crested, flexible bacteria.

Recent investigations show *Cristispira* to possess the characteristic ultrastructure (Ryter and Pillot, 1965) and locomotion (Jahn and Landman, 1965) of spirochetes. Unfavorable conditions cause the development of the crista, incurved state, spirochetal sphere, and bullae; such structures had earlier been mistaken for the undulating membrane, longitudinal division and sexual stages of protozoa (*cf.* notably Perrin, 1906).

Cristispira can be recognized readily in crystalline styles or the fluid of the digestive tract at low magnification in dark-field or phase-contrast illumination by its characteristic morphology, size and locomotion. It is advisable to search for *Cristispira* in mollusks from physiologically optimal conditions, for when the environment becomes adverse to their siphoning and feeding, the crystalline styles generally dissolve and the conspicuous populations of *Cristispira* disappear. Also, crystalline styles differ among species of mollusks and all types of crystalline styles are apparently not suitable environments for *Cristispira*.

In this Edition, only the description by Gross (1910) is presented for the proposed type species of the genus (Kuhn, 1970).

Numerous species have been named in the literature, but their descriptions are based solely on morphology and the commensals (mollusk symbionts or "hosts"). Until the organisms have been studied in pure culture and properties such as the extent of morphological variation, commensal range and commensal specificity are determined, these taxa are listed as *Species incertae sedis* (Table 5.1). To define the nomenclatural types (specific nomenifers), one commensal has been selected where more than one has been given in the original description, and the geographic source (type locality) has been recorded.

Spirochetes of smaller dimensions than *Cristispira* are also found in mollusks. They range in

TABLE 5.1

Species incertae sedis of the genus **Cristispira** *listed in chronological order of the original descriptions*

Nomenclatural type	Commensal[a]	Geographic source
C. balbianii (Certes) Gross 1910, 48 (*Trypanosoma balbianii* Certes 1882, 351; *Spirochaeta balbianii* (Certes) Swellengrebel 1907, 562; *Spirillum balbianii* (Certes) Macé 1913, 710)	*Ostrea edulis* Linnaeus 1766[a]	Arcachon, France (Marine)
C. anodontae (Keysselitz) Gross 1910, 57 (*Spirochaeta anodontae* Keysselitz 1906, 566; *Spirillum anodontae* (Keysselitz) Macé 1913, 711)	*Anodonta mutabilis* Clessin 1877 [*Anodonta cygnea* (Linnaeus 1758)]	?Hamburg, Germany (Fresh water)
C. pinnae (Gonder) Gross 1910, 57 (*Spirochaete* (sic) *pinnae* Gonder 1908, 491)	*Pinna squamosa* Gmelin 1791[a] [*Pinna nobilis* Linnaeus 1767]	?Gulf of Naples, Italy (Marine)
C. ostreae (Schellack) Noguchi 1918, 583 (*Spirochaeta ostreae* Schellack 1909, 409)	*Ostrea edulis* Linnaeus 1766	Rovingo, Adriatic Sea (Marine)
C. chamae (Schellack) Noguchi 1918, 583 (*Spirochaeta chamae* Schellack 1909, 409)	*Chama gryphoides* Linnaeus 1767[a]	Rovigno, Adriatic Sea (Marine)

TABLE 5.1—*Continued*

Nomenclatural type	Commensal[a]	Geographic source
C. spiculifera (Schellack) Gross 1910, 67 (*Spirochaeta spiculifera* Schellack 1909, 409)	*Anodonta mutabilis* Clessin 1877[a] [*Anodonta cygnea* (Linnaeus 1758)]	Berlin, Germany (Fresh water)
C. modiolae (Schellack) Gross 1910 46 (*Spirochaeta modiolae* Schellack 1909, 409)	*Modiola barbata* (Linnaeus 1767)	Rovigno, Adriatic Sea (Marine)
C. limae (Schellack) Gross 1910, 46 (*Spirochaeta limae* Schellack 1909, 410)	*Lima inflata* Lamarck 1819[a]	Rovigno, Adriatic Sea (Marine)
C. cardii - papillosi (Schellack) Ford 1927, 939 (*Spirochaeta cardii - papillosi* Schellack 1909 410)	*Cardium papillosum* Poli 1791	Rovigno, Adriatic Sea (Marine)
C. tapetos (Schellack) Gross 1910, 67 (*Spirochaeta tapetos* Schellack 1909, 410)	*Tapes decussatus* (Linnaeus 1767)	Rovigno, Adriatic Sea (Marine)
C. acuminata (Schellack) Gonder 1912, 495 (*Spirochaeta acuminata* Schellack 1909, 410)	*Tapes laeta* Wkff. [*Tapes aureus* (Gmelin 1791)]	Rovigno, Adriatic Sea (Marine)
C. saxicavae (Schellack) Ford 1927, 940 (*Spirochaeta saxicavae* Schellack 1909, 411)	*Saxicava arctica* (Linnaeus 1767)	Rovigno, Adriatic Sea (Marine)
C. gastrochaenae (Schellack) Ford 1927, 940 (*Spirochaeta gastrochaenae* Schellack 1909, 411)	*Gastrochaena dubia* (Pennant 1777)	Rovigno, Adriatic Sea (Marine)
C. mactrae (von Prowazek) Ford 1927, 940 (*Spirochaeta mactrae* von Prowazek 1910, 298)	*Mactra sulcataria* Deshayes 1853	Fukuoka, Japan (Marine)
C. veneris Dobell 1911, 508	*Venus casta* Chemnitz [*Meretrix casta* (Gmelin 1791)]	Tamblegam Lake, Ceylon (Salt water)
C. solenis (Fantham) *nov. comb.* (*Spirochaeta solenis* Fantham 1911, 480)	*Solen ensis* Linnaeus 1758 [*Ensis ensis* (Linnaeus 1758)]	(Marine)
C. helgolandica Collier 1921, 134	*Asterias rubens* Linnaeus 1767 (an echinoderm)	Helgoland, North Sea (Marine)
C. mina Dimitroff 1926, 159	*Ostrea virginiana* [*Ostrea (Crassostrea) virginica* Gmelin 1791]	Baltimore, U.S.A. (Marine)
C. tenua Dimitroff 1926, 160	*Ostrea virginiana* [*Ostrea (Crassostrea) virginica* Gmelin 1791]	Baltimore, U.S.A. (Marine)

[a] Commensals marked with an *a* were selected. The names of the commensals have been supplied with authorship and dating reference. Synonyms presently recognized by some experts on the taxonomy of mollusks are given in square brackets.

helical span from 6–25 μm, in width from 0.3–0.5 μm, and their ends taper into filaments. The ultra-structure of such a spirochete (Pillot and Ryter, 1965) confirms their morphological resemblance to treponemes and lends support to the reclassification of *Cristispira hartmanni* (Gonder) Gross 1910, 70 (*Spirochaete* (*sic*) *hartmanni* Gonder 1908, 493) to *Treponema hartmanni* (Gonder) Ford 1927, 990. Until the relationship of these organisms to *Cristispira* is further clarified, they are listed here as *species inquirendae* (Table 5.2).

Cristispira termitis (Leidy) Hollande 1922, 25 (*Vibrio termitis* Leidy 1881, 441; *Spirochaeta termitis* (Leidy) Dobell 1910, 81; *Treponema termitis* (Leidy) Dobell, 1912, 144) was observed in the intestine of termites and claimed to be different from *Cristispira* (Dobell, 1912). It may be related to the large insect spirochete described by Grimstone (1963) and discussed by Ryter and Pillot (1965).

TABLE 5.2

Species inquirendae of the genus **Cristispira** *listed in chronological order of the original description*

Nomenclatural type	Commensal[a]	Geographic source
C. hartmanni (Gonder) Gross 1910, 70 (*Spirochaete* (*sic*) *hartmanni* Gonder 1908, 493; *Treponema hartmanni* (Gonder) Ford 1927 990)	*Pinna squamosa* Gmelin 1791[a] [*Pinna nobilis* Linnaeus 1767]	Gulf of Naples, Italy (Marine)
C. pusilla (Schellack) Gross 1910, 44 (*Spirochaeta pusilla* Schellack 1909, 411)	*Anodonta mutabilis* Clessin 1877[a] [*Anodonta cygnea* (Linnaeus 1758)]	Berlin, Germany (Fresh water)
C. interrogationis Gross 1910, 5 (*Spirochaeta interrogationis* (Gross) Bosanquet 1911, 86)	*Pecten jacobaeus* (Linnaeus 1758)	Gulf of Naples, Italy (Marine)
C. parvula Dobell 1912, 142	*Venus casta* Chemnitz[a] [*Meretrix casta* (Gmelin 1791)]	Tamblegam Lake, Ceylon (Salt water)
C. polydorae Mesnil and Caullery 1916, 1120 (*Cristispirella polidorae* (*sic*) (Mesnil and Caullery) Hollande 1921, 1696)	*Polydora flava* Clpde. [a polychete annelid]	Cap de la Hague, Manche (Marine)
C. pachelabrae de Mello 1921, 241	*Pachelabra moesta* Reeve [probably *Pachylabra* Swainson 1840; a snail]	Nova Goa, India (Fresh water)

[a] See footnote to Table 5.1.

Description of type species of genus **Cristispira**

1. **Cristispira pectinis** Gross 1910, 44. (*Spirochaeta pectinis* (Gross) Bosanquet 1911, 86.)

pec′ti.nis. M.L. masc.n. *Pecten* a genus of mollusks; M.L. gen.n. *pectinis* of *Pecten*.

Cylindrical, helically coiled, flexible cells; span of helix 36–72 μm; average width of cell 1.5 μm; at most four complete turns. Ends rounded or tapered; no terminal appendages. 30–79 inclusions; cross-striations and polar granules in stained preparations. Transverse fission may be preceded by incurvation. In fixed preparations crista extends along side of cell; not readily seen on living specimens. Commensal.

Found in stomach and intestine, and at times in crystalline style of *Pecten jacobaeus* Linnaeus, 1758 from Gulf of Naples (marine environment).

Genus III. **Treponema** *Schaudinn 1905, 1728*

ROBERT M. SMIBERT

(*Spironema* Vuillemin 1905, 1568; *Microspironema* Stiles and Pfender 1905, 936.)
Tre.po.ne′ma. Gr. v. *trepo* turn; Gr. n. *nema* a thread; M.L. neut.n. *Treponema* a
turning thread.

Unicellular, helical rods 5–20 μm long, 0.09–0.5 μm wide, as determined by electron microscopy; with **tight regular or irregular spirals** (Plate 5.1 Fig. 3, 6–8). Cells have **one or more axial fibrils inserted at each end of the protoplasmic cylinder.**

Bullae and spirochetal spheres are seen in old cultures.

Motile. Gram-negative. Stain well with silver impregnation methods and Ryu's (1963) stain. Most species stain poorly with Giemsa's stain. Best **observed** under **dark-field or phase-contrast microscopy.** Most of the following part of the genus description refers only to the *in vitro* cultivatable species.

Chemoorganotrophs: **Metabolism fermentative,** using amino acids and/or carbohydrates.

Strict anaerobes. **Catalase and oxidase negative,** urease negative and Voges-Proskauer negative. Does not reduce nitrate. Not inhibited by iodoacetate.

Found in the oral cavity, intestinal tract and genital regions of man and animals. Some species are pathogenic.

The G + C content of the DNA (in the species examined) ranges from 32–50 moles % (T_m).

Type species: *Treponema pallidum* (Schaudinn and Hoffmann) Schaudinn 1905, 1728.

Further Comments

The classification of the genus *Treponema* has been unsatisfactory and it has been extremely dif-ficult to identify a species of the genus from the characteristics available.

The anaerobic technique of Hungate, as modified by Moore (1966), has allowed cultivation of many species and most of the information and characteristics given here have been compiled in the author's laboratory using the method of pre-reduced anaerobically sterilized media prepared by boiling and reducing the media to at least −100 MV, tubing in rubber-stoppered tubes under oxygen-free nitrogen, and autoclaving. All inoculations of media are made in the presence of a stream of sterile oxygen-free nitrogen or carbon dioxide, which flows into the tube by means of a bent hypodermic needle hooked over the lip of the culture tube.

The carbohydrates used in the study of all of the cultivated species of treponemes were adonitol, amygdalin, arabinose, cellobiose, dextrin, dulcitol, erythritol, fructose, galactose, galacturonic acid, gluconic acid, glucose, glucuronic acid, glycerol, glycogen, inositol, inulin, lactose, maltose, mannitol, mannose, melibiose, melezitose, mucin, pectin, raffinose, rhamnose, ribose, salicin, sorbitol, sorbose, sucrose, starch, tartaric acid, trehalose and xylose.

The classification of the cultivatable treponemes is based on a study of various labeled strains supplied by the Venereal Disease Research Laboratory of the National Communicable Disease Center, Atlanta, Georgia, U.S.A., and of strains isolated in the author's laboratory. The work was supported by National Institutes of Health Grant GMS-14604.

Key to the species of genus **Treponema**

I. Causes syphilis, yaws or pinta in man and animals.
  A. Cutaneous lesions in rabbits, no cutaneous lesions in hamsters or guinea pigs.
          1. *Treponema pallidum*
  B. Cutaneous lesions in rabbits and hamsters, but not in guinea pigs.
          2. *Treponema pertenue*
  C. Does not produce cutaneous lesions in rabbits, hamsters or guinea pigs.
          3. *Treponema carateum*
  D. Cutaneous lesions in rabbits and guinea pigs, not in hamsters.
          4. *Treponema paraluis-cuniculi*
II. Does not cause syphilis, yaws or pinta in man or animals.
  A. Ferments carbohydrates.

1. Produces *n*-butyric, acetic and propionic acids; ethanol, *n*-propanol and *n*-butanol. Ferments mannitol.

    5. *Treponema phagedenis*

2. Produces lactic, acetic and formic acids; traces of succinic acid. Does not ferment mannitol.

    6. *Treponema macrodentium*

B. Does not ferment carbohydrates.

1. Produces acetic acid and smaller amounts of succinic, lactic and formic acids.

    a. Cells 0.25–0.35 μm wide.

    7. *Treponema refringens*

    b. Cells 0.10–0.25 μm wide.

    8. *Treponema denticola*

2. Produces acetic and propionic acids.

    a. Uses lactate.

    9. *Treponema orale*

3. Produces acetic and traces of propionic and *n*-butyric acids.

    a. Does not use lactate.

    10. *Treponema scoliodontum*

4 Produces acetic and *n*-butyric acids.

    a. Does not use lactate.

    11. *Treponema vincentii*

Description of the species of genus **Treponema**
(Species 1–4 have not been cultivated.)

1. Treponema pallidum (Schaudinn and Hoffmann) Schaudinn 1905, 1728. (*Spirochaeta pallida* Schaudinn and Hoffmann 1905, 528; *Spironema pallidum* (Schaudinn and Hoffmann) Vuillemin 1905, 1567; *Microspironema pallidum* (Schaudinn and Hoffmann) Stiles and Pfender 1905, 936; *Spirillum pallidum* (Schaudinn and Hoffmann) Macé 1913, 668.)

pal'.li.dum. L. adj. *pallidum* pale, pallid.

Slender helical cells, tightly coiled, 6–20 μm long and 0.09–0.18 μm wide. The average length is 10–13 μm and the average width is 0.13–0.15 μm. The ends of the cells are pointed. Three axial fibrils are inserted into each end of the cell.

Motile with a sluggish drifting motion and graceful flexuous movements. It rarely rotates.

Pathogenic for man and monkeys. Virulent strains (Nichols pathogenic and Gand strains, etc.) are propagated by intratesticular inoculation of rabbits. Cutaneous inoculation of hamsters, mice and guinea pigs produces no apparent infection or visible lesions although a slight lesion is occasionally seen at the point of injection in guinea pigs.

Virulent strains have not been successfully cultivated *in vitro* although cells will remain motile for 4–7 days at 25 C under anaerobic conditions in a medium containing albumin, sodium bicarbonate, pyruvate, cysteine and serum ultrafiltrates.

The cause of venereal and congenital syphilis in man. Found in syphilitic lesions.

Reference strain: Rabbit-propagated Nichols.

For additonal information on *T. pallidum* see Willcox and Guthe (1966).

2. Treponema pertenue (Castellani) Castellani and Chalmers 1910, 310. (*Spirochaeta pertenuis* Castellani 1905, 54; *Spirochaete pertenuis* (*sic*) (Castellani) Lehman and Neumann 1912, 677; *Spironema pertenue* (Castellani) Gross 1912, 115; *Spirillum pertenue* (Castellani) Macé 1913, 693.)

per.ten'.u.e. L. neut.adj. *pertenue* very thin, slender.

Slender helical cells that are morphologically indistinguishable from *Treponema pallidum*. Motile.

Virulent strains are propagated by scarification of the skin of hamsters or rabbits. Cutaneous lesions are produced at the point of inoculation in hamsters and rabbits but not in guinea pigs or mice.

Virulent strains have not been cultivated *in vitro*.

The cause of yaws in man. Found in lesions from cases of yaws, a contagious disease that is spread by contact. The disease is common in tropical countries such as in Africa, Southeast Asia, the western Pacific islands, and tropical countries in the Americas.

Treponemes from lesions from cases of endemic syphilis produce cutaneous lesions in rabbits, hamsters and guinea pigs but not in mice. This organism, however, may not be related to *T. pertenue*.

3. **Treponema carateum** Brumpt 1939, 942. (*Treponema herrejoni* Léon and Blanco 1940, 5; *Treponema pictor* Pardo-Castello 1940, 118; *Treponema americanus* (*sic*) Léon 1940, 271; *Treponema pintae* Curbelo y Hernández 1941, 34.)

ca.ra'.te.um. M.L. n. *carate*, name of a South American disease, pinta. M.L. neut.adj. *carateum*, of carate.

Slender helical cells that are morphologically similar to *Treponema pallidum*. Motile.

Virulent strains have not been successfully grown *in vitro*. Experimental transmission of the disease has been accomplished in man as well as in chimpanzees by intradermal inoculation and by direct exposure of scarified areas of skin to abraded human lesions. Development of lesions took 35 and 70 days. Not proven virulent for rabbits, hamsters or guinea pigs.

The cause of pinta or carate in man. Found in the lymph fluid of cutaneous lesions of pinta. It is a contagious disease mainly restricted to Mexico, Central America, parts of subtropical South America, the West Indies and Cuba.

4. **Treponema paraluis-cuniculi** (Jacobsthal) Smibert *comb. nov.* (*Spirochaeta paraluis-cuniculi*

Jacobsthal 1920, 571; *Treponema pallidum* var. *cuniculi* Klarenbeek 1921, 211; *Treponema cuniculi* Noguchi 1921, 2052; *Spirochaeta cuniculi* Levaditi, Marie and Isaicu 1921, 51; *Spirochaeta pallida* var. *cuniculi* (Klarenbeek) Zuelzer 1925, 1765; *Spirochaeta paraluis* Pettit 1928, 91.)

par'a.lu.is-cu.ni'cu.li Gr pref. *para* resembling; L. n. *luis* pestilence, syphilis; L. gen.n. *cuniculi* of a rabbit; M.L. n. *paraluis-cuniculi* syphilis-like (disease) of rabbit.

Slender helical cells that are morphologically similar to *Treponema pallidum*. Motile.

Has not been cultivated *in vitro*.

The organism can be propagated by intratesticular inoculation of rabbits. Causes a latent infection in mice, guinea pigs and hamsters, treponemes are found in the lymph nodes. Cutaneous lesions are found only in guinea pigs and rabbits.

Produces benign venereal spirochetosis (rabbit spirochetosis) in rabbits. From the lesions in the genitoperineal area of rabbits. Primarily involves genitalia although cutaneous lesions often occur around the face, eyes, ears and nose.

For additional information on this organism see Smith and Persetsky (1967).

(Species 5–11 have been cultivated. Some of their identifying characteristics are given in Table 5.3.)

5. **Treponema phagedenis** (Noguchi) Brumpt 1922, 511. (*Spirochaeta phagedenis* Noguchi 1912, 261; *Spiroschaudinnia phagedenis* (Noguchi) Castellani and Chalmers 1913, 403; *Spironema phagedenis* (Noguchi) Bergey *et al.* 1923, 426; *Borrelia phagedenis* (Noguchi) Bergey *et al.* 1925, 435; *Treponema reiteri* Pillot 1965, 14.) Common or trivial names are Reiter treponeme, English Reiter treponeme, and Kazan treponemes including Kazan strains 2, 4, 5 and 8.

phag.e.den.is. Gr. *phagedena*, *phagedenis* of a cancerous sore.

Helical cells are 5–15 μm long and 0.24–0.40 μm wide. The average length is 6–12 μm and average width 0.25–0.30 μm. Widest cells, 0.40 μm, show double contours with a dark-field microscope. The ends of the cells are pointed. Three axial fibrils are inserted into each end of the cell. In old cultures axial fibrils may be seen trailing from the ends of the cells.

Motile with a jerky forward and backward motion. Rotates slowly.

On pre-reduced anaerobic peptone-yeast extract-serum agar (1.4%), small, white annular colonies, 0.5–1 mm in diameter, with dense centers appear in 2–5 days growing on the surface as well as in the agar. The appearance of colonies will vary with the concentration of agar.

Requires animal serum (inactivated at 56–60 C for 1 hr) for growth. Does not require rumen fluid or volatile fatty acids. Does not require cocarboxylase. Does not require carbohydrates as energy source for growth. Ferments carbohydrates and amino acids.

Ferments fructose, galactose (83%), glucose, lactose, mannitol, mannose and ribose (83%). Most (60%) ferment trehalose. The final pH of glucose fermentation ranges from 5.8–6.3. Dextrin, glycogen and starch are hydrolyzed but not fermented.

Does not ferment arabinose, xylose, sorbose, rhamnose, melibiose, sucrose, maltose, cellobiose, raffinose, melezitose, inulin, pectin, erythritol, sorbitol, dulcitol, adonitol, inositol, gluconic acid, galacturonic acid, glucuronic acid, amygdalin, glycerol, salicin, mucin or tartaric acid. A medium containing 1% mucin is blackened.

End-products of fermentation in peptone-yeast extract-serum medium without glucose are major amounts of *n*-butyric and acetic acids, moderate amounts of propionic and formic acids; and small amounts of succinic and lactic acids. Trace amounts of alcohols are also produced. In the same medium with glucose there is an increase in the amount of acetic and formic acids. Large

TABLE 5.3[a]

Some identifying characteristics of cultivatable treponemes

	T. phagedenis biotype Reiter	T. phagedenis biotype Kazan	T. macrodentium	T. refringens	T. refringens biotype calligyrum	T. denticola	T. denticola biotype comandoni	T. orale	T. scoliodontium	T. vincentii
Cells 0.1–0.25 μm wide	−	−	+	−	−	+	+	+	+	+
Cells 0.25–0.35 μm wide	+	+	−	+	+	−	−	−	−	−
No. of axial fibrils	3	3	1	?	?	2	?	1	?	?
Carbohydrates required as energy source	−	−	+	−	−	−	−	−	−	−
Carbohydrates fermented										
Fructose	+	+	+	−	−	−	−	−	−	−
Glucose	+	+	+	−	−	−	−	−	−	−
Lactose	+	+	−	−	−	−	−	−	−	−
Mannitol	+	+	−	−	−	−	−	−	−	−
Sucrose	−	−	+	−	−	−	−	−	−	−
Esculin hydrolysis	−	+	?	+	+	+	+	?	−	−
Indole	+	+	−	+	+	+	−	+	−	+
H₂S	d	d	+	+	+	+	+	+	−	+
1% glycine	+	+	?	−	+	d	d	?	−	−
Convert lactate	−	−	−	−	−	−	−	+	−	−

[a] Symbols: + = Acid production, a pH of 6.3 or lower or a positive test in 90% or more strains.

− = Most (90% or more) strains negative.

d = Character inconstant and in one strain may sometimes be positive, sometimes negative.

? = Information not available.

amounts of ethanol and *n*-butanol are produced. Smaller amounts of *n*-propanol are also produced.

Hydrolyzes hippurate and gelatin. Indole and hydrogen sulfide positive (80%). Cellulose is not digested. Lactate in a medium is not utilized, while pyruvate is utilized. Grows in a medium with 1% glycine. Very slight curd in skim milk. The Reiter and English Reiter strains of *T. phagedenis* do not hydrolyze esculin, while the Kazan strains do hydrolyze esculin. Hydrolysis of esculin is the only character that separates biotype Kazan from biotype Reiter.

It is weakly methyl red positive. Does not grow in 3% or 6.5% NaCl or 1% bile or at a pH of 6.0 or 9.6. Growth range for pH of medium is 6.5–8.0. Optimum pH is 7.0. Its growth range is from 30–42 C. Very slight growth of few strains at 25 C and 45 C. Optimum temperature is 37 C. It reduces neutral red and bromthymol blue. Chopped meat in a serum medium is neither blackened nor digested; a slight, fetid odor is produced.

A study of the antigenic relationship of the Reiter, Kazan and English Reiter treponemes by Meyer and Hunter (1967) showed them to be closely related. The Kazan strain contained an antigen not shared by the Reiter and English Reiter strains. The cultivated Nichols treponeme was placed in a different serologic group. Christiansen (1964) also reported that the Reiter and Kazan II strains were closely related but not identical. Dupouey (1963) reported that *T. phagedenis* and Reiter treponeme were closely related antigenically, sharing at least six common antigens; the Reiter treponeme has a large amount of an antigen that may be shared with a number of species. This broadly reacting protein antigen is probably the only one the Reiter biotype shares with *T. pallidum*.

Non-pathogenic. Isolated from phagedenic ulcer on human external genitalia. The Reiter treponeme was isolated from a case of primary syphilis in man. Also found on the anal and genital regions of normal male and female chimpanzees.

Although the Reiter treponeme was considered to be *Treponema pallidum*, it seems best to regard the non-virulent Reiter treponeme as a separate species. Thus, virulent *T. pallidum* remains uncultivated *in vitro*.

The G + C content of the DNA is 38–39 moles % (T_m). DNA from the Reiter strain of *T. phage-*

denis shows a very high homology with DNA from the English Reiter and Kazan strains. Shows very low homology with DNA from strains of *T. refringens* and *T. denticola* (chemical hybridization).

Reference strains: *Treponema phagedenis* (CIPP) or Reiter treponeme.

Additional information on this organism can be found in an excellent review by Wallace and Harris (1967).

6. Treponema macrodentium Noguchi 1912, 82. (*Spirochaeta macrodentium* (Noguchi) Pettit 1928, 182.)

mac.ro.den'.ti.um. Gr. adj. *macrus* long; L. n. *dens, dentis* tooth; M.L. gen.pl.n. *macrodentium* (*sic*) of large teeth.

Slender helical rods are 5–16 μm long and 0.1–0.25 μm wide. The ends of the cells are pointed. One axial fibril is inserted into each end of the cell.

Motile with a fairly rapid motion. Young cells rotate rapidly on their long axis.

Grows in peptone-yeast extract-medium or PPLO medium (BBL) containing 10% serum or ascitic fluid with cocarboxylase 5 μg/ml, glucose 1 mg/ml and cysteine 1 mg/ml. Requires animal serum for growth. This requirement can be replaced by isobutyric acid 20 μg/ml, spermine 150 μg/ml and nicotinamide 400 μg/ml. Will also grow in a medium supplemented with rumen fluid and cocarboxylase. Requires a fermentable carbohydrate as an energy source.

Carbohydrates are fermented. Acid but no gas is produced. Final pH in glucose broth is 5.0–5.4. Ferments fructose, glucose, maltose, ribose and sucrose. May ferment cellobiose, galactose and xylose. Does not ferment mannose, rhamnose, sorbose, lactose, arabinose, trehalose, mannitol, inulin, sorbitol or salicin. Starch is not hydrolyzed.

End-products of fermentation of glucose are major amounts of lactic acid, moderate amounts of acetic and formic acids and traces of succinic acid.

Gelatin is hydrolyzed. Indole negative. Hydrogen sulfide positive. Lactate not used. Ammonia not produced.

Optimum temperature 37 C. Grows at a pH of 7.0.

From the gingival crevice of man.

The G + C content of the DNA is 39 moles % (T$_m$).

Reference strain: TMI.

7. Treponema refringens (Schaudinn and Hoffmann) Castellani and Chalmers 1919, 461. (*Spirochaeta refringens* Schaudinn and Hoffmann 1905, 528; *Spironema refringens* (Schaudinn and Hoffmann) Gross 1912, 115; *Spiroschaudinnia refringens* (Schaudinn and Hoffmann) Castellani

and Chalmers 1913, 115; *Spirillum refringens* (Schaudinn and Hoffmann) Macé 1913, 697; *Treponema calligyrum* Noguchi 1913, 96; *Treponema minutum* Noguchi 1918, 671; *Treponema genitalis* Noguchi 1923, 260; *Spirochaeta calligyrum* (Noguchi) Zuelzer 1925, 1673; *Borrelia refringens* (Schaudinn and Hoffmann) Bergey *et al.* 1925, 436.) Common or trivial names: Nichols nonpathogenic and Noguchi treponemes.

re.frin'.gens. L. part.adj. *refringens* refringent, refractive.

Helical cells 5–16 μm long and 0.20–0.35 μm wide. The average cells are 5–8 μm long and 0.25–0.30 μm wide. Double contoured cells are 0.40 μm wide. Ends of the cells are pointed. Some cells may appear loosely coiled. Axial fibrils may be seen frequently in old cultures trailing from the ends of the cells.

Motile, with a slow, sluggish movement. Rotation of cells is rare. The cells usually rotate slowly.

Colonies on pre-reduced anaerobic peptone-yeast extract-serum agar (1.4%) are visible in 9–15 days. They are small, white, round, pinpoint surface colonies 0.5–1 mm in diameter. Some colonies after longer incubation are white, fluffy and up to 1.5 mm in diameter. Colonies grow on the surface of the medium as well as into the medium. Size and texture of colonies will vary with the concentration of agar in the medium.

Grows well in peptone-yeast extract-serum medium under anaerobic conditions. Requires animal serum (inactivated at 56–60 C for 1 hr) for growth. Does not require rumen fluid or volatile fatty acids for growth. Does not require cocarboxylase. Amino acids are fermented. Carbohydrates are not fermented and not required as an energy source for growth. Final pH of glucose broth is 6.5–6.7. Cellulose not hydrolyzed. Starch, dextrin and glycogen hydrolyzed.

End-products of fermentation of amino acids in peptone-yeast extract-serum medium are major amounts of acetic acid, moderate amounts of succinic, and smaller amounts of lactic acid. Trace amounts of propionic, *n*-butyric and formic acids produced by most strains. Trace amounts of ethanol, *n*-propanol and *n*-butanol are also produced by most strains. No additional end-products are produced in the presence of glucose.

Hydrolyzes esculin and gelatin. Hippurate is hydrolyzed by most strains. Indole positive. Strongly hydrogen sulfide positive. Methyl red negative. Skim milk is only slightly curdled. Lactate in a medium is not utilized. Pyruvate is used. Ammonia produced by most strains.

Does not grow in 3% or 6.5% NaCl or in 1% bile. Some strains do not grow in 1% glycine. Biotype calligyrum grows in medium with 1%

glycine in 5–6 days. Grows in a pH range of 6.5–8.0. Does not grow at a pH of 6.0 or 9.6. Optimum pH is 7.0. Grows in a temperature range of 30–42 C. Only very slight or no growth at 25 C or 45 C. Optimum temperature is 37 C. A strain labeled *T. minutum* had a temperature range of 34–40 C. Neutral red and bromthymol blue are reduced. Chopped meat serum medium is neither blackened nor digested. Only a very slight putrid odor is detectable.

Dupouey (1963) reported that strains labeled *T. refringens* and *T. calligyrum* were closely related antigenically, sharing 4–5 common antigens, but one labeled *T. minutum* was only slightly related to them antigenically. The three strains had only one antigen in common with *T. pallidum*.

Not pathogenic. Isolated from condyloma acuminata lesions, occasionally from syphilitic lesions, part of normal flora of male and female genitalia.

The G + C content of the DNA is 39–43 moles % (T_m). DNA from *T. refringens* shows a high homology with DNA from Nichols, Noguchi and strains labeled *T. calligyrum*, and a very low homology with DNA from strains of *T. denticola* and *T. phagedenis* (chemical hybridization).

Reference strains: *T. refringens*, Institute Pasteur, Paris or cultivated Nichols (non-pathogenic) treponeme.

8. Treponema denticola (Flügge) Brumpt 1922, 497. (*Spirochaete denticola* Flügge 1886, 390; *Spirochaete dentium* Miller 1889, 58; *Spirillum dentium* (Miller) Sternberg 1892, 694; *Spirochaeta dentium* (Miller) Migula 1895, 35; *Spironema dentium* (Miller) Gross 1912, 88; *Treponema dentium* (Miller) Dobell 1912, 158; *Treponema microdentium* Noguchi 1912, 82; *Spirochaeta orthodonta* Hoffmann 1920, 258; *Spirochaeta microdentium* (Noguchi) Heim 1922, 477; *Treponema orthodontum* (*sic*) (Hoffmann) Noguchi 1928, 481; *Treponema dentium-stenogyratum* Pettit 1928, 244; *Spirochaeta ambigua* Séguin and Vinzent 1936, 409; *Spirochaeta comandonii* Séguin and Vinzent 1936, 410; *Treponema ambiguum* (Séguin and Vinzent) Prévot 1940, 208; *Treponema comandoni* (Séguin and Vinzent) Prévot 1940, 208.)

den.ti′.co.la. L. masc.n. *dens, dentis* tooth; L. v.suff. *cola* from L. v. *colo* dwell; M.L. n. *denticola* tooth dweller.

Slender helical cells, 6–16 μm long and 0.10–0.25 μm wide. The ends of the cells are pointed and slightly bent. Two axial fibrils are inserted into each end of the cell. Very few axial fibrils are seen trailing from ends of cells.

Motile with a jerky but fairly rapid motion. Young cells rotate rapidly on their axis.

Surface and subsurface colonies, 0.3–1.0 mm in diameter, white, diffuse, appear after 2 weeks incubation.

Grows well in peptone-yeast extract-serum medium under anaerobic conditions. Requires animal serum for growth. Does not require rumen fluid or volatile fatty acids. Requires or greatly stimulated by cocarboxylase. Amino acids are fermented. Carbohydrates are not fermented and not required as an energy source for growth. Final pH of glucose broth 6.7. Mucin not fermented but medium with 1% mucin is blackened. Cellulose is not hydrolyzed. Starch, dextrin and glycogen are hydrolyzed.

End-products of fermentation of amino acids in peptone-yeast extract-serum medium are major amounts of acetic acid, moderate amounts of lactic acid and minor amounts of succinic and formic acids. Trace amounts of propionic acid, ethanol, *n*-propanol and *n*-butanol may occasionally be found. No additional end-products are produced in the presence of glucose. Hydrolyzes esculin and gelatin, but not hippurate. Strongly hydrogen sulfide positive. Most strains indole positive. Comandoni biotype is indole negative. Methyl red negative. Lactate in a medium is not utilized. Pyruvate is utilized. Produces ammonia from amino acids. Chopped meat-serum medium is neither blackened nor digested. No action on milk.

Does not grow in 3% or 6.5% NaCl or in 1% bile. Does not grow at a pH of 6.0 or 9.6. Grows at a pH range of 6.5–8.0. Optimum pH is 7.0. Only a culture labeled *T. ambiguum* failed to grow at a pH of 8.0. Grows in a temperature range of 30–42 C. Very slight or no growth of a few strains at 25 C and 45 C. Optimum temperature is 37 C. Both neutral red and bromthymol blue are reduced.

Some strains grow in a medium with 1% glycine.

Found in the oral cavity of man and chimpanzees, usually in the deposit at the juncture between the teeth and gums.

The G + C content of the DNA is 37–38 moles % (T_m). DNA from strains of *T. denticola* show a high homology with DNA from a culture labeled *T. comandonii* (chemical hybridization).

Reference strains: FM or TDI.

Comment: Examination of a culture labeled *T. ambiguum* showed that it was similar to cultures from the Pasteur Institute and the National Institute of Dental Health labeled *T. microdentium* and *T. microdentium* strain FM. Since the names *T. dentium* and *T. microdentium* are illegitimate (Rule 24b) the name of these organisms should be *Treponema denticola*.

9. Treponema orale Socransky, Listgarten, Hubersak, Cotmore and Clark 1969, 881. (*Tre-*

ponema oralis (*sic*) Socransky, Listgarten, Hubersak, Cotmore and Clark 1969, 881.)

or.al.e. L. n. *os, oris* the mouth; M.L. neut.adj. *orale* of the mouth.

Slender helical cells, 6–16 μm long and 0.10–0.25 μm wide. Occasional chains formed. One axial fibril is inserted into each end of the cell. Frequently end granules seen in broth cultures.

Motile with a jerky but fairly rapid motion.

Grows in either PPLO medium without crystal violet (BBL) or peptone-yeast extract medium. Each medium contains glucose 1 mg/ml, cysteine 1 mg/ml, nicotinamide 400 μg/ml, cocarboxylase 5 μg/ml, spermine tetrahydrochloride 150 μg/ml and sodium isobutyrate 20 μg/ml and each is further supplemented with 10% inactivated rabbit serum or ascitic fluid, or 0.05% α_2-globulin. Even turbidity in liquid media. Does not grow well on surface cultivation. Does not require carbohydrates as an energy source. Carbohydrates not fermented. Amino acids are fermented. Final pH in glucose broth is 6.8–7.2. End-products of fermentation of amino acids are acetic and propionic acids.

Hydrolyzes gelatin but not starch. Indole and hydrogen sulfide positive. Utilizes lactate. Does not produce ammonia in cultures.

Grows in a pH of 7.0 and at a temperature of 37 C.

Found in the gingival crevice of man.

The G + C content of the DNA is 37 moles %.

Reference strain: T01.

10. **Treponema scoliodontum** (Hoffmann) Noguchi 1928, 481. (*Spirochaeta skoliodonta* Hoffmann 1920, 258; *Spirochaeta acuta* Kritchevski and Séguin 1920, 618; *Treponema skoliodontum* (*sic*) (Hoffmann) Noguchi 1928, 481.)

sco.lio.dont.um. Gr. adj. *scolios* crooked, bent; Gr. n. *odous, odontos* tooth; M.L. neut.n. *scoliodontum* crooked tooth.

Helical cells 6–16 μm long and 0.10–0.24 μm wide. Very tightly coiled. The ends of the cells are pointed. Very few axial fibrils are seen trailing from the ends of cells.

Motile with a jerky but fairly rapid motion. Young cells rotate rapidly on their long axis.

Grows in peptone-yeast extract-serum medium under anaerobic conditions. Requires animal serum or ascitic fluid. Does not require rumen fluid or volatile fatty acids. Requires or greatly stimulated by cocarboxylase. Growth stimulated by the addition of TEM-4T (diacetyl tartaric acid ester of tallow monoglycerides) to culture media. Amino acids are fermented. Does not require carbohydrates as an energy source. Does not ferment carbohydrates.

Final pH of glucose broth is 6.9–7.1. A medium with 1% mucin is not blackened or fermented. Cellulose is not hydrolyzed. Starch, dextrin and glycogen are hydrolyzed.

End-products of fermentation of amino acids in a peptone-yeast extract-serum medium are moderate amounts of acetic acid, and small amounts of formic, succinic, lactic, propionic and *n*-butyric acids. No additional end-products are produced in the presence of glucose.

Does not hydrolyze esculin or hippurate. Gelatin is hydrolyzed. Indole and hydrogen sulfide negative Methyl red negative. Lactate and pyruvate are not used. Ammonia is not produced from amino acids. No action on milk. Chopped meat serum medium is neither blackened nor digested. Produces a slight fetid odor.

Does not grow in 3% or 6.5% NaCl, or in 1% bile or in 1% glycine media. Does not grow at a pH of 6.0 or 9.6. Grows at a pH range of 6.5–8.0. Optimum pH is 7.0. Grows in a temperature range of 30–42 C. Optimum temperature is 34 C. Neutral red and bromthymol blue not reduced.

From the oral cavity of man.

Reference strain: *T. scoliodontum* Institute Pasteur, Paris.

11. **Treponema vincentii** (Blanchard) Brumpt 1922, 515. (*Spirochaeta vincenti* Blanchard 1906, 3; *Spiroschaudinnia vincenti* (Blanchard) Castellani and Chalmers 1913, 402; *Spirillum vincenti* (Blanchard) Macé 1913, 914; *Spironema vincenti* (Blanchard) Park and Williams 1917, 506; *Borrelia vincenti* (Blanchard) Bergey *et al.* 1925, 435.)

vin.cen'.ti.i. M.L. gen.n. *vincentii* of Vincent; named for Dr. H. Vincent, a French bacteriologist.

Helical cells 5–16 μm long and 0.20–0.30 μm wide. The ends of the cells taper to a point. Cells may have shallow and irregular spirals. Axial fibrils may be seen trailing from the ends of the cells.

Actively motile with a rapid, jerky, vibratory motion. Young cells rotate rapidly on their long axis.

Colonies of strain N-9 visible after 2-week incubation. Small white colonies 12–15 mm in diameter, appearing as a slight haze in the agar.

Grows in a peptone-yeast extract medium under anaerobic conditions. Requires animal serum or ascitic fluid for growth. Greatly stimulated by TEM-4T (diacetyl tartaric acid ester of tallow monoglycerides). Does not require rumen fluid or volatile fatty acids for growth. Requires cocarboxylase. Amino acids are fermented. Does not require carbohydrates as an energy source for growth. Carbohydrates are not fermented. Final pH of glucose broth is 6.9–7.2. Medium with 1% mucin blackened. Cellulose is not hydrolyzed. Starch, dextrin and glycogen are hydrolyzed.

End-products of fermentation of amino acids in peptone-yeast extract-serum medium are major amounts of acetic and *n*-butyric acids, moderate amounts of lactic acid and smaller amounts of succinic and formic acids. Trace amounts of propionic acid may also be found. Ethanol, *n*-propanol and small amounts of *n*-butanol may also be found. No additional end-products are produced in the presence of glucose.

Does not hydrolyze esculin. Hydrolyzes gelatin. Weakly indole positive and strongly hydrogen sulfide positive. Methyl red negative. Skim milk is not changed. Lactate and pyruvate are not used. Ammonia is produced in cultures. Chopped meat is neither blackened nor digested. Slight fetid odor in cultures.

Does not grow in 3% or 6.5% NaCl. Does not grow in 1% glycine or 1% bile. Grows in a pH range of 6.5–7.5. Optimum pH is 7.0. Grows in a temperature range of 25–45 C. Optimum temperature is 37 C. Neutral red and bromthymol blue are reduced.

Meyer and Hunter (1967) reported that *T. vincentii* strain N-9 was antigenically distinct from *T. denticola* (FM) and the Nichols and Noguchi strains of *T. refringens*. Antigens were shared with *S. zuelzerae*, *T. phagedenis* (Reiter and Kazan strains).

Found in the oral cavity of man.

Reference strain: *T. vincentii* strain N-9.

Species incertae sedis

Addendum I. The following species are validly published and legitimate but their taxonomic relationships are not clear because of poor descriptions. Representative isolates have not been found in culture collections or have not been studied.

a. *Treponema penortha* (Beveridge) Prévot 1948, 269. (*Spirochaeta penortha* Beveridge 1936, 307; *Treponema pernortha* (*sic*) (Beveridge) Prévot 1948, 269.)

pen.orth.a. L. adv. *pen.* almost, nearly; Gr. adj. *orthos*, straight; M.L. adj. *penortha* almost straight.

The description is according to Beveridge (1936) and Prévot *et al.* (1967).

Helical cells 6–10 μm long and 0.25–0.3 μm wide. The ends of the cells are pointed. Spirochetal spheres are formed measuring 0.5–2.5 μm. Terminal fibrils were not observed trailing from the ends of the cells. Does not form spores.

Motile.

Colonies on anaerobic blood agar are convex and white with a pinkish sheen.

Grows in a medium containing inactivated

animal serum and 10% CO_2. Cultures produce gas and a putrid odor. Not hemolytic. Skim milk is coagulated.

Optimum temperature is 37 C. It is killed at 55 C.

Found associated with foot rot in sheep.

b. *Treponema mucosum* Noguchi 1912, 195. (*Spirochaeta mucosa* (Noguchi) Pettit 1928, 190.)

mu.co.sum. L. neut.adj. *mucosum* mucous or slimy.

The description is taken from the original article (Noguchi, 1912) and from Prévot *et al.* (1967).

Slender helical cells 5–10 μm long and 0.18–0.24 μm wide. The ends of the cells are pointed. According to Noguchi the characteristics of the organism are quite similar to those of *T. denticola*, the only differences being the ability of *T. mucosum* to produce "mucin" in cultures and to survive in rabbit testicles when inoculated along with agar.

Found in the oral cavity of man with alveolar pyorrhea.

c. *Treponema trimerodontum* (Hoffmann) Prévot 1940, 209. (*Spirochaeta trimerodonta* Hoffmann 1920, 258; *Leptospira dentium* Hoffmann 1920, 626; *Leptospira trimerodonta* (Hoffmann) Noguchi 1928, 487; *Treponema trimerodonta* (*sic*) (Hoffmann) Prévot 1940, 209.)

tri.mer.o.don'tum. Gr. adj. *trimeres* tripartite, 3-fold; Gr. masc.n. *odous, odontis* tooth; M.L. neut.adj. *trimerodontum* tripartite tooth, i.e. a molar.

The description is taken from Séguin and Vinzent (1941) and Prévot *et al.* (1967).

Helical cells 6–10 μm long and 0.20–0.30 μm wide. Ends of the cells pointed and hooked so as to resemble leptospires.

Motile with rapid rotational movement.

Grows in a medium containing animal serum. Gas is not produced in cultures. Hydrogen sulfide produced. Coagulated serum medium is not digested, and gelatin not liquefied. Neutral red is partly reduced and no odor produced by cultures.

From putrid bronchopleuropulmonary infections in man.

d. *Treponema enterogyratum* (Vinzent and Séguin) Prévot 1957, 332. (*Spirochaeta enterogyrata* Vinzent and Séguin 1939, 12; *Treponema enterogyrata* (*sic*) (Vincent and Séguin) Prévot 1957, 332.)

en.ter.o.gy.ra'ta. Gr. n. *enter, entero* intestine; L. neut.part.adj. *gyratum* spiral, whirl, turn; *enterogyrata* whirls in the intestine.

The description is after Prévot *et al.* (1967).

Helical cells 6–12 μm long and 0.20–0.30 μm wide with 3–15 spiral turns.

Actively motile.

Colonies appear in agar medium in the shape of bubbles.

Grows in a medium supplemented with animal serum. It is adaptable to a medium without serum. Gelatin is not liquefied and coagulated serum is not digested. Hydrogen sulfide is not produced. Found in human feces.

e. *Treponema buccale* (Steinberg) Swellengrebel 1907, 582. (*Spirillum buccale* Steinberg 1862, 434; *Spirochaete cohnii* Trevisan 1879, 149; *Spirochaete buccalis* (*sic*) (Steinberg) Schroeter 1886, 168; *Spirillum cohnii* (Trevisan) Trevisan 1889, 24; *Spirochaeta inaequalis* Gerber 1910, 512; *Spironema buccale* (Steinberg) Gross 1912, 84; *Spiroschaudinnia buccalis* (Steinberg) Castellani and Chalmers 1919, 450; *Borrelia buccalis* (Steinberg) Brumpt 1922, 495; *Treponema inequale* (Gerber) Brumpt 1922, 505; *Treponema undulatum* (Gerber) Brumpt 1922, 514.)

buc.ca.le. L. adj. *buccalis* buccal.

Some of the description is after Hampp (1954).

Helical cells 7–20 μm long and 0.30–0.40 μm wide. This is the largest of the oral treponemes.

Sluggishly motile with a jerky serpentine and flexuous movement. Cells rotate slowly.

This organism has probably neither been isolated in pure culture nor well described. Hampp (1954) did, however, report the isolation of this organism from the oral cavity.

Found in oral cavity of man. It probably invades lesions formed on the respiratory mucous membranes.

f. *Treponema hyos* (King and Drake) Smibert *comb. nov.* (*Spirochaeta suis* King, Baeslack and Hoffmann 1913, 253 not *Spirochaeta suis* Bosanquet 1911, 95; *Spirochaeta hyos* King and Drake 1915, 54; *Spironema hyos* (King and Drake) Bergey *et al.* 1923, 426; *Borrelia hyos* (King and Drake) Bergey *et al.* 1925, 436; *Spironema suis* (King and Drake) Ford 1927, 959.)

hy.os. Gr. n. *hys, hyos* of the hog.

Helical cells 5–7 μm long and 0.30–0.35 μm wide. Cells are short with only a few spirals. Ends of the cells are round.

Actively motile with rotational movement.

Grows in a medium containing serum or ascitic fluid.

From the intestinal ulcers of pigs with swine dysentery. Found in the blood, intestinal ulcers, crypts of the cecum and in external lesions of pigs with hog cholera. This organism is probably a part of the normal intestinal flora of pigs.

Addendum II. The following poorly defined species would be difficult to recognize from the published descriptions. Others may be found in *Index Bergeyana* (1966).

g. *Treponema aboriginalis* (*sic*) (Cleland) Brumpt 1922, 496.

h. *Treponema acuminatum* (Castellani) Brumpt 1922, 496.

i. *Treponema balanitidis* (Hoffmann and von Prowazek) Brumpt 1922, 496.

j. *Treponema bronchiale* (Castellani) Brumpt 1922, 496.

k. *Treponema bucco-pharyngei* (Macfie) Brumpt 1922, 497.

l. *Treponema bufonis* (Dobell) Ford 1927, 986.

m. *Treponema cotti* Duboscq and Lebailly 1912, 356.

n. *Treponema ctenocephali* (Patton) Ford 1927, 989.

o. *Treponema culicis* (Jaffé) Ford 1927, 989.

p. *Treponema eurygyratum* (Werner) Brumpt 1922, 500.

q. *Treponema forans* (Reiter) Brumpt 1922, 500.

r. *Treponema gadi* (Neumann) Duboscq and Lebailly 1912, 337.

s. *Treponema gallicolum* Lebailly 1913, 390.

t. *Treponema gracile* (Levaditi and Stanesco) Brumpt 1922, 501.

u. *Treponema grassi* (Döflein) Ford 1927, 988.

v. *Treponema hachaizae* (Kowalski) Brumpt 1922, 501.

w. *Treponema hilli* Duboscq and Grassé 1926, 34.

x. *Treponema intermedium* Dobell 1912, 160.

y. *Treponema intestinale* (Macfie and Carter) Brumpt 1922, 505.

z. *Treponema lari* Lebailly 1913, 390.

aa. *Treponema legeri* Duboscq and Lebailly 1912, 357.

bb. *Treponema lymphaticum* (Proescher and White) Brumpt 1922, 506.

cc. *Treponema microgyratum* (Loewenthal) Brumpt 1922, 506.

dd. *Treponema minei* (von Prowazek) Dobell 1912, 147.

ee. *Treponema minimum* de Beaurepaire-Aragão and Vianna 1913, 61.

ff. *Treponema mite* (Castellani) Brumpt 1922, 506.

gg. *Treponema noguchii* (Strong) Noguchi 1928, 483.

hh. *Treponema obtusum* (Castellani) Brumpt 1922, 508.

ii. *Treponema parvum* Dobell 1912, 151.

jj. *Treponema pavonis* Duboscq and Lebailly 1912, 356.

kk. *Treponema perexile* Duboscq and Lebailly 1912, 366.

ll. *Treponema podovis* Blaizot and Blaizot 1928, 912.

mm. *Treponema pseudopallidum* (Mulzer) Brumpt 1922, 511.

nn. *Treponema querquedulae* Lebailly 1913, 390.

oo. *Treponema rectum* (Gerber) Brumpt 1922, 511.

pp. *Treponema rigidum* Zinsser and Hopkins 1916, 489.

qq. *Treponema schaudinni* (von Prowazek) Brumpt 1922, 514.

rr. *Treponema spermiformis* (*sic*) Duboscq and Grassé 1927, 483.

ss. *Treponema squatarolae* Lebailly 1913, 390.

tt. *Treponema stenogyratum* (Werner) Brumpt 1922, 514.

uu. *Treponema stylopygae* Dobell 1912, 149.

vv. *Treponema tenue* (Gerber) Brumpt 1922, 514.

ww. *Treponema termitis* (Leidy) Dobell 1912, 144.

xx. *Treponema triglae* Duboscq and Lebailly 1913, 15.

yy. *Treponema urethrae* (Macfie) Brumpt 1922, 514.

zz. *Treponema urethrale* Castellani acc. to Castellani and Chalmers 1919, 1944.

aaa. *Treponema uretritis* Bacigalupo 1926, 1569.

bbb. *Treponema vaccinae* (Bonhoff) Brumpt 1922, 515.

ccc. *Treponema vaginalis* (*sic*) (Macfie) Brumpt 1922, 514.

ddd. *Spirochaeta suis* Bosanquet 1911, 95.

Genus IV. **Borrelia** *Swellengrebel 1907, 582*

Oscar Felsenfeld

Bor.rel′i.a. M.L. fem.n. *Borrelia* of Borrel; named after A. Borrel, a French bacteriologist.

Cells 0.2–0.5 by 3–20 μm, helical, with 3–10 or more coarse, uneven, often irregular coils, some of which may form obtuse angles (Plate 5.1, Fig. 4). Cells up to 1 μm wide and 25 μm long may occur; the measurements vary according to strain and the staining method used.

Electron microscopy shows a foamy, elastic envelope and a cytoplasmic membrane, between which there are 15–20 parallel fibrils coiled around the cell body. There is a central elongated fibrillar substance, no mitochondria and no undulating membrane. The fibrils constitute the locomotory apparatus, the coiled fibrils usually rotating in one direction, the cell body in the opposite direction but at the same rate. Motion is in forward and backward waves, laterally by bending and looping and corkscrew-like.

Gram-negative. Stain well with aniline dyes, particularly when phenol is used as a mordant.

Nutritional characteristics of most species unknown. Those studied possess fermentative metabolism. Growth of several species reported on complex media containing natural animal proteins or in developing chick embryos. Minimal nutritional requirements unknown.

Strict anaerobes. Growth observed at 20–37 C, optimum at 28–30 C. Preserved at −76 C.

Antigenic structure generally unstable resulting in considerable variations particularly in pathogenic strains during relapses.

No adequate system of species separation has yet been found.

Type species: *Borrelia anserina* (Sakharoff) Bergey *et al.* 1925, 435.

Further Comments

Borrelias are generally parasitic or living on mucous membranes. Some are pathogenic for man, other animals or birds. Those causing relapsing fever in man are transmitted by *Pediculus humanus* subsp. *humanus* or by *Ornithodoros* ticks. The latter also transmit borrelias to rodents. According to the unitarian acarine concept (Brumpt, 1937) each *Borrelia* species is carried by a different arthropod vector. This concept originated the trend to differentiate *Borrelia* species according to their vectors. Strains pathogenic for mammals are systemic parasites of argasid ticks and lice (Burgdorfer, 1951). Such species of *Borrelia* developed with acarinae, principally as their parasites, and evolved into different strains and species with the genetic changes that differentiated various *Ornithodoros* species (Baker and Wharton, 1951).

Infections with borrelias depend on the feeding habits and preferences of the vector. Some, like *Pediculus humanus* subsp. *humanus* and *Ornithodoros moubata* subsp. *anthropophilus* feed only on man; others are not discriminative, as *Ornithodoros tholozani*. When *Ornithodoros* takes its blood meal from several mammalian species, an animal (usually rodent) reservoir of the respective *Borrelia* develops. Several tick species have a long life span and the *Borrelia* carried by them remains viable in them for years. Hence ticks preserve several *Borrelia* strains without an animal reservoir.

Electron microscopy permits differentiation of *Borrelia* from related organisms. The number of fibrils coiled under cell wall is 2 in *Leptospira*, 3–7 in *Treponema* and around a hundred in *Cristispira*.

Phase and antigenic variations are the rule. Serologic tests are of limited value in the differentiation of *Borrelia* species even though the experimental animal was infected with a single *Borrelia* organism (Schuhardt and Wilkerson, 1951). Asymmetric (one-sided) immunity may occur in cross-protection tests. The immunity to *Borrelia* is of the premunition type.

Examination by dark-field microscopy, staining according to Giemsa and silver impregnation, tests for susceptibility of laboratory animals, borreliolysin and immobilizin titer determinations are used in routine diagnostic work; in specialized laboratories also xenodiagnosis. There is no record of serological testing of all known *Borrelia* species against each other, nor have all of them been examined in more recently introduced animal experiments which include the investigation of the influence of *Borrelia* on infections with *Trypanosoma* species in rodents, and the employment of the European hedgehog (*Erinaceus europaeus*) as a laboratory test animal. It is not yet possible, therefore, to tabulate differential diagnostic characteristics of *Borrelia* species.

The use of fluorescent-labeled antisera facilitates the finding of *Borrelia* but differentiation of species by this method has not yet been achieved to a satisfactory degree. Neither has growth in developing chick embryos proved valuable in separating the various species, principally because not all strains can be adapted to this medium.

The classification used in this and the 7th edition of THE MANUAL is based on the arthropod vectors. Borrelias causing disease in cattle, horses and birds are not carried by *Ornithodoros* or *Pediculus*. Transmission by contact has been established for *B. anserina* and borrelias of the human mucosae. These species, however, can be differentiated by other means described in the text.

Diseases caused by borrelias are frequently referred to, particularly in the veterinary literature, by names such as avian spirochetosis or bovine spirochetosis. The generic name *Spirochaeta* is also used for these organisms in recent literature outside North America.

Key to the species of genus **Borrelia**

I. From birds.
        1. *Borrelia anserina*
II. From animals other than birds.
 A. From man.
  1. Cause relapsing fever in man.
   a. Arthropod vector is the louse *Pediculus humanus* subsp. *humanus*.
        2. *Borrelia recurrentis*
   b. Arthropod vector is a tick of the genus *Ornithodoros*.
    b1. Transmitted by *Ornithodoros erraticus erraticus*.
        3. *Borrelia hispanica*
    b2. Transmitted by *Ornithodoros hermsi*.
        4. *Borrelia hermsii*
    b3. Transmitted by *Ornithodoros moubata*.
        5. *Borrelia duttonii*
    b4. Transmitted by *Ornithodoros parkeri*.
        6. *Borrelia parkeri*
    b5. Transmitted by *Ornithodoros rudis*.
        7. *Borrelia venezuelensis*
    b6. Transmitted by *Ornithodoros talaje*.
        8. *Borrelia mazzottii*
    b7. Transmitted by *Ornithodoros tholozani*.
        9. *Borrelia persica*
    b8. Transmitted by *Ornithodoros turicata*.
        10. *Borrelia turicatae*
    b9. Transmitted by *Ornithodoros verrucosus*.
        11. *Borrelia caucasica*

B. From animals other than man.
    1. From ticks (*Ornithodoros*).
        a. From *O. brasiliensis*.
                12. *Borrelia brasiliensis*
        b. From *O. dugesi*.
                13. *Borrelia dugesii*
        c. From *O. graingeri*.
                14. *Borrelia graingeri*
    2. From animals other than arthropods.
        a. From rodents.
           a1. Transmitted by *Ornithodoros erraticus sonrai*.
                15. *Borrelia crocidurae*
           a2. Transmitted by *O. tartakovskyi*.
                16. *Borrelia latyschewii*
           a3. Transmitted by *O. zumpti*.
                17. *Borrelia tillae*
        b. From ruminants and horses.
                18. *Borrelia theileri*
        c. From primates.
                19. *Borrelia harveyi*

Description of the species of genus **Borrelia**

1. Borrelia anserina (Sakharoff) Bergey *et al.* 1925, 435. (*Spirochaeta anserina* Sakharoff 1891, 565; *Spirochaete gallinarum* Stephens and Christopher 1904, 378; *Spiroschaudinnia anserina* (Sakharoff) Sambon 1907, 834.)

an.se.ri'na. L. adj. *anserina* pertaining to geese.

Cells 0.2–0.3 by 6–30 μm, usually 8–20 μm long, circular in cross-section; spirals with 5–8 turns, pitch of spiral usually 1.2–1.8 μm. Actively motile with lashing movements.

Grows in media containing natural proteins and tissue fragments. Some strains grow well in developing chick embryos.

Antigenically distinct. Several serologically different strains have been described but systematic immunological studies of the species are not available.

Pathogenic for geese, ducks, turkeys, pheasants, canaries, chickens and grouse; most strains also pathogenic for guinea fowl and pigeons. Mice, rats, rabbits, dogs, sheep and lizards not susceptible.

Causes avian borreliosis, also called range paralysis. Transmitted by the bites of ticks, principally *Argas miniatus*, *A. persicus* and *A. reflexus*, and possibly by other blood-sucking insects. Transmission may occur from bird to bird through feces and cannibalism. Found in infected birds and vector ticks in Africa, India, the Middle East, Europe, Australia, South America and the western United States.

Comment: *Spirochaete gallinarum* Stephens and Christopher 1904, 378, is considered identical with *Spirochaeta anserina* Sakharoff 1891, 565, which has precedence. When Swellengrebel introduced the name *Borrelia* he did not consider the earlier strain. On this basis *B. anserina* is accepted as the type species, as by Murray (Bergey *et al.*, 1948, 1058) and not *B. gallinarum* as in *Index Bergeyana* 1966, 421.

2. Borrelia recurrentis (Lebert) Bergey *et al.* 1925, 433. (*Spirochaete recurrentis* Lebert 1874, 273; *Spirochaete obermeieri* Cohn 1875, 196; *Spiroschaudinnia recurrentis* (Lebert) Sambon 1907, 833.)

re.cur.ren'tis. L. part.adj. *recurrens, recurrentis* recurring.

Cells cylindrical, sometimes flattened, 0.3–0.6 by 8–18 μm, with one or both ends pointed. Spirals with 3–8 turns, often inconstant; pitch of spiral 1.2–1.7 μm. Actively motile with lashing and corkscrew motion.

Grows in media with ascitic fluid or serum and either kidney fragments or coagulated egg white, at pH 7.2–7.4. Multiplies in developing chick embryos.

Pathogenic for man and monkeys. Symptoms vary according to *Borrelia* strain and species of monkey. Usually short and mild infection in mice but strains from Kenya often cause severe disease. Young rats are easily infected but are reportedly refractile to some North African strains. Guinea pigs are not susceptible.

The cause of epidemic relapsing fever in man. No animal reservoir known. The arthropod vector is *Pediculus humanus* subsp. *humanus*. After a meal of infected blood, the *Borrelia* penetrates into the coelomic fluid that is not connected with the gut.

The infection is transmitted by the coelomic fluid, not by the bite of the arthropod; this fluid is set free only when the *Pediculus* is crushed or mutilated. Scratching after the bite may damage the arthropod, and rub the *Borrelia* into the skin. The corkscrew-like motion of the organism assists its penetration through the abrasions of the skin or mucous membranes. The *Borrelia* is not transmitted by the transovarian route to subsequent generations of lice, whereas transovarial transmission is the rule in most *Ornithodoros* ticks. Found in infected lice and man.

Comment: More elongated, slender strains isolated from cases of relapsing fever in North Africa, but not fully identical with the reference strain of *Borrelia recurrentis*, were named *Borrelia berbera* (Sergent and Foley) Bergey *et al.* 1925, 435. Morphologically similar strains with considerable antigenic variability, isolated in India, were designated *Borrelia carteri* (Manson) Bergey *et al.* 1925, 435 (*Spirochaeta carteri* Manson 1907, 195).

Borrelia novyi (Schellack) Bergey *et al.* 1925, 434 isolated from a case of relapsing fever in New York, is most probably a rodent-adapted, laboratory strain of *B. recurrentis*. It may have lost many of its original antigenic characteristics but some serological cross-reactions with *B. recurrentis* are still present.

All three are probably strains of *B. recurrentis*.

3. Borrelia hispanica (de Buen) Steinhaus 1946, 453. (*Spirochaeta hispanica* de Buen 1926, 185.)

hi′spa′ni.ca. L. adj. *hispanica* Spanish.

Resembles *Borrelia recurrentis* morphologically.

Strains with somewhat different antigenic properties and of diverse geographical origin have been variously described as Moroccan (Baltazard, 1936), Normandian (Sautet, 1937), Peloponnesian or Greek (Caminopetros and Triantaphylopoulos, 1936) and Portuguese (Pinto, 1943).

Pathogenic for man. Monkeys develop disease usually with two or three relapses. Short and mild infections in mice and rats. Instillation into the eye of rabbits causes a keratitis similar to syphilitic keratitis. Mild illness in European hedgehogs (*Erinaceus europaeus*). No interference with *Trypanosoma brucei* infection in mice. Laboratory infections also in dogs, porcupines and bats. Cats and pigs are refractory.

Cause of endemic human relapsing fever in North Africa and countries bordering the Mediterranean.

The arthropod vector is *Ornithorodoros erraticus erraticus*, the large form of *O. erraticus*. This tick lives in animal burrows and stables and feeds on rodents, reptiles and amphibians in Africa and Mediterranean countries. Transovarial transmission is the rule. The infection is transmitted to mammals during the feeding of the tick. Found in cases of endemic relapsing fever, domestic and peridomestic rodents, dogs, jackals, foxes and porcupines in North Africa and countries bordering the Mediterranean.

4. Borrelia hermsii (Davis) Steinhaus 1946, 453. (*Spirochaeta hermsi* Davis 1942, 46.)

herm′si.i. M.L. gen.n. *hermsii* of *hermsi*, the specific epithet of the tick vector *O. hermsi*.

Morphologically resembles *Borrelia recurrentis* but shorter forms that appear wider when stained with aniline dyes are not infrequent.

Pathogenic for man. Monkeys are highly susceptible. Produces several relapses in mice and rats. Guinea pigs are susceptible.

Causes endemic relapsing fever in the western United States and Canada.

The arthropod vector is *Ornithodoros hermsi*. It lives at high elevations in western United States and Canada; feeds primarily on rodents and other ticks. *O. hermsi* is transported by rodents to mountain cabins. The transovarial transmission rate is low.

Found in *Ornithodoros hermsi*, squirrels (*Sciurus, Eutamias*), and infected humans.

5. Borrelia duttonii (Novy and Knapp) Bergey *et al.* 1925, 434. (*Spirillum duttoni* (*sic*) Novy and Knapp 1906, 296; *Spirochaeta duttoni* (Novy and Knapp) Breinl 1906, 1691.)

dut.to′ni.i. M.L. gen.n. *duttonii* of Dutton; named for J. E. Dutton.

Morphologically similar to *Borrelia recurrentis*.

Grown in media containing natural proteins and tissue fragments, and in developing chick embryos.

Pathogenic for man. Monkeys are readily infected; their mortality rate is high. Long-lasting infections in newborn and adult mice and rats. Guinea pigs are susceptible. Young rabbits may be infected but usually survive. Instillation into the conjunctival sack causes ulcerative keratitis in them. The European hedgehog (*Erinaceus europaeus*) is not susceptible. Protects mice against fatal *Trypanosoma brucei* infection.

Cause of endemic relapsing fever in Central and South Africa. *Ornithodoros moubata* is the only known reservoir.

The arthropod vector is *Ornithodoros moubata*. It is usually found in houses in which people and fowl live together, in Central and South Africa. The infection is transmitted through the saliva of nymphs and young ticks, and by the coxal fluid of adults. The hereditary transmission rate is high.

Found in human relapsing fever in Africa; *Ornithodoros moubata*.

Comment: *Borrelia kochii* (Novy) Bergey *et al.* 1925 is a strain of *Borrelia duttonii* from East Africa.

6. Borrelia parkeri (Davis) Steinhaus 1946, 453. (*Spirochaeta parkeri* Davis 1942, 46.)

par'ker.i. M.L. gen.n. *parkeri* the specific epithet of the tick vector *O. parkeri*.

Morphologically resembles *Borrelia recurrentis*.

Pathogenic for man but not neurotropic. Few organisms in the blood of infected animals. Monkeys develop two or three relapses. Mice and rats are susceptible but have few relapses. Guinea pigs are susceptible.

The arthropod vector is *Ornithodoros parkeri*. It transmits the infection by its bite. The vector lives with burrowing rodents in the western United States.

Found in relapsing fever in man, *Ornithodoros parkeri*, ground squirrels.

Comment: A serological variant *Borrelia parkeri* subsp. *hastingsi* was isolated from *Ornithodoros parkeri* var. *hastingsi* in the Hastings Reservation, California (Rafyi *et al.*, 1965, 631).

7. Borrelia venezuelensis (Brumpt) Brumpt 1922, 495. (*Treponema venezuelense* Brumpt 1921, 207; *Spirochaeta neotropicalis* Bates and St. John 1922, 575; *Borrelia neotropicalis* (Bates and St. John) Steinhaus 1946, 453.)

ve.ne.zue.len'sis. M.L. adj. *venezuelensis* the specific epithet of the tick vector *O. rudis* (*O. venezuelensis*).

Pathogenic for man. Monkeys, mice and rats are susceptible but the incubation time varies. Several strains are neurotropic in rodents. Guinea pigs are susceptible to some strains. Rabbits, dogs and fowls are not susceptible.

Cause of endemic relapsing fever in Panama, Colombia, Venezuela and Ecuador.

The arthropod vector is *Ornithodoros rudis*. It lives with monkeys and rodents but is acquiring the habits of bedbugs in Panama and the northern part of South America. The taxonomy of the vector was clarified by Dunn, 1927, 177, and Davis, 1942, 68.

8. Borrelia mazzottii Davis 1956, 17.

maz.zott'i.i. M.L. gen.n. *mazzottii* of L. Mazzotti.

Morphologically resembles *Borrelia recurrentis*.

Pathogenic for man. Mice and rats are susceptible but guinea pigs are not.

Causes endemic relapsing fever in Mexico and Guatemala.

The arthropod vector *Ornithodoros talaje* lives along the Pacific Coast of the Americas with rodents, armadillos and monkeys. It transmits the infection from animal to animal rather than from animal to man because only a few elect to feed on man.

Found in *Ornithodoros talaje*, and human relapsing fever in Guatemala.

Comment: Further characterization and elucidation of the relationship of this organism to its vector are desirable.

9. Borrelia persica (Dschunkowsky) Steinhaus 1946, 453. (*Spirochaeta persica* Dschunkowsky 1913, 419.)

per'si.ca. L. adj. *persica* Persian.

Morphologically resembles *Borrelia recurrentis*.

Pathogenic for man. Strongly neurotropic. Uzbek strains cause only mild disease. Monkeys and mice are highly susceptible. Rats are susceptible with a long incubation period, principally when infected with Uzbek, Jordanian and Tripoli strains. Guinea pigs are susceptible to most strains and develop hemoperitoneum. Severe illness in the European hedgehog (*Erinaceus europaeus*). Mice are not protected by this strain against fatal *Trypanosoma brucei* infection.

Causes endemic relapsing fever in man from Uzbekistan and Kashmir to Cyprus and Tripoli.

The arthropod vector is *Ornithodoros tholozani*. It ranges from Central Asia to the Mediterranean. Lives in caves, huts, stables and rodent burrows. The tick and its nymphal stages are long lived. The ticks transfer the infection to mammals rather by their bite than by their coxal fluid. Transovarial propagation is the rule.

Found in *Ornithodoros tholozani*, wild rodents and bats.

Comment: Strains from different geographic localities were described as *Borrelia babylonensis* in Iraq (Brumpt, 1939); *Borrelia sogdiana* in Central Asia (Nicolle and Anderson, 1927); and *Borrelia uzbekistanica* in Uzbekistan (Kassirsky, 1933).

10. Borrelia turicatae (Brumpt) Steinhaus 1946, 453. (*Spirochaeta turicatae* Brumpt 1933, 1369.)

tu.ri.ca'tae. M.L. gen.n. *turicatae* of *turicata*, the specific epithet of the tick vector, *O. turicata*.

Morphologically resembles *Borrelia recurrentis*.

Pathogenic for man. Monkeys develop 1–3 relapses. Mice and rats are susceptible. Guinea pigs are not susceptible to all strains. Dogs, foxes, cats, pigs and cotton rats were infected in the laboratory.

Cause of endemic relapsing fever in Mexico and the United States.

The arthropod vector is *Ornithodoros turicata*. It is found in the western United States, including Texas, Oklahoma, Kansas, and in Mexico. Lives with rodents and reptiles. It is becoming domesticated in Texas and Mexico. Infects mammals by its bite.

11. Borrelia caucasica (Kandelaki) Davis 1957, 901. (*Spirochaeta caucasica* Kandelaki acc. to Maruashvili 1945, 24.)

cau.ca′si.ca. M.L. adj. *caucasica* pertaining to the Caucasus.

Morphologically resembles *Borrelia recurrentis*. Causes mild relapsing fever in man in the Caucasus. Mice, rats and guinea pigs are susceptible. The arthropod vector is *Ornithodoros verrucosus*. It lives in the semi-desert areas of the Caucasus, principally in Azerbaijan, in burrows and caves inhabited by small rodents but readily attaches itself to man.

12. Borrelia brasiliensis Davis 1952, 476. bra.si.li.en′sis. M.L. adj. *brasiliensis* the specific epithet of the tick vector, *O. brasiliensis*.

Morphologically resembles *Borrelia recurrentis*. Mice, rats and guinea pigs are susceptible. The arthropod vector is *Ornithodoros brasiliensis*. It feeds on wild animals and birds but appears also in stables and houses in Brazil.

Source: *Ornithodoros brasiliensis* in Brazil.

13. Borrelia dugesii (Mazzotti) Davis 1957, 902. (*Spirochaeta dugesi* Mazzotti 1949, 278.)

du.ge′si.i. M.L. gen.n. *dugesii* of *dugesi* the specific epithet of the tick vector, *O. dugesi*.

Morphologically resembles *Borrelia recurrentis*. Pathogenic for mice and rats but not for guinea pigs.

The arthropod vector *Ornithodoros dugesi* is found in Mexico.

Isolated from *Ornithodoros dugesi* in Mexico.

14. Borrelia graingeri (Heisch) Davis 1957, 903. (*Spirochaeta graingeri* Heisch 1953, 133.)

grain′ger.i. M.L. gen.n. *graingeri* the specific epithet of the tick vector, *O. graingeri*.

Not reported from natural infection in man but circulates in the blood of persons injected with it. Present in the blood of mice and rats for a few days. Guinea pigs are not susceptible.

The arthropod vector is *Ornithodoros graingeri*, which lives in rodent caves and burrows in East Africa.

Isolated from *Ornithodoros graingeri* in Kenya.

15. Borrelia crocidurae (Leger) Davis 1957, 903. (*Spirochaeta crocidurae* Leger 1917, 281.)

cro.ci.du′rae. M.L. gen.n. *crocidurae* of *Crocidura*, a genus of Insectivora.

Resembles *B. recurrentis* morphologically.

A number of strains carried by the same tick vector are antigenically distinct, differ in behavior in animals and in geographical distribution (Bal-

tazard *et al.*, 1950). Four types, serologically different from each other, as well as from *B. duttonii*, *B. persica* and *B. recurrentis*, have been suggested. They are possibly serotypes of *B. crocidurae*.

The serotypes are mildly pathogenic for man and monkeys. Newborn mice, rats, rabbits and hamsters develop serious disease. Most strains, particularly from Egypt, circulate in the blood of guinea pigs without causing illness. Inapparent infection in the European hedgehog (*Erinaceus europaeus*).

The arthropod vector is *Ornithodoros erraticus sonrai* (the small form of *Ornithodoros erraticus*) which lives with small rodents in Africa, the Near East and Central Asia.

Originally isolated from the wild shrew (*Crocidura stampfii*) in West Africa.

16. Borrelia latyschewii (Sofiev) Davis 1948, 315. (*Spirochaeta latyschewi* (*sic*) Sofiev 1941, 271.)

la.ty.sche′wi.i. M.L. gen.n. *latyschewii* of Latyschew, named for Latyshew (Latyshev).

Morphologically resembles *Borrelia recurrentis*. Circulates in the blood of mice, and for a short time also in the blood of rabbits. Rats, guinea pigs and dogs are not susceptible.

The arthropod vector is *Ornithodorus tartakovskyi*. It lives with rodents and reptiles in burrows in Central Asia and Iran. Exudes coxal fluid after, not during feeding.

Isolated from gerbils and related rodents in Central Asia and Iran.

17. Borrelia tillae Zumpt and Organ 1961, 33. till′ae. M.L. gen.n. *tillae* of Till; named for Dr. W. Till.

Morphologically resembles *Borrelia recurrentis*. Serologically different from *Borrelia duttonii*.

Causes short, transient infection in monkeys. Neurotropic in mice. Rats are very susceptible but not guinea pigs.

The arthropod vector is *Ornithodoros zumpti*. It lives in rodent burrows in South Africa.

Originally isolated from *Ornithodoros zumpti* collected in nests of vlei rats (*Otomys saundersiae*) and from organs of *Rhabdomys pumilio* and *Rattus natalensis* in South Africa.

18. Borrelia theileri (Laveran) Bergey *et al.* 1925, 435. (*Spirochaeta theileri* Laveran 1903, 941.)

thei′le.ri. M.L. gen.n. *theileri* of Theiler; named for A. Theiler.

Cells slender, 0.25–0.3 by 20.0–30.0 μm when isolated from cattle. Smaller forms are seen in horses. The organism is actively motile with flexuous motion.

Causes cattle and horse borreliosis in South

Africa and Australia, respectively, a mild disease in cattle resembling tick fever, and a febrile disease in horses.

Transmitted by the ticks *Rhipicephalus decoloratus* and *R. evertsi* in Africa, by *Boophilus micropus* in Australia, where the small form is found also in cattle.

Isolated from the blood of infected cattle in Transvaal, South Africa.

19. **Borrelia harveyi** (Garnham) Davis 1948, 316. (*Spirochaeta harveyi* Garnham 1947, 49.)

har′vey.i. M.L. gen.n. *harveyi* of Harvey; named after A. E. C. Harvey.

Morphologically resembles *Borrelia recurrentis*. Serologically distinct from *Borrelia recurrentis* and *B. duttonii*.

Mild disease in man. Pathogenic for monkeys, mice and rats but not for guinea pigs.

Arthropod vector unknown.

Originally isolated from the blood of a grivet monkey (*Cercopithecus aethiops centralis*) in Kenya.

Species incertae sedis

a. *Borrelia glossinae* (Novy and Knapp) Bergey *et al.* 1923, 435. (*Spirillum glossinae* Novy and Knapp 1906, 385.)

Probably an organism accidentally acquired by a blood-sucking insect. Inadequately described.

b. *Borrelia dipodilli* (Heisch) Davis 1957, 903. (*Spirochaeta dipodilli* Heisch 1950, 271.)

Isolated from the pigmy gerbil (*Dipodillus* sp. = *Gerbillus* sp.) in East Africa.

c. *Borrelia merionesi* (Blanc and Maurice) Davis 1948, 316. (*Spirochaeta merionesi* Blanc and Maurice 1948, 141.)

Isolated from the gerbil (*Meriones shawi*) at Goulimine, Morocco.

Both *B. dipodilli* and *B. merionesi* may be related to or be serotypes of *B. crocidurae* and members of the "crocidurae subgroup."

d. *Borrelia queenslandica* Carley and Pope 1962, 261.

Isolated once in Queensland, Australia from *Rattus villosissimus.*

Genus V. **Leptospira** *Noguchi 1917, 755*

L. H. TURNER

Lep.to.spi′ra. Gr. adj. *leptos* thin, narrow, fine; Gr. n. *spira* a coil; M.L. fem.n. *Leptospira* a fine coil.

Cells **single, flexuous, helical,** 6–20 μm or more by 0.1 μm in diameter; coils 0.2–0.3 μm in overall diameter, pitch 0.3–0.5 μm (Plate 5.1, Fig. 5). **One or both ends may be bent**—more or less at right angles to the long axis—**or hooked.** Some strains may lose their hooked shape during subculture in certain media and retain the straight form in that medium.

Motile by means of axial filaments, comprising the axistyle, which are now regarded as flagellar analogues; no external flagella. **Movements are of three basic sorts: shunting** in either direction of the long axis (translation without polar differentiation); **rapid rotation** or oscillation about the long axis; and **flexion.** Leptospires may be motionless for a few moments and at such times the tightly coiled appearance, suggestive of a tiny rope, is usually visible as are the shape and size of the hooks if these are present. Translation is on a straight or gently curved course in liquid milieux; but in more viscous milieux (e.g. semi-solid agar) the course is usually sinuous or serpentine. When a hook-ended organism rotates rapidly around its long axis a variety of shapes may result, commonly like an hourglass, the figure 8, a tennis racquet or the capital letter T, depending on the number and form of the hooks. Flexion of the body may occur spontaneously when the leptospire is stationary and also when it meets an obstruction; it is usual during translation in viscous media; and it is also associated with transverse division of the organism, when it may be very vigorous.

Not readily visible in preparations treated with Giemsa stain or the Gram-reaction, but revealed, often distorted, by silver impregnation methods. **Readily visible by dark-field microscopy** at magnifications of 100 × or greater: less clearly observed by phase-contrast; not visible by ordinary illumination.

The structure of *Leptospira* as revealed by electron microscopy consists of a cellular body wound around an axistyle with a common external sheath enveloping both these elements. The cellular body is roughly circular in transverse section. Nuclear material, cytoplasm and a limiting cytoplasmic membrane can be differentiated. The axistyle seems to be a single structure, about 0.01–0.02 μm in diameter, comprised of two filaments each of which is inserted subterminally, by means of a disc or knob, at one end of the body with their free ends near the middle of the body.

Chemoorganotrophs: Metabolism respiratory. Terminal electron acceptors are cytochromes a, c and c_1 (for strains investigated; Baseman and Cox, 1969). Energy-yielding substrates should include fatty acids as the source of carbon. Parasitic strains require unsaturated fatty acids (containing 15–16 carbon atoms or more) or the corresponding polysorbate (Tween). This requirement is increased at temperatures above or below the optimum. Saturated fatty acids are not used unless suitable unsaturated fatty acids are also present (Johnson, Harris and Walby, 1969). Inorganic NH_4 salts are suitable as the source of nitrogen. Other minimal nutritional requirements include thiamine, vitamin B_{12}, and various metallic salts (see Ellinghausen and McCullough, 1965; Shenberg, 1967). Some strains seem to need other, undetermined, nutritional factors.

Can be cultured on various media which are based on a solution of inorganic salts, buffered with phosphates. Some media are supplemented with pooled rabbit serum, 7–20% v/v (Fletcher, 1928; Korthof, 1932; Stuart, 1948); others with bovine albumin, fraction V, and Tween 80 (Ellinghausen and McCullough, 1965; Johnson and Harris, 1967) usually without rabbit serum. Pooled normal rabbit serum added (2–5%, v/v) to bovine albumin Tween 80 medium enhances the rate of multiplication of certain strains—notably those which belong to serotypes which did not multiply in Shenberg's synthetic medium (Shenberg, 1967). Solid media are not recommended for routine isolation of strains because satisfactory multiplication is much less certain than in liquid or semi-solid (containing 0.2% agar, w/v) media.

Strict **aerobes** utilizing atmospheric oxygen.

Temperature optimum 28–30 C. More rapid multiplication may sometimes result from incubation at 37 C for 1–2 days. Exposure to 40 C is deleterious (Turner, 1966); and at 56 C it is lethal (for pathogenic strains investigated). Strains of free-living, presumed saprophytic, leptospires will multiply at lower temperatures (e.g. 13 C) than parasitic and pathogenic strains (Johnson and Harris, 1967).

Optimal pH requirement 7.2–7.4.

The G + C content of the DNA, for the few strains reported, ranges from 36–39 ± 1% (T_m and buoyant density, Haapala *et al.*, 1969).

Some strains parasitic or pathogenic in vertebrates; others free-living.

Type species: *Leptospira interrogans* (Stimson) Wenyon 1926, 1281.

Further Comments

The taxonomy of the genus is at present under consideration by the Subcommittee on Leptospira of the International Committee on Systematic Bacteriology. This subcommittee has recommended that until further study only one species be recognized, the type being the description by Stimson in 1907, which is the earliest.

In the 7th edition of THE MANUAL two species were recognized; the parasitic *L. icterohemorrhagiae* and the saprophytic *L. biflexa*. But there are not yet sufficient reliable taxonomic data for the circumscription of more than one species of *Leptospira* (see below).

For the general duty laboratory the determinative procedures cease, at present, with the recognition of *Leptospira* on morphological characters and movements as seen occasionally in fresh specimens of blood (more often after differential centrifugation, Wolff, 1954), cerebrospinal fluid, urine or in emulsions of organs from infected man or animals; in positive cultures of these materials; or in surface waters.

Isolates should be sent to a more specialized laboratory where they can be maintained and studied by techniques beyond the capacity of the referring agent.

Some strains of *Leptospira* are capable of persisting for long periods, even for the life of the host, in the proximal convoluted tubules in the renal cortex of various vertebrates, seemingly without detriment to the host. This is the reservoir or maintenance host condition. We do not know whether such organisms are parasitic on the epithelium of the renal tubules, as is commonly supposed, or whether they are commensal, living and multiplying in the urinary filtrate at that site. Whatever may be the precise nature of this relationship, it is presumed to be the last of a succession of phases in the course of infection—incubation period, septicemia (leptospiremia), leptospiruria and immunity (see e.g. Galton *et al.*, 1962; Turner, 1967).

Other strains of *Leptospira*, for which no vertebrate host is known, seem to be free-living or saprophytic in fresh surface waters and occasionally in domestic supplies or in saline waters.

Description of the species of genus **Leptospira**

1. **Leptospira interrogans** (Stimson) Wenyon 1926, 1281. (*Spirochaeta interrogans* Stimson 1907, 541; *Spirochaeta biflexa* Wolbach and Binger 1914, 25; *Spirochaeta nodosa* Hubener and Reiter 1916, 2; *Spirochaeta icterohaemorrhagiae* Inada, Ido, Hoki, Kaneko and Ito 1916, 379; *Spirochaeta icterogenes*

Uhlenhuth and Fromme 1916, 438; *Leptospira icteroides* Noguchi 1919, 581.)

in.ter′ro.gans. M.L. part. *interrogans* interrogation, here meaning shaped like a question mark.

Morphology and characteristics as for genus.

Further Comments

As the Subcommittee on Leptospira (ICSB) considers that strains of *Leptospira* cannot be accommodated satisfactorily and with confidence in taxa which are recognized by the Code, unofficial terms, such as "complex" and "serogroup" are used to denote tentative groupings.

Four genetically distinct groups of *Leptospira* have recently been demonstrated on the basis of G + C content of the DNA and of DNA annealing tests (Haapala *et al.*, 1969). In some "pathogenic" strains the G + C content was 36 ± 1%; in some other "pathogenic" strains and in the biflexa strains which were tested the G + C content was 39 ± 1%. The authors proposed that these four genetic groups be assigned the rank of species, but they did not propose names. Certain phenotypic biological characteristics did conform with these genetic groups and may be used to distinguish some of them.

For the time being the two long-recognized groups of pathogenic or parasitic strains, and of saprophytic, free-living or water leptospires are retained as "complexes," which have been named interrogans and biflexa.

Since 1960 a number of tests have been proposed to distinguish between these two complexes on biological grounds. However, none has proven very satisfactory and Kmety *et al.* (1966) concluded that none of the tests available at that time were sufficiently reliable for purposes of classification. Johnson and Harris (1968) using purine analogue sensitivity and lipase activity showed that the "interrogans" complex is comprised of at least two biological groups, 1 and 2, plus a unique group 4 which was unable to grow in the presence of DAP and 8-azaguanine and was lipase negative (*kabura*, Canicola serogroup). The "biflexa" complex comprised a biological group 3.

The genetic groups of Haapala *et al.* and the biological groups of Johnson and Harris conform quite well; they do not conform so neatly with the currently recognized serogroups based on agglutinogens. They conform quite well with the available information on cross immunity. However, much more study is required on these lines.

The enzymic properties of *Leptospira* strains, particularly the lipases, may be of some taxonomic value eventually. This aspect is discussed in the reports of Green *et al.* (1967) and Bakoss and Chorvath (1965).

For many years strains of leptospires have been compared and distinguished serologically, especially by means of the agglutination reaction and cross-agglutinin absorption studies carried out with antisera prepared in rabbits.

The arrangement by serogroups and serotypes which was proposed by Wolff and Broom (1954) has been revised and extended. The latest published list (World Health Organization, 1967) presents more than 130 serotypes under 18 serogroups. However, this list is not unanimously accepted: the results of different laboratories, when typing the same strain, have sometimes been discrepant. The causes of such discrepancies are being elucidated. There are not yet generally accepted techniques for culturing strains, for immunizing rabbits, for carrying out agglutinin-absorption, or for the microscopic agglutination reactions. The agglutinogenic composition of a strain, and its agglutinability, may be influenced by the medium in which it is cultured. The agglutinin factors of antisera depend on the agglutinogenic composition of cultures of the line of the strain which is used to immunize rabbits. Thus Borg-Petersen (1971) has already demonstrated that some strains possess an agglutinogenic component (or components) which is destroyed by heating the culture at 56 C—a procedure which is routinely adopted by at least one reference laboratory. Borg-Petersen studied strains belonging to the Icterohaemorrhagiae serogroup. Other serogroups have yet to be investigated in this respect. Some workers claim (unpublished information, Dr. M. Torten) that the results of typing a strain depend on the time elapsed, since the last immunizing inoculum, at which the antiserum is obtained.

Attempts have been made to analyze the agglutinogenic factors of leptospires (e.g. Kmety, 1967). Such studies, when extended to other serogroups, will enable the typing of strains to be based on qualitative criteria. At present the differentiation of strains must often be based on arbitrary quantitative criteria (Wolff and Broom, 1954).

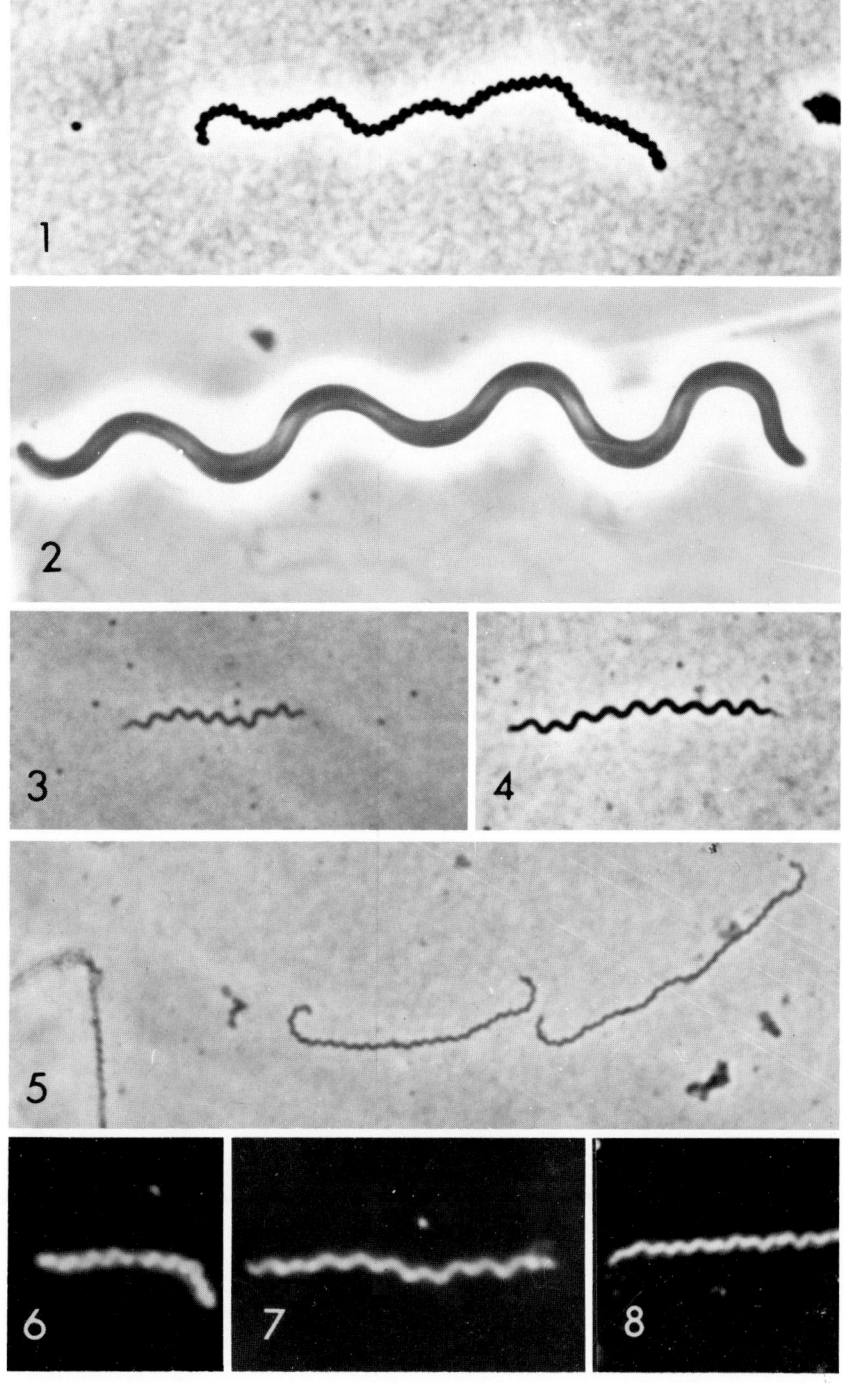

Plate 5.1

193

Plate 5.1

Plate 5.1. Spirochetes

Fig. 1. *Spirochaeta plicatilis* from activated sludge.

Fig. 2. *Cristispira* sp. from crystalline style of clam, *Cryptomya californica*.

Fig. 3. *Treponema pallidum* from rabbit testis.

Fig. 4. *Borrelia anserina* from chicken blood.

Fig. 5. *Leptospira interrogans* serotype icterohaemorrhagiae from Fletcher's medium.

Fig. 6. *Treponema phagedenis* biotype reiter.

Fig. 7. *Treponema refringens*.

Fig. 8. *Treponema denticola*.

Figs. 1–5. Phase-contrast photomicrographs by Dr. Daisy A. Kuhn, magnification ✕ 2200. Material for *Treponema* supplied by Dr. James N. Miller, for *Leptospira* and *Borrelia* by Dr. Ernst L. Biberstein; Figs. 6–8. Dark-field microscopy, magnification ✕ 2300, by Dr. R. M. Smibert.

Plate 6.1. *Bdellovibrio* and *Pelosigma*

Fig. 1. *Bdellovibrio bacteriovorus* ATCC 15143. Negatively stained preparation showing the vibrio-shaped cell with its thick ensheathed polar flagellum. From Burnham, J. C., Hashimoto, T. and Conti, S. F. 1968. J. Bacteriol. *96:* 1369.

Fig. 2. *Bdellovibrio bacteriovorus* ATCC 15143. Negatively stained flagellum showing blebs formed by sheath pulling away from the flagellar core. ✕34,300.

Fig. 3. *Bdellovibrio bacteriovorus* ATCC 15143. A bdellovibrio penetrating an *Escherichia coli* (ATCC 15144) host cell (p) and several *Escherichia coli* cells already infected (i). Phase-contrast. ✕1960.

Fig. 4. *Bdellovibrio bacteriovorus* ATCC 15143. Two *Escherichia coli* B/r cells being attacked by a number of *B. bacteriovorus* cells. Phase-contrast. ✕1764.

Fig. 5. *Bdellovibrio bacteriovorus* ATCC 15143. Schematic representation of the life cycle of parasitic, host-independent and facultative strains of the genus *Bdellovibrio*. Modified from Burnham, J. C., Hashimoto, T. and Conti, S. F. 1970. J. Bacteriol. *101:* 1004.

Fig. 6. *Pelosigma* sp. a coenobium from the mud of a fresh water lake (Plussee, Germany) growing attached to a submerged glass slide. ✕1254.

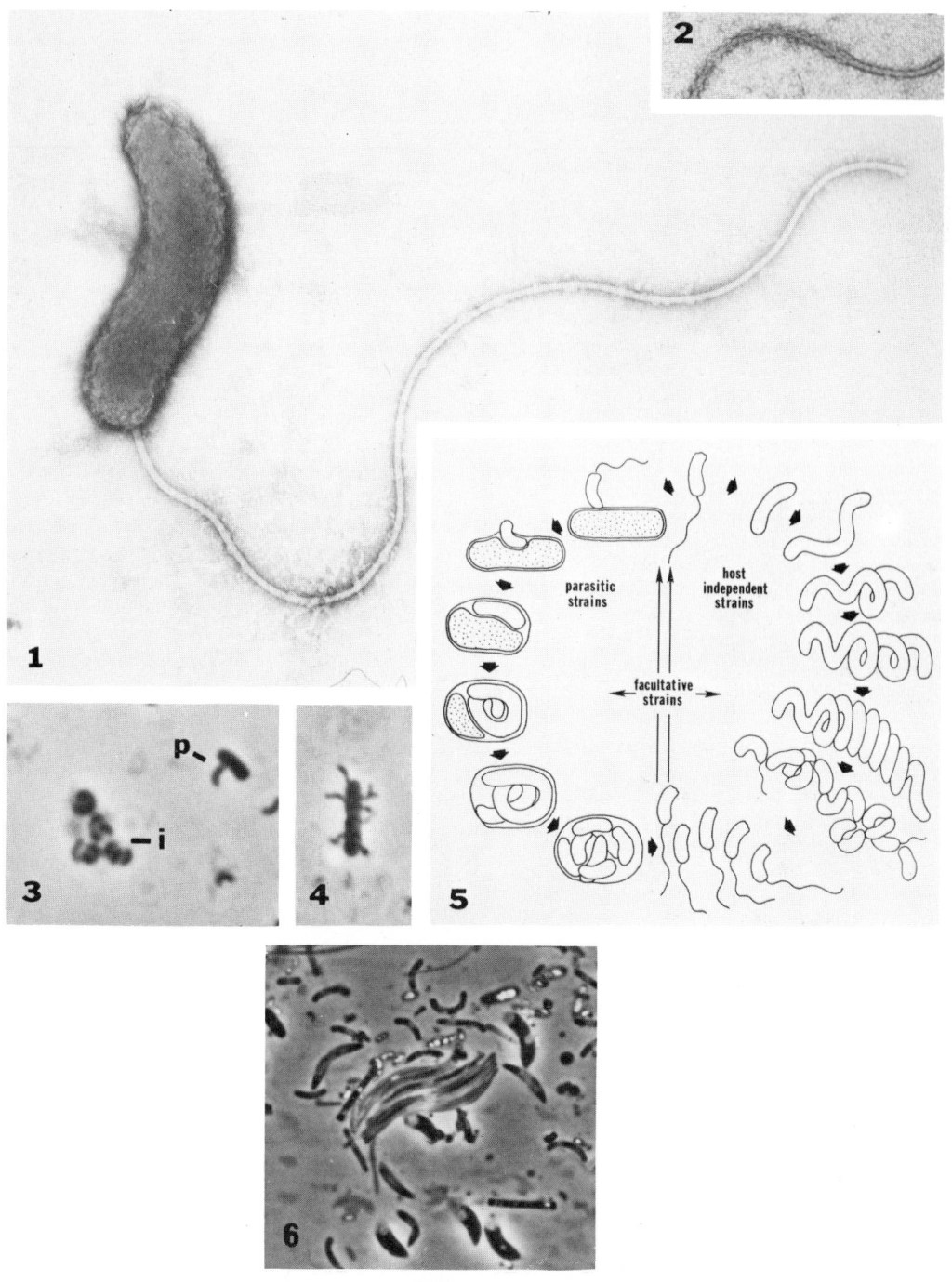

Plate 6.1

195

PART 6

SPIRAL AND CURVED BACTERIA

FAMILY I. **SPIRILLACEAE** MIGULA 1894, 237

N. R. KRIEG AND R. M. SMIBERT

Spi.ril.la′ce.ae. M.L. neut.n. *Spirillum* type genus of the family; *-aceae* ending to denote a family; M.L. fem.pl.n. *Spirillaceae* the *Spirillum* family.

Rigid, helically curved rods with less than one complete turn to many turns. Cell diameter, 0.2–1.7 µm; length of helix 0.5–60 µm. Motile, swimming in straight lines with a characteristic corkscrew-like motion. Cells may possess only a single polar flagellum or a fascicle of several polar flagella. Flagella may be at one or both poles. Some species are strictly aerobic or obligately microaerophilic, with oxygen required as the terminal electron acceptor. Others are anaerobic but can also grow under microaerophilic conditions. Chemoorganotrophs. Incapable of fermenting carbohydrates, although a few species can oxidize a limited variety. Usually no alcohols or fatty acids are detected by gas chromatography of culture supernatants; when detected they are present in only trace amounts. Oxidase positive. Indole negative. Some produce a yellowish green water-soluble fluorescent pigment. Some grow in a simple mineral salts medium containing a single carbon source and ammonium sulfate; other require a complex peptone-based medium. Some are free living in fresh water or marine environments. Others are saprophytic or parasitic. Some are pathogenic.

Key to the genera of family **Spirillaceae**

I. Cells 0.25–1.7 µm wide and 2–60 µm long. Polytrichous polar flagella, usually at both poles. Polyhydroxybutyrate granules usually present in cytoplasm. Strictly aerobic or obligately microaerophilic. The G + C content of the DNA ranges from 38–65 moles % (T_m).

Genus I. *Spirillum*

II. Cells 0.2–0.8 µm wide and 0.5–5 µm long. Single polar flagellum at one or both poles. Polyhydroxybutyrate granules not present. Microaerophilic to anaerobic. The G + C content of the DNA ranges from 30–35 moles % (T_m).

Genus II. *Campylobacter*

Genus I. **Spirillum** *Ehrenberg 1832, 38*

N. R. KRIEG*

Spi.ril′.lum. Gr. n. *spira* a spiral; M.L. dim.neut.n. *Spirillum* a small spiral.

* The assistance of Dr. P. B. Hylemon and Dr. J. S. Wells, Jr., is gratefully acknowledged. Without their assistance the thorough descriptions, DNA base compositions, etc., would not have been possible.

Rigid, helical cells, 0.25–1.7 μm in diameter, with less than one turn to many turns. Intracellular granules of polyhydroxybutyrate present in most species. Motile by means of polar polytrichous flagella. All but one species (*S. delicatum*) exhibit bipolar flagellation. Gram-negative.

Chemoorganotrophs, having a strictly respiratory metabolism with oxygen as the terminal electron acceptor. A few species can grow anaerobically with nitrate.

Acid reactions are usually not produced from sugars; in the few positive cases the reactions are evident only with a low peptone concentration. Sugars usually do not serve as sole carbon sources. Casein and hippurate not hydrolyzed. Indole, sulfatase, amylase and phenylalanine deaminase negative. Gelatin usually not attacked. Hydrogen sulfide usually produced from cysteine. A few species are urease positive.

Oxidase and catalase positive; the latter is sometimes weak. Phosphatase usually positive.

Yellowish green fluorescent water-soluble pigment produced by nearly half of the species.

The nutrition is simple; most species do not require amino acids, vitamins, purines or pyrimidines. Various organic acids, alcohols or amino acids can be utilized as sole carbon sources. Ammonium ions can usually be utilized as a sole nitrogen source. Incorporation of sea water into the media is required for marine species.

The type species is obligately microaerophilic. All other species can grow aerobically, although some may possibly prefer microaerophilic conditions. Optimum temperature 30 C, with no growth at 10 or 45 C. Fresh water species do not grow in NaCl concentrations greater than 3%, and some fail to grow in concentrations as low as 0.3%.

Found in fresh and salt waters containing organic matter.

The G + C content of the DNA ranges from 38–65 moles %.

Type species: *Spirillum volutans* Ehrenberg 1832, 38 (Stiles, 1905; Hylemon *et al.*, 1973).

Further Comments

Characteristics useful in differentiating the fresh water species are presented in Table 6.1; the marine species in Table 6.4. Other characteristics are summarized in Tables 6.2 and 6.3.

The descriptions are based largely on studies in the author's laboratory. With the exception of *S. volutans*, *S. anulus*, *S. itersonii*, *S. serpens* and *S. gracile*, only a single strain for each species is available for study, either the type or a reference strain. The degree of variation within most species remains to be defined, and consequent modifications may be expected in future editions. One limiting factor in this regard has been the diffi-

culty experienced by many investigators in isolating and preserving strains from natural sources.

The great range of G + C values, differences in size, nutrition, physiology and serology, strongly suggest that the genus could be divided into two or more genera. For the present it is useful to distinguish between five major groups: (1) the large obligately microaerophilic type species (G + C, 38 moles %); (2) the aerobic fresh water species (No. 2–13, with G + C = 50–65 moles %); (3) the marine species (No. 14) which attacks a relatively large number of carbohydrates (G + C, 63 moles %); (4) the marine species (No. 15–18) which do not attack carbohydrates (G + C, 42–48 moles %); (5) the obligately parasitic species (No. 19). (See Editorial Note below.)

In this edition of THE MANUAL, *S. atlanticum* is considered a subjective synonym of *S. linum*. *S. giesbergeri* and *S. graniferum* appear similar enough to warrant consideration as a single species, but additional information is needed.

Some species of spirilla produce spherical or oval coccoid bodies, 3–5 μm in diameter, which predominate usually in older cultures (1–6 weeks). These have been termed "microcysts" by Williams and Rittenberg (1957); however, since resistance to desiccation and the presence of thickened walls have not been demonstrated, it may be preferable to employ a term lacking these connotations. Williams and Rittenberg (1956) demonstrated that such coccoid body formation may occur from helical cells by (1) fusion of two entwined organisms to form one or more coccoid bodies; (2) the production of a protuberance at some point along the cell into which the entire organism is gradually absorbed; or (3) the gradual shortening and rounding of the organism to form an oval or spherical body. When coccoid bodies from an old culture are inoculated into fresh media, they germinate to form a helical cell. Germination occurs by either unipolar or bipolar emergence of the germ tube. Williams has demonstrated that fusion of helical cells followed by fusion and rearrangement of the chromatin bodies occurs in spirilla, suggesting a possible sexual mechanism.

McElroy and Krieg (1972) have shown that antisera prepared against heat-labile antigens of spirilla permit a rapid identification and also the serological distinction of most of the species from one another and from members of related genera (*Campylobacter* and *Pseudomonas*). Seventeen serogroups were established, and the value of serological identification was demonstrated by the rapid identification of some fresh isolates, later confirmed by other methods.

In order to provide a suitable basis for comparison with the descriptions, the media and methods described in the dissertations by Wells (1970) and Hylemon (1971) should be used for the determination of the characteristics of spirilla; the methods are summarized in Hylemon, Wells, Krieg and Jannasch in the October 1973 issue of the *International Journal of Systematic Bacteriology*.

Editorial Note. Hylemon, Wells, Krieg and Jannasch (The genus *Spirillum:* A taxonomic study. Int. J. Syst. Bacteriol. October, 1973) propose a division of the genus into three genera:

1. **Spirillum** with the obligately microaerophilic *S. volutans* as the type and only species.

2. **Aquaspirillum,** a new genus for the aerobic fresh water forms containing species 2, 4–13 recognized herein and two new species *A. dispar* (type strain: ATCC 27510) and *A. aquaticum* (type strain: ATCC 11330); *A. serpens* is the type species. *S. graniferum* (No. 3) is considered a synonym of *A. giesbergeri* (No. 2).

3. **Oceanospirillum,** a new genus for the marine forms that do not attack carbohydrates, with *O. linum* (No. 17) as the type species plus species 15, 16, 18 and a new species *O. maris* (type strain: ATCC 27509).

S. lunatum (No. 14) was not included in genus *Oceanospirillum* because of its higher G + C content and its ability to attack a variety of carbohydrates. The authors note that the authenticity of the type strain was in question; several characteristics are not consistent with the original description.

TABLE 6.1
Differential characteristics of the fresh water species of genus **Spirillum**

	1. S. volutans	2. S. giesbergeri	3. S. graniferum	4. S. anulus	5. S. meta-morphium	6. S. putridicon-chylium	7. S. itersonii	8. S. peregrinum	9. S. serpens	10. S. sinuosum	11. S. gracile	12. S. delicatum	13. S. poly-morphium
Microaerophilic	+	—	—	—	—	—	—	—	—	—	—	—	—
Cell diameter													
1.5 μm	+												
0.5–1.5 μm		+	+	+	+	+	+	+	+	+			
0.25–0.4 μm											+	+	+
Succinate as sole carbon source		—	—	—	+	+			—	—			
H₂S from cysteine		+	—	+	+	+			+	+			
Urease		+	+	—	—	—			d	+			
Gelatin liquefied		—	—	—	+	—			d	—			
Anaerobic growth from nitrate		—	—	—	—	—	+	—	—	—			
Fructose acid		—	—	—	—	—	+	+	—	—			
Glutamate as sole carbon source		—	—	+	+	+	+	+	+	—			
Glucose acid											+	—	—
Phosphatase											+	+	—
Krebs cycle acids as sole carbon source											—	+	—
G + C moles %	38	58	57	59	63	52	62	62	50	57	65	63	62

Description of the species of genus **Spirillum**

1. Spirillum volutans Ehrenberg 1832, 38. vo'lu.tans. L. v. *voluto* tumble about; L. part. adj. *volutans* tumbling about.

The largest of the spirilla. Cells observed by phase microscopy in 18-hr peptone-succinate-salts (PSS) broth are 1.4–1.7 μm in diameter; wave length 16–28 μm; diameter of helix 5–8 μm; length of helix 14–60 μm with less than one to a maximum of five turns. Exceptionally prominent dark granules of polyhydroxybutyrate in cytoplasm. Motile, possessing a fascicle of about 75 flagella at each pole easily seen in living cells by phase or dark-field microscopy. Front and rear fascicles appear to rotate at high speed, forming

TABLE 6.2
Other characteristics of species of genus **Spirillum**

Reactions are determined in peptone-succinate-salts (PSS) medium (see "Description of the species of genus *Spirillum*"). The peptone must be decreased to 0.2% to demonstrate acid reactions from sugars. Urease is determined using heavy washed suspensions in a BES-buffered phenol red medium. W = weak reaction; d = varies among strains.

Test	1. S. volutans	2. S. giesbergeri	3. S. graniferum	4. S. anulus	5. S. meta-morphum	6. S. putridiconchylium	7. S. itersonii	8. S. peregrinum	9. S. serpens	10. S. sinuosum	11. S. gracile	12. S. delicatum	13. S. polymorphum	14. S. lunatum	15. S. minutulum	16. S. beijerinckii	17. S. linum	18. S. japonicum
Catalase	W	+	+	+	+	W	+	+	+	+	+	+	+	+	+	W	W	W
Phosphatase	+	+	+	+	+	+	+	+	+	+	+	+	−	+	−	+	+	+
H₂S from cysteine	+	+	−	+	+	+	+	+	+	+	+	+	+	+	−	+	d	−
Urease	−	+	+	−	−	−	−	+	d	+	+	−	−	−	−	−	−	−
Esculin hydrolysis	−	−	−	−	−	−	+	+	−	−	−	−	+	−	−	−	−	−
Gelatin liquefaction	−	−	−	−	+	−	−	−	−	d	−	−	−	−	−	−	−	−
Nitrate reduction to nitrite or more reduced products	−	−	−	−	−	−	+	−	d	−	+*	+	+	+	+	−	−	−
Selenite reduction	−	−	−	−	−	−	d	+	−	−	−	−	W	+	+	−	W	W
Polyhydroxybutyrate granules	+	+	+	+	+	+	+	+	+	+	−	+	+	−	+	+	+	+
DNase	−	+	+	−	−	−	+	−	d	+	+	−	−	+	−	+	−	−
RNase	−	+	+	d	−	+	+	−	d	−	+	+	−	+	−	+	d	−
Acid reaction from sugars	−	−	−	−	−	−	+	+	−	−	+	−	−	+	−	−	−	−
Growth in presence of: a. 1% bile	−	+	+	−	+	+	+	+	+	+	+	−	+	+	+	+	+	+
b. 1% glycine	−	−	−	−	−	−	−	−	−	−	−	−	−	+	+	−	+	−
Growth on: a. Triple sugar iron slants	−	−	−	−	+	+	+	+	d	−	−	+	−	+	+	−	+	−
b. EMB agar	−	−	−	−	+	+	+	+	+	−	+	−	+	−	−	−	−	−
c. MacConkey agar	−	−	−	−	−	−	+	−	d	−	−	−	+	+	−	−	−	−
d. MRVP broth	−	+	+	−	−	−	+	−	d	−	−	+	−	+	−	−	+	−
e. Seller's slants	−	−	−	−	+	−	+	+	d	−	−	−	−	+	−	−	−	−
Pigment produced from: a. Phenylalanine	−	−	−	−	−	−	−	+	−	−	−	−	−	−	−	−	−	−
b. Tyrosine	−	−	−	−	−	−	+	−	−	−	−	−	−	+	−	+	+	−
c. Tryptophan	−	−	−	−	−	−	+	−	−	−	−	−	−	+	d	−	d	−
Alkaline reaction in litmus milk	−	−	−	−	−	−	−	+	−	−	−	−	−		−			
Water-soluble fluorescent pigment	−	−	−	−	+	−	+	+	d	−	d	−	+	+	−	−	+	−
Anaerobic growth with nitrate	−	−	−	−	−	−	+	−	−	−	−	−	−	+	−	−	−	−
Coccoid bodies predominant in older cultures	−	−	−	−	−	−	+	+	−	−	−	−	+	−	+	+	+	−

oriented cones of revolution. Reversal of orientation is accompanied by reversal of direction of cell
motion. The orientation of front and rear fascicles
is coordinated and the coordination can be reversibly affected by a variety of compounds (Krieg
et al., 1967).

Isolation is difficult, and as yet may be accomplished only by the capillary tube procedure of
Rittenberg and Rittenberg (1962).

Growth in nutrient broth scanty. Abundant,
cloudy growth with flocculation in peptone-succinate-salts (PSS) broth (grams/liter; Bacto-pep-
tone (Difco), 10.0; succinic-acid, 1.0; $(NH_4)_2SO_4$,
1.0; $MgSO_4 \cdot 7H_2O$, 1.0; $FeCl_3 \cdot 6H_2O$, 0.002; $MnSO_4 \cdot$
H_2O, 0.002). Colonies on PSS agar (0.7% agar) are
pinpoint with fimbriated edges.

As yet not cultivated in completely defined
media. Abundant growth occurs in a medium in
which 2.5 g of vitamin-free, salt-free acid-hydrolyzed casein (Nutritional Biochemicals Corp.) is
substituted for peptone in PSS media. Synthetic
casein hydrolysate fails to substitute for the acid
hydrolysate, even when supplemented with vitamins, synthetic peptides or trace minerals.

In PSS broth growth is inhibited by concentration of added phosphate greater than 0.003 м.

For other characteristics see Tables 6.1 and 6.2.

Obligately microaerophilic, requiring an atmosphere of 1–9% oxygen. Growth can be obtained
without such conditions by stratifying liquid
media with 0.15% agar. Microaerotactic.

Optimum temperature 30 C; scanty growth at
25 and 37 C; no growth at 10 or 42 C.

Has been isolated from stagnant pond water in
Virginia (U.S.A.) and from a mixed culture originally derived from cooling tower water of a beet
sugar refinery in England.

The G + C content of the DNA is 38 moles %
(T_m).

Proposed neotype strain: Wells; ATCC 19554
(Hylemon et al., 1973). The Pringsheim strain
(ATCC 19553) is identical in its characteristics.

2. **Spirillum giesbergeri** Williams and Rittenberg 1957, 86.

gies′ber.ger.i. M.L. gen.n. giesbergeri of Giesberger, the first investigator of the genus to define
certain physiological characteristics useful in
identification.

Cells observed by phase microscopy in 18-hr
PSS broth are 1.1–1.4 μm in diameter; diameter of
helix 2.0–5.0 μm; length of helix 4–40 μm with less
than one turn. Exceptionally prominent dark
granules of polyhydroxybutyrate in cytoplasm.
The tuft of flagella at each pole can be seen in
living cells by dark-field but not phase microscopy.

Growth in PSS broth moderate, with bottom
sediment and clear supernatant. Slow growing.

Colonies on PSS agar (1.5% agar) are circular,
white, convex, 1.0 mm.

Growth in defined mineral salts medium does
not occur for most of a large variety of sole carbon
sources; however, upon incorporation of 0.15%
agar into defined medium containing succinate, a
band of growth forms 3–6 mm below the surface
of the medium. Even after continued growth this
band fails to migrate to the surface.

This species appears to be serologically identical
with S. graniferum and S. sinuosum with regard
to heat-labile antigens.

For other characteristics see Tables 6.1, 6.2 and
6.3.

Capable of aerobic growth, but may possibly
prefer microaerophilic conditions.

Optimum temperature 30 C; moderate growth
at 25 C; scanty growth at 37 C; no growth at 10
or 42 C.

Isolated from pond water.

The G + C content of the DNA of the type
strain is 58 moles % (T_m).

Type strain: ATCC 11334.

3. **Spirillum graniferum** Williams and Rittenberg 1957, 92.

gra.ni′fer.um. L. n. granum a seed grain; L. v.
fero bear; M.L. neut.adj. graniferum grain bearing.

Cells observed by phase microscopy in 18-hr
PSS broth are 1.0–1.2 μm in diameter; wave length
7.0–8.4 μm; diameter of helix 3.5–4.2 μm; length of
helix 7–25 μm with one to three turns. Dark granules of polyhydroxybutyrate present in cytoplasm.
Motile, possessing a tuft of flagella at each pole
which can be seen in living cells by dark-field but
not phase microscopy.

Growth in PSS broth abundant, cloudy. Colonies on PSS agar are circular, convex, translucent,
1.0 mm.

This species appears to be serologically identical
with S. giesbergeri with regard to heat-labile
antigens.

For other characteristics see Tables 6.1, 6.2 and
6.3.

Strictly aerobic.

Optimum temperature 30 C; moderate growth
at 25 and 37 C; no growth at 10 or 42 C.

Source: fresh water.

The G + C content of the DNA of the type
strain is 57 moles % (T_m).

Type strain: ATCC 9875; NCIB 8230.

4. **Spirillum anulus** Williams and Rittenberg
1957, 86.

an′u.lus. L. masc.n. anulus a ring.

Cells observed by phase microscopy in 18-hr
PSS broth are 1.1–1.5 μm in diameter; wave length
5–13 μm; diameter of helix 1.7–4 μm; length of

TABLE 6.3

Sole carbon sources for species of genus **Spirillum**

The following are not used as sole carbon sources by any species of *Spirillum:* D-ribose, L-xylose, 2-ketogluconate, D-galactose, lactose, sucrose, maltose, sedoheptulose, D-raffinose, tryptophan, methionine, threonine. Propionate, ethanol, *n*-butanol, L-ornithine, L-citrulline, L-arginine, L-lysine, L-cysteine, L-leucine, L-isoleucine and L-valine are used by some strains of *S. itersonii.*

A washed standardized inoculum is used. + = the production of at least 10 Klett units of turbidity (blue filter) in the second 72-hr subculture in defined mineral salts basal medium containing carbon source, using 16-mm cuvettes.

Sole carbon source	2. S. giesbergeri	3. S. graniferum	4. S. anulus	5. S. metamorphum	6. S. patridicon chylium	7. S. itersonii	8. S. peregrinum	9. S. serpens	10. S. sinuosum	11. S. gracile	12. S. delicatum	13. S. polymorphum	14. S. lunatum	15. S. minutulum	16. S. beijerinckii	17. S. linum	18. S. japonicum
Citrate	−	−	−	−	−	−	−	−	−	−	−	−	+	−	−	−	−
Aconitate	−	−	−	−	+	+	+	−	−	−	+	−	+	−	−	−	−
Isocitrate	−	−	−	−	+	−	−	−	−	−	+	−	+	−	−	−	−
α-Ketoglutarate	−	−	−	+	+	+	+	d	−	−	+	−	+	d	−	−	−
Succinate	−	−	−	+	+	+	+	−	−	−	+	−	+	+	−	−	−
Fumarate	−	+	−	+	+	+	+	−	−	−	+	−	+	+	−	−	−
L-Malate	−	−	−	+	+	+	+	−	+	−	+	−	+	+	+	−	+
Oxaloacetate	−	−	−	+	+	d	+	d	+	−	+	−	+	+	−	−	+
Pyruvate	+	+	+	+	+	d	+	d	+	−	+	−	+	+	+	−	+
L-Lactate	−	−	−	+	+	d	+	−	−	−	+	−	+	+	+	−	+
Malonate	−	−	−	+	+	d	+	−	−	−	+	+	+	−	−	−	−
Acetate	−	−	−	−	+	−	+	−	−	−	+	+	+	+	−	d	−
β-Hydroxybutyrate	−	−	−	+	+	+	−	−	−	−	−	−	+	−	−	−	−
n-Propanol	−	−	−	−	−	+	−	−	−	−	−	−	+	−	−	−	−
L-Histidine	−	−	−	−	+	+	−	−	−	−	−	−	−	−	−	−	−
L-Tyrosine	−	−	−	−	−	−	−	−	−	−	−	−	+	−	−	−	−
L-Phenylalanine	−	−	−	−	−	+	−	−	−	−	−	−	−	−	−	−	−
L-Alanine	−	−	−	+	+	d	+	d	−	−	−	−	+	−	−	−	+
L-Glutamate	−	−	d	+	+	+	+	+	−	−	−	+	+	+	−	−	+
L-Aspartate	−	−	−	+	+	+	+	+	−	−	−	+	−	−	−	−	−
L-Asparagine	−	−	−	+	+	+	+	d	−	−	−	−	+	−	−	−	−
L-Proline	−	−	−	−	+	+	+	d	−	−	−	−	+	+	+	−	−
L-Hydroxyproline	−	−	−	−	−	d	+	−	−	−	−	−	−	−	−	−	−
L-Serine	−	−	−	−	−	−	−	d	−	−	−	−	−	−	−	−	−
Glycine	−	−	−	−	−	−	−	−	−	−	−	−	−	+	−	−	−
Putrescine	−	−	−	−	−	d	−	−	−	−	+	−	−	−	−	−	−
D-Glucose	−	−	−	−	−	−	−	−	−	−	−	−	−	+	−	−	−
D-Gluconate	−	−	−	−	−	−	−	−	−	−	−	−	−	+	−	−	−
D-Mannitol	−	−	−	−	−	−	−	−	−	−	−	−	−	+	−	−	−
D-Fructose	−	−	−	−	−	d	+	−	−	−	−	−	−	−	−	−	−
Glycerol	−	−	−	−	−	d	−	−	−	−	−	−	−	+	−	−	−

helix 4–52 μm with less than one to a maximum of six turns. Dark intracellular granules of poly-hydroxybutyrate present. Motile, possessing a tuft of flagella at each pole which can be seen in living cells by dark-field.

Growth in PSS broth abundant, cloudy. Colonies on PSS agar are white, pinpoint.

For other characteristics see Tables 6.1, 6.2 and 6.3.

Strictly aerobic.

Optimum temperature 30 C. Moderate growth at 25 and 37 C; no growth at 10 or 42 C.

Isolated from pond water.

The G + C content of the DNA of the type

strain is 59 moles % (T$_m$). The base composition of reference strain ATCC 19259 is identical.

Type strain: ATCC 11879; NCIB 9012.

5. Spirillum metamorphum Terasaki 1961, 220.

me.ta.mor'phum. Gr. neut.adj. *metamorphum* changing.

Cells observed by phase microscopy in 18-hr PSS broth are 1.1–1.3 μm in diameter; diameter of helix 2.8–3.0 μm; length of helix 4–9 μm with less than one turn. Dark granules of polyhydroxybutyrate present in cytoplasm.

Growth in PSS broth abundant, cloudy. Colonies on PSS agar are circular, convex, translucent, 2.0 mm.

For other characteristics see Tables 6.1, 6.2 and 6.3.

Strictly aerobic.

Optimum temperature 30 C. Moderate growth at 25 and 37 C; scanty growth at 42 C; no growth at 10 or 45 C.

Isolated from putrid infusion of a fresh water shellfish.

The G + C content of the DNA of the type strain is 63 moles % (T$_m$).

Type strain: ATCC 15280.

6. Spirillum putridiconchylium Terasaki 1961, 80.

pu'tri.di.con.chy.li.um. L. adj. *putridus* putrid, decayed; L. n. *conchylium* a shell fish; L. n. *putridiconchylium* decayed shellfish.

Cells observed by phase microscopy in 18-hr PSS broth are 0.9–1.2 μm in diameter; wave length 6–7 μm; diameter of helix 1.8–2.0 μm; length of helix 4–22 μm with less than one to a maximum of four turns (usually three). Dark granules of polyhydroxybutyrate present in cytoplasm. A tuft of flagella at each pole can be seen in living cells by dark-field but not phase microscopy.

Growth in PSS broth abundant, cloudy. Colonies on PSS agar are circular, convex, white, 1.5–2.0 mm.

For other characteristics see Tables 6.1, 6.2 and 6.3.

Strictly aerobic.

Optimum temperature 30 C. Moderate growth at 25 and 37 C; scanty growth at 42 C; no growth at 10 or 45 C.

Isolated from putrid infusion of a fresh water shellfish.

The G + C content of the DNA of the type strain is 52 moles % (T$_m$).

Type strain: ATCC 15279.

7. Spirillum itersonii Giesberger 1936, 68.

i.ter.so'ni.i. M.L. gen.n. *itersonii* of Iterson; named for G. Van Iterson, a Dutch bacteriologist.

Cells observed by phase microscopy in 18-hr PSS broth are 0.4–0.6 μm in diameter; wave length 3.0–3.5 μm; diameter of helix 1.2–1.6 μm; length of helix 2–7 μm with less than one to a maximum of two turns. Dark granules of polyhydroxybutyrate present in cytoplasm. Mitomycin C and ultraviolet light were found by Clark-Walker (1969) to induce the formation of coccoid bodies; the latter were found to contain phage tail parts, rhapidosomes and a granular substance not seen in normal cells. It was suggested that coccoid bodies might be formed as the result of induction of a defective phage.

Growth in PSS broth abundant, cloudy. Colonies on PSS agar are circular, convex, white, 0.8–1.5 mm. A brown water-soluble pigment formed in PSS agar containing either tryptophan or tyrosine (0.2%).

Acid reaction produced in complex media from glycerol and fructose but from no other sugars. When cultured in a fructose-casamino acids medium, the following enzyme activities have been detected: hexokinase, glucose 6-phosphate dehydrogenase, phosphogluconate dehydrase, 2-keto-3-deoxy-6-phosphogluconate aldolase, fructose 1,6-diphosphate aldolase and very low levels of 6-phosphogluconate dehydrogenase. When cultured on a succinate-casamino acids medium, the enzyme activities fall to a low basal level.

Other characteristics are to be found in Tables 6.1, 6.2 and 6.3.

Strictly aerobic.

Optimum temperature 30 C. Moderate growth at 25 and 37 C; no growth at 10 or 42 C.

Isolated from pond water.

The G + C content of the DNA of the type strain is 62 moles % (T$_m$).

Type strain: ATCC 12639.

8. Spirillum peregrinum Pretorius 1963, 407.

per.e.gri'num. L. neut.adj. *peregrinum* strange, foreign.

Cells observed by phase microscopy in 18-hr PSS broth are 0.5–0.7 μm in diameter; wave length 3.5–4.0 μm; diameter of helix 1.4–1.6 μm; length of helix 5–22 μm with four to five turns. Dark granules of polyhydroxybutyrate present in cytoplasm.

Growth in PSS broth abundant, cloudy. Colonies on PSS agar are circular, convex, white, 1.0–1.5 mm.

Yellow water-soluble pigment formed in PSS agar containing 0.2% phenylalanine. Acid reaction produced in complex media from fructose but no other sugar. When cultured in a fructose-casamino acids medium, hexokinase and fructose 1,6-diphosphate aldolase activities are detected. When cultured in a succinate-casamino acids medium, the enzyme activities fall to a low basal level.

Other characteristics can be found in Tables 6.1, 6.2 and 6.3.

Strictly aerobic.

Optimum temperature 30 C. Moderate growth at 25 and 37 C; no growth at 10 or 42 C.

The G + C content of the DNA of the type strain is 62 moles % (T_m).

Type strain: ATCC 15387.

9. **Spirillum serpens** (Müller) Winter 1884, 63. (*Vibrio serpens* Müller 1786, 48.)

ser'pens. L. v. *serpo* crawl or creep; L. part.adj. *serpens* creeping.

Cells observed by phase microscopy in 18-hr PSS broth are 0.6–1.1 μm in diameter; wave length 3.5–12 μm; diameter of helix 1.2–2.8 μm; length of helix 5–35 μm with less than one to a maximum of two turns. Dark granules of polyhydroxybutyrate present in cytoplasm. Motile, possessing a tuft of flagella at each pole. Visibility of the tufts in living cells by dark-field microscopy varies among strains. By electron microscopy the basal structure of the flagella appears similar to that of peritrichous flagella in *Proteus vulgaris* and other Gram-negative organisms. Each flagellum originates within the protoplast in an individual distinct vesicle and has a separate insertion site in the cell wall (Abram, 1969). The cell wall is composed of an inner peptidoglycan layer, a protein granule layer, a lipopolysaccharide layer and an external lipoprotein layer. The peptidoglycan contains alanine, glutamate and mesodiaminopimelate and is tightly knit, with 54% of the DAP molecules involved in cross-linkage between tetrapeptides (Kolenbrander and Ensign, 1968).

Growth in PSS broth abundant, cloudy. Colonies on PSS agar are circular, convex, white, 1.0–2.2 mm.

This species has been divided into several serological groups on the basis of heat-labile antigens.

For other characteristics see Tables 6.1, 6.2 and 6.3.

Strictly aerobic.

Optimum temperature 30 C. Moderate growth at 25 and 37 C; no growth at 10 or 42 C.

Isolated from fresh water.

The G + C content of the DNA ranges from 50–51 moles % (T_m).

Reference strain: ATCC 12638. Suggested neotype (ATCC Catalog of Strains, 9th ed., 1970).

10. **Spirillum sinuosum** Williams and Rittenberg 1957, 94.

sin.u.o'sum. L. neut.adj. *sinuosum* full of curves.

Cells observed by phase microscopy in 18-hr PSS broth are 0.6–0.9 μm in diameter; wave length 8.5–10.5 μm; diameter of helix 1.4–3.5 μm; length of helix 5–42 μm with one to two turns. Dark granules of polyhydroxybutyrate present in cytoplasm.

Growth in PSS broth abundant, cloudy. Colonies on PSS agar are circular, convex, white, 1.0 mm.

This species appears to be serologically identical with *S. giesbergeri* and *S. graniferum* with regard to heat-labile antigens.

For other characteristics see Tables 6.1, 6.2 and 6.3.

Strictly aerobic.

Optimum temperature 30 C. Moderate growth at 25 C; scanty growth of 37 C; no growth at 10 or 42 C.

Isolated from fresh water.

The G + C content of the DNA of the type strain is 57 moles % (T_m).

Type strain: ATCC 9786.

11. **Spirillum gracile** Canale-Parola, Rosenthal and Kupfer 1966, 114.

gra.cil'e. L. adj. *gracile* slender or thin.

Cells observed by phase microscopy in 18-hr PSS broth are 0.25–0.3 μm in diameter; wave length 2.8–3.5 μm; diameter of helix 0.5–2.0 μm; length of helix 4–14 μm with three to four turns. Polyhydroxybutyrate granules appear to be absent.

Growth in PSS broth moderate, cloudy. Colonies on PSS agar are white, pinpoint, with a large part of the colony embedded in the agar. When originally isolated, Canale-Parola *et al.* noted that all strains formed subsurface, spreading, semitransparent nonpigmented colonies, the spreading occurring within the agar. They noted that after prolonged subculturing some of the spirilla in each strain lost the ability to diffuse through agar and formed small, nonspreading colonies similar to those observed in recent analysis of these strains.

Isolation is accomplished by allowing the spirilla to pass through cellulose ester filter discs (0.45-μm pore size) to underlying agar.

An acid reaction is produced in complex media containing glucose, galactose and arabinose, but no other sugars. When cultured in a glucose-casamino acids-yeast extract medium, the following enzyme activities have been detected: glucokinase, glucose 6-phosphate dehydrogenase, phosphogluconate dehydrase, 2-keto-3-deoxy-6-phosphogluconate aldolase and fructose 1,6-diphosphate aldolase. Only low basal levels of activity occur when cells are cultured in succinate-casamino acids-yeast extract.

For other characteristics see Tables 6.1, 6.2 and 6.3.

Strictly aerobic.

Optimum temperature 30 C. Scanty growth at 25 and 37 C; no growth at 10 or 42 C.

Isolated from fresh water.

The G + C content of the DNA is 65 moles % (T_m).

Reference strains: ATCC 19624, ATCC 19625, ATCC 19626.

12. Spirillum delicatum Leifson 1962, 164.

del.i.ca'tum. L. neut.adj. *delicatum* delicate.

Cells observed by phase microscopy in 18-hr PSS broth are 0.3–0.4 μm in diameter; diameter of helix 0.4–0.7 μm; length of helix 3–5 μm with less than one turn. Cytoplasmic granules not evident, but tests for polyhydroxybutyrate are positive. Motile, with up to six polar flagella (usually one or two). This species constitutes an exception to the usual case of bipolar flagellation since the polar tuft is located at only one pole; however, for cells undergoing division, bipolar flagellation is frequently seen.

Growth in PSS broth abundant, cloudy. Colonies on PSS agar are circular, white, 0.5 mm.

For other characteristics see Tables 6.1, 6.2 and 6.3.

Strictly aerobic.

Optimum temperature 30 C. Moderate growth at 25 and 37 C; no growth at 10 or 42 C.

Isolated from distilled water.

The G + C content of the DNA of the type strain is 63 moles % (T_m).

Type strain: ATCC 14667.

13. Spirillum polymorphum Williams and Rittenberg 1957, 85.

po.ly.mor'phum. Gr. adj. *polymorphum* multiform.

Cells observed by phase microscopy in 18-hr PSS broth are 0.3–0.5 μm in diameter; diameter of helix 0.5–0.8 μm; length of helix 3.5–8.4 μm with less than one turn. Dark granules of polyhydroxybutyrate present in cytoplasm.

Growth in PSS broth abundant, cloudy. Colonies on PSS agar are circular, convex, translucent, pinpoint.

For other characteristics see Tables 6.1, 6.2 and 6.3.

Strictly aerobic.

Optimum temperature 30 C. Moderate growth at 25 and 37 C; no growth at 10 or 42 C.

Isolated from pond water.

The G + C content of the DNA of the type strain is 62 moles % (T_m).

Type strain: ATCC 11332.

14. Spirillum lunatum Williams and Rittenberg 1957, 83.

lu.na'tum. L. neut.adj. *lunatum* half-moon-shaped.

Cells observed by phase microscopy in 18-hr PSS broth are 0.5–0.6 μm in diameter; wave length 2.8–3.0 μm; diameter of helix 1.0–1.3 μm; length of helix 2.4–5.4 μm with less than one to a maximum of one turn. Dark granules occasionally present in cytoplasm, but tests for polyhydroxybutyrate are negative.

Growth in PSS broth abundant, cloudy. Colonies on PSS agar are circular, convex, translucent, 1.0 mm.

Dark brown water-soluble pigment formed in PSS agar containing 0.2% tryptophan. Light brown water-soluble pigment formed with 0.2% tyrosine. Acid reaction produced in complex media from glucose, fructose, galactose, mannose, mannitol and dextrin, but from no other carbohydrates. When cultured in a glucose-casamino acids or gluconate-casamino acids medium, the following enzyme activities have been detected: hexokinase, glucose 6-phosphate dehydrogenase, phosphogluconate dehydrase, 2-keto-3-deoxy-6-phosphogluconate aldolase and very low levels of 6-phosphogluconate dehydrogenase and fructose 1,6-diphosphate aldolase. Only low basal levels of activity occur when cells are cultured in succinate-casamino acids. A mannitol dehydrogenase is induced when cells are cultured in mannitol-Casamino acids.

Upon initial isolation, sea water was required for growth of the type strain (Williams and Rittenberg, 1957). The strain has since lost this requirement. Recent characterizations of the type strain using both fresh water and sea water media have yielded identical results.

For other characteristics see Tables 6.2, 6.3 and 6.4.

TABLE 6.4

Differential characteristics of sea water-requiring species of genus **Spirillum**

	14. S. lunatum	15. S. minululum	16. S. beijerinckii	17. S. linum	18. S. japonicum
Acid reaction from sugars	+	−	−	−	−
Phosphatase		−	+	+	+
Nitrate reduction		+	−	−	−
Selenite reduction		+	−	+	+
Krebs cycle acids and amino acids as sole carbon sources		+		−	+
G + C moles %	63	42	47	48	45

Strictly aerobic.

Optimum temperature 30 C. Moderate growth at 25 and 37 C; no growth at 10 or 42 C.

Isolated from coastal sea water.

The G + C content of the DNA of the type strain is 63 moles % (T_m).

Type strain: ATCC 11337.

15. Spirillum minutulum Watanabe 1959, 83.

mi.nu'tu.lum. L. dim.neut.adj. *minutulum* very little.

Cells observed by phase microscopy in 18-hr PSS broth are 0.3–0.4 μm in diameter; wave length 2.0–2.8 μm; diameter of helix 0.6–1.5 μm; length of helix 3–8 μm with one to two turns. Dark granules of polyhydroxybutyrate present in cytoplasm.

Growth in PSS broth abundant, cloudy. Colonies on PSS agar are circular, convex, white or translucent, less than 1.0 mm.

Natural or artificial sea water required for growth.

A dark brown water-soluble pigment is formed by one strain (ATCC 19192) in PSS agar containing 0.2% tryptophan.

For other characteristics see Tables 6.2, 6.3 and 6.4.

Strictly aerobic.

Optimum temperature 30 C. Moderate growth at 5, 25 and 37 C; no growth at 42 C.

Isolated from marine shellfish at Samugawa Beach.

The G + C content of the DNA is 42 moles % (T_m).

Reference strain: ATCC 19193.

16. Spirillum beijerinckii Williams and Rittenberg 1957, 90.

bei.jer.inck'i.i. M.L. gen.n. *beijerinckii* of Beijerinck; named for Prof. M. W. Beijerinck of Delft, Holland.

Cells observed by phase microscopy in 18-hr PSS broth are 0.7–1.0 μm in diameter; wave length 6.3–7.2 μm; diameter of helix 1.5–3.0 μm; length of helix 7–14 μm with one to two turns. Dark granules of polyhydroxybutyrate present in cytoplasm.

Growth in PSS broth abundant, cloudy. Colonies on PSS agar are circular, translucent, less than 0.5 mm.

Natural or artificial sea water required for growth.

A deep brown water-soluble pigment is formed in PSS agar containing 0.2% tyrosine.

For other characteristics see Tables 6.2, 6.3 and 6.4.

Strictly aerobic.

Optimum temperature 30 C. Moderate growth at 25 and 37 C; no growth at 10 or 45 C.

Isolated from coastal sea water of Long Island Sound.

The G + C content of the DNA of the type strain is 47 moles % (T_m).

Type strain: ATCC 12754.

17. Spirillum linum Williams and Rittenberg 1957, 82. (*Spirillum atlanticum* Williams and Rittenberg 1957, 91.)

li'num. L. n. *linum* flax, thread.

Cells observed by phase microscopy in 18-hr PSS broth are 0.4–0.6 μm in diameter; wave length 1.8–3.5 μm; diameter of helix 0.8–1.4 μm; length of helix 4–30 μm with two to six turns. Dark granules of polyhydroxybutyrate present in cytoplasm.

Growth in PSS broth abundant, cloudy. Colonies on PSS agar are circular, convex, translucent, 0.5–1.0 mm.

Natural or artificial sea water required for growth.

A dark brown water-soluble pigment is formed by one strain (ATCC 11336) in PSS agar containing 0.2% tyrosine or tryptophan.

Growth can be obtained in defined mineral salts media containing malate, succinate and methionine. No amino acid other than methionine can serve as nitrogen source.

For other characteristics see Tables 6.2, 6.3 and 6.4.

Strictly aerobic.

Optimum temperature 30 C. Moderate growth at 25 and 37 C; scanty growth at 5 C; no growth at 42 C.

Isolated from coastal sea water.

The G + C content of the DNA is 48 moles % (T_m).

Type strain: ATCC 11336.

18. Spirillum japonicum Watanabe 1959, 78.

ja.pon'i.cum. M.L. neut.adj. *japonicum* pertaining to Japan.

Cells observed by phase microscopy in 18-hr PSS broth are 1.0–1.2 μm in diameter; wave length 8–20 μm; diameter of helix 2–3 μm; length of helix 11–75 μm with less than one to a maximum of one turn. Dark granules of polyhydroxybutyrate present in cytoplasm. Motile, possessing a tuft of flagella at each pole which can be seen by dark-field microscopy.

Growth in PSS broth abundant, cloudy. Colonies on PSS agar are circular, convex, white, less than 1.0 mm.

Natural or artificial sea water required for growth.

For other characteristics see Tables 6.2, 6.3 and 6.4.

Strictly aerobic.

Optimum temperature 30 C. Moderate growth at 25 and 37 C; no growth at 5 or 42 C.

Isolated from marine shellfish at Samugawa Beach.

The G + C content of the DNA of the reference strain is 45 moles % (T_m).

Reference strain: ATCC 19191.

19. Spirillum minor Carter 1888, 47.

mi'nor. L. comp.adj. *minor* smaller.

Short thick cells with tapering ends, 0.2–0.5 by 3–5 μm, having two to three windings which are thick, regular and spiral. Actively motile in blood by means of bipolar tufts of flagella. Readily stained by ordinary aniline dyes and by Giemsa or Wright's stain. According to Robertson (1924) this organism is a spirillum, not a spirochaete.

Has not been cultivated on artificial media. Causes one type of rat bite fever in humans. Can be transferred to mice, rats, guinea pigs and monkeys. Appears to be a natural parasite of rats which act as healthy carriers.

Because of its habitat and widespread distribution this organism has been described under many different names. It is possible that some of these indicate varieties or even separate species. See Beeson (1943) for important literature.

Found in the blood of rats and mice.

APPENDIX

Included in this section are species whose taxonomic relationships are uncertain or for which no representative isolates are available.

a. *Spirillum virginianum* Dimitroff 1926, 19.

vir.gi.ni.a'num. M.L. neut.adj. *virginianum* pertaining to the State of Virginia.

A description is presented in the 7th edition of THE MANUAL. Strain NCIB 9075 was deposited by Marion A. Williams in 1960, and may be a lineal descendent of Dimitroff's culture. Thorough characterization of this strain has not yet been carried out.

b. *Spirillum undula* (Müller) Ehrenberg 1832, 38. (*Vibrio undula* Müller 1773, 46.)

un'du.la. L. n. *unda* a wave; M.L. dim.fem.n. *undula* a small wave.

A description is presented in the 7th edition of THE MANUAL. It is doubtful that Müller's *Vibrio undula* and Ehrenberg's *S. undula* were the same (Williams, 1959). Ehrenberg's description of *S. undula* could apply to any of several currently recognized species. Williams (1959) states that it is doubtful that any two investigators of the genus *Spirillum* have given the name *S. undula* to identical organisms. A type or reference strain does not appear to be available.

c. *Spirillum tenue* Ehrenberg 1838, 84.

te'nu.e. L. neut.adj. *tenue* thin.

A description is presented in the 7th edition of THE MANUAL. A type or reference strain does not appear to be available.

d. *Spirillum kutscheri* Migula 1900, 1024.

ku'tsche.ri. M.L. gen.n. *kutscheri* of Kutscher; named for K. H. Kutscher, the German bacteriologist who first isolated this organism.

A description is presented in the 7th edition of THE MANUAL. Representative isolates do not appear to be available.

e. *Spirillum curvatum* Williams and Rittenberg 1957, 84.

cur.va'tum. L. neut.adj. *curvatum* curved, bowed.

Cell diameter 0.4–0.6 μm. Possesses a tuft of flagella at each pole. Coccoid bodies predominate in older cultures.

Growth abundant in sea water yeast-autolysate nutrient broth; heavy turbidity with pellicle formation.

Grows well on pyruvate, succinate, fumarate and lactate. Grows less well on glucose and fructose. Malonate, citrate, fatty acids, ethanol and glycerol not used.

Catalase positive.

Optimum temperature 30 C. No growth at 40 C.

Requires sea water.

Isolated from coastal sea water.

A type or reference strain does not appear to be available.

f. *Spirillum mancuniense* Cayton and Preston 1955, 524.

man.cu.ni.en.se'. M.L. adj. *mancuniense* pertaining to the town of Manchester, England.

Cell diameter 0.7–1.0 μm. Possesses a tuft of flagella at one or both poles. Coccoid bodies present in older cultures.

Growth abundant in 0.5% peptone. No growth in the presence of 0.6% salt.

Although colonies on plates grow aerobically, diminution of air pressure to 60 mm Hg yields slightly better growth.

Gelatine slowly liquefied. No fermentation of glucose. Indol negative. Nitrate not reduced. Catalase negative.

Isolated from decaying grass at Manchester, England.

A type or reference strain does not appear to be available.

g. *Spirillum maritimum* Watanabe 1959, 82.

mar.i.ti'mum. L. neut.adj. *maritimum* belonging to the sea.

Cell diameter 0.6–0.7 μm. Slightly attenuated ends. By staining methods, said to possess a single helical flagellum at each pole. Formation of cyst-like body is conspicuous.

Growth in peptone-NaCl-calcium lactate-salts broth yields intense turbidity; crystals form a membrane on surface and a ring around tube wall.

Colonies on peptone-NaCl-calcium lactate-salts agar are round, pulvinate or convex, finely granular, smooth, milky, 0.5–1.0 mm. Production of small crystals occurs on or around colonies.

Gelatin not liquefied; indole negative; nitrates not reduced; acid produced from glucose and mannose.

Requires at least 1% NaCl.

Isolated from putrid infusions of marine shellfish collected at Samugawa Beach.

A type or reference strain does not appear to be available.

h. *Spirillum halophilum* Watanabe 1959, 80.

ha.lo.phi.lum Gr. n. *hals* salt; Gr. adj. *philus* loving; M.L. neut.adj. *halophilum* salt-loving.

Cell diameter 0.5–0.6 μm. Slender forms with one to three (seldom several) wavelike undulations. Motile with bipolar tufts of flagella. Growth in peptone-NaCl-calcium lactate-salts broth yields moderate turbidity. Calcium membrane is barely produced in old cultures. Clouded masses precipitate by slight stimulus.

Colonies on peptone-NaCl-calcium lactate-salts agar are round, entire, circular, 1.0–1.5 mm in diameter. Calcium crystals form as a scab.

Indol negative. Nitrates not reduced. Catalase negative. Little acid from glucose. No acid from other saccharides.

Requires at least 1.5% NaCl.

Isolated from putrid infusions of marine shellfish collected at Samugawa Beach.

Comment: ATCC strain 19192, said to be a strain of *S. halophilum*, does not fit the description of the species and should be placed in the species *S. minutulum*. A type or reference strain fitting the description of *S. halophilum* does not appear to be available.

Genus II. **Campylobacter** *Sebald and Véron 1963, 907*

ROBERT M. SMIBERT

Cam.py.lo.bac.ter. Gr. adj. *campylo* curved; Gr. n. *bacter* rod; M.L. masc.n. *Campylobacter* a curved rod.

Slender **non-spore-forming spirally curved rods,** 0.2–0.8 μm wide and 0.5–5 μm long. The rods may have one or more spirals and can be as long as 8 μm. They also appear S-shaped and gull-winged when two cells form short chains. Cells in old cultures form spherical or coccoid bodies.

Motile with a characteristic corkscrew-like motion. They have a **single polar flagellum at one or both ends of the cell.** Flagella may be 2–3 times the length of the cells. **Gram-negative.**

Chemoorganotrophs: Carbohydrates are neither fermented nor oxidized. Metabolism of those species studied is respiratory. No acid or neutral end products produced.

Do not require serum for growth. Energy from amino acids or tricarboxylic acid cycle intermediates not carbohydrates. Gelatin and urea not hydrolyzed. Methyl red and Voges Proskauer negative. No lipase activity. Oxidase positive. Pigments are not produced.

Microaerophilic to anaerobic. Some species are microaerophilic requiring an oxygen concentration of between 3 and 15%. Other species are anaerobic or can grow under either microaerophilic or anaerobic conditions. Occasionally a few strains may grow slightly under aerobic conditions.

Some species are pathogenic for man and animals. Found in the reproductive organs, intestinal tract and oral cavity of animals and man.

The G + C content of the DNA (of those species studied) ranges from 30–35 moles % (T_m).

Type species: *Campylobacter fetus* (Smith and Taylor) Sebald and Véron 1963, 907.

Further Comments

Most of the organisms in this genus have been studied using Albimi Brucella broth (Pfizer Diagnostics, Flushing, New York) or agar as the basal medium. Most tests were made in semisolid Brucella broth (0.16% agar) which provides microaerophilic conditions. For tests in liquid media or on agar plates, the oxygen content must be reduced to 3–15%. Casein, gelatin, DNA and RNA hydrolyses were made on Brucella agar plates incubated in an atmosphere of 5% oxygen and 95% nitrogen. Many of the methods are described in detail by Smibert (1965 and 1969).

Work in the author's laboratory was supported by National Institutes of Health Grant GM-14604.

Description of the species of genus **Campylobacter**

1. **Campylobacter fetus** (Smith and Taylor) Sebald and Véron 1963, 907. (*Vibrio fetus* Smith and Taylor 1919, 301; *Spirillum fetus* Lehmann and Neumann 1927, 552; *Vibrio fetus* var. *venerealis* Florent 1959, 955.)

fe'.tus. L. n. *foetus* fetus.

Slender curved rods that are 0.2–0.5 μm wide and 1.5–5 μm long. They appear comma-, S-, and gull-shaped. The ends of the cells are pointed. Loosely wound spiral filaments up to 8 μm long appear in old cultures. Spherical or coccoid forms are also found in old cultures especially when grown on agar plates.

Very actively motile with a characteristic darting and corkscrew-like motion. Motility and rotation of the cells are so rapid the curvation of the cells may be overlooked. Best observed with a phase-contrast microscope.

Whole cell hydrolysates contain small amounts of mesodiaminopimelic acid, while DAP was not found in isolated cell walls.

Several types of colonies are found on agar on primary isolation (Bryner *et al.*, 1962). Smooth colonies, the most frequently found, are small, 0.5 mm, round, slightly raised, smooth, colorless and slightly translucent. Cut glass colonies are 1 mm in diameter, round, raised, translucent and granular with reflecting facets. Rough colonies are rare and similar to smooth colonies with the exception of being granular and more opaque. Mucoid colonies are similar to smooth and cut glass colonies but are viscid. Sometimes on primary isolation colonies occur as a thin veil of confluent growth that is translucent and a very light gray or tan color. Colonies on blood agar are non-hemolytic, round, 1 mm in diameter, smooth, raised, convex and grayish white in appearance.

Slight even turbidity in broth.

Potassium gluconate is not oxidized. Final pH of glucose semisolid medium (0.16% agar) is 8.2–8.7 after 3 weeks.

Fatty acids are not produced from glucose as end products of fermentation. Small amounts of ethanol and *n*-propanol may be produced in cultures. Both pyruvate and lactate are utilized, but fatty acids do not accumulate in the medium as end products.

Hippurate, esculin, casein, ribonucleic acid and deoxyribonucleic acid are not hydrolyzed.

Glutamic acid, glutamine, aspartic acid and asparagine are deaminated and ammonia is produced in cultures; phenylalanine, tyrosine and tryptophan are not deaminated. Lysine, ornithine and arginine are not decarboxylated. Indole negative. Malonate is not utilized.

Catalase and oxidase positive. Hydrogen sulfide is not produced on triple sugar iron agar slants; there is only an alkaline reaction on this medium. Nitrate reduced to nitrite. Nitrite not reduced.

It does not grow in litmus milk. Peptone-supplemented litmus milk is reduced and turned slightly alkaline. Slight growth on Seller's medium (Sellers, 1964). The slant is slightly alkaline and the butt is not changed.

Grows in a semisolid medium containing 0.12–0.20% agar when incubated under aerobic conditions. Growth in a semisolid medium occurs only within the first few millimeters below the surface of the medium. There is little or no growth under strict anaerobic conditions. There is no growth in deep stab cultures. Only slight growth on MacConkey's agar.

Grows in a semisolid medium containing 1% bile but not in media containing 3.5% NaCl. Optimum temperature 37 C, optimum pH is 7.0.

Growth is not inhibited by 0.001 M iodoacetate or the vibriostatic agent 0/129.

TABLE 6.5

Some differential characteristics of species of the genus **Campylobacter**

	Catalase	Nitrite reduction	H₂S on TSI	H₂S, lead acetate strips	1% glycine	3.5% NaCl	25 C
1.a. *C. fetus* subsp. *fetus*	$+^a$	−	−	−	−	−	+
b. *C. fetus* subsp. *intestinalis*	+	−	−	+	+	−	+
c. *C. fetus* subsp. *jejuni*	+	−	−	+	+	−	−
2.a. *C. sputorum* subsp. *sputorum*	−	+	+	+	+	−	+
b. *C. sputorum* subsp. *bubulus*	−	+	+	+	+	+	d
3. *C. fecalis*	+	?	+	+	+	d	−

^{a} + = most (90% or more) strains positive for this characteristic; − = most (90% or more) strains negative; d = some (less than 90%) strains positive, some negative; ? = reaction not known.

The G + C content of the DNA ranges from 32–35 moles % (T_m).

1a. Campylobacter fetus subspecies fetus

subsp. nov. (NOT *Campylobacter fetus* subsp. *fetus* Véron and Chatelain 1973, 130.)

Morphology and characteristics as for species except as noted.

Cell walls contain galactose and either mannose or glucose.

Hydrogen sulfide usually not produced in a medium containing cysteine with lead acetate-impregnated paper strips as the detection system; a few isolates (3–4%) will be slightly H_2S positive after 5 days of incubation. These have been reported as subtype I (Bryner *et al.*, 1962) while these workers called the H_2S negative strains Type I. No growth on 0.1% sodium selenite medium and no reduction of selenite. No growth in semisolid media containing 1% glycine. No alkaline phosphatase activity and no aryl sulfatase activity.

Grows at 25 C but usually not at 42 or 45 C. A few strains have been reported that grow at 42 C.

Sensitive to chloramphenicol, 2 μg/ml; dihydrostreptomycin, 4 μg/ml; erythromycin, 2 μg/ml; neomycin, 8 μg/ml; oxytetracycline, 4 μg/ml; streptomycin, 4 μg/ml; and tetracycline, 0.5 μg/ml. Moderately sensitive to novobiocin, 64 μg/ml; and penicillin, 1 unit/ml. Resistant to bacitracin, 256 μg/ml, and polymyxin B, 512 units/ml.

One cause of abortion and infertility in cattle; transmitted venereally and found in the vaginal mucus of infected cows, the semen and prepuce of bulls and in the placenta and tissues of aborted bovine fetuses. Pathogenic for cattle, guinea pigs, hamsters and embryonated chicken eggs. Not pathogenic for rabbits, mice or rats when injected intraperitoneally. Will not multiply in the intestinal tract of man and animals.

Reference strain: UM. From bovine abortion.

1b. Campylobacter fetus subspecies intestinalis

(Florent) *comb. nov.* (*Vibrio fetus* var. *intestinalis* Florent 1959, 955; *Vibrio foetus-ovis* Buxton 1929, 47; *Campylobacter fetus* subsp. *fetus* Véron and Chatelain 1973, 130.)

in.test.in.al'.is. M.L. adj. *intestinalis* pertaining to the intestines.

Morphology and characteristics as for species except as noted.

Cell walls contain galactose only or galactose and mannose; or galactose and rhamnose or galactose, glucose and mannose.

Several types of colonies are found on agar on primary isolation (Bryner *et al.*, 1962). Smooth colonies are 1 mm in diameter, colorless to slightly cream-colored. Rough colonies are small, round, finely granular, opaque and white to cream- or tan-colored. They are 1–2 mm in diameter. Cut glass colonies do not develop in primary cultures. Smooth colonies incubated for 6–8 days become mucoid. Upon subculture smooth cut glass and rough cut glass colonies appear as well as smooth colonies. Frequently, on primary isolation colonies are low, flat, grayish to tan-colored and translucent with an irregular edge. They spread along the direction of the streak and coalesce. They may also form a thin veil of confluent growth on agar plates. Colonies on blood agar are non-hemolytic, round, 1–2 mm in diameter, smooth, convex and greyish white or light tan-colored.

Even turbidity in broth. A butyrous sediment may be seen in some broth cultures. Twenty-five per cent of the strains may be aryl sulfatase positive. No alkaline phosphatase activity.

Hydrogen sulfide is produced in a medium containing cysteine with lead acetate-impregnated paper strips as the method of detection. Grows on slants with 0.1% sodium selenite; selenite is reduced.

Grows in a semisolid medium containing 1% glycine.

Grows at 25 C, but usually not at 42 C. A few strains have been reported that grow at 25 and 42 C.

They are sensitive to chloramphenicol, 4 μg/ml; dihydrostreptomycin, 4 μg/ml; erythromycin, 2 μg/ml; neomycin, 8 μg/ml; oxytetracycline, 8 μg/ml; streptomycin, 4 μg/ml; and tetracycline, 1 μg/ml. Moderately sensitive to penicillin, 32 units/ml. Resistant to bacitracin, 128 μg/ml; novobiocin 128 μg/ml; and polymyxin B, 1024 units/ml.

Cause of abortion in sheep and sporadic abortion in cattle, as well as a cause of human infection; transmitted orally. Isolated from the placentas and stomach content of fetuses from aborted sheep and cattle; and from the blood, intestinal content and bile of infected ewes, cattle and man. This organism will grow in the intestinal tract and gall bladder of man and animals.

Reference strain: Montana 4440, serotype 2. From ovine abortion.

1c. Campylobacter fetus subspecies jejuni

subsp. nov. (*Vibrio jejuni* Jones, Orcutt and Little 1931, 861; *Vibrio hepaticus* Mathey and Rissberger 1964, 1339; *Campylobacter jejuni* Véron and Chatelain 1973, 128; *Campylobacter coli* Véron and Chatelain 1973, 127.)

je.ju'ni. L. adj. *jejunus* insignificant, meagre; M.L. gen.n. *jejuni* of the jejunum.

Morphology and characteristics as for species except as noted.

Cell walls contain galactose only, galactose and glucose, or galactose, glucose and mannose.

Two types of colonies are found on primary isolation (Smibert, 1965). One is low, flat, grayish, finely granular and translucent with an irregular edge. It spreads along the direction of the streak and tends to swarm and coalesce. The other is round, 1–2 mm in diameter, raised, convex, entire, smooth and glistening. It has a translucent edge with a darker, dirty brownish, slightly opaque center. Colonies on blood agar are non-hemolytic. Growth in broth usually has a butyrous sediment.

Casein, ribonucleic acid and deoxyribonucleic acid are hydrolyzed by about 60% of strains. Sixty-seven per cent have alkaline phosphatase activity and 6% are aryl sulfatase positive. Hydrogen sulfide is produced in a medium containing cysteine with lead acetate-impregnated paper strips as the method of detection. Grows on slants with 0.1% sodium selenite; selenite is reduced. Slight growth on MacConkey's agar. Grows in a medium with 1% glycine. Does not grow at 25 C. Grows at 42 C and usually at 45 C.

They are sensitive to chloramphenicol, 4 μg/ml; dihydrostreptomycin, 2 μg/ml; erythromycin, 8 μg/ml; neomycin, 8 μg/ml; oxytetracycline, 4 μg/ml; streptomycin, 2 μg/ml; and tetracycline, 1 μg/ml. Moderately sensitive to novobiocin, 64 μg/ml and penicillin 64 units/ml. Resistant to bacitracin, 128 μg/ml, and polymyxin B, 1024 units/ml.

Isolated from the placentas and stomach content of fetuses from aborted sheep. Also found in the intestinal tract of normal swine, cattle, sheep, goats, chickens, turkeys and wild birds. Causes disease in man. These are the "related vibrios" from human infection (King, 1957). Transmitted orally. Can grow in the intestinal tract of man and animals. Strains from aborted sheep are usually serotype 1 of Marsh and Firehammer (1953).

Reference strain: Montana 4849, serotype 1. From ovine abortion.

2. **Campylobacter sputorum** (Prévot) *comb. nov. (Vibrio sputorum* Prévot 1940, 85.)

spu.to′.rum. L. n. *sputum* spit, sputum; L. gen.pl.n. *sputorum* of sputa.

Slender curved non-spore-forming rods 0.3–0.6 μm wide and 2–4 μm long. They appear comma-shaped and gull-winged and occasionally occur as filaments up to 8 μm long. The ends of the cells are usually rounded. Motile. Ten- to fourteen-hour cultures have some motile cells that have a characteristic darting and corkscrew-like motion. Other cells appear non-motile. Best observed with a phase-contrast microscope.

Colonies on blood agar are gray, 1–2 mm in diameter, smooth, shiny, low convex and round with thin irregular spreading edges. Some may be weakly α-hemolytic. Growth in broth is light and evenly dispersed. Growth in a medium with 0.12–0.20% agar occurs in the upper third of the medium as a moderate evenly dispersed turbidity.

Final pH of glucose semisolid medium (0.16% agar) is 6.7–7.0 after 3 weeks. Fatty acids are not produced from glucose as end products or from peptone medium without glucose. No fatty acids or alcohols detected.

Catalase and indole negative. Hydrogen sulfide is produced in triple sugar iron agar or peptone iron agar. Also H₂S positive with lead acetate-impregnated paper strips as the detection system. Both nitrate and nitrite reduced. Nitrate in a medium enhances growth. A small amount of ammonia and CO_2 produced by cultures. No change in skim milk. Litmus reduced.

Microaerophilic to anaerobic. Growth on the surface of agar plates or in broth requires anaerobic conditions or oxygen concentration of the atmosphere to be reduced to 5% or below. No growth under aerobic conditions. Grows in deep stab cultures. Also grows in medium containing 0.12–0.20% agar when incubated under aerobic conditions.

2a. **Campylobacter sputorum** subspecies **sputorum** *subsp. nov.*

Morphology, colonies and other characteristics as for species except as noted.

Grows in semisolid medium containing 1 or 10% bile and in media with 2%, but not 3.5% NaCl. Grows at 25 C but not at 45 C.

Found in the gingival crevice flora of man. Represents about 5% of the flora.

Reference strain: Forsyth ER33, Forsyth Dental Center, Boston, Mass., U.S.A.

2b. **Campylobacter sputorum** subspecies **bubulus** (Loesche, Gibbons and Socransky) *comb. nov. (Vibrio bubulus* Thouvenot and Florent 1954, 237; *Campylobacter bubulus* (Thouvenot and Florent) Sebald and Véron 1963, 907; *Vibrio sputorum* var. *bubulum* (Prévot) Loesche, Gibbons and Socransky 1965, 1109.)

bub′.u.lus. L. adj. *bubulus* pertaining to cattle.

Morphology, colonies and other characteristics as for species except as noted.

Fatty acids and alcohols are usually not produced from glucose as end products of fermentation, although a few strains have been reported to produce a trace of acetic or acetic and lactic acids.

Grows in a medium with 2 or 3.5% NaCl. Growth in 1% bile medium is variable. Grows in medium with 0.1% selenite and selenite reduction is variable. Growth at both 25 C and 42 C is variable.

Sensitive to chloramphenicol, 1 μg/ml; dihydro-streptomycin, 8 μg/ml; erythromycin, 2 μg/ml; oxytetracycline, 4 μg/ml; penicillin, 0.25 unit/ml; polymyxin 4 units/ml; streptomycin, 8 μg/ml; tetracycline, 1 μg/ml. Moderately sensitive to neomycin, 16 μg/ml. Resistant to bacitracin, 64 μg/ml; and novobiocin, 256 μg/ml.

Found in the genital tract of male and female cattle and sheep. Can be isolated from semen, preputial and vaginal mucus of normal animals.

The G + C content of the DNA ranges from 29–31 moles % (T$_m$).

Reference strain: CIP 53103. Institute Pasteur, Paris.

3. Campylobacter fecalis comb. nov. (*Vibrio fecalis* Firehammer 1965. 493.)

fe.cal′is. L. n. *faex*, *faecis*, dregs; M.L. adj. *faecalis* pertaining to feces.

Slender, curved, non-spore-forming rods 0.3–0.6 μm wide and 2–4 μm long, ends of the cells are rounded. Morphology and motility otherwise similar to *C. fetus*.

Colonies on blood agar are pinpoint to 3.5 mm in diameter. They are shiny, smooth, convex and round, with entire edges.

Growth in a semisolid medium (0.12–0.20% agar) occurs in the upper third of the medium.

Indole negative. Catalase and oxidase positive. Hydrogen sulfide produced in peptone iron agar. Also H$_2$S positive with lead acetate strips as the method of detection. Nitrate reduced to nitrite.

Microaerophilic to anaerobic. Growth on the surface of agar plates or in broth requires the oxygen concentration to be reduced below 10%. Only traces of growth found on agar plates incubated under aerobic conditions. Poor growth under strict anaerobic conditions. Light growth in deep stab cultures.

Grows in semisolid medium containing 1% glycine. Growth in 1% bile is variable. Grows in medium with 0.1% selenite and reduces selenite. Grows in 2% NaCl. Some strains fail to grow in a medium with 4% NaCl. Some strains require at least 0.5% NaCl in the medium. Grows at 42 C but not at 25 C. Optimum temperature is 37 C.

Sensitive to chloramphenicol, 1 μg/ml; dihydro-streptomycin, 4 μg/ml; erythromycin, 4 μg/ml; oxytetracycline, 1 μg/ml; penicillin 0.25 unit/ml; streptomycin, 4 μg/ml; and tetracycline, 0.25 μg/ml. Resistant to bacitracin, 64 μg/ml; neomycin, 32 μg/ml; and novobiocin, 512 μg/ml. Moderately sensitive to polymyxin, 16 units/ml.

Isolated from sheep feces. May also be found in bovine semen and vagina.

Reference strain: from Montana State College, Bozeman, Montana, U.S.A.

Species incertae sedis

Representative strains of these species are not available for study. Although all are validly published and legitimate, their taxonomic relationships are not clear because of poor or sketchy descriptions. All are Gram-negative, motile rods which do not ferment carbohydrates, do not hydrolyze gelatin, change milk or reduce nitrates.

a. *Vibrio coli* Doyle 1948, 51.

co′li. Gr. n. *colum* or *colon*, the large intestine, colon; M.L. gen.n. *coli* of the colon.

Slender curved, non-spore-forming rods, 0.2–0.5 μm in diameter by 1.5–5 μm long. They appear comma-shaped and sometimes spiral-shaped. Occur singly or in chains of curved cells. Spherical or coccoid forms are found in old cultures. Single polar flagellum on one or both ends of cells.

Colonies on blood agar are small, round and dewdrop-like. Young colonies are translucent, becoming opaque and somewhat grayish as they age. Colonies may tend to elongate and follow the direction of streaking on an agar plate. Non-hemolytic.

Indole negative. Pigments not produced.

Microaerophilic. Growth on the surface of blood agar plates require 10–15% carbon dioxide. Little or no growth under aerobic conditions or anaerobic conditions. Said to cause swine dysentery but most strains isolated from the intestinal content of normal pigs and pigs with swine dysentery do not cause classic signs of the disease when fed to healthy pigs. The organisms usually found conform to the description of *C. fetus* subsp. *jejuni* and virulent strains of *V. coli* have probably not been isolated since the original description of the organism. Not pathogenic for rabbits, rats, mice, guinea pigs, cattle or chickens.

b. *Vibrio crassus* (Veillon and Repaci) Prévot 1940, 85. (*Spirillum crassum* Veillon and Repaci 1912, 306.)

cras.sus. L. adj. *crassus* thick, fat.

Curved, non-sporing rods, 0.6–0.8 μm wide and 2–3 μm long. Comma- and S-shaped. Cells have long peritrichous flagella.

Colonies in deep agar are circular and fluffy. Some strains may have reddish colonies. Slight turbidity in broth cultures.

Does not require serum for growth. Coagulated proteins not attacked.

Produces hydrogen sulfide, ammonia, volatile amines, aldehydes and acetic acid. Indole positive. Gas and fetid odor produced in glucose broth cultures.

Strict anaerobe. Neutral red and safranin are reduced. Optimum temperature 37 C.

Isolated from the buccal cavity and lungs of man.

c. *Zuberella rhinitis* (Tunnicliff) Prévot 1938, 293. (*Bacillus rhinitis* Tunnicliff 1915, 493.)

rhi.ni'tis. Gr. n. *rhis, rhinos* a nose; Gr. suff. *-itis* denoting an inflammatory disease; M.L. n. *rhinitis* inflammation of the nose.

Description from Sebald (1962, 52) and Prévot *et al.* (1967, 289).

Slender non-sporing rods, 0.3–0.5 µm wide and 5–8 µm long. Cells may be curved and have wavy filaments up to 30 µm long. Short rods may be in chains. Motile with undulating or rotary movement in young cultures. Subpolar flagella.

Colonies in deep agar (5 days to 1 month) are lenticular and yellowish. Surface colonies are circular with opaque yellowish center and transparent edges. Glucose broth cultures have soft crumbly sediment. Granular sediment in peptone broth cultures.

Requires serum for growth. Acetic and lactic acids, volatile amines, aldehydes and ketones produced.

Indole not produced. Hydrogen sulfide positive. Cultures do not produce gas or odor.

Optimum temperature is 37 C.

Thirty-one strains were isolated by Tunnicliff from secretions from head colds; one strain was isolated by Sebald from the posterior small intestine of a termite (*R. lucifugus*). No strains known to be extant.

GENERA OF UNCERTAIN AFFILIATION

Genus **Bdellovibrio** Stolp and Starr 1963, 243

JEFFREY C. BURNHAM AND JOHN ROBINSON

Bdel.lo.vib'ri.o. Gr. n. *bdella* leech, sucker; M.L. masc.n. *Vibrio* a generic name; M.L. masc.n. *Bdellovibrio* a leech-like vibrio.

Cells are single, small, curved, motile rods in the parasitic state; motility by means of a polar, sheathed flagellum, usually monotrichous. Cells elongate to helical (spiral) forms in the non-parasitic stage of their life cycle. Gram-negative.

May or may not be obligately parasitic on bacteria. Parasitic strains attach to and penetrate into the host; may exhibit a host-range specificity.

Chemoorganotrophs, metabolism respiratory, not fermentative. Parasitic strains require the presence of suitable host bacteria for intracellular multiplication; only facultative strains or host-independent (H-I mutants) derived from parasitic strains are able to grow in complex media.

Aerobic.

Temperature optimum generally 30 C; growth extremely poor above 37 C.

The G + C content of the DNA of most cultures of the type species is 50.4 ± 0.9 moles % (T_m, buoyant density). A second group of isolates of facultatively parasitic strains has a G + C content of 43 moles % (Seidler *et al.*, 1969, 1972).

Types species: *Bdellovibrio bacteriovorus* Stolp and Starr 1963, 243.

Further Comments

Some of the problems involved in determining the taxonomic position of the genus *Bdellovibrio* have been outlined by Shilo (1969). The DNA base composition of members of the genus *Bdellovibrio* (Seidler, Mandel and Baptist, 1972) is close to that of various members of the genera *Vibrio* and *Spirillum*. In addition to their basic morphological similarities, *Bdellovibrio* and *Campylobacter* have similar physiological characters such as their failure to utilize carbohydrates. They have, however, significantly different DNA base composition.

The outstanding characteristic of the genus *Bdellovibrio* is the ability of its members to attack, penetrate and develop within various Gram-negative and Gram-positive bacteria (Scherff *et al.*, 1966; Starr and Baigent, 1966; Burnham *et al.*, 1968; Burger *et al.*, 1968; Shilo, 1969). All the strains isolated from nature have been parasitic whereas host-independent strains have been derived in the laboratory from parasitic strains (Diedrich *et al.*, 1970; Seidler and Starr, 1969). The existence of independent strains questions the validity of the parasitic mode as a taxonomic criterion, although the identification of host-independent strains occurring in nature would be difficult to establish. The recent demonstration of a biochemical factor which is responsible for determining whether a bdellovibrio is host-dependent or host-independent (H-D) (M. Shilo, ASM Symposium, 1972, and personal communication) provides further evidence that the parasitic mode of bdellovibrio cannot be utilized in classification. In establishing the relationship of a bacterial strain to the genus *Bdellovibrio*, the important criterion should be the ability of the strain to revert to either the facultative or obligately parasitic state. Further data, such as bacteriophage typing, immunological techniques or physiological characteristics may eventually allow positive identification of questionable strains as members of the genus *Bdellovibrio*.

The electrophoretic migration patterns of specific enzymes isolated from an obligately parasitic bdellovibrio (109D) resembled those of similar enzymes isolated from the corresponding host-independent strain. The enzymes isolated from the host-independent cultures of strains A3.12 and UKi2 were distinct from each other and from all other host-independent bdellovibrios tested (Seidler *et al.*, 1972).

Nutritional and metabolic characterizations of known strains are incomplete and, therefore, presently are not valid taxonomic criteria.

Strains of *Bdellovibrio* have been determined on the basis of host susceptibility and on the location from which the parasite was isolated. Due to the variable and incompletely determined host specificity of many strains, this criterion generally should not be considered an accurate taxonomic character.

Bacteria resembling *B. bacteriovorus* have been shown to parasitize two species of *Clostridium* (Guelin *et el.*, 1968) and to lyse the green alga *Chlorella vulgaris* (Mamkaeva, 1966). Further data will show whether these bacteria should be included in the genus *Bdellovibrio*.

Key to the species of genus **Bdellovibrio**

I. High GC ratio (50.4 ± 0.9 moles %). Catalase positive; sensitive to vibriostat 0/129. Obligately parasitic or obligately host-independent. Relative low protease activity.

1. *Bdellovibrio bacteriovorus*

II. Low GC ratio (42–43 moles %). Facultatively parasitic or host-independent.

    1. Catalase positive; sensitive to vibriostat 0/129. High protease activity.

2. *Bdellovibrio stolpii*

    2. Catalase negative; resistant to vibriostat 0/129. Moderate protease activity; host-range restricted to pseudomonads.

3. *Bdellovibrio starrii*

Description of the species of genus **Bdellovibrio**

1. Bdellovibrio bacteriovorus Stolp and Starr 1963, 243.

bac.ter.i.o.vor'us. Gr. dim.n. *bakterion* a small rod; L.V. *voro* devour; M.L. adj. *bacteriovorus* bacteria devouring.

A marked and characteristic change in cell form occurs during the growth cycle (Plate 6.1, Fig. 5):

In the pre-infection stage of growth, cells are curved rods, 0.25–0.4 by 0.8–1.2 μm (Plate 6.1, Fig. 1), motile usually by means of a single, polar, sheathed flagellum (Plate 6.1, Fig. 2), although sometimes two flagella are present.

Cells penetrate the host bacterium (Plate 6.1, Figs. 3 and 4) and grow within the periplasmic space as curved rods or coiled filaments without flagella. The coiled filamentous forms divide into three or four daughter cells in *Escherichia coli* or into 17–20 daughter cells in a larger host cell such as *Spirillum serpens*. The flagellum has been observed to develop shortly after formation of the daughter cells (Burnham *et al.*, 1970).

Host-independent derivatives appear in culture as thin curved rods, coils or as filamentous cells, 0.25–0.4 μm wide and 2–40 μm long, average 10 μm (Plate 6.1, Fig. 5).

Susceptible hosts include many bacterial genera and species. Parasitic strains are unable to multiply in the absence of a host bacterium; growth of these strains may be seen as macroscopic plaques on a lawn of an appropriate host organism after 48–72 hrs at 30 C. Host-independent cultures have been obtained (Seidler and Starr, 1969) and are able to grow on media containing yeast extract and/or peptone and colonies appear as pinpoint, pale yellow or orange colonies after 5–6 days incubation.

Obligately parasitic and host-independent strains possess common physiological, morphological and genetic characteristics except that the latter are often yellow pigmented and non-motile (Seidler and Starr, 1969).

Because host-independent strains are more accessible for physiological measurement more information is available for this state rather than the host-dependent form. All H-I strains tested have been shown to be oxidase positive, can produce a non-specific protease and liquefy gelatin, can produce ammonia from peptone and do not ferment either carbohydrates or organic acids (Seidler and Starr, 1969).

Possesses a cytochrome spectrum and data suggest oxidation is coupled to phosphorylation. All strains tested contain enzymes of the tricarboxylic acid cycle (Simpson and Robinson, 1968; Seidler et al., 1972).

Low protease activity (3–10 units) (Seidler et al., 1972).

Growth is inhibited by the vibriostatic agent 0/129 (Seidler and Starr, 1969).

Obligately parasitic on bacterial hosts or obligately host-independent under laboratory conditions. No facultative forms of this species have yet been described. Has been isolated from a variety of habitats including soil, sewage, fresh water and marine environments (Shilo, 1969).

The G + C content of the DNA is 50.4 ± 0.9 moles % (buoyant density (Seidler et al., 1972).

Type strain: *Bdellovibrio* 100; ATCC 15356.

2. Bdellovibrio stolpii Seidler, Mandel and Baptist 1972, 216.

stolp'.i.i. M.L. gen.n. *stolpii;* named for Heinz Stolp, who discovered the bdellovibrios.

Morphology as for *B. bacteriovorus* depending upon the phase of the life cycle observed. May be either host-dependent or host-independent.

High protease activity (100 units) (Diedrich et al., 1970; Seidler et al., 1972)

Host range resembles *B. bacteriovorus.*

The G + C content of the DNA of strain UKi2 is 42.0 moles %. No homology between this strain and *B. bacteriovorus* strain 100.

Facultatively parasitic. Any cell may complete its life cycle either parasitically within a host bacterium or saprophytically on a complex medium.

Type strain: *Bdellovibrio* UKi2; ATCC 27111.

3. Bdellovibrio starrii Seidler, Mandel and Baptist 1972, 216.

star'.ri.i. M.L. gen.n. starrii; named for M.P. Starr, a contemporary investigator of bdellovibrios.

Morphology as for *B. bacteriovorus* depending upon the phase of the life cycle observed. May be either host-dependent or -independent. Facultatively parasitic.

Catalase negative.

Resistant to vibriostat 0/129.

Moderate protease activity (20–30 units) Seidler et al., 1972.

Host range restricted to pseudomonads.

The G + C content of the DNA of strain A3.12 is 43.5 moles % (Seidler et al., 1972). Strain A3.12 has a 16% DNA homology with *B. stolpii* and a 1% homology with *B. bacteriovorus.*

Type strain: *Bdellovibrio* A3.12; ATCC 27110.

Further Comments

Certain strains of *Bdellovibrio* possess characteristics that may eventually warrant the creation or further species. Examples are the strain W described by Burger et al. (1968) which is reported capable of attacking and penetrating *Rhodospirillum, Streptococcus* and *Lactobacillus* cells, and strain 6-5-S, which parasitizes *Spirillum serpens* and is facultatively parasitic resembling both *B. starrii* and *B. stolpii.*

Genus **Microcyclus** *Ørskov 1928, 183*

J. T. STALEY

Mi.cro.cy'clus. Gr. adj. *micrus* small; Gr. n. *cyclus* a circle; M.L. masc.n. *Microcyclus* a small circle.

Curved rods, 0.5–2.0 by 1.0 to more than 10 μm long. **Rings,** 1.5–10.0 μm outer diameter, **formed when cells elongate prior to division. Non-motile.** Resting stages not known. **Gram-negative.**

Chemoorganotrophs; **metabolism respiratory,** never fermentative. **Strict aerobes.**

Temperatures optimum 22–37 C; maximum usually below 37 C, for one strain 43 C; minimum 5 C.

Isolated from fresh water sources, including ponds, lakes and wells.

The G + C content of the DNA ranges from 39.5–68.4 moles % (buoyant density).

Type species: *Microcyclus aquaticus* Ørskov 1928, 183.

Further Comments

The wide range in DNA base composition values reported for the genus indicates that it is a hetero-

geneous collection of bacteria. Indeed, Claus et al. (1968) conclude from the base composition data that representatives of at least three distinct groups (i.e. genera) are included within the genus. Morphology (see descriptions of species) and other criteria support this contention and it must be concluded that the properties, ring formation by non-motile, heterotrophic, curved rods, are not sufficiently restrictive to circumscribe a single genus. As yet no formal taxonomic proposals have been published to resolve or clarify this problem, although Claus (1967) and Claus et al. (1968) have recommended that the genus *Spirosoma* Migula, 237, be revived and redescribed to contain *M. flavus* and two ring-forming strains that they have studied.

The medium of Hugh and Leifson (1953) is superior to other complex basal media for detecting aerobic acid production in isolated strains.

Description of the species of genus Microcyclus

1. **Microcyclus aquaticus** Ørskov 1928, 183.
a.qua'ti.cus. L. adj. *aquaticus* living in water.
Curved rods, 0.5–1.0 by 1.0–3.0 μm. Rings, 1.5–3.0 μm outer diameter are formed infrequently. Encapsulated. Gas vacuolated strains have been isolated.

Colonies are non-pigmented to cream-colored. Vacuolated strains form chalk-white colonies and in liquid media cells accumulate at the surface.

Glucose and gluconate catabolized via the Entner-Doudoroff and pentose phosphate pathways and tricarboxylic acid cycle. All strains produce acid aerobically on Hugh-Leifson medium from arabinose, fructose, galactose, glucose, inulin mannitol, mannose, melibiose, sorbitol, sucrose and xylose. Starch is not hydrolyzed.

Gelatin is not liquefied.

Growth occurs on a variety of complex media. Gas vacuolated strains use ammonium as sole nitrogen source and a number of carbon sources, including certain sugars, alcohols and sugar-alcohols, can be used as sole carbon source; biotin is stimulatory to the growth of these strains.

Temperature range 5–43 C; optimum 22–37 C.

Sensitive to neomycin, kanamycin, streptomycin, novobiocin and tetracycline; resistant to sulfonamides, chloromycetin, erythromycin and penicillin G.

Strains have been found in soil as well as in fresh waters.

The G + C content of the DNA is 66.3–68.4 moles %.

Type strain: ATCC 25396.

2. **Microcyclus flavus** Raj 1970, 62.
fla'vus. L. adj. *flavus* yellow.
Curved rods, 0.5–1.0 by 3.0–6.0 μm, tightly or loosely coiled helical forms and long slightly curved rods, 0.5–1.0 by 5.0–50.0 μm. Ring formation, outer diameter 1.5–3.0 μm, common. Encapsulated.

Colonies contain a water-insoluble, acetone-ethanol-soluble yellow pigment.

Glucose catabolized by the Embden-Meyerhof pathway, gluconate by the Entner-Doudoroff pathway; tricarboxylic acid cycle and pentose phosphate pathway are also present. Acid production is the same as for *M. aquaticus* except acid is produced aerobically from lactose and maltose but not from mannitol or sorbitol. Starch is not hydrolyzed. Gelatin not liquefied.

Optimum temperature 22–25 C; no growth at 37 C or below 7 C.

Sensitive to sulfonamides, kanamycin, streptomycin, penicillin G, novobiocin, erythromycin, tetracycline, chloromycetin; resistant to neomycin.

A marine strain was mentioned but not described by Raj (1970).

The G + C content of the DNA is 51.0–51.5 moles %.

Type strain: ATCC 23276.

3. **Microcyclus major** Gromov 1963, 733.
ma'jor. L. comp.adj. *major* larger.
Straight to curved rods, 1.0–2.0 by 5.0 to more than 50 μm long; helical forms in young cultures. Ring formation, 5–10 μm outer diameter, common. Capsules absent.

Colonies have a pale rose pigment.

Catabolic pathways not known. Acid formed aerobically from fructose, galactose, glucose, lactose, maltose, mannose, raffinose; acid not produced from rhamnose, xylose, dulcitol, glycerol or mannitol. Starch-hydrolyzed.

Slight liquefaction of gelatin.

The G + C content of the DNA is 39.5 moles %.
Type strain: BKM 859.

Genus **Pelosigma** *Lauterborn 1913, 100*

PETER HIRSCH

Pe.lo.sig'ma. Gr. adj. *pelos* dark-colored, hence anaerobic mud; Gr. n. *sigma* letter S; M.L. fem.n. *Pelosigma* S-shaped mud bacterium.

Cells S-shaped, slender filaments, 0.23–0.35 × 9–30 μm, colorless or pale gray, usually with a slight spiral twist. Generally arranged side by side in flat sigmoid aggregates of four (Plate 6.1 Fig. 6) or multiples of four; multiple aggregates of considerable thickness have been observed. The cells may be held together throughout their length by a mucoid substance, or only at one end forming a point, the other end being wider and spread out to reveal the individual filaments.

The aggregate may be motile by an organelle, visible with the light microscope and located at the pointed end. This organelle is interpreted by the author as a tuft of flagella.

Multiplication is presumably by synchronous cross-division of the aggregate and sudden rapid separation of the daughter aggregates.

Gram stain not recorded; have not been obtained in pure culture.

Found in and on mud in fresh and brackish waters.

Types species: *Pelosigma cohnii* (Warming) Lauterborn 1913, 100.

Description of the species of genus **Pelosigma**

1. **Pelosigma cohnii** (Warming) Lauterborn 1913, 100. (*Spiromonas cohnii* Warming 1875, 370.) coh'ni.i. M.L. gen.n. *cohnii* of Cohn; named for F. Cohn, a German microbiologist.

Cell morphology as for genus. Cells are wound around an imaginary axis in an S curve with 1¼ turns. Arranged in aggregates ranging in width from 1.2–4 µm and in length from 9–20 µm.

Originally found in actively decaying mud ("Faulschlamm") of brackish water habitats near Kalundborg, Vejle, Hofmansgave and from the Limfjord, Denmark.

2. **Pelosigma palustre** Lauterborn 1916, 418. pa.lus'tre. L. adj. *paluster, palustris, palustre* marshy, swampy.

Cell morphology as for genus. Aggregates are flat and bandlike, 8–10 µm wide and 20–25 µm long; one end is pointed, the other spread out like a fan.

The pointed end has a flagellum, probably a composite structure; movements are slow, trembly and connected with a rotation around the longitudinal axis; the aggregate may remain motionless for long periods.

Originally found in "Faulschlamm" mud of ponds with *Chara* spp. near Ludwigshafen, Germany.

Comment: Hirsch (unpublished) has observed *Pelosigma* aggregates frequently in mud of ponds and lakes in Northern Germany and in Lower Michigan, U.S.A. The aggregate length was surprisingly constant, 19.5 µm in all but one observation; the individual cell filaments were 0.23–0.31 µm wide and the aggregates 5–11 µm wide.

It is possible that *P. palustre* falls within the range of variability of *P. cohnii* and may indeed prove to be multiple aggregates of *P. cohnii*.

Genus **Brachyarcus** *Skuja 1964, 19*

PETER HIRSCH

Bra.chy.ar'cus. Gr. adj. *brachys* short; L. n. *arcus* a bow; M.L. masc.n. *Brachyarcus* a short bow.

Rod-shaped cells, bent like a bow, 1 by 1.5–2.5 µm, colorless, with several gas vesicles of reddish tinge arranged in the center, and occasionally with some minute sulfur granules. Cells arranged in groups (coenobia) as a result of polar growth and median cross-division combined with tight attachment to surfaces by means of a mucoid substance. Coenobia of two (rings) or four (clover leaf appearance) or more cells quite common. Delayed cell division resulting in the formation of pretzel-shaped cells. Division in the coenobia may or may not be synchronous. Coenobia of 2–10 cells may measure 3–6 µm in diameter, secondary families of

irregularly humped agglomerates up to 100 µm or more in size. The mucoid "capsule" thin and of uncertain (unsharp) limitation.

Non-motile. Microaerophilic or anaerobic.

Not obtained in pure culture.

Originally found in Lake Vuolep Njakajaure, Swedish Lappland, in April in 12–13.5 m depth. Hirsch (unpublished) has observed this organism in several Michigan marl lakes and ponds throughout the year, as well as in Lake Plussee (Germany).

Type species: *Brachyarcus thiophilus* Skuja 1964, 20.

Description of the species of genus **Brachyarcus**

1. **Brachyarcus thiophilus** Skuja 1964, 20. thi.o'phi.lus. Gr. n. *thium* sulfur; Gr. adj. *philus*

loving; M.L. adj. *thiophilus* sulfur loving.

Morphology and description as for the genus.

PART 7

GRAM-NEGATIVE AEROBIC RODS AND COCCI

FAMILY I. **PSEUDOMONADACEAE** WINSLOW, BROADHURST, BUCHANAN, KRUMWIEDE, ROGERS AND SMITH 1917, 555

Pseu.do.mo.na.da′ce.ae. M.L. fem.n. *Pseudomonas* type genus of family; *-aceae* ending to denote family; M.L. fem.pl.n. *Pseudomonadaceae* the *Pseudomonas* family.

Straight or curved rods, motile by polar flagella. Gram-negative.

Chemoorganotrophs: Metabolism respiratory, never fermentative. Do not fix nitrogen. Able to use other than single-carbon compounds as sole source of carbon.

Strict aerobes.

Catalase positive; oxidase usually positive.

Growth occurs from 4 or lower to 43 C.

The G + C content of the DNA ranges between 58–70 moles %.

Type genus: *Pseudomonas* Migula 1894, 237.

Genus I. **Pseudomonas** *Migula 1894, 237 Nom. cons.* Opin. 5, Jud. Comm. 1952, 121

M. DOUDOROFF AND N. J. PALLERONI[*]

(*Chlorobacterium* Guillebeau 1890, 32, *nom. rej.* Opin. 6, Jud. Comm. 1954, 143 (not *Chlorobacterium* Lauterborn 1916, 429); *Liquidomonas* Orla-Jensen 1909, 332; *Loefflerella* Holden 1935, 783.)

Pseu.do′mo.nas or Pseu.do.mo′nas. Gr. *pseudes* false; Gr. *monas* a unit, monad; M.L. fem.n. *Pseudomonas* false monad.

Cells single, **straight or curved rods,** but not helical. Dimensions, generally 0.5–1 μm by 1.5–4 μm. Motile by **polar flagella;** monotrichous or multitrichous. Do not produce sheaths or prosthecae. No resting stages known. **Gram-negative.**

Chemoorganotrophs: Metabolism **respiratory, never fermentative. Some are facultative chemolithotrophs,** able to use H_2 or CO as energy source. Molecular oxygen is the universal electron acceptor; **some can denitrify,** using nitrate as an alternate acceptor.

Strict aerobes, except for those species which can use denitrification as a means of anaerobic respiration. Catalase positive.

The G + C content of the DNA of those species examined ranges from 58–70 moles %.

Type species: *Pseudomonas aeruginosa* (Schroeter) Migula 1900, 884.

Further Comments

Members of the genus *Pseudomonas* are common inhabitants of soil, fresh water and marine environments, where their activities are important in the mineralization of organic matter. Some species cause diseases of plants, and exhibit varying degrees of host specificity. At least one species (*P. mallei*) appears to be a specialized mammalian parasite, while others are occasional animal pathogens (e.g. *P. pseudomallei*, *P. aeruginosa*).

Most species studied, including the parasitic ones, require no growth factors and can develop in mineral media with a single organic compound as sole source of carbon and energy. A few require additional amino acids or vitamins. Acetate can be used as the principal nutrient by all species that have been well characterized. Lactate, succinate and glucose can be used by the majority, but not by all species. An outstanding property of many members of the genus is their ability to

[*] With the assistance of Dr. Donald C. Hildebrand.

use a variety of organic compounds as sole or principal carbon sources for growth, some strains utilizing over 100 different substrates.

Some species produce acids oxidatively from alcohols and aldose sugars, especially when these are provided at high concentrations. Many species accumulate poly-β-hydroxybutyrate as an intracellular carbon reserve, particularly under conditions of nitrogen privation. A few species accumulate polysaccharides as reserve materials, while others appear to have no specialized reserve materials.

Most members of the genus are oxidase positive, but a few species give a weak or negative oxidase reaction.

Diffusible fluorescent pigments are produced by some species, particularly in iron-deficient media. The production of diffusible or insoluble blue, red, yellow or green phenazine pigments is also characteristic of certain species and varieties. At least four species are known to possess intracellular carotenoids. Colored by-products of metabolism may accumulate in certain media, particularly those containing aromatic or heterocyclic compounds.

Temperature relations vary considerably: growth occurs from 4–43 C. For most species the optimum lies near 30 C. All species can grow well at neutral or alkaline pH (7.0–8.5); most are incapable of growth at pH 6.0 or below. Marine species may require at least 1.0% NaCl for growth reflected either as a specific sodium or osmotic requirement, or both. Fresh water and soil species have no distinctive ionic or osmotic requirements.

Members of this genus are hosts to bacteriophages; the host ranges may be extremely narrow (confined to strains of a single species, characteristic of *P. aeruginosa* phages) or relatively wide (attacking as many as five different species). Genetic transfer by transduction has been shown for *P. aeruginosa* and one other species, *P. putida*; in *P. aeruginosa*, conjugation also occurs.

The morphological properties listed in the generic definition exclude the peritrichously or "degenerately peritrichously" flagellated aerobic organotrophic species assigned to the genera *Alcaligenes*, *Flavobacterium*, *Agrobacterium* and *Rhizobium*. It should be emphasized that it is often difficult to draw clear lines of distinction between polar, subpolar and "degenerately peritrichous" flagellation: both the number and mode of insertion of the flagella may vary on individual cells in a given culture; these characters often change with the conditions of cultivation. For this reason, a statistical analysis of flagellar distribution (Lautrop and Jessen, 1964) may be necessary to determine the flagellation of an unidentified motile strain. There are now many reasons to doubt the validity of mode of flagellar insertion as an indication of evolutionary relationships among bacterial species, and it is probable that some of the species presently assigned to *Pseudomonas* are less closely related to one another than they are to some peritrichously flagellated species.

The genus *Hydrogenomonas* Orla-Jensen 1909, 311 is not recognized in this edition of THE MANUAL (Davis *et al.*, 1969) and species with the property of oxidizing hydrogen are assigned to other genera as follows: *Pseudomonas facilis*, *P. saccharophila*, *P. ruhlandii*, *P. flava*, *P. palleronii* (no. 22–26), *Alcaligenes eutrophus* and *A. paradoxus*. *Hydrogenomonas carboxydovorans* is included in Addendum IV to *Pseudomonas* (p. 241).

Members of the genera *Zoogloea*, *Methylomonas*, *Halobacterium*, *Gluconobacter* and *Xanthomonas* possess most of the distinctive properties of *Pseudomonas* and can be differentiated from it only on grounds of certain special structural, physiological or ecological properties.

In the 7th edition of THE MANUAL, 160 *Pseudomonas* species were described; the majority were unrecognizable because the type strains had not been preserved or adequately characterized. Attempts to establish significant criteria for defining species by extensive comparative studies of large numbers of strains were initiated only recently, and cannot be considered either definitive or comprehensive. Some problems involved in the circumscription of the species were discussed by Stanier *et al.* (1966). The internal taxonomy of the genus is complicated because the various species can be differentiated from each other only on the basis of large constellations of phenotypic characters shared, in different combinations and permutations, by the individual strains which they comprise.

The following keys and species descriptions deal only with those species whose members have been compared in a large number of tests (at least 150) applied to all strains studied. Species recognizable from their descriptions and validly assignable to the genus, but which have not been included in the principal key are listed in Addendum I (p. 237).

None of the species in the principal list has a definitely vibrioid shape, is obligately halophilic, digests agar or cellulose, grows autotrophically with thiosulfate, utilizes methane or methanol as sole carbon source, or fixes nitrogen.

Wherever practicable, the information is presented in tabular form. It should be borne in mind that even important differential traits may be absent in some strains when isolated, or may be lost after laboratory cultivation. Dichotomous keys are used for the primary separation of species into groups that can be handled conveniently in separate sections. An attempt is made to arrange species in an order that may reflect their phylo-genetic relationships, but primary consideration is to practical determinative criteria.

The methodology employed in most of the studies is essentially that of Stanier *et al.* (1966), with only minor modifications by different authors.

Key to the sections of genus **Pseudomonas**

I. Growth factors not required. Capable of growth through several transfers in a purely mineral medium with the addition of an appropriate single carbon source.

(Acetate is used by all species, pyruvate (pure, filter-sterilized) and succinate by most. It is recommended that tests be also performed with glucose, lactate and L-alanine, all at 0.5% concentration.)

A. Poly-β-hydroxybutyrate is not accumulated as an intracellular carbon reserve.

(For this test DL-β-hydroxybutyrate is the preferred carbon source in a nitrogen-deficient medium (Stanier *et al.*, 1966). Most other species can use this substrate so, without evidence to the contrary, a strain incapable of growing with β-hydroxybutyrate but capable of growing with other carbon sources may be presumed not to accumulate the polymer.)

Section I

B. Poly-β-hydroxybutyrate is accumulated as an intracellular carbon reserve.

1. DL-Arginine and betaine are used as sole carbon sources. Grow in complex media at 40 C; most strains grow at 41 C.

Section II, p. 227

2. DL-Arginine is not used as sole carbon source and betaine is used by very few strains. Do not grow at 41 C, with the exception of *P. lemoignei*.

Section III, p. 230

II. Growth factors required; not capable of growth through several transfers in mineral media with a single carbon source.

Section IV, p. 235

Key to the species of Section I
(See also Tables 7.1 and 7.2)

A. Do not accumulate poly-β-hydroxybutyrate intracellularly. Do not hydrolyze poly-β-hydroxybutyrate extracellularly.

1. Fluorescent pigments produced by most strains.

a. Arginine dihydrolase present. Saprophytic.

1. *P. aeruginosa*
2. *P. putida*
3. *P. fluorescens*
4. *P. chlororaphis*
5. *P. aureofaciens*

b. Arginine dihydrolase absent. Phytopathogenic.

6. *P. syringae*
7. *P. cichorii*

2. Fluorescent pigments not produced.

a. Nitrate denitrified.

8. *P. stutzeri*
9. *P. mendocina*

b. Do not denitrify.

10. *P. alcaligenes*

Further Comments on Section I

Species grouped in this section include the type species *Pseudomonas aeruginosa*, all other saprophytic and phytopathogenic fluorescent pseudomonads, the non-pigmented denitrifying species of the "stutzeri group" (Palleroni *et al.* 1970) and the non-denitrifying non-pigmented *Pseudomonas alcaligenes*.

The "fluorescent" *Pseudomonas* species (species 1–7) are characterized by excretion of diffusible yellow-green pigments that fluoresce in ultraviolet light (wave length below 260 mμ). These pigments are produced particularly in iron-deficient and in special media (e.g. Medium B of King *et al.*, 1954) and should not be confused with non-fluorescent yellow or green pigments excreted by some other species of *Pseudomonas* (e.g. *P. cepacia*) or with fluorescent products of the oxidation of certain cyclic organic compounds. Some species of fluorescent pseudomonads also produce characteristic blue, green or orange phenazine pigments (particularly in Medium A of King *et al.*, 1954). Rare strains that do not produce the characteristic phenazine pigments can sometimes be identified with the appropriate species by their nutritional or physiological properties.

The saprophytic species of fluorescent pseudomonads can be distinguished from the phytopathogenic species, *P. syringae* and *P. cichorii*, by their positive arginine dihydrolase reaction, more rapid growth in most media and their ability to utilize certain organic substrates. The general outline of the taxonomic proposals for this group made by Stanier *et al.* (1966) is followed but the nomenclature is revised. The studies of Stanier *et al.* (1966) should be consulted for a discussion of the problems of the internal subdivision of the group and for a more complete characterization of the various species and biotypes than is presented here. A majority of strains encountered in nature and studied in the past fall into fairly well defined categories (species or biotypes); some

TABLE 7.1

Characteristics of species 1–10 of genus **Pseudomonas**

	1. P. aeruginosa	2. P. putida	3–5. See Table 7.2	6. P. syringae	7. P. cichorii	8. P. stutzeri	9. P. mendocina	10. P. alcaligenes
No. of flagella	1	>1	>1	>1	>1	1[a]	1[a]	1
Fluorescent pigments	d[b]	+[b]	+	+	+	−[b]	−	−
Pyocyanine	d	−	−	−	−	−	−	−
Carotenoids	−	−	−	−	−	−	+	−
Growth at 41 C	+	−	−	−	−	d	+	+
Levan formation from sucrose	−	−	d	d	−	−	−	−
Arginine dihydrolase	+	+	+	−	−	−	+	+
Oxidase reaction	+	+	+	−	+	+	+	+
Denitrification	+	−	d	−	−	+	+	−
Hydrolysis of:								
Gelatin	+	−	+	d	−	−	−	+
Starch	−	−	−	−	−	+	−	−
Poly-β-hydroxybutyrate	−	−	−	−	−	−	−	−
Carbon sources for growth:								
Glucose	+	+	+	+	+	+	+	−
Trehalose	+	+	+	−	−	−	−	−
2-Ketogluconate	+	+	+	−	−	−	−	−
meso-Inositol	−	−	+	d	d	−	−	−
Geraniol	+	−	−	−	−	−	+	−
L-Valine	+	+	+	−	−	+	+	+
β-Alanine	+	+	+	−	−	−	+	+
DL-Arginine	+	+	+	d	+	−	+	+

[a] Lateral flagella of short wave length may also be produced under certain conditions.

[b] + = positive for 90% or more of strains; − = negative for 90% or more of strains; d = positive for more than 10% but less than 90% of all strains studied.

strains cannot be assigned to any well character-ized species.

Phytopathogenic fluorescent pseudomonads present taxonomic problems similar to those en-countered in the classification of saprophytic species. The nomenclatorial confusion, illustrated by the listing of over 60 nomenspecies in the last edition of THE MANUAL can be largely attributed to the fact that early plant pathologists named many species isolated from different plant diseases on the assumption that each organism was highly specific for its host and in the type of lesion it produced. Most of the species are unrecognizable from their original descriptions, their host ranges were never examined and their type strains have been lost. In the present treatment, only two species are recognized, *P. syringae* and *P. cichorii*, which differ from each other mainly in their oxidase reaction. Future studies may reestablish

the validity of some other nomenspecies, which are here included in *P. syringae*.

Two species of non-fluorescent denitrifying *Pseudomonas* are included in Section I. One of these, *P. mendocina*, is a very homogeneous species. The other, *P. stutzeri*, which differs from *P. mendocina* in its characteristic colony mor-phology, pigmentation and several physiological and nutritional properties, is heterogeneous with respect to many phenotypic characters and DNA composition. The nomenclatural problem was discussed by Palleroni *et al.* (1970). The last species included in this section, *P. alcaligenes*, is represented by one well characterized strain. This species shows some resemblance to *P. pseudo-alcaligenes*, which is included in Section II.

The principal distinctive characters of the species included in this section are listed in Tables 7.1 and 7.2.

TABLE 7.2

Characteristics of species 3 (biotypes), 4 and 5 of genus **Pseudomonas**

	3. *P. fluorescens*					4. *P. chloro-raphis*	5. *P. aureo-faciens*
	Biotype I	Biotype II	Biotype III	Biotype IV	Mis-cella-neous strains		
P. fluorescens biotypes according to Stanier *et al.* (1966)[a]	A	B	C	F	G	D	E
Non-fluorescent pigments:							
Green (chlororaphin)	−[b]	−	−	−	−	+[b]	−
Orange (phenazine-1-carboxylate)	−	−	−	−	−	−	+
Blue, non-diffusible	−	−	−	+	−	−	−
Levan formation from sucrose	+	+	−	+	−	+	+
Denitrification	−	+	+	+	−	+	d[b]
Carbon sources for growth:							
L-Arabinose	+	+	−	+	d	−	+
Sucrose	+	+	−	+	d	+	d
Saccharate	+	+	−	+	d	+	+
Propionate	+	+	d	−	d	+	+
Butyrate	−	d	−	+	d	+	d
Sorbitol	+	+	d	+	d	−	−
Adonitol	+	d	+	−	d	−	−
Propylene glycol	−	+	d	−	d	−	−
Ethanol	−	+	d	−	d	+	−

[a] For more complete comparative tables, the original paper should be consulted.

[b] + = positive for 90% or more of strains; − = negative for all strains studied; d = positive for more than 10% but less than 90% of all strains studied.

Description of the species of Section I, genus **Pseudomonas**

1. **Pseudomonas aeruginosa** (Schroeter) Migula 1900, 884. (*Bacterium aeruginosum* Schroe-ter 1872, 126; *Bacterium aerugineum* Cohn 1872, 157; *Micrococcus pyocyaneus* Zopf 1884, 83; *Bacillus*

aeruginosus (Schroeter) Trevisan 1885, 101; *Bacillus pyocyaneus* (Zopf) Flügge 1886, 286; *Pseudomonas pyocyanea* (Zopf) Migula 1895, 29; *Bacterium pyocyaneum* (Zopf) Lehmann and Neumann 1896, 267. Probable synonym or variety, *Pseudomonas polycolor* Clara 1930, 704.)

ae.ru.gi.no'sa. L. fem.adj. *aeruginosa* full of copper rust or verdigris, hence green.

Rods, 0.5–0.8 by 1.5–3.0 μm, singly, in pairs or short chains. Motile with polar monotrichous flagellation (cells with two or more polar flagella may occur infrequently).

Diffusible fluorescent pigments and a soluble phenazine pigment, pyocyanin (blue in neutral or alkaline media, red in acid media), produced by most strains in suitable media (King *et al.*, 1954). Strains producing neither pigment extremely rare. Some strains also excrete a dark red pigment. Slime not produced in media with sucrose.

Weakly lipolytic. Egg yolk reaction negative.

Organic growth factors not required. Nutritionally versatile; individual strains can use from 76–82 or more different organic compounds for growth (Stanier *et al.*, 1966). The following combination of nutritional characters is particularly useful for distinguishing strains of this species that do not produce pyocyanin from other fluorescent *Pseudomonas* species: ability to use sebacate, geraniol, L-mandelate, acetamide and *n*-hexadecane and inability to use trehalose, mucate and meso-inositol.

Obligately aerobic, except in media with nitrate. Optimal temperature *ca.* 37 C. Growth at 41 C but not at 4 C.

Can be isolated from soil and water, particularly from enrichment cultures for denitrifying bacteria. Commonly isolated from clinical specimens (wound, burn and urinary tract infections). Causative agent of "blue pus," origin of the synonym "*pyocyaneus.*" Occasionally pathogenic for plants. Strains isolated from leaf spot of tobacco, identical with or similar to *P. aeruginosa* have been named *P. polycolor.*

The G + C content of the DNA is *ca.* 67 moles % (from buoyant density). Gene transfer by both conjugation and transduction demonstrated in this species (Holloway, 1969).

Neotype strain: ATCC 10145; NCIB 8295; NCTC 10332 Opin. 36, Jud. Comm.

2. **Pseudomonas putida** (Trevisan) Migula 1895, 29. (*Bacillus fluorescens putidus* Flügge 1886, 288; *Bacillus putidus* Trevisan 1889, 18; *Pseudomonas eisenbergii* Migula 1900, 913; *Pseudomonas convexa* Chester 1901, 325; *Pseudomonas incognita* Chester 1901, 323; *Pseudomonas ovalis* Chester 1901, 325; *Pseudomonas rugosa* (Wright) Chester 1901, 323; *Pseudomonas striata* Chester 1901, 325.)

pu'ti.da. L. fem.adj. *putida* stinking, fetid.

Rods, 0.7–1.1 by 2.0–4.0 μm. Some strains have oval-shaped cells (from which the name *Pseudomonas ovalis* is derived) but otherwise indistinguishable from other strains of normal cell shape. Motile with polar multitrichous flagellation.

Cultures produce diffusible fluorescent pigments, particularly in iron-deficient media. Other pigments not produced. Slime not produced on sucrose media.

Do not denitrify but may produce nitrite from nitrate.

Polyethylene sorbitan monooleate (Tween 80) very rarely hydrolyzed. Egg yolk reaction negative.

Organic growth factors not required. Nutritionally versatile; individual strains can use from 66 to more than 80 different carbon sources for growth (Stanier *et al.*, 1966). The following combination of nutritional characters distinguishes this species from other fluorescent *Pseudomonas* species: ability to use benzylamine and inability to use trehalose, meso-inositol and geraniol. Creatine used by the majority of strains but by only rare strains of other species. Sucrose used by few strains of *P. putida* and levan is never produced.

Obligately aerobic. Optimal temperature for growth *ca.* 25–30 C. No growth at 41 C. Some strains grow at 4 C or below.

Isolated from soil and water after enrichment in mineral media with various carbon sources. Soil appears to be better source than water; water samples more frequently yield *P. fluorescens* and *P. aeruginosa.*

The G + C content of the DNA is *ca.* 60–63 moles % (from buoyant density). Genetic transfer by transduction demonstrated (Holloway, 1969).

The majority of strains can be assigned to biotype A of Stanier *et al.* (1966), which is considered to be typical. This biotype has a DNA with a G + C content *ca.* 62.5 moles %. Biotype B, which has a G + C content of *ca.* 61 moles %, differs from biotype A only in a few phenotypic characters: all strains of this biotype utilize L-tryptophan, kynurenine and anthranilate, and most strains use D-galactose as carbon sources. None of the strains uses nicotinate.

Representative strain: ATCC 12633; NCIB 9494.

3. **Pseudomonas fluorescens** Migula 1895, 29. (*Bacillus fluorescens liquefaciens* Flügge 1886, 289; *Bacillus fluorescens* Trevisan 1889, 18; *Bacterium fluorescens* (Trevisan) Lehmann and Neumann 1896, 272; *Liquidomonas fluorescens* (Trevisan) Orla-Jensen 1909, 332.

flu.o.res'cens. L.n. *fluor* a flux; M.L. v. *fluoresco* fluoresce; M.L. part. adj. *fluorescens* fluorescing.

Rods, 0.7–0.8 by 2.3–2.8 μm during exponential

growth; may be shorter and thinner in old cultures. Occur singly and in pairs. Motile with polar multitrichous flagellation; occasionally non-motile.

Cultures produce diffusible fluorescent pigments, particularly in iron-deficient media. Some strains produce a non-diffusible blue pigment. On media containing 2–4% sucrose, colonies of strains of biotypes I, II and IV are slimy as a result of levan formation.

Some strains (biotypes II, III and IV) can denitrify.

Egg yolk reaction positive for biotypes I and III, variable in other strains. Biotype I generally lipolytic. Biotype II not lipolytic; other biotypes variable.

Organic growth factors not required. Nutritionally versatile; individual strains can use from 60 to more than 80 different carbon sources for growth (Stanier et al., 1966). Nutritional patterns that distinguish this species from other fluorescent pseudomonads and that are useful in distinguishing among the biotypes of P. fluorescens indicated in Tables 7.1 and 7.2.

Obligately aerobic, except for strains capable of denitrification and able to grow anaerobically in nitrate media. Optimal temperature for growth ranges from ca. 25–30 C. Most strains grow at 4 C or below ; no growth at 41 C.

Found in soil and water, from which it can be isolated after enrichment in media with various carbon sources, incubated aerobically. Denitrifying biotypes can be enriched in similar media containing nitrate, incubated under anaerobic conditions. Commonly associated with spoilage of foods (eggs, cured meats, fish and milk). Often isolated from clinical specimens. Some strains assigned to this species (biotype II) have been isolated from diseased plants (e.g. lettuce) and identified as P. marginalis.

The G + C content of the DNA ranges from 59.4–61.3 moles % (buoyant density). The lowest values found in biotype IV and the highest values in biotype II (Mandel, 1966).

Biotype I (biotype A of Stanier et al., 1966). Typical Pseudomonas fluorescens.

Biotype II (biotype B of Stanier et al., 1966). This biotype includes some strains believed to be phytopathogenic and identified as P. marginalis (Brown) Stevens, 1925, 30 and as P. tolaasii Paine 1919, 217. This biotype may deserve eventual recognition as a separate species, but a decision regarding its taxonomic and nomenclatural status must await further studies.

Biotype III (biotype C of Stanier et al., 1966). This biotype may eventually be found to deserve

specific rank and includes two subgroups that differ from each other in their ability to utilize higher dicarboxylic acids. No specific names other than P. fluorescens attached to well characterized members of this biotype.

Biotype IV (biotype F of Stanier et al., 1966). Contains the type strain of P. lemonnieri (Lasseur) Breed 1948, 178. Of the two thoroughly characterized strains, one has lost the characteristic blue pigment described by Starr, Blau and Cosens (1960).

The group of miscellaneous strains assigned to P. fluorescens biotype G by Stanier et al. (1966) is heterogeneous in the nutritional properties of its members, and may consist of strains that have lost one or more of the properties considered to be of diagnostic importance in differentiating among the better characterized biotypes. Among the non-authentic strains assigned to this group are strains labeled P. schuylkilliensis and P. geniculata (Stanier et al., 1966).

Neotype strain: ATCC 13525 (biotype I); NCIB 9046; NCTC 10038. Opin. 37 Jud. Comm. (IJSB 20: 15, 1970).

4. **Pseudomonas chlororaphis** (Guignard and Sauvageau) Bergey et al. 1930, 166. (Bacillus chlororaphis Guignard and Sauvageau 1894, 841.)

chlo.ro.ra'phis. Gr. adj. chlorus green; Gr. n. rhaphis a needle; M.L. fem.n. chlororaphis a green needle.

Rods, 0.7–0.8 by 1.5–3.6 μm during exponential growth, singly, in pairs or short chains. Motile with polar multitrichous flagellation.

Cultures produce diffusible fluorescent pigments, particularly in iron-deficient media. Green, insoluble phenazine pigment (chlororaphin) produced and crystallizes in the colony and surrounding medium. On media containing 2–4% sucrose, colonies slimy as a result of levan formation.

All strains can denitrify.

Egg yolk reaction positive. Lipolytic.

Organic growth factors are not required. Nutritionally versatile; individual strains can use from 65 to 80 different carbon sources for growth (Stanier et al., 1966). The nutritional pattern that distinguishes this species from other fluorescent pseudomonads indicated in Tables 7.1 and 7.2. All strains studied can use saccharate, valerate and ethanol as sole carbon sources, and none can use L-arabinose, sorbitol, propylene glycol or hydroxymethylglutarate.

Obligately aerobic except in media containing nitrate. All strains grow at 4 C. Optimal temperature for growth ca. 30 C.

Isolated from water.

The G + C content of the DNA is 63.5 moles % (buoyant density).

Suggested neotype strain: Stanier 30, NRRL-B-560; CCEB 292; ATCC 9446; NCIB 9392; IFO 3904 (Lysenko, 1961).

5. Pseudomonas aureofaciens Kluyver 1956, 406.

au.re.o.fa'ci.ens. L. adj. *aureus* golden; L. v. *facio* make; M.L. part.adj. *aureofaciens* making golden.

Rods, 0.7–0.8 by 1.9–2.8 μm during exponential growth, singly, in pairs or short chains. Motile with polar multitrichous flagellation.

Cultures produce diffusible fluorescent pigment, particularly in iron-deficient media. A diffusible orange-yellow pigment (phenazine-1-carboxylic acid) produced and may crystallize in the colony and in surrounding medium. On media containing 2–4% sucrose, colonies are slimy, as a result of levan formation.

Some strains can denitrify; all strains reduce nitrate to nitrite. Lipolytic.

Egg yolk reaction positive for most strains.

Organic growth factors not required. Nutritionally versatile. Individual strains can use 65–75 different carbon sources for growth. (One aberrant strain does not utilize glucose and a number of other substrates characteristically used not only by this species, but also by other fluorescent pseudomonads.) All strains studied can use L-arabinose, saccharate, valerate and phenylacetate as carbon sources, and none can use ethanol. The nutritional pattern that distinguishes this species from other fluorescent species indicated in Tables 7.1 and 7.2, and described by Stanier *et al.*, 1966.

Obligately aerobic. All strains grow at 4 C but not at 41 C. Optimal temperature for growth, *ca.* 30 C.

Occurs in soil and water. The type strain isolated from clay suspended in kerosene for 3 weeks.

The G + C content of the DNA is 63.6 moles % (buoyant density).

Type strain: ATCC 13985; NCIB 9030.

6. Pseudomonas syringae van Hall 1902, 141. (Provisionally included in this species are the following nomenspecies which may be synonyms, biotypes, pathotypes or varieties, and some of which may even deserve independent specific rank. *Phytomonas syringae* (van Hall) Bergey *et al.* 1930, 257; *Pseudomonas aptata* (Brown and Jamieson) Stevens 1925, 22; *Pseudomonas atrofaciens* (McCulloch) Stevens 1925, 22; *Pseudomonas avenae* Manns 1909, 133; *Pseudomonas barkeri* (Berridge) Clara 1934, 11; *Pseudomonas cerasi* Griffin 1911, 616; *Pseudomonas citrarefaciens* (Lee) Stapp 1928, 190; *Pseudomonas citriputealis* (Smith)

Stapp 1928, 190; *Pseudomonas coronafaciens* (Elliott) Stevens 1925, 27; *Pseudomonas delphinii* (Smith) Stapp 1928, 106; *Pseudomonas dysoxyli* Hutchinson 1949, 275; *Pseudomonas fraxini* (Brown) Škorić 1948, 78; *Pseudomonas garcae* do Amaral, Teixeira and Pinheiro 1956, 155; *Pseudomonas glycinea* Coerper 1919, 188; *Pseudomonas helianthi* (Kawamura) Săvulescu 1947, 11; *Pseudomonas hibisci* (Nakada and Takimoto) Stapp 1928, 203; *Pseudomonas holci* Kendrick 1926, 237; *Pseudomonas lachrymans* (Smith and Bryan) Carsner 1918, 201; *Pseudomonas matthiolae* (Briosi and Pavarino) Dowson 1943, 10; *Pseudomonas medicaginis* Sackett 1910, 553; *Pseudomonas mellea* Johnson 1923, 489; *Pseudomonas mori* (Boyer and Lambert) Stevens 1913, 30 (this name may have priority over *P. syringae*, but the organism is not sufficiently well described to identify it with the species); *Pseudomonas mors-prunorum* Wormald 1931, 251; *Pseudomonas nectarophila* (Doidge) Burkholder 1939, 193; *Pseudomonas panacis* (Nakata and Takimoto) Dowson 1943, 10; *Pseudomonas papulans* Rose 1917, 200; *Pseudomonas prunicola* Wormald 1930, 742; *Pseudomonas phaseolicola* (Burkholder) Dowson 1943, 10; *Pseudomonas pisi* Sackett 1916, 19; *Pseudomonas punctulans* (Brian) Săvulescu 1947, 12; *Pseudomonas rimaefaciens* Koning 1938, 11; *Pseudomonas savastanoi* (Smith) Stevens 1913, 33; *Pseudomonas savastanoi* var. *fraxini* (Brown) Dowson 1943, 11; *Pseudomonas sojae* (Wolf) Stapp 1928, 174; *Pseudomonas spongiosa* (Aderhold and Ruhland) Braun 1927, 2; *Pseudomonas tabaci* (Wolf and Foster) Stevens 1925, 36; *Pseudomonas tomato* (Okabe) Alstatt 1944, 530; *Pseudomonas tonelliana* (Ferraris) Burkholder 1948, 132; *Pseudomonas trifoliorum* (Jones, Williamson, Wolf and McCulloch) Stapp 1928, 177; *Pseudomonas utiformica* Clara 1932, 111; *Pseudomonas vignae* Gardner and Kendrick 1923, 275; *Pseudomonas viridifaciens* Tisdale and Williamson 1923, 150; *Pseudomonas xanthochlora* (Schuster) Stapp 1928, 231.)

sy.rin'gae. M.L. fem.n. *Syringa* generic name of lilac; M.L. fem.gen.n. *syringae* of the lilac.

Rods, *ca.* 0.7–1.2 by 1.5–3 or more μm. Long chains or filaments may occur. Motile with polar multitrichous flagellation.

Cultures produce diffusible fluorescent pigments, particularly in iron-deficient media. Most strains produce slime in media with 2–4% sucrose as a result of levan formation.

Cytochrome *c* not detectable in the spectrum.

Gelatin hydrolyzed by some strains.

Organic growth factors not required with rare exceptions. Growth of most strains slow in mineral media with a single carbon source and relatively slow in complex media. Nutritionally versatile; the nutritional spectrum similar to that of *P. cichorii*, and somewhat resembles that of saprophytic fluorescent *Pseudomonas* species, but is less extensive and more heterogeneous (i.e. more variable for different strains). Of 161 organic compounds tested, different strains found to utilize between 31 and 69 carbon sources for growth. Compounds that can be used as carbon sources include carbohydrates, organic acids and amino acids. Glucose, succinate and L-alanine are used by almost all strains. Trehalose, 2-ketogluconate, β-alanine, L-isoleucine, L-valine and spermine not used by any. DL-Lactate and DL-arginine used by only some strains. 2-Ketogluconate not produced from glucose (Lelliot *et al.*, 1966; Misaghi and Grogan, 1969; Sands *et al.*, 1970).

Obligately aerobic. Optimal temperature for growth *ca.* 25–30 C. No growth at 41 C. Some strains can grow at 4 C.

Isolated from various plants for which the various strains are pathogenic. Original strain isolated from lilac, and strains conforming to the original description are pathogenic on lilac, citrus, beans, cherries and other unrelated plants. Other nomenspecies or pathotypes in this species have been isolated from diseases of different plants, including peas, beans, soybean, blackberry, apple, tobacco, tomato, delphinium, passion flower, ash, sunflower, nasturtium, ginseng, cucumber, *Dysoxylum*, coffee bean plant, wheat and brom grass (*Bromus*). The host range of most strains of the various pathotypes or nomenspecies included in *P. syringae* has not been determined. When it has been studied, overlapping patterns have been found. Almost all strains produce a hypersensitive reaction in tobacco.

The G + C content of the DNA is *ca.* 59–61 moles % (buoyant density).

Future studies may help to clarify taxonomy and nomenclature of the nomenspecies and pathotypes included in *P. syringae*. Numerical analysis suggests that there may be a clustering of strains around certain nomenspecies (Sands *et al.*, 1970). At present, however, only a few of phenotypic characters examined offer some promise of being taxonomically useful. Among characters that may be useful in the subdivision of species are: production of levan from sucrose, and utilization of lactate, meso-inositol and sorbitol.

Proposed working type: ATCC 19310; NCPPB 281 (Sneath and Skerman, 1966).

7. Pseudomonas cichorii (Swingle) Stapp 1928, 291. (*Phytomonas cichorii* Swingle 1925, 730; *Bacterium cichorii* (Swingle) Elliott 1930, 112; *Pseudomonas endiviae* Kotte 1930, 609; *Bacterium formosanum* Okabe 1935, 65; *Chlorobacter cichorii* (Swingle) Patel and Kulkarni 1951, 80; probable

synonym: *Pseudomonas papaveris* Lelliott and Wallace 1955, 91.)

ci.cho'ri.i. Gr. *cichora* (pl.) succory, chicory; L. n. *cichorium* chicory; M.L. gen.n. *cichorii* of chicory.

Rods, *ca.* 0.8 by 0.2–3.5 μm. Motile with polar multitrichous flagellation.

Cultures produce diffusible fluorescent pigments, particularly in iron-deficient media. No slime produced on sucrose media.

Not lipolytic.

Organic growth factors not required. Growth slow in mineral media with single carbon sources, and relatively slow in complex media. Nutritionally versatile. Compounds that can be used as carbon sources for growth include carbohydrates, organic acids and amino acids. Nutritional spectrum is very similar to that of *P. syringae*. Glucose, succinate, L-alanine, DL-lactate and DL-arginine can be used as sole carbon sources. Trehalose, L-valine and β-alanine are not utilized. 2-Ketogluconate not produced from glucose and not utilized as carbon source (Lelliot *et al.*, 1966; Misaghi and Grogan, 1969; Sands *et al.*, 1970).

Obligately aerobic. Optimal temperature for growth *ca.* 30 C. No growth at 41 C.

Isolated from *Cichorium intybus* and *C. endivia*, for which it is pathogenic.

The G + C content of the DNA is *ca.* 59 moles % (buoyant density).

Suggested working type: ATCC 10857.

8. Pseudomonas stutzeri (Lehmann and Neumann) Sijderius 1946, 115. (*Bacillus denitrificans* II Burri and Stutzer 1895, 392; *Bacterium stutzeri* Lehmann and Neumann 1896, 237; *Bacillus nitrogenes* Migula 1900, 793; *Bacillus stutzeri* Chester 1901, 225; *Achromobacter sewerinii* Bergey *et al.* 1923, 140; *Achromobacter stutzeri* Bergey *et al.* 1930, 207; *Pseudomonas stanieri* Mandel 1966, 283. Not *Pseudomonas stutzeri* Migula 1900, 929.)

stut'ze.ri. Stutzer patronymic; *stutzeri* of Stutzer.

Motile with predominantly monotrichous polar flagellation. In some strains, lateral flagella of short wave length also produced, particularly in young cultures on complex solid media. These lateral flagella easily shed during manipulations incidental to flagella staining (Palleroni *et al.*, 1970).

Freshly isolated colonies adherent, have characteristic wrinkled appearance (van Niel and Allen, 1952), and reddish brown but not yellow color. After repeated transfers in laboratory media, colonies may become smooth, butyrous and pale. No diffusible pigments produced.

Metabolism respiratory with either oxygen or nitrate as electron acceptor. Capable of vigorous denitrification which may be delayed or require serial transfers in nitrate media under semi-aerobic conditions for its appearance. Characteristic color of freshly isolated colonies reflects high cytochrome content of the cells.

Starch hydrolyzed by most strains. Egg yolk reaction negative.

Organic growth factors not required. Nutritionally versatile and heterogeneous. Most strains can use at least 50 different organic compounds as sole carbon sources for growth, and some can use at least 65. Only five compounds (acetate, succinate, lactate, pyruvate and ethylene glycol) found to be universal substrates for all strains studied, and only 29 used by 90% of all strains. Among characteristic substrates for growth of most strains are starch, maltose, glycollate and ethylene glycol. Pentoses, hexoses other than glucose and fructose, disaccharides other than maltose, 2-ketogluconate, arginine, histidine and sarcosine are generally not used (Palleroni et al., 1970).

Obligately aerobic, except in media with nitrate. Most strains grow at 40 C and 41 C, some at 43 C. None can grow at 4 C. Optimum temperature for growth ca. 35 C.

Found in soil and water, from which it can be isolated after enrichment in media with nitrate under anaerobic conditions at 30 C, using various carbon sources, e.g. ethanol. L(+)-Tartrate gives excellent results in the enrichments (van Niel and Allen, 1952), although paradoxically, strains obtained in this manner may not grow with tartrate in pure culture. Many strains isolated from clinical specimens.

The G + C content of the DNA ranges from 60.7–66.3 moles % (from buoyant density). In vitro DNA hybridization experiments suggest that this species may be related to P. mendocina, and more distantly to fluorescent pseudomonads (Palleroni et al., 1970). The name P. stanieri Mandel (1966) has been proposed for strains with a G + C content around 62%. This species is not clearly differentiated from P. stutzeri on the basis of phenotypic characters (Palleroni et al., 1970).

Suggested neotype strain: Lautrop AB 201; Stanier 221; ATCC 17588 (Stanier et al., 1966).

9. Pseudomonas mendocina Palleroni in Palleroni, Doudoroff, Stanier, Solanes and Mandel 1970, 220.

men.do.ci'na. Spanish fem.n. mendocina native of Mendoza.

Rods, 0.7–0.8 by 1.4–2.8 μm, occurring singly and in pairs. Motile with predominantly monotrichous polar flagellation. In some strains, lateral flagella of short wave length also produced, particularly in young cultures on complex solid media. These lateral flagella are easily shed during the manipulations incidental to flagella staining.

Colonies yellowish as a result of production of intracellular carotenoid pigment; not adherent or wrinkled in appearance. No diffusible pigments produced.

Egg yolk reaction negative.

Organic growth factors not required. Individual strains can utilize from 56–65 or more different organic compounds as sole carbon sources for growth. These include arginine, geraniol, glycollate, ethylene glycol, propylene glycol and sarcosine. Pentoses, hexoses other than glucose and fructose, disaccharides and mannitol not used.

Obligately aerobic, except in media with nitrate. Growth at 41 C but not at 4 C. Optimal temperature for growth ca. 35 C.

Found in soil and water; isolated by enrichment in media with nitrate under anaerobic conditions especially at 40 C. Ethanol and L(+)-tartrate can be used as carbon sources for the enrichments.

The G + C content of the DNA, 62.8–64.3 moles % (buoyant density). In vitro DNA hybridization experiments suggest that this species may be related to P. stutzeri, and more distantly to fluorescent pseudomonads (Palleroni et al., 1970).

Type strain: ATCC 25411; NCIB 10541.

10. Pseudomonas alcaligenes Monias 1928, 332.

al.ca.li'ge.nes. M.L. adj. alcaligenes alkali-producing.

Rods, 0.5 by 2–3 μm. Motile with polar monotrichous flagellation.

Gelatin feebly hydrolyzed. Egg yolk reaction negative.

Organic growth factors not required. Very restricted nutritional spectrum, using only 23 of 146 organic compounds tested as sole carbon sources. Among the utilizable substrates are citrate, DL-arginine, β-alanine and spermine. Carbohydrates, including glucose and fructose, gluconate, β-hydroxybutyrate, glutarate, D-malate, ethanol, glycerol, mannitol, meta- and para-hydroxybenzoate, D-alanine, L-serine, L-aspartate, L-phenylalanine, putrescine and betaine not utilized.

Obligately aerobic. Growth at 41 C but not 4 C. Optimal temperature for growth ca. 35 C.

The neotype strain proposed by Hugh and Ikari (1964) isolated from swimming pool water is the only strain of the species so far characterized by the methods of Stanier et al., 1966.

The G + C content of the DNA is 66.3 moles % (buoyant density).

Proposed neotype strain: Hugh 1577; Stanier 142; NCTC 10367; ATCC 14909 (Hugh and Ikari, 1964); NCIB 9945.

Key to the species of Section II
(See also Table 7.3.)

A. Accumulate poly-β-hydroxybutyrate as a carbon reserve. Utilize arginine and betaine as sole carbon sources. Grow at 40 C in complex medium; most strains grow at 41 C.
   1. Arginine dihydrolase present.

               11. *P. pseudoalcaligenes*
               12. *P. pseudomallei*
               13. *P. mallei*
               14. *P. caryophylli*
   2. Arginine dihydrolase not present.

               15. *P. cepacia*
               16. *P. marginata*

Further Comments on Section II

Species included in this section are arbitrarily separated from species in Section III by their ability to use arginine and betaine as carbon sources for growth and to grow at 41 C in complex media. Some strains grow poorly at 41 C but all grow well at 40 C, a temperature generally unsuitable for the growth of species in Section III. The anaerobic "arginine dihydrolase" reaction is found only in some of the species. Most species are pathogenic for plants or for animals.

Description of the species of Section II, genus **Pseudomonas**

11. Pseudomonas pseudoalcaligenes Stanier in Stanier, Palleroni and Doudoroff 1966, 247.

pseu.do.al.ca.li'ge.nes. Gr. adj. *pseudes* false; M.L. adj. *alcaligenes* alkali-producing. M.L. adj. *pseudoalcaligenes* false alkali-producing.

Rods, 0.7–0.8 by 1.2–2.5 μm. Motile with polar monotrichous flagellation.

Denitrification does not occur.

Egg yolk reaction negative.

Organic growth factors not required. At least 43 different organic compounds can be used as sole sources of carbon. Fructose is the only carbohydrate utilized. Other utilizable substrates include glutarate, D-malate, mesaconate, ethanol, D-α-alanine, β-alanine, DL-arginine and betaine. Gluconate, 2-ketogluconate, mannitol, *meta*- and *para*-hydroxybenzoate and L-aspartate not utilized.

Obligately aerobic. Grow well at 40 C and poorly at 41 C. Do not grow at 4 C. Optimal temperature for growth *ca.* 35 C.

The type strain isolated from sinus discharge.

The G + C content of the DNA of type strain is 62.2 moles % (buoyant density).

The species as described by Stanier *et al.* (1966) appears to be heterogeneous and to include at least two groups of strains probably belonging to different species. Strains ATCC 12815 and 17443 do not accumulate poly-β-hydroxybutyrate, give a negative arginine dihydrolase reaction and share several nutritional characters not present in other strains. In some respects, these strains resemble *P. alcaligenes*, from which they differ in the arginine dihydrolase reaction, in possessing a broader nutritional spectrum and by having a lower G + C content of DNA, in the neighborhood of 62–63 moles %.

Type strain: ATCC 17440; NCIB 9946.

12. Pseudomonas pseudomallei (Whitmore) Haynes 1957, 100. (*Bacillus pseudomallei* Whitmore 1913, 9; *Bacterium whitmori* Stanton and Fletcher 1921, 196; *Malleomyces pseudomallei* Breed 1939, 300; *Loefflerella pseudomallei* Brindle and Cowan 1951, 574.)

pseu.do.mal'le.i. Gr. adj. *pseudes* false; L. n. *malleus* the disease glanders; M.L. gen.n. *pseudomallei* of false glanders.

Short rods, *ca.* 0.8 by 1.5 μm, singly and in short chains. Accumulate poly-β-hydroxybutyrate granules as intracellular carbon reserve, especially in nitrogen-deficient media (reported bipolar staining of the cells may be due to this phenomenon). Motile with polar multitrichous flagellation.

Colonies can range in structure from extreme rough to mucoid and in color from cream to bright orange.

Denitrification may occur. Almost all strains produce gas from nitrate under anaerobic conditions; those strains that do not produce gas grow well under the same conditions.

Polyethylene sorbitan monooleate (Tween 80) hydrolyzed.

No growth factors required. Nutritionally very versatile. Individual strains can use from 77–88 or more different organic compounds as sole carbon sources for growth. Among the characteristic carbon sources are many carbohydrates (including

TABLE 7.3

Characteristics of species 11–16 of genus **Pseudomonas**

	11. P. pseudo-alcaligenes	12. P. pseudo-mallei	13. P. mallei	14. P. caryo-phylli	15. P. cepacia	16. P. marginata
No. of flagella	1	>1	0	>1	>1	>1
Diffusible pigments[a]	−[b]	−	−	−	+[b]	d[b]
Arginine dihydrolase	+	+	+	+	−	−
Denitrification	−	+	+	+	−	−
Hydrolysis of:						
Starch	−	+	d	—	−	−
Extracellular poly-β-hydroxy-butyrate	−	+	d	−	−	−
Gelatin	−	+	+	−	d	+
Carbon sources for growth:						
Glucose	−	+	+	+	+	+
D-Xylose	−	−	+	+	d	+
D-Ribose	−	+	−	+	+	+
L-Rhamnose	−	−	−	+	d	−
Saccharate	−	−	−	+	+	+
Levulinate	−	+	−	+	+	+
Citraconate	−	−	−	+	+	+
Mesaconate	+	−	−	+	+	+
D(−)-Tartrate	−	−	−	+	+	+
Mesotartrate	−	−	−	+	+	+
Erythritol	−	+	−	−	+	+
Adonitol	−	d	−	−	+	+
2,3-Butylene glycol	−	−	−	+	+	−
meta-Hydroxybenzoate	−	−	−	−	+	−
Tryptamine	−	−	−	−	+	−
α-Amylamine	−	+	+	−	+	−

[a] Strains of *Pseudomonas cepacia* may produce non-fluorescent yellowish, greenish, brownish, violet or purple pigments that may be associated with the colony and/or diffuse into the medium. Diffusible yellow-green non-fluorescent pigments may also be produced by *Pseudomonas marginata* and *Pseudomonas caryophylli.*

[b] + = positive for 90% or more of strains; − = negative for 90% or more of strains; d = positive for more than 10% but less than 90% of all strains studied.

D-arabinose, D-fucose, trehalose, maltose, cellobiose, salicin and starch), saturated aliphatic dicarboxylic acids from C_4 to C_{10}, poly-β-hydroxybutyrate, erythritol, D-alanine and L-threonine. Compounds unsuitable as carbon sources include D-xylose, L-rhamnose, saccharate, 2,3-butyleneglycol, all three isomers of tartrate, citraconate, mesaconate, *meta*-hydroxybenzoate, glycine and nicotinate (Redfearn *et al.*, 1966).

Obligately aerobic except in media with nitrate. Optimal temperature for growth *ca.* 37 C. Growth occurs at 42 C but not at 4 C.

Isolated from human and animal cases of melioidosis and from soil and water in tropical regions, particularly Southeast Asia. Probably a soil organism and accidental pathogen, causing melioidosis (Redfearn *et al.*, 1966).

The G + C content of the DNA is *ca.* 69.5 moles % (buoyant density). *In vitro* DNA hybridization experiments suggest a close relationship to *P. mallei* and a distant relationship to *P. caryophylli*, *P. cepacia* and *P. marginata*.

Suggested neotype strain: WRAIR 286; NBL 121; ATCC 23343 (Redfearn *et al.*, 1966).

13. **Pseudomonas mallei** (Zopf) Redfearn, Palleroni and Stanier 1966, 305. (*Bacillus mallei* Zopf 1885, 89; *Pfeifferella mallei* (Zopf) Buchanan 1918, 54; *Malleomyces mallei* (Zopf) Pribram 1933, 93; *Actinobacillus mallei* (Zopf) Thompson 1933, 226; *Loefferella mallei* (Zopf) Holden 1935, 783;

Acinetobacter mallei (Zopf) Steel and Cowan 1964, 481.)

mal'le.i. L. n. *malleus* the disease glanders; L. gen.n. *mallei* of glanders.

Rods, ca. 0.5 by 1.5–4.0 μm, singly, in pairs and in groups. Accumulate poly-β-hydroxybutyrate granules as intracellular carbon reserve, especially in nitrogen-deficient media. Non-motile.

Denitrification commonly occurs. All strains can grow well anaerobically in the presence of nitrate, and some produce gas under these conditions. Most strains give a positive arginine dihydrolase reaction.

Polyethylene sorbitan monooleate (Tween 80) hydrolyzed. Most strains hydrolyze poly-β-hydroxybutyrate and some hydrolyze starch.

No growth factors required. Nutritionally versatile. Nutritional spectrum resembles that of *P. pseudomallei*, but is less extensive. Individual strains can use from 47–67 or more different organic compounds as sole carbon sources for growth. Compounds used by most strains of *P. mallei* that are not used by *P. pseudomallei* are D-xylose, glycine and DL-α-aminobutyrate. Among the compounds that are not used by *P. mallei* but are used by *P. pseudomallei* are D-ribose, erythritol and ethanolamine (Redfearn *et al.*, 1966).

Obligately aerobic except in media with nitrate. Optimal temperature for growth ca. 37 C. Growth at 42 C but not at 4 C.

Parasitic on horses and donkeys, in which it causes glanders and farcy. Infection is transmissible to man and other animal species.

The G + C content of the DNA is ca. 69 moles % (buoyant density). *In vitro* DNA hybridization experiments suggest a close relationship to *P. pseudomallei* and a distant relationship to *P. caryophylli*, *P. cepacia* and *P. marginata*.

Suggested neotype: NBL7, ATCC 23344 (Redfearn *et al.*, 1966).

14. **Pseudomonas caryophylli** (Burkholder) Starr and Burkholder 1942, 601. (*Phytomonas caryophylli* Burkholder 1942, 143.)

ca.ry.o'phyl.li. M.L. masc.n. *caryophyllus* specific epithet in *Dianthus caryophyllus*, carnation; M.L. gen.n. *caryophylli* of the carnation.

Rods, ca. 0.8 by 2.0 μm, (smaller and larger dimensions have been reported). Slightly curved cells may be observed. Accumulate poly-β-hydroxybutyrate granules as intracellular carbon reserve especially in nitrogen-deficient media. Motile with polar multitrichous flagellation.

Cultures of most strains produce a yellow-green non-fluorescent diffusible pigment, especially on media deficient in iron. Old cultures on potato glucose agar become brown. Slime not produced on sucrose media.

Capable of denitrification.

Some strains lipolytic and some give positive egg yolk reaction.

No growth factors required by most strains. Strains that can grow on mineral media with single carbon sources can use at least 61 different organic compounds. Among the characteristic carbon sources used by all strains studied are many carbohydrates (including D-ribose, D-xylose, D-arabinose, D-fucose, L-rhamnose, cellobiose and salicin), saccharate, glycollate, 2,3-butyleneglycol, mesotartrate, DL-arginine and betaine. Compounds unsuitable as carbon sources include maltose, saturated aliphatic dicarboxylic acids from C_3 to C_{10}, mesaconate, adonitol, *meta*-hydroxybenzoate, D-alanine, acetamide, nicotinate and trigonelline (Ballard *et al.*, 1970).

Obligately aerobic except in media with nitrate. Optimal temperature for growth ca. 30–33 C. All strains grow at 41 C and some can grow at 46 C; none can grow at 4 C.

Isolated from carnations, for which they are pathogenic.

The G + C content of the DNA is ca. 65.3 moles % (buoyant density). *In vitro* DNA hybridization experiments suggest that this species may be related to *P. pseudomallei* and *P. mallei* and more distantly to *P. cepacia* and *P. marginata* (Ballard *et al.*, 1970).

Suggested neotype strain: ICPB PC113; Ballard 720; ATCC 25418 (Ballard *et al.*, 1970).

15. **Pseudomonas cepacia** Burkholder 1950, 116. (*Pseudomonas multivorans* Stanier, Palleroni and Doudoroff 1966, 247.)

ce.pa'ci.a. L. fem.n. *caepa* or *cepa* onion; M.L. fem. adj. *cepacia* of or like onion.

Rods, 0.8–1.0 by 1.6–3.2 μm, singly or in pairs. Accumulate poly-β-hydroxybutyrate granules as intracellular carbon reserve, especially in nitrogen-deficient media. Motile with polar multitrichous flagellation (one to three flagella).

Non-fluorescent pigments produced by most strains in colonies and surrounding medium. On ordinary complex media the pigmentation usually yellow or greenish. Many strains also produce purple phenazine pigment, particularly in iron-deficient media. On chemically defined media such strains may show great variety of colors (green, brownish, red, violet, purple) depending on the carbon source used for growth, the pigments being sometimes exclusively associated with the organisms, sometimes also diffusible. Some strains produce slime on media containing 2–4% sucrose.

Denitrification does not occur, but nitrite produced from nitrate.

Gelatin hydrolyzed by some strains. Egg yolk reaction variable. Lipolytic.

No growth factors required. Nutritionally very versatile, most individual strains using from 95 to more than 105 different organic compounds as sole carbon sources for growth. Among characteristic carbon sources used are many carbohydrates (including D-ribose, D-arabinose, D-fucose, trehalose, cellobiose, salicin); mono- and dicarboxylic saturated aliphatic acids from C_2 to C_{10}; adonitol, 2,3-butylene glycol, levulinate, *meta*-hydroxybenzoate, tryptamine and α-amylamine. The following compounds are not utilized: maltose, D(−)-tartrate, mesaconate and erythritol (Ballard *et al.*, 1970).

Obligately aerobic. Optimal temperature for growth *ca.* 30–35 C. No growth at 4 C. Most strains grow at 41 C; almost all grow at 40 C.

Many strains isolated from rotted onions; others from soil and from clinical specimens (e.g. urinary tract infections). The species appears to be widely distributed in soil.

The G + C content of the DNA is *ca.* 67–68 moles % (buoyant density). *In vitro* DNA hybridization experiments indicate that this species may be related to *P. marginata*, *P. caryophylli*, *P. pseudomallei* and *P. mallei* (Ballard *et al.*, 1970).

Type strain: Burkholder 717; ICPB 25; ATCC 25416 (Ballard *et al.*, 1970).

16. Pseudomonas marginata (McCulloch) Stapp 1928, 56. (*Bacterium marginatum* McCulloch 1921, 115; *Phytomonas alliicola* Burkholder 1942, 146; *Pseudomonas alliicola* Starr and Burkholder 1942, 601; possible synonym: *Pseudomonas gladioli* Severini 1913; 420; this name may have priority over *P. marginata*, but no surviving strain is available.)

mar.gi.na'ta. L. fem.part.adj. *marginata* margined.

Rods, *ca.* 0.8 by *ca.* 2.0 μm (smaller and larger dimensions have been reported). Accumulate poly-β-hydroxybutyrate granules as intracellular carbon reserve, especially in nitrogen-deficient media. Motile with polar multitrichous flagellation.

A yellow-green non-fluorescent diffusible pigment produced by most cultures, especially in iron-deficient complex media. In complex media, old cultures of most strains designated as *P. alliicola* become dark brown. Slime produced on media containing 2–4% sucrose.

Denitrification does not occur. Oxidase reaction positive but very weak in some strains.

Egg yolk reaction positive. Lipolytic.

No growth factors required. Nutritionally versatile, individual strains using at least 78 different organic compounds as sole carbon sources for growth. Among characteristic carbon sources used by all strains are many carbohydrates (including D-ribose, D-xylose, D-arabinose, cellobiose, trehalose and salicin), meso- and D(−)-tartrate, 2,3-butyleneglycol, mesaconate, DL-arginine, betaine, nicotinate and trigonelline. Compounds unsuitable as carbon sources include maltose, L-rhamnose, levulinate, erythritol, *meta*-hydroxybenzoate, tryptamine and α-amylamine (Ballard *et al.*, 1970).

Obligately aerobic. Optimal temperature for growth *ca.* 30–35 C. All strains grow at 40 C and almost all grow at 41 C. No growth at 4 C.

Isolated from decayed onions, *Gladiolus* spp. and *Iris* spp., for which they are believed to be pathogenic.

The G + C content of the DNA is *ca.* 68.5 moles % (buoyant density). *In vitro* DNA hybridization experiments indicate that this species may be related to *P. cepacia*, *P. pseudomallei*, *P. mallei* and *P. caryophylli* (Ballard *et al.*, 1970).

Type strain: McCulloch Col. 3; NRRL B-792; ICPB PM107; NCPPB 1891; Ballard 704; ATCC 10248 (Ballard *et al.*, 1970).

Key to the species of Section III
(See also Table 7.4)

A. Accumulate poly-β-hydroxybutyrate as a carbon reserve. Arginine and betaine not utilized; arginine dihydrolase absent. Do not grow at 41 C, with the exception of *P. lemoignei*.
  1. Cannot grow autotrophically with hydrogen.
    17. *P. lemoignei*
    18. *P. testosteroni*
    19. *P. acidovorans*
    20. *P. delafieldii*
    21. *P. solanacearum*
  2. Can grow autotrophically with hydrogen.
    22. *P. facilis*
    23. *P. saccharophila*
    24. *P. ruhlandii*
    25. *P. flava*
    26. *P. palleronii*

Further Comments on Section III

None of the species known to be pathogenic for animals; one (*P. solanacearum*) is a plant pathogen. A number of species can grow autotrophically, using the oxidation of molecular hydrogen as an energy source. The carotenoid-containing species (*P. flava* and *P. palleronii*) may be easily confused with *Alcaligenes paradoxus* Davis (Davis *et al.*, 1969, 387).

The description of some species (*P. flava*, *P. ruhlandii* and *P. saccharophila*) based on studies of single representative strains, and diagnostic characters that have been selected may have to be revised when more strains become available for study.

Description of the species of Section III, genus **Pseudomonas**

17. **Pseudomonas lemoignei** Delafield, Doudoroff, Palleroni, Lusty and Contopoulou 1965, 1460.

le.moig'ne.i. M.L. gen.n. *lemoignei* of Lemoigne: named for M. H. Lemoigne, a French bacteriologist, who first identified the compound poly-β-hydroxybutyrate as a bacterial carbon reserve.

Straight to slightly curved rods, 0.75 by 1.6–2.6 μm singly and in pairs. In old liquid cultures cells tend to become swollen and ovoid, and to form clumps. Accumulate granules of poly-β-hydroxybutyrate as intracellular carbon reserve, especially in nitrogen-deficient media with DL-β-hydroxybutyrate as substrate. Motile with polar monotrichous flagellation.

Colonies small, very coherent and adherent to agar. On complex media or media containing tyrosine and suitable carbon source, colonies and surrounding media turn brown or black. No cellular or fluorescent pigments produced.

Nitrate not reduced to nitrite or to molecular nitrogen.

Organic growth factors not required. Extremely limited nutritional spectrum. In addition to poly-β-hydroxybutyrate, only six compounds of 146 tested supported good growth: acetate, butyrate, valerate, pyruvate, succinate and DL-β-hydroxybutyrate. Compounds commonly used by other species but unsuitable for *P. lemoignei* include DL-lactate, glucose and all other carbohydrates

TABLE 7.4
Characteristics of species 17–26 of genus **Pseudomonas**

	17. P. lemoig-nei	18. P. testos-teroni	19. P. acido-vorans	20. P. dela-fieldii	21. P. solana-cearum	22. P. facilis	23. P. sac-charo-phila	24. P. ruh-landii	25. P. flava	26. P. palle-ronii
No. of flagella	1	>1	>1	1	>1	1	1	1	1a	1a
Carotenoid pigments	−b	−	−	−	−	−	−	−	+b	+
Autotrophic growth with H₂	−	−	−	−	−	+	+	+	+	+
Hydrolysis of:										
Starch	−	−	−	−	−	−	+	−	−	−
Poly-β-hydroxybutyrate	+	db	−	+	−	+	−	−	−	−
Gelatin	−	−	−	−	−	+	−	−	−	−
Carbon sources for growth:										
DL-Lactate	−	+	+	+	d	+	+	+	+	+
Glucose	−	+	−	+	+	+	Mb	M	+	M
Fructose	−	−	+	+	+	+	M	−	+	−
L-Arabinose	−	−	+	+	+	−	+	−	+	−
Sucrose	−	−	−	−	+	−	+	−	+	−
Malonate	−	−	+	+	−	+	−	−	−	−
Ethanol	−	−	+	+	−	−	−	+	+	+
Glycollate	−	+	+	+	d	−	−	−	−	+
para-Hydroxybenzoate	−	+	+	+	d	−	−	−	−	+
Testosterone	+	+	−	−	−	−	−	−	−	−

a With a tendency toward multitrichous flagellation and subpolar flagellar insertion.

b + = positive for all strains studied; − = negative for all strains studied; d = positive for more than 10% but less than 90% of all strains studied; M = positive for all strains, but may require mutation in strains isolated from nature; > = one or more.

and sugar derivatives, polyols and all amino acids tested.

Obligately aerobic. Optimal temperature for growth, ca. 30 C. Growth at 41 C.

Isolated from soil by enrichment in mineral medium with poly-β-hydroxybutyrate as sole carbon source.

The G + C content of the DNA is 58.2 moles % (buoyant density).

Type strain: ATCC 17989; NCIB 9947.

18. Pseudomonas testosteroni Marcus and Talalay 1956, 661.

tes.tos.te.ro′ni. M.L. gen.n. *testosteroni* of testosterone, a chemical compound.

Rods, 0.7–0.8 by 2.1–2.9 μm, singly or in pairs. Motile with polar multitrichous flagellation.

Denitrification does not occur.

Lipolytic.

Organic growth factors not required. Individual strains can use from 53 to more than 60 different organic compounds as sole carbon sources. These include all saturated dicarboxylic acids from succinate to sebacate, glycollate, *para-* and *meta*-hydroxybenzoate, norleucine and testosterone. Carbohydrates including glucose and fructose, malonate, mannitol, L(+)-tartrate, ethanol, DL-arginine, L-tryptophan, betaine and acetamide not utilized (Stanier *et al.*, 1966).

Obligately aerobic.

Optimal temperature for growth, ca. 30 C. No growth at 4 C or 41 C.

Occurs in soil: isolated after aerobic enrichment with a variety of organic compounds as sole carbon sources, including testosterone.

The G + C content of the DNA is ca. 62 moles % (buoyant density).

Type strain: ATCC 11996; NCIB 8955.

19. Pseudomonas acidovorans den Dooren de Jong 1926, 106. (*Pseudomonas indoloxidans* Gray 1928, 263; *Pseudomonas desmolytica* Gray and Thornton 1928, 90.)

a.ci.do′vo.rans. L. neut.n. *acidum* acid; L. v. *voro* devour; M.L. part.adj. *acidovorans* acid-devouring.

Rods, 0.8–1.1 by 2.5–4.1 μm during exponential growth, singly and in pairs.

Motile with polar multitrichous flagellation.

Denitrification does not occur. Strains designated *P. indoloxidans* produce insoluble blue pigment, indigotin, in media containing indole and a suitable carbon source.

Lipolytic.

Organic growth factors not required. At least 68 different organic compounds can be used as sole carbon sources; include fructose but not glucose or any other sugars; gluconate but not 2-ketoglu-

conate. Other utilizable compounds are all saturated dicarboxylic acids from malonate to sebacate, ethanol, glycollate, *para-* and *meta*-hydroxybenzoate, norleucine and L-tryptophan and acetamide. DL-Arginine, betaine and testosterone are not utilized (Stanier *et al.*, 1966).

Obligately aerobic. Optimal temperature for growth ca. 30 C. No growth at 4 C or 41 C.

Occurs in soil; isolated after aerobic enrichment with a variety of organic compounds as sole carbon sources. Occasionally found in clinical specimens.

The G + C content of the DNA is ca. 67 moles % (buoyant density).

Type strain: den Dooren de Jong 7; Stanier 14; ATCC 15668 (Stanier *et al.*, 1966); NCIB 9681.

20. Pseudomonas delafieldii Davis in Davis, Stanier, Doudoroff and Mandel 1970, 12.

de.la.fiel′di.i. M.L. gen.n. *delafieldii* of Delafield; named for the American bacteriologist, F. P. Delafield, who first isolated the organism.

Rods, ca. 0.5 by 1.8–2.6 μm during exponential growth; ca. 0.5 by 1.4 μm in stationary phase. Motile with polar monotrichous flagellation.

Denitrification does not occur.

Organic growth factors not required. At least 48 different organic compounds can serve as sole carbon sources. These include D-ribose, D-xylose, L-arabinose, mannose, galactose, glycerol, mannitol, sorbitol, malonate, maleate, glutarate, adipate, suberate, poly-β-hydroxybutyrate, L(+)- and meso-tartrate, hydroxymethylglutarate, β-alanine, L-serine, δ-aminovalerate and histidine. Compounds unsuitable for growth include all common disaccharides, isobutyrate, valerate, 2,3-butylene glycol, ethanol, *n*-butanol, benzoate, *meta-* and *para*-hydroxybenzoate, citrate, itaconate, D-alanine, lysine, putrescine and testosterone (Davis *et al.*, 1970).

Obligately aerobic. Optimal temperature for growth ca. 30 C.

Isolated from soil by enrichment with poly-β-hydroxybutyrate as sole carbon source.

The G + C content of the DNA is ca. 65–66 moles % (buoyant density).

Type strain: Delafield FD-6; ATCC 17505.

21. Pseudomonas solanacearum (Smith) Smith 1914, 178. (*Bacillus solanacearum* Smith 1896, 10.)

so.la.na.ce.a′rum. M.L. fem.pl.n. *Solanaceae* the nightshade family; M.L. fem.pl.gen.n. *solanacearum* of the *Solanaceae*.

Rods, ca. 0.5–0.7 by 1.5–2.5 μm. Motile by one to four polar flagella.

At least two different types of colonies are produced on complex media; one type is smooth, fluid, elevated; the other is somewhat rough, dry and flat. Some strains produce a diffusible brown pigment in complex culture media.

Some strains capable of denitrification and most isolates are capable of growth under anaerobic conditions in media containing nitrate and an appropriate carbon source.

Polyethylene sorbitan monooleate (Tween 80) hydrolyzed by many strains. Egg yolk reaction negative.

Organic growth factors not required. Individual strains can use from 36 to more than 50 different organic compounds as sole carbon sources. In the nutritional spectrum there are no positive characters taxonomically useful for differentiation from other species of *Pseudomonas*, with the possible exception of the utilization of L-threonine, used by some strains. Negative nutritional characters of taxonomic importance are inability to use xylose, L-arabinose, 2-ketogluconate, isovalerate, heptanoate, caprylate, pelargonate, caprate, the common alcohols, most aromatic compounds, glycine, arginine and various amines (Palleroni and Duodoroff, 1971).

The species has been divided into four biotypes by Hayward (1964) mainly by acid formation from sugars and denitrification. The nutritional spectra of these biotypes are fairly similar, but biotypes I and II can be separated from III and IV as biotypes I and II are unable to use galactose, lactate, mannitol, sorbitol and *p*-hydroxybenzoate as carbon sources for growth (Palleroni and Doudoroff, 1971).

Important as a plant pathogen, particularly in warm and humid climates, causing wilt of many cultivated plants. Isolated from a wide variety of diseased plants, including several *Solanacea* (particularly potato, tomato, tobacco), *Casuarina*, *Strelitzia*, ginger, banana, *Heliconia*, peanut, *Pelargonium*. Believed to be transmitted through soil (for a discussion see Buddenhagen, 1965) and occasionally by insects (Buddenhagen and Elsasser, 1962). The isolated cultures easily lose their pathogenicity upon cultivation on laboratory media, and this phenomenon has been correlated with colony variation (Buddenhagen and Kelman, 1964).

The G + C content of the DNA is *ca.* 66.5–68 moles % (buoyant density). *In vitro* DNA experiments suggest that this species is unrelated to most other species of aerobic pseudomonads, with the possible exception of some strains of *P. cepacia* (Palleroni and Doudoroff, 1971).

Proposed neotype strain: NCPPB 325.

22. **Pseudomonas facilis** (Schatz and Bovell) Davis in Davis, Doudoroff, Stanier and Mandel 1969, 385. (*Hydrogenomonas facilis* Schatz and Bovell 1952, 88.)

fa'ci.lis. L. adj. *facilis* ready, quick.

Rods, 0.3–0.5 μm by 1.8–2.8 μm, singly, in pairs and short chains. Motile by one, rarely two, polar flagella.

No diffusible or cellular pigments produced.

Chemoorganotrophic and facultatively chemolithotrophic, using the oxidation of molecular hydrogen, but not of carbon monoxide or thiosulfate, as a source of energy for autotrophic growth. Denitrification does not occur.

Not lipolytic.

Organic growth factors not required. Autotrophic growth occurs in atmospheres containing H_2, CO_2 and O_2. At least 36 different organic compounds can serve as sole carbon sources for heterotrophic growth; include D-ribose, L-arabinose, malonate, mannitol, β-alanine, L-serine and L-leucine. Sucrose, trehalose, cellobiose, hydroxymethylglutarate, citrate, mesaconate, ethanol, *para*-hydroxybenzoate, L-tryptophan and nicotinate not utilized (Davis *et al.*, 1970).

Obligately aerobic. Optimal temperature for growth 28 C. No growth at 41 C.

Isolated from soil by enrichments in mineral media and atmosphere containing H_2, O_2 and CO_2.

The G + C content of the DNA is 61.7–63.8 moles % (buoyant density).

Type strain: ATCC 11228.

23. **Pseudomonas saccharophila** Doudoroff 1940, 59.

sac.cha.ro.phi'la. Gr. n. *sacchar* sugar; Gr. adj. *philus* loving; M.L. fem.adj. *saccharophila* sugar loving.

Rods, 0.5 by 3–4 μm, singly, in pairs and small clumps. Motile with polar monotrichous flagellation.

Young colonies colorless, but with aging the center becomes characteristically brownish red. Mucoid and non-pigmented clones may occur. No diffusible pigments produced.

Chemoorganotrophic and facultatively chemolithotrophic, using the oxidation of molecular hydrogen, but not of carbon monoxide or thiosulfate, as a source of energy for autotrophic growth. Denitrification does not occur, but nitrite produced from nitrate in organic media.

Not lipolytic.

Organic growth factors not required. Autotrophic growth occurs in atmospheres containing H_2, CO_2 and O_2. At least 47 different organic compounds can serve as sole carbon sources for heterotrophic growth; include L-arabinose, galactose, sucrose, trehalose, raffinose, cellobiose and quinate. The utilization of glucose, fructose, mannose and D-arabinose requires mutation. Mannitol and ethanol not utilized (Davis *et al.*, 1970).

Obligately aerobic. Optimal temperature for growth, 30 C. No growth at 41 C.

Sensitive to 10 units of penicillin/ml.

Isolated by enrichment in mineral medium inoculated with mud from a stagnant pool, and incubated in an atmosphere containing 83% hydrogen, 2% oxygen and 15% CO_2.

The G + C content of the DNA is 68.9 moles % (buoyant density).

Type strain: ATCC 15946.

24. **Pseudomonas ruhlandii** (Packer and Vishniac) Davis in Davis, Doudoroff, Stanier and Mandel 1969, 385. (*Hydrogenomonas ruhlandii* Packer and Vishniac 1955, 216.)

ruh.lan'di.i. M.L. gen.n. *ruhlandii* of Ruhland; named for the German microbiologist, W. Ruhland, who studied the physiology of the "hydrogen bacteria."

Rods, 0.4–0.75 by 0.75–2.0 μm (mean dimensions, 0.5 by 1.1 μm), singly and occasionally in small groups. Motile with polar monotrichous flagellation.

No diffusible or cellular pigments produced.

Chemoorganotrophic and facultatively chemolithotrophic, using oxidation of molecular hydrogen, but not of carbon monoxide or thiosulfate, as a source of energy for autotrophic growth. Denitrification does not occur, but nitrite produced from nitrate.

Not lipolytic.

Organic growth factors not required. Autotrophic growth occurs in atmospheres containing H_2, O_2 and CO_2 when oxygen concentration less than 20%. Autotrophic growth not initiated when inoculum grown in organic media with 20% oxygen (Packer and Vishniac, 1955). At least 36 different organic compounds can serve as sole carbon sources for heterotrophic growth. These include hydroxymethylglutarate, 2,3-butylene glycol, mesaconate, histidine and nicotinate. L-Arabinose, malonate, mannitol, L-serine and L-tryptophan not utilized. Glucose utilized only after a long lag period. Other pentoses, hexoses and disaccharides not utilized by the holotype strain according to Davis *et al.* (1970), although mannose, galactose and sucrose are reported as carbon sources in the original description.

Obligately aerobic. Microaerophilic under autotrophic conditions.

Optimal temperature 28 C. No growth at 41 C.

Isolated from soil by enrichment for autotrophic "hydrogen bacteria."

Type strain: Davis 368; ATCC 15749.

25. **Pseudomonas flava** (Niklewski) Davis in Davis, Doudoroff, Stanier and Mandel 1969, 385. (*Hydrogenomonas flava* Niklewski 1910, 123.)

fla'va. L. fem.adj. *flava* yellow.

Rods *ca.* 1.5 μm in length. Motile by one, more rarely two, flagella, inserted polarly or subpolarly.

Colonies pale yellow, turning darker yellow after prolonged incubation, from production of intracellular carotenoid pigments.

Chemoorganotrophic and facultatively chemolithotrophic, using oxidation of molecular hydrogen, but not of carbon monoxide or thiosulfate as source of energy for autotrophic growth. Denitrification does not occur.

Not lipolytic.

Organic growth factors not required. Autotrophic growth occurs in atmospheres containing H_2, CO_2 and O_2, with partial pressures of oxygen not exceeding 0.08 atmospheres. Potential for autotrophic growth may be lost after prolonged cultivation on ordinary laboratory media (Kluyver and Manten, 1942). At least 36 different organic compounds can serve as sole carbon sources for heterotrophic growth under ordinary aerobic conditions; include L-arabinose, L-rhamnose, most hexose sugars, sucrose, trehalose, maltose, cellobiose and mannitol. Saccharate, malonate, hydroxymethylglutarate, mesaconate, *para*-hydroxybenzoate and pantothenate not utilized (for more complete list of substrates see Davis *et al.*, 1970).

Obligately aerobic. Microaerophilic only for autotrophic growth with hydrogen. Optimal temperature, *ca.* 30 C. No growth at 41 C.

Isolated from soil by enrichment in mineral media and atmospheres containing H_2, CO_2 and O_2.

The G + C content of the DNA is 67.3 moles % (buoyant density).

The above description is based on the properties of the proposed neotype (Kluyver and Manten, 1942). Similar strains not sensitive to partial pressure of 0.2 atmospheres of oxygen during autotrophic growth occur in nature (Davis *et al.*, 1970). This species may be confused with *Alcaligenes paradoxus* (Davis *et al.*, 1969), some strains of which are initially microaerophilic after isolation. It can also be confused with *P. palleronii*. Nutritional characters useful in distinguishing the three species are included in the above description, and others are reported by Davis *et al.* (1970).

Proposed neotype: Strain designated as *Hydrogenomonas flava* by Kluyver and Manten (1942) in the collection of Lab. voor Mikrobiologie, Delft, The Netherlands (Davis *et al.*, 1970, 385).

26. **Pseudomonas palleronii** Davis in Davis, Stanier, Doudoroff and Mandel 1970, 11.

pal.le.ro'ni.i. M.L. gen.n. *palleronii* of Palleroni; named for the Argentine microbiologist, N. J. Palleroni, who first isolated the organism.

Rods, 0.4 by 1.5–2.6 μm during exponential

phase, *ca.* 0.5 by 2.9 μm in stationary phase. Motile by one flagellum, more rarely two, inserted polarly or subpolarly.

Colonies pale yellow, turning darker yellow after prolonged incubation, from production of a characteristic complement of intracellular carotenoid pigments.

Chemoorganotrophic and facultatively chemolithotrophic, using oxidation of molecular hydrogen but not of carbon monoxide or thiosulfate as source of energy for autotrophic growth. Denitrification does not occur.

Lipolytic.

Growth factors not required. Can grow autotrophically in atmospheres containing H_2, CO_2 and O_2. At least 46 different organic compounds can serve as sole carbon sources for growth; include meso-inositol, all three isomers of tartrate, phenol, *meta*- and *para*-hydroxybenzoate, L-mandelate and D-alanine. The utilization of glucose may require a mutation in cultures isolated by enrichment for autotrophic "hydrogen bacteria." Sugars other than glucose (pentoses, methyl pentoses, hexoses and disaccharides), mannitol and sorbitol not utilized as carbon sources (Davis *et al.*, 1970).

Obligately aerobic. All strains studied can tolerate a partial pressure of 0.2 atmosphere of oxygen for autotrophic growth with hydrogen. Optical temperature, *ca.* 30 C. No growth at 41 C.

Isolated from soil and water by enrichment in minimal media and atmospheres containing H_2, O_2 and CO_2.

The G + C content of the DNA is *ca.* 66.8 moles % (buoyant density).

The species easily confused with *Alcaligenes paradoxus* Davis *et al.* 1969, and with *P. flava* (Niklewski) Davis *et al.* (1970) and related strains. The distinguishing characters are listed by Davis *et al.* (1970).

Type: Davis 1-6-1; Stanier 366; ATCC 17724.

Key to the species of Section IV

(See also Table 7.5)

A. Require one or more growth factors. Do not accumulate poly-β-hydroxybutyrate as intracellular carbon reserve.

> 27. *P. maltophilia*
> 28. *P. vesicularis*
> 29. *P. diminuta*

Further Comments on Section IV

Two species (*P. diminuta* and *P. vesicularis*) appear to be closely related to each other but not to *P. maltophilia*. These two species have some similarities to the genus *Gluconobacter*.

It should be kept in mind that growth factors requiring strains of species included in the first three sections are occasionally encountered. Such strains can be sometimes allocated to the proper species on the basis of some outstanding phenotypic traits.

Description of the species of Section IV, genus **Pseudomonas**

27. **Pseudomonas maltophilia** Hugh and Ryschenkow 1960, 78.

mal.to.phi′li.a. Anglo-Saxon n. *malt;* Gr. n. *philia* friend. M. L. fem.n. *maltophilia* friend of malt.

Straight or slightly curved rods, 0.5 by 1.5 μm, singly or in pairs. Do not accumulate granules of poly-β-hydroxybutyrate as intracellular carbon reserve. Motile with polar multitrichous flagellation.

Colonies may be yellowish; the yellow color not due to carotenoid pigments.

Denitrification does not occur.

Strongly lipolytic.

Methionine required as the only organic growth factor. Limited in its nutritional spectrum. Individual strains used between 23 and 28 of 146 different organic compounds tested as principal carbon sources (Stanier *et al.*, 1966). These included glucose, mannose, sucrose, trehalose, maltose, cellobiose, lactose and salicin. Ethanol, glycerol, mannitol, aromatic compounds and amines not utilized. Under aerobic conditions acid readily produced in complex media with maltose but not with glucose (Hugh and Ryschenkow, 1960). Nitrate not used as nitrogen source.

Obligately aerobic. Do not grow at 4° C or 41° C. Optimal temperature for growth, *ca.* 35° C.

Most strains isolated from clinical specimens. Also found in water, milk and frozen food.

The G + C content of the DNA is *ca.* 67 moles % (buoyant density).

Type strain: Hugh 810-2; ATCC 13637; NCIB 9203; NCTC 10257; NRC 729.

TABLE 7.5

Characteristics of species 27–29 of genus **Pseudomonas**

	27. *P. maltophilia*	28. *P. vesicularis*	29. *P. diminuta*
Flagellation	Multitrichous	Monotrichous	Monotrichous
Color of colonies	Yellow	Yellow	None
Growth factors	Methionine	Pantothenate Biotin Cyanocobalamin	Pantothenate Biotin Cyanocobalamin Cystine
Oxidase reaction	Negative	Weak	Strong
Hydrolysis of:			
Starch	−[a]	−	−
Gelatin	+[a]	−	−
Tween 80	+	−	−
Poly-β-hydroxybutyrate	−	−	−
Carbon sources used for growth:			
Glucose	+	+	−
Cellobiose	+	+	−
DL-β-Hydroxybutyrate	−	+	+
Histidine	+	−	+
Pantothenate	−	−	+

[a] + = positive for 90% or more of strains; − = negative for 90% or more of strains.

28. **Pseudomonas vesicularis** (Büsing, Döll and Freytag) Galarneault and Leifson 1964, 167. (*Corynebacterium vesiculare* Büsing *et al.*, 1953, 76; *Pseudomonas vesiculare* (Büsing *et al.*) Galarneault and Leifson 1964, 167.)

ve.si.cu.la′ris. M.L. fem. adj. *vesicularis* pertaining to a vesicle.

Rods, *ca.* 0.5 by 1.0–4.0 μm. Accumulate poly-β-hydroxybutyrate granules as intracellular carbon reserve especially in nitrogen-deficient media with DL-β-hydroxybutyrate as carbon source. Motile by a single polar flagellum of very short wave length (0.6–1.0 μm).

Colonies are yellow because of intracellular carotenoid pigments.

Denitrification does not occur and nitrite not produced from nitrate. Oxidase reaction weakly positive. Cytochromes of *a*, *b* and *c* types present. The reduced/oxidized difference spectrum shows peak at 628 nm, characteristic of a_2 component.

Polyethylene sorbitan monooleate (Tween 80) not hydrolyzed. Egg yolk reaction negative.

Pantothenate, biotin, cyanocobalamin required as growth factors. Cystine not required. Growth slow in complex media and very slow in chemically defined media. Very few organic compounds can serve as principal carbon source for growth. Only

18 of 146 compounds tested are used by the two strains studied, and two other compounds used by one or other strain. Utilizable compounds include glucose, galactose, maltose, cellobiose, acetate, butyrate, succinate, DL-β-hydroxybutyrate, L-malate, D- and L-alanine and L-aspartate. Among compounds not utilized are fructose, gluconate, L-histidine and pantothenate. Can grow with 5% ethanol, producing acid. Isopropanol oxidized to acetone, but not used as carbon source. Ammonium salts, but not nitrates used as principal nitrogen source (Ballard *et al.*, 1968).

Obligately aerobic. Optimal temperature for growth *ca.* 30° C. Can grow at 37° C but not at 41° C. No growth at 4° C.

Type strain isolated from medicinal leech (*Hirudo medicinalis*). The other strain studied was isolated from a stream.

The G + C content of the DNA is 65.8 moles % (buoyant density).

Type strain: ATCC 11426; NCMB 1945.

29. **Pseudomonas diminuta** Leifson and Hugh 1954, 68.

di.mi.nu′ta. L. adj. *minutus* small; M. L. fem. adj. *diminuta* defective, minute.

Rods, *ca.* 0.5 by 1.0–4.0 μm. Accumulate poly-β-hydroxybutyrate granules as intracellular carbon reserve especially in nitrogen-deficient media with DL-β-hydroxybutyrate as carbon source. Motile by a single polar flagellum of very short wave length (0.6–1.0 μm).

Denitrification does not occur and nitrite rarely produced from nitrate. Cytochromes of a, b and c types present. The reduced/oxidized difference spectrum shows peak at 628 nm, characteristic of a_2 component.

Polyethylene sorbitan monooleate (Tween 80) not hydrolyzed. Egg yolk reaction negative.

Pantothenate, biotin, cyanocobalamin and cystine (or methionine) required as growth factors. Growth slow in complex media and very slow in chemically defined media. Few organic compounds can serve as principal carbon sources for growth. Only 11 of 146 compounds tested used by all strains studied, and 13 others by some strains.

The utilizable compounds include acetate, butyrate, DL-β-hydroxybutyrate, D- and L-alanine and pantothenate. Among compounds not utilized are all carbohydrates, gluconate, succinate, fumarate, L-malate and histidine. With few exceptions strains can utilize ethanol as sole carbon source. Such strains can grow with ethanol concentrations up to 5% in complex, but not in defined, media and produce acid (Ballard *et al.*, 1968).

Obligately aerobic. Optimal temperature for growth *ca.* 30° C. Can grow at 37° C and some strains at 41° C. No growth at 4° C.

Type strain was isolated from water; all other known strains, from clinical specimens.

The G + C content of the DNA, 66.3–67.3 moles % (buoyant density).

Type strain: RH342 ATCC 11568; NCTC 8545; NCIB 9393.

ADDENDA TO THE GENUS **PSEUDOMONAS**

General Comments

The preceding treatment of the genus *Pseudomonas* describes only those species of which the phenotypes have been extensively characterized. These include about 10% of the named species. The following four addenda list other nomenspecies which have been assigned to this genus, as well as certain nomenspecies originally assigned to other genera, but which appear to conform to the present definition of *Pseudomonas*, insofar as they have been characterized.

Addendum I lists nomenspecies of *Pseudomonas* which have been incompletely described, but which appear to conform to the generic definition. Strains (holotypes, cotypes, neotypes, proposed working types) of all these nomenspecies exist in culture collections. In principle, therefore, a determination of their taxonomic status is possible.

Addendum II lists nomenspecies of *Pseudomonas* which have been incompletely described, and for which no authentic strains appear to have been preserved. In our judgment, the great majority constitute *nomina nuda*.

Addendum III lists nomenspecies which heretofore have been assigned to the genus *Pseudomonas*, but which are definitely known to possess characters not in accord with the generic definition as used here.

Addendum IV lists nomenspecies heretofore assigned to other genera that are either not recognized in the present edition of THE MANUAL (e.g. *Cellvibrio*), or which have been redefined in a manner which eliminates the species in question (e.g. *Vibrio*). The descriptions of most of these species are incomplete; but, insofar as their properties have been determined, they appear to conform to the definition of *Pseudomonas* used here.

Finally, as an aid to the ready location of newly isolated *Pseudomonas* strains, we have included an Index to distinctive characters which have been described for all species included in the main taxonomic treatment of the genus, as well as in the four addenda.

Where cultures are available in culture collections, the strain numbers are indicated.

Species that were described in the 7th Edition of THE MANUAL are indicated by the page number (e.g. M. 121) immediately after the attribution.

Addendum I: Species represented in culture collections.

30. *Pseudomonas aceris* (Ark) Starr and Burkholder 1942, 601. (M. 121). Cotype ATCC 10853.

31. *Pseudomonas alboprecipitans* Rosen 1922, 383. (M. 149). ATCC 19860; NCPPB 1011.

32. *Pseudomonas aminovorans* den Dooren de Jong 1926, 161. NCIB 9039.

33. *Pseudomonas andropogonis* (Smith) Stapp 1928, 27. (M. 149). NCPPB 933; ATCC 23060.

34. *Pseudomonas apii* Jagger 1921, 186. (M. 123). ATCC 8722 and 9654. (Synonym: *Pseudomonas jaggeri* Stapp 1928.)

35. *Pseudomonas aromatica* Migula 1900, 880. NCIB 9043.

36. *Pseudomonas asplenii* (Ark and Tompkins) Săvulescu 1947, 11. (M. 124). Cotype ATCC 10203.

37. *Pseudomonas azotocolligans* Anderson 1955, 132. Type ATCC 12417; NCIB 9390.

38. *Pseudomonas azotogensis* Voets and Debacker 1956, 40. Type ATCC 15970.

39. *Pseudomonas beijerinckii* Hof 1935, 156. (M. 121). Cotype ATCC 19372; NCIB 9041.

40. *Pseudomonas berberidis* (Thornberry and Anderson) Stapp 1935, 407. (M. 124). ATCC 13454.

41. *Pseudomonas boreopolis* Gray and Thornton 1928, 92. (M. 101). ATCC 13476 and 15452.

42. *Pseudomonas cannabina* Sutić and Dowson 1959, 311. ATCC 13436; NCPPB 2069.

43. *Pseudomonas caryocyanea* (Dupaix) Dupaix 1933, 13. Type ATCC 19373; NCIB 9031.

44. *Pseudomonas cattleyae* (Pavarino) Săvulescu 1947, 11. (M. 148). ATCC 10200.

45. *Pseudomonas citronellolis* Seubert 1960, 428. Type ATCC 13674.

46. *Pseudomonas coenobios* ZoBell and Upham 1944, 272. ATCC 14402.

47. *Pseudomonas creosotensis* O'Neill, Drisko and Hochmann 1961, 473. Type ATCC 14582.

48. *Pseudomonas dacunhae* Gray and Thornton 1928, 90. (M. 113). ATCC 13261.

49. *Pseudomonas denitrificans* Bergey *et al.* 1923, 131. (M. 116). Proposed neotype ATCC 19244.

50. *Pseudomonas echinoides* Heumann 1962, 343. (See also Heumann and Marx, 1964.) Type ATCC 14820; NCIB 9420.

51. *Pseudomonas elongata* Humm 1946, 60. Cotype ATCC 10144; NCMB 1141.

52. *Pseudomonas eriobotryae* (Takimoto) Dowson 1943, 10. (M. 151). NCPPB 2331.

53. *Pseudomonas excibis* Steinhaus, Batey and Boerke 1956, 500. ATCC 12293.

54. *Pseudomonas extorquens* (Bassalik), Krasil'nikov 1949, 368. NCIB 9399.

55. *Pseudomonas flectens* Johnson 1956, 155. Type ATCC 12775.

56. *Pseudomonas fragi* (Eichholz) Gruber 1905, 122. (M. 110). Proposed neotype ATCC 4973.

57. *Pseudomonas hibiscicola* Moniz 1963, 177. NCPPB 1683; ATCC 19867.

58. *Pseudomonas huttiensis* Leifson 1962, 167. Type ATCC 14670; NCIB 9462.

59. *Pseudomonas indigofera* (Voges) Migula 1900, 950. ATCC 14036 and 19706; NCIB 9441.

60. *Pseudomonas lanceolata* Leifson 1962, 166. Type ATCC 14669; NCIB 9461.

61. *Pseudomonas lapsa* (Ark) Starr and Burkholder 1942, 601. (M. 130). ATCC 10859.

62. *Pseudomonas maculicola* (McCulloch) Stevens 1913, 28. (M. 125). ATCC 11781.

63. *Pseudomonas mangiferaindicae* Patel, Moniz and Kulkarni 1948, 189. (M. 126). Cotype ATCC 11637; NCPPB 490.

64. *Pseudomonas melanogenum* (*sic.*) Iizuka and Komagata 1963, 73. Type ATCC 17806.

65. *Pseudomonas melophthora* Allen and Riker 1932, 569. (M. 149). NCPPB 461, 462.

66. *Pseudomonas mildenbergii* Bergey *et al.* 1930, 172. (M. 109). CCM 233 (?).

67. *Pseudomonas mucidolens* Levine and Anderson 1932, 344. Cotype ATCC 4685; NCIB 9394.

68. *Pseudomonas myxogenes* Fuhrmann 1907, 462. (M. 103). CCM 221 (?).

69. *Pseudomonas nigrifaciens* White 1940, 640. (M. 117). (See also Norton and Jones, 1968.) Cotype ATCC 19375; NCIB 8614.

70. *Pseudomonas oleovorans* Lee and Chandler 1941, 378. (M. 113). Cotype ATCC 8062.

71. *Pseudomonas oryzicola* Klement 1954, 265. ATCC 19874. (Possibly synonym of *Pseudomonas marginalis;* see Goto, 1965.)

72. *Pseudomonas oxalaticus* Khambata and Bhat 1953, 507. Cotype ATCC 11883; NCIB 8642.

73. *Pseudomonas passiflorae* (Reid) Burkholder 1948, 138. (M. 144). ATCC 19876; NCPPB 224.

74. *Pseudomonas pastinaceae* Burkholder 1960, 280. Cotype NCPPB 806.

75. *Pseudomonas pavonaceae* Levine and Soppeland 1926, 41. (M. 105). Cotype ATCC 951.

76. *Pseudomonas perfectomarinus* (*sic*) ZoBell and Upham 1944, 277. ATCC 14405.

77. *Pseudomonas perlurida* Kellerman, McBeth, Scales and Smith 1913, 516. (M. 181). Cotype ATCC 490; NCIB 9325.

78. *Pseudomonas pictorum* Gray and Thornton 1928, 89. (M. 182). Type CCEB 322; NCIB 9152.

79. *Pseudomonas primulae* (Ark and Gardner) Starr and Burkholder 1942, 601. (M. 122). Cotype ATCC 19306; NCPPB 133.

80. *Pseudomonas reptilivora* Caldwell and Ryerson 1940, 335. (M. 100). Cotype ATCC 14836.

81. *Pseudomonas resinovorans* Delaporte, Raynaud and Daste 1961, 1075. Type ATCC 14235.

82. *Pseudomonas rhodos* Heumann 1962, 342. Type ATCC 14821; NCIB 9421.

83. *Pseudomonas ribicola* Bohn and Maloit 1946, 288. (M. 138). Cotype ATCC 13456; NCPPB 963.

84. *Pseudomonas rubrisubalbicans* (Christopher and Edgerton) Krasil'nikov 1949, 379. (Synonym: *Xanthomonas rubrisubalbicans* (Christoper and Edgerton) Săvulescu 1947, 55. See M. 176; possible synonym: *Xanthomonas floridana* Bourne 1970).

85. *Pseudomonas segnis* Goresline 1933, 452. (M. 183). JFCC IAM 1131.

86. *Pseudomonas septica* Bergey *et al.* 1930, 169. (M. 106). ATCC 14545.

87. *Pseudomonas sesami* Malkoff 1906, 665. (M. 135). ATCC 19879; NCPPB 1016.

88. *Pseudomonas setariae* (Okabe) Săvulescu 1947, 11. (M. 135). ATCC 19882; NCPPB 1392.

89. *Pseudomonas spinosa* Leifson 1962, 89. Type ATCC 14606.

90. *Pseudomonas stizolobii* (Wolf) Stapp 1935, 407. (M. 142). Suggested working type ATCC 19309; NCPPB 1024.

91. *Pseudomonas striafaciens* (Elliott) Starr and Burkholder 1942, 601. (M. 132). Cotype ATCC 10730; NCPPB 1898.

92. *Pseudomonas syncyanea* (Ehrenberg) Migula 1895, 29. (M. 106). ATCC 9979.

93. *Pseudomonas synxantha* (Ehrenberg) Breed 1948, 700. (M. 104). Proposed neotype ATCC 9890.

94. *Pseudomonas taetrolens* Haynes 1957, 108 (M. 108). ATCC 4683; NCIB 9396.

95. *Pseudomonas trifolii* Huss 1907, 68. (M. 180). ATCC 14537.

96. *Pseudomonas ulmi* Sutić and Tešić 1958, 23. ATCC 19883; NCPPB 632.

97. *Pseudomonas viburni* (Thornberry and Anderson) Stapp 1935, 407. (M. 141). Cotype ATCC 13458.

98. *Pseudomonas viridiflava* (Burkholder) Clara 1932, 111. (M. 133). Cotype NRRL B-895.

99. *Pseudomonas viridilivida* (Brown) Bergey *et al.* 1948, 115. (M. 122). Cotype ATCC 512.

100. *Pseudomonas vitiswoodrowii* Patel and Kulkarni 1951, 132. (M. 145). Cotype ATCC 11636.

101. *Pseudomonas washingtoniae* (Pine) Elliott 1951, 100. (M. 136). ATCC 13459; NCPPB 967.

102. *Pseudomonas woodsii* (Smith) Stevens 1925, 39. Suggested working type ATCC 19311.

103. *Pseudomonas xanthe* Zettnow 1915, 220. (M. 180). ATCC 8375; NCIB 9328.

Species incertae sedis

Addendum II: Species not known to be represented in culture collections.

104. *Pseudomonas agaramicus* Araki and Arai 1954, 12.

105. *Pseudomonas alboflava* Ueda, Ishikawa, Itami and Asai 1952, 37.

106. *Pseudomonas aleuritidis* (McCulloch and Demaree) Stapp 1935, 408. (M. 138).

107. *Pseudomonas aloes* (Passalacqua) Krasil'nikov 1949, 379.

108. *Pseudomonas ambigua* (Wright) Chester 1901, 308. (M. 112).

109. *Pseudomonas ananas* Serrano 1934, 355. (M. 134).

110. *Pseudomonas antimycetica* Thaysen and Thaysen 1955, 638. NCIB 8641.

111. *Pseudomonas arctica* (Issatchenko) Krasil'nikov 1949, 403.

112. *Pseudomonas arguta* McBeth 1916, 465. (M. 182).

113. *Pseudomonas arvilla* Gray and Thornton 1928, 90. (M. 113).

114. *Pseudomonas astatica* Kuchar 1954, 513.

115. *Pseudomonas astragali* (Takimoto) Săvulescu 1947, 11. (M. 146).

116. *Pseudomonas aucubicola* Trapp 1936, 265.

117. *Pseudomonas aurantiaca* Nakhimovskaya 1948, 58 NCIB 10066 and 10068.

118. *Pseudomonas beaufortensis* Humm 1946, 58.

119. *Pseudomonas bowlesiae* (Lewis and Watson) Dowson 1943, 9. (M. 134).

120. *Pseudomonas caesiae* (Kellerman *et al.*) Krasil'nikov 1949, 338.

121. *Pseudomonas calciprecipitans* Molisch 1925, 133. (M. 119).

122. *Pseudomonas calcis* (Drew) Kellerman and Smith 1914, 402. (M. 119).

123. *Pseudomonas calendulae* (Takimoto) Dowson 1943, 9. (M. 140).

124. *Pseudomonas castaneae* (Kawamura) Săvulescu 1947, 11. (M. 144).

125. *Pseudomonas caudata* (Wright) Conn in Conn and Bright 1919, 313. (M. 180).

126. *Pseudomonas caulivorae* (Prillieux and Delacroix) Krasil'nikov 1949, 356.

127. *Pseudomonas cerevisiae* Fuhrmann 1906, 309. (M. 181).

128. *Pseudomonas chitinovora* Sreenivasan 1955, 270.

129. *Pseudomonas chryothasia* Campbell and Williams 1951, 901.

130. *Pseudomonas cocovenenans* van Damme, Johannes, Cox and Berends 1960, 255.

131. *Pseudomonas cohaerens* (Wright) Chester 1901, 312. (M. 112).

132. *Pseudomonas colurnae* (Thornberry and Anderson) Burkholder 1948, 139. (M. 146).

133. *Pseudomonas corylli* (Brezezinski) Krasil'nikov 1949, 351.

134. *Pseudomonas cumini* (Kovachevski) Dowson 1943, 10. (M. 129).

135. *Pseudomonas cyanoides* (Angst) Krasil'nikov 1949, 338.

136. *Pseudomonas desaiana* (Burkholder) Săvulescu 1947, 11. (M. 129).

137. *Pseudomonas effusa* Kellerman, McBeth, Scales and Smith 1913, 515. (M. 102).

138. *Pseudomonas elastica* (Ellen) Krasil'nikov 1949, 361.

139. *Pseudomonas elegans* Araki and Arai 1954, 15.

140. *Pseudomonas ephemerocyanea* Fuller and Norman 1943, 274. (M. 102).

141. *Pseudomonas epstienii* (Peshkov) Krasil'nikov 1949, 403. NCIB 9538.

142. *Pseudomonas erodii* Lewis 1914, 231. (M. 129).

143. *Pseudomonas erythra* Fuller and Norman 1943, 276. (M. 110).

144. *Pseudomonas fabae* (Yu) Burkholder 1948, 139. (M. 145).

145. *Pseudomonas ferrugineum* Pillai 1938, 316.

146. *Pseudomonas fucicola* (Waksman *et al.*) Krasil'nikov 1949, 362.

147. *Pseudomonas gardeniae* (Burkholder and Pirone) Dowson 1943, 12. (M. 143).

148. *Pseudomonas gelatica* (Gran) Bergey *et al.* 1930, 175. (M. 119).

149. *Pseudomonas gelidicola* Kadota 1951, 58.

150. *Pseudomonas genevensis* Marca 1927, 15.

151. *Pseudomonas haemorrulcogenes* (Castellani) Castellani 1951, 77.

152. *Pseudomonas halestorga* Elazari-Volcani 1940, 82. (M. 120).

153. *Pseudomonas hypothermis* ZoBell and Upham 1944, 276.

154. *Pseudomonas ichthyodermis* (Wells and ZoBell) ZoBell and Upham 1944, 276. (M. 118).

155. *Pseudomonas iridicola* (Takimoto) Stapp 1935, 408. (M. 146).

156. *Pseudomonas jankei* Kuchar 1954, 517.

157. *Pseudomonas kyotoensis* Araki and Arai 1954, 14.

158. *Pseudomonas lacunogenes* Goresline 1933, 447. (M. 182).

159. *Pseudomonas lacustris* Marca 1927, 14.

160. *Pseudomonas lasia* Fuller and Norman 1943, 275. (M. 115).

161. *Pseudomonas lauracearum* Harvey 1952, 197.

162. *Pseudomonas levistici* Osterwalder 1909, 264. (M. 147).

163. *Pseudomonas lignicola* Westerdijk and Buisman 1929, 51. (M. 150).

164. *Pseudomonas ligustri* (d'Oliveira) Săvulescu 1947, 11. (M. 134).

165. *Pseudomonas limnophila* Kuchar 1954, 513.

166. *Pseudomonas marinodenitrificans* Venkataraman and Sreenivasan 1955, 32.

167. *Pseudomonas marinoglutinosa* (ZoBell and Allen) ZoBell 1943, 45 (M. 118).

168. *Pseudomonas marinopersica* ZoBell and Upham 1944, 275.

169. *Pseudomonas martyniae* (Elliott) Stapp 1928, 278. (M. 130).

170. *Pseudomonas maublancii* (Foex and Lansade) Săvulescu 1947, 11. (M. 147).

171. *Pseudomonas membranoformis* (ZoBell and Allen) ZoBell 1943, 45. (M. 118).

172. *Pseudomonas membranula* ZoBell and Upham 1944, 270.

173. *Pseudomonas mephitica* Claydon and Hammer 1939, 254. (M. 111).

174. *Pseudomonas minuscula* McBeth 1916, 467.

175. *Pseudomonas mira* McBeth 1916, 468. (M. 117).

176. *Pseudomonas mycolytica* Krasil'nikov 1949, 340.

177. *Pseudomonas mycophaga* Krasil'nikov 1949, 392.

178. *Pseudomonas neritica* ZoBell and Upham 1944, 255.

179. *Pseudomonas nivalis* Szilvinyi 1936, 217.

180. *Pseudomonas obscura* ZoBell and Upham 1944, 274.

181. *Pseudomonas oceanica* ZoBell and Upham 1944, 266.

182. *Pseudomonas panicimiliacei* (Ikata and Yamauchi) Săvulescu 1947, 11. (M. 151).

183. *Pseudomonas pellucidula* (Harrison) Krasil'nikov 1949, 402.

184. *Pseudomonas periphyta* ZoBell and Upham 1944, 276.

185. *Pseudomonas perolens* (Turner) Szybalski 1950, 733. (M. 111).

186. *Pseudomonas pestai* Kuchar 1953, 512.

187. *Pseudomonas petasitis* (Takimoto) Săvulescu 1947, 11. (M. 150).

188. *Pseudomonas pieris* Magrou 1953, 448.

189. *Pseudomonas pleomorpha* ZoBell and Upham 1944, 275.

190. *Pseudomonas polygoni* (Thornberry and Anderson) Burkholder 1948, 140. (M. 147).

191. *Pseudomonas propanica* Dostálek 1954, 169.

192. *Pseudomonas pseudoviolacea* Migula 1900, 943.

193. *Pseudomonas pseudozoogloeae* (Honing) Stapp 1928, 274. (M. 131).

194. *Pseudomonas pyri* (Djakova) Gorlenko 1961.

195. *Pseudomonas radiciperda* (Zhavoronkova) Săvulescu 1947, 12. (M. 147).

196. *Pseudomonas rathonis* Gray and Thornton 1928, 90. (M. 114).

197. *Pseudomonas roseola* Humm 1946, 62.

198. *Pseudomonas rubigenosa* (Catiano) Krasil'nikov 1949, 364.

199. *Pseudomonas rubra* (Zimmermann) Krasil'nikov 1949, 367.

200. *Pseudomonas rugosa* (Wright) Chester 1901, 323. (M. 109).

201. *Pseudomonas saliciperda* Lindeijer 1932, 23. (M. 151).

202. *Pseudomonas salopia* Gray and Thornton 1928, 91. (M. 114).

203. *Pseudomonas sapolactica* (Eichholz) de Rossi 1927, 693.

204. *Pseudomonas sessilis* ZoBell and Upham 1944, 259.

205. *Pseudomonas smaragdina* (Mez) Migula 1900, 890. (M. 107).

206. *Pseudomonas solaniolens* Paine 1923, 77.

207. *Pseudomonas sphingidis* (White) Krasil'nikov 1949, 410.

208. *Pseudomonas stereotropis* ZoBell and Upham 1944, 272.

209. *Pseudomonas subcreta* McBeth and Scales 1913, 37. (M. 182).

210. *Pseudomonas subrubra* Campbell and Williams 1951, 902.

211. *Pseudomonas tralucida* Kellerman, McBeth, Scales and Smith 1913, 517. (M. 115).

212. *Pseudomonas viciae* Uyeda 1915, 845. (M. 142).

213. *Pseudomonas villosa* (Keck) Pribram 1933, 49.

214. *Pseudomonas volfi* Schäperclaus 1930, 367.

215. *Pseudomonas wieringae* (Elliott) Săvulescu 1947, 11.

Species incertae sedis

Addendum III: Species heretofore assigned to *Pseudomonas* but possessing characters not in accord with the generic description. (Properties at variance with the generic description are given in each case.) Species included in the genus *Xanthomonas* in Breed *et al.* (1957) and in Elliott (1951) are excluded from this list.

216. *Pseudomonas atlantica* Humm 1946, 58. ATCC 19262. The G + C content of the DNA of this strain is *ca.* 43.5 moles % (from buoyant density) (Mandel, 1966).

217. *Pseudomonas betle* (Ragunathan) Săvulescu 1947, 11. (M. 137). ATCC 19861. Non-motile.

218. *Pseudomonas caviae* Scherago 1936, 83. (M. 101). ATCC 15468. Facultative anaerobe.

219. *Pseudomonas cissicola* (Takimoto) Burkholder 1948, 134. (M. 140). Non-motile.

220. *Pseudomonas corallina* Humm 1946, 59. Non-motile.

221. *Pseudomonas cruciviae* Gray and Thornton 1928, 91. (M. 114). ATCC 13262. The G + C content of the DNA is *ca.* 36–37.5 moles % (T_m and buoyant density) (Hill, 1966).

222. *Pseudomonas droebachensis* (Lundestad) Stanier 1941, 544. Non-motile.

223. *Pseudomonas floridana* Humm 1946, 60. Non-motile.

224. *Pseudomonas inertia* Humm 1946, 61. Non-motile.

225. *Pseudomonas iridescens* Stanier, 1941, 543. (M. 120). Non-motile.

226. *Pseudomonas iridis* van Hall 1902, 116. Facultative anaerobe.

227. *Pseudomonas musae* Gäumann 1921, 58. Gram-positive.

228. *Pseudomonas natriegens* Payne, Eagon and Williams 1961, 121 Type ATCC 14048. Capable of fermenting sugars (Eagon and Cho, 1965).

229. *Pseudomonas piscicida* (Bein) Buck, Myers and Leifson 1963, 1125. The G + C content of the DNA is 43.4–45.5 moles % (buoyant density) (Hill, 1966).

230. *Pseudomonas pomi* Cole 1959, 607. Non-motile.

231. *Pseudomonas putrefaciens* (Derby and Hammer) Long and Hammer 1941, 176. Cotype ATCC 8071. Capable of fermenting sugars (Levin, 1968).

The G + C content of the DNA is *ca.* 43.5–45.5 moles % (T_m and buoyant density) (Hill, 1966).

232. *Pseudomonas rhizoctonia* (Thomas) Stevens 1925, 34. Non-motile.

233. *Pseudomonas riboflavina* Foster 1944, 30. Holotype ATCC 9526. Non-motile.

234. *Pseudomonas rubescens* Pivnick 1954, 42. Cotype ATCC 12099. The G + C content of the DNA is *ca.* 46 moles % (melting point) (Rosypal and Rosypalová, 1966).

235. *Pseudomonas seminum* Cayley 1917, 461. (M. 145). Gram-positive; facultative anaerobe.

Species incertae sedis

Addendum IV: Species heretofore assigned to other genera but apparently conforming to the present description of *Pseudomonas*.

236. *Alginomonas alginica* (Waksman *et al.*) Kass, Lid and Molland 1945, 9. (M. 204).

237. *Alginomonas nonfermentans* Kass *et al.* 1945, 9. (M. 202).

238. *Alginomonas fucicola* (Waksman *et al.*) Kass *et al.* 1945, 9. (M. 203).

239. *Cellfalcicula fusca* Winogradsky 1929, 622. (M. 253).

240. *Cellfalcicula mucosa* Winogradsky 1929, 621. (M. 252).

241. *Cellfalcicula viridis* Winogradsky 1929, 616. (M. 252).

242. *Cellvibrio flavescens* Winogradsky 1929, 608. (M. 251).

243. *Cellvibrio fulvus* Stapp and Bortels 1934, 42. (M. 251). Cotype ATCC 12120; NCIB 8634.

244. *Cellvibrio ochraceus* Winogradsky 1929, 601. (M. 251).

245. *Cellvibrio vulgaris* Stapp and Bortels 1934, 44. (M. 252). Cotype ATCC 12209.

246. *Comamonas terrigena* (Gunther) Hugh 1962, 34. (See *Vibrio percolans* Mudd and Warren 1923, 447. (M. 244); and Stanier *et al.*, 1966.) (Paratype of *Vibrio percolans* ATCC 8461; NCIB 8193.)

247. *Hydrogenomonas carboxydovorans* Kistner 1954, 186.

248. *Vibrio agar-liquefaciens* (Gray and Chalmers) Bergey *et al.* 1934, 119. (M. 239).

249. *Vibrio agarlyticus* Cataldi 1940, 372.

250. *Vibrio andoii* Aoi and Orikura 1928, 331. (M. 240).

251. *Vibrio avidus* Humm 1946, 54.

252. *Vibrio beijerinckii* Stanier 1941, 539. (Not *Pseudomonas beijerinckii* Hof 1935, 156.) (M. 240). Suggested working type, strain 224 of Liston (1961).

253. *Vibrio cuneatus* Gray and Thornton 1928, 92. (M. 243). Holotype ATCC 6972.

254. *Vibrio cyclosites* Gray and Thornton 1928, 92. (M. 242). Holotype ATCC 14635.

255. *Vibrio fortis* Humm 1946, 55.

256. *Vibrio frequens* Humm 1946, 56.

257. *Vibrio fuscus* Stanier 1941, 540. (M. 241).

258. *Vibrio granii* (Lundestad) Stanier 1941, 538. (M. 241).

259. *Vibrio halonitrificans* Smith 1938, 33.

260. *Vibrio marinopraesens* ZoBell and Upham 1944, 256.

261. *Vibrio neocistes* Gray and Thornton 1928, 92. (M. 242). Cotype ATCC 14636; NCIB 2582.

262. *Vibrio notus* Humm 1946, 56.

263. *Vibrio oxaliticus* Bhat and Barker 1948, 360. (M. 242).

264. *Vibrio stanieri* Humm 1946, 57. (Not *Pseudomonas stanieri* Mandel 1966.)

265. *Vibrio turbidus* Humm 1946, 57.

INDEX TO SOME DISTINCTIVE CHARACTERS DESCRIBED FOR ORGANISMS LISTED IN THE MAIN TAXONOMIC TREATMENT AND ADDENDA

(The species are designated by their numbers.)

I. Special aspects of morphology and pigmentation.
- A. Cell morphology.
  1. Cell shape definitely vibrioid in exponentially growing cultures: 242, 243, 244, 245, 246, 248, 249, 250, 251, 252, 253, 254, 255, 256, 257, 258, 259, 260, 261, 262, 263, 264, 265.
  2. Spindle-shaped cells with pointed ends: 239, 240, 241.
  3. Cells polarly piliated, resulting in their aggregation to form rosettes: 50.
- B. Flagellar structure and insertion.
  1. Polar flagella of short wave length: 28, 29.
  2. Occasional lateral flagella, usually of short wave length, in addition to polar flagella: 8, 9, 218.
- C. Pigmentation.
  1. Diffusible pigments produced, or medium discolored.
     a. Fluorescent pigments: 1, 2, 3, 4, 5, 6, 7, 30, 34, 36, 40, 61, 62, 66, 68, 79, 80, 86, 87, 91, 92, 97, 98, 99, 106, 109, 116, 117, 119, 120, 123, 126, 133, 135, 136, 137, 138, 142, 163, 169, 193, 205, 212, 215, 219, 239.
     b. Non-fluorescent pigments:
        Yellow, orange or green: 4, 5, 14, 15, 16, 49, 83, 88, 93, 101, 117, 164, 192, 200, 217, 257.
        Blue, red, violet or purple: 1, 15, 43, 92, 140.
        Brown or black: 1, 16, 17, 21, 69, 75, 146, 147, 238.
  2. Pigments associated with the colonies.
        Yellow or orange: 5, 9, 12, 25, 26, 27, 28, 64, 66, 77, 78, 85, 95, 103, 112, 125, 127, 138, 153, 158, 222, 224, 225, 229, 242, 243.
        Green, blue, purple or violet: 4, 15, 39, 59, 75, 192, 238.
        Pink, red or brown: 8, 23, 54, 82, 143, 146, 148, 153, 197, 198, 199, 210, 220, 244, 257.

II. Special physiological and nutritional characters.
- A. Denitrification: 1, 3, 4, 8, 9, 12, 13, 14, 21, 49, 166, 259.
- B. Oxidase reaction negative or weakly positive: 6, 16, 27, 29.
- C. Autotrophic growth:
  1. With H_2: 22, 23, 24, 25, 26, 247.
  2. With CO: 246.
- D. Utilization or decomposition of special compounds.
  1. Starch: 8, 12, 13, 23, 31, 51, 63, 69, 76, 84, 103, 118, 125, 132, 136, 140, 148, 154, 155, 158, 160, 168, 209, 210, 216, 220, 222, 223, 225, 248, 249, 250, 251, 252, 255, 256, 262, 264, 265.
  2. Extracellular poly-β-hydroxybutyrate: 11, 13, 17, 18, 20, 22.
  3. Cellulose: 51, 105, 112, 120, 137, 140, 143, 160, 174, 175, 211, 239, 240, 241, 242, 243, 244, 245, 248, 257, 258, 264.
  4. Agar (agar-softening organisms included): 46, 51, 76, 85, 104, 118, 135, 139, 146, 148, 149, 153, 157, 158, 168, 197, 208, 209, 216, 222, 223, 224, 225, 248, 249, 250, 251, 252, 255, 256, 257, 258, 262, 264, 265.

5. Chitin: 51, 128, 129, 210, 223, 248, 256, 262.
6. Alginic acid: 51, 118, 146, 220, 223, 224, 236, 237, 238, 248, 251, 255, 256, 262, 264, 265.
7. Aliphatic hydrocarbons: 1, 15, 191.
8. Mineral oils: 70, 234.
9. Phenol and/or cresols: 2, 45, 78, 196, 221, 254.
10. Naphthalene: 2, 41, 113, 196, 202, 253, 261.
11. Isoprenoid compounds (e.g. citronellol, geraniol): 1, 9, 45.
12. Resins: 81.
13. Riboflavin: 233.
14. Oxalate: 54, 72, 263.
15. Methylamine: 32.
E. Production of special odors: 35, 67, 94, 171, 185, 205.
III. Special ecological properties:
  A. Parasites or occasional pathogens of plants: 1, 3, 6, 7, 14, 15, 16, 21, 30, 31, 33, 34, 36, 40, 42, 44, 52, 55, 57, 61, 62, 63, 65, 71, 73, 74, 79, 83, 84, 87, 88, 90, 91, 96, 97, 98, 99, 100, 101, 102, 106, *107*, 109, 115, *116*, 119, 123, 124, 126, 132, *133*, 134, 136, 142, 144, 147, 155, 161, 162, 163, 164, 169, 170, 182, 187, 190, 193, 195, 201, *206*, 212, 215, 217, 219, 226, 227, 230, 232, 235.
     (Numbers in *italics* may not be true plant pathogens; see Elliott, 1951.)
  B. Parasites or occasional pathogens of animals:
    1. Mammals: 1, 3, 12, 13, 15, 218.
    2. Fish and reptiles: 80, 154, 229.
    3. Insects: 53, 86, 188, 207.
  C. Associated with food spoilage:
    1. Milk and dairy products: 3, 56, 69, 92, 93, 94, 158, 173, 203, 231.
    2. Eggs: 3, 67, 94, 185.
    3. Other: 3, 39, 68, 92, 127, 130, 183.
  D. Isolated from sea water or brines; halophilic or salt tolerant: 39, 46, 51, 69, 76, 111, 118, 121, 122, 128, 129, 135, 146, 148, 149, 151, 153, 154, 157, 166, 167, 172, 173, 178, 180, 181, 184, 189, 197, 204, 208, 210, 216, 220, 222, 223, 224, 225, 229, 236, 238, 248, 250, 251, 252, 255, 256, 257, 258, 259, 260, 262, 264, 265.

Genus II. **Xanthomonas** *Dowson 1939, 187*

D. W. DYE AND R. A. LELLIOTT

(*Phytomonas* Bergey *et al.* 1923, 174.)
Xan.tho′mo.nas or Xan.tho.mo′nas Gr. adj. *xanthus* yellow; Gr. fem.n. *monas* unit, monad; M.L. fem.n. **X**anthomonas yellow monad.

Cells single, straight rods. Dimensions 0.2–0.8 by 0.6–2.0 μm, usually *ca.* 0.4–1.0 μm. Motile by means a **polar flagellum.** Do not produce sheaths. No resting stages known. **Gram-negative.** Growth on agar media usually yellow. Chemoorganotrophs. **Metabolism respiratory,** never fermentative. Molecular oxygen is the electron acceptor. **Oxidase reaction negative or weak.** Catalase positive. In a weakly buffered medium (Dye, 1962), acid is produced in small amounts from many carbohydrates but not from rhamnose, inulin, adonitol, dulcitol, inositol or salicin and rarely from sorbitol. **No acid produced in purple milk.** Acetate, citrate, malate, propionate and succinate are utilized but generally not benzoate, oxalate or tartrate; gluconate may be utilized after a delay. Most species hydrolyze starch and Tween 80 rapidly.

Nitrates not reduced. Hydrogen sulfide produced from cysteine and by most species from thiosulfate and peptone. Acetoin and indole not produced, sodium hippurate not hydrolyzed.

Minimal growth requirements are complex and usually include methionine, glutamic acid and nicotinic acid in various combinations. **Asparagine not utilized as a sole source of carbon and nitrogen. Growth** on nutrient agar **inhib-**

ited by 0.1% and **usually by 0.02% triphenyl-tetrazolium chloride.**

Strict aerobes.

Temperature optimum 25–27 C; none grow at 5 C, most grow at 7 C but some unable to grow below 9 C; all grow at 30 C, none grow at 40 C.

The G + C content of the DNA of 29 nomenspecies of the *X. campestris* group **ranges from 63.5–69.2 moles %** (mostly T_m).

Type species: *Xanthomonas campestris* (Pammel) Dowson 1939, 190.

Further Comments

Although the generic definition for the genus *Xanthomonas* does not differentiate it from the genus *Pseudomonas*, the two genera have not been combined because no extensive phenotypic comparisons of the two genera have yet been made.

All but four of the nomenspecies can be distinguished from the type species and/or each other only by their host range and may be regarded as members of a single taxospecies, here referred to as the *X. campestris* group.

X. fragariae and the *X. campestris* group produce a colorless polysaccharide slime on media containing glucose and colonies of these species are mucoid, domed and shining on glucose nutrient agar or glucose-yeast-chalk agar.

The yellow pigment of the 19 members of the *X. campestris* group that have been studied is a non-water-soluble carotenoid which in petroleum ether has an absorption spectrum with maxima at 418, 437 and 463 nm. Non-pigmented strains occur rarely; a diffusible brown pigment is produced by a few members of the *X. campestris* group.

All species recognized here are plant pathogens and so far as is known are found only in association with plants or plant materials. Reports of xanthomonads from other habitats have not been confirmed.

Hybridization studies of DNA from 27 nomenspecies of the *X. campestris* group and representative members of the genus *Pseudomonas* showed close homology within the *Xanthomonas* spp. and as close homology between these and *P. fluorescens* (45–75%) as between *P. fluorescens* and some other *Pseudomonas* spp. (De Ley et al., 1966). As a result of this work De Ley et al. (1966) proposed that the xanthomonads should be combined in a single species, transferred to the genus *Pseudomonas* and named *P. campestris;* they did not study strains of *Xanthomonas* spp. outside the *X. campestris* group.

Description of the species of genus **Xanthomonas**

1. **Xanthomonas campestris** (Pammel) Dowson 1939, 190. (*Bacillus campestris* Pammel 1895, 130; *Pseudomonas campestris* (Pammel) Smith 1897, 284; *Bacterium campestre* (Pammel) Smith 1897, 478; *Phytomonas campestris* (Pammel) Bergey et al. 1923, 176: subjective syn. *X. campestris* var. *aberrans* Knösel 1961, 5; *X. campestris* var. *armoraceae* (McCulloch) Starr and Burkholder 1942, 600.)

cam.pes′tris. L. gen.n. *campestris* specific epithet of *Brassica campestris* a host.

Grows at moderate rate on nutrient media.

Many strains of this nomenspecies and of other members of the group require amino acid(s), often methionine and/or glutamic acid and less frequently nicotinic acid for growth. Growth of most other strains is markedly enhanced by organic growth factors.

Causes a vascular and parenchymatous disease of many *Brassica* spp. Other hosts, including those determined experimentally, are: *Boerhavia erecta, Lepidium sativum, Capsella bursa-pastoris, Matthiola* spp., *Raphanus* spp., *Rorippa armoracia.*

A list of nomenspecies included with *X. campestris* in the *X. campestris* group is given in Addendum I (p. 245).

2. **Xanthomonas fragariae** Kennedy and King 1962, 875.

fra.gar′i.ae. M.L. n. *Fragaria* generic name of strawberry; M.L. gen.n. *fragariae* of strawberry.

Grows slowly on nutrient media.

Requires amino acid(s) for growth.

Causes a leaf spot disease of strawberry.

3. **Xanthomonas albilineans** (Ashby) Dowson 1943, 11. (*Bacterium albilineans* Ashby 1929, 135; *Phytomonas albilineans* (Ashby) Magrou 1937, 236; *Agrobacterium albilineans* (Ashby) Săvulescu 1947, 10; *Pseudomonas albilineans* (Ashby) Krasil′nikov 1949, 385; *Xanthomonas albilineans* var. *paspali* Orian 1962, 8.)

al.bi.li′ne.ans. L. adj. *albus* white; L. part.adj. *lineans* striping; M.L. adj. *albilineans* white-striping.

Grows poorly on nutrient agar; growth is better, but still slow, on 2% sucrose, 1% peptone agar and is yellowish.

Requires glutamic acid and methionine for growth.

A suitable inorganic basal medium for testing utilization of organic compounds is the YS medium of Dye (1962) without NaCl.

Causes leaf scald of sugar cane (*Saccharum officinale*). The following can be successfully inoculated experimentally but are probably not natural hosts: *Zea mays, Coix lacryma-jobi, Thysan-*

TABLE 7.6

Characters differentiating the species of genus **Xanthomonas**

	1. X. campestris	2. X. fragariae	3. X. albilineans	4. X. axonopodis	5. X. ampelina
Growth at 35°C	+[a]	+	+	+	−[a]
Hydrolysis of esculin[b]	+	+	+	+	−
Mucoid growth on nutrient glucose agar	+	+	−	−	−
Gelatin liquefaction	+	+	d[a]	−	−
Proteolysis of milk	+	−	−	−	−
H₂S from peptone	+	−	−	+	d
Urease	−	−	−	−	+
Tolerance of NaCl, %	2.0–5.0	0.5–1.0	<0.5	1.0	1.0
Acid production[c] from:					
Arabinose	+	−	−	−	+
Glucose	+	+	+	+	−
Mannose	+	+	+	−	−
Galactose	+	−	d	−	+
Trehalose	+	−	−	+	−
Cellobiose	+	−	−	−	−

[a] + = 90% or more of strains positive in this character; − = 90% or more of strains negative in this character; d = some (less than 90%) strains positive, some negative.

[b] Some reactions delayed.

[c] Within 21 days by the methods of Dye (1962).

olaena agrostis, Cymbopogon citratus, Paspalum scobiculatum var. commersonii, P. dilatatum, P. paniculatum, Sorghum halepense, Panicum maximum, Pennisetum purpureum and Bambusa vulgaris.

4. Xanthomonas axonopodis Starr and Garces 1950, 81. (*Xanthomonas axonoperis* (*sic*) Starr and Garces 1950, 81; an orthographic variant.)

ax.on.o′pod.is. M.L. n. *Axonopus* generic name of a grass; M.L. gen.n. *axonopodis* of *Axonopus*.

Grows slowly on nutrient media.

Requires amino acid(s) for growth.

Causes a gummosis of *Axonopus scoparius* and *A. micay*.

5. Xanthomonas ampelina Panagopoulos 1969, 75.

am.pel in′a. Gr. n. *ampelos* the grape vine; M.L. fem.adj. *ampelina* of the vine.

Grows very slowly; produces brown diffusible pigment on yeast-glucose-chalk agar on which growth is better than on most other nutrient media.

Requires glutamic acid for growth.

Maximum growth temperature 30 C.

Causes "Tsilik Marasi," a canker disease of grapevine (*Vitis vinifera*), which is similar to, or the same as, a group of diseases wrongly attributed to *Erwinia vitivora* (syn. *E. herbicola*).

Addendum I: Most of the nomenspecies listed alphabetically below can be distinguished from *X. campestris* or from each other, with certainty, only by plant host reactions. A few are, unlike *X. campestris*, non-pigmented (*X. manihotis*) or are known to have non-pigmented forms (*X. ricini*) but these are otherwise indistinguishable from *X. campestris* except by plant host reactions. The host plants that are given (in alphabetical order) for each nomenspecies include those determined as a result of inoculation tests as well as those on which they have been found naturally. Only subjective synonyms are given.

X. alangi Padhya and Patel 1962, 196. Host: *Alangium lamarckii*.

X. alfalfae (Riker, Jones and Davis) Dowson 1943, 11. Hosts: *Medicago sativa, Melilotus indica, Pisum sativum, Phaseolus vulgaris, Trigonella foenum-gracecum*.

X. alysicarpi Bhatt and Patel 1954, 165. Host: *Alysicarpus rugosus*.

X. alysicarpi var. *vaginalidis* Patel, Bhatt and Dhande 1954, 183. Host: *Alysicarpus vaginalis*.

X. amaranthicola Patel, Wankar and Kulkarni 1952, 347. Hosts: *Amaranthus* spp.

X. arecae Rao and Mohan 1970, 704. Host: *Areca catechu*.

X. argemoneae Srinivasan, Patel and Thirumalachar 1961, 106. Host: *Argemone mexicana*.

X. badrii Patel, Kulkarni and Dhande 1950, 104. Hosts: *Pisum sativum, Xanthium strumarium.*

X. barbarae Burkholder 1941, 348. Host: *Barbarea vulgaris.*

X. bauhiniae Padhya, Patel and Kotasthane 1965, 225. Host: *Bauhinia racemosa.*

X. begoniae (Takimoto) Dowson 1939, 190. (Syn. *X. flava-begoniae* (Wieringa) Magrou and Prévot 1948, 102; *X. flavozonatum* (*sic*) (McCulloch) Dowson 1939, 190.) Hosts: *Begonia* spp., and numerous cultivars and hybrids of tuberous and fibrous rooted begonias.

X. betlicola Patel, Kulkarni and Dhande 1951, 106. Hosts: *Piper betle, P. longum, P. hockeri.*

X. bilvae Patel, Allayyanavaramath and Kulkarni 1953, 217. Hosts: *Aegle marmelos, Citrus aurantifolia, Feronia elephantum.*

X. biophytii Patel, Chauhan, Kotasthane and Desai 1969, 274. Host: *Biophytum sensitivum.*

X. blepharidis Srinivasan, Patel and Thirumalachar in Srinivasan and Patel 1956, 367. Host: *Blepharis molluginifolia.*

X. brideliae Bhatt and Patel 1954, 165. Host: *Bridelia hamiltoniana.*

X. buteae Bhatt and Patel 1955, 94. Host: *Butea frendosa.*

X. cajani Kulkarni, Patel and Abhyankar 1950, 384. Host: *Cajanus cajan.*

X. carissa (*sic*) Moniz, Sabley and More 1964, 256. Hosts: *Carissa congesta, C. carandas, Cestrum nocturnum, Thevetia nerifolia.*

X. carotae (Kendrick) Dowson 1939, 190. Host: *Daucus carota.*

X. cassava Wiehe and Dowson 1953, 142. Hosts: *Manihot* spp.

X. cassiae Kulkarni, Patel and Dhande 1951, 47. Hosts: *Cassia tora, C. occidentalis, Cicer arientinum, Pisum sativum.*

X. celebensis (Gäumann) Dowson 1943, 11. (Syn. *X. musicola* Rangaswami and Rangarajan 1965, 1036.) Hosts: *Musa* spp.

X. citri (Hasse) Dowson 1939, 190. Hosts: *Aegle marmelos, Atalantia* spp., *Balsamocitrus paniculata, Casimiroa edulis, Chaetospermum glutinosa, Citropsis schweinfurthii, Citrus* spp. and hybrids, *Clausena lansium, Eremocitrus glauca, Euodia* spp., *Feroniella* spp., *Fortunella* spp., *Hesperethusa crenulata, Limonia* spp., *Melicope triphylla, Microcitrus* spp., *Murraya erotica, Paramigyna longipedunculata, Poncirrus trifoliata* and its hybrids, *Severinia buxifolia, Toddalia assiatica, Zanthoxylum* spp.

X. cleomei Abhyankar, Patel and Thirumalachar 1956, 93. Host: *Cleome monophylla.*

X. clerodendri Patel, Kulkarni and Dhande 1952, 74. Host: *Clerodendron phlomoides.*

X. coracanae Desai, Thirumalachar and Patel 1965, 386. Host: *Eleusine coracana.*

X. corianderi Srinivasan, Patel and Thirumalachar 1961, 301. Hosts: *Coriandrum sativum, Foeniculum vulgare.*

X. corylina (Miller, Bollen, Simmons, Gross and Barss) Starr and Burkholder 1942, 598. Hosts: *Corylus* spp.

X. cucurbitae (Bryan) Dowson 1939, 190. Hosts: *Cucurbita* spp., *Cucumis sativus, Citrullus vulgaris.*

X. cyamopsidis Patel, Dhande and Kulkarni 1953, 183. (Syn. *X. cyamophagus* Patel and Patel 1958, 258.) Host: *Cyamopsis tetragonoloba.*

X. cynodontis Desai, Patel, Gandhi and Kotasthane 1967, 213. Host: *Cynodon dactylon.*

X. desmodii Patel 1949, 213. Host: *Desmodium diffusum.*

X. desmodiigangeticii Uppal, Patel and Moniz in Patel and Moniz 1948, 140. Host: *Desmodium gangeticum.*

X. desmodiirotundifolii Desai and Shah 1960, 66. Host: *Desmodium rotundifolium.*

X. dieffenbachiae (McCulloch and Pirone) Dowson 1943, 12. Hosts: *Aglaonema robellinii, Anthurium andraeanum, Dieffenbachia* spp., *Dracaena fragrans.*

X. durantae Srinivasan, Patel and Thirumalachar in Srinivasan and Patel 1957, 90. Host: *Duranta repens.*

X. erythrinae Patel, Kulkarni and Dhande 1952, 346. Host: *Erythrina indica.*

X. glycines (Nakano) Magrou and Prévot 1948, 102. (Syn. *X. phaseoli* var. *sojensis* (Hedges) Starr and Burkholder 1942, 600.) Hosts: *Brunnichia cirrhosa, Dolichos uniflorus* (syn. *D. biflorus*) *Glycine* spp., *Lablab niger* (syn. *D. lablab*), *Phaseolus vulgaris.*

X. guizotiae Yirgou 1964, 1491. Host: *Guizotia abyssinica.*

X. gummisudans (McCulloch) Starr and Burkholder 1942, 600. Host: *Gladiolus* sp.

X. hederae (Arnaud) Dowson 1939, 190. Host: *Hedera helix.*

X. heliotropii Sabet, Ishag and Khalil 1969, 368. Host: *Heliotropium sudanicum.*

X. holcicola (Elliott) Starr and Burkholder 1942, 600. Hosts: *Sorghum halepense, S. vulgare.*

X. hyacinthi (Wakker) Dowson 1939, 188. Host: *Hyacinthus orientalis.*

X. incanae (Kendrick and Baker) Starr and Weiss 1943, 316. Host: *Matthiola incanae.*

X. ionidi Padhya and Patel 1963, 98. Host: *Ionidium heterophyllum.*

X. juglandis (Pierce) Dowson 1939, 190. Hosts: *Juglans* spp.

X. khayae Sabet 1959, 664. Hosts: *Khaya senegalensis, K. grandifoliola.*

X. lantanae Srinivasan, Patel and Thirumalachar in Srinivasan and Patel 1957, 90. Host: *Lantana camara* var. *aculeata.*

X. laureliae Dye 1963, 185. Host: *Laurelia novae-zelandiae.*

X. lawsoniae Patel, Bhatt and Kulkarni 1951, 326. Host: *Lawsonia alba.*

X. leeanum Patel and Kotasthane 1969, 519. Host: *Leea edgeworthii.*

X. lespedezae (Ayers, Lefebre and Johnson) Starr 1946, 136. Hosts: *Lespedeza* spp.

X. lochnerae Patel, Thirumalachar and Bhatt 1955, 21. Host: *Lochnera pusilla.*

X. maculifoliigardeniae (Ark and Barrett) Elrod and Braun 1947, 515. Hosts: *Gardenia* spp., *Ixora coccinea.*

X. malvacearum (Erw. Smith) Dowson 1939, 190. Hosts: *Gossypium* spp., *Thespesia lambas, Ceiba pentandra.*

X. manihotis (Arthaud-Berthet and Bondar) Starr 1946, 136. Non-pigmented. Hosts: *Manihot* spp.

X. marantae Zagatto and Pereira 1963, 36. Host: *Maranta arundinacea.*

X. martinicola Moniz and Patel 1958, 494. Host: *Martynia diandra.*

X. melhusi Patel, Kulkarni and Dhande 1952, 345. Host: *Tectona grandis.*

X. musacearum Yirgou and Bradbury 1968, 112. Hosts: *Ensete ventricosum, Musa* sp.

X. nakatae (Burkholder) Dowson 1943, 12. Host: *Corchorus capsularis.*

X. nakatae var. *fascicularis* Patel and Kotasthane 1969, 596. Host: *Corchorus fascicularis.*

X. nakatae var. *olitorii* Sabet 1957, 519. Host: *Corchorus olitorius.*

X. nakataecorchori Padhya and Patel 1963, 326. Host: *Corchorus acutangulus.*

X. nigromaculans (Takimoto) Dowson 1943, 12. Host: *Arctium lappa.*

X. nigromaculans f. sp. *zinniae* Hopkins and Dowson 1949, 253. Hosts: *Zinnia* spp.

X. oryzae (Uyeda and Ishiyama) Dowson 1943, 12. (Syn. *X. itoana* (Tochinai) Dowson 1943, 12; *X. kresek* Schure 1953, 3.) Hosts: *Isachne globosa, Leersia* spp., *Oryza sativa, Phalaris arundinacea, Phragmites communis, Zizania aquatica, Leptochloa chinensis, L. panacea.*

X. oryzae var. *dianthi* Thomas and Dickens 1953, 22. Host: *Dianthus barbatus.*

X. papavericola (Bryan and McWhorter) Dowson 1939, 190. Hosts: *Meconopsis baileyi, Papaver* spp.

X. passiflorae Pereira 1969, 169. Host: *Passiflora edulis.*

X. patelii Desai and Shah 1959, 378. Host: *Crotalaria juncea.*

X. pelargonii (Brown) Starr and Burkholder 1942, 600. (Syn. *X. geranii* (Burkholder) Dowson 1939, 190.) Hosts: *Geranium* spp., *Pelargonium* spp.

X. phaseoli (Erw. Smith) Dowson 1939, 190.

Hosts: *Lablab niger* (syn. *Dolichos lablab*), *Lupinus polyphyllus, Phaseolus vulgaris.*

X. phaseoli f. sp. *rhynchosiae* Sabet, Ishag and Khalil 1969, 368. Hosts: *Lupinus termis, Mucuna pruriens, Rhynchosia memnonia.*

X. phaseoli f. sp. *vigna-radiatae* Sabet, Ishag and Khalil 1969, 368. Hosts: *Lablab niger, Vigna radiata.*

X. phaseoli f. sp. *vignicola* (Burkholder) Sabet 1959, 329. (Syn. *X. vignicola* Burkholder 1944, 432.) Hosts: *Phaseolus vulgaris, Vigna unguiculata* (syn. *V. sinensis*), *V. pubigera.*

X. phaseolitrilobi Bhatt, Abhyankar and Patel 1956, 299. Host: *Phaseolus trilobus.*

X. phaseoli var. *fuscans* (Burkholder) Starr and Burkholder 1942, 600. Distinguishable from *X. phaseoli* f. sp. *vignicola* by the production of a diffusible brown pigment. Hosts: *Lablab niger, Phaseolus vulgaris, Vigna unguiculata.*

X. physalidicola Gota and Okabe 1958, 48. Host: *Physalis alkekengi* var. *francheti.*

X. physalidis Srinivasan, Patel and Thirumalachar 1962, 96. Hosts: *Physalis minima, P. peruviana.*

X. pisi Goto and Okabe 1958, 39. Host: *Pisum sativum.*

X. plantaginis (Thornberry and Anderson) Burkholder 1948, 161. Hosts: *Plantago* spp.

X. poinsettiaecola (sic) Patel, Bhatt and Kulkarni 1951, 327. Hosts: *Euphorbia pulcherrima, Manihot esculenta.*

X. pruni (Erw. Smith) Dowson 1939, 190. Hosts: *Prunus* spp., *Sorbus japonica.*

X. punicae Hingorani and Singh 1959, 47. Host: *Punica granatum.*

X. ricini (Yoshi and Takimoto) Dowson 1939, 190. (Syn. *X. ricinicola* (Elliott) Dowson 1939, 190; *X. anandensis* Desai and Shah 1963, 475.) Non-pigmented forms are common. Host: *Ricinus communis.*

X. ricini f. sp. *euphorbiae* Sabet, Ishag and Khalil 1969, 368. Hosts: *Euphorbia acalyphoides, E. pulcherrima, Ricinus communis.*

X. ricini f. sp. *phyllanthii* Sabet, Ishag and Khalil 1969, 368. Host: *Phyllanthus niruri.*

X. rubiidaei Sherengovyi 1968, 39. Host: *Rubus idaeus.*

X. sesami Sabet and Dowson 1960, 258. Host: *Sesamum orientale.*

X. sesbaniae Patel, Kulkarni and Dhande 1952, 75. Host: *Sesbania aegyptiaca.*

X. spermacoces Srinivasan, Thirumalachar and Patel in Srinivasan and Patel 1956, 366. Host: *Spermacoce hispida.*

X. stizolobiicola (sic) Patel, Kulkarni and Dhande 1951, 106. Host: *Stizolobium deeringeanum.*

X. tamarindi Patel, Bhatt and Kulkarni 1951, 327. Hosts: *Caesalpina sepiaria, Tamarindus indica.*

X. taraxaci Niederhauser 1943, 961. Host: *Taraxacum bicorne.*

X. tephrosiae Bhatt and Patel 1955, 94. Host: *Tephrosia purpurea.*

X. teramnii Bhatt, Pawar and Sukapure 1960, 181. Host: *Teramnus labialis.*

X. thirumalachari Padhya and Patel 1964, 342. Host: *Triumfetta pilosa.*

X. translucens (Jones, Johnson and Reddy) Dowson 1939, 190.

There are apparently a number of pathotypes, some with overlapping host ranges within this nomenspecies. Using the criteria of production of translucent, water-soaked lesions on unwounded plants and the data of Fang, Allen, Riker and Dickson (1950), Tominaga (1967) and Patel and Shekhawat (1971), six pathotypes are distinguishable as shown below.

Burkholder 1942, 600. Hosts: *Brassica* spp., *Capsicum annum, Lycopersicon esculentum, Nicotiana tabacum, Raphanus sativus.*

X. vitians (Brown) Dowson 1943, 13. (Syn. *X. lactucae* (Yamamoto) Dowson 1943, 12; *X. lactucae-scariolae* (Thornberry and Anderson) Săvulescu 1947, 12.) Hosts: *Lactuca* spp.

X. vitiscarnosae Moniz and Patel 1958, 495. Host: *Vitis carnosa.*

X. vitistrifoliae Padhya, Patel and Kotasthane 1965, 463. Host: *Vitis trifolia.*

Addendum II: *Species incertae sedis.* The following species have a doubtful taxonomic position for the reasons indicated.

a) cultures and/or description suggest inappropriate classification; b) inadequate description and no cultures available for further testing;

Pathotypes of *X. translucens* (using the *forma specialis* designation)	*Agropyron* spp.	*Bromus* spp.	*Dactylis glomerata*	*Hordeum* spp.	*Oryza sativa*	*Phleum pratense*	*Secale cereale*	*Triticum* spp.
cerealis Hagborg 1942, 317	+[a]	+	−[a]	a[a]	a	−	a/−	a/−
hordei Hagborg 1942, 317	a/−	−	+	+	a	−	a/−	−
oryzicola (Fang et al.[b]) Bradbury 1971, 72					+			
phleipratensis Wallin and Reddy 1945, 939	−			−		+	−	−
secalis (Reddy et al.[c]) Hagborg 1942, 317	−	−		a/−		−	+	a/−
undulosa (Smith et al.[d]) Hagborg 1942, 317	−	−		a/−		−	a	+

[a] + = natural infection; a = artificial infection; − = no infection.

[b] Fang *et al.* 1957, 119.

[c] Reddy *et al.* 1924, 1040.

[d] Smith *et al.* 1919, 48.

X. tribuli Srinivasan, Thirumalachar and Patel in Srinivasan and Patel 1956, 366. Host: *Tribulus terrestris.*

X. trichodesmae Patel, Kulkarni and Dhande 1952, 346. Host: *Trichodesma zeylanicum.*

X. uppalii Patel 1948, 67. Host: *Ipomoea muricata, Tropaeolum majus.*

X. vasculorum (Cobb) Dowson 1939, 190. Hosts: *Bambusa vulgaris, Branchiaria mutica, Cocos nucifera, Coix lacryma-jobi, Dictyosperma alba, Panicum maximum, Pennisetum purpureum, Saccharum officinale, Sorghum* spp., *Thysanolaena maxima, Zea mays.*

X. vernoniae Patel, Desai and Patel 1968, 221. Host: *Vernonia cinerea.*

X. vesicatoria (Doidge) Dowson 1939, 190. Hosts: *Capsicum* spp., *Datura stramonium, Hyoscyamus* spp., *Lycium* spp., *Lycopersicon* spp., *Nicotiana rustica, Physalis minima, Solanum* spp.

X. vesicatoria var. *raphani* (White) Starr and

c) not validly published; d) authors' cultures identified as *Erwinia herbicola* (q.v. Part 8).

X. acernea (Ogawa) Burkholder 1948, 165 (a).

X. annamalaiensis Rangaswami, Prasad and Eswaran 1961, 180 (d).

X. antirrhini (Takimoto) Dowson 1943, 11 (a).

X. arsenoxydans-quattor Turner 1954, 476 (a, c).

X. balsamivorum Rangaswami and Gowda 1963, 77 (d).

X. beticola (Smith, Brown and Townsend) Săvulescu 1947, 12 (a).

X. cannae (Bryan) Săvulescu 1947, 12 (a).

X. celebensis var. *gossypii* Ramalingam, Lewin, Sivapraksam and Krishnamurthy 1965, 53 (a).

X. conjac (Uyeda) Burkholder 1948, 171 (a).

X. cosmosicola Rangaswami and Gowda 1963, 74 (d).

X. eleusineae Rangaswami, Prasad and Eswaran 1961, 106 (a).

X. esculenti Rangaswami and Eswaran 1962, 1 (d).

X. fici (Cavara) Magrou and Prévot 1948, 102 (b).

X. gorlencovianum Tsilosani 1966, 324 (b).

X. gypsophilae (Brown) Magrou and Prévot 1948, 102 (a, c).

X. hemmiana (Yamamoto) Bu·kholder 1957, 162 (a).

X. heterocea (Vzorov) Săvulescu 1947, 12 (a).

X. hortoricola Hanada 1954, 257 (a).

X. indica Rangaswami, Prasad and Eswaran 1961, 106 (d).

X. jasminicola Rangaswami 1964, 286 (c).

X. jasminii Rangaswami and Eswaran 1961, 352 (a).

X. leersiae Fang, Ren, Chen, Chu, Faan and Wu 1957, 120 (b).

X. maydis Rangaswami, Prasad and Eswaran 1961, 393 (d).

X. necrosis (Kalinenko) Săvulescu 1947, 374 (b).

X. panici (Elliott) Săvulescu 1947, 13 (a).

X. penniseti Rajagopalan and Rangaswami 1958, 30 (d).

X. phormicola (Takimoto) Dowson 1943, 12 (a).

X. proteamaculans (Paine and Stansfield) Burkholder 1948, 169 (a).

X. radiciperda (Zhavoronkova) Magrou and Prévot 1948, 103 (a).

X. rubefaciens (Burr) Magrou and Prévot 1948, 103 (b).

X. rubrisorghi Rangaswami, Prasad and Eswaran 1961, 270 (d).

X. sambuci Tesic and Todorovic 1952, 9 (c).

X. suberfaciens (Burr) Magrou and Prévot 1948, 102 (b).

X. striaformans Thomas and Weinhold 1953, 23 (a, b).

X. tagetis Rangaswami and Gowda 1963, 77 (d).

X. tardicrescens (McCulloch) Dowson 1943, 12 (b).

X. zingiberi (Uyeda) Săvulescu 1947, 13 (b).

Genus III. **Zoogloea** Itzigsohn 1868, 30

KOBY CRABTREE AND ELIZABETH McCOY

(Not *Zoogloea* Cohn 1854, 123.)

Zo.o.gloe′ a. Gr. n. *zoon* animal; Gr. n. *gloea* jelly; M.L. fem.n. *Zoogloea* animal jelly.

Cells rod shaped, range 0.5–1.0 by 1.0–3.0 μm, **no spores or cysts. Young cells actively motile with polar monoflagellation.** In natural waters and certain media cells later **aggregate in macroscopic flocs,** free floating (Plate 7.1, Fig. 1*B*) or attached to surfaces (Plate 7.1, Fig. 2*A–C*). **On the flocs** there are **finger-like or dendritic outgrowths** (Plate 7.1, Fig. 1*A*). Capsules not demonstrable on the neotype strain but reported on one other strain. The flocs are firm, not soft or slimy; recent studies have shown **extracellular fibrils interlacing the cells in the flocs** (Plate 7.1, Fig. 3). **Pleomorphic with age;** old cells usually **distended with granules of poly-β-hydroxybutyric acid.** The amount of polymer and extent of flocculation are influenced by the age of culture and nature of medium (Crabtree *et al.*, 1965).

Gram-negative.

Cells occur naturally (in natural waters and sewage) in flocs as described above. In laboratory media, growth is **flocculent when the carbon: nitrogen ratio is** >10:1 and is **dispersed when ratio is** <5:1. In broth cultures containing 0.05% arginine and 0.5% glucose, there develops a fragile pellicle with lacy tapelike flocs adhering to the sidewalls and eventually extending to the bottom of the tube (Plate 7.1, Fig. 2*A–C*). In shake cultures, starlike flocs are formed (Plate 7.1, Fig. 1*B*).

Chemoorganotrophs. **Metabolism is respiratory, never fermentative. Molecular oxygen is the terminal electron acceptor.** Xylose, fructose, glucose, mannose, ethanol and glycerol and certain amino acids are oxidized. Dulcitol, melibiose, melezitose, raffinose, glycogen, soluble starch and inulin are not oxidized. Cellulose and native corn starch are not hydrolyzed.

Not proteolytic. Gelatin, coagulated egg albumin and casein not hydrolyzed. Litmus milk not peptonized. Indole and hydrogen sulfide not produced. Nitrates not reduced to nitrites or to gaseous nitrogen. Ammonia not produced from peptone and amino acids. Cytochrome oxidase, catalase positive. Citrate not utilized. Non-pigmented.

Vitamin B$_{12}$ required, biotin stimulatory. With vitamins provided, arginine, aspartic acid, glutamic acid, histidine, ornithine, citrulline, glutamine, pantothenate, glucosamine, adenine, uracil and thiamine may serve as sole source of carbon and nitrogen. Ammonia may also serve, when both vitamins are provided. Acetate, pyruvate, malate, fumarate and oxalate are utilized as carbon sources.

Strict aerobe. Optimum temperature 28–30 C;

slow growth at 10 C and none at 45 C. Optimum pH 7.0–7.5; no growth at pH 4.5 or 9.6. No distinctive ionic or osmotic requirement but will grow with 3% NaCl, not with 6.5% NaCl.

Base composition of DNA. Not determined.

Type species: *Zoogloea ramigera* Itzigsohn 1868, 30.

Further Comments

The nomenclature proposed by Crabtree and McCoy (1967) is adopted, although the proposal made to the Judicial Committee to declare *Zoogloea* Cohn 1854 a *nomen nudum* and to accept *Zoogloea* Itzigsohn, 1868 is still *sub judice*. The proposal was supported by Zvirbulis and Hatt (1967) and those interested in the nomenclature should refer to these two papers.

It should be noted that the genus *Zoogloea* bears some resemblance to *Pseudomonas* and to *Acetobacter*. The oxidative utilization of sugars and alcohols (including ethanol), as shown by O/F tests, is a prime characteristic, true of all three

genera. The extracellular fibrils in *Zoogloea* are reminiscent of the fibrils in *Acetobacter* (*A. xylinum*) but also in certain *Pseudomonas* cultures, i.e. *P. denitrificans* P95-5 (Friedman *et al.*, 1969). The chemical nature of the fibrils in *Zoogloea* is said to be cellulose-like but the evidence is less conclusive than in *Acetobacter* (Ohad *et al.*, 1962). *Zoogloea* differs from *Acetobacter* in inability to grow at pH 4.5 (acetic acid).

Chemically defined media for flocculation and physiological tests (modified from Crabtree *et al.*, 1965): 1. *Arginine medium* for general growth: arginine HCl, 0.05%; $MgSO_4 \cdot 7H_2O$, 0.02%; K_2HPO_4, 0.2%; KH_2PO_4, 0.1%; $FeSO_4 \cdot 5H_2O$, trace; vitamin B_{12}, 2 ng/ml; biotin (stimulatory), 2 ng/ml; distilled water.

2. *High C:N medium* for flocculation: arginine basal above plus suitable C source such as glucose 0.1–0.5%.

3. *Ammonium N medium:* ammonium salts can be substituted for arginine, if a suitable C source is provided.

Description of species of genus **Zoogloea**

1. **Zoogloea ramigera** Itzigsohn 1868, 30.

ram.i'ger.a. L. masc.n. *ramus* a branch; L. v. *gero* bear; M.L. adj. *ramigera* branch bearing.

Rods with round, blunt or pointed ends, 0.5–1.0 by 1.0–3.0 μm. Young cells are actively motile by a long straight flagellum at one or both poles.

Colonies tan to straw color, undulate, dry wrinkled, tough leathery with a distinctive dimple at the apex; at 72 hr usually 2–3 mm in diameter. Entire colony may be moved with a needle.

Additional carbohydrates oxidized by this species are: arabinose, galactose, sorbose, rhamnose, fucose, maltose, sucrose, lactose, trehalose, cellobiose, salicin, inositol, mannitol and sorbitol.

Arginine dihydrolase and tyrosine decarboxylase produced. Urea not hydrolyzed. Tyrosine may also serve as a sole source of carbon and nitrogen, if vitamin B_{12} is provided.

Originally isolated from flocs in activated sewage sludge. Found in some industrial wastes.

Proposed neotype strain: 1-16-M; ATCC 19623 (Crabtree and McCoy, 1967).

Comment: The above description is based on the proposed neotype and other strains of *Z. ramigera* (Crabtree, 1965). Other references useful in defining the species are: Blöch (1918), Butterfield (1935), Buterfield *et al.* (1937) and Wattie (1943). Presumably the latter had pure cultures but these cultures have not been kept. Recent investigators, Rich (1955), Dugan and Lundgren (1960), Dias and Bhat (1964) and Unz and Dondero (1967), have isolated and described numerous strains but there are many discrepancies *inter se* and between their concepts of the species and Itzigsohn's early brief

description. Nevertheless, the concept of *Z. ramigera* in its natural form in sewage flocs and in natural waters is widely accepted and is distinctive.

Editorial Note. Unz (1971) has proposed rejecting strain 1-16-M as the neotype strain and replacing it with a zoogloeal-forming strain 106 (ATCC 19544).

2. **Zoogloea filipendula** Beger 1928, 143.

fi.li.pen'du.la. L. neut.n. *filum* a thread; L. adj. *pendulus* hanging; M.L. adj. *filipendula* hanging thread.

Rods with round or pointed ends, 0.5–1.0 by 1.0–2.0 μm. Young cells are actively motile by a long wavy polar flagellum.

Does not oxidize arabinose, galactose, sorbose, rhamnose, fucose, maltose, sucrose, lactose, trehalose, cellobiose, salicin, inositol, mannitol or sorbitol.

Arginine dihydrolase and tyrosine decarboxylase are not produced. Urea is hydrolyzed.

Originally isolated by Beger from pump pistons and other submerged objects in waterworks near Berlin. Also found in sewage from activated treatment systems.

Species incertae sedis

a. *Zoogloea pulmonis equi* Bollinger 1870, 585. Found in slimy masses in pneumonitis in the horse. Not considered validly published.

b. *Zoogloea beigeliana* Eberth according to Trevisan 1879, 146. Not validly published.

Genus IV. **Gluconobacter** *Asai 1935, 689, emend. mut. char. Asai, Iizuka and Komagata 1964, 100*

J. DE LEY AND J. FRATEUR

(*Acetomonas* Leifson 1954, 109.)

Glu.con.o.bac′ter. M.L. n. *acidum gluconicum* gluconic acid; M.L. n. *bacter* masc. equivalent of Gr. neut.n. *bactrum* rod or staff; M.L. masc.n. *Gluconobacter* gluconate rod.

Cells ellipsoidal to rod-shaped, 0.6–0.8 by 1.5–2.0 μm, occurring singly, in pairs or in chains. Involution forms may occur in old cultures. **Motile with three to eight polar flagella**, rarely one flagellum, **or non-motile.** Endospores not formed. Gram-negative; weakly Gram-positive in older cultures.

Chemoorganotrophs: **Metabolism respiratory**, never fermentative; oxygen is terminal electron acceptor.

Characteristics differentiating the genus from *Acetobacter* and *Pseudomonas* are given in Table 7.7.

Oxidize ethanol to acetic acid, sometimes weakly, **at** neutral and **acid reactions (pH 4.5). Do not oxidize acetate or lactate to CO₂.** Usually pronounced **ketogenesis and acid formation from sugars.** Many strains produce 2- and

5-ketogluconic acids, water-soluble brown pigments and γ-pyrones. Usually strongly **catalase positive.**

Strict aerobes.

Temperature optimum 25–30 C; range tolerated 7–41 C, although many strains do not grow above 37 C. pH optimum 5.5–6.0; growth and acetic acid production occur at pH 4–4.5; slight growth in weakly alkaline media.

Occur in flowers, souring fruits, vegetables, beer, South African "kaffir beer," cider, wine, wine vinegar, baker's yeast, garden soil.

The G + C content of the DNA ranges from 60–64 moles %, with one exception of 57 (T_m, chemical analysis).

Type species: *Gluconobacter oxydans* (Henneberg) De Ley 1961, 47 (proposed by Asai, Iizuka and Komagata 1964, 119).

TABLE 7.7

Characteristics differentiating the genera **Gluconobacter, Acetobacter** *and* **Pseudomonas**

	Gluconobacter	Acetobacter	Pseudomonas
Flagellation	Polar or none	Peritrichous or none	Polar
Growth at pH 4.5	+	+	−
Oxidation of:			
Ethanol to acetic acid at pH 4.5	+(M)[a]	+(S)	−
Acetic acid to CO₂	−	+	d
Lactate to CO₂	−	+	+
Glucose to gluconate	+	d	d
Amino acids by resting cells	−	+	+
Krebs cycle	−	+	+
Production of 5-ketogluconate	+	d	−
Ketogenesis	+	d	−
Quinones Q₁₀	+	−	
Q₉	−	+	
Hydrolysis of:			
Lactose and starch	−	−	d
Gelatin	−/W	−	d
Greenish and/or fluorescent pigments	−	−	d

[a] M, moderate; S, strong; W, weak.

Description of the species of genus **Gluconobacter**

1. **Gluconobacter oxydans** (Henneberg) De Ley 1961, 47. (*Bacterium oxydans* Henneberg 1897, 224; *Bacillus oxydans* (Henneberg) Migula 1900, 800; *Acetobacter oxydans* (Henneberg) Bergey *et al.* 1923, 36; *Acetomonas oxydans* (Henneberg) Shimwell and Carr 1959, 363; *Gluconobacter oxydans*

(Henneberg) Asai in Asai, Iizuka and Komagata 1964, 118.)

ox.y.dans'. Gr. adj. *oxys* sharp, acid; L. part. *dans* giving; M.L. pres.part. *oxydans* acid-giving, oxidizing.

Morphology as for genus.

Colonies on wort agar and yeast extract-glucose agar circular, milky white to yellowish, sometimes becoming brownish in the center or yellowish at the periphery. Good growth on yeast extract-calcium-gluconate agar with or without formation of crystals of calcium 5-ketogluconate. On agar containing yeast extract, 3% ethanol and 2% CaCO₃ usually good growth and acetic acid production, dissolving the calcium carbonate.

Pantothenic acid, *p*-aminobenzoic acid, thiamine and other unidentified factor(s) in yeast extract required for growth; some strains require nicotinic acid.

Acid formed from L-arabinose, D-galactose, D-glucose, D-mannose, D-xylose, *n*-butanol, ethanol, isobutanol, propanol; usually produced from fructose. Frequently considerable 5-ketogluconate (crystallizing as Ca salt), 2-ketogluconate and other reducing compounds formed from glucose + CaCO₃ and gluconate. Oxalate is sometimes formed from fructose after a few weeks. Acid formation variable from cellobiose, maltose, melibiose, sorbose, sucrose, glycerol, inositol, mannitol, sorbitol, *n*-amylalcohol and isoamylalcohol. Little or no acid from methanol, isopropanol. Acid not produced from dextrin, dulcitol, glycogen, inulin, lactose, raffinose, rhamnose or trehalose.

Acetate, butyrate, formate, glycerate, lactate and propionate are not oxidized to CO₂.

Ketogenesis from polyols; usually considerable dihydroxyacetone, erythrulose and fructose from glycerol, erythritol and mannitol.

γ-Pyrones usually produced from fructose.

Form a distinct group serologically.

Type strain not extant. Suggested neotype: NCIB 9013.

1a. *Gluconobacter oxydans* subsp. *oxydans*. Morphology and description as for the genus and species.

1b. *Gluconobacter oxydans* subsp. *industrius* (Henneberg) *comb.nov.* (*Bacterium industrium* Henneberg 1898, 145; *Bacillus industrius* (Henneberg) Migula 1900, 803: *Bacterium aceti viscosum* Baker, Day and Hulton 1912, 662; *Acetobacter industrius* (Henneberg) Bergey *et al.* 1923, 36;? not *Gluconobacter viscosus* Asai 1935, 680; *Acetobacter capsulatum* (*sic*) Shimwell 1936, 585; *Acetobacter viscosum* (*sic*) (Baker, Day and Hulton) Shimwell 1936, 586; *Acetobacter suboxydans* var. *muciparum* (*sic*) Frateur 1950, 331; *Acetobacter suboxydans* var. *biourgianum* (*sic*) Frateur 1950, 331; *Gluconobacter*

industrius (Henneberg) Asai, Iizuka and Komagata 1964, 97; *Gluconobacter capsulatus* (Shimwell) Asai, Iizuka and Komagata 1964, 97.)

Little or no pigment formation on glucose CaCO₃ medium; a few strains produce a reddish to dark brown water-soluble pigment on media containing glucose, gluconate, beer, wort, variable on maltose.

Produces considerable amounts of crystalline calcium 5-ketogluconate on suitable media although this property may be lacking in a few strains or be lost on subculturing.

Does not produce γ-pyrones from galactose or glucose.

Viscous growth in beer and wort; most strains cause ropiness in beer. Frequently produce dextrans or levans on dextrin media or in media containing sucrose or raffinose.

1c. *Gluconobacter oxydans* subsp. *suboxydans* (Kluyver and de Leeuw) comb. nov. (*Acetobacter suboxydans* Kluyver and de Leeuw 1924, 179; *Bacterium gluconicum* Hermann 1928, 198; *Bacterium hoshigaki* var. *rosea* Takahashi and Asai 1930, 86; *Bacterium industrium* var. *hoshigaki* Takahashi and Asai 1930, 401; *Salmonella gluconica* (Hermann) Pribram 1933, 61; *Acetobacter hoshigaki* Bergey *et al.* 1934, 39; *Acetogluconobacter dioxyacetonicus* Asai 1935, 510; *Gluconoacetobacter cerinus* Asai 1935, 614; *Gluconoacetobacter rugosus* Asai 1935, 617; *Gluconoacetobacter nonoxygluconicus* Asai 1935, 676; *Gluconoacetobacter scleroideus* Asai 1935, 678; *Gluconoacetobacter opacus* Asai 1935, 677; *Gluconoacetobacter roseus* Asai 1935, 676; *Acetobacter roseum* (*sic*) (Asai) Vaughn 1942, 20; *Acetobacter gluconicum* (*sic*) (Hermann) Kelly and Vaughn 1948, 189; *Acetobacter suboxydans* var. *hoyerianum* Frateur 1950, 378; *Gluconobacter gluconicum* (*sic*) (Hermann) Asai and Shoda 1958, 294; *Gluconobacter cerinus* (Asai) Asai and Shoda 1958, 291; *Gluconobacter roseus* (Asai) Asai and Shoda 1958, 291; *Gluconobacter scleroideus* (Asai) Asai and Shoda 1958, 291; *Acetobacter dioxyacetonicus* (Asai) Kondô and Ameyama 1958, 370; *Acetobacter albidus* Kondô and Ameyama 1958, 370; *Gluconobacter albidus* (Kondô and Ameyama) Asai, Iizuka and Komagata 1964, 97; *Gluconobacter dioxyacetonicus* (Asai) Asai, Iizuka and Komagata 1964, 97; *Gluconobacter nonoxygluconicus* (Asai) Asai, Iizuka and Komagata 1964, 97.)

Similar to G. *oxydans* subsp. *industrius* but does not produce viscous growth in beer or polysaccharides from dextrin and sucrose.

Some strains produce a pink non-diffusible pigment (cytochromes).

1d. *Gluconobacter oxydans* subsp. *melanogenes* (Beijerinck) *comb. nov.* (*Acetobacter melanogenus* Beijerinck 1911, 171; *Acetobacter melanogenum* (sic) var. *maltovorans* Frateur 1950, 331; *Acetobacter melanogenum* (sic) var. *malto-saccharovorans* Frateur 1950, 331; *Acetomonas melanogena* (Beijerinck) Leifson 1954, 108; *Gluconobacter melanogenus* (Beijerinck) Asai and Shoda 1958, 294; *Acetobacter rubiginosus* Kondô and Ameyama 1958, 369; *Gluconobacter rubiginosus* (Kondô and Ameyama) Asai, Iizuka and Komagata 1964, 97.)

Colonies on yeast extract-glucose agar dark brown on aging. Considerable reddish to dark brown soluble pigments usually produced on glucose $CaCO_3$ media although in a few strains this property may be lacking or lost on subculturing.

May or may not produce calcium 5-ketogluconate; moderate amounts may be produced by some strains but this property is easily lost on subculturing; exceptionally this property may be gained. 2-5 Diketogluconate produced from glucose + $CaCO_3$ and gluconate.

γ-Pyrones produced from glucose and galactose.

A few strains produce slime on beer-gelatine and/or sucrose.

FAMILY II. **AZOTOBACTERACEAE** PRIBRAM 1933, 5

J. H. Becking

A.zo.to.bac.ter.a′ce.ae. M.L. masc.n. *Azotobacter* type genus of family; *-aceae* ending to denote family; M.L. fem.pl.n. *Azotobacteraceae* the *Azotobacter* family.

Cells **large,** predominantly **bluntly rod-shaped** to oval, but **changes dramatically in morphology** with time or changes in growth conditions. Cells often in pairs. **Motile by** peritrichous or polar **flagella** or non-motile.

Gram-negative but may be Gram-variable.

Endospores not produced but **some species form cysts.**

Heterotrophic. **Capable of fixing molecular nitrogen** in a nitrogen-free medium with an organic carbon source.

Normally fix (with 1–2% carbohydrate) **at least 10 mg of atmospheric nitrogen/g of carbohydrate consumed.** Under these conditions some *Beijerinckia* strains fix less (6–9 mg of N/g of carbohydrate consumed) because of copious polysaccharide (slime) production. Efficiency of nitrogen fixation appreciably increased at lower carbohydrate levels (Becking, 1971) or lower oxygen tension (Meyerhoff and Burk, 1928). Organic growth factors not required for growth but for nitrogen fixation trace elements, in particular molybdenum (a specific catalyst of nitrogen fixation) required.

Grows on media with or without combined nitrogen. However, some representatives (*Azomonas agilis, Beijerinckia* sp.) are unable to utilize nitrate as sole source of nitrogen or utilize it very poorly. Strains of *Beijerinckia* do not grow on peptone agar. Most strains capsulated and produce copious slime. Fluorescent pigments produced by some strains.

Strict aerobes, but also able to grow and fix nitrogen under reduced oxygen pressure. **Catalase positive or negative** (*Derxia*).

Inhabitants of soil, water and leaf surfaces (phyllosphere).

Type genus: *Azotobacter* Beijerinck 1901, 567.

Key to the genera of family **Azotobacteraceae**

I. Large ovoid cells; most species produce extracellular slime. Rapid growth; catalase positive.
  A. Cysts formed; G + C 63–66%.
               I. *Azotobacter*
  B. Cysts not formed; G + C 53–59%.
               II. *Azomonas*

II. Small rods producing very tenacious extracellular slime or gum, and conspicuous internal globular lipoid bodies. Slow growth; catalase may or may not be produced.
  A. Lipoid bodies bipolar; G + C 55–59%.
    Catalase positive.
               III. *Beijerinckia*
  B. Lipoid bodies numerous; G + C 70%.
    Catalase negative.
               IV. *Derxia*

Genus I. Azotobacter Beijerinck 1901, 567

D. B. JOHNSTONE

(*Parachromatium* Beijerinck 1903, 197; *Azotomonas* Orla-Jensen 1909, 328.)
A.zo.to.bac'ter. French n. *azote* nitrogen; M.L. masc.n. *bacter* the equivalent of Gr.
neut.n. *bactrum* a rod or staff; M.L. masc.n. *Azotobacter* nitrogen rod.

Large ovoid cells, 2 μm or more in diameter, of varying length down to coccoid morphology. Occur singly, in pairs or irregular clumps, and rarely in chains of more than four cells. Marked pleomorphism. Do not produce endospores, but **form thick-walled cysts.** May produce copious amounts of capsular slime. Motile with peritrichous flagella or non-motile. Gram-reaction negative, with marked variability.

Some strains elaborate a water-soluble fluorescent pigment which appears **green under ultraviolet light** (Table 7.8).

Generally fix at least 10 mg of atmospheric nitrogen non-symbiotically/g of carbohydrate consumed. Growth media contain carbohydrates (usually glucose), alcohols or organic acids. Molybdenum required for nitrogen fixation; may be replaced by vanadium.

Not proteolytic but can utilize nitrate, ammonia and amino acids as sources of nitrogen.

Catalase positive.

Grow well aerobically but can also grow under reduced oxygen tension.

Optimum temperature between 20 and 30 C. Growth range pH 5.5–8.5; optimum 7.0–7.5.

Found in soil and water.

The G + C content of the DNA ranges from 63–66 moles %.

Type species: *Azotobacter chroococcum* Beijerinck 1901, 567.

TABLE 7.8

Characters of the species of genera **Azotobacter** *and* **Azomonas**

	1. Azoto-bacter chroococcum	2. Azotobacter beijerinckii	3. Azoto-bacter vinelandii	4. Azoto-bacter paspali	1. Azomonas agilis	2. Azomonas insignis	3. Azomonas macro-cytogenes
Pigments							
Water-soluble fluorescence	None	None	Green	Green	White	None	White
Water-insoluble in cells	Black	Cinnamon					
Carbohydrate utilization							
Starch	+	−	−	−	−	−	−
Mannitol	+	−	+	−	−	−	+
Rhamnose	−	−	+	−	−	−	−
Motility	+	−	+	+	+	+	+
Flagella							
Peritrichous	+		+	+	+		
Lophotrichous						+	
Monotrichous							+
Cysts formed	+	+	+	+	−	−	−
Capsular slime produced	+	+	+	+	+	−	+
Habitat usually soil	+	+	+	+	−	−	+
Habitat limited to fresh water					+	+	
DNA base ratio GC%	65–66	66	66	63–65	53–54	57–58	58–59

Description of the species of genus Azotobacter

1. **Azotobacter chroococcum** Beijerinck 1901, 567. (*Bacillus azotobacter* Löhnis and Hanzawa 1914, 2; *Bacillus chroococcus* Buchanan 1925, 194.)

chro.o.coc'cum. Gr. n. *chroa* color; Gr. n. *coccus* a grain; M.L. neut.n. *chroococcum* colored coccus.

Large ovoid rods 2 by 5 μm, frequently in pairs and motile with peritrichous flagella (Plate 7.2, Fig. 1).

Cysts and capsular slime are formed (Plate 7.2, Fig. 2).

No water-soluble pigment produced, but growth on agar media characterized by a water-insoluble brown pigment which in some strains later becomes black.

Utilizes starch, which is unique among the species of this genus. Mannitol is utilized, but not rhamnose.

The cells are found in both soil and water.

The G + C content of DNA ranges from 65–66 moles %.

2. **Azotobacter beijerinckii** Lipman 1904, 248.
bei.jer.inck'ii. M.L. gen.n. *beijerinckii* of Beijerinck; named for M. W. Beijerinck of Delft, Holland.

Ovoid rods similar to *A. chroococcum* in size, but non-motile.

Cells occur singly, in pairs and sometimes in chains of several cells with large capsules. As cultures age, the cells become coccoid, form cysts, and turn yellow or cinnamon with a water-insoluble pigment.

More tolerant to acid than *A. chroococcum*.

Incapable of utilizing starch, mannitol or rhamnose.

The habitat is soil and water.

The G + C content of DNA is 66 moles %.

3. **Azotobacter vinelandii** Lipman 1903, 238.
vine.lan'di.i. M.L. gen.n. *vinelandii;* named for Vineland, New Jersey where first obtained.

Large ovoid rods, frequently in pairs; motile with peritrichous flagella.

Cysts are formed as well as a copious capsular slime. The latter contains a glucuronic acid polymer.

A water-soluble fluorescent pigment is produced which appears green under ultraviolet light.

Starch not utilized, but both mannitol and rhamnose are. This is the only species of this genus to use rhamnose.

The habitat is soil and water.

The G + C content of DNA is 66 moles %.

4. **Azotobacter paspali** Döbereiner 1966, 364.
pas.pal'i. M.L. gen.n. *paspali;* named for *Paspalum* generic name of a grass.

Large ovoid rods, so pleomorphic even in young cultures as to have coccoid cells of 2 μm to elongated filaments of 30 μm. Both coccoid and rod-shaped cells are motile with peritrichous flagella.

Cysts and capsular slime are formed.

A water-soluble fluorescent pigment produced which appears green under ultraviolet light and resembles that of *A. vinelandii*.

Starch, mannitol and rhamnose are not utilized.

The habitat is soil, particularly the root surface of *Paspalum notatum*.

The G + C content of DNA ranges from 63–65 moles %.

Genus II. **Azomonas** *Winogradsky 1938, 391*

D. B. JOHNSTONE

(*Azotococcus* Tchan 1953, 88.)

A.zo.mon'as. French n. *azote* nitrogen; Gr. n. *monas* a unit, monad; M.L. fem.n. *Azomonas* nitrogen monad.

Large ovoid cells 2 μm or more in diameter, of varying length down to coccoid morphology. Occur singly, in pairs or irregular clumps. Possess marked pleomorphism. **Do not produce** endospores or **cysts.** May produce copious amounts of capsular slime. Motile with polar or peritrichous flagella. Gram-negative or variable.

Strains that elaborate a water soluble fluorescent pigment **appear white under ultraviolet light** (see Table 7.8 for distinguishing characteristics).

Generally fix at least 10 mg of atmospheric nitrogen non-symbiotically/g of carbohydrate consumed. Growth media usually contain glucose or sucrose.

Not proteolytic. Catalase positive.

Grow well aerobically but can also grow under reduced oxygen tension.

Optimum temperature between 20 and 30 C. pH range 4.5–9.0; optimum 7.0–7.5.

Found in soil and water.

The G + C content of the DNA ranges from 53–59 moles %.

Type species: *Azomonas agilis* (Beijerinck) Winogradsky 1938, 400.

Description of the species of genus **Azomonas**

1. **Azomonas agilis** (Beijerinck) Winogradsky 1938, 400. (*Azotobacter agile* (*sic*) Beijerinck 1901, 577; *Azobacter agilis* Beijerinck 1901, 577; *Azotococcus agilis* (Beijerinck) Tchan 1953, 88.)

a'gi.lis. L. adj. *agilis* quick, agile.

Large ovoid rods, frequently in pairs and motile with peritrichous flagella (Plate 7.2, Fig. 3).

Cysts not formed, although large capsules produced containing a polymer of galactose and rhamnose.

A water-soluble fluorescent pigment elaborated which appears white under ultraviolet light. As cultures age, a pink pigment forms which turns dark blue, but this does not interfere with observing the characteristically white fluorescence under ultraviolet light.

Neither mannitol nor rhamnose utilized.

The habitat is limited to fresh water.

The G + C content of DNA ranges from 53–54 moles %.

2. **Azomonas insignis** (Derx) Baillie, Hodgkiss and Norris 1962, 118. (*Azotobacter insigne* (*sic*) Derx 1951, 344.)

in.sig'nis. L. adj. *insignis* distinguished by a mark.

Large ovoid rods, occasionally in pairs, but usually as single cells, very motile with lophotrichous flagella, two or more at one end. Motility characterized by frequent spinning of a cell on its own axis.

Cysts not formed. Little or no extracellular slime produced.

Cultures turn brownish with age. No soluble or fluorescent pigments.

Neither mannitol nor rhamnose utilized.

The habitat is limited to fresh water.

The G + C content of DNA ranges from 57–58 moles %.

3. **Azomonas macrocytogenes** (Jensen) Baillie, Hodgkiss and Norris 1962, 118. (*Azotobacter macrocytogenes* Jensen 1955, 280; *Beijerinckia macrocytogenes* (Jensen) Rubenchik 1959, 333.)

mac.ro.cy.tog'en.es. Gr. adj. *macrus* large: Gr.n. *kytos* cell; Gr. v. *gennaio* produce; M.L. part.adj. *macrocytogenes* large cell producing.

Large ovoid to coccoid cells singly and in pairs; motile with polar flagella, usually a single flagellum.

Cysts not formed, but profuse capsular slime produced.

Enormous cells have been observed, 8–10 µm, especially when utilizing ethanol.

A water-soluble pinkish pigment produced, and the cultures fluoresce white under ultraviolet light resembling *A. agilis*.

Mannitol utilized, but rhamnose is not.

Acid formation and acid tolerance are characteristic.

The habitat is soil.

The G + C content of DNA ranges from 58–59 moles %.

Genus III. **Beijerinckia** *Derx 1950, 145*

J. H. BECKING

Bei.je.rinck'ia. M.L. fem.n. *Beijerinckia;* named for M. W. Beijerinck, Dutch microbiologist.

Cells single, straight, slightly curved or pear-shaped rods, 0.5–1.5 by 1.7–4.5 µm with distinctly rounded ends. Sometimes large misshapen cells, 3 by 5–6 µm, that are occasionally branched or forked. Characteristic large, highly refractile, lipoid (poly-β-hydroxybutyrate) bodies occur at each end of the cell (Plate 7.3, Fig. 4). Under certain conditions some strains show rounded, coccoid cells without the terminal lipoid globules. Motile or non-motile (Table 7.9). Flagella peritrichous, but flagellation has been studied in only a few strains (Hofer, 1944).

Cysts (enclosing one cell) and capsules (enclosing several cells) found in some species (*B. mobile* and *B. fluminensis*) (Plate 7.3, Fig. 5). Gram-negative.

In liquid media no surface pellicle formed, but the whole medium becomes a homogenous, highly viscous semitransparent mass. In some species (*B. fluminensis*) the whole medium becomes opalescent and turbid, and adhering slime

is not produced. On agar medium copious slime is produced and giant colonies develop with a smooth, folded or plicated surface (Plate 7.3, Figs. 2 and 3). The slime is extremely tough, tenacious or elastic and makes it difficult to remove part of a colony with a loop. In some strains the slime is of a more granular consistency (*B. mobile, B. fluminensis*) resembling that of *Azotobacter* and therefore easier to remove. The polysaccharide slime consists of glucose, galactose, mannose, glucuronic acid and galacturonic acid, but no heptose. The main component, 51.8% of the dry weight of the polysaccharide, is glucose (López and Becking, 1968).

Atmospheric nitrogen is fixed in a nitrogen-deficient medium. **Molybdenum is required for nitrogen fixation, but, unlike Azotobacter, this cannot be replaced by vanadium.** Also in contrast to *Azotobacter*, calcium is not required for growth and nitrogen fixation.

Glucose, fructose and sucrose utilized by all

strains; a wide variability exists in the ability to attack other carbon substrates (see Table 7.10). Oxidize organic substrates to CO_2 and a small amount of acetic acid so that the pH of medium is lowered during growth.

No growth on peptone agar or broth. Glutamate utilized poorly or not at all.

Cells highly acid tolerant, growth between pH 3 and 9.5 or 10.0. In neutral and alkaline media acid is produced, in very acid media an alkaline substance which increases the pH of the medium is produced (Becking, 1961).

Cells contain cytochromes with absorption peaks at 415–424 nm (Sorêt); 480, 518 nm (β); 525–527 nm, 551–556 nm (αc); and 604, 630 nm ($a_1 a a_3 a_2$) (Moss and Tchan, 1958). Therefore, the cells contain cytochrome c (λ_{max} 524 and 552 nm) and cytochrome a (λ_{max} 590 and 630 nm).

Strict aerobe, but also grows and fixes nitrogen under reduced oxygen pressure.

Catalase positive.

Temperature range, 10–35 C; optimum for growth 20–30 C, no growth at 37 C. Cells frost resistant; no reduction of viability after storage for 3–4 months at −4 C (Becking, 1961).

The **G + C content** of DNA ranges from 54.7–60.7 moles % (T_m) (De Ley and Park, 1966). In DNA-hybridization experiments with ^{14}C-DNA fragments from *Pseudomonas fluorescens* and *P. putida* the percentage DNA bound relative to these two species is 28 and 19–29, respectively (De Ley and Park, 1966).

Type species: *Beijerinckia indica* (Starkey and De) Derx 1950, 146.

Further Comments

Although deserving separate generic rank (Derx, 1950), this genus is without doubt closely related to the genera *Azotobacter* and *Azomonas*.

Beijerinckia mobile is most probably the link between *Beijerinckia* and *Azotobacter*, because (1) the slime produced is less tenacious and of a consistency and color more like *Azotobacter;* (2) it produces good growth on benzoate, a property also present in most *Azotobacter* strains; (3) under certain conditions the cells are more rounded or spherical and lack the characteristic polar lipoid bodies of *Beijerinckia*. In this stage the cells resemble those of *Azotobacter*.

Beijerinckia derxii, which produces a water-soluble green fluorescent pigment in the medium is the counterpart of *Azotobacter vinelandii* in the genus *Azotobacter*.

Representatives of the genus *Beijerinckia* must also be phylogenetically closely related to *Azomonas agilis*. This view is based on the observation that both *Beijerinckia* species and *Azomonas agilis* (1) do not assimilate or very poorly assimilate nitrate; (2) vanadium cannot replace molybdenum in nitrogen fixation; (3) calcium is not required for nitrogen fixation; these properties all contrast with those of *Azotobacter vinelandii* and *A. chroococcum* (Becking, 1962). This opinion is supported by the G + C content of DNA in *Beijerinckia* and

TABLE 7.9

Differential characteristics of species of genus **Beijerinckia**

	B. indica	B. mobilis	B. fluminensis	B. derxii
Water-soluble green fluorescent pigment	−	−	−	+
Colony color after aging	Fulvus or pink	Amber brown	Fulvus or pink	Buff
Motility of cells	+^{a} or ±	+++	+^{a} or ±	−
Growth on casein agar	−/w	100^b	15	100
Growth with nitrate as N source	±^{c}	++	−/w	−/w
Growth when C and N source is asparagine	6	60	−/w	−/w
Growth on carbon sources:				
Lactose	70	0	20	100
Erythritol	0	45	0	0
Propanol	50	100	0	0
Acetate, butyrate, fumarate, lactate, malate	50–95	100	20–80	−/w
Benzoate	6	100	0	0

^{a} A minority of cells, mostly in young stages.
^{b} Numbers represent percentage of strains growing on the medium.
^{c} ± About 40–50% of strains tested positive; ++ good growth; −/w negative or weak reaction.

TABLE 7.10

Utilization of carbon compounds by **Beijerinckia** *species*

	1. B. indica	2. B. mobilis	3. B. fluminensis	4. B. derxii		1. B. indica	2. B. mobilis	3. B. fluminensis	4. B. derxii
	(48)[a]	(7)	(5)	(3)	Acetate[c]	50	100	20	−/w
Arabinose	85[b]	100	100	−/w	Butyrate	85	100	40	−/w
Galactose	85	45	100	100	Citrate	40	0	0	−/w
Inulin	65	15	0	−/w	Formate	0	70(w)	0	−/w
Maltose	90	70	100	100	Fumarate	80	100	60	−/w
Rhamnose	35	30	0	−/w	Lactate	95	100	80	−/w
Sorbose	70	100	100	100	Malate	85	100	20	−/w
Starch	55	30	20	−/w	Malonate	50	0	40	−/w
Xylose	70	45	100	100	Oxalate	2	0	0	−/w
					Propionate	0	0	0	−/w
Butanol	80	100	40	0	Succinate	80	100	40	100
Ethanol	90	100	80	100	Tartrate	15	0	0	−/w
Glycerol	70	100	100	0	Valerate	8	70(w)	0	−/w
Inositol	55	100	40	0					
Mannitol	90	85	40	100	Salicylate	6	0	0	0

[a] Figures in parentheses at top of table indicate number of strains tested.

[b] Numbers in the table indicate percentage of strains tested that utilize the compound. See also Table 7.9. All strains utilize glucose, fructose and sucrose; none can grow on amyl alcohol. With the exception of *B. indica* the percentages are indicative only, because of the small number of strains available for test. Tests made on nitrogen-free agar containing K_2HPO_4 0.8 g, KH_2PO_4 0.2 g, $MgSO_4 \cdot 7H_2O$ 0.5 g, $FeCl_3 \cdot 6H_2O$ 0.1 g, $Na_2MoO_4 \cdot 2H_2O$ 0.005 g, distilled water 1000 ml (pH 6.7). −/w, negative or weak reaction; (w), weak growth.

[c] Tested as Na- or Ca-salt; formate and oxalate tested only as Na-salt.

Azomonas agilis being in the same range but quite different from that of *Azotobacter vinelandii*, *A.* *chroococcum* and *A. beijerinckii* (De Ley and Park, 1966).

Description of the species of genus **Beijerinckia**

1. **Beijerinckia indica** (Starkey and De) Derx 1950, 146. (*Azotobacter* sp. Altson 1936, 268; *Azotobacter indicum* (*sic*) Starkey and De 1939, 337; *Azotobacter lacticogenes* Kauffmann and Toussaint 1951, 710; *Azotobacter acida* Roy 1958, 120; *Beijerinckia acida* (Roy) Peterson 1959, 73; *Beijerinckia congensis* Hilger 1965, 406.)

in'di.ca. L. fem.adj. *indica* of India.

Straight, or slightly curved, rods, 0.5–1.2 μm by 1.6–3.0 μm. Lipoid persists in aged cultures. No resting stages; cyst or ascococcus-formation in older cultures never observed.

Agar colonies raised. At first semitransparent, soon become uniformly turbid or opaque white. On aging colonies develop a light reddish, pink, cinnamon or fawn color on neutral or alkaline media; on acid media colonies remain colorless. On acid media slime more tenacious, tough and elastic than on alkaline media. Giant colonies may develop, first with a smooth surface, but later with a folded, wrinkled or plicated surface (Plate 7.3, Fig. 3).

Liquid media becomes viscous by slime production. On aging color produced, but less prominent than on agar.

Highly acid tolerant, growth between pH 3 and pH 10.0, optimum pH 4–10. Only a small reduction of growth at the higher pH values (Becking, 1961). Growth and utilization of nitrate-nitrogen are poor and molecular nitrogen is fixed in preference in the presence of nitrate in the medium (Becking, 1962). Weak growth on malt agar and no growth on plain broth or peptone agar. The latter two media can be used to check purity of strains.

Strict aerobe, but grows and fixes nitrogen under reduced oxygen pressure.

Temperature range about 10–35 C; no growth at 37 C.

Originally isolated from acidic soils of India (Dacca, pH 4.9) and of Burma (Insein, pH 5.2) (Starkey and De, 1939). Later found to be widely distributed in acidic tropical soils (Asia, North Australia, Africa and South America). Rare outside of tropics: Japan (Suto, 1954, 1957), Europe

(Becking, 1961), North America (Anderson, 1966) and non-tropical Australia (Thompson, 1968).

The G + C content of DNA is 54.7 ± 3.0 moles % (T_m) (De Ley and Park, 1966).

Type strain: Strain Starkey and De, ATCC 9039; Delft E.III.12.1.1.

Note. Derx (1950, 10) described *Beijerinckia indica* var. *alba* which can be distinguished from the type by the absence of pigmentation on aging, but under extreme (alkaline) conditions, it produced a pink pigment.

2. **Beijerinckia mobilis** Derx 1950, 10. (*Beijerinckia mobile* (*sic*) Derx 1950, 10.)

mo'bi.le. L. adj. *mobilis* movable, motile.

Straight, curved or pear-shaped rods, 0.6–1.0 µm by 1.6–3.0 µm. Sometimes misshaped or forked cells. Motility very conspicuous.

Agar colonies not as raised as *B. indica*. Older cultures do not show lipoid formation, but ascococcus-like clusters of cells often visible. The typical polar lipoid bodies may disappear in aging cells and the cells are then more rounded, and resemble *Azotobacter* cells.

Slime production less than by *B. indica;* slime formed neither elastic nor sticky; liquid media do not become viscous. There is a tendency to form a pellicle on the medium.

Older cultures on neutral or alkaline agar media show typical dark amber or deep reddish brown color.

Acid tolerant; growth between pH 3 and pH 10; optimal growth and nitrogen fixation at pH 4–5, and a sharp reduction in growth and nitrogen fixation toward the more alkaline pH values (Becking, 1961).

The composition of the slime has not yet been examined.

All strains tested produced good growth on nitrate and ammonium salts as nitrogen source (in contrast to *B. indica*). No or only weak growth on urea and the amino acids glycine and tyrosine. All strains grow on leucine and casein agar. Some growth on malt agar.

Strict aerobe, but also grows and fixes nitrogen under reduced oxygen pressure.

Temperature range: ca. 10–35 C. No growth at 37 C.

Originally isolated from Java, Indonesia (Derx, 1950). Common in Javanese soils; also isolated from soils of South America (Surinam) and tropical Africa.

The G + C content of DNA not yet examined.

Type strain: Derx; Delft E.III.12.2.1.

Note. The differences in levan production from sucrose observed by Derx (1950) are variable and cannot be used for the species differentiation of this species.

3. **Beijerinckia fluminensis** Döbereiner and Ruschel 1958, 269.

flu.mi.nen'sis. M.L. adj. *fluminensis;* named for the locality, "Baixada Fluminense," State Rio de Janeiro, Brazil from which soil it was first isolated.

Straight, or slightly curved rods, 1.0–1.5 µm by 3.0–3.5 µm. Motility slow, or absent, especially in older cells.

Agar colonies typical small granular, with an irregular, rough surface (Plate 7.3, Fig. 2). Moderately raised, slime not liquid, tenacious or elastic, but more granular and stiff. Colonies at first opaque white becoming, on neutral and alkaline media after 1–2 weeks, pink, reddish brown or fulvous (like *B. indica*).

Slime production in liquid media reduced. No pellicle or viscosity produced, but a bluish white turbidity of medium.

Older cultures show characteristic large capsules enclosing 2 up to 10 or more individual cells. Division of the cells within the capsules has been observed.

Considerably acid tolerant. Growth between pH 3.5 and 9.2.

The composition of the slime has not yet been examined.

Strict aerobe, but also growing and fixing nitrogen under reduced oxygen pressure.

Temperature range: ca. 10–35 C, optimum 26–33 C. No growth at 37 C.

Found in acidic soils of South America, Africa and Asia (China, Indonesia).

The G + C content of DNA is 56.2 ± 1.8 moles % (T_m) (type strain) (De Ley and Park, 1966). Hybridization with ^{14}C-DNA fragments from *Pseudomonas putida* showed 19% DNA bound relative to *P. putida* (De Ley and Park, 1966).

Type strain: Döbereiner and Ruschel CD10.

4. **Beijerinckia derxii** Tchan 1957, 315. (*Beijerinckia venezuelae* Materassi, Florenzano, Balloni and Favilli 1966, 210.)

derx.ii. M.L. gen.n. *derxii* of Derx; named for H. G. Derx, Dutch microbiologist.

Cells single, straight or curved rods, or rods with clavate extremities, 1.5–2.0 µm by 3.5–4.5 µm. Polar lipoid bodies very large and conspicuous. Cells non-motile. No cyst or capsule formation.

Agar colonies highly raised, slimy and smooth. Colonies at first semitransparent or opaque white, but after 2–3 weeks incubation a yellow-green, water-soluble fluorescent pigment produced, particularly in iron-deficient media. When pigment first appears, it remains within the colony, but later diffuses into the agar medium. In liquid media the whole medium turns uniformly turbid; pigment production less than on solid media, although under certain conditions pigment production on solid media may be very poor or absent.

Considerably acid tolerant. Growth between pH 4.0–9.0. Optimum pH between 6–7. There is no growth at pH 3 or 11 (Tchan, 1957).

The composition of the slime not yet examined. Temperature range not different from other *Beijerinckia* species. Growth between *ca.* 10–35 C. No growth at 37 C.

Isolated from soils from Queensland, Northern Australia. Up to now not isolated from soils of other tropical regions, but some strains isolated from Venezuelan soils most probably belong to this species (Materassi *et al.*, 1966; Florenzano, 1968).

The G + C content of DNA is 59.1 ± 1.6 moles % (T_m). In DNA-hybridization experiments with ^{14}C-DNA fragments from *Pseudomonas fluorescens* and *P. putida* the percentage DNA bound relative to these two species is 28 and 29%, respectively (De Ley and Park, 1966).

Type strain: Tchan Q13.

Genus IV. **Derxia** *Jensen, Petersen, De and Bhattacharya 1960, 193*

J. H. BECKING

Derx'i.a. M.L. fem.n. *Derxia;* named for H. G. Derx, a Dutch microbiologist.

Cells rod-shaped with rounded ends, 1.0–1.2 by 3.0–6.0 μm, occurring singly or in short chains. Cells rather pleomorphic, some cells occasionally becoming very large. Young cells have a homogenous cytoplasm; older cells show typical large refractive bodies throughout the whole cell. Motile by a short polar flagellum; motile cells numerous in liquid (glucose) medium containing combined nitrogen, but rare on nitrogen-deficient solid media. Resting stages not known. Gram-negative.

Agar colonies at first slimy and semitransparent; later massive and opaque, highly raised with a wrinkled surface (Plate 7.3, Fig. 1); older colonies develop a dark mahogany brown color. Liquid media turn into a gelatinous mass, but growth near the surface is more luxuriant forming a thick, tough pellicle.

A wide range of sugars, alcohols and organic acids are oxidized mostly to CO_2 and a small amount of acid, probably acetic, when growing in an alkaline medium.

Atmospheric nitrogen fixed in a nitrogen-deficient medium. Molybdenum required for nitrogen fixation; vanadium cannot replace molybdenum in this process.

Strict aerobe, also growing and fixing nitrogen under reduced oxygen pressure. Catalase negative.

Growth slow at 15 C, optimal at 25–35 C, feeble at 40 C, no growth at 50 C. Acid tolerant.

Found in tropical soils (Asia, Africa, South America).

The G + C content of the DNA is 70.4 moles %.

Type species: *Derxia gummosa* Jensen, Petersen, De and Bhattacharya 1960, 193.

Further Comments

In cell-morphology and type of flagellation—one polar flagellum—this genus is like the *Pseudomonaceae*, but in DNA-homology and DNA base composition it is widely different from *Pseudomonas* species (De Ley and Park, 1966).

There is some relation between *Derxia* and *Beijerinckia*, because (1) slime production, consistency of the slime, and colony shape are similar in both genera; and (2) in representatives of both genera vanadium cannot replace molybdenum in nitrogen fixation, in contrast to *Azotobacter* species, but like *Azomonas agilis*.

Description of the species of genus **Derxia**

1. **Derxia gummosa** Jensen, Petersen, De and Bhattacharya 1960, 193. (Derxia Indica Roy and Sen 1962, 605)

gum.mo'sa. L. fem.adj. *gummosa* slime (gum) producing.

Straight rods, 1.0–1.2 by 3.0–6.0 μm, singly or in short chains. Cells rather pleomorphic, depending on age and medium. In aging cultures cells often remain together forming long filaments of sometimes locally swollen or distorted cells. Some cells may assume enormous sizes (up to 30 μm).

Cells from young cultures on nitrogen-deficient agar appear as rods with rounded ends and homogeneous cytoplasm (Plate 7.2, Fig. 4). Older cells on sugar-rich media contain large refractive bodies throughout the whole cell. On glucose peptone agar especially, very elongated cells are produced containing many refractive bodies (Plate 7.2, Fig. 5). The refractive material is lipid (probably poly-β-hydroxybutyrate) as it stains with Sudan III and Sudan black, but some vacuoles which do not stain may also be involved. Under nitrogen-starvation conditions older cells undergo shrinkage (Plate 7.2, Fig. 6) and are finally enclosed by a slime envelope (Plate 7.2, Fig. 7).

Motile. However, motile cells are seldom seen in nitrogen-deficient solid media, but may become quite numerous in liquid glucose media with combined nitrogen (ammonia or glutamic acid). The motile cells have a rather short polar flagellum, often one at each cell pole.

No endospores, cysts or capsules. Gram-negative.

Agar colonies on nitrogen-deficient medium show at first a thin, whitish or semitransparent growth of scattered colonies. Later more massive, highly raised or dome-shaped colonies emerge, which gradually assume a diameter of 1 cm or more (giant colonies). These colonies are very reminiscent of those of *Beijerinckia* species.

Colonies at first whitish or dull yellow with a smooth surface, but the surface soon becomes coarse and wrinkled, while the color deepens to dark mahogany brown. The slime of these colonies is very tenacious and gum-like, but in other developmental stages more soft and smeary. Influence of oxygen concentration on colony type has been studied by Hill (1971).

Growth in liquid media usually starts as a ring at the glass-liquid interface and develops into a thick, wrinkled, tough pellicle. Shallow layers of medium change into a firm gelatinous mass after a couple of weeks. The color gradually becomes dark red-brown as in solid media.

Glucose, fructose, ethanol, glycerol, mannitol and sorbitol give good to excellent growth; mannose and lactate scant growth. Lactose, galactose, maltose, sucrose, formate, acetate, propionate, pyruvate, succinate, malate, fumarate, dulcitol and starch give no growth or a trace of growth. Butyrate, citrate, benzoate and xylose suppress growth.

Acid produced (probably acetic acid) from carbohydrates.

Growth with glutamic acid, ammonium acetate, alanine, sodium nitrate and urea decreasing from abundant to good in approximately the same sequence. Colonies on combined nitrogen change from pale yellow through rust brown to almost black (darkest with nitrate) and sometimes a light brown soluble pigment is produced in the agar. Growth with these combined nitrogen sources is much faster than with molecular nitrogen and is completely uniform, in contrast to the uneven growth on nitrogen-free agar. Aspartic acid, asparagine and peptone give a much slower growth, uneven and mostly confined to scattered colonies. Glycine seems to be toxic. Nitrate is not reduced to nitrite or gaseous nitrogen in a glucose nitrate medium. Indole is not formed.

The efficiency of the nitrogen fixation varies between 9 and 25 mg of N/g of glucose consumed, but in most strains it is distinctly lower than in *Azotobacter* or *Beijerinckia*. No requirement for amino acids, vitamins or growth factors. For nitrogen fixation trace elements, in particular molybdenum, required. Vanadium cannot replace molybdenum in nitrogen fixation.

Temperature range: growth very slow at 15 C (barely visible after 14 days), best at 25–35 C, feeble at 40 C and nil at 50 C. Cells heated in glucose medium are killed within 10 min at 60 C and 5 min at 70 C.

Acid tolerant. Growth from pH 5.5 to *ca.* pH 9.0. No growth at pH 4.4.

Originally isolated from a soil of India (West Bengal, Adisaptagram) of pH 6.5, but later also from slightly acidic or neutral soils of South America (Brazil, Surinam), South Africa and Java.

The G + C content of DNA is 70.4 ± 1.7 moles % (T_m) (type strain) (De Ley and Park, 1966). Hybridization with ^{14}C-DNA fragments from *Pseudomonas fluorescens* and *P. putida* the percentage DNA bound relative to both *Pseudomonas* species is 12 and 16, respectively (De Ley and Park, 1966).

Type strain: Bhattacharya, ATCC 15994.

FAMILY III. **RHIZOBIACEAE** CONN 1938, 321

D. C. JORDAN AND O. N. ALLEN

Rhi.zo.bi.a′ce.ae. M.L. neut.n. *Rhizobium*, type genus of the family; -*aceae*, ending to denote a family; M.L. fem.pl.n. *Rhizobiaceae*, the *Rhizobium* family.

Cells without endospores, normally **rod-shaped. Motile;** one polar or subpolar flagellum, or two to six peritrichous flagella. **Aerobic. Gram-negative. Many carbohydrates utilized.** Considerable extracellular slime usually produced during growth on carbohydrate-containing media. Some strains of rhizobia and agrobacteria show a close relationship in DNA base composition.

All species, with the exception of *Agrobacterium radiobacter*, **incite cortical hypertrophies on plants. Nodules** are **incited on roots of leguminous species** (family: *Leguminosae*) by strains of rhizobia (symbionts). **Gall hypertrophies are** produced **on roots and stems of** diverse **plant species** by strains of agrobacteria (tumorigenic phytopathogens). Can be isolated from nodules or galls; not easily identified when isolated from soil. Confirmation of isolates should be made by proper plant inoculation tests.

Key to the genera of the family Rhizobiaceae

I. Cells stimulate nodule production on roots of leguminous plants. Fix free nitrogen when in the symbiotic stage within root nodules.
Do not utilize citrate.
Do not produce 3-ketolactose.
Genus I. Rhizobium

II. Cells do not stimulate root nodule production on leguminous plants but most species do produce other types of hypertrophies on many plants.
Do not fix free nitrogen.
Utilize citrate.
Genus II. Agrobacterium, p. 264

Genus I. **Rhizobium** *Frank 1889, 338. Nom. gen. cons.* Opin 34, Jud. Comm. 1970, 11

D. C. JORDAN AND O. N. ALLEN

(*Phytomyxa* Schroeter 1886, 134; *Rhizobacterium* Kirchner 1896, 221.)
Rhi.zo'bi.um. Gr. n. *rhiza* a root; Gr. n. *bios* life; M.L. neut.n. *Rhizobium* that which lives in a root.

Rods 0.5–0.9 by 1.2–3.0 μm. **Commonly pleomorphic under adverse growth conditions.** Often contain granules of poly-β-hydroxybutyrate which are refractile under phase contrast, stainable with Sudan black B and soluble in chloroform. **Motile** by two to six peritrichous flagella or by a polar or subpolar flagellum. **Non-sporing. Gramnegative. Growth on carbohydrate media usually accompanied by** copious extracellular, polysaccharide **slime.**

Chemoorganotrophs: **Metabolism respiratory.** Molecular oxygen is the terminal electron acceptor. **3-Ketoglycosides not produced** (Bernaerts and DeLey, 1963). Gelatin is not liquefied or liquefied slowly within 2 months. Casein and agar not hydrolyzed. Produce little or no hydrogen sulfide on bismuth sulfite agar. Colonies colorless to white; exceptions are the pink to red colonies produced by strains from *Lotononis bainesii* Baker. Utilize a wide range of carbohydrates, without gas formation, but not cellulose or starch. Ammonium salts, nitrates and most amino acids can serve as nitrogen sources. Some strains will grow in a simple mineral-salts medium containing only vitamin-free casein hydrolysate as the sole source of both carbon and nitrogen. Some strains require water-soluble vitamins.

Aerobic; often able to produce excellent growth under oxygen tensions less than 0.01 atmosphere (Wilson, 1940). Temperature optima 25–30 C. pH range 5.0–8.5.

Members of this genus characteristically **able to invade root hairs of leguminous plants** (family: *Leguminosae*) **and incite production of root nodules,** wherein the bacteria occur as intracellular symbionts. All strains exhibit host range affinities (host "specificity"). Bacteria usually present in nodule as pleomorphic forms (bacte-

roids) enclosed singly or in small groups within plant-produced membranous sacs. **Nodule bacteroids** characteristically **involved in fixing molecular nitrogen** into combined forms utilizable by the host plant.

Plant tests are essential for identification.

The **G + C content** of the DNA ranges from **59.1–65.5** moles %, T_m (DeLey and Rassel, 1965).

Type species: *Rhizobium leguminosarum* (Frank) Frank 1889, 338.

Further Comments

Members of the genus *Rhizobium* are common soil inhabitants. Identification is relatively easy if isolated from host plant nodules but difficult if isolated from the soil or if a non-infective variant. The ability to cause nodule formation (infectiveness) is more stable than the ability to fix nitrogen in symbiosis (effectiveness). Strains tend to lose effectiveness after serial cultivation on media containing certain amino acids, especially DL- or D-forms, and after many years of storage on laboratory media. Infectiveness and effectiveness represent discrete phenomena and vary within wide limits, depending upon genetic factors present in both bacterial strain and host plant; occasional strains nodulate certain hosts but are ineffective.

Cells are commonly pleomorphic (swollen and either globular, ellipsoidal or branched) in old cultures, within root nodules, or when subjected to other conditions in which protein synthesis is partially inhibited. Such conditions vary with the strains but may include extremes of temperature and pH, low oxygen tension, and growth in laboratory media containing excessive amounts of various amino acids, or certain alkaloids, glycosides, dyes or antibiotics.

Most satisfactory growth is provided by media

containing yeast or other plant extracts. Yeast extract mineral salts media containing mannitol or glucose are among the most conventional. Pentoses are preferred by some species.

Species of *Rhizobium* can generally be distinguished from those of *Agrobacterium* by a combination of tests (see key to Rhizobiaceae and species description of *Agrobacterium radiobacter*). In addition, rhizobia have negative hypertrophy-initiating ability as shown by the Pinto bean (*Phaseolus vulgaris* L.) leaf test (Lippincott and Heberlein, 1965) and the carrot disc assay (Klein and Tenenbaum, 1955), as used by Lippincott and Lippincott (1969).

The taxonomic position of *Rhizobium* is controversial and the present classification can be regarded only as tentative. Studies involving transformation, numerical taxonomy, DNA base composition and DNA homology show a close relationship between some strains of the Group I species and certain strains of *Agrobacterium* species. This has led to the suggestion that these two genera should be amalgamated in part. Proposals also have been made relative to removal of the Group II species to a new genus or their consolidation into one species.

Admittedly, a classification of *Rhizobium* based on the plant affinity (cross-inoculation group) concept, complicated as it is by many instances of anomalous cross-infection, is not a particularly satisfying substitute for procedures designed to compare larger portions of the bacterial genome. Nevertheless at the present time it appears unwise to condemn any taxonomic treatment of this genus when only 8–9% of the 14,000 or so known species of leguminous plants have been examined for nodules and only about 0.3–0.4% studied with respect to their symbiotic relationships with nodule bacteria. The separation of *Rhizobium* from *Agrobacterium* and the present division of *Rhizobium* into two sections is in keeping with the general proposal of 't Mannetje (1967). However, any profound taxonomic revision must await comprehensive comparative studies involving large numbers of bacterial strains from a wide variety of leguminous plants.

Key to the groups of species of the genus **Rhizobium**

I. 2–6 peritrichous flagella; rapid growth on yeast extract media.

Group I.

1. *R. leguminosarum*
2. *R. phaseoli*
3. *R. trifolii*
4. *R. meliloti*

II. Polar or subpolar flagellum; slow growth on yeast extract media.

Group II.

5. *R. japonicum*
6. *R. lupini*

Description of the species of genus **Rhizobium**

(Group I)

Motile by two to six peritrichous flagella. Fimbriae described on a few strains. Some forms encapsulated.

Colonies circular, convex, semitranslucent, raised and mucilaginous; usually 2–4 mm in diameter within 3–5 days on yeast mannitol mineral salts agar. Pronounced turbidity develops after 2–3 days in agitated broth.

All species utilize a wide range of carbohydrates; glucose, mannitol or sucrose usually preferred. Some strains require biotin or other water-soluble vitamins.

Generally cause nodule formation on temperate zone leguminous plants.

The G + C content of the DNA ranges from 59.1–63.1 moles % (T$_m$).

1. **Rhizobium leguminosarum** (Frank) Frank 1889, 338. *emend.mut.char.* Baldwin and Fred 1929, 146. (*Schinzia leguminosarum* Frank 1879, 397; *Phytomyxa leguminosarum* (Frank) Schroeter 1886, 135; *Bacillus francki* Matzuschita 1902, 552; *Rhizobacterium leguminosarum* (Frank) Kirchner 1896, 221.)

le.gu.mi.no.sa'rum. M.L. fem.pl.n. *Leguminosae* old family name of the legumes; M.L. fem.gen.pl.n. *leguminosarum* of legumes.

Bacteroids in nodules commonly irregular with x-, y-, star- and club-shaped forms. Vacuolated forms predominate.

Litmus milk alkaline, serum zone formed.

Causes nodule formation on species of *Pisum* (pea), *Vicia* (vetch) and *Lens* (lentil).

Reference strain: ATCC 10004.

2. **Rhizobium phaseoli** Dangeard 1926, 197.

pha.se'o.li. M.L. masc.n. *Phaseolus* generic name of the bean; M.L. gen.n. *phaseoli* of *Phaseolus*.

Bacteroids in nodules usually rod-shaped and often vacuolated with a few branched forms.

Litmus milk alkaline; serum zone formed.

Causes nodule formation on temperate species of *Phaseolus* (*P. vulgaris*, kidney bean; *P. angustifolius*, bean; *P. multiflorus*, scarlet runner).

Reference strain: ATCC 14482.

3. **Rhizobium trifolii** Dangeard 1926, 191.

tri.fo′li.i. M.L. neut.n. *Trifolium* generic name of clover; M.L. gen.n. *trifolii* of clover.

Bacteroids in nodules are pear-shaped, swollen, vacuolated, rarely x- or y-shaped.

Litmus milk alkaline, serum zone formed.

Causes nodule formation on *Trifolium* spp.

Reference strain: ATCC 14480.

4. **Rhizobium meliloti** Dangeard 1926, 194.

me.li.lo′ti. M.L. masc.n. *Melilotus* generic name of sweet clover; M.L. gen.n. *meliloti* of *Melilotus*.

Bacteroids in nodules club-shaped, branched.

Litmus milk acid, serum zone formed.

Causes nodule formation on species of *Melilotus* (sweet clover), *Medicago* (alfalfa) and *Trigonella* (fenugreek).

Reference strain: ATCC 9930.

(Group II)

Motile by a polar or subpolar flagellum. Fimbriae have not been described.

Slow growth occurs on yeast extract media. Colonies circular, punctiform, opaque, rarely translucent, white, convex and granular in texture; do not exceed 1 mm in diameter within 5–7 days on yeast mannitol mineral salts agar. Moderate turbidity only after 3–5 days or longer in agitated broth. Faster growing strains on this medium are uncommon. Pentoses preferred as sources of carbon. Disaccharides and polysaccha-

rides generally not utilized. Usually no requirement for biotin.

These forms cause nodule formation on tropical zone leguminous plants. Certain strains may cause nodules on diverse hosts (Lange, 1961).

The G + C content of the DNA ranges from 61.6–65.5 moles % (T$_m$).

5. **Rhizobium japonicum** (Kirchner) Buchanan 1926, 90. (*Rhizobacterium japonicum* Kirchner 1896, 221.)

ja.po′ni.cum. M.L. adj. *japonicum* of Japan.

Bacteroids in nodules are longer than normal cells, slender and with only occasional branched and swollen forms.

Litmus milk alkaline, no serum zone.

Cause nodule formation on *Glycine* spp. Certain strains produce nodules on *Lupinus* spp. and plants in the cowpea miscellany.

Reference strain: ATCC 10324.

6. **Rhizobium lupini** (Schroeter) Eckhardt, Baldwin and Fred 1931, 273. (*Phytomyxa lupini* Schroeter 1886, 135.)

lu.pi′ni. M.L. masc.n. *Lupinus* generic name of lupine; M.L. gen.n. *lupini* of *Lupinus*.

Bacteroids in nodules are vacuolated rods, seldom branched.

Litmus milk alkaline, no serum zone, no reduction. An initial alkaline reaction followed by an acid reaction on rhamnose and xylose separates most strains from *R. japonicum* and also from strains isolated from plants in the cowpea miscellany.

Causes nodule formation on *Lupinus* spp. (lupines) and *Ornithopus* spp. (serradella). Limited infective ability shown for *Glycine* spp. and species in the cowpea miscellany.

Reference strain: ATCC 10319.

Genus II. **Agrobacterium** *Conn 1942, 359. Nom. gen. cons.* Opin. 33, Jud. Comm. 1970, 10

O. N. ALLEN AND A. J. HOLDING

(*Polymonas* Lieske 1928, 143, *nom. gen. rej.* Opin. 33, Jud. Comm. 1970, 10.)

Ag.ro.bac.te′ ri.um. Gr. n. *agrus* a field; Gr. dim.neut.n. *bakterion* a small rod; M.L. neut.n. *Agrobacterium* field rodlet or *bakterion*.

Rods 0.8 by 1.5–3.0 μm. Motile by one to four peritrichous flagella; if only one flagellum, lateral attachment is more common than polar. **Nonsporing. Gram-negative.** Fimbriae are common. **Growth on carbohydrate-containing media usually accompanied by** copious extracellular, polysaccharide **slime.** Colonies are non-pigmented and usually smooth tending to become striated with age, but rough colonies are produced by many strains.

Chemoorganotrophs: **Metabolism respiratory.** Molecular oxygen is the terminal electron acceptor. **3-Ketolactose commonly produced by** *A. tumefaciens* and *A. radiobacter* (DeLey *et al.*, 1966). Gelatin slowly liquefied in several weeks or not at all. Casein not hydrolyzed. Utilize a wide range of simple carbohydrates, organic acids and amino acids as energy sources but not cellulose, starch, agar or chitin. The release of either organic acids and/or CO$_2$ into the medium indicating utilization of carbohydrate usually produces an acidic reaction at the surface of synthetic media low in organic nitrogen compounds but the reaction may be hidden by the release of basic substances in media containing peptone or other materials rich in amino acids. Generally produce hydrogen sulfide on bismuth sulfite agar. Rapid rate of growth on meat extract or yeast extract peptone media; turbidity with pellicle usually within 24–36 hr; exception, strains of *A. rubi*. Ammonium salts, nitrates and some amino acids serve as sole nitrogen sources for some species; others require amino acids with additional growth factors (Starr, 1946). Normally oxidase positive (Table 7.11).

Aerobic, but able to grow under reduced oxygen tensions in plant tissue. Temperature optima 25–30 C. pH range 4.3–12.0. Optimum pH range 6.0–9.0.

With the exception of *A. radiobacter*, members of this genus **initiate stem hypertrophies on diverse plants** wherein the bacteria occur as intercellular parasites. **Bacteria enter host tissue through pre-existing lesions or abrasions; ability of strains to invade host plant tissue directly is lacking.** Host range affinities are not sharply defined. **Tissue puncture inoculation tests are essential to confirm strain identification** based on routine biochemical tests.

The **G + C content** of the DNA ranges **from 59.6–62.8** moles %, T_m (De Ley *et al.*, 1966).

Type species: *Agrobacterium tumefaciens* (Smith and Townsend) Conn 1942, 359.

Further Comments

Members of the genus *Agrobacterium* are soil inhabitants. *A. radiobacter* is common in many soils, while the other species probably occur mainly in soils previously contaminated with diseased plant material. Ease of isolating these organisms from soil is enhanced by using the medium of Schroth *et al.* (1965). Isolates from gall type growths are easily identified, but those from other sources, or non-infective variants, may be confused with members of the genus *Alcaligenes*. See *Further Comments* for the genus *Rhizobium* for interrelationships between the genera *Rhizobium* and *Agrobacterium*.

Key to the species of the genus **Agrobacterium**

A. Amino acids, nitrates and ammonium salts used as sole source of nitrogen. 3-Ketolactose produced.
  1. Produce galls.
    1. *A. tumefaciens*
  2. Do not produce galls.
    2. *A. radiobacter*
B. Do not utilize amino acids, nitrates and ammonium salts as sole source of nitrogen. 3-Ketolactose not produced.
  1. Produce hairy root of nursery stock.
    3. *A. rhizogenes*
  2. Produce galls on raspberries.
    4. *A. rubi*

Description of the species of genus **Agrobacterium**

1. Agrobacterium tumefaciens (Smith and Townsend) Conn 1942, 359. (*Bacterium tumefaciens* Smith and Townsend 1907, 672; *Pseudomonas tumefaciens* (Smith and Townsend) Duggar 1909, 114; *Phytomonas tumefaciens* (Smith and Townsend) Bergey *et al.* 1923, 189; *Polymonas tumefaciens* (Smith and Townsend) Lieske 1928, 143.)

tu.me.fa'ci.ens. L. part.adj. *tumefaciens* tumor-producing.

Growth rapid on meat extract or yeast extract peptone media.

Carbon and energy sources include a range of simple carbohydrates, organic acids and amino acids.

Mannitol nitrate glycerophosphate agar: mucoid growth with halo or browning and often with a white precipitate.

Litmus milk neutral to alkaline with serum zone that deepens with time; tan to grayish brown color after 2–3 weeks.

Causes galls of plants in more than 40 families. Paris daisy (*Chrysanthemum frutescens* L.) and tomato (*Lycopersicum esculentum* Mill.) are common test plants. Gall tissues are ill-defined consisting of disorganized masses of hyperplastic and hypertrophic tissues interspersed with badly organized groups of elements resembling trachae. Marked tumorigenic ability by the Pinto bean (*Phaseolus vulgaris* L.) leaf test (Lippincott and Heberlein, 1965) and the carrot root disc assay

TABLE 7.11

Differential characteristics of the species
of genus **Agrobacterium**

	1. A. tumefaciens,[a] 2. A. radiobacter	3. A. rhizogenes, 4. A. rubi
Asparagine used as sole source of carbon and nitrogen	+	−
Nitrites produced from nitrates	+	−
Ammonium salts, nitrates or amino acids utilized as sole nitrogen source	+	−
Growth factors and amino acids required	−	+
Calcium, or sodium, glycerophosphate mannitol nitrate agar:		
Growth	+	−/S[b]
Halo or browning	+	−
Starr's and Lippincotts' basal salts media, growth	+	−
Congo red and aniline blue mannitol agars:		
Growth	+	M
Dye absorbed	P	F
Sodium selenite-yeast water-glucose agar:		
Growth	+	S[c]
Selenite reduced	+	−
Litmus milk:		
Reaction	N/Alk	A
Serum zone formation	+	F
3-Ketolactose production	+	−

[a] Since this table was prepared it has been shown that there are two biotypes of these species, but the description given here applies only to biotype I (Keane, Kerr and New. 1970 Australian J. Biol. Sci. *23:* 585–595.

[b] S = scant growth; M = moderate growth; F = faint; A = acid; Alk = alkaline; N = neutral; P = pronounced.

[c] Some strains of *A. rubi* will not grow at all on this medium.

(Klein and Tenenbaum, 1955) as used by Lippincott and Lippincott (1969).
Reference strain: ATCC 4720.

2. **Agrobacterium radiobacter** (Beijerinck and van Delden) Conn 1942, 359. (*Bacillus radiobacter* Beijerinck and van Delden 1902, 3; *Bacterium radiobacter* (Beijerinck and van Delden) Löhnis 1905, 590; *Rhizobium radiobacter* (Beijerinck and van Delden) Pribram 1933, 53; *Achromobacter radiobacter* (Beijerinck and van Delden) Bergey *et al.* 1934, 230; *Alcaligenes radiobacter* (Beijerinck and van Delden) Conn 1939, 97; *Pseudomonas radiobacter* (Beijerinck and van Delden) Krasil'nikov 1949, 391.)

ra.di.o.bac'ter. L. n. *radius* a ray, beam; M.L. *bacter* masc. equivalent of Gr. neut.n. *bakterion* a rod or staff; M.L. masc.n. *radiobacter* ray rod.

The status of this species is controversial. Strains of this species are avirulent, otherwise their differentiation, morphologically, culturally, biochemically and serologically from strains of *Agrobacterium tumefaciens* is difficult. Whereas consideration has been given to designation of *Agrobacterium radiobacter* as either a prototype or variety of *Agrobacterium tumefaciens*, tentative acceptance of its status as a species here is believed wise until a greater number of strains from diverse sources has been examined along conventional lines. Protein electrophoretograms using polyacrylamide gel support this conclusion (Huisingh and Durbin, 1967).

Strains of this species also show superficial resemblances to species of Group I, genus *Rhizobium*. In addition to the reactions listed in the key to the *Rhizobiaceae*, strains of *Agrobacterium radiobacter*, as well as those of *A. tumefaciens*, are distinguished from *Rhizobium* spp. by producing halo formation, precipitates or browning on calcium glycerophosphate mannitol nitrate agar (Graham and Parker, 1964), raised, smooth, glistening colonies with halo on glycerol nitrate agar, turbidity with heavy ring or pellicle in peptone broths, turbidity with pellicle in citrate broth, pellicle after 8 days at 25 C in uric acid broth, H_2S production in mannitol tryptone medium, growth at pH 10.0–12.0 in yeast water mannitol broth, abundant growth with acid reaction in yeast water $CaCO_3$ mineral salts medium containing 1% dextrin, or inulin, or amygdalin, and in KNO_3 mineral salts medium without $CaCO_3$ containing 1% sodium 1-mannonate, or 1% sodium D-saccharate, and inability to initiate nodule formation (Allen and Allen, 1950).
Reference strain: NCIB 9042.

3. **Agrobacterium rhizogenes** (Riker *et al.*) Conn 1942, 359. (*Bacterium rhizogenes* Riker, Banfield, Wright, Keitt and Sagen 1930, 536.)

rhi.zo'ge.nes. Gr. n. *rhiza* a root; Gr. v. *gennaio* produce; M.L. adj. *rhizogenes* root-producing.

Causes the disease hairy root or woolly knot whereby masses of intertwined fleshy and fibrous roots are produced on nursery stock. Host tissue histology not explained. The Pinto bean leaf test shows a weak tumorigenic ability. Produces roots and tumors on carrot root discs.

Reference strain: ATCC 11325.

4. Agrobacterium rubi (Hildebrand) Starr and Weiss 1943, 316. (*Phytomonas rubi* Hildebrand 1940, 694.)

ru'bi. L. n. *Rubus* generic name of blackberry; L. gen.n. *rubi* of *Rubus.*

Causes the formation of small spherical growths, or elongated ridges, described as beading, coralling or knotting, on black and purple cane raspberries and to a lesser extent on red raspberries (*Rubus* spp.). Gall histology not described. Tumorigenic on Pinto bean leaves and on carrot discs.

Reference strain: ATCC 13334.

Species incertae sedis

a. *Agrobacterium pseudotsugae* (Hansen and Smith) Săvulescu 1947, 10. (*Bacterium pseudotsugae* Hansen and Smith 1937, 576; *Phytomonas pseudotsugae* (Hansen and Smith) Burkholder 1939, 209.)

A culture with the characteristics described in the 7th edition of THE MANUAL is not extant. The strain examined by De Ley *et al.* (1966) had peritrichous flagella and a G + C content of the DNA of 67.7 moles %, T_m.

b. *Agrobacterium gypsophilae* (Brown) Starr and Weiss 1943, 316. (*Bacterium gypsophilae* Brown 1934, 1109; *Pseudomonas gypsophilae* (Brown) Stapp 1935, 407; *Phytomonas gypsophilae* (Brown) Magrou 1937, 366; *Xanthomonas gypsophilae* (Brown) Magrou and Prévot 1948, 102.)

The cultures examined by Graham were yellow pigmented facultative anaerobes and could not be differentiated by routine biochemical tests from *Erwinia herbicola* (Graham and Quinn, 1974).

c. *Agrobacterium stellulatum* Stapp and Knösel 1954, 248.

This species, described in the 7th edition of THE MANUAL, possesses polar flagella, is non-pathogenic on plants tested, and the G + C content of the DNA falls outside the range for the genus adopted in this edition (*cf.* De Ley *et al.*, 1966).

d. *Agrobacterium polyspheroidum* Nikitin and Vasil'eva 1968, 444. Inadequately described.

Dr. Renate Ahrens stated, in a personal communication to O. N. Allen, her desire to withdraw the following names misplaced in the genus *Agrobacterium*:

e. *Agrobacterium ferrugineum* Ahrens and Rheinheimer 1967, 136.

f. *Agrobacterium luteum* Ahrens and Rheinheimer 1967, 136. (not *Agrobacterium luteum* (Babes) Hellmuth 1956, 506).

g. *Agrobacterium sanguineum* Ahrens and Rheinheimer 1967, 136.

h. *Agrobacterium agile* Ahrens 1968, 155.

i. *Agrobacterium gelatinovorum* Ahrens 1968, 155.

j. *Agrobacterium kielense* Ahrens 1968, 156.

k. *Agrobacterium aggregatum* Ahrens 1968, 157.

FAMILY IV. **METHYLOMONADACEAE** LEADBETTER 1974*

E. R. LEADBETTER

(*Methanomonadaceae* Breed 1957, 74.)

Meth.yl.o.mo.na'da.ce.ae. M.L. fem.n. *Methylomonas, -adis* a genus of bacteria; *-aceae* ending to denote a family; M.L. fem.pl.n. *Methylomonadaceae* the *Methylomonas* family.

Bacteria utilizing only one-carbon organic compounds, such as methane or methanol, as a carbon source. Gram-negative.

Further Comments

The rationale for using the generic prefix "Methylo" for organisms obligately dependent on either methane or methanol as a source of carbon and energy was developed by Foster and Davis (1966), who anticipated that " . . . methane dependence . . . may yet be found in . . . (diverse) . . . morphological groups." Although a recent extensive examination of methane-utilizing bacteria and their characteristics deliberately avoids making formal taxonomic proposals, it seems likely, indeed, that the prefix "Methylo" will be used later, at the generic level for several different groups (Whittenbury *et al.*, 1970). Another more limited study proposes a new genus *Methylovibrio* (Hazeu and Steenis, 1970) for two new isolates of

---

* Because of unexpected delays in publication, page numbers cannot be given. The paper should appear in the *International Journal of Systematic Bacteriology* during 1974, probably in April.

obligate methane-methanol bacteria. Considering then the sound reasoning behind Foster and Davis's proposal and its apparent adoption by other students of these organisms, Leadbetter (1974) proposed recognition of (1) the new genus *Methylococcus* suggested by Foster and Davis but regarded as not validly published (Hatt and Zvirbulis, 1967); (2) the genus *Methylomonas;* and (3) the family *Methylomonadaceae.*

It would be prudent to regard the genus *Methylovibrio* (Hazeu and Steenis) as *incertae sedis* until more extensive studies enable a careful comparison with the numerous organisms of diverse morphology characterized by Whittenbury *et al.* (1970).

Among the organisms able to use one-carbon organic compounds as carbon and energy sources those restricted to the utilization of only methane or methanol seem unique in their limited nutritional versatility. A number of bacteria are able to utilize multicarbon compounds as well as one-carbon compounds and are not obligately dependent on methane or methanol as a carbon and energy source. These have been considered as *Protaminobacter, Bacillus, Pseudomonas, Vibrio, Carboxydomonas* and *Brevibacterium.*

The basis for the existence of the genus *Protaminobacter* den Dooren de Jong 1927 seems doubtful and both species are considered *incertae sedis* pending additional and comparative studies. The type species, *P. alboflavus,* does not utilize one-

carbon compounds, is rather Gram-positive, and shares many properties of the soil mycobacteria-nocardia types (Leadbetter, unpublished; see also den Dooren de Jong, 1926, 158). *P. ruber* is probably identical with *Bacillus extorquens* Bassalik 1913, 255 (*Vibrio extorquens* (Bassalik) Bhat and Barker 1948, 367) as are also *Pseudomonas* AM1 (Peel and Quayle, 1961), *Pseudomonas* M-27 (Anthony and Zatman, 1964), *Pseudomonas* PRL-W4 (Kaneda and Roxburgh, 1959) and *Pseudomonas methanica* (Harrington and Kallio, 1960) (see Stocks and McCleskey, 1964).

Carboxydomonas oligocarbophila Orla-Jensen 1909, 311 was recognized in the last edition of THE MANUAL but Kistner's 1953 studies suggest that the descriptions of the organisms earlier described were inadequate. The place of *Hydrogenomonas carboxydovorans* Kistner 1954, 186 is also doubtful as a member of a non-extant genus (Davis *et al.,* 1969; see also comments under *Pseudomonas*). Both species are considered *incertae sedis.*

Other one-carbon-utilizing organisms, whose relationship with each other and with recognized organisms is not clear, include: *Pseudomonas aminovorans* den Dooren de Jong 1926, 161; several organisms regarded as *Pseudomonas* species, although unnamed (Kung and Wagner, 1970; Shaw *et al.,* 1966); and a Gram-negative diplococcus (Leadbetter and Gottlieb, 1967).

Organisms able to use single- as well as multi-carbon organic compounds require further study.

Genus I. Methylomonas Leadbetter 1974*

(*Methanomonas* Orla-Jensen 1909, 311.)

Me.thyl.o.mo'nas. M.L. n. *methyl* the methyl radical; Gr. *monas* a unit, monad; M.L. fem.n. *Methylomonas* methyl monad.

Cells **single, straight, curved** or **branched rods,** but **not helical.** Dimensions 0.5–1 by 1–4 μm. **Motile** by monotrichous polar flagellum. Sheaths or prosthecae not known. Resting stages not known. **Gram-negative.**

Chemoorganotrophs: **Metabolism is respiratory, using molecular oxygen as terminal electron acceptor. Methane and methanol are**

the **only known sources of carbon** and **energy.** Organic growth factors are not required.

Strict aerobes. Temperature range 20–35 C. The **G + C content** of the DNA of the type species examined is **52.1 moles** % (buoyant density).

Type species: *Methylomonas methanica* (Söhngen) Leadbetter 1974.*

Description of the species of genus Methylomonas

1. **Methylomonas methanica** (Söhngen) Leadbetter 1974.* (*Bacillus methanicus* Söhngen 1906, 515; *Methanomonas methanica* (Söhngen) Orla-Jensen 1909, 311; *Pseudomonas methanica* (Söhngen) Krasil'nikov 1949, 389; *Pseudomonas methanica* (Söhngen) *sensu* Dworkin and Foster 1956, 657 not *Pseudomonas methanica sensu* Harrington and Kallio 1960, 2.)

me.tha'ni.ca M.L. n. *methanum* methane; M.L. fem.adj. *methanica* relating to methane.

Straight rods 0.6 by 1.0 μm. Extracellular slime formed.

No growth on complex media (e.g. nutrient agar). The presence of organic compounds may inhibit growth on methane or methanol-mineral salts medium. Nitrate is a superior N source to ammonium ion; molecular nitrogen is not fixed. Poly-β-hydroxybutyric acid is not formed.

Catalase positive. Oxidase positive.

---

* See footnote on page 267.

Cells pink, reflecting the presence of carotenoid pigments.

Strict aerobes; grow best at oxygen concentrations of 20% or above. Optimum temperature 30 C; no growth at 37 C. pH optimum about 7, with range of 6.6–8.0 when sodium nitrate was N source.

The G + C content of the DNA of the strain examined is 52.1 moles % (buoyant density) (Foster and Davis, 1966).

Type strain not extant.

2. Methylomonas methanooxidans (Brown, Strawinski and McCleskey) Leadbetter 1974.* (*Methanomonas methanooxidans* Brown, Strawinski and McCleskey 1964, 795.)

me.than.o.ox′i.dans M.L. n. *methanum* methane; M.L. part.adj. *oxydans* oxidizing; M.L. part.adj. *methanooxidans* oxidizing methane.

Straight rods 1.0 by 1.5–3 μm. Rosettes may be formed. Some cells motile by monotrichous polar flagellum.

No growth on complex media (e.g. nutrient agar). Only microcolonies (less than 0.1 mm diameter) are formed on mineral salts-agar medium incubated in a methane-air atmosphere, although

copious growth is obtained in the same medium without agar.

No distinctive pigments are formed.

Strict aerobe. Temperature range 20–40 C; optimum 30–35 C. pH range 5–7; optimum 6.

Suggested neotype: ATCC 15573.

3. Methylomonas methanitrificans (Davis, Coty and Stanley) Leadbetter 1974.* (*Pseudomonas methanitrificans* Davis, Coty and Stanley 1964, 471.)

me.tha.ni.tri′fi.cans. M.L. n. *methanum* methane; M.L. v. *nitrifico* nitrify; M.L. part.adj. *metha-(no)nitrificans* nitrifying methane, actually, using nitrogen (gas) and methane.

Straight rods 1–2 by 2–4 μm. Motile, mode of flagellation not reported.

Methane is only reported source of carbon and energy. Growth on nutrient agar is meager. Molecular nitrogen can be used as sole N source. Poly-β-hydroxybutyric acid is formed.

Colonies of the best studied strain were a light yellow; the pigment was not water-soluble.

Presumably strictly aerobic.

Genus II. Methylococcus Leadbetter 1974*

Meth.yl.o.coc′cus. M.L. n. *methyl* the methyl radical; M.L. n. *coccus* a spherical cell; M.L. masc.n. *Methylococcus* methyl coccus.

Cells spherical, usually occurring in pairs. Non-motile. Resting stages not known. **Gram-negative.**

Chemoorganotrophs: **Metabolism is respiratory, using molecular oxygen as terminal electron acceptor. Methane and methanol are only known sources of carbon and energy.** No organic growth factors are required.

Strictly aerobic.

The **G + C content** of the DNA of the type strain is **62.5 moles** % (buoyant density) (Foster and Davis, 1966).

Type species: *Methylococcus capsulatus* Leadbetter 1974.*

Description of the species of genus Methylococcus

1. Methylococcus capsulatus Leadbetter 1974.*

cap.su.la′tus. M.L. adj. *capsulatus* encapsulated.

Cells 1.0 μm diameter. Capsules formed.

Poly-β-hydroxybutyric acid not formed. Nitrate superior to ammonium ion or casamino acids as N source. No organic growth factors required, al-

though colony formation on solid media is enhanced by complex extracts such as casamino acids.

Temperature optimum 37 C; range 30–50 C; no growth at 55 C.

Cells not pigmented.

Type strain: ATCC 19069.

FAMILY V. **HALOBACTERIACEAE** Fam. nov.

N. E. GIBBONS

Hal.o.bac.ter.i.ac′e.ae. M.L. n. *Halobacterium* type genus of family; -*aceae* ending to denote family; M.L. pl.n. *Halobacteriaceae* the *Halobacterium* family.

Rods and cocci which require high concentrations (>2 M or *ca.* 12%) of sodium chloride for growth. Reproduce by binary fission. Resting stages not known.

---

* See footnote on page 267.

Cell walls do not contain diaminopimelic acid or muramic acid. Cell walls largely lipoprotein, the proteins being acidic and probably more acidic than those of other Gram-negative organisms. Lipids mostly non-saponifiable phospholipid and glycolipid derivatives of a glycerol di-ether. The di-ether linkage is unusual and may be unique, as is the presence of di-hydrophytol groups rather than fatty acids (Kates et al., 1966; Kates, 1972) (ether-linked lipids have been reported in *Thermoplasma acidophilum* (Langworthy et al., 1972) and seem to be an adaptation to extreme environments). Lipid synthesis is by the mevalonic pathway, rather than the more usual malonate pathway (Kates et al., 1968). A fatty acid synthetase has been found in one species but is inhibited by the concentration of salt in the cell (Pugh et al., 1971).

Chemoorganotrophs: Metabolism respiratory, never fermentative. Energy is obtained from amino acids. Little is known about the utilization of carbohydrates.

Contain carotenoids; the main pigment is bacterioruberin (Kelly and Liaaen-Jensen, 1967). Colonies are of various shades of red: pink, orange-red, vermilion, mauve-red.

Found wherever NaCl and other required ions occur in adequate concentrations; salterns (where salt is prepared by solar evaporation of sea water), some salt lakes, the Dead Sea and proteinaceous material preserved with solar salts, such as fish, intestines (sausage casings) and hides.

The DNA shows a major and minor component. The minor component makes up 10–30% of the DNA; such a high content of a minor component has so far been found only in the family *Halobacteriaceae*.

The G + C content of the major component ranges from 66–68 moles %, of the minor component 57–60 moles % (buoyant density: Joshi et al., 1963; Moore and McCarthy, 1969).

Type genus: *Halobacterium* Elazari-Volcani 1957, 207.

Further Comments

Although *Halobacterium* is closely related to *Pseudomonas* in some respects, other characteristics, such as the high ionic requirement for growth and maintenance of cell structure, cell wall composition, lipid content, major and minor DNA components, link the red pigmented halophilic rods and cocci and they are here assigned to a separate family.

It should be emphasized that members of the family *Halobacteriaceae* require media containing at least 15% NaCl. Some physiological reactions differ depending on the salt concentration of the medium (Gibbons, 1957; Colwell and Gibbons, 1972) and may be quite independent of growth. Most tests reported are for media containing 20% NaCl.

Genus I. **Halobacterium** *Elazari-Volcani 1957, 207*

(Not *Halobacterium* Schoop 1935, 26 (nomen nudum); *Flavobacterium* (*Halobacterium*) Elazari-Volcani 1940, V and 59; *Halobacter* Anderson 1954, 66.)

Hal.o.bac.te′ri.um. Gr. n. *hals, halos* the sea, salt; Gr. n. *bakterion* a small rod; M.L. neut.n. *Halobacterium* the salt bacterium.

Cells rod-shaped, 0.6–1.0 by 1–6 μm, occurring singly; may be pleomorphic, especially in deficient media. Reproduce by binary fission. Motile by a tuft of polar flagella or non-motile. Resting stages not known. Gram-negative, although Gram-reaction and usual staining methods not satisfactory because of the high salt concentrations required for growth. Best observed by phase-contrast microscopy.

Mature cells retain their rod shape in 3.5 M NaCl; at lower concentrations pleomorphic forms appear and at 1.5 M the cells are spherical, because of loss of cell wall. At this concentration suspensions are viscous as a result of partial lysis and at 0.5 M few, if any, cells can be detected. Cell structure is maintained by MgCl₂ and CaCl₂.

Sodium, chloride and magnesium ions are required to maintain cell structure and rigidity. As their concentration is reduced the cell wall dissolves and the cell membrane breaks up into tiny fragments.

The internal ion is mainly potassium which is required to maintain the structural integrity of the ribosomes and for protein synthesis, as well as being one of the ions required to maintain activity of intracellular enzymes.

The cell wall does not contain diaminopimelic acid or muramic acid. The wall is largely lipoprotein, being acid and probably more acidic than that of other Gram-negative bacteria. The cell envelope contains about 20% lipid, mostly non-saponifiable derivatives of a glycerol di-ether. The di-ether linkage is unusual as is the presence of di-hydrophytol groups rather than fatty acids. Lipid synthesis seems to be by the mevalonic pathway, rather than the more usual malonate pathway.

Colonies on agar are small (<2 mm), round, convex, entire, translucent and pinkish mauve to bright red or vermilion in color.

Chemoorganotrophs: Metabolism respiratory, never fermentative. Energy obtained from amino acids. Carbohydrates are used only slightly, if at all; acid is not produced in sugar-containing media which become alkaline as the result of deamination and/or decarboxylation of amino acids. Some carbohydrates, such as glucose and glycerol, may be stimulatory under certain conditions and with 2% glycerol acid production has been noted (Gochnauer and Kushner, 1969). Some sugars may be utilized as indicated by manometric techniques. Starch not hydrolyzed.

Contain carotenoids; the main pigment is bacterioruberin, a 50-carbon carotenoid having four hydroxyl groups. This pigment is not restricted to the halophiles, since it occurs in *Corynebacterium poinsettiae*. A few colorless (white) strains have been isolated, probably mutants.

Squalene, dihydrosqualene, tetrahydroxysqualene and vitamin MK8 have been found (Tornabene *et al.*, 1969).

Proteins and proteoses preferred for growth in complex media (peptones not satisfactory), supplemented by at least 3 M NaCl plus Mg and K ions. A useful medium is 0.5% each of proteose-peptone and tryptone (Difco), 25% NaCl, 2% $MgSO_4 \cdot 7H_2O$ and 0.2% KCl. Will grow in media saturated with NaCl. The ions are required both for growth and maintenance of cell structure. Will not grow much below 3 M NaCl; will grow in 2.5 M NaCl + 1.5 M KCl. For details of cultivation see Eimhjellen (1965) or Gibbons (1969).

Oxidase and catalase positive. Gelatin usually liquefied. H_2S produced from peptone and thiosulfate, usually from cysteine. Indole produced in 25 and 30% NaCl. Arginine dihydrolase present. Urease negative. Some biochemical reactions affected by salt concentration in the growth medium.

Strict aerobes, growing on solid surfaces exposed to air, in shallow layers of liquid medium or in well aerated liquid cultures.

Grows well at 30–50 C; optimum for rate of growth 40 C, for cell production 35 C. pH range 5.5–8.0; optimum 7.2–7.4.

Most strains sensitive to polymyxin (300 units); not sensitive to aureomycin, chloromycetin, penicillin, terramycin.

Found where NaCl and other required ions occur in adequate concentrations: salterns, the Dead Sea and some other salt lakes, and heavily salted proteinaceous materials. Survives in solar salt for several years.

The DNA shows a major and a minor component, the latter making up 10–30% of the DNA. The G + C content of the major component ranges from 66–68 moles %, of the minor component 57–60 moles % (buoyant density; Joshi *et al.*, 1963; Moore and McCarthy, 1969).

Type species: *Halobacterium salinarium* (Harrison and Kennedy) Elazari-Volcani 1957, 208.

Further Comments

The extremely halophilic rod-shaped bacteria have been placed in *Pseudomonas* (Harrison and Kennedy, 1922) and in *Serratia* (Bergey *et al.*, 1923; Lochhead, 1934). Elazari-Volcani (1940) assigned them to subgeneric rank in the genus *Flavobacterium* because of their carotenoid pigments. In the 7th edition of THE MANUAL, Elazari-Volcani assigned them to full generic rank and recent work justifies this.

Since they are found on the surface of salt-evaporating ponds, exposed to semitropical sun, their descent from the phototrophic bacteria has been suggested (see Moore and McCarthy, 1969, 261). One halophilic photosynthetic species has been described (see *Ectothiorhodospira halophila*, p. 48). It also contains carotenoids but these differ from those of *Halobacterium*. Some strains of halophiles and some species of photosynthetic bacteria (purple and green) have gas vacuoles, which have some properties in common (Cohen-Bazire *et al.*, 1969; Walsby, 1972). The identification of the purple pigment as retinal and the suggestion that it may act as a photoreceptor (Oesterhelt and Stoeckenius, 1971) also indicates possible relations.

It has been suggested that all rod-shaped halophiles should be placed in one species. However, DNA homologies (Moore and McCarthy, 1969), while showing similarities, seem to justify some differentiation. Until more work has been done, two species are recognized.

Description of the species of genus **Halobacterium**

1. **Halobacterium salinarium** (Harrison and Kennedy) Elazari-Volcani 1957, 208. (*Pseudomonas salinaria* Harrison and Kennedy 1922, 121; *Serratia salinaria* (Harrison and Kennedy) Bergey *et al.* 1923, 93; *Serratia cutirubrum* Lochhead 1934, 275;

Flavobacterium (*Halobacterium*) *salinarium* (Harrison and Kennedy) Elazari-Volcani 1940, 59; *Halobacter salinaria* (Harrison and Kennedy) Anderson 1954, 68; *Halobacterium cutirubrum* (Lochhead) Elazari-Volcani 1957, 209.)

sal.in.ar'i.um. L. adj. *salinarium* belonging or pertaining to salt works.

Morphology and physiology as for genus.

Type strain not extant. Suggested reference strain: Lochhead 91-R6; NRC 34002.

Comments: Most biochemical data have been accumulated on *H. cutirubrum* (Lochhead) Elazari-Volcani 1957, 209, which is considered a subjective synonym of *H. salinarium*. When first isolated it was very proteolytic, completely clearing milk agar plates. Our culture has now lost this ability and is very similar to *H. salinarium;* this relation is borne out by numerical taxonomy studies (Colwell and Gibbons, 1972) and by DNA homology studies (Moore and McCarthy, 1969). For reference the type strain of *H. cutirubrum* is Lochhead 63-R2; NRC 34001.

Bacteriorhodopsin (bacterial visual purple) has been found in a membrane fraction of *H. halobium* and is said to occur only in this species (Oesterhelt and Stoeckenius, 1971; Blaurock and Stoeckenius, 1971). However, Gochnauer *et al.* (1972) find that in media containing 0.1% glycerol, carotenoids of *H. cutirubrum* are largely replaced by bacteriorhodopsin.

2. **Halobacterium halobium** (Petter) Elazari-Volcani 1957, 210. (*Bacillus halobius ruber* Klebahn 1919, 47; *Bacterium halobium* Petter 1931, 1417; *Flavobacterium (Halobacterium) halobium* (Petter) Elazari-Volcani 1940, V and 59.)

hal'o.bi.um. Gr. n. *hals, halos* salt; Gr. n. *bius* life; M.L. adj. *halobium* living on salt.

Rods motile by a tuft of polar flagella.

Many strains contain gas vacuoles; although under light microscopy the cells seem to be filled with a single vacuole, electron microscopy shows that there are numerous highly refractile vacuoles throughout the cytoplasm' (Cohen-Bazire *et al.*, 1969). Macroscopically the vacuoles give the colonies a purplish pink appearance; in liquid suspensions the color changes from pink to red when pressure is applied. Petter noted a wide range of color in her isolates.

Retinal plus a protein (bacteriorhodopsin) has been found in a strain of *H. halobium* (Oesterhelt and Stoeckenius, 1971).

Little biochemical data available. Gelatin hydrolysis and H_2S production in 25 and 30% salt have been reported (Gibbons, 1957). Venkataraman and Sreenivasan (1956) reported no sulfide production in 20% salt, but noted indole production.

Studies by Moore and McCarthy (1969) indicate the base sequence may be somewhat different than in *H. salinarium*, although the G + C content of the DNA is the same.

Type strain: Delft 9 (probably one of Petter's); NRC 34020. Sneath and Skerman (1966) suggested EV6.31.1; NCIB 8720 which is probably the same strain.

Species incertae sedis

a. *Halobacterium marismortui* Elazari-Volcani 1957, 210. (*Flavobacterium (Halobacterium) marismortui* Elazari-Volcani 1940, V and 48.)

As its name implies this organism was isolated from the Dead Sea. To our knowledge it has never been isolated since and no cultures are extant. Its main distinguishing features were the production of acid from fructose, glucose, mannose and glycerol, and the reduction of nitrate with the production of much gas.

b. *Halobacterium trapanicum* (Petter) Elazari-Volcani 1957, 211. (*Bacterium trapanicum* Petter 1931, 1419.)

Isolated from Trapani salt. Its main distinguishing characteristic was the orange color of its colonies.

Type strain: EV6.32.1 (Delft); NRC 34021.

c. *Amoebobacter morrhuae* Penso 1947, 593.

Isolated from salted cod and because of its extremely pleomorphic form placed in *Amoebobacter* (a phototrophic genus, p. 46). Pleomorphism was noted by most early workers with halophiles, partly at least as a result of the growth conditions used. However, Larsen (personal communication) indicates that this organism never assumes a regular rod shape as a result of varying nutritional and physical conditions of growth. While it undoubtedly belongs to the genus *Halobacterium*, its place and acceptance as a separate species must await further study.

Genus II. **Halococcus** *Schoop 1935, 817*

(Not *Halococcus* Sturges and Heideman 1924, 14; not *Halococcus* Hayashi, Kodaira, Baba and Kikuchi 1966, 635 and 639.)

Hal.o.coc'cus. Gr. n. *hals, halos* the sea, salt; Gr. n. *coccus* a berry; M.L. masc.n. *Halococcus* the salt coccus.

Cocci, 0.6–1.5 μm in diameter, occurring in pairs, tetrads and irregular clusters of tetrads. Nonmotile. Gram-negative.

Grow only in the presence of 2.5 M or higher concentrations of NaCl. Cells retain their shape in lower concentrations of salt and seem more resistant to osmotic damage although the cell walls resemble those of *Halobacterium* in not containing diaminopimelic acid or muramic acid.

Colonies small (<2 mm) opaque, convex, pink to red in color. Cells contain carotenoids.

Chemoorganotrophs: Metabolism respiratory, never fermentative.

Aerobic.

The G + C content of the DNA is 67 moles % for the major component and 59 moles % for the minor component, which makes up 31% of the total DNA (Moore and McCarthy, 1969).

Type species: *Halococcus morrhuae* (Farlow) Kocur and Hodgkiss 1973, 154.

Further Comments

The red pigmented halophilic cocci have not been studied as extensively as their rod-shaped counterparts. However, sufficient is known to place them in a separate genus, as proposed by Venkataraman and Sreenivasan (1956), Larsen (1962) and Kocur and Hodgkiss (1973). This is based on their Gram-reaction, high salt requirement for growth, the presence of di-ether linkages and dihydrophytol groups in their lipids and their cell wall composition.

These organisms have been placed in *Sarcina* and *Micrococcus*. *Sarcina* is now limited to anaerobic cocci. The organisms differ in many respects from *Micrococcus*.

The genus *Halococcus* Schoop was based on *Sarcina litoralis* Poulsen. Many authors have applied this name to the red halophilic cocci, although Poulsen's organism was not a halophile. In fact the description is so meager that the organism could not be recognized. It was found in decaying mud from the seashore, implying anaerobic, marine conditions. It was not grown and there is no mention of a pigment; also its dimensions were 2.66–3.99 μm.

The reddening of salted fish and meats was studied by many workers around the turn of the century but none used sufficient salt in their media until about 1915 (Kellerman, 15%). Organisms described previously were probably not the extreme halophiles considered here. It is possible, because of the greater resistance of the coccus to low salt concentrations, that the organism described by Farlow in 1880 from heavily salted codfish and named *Sarcina morrhuae* could have been the organism described herein.

Description of the species of genus **Halococcus**

1. **Halococcus morrhuae** (Farlow) Kocur and Hodgkiss 1973, 154. (Not *Sarcina litoralis* Poulsen 1879, 254; *Sarcina morrhuae* Farlow 1880, 974; *Micrococcus litoralis* (Poulsen) Kellerman 1915, 399; *Micrococcus morrhuae* (Farlow) Klebahn 1919, 47; *Halococcus litoralis* (Poulsen) Schoop 1935, 817.)

morr.hu'ae.. M.L. n. *morrhua* specific epithet of codfish, *Gadus morrhua;* M.L. gen.n. *morrhuae* of codfish.

Morphology as for genus.

Colonies pink to red in color.

Acid is not produced from glucose or other carbohydrates when tested by the usual methods. Glucose and glycerol utilized by some strains when tested manometrically. Starch may or may not be hydrolyzed and hydrolysis is dependent on salt concentration (20% NaCl optimum).

Gelatin may be hydrolyzed by some strains. Casein seldom hydrolyzed; litmus milk not changed. Indole produced in media containing 15–20% NaCl, in about ⅔ of tests in 25% NaCl. H₂S usually produced from sodium thiosulfate; frequently from cysteine. Nitrates reduced to nitrites, urease negative. Arginine dihydrolase usually negative.

Oxidase and catalase positive. Lipases produced.

Cells contain normal menaquinones (Jeffries, 1969).

Temperature range 30–37 C. pH optimum 7.2; some growth at pH 5.5; no growth at pH 8.

Some strains sensitive to tetracycline and polymyxin B. Not sensitive to aureomycin, chloromycetin, erythromycin, terramycin or penicillin.

Suggested neotype: L.D.3.1; CCM 537; NCMB 787; ATCC 17082 (Kocur and Hodgkiss, 1973, 154).

GENERA OF UNCERTAIN AFFILIATION

Genus **Alcaligenes** Castellani and Chalmers 1919, 936

A. J. HOLDING AND J. M. SHEWAN

Al.ca.li′ge.nes. Arabic *al* the; Arabic n. *galīy* the ash of saltwort; French n. *alcali* alkali; Gr. v. *gennaio* produce; M.L. masc.n. *Alcaligenes* alkali-producing (bacteria).

Cells rods, coccal rods or cocci 0.5–1.2 μm by 0.5–2.6 μm usually occurring singly. Motile with one to four (occasionally up to eight) peritrichous

or degenerate peritrichous flagella. No resting stages known. Gram-negative.

Chemoorganotrophs: Metabolism respiratory,

never fermentative. Molecular oxygen is the final electron acceptor.

Strict aerobes. Some strains capable of anaerobic respiration in the presence of nitrate or nitrite, which acts as an alternate electron acceptor.

Most strains (except *A. aquamarinus*) have simple nitrogenous nutritional requirements and produce very turbid growth in liquid media with ammonium or nitrate salts as the sole nitrogen source; some strains require organic nitrogen compounds (amino acids and/or vitamins). Do not fix gaseous nitrogen.

Not actively proteolytic in casein or gelatin media. Cellulose, chitin and agar not hydrolyzed. Oxidase positive (Kovacs' test).

Optimum temperatures between 20 and 37°C. All species grow quickly at pH 7.0.

The G + C content of the DNA ranges from 57.9–70 moles %.

Members of the genus *Alcaligenes* are common, apparently saprophytic, inhabitants of the intestinal tract of vertebrates. They occur in dairy products, rotting eggs and other foods and in fresh water, marine and terrestrial environments in which they are involved in decomposition and mineralization processes. Not known to enter into either pathogenic or symbiotic associations with plants or animals.

Type species: *Alcaligenes faecalis* Castellani and Chalmers, 1919, 936.

Further Comments

The generic description is very similar fundamentally to that of certain other bacteria that enter into special associations with plants or animals. The importance of these special associations should not be disregarded in any detailed consideration of these other organisms and any generic amalgamation at this time is considered to be undesirable. The bacteria that invade the root hairs of leguminous plants and incite the production of root nodules are placed in the genus *Rhizobium* and those bacteria producing disease-lesions in plants and their non-infective variants are allocated to the genus *Agrobacterium*. *Bordetella bronchiseptica* is an animal pathogenic organism with numerous properties in common with *Alcaligenes faecalis*. Because of the close similarity between previous descriptions of the genus *Achromobacter* and the present description of the genus *Alcaligenes*, a proposal to abandon the name *Achromobacter* has been published (Hendrie, Holding and Shewan, 1974*).

The general morphological characteristics of the genus exclude non-motile organisms which are considered to be members of the *Acinetobacter-Moraxella* group, gliding organisms that are simple myxobacteria, and motile bacteria with only polar flagella that are pseudomonads. More unusual morphological types such as spiralled, stalked and budding bacteria are also excluded.

The generic physiological characteristics exclude facultative anaerobes (i.e. those possessing both respiratory and fermentative metabolism), obligate anaerobes, photosynthetic and obligately chemolithotropic bacteria. Certain physiological characteristics including normally the inability to grow at pH 4.5 distinguish the organisms from the genus *Acetobacter*.

Description of the species of genus **Alcaligenes**

1. **Alcaligenes faecalis** Castellani and Chalmers 1919, 936. (*Bacillus faecalis alcaligenes* Petruschky 1896, 187; *Bacterium faecalis alcaligenes* Chester 1897, 73; *Bacterium alcaligenes* Mez 1898, 63; *Bacillus alcaligenes* (Mez) Migula 1900, 737; *Vibrio alcaligenes* (Mez) Lehmann and Neumann 1927, 548; *Bacterium faecale alcaligenes* Monias 1928, 330; *Pseudomonas alcaligenes* (Mez) Pribram 1933, 50; *Salmonella alcaligenes* (Mez) Pribram 1933, 58; *Achromobacter alcaligenes* (Mez) Brisou and Prévot 1954, 727; *Alcaligenes denitrificans* Leifson and Hugh 1954, 512; Hendrie, Holding and Shewan 1974*; *Achromobacter arsenoxydanstres* Turner 1954, 475; (not validly published) Hendrie, Holding and Shewan 1974*; *Achromobacter alcaligenes* (Mez) Brisou 1955, 173; *Lophomonas alcaligenes* (Mez) Galarneault and Leifson 1956, 102; *Alcaligenes odorans* (Malek and Kaz-dová-Kožiškova) Malek, Radochová and Lysenko 1963, 353.)

fae.ca'lis. L. n. *faex, faecis* dregs; M. L. adj. *faecalis* fecal.

Cocci or coccal rods 0.5 by 0.5–2.0 μm, usually occurring singly. Motile with one to eight peritrichous flagella. No special pigments produced. Characteristic strawberry-like odor produced by some strains.

Utilize as sole carbon and energy source, acetate, propionate, butyrate and some other organic acids; aspartic acid, asparagine, histidine, glutathione and some other organic nitrogenous compounds. Utilization of carbohydrates and chemolithotropic growth using H_2 gas has not been demonstrated. Oxidation of arsenite, which is very active in some strains, does not support chemolithotropic growth.

* Because of unexpected delays in publication, page numbers cannot be given. The paper should appear in the *International Journal of Systemic Bacteriology* during 1974.

Sole nitrogen sources include ammonium and nitrate salts and some amino acids and other organic nitrogenous compounds.

Some strains able to denitrify, by respiring anaerobically in the presence of nitrate or nitrite to produce nitrogen gas. May lose this ability on prolonged aerobic cultivation. Litmus milk alkaline.

Optimum temperature between 25° and 37°C.

The G + C content of the DNA is 58.9 moles %.

Reference strain: ATCC 8750; NCIB 8156.

2. **Alcaligenes aquamarinus** (ZoBell and Upham) Hendrie, Holding and Shewan 1974.* (*Achromobacter aquamarinus* ZoBell and Upham 1944, 264; *Bacterium aquamarinum* (ZoBell and Upham) Krasil'nikov 1949, 437.)

a.qua.ma.ri'nus. L. n. *aqua* water; L. adj. *marinus* of the sea; M.L. adj. *aquamarinus* pertaining to sea water.

Rods 0.7 by 2–4 μm. Motile with one to eight peritrichous flagella.

Oxidizes glucose, fructose, maltose and possibly other carbohydrates. Hydrolyzes starch.

Utilization of ammonium or nitrate salts as sole nitrogen source has not been demonstrated. Nitrate not reduced to nitrogen gas. Litmus milk unchanged.

Optimum temperature 20–25 C.

The G + C content of the DNA is 57.9 moles %.

Reference strain: ATCC 14400; NCMB 557.

3. **Alcaligenes eutrophus** Davis in Davis, Doudoroff, Stanier and Mandel 1969, 386. (*Hydrogenomonas eutropha* Wittenberger and Repaske 1958, 106 not validly published.)

eu.troph'us. Gr. prep. *eu* good, beneficial; Gr. n. *trophus* one who feeds; M.L. n. *eutrophus* good nutrition, well nourished.

Rods 0.5 by 1.8–2.6 μm; one to four peritrichous flagella. Colonies opaque, white or cream colored; after several days become brown.

Can use a variety of organic compounds as sole carbon source; these include glucose (mutant growth), fructose, testosterone, phenol, benzoate and many others. Cannot utilize ethanol, glycerol, mannitol, pentoses, disaccharides, D-tryptophan or acetamide.

Nitrates may be reduced to gaseous nitrogen but the ability may be lost on prolonged aerobic cultivation. Starch and gelatin not hydrolyzed.

Facultatively chemolithotrophic in an atmosphere containing H_2, O_2 and CO_2 gases.

Optimal temperature about 30°C.

Habitat: Soil and water.

The G + C content of DNA is 66.3–66.8 moles %.

Type strain: ATCC 17697.

* See footnote on page 274.

4. **Alcaligenes paradoxus** Davis in Davis, Doudoroff, Stanier and Mandel 1969, 387.

pa.ra.dox'us. Gr. prep. *para* amiss, contrary to; Gr. n. *doxus* an opinion; M.L. n. *paradoxus* contrary to expectation, in reference to the chemolithotropic and/or organotropic metabolism of the organism.

Rods 0.5 by 1.5–2.6 μm; singly or in pairs. Motile by one to two "degenerately peritrichous" flagella (subpolar or lateral insertion) which may be 4–6 times the length of the cell.

Colonies glistening and slimy, have a yellow carotenoid pigment.

Utilizes glucose, fructose, mannose, galactose and L-arabinose as carbon sources; sucrose and trehalose not used. Most strains can utilize pantothenate (may require mutant growth), malonate, sorbitol and many other organic compounds. In complex media incubated aerobically acid is produced from glucose.

Sorbitan mono-oleate (Tween 80) hydrolyzed; starch and gelatin not hydrolyzed.

Optimal temperature about 30 C.

Habitat: Soil.

G + C content of DNA is 68–70 moles %.

Type strain: ATCC 17713.

Two biotypes are recognized:

Biotype I. Facultatively autotrophic; reduce nitrate to nitrite in organic media.

Type strain: ATCC 17713.

Biotype II. Non-autotrophic; rarely reduce nitrate to nitrite in organic media.

Reference strain: ATCC 17549.

Species incertae sedis

The following extant species previously described as *Achromobacter* appear to be acceptable as *Alcaligenes*. They have not been compared with other species and their status is therefore uncertain.

a. *Achromobacter agile* (Jensen) Bergey *et al.* 1923, 138. NCIB 9986.

b. *Achromobacter cholinophagum* (*sic*) Shieh 1964, 839. NCMB 1501; ATCC 15918.

c. *Achromobacter pestifer* (Frankland and Frankland) Bergey *et al.* 1923, 140. ATCC 15445.

d. *Achromobacter starkeyi* Ruiz-Herrera 1970, 329. NCIB 10688.

Editorial Note: A list of many species previously assigned to *Achromobacter*, *Alcaligenes* and other genera, whose present status as to *Alcaligenes* is uncertain, has been compiled by the authors and will be published in the *International Journal of Systematic Bacteriology*.

Genus **Acetobacter** Beijerinck 1898, 215*

J. De Ley and J. Frateur

(*Ulvina* Kützing 1834, 1; *Mycoderma* Thompson 1852, 89; *Termobacterium* Lindner 1895, 243; *Acetobacterium* Ludwig 1898, 870; *Acetimonas* Orla-Jensen 1909, 312.)

A.ce.to.bac'ter. L. n. *acetum* vinegar; M.L. n. *bacter* masc. equivalent of Gr. neut.n. *bakterion* rod, staff; M.L. masc.n. *Acetobacter* vinegar rod.

Cells ellipsoidal to rod-shaped, straight or slightly curved, 0.6–0.8 by 1.0–3.0 μm, occurring singly, in pairs or in chains. Involution forms frequent in some species and may be spherical, elongated, swollen, club-shaped, curved, branched or filamentous. Motile by peritrichous flagella or

Oxidize ethanol to acetic acid at neutral and acid reactions (pH 4.5). Acetate and lactate are oxidized to CO_2 and H_2O ("overoxidizers"). Ethanol, lactate are good carbon sources. Hexoses and glycerol are used as carbon sources by most strains; mannitol and glutamate are used poorly

TABLE 7.12

Characteristics differentiating the species and subspecies of genus **Acetobacter**

	1. A. aceti				2. A. pasteurianus					3. A. peroxydans
	1a. subsp. aceti	1b. subsp. orle-anensis	1c. subsp. xylinum	1d. subsp. lique-faciens	2a. subsp. pasteur-ianus	2b. subsp. lovani-ensis	2c. subsp. estu-nensis	2d. subsp. ascendens	2e. subsp. para-doxus	
Catalase	(+)	(+)	(+)	(+)	(+)	(+)	(+)	(+)	−	−
Ketogenesis in glycerol or erythritol	+[a]	+[a]	+[a]	+[a]	−[b]	−[b]	−[b]	−[b]	−	−
Formation of										
5-Ketogluconate	(+)	(+)	(+)	(+)	−	−	−	−	−	−
2-Ketogluconate	(+)	(+)	(+)	(+)	d	−	−	−	−	−
Gluconate	+	+	+	+	(+)	+	+	−	−	−
Growth on ethanol[c]	+	−	(−)	+	(−)	+	+	−	−	+
Produces										
Cellulose	−	−	+	−	−	−	+	−	−	−
γ-Pyrones	−	−	−	+	−	−	−	−	−	−
Brown pigment	−	−	−	+	−	−	−	−	−	−
G + C, moles %	59–65	60–61	62–63	64	55–62		62		55–56	61–64

[a] Moderate to strong.

[b] Negative or very weak; ( ) usually positive or negative.

[c] In Hoyer-Frateur medium: ethanol, 3% (v/v); $(NH_4)_2SO_4$, 0.1%; K_2HPO_4, 0.01%; KH_2PO_4, 0.09%; $MgSO_4 \cdot 7H_2O$, 0.025%; $FeCl_3$, 0.0005%.

non-motile. Endospores not formed. Young cells Gram-negative, in older cultures some strains become Gram variable.

Chemoorganotrophs: Metabolism respiratory, never fermentative; oxygen is the terminal acceptor.

Characteristics differentiating the genus from *Gluconobacter* and *Pseudomonas* are given in Table 7.7.

or not at all; lactose, dextrin and starch are not hydrolyzed. γ-Pyrones usually not produced.

Many amino acids readily oxidized by resting cells.

Generally no pigments produced, but cell masses may be pink due to porphyrins; some strains produce a brown, water-soluble pigment.

Growth reported on simple and complex media; most strains do not require vitamins.

* The fact that Beijerinck did not formally propose the generic name *Acetobacter* is appropriately assessed by Kluyver (1940, 133) ". . . it is surprizing that neither Beijerinck nor Hoyer proposed in their publications the creation of a new genus for the acetic acid bacteria. In various papers which appeared shortly after 1898, the name *Acetobacter* is used without further explanation. There can be no doubt that in any case morally. . . Beijerinck is to be considered as the author of the genus *Acetobacter*. . .''

Strict aerobes.

Optimum temperature for growth *ca.* 30 C; temperature range 5–42 C. pH optimum 5.4–6.3; growth occurs at pH 4.0–4.5; little growth at pH 7–8.

Found on fruits and vegetables, souring fruit juices, vinegar, alcoholic beverages.

The G + C content of the DNA ranges from 55–64 moles % (T_m, chemical analysis).

Type species: *Acetobacter aceti* (Pasteur) Beijerinck 1898, 215.

Further Comments

Acceptance of the genus *Gluconobacter* Asai (p. 251) makes a redefinition of the genus *Acetobacter* necessary.

Some 50 specific epithets have been published

in naming species of the genus. In the 7th edition of The Manual five species were described: *A. aceti; A. xylinum; A. rancens; A. pasteurianus; A. kuetzingianus.* Frateur (1950) divided the genus into three groups: (a) peroxydans (*A. peroxydans, A. paradoxum*); (b) oxydans (*A. lovaniense, A. ascendens, A. rancens*); (c) mesoxydans (*A. aceti, A. xylinum, A. mesoxydans*). More recent investigations suggest that the number of accepted species should be reduced (Shimwell, 1957, 1959; De Ley, 1961; Scopes, 1962; Asai, Iizuka and Komagata, 1964).

The three groups of Frateur are here accepted as species with some of the former species epithets being accepted as subspecies.

Characteristics differentiating the three species and their subspecies are given in Table 7.12.

Description of the species of genus **Acetobacter**

1. **Acetobacter aceti** (Pasteur) Beijerinck 1898, 215. (*Mycoderma aceti* non visqueux (membraneux) Pasteur 1864, 125; *Bacterium aceti* (Pasteur) Lanzi 1876, 257; *Bacteriopsis aceti* (Pasteur) Trevisan 1885, 104; *Micrococcus aceti* (Pasteur) Maggi 1886, 81; *Bacillus aceticus* Flügge 1886, 313; *Bacterium hansenianum* Chester 1901, 126; *Acetimonas aceti* (Pasteur) Orla-Jensen 1909, 312; *Bacterium acetigenoideum* Krehan 1930, 496; *Acetobacter ketogenum* (*sic*) Walker and Thomas in Bousfield, Wright and Walker 1947, 258; *Acetobacter lafarianum* (*sic*) Janke 1950, 116; *Acetobacter aceti* var. *muciparum* (*sic*) (Hoyer) Frateur 1950, 330.)

a.ce′ti. L. n. *acetum* vinegar; L. gen.n. *aceti* of vinegar.

Morphology as for genus. Involution forms occur.

Moderate to good iridescent growth on agar containing yeast extract and 2% Ca acetate or Ca lactate. On yeast extract, 10% glucose, 3% $CaCO_3$ agar growth good, production of much gluconic acid, crystals of Ca 5-ketogluconate. With 10% fructose substituted, Ca oxalate is sometimes produced.

L-Arabinose, D-galactose, D-mannose and D-xylose oxidized to the corresponding acids. Acid formation from sucrose variable.

Pantothenic acid, *p*-aminobenzoic acid and nicotinic acid sometimes required.

Suggested neotype: NCIB 8621.

1a. *Acetobacter aceti* subsp. *aceti* comb. nov.

Morphology and description as for species (Table 7.12).

1b. *Acetobacter aceti* subsp. *orleanensis* (Henneberg) *comb. nov.* (*Mycoderma aceti* (Kützing)

Pasteur in Hansen 1879, 230; *Bacterium orleanense* Henneberg 1906, 106; *Bacillus orleanensis* (Henneberg) Macé 1913, 451; *Bacterium dihydroxyacetonicum* Virtanen and Bärlund 1926, 170; *Ulvina orleanense* (Henneberg) Pribram 1933, 75; *Acetobacter orleanense* (Henneberg) Frateur 1950, 315; *Acetobacter mesoxydans* Frateur 1950, 330.)

Description as for species but does not grow on ethanol (Table 7.12).

1c. *Acetobacter aceti* subsp. *xylinum* (Brown) *comb. nov.* (*Bacterium xylinum* Brown 1886, 439; *Bacillus xylinus* (Brown) Trevisan 1889, 16; *Acetobacterium xylinum* (Brown) Ludwig 1898, 870; *Bacterium acidi oxalici* Banning 1902, 396; *Bacterium xylinoides* Henneberg 1906, 113; *Bacillus xylinoides* (Henneberg) Macé 1913, 451; *Acetobacter xylinum* (Brown) Bergey *et al.* 1925, 37; *Acetobacter xylinoides* (Henneberg) Shimwell 1948, 31; *Acetobacter xylinum* var. *maltovorans* (*sic*) Frateur 1950, 330; *Acetobacter xylinum* var. *xylinoides* (*sic*) (Henneberg) Frateur 1950, 330; *Acetobacter bordeaux* Janke 1957, 732; *Acetobacter xylinum* var. *africanum* (*sic*) Kulka, Singh, Nattrass, Hall and Walker 1958, 490.)

Surface growth on beer tough and leathery.

Produces cellulose. Does not usually grow on ethanol (Table 7.12).

1d. *Acetobacter aceti* subsp. *liquefaciens* (Asai) *comb. nov.* (*Gluconoacetobacter liquefaciens* Asai 1935, 610; *Gluconobacter liquefaciens* Asai 1935, 679.)

Appears to have both peritrichous and polar flagella.

Produces brown pigment on yeast extract glucose $CaCO_3$ agar and on yeast extract gluconate agar. Produces γ-pyrones (Table 7.12).

2. **Acetobacter pasteurianus** (Hansen) Beijerinck 1916, 1199. (*Mycoderma pasteurianum* Hansen 1879, 230; *Bacterium pastorianum* (Hansen) Zopf 1883, 52; *Bacteriopsis pasteuriana* (Hansen) Trevisan 1885, 104; *Bacillus pasteurianus* (Hansen) Flügge 1886, 314; *Bacterium kützingianum* Hansen 1894, 289; *Bacterium rancens* Beijerinck 1898, 211; *Bacillus küttingianum* (*sic*) (Hansen) Takahashi 1906, 553; *Bacterium vini-acetati* Henneberg 1906, 122; *Acetobacter kützigianus* (*sic*) (Hansen) Bergey *et al.* 1923, 35; *Ulvina pasteuriana* (Hansen) Pribram 1933, 76; *Acetobacter turbidans* Cosbie, Tošić and Walker 1942, 82; *Acetobacter mobile* Tošić and Walker 1944, 296; *Acetobacter vini-acetati* (Henneberg) Shimwell 1948, 30; *Acetobacter agglutinans* Frateur 1950, 313; *Acetobacter acidum-mucosum* (*sic*) Tošić and Walker 1950, 192; *Acetobacter alcoholophilus* Kozulis and Parsons 1958, 47.)

pas.teur.i.a'nus. M.L. adj. *pasteurianus* of Pasteur; named for Louis Pasteur, French chemist and bacteriologist.

Morphology as for genus. Involution forms frequently produced. Some strains have capsules or produce slime.

Moderate to good growth on yeast extract agar plus 2% Ca acetate or lactate; usually with a pearl-like lustre. Many strains produce acetyl-methyl-carbinol from DL-lactate.

Some strains oxidize L-arabinose, D-galactose and D-xylose to their respective acids.

Acid not formed from D-arabinose, fructose, lactose, maltose, rhamnose, sucrose, raffinose or starch.

Amyloid substances formed in beer and on media containing fructose.

Pantothenic, *p*-aminobenzoic and nicotinic acids sometimes required for growth.

For other characteristics see Table 7.12.

Found in beer, wines and vinegars.

2a. *Acetobacter pasteurianus* subsp. *pasteurianus* comb. nov.

Description as for species.

2b. *Acetobacter pasteurianus* subsp. *lovaniensis* (Frateur) *comb. nov.* (*Acetobacter lovaniense* Frateur 1950, 328.)

Description as for species but grows on ethanol (Table 7.12).

Does not produce 2-ketogluconate.

2c. *Acetobacter pasteurianus* subsp. *estunensis* (Carr) *comb. nov.* (*Acetobacter estunenses* (*sic*) Carr 1958, 157.)

Description as for species but grows on ethanol and produces cellulose (Table 7.12).

2d. *Acetobacter pasteurianus* subsp. *ascendens* (Henneberg) *comb. nov.* (*Bacterium ascendens* Henneberg 1898, 145.)

Description as for species but does not produce gluconate from glucose.

2e. *Acetobacter pasteurianus* subsp. *paradoxus* (Frateur) *comb. nov.* (*Acetobacter paradoxum* (*sic*) Frateur 1950, 328.)

Description as for species but catalase negative. Does not produce gluconate from glucose. Grows on Ca gluconate agar but does not cause precipitation of $CaCO_3$. Does not grow on ethanol.

3. **Acetobacter peroxydans** Visser 't Hooft 1925, 98.

Rods, 0.5 by 2–3 µm, often in pairs or short chains. Involution forms do not occur. Motile.

Does not grow on or produce acid from D- or L-arabinose, D-fructose, D-galactose, D-glucose, lactose, maltose, rhamnose, sucrose, D-ribose, D-xylose, raffinose, starch, gluconate, sugar-alcohols. No or extremely weak ketogenesis on sugar-alcohols. No oxidation of butyrate, glycerate, 2- or 5-ketogluconate, or propionate. Amyl alcohol, *n*-butanol, isobutanol and *n*-propanol oxidized to the corresponding acids; isopropanol and *s*-butanol oxidized to corresponding ketones.

Several strains oxidize hydrogen.

Catalase negative, peroxidase positive.

Optimum temperature for growth 20–25 C; no growth at 40 C.

Found in wine, sugar beet pulp silage, ditch water, sewage.

The G + C content of the DNA ranges from 61–64 moles % (T_m, chemical analysis).

Type strain: NCIB 8618.

Genus **Brucella** *Meyer and Shaw 1920, 173*

W. J. BRINLEY-MORGAN AND N. B. McCULLOUGH[*]

Bru.cel'la. L. dim.ending -*ella*; M.L. fem.n. *Brucella* named after Sir David Bruce, who first recognized the organism causing undulant fever.

Coccobacilli or short rods, 0.5–0.7 by 0.6–1.5 µm, arranged singly, more rarely in short chains. No capsules. **Non-motile.** Do not form endo-spores. **Gram-negative.** Do not show bipolar staining.

Chemoorganotrophs: Metabolism respiratory.

[*] The authors are greatly indebted to Dr. Lois M. Jones, University of Wisconsin, for valuable comment and criticism.

Growth of some strains can occur in a chemically defined medium with ammonium ion as the sole source of nitrogen (McCullough and Dick, 1943); most strains show better growth, and from a smaller inoculum, using multiple amino acids (Rode, Oglesby and Schuhardt, 1958; Gerhardt, 1958).

The following vitamins are required for growth: thiamin, niacin and biotin; Ca pantothenate often stimulates growth; hemin (X factor) and coenzyme 1 (V factor) not required.

Catalase positive; oxidase usually positive but *B. neotomae* and *B. ovis* are oxidase negative. Urea hydrolyzed to a variable extent. Nitrates reduced to nitrites (except for *B. ovis*). Citrate not utilized; indole not produced; methyl red and Voges-Proskauer tests negative. Litmus milk no change.

Strict aerobes; some require 5–10% added CO_2 for growth especially on initial isolation.

Temperature range 20–40 C; optimum 37 C. Optimum pH 6.6–7.4.

Mammalian parasites and pathogens; facultatively intracellular with a relatively wide host range.

The G + C content of the DNA ranges from 56–58 moles % (buoyant density) (Hoyer and McCullough, 1968); members of the genus comprise a closely knit and sharply demarcated genetic group as defined by DNA hybridization studies.

Type species: *Brucella melitensis* (Hughes) Meyer and Shaw 1920, 179.

Further Comments

Most strains of *Brucella* grow poorly on peptone medium and cannot be grown from small inocula whereas peptones such as tryptose and trypticase soya provide for adequate growth; growth may be improved by addition of 1–5% (v/v) of serum. After 2 days of incubation, colonies pinpoint, reaching 2–3 mm after 4 days; non-hemolytic and non-pigmented, circular convex with a smooth, glistening surface and an entire edge. Growth in liquid medium shows moderate turbidity with slight sediment but no pellicle formation.

Members of the genus prone to spontaneous dissociation in laboratory media especially in static, liquid media. The three classical species—*B. melitensis*, *B. abortus* and *B. suis* almost invariably in the smooth (S) phase on primary isolation but intermediate (I), rough (R) and mucoid (M) colony types can occur. Colonial dissociation is best observed under obliquely transmitted light (Henry, 1933) or may be detected by agglutinability of cells in 1:1000 neutral acriflavine (Braun and Bonestell, 1947) or by their staining reaction with crystal violet (White and Wilson, 1951). If a culture normally produces S phase colonies, only colonies of

the S or I type should be used for serological identification. *B. ovis* and *B. canis*, as isolated in nature, lack S antigens.

Smooth species of *Brucella* show complete cross-reactivity by agglutination tests with unabsorbed anti-smooth Brucella sera. Different quantitative distribution of the M and A surface antigens of smooth *Brucella* can be shown by using monospecific antisera prepared by differential absorption (Wilson and Miles, 1932). Rough species of *Brucella* and rough variants of smooth *Brucella* show extensive cross-agglutination using unabsorbed anti-rough Brucella sera but show little cross-reaction with unabsorbed anti-smooth Brucella sera (Diaz, Jones and Wilson, 1967; Diaz *et al.*, 1968).

By using immune diffusion methods, soluble antigens of *Brucella* show extensive cross-reaction within the genus, but not with members of other genera. Most of the soluble antigens are identical in all the species and readily diffusible through agar but smooth cultures possess additional lipopolysaccharide M and A antigens which diffuse less readily (especially the A; Leong, Diaz and Wilson, 1968).

Brucella phage lyses smooth forms of some species of the genus but not members of other genera. *Brucella* not lysed by phages of other genera.

The differential characters of the species and biotypes of the genus *Brucella* are given in Table 7.13.

a. Use of "conventional" tests (Huddleson, 1929; Wilson and Miles, 1932). The requirement for added CO_2 for growth should be determined upon initial isolation as adaptation to non-dependence may occur quickly. The production of H_2S varies with the medium used and the determination of differential growth on media containing basic fuchsin and thionin is also influenced by the medium. Appropriate known reference cultures should always be included as controls. Unabsorbed anti-smooth, unabsorbed anti-rough sera as well as monospecific antisera should be used for serological identification.

b. Lysis by *Brucella* bacteriophage, using the Tbilisi (Tb) phage (Morgan, Kay and Bradley, 1960; Jones, 1960; Meyer, 1961; Ostrovskaya and Kaitmasova, 1966).

c. The use of manometric techniques (Meyer and Cameron, 1961; Meyer, 1964, 1966); the amino acids and carbohydrates oxidized are shown in Table 7.13. Rates of oxidation measured manometrically are especially of value in examining cultures which are non-typical or isolated from unusual hosts.

Details of techniques used for all these tests are available (Morgan and Gower, 1966; Alton and Jones, 1967).

TABLE 7.13

Differential characters of the species and biotypes of the genus **Brucella**

Biotype	CO_2 required	H_2S produced	Basic fuchsin b.	Thionin a.	Thionin b.	Mono A	Mono M	Anti-rough serum	Lysis by phage Tb, RTD[b]	L-Alanine	L-Asparagine	L-Glutamic acid	L-Arabinose	D-Galactose	D-Ribose	D-Glucose	L-Erythritol	D-Xylose	L-Arginine[c]	L-Lysine
1. B. melitensis																				
1	−	−	+	−	+	−	+	−	−	+	+	+	−	−	−	+	+	−	−	−
2	−	−	+	−	+	+	−	−	−	+	+	+	−	−	−	+	+	−	−	−
3	−	−	+	−	+	+	+	−	−	+	+	+	−	−	−	+	+	−	−	−
2. B. abortus																				
1	d	+	+	−	−	+	−	−	+	+	+	+	+	+	+	+	+	−	−	−
2	d	+	−	−	−	+	−	−	+	+	+	+	+	+	+	+	+	−	−	−
3	d	+	+	+	+	+	−	−	+	+	+	+	+	+	+	+	+	−	−	−
4	d	+	+	−	−	−	+	−	+	+	+	+	+	+	+	+	+	−	−	−
5	−	−	+	−	+	−	+	−	+	+	+	+	+	+	+	+	+	−	−	−
6	−	d	+	−	+	+	−	−	+	+	+	+	+	+	+	+	+	−	−	−
7	+	−	+	−	+	−	+	−	+	+	+	+	+	+	+	+	+	−	−	−
8	−	+	+	−	+	−	+	−	+	+	+	+	+	+	+	+	+	−	−	−
9	−	+	+	−	+	−	+	−	+	+	+	+	+	+	+	+	+	−	−	−
3. B. suis																				
1	−	+	−	+	+	+	−	−	−	−	−	−	+	+	+	+	+	+	+	+
2	−	−	−	+	+	+	−	−	−	−	+	+	+	+	+	+	+	+	+	−
3	−	−	+	+	+	+	−	−	−	−	−	+	−	−	+	+	+	+	+	+
4	−	−	+	+	+	+	+	−	−	−	−	+	−	−	+	+	+	+	+	+
4. B. neotomae	−	+	−	−	+	+	−	−	−	−	+	+	+	+	d	+	+	+	−	−
5. B. ovis	+	−	+	+	+	−	−	+	−	+	+	+	−	−	−	−	−	−	−	−
6. B. canis	−	−	−	+	+	−	−	+	−	−	−	−	−	−	+	+	d	−	+	+

[a] Certified dyes (National Aniline Division, Allied Chemical and Dye Co., New York) at concentrations a. = 1:25,000; b. = 1:50,000.

[b] RTD, routine test dilution

[c] Same reactions with DL-citrulline and DL-ornithine.

Description of the species of genus **Brucella**

1. **Brucella melitensis** (Hughes) Meyer and Shaw 1920, 179. (*streptococcus Miletensis* (*sic*) Hughes 1893, 325; *Micrococcus melitensis* (Hughes) Bruce 1893, 297.)

me.li.ten′sis. L. adj. *melitensis* of or pertaining to the Island of Malta.

Aerobic. Produce no H_2S or no more than a trace on peptone media. Usually Melitensis antigen predominant.

Usually pathogenic for goats and sheep but can also affect other species including cattle and man.

Three biotypes are recognized: See Table 7.13.

B. melitensis biotype 1—"classical" *B. melitensis* and most typical of the species.

Neotype strain of species and biotype reference strain: 16M; NCTC 10094; ATCC 23456 (Meyer and Morgan, 1973, 137).

B. melitensis biotype 2—(*B. intermedia* Renoux 1952) has Abortus antigen predominant. Reference strain: 63/9; NCTC 10508; ATCC 23457.

B. melitensis biotype 3—similar to biotype 1 but agglutinate in both abortus and melitensis monospecific sera (Wilson, 1933). Reference strain: Ether; NCTC 10509; ATCC 23458.

2. **Brucella abortus** (Schmidt and Weis) Meyer and Shaw 1920, 176. (Bacillus of abortion, Bang 1897, 250; *Bacterium abortus* Schmidt and Weis 1901, 266.)

a.bor′tus. L. n. *abortus*, abortion.

Usually require added CO_2 (5%) for growth especially on primary isolation. Usually produce moderate amounts of H_2S but may be negative. Usually have Abortus antigen predominant.

Usually pathogenic for cattle causing abortion but can also affect other species including man.

Nine biotypes are recognized:

B. abortus biotype 1—"classical" *B. abortus* and most typical of the species.

Neotype strain of species and biotype reference strain: 544; NCTC 10093; ATCC 23448 (Meyer and Morgan, 1973, 137).

B. abortus biotype 2—(Wilson, 1933) or "dyesensitive"; do not grow on media containing basic fuchsin or thionin; require serum or Tween 40 in media for growth. Reference strain: 86/8/59; NCTC 10501; ATCC 23449.

B. abortus biotype 3—thionin-resistant *B. abortus* (Bevan, 1930; Van der Schaaf and Rosa, 1940); may or may not require added CO_2 for primary isolation. Reference strain: TULYA; NCTC 10502; ATCC 23450.

B. abortus biotype 4—(Gilbert, 1930) a variety of *B. abortus* but agglutinated by monospecific melitensis rather than abortus serum. Reference strain: 292; NCTC 10503; ATCC 23451.

Biotypes 5–9 are typical of or similar to *B. melitensis* on conventional typing tests but are lyzed by Tbilisi Brucella phage at RTD and have the oxidative metabolic pattern of *B. abortus*.

B. abortus biotype 5—so-called British melitensis (Stableforth, 1959; Morgan *et al.*, 1960; Meyer, 1962). Reference strain: B3196; NCTC 10504; ATCC 23452.

B. abortus biotype 6—(*B. intermedia* Renoux 1952). Reference strain: 870; NCTC 10505; ATCC 23453.

B. abortus biotype 7—reference strain: 63/75; NCTC 10506; ATCC 23454.

B. abortus biotype 8—CO_2-requiring *B. melitensis* (Taylor *et al.*, 1932). Cultures not available for selection of reference strain.

B. abortus biotype 9—H_2S-producing *B. melitensis* (Taylor *et al.*, 1932). Reference strain: C68; NCTC 10507; ATCC 23455.

3. **Brucella suis** Huddleson 1929, 12. (Organism resembling *Bacillus abortus* Traum 1914, 86.)

su′is. L. n. *sus* the hog, swine; L. gen.n. *suis* of the hog.

Aerobic. Produce large amounts of H_2S or none at all. Usually have Abortus antigen predominant.

Usually pathogenic for pigs but can also affect hares, reindeer and other species including man.

Four biotypes are recognized:

B. suis biotype 1—typical of the species.

Neotype strain of species and biotype reference strain: 1330; NCTC 10316; ATCC 23444 (Meyer and Morgan, 1973, 137).

B. suis biotype 2—isolated by Thomsen (1929) and characterized by Kristensen (1931). Appears to be of low pathogenicity for man. Reference strain: Thomsen; NCTC 10510; ATCC 23445.

B. suis biotype 3—cultures isolated from hogs and man in U. S. A. (Huddleson, 1957). Reference strain: 686; NCTC 10511; ATCC 23446.

B. suis biotype 4—(*B. rangiferi tarandi* Davydov 1961, 31); originally isolated from reindeer in the far north of Russia; also found in caribou and Eskimos in Alaska (Huntley, Phillip and Maynard, 1963; Meyer, 1966). Reference strain: 40; ATCC 23447.

4. **Brucella neotomae** Stoenner and Lackman 1957, 947.

neo.tom′ae. M.L. fem.n. *Neotoma* generic name of the desert wood rat of the western U. S. A., *Neotoma lepida* Thomas; M.L. fem.gen.n. *neotomae* of the desert wood rat, the host from which the organism was first isolated.

Aerobic; Abortus antigen predominant. Not lysed by Brucella phage Tb at RTD, lysed at 10,000 × RTD.

Occurs in the desert wood rat (*Neotoma lepida* Thomas).

Type strain: 5K33; NCTC 10084; ATCC 23459.

5. Brucella ovis Buddle 1956, 362. (Not *Brucella ovis* van Drimmelen 1953, 302.)

o'vis L. n. *ovis* a sheep.

Require added CO_2 for growth. Even on primary isolation, cultures are in the non-smooth phase and do not possess Abortus and Melitensis antigens of smooth brucellae.

Pathogenic for sheep causing epididymitis in rams and may cause abortion in ewes. Cross-agglutinates with *B. canis* and rough variants of the other species.

B. ovis polynucleotides are similar to those of the other *Brucella* species, but *B. ovis* DNA lacks some of the polynucleotide sequences present in *B. suis* DNA (Hoyer and McCullough, 1968).

Neotype strain: 63/290; NCTC 10512; ATCC 25840 (Meyer and Morgan, 1973, 137).

6. Brucella canis Carmichael and Bruner 1968, 579.

can'is. L. n. *canis* a dog.

Even on primary isolation, cultures are in the non-smooth phase and do not possess Abortus and Melitensis antigens of smooth brucellae. Cross-agglutinates with *B. ovis* and rough variants of the other species (Diaz, Jones and Wilson, 1968; Jones *et al.*, 1968).

Pathogenic for dogs causing epididymitis and abortion.

Reference strain: RM 6/66; ATCC 23365.

The strains of *B. canis* have some characteristics in common with *B. suis*. Their acceptance as a new species is provisional.

Genus **Bordetella** *Moreno-López 1952, 178*

MARGARET PITTMAN

Bor.de.tel'la. M.L. dim.ending *-ella;* M.L. fem.n. *Bordetella* named for Jules Bordet, who with O. Gengou, first isolated the organism causing pertussis.

Minute coccobacilli, 0.2–0.3 μm by 0.5–1.0 μm, arranged singly or in pairs, more rarely in short chains. **Non-motile** or **motile** by lateral polytrichous flagella. **Gram-negative,** bipolar.

Colonies on potato-glycerol-blood agar (Bordet and Gengou, 1906) medium, smooth, convex, pearly, glistening, nearly transparent, surrounded by zone of hemolysis without definite periphery.

Chemoorganotrophs: **Metabolism respiratory,** never fermentative. Contain catalase. No indole reaction. No gelatin liquefaction. Litmus milk made alkaline. **Require nicotinic acid, cysteine** and **methionine** (most strains); **utilize oxidatively** alanine, asparagine, glutamic acid, proline and serine (Rowatt, 1957; Meyer and

Cameron, 1957; **hemin** (X factor) and **coenzyme I** (V factor) **not required.**

Strict aerobes.

Temperature optimum 35–37 C.

Mammalian parasites and **pathogens** of the **respiratory tract:** cause whooping cough or whooping cough-like disease in man; one species causes bronchopneumonia in dogs, guinea pigs and other animal species.

Have a **genus specific heat stable O antigen, antigenic heat-labile dermonecrotic toxin,** and the **common heat-labile agglutinogen no. 7. Each species** has a **specific agglutinogen;** 10 other agglutinogens occur among the species.

Type species: *Bordetella pertussis* (Bergey *et al.*) Moreno-López 1952, 178.

Description of the species of genus **Bordetella**

(Differential characteristics of the species are summarized in Table 7.14.)

1. Bordetella pertussis (Bergey *et al.*) Moreno-López 1952, 178. (There is question concerning the valid publication of the specific epithet *pertussis*. The Judicial Commission has been asked for a ruling.) (Microbe de coqueluche, Bordet and Gengou 1906, 731; *Hemophilus pertussis* Bergey *et al.* 1923, 269; *Bacterium tussis-convulsivae* Lehmann and Neumann 1927, 317; *Haemophilus pertussis* (Bergey *et al.*) Pribram 1933, 68.)

per.tus'sis. L. pref. *per* very, severe; L. n. *tussis* cough; M.L. gen.n. *pertussis* of a severe cough, of whooping cough.

Has a capsule-like sheath that does not swell in antiserum. Primary isolation requires inhibitors of unsaturated fatty acids and collodial sulfur or sulfides such as 15% or more blood, charcoal or ion-exchange resins. Potato-glycerol-blood agar of Bordet and Gengou is highly suitable.

Has marked propensity to modulate culturally and serologically. Adapts to growth on plain agar with loss of hemolytic, dermonecrotic, infectious, immunogenic, histamine sensitization factor, serotype factors and other characters (Pittman, 1970, 247).

TABLE 7.14
Differential characteristics of species of genus **Bordetella**

	1. B. pertussis	2. B. parapertussis	3. B. bronchiseptica
Flagella	$-^a$	−	$+^a$
Nitrate → nitrite	−	−	+
Citrate utilized	−	+	+
Urease produced	−	+	+
Peptone agar			
Growth	−	+	+
Browning	−	+	−
Bordet-Gengou agar			
Growth in 1–2 days	−	+	+
Growth in 3–4 days	+		
Litmus milk alkaline			
In 1–4 days	−	+	+
In 12–14 days, M^b	+		
Sensitize mice to histamine	+	−	−
Specific heat-labile antigen			
Factor 1	+	−	−
Factor 12	−	−	+
Factor 14	−	+	−
G + C content, moles %, T$_m$	61	61	66

^{a} − = All strains negative; + = all strains positive.
^{b} M = modulated from Phase I to Phase II–IV.

Experimentally in mice causes localized respiratory and brain but not peritoneal or blood infections. Large numbers of bacteria (10^9) intraperitoneally cause fatal toxemia.

Type strain not extant. Proposed reference strain: Kendrick-Eldering 10536; ATCC 10380; serotype 1.2.3.5.6.

2. **Bordetella parapertussis** (Eldering and Kendrick) Moreno-López 1952, 178. *Bacillus parapertussis* Eldering and Kendrick 1938, 571; *Haemophilus parapertussis* (Eldering and Kendrick) Wilson and Miles 1946, 809; *Acinetobacter parapertussis* (Eldering and Kendrick) Steel and Cowan 1964, 481.)

pa.ra.per.tus'sis. Gr. prep. *para* resembling; M.L. n. *pertussis* a specific epithet; M.L. adj. *parapertussis* resembling (*Bordetella*) *pertussis*.

Brown coloring of medium is attributed to tyrosinase (Ensminger, 1953.)

Type strain: Eldering and Kendrick 522; NCTC 5952 (Steel and Cowan, 1964, 482, under *Acinetobacter parapertussis*).

3. **Bordetella bronchiseptica** (Ferry) Moreno-López 1952, 178. (*Bacillus bronchicanis* Ferry 1911, 404; *Bacillus bronchisepticus* Ferry 1912, 377; *Bacterium bronchisepticus* (Ferry) Evans 1918, 582; *Alcaligenes bronchisepticus* (Ferry) Bergey *et al.* 1925, 257; *Brucella bronchiseptica* (Ferry) Topley and Wilson 1929, 517; *Alcaligenes bronchicanis* (Ferry) Haupt 1935, 188; *Haemophilus bronchisepticus* (Ferry) Wilson and Miles 1946, 787.)

bron.chi.sep'ti.ca. Gr. n. *bronchus* the trachea; Gr. adj. *septicus* putrefactive, septic; M.L. fem. adj. *bronchiseptica* intended to mean with an infected bronchus.

Grows more rapidly than *B. pertussis* and *B. parapertussis*. Utilizes citrate as the sole source of carbon. Splits urea in 4 hr. Splits asparagine.

Type strain: Dog 71 (Ferry, 1912, 79); NCTC 452; ATCC 19395.

ATCC 4617 is used for assay of polymyxin and sodium colistimate.

Further Comments

B. bronchiseptica possesses some properties in common with genus *Alcaligenes*. Classification in genus *Bordetella* is based on special common properties with other species in this genus; ecology (respiratory tract), common heat-stable O antigen, common antigenic heat-labile dermonecrotic toxin and some of the 14 heat-labile agglutinogens among the species; and identical amino acid utilization with that of *B. pertussis* (Meyer and Cameron, 1957, 158); it differs in G + C content of DNA from both *Alcaligenes faecalis*, 62 moles % (Hoyer and McCullough, 1968, 1785) and other species of *Bordetella*, 61 moles % (B. H. Hoyer, personal communication) and the historically related genus *Brucella*, 57–58 mole % (Hoyer and McCullough, 1968, 446).

Genus **Francisella** Dorofe'ev 1947, 176

CORA R. OWEN*

Fran.cis.el'la. M.L. dim.ending -*ella;* M.L. fem.n. *Francisella* named for Edward Francis, an American bacteriologist who studied these organisms.

---

* With the assistance of Drs. W. L. Jellison, C. B. Philip, G. M. Kohls and W. Burgdorfer of the Rocky Mountain Laboratory.

Very small, coccoid to ellipsoidal pleomorphic rods, which often show bipolar staining by special methods; many pass through Berkefeld filters. **Non-motile. Gram-negative.**

Acid without gas produced in media containing some carbohydrates. Catalase negative, H₂S produced. Cells soluble in sodium lauryl sulfate and sodium ricinoleate.

No growth on ordinary media without enrichment.

Strictly aerobic, no growth anaerobically. Optimum temperature 37 C.

Frequently found in natural waters; can be parasitic on man, other mammals, birds and arthropods.

Type species: *Francisella tularensis* (McCoy and Chapin) Dorofe'ev 1947, 178.

Further Comments

Two species are recognized, distinguished by the production of acid from sucrose, serological tests and pathogenicity. Skin-test antigens cross-react; no cross-agglutination or other serological cross-reactions; non-living vaccines of *F. novicida* do not confer immunity to *F. tularensis* (Owen *et al.*, 1964); living vaccines of one species confer some immunity to the other species.

Description of the species of genus **Francisella**

1. **Francisella tularensis** (McCoy and Chapin) Dorofe'ev 1947, 178. (*Bacterium tularense* McCoy and Chapin 1912, 61; *Pasteurella tularensis* (McCoy and Chapin) Bergey *et al.* 1923, 267; *Brucella tularensis* (McCoy and Chapin) Topley and Wilson 1929, 509; *Francisella tularense* (*sic*) (McCoy and Chapin) Dorofe'ev 1947, 178.)

tu.la.ren'sis. M.L. adj. pertaining to Tulare County, California, in which tularemia was first observed.

Rods and cocci in equal numbers; 0.2 by 0.2–0.7 μm; occurring singly. Extremely pleomorphic. Capsules rare or absent.

No growth on ordinary media without enrichment. Growth occurs on coagulated egg yolk, blood-glucose-cystine agar, glucose-blood agar, glucose-serum agar and blood agar; addition of fresh sterile animal tissues, especially rabbit spleen, favors growth on media without cystine. Grows in thioglycollate broth. On blood cystine media, colonies gray and may attain a diameter of 4 mm after 2–5 days; blood discolored and may become greenish. Growth easily emulsified. Rough variants occasionally occur; mucoid variants not reported.

Slight acid without gas produced from glucose, maltose, mannose, fructose and dextrin. No acid from sucrose.

Growth slight in litmus milk, slight acid may be produced.

Optimum temperature 37 C. Thermal death point 56 C for 10 min. Survives best at low temperatures, including storage at −70 C. Remains viable for years lyophilized in skim milk and stored under vacuum at +4 C.

The cause of tularemia in man and many other warm-blooded animals transmitted by blood-sucking arthropods, inhalation, ingestion and contact. The last mode of infection is very infrequent between human beings; infection of man from contact with animals common. The organism penetrates unbroken skin and mucous membranes to cause infection. Buboes and areas of necrosis produced in organs and tissues of man and animals. Unlocalized forms of disease occur after inhalation infection; anginal forms after digestion. Highly infectious for white mice and guinea pigs; can cause disease in sheep, horses (Claus *et al.*, 1959), cattle and birds; can infect blood-sucking arthropods which feed on bacteremic hosts and may serve as vectors.

Originally isolated from California ground squirrels but since found in many species of wild animals throughout North America, Europe (except British Isles, Spain, Portugal), Russia, Turkey and Japan. Also found in natural waters.

Immunologic types have not been found.

Reference strain: B38; ATCC 6223.

1a. *Francisella tularensis* var. *tularensis* Olsufiev, Emelyanova and Dunaeva 1959, 146. (*Francisella tularensis* var. *nearctica* Olsufiev 1968, 5.)

Morphology and characteristics as for species.

Utilizes glycerol (Olsufiev *et al.*, 1959) and produces citrulline ureidase (Marchette and Nicholes, 1961).

As virulent for the domestic rabbit as for white mouse and guinea pig (Bell *et al.*, 1955). Has been isolated in nature in North America only.

1b. *Francisella tularensis* var. *palaearctica* Olsufiev, Emelyanova and Dunaeva 1959, 148.

pa.lae.arc'ti.ca. M.L. adj. *palaearctica* of the Old World, Northern Hemisphere; named for the part of the world in which it occurs.

Does not utilize glycerol or produce citrulline ureidase.

Requires at least 5 log₁₀ more cells to kill domestic rabbit than to kill white mouse or guinea pig.

Found wherever tularemia occurs throughout northern hemisphere.

(A third variety, *F. tularensis* var. *mediaasiatica*,

THERMUS 285

has been proposed (Aikimbaev, 1966, 42) but this has not yet been validated.)

2. Francisella novicida (Larson *et al.*) Olsufiev, Emelyanova and Dunaeva 1959, 146. (*Pasteurella novicida* Larson, Wicht and Jellison 1955, 253.)

no.vi′ci.da. L. adj. *novus* new; L. v.suff. -*cida* from L. v. *caedo* cut, kill; M.L. n. *novicida* new killer.

Coccoid to ovoid or short rod-shaped cells, 0.2–0.3 by 0.3 μm in tissues, 0.7 by 1.7 μm in liquid media and 0.5 by 0.5–0.9 μm on solid media. Capsules not observed.

Grows in gelatin (without added cystine); no liquefaction.

On glucose-cystine agar colonies translucent, 6–7 mm diameter. On glucose-cystine-blood agar, colonies smooth, glistening, entire, amorphous, elevated, gray with blue cast, 8 mm diameter in 3 days at 37 C; green discoloration of blood. Good growth on primary isolation. Growth easily emulsified. Similar colonies on blood agar except gray color less pronounced, attain 4 mm diameter; no hemolysis.

Moderate uniform turbidity in peptone broth; no pellicle or ring; slight sediment which disintegrates on shaking.

Litmus milk acid and soft coagulation in 2 weeks.

Acid but no gas from glucose, sucrose, fructose and mannose. Utilizes glycerol.

Methyl red test negative, acetylmethylcarbinol not produced. Indole and ammonia not produced; nitrate not reduced to nitrite. Methylene blue reduced.

Optimum temperature 37 C. Thermal death point 60 C for 10 min. Remains viable for years when lyophilized in skim milk and stored under vacuum at 4 C.

Experimentally pathogenic for white mice, guinea pigs and hamsters, producing lesions similar to those of tularemia; rabbits, white rats and pigeons resistant. Not known to infect man.

Isolated from a water sample taken from Ogden Bay, Utah, 1951.

Type strain: Utah 112; ATCC 15482.

Genus **Thermus** *Brock and Freeze 1969, 295*

T. D. BROCK

Ther.mus. Gr. adj. *thermus* hot; M.L. masc.n. *thermus* to indicate an organism living in hot places.

Rods and filaments, 0.5–0.8 μm in diameter. Rods 5–10 μm in length, filaments variable length from 20 to greater than 200 μm. **Large spheres** 10–20 μm in diameter derived from the association of individual cells usually formed in old cultures. **Gram-negative. Non-motile. Flagella and endospores absent.**

Compact slowly spreading colonies. Many strains produce **yellow to bright orange colonies** although non-pigmented strains have also been isolated (Ramaley and Hixson, 1970). Surface pellicle often formed in unshaken liquid cultures.

Chemoorganotrophs. **Obligately aerobic.** Growth on amino acids, sugars, organic acids.

Temperature optimum 70–72 C; maximum 79 C, minimum 40 C. pH optimum 7.5–7.8, maximum 9.5, minimum 6.

The **G + C content** of the DNA ranges from **64–67 moles** % (buoyant density) among five strains tested.

Type species: *Thermus aquaticus* Brock and Freeze 1969, 295.

Further Comments

Members of the genus *Thermus* are common inhabitants of aquatic thermal habitats, both natural and man-made. The organism has been isolated from hot springs, hot water tanks, and thermally polluted rivers. In the drainways of some hot springs at 75–80 C, bright orange tufts composed of intertwined filaments of *Thermus* can be seen macroscopically. Detailed taxonomic studies of the genus have not been made, and more than one species may ultimately be recognized.

Studies on one strain of *Thermus aquaticus* have shown that the organism produces unusually thermostable enzymes, ribosomes, transfer RNA and plasma membrane.

Description of the species of genus **Thermus**

1. Thermus aquaticus Brock and Freeze 1969, 295.

a.qua′ti.cus. L. adj. *aquaticus* living in water.

Morphology and physiology as for genus.
Type species: YT1; ATCC 25104.

Plate 7.1. *Zoogloea* (page 287)

Fig. 1. *A*, finger-like outgrowth of floc, showing cells of *Z. ramigera*. Bright phase-contrast. × 833. *B*, Flocs of *Z. ramigera* at low magnification, showing star-like masses. × 25.

Fig. 2. *A*, tapelike growth in broth of *Z. ramigera*, strain I-16-M. *B* and *C*, lacy hanging growth in broth of *Z. filipendula*, strain P-8-4 and P-77-4 respectively.

Fig. 3. Fibrils interlacing cells of *Z. ramigera*, strain I-16-M. Electron micrograph, shadow cast with carbon platinum. From Friedman, B. A., Dugan, P. R., Pfister, R. M. and Remsen, C. C. 1969 J. Bacteriol. *98:* 1330.

Plate 7.2. *Azotobacter, Azomonas and Derxia* (page 288)

Fig. 1. *Azotobacter chroococcum*. Strain with typical rod-shaped cells; 3-day-old cells on nitrogen-free glucose agar.

Fig. 2. *Azotobacter chroococcum*. Cysts and germinating cysts; 14-day-old culture on nitrogen-free glucose agar.

Fig. 3. *Azomonas agilis*. Large, almost spherical yeast-like cells; 3-day-old culture on nitrogen-free glucose agar.

Fig. 4. Seven-day-old cells of *Derxia gummosa* on nitrogen-free agar containing 2% glucose.

Fig. 5. Ten-day-old cells of *Derxia gummosa* on peptone agar with 2% glucose.

Fig. 6. Three-week-old cells of *Derxia gummosa* on nitrogen-free glucose agar. The cells show shrinkage.

Fig. 7. Three-month-old cells of *Derxia gummosa* on nitrogen-free glucose agar. The cells show shrinkage and are enclosed by a thick slime envelope.

All phase-contrast photographs of living cells. Figures 1–3: × 1500, Figures 4–7: × 950.

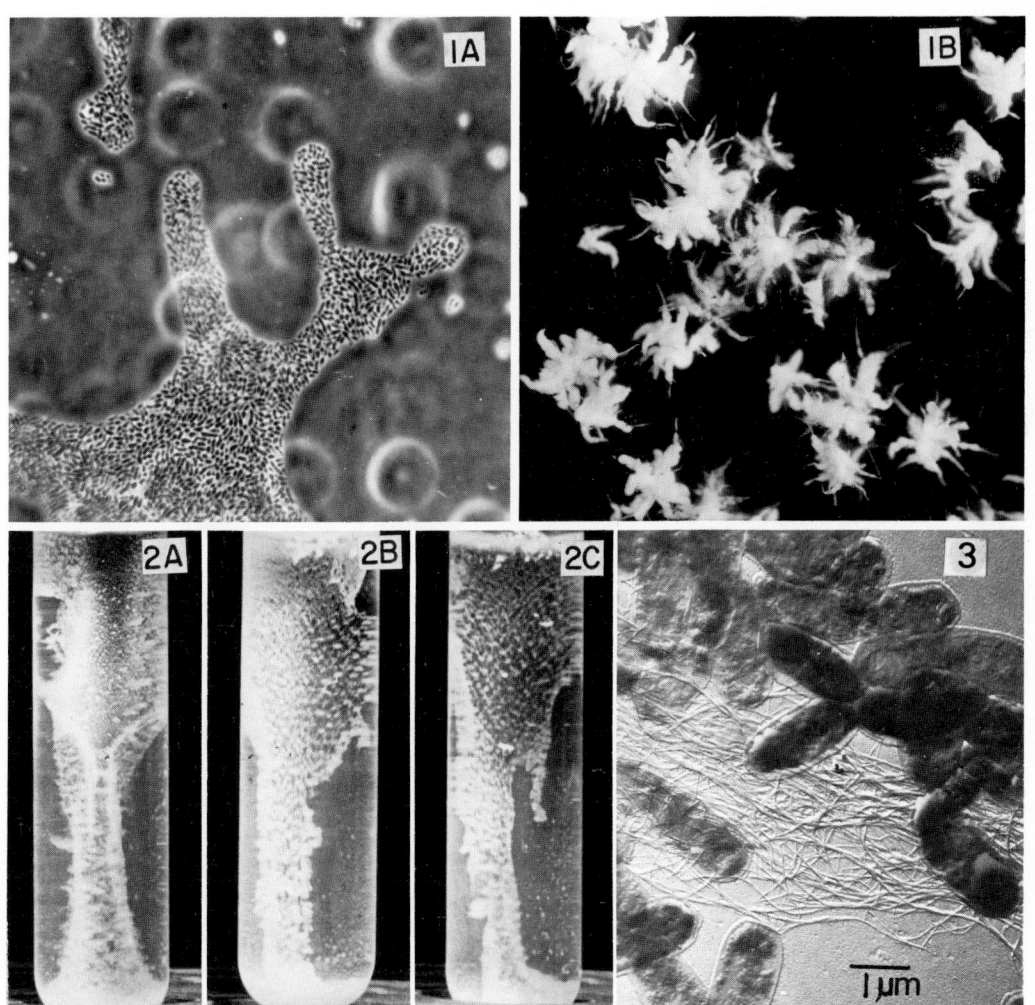

Plate 7.1

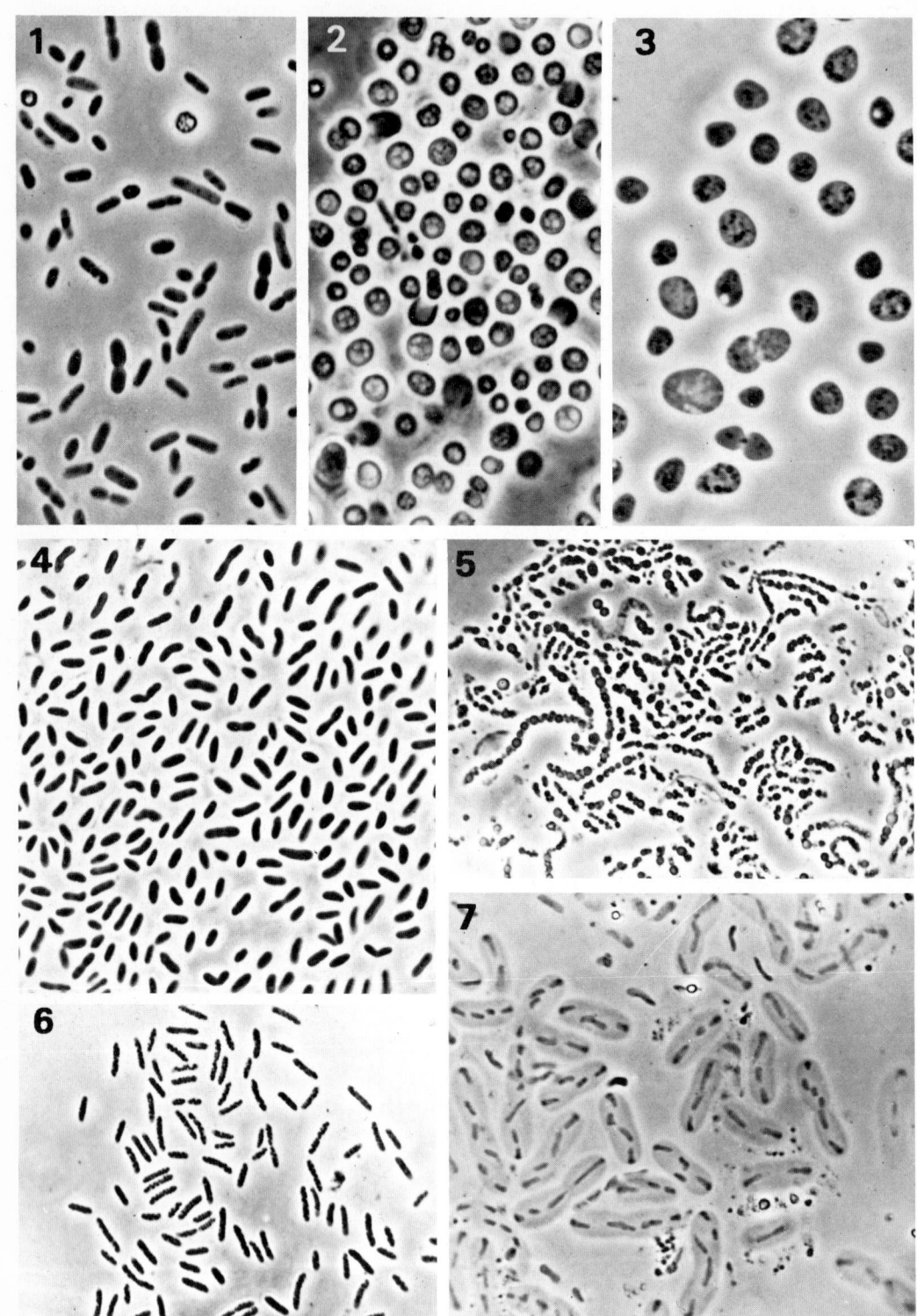

Plate 7.2.

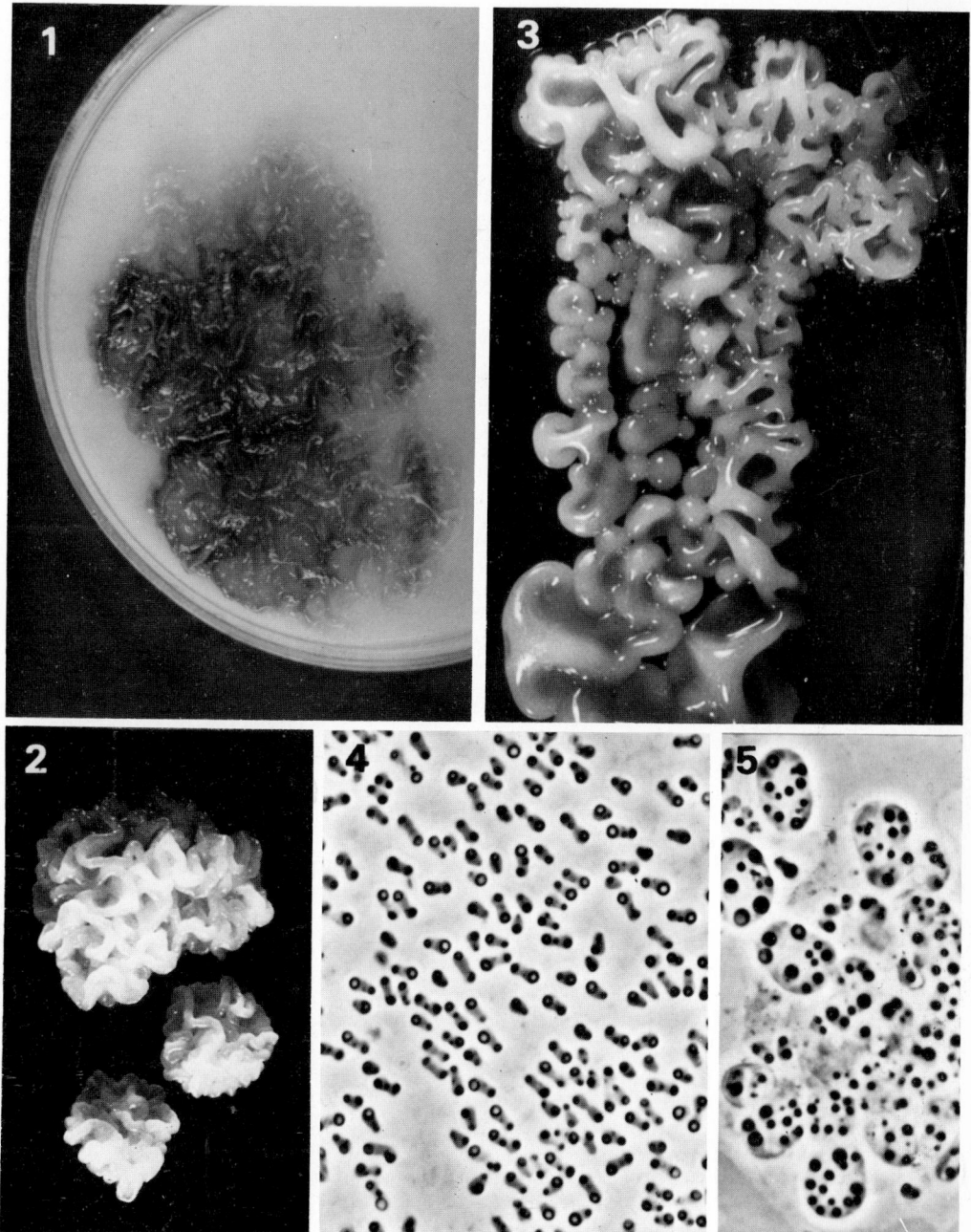

Plate 7.3. *Derxia, Beijerinckia*

Fig. 1. Colony-type of *Derxia gummosa* on nitrogen-free glucose agar with calcium carbonate. × 1.0.

Fig. 2. Typical colony of *Beijerinckia fluminensis* showing very plicated surface, while the slime is of more granular consistency. × 4.0.

Fig. 3. *Beijerinckia indica* colony on nitrogen-free glucose agar without calcium carbonate. × 2.0.

Fig. 4. Typical *Beijerinckia* cells (*B. indica*) showing characteristic polar lipoid bodies. × 1500.

Fig. 5. Capsule formation in *Beijerinckia fluminensis* on nitrogen-free glucose agar. × 1500.

PART 8

GRAM-NEGATIVE FACULTATIVELY ANAEROBIC RODS

FAMILY I. **ENTEROBACTERIACEAE** RAHN 1937, 281 *Nom. gen. cons.* Opin. 15, Jud. Comm. 1958, 73

S. T. COWAN*

Small **Gram-negative rods; motile by peritrichate flagella** or non-motile. Capsulated or non-capsulated. Not spore-forming; not acid-fast. Aerobic and **facultatively anaerobic.** Grow readily on meat extract media but some members have special growth requirements. **Chemoorganotrophic; metabolism respiratory and fermentative. Acid is produced** from the fermentation of **glucose,** other carbohydrates and alcohols; **usually aerogenic** but anaerogenic groups and mutants occur. **Catalase positive** with the exception of one serotype of *Shigella;* **oxidase negative. Nitrates are reduced to nitrites** except by some strains of *Erwinia.* G + C content of DNA: 39–59 moles %.

Type genus: *Escherichia* Castellani and Chalmers 1919, 941. Designated type genus Opinion 15 Jud. Comm. 1958, 73.

Further Comments

Circumscription. The definition circumscribes a large, apparently natural, group of many smaller interrelated groups. The delimitation of this large group, the family *Enterobacteriaceae*, from other families is in most cases clear-cut, but in some instances, to be mentioned below, the borderline is ill defined.

The genera *Aeromonas, Vibrio* and *Photobacterium* are distinguished from members of the family *Enterobacteriaceae* by their polar flagellation and often by the oxidase reaction. However, in recent studies it has been questioned whether these characters deserve the classificatory weight

accorded them at present, since genetic and other data suggest a fairly close relationship between these three genera and the present family.

The borderline toward the two genera *Alcaligenes* (*Achromobacter*) and *Flavobacterium* is ill defined because of the uncertainty of the circumscription of these genera. The tendency today is to require a strictly aerobic metabolism of *Alcaligenes* and *Flavobacterium* and then the uncertainty is formally removed. There remains, however, the problem that some bacteria are left which cannot be satisfactorily fitted into the (present) system.

From a practical point of view, and with identification in mind, it should be noted that some spore-forming species such as *Bacillus polymyxa* and *B. macerans*, if spores are not present or not discovered, may masquerade as members of the family because the Gram reaction usually comes out as negative in these species.

Subdivision of the Family. The family lends itself to an apparently non-arbitrary subdivision into a few major groups (here, without prejudice, called Tribes) on the basis of a few metabolic characteristics and the moles % G + C (Table 8.1).

By using, separately or in combination, morphological characteristics, further biochemical characteristics, antigenic differences and differences in bacteriophage or bacteriocin susceptibility, it is possible to split each major division into a large number of more-or-less distinct subgroups (Tables 8.2 and 8.3). However, there is evidence that the end products of metabolism may not be

---

* With the assistance of the authors for the genera of the family.

TABLE 8.1

Distinguishing characteristics of the five primary groups (tribes)

	Tribe I. Escherichieae	Tribe II. Klebsielleae	Tribe III. Proteeae	Tribe IV. Yersinieae	Tribe V. Erwinieae
Fermentation pattern	Mixed acid	2,3-Butanediol		Mixed acid	Mixed acid & 2,3-butanediol
M.R.	+	D	+	+	
V.P.	–	D	D	–	D
Phenylalanine deamination	–	–	+	–	D
Nitrate reduction	+	+	+	+	D
Urease	–	D	D	D	–
KCN, growth in	D	+	+	–	D
Optimal temp. for growth	37 C	37 C	37 C	30–37 C	27–30 C
G + C, %	50–53	52–59	39–42	45–47	50–58

satisfactory criteria for distinguishing *Erwinia* species from each other and from other enterobacteria (White and Starr, 1971).

Because of the diagnostic and epidemiological interest in a system of refined pinpointing of strains from human illnesses and the possibilities for extensive splitting, the formal subdivision of the family has been carried further than in most other bacterial families. The result is that the ranking within the family has come out of register with that of other areas of bacterial classification, so that taxa which in other families are ranked as species, in this family occur as genera or in some cases as tribes. A consequence of the extensive formal subdivision is a profusion of scientific names and, as some of the names are old and ingrained, they become the main obstacle for a necessary change of rank within the family as a first step toward a revision of the classification.

Although a revision of the family is already due and is unavoidable in the foreseeable future, the present edition of THE MANUAL has adopted a semiconservative policy; some modifications and some additions have been made to bring the classification into line with current usage, but the basis of the traditional system and its names are retained. Thus we have followed traditional lines and treated the five "tribes" as a family and the main subdivisions of the tribes as genera, with the clear understanding that these cannot be equated with the genera of other families. There is increasing evidence of the sharing of characters, including the antigens responsible for the serological subdivisions, of experimental transfer of genetic material and apparent intra- and intergeneric transformation, so that taxonomic opinion is moving toward a reduction in the number of genera. The various authors have retained those groups that have been generally accepted as genera, but it is easy to forecast that within the working life of this edition of THE MANUAL there will be fusion of several of them, perhaps to the extent that those collections of genera now called tribes will themselves become the genera of the future.

Note Added in Proof. The contributions to this family were written before 1970; during the delay inevitable in the production of THE MANUAL various proposals have been made for the creation of a new genus, new combinations and some new species. At this stage opinion on the worthiness of these new taxa has not crystallized; some may be warranted, most are probably not needed. It is sufficient to say that the state of the enterobacteria is still fluid.

Further Notes

Although often referred to as the Coli-Typhoid Group, the members of the *Enterobacteriaceae* are not all intestinal parasites, and certainly not all are pathogenic to their hosts. The group is of wide interest because it includes important pathogens, some for man and other animals, others for plants; they are common causes of ill health and severe economic loss in all parts of the world. In many cases it is possible to characterize strains in such detail that the source of an epidemic or epizootic can be determined with great accuracy. The ability to characterize strains in detail depends on biochemical differentiation, serological analysis, sometimes supported by phage typing and further biochemical differentiation.

Most important in epidemiological investigations is the serological analysis of a series of different surface antigens, some of polysaccharide and some of protein nature. The best characterized are (1) the O (somatic) antigens which are lipo-polysaccharides embedded in the cell wall; (2) the K (capsular) antigens most of which are poly-saccharides; and (3) the H (flagellar) antigens which are proteins. Other immunogenic structures such as fimbriae (proteins) and the M (mucoid, polysaccharide) antigen may be of importance for serological analysis. For a proper evaluation of serological analysis it is important to understand the genetic variation that regulates these antigens. For a more detailed description see Kauffmann (1966) and Edwards and Ewing (1962).

Escherichia is now regarded as a genus with only one species in which there are several hundred different antigenic specificities; together these specificities produce by different combination of the O, K and H antigens several thousand sero-types. Two species that were in the genus *Esche-richia* in the seventh edition of THE MANUAL will now be found in *Citrobacter*, while a fourth species, *E. aurescens*, is thought to be a pigmented form of *E. coli* and is denied specific status.

A new genus *Edwardsiella* appears in this first major subdivision of the family but not all workers are willing to give it separate generic recognition, and some would place it as a species of *Escherichia*.

Like the other genera in the family the genus *Salmonella* has been further subdivided. Certain kinds of *Salmonella* were originally singled out as separate species on the basis of striking, although small, deviations from the common biochemical pattern and/or on the basis of their ability to produce a clinically recognizable specific disease, e.g. enteric (typhoid) fever. The development of refined antigenic analysis, now embodied in the Kauffmann-White scheme, led to the recognition that the earlier established species were also recog-nizable serologically, and as an extension of this observation the notion was introduced that all the different serotypes represented separate spe-cies, although it was realized that many of the serotypes were biochemically heterogeneous. This concept was later carried to an extreme by Kauff-mann (1961) in his definition of a species as "a group of related serofermentative phage types," and his monograph on *Salmonella* species.

As a supplement to this extreme formal sub-division Kauffmann introduced a primary sub-division of the genus into four subgenera; he did not give formal names to these taxa but labeled them I to IV. The subgenera are differentiated on the basis of minor biochemical differences. Subgenus I comprises the majority of the species and subgenus III comprises the species which for historical reasons are usually placed in the sep-arate Arizona group (or genus), although it is

generally recognized that they are slightly deviat-ing salmonellas.

In contrast to Kauffmann's ideas on the sub-division of the genus, many taxonomists are of the opinion that the genus *Salmonella* does not deserve the rank of genus, but more properly corresponds to a species as this concept is used within other areas of bacteriology, and favoring the view that the biochemical variants and the serotypes should be recognized as infrasubspecific groupings outside the formal nomenclatural sys-tem. The problem whether or not within this concept a few formally named species should be recognized as a concession to tradition is open to debate, and the same is true of the name of the species to which the many thousand serotypes should be allocated; this problem will be discussed by Le Minor and Rohde.

A reconciliation of the two extreme concepts is impossible and even a partial reversal of the de-velopment toward splitting is difficult taking re-gard of the priority rule of the Code and the es-tablished usage in clinical medicine of certain specific names such as *S. typhi*, *S. paratyphi-B* and *S. typhimurium*. For these reasons the present edition of THE MANUAL leaves matters more or less as they stand, recognizing that this is neither logical nor satisfactory, and hoping for a clarifica-tion before the next edition is due.

Shigella is another group (or genus) subdivided mainly on the basis of antigenic structure. Al-though there are differences in biochemical char-acters that justify the recognition of subgroups, two of them, *S. flexneri* and *S. boydii*, can only be distinguished satisfactorily by serological means.

The second division (tribe) includes *Klebsiella*, *Enterobacter*, *Hafnia* and *Serratia*. The character-istics differentiating the genera are given in Table 8.11 and further information in Table 8.3.

Editors of earlier editions of THE MANUAL fore-cast the fusion of *Klebsiella* and *Aerobacter*, and this now seems to have taken place (see "Adden-dum I," p. 339). Of the bacteria that have been mislabeled *A. aerogenes*, those that grow at 37 C and are non-motile should be placed in the genus *Klebsiella*, while those motile by peritrichous flagella should be included in *Enterobacter* as *E. aerogenes*. In the genus *Enterobacter* Hormaeche and Edwards is the species *E. cloacae* (Jordan's *B. cloacae*), the type of the genus. Another species, *E. liquefaciens*, which grows at temperatures below 37 C should probably be included in Grimes' re-defined *Aerobacter*, but some workers regard it as a member of the genus *Serratia*. There have been various suggestions for the combination of one or more of the genera *Klebsiella*, *Enterobacter*, *Hafnia* and *Serratia*, and although some of these sugges-

TABLE 8.2

Distinguishing characteristics of the genera of the tribe **Escherichieae** *(group I)*

	Escherichia	Edwardsiella	Citrobacter	Salmonella	Shigella
Motility	+	+	+	+	−
Indole	+	+	D	−	d
H₂S on TSI	−	+	D	+	−
β-Galactosidase	+	−	+	D	d
Lactose	+/×	−	+/×	D	−/×
Mucate	+	−	+	D	−
KCN	−	−	d	D	−
Gelatin	−	−	−	D	−
Malonate	−	−	d	D	−
d-Tartrate	d	−	−	D	−
Dulcitol	d	−	d	D	d

tions are likely to be adopted in the foreseeable future, we have not made any new combinations for this edition of THE MANUAL.

Some workers would split *Proteus* into three or four subgenera, but here Lautrop has treated them as one genus. *Proteus* is a group distinguished from all other members of the family by having the ability to deaminate phenylalanine, but differences in the G + C base ratios suggest that there may be two distinct subgroups sufficiently different to be worthy of separation into two genera, one of which would be monotypic and include Morgan's bacillus. The Providence subgroup, in spite of its lack of urease activity, is not thought to justify separate generic recognition.

For the first time in any comprehensive classification of bacteria, the genus *Yersinia* is placed in the family *Enterobacteriaceae*. It is a genus created by the division of *Pasteurella* and includes three species, *Yersinia pestis, Y. pseudotuberculosis* and *Y. enterocolitica*, which produce infections in man and animals. *Yersinia pestis*, probably because it is a particularly dangerous organism to work with, has been less well characterized than the other species of the genus.

Erwinia is a genus of bacteria that normally grow at temperatures below 37 C and may not grow at that temperature. They are associated with plants, as saprophytes, epiphytes and/or pathogens; at least one has also been isolated from animals including man. Many species that have been recognized have been collected into three groups, and details of the included nomenspecies will be found in the section by Lelliott.

Genus I. **Escherichia** *Castellani and Chalmers 1919, 491 Nom. cons.* Opin. 15, Jud. Comm. 1958, 73

F. ØRSKOV

Esch.er.i′chi.a. M.L. fem.n. *Escherichia* named after Theodor Escherich who isolated the type species of the genus.

Straight rods, 1.1–1.5 by 2.0–6.0 μm (living) or 0.4–0.7 by 1.0–3.0 μm (dried and stained) (Luria, 1960). Singly or in pairs. Motile by peritrichous flagella or non-motile.

Grow readily on simple nutrient media. Colonies on nutrient agar may be smooth (S), low convex, moist, shiny surface, entire edge, gray and easily emulsified in saline, or rough (R), dry and do not emulsify well in saline. There are intergrading forms between these extremes. Mucoid forms may occur.

Acetate can be used but citrate cannot be used as sole carbon source. Glucose and other carbohydrates are fermented with the production of pyruvate, which is further converted into lactic, acetic and formic acids. Part of the formic acid is split by a complex hydrogenlyase enzyme system into equal amounts of CO_2 and H_2 (de Ley, 1962). Some strains are anaerogenic. Lactose is fermented by most strains, but this may be delayed or absent.

Other biochemical characteristics are shown in Tables 8.2 and 8.3.

G + C content of DNA: 50–51 moles % (buoyant density and T_m).

Type species: *Escherichia coli* (Migula) Castellani and Chalmers 1919, 941.

Further Comments

In morphology and colony form on nutrient media the organisms of this genus cannot be distinguished from those of other genera of the fam-

TABLE 8.3

Main biochemical characters of primary groups I to IV

	Group I					Group II				Group III. Proteus	Group IV. Yersinia
	Escherichia	Edwardsiella	Citrobacter	Salmonella	Shigella	Klebsiella	Enterobacter	Hafnia	Serratia		
Catalase	+	+	+	+	Dᵃ	+	+	+	+	+	+
Oxidase	-	-	-	-	-	-	-	-	-	-	-
β-Galactosidase	+	(-)	+	D	d	+	+	+	+	(-)	+
Gas from glucose at 37 C	+	+	+	+	(-)	d	+	+	d	D	-
KCN (growth on)	-	-	+	D	-	+	+	+	+	+	-
Mucate (acid)	+	-	+	D	-	d	d	-	-		
Nitrate reduced	+	+	+	+	+	+	+	+	+	+	+
G + C, moles %	50–51			50–53		52–56	52–59	52–57	53–59	39–42 (one species = 50)	45–47
Carbohydrates (acid from)											
Adonitol	-	-	-	-	-	d	+	-	d	D	D
Arabinose	+	-	+	+	-	+	+	+	-	-	+
Dulcitol	d	-	d	D	d	d	-	-	-	-	+
Esculin	d	-	d	-	-	-	D	-	-	d	D
Inositol	-	-	-	d	-	+	D	-	d	D	-
Lactose	+ or X	-	+ or X	D	D	+	+	-	-	-	-
Maltose	+	+	+	+	-	+	+	+	+	D	+
Mannitol	+	-	+	+	D	+	+	+	+	D	+
Salicin	d	-	d	-	-	+	+	-	+	d	D
Sorbitol	+	-	+	+	-	+	+	-	-	-	D
Sucrose	d	-	d	-	D	(+)	+	- or X	+	D	D
Trehalose	+	-	+	+	-	+	+	+	+	d	+
Xylose	d	-	+	+	D	+	+	+	d	D	D
Related C sources											
Citrate	-	-	+	+	-	d	+	+	+		-
Gluconate		-		-	-	+	+	-	-		
Malonate	-	-	d	D	-	D	+	-			
d-Tartrate	d	-	+	D	-	d	-	-	-		D
M.R.	+	+	+	+	+	D	-	-	-	+	+
V.P.	-	-	-	-	-	D	+	+	D	d	-
Protein reactions											
Arginine	d	-	d	+			D	-	-	-	-
Gelatin hydrolysis	-	-	-	D	-	(d)	(+)	-	+	D	D
H₂S from TSI	-	+	D	+	-	-	-	-	+	D	D
Indole	+	+	D	-	D	d	-	-	-	D	D
Lysine decarboxylated	+	+	-	+	-	d	D	+	+	d	-
Ornithine	d	+	d	+	d	-	+	+	+	D	D
Urea hydrolyzed	-	-	(+)	-	-	d	(d)	-	-	D	D
Glutamic acid	-	-	-	-	-	-	-	-	-	+	-
Phenylalanine	-	-	-	-	-	-	-	-	-	+	-

ᵃ D = different reactions given by different species of a genus; d = different reactions given by different strains of a species or serotype; X = late and irregularly positive (mutative).

ily. The characters recorded are taken from recent work by Ewing and his colleagues (Ewing, 1966; Ewing, Davis and Martin, 1967) and from the results of extensive work in Copenhagen.

Wild strains are sensitive *in vitro* to most antibiotics and to sulfonamides; sensitivity to penicillin is low.

In the older literature many varietal names can be found; most of the varieties were based on differences in sugar fermentation patterns. Those interested are referred to the seventh edition of THE MANUAL. One such name often found in the literature was the Alkalescens-Dispar group, which was distinguished from *E. coli* by the absence of gas production in the fermentation of glucose, non-motility and in the delayed, or absence of, fermentation of lactose; these strains are now included in *Escherichia coli*.

Leclerc (1962) collected strains from water which were negative in the lysine, ornithine and glutamic acid decarboxylase tests and named them *Escherichia adecarboxylata:* they further differ from typical *E. coli* in being malonate positive, KCN positive, sorbitol negative, cellobiose positive and often produce urease; they liquefy gelatin and produce a yellow pigment. On the knowledge available they are not thought to justify the recognition of a second species of *Escherichia*.

Description of the species of genus Escherichia

1. **Escherichia coli** (Migula) Castellani and Chalmers 1919, 941. (*Bacterium coli commune* Escherich 1885, 518; *Bacillus coli* Migula 1895, 27; *Bacterium coli* (Migula) Lehmann and Neumann 1896, 224.)

co'li. Gr. n. *colon* large intestine, colon; M.L. gen.n. *coli* of the colon.

Many strains have capsules or similar less well developed structures (microcapsules). Fimbriae (pili) on many strains; subdivided by their direct hemagglutinating capacity or differences in morphology into several fimbrial types (Duguid *et al.*, 1955; Brinton, 1965); the sex (or F) fimbrial type can be detected by its affinity for special male phages and by its antigenic properties.

In broth, growth is shown by a general turbidity and a heavy deposit which disperses completely on shaking (S form); the extreme R form shows a clear supernatant and a granular deposit which does not disintegrate completely on shaking.

Found in the lower part of the intestine of warm-blooded animals. Many, if not all, members may show opportunistic pathogenicity (urinary tract infections in man, mastitis in cows, to mention two examples). A limited number of well defined serotypes is closely associated with certain infectious enteric diseases in human infants and young of other animals (Kauffmann and Ørskov, 1956; Ewing, Davis and Montague, 1963; Sojka, 1965); some of these strains from infections in young pigs produce toxins genetically determined by plasmids (Smith and Halls, 1967). Hemolytic strains are found in high frequency in pigs; this hemolytic character can likewise be determined by a plasmid (Smith and Halls, 1968).

Neotype: U5/41 (serotype O1:K1:H7); NCTC 9001; ATCC 11775 (Jud. Comm. 1963, Opin. 26).

Further Comments

Subdivision of *Escherichia* into biotypes can be made using the fermentation and other biochemical tests regularly carried out on the *Enterobacteriaceae;* these biotypes are quite stable within the same clones when not exposed to selection pressures, but the enzymic properties change with a frequency of other mutations. Experience shows that the fermentation pattern of a particular strain has no greater tendency to change than its antigenic properties. No correlation has been found between fermentative properties and single antigens; however, a well characterized serotype such as O111:K58:H2 (from infantile diarrhea) is found all over the world associated with practically the same biotype, and it is probably one clone that has become widely distributed.

Other subdivisions can be made on the activity of phages and/or colicins, based either on the sensitivity of strains to a set of known phages/colicins or on the carriage of phages/colicins active toward a known set of test strains. Special phage-typing systems have been developed for serotypes from human infantile diarrhea.

Serology is probably the most useful way of subdividing *E. coli;* it is based on the properties of the different surface structures, which in serological terms are expressed as O (somatic), K (capsular or microcapsular) and H (flagellar) antigens (Kauffmann, 1966). The antigenic pattern of a strain is recorded as the number of the particular O, K and H antigens, as O111:K58:H2; there are 150 recognized O antigens, 90 K and 50 H. The *E. coli* O antigens are not yet as well characterized in single factors as the O antigens in the *Salmonella* group. Although the fundamental characters of the O antigens are like those of the Salmonella O antigens, the *E. coli* O antigens are better compared with the Salmonella antigenic complexes forming the groups A, B, C, etc.

The O antigen-determinant groups are found in the polysaccharide part of the lipopolysaccharide molecule in the outer layer of the bacterial cell wall; for a general description see Lüderitz *et al.* (1966), and for details of 100 polysaccharides from *E. coli* O antigens see Ørskov *et al.* (1967). The K antigens were defined by Kauffmann and his associates (for a review see Kauffmann, 1966). K antigens have chemical compositions comparable to those found in other polysaccharide capsules (Lüderitz *et al.*, 1968), but one, at least, is protein (Stirm *et al.*, 1967), and determined by a plasmid (Ørskov *et al.*, 1961). Some strains have two well defined K antigens, a polysaccharide and a protein K antigen. The complexity of the K antigens may be resolved by further research, and this may require revision of the present ideas on them. The H antigens, like those of *Salmonella*, are protein in nature; phase variation has not been detected in *E. coli* strains.

With 150 O test sera it is possible to determine the O group of about 90% of human strains in temperate zones (Ewing and Davis, 1961), but in other climates and from other animal sources the proportion identified will probably be lower; thus a considerable number of unknown O groups exists and the same applies to the K and H antigens. The large numbers of O, K and H antigens, and the possible permutations and combinations indicate that the total number of serotypes of *E. coli* is very high.

Genus II. **Edwardsiella** *Ewing and McWhorter 1965, 37*

RIICHI SAKAZAKI

(Asakusa group: Sakazaki and Murata 1962, 616; Bartholomew group: King and Adler 1964, 230.)

Ed.ward.si.el′la. M.L. dim. ending *-ella;* M.L. fem.n. *Edwardsiella;* named after P. R. Edwards, an American bacteriologist.

Motile peritrichously flagellated **rods:** not encapsulated.

Citrate and malonate cannot be used as a sole C source. Late fermentation sometimes occurs in citrate.

H₂S is produced abundantly **from TSI agar,** but poorly producing biotypes may occur. Indole is produced. Lysine and ornithine decarboxylases are present.

Mannitol is not fermented by most strains. No action on *d*-tartrate and mucate.

Type species: *Edwardsiella tarda* Ewing and McWhorter 1965, 37.

Further Comments

This group was originally described by Sakazaki and Murata (1962) and named the Asakusa group in a paper published in Japanese (see also Sakazaki, 1965); it was also named the Bartholomew group by King and Adler in 1964. Members of the group seem to resemble *Escherichia, Citrobacter, Proteus* and *Salmonella;* some of the differences and similarities are shown in Table 8.2.

Description of the species of genus **Edwardsiella**

1. **Edwardsiella tarda** Ewing and McWhorter in Ewing *et al.* 1965, 37.

tar′da. L. fem.adj. *tarda* slow, here it implies inactivity.

Some strains may ferment sucrose rapidly, and these strains may also ferment mannitol and salicin slowly; the majority of strains do not ferment these carbohydrates.

Alginate is not utilized; pectate is not decomposed; lipase is not produced.

Serology: the antigenic formulae of 18 serotypes have been described by Sakazaki (1967). Antigenic relationships have not been observed with other members of the *Enterobacteriaceae.*

Probably a normal intestinal inhabitant of snakes. Occasionally isolated from the stool of humans with diarrhea or those in good health, from blood of man and animals and from urine of man. Also found in water.

Type strain: 148359; ATCC 15947.

Genus III. **Citrobacter** *Werkman and Gillen 1932, 173*

JIRI SEDLÁK

Cit.ro.bac′ter. L. n. *citrus* lemon; M.L. n. *bacter* a small rod; M.L. masc.n. *Citrobacter* a citrate using rod.

Motile peritrichously flagellated **rods,** not encapsulated. Grow readily on ordinary media and **can use citrate as sole C source.** Glucose and other carbohydrates fermented with the production of acid and gas (CO₂:H₂ in the ratio 1:1). Lactose fermentation may be delayed or absent. **Trimethyleneglycol formed from glycerol.** Growth not inhibited by KCN.

Able to grow on Muller's tetrathionate broth, Leifson's selenite broth, sodium deoxycholate-citrate agar, Wilson and Blair's bismuth sulfite medium and Kristensen's brilliant green-phenol red agar, all of which inhibit or retard the growth of *E. coli.*

Other characteristics are given in Tables 8.2, 8.3 and 8.4.

Found in water, food, feces and urine, but their pathogenicity is problematical. They seem to be normal intestinal inhabitants and are found constantly in healthy people. Certain serotypes seem to occur in sporadic and mass alimentary infections, and in infections of the urinary tract, gall bladder, middle ear and meninges.

Type species: *Citrobacter freundii* (Braak) Werkman and Gillen 1932, 173.

Further Comments

The genus was created for the citrate-utilizing, lactose-fermenting coliform bacteria, but it did not readily gain acceptance and its species were often included in *Escherichia* (Vaughn and Levine, 1942). Citrate-utilizing, slow lactose-fermenting 'paracolons,' at one time known as the Bethesda-Ballerup group (Edwards, West and Bruner, 1948, 711), were found to share a similar series of antigens with the lactose-fermenting citrobacters, and all are now included in the genus *Citrobacter* without distinguishing between prompt and slow lactose fermentation. More detailed biochemical characterization of *Salmonella coli* 1 (Kauffmann, 1941) and some of the *Escherichia coli* serotypes showed that these bacteria had been mislabeled and were, in fact, Citrobacter serotypes, and these, too, are now included in the genus.

Of the six species described by Werkman and Gillen few are now accepted. The whole Citrobacter group is in the melting pot and it has been broken down into other species (*C. koseri, C. diversus*) and other genera (*Levinea, Padlewskia*). The difficulties of trying to rationalize the present position can be gauged by consulting papers by Frederiksen (1970) and by Ewing and Davis (1971).

Description of the species of genus **Citrobacter**

1. **Citrobacter freundii** (Braak) Werkman and Gillen 1932, 1973. (*Bacterium freundii* Braak 1928, 162; *Escherichia freundii* (Braak) Yale 1939, 394.)

freun'di.i. M.L. gen.n. *freundii* of Freund; named after A. Freund, the bacteriologist who first observed that trimethyleneglycol was a product of fermentation.

In addition to the characters common to all species of the genus (Table 8.2) the following are characters of the species.

The morphology of the bacterium and the colony form resemble those of *Escherichia coli*, but growth occurs on some media that inhibit *E. coli*.

H_2S produced; indole not produced. Lysine is not decarboxylated. Malonate is not utilized.

Neotype: NCTC 9750; ATCC 8090 (Jud. Comm. 1963; Opin. 26).

2. **Citrobacter intermedius** Werkman and Gillen 1932, 178. (*Escherichia intermedia* (Werkman and Gillen) Vaughn and Levine 1942, 498.)

in.ter.me'di.us. L. adj. *intermedius* intermediate.

Differs from the type species in the following characters.

H_2S is not produced; some strains produce indole, decarboxylate lysine and utilize malonate. At least two biotypes can be distinguished (Table 8.4).

Serology of Citrobacter Species

Citrobacter organisms have a mosaic of O, K and H antigens that are found in the *Enterobacteriaceae*. An antigenic diagnostic scheme was first made for the so-called Bethesda-Ballerup group by West and Edwards (1954) and had 32 O groups and 87 H antigens. This was supplemented by Sedlák and Slajsova (1966, 1967) to include 10 strains earlier designated *Salmonella coli* 1 (Kauffmann, 1941), *E. coli* O groups 67, 72, 94 and 122 (Ørskov, 1956) and the type strain of *Citrobacter freundii* which had not hitherto been included in the antigenic scheme. By 1969 this scheme had 42 O groups and more than 90 H antigens. Crossreactions with antisera of other members of the *Enterobacteriaceae* suggest that there are relations

TABLE 8.4

Distinguishing characters between **Citrobacter freundii** *and biotypes of* **C. intermedius**

	H_2S	Indole	Malonate	KCN	Other designations
1. *C. freundii*	+	−	−	+	*Salmonella coli*
2. *C. intermedius* biotype a	−	+	−	+	*Padlewskia, Levinea amalonatica*
C. intermedius biotype b	−	+	+	−	*C. koseri, Levinea malonatica*

between *Citrobacter* on the one hand, and *Salmonella*, *Arizona* and *Escherichia* on the other

TABLE 8.5

Relations between O antigens of **Citrobacter** *and* **Salmonella**

Citrobacter O group	Salmonella O antigen	
Ci 3	Sa 41	*S. waycross*
Ci 9	Sa 30	*S. urbana*
Ci 11	Sa 40	*S. riogrande*
Ci 12	Sa 44	*S. niarembe*
	Sa 57	*S. locarno*
Ci 14	Sa 38	*S. inverness*
Ci 19	Sa 6, 14, 25	*S. onderstepoort*
Ci 20	Sa 17	*S. kirkee*
Ci 21	Sa 6, 14, 24	*S. carrau*
Ci 22	Sa 4, 5	*S. paratyphi-B*
Ci 23	Sa 18	*S. cerro*
Ci 26	Sa 21	*S. minnesota*
Ci 37	Sa 48	*S. djakarta*
Ci 38	Sa 8, 20	*S. kentucky*
Ci 39	Sa 3, 10	*S. anatum*
Ci 40	Sa 57	*S. locarno*
Ci 41	Sa 55	*S. tranoroa*
Ci 42	Sa 54	*S. uccle*
Ci 43	Sa 28	*S. dakar*
Ci 44	Sa 35	*S. adelaide*

(Edwards and Ewing, 1955, 1962; Sedlák and Slajsova, 1966, 1967; and unpublished data of van Oye, Hoffman, Rohde and Harsova).

The O and H antigens are either simple or are a complex made up of two or three partial antigens; several are common to many groups and this suggests relationships between these groups. Several strains of the serogroups O5 and O29 have a Vi antigen serologically identical with the Vi antigen of *Salmonella typhi* and *S. paratyphi-C;* examples from the supplemented antigenic scheme are shown in Table 8.5.

Immunochemistry

The fundamental immunochemical studies were made by Westphal *et al.* (1960); more recently it has been found that serologically identical or closely related strains of *Citrobacter*, *Salmonella* and *Arizona* (*Salmonella arizonae*) have identical or very similar sugar constituents of their O antigens. The polysaccharide complex of most members of the *Enterobacteriaceae* has the following carbohydrates in common: glucose, galactose, glucosamine, heptose and KDO (2-keto-3-deoxyoctonic acid). Other carbohydrates, alone or in combination, confer specificity; these include galactosamine, mannose, rhamnose, aquebose, colominic acid and a new dideoxy-sugar (Sedlak *et al.*, 1971; Keléti *et al.*, 1971).

Genus IV. **Salmonella** *Lignières 1900, 389*

L. LE MINOR AND R. ROHDE

Sal.mon.el'la. M.L. dim. ending *-ella*. M.L. fem.n. *Salmonella;* named after D. E. Salmon, an American bacteriologist.

Rods, usually motile by peritrichous flagella; non-motile mutants may occur and one type (*S. gallinarum*, *S. pullorum*) is always non-motile. Colonies are generally 2–4 mm in diameter but certain types (*S. abortus-equi*, *S. typhi-suis* and *Salmonella abortus-ovis*) produce colonies of about 1 mm. Most strains will grow on defined media without special growth factors, and they **can use citrate as C source. Most strains are aerogenic** but *S. typhi*, an important exception, never produces gas. Anaerogenic variants of normally gas-producing serotypes are found in nature, and this is particularly common with *S. dublin*.

The common biochemical characters are shown in Table 8.3. Subdivision by biochemical characters into the so-called subgenera of Kauffmann (1960, 1963, 1964) is shown in Table 8.6. These subdivisions correspond more closely to species or subspecies in other groups of bacteria. Whatever rank is assigned to them, the worthiness of the

subdivisions was confirmed by Rohde (1965, 1966, 1967).

DNase and lipase are not produced.

G + C content of DNA: 50–53 moles % (by chemical analysis, T_m and buoyant density (Hill, 1966)).

Type species: *Salmonella cholerae-suis* (Smith) Weldin.

Further Comments

Most strains of the genus are susceptible to the O1 phage of Felix and Callow (1943); this phage is highly specific for *Salmonella*, lysing 99.5% of strains tested and only 0.3% of strains belonging to other members of the family (Thal and Kallings, 1955).

The names given to the different salmonellas do not follow the usual rules of nomenclature. Because of their importance in pathology, the first salmonellas were given names which indicated the disease and the animal from which they were isolated, and names of this kind (such as *S. typhi*, *S. paratyphi-A*, *S. paratyphi-B*, *S. cholerae-suis*, *S. typhi-murium* and *S. abortus-ovis*) continue to be used in clinical bacteriology. This nomenclature was abandoned by the more systematically minded, for the names implied the limitation of pathogenicity to definite host species, whereas *S. typhi-murium* and *S. bovis-morbificans* are frequently isolated from human infections. New types are now given the name of the town, region or country in which the first strain was isolated, as *S. london*, *S. panama*, *S. stanleyville*. New types of subgenera II, III and IV described after 1966 are designated simply by formula; this allows the *Arizona* group (subgenus III or *S. arizonae*) of Edwards, Fife and Ramsey (1959) to be included in the Kauffmann-White scheme, simplifies the terminology of the antigenic factors, and allows the same sera to be used to establish the antigenic formulae (Kauffmann and Rohde, 1962; Kauffmann, 1965; Rohde, 1967). With few exceptions, the formulae of Arizona serotypes published by Edwards, Fife and Ewing (1965) may be translated into Salmonella formulae and included in the Kauffmann-White scheme.

The International Enterobacteriaceae Subcommittee has not given clear guidance on the naming of the different types; it is paradoxical that serotypes of subgenus I bear species-like epithets, while those of *Escherichia* and *Salmonella* subgenus III (*S. arizonae*) do not. *S. typhi* owes its name to the importance of the bacterium in human pathology, but when these infection-syndrome names were first applied no one could have imagined that by 1968 there would be 1200 closely related types of bacteria each serologically characterized. Borman, Stuart and Wheeler (1948) proposed the subdivision of the genus into three species, *S. cholerae-suis* (the type species), *S. typhosa* and *S. kauffmannii*, the last to serve as a species for all the serological types. Kauffmann and Edwards (1952) made a similar proposal, but designated the all-embracing species *Salmonella enterica*. Ewing's (1966) proposal based the multiserotype species on *S. enteritidis* so that *S. paratyphi-A* would, by his scheme, become *S. enteritidis* serotype Paratyphi-A, an awkward and clumsy nomenclature for an organism that is frequently isolated.

Scientifically none of the present methods of nomenclature of *Salmonella* is satisfactory, and so without prejudice as to what constitutes a species, the *Enterobacteriaceae* Subcommittee considers the diagnostic use of the Kauffmann-White scheme to be overridingly important, and that the nomenclatural practice of giving names to the serotypes

of subgenus I should continue, but that new serotypes of the other subgenera should be designated only by formulae.

Division into Serotypes

The Kauffmann-White scheme, in which organisms are represented by the numbers and letters given to the different O, Vi and H antigens, indicates only those antigens of primary diagnostic importance and is not a complete record of the antigenic complement or its complexity (Kauffmann, 1966). The scheme, expanded to include all four subgenera, is given in Table 8.7. The original Arizona antigens (given in brackets) have been converted to the Salmonella designations.

Antigenic formulae (e.g. 6.7:r:1,7) represent the O antigens: the phase 1 antigen(s): the phase 2 antigen(s), respectively. Those with particular O antigens in common are collected into an O group and arranged alphabetically by H antigens within the group.

Subdivision of Serotypes

Biotypes are different sugar patterns shown by strains of the same serotype; they are determined by the presence or absence of enzymes and hence are genetically determined. Biotypes may serve as markers and be of interest epidemiologically, as the xylose$^+$ and xylose$^-$ character of *S. typhi*.

Phage types depend on the sensitivity of cultures to a series of bacteriophages at appropriate dilutions. Phage typing of *S. typhi* and other salmonellas which possess the Vi antigen (*S. paratyphi-C* and rarely *S. dublin*) is based on a series of adapted phages from phage Vi II of Craigie and Yen (1937, 1938). Phage typing of *S. paratyphi-B* (Felix and Callow, 1943) and *S. typhimurium* (Anderson, 1964) use different series of phages, and analogous methods have been proposed for other serotypes of *Salmonella*, some of them making use of the lysogenicity of the strains.

Other subdivisions of the serotypes may be made on the production of, or the sensitivity to, bacteriocins, on the resistance to antibiotics and on the resistance transfer factors.

Stability and Specificity of Antigens

Lysogenization by certain converting phages may produce changes in the O antigenic formulae of salmonellas. In groups A, B and D the presence of factor O1 is associated with lysogenization (Iseki and Kashiwagi, 1955, 1957; Stocker, 1958; Zinder, 1957), but the presence or absence of this

TABLE 8.6

*Distinguishing characters of the subgenera
of* **Salmonella**

	I	II	III	IV
Dulcitol	+	+	−	−
Lactose	−	−	+ or ×[a]	−
β-Galacto-sidase	−	×	+	−
d-Tartrate	+	−	−	−
Mucate	+	+	d	−
Malonate	−	+	+	−
Gelatin	−	(+)	(+)	(+)
KCN	−	−	−	+

[a] × = late and irregularly positive; (+) = late, but always positive.

factor in strains of these groups does not change the name of the organism; thus, *S. typhimurium* designates both the 1+ and 1− strains. On the contrary, in group E the name is changed. Phage ϵ_{15} (Iseki and Sakai, 1953) alters the O antigen 3, 10 to 3, 15; that is, it makes *S. anatum* become *S. newington;* in a similar way phage ϵ_{34} changes *S. newington* to *S. minneapolis.* The same applies to *S. cerro* and *S. siegburg,* the latter being simply the lysogenic variant of the former (Le Minor, 1965), and to all the strains of group C_4 (6, 7, 14)

which are lysogenic variants of strains of group C_1 (6, 7) although they bear different names. For this reason, all the factors connected with the conversion are *italicized* in the joint Kauffmann-White table which for the first time includes all *Arizona* serotypes with their corresponding *Salmonella* formulae. The converting phages of *Salmonella* are identical in morphology (Vieu *et al.*, 1965) but their action is limited to certain O groups; they are serologically different from one another (Le Minor, 1968).

The specificities of the O factors of *Salmonella* are determined by the composition and structure of the polysaccharides, and they are modified during S → R mutation and bacteriophage conversions (see review by Staub and Westphal, 1964). Thus, the only difference known between the 4, 12 and the 9, 12 O-specific repeat units is in the dideoxyhexose branch unit attached to the mannose, which is abequose in 4, 12 and tyvelose in 9, 12. In the conversion of 3, 10 → 3, 15 the terminal acetyl radical of the chain is suppressed and the α linkage between galactose and mannose is transformed into β.

Other modifications of the specificity of somatic antigens may occur after a mutation, resulting in new specificities called T_1 and T_2 by Kauffmann (1956) and in different R types (reviewed by Stocker, Wilkinson and Mäkelä, 1966; and Sarvas, 1967).

Description of some **Salmonella** *types*

SUBGENUS I

1. **Salmonella cholerae-suis** (Smith) Weldin 1927, 155. (*Bacillus cholerae-suis* Smith 1894, 9.)

chol.er.ae su′is. Gr. n. *cholera* cholera; L. n. *sus* swine, hog; M.L. gen.n. *suis* of a hog; M.L. gen.n. *choleraesuis* of hog cholera.

Antigenic formula: 6,7:c:1,5.

The detailed O formula is normally 6_2, 7, and this may be transformed by lysogenization into 6_1, 7 or 6_2, 7, 14.

Arabinose and trehalose are not fermented; dulcitol is slowly and irregularly fermented.

H_2S is not produced.

Pathogenic for man and other animals.

Neotype: NCTC 5735; ATCC 13312 (Opin. 26 Jud. Comm. 1963).

1a. *Salmonella cholerae-suis* var. *kunzendorf* Salmonella Subcommittee 1934, 341.

Description as for *S. cholerae-suis* except H_2S is produced.

2. **Salmonella hirschfeldii** Weldin 1927, 161. (Paratyphoid C bacillus, Hirschfeld 1919, 296;

Salmonella paratyphi-C (Salmonella Subcommittee 1934.)

hirsch.fel′di.i. M.L. gen.n. *hirschfeldii* of Hirschfeld; named after Hirschfeld who first called it the paratyphoid C bacillus, a name still in use today.

Antigenic formula: 6,7[Vi]:c:1,5.

Ferments dulcitol and trehalose; produces H_2S. Arabinose fermentation is variable.

Editorial Note. In 1934 the *Salmonella* Subcommittee of the International Committee on Bacteriological Nomenclature recommended the names *Salmonella paratyphi-A, S. paratyphi-B* and *S. paratyphi-C;* these names, although not in conformity with the rules of botanical nomenclature, were adopted by medical bacteriologists and by the WHO International *Salmonella* Center. The Bacteriological Code, adopted in 1947, was retroactive, thus making illegitimate these names which had been accepted and which are still used in medical work.

3. **Salmonella typhi** (Schroeter) Warren and Scott 1930, 416 *Nom. cons.* Opin. 18, Jud. Comm. 1963, 32. (*Bacillus typhi* Schroeter 1886, 165; *Bac-*

TABLE 8.7

Kauffmann-White scheme including subgenus III (S. arizonae) and serotypes described up to December 1970

Species or serotype	O	Phase 1	Phase 2	Species or serotype	O	Phase 1	Phase 2
Group A				*S. caledon	4,12	g,m,t	e,n,x
S. paratyphi A	1,2,12	a	—	S. hato	4,[5],12	g,m,s	—
S. nitra	2,12	g,m	—	S. california	4,12	g,m,t	—
S. kiel	1,2,12	g,p	—	S. kingston	1,4,[5],12,27	g,s,t	—
Group B				S. budapest	1,4,12	g,t	—
*S. makoma	4,12	a	—	*S. bechuana	4,12,27	g,t	—
S. kisangani	1,4,[5],12	a	1,2	S. travis	4,[5],12	g,z51	1,7
S. hessarek	4,12,27	a	1,5	S. banana	4,[5],12	m,t	—
S. fulica	4,[5],12	a	1,5	S. typhimurium	1,4,[5],12	i	1,2
S. arechavaleta	4,[5],12	a	[1,7]	S. lagos	1,4,12	i	1,5
S. bispebjerg	1,4,[5],12	a	e,n,x	S. agama	4,12	i	1,6
S. abortus-equi	4,12	—	e,n,x	S. gloucester	1,4,12,27	i	l,w
S. tinda	1,4,12,27	a	e,n,z15	S. massenya	1,4,12,27	k	1,5
S. nakuru	1,4,12,27	a	z6	S. neumuenster	1,4,12,27	k	1,6
S. schottmuelleri	1,4,[5],12	b	1,2	*Salmonella	1,4,12,27	k	1,6
(S. paratyphi B)				S. ljubljana	4,12,27	k	e,n,x
S. java1	1,4,[5],12	b	[1,2]	S. texas	4,[5],12	k	e,n,z15
S. limete	1,4,12,27	b	1,5	S. azteca	4,[5],12	l,v	1,5
S. canada	4,12	b	1,6	S. clackamas	4,12	l,v	1,6
S. uppsala	4,12,27	b	1,7	S. bredeney	1,4,12,27	l,v	1,7
*S. sofia	1,4,12,27	b	[e,n,x]	S. fyris	4,5,12	l,v	1,2
S. abony	1,4,[5],12	b	e,n,x	S. kimuenza	1,4,12,27	l,v	e,n,x
S. abortus-bovis	1,4,12,27	b	e,n,x	S. brandenburg	1,4,12	l,v	e,n,z15
S. wagenia	1,4,12,27	b	e,n,z15	S. togo	4,12	l,w	1,6
S. schleissheim	4,12,27	b	—	*S. kilwa	4,12	l,w	e,n,x
S. wien	1,4,12,27	b	l,w	S. ayton	1,4,12,27	l,w	z6
S. legon	4,12	c	1,5	S. vom	4,12,27	l,z13,z28	e,n,z15
S. abortus-ovis	4,12	c	1,6	S. kunduchi	1,4,[5],12,27	l,z28	1,2
S. altendorf	4,12,27	c	1,7	S. tyresoe	4,12	l,z28	1,5
S. jericho	1,4,12,27	c	e,n,z15	S. heidelberg	1,4,[5],12	r	1,2
S. bury	4,12,27	c	z6	S. bradford	4,12,27	r	1,5
S. stanley	1,4,[5],12,27	d	1,2	S. remo	1,4,12,27	r	1,7
S. eppendorf	1,4,12,27	d	1,5	S. bochum	4,[5],12	r	l,w
S. schwarzengrund	1,4,12,27	d	1,7	S. africana	4,12	r,i	l,w
*S. kluetjenfelde	4,12	d	e,n,x	S. coeln	4,[5],12	y	1,2
S. sarajane	4,12,27	d	e,n,x	S. trachau	4,12,27	y	1,5
S. duisburg	1,4,12,27	d	e,n,z15	S. teddington	4,12,27	y	1,7
S. salinatis	4,12	d,e,h	d,e,n,z15	S. ball	1,4,[5],12,27	y	e,n,x
S. mons	1,4,12,27	d	l,w	S. jos	1,4,12,27	y	e,n,z15
S. ayinde	4,12,27	d	z6	S. kamoru	4,12,27	y	z6
S. saint-paul	1,4,[5],12	e,h	1,2	S. shubra	4,[5],12	z	1,2
S. reading	4,[5],12	e,h	1,5	*Salmonella	1,4,12,27	z	1,5
S. kaapstad	4,12	e,h	1,7	S. kiambu	4,12	z	1,5
S. chester	4,[5],12	e,h	e,n,x	S. indiana	1,4,12	z	1,7
S. san-diego	4,[5],12	e,h	e,n,z15	*S. nordenham	1,4,12,27	z	e,n,x
*S. makumira	1,4,12,27	e,n,x	1,7	S. koenigstuhl	4,12	z	e,n,z15
*Salmonella	4,12	(f),g	—	S. preston	1,4,12	z	l,w
S. derby	1,4,[5],12	f,g	[1,2]	S. entebbe	1,4,12,27	z	z6
S. agona	1,4,12	f,g,s	—	S. stanleyville	1,4,[5],12,27	z4,z23	[1,2]
S. essen	4,12	g,m	—	S. kalamu	4,12	z4,z24	[1,5]
				S. haifa	1,4,[5],12	z10	1,2

This table was prepared by the World Health Organization International Reference Centre for Salmonella, Pasteur Institute, Paris.

Definition of symbols. * Subgenus II, † Subgenus III, original Arizona formula (Edwards *et al.*, 1965) given in brackets, ‡ Subgenus IV; ( ), atypical, weakly agglutinable by ordinary sera; [ ], may be lacking. O antigens numbered 1-65; when italicized present only when organism is lysogenized by converting bacteriophage. H antigens: Phase 1 lettered a–z, z_1–z_{59}; Phase 2 originally numbered but because of crossed reactions now includes many with Phase 1 designations. Groups: A group contains one or more O antigens in common. Listing within groups is alphabetically by Phase 1 and/or numerically by Phase 2 antigens.

[1] d-tartrate positive contrary to *S. paratyphi* B.

TABLE 8.7—*Continued*

Species or serotype	O	H Phase 1	H Phase 2	Species or serotype	O	H Phase 1	H Phase 2
S. ituri	1,4,12	z_{10}	1,5	*S. norton*	6,7	i	1,w
S. tudu	4,12	z_{10}	1,6	*S. galiema*	6,7	k	1,2
S. albert	4,12	z_{10}	e,n,x	*S. thompson*	6,7	k	1,5
S. tokoin	4,12	z_{10}	e,n,z_{15}	*S. daytona*	6,7	k	1,6
S. mura	1,4,12	z_{10}	l,w	*S. baiboukoum*	6,7	k	1,7
S. fortune	1,4,12,27	z_{10}	z_6	*S. singapore*	6,7	k	e,n,x
S. vellore	1,4,12,27	z_{10}	z_{35}	*S. escanaba*	6,7	k	e,n,z_{15}
S. brancaster	1,4,12,27	z_{29}		†*Salmonella arizonae*	6,7	(k)	z
**S. helsinki*	1,4,12	z_{29}	[e,n,x]	(Ar.27:22:31)			
S. tafo	1,4,12,27	z_{35}	1,7	*Salmonella	6,7	k	[z_6]
S. sloterdijk	1,4,12,27	z_{35}	z_6	*S. concord*	6,7	l,v	1,2
S. yaounde	1,4,12,27	z_{35}	e,n,z_{15}	*S. irumu*	6,7	l,v	1,5
S. tejas	4,12	z_{36}	—	*S. bonn*	6,7	l,v	e,n,x
S. wilhelmsburg	1,4,[5],12,27	z_{38}	—	*S. potsdam*	6,7	l,v	e,n,z_{15}
**S. durbanville*	1,4,12,27	z_{39}	1,[5],7	†*Salmonella arizonae*	6,7	l,v	z
				(Ar.27:23:31)			
Group C₁				*S. gdansk*	6,7	l,v	z_6
S. san-juan	6,7	a	1,5	†*Salmonella arizonae*	6,7	l,v	z_{53}
S. umhlali	6,7	a	1,6	(Ar.27:23:25)			
S. austin	6,7	a	1,7	*S. gabon*	6,7	l,w	1,2
S. oslo	6,7	a	e,n,x	*S. colorado*	6,7	l,w	1,5
S. denver	6,7	a	e,n,z_{15}	*S. ness-ziona*	6,7	l,z_{13}	1,5
S. coleypark	6,7	a	l,w	*S. kenya*	6,7	l,z_{13}	e,n,x
*Salmonella	6,7	a	z_6	*S. neukoelln*	6,7	l,z_{13},z_{28}	e,n,z_{15}
**S. calvinia*	6,7	a	z_{42}	*S. makiso*	6,7	l,z_{13},z_{28}	z_6
S. brazzaville	6,7	b	1,2	**S. heilbron*	6,7	l,z_{28}	1,5:[z_{42}]
S. edinburg	6,7	b	1,5	*S. virchow*	6,7	r	1,2
S. koumra	6,7	b	1,7	*S. infantis*	6,7	r	1,5
S. georgia	6,7	b	e,n,z_{15}	*S. nigeria*	6,7	r	1,6
S. ohio	6,7	b	l,w	*S. colindale*	6,7	r	1,7
S. leopoldville	6,7	b	z_6	*S. papuana*	6,7	r	e,n,z_{15}
S. kotte	6,7	b	z_{35}	*S. richmond*	6,7	y	1,2
**S. bloemfontein*	6,7	b	[e,n,x]:z_{42}	*S. bareilly*	6,7	y	1,5
S. hirschfeldii (*S. paratyphi C*)	6,7,[Vi]	c	1,5	*S. gatow*	6,7	y	1,7
				S. hartford	6,7	y	e,n,x
S. cholerae-suis	6,7	[c]	1,5	*S. mikawasima*	6,7	y	e,n,z_{15}
S. typhi-suis	6,7	c	1,5	**S. tosamanga*	6,7	z	1,5
S. birkenhead	6,7	c	1,6	*S. oakland*	6,7	z	1,6,7
S. kisii	6,7	d	1,2	*S. businga*	6,7	z	e,n,z_{15}
S. isangi	6,7	d	1,5	*Salmonella	6,7	z	z_6
S. kivu	6,7	d	1,6	**S. oysterbeds*	6,7	z	z_{42}
S. amersfoort	6,7	d	e,n,x	*S. goma*	6,7	z_4,z_{23}	z_6
S. gombe	6,7	d	e,n,z_{15}	*S. obogu*	6,7	z_4,z_{23}	1,5
S. livingstone	6,7	d	l,w	*S. aequatoria*	6,7	z_4,z_{23}	e,n,z_{15}
S. wil	6,7	d	l,z_{13},z_{28}	‡*S. roterberg*	6,7	z_4,z_{23}	—
S. larochelle	6,7	e,h	1,2	*S. somone*	6,7	z_4,z_{24}	—
S. lomita	6,7	e,h	1,5	‡*S. kralendyk*	6,7	z_4,z_{24}	—
S. norwich	6,7	e,h	1,6	†*Salmonella arizonae*	6,7	z_4,z_{32}	—
S. braenderup	6,7	e,h	e,n,z_{15}	(Ar.27:1,6,7:−)			
S. rissen	6,7	f,g	—	**S. cape*	6,7	z_6	1,7
S. afula	6,7	f,g,t	e,n,x	*S. menden*	6,7	z_{10}	1,2
S. montevideo	6,7	g,m,s,[p]	—	*S. inganda*	6,7	z_{10}	1,5
S. othmarschen	6,7	g,m,[t]	—	*S. eschweiler*	6,7	z_{10}	1,6
S. menston	6,7	g,s,t	—	*S. ngili*	6,7	z_{10}	1,7
S. riggil	6,7	g,t	—	*S. djugu*	6,7	z_{10}	e,n,x
*Salmonella	6,7	g,t	e,n,x:z_{42}	*S. mbandaka*	6,7	z_{10}	e,n,z_{15}
S. alamo	6,7	g,z_{51}	1,5	*Salmonella	6,7	z_{10}	z_{35}
S. haelsingborg	6,7	m,p,t,[u]	—	*S. tennessee*	6,7	z_{29}	—
S. oranienburg	6,7	m,t	—	*Salmonella	6,7	z_{29}	—
S. augustenborg	6,7	i	1,2	‡*S. argentina*	6,7	z_{36}	—
S. oritamerin	6,7	i	1,5	**S. bacongo*	6,7	z_{36}	z_{42}
S. garoli	6,7	i	1,6	*S. lille*	6,7	z_{38}	—

TABLE 8.7—Continued

Species or serotype	O	H Phase 1	H Phase 2	Species or serotype	O	H Phase 1	H Phase 2
*S. gilbert	6,7	z_{39}	1,5,7	S. tallahassee	6,8	z_4, z_{32}	—
S. hillsborough	6,7	z_{41}	l,w	S. zerifin	6,8	z_{10}	1,2
S. tamilnadu	6,7	z_{41}	z_{35}	S. mapo	6,8	z_{10}	1,5
*S. sullivan	6,7	z_{42}	1,7	S. cleveland	6,8	z_{10}	1,7
*Salmonella	6,7	z_{42}	e,n,x:1,6	S. hadar	6,8	z_{10}	e,n,x
				S. glostrup	6,8	z_{10}	e,n,z_{15}
Group C₂				S. wippra	6,8	z_{10}	z_6
S. doncaster	6,8	a	1,5	S. uno	6,8	z_{29}	—
S. curacao	6,8	a	1,6	S. yarm	6,8	z_{35}	1,2
S. nordufer	6,8	a	1,7				
S. narashino	6,8	a	e,n,x	**Group C₃**			
S. leith	6,8	a	e,n,z_{15}	S. djelfa	8	b	1,2
*S. tulear	6,8	a	z_{52}	S. korbol	8,20	b	1,5
S. nagoya	6,8	b	1,5	S. sanga	8	b	1,7
S. stourbridge	6,8	b	1,6	S. shipley	8,20	b	e,n,z_{15}
S. gatuni	6,8	b	e,n,x	S. alexanderpolder	8	c	l,w
S. presov	6,8	b	e,n,z_{15}	S. virginia	8	d	1,2
S. bukuru	6,8	b	l,w	S. labadi	8,20	d	z_6
S. banalia	6,8	b	z_6	S. atakpame	8,20	e,h	1,7
S. wingrove	6,8	c	1,2	S. bardo	8	e,h	1,2
S. utah	6,8	c	1,5	S. ferruch	8	e,h	1,5
S. bronx	6,8	c	1,6	S. rechovot	8,20	e,h	z_6
S. belfast	6,8	c	1,7	S. emek	8,20	g,m,s	—
S. belem	6,8	c	e,n,x	S. yokoe	8	m,t	—
S. quiniela	6,8	c	e,n,z_{15}	S. kentucky	8,20	i	z_6
S. muenchen	6,8	d	1,2	S. kentucky var. jerusalem	8	i	z_6
S. manhattan	6,8	d	1,5	S. haardt	8	k	1,5
S. sterrenbos	6,8	d	e,n,x	S. pakistan	8	l,v	1,2
S. herston	6,8	d	e,n,z_{15}	S. amherstiana	8	l,v	1,6
S. newport	6,8	e,h	1,2	S. hindmarsh	8	r	1,5
S. kottbus	6,8	e,h	1,5	S. pikine	8,20	r	z_6
S. tshiongwe	6,8	e,h	e,n,z_{15}	S. altona	8,20	r,i	z_6
S. sandow	6,8	f,g	e,n,x	S. cocody	8,20	r,i	e,n,z_{15}
S. chincol	6,8	g,m,s	e,n,x	S. giza	8	y	1,2
*Salmonella	6,8	g,m,t	[e,n,x]	S. brunei	8,20	y	1,5
*S. baragwanath	6,8	m,t	1,5	S. alagbon	8	y	1,7
*S. germiston	6,8	m,t	e,n,x	S. sunnycove	8	y	e,n,x
S. lindenburg	6,8	i	1,2	S. kralingen	8,20	y	z_6
S. takoradi	6,8	i	1,5	S. corvallis	8,20	z_4, z_{23}	—
S. warnow	6,8	i	1,6	S. albany	8,20	z_4, z_{24}	—
S. bonariensis	6,8	i	e,n,x	S. paris	8,20	z_{10}	1,5
S. aba	6,8	i	e,n,z_{15}	S. molade	8,20	z_{10}	z_6
S. blockley	6,8	k	1,5	S. istanbul	8	z_{10}	e,n,x
S. schwerin	6,8	k	e,n,x	S. tamale	8,20	z_{29}	[e,n,z_{15}]
S. litchfield	6,8	l,v	1,2	S. angers	8,20	z_{35}	z_6
S. loanda	6,8	l,v	1,5	S. diogoye	8,20	z_{41}	z_6
S. manchester	6,8	l,v	1,7	S. apeyeme	8,20	z_{38}	—
S. holcomb	6,8	l,v	e,n,x				
S. edmonton	6,8	l,v	e,n,z_{15}	**Group C₄** (Salmonella of group C₁ lysogenized by "phage 14")			
S. fayed	6,8	l,w	1,2				
S. breukelen	6,8	l,z_{13}	e,n,z_{15}				
S. bovis-morbificans	6,8	r	1,5	S. nienstedten	6,7,14	b	[l,w]
S. akanji	6,8	r	1,7	S. kaduna	6,7,14	c	e,n,z_{15}
S. hidalgo	6,8	r	e,n,z_{15}	S. hissar	6,7,14	c	1,2
S. gold-coast	6,8	r	l,w	S. omderman	6,7,14	d	e,n,x
S. tananarive	6,8	y	1,5	S. eimsbuettel	6,7,14	d	l,w
S. praha	6,8	y	e,n,z_{15}	S. nieukerk	6,7,14	d	z_6
S. mowanjum	6,8	z	1,5	S. ardwick	6,7,14	f,g	—
S. kuru	6,8	z	l,w	S. thielallee	6,7,14	m,t	—
S. lezennes	6,8	z_4, z_{23}	1,7	S. gelsenkirchen	6,7,14	l,v	z_6
S. chailey	6,8	z_4, z_{23}	e,n,z_{15}				
S. duesseldorf	6,8	z_4, z_{24}	—				

TABLE 8.7—Continued

Species or serotype	O	H Phase 1	H Phase 2	Species or serotype	O	H Phase 1	H Phase 2
S. jerusalem	6,7,14	z_{10}	l,w	S. jamaica	9,12	r	1,5
S. bornum	6,7,14	z_{38}	—	S. lomé	9,12	r	z_6
				S. lawndale	1,9,12	z	1,5
Group D₁				*S. stellenbosch	1,9,12	z	1,7
S. sendai	1,9,12	a	1,5	*S. angola	1,9,12	z	z_6
S. miami	1,9,12	a	1,5	*S. hueningen	9,12	z	z_{39}
*Salmonella	9,12	a	1,5	S. wangata	1,9,12	z_4,z_{23}	[1,7]
S. os	9,12	a	1,6	S. portland	9,12	z_{10}	1,5
S. saarbruecken	1,9,12	a	1,7	*S. canastel	9,12	z_{29}	1,5
S. loma-linda	9,12	a	e,n,x	S. penarth	9,12	z_{35}	z_6
S. durban	9,12	a	e,n,z_{15}	S. elomrane	1,9,12	z_{38}	—
S. onarimon	1,9,12	b	1,2	*S. wynberg	1,9,12	z_{39}	1,7
S. frintrop	1,9,12	b	1,5	S. gallinarum	1,9,12	—	—
*S. mjimwema	1,9,12	b	e,n,x				
*S. blankenese	1,9,12	b	z_6	Group D₂[1]			
*S. suederelbe	1,9,12	b	z_{39}	S. baildon	9,46	a	e,n,x
S. goeteborg	9,12	c	1,5	S. zadar	9,46	b	1,6
S. ipeko	9,12	c	1,6	*S. lundby	9,46	b	e,n,x
S. alabama	9,12	c	e,n,z_{15}	S. bamboye	9,46	b	l,w
S. ridge	9,12	c	z_6	S. itutaba	9,46	c	z_6
S. typhi	9,12[Vi]	d	—	S. strasbourg	9,46	d	1,7
S. ndolo	1,9,12	d	1,5	S. plymouth	9,46	d	z_6
S. tarshyne	9,12	d	1,6	S. bergedorf	9,46	e,h	1,2
*S. rhodesiense	9,12	d	e,n,x	S. wernigerode	9,46	f,g	—
S. zega	9,12	d	z_6	*S. duivenhoks	9,46	g,m,s,t	e,n,x
S. jaffna	1,9,12	d	z_{35}	S. gateshead	9,46	g,s,t	—
S. bournemouth	9,12	e,h	1,2	S. mathura	9,46	i	e,n,z_{15}
S. eastbourne	1,9,12	e,h	1,5	S. potto	9,46	i	z_6
S. israel	9,12	e,h	e,n,z_{15}	S. marylebone	9,46	k	1,2
*Salmonella	9,12	e,n,x	1,6	S. ceyco	9,46	k	z_{15}
*S. lindrick	9,12	e,n,x	1,[5],7	S. india	9,46	l,v	1,5
S. berta	1,9,12	f,g,t	—	S. geraldton	9,46	l,v	1,6
S. enteritidis	1,9,12	g,m	—	S. shoreditch	9,46	r	e,n,z_{15}
S. blegdam	9,12	g,m,q	—	S. mayday	9,46	y	z_6
*S. muizenberg	9,12	g,m,s,t	1,5	S. benin	9,46	y	1,7
*S. kuilsrivier	1,9,12	g,m,s,t	e,n,x	*S. haarlem	9,46	z	e,n,x
S. manica	1,9,12	g,m,s,t	z_{42}	S. ekotedo	9,46	z_4,z_{23}	—
*S. hamburg	1,9,12	g,m,t	—	*S. maarssen	9,46	z_4,z_{24}	$z_{39}:z_{42}$
S. dublin	1,9,12[Vi]	g,p	—	S. lishabi	9,46	z_{10}	1,7
S. naestved	1,9,12	g,p,s	—	S. inglis	9,46	z_{10}	e,n,x
S. rostock	1,9,12	g,p,u	—	*Salmonella	9,46	z_{10}	z_6
S. moscow	9,12	g,q	—	S. ouakam	9,46	z_{29}	—
*S. neasden	9,12	g,s,t	e,n,x	S. hillegersberg	9,46	z_{35}	1,5
S. new-mexico	9,12	g,z_{51}	1,5	S. fresno	9,46	z_{38}	—
S. pensacola	1,9,12	m,t	—				
S. seremban	9,12	i	1,5	Group D₃			
S. claibornei	1,9,12	k	1,5	*S. zuerich	1,9,12,(46),27	c	z_{39}
S. goverdhan	9,12	k	1,6	*Salmonella	1,9,12,(46),27	y	z_{39}
S. mendoza	9,12	l,v	1,2				
S. panama	1,9,12	l,v	1,5	Group E₁			
S. kapemba	9,12	l,v	1,7	S. aminatu	3,10	a	1,2
*Salmonella	9,12	l,v	e,n,x	S. goelzau	3,10	a	1,5
S. goettingen	9,12	l,v	e,n,z_{15}	S. oxford	3,10	a	1,7
S. victoria	1,9,12	l,w	1,5	*S. matroosfontein	3,10	a	e,n,x
†S. dar-es-salaam	1,9,12	l,w	e,n,x	S. galil	3,10	a	e,n,z_{15}
(S. salamae)				S. kalina	3,10	b	1,2
S. miyazaki	9,12	l,z_{13}	1,7	S. butantan	3,10	b	1,5
S. napoli	1,9,12	l,z_{13}	e,n,x	S. allerton	3,10	b	1,6
S. javiana	1,9,12	l,z_{28}	1,5	S. huvudsta	3,10	b	1,7

[1] The serotypes of this group also contain the factors O:3 and (10) the latter not very well developed. They can be lysogenized by phages ϵ_{15} and ϵ_{34} and in the case of double lysogenization become strongly agglutinable, like strains of group E₄, by antisera against O:34 and O:12₂.

TABLE 8.7—*Continued*

Species or serotype	O	H Phase 1	H Phase 2	Species or serotype	O	H Phase 1	H Phase 2
S. benfica	3,10	b	e,n,x	*S. clerkenwell*	3,10	z	l,w
S. yaba	3,10	b	e,n,z15	**S. tafelbaai*	3,10	z	z39
S. epicrates	3,10	b	l,w	*S. adabraka*	3,10	z4,z23	[1,7]
S. pramiso	3,10	c	1,7	*S. florian*	3,10	z4,z24	—
S. agege	3,10	c	e,n,z15	*S. okerara*	3,10	z10	1,2
S. anderlecht	3,10	c	l,w	*S. lexington*	3,10	z10	1,5
S. okefoko	3,10	c	z6	*S. coquilhatville*	3,10	z10	1,7
S. stormont	3,10	d	1,2	*S. kristianstad*	3,10	z10	e,n,z15
S. shangani	3,10	d	1,5	*S. biafra*	3,10	z10	z6
S. onireke	3,10	d	1,7	*S. jedburgh*	3,10	z29	—
S. souza	3,10	d	e,n,x	*S. cairina*	3,10	z35	z6
S. madjorio	3,10	d	e,n,z15	*S. macallen*	3,10	z36	—
S. birmingham	3,10	d	l,w	*S. bolombo*	3,10	z38	—
S. weybridge	3,10	d	z6	**S. mpila*	3,10	z38	z42
S. maron	3,10	d	z35	**S. winchester*	3,10	z39	1,7
S. vejle	3,10	e,h	1,2				
S. muenster	3,10	e,h	1,5	Group E2 (*Salmo-*			
S. anatum	3,10	e,h	1,6	*nella* of group E1			
S. nyborg	3,10	e,h	1,7	lysogenized by			
S. newlands	3,10	e,h	e,n,x	phage ε15)			
S. meleagridis	3,10	e,h	l,w	*S. clichy*	3,15	a	1,5
S. sekondi	3,10	e,h	z6	*S. rosenthal*	3,15	b	1,5
*S. chudleigh	3,10	e,n,x	1,7	*S. pankow*	3,15	d	1,5
S. regent	3,10	f,g	—	*S. eschersheim*	3,15	d	e,n,x
S. suberu	3,10	g,m	—	*S. goerlitz*	3,15	e,h	1,2
S. amsterdam	3,10	g,m,s	—	*S. new-haw*	3,15	e,h	1,5
S. westhampton	3,10	g,s,t	—	*S. newington*	3,15	e,h	1,6
*S. islington	3,10	g,t	—	*S. selandia*	3,15	e,h	1,7
S. southbank	3,10	m,t	—	*S. cambridge*	3,15	e,h	l,w
*S. stikland	3,10	m,t	e,n,x	*S. drypool*	3,15	g,m,s	—
S. amounderness	3,10	i	1,5	**S. parow*	3,15	g,m,s,t	—
S. falkensee	3,10	i	e,n,z15	*S. halmstad*	3,15	g,s,t	—
S. yeerongpilly	3,10	i	z6	*S. nancy*	3,15	l,v	1,2
S. wimborne	3,10	k	1,2	*S. portsmouth*	3,15	l,v	1,6
S. zanzibar	3,10	k	1,5	*S. new-brunswick*	3,15	l,v	1,7
S. marienthal	3,10	k	e,n,z15	*S. kinshasa*	3,15	l,z13	1,5
S. new-rochelle	3,10	k	l,w	*S. lanka*	3,15	r	z6
S. nchanga	3,10	l,v	1,2	*S. tuebingen*	3,15	y	1,2
S. sinstorf	3,10	l,v	1,5	*S. binza*	3,15	y	1,5
S. london	3,10	l,v	1,6	*S. tournai*	3,15	y	z6
S. give	3,10	l,v	1,7	*S. manila*	3,15	z10	1,5
S. ruzizi	3,10	l,v	e,n,z15				
*S. fuhlsbuettel	3,10	l,v	z6	Group E3 (*Salmonella*			
S. uganda	3,10	l,z13	1,5	of group E1 lysogen-			
S. fallowfield	3,10	l,z13,z28	e,n,z15	ized by phages ε15			
S. joal	3,10	l,z28	1,7	and ε34)			
*S. westpark	3,10	l,z28	e,n,x	*S. khartoum*	3,15,34	a	1,7
*Salmonella	3,10	l,z28	z39	*S. arkansas*	3,15,34	e,h	1,5
S. seegefeld	3,10	r,i	1,2	*S. minneapolis*	3,15,34	e,h	1,6
S. ughelli	3,10	r	1,5	*S. wildwood*	3,15,34	e,h	l,w
S. elisabethville	3,10	r	1,7	*S. canoga*	3,15,34	g,s,t	—
S. simi	3,10	r	e,n,z15	*S. menhaden*	3,15,34	l,v	1,7
S. weltevreden	3,10	r	z6	*S. thomasville*	3,15,34	y	1,5
S. amager	3,10	y	1,2	*S. illinois*	3,15,34	z10	1,5
S. orion	3,10	y	1,5	*S. harrisonburg*	3,15,34	z10	1,6
S. mokola	3,10	y	1,7				
S. ohlstedt	3,10	y	e,n,x	Group E4			
S. bolton	3,10	y	e,n,z15	*S. gwoza*	1,3,19	a	e,n,z15
S. langensalza	3,10	y	l,w	*S. gnesta*	1,3,19	b	1,5
S. stockholm	3,10	y	z6	*S. visby*	1,3,19	b	1,6
*S. alexander	3,10	z	1,5	*S. broughton*	1,3,19	b	l,w
*S. finchley	3,10	z	e,n,x	*S. accra*	1,3,19	b	z6

TABLE 8.7—*Continued*

Species or serotype	O	H Phase 1	H Phase 2	Species or serotype	O	H Phase 1	H Phase 2
S. madiago	1,3,19	c	1,7	*S. colobane*	11	k	1,7
S. ahmadi	1,3,19	d	1,5	*S. kisarawe*	11	k	e,n,x
S. liverpool	1,3,19	d	e,n,z_{15}	*S. amba*	11	k	1,z_{13},z_{28}
S. tilburg	1,3,19	d	1,w	†*Salmonella arizonae*	11	k	z_{53}
S. niloese	1,3,19	d	z_6	(Ar.17:29:25)			
S. vilvoorde	1,3,19	e,h	1,5	*S. stendal*	11	l,v	1,2
S. sankt-marx	1,3,19	e,h	1,7	*S. maracaibo*	11	l,v	1,5
S. sao	1,3,19	e,h	e,n,z_{15}	*S. fann*	11	l,v	e,n,x
S. calabar	1,3,19	e,h	1,w	*S. bullbay*	11	l,v	e,n,z_{15}
S. rideau	1,3,19	f,g	—	*S. glidji*	11	l,w	1,5
S. maiduguri	1,3,19	f,g,t	e,n,z_{15}	*S. osnabrueck*	11	l,z_{13},z_{28}	e,n,x
S. senftenberg	1,3,19	g,[s],t	—	*S. huila*	11	l,z_{28}	e,n,x
S. cannstatt	1,3,19	m,t	—	*S. senegal*	11	r	1,5
S. stratford	1,3,19	i	1,2	*S. rubislaw*	11	r	e,n,x
S. machaga	1,3,19	i	e,n,x	*S. volta*	11	r	l,z_{13},z_{28}
S. avonmouth	1,3,19	i	e,n,z_{15}	*S. solt*	11	y	1,5
S. zuilen	1,3,19	i	1,w	*S. herzliya*	11	y	e,n,x
S. taksony	1,3,19	i	z_6	*S. nyanza*	11	z	z_6
S. ngor	1,3,19	l,v	1,5	*S. soutpan*	11	z	z_{29}
S. westerstede	1,3,19	l,z_{13}	[1,2]	†*Salmonella arizonae*	11	z_4,z_{23}	—
S. lokstedt	1,3,19	l,z_{13},z_{28}	1,2	(Ar.17:1,2,5:−)			
S. bedford	1,3,19	l,z_{13},z_{28}	e,n,z_{15}	‡*S. parera*	11	z_4,z_{23}	—
S. yalding	1,3,19	r	e,n,z_{15}	*S. remete*	11	z_4,z_{23}	1,6
S. fareham	1,3,19	r,i	1,w	*S. etterbeek*	11	z_4,z_{23}	e,n,z_{15}
S. krefeld	1,3,19	y	1,w	*S. yehuda*	11	z_4,z_{24}	—
S. korlebu	1,3,19	z	1,5	‡*Salmonella*	11	z_4,z_{32}	—
S. schoeneberg	1,3,19	z	e,n,z_{15}	*S. wentworth*	11	z_{10}	1,2
S. carno	1,3,19	z	1,w	*S. straengnaes*	11	z_{10}	1,5
S. dallgow	1,3,19	z_{10}	e,n,z_{15}	*S. tel-hashomer*	11	z_{10}	e,n,x
S. llandoff	1,3,19	z_{29}	—	*S. lene*	11	z_{38}	—
S. chittagong	(1),3,10,(19)	b	z_{35}	*S. maastricht*	11	z_{41}	1,2
S. bilu	(1),3,10,(19)	f,g,t	1,(2),7	*Salmonella*	11	—	1,5
S. ilugun	(1),3,10,(19)	z_4,z_{23}	z_6				
S. dessau	(1),3,*15*,(19)	g,s,t	—	**Group G$_1$**			
S. cannonhill	(1),3,*15*,(19)	y	e,n,x	†*Salmonella arizonae*	13,22	—	—
				(Ar.18:−:−)			
Group F				*S. mim*	13,22	a	1,6
S. toowong	11	a	1,7	*S. ibadan*	13,22	b	1,5
S. marseille	11	a	1,5	*S. vaertan*	13,22	b	e,n,x
S. luciana	11	a	e,n,z_{15}	*S. bahati*	13,22	b	e,n,z_{15}
S. glencairn	11	a	z_6:z_{42}	*S. haouaria*	13,22	c	e,n,x,z_{15}
S. leeuwarden	11	b	1,5	*S. friedenau*	13,22	d	1,6
Salmonella	11	b	1,7	*S. diguel*	*1*,13,22	d	e,n,z_{15}
S. srinagar	11	b	e,n,x	*S. willemstad*	13,22	e,h	1,6
S. pharr	11	b	e,n,z_{15}	*S. raus*	13,22	f,g	e,n,x
S. chandans	11	d	e,n,x	*Salmonella*	13,22	(f),g,t	—
S. montgomery	11	d(a)	d,e,n,z_{15}	*S. bron*	13,22	g,m	[e,n,z_{15}]
S. findorff	11	d	z_6	*S. limbe*	*1*,13,22	g,m,t	[1,5]
S. chingola	11	e,h	1,2	*S. rotterdam*	*1*,13,22	g,t	1,5
S. adamstua	11	e,h	1,6	*S. lovelace*	13,22	l,v	1,5
S. redhill	11	e,h	l,z_{13},z_{28}	*S. borbeck*	13,22	l,v	1,6
S. grabouw	11	g,m,s,t	z_{39}	*S. tanger*	*1*,13,22	y	1,6
‡*S. mundsburg*	11	g,z_{51}	—	*S. poona*	*1*,13,22	z	1,6
S. lincoln	11	m,t	e,n,x	*S. bristol*	13,22	z	1,7
S. aberdeen	11	i	1,2	*S. roodepoort*	*1*,13,22	z_{10}	1,5
S. brijbhumi	11	i	1,5	*S. agoueve*	13,22	z_{29}	—
S. heerlen	11	i	1,6	*S. clifton*	13,22	z_{29}	1,5
S. veneziana	11	i	e,n,x	*S. goodwood*	13,22	z_{29}	e,n,x
S. pretoria	11	k	1,2	*S. mampong*	13,22	z_{35}	1,6
S. abaetetuba	11	k	1,5	*S. leiden*	13,22	z_{38}	—
S. sharon	11	k	1,6	*Salmonella*	13,22	z_{39}	1,5,(7)

TABLE 8.7—Continued

Species or serotype	O	H Phase 1	H Phase 2	Species or serotype	O	H Phase 1	H Phase 2
Group G₂				S. finkenwerder	[1],6,14,[25]	d	1,5
S. chagoua	1,13,23	a	1,5	S. florida	1,6,14,25	d	1,7
*S. tygerberg	1,13,23	a	z42	S. lindern	6,14,24	d	e,n,x
S. atlanta	13,23	b	—	S. charity	1,6,14,25	d	e,n,x
S. mississippi	1,13,23	b	1,5	S. teko	1,6,14,25	d	e,n,z15
*S. acres	1,13,23	b	[1,5]:z42	S. encino	1,6,14,25	d	l,z13,z28
S. bracknell	13,23	b	1,6	S. albuquerque	1,6,14,24	d	z6
S. ullevi	1,13,23	b	e,n,x	S. bahrenfeld	6,14,24	e,h	1,5
S. durham	13,23	b	e,n,z15	S. onderstepoort	1,6,14,[25]	e,h	1,5
S. mishmar-haemek	1,13,23	d	1,5	S. magumeri	1,6,14,25	e,h	1,6
S. grumpensis	13,23	d	1,7	S. beaudesert	[1],6,14,[25]	e,n	1,7
S. tel-el-kebir	13,23	d	e,n,z15	S. warragul	1,6,14,25	g,m	—
*Salmonella	13,23	d	e,n,x	S. caracas	1,6,14,25	g,m,s	—
S. putten	13,23	d	l,w	S. kaitaan	1,6,14,25	m,t	—
S. isuge	13,23	d	z6	*S. rooikrantz	1,6,14	m,t	1,5
S. wichita	13,23	d	[z37]	*S. emmerich	6,14	[m,t]	e,n,x
S. tschangu	1,13,23	e,h	1,5	S. mampeza	1,6,14,25	i	1,5
*S. epping	1,13,23	e,n,x	1,7	S. buzu	1,6,14,25	i	1,7
S. havana	1,13,23	f,g,[s]	—	S. schalkwijk	6,14,24	i	e,n,..
S. agbeni	13,23	g,m	—	S. harburg	1,6,14,25	k	1,5
*S. kraaifontein	1,13,23	g,m,t	[e,n,x]	*Salmonella	6,14	k	[e,n,x]
S. okatie	13,23	g,s,t	—	†Salmonella arizonae	6,14	k	z
*S. luanshya	13,23	g,m,s,t	—	(Ar.7a,7c:29:31)			
S. congo	13,23	g,m,s,t	—	*Salmonella	1,6,14	k	z6:z42
*S. gojenberg	1,13,23	g,t	1,5	†Salmonella arizonae	6,14	k	z51
†Salmonella arizonae	1,13,23	g,z51	—	(Ar.7a,7c:29:25)			
(Ar.18:13,14:-)				S. boecker	[1],6,14,[25]	l,v	1,7
S. kintambo	13,23	m,t	—	S. horsham	1,6,14,25	l,v	e,n,x
*S. katesgrove	1,13,23	m,t	1,5	S. aflao	1,6,14,25	l,z28	e,n,x
*S. worchester	1,13,23	m,t	e,n,x	S. surat	1,6,14,25	[r],i	e,n,z15
*S. boulders	13,23	m,t	z42	S. carrau	6,14,24	y	1,7
S. kedougou	1,13,23	i	l,w	S. madelia	1,6,14,25	y	1,7
S. idikan	13,23	i	1,5	S. fischerkietz	1,6,14,25	y	e,n,x
S. jukestown	13,23	i	e,n,z15	S. homosassa	1,6,14,25	z	1,5
*Salmonella	13,23	l,z28	z6	S. soahanina	6,14,24	z	e,n,x
*S. vredelust	1,13,23	l,z28	z42	S. sundsvall	1,6,14,25	z	e,n,x
S. adjame	13,23	r	1,6	S. poano	1,6,14,25	z	l,z13,z28
S. linton	13,23	r	e,n,z15	S. bousso	1,6,14,25	z4,z23	[e,n,z15]
S. yarrabah	13,23	y	1,7	S. uzaramo	1,6,14,25	z4,z24	—
S. ordonez	1,13,23	y	l,w	S. nessa	1,6,14,25	z10	1,2
S. tunis	1,13,23	y	z6	*S. simonstown	1,6,14	z10	1,5
*S. nachshonim	1,13,23	z	1,5	*S. bornheim	1,6,14,25	z10	1,(2),7
S. farmsen	13,23	z	1,6	†Salmonella arizonae	6,14	z10	z
S. worthington	1,13,23	z	l,w	(Ar.7a,7c:27:31)			
S. ajiobo	13,23	z4,z23	—	†Salmonella arizonae	6,14	z10	z:z56
S. romanby	13,23	z4,z24	—	(Ar.7a,7c:27:31:38)			
†Salmonella arizonae	1,13,23	z4,z24	—	*S. slangkop	1,6,14	z10	z6:z42
(Ar.18:1,3,11:-)				†Salmonella arizonae	6,14	z10	z56
S. demerara	13,23	z10	l,w	(Ar.7a,7c:27:38)			
S. cubana	1,13,23	z29	[z37]	S. sara	1,6,14,25	z38	[e,n,x]
S. fanti	13,23	z38	—	*Salmonella	1,6,14	z42	1,6
*S. stevenage	1,13,23	[z42]	1,7				
Group H				**Group I (O 16)**			
S. garba	1,6,14,25	a	1,5	S. hannover	16	a	1,2
S. ferlac	1,6,14,25	a	e,n,x	S. brazil	16	a	1,5
S. tucson	[1],6,14,[25]	b	[1,7]	S. amunigun	16	a	1,6
*Salmonella	6,14	b	e,n,x,z15	S. nyeko	16	a	1,7
S. blijdorp	1,6,14,25	c	1,5	S. fischerhuette	16	a	e,n,z15
S. kassberg	1,6,14,25	c	1,6	S. heron	16	a	z6
S. minna	1,6,14,25	c	l,w	S. hull	16	b	1,2
S. heves	6,14,24	d	1,5	S. wa	16	b	1,5
				S. glasgow	16	b	1,6

TABLE 8.7—*Continued*

Species or serotype	O	H Phase 1	H Phase 2	Species or serotype	O	H Phase 1	H Phase 2
S. hvittingfoss	16	b	e,n,x	*S. kibi*	16	z_4,z_{23}	—
S. sangera	16	b	e,n,z_{15}	*$S. haddon*	16	z_4,z_{23}	—
S. malstatt	16	b	z_6	‡*S. ochsenzoll*	16	z_4,z_{23}	—
Salmonella	16	b	z_{42}	‡*S. chameleon*	16	z_4,z_{32}	—
S. vancouver	16	c	1,5	†*Salmonella arizonae*	16	z_{10}	1,5
S. shamba	16	c	e,n,x	(Ar.25:27:30)			
S. oldenburg	16	d	1,2	*S. lisboa*	16	z_{10}	1,6
S. gaminara	16	d	1,7	†*Salmonella arizonae*	16	z_{10}	e,n,x,z_{15}
S. barranquilla	16	d	e,n,x	(Ar.25:27:28)			
S. nottingham	16	d	e,n,z_{15}	*S. redlands*	16	z_{10}	e,n,z_{15}
S. barmbek	16	d	z_6	*$S. jacksonville*	16	z_{29}	—
S. malakal	16	e,h	1,2	*$S. woodstock*	16	z_{42}	1,(5),7
S. saboya	16	e,h	1,5	*$S. elsiesrivier*	16	z_{42}	1,6
S. weston	16	e,h	z_6	**Group J (O 17)**			
$S. bellville	16	e,n,x	1,7	*S. bonames*	17	a	1,2
S. tees	16	f,g	—	*S. jangwani*	17	a	1,5
S. adeoyo	16	g,m	—	*S. kinondoni*	17	a	e,n,x
S. nikolaifleet	16	g,m,s	—	*S. kirkee*	17	b	1,2
$S. mobeni	16	g,m,[s],t	—	*$S. hillbrow*	17	b	e,n,x,z_{15}
$S. merseyside	16	g,t	1,5	*S. victoriaborg*	17	c	1,6
$S. rowbarton	16	m,t	[z_{42}]	*$S. woerden*	17	c	z_{39}
S. amina	16	i	1,5	*S. berlin*	17	d	1,5
S. wisbech	16	i	1,7	*S. niamey*	17	d	l,w
S. frankfurt	16	i	e,n,z_{15}	*$S. verity*	17	e,n,x,z_{15}	1,6
S. abobo	16	i	z_6	*$S. bleadon*	17	(f),g,t	[e,n,x,z_{15}]
†*Salmonella arizonae*	16	i	z_{35}	*Salmonella*	17	k	—
(Ar.25:33:21)				*S. irenea*	17	k	1,5
S. szentes	16	k	1,2	*S. matadi*	17	k	e,n,x
S. nuatja	16	k	e,n,x	*S. morotai*	17	l,v	1,2
S. orientalis	16	k	e,n,z_{15}	*S. michigan*	17	l,v	1,5
†*Salmonella arizonae*	16	k	z	*S. carmel*	17	l,v	e,n,x
(Ar.25:29:31)				†*Salmonella arizonae*	17	l,v	e,n,x,z_{15}
†*Salmonella arizonae*	16	k	z_{53}	(Ar.12:23:28)			
(Ar.25:29:25)				†*Salmonella arizonae*	17	l,v	z_{35}
†*Salmonella arizonae*	16	(k)	z_{35}	(Ar.12:23:21)			
(Ar.25:22:21)				*S. gori*	17	z	1,2
†*Salmonella arizonae*	16	l,v	1,5,7	*$S. constantia*	17	z	l,w:z_{42}
(Ar.25:23:30)				†*Salmonella arizonae*	17	z_4,z_{23}	—
S. shanghai	16	l,v	1,6	(Ar.12:1,2,5:-)			
S. welikade	16	l,v	1,7	(Ar.12:1,2,6:-)			
S. salford	16	l,v	e,n,x	†*Salmonella arizonae*	17	z_4,z_{24}	—
S. burgas	16	l,v	e,n,z_{15}	(Ar.12:1,3,11:-)			
†*Salmonella arizonae*	16	l,v	z	†*Salmonella arizonae*	17	z_4,z_{32}	—
(Ar.25:23:31)				(Ar.12:1,6,7:-)			
S. los-angeles	16	l,v	z_6	(Ar.12:1,6,7,9:-)			
†*Salmonella arizonae*	16	l,v	z_{35}	(Ar.12:1,7,8:-)			
(Ar.25:23:21)				*S. djibouti*	17	z_{10}	e,n,x
†*Salmonella arizonae*	16	l,v	z_{53}	†*Salmonella arizonae*	17	z_{10}	e,n,x,z_{15}
(Ar.25:23:25)				(Ar.12:27:28)			
S. lomnava	16	l,w	e,n,z_{15}	†*Salmonella arizonae*	17	z_{10},(a)	e,n,x,z_{15}
$S. noordhoek	16	l,w	z_6	(Ar.12:27,(35):28)			
S. mandera	16	l,z_{13}	e,n,z_{15}	†*Salmonella arizonae*	17	z_{10}	z
S. enugu	16	l,z_{13},z_{28}	—	(Ar.12:27:31)			
$S. sarepta	16	l,z_{28}	z_{42}	*S. kandla*	17	z_{29}	—
S. annedal	16	r,i	e,n,x	†*Salmonella arizonae*	17	z_{36}	—
S. zwickau	16	r,i	e,n,z_{15}	(Ar.12:17,20:-)			
S. avignon	16	y	e,n,z_{15}	**Group K (O 18)**			
S. saphra	16	y	1,5	*S. fluntern*	6,14,18	b	1,5
S. akuafo	16	y	1,6	*S. usumbura*	18	d	1,7
S. kikoma	16	y	e,n,x	†*Salmonella arizonae*	18	g,z_{51}	—
S. lingwala	16	z	1,7	(Ar.7a,7b:13,14:-)			
$S. louwbester	16	z	e,n,x				

TABLE 8.7—Continued

Species or serotype	O	H Phase 1	H Phase 2
S. langenhorn	18	m,t	—
*Salmonella	18	m,t	1,5
S. memphis	18	k	1,5
†Salmonella arizonae (Ar.7a,7b:22:34)	18	(k)	z_{54}
†Salmonella arizonae (Ar.7a,7b:24:31)	18	r	z
*Salmonella	18	y	e,n,x,z_{15}
S. cerro	6,14,18	z_4,z_{23}	[1,5]
†Salmonella arizonae (Ar.7a,7b:1,2,5:−) (Ar.7a,7b:1,2,6:−)	18	z_4,z_{23}	—
S. blukwa	18	z_4,z_{24}	—
†Salmonella arizonae (Ar.7a,7b:1,7,8:−)	18	z_4,z_{32}	—
S. carnac	18	z_{10}	z_6
*S. zeist	18	z_{10}	z_6
*S. beloha	18	z_{36}	—
S. sinthia	18	z_{38}	—
Group L (O 21)			
S. assen	21	a	—
S. ghana	21	b	1,6
S. minnesota	21	b	e,n,x
S. rhône	21	c	e,n,x
*Salmonella	21	c	e,n,x
†Salmonella arizonae (Ar.22:32:28)	21	c	e,n,x,z_{15}
S. spartel	21	d	1,5
S. magwa	21	d	e,n,x
S. good	21	f,g	e,n,x
†Salmonella arizonae (Ar.22:13,14:−)	21	g,z_{51}	—
S. diourbel	21	i	1,2
†Salmonella arizonae (Ar.22:33:28)	21	i	e,n,x,z_{15}
†Salmonella arizonae (Ar.22:29:28)	21	k	e,n,x,z_{15}
†Salmonella arizonae (Ar.22:23:31)	21	l,v	z
S. ruiru	21	y	e,n,x
S. baguida	21	z_4,z_{23}	—
‡S. soesterberg	21	z_4,z_{23}	—
*S. gwaai	21	z_4,z_{24}	—
†Salmonella arizonae (Ar.22:27:28)	21	z_{10}	e,n,x,z_{15}
†Salmonella arizonae (Ar.22:27:31)	21	z_{10}	z
*S. wandsbek	21	z_{10}	z_6
†Salmonella arizonae (Ar.22:16,17,18:−)	21	z_{29}	—
S. gambaga	21	z_{35}	e,n,z_{15}
Group M (O 28)			
S. solna	28	a	1,5
S. dakar	28	a	1,6
S. seattle	28	a	e,n,x
S. honelis	28	a	e,n,z_{15}
S. moero	28	b	1,5
S. ashanti	28	b	1,6
S. bokanjac	28	b	1,7

Species or serotype	O	H Phase 1	H Phase 2
S. langford	28	b	e,n,z_{15}
*S. kaltenhausen	28	b	z_6
S. hermannswerder	28	c	1,5
S. eberswalde	28	c	1,6
S. halle	28	c	1,7
S. dresden	28	c	e,n,x
S. wedding	28	c	e,n,z_{15}
S. techimani	28	c	z_6
S. amoutive	28	d	1,5
S. mundonobo	28	d	1,7
S. mocamedes	28	d	e,n,x
S. patience	28	d	e,n,z_{15}
S. kpeme	28	e,h	1,7
*Salmonella	28	e,n,x	1,7
S. friedrichsfelde	28	f,g	—
S. abadina	28	g,m	[e,n,z_{15}]
S. croft	28	g,m,s	—
S. ona	28	g,s,t	—
*S. llandudno	28	g,s,t	1,5
S. vinohrady	28	m,t	—
*Salmonella	28	m,t	—
S. cotham	28	i	1,5
S. volkmarsdorf	28	i	1,6
S. kuessel	28	i	e,n,z_{15}
S. guildford	28	k	1,2
S. ilala	28	k	1,5
S. adamstown	28	k	1,6
S. taunton	28	k	e,n,x
S. ank	28	k	e,n,z_{15}
S. leoben	28	l,v	1,5
S. vitkin	28	l,v	e,n,x
S. nashua	28	l,v	e,n,z_{15}
S. chicago	28	r	1,5
S. bassadji	28	r	1,6
S. kibusi	28	r	e,n,x
*S. oevelgoenne	28	r	e,n,z_{15}
S. sankt-georg	28	r,i	e,n,z_{15}
S. oskarshamn	28	y	1,2
S. nima	28	y	1,5
S. pomona	28	y	1,7
S. kitenge	28	y	e,n,x
S. tel-aviv	28	y	e,n,z_{15}
S. shomolu	28	y	l,w
S. ezra	28	z	1,7
S. brisbane	28	z	e,n,z_{15}
*S. ceres	28	z	z_{39}
S. teltow	28	z_4,z_{23}	1,6
S. babelsberg	28	z_4,z_{23}	[e,n,z_{15}]
S. rogy	28	z_{10}	1,2
S. umbilo	28	z_{10}	e,n,x
S. luckenwalde	28	z_{10}	e,n,z_{15}
S. moroto	28	z_{10}	l,w
S. djermaia	28	z_{29}	—
S. aderike	28	z_{38}	—
Group N (O 30)			
S. doulassame	30	a	e,n,z_{15}
S. overvecht	30	a	1,2
S. zehlendorf	30	a	1,5
*S. odijk	30	a	z_{39}
S. louga	30	b	1,2
S. aschersleben	30	b	1,5

TABLE 8.7—*Continued*

Species or serotype	O	H Phase 1	H Phase 2
S. urbana	30	b	e,n,x
S. messina	30	d	1,5
*S. slatograd	30	f,g,t	—
S. godesberg	30	g,m	—
S. giessen	30	g,m,s	—
S. sternschanze	30	g,s,t	[z_{59}]
S. wayne	30	g,z_{51}	—
S. landau	30	i	1,2
S. morehead	30	i	1,5
S. soerenga	30	i	l,w
S. hilversum	30	k	1,2
S. ramat-gan	30	k	1,5
S. aqua	30	k	1,6
S. angoda	30	k	e,n,x
S. odozi	30	k	e,n,x,z_{15}
*Salmonella	30	k	e,n,[x],z_{15}
S. ligeo	30	l,v	1,2
S. donna	30	l,v	1,5
S. morocco	30	l,z_{13},z_{28}	e,n,z_{15}
S. gege	30	r	1,5
S. matopeni	30	y	1,2
S. steinplatz	30	y	1,6
S. baguirmi	30	y	e,n,x
S. bietri	30	y	1,5
S. bodjonegoro	30	z_4,z_{24}	—
S. kumasi	30	z_{10}	e,n,z_{15}
S. aragua	30	z_{29}	—
S. ago	30	z_{38}	—
*Salmonella	30	z_{39}	1,7
Group O (O 35)			
S. umhlatazana	35	a	e,n,z_{15}
S. tchad	35	b	—
S. yolo	35	c	—
S. dembe	35	d	l,w
S. gassi	35	e,h	z_6
S. adelaide	35	f,g	—
S. ealing	35	g,m,s	—
*Salmonella	35	g,m,s,t	—
S. ebrie	35	g,m,t	—
S. anecho	35	g,s,t	—
S. agodi	35	g,t	—
†*Salmonella arizonae* (Ar.20:13,14:−)	35	g,z_{51}	—
S. monschaui	35	m,t	—
†*Salmonella arizonae* (Ar.20:33:28)	35	i	e,n,x,z_{15}
S. gambia	35	i	e,n,z_{15}
S. bandia	35	i	l,w
†*Salmonella arizonae* (Ar.20:33:31)	35	i	z
†*Salmonella arizonae* (Ar.20:29:31)	35	k	z
†*Salmonella arizonae* (Ar.20:29:25)	35	k	z_{53}
†*Salmonella arizonae* (Ar.20:22:31)	35	(k)	z
†*Salmonella arizonae* (Ar.20:23:30)	35	l,v	1,5,7
†*Salmonella arizonae* (Ar.20:23:21)	35	l,v	z_{35}
*Salmonella	35	l,z_{28}	—

Species or serotype	O	H Phase 1	H Phase 2
†*Salmonella arizonae* (Ar.20:24:28)	35	r	e,n,x,z_{15}
S. massakory	35	r	l,w
†*Salmonella arizonae* (Ar.20:24:21)	35	r	z_{35}
†*Salmonella arizonae* (Ar.20:24:41)	35	r	z_{53}
S. alachua	35	z_4,z_{23}	—
†*Salmonella arizonae* (Ar.20:1,2,6:−)	35	z_4,z_{23}	—
S. westphalia	35	z_4,z_{24}	—
S. camberene	35	z_{10}	1,5
S. enschede	35	z_{10}	l,w
S. ligna	35	z_{10}	z_6
†*Salmonella arizonae* (Ar.20:27:21)	35	z_{10}	z_{35}
†*Salmonella arizonae* (Ar.20:16,17,18:−)	35	z_{29}	—
*S. utbremen	35	z_{29}	e,n,x
†*Salmonella arizonae* (Ar.20:26:30)	35	z_{52}	1,5,7
†*Salmonella arizonae* (Ar.20:26:28)	35	z_{52}	e,n,x,z_{15}
†*Salmonella arizonae* (Ar.20:26:21)	35	z_{52}	z_{35}
Group P (O 38)			
S. sheffield	38	c	1,5
S. kidderminster	38	c	1,6
*S. carletonville	38	d	1,5
S. thiaroye	38	e,h	1,2
S. kasenyi	38	e,h	1,5
S. korovi	38	g,m,s	—
*S. foulpointe	38	g,t	—
†*Salmonella arizonae* (Ar.16:13,14:−)	38	g,z_{51}	—
S. mgulani	38	i	1,2
S. lansing	38	i	1,5
†*Salmonella arizonae* (Ar.16:33:25)	38	i	z_{53}
S. njala	38	k	e,n,x
S. echa	38	k	1,2
S. inverness	38	k	1,6
†*Salmonella arizonae* (Ar.16:29:31)	38	k	z
†*Salmonella arizonae* (Ar.16:22:30)	38	(k)	1,5,7
†*Salmonella arizonae* (Ar.16:22:31)	38	(k)	z
†*Salmonella arizonae* (Ar.16:22:21)	38	(k)	z_{35}
†*Salmonella arizonae* (Ar.16:22:34)	38	(k)	z_{54}
†*Salmonella arizonae* (Ar.16:22:35)	38	(k)	z_{55}
S. alger	38	l,v	1,2
S. kimberley	38	l,v	1,5
S. roan	38	l,v	e,n,x
†*Salmonella arizonae* (Ar.16:23:31)	38	l,v	z
†*Salmonella arizonae* (Ar.16:23:21)	38	l,v	z_{35}

TABLE 8.7—Continued

Species or serotype	O	H Phase 1	H Phase 2	Species or serotype	O	H Phase 1	H Phase 2
†Salmonella arizonae (Ar.16:23:25)	38	l,v	z_{53}	*S. sunnydale	1,40	k	e,n,x,z_{15}
†Salmonella arizonae (Ar.16:23:34)	38	l,v	z_{54}	†Salmonella arizonae (Ar.10a,10b:29:31:40a,40c)	40	k	z:z_{57}
S. lindi	38	r	1,5	S. millesi	1,40	l,v	1,2
S. emmastad	38	r	1,6	†Salmonella arizonae (Ar.10a,10b:23:25)	40	l,v	z_{53}
†Salmonella arizonae (Ar.16:24:31)	38	r	z	S. bukavu	1,40	l,z_{28}	1,5
†Salmonella arizonae (Ar.16:24:31:40a,40b)	38	r	z:z_{57}	S. santhiaba	40	l,z_{28}	1,6
†Salmonella arizonae (Ar.16:24:21)	38	r	z_{35}	*S. bulawayo	1,40	z	1,5
S. freetown	38	y	1,5	S. nowawes	40	z	z_6
S. colombo	38	y	1,6	†Salmonella arizonae (Ar.10a,10b:1,2,6:-)	40	z_4,z_{23}	—
S. perth	38	y	e,n,x	‡S. sachsenwald	1,40	z_4,z_{23}	—
‡Salmonella	38	z_4,z_{23}	—	†Salmonella arizonae (Ar.10a,10b:1,3,11:-)	40	z_4,z_{24}	—
S. yoff	38	z_4,z_{23}	1,2	‡Salmonella	40	z_4,z_{24}	—
†Salmonella arizonae †(Ar.16:27:31)	38	z_{10}	z	*S. degania	40	z_4,z_{24}	z_{39}
†Salmonella arizonae (Ar.16:27:25)	38	z_{10}	z_{53}	‡S. bern	1,40	z_4,z_{32}	—
Salmonella arizonae (Ar.16:39:25)	38	z_{47}	z_{53}	†Salmonella arizonae (Ar.10a,10b:1,7,8:-)	40	z_4,z_{32}	—
Group Q (O 39)				‡Salmonella	40	z_4,z_{32}	—
S. wandsworth	39	b	1,2	*Salmonella	1,40	z_6	1,5
S. logone	39	d	1,5	S. trotha	40	z_{10}	z_6
S. mara	39	e,h	1,5	†Salmonella arizonae (Ar.10a,10c:27:21)	40	z_{10}	z_{35}
S. hofit	39	i	1,5	S. omifisan	40	z_{29}	—
S. champaign	39	k	1,5	†Salmonella arizonae (Ar.10a,10b:16,17,18:-)	40	z_{29}	—
S. kokomlemle	39	l,v	e,n,x	*S. fandran	1,40	z_{35}	e,n,x,z_{15}
*S. mondeor	39	l,z_{28}	e,n,x	†Salmonella arizonae (Ar.10a,10b:17,20:-)	40	z_{36}	—
S. anfo	39	y	1,2	*S. grunty	1,40	z_{39}	1,6
S. windermere	39	y	1,5	S. karamoja	1,40	z_{41}	1,2
Group R (O 40)				*Salmonella	40	—	1,7
S. shikmonah	40	a	1,5				
S. greiz	40	a	z_6	**Group S (O 41)**			
*S. springs	40	a	z_{39}	S. vietnam	41	b	—
*Salmonella	40	b	—	*Salmonella	41	b	—
S. riogrande	40	b	1,5	S. egusi	41	d	[1,5]
S. johannesburg	1,40	b	e,n,x	*S. hennepin	41	d	z_6
S. duval	1,40	b	e,n,z_{15}	*S. lethe	41	g,t	—
S. benguella	40	b	z_6	†Salmonella arizonae (Ar.13:13,14:-)	41	g,z_{51}	—
*S. suarez	1,40	c	e,n,x,z_{15}	*Salmonella	41	k	—
*S. ottershaw	40	d	—	†Salmonella arizonae (Ar.13:22:-)	41	(k)	—
S. driffield	1,40	d	1,5	*S. dubrovnik	41	z	1,5
S. tilene	1,40	e,h	1,2	‡Salmonella	41	z_4,z_{23}	—
*Salmonella	1,40	(f),g	e,n,x,z_{15}	S. waycross	41	z_4,z_{23}	—
*S. boksburg	40	g,m,s,t	e,n,x	†Salmonella arizonae (Ar.13:1,2,5:-)	41	z_4,z_{23}	—
*S. alsterdorf	1,40	g,m,t	—	S. ipswich	41	z_4,z_{24}	—
†Salmonella arizonae (Ar.10a,10b:13,14:-)	40	g,z_{51}	—	†Salmonella arizonae (Ar.13:1,3,11:-)	41	z_4,z_{24}	—
‡S. seminole	1,40	g,z_{51}	—				
*Salmonella	1,40	m,t	z_{42}				
†Salmonella arizonae (Ar.10a,10b:33:30)	40	i	1,5,7				
S. goulfey	1,40	k	1,5				
S. allandale	1,40	k	1,6				
S. hann	40	k	e,n,x				

TABLE 8.7—*Continued*

Species or serotype	O	H Phase 1	H Phase 2
†*Salmonella arizonae* (Ar.13:1,6,7:−) (Ar.13:1,7,8:−)	41	z_4,z_{32}	—
S. negev	41	z_{10}	1,2
S. leipzig	41	z_{10}	1,5
S. landala	41	z_{10}	1,6
S. inpraw	41	z_{10}	e,n,x
S. lurup	41	z_{10}	[e,n,x,z_{15}]
S. lichtenberg	41	z_{10}	z_6
†*Salmonella arizonae* (Ar.13:16,17,18:−)	41	z_{29}	—
†*Salmonella arizonae* (Ar.13:17,20:−)	41	z_{36}	—
S. offa	41	z_{38}	—
Group T (O 42)			
S. faji	1,42	a	e,n,z_{15}
S. chinovum	42	b	1,5
S. uphill	42	b	e,n,x,z_{15}
S. egusitoo	1,42	b	z_6
S. kampala	1,42	c	z_6
S. fremantle	42	(f),g,t	—
†*Salmonella arizonae* (Ar.15:13,14:−)	42	g,z_{51}	—
S. maricopa	1,42	g,z_{51}	1,5
S. waral	1,42	m,t	—
Salmonella	42	m,t	e,n,x,z_{15}
S. kaneshie	1,42	i	l,w
S. middlesbrough	1,42	i	z_6
†*Salmonella arizonae* (Ar.15:29:−)	42	k	—
S. haferbreite	42	k	1,6
†*Salmonella arizonae* (Ar.15:22:21)	42	(k)	z_{35}
†*Salmonella arizonae* (Ar.15:23:30)	42	l,v	1,5,7
S. portbech	42	l,v	e,n,x,z_{15}
S. sipane	1,42	r	e,n,z_{15}
S. nairobi	42	r	—
†*Salmonella arizonae* (Ar.15:24:31)	42	r	z
†*Salmonella arizonae* (Ar.15:24:42)	42	r	z_{50}
†*Salmonella arizonae* (Ar.15:24:25)	42	r	z_{53}
S. harvestehude	1,42	y	z_6
S. detroit	42	z	1,5
S. ursenbach	42	z	1,6
S. rand	42	z	e,n,x,z_{15}
S. nuernberg	42	z	z_6
†*Salmonella arizonae* (Ar.15:1,2,6:−)	42	z_4,z_{23}	—
S. gera	1,42	z_4,z_{23}	1,6
S. toricada	1,42	z_4,z_{24}	—
†*Salmonella arizonae* (Ar.15:1,3,11:−)	42	z_4,z_{24}	—
†*Salmonella arizonae* (Ar.15:27:28)	42	z_{10}	e,n,x,z_{15}
†*Salmonella arizonae* (Ar.15:27:31)	42	z_{10}	z
S. loenga	1,42	z_{10}	z_6
S. kahla	1,42	z_{35}	1,6

Species or serotype	O	H Phase 1	H Phase 2
S. weslaco	42	z_{36}	—
Salmonella	42	—	1,6
S. taset	1,42	z_{41}	—
Group U (O 43)			
S. graz	43	a	1,2
S. berkeley	43	a	1,5
S. kommetje	43	b	z_{42}
Salmonella	43	e,n,x,z_{15}	1,(5),7
Salmonella	43	e,n,x,z_{15}	1,6
S. milwaukee	43	f,g	—
Salmonella	43	f,g,t	1,5
S. mosselbay	43	g,m,s,t	z_{42}
S. veddel	43	g,t	—
S. mbao	43	i	1,2
S. ahuza	43	k	1,5
†*Salmonella arizonae* (Ar.21:29:31)	43	k	z
†*Salmonella arizonae* (Ar.21:23:25)	43	l,v	z_{53}
†*Salmonella arizonae* (Ar.21:24:28)	43	r	e,n,x,z_{15}
S. farcha	43	y	1,2
S. kingabwa	43	y	1,5
Salmonella	43	z	1,5
‡S. houten (S. houtenae)	43	z_4,z_{23}	—
†*Salmonella arizonae* (Ar.21:1,3,11:−)	43	z_4,z_{24}	—
‡S. tuindorp	43	z_4,z_{32}	—
‡*Salmonella*	43	z_{29}	—
Salmonella	43	z_{29}	z_{42}
S. ahepe	43	z_{35}	1,6
†*Salmonella arizonae* (Ar.21:17,20:−)	43	z_{36}	—
‡S. volksdorf	43	z_{36},z_{38}	—
S. irigny	43	z_{38}	—
S. bunnik	43	z_{42}	1,5,7
Group V (O 44)			
S. niakhar	44	a	1,5
S. niarembe	44	a	1,w
S. sedgwick	44	b	e,n,z_{15}
S. madigan	44	c	1,5
S. bobo	44	d	1,5
S. fischerstrasse	44	d	e,n,z_{15}
S. vleuten	44	f,g	—
S. gamaba	44	g,m,s	—
‡*Salmonella*	44	g,z_{51}	—
S. muguga	44	m,t	—
S. lawra	44	k	e,n,z_{15}
S. uhlenhorst	44	z	l,w
S. kua	44	z_4,z_{23}	—
Salmonella	44	z_4,z_{23}	—
†*Salmonella arizonae* (Ar.13:1,2,5:−) (Ar.13:1,2,6:−)	44	z_4,z_{23}	—
S. christiansborg	44	z_4,z_{24}	—
‡*Salmonella*	44	z_4,z_{24}	—
†*Salmonella arizonae* (Ar.13:1,6,7,9:−) (Ar.13:1,7,8:−) (Ar.13:1,2,10:−)	44	z_4,z_{32}	—

TABLE 8.7—*Continued*

Species or serotype	O	Phase 1	Phase 2	Species or serotype	O	Phase 1	Phase 2
‡*S. lohbruegge*	44	z_4, z_{32}	—	*S. lyon*	47	k	e, n, z_{15}
S. guinea	44	z_{10}	[1,7]	†*Salmonella arizonae* (Ar.28:29:31)	47	k	z
‡*Salmonella*	44	$z_{36}, [z_{38}]$	—	†*Salmonella arizonae* (Ar.23:23:30:42)	47	l,v	$1,5,7:[z_{50}]$
S. koketime	44	z_{38}	—				
S. clovelly	1,44	z_{39}	e, n, x, z_{15}	†*Salmonella arizonae* (Ar.28:23:28)	47	l,v	e, n, x, z_{15}
Group W (O 45)				†*Salmonella arizonae* (Ar.28:23:21)	47	l,v	z_{35}
S. vrindaban	45	a	e, n, x, z_{15}	†*Salmonella arizonae* (Ar.28:23:25)	47	l,v	z_{53}
S. ejeda	45	a	z_{10}				
S. riverside	45	b	1,5	*S. teshie*	1,47	l, z_{13}, z_{28}	e, n, z_{15}
S. deversoir	45	c	e, n, x	†*Salmonella arizonae* (Ar.23:24:31)	47	r	z
S. dugbe	45	d	1,6	†*Salmonella arizonae* (Ar.23:24:25)	47	r	z_{53}
S. karachi	45	d	e, n, x				
S. suelldorf	45	f,g	—	*Salmonella*	47	r	z_{53}
S. tornow	45	g,m	—	*S. moualine*	47	y	1,6
S. bremen	45	g,m,s,t	e, n, x	*S. mount-pleasant*	47	z	1,5
S. windhoek	45	g,t	1,5	*S. kaolack*	47	z	1,6
†*Salmonella arizonae* (Ar.11:13,14:−)	45	g, z_{51}	—	*S. chersina*	47	z	z_6
				S. bere	47	z_4, z_{23}	z_6
‡*Salmonella*	45	g, z_{51}	—	*Salmonella*	47	z_6	1,6
S. apapa	45	m,t	—	†*Salmonella arizonae* (Ar.28:27:30)	47	z_{10}	1,5,7
S. perinet	45	m,t	e, n, x, z_{15}				
S. casablanca	45	k	1,7	†*Salmonella arizonae* (Ar.28:27:31)	47	z_{10}	z
S. cairns	45	k	e, n, z_{15}	†*Salmonella arizonae* (Ar.28:16,17,18:−)	47	z_{29}	—
S. klapmuts	45	z	z_{39}				
‡*Salmonella*	45	z_4, z_{23}	—	*S. alexanderplatz*	47	z_{38}	—
S. jodhpur	45	z_{29}	—	*S. quinhon*	47	z_{44}	—
†*Salmonella arizonae* (Ar.11:16,17,18:−)	45	z_{29}	—	†*Salmonella arizonae* (Ar.28:26:30)	47	z_{52}	1,5,7
S. lattenkamp	45	z_{35}	1,5	†*Salmonella arizonae* (Ar.28:26:28)	47	z_{52}	e, n, x, z_{15}
Group X (O 47)				†*Salmonella arizonae* (Ar.28:26:31)	47	z_{52}	z
S. bilthoven	47	a	1,5				
S. saka	47	b	—	†*Salmonella arizonae* (Ar.28:26:21)	47	z_{52}	z_{35}
S. phoenix	47	b	1,5				
S. khami	47	b	$[e, n, x, z_{15}]$	**Group Y (O 48)**			
†*Salmonella arizonae* (Ar.23:32:28) (Ar.28:32:28:[40ₐ, 40ₑ])	47	c	$e, n, x, z_{15}:$ $[z_{57}]$	*S. hisingen*	48	a	1,5,7
				Salmonella	48	a	z_6
†*Salmonella arizonae* (Ar.28:32:31)	47	c	z	†*Salmonella arizonae* (Ar.5:35:21)	48	a	z_{35}
S. stellingen	47	d	e, n, x	*S. hagenbeck*	48	d	z_6
S. quimbamba	47	d	z_{39}	*S. fitzroy*	48	e,h	1,5
S. sljeme	1,47	f,g	—	*S. hammonia*	48	e, n, x, z_{15}	z_6
S. luke	1,47	g,m	—	*S. erlangen*	48	g,m,t	—
S. mesbit	47	m,t	e, n, z_{15}	†*Salmonella arizonae* (Ar.5:13,14:−)	48	g, z_{51}	—
†*Salmonella arizonae* (Ar.23:33:28:[42])	47	i	$e, n, x, z_{15}:$ $[z_{50}]$	‡*S. marina*	48	g, z_{51}	—
				S. sydney	48	i	z
S. bergen	47	i	e, n, z_{15}	†*Salmonella arizonae* (Ar.5:33:25)	48	i	z_{53}
†*Salmonella arizonae* (Ar.28:33:31)	47	i	z				
†*Salmonella arizonae* (Ar.23:33:21) (Ar.28:33:21)	47	i	z_{35}	†*Salmonella arizonae* (Ar.5:29:30)	48	k	1,5,7
				†*Salmonella arizonae* (Ar.5:29:28)	48	k	e, n, x, z_{15}
†*Salmonella arizonae* (Ar.23:33:25) (Ar.28:33:25:[40ₐ, 40ₑ]	47	i	$z_{53}:[z_{57}]$	*S. dahlem*	48	k	e, n, z_{15}
S. bootle	47	k	1,5	†*Salmonella arizonae* (Ar.5:29:21)	48	k	z_{35}
S. dahomey	47	k	1,6				
†*Salmonella arizonae* (Ar.28:29:28)	47	k	e, n, x, z_{15}				

TABLE 8.7—*Continued*

Species or serotype	O	Phase 1	Phase 2
*S. sakaraha	48	k	z_{39}
†Salmonella arizonae (Ar.5:29:25)	48	k	z_{53}
†Salmonella arizonae (Ar.5:22:25)	48	(k)	z_{53}
†Salmonella arizonae (Ar.5:23:30:[39])	48	l,v	1,5,7:[z_{47}]
†Salmonella arizonae (Ar.5:24:28)	48	r	e,n,x,z_{15}
S. toucra	48	z	1,5:[z_{58}]
†Salmonella arizonae (Ar.5:1,2,5:−) (Ar.5:1,2,5,6:−) (Ar.5:1,6:−)	48	z_4,z_{23}	—
S. djakarta	48	z_4,z_{24}	—
†Salmonella arizonae (Ar.5:1,3,11:−)	48	z_4,z_{24}	—
†Salmonella arizonae (Ar.5:1,6,7:−) (Ar.5:1,7,8:−)	48	z_4,z_{32}	—
†Salmonella	48	z_4,z_{32}	—
*S. ngozi	48	z_{10}	[1,5]
†Salmonella arizonae (Ar.5:27:28)	48	z_{10}	e,n,x,z_{15}
†Salmonella arizonae (Ar.5:16,17,18:−)	48	z_{29}	—
S. bongor	48	z_{35}	—
†Salmonella arizonae (Ar.5:17,20:−)	48	z_{36}	—
†Salmonella arizonae (Ar.5:26:31)	48	z_{52}	z
Group Z (O 50)			
*Salmonella	50	b	z_6
S. rochdale	50	b	e,n,x
*S. krugersdorp	50	e,n,x	1,7
*S. namib	50	g,m,s,t	1,5
†Salmonella arizonae (Ar.9a,9b:13,14:−)	50	g,z_{51}	—
‡S. wassenaar	50	g,z_{51}	—
*S. atra	50	m,t	z_6:z_{42}
†Salmonella arizonae (Ar.9a,9c:33:30)	50	i	1,5,7
†Salmonella arizonae (Ar.9a,9c:33:28)	50	i	e,n,x,z_{15}
†Salmonella arizonae (Ar.9a,9c:33:31)	50	i	z
†Salmonella arizonae (Ar.9a,9c:29:30)	50	k	1,5,7
†Salmonella arizonae (Ar.9a,9c:29:28)	50	k	e,n,x,z_{15}
†Salmonella arizonae (Ar.9a,9b:29:31) (Ar.9a,9c:29:31)	50	k	z
†Salmonella arizonae (Ar.9a,9b:22:31)	50	(k)	z
*S. seaforth	50	k	z_6
†Salmonella arizonae (Ar.9a,9b:29:21)	50	k	z_{35}
†Salmonella arizonae (Ar.9a,9b:22:21)	50	(k)	z_{35}
†Salmonella arizonae (Ar.9a,9c:29:25)	50	k	z_{53}

Species or serotype	O	Phase 1	Phase 2
†Salmonella arizonae (Ar.9a,9b:23:28)	50	l,v	e,n,x,z_{15}
†Salmonella arizonae (Ar.9a,9c:23:31)	50	l,v	z
†Salmonella arizonae (Ar.9a,9c:23:21)	50	l,v	z_{35}
*Salmonella	50	l,w	e,n,x,z_{15}: z_{42}
*Salmonella	50	l,z_{28}	z_{42}
†Salmonella arizonae (Ar.9a,9b:24:30)	50	r	1,5,7
†Salmonella arizonae (Ar.9a,9b:24:31)	50	r	z
†Salmonella arizonae (Ar.9a,9b:24:21)	50	r	z_{35}
†Salmonella arizonae (Ar.9a,9b:24:25)	50	r	z_{53}
S. dougi	50	y	1,6
*S. greenside	50	z	e,n,x
†Salmonella arizonae (Ar.9a,9b:1,2,5:−) (Ar.9a,9b:1,2,6:−)	50	z_4,z_{23}	—
‡S. flint	50	z_4,z_{23}	—
‡Salmonella	50	z_4,z_{24}	—
†Salmonella arizonae (Ar.9a,9b:1,3,11:−)	50	z_4,z_{24}	—
†Salmonella arizonae (Ar.9a,9b:1,6,7:−) (Ar.9a,9b:1,7,8:−)	50	z_4,z_{32}	—
‡S. bonaire	50	z_4,z_{32}	—
†Salmonella arizonae (Ar.9a,9c:27:31)	50	z_{10}	z
*S. hooggraven	50	z_{10}	z_6:z_{42}
†Salmonella arizonae (Ar.9a,9b:16,17,18:−)	50	z_{29}	—
†Salmonella arizonae (Ar.9a,9b:17,20:−)	50	z_{36}	—
*S. faure	50	z_{42}	1,7
†Salmonella arizonae (Ar.9a,9b:26:30)	50	z_{52}	1,5,7
†Salmonella arizonae (Ar.9a,9b:26:31) (Ar.9a,9c:26:31)	50	z_{52}	z
†Salmonella arizonae (Ar.9a,9b:26:21) (Ar.9a,9c:26:21)	50	z_{52}	z_{35}
†Salmonella arizonae (Ar.9a,9b:26:25) (Ar.9a,9c:26:25)	50	z_{52}	z_{53}
Group 51			
S. tione	51	a	e,n,x
S. gokul	1,51	d	[1,5]
S. meskin	51	e,h	1,2
†Salmonella arizonae (Ar.1,2:13,14:−)	51	g,z_{51}	—
S. kabete	51	i	1,5
†Salmonella arizonae (Ar.1,2:33:21)	51	i	z_{35}
S. dan	51	k	e,n,z_{15}
S. overschie	51	l,v	1,5

TABLE 8.7—Continued

Species or serotype	O	Phase 1	Phase 2
†*Salmonella arizonae* (Ar.1,2:23:31)	51	l,v	z
**S. askraal*	51	l,z_{28}	—
S. antsalova	51	z	1,5
S. treforest	1,51	z	1,6
†*Salmonella arizonae* (Ar.1,2:1,2,5:−)	51	z_4,z_{23}	—
‡*S. harmelen*	51	z_4,z_{23}	—
†*Salmonella arizonae* (Ar.1,2:1,3,11:−)	51	z_4,z_{24}	—
*Salmonella	51	z_{29}	e,n,x,z_{15}
†*Salmonella arizonae* (Ar.1,2:26:31)	51	z_{52}	z
**S. roggeveld*	51	—	1,7
Group 52			
S. flottbek	52	b	—
S. utrecht	52	d	1,5
*Salmonella	52	d	e,n,x,z_{15}
S. butare	52	e,h	1,6
S. sainte-marie	52	g,t	—
†*Salmonella arizonae* (Ar.31:29:21)	52	k	z_{35}
†*Salmonella arizonae* (Ar.31:23:25)	52	l,v	z_{53}
*Salmonella	52	z_{44}	1,5,7
Group 53			
†*Salmonella arizonae* (Ar.1,4:13,14:−)	53	g,z_{51}	—
†*Salmonella arizonae* (Ar.1,4:33:31)	53	i	z
†*Salmonella arizonae* (Ar.1,4:29:28)	53	k	e,n,x,z_{15}
†*Salmonella arizonae* (Ar.1,4:29:31)	53	k	z
†*Salmonella arizonae* (Ar.1,4:22:21)	53	(k)	z_{35}
†*Salmonella arizonae* (Ar.1,4:23:28)	53	l,v	e,n,x,z_{15}
**S. midhurst*	53	l,z_{28}	z_{39}
†*Salmonella arizonae* (Ar.1,4:24:31)	53	r	z
†*Salmonella arizonae* (Ar.1,4:24:21)	53	r	z_{35}
†*Salmonella arizonae* (Ar.1,4:24:38)	53	r	z_{56}
†*Salmonella arizonae* (Ar.1,4:31:30)	53	z	1,5,7
*Salmonella	53	z	z_6
†*Salmonella arizonae* (Ar.1,4:1,2,5:−) (Ar.1,4:1,2,6:−)	53	z_4,z_{23}	—
**S. humber*	53	z_4,z_{24}	—
†*Salmonella arizonae* (Ar.1,4:1,3,11:−)	53	z_4,z_{24}	—
†*Salmonella arizonae* (Ar.1,4:1,6,7:−) (Ar.1,4:1,6,7,9:−)	53	z_4,z_{32}	—
†*Salmonella arizonae* (Ar.1,4:27:21)	53	z_{10}	z_{35}
†*Salmonella arizonae* (Ar.1,4:16,17,18:−)	53	z_{29}	—

Species or serotype	O	Phase 1	Phase 2
‡*S. bockenheim*	1,53	z_{36},z_{38}	—
†*Salmonella arizonae* (Ar.1,4:26:21)	53	z_{52}	z_{35}
†*Salmonella arizonae* (Ar.1,4:26:25)	53	z_{52}	z_{53}
Group 54			
S. tonev	54	b	e,n,x
S. rossleben	54	e,h	1,6
S. uccle	54	g,s,t	—
S. poeseldorf	54	i	z_6
S. ochsenwerder	54	k	1,5
S. steinwerder	(3),(15),54	y	1,5
Group 55			
**S. tranoroa*	55	k	z_{39}
Group 56			
**S. artis*	56	b	—
*Salmonella	56	d	—
*Salmonella	56	e,n,x	1,7
†*Salmonella arizonae* (Ar.14:1,2,5:−) (Ar.14:1,2,6:−)	56	z_4,z_{23}	—
†*Salmonella arizonae* (Ar.14:1,6,7,9:−)	56	z_4,z_{32}	—
Group 57			
*Salmonella	57	g,m,s,t	z_{42}
†*Salmonella arizonae* (Ar.34:33:28)	57	i	e,n,x,z_{15}
†*Salmonella arizonae* (Ar.34:33:31)	57	i	z
**S. locarno*	57	z_{29}	z_{42}
**S. manombo*	57	z_{39}	e,n,x,z_{15}
**S. tokai*	57	z_{42}	1,6:z_{53}
Group 58			
*Salmonella	58	a	—
†*Salmonella arizonae* (Ar.1,33:23:28)	58	l,v	e,n,x,z_{15}
†*Salmonella arizonae* (Ar.1,33:23:21)	58	l,v	z_{35}
**S. basel*	58	l,z_{13},z_{28}	1,5
†*Salmonella arizonae* (Ar.1,33:24:28)	58	r	e,n,x,z_{15}
†*Salmonella arizonae* (Ar.1,33:24:25: [39]:[40$_a$,40$_c$]	58	r	z_{53}:[z_{47}]:[z_{57}]
†*Salmonella arizonae* (Ar.1,33:26:31)	58	z_{52}	z
†*Salmonella arizonae* (Ar.1,33:26:21)	58	z_{52}	z_{35}
Group 59			
†*Salmonella arizonae* (Ar.19:32:28)	59	c	e,n,x,z_{15}
†*Salmonella arizonae* (Ar.19:33:31)	59	i	z
†*Salmonella arizonae* (Ar.19:33:21)	59	i	z_{35}
**S. betioky*	59	k	(z)
†*Salmonella arizonae* (Ar.19:29:25)	59	k	z_{53}

TABLE 8.7—*Continued*

Species or serotype	O	H Phase 1	H Phase 2
†*Salmonella arizonae* (Ar.19:22:28)	59	(k)	e,n,x,z_{15}
†*Salmonella arizonae* (Ar.19:22:31)	59	(k)	z
†*Salmonella arizonae* (Ar.19:22:21)	59	(k)	z_{35}
†*Salmonella arizonae* (Ar.19:1,2,5,:-) (Ar.19:1,2,6:-)	59	z_4,z_{23}	—
†*Salmonella arizonae* (Ar.19:27:25)	59	z_{10}	z_{53}
†*Salmonella arizonae* (Ar.19:27:40a,40c)	59	z_{10}	z_{57}
†*Salmonella arizonae* (Ar.19:16,17,18:-)	59	z_{29}	—
†*Salmonella arizonae* (Ar.19:17,20:-)	59	z_{36}	—
†*Salmonella arizonae* (Ar.19:26:-)	59	z_{52}	—
Group 60			
*S. setubal	60	g,m,t	z_6
†*Salmonella arizonae* (Ar.24:33:21)	60	i	z_{35}
†*Salmonella arizonae* (Ar.24:29:28)	60	k	e,n,x,z_{15}
†*Salmonella arizonae* (Ar.24:29:31)	60	k	z
†*Salmonella arizonae* (Ar.24:29:21)	60	k	z_{35}
†*Salmonella arizonae* (Ar.24:22:25)	60	(k)	z_{53}
†*Salmonella arizonae* (Ar.24:23:31)	60	l,v	z
†*Salmonella arizonae* (Ar.24:24:28)	60	r	e,n,x,z_{15}
†*Salmonella arizonae* (Ar.24:24:31)	60	r	z
†*Salmonella arizonae* (Ar.24:24:25)	60	r	z_{53}
*S. luton	60	z	e,n,x
†*Salmonella arizonae* (Ar.24:27:31)	60	z_{10}	z
†*Salmonella arizonae* (Ar.24:26:21)	60	z_{52}	z_{35}
†*Salmonella arizonae* (Ar.24:26:25)	60	z_{52}	z_{53}
Group 61			
†*Salmonella arizonae* (Ar.26:32:30)	61	c	1,5,7
†*Salmonella arizonae* (Ar.26:32:21)	61	c	z_{35}
†*Salmonella arizonae* (Ar.26:33:28)	61	i	e,n,x,z_{15}
*S. eilbek	61	i	z
†*Salmonella arizonae* (Ar.26:33:31)	61	i	z
†*Salmonella arizonae* (Ar.26:33:21)	61	i	z_{35}
†*Salmonella arizonae* (Ar.26:33:25)	61	i	z_{53}

Species or serotype	O	H Phase 1	H Phase 2
†*Salmonella arizonae* (Ar.26:29:30)	61	k	1,5,7
†*Salmonella arizonae* (Ar.26:22:25)	61	(k)	z_{53}
†*Salmonella arizonae* (Ar.26:23:30:[40a, 40b])	61	l,v	$1,5,7:[z_{57}]$
†*Salmonella arizonae* (Ar.26:23:31)	61	l,v	z
†*Salmonella arizonae* (Ar.26:23:21)	61	l,v	z_{35}
†*Salmonella arizonae* (Ar.26:23:25)	61	l,v	z_{53}
†*Salmonella arizonae* (Ar.26:24:30)	61	r	1,5,7
†*Salmonella arizonae* (Ar.26:24:25)	61	r	z_{53}
†*Salmonella arizonae* (Ar.26:26:25)	61	z_{52}	z_{53}
Group 62			
†*Salmonella arizonae* (Ar.6:13,14:-)	62	g,z_{51}	—
†*Salmonella arizonae* (Ar.6:1,2,5:-)	62	z_4,z_{23}	—
†*Salmonella arizonae* (Ar.6:1,7,8:-)	62	z_4,z_{32}	—
Group 63			
†*Salmonella arizonae* (Ar.8:13,14:-)	63	g,z_{51}	—
†*Salmonella arizonae* (Ar.8:1,7,8:-)	63	z_4,z_{32}	—
†*Salmonella arizonae* (Ar.8:17,20:-)	63	z_{36}	—
Group 64			
†*Salmonella arizonae* (Ar.29:32:31)	64	c	z
†*Salmonella arizonae* (Ar.29:33:31)	64	i	z
†*Salmonella arizonae* (Ar.(5),29:33:[21]:[40a,40b])	64	i	$[z_{35}]:[z_{57}]$
†*Salmonella arizonae* (Ar.(5),29:29:30)	64	k	1,5,7
*Salmonella	64	k	e,n,x,z_{15}
†*Salmonella arizonae* (Ar.29:29:25)	64	k	z_{53}
†*Salmonella arizonae* (Ar.29:23:31)	64	l,v	z
†*Salmonella arizonae* (Ar.29:24:31) (Ar.(5),29:24:31)	64	r	z
†*Salmonella arizonae* (Ar.29:27:31)	64	z_{10}	z
*Salmonella	64	z_{29}	—
†*Salmonella arizonae* (Ar.(5),29:17,20:-)	64	z_{36}	—
†*Salmonella arizonae* (Ar.29:26:28)	64	z_{52}	e,n,x,z_{15}

TABLE 8.7—*Continued*

Species or serotype	O	Antigen H Phase 1	Antigen H Phase 2	Species or serotype	O	Antigen H Phase 1	Antigen H Phase 2
Group 65				†Salmonella arizonae (Ar.30:23:31)	65	l,v	z
†Salmonella arizonae (Ar.30:32:25)	65	c	z_{53}	†Salmonella arizonae (Ar.30:23:25)	65	l,v	z_{53}
†Salmonella arizonae (Ar.30:22:31)	65	(k)	z	†Salmonella arizonae (Ar.30:27:28)	65	z_{10}	e,n,x,z_{15}
†Salmonella arizonae (Ar.30:22:21)	65	(k)	z_{35}	†Salmonella arizonae (Ar.30:26:31)	65	z_{52}	z
†Salmonella arizonae (Ar.30:22:25)	65	(k)	z_{53}	†Salmonella arizonae (Ar.30:26:21)	65	z_{52}	z_{35}
†Salmonella arizonae (Ar.30:23:30)	65	l,v	1,5,7	†Salmonella arizonae (Ar.30:26:25)	65	z_{52}	z_{53}

terium (*Eberthella*) *typhi* (Schroeter) Buchanan 1918, 53.)

For a discussion of the name see Int. Bull. Bact. Nom. Tax. 1958, **8**, 155–157.

ty'phi. Gr. n. *typhus* a stupor; M.L. gen.n. *typhi* of typhoid.

Antigenic formula: 9,12,[Vi]:d:-.

Does not grow on Simmons' citrate medium or on a minimal defined medium; needs tryptophan as growth factor.

Does not produce gas from the fermentation of glucose or other sugar. Fermentation of xylose is variable.

Many strains are agglutinated by Vi and are inagglutinable by O9 antisera; their colonies are opaque and have an iridescent appearance when examined by transmitted light. Colonies of intermediate appearance, agglutinable by both Vi and O antisera, may occur (VW colonies).

In nature pathogenic only in man, causing typhoid (enteric) fever, transmitted by water or food contaminated by human excreta.

4. **Salmonella paratyphi-A** (Brion and Kayser) Castellani and Chalmers 1919, 939. (*Bacterium paratyphi* Kayser 1902, 426; *Bacterium paratyphi* typus A Brion and Kayser 1902, 613.)

pa.ra.ty'phi. Gr. prep. *para* alongside of; Gr. n. *typhus* a stupor; M.L. n. *paratyphus* paratyphoid.

Antigenic formula: 1,2,12:a:-. As with other strains of groups A, B and D, the presence of factor 1 is connected with lysogenization.

Aerogenic; ferments arabinose but not xylose.

Cannot use citrate as sole C source.

The majority of strains do not produce H₂S, and in this character it is unlike most other salmonellas.

Lysine decarboxylase is weak or inactive.

Pathogenic only for man.

5. **Salmonella schottmuelleri** (Winslow, Kliger and Rothberg) Bergey *et al.* 1923, 213. (*Bacterium paratyphi* typus B Brion and Kayser 1902, 613; *Bacillus schottmuelleri* Winslow, Kliger and Rothberg 1919, 479; *Salmonella paratyphi-B* (Brion and Kayser) Castellani and Chalmers 1919, 939; *Salmonella paratyphi-B* Salmonella Subcommittee 1934.) (See "Editorial Note" to *S. hirschfeldii*.)

schott.muel'ler.i. M.L. gen.n. *schottmuelleri* of Schottmüller, named after Prof. R. Schottmüller, who isolated the organism in 1899.

Antigenic formula: 1,4,5,12:b:1,2.

Produces a slime wall when grown on a medium containing 0.5% glucose, 0.2 M sodium phosphate, pH 7 (Anderson, 1961).

Is *d*-tartrate negative.

Causes enteric fever in man and very rarely infects animals.

A variant known as *S. java* is *d*-tartrate positive, fails to produce a slime wall, usually causes enteritis in man and not uncommonly infects animals (Kauffmann, 1941).

Some strains are intermediate between these two extremes.

6. **Salmonella typhimurium** (Loeffler) Castellani and Chalmers 1919, 939. (*Bacillus typhi murium* Loeffler 1892, 134.)

ty.phi.mur'i.um. Gr. n. *typhus* a stupor; L.n. *mus* mouse; L. gen.pl.n. *murium* of mice; M.L. gen.pl.n. *typhimurium* typhoid of mice.

Antigenic formula: 1,4,5,12:i:1,2.

The presence of factor 1 follows lysogenization by a converting phage named iota or PLT₂₂.

Ubiquitous and frequently the cause of infections in man and animals; the most frequent agent of food-poisoning in man.

The well known chromosome map of the Salmonella is that of *S. typhimurium* strain LT₂ (for a review see Sanderson, 1967).

Neotype: NCTC 74; ATCC 13311 (Jud. Comm. 1963; Opin. 26).

7. Salmonella enteritidis (Gaertner) Castellani and Chalmers 1919, 939. (*Bacillus enteritidis* Gaertner 1888, 573).

en.ter.it′id.is. Gr. n. *enteron* gut, intestine; M.L. n. *enteritis* enteritis; M.L. gen.n. *enteritidis* of enteritis.

Antigenic formula: 1,9,12:g,m:-.

Frequent in man and animals.

8. Salmonella gallinarum (Klein) Bergey *et al.* 1925, 236. (*Bacillus gallinarum* Klein 1889, 689; *Bacterium pullorum* Rettger 1909, 123; *Salmonella gallinarum-pullorum* Taylor *et al.* 1952, 140.)

gal.lin.ar′um. L. n. *gallina* hen; L. gen.pl.n. *gallinarum* of hens.

Antigenic formula: 1,9,12:-:-.

Always non-motile. May be subdivided into biotypes on fermentation characteristics, production of gas and H₂S.

Does not grow on a minimal defined medium.

Isolated chiefly from chickens and other birds. Causative agent of fowl typhoid.

SUBGENUS II*[1]

9. Salmonella salamae Le Minor, Rohde and Taylor 1970, 209. (*Salmonella dar-es-salaam* Salmonella Subcommittee 1934, 346; *Salmonella daressalamensis* Haupt 1932, 674.)

sal.am′ae. M.L. gen.n. *salamae* of (Dar-es) salaam.

Antigenic formula: 1,9,12:1,w:e,n,x.

Mucate and malonate positive; gelatin liquefaction slow.

Isolated in 1922 from the urine of a patient in Dar-es-Salaam (Tanzania) and the antigenic structure was determined by White (1926). The biochemical characteristics differed from previously identified salmonellas (Table 9.5) and the organism became the type species of subgenus II.

Type strain: NCTC 5773; ATCC 6959.

SUBGENUS III†

10. Salmonella arizonae Kauffmann in van Oye 1964, 37. (*Arizona arizonae* Kauffmann and Ed-

wards 1952, 6; *Paracolobactrum arizonae* (Kauffmann) Borman 1957, 347; *Arizona hinshawii* (Ewing 1969).)

a.ri.zon′ae. M.L. gen.n. *arizonae* of Arizona, a state in the U.S.A.

The type species of Salmonella subgenus III (see Table 8.5).

Antigenic formula: 51:z₄,z₂₃:- (the corresponding Arizona formula is :1,2:1,2,5:-).

The original strain isolated from reptiles was designated Dar-es Salaam type var. from Arizona (Caldwell and Ryerson, 1939). The antigenic formula was worked out by Kauffmann (1941) as 33:z₄,z₂₃,z₃₆:- and he gave it the name *Salmonella* sp. (ser) *arizona*. After Edwards, West and Bruner (1947) established Arizona as an independent group, the O antigen 33 was deleted from the Kauffmann-White scheme. O group 51 is identical with the old O:33 and with the Arizona antigen designated 1,2 by Edwards *et al.* The H antigens z₄, z₂₃, z₃₆ (simplified to z₄, z₂₃) correspond to H 1,2,5 of Edwards *et al.*

Type strain: NCTC 9297; ATCC 13314.

SUBGENUS IV‡

11. Salmonella houtenae Le Minor, Rohde and Taylor 1970, 209. (*Salmonella houten* Kauffmann 1962, 352.)

hou.ten.ae. M.L. gen.n. *houtenae* of Houten, a town in Holland.

Antigenic formula: 43:z₄,z₂₃:-.

The type species of Salmonella subgenus IV; it is the oldest known member of the subgenus (see discussion in Kauffmann, 1966, 244, on *S. delplata*, a mixed culture from which the strain *S. houtenae* was obtained).

Type strain: NCTC 10401.

Geographical Distribution of Salmonella Types

While certain types are ubiquitous (e.g. *S. typhimurium*), others seem to be localized in different regions of the globe, as *S. berta* in North America and *S. sendai* in the Far East. The essential bibliography of *Salmonella* and its epidemiology will be found in the books by Kauffmann (1966), van Oye (1964) and Kelterborn (1967).

Genus V. **Shigella** *Castellani and Chalmers 1919, 936*

K. PATRICIA CARPENTER†

Shi.gel′la. M.L. dim. ending -*ella;* M.L. fem.n. *Shigella;* named after K. Shiga, the Japanese bacteriologist who first discovered the dysentery bacillus.

---

[1] In this section on *Salmonella* and in the Kauffmann-White scheme (Table 8.7), one asterisk indicates a member of subgenus II; †, subgenus III; ‡, subgenus IV.

† Deceased.

Non-motile rods, not encapsulated. Grow well on nutrient media and do not need special growth factors; growth inhibited by bismuth sulfite in Wilson and Blair's medium. **Cannot use citrate or malonate as sole C source. Growth inhibited by KCN.** H_2S is not produced.

Glucose and other carbohydrates are **fermented with the production of acid but not gas** (except for aerogenic biotypes of one serotype). Adonitol, inositol and salicin are not fermented.

Catalase is usually produced, but exceptions occur in one species.

Other biochemical characters are shown in Tables 8.3 and 8.8.

Type species: *Shigella dysenteriae* (Shiga) Castellani and Chalmers.

Further Comments

The normal habitat of all species is the intestinal tract of man and higher monkeys, and all species produce dysentery. The four species are often referred to as subgroups A, B, C and D; subgroup A consists essentially of mannitol non-fermenters (but exceptional strains occur), the other three subgroups are fermenters of mannitol.

Each member of the genus has a distinctive antigenic structure by which it can be recognized, and the four subgroups are differentiated by a combination of biochemical (Table 8.8) and serological (Table 8.9) characteristics. Subgroup B (*S. flexneri*) is characterized by the fact that all members of the subgroup are inter-related serologically. In subgroup C (*S. boydii*) the members are not related serologically to each other or to the other subgroups. Subgroup D (*S. sonnei*) strains generally ferment lactose and sucrose after some days of incubation.

The main biochemical characters of the genus are shown in Table 8.3, and those that distinguish the subgroups (or species) are shown in Table 8.8.

Description of the species of genus **Shigella**

1. **Shigella dysenteriae** (Shiga) Castellani and Chalmers 1919, 935. *Epit. spec. cons.* Opin. 11, Jud. Comm. 1954, 149. (*Bacillus dysenteriae* Shiga 1898, 817; *Bacillus shigae* Chester 1901, 228; *Eberthella dysenteriae* (Shiga) Bergey *et al.* 1925, 250; *Shigella shigae* (Chester) Wilson and Miles 1946, 685.)

dys.en.te'ri.ae. Gr. n. *dysenteria* dysentery; M.L. gen.n. *dysenteriae* of dysentery.

Colonies of serotype 1 often have a pinkish tinge on Leifson's deoxycholate citrate agar. Catalase is not produced by serotype 1, but is usually produced by strains of other serotypes.

Mannitol is not fermented. Dulcitol is fermented by strains of serotype 5. Indole is not produced by serotype 1 but is always produced by strains

TABLE 8.8

Distinguishing characters of the Shigella species

	1. S. dysenteriae	2. S. flexneri	3. S. boydii	4. S. sonnei
Lactose (acid)	−	−	−	(+)
Mannitol	−	+	+	+
Sucrose	−,	+	−	(+)
Dulcitol	d	−	d	−
Xylose	−	−	d	d
Indole	d	d	d	−
Ornithine	−	−	−	+
Arginine	−	−	−	d

of serotype 2; strains of other serotypes vary in indole production.

Each of the ten serotypes has a distinctive antigen by which it can be recognized; there are few cross-reactions, either within the subgroup or between subgroups. All the serotypes have, at one time or another, been known by other designations, and the main ones are shown in Table 8.9. Serotype 1 (Shiga's bacillus) is peculiar in that it produces a potent exotoxin (Shiga toxin).

Neotype strain: Newcastle 1934; NCTC 4837; ATCC 13313 (Jud. Comm. 1963, Opin. 26).

2. **Shigella flexneri** Castellani and Chalmers 1919, 937. *Epit. spec. cons.* Opin. 11, Jud. Comm. 1954, 149. (*Shigella paradysenteriae* Weldin 1927, 178.)

flex'ner.i. M.L. gen.n. *flexneri* of Flexner; named after Simon Flexner, the American bacteriologist who studied the bacteria of dysentery.

Catalase is produced.

Mannitol is fermented, except by biotype Newcastle, serotype 6 and a mannitol negative, xylose positive biotype of serotype 4a (sometimes known as *S. rabaulensis*): dulcitol is fermented by certain biotypes of serotype 6 (see Table 8.10), some of which produce gas from fermentable sugars.

Indole is not produced by serotype 6; in other serotypes production is variable.

All the serotypes are antigenically related but each has a qualitatively distinct major (or type) antigen; the antigenic complex (the group anti-

TABLE 8.9

Earlier designations and antigenic formulae of **Shigella** *species*

Subgroup and Species	Serotype	Subserotype	Formula	Main earlier designations or synonyms
Subgroup A				
S. dysenteriae	1			*S. shigae*
	2			*S. schmitzii, S. ambigua*
	3			*S. largei* Q771, *S. arabinotarda* A
	4			*S. largei* Q1167, *S. arabinotarda* B
	5			*S. largei* Q1030
	6			*S. largei* Q454
	7			*S. largei* Q902
	8			Serotype 599-52 (Ewing *et al.*)
	9			Serotype 58 (Cox and Wallace)
	10			Serotype 2050 (Ewing)
Subgroup B				
S. flexneri	1	1a	I :2,4	V (Andrewes and Inman)
		1b	I :'S':6:2,4	VZ (Andrewes and Inman)
	2	2a	II :3,4	W (Andrewes and Inman)
		2b	II :7,8	WX (Andrewes and Inman)
	3	3a	III :6:7,8	Z (Andrewes and Inman)
		3b	III :6:3,4	
		3c	III :6:	
	4	4a*	IV :'B':3,4	103 (Boyd)
		4b	(IV) :'B':6:3,4	103Z (Rewell and Bridges)
	5		V :7,8	P119 and P119X (Boyd), (Bridges)
	6		VI :(2),4	*S. newcastle;* Manchester bacillus; Boyd 88 (Newcastle and Manchester—aerogenic; Newcastle—mannitol-negative)
	X		— :7,8	X (Andrewes and Inman)
	Y		— :3,4	Y (Andrewes and Inman)
Subgroup C				
S. boydii	1			170 (Boyd)
	2			P288 (Boyd)
	3			D1 (Boyd)
	4			P274 (Boyd)
	5			P143 (Boyd)
	6			D19 (Boyd)
	7			Lavington I; *S. etousae*
	8			Serotype 112 (Cox and Wallace)
	9			Serotype 1296/7 and 1320 (Francis)
	10			Serotype 430 (Ewing); D15 (Szturm *et al.*)
	11			Serotype 34 and 732 (Ewing)
	12			Serotype 123 (Ewing and Hucks)
	13			Serotype 425 (Ewing and Hucks)
	14			Serotype 2770-51 (Ewing and Hucks)
	15			Serotype 703 (Ewing *et al.*)
Subgroup D				
S. sonnei				Duval's bacillus; *B. ceylanensis* A

* The group phase of this subserotype, corresponding to Boyd's 103B organism, has the formula — :'B':3,4.

gens) is shared by other members of the subgroup (see Table 8.9). Because of the important intragroup relations, highly absorbed sera are needed for the detailed serotyping of *S. flexneri*, and these are normally only available in specialized reference laboratories.

Variation may occur in which the type antigen is lost and the organism assumes a phase in which

TABLE 8.10

Biochemical varieties of **Shigella flexneri** *6*

Old designation	Glucose	Mannitol	Dulcitol
Type 88 Boyd	A[a]	A	—
Type 88 Boyd	A	A	A
Newcastle bacillus	AG/A	—	AG/A/—
Manchester bacillus	AG	AG	AG/—

[a] A = acid produced; AG = acid and gas produced. The fermentation of dulcitol may be delayed.

only the group antigen factors are found. Boyd first described this phenomenon as A → B variation; these variants occur in nature as well as in laboratory cultures.

3. Shigella boydii Ewing 1949, 634. *Epit. spec. cons.* Opin. 11, Jud. Comm., 1954, 149.

boy'di.i. M.L. gen.n. *boydii* of Boyd; named after Sir John Boyd, the British bacteriologist who made a study of dysentery bacilli.

Catalase is produced.

Mannitol is fermented; dulcitol is usually fermented by serotypes 2, 3, 4, 6 and 10, but this may be delayed. Xylose fermentation is variable.

Indole may or may not be produced.

Each serotype has a qualitatively distinct antigen; although there may be some cross-reactions with antisera to other *Shigella* species, these

seldom interfere with diagnosis. Serotypes 10 and 11 share a major antigen, although each possesses a specific antigen.

4. Shigella sonnei (Levine) Weldin 1927, 182. *Epit. spec. cons.* Opin. 11, Jud. Comm., 1954, 149. (*Bacterium sonnei* Levine 1920, 31.)

son'ne.i. M.L. gen.n. *sonnei* of Sonne; named after Carl Sonne who worked with this organism.

On deoxycholate citrate agar colonies are at first colorless but after a few days show bright pink papillae consisting of lactose-fermenting mutant cells. On MacConkey's taurocholate lactose agar, phase I colonies are indistinguishable from colonies of other shigellas, but phase II colonies are larger, flatter and more translucent and have an irregular edge. On subculture phase I colonies produce both phase I and phase II colonies, but phase II colonies give rise to phase II colonies only.

Mannitol is fermented rapidly, lactose and sucrose more slowly; some strains may ferment xylose.

Catalase is produced; indole is not produced.

Ornithine is decarboxylated; arginine may be decarboxylated (or dihydrolyzed).

The one serotype exists in two phases, I and II, and each has a distinctive antigen. Phase II is regarded as a loss variation, but organisms in that phase may be isolated from patients, usually during convalescence and toward the end of an outbreak of Sonne dysentery. An antiserum containing agglutinins for both phases should be used for identification.

Genus VI. **Klebsiella** *Trevisan 1885, 105 Nom. cons.* Opin. 13, Jud. Comm. 1954, 152

IDA ØRSKOV

(*Hyalococcus* Schroeter 1886, 152.)

Kleb.si.el'la. M.L. dim. ending -*ella*. M.L. fem.n. *Klebsiella;* named after Edwin Klebs (1834–1913), a German bacteriologist.

Non-motile, capsulated **rods,** 0.3–1.5 μm by 0.6–6.0 μm, arranged singly, in pairs or short chains. Grow on meat extract media producing more-or-less dome-shaped, glistening colonies of varying degrees of stickiness, depending on the strain and the composition of the medium.

No special growth requirements and most strains **can use citrate and glucose as sole C source,** and ammonia as N source. Glucose is fermented with the production of acid and gas (more CO_2 than H_2), but anaerogenic strains occur. Most strains produce 2,3-butanediol as a major end product of the fermentation of glucose and the V.P. reaction is usually positive; lactic, acetic

and formic acids are formed in smaller amounts and ethanol in larger amounts than in a mixed acid fermentation. H_2S is not produced from TSI; gelatinase and indole are usually not produced.

Other characters are shown in Table 8.3.

Optimal temperature for growth 35–37 C; optimal pH about 7.2.

G + C content of DNA: 52–56 moles % (T_m and buoyant density).

Klebsiella strains are resistant to penicillin in standard doses but may be sensitive to high concentrations; sensitivity to the following drugs may vary: ampicillin, cephalosporins, streptomycin, chloramphenicol, tetracycline, neomycin, kanamycin, polymyxin B, sulfonamides and nitro-furantoin. The proportion of resistant strains is increasing steadily, which may be due to mutations but in many cases is probably caused by transfer of plasmids carrying resistance determinants to a varying number of drugs.

Type species: *Klebsiella pneumoniae* (Schroeter) Trevisan.

Further Comments

Non-motile bacteria listed in the 7th edition of THE MANUAL as *Aerobacter aerogenes* are, in this edition, included in *Klebsiella pneumoniae;* this is a consequence of the demonstration by Edwards (1928, 1929) and Kauffmann (1949) that strains of *Aerobacter aerogenes* able to grow at 37 C were indistinguishable from *Klebsiella*. All *Klebsiella* strains not belonging to *K. ozaenae* or *K. rhinoscleromatis* (see below and Table 8.11) are considered to be *K. pneumoniae*, which agrees with Ewing's (1966) ideas on the classification of the group.

Some strains produce acetoin and 2,3-butanediol very slowly and in such small amounts that the methyl red reaction remains positive. In other strains the acetoin will disappear before the Voges-Proskauer reaction is tested. Seemingly paradoxical M.R. and V.P. reactions, ++ and − −, therefore occur. *Klebsiella pneumoniae* capsule type 3 is M.R. positive and V.P. negative, and the same reactions may occur with capsule type 1 strains. Strains from the upper respiratory tract may show more aberrant biochemical characters than strains from the intestinal canal, e.g. the formation of little or no gas from glucose and/or late fermentation of lactose or inositol.

Cowan *et al.* (1960) had no indole-positive strains in their study and excluded gelatin-liquefying strains from the genus, but in Denmark Brooke (1951) and Ørskov (1955) found that the percentage of *K. pneumoniae* strains positive in these two characters was so high that the characters must be regarded as variable. Liquefaction of gelatin is slow and usually but not always correlated with indole production. Lautrop (1956) revived the name *Bacterium oxytocum* Flügge for

these indole positive, gelatin liquefying strains many of which fermented dulcitol, melezitose and inulin (late). *Aerobacter oxytocum* was recognized in early editions of THE MANUAL but not in the last two. Korth *et al.* (1969) showed that the 'oxytoca' variants of *Klebsiella* produce a dark brown pigment when grown on a defined medium containing gluconate and ferric citrate.

Henriksen (1954) considered that there were two subgroups of the *Klebsiella* group; the first consisted of the rhinoscleroma and ozaena organisms and *K. pneumoniae* type 3, characterized by being M.R. positive, V.P. negative, indole negative, gelatin negative and variable in the citrate and urease characters. The second subgroup comprised the *Aerobacter* organisms, and *Klebsiella pneumoniae* types 1 and 2; these were characterized as M.R. negative, V.P. positive, citrate positive, urease positive, gelatin variable and indole variable.

On the basis of other characters, Cowan *et al.* (1960) divided 176 strains of the *Klebsiella* group into six subgroups of five species and one variety. Durlakowa *et al.* (1967) and Slopek and Durlakowa (1967) examined 851 strains and divided them into the six taxa of Cowan *et al.;* the names and the ranks of the subgroups were, however, changed.

Bascomb *et al.* (1971) divided the genus *Klebsiella* into six taxa; two of these, i.e. *K. aerogenes/ oxytoca/edwardsii* and *K. pneumoniae* correspond to what in this edition of THE MANUAL is called *K. pneumoniae*. Taxon II of Bascomb *et al.*, which included four capsulated strains, was described as the most homogeneous group. It was not stated, but it seemed probable, that all four strains belonged to capsule type 3 as were the strains named *K. pneumoniae* by Cowan *et al.* (1960). If that is so, the homogeneity is more explicable.

Strains of klebsiella-like organisms have been reported in plants and in soils, where they seem to be responsible for nitrogen fixation; at least two new species have been named, *K. rubiacearum* Centifanto and Silver, 1964, 780, and *Bacterium nodoantrum* Skripal 1971, 51. Comparative studies will be required before a judgment can be made on these and other klebsiella-like strains.

Description of the species of genus **Klebsiella**

1. **Klebsiella pneumoniae** (Schroeter) Trevisan 1887, 94 (includes *Aerobacter aerogenes* as described in the 7th edition of THE MANUAL; see p. 339). (*Bacterium pneumoniae crouposae* Zopf 1885, 66; *Hyalococcus pneumoniae* Schroeter 1886, 152; *Bacillus pneumoniae* (Schroeter) Flügge 1886, 204.)

pneu.mo′ni.ae. Gr. n. *pneumonia* pneumonia,

inflammation of the lungs; M.L. gen.n. *pneumoniae* of pneumonia.

Fimbriae are present in most strains; two types are recognized (Duguid, 1959): the thick type adheres to red cells of the guinea pig and other animals except the ox; the reaction is inhibited by D-mannose (MS = mannose-sensitive). The thin type adheres to cells treated with tannic acid (ox

TABLE 8.11

Differentiation within the genus **Klebsiella**

Character or test	1. K. pneumoniae	2. K. ozaenae	3. K. rhinoscleromatis
Gas from glucose	+	d	−
Lactose (acid)	+	(+)	−
Dulcitol (acid)	d	−	−
Methyl red	−	+	+.
Voges-Proskauer	+	−.	−
Citrate utilization	+	d	−
Urease	+	d	−
Malonate	+	−	+
Lysine decarboxylase	+	d	−
Mucate	+	d	−
Organic acids			
Citrate	d	d	−
d-Tartrate	d	d	−

cells most suitable) and is not influenced by mannose (MR = mannose-resistant). MS fimbriae are 6.5–7.0 nm as shown by negative staining (Thornley and Horne, 1962).

The biochemical characters are shown in Table 8.11.

Klebsiella pneumoniae can be subdivided into many biotypes. Ørskov (1957) subdivided 226 *K. pneumoniae* strains into 23 biotypes using dulcitol, adonitol, inositol, sorbose, urease and organic acids (d-tartrate and sodium citrate) and into 32 biotypes when indole and gelatin were added.

K. pneumoniae is widely distributed in nature, in soil, water, grain, etc. and is normally found in the intestinal canal of man and animals. It may be isolated in association with several pathological conditions in man, e.g. infection of the urinary or the respiratory tract. Capsule types 1, 2 and 3 may be the causative agent in pneumonia. In animals *K. pneumoniae* may be isolated from e.g. metritis in mares.

K. pneumoniae capsule types 1 and 2 are highly virulent for mice; all other strains are either nonpathogenic or of low pathogenicity for mice.

Suggested neotype strain: CDC 298/53; NCTC 9633; ATCC 13883 (Cowan *et al.*, 1960).

2. **Klebsiella ozaenae** (Abel) Bergey *et al.* 1925, 266. (*Bacillus mucosus ozaenae* Abel 1893, 167; *Bacillus ozaenae* Abel 1893, 172; *Bacterium ozaenae* (Abel) Lehmann and Neumann 1896, 204.)

o.zae'nae. L. fem.n. *ozaena* ozena; L. gen.n. *ozaenae* of ozena.

Neither of the fimbrial types described in *K. pneumoniae* is found in *K. ozaenae*. In a few strains

some very long (up to 10 μm) and thick (10 nm) fimbriae with a very weak hemagglutinating activity of the MS type have been found (Duguid *et al.*, 1966).

It may be difficult to classify a strain as *K. ozaenae* because of great variability in many biochemical characteristics. The majority of strains will grow with ammonia as N source and glucose as C and energy source.

Capsule type 4 is the most common type; types 3, 5, 6 and 1/5 are also found.

K. ozaenae occurs in ozena and other chronic diseases of the respiratory tract.

Suggested neotype strain: D5050; NCTC 5050.

3. **Klebsiella rhinoscleromatis** Trevisan 1887, 95. (*Bacterium rhinoscleromatis* (Trevisan) Migula 1900, 352.)

rhi.no.scle.ro'ma.tis. M.L. adj. *rhinoscleromatis* pertaining to rhinoscleroma.

Fimbriae have not been demonstrated in this species.

Does not grow with ammonia as N source or with glucose as C source.

Serologically strains of this species belong to capsule type 3.

K. rhinoscleromatis is found constantly and exclusively in patients with rhinoscleroma and their contacts.

Suggested neotype strain: R.70; NCTC 5046; ATCC 13884 (Cowan *et al.*, 1960).

Further Comments

Serology: *Klebsiella* cultures are classified serologically on the basis of their K (capsular) and O (somatic) antigens. Capsular types A–C of Julianelle (1926) and D–F of Goslings and Snijders (1936) were redesignated 1–6 by Kauffmann (1949), who also established eight new types; other workers have brought the total of K types to 80 (for a review see Slopek, 1968) and there are 11 different O types. As the number of O types is small compared with the K types, and as their determination is difficult because of the heat-stable K antigens, determination of the serological type is primarily based on the K determination. All the K types may be found in *K. pneumoniae*; *K. ozaenae* contains types 4, 5, 6, 1/5 and occasionally 3, but most belong to capsule type 4. *K. rhinoscleromatis* strains are invariably members of K type 3.

Many relationships exist between the K antigens of *Klebsiella* and some K antigens of other bacteria such as *Streptococcus pneumoniae*, *Escherichia coli*, and *Salmonella paratyphi-B* (M antigen). When examined, the O antigens are found to cross-react to a great extent with antisera of the *Escherichia coli* group.

The K antigens of K types 1–72 are acidic polysaccharides containing hexuronic acids (glucuronic or galacturonic acid) in almost all cases, and also two to four of the following sugars: galactose, glucose, mannose, fucose and rhamnose (Nimmich, 1968, 1971). See Lüderitz *et al.* (1968) for a review of qualitative and quantitative analyses of *Klebsiella* capsular substances. For the lipopolysaccharides of the O antigens see Nimmich (1968, 1971).

There are no reports on the serology of R antigens (from strains devoid of O antigen with or without K antigen) or the fimbrial antigens. A group-specific antigen present in almost all *Klebsiella* strains has been reported (Pickett and Cabelli, 1953).

Differentiation by Bacteriocins and Phages

Many *Klebsiella* strains produce bacteriocins and there is some evidence of species specificity (Slopek and Maresz-Babczyszyn, 1967). Differentiation by phages isolated from stools has also been attempted and a typing system of 15 selected phages has been proposed (Slopek *et al.*, 1967).

Genetics: Successful chromosomal recombination has been reported by Matsumoto and Tazaki (1970). Transfer of plasmids through conjugation does take place, and the lac character has been transferred from *E. coli* K-12 by the F-lac factor; resistance to antibiotics is transmissible by R factors to *Klebsiella* cultures from other members of the *Enterobacteriaceae* and *vice versa*. Furthermore, transfers of bacteriocinogenic factors have been demonstrated in mixed cultures of *Klebsiella*.

Genus VII. **Enterobacter** Hormaeche and Edwards 1960, 72 *Nom. cons.* Opin. 28, Jud. Comm. 1963, 38

RIICHI SAKAZAKI

(*Cloaca* Castellani and Chalmers 1919, 941; *Aerobacter* Hormaeche and Edwards 1958, 113; not *Aerobacter* Beijerinck 1900, 193; not *Enterobacter* Rahn 1937, 273.)

En.te.ro.bac′ter. Gr. neut.n. *enteron* intestine; M.L. masc.n. *bacter* equivalent of bacterium, a small rod; M.L. masc.n. *Enterobacter* intestinal small rod.

Motile rods, peritrichously flagellated; some strains encapsulated. **Citrate and acetate can be used as sole C source.**

At 37 C glucose is fermented with the production of acid and gas (CO_2:H::2:1). At 44.5 C gas is not produced from glucose. Dulcitol is usually not attacked. The Voges-Proskauer test is usually positive; the methyl red test is negative.

Alginate is not utilized; pectate is not decomposed.

Gelatin is liquefied slowly by most strains; H_2S is not produced on TSI agar.

Other characteristics are given in Tables 8.3, 8.12 and 8.13.

G + C content of DNA: 52–59 moles %.

Type species: *Enterobacter cloacae* (Jordan) Hormaeche and Edwards 1960, 72.

Further Comments

Aerobacter Beijerinck was a genus composed of motile (peritrichously and polarly flagellated) and non-motile bacteria that grew at temperatures below 37 C but feebly or not at all at 37 C. Castellani and Chalmers (1919) separated from it the genus *Cloaca*, mainly on the basis of slow liquefaction of gelatin; the description of this genus was amplified by Hormaeche and Edwards (1958) but they retained the name *Aerobacter;* this led to confusion and they later changed the name to

Enterobacter Hormaeche and Edwards 1960, 72. In 1963 the generic name *Enterobacter* Hormaeche and Edwards was conserved against *Cloaca* Castellani and Chalmers (Jud. Comm., Opinion 28).

Sakazaki (1961) and Ewing (1963) suggested that *Hafnia alvei* Møller 1954 should be included in the genus *Enterobacter*, and it has also been suggested that the *Serratia* group may be related. *E. liquefaciens* may be a non-pigmented species of *Serratia* (Bascomb *et al.*, 1971; Ewing *et al.*, 1973). Differences and similarities of these genera and *Klebsiella*, a genus from which *Enterobacter* has to be distinguished, are given in Table 8.12.

TABLE 8.12

Differentiation between **Enterobacter** *and related genera*

	Klebsiella	Enterobacter	Hafnia	Serratia
Motility	−	+	+	+
Ornithine decarboxylase	−	+	+	+
Sorbitol	+	+	−	−
Red pigment	−	−	−	D
DNase	−	−	−	+
Arginine dihydrolase	−	D	−	−

Description of the species of genus Enterobacter

1. **Enterobacter cloacae** (Jordan) Hormaeche and Edwards 1960, 72. (*Bacillus cloacae* Jordan 1890, 836; *Bacterium cloacae* (Jordan) Lehmann and Neumann 1896, 236; *Cloaca cloacae* (Jordan) Castellani and Chalmers 1919, 958; *Aerobacter cloacae* (Jordan) Bergey *et al.* 1923, 207; *Aerobacter cloacae* (Jordan) Hormaeche and Edwards 1958, 112.)

clo.a'cae. L. n. *cloaca* a sewer; L. gen.n. *cloacae* of a sewer.

Gas is produced in the fermentation of most carbohydrates at 37 C. Gas is not produced in the fermentation of glycerol or inositol; lactose fermentation may be slow; adonitol is fermented by some strains. Malonate may or may not be used as a C source.

Gelatin is liquefied slowly but this property may be lost (Sakazaki and Namioka, 1957). Esculin is not hydrolyzed; lysine is not decarboxylated. Arginine dihydrolase is present.

A non-diffusible yellow pigment may be produced by some strains.

The antigenic formulae of 79 serotypes have been recorded (Sakazaki and Namioka, 1960).

Found in feces of man and other animals, sewage, soil and water. Occasionally found in urine, pus and other pathological material from animals.

Neotype strain: CDC 279-56; NCTC 10005; ATCC 13047.

2. **Enterobacter aerogenes** Hormaeche and Edwards 1960, 74. (*Aerobacter aerogenes* Hormaeche and Edwards 1958, 114; not *Bacillus aerogenes* Miller 1896, 119; not *Bacterium aerogenes* (Kruse) Chester 1897, 53; not *Aerobacter aerogenes* (Kruse) Beijerinck 1900, 193.)

a.e.ro'gen.es. Gr. masc.n. *aer* air; Gr. v. *gennaio* produce; M.L. adj. *aerogenes* gas producing.

Gas produced from most carbohydrates at 37 C, including adonitol, glycerol and sometimes inosi-

tol. Malonate is used as sole C source. The ability to produce gas at 44.5 C varies; of the two biotypes, one (negative at 44.5) seems to be of fecal origin; the other (positive at 44.5) is from soil or vegetation free from fecal contamination (Hendricks, 1970).

Gelatin is very slowly liquefied by most strains. Indole is usually not produced, but some strains may do so (Sakazaki and Namioka, 1957). Esculin is hydrolyzed; lysine is decarboxylated; arginine dihydrolase is not present.

Most of the motile strains labeled *Aerobacter aerogenes* (*Enterobacter aerogenes*) were found by Edwards and Ewing (1962) to be agglutinable by and give quellung reactions with *Klebsiella* capsular antisera.

Found in feces of man and other animals, sewage, soil, water and dairy products.

Type strain: CDC 819-56; NCTC 10006; ATCC 13048.

TABLE 8.13

Differentiation between **Enterobacter cloacae** *and* **E. aerogenes**

	1. E. cloacae	2. E. aerogenes
Gas produced from		
Glycerol	−	+
Inositol	−	d
Adonitol	d	+
Malonate as C source	d	+
Gelatin liquefaction	(+)[a]	((+))
Esculin hydrolysis	−	+
Lysine decarboxylase	−	+
Arginine dihydrolase	+	−

[a] (+) slow liquefaction; ((+)) very slow liquefaction.

Genus VIII. Hafnia Møller 1954, 272

RIICHI SAKAZAKI

Haf'ni.a. O.L. fem.n. *Hafnia* the old name for Copenhagen.

Motile, peritrichously flagellated **rods;** not encapsulated. **Citrate and acetate can be used as sole C source.**

Glucose is fermented with the production of acid and gas. Acid is not produced from citrate, *d*-tartrate, and mucate. The M.R. test is usually negative at 37 C; the V.P. test is usually positive at 22 C, but may be negative at 37 C.

Alginate is not utilized; pectate is not decomposed. H_2S is not produced from TSI agar, but is

produced weakly in peptone iron agar by many strains. Lipase is not produced; DNase is not produced.

Other characteristics are given in Tables 8.3 and 8.11.

G + C content of DNA is 52–57 moles %.

(See Sakazaki and Namioka (1957) for a review of the literature.)

Type species: *Hafnia alvei* Møller 1954, 272.

Further Comments

Sakazaki (1961) suggested that the Hafnia group of bacteria should be placed in the genus *Enterobacter* and proposed the combination *Enterobacter alvei* as the name of the organism. Møller

(1954) stated that the name *Hafnia alvei* was derived from the name *Bacillus paratyphi-alvei* Bahr 1919, 45; Ewing and Fife (1968) found that the characters of *Hafnia alvei* were not those described by Bahr, so that Møller's name stands as a new name for a hitherto undescribed species.

Description of the species of genus Hafnia

1. **Hafnia alvei** Møller 1954, 272. (*Enterobacter alvei* (Møller) Sakazaki 1961, 238; *Enterobacter aerogenes* subsp. *hafniae* Ewing 1963, 110; not *Bacillus paratyphi-alvei* Bahr 1919, 45.)

al've.i. L. n. *alveus* a beehive; L. gen.n. *alvei*, of a beehive.

Most strains fail to ferment sucrose, but some strains slowly and irregularly attack the sugar. A few strains attack salicin, but the majority do not.

Many strains can utilize malonate as a C source.

Serology: the antigenic formulae of 49 serotypes were recorded by Sakazaki (1961) and Matsumoto (1963, 1964) extended the number to 197.

Found in feces of man and other animals, sewage, soil, water and dairy products.

Type strain: Stuart 32011; NCTC 8106; ATCC 13,337.

Genus IX. **Serratia** Bizio 1823, 288

RIICHI SAKAZAKI

Ser.ra'ti.a. M.L. fem.n. *Serratia;* named after Serafino Serrati, the Italian physicist who invented a steamboat at Florence before 1787.

Motile peritrichously flagellated **rods;** some strains are capsulated.

Citrate and acetate can be used as sole C source. **Many strains produce pink, red or magenta pigment.**

Glucose is fermented with or without the production of a small volume of gas; cellobiose, inositol and glycerol are fermented without gas production. The fermentation of xylose and adonitol are variable characters. The M.R. test is negative, the V.P. is usually positive.

Malonate and alginate are not utilized; pectate is not decomposed. Urea is usually not decomposed, but some strains attack it weakly. **DNase is produced.** For other characteristics see Table 8.12.

When it is present, the characteristic feature of the genus is a non-diffusible pigment, prodigio-

sin. But pigmentation is a variable characteristic of strains and it is influenced by cultural conditions and medium; many strains fail to produce it.

$G + C$ content of DNA: 53–59 moles %.

Type species: *Serratia marcescens* Bizio 1823, 288.

Further Comments

In the 4th edition of THE MANUAL (1934) 27 species of the genus were listed; later, Breed *et al.* (1957) treated many of the epithets as synonyms and in the 7th edition listed only five species, *S. marcescens, S. kiliensis, S. indica, S. plymuthica* and *S. piscatorum.* However, Davis and Ewing (1957) thought that only one species was needed, while Ewing, Davis and Reavis (1959), Ewing, Davis and Johnson (1962) and Martinec and Kocur (1960, 1961) proposed that the one species should have a subspecies for the V.P.-negative strains.

Description of the species of genus Serratia

1. **Serratia marcescens** Bizio 1823, 288. (*Bacillus marcescens* (Bizio) Trevisan in de Toni and Trevisan 1889, 976.)

mar.ces'cens. L. v. *marcesco* fade; L. part.adj. *marcescens* fading away.

Cellobiose, inositol and glycerol are fermented without gas production. Arabinose is not fermented. The fermentation of xylose and adonitol are variable characters.

V.P. reaction positive; rarely negative.

Serology: 46 serotypes were found by Davis and Woodward (1957) and by Ewing, Davis and Reavis (1959).

Found in water, soil and food, presumably widely distributed. Occasionally found in pathological specimens, in which it may grow after collection and e.g. in sputum suggests hemoptysis (pseudohemoptysis).

Neotype strain: BS 303; ATCC 13880.

Genus X. **Proteus** Hauser 1885, 12

HANS LAUTROP

(*Liquidobacterium* Orla-Jensen 1908, 318 (Danish), 1909, 338 (German).)
Pro'te.us. Gr. n. *Proteus;* a god able to transform himself into many shapes.

Usually straight rods, 0.4–0.6 by 1.0–3.0 μm; coccoid and irregular involution forms, filaments and spheroplasts are frequent under certain conditions. May occur in pairs or chains. Not encapsulated. Motile by peritrichous flagella; swimming motility most pronounced at 20 C and often absent at 37 C. Non-pigmented.

Acid is formed regularly and rapidly from glucose, and more slowly and less regularly from fructose, galactose and glycerol; the methyl red test is usually positive. Gas in small amounts may or may not be formed from glucose. With rare exceptions, acid is not formed from L-arabinose, dextrin, dulcitol, glycogen, inulin, lactose, melibiose, raffinose, sorbitol, sorbose or starch. Acid formation from other sugars varies (Table 8.15). Pectolytic and alginolytic activity not present.

Nitrates reduced to nitrites; Pichinoty's type A reductase present in all species but some strains of *P. inconstans* have type B reductase (Pichinoty *et al.*, 1966). Sulfite reductase not present.

Phenylalanine deaminated to phenylpyruvic acid. Indole formed except by one species. Glutamic acid decarboxylated; arginine not decarboxylated. Growth not inhibited by KCN in Møller's test.

Some species grow in a mineral medium with ammonium ion as sole N source and glucose as sole energy and C source; other species require growth factors (Table 8.14).

Temperature range for growth about 10–43 C.

The G + C content of the DNA ranges from 38–42 moles % (T_m) for four species and is about 50 for *P. morganii.*

Type species: *Proteus vulgaris* Hauser 1885, 12.

Further Comments

The ranking of the five included taxa as species is not universally accepted. *Proteus morganii* stands clearly apart from the four others and might deserve generic rank; indeed, Fulton proposed the generic name *Morganella* for this taxon (Fulton, 1943, 81). *Proteus inconstans* has also been given generic rank as *Providencia* and divided into two species, *P. alcalifaciens* and *P. stuartii*, corresponding to the subgroups A and B, respectively (Ewing, 1958, 1962). An arrangement of the five species in four different genera, *Proteus, Morganella, Rettgerella* and *Providencia*, has also been used (Kauffmann, 1966, 359).

Proteus vulgaris and *P. mirabilis* show a high degree of phenotypic similarity and might be considered as two biotypes of one species.

Proteus zenkeri Hauser 1885, 44 appears from the original description to be a non-liquefying variant of *P. vulgaris*, but it has also been classified as a species of *Kurthia. Proteus myxofaciens* Cosenza and Podgwaite 1966, 190, is probably not a *Proteus* species, but identical with *Erwinia herbicola.*

P. hydrophilus and *P. ichthyosmius* are now classified in the genus *Aeromonas.*

Spontaneous swarming on the surface of solid media is a striking characteristic of most strains of *P. vulgaris* and *P. mirabilis;* the surface is covered by a thin layer of moving cells arranged at the edge in changing patterns of whirls and bands; during active swarming the cells are elongated and provided with an enormous number of flagella (Hughes, 1957; Morrison and Scott, 1966; Jones and Park, 1967). On a surface two swarming cultures may either merge completely or come to a halt a few millimeters apart (Dienes phenomenon) (Dienes, 1946). In some strains of the other three species a kind of surface spreading may be induced by growth at low temperature on plates with a low agar concentration (Rauss, 1936, 1962). It is still unknown whether the spontaneous swarming and the induced spreading are basically the same character.

Both normal and curly flagella occur; their number varies from very few to very many. Mean wave length 2.25 μm (Leifson, 1960). Curly flagella predominate in cultures with a pH below 6.0–6.5 (Hoeniger, 1965). Flagella diameter 11–14 nm (Lowy and Hanson, 1965).

Sensitivity to antibiotics highly variable and rarely of diagnostic significance.

Thermostable somatic (O) antigens and thermolabile flagellar (H) antigens are usually present, and differences in the serological specificities of both antigens form the basis for the recognition of a large number of serotypes within each species. An antigenic scheme common to the whole genus has not been developed.

Virulent and temperate phages are known for all species. There is extensive cross-susceptibility inside the genus, with only *P. morganii* in a relatively isolated position; outside the genus only a few strains of *E. coli* have been found susceptible. Bacteriocins are produced by all species except *P. rettgeri;* bacteriocin susceptibility is limited to

TABLE 8.14

Differential characteristics of species within the genus **Proteus**

	1. *P. vulgaris*	2. *P. mirabilis*	3. *P. morganii*	4. *P. rettgeri*	5. *P. inconstans*
Gas from glucose	+[a]	+	+	d[b]	d
Urease	+	+	+	+	+
Indole production	+	−[c]	+	+	+
Gelatin liquefaction	+	+	−	−	−
H₂S production	+	+	−	−	−
Ornithine decarboxylase	−	+	+	−	−
Growth requirement					
Nicotinic acid	+	+	+	−	−
Pantothenic acid	−	−	+	−	−
Moles % G + C	39.3	39.3	50.0	39.0	41.5
±	1.2	1.4	0.7	1.5	0.6

[a] + = more than 90% strains have this character.
[b] d = 89–11% positive.
[c] − = less than 10% of strains positive for this character.

strains of the same species, except in the case of *P. vulgaris* and *P. mirabilis* where extensive cross-susceptibility occurs.

Generalized transduction is known to occur in all species. Transference of episomes from *E. coli* to *P. mirabilis* is known, but the episomes persist in a non-integrated state. Transformations and lysogenic conversions are not known to occur.

Description of the species of genus **Proteus**

The most important distinguishing characters are shown in Table 8.14, and Table 8.15 shows the different behavior in carbohydrate fermentation.

1. Proteus vulgaris Hauser 1885, 12.

vul.ga′ris. L. adj. *vulgaris* common.

Fimbriae may be present. Strong tendency to spontaneous swarming. Strong putrefactive odor in protein-containing media. Some strains hemolytic on blood agar. Acetoin may be formed, but rarely and only at 22 C. Extracellular deoxyribonuclease formed by some strains. Phosphatase and lipase regularly present. Regularly produces penicillinase.

Serology: The combined antigenic scheme of *P. vulgaris* and *P. mirabilis* lists 110 serotypes (Kauffmann, 1966, 340–346). O antigens 1, 2 and 3 of the scheme react with antibodies formed in man during some rickettsial infections, and are used in the Weil-Felix reaction for the diagnosis of typhus fever.

Found in fecal matter of many animals, sewage and soil, especially where animal protein is decomposing. Less frequently found in clinical specimens than *P. mirabilis*.

Neotype strain: Strain Lehman; NCTC 4175; ATCC 13315; NCIB 4175; Jud. Comm. 1963, Opin. 26.

2. Proteus mirabilis Hauser 1885, 34.

mi.ra′bi.lis. L. adj. *mirabilis* wonderful, surprising.

Fimbriae may be present. Pronounced tendency to spontaneous swarming. Strong putrefactive odor in protein-containing media. Some strains hemolytic on blood agar. About 50% of the strains form acetoin during glucose fermentation at 22 C. When it occurs, acid formation from sucrose is delayed. Acid formation from maltose absent due to lack of the specific permease. Phosphatase and lipase regularly present. Only some strains form penicillinase. Fairly uniform in its sensitivity pattern to antibiotics; in contrast to other species usually sensitive to penicillin, ampicillin and cephalosporin.

Serology: see Kauffmann and Perch for antigenic scheme (Kauffmann, 1966, 340–346).

Found in fecal matter of many animals, sewage and soil, especially where animal protein is decomposing. It is the most frequent *Proteus* species found in human clinical material.

Type strain not extant; neotype has not been proposed.

Hauser in 1892 stated that he then considered *P. mirabilis* as a modification of *P. vulgaris*. The present taxon *P. mirabilis* was separated formally from *P. vulgaris* on the basis of no fermentation of maltose and delayed sucrose fermentation (Wenner and Rettger, 1919, 350).

TABLE 8.15

Fermentation reactions within the genus **Proteus** (*all known sugars giving differential reactions are included*)

	1. *P. vulgaris*	2. *P. mirabilis*	3. *P. morganii*	4. *P. rettgeri*	5. *P. inconstans* A	B
Adonitol	−	−	−	+	+	−
Amygdalin[a]	+	−	−	−	(d)[b]	(d)
Arabitol[a]	−	−	−	(+)	(d)?	(d)?
Arbutin[a]	d	−	−	+	−	−
Cellobiose	(d)	(d)	−	(d)	−	(d)
Erythritol	−	−	−	+	−	−
Esculin	d	−	−	+	−	−
Inositol	−	−	−	+	−	+
Maltose	+	−	−	−	−	−
Mannitol	−	−	−	+	−	−
Mannose[a]	−	−	+	+	+	+
Melezitose[a]	(+)	−	−	−	−	−
Rhamnose	−	−	−	d	−	−
Salicin	d	(d)	−	d	−	−
Sucrose	+	(+)	−	d	(d)	(+)
Trehalose	(+)	+	d	−	−	d
Xylose	d	+	−	d	−	−

[a] Results for these sugars are based on a small number of strains.

[b] Results in parentheses mean that the reaction is usually delayed, i.e. occurring later than 48 hr.

3. **Proteus morganii** (Winslow, Kligler and Rothberg) Yale 1939, 435. (Organism No. 1 Morgan 1906, 908; *Bacillus morgani* (sic) Winslow, Kligler and Rothberg 1919, 481; *Salmonella morgani* (Winslow *et al.*) Castellani and Chalmers 1919, 939; *Morganella morganii* (Winslow *et al.*) Fulton 1943, 81.)

mor.ga′ni.i. M.L. gen.n. *morganii* of Morgan; named after H. de R. Morgan, the bacteriologist who first studied the organism.

Fimbriae may be present. Does not swarm spontaneously but may be induced to spread on solid surfaces by reducing the concentration of agar. Putrefactive odor in protein-containing media. About one-third of the strains hemolytic on blood agar. Acetoin not formed. Lysine weakly decarboxylated by some strains. H_2S formed, but in small amount, in ferrous chloride gelatin.

The urease and phenylalanine deaminase are serologically distinct from the similar enzymes in the other *Proteus* species.

Many strains are sensitive to nitrofurantoin and tetracyclines, antibiotics toward which most other *Proteus* strains are resistant.

Serology: The antigenic scheme lists 66 serotypes (Rauss and Vörös, 1959, 1967).

Habitat not known; most frequently found in the feces of man.

G + C content of DNA about 50 moles % (T_m)

is different from that of the other *Proteus* species (Falkow *et al.*, 1962).

Proposed neotype: ATCC 25830; ATCC 8076h; NCTC 235; NCIB 235 (Lessel, 1971).

4. **Proteus rettgeri** (Hadley, Elkins and Caldwell) Rustigian and Stuart 1943, 243. (*Bacterium rettgeri* Hadley, Elkins and Caldwell 1918, 180; *Shigella rettgeri* (Hadley *et al.*) Weldin 1927, 181; *Proteus entericus* (Hadley *et al.*) Rustigian and Stuart 1943, 199.)

rett′ge.ri. M.L. gen.n. *rettgeri* of Rettger; named after L. F. Rettger, the American bacteriologist who first isolated this organism in 1904.

Does not swarm spontaneously, but spreading may be induced in some strains.

Serology: The antigenic scheme lists 45 serotypes (Namioka and Sakazaki, 1958).

Bacteriocin-producing strains not described.

Habitat not known; most strains isolated from a variety of human clinical specimens. Also isolated from chicken feces.

Type strain not extant; neotype has not been proposed.

5. **Proteus inconstans** (Ornstein) Shaw and Clarke 1955, 155. (*Bacillus inconstans* Ornstein 1921, 166; Anaerogenic paracolon type 29911

Stuart *et al.* 1943, 111; *Proteus stuartii* Buttiaux *et al.* 1954, 385; *Providencia inconstans* (Ornstein) Ewing 1958, 18; *Providencia alcalifaciens* (De Salles Gomes) Ewing 1962, 96; *Providencia stuartii* (Buttiaux *et al.*) Ewing 1962, 96.)

in.con'stans. L. adj. *inconstans* inconstant, changeable.

Spontaneous swarming does not occur.

Resistant to more antibiotics than the other four species but usually sensitive to kanamycin and gentamycin.

Serology: The antigenic scheme lists 156 serotypes (Ewing *et al.*, 1954; Edwards and Ewing, 1962, 230).

Habitat not known. Usually isolated from human clinical specimens, especially urine and feces.

May be divided into two subgroups biochemically. Subgroup A: glucose gas, adonitol acid, inositol no acid. Subgroup B: glucose no gas, adonitol no acid, inositol acid. A few strains occur that do not fit into either of these subgroups.

ATCC 9886 is a typical strain of subgroup A and was designated as the type strain of *Providencia alcalifaciens* by Ewing, 1962.

Genus XI. **Yersinia** van Loghem 1944, 15

H. H. MOLLARET AND E. THAL

Yer.sin'i.a. M.L. fem.n. *Yersinia* named after the French bacteriologist A. J. E. Yersin who first isolated the causal organism of plague in 1894.

Cells ovoid or rods, 0.5–1.0 by 1–2 μm; nonmotile at 37 C; at temperatures below 37 C two species motile with peritrichous flagella. Not encapsulated.

Lactose usually not fermented; β-galactosidase produced. Fructose, glucose, glycerol, maltose, mannitol, mannose and trehalose fermented without the production of gas. Dulcitol, erythritol, fucose, glycogen, inositol, melezitose and raffinose not fermented. Methyl red test positive; Voges-Proskauer negative at 37 C for *Y. pestis* and *Y. pseudotuberculosis*.

Grows well on meat extract agar (1–2 days); cannot utilize citrate as sole source of C. Malonate not utilized. Tetrathionate not reduced.

Gelatin not hydrolyzed; nitrates reduced, except by one variety. Lysine not decarboxylated, phenylalanine not deaminated. Arginine dihydrolase reaction not present. Indole usually not produced by *Y. pestis* and *Y. pseudotuberculosis*.

Temperature range said to be between −2 C and 45 C; optimal 30–37 C.

The G + C content of the DNA ranges from 45.8–46.8 moles % (T_m).

Type species: *Yersinia pestis* (Lehmann and Neumann) van Loghem 1944, 15.

Further Comments

The genus was proposed by van Loghem (1944) to separate from the genus *Pasteurella* those bacteria that differed from the septicemic group (Pasteurella *sensu stricto*). In 1954 Thal proposed that the genus *Yersinia* should be included in the family *Enterobacteriaceae*. Numerical methods seem to justify the division of *Pasteurella* (Smith and Thal, 1965) and the relation of *Yersinia* to the *Enterobacteriaceae* (Talbot and Sneath, 1960).

The main characteristics that distinguish the species are shown in Table 8.16, which also compares them with *Salmonella* and *Pasteurella multocida;* other characteristics are given in Table 8.17.

TABLE 8.16

Distinguishing characters between the species of **Yersinia,** *and a comparison of these characters with those of* **Salmonella** *and* **Pasteurella multocida**

	1. *Y. pestis*	2. *Y. pseudotuberculosis*	3. *Y. enterocolitica*	*Salmonella*	*Pasteurella*
Motility					
22 C	−	+	+	+	−
37 C	−	−	−	+	−
Urea hydrolyzed	−a	+	+	−	−
Esculin	+	+	−	−	−
Rhamnose	(d)	+	−	+	−
Salicin	d	+	−	−	−
Sucrose	−	−	+	−	+
Gas from glucose	−	−	−	+	−
β-Galactosidase	+	+	+	d	−
Ornithine decarboxylase	−	−	+	+	
Citrate utilized	−	−	−	+	−
Indole	−	−	d	−	d

a = sometimes positive in freshly isolated strains.

Description of the species of genus Yersinia

1. **Yersinia pestis** (Lehmann and Neumann) van Loghem 1944, 15. (*Bacterium pestis* Lehmann and Neumann 1896, 194; *Bacillus pestis* (Lehmann and Neumann) Migula 1900, 749; *Pasteurella pestis* (Lehmann and Neumann) Bergey *et al.* 1923, 267; *Pestisella pestis* (Lehmann and Neumann) Dorofeev 1947, 177.)

pes'tis. L. n. *pestis* plague, pestilence.

Non-motile cells, coccoid, oval or rod-shaped; involution forms common.

Growth on meat extract agar from a heavy inoculum. Grows uniformly in peptone water over a temperature range from −2 C to 40 C, optimum 27–28 C. On first isolation growth is slower and does not produce turbidity in broth. In older cultures there is pellicle formation with stalactites; sedimentation occurs later.

The biochemical characteristics are shown in Table 8.17, but most are based on the examination of very few strains; the more important distinguishing characteristics, based on a larger number of strains are shown in Table 8.16.

Y. pestis produces alkali from nitrogenous constituents more slowly than *Y. pseudotuberculosis*, and this property is useful in differentiating the two species, especially on deoxycholate citrate agar (Thal and Chen, 1955).

Causative organism of plague in man, rats, ground squirrels and other rodents; transmitted from rat to rat and from rat to man by the rat flea. Isolated from buboes, blood, sputum and lung exudate; it has also been isolated from the throats of healthy carriers. Experimentally infective for the rat and guinea pig; the mouse is more susceptible to the toxin.

Reference strain: Soemedang, NCTC 5923; ATCC 19428 (Sneath and Skerman, 1966, 88).

Further Comment

Separate envelope and somatic antigens were shown by Schütze (1932). Baker *et al.* (1947, 1952) extracted several fractions of which fraction 1 protected the bacteria from phagocytosis and specific antibodies, and was related to the R antigen complex of *Y. pseudotuberculosis*.

All strains are sensitive to the same phages as *Y. pseudotuberculosis;* none has been found to be lysogenic. Some strains produce pesticins (bacteriocins) active on both *Y. pestis* and *Y. pseudotuberculosis* (Ben Gurion and Hertman, 1958; Brubaker and Surgalla, 1962).

Ecology and geographical distribution: human plague is much less widely distributed than plague-infected rats. The organism is found in infected dead rats and in their fleas; it can persist in the soil (Mollaret, 1963) and even multiply there (Domaradskij *et al.*, 1968). For the geographical distribution of *Y. pestis* see Rodenwaldt and Jusatz (1956) and Mollaret (1969).

2. **Yersinia pseudotuberculosis** (Pfeiffer) Smith and Thal 1965, 220. (*Bacillus pseudotuberkulosis* (*sic*) Pfeiffer 1889, 5; *Pasteurella pseudotuberculosis* (Pfeiffer) Topley and Wilson 1929, 825; *Bacterium pseudotuberculosis* (Pfeiffer) Migula 1900, 374; *Shigella pseudotuberculosis* (Pfeiffer) Haupt 1935, 216.)

pseu.do.tu.ber.cu.lo'sis. Gr. adj. *pseudes* false; M.L. fem.n. *tuberculosis;* M.L. gen.n. *pseudotuberculosis* pseudotuberculosis, false tuberculosis.

Cells coccoid, oval or rod-shaped; 0.8–6.0 μm by 0.8 μm; when grown below 30 C motile by peritricate flagella; not encapsulated.

Grow easily on meat extract media; in broth the S phase produces a uniform turbidity and after several days a sediment; the R phase strains grow without producing a turbidity.

Esculin and rhamnose fermented; urea hydrolyzed. Other biochemical characters are shown in Tables 8.3, 8.16 and 8.17.

Isolated from animals with pseudotuberculosis lesions, generally in the mesenteric glands; also from the intestines and lymph nodes of healthy carriers. Also found in man. The guinea pig is highly susceptible to natural and experimental infection, which can be produced per os; the mouse is most susceptible to the toxin.

Serology: 15 somatic (O) and 5 flagellar (H) antigens have been identified and a diagnostic antigenic scheme of 10 serotypes has been produced (Thal and Knapp, 1969). In diagnostic work it is generally sufficient to identify the O group, but analysis of the H and O antigens is of epidemiological interest.

Serological cross-reactions with other organisms are common.

a. Antigen (1), which represents the R antigen complex, is related to a similar antigen complex of *Y. pestis* (Schütze, 1932) and accounts for the capacity of strains of *Y. pseudotuberculosis* to immunize against plague (Thal, 1955, 1956, 1962; Thal *et al.*, 1967).

b. With *Salmonella;* the O antigens of group II of *Y. pseudotuberculosis* and the O antigen 4 of *Salmonella* group B cross-react (Kauffmann, 1933), as do the O antigens of group IV of *Y. pseudotuberculosis* with those of O antigen 9 of *Salmonella* group D (Knapp, 1955). The polysaccharide abequose, immunologically determinant for the O antigens of *Salmonella* group B, is also present

TABLE 8.17

Characters, other than those of special value in differentiating species **of Yersinia**

	1. Y. pestis[a]	*2. Y. pseudo-tuberculosis*	*3. Y. entero-colitica*
Adonitol	−	+	−
Amygdalin	(+)[b]	−	−
Arabinose	+	+	d
Arabitol	−	−	−
Arbutin	−	+	+
Cellobiose	−	−	+
Dextrin	+	+	d
Galactose	(+)	+	+
Inulin	w[c]	−	−
Lactose	−	−	−
Melibiose	d	+	−
Sorbitol	d	−	+
Sorbose	−	−	+
Starch	w	+	d
Xylose	+	−	d
KCN (growth in)	−	−	d
d-Tartrate	+	−	−
H₂S production	+	−	−

[a] The reactions given for *Y. pestis* are based on the examination of only a small number of strains.
[b] ( ) = delayed reaction.
[c] w = weak reaction

in the O antigens of *Y. pseudotuberculosis* group II; similarly *Salmonella* group D and *Y. pseudotuberculosis* group IV have the polysaccharide tyvelose in common (Westphal *et al.*, 1960).

c. Cross-reactions occur between *Y. pseudotuberculosis* group IVA and *E. coli* O antigens 77 and 17, and *Enterobacter cloacae* (Knapp, 1968).

An exotoxin, independent of the O group, has been found, mostly in strains of group III but also in other groups. Toxins from the different O groups are immunologically identical (Thal, 1954, 1966). Some bacteriophages are active on both *Yersinia pestis* and *Y. pseudotuberculosis*, and under suitable conditions of temperature and test dilution, one such as 4FLU 1927, will distinguish between the species (Gunnison *et al.*, 1951).

Reference strain: Strain 14-1 Thal; CIP Y. pst 48; NCTC 10275.

3. Yersinia enterocolitica Frederiksen 1964, 104. (*Bacterium enterocoliticum* Schleifstein and Coleman 1943, 56.)

en'ter.o.col.it'ic.a. Gr. n. *enteron* intestine; Gr. n. *colon* the colon; Gr. suff. *-iticos* pertaining to; M.L. fem.adj. *enterocolitica* pertaining to the intestine and colon.

Cells ovoid or rod-shaped, motile at temperatures below 30 C. In broth there may be turbidity or a pellicle with a clear supernatant and a sediment.

On first isolation a smooth colony form, 0.5–2.0 mm in diameter; in subculture dissociation occurs and R forms are frequent. V.P. test negative at 37 C, sometimes positive at 22–30 C.

Biochemical characters are shown in Tables 8.16 and 8.17.

Optimal temperature for growth 30–37 C.

Y. enterocolitica is ubiquitous; it has been isolated from the feces and lymph nodes of both sick and healthy animals and man, and from the cadavers of cattle, hare, dog, guinea pig, horse, monkey and sheep. Since 1955 the species has been isolated with increasing frequency from man and animals, e.g. pigs, rabbits and chinchillas. It has also been isolated from material likely to be contaminated by feces, as milk and ice cream. It is not pathogenic for the usual laboratory animals (Knapp and Thal, 1963; Mollaret and Guillon, 1965).

Reference strain: Hässig 3/24; CIP Y.e. 160.

Further Comments

The organism that is now named *Yersinia enterocolitica* was isolated by Hässig from septicemia in man (Hässig *et al.*, 1949), but at least one of his strains (2/15) does not conform to the definition of the species.

The biochemical differentiation from other enterobacteria and the serology of the species is being developed; at present both O and H antigens have been shown. Winblad (1968) identified eight O antigens, and some strains have another O antigen related to *Brucella abortus* (Ahvonen and Jansson, 1968; Ahvonen, Jansson and Aho, 1969). The H and O antigens are common to strains from man and animals (Knapp and Thal, 1963) and are not related to those of *Y. pseudotuberculosis*.

Genus XII. **Erwinia** *Winslow, Broadhurst, Buchanan, Krumwiede, Rogers and Smith 1920, 209*

R. A. LELLIOTT*

(Includes *Pectobacterium* Waldee 1945, 469.)
Er.wi'ni.a. M.L. fem.n. *Erwinia*; named for Erwin F. Smith.

* With the assistance of D. W. Dye, D. C. Graham and W. Frederiksen.

Cells predominantly single, straight rods, 0.5–1.0 by 1.0–3.0 μm. **Motile** (one exception) by peritrichous flagella. Gram-negative.

Produce **acid from** fructose, glucose, galactose, **β-methylglucoside and sucrose,** usually from mannose and ribose, but **rarely from adonitol, dulcitol or melezitose.** Utilize acetate, fumarate, gluconate, malate and succinate but **not benzoate, oxalate or propionate, as carbon- and energy-yielding sources. Gas production comparatively weak or absent.** Fermentation end products from glucose are CO_2 and different combinations of succinate, lactate, formate and acetate; some form 2,3-butanediol and some ethanol (White and Starr, 1971). **Do not hydrolyze starch** beyond dextrins.

Do not decarboxylate glutamic acid. Decarboxylases for arginine, lysine or ornithine can be detected by Møller's method (Møller, 1955) in only a few (less than 5%) of strains of the *E. carotovora* group and in none of other species.

Rarely produce urease or lipases.

Optimum temperature for growth 27–30 C; maximum varies between 32 C and at least 40 C.

Facultatively anaerobic but anaerobic growth by some species is weak. Oxidase negative, catalase positive.

Associated with plants as pathogens, saprophytes or as constituents of the epiphytic flora. One species has also been isolated from animal and human hosts.

The G + C content of DNA ranges from 50–58 moles % (buoyant density and T_m).

Type species: *Erwinia amylovora* (Burrill) Winslow *et al.* 1920, 209.

Further Comments

Many workers have supported, and some may still support, the suggestion by Waldee (1945) that *Erwinia* should be restricted to the pathogens (*E. amylovora*, *E. salicis* and *E. tracheiphila*) that cause necrotic or wilt diseases, utilize a restricted range of carbon compounds and usually require organic nitrogen compounds for growth; and that the biochemically more active soft rotting pathogens (*E. carotovora* and *E. chrysanthemi*) should be placed in a separate genus *Pectobacterium*. The realization that there are species taxonomically intermediate between these two groups, and that there are pathogens that resemble *E. carotovora* in most of their characters but do not cause rots has seriously weakened this argument. A compromise has been adopted here by retaining one genus but recognizing three clusters of organisms within it: the (A) amylovora, (B) herbicola and (C) carotovora groups. This may be the most practical arrangement possible at present but recent evidence of taxonomically unacceptable heterogeneity within the genus, or suggestions (Starr and Mandel, 1969; White and Starr, 1971) that its members need to be shuffled with other members of the *Enterobacteriaceae* into new groupings when these can be more precisely defined, should not be ignored.

Some nomenspecies that were combined by Martinec and Kocur (1963) or Dye (1968, 1969) have been retained; this seems justifiable until more isolates of these species have been studied.

Additional characters of the species and subspecies of the genus are given in Tables 8.18, 8.19 and 8.20; small numbers of isolates of *E. salicis*, *E. tracheiphila*, *E. nigrifluens*, *E. quercina*, *E. rubrifaciens*, *E. uredovora*, *E. cypripedii* and *E. rhapontici* have been studied and data given for these species should be treated with reserve.

Description of the species of genus **Erwinia**

A. Amylovora Group

1. **Erwinia amylovora** (Burrill) Winslow, Broadhurst, Buchanan, Krumwiede, Rogers and Smith 1920, 209. (*Micrococcus amylovorus* Burrill 1882, 134; *Bacillus amylovorus* (Burrill) Trevisan 1889, 19; *Bacterium amylovorus* (Burrill) Chester 1897, 127.)

a.my.lo'vo.ra. Gr. n. *amylum* starch; L. v. *voro* devour; M.L. fem.adj. *amylovora* starch-destroying.

Colonies on 5% sucrose nutrient agar are typically white, domed, shining, mucoid (levan-type) with radial striations and a dense flocculent center or central ring after 2–3 days at 27 C. Non-levan forms are isolated rarely.

Agglutination with *E. amylovora* antiserum is the most rapid and accurate method of determination (Lelliott, 1968); the species is serologically homogeneous and has few agglutinogens in common with related species or with the saprophytes found in diseased material. Bacteriophage typing cannot always differentiate this species from *E. herbicola*.

Causes a necrotic disease (Fireblight) of most species of the *Pomoideae* and of some species in other subfamilies of the *Rosaceae*. A *forma specialis* has been described from raspberry (*Rubus idaeus*) by Starr *et al.* (1951).

The G + C content of the DNA of seven strains ranges from 53.6–54.1 moles % (buoyant density).

TABLE 8.18

Acid production from organic compounds[a] by **Erwinia** *species and their varieties[b]*

	Amylovora group						Herbicola group				Carotovora group				
	1. E. amylovora	2. E. salicis	3. E. tracheiphila[c]	4. E. nigrifluens	5. E. quercina	6. E. rubrifaciens	7a. E. herbicola var. herbicola	7b. E. herbicola var. ananas	8. E. stewartii	9. E. uredovora	10a. E. carotovora var. carotovora	10b. E. carotovora var. atroseptica	11. E. chrysanthemi[d]	12. E. cypripedii	13. E. rhapontici
Arabinose	d[e]	−	−	+	−	+	+	+	+	+	+	+	+	+	+
Mannitol	−	+	−	+	+	+	+	+	+	+	+	+	+	+	+
Salicin	−	+	−	+	+	−	d	+	−	d	+	+	+	+	+
α-Methyl glucoside	−	−	−	−	+	+	−	−	−	−	−	+	−	−	d
Xylose	−	−	−	+	−	−	+	+	+	+	+	+	+	+	−
Raffinose	−	+	−	+	−	−	d	+	+	+	+	+	d	−	+
Dulcitol	−	−	−	−	−	−	−	−	−	−	−	−	−	−	+
Inositol	−	+	−	+	−	−	−	+	−	+	d	−	−	+	+
Lactose	−	−	−	−	−	−	d	+	+	+	+	+	−	−	+
Melezitose	−	−	−	−	−	−	−	−	−	+	−	−	−	−	+
Melibiose	−	+	−	+	−	−	−	+	+	+	+	+	d	d	+
Maltose	−	−	−	−	−	−	+	+	−	−	d	+	−	d	+
Adonitol	−	−	−	−	−	−	−	−	−	−	−	−	−	−	−
Cellobiose	−	−	−	−	−	−	−	−	−	−	+	+	+	+	+
Dextrin	−	−	−	−	−	−	+	−	−	+	−	−	−	−	−
Esculin	−	+	−	+	+	−	d	d	−	d	+	+	+	+	+
Glycerol	−	d	−	+	+	d	−	+	−	+	+	+	+	+	+
Mannose	−	+	−	+	+	+	+	+	+	+	+	+	+	+	+
Rhamnose	−	−	−	+	−	−	+	d	−	−	+	+	+	+	+
Ribose	+	+	−	+	+	+	+	+	+	+	+	+	+	+	+
Sorbitol	d	+	−	+	+	+	+	+	+	+	+	+	+	+	+

[a] After 7 days growth at 27 C in unshaken aqueous solution of 1% organic compound, 1% peptone with bromcresol purple as an indicator. Data mostly from Dye (1968, 1969).

[b] Characters in boxes are those most useful for differentiation within groups. For invariant characters see generic description.

[c] *E. tracheiphila* grows very slowly in the above medium.

[d] Many strains of *E. chrysanthemi* are delayed lactose fermenters.

[e] + = 80% or more of strains positive; − = 20% or less of strains positive; d = 21–79% of strains positive.

2. **Erwinia salicis** (Day) Chester 1939, 406. (*Bacterium salicis* Day 1924, 14; *Phytomonas salicis* (Day) Magrou 1937, 408; *Pseudobacterium salicis* (Day) Krasil'nikov 1949, 226; *Erwinia amylovora* var. *salicis* Martinec and Kocur 1963, 142; (subj. syn. *Pseudomonas saliciperda* Lindeijer 1932, 23; (see Gremmen and Kam 1970, 249).)

sa'li.cis. L. n. *salix* the willow; L. gen.n. *salicis* of the willow.

Grows poorly on nutrient agar but moderately well on yeast extract-glucose-chalk agar or on glucose nutrient agar.

Colonies on 0.5% starch potato agar (pH 6.5) are yellowish in 2–3 days. A bright yellow pigment is produced on autoclaved potato tissue. Craters form round colonies on the pectate gel of Paton (1959).

Causes a vascular wilt of *Salix* spp.

The G + C content of the DNA of two strains is 51.3 and 51.5 moles % (buoyant density).

3. **Erwinia tracheiphila** (Smith) Bergey, Harrison, Breed, Hammer and Huntoon 1923, 173. (*Bacillus tracheiphilus* (sic) Smith 1895, 364; *Bacterium tracheiphilus* (Smith) Chester 1897, 72; *Erwinia amylovora* var. *tracheiphila* (Smith) Dye 1968, 605.)

tra.che.i'phi.la. L. n. *trachia* the windpipe: Gr. adj. *philus* loving; M.L. adj. *tracheiphila* trachea-

TABLE 8.19
Biochemical, physiological and cultural characters of Erwinia species and their varieties[a]

	Amylovora group						Herbicola group				Carotovora group				
	1. E. amylovora	2. E. salicis	3. E. tracheiphila	4. E. nigrifluens	5. E. quercina	6. E. rubrifaciens	7a. E. herbicola var. herbicola	7b. E. herbicola var. ananas	8. E. stewartii	9. E. uredovora	10a. E. carotovora var. carotovora	10b. E. carotovora var. atroseptica	11. E. chrysanthemi	12. E. cypripedii	13. E. rhapontici
Anaerobic growth	±[b]	±	±	+	+	+	+	+	+	+	+	+	+	+	+
Growth factors required[c]	+	+	+	−	+	−	−	−	−	−	−	−	−	−	−
H₂S from cysteine[d]	−	+	+	+	+	+	+	d	−	−	+	+	+	+	+
Urease[d]	−	−	−	+	−	−	−	−	−	−	−	−	−	−	−
Growth at 36 C	−	−	−	+	+	+	+	+	d	+	+	+	+	+	+
Gluconate oxidation[e]	−	−	+	−	−	−	d	+	d	+	−	−	+	−	−
Sucrose, reducing compounds[f]	+	+	d	−	+	−	d	+	d	+	−	+	d	−	−
Pectate degradation[g]	−	+	−	−	−	+	−	+	−	−	+	+	+	−	−
Pink diffusible pigment[h]	−	−	−	−	−	+	−	−	−	−	−	−	−	−	+
Acetoin[d]	+	+	d	+	+	−	+	+	−	+	+	+	+	−	−
Mucoid growth[i]	+	+	−	−	+	+	d	+	+	−	d	−	d	d	d
Symplasmata[j]							d	d	−	d	·	·	·	·	
Motility	+	+	+	+	+	+	+	+	−	+	+	+	+	+	+
Nitrate reduction[d]	−	−	−	−	−	−	+	−	−	+	+	+	+	+	+
DNase[k]							−	−	−	−	−	−	−	−	−
Gelatin liquefaction[d]	+	−	−	−	−	−	+	−	−	+	+	+			
Phenylalanine deaminase[l]	−	−	−	−	−	−	+	−	−	−	−	−	−	+	d
Indole[m]	−	−	−	−	−	−	−	+	−	−	−	−	+	−	−
Gas from glucose[n]	−	−	−	−	−	−	−	−	−	−	d	d	+	+	−
Casein hydrolysis[d]	−	−	−	−	−	−	−	−	−	−	+	d	d	−	−
Blue pigment[o]	−	−	−	−	−	−	−	−	−	−	−	−	+	−	−
Yellow pigment[p]	−	−	−	−	−	−	+	+	+	+	−	−	−	−	−
Cottonseed oil hydrolysis[d]	−	−	−	−	−	−	−	−	−	−	d	−	d	−	d
KCN inhibition	+	+	+	+	+	+	+	+	+	+	d	d	d	+	−
Maximum growth temperature[q]	33–34	35	32–34	38	38	38	37–40	40	35–39	37–39	37–40	35	37–40	38	34
Growth in 5% NaCl											+	+	−	+	+
Lecithinase[r]											−	+	−	+	+
Phosphatase[s]											−	−	+	+	+
Sensitivity to erythromycin (50 µg)											−	−	+	+	+

[a] Characters in boxes are those most useful for differentiation within groups. For invariant characters see generic description.

[b] ± = weak growth; + = 80% or more of strains positive; − = 20% or less of strains positive; d = 21–79% of strains positive.

[c] E. amylovora requires nicotinic acid. Other growth factor requiring species will grow in an inorganic salts medium with utilizable C source and yeast extract; their exact requirements are not known.

[d] By the methods of Dye (1968).

[e] After 4 days shake culture at 27 C in the medium of Shaw and Clarke (1955). The production of an orange or brown color (with or without precipitate) with an equal volume of Benedict's quantitative reagent after 10 min in a boiling water bath constitutes a positive reaction.

[f] After 2 days shake culture at 27 C in 4% sucrose, 1% peptone, 0.5% beef extract broth and tested and read as footnote e above.

[g] In 3 days at 27 C on Paton's medium (Paton, 1959).

[h] On 1% yeast extract, 1% D-glucose, 2% ppt. chalk agar (YDC). Pigment production by E. rhapontici is more consistent on media containing (%): sucrose, 2; peptone, 0.5; K₂HPO₄, 0.05; MgSO₄, 0.025; agar, 2; pH 7.2–7.4.

[i] On 5% sucrose nutrient agar.

[j] See Graham and Hodgkiss (1967).

[k] On DNase test agar after 2 days at 27 C (Graham and Hodgkiss, 1967).

[l] On phenylalanine agar after 2–3 days at 27 C (Anon, 1958). The reaction is weaker than that given by Proteus spp.

[m] After 2 and 5 days at 27 C in 1% tryptone, 0.1% tryptophan broth and tested with Kovacs' reagent. E. chrysanthemi probably converts tryptophan to α-methyl indole and not indole (Lelliott, 1956).

[n] In the sealed tube of Hugh and Leifson's (1953) O/F medium. E. quercina and E. rubrifaciens produce small amounts of gas (possibly from peptone) on some other media.

[o] On YDC (see footnote h above) after 5–10 days at 27 C.

[p] On nutrient agar. Non-pigmented strains of E. herbicola occur (Billing and Baker, 1963) but their frequency in relation to pigmented forms is not known.

[q] Data from Dye (1968, 1969); not tested for growth above 40 C. Range indicates strain variation.

[r] On egg yolk agar after 7 days at 27 C.

[s] As Cowan and Steel (1965) but on 0.05% sodium phenolphthalein diphosphate agar after 2 days at 27 C.

TABLE 8.20

Utilization of some organic compounds[a] as a source of carbon and energy by **Erwinia** *species and their varieties[b]*

	Formate	Malonate	Tartrate	Lactate	Citrate	Galacturonate
1. *E. amylovora*	+[c]	−	−	+	+	−
2. *E. salicis*	−	−	−	−	−	−
3. *E. tracheiphila*	d	−	−	−	d	−
4. *E. nigrifluens*	+	−	+	+	−	−
5. *E. quercina*	+	−	−	+	+	−
6. *E. rubrifaciens*	+	−	+	+	+	−
7a. *E. herbicola* var. *herbicola*	+	d	d	+	+	−
7b. *E. herbicola* var. *ananas*	+	−	+	+	+	d
8. *E. stewartii*	+	−	+	+	+	−
9. *E. uredovora*	+	−	+	+	+	−
10. *E. carotovora*	+	−	−	+	+	d
11. *E. chrysanthemi*	+	+	+	+	+	d
12. *E. cypripedii*	+	d	+	+	+	+
13. *E. rhapontici*	+	+	−	+	+	−

[a] In 21 days at 27 C in OY medium (Dye, 1968).

[b] For invariant characters see generic description.

[c] + = 80% or more of strains positive; − = 20% or less of strains negative; d = 21–79% of strains positive. Data mostly from Dye (1968, 1969).

loving, i.e. growing in the tracheae of the vascular bundles.

Grows very poorly on nutrient agar but moderately well in yeast extract-glucose-chalk agar or glucose-nutrient agar.

Causes a vascular wilt of *Cucurbita* spp.

The G + C content of the DNA of three strains ranges from 50–52 moles % (buoyant density).

4. Erwinia nigrifluens Wilson, Starr and Berger 1957, 673. (*Erwinia amylovora* var. *nigrifluens* Dye 1968, 605.)

ni.gri.flu′ens. L. adj. *niger, nigra* black; L. v. *fluo* flow; M.L. part.adj. *nigrifluens* black flowing.

Colonies on Bacto EMB agar are dark violet with a green metallic sheen.

Growth media should contain yeast extract and should be at pH 7–8.

Causes a bark necrosis of the Persian walnut (*Juglans regia*).

The G + C content of the DNA of one strain is 56.1 moles % (buoyant density).

5. Erwinia quercina Hildebrand and Schroth 1967, 253. (*Erwinia amylovora* var. *quercina* (Hildebrand and Schroth) Dye 1968, 605.)

quer.ci′na. L. n. *quercus* oak; L. suff. *-ina* belonging to; M.L..part.adj. *quercina* oak belonging.

Growth on potato-glucose-peptone-calcium carbonate (PGPC) agar is luxuriant and after 24 hrs colonies are white, circular and raised with entire margins.

Small amounts of gas are produced (possibly from peptone) in a glucose peptone medium and in PGPC.

Superficially rots onion (but not potato) slices and induces profuse lateral root development in 3–4 days on slices of carrot, turnip or beet.

Causes copious oozing of sap from acorns and, by artificial inoculation, shoot blight of *Quercus agrifolia* and *Q. wislizeni*.

The G + C content of the DNA of two strains is 54.6 and 55.1 moles % (buoyant density).

6. Erwinia rubrifaciens Wilson, Zeitoun and Fredrickson 1967, 621. (*Erwinia amylovora* var. *rubrifaciens* Dye 1968, 605.)

rub.ri.fac′i.ens. *ruber* red; L. v. *facio* make; M.L. part.adj. *rubrifaciens* red producing.

Grows poorly on nutrient agar, but well on yeast extract-glucose-chalk agar on which colonies are cream to yellow, low convex, smooth, shining with entire margins. Small craters form around colonies on the pectate gel of Paton (1959).

Causes a phloem necrosis of Persian walnut trees (*Juglans regia*).

The G + C content of the DNA of three strains ranges from 52.0–52.6 moles % (buoyant density).

B. Herbicola Group

7. Erwinia herbicola (Löhnis) Dye 1964, 268. (*Bacterium herbicola* Löhnis 1911, 470; *Bacterium herbicola* Geilinger 1921, 105; *Pseudomonas herbicola* (Geilinger) de′Rossi 1927, 369.) (Subj. syns. *Agrobacterium gypsophilae* (Brown) Starr and Weiss 1943, 316; *Bacterium typhi flavum* Smith 1948, 533 *sensu* Dresel and Stickl 1928, 517 and Cruickshank 1935, 354; *Enterobacter agglomerans* (Beijerinck) Ewing and Fife 1972, 10; *Erwinia cassavae* (Hansford) Burkholder 1948, 466; *Erwinia citrimaculans* (Doidge) Magrou 1937, 203; *Erwinia lathyri* (Manns and Taubenhaus) Magrou 1937, 209; *Erwinia mangiferae* (Doidge) Bergey *et al.* 1923, 173; *Erwinia milletiae* (Kawakami and Yoshida) Magrou 1937, 213; *Erwinia vitivora* (Baccarini) du Plessis 1940, 58; *Flavobacterium rhenanum* (Migula) Bergey and Breed 1948, 433; *Xanthomonas trifolii* (Huss) James 1955, 484.)

her.bic′o.la. L. n. *herba* grass, green plants; L. suff. *cola* dweller; M.L. n. *herbicola* grass dweller.

The yellow (YC) and non-pigmented (DC)

Erwinia-like organisms from plant sources described by Billing and Baker (1963) are included in this species.

Colonies of most strains are yellow; non-pigmented forms have been isolated and may be common.

Exists on plant surfaces and as secondary organisms in lesions caused by many plant pathogens. Some strains (syn. *Erwinia milletiae*) are said to cause galls on *Milletia japonica* and some (syn. *Agrobacterium gypsophilae*) to cause galls on *Gypsophila paniculata*. Has been isolated from water (syn. *Flavobacterium rhenanum*), the enteric tract of man (syn. *Bacterium typhi flavum*, *Enterobacter* pigmentées anaérogènes LeClerc, 1962, and see Gilardi *et al.*, 1970), from septic tonsils of man and the spleen and liver of symptomless deer (Muraschi *et al.*, 1965).

The G + C content of the DNA of 30 strains ranges from 52.6–57.7 moles % (buoyant density).

7a. *E. herbicola* var. *herbicola* Dye 1969, 223.
Characteristics as for species.

7b. *E. herbicola* var. *ananas* (Serrano) Dye 1969, 223. (*Bacillus ananas* Serrano 1928, 271; *Erwinia ananas* Serrano 1928, 271; *Bacterium ananas* (Serrano) Burgvits 1935, 44; *Pectobacterium ananas* (Serrano) Patel and Kulkarni 1951, 80.)

The description by Serrano (1928) is indistinguishable from *E. herbicola* var. *herbicola* but studies of recent isolates of this pineapple pathogen indicate that it should be regarded as a distinct variety (see Tables 8.18 and 8.19).

Causes rot of pineapple (*Ananas sativus*) fruitlets.

The G + C content of the DNA of four strains is 53.1–54.1 moles % (buoyant density).

8. **Erwinia stewartii** (Smith) Dye 1963, 504. (*Pseudomonas stewarti* Smith 1898, 422; *Bacterium stewarti* (Smith) Smith 1911, 66; *Aplanobacter stewarti* (Smith) McCulloch 1918, 440; *Phytomonas stewarti* (Smith) Bergey *et al.* 1923, 192; *Xanthomonas stewarti* (Smith) Dowson 1939, 190; *Pseudobacterium stewarti* (Smith) Krasil'nikov 1949, 240.)

stewart'i.i. M.L. gen.n. *stewartii* of Stewart; named for F. C. Stewart.

Growth slow; but better on nutrient media with a utilizable carbohydrate such as glucose or sucrose than without.

Causes a vascular wilt of corn (*Zea mays*) and some related plants, and exists in its insect vector, *Chaetocnema pulicaria*.

The G + C content of the DNA of two strains is 54.6 and 55.1 moles % (buoyant density).

9. **Erwinia uredovora** (Pon, Townsend, Wessman, Schmitt and Kingsolver) Dye 1963, 149. (*Xanthomonas uredovorus* Pon *et al.* 1954, 710.)

ur.e.dov'or.a. L. n. *uredo* blight; L. v. *voro* devour; M.L. n. *uredovora* blight devourer (eats uredospores and uredia).

Attacks uredia of *Puccinia graminis* and can exist in soil.

The G + C content of the DNA of five strains ranges from 53.0–54.4 moles % (buoyant density and T_m).

C. Carotovora Group

10. **Erwinia carotovora** (Jones) Bergey, Harrison, Breed, Hammer and Huntoon 1923, 171. (*Bacillus carotovorus* Jones 1901, 12; *Bacterium carotovorum* (Jones) Lehmann and Neumann 1927, 446; *Pectobacterium carotovorum* (Jones) Waldee 1945, 469.)

ca.ro.tov'or.a. L. n. *carota* carrot. L. v. *voro* devour. M.L. n. *carotovora* carrot devourer.

Gas production from carbohydrates is erratic; some strains (syn. *E. aroideae*) are anaerogenic when isolated, others produce moderate or small amounts and often become anaerogenic after prolonged artificial culture.

10a. *E. carotovora* var. *carotovora* Dye 1969, 81. (Subj. syns. *Bacillus apiovorus* Wormald 1914, 217; *Bacillus cepivorus* Delacroix 1906, 368; *Bacillus oleraceae* Harrison 1904, 46; *Bacillus omnivorus* van Hall 1902, 123; *Erwinia aroideae* (Townsend) Bergey *et al.* 1923, 171; *Erwinia betivora* (Takimoto) Magrou 1937, 200; *Erwinia croci* (Mizusawa) Magrou 1937, 204; *Erwinia cytolytica* Chester 1938, 431; *Erwinia dahliae* (Hori and Bokura) Magrou 1937, 205; *Erwinia melonis* (Giddings) Bergey *et al.* 1923, 170; *Erwinia papaveris* (Ayyar) Magrou 1937, 214; *Erwinia solanisapra* (Harrison) Bergey *et al.* 1923, 170; *Pectobacterium delphinii* Waldee 1945, 471.)

Causes rotting, particularly of storage tissue, of a wide range of plants.

The G + C content of the DNA of 11 strains ranges from 50.5–53.1 moles % (buoyant density).

10b. *E. carotovora* var. *atroseptica* (Hellmers and Dowson) Dye 1969, 81. (*Bacillus atrosepticus* van Hall 1902, 134; *Erwinia atroseptica* (van Hall) Jennison 1923, 43; *Bacterium atrosepticum* (van Hall) Lehmann and Neumann 1927, 446; *Pectobacterium atrosepticum* (van Hall) Patel and Kulkarni 1951, 80; *Bacterium carotovorum* (Jones) var. *atrosepticum* Hellmers and Dowson 1953, 110; *Pectobacterium carotovorum* var. *atrosepticum* Hellmers and Dowson 1957, 171.) (Subj. syns. *Bacillus melanogenes* Pethybridge and Murphy 1911, 31; *Erwinia phytophthora* (Appel) Magrou 1937, 215.)

Causes a vascular and parenchymatal disease (blackleg) of potato (*Solanum tuberosum*) plants and a storage rot of potato tubers.

The G + C content of the DNA of two strains is 51.3 and 53.1 moles % (buoyant density).

11. Erwinia chrysanthemi Burkholder, McFadden and Dimock 1953, 526. (*Erwinia carotovora* var. *chrysanthemi* (Burkholder, McFadden and Dimock) Dye 1969, 93; *Pectobacterium carotovorum* f. sp. *chrysanthemi* Dowson 1957, 178; *Pectobacterium carotovorum* var. *chrysanthemi* (Burkholder, McFadden and Dimock) Graham and Dowson 1960, 56; *Pectobacterium parthenii* Hellmers 1958, 128; *Pectobacterium parthenii* var. *chrysanthemi* (Burkholder, McFadden and Dimock) Hellmers 1958, 129; *Pectobacterium parthenii* var. *dianthicola* Hellmers 1958, 129.) (Subj. syns. *Erwinia carotovora* f. sp. *parthenii* Starr 1947, 299; *Erwinia carotovora* var. *zeae* Sabet 1954, 66; *Erwinia dieffenbachiae* McFadden 1961, 663; *Pectobacterium carotovorum* var. *graminarum* Dowson and Hayward 1960, 275.)

chrys.an'the.mi. M.L. n. *Chrysanthemum* generic name; M.L. gen.n. *chrysanthemi* of chrysanthemums.

Colonies on potato-glucose-agar (pH 6.5) are characteristically umbonate with undulate to coralloid margins ('fried egg') at 3–6 days growth.

Causes vascular wilts or parenchymatal necroses of *Begonia bertini*, *Chrysanthemum morifolium*, *Dahlia* spp., *Dianthus caryophyllus*, *Parthenium argentatum*, *Philodendron* spp., *Dieffenbachia* spp., *Zea mays*. There is evidence of differentiation into pathotypes; the 'carnation' strain can infect chrysanthemum but isolates from other hosts are not known to infect carnation; and chrysanthemum and parthenium isolates are only weakly pathogenic to philodendron.

The G + C content of the DNA of six strains is 55.1–57.1 moles % (buoyant density).

12. Erwinia cypripedii (Hori) Bergey, Harrison, Breed, Hammer and Huntoon 1923, 171. (*Bacillus cypripedii* Hori 1911, 91; *Erwinia carotovora* var. *cypripedii* (Hori) Dye 1969, 93.)

cyp.ri.ped'i.i. M.L. n. *Cypripedium* generic name; M.L. gen.n. *cypripedii* of cypripedium orchids.

Causes a brown rot of cypripedium orchids (*Cypripedium* spp.).

The G + C content of the DNA of two strains is 54.1 and 54.6 moles % (buoyant density).

13. Erwinia rhapontici (Millard) Burkholder 1948, 475. (*Phytomonas rhapontica* (sic) Millard 1924, 11; *Bacterium rhaponticum* Millard 1924, 11; *Aplanobacter rhaponticum* (Millard) White 1936, 42; *Xanthomonas rhapontica* (Millard) Savulescu 1947, 13; *Pectobacterium rhapontici* (Millard) Patel and Kulkarni 1951, 80; *Pseudobacterium rhapontici*

(Millard) Krasil'nikov 1949, 241; *Erwinia carotovora* var. *rhapontici* (Millard) Dye 1969, 93.)

rha.pon'ti.ci. M.L. n. *rhaponticum* specific epithet; M.L. gen.n. *rhapontici* of *rhaponticum*, of rhubarb.

Rots onion and cucumber slices; potato slices are rotted slightly or not at all.

Causes a crown rot of rhubarb (*Rheum rhaponticum*).

The G + C content of the DNA of three strains ranges from 51.0–53.1 moles % (buoyant density).

Addendum I. The taxonomic position of the following species is doubtful; most are excluded from the genus *Erwinia* as described here.

a. *Erwinia cancerogena* Urosevic 1966, 500.

Causes a canker disease of poplar (*Populus* spp.).

This species is excluded from the genus by its positive arginine and ornithine decarboxylase reactions. It is probably a species of *Enterobacter*.

b. *Erwinia carnegieana* Standring 1942, 310.

In the original description *E. carnegieana* is described as, *inter alia*, a Gram-positive organism which does not ferment lactose, produces a necrotic disease of *Carnegiea gigantea* and does not attack *Opuntia* spp. or rot carrots. Later, Boyle (1949) with other isolates showed that the Gram reaction became nearly negative with continued culture, confirmed the lactose reaction and showed that they were not agglutinated by *E. carotovora* antiserum. Burkholder (1957) emended the description to, *inter alia*, Gram-negative with Gram-positive granules in the cells of old cultures and lactose-positive. Alcorn (1961) obtained isolates from *C. gigantea* and *Opuntia* spp. that would cross-infect.

A co-type (NCPPB 439) is Gram-negative, lactose-positive, does not rot carrot or liquefy pectate gel and produces lysine decarboxylase; two of Alcorn's isolates (NCPPB 671 and 672) are typical of *E. carotovora* (Lelliott and Graham, unpublished). There may therefore be two pathogens of *C. gigantea*: *E. carnegieana* and *E. carotovora*, both of which cause a similar disease.

c. *Erwinia dissolvens* (Rosen) Burkholder 1948, 472. (*Pseudomonas dissolvens* Rosen 1922, 497; *Bacterium dissolvens* Rosen 1922, 499; *Phytomonas dissolvens* (Rosen) Rosen 1926, 264; *Aplanobacter dissolvens* (Rosen) Rosen 1926, 264; *Aerobacter dissolvens* (Rosen) Waldee 1945, 473.)

Isolated from rotting corn stalks (*Zea mays*).

Taxonomically nearer to *Klebsiella* than *Erwinia;* non-motile, produces large amounts of gas from many carbohydrates and decarboxylates arginine and/or lysine (Dye, 1969).

d. *Erwinia nimipressuralis* Carter 1945, 423.

Isolated from wet wood of elms (*Ulmus* spp.) but its pathogenicity is doubtful.

Probably an *Enterobacter* sp.; it is excluded from *Erwinia* because co-types produce large amounts of gas from many carbohydrates (including lactose), decarboxylate arginine and produce lipase.

e. *Erwinia proteamaculans* (Paine and Stansfield) Dye 1966, 850. (*Pseudomonas proteamaculans* Paine and Stansfield 1919, 33; *Phytomonas proteamaculans* (Paine and Stansfield) Bergey *et al.* 1930, 247; *Bacterium proteamaculans* (Paine and Stansfield) Elliott 1930, 186; *Enterobacter proteamaculans* (Paine and Stansfield) Dye 1969, 838.)

Said to cause a leaf spot of *Protea cynaroides*.

Possibly an *Enterobacter* sp.; excluded from *Erwinia* because the only known extant co-type produces lipase, and lysine and ornithine decarboxylases.

f. *Erwinia salmonis* Brisou, Tysset and Vacher 1959, 244.

Isolated from heart blood of common trout (*Salmo fario*) and causes fatal septicemia experimentally in fish and some mammals.

No direct comparison with other *Erwinia* spp. has been reported. Differs from other *Erwinia* spp. in failing to produce acid from sucrose, in hydrolyzing starch and utilizing benzoate.

g. *Erwinia sinocalami* Lo, Chen and Huang 1966, 20.

Description summarized from Lo *et al.* (1966).

Single straight rods, 0.7–0.9 by 1.5–1.8 μm, motile by two to eight peritrichous flagella. Non-spore-forming, capsulated.

Gram-negative.

Produces acid from arabinose, rhamnose, xylose, glucose, fructose, galactose, lactose, sucrose, maltose and salicin within 20 days in inorganic basal medium with bromcresol purple as an indicator. No acid produced from rhamnose, starch and inulin. Variable results from dextrin, glycerol, mannitol and sorbitol. No gas produced.

Starch not hydrolyzed. Nitrate not reduced to nitrite. Gelatin liquefied. H_2S not produced. Indole production small (only detected after 20 days). Litmus milk reduced. Not pigmented.

Aerobic. Optimum temperature for growth 32 C, maximum 44 C.

Causes a shoot blight and culm-spot of the bamboos *Sinocalamus latiflorus* and *Leleba oldhami.*

No direct comparison with other *Erwinia* spp. has been reported but it is probably a member of the high temperature section of the amylovora group (q.v.). Although it is described as aerobic

the method used to test this character does not preclude the weak anaerobic growth characteristic of some of the amylovora group.

Addendum II. The description of the following species is such that they are unlikely to be recognized if isolated again: *Erwinia araliavora* (Uyeda) Magrou, 1937, 197; *E. asteracearum* (Pavarino) Magrou 1937, 199; *E. bussei* (Migula) Magrou 1937, 200; *E. cacticida* (Johnston and Hitchcock) Magrou 1937, 201; *E. edgeworthiae* (Hori and Bokura) Magrou 1937, 206; *E. erivanensis* (Kalantarian) Bergey *et al.* 1930, 239; *E. flavida* (Fawcett) Magrou 1937, 207; *E. ixiae* (Severini) Magrou 1937, 208; *E. lilii* (Uyeda) Magrou 1937, 210; *E. nelliae* (Welles) Magrou 1937, 213; *E. nicotianae* (Uyeda) Bergey *et al.* 1923, 172; *E. papayae* (Rant) Magrou 1937, 214; *E. sacchari* Roldan 1931, 256; *E. scabiegena* (von Faber) Magrou 1937, 217; *E. serbinowi* (Potebnia) Magrou 1937, 217; *E. uvae* (Kruse) Magrou 1937, 220.

Addenda to the family **Enterobacteriaceae**

Addendum I: The generic name "Aerobacter" Beijerinck 1900, 193 has been rejected (Judicial Commission Opinion 46). It is agreed by all the authors who have contributed sections on *Enterobacteriaceae* to this edition of THE MANUAL that the name *Aerobacter* was and may remain a source of confusion, and that most bacteria named *Aerobacter aerogenes* are mislabeled. The wide use and misuse of the taxon *Aerobacter* in the past and the need for explanation has prompted this inclusion in an addendum to the family as a matter for historians rather than for practical bacteriologists.

Beijerinck, in his original paper, stated that the genus *Aerobacter* consisted of motile and nonmotile organisms, some peritrichously and others polarly flagellated, *Aerobacter aerogenes* was said to be peritrichously flagellated but might occur as a non-motile organism; it grew best at about 30 C, and only feebly or not at all at 37 C. This statement on the temperature limit for growth has been overlooked by most workers, and many strains misidentified and mislabeled as *Aerobacter aerogenes* grow readily at 37 C and at higher temperatures. If *Aerobacter* is recognizable at all it is in what Grimes (1961) describes as the cold-tolerant mesophilic *Aerobacter* types. The organism should probably be renamed to avoid even further confusion but its main characters, including those most useful for distinguishing the group from *Enterobacter* and *Klebsiella*, are repeated here in order to provide a place for valid organisms otherwise dislocated by the decision.

Aerobacter Beijerinck, emend. Grimes 1961, 12.

Motile rods; M.R. negative; V.P. positive. Uric acid and citrates can be used as sole C sources.

Grows in KCN media. Carbohydrates are fermented with the production of acid and gas.

Arginine dihydrolase positive; lysine and ornithine decarboxylases present.

Optimal temperature 20–30 C; good growth at 15 C; poor or no growth at 37 C.

Isolated from soil and plant grains.

Addendum II. *Genera that were nomenclaturally superfluous when published (Rule 24a).*

1. **Colobactrum** Borman, Stuart and Wheeler 1944, 357.

2. **Paracolobactrum** Borman, Stuart and Wheeler 1944, 361.

These names for lactose-fermenting (*Colobactrum*) and late or non-lactose-fermenting (*Paracolobactrum*) coliform bacteria were proposed because the authors thought that lactose fermentation was of overriding importance, and that the term "paracolon" was worthy of generic recognition.

Research on β-galactosidase has shown that a differentiation based on the rapidity of lactose fermentation is not justified. Organisms that were named *Colobactrum* or *Paracolobactrum* should have been placed in genera such as *Escherichia*, *Citrobacter* or *Klebsiella*, and for this reason the Ballerup-Bethesda group are placed in the genus *Citrobacter*.

FAMILY II. **VIBRIONACEAE** VÉRON 1965, 5245

Rigid Gram-negative rods, straight or curved; usually motile by polar flagella but some cells may have, in addition, lateral flagella produced under certain growth conditions.

Chemoorganotrophs, metabolism both fermentative and respiratory. Oxidase positive. Several species produce butylene glycol from glucose, some are proteolytic, and some produce indole.

Facultative anaerobes without exacting nutritional requirements.

Usually found in fresh or sea water, occasionally in fish or man.

The G + C content of the DNA ranges from 39 to about 63 moles %.

Type genus: *Vibrio* Pacini 1854, 411.

Genus I. **Vibrio** *Pacini 1854, 411*

J. M. SHEWAN AND M. VÉRON

(*Pacinia* Trevisan 1885, 83; *Microspira* Schroeter 1886, 168.)

Vib′ri.o. L. v. *vibro* move rapidly to and fro, vibrate; M.L. masc. n. *Vibrio* that which vibrates.

Short asporogenous **rods, axis curved or straight,** 0.5 by 1.5–3.0 μm, single or occasionally united into S shapes or spirals. **Motile by a single polar flagellum,** or, in some species, two or more flagella in one polar tuft; very occasionally non-motile. In some species the flagellum has a central core with an outer sheath (visible in electron microscope preparations). Spheroplasts frequently present, usually formed in adverse environmental conditions. **Gram-negative.** Not acid-fast. No capsules. Grow well and rapidly on standard nutrient media.

Chemoorganotrophs, **metabolism is both respiratory** (oxygen is utilized) **and fermentative.** Metabolism of carbohydrates is fermentative with mixed products but **no CO₂ or H₂. Oxidase-positive.** Non-pigmented or yellow. Generally able to grow on simple mineral ammonium media with a simple carbon source; glutamate and suc-

cinate are oxidizable substrates, probably universal within the genus, but the range of substrates utilized is relatively limited. Frequently V.P. positive. Nitrites usually formed from nitrates. Acid but no gas formed from glucose. Urease negative.

Facultatively anaerobic. Temperature optima range from 18–37 C. pH range 6.0–9.0. Optimum NaCl requirement usually 3.0%, some strains fail to grow in the absence of sodium chloride. Usually **sensitive to 2,4-diamino-6,7-diisopropyl pteridine** (0/129) and novobiocin.

The G + C content of the DNA (of those species examined) ranges from 40–50 moles %.

Found in fresh and salt water, and in the alimentary canal of man and animals; some species are pathogenic for man and other vertebrates (fish).

Type species: *Vibrio cholerae* Pacini 1854, 411.

TABLE 8.21

Features which distinguish the genus **Vibrio** *from related genera*

	Vibrio	Aeromonas	Photobacterium	Lucibacterium	Pseudomonas
Flagella	Polar	Polar	Polar	Usually peri-trichous	Polar
Oxidase +	+	+	− or +	+	+ (Occasionally −)
O/129 sensitivity	+	−	+	−	−
Luminescence	d	−	+	+	−
Carbohydrate metabo-lism (O/F medium)	Ferment-ative	Ferment-ative	Ferment-ative	Fermentative	Respiratory or not metabolized
Gas production	−	d	+	−	
Diffusible pigment	−	− or brown	−	−	Green fluorescent or − (occasionally red, brown or orange pigments produced)

Further Comments

It has been suggested (Hendrie *et al.*, 1971) that the genus should include the C27 organisms (Ferguson and Henderson, 1947), previously assigned to *Aeromonas* (Ewing *et al.*, 1961), *Plesiomonas* (Habs and Schubert, 1962; Eddy and Carpenter, 1964) and *Fergusonia* (Sebald and Véron, 1963).

Heiberg (1934) described a biochemical classification of vibrios based on the fermentation of mannose, sucrose and arabinose. Almost all cholera vibrios belonging to serogroup O:1 can be classified into Heiberg's group I (mannose and sucrose fermented, arabinose not fermented). The non-cholera vibrios (NCV) belong either to his group I or to group II, which differs from group I by the non-fermentation of mannose. The Heiberg groups III to VI are very seldom found in the true vibrios.

DNA hybridization tests, performed on DNA agar, between different species and strains show that there is a high percentage binding between some species, but also a wide variation in percentage binding between other species and strains within species (Hanaoka, Kato and Amano, 1969).

The microaerophilic species are now excluded and transferred to the genus *Campylobacter* (Sebald and Véron, 1963).

Strictly anaerobic strains of a *Vibrio* sp. are not known.

Features distinguishing the genus from other genera with polar flagellated, oxidase positive, Gram-negative rods are given in Table 8.21.

Description of the species of genus **Vibrio**

The characters which distinguish the species are given in Table 8.22, and further information on the biotypes of *V. cholerae* is in Table 8.23.

1. **Vibrio cholerae** Pacini 1854, 411. (*Vibrio cholera* Pacini 1854, 411; *Kommabacillus* Koch 1884, 479; *Bacillo virgola del* Koch Trevisan 1884, 373; *Bacillus cholerae* (Pacini) Trevisan 1884, 374; *Bacillus cholerae-asiaticae* Trevisan 1884, according to Trevisan 1885, 84; *Pacinia cholerae-asiaticae* (Trevisan) Trevisan 1885, 84; *Spirillum cholerae-asiaticae* (Trevisan) Zopf 1885, 69; *Microspira comma* Schroeter 1886, 168; *Spirillum cholerae* (Pacini) Macé 1889, 606; *Vibrio cholerae-asiaticae* (Trevisan) Pfeiffer 1896, 527; *Vibrio comma* (Schroeter) Blanchard 1906, 1; *Liquidovibrio cholerae* (Pacini) Orla-Jensen 1909, 333.)

chol'er.ae. Gr. n. *cholera* the cholera, an intestinal disease; M.L. gen.n. *cholerae* of cholera.

On first isolation from patients colonies are smooth, glistening and transparent and by oblique light show a typical greenish to red-bronze iridescence and fine granular transparency (Lankford, 1959). Colonial variants are either opaque and corrugated (rugose variants) or rough. The rugose variant is seen especially in old cultures.

On blood agar most strains show a zone of hemolysis (hemodigestion); except with the *cholerae* biotype and some *eltor* biotype strains, a soluble hemolysin can be shown in the test tube with washed sheep or goat red cells by the technique of Feeley and Pittman (1963). Hemolysin is differentiated from hemodigestion by anaerobic incuba-

TABLE 8.22

Characteristics of the species of genus **Vibrio**

	Vibrio cholerae	Vibrio para-haemolyticus	Vibrio anguillarum	Vibrio fischeri	Vibrio costicola
Flagella					
1 polar	+[a]	+	+	d	+
lophotrichous	−	−	−	d	−
Indole	+	+	+	+	−
Methyl red	d	d	d	+	d
Voges-Proskauer	d	d	d	−	−
Citrate utilization	d	d	−	−	−
O/129 sensitivity	+	(+)	+	+	d
Novobiocin sensitivity	+	(+)	+	+	+
Gelatin liquefaction	+	+	+	+	d
Casein hydrolysis	+	+	+	+	−
Tween 80 hydrolysis	+	+	d	+	d
Lecithinase (egg yolk)	+	+			d
Hydrogen sulfide	−	−	−	−	−
Acid from arabinose	−	d	−	−	−
Acid from mannose	d	+	+	+	+
Acid from sucrose	+	d	+	+	+
Acid from mannitol	+	+	+	d	d
Acid from inositol	−	−	−	d	−
Acid from salicin	−	−	−	d	−
Hydrolysis of starch	+	+	+	d	
Growth without added NaCl	+	−	+	−	−
Growth in 7.0% NaCl	d	+	d	+	+
Growth in 10.0% NaCl	−	d	−	−	+
Growth at 5 C	−	−	+	+	+
Growth at 37 C	+	+	d	−	d
Growth at 42 C	d		−		−
Møller's media:					
Arginine–alkaline reaction	−	−	+	−	+
Lysine decarboxylase	+	+	−	+	−
Ornithine decarboxylase	+	+	−	−	−
Hemolysis: horse blood agar	d	d	d	−	−
Luminescence	−[b]	−	−	d	−
Pathogenicity	+	+	+	−	−
DNA base ratio–moles % G + C	45–49	42–45	44–45	40–46	50

[a] + positive; − negative; (+) weak positive; d more than 10% of strains positive or negative.

[b] *Vibrio cholerae* biotype *albensis* is luminous, but only one strain is known and has been examined.

tion (de Moor, 1963) when only hemolysis can occur.

In addition to the carbohydrate reactions given in Tables 8.22 and 8.23, fructose, galactose, maltose and trehalose are fermented without gas. Lactose is usually fermented with a delayed reaction. Xylose, rhamnose, melibiose, raffinose, melezitose, adonitol, dulcitol and esculin are not fermented.

Glucose, fructose, galactose; acetate, lactate, pyruvate, citrate, malate, succinate; glycerol, mannitol; glutamate can be utilized as the sole source of carbon by most strains.

Catalase positive. Lecithinase positive.

More than 90% of strains are sensitive to 0/129 vibriostatic compound (2,4-diamino-6,7-diisopropyl pteridine) (Shewan *et al.*, 1954) and to novobiocin (disc with 10 μg). Growth does not occur in KCN broth, or at pH 5 and 11, but growth does occur at pH 10.

The majority of strains may be classified into six antigenic O groups. The cholera vibrios (biotypes *cholerae* and *eltor*) belong to the O:1 group; strains belonging to other groups are usually called 'non-cholera vibrios.'

Three antigenic factors, A, B and C, have been

observed in strains belonging to serogroup O:1, which in combination give three serotypes.

Serotypes	O Factors	Agglutination by specific sera	
		Anti-B	Anti-C
Ogawa	AB	+	−
Inaba	AC	−	+
Hikojima	ABC	+	+

A fourth serotype, having only the A factor, has been suggested.

Found in the intestinal contents of man (in health and disease) and of other animals. Also found in water. While classical cholera is always associated with strains of serogroup O:1, strains of other groups may produce in man and animals illnesses similar to the milder forms of cholera (McIntyre *et al.*, 1965: Aldova *et al.*, 1968).

Cholera vibrios are pathogenic experimentally to guinea pigs, infant rabbits and for adult rabbits by the ligated ileal loop test of De and Chatterjee (1953), and also for adult dogs (Sack and Carpenter, 1969).

The average G + C content of the DNA of *V.*

cholerae is 47 moles % and the majority of strains have values between 45 and 49 moles %, whatever their serogroup.

1a. *Vibrio cholerae* biotype *cholerae*.

This biotype belongs to any one serotype of serogroup O:1.

Tube hemolysis test negative. Chicken erythrocyte agglutination (slide test of Finkelstein and Mukerjee, 1963) negative or delayed. Voges-Proskauer test sometimes positive at 37 C, but negative or weakly positive at 22 C. Sensitive to 50 I.U. polymyxin B (disc test, Gan and Tjia, 1963). Sensitive to group IV cholera phages of Mukerjee (1963).

Found in intestinal tract of human beings and animals, suffering from cholera or simple diarrhea only and in healthy carriers. Also found in water and occasionally in foods.

Neotype strain: Ogawa strain 3; NCTC 8021; ATCC 14035, serogroup O:1, serotype Ogawa.

1b. *Vibrio cholerae* biotype *eltor* Pribram 1919, 121. (El Tor vibrio, Gotschlich 1906, 281; *Vibrio El Tor* Pribram 1919, 121; *Vibrio El Tor* Heim 1911, 384.)

el.tor′. Named for the El Tor lazaret (Sinai) in

TABLE 8.23

Characters that distinguish the biotypes of **Vibrio cholerae**

	cholerae	eltor	proteus	albensis
Number of flagella	1	1	1	1 or more
Hemolysis (in tube)	−	d	+	−
Fermentation of arabinose	−	−	−	−
Fermentation of mannose	+	+	+	−
Fermentation of sucrose	+	+	+	+
V.P. at 22 C	−	+	d	+
V.P. at 37 C	d	+	d	+
M.R. at 22 C	+	d	d	−
M.R. at 37 C	+	+	d	d
Nitrates reduced to nitrites	+	+	−	+
Indole produced at 22 C	+	+	+	(+)[a]
Indole produced at 37 C	d	d	(+)	−
Cholera red reaction	+	+	−	−
Lysine decarboxylated	+	+	−	+
Growth at 5 C	−	−	+	−
Growth at 37 C	+	+	+	−
Growth at 42 C	d	d	+	+
Growth in NaCl 0%	+	+	−	−
Growth in NaCl 4%	+	+	−	+
Growth in NaCl 7%	d	d	+	+
Growth in NaCl 8%	−	−	(+)	−
Growth at pH 10	+	+	+	−
Growth at pH 11	d	d	+	+
Agglutinated by O:1 antiserum	+	+	−	−
Luminescence	−	−	−	(+)

[a] (+) = delayed.

which the organism was first isolated from healthy pilgrims returning from Mecca.

This biotype also belongs to any one serotype of serogroup O:1.

The morphology of colonies is similar to that of *cholerae* biotype.

Tube hemolysin test variable; until 1962 most *eltor* strains were hemolytic when tested soon after isolation; recent isolations are non-hemolytic or become hemolytic only after several subcultures.

Chicken erythrocyte agglutination test positive. Voges-Proskauer test positive at 37 and 22 C. Not sensitive to 50 I.U. polymyxin B. Not susceptible to group II and group IV cholera phages of Mukerjee (1963) at suitable dilutions.

Same habitat as *cholerae* biotype. Can survive long periods in foods.

Reference strain: NCTC 8457; ATCC 14033, serogroup O:1, serotype Inaba.

1c. *Vibrio cholerae* biotype *proteus* (Buchner) *comb. nov.* (Kommabacillus der cholera nostras, Finkler and Prior 1884, 632; *Vibrio proteus* Buchner 1885, 10 (not *Vibrio proteus* Müller 1773, 45); *Pacinia finkleri* Trevisan 1885, 84; *Microspira finkleri* (Trevisan) Schroeter 1886, 169; *Spirillum finkleri* (Trevisan) Crookshank 1887, 258; *Vibrio metschnikovii* Gamaléia 1888, 485; *Pacinia metschnikoffi* (Gamaléia) Trevisan 1889, 23; *Spirillum metschnikovi* (Gamaléia) Sternberg 1892, 511; *Microspira metschnikofii* (Gamaléia) Migula 1895, 33; *Microspira protea* (Buchner) Chester 1901, 338; *Vibrio nordhafen* Bergey, Haynes and Hitchins 1948, 196.)

pro'te.us. Gr. n. *Proteus*, a sea god who could change his form; M.L. masc.n. *proteus* in reference to changeable form of organism.

Pleomorphic forms observed very frequently.

Some strains are oxidase negative.

Although formerly described as indole negative, indole is produced at 30 C in 1 day and at 37 C in 2 days in 1% peptone broth with 0.5% NaCl.

Milk is acidified after some days; coagulation and peptonization may occur late, if at all.

No growth in peptone broth without NaCl, growth abundant in 7% NaCl, slight in 8% NaCl.

Found in intestinal tract or feces of man and other animals (birds). Causes gastroenteritis ("Cholera nostras").

Proposed reference strain: NCTC 8563.

1d. *Vibrio cholerae* biotype *albensis* (Lehmann and Neumann) *comb. nov.* (Vibrionenart aus dem Elbestrom, Dunbar 1893, 799; *Vibrio albensis* Lehmann and Neumann 1896, 340; *Microspira albensis* (Lehmann and Neumann) Mez 1898, 65; *Microspira dunbari* Migula 1900, 1013; *Photospirillum dunbari* (Migula) Miquel and Cambier 1902, 886; *Vibrio phosphorescens* Matzuschita 1902, 108; *Spirillum phosphorescens* (Matzuschita) Macé 1913, 656; *Photobacterium dunbari* (Migula) Ford 1927, 621.)

al.ben'sis. M.L. adj. *albensis*, pertaining to the (river) Elbe.

Flagellation usually monotrichous but occasionally lophotrichous.

Luminescent on media containing up to 1.2% NaCl (Warren, 1945), but this property may be lost after repeated subculturing.

Found in fresh water, in human feces and in bile. Originally described as pathogenic to guinea pigs and pigeons.

Proposed reference strain: NCMB 41; ATCC 14547. This strain is actually not pathogenic for guinea pigs or pigeons.

2. **Vibrio parahaemolyticus** (Fujino *et al.*) Sakazaki, Iwanami and Fukumi 1963, 181. (*Pasteurella parahaemolytica* Fujino, Okuno, Nakada, Aoyama, Fukai, Mukai and Ueho 1951, 11; *Pseudomonas enteritis* Takikawa 1958, 322; *Oceanomonas enteritidis* (Takikawa) Miyamoto, Nakamura and Takizawa 1961, 480; *Oceanomonas parahaemolytica* (Fujino *et al.*) Miyamoto *et al.* 1961, 481; *Oceanomonas alginolytica* Miyamoto *et al.* 1961, 481; *Vibrio alginolyticus* (Miyamoto *et al.*) Sakazaki 1968, 360; *Beneckea alginolytica* (Miyamoto *et al.*) Baumann, Baumann and Mandel 1971, 289; *Beneckea parahaemolytica* (Fujino *et al.*) Bauman *et al.* 1971, 291.)

para.hae.mo.ly'ti.cus. Gr. prep. *para* by the side of, beside; Gr. n. *haema* blood; M.L. adj. *haemolytica* a specific epithet M.L. adj. *parahaemolyticus* similar to (*Pasteurella*) *haemolytica*.

The two biotypes of *V. parahaemolyticus* can be distinguished as follows.

	Biotype 1 (parahae-molyticus)	Biotype 2 (algino-lyticus)
Growth in 10% NaCl	−	+
V.P.	−	+
Methyl red	+	−
Acid from arabinose	d	−
Acid from sucrose	−	+

Habitat: The marine environment, sea foods and feces of patients with acute enteritis.

Suggested working types: ATCC 17802; ATCC 17803; ATCC 17749; ATCC 17750.

3. **Vibrio anguillarum** Bergman 1909, 28. (*Vibrio piscium* David 1927, 55; *Achromobacter ichthyodermis* Wells and ZoBell 1934, 123; *Pseudomonas ichthyodermis* (Wells and ZoBell) ZoBell and Upham 1944, 246; *Vibrio piscium* var. *japonicus* Hoshina 1957, 66; *Vibrio ichthyodermis* (Wells

and ZoBell) Shewan, Hobbs and Hodgkiss 1960, 384.)

an.guil.lar'um. L. n. *anguilla* eel; L. gen.pl.n. *anguillarum* of eels.

Cells usually have a single polar flagellum, but in the original description of *Achromobacter ichthyodermis* lophotrichous flagella were indicated.

Strain to strain variations occur in the results of the methyl red and Voges-Proskauer tests; salt and temperature tolerance; hemolysis of horse blood agar.

Habitat: water, usually isolated from diseased conditions in fish, both marine and fresh water.

Suggested working strains: NCMB 6, ATCC 19264; NCMB 407; NCMB 571; NCMB 829, ATCC 14181 (Hendrie *et al.*, 1971).

4. Vibrio fischeri (Beijerinck) Lehmann and Neumann 1896, 342. (Einheimischer Leuchtbacillus Fischer 1888, 107; *Photobacterium fischeri* Beijerinck 1889, 402; *Bacillus fischeri* (Beijerinck) Trevisan 1889, 18; *Bacillus phosphorescens indigenus* Eisenberg 1891, 124; *Spirillum marinum* Russell 1892, 198; *Bacterium phosphorescens indigenus* (Eisenberg) Chester 1897, 121; *Microspira marina* (Russell) Migula 1900, 1002; *Microspira*

fischeri (Beijerinck) Chester 1901, 333; *Vibrio marinus* (Russell) Ford 1927, 347; *Achromobacter fischeri* (Beijerinck) Bergey *et al.* 1930, 220; *Vibrio noctiluca* Weisglass and Skreb 1963, 9.)

fisch'er.i. M.L. gen.n. *fischeri* of Fischer; named for Bernhard Fischer, one of the earliest students of luminescent bacteria.

Flagellation is usually lophotrichous, but some strains are monotrichous.

Variable results between strains are found in production of acid from sucrose and salicin, hydrolysis of starch and luminescence.

Luminescence is not a constant feature, and the ability to emit light may be irretrievably lost during prolonged maintenance in artificial culture.

Habitat: sea water and marine animals.

Neotype: ATCC 7744, NCMB 1281 (Hendrie *et al.*, 1971).

5. Vibrio costicola Smith 1938, 29.

cost.ti'co.la. L. n. *costa* rib; L. subst. *cola* dweller; M.L. n. *costicola* rib dweller (from bacon).

This species tolerates salt concentrations from 2–23% with the optimum concentration 6–12%.

Habitat: cured meats and brines.

Type strain: NCMB 701.

Genus incertae sedis

The genus *Beneckea*, Campbell 1957, 328 originally described as having peritrichous flagella, seems to be composed of many vibrio-like organisms, and one species, *B. parahaemolytica*, is in this edition of THE MANUAL placed in the genus *Vibrio*. Baumann *et al.* (1971) recently described the genus *Beneckea* in detail and listed several new species; the relationship of these species to the earlier described peritrichously flagellated *Beneckea* remains to be determined. Meanwhile the genus and species listed by Baumann *et al.* should be regarded as a genus and species of uncertain taxonomic position.

Beneckea campbellii Baumann, Baumann and Mandel 1971, 288.

Beneckea neptuna Baumann *et al.* 1971, 289.

Beneckea nereida Baumann *et al.* 1971, 289.

Beneckea pelagia Baumann *et al.* 1971, 291.

Beneckea nigrapulchrituda Baumann, Baumann, Mandel and Allen 1971, 1383.

Beneckea natriegens (Payne *et al.*) Baumann, Baumann and Mandel 1971, 291. (*Pseudomonas natriegens* Payne, Eagon and Williams 1961, 121; *Vibrio natriegens* Webb and Payne 1971, 1080.)

Genus II. **Aeromonas** Kluyver and van Niel 1936, 398

RALPH H. W. SCHUBERT

Ae.ro.mo'nas. Gr. masc.n. *aer* air, gas; Gr. fem.n. *monas* unit, monad; M.L. fem.n. *Aeromonas* gas (-producing) monad.

Cells straight, rod-shaped with rounded ends to coccoid 1.0–4.4 μm, occasionally forming filaments up to 8 μm long; occurs singly, in pairs or chains; **motile by polar flagella, generally monotrichous;** some species non-motile. No resting stages known. **Gram-negative.**

Chemoorganotrophs: **metabolism both respiratory and fermentative;** carbohydrates broken down to acid or acid and gas (CO₂ and H₂). Some

species show a 2,3-butanediol fermentation (produce acetoin from glucose).

Some species grow on a mineral medium with ammonia as the sole source of nitrogen and one of the following as the sole source of carbon: glucose, arginine, asparagine or histidine; all species grow in a mixture of L-arginine, asparagine, leucine and methionine.

Glucose, fructose, maltose and trehalose are

broken down; adonitol, dulcitol, inositol, inulin, melezitose, sorbose and xylose are not fermented. Starch, dextrin and glycerol hydrolyzed.

Casein hydrolyzed; gelatin liquefied; deoxyribonuclease, arginine dehydrogenase and phosphatase produced; glutamic acid not decarboxylated; urea not broken down. Nitrate reduced to nitrite. Cytochrome oxidase, **oxidase** and catalase **positive.**

Facultative anaerobes.

Some species show no growth at 37 C, for others the maximum growth temperature is 38–41 C. Minimum growth temperature 0–5 C. Most biochemical tests are performed at the optimum temperature of 30 C, although some are best performed at 20 C. Growth is restricted to the pH range 5.5–9.0.

Not sensitive to the vibriostatic agent 2,4-diamino-6,7-diisopropyl pteridine (O/129).

The G + C content of the DNA ranges from 57–63 moles %.

Proposed type species: *Aeromonas hydrophila* (Chester) Stanier 1943, 213.

Further Comments

Aeromonas species share some properties with members of the *Enterobacteriaceae* and some with members of the genera *Pseudomonas* and *Vibrio*. Pertinent methods are given in the Report of the Enterobacteriaceae Subcommittee (1958) and in Schubert 1960, 1964, 1969; Eddy, 1960, 1962.

When they proposed the genus, Kluyver and van Niel recognized *Aeromonas liquefaciens* (Beijerinck) Kluyver and van Niel 1936, 398, as the type species. Schubert (1967, 1968) claims that the species is unrecognizable and should be rejected. In a request to the Judicial Commission, Schubert (1971, 90) has proposed *Aeromonas hydrophila* (Chester) Stanier 1943, 213 as type species of the genus as now defined.

Editorial Note. After a study of over 90 named strains of *Aeromonas*, excluding *A. salmonicida*, in a computer study using more than 80 tests plus immunodiffusion serological analysis and enterase isozyme patterns, D. H. McCarthy (thesis, Institute of Biology, London, 1973) suggests that *Aeromonas punctata* (Zimmerman 1890, *emend.* Lehmann and Neumann 1896) Snieszko 1957, 190 is the correct name with *A. formicans*, *A. liquefaciens* and *A. hydrophila* as synonyms. McCarthy doubts the need of a subspecies for the partly anaerogenic, gluconate and acetylmethycarbinol negative strains, but if one is needed the name should be *Aeromonas punctata* subsp. *caviae* Scherago 1936. He also suggests that *A. hydrophila* subsp. *proteolytica* does not belong in the genus *Aeromonas* on the basis of the G + C ratio (50.5 moles %) and other characteristics. This work will be published as "Studies on the genus *Aeromonas*" *Fish Diseases Technical Reports* Series No. 1. Ministry of Agriculture, Fisheries and Food, Fish Disease Laboratories, Weymouth, Dorset, probably in 1973, and a summary in the *International Journal of Systematic Bacteriology* in 1974.

However, the Judicial Commission has conserved the name *Aeromonas* Stanier with *A. hydrophila* as the type species (Opinion 48).

Description of the species of genus **Aeromonas**

For differential characteristics see Table 8.24.

1. Aeromonas hydrophila (Chester) Stanier 1943, 213. (*Bacillus hydrophilus fuscus* Sanarelli 1891, 197; *Bacillus hydrophilus* Chester 1901, 235; *Proteus hydrophilus* (Chester) Bergey *et al.* 1923, 211; *Bacterium hydrophilum* (Chester) Weldin and Levine 1923, 14; *Pseudomonas hydrophila* (Chester) Breed *et al.* 1948, 102; *Bacillus ichthyosmius* Hammer 1917, 243; *Escherichia ichthyosmia* (Hammer) Bergey *et al.* 1923, 201; *Proteus ichthyosmius* (Hammer) Bergey *et al.* 1934, 364; *Pseudomonas ichthyosmia* (Hammer) Breed *et al.* 1948, 103; *Pseudomonas fermentans* von Wolzogen-Kühr 1932, 228; *Flavobacterium fermentans* (von Wolzogen-Kühr) Bergey *et al.* 1934, 155; *Proteus melanovogenes* Miles and Halnan 1937, 91; *Vibrio jamaicensis* Caselitz 1955, 62.)

hy.dro'phi.la. Gr. n. *hydro* water; Gr. adj. *philos* loving; M.L. adj. *hydrophila* water loving.

Morphology and physiology as for genus and Table 8.24.

1a. *Aeromonas hydrophila* subspecies *hydrophila*.

Produces gas from glucose and glycerol; Voges-Proskauer reaction positive; oxidizes gluconic acid.

Cause of red leg disease in frogs. Pathogenic for snakes causing septicemia and stomatitis (Camin, 1948; Page, 1961). May also cause infections in fresh water fish and may be a secondary invader in virus-infected fish (Heuschmann-Brunner, 1965). Mice may be infected experimentally (Schubert, 1964). Human infection may be caused by biotype I; the pathology of infection in man and animals has been reviewed by Caselitz, 1966.

Found in fresh, uncontaminated water.

Neotype strain: ATCC 7966; NCTC 8049, NCMB 86.

1b. *Aeromonas hydrophila* subspecies *anaerogenes* Schubert 1964, 350.

an.ae.ro'gen.es. Gr. pref. *an* not; Gr. masc.n

TABLE 8.24
Characteristics differentiating the species, subspecies and biotypes of genus **Aeromonas**

	1. *A. hydrophila* subsp.			2. *A. punctata* subsp.		3. *A. salmonicida* subsp.		
	hydrophila	anaerogenes	proteo-lytica	punc-tata	caviae	sal-moni-cida	ach-romo-genes	maso-ucida
Growth in nutrient broth at 37 C (water bath)	+[a]	+	+	+	+	−	−	−
Brown pigment (water-soluble) on trypticase soy agar	−	−	−	−	−	+	−[b]	−[b]
Ammonium ion and glucose as sole source of nitrogen and carbon	+[c]	+[c]	+	+	+[c]	−	−	−
Amino acids as sole source of carbon								
L-Arginine	+	+	+	+	+	−	−	−
L-Asparagine	+	+	+	+	+	−	−	−
L-Histidine	+	+	+	+	+	−	−	−
L-Glutamic acid	+	+[c]	+	+	+	−	−	−
L-Serine	+[c]	d	+	+	+	−	−	−
L-Alanine	d	d	+	d	+	−	−	−
Growth in nutrient broth containing 7.5% NaCl	−	−	+	−	−			
Butanediol dehydrogenase	+	+	+			+	+	+
Production of:								
Gas from glycerol	+	−	−	−	−	+	−	+
Gas from glucose	+	−	−	+	−	+	−	+
Indole in tryptone broth containing 0.1% tryptophan	+[c]	+[c]	+	+	+	+	+	+
Hydrogen sulfide from 2.5% peptone water	+	+	−	+	+	−	−[b]	+
KCN broth (Møller technique)	+	+	+	+	+	−	−	−
Lysine decarboxylase reaction (Møller technique)	−	−	+			−	−	+
	biotypes							
	1	2						
Voges-Proskauer reaction	+	−	+	+	−	−	−	+
Gluconate oxidase test	+	−	+	−	−	−	−	+
Breakdown of								
Galactose	+	+	−	+	+	+	+	+
Sucrose	d	+	−	d	+	−	+	+
Mannitol	+	+	+	d	+	+	−	+
Arabinose	d	+	−	d	+	+	+	+
Esculin	d	d	−	d	+	+	−	+

[a] + = positive; − = negative; d = some strains positive, some negative.
[b] May be a delayed positive reaction.
[c] Aberrant strains occur.

...er air, gas; Gr. v. *gennaio* produce; M.L. adj. *anerogenes* not gas producing.

Does not produce gas from glucose or glycogen; Voges-Proskauer reaction positive; gluconic acid oxidized.

Found in sewage and contaminated fresh water. Not pathogenic to frogs.

Type strain: ATCC 15467.

A biotype is recognized which is V.P. negative and does not oxidize gluconic acid.

1c. *Aeromonas hydrophila* subspecies *proteolytica* (Merkel *et al.*) Schubert 1969, 412. (*Aeromonas proteolytica* Merkel, Traganza, Mukherjee, Griffin and Prescott 1965, 1230.)

pro.te.o.ly′ti.ca. M.L. fem.adj. *proteolytica* proteolytic.

Originally isolated from the alimentary canal of the marine borer *Limnoria tripunctata*.

Type strain: ATCC 15338; NCMB 1326.

2. **Aeromonas punctata** (Zimmermann) Snieszko 1957, 190. (*Bacillus punctatus* Zimmermann 1890, 86; *Bacterium punctatum* Lehmann and Neumann 1896, 238; *Pseudomonas punctata* (Zimmermann) Chester 1901, 313.)

punc.ta′ta. L. n. *punctum* a point; M.L. fem. adj. *punctata* full of points.

Morphology and physiology as for genus and Table 8.24.

2a. *Aeromonas punctata* subspecies *punctata*.
Produces gas from glucose.

Found mostly in fresh uncontaminated brooks and rivers.

May induce red leg disease in frogs under experimental conditions.

Neotype strain: NCMB 74.

2b. *Aeromonas punctata* subspecies *caviae* (Scherago) Schubert 1964, 350. (*Pseudomonas caviae* Scherago 1936, 83; *Pseudomonas formicans* Crawford 1954, 734; *Aeromonas formicans* (Crawford) Pivnick and Sabina 1957, 251.)

ca′vi.ae. M.L. fem.n. *Cavia* generic name of guinea pig; M.L. gen.n. *caviae* of guinea pig.

No gas produced from glucose.

Found in sewage and sewage-contaminated water.

Originally isolated from epizootic in young guinea pigs. Not pathogenic for frogs.

Type strain: ATCC 15468.

3. **Aeromonas salmonicida** (Lehmann and Neumann) Griffin, Snieszko and Friddle 1953, 138. (*Bacterium salmonicida* Lehmann and Neumann 1896, 240; *Bacillus salmonicida* (Lehmann and Neumann) Kruse 1896, 322; *Proteus salmonicida* (Lehmann and Neumann) Pribram 1933, 73.)

sal.mon.ic′i.da. L. n. *salmo, salmonis* salmon; L. suff. *-cida* from L. v. *caedo* cut or kill; M.L. n. *salmonicida* salmon killer.

Morphology and physiology as for genus and Table 8.24. Not motile.

Found on and pathogenic to salmonid fishes, causing furunculosis. May also cause serious infections in other fish. Not found in surface waters.

3a. *Aeromonas salmonicida* subspecies *salmonicida*.
Proposed neotype: NCMB 1102 (Schubert 1967, 30).

3b. *Aeromonas salmonicida* subspecies *achromogenes* (Smith) Schubert 1967, 278. (*Necromonas achromogenes* Smith 1963, 273.)

a.chro.mo.gen′es. Gr. adj. *achromos* colorless; Gr. v. *gennaio* produce; M.L. adj. *achromogenes* not producing color.

Does not produce the brown pigment typical of the primary subspecies.

May produce indole.

Type strain: NCMB 1110.

3c. *Aeromonas salmonicida* subspecies *masoucida* Kimura 1969, 52.

ma.sou.ci′da. Japanese n. *masou* specific epithet of *Oncorhynchus masou;* L.v. suff. *-cida* from L. v. *caedo* cut, kill; M.L. fem.n. *masoucida Oncorhynchus masou* killer.

Does not produce brown pigment. Is aerogenic and differs in several respects from other subspecies (Table 8.24).

Isolated from Japanese salmon.

Type strain: Kimura 1-a-1.

Genus III. **Plesiomonas** *Habs and Schubert 1962, 324*

RALPH H. W. SCHUBERT

(*Fergusonia* Sebald and Véron 1963, 907; "C 27 Group" Ferguson and Henderson 1947, 179.)

Ple.si.o.mo′nas. Gr. masc.n. *plesios* neighbor; Gr. fem.n. *monas* unit, monad; M.L. fem.n. *Plesiomonas* neighbor monad (to Aeromonas).

Cells round-ended, straight, rod-shaped 0.8–1.0 by 3.0 μm; growing singly in pairs or short chains, motile by polar flagella, generally lophotrichous. Resting stages not known. Gram-negative.

Chemoorganotrophs: **metabolism** both **respiratory and fermentative;** breakdown of carbohydrates by **production of acid but without gas.** Most strains grow on mineral media containing ammonia as sole source of nitrogen and glucose as sole source of carbon. Cytochrome oxidase, **oxidase** and **catalase** reactions **positive. Absence of exoferments** like diastase, lipase, deoxyribonuclease and proteinases; breakdown of inositol; positive lysine, ornithine decarboxylase and **arginine dehydrogenase** reactions (Møller technique).

Facultative anaerobe. Optimum growth at 30 C, good growth at 37 C; maximum growth temperature ranges in between 39–41 C. No growth in nutrient broth containing 7.5% sodium chloride.

Most strains are **sensitive to** the vibriostatic agent **2,4-diamino-6,7-diisopropyl pteridine** (0/129).

The G + C content of the DNA is 51 moles %.

Type species: *Plesiomonas shigelloides* (Bader) Habs and Schubert 1962, 324.

Further Comments

As *Plesiomonas* is a monotypic genus, the genus definition obviously emphasizes those characters by which *Plesiomonas* is differentiated from other genera. Characters not mentioned in the genus definition are given in the species definition. For the demonstration of the properties of *Plesiomonas shigelloides* strains the same methods are used as in the taxonomy of *Aeromonas*. Ferguson and Henderson (1947, 179) assigned their strain, "C$_{27}$", without naming it, to "Paracolon," family *Entero-*

bacteriaceae. The specific epithet 'shigelloides' is due to Bader (1954, 455) who, because of its polar flagellation, classified the species in the genus *Pseudomonas*. The transfer of the species to the genus *Aeromonas* (Ewing *et al.*, 1961) was followed by its transfer to the newly created genus *Plesiomonas* Habs and Schubert 1962, 324. The creation of the new genus *Plesiomonas* was inevitable because *P. shigelloides* strains do not exhibit some essential features of either *Aeromonas* or *Vibrio*, the hitherto known genera of oxidase-positive, fermentative, polarly flagellated, Gram-negative rods. *Plesiomonas* is characterized by lophotrichous polar flagella (two to seven flagella) whereas *Vibrio* and *Aeromonas* have only one polar flagellum. *Plesiomonas* lacks exoenzymes whereas *Vibrio* and *Aeromonas* produce diastase, lipase, DNase and various proteinases (gelatinase, caseinase). *Plesiomonas* ferments inositol, whereas the two other genera do not attack this compound; on the other hand *Plesiomonas* differs from members of *Aeromonas* and *Vibrio* in the restricted range of carbohydrates fermented (Eddy and Carpenter, 1964, 107).

Description of the species of genus **Plesiomonas**

1. **Plesiomonas shigelloides** (Bader) Habs and Schubert 1962, 324. (*Pseudomonas shigelloides* Bader 1954, 455; *Aeromonas shigelloides* (Bader) Ewing, Hugh and Johnson 1961, 32; *Fergusonia shigelloides* (Bader) Sebald and Véron 1963, 908.)

shi.gel.loi'des. M.L. fem.n. *Shigella*, a generic name; Gr. suf. *eides* similar; M.L. adj. *shigelloides*, Shigella-like.

Maltose, trehalose and glycerol fermented. Starch, dextrin, glycogen, mannitol, fructose, sucrose, arabinose, esculin, raffinose, cellobiose, salicin, sorbitol, inulin, melezitose, rhamnose, xylose, dulcitol and adonitol not fermented.

M.R. reaction variable; V.P. negative. Gluconate negative; malonate negative. Cannot use citrate as sole carbon source.

KCN negative; H$_2$S negative.

Nitrates reduced; gelatin not liquefied; indole produced. Phosphatase produced. Phenylalanine deaminase negative.

Plesiomonas shigelloides has been isolated from feces of humans and monkeys (Schmid *et al.*, 1954), from canine lymph nodes and dead chicken embryos; Japanese authors reported on outbreaks of infectious gastroenteritis in humans due to *Plesiomonas shigelloides* (see review by Sakazaki *et al.*, 1959).

Only some *Plesiomonas shigelloides* strains share a common O antigen with *Shigella sonnei* (Sakazaki *et al.*, 1959). Fifty-seven strains studied serologically were distributed in 16 O groups by Quincke (1957); only one of these groups showed antigenic relationships with *S. sonnei*.

Type strain: ATCC 14029; NCIB 9242.

Genus IV. **Photobacterium** Beijerinck 1889, 401

MARGARET S. HENDRIE AND J. M. SHEWAN

(*Photobacter* Beijerinck 1900, 354; *Photomonas* Orla-Jensen 1921, 271.)

Pho.to.bac.te'ri.um. Gr. n. *phos* light; Gr. neut.dim.n. *bakterion*, a small rod; M.L. neut.n. *Photobacterium* light (-producing) bacterium.

Cells **coccobacilli** or occasionally rods, 1.0–2.5 by 0.4–1.0 μm, **asporogenous**, axis straight, occurring singly and occasionally in pairs.

Pleomorphic forms occur in adverse conditions of growth. **Motile** by one or more **polar flagella,** or occasionally non-motile. Strains may appear non-

TABLE 8.25

*Features which distinguish **Photobacterium** from other heterotrophic Gram-negative bacteria*

	Photobacterium	Lucibacterium	Vibrio	Aeromonas	Pseudomonas	Alcaligenes	Enterobacteriaceae
Flagella	Polar (occasionally nonmotile, no flagella)	Usually peritrichous	Polar (one or more per cell)	Polar (usually one per cell)	Polar (one or more per cell)	Peritrichous	Peritrichous
Oxidase	d	+	+	+	+	+	—
Sensitivity to 2,4-diamino-6,7-diisopropyl pteridine	+	—	+	—	—	—	—
Luminescence	+	+	d	—	—	—	—
Carbohydrate metabolism	Fermentative	Fermentative	Fermentative	Fermentative	Respiratory, or not metabolized	Respiratory, or not metabolized	Fermentative
Gas production	+	—	—	d	—	—	d
DNA base ratio, moles % G + C	39–42	45–46	40–50	57–63	58–70	58–70	39–59

motile as only a small proportion of the cells in a culture are seen to move actively and detection of flagella is difficult. **Gram-negative.** No capsules. Some strains do not grow well on nutrient media unless the NaCl concentration is 2.0–3.0%.

Chemoorganotrophs, metabolism is both respiratory and fermentative. **Carbohydrate metabolism is fermentative and gas is usually produced.** In peptone media **acid** and usually gas produced from **hexoses** but pentoses and more complex carbohydrates are not generally attacked. Nitrites produced from nitrates. Ammonia produced from peptone. **Indole not produced.** Methyl red positive. **Acetylmethylcarbinol usually produced.** Trimethylamine oxide generally reduced to trimethylamine. Possess a lysine decarboxylase but not an ornithine decarboxylase, alkaline products usually produced from arginine. **Starch not hydrolyzed.** Chitin may be digested. Gelatin may occasionally be slowly liquefied.

Sensitive to chloramphenicol, streptomycin, polymyxin B and **2,4-diamino-6,7-diisopropyl pteridine.** Insensitive to penicillin.

Facultatively anaerobic. Optimum temperature usually 20–30 C. Optimum NaCl concentration usually 3.0%, growth occurs in nutrient media with 0.5–5.0% NaCl, no growth without NaCl. Growth at pH 6–9. Usually luminescent in suitable conditions, but luminescence may be lost after long periods in artificial culture.

The G + C content of the DNA (in the strains examined) ranges from 39–42 moles %.

Habitat: Sea water, surface and alimentary tract of some marine fishes and the luminous organs of some fish and cephalopods.

Type species: *Photobacterium phosphoreum* (Cohn) Ford 1927, 615.

Features which distinguish the genus from other Gram-negative, motile heterotrophic bacteria are given in Table 8.25.

Description of the species of genus **Photobacterium**

Features distinguishing the species are given in Table 8.26.

1. **Photobacterium phosphoreum** (Cohn) Ford 1927, 615. (*Micrococcus phosphoreus* Cohn 1878, 126; *Bacillus phosphorescens* II Baumgarten 1888, 344; *Bacterium phosphorescens* Fischer 1888, 107; *Photobacterium phosphorescens* Beijerinck 1889, 401; *Bacillus hermesi* Trevisan 1889, 31; *Streptococcus phosphoreus* (Cohn) Trevisan 1889, 31; *Bacillus phosphoreus* (Cohn) Macé 1901, 995; *Photobacter phosphorescens* Beijerinck 1901, 46; *Pseudomonas lucifera* Molisch 1904, 721; *Bacterium phosphoreum* (Cohn) Molisch 1912, 66; *Photobacter phosphoreum* (Cohn) Beijerinck 1916, 15; *Micrococcus physiculus* Kishitani 1930, 813; *Coccobacillus acropoma* Yasaki and Haneda 1936, 57; *Acinetobacter phosphorescens* Brisou 1955, 67; *Photobacterium profundum* Weisglass and Gavrilović 1963, 69.) phos.pho're.um. Gr. v. *phosphoreo* bring light; M.L. neut.adj. *phosphoreum* light-bearing.

Suggested working types: ATCC 11040 (NCMB 1282); NCMB 844; NCMB 1275.

From the descriptions in the literature it is likely that the following named species could belong to *Photobacterium phosphoreum: Coccobacillus loligo* Kishitani 1928, 611; *Micrococcus sepiola* Kishitani 1928, 396; *Pseudomonas euprymna*

Kishitani 1928, 308; *Pseudomonas luminescens* Kishitani 1928, 69; *Pseudomonas photogena* Kishitani 1928, 70; *Pseudomonas phosphorescens* Kishitani 1928, 71; *Microspira asamushiensis* Kishitani 1930, 813.

2. **Photobacterium mandapamensis** Hendrie, Hodgkiss and Shewan 1970, 165. man.da.pam.en'sis. M.L. gen.n. *mandapamensis* from (sea water near) Mandapam (a town in southern India).

Glycerol may be slowly fermented.

Suggested working types: NCMB 391; NCMB 1198.

TABLE 8.26
Characters that distinguish species of genus **Photobacterium**

	P. phos-phoreum	P. manda-pamensis
Oxidase (Kovacs)	+	−	+
Maltose fermentation	+	−
Voges-Proskauer	+	+
2,3-Butanediol (Bullock's method)	+	−
Growth at 5 C	+	−
Growth at 37 C	−	+

Genus V. **Lucibacterium** Hendrie, Hodgkiss and Shewan 1970, 166

MARGARET S. HENDRIE AND J. M. SHEWAN

Lu'ci.bac.te'ri.um. L. fem.n. *lux, lucis* light, radiance; Gr. neut.dim.n. *bakterion* a small rod; M.L. neut.n. *Lucibacterium* light (-emitting) bacterium.

Cells **asporogenous rods,** axis straight or slightly curved, 0.5–1.0 by 1.2–2.5 μm. **Motile by peritrichous flagella.** Some strains also have a thick, sheathed polar flagellum in addition to the finer lateral flagella. Fimbriae may also be present. Occasionally only the polar flagellum can be detected. Pleomorphic forms occur in adverse cultural conditions. No capsules. **Gram-negative.**

Chemoorganotrophs, metabolism is both respiratory and fermentative. **Carbohydrate metabolism is fermentative without gas** production. A wide range of carbohydrates are fermented but pentoses, lactose, raffinose, cellulose, inositol, dulcitol and inulin are not usually attacked. Nitrites produced from nitrates. **Indole produced.** Methyl red positive. **Acetylmethylcarbinol not produced.** Trimethylamine oxide reduced to trimethylamine. Ammonia produced from peptone. **Oxidase** (Kovacs') and catalase **positive. Starch hydrolyzed. Gelatin liquefied.** Chitin may be digested. Possesses lipases and lecithinase. Usually possesses decarboxylases for lysine and ornithine but does not generally give rise to alkaline products from arginine.

Sensitive to chloramphenicol, streptomycin and polymyxin B. **Insensitive** to penicillin and **2,4-diamino-6,7-diisopropyl pteridine.**

Facultatively anaerobic. Optimum temperature 25–30 C. Growth in nutrient media containing 0.5–5.0% NaCl, optimum NaCl concentration 2.0–3.0%, **no growth without NaCl.** Growth at pH 6–9. Usually luminescent in suitable growth conditions. In some strains luminescence is transient (i.e. it fades rapidly).

The G + C content of the DNA (of the strains examined) is 45–46 moles %.

Habitat: Sea water. Also found on the surfaces of dead marine animals.

Type species: *Lucibacterium harveyi* (Johnson and Shunk) Hendrie, Hodgkiss and Shewan 1970, 166.

Features which distinguish the genus from other heterotrophic, motile, Gram-negative bacteria are given in Table 8.25.

Description of the species of genus **Lucibacterium**

1. **Lucibacterium harveyi** (Johnson and Shunk) Hendrie, Hodgkiss and Shewan 1970, 166. (*Photobacter splendidum* Beijerinck 1900, 362; *Vibrio splendidus* (Beijerinck) Lehmann and Neumann 1927, 543; *Achromobacter harveyi* Johnson and Shunk 1936, 587; *Photobacterium splendidum* (Beijerinck) Eymers and van Schouwenburg 1937, 236; *Photobacterium sepiae* Kluyver 1938 in Doudoroff 1942, 451; *Pseudomonas harveyi* (Johnson and Shunk) Breed 1948, 110; *Photobacterium harveyi* (Johnson and Shunk) Breed and Lessel 1954, 61.)

har'vey.i. M.L. gen.n. *harveyi* of Harvey; named for E. N. Harvey.

Although growth usually occurs at 37 C and not at 5 C, strains do occur with a lower temperature range.

Occasionally strains do not possess decarboxylase for lysine and ornithine, but may produce alkaline products from arginine.

Some strains tend to have a spreading growth on solid media.

Type strain: ATCC 14126 (NCMB 1280).

GENERA OF UNCERTAIN AFFILIATION

Genus **Zymomonas** *Kluyver and van Niel 1936, 399*

J. Geoffrey Carr

Zy.mo'mo.nas or Zy.mo.mo'nas. Gr. n. *zyme* leaven, ferment; Gr. n. *monas* a unit, monad; M.L. fem.n. *Zymomonas* fermenting monad.

Plump rod-shaped cells with rounded ends, occurring singly or in pairs, motile by means of lophotrichous flagella. No resting stages known. Gram-negative.

Chemoorganotrophs showing vigorous fermentative metabolism of glucose or fructose which yields nearly equimolar quantities of ethanol and CO_2 by the Entner-Doudoroff pathway. Although not an end product, acetaldehyde accumulates in the culture medium. Arabinose, rhamnose, xylose, galactose, lactose, maltose, raffinose, mannitol and dulcitol are not fermented.

Produces catalase and H_2S. Gelatin is not liquefied, indole not produced, nitrate is not reduced to nitrite, and the methyl red test is negative (Millis, 1956).

Anaerobes, but able to tolerate some oxygen. Optimum temperature 30 C. The G + C content of the DNA is 47–48 moles %.

Type species: *Zymomonas mobilis* (Lindner) Kluyver and van Niel 1936, 399.

Description of the species of genus **Zymomonas**

1. **Zymomonas mobilis** (Lindner) Kluyver and van Niel 1936, 399. (*Termobacterium mobile* Lindner 1928, 253; *Pseudomonas lindneri* Kluyver and Hoppenbrouwers 1931, 259; *Zymomonas mobile* (*sic*) (Lindner) Kluyver and van Niel 1936, 399; *Saccharomonas lindneri* (Kluyver and Hoppenbrouwers) Shimwell 1950, 182.)

mo'bi.lis. L. adj. *mobilis* movable, motile.

Short rods, 1.4–2.0 by 4.0–5.0 μm, usually in pairs, less frequently in short chains.

One mole of glucose is fermented with the production of 1.6 moles of ethyl alcohol and 1.8 moles of CO_2, with traces of lactate (Belaich and Senez, 1965). Fructose and sucrose are also fermented; the latter is metabolized more slowly than the monosaccharides and levan is produced (Dawes *et al.*, 1966). The fermentation of sucrose is the main characteristic distinguishing this species from *Z. anaerobia*. Growth but no gas production from sorbitol. Ethyl alcohol is metabolized to acetate in aerated cultures (Belaich and Senez, 1965).

Pantothenate is the only growth factor required. Growth occurs in a synthetic medium plus pantothenate but is about half of that obtained in yeast extract medium. Good growth is obtained with ammonium ion as the sole source of nitrogen (Bexon and Dawes, 1970). Cells grown aerobically contain cytochromes of the *c* and a_2 types; those grown anaerobically also contain a cytochrome of the *b* type (Belaich and Senez, 1965).

Originally isolated from pulque, the fermenting sap of *Agave americana*. Also found in the fermenting juice of the Gomuti palm, *Arenga pinnata* (*A. sacchifera*) in central Java (Roelofsen, 1941). Apparently common in fermenting plant juices in tropical countries.

The G + C content of the DNA (strain NCIB 8938) is 48.3 moles %.

2. **Zymomonas anaerobia** (Shimwell) Kluyver 1957, 199. (*Achromobacter anaerobium* (*sic*) Shimwell 1937, 509; *Saccharomonas anaerobia* (Shimwell) Shimwell 1950, 181.) Vernacular name: 'Cider sickness organism' Barker and Hillier 1912, 78.

an.ae.ro'bi.a. Gr. pref. *an* not; Gr. n. *aer* air: Gr. n. *bius* life; M.L. adj. *anaerobius* not living in air.

Plump rods, 1.0–1.5 by 2.0–3.0 μm occurring singly but more commonly as diplobacilli. Two strains are found in cider; both motile by lophotrichous flagella; in one the cells remain separated; in the other they form motile clumps or rosettes. Similar clumps are found in strains from beer.

Ferments only glucose and fructose; 1 mole of glucose yields 1.8 moles of ethyl alcohol and 1.9 moles CO_2, with some acetaldehyde; 1 mole of fructose produces 1.5 moles each of ethyl alcohol and CO_2 with some glycerol (McGill *et al.*, 1965). Slight growth but no gas production in the presence of sorbitol. Sucrose and ethyl alcohol are not metabolized.

Acetylmethylcarbinol and diacetyl are not produced. Cytochrome *b* and traces of *c* are found in cells grown aerobically or anaerobically.

Optimum temperature 30 C; no growth at 41 C. Killed by exposure to 60 C for 5 min. Optimum pH range 4.5–6.5; slight growth at pH 3.5, vigorous growth up to pH 8.0.

Found in tainted cider, known in England as 'cider sickness' and in France as 'framboisé'; the unpleasant flavor is caused by the accumulation of acetaldehyde. Also found in tainted beer where the unpleasant flavor is enhanced by the release of H_2S. This species has been isolated from brewery yards and brushes of cask-washing machines.

The G + C content of the DNA (strain NCIB 8227) is 47.6 moles %.

2a. *Zymomonas anaerobia* var. *anaerobia*. Description as for species.

2b. *Zymomonas anaerobia* var. *immobilis* (Shimwell) *comb. nov.* (*Saccharomonas anaerobia* var. *immobilis* Shimwell 1950, 182.)

Does not have flagella and is not motile at any stage.

Usually found in beer and its colonies may be distinguished from those of the primary variety (Shimwell, 1950).

2c. *Zymomonas anaerobia* var. *pomaceae* Millis 1956, 527.

Description as for species. Able to grow slowly in the presence of 500 p.p.m. SO_2 but is inhibited by 750 p.p.m.

Isolated from brewing yards, brushes of cask-washing machines, fermenting beers and sick ciders.

Genus **Chromobacterium** *Bergonzini 1881, 153. Nom. gen. cons.* Opin 16, Jud. Comm., 1958, 152

P. H. A. SNEATH

(*Cromobacterium* (*sic*) Bergonzini 1881, 153, spelling emended Buchanan 1918, 52; not *Cromobacterium* Bergonzini 1879, 39 *nom. rej.* Opin. 16, Jud. Comm. 1958, 152.)

Chro.mo.bac.te′rium. Gr. n. *chroma* color; Gr. n. *bakterion* a small rod; M.L. neut.n. *Chromobacterium* a colored rod.

Rods with rounded ends, sometimes slightly curved, occurring singly, occasionally with some pairs or short chains. Dimensions 0.6–1.2 by 1.5–6 μm, without definite capsules although sometimes with intercellular slime. Motile by both one **polar flagellum** and usually one to four **subpolar or lateral flagella. No resting stages known. Gram-negative,** often with barred or bipolar staining and lipid inclusions.

Produce low convex round **violet colonies,** and in nutrient broth a **violet ring** at the junction of the liquid surface and the container wall.

Chemoorganotrophs: **metabolism either respiratory (oxidative) or fermentative,** yielding organic acids from glucose and some other carbohydrates, but no visible gas. Oxidize lactate to CO_2.

Usually oxidase positive by the method of Kovacs (1956), although pigment may interfere with reading. Catalase positive, but **highly sensitive to hydrogen peroxide.** Indole negative, V.P. negative. Nitrate and nitrite reduced, sometimes with visible gas production. Produce ammonia from peptone. Phosphatase positive, arylsulfatase negative.

Violet pigment (violacein) produced on suitable media, not diffusing readily, **soluble in ethanol but not in water or chloroform.** Violacein has antibiotic properties (DeMoss 1967).

Grow on ordinary peptone media. Utilize citrate and NH_3 as sole carbon and nitrogen sources for growth. Without distinctive organic growth factor requirements.

Strict aerobes or facultative anaerobes.

Grow at 25 C, but species differ in optimum and in maximum and minimum temperatures. Do not grow below pH 5, optimum pH 7–8. Do not grow in media containing 6% or more of NaCl.

Resistant to benzylpenicillin, 10 μg/ml; sensitive to tetracycline, 30 μg/ml. **Resistant to** 0/129 (2,4-diamino-6,7-diisopropyl pteridine) disc method, 30 μg/disc.

Soil and water organisms, occasionally causing infections of mammals or food spoilage.

The **G + C content** of the DNA ranges from **63–72 moles** % (T_m method).

Type species: *Chromobacterium violaceum* Bergonzini 1881, 153.

Further Comments

The nomenclature of the genus has been much confused, and this should be noted in consulting the literature. The major synonyms are noted in the species descriptions, and a full synonymy may be found in Sneath (1960). There are two well established species, one growing at a temperature of 37 C but not at 4 C (mesophilic), and one growing at 4 C but not at 37 C (psychrophilic). The name *Chromobacterium violaceum* has been widely used for both species, and since there is doubt over the identity of Bergonzini's original strain any decision is necessarily arbitrary. The Judicial Commission (1958) fixed the type species as a mesophil, and as a consequence the following changes have been made from the 7th edition of THE MANUAL. The mesophilic organism there referred to (p. 295) as *C. janthinum* is here described under the name *C. violaceum,* while the psychrophilic organisms there named (p. 294) *C. violaceum* and *C. amethystinum* have been united under the name *C. lividum,* which has priority.

The descriptions are based in the main on Leifson (1956), Sneath (1956, 1960), Steel and Midley (1962) and Moffett and Colwell (1968). The earlier literature is summarized in Sneath (1960)

Chromobacterium violaceum and *C. lividum* a well defined and reasonably distinct species though the possible occurrence of intermediate strains requires further investigation. Within them there are no well marked varieties. Thus those strains of *C. violaceum* that do not acidify carbohydrates anaerobically (*C. laurentium* Leifson, 1956) do not differ in most other respects from the fermentative strains, and extensively cross-react serologically with them; they are here considered as belonging to the same species. The membrane-forming (gelatinous) strains of *lividum* earlier called *C. amethystinum* and *membranaceum* likewise cross-react with other strains of *C. lividum* and do not differ notably other ways. The two species, however, show little or no serological cross-reaction.

The pigment violacein is readily identified spectrophotometrically, showing an absorption maximum in ethanolic solution at 579 nm and minimum at 430 nm. In 10% (v/v) H_2SO_4 in eth nol the pigment gives a green solution with absorption maximum at 700 nm. If NaOH is added to an ethanolic solution it becomes green, then reddish brown. Pigment is only freely produced on media containing tryptophan and may be suppressed by certain brands of peptone. Both species may give rise to unpigmented variants.

The characteristic flagellar arrangement is best seen in young cultures on solid media. The single polar flagellum is inserted at the tip of the cell, shows long shallow waves and often stains faintly. The lateral flagella (usually long, and 1–4 but up to 8 in number) may be inserted subpolarly or laterally, and usually show deep short waves and stain readily. The two forms of flagella are antigenically distinct. Old cultures and those in liquid media show few lateral flagella. Occasional strains lack the lateral flagella.

The testing methods for hydrogen cyanide production, carbohydrate reactions, and hydrolysis of casein and esculin, together with characteristics of violacein are described in Sneath (1960, 1966).

Members of the genus die quickly at 0 C (Efthimion and Corpe, 1969), although C. lividum grows slowly at 4 C. C. lividum survives lyophilization (Fisher, 1963) but the proportion of viable cells is low. The genus is of interest to biochemists in four areas, although it is not always clear which species was studied: indole metabolism and biosynthesis of violacein; production of HCN; occurrence of unusual sugar compounds; and the production of extracellular polysaccharide. Recent publications in these areas are those of DeMoss (1967), Brysk et al. (1969), Stevens et al. (1963) and Corpe (1964). Although the organisms are indole negative by the usual testing methods they may under certain conditions accumulate pyrrole compounds that give a positive reaction with indole reagents (Corpe, 1961, 1963; Sebek and Jäger, 1962). Violacein production has been used as an assay for L-tryptophan (Sebek, 1965).

Although commonly grouped with Rhizobium and Agrobacterium, the genus Chromobacterium is not particularly close to them or to any other genus so far as is known (Heberlein et al., 1967; Moffett and Colwell, 1968). It is possible that the genus is not a very natural taxon, although the species share certain unusual features (the pigment violacein, flagellar arrangement, sensitivity to hydrogen peroxide). The species are almost as different from one another as are many genera. Unpigmented strains may be difficult to identify.

These organisms are common in soil and fresh water, and it is likely that C. violaceum predominates in the tropics and C. lividum in temperate regions. The enrichment method of Corpe (1951) appears to yield mainly C. lividum. They occasionally cause violet discoloration of foodstuffs (usually C. lividum, e.g. Seitz et al., 1961). C. lividum has been reported from the leaf nodules of the plants Psychotria nairobiensis and Ardisia crispa (Bettelheim et al., 1968; these strains were reported to fix nitrogen). C. violaceum is occasionally responsible for pyogenic infections of man and animals. Bacteriophages have occasionally been reported for the genus.

Characteristics differentiating the species are given in Table 8.27.

The figures in parentheses in the descriptions indicate the proportion of strains possessing the property if it is variable (e.g. 0.9 means 90% of strains).

Description of the species of genus Chromobacterium

1. **Chromobacterium violaceum** Bergonzini 1881, 153, emend. mut. char. Sneath 1956, 79. (Subjective synonyms: Chromobacterium janthinum (Zopf) Ford 1927, 474; Chromobacterium manilae (Krasil'nikov) Leifson 1956, 399; Chromobacterium laurentium (Migula) Leifson 1956, 399; not Chromobacterium violaceum Leifson 1956, 399; not Chromobacterium violaceum as described by Breed et al. 1957, 294.)

vi.o.la′ce.um. L. neut.adj. violaceum, violet-colored.

Rods, 0.6–0.9 by 1.5–3 µm, often coccobacillary. Rarely contain metachromatic granules. Usually (0.8) contain poly-β-hydroxybutyrate inclusions.

Colonies smooth although rough and non-pigmented variants may occur, not gelatinous. Violet ring in nutrient broth at surface with fragile pellicle. Gelatin liquefied in 7 days at 20 C (infundibuliform liquefaction with violet pellicle).

Attack on carbohydrates usually fermentative (0.8), sometimes oxidative (0.2). Acidity from carbohydrates is detectable in ordinary peptone water media, but the medium of Hugh and Leifson (1953) is preferable. In this medium acid without visible gas is produced from glucose, fructose, trehalose; often from maltose (0.5), mannose (0.8), sorbitol (0.6) and rhamnose (0.5); rarely from sucrose (0.25), starch (0.2), dextrin (0.2), glycogen (0.2), salicin (0.1) and glycerol (0.1). No acid from adonitol, L-arabinose, cellobiose, dulcitol, galactose, m-inositol, inulin, lactose, mannitol, melibiose, melezitose, raffinose and xylose. Inulin may yield acid if heated during preparation.

3-Ketolactose is not produced from lactose.

2-Ketogluconate not produced from gluconate. Chitin is usually digested (0.95). Starch digestion weak or negative. Agar and pectate not digested. Acetic acid not produced from ethanol. Esculin is not hydrolyzed.

Casein is hydrolyzed, usually strongly. Usually hemolytic (0.95) on horse blood. Urease negative or weak. Reduces nitrate to nitrite (0.95) and nitrite usually (0.8) reduced further without visible gas. H_2S production negative or weak. Arginine dihydrolase variable (0.5); L-arginine decarboxylase positive; L-ornithine and L-lysine decarboxylases negative. Melanin often produced from phenylalanine (0.8), phenylpyruvate not produced.

Oxidase (Kovacs' method) usually positive (0.7), although pigment may interfere. Produces HCN and cultures smell of ammonium cyanide (0.95). Produces turbidity from egg yolk. Methylene blue reduced, often weakly. Penicillinase is sometimes produced (0.3).

Grows on ordinary peptone media. Without distinctive organic growth factor requirements. Utilizes citrate and NH₃ as sole carbon and nitrogen sources (although slowly) and also the following: L-alanine, L-glutamic acid, L-histidine, L-lysine, L-ornithine, L-phenylalanine, L-tyrosine, less often L-arginine (0.6), L-serine (0.6), L-proline (0.6) but not L-cystine or L-leucine. Malonate and acetate are not utilized as sole carbon source, and formate seldom (0.3).

Grows in KCN medium (Steel and Midgley, 1962).

Facultatively anaerobic, although the oxidative strains grow slowly anaerobically. Growth is best at 30–35 C, minimum 10–15 C, maximum 40 C (a few strains (0.2) grow at 44 C).

Soil and water organisms, common in tropical countries, occasionally causing serious pyogenic or septicemic infections of mammals, including man (reviewed in Sneath, 1960).

TABLE 8.27

Main differential characters of species

	C. violaceum	C. lividum
Growth at 4 C (7 days)	−	+
Growth at 37 C (7 days)	+[a]	−
HCN production	+	−
Turbidity from egg yolk	+	−
Acid from trehalose	+	−
Acid from arabinose	−	+
Acid from xylose	−	+
Casein hydrolysis	+	− or w
Esculin hydrolysis	−	+

[a] + = positive; w = weak; − = negative.

The G + C ratio ranges from 63–68 moles % (T_m). Neotype strain: MK; NCTC 9757; NCIB 9131; ATCC 12472; D 252 (Editorial Board 1958, 152).

2. **Chromobacterium lividum** (Eisenberg) Bergey *et al.* 1923, 119. (Chromogene Bacterienart, Plagge and Proskauer 1887, 463; *Bacillus lividus* Eisenberg 1891, 81 (Bergey *et al.* 1923, attributed the name erroneously); *Bacillus violaceus berolinensis* Kruse 1896, 311; *Bacterium lividus* (*sic*) (Eisenberg) Chester 1897, 117; *Bacillus berolinensis* Chester 1901, 305; *Chromobacterium violaceum* Ford 1927, 469; *Chromobacterium amethystinum* Breed *et al.* 1957, 294; *Chromobacterium violaceum* Leifson 1956, 399.)

li'vi.dum. L. neut.adj. *lividum*, leaden colored, dark blue.

Cells 0.8–1.2 by 2.5–6 μm, occasionally in short chains, contain little lipid. Intercellular slime may be present.

Growth is often gelatinous or rubbery (0.7) and a tough pellicle is then seen on broth cultures. The violet ring in broth is usually viscous and frondlike. Violet pigmentation is often less intense and is produced more slowly, sometimes as concentric rings in the colonies. A few strains show a pale yellow diffusing fluorescent pigment in young cultures. The cultures do not smell of ammonium cyanide.

Attack on carbohydrates is oxidative; acidity is often not detectable in the usual peptone water media, and rarely (0.05) is not detectable in Hugh and Leifson's medium. In Hugh and Leifson's medium acid is almost always (0.95) produced from L-arabinose, fructose, galactose, glucose, *m*-inositol, maltose, mannose, sorbitol, xylose; often from lactose (0.9), glycerol (0.8), mannitol (0.8), sucrose (0.8), cellobiose (0.6) and inulin (0.4); rarely from salicin (0.1), not from dulcitol, starch or trehalose. Esculin hydrolysis positive (0.95). Chitin rarely digested (0.05). Reducing substances sometimes produced from gluconate (0.3).

Gelatin stab cultures seldom show liquefaction (7 days at 20 C). Casein hydrolysis rarely positive (0.1), sometimes weak. Rarely shows weak hemolysis on horse blood agar (0.2). Arginine decarboxylase rarely positive (0.2); arginine dihydrolase negative. Melanin usually not produced from phenylalanine. Visible gas may be produced from nitrate (0.4).

No turbidity produced from egg yolk. Hydrogen cyanide not produced. Methylene blue seldom reduced, even weakly.

Growth on citrate and NH₃ as sole carbon and nitrogen sources is usually rapid, and growth usually occurs on L-leucine but not on L-ornithine or L-phenylalanine. Malonate is sometimes utilized for growth.

Strictly aerobic. Grows best at about 25 C, minimum for most strains about 2 C (occasionally (0.1) 10 C), maximum usually about 32 C (occasionally 30 or 35 C).

Common in soil and water in temperate regions and recorded from leaf nodules of certain plants.

The G + C ratio ranges from 65–72 moles % (T_m).

Neotype strain: HB; NCTC 9796; NCIB 9130; ATCC 12473; D 303 (Sneath, 1956).

Species incertae sedis

a. *Chromobacterium marismortui* Elizari-Volcani 1940, 76.

Although included in the 7th edition of THE MANUAL, it has been omitted as it does not produce violacein or show the characteristic flagellar arrangement; optimum NaCl concentration 12%; its systematic position is uncertain. (See also p. 272.)

b. *Chromobacterium marinum* Hamilton and Austin 1967, 262.

A marine organism whose taxonomic position is still uncertain.

The following organisms which produce blue or violet pigments have at one time been placed in the genus.

c. *Chromobacterium viscosum* Grimes 1927, 367.

d. *Chromobacterium iodinum* Davis 1939, 273.

e. *Chromobacterium chocolatum* Knutsen in Lasseur *et al.* 1944, 293.

Strains of *Pseudomonas cepacia* misidentified as *C. janthinum*. (*cf.* Gilman, 1953; Stanier *et al.*, 1966, 247.)

Genus **Flavobacterium** Bergey et al. 1923, 97

OWEN B. WEEKS

Fla.vo.bac.te′ri.um. L. adj. *flavus* yellow; Gr. neut.n. *bakterion* a small rod; M.L. neut.n. *Flavobacterium* a yellow bacterium.

Cells vary from coccobacilli to slender rods. **Motile with peritrichous flagella or non-motile.** The latter **do not show gliding movement** (flexibacterial) **or swarming** (cytophagal) **growth on a nutrient agar.** Endospores not formed. Gram-negative.

Growth on solid media is pigmented yellow, orange, red or brown and **hue may vary with media and temperature.** Color most pronounced on potato, gelatin or milk-containing media and at lower temperatures (15–20 C). Often light is required for maximum pigmentation. **Pigments not soluble in media** and have not been characterized but generally presumed to be carotenoid; however, non-carotenoid pigmentation is probable for at least two of the species. Colonies typically translucent, smooth and entire, or occasionally opaque.

Chemoorganotrophs. Cultures often difficult to maintain following primary isolation and fastidiousness has been associated with nitrogen requirements, need for exogenous metabolites such as B complex vitamins or physical conditions such as concentration of agar in the medium.

Metabolism respiratory. Fermentation behavior, not conspicuous and **neither acid nor gas reactions common** in carbohydrate-containing broth. Acid reactions are common when such cultures are shaken during incubation and concentrations of peptone in media are low. A few cultures are facultatively anaerobic.

Characteristics for the species are given in Table 8.28.

Incubation temperatures below 30 C preferable and growth may be inhibited by higher temperatures. A few species grow at 37 C.

Widely distributed in soil and fresh and marine waters. Commonly found on vegetables during commercial processing and in dairy products. Unidentified pigmented isolates from human infections have been assigned to the genus and one pathogenic species has been included.

Two distinct ranges of DNA nucleotide base ratios (G + C moles %) have been reported: 30–42 and 63–70.

Type species: *Flavobacterium aquatile* (Frankland and Frankland) Bergey *et al.* 1923, 100.

Further Comments

Flavobacterium had its inception as a color genus (Bergey *et al.*, 1923) and has been taxonomically heterogeneous from the start. Heterogeneity was lessened by the exclusion of Gram-positive species (Breed *et al.*, 1957) and this continues in the present arrangement with exclusion of non-motile species which show cytophagal-like swarming on agar media. Differentiation of non-motile flavobacteria from *Cytophaga* is a major, unresolved problem (Mitchell *et al.*, 1969; Weeks, 1969). For the present non-motile cultures of flavobacterial species which show the swarming are assumed to be *Cytophaga*. Cultures which do not exhibit the property have been retained in *Flavobacterium* even though gliding motility can be demonstrated by careful study and special technique (L. R. Perry, Torry Research Station, unpublished studies). A similar reservation has been made for cultures which are reported to show swarming behavior under unusual experimental conditions (Mitchell *et al.*, 1969). Further studies may lead to the decision that gliding motility and swarming on agar media are unique to *Cytophaga* and that the flavobacteria exhibiting only the former character, or the latter, under special test conditions, are aberrant *Cytophaga*. Whatever the decision the actual circumstances suggest that *Flavobacterium* is an uncertain taxonomic concept (*genus incertae sedis*) since most of the species are non-motile and phenotypically similar to *Cytophaga*. Taxonomic uncertainty is indicated also by the two disparate DNA nucleotide base ratio ranges into which the species may be sorted.

The genus has been arranged into two sections: Section I contains the non-motile flavobacteria characterized by a G + C content of the DNA ranging from 30–42 moles %; Section II contains the non-motile and motile by peritrichous flagella species having a G + C content of the DNA ranging from 63–70 moles %.

The genus includes only Gram-negative species and cultures. A number of the reference strains extant, reportedly Gram-negative, are Gram-positive (see "*Species incertae sedis,*" II) which suggests investigators should pay special attention to this property. The Kopeloff-Beerman modification of Gram's procedure (Conn, 1957) is the preferred technique. It is useful to supplement staining with complementing tests such as the ability of Gram-negative bacteria to grow upon an agar medium containing crystal violet (2.5 × 10⁻⁵ and 1 × 10⁻⁶ concentrations) but not upon a medium containing 1% sodium dodecyl sulfate. Gram-positive bacteria usually give the antithetical results.

Description of the species of genus **Flavobacterium**

1. **Flavobacterium aquatile** (Frankland and Frankland) Bergey *et al.* 1923, 100. (*Bacillus aquatilis* Frankland and Frankland 1889, 381; *Bacterium aquatilis* (*sic*) (Frankland and Frankland) Chester 1897, 961; *Flavobacterium aquatilis* (*sic*) (Frankland and Frankland) Bergey *et al.* 1923, 100; *Chromobacterium aquatilis* (*sic*) (Frankland and Frankland) Topley and Wilson 1929, 404; *Empedobacter aquatile* (Frankland and Frankland) Brisou *et al.* 1960, 359.)

a.qua'ti.le. L. neut.adj. *aquatile* living in water.

Rods, 0.5–0.7 by 1.0–3.0 μm, approaching coccobacillary form in young cultures; filamentous forms, 10–40 μm long in both liquid and solid media. No swarming on solid media at 25–30 C, i.e. discrete colonies form, but swarming has been reported in cultures continuously propagated at 15 C (Mitchell *et al.*, 1969) and, more recently, a gliding movement has been demonstrated using a special technique (L. B. Perry, Torry Research Station, unpublished studies).

Pigments are carotenoids, not characterized, resembling β-carotene, cryptoxanthin and zeaxanthin. Carotenogenesis is temperature regulated, light yellow-brown at 30 C, bright orange at 15–20 C. Colonies 1–3 mm, smooth, entire, glistening, transparent but more mucoid and spreading at 15 C.

Poor growth on nutrient agar, or in nutrient broth, containing meat extracts and peptones. Excellent growth when organic nitrogenous source is an enzymic digest of casein, or gelatin, supplemented with glucose and yeast extract. Thiamine will replace yeast extract and *F. aquatile* strain Taylor ATCC 11947 may be used in bioassay of thiamine (Weeks and Beck, 1960). Acid-hydrolyzed casein will not supply nitrogen requirements.

Slow growth in litmus milk with peptonization. No growth under anaerobic conditions, at 37 C and in media containing more than 1% added NaCl.

The G + C content of the DNA is 32 moles % (T_m).

Isolated originally from deep wells in the chalk region of Kent, England. The bacterium has not been isolated from marine environments.

Neotype strain: Taylor 36; ATCC 11947; NCIB 8694.

2. **Flavobacterium breve** (Lustig) Bergey *et al.* 1923, 116. (*Bacillus canicolis brevis* Cornil and Babes 1890, 292; *Bacillus canalis parvus* Eisenberg 1891, 362; *Bacillus brevis* Lustig 1890, 52; *Bacterium canalis parvus* (*sic*) (Eisenberg) Chester 1897, 130; *Bacterium canale* Mez 1898, 55; *Bacterium breve* (Lustig) Chester 1901, 172; *Flavobacterium brevis* (*sic*) (Lustig) Bergey *et al.* 1923, 116; *Pseudobacterium brevis* (*sic*) (Lustig) Krasil'nikov 1949, 239; *Empedobacter breve* (Lustig) Prévot 1961, 181.) German vernacular name: Der kurze Canalbacillus Mori, 1888, 53.

bre've. L. neut.adj. *breve* short.

Rods, 0.8–1.0 by 1.0–2.5 μm, purportedly show polar staining. Non-motile and no swarming growth has been reported.

Pigmentation is light yellow and hue does not change with variation of medium and temperature.

Does not grow well on nutrient agar, potato or gelatin. Colonies minute and develop slowly.

Pathogenic for laboratory animals (Lustig).

Isolated originally from sewage in Berlin.

The G + C content of the DNA is 26 moles % (T_m).

Reference strain: ATCC 14234.

3. **Flavobacterium meningosepticum** King 1959, 247.

me.nin.go.sep'ti.cum. Gr. n. *meninx, meningos*

TABLE 8.28
Characteristics of the species of genus **Flavobacterium**

Character	Section I						Section II					
	1. F. aquatile	2. F. breve	3. F. meningosepticum	4. F. ferrugineum	5. F. halmephilum	6. F. uliginosum	7. F. capsulatum	8. F. lutescens	9. F. rigense	10. F. indoltheticum	11. F. tirremicum	12. F. devorans
Motile	−	−	−	−	−	−	−	−	+	+	+	+
Pigmentation	yo^a	y	y	yo	y	yo	y	y	yo	yo	r	y
Specific attribute basis for isolation	−	−	−	+	+	−	+	−	−	+	+	+
Grows at 37 C	−	0	+	+	−	−	−	+	+	−	−	−
Halotolerant	−	+	+	+	+	−	+	+	+	+	+	+
Require added NaCl	−	−	−	−	+	+	−	−	−	+	+	−
Hydrolysis of												
Gelatin	+	−	+	+	+	+	−	+	+	+	−	+
Casein	+	−	+	0	0	+	0	+	−	+	0	+
Starch	−	0	0	+	−	−	−	+	0	+	−	−
Agar	−	0	0	0	0	+	0	0	0	0	0	−
Cellulose	−	0	0	0	0	0	+	0	0	0	0	−
Chitin	0	0	0	0	0	0	0	0	0	+	0	−
Acid from												
Glucose	+	+	+	+	−	+	+	+	+	−		+
Lactose^b	−	−	+	+	−	+	+	−	−	−	−	−
Sucrose	+	0	−	+	−	+	+	−	0	+	−	+
Maltose	+	+	+	+	−	+	+	−	0	+	−	+
Methyl red	−	−	0	0	0	0	0	−	0	0	0	−
Citrate	−	−	V	0	0	0	−	+	0	−	0	+
Indole	−	−	+	0	−	−	−	0	−	+	−	−
H$_2$S	−	0	V	0	0	0	0	0	0	+	0	0
Urease	−	0	−	0	0	−	0	−	0	−	0	−
NO$_3$ to NO$_2$	−	0	−	0	0	+	+	+	+	−	−	−
Catalase	+	0	+	0	0	0	+	0	0	0	0	+
Isolated from marine environment	−	−	−	−	+	+	−	+	−	+	+	−
Isolated from soil or fresh water	+	+	−	+	−	−	+	−	+	−	−	+
Human pathogen	−	−	+	−	−	−	−	−	−	−	−	−
Moles % G + C	33	26	36	43	0	32	63	65	69	70	0	69

^{a} Pigmentation: y = yellow; yo = yellow to orange; r = red.
^{b} Slowly; V = variable; 0 = no data recorded.

meninges, membrane covering the brain, Gr. adj. *septikos* putrefactive. M.L. adj. *meningosepticum* apparently referring to association of the bacterium with both meningitis and septicemia and not to septic meningitis as the name infers.

Rods, slender and slightly curved or short with rounded ends. Filaments common. Some strains encapsulated, especially following animal passage (mice).

Neither cellular gliding movement nor swarming growth reported.

Pigmentation is yellow and hue constant on different media. No pigmentation apparent at 37 C. Pigmentation develops slowly (1 week) in cultures held at room temperatures.

Colonies on heart infusion, 5% rabbit blood, 37 C, vary from punctiform to 2 mm diameter, usually 1–1.5 mm. Smooth, entire, glistening, translucent, butyrous, gray-white. Blood not hemolyzed but medium may show green discoloration. Colonial variation has been related to antigenicity which divides strains into groups A, B and C.

Litmus milk alkaline and peptonized.

H_2S test is positive (lead acetate paper, negative with TSI medium).

Grows poorly on MacConkey's agar and not upon SS agar.

Acid reactions (Table 8.28) from tests with nutrient broth basal medium containing bromcresol purple indicator. Complete data for C sources tested: acid from glucose, mannose, fructose, maltose, lactose, trehalose and mannitol; acid not produced from arabinose, xylose, rhamnose, galactose, sucrose, raffinose or salicin. Gas production not observed.

Pathogenic for man, primarily the newly born; meningitis and septicemia are associated with the infections. Not pathogenic for rabbits, hamsters and doubtfully pathogenic for mice. Isolated from spinal fluid, blood, ventricular fluid and throats of human infants, especially those born prematurely, and also from throats of healthy adults and adults with meningitis.

The G + C content of the DNA is 38.3 moles % for serogroup A and 36.4 moles % for serogroups B and C (Mitchell et al., 1969).

Type strain: King 14, Group A; ATCC 13253; NCTC 10016.

Reference strains: King 422, Group B; ATCC 13254; NCTC 10585: King 3375, Group C; ATCC 13255; NCTC 10586.

4. Flavobacterium ferrugineum Sickles and Shaw 1934, 429. (Pseudobacterium ferrugineum (Sickles and Shaw) Krasil'nikov 1949, 234; Empedobacter ferrugineum (sic) (Sickles and Shaw) Prévot 1961, 181.)

fer.ru.gi'ne.um. L. neut.adj. ferrugineum resembling iron rust, dark red.

Rods, small and slender, 0.5–0.7 by 1.0 μm, occuring singly and in pairs.

No gliding movement or swarming growth reported.

Pigment yellow to bright orange, depending upon medium. Blood agar colonies dull, rustcolored, entire, umbilicate, dry consistency, 1 cm diameter. Nutrient agar similar but light yellow. Potato, bright orange.

Poor growth in infusion broth, moderate growth in meat-extract broth, uniform turbidity. Growth enhanced by addition of pneumococcal polysaccharide. Poor growth on nutrient agar slopes.

Acid reactions from dextrin and inulin as well as the substrates in Table 8.28. Hydrolyzes nonspecific polysaccharide from Type I pneumococcus.

Facultatively anaerobic.

Isolated from swamp soil.

The G + C content of the DNA is 42.6 moles % (T_m).

Type strain: ATCC 13524.

5. Flavobacterium halmephilum Elazari-Volcani 1940, 85. (Empedobacter halmephilum (sic) (Elazari-Volcani) Prévot 1961, 182.)

hal.me'phil.um. Gr. n. halme brine, sea water; Gr. adj. philus loving, M.L. neut.adj. halmephilum sea water-loving.

Rods, 0.5–0.7 by 0.7–2 μm; morphology and size unchanged by salt concentrations. Occurs singly and in pairs. No report of gliding cell movement or swarming growth.

Yellow pigmentation not varying with medium and temperature. Pigments not characterized.

Halophilic; does not grow in media containing 0.5% salt but grows well in concentrations from 3–24% and moderately in 30% and in water of the Dead Sea.

Agar colonies (12% NaCl, 1% proteose peptone, 2% KNO_3) are circular, entire, smooth, convex, glistening, opaque and yellow. Moderate growth on sloped media, smooth or slightly roughened, opaque. Broth (12% NaCl) supports moderate growth, uniformly turbid, delicate pellicle, yellow sediment and broth becomes yellow.

No acid reactions in 12% salt-peptone broth containing glucose, fructose, mannose, galactose, lactose, sucrose, maltose, arabinose, xylose, raffinose, inulin, dextrin, glycerol, mannitol or salicin.

Growth temperature optimum, 30 C.

Isolated from water of the Dead Sea.

DNA nucleotide base ratio not known.

Reference strain: ATCC 19717; NCIB 8718.

6. Flavobacterium uliginosum ZoBell and Upham 1944, 263. (Agarbacterium uliginosum (ZoBell and Upham) Breed 1957, 326.)

u.li.gi.no'sum. L. neut.adj. uliginosum wet, damp.

Rods, 0.4–0.6 by 1.2–4 μm, sometimes slightly curved. Gliding movement or swarming growth not reported.

Pigmentation is yellow-orange, no variation in hue reported. Pigments not characterized.

Marine, no growth in fresh water broth or in any non-sea water medium except litmus milk in which growth occurred. All cultural properties are for populations grown in media prepared with sea water unless otherwise specified. Colonies sunken, irregular, gummy, adherent, orange-yellow. Agar becomes discolored and slowly liquefied. Colonies on gelatin, 1 mm. Moderate growth in broth, yellow pellicle.

Acid reactions from xylose and salicin as well as the substrates in Table 8.28. Acid not produced from glycerol or mannitol. Ability to hydrolyze agar is lost during laboratory cultivation.

Ammonia produced from peptone but not urea.

Grows best at temperatures between 20 and 30 C.

Isolated from marine bottom deposits.

The G + C content of the DNA is 32 moles % (T_m).

Type strain: 553; ATCC 14397 (listed as *Agarbacterium*); NCMB 1863.

7. Flavobacterium capsulatum Leifson 1962, 163.

cap.su.la′tum. L. n. *capsula* a small chest, capsule; M.L. neut.adj. *capsulatum* encapsulated.

Rods usually 0.6 by 2–5 µm, short and long chains common. Conspicuous capsule formed under all conditions studied. Neither gliding movement nor swarming growth reported.

Yellow pigmentation on agar media and yellow pellicle in broth. Pigment is not soluble in media, presumably carotenoid.

Colonies on agar fairly large. Growth on agar slopes heavy and mucoid. Broth cultures uniformly turbid with a pellicle.

Nutritionally not fastidious, growing well in glucose, inorganic salts medium, peptone media and with vitamin-free casamino acids.

Acid reactions from xylose, D-sorbitol and raffinose as well as the substrates in Table 8.28.

Described as mesophilic.

Isolated from distilled water in several separate locales.

The G + C content of the DNA is 63 moles % (T_m) (Mitchell *et al.*, 1969).

Type strain: ATCC 14666; NCIB 9890.

8. Flavobacterium lutescens (Migula) Bergey *et al.* 1923, 114. (*Bacterium lutescens* Migula 1900, 476; *Pseudobacterium lutescens* (Migula) Krasil′nikov 1949, 238; *Aplanobacter lutescens* (Migula) Brisou 1955, 170; *Empedobacter lutescens* (Migula) Brisou *et al.* 1960, 492.) Italian vernacular name: Bacillo giallo Lustig 1890, 91.

lu.tes′cens. L. part.adj. *lutescens* becoming muddy.

Rods, 0.5 by 2 µm. Gliding movement or swarming growth not reported.

Pigmentation yellow and no change in hue reported. Pigments not characterized.

Colonies on nutrient agar, 1–2 mm, circular, entire, raised, mucoid, opaque. Broth cultures turbid.

Nutritionally not fastidious.

Litmus milk, poor growth and sometimes an alkaline reaction. Cultures grow well in media containing 4% added NaCl.

Growth temperature range is 15–37 C.

Facultatively anaerobic. Oxidase negative.

Isolated from water, apparently both marine and fresh water.

The G + C content of the DNA is 65 moles % (T_m).

Reference strain: Brisou 2611 (designated *Empedobacter lutescens*).

9. Flavobacterium rigense Bergey *et al.* 1923, 100. (*Bacillus brunneus rigensis* Bazarewski 1905, 1; *Flavobacterium rigensis* (sic) Bergey *et al.* 1923, 100; *Chromobacterium rigense* (Bergey *et al.*) Krasil′nikov 1949, 499.)

ri.gen′se. M.L. neut.adj. *rigense* pertaining to Riga, the city where this species was isolated.

Rods, 0.8 by 1–2.5 µm. Cells commonly encapsulated. Motile by means of peritrichous flagella.

Pigmentation yellow, orange or brown and develops slowly. Noticeable change in hue with variations in incubation temperature. Pigments are reported to be soluble in water and ethanol and insoluble in diethyl ether and may not be carotenoid.

Colonies on nutrient agar and gelatin become pigmented slowly. Colony form is circular, entire to undulate, moderate size. Broth cultures turbid with pellicle and brown sediment.

Nutritionally not fastidious.

Litmus milk unchanged.

Facultatively anaerobic.

Grows well at both 30 and 37 C.

Isolated from soil and fresh water.

The G + C content of the DNA is 69 moles % (T_m).

Reference strain: F18 (O. B. Weeks), isolated by Fred Mindach, Indianapolis, Indiana.

10. Flavobacterium indoltheticum Campbell and Williams 1951, 903. (*Beneckea indolthetica* (Campbell and Williams) Campbell 1957, 331.)

in.dol.the′ti.cum. M.L. n. *indolum* indole; Gr. adj. *theticus* positive; M.L. neut.adj. *indoltheticum* indole-positive.

Rods, 0.5–1.0 by 1.0–1.5 µm occurring singly and in pairs. Motile by peritrichous flagella.

Pigmentation is yellow to yellow-orange, no change in hue with cultural conditions. Preliminary studies indicate the pigment is not carotenoid.

Colonies on nutrient agar, circular, smooth, entire and opaque. Nutrient broth, turbid, slight pellicle and yellow sediment.

Not nutritionally fastidious. Ammonium chloride serves as nitrogen source and a variety of organic compounds as carbon sources.

Acid reactions in nutrient broth containing: dextrin, mannose, raffinose, trehalose or cellobiose and the substrates listed in Table 8.28. Acid not produced from lactose, arabinose, rhamnose, xylose, inulin, mannitol, salicin, dulcitol or inositol.

Ammonia is produced from peptone.

Various carbohydrates, and some fatty acids,

serve as sole carbon sources in an inorganic medium containing ammonium chloride as a nitrogen source including: glucose, maltose, sucrose, dextrin, mannose, raffinose, salicin, cellobiose, trehalose, starch, glycogen, glucosamine, pyruvate and acetate. Carbon and energy needs are not met by: lactose, arabinose, rhamnose, xylose, inulin, mannitol, dulcitol, galactose, inositol, adonitol, succinate, malonate, tartrate, β-alanine, asparagine, propionate, salicylate, valerate, oxalate, butyrate, mandelate, ethanol and propanol.

Will grow in media containing up to 10% NaCl but does not require added salt.

Facultatively anaerobic.

Isolated from marine mud and found in marine environments.

The G + C content of the DNA ranges from 69–70 moles % (T_m).

Type strain: F37 (O. B. Weeks), received from O. B. Williams.

11. **Flavobacterium tirrenicum** Marini and Spalla 1964, 37.

tir.ren'i.cum. L. adj. *tyrrhenus* of Tyrrhen, Tirrene; M.L. adj. *tirrenicum* of or pertaining to Tuscany.

Rods, 0.6–0.7 by 2–6 µm, may be slightly curved.

Motile but type of flagellation not reported.

Pigmented pink or light red (rosato). Pigment not characterized.

Colonies small, glossy and translucent. Scanty growth in broth, uniformly turbid.

Nutritionally fastidious requiring an iron-containing peptide (FT factor) for growth (Marini and Merli, 1964); fish meal, liver extracts and culture filtrates of a *Streptomyces* sp. contain the sideramine which is related to the terregenes factor (*Arthrobacter terregenes*). *F. tirrenicum* seems to represent a type of nutritionally fastidious, pigmented bacterium commonly found in marine habitats.

No acid reactions in sea water media containing common carbohydrates.

Halotolerant.

Optimum temperature for growth, 30 C.

Isolated from deep water of Gulf of Naples.

DNA nucleotide base ratio not known.

Type strain: Na 540; ATCC 15997.

12. **Flavobacterium devorans** (Zimmermann) Bergey *et al.* 1923, 102. (*Bacillus devorans* Zimmermann 1890, 96; *Chromobacterium devorans* (Zimmermann) Krasil'nikov 1949, 499.)

de'vor.ans. L. part.adj. *devorans* consuming, devouring.

Rods, 0.5–1.0 by 1–2 µm, may occur in short chains. Motile by peritrichous flagella and a thin, spreading growth on nutrient agar is common.

Pigmented yellow but not especially pronounced. Pigments not characterized, presumably carotenoid.

Colonies on nutrient agar are 1 mm, or more, circular, convex, entire, opaque and light yellow. Growth may spread upon moist agar. Broth, uniformly turbid with white sediment.

Not nutritionally fastidious.

Litmus milk, unchanged.

Growth temperature range, 15–30 C. No growth on nutrient agar made anaerobic by pyrogallol technique. Growth in media containing up to 8% added NaCl.

Isolated from fresh water.

The G + C content of the DNA is 69 moles % (T_m).

Reference strain: ATCC 10829; NCIB 8195.

Species incertae sedis

I. Species names which have been published validly but which are not well enough described to permit an arrangement within the genus. Reference cultures exist.

a. *Flavobacterium heparinum* Payza and Korn 1956, 854. Small, Gram-negative, motile rod growing at 25 C but not at low temperature or at 37 C. Presumably pigmented but this is not recorded. Known primarily from its ability to degrade heparin which will serve as a sole carbon and nitrogen source. DNA contains 42.2 moles % G + C (T_m) according to Mitchell *et al.* (1969) who regard the bacterium as a *Cytophaga*. Type strain: ATCC 13125; NCIB 9290.

b. *Flavobacterium buccalis* (Chester) Bergey *et al.* 1923, 113. (*Bacillus buccalis minutus* Sternberg 1892, 643; *Bacterium buccalis minutus* (Sternberg) Chester 1897, 108; *Bacterium vignali* Migula 1900, 443; *Bacterium buccalis* Chester, 1901, 167.) The original description is too general to allow any arrangement within the genus. Recently a bacterium designated *F. buccalis* has been studied for its ability to dehydrogenate 3β-ol-Δ⁵-steroids to the corresponding Δ⁴-3-ketones (Protiva *et al.*, 1964) but additional characters of the bacterium are not reported. The steroid-transforming abilities are similar to *F. dehydrogenans* which is a Gram-positive bacterium. DNA nucleotide base ratio is not known. Reference cultures do not exist except that of Protiva *et al.* (1964) No. 38/3, Res. Inst. Natural Drugs, Prague, Czechoslovakia.

II. Species names which have been validly published, adequately described and which are represented by reference cultures but which should not be considered *Flavobacterium* for the reasons specified.

A. Reference cultures are Gram-positive.

   c. *Flavobacterium acidificum* Steinhouse 1941, 757. Type strain: ATCC 8366.

   d. *Flavobacterium arborescens* (Frankland and Frankland) Bergey *et al.* 1923, 113. (*Bacillus arborescens* Frankland and Frankland 1889, 379.) DNA contains 71 moles % G + C (*T_m*). Reference strain: ATCC 4358.

   e. *Flavobacterium dehydrogenans* (Arnaudi) Arnaudi 1942, 358. (*Micrococcus dehydrogenans* Arnaudi 1939, 211.) The species is very well studied and has been known to be Gram-positive since it was first described. For characterization see: Ferarri and Zannini (1958), Ferarri (1963). For definitive characterization of the carotenoid pigments see: Jensen *et al.* (1968), Weeks *et al.* (1969). DNA contains 72–74 moles % G + C (*T_m*). Type strain: ATCC 13930.

   f. *Flavobacterium esteraromaticum* (Omelianski) Bergey *et al.* 1930, 149. (*Bacterium esteraromaticum* Omelianski 1923, 407.) DNA contains 69 moles % G + C (*T_m*). Reference strain: ATCC 8091.

   g. *Flavobacterium flavescens* (Pohl) Bergey *et al.* 1923, 107. (*Bacillus flavescens* Pohl 1892, 141.) DNA contains 71 moles % G + C (*T_m*). Reference strain: ATCC 8315.

   h. *Flavobacterium harrisonii* (Buchanan and Hammer) Bergey *et al.* 1923, 104. (Variety No. 6 Harrison 1905, 129; *Bacillus lactis harrisonii* Conn *et al.* 1907, 169; *Bacillus harrisonii* Buchanan and Hammer 1915, 257.) DNA contains 72 moles % G + C (*T_m*). Reference strain: ATCC 14589.

   i. *Flavobacterium marinotypicum* ZoBell and Upham 1944, 268. DNA contains 69 moles % G + C (*T_m*). Reference strain: ATCC 19260.

   j. *Flavobacterium resinovorum* Delaporte and Daste 1956, 834. DNA contains 70 moles % G + C (*T_m*). Type strain: ATCC 12524.

   k. *Flavobacterium suaveolens* Soppeland 1924, 276. DNA contains 70 moles % G + C (*T_m*). Reference strain: ATCC 958.

B. Reference cultures Gram-negative but not *Flavobacterium* for other reasons.

   l. *Flavobacterium pectinovorum* Dorey 1959, 94. Reference culture is said to show gliding movement (Lund, 1969) and Mitchell *et al.* (1969) consider the species to be a *Cytophaga*. DNA contains 32–33 moles % G + C (*T_m*). Type strain: NCIB 9059; ATCC 19366.

   m. *Flavobacterium proteus* Shimwell and Grimes 1936, 348. (*Obesumbacterium* Shimwell 1963, 759.) The species is reported to produce acid and gas from glucose and maltose in the original description and should not be considered a *Flavobacterium* because of its aerogenic character. DNA contains 30 moles % G + C (*T_m*). Reference strain: NCIB 8771; ATCC 12841.

III. Species validly described but for which no reference cultures are known to exist.

   n. *Flavobacterium amocontactum* ZoBell and Allen 1935, 246. (*Agarbacterium amocontactum* Breed 1957, 325.)

   o. *Flavobacterium balustinum* Harrison 1929, 234.

   p. *Flavobacterium diffusum* (Frankland and Frankland) Bergey *et al.* 1923, 100. (*Bacillus diffusus* Frankland and Frankland 1889, 396.)

   q. *Flavobacterium dormitator* (Wright) Bergey *et al.* 1923, 115. (*Bacillus dormitator* Wright 1895, 442.)

   r. *Flavobacterium fucatum* Harrison 1929, 229.

   s. *Flavobacterium halohydrium* ZoBell and Upham 1944, 278.

   t. *Flavobacterium invisibile* (Vaughan) Bergey *et al.* 1923, 109. (*Bacillus invisibilis* Vaughan 1892, 191; *Flavobacterium invisibilis* (*sic*) Bergey *et al.* 1923, 109.)

   u. *Flavobacterium lactis* (Grimm) Bergey *et al.* 1923, 108.

   v. *Flavobacterium marinum* Harrison 1929, 234.

   w. *Flavobacterium marinovirosum* ZoBell and Upham 1944, 271.

   x. *Flavobacterium neptunium* ZoBell and Upham 1944, 278.

   y. *Flavobacterium okeanokoites* ZoBell and Upham 1944, 270.

   z. *Flavobacterium peregrinum* Stapp and Spicher 1954, 119.

aa. *Flavobacterium solare* (Lehmann and Neumann) Bergey *et al.* 1923, 116.

(*Bacterium solare* Lehmann and Neumann 1896, 258.)

Genus **Haemophilus** Winslow, Broadhurst, Buchanan, Krumwiede, Rogers and Smith 1917, 561

K. ZINNEMANN AND E. L. BIBERSTEIN

Hae.mo′phi.lus or Haem.oph′il.us Gr. n. *haema* blood; Gr. adj. *philos* loving; M.L. masc.n. *Haemophilus* the blood lover.

Minute to medium-sized, **coccobacillary to rod-shaped** cells which sometimes form **threads and filaments,** and may show marked **pleomorphism.** Non-motile. Gram-negative. Strict parasites, **requiring provision of growth factors present in blood, especially X and/or V factors.** Aerobic, facultatively anaerobic. May or may not be pathogenic. Optimum temperature usually 37 C; range usually 25–43 C. Occur in various lesions and secretions as well as on normal mucous membranes of vertebrates.

The G + C content of the DNA (in the species examined) is from 38–42 moles %.

Type species: *Haemophilus influenzae* (Lehmann and Neumann) Winslow *et al.* 1917, 561.

Further Comments

The genus *Haemophilus* is fairly well circumscribed. Some difficulty may be experienced in the demarcation from pasteurellae because on first isolation the latter often do not grow on media without X factor, and because colonial and microscopical morphology can be similar. It is advisable always to use an optimum (chocolate, Levinthal, 1918, or Fildes' agar, 1920) and a fresh blood medium in conjunction, for primary isolation as well as for screening and determination of additional characters. A non-hemolytic *Haemophilus* species would be suggested by good growth on an optimum medium with colonies 1–3 mm in size and by poor or no growth on fresh blood agar. Small or medium-sized hemolytic colonies on fresh blood agar with larger *Haemophilus*-like colonies on an optimal medium suggest one of the hemolytic *Haemophilus* species.

Non-hemolytic and hemolytic species dependent on, or preferring increased CO_2 tension, are recognized by either complete absence of, or very poor growth or hemolysis in air, while luxurious growth is seen in air with at least 5% CO_2.

Good growth on optimum solid media distinguishes *Haemophilus influenzae* and *Haemophilus parainfluenzae* from the poorly growing *Haemophilus suis* and *Haemophilus parasuis*.

Methods of screening for, and determination of V and X requirements are given by Zinnemann, 1960; Zinnemann *et al.*, 1968; and White 1963. The δ-aminolevulinic acid (ALA) test is described by Biberstein *et al.*, 1963. For the growth factor requirements of species see Table 8.29.

Where available, production of acid from sugars is recorded, but the usefulness of sugar reactions in characterizing *Haemophilus* species is disputed.

Description of the species of genus **Haemophilus**

1. **Haemophilus influenzae** (Lehmann and Neumann) Winslow *et al.* 1917, 561. (Influenzabacillus Pfeiffer 1892, 28; *Bacterium influenzae* Lehmann and Neumann 1896, 187; *Mycobacterium influenzae* (Lehmann and Neumann) Chester 1901, 351; *Coccobacillus pfeifferi* Neveu-Lemaire 1921, 20; *Haemophilus meningitidis* (Martins) Hauduroy *et al.* 1937, 254.)

in.flu.en′zae. Italian n. *influenza* influenza; M.L. gen.n. *influenzae* of influenza.

Small rods or filaments 0.2–0.3 by 0.5–2.0 µm. Capsulated strains show iridescence on transparent media in oblique light.

Grows satisfactorily on chocolate agar in air or air with raised carbon dioxide content.

Abundance of growth on blood agar is never great but directly related to the degree of hemolysis and inversely to the age of the blood (V factor).

Odor of cultures often suggestive of indole.

Produces acid from glucose and sucrose and, less consistently, galactose, fructose, maltose and xylose.

Killed after 30 min at 55 C.

Six serotypes based on capsular polysaccharide antigens. Noncapsulated strains are serologically heterogeneous.

Isolated originally from cases of pandemic influenza and regarded as its causative agent at the time; found normally in the nasopharynx of man in small numbers and in clinical conditions in sputum, paranasal sinuses, conjunctivae, cerebrospinal fluid, blood, pus from joints or osteomyelitis, otitis, obstructive epiglottitis and various other sites in rare septic conditions.

Reference strain: NCTC 4560; ATCC 19418 (Sneath and Skerman 1966).

2. **Haemophilus suis** Hauduroy, Ehringer, Urbain, Guillot and Magrou, 1937, 258. (*Haemophilus influenzae suis* Lewis and Shope 1931, 364.)

TABLE 8.29
Differential characteristics of the species of genus Haemophilus[a]

	1. H. influenzae	2. H. suis	3. H. haemolyticus	4. H. parainfluenzae	5. H. parasuis	6. H. para-haemolyticus	7. H. haemo-globinophilus	8. H. influ-enzaemurium	9. H. gallinarum	10. H. paraphro-philus	11. H. para-gallinarum	12. H. para-phrohaemolyticus	13. H. aphrophilus	14. H. ducreyi	15. H. ovis	16. H. pulo-riorum	17. H. citreus	18. H. piscium	19. H. aegyptius
															\multicolumn{5}{ }{Species incertae sedis}				
V requirement	+	+	+	+	+	+	−	−	+	+	+	+	−	−	−	?[b]	?	−†	+
X requirement	+	+	+	−	−	−	+	+	+	−	−	−	+	+	w	?	?	−†	+
Hemolysis	−	−	+	−	−	+	−	−	−	−	−	+	−	(+)	−	−	d	+	−
Increased CO₂ requirement	−	−	−	−	−	−	−	+	+	+	+	+	+	+	−	−	−	−	−
Serum or other additional enrichment required	−	+	−	−	+	−	−	−	+	−	+	−	−	+	−	−	−	−	−
Nitrites from nitrates	+	+	+	+	+	+	+	?	+	+	+	+	+	?	+	?	+	−	+
Indole	d	−	d	d	−	−	+	−	−	−	−	d	−	?	−	?	+	−	+
Catalase	+	+	+	+	+	+	+	−	?	+	−	+	(+)	?	?	?	?	?	?
δ-ALA utilization	−	−	?	+	+	+	?	?	?	?	?	+	?	?	?	?	?	?	?

[a] Species may be identified by the first five requirements listed.

[b] ?, not determined, not known; †, H. piscium requires diphosphothiamine or ATP; w, weak reaction or poor growth, in H. ovis X requirement is demonstrable in freshly isolated cultures; (+), delayed; d, some (less than 90%) strains positive, some negative.

su'is. L. n. sus hog, swine; L. gen.n. suis of swine.

Morphology coccobacillary or filamentous.

Growth on solid media is never abundant. Requires complex organic constituents in media in addition to growth factors. Chocolate agar satisfactory, blood agar less so.

Weak acid from maltose and sucrose.

Three serotypes based on capsular polysaccharides. Noncapsulated cultures are antigenically heterogeneous.

Isolated from all parts of the respiratory tract, heart blood, meninges, serous cavities and joints of swine. When in association with swine influenza virus it is found mainly in the respiratory tract.

Reference strain: NCTC 4557; ATCC 19417 (Sneath and Skerman 1966).

3. Haemophilus haemolyticus Bergey et al. 1923, 269. (Bacillus X, Pritchett and Stillman 1919, 260; also Stillman and Bourn 1920, 665.)

hae.mo.ly'ti.cus. Gr. n. haema blood; Gr. adj. lyticus loosening, dissolving; M.L. adj. haemolyticus blood-dissolving.

Bacillary forms predominate, sometimes with tangled filaments.

Grows quite well on either chocolate or fresh blood agar in air or air with increased carbon dioxide content.

Hemolysis described in liquid and solid blood media.

Acid production from sugars variable.

In the upper respiratory tract of man; usually as a commensal but may cause sore throat when present in profusion.

Sneath and Skerman (1966), in following the Type Culture Collections' catalogues, designated strain NCTC 8479, ATCC 10014 as authentic. In fact, this strain is V-dependent and the strain on which M. Pittman based her description of H. parahaemolyticus.

4. Haemophilus parainfluenzae Rivers 1922, 431.

pa.ra.in.flu.en'zae. Gr. prep. para alongside of, resembling; M.L. n. influenzae specific epithet; M.L. adj. parainfluenzae (Haemophilus)influenzae-like.

Rods sometimes with pointed ends.

Grows well on chocolate agar, less consistently so on fresh blood agar.

Capsulated strains have type-specific antigen and are homogeneous; non-capsulated cultures are heterogeneous.

Acid from glucose, galactose, fructose and sucrose; occasionally also from maltose, xylose and dextrin.

Frequently non-pathogenic in upper respiratory tract of man and cats; organisms indistinguishable from Haemophilus parainfluenzae have also been reported in cattle, sheep and fowl.

5. Haemophilus parasuis Biberstein and White 1969, 77.

pa.ra.su'is. Gr. prep. *para* alongside of, resembling; M.L. n. *suis* specific epithet; M.L. adj. *parasuis* (*Haemophilus*) *suis*-like.

Except for absence of X factor requirements, identical with *H. suis* q.v.

Most strains in culture collections require V factor only and must be regarded, therefore, as *H. parasuis*.

6. **Haemophilus parahaemolyticus** Pittman 1953, 750. (Haemolytic influenza bacillus Fildes 1924, 69; hemolytic strain *Haemophilus parainfluenzae* Valentine and Rivers 1927, 1001; *Haemophilus pleuropneumoniae* White, Leidy, Jamieson and Shope 1964, 1.)

pa.ra.hae.mo.ly'ti.cus. Gr. prep. *para* alongside of, resembling; M.L. n. *haemolyticus* specific epithet; M.L. adj. *parahaemolyticus* (*Haemophilus*)-*haemolyticus*-like.

Stout rods or tangled filaments.

Grows well on chocolate or blood agar.

Acid from glucose and maltose.

Found in the upper respiratory tract of man; frequently associated with acute pharyngitis; causes occasional endocarditis in man and pleuropneumonia as well as septicemia in swine; has been found sporadically in septic conditions of cattle and sheep.

Reference strain (holotype): NCTC 8479; ATCC 10014.

7. **Haemophilus haemoglobinophilus** (Lehmann and Neumann) Murray 1939, 309. (*Bacillus haemoglobinophilus canis* Friedberger 1903, 406; *Bacterium haemoglobinophilus* Lehmann and Neumann 1907, 270; *Hemophilus* (*sic*) *canis* Rivers 1922, 581.)

hae.mo.glo.bi.no'phi.lus. M.L. n. *haemoglobinum* hemoglobin; Gr. adj. *philos* loving; M.L. adj. *haemoglobinophilus* hemoglobin-loving.

Small to long rods and filaments measuring 0.2–0.3 by 0.5–2.0 μm.

Grows well on chocolate agar but some strains will not grow on peptone agar media with X factor added without 0.5% sodium chloride.

Occurs in prepuce and vagina of dogs. No known pathogenicity.

Reference strain: NCTC 1659; ATCC 19416 (Sneath and Skerman 1966).

8. **Haemophilus influenzae-murium** (Kairies and Schwartzer) Lwoff 1939, 171. (*Bacterium influenzae murium* (*sic*) Kairies and Schwartzer 1936, 351; *Haemophilus influenzae murium* (*sic*) Lwoff 1939, 171; *Hemophilus muris* (*sic*) Prévot in Hauduroy *et al.* 1953, 272.)

in.flu.en'zae-mu'ri.um. Ital. n. *influenza* influenza; L. n. *mus, muris* mouse; M.L. gen.pl.n. *influenzae-murium* influenza of mice.

Stout rods or filaments, sometimes with pointed ends.

Grows quite well on chocolate agar. Colonies turn white and opaque with age.

Acid from glucose, lactose, maltose, fructose and sucrose.

Isolated from nose and pharynx of mice ill with respiratory infection and conjunctivitis during epidemics in stock colonies.

9. **Haemophilus gallinarum** Delaplane, Erwin and Stuart 1934, 11. (*Bacillus haemoglobinophilus coryzae gallinarum* de Blieck 1931, 313.)

gal.li.na'rum. L. n. *gallina* hen; L. gen.pl.n. *gallinarum* of hens.

Small rods occurring singly, in pairs, chains or filaments.

Requires approximately 1% sodium chloride in medium and raised CO_2 tension for growth.

Acid from glucose; variable from galactose, mannose, fructose, maltose, sucrose and dextrin.

Killed at 55 C in 4–6 min.

Found in nasal and paranasal sinus exudate of fowl suffering from coryza.

10. **Haemophilus paraphrophilus** Zinnemann, Rogers, Frazer and Boyce 1968, 418.

pa.ra.phro'phi.lus. Gr. prep. *para* alongside of, resembling; M.L. adj. *aphrophilus* specific epithet; M.L. adj. *paraphrophilus* (*Haemophilus*) *aphrophilus*-like.

Organisms are bacillary or filamentous. Involution forms occur under fully aerobic incubation.

Requires increased CO_2 tension for satisfactory growth.

Acid from glucose, galactose, fructose, maltose, sucrose; variable from lactose.

Killed at 50 C after 120 min and at 60 C in less than 10 min.

Occurs in fauces of man; may cause subacute endocarditis, osteomyelitis of maxilla, paronychia, brain abscess and has been isolated from inflamed appendix, from urine of children with congenital malformation of urogenital tract and from vagina of mature women.

Holotype strain: NCTC 10557.

11. **Haemophilus paragallinarum** Biberstein and White 1969, 77.

pa.ra.gal.li.na'rum. Gr. prep. *para* alongside of, resembling; M.L. gen.pl.n. *gallinarum* specific epithet, M.L. adj. *paragallinarum* (*Haemophilus*)-*gallinarum*-like.

Most strains in culture collections labeled *H. gallinarum* do not require X but only V factor and must, therefore, be regarded as *H. paragallinarum*.

Does not require more than 0.5% sodium chloride in media. Two agglutination groups identified.

Other characteristics as *H. gallinarum.* Occurs in nasal exudate of fowls.

12. Haemophilus paraphrohaemolyticus

Zinnemann, Rogers, Frazer and Devaraj 1971, 143.

par.aph.ro.hae.mo.ly'ti.cus. M.L. adj. *paraphro-* resembling *H. aphrophilus;* M.L. adj. *haemolyticus* blood dissolving M.L. adj. *paraphrohaemolyticus* like *H. aphrophilus* but hemolytic.

Growing in air with 10% CO_2, short to medium length rods 0.75–2.5 μm and 0.4–0.5 μm in width with occasional short filament; in air without added CO_2 short to long, coarse rods with involution forms and twisted filaments.

Requires increased CO_2 tension for satisfactory growth, microscopical morphology and good hemolysis.

Acid from fructose, glucose and maltose and variably, although consistently for each strain, from dextrin, galactose, lactose, mannitol, sucrose and xylose.

Gelatin not liquefied; urease reaction positive but may be slow.

No inhibition of growth by optochin.

Killed at 50 C after 10–60 min, at 55 C in 2–18 min, at 60 C in less than 2 min. Survives for 7–8 days at room temperature, for 10 days on incubator shelf and for 21 days at 37 C in jar filled with air with added CO_2.

Occurs in mouth of man, may cause acute sore throat, has been found in ulcer of mouth, in sputum and in urethral discharge of adult males.

Holotype strain: NCTC No. 10670.

13. Haemophilus aphrophilus Khairat 1940, 505.

a.phro'phi.lus. Gr. n. *aphros* foam; Gr. adj. *philos* loving; M.L. adj. *aphrophilus* foam-loving.

Small rods and filaments measuring 0.4 by 1.5–2.0 μm.

Grows well on chocolate agar in 5% CO_2. Requires increased CO_2 tension but produces a comparatively high rate of X- and/or CO_2-independent variants.

Growth is not markedly affected by the presence or absence of NaCl in proteose peptone media with X factor.

Acid from glucose, galactose, fructose, maltose, lactose, sucrose, dextrin.

Killed at 60 C after 10 min. Serologically not homogeneous.

Isolated from blood during life and from heart valves after death from endocarditis in man. Has been reported to occur in pharynx of pet dogs.

Holotype strain: NCTC 5886; ATCC 19415 (Sneath and Skerman 1966); challenged by Boyce *et al.* (1969) who designate NCTC 5906 as holotype.

14. Haemophilus ducreyi (Neveu-Lemaire)

Bergey *et al.* 1923, 271. (*Bacillus ulceris cancrosi* Kruse 1896, 456; *Coccobacillus ducreyi* Neveu-Lemaire 1921, 20.)

du.crey'i. M.L. gen.n. *ducreyi* of Ducrey; named after Ducrey, the bacteriologist who first isolated this organism.

Slender rods in pairs or chains, measuring 0.5 by 1.5–2.0 μm.

Grows moderately well on chocolate agar with increased CO_2 tension, sparingly on blood agar.

No information on acid production from sugars and serology.

Isolated from soft chancre in man. Has become comparatively rare in many parts of the world during last two to three decades.

Species incertae sedis. A

a. *Haemophilus ovis* Mitchell 1925, 12.

o'vis. L. fem.n. *ovis* sheep; L. gen.n. *ovis* of a sheep.

Small pleomorphic rods sometimes in short chains.

Grows on chocolate agar in air.

Acid produced from glucose, galactose, maltose, mannitol, fructose, mannose, raffinose, sorbitol, sucrose and feebly from lactose and xylose.

Optimum temperature 37 C, slight growth at 28 C.

No information on serology.

Isolated from lungs of sheep suffering from bronchopneumonia and generalized hemorrhagic involvement. Pathogenic also for guinea pigs. Has not been recovered since original description. No strains available from culture collections.

b. *Haemophilus putoriorum* Hauduroy *et al.* 1937, 258. (*Bacterium influenzae putoriorum multiforme* Kairies 1935, 17.)

pu.to.ri.o'rum. M.L. masc.n. *Putorius* generic name of ferret; M.L. gen.pl.n. *putoriorum* of ferrets.

Pleomorphic small rods.

Grows on blood media in air. Odor like that of *Haemophilus influenzae.*

No data on acid production from sugars.

Antigenically distinct from *H. influenzae* and *H. suis.*

Different isolates cross-react serologically but are not identical.

Isolated from respiratory tract of ferrets. Nonpathogenic for laboratory animals when inoculated in pure culture but intracutaneous inoculation produces a marked hemorrhagic lesion.

No strain available from culture collections.

c. *Haemophilus citreus* Diernhofer 1949, 588.

cit're.us. M.L. adj. *citreus* lemon-colored.

Rods occurring in short chains or filaments, 0.5 by 1.0–2.0 μm.

Grows well on blood agar in air.

Colonies show yellow pigmentation.

Weak acid production from glucose.

Killed after 30 min at 50 C.

No information on antigenic properties.

Isolated in acute and chronic cases of coital exanthema of cattle but is non-pathogenic for mice, guinea pigs and calves. Does not induce the disease in genital tracts of cattle.

d. *Haemophilus piscium* Snieszko, Griffin and Friddle 1950, 699.

pis'ci.um. L. n. *piscis* fish; gen.pl.n. *piscium* of fishes.

Rods, singly, in pairs or in filaments and measuring 0.5 by 1.0–3.0 μm.

Can be cultured only in media containing fish extract.

Temperature range 3–34 C, optimum 20–25 C.

Acid but no gas from glucose, fructose, sucrose. Delayed acid production from maltose, trehalose, starch. Weak and slow acidification in galactose, mannose, cellobiose, dextrin. Is methyl red positive, Voges-Proskauer negative.

Pathogenic for trout and a cause of ulcer disease in trout.

Reference strain: ATCC 10801 (Sneath and Skerman 1966).

e. *Haemophilus aegyptius* (Trevisan) Pittman and Davis 1950, 413. (*Bacillus aegyptius* Trevisan 1889, 13; *Bacillus conjunctivitidis* Kruse 1896, 440; *Bacterium conjunctivitis* (*sic*) Chester 1897, 67; *Bacterium pseudo conjunctivitidis* (Kruse) Chester 1897, 108; *Bacterium aegyptiacum* Lehmann and Neumann 1899, 191; *Hemophilus* (*sic*) *conjunctivitidis* (Kruse) Bergey *et al.* 1923, 270.)

ae.gyp'ti.us. L. adj. *aegyptius* Egyptian.

Common name: Koch-Weeks Bacillus.

Small rods to filaments measuring 0.25–0.5 by 1.0–2.5 μm.

Grows moderately well on chocolate agar in air.

Forms comet-like colonies in semi-solid agar media.

Produces feeble acidity in glucose, galactose, fructose, appears to be consistently negative on xylose (*cf. H. influenzae*).

Temperature range 25–40 C; optimum 34–37 C.

Serologically heterogeneous. Some antigens "species"-specific, others shared with *H. influenzae*.

The features distinguishing it from *H. influenzae* are minor ones of which one, hemagglutination, is reported to be also a property of some *H. influenzae* strains (Ivler *et al.*, 1963). The differences may justify its designation as a biotype.

Causes acute or subacute infectious conjunctivitis in hot climates.

Reference strain: NCTC 8502; ATCC 11116 (Sneath and Skerman 1966).

Species incertae sedis. B

Haemophilus vaginalis* Gardner and Dukes 1955, 963

S. P. Lapage

(*Hemophilus hemolyticus vaginalis* Lutz, Grootten and Wurch 1956, 132; *Corynebacterium vaginale* Zinnemann and Turner 1963, 214; Leopold (1953) described but did not name the organism; not *Haemophilus vaginalis* Amies and Jones 1957, 582, whose description was based on strains of true *Haemophilus* (Lapage, 1961; Redmond and Kotcher, 1961; Zinnemann and Turner, 1963).)

vag.in.al'is. L. n. *vagina* sheath; L. suf. *-alis* pertaining to; L. adj. *vaginalis* of the vagina.

Description from authors mentioned in text and from: Dukes and Gardner, 1961; Lutz, Wurch and Grootten, 1956; Lutz, Grootten and Wurch, 1956; Lapage, 1961; Redmond and Kotcher, 1961; Park *et al.*, 1968; Dunkelberg and McVeigh, 1969.

Bacilli and coccobacilli, 0.3–0.6 by 1–2 μm. Pleomorphism not marked, filaments not seen. Arranged singly or in pairs lying end to end; palisades, cells lying at sharp angles and clubs may be present. Arrangement suggestive of *Corynebac-*

terium rather than *Haemophilus*. No capsules, spores or flagella.

Gram-variable; uneven staining of cells, some cells retain Gram stain which may be more marked in young (8–12 hrs) cultures. Frankly Gram-positive on inspissated serum (Zinnemann and Turner, 1963). Metachromatic granules with Albert's or Neisser's stain. Sudanophilic inclusions found. Not acid-fast.

Electron microscopy shows cell walls and septa

* **Editorial Note.** This species does not belong in the genus *Haemophilus* but its taxonomic position has not yet been settled. It is put here as the place in which readers will most probably look for it.

to be of Gram-positive type (Reyn *et al.*, 1966). Vickerstaff and Cole (1969) found 6-deoxytalose and no arabinose in the wall of the type strain.

Type strain contained lysine, glutamic and aspartic acids but diaminopimelic acid not demonstrated (Lapage, unpublished).

No growth on usual diagnostic media. No growth on nutrient agar, and poor if any growth on blood agar; on Casman's agar (1947) colonies hardly visible in 24 hrs, 0.1–0.2 mm diameter in 24–48 hrs; convex, circular, entire, homogeneous, smooth, translucent and colorless. On further incubation, become dull and more opaque and may reach 1 mm diameter, no further differentiation occurs.

Reports on hemolysis vary from α,β to nonhemolytic. Differs with species of origin of blood. On Casman's medium with human or rabbit blood, present author finds α-lysis at first then β, followed by conversion of the medium to a chocolate color. Little action on horse or sheep blood.

In stabs in semi-solid media, no growth on surface or in the top few millimeters of the medium, then growth to the base of the medium. Light inocula may form discrete colonies.

In liquid media turbidity and deposit. May be discrete colonies.

Chemoorganotroph, fermentative. Principal product of fermentation acetic acid (Moss and Dunkelberg, 1969). Methyl red test may be positive in glucose-containing media. There are differences in the published results on the fermentation patterns, perhaps due to the different media used. Acid is produced but no gas, and fermentation may be weak especially of the sugars less frequently attacked. Most strains produce acid but no gas from starch, glycogen and dextrin; glucose and maltose are commonly fermented; arabinose, fructose, galactose, rhamnose and xylose may be attacked. Some authors report strains which ferment sucrose, glycerol and inulin. Do not ferment dulcitol, inositol, lactose, mannitol, raffinose, salicin, sorbitol or trehalose.

Catalase and cytochrome oxidase not produced, nitrates not reduced. Peroxidase production variable (Edmunds, 1962). Does not produce acetylmethylcarbinol, H_2S, indole or urease. Fails to liquefy gelatin or coagulated serum; litmus milk no change. Sensitive to hydrogen peroxide by inhibition test on an agar plate (Dunkelberg *et al.*, 1970). Does not decarboxylate lysine, ornithine or arginine, oxidize ethanol, digest cellulose or form starch from maltose (Dunkelberg *et al.*, 1970).

Exacting growth needs, but does not need hemin (X factor), niacin adenine nucleotide (V factor) or similar coenzyme-like substances. A need for blood has been found by most workers and Edmunds (1960) found that a factor in red cell stroma was necessary. Dunkelberg *et al.* (1970)

have grown the organism on a semi-synthetic medium containing enzymic digest of casein, carbohydrates, vitamins, bases, salts and trace metals. They considered that the strains tested required particular commercial proteose peptones as nitrogen, vitamin and purine pyrimidine base sources, five B vitamins: thiamine, riboflavin, nicotinic acid, folic acid and biotin, and one or more nucleic acid bases; the type strain required adenine; the amino acid requirements are not known. Growth improved by fermentable carbohydrates (Zinnemann and Turner, 1963; Dunkelberg *et al.*, 1970).

Facultative anaerobe. Optimum temperature 36–37 C (Dukes and Gardner, 1961). Optimum pH 6–6.5, little growth at pH 4.5, none at pH 8 (Edmunds, 1960). Added CO_2 may improve growth on some media, e.g. chocolate blood agar, but have little effect on others, e.g. Casman's medium. Viability poor, cultures on solid media may die after 24 hrs but semi-solid cultures may live 7 days or longer. Viability good when freeze-dried.

Reports vary on the presence of growth on tellurite media, grows in the presence of 0.01% potassium cyanide; but does not grow in the presence of 0.5% NaCl or 0.5% bile in peptone starch glucose medium (Dunkelberg *et al.*, 1970).

Reports vary but most strains seem to be sensitive to penicillin, chloramphenicol, tetracycline, bacitracin, erythromycin, neomycin. Sensitivity to streptomycin varies.

Seven serological groups by precipitin tests (Edmunds, 1962), common antigen by tube agglutination and fluorescent antiserum (Redmond and Kotcher, 1963). Some strains show hemagglutination (cold) probably not fimbrial, of human, horse, ox and fowl red cells, but not sheep or rabbit (Edmunds, 1962).

Found in the human genital tract, and has been isolated in America, Europe and Australia. Reported as cause of 'non-specific' vaginitis and urethritis. Masses of bacterial cells may be seen to cover the epithelial cells in the discharge, the so-called 'clue-cells.' Transmission has been claimed in human volunteer experiments (Gardner and Dukes, 1955). Not pathogenic for laboratory animals.

The G + C content of the DNA is 42 ± 1 moles % (strain Dukes 594).

Type strain (holotype): 594, Gardner and Dukes; NCTC 10287; ATCC 14018.

Further Comments

H. vaginalis is a perfectly good species although reported characteristics vary in detail partly due to the use of different media, difficulty in obtaining pure cultures and confusion with other organisms.

For a time there was some doubt as to whether strains distributed as 594 were authentic (unpublished). Strains NCTC 10287 and ATCC 14018, both listed as strain 594 in the respective catalogues, have, however, a G + C content of 42 ± 1 moles % (Lapage, 1973, unpublished).

It has been suggested that the species belongs to *Corynebacterium* (Zinnemann and Turner, 1963; Dunkelberg *et al.*, 1970) but this is unlikely since its cell wall does not contain DAP and contains 6-deoxytalose in place of arabinose. It differs from *Corynebacterium cervicis* (Vickerstaff and Cole, 1969) and from *C. parvum* (Zinnemann and Turner, 1963). *Lactobacillus* has been proposed (Amies and

Garabedian, 1963; Garabedian, 1969) but it probably does not belong to that genus or to *Eubacterium* or *Propionibacterium* since acetic acid is the principal acid produced rather than lactic, butyric or propionic (Moss and Dunkelberg, 1969). The finding in the cell wall of 6-deoxytalose (Vickerstaff and Cole, 1969) and of lysine, glutamic acid and aspartic acid but no DAP (Criswell *et al.*, 1971; Lapage, unpublished) suggests a possible relationship to *Actinomyces bovis*.

The consensus is that the organism represented by strain 594 does not belong in the genus *Haemophilus* but further study is required before it can be assigned to a definite genus.

Genus **Pasteurella** *Trevisan 1887, 94 Nom. cons.* Opin. 13, Jud. Comm. 1954, 153

J. E. SMITH

Pas.teu.rel'la. M.L. dim.fem.n. *Pasteurella;* named for Louis Pasteur.

Cells ovoid or rod-shaped, 1.4 ± 0.4 by 0.4 ± 0.1 μm, singly or less frequently in pairs or short chains. **Non-motile;** do not form endospores. **Gram-negative; bipolar staining** common, especially in preparations made from infected animal tissues stained with Giemsa or methylene blue.

Chemoorganotrophic, growing best on media containing blood; minimal nutritional requirements variable. **Metabolism fermentative;** glucose and other fermentable substances including fructose attacked with the production of small amounts of acid but not gas.

Catalase positive; almost always **oxidase positive;** reduce nitrates; gelatinase negative; M.R. and V.P. negative. Aerobic, **facultatively anaerobic;** temperature range 22–42 C, optimum 37 C.

Parasitic on mammals (including man) and birds.

G + C content of DNA of those members tested is in the range 36.5–43.0 moles %.

Type species: *Pasteurella multocida* (Lehmann

and Neumann) Rosenbusch and Merchant 1939, 85 (see "Further Comments," below).

Further Comments

The genus is closely associated with the genus *Actinobacillus*, to which the species *P. haemolytica* shows the strongest relationship.

The genus is a heterogeneous group of usually bipolar bacteria, parasitic on a wide range of mammals and birds, of considerable importance in diseases of domesticated animals and also occurring in man. Characteristics differentiating the species are given in Table 8.30.

The type species presents a problem in that the Judicial Commission, by Opinion 13, designated *P. cholerae-gallinarum*, an organism now included in the umbrella species, *P. multocida*. However, for many years bacteriologists in all parts of the world have regarded *P. multocida* as the type and, as in the seventh edition of THE MANUAL, that practice is followed here.

Description of the species of genus **Pasteurella**

1. **Pasteurella multocida** (Lehmann and Neumann) Rosenbusch and Merchant 1939, 85. (*Pasteurella cholerae-gallinarum* Trevisan 1887, 94; *Pasteurella bollingeri* Trevisan 1889, 21; *Bacterium bipolare multocidum* Kitt 1893, 304; *Bacterium multocidum* Lehmann and Neumann 1899, 196; *Bacillus polaris septicus* Hutyra 1913, 67; *Pasteurella septica* Topley and Wilson 1929, 488.)

This summary gives only the principal landmarks in the lengthy development of the nomenclature of this species. Hussaini (1966) lists completely the large number of synonyms used between 1878 and 1966, comprising varied combina-

tions of seven generic names (*Bacillus, Bacterium, Coccobacillus, Eucystia, Micrococcus, Pasteurella, Octopsis*) with 39 specific epithets. It was long considered that isolates derived from different species of host animal represented different species of the organism, largely because naturally occurring outbreaks of disease rarely crossed to different host forms, and the period 1880–1930 produced a great number of specific epithets coined to indicate these special forms, such as *bovicida, bubalseptica, cuniculicida, equiseptica, ferarum, gallinae, lepiseptica, oviseptica, muricida, suilla, vituliseptica* and others. It became apparent that most of these forms were probably no more than minor biotypes of a single species for which the name *Pasteurella septica* was suggested by Topley and Wilson (1929) and *Pasteurella multocida* by Rosenbusch and Merchant (1939); the latter name is now universally employed.

mul.to′ci.da. L. adj. *multus* many; L. adj.suf. *-cidus* from L. v. *caedo* cut, kill; M.L. adj. *multocidus* many-killing, i.e. pathogenic for many (species of animal).

Cells usually coccobacillary or short rods in cultures from diseased tissues; strains from healthy animals often pleomorphic with longer bacillary forms and occasional short filaments. Bipolar staining usual in the coccobacillary and rod forms. Capsules, when present, best demonstrated in cultures by negative staining methods; in infected tissues better seen in Giemsa-stained preparations.

Growth from small inocula poor except on media containing blood or hematin. Non-hemolytic, but most strains produce a brownish discoloration of blood media in regions of confluent growth. Mucoid, smooth and rough variants occur (Carter and Bain, 1960); colonies of capsulated smooth strains are iridescent in obliquely transmitted light (Smith, 1958). Cultures on blood agar have a faint but distinctive smell of value in recognition.

Acid produced, usually within 48 hrs, from galactose, mannose and sucrose. Most strains ferment mannitol, sorbitol and xylose (Table 8.30), although strains from cats and dogs may be negative. Adonitol, amygdalin, erythritol, esculin, glycogen, inulin, inositol, lactose, melezitose, melibiose, rhamnose, salicin and sorbose generally not attacked. Arabinose and dulcitol fermentation most common in strains from birds. Dextrin, glycerol, maltose and raffinose occasionally fermented; attack of maltose is characteristic of strains from dogs and cats.

Indole, ammonia and H_2S produced. Urease negative. Ornithine decarboxylase usually produced; arginine dihydrolase, lysine and glutamic acid decarboxylases negative. Malonate and β-galactosidase negative. Gluconate oxidation variable. Most strains fail to grow on MacConkey's agar; a few are oxidase negative when tested by Kovacs' method. Most strains grow in KCN.

Growth occurs between 25–40 C; most avian strains and a few others grow at 42 C. Optimum temperature 37 C.

Most strains highly sensitive *in vitro* to penicillin and generally show an antibiotic sensitivity pattern more resembling that of a Gram-positive than a Gram-negative bacterium.

Highly pathogenic for mice and rabbits; injection of one to ten cells of a virulent strain often produces a rapidly fatal septicemia. Strains from birds also pathogenic for pigeons.

Found in Asian and African hemorrhagic septicemia of ruminant animals, in pneumonic conditions in cattle, calves, sheep and pigs, in some cases of transit fever (transit pneumonia) of young cattle, in cat- and dog-bite wounds in cats, dogs and humans, in occasional respiratory and other infections of man, in fowl, duck and turkey cholera. It may be a commensal in the throats and noses of domesticated and wild mammals and birds of a wide variety of species. There is a marked degree of specificity for the different host species and some characteristics may vary according to origin; strains from pigs are often mucoid and strains from poultry often ferment arabinose.

Four main antigenic groups, designated I to IV, were distinguished by Roberts (1947) using mouse protection tests. The serological classification in current use is that of Carter (1952, 1957, 1961) in which the five types depend on capsular or surface polysaccharides. Carter's types A and B are probably the same as Roberts' II and I, respectively; C and D may correspond to Roberts' III and IV (Carter, 1955). Serotyping is at present carried out by a simplified form of the passive hemagglutination technique described by Carter (1955) or by a slide capsular agglutination technique using young (6-hr) cultures (Namioka and Murata, 1961). Capsular types show broad patterns of host specificity. The mucoid material found in strains from pigs and poultry appears to be a form of hyaluronic acid and may interfere with the antigenicity of the

TABLE 8.30

Characters of the species of genus **Pasteurella**

Characteristic	1. P. multocida	2. P. pneumotropica	3. P. haemolytica	4. P. ureae
Hemolysis on blood agar	−[a]	−	+	+
Growth on MacConkey's agar	−	−	+	−
Indole	+	+	−	−
Urease	−	+	−	+
H_2S	+	+	d	−
Ornithine decarboxylase	+	d	d	−
Mannitol	+	−	+	+
Sorbitol	+	−	d	+
Trehalose	d	+	d	−
Xylose	d	d	+	−
Lactose	−	d	d	−

[a] − = most (90% or more) strains negative; + = most (90% or more) strains positive; d = some (less than 90%) strains positive, some negative.

type-specific polysaccharide and its reactions in capsule-typing techniques.

The G + C content of DNA is in the range 36.5–40.05 moles % (Belozersky and Spirin, 1960; Bailie, 1969).

Suggested working strain; NCTC 3195 (Sneath and Skerman 1966, 88).

2. **Pasteurella pneumotropica** Jawetz 1950, 179.

pneu.mo.tro′pi.ca. Gr. n. *pneumon* lung; Gr. n. *tropicus* tropic, circle; M.L. fem.adj. *pneumotropica* having an affinity for the lungs.

Cells rod-shaped, about 1.2 by 0.5 μm with occasional longer forms. Bipolar staining not common.

Colonies on blood agar 1.6–2.0 mm in diameter after 48 hrs at 37 C, smooth, grayish translucent, butyrous, with a characteristic odor. Not hemolytic.

Acid produced only in small amounts from fermentable substrates: dextrin, galactose, glycerol, maltose, mannose, starch and trehalose are fermented by most strains; lactose by many, amygdalin and raffinose by some. Adonitol, arabinose, cellobiose, dulcitol, glycogen, inositol, erythritol,

mannitol, rhamnose, salicin, sorbitol and xylose are usually not attacked.

Indole is produced and, usually, H_2S. Arginine dihydrolase and lysine decarboxylase not produced; a few strains produce ornithine decarboxylase. Most strains fail to grow on MacConkey's agar, but some grow in KCN.

Aerobic, facultatively anaerobic. Growth occurs between 22–40 C; optimum 37 C.

Some degree of sensitivity to penicillin *in vitro*, but this is less marked than that of *P. multocida*.

Serological cross-reactions may occur with *P. multocida* due to the sharing of minor somatic antigens.

P. pneumotropica occasionally causes enzootic disease of mice, rabbits and other laboratory rodents; these include respiratory infections and abscesses. Although of low experimental pathogenicity, it is considered to be a potential pathogen of laboratory animal stock and is required to be excluded from rodent colonies designated as within the higher grades of specific pathogen-free (SPF) stock. Frequently present in the nasopharynx of non-SPF rats, mice and hamsters; it occurs in the mouth and throat of healthy dogs. Occurs occasionally in infections of man following dog bites (Winton and Mair, 1969) although *P. multocida* is more common; occasionally present in the human respiratory tract (Henriksen and Jyssum, 1961; Henriksen, 1962). Winton and Mair considered this bacterium to be a variant of *P. multocida* and suggested the name *P. multocida* var. *ureae*.

3. **Pasteurella haemolytica** Newsom and Cross 1932, 715.

hae.mo.ly′ti.ca. Gr. n. *haema* blood; Gr. adj. *lyticus* dissolving; M.L. fem.adj. *haemolytica* blood-dissolving.

Cells rod-shaped, somewhat larger than *P. multocida;* slight pleomorphism and bipolar staining seen, but capsular material not usually demonstrable by microscopical methods.

Zone of hemolysis surrounds colonies of freshly isolated strains but may be reduced or lost after a few subcultures. A double zone of hemolysis on lamb's blood agar is characteristic; no effect on ovine erythrocytes suspended in broth.

Fermentation tests are carried out in peptone water with 1% substrate, 10% nutrient broth, and bromthymol blue as indicator. Dextrin, glycogen, inositol, maltose, mannitol, raffinose, sorbitol, and starch attacked, but not adonitol, erythritol, or inulin; results vary with dulcitol, glycerol and rhamnose.

Two biotypes, A and T, recognized (Smith, 1961); the differences are summarized in Table 8.31 (Biberstein and Francis, 1968; Shreeve, Ivanov and Thompson, 1970).

Serotypes (1–12) distinguished on the basis of

TABLE 8.31

Characteristics of the biotypes of
Pasteurella haemolytica

	A	T
Acid (within 10 days) from		
Arabinose	+	−
Xylose	+	−
Lactose	d	−
Mannose	−	d
Salicin	−	d
Trehalose	−	+
Cultures	Die out rapidly	Survive longer
Penicillin	Sensitive	Resistant
Capsule serotypes	1, 2, 5, 6, 7, 8, 9, 11, 12	3, 4, 10
Somatic serotypes	A, B	C, D
Found as		
Pathogen	Pneumonia in cattle and sheep Septicemia in newborn lambs	Septicemia in lambs over 3 months
Commensal	Nasopharynx of cattle and sheep	−

indirect hemagglutination tests using soluble surface (capsule) antigens (Biberstein, Gills and Knight, 1960; Biberstein and Gills, 1962; Biberstein and Thompson, 1966); these correlate with the Smith biotypes but non-serotypable strains occur (Åarslaff, Biberstein, Shreeve and Thompson, 1970).

Pathogenic for mice only when inoculated intracerebrally or with gastric mucin.

Found in a common form of enzootic pneumonia of sheep and in septicemia of lambs; occurs, sometimes in mixed culture with *P. multocida*, in pneumonia in cattle. Found in disease of fowls, including a condition resembling fowl cholera.

4. **Pasteurella ureae** Jones 1962, 150. (*Pasteurella haemolytica* var. *ureae* Henriksen and Jyssum 1960, 443.)

u.re′ae. Gr. n. *urum* urine; M.L. gen.n. *ureae* of urine.

Cells rod-shaped, pleomorphic, depending on growth medium. Bipolar staining occasionally seen. Growth best on medium containing blood or serum; on blood agar colonies surrounded by zone of hemolysis.

Attacks fructose, mannitol and sorbitol but not usually dulcitol, glycerol, lactose, salicin or xylose.

Occurs infrequently in noses of healthy persons and in occasional cases of ozaena and other infections of the nose.

Although originally described as a variant of *P. haemolytica* its affinities with that organism are doubtful (Jones, 1962). It appears to be phenetically closer to *P. pneumotropica* (Smith and Thal, 1965) and probably represents only one of an undescribed range of *Pasteurella* intermediates.

Type strain: Henriksen 3520/59; NCTC 10219; ATCC 25976.

Species incertae sedis

a. *Pfeifferella anatipestifer* Hendrickson and Hilbert 1932, 249. (*Pasteurella anapestifer* (*sic*)

Hauduroy *et al.* 1953, 367; *Pasteurella anatipestifer* (Hendrickson and Hilbert) Merchant 1957, 397.)

Cells rod-shaped, about 0.5 μm by 1.0–2.0 μm, occurring singly and in pairs; frequently pleomorphic on subculture, with filaments developing in some strains.

Colonies on blood agar small, transparent, with a distinct glutinous consistency; no hemolysis. Slight turbidity in broth and peptone water with a surface ring; growth in liquid media improved by horse serum. Thiamin required for growth, but low concentrations of pyrithiamin and amprolium are inhibitory. With moist incubation conditions, CO_2 not required (Harry, 1969).

No growth on MacConkey's agar or on yeast extract agar containing 0.05% sodium taurocholate.

No acid from any carbohydrate or other substrate; liquefies gelatin, coagulated serum and egg; no change in litmus milk; nitrates not reduced; indole, citrate, malonate, phenylalanine deaminase, KCN and urease negative; catalase, oxidase and arginine dihydrolase positive.

Isolated from septicemic disease in ducks and ducklings in U. S. A. and U. K.; also from septicemia in geese, turkeys and waterfowl. Agglutination tests indicate eight serotypes of which type A, the most common, is associated with the highest flock mortality and the only one to reproduce the disease experimentally (Harry, 1969).

Appears to be almost identical with *P. septicaemiae*, described in the seventh edition of THE MANUAL and isolated from septicemia in young geese. It should probably be placed in the genus *Moraxella*.

b. *Pasteurella pfaffi* (Hadley) Hauduroy, Ehringer, Guillot, Magrou, Prévot, Rosset and Urbain 1953, 377. (*Bacillus pfaffi* Hadley 1918, 204.)

Isolated from septicemia in canaries. Although no cultures now exist, this organism was probably a minor variant of *Pasteurella multocida*.

Genus **Actinobacillus** Brumpt 1910, 849

J. E. PHILLIPS

Ac.ti.no.bac.il′lus. Gr. n. *actis, actinis* a ray; L. dim.masc.n. *bacillus* a small staff or rod; M.L. masc.n. *Actinobacillus* ray bacillus or rod.

Cells spherical, oval or rod-shaped, 0.4 ± 0.1 by 1.0 ± 0.4 μm with mostly **bacillary** but **interspersed with coccal elements** which often lie at the pole of a bacillus **giving a characteristic Morse code form;** occasionally longer forms up to 6 μm, more common on media containing glucose or maltose; arranged singly, in pairs or more rarely in chains. **Non-motile.** Do not form endospores.

Gram-negative. Irregular staining. Small amounts of extracellular slime may be demonstrated in wet india ink preparations.

Cultures very sticky especially on primary isolation; colonies may be difficult to remove completely from the agar surface. Surface cultures are of low viability, dying out in 5–7 days.

Chemoorganotrophs: metabolism fermentative.

Acid but no gas within 24 hrs from glucose, fructose and xylose. Dulcitol, inositol and inulin are not fermented. Other carbohydrates may be fermented with acid but no gas. β-**Galactosidase positive.** Methyl red test negative.

Hydrogen sulfide produced by most strains (tested by lead acetate paper). **Nitrates reduced to nitrites. Indole not produced. Urease positive.**

Growth reported only on complex media; minimal nutritional requirements unknown. Grows on MacConkey's medium.

Facultative anaerobes.

Temperatures range 20–42 C; optimum 37 C.

Not inhibited by novobiocin or oleandomycin.

The G + C content of the DNA ranges from 40.6–42.0 moles % (T_m).

Type species: *Actinobacillus lignieresii* Brumpt 1910, 849.

Further Comments

Three species included in the genus in the last edition of THE MANUAL are excluded from it in the present edition. *Actinobacillus mallei* will be found as *Pseudomonas mallei* on page 228; *A. actinomycetemcomitans* and *A. actinoides* are listed as "Species incertae sedis", together with several other species, assigned to the genus by the workers describing them, whose taxonomic position is still doubtful.

Failure to ferment any of a wide range of carbohydrates should exclude from this genus *Actinobacillus seminis* Baynes and Simmons 1960, 459, isolated from cases of ovine epididymitis.

Description of the species of genus **Actinobacillus**

The main differences between *Actinobacillus lignieresii* and *A. equuli* are shown in Table 8.32.

1. **Actinobacillus lignieresii** Brumpt 1910, 849. (Actinobacilo Lignières and Spitz 1902, 169; l'actinobacille Lignières and Spitz 1902, 487; *Actinobacillus lignieresi* Brumpt 1910, 849; *Bacterium purifaciens* Christiansen 1917, 458.)

lig.ni.e.re.'si.i. M.L. gen.n. lignieresii of Lignières; named for J. Lignières, one of the bacteriologists who first isolated this organism.

Cells usually rod-shaped, but marked variability depending upon the growth medium. Long bacillary forms most frequently seen on media containing glucose and maltose; shorter bacillary and coccobacillary forms more usual on media containing blood or serum. Extracellular slime cannot be demonstrated in stained smears.

Stickiness of colonies lost on repeated subculture. Fluorescent, granular and dwarf colonial variants have been described. Broth cultures uniformly turbid; little deposit.

Acid but no gas within 24 hrs from maltose, mannitol, mannose and dextrin; most strains attack galactose and sucrose promptly but some do so more slowly. Lactose is fermented slowly (3–7 days) by most strains, but some fail to do so. Arabinose and glycerol are fermented by most strains. Salicin, rhamnose and trehalose are not fermented; raffinose is usually not attacked but a small number of strains may give positive results.

Most strains utilize glucose and maltose to synthesize starch. Sodium gluconate not oxidized.

Most strains give a weakly positive catalase test. The oxidase test may be positive or negative. Weak positive reaction in methylene blue test.

Aerobic, facultatively anaerobic.

Growth occurs between 20 and 39 C. Optimum temperature 37 C. Does not grow at 44 C.

Pathogenic for cattle and sheep. Found in granulomatous lesions of the upper alimentary tract of cattle, particularly in the tongue (wooden tongue). Associated with suppurative lesions in the skin and lungs of sheep. May also be associated with granulomatous lesions in other sites in both these species. Christiansen (1917) first described the organism in sheep as *Bacterium purifaciens* but Tunnicliff (1941) compared it with *A. lignieresii* and concluded that they were identical. Has been recovered from ruminal contents of cattle and sheep (Phillips, 1961) and from the lining membrane of the mouth (Phillips, 1964). Has been described as an etiological agent in disease in man (Thompson and Willius, 1932) and in the dog (Kemenes and Markoi, 1959).

Not pathogenic for rabbits, mice or guinea pigs.

TABLE 8.32
Differences between **Actinobacillus lignieresii** *and* **Actinobacillus equuli**

	A. lignieresii	A. equuli
Broth deposit	$-^a$	+
Methylene blue reduction	+	−
Voges-Proskauer test	d	−
Sodium hippurate hydrolysis	−	+
Gelatinase	−	+
Fermentation of		
Raffinose	−	+
Trehalose	−	+

a − = most (90% or more) strains negative in this test; + = most (90% or more) strains positive in this test; d = some (less than 90%) strains positive, some negative.

Six antigenic types have been described (Phillips, 1967) which can be distinguished by differences in heat-stable antigens. There is evidence of host specificity with antigenic types 3 and 4 which have been recovered only from sheep. Heat-labile antigens associated with extracellular slime can be demonstrated.

The G + C content of the DNA ranges from 41.8–42.6 moles % (T_m) (Boháček and Mráz, 1967).

2. **Actinobacillus equuli** (van Straaten) Haupt 1934, 513. (*Bacillus nephritidis equi* Meyer 1910, 154; *Bacterium viscosum equi* Magnusson 1917, 146; *Bacillus equuli* van Straaten 1918, 75; *Bacillus pyosepticus equi* de Blieck and van Heelsbergen 1919, 496; *Bacillus equirulis* de Blieck and van Heelsbergen 1919, 496; *Bacterium pyosepticum viscosum* Miessner 1921, 186; *Bacterium pyosepticum viscosum equi* Lütje 1921, 468; *Bacterium pyosepticum* Miessner and Berge 1922, 481; *Bacterium equi* Weldin and Levine 1923, 16; *Bacillus pyosepticus* Clarenburg 1925, 196; *Eberthella viscosa* Snyder 1925, 481; *Shigella equi* (Weldin and Levine) Weldin 1927, 183; *Shigella viscosa* (Snyder) Bergey 1930, 363; *Shigella equirulis* (de Blieck and van Heelsbergen) Edwards 1931, 491; *Shigella equuli* (van Straaten) Dimock, Edwards and Bruner 1947, 16).

e.quu'li. L. n. *equulus* a foal; L. gen.n. *equuli* of a foal.

Cells usually rod-shaped, but showing marked variability depending upon the growth medium. Long bacillary forms similar to those seen with *Actinobacillus lignieresii* occur on media containing glucose or maltose.

Capsules are not produced, but considerable amounts of extracellular slime may be seen in wet india ink preparations and as faintly staining interstitial substance in stained smears.

Stickiness of colonies not lost on repeated subculture. Broth cultures show uniform turbidity at first which is followed later by the development of a marked deposit. The medium increases in viscosity so that a strand of sticky material may be drawn out between the surface of the broth and an inoculating loop. Low viability of fluid cultures in media containing even small amounts of fermentable substrates.

Acid but no gas within 24 hrs from raffinose, sucrose and trehalose; all strains ferment galactose, lactose, maltose and mannitol, most within 24 hrs but a few give a delayed reaction. All strains give a delayed fermentation of dextrin and mannose. Arabinose, glycerol, rhamnose and salicin may or may not be fermented. Most strains utilize glucose and maltose to synthesize starch but the species is not as active in this respect as *A. lignieresii*.

Sodium gluconate not oxidized.

Most strains are catalase and oxidase positive. Aerobic, facultatively anaerobic.

Growth occurs between 20 and 39 C; some strains will grow at 44 C.

Pathogenic for horses and pigs. Causes suppurative lesions especially in the kidneys and joints of foals and piglets, and endocarditis in pigs. Also occurs in normal horses in the mouth (Cottew and Francis, 1954), the tonsillar region (Dimock *et al.*, 1947) and the intestinal tract (Laudien, 1923). Not pathogenic for rabbits, guinea pigs or rats. Cultures injected by the intraperitoneal route may show slight virulence for guinea pigs (Haupt, 1934), but this is a non-specific reaction which may be induced by killed organisms (Dimock *et al.*, 1947).

There is considerable serological heterogeneity between strains. Evidence of an antigenic relationship with *Actinobacillus lignieresii* has been shown (Haupt, 1934; Vallée, Thibault and Second, 1963). Haupt (1934) reported cross-agglutination reactions with *Pseudomonas mallei*.

The G + C content of the DNA ranges from 40.0–41.8 moles % (T_m) (Boháček and Mráz, 1967).

Species incertae sedis

a. *Bacterium actinomycetem comitans* Klinger 1912, 198. (*Bacterium acetinomycetum comitans* (*sic*) (Klinger) Colebrook 1920, 208; *Bacterium comitans* Lieske 1921, 233; *Actinobacillus actinomycetem-comitans* (Klinger) Topley and Wilson 1929, 256.)

Cells spherical and rod-shaped 0.7 ± 0.1 by 1.0 ± 0.4 μm. Bacillary forms seen more frequently in agar cultures than in broth or gelatin cultures. Non-motile. Gram-negative.

Agar colonies 1 mm in diameter after 2–3 days. Adherent to the medium. Difficult to break up. Colonies described as star-like (Colebrook, 1920) or like "crossed cigars" (Heinrich and Pulverer, 1959).

Growth in broth in the form of granules at the bottom and up the sides of the tube. On repeated subculture strains may produce a uniformly turbid growth in broth.

Attack of carbohydrates variable. Glucose fermented by all strains with acid but no gas produced. Eight biotypes recognizable on basis of reactions with mannitol, galactose and xylose (Pulverer and Ko, 1970). Variable result with lactose, maltose, sucrose and dulcitol. No fermentation of arabinose, inositol, glycerol, inulin, raffinose, salicin and xylose.

Hydrogen sulfide not produced. Nitrates may or may not be reduced. Indole not produced. Urease negative. Grows under aerobic and slightly better under anaerobic conditions, (Thjøtta and Sydnes, 1951). Microaerophilic (Heinrich and Pulverer, 1959). Growth is improved by increased CO₂ tension (Holm, 1954).

No growth at 22 C. Optimum growth at 37 C (Heinrich and Pulverer, 1959).

Serologically uniform (Heinrich and Pulverer, 1959).

Non-pathogenic for laboratory animals. Pathogenicity for man is doubtful since it occurs in conjunction with actinomycetes in lesions of actinomycosis. Thjøtta and Sydnes (1951) reported a jaw infection in a woman due solely to *Bacterium actinomycetemcomitans*. Isolated from abscesses in a naturally occurring infection in laboratory mice (Vallée and Gaillard, 1953).

b. *Bacillus actinoides* Smith 1918, 342. (*Actinobacillus actinoides* (Smith) Topley and Wilson 1929, 256.)

Cells rod-shaped arranged in groups in tissues. In culture coccoid and bacillary forms occur. In the condensation water of coagulated serum, small flakes up to 1 mm in diameter occur consisting of sheathed filaments each of which terminates in a club-shaped expansion. The sheaths and clubs cannot be stained and enclose chains of minute bacilli. Capsules and clubs are not formed on agar media containing blood or milk. Non-motile. Gram-negative.

Growth on nutrient agar poor. No growth in nutrient broth. On coagulated serum whitish flocculi appear after 3 days but surface colonies do not make their appearance for several weeks. Growth may occur on agar to which a piece of guinea pig spleen has been added (Topley and Wilson, 1929).

Growth only occurs under raised CO₂ tension.

Optimum temperature 37 C.

Not pathogenic for laboratory animals. Intratracheal injection in calves causes lesions in the lungs similar to those seen in the natural disease. Pathogenic for goats by the intratracheal route. Isolated from cases of chronic pneumonia in calves, but its etiological significance is not convincing. Comment: Smith (1918) commented on the similarity of this organism to *A. lignieresii* and Topley and Wilson (1929) included it in the genus. Dienes and Edsall (1937) noted the resemblance between *B. actinoides* and *Streptobacillus moniliformis* but Levi and Cotchin (1950) concluded that the organisms were distinct, but thought that they should be placed in the same group.

c. *Actinobacillus capsulatus* Arseculeratne 1962, 38.

Cells rod-shaped, 0.6 by 1.2 μm. Old cultures show filamentous bacilli with fragmentation into minute coccoid bodies. Capsules present. Monili-

form bodies produced in 5-day cultures on Loeffler's serum. Do not form endospores. Non-motile. Gram-negative. Not acid-fast.

Primary cultures will not grow on nutrient agar or in nutrient broth, but subcultures grow as pinpoint colonies or a faint turbidity.

Colonies on sheep blood agar very sticky. "Flower Head" colonies produced on rabbit blood agar. Non-hemolytic.

Growth on Loeffler's serum scanty, but profuse on Dorset egg.

Small discrete mural colonies in serum broth.

Acid but no gas from glucose, lactose, maltose, sucrose, mannitol and galactose. Arabinose not fermented. Gelatinase negative.

Weakly positive catalase test. Nitrites produced from nitrates. Citrate not utilized. Urease produced.

Aerobic, facultatively anaerobic. Growth favored on primary isolation only by the addition of 10% CO₂.

Optimum temperature 37 C. No growth at 22 C. Killed by heat at 60 C for 10 min.

Viability best on Dorset egg on which it survives no longer than 10 days. Resistant to penicillin; sensitive to streptomycin, tetracycline and chloramphenicol.

Pathogenic for rabbits, guinea pigs less susceptible and in rats and mice only transient infections after subcutaneous injection. Isolated from lesions in the region of tarsal joints in three rabbits.

d. *Actinobacillus suis* van Dorssen and Jaartsveld 1962, 456.

Cells rod-shaped, 0.3–1.0 μm in length, occurring singly, in chains and in masses. Non-motile. Gram-negative.

Colonies on serum agar adherent to medium but not markedly stringy. Zone of complete hemolysis on blood agar. Viscous growth in serum broth.

Acid but no gas produced from glucose, fructose, xylose, lactose, maltose, sucrose, arabinose and salicin. Rhamnose, dulcitol, sorbitol, inositol, inulin and adonitol not fermented.

Methyl red test positive. Voges-Proskauer test negative.

Hydrogen sulfide not produced (in TSI agar). Nitrites produced from nitrates. Indole not produced. Catalase test negative. Urease positive. Citrate not utilized.

Aerobic, facultatively anaerobic. Optimum temperature 37 C.

Non-pathogenic for rabbits and guinea pigs. Pathogenic for mice only after intraperitoneal injection. Isolated from a variety of lesions in piglets.

e. *Actinobacillus suis* Zimmermann 1964, 460.

Cells short oval rods up to 3–4 μm in length. Occasional filaments.

Colonies mucoid and sticky and on blood agar surrounded by zones of complete hemolysis. Cultures in nutrient broth uniformly turbid and viscous after 18 hrs, but later the medium clears and a sediment is produced.

Acid but no gas from glucose and lactose. Man-nitol, rhamnose, dulcitol, sucrose and arabinose not fermented.

Hydrogen sulfide produced. Nitrites produced from nitrates. Indole not produced.

Low viability of strains. Nutrient agar and nutrient broth cultures die out within 15 days at 4 C.

Pathogenic for mice. Isolated from six cases of septicemic disease in piglets.

Genus **Cardiobacterium** Slotnick and Dougherty 1964, 271

S. P. LAPAGE

(Group II D, Tucker *et al.* 1962.)

Car.di.o.bac.te′ri.um. Gr. n. *cardia* heart; Gr. n. *bakterion* small rod; M.L. neut.n. *Cardiobacterium* bacterium of the heart.

Rods, 0.5–0.75 by 1–3 μm, arranged singly, in pairs, short chains and clusters. Pleomorphic. No flagella, endospores or capsules demonstrated. **Gram-negative with possible retention of stain in parts of the cells.**

Chemoorganotroph, fermentative. Acid but no gas produced from fructose, glucose, mannose, sorbitol and sucrose. Glucose fermented largely to lactate and smaller amounts of pyruvate, formate and propionate.

Cytochrome oxidase and small amounts of indole produced. Catalase not produced, nitrates not reduced. Does not liquefy gelatin or serum; does not produce acetylmethylcarbinol, urease, ornithine or lysine decarboxylase, arginine dihydrolase or phenylalanine deaminase.

Facultative anaerobe. Carbon dioxide needed by some strains on isolation. Favored by high humidity, may not grow in dry incubator. Grows at 30–37 C, no growth at 22 or at 42 C. Optimum pH 7–7.2.

Type species: *Cardiobacterium hominis* Slotnick and Dougherty 1964, 271.

Further Comments

The genus was created after a study of eight strains and is probably justified as there are no serological cross-reactions with 12 "related" genera (Slotnick and Dougherty, 1964). The main differences between *Cardiobacterium* and similar genera are given in Table 8.33.

Description of the species of genus **Cardiobacterium**

1. **Cardiobacterium hominis** Slotnick and Dougherty 1964, 271. (Group II D, Tucker *et al.* 1962).

ho′min.is. L. gen.n. *hominis* of man.

Description taken from Slotnick and Dougherty (1964, 1965); Midgley *et al.* (1970). Only a few strains have been studied so far.

Pleomorphic; rods to ovals, cells with swollen ends and occasional filaments. Slotnick and Dougherty (1964) and Tucker *et al.* (1962) describe a high degree of pleomorphism and metachromatic and sudanophilic granules; these were not found by Midgley *et al.* (1970). Not acid-fast. Cell wall of Gram-negative three-layered type (Reyn *et al.*, 1969).

Poor growth on nutrient agar; on horse blood agar colonies 0.5 mm diameter after 24–28 hrs, 1–2 mm diameter in 3–4 days. Circular, low convex, smooth, entire, shiny, opaque and butyrous. May later differentiate to whiter, more opaque center. No hemolysis but heavy growth may turn the blood slightly green.

Strains differ in fermentation of maltose and mannitol. Adonitol, arabinose, cellobiose, dulcitol, erythritol, galactose, glycerol, inositol, lactose, melibiose, melezitose, rhamnose, trehalose and xylose not fermented. β-Galactosidase not produced (ONPG test).

Grows in peptone water and the usual laboratory test media but growth may be better if serum added. Does not grow in Koser's citrate, Møller's cyanide medium, on MacConkey's or on tellurite agar. Some strains produce H₂S (lead acetate paper). Litmus milk no change or less frequently weak acid. Gelatin and serum not liquefied. Does not utilize malonate, or oxidize gluconate.

Grows on a synthetic medium containing glucose, various amino acids, pantothenic acid, nicotinamide, pyridoxine, thiamine, biotin and a basal salts mixture; inhibited by low concentrations of riboflavin and flavine nucleotide (Slotnick and Dougherty, 1965).

Should survive several days in culture. Stands storage in the frozen state, and survives freeze-

TABLE 8.33

Distinguishing characteristics of **Cardiobacterium** *and similar genera*

	Catalase	Oxidase	Indole	Nitrate reduced	Acid from glucose	Hugh and Leifson O/F test	% G + C
Cardiobacterium hominis	−[a]	+	+	−	+	F	61.7[b]
Bacteroides corrodens	−	+	−	+	−	N	56.2–58.2[b]
Moraxella kingii	−	+	−	d	+	F	44.5[c]
Other *Moraxella*	+	+	−	D	−	N	41–46[c,d]
Haemophilus aphrophilus	−	−	−	+	+	F	40[e]
Actinobacillus actinomycetem-comitans	+	−	−	+	+	F	39[f]
Other *Haemophilus*	+	+	D	+	+	F	38–42[g]
Streptobacillus moniliformis	−	−	−	−	+	NG	23.9[h]
Pasteurella sensu stricto	+	+	D	+	+	F	33–46[g,i,j]
Brucella	+	+	−	+	−	N	56–58[g,i,k,m]
Bordetella (excluding *B. bronchiseptica*)	+	D	−	−	−	N	61–66[g,l]

[a] − = most (90% or more) strains negative in this character; + = most (90% or more) strains positive in this character; F = fermentative; N = no reaction or an alkaline reaction; d = some (less than 90%) strains positive, some negative; D = different species vary in their results; NG = fails to grow in medium.

[b] Hill, *et al.*, 1970.
[c] Bøvre *et al.*, 1969.
[d] Bøvre, 1967.
[e] Lapage, unpublished.
[f] Goodman, 1972.
[g] Hill, 1966.
[h] Williams *et al.*, 1969 (L-form only).
[i] Bohaček and Mraz, 1967.
[j] Bailie, 1969.
[k] Hoyer and McCullough, 1968, 444.
[l] Bacon *et al.*, 1967.
[m] Hoyer and McCullough, 1968, 1783.

drying. Sensitive to penicillin, ampicillin, streptomycin, tetracycline, chloramphenicol and neomycin.

Common and specific antigens among four strains by agglutination and immunofluorescent studies (Slotnick and Dougherty, 1964).

Found in the human nose and throat (Slotnick *et al.* 1964). Produces endocarditis in man. Not shown to be pathogenic to the laboratory animals tested.

The G + C content of the DNA is 61.7 moles % (T_m, Hill *et al.*, 1970).

Type strain: not designated; strain 6573, a co-type, is suggested as a reference strain.

(Genus **Streptobacillus** Levaditi, Nicolau and Poincloux 1925, 1188)

RUTH G. WITTLER AND SYLVIA G. CARY

(Not *Streptobacillus* Ucke 1898, 1000; *Haverhillia* Parker and Hudson 1926, 358.)
Strep.to.ba.cil′lus. Gr. adj. *streptos* twisted, curved; L. dim.n. *bacillum* a small rod; M.L. n. *bacillus* a small rod; M.L. masc.n. *Streptobacillus* a twisted or curved small rod.

Rods, 0.3–0.7 by 1–5 μm long, with rounded or pointed ends, **frequently in chains** and **filaments** 0.5–0.9 μm by 10–150 μm long. Rods may show a central thickening; filaments often have a series of oval to elongated **bulbous swellings,** 1–3 μm in diameter, giving the appearance of a string of beads. True branching does not occur.

Non-encapsulated. Non-motile. No resting stages known. **Gram-negative.** Non-acid-fast.

Spontaneous conversion to **L phase** or to

transitional phase variant (McGee and Wittler, 1969) occurs during cultivation *in vitro* and may occur *in vivo*.

Chemoorganotrophs: **metabolism fermentative.**

Serum, ascitic fluid or blood required for growth. Neopeptone not recommended for optimal cultivation (Brown and Nunemaker, 1942). Minimal nutritional requirements unknown.

Facultative anaerobe. CO_2 and moisture required for fresh isolates and for some laboratory strains.

Temperature optimum 35–37 C; range 30–38 C; no growth at 23 C. pH optimum 7.4–7.6; range 7.0–8.0.

Parasitic to **pathogenic** for **rats** and **other mammals.**

The G + C content of the DNA of the bacillary phase of the organism (three strains) is 24–25 moles %, of the L phase variant strains L₁ Rat 30 and MLS, 24–26 moles % (T_m) (Williams *et al.*, 1969 and unpublished data of Williams and Wittler, 1970).

Type species: *Streptobacillus moniliformis* Levaditi *et al.* 1925, 1188.

Further Comments

The genus *Streptobacillus* was transferred from the family *Parvobacteriaceae*, tribe *Haemophileae* (Breed *et al.*, 1948) to the family *Bacteroidaceae* (Breed *et al.*, 1957).

The generic name, *Streptobacillus* Levaditi *et al.*, 1925, 1188 is illegitimate (Buchanan *et al.*, 1966). However, *Streptobacillus* has come into common use, and is retained in this edition of THE MANUAL pending studies to determine the taxonomic position of the genus.

The genus *Haverhillia*, type species *Haverhillia multiformis*, and the genus *Streptobacillus*, type species *Streptobacillus moniliformis*, have been regarded as synonymous on the basis of published descriptions without experimental bacteriological comparisons ever having been made on strains of the type species (Heilman, 1941; Freundt, 1957; Roughgarden, 1965).

Description of the species of genus **Streptobacillus**

1. **Streptobacillus moniliformis** Levaditi, Nicolau and Poincloux 1925, 1188. (*Streptothrix muris ratti* Schottmüller 1914, 87; *Nocardia muris* de Mello and Pais 1918, 183; *Actinomyces muris ratti* (Schottmüller) Lieske 1921, 31; *Haverhillia multiformis* Parker and Hudson 1926, 358; *Actinomyces muris* (de Mello and Pais) Topley and Wilson 1936, 274; *Asterococcus muris* (de Mello and Pais) Heilman 1941, 32; *Proactinomyces muris* (de Mello and Pais) Krasil'nikov 1941, 76; *Haverhillia moniliformis* (Levaditi, Nicolau and Poincloux) Prévot 1948, 25; *Actinobacillus muris* (de Mello and Pais) Wilson and Miles 1955, 475.)

mo.ni.li.for'mis. L. n. *monile* necklace; L. n. *forma* shape or form; M.L. adj. *moniliformis* necklace-shaped.

Morphology as for genus.

Arrangement and morphology of cells are extremely variable and dependent on environmental or cultural conditions and on age of growth (Heilman, 1941). In smears of blood or exudate from infected hosts and in recent isolates maintained under optimal cultural conditions, small rods and short filaments predominate (Freundt, 1956).

In 20% horse serum infusion agar cultures of laboratory strains, a variety of morphological forms may develop. In 1- to 4-hr-old cultures, filaments are short and consist of single cells that may or may not contain granules. After 6–12 hrs, filaments elongate and develop multiple rounded or fusiform swellings. The filaments lie in wavy loops and often in a parallel or twisted arrangement. After 12–18 hrs, the filaments fragment into seg-

ments forming chains of rods and of round, oval or fusiform bodies. Between 24 and 30 hrs, tiny coccoid and ring forms appear accompanied by masses of bubbles and large irregular shaped droplets of cholesterol or cholesterol-like material. After 3–5 days, the cells consist of very small, pale granulated rods.

In 20% horse serum infusion broth cultures, the growth for the first 24 hrs consists of chains of short rods with rounded ends and one to four granules located centrally, subterminally or terminally. Longer granulated rods with tapered ends and long, tangled filaments with granules may or may not be present. After 48–72 hrs, small, slender, granulated rods predominate or, if filaments were abundant earlier, the tangled filamentous form tends to predominate. Masses of bubbles are numerous. Tiny coccoid forms and round bodies may or may not be present.

The cell envelope of one strain (ATCC 14647) was found to have a significantly higher free lipid and phospholipid content and a lower glucosamine and muramic acid content than *E. coli* (Knipp and Sokatch, 1969). The low hexosamine content may be related to the tendency of *S. moniliformis* to convert to transitional phase or L phase variants during cultivation.

Colonies on serum or ascitic fluid agar, 1–3 mm in diameter after 48–72 hrs of growth; circular with smooth or slightly irregular edge; convex or low cone with flattened surface facets; colorless, gray or whitish gray; slightly translucent to opaque; glistening surface; butyrous consistency.

In serum or ascitic fluid broth, whitish granules or flocculent balls of growth after 24 hrs; growth sediments on bottom and along side of tube leaving a clear supernatant fluid; occasional strains produce a slight turbidity. No odor.

On sheep blood agar, colonies barely visible at 24 hrs; after 48–72 hrs, colonies smooth, gray, translucent and non-hemolytic; slight α-hemolysis of sheep and horse erythrocytes demonstrable by blood cell-agar overlay technique (Aluotto et al., 1970).

Growth on Loeffler's serum agar slants poor.

Acid but no gas from dextrin, fructose, galactose, glucose, glycogen, maltose, mannose and starch. Acid may or may not be produced from arabinose, lactose, salicin, sucrose and xylose. No acid from dulcitol, glycerol, inositol, inulin, mannitol, raffinose, rhamnose, sorbitol and trehalose.

Poor or no growth on milk, no coagulation. Gelatin not liquefied, casein and coagulated serum not digested. Hydrogen sulfide produced in slight amounts, indole not produced. Arginine hydrolyzed, phenylalanine not deaminated. Urea not hydrolyzed. Nitrate not reduced to nitrite. Esculin hydrolyzed slightly. Catalase and oxidase negative, phosphatase positive. Benzidine negative. Gluconate negative. Methylene blue chloride reduced anaerobically, potassium tellurite and 2,3,5-triphenyltetrazolium chloride reduced aerobically and anaerobically. Not sensitive to optochin (ethyl hydrocuprein hydrochloride).

Film and spots produced on 20% horse serum agar by bacillary phase, similar to those produced by Mycoplasma strains (cf. Fabricant and Freundt, 1967).

At 3–4 C serum broth cultures are viable 7–10 days; sealed serum agar cultures viable 14–15 days. At 37 C serum broth and agar cultures are viable 2–4 days. Organisms in infected tissues, body fluids and cultures remain viable several years at −25 to −70 C. Killed by exposure to 55 C for 30 min (Wilson and Miles, 1955).

Isolated from joint fluid and blood in spontaneous cases of polyarthritis of mice (Freundt, 1956). Isolated from hock joint and sternal bursa in cases of arthritis of turkeys (Boyer et al., 1958; Yamamoto and Clark, 1966).

Inhabitant of nasopharynx of wild and laboratory rats. Cause of streptobacillary-rat-bite fever of man (Roughgarden, 1965) and streptobacillary arthritis of mice and turkeys (Freundt, 1956; Boyer et al., 1958; Yamamoto and Clark, 1966). Similar or possibly identical organisms were isolated from the blood and joint fluid of patients with Haverhill fever (termed by Place et al., 1926, "erythema arthriticum epidemicum") during a milk-borne epidemic that occurred in Haverhill, Mass., in 1926 (Parker and Hudson, 1926). (Distinct from Spirillum minor Carter, 1888, 47, cause of spirillary-rat-bite fever or sodoku (Brown and Nunemaker, 1942).)

Type strain not extant. Reference strain: ATCC 14647.

Further Comments

Transitional Phase Variant. All strains of *S. moniliformis* undergo a reversible conversion to transitional phase variants. These variants have weakened or defective cell walls and during serial culture show alterations in cellular and colonial morphology and biology.

L Phase Variant. The designation "L₁" was first used by Klieneberger (1935) to refer to the "pleuropneumonia-like organism" associated with *S. moniliformis* in cultures. The term "L" (standing for Lister Institute) was later applied to bacterial variants that reproduced in the form of very small cells that lacked rigid walls and that produced colonies with central cores, which penetrated the agar.

The following names applied to the L phase variant have no standing in nomenclature as they were given to an infrasubspecific form: *Murimyces streptobacilli-moniliformis* (Levaditi, Nicolau and Poincloux) Sabin 1941, 57; *Bactepneumonia moniliformis* (Levaditi, Nicolau and Poincloux) Tulasne and Brisou 1955, 239.

S. moniliformis can undergo conversion to the L phase variant spontaneously in media containing serum or ascitic fluid. The proportion of cells that undergo conversion varies from strain to strain and with the cultural conditions. The L phase variant differs from the bacillus in cellular and colonial morphology, and in degree of resistance to those antibiotics that act primarily on the bacterial cell wall.

Individual strains differ in degree of stability in the variant L phase. Conversion of *S. moniliformis* to the L phase is ordinarily a transient phenomenon, and the variant obtained is termed an "unstable" L phase organism. Occasionally, permanent conversion may be achieved by laboratory manipulation and such a variant is termed a "stable" L phase organism. The *S. moniliformis* L phase variant strain, L₁ Rat 30, was stabilized by Klieneberger in 1938, and reversion to the bacillus has never since been obtained; the following description is based on this strain.

The individual cells vary from granule-like bodies 0.3 μm to large bodies 3.0 μm in diameter, average 1.0 ± 0.5 μm. The smallest bodies are generally coccoid in shape. The larger bodies may be distorted from the normally round or spherical configuration to highly pleomorphic shapes and protoplasmic threads.

The cell membrane of strain L₁ Rat 30 was found

to contain 53% protein and 40% lipid. Approximately one-third of the total lipid consisted of cholesterol in the nonesterified form (Razin and Boschwitz, 1968).

On serum or ascitic fluid agar, colonies of the L phase variant range from 10–2000 μm in diameter, averaging 300–500 μm after 24 hrs of growth. Viewed at magnifications of 10–200 times, 24- to 48-hr-old colonies show coarse surface tracings due to the presence of large oil-like droplets of cholesterol or cholesterol-like substance (Partridge and Klieneberger, 1941). The peripheral portion of the colony appears translucent and lacy due to the accumulation of oil-like droplets as large as 30–40 μm in diameter, and to the presence of large, swollen bodies and amorphous material interspersed with small coccoid bodies. The central portion of the colony penetrates the agar to a depth of 30–50 μm, and has a round or irregular shape and a deep brownish color. It is composed of densely packed large and small bodies, granules, amorphous material and smaller oil-like droplets. In very young, actively growing colonies, the predominant cell form both in the core and at the edge is the small coccoid body, 0.3–1.0 μm in diameter; bubbles and oil-like droplets are located near the edge.

In serum or ascitic fluid broth, growth of the L phase variant resembles that of the bacillus. Whitish, fluffy or granular balls of growth settle on the bottom or along the side of the tube leaving the supernatant clear. Upon shaking, the balls of growth break into smaller aggregates, but the suspension retains a granular consistency amid general turbidity.

Nutritional requirements for the L phase variant are less exacting than for the bacillus. Biochemical characters are generally similar to, but somewhat weaker than, those of the bacillus (Cohen et al., 1968). The L phase variant does not produce film and spots on 20% horse serum agar.

Temperature, pH and environmental requirements are similar to those of the bacillary phase.

The L phase variant is resistant to approximately 10,000 times higher concentration of penicillin than is the bacillus.

The L phase variant is generally encountered under laboratory conditions. However, *S. monili-formis* L phase variant has been isolated from the lung lesions of a laboratory rat (Klieneberger, 1938) and from the blood of patients with strepto-bacillary-rat-bite fever following penicillin therapy (Dolman et al., 1951).

Inoculated experimentally into mice, the L phase variant is non-pathogenic. However, it frequently reverts *in vivo* to the bacillary phase and the reverted bacillus retains the pathogenic potential of the original bacillary phase (Freundt, 1956).

The L phase variant and the bacillus share a common antigen, but the L phase lacks another antigen present in the bacillus (Klieneberger, 1942). The L phase variant protects against experimental challenge with the homologous L phase variant, but does not protect against challenge with the bacillus (Freundt, 1956).

Genus **Calymmatobacterium** *Aragao and Vianna 1913, 221*

R. B. DIENST

(*Donovania* Anderson, De Monbreun and Goodpasture 1945, 37.)
Ca.lym.ma.to.bac.te'ri.um. Gr. n. *calymma* mantel, sheath; Gr. dim.neut.n. *bakterion* a small rod; M.L. neut.n. *Calymmatobacterium* the sheathed rodlet.

Pleomorphic rods, 1–2 μm in length, **with rounded ends,** occurring singly and in clusters. Exhibit **single or bipolar condensation of chromatin,** the latter giving rise to the characteristic "safety pin" forms. Usually **encapsulated** and readily demonstrated by Wright's stain as blue bacillary bodies surrounded by well defined, dense pinkish capsules. Non-motile. **Gram-negative.**

May be **isolated and grown on fresh egg yolk medium** (Dienst, 1948). **After adaptation grows on artificial media.** Colonies on Levinthal beef heart infusion agar shiny and translucent, up to 1.5 mm diameter; gradually become gray, later brownish.

Type species: *Calymmatobacterium granulomatis* Aragao and Vianna 1913, 221

Description of the species of genus **Calymmatobacterium**

1. **Calymmatobacterium granulomatis** Aragao and Vianna 1913, 221. (*Donovania granulomatis* (Aragoa and Vianna) Anderson, De Monbreun and Goodpasture 1945, 37.)
gran.u.lo'ma.tis. L. dim.n. *granulum* a small grain; Gr. suff. *oma* a swelling or tumor; M.L. n.

granuloma a granuloma; M.L. gen.n. *granulomatis* of a granuloma.

Description as for genus.

Cause of granuloma inguinale and other granulomatous lesions of man. Not pathogenic for laboratory animals.

ADDENDUM TO PART 8

Bacteria Symbiotic or Parasitic in Protozoa

Protozoologists have occasionally found various species of protozoa to be infected with organisms, many of which are of the size of ordinary bacteria. Some are found within the host, others on the surface; some have been reported to be parasitic, others are known to be symbionts. Some have the morphology of cocci, others are spirally shaped and still others appear to be spore-bearing rods. Unfortunately, because of their specialized ecological niche, only a few of these organisms have been cultured away from the eukaryotic cells they inhabit and then only with great difficulty. Consequently, little information has been obtained that would permit assessment of the taxonomic affinities of these organisms. In the 6th edition of THE MANUAL, these bacterial inhabitants of protozoa were placed in an appendix to the Rickettsiales, but in the seventh edition, because of their greater similarities to true bacteria, they were described in an addendum to the Class Schizomycetes. Information is still too scanty to warrant any taxo-nomic positioning of these organisms and may in fact be quite diverse. However, as they are all Gram-negative they are now placed as an Addendum to Part 8. It is important that bacteriologists recognize their existence, for they represent an interesting group worthy of more detailed study.

The best information comes from the work of protozoologists who, until recently, have not been too concerned with the taxonomic relationships of the bacteria living in protozoa to well studied groups of bacteria. Dr. John R. Preer, an expert on the intracellular bacteria of *Paramecium* has provided a brief summary of the knowledge available about these organisms. It is presented here for the information of bacteriologists.

For information on other genera and species of bacterial parasites of protozoa, the reader is referred to pages 927–930 of the 7th edition of THE MANUAL.

The Editors

The Intracellular Bacteria of *Paramecium aurelia*

JOHN R. PREER, JR.

Many kinds of bacteria live within *Paramecium* and other ciliated protozoa; those associated with *P. aurelia* are best known. In many localities 50% or more of the individuals in natural populations are infected.

All are Gram-negative rods and show typical bacterial structure in the electron microscope. Limited growth free of *Paramecium* has been reported for two of the forms.

The bacteria are highly restricted in the strains of paramecia they can infect. Infection generally depends on the presence of one or two specific genes

TABLE 8.34
Main characteristics of intracellular bacteria in **Paramecium aurelia**

	Morphology	Flagellation	Syngen or varieties infected	Action on sensitives
Kappa	Rods with R bodies	None	2, 4	Paralysis, spinning, vacuolization, hump formation, depending upon strain of Kappa
Mu[a]	Rods	None	1, 2, 8	Mate killing
Lambda[a]	Large straight rods	Peritrichous	4, 8	Rapid lysis
Sigma	Large curved rods	Peritrichous	2	Rapid lysis
Delta	Rods thick cell wall	Sparse	1, 2, 4, 6	Weak killer or no action
Gamma	Small rods; enclosed in extra set of membranes	None	8	Strong vacuolization
Tau	Rods	None		Weak paralysis
Nu	Rods	None	1, 2, 5	No toxic action
Alpha	Spiral	None	2	No toxic action

[a] Restricted *in vitro* culture in a complex medium has been reported for these forms (van Wagtendonk *et al.*, 1963; Williams, 1971).

in the protozoon. Under certain conditions in the laboratory the infected individuals have a competitive advantage over uninfected paramecia, while under other conditions the opposite is true. Most of the bacteria convert their hosts into killers, which liberate toxic particles (generally, if not always, the bacteria themselves) into the medium and thereby kill sensitive strains of paramecia. Although each host paramecium is made resistant to the toxin by the bacterial strain it harbors, organisms free of the bacteria are generally sensitive.

The first form discovered was recognized genetically and given the designation kappa. The practice of using Greek letters has been continued for other forms (Table 8.34).

Kappas are of many kinds and found only in syngens (varieties) 2 and 4 of *P. aurelia*. They are distinguished by the presence of refractile (R) bodies in 5–30% of each population. R bodies are unique and are seen as very distinctive rolls of "ribbon," which may suddenly unroll; they arise in associa-

tion with phage-like particles which infect kappa and are involved in the toxic action of kappa-bearing (killer) strains of paramecia. The cytochromes of kappa are completely different from those of the host organisms and very similar to those of many of the Enterobacteriaceae and of *Brucella*.

Mus are found in syngens 1, 2 and 8 and lead to mate-killing (death of mates at conjugation); no active toxin is liberated into the medium. Diaminopimelic acid has been demonstrated in mu.

The spiral form alpha is restricted to the macronucleus and is apparently closely related to *Drepanospira* and *Holospora*.

Recent reviews are by Beale, Jurand and Preer (1969) and Preer (1971).

Editorial Note. A review by Preer, Preer and Jurand is expected to appear in *Bacteriological Reviews* in late 1973 or early 1974; it includes DNA base ratios, a discussion of the relationships of the symbionts with known bacteria and assigns binomial names.

Important Notes

for

Users of this Edition

1. Always read both generic and species descriptions because characters listed in the generic description are not usually listed in the species descriptions.

2. In tables, characters common to all taxa are not shown but may be listed in footnotes.

3. Generally in tables (exceptions are clearly indicated in footnotes, q.v.) the meanings of symbols are as follows:

+	more than 90 % strains positive
−	more than 90 % strains negative
d	11–89 % strains positive
( )	delayed reaction
w	weak reaction
D	Different reactions in different taxa (species of a genus or genera of a family)
v	strain instability (NOT differences between strains)

PART 9

GRAM-NEGATIVE ANAEROBIC BACTERIA

FAMILY I. **BACTEROIDACEAE** PRIBRAM 1933, 10

Lillian V. Holdeman and W. E. C. Moore

Bac.te.ro.i.da'ce.ae or Bac.te.roi.da'ce.ae. M.L. n. *Bacteroides* type genus of the family; *-aceae* ending to denote a family; M.L. fem.pl.n. *Bacteroidaceae* the *Bacteroides* family.

Gram-negative, uniform or pleomorphic rods. Non-motile or motile with peritrichous flagella. Non-spore-forming.

Chemoorganotrophs.

Obligate anaerobes (require reduced oxygen tension).

Isolated from the natural cavities of man and other animals and the intestinal tract of insects; also isolated from infections of man and other animals. Some species are pathogenic.

Key to the genera of family **Bacteroidaceae**

I. Produce (from peptone or glucose) mixtures of acids including succinic, acetic, formic, lactic, propionic; butyric acid usually not a major product. Some species produce a mixture of butyric, isobutyric and isovaleric acids, along with major amounts of succinic acid.

Genus I. *Bacteroides*

II. Produce (from peptone or glucose) butyric acid (without isobutyric and isovaleric acids) as a major product.

Genus II. *Fusobacterium*

III. Produce (from peptone or glucose) lactic acid as the only major fermentation acid.

Genus III. *Leptotrichia*

Note. We wish to thank the many people and organizations who assisted in the research and preparation of descriptions of the genera *Bacteroides, Fusobacterium,* and *Propionibacterium* and *Eubacterium* (Part 17). We are especially indebted to other members of the International Subcommittee on Anaerobic Gram-negative Rods (Drs. Ella Barnes, H. Beerens, J. Bittner, M. P. Bryant, L. Fievez, S. M. Finegold, H. S. Goldberg, L. Reinhold, A. C. Sonnenwirth, S. Suzuki and H. Werner), the American Type Culture Collection and Dr. A. R. Prévot for cultures and descriptive information; the National Institutes of Health, General Medical Sciences, for support of the characterization studies; Dr. L. S. McClung for library and reference information; and the staff of our laboratory, particularly Elizabeth P. Cato, who assisted with computation of data and proofing this manuscript. In addition to original and emended descriptions, two references were especially valuable: *Étude sur les Bactéries Anaérobies Gram-négatives Asporulées,* Madeleine Sebald, 1962, and *Les Bactéries Anaérobies,* A. R. Prévot, A. Turpin and P. Kaiser, 1967.

Genus I. **Bacteroides** Castellani and Chalmers 1919, 959

LILLIAN V. HOLDEMAN AND W. E. C. MOORE

(*Ristella* Prévot 1938, 290.)

Bac.te.ro.i'des or Bac.te.roi'des. M.L. n. *bacter* the masc. equivalent of Gr. neut.n. *bacterum* a staff or rod; Gr. n. *idus* form, shape; M.L. masc.n. *Bacteroides* rod-like.

Gram-negative, non-spore-forming rods. Non-motile or motile with peritrichous flagella. Chemoorganotrophs. **Metabolize carbohydrates or peptone. Fermentation products** of saccharoclastic species **include combinations of succinic, lactic, acetic, formic or propionic acids,** sometimes with short-chained alcohols; **butyric acid is usually** not a **major product.** When *n*-butyric acid is produced, isobutyric and isovaleric acids are also present. From peptone non-saccharoclastic species produce either (a) trace to moderate amounts of succinic, formic, acetic and lactic acids or (b) major amounts of acetic and butyric acids (sometimes also succinic acid) with moderate amounts of alcohols and isovaleric, propionic and isobutyric acids. Carbon dioxide utilized or required by strains of several intestinal species and is incorporated into succinic acid (Caldwell *et al.*, 1969).

Obligately anaerobic. The maximum Eh that will allow initiation of growth varies among species and, within species, differs with the size of inoculum used. Eh sensitivity ranges from those that can initiate growth in partially oxidized media or in a candle jar to those that have not yet been cultured in media with an initial oxidation-reduction potential above minus 100 mV, even when large inocula are used. Oxidized media are inhibitory for some species. Addition of serum or ascitic fluid to such media may enable them to support growth.

Catalase usually not produced, or only in trace amounts. Hippurate usually not hydrolyzed. Lecithinase and lipase usually not produced on McClung-Toabe egg yolk agar. Meat usually not digested.

Growth usually most rapid at 37 C and pH near 7.0.

The G + C content of the DNA (in the species examined) ranges from 40–55 moles % (chromatographic separation (Sebald, 1962) and T_m (J. L. Johnson, personal communication)).

Found in cavities of man and other animals, infections of soft tissue, and sewage. Some species may be pathogenic.

Characteristics by which species in the genus can be differentiated are given in the Key and in Tables 9.1, 9.2, 9.4, 9.5, 9.6 and 9.7. Additional characteristics are given in the text concerning each species. The drawings of each species (Figs. 9A–9D) are composites of numerous cultures. Individual cultures usually show somewhat less variation in cellular morphology than is depicted. Cells may be more pleomorphic when grown in media that are not highly reduced (Doetsch *et al.*, 1957). Terminal or central swellings, vacuoles, or filaments are common.

The type species: *Bacteroides fragilis* (Veillon and Zuber) Castellani and Chalmers 1919, 959.

Key to species in the genus **Bacteroides**

I. Black pigment not produced.
  A. Carbohydrates strongly fermented.
    1. Succinic acid a major product.
      a. Good growth in 20% bile (2% ox gall).
        b. Indole not produced.
          c. Trehalose not acid.
            d. Rhamnose not acid.
              1a. *B. fragilis* subsp. *fragilis*
            dd. Rhamnose acid.
              1b. *B. fragilis* subsp. *vulgatus*
          cc. Trehalose acid. 1c. *B. fragilis* subsp. *distasonis*
        bb. Indole produced.
          c. Mannitol acid. 1d. *B. fragilis* subsp. *ovatus*
          cc. Mannitol not acid.
              1e. *B. fragilis* subsp. *thetaiotaomicron*

 aa. Growth inhibited by 20% bile (2% ox gall).
  b. Mannose acid.
   c. Hemin required for growth.
                    2a. *B. ruminicola* subsp. *ruminicola*
   cc. Hemin not required for growth.
    d. Arabinose acid.
                    2b. *B. ruminicola* subsp. *brevis*
    dd. Arabinose not acid.
     e. Long, thin cells present.
                    3. *B. ochraceus*
     ee. Long, thin cells not present.
                    4. *B. oralis*
   bb. Mannose not acid.   5. *B. amylophilus*
 2. Succinic acid not a major product.
  a. Propionic acid a major product.
   b. Not motile.   6. *B. hypermegas*
   bb. Motile.   7. *B. serpens*
  aa. Propionic acid not a major product.
   b. Not motile.
    c. Lactose not acid.   8. *B. termitidis*
    cc. Lactose acid.
     d. Melezitose acid.
                    9. *B. biacutus*
     dd. Melezitose not acid.
          10a. *B. clostridiiformis* subsp. *clostridiiformis*
(Also see *Clostridium ramosum*, which now includes organisms formerly identified as *Bacteroides trichoides* (*B. terebrans*) and in which spores usually are not seen.)
    bb. Motile.
     c. Abundant gas produced.
                    10b. *B. clostridiiformis* subsp. *girans*
     cc. Little, if any, gas produced.
                    11. *B. constellatus*
B. Carbohydrates not fermented, or only weakly so.
 1. Indole produced.
  a. Propionic acid produced.
                    12. *B. putredinis*
  aa. Propionic acid not produced.
                    13. *B coagulans*
 2. Indole not produced.
  a. Nitrate reduced.
   b. Butyric acid produced.
                    14. *B. praeacutus*
   bb. Butyric acid not produced.
                    15. *B. corrodens*
  aa. Nitrate not reduced.
   b. Gelatin completely liquefied.
                    16. *B. nodosus*
   bb. Gelatin not completely liquefied.
    c. Lactic acid a major product.
                    17. *B. furcosus*
    cc. Lactic acid not a major product.
     d. Esculin hydrolyzed.
                    18. *B. capillosus*

dd. Esculin not hydrolyzed.
   e. Cellobiose weak acid.
      19. *B. succinogenes*
  ee. Cellobiose not acid.
      20. *B. pneumosintes*
II. Black pigment on laked blood agar.
  A. Glucose acid.
    1. Highly saccharolytic, non-proteolytic.
      21a. *B. melaninogenicus* subsp. *melaninogenicus*
    2. Moderately saccharolytic, moderately proteolytic.
      21b. *B. melaninogenicus* subsp. *intermedius*
  B. Glucose not acid.
      21c. *B. melaninogenicus* subsp. *asaccharolyticus*
III. Black pigment in gelatin culture after incubation for several days.
      22. *B. niger*

Description of the species of genus **Bacteroides**

1. Bacteroides fragilis (Veillon and Zuber) Castellani and Chalmers 1919, 959. (*Bacillus fragilis* Veillon and Zuber 1898, 536; *Fusiformis fragilis* Topley and Wilson 1929, 303; *Ristella fragilis* Prévot 1938, 290; *Pseudobacterium fragilis* Krasil'nikov 1949, 248; *Bacteroides inaequalis* Eggerth and Gagnon 1933, 407; *Sphaerophorus inaequalis* (Eggerth and Gagnon) Prévot 1938, 298; *Pseudobacterium inaequalis* (Eggerth and Gagnon) Krasil'nikov 1949, 248; *Bacteroides incommunis* Eggerth and Gagnon 1933, 402; *Ristella incommunis* (Eggerth and Gagnon) Prévot 1938, 291; *Pseudobacterium incommunis* (Eggerth and Gagnon) Krasil'nikov 1949, 248; *Bacteroides uncatus* Eggerth and Gagnon 1933, 404; *Ristella uncata* (Eggerth and Gagnon) Prévot 1938, 291; *Pseudobacterium uncatum* (Eggerth and Gagnon) Krasil'nikov 1949, 245; *Sphaerophorus intermedius* Bergan and Hovig 1968, 429.)

fra'gi.lis. L. adj. *fragilis* fragile.

Description based on study of 326 strains, including NCTC 9343 and other labeled strains cited in descriptions of subspecies (see Tables 9.1, 9.2 and Fig. 9A).

The original description of *B. fragilis* was based on strains isolated from abdominal abscesses following appendicitis. The strains isolated were anaerobic Gram-negative rods that produced abscesses in test animals. It is generally accepted and reasonable to assume that these organisms were similar to those subsequently described as *B. fragilis*. Of the 326 strains we tested, 6 strains (including NCTC 9343) fit the original description for *B. convexus* (Eggerth and Gagnon), 7 fit the original description for *B. vulgatus* (Eggerth and Gagnon), 1 fits the original description for *B. distasonis* (Eggerth and Gagnon), and many fit the more general description of

Ristella pseudoinsolita (Beerens). The majority, however, did not exactly fit the reactions given in the literature for the species listed as synonyms or subspecies. Rather, there is a continuum of variants. However, analysis of certain sets of characteristics showed clusters of strains; these clusters have been designated subspecies. The subspecies conform well to different serotypes reported by Beerens *et al.* (1971) and Lombard and Dowell (1972). Additional study is needed to determine if the subdivision into subspecies is warranted and to determine the position of the many strains, now designated as *B. fragilis*, that have the characteristics of the complex but not of any of the designated subspecies.

Surface colonies on horse blood agar plates 1–3 mm, circular, entire, low convex, translucent to semi-opaque often with an internal structure of concentric rings when viewed by obliquely transmitted light. In general, strains produce no hemolysis on horse blood or rabbit blood agar; a few strains may be slightly hemolytic, particularly in the area of confluent growth. A very few strains (less than 1%) are frankly beta-hemolytic.

Glucose broth cultures are turbid with sediment. Sediment usually is smooth, but may be stringy or ropy. Terminal pH in glucose (1%) broth is between 5.0 and 6.0.

Growth is often enhanced by 20% bile. For some strains, typical positive reactions are not obtained unless heme is added to the differential media. Growth may occur at 25 C or 45 C. Upper and lower temperatures permitting growth have not been determined for most strains, but they vary with growth conditions and size of inoculum. Most strains grow at pH 8.5, but more slowly and less luxuriantly than at pH 7.0.

Glutamic acid not decarboxylated (Werner,

1970). A few of the strains examined reduce nitrate to nitrite. About half the strains partially digest gelatin in 3 weeks; very few completely digest gelatin.

Most strains are resistant (Finegold and Miller, 1968) to penicillin (1 μg/ml), kanamycin (3200 μg/ml), and neomycin (3200 μg/ml); 98% of the strains are sensitive to erythromycin (50 μg/ml). Most strains are sensitive (Wilkins *et al.*, 1972) to clindamycin (0.8 μg/ml); 80% of the strains are sensitive to tetracycline (3.125 μg/ml).

Isolated from contents of the lower intestinal tracts of man and other animals, where it occurs as a part of the predominant flora. Occasionally isolated from the mouth. Also isolated, either with other organisms or as the only organism present, from specimens from appendicitis, peritonitis, heart valve infections, blood, rectal abscesses, pilonidal cysts, surgical wounds and lesions of the urogenital tract. *B. fragilis* is the most common species of anaerobic bacteria isolated from human soft tissue infections.

1a. *Bacteroides fragilis* subsp. *fragilis* (Veillon and Zuber) Castellani and Chalmers 1919, 959. (*Bacillus fragilis* Veillon and Zuber 1898, 536; *Bacteroides convexus* Eggerth and Gagnon 1933, 406; *Pasteurella convexa* (Eggerth and Gagnon) Prévot 1938, 292; *Pseudobacterium convexum* (Eggerth and Gagnon) Krasil'nikov 1949, 246; *Ristella pseudoinsolita* Beerens 1949, 4; *Eggerthella convexa* (Eggerth and Gagnon) Beerens, Castel and Fievez 1962, 120.)

Description based on study of NCTC strains 9343, 9344, 8560 and 106 other strains with similar characteristics (Fig. 9A).

G + C content of DNA is 43 moles % (chromatographic separation, Sebald, 1962).

Reference strain: NCTC 9343.

1b. *Bacteroides fragilis* subsp. *vulgatus* (Eggerth and Gagnon) Holdeman and Moore 1970, 35.* (*Bacteroides vulgatus* Eggerth and Gagnon 1933, 401; *Pasteurella vulgata* (Eggerth and Gagnon) Prévot 1938, 292; *Pseudobacterium vulgatum* (Eggerth and Gagnon) Krasil'nikov 1949, 245.)

vul.ga'tus. L. adj. *vulgatus* common.

Description based on study of type strain and 34 other strains with similar characteristics (Fig. 9A).

Type strain: ATCC 8482.

1c. *Bacteroides fragilis* subsp. *distasonis* (Eggerth and Gagnon) Holdeman and Moore 1970, 35.

(*Bacteroides distasonis* Eggerth and Gagnon 1933, 403; *Ristella distasonis* (Eggerth and Gagnon) Prévot 1938, 291; *Pseudobacterium distasonis* (Eggerth and Gagnon) Krasil'nikov 1949, 248.)

dis.ta.so'nis. M.L. gen.n. *distasonis* of Distaso; named for A. Distaso, a Romanian bacteriologist.

Description based on study of type strain and 25 other strains with similar characteristics (Fig. 9A).

Type strain: ATCC 8503.

1d. *Bacteroides fragilis* subsp. *ovatus* (Eggerth and Gagnon) Holdeman and Moore 1970, 35. (*Bacteroides ovatus* Eggerth and Gagnon 1933, 405; *Pasteurella ovata* (Eggerth and Gagnon) Prévot 1938, 292; *Pseudobacterium ovatum* (Eggerth and Gagnon) Krasil'nikov 1949, 245.)

o.va'tus. L. adj. *ovatus* ovate, egg-shaped.

Description based on study of type strain and eight other strains with similar characteristics (Fig. 9A).

Type strain: ATCC 8483.

1e. *Bacteroides fragilis* subsp. *thetaiotaomicron* (Distaso) Holdeman and Moore 1970, 35. (*Bacillus thetaiotaomicron* Distaso 1912, 444; *Bacteroides thetaiotaomicron* (Distaso) Castellani and Chalmers 1919, 960; *Sphaerocillus thetaiotaomicron* (Distaso) Prévot 1938, 300—Prévot placed this organism in the genus *Sphaerocillus* because Distaso thought the strains he examined were motile, but Eggerth and Gagnon (1933, 399) believe that Distaso "observed Brownian movement, which is very active"; *Pseudobacterium thetaiotaomicron* (Distaso) Krasil'nikov 1949, 248; *Bacillus variabilis* Distaso 1912, 441 (capsulated strains); *Bacteroides variabilis* (Distaso) Castellani and Chalmers 1919, 960; *Capsularis variabilis* (Distaso) Prévot 1938, 293; *Pseudobacterium variabilis* (Distaso) Krasil'nikov 1949, 244.)

the.ta.i.o.ta.o'mi.cron. M.L. n. *thetaiotaomicron* a combination of the Greek letters theta, iota and omicron.

Description based on study of ATCC 8492 and 64 other strains with similar characteristics (Fig. 9A).

Reference strain: ATCC 8492; Eggerth.

Further Comments

A number of species have been described which probably are synonyms of the *B. fragilis* group but the original strains have been lost.

i. *Bacteroides exiguus* Eggerth and Gagnon 1933, 407 differed from present descriptions of *B.*

---

* **Editorial Note.** About half of the names in the genus *Bacteroides* are new combinations attributed to Holdeman and Moore in 1970 and 1972 editions of laboratory manuals of the Virginia Polytechnic Institute Anaerobe Laboratory. The authorities for these new combinations are confirmed in Moore and Holdeman (1973).

TABLE 9.1

Fermentation reactions of the species of genus **Bacteroides** *species 1–5[a]*

	1a. *B. fragilis* ss. *fragilis*	1b. *B. fragilis* ss. *vulgatus*	1c. *B. fragilis* ss. *distasonis*	1d. *B. fragilis* ss. *ovatus*	1e. *B. fragilus* ss. *thetaiotaomicron*	2a. *B. ruminicola* ss. *ruminicola*	2b. *B. ruminicola* ss. *brevis*	3. *B. ochraceus*	4. *B. oralis*	5. *B. amylophilus*
Products from peptone yeast extract glucose broth[a]	SAp(l ibivf)	SA(Lfp ibiv)	SAp(fl ibiv)	SA(Lfp ibiv)	SA(Lpf ibiv)	SA(flp ibbiv)	SA(lfp ibivv)	SA(l)	SA(lf ibiv)	Sfa (pl)
Products from lactate	−p	−	−	−p	−p	−	−	−	−	−
Acid produced from:										
Adonitol	−	−	−w	−	−	−w	−w	−	−w	v
Amygdalin	+	v	w+	w+	+w	+−	+w	+	+w	−
Arabinose	−	+−	+−	w+	+w	v	+w	−w	−	−
Cellobiose	−w	−w	v	w+	+−	v	+w	w+	+−	−
Dulcitol	−	−	−	−	−	−	−w	−	−+	−
Erythritol	−	−	−w	−	−	−w	−+	−	−w	−
Esculin	−w	−w	w−	w−	v	d	v	−w	−+	−
Esculin hydrolyzed	+	+−	+	+	+	+	+	+	+−	−
Fructose	+w	+w	+w	+w	+w	+w	+w	w+	+w	−w
Galactose	w+	+w	+w	+w	+w	w+	+w	w+	+w	−
Glucose	+w	+w	+w	+w	+w	+w	+	+	+w	−
Glycerol	−	−	−	−	−	−w	−+	−	−w	−
Glycogen	w+	+−	−+	v	+w	+−	+−	+w	+w	+w
Inositol	−	−	−	−	−	−w	−w	−	−w	−
Inulin	v	v	v	w+	+−	−+	+−	+w	d	−
Lactose	w+	+w	+w	+w	+w	+w	+w	+w	+w	−
Maltose	+w	+w	+w	+w	+w	−+	+w	v	+−	+
Mannitol	−	−	−	w	−	−w	−	−	−	−
Mannose	+w	+w	+w	+w	+w	−	+	+w	+w	−
Melezitose	−	−	v	−w	v	−	−w	−	−w	−
Melibiose	w+	+−	+w	w+	+w	−+	+−	−w	v	−
Raffinose	+w	+w	+w	+−	+w	−+	+w	w+	+w	−
Rhamnose	−	+w	v	v	v	−w	−+	−w	−+	−
Ribose	−w	d	v	v	v	−w	v	−w	v	−+
Salicin	−	−	−+	v	d	−+	v	−w	−+	−
Sorbitol	−	−	−	−w	−	−+	−w	−	−w	·
Sorbose	−	−	−w	−w	−	−w	−w	−w	−w	−
Starch	+w	+w	w+	w+	+w	d	+w	+w	+	+
Starch hydrolyzed	+	+−	d	+	+	+−	+	+−	+	+
Sucrose	+w	+w	+w	+w	+w	+−	+	+w	+w	−w
Trehalose	−	−	w	v	+−	−w	−+	−	−w	−
Xylose	+w	+w	+w	+w	−+	+−	+−	−w	−w	−

[a] Products from PYG broth: Capital letters indicate >1 meq acid/100 ml broth; small letters <1 meq/100 ml. A, acetic; B, butyric; F, formic; iB, isobutyric; iV, isovaleric; L, lactic; P, propionic; S, succinic; v, valeric; 2, ethanol; ( ) products of occasional strains. +, strong acid, i.e. pH 5.5 or below in 90–100% of strains. w, weak acid, i.e. pH 5.5–6.0 in 90–100% of strains. −, acid not produced, i.e. pH 6.0 or above in 90–100% of strains. (Uninoculated xylose and arabinose, under CO_2 (pre-reduced media), are often pH 5.9. For cultures in these media, we consider pG 5.4 to 5.7 weak acid and pH below 5.4 strong acid.) d, reaction positive 40–60% strains. v, reaction variable within a strain. When any of the above used as superscripts—reaction of 10–40% of strains. ·, not tested. Acid products determined by procedures given in Holdeman and Moore, 1972.

fragilis only in that it had "*slender* cells 0.5–1.0 microns long."

ii. *B. gulosus* Eggerth and Gagnon 1933, 398. The description is indistinguishable from *B. fragilis* subsp. *ovatus*. Although no labeled strains are now extant, Sebald (1962) reported a G + C % of 41.5 and 43.4 of two labeled strains, which also suggests identity with *B. fragilis*.

iii. *Ristella glycolytica* (Tardieux and Ernst) Sebald 1962, 149. From the description this species cannot be differentiated from *B. fragilis*. Two labeled strains from the Pasteur Institute were *B. fragilis*. As was also reported by Sebald, they differed from the original description by producing propionic rather than butyric acid. Sebald (1962) found the G + C % of these strains to be 45.

TABLE 9.2

Biochemical reactions of species of genus **Bacteroides,** *species 1–5*
(not motile, do not produce black colonies)[a]

	1a. B. fragilis ss. fragilis	1b. B. fragilis ss. vulgatus	1c. B. fragilis ss. distasonis	1d. B. fragilis ss. ovatus	1e. B. fragilis ss. thetaiotaomicron	2a. B. ruminicola ss. ruminicola	2b. B. ruminicola ss. brevis	3. B. ochraceus	4. B. oralis	5. B. amylophilus
Gelatin digested	−w	−+	−w	−w	−w	−+	+w	w−	+−	+w
Milk	c	c	c	c	c	c−	c^d	c	c−	.
Indole produced	−	−	−	+	+	−	−+	−	−	−
H₂S produced	−+	−+	−+	−+	d	−	−	−	−	−
Nitrate reduced	−	−	−	−	−	−+	−	−	−	−
Gas	2−	+4	3−	+3	+3	−	−	−	−2	−
Acetoin	−	−	−w	−	−w	−w	−+	−+	−+	−+
Growth in 20% bile	s4	s4	s4	s4	3s	−	−2	2−	−2	−
Hemolysis	−	−b	−w	−b	−ba	−+	−b	−	b−	−
Propionate ← threonine	−	−	−	−	−	+−	−+	−+	−	−
Growth enhanced by	h	h	h	h	h	hsr	rt	th	hrt	CO₂

[a] +, positive reaction—90–100% of strains. −, negative reaction—90–100% of strains. Hemolysis, a = alpha; b = beta; w = weak. Gas and growth, relative amounts indicated by +, 2, 3, 4; s = stimulated. Growth enhanced by b, bile; h, hemin; r, rumen fluid; s, serum; t, Tween 80. Milk—c, curd; d, digested. Gelatin—w (weak) after chilling until uninoculated controls are solidified and returning to room temp., liquefaction in less than half the time of the controls. Any of the above used as superscript indicates reaction of 10–40% of strains. Reactions determined by procedures given in Holdeman and Moore, 1972.

iv. *B. insolitus* Eggerth and Gagnon 1933, 408. The original description of this species is typical of many strains of *B. fragilis* when cultured without heme. Five labeled strains from the Pasteur Institute (Paris) gave typical reactions of *B. fragilis* when heme was added to the media. One of these (Ni2) has a G + C % of 42.6 (Sebald, 1962) which is also typical of *B. fragilis*. Strain 1617 (G + C % = 39.6, Sebald, 1962) was not available for study.

v. *B. pyogenes* Hauduroy *et al.* 1937, 69. Six of six strains from the Pasteur Institute (Paris) are *B. fragilis* as defined. Strain 1748B (not now available) was reported to have a G + C % of 34 (Sebald, 1962) and to produce formic, acetic and butyric acids (Prévot *et al.*, 1967). It probably was not representative of strains of this species in the Pasteur Institute collection.

vi. *Pasteurella serophila* Bokkenheuser 1951, 551. Pasteur Institute strain 599 (possibly the type strain) had characteristics conforming to those recorded for that strain (except that it grew in our laboratory without serum added to the medium) and was identical to *B. fragilis* as defined. No other labeled strains are known to be extant.

vii. *Ristella tumida* (Eggerth and Gagnon) Prévot 1938, 292. Reported to produce formic,

acetic, propionic, butyric and lactic acids (Prévot *et al.*, 1967). However, strain "Meunier II," found by Sebald (1962) to have a G + C % of 45, produced succinic, acetic and lactic acids and had other characteristics described here for *B. fragilis*.

viii. *Bacteroides uniformis* Eggerth and Gagnon 1933, 400. According to the original description this species would be identical to *B. fragilis* as defined. However, Sebald (1962) reported a G + C % of 37 for two labeled strains. One of these (1206A) was reported as microaerophilic by Dr. Prévot and grew on the surface of aerobic blood agar plates in our laboratory. Other reactions were comparable between the two laboratories. This strain is not a member of the genus *Bacteroides*. The other strain, TE 105B, was not available for study. No other labeled strains are known to be extant.

2. Bacteroides ruminicola Bryant, Small, Bouma and Chu 1958, 18. (*Ruminobacter ruminicola* (Bryant *et al.*) Prévot 1966, 121.)

ru.mi.ni′co.la. M.L. adj. *rumin-* of or relating to the rumen; L. substantive ending *-cola* inhabitant; M.L. n. *ruminicola* inhabitant of the rumen.

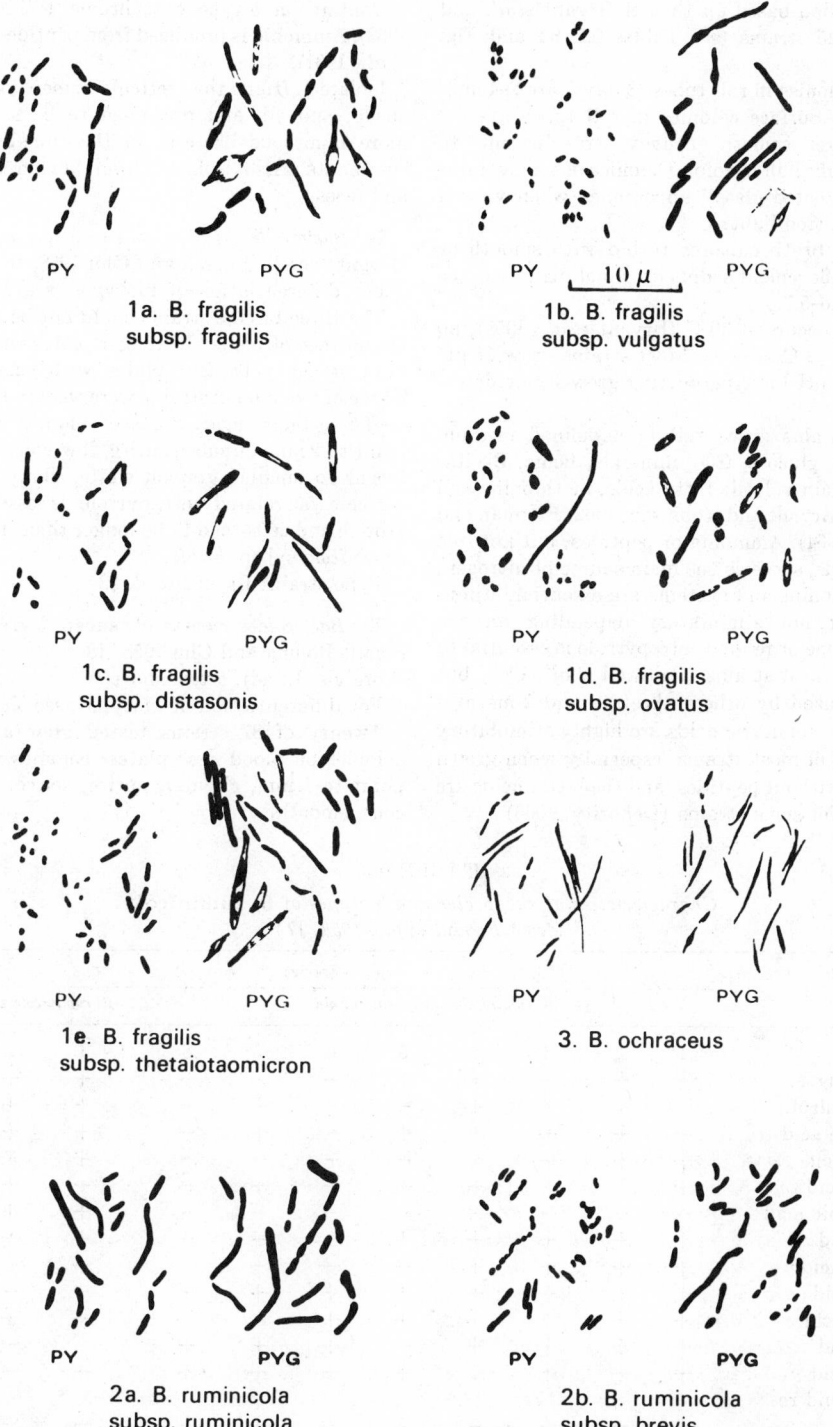

1a. B. fragilis
subsp. fragilis

1b. B. fragilis
subsp. vulgatus

1c. B. fragilis
subsp. distasonis

1d. B. fragilis
subsp. ovatus

1e. B. fragilis
subsp. thetaiotaomicron

3. B. ochraceus

2a. B. ruminicola
subsp. ruminicola

2b. B. ruminicola
subsp. brevis

Fig. 9A. *Bacteroides*. The line drawings are composites of several cultures and individual cultures usually show somewhat less variation in cellular morphology. PY, peptone-yeast extract broth; PYG, PY-glucose broth. Scale: 9/16 inch = 10 μm.

Description based on that of Bryant et al. and study of 45 strains (see Tables 9.1, 9.2 and Fig. 9A).

Deep colonies in roll tubes (3 days) are 2–4 mm, lenticular. Surface colonies in roll tubes are 1–2 mm, entire, smooth, convex, transluscent to opaque, light buff in color. Colonies of some strains have a "frosted glass" appearance when viewed by transmitted light.

Glucose broth cultures turbid with smooth or stringy to flocculent sediment. Final pH in glucose broth is 4.6–5.7.

Growth occurs at 30 C (Bryant et al., 1958), no growth at 22 C or 45 C. Most strains grow at pH 8.5; lower pH limit permitting growth not determined.

Most strains grow well in a defined medium containing glucose, CO_2, minerals, heme, B-vitamins, certain volatile fatty acids, methionine and cysteine (Bryant and Robinson, 1962; Pittman and Bryant, 1964). Ammonia or peptides, but not free amino acids, serve as the main source of nitrogen, and methionine and cysteine are essential, stimulatory, or not stimulatory depending on the strain. Heme or related tetrapyrrole is essential to growth of most strains (Caldwell et al., 1965) but is synthesized by others. Acetate and 2-methylbutyric or isobutyric acids are highly stimulatory to growth of most strains, especially when grown in media without peptides, and the latter acids are essential for some strains (Dehority, 1966).

Contains a b-type cytochrome (White et al., 1962). Ammonia is produced from peptides (Bladen et al., 1961).

Isolated from the reticulo-rumen of cattle, sheep, and elk and presumed to be among the more numerous bacteria in the rumen of most ruminants. Also isolated from human abscesses and feces.

2a. *Bacteroides ruminicola* subsp. *ruminicola* Bryant, Small, Bouma and Chu 1958, 18.

For differentiation of biotypes, see Table 9.3.

The three bovine strains could not be grown on the surface of horse blood agar plates with rumen fluid or egg yolk agar plates with rumen fluid. Three of five human strains were weakly hemolytic on horse blood agar. Surface colonies were pinpoint to 2 mm, circular, entire, low convex, translucent to opaque, grayish white, shiny, smooth.

Heme or related tetrapyrrole is essential for growth and cells tend to be longer than those of *B. ruminicola* subsp. *brevis*.

Type strain: Bryant et al. 23.

2b. *Bacteroides ruminicola* subsp. *brevis* Bryant, Small, Bouma and Chu 1958, 18.

bre'vis. L. adj. *brevis* short.

For differentiation of biotypes, see Table 9.3.

Twenty of 37 strains tested grew as surface colonies on blood agar plates; colonies were pinpoint to 1 mm, circular, entire, convex, translucent, smooth.

TABLE 9.3

Characteristics of subspecies and biotypes of **B. ruminicola**
(from Bryant et al., 1958, 17)

| | Species | | | | | | | | | | |
| | B. ruminicola subsp. ruminicola | | | | | | | | B. ruminicola subsp. brevis | | |
Biotype	1	2	3	4	5	6	7	8	1	2	3
Gelatin dig.	+	+	−	+	+	−	−	−	+	+	+
Starch hydrol.	+	+	+	+	−	−	−	−	+	+	+
Arabinose acid	+	+	+	+	+	−	+	−	+	+	+
Dextrin acid	+	+	+	+	−	−	−	−	+	+	+
Esculin acid	+	+	+	+	+	+	+	−	+	+	+
Gum arabic acid	−	−	−	−	−	−	−	−	+	+	−
Inulin acid	+	+	+	+	+	−	−	−	+	+	+
Maltose acid	+	+	+	+	−	−	+	−	+	+	+
Salicin acid	+	+	+	+	+	+	+	+	+	−	+
Sucrose acid	+	+	+	+	+	+	−	−	+	+	+
Xylan acid	+	+	+	+	+	+	+	+	−	−	−
Xylose acid	+	+	+	+	+	+	+	+	−	−	−
Rumen fluid required[a]	+	+	+	−	+	+	+	+			
H₂S prod.[b]	−	±	−	±	−	−	−	−	+	−	+

[a] Hemin replaces the rumen fluid requirement of the majority of strains (Caldwell et al., 1965).

[b] Medium of Bryant and Small, 1956.

Although this subspecies does not require heme or related tetrapyrrole in the growth medium (i.e. it synthesizes heme (White *et al.*, 1962)), heme may enhance growth of some strains. Most cells are much shorter (mainly coccoid to oval) than in *B. ruminicola* subsp. *ruminicola*.

Type strain: Bryant *et al.* GA 33.

3. **Bacteroides ochraceus** (Prévot) Holdeman and Moore 1972, 27. (*Fusiformis nucleatus* var. *ochraceus* Prévot, Joubert, Tardieux and de Cadore 1956, 793; *Ristella ochraceus* (Prévot) Sebald 1962, 77; *Bacteroides oralis* subsp. *elongatus* Loesche, Socransky and Gibbons 1964, 1335.)

o.chra′ce.us. Gr. n. *ochra* yellow ochre; M.L. adj. *ochraceus* of the color of ochre.

Surface colonies (2–3 days) on horse blood agar are 0.5–2.0 mm, circular, entire, convex, shiny, smooth, translucent, often yellow to light orange. Rhizoid colonies with growth penetrating into the agar may be observed when inoculated with broth cultures in which long cells are predominant. On subculture, the more regular colonial type returns (Loesche *et al.*). Laboratory strains grow as surface colonies on blood agar plates incubated in a candle jar. For morphology see Figure 9A.

Growth in glucose broth is turbid, usually with granular sediment.

Growth of most strains stimulated by hemin (5 μg/ml). No growth at 45 C; may grow at 25 C, but more slowly than at 37 C.

Of five strains, one fer mented arabinose another xylose (Loesche *et al.*) (Table 9.1). Gelatin not liquefied (Loesche *et al.*); gelatin partially or completely liquefied in our laboratory (Table 9.2).

Isolated from the gingival crevice and infections of man.

Type strain: Loesche *et al.* R-42.

4. **Bacteroides oralis** Loesche, Socransky and Gibbons 1964, 1334. (*Ristella oralis* (Loesche *et al.*) Prévot *et al.* 1967, 264.)

o.ra′lis. M.L. adj. *oralis* of the mouth.

Description based on that of Loesche *et al.* and study of the type strain and 17 similar strains. (Tables 9.1, 9.2 and Fig. 9B).

Surface colonies (2 days) are 0.5–2.0 mm, circular, entire, convex, shiny, smooth, translucent. Non-hemolytic on horse blood infusion agar (Loesche *et al.*); some strains beta-hemolytic on horse blood brain heart infusion agar (our laboratory).

In 24 hr, glucose broth cultures moderately turbid with no sediment or with smooth or stringy sediment. Final pH is 4.8–5.4.

One of 18 strains produced acid from arabinose and xylose. Gelatin not liquefied (Loesche *et al.*); 11 of 18 strains completely liquefied gelatin after incubation for 3 weeks (our laboratory). Meat not

digested. No decarboxylase for L-lysine, L-ornithine and L-arginine (Loesche *et al.*), or glutamic acid (Werner, 1970). Urea not hydrolyzed. No hydrogen sulfide detected in Peptone Iron Agar or in semi-solid agar with 0.02% ferrous sulfate and 0.03% sodium hyposulfite. Slight blackening of lead acetate paper suspended over culture in trypticase broth.

Isolated from the gingival crevice area of man and from infections, usually of the oral cavity and upper respiratory and genital tracts. Significance of this organism in infection is unknown.

Type strain: Loesche *et al.* 7CM.

5. **Bacteroides amylophilus** Hamlin and Hungate 1956, 552. (*Ruminobacter amylophilum* (Hamlin and Hungate) Prévot 1966, 121.)

am.y.lo′phi.lus. Gr. n. *amylo* starch; Gr. part. *philo* loving; M.L. adj. *amylophilus* starch-loving.

Description from Hamlin and Hungate and from study of five strains (Tables 9.1, 9.2 and Fig. 9B).

Surface colonies (2 days) on rumen fluid-glucose-cellobiose agar roll tubes are 1 mm, circular, entire, slightly convex, translucent, smooth, glistening, white to tan. The two bovine strains tested did not grow on the surface of plates incubated in an anaerobe jar. Colonies in deep agar are 0.8 mm, lenticular, entire or irregular, white, soft butyrous.

Starch broth cultures are turbid and have a final pH of 5.5–5.7.

Carbon dioxide required for growth; CO_2 is fixed, ammonia is assimilated. Most rapid growth at 39–45 C, grows at 35 C, does not grow at 30 or 55 C. The one strain tested grew best at pH 6.8–7.8 and failed to grow at pH 6.0 and 8.8.

Occurs sporadically in rumen contents of cattle but, when present, may be the predominant starch digester and constitutes 10% of the bacterial population of the rumen (Bryant and Robinson, 1961; Blackburn and Hobson, 1962).

6. **Bacteroides hypermegas** Harrison and Hansen 1963, 28. (*Bacteroides saucissus* Stevens 1956, 101 (not validly published); *Sphaerophorus hypermegas* (Harrison and Hansen) Prévot 1966, 145.)

hy.per.meg′as. Gr. pref. *hyper* excessive; Gr. adj. *megas* great; M.L. adj. *hypermegas* excessively great.

Description from Harrison and Hansen (1963), Goldberg *et al.* (1964), and study of five strains, including ATCC 25560 (Tables 9.4, 9.5 and Fig. 9B).

Cells contain volutin but no fat or glycogen (Harrison and Hansen).

Surface colonies (2–3 days) on horse blood agar are 1–2 mm, circular, entire to erose, low convex, translucent, often with patterned structure when

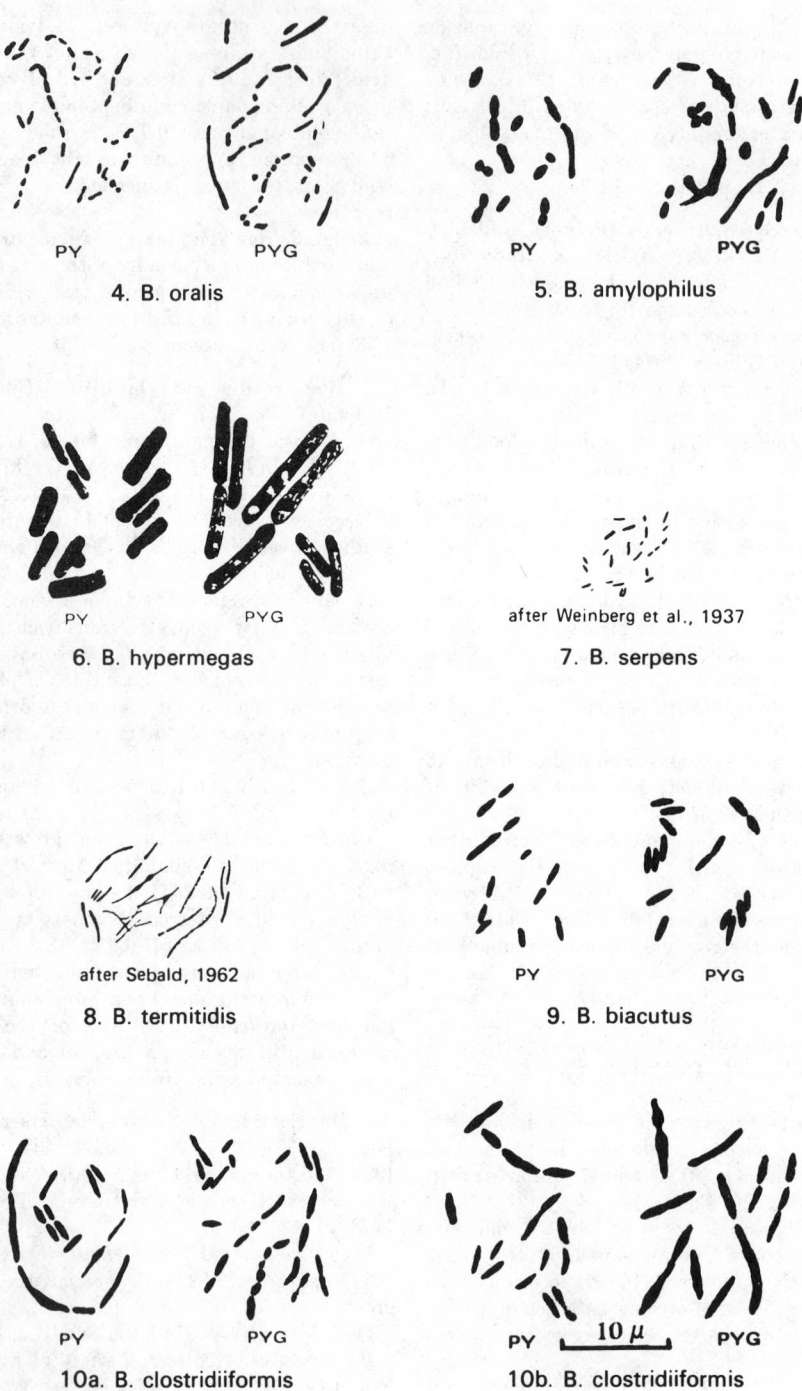

PY   PYG
4. B. oralis

PY   PYG
5. B. amylophilus

PY   PYG
6. B. hypermegas

after Weinberg et al., 1937
7. B. serpens

after Sebald, 1962
8. B. termitidis

PY   PYG
9. B. biacutus

PY   PYG
10a. B. clostridiiformis
subsp. clostridiiformis

PY ⌊ 10 μ ⌋ PYG
10b. B. clostridiiformis
subsp. girans

Fig. 9B. *Bacteroides*. The line drawings are composites of several cultures and individual cultures usually show somewhat less variation in cellular morphology. PY, peptone-yeast extract broth; PYG, PY-glucose broth. Scale: $9/16$ inch = 10 μm.

viewed by obliquely transmitted light. Surface colonies (2 days) on Reinforced Clostridial Medium Agar (BBL) are 3–5 mm, convex, gray, opaque, smooth.

Glucose broth cultures uniformly turbid, sometimes with flocculent sediment; final pH is 4.6–5.5.

Growth stimulated by fermentable carbohydrate. Grows poorly, if at all, at temperatures below 25 C or above 45 C. Does not grow in media with pH below 4.8 or above 8.6.

Trace to moderate amounts of gas produced in glucose agar deep cultures (our laboratory); no gas produced (Harrison and Hansen). Growth in medium with 1:100,000 brilliant green is variable. Growth inhibited in medium with NaCl concentration much above 1.5%. Glutamic acid not decarboxylated (Werner, 1970).

Isolated from the intestinal tracts of poultry, particularly turkeys and chickens, where it is present at levels of 10^7–10^8/g in the cecum.

Reference strain: ATCC 25560; V.P.I. 2366; Barnes F/61/42.

Further Comments

One of the strains (21-28) of *B. hypermegas* on which the description by Harrison and Hansen was based was previously described by Stevens (1956) and named *Bacteroides saucissus* with 21-28 designated the type strain. However, publication of a new species in a thesis not printed for distribution is thought not to constitute a valid publication (see Rule 11, Int. J. Syst. Bacteriol., *16*: 470, 1966).

The large amounts of non-volatile acid present in glucose cultures reported by Harrison and Hansen have not been detected as succinic, lactic or pyruvic acids in the strains of Goldberg *et al.*, (strains of Harrison and Hansen have been lost). This non-volatile acid, "calculated as lactic acid," may be some non-volatile acid other than succinic, lactic or pyruvic.

7. **Bacteroides serpens** (Veillon and Zuber) Hauduroy *et al.* 1937, 74. (*Bacillus serpens* Veillon and Zuber 1898, 532; *Bacillus radiiformis* Rist and Guillemot in Rist 1898, 162; *Zuberella serpens* (Veillon and Zuber) Prévot 1938, 293; *Pseudobacterium serpens* (Veillon and Zuber) Krasil'nikov 1949, 246.)

ser'pens. L. part.adj. *serpens* creeping.

No strains conforming to the descriptions were available for study. Description from Prévot *et al.* (1967) and Sebald (1962) (Tables 9.4, 9.5 and Fig. 9B).

Cells have peritrichous flagella.

Colonies in deep agar (2 days) are pinpoint, lenticular or mulberry shaped; upon continued incubation, divergent offshoots develop and the original colony degenerates. Glucose broth turbid with white sediment, gas and fetid odor. Coagulated proteins are not attacked.

G + C content of the DNA is 36 moles % (chromatographic separation, Sebald, 1962).

Isolated from infections of the intestinal tract, respiratory tract, middle ear, blood; and from contaminated sea water.

8. **Bacteroides termitidis** (Sebald) Holdeman and Moore 1970, 33. (*Sphaerophorus siccus* var. *termitidis* Sebald 1962, 124.)

ter.mi'ti.dis. L. n. *tarmes, tarmit-* (L. L. var. *termes, termit-*) worm that eats wood; M.L. adj. *termitidis* pertaining to the termite.

Description from Sebald (1962, 55) (Tables 9.4, 9.5 and Fig. 9B).

Surface colonies are 1–2 mm, circular, transparent to opaque. Colonies in deep agar are lenticular and not pigmented.

Glucose broth culture slightly turbid with sediment and gas (H_2). Final pH in glucose broth is 3 (probably comparable to pH 4.5–5 as determined by methods used for pH analysis in our laboratory).

No growth at 56 C. Growth not inhibited by 1:100,000 gentian violet or by brilliant green-N_3Na medium (Beerens and Tahon-Castel, 1965).

Coagulated proteins not attacked. Neither urease, chitinase nor toxin produced. Not pathogenic for mice or guinea pigs.

G + C content of the DNA is 34 moles % (chromatographic separation, Sebald, 1962).

Isolated from posterior intestinal contents of termites (*Reticulitermes lucifugus*) where they are part of the predominant bacterial flora. Thought to be important by fermenting products of cellulose digestion.

9. **Bacteroides biacutus** (Weinberg and Prévot) Holdeman and Moore 1970, 33. (*Fusobacterium biacutum* Weinberg and Prévot 1926, 522; *Fusiformis biacutus* (Weinberg and Prévot) Hauduroy *et al.* 1937, 238; *Ristella biacuta* (Weinberg and Prévot) Sebald 1962, 148.)

bi.a.cu'tus. L. adv. *bis* twice; L. adj. *acutus* sharped, pointed; L. adj. *biacutus* two-pointed.

Description based on study of CIPP (Prévot) strain 132 II and one other strain and, in part, on descriptions of Prévot *et al.* (1967) and Sebald (1962) (Tables 9.4, 9.5 and Fig. 9B).

Surface colonies (3 days) 0.5 mm, circular, entire, low convex, translucent, mottled, white, shiny, smooth.

Glucose broth cultures turbid with smooth sediment and final pH of 5.0–5.5.

Grows well at 45 C, less rapidly at 25 C and pH 7.6. Does not grow in 6.5% NaCl.

TABLE 9.4

Fermentation reactions of the species of genus **Bacteroides** *species 6–14*[a]

	6. B. hypermegas	7. B. serpens	8. B. termitidis	9. B. biacutus	10a. B. clostridiiformis ss. clostridiiformis	10b. B. clostridiiformis ss. girans	11. B. constellatus	12. B. putredinis	13. B. coagulans	14. B. praeacutus
Products from peptone yeast extract glucose broth	PA1 (fs)	AP	FALs2	AF2 (Lsp)	Af(sl 2)	Afls (2)	A2	aspiv bib(lf)	alsf	ABiViB p(slf)
Products from lactate	p⁻	·	·	−	−	−	·	−	−	−
Acid produced from:										
Adonitol	−	·	·	−		−w	·	−	−	−
Amygdalin	−w	·	·	−	−w	−w	·	−	−	−
Arabinose	+w	·	−	w⁻	w⁻	+w	·	−	−	−
Cellobiose	+⁻			+	d	−⁺	·	−	−	−
Dulcitol	v			−	−	−		−	−	−
Erythritol	−	·	·	−	−	−		−	−	−
Esculin	−	·	·	−	−w	−	·	−	−	−
Esculin hydrolyzed	+	·	·	+	+	+	·	−	−	−
Fructose	+	+	+	+	w⁺	+w	w	−	−	−
Galactose	+	+	+	+	w⁺	+w	w	−	−	−
Glucose	+	+	+	+	+w	+	+	−w	−	−
Glycerol	−⁺	·	−	−	−	−	w	−	−	−
Glycogen	−	·	·	−	−w	−w	·	−	−	−
Inositol	+⁻	·	·	−	−	−w	·	−	−	−
Inulin	+⁻	·	·	·	−	−	·	−	−	−
Lactose	+	+	−	+	w⁺	v	w	−	−	−
Maltose	+	+	+	w⁺	v	+w	+	−	−	−
Mannitol	+	·	·	−	−	−w	·	−	−	−
Mannose	+	·	·	+	v	+	·	−w	−	−
Melezitose	−w	·	·	w⁺	−	v	·	−	−	−
Melibiose	+	·	·	w⁻	−⁺	v	·	−	−	−
Raffinose	+	·	·	+	v	+w	·	−	−	−
Rhamnose	+⁻	·	·	+	d	+w	·	−	−	−
Ribose	+w	·	·	w⁻	d	+⁻	·	−	−	−
Salicin	+⁻	·	·	w⁺	v	−⁺	·	−	−	−
Sorbitol	+	·	·	−	−	v	·	−	−	−
Sorbose	−w	·	·	−	−w	−w	·	−	−	−
Starch	−w	·	·	w⁻	−w	w⁻	+	−	−	−
Starch hydrolyzed	−	·	·	−	−⁺	−	·	−	'	−
Sucrose	+	·	+	+	+w	+	w	−	−	−
Trehalose	+w	·	·	w	−w	w⁻	·	−	−	−
Xylose	+	·	+	+	d	+w	·	−	−	−

[a] For symbols see Table 9.1.

Cell walls contain meso-diaminopimelic acid (C. S. Cummins, personal communication).

G + C content of the DNA is 40 moles % (chromatographic separation, Sebald, 1962).

Isolated from infected appendix and abdominal wound.

Further Comments

Some previous descriptions of *B. biacutus* were based on heterogeneous groups of organisms. Strains were reported to have a % G + C ranging from 41–55 (Sebald, 1962) and to produce two or three different sets of fermentation acids: acetic-propionic (Prévot and Kirchheiner, 1938); acetic-butyric, sometimes with formic; or acetic-formic (Sebald, 1962).

Strains formerly designated *Ristella* (*Eggerthella*) *clostridiiformis* and *Fusobacterium girans* are similar to *Bacteroides biacutus* morphologically and are described as having cultural characteristics similar to the acetic-formic fermentation group of *B. biacutus*. Type strains are not available. Therefore, we proposed (Moore and Holdeman, 1970) that of the strains producing acetic and formic acids: (1) non-motile strains having a G + C % of about 40 be considered *B. biacutus*;

TABLE 9.5

Biochemical reactions of species of genus **Bacteroides,** *species 6–14*[a]

(*do not produce black colonies*)

	6. B. hyper-megas	7. B. serpens	8. B. termitidis	9. B. biacutus	10a. B. clostridii-formis ss. clostridii-formis	10b. B. clostridii-formis ss. girans	11. B. con-stellatus	12. B. putredinis	13. B. coagulans	14. B. prae-acutus
Gelatin digested	−	+	−	w⁻	−ʷ	−	−	+	+	+
Milk	c	c	−	c	−ᶜ	c	c	dᶜ	cᵈ	−
Indole produced	−	·	−	−	−⁺	−	·	+	+	−
H₂S produced	−	+	·	−	−⁺	+	·	+⁻	−	+
Nitrate reduced	−	−	−	+	d	+	+	−	−	+
Gas	2⁺	+³	4	4	4	4	+⁻	4⁻	−²	+²
Acetoin	w⁻	·	−	+	+⁻	+⁻	·	−	−ʷ	
Growth in 20% bile	+⁴	·	·	3⁴	4⁻	4	·	3⁻	+ˢ	+
Hemolysis	−	−	−	−	−ᵇ	−ᵃ	+	−	−	−
Motility	−	+	−	−	−	+	+	−	−	+
Propionate ← threonine	−	·	·	−	−⁺	−	−	−⁺	−	−
Growth enhanced by	−	·	·	−	−	−	s	−	r	

[a] For symbols see Table 9.2.

(2) non-motile strains having a G + C % of 53–55 be considered *B. clostridiiformis* subsp. *clostridiiformis*; (3) motile strains be considered *B. clostridiiformis* subsp. *girans*; and that strains producing butyric acid as a major product be considered fusobacteria.

We have not detected spores in any of the strains listed above or in nine other strains that are similar to *B. clostridiiformis* subsp. *clostridiiformis*. Reinhold *et al.* (1967) reported spores in *Ristella biacuta* (*B. biacutus*) strain 139 I (CIPP (Prévot)). We also have detected spores in this strain and in numerous strains labeled *R. biacuta*, *R. clostridiiformis* or *Sphaerophorus clostridiiformis*, *Zuberella clostridiiformis* and *Fusocillus girans*. Some of the spore-forming strains are motile and some are not; all belong in the genus *Clostridium*. Spores are often difficult to detect in the spore-forming strains and could easily be overlooked because they are quite small and often are seen only in cultures (in chopped meat) that are 3–4 weeks old. The spores resist heating at 80 C for 10 min.

10. Bacteroides clostridiiformis (Burri and Ankersmit) Holdeman and Moore 1970, 33.

clos.tri.di.i.for′mis. Gr. n. *closter* a spindle; M.L. n. *clostridium* a small spindle; L. adj.comb.form *-formis* in the form of; M.L. adj. *clostridiiformis* in the form of a small spindle, spindle shaped.

Glucose broth cultures turbid with smooth sediment. Final pH in glucose broth is 5.0–5.5.

All strains grow at 30 C, most grow at 45 C and 25 C. Some strains grow at pH 7.5 but none grows at pH 8.5. No growth in 6.5% NaCl.

Further Comments

See *B. biacutus, Further Comments.*

10a. *Bacteroides clostridiiformis* subsp. *clostridiiformis* (Burri and Ankersmit) Holdeman and Moore 1970, 33. (*Bacterium clostridiiforme* Burri and Ankersmit 1906, 115; *Eggerthella clostridiiformis* (Burri and Ankersmit) Beerens, Castel and Fievez 1962, 120; *Ristella clostridiiformis* (Burri and Ankersmit) Prévot 1938, 291.)

Description based on Prévot *et al.* (1967), Sebald (1962), and a study of 10 strains (Tables 9.4, 9.5 and Fig. 9B).

Surface colonies on blood agar (2–3 days) are 1–2 mm, circular, entire to slightly scalloped, convex, translucent. Colonies of most strains have mottled or mosaic appearance when viewed by transmitted light.

G + C content of DNA is 53 moles % (chromatographic separation, Sebald, 1962).

Isolated from tissue and intestinal tract abscesses, turkey liver lesions and rumen contents of calves.

10b. *Bacteroides clostridiiformis* subsp. *girans* (Prévot) Holdeman and Moore 1970, 33. (*Bacterium clostridieformis* (*sic*) Choukévitch 1911, 350; *Zuberella clostridiiformis mobilis* Prévot 1938, 293; *Fusocillus girans* Prévot 1940, 249; *Fusobacterium girans* (Prévot) Macdonald 1953, 52; *Zuberella girans* (Prévot) Sebald 1962, 148.)

gi'rans. L. v. *gyro* turn around in circles; M.L. pres.part. *gyrans, girans* turning around in circles.

Description based on study of three strains (Tables 9.4, 9.5 and Fig. 9B).

Isolated from stomach contents, peritoneal fluid and colon abscess.

11. Bacteroides constellatus (Martres, Brygoo and Thouvenot) Holdeman and Moore 1970, 33. (*Zuberella constellata* Martres *et al.* 1952, 141.)

con.stel.la'tus. L. adj. *constellatus* studded with stars.

Description from Martres *et al.* (Tables 9.4 and 9.5).

Rods 0.5 by 1.8 μm with gyratory motility; presence or type of flagella not determined.

Colonies in deep agar are 1–2 mm, lenticular with an opaque center and surrounded by smaller colonies.

Peptone-serum-glucose broth cultures turbid with sediment and gas. No toxin or hemolysin in broth culture fluids.

Serum required for growth.

Lipids hydrolyzed. Coagulated protein not attacked. Sulfites not reduced.

Isolated from inflamed lacrymal sac.

12. Bacteroides putredinis (Weinberg, Nativelle and Prévot) Kelly 1957, 430. (*Bacillus putredinis* Weinberg *et al.* 1937, 755; *Ristella putredinis* (Weinberg *et al.*) Prévot 1938, 291; *Pseudobacterium putredinis* (Weinberg *et al.*) Krasil'nikov 1949, 243.)

put.re'di.nis. L. n. *putredo* putridity; M.L. gen.n. *putredinis* of putridity.

Description based on previous literature descriptions and study of 16 strains (Tables 9.4, 9.5 and Fig. 9C).

Surface colonies (2 days) are pinpoint to 0.5 mm, circular to slightly irregular with entire to erose margins, low convex, translucent, gray, dull, smooth.

Broth cultures are lightly turbid with smooth or stringy sediment.

Most strains grow at 25–45 C.

Some strains digest chopped meat, most strains produce trace amounts of catalase. Glutamic acid decarboxylated (Werner, 1970).

Isolated from feces, abdominal and rectal abscesses, from cases of acute appendicitis and from foot rot in sheep.

13. Bacteroides coagulans Eggerth and Gagnon 1933, 409. (*Pasteurella coagulans* (Eggerth and Gagnon) Prévot 1938, 292; *Pseudobacterium coagulans* (Eggerth and Gagnon) Krasil'nikov 1949, 246.)

co.a'gu.lans. L. part.adj. *coagulans* curdling, coagulating.

Description from Eggerth and Gagnon and study of three strains (Tables 9.4, 9.5 and Fig. 9C).

Surface colonies on horse blood agar (2 days) punctate, circular, entire, slightly raised, translucent; non-hemolytic.

Glucose broth cultures turbid; no acid produced.

Grows poorly, if at all, at 25 and 45 C; does not grow at pH 8.5.

Isolated from human feces and human lung.

14. Bacteroides praeacutus (Tissier) Holdeman and Moore 1970, 33. (*Coccobacillus praeacutus* Tissier 1908, 193; *Zuberella praeacuta* (Tissier) Prévot 1938, 293; *Fusobacterium praeacutum* (Tissier) Hoffman 1957, 439.)

prae.a.cu'tus. L. pref. *prae* very, quite; L. adj. *acutus* sharp; M.L. adj. *praeacutus* quite sharp.

Description from Prévot *et al.* (1967) and study of three strains including ATCC 25539 (Tables 9.4, 9.5 and Fig. 9C).

Motile with peritrichous flagella.

Surface colonies on horse blood agar (2 days) are 0.5 mm, circular, flat, scalloped to diffuse edge, grayish, dull, smooth, translucent with "mosaic" appearance when viewed by obliquely transmitted light.

Glucose broth cultures (2 days) are moderately turbid with sediment. Nitrates completely reduced by one of three strains. Hippurate hydrolyzed by two of three strains.

Isolated from intestinal tracts of infants, from gangrenous lesions and from blood.

Reference strain: ATCC 25539; V.P.I. 0563; CIPP (Prévot) 3722 B.

Further Comments

The type strain is not extant. Cultural reactions of the reference strain are similar to those originally described and morphological characteristics are similar to those depicted in Weinberg *et al.* (1937, 786) for this species.

15. Bacteroides corrodens Eiken (in part) 1958, 415. (*Ristella corrodens* (Eiken) Prévot 1966, 118.)

cor.ro'dens. L. v. *corrodo* gnaw; M.L. part.adj. *corrodens* gnawing; the gnawing bacillus.

Description from Eiken (1958), Khairat (1967), Jackson *et al.* (1971), and study of 13 anaerobic strains that failed to grow as surface colonies in a candle jar (Tables 9.6, 9.7 and Fig. 9C).

Surface colonies on blood agar (4–5 days) are 1 mm, circular with entire to slightly undulating margins, convex to slightly umbonate, gray-white. Colonies grow in depressions in the agar or produce slight zones of depression in the agar around the colonies. Colonies may have thin spreading edges.

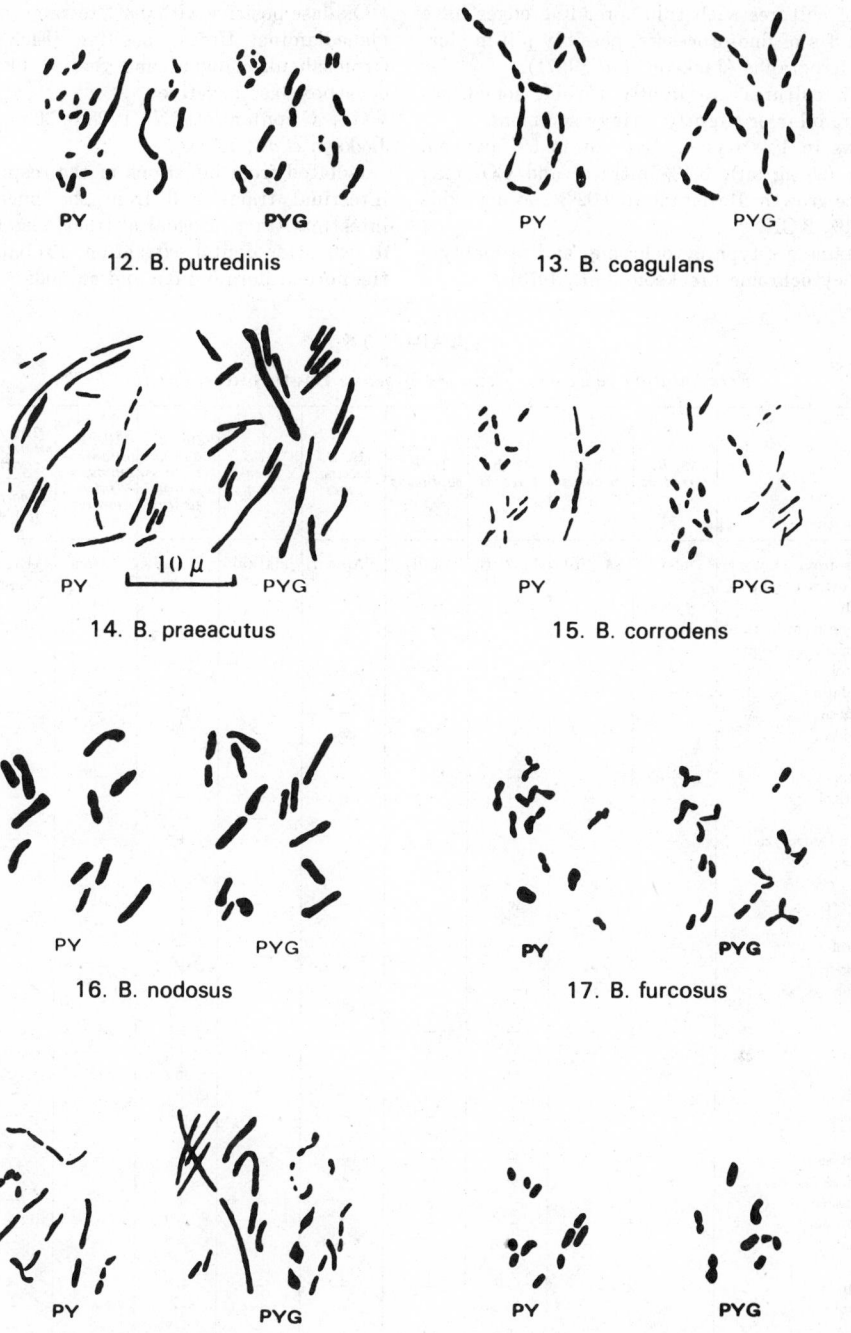

12. B. putredinis

13. B. coagulans

14. B. praeacutus

15. B. corrodens

16. B. nodosus

17. B. furcosus

18. B. capillosus

19. B. succinogenes

Fig. 9C. *Bacteroides*. The line drawings are composites of several cultures and individual cultures usually show somewhat less variation in cellular morphology. PY, peptone-yeast extract broth; PYG, PY-glucose broth. Scale: 9/16 inch = 10 μm.

Cells of cultures with thin spreading edges have polar tufts of fine processes, possibly pili in electron micrographs (Jackson et al., 1971).

Broth cultures are lightly turbid, sometimes with granular or slightly stringy sediment.

Grows in 1% oxygen, but not in 5% oxygen. Hemin (25 μg/ml), 0.1% nitrate, and CO_2 may enhance growth. Resistant to 0.02% sodium azide and 0.1% KCN.

Contains a c-type cytochrome, and probably a b-type cytochrome (Jackson et al., 1970).

Oxidase positive with 0.3% tetramethyl-p-phenylenediamine. Urease positive (Elek's method; Cruickshank, Duguid and Swain, 1968). Lysine decarboxylase negative.

G + C content of DNA is 28 to 30 moles % (T_m, Jackson et al., 1971).

Isolated from infections of the respiratory and intestinal tracts and from the buccal cavity, intestinal tract, urogenital tract; also from blood drawn after dental extraction. Probably part of the normal flora of man and animals.

TABLE 9.6

Fermentation reactions of species of genus **Bacteroides**, *species 15–22*[a]

	15. B. corrodens	16. B. nodosus	17. B. furcosus	18. B. capillosus	19. B. succinogenes	20. B. pneumosintes	21a. B. melaninogenicus ss. melaninogenicus	21b. B. melaninogenicus ss. intermedius	21c. B. melaninogenicus ss. asaccharolyticus	22. B. niger
Products from peptone yeast extract glucose broth	alsf	sa (lfp)	Lsa2 (f)	Sa(lf)	SApiv (f)	s(afliv)	Sl (AF ivib)	SAibiv	ABibiv (splf)	AFL
Products from lactate	−	−	−	−	−	−	−	−P	−P	.
Acid produced from:										
Adonitol	−	−	−	−	.	−	−w	−	−	.
Amygdalin	−	−	−	−w	.	−	+−	w−	−	.
Arabinose	−	−	−	−	−	−	w−	−w	−	.
Cellobiose	−	−	−	−w	w	−	+−	−	−	.
Dulcitol	−	−	−	−	.	−	−w	−w	−	.
Erythritol	−	−	−	−	−	−	−	−	−	.
Esculin	−	−	−	−	−	−	−	−	−	.
Esculin hydrolyzed	−	−	+−	+	−	−	+−	−	−	.
Fructose	−	−	w	−w	−	−	+	v	−	−
Galactose	−	−	w	−w	−	−	+	d	−	.
Glucose	−	−	w+	+−	w	−	+	+w	−	−
Glycerol	−	−	−	−	.	−	−+	−+	−	.
Glycogen	−	−	−	−	.	−	+	+−	−	.
Inositol	−	−	−w	−	.	−	−	−w	−	.
Inulin	−	−	−	−	.	−	+−	−w	−	.
Lactose	−	−	−	−w	v	−	+	−+	−	−
Maltose	−	−	−	−w	v	−	+	v	−	−
Mannitol	−	−	−	−	−	−	−	−	−	.
Mannose	−	−	−	−w	−	−	+	d	−	.
Melezitose	−	−	−	−	−	−	−	−	−	.
Melibiose	−	−	−	−	−	−	w−	−	−	.
Raffinose	−	−	−	−	−	−	+	−+	−	.
Rhamnose	−	−	−	−	−	−	−w	−	−	.
Ribose	−	−	−	−w	.	−w	−w	−w	−	.
Salicin	−	−	−w	−	−	−	−w	−	−	.
Sorbitol	−	−	−	−	−	−	−	−	−	.
Sorbose	−	−	−	−	.	−	−w	−	−	.
Starch	−	−	−	−w	v	−	+	+w	−	.
Starch hydrolyzed	−	−	−	−+	v	−	+	+−	−	.
Sucrose	−	−	v	−w	−	−	+	v	−	−
Trehalose	−	−	−	−w	v	−	−	−w	−	.
Xylose	−	−	−	−w	−	−	−w	−w	−	.

[a] For symbols see Table 9.1.

TABLE 9.7

Biochemical reactions of species of genus **Bacteroides** *species 15–22[a]*

	15. B. corrodens	16. B. nodosus	17. B. furcosus	18. B. capillosus	19. B. succinogenes	20. B. pneumosintes	21a. B. melaninogenicus ss. melaninogenicus	21b. B. melaninogenicus ss. intermedius	21c. B. melaninogenicus ss. asaccharolyticus	22. B. niger
Black colonies	–	–	–	–	–	–	+	+	+	+
Gelatin digested	–	+	–w	–w	–	–w	+	+	+	–
Milk	–	d^c	–	–^{c}	–	–	c	d^{c-}	d$^-$	–
Indole produced	–	–	–	–	–	–	–	+$^-$	+$^-$	–
H$_2$S produced	–	+$^-$	–	–$^+$	–	–	–	–$^+$	+$^-$	+
Nitrate reduced	+	–	–	–	–	–	–	–	–$^+$	+
Gas	–	+$^-$	–^{2}	v	–	–	–$^{-+}$	+$^-$	2$^-$	–
Acetoin	–	–	–	–	–	–	v	–	–$^+$	.
Growth in 20% bile	+	–	+$^-$	+^{s}	–	–$^+$	–^{3}	2$^-$	–$^+$	.
Hemolysis	–	–	–	–	.	–^{a}	b	b^{a-}	–^{b}	.
Motility	–	–	–	–	–	–	–	–	–	–
Propionate ← threonine	–	–$^+$	–	–	–	–	–	–	–	–
Growth enhanced by	–	–	t	bth	CO$_2$	–	h	h	hr	s

[a] For symbols see Table 9.2.

Further Comments

Bacteroides corrodens as originally described by Eiken (1958) included both obligately anaerobic and facultative strains. ATCC 23834, NCTC 10596, Henricksen 333/54–55 (Eiken), deposited by Henricksen (1969) to represent *B. corrodens*, is a facultative strain and the type strain of *Eikenella corrodens* (Eiken) Jackson and Goodman 1972, 73. No reference strain has been designated for *Bacteroides corrodens* (anaerobic).

16. **Bacteroides nodosus** (Beveridge) Mráz, Tesarčik and Vařejka 1963, 85. ("Organism K," Beveridge 1938, 1; *Fusiformis nodosus* Beveridge 1941, 23; *Ristella nodosa* (Beveridge) Prévot 1948, 82.)

no.do′sus. L. adj. *nodosus* knotty or swollen.

Description from Beveridge (1941) and study of ATCC 25549 and two other strains (Tables 9.6, 9.7 and Fig. 9C).

Terminal enlargements of cells are more pronounced in lesions than in cultures.

Surface colonies are 0.5–2 mm, smooth, convex, translucent or semi-opaque. Colonies often etch into the surface of the medium immediately under the colony, producing a sunken appearance.

Broth cultures are lightly turbid, sometimes granular.

Growth enhanced by at least 10% CO$_2$ and 10% horse serum. Growth is most rapid at pH 7.4–7.6; does not grow at pH 4–6; grows at pH 8–9.

Hoof powder is digested. Tyrosine crystals develop in chopped meat after incubation for 4–6 weeks.

Causative agent of foot rot in sheep (and possibly in goats); infected hoofs apparently the only natural habitat.

Reference strain: ATCC 25549; V.P.I. 2340; Beveridge 11342.

17. **Bacteroides furcosus** (Veillon and Zuber) Hauduroy *et al.* 1937, 61. (*Bacillus furcosus* Veillon and Zuber 1898, 541; *Fusiformis furcosus* (Veillon and Zuber) Topley and Wilson 1929, 302; *Ristella furcosa* (Veillon and Zuber) Prévot 1938, 291; *Pseudobacterium furcosum* (Veillon and Zuber) Krasil'nikov 1949, 247.)

fur.co′sus. L. adj. *furcosus* forked.

Description based on Prévot *et al.* (1967, 220) and study of ATCC 25662 and two other strains (Tables 9.6, 9.7 and Fig. 9C).

Surface colonies (2 days) are 0.5 mm, circular, entire, convex, translucent to semi-opaque, grayish white, shiny, smooth.

Glucose broth cultures are turbid with smooth sediment. Final pH in glucose broth is 5.7–5.9.

Grows at 25 C but not at 45 C. Grows at pH 8.0.

Reported to ferment maltose and mannitol (Prévot et al., 1967, 221); maltose and mannitol not fermented in our laboratory.

Isolated from infected appendix, lung abscesses and feces.

Reference strain: ATCC 25662; V.P.I. 3253; Suzuki T-301-2a.

18. **Bacteroides capillosus** (Tissier) Kelly 1957, 433. (*Bacillus capillosus* Tissier 1908, 193; *Ristella capillosa* (Tissier) Prévot 1938, 292; *Pseudobacterium capillosum* (Tissier) Krasil'nikov 1949, 247.)

ca.pil.lo'sus. L. adj. *capillosus* very hairy.

Description from Sebald (1962, 51) and study of 18 strains (Tables 9.6, 9.7 and Fig. 9C).

Surface colonies are minute to 1 mm, circular, entire, convex, translucent, smooth. Two strains did not grow on agar plates incubated in an anaerobe jar.

Broth cultures are lightly turbid, sometimes with sediment (smooth to slightly stringy). Development of black pigment in gelatin cultures reported by Sebald.

Growth of most strains enhanced by hemin, rumen fluid, or Tween 80. Strains are generally non-fermentative unless Tween 80 is added to the medium, in which case they may be slightly fermentative.

Isolated from cysts and wounds, human feces, intestinal tracts of hogs, mice, and *R. lucifugus* (termite) and from sludge.

19. **Bacteroides succinogenes** Hungate 1950, 13. (*Ruminobacter succinogenes* (Hungate) Prévot 1966, 122.)

suc.ci.no'ge.nes. M.L. n. *acidum succinicum* succinic acid; Gr. v. *gennaio* produce; M.L. adj. *succinogenes* succinic acid-producing.

Description from Hungate (1950, 1966), Bryant and Doetsch (1954) and study of one strain (Tables 9.6, 9.7 and Fig. 9C).

Surface colonies on rumen fluid glucose cellobiose agar (RGCA) in roll tube are entire, slightly convex, translucent to opaque, non-pigmented or sometimes yellow, often with "frosted glass" appearance. One strain tested did not grow on the surface of agar plates incubated in an anaerobe jar. Deep colonies (RGCA) are 1–3 mm, lenticular. Colonies in cellulose agar roll tubes are not visible macroscopically, but are surrounded by definite clear zones of cellulose digestion; rods may be observed microscopically at the periphery of the area cleared of cellulose. Cells migrate through agar.

Glucose broth cultures are turbid with smooth sediment. Broth clears in older cultures as cells lyse. Final pH in glucose broth is about 5.5.

Volatile fatty acids other than acetate and NH_4^+ (but little or no organic nitrogen) are required for growth (Bryant and Robinson, 1962). Good growth and cellulose digestion occur in media containing only cellulose, p-aminobenzoic acid, biotin, cysteine, alanine, phenylalanine, valerate, isobutyrate, a carbonic acid-bicarbonate buffer, resazurin, and minerals (Bryant et al., 1959).

No acid from gum arabic or xylan. Cultures show CO_2 uptake in fermentation of cellulose or cellobiose.

Optimum temperature for growth about 40 C; grows well in glucose medium at 30–38 C; does not grow at 22 or 45 C. Grows in glucose medium with an initial pH of 6.05–7.7, but not at pH 5.5 (Bryant and Doetsch, 1954).

Isolated from rumen contents of cattle, sheep, and several wild African antelopes; presumed to be among the more numerous bacteria in the rumen of most ruminants.

20. **Bacteroides pneumosintes** (Olitsky and Gates) Holdeman and Moore 1970, 33. (*Bacterium pneumosintes* Olitsky and Gates 1921, 727; *Dialister pneumosintes* (Olitsky and Gates) Bergey et al. 1923, 271; *Bacillus pneumosintes* (Olitsky and Gates) Ford 1927, 634; *Dialister pneumosintes* var. *septicemiae* (sic) Hauduroy et al. 1953, 195.)

pneu.mo.sin'tes. Gr. n. *pneuma* air; Gr. n. *sintes* a spoiler, thief; M.L. adj. *pneumosintes* breath-destroying.

Description from Hitchens (1957, 441), Prévot et al. (1967, 190), and examination of CIPP (Prévot) Bi2 and five other strains (Tables 9.6, 9.7 and Fig. 9D).

Cells from naso-pharynx 0.15–0.3 μm long and one-half to one-third as wide; filterable through Berkfeld V and N filters. Cells from glucose broth are 0.5–1 μm long.

Deep agar colonies are punctiform, granular, white, no gas. Surface colonies on horse blood agar (2 days) are punctiform, circular, entire, convex, clear, transparent, shiny, smooth.

Glucose broth cultures are lightly turbid without sediment; no acid produced.

Grows at pH 8.5 and at 25–45 C.

Isolated from naso-pharyngeal washings from normal individuals (may be involved in secondary infections of the upper respiratory tract), blood, and lung and brain abscesses (in association with *Peptostreptococcus intermedius*).

Further Comments

The type strain of *D. pneumosintes* var. *septicemia* (CIPP (Prévot) Bi2) was originally described as weakly fermenting glucose, fructose, galactose, arabinose, maltose, lactose and as

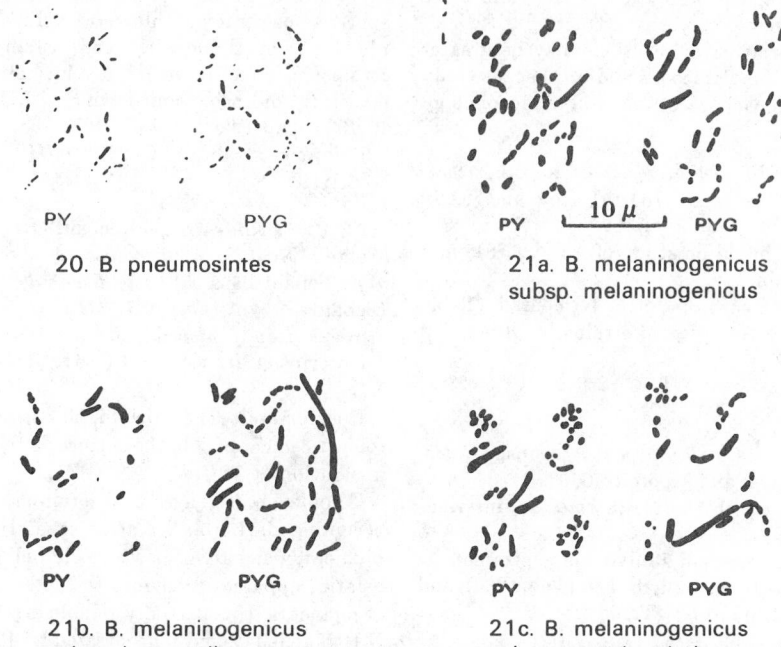

PY   PYG

20. B. pneumosintes

PY   10 μ   PYG

21a. B. melaninogenicus
subsp. melaninogenicus

PY   PYG

21b. B. melaninogenicus
subsp. intermedius

PY   PYG

21c. B. melaninogenicus
subsp. asaccharolyticus

Fig. 9D. *Bacteroides.* The line drawings are composites of several cultures and individual cultures usually show somewhat less variation in cellular morphology. PY, peptone-yeast extract broth; PYG PY-glucose broth. Scale: $\frac{9}{16}$ inch = 10 μm.

coagulating serum and milk (Prévot and Raynaud, 1947). Chen and Cleverdon (1963) reported that the strain fermented only glucose in conventional diagnostic media. Carbohydrates were not fermented when tested by our methods. This strain was deposited as ATCC 25558, but has been lost by us and ATCC.

Prévot *et al.* (1967) described *D. pneumosintes* var. *septicus*, which is an anaerobic microaerophile that utilizes urea and ferments glucose, fructose, and mannitol. Lactose, sucrose, raffinose and glycerol are weakly fermented. Milk is coagulated. Isolated in 1954 by Weiglass from blood of a young man with chronic polyarthritis and endocarditis.

21. Bacteroides melaninogenicus (Oliver and Wherry) Roy and Kelly 1939, 569. (*Bacterium melaninogenicum* Oliver and Wherry 1921, 341; *Hemophilus melaninogenicus* (*sic*) (Oliver and Wherry) Bergey *et al.* 1930, 314; *Ristella melaninogenica* (Oliver and Wherry) Prévot 1938, 290; *Fusiformis nigrescens* Schwabacher, Lucas and Rimington 1947, 109.)

me.la.ni.no.ge'ni.cus. Gr. adj. *melas* black; M.L. n. *melaninum* melanin; M.L. adj. *genicus* producing, probably derived from Gr. n. *genetes* a producer; M.L. adj. *melaninogenicus* melanin-producing. (Black pigment is due to a hematin

derivative (Schwabacher, Lucas and Rimington, 1947) and not to melanin, as originally thought.)

Surface colonies on blood agar (2–3 days) are 0.5–3.0 mm, circular, entire, convex to pulvinate, opaque. After incubation for 2–3 days, colonies may be gray or brown or black. Colonies become darker upon continued incubation (5–14 days). Dark pigment develops more rapidly when laked blood, rather than blood containing whole red blood cells, is used.

Glucose broth cultures usually turbid with smooth or stringy sediment.

Hemin (1.0 μg/ml), menadione (0.1 μg/ml), or both are required for or enhance growth of most strains. Most strains grow at 25 C and pH 8.5; some grow at 45 C.

Isolated from the mouth, urine, feces, and infections of the mouth, soft tissue, respiratory tract, urogenital tract and intestinal tract. Pathogenic, but usually in association with other kinds of organisms.

Further Comments

Sawyer *et al.* (1962) grouped 31 strains of *B. melaninogenicus* as (1) strongly fermentative; (2) weakly fermentative; and (3) non-fermentative. Three subspecies are recognized on the basis of saccharolytic and proteolytic activity, and acids produced in peptone yeast extract glucose cultures. The variation in reactions may be greater than indicated in Tables 9.6 and 9.7, because only a few strains of each were studied. For morphology see Figure 9D.

21a. *Bacteroides melaninogenicus* subsp. *melaninogenicus* (Oliver and Wherry) Roy and Kelly 1939, 569.

Description based on study of ATCC 25845 and one other strain.

A number of carbohydrates fermented (Table 9.6), similar to the original strains of Oliver and Wherry (1921).

Reference strain: ATCC 25845; VPI 2381; Finegold B282.

21b. *Bacteroides melaninogenicus* subsp. *intermedius* Holdeman and Moore 1970, 33.

in.ter.me'di.us. L. adj. *intermedius* intermediate.

Description based on study of nine strains.

Ferments some carbohydrates (Table 9.6) and is somewhat proteolytic (Table 9.7).

21c. *Bacteroides melaninogenicus* subsp. *asaccharolyticus* Holdeman and Moore 1970, 33.

a.sac.cha.ro.ly'ti.cus. Gr. pref. *a* not; Gr. n. *sacchar* sugar; Gr. adj. *lyticus* able to loosen; M.L. adj. *asaccharolyticus* not digesting sugar.

Description based on ATCC 25260 and 14 similar strains.

Carbohydrates not fermented (Table 9.6); proteolytic (Table 9.7). Does not decarboxylate glutamate (Werner, 1970; Werner *et al.*, 1971).

Six strains sensitive to penicillins, cephalosporins, bacitracin, chlortetracycline, chloramphenicol, erythromycin, and rifampicin. Six strains resistant to streptomycin, colistin, polymyxin B, and neomycin, having M.I.C. values of 10–100 µg/ml (Werner *et al.*, 1971).

Reference strain: ATCC 25260; VPI 4198; Finegold B440.

22. **Bacteroides niger** (Sebald) Holdeman and Moore 1970, 33. (*Sphaerophorus abscedens* var. *niger* Sebald 1962, 55; *Ristella abscedens* var. *niger* (Sebald) Prévot *et al.* 1967, 252.)

ni'ger. L. adj. *niger* black.

Description from Sebald (1962) (Tables 9.6 and 9.7).

Rods 0.5 by 2–6 µm with bipolar staining; spheroids 2–3 µm appear in the cultures after prolonged incubation at 37 C.

Colonies in deep agar plus serum are lenticular, becoming dark (black) after several days. The pigment, which forms particularly well on nutrient gelatin, apparently belongs to the melanin group of pigments. It is partially soluble in 0.1 N NaOH, N HCl at 100 C, and concentrated HCl in the cold; it is completely soluble in concentrated HCl and H_2SO_4 at 100 C; it is insoluble in water, alcohol, chloroform, ether, or acetone.

Cultures in serum-glucose broth have a granular sediment.

Growth enhanced by serum, serum globulin, or brain extract. Coagulated proteins not attacked.

One strain isolated from rectum of a termite, *R. lucifugus*.

Genus II. **Fusobacterium** *Knorr 1922, 4*

W. E. C. MOORE AND LILLIAN V. HOLDEMAN

Fu.so.bac.te'ri.um. L. n. *fusus* a spindle; Gr. dim.n. *bakterion* a small rod; M. L. neut.n. *Fusobacterium* a small spindle-shaped rod.

Gram-negative; non-spore-forming rods. Nonmotile or motile with peritrichous flagella.

Chemoorganotrophs: Metabolize carbohydrates or peptone. **Major products from carbohydrate or peptone include butyric acid,** often with acetic and lactic acids, and lesser amounts of propionic, succinic, and formic acids and short-chained alcohols. Strains of some species convert threonine to propionate. Pyruvate converted to acetate and butyrate and sometimes also to formate, succinate and lactate.

Adonitol, dulcitol, erythritol, glycerol, inositol, melezitose, rhamnose, ribose, sorbitol and sorbose usually not attacked. Catalase usually not produced.

Obligately anaerobic. Will not grow on the surface of agar plates incubated aerobically. The maximum Eh permitting growth varies, depending upon the species, the size of inoculum, and the medium. Oxidized media are inhibitory for some species; addition of serum or ascitic fluid to such media may enable them to support growth.

Growth usually most rapid at 37 C and pH near 7.

The G + C content of the DNA in most of the species examined ranges from 26–34 moles % (chromatographic separation, Sebald, 1962 and T_m, Johnson, personal communication).

Found in cavities of man and other animals. Some species are pathogenic and occur in various purulent or gangrenous infections and in organ infarcts.

Characteristics by which species in the genus can be differentiated are given in the key and in Tables 9.8–9.11. Other characteristics of the species in the genus are given in the text.

Type species: *Fusobacterium nucleatum* Knorr 1922, 17.

Further Comments

Knorr (1922) described the genus *Fusobacterium* and three species: *F. plauti-vincenti*, *F. nucleatum*, and *F. polymorphum;* no type species was designated. In the 5th edition of *Bergey's Manual of Determinative Bacteriology* (1939, 586), Roy and Kelly designated *F. plauti-vincenti* the type species of *Fusobacterium* Knorr. Breed (6th edition of THE MANUAL, 1948, 581) also lists *F. plauti-vincenti* as the type species. Hoffman (7th edition of THE MANUAL, 1957, 437) lists the type species of *Fusobacterium* as *F. fusiforme* (Veillon and Zuber) Hoffman and designates *F. plauti-vincenti* as synonymous. *F. plauti-vincenti* Knorr and *F. fusiforme* (Veillon and Zuber) Hoffman are considered later synonyms of *Leptotrichia buccalis* (Robin) Trevisan 1879, 147 (q.v.). Furthermore, *F. fusiforme* (Veillon and Zuber) Hoffman probably is not the organism described by Vincent and named by Veillon and Zuber and is certainly not the organism commonly recognized as *Fusiformis fusiformis*.

If Rule 9c(3) is followed precisely, the genus *Fusobacterium* Knorr has no nomenclatural standing because the first type species designated (*F.*

plauti-vincenti) is synonymous with the type species of a prior genus (*Leptotrichia* Trevisan).

We believe that Principle 2 of the Code of Nomenclature applies to this case and pending Judicial Commission Opinion, suggest that *F. nucleatum* now be designated the type species of the genus *Fusobacterium* Knorr to comply with Principle 1(1).

According to Rule 9c(3b), the type species of a genus must be one of those included when the genus was originally described. *F. nucleatum* Knorr and *F. polymorphum* Knorr are thought to be synonymous by later workers (Spaulding and Rettger, 1937; Omata and Braunberg, 1960) and the name *F. nucleatum* is usually used. Sebald (1962) reported a G + C content of 27 moles % for CIPP (Prévot) strains of *Fusiformis fusiformis* and of 32 moles % (chromatographic separation) for CIPP (Prévot) strains of *F. nucleatum*. Since these species were closely related by G + C content to *Sphaerophorus*, she suggested they be considered *Sphaerophorus fusiformis* and *S. nucleatus*. We retested all available strains of these species examined by Sebald. *S. fusiformis* (Vincent) Sebald and *S. nucleatus* (Knorr) Sebald could not be differentiated by their phenotypic characteristics. The differences she noted in the nucleic acid properties of the two species could not be verified (T_m, J. L. Johnson, personal communication). On the basis of these data, we consider the species synonymous, bearing the name *Fusobacterium nucleatum* Knorr, which, pending referral to the Judicial Commission, we consider the type species of *Fusobacterium* Knorr.

Key for some species in the genus **Fusobacterium**

I. Indole produced.
  A. Esculin not hydrolyzed.
    1. Propionate not formed from lactate.
      a. Lactose not acid.
        b. Mannose not acid.
          c. Propionate formed from threonine.
            d. Little or no gas formed in glucose agar deeps.
              1. *F. nucleatum*
            dd. Much gas formed in glucose agar deeps.
              2. *F. gonidiaformans*
          cc. Propionate not formed from threonine.
            3. *F. naviforme*
        bb. Weak acid from mannose.
          4. *F. varium*
      aa. Lactose acid.
        5. *F. glutinosum*
    2. Propionate formed from lactate.
      6. *F. necrophorum*
  B. Esculin hydrolyzed.
    7. *F. aquatile*

II. Indole not produced.
    A. Abundant gas produced in glucose agar deeps.
        1. Maltose and mannitol acid.

                        8. *F. stabile*

        2. Maltose and mannitol not acid.
          a. Lactose acid.
            b. Esculin hydrolyzed.

                        9. *F. mortiferum*

            bb. Esculin not hydrolyzed.

                     10. *F. symbiosum*

          aa. Lactose not acid.
            b. Sucrose acid.
              c. Elongate cells present.

                     11. *F. necrogenes*

              cc. Short ovoid cells.

                     12. *F. perfoetens*

            bb. Sucrose not acid.
              c. Esculin hydrolyzed.

                     11. *F. necrogenes*

              cc. Esculin not hydrolyzed.
                d. Threonine converted to propionate.

                     4. *F. varium*

              dd. Threonine not converted to propionate.

                     10. *F. symbiosum*

    B. Little or no gas produced in glucose agar deeps.
        1. Esculin hydrolyzed.

                     13. *F. prausnitzii*

        2. Esculin not hydrolyzed.
          a. Nitrate reduced.

                     14. *F. plauti*

          aa. Nitrate not reduced.
            b. Motile.

                     15. *F. bullosum*

            bb. Not motile.

                     16. *F. russii*

Description of the species of genus **Fusobacterium**

1. Fusobacterium nucleatum Knorr 1922, 17. (*Bacillus fusiformis* Veillon and Zuber 1898, 540; *Corynebacterium fusiforme* (Veillon and Zuber) Lehmann and Neumann 1907, 529; *Fusiformis nucleatus* (Knorr) Bergey *et al.* 1930, 514; *Fusiformis fusiformis* (Veillon and Zuber) Topley and Wilson 1936, 357; Group I, Spaulding and Rettger 1937, 535; Group III (and probably *Fusobacterium polymorphum*) Baird-Parker 1960, 458; *Sphaerophorus fusiformis* (Veillon and Zuber) Sebald 1962, 149; **NOT** *Fusobacterium plauti-vincenti* Knorr 1922, 5; **NOT** the organism described as *F. fusiforme* by Hoffman in 7th edition of THE MANUAL.)

nu.cle.a'tum. L. neut.adj. *nucleatum* having a kernel, nucleated.

Description based on Knorr (1922) and study of ATCC 25586 and CIPP (Prévot) strains 1599A, 679, 1210, 1789, and 1734D and 71 other strains with similar characteristics (Tables 9.8, 9.9 and Fig. 9E).

Surface colonies on horse blood agar (2 days) are 1–2 mm, circular to slightly irregular, convex to pulvinate, translucent, often with "flecked" appearance when viewed by transmitted light; usually non-hemolytic, but may be slightly hemolytic under area of confluent growth or may produce greenish discoloration of the blood agar upon exposure to oxygen.

Glucose broth cultures have flocculent or granular sediment, with or without turbidity, and final pH of 5.8–6.2.

G + C content of the DNA is 27 to 28 moles %

(chromatographic separation, Sebald 1962; T_m, J. L. Johnson, personal communication).

Isolated from the mouth and infections of the upper respiratory tract and pleural cavity, occasionally from wounds and other kinds of infections.

Reference strain: ATCC 25586; V.P.I. 4355; CIPP (Prévot) 1612A, isolated from cervico-facial lesion.

2. **Fusobacterium gonidiaformans** (Tunnicliff and Jackson) Moore and Holdeman 1970, 45.* (*Bacillus gonidiaformans* Tunnicliff and Jackson 1925, 430; *Actinomyces gonidiaformis* (Tunnicliff and Jackson) Bergey *et al.* 1930, 469; *Sphaerophorus gonidiaformans* (Tunnicliff and Jackson) Prévot 1938, 299; *Pseudobacterium gonidiaformans* (Tunnicliff and Jackson) Krasil'nikov 1949, 243.)

go.ni.di.a.for'mans. Gr. n. *gone* offspring, seed; M.L. n. *gonidium* gonidium; L. part.adj. *formans* forming; M.L. part.adj. *gonidiaformans* gonidiaforming.

Description supplemented by study of ATCC 25563 and four other similar strains (Tables 9.8, 9.9 and Fig. 9E).

The spheroid or gonidial forms implied by the name of this organism are seen most often in old cultures or in media that are not highly reduced.

Surface colonies on horse blood agar plates minute to 0.5 mm, circular, entire, translucent.

Glucose broth cultures turbid with smooth sediment and final pH of 5.9–6.2.

Isolated from human infections of the respiratory tract, urogenital tract, and intestinal tract; also from a lamb with pneumonia.

Reference strain: ATCC 25563; V.P.I. 0482A; CIPP (Prévot) 3554A.

3. **Fusobacterium naviforme** (Jungano) Moore and Holdeman 1970, 45. (*Bacillus naviformis* Jungano 1909, 123; *Ristella naviformis* (Jungano) Prévot 1938, 291; *Pseudobacterium naviformis* (Jungano) Krasil'nikov 1949, 247.)

na.vi.for'me. L. n. *navis* ship; L. n. *forma* shape; M.L. neut.adj. *naviforme* in the shape of a ship.

Description based on study of 15 strains (Tables 9.8, 9.9 and Fig. 9E).

Surface colonies on horse blood agar (2 days) are minute to 2.0 mm, circular, entire, low convex, gray-white, translucent with mottled appearance when viewed by obliquely transmitted light.

Glucose broth cultures are lightly turbid with smooth sediment and final pH of 5.8–6.4.

Isolated by Jungano from the large intestine of a laboratory rat. Other strains isolated from abscesses and pilonidal cyst.

4. **Fusobacterium varium** (Eggerth and Gagnon) Moore and Holdeman 1969, 12. (*Bacteroides varius* Eggerth and Gagnon 1933, 409; *Sphaerophorus varius* (Eggerth and Gagnon) Prévot 1938, 299; *Pseudobacterium varium* (Eggerth and Gagnon) Krasil'nikov 1949, 242; *Sphaerophorus varius* var. *sulfitoreductans* Sebald 1962, 124.)

va'ri.um. L. neut.adj. *varium* diverse, varied.

Description from study of ATCC 8501 and 28 similar strains, including CIPP (Prévot) BL1, 700, CSA, CSB, B1-7-In, 2333, Te53, and 1669 (Tables 9.8, 9.9 and Fig. 9E).

Surface colonies on horse blood agar (2 days) are minute to 1 mm, circular with entire edges, flat to low convex, translucent, usually with gray-white centers and colorless edges.

Glucose broth cultures are turbid with smooth sediment and final pH of 5.6–6.0.

Isolated from human feces, cecal contents of mice, intestinal contents of *Blatta orientalis* (roach), posterior intestinal tract of *R. lucifugus* (termite), and purulent infections of man (surgical wounds, peritonitis, pleurisy, sinusitis, etc.). A bacteriophage active against four strains was isolated from filtrate of feline feces (Huet and Thouvenot, 1964).

Type strain: ATCC 8501 (Eggerth); NCTC 10560.

Further Comments

Strains which have a high degree of nucleic acid similarity (J. L. Johnson, personal communication) with ATCC 8501 all weakly ferment mannose. Several strains that do not ferment mannose, but are similar in other phenotypic characteristics, may be similar to CIPP (Prévot) strain 1802, *Sphaerophorus siccus* (Eggerth and Gagnon) Prévot 1938, which is described as fermenting glucose and fructose, producing gas, H_2S, and formic, acetic, butyric, lactic, succinic and valeric acids. G + C content of strain 1802 is 31 moles % (chromatographic separation, Sebald, 1962).

5. **Fusobacterium glutinosum** (Hauduroy *et al.*) Moore and Holdeman 1970, 45. (*Bacillus glutinosus* (Guillemot, Hallé and Rist) 1904, 599 (illeg.); *Bacteroides glutinosus* Hauduroy, Ehringer, Urbain, Guillot and Magrou 1937, 61; *Ristella glutinosa* (Hauduroy *et al.*) Prévot 1938, 292.)

* **Editorial Note.** The majority of the names in the genus *Fusobacterium* are new combinations attributed to Moore and Holdeman 1969, 1970 and 1972 in several editions of laboratory manuals of the Virginia Polytechnic Institute Anaerobe Laboratory. The authorities for these new combinations are confirmed in Moore and Holdeman (1973).

TABLE 9.8

Fermentation reactions of species of genus **Fusobacterium**, *species 1–8[a]*

	1. F. nucleatum	2. F. gonidia- formans	3. F. naviforme	4. F. varium	5. F. glutinosum	6. F. necro- phorum	7. F. aquatile	8. F. stabile
Products from peptone- yeast extract-glu- cose broth	Baps (Lf)	Bap (lfs)	BLap (fs)	BLA (p)	Bap (Lfs)	Bap (Lsf)	Ba(ls)	FBL
Lactate → propionate	−	−	−	−	−	P	−	·
Acid produced from:								
Amygdalin	−	−	−	−	−	−	w	·
Arabinose	−	−	−	−	−	−	w+	·
Cellobiose	−	−	−	−	−	−	+	·
Dextrin	−	−	−	−	−	−	d	·
Fructose	w−	−	−	w	+	−w	+	+
Galactose	−	−	−	−w	w	−	w	·
Glucose	−w	w−	−+	w+	+	−w	+w	+
Glycogen	−	−	−	−	−	−	+w	·
Inulin	−	−	−	−	−	−	w	·
Lactose	−	−	−	−	+	−	+w	+
Maltose	−	−	−	−	w	−	+w	+
Mannitol	−	−	−	−	−	−	−	+
Mannose	−	−	−	w	−	−	+	·
Melibiose	−	−	−	−	−	−	−w	·
Raffinose	−	−	−	−	−	−	w	·
Salicin	−	−	−	−	−	−	+w	·
Starch	−	−	−	−	−	−	w	+
Starch hydrolyzed	−	+−	−	−	−	−	+	·
Sucrose	−	−	−	−	+	−	+	+
Trehalose	−	−	−	−	−	−	−	·
Xylose	−	−	−	−	−	−	w+	·

[a] Products from PYG broth: Capital letters indicate >1 meq acid/100 ml broth; small letters <1 meq/100 ml. A, acetic; B, butyric; F, formic; iV, isovaleric; L, lactic; P, propionic; S, succinic; v valeric; 2, ethanol; ( ) products of occasional strains. +, strong acid, i.e. pH 5.5 or below in 90–100% of strains. w, weak acid, i.e. pH 5.5–6.0 in 90–100% of strains. (Uninoculated xylose and arabinose, under CO_2 (pre-reduced media), are often pH 5.9. For cultures in these media, pH 5.4–5.7 are considered weak acid, below pH 5.4 strong acid.) −, acid not produced, i.e. pH 6.0 or above in 90–100% of strains. d, reaction positive 40–60% strains. v, reaction variable within a strain. When any of the above used as superscripts, reaction of 10–40% of strains. ·, not tested. Acid products determined by procedures given in Holdeman and Moore, 1972.

glu.ti.no′sum. L. neut.adj. *glutinosum* glutinous.

Literature descriptions supplemented by results from study of one strain (V.P.I. 4865) (Tables 9.8, 9.9 and Fig. 9E).

Surface colonies minute, irregular with scalloped edge and bumpy surface, opaque, white, shiny.

Glucose broth cultures have granular or crumby sediment, no turbidity; final pH is 4.9–5.3.

Isolated from purulent pleurisy, pulmonary gangrene, fatal perinephritis with metastatic abscess in the brain.

Further Comments

No labeled strains are known to be extant. A collection culture, obviously not what it was labeled (differing in reactions from several others bearing the same label as well as in reactions originally recorded for the culture), is very similar, morphologically, to *B. glutinosus* as depicted in Weinburg *et al.*, 1937, 766. Cultural reactions are similar to those given for this species under *Ristella glutinosa* in Prévot *et al.*, 1967, 242.

TABLE 9.9

Biochemical reactions of species of genus **Fusobacterium,** *species 1–8[a]*

	1. *F. nucleatum*	2. *F. gonidia- formans*	3. *F. naviforme*	4. *F. varium*	5. *F. gluti- nosum*	6. *F. necro- phorum*	7. *F. aquatile*	8. *F. stabile*
Gelatin	—ᵂ	w⁻	—	—ᵂ	—	v	—	d
Milk	—	—	—	—	c	cd⁻	c	c
Indole produced	+	+	+	+⁻	+	+	+	—
Nitrate reduced	—	—	—	—	—	—	—	—
Esculin hydrolyzed	—	—	—	—	—	—	+	·
Gas	—³	4²	—²	4	2	4²	4²	4
Acetoin	—⁺	—	—	—	—	—	+	·
Lecithinase	—	—	—	—	—	—	—	·
Lipase	—	—	—	—	—	+⁻	—	·
Hemolysis	—ᵃᵇ	a⁻	—ᵇ	—ᵃ	w	bᵃ⁻	—	·
Motility	—	—	—	—	—	—	+	—
Hippurate hydrolyzed	—⁺	y	—⁺	—⁺	+	—	d	·
20% Bile — growth	—²	—	—²	4	4	—⁴	s	·
Propionate ← threonine	+	+	—	+	+	+	d	·
Gr. enhanced by	t	—	tr	—	—	—	b	

[a] +, positive reaction—90–100% of strains. —, negative reaction—90–100% of strains. Hemolysis— a = alpha; b = beta; w = weak. Gas and growth—relative amounts indicated by +, 2, 3, 4; s = stimulated; growth enhanced by b, bile; h, hemin; r, rumen fluid; t, Tween 80. milk—c, curd, d, digested. gelatin—w (weak) after chilling until uninoculated controls are solidified and returning to room temp., liquefaction in less than half the time of the controls. Any of the above used as superscripts indicates reaction of 10–40% of strains. Reactions determined by procedures given in Holdeman and Moore (1972).

This strain was deposited as ATCC 25580, but has since been lost by both ATCC and us.

6. Fusobacterium necrophorum (Flügge) Moore and Holdeman 1969, 12. (Bacillus der Kälberdiphtherie Loeffler 1884, 493; *Bacillus necrophorus* Flügge 1886, 273; *Bacillus necroseos* Jensen 1897, 123; *Bacillus funduliformis* Hallé 1898, 30; *Streptothrix necrophorus* (Flügge) Kitt 1899, 42; *Actinomyces necrophorus* (Flügge) Lehmann and Neumann 1899, 434; *Corynebacterium necrophorum* (Flügge) Lehmann and Neumann 1907, 531; *Bacterium necrophorum* (Flügge) Lehmann and Neumann 1927, 504; *Actinomyces pseudonecrophorus* Harris and Brown 1927, 208; *Fusiformis necrophorus* (Flügge) Topley and Wilson 1929, 299; *Streptothrix necupthora* (*sic*) (Flügge) Bergey *et al.* 1930, 468; *Bacteroides fundibuliformis* (*sic*) (Bergey *et al.*) Bergey *et al.* 1930, 373; *Bacillus fundibuliformis* (*sic*) Bergey *et al.* 1930, 373; *Bacillus necrosus* (*sic*) (Jensen) Weinberg *et al.* 1937, 660; *Sphaerophorus necrophorus* (Flügge) Prévot 1938, 298; *Sphaerophorus funduliformis* (Hallé) Prévot 1938, 298; *Bacterium funduliforme* (Hallé) Dack, Dragstedt, Johnson and McCullough 1938, 169; *Bacteroides funduliformis* (Hallé) Bergey *et al.* 1939, 559; *Sphaerophorus pseudonecrophorus* (Harris

and Brown) Prévot 1940, 178; *Proactinomyces necrophorus* (Flügge) Krasil'nikov 1941, 68; *Necrobacterium necrophorus* (*sic*) (Flügge) Lahelle and Thjøtta 1945, 321; *Pseudobacterium funduliformis* (Hallé) Krasil'nikov 1949, 243; *Fusiformis hemolyticus* (*sic*) Beerens and Gaumont 1953, 117; *Bacteroides necrophorus* (Flügge) Beerens and Tahon-Castel 1965, 45.)

ne.cro′pho.rum. Gr. adj. *necros* dead; Gr. adj. *phorum* bearing; M.L. neut.adj. *necrophorum* necrosis-producing.

Description from study of ATCC 25286 and 56 other strains with similar characteristics (Tables 9.8, 9.9 and Fig. 9E).

Surface colonies on horse blood agar are 1–2 mm; circular with scalloped to erose edges, convex to umbonate, often with bumpy, ridged, or uneven surface; translucent to opaque, often with mosaic internal structure when viewed by transmitted light.

Glucose broth cultures have smooth, flocculent, granular, or stringy sediment and usually are turbid; final pH is 5.8–6.3.

G + C content is 31 to 34 moles % (chromatographic separation, Sebald, 1962).

Isolated from the natural cavities of man and other animals and from necrotic lesions, abscesses,

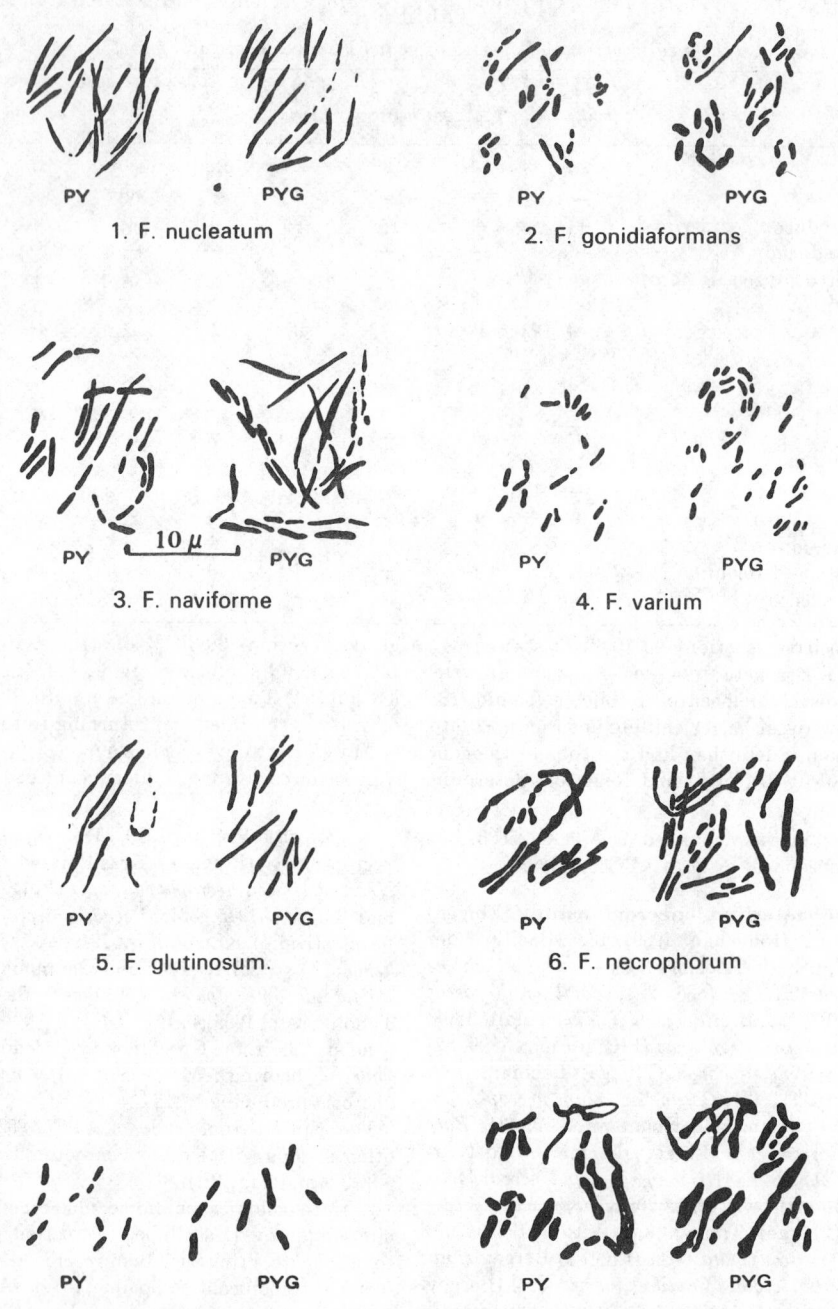

1. F. nucleatum

2. F. gonidiaformans

3. F. naviforme

4. F. varium

5. F. glutinosum.

6. F. necrophorum

7. F. aquatile

9. F. mortiferum

Fig. 9E. *Fusobacterium*. The line drawings are composites of several cultures and individual cultures usually show somewhat less variation in cellular morphology. PY, peptone-yeast extract broth; PYG, PY-glucose broth. Scale: 9/16 inch = 10 μm.

and blood of man and other animals, particularly liver abscesses of cattle and pigs. For a review of natural and experimental pathogenicity, see Prévot *et al.*, 1967, 307–321 and 335–343.

Reference strain: ATCC 25286; V.P.I. 2891; Fievez 2358.

Further Comments

Beerens, Fievez and Wattre (1971) suggest that this species may exist in three phases. Phase A corresponds to organisms previously recognized as *S. necrophorus* and is hemolytic, has a hemagglutinin, and is pathogenic for mice. Phase B corresponds to organisms previously recognized as *S. funduliformis* and is hemolytic, has no hemagglutinin, and demonstrates little, if any, pathogenicity for mice. Phase C corresponds to organisms previously recognized as *S. pseudonecrophorus* and is not hemolytic, has no hemagglutinin, and is not pathogenic for mice. A tendency toward mutation to penicillin resistance (to 500 units/ml) also is reported to accompany Phase A to C mutation. Phase A organisms are isolated more often from cattle than from man and Phase B and C organisms from man more often than from other animals.

Strains of *F. necrophorum* that produce alphahemolysis or are non-hemolytic may be differentiated from *F. nucleatum* by their conversion of lactate to propionate and by their production of large amounts of gas in peptone yeast extract glucose agar deeps. The gas splits the agar and often raises parts of the agar to the top of the tube.

Stevens (1956, 99 and 105) proposed ATCC 9817 (Dienes strain) as the type (*sic*) strain for *Bacteroides funduliformis* (Hallé) Bergey *et al.* This strain does not have the cultural reactions of *F. necrophorum* (or of Phase B) and has a G + C content of 27 moles % (J. L. Johnson, unpublished data); it therefore cannot be the neotype strain of this species.

7. Fusobacterium aquatile (Prévot) Moore and Holdeman 1973, 72. (*Zuberella aquatilis* Prévot 1938, 293; *Zuberella nova* Prévot 1947, 411; *Fusobacterium novum* (Prévot) Moore and Holdeman 1970, 45.)

a.qua′ti.le. L. neut.adj. *aquatile* living in or near water.

Literature description supplemented with information obtained from study of one strain (Tables 9.8, 9.9 and Fig. 9E).

Motile, probably with peritrichous flagella. When stained with Leifson method, most cells appear to have only one to three flagella, arranged at random, probably because the flagella are easily separated from the cell during centrifugation and resuspension.

Surface colonies on horse blood agar (2 days) 1–1.5 mm, circular, entire, low convex, opaque, white, shiny, smooth, non-hemolytic. Colonies in deep agar lenticular, later surrounded by a cottony halo.

Glucose broth cultures turbid with smooth sediment and final pH of 5.3–5.6.

Isolated from river water by Spray and Laux (1930), appendicitis (Prévot strain) and from human feces.

8. Fusobacterium stabile (Patočka and Prévot) Moore and Holdeman 1973, 72. (*Capsularis stabilis* Patočka and Prévot 1947, 838.)

sta′bi.le. L. neut.adj. *stabile* steady, standing firm.

Description from Prévot *et al.*, 1967, 275 (Tables 9.8 and 9.9).

Cells are 0.5–0.6 by 1.4 μm or more in length with rounded ends and clearly visible capsules.

In deep agar, colonies are punctiform; slightly fetid odor produced.

Peptone broth cultures are slightly turbid; glucose broth cultures are uniformly turbid.

Growth enhanced by serum. Growth most rapid at 37 C and pH 7.4.

Neutral red and phenosafranin not reduced. Coagulated proteins not attacked. Produce volatile amines and acetone.

Originally isolated from salpingitis; later isolated from peritonitis and septicemia following appendicitis.

No labeled strains are known to be extant. No strains conforming to the description were available for study. However, its description is sufficiently complete to permit recognition if it is isolated again.

9. Fusobacterium mortiferum (Harris) Moore and Holdeman 1970, 45. (*Bacillus mortiferus* Harris 1901, 546; *Bacteroides freundii* Hauduroy *et al.* 1937, 57; *Bacillus necroticus* Nativelle 1936 in Weinberg *et al.* 1937, 693; *Spherophorus mortiferus* (*sic*) (Harris) Prévot 1938, 299; *Spherophorus freundi* (*sic*) (Hauduroy *et al.*) Prévot 1938, 299; *Spherophorus necroticus* (*sic*) (Nativelle) Prévot 1938, 298; *Spherophorus ridiculosus* (*sic*) Prévot 1948, 387; *Pseudobacterium mortiferum* (Harris) Krasil′nikov 1949, 243; *Pseudobacterium freundii* (Hauduroy *et al.*) Krasil′nikov 1949, 244; *Pseudobacterium necroticum* (Nativelle) Krasil′nikov 1949, 242; *Fusobacterium ridiculosum* (Prévot) Moore and Holdeman 1969, 12.)

mor.ti′fer.um. L. n. *mors, mortis* death; L. v. *fero* bear; M.L. neut.adj. *mortiferum* death-bearing.

Description from Prévot *et al.* (1967, 363), Prévot (1966, 142) and study of 19 strains, including ATCC 9817 (Tables 9.10, 9.11 and Fig. 9E).

Surface colonies on horse blood agar (2 days) are 1–2 mm, circular with entire, diffuse, or slightly scalloped edge; convex or slightly umbonate; translucent; smooth. Colonies in glucose agar deeps are lenticular or irregular and may be surrounded by smaller colonies. Gas breaks the agar.

Glucose broth cultures uniformly turbid with smooth or semi-viscous sediment.

G + C content is 26 to 28 moles % (chromatographic separation, Sebald, 1962; T_m, J. L. Johnson, personal communication).

Isolated from necrotic abscesses of human liver and rectum, septicemia, purulent pleurisy, and urinary tract; from buccal cavity, intestinal tract and feces; once from sheep, and once from irradiated mice.

Neotype strain: ATCC 25557; V.P.I. 4123; CIPP (Prévot) 350A (type strain of *Sphaerophorus ridiculosus* Prévot 1948).

Further Comments

Nucleic acid studies (J. L. Johnson, personal communication) show a high degree of similarity between the neotype strain of *F. mortiferum* (type strain of *F. ridiculosus*) and the neotype strain of *Sphaerophorus freundii* (Hauduroy *et al.*) Prévot 1938, ATCC 9817, suggested by Werner (1972).

10. Fusobacterium symbiosum (Stevens) Moore and Holdeman 1972, 21. (*Bacteroides symbiosus* Stevens 1956, 100; *Zuberella pedipedis* (Prévot) Sebald 1962, 92; **NOT** *Fusocillus pedipedis* Prévot 1948, 246.)

sym.bi.o′sum. Gr. adj. *symbiosum* living together with, symbiotic.

Description based on study of ATCC 14940 and eight other strains with similar characteristics (Tables 9.10, 9.11 and Fig. 9F).

Surface colonies on horse blood agar (2 days) are minute to 1 mm, circular, entire, low convex, translucent, smooth; usually with mottled appearance when viewed by transmitted light.

Glucose broth cultures are turbid with smooth or ropy sediment and final pH of 5.3–5.7.

G + C content of the DNA is 48 moles % (chromatographic separation, Sebald, 1962).

Isolated from feces, blood and soft tissue infections.

Reference strain: ATCC 14940; Reeves.

Further Comments

Sebald (1962) studied two strains ("Reeves" and "ATCC") of motile Gram-negative rods, sent to the Pasteur Institute (Paris) as *Bacteroides symbiosus* and identified by Dr. Prévot as *Fusocillus pedipedis*. These strains were reclassified by Sebald (1962) as *Zuberella pedipedis* and have

cultural characteristics that are not at variance with the descriptions of *F. pedipedis*. However, the morphological characteristics of the cells of these two strains are not like those depicted by Beveridge for his motile fusiform (Beveridge, 1941). They are so dissimilar in appearance, even when one makes allowances for differences in culture medium and methods, that it is highly improbable that these two strains are representatives of the Beveridge organism and they therefore are retained here as *F. symbiosum*.

Other strains, with similar morphological and phenotypic characteristics, produce heat-resistant spores and are clostridia.

11. Fusobacterium necrogenes (Weinberg *et al.*) Moore and Holdeman 1970, 45. (*Bacillus necrogenes* Weinberg, Nativelle and Prévot 1937, 681; *Proactinomyces gedanensis* var. *necrogenes* Krasil'nikov 1949, 120; *Spherophorus necrogenes* (*sic*) (Weinberg *et al.*) Prévot 1938, 298; *Pseudobacterium necrogenes* (Weinberg *et al.*) Krasil'nikov 1949, 244.)

ne.cro′ge.nes. Gr. adj. *necros* dead; Gr. v. *gennaio* produce; M.L. adj. *necrogenes* necrosis-producing.

Description based on study of ATCC 25556 and three other strains (Tables 9.10, 9.11 and Fig. 9F).

Surface colonies on horse blood agar (2 days) minute to 0.5 mm, circular, flat to low convex, entire, translucent, white, smooth, shiny.

Glucose broth cultures moderately turbid with pH of 5.7–6.0.

Originally isolated by Kawamura from necrotic abscess of a chicken. Barnes strain isolated from cecal contents of a duck. Other strains isolated from human feces.

Reference strain: ATCC 25556; V.P.I. 2368; Barnes EB/D/1/4a.

12. Fusobacterium perfoetens (Tissier) Moore and Holdeman 1973, 72. (*Cocco-Bacillus anaerobius perfoetens* Tissier 1900, 70; *Coccobacillus perfoetens* Tissier 1905, 110; *Bacterium perfoetens* (Tissier) Weinberg *et al.* 1937, 790; *Bacteroides perfoetens* (Tissier) Hauduroy *et al.* 1937, 67; *Ristella perfoetens* (Tissier) Prévot 1938, 291; *Pseudobacterium perfoetens* (Tissier) Krasil'nikov 1949, 247; *Sphaerophorus perfoetens* (Tissier) Sebald 1962, 149.)

per.foe′tens. L. pref. *per* very; L. part.adj. *foetens* stinking; M.L. part.adj. *perfoetens* very stinking.

Description from Prévot *et al.* (1967, 381) and Weinberg *et al.* (1937, 790) (Tables 9.10 and 9.11).

Cells are 0.6–0.8 by 0.8–1 μm, oval, never elongated, occurring singly, in pairs, in chains of no more than three cells, or in masses. No flagella or capsule.

TABLE 9.10

Fermentation reactions of species of genus **Fusobacterium**, *species 9–16[a]*

	9. F. mortiferum	10. F. symbiosum	11. F. necrogenes	12. F. perfoetens	13. F. prausnitzii	14. F. plauti	15. F. bullosum	16. F. russii
Products from peptone-yeast extract-glucose broth	BAp(LF ivs4)	BA (L24f)	Ba(lf psv4)	ABL	B(AF lsp2)	LBas	Abp4	Ba(lf s24)
Lactate → propionate	−	−	−	·	−	−	−	−
Acid produced from:								
Amygdalin	−w	−	−	·	−	−	−	−
Arabinose	−	−+	−w	·	−	w−	−	−
Cellobiose	w−	−+	w−	·	−w	−w	−	−
Dextrin	−w	−	−w	·	−w	+−	−	−
Fructose	w+	+w	w+	·	−w	−w	−	−
Galactose	v	+−	w+	·	w−	−	−	−
Glucose	+w	+w	+w	+	v	w+	−	−w
Glycogen	−	−	−	·	−	w+	−	−
Inulin	−	−	−	·	−w	−w	−	−
Lactose	w+	−w	−	·	−w	−	−	−
Maltose	w−	−	−	·	w−	−	−	−
Mannitol	−	w−	−·	·	−	−	−	−
Mannose	w+	+−	w+	·	−w	−w	−	−
Melibiose	v	−	−w	·	−	−	−	−
Raffinose	v	−	−w	·	−	−	−	−
Salicin	w−	−	−w	·	−w	−	−	−
Starch	−w	−	−	·	−	+	w	−
Starch hydrolyzed	−+	−	−	·	−	+−	+	−
Sucrose	v	−	w−	+	−w	−	−	−
Trehalose	−w	−w	w−	·	−w	−	−	−
Xylose	−+	−	−	·	−	−	−	−

[a] For symbols see Table 9.8; 4, butanol.

Colonies in deep agar are 1 mm, lenticular.

Growth is rapid in glucose broth cultures, producing gas and fetid odor.

Coagulated proteins not attacked. Produces H_2S, CO_2, NH_3.

G + C content of DNA is 30 moles % (chromatographic separation, Sebald, 1962).

Isolated by Tissier in 1900 from an infant with diarrhea and in 1905 from nursing infants. Strain CC1 isolated in 1947 by Prévot from the cecum of a horse and studied by Sebald (1962) has been lost. No strains conforming to the description are known to be extant. However, strains of this species could be recognized by the characteristics cited.

Further Comments

In 1905 Tissier also described a lactose-fermenting variant (*Sphaerophorus perfoetens* var. *lacticus*, Prévot *et al.* 1967), which is not known to have been isolated since. From the relatively poor description of *S. perfoetens* var. *lacticus*, it would be difficult to recognize the variety.

13. **Fusobacterium prausnitzii** (Hauduroy *et al.*) Moore and Holdeman 1970, 45. (*Bacillus mucosus anaerobius* Prausnitz 1922, 126; *Bacteroides praussnitzii* (sic) Hauduroy *et al.* 1937, 68; *Bacterium zoogleiformans* Weinberg, Nativelle and Prévot 1937, 725; *Capsularis zoogleiformans* (Weinberg *et al.*) Prévot in Hauduroy *et al.* 1953, 122. Probable synonym: *Ristella abscedens* (Tardieux and Monteverde) Sebald 1962, 149.)

praus.nit′zi.i. M.L. gen.n. *prausnitzii* of Prausnitz; named for C. Prausnitz, the bacteriologist who first isolated this organism.

Literature description supplemented with re-

TABLE 9.11
Biochemical reactions of species of genus **Fusobacterium**, *species 9–16*[a]

	9. F. morti- ferum	10. F. sym- biosum	11. F. necro- genes	12. F. perfoe- tens	13. F. praus- nitzii	14. F. plauti	15. F. bullosum	16. F. russii
Gelatin	−	−w	−	·	−w	−	−	−w
Milk	−	−c	−	−	−	−	−	−
Indole produced	−	−	−	−	−	−	−	−
Nitrate reduced	−	−	−	·	−	+	−	−
Esculin hydrolyzed	+	−	+	·	+	−	−	−
Gas	4	4²	4	4	−²	−+	2	2⁻
Acetoin	−	v	−	·	−w	−	−	−
Lecithinase	−	−	−	·	−	−	−	−
Lipase	−	−	−	·	−	−	−	−
Hemolysis	−	−b	−b	·	−a	−	−	−
Motility	−	d	−	−	−	+⁻	+	−
Hippurate hydrolyzed	−	−	−	·	−	−	+	−+
20% Bile − growth	4s	v	4	·	4+	+²	2	3⁻
Propionate ← threonine	+	−	+	·	−	−	−	−
Gr. enhanced by	b	−	−	·	rh	−	−	−

[a] For symbols see Table 9.9.

sults from study of 54 strains (Tables 9.10, 9.11 and Fig. 9F).

Surface colonies on horse blood agar plates 2 mm, irregular, slightly scalloped, translucent, iridescent, smooth.

Glucose broth cultures turbid with ropy sediment and final pH of 5.7–5.9.

First isolated from purulent pleurisy. Also isolated from human and animal feces.

14. **Fusobacterium plauti** Séguin 1928, 439. (*Bacille* de Plaut, Kritchevsky and Séguin 1921, 722; *Bacillus plauti* Séguin 1928, 439; *Fusocillus plauti* (Séguin) Prévot 1938, 300; *Zuberella plauti* (Séguin) Sebald 1962, 149.)

plau′ti. M.L. gen.n. *plauti* of Plaut; named for R. Plaut, the bacteriologist who first described this organism.

Literature description supplemented by study of four strains, CIPP (Prévot) S1, S2, S3, S4 (Tables 9.10, 9.11 and Fig. 9F).

Motile with peritrichous flagella.

Surface colonies on horse blood agar (2 days) 0.5 mm, circular with diffuse edges, gray-white, dull, smooth, translucent with mottled appearance when viewed by obliquely transmitted light.

Glucose broth cultures moderately turbid with smooth (occasionally flocculent) sediment and final pH of 5.3–5.9.

Original strains isolated from the buccal cavity. Extant strains isolated from cultures of *Entamoeba histolytica*.

Further Comments

In comparing stains of the extant strains with drawings of the oral organisms described by Séguin (1928, 440), there is some doubt that the organisms are the same. The drawings of the oral organisms show long rods with several curves (like spirochetes) and peritrichous flagella. It is possible that this appearance would be observed in the extant strains if the individual cells in a chain were not seen. It is unfortunate that no strains isolated from the buccal cavity are extant.

15. **Fusobacterium bullosum** (Distaso) Moore and Holdeman 1970, 45. (*Bacillus bullosus* Distaso 1912, 443; *Bacteroides bullosus* (Distaso) Castellani and Chalmers 1919, 960; *Sphaerocillus bullosus* (Distaso) Prévot 1938, 300.)

bul.lo′sum. L. n. *bulla* a knob; M.L. neut.adj *bullosum* knobbed.

Literature descriptions supplemented with results of study of one strain (Tables 9.10, 9.11 and Fig. 9F).

Motile with peritrichous flagella. Flagella difficult to stain by usual methods and cells may appear to have only one, or no, flagellum.

Surface colonies on horse blood agar (2 days) 0.5–1 mm, circular or slightly irregular, erose, flat or low convex, gray, translucent with mosaic appearance when viewed by obliquely transmitted light.

Glucose broth cultures moderately turbid with small smooth sediment; final pH is 6.0–6.5.

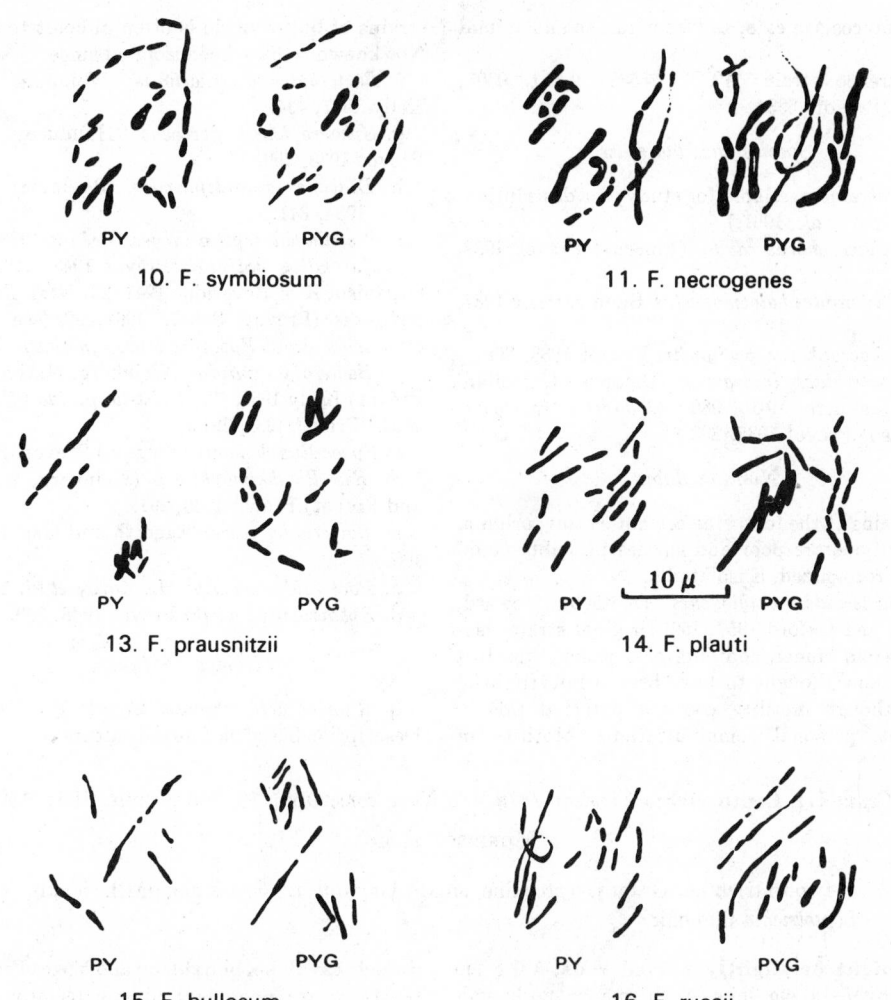

10. F. symbiosum   11. F. necrogenes

13. F. prausnitzii   14. F. plauti

15. F. bullosum.   16. F. russii

Fig. 9F. *Fusobacterium*. The line drawings are composites of several cultures and individual cultures usually show somewhat less variation in cellular morphology. PY, peptone-yeast extract broth; PYG, PY-glucose broth. Scale: 9/16 inch = 10 μm.

Isolated from human intestinal contents, actinomycosis (along with *A. israelii*), bone marrow.

Further Comments

Other strains with similar characteristics produce heat-resistant spores and are therefore clostridia.

16. **Fusobacterium russii** (Hauduroy *et al.*) Moore and Holdeman 1970, 45. (Influenzabacillenähnliches anaërobes Stäbchen, Russ 1905, 357; *Bacteroides russii* Hauduroy, Ehringer, Urbain, Guillot and Magrou 1937, 73; *Sphaerophorus influenzaeformis* (Russ) Prévot 1938, 299; *Bacillus influenzaeformis* Bergey *et al.* 1939, 583.)

rus'si.i. M.L. gen.n. *russii* of Russ, the bacteriologist who first cultured this organism.

Description based on Prévot *et al.* (1967) and study of ATCC 25533 and 13 similar strains (Tables 9.10, 9.11 and Fig. 9F).

Surface colonies on horse blood agar (2 days) 0.5–1 mm, circular, smooth, shiny, entire, convex, translucent with mottled appearance when viewed by obliquely transmitted light.

Glucose broth cultures turbid, often with stringy sediment; final pH is 5.9–6.1.

G + C content of DNA is 30 moles % (Sebald, 1962).

Isolated by Russ from perianal abscess; extant strains isolated from infections of cats, including actinomycosis of cats, and from human and animal feces.

Reference strain: ATCC 25533; V.P.I. 0307; CIPP (Prévot) 593A.

Species incertae sedis

(No strains available for study; for description, see Prévot et al., 1967.)

a. *Sphaerophorus caviae* (Vincent) Prévot 1938, 299.

b. *Bacteroides halosmophilus* Baumgartner 1937, 323.

c. *Sphaerophorus peritonitis* Prévot 1938, 298.

d. *Bacteroides variegatus* (Distaso) Castellani and Chalmers 1919, 960. (*Zuberella variegatas* (Distaso) Prévot 1938, 293.)

Nomina dubia

(Strains of the following are not extant; original descriptions are poor and species probably could not be recognized, if isolated.)

e. *Bacteroides amylogenes* Doetsch, Howard, Mann, and Oxford 1957, 156. Original strain, isolated from rumen contents of a sheep, was lost but is now thought to have been a butyrivibrio, even though motility was not detected (M. P. Bryant, personal communication). Motility of strains of butyrivibrio is often difficult to detect. Not known to have been isolated since.

f. *Bacteroides destillationis* (Weinberg et al.) Kelly 1957, 434.

g. *Sphaerophorus floccosus* (Hauduroy et al.) Prévot 1938, 299.

h. *Dialister granuliformans* (Pavlovic) Bergey et al. 1934, 341.

i. *Ristella haloseptica* (Wyss) Prévot 1938, 291.

j. *Fusocillus pedipedis* Prévot 1948, 246. (NOT Organism K of Beveridge 1941, 23; NOT *Zuberella pedipedis* (Prévot) Sebald 1962, 92 (see *Further Comments* about *Fusobacterium symbiosum*).)

k. *Bacteroides putidus* (Weinberg, Nativelle and Prévot) Kelly 1957, 429. (*Ristella putida* (Weinberg et al.) Prévot 1938, 291.)

l. *Fusocillus shmamini* (Séguin) Prévot 1938, 300.

m. *Ristella thermophila* β (Weinberg, Nativelle and Prévot) Prévot 1938, 291.

n. *Bacteroides vescus* Eggerth and Gagnon 1933, 406.

o. *Bacteroides viscosus* Hauduroy et al. 1937, 81.

p. *Sphaerocillus wirthi* Prévot 1938, 300.

Nomen confusum

q. *Ruminobacter parvum* Sijpesteijn 1949, 114. Description based on a mixed culture.

Genus III. **Leptotrichia** *Trevisan 1879, 138 Nom. cons.* Opin. 13, Jud. Comm. 1954, 152

MORRISON ROGOSA

Lep.to.trich′i.a. Gr. adj. *leptus* fine, small; Gr. n. *thrix, thricis* hair; M.L. fem.n. *Leptotrichia* fine hair.

Straight or slightly curved rods, 1–1.5 μm wide by 5–15 μm long, with one or both ends rounded or pointed; **many cells with pointed ends are generally found** in every culture. Two or more cells arranged in septate filaments of varying length, in older cultures filaments up to 200 μm long may twist around each other and large coccoid bodies or bulbous swellings within a filament may sometimes be found as cells lyse.

There is **no club formation or branching. Cells are non-motile.** No resting stages known; cells do not survive 60 C for 10 min. **Gram-nega-tive.** Granules, which may be Gram-positive, are distributed rather evenly along the long axis but one or more large granules may localize near the end of the cell.

Heterotrophic with complex nutritional requirements. **Acid without gas is produced from glu-cose;** the **major products are DL-lactic acid,** about 90%, and acetic acid, about 10%; **butyric acid is not a product.**

Hydrogen sulphide and indole are not pro- duced; **catalase, benzidine** and Voges Proskauer reactions are **negative;** nitrate usually not reduced; gelatin is not liquefied.

Anaerobic; 5% carbon dioxide is essential for successful isolation and optimal growth.

Good growth at 35–37 C; little or no growth at 25 C, no growth at 45 C.

Found in the oral cavity of man.

The **G + C content** of one strain examined was **31.5 moles %,** chromatographic analysis (Lee et al., 1956); a second strain had a **G + C content of 34 moles %** (T_m) (Hofstad, 1970).

Type species: *Leptotrichia buccalis* (Robin) Trevisan 1879, 147.

Further Comments

The Judicial Commission and the ICSB conserved the name *Leptotrichia* Trevisan and designated *L. buccalis* as the type species (Opinion 13 Jud. Comm. 1954, 153). However, in the 7th edition of THE MANUAL, p. 437, Hoffman confused *L buccalis, Bacillus fusiformis* Veillon and Zuber and

TABLE 9.12

Characteristics differentiating **Leptotrichia, Fusobacterium,** *and* **Bacterionema**

	Leptotrichia buccalis	Fusobacterium nucleatum	Bacterionema matruchotii
Morphology	Unbranched rods, pointed ends present	Unbranched rods, pointed ends present	Branched filaments, bacillus-like body at one end
Gram-reaction and cell structure	Gram-negative	Gram-negative	Gram-positive
Relation to oxygen	Anaerobic	Anaerobic	Preferentially aerobic
G + C moles %	31.5–34	27–28	?
Catalase	−	−	+
H₂S	−	+	−
Indole	−	+	−
Nitrate reduction	−	±	+
Heterofermentative	−	+	+
Butyric acid produced	−	+	−
Propionic acid produced	−	−	+
Ethanol produced	−	+	?
Ammonia produced	−	+	?
CO₂ produced	−	+	+
Vosges Proskauer	−	?	+
Foul odor	−	+	−

Fusobacterium plauti-vincenti Knorr. The last is a later synonym of *L. buccalis*; *B. fusiformis* is a synonym of *Fusobacterium nucleatum*.

Many recent authors (Hamilton and Zahler, 1957; Kasai, 1961, 1965) have described *L. buccalis* as Gram-positive and related to *Lactobacillus*. Thjøtta, Hartman and Bøe (1939) and Bøe and Thjøtta (1944), who were the first to give an adequate description of this organism, considered *Leptotrichia* to be Gram-negative and related to *Fusobacterium*. The latter stated "Leptotrichia is Gram-positive in very young cultures but it looses (sic) this character very soon, and thus older cultures will show negative rods." Authors who would include *L. buccalis* in *Lactobacilleae* also overlooked the finding of Bøe and Thjøtta that this organism elicited the Schwartzman reaction in rabbits.

There is now agreement that the cells are Gram-negative in 24 hr. or less (usually 6 hr.); that the fine structure is that of typical Gram-negative organisms (Hofstad and Selvig, 1969; Hofstad, 1967, 1970); that they contain lipopolysaccharides characteristic of Gram-negative organisms, which also have the characteristics of endotoxins causing elevation of temperature and the Schwartzman reaction, and which have a serologically specific O-antigen (Araujo *et al.*, 1963; Gustafson *et al.*, 1966).

Earlier workers have confused *L. buccalis* with *Bacterionema matruchotii* q.v. (Actinomycetaceae) and various members of *Fusobacterium* such as *Fusobacterium nucleatum*.

The characteristics differentiating the three oral organisms are given in Table 9.12.

Description of the species of genus **Leptotrichia**

1. **Leptotrichia buccalis** (Robin) Trevisan 1879, 147. (*Leptothrix buccalis* Robin 1853, 345; *Fusobacterium plauti-vincenti* Knorr 1922, 5; *Fusobacterium fusiforme* (Veillon and Zuber) Hoffman 1957, 437.)

buc.ca'lis. L. adj. *buccalis* buccal, pertaining to the mouth.

Typical sub-surface colonies are lobate, convoluted (Medusa head) and iridescent in media containing crystal violet. Surface colonies vary in consistency from butyrous to brittle.

In primary isolation 5% CO₂ is essential and supplements of serum, ascitic fluid or starch are very useful. An initial pH of 7.2–7.4 favors growth. Growth in deep agar tubes usually begins about 1 cm below the surface. Thioglycollate may be inhibiting for some strains and there is no growth with 2% NaCl. Good growth occurs in media containing 0.001% crystal violet or 10 μg/ml of streptomycin.

Fructose, glucose, maltose, mannose and sucrose are fermented with production of acid but no gas; cellobiose, salicin and trehalose are usually fermented; galactose, lactose, raffinose and starch may be fermented. The final pH ranges from about 4.6–5.2. Acid is not produced from arabinose, dulcitol, glycerol, inulin, mannitol, rhamnose, sorbitol or xylose. Lactate is not fermented.

In media containing amino acids or peptones and a fermentable carbohydrate, lactic and acetic acids are the only significant products (Jackins and Barker, 1951); butyric acid, butanol, ammonia, carbon dioxide and ethanol are not formed.

Washed cell suspensions incubated anaerobically without CO_2 in M/30 phosphate buffer, pH 7.2, ferment glucose exclusively to lactic acid; pyruvate is fermented with low carbon recovery chiefly to acetic and formic acids, with lesser amounts of CO_2, and only 5% of the total products is lactate (Jackins and Barker, 1951).

Cultures die rapidly in 0.5–1% glucose media at 4 C; survival is improved in 0.2% or less glucose. Cells have been lyophilized successfully in skim milk and have survived at least 2 years at 4 C.

Cells maintained at −70 C are viable for at least 6 months but do not survive at −20 C.

Species incertae sedis

a. *Leptotrichia innominata* (Miller) Nannizz 1934, 56.

b. *Leptotrichia innominata* var. *ochracea* (sic) Prévot 1957, 310. Une variété pigmentée de *Leptotrichia innominata* Mazurek 1955, 208.

c. *Leptotrichia tenuis* Trevisan 1889, 10.

d. *Leptotrichia vaginalis* Patočka and Reyne 1947, 600. Not *Leptotrichia vaginalis* (Donné) De Toni and Trevisan 1889, 935.

e. *Leptotrichia haemolytica* Bezjak 1952, 100.

The above organisms (cited by Prévot, 1957) cannot be identified from published descriptions. Some are probably members of the genus *Fusobacterium*. Existing data for three of them indicate that their fermentations are clearly heterofermentative and thus inconsistent with the genus and type as presented here.

GENERA OF UNCERTAIN AFFILIATION

Genus **Desulfovibrio** *Kluyver and van Niel 1936, 397*

JOHN R. POSTGATE

(*Sporovibrio* Starkey 1938, 300.)

De.sul.fo.vib′ri.o. L. pref. *de* from; L. n. *sulfur* sulfur; L. v. *vibro*. vibrate; M. L. masc.n. *Vibrio* that which vibrates, a generic name; M.L. masc.n. *Desulfovibrio* a vibrio that reduces sulfur compounds.

Curved rods, sometimes sigmoid or spirilloid. Morphology influenced by age and environment. Descriptions refer to freshly grown cultures in Baars's medium (1930) containing 0.02–0.1% NaCl. **Motile** by means of **polar flagella. Do not form endospores. Gram-negative.**

Chemoorganotrophs which obtain energy by anaerobic respiration reducing sulfates or other reducible sulfur compounds to H₂S. Lactate, pyruvate and usually malate, are oxidized to acetate and CO_2. Carbohydrates rarely utilized, gas never formed from carbohydrates.

Cells contain C₃ cytochromes and desulfoviridin, absorption band 630 nm; the latter is responsible for the **characteristic red fluorescence of cells** if inspected in light of 365 nm immediately after the addition of a few drops of 2.0 N NaOH; this reaction releases the chromophore of the pigment. Organic growth factors not required.

Hydrogenase usually present. Gelatin not liquefied, nitrates not reduced. Nitrogen sometimes fixed.

Strict anaerobes.

Upper temperature limit for growth 44 C. Optimum influenced by origin and previou history, usually 25–30. Some strains can grow a temperatures at or below 0 C.

Some species and subspecies moderately halophilic.

Species generally show some degree of antigeni cross-reaction. Pathogenicity not recorded.

Type species: *Desulfovibrio desulfuricans* (Beijerinck) Kluyver and van Niel 1936, 397.

Further Comments

The rationale of the classification given here which has some tentative features but which ha proved to be a useful working system, is given by Postgate and Campbell (1966). Its official statu and permanence is indicated by Postgate (1967). Some comments on the use of the criteria wer made by Miller *et al.* (1968). In particular, th hibitane resistance test seems to be of limite value and some strains of *D. salexigens* do no have the high resistance quoted; one strain of *D desulfuricans* lacking desulfoviridin has been re ported by Miller and Saleh (1964). Postgate and Campbell (1966) proposed a nitrogen-fixing variety, *D. desulfuricans* subsp. *azotovorans*, but

Reiderer-Henderson and Wilson (1970) have shown that nitrogen fixation is widespread within the genus.

Description of the species of genus **Desulfovibrio**

Differential reactions of the species and subspecies are given in Table 9.13.

1. **Desulfovibrio desulfuricans** (Beijerinck) Kluyver and van Niel 1936, 397. (*Spirillum desulfuricans* Beijerinck 1895, 113; *Bacillus desulfuricans* (Beijerinck) Saltet 1900, 669; *Microspira desulfuricans* (Beijerinck) Migula 1900, 1016; *Vibrio cholinicus* Hayward and Statman 1959, 560.)

de.sul.fur'i.cans. L. pref. *de* from; L. n. *sulfur* sulfur; M.L. *desulfuricans* reducing sulfur compounds.

Sigmoid forms may occur.

Usually require special media containing sulfates; media containing iron salts become blackened and bacteria often associate with precipitated FeS. In lactate sulfate agar containing excess of ferrous salt produces round wholly black colonies; in peptone glucose sulfate agar colonies similar but show a golden sheen at earliest stages of development.

Resistant to 10–25 mg of hibitane/liter, but not half of these concentrations (Saleh, 1964).

Found in fresh water, particularly polluted waters showing blackening and sulfide formation, in soils, particularly anaerobic or water-logged soils rich in organic materials, in marine and brackish waters. Fresh water strains readily adapt to sea water and *vice versa*.

1a. *Desulfovibrio desulfuricans* subsp. *desulfuricans*.

Description as above.

Type not extant. Suggested neotype strain: Essex 6, NCIB 8307.

1b. *Desulfovibrio desulfuricans* subsp. *aestuarii* Postgate and Campbell 1966, 734.

Description as for *D. desulfuricans* except strains of marine or brackish origin incapable of adapting to fresh water environments. Definite NaCl requirement, usually 2.5%.

Reference strain: Sylt 3, NCIB 9335.

2. **Desulfovibrio vulgaris** Postgate and Campbell 1966, 734. (Includes strains earlier accepted as *D. desulfuricans* as defined by ZoBell, 1953).

vul.gar'is. L. adj. *vulgaris* common.

Carbon sources restricted to lactate, pyruvate, formate and certain simple primary alcohols including methanol, ethanol, propanol, butanol.

Similar to *D. desulfuricans* except for characteristics noted in Table 9.13.

2a. *Desulfovibrio vulgaris* subsp. *vulgaris*. Description as for *D. vulgaris*.
Type strain: Hildenborough; NCIB 8303.

2b. *Desulfovibrio vulgaris* subsp. *oxamicus* Postgate and Campbell 1966, 734.

Description as for *D. vulgaris* except organism grows in choline or pyruvate without sulfate and is capable of metabolizing oxamate and oxalate.
Type strain: Monticello 2; NCIB 9442.

3. **Desulfovibrio salexigens** Postgate and Campbell 1966, 735. (*Microspira aestuarii* van Delden 1903, 81.)

sal.ex'i.gens. L. n. *sal* salt; M.L. *exigens* demanding; M.L. *salexigens* salt-demanding

Morphology as for *D. desulfuricans*.

Requires chloride ion for growth, supplied as NaCl >0.6% (usually 2.5–5%).

Hibitane resistance variable, often more than 1 g/liter. Some strains with resistance of 250 mg/liter have been reported (Miller *et al.*, 1968).

Found in sea water, marine and estuarine muds, pickling brines.

Type strain: British Guiana; NCIB 8403; ATCC 14822.

4. **Desulfovibrio africanus** Campbell, Kasprzycki and Postgate 1966, 1127.

af.ric.an'us. L. adj. *africanus* pertaining to Africa.

Long sigmoid rods 0.5 by 5–10 μm. Lophotrichous polar flagella provide rapid progressive motility.

Except as noted in Table 9.13, similar to *D. vulgaris*.

Isolated from salt and fresh waters from Africa. Has wide salt tolerance.

Type strain: Benghazi NCIB 8401.

5. **Desulfovibrio gigas** LeGall 1963, 1120.

gi'gas. L. n. *gigas* giant.

Large curved rods 1.2–1.5 by 5–10 μm, often in chains appearing as spirilla. Motility slow by means of lophotrichous polar flagella. Young organisms show areas of low contrast when examined by phase microscopy.

An Eh of about 80 mv (ascorbate at pH 7) seems most suitable for growth which is slower than other species.

Isolated from Etang de Berre, near Marseilles, France. Despite its salt water origin, saline media was not used for its culture.

Type strain: NCIB 9332.

TABLE 9.13

Differential characteristics of species of genus **Desulfovibrio**

	1a. D. desulfuricans subsp. desulfuricans	1b. D. desulfuricans subsp. aestuarii	2a. D. vulgaris subsp. vulgaris	2b. D. vulgaris subsp. oxamicus	3. D. salexigens	4. D. africanus	5. D. gigas
Dimensions μm	0.5–1 by 3–5		0.5–1 by 3–5		0.5–1 by 3–5	0.5 by 5–10	1.2–1.5 by 5–10
Flagella polar							
Single	+	+	+	+	+	−	−
Lophotrichous	−	−	−	−	−	+	+
Thickness of flagellum (nm)	20–25	20–25	20–25		20–25	12	9
Growth in[a]							
Pyruvate minus sulfate	+	+	−	+	−	−	−
Malate plus sulfate	+	+	−	−	+	+	−
Malate minus sulfate	−	−	−	−	−	−	−
Choline plus sulfate	+	+	−	+	−	−	−
Choline minus sulfate	+	+	−	+	−	−	−
NaCl requirement	−	+	−	−	+	−	−
Hibitane resistance mg/l	10–25	10–25	2.5	2.5	1000	2.5	2.5
G + C mole %	55.3 ± 1		61.2 ± 1		46.1 ± 1	61.2 ± 1	60.2
Eh requirement (mv) for growth at pH 7.2	−100		−100		−100	−100	80

[a] All species grow in lactate with sulfate and in pyruvate with sulfate.

Species incertae sedis

The status of the following strains has been discussed by Postgate and Campbell (1966), and the Subcommittee on Sulfate-reducing Bacteria (ICSB) has not accepted them (Postgate, 1967).

a. *Desulfovibrio rubentschikii* (Baars) ZoBell 1948, 248. (*Vibrio rübentschickii* (sic) Baars 1930, 89; *Sporovibrio rubentschickii* (sic) (Baars) Brisou 1955, 227.)

Baars described a strain similar to *D. desulfuricans* but distinguished by its ability to oxidize acetate to CO_2 and water. No strains are extant and there is some doubt as to the purity of Baars's culture.

b. *Desulforistella hydrocarbonoblastica* Hvid-Hansen 1951, 332.

Short, pointed, non-motile, non-sporulating rods (in a formate-sulfate medium) from subterranean water in Sjaelland, Denmark. Described as an obligate anaerobe, facultative autotroph; grew best at 30 C, did not grow at 37 C; utilized formate, lactate, propionate or acetate; also other organic compounds. Old cultures sometimes contained an ether-soluble bituminous material.

The original culture has been lost and it has not been re-isolated in Denmark or elsewhere.

Genus **Butyrivibrio** *Bryant and Small 1956, 18*

M. P. BRYANT

Bu.ty.ri.vib′ri.o. M.L. adj. *butyricus* butyric; L. v. *vibro* vibrate; M.L. n. *Vibrio* that which vibrates; a generic name; M.L. masc.n. *Butyrivibrio* a butyric vibrio.

Curved rods, single or in chains or filaments which may or may not be helical. **Motile** with **monotrichous, polar** or sub-polar **flagella.** No resting stages known. **Gram-negative.**

Chemoorganotrophs: **Metabolism fermentative, carbohydrates** being the main fermentable substrates. **Glucose is fermented with production of butyrate** as one of the important products. Some strains in some media produce much lactic acid and little butyric acid (Gill and King, 1958; Lee and Moore, 1959).

Strict anaerobe.

Type species: *Butyrivibrio fibrisolvens* Bryant and Small 1956, 19.

Further Comments

Members of this genus constitute one of the most numerous and biochemically versatile groups of bacteria present in the rumen; and there is great variation among the features of strains included in the genus. For example, great variation occurs in energy sources, fermentation products (Bryant and Small, 1956; Shane *et al.*, 1969), serology (Margherita and Hungate, 1963), and nutrition (Bryant and Robinson 1962; Shane *et al.*, 1969). A few non-motile strains are reported. Hungate named a new species, *Butyrivibrio alactacidigens*,

Hungate 1966, 43, which was separated from *B. fibrisolvens* because lactate was not produced. Shane *et al.* (1969) placed most strains in two groups. One group produced appreciable amounts of lactate, low formate and removed acetate from the medium during cellobiose fermentation, while a second group produced acetate, more formate but little or no lactate; the latter group was more exacting in nutritional requirements. However, both groups varied greatly in other features and strains in the second group producing little or no lactate usually differed greatly from strains of Hungate which did not produce lactic acid. Until more definitive studies are carried out, it seems futile to assign more than one species to the genus.

Description of the species of genus **Butyrivibrio**

1. **Butyrivibrio fibrisolvens** Bryant and Small 1956, 19.

fi.bri.sol′vens. L. n. *fibra* fiber; L. part.adj. *solvens* dissolving; M.L. part.adj. *fibrisolvens* fiber-dissolving.

Curved rods, 0.3–0.8 by 1–5 μm with tapered and rounded ends. Some strains are almost straight. Some strains have bluntly pointed ends and are almost spindle shaped. Cells occur singly, in pairs and chains and often long chains and filaments are seen. Chains may or may not show helical arrangement. Rapid, vibrating motility, often progressive, but often only a few cells show motility. A few non-motile nonflagellated strains are reported.

In rumen fluid-glucose-cellobiose agar, surface colonies are usually smooth, entire, slightly convex, translucent, light tan in color and 2–4 mm in diameter. Some strains have rough colonies that are more flat, lighter in color and have filamentous margins. Deep colonies are usually lenticular or Y shaped but some form compound lenticular colonies.

In cellulose agar, colonies of cellulolytic strains vary from simple lens-shaped to triangular to compound lenticular, and some strains form branched rhizoidal colonies. Zones of cellulose digestion around colonies vary from very narrow with slow and indistinct digestion to broad zones with rapid and complete digestion.

Growth in glucose liquid medium varies from flocculent to granular sediment, some of which may adhere to the wall of tubes, to evenly turbid.

Energy-yielding metabolism is fermentative and glucose is fermented with production of butyric and formic acids, and, usually, lactic acid, hydrogen and carbon dioxide. There can be either net

production or utilization of acetic acid. Under some conditions of culture, lactic acid production may be increased and only a small amount of butyric acid is formed (Gill and King, 1958). Some ethanol and propionic acid may be produced, but succinic acid is not produced.

D-Xylose, fructose, glucose, cellobiose, lactose, maltose, sucrose, esculin, salicin and inulin are fermented. Trehalose, dextrin, pectin, starch and xylan are often fermented. Gum arabic, glycerol, inositol, mannitol and lactate are not fermented. Cellulose is fermented by some strains but usually not by strains isolated using media without cellulose as energy source.

Indole and catalase are not produced. Nitrate is usually not reduced. Hydrogen sulfide and acetoin production and gelatin liquefaction are variable.

Minimal nutritional requirements are quite variable but most strains grow in defined media and many strains grow well in media containing glucose, minerals, B-vitamins, cysteine or sulfide, and with ammonia as the main nitrogen source. Amino acids, acetate and certain other volatile acids may be quite stimulatory and ammonia may be essential for some strains (Bryant and Robinson, 1962; Gill and King, 1958).

Good growth occurs at 30 and 37 C and, usually, at 45 C but not at 22 C. No growth occurs at 50 C.

Final pH in poorly buffered glucose medium is usually 5.0–5.6, but for some strains, as low as 4.6.

Found in the rumen of most, if not all, ruminants; and closely related if not identical species are present in the intestinal tract of other mammals (Brown and Moore, 1960).

Type strain: D1; ATCC 19171.

Genus **Succinivibrio** *Bryant and Small 1958, 22*

M. P. BRYANT

Suc.cin.i.vib'ri.o. M.L. n. *acidum succinicum* succinic acid; M.L. masc.n. *Vibrio* that which vibrates, a generic name; M.L. masc.n. *Succinivibrio* the succinic acid vibrio.

Curved rods with **pointed ends. Cells** are **helical** with less than one to three or more coils per cell and may become straight to slightly curved with pointed ends after maintenance on artificial media. Rapid progressive vibrating motility with **monotrichous** and **polar flagellation.** No resting stages known. **Gram-negative.** Chemoorganotrophs: **Metabolism fermenta-** tive, **carbohydrates** being the fermentable substrates. **Major products** from glucose are **succinate, acetate,** formate and sometimes lactate. **Butyrate or hydrogen are not formed.** A large net uptake of CO_2 may occur.

Strict anaerobe.

Type species: *Succinivibrio dextrinosolvens* Bryant and Small 1958, 22.

Description of the species of genus **Succinivibrio**

1. **Succinivibrio dextrinosolvens** Bryant and Small 1958, 22.

dex.trin.o.sol'vens. M.L. n. *dextrinosum* dextrin; L. part.adj. *solvens* dissolving; M.L. part.adj. *dextrinosolvens* dextrin-dissolving.

Helical rods with pointed ends, 0.3–0.7 by 1–7 μm, many cells are short with one or less coils but longer cells containing two or three coils are present. Cells are arranged as singles with a few short chains. After culture in artificial media cells of some strains become straight or slightly curved. Swollen forms and round bodies may be present.

On rumen fluid-glucose-cellobiose agar, surface colonies are 1–2 mm, entire, translucent and light tan in color. Deep colonies are lenticular.

In rumen fluid-glucose liquid medium, heavy flocculent sediment with some turbidity.

Energy-yielding metabolism is fermentative and large amounts of succinic and acetic acids and, often, formic and lactic acids are produced from glucose.

Amino acids and peptides are not fermented.

D-Xylose, galactose, glucose, maltose and dextrin are fermented with acid production. L-Arabinose, fructose, cellobiose, sucrose, esculin, salicin and mannitol are fermented by some strains. Lactose, trehalose, cellulose, inulin, starch, xylan, glycerol and inositol are not fermented (however, see Hungate 1966, 72). Starch is not hydrolyzed.

Nitrate is not reduced. Gelatin is not liquefied. Catalase, acetoin, indole and hydrogen sulfide are not produced.

Growth of most strains is good in glucose media containing peptone and yeast extract in place of rumen fluid (however, see Scardovi, 1963). Addition of carbon dioxide-bicarbonate to growth media allows better growth.

Growth occurs at 30–37 C but not at 22 or 45 C.

Final pH in lightly buffered glucose medium is 4.8–5.2.

Found in rumen contents of cattle and sheep (Wilson, 1953; Scardovi, 1963). Sometimes the most numerous bacterium cultured from cattle fed diets high in grain (Bryant *et al.*, 1961). Presumably, it is present in rumens of other animals.

Type strain: 24; ATCC 19716.

Genus **Succinimonas** *Bryant, Small, Bouma and Chu 1958, 21*

M. P. BRYANT

Suc.cin.i.mon'as. M.L. n. *acidum succinicum* succinic acid; Gr. n. *monas* a unit, monad; M.L. fem.n. *Succinimonas* succinic acid monad.

Straight rods with rounded ends, usually short to coccoid. Motile, with monotrichous polar flagella. No spores or other resting stages known. **Gram-negative.**

Chemoorganotrophs: **Metabolism fermentative, carbohydrates** being the **fermentable sub-** strates. **Large amounts of succinate** and some acetate are the major products of glucose fermentation. **No butyrate or gas is formed.**

Strict anaerobe. Catalase is not produced.

Type species: *Succinimonas amylolytica* Bryant *et al.* 1958, 21.

Description of the species of genus **Succinimonas**

1. **Succinimonas amylolytica** Bryant, Small, Bouma and Chu 1958, 21.

am.y.lo.ly'ti.ca. Gr. n. *amylum* fine meal, starch; Gr. adj. *lyticus* loosening, dissolving; M.L. fem.adj. *amylolytica* starch-dissolving.

Straight rods with rounded ends, 1.0–1.5 by 1.0–3 μm, arranged as singles, pairs and clumps. Capsules are not evident. Relatively slow progressive motility which may be rapidly lost on exposure to air.

Surface colonies on rumen fluid glucose cellobiose agar are smooth, entire, convex, translucent, light tan and 0.7–1.5 mm in diameter after 3 days at 37 C; deep colonies are lenticular and 0.7–1 mm in diameter.

Growth in rumen fluid-glucose liquid medium is evenly turbid and relatively light.

Energy-yielding metabolism is fermentative with glucose being fermented mainly to succinic and acetic acids concurrent with a large net uptake of carbon dioxide.

Acid but no gas is produced from glucose, maltose, dextrin and starch. Arabinose, xylose, fructose, cellobiose, lactose, sucrose, esculin, salicin, cellulose, inulin, xylan, glycerol, mannitol and lactate are not fermented.

Amino acids or peptides are not fermented (Bladen *et al.*, 1961). H_2S, catalase, and indole are not produced. Gelatin is not liquefied. Voges-Proskauer test is variable.

Growth is good in glucose medium with trypticase and yeast extract replacing rumen fluid; carbon dioxide is required. The type strain grows well in a defined medium containing glucose, carbon dioxide-bicarbonate buffer, minerals, B-vitamins and acetate. Ammonia serves as the nitrogen source and is not replaced by amino acids or peptides; sulfide serves as the reducing agent and sulfur source. Acetate (30 mM) is highly stimulatory in the presence or absence of amino acids even though it is produced from glucose (unpublished data of Roberton and Bryant; Bryant and Robinson, 1962).

Growth occurs at 30 and 37 C but not at 22 or 45 C.

Final pH in poorly buffered glucose medium is 5.2–5.8. Good growth at pH 6.5–7.0 but higher limit not determined.

Found in rumen contents of cattle fed forage and grain, but usually constitutes less than 6% of total viable bacteria.

Presumably, also occurs in the rumen of other animals.

Type strain: $B_2$4; ATCC 19206.

Genus **Lachnospira** *Bryant and Small 1956, 24*

MORRISON ROGOSA AND M. P. BRYANT

Lach.no.spir'ra. Gr. n. *lachnos* woolly hair, down; L. n. *spira* a coil; M.L. fem.n. *Lachnospira* woolly (colony-producing) spiral.

Curved rods, motile with lateral to subterminal monotrichous flagella. Weakly Gram-positive.

Ferments glucose producing large amounts of ethanol, lactic, formic and acetic acids, and CO_2; small amounts of hydrogen also produced. Not proteolytic.

Anaerobic.

Found in bovine rumen in significant numbers and probably present in rumen of other animals.

Type species: *Lachnospira multiparis* Bryant and Small 1956, 24.

Description of the species of genus **Lachnospira**

1. **Lachnospira multiparis** Bryant and Small, 1956, 24.

Curved rods, 0.4–0.6 μm wide and 2–4 μm long with bluntly pointed ends. Cells generally singly or in pairs. Single cells appear as curved or helicoidal rods; occasionally a few very long chains of cells only slightly curved and with more rounded ends are observed; in liquid media without rumen fluid they may be arranged mainly as very long chains and filaments. Many Gram-positive cells in young culture which rapidly become Gram-negative.

Surface colonies in agar relatively large, flat, characteristically filamentous, white. Deep colonies appear as woolly balls.

Fermentation products from glucose chiefly carbon dioxide, ethanol, and acetic, formic, and lactic acids; also a small amount of hydrogen.

Glucose, fructose, cellobiose, esculin, pectin, salicin and sucrose fermented. D-Xylose variably or weakly fermented. L-arabinose, cellulose, dextrin, galactose, glycerol, gum arabic, inositol, inulin, lactose, maltose, mannitol, trehalose and xylan not fermented. Starch not hydrolyzed.

Catalase, indole, hydrogen sulfide negative. Nitrate not reduced. Gelatin not liquefied. Acetylmethylcarbinol (Voges-Proskauer) reaction usually weakly positive although varying from negative to strongly positive.

Growth is good in media without added CO_2 or bicarbonate, or in a medium in which yeast extract and trypticase replace rumen fluid. In glucose rumen fluid media growth results in a heavy flocculent sediment. All strains so far studied grow well in a defined medium containing B-vitamins, cysteine, glucose, ammonia as nitrogen source, minerals, acetate, bicarbonate and CO_2 (Bryant and Robinson, 1962).

Good growth at 30–45 C; no growth at 22 or 50 C.

Isolated from the reticulo-rumen of cattle; probably also occurs in rumen of other animals.

Type strain: D32; ATCC 19207.

Further Comments

The organism was originally described as polarly flagellated. However, Leifson (1960) demonstrated lateral flagellation and Tripathy, Wolin and Bryant confirmed this by electron microscopy (unpublished data, 1967).

Several isolates from sheep rumen, which resembled *Lachnospira* and were tentatively identified as such (Blackburn and Hobson, 1962; Akkada and Blackburn, 1963), have quite different biochemical properties. Apparently organisms with similar morphology, cultural properties and habitat exist, but their relationship to the genus or to *L. multiparis* is not known.

Genus **Selenomonas** *von Prowazek 1913, 36 Nom. cons.* Opin. 21. Jud. Comm. 1958, 163

M. P. BRYANT

Se.le.no.mo′nas. Gr. n. *selene* the moon; Gr. n. *monas* a unit, monad; M.L. fem.n. *Selenomonas* moon (-shaped) monad.

Curves **curved to helical rods,** ends usually tapered and rounded to give **kidney- to crescent-shaped short cells.** Long cells and chains of cells are helical. Strains differ markedly in size; pure cultures are often smaller than those in the natural habitat. **Motile with active tumbling;** up to 16 **flagella present as a tuft often near the center of the concave side** at or near the point of cell fission. Long cells with several tufts often seen. Spheroplasts with flagella attached are sometimes seen. Capsules not formed. Resting stages not known. Gram-negative.

Chemoorganotrophs: **Metabolism fermentative, carbohydrates and, sometimes, certain amino acids and lactic acid being the fermentable substrates. Fermentation of glucose yields chiefly acetate, propionate, CO_2 and/or lactate.** Organic growth factors required but not adequately studied. Usually grown in complex media.

Strict anaerobes, catalase not produced.

Temperature optimum 35–40 C; maximum about 45 C; minimum 20–30 C. Grow at lower pH (4.5–5.0) than most anaerobic, Gram-negative, motile rods.

Found mainly in the gastrointestinal tract of mammals but seen in dirty river water.

The **G + C content** of the DNA ranges from **53–61 moles %** (buoyant density).

Type species: *Selenomonas sputigena* (Flügge) Boskamp 1922, 70 Opin. 21, Jud. Comm. 1958, 163.

Further Comments

In the 7th edition of THE MANUAL, three species were described, the separation based primarily on habitat. *Selenomonas palpitans*, found in the ceca of guinea pigs, was the type species; however, this organism has not been obtained in pure culture and might be identical with one of the species described below (see Lessel and Breed, 1954, 165).

More recent important studies on the characteristics of the genus and species include those of Macdonald (1953); Bryant (1956); Macdonald *et al.* (1959); Bryant and Robinson (1962); Hobson *et al.* (1962); Loesche and Gibbons (1965); Kanegasaki and Takahashi (1967); Kingsley (1968); and Prins (1971).

Key to the species of genus **Selenomonas**

I. Found in the human buccal cavity. Cellobiose, salicin and dulcitol are not fermented; H_2S is usually not produced. G + C content of DNA is 60.6 ± 0.8 moles %.

1. *Selenomonas sputigena*

II. Found in rumen contents of ruminants. Cellobiose, salicin and dulcitol are fermented; H_2S is usually produced. G + C content of DNA is 54.0 ± 0.8 moles %.

2. *Selenomonas ruminantium*

Description of the species of genus **Selenomonas**

1. **Selenomonas sputigena** (Flügge) Boskamp 1922, 58. (*Spirillum sputigenum* Flügge 1886, 387; *Vibrio sputigenus* Prévot 1940, 85.)

spu.ti'ge.na. L. n. *sputum* spit, sputum; L. v. *gigno* produce; M.L. fem.adj. *sputigena* sputum-produced.

Curved, helical rods, rounded or bluntly tapered ends, 0.9–1.1 by 3–5.5 μm; some strains only 0.5 μm wide. Frequently two to five curves giving cells a spirillar appearance, sometimes long spirals. Cells occur as singles, pairs and short chains.

Colonies on blood agar are smooth, convex, grayish yellow, sometimes mottled and sometimes with a more opaque center. Diameter less than 0.5 mm except sometimes a smooth granular, flat, grayish translucent, irregular border and up to 2 mm diameter (Macdonald, 1953).

Heavy floccules and coarse granules in Brewer's thioglycollate broth (Difco).

Litmus milk acid and coagulated.

Fermentative, producing propionic and acetic acids as the acid products from glucose (Loesche and Gibbons, 1965). Carbon dioxide not determined.

Acid but no visible gas produced from arabinose, xylose, galactose, glucose, lactose, maltose, esculin, dextrin, inulin, glycerol and mannitol. Cellobiose, salicin, cellulose, xylan, dulcitol, inositol and sorbitol are not fermented (Kingsley, 1968). Starch is hydrolyzed. Gelatin is not liquefied. Indole is not produced. Nitrite produced from nitrate. H_2S usually not produced.

Temperature range, minimum about 20–30 C, maximum about 45 C. No growth at 50 C.

The pH range is 4.5–8.6.

Found in the human buccal cavity.

G + C content of the DNA is 60.6 ± 0.8 moles % (buoyant density).

Type strain not designated.

2. **Selenomonas ruminantium** (Certes) Wenyon 1926, 311. (*Ancyromonas ruminantium* Certes 1889, 70; *Selenomastix ruminantium* Woodcock and Lapage 1913–1914, 433; *Spirillum ruminantium* Macdonald *et al.* 1959, 559; *Selenomonas lactilytica* Hungate 1966, 69.)

ru.mi.nan'ti.um. M.L. pl.n. *ruminantia* ruminants; M.L. pl.gen.n. *ruminantium* of ruminants.

Curved or helical rods, rounded or bluntly tapered ends, frequently crescentic in shape, usually 0.8–1 by 2–7 μm. Cells arranged as singles and pairs with occasional short chains. A few helical filaments up to 20 μm in length are sometimes present. Some strains are smaller, 0.4–0.6 by 1.8–3 μm, and some are bigger, to about 3 μm in diameter.

Cells in rumen contents tend to be larger than those in pure culture (Purdom, 1963).

On rumen fluid glucose-cellobiose agar surface colonies entire, slightly convex, translucent and light tan, 2–6 mm diameter in 3 days; deep colonies thin and lenticular.

Growth in rumen fluid-glucose liquid medium is heavily turbid, often with some lightly flocculent sediment after 1–2 days.

Fermentative, producing propionic and acetic acids, carbon dioxide and/or DL-lactic acid as major products of glucose fermentation.

Acid produced from arabinose, xylose, fructose, galactose, glucose, cellobiose, lactose, maltose, esculin, salicin and dulcitol, and, usually, from sucrose and mannitol. Strains vary on trehalose, dextrin and inulin. No acid from cellulose, gum arabic, pectin, xylan or glycerol.

Starch is hydrolyzed by some strains. Gelatin not liquefied. Indole not produced. H_2S usually produced from cysteine. Nitrite produced from nitrate by some strains.

Grows in chemically defined medium containing B-vitamins, glucose, minerals, carbon dioxide, acetate and certain other volatile fatty acids (Bryant and Robinson, 1962). Fatty acid such as *n*-valerate is necessary for good growth of some strains even in a glucose medium containing trypticase and yeast extract (Kanegasaki and Takahashi, 1967). Ammonia, cysteine, serine or aspartic acid may serve as the sole source of nitrogen and cysteine or sulfide, as the source of sulfur.

Grow at 30–37 C, variable at 45 C, no growth at 22 or 50 C.

Final pH in poorly buffered glucose medium is 4.3–4.6.

Usually quite abundant in rumen contents of almost any ruminant as determined in direct microscopic studies. It has been isolated from rumen contents of cattle, sheep and elk.

G + C content of the DNA is 54.0 ± 0.8 moles % (buoyant density).

No type strain designated. Reference strain GA 192 available from M. P. Bryant.

2a. *Selenomonas ruminantium* subsp. *ruminantium* comb. nov.

Description as for species

2b. *Selenomonas ruminantium* subsp. *lactilytica* Bryant 1956, 165. (*Selenomonas lactilytica* Hungate 1966, 68.)

lac.ti.ly'ti.ca. L. n. *lac, lactis* milk; Gr. adj. *lyticus* dissolving; M.L. adj. *lactilyticus* milk dissolving.

This variety possesses the same characteristics as *Selenomonas ruminantium* except lactic acid and glycerol are fermented and possible antigenic differences. Lactic acid is fermented with production of propionic and acetic acid and, probably, CO_2 (Bryant, 1956). Glycerol is fermented mainly to propionic acid with small amounts of lactic, succinic and acetic acids (Hobson and Mann, 1961).

Hobson, Mann and Smith (1962) found that "O" antisera from sheep strains of *S. ruminantium* subsp. *lactilytica* reacted with a bovine strain of the same subspecies but not with a bovine strain of *S. ruminantium*. It was suggested that the two subspecies may differ in "O" antigens.

Reference strain: ATCC 19205; PC18 (Bryant 1956, 164).

2c. *Selenomonas ruminantium* subsp. *bryanti* Prins 1971, 825.

bry.ant'i. M.L. gen.n. *bryanti* of Bryant.

This subspecies possesses the same characteristics as *Selenomonas ruminantium* subsp. *ruminantium* except that cells in pure cultures are larger, measuring 2–3 by 5–10 μm, H_2S is usually not produced from cysteine and arabinose, xylose, galactose, lactose and dulcitol are usually not fermented.

Important Notes

for

Users of this Edition

1. Always read both generic and species descriptions because characters listed in the generic description are not usually listed in the species descriptions.

2. In tables, characters common to all taxa are not shown but may be listed in footnotes.

3. Generally in tables (exceptions are clearly indicated in footnotes, q.v.) the meanings of symbols are as follows:

+   more than 90 % strains positive

−   more than 90 % strains negative

d   11–89 % strains positive

( )   delayed reaction

w   weak reaction

D   Different reactions in different taxa (species of a genus or genera of a family)

v   strain instability (NOT differences between strains)

PART 10

GRAM-NEGATIVE COCCI AND COCCOBACILLI

FAMILY I. **NEISSERIACEAE** PRÉVOT 1933, 119

ALICE REYN

Neis.se.ri.a′ce.ae. M.L. fem.n. *Neisseria* type genus of the family; *-aceae* ending to denote family; M.L. fem.pl.n. *Neisseriaceae* the *Neisseria* family.

Organisms spherical in pairs or in masses with adjacent sides flattened, or rod-shaped in pairs or short chains. Not flagellated. Some species show twitching motility. Gram-negative.

Some species form xanthophyll pigment. Some species have complex growth requirements immediately after isolation, but may later grow in simple defined media. Catalase and cytochrome-oxidase produced by most species.

Aerobic.

Optimum temperature about 32–37 C.

Neisseria, Branhamella and *Moraxella* parasitic; *Acinetobacter* saprophytic or an opportunistic pathogen. Other characteristics differentiating the genera are given in Table 10.1.

The G + C content of the DNA ranges from 39–52 moles %.

Type genus: *Neisseria* Trevisan 1885, 105.

TABLE 10.1

Differential characteristics of the genera of family **Neisseriaceae**[a]

	I. *Neisseria*	II. *Branhamella*	III. *Moraxella*	IV. *Acinetobacter*[b]
Cells	Coccal	Coccal	Plump rods in pairs or short chains	Rod-shaped in exponential phase but nearly spherical in stationary phase
Division	Two planes	Two planes	One plane	One plane
Growth in simple defined media	d	−	d	+
Penicillin	Originally sensitive	Originally sensitive	Originally sensitive	Originally resistant
Oxidase	+	+	+	−
Reduction of nitrate	− or d	+	d	−
G + C moles %	47–52	40–45	40–46	39–47

[a] +: Most strains positive (≧90%). −: Most strains negative (≧90%). d: Some strains positive, some negative.

[b] Temporarily associated with *Neisseriaceae* (Report Subcommittee on Taxonomy of *Moraxella* and Allied Bacteria 1971. Int. J. Syst. Bacteriol., *21*: 213. Min. 13.

Genus I. **Neisseria** *Trevisan 1885, 105. Nom. cons.* Opin. 13, Jud. Comm. 1954, 153

Alice Reyn

(*Merismopedia* Zopf 1885, 51; *Gonococcus* Lindau 1898, 100.)
Neis.se′ri.a. M.L. fem.n. *Neisseria* named for Dr. Albert Neisser, who discovered the organism of gonorrhea in 1879.

Cocci, 0.6–1.0 μm in diameter, occurring singly but often **in pairs** with **adjacent sides flattened.** Division in two planes at right angles to each other sometimes resulting in the formation of tetrads. Cell wall of the Gram-negative type (Plate 10.1, Fig. 1). No formation of endospores; non-motile. Capsules and fimbriae may be present.

Gram-negative.

Yellow-greenish pigment produced by two species.

Complex growth requirements. Some species are hemolytic.

Chemoorganotrophic. **Few carbohydrates utilized** (Table 10.2).

Aerobic or facultatively anaerobic. Temperature optimum about 37 C.

Catalase and cytochrome-oxidase produced.

Parasites of mucous membranes of mammals.

The G + C content of the DNA (reported for all species) ranges **from 47.0–52.0 moles %.** Few or no genetic affinities with *Branhamella.*

Type species: *Neisseria gonorrhoeae* (Zopf) Trevisan 1885, 106.

Further Comments

Members of the genus *Neisseria* are susceptible to drying and exposure to sunlight. *N. gonorrhoeae* and *N. meningitidis* autolyse easily; they are also less resistant than other *Neisseria* to disruption by pressure in a Ribi Cell Fractionator (Martin *et al.*, 1969).

Generally, the sensitivity to various antibiotics is taxonomically not very important since this quality varies from one species to another and also within species. The tendency to *in vivo* and *in vitro* development of decreased sensitivity and complete resistance should be borne in mind. However, most *Neisseria* are sensitive to penicillin, streptomycin, tetracyclines, the macro-lide group and polymyxin B sulfate and related drugs, (*N. gonorrhoeae* and *N. meningitidis* are relatively resistant to polymyxin B sulfate); they are relatively resistant to vancomycin and related drugs. *N. mucosa* is very sensitive to chloramphenicol (Véron *et al.*, 1959; Berger and Miersch, 1970). Exact inhibitory concentrations depend upon the method used.

The differential diagnosis between *N. gonorrhoeae* and *N. meningitidis* is sometimes difficult as acid production from glucose and/or maltose may be weak or absent. After 24 hr. incubation on 10% horse blood agar plates single colonies of *N. meningitidis* are generally larger than those of *N. gonorrhoeae*, and the occurrence of β-hemolysis on 5% horse blood agar after 48–72 hr. incubation points to *N. meningitidis*. Lack of acid production from glucose and maltose, and growth at 22 C points to *Branhamella catarrhalis* (see p. 432), especially if nitrates are reduced.

The former *N. subflava, N. flava* and *N. perflava* are here united into one species, *N. subflava*, because the diversities of cultural and biochemical reactions of the chromogenic, saccharolytic *Neisseria* make differentiation between species most unsatisfactory. This action has often been recommended in varying degree (Wilson, 1928; Wilson and Smith, 1928; Wilson and Miles, 1955; Berger and Brunhoeber, 1961; Berger, 1962). *N. perflava* is by far the most common of the three (Berger and Wulf, 1961), and it seems likely that *N. subflava* and *N. flava* may be considered as defective variants of one species. Genetic data give strong support to this view (Catlin and Cunningham, 1961; Kingsbury, 1967; Henriksen and Bøvre, 1968).

The former *N. catarrhalis* has been transferred to the genus, *Branhamella* (p. 432), justified by the biochemical, physiological and genetical differences. *N. caviae* Pelczar 1953 and *N. ovis* Lindquist 1960 are discussed under *Branhamella.*

Description of the species of genus **Neisseria**

1. Neisseria gonorrhoeae (Zopf) Trevisan 1885, 106. (*Gonococcus* Bumm 1885, 16; *Merismopedia gonorrhoeae* Zopf 1885, 54; *Micrococcus gonorrhoeae* (Zopf) Flügge 1886, 156; *Micrococcus gonococcus* Schroeter 1886, 147; *Diplococcus gonorrhoeae* (Zopf) Lehmann and Neumann 1896, 150; *Gonococcus neisseri* Lindau 1898, 100: Micrococcus der gonorrhoe Neisser 1879, 497.)

go.nor.rhoe′ae. Gr. n. *gonorrhoea* gonorrhea; M.L. gen.n. *gonorrhoeae* of gonorrhea.

Common name: Gonococcus.

Acid produced from glucose but may be absent or weak even on adequate medium. Repeated subculturing will often result in normal acid production.

Nutritional requirements complex, varying from

TABLE 10.2

Characteristics differentiating the species of genus **Neisseria**[a]

	1. N. gonorrhoeae	2. N. meningitidis	3. N. sicca	4. N. subflava	5. N. flavescens	6. N. mucosa
Capsules	—	v	v	+	—	+
Acid from:						
Glucose	+	+	+	+	—	+
Maltose	—	+	+	+	—	+
Fructose	—	—	+	v	—	+
Sucrose	—	—	+	v	—	+
Starch	—	—	v	v	—	+
Polysaccharide produced from 5% sucrose	0	0	+	d	+	+
Production of H_2S	—	—	+	+'	+	+
Reduction of:						
Nitrate	—	—	—	—	—	+
Nitrite	—	d	+	+	+	+
Pigment	—	—	d	+	+	—
Extra CO_2	//	/	*	*	*	*
Growth at 22 C	—	—	d	d	+	+
G + C moles %	49.5–49.6	50.0–51.5	49.0–51.5	48.0–50.5	46.5–50.1	50.5–52.0

[a] +: Most strains positive ($\geq 90\%$). —: Most strains negative ($\geq 90\%$). d: Some strains positive, some negative. v: Character inconstant and in one strain may sometimes be positive, sometimes negative. 0: No growth on medium with 5% sucrose. G + C: Guanine + cytosine in the deoxyribonucleic acid. ?: Not tested. //: Very important. /: Important. *: not necessary.

strain to strain. Glutamine necessary on primary isolation for about 20% of strains, cocarboxylase for about 1%. A need for gluthathione is sometimes acquired by old laboratory strains. Iron is an essential growth factor (Kellogg et al., 1963). Solid synthetic media have been described but growth is sparse. A chemical supplement, containing glucose, glutamine and cocarboxylase was described by Lankford, 1950, and a fluid protein-free synthetic medium by Kenny et al. (1967). Growth in this medium is sparse in comparison to media enriched with serum or ascitic fluid. Protective substances incorporated in solid media, as starch, cholesterol, albumin or carbon particles promote growth. Addition of Ca^{++} to transport media enhances viability.

Growth on solid media may be obtained within 30–38.5 C, optimum at 35–36 C. A certain degree of moisture is important and extra CO_2 is necessary on primary isolation.

Primarily found in purulent venereal discharges. Also found in the blood, the conjunctiva, joints, petechiae in the skin and cerebrospinal fluid. Causes gonorrhea and other infections in man. Not found in other animals.

Genetically, closely related to *N. meningitidis* as demonstrated in transformation experiments (Catlin, 1967), by direct hybridization and thermal stability examination of DNA duplexes (Kings-

bury, 1967; Kingsbury et al., 1969). By chromatography the G + C content of the DNA has been found to be 49.6 (Lee et al., 1956) and 49.5 (Belozersky and Spirin, 1960) moles %.

Antigenic composition complicated. Recognized groups or international types do not exist. Differing degrees of cross-reactivity with other *Neisseria* species, most strongly with *N. meningitidis;* weak cross-reactivity with *Branhamella.*

Reference strain: J. F. Wilson B. 5025; ATCC 19424; NCTC 8375.

2. Neisseria meningitidis (Albrecht and Ghon) Murray 1929, 8. *Nom. cons.* Opin. 35, Jud. Comm. 1970, 13. (*Diplokokkus intracellularis meningitidis (sic)* Weichselbaum 1887, 583; *Neisseria weichselbaumii* Trevisan 1889, 32; *Micrococcus intracellularis* (Jaeger) Migula 1900, 189; *Micrococcus meningitidis cerebrospinalis* Albrecht and Ghon 1901, 988; *Micrococcus meningitidis* Albrecht and Ghon 1903, 498.)

me.nin.gi'ti.dis. Gr. fem.n. *meninx, meningis* the membrane enclosing the brain; M.L. fem.n. *meningitis, meningitidis* inflammation of the meninges.

Vernacular name: Meningococcus.

Most serogroups encapsulated in fresh isolates.

Acid produced from glucose and maltose, but strains that fail to produce acid on either or both do occur. Repeated subculturing often results in normal acid production. The DNA of such strains has been shown to contain a mutator gene (Jyssum and Jyssum, 1968).

Nutritional requirements complex upon primary isolation but many laboratory strains grow in simple synthetic media. Glutamic acid required as carbon source. Strains can be adapted to grow with NH_4 as the sole nitrogen source (Jyssum, 1959).

Extra CO_2 is not essential on primary isolation but stimulates growth. A certain degree of moisture is desirable.

The NaCl concentration in medium should not be too high; Ca^{++} stimulate growth, and Mg^{++} diminish the lag phase. Starch acts as a growth-promoting protective substance.

Originally found in the cerebrospinal fluid. Cause of epidemic cerebrospinal fever. Also found in the nasopharynx, in blood, in conjunctiva, in joints and petechiae in skin. Occasionally found in venereal discharges. Found in the nasopharynx of man (healthy carriers).

Genetically closely related to *N. gonorrhoeae* (Catlin, 1967). Less related to *N. subflava*, *N. sicca* and *N. flavescens;* not related to *Branhamella catarrhalis* and *N. caviae* (Kingsbury, 1967; Kingsbury *et al.*, 1969). By chromatography, the G + C content of the DNA was found to be 50.5 (Belozersky and Spirin, 1960) and 51.3 moles % (Catlin and Cunningham, 1961). By analysis of the thermal denaturation temperature Marmur and Doty (1962) found 51.5 moles %, and Schildkraut *et al.* (1962) found 50.0 moles % by buoyant density.

Four main groups of *Neisseria meningitidis* have been differentiated on the basis of agglutination

TABLE 10.3

Relationships among the various classifications of meningococci

In common use, 1940	Recommended by Committee 1950	Slaterus, 1961, 1963	Hollis *et al.*, 1968
I	A	A	A
II	B	B	B
IIα	C	C	C[a]
IV	D	D	D
		X	E
		Y	F
		Z	G

[a] Vedros and Culver 1968, described a new "encapsulated" serogroup (called E) among nongroupable strains isolated in California; it was antigenically related to group C, but antisera to E did not protect mice challenged with groups A, B or C organisms. The new group E strains were not compared with the X, Y and Z strains of Slaterus.

reactions with immune serum. The Subcommittee on *Neisseria* of the International Committee on Bacteriological Nomenclature has suggested (Int. Bull. Bacteriol. Nomencl. Taxon., *4:* 1954, 95) that these groups be designated as A, B, C and D. Relationships of these groups to those used previously and to recent results obtained by Slaterus (1961), Slaterus *et al.* (1963) and Hollis *et al.* (1968) are shown in Table 10.3. Slaterus (1961) described three new serological groups, X, Y and Z and these were confirmed by Hollis *et al.* (1968) who suggested the terms E, F and G to follow the established groups: A, B, C and D. Serogrouping can also be made by the capsule swelling method treating fresh smooth cultures (or cultures preserved in the smooth stage) with specific antibody. Capsular swelling has never been demonstrated for group B organisms. Specific B polysaccharide (and other specific polysaccharides) can be shown by spot-inoculation on agar containing specific antibody (Branham, 1953, 1958).

Differing degrees of serological cross-reactivity with other *Neisseria* species, strongest with *N. gonorrhoeae*.

Neotype strain: Sara E. Branham M 1027; ATCC 13077; NCTC 10025 (Int. J. Syst. Bacteriol., 1970, 14).

3. **Neisseria sicca** (von Lingelsheim) Bergey *et al.* 1923, 43. (*Diplococcus pharyngis siccus* von Lingelsheim 1906, 409; *Diplococcus siccus* von Lingelsheim 1908, 476.)

sic'ca. L. fem.adj. *sicca* dry.

Sometimes encapsulated (Berger, 1963).

Colony morphology on blood agar variable, both dry grayish and slimy white or yellow colonies are described.

Acid production from carbohydrates varies with the medium. Polysaccharide produced on medium with 5% sucrose; some strains produce slime. Xanthophyll pigment produced by some strains (Berger, 1963). Nutritional requirements complex.

Found in nasopharynx, saliva and sputum of man.

The G + C content of the DNA is 51.5 moles % by chromatography (Catlin and Cunningham, 1961), 49.0 by thermal denaturation temperature (Marmur and Doty, 1962) and 51.0 by buoyant density (Schildkraut *et al.*, 1962). Results of interspecific transformation indicate genetic affinities with other *Neisseria* (Catlin and Cunningham, 1961).

Direct interspecies hybridization experiments with DNA (Kingsbury, 1967) show most relation with *N. subflava*, less with *N. meningitis* and little with *B. catarrhalis*. Examination of thermal stability of interspecies duplexes between *N. men-*

ingitidis DNA and *N. sicca* DNA show that the relatedness is far less than judged by the previous hybridization experiments; interspecies examination with other *Neisseria* not performed (Kingsbury *et al.*, 1969).

Spontaneously agglutinable in saline. Reports on antigenic analyses questionable. Possibly serologically distinct from *N. flava*, *N. perflava* and *N. flavescens* (Berger and Wulf, 1961; Berger and Brunhoeber, 1961). Serologically related to *N. mucosa*. (Véron *et al.*, 1961).

Type strain not designated.

4. **Neisseria subflava** (Flügge) Trevisan 1889, 32. (*Micrococcus subflavus* Flügge 1886, 159; *Neisseria flava* Bergey *et al.* 1923, 43; *Neisseria perflava* Bergey *et al.* 1923, 43.)

sub.fla′va. L. pref. *sub* less than, somewhat; L. adj. *flavus* yellow; L. fem.adj. *subflava* yellowish.

Encapsulated (Berger, 1963).

Most strains produce acid from glucose and maltose, and many strains also from fructose, sucrose and starch (Table 10.2). The production of acid varies with the medium and also from time to time using the "same" medium (Wilson, 1928; Hajek *et al.*, 1950; Berger, 1961).

Found in secretion from human nasopharynx and rarely in cerebrospinal fluid in cases of meningitis (Noguchi *et al.*, 1963).

The G + C content of the DNA ranges from 49.2–50.5 moles % as determined by chromatography (Catlin and Cunningham, 1961; La Macchia and Pelczar, 1966); 49 moles % by thermal denaturation (Marmur and Doty, 1962) and 48 moles % by buoyant density (Schildkraut *et al.*, 1962). Experiments with DNA indicate some relation to the DNA of *N. sicca*, *N. flavescens* and *N. meningitidis* but little or no relation to that of *B. catarrhalis* or *N. caviae* (Catlin and Cunningham, 1961; Kingsbury, 1967; Kingsbury and Duncan, 1967). Determination of the thermal stability of interspecies DNA duplexes is lacking.

Often agglutinated in saline. Serological results questionable. Presumably unrelated to *N. flavescens* and *N. sicca*. Some relation to *N. gonorrhoeae*.

Type strain not designated.

5. **Neisseria flavescens** Branham 1930, 849.

fla.ves′cens. L. v. *flavesco* become golden yellow; L. part.adj. *flavescens* becoming golden yellow.

Normally, no acid production from glucose, maltose, fructose and sucrose; occasionally weak acid production from glucose, maltose and fructose (Hajek *et al.*, 1950; Berger, 1963). Xanthophyll pigment produced but absorption spectra vary (Ellinghausen and Pelczar, 1955; Berger, 1961). Complex nutritional requirements. Found in cerebrospinal fluid from patients with meningitis and in

blood from cases of septicemia. Rare (Branham, 1930; Wertlake and Williams, 1968).

The G + C content of the DNA is 50.1 moles % by chromatography (Catlin and Cunningham, 1961); 49 moles % by thermal denaturation (Marmur and Doty, 1962) and from 46.5–47.5 by buoyant density (Schildkraut *et al.*, 1962; Bøvre, 1967). Experiments with DNA indicated a close relation to *N. meningitidis*, *N. perflava*, *N. subflava*, *N. flava* and *N. sicca* (Catlin and Cunningham, 1961; Kingsbury, 1967) but much less relation to *B. catarrhalis* and *N. caviae*. Determination of thermal stability of interspecies DNA duplexes is lacking.

Type strain: Sara Branham N 155; NCTC 8263; ATCC 13120.

6. **Neisseria mucosa** Véron, Thibault and Second 1959, 508. (*Diplococcus mucosus* von Lingelsheim 1906, 395.)

mu.co′sa. L. fem.adj. *mucosa* slimy.

The cocci are either surrounded by individual distinct capsules or they are together in masses surrounded by slime. Capsules vary in size within strains, and production of slime varies with the medium.

Most strains are non-pigmented (Berger and Miersch, 1970), but one of the strains described by Véron *et al.*, 1959, was slightly yellow. Growth is favored by a certain degree of humidity and by serum, blood or ascitic fluid in the medium. Optimal growth between 30 and 37 C; no growth below 22 C. A few thermoresistant organisms can be picked at 42 C.

Found in rhinopharynx of man. Occasionally pathogenic for man. Pathogenic for mice under certain conditions.

The G + C content of the DNA is 51.0 ± 1.3 moles % as determined by buoyant density (Mandel, 1967; three strains examined). Hybridization experiments not performed.

Most strains are serologically related to *N. sicca*. (Véron *et al.*, 1961).

Type strain not designated.

Species incertae sedis

a. *Neisseria animalis* Berger 1960, 160.

b. *Neisseria canis* Berger 1962, 455.

c. *Neisseria caviae* Pelczar 1953, 744.

d. *Neisseria cinerea* (von Lingelsheim) Murray 1939, 283; (*Micrococcus cinereus* von Lingelsheim 1906, 396).

e. *Neisseria cuniculi* Berger 1962, 455; *Neisseria cuniculi* var. *giganta* Berger 1962, 456.

f. *Neisseria denitrificans* Berger 1962, 455.

g. *Neisseria elongata* Bøvre and Holten 1970, 73.

h. *Neisseria lactamicus* (*sic*) Hollis, Wiggins and Weaver 1969, 72 (*Neisseria meningococcoides* Berger and Schlez 1970, 59).

i. *Neisseria ovis* Lindquist 1960, 165.

j. *Neisseria suis* Gallo, Ivanov and Morris 1961/ 1962, 57.

These species (except *N. cinerea*, *N. caviae* and *N. ovis*, see below) have not been analyzed for moles % G + C of the DNA. The moles % G + C of *N. cinerea* is 49.0 (Bøvre *et al.*, 1969). Bøvre (1965) demonstrated by transformation of resist-

ance to streptomycin that it probably belongs to the same group as the pigmented, asaccharolytic *N. flavescens*. Bøvre did not find any compatibility between *N. cinerea* and *N. flavescens* and *B. catarrhalis*, *N. caviae* and *N. ovis*.

Because of the lack of genetic information the species listed above are provisionally considered as *species incertae sedis*.

Genus II. **Branhamella** *Catlin 1970, 157*

ALICE REYN

Bran.ham.el′la. M.L. fem.n. *Branhamella* named in honor of Sara Branham, who contributed to the knowledge of the *Neisseria* family.

Cocci commonly arranged in pairs with adjacent sides flattened; division in two planes at right angles to each other. Cell wall of the Gram-negative type (Plate 10.2, Fig. 1). No formation of endospores; non-motile. **Gram-negative** with a tendency to resist decoloration.

Chemoorganotrophic. **Acid not produced from carbohydrates.** No xanthophyll pigment.

Growth occurs on customary media without blood. Aerobic. Temperature optimum about 37 C. **Catalase and cytochrome-oxidase produced; nitrates usually reduced.**

Parasites of mucous membranes of mammals.

The G + C content of the DNA ranges **from 40 to about 45 moles** %. Differs from *Neisseria* in DNA base content and fatty acid composition; few or no genetic affinities with *Neisseria*.

Type species: *Branhamella catarrhalis* (Frosch and Kolle) Catlin.

Further Comments

N. caviae and *N. ovis* do not form xanthophyll pigments (Ellinghausen and Pelczar, 1955; Berger,

1961), and are not saccharolytic; it is therefore tempting to group them with *B. catarrhalis*. However, both oxidize butyrate and *N. caviae* does not contain decanoic acid, a component of *B. catarrhalis* (Lewis *et al.*, 1968). *N. ovis* also attacks caproate (Baumann *et al.*, 1968).

The G + C content of their DNA as determined by buoyant density of two *N. caviae* and three *N. ovis* strains is about 44.5 moles % (Bøvre, 1967; Bøvre *et al.*, 1969). By chromatography of 11 strains of *N. caviae* La Macchia and Pelczar (1966) found a range from 47.7–50.4 moles %. DNA-DNA hybridization (Kingsbury, 1967) indicates a closer relationship between *N. caviae* and *B. catarrhalis* than between *N. caviae* and the *Neisseria* of the equimolar DNA base type. Baumann *et al.* (1968) showed that the nutritional and physiological properties of *B. catarrhalis*, *N. caviae* and *N. ovis* were similar though not identical. The thermal stability of the DNA duplexes between *B. catarrhalis*, *N. caviae* and *N. ovis* has not yet been examined. Thus, final classification of these species must await further investigation.

Description of the species of genus **Branhamella**

1. **Branhamella catarrhalis** (Frosch and Kolle) Catlin 1970, 157. (*Mikrokokkus catarrhalis* (*sic*) Frosch and Kolle in Flügge 1896, 154.)

ca.tar.rha′lis. Gr. adj. *catarrhus* downflowing, catarrh; M.L. adj. *catarrhalis* of catarrh.

Capsules visible, both by light and electron microscopy (Plate 10.2, Fig. 1); Gram-negative with a tendency to resist decoloration.

No acid from carbohydrates; polysaccharide not produced from sucrose. Growth occurs on nutrient agar and in a medium composed of amino acids, mineral salts, biotin and with lactate or succinate as principal carbon and energy source. Butyrate is not attacked.

No production of extracellular urease, hydrogen sulfide or indole. Production of lipase, cytochrome-oxidase, catalase and extracellular deoxyribonu-

clease. Nitrates reduced to nitrites and beyond without gas production.

Aerobic. Temperature optimum 37 C, but grows well at 22 C.

Susceptible to penicillin, streptomycin, tetracyclines and polymyxin B sulfate and similar antibiotics. Relatively resistant to vancomycin and related drugs.

Spontaneously agglutinated in saline. Serologically different from *N. flavescens*, *N. flava*, *N. perflava* and *N. sicca* (Warner *et al.*, 1952). Also different from *N. cinerea* (Berger and Paepcke, 1962).

Parasite of mucous membranes of mammals. Occasionally found in venereal discharges. May be responsible for inflammations of mucous membranes, alone or in association with other organisms. Has been reported as causing meningitis. Rare (Cocchi and Ulivelli, 1968).

The G + C content of the DNA is 40–44 moles %. Proposed neotype: Catlin Ne 11; ATCC 25238.

Genus III. **Moraxella** *Lwoff 1939, 173 Nom. cons.* Opin. 41, Jud. Comm. 1971, 106

HANS LAUTROP

(*Diplobacillus* McNab 1904, 62.)

Mo.rax.el′la. M.L. dim.ending -*ella*. M.L. fem.n. *Moraxella* named for V. Morax, a Swiss ophthalmologist who isolated the type species.

Rods, usually very short and plump, typically 1.0–1.5 by 1.5–2.5 μm; often approaching coccus shape, predominantly in pairs and short chains. Some cultures fairly uniform, others pleomorphic showing variation in cell size and shape, filaments and long chains. Pleomorphism enhanced by lack of oxygen and above optimum temperatures. No spores formed; flagella not present. Some strains show under special conditions a "twitching" motility on solid surfaces. Capsules and fimbriae may or may not be present. Gram-negative.

Chemoorganotrophic with oxidative metabolism. Most strains more or less fastidious, but specific growth requirements not known. A limited number of organic acids, alcohols and amino acids serve as carbon and energy sources. **Carbohydrates not utilized.**

Oxidase positive, catalase usually positive. Indole, acetoin and H₂S not produced.

Strict aerobes. Optimum temperature 32–35 C. Optimum pH 7–7.5. **Highly sensitive to penicillin.**

Parasites on the mucous membranes of man and warm-blooded animals and possibly also saprophytes.

The G + C content of the DNA ranges from 40–46 moles % (buoyant density).

Type species: *Moraxella lacunata* (Eyre) Lwoff 1939, 173.

Further Comments

The present circumscription excludes members of the more or less similar genera *Neisseria, Branhamella* and *Acinetobacter*.

Moraxella and *Branhamella* are phenotypically and genetically so similar that a fusion of the two

genera into one has been proposed (Henriksen and Bøvre, 1968). The only reasonable objection is that *Moraxella* is rod shaped and *Branhamella* coccus shaped and this difference is considered so important that the proposal has not been followed here.

The G + C content of the DNA and transformation experiments show that *Moraxella* and the revised genus *Neisseria* are not as closely related as might at first be suspected.

Some French authors (Audureau, 1940; Piéchaud *et al.* 1951, 1956; Lwoff, 1964) have considered members of the genus *Acinetobacter*, as defined here, as members of the genus *Moraxella*. Although the G + C content of their DNA is about the same and they are phenotypically close, transformation and hybridization experiments have given no definite evidence of a close relationship. They also differ in oxidase reaction, cell wall composition, nutritive versatility, penicillin sensitivity and habitat. For these reasons they are considered as separate genera. It is important to note that, if only ordinary routine tests are employed, the only regular difference between some strains from these genera is the oxidase test.

In the present definition, *Moraxella* is characterized by its inability to form acid from glucose. It does not seem advisable to change the generic concept to include saccharolytic *Moraxella*-like taxa; they are listed as *species incertae sedis* until further studies have clarified their taxonomic position. *Moraxella saccharolytica* Flamm 1956, 500, seems to belong in the genus *Flavobacterium*.

Five species are recognized. Some would divide the type species into two subspecies and *M. bovis* is so closely related to the type species that it could be considered a third subspecies of *M. lacunata* rather than a separate species.

Key to the species of genus **Moraxella**

I. Do not grow in mineral medium with acetate and ammonium salts.
   A. Coagulated serum liquefied. Hemolysis on chocolate agar.
      1. Nitrite usually produced. Blood agar not hemolyzed.
            1. *M. lacunata*
      2. Nitrite not produced. Blood agar usually hemolyzed.
            2. *M. bovis*

B. Coagulated serum not liquefied. No hemolysis on chocolate agar.
  1. Phenylalanine deaminase absent.
                                              3. *M. nonliquefaciens*
  2. Phenylalanine deaminase present.
                                              4. *M. phenylpyruvica*
II. Grow in mineral medium with acetate and ammonium salts.
                                              5. *M. osloensis*

Description of the species of genus **Moraxella**

1. Moraxella lacunata (Eyre) Lwoff 1939, 173. (Diplobacille de la conjunctivite subaigue Morax 1896, 337; Die chronische Diplobacillen conjunctivitis Axenfeld 1897; 1; *Bacillus lacunatus* Eyre 1900, 5; *Diplobacillus moraxaxenfeld* McNab 1904, 62.)

la.cu.na′ta. L. n. *lacuna* a shallow depression; M.L. fem.adj. *lacunata* pitted.

Cells vary from small lanceolate coccoids to plump rods, 0.8–1.2 by 1.5–3.0 μm, arranged predominantly in pairs and short chains. Aberrant cell forms frequent. Capsules, if present, very narrow.

Typical colonies on blood agar 0.1–0.3 mm, but some strains may produce colonies up to 3 mm in diameter. Smooth and rough colonies occur. Occasionally pitting of the agar under the colony. No hemolysis on horse blood agar (Oag, 1942 reported hemolysis by some strains, kind of blood not stated). On chocolate agar large dark zones around the very small colonies.

Coagulated horse serum liquefied more or less rapidly. Growth in gelatin stab usually too poor to test liquefaction by this method. Nitrates reduced to nitrites by the majority of strains.

Requires complex media often with the addition of serum or oleic acid. In complex media utilizes butyrate and usually caproate; ethanol utilized by some strains; propionate not utilized.

Commonly found in the conjunctiva of guinea pigs; probably causes conjunctivitis in humans living under poor hygienic conditions, formerly reported frequently, now rarely.

The G + C content of the DNA ranges from 42–43 moles % (Bøvre, 1967; Baumann *et al.*, 1968).

Neotype strain: Morax 260; ATCC 17967.

Further Comments

The species as here described includes both *M. lacunata* Lwoff and *M. duplex* Lwoff 1939, 173 (*M. liquefaciens* (McNab) Murray 1948, 591), as no definite distinguishing characteristics have been found. Strains previously labeled *M. liquefaciens* (reference strain ATCC 17952; NCTC 7911) are usually less fastidious, produce large colonies, liquefy coagulated serum more slowly and are more rod-like than coccobacillary and may be classified as *M. lacunata* subsp. *liquefaciens*, in contrast to *M. lacunata* subsp. *lacunata*.

2. Moraxella bovis (Hauduroy *et al.*) Murray 1948, 591. (*Haemophilus bovis* (*sic*) Hauduroy, Ehringer, Urbain, Guillot and Magrou 1937, 247; *Moraxella duplex* des bovides Lwoff 1939, 174.)

bo′vis. L. gen.n. *bovis* of a cow or ox.

Morphology of fresh isolates as for genus; collection strains dominated by rods and filaments.

Colonies on blood agar at 48 hr. about 3 mm; fresh isolates usually smooth, older strains frequently rough. With very few exceptions, broad hemolytic zones present. Pitting of substrate may occur. Dark hemolytic zones on chocolate agar.

Liquefies gelatin and coagulated serum, sometimes slowly. Nitrates usually not reduced; may be very late in a few strains. Catalase present in about half the strains according to Pugh *et al.*, 1966; others report presence of catalase regularly.

Less fastidious than the type species. In complex media utilizes butyrate, caproate and ethanol, but not propionate.

Primary cause of "pink-eye" in cattle, a kind of epizootic keratoconjunctivitis.

The G + C content of the DNA 41–43 moles %.

Suggested working strain: Hawksley's 36; NCTC 8561; ATCC 17947.

Further Comments

See *Further Comments* to *M. lacunata*.

Moraxella caprae Pande and Sekariah 1960, 277, isolated from keratitis in goats, is indistinguishable from *M. bovis*, except serologically, according to the description.

Moraxella equi Hughes and Pugh 1970, 462, isolated from conjunctivitis in horses, is non-hemolytic but otherwise similar to *M. bovis*, except perhaps in host species predeliction and pathologic changes produced.

3. Moraxella nonliquefaciens Lwoff 1939, 171. (*Bacillus duplex non liquefaciens* Scarlett 1916, 107.)

non.li.que.fa′ci.ens. L. prefix *non* not; L. part.-adj. *liquefaciens* dissolving; L. part.adj. *nonliquefaciens* not liquefying.

Morphology as described for genus.

Colonies on blood agar after 48 hr. 2–3 mm; some very mucoid. No hemolysis on blood or chocolate agar.

Gelatin and coagulated serum not liquefied. Nitrates usually reduced to nitrites. Urease present in a few isolates.

Growth requirements variable; some strains very fastidious requiring complex media with added serum or oleic acid, others satisfied with complex media without additions; none can grow in a simple mineral medium.

Some strains very restricted in utilization of carbon sources; utilizes acetate and lactate but not propionate, butyrate, caproate or ethanol.

Found in upper respiratory tract, especially the nose. Pathogenicity uncertain but may be a secondary invader in the respiratory tract.

The G + C content of the DNA about 40–44 moles %.

Neotype strain: 4663/62 of Bøvre and Henriksen; ATCC 19975; NCTC 10464.

Further Comments

See *Further Comments* to *M. lacunata*.

Earlier this species was no doubt often confused with *M. osloensis*.

4. Moraxella phenylpyruvica Bøvre and Henriksen 1967, 344; *Epit. spec. cons.* Opin. 42, Jud. Comm. 1971, 107. (*Moraxella polymorpha* Flamm 1957, 266; *Moraxella phenylpyrouvica* (sic) Bøvre and Henriksen 1967, 344.)

phe.nyl.py.ru'vi.ca. From phenylpyruvic acid, the product of deamination of phenylalanine by this organism.

Morphology as for genus.

Colonies on blood agar 0.5–1.0 mm after 48 hr. No hemolysis on blood agar or chocolate agar.

Gelatin and coagulated serum not liquefied. Nitrates usually reduced to nitrites. Urease usually present. Distinctive property is deamination of phenylalanine to phenylpyruvic acid; tryptophan also deaminated.

Very fastidious, growing poorly even in complex media; addition of serum improves growth but is not required.

The G + C content of the DNA is about 43–43.5 moles %.

Isolated from different kinds of pathological specimens. Pathogenicity not known.

Neotype strain: 2863 of Bøvre and Henriksen; ATCC 23333; NCTC 10526.

Further Comments

There are phenylalanine deaminase-positive *Moraxella* strains which do not belong in this species, but they have not yet been studied in detail.

5. Moraxella osloensis Bøvre and Henriksen 1967, 131.

os.lo.en'sis. M.L. adj. *osloensis* from Oslo, the Norwegian capital where the species was first recognized.

Morphology as for genus, except that poly-β-hydroxybutyrate inclusions are regularly present after growth in appropriate medium (for technique, see Stanier *et al.*, 1966).

Colonies on blood agar 2.0–2.5 mm after 48 hr., in some cultures in addition a class of smaller colonies. No hemolysis.

Serum and gelatin not liquefied. Nitrite produced by about half of the strains. Urease rarely present. Phenylalanine not deaminated. Acid formed regularly from ethanol in O/F medium (Hugh and Leifson, 1953).

No specific growth requirement. Fatty acids and alcohols, but not carbohydrates, utilized as carbon and energy sources. Citrate rarely utilized.

Found in the upper respiratory tract of man. Pathogenicity uncertain.

The G + C content of the DNA about 43–43.5 moles %.

Type strain: A 1920 of Bøvre and Henrikens; ATCC 19976; NCTC 10465.

Further Comments

Rare strains having unknown growth requirements may be indistinguishable from *M. nonliquefaciens*. See also *Further Comments* to *Moraxella urethralis*.

Species incertae sedis

a. *Moraxella kingii* Henriksen and Bøvre 1968, 383.

king'i.i. M.L. gen.n. *kingii* named for Elisabeth O. King, the American bacteriologist who did the pioneer work with this species.

Differs from the genus in the following characteristics:

Catalase absent. Acid produced from glucose and maltose on ascites agar medium. In semisolid media some tendency to grow in the depth of the tube, but no growth under strict anaerobic conditions. Cells coccoid to rod shaped with square ends; somewhat smaller than most *Moraxella*.

Colonies on blood agar 1.2–1.5 mm after 48 hr. Some strains mucoid. Pitting of the surface may occur. β-Hemolytic zones on blood agar but no hemolysis on chocolate agar. Very poor survival of blood agar cultures kept at room temperature.

No liquefaction of coagulated serum. Some strains reduce nitrate to nitrite. Urease or phenylalanine deaminase not present.

Requires complex media but serum does not improve growth.

The G + C content of the DNA is 44.5 moles %.

Strains have been isolated from a variety of human pathological specimens. Pathogenicity not known.

Type strain: 4177/66 of Henriksen and Bøvre; ATCC 23330.

b. "*Moraxella urethralis*" Lautrop *et al.* 1970, 255. u.re.thra'lis. M.L. adj. *urethralis* from Greek *ourethra* the distal part of the urinary tract which is probably the habitat of this species.

Cells smaller than typical for the genus, 0.6 by 0.1–1.5 μm and without the plump coccoid appearance of typical *Moraxella*. Intracellular poly-β-hydroxybutyrate inclusions regularly present but may be difficult to recognize in the small cells. Strong tendency to aggregate into small granules in fluid mineral media. Capsules regularly absent.

Colonies on blood agar 1.5–2.0 mm after 48 hr., more white in color than other *Moraxella*. No hemolysis.

Serum and gelatin not liquefied. Nitrates not reduced. Urease and phenylalanine deaminase not present. Weak and late acid formation from ethanol in O/F medium.

No specific growth requirements, but growth in mineral media slow. Utilizes acetate and butyrate as only carbon and energy sources. Citrate used by about half of the strains.

Found in urines and secretions from the genital tract of man. Pathogenicity unknown.

The G + C content of the DNA about 46 moles %.

Type strain not designated; ATCC 17960 is a typical strain.

Comment: A recognizable species, but generic assignment uncertain. Has not yet been proposed formally as a species of *Moraxella*. Some of the strains formerly designated *Mima polymorpha* var. *oxidans* belong in this species.

Distinction from *M. osloensis* may be difficult. The smaller cells and the smaller, white colonies and, in some cases, nitrite production or citrate utilization are the criteria on which a phenotypic differentiation is based.

c. *Other Moraxella-like Taxa*. Oxidase positive, saccharolytic strains showing *Moraxella* morphology (Thornley, 1967; Bijsterveld, 1970; Shewan, 1971; and possibly Ming-ching *et al.*, 1966).

Strains with the above-mentioned characteristics but otherwise conforming to the generic descriptions of both *Moraxella* and *Acinetobacter* are found in cold-stored poultry and fish and from a variety of human pathological specimens (Lautrop, unpublished). May or may not grow at 37 C. The kinds of sugars from which acid is produced oxidatively vary. Sensitivity to penicillin varies, usually intermediate between *Moraxella* and *Acinetobacter*.

The G + C content of the DNA of Thornley's strains ranges from 44–46 moles %.

Comment: The classificatory problem of these strains is as yet unsolved. A new genus may be required.

Genus IV. **Acinetobacter** *Brisou and Prévot 1954, 727*

HANS LAUTROP

A.ci.ne'to.bacter. Gr. adj. *akinetos* unable to move; M.L. masc.n. *bacter* a rod or staff; M.L. masc.n. *Acinetobacter* non-motile rod.

Rods, usually very short and plump, typically 1.0–1.5 by 1.5–2.5 μm in logarithmic phase, approaching coccus shape in stationary phase; predominantly in pairs and short chains. Large, irregular cells and filaments in small numbers in all cultures, but may dominate. No spores formed; flagella not present. Some strains show under special conditions a "twitching" motility on solid surfaces. Capsules and fimbriae may or may not be present. Intracellular poly-β-hydroxybutyrate inclusions do not occur. Gram-negative.

Chemoorganotrophic with oxidative metabolism. Usually versatile in the utilization of organic compounds as carbon and energy source. No specific growth requirements. Acid may or may not be formed from sugars.

Oxidase negative, catalase positive. Acetoin, indole and H₂S not produced.

Strict aerobes. Optimum temperature 30–32 C. Optimum pH about 7. **Resistant to penicillin.** Free-living ubiquitous saprophytes.

The G + C content of the DNA ranges from 40–47 moles % (thermal melting point).

Type species: *Acinetobacter calcoaceticus* (Beijerinck) Baumann, Doudoroff and Stanier 1968, 1538 (designated by Baumann *et al.*, 1968, 1538).

Further Comments

The tribe *Mimeae* de Bord 1942, used for the present taxon lost its standing in nomenclature by Opinion 40, Jud. Comm. 1971, 105. The adoption of Brisou and Prévot's genus *Acinetobacter* 1954 is a consequence of Brisou's designation of *Acinetobacter anitratum* as the type species of that genus in 1957, 401. Prévot's later designation of *Acineto-*

bacter stenohalis as the type species in 1961 must be disregarded according to international rules. The requirement for strict aerobiosis in the present circumscription of the genus excludes the facultatively anaerobic members of the genus as originally conceived by Brisou and Prévot.

The requirement for a negative oxidase reaction provides for an easy and clear-cut separation from the oxidase positive genus *Moraxella*. Phenotypically the genera *Acinetobacter* and *Moraxella* are otherwise so similar that strains of *Acinetobacter* have been included as species in the genus *Moraxella* (Audureau, 1940; Piéchaud *et al.*, 1951, 1956; Lwoff, 1964). There are however differences as regards chemical structure of the wall, metabolic capacity, penicillin sensitivity and habitat, and genetic studies have not supported the existence of a close relationship, although the G + C mole % of the DNA is the same.

The requirement for a negative oxidase test also serves to distinguish the genus from *Branhamella* with which otherwise it has considerable phenotypic similarity including morphology when *Acinetobacter* cells approach the coccus shape.

An unfortunate side effect of this requirement is that taxa which have the general characters of both *Acinetobacter* and *Moraxella*, but do not fit either as here circumscribed, are left without generic assignment (Thornley, 1967). For these taxa, see Appendix to the genus *Moraxella*.

Although several proposals exist for a subdivision of the genus (Piéchaud *et al.*, 1956; Mannheim and Stenzel, 1962; Baumann *et al.*, 1968), all members are here united in a single species. The previous subdivisions were based on acid formation, hemolysis and gelatin liquefaction (which are strongly correlated) and degree of nutritional versatility, but the resulting subgroups were neither homogeneous nor known to be biologically significant.

The one subgroup within the complex which appears to come closest to deserving status as a separate species is distinguished by the following set of characters: (1) restricted metabolic capacity usually including inability to utilize citrate; (2) a moderate or low degree of resistance to penicillin; (3) especially high degree of sensitivity to a bacteriocin which lyses most *Acinetobacter* strains; (4) a neutral or very weak acid reaction in the O/F glucose tube in contrast to either a clear-cut acid or alkaline reaction; (5) absence of hemolysis and gelatin liquefaction; (6) G + C content of DNA above 43 moles %. In DNA homology studies these strains come out as a distinct group (Johnson *et al.*, 1970). Such strains belong in the French "lwoffi" subgroup but only constitute part of it; in Baumann *et al.*'s subdivision they correspond to the B₂ subgroup, and in Mannheim and Stenzel's subdivision they are within the species *Achromobacter metalcaligenes*.

Description of the species of genus **Acinetobacter**

1. **Acinetobacter calcoaceticus** (Beijerinck) Baumann, Doudoroff and Stanier 1968, 1538. (*Bacterium viscolactis* Mez 1898, 61; *Micrococcus chinicus* Emmerling and Abderhalden 1903, 338; *Micrococcus calco-aceticus* Beijerinck 1911, 1067.)

Synonyms for acid-forming strains only: *Herellea vaginicola* de Bord 1942, 476; *Bacterium anitratum* Schaub and Hauber 1948, 385; *Neisseria winogradskyi* Lemoigne, Gerard and Jacobelli 1952, 395; *Achromobacter anitratum* (Schaub and Hauber) Brisou 1953, 813; *Acinetobacter anitratum* (Schaub and Hauber) Brisou and Prévot 1954, 727; *Moraxella glucidolytica* Piéchaud, Piéchaud and Second 1956, 522; *Micrococcus cerificans* Finnerty, Hawtrey and Kallio 1962, 176; *Achromobacter conjunctivae* Mannheim and Stenzel 1962, 77; *Achromobacter haemolyticus* var. *glucidolytica* Mannheim and Stenzel 1962, 78; *Lingelsheimia anitrata* (Schaub and Hauber) Seeliger, Schubert and Schlieber 1966, 257.

Synonyms for non-acid-forming strains only: *Alcaligenes haemolysans* Henriksen 1937, 167; *Moraxella lwoffi* Audureau 1940, 150; *Mima polymorpha* de Bord 1942, 475; *Acinetobacter lwoffi*

(Audureau) Brisou and Prévot 1954, 727; *Achromobacter haemolyticus* var. *alcaligenes* Mannheim and Stenzel 1962, 78; *Achromobacter citroalcaligenes* Mannheim and Stenzel 1962, 80. Formerly strains of this species have informally been designated: "B5W" (Stuart *et al.*, 1949) and "Vibrio O1" (see Fewson, 1967).)

cal.co.a.ce'ti.cus. L. n. *calx* chalk; L. n. *acetum* acetic acid. M.L. n. *calcoaceticum* calcium acetate which was used by Beijerinck in the enrichment medium from which he isolated the organism.

Morphology as for genus.

Colonies 2–3 mm after 24 hr., usually smooth, sometimes extremely mucoid and adherent to substrate. Surface spreading, indicating twitching motility, rare on usual media, common on plates with tryptone 0.5%, acetate 0.5% and yeast extract 0.5%. On sheep or ox blood agar plates 10–20% of the strains show broad hemolytic zones. Some strains produce a special, disagreeable odor.

Acid may or may not be produced. If produced it is due to a single nonspecific aldose dehydrogenase active on L-arabinose, D-xylose, D-ribose, D-glucose, D-galactose, D-mannose, lactose, cellobiose and late and irregularly on maltose and L-rhamnose. Nitrates not reduced except in extremely rare cases, but a repressed nitrate-reductase regularly present according to Jyssum and Joner (1965).

Gelatin liquefied by 10–20% of the strains. Urease present in about half of the strains. Some strains grow in KCN medium. Phenylalanine not deaminated and ornithine, lysine and arginine not decarboxylated.

All strains grow in a simple mineral medium with ammonium salts and acetate, butyrate, or pyruvate as carbon and energy source. With very few exceptions hexoses, disaccharides and alcohol sugars cannot be utilized. Citrate and many other organic compounds may or may not be utilized.

Usually resistant to 5 units of penicillin and many strains highly resistant due to production of penicillinase.

Present in soil and water and frequently isolated from healthy or diseased animals and man. Pathogenicity uncertain but probably of some significance in otherwise debilitated individuals.

The G + C content of DNA about 40–47 moles %.

Type strain: Delft 1 of Beijerinck; ATCC 23055. Morphologically the type strain is dominated by aberrant cell forms including filaments and metabolically it deviates in certain respect from the typical pattern of other acid-producing strains.

Further Comments

Non-acid-producing strains of the species are phenotypically very similar to *Moraxella osloensis*, but besides the positive oxidase test the latter is characterized by having intracellular poly-β-hydroxybutyrate inclusions.

GENERA OF UNCERTAIN AFFILIATION

Genus **Paracoccus** Davis in Davis, Doudoroff, Stanier and Mandel 969, 384

M. DOUDOROFF

Pa.ra.coc′cus. Gr. prep. *para* like, alongside of; Gr. n. *coccus* a grain, berry; M.L. masc.n. *Paracoccus* like a coccus.

Cells **spherical** or nearly spherical, occurring singly, in pairs or aggregates. Dimensions, 0.5–1.1 μm in diameter. (Short rod-shaped cells may occur in young cultures.) **Non-motile.** Accumulate granules of poly-β-hydroxybutyrate as intracellular carbon reserve, especially in nitrogen-deficient media. No resting stages are known. **Gram-negative.**

Chemoorganotrophs: Metabolism **respiratory**, never fermentative. One species is **facultatively chemolithotrophic,** able to use the oxidation of H₂ as energy source for autotrophic growth. **Molecular oxygen or nitrate** can serve as alternative electron acceptors. Nitrate is reduced to nitrous oxide and **molecular nitrogen** under anaerobic conditions. Oxidase and catalase reactions positive.

Aerobic, except in media containing nitrate.

The G + C content of the DNA, *ca.* 64–67 moles % (from buoyant density).

Type species: *Paracoccus denitrificans* (Beijerinck) Davis.

Further Comments

In many respects, the type species of the genus resembles some facultatively chemolithotrophic species belonging to the genera *Alcaligenes* and

Pseudomonas, from which it differs mainly in its coccoid shape during exponential growth and the absence of flagellation (Davis *et al.*, 1969, 1970). For instance, it shares the following properties with the peritrichously flagellated species *Alcaligenes eutrophus* Davis: Gram-negative staining, ability to grow autotrophically in atmospheres containing hydrogen and oxygen, accumulation of poly-β-hydroxybutyrate as reserve material, growth and denitrification in organic but not in inorganic media, inability to use arginine aerobically or anaerobically as carbon or energy source, and the identical range of DNA base composition (66.3–66.8 moles % G + C). Furthermore, the cells of *A. eutrophus* become almost spherical in the stationary phase of growth.

P. denitrificans also resembles in many respects some facultatively organotrophic non-motile strains of *Thiobacillus*. The claims (THE MANUAL, 7th edition, p. 43) that *P. denitrificans* may be identical with or closely related to *T. novellus* have not been substantiated (Kocur *et al.*, 1968; Taylor and Hoare, 1969). The denitrifying strain of *Thiobacillus* studied by Taylor and Hoare does not grow with hydrogen, while *P. denitrificans* cannot grow with thiosulfate as source of energy.

On many grounds, it seems likely that the genera

Paracoccus, Pseudomonas and *Alcaligenes* are very closely related to each other and that at least the facultatively organotrophic species of the genus *Thiobacillus* belong in the same generic cluster.

Key to the species of genus **Paracoccus**

1. Facultatively chemolithotrophic, not halophilic, no growth factor requirements, gelatin not hydrolyzed.

1. *Paracoccus denitrificans*

2. Not chemolithotrophic, halophilic, complex media required, gelatin hydrolyzed.

2. *Paracoccus halodenitrificans*

Description of the species of genus **Paracoccus**

1. **Paracoccus denitrificans** (Beijerinck) Davis in Davis, Stanier, Doudoroff and Mandel 1969, 384. (*Micrococcus denitrificans* Beijerinck 1910, 53.)

de.ni.tri'fi.cans. L. prep. *de* away from; L. n. *nitrum* soda; M.L. n. *nitrum* nitrate; M.L. v. *denitrifico* denitrify; M.L. part.adj. *denitrificans* denitrifying.

Cocci or very short coccobacilli during exponential growth, with the average dimensions of 1.1 by 1.3 μm; cells spherical in the stationary phase, 1 micron in diameter. Occur singly, in pairs and aggregates.

Chemoorganotrophic and facultatively chemolithotrophic, capable of using the oxidation of molecular hydrogen as a source of energy. Either oxygen or nitrate can serve as electron acceptor, nitrate being reduced to nitrous oxide and molecular nitrogen. Arginine dihydrolase reaction is negative.

Gelatin, starch and extracellular poly-β-hydroxybutyrate are not hydrolyzed.

Organic growth factors are not required for aerobic or anaerobic growth with organic substrates, or for autotrophic growth in inorganic media exposed to atmospheres containing H_2, O_2 and CO_2. Anaerobic growth in inorganic media containing nitrate and exposed only to H_2 and CO_2 is, however, abortive unless organic substances such as yeast extract are added. This appears to be due to the inhibition of chemolithotrophic metabolism by nitrite, which is the first product of denitrification (Bovell, 1967).

At least 64 different organic compounds can be used as sole carbon sources for growth by different strains of the species. These include sucrose, trehalose, maltose, malonate, mesaconate, mannitol, *meso*-inositol, 2,3-butylene glycol, *para*-hydroxybenzoate, leucine, histidine, sarcosine and creatine. Among the compounds that are not utilized are D-arabinose, *meso*-tartrate, arginine and nicotinate (Davis *et al.*, 1970). Both ammonium salts and nitrate can serve as source of nitrogen.

Aerobic except in media containing organic compounds and nitrate. Optimal temperature, *ca.* 30 C.; temperature range, 5–37 C.

Isolated from soil by enrichment for denitrifying bacteria and by enrichment for autotrophic "hydrogen bacteria."

The G + C content of the DNA is *ca.* 66.5 moles % (buoyant density).

Holotype: ATCC 17741.

2. **Paracoccus halodenitrificans** (Robinson and Gibbons) Davis in Davis *et al.* 1969, 386. (*Micrococcus halodenitrificans* Robinson and Gibbons 1952, 154; *Micrococcus denitrificans* var. *halodenitrificans* Kocur and Martinec 1962, 89.)

ha.lo.de.ni.tri'fi.cans. Gr. n. *hals, halis* salt; M.L. v. *denitrifico* denitrify; M.L. part.adj. *halodenitrificans* salt (requiring) denitrifying.

Cocci, *ca.* 0.5 μm in diameter, occurring singly or in pairs. Non-motile. Accumulate granules of poly-β-hydroxybutyrate, especially in nitrogen-deficient media. Cells may stick together due to a viscous surface slime, which is either excreted or adsorbed by the cells and which can be digested with deoxyribonuclease (Smithies and Gibbons, 1955). Gram-negative.

No growth in media containing less than 0.4 M NaCl. Good growth and survival of cultures requires media containing 0.8 M–1.6 M NaCl, and slow growth occurs in media with up to 4 M NaCl.

Chemoorganotrophic. Metabolism is respiratory, not fermentative. Either oxygen or nitrate can serve as electron acceptor, nitrate being reduced to nitrous oxide and molecular nitrogen.

Nutritional requirements appear to be complex and have not been elucidated. No growth occurs in mineral media with simple carbon sources, but peptone media support good growth. Acetate, pyruvate and glycerol are metabolized when added to complex media. No acid is produced oxidatively in media containing glucose.

Aerobic except in media containing nitrate. Optimal temperature, *ca.* 20 C; temperature range, 0–32 C.

Isolated from meat-curing brines. Presumably widely distributed in natural and artificial brines.

The G + C content of the DNA is 64–66 moles % (from buoyant density).

Holotype: ATCC 13511.

Genus **Lampropedia** *Schroeter 1886, 151*

HARRY W. SEELEY, JR.

Lam.pro.ped'i.a. Gr. adj. *lamprus* bright, radiant; Gr. n. *pedia* a plain, flat country; M. L. fem.n. *Lampropedia* a shining flat sheet (of cells).

Cells rounded or almost cubical when packed together, 1.0–1.5 by 1.0–2.5 μm (Plate 10.3, Fig. 1), occurring in **pairs, tetrads and regular, squared tablets of cells.** Division takes place alternately (and synchronously) in two planes. The cells forming a tablet are enclosed within a **complex, structured envelope** (Plate 10.3, Figs. 2–4); each cell is enclosed by a conventional wall of **Gram-negative** type. **Not motile** and non-flagellated but exhibiting a **flickering movement of groups of cells** in rapidly growing cultures. No resting stages known.

Growth forms a thin, dry, wrinkled pellicle on the surface of both solid and liquid media.

Chemoorganotroph: Respiratory energyyielding pathway. Energy sources limited to Krebs cycle intermediates; carbohydrates, alcohols, glucosides, fatty acids and many related compounds not utilized. Limited exoenzyme production and limited amino acid utilization. **Carotenoid or photosynthetic pigments not formed.** Prominent **inclusions of poly-β-hydroxybutyrate. Do not contain gas vacuoles or sulfur inclusions.** Vitamins may be required.

Obligate aerobe. Grows over temperature range 10–35 C and at pH 6–8.5.

Ecological niche unknown but rich, organic environments likely.

The G + C content is in region of 61 moles % (buoyant density).

Type species: *Lampropedia hyalina* Schroeter 1886, 151.

Further Comments

This genus is sufficiently distinctive to be im-mediately recognizable under the microscope (Pringsheim, 1955), but it is assumed that the morphological characters are stable and that the few extant cultures of *L. hyalina* are representative. Doubts may be based on two observations: (1) Pringsheim (1955) noted that motility, observed in the original material, was lost on culturing; no other author has mentioned true motility (with the possible exception of the synonomous *Pedioplana* Wolff 1907, 10). (2) A strain derived from Pringsheim's original isolation was found to have lost the capability of forming tablets of cells and, even though some component of its structured surface was missing, it retained enough characters to be recognizable (Murray, 1963). It is not known if these or more extreme variants occur in nature; if so, it would be difficult to recognize the genus.

Despite morphological similarities, i.e. growth in perpendicular planes forming tablets of (about) 4–256 individual cells, the physiological characteristics of *Lampropedia* (strict aerobes, absence of photosynthetic pigments, limited largely to Krebs cycle intermediates as energy sources) clearly set it aside from the anaerobic, photosynthetic, carotenoid-containing *Thiopedia* and *Thiocapsa*.

Merismopedia bears a grossly similar superficial resemblance to *Lampropedia* (Murray, 1963) but individual cells of this alga are several times larger, and as well, contain chromatophores and gas vacuoles.

The circumscription of the genus is changed from that adopted in the 6th edition of THE MANUAL, the last in which the genus was recognized, to avoid confusion with the sulfur-purple organisms, *Thiopedia*.

Description of the species of genus **Lampropedia**

1. **Lampropedia hyalina** (Ehrenberg) Schroeter 1886, 151. (*Gonium hyalinum* Ehrenberg 1832, 63; *Merismopedia hyalina* (Ehrenberg) Kützing 1849, 471; *Sarcina hyalina* (Ehrenberg) Winter 1884, 51; *Pediococcus hyalinus* (Ehrenberg) Trevisan 1889, 29; *Micrococcus hyalinus* (Ehrenberg) Migula 1900, 195.)

hy.a.li'na. Gr. adj. *hyalinus* glassy, shiny; M.L. fem.adj. *hyalina* hyaline.

Cells enclosed in a complex structured envelope (Chapman and Salton, 1962; Chapman *et al.*, 1963; Murray, 1963; Kuhn *et al.*, 1965; Pangborn and Starr, 1966); unique motion in growing culture characterized by detachment of small groups of cells from larger groups and reattachment to other large groups (Puttlitz and Seeley, 1968).

(Nutrition, physiology and some cultural characteristics taken from Puttlitz and Seeley, 1968.)

Utilizes pyruvate, lactate, butyrate, fumarate, malate, succinate (and acetate in the presence of catalytic levels of pyruvate) as sole energy sources.

Non-hemolytic. Does not hydrolyze casein, gelatin, fats, starch, DNA, sodium hippurate, urea. Does not produce indole or acetylmethylcarbinol.

Catalase positive. Absorption peaks detected in reduced intact cell preparations at 426, 530, 560 and 590 nm.

Utilizes NH$_4$Cl, alanine, arginine and tyrosine as sole nitrogen sources. Requires biotin and thiamine for growth.

Strict aerobe. Temperature optimum for growth 30 C, limits 10–35 C. Optimum pH 7.0, range 6.0–8.6. Tolerates 1.0 but not 1.5% NaCl and 2.0 but not 4.0% sucrose. Does not tolerate 0.5% bile.

The microorganism originally designated as *Lampropedia hyalina* (Schroeter, 1886) was found in swamp water and decomposing materials from sugar refineries. Pringsheim's (1955) isolation was from a barnyard puddle. Cell groupings characteristic of *Lampropedia* (but species not designated) were reported in the rumen of sheep and cattle (Hungate, 1960; Eadie, 1962) and in gut flora of herbivorous reptiles and nematodes (Schad *et al.*, 1964). The sensitivity of *L. hyalina* to 0.5% bile and its failure to grow below pH 6 and at temperatures above 35 C suggest that the organism was a transient in some samples noted or that species with other characteristics may exist (see *species incertae sedis* following).

The G + C content of the DNA of those strains examined ranges from 60.7–61.2 moles % buoyant density (Mandel, 1970).

Species incertae sedis

a. *Lampropedia reitenbachii* (Caspary) de Toni and Trevisan 1889, 1048.

b. *Lampropedia violacea* (Brébisson) de Toni and Trevisan 1889, 1048.

c. *Lampropedia ochracea* (Mettenheimer) de Toni and Trevisan 1889, 1049.

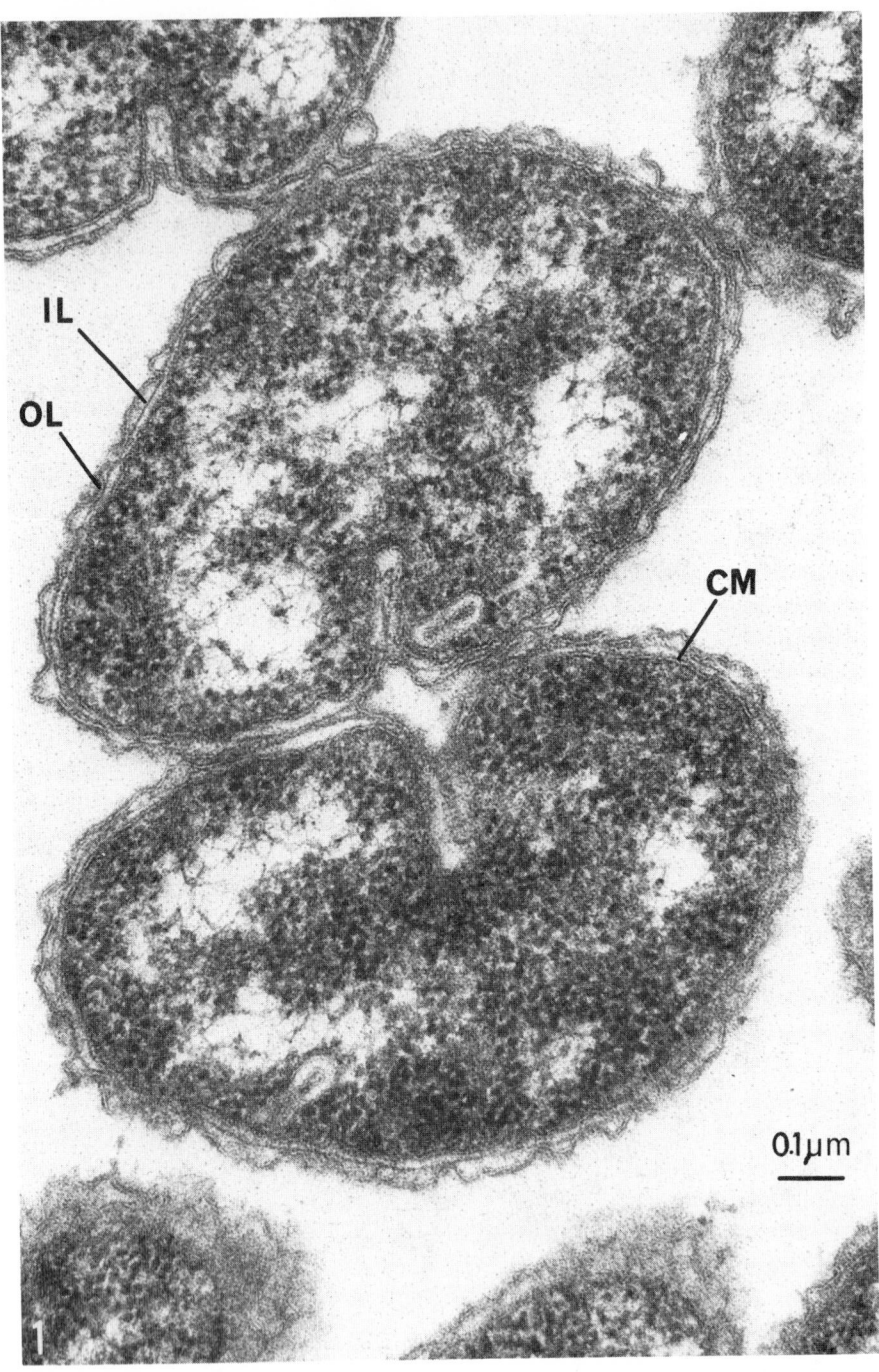

Plate 10.1. *Neisseria*

Fig. 1. Section of dividing cells of *Neisseria gonorrhoeae* (Statens Seruminstitut, Copenhagen, 87115/1961). Note outer layer (*OL*) and intermediate layer (*IL*) of cell wall and cytoplasmic membrane (*CM*). Stained with magnesium uranyl acetate and lead citrate. × 90,000. From unpublished work of A. Reyn and A. Birch-Andersen.

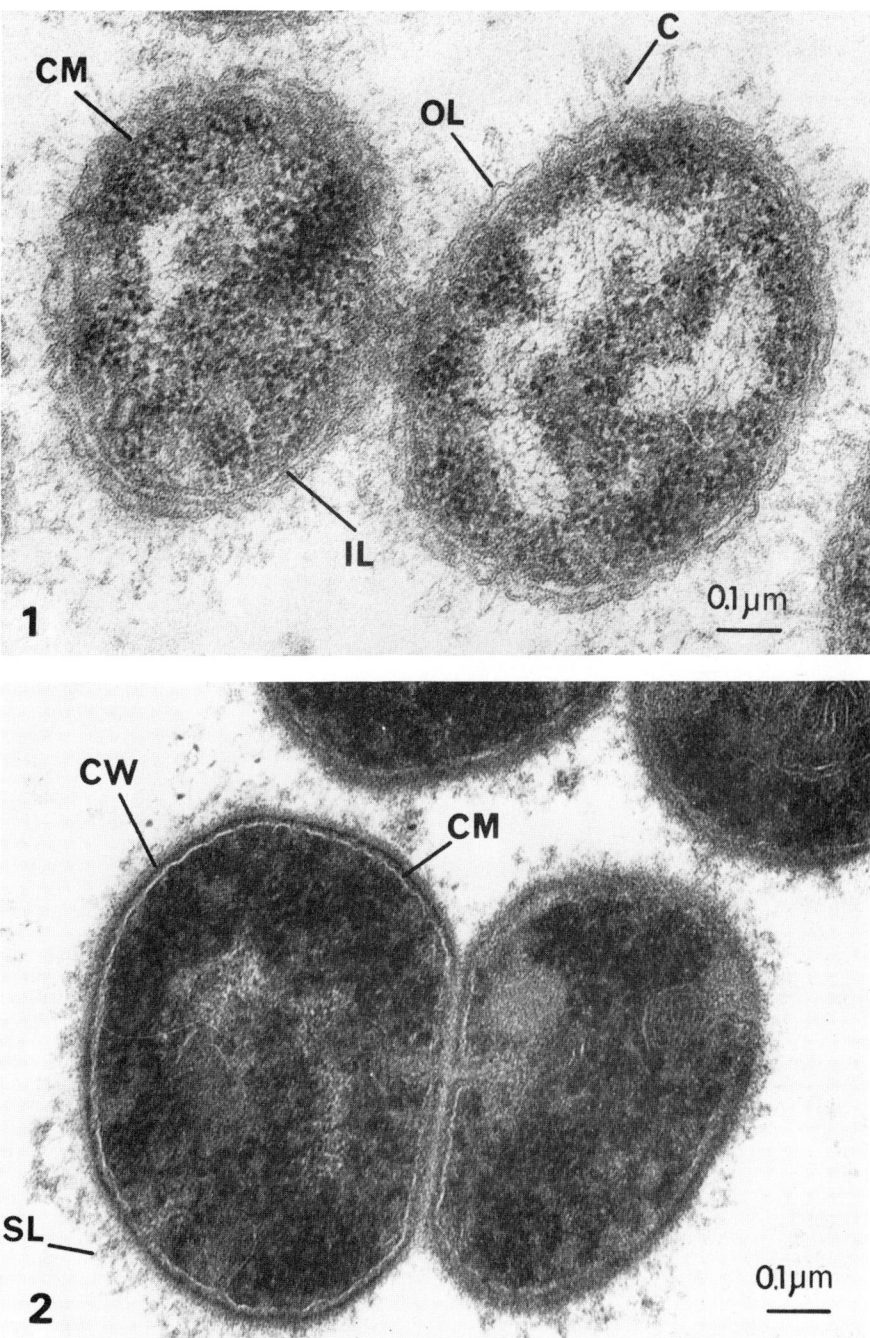

Plate 10.2. *Branhamella* and *Gemella*

Fig. 1. Section of dividing cell of *Branhamella catarrhalis* (Statens Seruminstitut, Copenhagen, 87895/1959). Note outer layer (*OL*) and intermediate layer (*IL*) of cell wall, cytoplasmic membrane (*CM*) and capsule (*C*). Stained with magnesium uranyl acetate and lead citrate. × 90,000. From unpublished work of A. Reyn and A. Birch-Andersen.

Fig. 2. Section of dividing cells of *Gemella haemolysans* (ATCC 10379). Note slime layer (*SL*), cell wall (*CW*) and cytoplasmic membrane (*CM*). Stained with magnesium uranyl acetate and lead citrate. × 90,000. From unpublished work of A. Reyn and A. Birch-Andersen. (See genus *Gemella*, Part 14.).

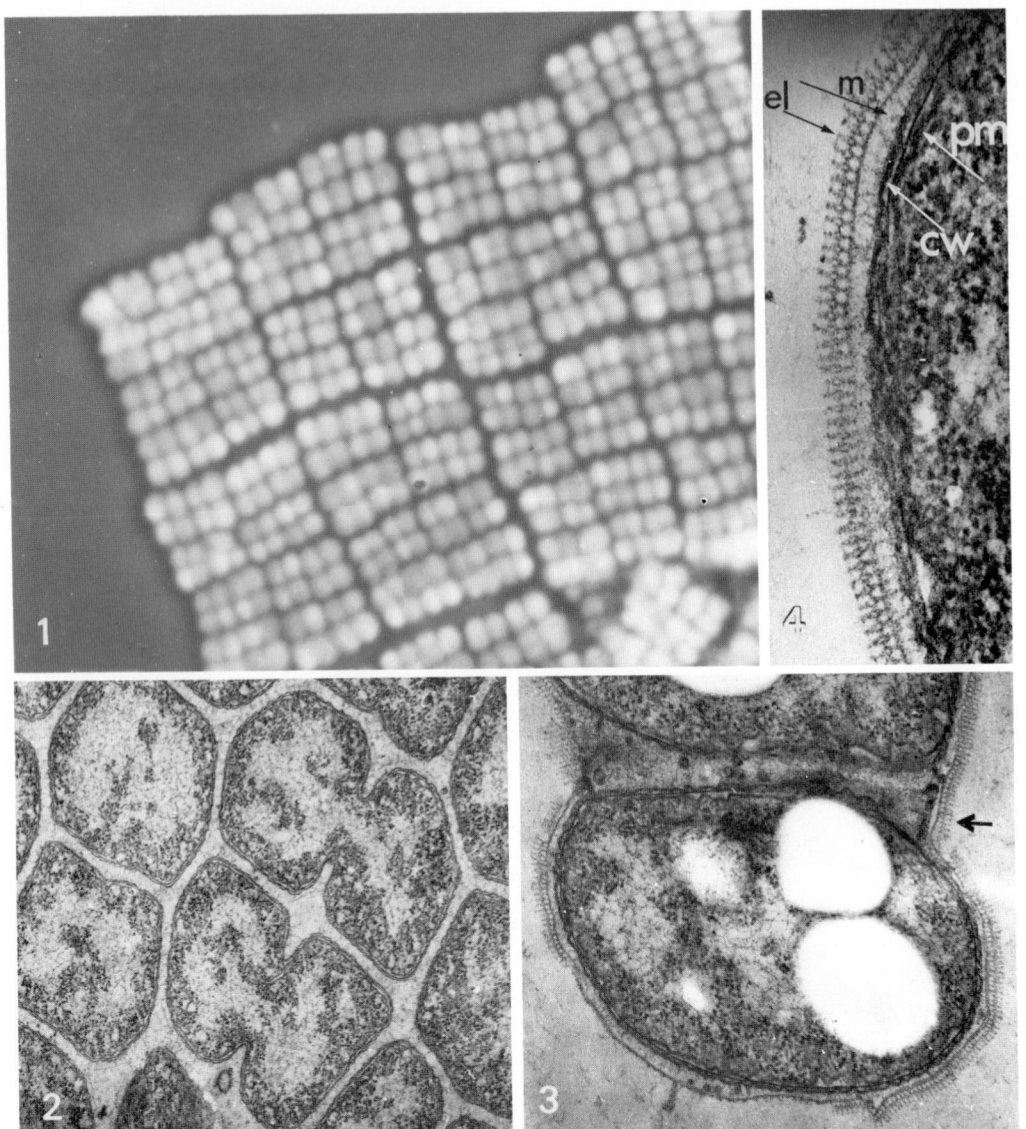

Plate 10.3. *Lampropedia hyalina*

Fig. 1. Light micrograph of a nigrosin preparation showing a corner of a sheet of actively growing tablets of cells. Adjacent tablets are almost synchronized in division. × 2880.

Fig. 2. Electron micrograph of a section in the plane of a tablet to show division and the relatively large volume of nucleoplasm. A matrix substance separates the cell walls (of usual Gram-negative character, see Fig. 4) of adjacent cells. Divisions always show a "constrictive" form. × 25,200.

Fig. 3. Section transverse to the edge of a tablet of cells showing a bipartite external structured envelope. The inner layer is obvious in an area exposed by stripping. The envelope encloses the tablet and bridges over the matrix separating adjacent cells (*arrow*). The low density vesicles represent poly-β-hydroxybutyrate granules. × 42,300.

Fig. 4. A high magnification of a section showing the complex structured envelope (*el*), the matrix (*m*), the cell wall (*cw*) and plasma membrane (*pm*). × 99,000. (Micrographs by R. G. E. Murray and S. Lanys.)

PART 11

GRAM-NEGATIVE ANAEROBIC COCCI

FAMILY I. **VEILLONELLACEAE** ROGOSA 1971, 232

MORRISON ROGOSA

Veil.lo.nel.la′ce.ae. M.L. fem.n. *Veillonella* type genus of the family; *-aceae* ending to denote a family; M.L. fem.pl.n. *Veillonellaceae* the *Veillonella* family.

Cocci, varying in diameter from small (*ca.* 0.3–0.5 μm) to large (*ca.* 2.5 μm); occur characteristically in pairs; single cells, masses or chains may also occur although the chains may show gaps, illustrating the basic diplococcal arrangement. Adjacent sides of cell pairs may be flattened. No endospores. Non-motile; no flagella. Gram-negative but tend to resist decolorization.

Chemoorganotrophic. Possess complex nutritional requirements. Gas is produced, frequently abundantly. Carbohydrates may or may not be fermented. Lactic acid may not be produced and if present is not a major product; lactate is fermented by some genera with the production of CO_2, H_2 and various lower volatile fatty acids containing 2–6 C atoms.

Anaerobic. Cytochrome oxidase negative. Catalase negative, but some strains decompose peroxide by a pseudocatalase (non-heme-containing).

Parasites of homothermic animals such as man, ruminants, rodents and pigs, particularly found in the alimentary tract.

The type genus: *Veillonella* Prévot 1933, 118.

The key differentiating characteristics of the genera are shown in Table 11.1.

TABLE 11.1

Characteristics of the genera of family **Veillonellaceae**[a]

	Cell diameter	G + C	Carbohydrates femented	Lactate fermented	Amino acids main energy source	Pyruvate utilized	Succinate decarboxylated	Products in growth media						
								Gases		Volatile fatty acid				
								CO_2	H_2	C_2	C_3	C_4	C_5	C_6
	μm	*moles %*												
Veillonella	0.3–0.5	40.3–44.4	−	+	−	+	+	+	+	+	+	−	−	−
Acidamino-coccus	0.6–1.0	56.6	−[b]	∓	+	−	−	+	−	+	−	+	−	−
Megasphaera	1.7–1.9 2.4–2.6	53.1–54.1	+	+	−	+	−	+	w	+	+	+	+	+

[a] + = positive reaction; − = negative reaction; w = slight.
[b] Generally negative, slight or variable.

445

Genus I. **Veillonella** *Prévot 1933, 118, emend. mut. char. Rogosa 1965, 706*

MORRISON ROGOSA

Veil.lo.nel′la. M.L. dim.ending -*ella*; M.L. fem.n. *Veillonella* named for A. Veillon, the French bacteriologist who isolated the type species.

Small cocci, 0.3–0.5 μm in diameter. By light microscopy appear as spherical **diplococci, masses** and short chains. By electron microscopy, logarithmic-phase single cells appear spherical, whereas diplococci have a flattening at the cell junctions. Non-motile. Non-sporulating. **Gram-negative.**

Lactate agar media containing 7.5 μg/ml of vancomycin favor isolation; colonies in poured plates 1–3 mm in their greatest dimension; smooth; entire; lens, diamond, or heart shaped; opaque; grayish white and butyrous.

Carbohydrates and **polyols not** fermented. D-Ribose not fermented but is incorporated into nucleic acids. **Glucokinase, fructokinase** and **glucose permease not** detectable.

Cultures produce **acetate, propionate, CO_2** and **H_2 from lactate.** Hypoxanthine may be fermented.

Resting cells respire on lactate; attack oxaloacetate aerobically producing CO_2 and H_2; and anaerobically metabolize pyruvate, oxaloacetate, malate, fumarate and succinate. **CO_2** and **propionate** produced **from succinate.** Formate, citrate, isocitrate and malonate not attacked.

H_2S produced from reduced glutathione, cysteine, cystine, thiosulfate, thiocyanate and thioglycolate. Gelatin not liquefied. Non-hemolytic. Indole negative. Nitrate reduced.

Cytochrome oxidase and **benzidine negative;** but some species produce an atypical catalase lacking porphyrin.

Chemoorganotrophic. Complex nutritional requirements. CO_2 required. No growth in amino acid media containing necessary vitamins and other co-factors unless supplemented with pyruvate, which supports growth best, or with lactate, malate, fumarate and oxaloacetate; no growth with added succinate, carbohydrates, polyols, phosphorylated hexoses or trioses. Riboflavin and folic acid not required; niacin and calcium pantothenate often stimulatory but dispensable; biotin and *p*-aminobenzoic acid frequently stimulatory and sometimes indispensable; pyridoxal and thiamine indispensable; some organisms require putrescine or cadaverine.

Growth in 1% Tween 80, variable or inhibited by 10% bile, 0.0002% crystal violet or brilliant green and 1% NaCl; no growth in 4% NaCl, 0.25% phenethyl alcohol or 0.001% potassium tellurite. A variety of **inorganic compounds, dyes** and **indicators** are **reduced;** 2,3,5-triphenyl tetrazolium chloride not reduced.

Sensitive to penicillin G (0.4–3.1 units/ml); chloramphenicol, chlortetracycline, oxytetracycline, polymyxin B (all <1.0 μg/ml), and erythromycin (1.3–5.0 μg/ml); resistant to streptomycin (>25 μg/ml), neomycin (>50 μg/ml) and vancomycin (500 μg/ml).

Anaerobic.

Growth good from 30–37 C at pH 6.5–8.0, poor at 40 C, slow at 24 C, and negative at 18 and 45 C; cells killed at 60 C for 30 min. Cells suspended in skim milk can be successfully lyophilized.

Serologically specific **endotoxins** (lipopolysaccharides) induce pyrogenicity and the Schwartzman reaction in rabbits.

Parasitic in the mouth and in the intestinal and respiratory tracts of man and other animals.

The G + C content of DNA ranges from 40–44 moles % (buoyant density). Values of 34–36.5 moles % (chromatographic) cited by Rosypal and Rosypalova (1966) for *V. parvula* are incorrect.

Type species: *Veillonella parvula* (Veillon and Zuber) Prévot 1933, 119.

Description of the species of genus **Veillonella**

1. **Veillonella parvula** (Veillon and Zuber) Prévot 1933, 119. (*Staphylococcus parvulus* Veillon and Zuber 1898, 542.)

par′vu.la. L.fem.dim.adj. *parvula* very small.

Prévot (1933) and THE MANUAL (7th ed.) describe the species as fermenting glucose and producing indole. Study of the type strain demonstrated that glucose is not fermented and indole is not produced (Rogosa, 1964, 1965).

Catalase negative. Most strains do not require putrescine or cadaverine for growth.

The G + C content of DNA is 41.3 moles %. Type strain: Prévot Te 3; ATCC 10790.

1a. *Veillonella parvula* subsp. *parvula* Rogosa 1965, 707. Description as for the species. Antigenic group VI (Rogosa, 1965). All of human origin, from mouth, respiratory or intestinal tracts.

1b. *Veillonella parvula* subsp. *rodentium* Rogosa 1965, 707. One-third of strains require putrescine and cadaverine. Antigenic group II.

All of buccal or intestinal origin, from hamster, rat and rabbit.

The G + C content of DNA is 44.4 moles %.
Type strain: HV19; ATCC 17743.

1c. *Veillonella parvula* subsp. *atypica* Rogosa 1965, 707. Does not require putrescine or cadaverine for growth. Antigenic group V.

Isolated from buccal cavity of man and the rat.
Type strain: KON (Langford *et al.*, 1950); ATCC 17744.

2. **Veillonella alcalescens** Prévot 1933, 127. (*Micrococcus gazogenes alcalescens anaerobius* Lewkowicz 1901, 633; *Micrococcus gazogenes* Hall and Howitt 1925, 113; Not *Micrococcus gazogenes* Choukévitch 1911, 356; *Veillonella gazogenes* (Hall and Howitt) Murray 1939, 287; *Micrococcus lactilyticus* Foubert and Douglas 1948, 30.)

al.ca.les'cens. M.L. v. *alcalesco* make alkaline; M.L. part.adj. *alcalescens* alkaline-making.

Decomposes hydrogen peroxide; requires putrescine or cadaverine for growth; growth often rough, beginning at bottom of tube. Isolated primarily from the human mouth, occasionally from the mouths of rat and rabbit. Has been isolated from blood of man after surgical procedures.

The G + C content of DNA is 42.3 moles %.

Type strain: Prévot 259; ATCC 17745.

2a. *Veillonella alcalescens* subsp. *alcalescens* Rogosa 1965, 707. Description as for the species. Antigenic group IV.

2b. *Veillonella alcalescens* subsp. *ratti* Rogosa 1965, 708. Does not require putrescine or cadaverine for growth. Antigenic group III. Isolated from the mouth and intestinal contents of rats.

The G + C content of DNA is 44.4 moles %.

Type strain: RV-12X; ATCC 17746.

2c. *Veillonella alcalescens* subsp. *criceti* Rogosa 1965, 708. Two-thirds of strains require putrescine or cadaverine for growth. Antigenic group I. Isolated from the mouth of the hamster.

The G + C content of DNA is 40.3 moles %.

Type strain: HV-1; ATCC 17747.

2d. *Veillonella alcalescens* subsp. *dispar* Rogosa 1965, 708. Requires putrescine or cadaverine for growth. Antigenic group VII. Isolated from mouth and respiratory tract of man.

The G + C content of DNA is 42.3 moles %.
Type strain: ERN (Langford *et al.*, 1950); ATCC 17748.

Species incertae sedis

In the following only one strain was isolated or inadequately described; none is extant or is recognizable.

a. *Veillonella parvula* subsp *minima* Prévot 1933, 125. (*Staphylococcus minimus* Gioelli 1907, 164.)

b. *Veillonella parvula* subsp. *branhamii* Prévot 1933, 126. (Anaerobic micrococcus Branham 1927, 203; *Micrococcus branhamii* Bergey *et al.* 1930, 92.)

c. *Veillonella parvula* subsp. *thomsonii* Prévot 1933, 126. (Anaerobic diplococcus Thomson, 1923, 227.)

d. *Veillonella alcalescens* subsp. *gingivalis* Prévot 1933, 133. (Kleiner Micrococcus Ozaki 1912, 83; *Micrococcus gingivalis* Bergey *et al.* 1923, 69.)

e. *Veillonella alcalescens* subsp. *minutissima* Prévot 1933, 134. (*Micrococcus minutissimus* Oliver and Wherry 1921, 342.)

f. *Veillonella alcalescens* subsp. *syzygios* Prévot 1933, 134. (*Syzygiococcus scarlatinae* Herzberg 1928, 575; *Micrococcus syzygios scarlatinae* Herzberg 1929, 383; *Micrococcus syzygios* Bergey *et al.* 1930, 92.)

The following species described in the 7th edition of THE MANUAL also differ widely from the present definition of the genus.

g. *Veillonella discoides* (Prévot) Pelczar 1957, 488. (*Neisseria discoides* Prévot 1933, 106.)

h. *Veillonella reniformis* (Cottet) Pelczar 1957, 489. (*Diplococcus reniformis* Cottet 1900, 42, *Micrococcus reniformis* (Cottet) Oliver and Wherry 1921, 341; *Neisseria reniformis* (Cottet) Prévot 1933, 102.)

i. *Veillonella orbiculus* (Tissier) Pelczar 1957, 489. (*Diplococcus orbiculus* Tissier 1908, 204; *Neisseria orbiculata* (Tissier) Prevot 1933, 109.)

j. *Veillonella vulvo-vaginitidis* (Reynes) Pelczar 1957, 489. (*Neisseria vulvo-vaginitis* Reynes 1947, 601.)

Genus II. **Acidaminococcus** Rogosa 1969, 765

MORRISON ROGOSA

A.cid.a.min.o.coc'cus. M.L. n. *acidum* acid; M.L. adj. *amino* amino; M.L. masc.n. *Acidaminococcus* the amino acid coccus.

Cocci, 0.6–1.0 μm in diameter, often as oval or kidney-shaped **diplococci. Gram-negative.** Lipopolysaccharide (endotoxin) outer cell wall layer demonstrable in thin sections by electron microscopy, and by the Schwartzman reaction in rabbits. Flagella, motility and spores not demonstrable.

Weak saccharoclastic activity; glucose is not

fermented by about 60% of strains and only weakly catabolized by the remainder. **Lactate, fumarate, malate, succinate, citrate and pyruvate are not used** as energy sources; pyruvate suppresses growth completely.

Amino acids, of which glutamic acid is the most important, **can serve as sole energy source for growth;** in such media acetic and butyric acids accumulate in a molar ratio of 2:1 and CO_2 is also formed; hydrogen and propionic acid are not detectable.

Nutritional requirements multiple. Tryptophan, glutamic acid, valine and arginine are required; cysteine and histidine are required by 93% of strains, tyrosine by 79%, phenylalanine and serine by 50%. Glycine is sometimes stimulatory. Alanine, leucine, isoleucine, proline, threonine, methionine, lysine and aspartic acid are not required for growth. In amino acid media, vitamin B_{12}, pyridoxal, pantothenate and biotin are indispensable for growth; p-aminobenzoic acid is essential or highly stimulatory; exogenous putrescine, folic acid, folinic acid, thiamine, niacin and riboflavin are not required. No growth in lactate or pyruvate media which support growth of *Veillonella*.

Ammonia produced. Gelatin generally not liquefied although slow and partial liquefaction may sometimes occur. H_2S and indole not produced. Cytochrome oxidase, catalase, benzidine and nitrate reduction tests negative. Sulfonthalein indicators not reduced.

Anaerobic. No growth on surface of agar media incubated in air.

Good growth at 30–37 C; poor or absent at 25 and 45 C. Cells do not survive 60 C for 30 min. Growth occurs at initial pH values between 6.2–7.5, although best growth at neutral reaction; final pH values in media initially at pH 7.5 range from about 6.1–6.7.

No serological cross-reactions between strains of *Acidaminococcus* and either *Veillonella* serotypes or *Peptococcus aerogenes*.

Resistant to vancomycin (7.5 µg/ml).

The G + C content of the DNA of 15 strains was 56 ± 0.9 moles % (buoyant density).

Type species: *Acidaminococcus fermentans* Rogosa 1969, 765.

Further Comments

Fuller (1966) isolated 49 strains from the intestinal tract of pigs and suggested a possible relationship to *Veillonella reniformis*. However, the present description of the genus *Veillonella* makes this highly improbable. Rogosa suggested a new genus *Acidaminococcus* 1969, 765. Although *Acidaminococcus* and *Peptococcus aerogenes* both ferment glutamate to very similar products, the latter is differentiated nutritionally, serologically, by Gram reaction and a widely different G + C content of the DNA.

Description of the species of genus **Acidaminococcus**

1. **Acidaminococcus fermentans** Rogosa 1969, 765.

fer.men'tans. M.L. part.adj. *fermentans* fermenting.

Morphology as for genus.

Surface colonies on complex media incubated in 95% H_2 + 5% CO_2 0.1–0.2 mm in diameter in 48 hr; round, entire, slightly raised, whitish gray or nearly transparent. In peptone + yeast extract broth growth starts at the bottom of the tube and the broth becomes evenly turbid. Supplements of 0.4–0.5% sodium glutamate enhance growth and gas (CO_2) production. Derivative products from glucose autoclaved in amino acid media are necessary or highly stimulatory for growth. Cells suspended in skim milk may be lyophilized successfully.

Polyols including adonitol, dulcitol, erythritol, glycerol, inositol, mannitol and sorbitol not attacked; amygdalin, arabinose, fructose, galactose, inulin, maltose, mannose, melezitose, α-methyl-D-glucoside, α-methyl-D-mannoside, raffinose, salicin, sorbose, sucrose, trehalose, xylose, erythrose and esculin not attacked; ambiguous, extremely weak, or negative reactions with cellobiose, fucose, lactose, melibiose, rhamnose and ribose.

Isolated from intestinal tract of the pig and man; probably not pathogenic because all isolations were made from normal animals; may be widespread in the intestinal tracts of homothermic animals.

Type strain: VR4; ATCC 25085; this strain does not ferment glucose. Reference strains fermenting glucose weakly: VR7; ATCC 25086 and VR11; ATCC 25087.

Genus III. **Megasphaera** Rogosa 1971, 187

MORRISON ROGOSA

Me.ga.sphae'ra. Gr. adj. *megas* big; Gr. n. *sphaera* a sphere; M.L. fem.n. *Megasphaera* big sphere.

Relatively large cocci, 2 μm or more in diameter, in pairs; occasionally the diplococci are arranged in chains. Non-motile; non-sporing. Gram-negative.

Chemoheterotrophic. Gas produced. Lactate and glucose are fermented with the production of lower fatty acids, CO_2 and some H_2. Succinate, fumarate and malate are not attacked.

Complex nutritional requirements.

Obligately anaerobic.

Types species: *Megasphaera elsdenii* (Gutierrez *et al.*) Rogosa 1971, 189.

Further Comments

Differs from *Veillonella*, which produces propionate from succinate, by producing propionate by an acrylic pathway as in *Clostridium propionicum*.

Description of the species of genus **Megasphaera**

1. **Megasphaera elsdenii** (Gutierrez *et al.*) Rogosa 1971, 189. (Organism LC Elsden and Lewis 1953, 183; rumen organism LC Elsden *et al.* 1956, 686; *Peptostreptococcus elsdenii* Gutierrez, Davis, Lindahl and Warwick 1959, 20.)

els.den'i.i. M.L. gen.n. *elsdenii* of Elsden; named after S. R. Elsden who first isolated the organism.

In wet mounts, cells spherical, 2.4–2.6 μm; in stained or fixed preparations adjacent sides of diplococci flattened and diameter ranges from 1.2–1.9 μm. Occasionally 8–10 diplococci are arranged in a chain. Thin smears are Gram-negative even in 4-hr cultures.

Surface colonies after 4 days are 0.5–2.0 mm in diameter, slightly raised, circular and entire; surface glistening to slightly rough and adherent to butyrous. Deep colonies are thin and disc-shaped, up to 4 mm in diameter and greenish yellow or honey colored.

Grows well in a 0.4% yeast extract medium containing 0.05% KH_2PO_4, 0.05% NH_4Cl, 0.03% $MgCl_2·6H_2O$, 0.03% thioglycolic acid, 2% soluble starch and 1.3% DL-sodium lactate at pH 7.4. Starch in the medium and 5% or more CO_2 facilitate isolation.

In a medium containing 0.4% yeast extract, 0.03% thioglycolic acid and 1% substrate (pH 7.4) and in an atmosphere of 95% H_2 + 5% CO_2 at 35–38 C, there is good growth and gas production with lactate, glucose and fructose; variable growth and fermentation with glycerol, maltose, mannitol, sorbitol and sucrose; no growth with arabinose, cellobiose, dextrin, esculin, galactose, inulin, lactose, mannose, raffinose, rhamnose, salicin, starch, trehalose or xylose. Final pH on glucose or fructose is 4.0–5.0; on lactate, 7.8–8.0. Products from lactate are acetate, propionate, C_4 straight and branched acids, valerate, little or no caproic acid, a large quantity of CO_2, and small amounts of H_2. Products from glucose are different from those from lactate in that some formate is produced, less acetate, propionate, butyrate and valerate are formed, and caproate (about 60% or more) is the most copious product.

H_2S produced. Gelatin not liquefied. Catalase and indole not produced. Nitrate not reduced.

Growth occurs from 25–40 C but generally not at 45 C; some strains grow at 45 but not at 50 C. Cultures stored at 4 C must be transferred at least fortnightly; may not survive freeze-drying procedures; survives in liquid nitrogen.

Found in rumen of cattle and sheep and in cecum of pigs fed significant amounts of starch.

The G + C content of the DNA of type and two similar strains is 53.6 ± 0.5 moles % (buoyant density).

Type strain: LC1; NCIB 8927; ATCC 25940 (Elsden *et al.*, 1956) from rumen of a sheep.

Further Comments

Strains from the cecum of pig appear to be different from rumen strains in having smaller cells, in fermenting raffinose and variably fermenting xylose. The G + C content of pig strains has not been determined.

PART 12

GRAM-NEGATIVE CHEMOLITHOTROPHIC BACTERIA

Key to the gram-negative chemolithotrophic bacteria

I. Organisms oxidizing ammonia or nitrite.
                         Family *Nitrobacteraceae*

II. Organisms metabolizing sulfur and sulfur compounds.
                                         p. 456

III. Organisms depositing iron and/or manganese oxides.
                         Family *Siderocapsaceae*, p. 464

FAMILY I. **NITROBACTERACEAE** BUCHANAN 1917, 349, *emend. mut. char.*
STARKEY 1948, 69; WATSON 1971, 262

Ni.tro.bac.te.ra′ce.ae. M.L. n. *Nitrobacter* type genus of the family; *-aceae* ending to denote a family; M.L. pl.n. *Nitrobacteraceae* the *Nitrobacter* family.

Rod-shaped, ellipsoidal, spherical, spirillar and lobular cells without endospores. Flagella subpolar or peritrichous and often absent. Gram-negative. Cells derive energy from the oxidation of ammonia or nitrite and satisfy their carbon needs by the fixation of CO_2. Only a few strains of one species, *Nitrobacter winogradskyi*, shown to be facultatively chemoorganotrophic. Not parasitic. Commonly found in soil, fresh water and sea water. All of the organisms placed in this family are obligate aerobes and none requires organic growth factors. Cells are rich in cytochromes but no other pigments have been demonstrated.

The type genus of the family is *Nitrobacter* Winogradsky 1892, 127 (*gen. cons.* Opin. 23. Jud. Comm. 1958, 169).

Further Comments

The organisms placed in this family are commonly referred to as nitrifying bacteria and represent a physiological community of morphologically different organisms, both terrestrial and aquatic.

Two physiological groups exist; the first group oxidizes ammonia while the second group oxidizes nitrite. All of these organisms are chemolithotrophs fulfilling their energy needs by the oxidation of ammonia or nitrite and their carbon needs by the fixation of carbon dioxide. With the exception of *Nitrobacter winogradskyi* all of the nitrifying bacteria are obligate chemolithotrophs.

Most of our previous knowledge concerning the morphology and taxonomy of the organisms in this family stemmed from the work of Helene and Sergei Winogradsky. The classification in the 7th edition of THE MANUAL was based primarily on their studies and the nitrifying bacteria were placed in 14 species in seven genera: *Nitrobacter, Nitrosomonas, Nitrosococcus, Nitrosospira, Nitrosocystis, Nitrosogloea* and *Nitrocystis*. Recently, Watson (1971), in revising the family, accepted the first four genera, combined *Nitrocystis* with *Nitrobacter*, placed *Nitrosocystis* and *Nitrosogloea* as *genera incertae sedis* and added three new genera: *Nitrospina, Nitrococcus* and *Nitrosolobus*.

In most cases multiplication is by binary fission but *Nitrobacter winogradskyi* has a polar type of cell growth and multiplies by budding. Cytomembranes are common in several morphological types of nitrifying bacteria. In the larger spherical- and lobular-shaped cells the cytomembranes intrude into the inner region of the cells but in the short rod-shaped cells the cytomembranes are restricted to the peripheral regions. No cytomembranes are evident in the elongated nitrifying bacteria.

The present classification is based entirely, with the exception of *Nitrosococcus nitrosus*, on the cellular morphology of strains grown in pure culture.

Key to the genera of family Nitrobacteraceae

I. Nitrite oxidized to nitrate.
  A. Cells rod shaped.
    1. Short rods, often wedge or pear shaped, with a polar cap of cytomembranes.
           Genus I. *Nitrobacter*
    2. Long, slender rods with no extensive cytomembrane system.
           Genus II. *Nitrospina*
  B. Cells spherical (diameter 1.54 μm or more) with cytomembranes forming a branched, tubular network in the cytoplasm.
           Genus III. *Nitrococcus*
II. Ammonia oxidized to nitrite.
  A. Cells are straight rods with *peripheral* membranes occurring as flattened lamellae.
           Genus IV. *Nitrosomonas*
  B. Cells are not straight rods and do not contain flattened lamellae in the peripheral regions.
    1. Cells are spirals with no evident cytomembranes.
           Genus V. *Nitrosospira*
    2. Cells are not spirals but do contain cytomembranes.
      a. Cells spherical (diameter 1.5 μm or more) with cytomembranes forming flattened lamellae in center of cells.
           Genus VI. *Nitrosococcus*
      b. Cells lobular and partially compartmentalized by cytomembranes.
           Genus VII. *Nitrosolobus*

Genus I. Nitrobacter *Winogradsky 1892, 127, nom. cons.* Opin. 23. Jud. Comm. 1958, 169

Ni.tro.bac'ter. L. n. *nitrum* nitrate; M.L. n. *bacter* the masculine form of the Gr. neut.n. *bactrum* a rod; M.L. masc.n. *Nitrobacter* nitrate rod.

Cells **short rods, often wedge** or **pear shaped. Reproduce by budding.** Cells possess a **polar cap of cytomembranes. Usually non-motile.** No resting stages known. **Gram-negative.** Cells rich in cytochromes imparting a yellowish color to cell suspensions; void of other pigments. Some strains are **obligate chemolithotrophs** which oxidize nitrite to nitrate and fix CO_2 to fulfill energy and carbon needs. The autotrophic medium consists of fresh water or sea water enriched with nitrite and other inorganic salts; no organic growth factor being required. **Some strains** can be grown **heterotrophically,** but the growth rate is slower than when the cells are grown autotrophically. **Strictly aerobic** using oxygen as terminal electron acceptor. **pH range for growth, 6.5–8.5. Temperature range for growth, 5–40 C.**

The **G + C content** of the DNA ranges from 60.7–61.7 moles % (buoyant density; one species, six strains).

Habitat: **Soils, fresh water** and **sea water.**

Type species: *N. winogradskyi* Winslow *et al.* 1917, 552, *nom. cons.* Opin. 23. Jud. Comm. 1958, 169.

Comment: Hirsch (1970) pointed out the similarity of *Nitromicrobium germinans* Stutzer and Hartleb 1899, 197, a nitrite-oxidizing, budding microorganism, to *Nitrobacter spp.* and its possible synonymy with *Nitrobacter winogradskyi* Winslow *et al.* 1917.

Description of the species of genus Nitrobacter

1. **Nitrobacter winogradskyi** Winslow, Broadhurst, Buchanan, Krumweide, Rogers and Smith 1917, 552, *emend. mut. char.* Watson 1971, 264. (*Bac-terium nitrobacter* Lehmann and Neumann 1899, 187; *Bacillus nitrobakter* (*sic*) Matzuschita 1902, 536; *Nitrobacterium nitrobacter* Castellani and

Chalmers 1919, 933: Subj. syn. *Nitrobacter agilis* Nelson 1931, 287; *Nitrocystis sarcinoides* Winogradsky 1937, 336; *Nitrocystis micropunctata* Winogradsky 1937, 336; *Nitrogloea micropunctata* Winogradsky 1935, 1888.)

wi.no.grad'sky.i. M.L. gen.n. *winogradskyi* of Winogradsky; named for S. Winogradsky, the microbiologist who first isolated these bacteria.

Cells short rods, often wedge or pear shaped, 0.6–0.8 by 1.0–2.0 μm (Plate 12.1, Fig. 1). Cells possess a polar cap of cytomembranes arranged to form flattened vesicles (Plate 12.1, Fig. 3). Usually non-motile but when grown in continuous culture motile cells with a single subterminal flagellum are produced (Plate 12.1, Fig. 2). In liquid media grow either free or in small clumps of a hundred or more cells embedded in a slime matrix.

Morphology and physiology as for genus unless otherwise noted.

Optimum growth temperature, 25–30 C. Optimum growth pH, 7.5–8.0.

Many strains will grow either in a fresh or sea water medium. Marine strains are morphologically indistinguishable from terrestrial strains.

Cell suspensions show characteristic (oxidized minus dithionite reduced) absorption peaks at 420, 440, 522, 550, 587 and 600 nm.

Storage materials: Glycogen, poly-β-hydroxy-butyrate and polyphosphates.

The G + C content of the DNA ranges from 60.7–61.7 moles % (buoyant density; six strains).

Widely distributed in soils.

Neotype strain: ATCC 25391 (isolated from soil), Watson 1971, 264.

Genus II. **Nitrospina** *Watson and Waterbury 1971, 225*

Ni.tro.spi'na. L. n. *nitrum* nitre, nitrate; L. n. *spina* spine; M.L. masc.n. *Nitrospina* nitrate spine.

Cells are **straight, slender rods; spherical forms,** are found in senescent cultures. There is **no extensive cytomembrane system. Gram-negative. Non-motile.** Cells have cytochromes but no other pigments.

Obligately chemolithotrophic bacteria which **oxidize nitrite to nitrate** and **fix CO₂** to fulfill energy and carbon needs. **Grow in sea-water** enriched with nitrite and inorganic salts; **no organic growth factors** are **required.**

Strictly aerobic using oxygen as a terminal electron acceptor.

Temperature range for growth, 20–30 C. pH range for growth, 7.0–8.0. Optimum growth in 70–100% sea water; no growth in distilled water mineral salts medium even if NaCl is included.

The **G + C content** of the DNA is **57.7 moles %** (buoyant density; one species, one strain).

Habitat: South Atlantic Ocean.

Type species: *N. gracilis* Watson and Waterbury, 1971, 225.

Description of the species of genus **Nitrospina**

1. **Nitrospina gracilis** Watson and Waterbury 1971, 225.

gra'ci.lis. L. adj. *gracilis* slender.

Cells are long, slender rods, 0.30–0.40 by 2.7–6.5 μm (Plate 12.1, Fig. 4); spherical forms 1.35–1.45 μm in diameter are found in old cultures. Cells lack extensive cytomembrane system, but occasional bleb-like invaginations of the plasma membrane occur (Plate 12.1, Fig. 5). Occur singly or in pairs. Grow free in liquid media but will adhere to culture vessel walls in old cultures.

Morphology and physiology as for genus unless otherwise noted. Will not grow on organic media,

and no organic growth factors are required; some organic compounds inhibit growth. Grow in 70–100% sea water enriched with nitrite and other mineral salts. Optimum growth temperature, 25–30 C. Optimum growth pH, 7.5–8.0.

Cell suspensions show characteristic (oxidized minus dithionite reduced) absorption peaks at 425, 532 and 553 nm.

Storage material: Glycogen.

The G + C content of the DNA is 57.7 moles % (buoyant density).

Type strain: ATCC 25379 (isolated from the South Atlantic Ocean).

Genus III. **Nitrococcus** *Watson and Waterbury 1971, 224*

Ni.tro.coc'cus. L. n. *nitrum* nitre, nitrate; Gr. n. *coccus* grain, berry; M.L. masc.n. *Nitrococcus* nitrate sphere.

Cells **spherical, 1.5 μm or larger. Gram-negative. Motile** by means of **one to two subterminally inserted flagella.** Cells are rich in cyto-

chromes imparting a yellowish to reddish color to cell suspensions; void of other pigments.

Obligately chemolithotrophic bacteria which

oxidize nitrite to nitrate and fix CO_2 to fulfill energy and carbon needs. **Growth medium is sea water** enriched with nitrite and other inorganic salts, **no organic growth factor** being **required**.

Strictly aerobic using oxygen as a terminal electron acceptor.

Temperature range for growth, 15–30 C. pH range for growth, 6.8–8.0.

The **G + C content** of the DNA is 61.2 moles % (buoyant density; one species, one strain).

Habitat: South Pacific Ocean.

Type species: *N. mobilis* Watson and Waterbury 1971, 224.

Description of the species of genus **Nitrococcus**

1. **Nitrococcus mobilis** Watson and Waterbury 1971, 224.

mo'bi.lis. L. adj. *mobilis* movable, motile.

Cells spherical, 1.5–1.8 μm in diameter, following division but elongating with dimensions of 1.8 by 3.5 μm just prior to division. Cells occur singly or in pairs (Plate 12.1, Fig. 6). Cells have an extensive cytomembrane system which forms tubes randomly arranged in the cytoplasm (Plate 12.1, Fig. 8). Motile by means of one or two flagella inserted subpolarly in elongated cells; the flagella are 12.5 nm wide and 3–4 μm long (Plate 12.1, Fig. 7). When motile, cells have a turning, twisting, erratic motion. Grow either free in liquid media or in small clumps of a hundred or more cells embedded in a slime matrix; this aggregate of cells is not surrounded by any type of limiting membrane.

Morphology and physiology as for genus unless otherwise noted. Optimum growth in 70–100% sea water; will not grow in fresh water even if NaCl is included. Optimum temperature range, 25–30 C. Optimum pH range, 7.5–8.0.

Cell suspensions show characteristic (oxidized minus dithionite reduced) absorption peaks at 420, 440, 522, 550, 587 and 600 nm.

Storage materials: Glycogen and poly-β-hydroxybutyrate.

The G + C content of the DNA is 61.2 moles % (buoyant density).

Type strain: ATCC 25380 (isolated from South Pacific Ocean).

Genus IV. **Nitrosomonas** *Winogradsky 1892, 127, nom. cons.* Opin. 23. Jud. Comm. 1958, 169

(*Nitromonas* Winogradsky 1890, 258 (*nom. rej.* Opin. 23); Not *Nitromonas* Orla Jensen 1909, 312 (*nom. rej.* Opin. 23).)

Ni.tro.so.mo'nas. M.L. adj. *nitrosus* nitrous; Gr. fem.n. *monas*, a unit, monad; M.L. fem.n. *Nitrosomonas* nitrite monad, i.e. the monad producing nitrite.

Cells **ellipsoidal** or **short rods, motile or nonmotile, occurring singly**, in **pairs** or as **short chains. Gram-negative.** Possess **cytomembranes** which **occur as flattened vesicles in the peripheral regions of the cytoplasm.** Cells grow free in the medium or are embedded in a slime matrix. Cells are rich in cytochromes which impart a yellowish to reddish color to cell suspensions; void of other pigments.

Obligately chemolithotrophic bacteria which **oxidize ammonia to nitrite** and **fix CO_2** to fulfill energy and carbon needs. **Grow in fresh water or sea water** enriched with ammonia and inorganic salts, **no organic growth factors** being **required. Strictly aerobic**, using oxygen as a terminal electron acceptor. **Temperature range for growth, 5–30 C. pH range for growth, 5.8–8.5.**

The **G + C content** of the DNA ranges from **47.4–51.0** moles % (buoyant density; 13 strains).

Habitat: **Soils, fresh water and sea water.**

Type species: *N. europaea* Winogradsky 1892, 127, *nom. cons.* Opin. 23. Jud. Comm. 1958, 169.

Description of the species of genus **Nitrosomonas**

1. **Nitrosomonas europaea** Winogradsky 1892, 127, *emend. mut. char.* Watson 1971, 266. (*Pseudomonas europaea* (Winogradsky) Migula 1895, 29; *Bacterium nitrosomonas* Lehmann and Neumann 1899, 187; *Planococcus europaeus* (Winogradsky) Vuillemin 1913, 525: Subj. syn. *Nitrosomonas monocella* Nelson 1931, 287.)

eu.ro.pae'a. Gr. adj. *europaeus* of Europe, European.

Rods, 0.8–0.9 by 1.0–2.0 μm, occurring singly, rarely in chains (Plate 12.2, Fig. 1). Gram-negative. When motile, possess one to two subpolar flagella 3–4 times the length of the rod (Plate 12.2, Fig. 2). Cells have cytomembranes which form flattened lamellae in the peripheral regions of the cytoplasm (Plate 12.2, Fig. 3).

Cells are obligate chemolithotrophs oxidizing ammonia and hydroxylamine to nitrite. Growth occurs with ammonia but not with hydroxylamine. Grow in particulate-free liquid salts media as individual cells or in small aggregates embedded in a slime. Can oxidize ammonia at reduced oxygen tensions. Optimum growth temperature, 25–30 C. Optimum growth pH, 7.5–8.0. No sodium chloride requirement. Cannot be grown in sea water.

Cell suspensions show characteristic (oxidized minus dithionite reduced) absorption peaks at 423, 465, 522, 552 and 605 nm.

Storage materials: Polyphosphates.

The G + C content of the DNA ranges from 50.5–51.0 moles % (buoyant density; three strains). Widely distributed in soils.

Neotype strain: ATCC 25978 (isolated from soils in the United States), Watson 1971, 266.

Genus V. **Nitrosospira** *Winogradsky and Winogradsky 1933, 406*

Ni.tro.so.spi'ra. M.L. adj. *nitrosus* nitrous; Gr. n. *spira* a coil, spiral; M.L. fem.n. *Nitrosospira* nitrous spiral.

Cells **spiral shaped. Gram-negative. Cells lack cytomembranes** (Plate 12.2, Fig. 6). **Non-motile or motile** by means of **peritrichous flagella.** Grow free in liquid medium. Cells are rich in cytochromes which impart a yellowish to reddish color to cell suspensions; void of other pigments.

Obligately chemolithotrophic bacteria which **oxidize ammonia to nitrite** and **fix CO₂** to fulfill energy and carbon needs. Grow in **fresh water** enriched with ammonia and inorganic salts; **no** organic growth factors are **required. Strictly aerobic** using oxygen as a terminal electron acceptor. **Temperature range for growth, 15–30 C. pH range for growth, 6.5–8.5.**

The **G + C content** of the DNA is **54.1** moles % (buoyant density; one species, one strain).

Isolated from soils from Brie, France; Island of Crete; the summit of Mt. Pilatus in Switzerland; and the Parthenon in Athens, Greece.

Type species: *N. briensis* Winogradsky and Winogradsky 1933, 407.

Description of the species of genus **Nitrosospira**

1. **Nitrosospira briensis** Winogradsky and Winogradsky 1933, 407. (Subj. syn. *Nitrosospira antarctica* Winogradsky and Winogradsky 1933, 407.)

bri.en'sis. *Brie* French place name; M.L. adj. *briensis* of Brie.

Cells are tightly wound spirals with 3–20 turns (Plate 12.2, Fig. 5). Short spirals have the appearance of short rods and ellipsoidal cells (Plate 12.2, Fig. 4); the spiral nature of the cells may not be evident when examined with a light-field or a phase-contrast microscope. Width of spiral filament, 0.3–0.4 μm; amplitude of the spiral, 0.8–1.0 μm. Small pseudococci are observed in senescent cultures. Motile or non-motile; when motile, cells are propelled by one to six peritrichous flagella 3–5 μm in length (Plate 12.2, Fig. 5). Grow in particulate-free liquid mineral salts medium as individual cells.

Cells are obligate chemolithotrophs oxidizing ammonia and hydroxylamine to nitrite and fixing CO_2. Growth occurs with ammonia but not with hydroxylamine. Strict aerobes but ammonia can be oxidized at reduced oxygen tensions. Optimum growth temperature, 25–30 C. Optimum growth pH, 7.5–8.0. No sodium chloride requirement; cannot be grown in sea water.

Cell suspensions show characteristic (oxidized minus dithionite reduced) absorption peaks at 423, 465, 522, 552 and 605 nm.

The G + C content of the DNA is 54.1 moles % (buoyant density).

Widely distributed in soils.

Neotype strain: ATCC 25971 (isolated from soil from the Island of Crete), Watson 1971, 267.

Genus VI. **Nitrosococcus** *Winogradsky 1892, 127, nom. cons.* Opin. 23. Jud. Comm. 1958, 169

Ni.tro.so.coc'cus. M.L. adj. *nitrosus* nitrous; Gr. n. *coccus* grain, berry; M.L. masc.n. *Nitrosococcus* nitrous sphere.

Cells **spherical. Motile or non-motile. Gram-negative.** Grow **singly, in pairs** or in **tetrads** either suspended free in liquid medium or embedded in slime to form aggregates which are either attached to vessel walls or are suspended in the liquid medium.

Obligately chemolithotrophic bacteria which **oxidize ammonia to nitrite** and **fix CO₂** to fulfill energy and carbon needs. Grow in **fresh water** or **sea-water** enriched with ammonia and inorganic salts, **no organic growth factors** being **required. Strictly aerobic** using oxygen as a terminal electron acceptor. **Temperature range for growth, 2–30 C. pH range for growth, 6.0–8.0.**

The **G + C content** of the DNA ranges from 50.5–51.0 moles % (buoyant density; one species, three strains).

Habitat: **Soils** and **sea water.**
Type species: *N. nitrosus* (Migula) Buchanan 1925, 402. (Type species not presently in culture.)

Description of the species of genus **Nitrosococcus**

1. **Nitrosococcus nitrosus** (Migula) Buchanan 1925, 402, *nom. cons.* Opin. 23. Jud. Comm. 1958, 169. (*Micrococcus nitrosus* Migula 1900, 194; *Nitrosocystis coccoides* Starkey ·1948, 72.)

ni.tro'sus. M.L. adj. *nitrosus* nitrous.

Large spheres, 1.5–1.7 µm in diameter, with thick cell membranes. Motility has not been demonstrated. Stain readily with aniline dyes. Zoogloea formation not observed.

Growth medium consists of fresh water, inorganic salts and ammonia. Aerobic cells use oxygen as a terminal electron acceptor. Optimum temperature, 20–25 C.

Isolated from soils from Quito, Ecuador; Campinas, Brazil; and Melbourne, Australia. Presumably widely distributed in soil.

2. **Nitrosococcus oceanus** (Watson) Watson 1971, 267. (*Nitrosocystis oceanus* Watson 1965, R279.)

o.ce.an'us. L. n. *oceanus* ocean.

Cells spherical to ellipsoidal, 1.8–2.2 µm in size (Plate 12.2, Fig. 7). Occur singly, in pairs and occasionally as tetrads either growing free in a liquid medium or as aggregates suspended in the medium. Cysts were occasionally observed in mixed cultures, but have not been found in pure cultures. Motile or non-motile; when motile, propelled by a single flagellum or a small tuft of peritrichous flagella (Plate 12.2, Fig. 8). Cells have cytomembranes which are arranged as flattened lamellae in the central region of the cell (Plate 12.2, Fig. 9).

Cells are obligate chemolithotrophs oxidizing ammonia and hydroxylamine to nitrite and fixing CO_2 to fulfill their energy and carbon needs. Growth occurs with ammonia but not with hydroxylamine. Optimum growth temperature, 25–30 C. Optimum growth pH, 7.5–8.0. Growth in 70–100% sea water; will not grow in fresh water even if NaCl is included.

Cell suspensions show characteristic (oxidized minus dithionite reduced) absorption peaks at 423, 465, 522, 552 and 605 nm.

Storage materials: Glycogen and polyphosphates.

The G + C content of DNA ranges from 50.5–51.0 moles % (buoyant density; three strains).

Habitat: Atlantic and Pacific Oceans.

Type strain: ATCC 19707 (isolated from North Atlantic sea water).

Genus VII. **Nitrosolobus** *Watson, Graham, Remsen and Valois 1971, 200*

Ni.tro.so.lob'us. M.L. *nitrosus* nitrous; M.L. n. *lobus* a lobe; M.L. fem.n. *Nitrosolobus* nitrous lobe, i.e. a lobe producing nitrite.

Cells **pleomorphic** and **lobate,** 1.0–1.5 µm in diameter; **division by constriction. Gramnegative.** Cells partially **compartmentalized by the invagination of the plasma membrane** and other segments of the cell envelope into the cytoplasm forming vesicular regions. **Motile by** means of **peritrichous flagella.** Grow free in liquid medium. Cells are rich in cytochromes which impart a yellowish to reddish color to cell suspensions; void of other pigments.

Obligately chemolithotrophic bacteria which **oxidize ammonia to nitrite** and **fix CO_2** to fulfill energy and carbon needs. Grow in fresh water enriched with ammonia and inorganic salts; **no organic growth factors are required. Strictly aerobic** using oxygen as a terminal electron acceptor. **Temperature range for growth, 15–30 C. pH range for growth, 6.0–8.2.**

The **G + C content** of DNA ranges from **53.6–55.1** moles % (buoyant density; one species, three strains).

Habitat: Soils from Surinam, South America; Galapagos Archipelago; Southwest Africa; and Russia.

Type species: *N. multiformis* Watson *et al.* 1971, 200.

Description of the species of genus **Nitrosolobus**

1. **Nitrosolobus multiformis** Watson, Graham, Remsen and Valois 1971, 200.

mul.ti.for'mis. L. adj. *multus* many; L. n. *forma* shape. M.L. adj. *multiformis* many shapes.

Cells composed of multiple ellipsoidal units, randomly arranged, forming lobular pleomorphic cells (Plate 12.2, Fig. 10) which are internally compartmentalized, having one to five membrane-bounded central areas surrounded by numerous smaller membrane-bounded areas rich in glycogen deposits (Plate 12.2, Fig. 12). Cells brain-like in appearance and 1.0–1.5 by 1.0–2.5 μm. Motile by means of 1–20 flagella randomly arranged; the flagella are 15.0 nm wide and 2.2–5.0 μm long (Plate 12.2, Fig. 11). Grow in particulate-free liquid mineral salts medium as individual cells.

Cells are obligate chemolithotrophs oxidizing ammonia, hydroxylamine and biuret to nitrite and fixing CO_2. Growth occurs with ammonia but not with hydroxylamine or biuret. Cells have urease system. Strict aerobes, although cells can oxidize ammonia at reduced oxygen tensions. Optimum growth temperature, 25–30 C. Optimum growth pH, 7.5–7.8. No sodium chloride requirement; cannot be grown in sea water.

Cell suspensions show characteristic (oxidized minus dithionite reduced) absorption peaks at 423, 465, 522, 552 and 605 nm.

Storage materials: Glycogen and polyphosphates.

The G + C content of the DNA is 54.6 moles % (buoyant density).

Widely distributed in soils.

Type strain: ATCC 25196 (isolated from soils from Surinam, South America).

Genera and species incertae sedis

a. *Nitrosogloea merismoides* Winogradsky 1935, 1887.

b. *Nitrosogloea membranacea* Winogradsky 1935, 1887.

c. *Nitrosogloea schizobacteroides* Winogradsky 1935, 1887.

d. *Nitrosocystis javanensis* (Winogradsky) Starkey 1948, 72. (*Nitrosomonas javanensis* Winogradsky 1892, 127; *Pseudomonas javanensis* (Winogradsky) Migula 1895, 30.)

These genera were based on the formation of cell aggregations, a property shown to depend on cultural conditions. Also the four species were described briefly, no cultures were maintained and they have not been reisolated. Since the organisms could not be identified from the original brief description, these taxa are considered *incertae sedis*.

ORGANISMS METABOLIZING SULFUR AND SULFUR COMPOUNDS

Key to genera of organisms metabolizing sulfur

I. Organisms metabolizing sulfur and sulfur compounds.

(Although the genus *Thiobacillus* could be placed in a monogeneric family Thiobacillaceae, with a description identical to that of the genus, this seems to serve no useful purpose and until all genera have been grown in pure culture it seems more appropriate to consider them all as genera of uncertain affiliation.)

A. Obtain energy from oxidization of reduced or partially reduced sulfur compounds.

1. Rods, motile.

1. *Thiobacillus*

2. Cells spherical with lobes.

2. *Sulfolobus*, p. 461

B. Have not been grown in pure culture, cells contain sulfur granules.

1. Rods, embedded in gelatinous mass, non-motile.

3. *Thiobacterium*, p. 462

2. Cells not embedded in gelatinous mass, motile.

a. Cylindrical cells, flagella polar.

4. *Macromonas*, p. 462

b. Ovoid cells, flagella peritrichous.

5. *Thiovulum*, p. 463

c. Spiral cells, flagella polar.

6. *Thiospira*, p. 464

Genus 1. **Thiobacillus** Beijerinck 1904, 597

W. V. Vishniac

(*Sulfomonas* Orla-Jensen 1909, 314; Not *Thiobacillus* Ellis 1932, 130.)

Thi.o.ba.cil'lus. Gr. n. *thium* sulfur; L. n. *bacillus* a small rod; M.L. masc.n. *Thiobacillus* sulfur rodlet.

Small rod-shaped cells. Motile by means of a single polar flagellum: two species non-motile. No resting stages known. Gram-negative.

Energy derived from the oxidation of one or more reduced or partially **reduced sulfur compounds,** including sulfides, elemental sulfur, thiosulfate, polythionates and sulfite. The final oxidation product is sulfate, but sulfur and polythionates accumulate, sometimes transiently, under certain conditions. *T. ferrooxidans* also utilizes ferrous compounds as electron donors.

The genus includes strictly autotrophic species which derive their carbon from carbon dioxide, facultative autotrophs and at least one species which requires both a partially reduced sulfur compound and organic matter for optimal growth.

Obligate aerobes, except *T. denitrificans* which grows anaerobically with nitrate as electron acceptor.

Optimum temperature about 28–30 C. pH range wide; some strains prefer strongly acid conditions, others mildly alkaline conditions.

Found in sea water, marine mud, soil, fresh water, acid mine waters, sewage, sulfur springs and in or near sulfur deposits. Especially frequent in environments in which hydrogen sulfide is produced or sulfur deposited.

The G + C content of the DNA, for the species tested, ranges from 50–68 moles %.

Type species: *Thiobacillus thioparus* Beijerinck 1904, 597.

Further Comments

The utilization of various sulfur compounds and the formation of products other than sulfate are dependent on many external factors, including oxygen concentration, pH, rate of growth, phosphate concentration, trace metals and possibly other unidentified factors. Therefore, the ability to use certain sulfur compounds or the deposit of intermediate products is not a useful diagnostic feature.

The type species is strictly autotrophic, in the sense that it requires an inorganic electron donor, an inorganic electron acceptor and uses only carbon dioxide as its major carbon source. However, many, and probably all, species are capable of assimilating certain organic compounds provided that thiosulfate or another inorganic electron donor is available as an energy source. It has been claimed that some species, commonly assumed to be strict autotrophs, grow on sucrose under specified conditions. These observations are not sufficiently well established to be useful as diagnostic features (Kelly, 1971). Many heterotrophic bacteria have the ability to oxidize thiosulphate, but do not derive energy from the process (Guittonneau, 1925; Starkey, 1935). These organisms bear no relationship to the thiobacilli; therefore, the species previously listed as *T. coproliticus* Lipman and McLees 1940, 430 and *T. trautweinii* Bergey *et al.* 1925, 39 are not included in this description.

Among the thiobacilli it has been possible to establish three types which differ in their lipid content determined as fatty acid methyl esters

TABLE 12.1
Types of fatty acid methyl ester (FAME) profiles found in genus **Thiobacillus**[a]

Fatty acid[b]	Total extracted fatty acids in:		
	Type I	Type II	Type III
	%	%	%
C_6			7
C_8	6	4	6
C_9			7
C_{10}	6		3
C_{11}	2		10
C_{12}	4		14
C_{13}			7
$C_{14:1}$	30	2	16
C_{14}	12	3	21
C_{15}	2	32	
$C_{16:1}$	15	2	
C_{16}	9	7	4
$C_{17:1}$		15	
$C_{:7}$		20	

[a] Levin (1971) obtained a fatty acid profile for *T. thiooxidans* (ATCC 8085) different from Type III above (obtained from two strains maintained by D. G. Lundgren, one originally ATCC 8085) and different profiles from Types I and II. Its main feature was the predominance of a C_{19}-cyclopropane acid.

[b] Subscript gives length of carbon chain; number to right of colon indicates number of double bonds.

(FAME profiles, Agate and Vishniac, 1973) (Table 12.1), and also three groups which differ in the G + C content of their DNA (Jackson *et al.*, 1968). While the information is not available for all species, the beginning of a useful matrix for the classification of the thiobacilli has been established (Table 12.2).

The isolation of halophilic and thermophilic strains of *Thiobacillus* has been mentioned repeatedly (see e.g. Egorova and Deryugina, 1963 for a thermophilic and reportedly sporulating strain); halophilic strains are discussed under *T. neapolitanus*. Furthermore, strains resistant to a variety of heavy metals have been found in experiments designed to test the ability of *Thiobacillus* strains to oxidize mineral sulfides for the purpose of extracting metals from their ores by the resulting sulfuric acid. For a review of various *Thiobacillus* species and strains, and their role in nature, see Sokolova and Karavaiko, 1964. A now partly outdated survey from a physiological point of view is Vishniac and Santer, 1957.

Finally, it is well to remember the words of Bunker (1936): "It may . . . be . . . that there is . . . a . . . gradation of types, from the strictly

TABLE 12.2

Classification of genus **Thiobacillus** *based on lipid and nucleotide content*

FAME profile	G + C fraction of DNA			
	50–52% moles %	56–57 moles %	62–68 moles %	Not determined
Type I		T. neapolitanus T. ferrooxidans		
Type II			T. thioparus T. novellus	T. denitrificans T. intermedius
Type III	T. thiooxidans			
Not determined				T. perometabolis

autotrophic organisms for which the presence of sulphur is an absolute essential for metabolism, through facultative autotrophs, for some of which sulphur is a vital necessity, to . . . types which may be little more than incidental oxidizers of sulphur and sulphur compounds.''

Key to the species of genus **Thiobacillus**

I. Strictly autotrophic
  A. Oxidize sulfur compounds only.
    1. Strictly aerobic.
      a. Grows optimally in neutral medium (pH 6.0–8.0).
        aa. FAME profile II (Table 12.1), G + C fraction 62–68%.
          1. *Thiobacillus thioparus*
        aaa. FAME profile I (Table 12.1), G + C fraction 56–57%.
          2. *Thiobacillus neapolitanus*
      b. Grows optimally in acid medium (pH 1.0–3.5).
        3. *Thiobacillus thiooxidans*
    2. Facultatively anaerobic, uses nitrate in absence of oxygen.
      4. *Thiobacillus denitrificans*
  B. Oxidizes ferrous compounds or sulfur compounds.
    5. *Thiobacillus ferrooxidans*
II. Facultatively autotrophic.
  A. Oxidation of thiosulfate repressed by organic substrates.
    6. *Thiobacillus novellus*
  B. Simultaneous oxidation of thiosulfate and organic substrates occurs.
    7. *Thiobacillus intermedius*
III. Heterotrophic, but requires simultaneous utilization of sulfur compounds and organic substrates for optimal growth.
    8. *Thiobacillus perometabolis*

Description of the species of genus **Thiobacillus**

1. **Thiobacillus thioparus** Beijerinck 1904, 597. (*Bacterium thioparum* (Beijerinck) Schoenichen 1925, 44; *Bacterium thiocyanoxidans* Happold and Key 1937, 1328. *Thiobacillus thiocyanoxidans* (Happold and Key) Happold, Johnstone, Rogers and Youatt 1954, 265.)

thi.o′par.us. Gr. n. *thium* sulfur; L. v. *paro* produce; M.L. adj. *thioparus* sulfur producing.

Thin, short rods, 0.5 by 1.0–3.0 μm, averaging 0.5 by 1.7 μm. Motile with polar flagellum; occasionally embedded in slime which renders them non-motile (unpublished observation).

Colonies on thiosulfate agar are small (1–2 mm in diameter), circular, whitish yellow due to precipitated sulfur. Turn pink, then brown in old cultures, especially in the center.

Liquid thiosulfate medium becomes uniformly turbid. At the usual thiosulfate concentration,

1%, a pellicle forms consisting of cells and free sulfur. No sulfur is formed at lower thiosulfate concentrations or in a chemostat even at 1% concentration under optimum growth conditions. In stagnant cultures a slimy pellicle with strands trailing to the bottom is formed occasionally. pH drops to 4.5.

Strictly autotrophic. Energy derived by the oxidation of thiosulfate to sulfate. Sulfur granules and polythionates may accumulate depending on culture conditions. Elemental sulfur is slowly oxidized. Also oxidizes other partially reduced sulfur compounds, including hydrogen sulfide (Jacobsen, 1914), tetrathionate, and in some strains, thiocyanate (Happold et al., 1954).

Nitrates and ammonium salts utilized as nitrogen sources.

Strictly aerobic.

Optimum temperature 28 C. Optimum pH between 6.6–7.2. Growth occurs between pH 4.5–7.8; in some strains to pH 10.0.

Found in mud, soil, canal water and other fresh water sources. Presumably widely distributed.

2. **Thiobacillus neapolitanus** Parker 1957, 86. (*Thiobacillus* X Parker and Prisk 1953, 352.)

ne.a.po.li.ta'nus. L. adj. *neapolitanus* pertaining to Naples.

Short rods, 0.5 by 1.0–1.5 μm. Non-motile; some freshly isolated marine strains motile (Vishniac, 1952).

Colonies on thiosulfate agar are small (1–2 mm) circular, convex, glistening, whitish yellow due to precipitated sulfur. Center of old colonies turns pink.

In liquid thiosulfate medium, uniform turbidity with pellicle which contains free sulfur and a sulfur precipitate. pH drops to 3.5. No sulfur accumulates at low thiosulfate concentrations, or in chemostat or well aerated cultures under optimum growth conditions. Accumulation of sulfur and polythionates is a function of oxygen and thiosulfate concentration.

Strictly autotrophic, but will assimilate organic compounds in the presence of an oxidizable sulfur compound (Kelly, 1969; Johnson and Vishniac, 1970). Derives energy by the oxidation of thiosulfate to sulfate. Elemental sulfur, hydrogen sulfide and other partially reduced sulfur compounds are also oxidized.

Uses ammonium salts and nitrates as nitrogen source. No growth on organic media.

Strictly aerobic.

Optimum temperature 28 C; growth range 8–37 C. Optimum pH 6.2–7.0; growth range between 3.0–8.5.

Originally isolated by Nathansohn (1902) from sea water at Naples, Italy. Isolated by Parker from corroding concrete sewers. Frequently found in marine mud and sea water. Presumably widely distributed in soil, fresh water and marine environments.

Further Comments

The difference between *Thiobacillus thioparus* and *T. neapolitanus* was recognized by Parker (1953). The separation has been supported by the FAME profiles and G + C content (Table 12.2).

The marine and halophilic strains isolated by Issatchenko (Issatchenko and Salimovakaya, 1928) and later described as *Thiobacterium issatchenkoi* Zaslavskii 1952, 35, may be related. Lange-Posdeeva (1930) described a strain which grew in 40% sodium thiosulfate.

3. **Thiobacillus thiooxidans** Waksman and Joffe 1922, 239. (*Thiobacterium thiooxydans* (*sic*) (Waksman and Joffe) Lehmann and Neumann 1927, 517; *Thiobacillus thermitanus* Emoto 1928, 422; *Thiobacillus lobatus* Emoto 1929, 148; *Thiobacillus crenatus* Emoto 1929, 149; *Thiobacillus umbonatus* Emoto 1929, 150; *Thiobacillus concretivorus* Parker 1945, 82.)

thio.ox'i.dans. Gr. n. *thium* sulfur; M.L. v. *oxido* make acid, oxidize; M.L. part.adj. *thiooxidans* oxidizing sulfur.

Short rods, 0.5 by 1.0–2.0 μm, occurring singly, in pairs or in short chains.

Colonies on thiosulfate agar are minute (usually less than 1.0 mm), transparent or whitish yellow which clears on prolonged growth, entire edges but variations in colony shape have been observed (Emoto, 1929).

Uniform turbidity in liquid media. Sulfur broth becomes more acid than pH 1.0. In thiosulfate medium sulfur may precipitate transiently; polythionates may appear transiently.

Ammonium salts serve as nitrogen sources, but use of nitrates reported for two strains.

Strictly autotrophic; remarkable for its ability to carry out a rapid oxidation of elemental sulfur. Other partially reduced sulfur compounds also utilized (London and Rittenberg, 1964).

Strictly aerobic.

Optimum temperature 28–30 C; growth range from 10–37 C. Optimum pH between 2.0–3.5; upper limit of growth near pH 6.0, lower limit usually near pH 0.5, but a negative pH has been reported (quoted by Sijderius, 1946).

Originally isolated from soil containing flowers of sulfur. Found in sulfur springs (Emoto, 1928, 1929; Baudisch, 1935). Also isolated from composts of soil, acid mine waters and corroding steel and concrete.

Reference strain: K.R.Butlin 3/TA; NCIB 8343; ATCC 19377.

4. **Thiobacillus denitrificans** Beijerinck 1904, 153. (*Sulfomonas denitrificans* (Beijerinck) Orla-Jensen 1909, 314.)

de.ni.tri'fi.cans. M.L. v. *denitrifico* denitrify; M.L. part.adj. *denitrificans* denitrifying.

Short rods, 0.5 by 1–3 μm.

Colonies on the surface of thiosulfate agar thin, clear or weakly opalescent. Deep agar colonies in the presence of nitrate are opalescent or glistening and star shaped as a result of nitrogen splitting the agar.

Uniform turbidity in thiosulfate broth. In air sulfur and polythionates may accumulate, usually transiently. No intermediates accumulate under optimum growth conditions. With nitrate present and incubation in glass stoppered bottles foam results from vigorous nitrogen production.

Strictly autotrophic. Facultatively anaerobic. Oxidizes thiosulfate to sulfate, but can also utilize other partially reduced sulfur compounds including polythionates and sulfide. Elemental sulfur is slowly oxidized. Under anaerobic conditions nitrate is required and gaseous nitrogen is evolved. A transient formation of NO has been observed (Baalsrud and Baalsrud, 1954).

Found in canal water, mine waters, marine sources and soil; presumably widely distributed.

Further Comments

A variety of strains have been observed, including some which oxidize mineral sulfides and are resistant to a variety of heavy metals (Kramarenko and Prisrenova, 1961).

5. **Thiobacillus ferrooxidans** Temple and Colmer 1951, 605. (*Ferrobacillus ferrooxidans* Leathen and Braley 1954, 44; *Ferrobacillus sulfooxidans* Kinsel 1960, 631.)

fer.ro.ox'i.dans. L. n. *ferrum* iron; M.L. v. *oxido* oxidize, make acid; M.L. part.adj. *ferrooxidans* iron-oxidizing.

Short rods, 0.5 by 1.0 μm, with rounded ends, occurring singly or in pairs, rarely in short chains.

Colonies on thiosulfate agar very thin and small with irregular margins, becoming whitish in center upon aging.

Uniform turbidity in thiosulfate and in sulfur liquid media; in the former a delicate pellicle after 2–3 weeks.

On ferrous agar (agarose is a more suitable solidifying agent; Kelly, unpublished data) colonial appearance varies with the ferrous-iron content. With low to moderate iron concentrations, an amber zone reveals the presence of microscopic colonies which become lobed and coated with hydrated ferric oxide; with high ferrous iron concentrations, growth is abundant becoming heavily encrusted with hydrated ferric oxide.

Ferrous liquid medium remains clear, rapidly turning amber to reddish brown due to production of ferric iron; if pH rises above 1.9, ferric hydrate is precipitated and a pellicle composed of ferric hydrate and cells is formed.

Utilizes ammonia as a nitrogen source; nitrates are used slowly.

Oxidizes sulfur and presumably other partially reduced sulfur compounds in addition to thiosulfate. Maintenance on sulfur-containing substrates does not lead to loss of ability to use ferrous compounds, and prolonged cultivation on iron media does not lead to a loss of ability to oxidize sulfur compounds (Kelly and Touvinen, 1972).

Strictly aerobic.

Considered to be strictly autotrophic, although two out of eight strains have been reported to grow on sucrose after a period of adaptation (Shafia and Wilkinson, 1969).

Optimum temperature reported as 15–20 C, but has been cultivated at 25 C.

The optimum pH lies between 2.5–5.8; growth occurs between pH 1.4–6.0, the lower range may extend below 1.4 (Tuovinen and Kelly, unpublished data).

Isolated from bituminous coal mine drainage waters, strongly acid and high in ferrous iron, in West Virginia and Pennsylvania. Found in acid waters of high iron content and in soils containing pyrite or marcasite.

6. **Thiobacillus novellus** Starkey 1934, 365.

no.vel'lus. L. dim.adj. *novellus* new.

Short rods, coccoidal or ellipsoidal cells, 0.4–1.0 by 0.6–4.0 μm, occurring singly, occasionally in pairs. Non-motile.

Slow growth on nutrient agar plates, colonies colorless, moist, raised, circular, 1 mm in diameter. Colonies and growth similar on thiosulfate agar but become white with precipitated sulfur.

Nutrient broth becomes slightly turbid with a gelatinous pellicle, which forms a long streamer-like network extending to the bottom of the tube. Some sediment.

Thiosulfate broth becomes uniformly turbid with no pellicle; a whitish sediment with thin incomplete membrane forms on the bottom of the flask. Reaction changes slowly from pH 7.8–5.8 with decomposition of only a fraction of the thiosulfate; less than one third is decomposed according to Sokolova and Karavaiko (1964).

No visible growth in elemental sulfur media.

Slow liquefaction of gelatin.

Although in a thiosulfate medium growth proceeds until only a fraction of the thiosulfate is consumed, resting cell suspensions will oxidize thiosulfate stoichiometrically to sulfate, as measured manometrically (Santer *et al.* 1959). In synthetic organic media only glutamic and aspartic acids support growth well. Citrate is used slowly, but other members of the tricarboxylic acid cycle and carbohydrates are not used. Utilizable organic compounds repress the oxidation of thiosulfate.

Strictly aerobic; facultatively autotrophic. In transferring cells from autotrophic to heterotrophic conditions and *vice versa* each cell is capable of adaptation after a lag period as shown by replicate plating.

Optimum temperature 30 C. Optimum pH range 7.8–9.0; growth range pH 5.0–9.2.

Isolated from soil; presumably widely distributed.

Reference strain (co-type): ATCC 8093.

7. Thiobacillus intermedius London 1963, 335.

in.ter.me′di.us. L. prep. *inter* between, among; L. adj. *medius* middle; M.L. adj. *intermedius* in between, intermediate.

Thin, short rods, 0.5 by 1.0–2.0 μm.

On yeast extract agar, colonies small, thin, clear and spreading. Growth in yeast extract media without thiosulfate is greatly increased by the addition of glucose or glutamate.

On thiosulfate agar, colonies are small (less than 1.0 mm), yellow-opaque with raised centers and flat and veil-like fringe. Opacity results from sulfur deposition, but less sulfur is precipitated than in colonies of *T. thioparus*. Sulfur precipitation may extend beyond the edge of the colony by diffusion of acid which decomposes thiosulfate.

Rate and extent of growth in thiosulfate medium are increased by addition of yeast extract, glucose, fructose, sucrose, maltose, aspartate or glutamate. Many other carbohydrates, alcohols and acids have no effect.

Special characteristic: While either thiosulfate or organic compounds may support the growth of this organism, optimal growth occurs in the presence of glutamic acid and glucose. In media containing both sulfur compounds and organic compounds both are oxidized simultaneously in contrast to *T. novellus*.

Strictly aerobic.

Optimum temperature 30 C.

Optimum pH not recorded, presumably between 6.0–7.0; growth range between 1.9–7.0.

Isolated from fresh water mud; presumably widely distributed.

Type strain: ATCC 15466.

8. Thiobacillus perometabolis London and Rittenberg 1967, 218.

pe.ro.me.ta′bo.lis. Gr. adj. *peros* maimed, crippled; Gr. v. *metabole* alter, change; M.L. part.adj. *perometabolis* with a maimed metabolism.

Thin, short rods, 0.5 by 1.0–2.0 μm.

Colonies on thiosulfate agar barely visible. Upon supplementation with yeast extract, colonies are 1.0–3.0 mm in diameter, smooth, entire, creamy. Some sulfur deposition, but less than in cultures of either *T. thioparus* or *T. intermedius*. Old colonies pink to orange in center. No growth in strictly mineral liquid medium but vigorous growth occurs with uniform turbidity upon the addition of yeast extract or of one of fructose, arabinose, ribose or xylose.

Slow growth with uniform turbidity in casein hydrolysate or yeast extract broth. Upon supplementation with thiosulfate the yield doubles and the generation time halves.

Obligately aerobic. Although suspensions of the organism are capable of vigorous oxidation of yeast extract, thiosulfate, tetrathionate, elemental sulfur and sulfide, it is not capable of autotrophic growth. Its growth on organic media is limited in rate and yield. Optimal growth conditions require the presence of a partially reduced sulfur compound and organic matter. The organism might be described as an *obligate mixotroph*.

Optimum temperature 30 C. pH range 2.8–6.8.

Isolated from soil; presumably widely distributed.

Genus 2. **Sulfolobus** *Brock, Brock, Belly and Weiss 1972, 54*

T. D. Brock

Sulf.o.lo′bus. L. n. *sulfur* sulfur; L. n. *lobus* a lobe; M.L. masc.n. *Sulfolobus* a lobed sulfur-oxidizing organism.

Spherical cells with lobes, 0.8–1 μm in diameter. Resemble mycoplasmas but are more refractile and more regular in diameter. Not motile; **flagella and endospores absent. Gram-negative. Cell wall devoid of peptidoglycan.** Subunit structure of cell wall visible upon negative staining and electron microscopy.

Colonies smooth, glistening, non-pigmented.

Facultative autotrophs. **Use elemental sulfur as energy source.** May also use yeast extract, glutamate or ribose as a carbon and energy source. Aerobic.

Temperature optimum 70–75 C; maximum 85 C; minimum 55 C.

pH optimum 2–3; maximum 5.8; minimum 0.9. The G + C content of the DNA ranges from 60–68 moles % amongst five strains tested (buoyant density).

Found in solfatara areas containing hot acid environments, both soil and water.

Type species: *Sulfolobus acidocaldarius* Brock, Brock, Belly and Weiss 1972, 66.

Description of the species of genus **Sulfolobus**

1. **Sulfolobus acidocaldarius** Brock, Brock, Belly and Weiss 1972, 66.

a.ci.do.cal.dar'i.us. M.L. neut.n. *acidum* acid; L. adj. *caldarius* pertaining to warm or hot; M.L. adj. *acidocaldarius* organism living in acid and hot environments.

Morphology and description as for the genus. Type strain: 98-3.

Genus 3. **Thiobacterium** *Janke 1924, 68*

J. W. M. LA RIVIÈRE

(*Thiodendron* Lackey and Lackey 1961, 36.) Thi.o.bac.te'ri.um. Gr. n. *thium* sulfur; Gr. dim.n. *bakterion* a small rod; M.L. neut.n. *Thiobacterium* small sulfur rod.

Rod-shaped cells, each containing one or more sulfur droplets. Cells embedded in gelatinous masses, which are spherical when free floating **or dendroid** when attached to a solid substrate. **Non-motile.** No resting stages known.

Has not been grown in pure culture.

Type species: *Thiobacterium bovista* (Molisch) Janke 1924, 68.

Further Comments

The present description restricts the genus to non-motile, colorless, rod-shaped bacteria which contain sulfur inclusions and are embedded in gelatinous material; it thus excludes *Thiobacterium cristalliferum* and *Thiobacterium retiformans*, which, in the 7th edition of THE MANUAL

(pp. 79–80), were characterized *inter alia* by the presence of external sulfur globules. The above circumscription seems warranted until pure culture studies make it possible to relate the observed sulfur granules unequivocally to the bacteria in question; chemical oxidation is one possible alternative explanation.

At the same time, the description has been widened to include the organism described by Lackey and Lackey (1961) under the name *Thiodendron mucosum*. It is possible that the larger cell size and dendroid shape of the embedding material are the result of growth in flowing water, the puff-ball-shaped colonies being the stagnant water form.

Pure culture studies may warrant the recognition of these and other species.

Description of the species of genus **Thiobacterium**

1. **Thiobacterium bovista** (Molisch) Janke 1924, 68. (*Bacterium bovista* Molisch 1912, 59; *Thiodendron mucosum* Lackey and Lackey 1961, 36.)

bo.vist'ta. M.L. n. *Bovista* a genus of puff balls; M.L. fem.n. *bovista* puff ball.

Rods, usually 0.6–1.5 by 2.5 μm, but cells 2.5 by 9 μm occur. When sulfur inclusions are present,

cell masses are white in reflected light, black or bluish in transmitted light.

In the spherical colonies, the cells are embedded in the gelatinous walls of a bladder-like structure, filled with water; such colonies occur near the surface of sulfurous brackish and marine waters and have the appearance of groups of puff balls of different sizes. Dendroid colonies show extensive branching and may reach a size of 2–3 mm.

Genus 4. **Macromonas** *Utermöhl and Koppe in Koppe 1924, 632*

J. W. M. LA RIVIÈRE

Mac.ro.mo'nas. Gr. adj. *macrus* large; Gr. n. *monas* a unit, monad; M.L. fem.n. *Macromonas* a large monad.

Cylindrical to bean-shaped cells, motile by a polar flagellum. Inclusions of calcium carbonate sometimes accompanied by sulfur globules. Multiplication by constriction followed by fission. No resting stages known. **Microaerophilic;** found in fresh water environments with low oxygen concentrations, e.g. hypolimnia and upper layers of bottom muds. Has not been grown in pure culture.

Type species: *Macromonas mobilis* (Lauterborn) Utermöhl and Koppe in Koppe 1924, 632.

Further Comments

Until further studies show that observed morphological differences like size warrant further differentiation, the number of species is restricted to two.

Description of the species of genus **Macromonas**

1. Macromonas mobilis (Lauterborn) Utermöhl and Koppe in Koppe 1924, 632. (*Achromatium mobile* Lauterborn 1915, 413.)

mo'bi.lis. L. fem.adj. *mobilis* motile.

Single, slightly curved elongated ellipsoidal or cylindrical cells, usually 9 by 20 μm, sometimes 6–14 by 10–30 μm, with broad hemispherical ends. Motile by a single polar flagellum; 20–40 μm long and distinctly visible by light microscopy; may consist of a tuft of flagella. Rate of movement sluggish, *ca.* 800 μm/min, probably as result of high specific gravity of cells.

Normally contain small sulfur inclusions and one to four large spherical inclusions of calcium carbonate, one or two such bodies being found in each newly divided cell. The calcium carbonate inclusions may almost fill the cell. They may also disappear, probably as a result of decrease of pH

of the surrounding water; their disappearance usually precedes the disappearance of sulfur granules.

2. Macromonas bipunctata (Gicklhorn) Utermöhl and Koppe in Koppe 1924, 632. (*Pseudomonas bipunctata* Gicklhorn 1920, 425.)

bi.punc.ta'ta. L. *bis* twice; L. part.adj. *punctatus* punctate, dotted; M.L. fem.adj. *bipunctata* twice punctate.

Single, cylindrical cells with hemispherical ends, often pear shaped just after cell division. Motile by a single polar flagellum, 10–15 μm long, visible only after staining. Rate of movement sluggish, *ca.* 600 μm/min, probably as a result of the high specific gravity of the cells. Normally contain one to two large calcium carbonate inclusions, which almost fill the cell; sulfur globules may also be present.

Genus 5. **Thiovulum** *Hinze 1913, 195*

J. W. M. LA RIVIÈRE

Thi.o'vu.lum. Gr. n. *thium* sulfur; L. n. *ovum* egg; M.L. neut.dim.n. *Thiovulum* small sulfur egg.

Cells **round** to **ovoid,** 5–25 μm in diameter. Cytoplasm often concentrated at one end of cell, the remaining space being occupied by a large vacuole. Cytoplasm normally **contains orthorhombic sulfur inclusions,** sometimes concentrated at one end, sometimes filling cells almost completely. **Strongly motile** by **peritrichous flagella,** forward movement accompanied by **rotation** around the long axis. Cells characterized by presence of **one polar fibrillar organelle,** visible in thin sections by electron microscopy; its function is not known. No resting stages observed.

Multiplication by constriction followed by fission.

Gram-negative.

Microaerophilic. Catalase negative.

Type species: *Thiovulum majus* Hinze 1913, 195.

Further Comments

Although advances have been made towards pure culture study (La Rivière, 1963), no results are at present available that warrant recognition of different species. Present knowledge of the organism has been summarized by Starr and Skerman (1965).

Description of the species of genus **Thiovulum**

1. Thiovulum majus Hinze 1913, 195. (*Monas muelleri* Warming 1875, 363; *Thiovulum minus* Hinze 1913, 195; *Thiovulum muelleri* (Warming) Lauterborn 1916, 414.)

ma'jus L. comp.adj. *major* larger.

Morphology and physiology as for genus.

Strongly chemotactic with respect to oxygen. Grows only in liquid cultures and in characteristic

webs at the interface of oxygen- and of sulfide-bearing water layers. Further study is required to establish metabolic relationship with sulfide.

Isolation in pure culture difficult. Found in fresh water and marine environments where sulfide-containing water layers are in contact with overlaying oxygen-containing water.

Genus 6. **Thiospira** *Visloukh 1914, 48*

J. W. M. LA RIVIÈRE

(*Sulfospirillum* Kluyver and van Niel 1936, 396; not *Thiospirillum* Winogradsky 1888, 104.)

Thi.o.spi′ra. Gr. n. *thium* sulfur; Gr. n. *spira* a coil; M.L. fem.n. *Thiospira* sulfur coil or spiral.

Spirilla, usually with pointed ends, with **sulfur inclusions. Motile** by mono- or polytrichous **polar flagella. No resting stages known.** Has not been cultivated in pure culture.

Type species: *Thiospira winogradskyi* (Omelianski) Visloukh 1914, 48.

Further Comments

Pure culture studies will have to determine whether further differentiation into more than two species is warranted.

Description of the species of genus **Thiospira**

1. **Thiospira winogradskyi** (Omelianski) Visloukh 1914, 48. (*Thiospirillum winogradskii* Omelianski 1905, 770.)

wi.no.grad′sky.i. M.L. gen.n. *winogradskyi* of Winogradsky; named for S. N. Winogradsky, a Russian microbiologist.

Colorless spirilla, somewhat pointed at the ends 2–2.5 by 50 μm. Cells contain numerous sulfur globules. Strongly motile with one or two polar flagella. Probably microaerophilic and strongly chemotactic with respect to oxygen.

Found in waters overlaying sulfurous muds.

2. **Thiospira bipunctata** (Molisch) Visloukh 1914, 48. (*Spirillum bipunctatum* Molisch 1912, 59.)

bi.punc.ta′ta. L. *bis* twice; L. part.adj. *punctatus* punctate, dotted; M.L. fem.adj. *bipunctata* twice punctate.

Colorless spirilla, markedly pointed at the ends, 1.7–2.4 by 6.6–14 μm. Both ends contain large volutin granules; several small sulfur globules are present in the clear center and sometimes at the ends. Polar flagella at one or both ends.

Found in sulfurous marine and brackish waters.

FAMILY **SIDEROCAPSACEAE*** PRIBRAM 1929, 377

G. A. ZAVARZIN

Si.de.ro.cap.sa′ce.ae. M.L. fem.n. *Siderocapsa* type genus of family; M.L. suff. *-aceae* denoting family; M.L. fem.pl.n. *Siderocapsaceae* the *Siderocapsa* family.

Cells spherical, ellipsoidal or rod-shaped. May have thick or thin capsules.

Able to deposit iron and/or manganese oxides on or in capsules when present, or on extracellular material.

Found in iron-bearing waters.

Further Comments

Few of the organisms in this family have been cultivated, and the validity of a taxon based on deposition of iron and manganese oxides is doubtful, as it is now known that many other bacteria have this ability. The four genera recognized have been observed and described by numerous investigators and they are of geological importance. Differentiation is mainly on morphological features.

* *Editorial Note.* The author has questioned the validity of the family and to the editors, its placement has presented a problem.

The four genera included have only one feature in common—the ability to deposit or to cause the deposition of iron and/or manganese oxides. Most are organotrophs. *Siderococcus* reproduces by budding and could be placed in Part 4. Dr. Dubinina has recently obtained *Siderocapsa* in pure culture and found it indistinguishable from *Arthrobacter globiformis*. A paper is expected during 1974 in *International Journal of Systematic Bacteriology. Ochrobium* may be an alga.

As it is impossible to place the family, as here constituted, in a more logical place it is included with the chemolithotrophs, even though it is known that in many instances the deposition of iron or manganese is the result of the metabolism of the organic ion of the compound.

Genera incertae sedis

a. *Ferribacterium* Brussoff 1916, 548.
b. *Siderobacter* Naumann 1921, 55.

c. *Sideromonas* Cholodny 1922, 334.
d. *Sideronema* Beger 1941, 323.
e. *Siderosphaera* Beger 1949, 7.

Key to the genera of family **Siderocapsaceae**

I. Iron or manganese oxides deposited.
   A. Oxides deposited on or in capsular material which surrounds groups of spherical or ovoid cells.

Genus I. *Siderocapsa*

   B. Oxides form a delicate sheath (torus) which emphasizes the margin of the cell; cells usually rod-shaped.
      1. Torus completely surrounds the cell.

Genus II. *Naumanniella*

      2. Torus is open at one end, resembling a horseshoe.

Genus III. *Ochrobium*

II. Iron but not manganese oxide deposited in the medium. Cells coccoid.

Genus IV. *Siderococcus*

Genus I. **Siderocapsa** *Molisch 1910, 30*

H. SKUJA†

    Si.de.ro.cap′sa. Gr. n. *siderus* iron; L. fem.n. *capsa* box; M.L. fem.n. *Siderocapsa* iron box.

One to many **spherical to ovoid cells embedded in a common capsule,** partially **encrusted with iron and/or manganese compounds.** Cells colorless but outer portion of capsule yellowish to brownish from metal compounds, especially when old. Motility not known.

Common in fresh water; may be attached to surface of water plants or other substrates and in plankton or free floating. Generally aerobic but can grow under reduced oxygen tensions. Develop under neutral or slightly alkaline conditions.

As they have not been grown in pure culture, differentiation is based on morphological characteristics.

Type species: *Siderocapsa treubii* Molisch 1910, 30.

Key to the species of genus **Siderocapsa**

I. Grow attached to substrates, especially leaves of fresh water plants.
   A. Many cells embedded in a relatively large mass of capsular material.
      1. Cocci, 0.4–0.6 μm in diameter.

1. *Siderocapsa treubii*

      2. Cocci, 0.7–1.8 μm in diameter.

2. *Siderocapsa major*

   B. Single cocci in thin, closely fitting capsule, 0.5–0.75 μm.

3. *Siderocapsa monoica*

II. Grow in plankton or free floating in surface layers of fresh water.
   A. Capsules thin, closely fitting and more or less homogeneous.
      1. Single coccus, 0.2–0.5 μm in diameter, per capsule.

4. *Siderocapsa anulata*

   B. Capsular material more or less abundant, gelatinous and may have stratified, indistinctly radial or granular structure.
      1. More or less ovoid cells, 0.5–0.6 by 0.8 μm, usually in pairs in capsule.

5. *Siderocapsa geminata*

---

† Deceased.

2. Generally many spherical cells in each capsule (colony).
   a. Generally neustic forms, up to 90 colonies aggregate in larger zoogloeal masses. Cells 1.0–1.2 µm in diameter.
             6. *Siderocapsa coronata*
   b. Generally planktonic forms, especially in hypolimnion; colonies do not aggregate.
    bb. Cells 0.4–1.0 µm in diameter, one to four per colony.
             7. *Siderocapsa arlbergensis*
    bbb. Cells 1–2 µm in diameter, usually up to 60 cells per colony, rarely one or two.
             8. *Siderocapsa eusphaera*

Description of the species of genus **Siderocapsa**

1. **Siderocapsa treubii** Molisch 1910, 30.

treu′bi.i. M.L. gen.n. *treubii* of Treub; named for Professor Treub, an early director of the tropical garden, Buitenzorg, Java.

Capsules surrounded by ferric hydroxide, which is also deposited on surface of water plants. Found attached to leaves of *Elodea, Nymphaea, Sagittaria, Salvinia* etc. Widely distributed.

2. **Siderocapsa major** Molisch 1910, 13.

ma′jor. L. comp.adj. *major* larger.

Similar to *S. treubii*, cells larger and capsule less sharply defined.

Originally isolated from *Spirogyra* sp. Widely distributed on fresh water plants.

Comment: As the difference between *S. treubii* and *S. major* is mainly size and intermediate forms occur, some authors (Hartman and Henrici, 1939) consider the latter taxon an extreme form of the former. This opinion is supported by Drake (1965) who doubts the existence of *S. major* and *S. monoica*.

3. **Siderocapsa monoica** Naumann 1921, 49.

mo.noi′ca. Gr. adj. *monos* single; Gr. n. *oicus* house; M.L. fem.adj. *monoica* solitary dwelling.

A single cell per capsule; capsule relatively thin but older forms are encrusted with ferric hydroxide and dark brown in color.

Found on *Potamogeton natans* in Sweden.

4. **Siderocapsa anulata** Kalbe, Keil and Thiele 1965, 35.

a.nu.la′ta. L. fem.adj. *anulata* with a ring.

Capsule moderately thick when embedded with iron compounds, outer diameter 1.2–1.9 µm, inner diameter 0.65–1.9 µm; edge often has a cog-wheel appearance; yellow to brown color.

Found in plankton of Kummerower See, Mecklenburg, Germany.

5. **Siderocapsa geminata** Skuja 1956, 19.

gem.in.a′ta. L. v. *gemino* double; M.L. fem.-part.adj. *geminata* doubled, paired.

Cells in pairs, although single cells often occur especially after division of a colony. Capsule round, 7–11 µm in diameter; at first homogeneous and almost colorless, the outer portion becomes stratified concentrically and brownish colored due to impregnation with iron or manganese compounds.

Found in hypolimnion of some Swedish lakes; also in epilimnion at time of vernal and autumnal circulation.

6. **Siderocapsa coronata** Redinger 1931, 413.

co.ro.na′ta. L. fem.part.adj. *coronata* crowned.

Two to eight cells in an irregularly spherical capsule, up to 24 µm in diameter, more or less homogeneous or indistinctly radially structured; at first nearly colorless, the outer portion later becoming impregnated with iron or manganese compounds.

Generally a neustic form of surface layers in alpine or subalpine lakes and pools. Ruttner (1937) reported maximum distribution at reduced oxygen tension and at a depth of 17.5–27.5 meters.

7. **Siderocapsa arlbergensis** Wawrik 1956, 21.

arl.berg.en′sis. L. adj.suff. *-ensis* belonging to; M.L. adj. *arlbergensis* from Arlberg, a town in Austria.

Capsule irregularly spherical, 6–15 µm in diameter, uniformly granular; at first colorless subsequently outer portions more or less ochraceous. Usually single colony, a few may aggregate in irregular, small zoogloeal masses.

Isolated from alpine pools at Arlberg, Tyrol, Austria.

8. **Siderocapsa eusphaera** Skuja 1948, 13.

eu.sphae′ra. Gr. prep. *eu* true, nice, beautiful; Gr. n. *sphaera* ball, sphere; M.L. n. *eusphaera* a beautiful sphere.

Older cells may have a central gas vacuole. Capsules regular spheres, 10–50 µm in diameter, usually stratified with two to three concentric brownish layers due to impregnation with iron compounds, inner layers darker colored. Occurs singly, free floating in plankton.

Found in hypolimnion, sometimes in epilimnion of some larger meso- and eutrophic lakes in Sweden.

Genus II. **Naumanniella** Dorff 1934, 19

G. A. ZAVARZIN

Nau.man.ni.el'la. M.L. dim.ending -*ella;* M.L. fem.n. *Naumanniella* named for Einar Naumann, a Swedish limnologist.

Rod-shaped cells, each surrounded by a delicate, very regular capsule. Capsule becomes golden yellow from deposits of iron compounds, which **emphasizes the margin and gives the cell the appearance of a minute diatom** or link of a chain ("torus"). Manganese oxide stains the capsule black and renders it opaque.

Complex organic compounds of iron and manganese are decomposed.

Widely distributed in iron-bearing waters.

Type species: *Naumanniella neustonica* Dorff 1934, 20.

Description of species of genus **Naumanniella**

1. **Naumanniella neustonica** Dorff 1934, 20.
neus.to'ni.ca. Gr. adj. *neustus* swimming, floating; M.L. adj. *neustonica* of the neuston or surface film.

Cells rod shaped, 1.8–3.3 by 4.9–10.0 μm (cell + torus), straight, occurring singly; may be constricted. Multiplication by binary fission.

Growth in ferrous citrate solutions produces a golden brown surface film and voluminous cloudy deposits of iron hydroxide. Growth in other media has not been reported.

2. **Naumanniella minor** Dorff 1934, 21.
mi'nor. L. comp.adj. *minor* smaller.

Cells frequently curved or spiral shaped, occur singly; cells + torus 1.2–1.5 by 3.1–3.6 μm.

3. **Naumanniella catenata** Beger 1941, 321.
ca.te.na'ta. L. part.adj. *catenatus* in chains.

Cells occur as chains of slightly curved rods; cells + torus 1.0–1.2 by 4.9–5.5 μm.

4. **Naumanniella pygmaea** Beger 1949, 65.
pyg.mae'a. Gr. adj. *pygmaea* dwarfish.

Cells straight rods with rounded ends, occurring singly; cell + torus 1.0 by 2.0 μm. Multiplication by budding has been claimed.

Grows slowly on soil extract agar producing minute brownish colonies (Mun, 1967).

5. **Naumanniella elliptica** Beger 1949, 66.
el.lip'ti.ca. Gr. adj. *elliptica* elliptical.

Cells ellipsoidal with pronounced torus; cell + torus 2.0 by 2.5–3.0 μm.

6. **Naumanniella polymorpha** Ten 1969, 698.
po.ly.mor'pha. Gr. adj. *poly* many; Gr. n. *morphus* form, shape; M.L. adj. *polymorpha* of many shapes.

Cells ellipsoidal, sometimes coccoid, 0.7–1.0 by 1.0–2.0 μm. Multiplication by budding; the bud is motile and free from metal oxides.

Colonies in manganese carbonate or manganese acetate agar are dark brown, minute and swarm on the surface. No growth in ordinary organic media. Oxidizes manganous but not ferrous compounds.

Originally isolated from soils of Sachalin with high manganese content.

Genus III. **Ochrobium** Perfil'ev in Visloukh 1921, 88

G. A. ZAVARZIN

O.chro'bi.um. Gr. n. *ochra* yellow ochre; Gr. n. *bios* life; M.L. neut.n. *Ochrobium* yellow life referring to color imparted by iron oxides.

Ellipsoidal to rod-shaped cells, 0.5–3.0 by 1.5–5.0 μm, partially surrounded by a marginal thickening (torus) that is heavily impregnated with iron. The torus remains open at one end, resembling a horseshoe; pairs of cells may unite with the open ends together. The cells are surrounded by a delicate transparent capsule that contains a small amount of iron; clumps of several cells may form. May be motile by means of two unequal polar flagella.

Widely distributed in iron-bearing fresh water.

Type species: *Ochrobium tectum* Perfil'ev in Visloukh 1921, 88.

Further Comment

The cells are much like those of the algal genus *Pteromonas*, but smaller. Thus the organism may not be a bacterium and the genus should be considered *incertae sedis*.

Description of the species of genus **Ochrobium**

1. **Ochrobium tectum** Perfil'ev in Visloukh 1921, 88.

tec′tum. L. past part.neut *tectum* of L. v. *tego* covered.

Morphology as for genus.

Originally found in the region around Leningrad; later found independently in Sweden by Naumann (1929) and by Beger (1949) in wells near Berlin.

Genus IV. **Siderococcus** *Dorff 1934, 9*

G. A. ZAVARZIN

Si.de.ro.coc′cus. Gr. n. *siderus* iron; Gr. n. *coccus* a berry, sphere; M.L. masc.n. *Siderococcus* iron coccus.

Cells spherical 0.2–0.5 μm in diameter; occur singly or in small motile colonies of 6–15 cells resembling an ear of corn (*Zea mays*). **Motile** but type of flagellation is not known. Cells have filamentous appendages **but no capsules. Multiplication is apparently by budding,** the cells becoming pear shaped.

The organisms usually develop as a homogeneous, continuous accretion although sometimes chains of coccoid cells are arranged in a net-like structure. **Cells are not encrusted with iron compounds** although older colonies may become covered with iron oxides and the cells themselves are hardly recognizable. The orange-yellow deposits contain only ferric hydroxide.

So far growth has been observed only in the natural habitat and in isolated mud samples; *Siderococcus* microzones are found in mud horizons with low concentrations of oxygen and a neutral pH (Perfil'ev and Gabe, 1964). Widely distributed in fresh waters and bottom deposits.

Type species: *Siderococcus limoniticus* Dorff 1934, 9.

Description of species of genus **Siderococcus**

1. **Siderococcus limoniticus** Dorff 1934, 9. li.mo.ni′ti.cus. M.L. n. *limonitum* limonite, a

mineral, ferrous oxide; M.L. adj. *limoniticus* of limonite.

Description as for genus.

Important Notes

for

Users of this Edition

1. Always read both generic and species descriptions because characters listed in the generic description are not usually listed in the species descriptions.

2. In tables, characters common to all taxa are not shown but may be listed in footnotes.

3. Generally in tables (exceptions are clearly indicated in footnotes, q.v.) the meanings of symbols are as follows:

+ more than 90 % strains positive
− more than 90 % strains negative
d 11–89 % strains positive
( ) delayed reaction
w weak reaction
D Different reactions in different taxa (species of a genus or genera of a family)
v strain instability (NOT differences between strains)

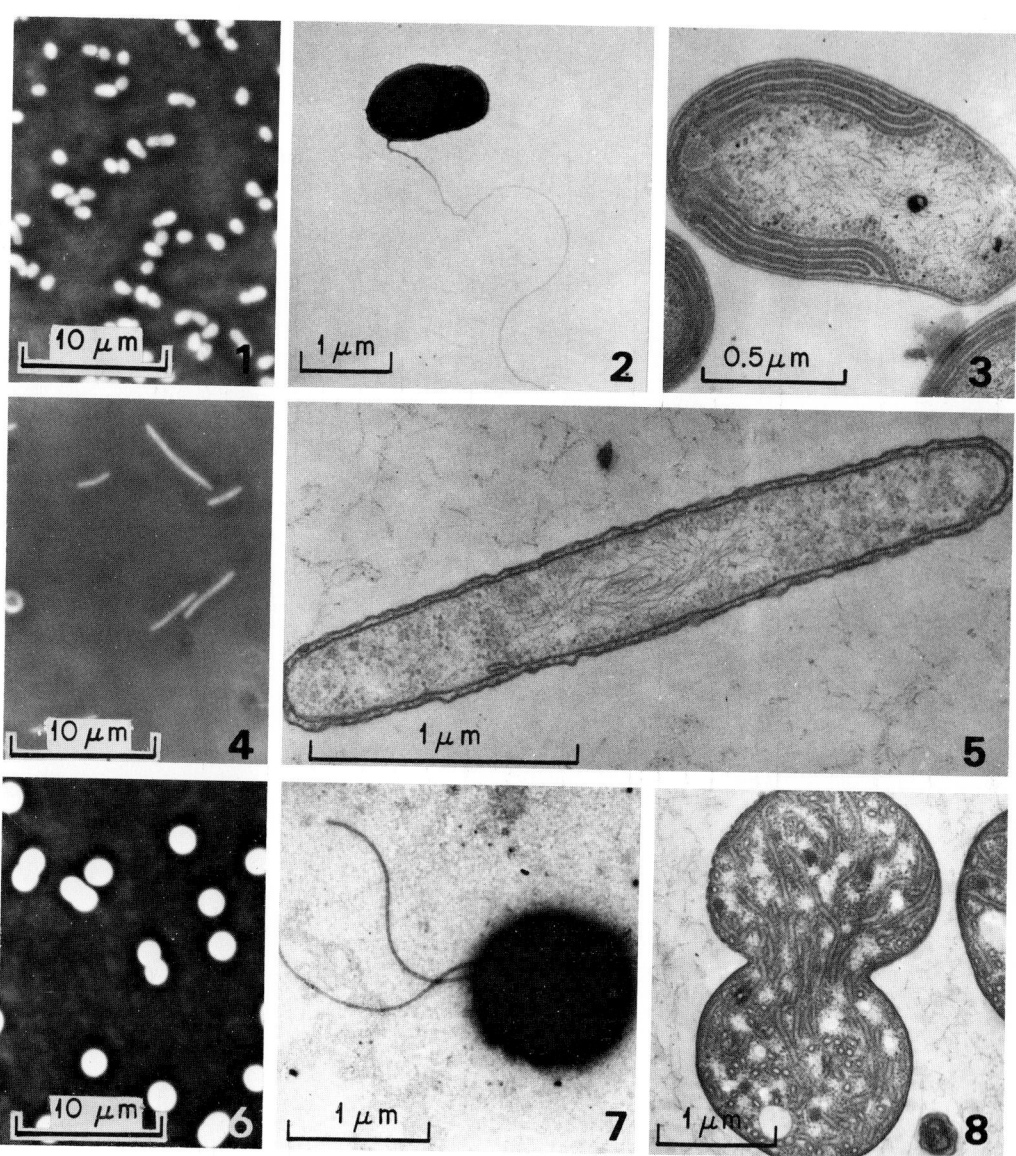

Plate 12.1. *Nitrobacteraceae*

Fig. 1. Phase-contrast photomicrograph of *Nitrobacter winogradskyi* showing pear-and wedge-shaped cells. ✕ 2000.

Fig. 2. Electron micrograph of a negatively stained cell of *Nitrobacter winogradskyi* showing single subpolar flagellum. ✕ 12,800.

Fig. 3. Electron micrograph of a section of *Nitrobacter winogradskyi* showing polar cap of peripheral cytomembranes. ✕ 41,000.

Fig. 4. Phase-contrast photomicrograph of *Nitrospina gracilis* showing the shape and size of cells. ✕ 2000.

Fig. 5. Electron micrograph of a section of *Nitrospina gracilis* showing the lack of extensive cytomembranes found in most other nitrifying bacteria but shows small bleb-like intrusion of plasma membrane. ✕ 37,500.

Fig. 6. Phase-contrast photomicrograph of *Nitrococcus mobilis* showing size and shape of cells. ✕ 2000.

Fig. 7. Electron micrograph of a negatively stained cell of *Nitrococcus mobilis* showing pair of flagella. ✕ 19,300. Figures 5 and 7 from Watson, S. W., Waterbury, J. B.: Characteristics of Two Marine Nitrite-Oxidizing Bacteria, *Nitrospina gracilis* nov. gen. nov. sp. and *Nitrococcus mobilis* nov. gen. nov. sp. Arch Mikrobiol. *77:* 203–230, 1971. Berlin-Heidelberg-New York: Springer.

Figure 8. Electron micrograph of a section of *Nitrococcus mobilis* showing tubular type cytomembranes extending throughout the cytoplasm. ✕ 16,000. With permission of the American Society for Microbiology. Journal of Bacteriology, *107;* 563–569, 1971.

Plate 12.2. *Nitrobacteraceae*

Fig. 1. Phase-contrast photomicrograph of *Nitrosomonas europaea* showing shape and size of cells. ✕ 2000.

Fig. 2. Electron micrograph of a negatively stained cell of *Nitrosomonas europaea* showing pair of flagella with subpolar insertion. ✕ 13,600.

Fig. 3. Electron micrograph of a section of *Nitrosomonas europae* showing peripheral cytomembranes. ✕ 35,200.

Fig. 4. Phase-contrast photomicrograph of *Nitrosospira briensis* showing tightly coiled spirals. ✕ 2000.

Fig. 5. Electron micrograph of a negatively stained cell of *Nitrosospira briensis* showing peritrichous flagella. ✕ 6,000.

Fig. 6. Electron micrograph of a section passing through a spiral of a single cell of *Nitrosospira briensis*, note the lack of cytomembranes. ✕ 32,500. Figures 5 and 6 from Watson, S. W.: Reisolation of *Nitrosospira briensis* S. Winogradsky and H. Winogradsky, 1933. Arch Mikrobiol *75:* 179–188, 1971. Berlin-Heidelberg-New York: Springer.

Fig. 7. Phase-contrast photomicrograph of *Nitrosococcus oceanus* showing shape and size of cells. ✕ 2000.

Fig. 8. Electron micrograph of a shadowed cell of *Nitrosococcus oceanus* showing peritrichous flagella. ✕ 10,000.

Fig. 9. Electron micrograph of a section of *Nitrosococcus oceanus* showing cytomembranes arranged as flattened lamellae in the central region of the cell. ✕ 22,500.

Fig. 10. Phase-contrast photomicrograph of *Nitrosolobus multiformis* showing the lobular shape of cells. ✕ 2000.

Fig. 11. Electron micrograph of a negatively stained cell of *Nitrosolobus multiformis* showing peritrichous flagella. ✕ 8900.

Fig. 12. Electron micrograph of a section of *Nitrosolobus multiformis* showing its lobular shape and cytomembranes which partially compartmentalize the cell. ✕ 22,500. Figures 11 and 12 from Watson, S. W., Graham, L. B., Remsen, C. C., Valois, F. W.: A Lobular, Ammonia-Oxidizing Bacterium, *Nitrosolobus multiformis* nov. gen. nov. sp. Arch Mikrobiol *76:* 183–203, 1971. Berlin-Heidelberg-New York: Springer.

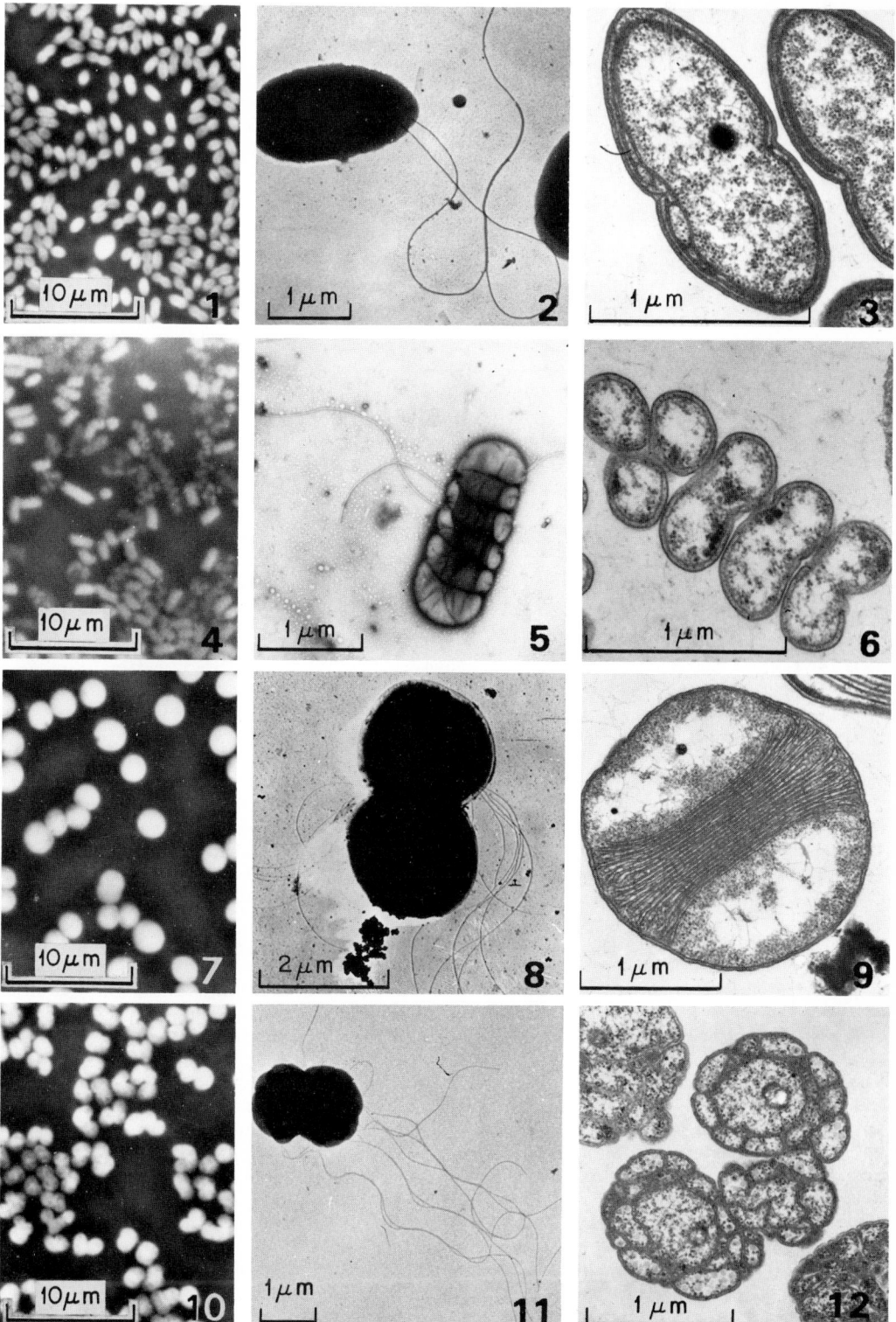

PART 13

METHANE-PRODUCING BACTERIA

MARVIN P. BRYANT

FAMILY I. **METHANOBACTERIACEAE** BARKER 1956, 15

Me.tha.no.bac.ter.i.a′ce.ae. M.L. neut.n. *Methanobacterium* type genus of the family; *-aceae* ending to denote a family; M.L. fem.pl.n. *Methanobacteriaceae* the *Methanobacterium* family.

Cells are **rods or cocci, motile** or **non-motile,** Gram-positive or Gram-negative. Spores have not been found in any pure culture. Very strict **anaerobes** which obtain energy for growth via **formation of methane** by reduction of CO_2, utilizing electrons generated in the oxidation of compounds such as hydrogen and formate, or via fermentation of compounds such as acetate and methanol with formation of methane and CO_2 as the products. A **highly specialized physiological group** which does not utilize carbohydrate or proteinaceous materials or organic compounds, other than those listed above, as energy sources.

Further Comments

Although previous classifications, including the 7th edition of THE MANUAL, placed primary emphasis on morphology and dispersed the methanogenic bacteria among the genera and families of better known bacteria of similar form, Barker (1956, 14) emphasized that "those investigators who have had personal experience with the methane bacteria have been impressed by the striking physiological characteristics of all members of the group, and have preferred to think of them as belonging to a physiological family. . . ."

Although much remains to be done to understand the biology of these bacteria adequately, their physiological uniqueness is even more apparent in 1972 than it was in 1956. Because of their high degree of specialization in regard to energy sources, their growth requirement for very strict anaerobic conditions and the probability that they are killed more rapidly by relatively short times of exposure to air than other anaerobic forms, they are not encountered except by workers specifically searching for them.

The energy sources utilized for their growth appear at the present time to be even fewer than previously supposed. Bryant *et al.* (1967) showed that the one supposedly authentic pure culture, which fermented ethanol and certain aliphatic alcohols other than methanol, was an association of two species. *Methanobacterium* strain M.O.H., isolated from the mixed culture, utilizes only hydrogen gas in the reduction of CO_2 to methane and, in the association, the hydrogen is produced by the second non-methanogenic species during the oxidation of ethanol to acetate. This and further studies (Bryant, 1969; Reddy *et al.*, 1972) indicate that methanogenic species probably do not utilize alcohols other than methanol. Based on the above studies on ethanol fermentations and on the fact that authentic pure cultures of methanogenic species reported to attack any aliphatic, organic, carboxylic acids other than acetate and formate have not been obtained, Bryant *et al.* (1967) suggested that fatty acids other than acetate and formate may not be attacked by methanogenic bacteria as such but may be attacked by associations similar to those which attack ethanol. Thus, the only energy sources definitely known to be utilized by methanogenic bacteria include hydrogen-carbon dioxide, formate, acetate, methanol and carbon monoxide. All methane bacteria so far studied for the feature utilize hydrogen.

The classification of methanogenic bacteria is in drastic need of modification. For example, none of the type species of any of the genera described herein has ever been obtained in authentic pure culture and neither type nor reference strains are extant. Thus, one must interpret the described features of the type species with caution; they are included here only because they are the type species now recognized.

The genus *Methanobacterium*, in which the type species is a Gram-negative, non-motile, cylindrical rod, now includes other species which should probably be placed in new genera. *Methanobacterium mobile* is a monotrichously flagellated motile organism. *Methanobacterium ruminantium* is strongly Gram-positive and resembles some streptococci in shape, and most of the presently available strains of *Methanobacterium formicicum* are definitely Gram-positive (Langenberg *et al.*, 1968).

Some new species and genera are known but names have not been validly published. For example, Smith (1966) has isolated from sludge and briefly described a *"Methanospirillum."* This organism is abundant in sludge and has also been isolated in the author's laboratory and by James Ferry in the laboratory of R. S. Wolfe. It is Gram-negative, has tufts of polar flagella and ferments hydrogen-carbon dioxide and formate (Wolfe, 1971). Smith (1966) has also isolated from sludge strains of *Methanococcus* sp. which are Gram-positive, non-motile diplococci which ferment hydrogen-carbon dioxide and formate and do not require high pH for growth. Another unnamed species fermenting these same materials was a non-motile, Gram-negative, heavy-bodied, short rod (unpublished data of C. A. Hollowell, N. R. Crabill and M. J. Wolin, 1969). Excellent photographs of cells of many of these species as well as descriptions of culture methods and biochemical studies are given by R. S. Wolfe (1971, 114).

Key to the genera of family **Methanobacteriaceae**

I. Rods or chain-forming lancet-shaped coccoids.
Genus I. *Methanobacterium*
II. Cocci other than chain-forming lancets.
    A. Large cocci in packets.
Genus II. *Methanosarcina*
    B. Cocci occurring singly, in pairs or clumps.
Genus III. *Methanococcus*

Genus I. **Methanobacterium** Kluyver and Van Niel 1936, 399

Me.tha.no.bac.ter′i.um. M.L. n. *methanum* methane; Gr. n. *bakterion* a small rod; M.L. neut.n. *Methanobacterium* the methane (-producing) rodlet.

Curved, crooked, to straight **non-sporing rods,** long and filamentous to **coccoid,** about 0.5–1.0 μm in width, Gram-positive to Gram-negative, non-motile or motile with monotrichous polar flagellation.

Chemolithotrophs: **Energy for growth** obtained via **reduction of CO₂ to methane utilizing H₂** and sometimes formate as the electron donor. The type species supposedly utilized acetate and butyrate as energy sources but it was never obtained in pure culture (see *Further Comments* under the description of the family).

Autotrophic to heterotrophic in nutrition.

Very strict **anaerobes.**

Mesophiles to extreme thermophiles.

Widely distributed in nature being found in anaerobic habitats such as sediments of natural waters, soil, anaerobic sewage digestors and the gastrointestinal tract of animals.

Type species: *Methanobacterium soehngenii* Barker 1936, 437.

Key to the species of the genus **Methanobacterium**

I. Rods which produce methane from acetate.
1. *Methanobacterium soehngenii*
II. Rods or chained coccoids which do not produce methane from acetate.
    A. Slender, cylindrical, straight to irregularly crooked, non-motile rods.
        1. Grow at mesophilic temperatures.
2. *Methanobacterium formicicum*

2. Grow at thermophilic temperature, 40–75 C.

3. *Methanobacterium thermoautotrophicum*

B. Gram-positive coccoids to short, lancet-shaped rods usually in chains.

4. *Methanobacterium ruminantium*

C. Gram-negative, motile, short, straight or slightly curved rods.

5. *Methanobacterium mobile*

Description of the species of genus **Methanobacterium**

1. Methanobacterium soehngenii Barker 1936, 437.

soehn.ge′ni.i. M.L. gen.n. *soehngenii* of Söhngen; named for N. L. Söhngen who first studied this organism.

Rods straight to slightly curved, moderately long and characteristically joined into long chains which often lie parallel to one another so as to form bundles. Non-motile. Gram-negative.

Acetate and butyrate are fermented with the production of methane and carbon dioxide. Ethanol and butanol are not fermented. As this organism was never obtained in pure culture, it appears quite doubtful that the methanogenic organism *per se* utilized butyrate (see *Further Comments* on the family).

Obtained only from enrichment cultures containing acetate or butyrate as the only organic compound and inoculated with canal mud. Numbers in the source material were not estimated.

2. Methanobacterium formicicum Schnellen 1947, 85.

for.mi′ci.cum. M.L. n. *acidum formicum;* M.L. adj. *formicicum* pertaining to formic acid.

Original description supplemented by material from Mylroie and Hungate (1954, 58) and Smith (1966, 159).

Cells are slender and cylindrical with blunt rounded ends. Some chains and filaments are commonly seen and some cells are unevenly crooked. Cell width is from 0.4–0.8 μm depending on the strain and length, in a single strain, from 2–15 μm. Mylroie and Hungate (1954) indicate that cells are Gram-negative but strains of Smith (1966) and most from author's laboratory (unpublished data) are Gram-positive. Electron micrographs of both negatively stained cells and thin sections indicated that two strains were Gram-positive and contain characteristic intracytoplasmic membranous elements (Langenberg *et al.*, 1968). Non-motile and non-sporing.

Surface colonies are white to gray, flat and filamentous. Deep colonies appear as profusely filamented spheroids. Colonies in agar roll tubes containing H_2-CO_2 gas phase appear after 3–5 days at 37 C and in about 2 weeks attain a diameter of 2–5 mm, depending on the number of colonies in relation to the H_2 energy source.

Type of growth in liquid medium with CO_2-H_2 gas phase depends on the strain and results either in even turbidity or highly granular clumps which do not break up even with vigorous agitation (Langenburg *et al.*, 1968).

Energy-yielding metabolism involves reduction of CO_2 to methane with H_2 or formate serving as the electron donor. Formate may also replace CO_2 as the source of carbon for methane but is probably first converted to CO_2 via a formate hydrogenlyase system. Carbohydrates, amino acids, ethanol, methanol, acetate, propionate, butyrate and lactate are not fermented. Carbon monoxide may be fermented by some strains (Kluyver and Schnellen, 1947) and some strains which probably belong to the species do not utilize formate (Bryant *et al.*, 1968; Langenberg *et al.*, 1968). Nitrate, sulfate or fumarate are not utilized as electron acceptors in place of CO_2.

At least some strains are probably capable of autotrophic growth (Mylroie and Hungate, 1954; Bryant *et al.*, 1971); ammonia may be essential as the main nitrogen source and acetate and cysteine may be quite stimulatory to growth (Bryant *et al.*, 1971).

Good growth at 38 and 45 C and no growth at 55 C.

Numerous in sludge from anaerobic sewage digestors and found in anaerobic sediments of natural waters and, in low numbers, in the rumen of cattle (Bryant, 1965).

References strains available from M. P. Bryant or P. H. Smith.

Further Comments

The culture known as *Methanobacterium* M.O.H. and isolated from the mixed culture previously known as *Methanobacillus omelianskii* appears to be identical with *M. formicicum* but it does not utilize formate (Langenberg *et al.*, 1968). The G + C content of the DNA of this strain is 38 moles % (buoyant density) (Zeikus and Wolfe, 1972).

It is possible that further studies will show that two species actually are included in the above description; one, a Gram-negative slender rod and the other, a Gram-positive slender rod.

3. Methanobacterium thermoautrophicum Zeikus and Wolfe 1972, 712. (*Methanobacterium* *thermoautrophicus* (*sic*) Zeikus and Wolfe 1972, 712.)

ther.mo.au.to.tro'phi.cum. Gr. adj. *thermus* hot; Gr. pref. *auto* self; Gr. n. *trophos* one who feeds; M.L. neut.adj. *thermoautotrophicum* thermo(philic) and autotrophic.

Cells are slender, cylindrical rods 0.35–0.6 μm in width and 3–7 μm long but often with filaments from 10–120 μm in length. Cells are often irregularly crooked. Gram-positive, non-motile and non-sporing. Electron micrographs of thin sections reveal the characteristic intracytoplasmic membranous elements similar to those of *Methanobacterium formicicum* (personal communication, G. Zeikus).

Deep colonies in roll tubes are tannish white, roughly spherical, diffuse and filamentous.

Energy-yielding metabolism involves reduction of CO_2 to methane with H_2 serving as the electron donor. Neither growth nor methane formation occurred within 24 hr when the H_2 electron donor was replaced by ethanol, methanol, formate, acetate or pyruvate. Other possible electron donors and acceptors and longer incubation times were not studied.

The organism grows rapidly under autotrophic conditions with CO_2 as sole carbon source and ammonia as sole nitrogen source.

Optimum temperature for growth is 65–70 C; maximum, 75 C; minimum, 40 C.

pH optimum is about 7.2–7.6. Little or no growth below pH 6.0 or above pH 8.8.

The G + C content of the DNA is 52 moles % (buoyant density).

Isolated from sludge from anaerobic sewage digestor.

Reference culture available from Dr. J. G. Zeikus, University of Wisconsin, Madison.

4. Methanobacterium ruminantium Smith and Hungate 1958, 717.

ru.mi.nan'ti.um. L. part.adj. *ruminans, ruminantis* ruminating; M.L. neut.pl.n. *ruminantia* ruminants; M.L. pl.gen.n. *ruminantium* of ruminants.

Short lancet-shaped to oval rod or coccus 0.7 μm in width and 0.8–1.8 μm in length. May occur predominantly in pairs but usually in chains which may contain up to 20 or more cells resembling streptococci. Very strongly Gram-positive, even in relatively old cultures. Non-motile. The coccoid appearance and large number of cross walls are emphasized in electron micrographs (Langenberg *et al.*, 1968).

Surface colonies in roll tube cultures with H_2-CO_2 gas phase are translucent, convex, circular with entire margins and usually off-white to yellow in color. They may be visible after about 3 days of incubation at 37 C and may reach a diameter of 3–4 mm depending on the number of colonies and amount of energy source available. Colonies in deep agar are lenticular.

Growth in liquid medium with H_2-CO_2 gas phase is evenly turbid or may floc. In the latter case vigorous shaking will result in breakdown of the floc and even suspension.

Energy-yielding metabolism involves reduction of CO_2 to methane with H_2 or formate serving as the electron donor. Utilization of formate may be slow. Formate also replaces CO_2 as the source of carbon for methane but is probably first converted to CO_2 via a formate hydrogenlyase system. Carbohydrates, amino acids, methanol, ethanol, acetate, isobutyrate, propionate, valerate, caproate, succinate and pyruvate are not utilized as electron donors. Pyruvate does not replace CO_2 as carbon source for methane formation because cells are impermeable to pyruvate.

In addition to the CO_2 and H_2 required as energy source, acetate is essential as a major source of cell carbon, ammonia as the major source of cell nitrogen and sulfide may be essential as the main source of cell sulfur. One or more B vitamins are required. Amino acids, peptides and organic materials in yeast extract are not used effectively as major sources of cell carbon, nitrogen or sulfur. Some rumen strains also require for growth 2-methyl-*n*-butyrate, a few amino acids and an as yet unidentified vitamin-like growth factor present in rumen fluid and sewage sludge and produced by other methanogenic bacteria (Bryant *et al.*, 1971).

Very strict anaerobes. As little as 0.8 μl O_2/5 ml of medium may inhibit growth.

Temperature optimum for growth is about 37–43 C. Little or no growth at 47 C and some rumen strains do not grow at 33 C.

Growth does not occur above pH 8.0 nor below pH 6.0.

Widely distributed in nature. It is the major rumen species and is the only methanogen so far isolated from the gastrointestinal tract of man (Nottingham and Hungate, 1968). It is one of the most numerous species in anaerobic sewage digestors (Smith, 1966) and in the stomach of langur monkeys (Bauchop and Martucci, 1968).

Reference strains available from M. P. Bryant or P. H. Smith.

5. Methanobacterium mobile Paynter and Hungate 1968, 1951. (*Methanobacterium mobilis* (*sic*) Paynter and Hungate 1968, 1951.)

mo'bi.le. L. neut.adj. *mobile* movable, motile.

Straight to slightly curved rods, 0.7 μm wide and 1.5–2.0 μm long, with rounded ends and usually occurring as singles or pairs but not chains. Motile with monotrichous polar flagellation. Gram-negative.

Colonies are small, translucent, entire, convex

and colorless to pale yellow. They become visible after 4 days at 39 C and reach a maximum diameter of 0.7–1.0 mm after 15 days. Deep colonies are lenticular and 0.5–0.7 mm in diameter in 15 days.

Energy-yielding metabolism involves reduction of CO_2 to methane with H_2 or formate serving as the electron donor. Formate, it is presumed, can replace CO_2 as the electron acceptor. Compounds not utilized as electron donors include acetate, propionate, butyrate, isobutyrate, valerate, isovalerate, caproate, succinate, glucose, pyruvate, methanol, ethanol, propanol, isopropanol and butanol.

The organism has not been grown satisfactorily in media devoid of rumen fluid or extracts of the mixed ruminal bacteria. An unidentified factor similar to that required by rumen strains of *Methanobacterium ruminantium* is required (Bryant *et al.*, 1971).

Temperature of maximum gas uptake by growing cultures is about 40 C. Uptake was very slow at 30 and 45 C and did not occur within 28 days at 28 or 50 C.

The maximal rate of gas uptake during growth was at pH 6.1–6.9. Growth occurred at pH 5.9 and 7.7 but not at pH 5.6.

Isolated repeatedly from rumen contents of one cow.

Reference cultures available from P. H. Smith, University of Florida, Gainesville.

Genus II. **Methanosarcina** *Kluyver and van Niel 1936, 400, emend. mut. char. Barker 1956, 14*

Me.tha.no.sar.ci'na. M.L. n. *methanum* methane; L. n. *Sarcina* a generic name; M.L. fem.n. *Methanosarcina* methane sarcina.

Large spherical cells, 1.5–2.5 μm in diameter, generally occurring in regular **packets.** Gram-variable to Gram-positive. Non-motile.

The energy metabolism involves **formation of methane from acetate** and sometimes from **methanol,** CO and possibly, from butyrate. H_2 may also be utilized in reduction of methanol.

Strict anaerobes.

Good growth at 30–37 C.

Found in muddy sediments of natural waters and in anaerobic sewage digestors.

Type species: *Methanosarcina methanica* (Smit) Kluyver and van Niel 1936, 400.

Description of the species of genus **Methanosarcina**

1. **Methanosarcina methanica** (Smit) Kluyver and van Niel 1936, 400. (Methaansarcine Söhngen 1906, 104; *Zymosarcina methanica* Smit 1930, 25; *Sarcina methanica* (Smit) Weinberg, Nativelle and Prévot 1937, 1032.)

me.tha'ni.ca. M.L. n. *methanum* methane; M.L. fem.adj. *methanica* pertaining to methane.

Description taken from Söhngen (1906, 104), Smit (1930, 25) and Barker (1936, 428).

Spheres, 2.0–2.5 μm in diameter, occurring in characteristic packets, usually of eight or more but occasionally in smaller groups. Gram-variable and non-motile.

In acetate agar colonies are very small, 50–100 μm, and show gas splits.

Energy sources include acetate and, perhaps, butyrate which are fermented with formation of methane and CO_2. Ethanol, peptones and carbohydrates are not fermented. Methanol, CO and H_2-CO_2 apparently were not tested.

Nutrition is simple in that "purified" cultures grow in media with acetate as the sole organic compound, ammonium salts as nitrogen source and sulfide as sulfur source and reducing agent.

Found in swamp waters and mud and anaerobic sewage digestor sludge.

No reference strains are available.

Further Comments

Strains identified as this species have never been isolated in pure culture and have not been studied adequately in comparison with strains identified as *Methanosarcina barkeri*, see below, which have been obtained in pure culture and more adequately studied. The two species are probably identical.

2. **Methanosarcina barkeri** Schnellen 1947, 73. (*Methanosarcina barkerii* (*sic*) Schnellen 1947, 73; *Sarcina barkeri* Breed and Smit 1957, 470.)

bar'ker.i. M.L. gen.n. *barkeri* of Barker; named for H. A. Barker who has made many of the definitive studies on this and other methanogenic bacteria.

Original material supplemented with material from Kluyver and Schnellen (1947, 60) and Stadtman and Barker (1951, 81).

Spheres, 1.5–2.0 μm in diameter, occurring mostly in packets of eight or less but sometimes in large masses. Non-motile. Gram-positive.

Deep colonies in methanol agar with inorganic salts are whitish and 0.5–1.0 mm.

In liquid methanol medium growth may occur as zoogloeal masses or as flocculent sediment with active gas formation.

Energy-yielding metabolism involves reduction of CO_2 with H_2 yielding methane. Methanol, acetate and CO also are fermented with formation of methane and CO_2. The methyl group of methanol and acetate is reduced to methane without intermediate conversion to CO_2. Carbohydrates, amino acids, formate, ethanol, propionate and butyrate are not fermented.

Growth occurs in media with methanol or acetate as the sole organic compound and with ammonia as nitrogen source and sulfide as sulfur source. Growth and methane formation are much more rapid in methanol than in acetate media.

Growth good at pH 7.0.

Catalase negative. Nitrate is not reduced.

Found in mud and in sludge from anaerobic sewage digestors.

Reference strains available from P. H. Smith, University of Florida, Gainesville; or M. P. Bryant.

Genus III. **Methanococcus** *Kluyver and van Niel 1936, 400, emend. mut. char. Barker 1936, 430*

Me.tha.no.coc'cus. M.L. n. *methanum* methane; M.L. n. *coccus* a spherical cell; M.L. masc.n. *Methanococcus* methane coccus.

Spherical cells, occurring **singly,** in **pairs** or in **masses.** Motile or non-motile. Gram-positive or -negative. (It should be noted that the genus *Methanobacterium* includes Gram-positive, chain-forming coccoids.)

The **energy metabolism involves the formation of methane** from H_2 and CO_2, formate, acetate or butyrate. Authentic pure cultures have not been shown to utilize acetate or butyrate in methane formation and it is quite possible that butyrate is not fermented by methanogenic bacteria *per se* (see *Further Comments* on the family).

Strict anaerobes.

The type species: *Methanococcus mazei* Barker 1936, 433.

Descriptions of this genus and two species are only slightly modified from those of Barker (1957).

Description of the species of genus **Methanococcus**

1. **Methanococcus mazei** Barker 1936, 433. (*Pseudosarcina* Mazé 1915, 398.)

ma'ze.i. M.L. gen.n. *mazei* of Mazé; named for P. Mazé, the French bacteriologist who first studied the organism.

Small spherical cells occurring singly, in large irregular masses or in regular cysts of various sizes and forms. Non-motile. Gram-variable.

Ferments acetate and butyrate with the production of methane and CO_2. Ethanol and butanol are not attacked.

Grows in a medium containing acetate as the sole organic constituent and utilizes ammonia but not nitrate as the nitrogen source. Yeast extract is not fermented.

Optimum temperature between 30 and 37 C.

Isolated from garden soil, sewage sludge, black mud, and feces of herbivorous animals.

No type or reference strains are available.

Further Comments

This species was never obtained in pure culture and the above features should be interpreted in this light. It should especially be emphasized that it is quite possible that methanogenic bacteria *per se* do not ferment butyrate or other acids with the exceptions of acetate and formate (see *Further Comments* on the family).

2. **Methanococcus vannielii** Stadtman and Barker 1951, 269.

van.niel'i.i. M.L. gen.n. *vannielii* of van Niel; named for C. B. van Niel, the bacteriologist who developed the carbon dioxide reduction theory of methane formation.

Cocci, often slightly ellipsoidal, which vary from $0.5-4.0$ μm in diameter and which frequently occur in pairs. Different sizes sometimes suggest budding. Motile. Cells disintegrate on drying so that Gram-reaction and type of flagellation were not recorded.

Agar deep colonies $0.5-1.0$ mm in diameter, lenticular and light brown.

Formate is fermented with production of methane and CO_2 and, under some conditions, H_2. Methane is also formed from CO_2 and H_2. Acetate, propionate, butyrate, succinate, glucose, ethanol and methanol are not fermented.

Grows well in a medium containing formate as the sole organic compound and with ammonia as the nitrogen source.

Grows well at 30–40 C.

pH range for growth is 7.4–9.2, with the optimum being about 8.0.

Only one strain has been isolated: from black mud from the shore of San Francisco Bay.

Reference strain available from Dr. Theresa Stadtman, National Heart Institute, Bethesda, Maryland, or Dr. P. H. Smith, University of Florida, Gainesville.

PART 14

GRAM-POSITIVE COCCI

FAMILY I. **MICROCOCCACEAE** PRIBRAM 1929, 385

A. C. BAIRD-PARKER

Mi.cro.coc.ca'ce.ae. M.L. masc.n. *Micrococcus* type genus of the family; *-aceae* ending to denote family; M.L. fem.pl.n. *Micrococcaceae* the *Micrococcus* family.

Cells spherical, 0.5–3.5 μm in diameter, characteristically dividing in more than one plane to form regular or irregular clusters or packets. Motile or non-motile. Resting stages not produced. Gram-positive.

Chemoorganotrophs; metabolism respiratory or fermentative. Acid without gas produced from glucose, when attacked.

Nutritional requirements variable. All strains grow in presence of 5% NaCl; many grow in 10–15%.

Catalase positive.

Aerobic or facultatively anaerobic.

The G + C content of the DNA ranges from about 30–75 moles %.

Type genus: *Micrococcus* Cohn 1872, 151.

Further Comments

As now conceived the family consists of three genera, *Micrococcus*, *Staphylococcus* and *Planococcus*. *Gaffkya* is unrecognizable and the name has been rejected (Opinion 39, Jud. Comm. 1971, 104). *Sarcina*, as suggested by Shaw *et al.* (1951), is limited to anaerobic forms and will be found later in the Part in *Peptococcaceae; Methanococcus* will be found in Part 13. *Aerococcus* is regarded as an intermediate rather closer to *Streptococcus* than to *Micrococcus* and *Staphylococcus* and will be found in the *Streptococcaceae*.

Distinguishing features of the genera *Micrococcus, Staphylococcus, Planococcus* and *Aerococcus* are shown in Table 14.1.

Genus I. **Micrococcus** *Cohn 1872, 151*

A. C. BAIRD-PARKER

Mi.cro.coc'cus. Gr. adj. *micros* small; Gr. n. *coccus* a grain or berry; M.L. masc.n. *Micrococcus* small coccus.

Cells **spherical,** 0.5–3.5 μm in diameter, occurring singly, in pairs and characteristically dividing in more than one plane to form irregular clusters, tetrads or cubical packets. Usually non-motile. No resting stages known. Gram-positive. **Chemoorganotrophs: Metabolism strictly respiratory.** Menaquinones and cytochromes form the electron transport system; carotenoid pigments may be present. **Oxygen is the universal electron acceptor.** Carbon-containing compounds such as pyruvate, acetate, lactate, succinate, glutamate and carbohydrate are usually utilized. **Glucose oxidized** to mainly acetate or

completely oxidized to carbon dioxide and water. Glucose is metabolized by hexose monophosphate pathway and citric acid cycle enzymes; a functional glycolysis cycle may also be present.

Indole not produced. Teichoic acids are uncommon.

Catalase produced.

Nutritional requirements are variable. Members of some species can grow with ammonium phosphate as sole nitrogen source or with glutamic acid as carbon, nitrogen and energy source and thiamine and/or biotin as growth factors.

Aerobes. Growth optimum 25–30 C. All grow in

478

TABLE 14.1

Differential characteristics of the genera of family **Micrococcaceae** *and of genus* **Aerococcus** *(Streptococcaceae)[a]*

	Micrococcus	Staphylococcus	Planococcus	Aerococcus
Cells, spherical, Gram-positive	+	+	+	+
Arrangement: Irregular clusters	±	+	−	−
Tetrads	v	−	+	+
Glucose fermentation[b]	−	+	−	d
Cytochromes	+	+	+	−
Catalases: Heme	+	+	+	−
Non-heme	−	−	−	v
Hydrogen peroxide formation	−	−	−	+
Motility	−	−	+	−
Yellow-brown pigment	−	−	+	−
G + C content of DNA (moles %)	66–75	30–40	39–52	37–41

[a] + = most (90% or more) strains positive; − = most (90% or more) strains negative; v = characters inconstant and in one strain may sometimes be positive, sometimes negative; d = some (less than 90%) strains positive.

[b] Growth and acid anaerobically from glucose in the Standard Medium proposed by the I.C.S.B. Subcommittee on the Taxonomy of Staphylococci and Micrococci (1965), with exception of strains of *S. saprophyticus* which only weakly ferment glucose (see species description).

the presence of up to 5% sodium chloride. Cells not lysed by lysostaphin endopeptidase (1 unit/ml).

Common inhabitant of soils and fresh water. Frequently found on the skin of man and other animals.

The G + C content of the DNA ranges from 66–75 moles % (T_m; buoyant density; chromatography; E_{260}/E_{280} at pH 3.0).

Type species: *Micrococcus luteus* (Schroeter) Cohn 1872, 153.

Further Comments

In scope and content the genus *Micrococcus* is very different from that presented in previous editions of THE MANUAL. The number of species has been reduced from 16 in the 7th edition to 3 (Table 14.2).

The ability to form cubical packets of cells is by no means constant among the aerobic sarcinae and these organisms are indistinguishable from *M. luteus* on the basis of their metabolism (Kocur and Martinec, 1962; Rosypal and Kocur, 1963; Baird-Parker, 1965; Blevins et al., 1969; Kocur et al., 1972), their cell wall composition (Baird-Parker, 1965, 1970; Campbell et al., 1969) and their DNA composition (Rosypalova et al., 1966). The DNA of *Sarcina lutea* has been shown to be able to transform with that of *M. luteus* (Kloos and Schultes, 1969; Schleifer et al., 1972).

The genus *Sarcina* is retained for the anaerobic packet-forming cocci. *Sarcina litoralis* may be found in the genus *Halococcus*. *Sarcina hansenii* is probably identical to *Sarcina urea*, which is now placed in the genus *Sporosarcina*.

Description of the species of genus **Micrococcus**

1. **Micrococcus luteus** (Schroeter) Cohn 1872, 153. (*Bacteridium luteum* Schroeter 1872, 126; *Sarcina lutea* (Schroeter) Schroeter 1886, 154; *Sarcina aurantiaca* Flügge 1886, 180; *Sarcina flava* de Bary 1887, 185; *Micrococcus flavus* Trevisan 1889, 34; *Sarcina marginata* Gruber 1895, 32; *Sarcina subflava* Ravenel 1896, 10; *Sarcina variabilis* Stubenrath in Lehmann and Neumann 1896, 143; Not *Micrococcus luteus* Lehmann and Neumann 1896, 161; *Micrococcus lysodeikticus* Fleming 1922, 306; *Sarcina citrea* (Migula) Bergey et al. 1923, 74; *Micrococcus afermentans* Castellani 1928,

537; *Staphylococcus flavocyaneus* Knaysi 1942, 368; *Sarcina pelagia* ZoBell and Upham 1944, 279; *Staphylococcus afermentans* Castellani in Shaw et al. 1951, 1022; *Micrococcus sodonensis* Aaronson 1955, 67; *Sarcina exigua* Müller 1961, 526.)

lu'te.us. L. adj. *luteus* golden-yellow.

Spheres, 1.0–2.0 μm in diameter occurring singly, in pairs and dividing in more than one plane to form tetrads, irregular clusters or regular packets of cells. Non-motile.

Cell wall peptidoglycan contains the unique L-lysine peptide subunit as well as glucosamine,

TABLE 14.2

Differential characteristics of the species
of genus **Micrococcus**[a]

	1. M. luteus	2. M. roseus	3. M. varians
Pigment: Yellow	+	−	+
Red	−	+	−
Novobiocin sensitivity[b]	S/R	S	R
Glucose: acid[c]	−	d	+
Xylose: acid[d]	−	d	+
Fat: fatty acids[d]	+	−	d
Arginine: ammonia[d]	−	−	d
Cell wall: L-lys peptide	+	−	−
G + C content of DNA (moles %)	71–75	66–75	66–72

[a] + = most (90% or more) strains positive;
− = most (90% or more) strains negative; d = some (less than 90%) strains positive.

[b] R = M.I.C. > 2.0 µg/ml; S = M.I.C. < 0.6 µg/ml (Mitchell and Baird-Parker, 1967; Jeffries, 1969).

[c] Standard method of I.C.S.B. Subcommittee on the Taxonomy of Staphylococci and Micrococci (1965).

[d] Baird-Parker (1963).

muramic acid, alanine, glutamic acid, glycine (Schleifer *et al.*, 1972). Teichoic acids are absent.

Strains growing as tetrads or irregular cell clusters form smooth, convex colonies with a regular edge. Colonies of strains forming cubical packets usually have a granular surface and a matt appearance. Colonies are yellow, yellowish green or orange colored: some strains form a violet pigment which diffuses into the medium.

In liquid media uniform turbidity followed by clearing, with a fine or somewhat granular or mucoid deposit.

Metabolism strictly respiratory. Membrane-bound electron transport system contains cytochromes a, b and c (Lukyanova and Taptykova, 1968), hydrogenated menaquinones (Jeffries *et al.*, 1968), carotenoid pigments (Thirkell and Strang, 1967; Thirkell and Hunter, 1969). NADH, malate and succinate dehydrogenases present (Erickson and Parker, 1969). Enzymes of the glycolysis, hexose monophosphate and citric acid cycles present; glycolysis cycle functions only under aerobic conditions (Dawes and Homes, 1958). Hexose monophosphate and citric acid cycles main pathways for energy requirements (Dawes and Homes, 1958; Perry and Evans, 1967; Blevins *et al.*, 1969). Glyoxylate cycle enzymes demonstrated in acetate grown cells (Perry and Evans, 1966). Many strains show oxidase activity (Boswell *et al.*, 1972; Kocur *et al.*, 1972).

Usually no detectable acid from carbohydrates; final pH in glucose broth is 6.4–8.2. Carbon-containing compounds, oxidized to carbon dioxide and water, include acetate, lactate, pyruvate, succinate, fructose, galactose, glucose, glycerol, maltose, sucrose. Variable oxidation of mannitol, sorbitol, arabinose, rhamnose, ribose, xylose and starch; dulcitol not oxidized (Rosypal and Kocur, 1963; Perry and Evans, 1966).

Proteins, polypeptides and fats may be hydrolyzed (Baird-Parker, 1965). Arginine usually not hydrolyzed. Nitrates usually not reduced.

Some strains will grow on a defined medium containing pyruvic acid as carbon and energy source, glutamic acid, biotin and mineral salts (Perry and Evans, 1967). Other strains require a more complex medium containing a number of amino acids (Grula, 1962).

Strict aerobe. Optimum growth temperature about 30 C. Most strains grow at 10 and some at 45 C. Growth in 5% NaCl but not usually in 10 or 15% (Baird-Parker, 1965; Kocur *et al.*, 1972). Resistant to lysis by lysostaphin (1 unit/ml) (Schindler and Schuhardt, 1964; Thomas, 1964). Most strains sensitive to lysozyme (lysed by 100 µg/ml lysozyme; MIC < 25 µg/ml; (Jeffries, 1969) and to novobiocin (MIC < 1 µg/ml) (Mitchell and Baird-Parker, 1967; Jeffries, 1969).

Not pathogenic to plants and animals. Common in soil, dust, water, skin of man and other animals.

G + C content of the DNA is 70.7–75.5 moles % (Kocur *et al.*, 1972).

Proposed neotype strain: CCM 169; ATCC 4698; NCTC 2665.

Further Comments

Micrococcus luteus, when defined on the basis of biochemical characters, was a poor species containing organisms placed together primarily on their inability to form detectable acid when grown aerobically in a complex media containing glucose (Baird-Parker, 1963; ICSB Subcommittee on the Taxonomy of Staphylococci and Micrococci, 1965). The species has been better circumscribed by Kocur *et al.* (1972) and Schleifer *et al.* (1972).

The main component of the cell wall is a peptidoglycan which is composed of a glycan containing repeating units of β-1,4 linked N-acetyl glucosamine and N-acetyl muramic acid residues attached to peptide subunits possessing the structure and composition of N^{α}-[L-alanyl-γ(α-D-glutamyl glycine)]-L-lysyl-D-alanine (Campbell *et al.*, 1969). The peptide subunits are cross-linked by a variety of bridges; the main type is the L-lysyl tetrapeptide described by Schleifer and Kandler (1970) and Schleifer *et al.* (1972).

Strain ATCC 398 (NCTC 8512), proposed as the neotype by Breed (1952), is not a typical member of the species. It has a G + C content of 66 moles %, a different cell wall structure (Schleifer and Kandler, 1970), a different menaquinone pattern (Jeffries *et al.*, 1968), contains glucosaminyl ribitol teichoic acid (Partridge *et al.*, 1973), and does not transform with *M. luteus* strain ISU (Kloos, 1969) or strain ATCC 27141 (Schleifer *et al.*, 1972). Kocur *et al.* (1972) suggest ATTC 398 is another species and propose strain CCM 169; ATCC 4698 (*M. lysodeikticus*); NCTC 2665 (*Staphylococcus afermentans*) as the neotype.

2. **Micrococcus roseus** Flügge 1886, 183. (*Micrococcus tetragenus ruber* Schneider 1894, 215; *Micrococcus agilis* Ali-Cohen 1889, 36; *Staphylococcus roseus* (Flügge) Tavel acc. to Lehmann and Neumann 1896, 177; *Micrococcus rubens* Migula 1900, 177; *Micrococcus roseofulvus* Hucker 1928, 27.)

ro.se.us L. adj. *roseus* rose-colored.

Spheres (1.0–2.5 μm in diameter) occurring singly, in pairs and dividing in more than one plane to form irregular clusters, tetrads or cubical packets. Some strains motile by one or two flagella.

Cell walls contain galactose, glucose, mannose, alanine, glutamic acid, lysine, glucosamine and muramic acid (Baird-Parker, 1965). The main component is a peptidoglycan consisting of a glycan composed of *N*-acetyl glucosamine and *N*-acetyl muramic acid residues and peptide subunits containing alanine, glutamic acid and lysine (Ghuysen, 1968). The peptide subunits are joined by a peptide bridge containing L-alanine or L-alanine and L-threonine (Petit, Munoz and Ghuysen, 1966; Schleifer and Kandler, 1970). Teichoic acids absent.

Pink or red, smooth, slightly convex colonies with regular margins.

In broth, slight turbidity with a pink or colorless, fine, slightly mucoid or granular deposit; some strains form slight pellicle with ring.

Chemoorganotrophs: Metabolism strictly respiratory. Cytochromes, menaquinones and carotenoid pigments present. The main pigment has been identified as canthaxanthin; further pigments are α or β carotene derivatives (Ungers and Cooney, 1968).

Acid production from carbohydrates is variable. Small amounts of acid may be produced from any number of carbohydrates including glucose, cellobiose, dextrin, galactose, glycerol, maltose, mannose, mannitol, rhamnose and xylose (Baird-Parker, 1965). Terminal pH in unbuffered glucose broth (5.2–6.5; mean pH 6.3). No acid in standard broth (Subcommittee, 1965).

Nitrates usually reduced to nitrite.

Arginine not hydrolyzed.

Fats (butter and lard) not hydrolyzed.

Casein and egg yolk not hydrolyzed. Gelatin may be weakly hydrolyzed (Baird-Parker, 1965).

Differential characters are given in Table 14.2.

Strains studied by Eisenberg and Evans (1963) grew with glutamate as the sole carbon, nitrogen and energy source and biotin and thiamine as growth factors (see also Cooney and Thierry, 1966). Many carbon compounds stimulate growth: glucose, xylose, succinate, lactate and pyruvate (Eisenberg and Evans, 1963). No growth on Simmons citrate agar (Kocur and Páčová, 1970).

Strict aerobe. Optimum growth temperature 25 C; grows well at 10 C. Grows in 5% NaCl but not in 15%. Usually resistant to lysozyme (MIC > 25 μg/ml) and to lysostaphin and sensitive to novobiocin (MIC < 0.6 μg/ml) (Jeffries, 1969).

Not pathogenic to man, plants and animals.

Isolated from dust, water and salt-containing foods. Not common.

The G + C content of the DNA is 66–74 moles % (T_m; E_{260}/E_{280} at pH 3).

Suggested neotype: ATCC 186; NCTC 7523; CCM 679, Kocur and Páčová 1970, 237. (See *Further Comments* on the species.)

Further Comments

Strain ATCC 186 is the holotype strain of *Micrococcus rubens* Migula (Migula → Kral → Breed → ATCC). Hill 1959, 282 suggested it as the neotype. However, Kocur and Martinec (1962) suggested NCTC 7520, ATCC 418 as the neotype and this was accepted by the Subcommittee 1971, 162. This was not one of the strains studied by Kocur and Páčová, 1970. Shaw *et al.* 1951, 1022 indicate NCTC 7520 is the type strain of *Staphylococcus roseus* Tavel, although the ATCC catalogue (1972) states there is no evidence that it was one of Tavel's original strains.

3. **Micrococcus varians** Migula 1900, 135. (*Staphylococcus lactis* Shaw, Stitt and Cowan 1951, 1021; *Micrococcus pulcher* Müller 1961, 534; *Micrococus* subgroups 5 and 6 Baird-Parker 1963, 417.)

va'ri.ans. L. part.adj. *varians* varying.

Spheres, 1.0–1.5 μm in diameter, singly, in pairs, tetrads and commonly irregular clusters; some strains may divide regularly in three perpendicular planes to form regular packets of cells. Non-motile.

Cell walls contain glucosamine, muramic acid, lysine, alanine, glutamic acid, serine, glycine and glucose together with mannose, galactose and galactosamine in some strains (Baird-Parker, 1965). The peptidoglycan is of the Lys-Ala 3–4 type (Schleifer and Kandler, 1970). Do not contain teichoic acid (Kocur, personal communication, 1971) but may contain *N*-acetylglucosamine 1-phosphate polymers (Partridge *et al.*, 1973).

Colonies yellow, smooth, convex with a regular edge.

Chemoorganotrophs: Metabolism strictly respiratory. Menaquinones and cytochromes form the electron transport system (Baird-Parker, unpublished; Jeffries, 1969).

Acid but not gas produced from fructose and glucose; terminal pH in unbuffered glucose broth 4.5–5.9. Acid production variable in galactose, lactose, maltose, sucrose.

Nitrates and nitrites usually reduced. Ammonia usually not produced from arginine. Hippurate usually hydrolyzed. Fats, and casein sometimes hydrolyzed. Gelatin usually hydrolyzed. Tweens not hydrolyzed (Baird-Parker, 1963, 1965). Growth reported only on complex media; minimal nutritional requirements not known. Some strains grow on Simmons's citrate agar.

Strict aerobe. Good growth between 22–37 C; most strains grow at 10 but not usually at 45 C. All strains grow on solid media with 7.5% NaCl but not with 15%. Resistant to novobiocin (M.I.C. > 2.0 μg/ml) (Mitchell and Baird-Parker, 1967; Jeffries, 1969). Not lysed by lysozyme (100 μg/ml) or lysostaphin (1 unit/ml) (Thomas, 1964).

Not pathogenic. Originally isolated from milk. Common in milk and dairy products, animal carcasses, dust and soils.

The G + C content of the DNA is 66–72 moles % (T_m).

Proposed neotype strain: CCM 884; ATCC 15306; NCTC 7564 (Kocur and Martinec, 1972).

Further Comments

It is possible to distinguish strains of *M. varians* (G + C content of DNA 66–72 moles %) from weakly anaerobic staphylococci (G + C 30–35 %) that have similar biochemical and physiological characteristics by the test proposed by Evans and Kloos (1972).

Species incertae sedis

a. *Micrococcus colpogenes* Campbell and Williams 1951, 904.

col.po'ge.nes. Gr. n. *colpus* bosom, fold; Gr. v. *gennaio* bear, produce; M. L. adj. *colpogenes* fold producing.

Description based on Campbell and Williams (1951).

Spheres occurring in pairs or clumps. Nonmotile. Gram-positive.

Colonies, circular, raised, smooth, glistening, yellow with entire edge.

Moderate turbidity in broth, slight granular sediment.

Chemoorganotroph: Metabolism respiratory. Oxygen is the terminal oxygen acceptor. Glucose, maltose, mannitol, acetate, malate, pyruvate, succinate, glycogen, glucosamine and chitin are utilized as carbon sources.

No acid or gas from carbohydrates.

Chitin hydrolyzed; ammonia and reducing sugars produced in 8 days. Casein and urea hydrolyzed. Indole and H_2S not formed. Nitrates reduced to nitrites. Not lipolytic.

Will grow on medium containing ammonium chloride as the nitrogen source in the presence of a utilizable carbon source.

Aerobe. Optimum growth temperature 20–30 C; no growth at 4 C. Growth in the presence of 4% NaCl but not in presence of 10%.

Originally isolated from marine muds. All original isolates have been lost.

b. *Micrococcus cryophilus* McLean, Sulzbacher and Mudd 1951, 723.

cry.oph'il.us. Gr. n. *cryos* cold, frost; Gr. adj. *philus* loving; M. L. adj. *cryophilus* cold loving.

Description from McLean *et al.* (1951) and Kocur and Martinec (1962).

Spheres 1.0–3.6 μm in diameter, usually 1.2–1.8 μm, singly, in pairs, chains and clusters. Thick capsule observed in sections examined in the electron microscope. Not motile. Gram-positive.

The cell wall peptidoglycan contains glucosamine, muramic acid, diaminopimelic acid, alanine, glutamic acid and has a structure similar to the peptidoglycan of many species of *Bacillus* (Professor O. Kandler, personal communication, 1969). The cell wall is surrounded by a capsule and in cross-section is quite unlike that of micrococci or staphylococci (Mazanec *et al.*, 1966).

Colonies glistening, smooth, slightly convex, creamy white with regular margin. Turbidity at first in broth, later slight sediment and ring at surface.

Chemoorganotroph: Metabolism strictly respiratory. Oxygen is the universal electron acceptor.

No acid from carbohydrates in peptone media. Starch and esculin not hydrolyzed.

Gelatin and casein are not hydrolyzed. Nitrates and nitrites not reduced. Urease and lipases may be produced. L-Glutamic acid used as sole carbon, nitrogen and energy source (Tai and Jackson, 1969).

Aerobe. Optimum temperature 10 C, growth range −4 to +24 C. Mutants with an optimum temperature of 20 C have been isolated (Tai and Jackson, 1969). pH range 5.5–9.5, optimum 6.8–7.2. Grows in presence of 5% NaCl. Not lysed by lysozyme. Isolated from pork sausage.

The G + C content of the DNA is 41 moles % (T_m).

Comment: This species is similar to the so-called false *Neisseria* spp. (Sleytr and Kocur, 1971). It might be accommodated in the genus *Branhamella*.

c. *Micrococcus radiodurans* Raj, Duryee, Deeney, Wang, Anderson and Elliker 1960, 289.

This U.V.- and γ-radiation-resistant coccus is similar to *M. roseus* with respect to its biochemical and physiological characters, G + C content and its red pigment (Anderson *et al.*, 1956; Raj *et al.*, 1960; Duryee *et al.*, 1961; Moseley and Schein, 1964; Krabbenhaft, 1966; Thirkell, 1969). It differs from *M. roseus* in the structure and composition of its cell wall which is completely different from all accepted members of the genus *Micrococcus* (Work and Griffith, 1968; Schleifer and Kandler, 1970; Girard, 1970). Similar

radiation resistant cocci were reported by Davis *et al.* (1963) and Lewis (1971).

d. *Micrococcus freudenreichii* Guillebeau 1891, 135.

The strain Hucker 71; ATCC 407; CCM 764 does not resemble the original description, but Bergan *et al.* (1970) suggested it be considered the neotype. McDonald (1971) suggests the name should be rejected as *nomen dubium*. The organism as now known is interesting in that it utilizes maltose, maltotriose and maltotetraose but not glucose. It may be a synonym of *M. luteus*.

e. *Micrococcus mucilaginosus* Migula 1900, 119.

f. *Micrococcus halobius* Onishi and Kamekura 1972, 235.
Type strain: 28-3; ATCC 21727.

Genus II. **Staphylococcus** *Rosenbach 1884*, 18 nom. cons. Opin. 17 Jud. Comm. 1958, 153

A. C. BAIRD-PARKER

(*Aurococcus* Winslow and Rogers 1906, 540.)
Staph.y.lo.coc′cus. Gr. n. *staphylo* bunch of grapes; Gr. n. *coccus* a grain or berry; M.L. masc.n. *Staphylococcus* the grape-like coccus.

Cells spherical, 0.5–1.5 μm in diameter occurring singly, in pairs and characteristically dividing in more than one plane to form irregular clusters. Non-motile. No resting stages known. Gram-positive. **Cell wall contains two main components; a peptidoglycan and its associated teichoic acids.** The peptidoglycan consists of glycan composed of repeating units of β-1,4 linked *N*-acetyl-glucosamine and *N*-acetyl-muramic acid residues that are linked through *N*-acetyl-muramyl-L-alanine linkages to peptide subunits consisting of *N*α-(L-alanyl-D-isoglutamyl)-L-lysyl-D-alanine. The peptide subunits are cross-linked by penta-peptide bridges containing solely or mainly glycine. These bridges extend from the *N*ε-lysine residue of one peptide subunit to the C-terminal D-alanine of a neighboring peptide subunit. Attached to or in close proximity to the peptidoglycan are teichoic acids which, depending on the species, contain either ribitol or glycerol linked to a sugar or an aminosugar.

Chemoorganotrophs: **Metabolism respiratory and fermentative.** Catalase produced. Menaquinones and cytochromes a, b₁ and O form the electron transport system; carotenoid pigments may be present. Oxygen is the universal terminal electron acceptor.

A wide range of **carbohydrates may be utilized**, particularly in the presence of air, **with the production of acid** but gas not detectable by standard procedures. Under anaerobic conditions the main product of glucose fermentation is lactic acid; in the presence of air the main product is acetic acid with small amounts of CO₂. Acetoin is usually formed as an end-product of glucose metabolism. Terminal pH in unbuffered glucose broth is 4.2–5.4. Acid usually not produced from arabinose, cellobiose, inositol, inulin or raffinose. Starch and esculin usually not hydrolyzed.

Extracellular enzymes and toxins are formed. A variety of protein and fat-containing substrates are hydrolyzed. Hippurate and arginine may be broken down. Indole not formed.

Amino acids and vitamins required for aerobic growth. In addition, uracil and a fermentable carbon source are required for anaerobic growth.

Facultative anaerobes, growth more rapid and abundant under aerobic conditions. Temperature optimum 35–40 C; growth range 6.5–46 C. pH optimum 7.0–7.5; growth range pH 4.2–9.3.

Most strains grow in the presence of 15% sodium chloride or 40% bile. **Sensitive to lysis by lysostaphin endopeptidase (1 unit/ml). Resistant to lysis by lysozyme (100 μg/ml).**

Usually **sensitive to antibiotics** such as the β-lactam and macrolide antibiotics, tetracylines, novobiocin and chloramphenicol but **resistant** to polymyxin and polyenes. Sensitive to antibacterials such as phenols and their derivatives, surface active compounds, salicylanilides, carbanilides, halogens (chlorine and iodine) and their derivatives such as chloramines and iodophors. Usually sensitive to heat ($D_{60\,C}$ value 3 min or less in buffer, pH 7.0). Moderately resistant to γ-radiation (D value *ca.* 20 Krads in buffer, pH 7.0).

Host for a wide range of bacteriophages which may have a narrow or wide host range. Transfer of characters by transduction has been shown for *S. aureus*.

Mainly associated with skin, skin glands and mucous membranes of warm-blooded animals. Their host range is wide and many strains are potential pathogens.

The G + C content of DNA ranges from 30–40 moles % (T_m, buoyant density, chromatography).

Type species: *Staphylococcus aureus* Rosenbach 1884, 18.

Further Comments

Three species are recognized in the genus *Staphylococcus*; characters differentiating these species are shown in Table 14.3. It is separated from the genus *Micrococcus* by the G + C content of the DNA, cell wall composition and an ability to grow anaerobically, and under these conditions to ferment glucose. The basis for separation on this latter character has been criticized (Mortensen and Kocur 1967; Gibson, 1967; Klesius and Schuhardt, 1968) for there exist strains of Gram-positive, catalase positive cocci with a primarily respiratory metabolism but which may grow slowly anaerobically in complex media and under these conditions form small amounts of acid from glucose (Evans and Kloos, 1972). Where information is available these organisms are similar to staphylococci in the G + C content of their DNA (Kocur, 1971) and the sensitivity of their cells to lysostaphin (Lachica *et al.*, 1971). They differ from other staphylococci in resistance to novobiocin (Mitchell and Baird-Parker, 1967, Mortensen, 1969), presence of generally different menaquinones in cell membranes (Jeffries *et al.*, 1968), often differences in growth requirements (Baird-Parker, 1965), in cell wall components (Oeding and Hasselgren, 1972) and in their inability to form detectable amounts of acid anaerobically in the standard medium proposed by the I.C.S.B. Subcommittee on the Taxonomy of Staphylococci and Micrococci (hereinafter referred to as Subcommittee) (1965) for the determination of anaerobic growth and fermentation of glucose. These organisms were recently classified as *Micrococcus saprophyticus* (Subcommittee, 1971) However, in view of their DNA and cell wall composition and their ability to grow slowly anaerobically in certain media, they appear to be more closely related to the staphylococci and are here transferred to the genus *Staphylococcus* as *S saprophyticus*.

Description of the species of genus **Staphylococcus**

1. **Staphylococcus aureus** Rosenbach 1884, 18. (*Staphylococcus pyogenes aureus* Rosenbach 1884, 19; *Staphylococcus pyogenes albus* Rosenbach 1884, 21; *Staphylococcus albus* Rosenbach 1884, 18; *Staphylococcus pyogenes citreus* Passet 1885, 9; *Micrococcus aureus* (Rosenbach) Zopf 1885, 57; *Micrococcus pyogenes* Lehmann and Neumann 1896, 165; *Micrococcus citreus* Migula 1900, 147; *Micrococcus albus* (Rosenbach) Buchanan 1911, 196; *Staphylococcus citreus* (Migula) Bergey *et al.* 1923, 55; *Micrococcus pyogenes var. aureus* Hucker 1948, 241.)

au're.us. L. adj. *aureus* golden.

Spheres, 0.8–1.0 μm in diameter. Cells occur singly, in pairs and divide in more than one plane to form irregular clusters. Some strains possess a capsule or slime layer.

Cell walls contain organic phosphorus, ribitol, glucosamine, muramic acid, glycine, lysine, aspartic acid, serine, glutamic acid, alanine and small amounts of threonine, proline, valine and leucine. These compounds can be accounted for in three of the main components of the wall which are the peptidoglycan, the ribitol teichoic acids and the species-specific precipitinogen Protein

A. For details on the biochemistry of the cell wall see: Ghuysen and Strominger, 1963; Oeding, 1965; Tipper and Strominger, 1966; Grov and Rude 1967; Marandon and Oeding, 1967; Ghuysen 1968; Tipper and Berman, 1969; Schleifer and Kandler, 1970.

Cell membranes contain the glycolipids, mono- and diglucosyl-diglyceride and the phospholipids lysyl-phosphatidyl-glycerol, phosphatidyl-glycerol and cardiolipin (White and Frerman, 1967).

Colonies are smooth, low-convex, glistening butyrous and with an entire edge. Under conditions inhibitory to normal growth, R (rough) or G (dwarf) colony variants may be produced Colonial pigmentation is extremely variable, hence the variety of specific epithets such as *aureus* *albus* and *citreus* that have been applied to this species (see 7th edition of THE MANUAL). Colonies of most strains are orange in color although certain antibiotic-resistant strains and strains from bovine sources are more commonly yellow pigmented (Willis *et al.*, 1964).

Growth on agar slants abundant, opaque smooth, flat, moist and white, yellow or orange in color.

TABLE 14.3

Characteristics differentiating species of genus **Staphylococcus**[a]

	1. S. aureus	2. S. epidermidis[b]	3. S. saprophylicus
Coagulases	+	−	−
Mannitol:			
Acid aerobically	+	d	d
Acid anaerobically	+	−	−
α-Toxin	+	−	−
Heat-resistant endonucleases	+	−	−
Biotin for growth	−	+	NT
Cell wall:			
Ribitol	+	−	+
Glycerol	−	+	d
Protein A	+	−	−
Novobiocin sensitivity[c]	S	S	R

[a] + = most (90% or more) strains positive; − = most (90% or more) strains negative; d = some (less than 90%) strains positive, some negative; NT, not tested.

[b] Four biotypes recognized: see Table 14.4.

[c] R = MIC > 2.0 μg/ml; S = MIC < 0.6 μg/ml (Mitchell and Baird-Parker, 1967; Jeffries, 1969).

In broth growth first results in turbidity, later the broth becomes clear with a fine, easily suspended deposit; frequently a ring pellicle.

Chemoorganotrophs: Metabolism respiratory and fermentative. Menaquinones and cytochromes a, b₁ and O form the membrane-bound electron transport system. Menaquinones have the characteristic pattern of MK-8 as the major isoprenologue and MK-9 and MK-7 as minor isoprenologues (Jeffries *et al.*, 1968). Carotenoid pigments (Hammond and White, 1970) are produced by most strains giving cells ranging in color from deep orange to pale yellow; the production of pigments depends on growth conditions and may be variable within a single strain (O'Connor, Willis and Smith, 1966). Energy is obtained via glycolysis, the hexose monophosphate pathway and citric acid cycles (Strasters and Winkler, 1963; Montiel and Blumenthal, 1965). The operation of these cycles is dependent on growth conditions (Collins and Lascelles, 1962). Endogenous respiration occurs either by the utilization of free amino acids within the cell pool or by the utilization of poly-β-hydroxybutyrate (Ivler, 1965).

Catalase produced by aerobically grown cells; may be absent in respiratory deficient mutants (Jensen, 1963).

Glucose is fermented with the production of optically inactive or L-lactic acid (Orla-Jensen, 1919). In the presence of air, mainly acetate and small amounts of carbon dioxide are produced (Gardner and Lascelles, 1962).

Acid produced aerobically and anaerobically from glucose, lactose, maltose and mannitol. In air a wider range of carbohydrates are used as carbon and energy sources and hexoses, pentoses, disaccharides and sugar alcohols are metabolized with the production of acid (Baird-Parker, 1965). No acid from arabinose, cellobiose, dextrin, inositol, raffinose, rhamnose or xylose. Esculin and starch not usually hydrolyzed. Hyaluronic acid may be hydrolyzed by hyaluronic acid lyase and cell walls of *M. luteus* by muramidase (Hawiger, 1968; Arvidson *et al.*, 1970).

Acetoin is produced as an end-product of glucose metabolism; terminal pH in glucose broth 4.4–5.0.

Nitrates reduced by nitratases. Ammonia produced from arginine by arginine dihydrolase. Trace amounts of hydrogen sulfide may be formed from cysteine; glutamic acid and lysine not decarboxylated (Baird-Parker, 1963; Hugh and Ellis, 1968).

Proteases, lipases, phospholipases, lipoprotein lipases, esterases and lyases are produced. Most strains will hydrolyze native animal proteins e.g. hemoglobin, fibrin, egg white and casein and polypeptides such as gelatin. Lipids, Tweens, Spans and phospholipoproteins are hydrolyzed with the release of fatty acids. Some strains produce lecithinase A but none lecithinase C (Nygrén *et al.*, 1966); beta-hemolysin possesses phospholipase C activity against sphingomyelin (Maheswaran and Lindorfer, 1967). Mucopolysaccharides may be degraded by hyaluronic acid lyase and cell wall glycans by lysozyme (Hawiger, 1968). Strains from animal sources and antibiotic resistant strains generally produce fewer of these enzymes (Willis *et al.*, 1966).

Coagulases are produced by virtually all strains; several antigenically distinct and substrate specific coagulases are produced. For further information on the coagulase test see *Further Comments* at the end of species description.

At least three hemolysins are produced (alpha, beta and delta), distinguished by type and range of hemolysis on sheep, rabbit and human erythrocytes (Elek and Levy, 1950). A single strain frequently produces more than one hemolysin; beta-hemolysin is more commonly produced by strains from animal sources.

Acid and/or alkaline phosphatases are produced (Melveaux and San Clemente, 1967; Tirunarayanan, 1968) and heat-resistant nucleases (phosphodiesterases) (Chesbro and Auborn, 1967; Wadstrom, 1967; Lachica *et al.*, 1969, 1971).

Up to 12 amino acids, adenine and thiamine

are required for aerobic growth (Files *et al.*, 1936; Mah *et al.*, 1967); for anaerobic growth, uracil and a fermentable carbon source are also required (Richardson, 1936).

Facultative anaerobes: growing best under aerobic conditions. Most strains grow between 6.5 and 46 C; optimum 30–37 C; pH values between 4.2 and 9.3 (optimum pH 7.0–7.5) and in 15% sodium chloride or 40% bile. Minimum water activity permitting growth of aerobically grown cells is 0.86 (Scott, 1953).

Originally isolated from pus in wounds; found in nasal membranes, hair follicles, skin and perineum of warm-blooded animals. Potential pathogens causing a wide range of infections and intoxications; boils, abscesses, meningitis, furunculosis, pyemia, osteomyelitis, suppuration of wounds and food poisoning; see *Further Comments* at end of species description.

The G + C content of the DNA is 30.7–39 moles % (T_m; chromatography; buoyant density); according to Garrity, Detrick and Kennedy (1969) most likely range is 32–35 moles %.

Neotype strain: S33R$_4$; NCTC 8532; ATCC 12600; CCM 885. Opin. 17, Jud. Comm. 1958, 153.

Further Comments

The main diagnostic characters of *Staphylococcus aureus* are listed in Table 14.3. The main diagnostic test for *S. aureus* is the detection of coagulases; these enzymes should be tested for under carefully standardized conditions using rabbit plasma and the tube test; see standard methods proposed by the I.C.S.B. Subcommittee (1965). For routine diagnostic work the "slide coagulase test" is frequently used and usually correlates with the test tube. According to Tager and Drummond (1965) the "slide coagulase test" detects the "clumping factor" generally regarded to be distinct from coagulase, but may be closely related (Blackstock, Hyde and Kelly, 1968). Rabbit, human, horse and pig plasmas are usually coagulated by strains from both human and animal sources; but canine strains do not coagulate human plasma (Hájek and Maršálek, 1971); bovine plasma (Meyer, 1967; Grün, 1968) and sheep plasma (Hájek, Maršálek and Černá, 1968), are usually only coagulated by strains from animal sources. Coagulase, although produced, may not be detected; for reasons see Baird-Parker (1965). Coagulase negative mutants of *S. aureus* occur and if an organism is suspected to be a strain of *S. aureus* it should be investigated for the characters outlined in Table 14.3 and by Baird-Parker (1965). Citrate utilizing bacteria may cause false clotting of citrated plasma (Evans, Buettner and Niven, 1952); this can be avoided by the use of plasma containing 0.1% ethylene diaminetetraacetic acid (EDTA) (Baer, 1968). Bacteria other than staphylococci may produce coagulase and morphology should be checked. Brown *et al.* (1967) isolated a staphylococcus which, although producing coagulases, possessed biochemical characters more like *Staphylococcus epidermidis*; see also Jeffries *et al.* (1968).

Diagnostic and selective media were reviewed by Baird-Parker (1969). The most generally accepted diagnostic character is the ability of strains of *S. aureus* to clear egg yolk by a phospholipoprotein lipase which splits the lipid moiety from lipovitellenin (Tirunarayanan and Lundbeck, 1967). Selective agents are based on tolerance of *S. aureus* to salts such as sodium chloride, lithium chloride and potassium thiocyanate and sodium azide, amino acids such as glycine and antibiotics such as polymyxin.

Differentiation of strains of *S. aureus* by phage typing is commonly used; it is well developed for typing human strains (Parker, 1962; Wentworth, 1963) and progress is being made in developing an equivalent system for strains from animal sources (Blair and Parker, 1967; Frost, 1967; Meyer, 1967). The antigenic structure of *S. aureus* is complex, the demonstration of a particular antigen may be difficult and the production of specific antisera complicated (Oeding, 1957; Torres Pereira, 1961; White *et al.*, 1962; Haukenes, 1967). Although of proven value antigenic typing has yet to find international acceptance. Continued cultivation in the laboratory leads to antigenic changes which complicate the serological picture (Torres Pereira, 1961).

All strains are potential pathogens. Under suitable conditions strains produce a variety of enzymes believed by some to play a role in initiating infection; these include hyaluronic acid lyase, fibrinolysin, nucleases and coagulases. Certain antigenic factors are thought to play a part in enabling a strain to set up an infection (Torres Pereira, 1967). The symptoms may be caused by a variety of toxins, the most important of which is probably the α-toxin which in animal tests is lethal, dermonecrotic, hemolytic, leucocidal and causes platelet damage. Other toxins are the beta-hemolysin (disrupts the red cell membranes) and the Panton-Valentine leucocidin (destroys human leucocytes). For further information on pathogenicity see Elek (1959); Ivler (1965); Bernheimer (1968); Florman (1968); Williams (1969). Food poisoning strains of *S. aureus* produce enterotoxins of which six antigenic types have been reported (Bergdoll, 1967; Casman, 1967; Minor and Marth, 1972). Rabbits and mice are susceptible to the α-toxin.

The division of *S. aureus* into subspecies or

varieties was considered by the I.C.S.B. Sub-committee (1967), which decided at that time there was insufficient evidence to propose sub-division of the species. However, the evidence now available from studies by Meyer (1967), Grün (1968), Scharmann and Blobel (1968), Oeding *et al.* (1971), Hájek and Maršálek (1971), Live (1972), Tschäpe and Rische (1972) have shown that it is possible to distinguish between strains from human and different animals by differences in susceptibility to phages, antigenic components of the cell wall, nutritional requirements, bio-chemical characters such as coagulation of bovine plasma, beta-hemolysin, fibrinolysin and serologi-cal differences in nucleases. Torres Pereira (1967) proposed the subdivision of human isolates into two varieties based on a general correlation be-tween antigenic structure, susceptibility to phages and antibiotics and differences in virulence.

2. **Staphylococcus epidermidis** (Winslow and Winslow) Evans 1916, 449. (*Staphylococcus epi-dermidis albus* Welch 1891, 441; *Albococcus epi-dermis* Winslow and Winslow 1908, 201; *Mi-crococcus epidermidis* (Winslow and Winslow) Hucker 1924, 21; *Staphylococcus saprophyticus* Fairbrother 1940, 88; *Micrococcus hyicus* Som-polinsky 1953, 307; *Micrococcus violagabriellae* Castellani 1955, 477; *S. epidermidis* subsp. *vio-lagabriella* Marples 1969, 49.)

e.pi.der′mi.dis. Gr. n. *epidermidis*, the outer skin; M.L. gen.n. *epidermidis* of the epidermis.

Spheres, 0.5–1.5 μm in diameter, occurring singly, in pairs and dividing in more than one plane to form irregular clusters; occasionally tetrads ob-served.

Cell walls are similar in composition to those of *S. aureus* but contain glycerol in place of ribitol. Most also contain glucose and some galactosamine (Baird-Parker, 1965). Protein A is absent from this species (Oeding, 1960). Detailed structure of the peptidoglycan has been reported for only a few strains; in strains studied, the glycan and peptide subunits are identical with those of *S. aureus*: the peptide subunits are bridged by a pentapeptide containing glycine together with usually L-serine or sometimes L-alanine (Schleifer *et al.*, 1968; Tipper, 1969; Tipper and Berman, 1969). Four types of teichoic acid have been de-scribed in which glycerol is either α- or β-linked to glucose or α-linked to glucosamine (Davison and Baddiley, 1964; Davison *et al.*, 1964; Oeding *et al.*, 1967; Archibald *et al.*, 1968; Aasen and Oed-ing, 1971). The teichoic acids are serologically distinct. The α-linked glucosyl-glycerol teichoic acid probably corresponds to the group specific precipitin (Carbohydrate B) of Julianelle and Wieghard (1935) (Losnegard and Oeding, 1963).

Colonies are circular, convex with a smooth or slightly granular surface, and an entire or slightly irregular edge. They are usually white or yellow, occasionally orange and rarely purple. Purple-pigmented strains, named *Micrococcus viologabriellae* by Castellani (1955), were renamed *S. epidermidis* subsp. *violagabriellae* by Marples (1969).

Growth in broth is first turbid, later becoming clear with a fine or slightly mucoid deposit.

Chemoorganotrophs: Metabolism respiratory and fermentative. Menaquinones and cytochromes a, b₁, and O form the electron transport system; carotenoid pigments may be present. Nitrate functions as an electron acceptor in nitrate-reduc-ing strains (Jacobs *et al.*, 1969).

Acid production from carbohydrates is variable and forms the basis for recognizing biotypes within this species; (Baird-Parker, 1965) (Table 14.4). Acid from glucose and usually lactose and maltose in presence or absence of air, acid from glycerol and mannitol usual only in air. No acid from arabinose, cellobiose, inulin, raffinose, rhamnose, salicin, sorbitol or xylose. Starch and esculin not hy-drolyzed (Baird-Parker, 1965; High and Ellis, 1968). Some strains produce a muramidase (lyso-zyme) (Hawiger, 1968) and a heat-sensitive nuclease (Lachica *et al.*, 1971). Acetoin is produced as an end product of glucose metabolism; terminal pH in glucose broth 4.2–5.8.

Nitrates reduced by nitratases and may be further reduced by nitritases (Jones *et al.*, 1963). Ammonia produced from arginine by arginine dihydrolase. Hippurate hydrolyzed by most strains.

TABLE 14.4

Characteristics of biotypes of
Staphylococcus epidermidis[a]

	Biotype			
	1	2	3	4
Acetoin	+	−	+	+
Phosphatase	+[b]	+	−	−
Acid aerobically from:				
Lactose	+	d	−	d
Maltose	+	−	d	d
Mannitol	−	−	−	+

[a] + = most (90% or more) strains positive; − = most (90% or more) strains negative; d = some (less than 90%) strains positive some nega-tive. For methods see Baird-Parker (1963).

[b] On the plate test (Baird-Parker, 1963) some of these organisms are phosphatase negative but by the more sensitive tube test of Pennock and Huddy (1967) they are phosphatase positive.

Traces of H₂S may be produced from cysteine. No decarboxylation of glutamic acid or ornithine (Hugh and Ellis, 1968). Proteases, lipases, phospholipases, lipoprotein lipases, esterases and lyases may be produced (Baird-Parker, 1963, 1965).

Many strains produce a thermostable hemolysin that causes complete hemolysis of guinea pig, rabbit, horse, human and sheep erythrocytes (Kleck and Donahue, 1968). This hemolysin, probably identical with the delta-hemolysin of *S. aureus*, is claimed by Elek (1959) and others to be distinct and has been labeled epsilon-hemolysin. Some strains produce small quantities of hemolysins with patterns corresponding to the alpha- and beta-hemolysins of *S. aureus* (Kocur, Přecechtěl and Martinec, 1966), but these may be antigenically distinct from the hemolysins produced by *S. aureus* (Abramson and Friedman, 1967).

Unable to grow with ammonium phosphate and glucose as sole carbon, nitrogen and energy sources (Baird-Parker, 1963, 1965); no growth in Koser's or Simmons' citrate (Hugh and Ellis, 1968). Requires an organic nitrogen source (amino acids) and a number of B group vitamins, including biotin, for growth (Gretler *et al.*, 1955). For anaerobic growth most strains also require uracil and a fermentable carbon source (Jones *et al.*, 1963).

Facultative anaerobe. Grows at 45 C, often at 10 C (optimum 30–37 C). Most grow in 10–15% sodium chloride and many in 40% bile. Sensitive to novobiocin (M.I.C. < 0.6 µg/ml), usually sensitive to lysis by lysostaphin (1 unit/ml) and resistant to lysis by lysozyme (100 µg/ml) (Thomas, 1964; Mitchell and Baird-Parker, 1967; Klesius and Schuhardt, 1968; Jeffries, 1969).

Isolated from small stitch abscesses and other wounds. Very common, but mainly found on the skin and mucous membranes of warm-blooded animals. Commensals or parasites, many strains may be primary pathogens or secondary invaders. *Micrococcus hyicus*, a strain belonging to *S. epidermidis* biotype 2 (Table 14.4), is a specific pig pathogen causing contagious impetigo in swine (Sompolinsky, 1953; Hunter *et al.*, 1970). *Micrococcus violagabriellae* was believed by Castellani (1955) to cause chronic skin infections in man; this organism is a member of *S. epidermidis* biotype 1. Subacute bacterial endocarditis may be caused by penicillin-sensitive strains of *S. epidermidis* and post-cardiotomy endocarditis by penicillin-resistant strains (Quinn, Cox and Drake, 1966). *Staphylococcus epidermidis* biotype 1 frequently causes infection following the insertion of ventriculo-atrial shunts in the treatment of hydrocephalus (Holt, 1969, 1972). Some biotypes of *S. epidermidis* may be further subdivided by phage typing (Verhoef *et al.*, 1972).

The G + C content of the DNA is 30–37 moles % (Tₘ, buoyant density).

Neotype strain: ATCC 14990 (Hugh and Ellis 1968, 237).

3. **Staphylococcus saprophyticus** (Fairbrother) *emend. mut. char.* Shaw, Stitt and Cowan 1951, 1021. (Not *Staphylococcus saprophyticus* Fairbrother 1940, 88; Micrococcus subgroups 1, 2, 3 and 4 Baird Parker 1961, 417.)

sa.pro.phy'tic.us. Gr. adj. *sapros* putrid; Gr. n. *phyton* plant; M.L. adj. *saprophyticus* saprophytic, growing on dead tissues.

Spheres, 0.5–1.5 µm in diameter, occurring singly, in pairs and occasionally tetrads or cubical packets, but usually forming compact or loose irregular clusters. Not motile.

Cell walls contain organic phosphorus, glucosamine, muramic acid, lysine, alanine, glutamic acid, serine, glycine, ribitol and/or glycerol. The teichoic acid of most strains contains ribitol that is β-linked to *N*-acetyl glucosamine (Davison *et al.*, 1964; Oeding and Haselgren, 1972).

Smooth, convex colonies with regular or slightly irregular margins. Mostly white but occasionally yellow or orange in color.

Uniform turbidity in broth followed by clearing with a fine or sometimes a granular or mucoid deposit.

Chemoorganotrophs: Metabolism mainly respiratory. Menaquinones and cytochromes form the electron transport system. Slight growth anaerobically in certain media and glucose fermented weakly (Auletta and Kennedy, 1966; Mortensen and Kocur, 1967; Mortensen, 1969; Evans and Kloos, 1972).

Acid from glucose, glycerol, lactose and sucrose. Some strains form acid from maltose and mannitol and on the basis of differences in carbohydrate utilized four biotypes can be recognized (Baird-Parker, 1963). Acid usually not produced from arabinose, cellobiose, dextrin, inositol, raffinose, rhamnose, salicin or xylose. Acetoin produced as an end-product of glucose metabolism. Terminal pH in unbuffered glucose broth is 4.4–5.8 (mean pH 4.9).

Nitrate usually reduced with the formation of nitrite or ammonia. Ammonia usually produced from arginine. Casein, gelatin, egg yolk, fats, Tweens and urea usually hydrolyzed. Some strains produce muramidase and/or nuclease (Holt, 1972).

Growth reported only on complex media; minimal nutritional requirements unknown.

Facultative anaerobe showing slight anaerobic growth in suitable media (Evans and Kloos, 1972). Optimum growth temperature 30–37 C; most strains grow at 10 and some at 45 C. Grows in at least 5% NaCl.

Resistant to novobiocin (M.I.C. > 2.0 µg/ml) Mitchell and Baird-Parker, 1967; Jeffries, 1969) and lysozyme (M.I.C. > 300 µg/ml) (Jeffries, 1969). Two strains tested by Klesius and Schuardt (1968) were lysed by 1 unit/ml of lysostaphin; Kloos (personal communication, 1972) found all strains studied were sensitive to lysostaphin.

Usually regarded as nonpathogenic but some strains cause infections of the urinary tract, i.e. strains of Baird-Parker's *Micrococcus* subgroup 3 Torres Pereira, 1962; Mitchell, 1964, 1968; Roberts, 1967).

Isolated from urine; common in air, soil, dust, dairy products and surface of animal carcasses.

The G + C content of the DNA is 30–37 moles % (T_m, buoyant density).

Suggested neotype strain: S.41; NCTC 7292; ATCC 15305 (Shaw, Stitt and Cowan 1951, 1021).

Species incertae sedis

a. *Staphylococcus salivarius* Andrewes and Gordon 1907, 558.

The organism assigned to this taxon by Gordon (1957) is not a member of this genus. It is a coccus (1.5–1.8 µm in diameter); grows poorly and only weakly ferments glucose in the standard medium proposed by the I.C.S.B. Subcommittee (1965). Its DNA has a G + C content of 55.4–58.5 moles %, it does not grow in 5% sodium chloride or 40% bile, is oxidase positive and resistant to lysostaphin. It also has a cell wall composition different from that reported for staphylococci (Bowden, 1969). This organism appears to be an important part of the oral flora. Further study is required to find its correct taxonomic position. Bergan *et al.* (1970) suggest it is a micrococcus and should be called *Micrococcus mucilaginosus* Migula 1900, 119.

Genus III. **Planococcus** *Migula 1894, 236*

MILOSLAV KOCUR

Plan.o.coc'cus. Gr. comb.form *planos* wandering; Gr. n. *coccus* a grain, berry; M.L. masc.n. *Planococcus* motile coccus.

Spheres, 1.0–1.2 µm in diameter, **occurring singly, in pairs, in threes or in tetrads. Motile,** each cell usually possessing one or two flagella, occasionally three to four. Non-spore-forming. **Gram-positive.**

Chemoorganotrophs: **Metabolism respiratory, never fermentative.** Do not produce acid or gas from glucose, maltose, lactose or sucrose in standard medium (Subcommittee, 1965). Produce acid but no gas from glucose, maltose and mannitol but not from lactose or sucrose in MOF medium (Leifson, 1963). Final pH in glucose broth 6.6–7.0. Do not produce acetylmethylcarbinol, coagu-

lase, phosphatase; do not reduce nitrate or nitrite, or hydrolyze arginine. Gelatin is hydrolyzed.

Strict aerobes.

Catalase positive; benzidine test for porphyrins is positive.

Good growth between 20 and 37 C.

Found in sea water, but grows equally well in sea water medium and in media without added salt.

The G + C content of the DNA ranges from 48–52 moles % (T_m) (Bohacek *et al.*, 1967, 1968).

Type species: *Planococcus citreus* Migula 1894, 236.

Description of the species of genus **Planococcus**

1. **Planococcus citreus** Migula 1894, 236. (*Micrococcus aquivivus* ZoBell and Upham 1944, 239; *Sarcina citrea* (Migula) Bergey *et al.* 1948, 288; *Micrococcus eucinetus* Leifson 1964, 41.)

cit're.us. M.L. adj. *citreus* lemon-yellow.

Morphology as for genus.

Agar colonies are round, 2–3 mm in diameter, slightly convex, smooth and glistening. A water-insoluble yellowish orange pigment is produced.

In sea water or nutrient broth, moderate turbidity with sediment but no pellicle.

Methyl red test negative. Indole and hydrogen sulfide not produced. Starch and esculin not hydrolyzed. No growth on Simmons' citrate agar. Phenylalanine deaminase, oxidase or urease not

produced. Egg yolk reaction is negative. Tween 80 is not split. Human and rabbit plasma is not coagulated. Does not produce hemolysis.

Further Comments

Pending further study, other strains with very similar biochemical characteristics and a G + C content of 39–42 moles %, have not yet been given a species designation.

Species incertae sedis

a. *Planococcus roseus* (Winogradsky) Migula 1895, 19.

b. *Planococcus casei* Migula 1900, 270.

c. *Planococcus ochroleucus* (Prove) Migula 1900, 272.

d. *Planococcus löffleri* (*sic*) Migula 1900, 273.

e. *Planococcus luteus* (Adametz) Migula 1900, 274.

f. *Planococcus agilis* (Ali-Cohen) Chester 1901, 115.

g. *Planococcus europaeus* (Winogradsky) Vuillemin 1913, 525.

The above are found in the literature. Those studied have similar biochemical characteristics and a G + C content of 39–42 moles %. Until further studies have been made they are considered *incertae sedis*.

FAMILY II. **STREPTOCOCCACEAE** *fam. nov.*

R. H. Deibel and H. W. Seeley, Jr.

Strep.to.coc.ca'ce.ae. M.L. masc.n. *Streptococcus* type genus of family; *-aceae* ending to denote family; M.L. gen.pl.n. *Streptococcaceae* the *Streptococcus* family.

Cells spherical or ovoid, in pairs or chains of varying length or in tetrads. Non-motile or rarely motile. Endospores not formed. Gram-positive.

Chemoorganotrophs. Metabolism fermentative; lactic, acetic and formic acids, and ethanol and CO_2 formed from carbohydrates.

Nutritional requirements complex and variable. Catalase test is variable, benzidine negative. Facultatively anaerobic.

The G + C content of DNA ranges from 33–44 moles %.

Type genus: *Streptococcus* Rosenbach 1884, 22.

Key to the genera of family **Streptococcaceae**

I. Glucose is fermented by the hexose diphosphate pathway yielding chiefly dextrorotatory lactic acid (homofermentative). Cell division in one plane resulting in pairs and chains. Catalase negative.

Genus I. *Streptococcus*

II. Glucose is fermented by the hexose monophosphate pathway yielding levorotatory lactic acid, CO_2, ethanol and/or acetic acid (heterofermentative). Cell division in one plane resulting in pairs and chains. Catalase negative.

Genus II. *Leuconostoc*, p. 510

III. Glucose is fermented yielding inactive lactic acid (homofermentative). Cell division in two planes resulting in pairs and tetrads. Catalase variable.

Genus III. *Pediococcus*, p. 513

IV. Glucose is fermented yielding dextrorotatory lactic acid (homofermentative). Cell division in two planes resulting in pairs and tetrads. Catalase variable.

Genus IV. *Aerococcus*, p. 515

V. Glucose is fermented but the products of fermentation have not yet been reported. Gram-indeterminate, but cell wall structure is of the Gram-positive type. Catalase negative.

Genus V. *Gemella*, p. 516

Genus I. **Streptococcus** *Rosenbach 1884, 22*

R. H. Deibel and H. W. Seeley, Jr.

Strep.to.coc'cus. Gr. adj. *streptus* pliant; Gr. n. *coccus* a grain, berry; M.L. masc.n. *Streptococcus* pliant coccus.

Cells spherical to ovoid, less than 2 μm in diameter (varies with species), occurring in **pairs or chains** when grown in liquid media. Occasional motile strains in serological group D. **Gram-positive.**

Chemoorganotrophs: **Metabolism fermentative,** predominant end product of glucose fermentation is **dextrorotatory** lactic acid (**homofermentative**), some species ferment organic acids (malic and citric) and amino acids (serine and arginine). Pigments, found only in groups B and D, usually red or yellow.

Pellicles are never formed. Do not contain heme compounds; benzidine negative and **catalase negative.**

Facultative anaerobes. Oxygen, or other hydrogen acceptors, may alter end products of carbohydrate metabolism. Hydrogen peroxide may accumulate in presence of oxygen.

Minimal nutritional requirements are generally complex (but variable) and may involve amino acids, purines, pyrimidines, peptides, vitamins and occasionally fatty acids and elevated CO_2 tension. In contrast, strains of one species (*S. bovis*) may only require glucose and ammonium and inorganic salts for growth.

Temperature optimum, about 37 C unless noted otherwise; maximum and minimum temperatures vary with the species.

The G + C content of the DNA (reported for 15 species) ranges from 33–42 moles % (most used method was chemical analysis).

Type species: *Streptococcus pyogenes* Rosenbach 1884, 23.

Key to the species of genus **Streptococcus**

I. Does not grow at 10 or 45 C (some exceptions in *S. sanguis*). Does not grow in 6.5% NaCl broth, at pH 9.6, or in 0.1% methylene blue milk.
  A. Does not hydrolyze sodium hippurate (some exceptions in *S. dysgalactiae*).
    1. Does not require a high CO_2 tension for rapid growth on blood agar. Cells and colonies not "minute." Ammonia produced from arginine.
      a. Does not ferment inulin.
        b. Fibrinolytic.
          c. Beta-hemolytic. Ferments trehalose but not sorbitol.
            d. Does not ferment glycerol. Lancefield GROUP A.
              1. *Streptococcus pyogenes*
            dd. Ferments glycerol aerobically. Lancefield GROUP C.
              2. *Streptococcus equisimilis*
        bb. Not fibrinolytic.
          c. Beta-hemolytic.
            d. Ferments sorbitol but not trehalose. Ferments glycerol aerobically. Lancefield GROUP C.
              3. *Streptococcus zooepidemicus*
            dd. Does not ferment trehalose, sorbitol or glycerol. Lancefield GROUP C.
              4. *Streptococcus equi*
          cc. Not beta-hemolytic. Ferments trehalose. Usually ferments sorbitol. Does not ferment glycerol. Lancefield GROUP C.
            5. *Streptococcus dysgalactiae*
      aa. Inulin usually fermented.
        b. May or may not be beta-hemolytic. Not bile soluble. Viscous polysaccharide may be produced in 5% sucrose broth. Lancefield GROUP H.
          6. *Streptococcus sanguis*
        bb. (Viscous polysaccharide not produced in 5% sucrose broth.) Alpha-reaction on blood agar. Bile soluble. No group antigen demonstrated.
          7. *Streptococcus pneumoniae*
    2. High CO_2 tension required for rapid growth on blood agar. "Minute" cells and colonies. Ammonia produced from arginine. Lancefield GROUP F and Type 1. GROUP G.
      8. *Streptococcus anginosus*
  B. Hydrolyzes sodium hippurate.
    1. Produces ammonia from arginine. Ferments glycerol aerobically. May be beta-hemolytic. Final pH in glucose broth 4.2–4.8. Lancefield GROUP B.
      9. *Streptococcus agalactiae*
    2. Does not produce ammonia from arginine. Does not ferment glycerol. Not hemolytic. Final pH in glucose broth 5.6–6.5. No group antigen demonstrated.
      10. *Streptococcus acidominimus*

II. Does not grow at 10 C. Grows at 45 C (exceptions in *S. mitis*). Does not grow in 6.5% NaCl broth, at pH 9.6 or in 0.1% methylene blue milk. Does not produce ammonia from arginine. (Exceptions in *S. mitis*.) Not beta-hemolytic.

  A. Does not grow at 50 C. Grows in 2% NaCl broth.

    1. Does not hydrolyze starch. Does not grow on 40% bile blood agar.

      a. Indifferent gamma-reaction on blood agar. Ferments raffinose and inulin. Produces large mucoid colonies on 5% sucrose and raffinose agar. Lancefield GROUP K.

        11. *Streptococcus salivarius*

      aa. Alpha-reaction on blood agar. Does not ferment inulin. May or may not ferment raffinose. Does not produce mucoid colonies on 5% sucrose agar. No group antigen demonstrated.

        12. *Streptococcus mitis*

    2. Hydrolyzes starch. Grows on 40% bile blood agar.

      a. Ferments lactose. Lancefield GROUP D.

        13. *Streptococcus bovis*

      aa. Does not ferment lactose. Lancefield GROUP D.

        14. *Streptococcus equinus*

  B. Grows at 50C. Does not grow in 2% NaCl broth. No group antigen demonstrated.

        15. *Streptococcus thermophilus*

III. Grows at 10 and 45 C.

  A. Grows in 6.5% NaCl, at pH 9.6 and in 0.1% methylene blue milk.

    1. Uses arginine as a source of energy; ferments sorbitol, does not ferment arabinose. Does not require folic acid for growth in simplified media. Lancefield GROUP D.

        16. *Streptococcus faecalis*

    2. Does not utilize arginine as a source of energy but hydrolyzes it to produce ammonia. Ferments arabinose, does not ferment sorbitol. Requires folic acid for growth in simplified media. Lancefield GROUP D.

        17. *Streptococcus faecium*

  B. Grows in 6.5 % NaCl broth and at pH 9.6. Does not grow in 0.1% methylene blue milk. Ferments arabinose and sorbitol. Requires folinic acid for growth in simplified media. Does not produce ammonia from arginine. Contains Lancefield GROUP Q and GROUP D antigens.

        18. *Streptococcus avium*

  C. Does not grow in 6.5% NaCl broth, at pH 9.6 or in 0.1% methylene blue milk. Ferments sorbitol. Does not ferment arabinose. Usually requires folic acid for growth, in simplified media. Produces ammonia from arginine.

        19. *Streptococcus uberis*

IV. Does not grow at 45 C; grows at 10 C. Does not grow in 6.5% NaCl broth, or at pH 9.6. Grows in 0.1% methylene blue milk. Lancefield GROUP N.

  A. Produces ammonia from arginine.

        20. *Streptococcus lactis*

  B. Does not produce ammonia from arginine.

        21. *Streptococcus cremoris*

Further Comments

The majority of the streptococci possess a dominant serologically-active carbohydrate (termed the "C" substance). Employing the precipitin technique, Lancefield (1933) established serological groups which reflected the distinctive antigenicity of these carbohydrates. Subsequently the serological and physiological characteristics were used to delineate the various species. In the past (Sherman, 1937), the streptococci were sep-arated into physiological divisions on the basis of growth at 10 and 45 C (Sherman criteria); in this edition of THE MANUAL, designations of the physiological divisions (pyogenic, viridans, lactic, enterococcus) have been omitted from the key since newly recognized species may cut across the broad lines of Sherman's major divisions. Nevertheless, Table 14.6 demonstrates that growth and tolerance tests as proposed by Sherman for the separation of his divisions still have some usefulness.

TABLE 14.5

Some chemical characteristics and cellular location of streptococcal antigens

Group antigen (hapten)			Type antigen (hapten)			Species
Desig-nation	Chemical nature	Cellular location	Designation	Chemical nature	Cellular location	
A	Rhamnose-*N*-acetyl-glucosamine poly-saccharide	Wall	M R T	Protein Protein Protein	Envelope Envelope Envelope	*S. pyogenes*
B	Rhamnose-glucos-amine polysacchar-ide	Wall	S (5 types)	Glucose-galactose-*N*-acetyl-glucosamine polysaccharide	Envelope	*S. agalactiae*
C	Rhamnose-*N*-acetyl-galactosamine polysaccharide	Wall	(8 types described)	Protein		*S. equisimilis*
			(8 types)	Protein		*S. zooepidemicus*
			(Only 1 type de-scribed)	Protein	Envelope	*S. equi*
			(3 types)	Protein		*S. dysgalactiae*
D	Glycerol teichoic acid containing D-alanine and glucose	"Intracellular" be-tween wall and membrane	(11 types established)	Rhamnose-glucos-amine-glucose polysaccharide	Cell wall	*S. faecalis*
D			(19 types described; many more exist)			*S. faecium*
D, Q						*S. avium*
D			(Many types exist)	Carbohydrate?	Capsule	*S. bovis*
D						*S. equinus*
E	Rhamnose polysac-charide	Wall	I to V	Polysaccharide		*Streptococcus* sp.
F	Rhamnose and a glu-copyranosyl-*N*-ace-tyl-galactosamine tetrasaccharide	Wall	I to V	Carbohydrates—some types contain glu-cose, galactose and rhamnose		*S. anginosus*
G	Rhamnose-galactos-amine polysacchar-ide	Wall	(3 types described)			*Streptococcus* sp. (large colony)
H	Rhamnose polysac-charide	Wall	(5 types described)			*S. sanguis*
K	Rhamnose polysac-charide	Wall	I and II, and I-II	Galactose, glucose rhamnose Type I—*O*-β-D galac-topyranosyl-(1->6)-D-galactose	Cell wall	*Streptococcus* sp., *S. salivarius*
N	Glycerol teichoic acid containing D-alanine and galactose phos-phate	"Intracellular" between wall and membrane	Many types exist			*S. lactis* *S. cremoris*
D			(Only 1 type)	Contains glucose, glucosamine, galac-tosamine	Capsule	*S. suis*
None	Ribitol teichoic acid with choline phos-phate	Wall	(80 types) M	Carbohydrate Protein	Capsule Envelope	*S. pneumoniae*

The similarity of the pneumococci to the streptococci has been recognized for many years and in this edition of THE MANUAL the organism is classified as *Streptococcus pneumoniae* (Klein) Chester 1901, 63.

Twenty-one species are now recognized: five do not possess demonstrable group antigens. Serological groups from A to H and K to T have been worked out over the years; but no species designations have been suggested for many of the groups (i.e. E, G, L, M *etc*.). In each of the groups C, D and N more than one species has been established and there may be reason in the future to separate the group D streptococci as a new genus. A listing of the serological groups and some characteristics of the group and type-specific antigens are presented in Table 14.5.

The rhamnosyl residues in the wall of *S. agalactiae* contain a 1-2 linkage while in groups A, C,

E and H the linkage is 1-3; and in groups F, G and K the linkage is 1-4 (Chionglo and Hayashi, 1969). Although qualitative and quantitative differences exist in the sugar moieties of the cell wall and the associated antigens, their utility for determinative purposes is minimal (Slade and Slamp, 1962).

Identification of the individual species is based on serological and physiological tests. Species lacking a recognizable group antigen (e.g. *S. mitis*, *S. thermophilus*) are identified by physiological tests. A strain that lacks a group antigen can often be placed in a serological group by its physiological characteristics. Identification by both physiological and serological procedures offers the most definitive approach.

Ever increasing numbers of strains from different serological groups are being reported to possess type antigens that cross-react with those in other groups. Consequently, the specificity of type re-

TABLE 14.6

Tolerance tests used to differentiate streptococcal species[a]

Streptococcal species	Growth at:		Growth in media containing:			Growth initiation at pH 9.6	Heat tolerance (60 C for 30 min)
	10 C	45 C	Methylene blue (0.1% in milk)	Sodium chloride (6.5%)	Bile (40%)		
1. *S. pyogenes*	–	–	–	–	–	–	–
2. *S. equisimilis*	–	–	–	–	–	–	–
3. *S. zooepidemicus*	–	–	–	–	–	–	–
4. *S. equi*	–	–	–	–	–	–	–
5. *S. dysgalactiae*	–	–	–	–	–	–	–
6. *S. sanguis*	–	d	–	–	+	ND	d
7. *S. pneumoniae*	–	–	ND	–	–	–	–
8. *S. anginosus*	–	d	–	–	d	–	–
9. *S. agalactiae*	–	–	–	–	+	–	–
10. *S. acidominimus*	–	–	–	–	ND	–	–
11. *S. salivarius*	–	+	–	–	–	–	–
12. *S. mitis*	–	d	–	–	–	–	–
13. *S. bovis*	–	+	–	–	+	–	+
14. *S. equinus*	–	+	–	–	+	–	–
15. *S. thermophilus*	–	+	–	–	–	–	+
16. *S. faecalis*	+	+	+	+	+	+	+
17. *S. faecium*	+	+	+	+	+	+	+
18. *S. avium*	+	+	–	+	ND	+	ND
19. *S. uberis*	+	+	–	–	–	–	+
20. *S. lactis*	+	–	+	–	+	–	d
21. *S. cremoris*	+	–	d	–	+	–	d
a. *S.* sp., group G	–	–	–	–	–	–	–
b. *S.* sp., group E	–	–	–	–	–	–	–
c. *S.* sp., group L	ND	ND	ND	ND	ND	ND	ND
d. *S.* sp., group M	ND	ND	ND	–	–	ND	ND
e. *S.* sp., group O	ND	ND	ND	ND	–	ND	–
f. *S.* sp., group R	ND	–	–	–	+	–	–
g. *S. suis*, group D	ND	ND	ND	ND	d	ND	–

[a] – = most (90–100%) strains negative; + = most (90–100%) strains positive; ND = no data or insufficient data; d = some strains positive, some negative.

actions cannot be equated with the specificity of the group antigens. Cross-reactions with "group" antisera should be viewed with suspicion for cross-reactivity may reside in common type-specific antigens. The use of strains lacking type-specific antigens to prepare grouping antisera should prevent these cross-reactions. Strains lacking type antigens have been observed in groups E and F (Yao *et al.*, 1964; Ottens and Winkler, 1962).

The characteristic morphology associated with a *Streptococcus* isolate is demonstrated in broth cultures. Smears prepared from agar cultures or certain natural materials may and usually do present a radically different morphological appearance. Although the streptococci are Gram-

positive, cultures beyond the logarithmic phase of growth and smears of natural materials may show Gram-negative cells as well as cellular and staining aberrations. Young broth cultures offer the best material for Gram-staining.

In general, the cell structure of streptococci does not differ from that of other Gram-positive cocci (Cole, 1968). The cell wall is homogenous in structure with each cell in the chain showing cross walls at right angles to the plane of division as viewed in electron micrographs. The mesosomes and tripled-layered cytoplasmic membrane are similar also to other Gram-positive organisms.

True capsule formation in certain of the streptococci is not uncommon. In some instances (*S.*

TABLE 14.7

Hemolysin and fibrinolysin production and key hydrolytic reactions of streptococci[a]

Streptococcal species	Soluble hemolysin production			Fibrinolysin production	Hydrolysis of:				
	O	S	Other		Gelatin	Starch	Hippurate	Esculin	Arginine
1. *S. pyogenes*	+	+	−	+	d	−	−	+	+
2. *S. equisimilis*	+	−	+	+[b]	−	d	−	d	+
3. *S. zooepidemicus*	−	−	+	−	−	+	−	+	+
4. *S. equi*	−	−	+	−	−	ND	−	−	+
5. *S. dysgalactiae*	−	−	−	d[c]	−	ND	d	d	+
6. *S. sanguis*	ND	ND	−	−	−	−	−	+	+
7. *S. pneumoniae*	+	−	−	−	−	ND	ND	ND	ND
8. *S. anginosus*	−	−	d	−	−	−	−	+	+
9. *S. agalactiae*	−	−	+	−	−	−	+	−	+
10. *S. acidominimus*	ND	ND	−	−	−	−	(+)[d]	−	−
11. *S. salivarius*	−	−	−	−	−	−	−	+	−
12. *S. mitis*	−	−	ND	−	−	−	−	−	d
13. *S. bovis*	−	−	−	−	−	+	d	+	−
14. *S. equinus*	−	−	−	−	−	+[e]	−	+	−
15. *S. thermophilus*	−	−	−	−	−	+	−	−	−
16. *S. faecalis*	−	−	ND	−	−[e]	−	+	+	+
17. *S. faecium*	−	−	ND	−	−	−	ND	+	+
18. *S. avium*	ND	ND	ND	ND	−	−	−	+	−
19. *S. uberis*	−	−	−	ND	−	d	(+)[d]	+	+
20. *S. lactis*	−	−	−	−	−	−	d	d	+
21. *S. cremoris*	−	−	−	−	−	−	−	d	−
a. *S.* sp., group G	+	ND	−	d	−	+	−	+	+
b. *S.* sp., group E	ND	ND	ND	−	d	−	−	+	+
c. *S.* sp., group L	−	−	−	ND	ND	ND	−	ND	ND
d. *S.* sp., group M	ND	ND	d	ND	ND	ND	−	−	d[b]
e. *S.* sp., group O[b]	−	−	−	−	ND	ND	−	ND	ND
f. *S.* sp., group R[b]	ND	ND	d	ND	ND	ND	−	+	+
g. *S. suis*, group D	ND	ND	ND	ND	ND	+	ND	+	ND

[a] See Table 14.6 for key to symbols.
[b] See species description.
[c] For animal fibrin only.
[d] ( ) Delayed reaction.
[e] See description of varieties.

pyogenes) the capsule has no serologic importance and superficial antigens (i.e. M, R and T) may impart serological specificity or virulence. In other instances the capsule *per se* may play an important role in determining serological specificity and/or virulence (e.g. *S. equi*, *S. pneumoniae*, *S. suis* and *S. bovis*). Attempts have been made to distinguish between the superficial antigens that are closely associated with the cell wall as compared to the true capsular antigens. Terms such as microcapsular, K and envelope antigens have been employed. In this part of THE MANUAL, the term "envelope" has been employed to denote the superficial antigens such as the M, type-specific antigens of *S. pyogenes* and the distinction is made between envelope and true capsular antigens (see Table 14.5).

The cardinal physiological reactions employed to differentiate the streptococci are presented in Tables 14.6 and 14.7. These characteristics and the production of hemolysins and fibrinolysin have been tabulated to facilitate comparison and to present data. For the most part, this material is not repeated in the individual descriptions of the species. Generally the reactions of litmus milk cultures are omitted; aside from the reduction of litmus prior to acid curd formation (as observed in group N and most group D species) and in certain instances where peptonization occurs, the taxonomic value of this set of characteristics is limited.

As yet, no reports dealing with streptococcal transduction or conjugation have appeared in the literature. A study involving DNA-RNA homology among the streptococci indicated a relationship between the various serologic groups (including *S. pneumoniae*), except for *S. faecalis* which was only remotely related (Weissman *et al.*, 1966).

Description of the species of genus **Streptococcus**

1. **Streptococcus pyogenes** Rosenbach 1884, 23. (*Streptococcus erysipelatos* (*sic*) Rosenbach 1884, 22; *Micrococcus scarlatinae* Klein 1884, 332; *Streptococcus scarlatinae* Klein 1887, 391; *Streptococcus hemolyticus* Rolly 1911, 87.)

py.og′en.es. Gr. n. *pyum* pus; Gr. v. *gennaio* beget; M.L. adj. *pyogenes* pus-producing.

Spherical or ovoid cells, 0.6–1.0 μm in diameter. Occur as pairs and short to moderate chains in clinical material; long chains frequent in broth cultures. Some strains form a capsule of hyaluronic acid demonstrable in 2–4 hr cultures. Hyaluronidase production by many strains precludes capsule formation.

The peptide subunit in the peptidoglycan is in the following sequence: L-alanine, D-isoglutamine, L-lysine, D-alanine. The cross-bridge dipeptide L-alanine-L-alanine links the peptide subunits *via* D-alanine and L-lysine.

An arrangement for cell wall construction has been proposed (Ghuysen, 1968). Proceeding distally the following polymeric constituents occur: the "C" polysaccharide backbone is linked by trirhamnose groups to the peptidoglycan which is in turn linked *via* phosphodiester bridges to a polysaccharide. The latter serves to attach the T, R and M antigens.

Although the literature abounds with descriptions of mucoid, matt and glossy colony forms, only mucoid and glossy exist; Wilson (1959) has demonstrated that matt is merely a dehydrated mucoid form. Culture of either matt or mucoid on hyaluronidase-containing agar yields the glossy form and plate cultures containing a matt-type strain, sealed to prevent moisture loss, give rise to a mucoid colony. Colony form is related directly to hyaluronic-acid-capsule formation and not to M protein production, as originally thought.

Constitutes Lancefield's group A; the A antigenic determinant consists of a branched rhamnose chain with terminal *N*-acetylglucosamine residues. It is an integral part of the cell wall (Krause and McCarty, 1961). Using the precipitin test, about 55 M antigens described; some strains lack an M antigen. M antigens, serologically identical to group A types, may be found in other groups (Maxted and Potter, 1967); group A strains may also possess type-specific antigens of other groups (Jablon *et al.*, 1965). Some strains produce a proteinase that destroys the M antigen but if grown at 22 C rather than 37 C, the type-specific antigen may be demonstrated. Nutritionally, peptides, eight amino acids and an energy source (glucose) are needed for M protein synthesis. Growth media (Kremers and Quinn, 1965) and extraction and testing procedures (Swift *et al.*, 1943) have been described. Incubation of cells with trypsin destroys M protein without affecting viability. Protoplasts and L forms synthesize M protein which can diffuse into the medium. Each M type may be produced in multiple molecular forms (Fox and Wittner, 1965; Johnson and Vosti, 1968). Marked heat stability (boiling for 0.5 hr or several min at pH 2), acid resistance and alcohol solubility are common properties of the M proteins. Although they are heat stable in acid, extracted proteins are poor antigens even though they still react with homologous antisera (Besdine and Pine, 1968). Strains lacking these proteins are avirulent. M types I, IV, XII, XVIII, XXV and

[L are associated with glomerulonephritis and referred to as nephrogenic strains. All other group A streptococci regardless of type may be associated with rheumatic fever, subsequent to the initial infection.

The T antigens, resistant to pepsin and trypsin, are not heat-stable or markedly resistant to acid. The Griffith (1934) slide-agglutination technique utilizes trypsinized suspensions and is used when strains cannot be typed by the M antigens (especially useful in epidemiologic studies of skin isolates; Parker, 1969). For innovations and general methodology of T-agglutination typing see Padula et al. (1969).

The R antigen was first observed in a standard Griffith strain but later in some strains of groups B, C and G. It is sensitive to pepsin but not trypsin. Heat in the presence of acid destroys it but it tolerates mild heat in slightly alkaline solutions.

Energy-yielding metabolism is fermentative and the final pH range in glucose broth is 4.8–6.0. Acid produced from glucose, maltose, lactose, sucrose and salicin. No acid from inulin, raffinose, arabinose or common polyols.

Although many group A strains elaborate a proteinase in the form of a zymogen (Elliott, 1945; Elliott and Dole, 1947), many descriptions include the general inability of S. pyogenes to degrade proteins (especially gelatin) in laboratory cultures. However, using anaerobiosis and the plate method of detection, many strains evidence significant activity with gelatin, wheat gluten, beef protein, casein and denatured pepsin. Bovine serum albumin, lactalbumin and denatured lysozyme are not degraded (Deibel, 1963).

Group A streptococci elaborate four serologically distinct nucleases (designated A-D; Wannamaker, 1962). All possess DNase activity and types B and D also possess RNase activity (Wannamaker et al., 1967). Most group A strains hydrolyze both DNA and RNA when tested by the plate method and unlike proteolytic activity, nucleolytic activity is independent of oxygen tension (Deibel, 1963).

Most group A strains produce an extracellular enzyme that cleaves diphosphopyridine nucleotide (DPNase or NADase) at the nicotinamide-ribose linkage (Carlson et al., 1957). Ayoub and Ferretti (1966) have described a procedure to measure this activity. Groups C and G also produce the enzyme.

A majority of strains produce an enzyme, fibrinolysin, that activates a protease in normal human plasma and causes lysis of clots. It is trypsin sensitive and withstands boiling for 1 hr.

Two soluble, antigenic hemolysins are produced in fluid culture. Streptolysin O (probably identical to pneumolysin, listeriolysin etc.) is reversibly oxygen labile, susceptible to proteolytic enzymes and thermolabile. Streptolysin S effects hemolysis on blood agar plates whether they are incubated aerobically or anaerobically. It is thermolabile but stable to oxygen. However, its unstable nature leads to rapid loss of activity in the laboratory. In cultures, its production is greatly enhanced by nucleic acid fractions (see Okamoto, 1962). Most strains produce a strong beta-hemolytic reaction within 24 hr when cultured on blood agar.

The heat-stable erythrogenic toxin is produced by most strains. Three serologic types have been described (A, B and C). Zabriskie (1964) showed that toxin production is a transmissible character mediated by a temperate phage and thus confirmed the early observation of Frobisher and Brown (1927).

Facultative anaerobes. Optimum temperature for growth 37 C. Nutritionally fastidious, especially upon primary isolation; using laboratory strains and huge inocula, only moderate growth obtained in a synthetic medium in serial transfer (Mickelson, 1964). Adaptation is also necessary for growth of L-phase variants (van Boven et al., 1967).

Source: human mouth, throat, respiratory tract, blood, various lesions and inflammatory exudates.

The G + C content of DNA ranges from 34.5 and 38.5 (T_m) to 38–40 (method not stated) moles %.

2. **Streptococcus equisimilis** Frost and·Engelbrecht 1936, 3. (Type B, Ogura 1929, 174; Human C, Sherman 1937, 35.)

e.qui.si′mi.lis. M.L. gen.n. equi a species of Streptococcus; L. adj. similis resembling; M.L. adj. equisimilis resembling S. equi.

Cellular and colonial morphology are similar to S. pyogenes.

Lancefield's group C. The antigenic determinant is an integral part of the cell wall (Krause and McCarty, 1962). In Griffith's agglutinative scheme of the 27 original types, three (7, 20 and 21) are now classified in this species. Various schemes utilizing agglutination, precipitin and biochemical characters have designated eight "serotypes" but the individuality of the types may be questionable. The type-specific antigens are protein in nature; some digestible with trypsin. Some contain the T or R antigens of S. pyogenes types (Stableforth, 1959).

Final pH range in glucose broth 4.6–5.4. Acid is produced from glucose, maltose, sucrose and trehalose. Acid from glycerol only under aerobic conditions. Arabinose, raffinose, inulin, mannitol and sorbitol are not fermented. In contrast to strains from cattle, swine and man, the equine strains do not ferment lactose.

Beta-hemolytic on blood agar. Streptolysin O and/or other unidentified lysins are produced in broth cultures. Usually fibrinolytic but animal strains may not lyse human fibrin.

Growth reported only on complex media; minimal nutritional requirements unknown.

Source: upper respiratory tract of normal and diseased humans and animals. Occasionally associated with erysipelas and puerperal fever.

3. **Streptococcus zooepidemicus** Frost and Englebrecht 1936, 3. (Animal pyogenes, type A Edwards 1934, 527; *Streptococcus pyogenes animalis* Seelemann 1942, 8.)

zo.o.e.pi.dem'i.cus. Gr. n. *zoon* animal; Gr. adj. *epidemios* dwelling among, prevalent, epidemic; M.L. adj. *zooepidemicus* prevalent among animals.

Cellular and colonial morphology resembles that of *S. pyogenes*. Many strains produce capsular hyaluronic acid and mucoid colonies are common.

Lancefield's group C (see *S. equisimilis* for description of group antigen). Eight serotypes defined by nine antigens; eight of antigens trypsin sensitive; remaining one trypsin resistant and pepsin sensitive (Moore and Bryans, 1969). Relationship of serotypes from humans and animals unknown.

Final pH range in glucose broth is 4.6–5.0. Acid from glucose, sucrose, sorbitol and salicin. Lactose and maltose fermentations are variable. No acid from xylose, arabinose, trehalose, raffinose, inulin, glycerol or mannitol.

Wide zones of beta-hemolysis produced on blood agar. A soluble hemolysin, unrelated to streptolysins O and S, is produced.

Growth reported only on complex media; minimal nutritional requirements unknown.

Source: blood, inflammatory exudates and lesions of diseased animals. Not a human pathogen. Septicemia of cows, rabbits and swine. Almost always isolated from wound infections of horses (Stableforth, 1959). Not infrequently associated with various avian disease processes (Peckham, 1966). Isolated from the fetus and metritis of horses and cows.

Acid from sorbitol but not from trehalose and the inability to produce fibrinolysin characterize this species.

4. **Streptococcus equi** Sand and Jensen 1888, 436. (*Bacillus adenitis equi* Baruchello 1887, 248.)

e'qui. L. n. *equus* horse; L. gen.n. *equi* of a horse.

Ovoid or spherical cells 0.6–1.0 μm in diameter; sometimes in pus the long axis of the cells is transverse to the long axis of the chain and at other times parallel with the long axis of the chain, in the latter case resembling streptobacilli; bacillary forms are not rare; occur in pairs, short or long

chains; very long chains common in broth. Capsules are demonstrable in some strains when young cultures are examined and when serum is added to the growth medium.

Lancefield's group C (see *S. equisimilis* for description of group antigen). Only one serotype; the type-specific antigen is a protein, but apparently not associated with immunity to infection. A protective antigen distinct from the envelope type antigen demonstrated in capsular material. Among the streptococci this situation is unparalleled except for *S. pneumoniae* (see Englebrecht, 1969).

Final pH range in glucose broth is 4.8–5.5; a restricted fermentative pattern: acid from glucose, maltose, sucrose, salicin but no acid from arabinose, lactose, trehalose, raffinose, inulin, glycerol, mannitol or sorbitol.

Wide zones of beta-hemolysis observed on blood agar where the small, watery colony dries out rapidly and ultimately leaves a flat, glistening colony. A soluble hemolysin, distinct from streptolysins O and S, produced in serum-fortified broth cultures.

Growth in common laboratory media poor unless fortified with serum. The minimum nutritional requirements of some strains have been identified partially; in a casein-hydrolysate medium, at least two B vitamins and uracil are required. The amino acid requirements have not been determined.

The minimum temperature is approximately 20 C.

The cause of equine strangles. Isolated from abscesses in submaxillary glands and mucopurulent discharges of the upper respiratory system of horses and their immediate environment. Rarely isolated from other animals.

This species is distinctive by inability to ferment trehalose, lactose, sorbitol and the other common polyols.

5. **Streptococcus dysgalactiae** Diernhofer 1932, 369. (Group II, Minett 1934, 511; *Streptococcus pseudoagalactiae* Plastridge and Hartsell 1937, 110.)

dys.ga.lac'ti.ae. Gr. pref. *dys* ill, hard; Gr. n. *galactia* pertaining to milk; M.L. n. *dysgalactia* loss or impairment of milk secretion, dysgalactia; M.L. gen.n. *dysgalactiae* of dysgalactia.

Spherical or ovoid cells occurring in chains of medium length.

Colony forms may be similar to those of *S. pyogenes*.

Lancefield's group C (see *S. equisimilis* for description of group antigen). At least three serotypes; type specificity associated with a protein antigen. Strains from lambs suffering from polyarthritis may be a distinct serotype (Blakemore et al., 1941).

Final pH ranges in glucose broth: bovine strains, 5.0–5.2; ovine strains, 4.4–4.9. Acid from glucose, maltose, sucrose and trehalose. No acid from raffinose, inulin, glycerol or mannitol. Results with lactose, sorbitol and salicin are variable. Most bovine strains produce hyaluronidase. May produce a fibrinolysin for bovine fibrin but not human fibrin.

This species produces a strong alpha-reaction on blood agar.

Growth only on complex media; minimal nutritional requirements unknown.

Isolated from milk and udders of cows with acute but mild mastitis. Also from blood and tissues of lambs with polyarthritis (joint-ill).

Further study is needed to establish the species' serologic relationship to group C and the relationship of ovine and bovine strains. In contrast to other group C species, pathogenicity of *S. dysgalactiae* for mice is very low.

6. **Streptococcus sanguis** White in White and Niven 1946, 722. (Serolgical group H Hare 1935, 509; *Streptococcus* s. b. e. Loewe, Plummer, Niven and Sherman 1946, 257.)

san'guis. L. n. *sanguis* blood.

Spherical or ovoid cells 0.8–1.2 μm in diameter occurring in medium to long chains. Cultures grown aerobically may have an occasional rod-shaped cell.

"Matt" (use of this term is questionable—see description of *S. pyogenes*) or glossy-type colonies may be produced, usually 0.7–0.9 mm in diameter on blood agar.

Lancefield's group H; five serotypes described (Dodd, 1949); some strains have more than one type-specific antigenic determinant (Washburn et al., 1946).

Final pH range in glucose broth is 4.6–5.2. Acid produced from glucose, maltose, lactose, sucrose, trehalose, salicin and usually inulin. Arabinose, xylose, glycerol, mannitol, sorbitol and raffinose usually not fermented. Nearly all cultures synthesize a polysaccharide (dextran) from sucrose in broth culture (Niven, Kiziuta and White, 1946) or on sucrose agar when incubated anaerobically (Hehre and Neill, 1946); this polysaccharide cross-reacts with type II pneumococcus antiserum.

Recent studies indicate that most strains produce an alpha-reaction on blood agar. In earlier investigations, alpha-, weak beta- and gamma-reactions were reported.

Growth only on complex media; minimal nutritional requirements unknown. However, synthetic media for competence development has

been reported; see review by Lawson and Gooder (1970).

May or may not grow at 45 C.

Consistently found in dental plaques where it constitutes a significant part of the predominant flora (Carlsson, 1965, 1967). In blood and on heart valves in subacute endocarditis. Infrequently encountered in saliva and throat specimens. L-forms are associated with recurrent aphthous stomatitis (Barile et al., 1968).

Although the strains isolated from cases of endocarditis react with group H antisera, attempts to prepare group H sera with these strains have failed. Further serological studies needed to establish the validity of the physiological grouping.

The G + C content of the DNA ranges from 38–40 moles % (method not stated). The G + C content of *S. sanguis* strain Challis is 41.8 (T_m) moles %.

Reciprocal intraspecies, intrageneric (*S. anginosus*, *S. pneumoniae* and group O streptococci) and non-reciprocal intergeneric (low frequency with *Staphylococcus aureus*) transformations reported (Pakula, 1963; Perry and Slade, 1964; Willers et al., 1968). The *Staphylococcus* transformation has been verified (Dobrzanski et al., 1968).

7. **Streptococcus pneumoniae** (Klein) Chester 1901, 63. (*Micrococcus pneumoniae* Klein 1884, 329; *Diplococcus pneumoniae* (Klein) Weichselbaum 1886, 506.)

pneu.mo.ni.ae. Gr. n. *pneumon* the lungs; M.L. fem.n. *pneumonia* pneumonia. M.L. gen.n. *pneumoniae* of pneumonia.

Oval or spherical, coccal-like forms 0.5–1.25 μm typically in pairs, sometimes singly or in short chains. The distal ends of each pair of organisms tend to be pointed or lance shaped. Generally, upon primary isolation, heavily encapsulated with polysaccharide (termed SSS or specific soluble substance) that is distinct from envelope, M-protein antigens. Continued growth in laboratory media promotes chain formation. Gram-positive reaction of young cells may be lost as culture ages and subsequently stains Gram-negatively.

A tentative structure of the peptide subunit of the peptidoglycan (Mosser and Tomasz, 1970) indicates identity with that of *S. pyogenes*. No bridge amino acid or peptide described. The major polymeric components of the cell wall are the peptidoglycan and the choline-ribitol-teichoic-acid complex (Mosser and Tomasz, 1970) which constitutes the species specific substance.

Mucoid colonies result from copious capsular polysaccharide synthesis. Smooth colonies are glistening and dome shaped, and reflect decreased capsular polysaccharide. Rough colonies occur rarely and have a wrinkled, mycelium-like appearance. "Phantom" colonies reflect early and rapid partial autolysis of a mucoid colony (averted by incubation under increased CO_2 tension).

A species-specific, somatic carbohydrate distinct from other streptococcal species has been demonstrated (Tillett and Francis, 1930) and consists of a ribitol teichoic acid with choline phosphate (Brundish and Baddiley, 1968). M proteins, analogous to those of *S. pyogenes* but immunologically distinct, occur but specific types are not related to the types based on capsular polysaccharides. Serology of the capsular polysaccharide has defined some 80 types. Polysaccharides from other organisms frequently cross-react with various type-specific pneumococcal antisera. The structure of types III, VI and VIII is known and pathway for synthesis of type III has been described (Austrian, 1953).

The energy-yielding metabolism is fermentative yielding primarily lactic acid but low in quantity unless the culture is periodically neutralized. Optimum pH is 7.8 with an optimal range of 6.5–8.3. Final pH in glucose broth approximately 5.0. Aerobically, a significant quantity of hydrogen peroxide accumulates as well as acetic and formic acids. Glucose, galactose, fructose, sucrose, lactose, maltose, raffinose, glycogen and inulin are fermented. Slow acid production from glycerol (aerobic?), xylose, arabinose and erythritol. Some strains may ferment mannitol. No acid from dulcitol and sorbitol.

The addition of bile or bile salts to a neutralized culture activates an autolytic amidase which cleaves the bond between alanine and muramic acid in the peptidoglycan (Mossler and Tomasz, 1970); thus, bile soluble.

Growth inhibited by approximately 1:400 ethyl hydrocuprein hydrochloride (optochin) and this sensitivity test is as reliable as the bile-solubility test (Bowers and Jeffries, 1955).

A strong alpha-reaction is noted on blood agar cultures incubated aerobically. Anaerobic incubation results in a beta-hemolytic reaction due to pneumolysin O (identical to streptolysin O) activity.

The addition of blood, serum or ascitic fluid to media enhances growth especially upon primary isolation.

In contrast with other streptococci, the pneumococci require choline for growth in defined media. Ethanolamine replaces choline but not on a molar basis (Badger, 1944). Reducing agents are practically essential. At least 4 of the B vitamins; adenine, guanine and uracil; and 7 to 10 amino acids are required for growth by most strains.

Facultative anaerobe with marked tendency to accumulate hydrogen peroxide upon aerobic culture. Temperature range 25–42 C.

Source: upper respiratory tract, inflammatory exudates and various body fluids of diseased humans, and rarely, domestic animals. Upper respiratory tract of normal humans and domestic animals.

The G + C content of DNA ranges from 38.5 (chemical analysis) to 39 (T_m) and 42 (buoyant density) moles %.

Many genetic markers have been transformed among various strains of pneumococci. Reciprocal intraspecies and intrageneric transformations have been demonstrated (Bracco *et al.*, 1957; Pakula *et al.*, 1958; Ravin and De Sa, 1964; Chen and Ravin, 1966).

8. **Streptococcus anginosus** Andrewes and Horder 1906, 713. (Minute beta-hemolytic *Streptococcus* Long and Bliss 1934, 619; *Streptococcus* MG. Mirick, Thomas, Curnen and Horsfall 1944, 391.)

an.gi.no'sus. L. adj. *anginosus* pertaining to angina.

The majority of the hemolytic strains are minute cocci that are approximately one-half the size of a typical *S. pyogenes* cell. Generally occur in pairs and short chains. The alpha- and gamma-reacting strains (on blood agar as represented by the previous designation of *Streptococcus* MG) present a cell morphology more consistent with *S. pyogenes* in size but also occur in pairs and short chains.

The species consists of Lancefield's group F and group G, type I. Organisms known as *Streptococcus* MG are held to be distinct but Willers *et al.* (1964) demonstrated that cultures bearing this designation were indeed group F, type III streptococci. The striking similarity in physiological characteristics reinforces the serological studies. The predominant determinant in the group antigen is glucosyl-*N*-acetyl-D-galactosamine (Michel and Willers, 1964).

In group F, of five serotypes, types I, II and III are relatively common while types IV and V are infrequent. Strains with altered group specificity, but retaining type specificity, have been described and the use of type-less and group-less strains for antiserum production has been suggested (Willers and Alderkamp, 1967). Type antigens of group F occur in other serological groups aside from type I, group G (groups C, T, A, L and type III cross-reacting with type I of *S. salivarius* (Willers and Alderkamp, 1967). Type I, group G is included in this species because of its similarity in physiological and nutritional characteristics. Type I antigen contains *N*-acetylgalactosamine; type II, rhamnose, glucose, galactose, galactosamine (ratio 2:2:1:1; Michel and Krause, 1967) and type III, glucose, galactose and rhamnose (5:3:1; Willers and Alderkamp, 1967).

Final pH range in glucose broth *ca.* 4.5–5.2. Acid from glucose, maltose, salicin, sucrose and usually lactose and trehalose. Inulin, xylose, arabinose, glycerol, mannitol and sorbitol are not fermented. Raffinose usually fermented by the type I, group G strains.

Some strains produce either an alpha- or gamma-reaction on blood agar and a soluble hemolysin usually is lacking in filtrates of these cultures; generally, colonies of these strains are moderate in size. The majority of the group F and type I, group G strains form minute colonies on blood agar with relatively large hemolytic zones after incubation for 48 to 96 hr. The hemolytic zones may appear before the colonies are visible to the unaided eye. Incubation under 10% CO_2 stimulates hemolysis and growth significantly (Deibel and Niven, 1955). A soluble hemolysin is produced by certain strains but it may be difficult to demonstrate (Long and Bliss, 1934).

For growth in synthetic medium the cultures must contain oleic acid or be incubated under increased CO_2 tensions (Deibel and Niven, 1955). Folinic acid and four other B vitamins are required. Other vitamins, a reducing substance and a peptide factor are stimulatory (Niven *et al.*, 1946; Deibel and Niven, 1955).

No growth at 10 C and only rarely at 45 C.

Source: human throat, sinuses, abscesses, vagina, skin and feces. The alpha- and gamma-reacting strains previously designated as *Streptococcus* MG are associated with cases of primary atypical pneumonia.

All streptococci classified here as *Streptococcus anginosus* do not conform to the description as given by Andrewes and Horder. The grouping is based upon substantial serological, physiological and nutritional characteristics. The nutritional characteristics of strains previously designated as *Streptococcus* MG have not been reported.

Reciprocal intraspecies and interspecies (*S. sanguis*) transformations are reported.

9. **Streptococcus agalactiae** Lehmann and Neumann 1896, 126, *nom. cons.* Opin. 8 Jud. Comm. 1954, 152. (*Streptococcus de la mammite* Nocard and Mollereau 1887, 115; *Streptococcus agalactiae contagiosae* Kitt 1893, 322; *Streptococcus mastitidis* Migula 1900, 19.)

a.ga.lac'ti.ae. Gr. *agalactia* want of milk, agalactia; M.L. gen.n. *agalactiae* of agalactia.

Spherical or ovoid cells 0.6–1.2 µm in diameter, occurring in chains of seldom less than four cells and frequently very long. Chains may appear to be composed of paired cocci.

Lancefield's group B; subdivided into at least four serotypes on the reactivity of carbohydrate haptens (S substance) in the cell envelope. Some strains lack S substance. These substances appear to have a direct relationship to virulence and are concerned with the protective action of immune sera. Weak group cross-reactions between groups B and G may occur due to the similarity of their rhamnose-hexosamine antigens (Curtis and Krause, 1964).

Energy-yielding metabolism is fermentative with lactic acid constituting the chief end-product. Final pH range in glucose broth is 4.2–4.8. Acid produced from glucose, maltose, sucrose and trehalose. Glycerol fermented only aerobically. Lactose usually fermented by strains from bovine sources but this characteristic may be variable in strains isolated from other animals and humans. Xylose, arabinose, raffinose, inulin, mannitol and sorbitol are not fermented. About half of the isolates from bovine sources produce a narrow but definitive zone of beta-hemolysis on blood agar; other strains show typical alpha-, beta- or gamma-reactions. Most hemolytic strains produce a soluble hemolysin, not antigenic, moderately sensitive to heat and acid and distinct from streptolysins O and S. Christie *et al.* (1944) noted that most group B strains would lyse sheep or ox cells when grown in the proximity of beta-lysin-producing *Staphylococcus aureus* strains under anaerobic conditions (the CAMP test). About 80% of the group A strains also give this reaction but only rare strains of groups C, F and G.

Some strains produce a yellow, orange or brick red pigment and production may be enhanced by addition of starch or anaerobiosis. Most strains produce hyaluronidase (Gochnauer and Wilson, 1951).

At least six vitamins and nine amino acids are required for growth (Willett *et al.*, 1967).

Source: milk and udder tissues of cows with mastitis. Also associated with various human infections and rarely as the etiologic agent of epizootics involving fish (Robinson and Meyer, 1966).

Neotype strain: ATCC 13813; NCTC 8181.

10. **Streptococcus acidomininus** Ayers and Mudge 1922, 49.

a.ci.do.mi'ni.mus. L. adj. *acidus* sour, acid; L. sup.adj. *minimus* very least; M.L. adj. *acidominimus* literally acid least, probably intended to mean that this organism produces the least amount of acid.

Spherical cells occurring in short chains.

No group-specific antigen has been demonstrated.

Weakly fermentative; most strains fail to decrease pH value of carbohydrate-containing media below 6.0. Acid produced from glucose, lactose and sucrose and generally maltose and trehalose. Glycerol, raffinose, inulin, arabinose and xylose are not fermented. The high limiting pH value offers difficulty in determining fermentation reactions.

Alpha-reaction on blood agar.

Growth only on complex media; minimal nutritional requirements unknown.

Source: common in bovine vagina, occasionally found on skin of calves and in raw milk.

This species may be confused with *S. agalactiae* because of its slow hydrolysis of hippurate. Differentiation is effected by serological reactions and the inability of *S. acidominimus* to ferment glycerol or to hydrolyze arginine.

11. Streptococcus salivarius Andrewes and Horder 1906, 712.

sa.li.va′ri.us. L. adj. *salivarius* salivary, slimy.

Description based on Sherman *et al.* (1943).

Spherical or ovoid cells 0.8–1.0 μm in diameter. Chain length may vary from short to very long.

Smooth and rough variants occur with the rough variant often reverting upon subculture in broth.

The serology of this species has not been resolved completely. Serotypes designated as I and II were established and additional but less frequently encountered types exist. Type I and *Streptococcus* MG (*Streptococcus anginosus*) cross-react but they differ substantially in their physiological reactions. Also, only type I strains react with group K antiserum (Montague and Knox, 1968). The specificity of the group K antigen has been questioned (Williams, 1956).

Final pH range in glucose broth is 4.0–4.4. Acid from glucose, sucrose, maltose, raffinose, inulin, salicin and usually trehalose and lactose. No acid from glycerol, mannitol, sorbitol, xylose or arabinose. Some strains ferment only the terminal fructofuranose portion of raffinose (forming polysaccharide) and melibiose accumulates.

On sucrose agar most isolates produce soluble levans which result in large mucoid colonies. Some strains produce insoluble dextrans. Although serologically active, these substances are not related to the type specific polysaccharides. Colonies on sucrose agar may vary from "smooth" to "rough" depending upon relative proportions of levan and dextran synthesized. These variations are not related to the smooth-rough variants observed in other media.

Most *S. salivarius* strains uniformly yield a gamma-reaction on blood agar. Various studies have reported strains yielding the alpha-reaction. Occasionally beta-hemolytic "group K streptococci" also have been reported.

The minimal nutritional requirements were determined by Smiley *et al.* (1943) and nine amino acids, five vitamins and uracil were required for growth.

Growth at 45 but not at 47 C.

Source: tongue, saliva and feces of humans.

The G + C content of the DNA ranges from 39–42 (method not stated) moles %.

A more complete definition of this species and its relationship with group K and other closely related slime-forming and non-slime-forming streptococci appears to be warranted. The studies of Williams (1956) and Stewart and McKeever (1963) indicate that knowledge of the physiological and serological relationships between group K and *S. salivarius* is far from satisfactory. Carlsson (1965) has suggested a relation between dental caries and *Streptococcus mutans* (Clark 1924, 142), an organism quite similar to *S. salivarius*. *S. mutans* has not yet been extensively studied and compared with *S. salivarius*.

12. Streptococcus mitis Andrewes and Horder 1906, 712.

mi′tis. L. adj. *mitis* mild.

Description based on Sherman *et al.* (1943).

Spherical or ellipsoidal cells 0.6–0.8 μm in diameter. Long chains in broth cultures.

Smooth and rough colony variants occur with frequent reversion of rough to smooth upon subculture in broth.

No group antigen has been shown but many serological types are demonstrable by the precipitin test. Using the cell wall agglutination procedure two "groups" have been defined (Kalonaros and Bahn, 1965). One group was agglutinated by group O antiserum while the other agglutinated by group O and N antisera. Only slight reactions occurred in the precipitin test. Apparently, serological reactions are of little value in identifying this species.

Final pH range in glucose broth is 4.2–5.8, average about 4.5. Acid produced from glucose, maltose sucrose and usually from lactose and salicin. Occasional strains ferment raffinose and trehalose. No acid from inulin, mannitol, sorbitol, glycerol, arabinose or xylose. Rarely a strain produces a typical mucoid colony on sucrose agar or a colony resembling small bits of broken glass. Some strains oxidize butyric acid (Wolin *et al.*, 1952) with the accumulation of hydrogen peroxide in the medium.

On blood agar incubated aerobically a pronounced alpha-reaction occurs.

Only limited nutritional studies have been conducted. The requirements for some strains include four B vitamins. Most strains require unidentified factors in yeast extract (Nevin, 1954).

Growth may or may not occur at 45 C.

Source: human saliva, sputum and feces.

This species comprises a heterogeneous group of alpha-hemolytic streptococci associated with the human respiratory tract. It has no unique identifiable characteristic and some non-hemolytic varieties of typically hemolytic species may be confused with it.

13. Streptococcus bovis Orla-Jensen 1919, 137; *emend. mut. char.* Sherman 1937, 57.

bo'vis. L. n. *bos* a cow; L. gen.n. *bovis* of a cow.

Spherical or ovoid cells 0.8–1.0 μm in diameter in pairs, moderate chains and occasionally in long chains.

By disintegrating the cell preparation used to inoculate rabbits, Shattock (1949) demonstrated the group D antigen in this species. At least 11 serotypes distinguished and a multiplicity of types indicated (Medrek and Barnes, 1962). Large capsules surround the cells and type specificity is associated with the capsule. Type-specific cell wall antigens also occur. The type specificity in one strain was shown to be associated primarily with a *N*-acetyl glucosamine determinant (Kane and Karakawa, 1969). Cross-type reactions occur with other streptococci of groups E and N (Perry *et al.*, 1958).

The structure of the mucopeptide is similar to that of other streptococci except there is evidence that the crossbridge compound is threonine and this may be a useful taxonomic tool (Kane *et al.*, 1969).

Final pH range in glucose broth of 4.0–4.5. Acid produced from glucose, fructose, mannose, lactose, galactose, maltose, sucrose, raffinose and salicin. The fermentation of arabinose, xylose, mannitol, sorbitol, trehalose or inulin is variable. No acid from glycerol. Starch is usually hydrolyzed and also fermented with the production of acid. This character may be lost upon subculture (see Medrek and Barnes, 1962). Large amounts of polysaccharide are produced in sucrose broth (Niven *et al.*, 1948). When cultivated on sucrose-containing agar media under increased carbon dioxide tensions, the majority of the strains form copious quantities of a dextran (Dain *et al.*, 1956). Physiological patterns vary with source of strains.

On blood agar, most strains produce small colonies with an alpha-reaction of varying intensity. Some strains give a gamma-reaction.

Among the streptococci, this is the least nutritionally fastidious species. Niven *et al.* (1948) could not show a requirement for any specific amino acid. Wolin *et al.* (1959) showed that many isolates could utilize ammonium salts as a source of nitrogen. Generally biotin (Niven *et al.*, 1948) or oleate is required for growth and for dextran production (Barnes *et al.*, 1961).

Source: alimentary tract of cows, sheep and other ruminants; occasionally occurring in large numbers in human feces. Occasionally encountered in cases of human endocarditis (Niven *et al.*, 1948).

The G + C content of DNA ranges from 38–40 (method not stated) to 42 (T_m) moles %.

Although this species lacks precise definition, the results of tolerance tests, tests for starch hydrolysis and dextran production and simple nutritional requirements offer approaches to classification. In a limited study Sims (1964) observed characteristic growth initiation of this species at a pH value of 5.0, provided sufficient CO_2 (*ca.* 5%) was present. This was in contrast to *S. equinus*.

14. Streptococcus equinus Andrewes and Horder 1906, 712.

e.qui'nus. L. adj. *equinus* pertaining to a horse.

Spheres occurring in moderately long chains. Chaining pronounced in broth cultures.

Lancefield's group D; demonstration of group antigen necessitates cellular disruption of preparation used to inoculate rabbits (Smith and Shattock, 1962; Fuller and Newland, 1963).

Final pH range of 4.0–4.5 in glucose broth. Acid from glucose, fructose, galactose, maltose and usually from sucrose and salicin. Raffinose and inulin are seldom fermented. No acid from arabinose, xylose, lactose, mannitol or glycerol.

In the presence of small concentrations of a fermentable monosaccharide, this species hydrolyzes starch but not to the level of reducing sugars. Anaerobic incubation inhibits hydrolysis. In contrast, for *S. bovis* strains no monosaccharide is required, reducing sugars accumulate and anaerobiosis is not inhibitory (Dunican and Seeley, 1962).

Inhibited by 0.04% tellurite. No extracellular polysaccharide synthesized from sucrose.

Weak alpha-reaction observed on blood agar.

Growth only on complex media, minimal nutritional requirements unknown.

Grows at 45 but not at 50 C.

Predominant streptococcus in alimentary tract of the horse.

Although this species is ill defined and lacks specific characteristics for identification, the overall physiological characteristics indicate a distinctiveness that merits species status.

15. Streptococcus thermophilus Orla-Jensen 1919, 136.

ther.mo'phil.us. Gr. n. *therme* heat; Gr. adj. *philus* loving; M.L. adj. *thermophilus* heat-loving.

Spherical or ovoid cells, 0.7–0.9 μm in diameter in pairs to long chains.

The structure of the peptidoglycan is identical to that of *S. faecalis* (Schleifer and Kandler, 1967).

No group-specific antigen demonstrated.

Final pH range in glucose broth is 4.0–4.5. The preferential fermentation of the disaccharides, sucrose and lactose, may result in a lower pH value as compared to glucose. Acid is produced from glucose, fructose, lactose and sucrose. No acid from trehalose, maltose, inulin, glycerol, mannitol, sorbitol or salicin and rarely from raffinose, xylose or arabinose.

Most strains produce a weak alpha-reaction on blood agar.

Six B vitamins are required for serial transfer in synthetic media. Purines and pyrimidines are not required. The amino acid requirements have not been resolved completely (Guss and Delwiche, 1954; Nurmikko and Karha, 1962).

Optimum temperature is between 40 and 45 C. Growth occurs at 50 but not at 53 C. No growth at temperatures below 20 C. Heat tolerance marked, survives 65 C for 30 min.

Source: milk and milk products such as Swiss cheese and yogurt. Often used as a starter culture for these products.

This species easily recognized by its high temperature limit of growth, thermal tolerance, inability to ferment maltose and inability to grow in media containing 2.0% sodium chloride.

Suggested "working type" strain: ATCC 19258; NCDO 573 (Sneath and Skerman, 1966).

16. Streptococcus faecalis Andrewes and Horder 1906, 713 *nom. cons.* Opin. 30, Jud. Comm. 1963, 167. (*Micrococcus ovalis* Escherich 1886, 89; *Enterocoque* Thiercelin 1902, 1082; *Enterococcus proteiformis* Thiercelin and Jouhaud 1903, 686; *Streptococcus glycerinaceus* Orla-Jensen 1919, 140.)

fae.cal'is. L. n. *faex, faecis* dregs; M. L. adj. *faecalis* relating to feces.

Ovoid cells elongated in direction of chain; 0.5–1.0 μm in diameter, mostly in pairs or short chains. Generally non-motile. The peptide subunit of the peptidoglycan consists of L-alanine, D-glutamine, L-lysine and D-alanine. A bridge tripeptide of L-alanine joining peptide subunits through L-lysine and D-alanine was described by Kandler *et al.* (1968) who obtained an excellent correlation between peptidoglycan structure and physiological characteristics.

In rich media such as APT agar (Evans and Niven, 1951), colonies of most group D species are larger than usual and may be confused with staphylococci, micrococci, *etc.* Colonies are smooth and entire; rarely pigmented (false pigmentation may be due to precipitation of metal ions; Jones *et al.*, 1963).

Lancefield's group D. The group specific antigenic determinant is a glycerol teichoic acid (Elliott, 1962) in which the glycerol phosphate contains glucose in glucosidic linkage to glycerol and D-alanine is esterified to hydroxyl groups of glucose (Wicken and Baddiley, 1963; Wicken *et al.*, 1963). Slight variations in glucose content exist. Jones and Shattock (1960) demonstrated an intracellular location of the polymer and Slade and Shockman (1963) associated the polymer with the cytoplasmic membrane. The substance is released in conversion of cells to protoplasts (Shattock and Smith, 1963) and L forms do not contain it (Hijmans, 1962). Antigen production enhanced in buffered-glucose medium (Medrek and Barnes, 1962). The type-specific antigens are polysaccharide in nature, contain N-acetylhexosamines (Willers and Michel, 1966) and are located in the cell wall (Elliott, 1960; Sharpe, 1964). A numerical series and type cultures for the 11 type-specific antigens established by Sharpe (1964). Stewart and McLoughlin (1965) described a red cell-sensitizing antigen (referred to as D') which paralleled the occurrence of the group D antigen.

Glucose is fermented yielding primarily lactic acid; however, if neutrality of culture is maintained, greater amounts of formate, acetate and ethanol are formed (Gunsalus and Niven, 1942). Gluconate fermentation involves a mixed pathway (Goddard and Sokatch, 1964). Aerobically, glycerol phosphorylated and oxidized to lactate with oxygen serving as hydrogen acceptor. Anaerobically, a pyridine-linked glycerol dehydrogenase couples with fumarate as hydrogen acceptor (Jacobs and VanDemark, 1960). Aerobic incubation alters the end-products of carbohydrate metabolism (London and Appleman, 1962).

Pyruvate utilized as an energy source and fermented *via* the phosphoroclastic and dismutation pathways. Citrate, serine and malate are also fermented and the energy-yielding processes are linked apparently to pyruvate metabolism as lipoate is required with these energy sources (Deibel and Niven, 1964; Deibel, 1964). Arginine (and often agmatine) utilized as an energy source but the mechanism has not been elucidated. Similar or identical hydrolytic reactions are effected by *S. faecium* but the energy released is not available for growth (Deibel, 1964). Some characteristics of the fumarate reductase (Aue and Deibel, 1967) and its role in radically altering energy production and fermentation products of glucose are described (Deibel and Kvetkas, 1964). Rarely strains of the species possess catalase activity maintained and enhanced by aerobic serial transfer and cation fortification of medium (Jones *et al.*, 1964). The activity is not associated with iron-porphyrin compounds. Some strains, when grown aerobically, possess a potent, adaptive peroxidase which precludes demonstrable peroxide formation in the culture (Seeley and VanDemark, 1951).

The activity is more frequently associated with this group D species.

Final pH in glucose broth between 4.1 and 4.6. Most strains produce acid from glucose, sucrose, mannose, fructose, galactose, maltose, cellobiose, trehalose, lactose, melezitose, sorbitol, glycerol and mannitol (polyol fermentation is enhanced by fumarate addition). Generally, arabinose, inulin, melibiose and raffinose are not fermented.

Characteristically grows in presence of 0.04% tellurite reducing it to tellurium (Skadhauge, 1950). Growth occurs at pH 10–10.5 in carbonate-buffered media; the use of glycine-buffered media for this tolerance test is not recommended (Chesbro and Evans, 1959). The heat-tolerance test (60 C for 30 min) is markedly affected by the pH value of the menstruum (White, 1963).

Most strains decarboxylate tyrosine to tyramine plus carbon dioxide. Gelatin is not hydrolyzed.

Growth occurs in presence of 0.1% thallous acetate, or 0.02% sodium azide or 0.5–1.0 units penicillin/ml.

A gamma-reaction is observed on blood agar (Deibel et al., 1963).

Some strains produce hyaluronidase (Rosan and Williams, 1966). Citrate utilization (in the plasma preparation) may lead to a false positive coagulase test (Evans et al., 1952).

Variations occur depending upon the strain but generally 7–13 amino acids and five of the B vitamins necessary for serial subculture in synthetic media. Purines and pyrimidines stimulatory. Folate not required (Deibel et al., 1963). Lipoate required when energy source consists of pyruvate, serine, citrate or malate (Deibel, 1964).

Growth at 47 but not at 50 C.

Source: feces of humans and warm-blooded animals. Occasionally in urinary tract infections and subacute endocarditis. Common in many food products, often unrelated to direct fecal contamination (Deibel, 1964). Relationship with food poisoning is questionable (Deibel and Silliker 1963). Occurs frequently in plants where an epiphytic relationship exists (Mundt et al., 1962).

The G + C content of DNA ranges from 33.5 (chemical analysis) to 38 (buoyant density) moles %.

Reference strain: ATCC 19433; NCTC 775; NCDO 581.

16a. Streptococcus faecalis subsp. **faecalis** subsp. nov.

Description as for the species.

Milk acid and curdled but not peptonized. Not hemolytic.

16b. Streptococcus faecalis subsp. **liquefaciens** Mattick 1949, 519. (*Streptococcus liquefaciens* Sternberg 1892, 613.)

Milk acid, curdled and finally peptonized. Not hemolytic.

Gelatin liquefied; however, not all strains that liquefy gelatin cause extensive peptonization of milk. A quantitative difference for gelatin liquefication exists and some doubt can be cast upon the validity of this subspecies (Deibel et al., 1963).

16c. Streptococcus faecalis subsp. **zymogenes** Mattick 1947, 519. (*Micrococcus zymogenes* MacCallum and Hastings 1899, 521.)

Litmus milk is acidified and curdled and may be peptonized. Gelatin may or may not be liquefied.

Broad zones of hemolysis occur on blood agar but this character may be lost upon serial transfer in the laboratory. Irwin and Seeley (1958) studied some characteristics of the hemolysin and Brock et al. (1963) observed that the hemolysin was also a bacteriocin.

17. Streptococcus faecium Orla-Jensen 1919, 139. (*Streptococcus durans* Sherman and Wing 1937, 165.)

fae.ci′um. L. n. *faex, faecis* dregs; L. gen.pl.n. *faecium* of the dregs, of feces.

Ovoid cells elongated in direction of chain, chiefly in pairs and occasionally in short chains. Motile varieties often encountered.

The peptidoglycan structure similar to that of S. faecalis but differs in that the bridge compound consists of D-asparagine (Kandler et al., 1968).

Colonial morphology approximates that of S. faecalis. Yellow-pigmented varieties are encountered.

Lancefield's group D. Chemistry and location of group antigen parallels that of S. faecalis. At least 19 type-specific antigens are described and many more exist (Barnes, 1964). Type antigens are located in the cell wall, are resistant to trypsin and one has been characterized as a polysaccharide (Barnes, 1964; Elliott, 1960).

In glucose broth the final pH is 4.0–4.4. Aerobic incubation radically alters products of carbohydrate metabolism (London and Appleman, 1962). Glycerol utilized only under aerobic conditions. Most strains produce acid from fructose, galactose, melibiose, trehalose, maltose, cellobiose and lactose. Barnes (1964) observed high variability in the fermentation of sucrose, mannitol, sorbitol, arabinose and raffinose. Does not ferment inositol, melezitose or inulin. Pyruvate, gluconate, citrate, serine and malate not used as an energy source. Ammonia produced from arginine but is not used as an energy source. Tyrosine decarboxylation is variable. Rarely strains reduce nitrate (Langston and Williams, 1962).

No growth in media containing 0.04% tellurite. Growth occurs in media containing 0.02% sodium azide, 0.1% thallous acetate or 0.5 to 1.0 units penicillin/ml.

A moderate to strong alpha-reaction on blood agar is observed.

Ten to 14 amino acids and 5–6 B vitamins are required for growth in synthetic media. Most strains require folic acid (or thymine) for growth. Many strains require lipoate for aerobic growth in defined media regardless of the energy source (Deibel, 1964). Purines and pyrimidines not essential.

Rapid growth at 50 C. Many strains fail to grow at 52 C.

Source and habitat parallels that of *S. faecalis*.

The G + C content of the DNA ranges from 34–38 (method not stated) to 40 (T_m) moles %.

A strain of this species has been transformed with DNA from *S. sanguis* strain Challis (Mehta et al., 1967). A chromosomal map with several markers is reported (Stonehill and Hutchison, 1966).

Reference strain: ATCC 19434; NTCC 7171; NCDO 942.

18. **Streptococcus avium** Nowlan and Deibel 1967, 295. (Group Q *Streptococcus* Guthof 1955, 60.)

av.i'um. L. n. *avis* bird; L. gen. pl. *avium* of birds.

Cell and colonial morphology similar to that of *S. faecalis* and *S. faecium*. No motile, pigmented or slime-producing varieties described.

Lancefield's group Q. Group antigen is associated with the cell wall; this antigen is not demonstrable in all strains; when it is not demonstrable, identification rests on physiological characteristics. Serologically, these streptococci are peculiar in that the majority of the strains encountered also contain group D antigen; as in the established group D species, antigen is not an integral part of the cell wall, and is located between the wall and the membrane (Smith and Shattock, 1964).

An average final pH value of 4.2 in glucose broth, and acid is produced from ascorbate, arabinose, xylose, glucose, fructose, galactose, maltose, mannose, lactose, sucrose, trehalose, melezitose, cellobiose, inulin, dextran, glycerol (aerobically and anaerobically), arabitol, dulcitol, ribitol, sorbitol, xylitol and sorbose. Variable results obtained with melibiose. Erythritol, inositol, raffinose not fermented. Starch not hydrolyzed. Gluconate utilized as an energy source by most strains but pyruvate, malate, citrate, arginine and serine are not. Acetylmethylcarbinol not produced. Nitrate is not reduced. Isolation readily effected by culture at pH 10 with sorbose as the energy source in a medium containing 0.02% sodium azide. Slow growth may be observed in 6.5% sodium chloride broth. Will not grow in

presence of 0.04% tellurite or in 0.10% methylene blue milk.

Most strains produce an alpha-reaction on blood agar.

No requirement for riboflavin or pyridoxal in a casein-hydrolysate medium. Folinic acid (or folic acid plus thymine) is required for growth (Nowlan and Deibel, 1967). Amino acid requirements are unknown.

Growth at 45 C but response may be slow (48 hr) with some strains.

Source: feces of chickens characteristically, and occasionally from man, dogs and pigs.

This species has a marked, broad fermentation pattern. Sorbose fermentation, inability to hydrolyze arginine, no growth in milk with 0.1% methylene blue when sterilized separately and added aseptically and the response to folinic acid characterize this species.

19. **Streptococcus uberis** Diernhofer 1932, 370.

u'ber.is. L. n. *uber*, udder, teat; L. gen.n. *uberis* of an udder.

Description from Seeley (1951).

Spheres occurring in pairs to chains of moderate length.

The serology of this species has not been resolved. Between a third to a half of the strains isolated in various investigations react with group E antiserum. To a lesser extent, reactions with groups C, G and D antisera have been reported (Roguinsky, 1969). These reactions appear to be due to similar type-specific antigens. Stableforth (1959) was unable to obtain group E antibody by injection of *S. uberis* strains that cross-reacted with group E antiserum.

Final pH range in glucose broth 4.6–4.9. Acid from glucose, fructose, maltose, lactose, sucrose, trehalose, mannitol, sorbitol and salicin. Most strains ferment inulin but fail to ferment raffinose. Xylose, arabinose and melibiose not fermented. Glycerol fermented aerobically but not anaerobically.

Growth on blood agar may be characterized by a weak alpha- or gamma-reaction.

At least six amino acids, and usually seven B vitamins are essential for growth in synthetic media. Uracil is required.

Optimum temperature 35–37 C.

Source: skin and lips of cows (Cullen, 1966); raw milk and udder tissue of cows with mastitis.

A definitive identification scheme based on either physiological or serological procedures is not available. Similarity to the group D streptococci exists in that this species grows at 10 and 45 C and generally survives 60 C for 30 min. Furthermore, the folic acid requirement reflects a relationship with *S. faecium*. However, the overall nutritional and physiological pattern is definitively different and when considered in light of serological reactions, it is difficult to relate this species to the group D streptococci.

Reference strain: ATCC 19436; NCTC 3858.

20. Streptococcus lactis (Lister) Löhnis 1909, 554. (*Bacterium lactis* Lister 1873, 408.)

lac'tis. L. n. *lac* milk; L. gen.n. *lactis* of milk.

Ovoid cells elongated in direction of the chain; 0.5–1.0 μm in diameter. Mostly in pairs or short chains. Some cultures produce long chains.

The peptidoglycan is similar to that of *S. pyogenes* with the exception that the crossbridge consists of D-isoasparagine (Schleifer and Kandler, 1967).

Lancefield's group N. The group antigenic determinant is a glycerol teichoic acid containing galactose phosphate (Elliott, 1963). It is not a wall constituent and it occurs intracellularly like the group D teichoic acid (Smith and Shattock, 1964). The group N teichoic acid cross-reacts with certain type-specific antipneumococcal sera (Heidelberger and Elliott, 1966).

Many serological types are known.

Final pH range of 4.0–4.5 in glucose broth. Acid from glucose, maltose, lactose. Xylose, arabinose, sucrose, trehalose, mannitol and salicin may or may not be fermented. No acid from raffinose, inulin, glycerol or sorbitol. Tyrosine is not decarboxylated. Some strains produce an antibiotic, nisin (Hirsch, 1951), that inhibits many Grampositive organisms. Growth in media containing 4.0 but not 6.5% sodium chloride. Growth initiated at pH 9.2 but not 9.6. Variations in the latter two characters occur. Grows in 0.3% methylene blue in milk. Some strains may metabolize leucine to produce 3-methylbutanal which gives a malty-flavor defect in dairy products (Jackson and Morgan, 1954).

On blood agar a weak alpha- or gamma-reaction is observed.

The nutrition of this species is comparatively complex. Generally, 4 or 5 of the B vitamins, 10 to 13 amino acids (Niven, 1944) and acetate and oleate or lipoate (Collins *et al.*, 1950) are required for growth in synthetic media. Purines and pyrimidines are not required but may be stimulatory.

Optimum temperature *ca.* 30 C. Some strains fail to grow at 41 C. No growth at 45 C.

A common contaminant in milk and dairy products (Stark and Sherman, 1935).

The G + C content of DNA ranges from 38.4–38.6 moles % (T_m).

Intraspecies transformation does not occur. DNA from *S. lactis* can transform *S. sanguis*, strain Challis.

Reference strain: ATCC 19435; NCTC 6681.

20a. Streptococcus lactis subsp. **diacetylactis** subsp. nov. (*Streptococcus diacetilactis* (*sic*) Matuszewski, Pijanowski and Supinska 1936, 23).

di.acety.lac'tis. Reference to a *Streptococcus lactis* isolate that produces diacetyl.

This variety possesses the same characteristics as *S. lactis* except that it is capable of fermenting citrate (in conjunction with a fermentable carbohydrate) with the production of carbon dioxide, acetoin and diacetyl.

21. Streptococcus cremoris Orla-Jensen 1919, 132. (*Streptococcus hollandicus* Scholl 1891, 51; *Streptococcus lactis* B Ayers, Johnson and Mudge 1924, 39.)

cre.mo'ris. L. n. *cremor* juice, cream; L. gen.n. *cremoris* of cream.

Spheres or ovoid cells elongated in direction of the chain; 0.6–1.0 μm in diameter (often larger than *Streptococcus lactis*); form long chains, especially in milk, but in some cultures predominantly as pairs.

The peptidoglycan is similar to that of *S. pyogenes* except for differences in crossbridge compounds. Two types of crossbridges have been reported for *S. cremoris*: one is identical to *S. lactis* (D-isoasparagine); the other, a dipeptide consisting of L-alanyl-threonine (Schleifer and Kandler, 1967).

Lancefield's group N. The antigenic determinant is the same as that described for *S. lactis*. Many serological types are known to exist.

Final pH range of 4.0–4.5 in glucose broth. Acid produced from glucose and lactose. May or may not ferment trehalose and salicin. Rarely ferments maltose, sucrose, raffinose or mannitol. Arabinose, xylose, inulin, glycerol and sorbitol not fermented. In the presence of a fermentable sugar, some strains degrade citrate producing carbon dioxide, acetic acid and diacetyl. Some strains also produce antibiotic-like substances (Oxford, 1944).

On blood agar a weak alpha- or gamma-reaction is observed.

The nutritional characteristics closely parallel those of *S. lactis*.

Optimum temperature is about 30 C. No growth at 40 C.

Source: raw milk and milk products.

Differentiated from *S. lactis* which produces ammonia from arginine, generally grows in broth containing 4.0% sodium chloride, usually initiates growth in broth adjusted to pH 9.2 and grows in the presence of 0.3% methylene blue when added to milk. *S. cremoris* gives a negative reaction in these tests.

The G + C content of DNA ranges from 38–40 moles % (method not stated).

Suggested "working type" strain: ATCC 19257; NCDO 607 (Sneath and Skerman, 1966).

Species incertae sedis

Descriptions of species to which no name has been given or of poorly defined species, the taxonomic relationships of which are not clear.

a. *Streptococcus* sp. (group G, all serotypes except type I; commonly referred to as large colony, group G) Lancefield and Hare 1935, 346.

Spherical or ovoid cells 0.6–1.0 μm in diameter in medium to long chains.

Characteristic, "matt"-type colonies (indistinguishable from *S. pyogenes*).

Lancefield's group G; at least three serological types (Simmons and Keogh, 1940) described. Some strains have a protein antigen in common with *S. equisimilis* that may give confusing cross-reactions.

Final pH range in glucose broth 4.8–5.2. Acid produced from glucose, lactose, sucrose, trehalose and glycerol (aerobically only). Salicin fermentation is variable and occasional strains ferment inulin. No acid from mannitol, sorbitol or raffinose.

The broad zones of beta-hemolysis on blood agar may be larger than those of *S. pyogenes*. Streptolysin O is produced.

The minimal nutritional requirements have not been thoroughly investigated. The organism grows in a casein-hydrolysate medium and in contrast to *S. anginosus* (including the group G, type I strains) folinic acid, oleic acid or increased CO_2 tensions are not required.

Obtained from the human throat, nose, skin, vagina and feces. Also found in the throat of domestic animals, especially the dog.

This group deserves species recognition, but no suitable name has been proposed. More than one variety or species may be included (Sherman, 1937). Some strains, especially those that are fibrinolytic, are very difficult to differentiate from *S. pyogenes* except by serological methods.

The G + C content of DNA is 41 moles % (buoyant density).

b. *Streptococcus* sp. (group E) Brown, Frost and Shaw 1926, 381; Lancefield 1933, 571.

Description from Deibel et al., 1964; Yao et al., 1964.

Spherical to ovoid cells generally in chains of small to medium length.

Occasional strains give colonial variants that are large, flat and translucent but the character is not stable and reversion occurs to the typical small, elevated and entire colonial morphology.

This group consists of at least five serotypes (I–V). Frequently, strains from swine are encountered that lack a type antigen and the existence of types I and III is questioned (Yao et al., 1964; Payne and Armstrong, 1970).

Energy-yielding metabolism is fermentative and a final pH range of 4.2–5.6, average 4.9 is produced in glucose broth. The porcine strains produce acid from glucose, glycerol, mannose, fructose, trehalose, cellobiose, salicin and usually galactose, maltose, mannitol and sorbitol. Arabinose, xylose, rhamnose, melibiose, lactose, raffinose and inulin are not fermented. The fermentation pattern of isolates from dairy sources may deviate. Most strains utilize pyruvate as an energy source. Deoxyribonucleic (but not ribonucleic) acid is hydrolyzed but only if the cultures are incubated anaerobically. Approximately one-half of the porcine isolates hydrolyze gelatin but again only if cultured anaerobically. Most of these strains will also form a rennet-like curd in litmus milk with subsequent digestion if the medium is fortified with yeast extract.

On blood agar, beta-hemolysis is rapidly produced by dairy strains but an equivalent reaction by the porcine strains requires 48 hr.

Growth reported only on complex media, minimal nutritional requirements unknown.

Group E streptococci (most frequently type IV) are the etiologic agent of cervical lymphadenitis of swine.

In the United States responsible for considerable economic loss and neither preventive nor practical control measures are available. Minor physiological differences exist in strains from the dairy and from pigs. Cross-reactions with group E antisera and other streptococcal species have been reported but the specificity of these reactions has not been investigated sufficiently. Various epithets have been proposed such as *infrequens* and *subacidus*. However, it is neither infrequent in occurrence nor can differences of acid production in milk be considered a valid character when the vast majority of the isolates does not ferment lactose. In addition, litmus milk is nutritionally inadequate for these organisms (Deibel et al., 1964).

c. *Streptococcus* sp. (group L) (Fry) Hare and Fry 1938, 1537. (See Laughton (1948) for additional information.)

Spherical or ovoid cells in long chains.

Serological group L.

Acid from maltose, lactose, sucrose, trehalose and salicin. Glycerol and sorbitol may or may not be fermented. Final pH range in glucose broth is 4.7–5.2.

Glossy- and intermediate-type colonies on blood agar evidencing a beta-hemolytic reaction.

Isolated from miscellaneous infections of the dog.

d. *Streptococcus* sp. (group M) Fry 1941, 676. (Description taken primarily from Skadhauge and Perch, 1959.)

Spherical or ovoid cells in long chains.

Belongs to serological group M. Three biotypes distinguished by physiological and serological tests. Biotype I contains a "group" antigen that is heat stable (127 C for 2 hr) and insensitive to pepsin. Biotype III contains a "group" antigen that is heat labile and sensitive to pepsin. Biotype II has both antigens. Some biotype I strains cross-react with group K antiserum, probably reflecting common type-specific antigens.

Acid from glucose, maltose, lactose, sucrose and no acid from melibiose, melezitose, inositol, rhamnose, dulcitol, xylose or adonitol. Generally, no growth in 40% bile-blood agar, in media containing either 6.5% sodium chloride or 0.04% tellurite.

Biotype I consists of alpha-hemolytic human strains that fail to hydrolyze arginine and have a final pH range in glucose broth of 4.6–5.2. Biotype II strains are of animal origin; they are beta-hemolytic, hydrolyze arginine and attain a final pH range of 6.3–7.2. Biotype III strains are also of animal origin, beta-hemolytic, hydrolyze arginine but produce more acid from glucose (5.9–6.7). Only three strains of the latter biotype have been isolated, however.

Isolated from cases of human subacute endocarditis. Other human sources include abscesses, nasopharynx and vagina (Rifkind and Cole, 1962). Isolated from the urethra, vagina and tonsillar area of dogs.

e. *Streptococcus* sp. (group O) Boissard and Wormald 1950, 37.

Spherical or ovoid cells in long chains in broth culture. Rhamnose lacking in cell wall (Slade and Slamp, 1962).

Serological group O. Extraction of group polysaccharide by Fuller's formamide procedure destroys specificity; must use Lancefield's acid procedure.

Acid from glucose and lactose. May or may not ferment trehalose and salicin. No acid from mannitol or sorbitol. Final pH range in glucose broth is 4.5–5.1.

Reaction on blood agar ranges from strong alpha-reaction to moderately wide zones of beta-hemolysis similar to that produced by *S. pyogenes*. Growth under anaerobic conditions reduces or completely inhibits the beta-hemolysis. Surface colonies on blood agar are 0.4–0.8 mm in diameter after 18 hr at 37 C with flattened margin and raised center. The margin is radically striated and has a beaded or pleated edge. Colony has rubbery and coherent consistency.

Isolated from nasopharynx of normal humans; occasionally from throats of individuals suffering from tonsilitis.

Reciprocal intraspecies and intrageneric (*S. sanguis*) transformations have been reported.

f. *Streptococcus* sp. (group R) de Moor 1957, 174. (Description taken from de Moor (1963) and Perch *et al.* (1968).)

Small ovoid cocci, singly, in pairs or rarely in short chains. Some tendency toward rod formation noted.

Serological group R. Extracts made by Lancefield's hot-acid procedure destroys the reactivity. Decreasing the acid content to 0.66 N affords extraction without destruction (de Moor, 1963). Fuller's formamide procedure also destroys the antigen (Perch *et al.*, 1968).

Acid produced from glucose, sucrose, lactose, maltose, salicin, trehalose, raffinose, inulin and melibiose. No acid from arabinose, mannitol, sorbitol, glycerol and melezitose. Glycogen is hydrolyzed. No growth in a medium with 0.04% tellurite. No polysaccharide formed on sucrose agar.

Narrow, weak zones of beta-hemolysis or gamma-reaction noted on horse blood agar. Gamma-reaction on sheep blood agar. A weak CAMP test is observed with some strains (see *S. agalactiae*).

Isolated from septicemia of pigs; meningitis of man.

g. *Streptococcus suis* Elliott 1966, 211. (Group S *Streptococcus* de Moor 1963, 272.)

Cocci occurring singly, in pairs and rarely short chains as examined in clinical material (Field *et al.*, 1954).

Lancefield's group D. Strains from various investigations are identical serologically. Group antigen contains small amount of xylose. Type specificity associated with a capsular polysaccharide composed of glucose, glucosamine and galactosamine. Only one serotype (type I) described (Elliott, 1966).

Acid from trehalose, salicin, inulin, lactose, dextrose, maltose, fructose, sucrose, galactose and no acid from sorbitol, mannitol, raffinose, arabinose, dulcitol, glycerol, inositol, rhamnose or xylose. Dextrans are not produced. Slow growth in media containing 40% bile.

Source: bacteremia of young pigs frequently involving the brain and joints; throat cultures of diseased pigs and their healthy littermates.

The hydrolysis of starch and esculin, intolerance to heat and the capsular location of the type specific antigen tend to relate this species most closely to *S. bovis* rather than other group D streptococci.

Genus II. **Leuconostoc** van Tieghem 1878, 198, emend. mut. char.
Hucker and Pederson 1930, 66

ELLEN I. GARVIE

(*Betacoccus* Orla-Jensen 1919, 146.)

Leu.co.nos'toc. Gr. adj. *leucus* clear, light; M.L. neut.n. *Nostoc* algal generic name; M.L. neut.n. *Leuconostoc* colorless nostoc.

Cells may be **spherical but often lenticular** particularly on agar, **usually in pairs and chains.** Non-motile. **Gram-positive.** Spores not formed.

Colonies are small, usually less than 1 mm diameter, smooth, round, grayish white.

Chemoorganotrophs requiring rich media; **often having complex growth factor and amino acid requirements.** Nicotinic acid, thiamine, pantothenic acid and biotin are required by all species but cobalamin and *p*-amino benzoic acid by none.

Growth is dependent on the presence of a fermentable carbohydrate; glucose is fermented with the production of D(−) lactic acid, ethanol and CO₂. Some strains have an oxidative mechanism and acetic acid is formed in place of ethanol. Rhamnose, melezitose, inulin, starch, glycerol, sorbitol and inositol are not fermented. Dextrin is seldom fermented. Sorbose, erythritol, adonitol and dulcitol are not fermented by species 1–5 and not recorded for 6. Other reactions on carbohydrates and growth requirements are given in Table 14.8.

Some species produce a characteristic slime from sucrose. **Mannitol is formed from fructose.**

The amino acid composition of the cross-linking peptide of the cell wall murein consists of alanine, serine and lysine. Peptides from different species are not however identical (Kandler *et al.*, 1967).

Catalase negative although some strains will destroy H_2O_2 when grown in the presence of heme and others have peroxidase activity. Orthotoluidene is not attacked. Cytochromes are absent.

Arginine is not hydrolyzed. Milk is rarely acidified and curdled unless supplemented with a fermentable carbohydrate and yeast extract.

Non-proteolytic. Indole is not formed. Do not reduce nitrates. Non-hemolytic.

Facultative anaerobes.

Optimum temperature between 20 and 30 C.

Non-pathogenic to animals and man.

The G + C content of the DNA for *L. lactis* is 43–44 moles % and for all other species from 38–42 moles % (T_m, buoyant density).

Type species: *Leuconostoc mesenteroides* (Tsenkovskii) van Tieghem 1878, 191.

Further Comments

The genus can be divided into two groups: the first comprises five species which may be closely related and difficult to separate from each other, and which initiate growth between pH 5.5 and 6.5, rarely below 5.0 (Garvie, 1967). Their NAD-dependent lactate dehydrogenases (EC 1.1.1.28) have the same electrophoretic mobility (Garvie, 1969). They are found in slimy sugar solutions, fermenting vegetables and in milk and dairy products. The second group consists of organisms found in wine which comprise one species *L. oenos* which will grow in media with a pH between 4.8 and 4.2 or even lower; they may not grow at higher pH values. The electrophoretic mobility of the lactate dehydrogenase of this group is markedly different from that of the first group.

Leuconostoc oenos usually destroys a factor found in tomato juice (TJF), specific for many strains of the species (Garvie and Mabbitt, 1967).

Separation of species of the genus *Leuconostoc* from gas-forming lactobacilli is sometimes difficult. No known leuconostocs hydrolyze arginine and all form only D(−)-lactic acid. Gas-forming lactobacilli form DL lactic acid although D(−)-lactic acid may be a major product under some circumstances (Garvie, 1967), and many hydrolyze arginine. Some common and widely distributed gas-forming lactobacilli form dextran and share other properties with leuconostocs (Sharpe, Garvie and Tilbury, 1972).

The nomenclature used differs from that in earlier editions. Dextran-forming strains fermenting xylose but not arabinose are grouped with the strains fermenting neither pentose as *Leuconostoc dextranicum*. *Lactococcus dextranicus* (Beijerinck) hydrolyzed esculin, a property more often associated with the xylose-fermenting strains than with those which do not ferment xylose. The distinction between the active *Leuconostoc mesenteroides* and less active *Leuconostoc dextranicum* is blurred.

Leuconostoc paramesenteroides was previously included in *Leuconostoc mesenteroides* as a non-dextran-forming variant. The species has been isolated from sausages (Niven *et al.*, 1949), fermenting cucumbers (Pederson and Ward, 1949), and herbage (Whittenbury, 1966); strains from sausages and fermenting cucumbers have a high salt tolerance.

Leuconostoc lactis was first described by Abd-el-Malek and Gibson (1948) who tentatively used the name *Streptococcus kefir* for a species which they had found to be "the prevalent type of heterofermentative streptococcus in milk." They recognized that Hucker and Pederson (1930) regarded *Streptococcus kefir* as a synonym of *Streptococcus paracitrovorus* (Hammer) and *Lactococcus dextranicus* (Beijerinck). *Streptococcus kefir* (Abd-el-Malek and Gibson) differed from the organism described by Beijerinck, a difference noted by Abd-el-Malek and Gibson. Hucker and Pederson (1930) examined a strain of *Streptococcus kefir* deposited by Evans in the ATCC; it formed dextran from sucrose.

Bacteria with the properties of *Leuconostoc lactis* have not been widely isolated from milk and dairy products. There appears little justification for the argument used by Abd-el-Malek and Gibson for retaining the name *Streptococcus kefir* for the species they isolated. The conclusions of Hucker and Pederson that *Streptococcus kefir* (Evans) and *Leuconostoc dextranicum* are synonyms has not been challenged.

In the 6th edition of THE MANUAL it was suggested that a strain of *Bacterium gracile* (Müller-Thurgau and Osterwalder, 1913) was a leuconostoc. This strain has not been traced but Charlton *et al.* (1934) examined a culture of *Lactobacillus gracilis* obtained from Dr Osterwalder and found it formed DL-lactic acid. They considered that morphologically the culture was more coccoid than rod-like and was a leuconostoc. It would not however be included in the present genus *Leuconostoc*, the species of which form only D(−)-lactic acid. Bidan (1956) used the name *Leuconostoc gracile* for a strain, isolated from wine, which fermented both lactose and raffinose; it was unlike *Leuconostoc oenos*. Peynaud (1968) has suggested that *Leuconostoc oenos* should be divided into two species on the basis of pentose fermentation. More information is required to know if this is justified; if a split is desirable the name *Leuconostoc oenos* will apply to the non-pentose-fermenting group, which includes the type strain.

Description of the species of genus Leuconostoc

1. **Leuconostoc mesenteroides** (Tsenkovskii) van Tieghem 1879, 198. (*Ascococcus mesenteroides* Tsenkovskii 1878, 159; *Betacoccus arabinosaceus* Orla-Jensen 1919, 152.)

me.sen.ter.oi'des. Gr. n. *mesenterium* the mesentery; Gr. n. *oides* form, shape; M.L. adj. *mesenteroides* mesentery-like.

Spheres or lenticular cells, 0.5–0.7 by 0.7–1.2 μm; in pairs or chains, usually short.

A characteristic slime of dextran formed from sucrose, favored by a temperature of 20–25 C. Different colonial types are formed on sucrose agar depending on the characteristics of the dextran formed. Some strains, particularly those from dairy sources, produce little dextran.

Glucose broken down anaerobically by a pentose pathway yielding 1 mole each of D(−)-lactic acid, ethanol and CO_2. Phosphoketolase is present; fructose 1,6-diphosphate aldolase is absent. Some strains also have an aerobic oxidative metabolism using 1 mole oxygen/mole glucose and yielding equimolar quantities of CO_2, lactate and acetate. These strains also have peroxidase activity.

Acetate, lactate and tartrate are not used as sole carbon source.

The number of amino acids essential for growth is small; only valine and glutamic acid are required by all strains.

Do not withstand heating to 55 C for 30 min, although slimy cultures in sugar factories may withstand heating to 80–85 C.

Temperature range 10–37 C; optimum 20–30 C.

Found in slimy sugar solutions, on fruit and vegetables, in milk and dairy products.

G + C content of DNA for two strains is 39–42 moles % (Marmur and Doty, 1962; Schildkraut, Marmur and Doty, 1962; Garvie, unpublished).

Suggested reference strains: NCDO 523; ATCC 8293; NCDO 768; ATCC 12291.

2. **Leuconostoc dextranicum** (Beijerinck) Hucker and Pederson 1930, 67. (*Lactococcus dextranicus* Beijerinck 1912, 27; *Betacoccus bovis* Orla-Jensen 1919, 152; *Streptococcus paracitrovorus* Hammer 1920, 93; *Betacoccus cremoris* (A form) Knudsen and Sørensen 1929, 82; *Streptococcus kefir* Migula 1900, 44.)

dex.tra'ni.cum. M.L. n. *dextranum* dextran; M.L. neut.adj. *dextranicum* relating to dextran.

Spheres or lenticular cells, 0.5–0.7 by 0.7–1.2 μm; in pairs or short chains.

Dextran is formed from sucrose but not as actively as with most strains of *L. mesenteroides*.

The carbohydrate breakdown pathway has not been studied.

Amino acid requirements are more varied than with *L. mesenteroides* but only a few amino acids affect the growth of any one strain (Garvie, 1967).

Temperature range 10–37 C; optimum 20–30 C.

TABLE 14.8

Differentiating characteristics of the species of genus **Leuconostoc**[a]

	1. *L. mesenteroides*	2. *L. dextranicum*	3. *L. paramesenteroides*	4. *L. lactis*	5. *L. cremoris*	6. *L. oenos*
Acid from:						
Amygdalin	d	d	(d)	−	−	.
Arabinose	+	−	d	−	−	d
Arbutin	d	−	−	−	−	.
Cellobiose	d	d	(d)	−	−	d
Fructose	+	+	+	+	−	+
Galactose	+	d	+	+	d	d
Glucose	+	+	+	+	+	+
Lactose	(d)	+	(d)	+	+	−
Maltose	+	+	+	+	d	−
Mannitol	d	d	(d)	−	−	−
Mannose	+	d	+	d	−	d
Melibiose	d	d	+	d	−	d
Raffinose	d	d	d	d	−	−
Ribose	+	.	.	.	.	.
Salicin	d	d	−	d	−	d
Sucrose	+	+	+	+	−	−
Trehalose	+	+	+	−	−	+
Xylose	d	d	d	−	−	d
Hydrolysis of esculin	d	d	d	−	−	+
Required for growth:						
Uracil	−	−	−	−	+	−
Guanine + adenine + xanthine + uracil	−	d	d	−	+	+
Riboflavin	d	d	+	+	+	+
Pyridoxal	d	d	+	−	+	+
Folic acid	d	d	+	−	+	+
Tomato juice factor	−	−	−	−	−	d
Destruction of tomato juice factor	−	−	−	−	−	d
Dextran formation	+	+	−	−	−	−
Utilize citrate (carbohydrate present)	d	d	d	d	+	d
Utilize malate:						
No carbohydrate	d	−	d	−	−	.
+ carbohydrate	d	−	d	−	−	+
Yeast glucose litmus milk acid	+	+	+	+	+	d
Clot	d	d	d	d	d	−
Reduction	d	d	d	(d)	−	d
Gas	d	d	−	−	−	−
Growth in: 3.0% NaCl	+	d	d	d	−	.
6.5% NaCl	d	−	d	−	−	.
Growth at pH:						
4.8 (initial)	−	−	d	−	−	+
6.5 (initial)	+	+	+	+	+	d
Growth at 37 C	d	+	d	+	−	d
Final pH in glucose broth	4.5	4.5	4.4	4.7	5.0	.

[a] + = > 90% strains positive; d = 10–90% strains positive; − = > 90% strains negative; ( ) = delayed reaction; . = not known.

Found on fruit and vegetables and in milk and milk products.

The G + C content of the DNA is 38–39 moles % (two strains; Garvie unpublished).

Suggested reference strains: NCDO 812 (ferments xylose); NCDO 529; ATCC 19255 (does not ferment xylose).

3. **Leuconostoc paramesenteroides** Garvie 1967, 446.

pa.ra.me.sen.ter.oi′des. Gr. prep. *para* resembling; M.L. *mesenteroides* a specific epithet; M.L. adj. *paramesenteroides* resembling *L. mesenteroides*.

Spheres or lenticular cells 0.5–0.7 by 0.7–1.0 μm; in pairs or chains.

Dextran is not formed from sucrose.

The carbohydrate breakdown pathway has not been studied.

Amino acid requirements are complex and variable. More amino acids influence growth of most strains than for strains of *L. mesenteroides*.

Facultative anaerobe although some strains prefer reducing conditions for growth (Garvie, 1967); these strains also have a low optimum temperature, 18–24 C.

Whittenbury (1964) found that some strains formed an enzyme which destroyed hydrogen peroxide when grown on a medium containing heme.

Tolerant of higher concentrations of sodium chloride than other species.

Temperature range 10–37 C; optimum 20–30 C.

Found on herbage, in fermenting vegetables, milk and milk products. Apparently widely distributed.

The G + C content of the DNA is 38–39 moles % (two strains; Garvie unpublished).

Type strain: NCDO 803.

4. Leuconostoc lactis Garvie 1960, 290. (*Streptococcus kefir* Abd-el-Malek and Gibson 1948, 246; Not *Streptococcus kefir* Evans 1918, 243.)

lac'tis L. n. *lac* milk; L. gen.n. *lactis* of milk.

Spheres or lenticular cells, 0.5–0.7 by 0.7–1.2 μm.

Amino acid requirements complex and similar to *L. paramesenteroides*.

More heat resistant than other species and normally survives heating at 60 C for 30 min.

Temperature range 10–40 C; optimum 25–30 C.

Found in milk and dairy products, not widely recorded.

The G + C content of the DNA is 43–44 moles % (two strains; Garvie unpublished).

Type strain: NCDO 533; ATCC 19256.

5. Leuconostoc cremoris (Knudsen and Sørensen) Garvie 1960, 289. (*Betacoccus cremoris* (X form) Knudsen and Sørensen 1929, 81; *Leuconostoc citrovorum* (Hammer) Hucker and Pederson 1930, 67, *nom. rej.* Opin. 45, Jud. Comm. 1971.)

cre.mor'is. L. n. *cremor* cream; L. gen.n. *cremoris* of cream.

Spheres or lenticular cells, 0.8–1.2 μm; generally in long chains in which the cells appear to be in pairs.

Normally sucrose is not fermented but "mutant colonies" formed in soft agar may form acid from sucrose (Whittenbury, 1966).

Citrate is broken down when a fermentable carbohydrate is present. Acetate, pyruvate and CO_2 are formed; pyruvate is converted to acetoin and diacetyl (Speckman and Collins, 1968). There is no citric acid cycle.

Amino acid requirements are complex as omission of any one of a number of them will prevent growth.

Facultative anaerobe but prefers reducing conditions (Garvie, 1967).

Temperature range 10–30 C; optimum 18–25 C.

The least active species, characterized by limited fermentation ability and complex growth factor requirements.

Found in starter cultures, milk and dairy products. No "wild" source is known.

The G + C content of the DNA is 39–42 moles % (two strains; Garvie unpublished).

Suggested reference strain: NCDO 543; ATCC 19254.

6. Leuconostoc oenos Garvie 1967, 431.

oe.nos'. Gr. n. *oinos* wine; Gr. gen.n. *oenos* of wine.

Spheres or lenticular cells, 0.5–1.0 by 0.7–1.5 μm.

Growth on agar surface poor even anaerobically. Grow best in medium of pH 4.2–4.8 containing tomato juice. Growth is slow.

The carbohydrate breakdown pathway has not been studied. Aldolase is absent.

Amino acid requirements are complex and omission of any of a number of amino acids will prevent growth. Studies of growth requirements are complicated by the requirement by some strains of a growth factor in tomato juice (Garvie and Mabbitt, 1967), probably L'-O-(β-glucopyranosyl-D(R)-pantothenic acid (Amachi *et al.*, 1971).

Facultative anaerobe—preferring reducing conditions.

Grow slowly in 10% ethanol at pH 4.8.

Temperature range 10–35 C; optimum 18–24 C.

Found in wine.

The G + C content of the DNA is 39–40 moles % (two strains; Garvie unpublished).

Type strain: NCDO 1674; ATCC 23279.

Genus III. **Pediococcus** *Balcke 1884, 257*

KAKUO KITAHARA

Pe.di.o.coc'cus. Gr. n. *pedium* a plane surface; Gr. n. *coccus* a berry; M.L. masc.n. *Pediococcus* plane coccus.

Cocci occurring **in pairs or in tetrads** as the result of alternate division along the two perpendicular planes; single cells or chains are rare.

Non-motile, do not form endospores. Gram-positive.

Chemoorganotrophs: **metabolism fermenta-**

tive. The fermentation is **homolactic, producing DL-lactic acid;** the L(+) enantiomorph generally predominates. Acid but no gas produced from glucose, fructose and mannose; sorbitol and starch not fermented.

Gelatin not liquefied. Nitrates not reduced to nitrites.

Nutritional requirements complex.

Microaerophilic; surface growth is poor. Gen-

erally **catalase negative** although may possess some non-heme catalase activity.

Saprophytes found in fermenting plant materials, especially in spoiled beer; rare in milk and dairy products.

The G + C content of the DNA ranges from 34–44 moles %.

Type species: *Pediococcus cerevisiae* Balcke 1884, 257.

Key to the species of genus **Pediococcus**

I. Grow at pH 5.0 but not at pH 9.0.
    A. No growth at pH 7.0 or at 35 C; prefers anaerobic conditions.
        1. *Pediococcus cerevisiae*
    B. Growth at pH 7.0 and at 35 C; microaerophilic.
        1. Can grow at 50 C.
            2. *Pediococcus acidilactici*
        2. Can not grow at 50 C.
            3. *Pediococcus pentosaceus*
II. Do not grow at pH 5.0 but grow at pH 9.0; microaerophilic.
    A. Halophilic.
            4. *Pediococcus halophilus*
    B. Not halophilic
            5. *Pediococcus urinae-equi*

Description of the species of genus **Pediococcus**

1. Pediococcus cerevisiae Balcke 1884, 257. (Ferment no. 7 Pasteur 1876, 6; Sarcina from beer Hansen 1879, 234; *Pediococcus damnosus* Claussen 1903, 68; *Pediococcus perniciosus* Claussen 1903, 68; *Pediococcus damnosus* var. *salicinaceus* Mees 1934, 96; *Pediococcus damnosus* var. *diastaticus* Andrews and Gilliland 1952, 195; *Pediococcus mevalovorus* Kitahara and Nakagawa 1958, 28.)

ce.re.vi'si.ae. L. n. *cerevisia* beer; L. gen.n. *cerevisiae* of beer.

Spheres, 0.6–1.0 μm in diameter.

Colonies on glucose peptone yeast extract gelatin white becoming yellowish brown; on wort gelatin colonies white 2–3 mm in diameter. Scant beaded growth on agar slants. In agar stabs filiform growth along stab, no surface growth.

Acid but no gas is produced from maltose, sometimes from galactose and salicin. No acid from pentoses, lactose, sucrose or mannitol.

Diacetyl is generally produced and is responsible for the "sarcina odor" of spoiled beer.

Niacin and biotin are essential for growth; pyridoxine is stimulatory. Ascorbic acid is a stimulant for some strains.

The strain originally described as *P. mevalovorus* requires mevalonic acid, which is present in beer but not in wort; it also requires Tween 80 in place of biotin (Nakagawa and Kitahara, 1959).

Requires CO_2 for growth (Weinfurtner *et al.*, 1955).

Optimum temperature 25 C, no growth at 35. Thermal death point 60 C for 10 min.

Highly tolerant of hop antiseptics; hop-humulone may cause formation of giant cells 5–15 μm in diameter (Nakagawa and Kitahara, 1962).

pH range: Growth between pH 3.5–6.2, optimum about 5.5.

Found in spoiled beer and brewer's yeasts.

2. Pediococcus acidilactici Lindner 1887, 440. (*Pediococcus lindneri* Henneberg 1926, 105.)

a.ci.di.lac.ti'ci. M.L. n. *acidum lacticum* lactic acid; M.L. gen.n. *acidilactici* of lactic acid.

Spheres, 0.6–1.0 μm in diameter.

Small white colonies on glucose peptone yeast extract gelatin. In stabs filiform growth along the stab; scant leafy or no growth on surface. Grows in unhopped wort but not in hopped wort or beer.

Acid from galactose, arabinose, xylose, salicin and trehalose; some strains produce slight acid from sucrose and lactose. No acid from maltose, mannitol, α-methyl glucoside or dextrin. Diacetyl is produced.

Nearly all amino acids except methionine are required for growth. Riboflavin, pyridoxine, pantothenic acid, niacin and biotin are also essential (Sakaguchi, 1960).

Optimum temperature 40 C, maximum 52 C. Thermal death point 70 C for 10 min; some strains, especially when newly isolated, are more heat tolerant.

Found in sauerkraut and fermenting mashes.

The G + C content of the DNA is 44 moles % (T_m) (Sakaguchi and Mori, 1969).

3. Pediococcus pentosaceus Mees 1934, 96.

(*Tetracoccus* No. 2 Orla-Jensen 1919, 78; *Pediococcus hennebergii* Sollied 1903, 491; *Pediococcus parvulus* Günther and White 1961, 195.)

pen.to.sa'ce.us. M.L. neut.n. *pentosum* a pentose sugar; M.L. adj. *pentosaceus* relating to a pentose.

Spheres, 0.8–1.0 μm in diameter.

On glucose peptone yeast extract gelatin, surface colonies are white and pinpoint; in stabs white growth along stab but no surface growth.

Acid is produced from galactose, maltose and usually from arabinose, xylose, lactose, salicin and α-methyl glucoside; sometimes slight acid from sucrose; ordinarily no acid from rhamnose, trehalose, mannitol, dextrin or inulin. Diacetyl is produced.

All amino acids are required for growth; among them, however, serine, methionine and lysine are stimulatory.

Folinic acid, niacin and pantothenic acid are essential; biotin is stimulatory. A few strains also require riboflavin.

Usually catalase negative but may be weakly positive in media of low sugar content (Felton *et al.*, 1953). The catalase is of the non-heme type (Delwiche, 1961).

Optimum temperature 35 C; maximum 42–45 C. Killed at 65 C in 8 min.

Sensitive to hop-antiseptics.

Originally isolated from malt mash; widely distributed in fermenting materials such as sauerkraut, pickles, silages and cereal mashes.

The G + C content of the DNA is 38 moles % (Suzuki and Kitahara, 1964).

4. Pediococcus halophilus Mees 1934, 96.

(*Tetracoccus* No. 1 Orla-Jensen 1919, 77; *Sarcina hamaguchiae* Saito 1907, 155; *Pediococcus soyae* Sakaguchi 1958, 353.)

hal.o.phi'lus. Gr. n. *halos* salt; Gr. adj. *philus* loving; M.L. adj. *halophilus* salt loving.

Spheres, 0.6–0.8 μm in diameter.

Little or no growth on surface of solid media; white filiform growth along agar stab.

Acid is produced from arabinose, galactose, maltose, mannitol, α-methyl glucoside, dextrin and usually from sucrose, raffinose and salicin. No acid from rhamnose, glycerol or inulin.

Uracil, riboflavin, pyridoxine, pantothenic acid,

niacin, folinic acid, biotin and either betain or carnitine are essential for growth. Glutamic acid, arginine, histidine, isoleucine, leucine, methionine, serine, tryptophan and an undetermined peptidic "P-factor" are also required (Sakaguchi, 1960; 1962).

Grows best in presence of 6–8% NaCl; tolerates more than 15%.

Temperature optimum 25–30 C; maximum 40 C.

Isolated from anchovies and soy mash. It is active in the ripening of soy mash and soy bean cheese *miso*.

The G + C content of the DNA is 34–35 moles % (T$_m$) (Sakaguchi and Mori, 1969).

Comment added in proof: *Tetracoccus soyae* Ueno and Omata 1961, 369, a homofermentative, halophilic tetrad isolated from soy mash, differs from species of *Pediococcus* in being definitely catalase positive, producing surface growth and reducing nitrate, thus resembling *Tetracoccus* Orla-Jensen 1919, 77. Its taxonomic position is still uncertain.

5. Pediococcus urinae-equi Mees 1934, 97.

u.ri'nae.e.qui. L. fem.n. *urina* urine; L. masc.n. *equus* horse; M.L. gen.n. *urinae-equi* of urine of horse.

Spheres, 0.8–1.0 μm in diameter.

Agar colonies are circular, 1–2 mm in diameter, grayish white and raised. White growth along agar stab with limited surface growth.

Acid is produced from galactose, maltose, sucrose, lactose, salicin, dextrin and glycerol; some strains produce acid from pentoses, raffinose and mannitol. No acid from dulcitol or starch. Lactic acid produced is exclusively L(+) form.

Glutamic acid, arginine, histidine, isoleucine, leucine, methionine, phenylalanine and tryptophan are required for growth as well as biotin, pantothenic acid, niacin, riboflavin, pyridoxine, betaine and folinic acid.

Temperature optimum between 25 and 30 C; maximum 42 C.

Tolerates 10% but not 15% NaCl.

Frequently found in brewer's yeast. Originally isolated from horse urine.

The G + C content of the DNA is 40 moles % (T$_m$) (Sakaguchi and Mori, 1969).

Genus IV. Aerococcus *Williams, Hirch and Cowan 1953, 475*

JAMES B. EVANS

A.ë.ro.coc'cus. Gr. masc.n. *aër* air, gas; Gr. n. *coccus* a grain, berry; M.L. masc.n. *Aërococcus* air coccus.

Spheres, 1.0–2.0 μm in diameter, with a strong tendency toward tetrad formation when grown in suitable liquid media (Deibel and Niven, 1960). Non-motile. **Gram-positive.**

Growth on solid media is generally sparse and beaded (small discrete colonies).

Chemoorganotrophic: Acid but no gas is produced from glucose, fructose, galactose, mannose, maltose and sucrose; acid is usually produced from lactose, trehalose and mannitol. Final pH in glucose broth 5.0–5.5.

Microaerophilic. In shake cultures or soft sugar agar a heavy band of discrete colonies is produced just beneath the surface. Anaerobic growth frequently absent and when it does occur is delayed and often consists of only a few discrete colonies.

Catalase activity is absent or weak and when present is a non-heme pseudocatalase. Porphyrin respiratory enzymes are absent (method of Deibel and Evans, 1960). Hydrogen peroxide is produced during aerobic growth.

The cell wall murein contains no interpeptide bridge. The carboxyl group of D-alanine is directly bound to the ε-amino group of L-lysine of an adjacent peptide subunit (Kandler et al., 1970).

Widely distributed in air, in meat brines and on raw and processed vegetables.

The G + C content of DNA of the type species ranges from 36–40 moles % (T_m) (Schultes and Evans, 1971).

Type species: *Aerococcus viridans* Williams, Hirch and Cowan 1953, 477.

Description of species of genus **Aerococcus**

1. **Aerococcus viridans** Williams, Hirch and Cowan 1953, 477. (*Gaffkya homari* Hitchner and Snieszko 1947, 48; *Pediococcus homari* Deibel and Niven 1960, 178.)

vi.ri.dans. L. part.adj. *viridans* producing a green color.

Morphology as for genus. Cell wall contains glucose, galactose and galactosamine but no rhamnose, glycerol or ribitol.

Agar colonies are round, 0.5–1.0 mm in diameter, semitransparent, white or gray. On blood agar the colonies are larger and are surrounded by a zone of greening, presumably the result of hydrogen peroxide production.

In broth there is moderate uniform turbidity that tends to settle rather quickly into a packed sediment.

Acetylmethylcarbinol is not produced. Gelatin not liquefied. Ammonia not produced from arginine. Nitrates not reduced to nitrites.

The vitamins pantothenic acid, nicotinic acid and biotin are either absolutely required or are markedly stimulatory. Guanine or another purine base is required. Amino acids are required but the requirement is not specific or sharply defined (Miller and Evans, 1970).

Growth is not prevented by 40% bile, 10% NaCl, 0.01% potassium tellurite or a pH of 9.6.

Growth at 10 but not at 45 C (exceptions may be common).

May be pathogenic for lobsters. Has been isolated from human infections of the urinary tract and from endocarditis.

Further Comments

This seems to be a relatively homogeneous, identifiable species but considerable confusion has existed regarding its nomenclature. It appears to be indistinguishable from *Gaffkya homari*, the lobster pathogen; but *Gaffkya* is a *nomen rejiciendum* (Opinion 39 Jud. Comm., 1971). Deibel and Niven suggested placing it in *Pediococcus* but the majority of other workers favor accepting the name *Aerococcus viridans*. The relationship of these organisms to the genera *Pediococcus*, *Streptococcus* (enterococci) and *Staphylococcus* (respiratory deficient) must await further studies, such as the determination of DNA homologies.

Genus V. **Gemella** *Berger 1960, 253*

ALICE REYN

Ge.mel′la. M.L. dim.n. *gemellus* a twin; M.L. fem.n. *Gemella* a little twin.

Cocci singly or in pairs with adjacent sides flattened. Gram-indeterminate but cell wall is of the Gram-positive type (Plate 10.2, Fig. 2). Endospores not formed; non-motile.

On blood agar small smooth colonies resemble those of beta-hemolytic streptococci; the hemolysin is produced on rabbit or horse blood agar. Pigment is not produced.

Chemoorganotroph: Fermentative, acid is produced from several carbohydrates. Growth requirements complex.

Nitrates not reduced; nitrites reduced by some strains.

Aerobic or facultatively anaerobic. Respiratory system resistant to KCN.

Optimal temperature 37 C, grows at 22 C.

Parasites of mammals.

Mean G + C content of DNA 33 ± 1.6 moles %.

Type species: *Gemella haemolysans* (Thjøtta and Bøe) Berger 1960, 253.

Further Comments

By gas chromatography Yamakawa and Ueta (1964) demonstrated that the fatty acid content and sugars of *Neisseria haemolysans* differed from those of other *Neisseria* species studied.

Smears stained by Gram's method contain distinct Gram-positive, Gram-negative and Gram-variable cells. The fine structure resembles that of many Gram-positive organisms, though the cell wall is comparatively thin (about 10 nm) and of varying thickness; this probably accounts for the variability in decolorization during the Gram-staining method (Reyn *et al.*, 1966, 1970). By comparison, the cell wall structure of *Veillonella* is of the complex Gram-negative type (Bladen and Mergenhagen, 1964) as is that of *Neisseria* species.

Berger (1960, 1961) proposed *Gemella* as an aerobic, oxidase negative, catalase negative genus within the *Neisseriaceae*, but recent studies have indicated that it does not belong in this family of Gram-negative cocci (Reyn *et al.*, 1966; Reyn, 1970; Reyn *et al.*, 1970).

The G + C content of the DNA is 33 ± 1.6 moles % (buoyant density) compared to *Neisseriaceae* (47–52) and *Veillonella* (36.5) (Lee *et al.*, 1956). The range observed corresponds to the peroxidase negative, catalase negative family *Streptococcaceae*, but although it has many physiological features in common, as for example, resistance to H_2O_2 and KCN (Berger, 1960), *Gemella* does not fit completely into any of the described genera.

Description of the species of genus **Gemella**

1. **Gemella haemolysans** (Thjøtta and Bøe) Berger 1960, 253. (*Neisseria haemolysans* Thjøtta and Bøe 1938, 531.)

hae.mo.ly'sans. Gr. n. *haema* blood; Gr. v. *lyo* loosen; M.L. part.adj. *haemolysans* dissolving blood.

Cocci 0.5 by 0.5–0.6 μm singly or in pairs with adjacent sides flattened. Considerable variation in size; division in several planes. Thin sections observed by electron microscopy reveal many giant cells with several ingrowing septa at different angles to each other.

Stearic acid and galactose are the predominant components of whole cells (Yamakawa and Ueta, 1964).

Beta-hemolysis on solid medium with rabbit or horse blood; poor growth on media without protein. Growth on medium with tellurite very weak or absent. Pigment not formed.

Acid produced fermentatively from glucose, maltose, fructose, sucrose, glycogen, starch and dextrin and occasionally from mannitol, arabinose, sorbitol and inulin. Polysaccharide not produced on 5% sucrose. Very feeble growth in Hugh and Leifsen's (1953) medium under anaerobic conditions. Esculin not broken down.

Arginine not hydrolyzed. Indole, H_2S, gelatinase, urease, cytochrome oxidase, catalase and peroxidase not produced. H_2O_2 formed under aerobic conditions. Resistant to optochin.

Nitrates not reduced; nitrites reduced by some strains.

Aerobic, facultatively anaerobic. Weak growth under completely anaerobic conditions. Resistant to KCN.

Temperature optimum 37 C; grows at 22 but not at 10 or 45 C.

Sensitive to penicillin, streptomycin, tetracycline, sulfathiazole, chloramphenicol, macrolides, vancomycin and related drugs.

Serologically different from *N. perflava* and *N. sicca*.

Found in bronchial secretions and slime from the respiratory tract.

The G + C content of DNA is 33.5 moles % (Mandel, personal communication).

FAMILY III. **PEPTOCOCCACEAE** ROGOSA 1971, 235

MORRISON ROGOSA

Pep.to.coc.ca'ce.ae. M.L. masc.n. *Peptococcus* type genus of the family; -*aceae* ending to denote a family; M.L. fem.pl.n. *Peptococcaceae* the *Peptococcus* family.

Cocci varying in diameter (0.5–2.5 μm). Occur singly, in pairs, tetrads and irregular masses, occasionally in three-dimensional cubic packets. Some organisms may form short or long chains with some elongation of individual elements. Flagella, motility and endospores are absent.

Gram-positive but may stain equivocally. Anaerobic. Chemoorganotrophic with complex nutritional requirements. Gas consisting principally of CO_2 and usually H_2 is produced in cultures from amino acids or carbohydrates or both; H_2S may also be detectable. Carbohydrates may or may not be fermented. In complex media the major non-gaseous products are lower fatty acids; in some cases succinate or ethanol may also be major products. Lactate, if produced, is generally not a major product and is always associated with a considerable heterofermentation. Found in the mouth and intestinal and respiratory tracts of man and other animals; and frequently in the normal and pathological human female urogenital tracts. Packet formers are found in soil, on the surface of cereal grains and in the alimentary tract of man and animals.

The characteristics differentiating the genera of the family are shown in Table 14.9.

Genus I. **Peptococcus** Kluyver and van Niel 1936, 400

MORRISON ROGOSA

Pep.to.coc'cus. Gr. v. *pepto* cook, digest; Gr. n. *coccus* a grain, berry; M.L. masc.n. *Peptococcus* the digesting-coccus.

Spherical bacteria about 0.5–1.0 μm or exceptionally 1.6 μm in diameter. Occur **singly,** in **pairs, tetrads** or **irregular masses** or rarely in short chains but not in three-dimensional cubical packets; long chains never formed. Non-flagellated and non-motile. No spores. **Gram-positive.**

Chemoorganotrophic. Can use protein decomposition products (peptones, amino acids) as sole energy source. Carbohydrates not required and saccharoclastic activity often limited or negative but carbohydrates may enhance growth or required fatty acids may stimulate carbohydrate fermentation of some species. Where tested, such nitrogenous compounds as **amino acids,** purines or pyrimidines are fermented. Gas produced in cultures by some species with or without carbohydrates. Ammonia produced. Depending on the substrate, acetate, propionate, butyrate, formate, succinate, lactate, CO_2, ammonia and occasionally hydrogen are produced. In complex media, various combinations of the **C_1 to C_4 lower volatile fatty acids, CO_2, H_2 and ammonia** tend to be the **major products** from amino acids; however, lactate is also produced from histidine. Generally **lactate is not** a **major** product in glucose complex media and it is **not formed homofermentatively** as in *Streptococcus* but is always associated with a very considerable heterofermentation. Lactate and malate are not fermented; a number of species produce gas from pyruvate, but most species do not produce gas from citrate or tartrate.

Weak or variable catalase reaction. Rarely hemolytic. Coagulase negative. Indole sometimes produced.

Temperature range generally 25–37 C; optimal temperature 35–37 C.

Anaerobic. pH range generally 6–8; optimal pH 7–8.

Sensitive to penicillin G (1–10 units/ml; often to less than 1 unit/ml).

Isolated from the human female urogenital tract, the human intestinal and respiratory tract, inflamed appendices, cystitis, pleurisy, the gums, postpartum septicemia, tonsils, plasma, skin, caseous lymphadenitis in sheep, and tidal bay mud.

TABLE 14.9
Characteristics differentiating the genera of family **Peptococcaceae**[a]

| | Growth at pH 2–2.5 | Arrangement | | G + C moles % | Carbo-hydrates fermented | Cellulose digested | Major products | | | Peptones or amino acids main N and energy source |
		Chains	Cubic packets				Succinate	Ethanol	Volatile acids with >3C atoms	
Peptococcus	−	−	−	35.7–36.7	d	−	−	−	+	+
Peptostrepto-coccus	−	+	−	33.5	d	−	−	d	+	+
Ruminococcus	−	+ or −	−	39.8–45.4	+	+	+ or −	+ or −	−	−
Sarcina	+	−	+	28.6–30.6	+	−	−	+ or −	+ or −	−

[a] + = positive; − = negative; d = usually (90–100%) positive.

Pathogenicity is uncertain because isolations are very frequently from sites where other potential pathogens are present.

Type species: *Peptococcus niger* (Hall) Kluyver and van Niel 1936, 400.

Further Comments

Kluyver and van Niel (1936) proposed the genus *Peptococcus* to include anaerobic, Gram-positive cocci, not regularly forming chains, which were chemoheterotrophic and "capable of fermenting protein decomposition products." The present descriptions conform to this concept.

For this reason, a number of frankly saccharoclastic organisms, which use amino acids chiefly or only for biosynthetic purposes, or whose nitrogen metabolism is little understood are excluded from *Peptococcus*. Among these are the following included in THE MANUAL, 7th edition: (1) *Peptococcus grigoroffii* (Prévot) Douglas 1957, 477; (2) *Peptococcus saccharolyticus* (Foubert and Douglas) Douglas 1957, 478. The latter ferments glucose to CO_2, ethanol, acetic and formic acids, with traces of lactic acid (Foubert and Douglas, 1948); nothing is known of the glucose fermentation of *P. grigoroffii*.

Gaffkya anaerobius (*sic*) (Choukévitch) Prévot 1933 (*Tetracoccus anaerobicus* Choukévitch, 1911; *Micrococcus tetragenes anaerobius* Hamm 1912) was described in THE MANUAL (6th edition) under the name *Gaffkya anaerobia* (Choukévitch) Prévot but was omitted from the 7th edition. Recently, *Gaffkya* Trevisan was rejected as a name (Opinion 39, Jud. Comm. 1971, 104). Although Prévot was unable to isolate the organism described by Choukévitch, he continued to describe it through a number of publications including Prévot and Fredette (1966). Recently, Moore and Holdeman (personal communication) isolated from the human vagina three strains consistent with the general descriptions of Choukévitch (1911) and Prévot (1933); however, Moore and Holdeman add the information that lactate and butyrate are the principal fermentation products in peptone-yeast extract-glucose media; the lactate is derived exclusively from the glucose, whereas the butyrate results from the fermentation of both peptone and glucose. This appears to be an unusual distribution of products for anaerobic cocci.

Peptococcus prevotii Douglas 1957 was originally described under the name *Micrococcus prevotii* (Foubert and Douglas, 1948). Different strains were described; one group was morphologically like the genus with the rejected name *Gaffkya* whose cells were arranged in "contiguous plates of four arranged in many planes;" another group was not characterized by modal tetrad formation; two biochemical groups, one fermenting glucose and five other carbohydrates, and one group not fermenting any carbohydrates, were also described. From study of extant strains, there is no doubt that the descriptions of *Micrococcus prevotii* Foubert and Douglas 1948, 31 and *Peptococcus prevotii* Douglas 1957, 477 are composite descriptions of two organisms. One of the discordant elements has cells arranged typically as tetrads, ferments glucose and certain other carbohydrates, and produces lactate and butyrate as major products; in these and in about 100 other characteristics this element of *P. prevotii* is indistinguishable from the organism with the rejected name *Gaffkya anaerobia*. Except for a negative indole reaction, the second element is indistinguishable from *P. asaccharolyticus*. Therefore, according to Rule 14a (3) and Provision 3 of the International Code of Nomenclature, *Peptococcus prevotii* and its basionym *Micrococcus prevotii* should be considered as *nomina confusa* and placed on the list of *nomina rejicienda*.

Group VII b of Hare (1967) has the general characteristics of *Peptococcus* and is similar to or identical with *Micrococcus abscedens ovis* Morel 1911 isolated from caseous lymphadenitis in sheep and the human nose and skin (Joubert, 1958).

Present judgments of the species of *Peptococcus* will doubtless require revision as urgently needed modern data become available.

Description of the species of genus **Peptococcus**

1. **Peptococcus niger** (Hall) Kluyver and van Niel 1936, 400. (*Micrococcus niger* Hall 1930, 409.) ni'ger. L. adj. *niger* black.

Small spheres about 0.6 μm in diameter, occurring in irregular masses or occasional pairs but not in chains. No growth less than 3 cm from the surface in tubes of deep agar incubated in air.

Growth is slow in all agar media tested (Hall, 1930) requiring 2–4 days at 37 C for colony appearance or about 7 days at 25–30 C. In agar deeps, colonies are initially without distinctive color but turn brown to black in 4 days or more with production of a non-diffusible pigment. Small gas bubbles sometimes appear. Colonies are generally never more than 0.5 mm; colonies are irregularly globular, smooth and dense. On blood agar, colonies are first like tiny black pearls, round, smooth and glistening, reaching a maximal diameter of 0.5 mm; non-hemolytic. On coagulated blood serum growth is slower, requiring 8 days for the appearance of minute brown colonies. In gelatin media, a sediment, gradually turning black, is noticeable after 5 or more days. In brain medium growth is slow; gas is uniformly produced after 6 days. No growth or change in milk. Ascitic fluid does not enhance growth in infusion agar with 2% peptone. In broth, growth is slow, becoming visible in 4–5 days.

Probably no carbohydrates or polyols are fermented. Gas is observable and a black sediment collects at the bottom of the tube; all cultures, including those with carbohydrates, become more alkaline; ammonia and carbon dioxide are presumably produced from non-carbohydrate nitrogenous substrates (e.g. amino acids). The original descriptions (Hall, 1930) did not indicate the nature of the gas.

Gelatin and coagulated blood serum are not liquefied. H_2S is produced in brain medium.

The organism was non-pathogenic for one guinea pig (subcutaneously) and one rabbit (intravenously) (Hall, 1930).

Isolated from the urine of an aged woman. No strains are extant and it has never been reisolated.

TABLE 14.10

Characteristics of the species of genus
Peptococcus[a]

	1. P. niger	2. P. a-sacchar-olyticus	3. P. aero-genes	4. P. activus	6. P. anaero-bius
Sugars fermented	−	−	w	+	w
Visible gas	+	+	+	+	−
Amino acids fermented					
Glycine	NT	−	−	−	+
Serine	NT	+	+	+	−
Threonine	NT	+	+	+	−
Glutamate	NT	+	+	+	−
Histidine	NT	+	+	+	−
Purines fermented	NT	+	+	+	−
Nitrate reduction	NT	+	+	+	−
Indole	NT	+	+	+	−
Gelatin	−	−	−	+	−
H_2S	+	+	+	+	+

[a] NT = not tested; + = positive; − = negative; w = negative or feebly positive. No other amino acids than those listed are fermented. Products from glycine are CO_2, NH_3 and acetate; from serine CO_2, NH_3, H_2 and acetate; from threonine CO_2, NH_3, H_2 and propionate; from glutamate CO_2, NH_3, H_2, acetate and butyrate.

Further Comments

Dark or black colony formation is not an unusual feature among distinctly different species of *Peptococcus*, as well as *Peptostreptococcus*. The black colonies appear as variants among the great non-pigmented population and often revert; there is little doubt that they are mutants. Thus, Hall's original isolate, on which the type species concept is based, unfortunately was most probably such a mutant strain of vague origin.

2. **Peptococcus asaccharolyticus** (Distaso) Douglas 1957, 476. (*Staphylococcus asaccharolyticus* Distaso 1912, 445; *Micrococcus asaccharolyticus* (Distaso) Hall 1948, 246.)

a.sac.cha.ro.ly′ti.cus. Gr. pref. *a* not; Gr. n. *sacchar* sugar; Gr. adj. *lyticus* able to loose; M.L. adj. *asaccharolyticus* not digesting sugar.

Spheres of varying diameter, ranging from 0.6–1.5 μm within a single culture or in different strains, and occurring singly, in pairs, tetrads, irregular masses and short chains.

Colonies are generally circular, 0.5–2 mm in diameter, smooth, entire, low convex, opaque or translucent, grayish white, butyrous. Deep agar colonies tend to be lenticular. In usual infusion agar media colonies may be pinpoint and thin and gas bubbles may be observed; in peptone (2%) and yeast extract (1%) reduced media supplemented with 0.2–0.5% fermentable amino acids (Table 14.10) growth and gas production (CO_2 and H_2) are abundant and broth media develop an even or granular turbidity. Growth is not enhanced by glucose, blood or serum.

Egg albumin, serum and casein are not digested.

Litmus milk is unchanged or reduced. Fatty acids do not stimulate carbohydrate fermentation. Gas is produced from pyruvate, citrate and tartrate.

Amino acids fermented by cell suspensions and the products formed are shown in Table 14.10. An initial deamination of these amino acids very probably occurs because pyruvate and α-ketobutyrate yield, except for ammonia, the same quantities of products as from serine and threonine, respectively. Also, mixtures of serine and pyruvate, or of threonine and α-ketobutyrate are decomposed at the same rates as the keto acids alone (Whiteley, 1957).

Isolated from human large intestine, buccal cavity, pleura, uterus, vagina.

The G + C content of the DNA of one authentic strain is 36 moles % (buoyant density).

Further Comments

Group II of Hare *et al.* (1952) and Hare (1967) is very probably *P. asaccharolyticus*.

3. **Peptococcus aerogenes** (Schottmüller) Douglas 1957, 476. (*Staphylococcus aerogenes* Schottmüller 1912, 270; *Micrococcus aerogenes* (Schottmüller) Bergey *et al.* 1923, 70; Not *Micrococcus aerogenes* Miller 1886, 119.)

a.e.ro'ge.nes. Gr. n. *aër* air; Gr. v. *gennaio* produce; M.L. part.adj. *aerogenes* gas-producing.

Microscopic appearance, general cultural characteristics, amino acid metabolism and G + C content (36 moles %) of the DNA are indistinguishable from *P. asaccharolyticus*. However, *P. aerogenes* may differ in fermenting certain carbohydrates and in not producing gas from citrate or tartrate.

Only small amounts of glucose and fructose may be fermented in media without fatty acid supplementation but addition of sodium oleate or certain other fatty acids stimulates active fermentation of glucose and fructose with acid and gas production (Wildy and Hare, 1953; Hare, 1967). Variable fermentation of sucrose, maltose and galactose.

Isolated from cases of puerperal fever, female genital tract, tonsils, and the nose.

The G + C content of the DNA of two strains was 35.7 and 36.7 moles % (buoyant density) (Rogosa, 1969).

Further Comments

Group VIII (Hare, 1967) appears to be *P. aerogenes*.

4. **Peptococcus activus** (Prévot and Taffanel) Douglas 1957, 475. (*Staphylococcus activus* Prévot and Taffanel 1945, 152; group III Hare *et al.* 1952.)

ac'ti.vus. L. adj. *activus* active.

Spheres, 0.8–1.0 μm in diameter. Colonies may attain 4 mm in 4 days; smooth to rough surface, entire, convex, opaque, grayish white, slightly yellow, cream or light brown, butyrous; deep colonies lenticular.

In peptone yeast extract or infusion broth, growth is highly turbid with abundant gas within 48–72 hr.

Acid and gas are produced from glucose, fructose, mannose, maltose, sucrose and generally from galactose; lactose, L-arabinose, or D-xylose may be variably and generally feebly fermented; no fermentation of raffinose, starch, inulin, salicin, glycerol or mannitol. Fatty acids, particularly sodium oleate, stimulate carbohydrate attack.

Gas produced from pyruvate but not from citrate or tartrate.

Coagulated serum and egg albumin, fibrin, brain and casein are at least partially digested with variable blackening.

Litmus milk is generally reduced and some strains partially coagulate whole milk with some appearance of digestion.

Nitrate reduced; nitrite may also be reduced.

Isolated from puerperal septicemia and the female genital tract.

5. **Peptococcus constellatus** (Prévot) Douglas 1957, 477. (*Diplococcus constellatus*, Prévot 1924, 426.)

con.stel.la'tus. L. adj. *constellatus* studded with stars.

Spheres, 0.5–0.6 μm in diameter, occurring in pairs and tetrads, rarely in very short chains, never in clusters (Prévot, 1924); spheres 0.8–1.0 μm in diameter arranged in masses (Hare, 1967).

Deep agar lenticular colonies are biconvex, thick, opaque, yellowish, at first very small, attaining a maximum size of 0.5–1.5 mm in 5 days. Colonies are surrounded by many very small satellite colonies visible microscopically (Prévot, 1924); no satellite colonies (Hare, 1967).

Growth poor in media without fermentable carbohydrate; good growth in glucose infusion media. Gas is not produced.

Added fatty acids in complex media not required for stimulation of carbohydrate fermentation. Acid without gas from glucose, fructose, galactose, maltose, sucrose and mannose; lactose fermentation negative or variable; cellobiose, dextrin, coniferin and esculin not fermented. Original report (Prévot, 1924; Prévot and Fredette, 1966; THE MANUAL, 7th edition) of a positive arabinose fermentation and a negative trehalose fermentation are in error; modern techniques show arabinose is not fermented and trehalose is fermented.

Products from glucose in complex media are mainly lactate and some acetate.

No gas from pyruvate, citrate or tartrate.

Litmus milk unchanged; coagulated proteins not attacked; gelatin not liquefied. Indole not formed.

Isolated from tonsils, purulent pleurisy, appendix, the nose and throat, gums and infrequently from skin and the vagina.

Further Comments

Prévot's (1924) observation of satellite colonies has never been verified and must have been due to some very special strain or cultural and medium characteristics. Group VI b (Hare, 1967) is *P. constellatus* in all other major characteristics. The nitrogen metabolism of *P. constellatus* has never been studied and is therefore not included in Table 14.10. More sophisticated studies than previously are required to resolve its present doubtful taxonomic allocation.

6. **Peptococcus anaerobius** (Hamm) Douglas 1957, 479. (*Diplococcus magnus anaerobius* Tissier and Martelly 1902, 885; *Staphylococcus anaerobius*

Hamm 1912, 79; not *Staphylococcus anaerobius* Heurlin 1910, 130; *Diplococcus magnus* Prévot 1933, 140; *Diplococcus glycinophilus* Cardon and Barker 1946, 634; *Micrococcus variabilis* Foubert and Douglas 1948, 32; *Micrococcus anaerobius* (Hamm) Hall 1948, 247; *Peptostreptococcus magnus* (Prévot) Smith 1957, 539; *Peptococcus glycinophilus* (Cardon and Barker) Douglas 1957, 478; *Peptococcus variabilis* (Foubert and Douglas) Douglas 1957, 479.) French vernacular name: Staphylocoque anaérobie, Jungano 1907, 707.

an.a.e.ro'bi.us. Gr. pref. *an* not; Gr. n. *aër* air; Gr. n. *bius* life; M.L. adj. *anaerobius* not living in air.

Spheres, 0.6–1.0 μm mean diameter, occurring singly, in pairs, tetrads and masses. Agar colonies are 0.5–3.0 mm in diameter, circular, smooth to rough surface, entire, low convex, opaque, grayish to light brown, opalescent, butyrous. Deep agar colonies are lenticular. In peptone yeast extract broth moderate growth is turbid or granular without visible gas production. Growth is not enhanced by glucose, blood, or serum.

Action on carbohydrates is negative, feeble or questionable.

Gelatin is not liquefied, and egg albumin, beef serum and casein are not obviously attacked.

Reported catalase positive but, if present, this reaction is weak; nitrate not reduced; indole is not produced; hydrogen sulfide is produced. Litmus milk is unchanged or reduced.

Cell suspensions and growing cultures decompose glycine to CO_2, NH_3 and acetic acid. Other amino acids tested (23 compounds) are not fermented (Whiteley, 1957) (Table 14.10).

Isolated from appendices, cystitis, female genital tract, draining sinus, and tidal bay mud.

Further Comments

P. glycinophilus, P. variabilis and *P. anaerobius* (THE MANUAL, 7th edition) are here considered as synonyms of *P. anaerobius*. At best, they are probably varieties or subspecies of *P. anaerobius*. All three are generally non-saccharoclastic, ferment glycine but no other amino acid with the production of CO_2, NH_3 and acetic acid (Cardon and Barker, 1946; Whiteley, 1957; Douglas, 1951).

Reported differences in cell sizes, in nitrate reduction, H_2S production and gelatin liquefaction were of very few strains and the methods frequently give equivocal or variable results. However, named strains of *P. variabilis* liquefy gelatin. Uniform and smaller cell sizes are obtained only in media supplemented with certain fatty acids (Wildy and Hare, 1953).

Hare (1967) described 52 strains of Group IX which, like *P. anaerobius*, did not ferment carbohydrates. However, glycine was not fermented. Unfortunately, other amino acids were not tested as possible substrates. Thus, the taxonomic status of his Group IX, although probably different from *P. anaerobius*, remains uncertain.

By previous historical definitions, *P. anaerobius* and *P. asaccharolyticus* are identical in their negative or feeble action on carbohydrates but the former does not produce gas in usual media whereas *P. asaccharolyticus* does. Furthermore, their amino acid metabolism is very different (Table 14.10).

Editorial Note. West and Holdeman (1973) suggest that *Peptococcus anaerobius* is a *nomen confusum* and should be rejected.

Genus II. **Peptostreptococcus** *Kluyver and van Niel 1936, 401*

MORRISON ROGOSA

Pep.to.strep.to.coc'cus. Gr. v. *pepto* cook, digest; M.L. masc.n. *Streptococcus* a generic name; M.L. masc. n. *Peptostreptococcus* the digesting streptococcus.

Spherical to **ovoid** cells, usually 0.7–1.0 μm, but occasionally 0.3–0.5 μm in diameter, occurring in **pairs, short or long chains.** Non-flagellated and non-motile. No spores. **Gram-positive.**

Chemoorganotrophic. With one exception, **carbohydrates** are **fermented** with production **of acid, gas or both.** Gas production in carbohydrate-free peptone media by some species. **Pyruvate frequently fermented producing acid or gas or both;** malate, citrate, and tartrate **not fermented. Ammonia** is **produced** but an alkaline reaction may not result when acid is produced in carbohydrate medium. **Acidic products** may be any one or combinations of the following:

acetic, formic, propionic, butyric, isobutyric, valeric, isovaleric, isocaproic, caproic and **succinic.** Amines and various alcohols may also be produced. **Lactate not produced. Catalase negative.** Nitrates not reduced. Indole generally not produced. Gelatin generally not liquefied and coagulated proteins not attacked. Very rarely hemolytic.

Anaerobic.

Initial pH growth range about 6.0–8.0; optimum 7.0–7.5. Growth generally from 25–38 C; optimum about 35–37 C. Complex media required.

Isolated from the normal and pathological female genital tract and blood in puerperal fever, from respiratory and intestinal tracts in normal humans and animals, from the oral cavity, from pyogenic infections, septic war wounds and appendicitis. May be pathogenic. Generally sensitive to penicillins (10 unit/ml) and often <1 unit/ml.

Type species: *Peptostreptococcus anaerobius* (Krönig) Kluyver and van Niel 1936, 401.

Further Comments

The species listed in the 7th edition of THE MANUAL may be divided into three groups. 1) *P. anaerobius, P. foetidus, P. putridus, P. productus* and *P. lanceolatus;* 2) *P. micros, P. parvulus, P. intermedius* and *P. evolutus;* 3) *P. magnus, P. paleopneumonia, P. plagarumbelli* and *P. morbillorum.*

Three species of group 1 are accepted in this edition; *P. foetidus* and *P. putridus* are considered synonyms of *P. anaerobius.*

Only the first two species of group 2 are retained provisionally since they do not appear to depend mainly on carbohydrates for energy (even though they produce lactate unlike the generic description which applies strictly only to group 1). Whether they are different species is conjectural.

P. intermedius is probably a strain variant of *P. evolutus.* Since it is not always strictly anaerobic, grows moderately well in air after several passages, is primarily saccharoclastic with lactic acid as the major product, it should be classified with *Streptococcus* (see Prévot, 1924, 1925).

Group 3 does not form long chains, gas or fetid odors. *P. magnus* cannot be distinguished from *Peptococcus anaerobius* (see p. 521). *P. paleopneumoniae* may become aerotolerant, is bile soluble, produces lactate homofermentatively and otherwise resembles *Streptococcus pneumoniae* (Smith, 1936; Rist, 1902; Bolognesi, 1907). Strains of *P. plagarumbelli* and *P. morbillum* are not available, their descriptions are poor and based on now unacceptable methods; they are therefore considered as *Species incertae sedis.*

Organism LC (Elsden and Lewis, 1953; Elsden *et al.,* 1956), isolated from the rumen of sheep and cattle, was later named *Peptostreptococcus elsdenii* by Guitierrez *et al.* (1959). This name is inappropriate because this organism does not have the characteristics of *Peptostreptococcus* Kluyver and van Niel, 1936. Accordingly, Rogosa (1971) transferred *P. elsdenii* to a new genus, *Megasphaera*, which he included in a new family, *Veillonellaceae* (1971); these are described elsewhere in THE MANUAL.

Clearly, this is not a homogeneous genus and present and past descriptions reflect this. Isolations were frequently from sources containing a teeming mixed flora of other anaerobic and facultative organisms and where purification of cultures is tedious. With added information and restudy of extant historical strains by modern approaches and techniques, more strains can now be identified by the criteria presented here.

Description of the species of genus **Peptostreptococcus**

1. **Peptostreptococcus anaerobius** (Krönig) Kluyver and van Niel 1936, 401. (*Micrococcus foetidus* Flügge 1886, 172; Not *Micrococcus foetidus* Eisenberg 1891, 22; *Streptococcus anaerob (sic)* Krönig 1895, 409; *Streptococcus anaerobius* (Krönig) Natvig 1905, 724; *Streptococcus putridus* Schottmüller 1910, 450; *Streptococcus foetidus* (Veillon) Prévot 1933, 189; Not *Streptococcus foetidus* Migula 1900; 38; *Peptostreptococcus putridus* (Schottmüller) Smith 1957, 535; *Peptostreptococcus foetidus* (Veillon) Smith 1957, 535; German vernacular name: Stinkcoccus, Klamann 1887, 1346.)

an.a.e.ro.bi'us. Gr. pref. *an* not; Gr. n. *aër* air; Gr. n. *bius* life; M.L. adj. *anaerobius* not living in air, anaerobic.

Cocci, averaging 0.8 μm in diameter, occurring in chains.

Colonies are translucent, smooth pearly gray domes with an entire margin; mean diameter about 1.5–2.0 mm in 4 days. Deep agar colonies are lenticular, about 1–2 mm after 48–72 hr.

Growth is abundant in peptone, meat or liver infusion broths; gas produced in all media (large amount of CO_2 and some H_2). pH in peptone medium not decreased. Fetid odor in all media.

Coagulated proteins not attacked. Fresh fibrin and fresh organs appear partially disintegrated and blackened from H_2S produced; abundant gas produced and marked fetid odor. H_2S also produced in media without fresh tissues provided thioglycollate (0.1%), or cystine (0.1%), or sodium thiosulfate (1%), or flowers of sulfur (1%) are present.

Gelatin not liquefied. Milk unchanged. Indole not produced.

Weak fermentation of fructose, glucose and maltose only (Table 14.11); Prévot and Fredette (1966) report acid and additional gas from glucose and fructose; Smith (1957) fermentation of fructose, glucose, galactose, maltose, sucrose and sometimes arabinose and mannitol. Carbohydrate fermentation stimulated by the sulfur compounds named above but not by fatty acids.

Fermentation products in complex media include acetic and formic acids, ethanol and amines. Lactate is not produced or is only a very minor product.

May be pathogenic.

Isolated in cases of putrefactive gangrene, war wounds, osteomyelitis, normal and pathological

TABLE 14.11

Characteristics of the species of genus **Peptostreptococcus**[a]

	1. *P. anaerobius*	2. *P. productus*	3. *P. lanceolatus*	4. *P. micros*	5. *P. parvulus*
Gas production from complex media and from pyruvate	+	+	+	−	−
Fetid odor	+	+	+	−	−
Propionate from threonine	+	−	−	−	−
Major acids produced					
Lactate	−	− or w	−	+	+
Acetate	+	+	+	+	+
C_3–C_6	+	+	+	−	−
Carbohydrates fermented					
Arabinose	−	w	−	−	d
Fructose	w	+	−	−	−
Galactose	−	+	−	−	−
Glucose	w	+	+	−	+
Lactose	−	+	−	−	+
Maltose	w	+	−	−	−
Sucrose	−	+	+	−	−
Xylose	−	+	−	−	−
Litmus milk acid and coagulated	−	+	−	−	+

[a] + = 90% or more strains positive; − = 90% or more strains negative; d = variable; w = weak positive or trace. All species produce chains of cells. None ferments adonitol, cellulose, erythritol or glycogen. None produces lecithinase, lipase or acetylmethylcarbinol.

genital secretions, tissues and blood in puerperal fever, appendicitis, the intestinal tract, pleurisy, sinusitis and the mouth.

2. Peptostreptococcus productus (Prévot) Smith 1957, 536. (*Streptococcus productus* Prévot 1941, 105.)

pro.duc′tus. L. adj. *productus* produced.

Spheres, 0.7–1.2 μm in diameter, occurring in chains of about 6–20 cells.

In addition to the carbohydrates listed in Table 14.11, the following are fermented with acid and gas: amygdalin, cellobiose, mannose, melezitose, melibiose, pectin, raffinose, ribose, sorbitol, trehalose; variable or weak fermentation occur with arabinose, dextrin, dulcitol, esculin (hydrolyzed), glycerol, inositol, inulin, mannitol, salicin, sorbose and starch. Major products are acetate and succinate with some lactate and traces of formate. General characteristics in Table 14.11.

Isolated from a subacute case of pulmonary gangrene, intestinal tract, brain and pelvic abscesses, urine, blood.

3. Peptostreptococcus lanceolatus (Prévot) Smith 1957, 537. (*Coccus lanceolatus anaerobius* Tissier 1926, 447; *Streptococcus lanceolatus* (Tissier) Prévot 1933, 193; not *Streptococcus lanceolatus* Gamaleia 1888, 442.)

lan.ce.o.la′tus. L. adj. *lanceolatus* lancet-shaped.

Described as having large ovoid cells, 1.2–1.4 μm in diameter, with pointed ends, occurring in short chains in culture and in pairs in exudates.

Unlike *P. anaerobius* does not produce H_2. General characteristics in Table 14.11.

Has been isolated from putrid diarrhea, dental infection, vulvovaginitis, arthritic and other abscesses.

4. Peptostreptococcus micros (Prévot) Smith 1957, 537. (*Streptococcus anaerobius micros* Lewkowicz 1901, 645; *Streptococcus micros* Prévot 1933, 195.)

mi′cros. Gr. adj. *mikros* small.

Small spheres, 0.3–0.5 μm in diameter, occurring in long chains or in pairs; dimensions may become somewhat larger on transfers.

Described by Prévot (1933) as fermenting glucose, fructose, galactose, sucrose and maltose with the production of acid consisting of propionic, formic and lactic acids. No gas is formed. Studies of 22 strains by Moore and Holdeman (personal communication), including 19 original Prévot strains, do not confirm Prévot; their strains do not ferment any carbohydrates or polyols and propionic acid is not formed but rather lactic acid is the major product (50–90%) with acetate and succinate as significant products and formate present in lesser amounts. General characteristics in Table 14.11.

Isolated from purulent pleurisy, lochia and

uterus in puerperal sepsis, appendicitis, brain and dental abscess, actinomycosis.

5. Peptostreptococcus parvulus (Weinberg *et al.*) Smith 1957, 538. (*Streptococcus parvulus non liquefaciens* Repaci 1910, 293; *Streptococcus parvulus* Weinberg *et al.* 1937, 1011; Not *Streptococcus parvulus* Levinthal 1928, 200.)

par'vu.lus. L. dim.adj. *parvulus* somewhat small.

Small spheres, 0.3–0.4 μm in diameter, occurring in short chains or occasional pairs. Colonies may become black. Feeble and variable fermentation of carbohydrates (Table 14.11) may be due to poor growth. Fermentation products are similar to those of *P. micros* and these species may be strain variants of each other.

Isolated once from the respiratory tract and once from the mouth.

Genus III. Ruminococcus Sijpesteijn 1948, 152

MORRISON ROGOSA

Ru.min.o.coc'cus. L. adj. *ruminalis* of the rumen; Gr. n. *coccus* a grain, berry; M.L. masc.n. *Ruminococcus* coccus of the rumen.

Spherical to **elongated coccoid** cells. When elongated, ends of cells rounded or flat rather than pointed or lancet shaped. Sides of cells may be flattened when in contact with other cells, particularly with diplococci which resemble neisseria and pediococci. Occur as single cells, pairs or short chains of 3–4 cells (*R. albus*) or chains of 8–50 or more cells (*R. flavefaciens*). Average diameter varies from 0.7–1.2 μm with a modal range of 0.8–1.0 μm.

Gram-positive anatomy but may decolorize relatively easily. Some cells of all strains in young culture contain an **iodophilic** reserve **polysaccharide** (a glucose polymer) which may rapidly disappear in the stationary phase of growth. There are no endospores and the cells are non-motile.

Roll tube surface colonies after 3 days in rumen fluid glucose cellobiose agar are entire, smooth and slightly convex. In transmitted light, they are translucent to opaque but occasionally have a fluorescent to "frosted glass" appearance. Colonies may be white, light tan or yellow to orange and are generally 2–4 mm in diameter. Deep colonies are lenticular.

Cellulose is **usually digested** and reducing sugar is produced in media containing excess cellulose. **Cellobiose** is always **fermented**. Xylan is fermented by about 85% of strains. A given strain may not ferment any other sugars or may attack one or more of the following: glucose, D-xylose, esculin, fructose, sucrose, lactose and L-arabinose. A glucose permease is lacking where whole cells do not ferment glucose; a glucokinase is present in cell extracts (Ayers, 1958). No acid is produced from a wide variety of other carbohydrates and polyols.

In media **containing bicarbonate** and **gassed** with **CO₂, cellobiose is fermented; major products are acetic** and **formic** acids; **ethanol** and/or **succinic** acid may also be major products; but **when ethanol** is a **major product**, there is **more** hydrogen and little or **no succinic acid** (*R. albus*); and **when succinate** is a **major product** there is **little hydrogen** and **little** or **no ethanol** (*R. flavefaciens*). Lactic acid is usually a minor product, if produced; methane and fatty acids containing three or more carbon atoms are not formed. Succinate production involves CO_2 fixation. Organisms intermediate between *R. albus* and *R. flavefaciens* in colony color, morphology and fermentation products have been isolated (Hungate, 1957).

Chemoorganotrophic and **nutrient requirements** may be **complex** (Ayers, 1958). Rumen fluid or fecal extracts may be required for growth in usual media (Hungate, 1957); however, more than half of strains can be grown in defined media containing B vitamins, cellobiose, minerals, ammonia, sulfide, carbon dioxide, acetate and one or more of the branch-chained acids, isovalerate, isobutyrate and 2-methyl-butyrate. **Ammonia is** essential as the **main nitrogen source** for all strains adequately studied and exogenous amino acid carbon and nitrogen are very poorly utilized although a few strains require methionine or vitamin B₁₂ (Allison *et al.*, 1962; Bryant and Robinson, 1961; Dehority *et al.*, 1967).

Growth in rumen fluid cellobiose broth is evenly turbid within 1 day. There is good growth from 30–37 C and no growth at 15 or 50 C. Occasional strains grow at 22 or 45 C.

Catalase, indole and **H₂S** are **not produced.** Starch is not hydrolyzed and nitrate is not reduced. Ammonia is not produced from amino acids or peptides. Voges-Proskauer reaction is variable and a small minority of strains liquefy gelatin.

Strictly anaerobic, requiring technique of Hungate (1950) or similar stringent anaerobic method for successful isolation. An initial pH range of 6.0–7.0 favors growth; a pH of about 6.5 is optimal.

There is a multiplicity of serological types (Jarvis, 1967).

Isolated from bovine and ovine rumen and from ceca of rabbits and guinea pigs; probably widely distributed in the rumen of *Ruminantia* and in the cecum and colon of herbivores and thought to play an important role in the fermentation of cellulose in these organs.

The G + C content of the DNA ranges from 39.8–45.4 moles % (T_m).

Type species: *Ruminococcus flavefaciens* Sijpesteijn 1948, 152.

Further Comments

A few non-cellulolytic strains which otherwise resemble *R. albus* (Bryant and Robinson, 1961) have also been isolated. Pure cultures have lost their cellulolytic activity when carried in cellulose or cellobiose agar medium; some regain it when grown on liquid rumen fluid medium. However, other reported losses of activity when transfers were made in cellobiose medium, may be attributed, as Hungate (1957) has emphasized, to the study of impure cultures and the selection of a non-cellulolytic contaminant in the purification process.

Isolation of pure cultures from cellulose media is difficult, particularly when rumen fluid is necessarily used. Obviously, rumen fluid helps support a wide diversity of organisms in the habitat. Based on pertinent investigations, it may be said that the presence of methane, propionate, butyrate or volatile acids of greater carbon number should serve as a warning light for the investigator to pursue the purification of cultures with great diligence and patience.

Animal rations may significantly affect the rumen microbiota. Sijpesteijn (1948) conducted most of her investigations during war years and many of her samples were from animals in poor condition. She found *R. flavefaciens* but not *R. albus*. Kistner and Gouws (1964) found only *R.*

albus in the rumen of sheep fed a good ration, but *R. flavefaciens* was also isolated when a poor ration was fed.

A smaller coccus, *Ruminobacter parvum*, was observed by Sijpesteijn (1948, 1949) but was not isolated in pure culture. Henneberg (1922, 1926) also observed an iodophilic coccus, *Micrococcus ruminantium* syn. *Streptococcus jodophilus* in the microscopic examination of rumen contents. This organism was also not isolated. Further historical discussion of cocci in the rumen is found in Bryant (1959), Sijpesteijn (1948) and Hungate (1950).

Previous fermentation balance studies of *Ruminococcus* show incomplete utilization of cellulose or cellobiose and low recovery of products; CO_2 production or uptake being generally neglected (Sijpesteijn, 1951; Bryant *et al.*, 1958; Kistner and Gouws, 1964; Jarvis and Annison, 1967) and large polysaccharide synthesis (Hungate, 1963) not considered. Moreover, because little exogenous amino acid carbon or nitrogen is incorporated into cells, much of the carbon of protein and other cellular constituents is probably derived from the carbohydrate energy source (Allison *et al.*, 1962; Bryant and Robinson, 1963).

The question of CO_2 uptake deserves some mention. Hopgood and Walker (1967) found that with a glucose-fermenting strain of *R. flavefaciens* there was a net uptake of $^{14}CO_2$ with the formation of carboxyl-labeled succinate. Without CO_2 and bicarbonate the glucose fermentation was essentially homolactic with a yield of 84% lactic acid. Whether such shifts in product accumulation occur in fermentations of cellulose, cellobiose or xylan is unknown.

The position of *Ruminococcus* in supra-generic taxa is conjectural. Sijpesteijn (1948) suggested that the genus be included in *Streptococceae* although she recognized the attendant difficulties. In the past *Streptococceae* was included in *Lactobacillaceae*, but these rumen cocci are not lactic acid bacteria. Lactate is not a major product and is present only in very minor or trace amounts.

Description of the species of genus **Ruminococcus**

1. **Ruminococcus flavefaciens** Sijpesteijn 1948, 152.

fla.ve.fac′i.ens. L. adj. *flavus* yellow. L. pres.-part. *faciens* producing; M.L. part.adj. *flavefaciens* yellow producing.

Generally appear as spherical iodophilic streptococci whose chain length varies with cultural conditions and may consist of 8–50 or more cells. Average cell diameter is 0.8–0.9 μm. Some strains appear Gram-negative, others Gram-variable; electron microscopy of thin sections of cells shows gram positive anatomy. In a good rumen fluid-

cellobiose-agar medium colonies are 2–4 mm in diameter in 3 days and vary in pigmentation from typical yellow or orange to occasional white.

Final pH in cellobiose broth is 5.0–5.6. Gas splits are not observed in deep tubes of cellobiose agar but a small amount of hydrogen is produced (Kistner and Gouws, 1964). Xylan is usually fermented. Glucose and D-xylose are generally not fermented; L-arabinose, dextrin, esculin, fructose, glycerol, gum arabic, inulin, lactose, maltose, mannitol, salicin and sucrose are not fermented (Bryant *et al.*, 1958).

G + C content of the DNA is 39.8–43.5 moles %, based on T_m measurements on seven strains (Sharma, 1968).

Found in the rumen of cattle and sheep, cecum and colon of rabbits and probably other herbivores.

2. Ruminococcus albus Hungate 1957, 307.

al'bus. L. adj. *albus* white.

Coccoid cells, 0.8–2.0 μm in diameter, singly, in pairs or short chains, never in tetrads or cubical packets, often slightly elongated prior to division. Some strains appear Gram-negative, others Gram-variable. Often capsulated. Iodophilic. The original description stated that the species was usually not iodophilic (Hungate, 1957) but this was later corrected (Hungate, 1963; Bryant *et al.*, 1958).

The colonies are usually white but varying degrees of pigmentation occur. Surface colonies on cellobiose rumen fluid agar are circular and convex; deep colonies are usually lens shaped but may be more complex.

Final pH in cellobiose medium is 5.1–5.6; the majority of strains also ferment cellulose and xylan. Fructose, glucose, lactose and mannose are fermented by some strains; L-arabinose, dextrin, esculin, galactose, glycerol, inulin, lactate, maltose, mannitol, raffinose, sucrose, trehalose and xylose are not fermented.

In the fermentation of cellulose, CO_2, H_2, acetate and ethanol were increased and lactate was decreased as compared to products from cellobiose (Hungate, 1950).

The G + C content of the DNA ranges from 42.6–45.4 moles % (T_m).

Found in large numbers in the rumen of cattle and sheep.

Genus IV. Sarcina Goodsir 1842, 434

E. Canale-Parola

Sar.ci'na. L. n. *Sarcina* a package, bundle.

Nearly spherical cells, 1.8–3 μm in diameter, occurring in packets of eight or more. Some of the cells in cultures may be present singly, or as groups of fewer than eight cells. Generally the cells are flattened in the areas of contact with adjacent cells. Division occurs in three perpendicular planes. Non-motile. A report of spore formation by these organisms has been published (Knöll, 1965). Gram-positive.

Chemoorganotrophs: Strictly fermentative metabolism, carbohydrates are the fermentable substrates. The main products of glucose fermentation are CO_2, H_2, acetic acid as well as ethanol for one species and butyric acid for the other. Not pigmented. Catalase not present. The minimal growth requirements include numerous amino acids and few vitamins.

Strict anaerobes. Grow from pH values near 1 to 9.8.

The G + C content of the DNA ranges from 28–31 moles % (buoyant density).

Type species: *Sarcina ventriculi* Goodsir 1842, 437.

Further Comments

In addition to *Sarcina ventriculi* and *Sarcina maxima*, the following species were included in the genus *Sarcina* in the 7th edition of THE MANUAL: *S. methanica, S. barkeri, S. lutea, S. flava, S. aurantiaca, S. litoralis, S. hansenii* and *S. ureae*. In the present edition the methanogenic sarcinae are included in the genus *Methanosarcina*, the aerobic spore-forming sarcinae in the genus *Sporosarcina*, and the non-spore-forming aerobes in *Micrococcus* and *Halococcus*. Discussions of the classification of packet-forming cocci have been published (Canale-Parola *et al.*, 1967; Canale-Parola, 1970).

Description of the species of genus Sarcina

1. Sarcina ventriculi Goodsir 1842, 437. (*Zymosarcina ventriculi* (Goodsir) Smit 1930, 26.) ven.tri'cu.li. L. n. *ventriculus* the stomach; L. gen.n. *ventriculi* of the stomach.

Nearly spherical cells, 1.8–2.4 μm in diameter, occurring in packets of eight to several hundred or more. Large packets (e.g. consisting of about 60 or more cells) tend to have an irregular or distorted appearance. Frequently cells in these packets exhibit flattened shapes and are irregularly arranged. Generally a fibrous layer 150–200 nm thick, composed either totally or in great part of cellulose (or a cellulose-like material) is present on the outer surface of the cell wall. This layer is absent in some strains. Reported to form spherical spores (Knöll, 1965). Subsurface colonies in agar media are star shaped or irregularly cubical, and measure up to several mm in diameter. Surface colonies (anaerobic) are roundish, often with rugged edges.

Products of glucose fermentation (growing cells, μmoles/100 μmoles glucose): ethanol, 100;

acetic acid, 60; lactic acid, 10; CO_2, 190; H_2, 140; acetoin, trace amounts (strain EC-1, see Canale-Parola and Wolfe, 1960). D-Fructose, D-galactose, D-glucose, D-mannose, lactose, maltose and sucrose are fermented. L-Arabinose, D-ribose, D-xylose, cellobiose, raffinose, dextrin, starch, dulcitol, g. /cerol, mannitol and citric, gluconic, succinic acids are not fermented. Utilize the Embden-Meyerhof pathway in the fermentation of glucose (Milhaud et al., 1956). May possess both a yeast pyruvate decarboxylase and a coliform-type pyruvate phosphoroclastic system (Arbuthnott et al., 1960). Two strains studied required biotin, nicotinic acid, arginine, glutamic acid, histidine, isoleucine, leucine, methionine, phenylalanine, serine, tryptophan, tyrosine and valine. Both strains grew in a chemically defined medium containing the vitamins and amino acids listed above, as well as glucose and inorganic salts (Canale-Parola and Wolfe, 1960; Knöll and Horschak, 1964).

Temperature optimum: 30–37 C.

Isolated from soil, mud, contents of diseased human stomach, rabbit and guinea pig stomach contents and the surface of cereal seeds.

The G + C content of the DNA is 30.6 ± 1 moles % (buoyant density).

Phase contrast photomicrographs and electron micrographs of this species have been published (Canale-Parola and Wolfe, 1960; Canale-Parola et al., 1961; Holt and Canale-Parola, 1967).

Reference strain: AL; ATCC 19633.

2. **Sarcina maxima** Lindner 1888, 54. (*Zymosarcina maxima* (Lindner) Smit 1930, 26.)

max'i.ma. L. sup.adj. *maximus* greatest, largest.

Nearly spherical cells, 2–3 μm in diameter, occurring in packets of eight or more. The cells lack a cellulose outer layer. Reported to form oval spores (Knöll, 1965). Subsurface colonies in agar media are cuboid with protuberances or unevenly spherical. Surface colonies (anaerobic) are roundish, often with rugged edges.

Products of glucose fermentation (growing cells, μmoles/100 μmoles glucose): butyric acid, 77; acetic acid, 40; CO_2, 197; H_2, 223 (strain 11, see Kupfer and Canale-Parola, 1967). D-Fructose, D-galactose, D-glucose, lactose, maltose and sucrose are fermented. L-Arabinose, D-xylose and raffinose are fermented poorly or not at all. Dextrin, starch, dulcitol, glycerol and mannitol are not fermented. Utilize the Embden-Meyerhof pathway for the fermentation of glucose (Kupfer and Canale-Parola, 1968). Pyruvate is metabolized via a phosphoroclastic system similar to that of saccharolytic clostridia; a coliform-type pyruvate clastic system, with formate as an intermediate, may also be present (Kupfer and Canale-Parola, 1967). A strain studied required biotin, nicotinic acid, thiamine, alanine, arginine, aspartic and glutamic acids, histidine, isoleucine, leucine, methionine, phenylalanine, serine, threonine, tryptophan, tyrosine and valine (Knöll and Horschak, 1964).

Temperature optimum: 30–35 C.

Isolated from the hull or outer coat of cereal grains, such as wheat, oat, rice and rye. Fresh wheat bran has been used as a source. Also isolated from horse manure and soil.

The G + C content of the DNA is 28.6 ± 1 (buoyant density).

Phase contrast photomicrographs and electron micrographs of this species have been published (Smit, 1930; Holt and Canale-Parola, 1967).

Reference strain: 11 (Canale-Parola, Mandel and Kupfer, 1967).

PART 15

ENDOSPORE-FORMING RODS AND COCCI

FAMILY I. **BACILLACEAE** FISCHER 1895, 139

Ba.cil.la'ce.ae. M.L. n. *Bacillus* type genus of the family; *-aceae* ending to denote family; M.L. fem.pl.n. *Bacillaceae* the *Bacillus* family.

Cells rod-shaped, in one genus spherical. Mycelium not produced. Endospores formed. Endospores differ from vegetative cells in being more refractive and less susceptible to staining, in having greater resistance to heat and other destructive agents, and in containing dipicolinic acid (5–15% of dry weight). The spore contains a central cell (the core) which is enclosed by a cortex of peptidoglycan and an outer spore coat.

Majority Gram-positive. Motile by lateral or peritrichous flagella, or non-motile.

Aerobic, facultative or anaerobic.

Key to the genera of family **Bacillaceae**

I. Cells rod-shaped
  A. Aerobic or facultative, catalase usually produced.

  Genus I. *Bacillus*

  B. Microaerophilic, catalase not produced.

  Genus II. *Sporolactobacillus*, p. 550

  C. Anaerobic.
    1. Sulfate not reduced to sulfide.

    Genus III. *Clostridium*, p. 551

    2. Sulfate reduced to sulfide.

    Genus IV. *Desulfotomaculum*, p. 572

II. Cells spherical, in packets.

    Genus V. *Sporosarcina*, p. 573

Genus I. **Bacillus** *Cohn 1872, 174; nom.gen.cons.* Nomencl. Comm. Intern. Soc. Microbiol. 1937, 28; Opin. A. Jud. Comm. 1955, 39

T. GIBSON† AND RUTH E. GORDON

Ba.cil'lus. L. dim. n. *bacillus* a small rod; M.L. n. *Bacillus* a rodlet.

Cells rod-shaped, straight or nearly so, 0.3–2.2 by 1.2–7.0 μm. **Majority motile;** flagella typically lateral. **Heat-resistant endospores formed; not more than one in a sporangial cell.** Sporulation not repressed by exposure to air. Gram reaction: positive, or positive only in early stages of growth, or negative.

Chemoorganotrophs; metabolism strictly res-

piratory, strictly fermentative or both respiratory and fermentative, using various substrates. The terminal electron acceptor in respiratory metabolism is molecular oxygen, replaceable in some species by nitrate. Catalase formed by most species.

Strict aerobes or facultative anaerobes.

The G + C content of the DNA of those strains

† Deceased.

examined ranges from 32–62 moles % (T_m and buoyant density).

Type species: *Bacillus subtilis* (Ehrenberg) Cohn 1872, 174.

Further Comments

A feature of the genus *Bacillus*, in contrast to other bacterial genera, is the great diversity in the properties of the organisms it embraces. In the fermentation of glucose there are several variants: (1) *B. coagulans* produces a homolactic fermentation; (2) *B. subtilis*, *B. licheniformis* and *B. cereus* form 2,3-butanediol and glycerol as major products; (3) *B. polymyxa* forms 2,3-butanediol, ethanol and H_2 as the main products; (4) *B. macerans* produces chiefly ethanol, acetone and acetic and formic acids. *B. pycnoticus* Ruhland and Grohmann (in Grohmann 1924, 261) and some other briefly characterized species have been described as facultative chemolithotrophs which use H_2 as an energy source in the absence of organic compounds (Ruhland, 1924).

Nutritional requirements for vegetative growth of the chemoorganotrophs vary between a single source of carbon and energy with inorganic nitrogen but no growth factors at one extreme, to distinctly complex, as yet unidentified, needs at the other. One species, *B. fastidiosus*, grows with uric acid or allantoin as the sole source of both carbon and nitrogen. Many strains of *B. polymyxa* and *B. macerans* assimilate molecular nitrogen.

Vegetative cell walls differ widely in structure, composition and susceptibility to lytic enzymes in those species examined (chiefly *B. cereus*, *B. megaterium* and *B. polymyxa*). Flagella have been reported to be lateral except in *Bacillus glycinophilus* Rippel 1937, 42, *Bacillus pacificus* Delaporte 1967, 3071 and *Bacillus uniflagellatus* Mann 1968, 351 in which the insertion was described as polar or almost so.

The maximum temperature for vegetative growth ranges from about 25 to above 75 C; the minimum from about −5 to about 45 C. The minimum pH value for growth varies from 7.5–8 to about 2. Salt tolerance ranges from less than 2–25% NaCl.

Proposals have been made to simplify this genus by dividing it into several genera. The following genera, which have been introduced or formally proposed, comprise species usually included in the genus *Bacillus*:

1. *Urobacillus* Miquel 1889, 519. Type sp. (monotype): *Urobacillus pasteurii* Miquel 1889, 519.

2. *Aerobacillus* Donker 1926, 138. Type sp. (subs. des. Kluyver and van Niel 1936, 402): *Aerobacillus polymyxa* (Prazmowski) Donker 1926, 138.

3. *Zymobacillus* Kluyver and van Niel 1936, 402.

Type sp.: *Zymobacillus macerans* (Schardinger) Kluyver and van Niel 1936, 402.

4. *Denitrobacillus* Verhoeven 1952, 133. Type sp.: *Denitrobacillus licheniformis* (Weigmann) Verhoeven 1952, 134.

5. *Thermobacillus* Feirer 1927, 49. Type sp.: Not designated but ten new thermophilic species named.

Each of these arrangements has its problems. One, on which available information does not give a complete answer, is whether the genera could be circumscribed so that difficulties would not arise owing to intermediate strains.

Spore formation has proved to be a generally reliable diagnostic character of the genus *Bacillus*. It is controlled by a number of genes which occur in different parts of the chromosome and, under laboratory conditions, mutants may be obtained in which sporulation is diminished or does not occur. Nevertheless, asporogenous mutants of these organisms appear to have little capacity to survive in nature and their taxonomic significance is possibly slight. Sporulation by normal strains is dependent on culture conditions; some media may provide insufficient manganese.

Several species of *Bacillus* produce antibiotics of the polypeptide class; a single strain may form more than one. Antibiotics appear in a culture when the stage of sporulation is reached, thus suggesting that their essential function is in that process. Another sort of antimicrobial action by strains of several species is a lysis of the vegetative cells of other organisms. In several instances an enzymatic degradation of constituents of cell walls has been demonstrated.

Serological information on this genus is not extensive. The spore possesses antigens different from those of the vegetative cell, and in parts of the genus spore antigens have provided a separation of species corresponding with that based on other criteria. Vegetative H-antigens have been utilized for the recognition of serotypes of the insect-pathogenic *B. thuringiensis*.

Genetic recombination by transformation and transduction and DNA-DNA hybridization have been used for indicating relationships in one group of species, that of *B. subtilis*, *B. licheniformis* and similar organisms.

All species of *Bacillus* appear to have a mode of life based on accumulation of their spores in nature. If germination of a spore is initiated by contact with nutrient material, multiplication occurs and if nutrients become exhausted a fresh crop of spores may be formed. This pattern indicates that few species have distinctive habitats.

Taxonomy within the genus *Bacillus* has developed unevenly. The most completely characterized organisms are those that grow rapidly and abundantly on nutrient agar and those that have attracted attention because of their pathogenicity or industrial importance. Other species have been examined less intensively. There are indications of further distinguishable groups which have not yet been defined and named.

Since endospore formation is the dominant feature in the characterization of *Bacillus*, it is essential that there is a boundary separating this genus from other genera in which endospores are produced. The genus *Clostridium* (see p. 551) is distinguished from *Bacillus* by inability to grow on the surface of agar media in air or, if growth does occur under these conditions, it is slight and does not lead to sporulation. There is also little or no catalase activity. *Sporosarcina* (see p. 573) is sharply separated from *Bacillus* by the coccal form of its vegetative cells. *Sporolactobacillus* (see p. 550) has the physiological properties of the genus *Lactobacillus* including the inability to form catalase.

Endospore-forming Bacteria of Uncertain Taxonomic Position

Various exceptionally large organisms which produce endospores occur in the alimentary tract of animals. They have been described solely by microscopical features; their response to oxygen and the physiology and the finer structure of their spores are still unknown. These large organisms have been placed in the following genera.

Genus *Bacillus*. Collin (1913), Hollande (1934) and Delaporte (1964) have described and named 15 species in the intestine of tadpoles. The spore in most of the species is larger and much more elongate than it is in other species of *Bacillus*. In seven of the species two spores may be formed in a single nonseptate cell, and Delaporte (1963) has shown that in *Bacillus camptospora* Collin 1913, 60

and *Bacillus enterothrix* Collin 1913, 60 one of the two spores may move rapidly through much of the interior of the long vegetative cell. The microscopical evidence thus suggests that the spores of these organisms are markedly different from typical endospores.

The genus *Fusosporus* Delaporte 1964, 857, in which two spores occur per cell, was separated from *Bacillus* because the vegetative cell tapers to pointed ends.

The genera *Arthromitus* Leidy 1850, 227, *Anisomitus* Grassé 1925, 343, *Entomitus* Grassé 1924, 30 and *Coleomitus* Duboscq and Grassé 1930, 28 comprise organisms which grow as septate filaments. In the first two genera the filaments occur attached by the basal end to a solid surface. The endospore in most of these organisms is morphologically not unlike the spore of cultivable species of *Bacillus*. In the 7th edition of THE MANUAL *Arthromitus* and *Coleomitus* were placed in the order *Caryophanales*.

The genera *Bacillospira* Hollande 1933, 1830 and *Sporospirillum* Delaporte 1964, 257, identified in batrachian gut contents, are distinguished from *Bacillus* by a rigid spiral form. Most produce two endospores in the vegetative cell. *Bacillospira* possesses an axial filament and was regarded as a spirochete by Hollande (1933, 1934).

Two specially large spore formers which occur in the alimentary tract of mammals appear to differ significantly from *Bacillus* in morphology. They are *Oscillospira* (see p. 574) and *Metabacterium* Chatton and Pérard 1913, 1234 which forms up to eight spores in a nonseptate cell 5 by 10–25 μm in size.

Interrelationships among these large intestinal organisms remain to be elucidated, as do their relationships to other genera and problems concerning the properties of their endospores.

Description of the species of genus **Bacillus**

The species in the following account are divided arbitrarily into two groups.

Group I comprises 22 species which are widely accepted as distinct entities. Their precise outlines may not be entirely clear, but at least some information is available on the internal variation in each and also on organisms which are distinguishable yet not markedly different.

Group II, which begins on page 545, contains 26 species which so far have received less widespread recognition.

Characteristics for primary differentiation of species in Group I are presented in Table 15.1. This table serves as a key and may be used as a lead to the appropriate description in the text or in Tables 15.2–15.7.

Group I

1. **Bacillus subtilis** (Ehrenberg) Cohn 1872, 174; *Nom. cons.* Nomencl. Comm. Intern. Soc. Microbiol. 1937, 28; Opin. A. Jud. Comm. 1955, 39. (*Vibrio subtilis* Ehrenberg 1835, 279.)

sub'ti.lis. L. adj. *subtilis* slender.

See Tables 15.1 and 15.2.

Rods seldom in chains; stain uniformly. Flagella lateral. Endospores 0.8 by 1.5–1.8 μm; surface of free spore stains faintly. On germination, the spore coat breaks equatorially. After emergence of vegetative cell the lysis of spore coat is distinctly slow.

Colonies on agar media round or irregular; surface dull; become thick and opaque; may be

TABLE 15.1

Differential characteristics of the species of genus **Bacillus** *(Group I)*

The symbols used are: E, elliptical or cylindrical; S, spherical or nearly so; C, central; T, terminal or subterminal; CT, central to terminal, variation within or between strains; W, weakly positive.

	Spore			Products of action on glucose		
	Shape	Distends sporangium distinctly	Dominant position	Acid[a]	Gas[b]	Acetoin[c]
1. *B. subtilis*	E	−	C	+	−	+
2. *B. pumilus*	E	−	C	+	−	+
3. *B. licheniformis*	E	−	C	+	W or −	+
4. *B. cereus*	E	−	C	+	−	+
5. *B. anthracis*	E	−	C	+	−	+
6. *B. thuringiensis*	E	−	C	+	−	+
7. *B. megaterium*	E	−	C	+	−	−
8. *B. polymyxa*	E	+	CT	+	+	+
9. *B. macerans*	E	+	T	+	+	−
10. *B. circulans*	E	+	CT	+	−	−
11. *B. stearothermophilus*	E	+ or −	T	+	−	
12. *B. coagulans*	E	+ or −	CT	+	−	+ or −
13. *B. alvei*	E	+	CT	+	−	+
14. *B. firmus*	E	−	C	W	−	−
15. *B. laterosporus*	E	+	C	+	−	−
16. *B. brevis*	E	+	CT	+ or −	−	−
17. *B. sphaericus*	S	+	T	−	−	−
18. *B. pasteurii*[d]	S	+	T	−	−	−
19. *B. fastidiosus*[d]	E	−	C	−	−	
20. *B. larvae*[d]	E	+	CT	+	−	
21. *B. popilliae*[d]	E	+	C	+	−	
22. *B. lentimorbus*[d]	E	+	C	+	−	

[a] Medium 1 (grams per liter): $(NH_4)_2HPO_4$, 1.0; KCl, 0.2; $MgSO_4 \cdot 7H_2O$, 0.2; yeast extract, 0.2; agar, 15; bromcresol purple, 0.008; glucose (added after sterilization), 5; used as slants.

[b] Medium 2 (grams per liter): peptone, 5; yeast extract, 3; NaCl, 5; agar, 3; bromcresol purple, 0.008; glucose (added after sterilization), 10; pH 7.0; used for stab cultures; may show acid formation if growth fails on medium 1.

[c] Medium 3 (grams per liter): proteose peptone, 7; glucose, 5; NaCl, 5.

[d] Little or no growth in medium 1, 2 or 3; in media that support growth the results are as shown.

wrinkled and may become cream-colored or brown. Features of colonies vary greatly with composition of the medium. Active spreading occurs on agar with a moist surface. Cell material grown on agar does not disperse readily in liquids.

In 1% glucose nutrient agar stab surface growth becomes thick, often rugose and brown. A disc of reddish pigment may form below the growth. Deep growth starts but soon comes to a standstill. Weak acid formation (to bromcresol purple) occurs to the bottom and neutralization proceeds from the surface.

In broth dull, wrinkled, coherent pellicle; little or no turbidity.

Energy-yielding metabolism is predominantly respiratory, oxygen being the terminal electron acceptor. Anaerobically, in complex media containing glucose, growth and fermentation are weak; admission of oxygen permits abundant growth with the formation of 2,3-butanediol, acetoin and CO_2 as major products.

Pectin and polysaccharides of plant tissues are decomposed, and some strains produce a rot in live potato tubers.

Levan is formed extracellularly from sucrose and raffinose, the yield varying with the strain. The chief endocellular storage product is a glycogen-like carbohydrate.

Pigments, which in particular cases have been identified as pulcherrimin or melanins, may be produced in colonies or the adjacent medium. In many strains they are brown or red, in fewer orange or black. Occurrence of each pigment is dependent on composition of medium.

Nutrient gelatin (22 C) liquefied to 1 cm or more within 7 days.

TABLE 15.2
Differential characteristics of **Bacillus** *species 1–3*

Common characteristics: Gram-positive; motile; catalase positive; intracellular globules unstainable by fuchsin in cells grown on glucose agar not observed; acid produced from arabinose, xylose and mannitol in Medium 1 (Table 15.1) with glucose replaced; egg yolk reaction negative; no growth in 0.02% azide by strains growing at 55 C; growth in 7% NaCl and in Sabouraud dextrose broth and/or agar; variable growth in 0.001% lysozyme; alkali on citrate-salts agar; casein hydrolyzed; tyrosine not decomposed. Data are from Gordon *et al.* (1973).

	1. *B. subtilis*	2. *B. pumilus*	3. *B. licheniformis*
Rods			
Width, μm	0.7–0.8	0.6–0.7	0.6–0.8
Length, μm	2–3	2–3	1.5–3
Hydrolysis of			
Starch	+[a]	−	+
Hippurate	−	+	−
NO_3^- to NO_2^-	+	−	+
Growth in anaerobic agar	−	−	+
Temperature for growth, C			
Maximum	45–55	45–50	50–55
Minimum[b]	5–20	5–15	15

[a] + = positive for 90–100% of strains; − = negative for 90–100% of strains.

[b] Lowest temperature tested 3 C.

Litmus milk reduced; rapid digestion of casein without much clotting.

Arginine dihydrolase absent. Hemolysis variable. Lecithinase not produced.

Polypeptide antibiotics are produced; a single strain may form several. Enzymes are liberated which have a lytic action on live bacterial cells.

A minimal medium for vegetative growth has no vitamins and contains glucose, citrate and an ammonium salt as the sole sources of carbon and nitrogen.

Aerobic, excepting that glucose and, less effectively, nitrate permit a much restricted anaerobic growth in complex media. Growth active at pH 5.5–8.5; pH limits not recorded.

Endospores widespread; occur in many heat-treated materials. Vegetative growth, with participation in early stages of breakdown, in various materials of plant and animal origin. Growth in non-acid foods if oxygen available. Causative agent of slimy bread.

The G + C content of the DNA has been most frequently found to be 42–43 moles % (by analysis, T_m and buoyant density).

Neotype strain: Marburg strain; ATCC 6051; NCIB 3610; NCTC 3610.

Comments: The foregoing description applies to the neotype strain of *B. subtilis* and to other strains which appear to be essentially similar. Among freshly isolated organisms, derived more especially from severely heated materials (surgical dressings, canned foods, etc.), some may have properties that link them to *B. subtilis* yet they diverge from the typical form of that species in several of the following directions: cells larger (about 1 μm thick); surface of spore more stainable; diminished capacity to sporulate; colonies smooth, glossy, lacking pigments and easily dispersed in liquids; in some, colonies thin and translucent; turbidity without a pellicle in broth; little acetoin formed; no action on nitrate; more complex nutrient requirements. In laboratory work mutants showing most of these features have been obtained from typical strains of *B. subtilis*. This evidence, together with a gradation in the properties of fresh isolates, has led to the inclusion of the aberrant organisms in that species.

Subspecies most frequently recognized are:

a. *Bacillus subtilis* var. *aterrimus* Smith, Gordon and Clark 1946, 64 (*Bacillus aterrimus* Lehmann and Neumann 1896, 303), which produces a black pigment exclusively on carbohydrate media.

b. *Bacillus subtilis* var. *niger* Smith, Gordon and Clark 1946, 66 (*Bacillus niger* Migula 1900, 636), which forms a black pigment on media containing tyrosine. In neither is pigment formation fully stable.

Probable synonyms: Among the species of *Bacillus* that appear to lack a type culture, several as judged by description are indistinguishable from *B. subtilis*. Notable are *Bacillus mesentericus* Trevisan 1889, 19, *Bacillus vulgatus* Trevisan 1889, 19 and *Bacillus panis* Migula 1900, 576. Strains of *B. subtilis* that form a red pigment have sometimes been identified (possibly misidentified) as *Bacillus globigii* Migula 1900, 554. Original cultures of *Bacillus natto* Sawamura 1906, 109 were found to be identical with *B. subtilis* by Smith *et al.* (1946).

2. **Bacillus pumilus** Meyer and Gottheil in Gottheil 1901, 681.

pu′mi.lus. L. adj. *pumilus* little.

See Tables 15.1 and 15.2.

Differentiation from *B. subtilis* is the main problem in identification. Apart from the differences shown in Table 15.2, there are two significant distinguishing properties of *B. pumilus:* (1) the colony of most strains on nutrient agar is smooth and becomes slightly yellowish, and (2) there is a requirement for biotin with, in some strains, a further requirement for amino acids. The characters of this species approach those of certain variants of *B. subtilis* (above). Serological tests and the method of spore germination have not provided a reliable separation of the two species.

Spores ubiquitous; occur in soil more frequently than those of *B. subtilis*.

The G + C content of the DNA has been reported in the range 39–43 moles % (T_m and buoyant density).

Type strain: ATCC 7061; NCIB 9369; NCTC 10337.

3. Bacillus licheniformis (Weigmann) Chester 1901, 287. (*Clostridium licheniforme* Weigmann 1898, 822; *Denitrobacillus licheniformis* (Weigmann) Verhoeven 1952, 134.)

li.che.ni.for'mis. Gr. n. *lichen* lichen; L. n. *forma* shape; M.L. adj. *licheniformis* lichen-shaped.

See Tables 15.1 and 15.2.

Spore antigens are distinguished from those of *B. subtilis* in agglutination and precipitation tests.

On germination of spore the site of emergence of the vegetative cell varies from equatorial to polar; the spore coat is not split widely and does not undergo rapid lysis.

Colonies on agar become opaque with dull to rough surface; hair-like outgrowths common; usually attached strongly to agar; mounds and lobes consisting largely of slime often accumulate on colony, especially on glucose agar or glutamate-glycerol agar. Liquefaction of nutrient gelatin (22 C) slow, saucer-shaped in 7 days.

Glucose fermented anaerobically to various products among which 2,3-butanediol and glycerol are the most characteristic.

Glutamyl polypeptide formed as an extracellular amorphous slime. Levan produced extracellularly from sucrose and raffinose.

Red pigment (presumably pulcherrimin) formed by many strains on carbohydrate media containing sufficient iron. Aged cultures may become brown.

Arginine dihydrolase positive in most strains.

Penicillinase inducible.

Polypeptide antibiotics produced.

Freshly isolated strains grow with ammonia as the sole source of nitrogen in the absence of growth factors.

Anaerobic growth occurs in complex media containing glucose or nitrate. CO_2 is produced from glucose, N_2 and N_2O from nitrate, but in each case visual gas formation is erratic.

Spores occur in soil; may survive severe heat treatment. Vegetative growth in many foods, especially if held at 30–50 C.

The G + C content of the DNA of several strains has been reported in the range 43–47 moles % (T_m and buoyant density). Transformation and transduction have been utilized in genetic studies.

Neotype strain: ATCC 14580; NCIB 9375; NCTC 10341.

Probable synonyms: The following have not been differentiated from *B. licheniformis* by any property: *Semiclostridium commune* Maassen 1907, 6; *Bacillus tinakiensis* Kurochkin 1958, 224.

Strains that produce a red pigment have sometimes been identified as *Bacillus globigii* Migula 1900, 554.

4. Bacillus cereus Frankland and Frankland 1887, 279.

ce're.us. L. adj. *cereus* waxen, wax-colored.

See Tables 15.1 and 15.3.

Cells in early stages of growth on glucose agar contain globules of lipid, largely poly-β-hydroxybutyrate. Volutin cell-inclusions also occur. The spore when liberated from the sporangium is encased in a loose fitting exosporium. On germination the spore coat undergoes rapid lysis while the vegetative cell is emerging.

The bacilli tend to occur in chains; the stability of the chains determines the form of the colony, which varies greatly in different strains. At one extreme the colony has a dull or frosted glass appearance and an undulate margin from which outgrowths do not develop. At the other extreme the colony forms root-like outgrowths which spread widely over the surface of agar. The outgrowths are irregularly tangled or they curve predominantly clockwise or counterclockwise in different strains.

Products of fermentation of glucose include 2,3-butanediol, acetoin, glycerol, lactic, succinic, formic and acetic acids and CO_2, with variations determined by the conditions.

Extracellular products include hemolysin, soluble toxin lethal for mice, enzymes lytic for bacterial cells, proteolytic enzymes and phospholipase C. Organism has been incriminated in food poisoning when extensive multiplication had occurred in foods.

The red pigment pulcherrimin is produced by some strains in starch media containing sufficient iron. Some strains produce a yellow-green fluorescent pigment in various media. On nutrient agar some strains darken the medium slightly, and some produce a pinkish brown diffusible pigment.

Has an absolute requirement for one or several amino acids; strains differ in those needed. Vitamins are not required.

Anaerobic growth occurs in complex media; it is promoted by glucose or nitrate.

Not susceptible to γ-phage. Does not react with anthrax fluorescent antibody conjugate (Weaver *et al.*, 1970).

Spores widespread. Multiplication has been observed chiefly in foods.

The G + C content of the DNA is reported to

be 32–33 moles % (T_m) and 33–37 moles % (analysis) for several strains.

Neotype strain: ATCC 14579; NCIB 9373; NCTC 2599.

Subspecies that have been designated include:

a. *Bacillus cereus* var. *fluorescens* Laubach 1916, 508. Distinctive property: produces yellow-green fluorescent pigment. Original strain: ATCC 13824; NCIB 2600; NCTC 2600.

b. *Bacillus cereus* var. *albolactis* (Migula) de Soriano 1935, 544 (*Bacillus albolactis* Migula 1900, V). Distinctive property: acid formed from lactose.

c. *Bacillus cereus* var. *mycoides* (Flügge) Smith, Gordon and Clark 1946, 54 (*Bacillus mycoides* Flügge 1886, 324). Distinctive properties: colony rhizoid; non-motile; acetoin formation variable; few strains grow at 37 C; some strains give rise to variants which form compact colonies but retain other properties. The G + C content of the DNA has been given as 39 moles % (T_m) for one strain.

Synonyms: possibly there are many which cannot be certainly identified by the descriptions or by preserved cultures. Of 15 accepted by Smith *et al.* (1952) only one, *B. mycoides* (see above), has appeared frequently in the literature. Prior to about 1940 organisms which have the properties of the neotype strain of *B. cereus* were widely regarded as typical representatives of the species *Bacillus subtilis* (Ehrenberg) Cohn. There is fairly general agreement that *Bacillus anthracoides* Trevisan 1889, 20 and *Bacillus pseudanthracis* Wahrlich 1890–1891, 26 are synonyms of *Bacillus cereus*.

5. Bacillus anthracis Cohn 1872, 177. (*Bacillus cereus* var. *anthracis* Smith, Gordon and Clark 1946, 55; *Bacteridium anthracis* (Cohn) Hauduroy *et al.* 1953, 85.)

an'thra.cis. Gr. n. *anthrax* charcoal, a red precious stone, a carbuncle; M.L. n. *anthrax* the disease anthrax; M.L. gen.n. *anthracis* of anthrax.

This species is similar to *B. cereus* but differs in the characteristics shown in Table 15.3 and the following.

Thiamine and a greater number of amino acids are required in a minimal medium.

Growth fails on agar + 10 units of penicillin G per ml. On agar + 0.05 or 0.5 unit of penicillin G per ml, cells in young cultures become large and spherical; such cells occur in chains which, as seen on the surface of the agar, are likened to strings of pearls.

No hemolysis of sheep cells in 24 hrs; spreads little.

TABLE 15.3
Differential characteristics of **Bacillus** *species 4–7*

Common characteristics: Gram-positive, catalase positive, cells grown on glucose agar contain intracellular globules unstainable by fuchsin, grow in 7% NaCl and in Sabouraud dextrose broth and/or agar, hydrolyze starch and casein.

	4. B. cereus	5. B. anthracis	6. B. thuringiensis	7. B. megaterium
Rods				
Width, μm	1.0–1.2	1.0–1.2	1.0–1.2	1.2–1.5
Length, μm	3–5	3–5	3–5	2–5
Intracellular protein crystals	−[a]	−	+	−
Motility	d	−	d	d
Acid from[b] arabinose, xylose and				
mannitol	−	−	−	d
NO_3^- to NO_2^-	+	+	+	d
Egg yolk reaction	+	+	+	−
Growth in				
Anaerobic agar	+	+	+	−
0.001% lysozyme	+	+	+	−
Alkali on citrate-salts agar	+	d	+	+
Temperature for growth, C				
Maximum	35–45	40	40–45	35–45
Minimum[c]	10–20	15–20	10–15	3–20
Tyrosine decomposed	+	d	d	d

[a] + = positive for 90–100% of strains; − = negative for 90–100% of strains; d = reactions differ, positive for 11–89% of strains.

[b] Medium 1 (Table 15.1) with glucose replaced.

[c] Lowest temperature tested 3 C.

Egg yolk reaction on agar weak; seldom proceeds beyond edge of colony.

Susceptible to γ-phage. Reacts with anthrax fluorescent antibody conjugate (Weaver et al., 1970).

Virulent strains of B. anthracis form capsules of glutamyl polypeptide during in vivo multiplication; they also form capsules on agar plus bicarbonate under CO_2, and their colonies are then mucoid.

The causative agent of the disease anthrax in man and animals. Spores persist for long periods in contaminated materials.

Neotype strain: ATCC 14578; NCIB 9388; NCTC 10340.

6. **Bacillus thuringiensis** Berliner 1915, 39. (Bacillus cereus var. thuringiensis Smith, Gordon and Clark 1952, 67.)

thur.in.gi.en'sis. M.L. gen.n. thuringiensis of Thuringia; named for Thuringia, a German province.

See Tables 15.1 and 15.3.

This species is distinguished from B. cereus by pathogenicity for larvae of Lepidoptera and by the production of a crystalline protein body, rarely two or three, in the cell during the phase of spore formation. This body stains like other cell material; it is formed outside the exosporium and it separates readily from the liberated spore. In the larval gut toxin is released from the crystal by enzymatic action. The capacity to form crystals may be lost by laboratory cultures.

B. thuringiensis has been divided on the basis of H-antigens into 11 serotypes, in one of which there are two subtypes. Proposals have been made to give the serotypes the rank of variety, and several have been divided into two "varieties" on the grounds of biochemical properties or pathogenicity. The following table gives a summary.

Serotype	Subspecific (varietal) epithets
1	thuringiensis Heimpel and Angus 1958, 538 (berliner de Barjac and Bonnefoi 1967, 1812); amuscatoxicus Heimpel 1967, 290.
2ᵃ	finitimus de Barjac and Bonnefoi 1967, 1812.
3	alesti Heimpel and Angus 1958, 538.
4a, 4b	sotto Heimpel and Angus 1958, 538; dendrolimus Krieg 1961, 233.
4a, 4c	kenyae de Barjac and Bonnefoi 1967, 1812.
5	galleriae Heimpel 1967, 291.
6	entomocidus Heimpel 1967, 292; subtoxicus Heimpel 1967, 292.
7	aizawai Heimpel 1967, 290; pacificus Heimpel 1967, 290.

Serotype	Subspecific (varietal) epithets
8	anagastae Heimpel 1967, 291 (morrisoni de Barjac and Bonnefoi 1968, 343).
9	tolworthi de Barjac and Bonnefoi 1968, 343.
10	darmstadiensis Krieg, de Barjac and Bonnefoi 1968, 430.
11	toumanoffii Krieg 1969, 281.

ᵃ See note on serotype 2.

Separation of the several molecular forms of esterases extractable from vegetative cells has demonstrated groups of strains corresponding closely with the serological groups (Norris, 1964). A discrepancy is that serotypes 5 and 7 have shown the same esterase pattern.

Among strains of B. thuringiensis there are variations in physiological properties such as in the production of acetoin, lecithinase, pulcherrimin, proteolytic enzymes and urease and in acid formation from sucrose and salicin. These properties have been utilized for the recognition of strains; their association with particular serotypes has diminished as the number of isolates increased (de Barjac and Bonnefoi, 1968; Krieg, 1968). All the above-mentioned physiological variations occur in B. cereus. Some strains of that species share spore and flagellar antigens with strains of B. thuringiensis but information on these links is fragmentary.

Serotype 2 was originally given the status of an independent species, Bacillus finitimus Heimpel and Angus 1958, 539. Its source was dead larvae of Malacosoma disstria. Subsequently it has frequently been treated as a serotype, or variety, of B. thuringiensis, although it does not conform to the accepted circumscription of that species. In particular the intracellular inclusion body of B. finitimus is formed between the spore and the exosporium, remains strongly attached to the liberated spore and it has shown little toxicity for insect larvae. Compare B. medusa (Species No. 24).

7. **Bacillus megaterium** de Bary 1884, 499.

me.ga.te'ri.um. Gr. adj. mega large; Gr. n. teras, teratis monster, beast; M.L. n. megaterium big beast.

See Tables 15.1 and 15.3. The notes immediately following apply to strains which may be regarded as typical of B. megaterium. Other strains which have been allocated to this species possess certain divergent properties; they will be discussed under "Possible Synonyms."

On nutrient agar, cells cylindrical to oval or pear-shaped; diameter mainly about 1.5 μm; tend to occur in short twisted chains. In carbohydrate media cells larger; diameter may reach 3.0 μm or more in some strains.

Lipid, largely poly-β-hydroxybutyrate, appears as unstained intracellular globules in weakly stained organisms, especially in those grown on glucose media.

Motility occurs in most strains; it is slow and requires free aeration.

Spores vary from short-oval to elongate. In some strains the spore coat is stainable with fuchsin.

On nutrient agar growth heaped and non-spreading, glossy or moderately dull, sometimes slightly rugose; on aging, usually some shade of yellow; on long incubation, growth and medium may become brown or black.

Growth on glucose agar mucoid to various degrees.

Acid usually formed from arabinose, xylose and mannitol.

Active liquefaction of nutrient gelatin (22 C). Casein actively digested. Phenylalanine deaminated. Nitrate assimilated but no accumulation of nitrite in the medium.

Multiplies without growth factors on ammonium salt or nitrate and glucose as sole sources of nitrogen and carbon.

Aerobic.

Spores occur in soil.

The G + C content of the DNA has been reported as 36–38 moles % (analysis, T_m and buoyant density).

Neotype strain: ATCC 14581; NCIB 9376; NCTC 10342.

Possible synonyms: 1. There is general agreement that the following species are indistinguishable from *B. megaterium*: *Bacillus tumescens* Zopf 1885, 82; *Bacillus oxalaticus* Migula 1894, 139; *Bacillus ruminatus* Meyer and Gottheil in Gottheil 1901, 485; *Bacillus graveolens* Meyer and Gottheil in Gottheil 1901, 496; *Bacillus petasites* Meyer and Gottheil in Gottheil 1901, 535; *Bacillus silvaticus* Meyer and Neide in Neide 1904, 25; *Bacillus malabarensis* Löhnis and Pillai 1907, 91; *Bacillus danicus* Löhnis and Westermann 1909, 253.

2. *Bacillus carotarum* Koch 1888, 279 was accepted as a synonym of *B. megaterium* by Smith *et al.* (1946) but differs from typical members of that species in:

Organisms smaller, mainly 1.0–1.2 μm in diameter (range 0.8–1.7 μm). Many strains non-motile (despite good aeration). The chief intracellular storage product gives the iodine color of glycogen. Poly-β-hydroxybutyrate has not been reported. Agar cultures lack yellow and black pigments.

Acid formation from arabinose, xylose, mannitol variable. Starch weakly hydrolyzed; colony on starch agar may be surrounded by a zone giving a reddish color with iodine.

Liquefaction of nutrient gelatin (22 C) slow, little at 7 days. Litmus milk alkaline; digestion of casein slow, starting in 1–3 weeks, often without clot formation. Phenylalanine not deaminated. Nitrate reduced to nitrite by most strains (>90%).

More numerous than typical *B. megaterium* in soil.

Possible synonyms of *B. carotarum*: *Bacillus simplex* Meyer and Gottheil in Gottheil 1901, 685; *Bacillus cohaerens* Meyer and Gottheil in Gottheil 1901, 689; *Bacillus cobayae* Meyer and Stapp in Stapp 1920, 10; *Bacillus capri* Meyer and Stapp in Stapp 1920, 19; *Bacillus musculi* Meyer and Stapp in Stapp 1920, 39. These five have been distinguished by properties which might be regarded as within the range of variation in *B. carotarum*.

3. *Bacillus flexus* Batchelor 1919, 32 has many of the properties of a typical *B. megaterium*, including morphology and abundant intracellular storage of lipid, but differs from that species in failing to produce acid from pentoses or to deaminate phenylalanine and, although it digests casein in milk agar it fails to do so in litmus milk within 4 weeks.

Originally isolated from feces.

8. **Bacillus polymyxa** (Prazmowski) Macé 1889, 588. (*Clostridium polymyxa* Prazmowski 1880, 37; *Granulobacter polymyxa* (Prazmowski) Beijerinck 1893, 9; *Aerobacillus polymyxa* (Prazmowski) Donker 1926, 138.)

po.ly.my'xa. Gr. pref. *poly-* much, many; Gr. n. *myxa* slime or mucus; M.L. n. *polymyxa* much slime.

See Tables 15.1 and 15.4.

Spore has parallel, longitudinal surface ridges so that it is star-like in cross-section.

Colonies on nutrient agar thin; often with amoeboid spreading. On glucose agar usually heaped, mucoid with matt surface. Growth generally adherent to any agar medium.

Glucose fermented; chief products are 2,3-butanediol, ethanol, CO_2 and H_2. Many carbohydrates, polyols and other substances are fermented. Pectin and polysaccharides of plant tissues decomposed. Action on cellulose is weak or lacking.

Levan formed from sucrose; accumulates in large capsules, not in surrounding medium.

Nitrogen is fixed under anaerobic conditions by the majority of strains tested. A minimal medium consists of a carbon-energy source, NH_4-N and biotin. Selective conditions for growth are a carbon source (lactose specially selective), nitrogen as NH_4^+ or N_2, biotin or trace of yeast extract, pH 6–7, 30 C, oxygen excluded.

Anaerobic growth vigorous in the presence of a fermentable carbohydrate.

Spores widespread; multiplication chiefly in

TABLE 15.4

Differential characteristics of **Bacillus** *species 8–10*

Common characteristics: Gram-positive, -variable or -negative; catalase positive; acid produced from arabinose, xylose and mannitol in Medium 1 (Table 15.1) with glucose replaced; starch hydrolyzed; indole not produced; phenylalanine not deaminated; tyrosine not decomposed.

	8. *B. polymyxa*	9. *B. macerans*	10. *B. circulans*
Rods			
Width, μm	0.6–0.8	0.5–0.7	0.5–0.7
Length, μm	2–5	2.5–5	2–5
Motility	+[a]	+	d
Temperature for growth, C			
Maximum	35–45	40–50	35–50
Minimum	5–10	5–20	5–20
Gas from arabinose, xylose and mannitol	+	+	−
Growth in			
Anaerobic agar	+	+	d
Sabouraud dextrose broth and/or agar	+	+	d
5% NaCl	−	−	d
7% NaCl	−	−	d
0.001% lysozyme	d		d
Production of			
Crystalline dextrins	−	d	−
Dihydroxyacetone	+	−	−
NO$_3^-$ to NO$_2^-$	+	+	d
Decomposition of casein	+	−	d

a + = positive for 90–100% of strains; − = negative for 90–100% of strains; d = reactions differ, positive for 11–89% of strains.

decomposing vegetation. Participates in retting of flax.

The G + C content of the DNA has been reported to be 43–46 moles % (T_m) in five strains.

Neotype strain: ATCC 842; NCIB 8158; NCTC 10343.

Comments: *B. polymyxa* has been widely regarded as a clearly demarcated species. One reported aberration is the strain Hino which has some of the special properties of *B. polymyxa* and some of those of *B. macerans* (Hino and Wilson, 1958; Ouellette *et al.*, 1969).

Synonyms recognized by Porter *et al.* (1937) on the evidence of comparative studies: *Astasia asterospora* Meyer 1897, 185; *Bacillus asterosporus*

(Meyer) Migula 1900, 528; *Aerobacillus asterosporus* (Meyer) Donker 1926, 141; *Bacillus aerosporus* Greer 1928, 508.

9. Bacillus macerans Schardinger 1905, 772. (*Aerobacillus macerans* (Schardinger) Donker 1926, 139; *Zymobacillus macerans* (Schardinger) Kluyver and van Niel 1936, 402; *Bactrillum macerans* (Schardinger) Pribram 1933, 80.)

ma'ce.rans. L. part.adj. *macerans* softening by steeping, retting.

See Tables 15.1 and 15.4.

Spore surface has marked longitudinal ridges as in *B. polymyxa* (shown for three strains by carbon replica technique).

Colonies on nutrient agar thin, round to spreading. On glucose agar may be more opaque; not mucoid.

Products of the fermentation of carbohydrates are ethanol, acetone, formic and acetic acids, CO_2 and H_2. A wide range of carbohydrates and other substances are fermented. Pectin and polysaccharides of plant tissues decomposed. Weak or no action on cellulose.

A minimal medium for typical strains contains a carbon-energy source, NH$_4$-N, biotin and thiamine. N$_2$ is fixed under anaerobic conditions by the majority of strains examined.

Anaerobic growth vigorous in media containing fermentable sugar.

Spores relatively scarce in soil; isolation from that source usually requires enrichment culture. Multiplication in plant materials at elevated temperatures. Participates in retting of flax. Growth has been recorded in canned fruits initially at pH 3.8–4.

The G + C content of the DNA has been found to be 49–51 moles % (T_m) in five strains.

Type strain: ATCC 8244; NCIB 9368; NCTC 6355.

Comments: *B. macerans* is positively distinguishable from the other active gas-forming species, *B. polymyxa*, by the properties shown in Tables 15.1 and 15.4 (for an exception see above). The separation from *B. circulans* is much less precise. Some organisms which have been placed in the latter species form crystalline dextrins from starch, thus leaving gas formation as the principal differential character. In typical strains of *B. macerans* some free gas may be formed from most of the substances which are fermented with acid production, but other strains have produced gas from few of the fermented compounds and these did not always include glucose. The distinction from *B. circulans* thus deserves further study.

Synonyms based upon comparative studies: *Bacillus acetoethylicum* (*sic*) Northrop, Ashe and Senior 1919, 2 (see Porter *et al.*, 1937); *Aerobacillus acetoethylicus* (Northrop *et al.*) Donker 1926, 139 (see Porter *et al.*, 1937); *Bacillus polymyxa* var. *acetoethylicum* de Soriano 1935, 557; *Bacillus betanigrificans* Cameron, Esty and Williams 1936, 74 (see Smith *et al.*, 1946); *Aerobacillus schuylkilliensis* Eisenberg 1942, 365 (see Smith *et al.*, 1952); *Bacillus vagans* Alarie and Gray 1947, 236 (see Smith *et al.*, 1952); *Bacillus soli* Alarie and Gray 1947, 237 (see Smith *et al.*, 1952).

10. **Bacillus circulans** Jordan 1890, 831.

cir'cu.lans. L. part.adj. *circulans* circling.

See Tables 15.1 and 15.4.

In most strains the spore is terminal to subterminal; it is central in a spindle-shaped sporangium if the bacillus is short. In many strains deeply stainable material persists on the surface of free spores.

Growth on nutrient agar is generally thin; in some strains it spreads actively and may give rise to "motile colonies."

Acid without gas is formed from many carbohydrates and other substances. Cellulose is degraded weakly by some strains.

Proteolytic action in nutrient gelatin or litmus milk is feeble or lacking.

Strains placed in this species by Knight and Proom (1950) and Proom and Knight (1955) showed a wide spectrum in minimal requirements from those that grew with NH_4-N without growth factors at one extreme to those that required complex undefined media at the other.

Anaerobic growth in glucose media may be good, weak or none depending on the strain.

Spores numerous in soil; no special conditions which promote multiplication have been recognized.

The G + C content of the DNA in moles per cent has been reported to be 35 (T_m) in the neotype strain, 47 (analysis) in strain IAM 1165, 50 (analysis) in strain ATCC 9966.

Neotype strain: ATCC 4513; NCIB 9374; NCTC 2610.

Comments: This species, as currently delineated, is a markedly heterogeneous collection. Table 15.4 shows relatively few properties which are either positive or negative in over 90% of the strains that provided the data; nutritional requirements and the G + C content of the DNA both vary widely. The obstacle to a division of this assemblage is that the variant features occur in many combinations without any obvious association in discrete groups. Owing to the internal heterogeneity, the boundary around the species is not well defined.

Possible synonyms: Original strains of the following species were identified as *Bacillus circulans* by Smith *et al.* (1946, 1952): *Bacillus amylolyticus* Kellerman and McBeth 1912, 490; *Bacillus clo-*

steroides Gray and Thornton 1928, 93; *Bacillus krzemieniewski* Kleczkowska, Norman and Snieszko 1940, 185; *Bacillus effluens* Alarie and Gray 1947, 238; *Bacillus kellermanii* Alarie and Gray 1947, 237; *Bacillus torquens* Alarie and Gray 1947, 237.

Other species which have the general character of *B. circulans* appear to have been established chiefly on the ground of a single property, as follows: *Bacillus latvianus* Kalnins 1930, 265; decomposes cellulose. *Bacillus palustris* Sickles and Shaw 1934, 429; decomposes specific polysaccharide of type 3 pneumococcus. *Bacillus aporrhoeus* Fuller and Norman 1943, 277; decomposes cellulose.

11. **Bacillus stearothermophilus** Donk 1920, 373.

ste.a.ro.ther.mo'phi.lus. Gr. n. *stear* fat; Gr. n. *thermus* heat; Gr. adj. *philus* loving; M.L. adj. *stearothermophilus* (presumably intended to mean) fat- and heat-loving.

See Tables 15.1 and 15.5. The most distinctive diagnostic characters are capacity to grow at 65 C, sensitivity to azide and a limited tolerance to acid.

In glucose media most strains grow actively without oxygen until the pH level reaches 5.3–4.8. Other strains fail to grow anaerobically. Products of anaerobic fermentation are chiefly L(+)-lactic acid plus small amounts of formic and acetic acids and ethanol in the ratio 2:1:1 (reported for 10 strains by McKray and Vaughn, 1957).

Nitrate supports anaerobic growth with gas production in some strains; other strains have no action or reduce to nitrite only.

Minimal nutritional requirements vary greatly among strains and range from a carbon source and NH_4-N, without growth factors, to requirements for various amino acids and vitamins. Some strains have shown increased needs when the incubation temperature approached the lower or the higher limit for growth.

The spores are more resistant to heat than those of any mesophilic species in the genus. In contrast, the vegetative cells are markedly sensitive to unfavorable conditions. If cooled to room temperature their viability may be lost immediately. Initiation of growth in diagnostic media may be erratic.

Sulfonamide resistance is weaker than in other species of *Bacillus* except *B. alvei* and *B. laterosporus* (Kundrat, 1963).

Spores occur in soil in all climatic zones; have been found in deep cores of ocean sediments estimated to have an age of thousands of years (Bartholomew and Paik, 1966). Vegetative growth is rapid in many foods of pH above 5.0 if held at an appropriate elevated temperature; also in heaps of heating compost. Growth in canned foods results in a "flat sour" condition.

The G + C content of the DNA in 16 strains has been found to be in the range 49–53 moles % (T_m, buoyant density and analysis). In three other strains the values were in the range 44–46% (T_m and buoyant density).

Type strain: ATCC 12980; NCIB 8923; NCTC 10339.

Comments: In the genus *Bacillus* many species of thermophiles have been proposed. Of those capable of vigorous growth at 65 C and pH 7 only *B. stearothermophilus* has been widely accepted, and this species is frequently regarded as embracing all strains adapted to these conditions. Two weaknesses of this situation are: (1) *B. stearothermophilus* is markedly heterogeneous and there are uncertainties about its demarcation; (2) the emphasis on ability to grow at 65 C has the effect of excluding organisms that have temperature maxima between 55 and 65 C although they have not so far been distinguished from *B. stearothermophilus* by any other property. As yet, there has been no agreement on how classification in this part of the genus might be improved.

12. Bacillus coagulans Hammer 1915, 129.

co.a'gu.lans. L. part.adj. *coagulans* curdling, coagulating.

See Tables 15.1 and 15.5.

Morphological variation is considerable; in a diagram of species of *Bacillus*, Smith *et al.* (1952) placed *B. coagulans* on the line dividing Group 1 (sporangia not appreciably swollen by oval or cylindrical spores) from Group 2 (sporangia swollen by oval spores).

Products of fermentation of glucose are mainly L(+)-lactic acid plus small amounts of 2,3-butanediol, acetoin, acetic acid and ethanol. Final pH 4.0–5.0, varying with strain, medium and conditions. Fermentable sugars increase the amount of aerobic growth on agar, and they support anaerobic growth.

Minimal nutritional requirements are diverse, varying with strain and incubation temperature. Several amino acids and several vitamins fall within the range of requirements. Aciduric; for initiation of growth optimum pH level is close to 6; minimum 4.0–5.0 in different strains.

Spores relatively scarce in soil. Selective conditions for vegetative growth are presence of fermentable sugar, pH 4.5–5.5 initially or resulting from fermentation, incubation at 45–55 C aerobically or anaerobically. May multiply in acid foods such as canned tomato juice and silage.

The G + C content of the DNA recorded for three strains is 47–48 moles % (T_m and analysis).

Type strain: ATCC 7050; NCIB 9365; NCTC 10334.

TABLE 15.5

Differential characteristics of **Bacillus** *species 11–13*

Common characteristics: Starch hydrolyzed, no growth in 7% NaCl, phenylalanine not deaminated.

	11. B. stearo-thermophilus	12. B. coagulans	13. B. alvei
Rods			
Width, μm	0.6–1	0.6–1	0.5–0.8
Length, μm	2–3.5	2.5–5	2–5
Gram reaction	d and v[a]	+	d and v
Motility	+	+	d
Catalase	d	+	+
Temperature for growth, C			
Maximum	65–75	55–60	35–45
Minimum	30–45	15–25	15–20
Acid from[b] arabinose, xylose and mannitol	d	d	–
Growth in			
Anaerobic agar	–	+	+
Sabouraud dextrose broth and/or agar	–	+	–
5% NaCl	d		d
0.02% azide	–	+	
0.001% lysozyme	–	–	+
Production of			
Dihydroxyacetone	–	d	+
Indole	–	–	+
NO_3^- to NO_2^-	d	d	–
Decomposition of			
Casein	d	d	+
Tyrosine	–	–	d

[a] + = positive for 90–100% of strains; – = negative for 90–100% of strains; d = reactions differ, positive for 11–89% of strains; v = character inconstant in one strain.

[b] Medium 1 (Table 15.1) with glucose replaced.

Comments: Some problems have arisen in the classification of organisms that diverge from the typical *B. coagulans* in one or more of the following respects: growth at 65 C, failure to grow at pH 5 or to grow anaerobically in glucose media, no acetoin accumulated. Such organisms appear to obscure the distinctions between *B. coagulans* and *B. stearothermophilus* or *B. circulans*.

Wolf and Barker (1968) divided 130 strains of *B. coagulans* into two major groups in one of which acetoin was not formed and growth occurred at 65 C on an agar medium at pH 6.2. Smith *et al.* (1952), using 77 strains, had different results: 27%

failed to accumulate acetoin but none showed growth at 65 C.

Possible synonyms: On the basis of a study of original strains Smith *et al.* (1952) accepted: *Bacillus thermoacidurans* Berry 1933, 72; *Bacillus dextrolacticus* Andersen and Werkman 1940, 187; *Bacillus thermoacidificans* Renco 1942, 109; *Lactobacillus cereale* Olsen 1944, 125.

13. Bacillus alvei Cheshire and Cheyne 1885, 592.

al′ve.i. L. n. *alveus* a beehive; L. gen.n. *alvei* of a beehive.

See Tables 15.1 and 15.5.

Typical strains spread vigorously on agar and may show "motile colonies." Their free spores may lie side by side in long rows on the agar.

The separation of this species from *B. circulans* relies principally on the production of dihydroxyacetone and indole. Some organisms give various intermediate results in tests for acetoin formation and action on pentoses, mannitol and proteins.

Minimal nutritional requirements are several amino acids plus thiamine or thiamine and biotin or more complex unidentified requirements.

Anaerobic growth occurs in complex media containing glucose.

Isolated from soil and from honeybee larvae suffering from European foulbrood.

The G + C content of the DNA in the type strain is 33 moles % (T_m).

Neotype strain: ATCC 6344; NCIB 9371; NCTC 6352.

Possible synonym: Smith *et al.* (1946) found original strains of *Bacillus para-alvei* Burnside and Foster 1935, 579 to be indistinguishable from *B. alvei.*

14. Bacillus firmus Bredemann and Werner in Werner 1933, 470.

fir′mus. L. adj. *firmus* strong, firm.

See Tables 15.1 and 15.6.

Especially useful diagnostic characters: morphology similar to that of *B. subtilis;* acid formed oxidatively from glucose; acid-sensitive, grows only at pH values above 6; addition of glucose to media that support growth results in early inhibition and the pH value does not fall below 6.

Minimal nutritional requirements are a mixture of amino acids and biotin or both biotin and thiamine.

Has been isolated chiefly from soil.

The G + C content of the DNA of one strain has been reported to be 41 moles % (T_m).

Type strain: ATCC 14575; NCIB 9366; NCTC 10335.

Possible synonym, on the basis of a comparative study by Smith *et al.* (1952), is *Bacillus imomarinus* ZoBell and Upham 1944, 265.

15. Bacillus laterosporus Laubach 1916, 511.

la.te.ro.spor′us. L. n. *latus, lateris* the side; M.L. n. *spora* spore; M.L. adj. *laterosporus* with lateral spore.

Tables 15.1 and 15.6.

An important diagnostic feature is the production of a canoe-shaped body attached to the side of the spore. As a consequence the spore occupies a lateral position in a spindle-shaped sporangium. The parasporal body, which is stainable with fuchsin, remains firmly adherent to the spore after lysis of the sporangium. Fitz-James and Young (1958) give further details. This feature has limitations in diagnosis as the proportion of sporangia which contain parasporal bodies varies with the strain and with the growth medium; moreover, lateral spores may occur in other species of *Bacillus.*

Moderate growth on nutrient agar, becoming dull and opaque; may spread actively if surface is moist. Growth thicker on glucose nutrient agar; may become wrinkled.

Anaerobic growth occurs at expense of glucose.

Has been isolated, only rarely, from dead honeybee larvae, soil and water.

The G + C content of the DNA of the type strain is 40 moles % (T_m).

Type strain: ATCC 64; NCIB 9367; NCTC 6357.

Probable synonym: On the evidence of a comparative study, Smith *et al.* (1946) reported that *Bacillus orpheus* McCray 1917, 410 is identical.

16. Bacillus brevis Migula 1900, 583.

bre′vis. L. adj. *brevis* short.

See Tables 15.1 and 15.6.

The recognition of this species depends largely on: (1) the elliptical spore which distends the sporangium into a spindle-shaped or clavate body; when liberated, its surface shows considerable stainability; (2) acid formation from glucose may be positive, weak or absent; it is not detectable in glucose peptone water.

Minimal nutritional requirement of most strains is a mixture of amino acids without vitamins.

Aerobic.

Has been isolated chiefly from soil and foods.

The G + C content of the DNA in moles per cent is reported to be 43 (T_m) or 45 (buoyant density) in strain ATCC 9999; 44 (analysis) in strain ATCC 8185.

Neotype strain: ATCC 8246; NCIB 9372; NCTC 2611.

Possible synonyms: Smith *et al.* (1952), on the basis of comparative work, accepted: *Bacillus centrosporus* Ford 1916, 524; *Bacillus hollandicus* Meyer and Stapp in Stapp 1920, 47.

TABLE 15.6

Differential characteristics of **Bacillus** *species 14–17*

Common characteristics: Catalase positive; dihydroxyacetone not produced.

	14. B. firmus	15. B. laterosporus	16. B. brevis	17. B. sphaericus
Rods				
Width, μm	0.6–0.9	0.5–0.8	0.6–0.9	0.6–1
Length, μm	1.2–4	2–5	1.5–4	1.5–5
Gram reaction	$+^a$	d and v	d and v	d and v
Easily stainable body attached to one side of spore	−	+	−	−
Motility	d	+	+	+
Temperature for growth, C				
Maximum	40–45	35–50	40–60	30–45
Minimum	5–20	15–20	10–35	5–15
Acid from^b				
Arabinose and xylose	d	−	−	−
Mannitol	±	+	d	−
Hydrolysis of starch	+	−	−	−
Growth in				
Anaerobic agar	−	+	−	−
0.02% azide^c			−	
0.001% lysozyme	−	+	d	d
7% NaCl	+	−	$−^d$	d
Sabouraud dextrose broth and/or agar	−	−	d	d
Production of				
Alkaline reaction in V-P broth	−	−	+	+
Indole	−	d	−	−
NO_3^- to NO_2^-	+	+	d	−
Decomposition of				
Casein	+	+	+	d
Tyrosine	d	+	+	−
Deamination of phenylalanine	d	−	−	+

^{a} + = positive for 90–100% of strains; − = negative for 90–100% of strains; d = reactions differ, positive for 11–89% of strains; v = character inconstant in one strain.

^{b} Medium 1 (Table 15.1) with glucose replaced.

^{c} Only strains that grow at 55 C were tested.

^{d} Cultures of *B. brevis* did not grow in 5% NaCl.

17. Bacillus sphaericus Meyer and Neide in Neide 1904, 350.

sphae′ri.cus. Gr. adj. *sphaericus* spherical.

Tables 15.1 and 15.6.

Growth on nutrient agar varies in different strains from compact and heaped to a wide spreading over the surface. Uncommon strains produce pink colonies.

Proteolytic action on nutrient gelatin (22 C) and litmus milk shows gradation from obvious by 7 days to none in 3 weeks.

Growth has been reported on casein hydrolysate, some strains requiring thiamine or thiamine plus biotin in addition. Growth in complex media is not promoted by ammonia or urea.

Aerobic in all media.

Spores occur in soil.

The G + C content of the DNA reported for two strains is 37 moles % (by T_m) and 43 moles % (by analysis).

Type strain: ATCC 14577; NCIB 9370; NCTC 10338.

Comments: The boundary around this species is diffuse and arbitrary. Ureaclastic strains intergrade without any clear line of separation with organisms that require for optimum growth alkaline media containing NH_3 (see comment on *B. pasteurii,* below). Other organisms close in character to *B. sphaericus* diverge from that species in one or more of the following properties: mature spore slightly oval, weak acid production from sugars or polyols, weak hydrolysis of starch, nitrate reduced to nitrite, maximum temperature for growth below 30 C, no growth at pH 6 or in 5% NaCl broth. These characters occur in various combinations, apparently not in distinct groups. The special significance of the organisms that cluster around *B. sphaericus* is that they are among the most numerous representatives of the genus *Bacillus* in soil.

Bacillus rotans Roberts 1935, 234 which has been distinguished by degradation of glucose without appreciable acid accumulation and by failure to grow at pH 6 or in 4% NaCl broth seems to stand close to *B. sphaericus*. It has even been included in that species as *Bacillus sphaericus* var. *rotans* Smith, Gordon and Clark 1946, 96.

Synonyms of *B. sphaericus* recognized in comparative studies by Smith *et al.* (1952): *Bacillus fusiformis* Meyer and Gottheil in Gottheil 1901, 724 (*Bacillus sphaericus* var. *fusiformis* Smith, Gordon and Clark, 1946, 97); *Bacillus lactimorbi* Jordan and Harris 1908, 1669; *Bacillus serositidis* Lacorte 1932, 3.

18. **Bacillus pasteurii** (Miquel) Chester 1898, 110. (*Urobacillus pasteurii* Miquel 1889, 519; *Bacillus probatus* Meyer and Viehoever in Viehoever 1913, 209.)

pas.teur'i.i. M.L. gen.n. *pasteurii* of Pasteur; named for Louis Pasteur, French chemist and bacteriologist.

Rods, 0.5–1.2 by 1.3–4 μm, shape varying in different strains; little tendency to occur in chains; motile. Gram-positive but easily decolorized.

Endospores spherical or slightly oval, 0.8–1.3 μm in diameter; usually distend the sporangium in a terminal or subterminal position.

Colonies on agar not distinctive; usually circular and glossy; size and opacity vary on different media.

Liquid media turbid; deposit slimy; rarely a fragile pellicle.

Energy-yielding metabolism respiratory, oxygen being the terminal electron acceptor. Oxidizable substrates include amino acids.

Glucose weakly oxidized by some strains, apparently not by others. Starch not hydrolyzed.

Gelatin and casein slowly digested by some strains; unaffected by others in 3 weeks. Nitrate reduced to nitrite by some strains.

Urea converted to ammonium carbonate more actively than by any other known bacterium. Typical strains decompose 3 g of urea per liter of culture per hr during the phase of maximum activity, and they complete the decomposition of 10 g of urea in 100 ml of broth (Miquel, 1889). Ureaclastic ability commonly decreases during maintenance on artificial media.

Alkaline media (optimum about pH 9) containing NH₃ (optimum about 1% NH₄Cl) are required.

A suitable basis is peptone and extract of meat or yeast. The ammonium salt may be replaced by urea which is partially converted to ammonium carbonate during heat sterilization, permitting growth to start. Filter-sterilized urea is less favorable. A delayed growth may occur from a large inoculum in alkaline media containing neither ammonia nor urea. The function of free NH₃ appears to be in facilitating the transport of nutrients across the cell membrane (Wiley and Stokes, 1963).

A defined medium which supports growth contains casein hydrolysate at pH 8.5–9.5, ammonia, thiamine and, for some strains, biotin and nicotinic acid.

Aerobe: slight growth may occur anaerobically in the absence of sugars.

Maximum temperature 33–42 C in different strains.

Isolated by Miquel from soil, water, sewage and incrustations on urinals. Repeatedly isolated from soil.

The G + C content of the DNA of the neotype strain has been reported to be 42 moles % (T_m).

Neotype strain: ATCC 11859; NCIB 8841; NCTC 4822.

Comments: Although *B. pasteurii* as originally characterized is a unique and exclusive species, its boundary has been difficult to define in precise terms. Most of the urea-decomposing bacteria that occur in soil appear to fall into a continuous series linking the typical *B. pasteurii* to *B. sphaericus* (Gibson, 1934; 1935). In this series there is a gradual decrease to zero in ureaclastic activity and in the requirement for NH₃ and alkaline media. As yet, any division of this assemblage would be arbitrary. One attempt at division was the proposal of the species *Bacillus loehnisii* Gibson 1935, 495 to comprise the strains that do not complete the decomposition of 2% urea in broth within 48 hours and are less dependent than *B. pasteurii* on NH₃ and alkalinity. With a slight change of emphasis on differential properties, a similar group was later classified as *Bacillus sphaericus* var. *loehnisii* Smith, Gordon and Clark 1946, 98.

19. **Bacillus fastidiosus** den Dooren de Jong 1929, 349.

fas.tid'i.os.us. L. adj. *fastidiosus* disdainful, fastidious.

Rods large, 1.5–2.5 by 3–6 μm; stain uniformly; often in chains. Motile; flagella lateral. Gram-positive in early stages of growth.

Endospores oval to cylindrical, 1.4–1.7 by 1.8–3 μm; occupy most of the interior of the shorter sporangia; terminal or subterminal in longer rods; may lie obliquely to the axis of closely septate filaments; produce little or no swelling of the sporangium; have stainable surface after release.

Uric acid utilized as the only source of energy, carbon and nitrogen; the products are solely CO_2, NH_3 and cell materials. Allantoin likewise supports growth. Of many other organic compounds tested only allantoic acid and glyoxylate were utilized. Peptone and glucose do not repress growth on uric acid.

Colonies on uric acid (1%) agar become opaque but unpigmented; often have a ragged outline and hair-like outgrowths; may grow into the agar and resist removal by scraping. Uric acid is cleared although separated from the colony. Reaction becomes strongly alkaline.

Nitrate not reduced. Catalase positive.

Strictly aerobic.

Active growth at 20–37 C.

Originally isolated from soil. Subsequent isolations from soil and poultry litter.

20. **Bacillus larvae** White 1906, 40.

lar'vae. L. n. *larva* ghost; M.L. n. *larva* a larva; M.L. gen.n. *larvae* of a larva.

See Tables 15.1 and 15.7.

Description based largely on Gordon *et al.*, 1973.

Cultures require thiamine and certain amino acids for growth (Lochhead, 1942; Katznelson and Lochhead, 1948) and will not survive serial transfer in nutrient broth. Growth and sporulation are satisfactory on a tryptone-glucose-yeast extract medium (J-medium of Haynes and Rhodes, 1963) and on the soluble starch medium recommended by Bailey and Lee (1962).

Cause of American foulbrood of honeybees (*Apis mellifera*).

Comment: In their physiological properties, the strains of *Bacillus pulvifaciens* Katznelson 1950, 155 were similar to strains of *B. larvae* except for their growth upon serial transfer in nutrient broth and their growth at 20 C. Although the strains of *B. pulvifaciens* were isolated from dead larvae of honeybees, they were not pathogenic for *Apis mellifera* (Katznelson and Jamieson, 1952). Morphologically *B. pulvifaciens* has been separated from *B. larvae* only by the surface contour of their spores; the surfaces of spores of *B. pulvifaciens* (three strains) had four longitudinal ribs and a fine groove on either side of the base of each rib, the spores of *B. larvae* (one strain) were smooth. Serologically two strains of *B. pulvifaciens* were differentiated from two strains of *B. larvae* (Hrubant and Rhodes, 1968).

21. **Bacillus popilliae** Dutky 1940, 57.

po.pil'li.ae. M.L. n. *Popillia* generic name of the Japanese beetle; M.L. gen.n. *popilliae* of *Popillia*.

See Tables 15.1 and 15.7.

Description based largely on Gordon *et al.*, 1973.

Thiamine is essential for growth (Dutky, 1947);

TABLE 15.7

Differential characteristics of **Bacillus** *species 20–22[a]*

Common characteristics: Acid produced from trehalose, acid not produced from arabinose and xylose, starch not hydrolyzed, alkali not produced in citrate-salts medium, catalase negative, dihydroxyacetone and indole not produced, phenylalanine not deaminated, tyrosine not decomposed, growth in anaerobic agar and in 0.001% lysozyme, no growth on serial transfer in nutrient broth or in 5% NaCl.

	20. B. larvae	21. B. popilliae	22. B. lentimorbus
Rods			
Width, μm	0.5–0.6	0.5–0.8	0.5–0.7
Length, μm	1.5–6	1.3–5.2	1.8–7
Parasporal body in sporangium	–[b]	+	–
Gram reaction	+	–[c]	–[c]
Motility	d	d	–
Temperature for growth, C			
Maximum	40	35	35
Minimum	25	20	20
Acid from mannitol	d	–	–
NO_3^- to NO_2^-	d	–	–
Growth in 2% NaCl	+	+	+
Decomposition of			
Casein	+	–	–
Gelatin	+	–	–

[a] All cultures grown in J-medium or modifications of J-medium (Gordon *et al.*, 1973).

[b] + = positive for 90–100% of strains; – = negative for 90–100% of strains; d = reactions differ, positive for 11–89% of strains.

[c] Sporangia and presporal forms are Gram-positive.

biotin, myoinositol and niacin are stimulatory (Sylvester and Costilow, 1964). Cultures will grow indefinitely upon serial transfer in liquid (shaken) or semi-solid J-medium (Haynes and Rhodes, 1963) but will not survive more than four serial transfers in nutrient broth.

This species causes the more widespread of two milky diseases of the Japanese beetle (*Popillia japonica* Newman), and together with *B. lentimorbus* is a prospective weapon for the control and elimination of the Japanese beetle and the European chafer (*Amphimallon majalis* Razoumowsky). The larvae become milky white because of the prolific production of spores by the bacilli in the hemolymph.

With a few exceptions, strains will sporulate readily when injected into the hemolymph or fed to susceptible insects (Pridham *et al.*, 1964). In laboratory media, only selected strains have been induced to sporulate (Haynes and Rhodes, 1966; Sharpe *et al.*, 1970). Spores formed *in vitro* caused milkiness in larvae infected by injection but not by feeding (Lüthy, 1968; Schwartz and Sharpe, 1970).

The parasporal body that distinguishes strains of *B. popilliae* from strains of *B. lentimorbus* was described by different authors as hemispherical or subconical, triangular, rhombohedral or indefinite in shape.

Krieg (1961, 244) recognized two subspecies of this species: (1) *Bacillus popilliae* subsp. *new zealand*, described by Dumbleton (1945) as the cause of a milky disease in *Odontria zealandica* White, and (2) *Bacillus popilliae* subsp. *fribourgensis*, first characterized as the cause of milky disease of *Melolontha melolontha* Linnaeus and named *Bacillus fribourgensis* by Wille 1956, 274.

22. **Bacillus lentimorbus** Dutky 1940, 57.

len.ti.mor'bus. L. adj. *lentus* slow; L. n. *morbus* disease; M.L. n. *lentimorbus* the slow disease.

See Tables 15.1 and 15.7.

Description based largely on Gordon *et al.*, 1973.

This species, which is more fastidious nutritionally and less widespread than *B. popilliae*, also infects the larvae of the Japanese beetle (*Popillia japonica* Newman) and the European chafer (*Amphimallon majalis* Razoumowsky).

Pure cultures can be isolated most readily from dry films of hemolymph of infected larvae. They can be maintained indefinitely by serial transfer in diphasic J-medium (Haynes and Rhodes, 1963). Spores are produced by injection of vegetative cells or spores into susceptible grubs (Haynes *et al.*, 1961). Sporulation by cultures *in vitro*, however, has been reported only by Steinkraus and Tashiro (1955); other methods producing sporulation *in vitro* by *B. popilliae* have been unsuccessful with *B. lentimorbus* (Rhodes *et al.*, 1965; Haynes and Rhodes, 1966; Sharpe *et al.*, 1970). Strains of *B. lentimorbus* have been separated from strains of *B. popilliae* by cross-agglutination reactions (Hrubant and Rhodes, 1968) and by the surface topography of their spores (Bulla *et al.*, 1969).

Beard (1956, 641) isolated strains in Australia that differed from strains of *B. lentimorbus* in having a somewhat different host range and being less infective for the Japanese beetle. He designated his strains as *Bacillus lentimorbus* subsp. *australis*.

Group II

The 26 species in this group are placed together because it might be considered that they need further investigation to show how they differ from other species or from closely similar but unnamed organisms. Several have the severe limitation that the original description was based on very few strains, while not more than a single culture may now be available for study.

Some differential properties of the species in Group II are given in Table 15.8.

Editorial Note: The status of the species in this section has been the subject of considerable correspondence between the authors and editors. In many genera, most, if not all of the species in this group would have been listed as *species incertae sedis*, and one author agrees. It is evident that more work is needed to establish their status as distinct species. *B. acidocaldarius* seems quite distinct and is placed at the end because its description appeared after the manuscript and tables had been written.

23. **Bacillus amyloliquefaciens** Fukumoto 1943, 488. (*Bacillus amyloliguifaciens* (*sic*) Fukumoto 1943, 488.)

am.yl.o.li.que.fac'i.ens. L. n. *amylum* starch; M.L. part.adj. *amyloliquefaciens* starch-digesting.

This species has been separated from *B. subtilis* by Welker and Campbell (1967) on the grounds that the G + C content of the DNA is 43.5–44.9 moles % (by analysis, T_m and buoyant density), DNA-DNA hybridization shows only about 15% homology and the α-amylase has different properties and is produced in larger amounts. Other differences were hitherto accepted among the variants of *B. subtilis*.

Neotype strain: ATCC 23350.

24. **Bacillus medusa** Delaporte 1969, 1131.

me.du'sa. L. n. *medusa* Gorgon with serpent hair.

Distinguished from *B. cereus* by forming a large spherical to elliptical parasporal body which consists mainly of protein and is refractile and deeply stainable.

Source: Cow dung.

Type strain not designated. Original strain: NCIB 10437.

Comment: It is not clear how this species differs from *Bacillus finitimus* Heimpel and Angus 1958, 539, more especially from the non-motile *Bacillus finitimus* var. *fowleri* Heimpel 1967, 292. See page 536.

25. **Bacillus maroccanus** Delaporte and Sasson 1967, 2346.

mar.oc.can'us. M.L. adj. *maroccanus* of Morocco.

TABLE 15.8

Differential characteristics of the species of genus **Bacillus** *Group II*

	Shape	Spore Distends sporangium distinctly	Dominant position	Acid from glucose	Starch hydrolysis	Anaerobic growth	Growth at 3 C
23. *B. amyloliquefaciens*	E[a]	−	C	+	+	−	−
24. *B. medusa*	E	−	C	+	+	+	−
25. *B. maroccanus*	E	−	CT	+	+	−	−
26. *B. pacificus*	E	+	C	+	+	−	−
27. *B. lentus*	E	−	C	+	+	−	−
28. *B. epiphytus*	E and S	−	C	+	+	−	−
29. *B. apiarius*	E	+	C	+	+	+	−
30. *B. psychrosaccharolyticus*	E	+	C	+	+	+	+
31. *B. macquariensis*	E	+	T	+	+	+	+
32. *B. laevolacticus*	E	+	T	+	+	+	−
33. *B. racemilacticus*	E	+	T	+	+	+	−
34. *B. filicolonicus*	E	+	T	+	+	+	−
35. *B. pantothenticus*	E and S	+	T	+	+	+	−
36. *B. thiaminolyticus*	E	+	T	+	+	+	−
37. *B. pulvifaciens*	E	+	CT	+	−	+	−
38. *B. cirroflagellosus*	E	+	C	−	+	−	−
39. *B. freudenreichii*	E	−	C	−	−	−	−
40. *B. alcalophilus*	E	−	T	−	+	−	−
41. *B. badius*	E	−	CT	−	−	−	−
42. *B. aneurinolyticus*	E	v	T	−	−	−	−
43. *B. macroides*	E	v	T	−	−	−	−
44. *B. aminovorans*	S	−	C	+	+	−	−
45. *B. insolitus*	S	−	CT	−	−	−	+
46. *B. globisporus*	S	+	T	+	d	−	+
47. *B. psychrophilus*	S	v	T	+	+	−	+
48. *B. acidocaldarius*	E	+	T	−	+	−	−[b]

[a] E = elliptical or cylindrical; S = spherical or nearly so; C = central; T = terminal or subterminal; CT = central to terminal; variation within or between strains; + = positive for 90–100% of strains; − = negative for 90–100% of strains; d = reactions differ, positive for 11–89% of strains; v = character inconstant in one strain.

[b] Minimum temperature for growth 45 C.

Relatively large organisms distinguished from *B. megaterium* chiefly by the production of lecithinase and a failure to store poly-β-hydroxybutyrate.
Source: Desert soil.
Type strain not designated. Original strain: NCIB 10500.

26. Bacillus pacificus Delaporte 1967, 3071.
pa.ci′fic.us. L. adj. *pacificus* pacific.
Cells oval, exceptionally large, measuring 1.5–2.1 by 2.7–3.4 μm; have capsules and lipid inclusions; motile; one or two flagella inserted at or near one pole or both poles. Spores 1.3–1.5 by 2.7–3.4 μm in size.
Best medium reported is 0.1% tryptone in sea water. No growth on ordinary nutrient agar. Glucose broth reaches pH 6 in 10 days; no acetoin formed. Gelatin slowly liquefied. Nitrate reduced to nitrite. Catalase formed. Grows in 10% NaCl. Growth good at 28–40 C; none at 4 C.
Isolated from sand of shore, Pacific Ocean, California.
Type strain not designated. Original strain: NCMB 1862.

27. Bacillus lentus Gibson 1935, 368.
len.tus. L. adj. *lentus* slow.
Originally distinguished from *B. firmus* (see p. 541) by the production of a distinct titratable alkalinity in urea broth and by a lack of action on gelatin and casein. The G + C content of the DNA of one strain has been reported to be 37 moles % (T_m). Strains intermediate between the two species in physiological properties have now been isolated so that the status of *B. lentus* as a clearly delimited species has become questionable.

Source: Soil.
Type strain: ATCC 10840; NCIB 8773; NCTC 4824.

28. Bacillus epiphytus ZoBell and Upham 1944, 266.

e.pi.phyt'us. Gr. prep. *epi* upon; Gr. n. *phyton* a plant; M.L. adj. *epiphytus* growing, as an epiphyte, upon a plant.

Has the properties given for *B. firmus* in Tables 15.1 and 15.6, with the exception that the spore is oval to spherical.

Grows in 10% NaCl broth; no action on tyrosine; phenylalanine deaminated.

Source: Marine phytoplankton.
Original strain: ATCC 14412; NCMB 444.

29. Bacillus apiarius Katznelson 1955, 636.

a.pi.ar'i.us. L. adj. *apiarius* relating to bees.

A special feature is the nature of the spore coat, which is ridged, rectangular in outline and unusually thick. The coat remains covered by stainable remnants of the sporangium for a considerable time. Diameter of vegetative cells 0.6–0.8 μm, often less at poles.

In other properties similar to *B. laterosporus* as described on page 541 and in Tables 15.1 and 15.6, except that a parasporal body has not been described, growth occurs in Sabouraud dextrose media, acid is not formed from mannitol, starch is hydrolyzed and phenylalanine is deaminated.

Source: Dead larvae of honeybee.
Type strain not designated.
Comment: Spores with rectangular shape are also formed by *Bacillus brachysporus* (Burchard) Mez 1898, 33 (*Bacterium brachysporum* Burchard 1898, 20), which differs from *B. apiarius* in having nonmotile and larger (1.2 μm diameter) vegetative cells.

30. Bacillus psychrosaccharolyticus Larkin and Stokes 1967, 890.

psy.chro.sac.char.o.lyt'i.cus. Gr. adj. *psychros* cold; Gr. n. *saccharon* sugar; Gr. adj. *lytos* dissolvable; M.L. adj. *psychrosaccharolyticus* cold (adapted), sugar-fermenting.

Distinctly pleomorphic; vary from coccal to elongate; on glucose media may contain globules unstainable with fuchsin; if sporulation does not occur, organisms may swell and become faintly stainable, often pear-shaped bodies up to 2 μm in diameter. The spore frequently fills most of the sporangium; it may form in a lateral position.

Overgrowth of laboratory cultures by asporogenous mutants appears to occur frequently.

On agar media relatively thick opaque growth without spreading or outgrowths.

Glucose promotes anaerobic growth only slightly.

Proteolysis relatively weak.
Growth and sporulation occur at 0 C.
Sources: Soil and marshes.
Type strain: ATCC 23296.
Comments: Direct plating of soil frequently yields organisms which have the characteristics of *B. psychrosaccharolyticus*, except that some of them may diverge from that species in action on nitrate (none or denitrification), proteins, starch or particular sugars, or in the utilization of glucose for anaerobic growth. These organisms, which do not appear to have been named, have not been subjected to comparative studies and their possible relationship to *B. psychrosaccharolyticus* remains to be examined.

31. Bacillus macquariensis Marshall and Ohye 1966, 45.

mac.qua'ri.en.sis. M.L. adj. *macquariensis* pertaining to Macquarie Island.

Special property: Grows and sporulates at 0 C. In other properties similar to *B. circulans* (p. 539 and Tables 15.1, 15.4).

Isolated from soil from Macquarie Island (subantarctic).

Type strain: NCTC 10419; NCIB 9934.

32. Bacillus laevolacticus Nakayama and Yanoshi 1967, 149.

lae.vo.lac'tic.us. M.L. adj. *laevolacticus* pertaining to levolactic acid.

Cells 0.4–1 μm in diameter; motile; flagella lateral and polar. Spores oval, terminal or nearly so; swell sporangium slightly.

Glucose fermented to D(−)-lactic acid equivalent to 94–99% of sugar consumed; final pH 3.8–3.2. No growth in carbohydrate-free media. Glucose supports active anaerobic growth. Catalase formed. Inulin and starch fermented. Glucose-gelatin slowly liquefied. Nitrate not reduced.

Maximum temperature 45–50 C. Grow in 2% NaCl broth; slight or no growth in 5%.

Sources: Rhizosphere of various plants.
Type strain: ATCC 23492; NCIB 10269.

33. Bacillus racemilacticus Nakayama and Yanoshi 1967, 150.

ra.ce.mi.lac'tic.us. M.L. adj. *racemilacticus* relating to racemic lactic acid.

Produces DL-lactic acid in a homolactic fermentation of glucose; otherwise similar in properties to *B. laevolacticus* (above).

Sources: Rhizosphere of various plants.
Type strain: ATCC 23496; NCIB 10274.

34. Bacillus filicolonicus ZoBell and Upham 1944, 270.

fi.li.co.lon'i.cus. L. n. *filum* a thread; L. adj. *colonicus* pertaining to a colony; M.L. adj. *filicolonicus* with thread (-like) colonies.

Has the properties given for *B. circulans* in Tables 15.1 and 15.4 except that action on pentoses is lacking and growth occurs in 10% NaCl broth. Remains to be compared in detail with *B. pantothenticus* (below) which is also salt-tolerant.

Sources: Sea water and marine mud.

Original strain: ATCC 14413; NCMB 445.

35. Bacillus pantothenticus Proom and Knight 1950, 539.

pan.to.then'tic.us. M.L. n. *acidum pantothenicum* panthothenic acid; M.L. adj. *pantothenticus* relating to pantothenic (acid).

Originally this species was delimited chiefly by: (1) a nutritional requirement, apparently unique in the genus *Bacillus*, for pantothenic acid. Also required are amino acids, thiamine and biotin. (2) Growth is improved by addition of 4% NaCl to media, and is good in 10% NaCl broth. (3) The spore is terminal and oval or sometimes spherical. Other properties are similar to those given for *B. circulans* in Tables 15.1 and 15.4 except that action on mannitol is lacking and phenylalanine is deaminated.

Growth in the presence of glucose is usually arrested at an early stage. Anaerobic growth in glucose media is weak, but is increased if medium is initially alkaline and strongly buffered.

Source: Soil.

Type strain: ATCC 14576; NCIB 8775; NCTC 8162.

36. Bacillus thiaminolyticus Kuno 1951, 364. (Clostridium thiaminolyticum (Kuno) Kimura and Liao 1953, 133.)

thi.am.in.o.ly'tic.us. M.L. n. *thiamina* thiamine; M.L. adj. *lyticus* dissolving; M.L. adj. *thiaminolyticus* decomposing thiamine.

Decomposes thiamine actively.

In his original description, Kuno recognized the similarity of this species to *B. alvei* (see p. 541). After a study of 44 strains of *B. thiaminolyticus*, Hayashi and Nakayama (1953) reported that it was variable in some of the characteristics by which it was distinguished from *B. alvei*. The need for further study of more strains of *B. alvei* and *B. thiaminolyticus* is thus indicated. Decomposition of thiamine by *B. alvei* has not been reported.

Source: Human feces.

37. Bacillus pulvifaciens Katznelson 1950, 155.

pul.vi.fac'i.ens. L. n. *pulvis* powder, dust; L. pres.part. *faciens* making; M.L. adj. *pulvifaciens* making powder or dust.

Growth on agar media unpigmented or buff to red; thin and non-spreading on nutrient agar; heavier if glucose is added. Anaerobic growth in glucose media is moderate.

Minimal nutritional requirement is a mixture of amino acids plus biotin.

Catalase negative on nutrient agar. Production dependent on medium (positive on J-medium).

Closely resembles *Bacillus larvae* (page 544 and Tables 15.1, 15.7) from which it is distinguished by an ability to grow at 20 C and also in ordinary nutrient broth on serial transfer.

Source: Dead larvae of honeybee.

Type strain not designated. Original strain: ATCC 13537.

38. Bacillus cirroflagellosus ZoBell and Upham 1944, 266.

cir.ro.flag.el.los'us. L. n. *cirrus* a tuft; L. dim.n. *flagellum* a small whip; M.L. adj. *cirroflagellosus* with tufts of flagella.

Neither close resemblance to another species nor a specially distinctive property has been reported.

Has no action on casein or tyrosine; phenylalanine is deaminated; nitrate reduced to nitrite. No growth in Sabouraud dextrose broth or agar. Despite a reputed marine origin, 5% NaCl prevents growth. Maximum temperature 30 C.

Source: Marine mud.

Original strain: ATCC 14411; NCMB 1044.

39. Bacillus freudenreichii (Miquel) Chester 1898, 110. (Urobacillus freudenreichii Miquel 1890, 367.)

freud.en.reich'i.i. M.L. gen.n. *freudenreichii* of Freudenreich; named for E. von Freudenreich, a Swiss bacteriologist.

This species is close to *B. brevis* (see p. 541 and Tables 15.1, 15.6) in morphology and physiology except in the difference that it produces a considerable titratable alkalinity in urea broth and it is less tolerant of acid. Additionally, growth occurs in 5% NaCl broth and phenylalanine is deaminated. In nutritional requirements it appears to be similar to or identical with *B. brevis* (Bornside and Kallio, 1956). It might be regarded as an intermediate between *B. brevis* and *B. pasteurii*.

Has been isolated from soil, river water and sewage.

Type strain: Not designated. No original strain exists.

40. Bacillus alcalophilus Vedder 1934, 141.

al.cal.o.phil'us. M.L. *alcali* Eng. alkali from the Arabic *al* the end; *qaliy* soda ash; Gr. adj. *philus* loving; M.L. adj. *alcalophilus* liking alkaline (media).

Originally characterized chiefly by marked tolerance to alkali and inability to grow on media at pH 7.

Rods 0.7–0.9 µm thick. Spores oval, mainly subterminal; produce little swelling of sporangium.

Aerobic. Action on glucose slow and not detected by acid formation in cultures. Proteolytic activity weak. Starch digested. Nitrate not reduced.

Unlike *B. pasteurii*, lacks urease activity and has no requirement for ammonia in addition to alkalinity.

Isolated from various materials using preliminary enrichment in broth at pH 10.

Type strain not designated. Original strains: NCTC 4553; NCIB 10436; NCTC 4554; NCIB 10438.

41. Bacillus badius Batchelor 1919, 25.

ba.di′us. L. adj. *badius* chestnut brown.

Distinguished from *B. brevis* as described on page 541 and in Tables 15.1 and 15.6 by growth in 5% NaCl broth; rods are greater in diameter (0.8–1.2 μm) and are not distended by the spore; free spores show little surface stainability.

The type strain of *B. badius* grows as chains of rods with blunt or flat ends, and its colony has a folded hair structure and rhizoid outgrowths. Other strains appear to differ from the type culture only in the absence of chains and the production of smooth colonies.

Has been isolated infrequently from feces, dust and foods.

Type strain: ATCC 14574; NCIB 9364; NCTC 10333.

42. Bacillus aneurinolyticus Kimura and Aoyama in Aoyama 1952, 127.

an.eur.in.o.lyt′ic.us. M.L. n. *aneurinum* thiamine; L. adj. *lyticus* dissolving; M.L. adj. *aneurinolyticus* digesting thiamine.

The only definite distinction from the characters of *Bacillus brevis* as given in Tables 15.1 and 15.6 is the lack of action on casein by *B. aneurinolyticus*. The significance of this feature might be questioned since strains that have been classified in *B. brevis* show a gradation in proteolytic activity from moderately active to none. The special property of *B. aneurinolyticus*, the decomposition of thiamine (aneurine), has not been reported in *B. brevis*.

Source: Human feces.

The G + C content of the DNA is reported to be 42 moles % (analysis).

Type strain not designated. Strain: IAM 1077; ATCC 12856 has been distributed as representative by Kimura and Aoyama.

43. Bacillus macroides Bennett and Canale-Parola 1965, 204. (*Lineola longa* Pringsheim 1950, 209.)

mac.roi′des. M.L. adj. *macroides* rather long, elongated.

Characters conform to those of *B. sphaericus* given in Tables 15.1 and 15.6 with one exception: the spore is frankly oval and scarcely distends the

sporangium. The spherical to slightly oval spores formed by many strains of *B. sphaericus* have so far been distinguishable. The properties of *B. macroides* are also not greatly different from those of *B. badius* (Species No. 41).

Minimal nutritional requirements: a carbon-energy source, NH₄-N, thiamine, biotin and, in one strain, guanine.

Carbon sources: various amino acids and C₂ to C₅ *n*-fatty acids but not sugars. Proteolytic action is not detected within 3 weeks.

Sources: Cow dung, plant material decaying in water.

The G + C content of the DNA is reported to be 42 moles % (T_m).

Type strain: ATCC 12905; NCIB 8796 (possibly different) were deposited as type cultures by Pringsheim.

44. Bacillus aminovorans den Dooren de Jong 1926, 157.

am.in.o.vor′ans. M.L. n. *aminum* amine; L. pres.part. *vorans* devouring, digesting; M.L. pres. part. *aminovorans* amine digesting.

Vegetative cells are thick (1.1–1.5 μm) and often coccal or rectangular in outline.

Glucose utilized as sole source of carbon.

Urea decomposed. No action on casein.

Special feature: Utilizes methylamine as source of both carbon and nitrogen.

Maximum temperature for growth 35 C.

Source: Soil.

Type strain: ATCC 7046; NCIB 8292; NCTC 2870.

45. Bacillus insolitus Larkin and Stokes 1967, 891.

in.so.li′tus. L. adj. *insolitus* unusual.

Data from original description: Growth and sporulation occur at 0 C. Spores are apparently unique in that they vary in shape from round to cylindrical, and in size from 0.7–1.4 μm in diameter and up to 2.4 μm in length, depending on the medium on which they are produced.

Vegetative cells are stout and often short like those of *B. aminovorans* (above). Differs from that species in: maximum temperature for growth 25 C; no growth in 5% NaCl broth; no action on glucose, starch or urea.

Source: Soil.

Type strain: ATCC 23299.

46. Bacillus globisporus Larkin and Stokes 1967, 892.

glo.bis′por.us. L. n. *globus* a sphere; M.L. n. *spora* a spore; M.L. adj. *globisporus* with spherical spores.

Seems to be a segment of the assemblage that shows affinities to *B. sphaericus*. Has properties of that species (see p. 542, Tables 15.1 and 15.6) but diverges in maximum temperature for growth 25–30 C; grows and sporulates at 0 C; glucose weakly oxidized; urea decomposed; no growth in 5% NaCl broth or, from light inoculum, on nutrient agar at pH 6. Characterization has been hindered by weak and erratic responses to some of the commonly used diagnostic tests.

Sources: Soil, river water.

Type strain: ATCC 23301.

47. **Bacillus psychrophilus** Larkin and Stokes 1967, 894.

psy.chro'phil.us. Gr. adj. *psychros* cold; Gr. adj. *philus* liking, preferring; M.L. adj. *psychrophilus* preferring cold.

Similar to *B. sphaericus* as described in Tables 15.1 and 15.6 but differs in maximum temperature for growth 25–30 C; grows and sporulates at 0 C; glucose weakly oxidized; urea decomposed; phenylalanine not deaminated; nitrate reduced to nitrite; no growth from light inoculum on nutrient agar at pH 6. Results of tests for physiological properties may be weak and variable.

Sources: Soil, river water.

Type strain: ATCC 23304.

48. **Bacillus acidocaldarius** Darland and Brock 1971, 9.

a.ci.do.cal.dar'i.us. M.L. n. *acidum* acid; L. adj. *caldarius* pertaining to warm or hot; M.L. adj. *acidocaldarius* pertaining to acid thermal (habitats).

Carbon-energy sources include glucose, galactose, starch, glycerol and casamino acids. No growth with ethanol, sorbitol, acetate, succinate or citrate as sole C source. Growth occurs with NH_4^+ but not with NO_3^- as the sole N source. Growth factors are not required.

Aerobic in media containing glucose or nitrate.

Temperature limits for growth: 45 and 65 or 70 C.

Limits of pH for growth: 2 and 5–6 except in 2 of 15 strains which have a lower limit of pH 3.

Spores have relatively weak heat resistance, half-time death at 86 C being 10–12 min.

Sources: Thermal, markedly acid water and soil. Enrichment from more nearly neutral soils has failed.

The G + C content of the DNA in three strains is 61–62 moles % (buoyant density).

Type strain: 104-1A; ATCC 27009.

Comment: This species is separated widely from all others in the genus *Bacillus*. From the ubiquitous thermophiles, *B. stearothermophilus* and the assumptive variants of that species, it is distinguished most decisively by its acidophilic nature and the composition of its DNA.

Genus II. **Sporolactobacillus** *Kitahara and Suzuki 1969, 69*

KAKUO KITAHARA

Spo.ro.lac.to.ba.cil'lus. Gr. n. *spora* seed; L. n. *lactis* milk; L. dim. n. *bacillus* a small rod; M.L. masc.n. *Sporolactobacillus* sporing milk rodlet.

Cells straight rods, 0.7–0.8 by 3–5 μm, occurring singly, in pairs, rarely in short chains. Motile by means of a small number of long peritrichous flagella. **Endospores formed.** Gram-positive.

Chemoorganotrophs; metabolism fermentative, hexose sugars are decomposed through the typical homo-fermentative pathway producing only lactic acid. **Do not contain heme compounds such as catalase or cytochromes.**

Microaerophilic.

Does not grow below 10 C or above 45 C. Grows well between 15 and 40 C. Optimum for growth 35 C, for fermentation 30 C.

Type species: *Sporolactobacillus inulinus* (Kitahara and Suzuki) Kitahara and Lai 1967, 197.

Further Comments

Resembles *Lactobacillus* in many characteristics; cannot grow in sugar-free media; produces pinpoint colonies on agar plates; uniform growth along agar stab but no surface growth. On the other hand, spore formation and the presence of diaminopimelic acid in the cell wall murein (Kandler, 1967) reveals a relation to *Bacillus*.

Description of the species of genus **Sporolactobacillus**

1. **Sporolactobacillus inulinus** (Kitahara and Suzuki) Kitahara and Lai 1967, 197. (*Sporolactobacillus* (*Lactobacillus*) *inulinus* Kitahara and Suzuki 1963, 69.)

in.u.lin'us. M.L. n. *inulum* inulin; M.L. adj. *inulinus* pertaining to inulin.

Morphology as for genus. Tadpole-like cells appear under certain conditions (Kitahara and Lai, 1967). Elliptical endospores, 0.8 by 1.0 μm, are formed in some cells in a terminal position, the sporangium swelling with maturation of the endospore. Spores tolerate heating for 10 min at 85 C; some survive 10 min at 95 C.

Good growth in glucose-yeast extract-peptone

(GYP) media. On GYP-agar colonies pinpoint; somewhat larger on or in semi-solid agar and show motility. In deep agar, small colonies distributed equally throughout the agar except on the surface. Broth becomes turbid and the growth gradually precipitates.

A strictly homolactic fermentation produces D(−)-lactic acid. Acid but no gas from fructose, glucose, inulin, maltose, mannose, raffinose, sucrose, trehalose, mannitol, sorbitol and α-methyl glucoside. Slight or no acid from arabinose, xylose, galactose, lactose, melibiose, cellobiose, melezitose, dextrin, starch, glycerol, erythritol, adonitol and salicin. Limiting pH is 4.0; however, when calcium carbonate is present 20% or more glucose may be fermented completely.

Gelatin not liquefied. Indole not formed. Nitrates not reduced to nitrites. Litmus milk unchanged.

Composition of cell wall murein: GuNAc: MurNAc:D-Glu:DAP:L-Ala:D-Ala:NH₃ = 1:1:1: 1:1:0.5:2. Teichoic acid not detected (Weiss *et al.*, 1967) Dipicolinic acid content of spore 5.16% (Kitahara and Lai, 1967).

Leucine, valine, biotin and pantothenic acid are essential nutrients.

The G + C content of the DNA 39.3 moles % (chemical analysis, Suzuki and Kitahara, 1964; 47.3 moles % (T_m, Miller *et al.*, 1970)).

Originally isolated from a chicken feed.

Type strain: ATCC 15538.

Editorial Note. The sole species was originally named *Sporolactobacillus* (*Lactobacillus*) *inulinus* thus inadvertently reducing *Lactobacillus* to subgeneric rank. This presumably was a *lapsus calami* for *Lactobacillus* (*Sporolactobacillus*) *inulinus*. However, in subsequent papers, Kitahara and his co-workers referred to the organism as *Sporolactobacillus inulinus* and the recognition of generic rank is here attributed to Kitahara and Lai (1967). Kitahara and Toyota (1972) in a footnote state: "The name *Sporolactobacillus* is considered more appropriate to be raised to a genus of the family Bacillaceae from a subgenus of the genus *Lactobacillus*..."

Genus III. **Clostridium** *Prazmowski 1880, 23*

LOUIS DS. SMITH AND GEOFFREY HOBBS

Clos.tri′di.um. Gr. n. *closter* a spindle; M.L. dim.n. *Clostridium* a small spindle.

Rods, usually motile by means of peritrichous flagella; occasionally non-motile. **Form ovoid to spherical spores that usually distend the bacilli.** Generally Gram-positive, at least in the early stages of growth.

Chemoorganotrophs. Some species are saccharolytic, some proteolytic, some both, some neither. Ferment sugars, polyalcohols, amino acids, organic acids, purines and other organic compounds. Some species fix nitrogen. **Do not reduce sulfate.**

Most strains are strictly anaerobic, although some may grow in the presence of air at atmospheric pressure. Catalase is usually not produced, but when it is, in small amounts.

Commonly found in soil, marine and fresh water sediments, and in the intestinal tract of man and other animals.

The G + C content of the DNA ranges from 23–43 moles %.

Type species: *Clostridium butyricum* Prazmowski 1880, 24.

Further Comments

For convenience in identification the genus is divided into four groups on the basis of spore posi-

tion and gelatin liquefaction. Tests by which the species in each of the four groups may be distinguished are given in Tables 15.9, 15.12, 15.15 and 15.18. For some species and subspecies, toxin-antitoxin reactions are required for dependable identification; these species and subspecies are marked in the tables. As far as possible only stable characteristics are used in the tables; those marked variable are not essential for recognition of that species.

More than 300 species of spore-forming anaerobic bacilli have been described and placed, by various authors at various times, in some 40 genera. The taxonomy of these bacteria is confused, primarily because sufficient data are not available to enable a thorough taxonomic study to be carried out. While there may be justification for separating these bacteria into several genera when adequate data become available, for the present, the anaerobic bacilli, with the exception of *Desulfotomaculum*, are retained in one genus.

Five species, that require special media or conditions and therefore could not be compared directly with the others, are described in Group V.

Species for which strains were not available for study are listed as *species incertae sedis*.

Key to the groups of genus **Clostridium**

I. Spores subterminal.
   A. Gelatin not hydrolyzed.

                Group I. Species 1–11

   B. Gelatin hydrolyzed.

                Group II. Species 12–31

II. Spores terminal.
   A. Gelatin not hydrolyzed.

                Group III. Species 32–50

   B. Gelatin hydrolyzed.

                Group IV. Species 51–56

III. Species with special growth requirements.

                Group V. Species 57–61

Description of the species of genus **Clostridium**

Group I

1. Clostridium butyricum Prazmowski 1880, 24. (*Bacillus amylobacter* van Tieghem 1877, 128; *Metallacter amylobacter* (van Tieghem) Trevisan 1879, 147; *Bacterium navicula* Reinke and Berthold 1879, 21; *Bacillus butyricus* (Prazmowski) Flügge 1886, 295; *Bacillus navicula* (Reinke and Berthold) Chester 1898, 128; *Amylobacter navicula* (Reinke and Berthold) Wehmer 1898, 696; *Clostridium naviculum* (Reinke and Berthold) Prévot 1938, 78.)

bu.ty′ri.cum. Gr. n. *butyrum* butter; M.L. neut.adj. *butyricum* related to butter, butyric.

Straight or slightly curved rods, 0.6–1.2 by 3.0–7.0 μm, with rounded ends; occurring singly, in pairs, in short chains and occasionally long filaments. Motile with peritrichous flagella. Spores are oval and eccentric to subterminal, with no exosporium and no appendages. Gram-positive becoming negative in old cultures; often granulose positive. Cell wall contains DL-diaminopimelic acid; glucose is the only cell wall sugar.

Little or no growth on nutrient agar; good growth on glucose agar. Surface colonies circular to slightly irregular, 1–3 mm in diameter, slightly raised, white to cream color, glossy to matt surface.

Little or no growth in cooked meat broth; good growth and gas produced in broth media with a fermentable carbohydrate added.

Lactate is utilized by some strains. Fermentation products include acetic acid, butyric acid and butanol.

Casein and gelatin are not hydrolyzed. Milk becomes acid with early coagulation and often with stormy fermentation; the clot is fragmented but not digested.

Limited fixation of atmospheric nitrogen.

Does not require amino acids or vitamins, other than biotin, for growth.

Optimum temperature for growth is 25–37 C.

Has been found in soil, animal feces, cheese, naturally soured milk.

TABLE 15.9

Distinguishing characteristics of species of **Clostridium** *Group I*
Spores subterminal, gelatin not hydrolyzed

	Maltose	Mannose	Raffinose	Lactose	Ribose	Starch	Dulcitol	Sorbitol	Nitrate
1. *C. butyricum*	+	+	+	+	+	+	−	−	d
2. *C. beijerinckii*	+	+	+	+	−	−	−	−	−
3. *C. oroticum*	+	+	+	+	+	−	+	−	+
4. *C. rectum*	+	+	+	+	−	−	−	−	+
5. *C. paraperfringens*	+	+	−	+	V[a]	+	−	−	V
6. *C. rubrum*	+	+	+	+	V	+	−	V	−
7. *C. fallax*	+	+	−	−	d	+	−	d	−
8. *C. pasteurianum*	+	+	+	−	−	−	−	−	d
9. *C. sticklandii*	+	−	−	−	+	−	−	−	−
10. *C. tyrobutyricum*	−	+	−	−	−	−	−	−	−
11. *C. propionicum*	−	−	−	−	−	−	−	−	−

[a] V = variable.

The G + C content of the DNA is 27–28 moles %. Neotype strain: ATCC 19398; NCTC 7423; NCIB 7423.

2. Clostridium beijerinckii Donker, 1926, 145.
bei.jer.inck'i.i. M.L. gen.n. *beijerinckii*, named for M. W. Beijerinck, Dutch bacteriologist.

Straight to slightly curved rods 1.5–7.5 μm, with rounded ends. Motile with peritrichous flagella. Spores are oval and eccentric to subterminal with no exosporium or appendages. Gram-positive, becoming negative in old cultures. Cell wall contains DL-diaminopimelic acid; cell wall sugars are glucose, galactose.

Little or no growth on nutrient agar; good growth on agar with fermentable carbohydrate. Surface colonies are irregular, circular, 2 mm in diameter, raised with entire edge, translucent, gray with glossy surface.

Poor growth in nutrient broth; moderate growth in broth media with fermentable carbohydrate. Fermentation products include acetic acid, butyric acid.

Milk becomes acid; no other change.

Will not grow with biotin as the only vitamin, or ammonia as nitrogen source.

Optimum temperature for growth is 30 C.

Has been found in soil, infected wounds, fermenting olives, spoiled candy.

The G + C content of the DNA is 26–28 moles %. Reference strain: NCIB 9362; ATCC 25752.

3. Clostridium oroticum (Wachsman and Barker) Cato, Moore and Holdeman 1968, 9. (*Zymobacterium oroticum* Wachsman and Barker 1954, 400.)

o.ro'ti.cum. M.L. n. *acidum oroticum* orotic acid; M.L. neut.adj. *oroticum* pertaining to orotic acid.

Rods, 0.6 by 1.3–1.9 μm with rounded ends occurring in long tangled chains. Not motile. Spores are oval, subterminal. Gram-positive.

Surface colonies circular, 1 mm in diameter, slightly raised, translucent, opaque white with matt surface.

Good growth in most liquid media, forming a granular deposit.

Fermentation products include large quantities of acetic acid. Characteristically ferments orotic acid.

Milk made acid, otherwise unchanged.

Optimum temperature for growth 30–40 C.

Isolated from black mud from San Francisco Bay.

The G + C content of the DNA is 44 moles %. Type strain: ATCC 25750; NCIB 10650.

4. Clostridium rectum (Heller) comb. nov. (*Hiberillus rectus* Heller 1922, 17; *Inflabilis rectus* (Heller) Prévot 1938, 77.)

rec'tus. L. neut.adj. *rectum* straight.

Straight rods, 0.5–1.1 by 1.6–3.1 μm, with squared ends, occurring singly and in pairs. Not motile. Spores are oval, eccentric to subterminal. Gram-positive.

Surface colonies circular, 1–2 mm in diameter with entire edge, convex, translucent, with a glossy surface.

Moderate growth in broth media with added carbohydrate. Diffuse turbidity, sediment.

Fermentation products include major amounts of butyric and smaller amounts of acetic, propionic and valeric acids.

Proteins not attacked. Milk is made acid, otherwise unchanged.

Optimum temperature for growth is 37–45 C.

Has been found in beet rhizosphere, horse manure.

The G + C content of the DNA is 26 moles %. Reference strain: ATCC 25751; NCIB 10651.

5. Clostridium paraperfringens Nakamura, Tamai and Nishida 1970, 137.

pa.ra.per.fring'ens. Gr. pref. *para* beside, M.L. n. *perfringens* a specific epithet; M.L. adj. *paraperfringens* resembling (*Clostridium*) *perfringens*.

Straight rods, 0.9–1.3 by 1.6–3.9 μm, with rounded ends. Not motile. Spores round to oval, eccentric to subterminal. Gram-positive; often granulose positive. Cell wall contains DL-diaminopimelic acid.

Surface colonies are circular, 0.5–1 mm in diameter, slightly raised, translucent with a dull smooth surface.

Moderately good growth in liquid media, diffuse turbidity and ropy sediment.

Fermentation products include acetic and butyric acids, with or without alcohols.

Milk becomes acid, otherwise unchanged.

Optimum temperature for growth is 30–45 C.

Has been found in wounds, feces.

The G + C content of the DNA is 28 moles %. Reference strain: ATCC 25753; NCIB 10652.

6. Clostridium rubrum Ng and Vaughn 1963, 1111.

ru'brum. L. neut.adj. *rubrum* reddish.

Straight or slightly curved rods, 0.5–0.8 by 2.5–6.7 μm, occurring singly, in pairs, in short chains. Motile with peritrichous flagella. Gram-positive. Spores are oval, eccentric to subterminal with no exosporium and no appendages. Cell wall contains DL-diaminopimelic acid.

Surface colonies circular with entire margin, 0.5–1.0 mm in diameter, convex, translucent, white with pink to red centers, glossy surface. Pigment best produced on high carbohydrate media.

Moderate growth in nutrient broth, cooked meat broth; better growth in broth with fermentable carbohydrate.

TABLE 15.10
Carbohydrates fermented by species of genus **Clostridium** Group I

Adonitol, cellulose and erythritol not fermented.

	1. C. butyricum	2. C. beijerinckii	3. C. oroticum	4. C. rectum	5. C. paraperfringens	6. C. rubrum	7. C. fallax	8. C. pasteurianum	9. C. sticklandii	10. C. tyrobutyricum	11. C. propionicum
Amygdalin	d	−	−	−	−	+	−	−	−	−	−
Arabinose	d	+	+	−	−	+	−	+	−	−	−
Cellobiose	+	+	+	+	+	+	−	−	−	−	−
Dulcitol	−	−	+	−	−	−	−	−	−	−	−
Esculin	d	−	−	−	−	+	+	−	−	−	−
Fructose	+	+	+	+	+	+	+	+	−	+	−
Galactose	+	+	+	+	+	+	+	+	−	−	−
Glucose	+	+	+	+	+	+	+	+	w	+	−
Glycerol	d	−	−	−	−	−	−	+	−	−	−
Glycogen	+	−	−	−	+	+	−	−	−	−	−
Inositol	d	+	+	−	−	−	−	+	−	−	−
Inulin	d	+	+	−	−	+	−	+	−	−	−
Lactose	+	+	+	+	+	+	−	−	−	−	−
Maltose	+	+	+	+	+	+	+	+	w	−	−
Mannitol	d	−	−	−	−	d	−	+	−	+	−
Mannose	+	+	+	+	+	+	+	+	−	+	−
Melezitose	−	+	+	−	−	+	−	+	−	−	−
Melibiose	+	+	−	−	−	−	−	−	−	−	−
Raffinose	+	+	+	+	−	+	−	+	−	−	−
Rhamnose	d	−	+	−	−	d	−	−	−	−	−
Ribose	+	−	+	−	d	d	d	d	w	−	−
Salicin	+	−	+	−	+	+	−	d	−	−	−
Sorbitol	−	−	−	−	−	d	d	+	−	−	−
Sorbose	−	−	−	−	−	−	d	+	−	−	−
Starch	+	−	−	−	+	d	+	−	−	−	−
Sucrose	+	+	+	+	+	+	−	+	−	−	−
Trehalose	+	+	+	−	d	d	−	+	−	−	−
Xylose	+	+	+	−	−	+	+	−	−	+	−

Fermentation products include acetic and butyric acids.

Milk becomes acid but no other change.

Optimum temperature for growth not determined. Grows well from 30–37 C.

Has been found in soil.

The G + C content of the DNA is 28 moles %. Type strain: ATCC 14949; NCIB 9503.

7. Clostridium fallax (Weinberg and Séguin) Bergey *et al.* 1923, 325. (*Bacillus fallax* Weinberg and Séguin 1915, 686; *Vallorillus fallax* (Weinberg and Séquin) Heller 1922, 16.)

fal′lax. L. adj. *fallax* deceptive.

Straight to curved rods, 0.6–0.9 by 2.8–10.7 μm, with rounded ends. Motile with peritrichous flagella. Spores are oval, eccentric to subterminal.

Gram-positive. Cell wall contains DL-diaminopimelic acid; cell wall sugar is glucose.

Surface colonies are circular to slightly irregular, 4–5 mm in diameter, entire to slightly erose margins, raised, translucent, grayish white, glossy surface.

Poor growth in nutrient broth; good growth in broth with fermentable carbohydrate.

Fermentation products include large amounts of acetic, butyric and lactic acids.

Milk is unchanged.

Optimum temperature for growth is 37–45 C.

Has been found in soil, wounds.

Type strain: ATCC 19400; NCIB 10634.

8. Clostridium pasteurianum Winogradsky 1895, 330. (*Clostridium pastorianus* (*sic*) Winograd

TABLE 15.11

Biochemical reactions of species of genus **Clostridium** *Group I*

Gelatin and casein are not hydrolyzed, indole or acetylmethylcarbinol not produced, urease or lipase not formed, toxin not produced, not pathogenic for laboratory animals.

	1. C. butyricum	2. C. beijerinckii	3. C. oroticum	4. C. rectum	5. C. paraperfringens	6. C. rubrum	7. C. fallax	8. C. pasteurianum	9. C. sticklandii	10. C. tyrobutyricum	11. C. propionicum
H₂S produced	−	−	−	+	+	−	+	−	−	−	+
Lecithinase on egg-yolk agar	−	−	−	−	+	−	−	−	−	−	−
Blood agar hemolyzed	−	+	−	−	d	−	−	−	−	−	−
Nitrate reduced	d	−	+	+	d	−	−	−	−	−	−

sky 1902, 43; *Bacillus pasteurianus* (Winogradsky) Lehmann and Neumann 1907, 82; *Bacillus pastorianus* (Winogradsky) Lehmann and Neumann 1907, 462; *Bacillus winogradsky* Matzuschita 1902, 548; *Butyribacillus pasteurianus* (Winogradsky) Orla-Jensen 1909, 342.)

pas.teu.ri.a′num. M.L. neut.adj. *pasteurianum* pertaining to Louis Pasteur, French bacteriologist.

Straight to slightly curved rods 0.5–0.8 by 3.4–13.2 μm. Motile with peritrichous flagella. Spores are oval, subterminal with no exosporium and no appendages. Gram-positive. Cell wall contains DL-diaminopimelic acid; cell wall sugars are glucose, galactose, rhamnose, mannose.

Surface colonies are circular with slightly rhizoid margin, 1–3 mm in diameter, translucent, gray, glossy surface.

Moderate growth in nutrient broth, cooked meat broth; better growth in media with fermentable carbohydrate.

Fermentation products include major amounts of acetic and butyric acids.

Milk is unchanged.

Optimum temperature for growth is 37 C.

Has been found in soil.

Fixes atmospheric nitrogen.

The G + C content of the DNA is 26–28 moles %.

Reference strain: ATCC 6013; NCIB 9486.

9. **Clostridium sticklandii** Stadtman and McClung 1957, 218.

stick′lan.di.i. M.L. gen.n. *sticklandii*, pertaining to L. H. Stickland, British biochemist.

Rods slender, straight or curved, 0.3–0.5 by 1.3–3.8 μm. Motile with peritrichous flagella. Spores are oval, subterminal. Gram-positive.

Surface colonies are circular with an entire margin, 2 mm in diameter, convex, white, translucent, glossy surface.

Good growth in nutrient broth, diffuse turbidity.

Fermentation products include acetic, butyric, isovaleric and smaller amounts of propionic and isobutyric acids, isobutyl and butyl alcohols.

Milk is unchanged.

Optimum temperature for growth is 30–35 C.

Found in black mud from San Francisco Bay.

The G + C content of the DNA is 26 moles %.

Type strain: ATCC 12662; NCIB 10654.

10. **Clostridium tyrobutyricum** van Beynum and Pette 1935, 208.

ty.ro.bu.ty′ri.cum. Gr. n. *tyrus* cheese; M.L. n. *acidum butyricum* butyric acid; M.L. neut.adj. *tyrobutyricum* the butyric acid producing organism from cheese.

Straight rods 1.5 by 4.5 μm. Motile with peritrichous flagella. Spores are oval, subterminal. Gram-positive. Cell wall contains DL-diaminopimelic acid; glucose is the only cell wall sugar.

Surface colonies are circular, 0.5 mm in diameter, convex with entire margin, gray, translucent with glossy surface.

Good growth in broth when fermentable carbohydrate is present.

Fermentation products include large amounts of acetic and butyric acids.

Milk is made acid, otherwise unchanged.

Optimum temperature for growth is 37 C.

Has been found in silage and cheese.

The G + C content of the DNA is 28 moles %.

Reference strain: ATCC 25755; NCIB 10635.

11. **Clostridium propionicum** Cardon and Barker 1946, 631.

pró.pi.o'ni.cum. M.L. neut.adj. *propionicum* pertaining to propionic acid.

Straight to slightly curved rods, 0.5–0.8 by 1.8–5.0 μm. Not motile. Spores oval. Gram-positive.

Surface colonies are circular, 0.5–1.0 mm in diameter, convex, entire margins, translucent, grayish, glossy surface.

Slight to no growth in nutrient broth not containing fermentable carbohydrate.

Fermentation products include propionic acid, with smaller amounts of acetic, isobutyric, butyric and isovaleric acids.

Milk is unchanged.

Optimum temperature for growth is 25 C.

Found in black mud from San Francisco Bay. Reference strain: ATCC 25522; NCIB 10656.

Group II

12. Clostridium ghoni Prévot 1938, 83.

gho'ni. M.L. gen.n. *ghoni*, pertaining to Professor Ghon, German bacteriologist.

Straight rods, 0.5–0.6 by 1.6–3.8 μm. Motile with peritrichous flagella. Spores are oval, central to subterminal. Gram-positive. Cell wall contains DL-diaminopimelic acid.

Surface colonies are circular with slightly irregular margin, 1–2 mm in diameter, gray, translucent with matt surface.

TABLE 15.12

Distinguishing characteristics of species of genus **Clostridium** *Group II*
Spores subterminal, gelatin hydrolyzed

	Casein digested	Lecithinase	Indole	Lipase	Glucose	Mannose	Maltose	Lactose	Salicin
12. *C. ghoni*	+	+	+	+	−	−	−	−	−
13. *C. bifermentans*[a]	+	+	+	−	+	V[g]	+	−	−
14. *C. sordellii*[a]	+	+	+	−	+	V	+	−	−
15. *C. lituseburense*	+	+	−	−	+	−	+	−	−
16. *C. limosum*[b]	+	+	−	−	−	−	−	−	−
17. *C. subterminale*[b, c]	+	V	−	−	−	−	−	−	−
18. *C. mangenotii*	+	−	+	−	−	−	−	−	−
19. *C. sporogenes*[d]	+	−	−	+	+	−	V	−	d
20. *C. botulinum*									
Types A, B, C, D, F[d]	+	−	−	+	+	−	V	−	V
Types B, C, D, E, F[d]	−	V	−	+	+	+	V	−	−
Type G[d]	+	−	−	−	−	−	−	−	−
21. *C. plagarum*	+	−	−	+	+	+	+	+	
22. *C. acetobutylicum*	d	−	−	+	+	+	+	+	+
23. *C. histolyticum*[c]	+	−	−	−	−	−	−	−	−
24. *C. aurantibutyricum*	−	−	−	+	+	+	+	+	+
25. *C. novyi*									
Type A	−	+	−	+	+	−	d	−	−
Type B	V	+	−	−	+	d	+	−	−
Type C	−	−	d	−	+	−	−	−	−
26. *C. perfringens*	−	+	−	−	+	+	−	+	V
27. *C. haemolyticum*	−	+	+	−	+	V	−	−	−
28. *C. felsineum*[e]	−	−	−	−	+	+	V	+	+
29. *C. chauvoei*	−	−	−	−	+	+	+	+	−
30. *C. septicum*[e]	−	−	−	−	+	+	+	+	V
31. *C. difficile*[f]	−	−	−	−	+	+	−	−	d

[a] *C. sordellii* produces urease; *C. bifermentans* does not.

[b] *C. limosum* produces collagenase; *C. subterminale* does not.

[c] *C. histolyticum* is aerotolerant; *C. subterminale* is not.

[d] Toxin neutralization tests are necessary for identification.

[e] *C. felsineum* ferments sucrose; *C. septicum* does not.

[f] *C. difficile* hydrolyzes gelatin very slowly.

[g] V = variable.

Moderate growth in nutrient broth and cooked meat medium.

Fermentation products include acetic, propionic, isobutyric, butyric, isovaleric, isocaproic acids and isobutanol.

Milk is coagulated and digested.

Optimum temperature for growth is 30–37 C.

Has been found in soil and the female genital tract.

Reference strain: ATCC 25757; NCIB 10636

13. Clostridium bifermentans (Weinberg and Séguin) Bergey *et al.* 1923, 323. (*Bacillus bifermentans sporogenes* Tissier and Martelly 1902, 894; *Bacillus bifermentans* Weinberg and Séguin 1918, 128; *Martellillus bifermentans* (Weinberg and Seguin) Heller 1922, 25.)

bi.fer.men'tans. L. pref. *bis* twice; L. part.adj. *fermentans* leavening; M.L. adj. *bifermentans* doubly fermenting.

Rods, 0.5–0.6 by 1.0–3.0 μm. Singly and in short chains. Sluggishly motile with peritrichous flagella. Spores are oval, central to subterminal, with a thick exosporium; some strains have appendages on spores. Gram-positive. Cell wall contains DL-diaminopimelic acid; cell wall sugars are glucose, rhamnose, mannose, galactose with variation from strain to strain.

Surface colonies are circular with irregular margins, 2–4 mm in diameter, convex, translucent to opaque depending upon degree of sporulation, white to yellowish with glossy surface.

Good growth in nutrient broth, cooked meat broth. Viscid sediment.

Fermentation products include acetic, isobutyric, isovaleric, isocaproic acids, sometimes butyric acid and ethyl, propyl and isobutyl alcohols.

Milk is digested. Nitrate reduction variable depending upon basal medium.

Optimum temperature for growth is 30–37 C.

Has been found in soil, fresh water and marine sediments, feces.

The G + C content of the DNA is 26 moles %.

Reference strain: ATCC 638; NCIB 10716.

14. Clostridium sordellii (Hall and Scott) Prévot 1938, 83. (*Bacillus oedematis sporogenes* Sordelli 1923, 55; *Bacillus sordelli* Hall and Scott 1927, 330.)

sor.del'li.i. M.L. gen.n. *sordellii* pertaining to Professor Sordelli, Argentinian bacteriologist.

Straight rods, 1.1–1.6 by 3.1–4.5 μm. Motile with peritrichous flagella. Spores are oval, central to subterminal with a thick exosporium, some with appendages. Gram-positive. Cell wall contains DL-diaminopimelic acid; cell wall sugar is glucose, trace of galactose.

Surface colonies circular to irregular, 1–2 mm in diameter, translucent to opaque depending upon degree of sporulation, grayish, matt surface.

Good growth in nutrient broth, cooked meat broth, glucose broth.

Fermentation products include major amounts of acetic, isobutyric and isovaleric acids, with smaller amounts of propionic and isocaproic acids and ethyl, propyl, isobutyl and isoamyl alcohols.

Milk is slowly digested. Nitrate reduction variable, depending upon basal medium.

Optimum temperature for growth is 37 C.

Found in infections of man and animals.

The G + C content of the DNA is 26 moles %.

Reference strain: ATCC 9714; NCIB 10717.

15. Clostridium lituseburense (Laplanche and Saissac) McClung and McCoy 1957, 664. (*Inflabilis litus-eburense* Laplanche and Saissac in Prévot 1948, 276.)

li'tus.e.bu.ren'se. L. n. *litus* coast; L. n. *ebur* ivory; M.L. adj. *litus-eburense* pertaining to the Ivory Coast.

Straight or slightly curved rods, 1.4–1.7 by 3.1–6.3 μm. Motile with peritrichous flagella. Oval subterminal spores. Gram-positive.

Surface colonies 2 mm in diameter, irregular margins, raised center, translucent, pinkish to white, glossy surface.

Moderately good growth in nutrient broth, mucoid sediment.

Fermentation products include acetic, butyric and isovaleric acids and smaller amounts of propionic and isobutyric acids and ethyl, propyl, isobutyl alcohols.

Milk is digested.

Optimum temperature for growth is 30 C; grows 25–45 C.

Has been found in soil from Ivory Coast.

The G + C content of the DNA is 26 moles %.

Reference strain: ATCC 25759; NCIB 10637.

16. Clostridium limosum Andre in Prévot 1948, 165.

li.mo'sum. L. adj. *limosum* muddy or slimy.

Rods, 0.8–1.1 by 1.7–3.1 μm. Motile with peritrichous flagella. Spores are oval, subterminal. Gram-positive. Cell walls contain DL-diaminopimelic acid; cell wall sugars are glucose, galactose, rhamnose, mannose.

Surface colonies circular, 1–2 mm in diameter, low convex, white to gray, translucent, glossy surface.

Grows well in nutrient broth with uniform turbidity, moderate sediment.

Fermentation products from peptone, yeast extract medium consist largely of acetic acid.

Milk is digested.

Toxin is produced but usually in small amounts; sometimes pathogenic for laboratory animals.

Optimum temperature for growth is 37 C; grows at 25–45 C.

Has been found in a variety of animal infections, soil.

The G + C content of the DNA is 26 moles %.

Reference strain: ATCC 25760; NCIB 10638.

17. Clostridium subterminale (Hall and Whitehead) Spray 1948, 786. (*Bacillus subterminalis* Hall and Whitehead 1927, 67.)

sub.ter.min.na′le. L. pref. *sub* under; L. adj. *terminalis* terminal; M.L. neut.adj. *subterminale* near the end, subterminal.

Rods, 0.5–1.0 by 1.9–4.4 µm. Most strains motile with peritrichous flagella. Spores are oval, subterminal with thick exosporium, no appendages. Gram-positive.

Surface colonies are 1–2 mm in diameter, irregularly circular, lobate margin, low convex, translucent, gray, matt surface.

Slight to moderate growth in nutrient broth; moderate growth in cooked meat broth.

Fermentation products from peptone, yeast extract medium include acetic, butyric, isobutyric, isovaleric acids, and ethyl, isobutyl, butyl and isoamyl alcohols.

Milk is coagulated, digested. A small amount of lecithinase may be produced on egg yolk agar.

Optimum temperature for growth is 37 C; grows at 25–45 C.

Has been found in soil, wounds.

The G + C content of the DNA is 28 moles %.

Reference strain: ATCC 25774; NCIB 9384.

18. Clostridium mangenotii (Prévot and Zimmès-Chaverou) McClung and McCoy 1957, 664. (*Inflabilis mangenoti* Prévot and Zimmès-Chaverou 1947, 603.)

man.ge.no′ti.i. M.L. gen.n. *mangenotii* pertaining to Professor Mangenot, Italian bacteriologist.

Rods, 0.6–0.9 by 3.1–8.2 µm. Not motile. Spores are oval, subterminal. Gram-positive.

Surface colonies are 0.5–1.0 mm in diameter, convex, translucent, grayish white, matt surface.

Moderate to poor growth in nutrient or cooked

TABLE 15.13
*Carbohydrates fermented by species of genus **Clostridium** Group II*

	12. C. ghoni	13. C. bifermentans	14. C. sordellii	15. C. lituseburense	16. C. limosum	17. C. subterminale	18. C. mangenotii	19. C. sporogenes	20. C. botulinum A, B, C, D, F	B, C, D, E, F	G	21. C. plagarum	22. C. acetobutylicum	23. C. histolyicum	24. C. aurantibutyricum	25. C. novyi, Type A	Type B	Type C	26. C. perfringens	27. C. haemolyticum	28. C. felsineum	29. C. chauvoei	30. C. septicum	31. C. difficile
Amygdalin	−	−	−	−	−	−	−	−	−	d	−	−	d	−	−	−	−	−	−	−	d	−	−	−
Arabinose	−	d	−	−	−	−	−	−	−	d	−	−	+	−	−	+	d	−	−	−	−	+	−	d
Cellobiose	−	−	−	−	−	−	−	−	−	−	−	−	−	−	−	−	−	−	−	−	+	d	−	−
Dulcitol	−	−	−	−	−	−	−	−	−	−	−	+	+	−	+	−	−	−	d	−	+	d	+	d
Esculin	−	−	−	−	−	−	−	−	−	−	−	−	d	−	−	−	−	−	−	−	d	−	d	−
Fructose	−	d	d	+	−	−	−	+	+	+	−	+	+	−	+	d	−	−	+	+	+	+	+	+
Galactose	−	−	−	−	−	−	−	−	d	−	−	+	+	−	+	−	+	−	+	−	+	+	d	−
Glucose	w	+	+	−	−	+	+	+	+	+	−	+	+	−	+	+	+	w	+	+	+	+	+	+
Glycerol	−	d	d	−	−	−	−	d	d	d	−	d	−	−	−	d	−	−	d	d	−	−	−	−
Glycogen	−	−	−	−	−	−	−	−	−	−	−	+	+	−	+	−	−	−	d	−	d	−	−	−
Inositol	−	−	−	−	−	−	−	−	−	d	−	−	d	−	−	d	+	w	d	−	d	−	−	−
Inulin	−	−	−	−	−	−	−	−	−	−	−	−	d	−	−	−	−	−	−	−	d	−	−	d
Lactose	−	−	−	−	−	−	−	−	−	−	−	+	+	−	+	−	−	−	+	−	d	+	+	−
Maltose	w	+	+	+	−	−	d	d	d	−	+	+	+	−	+	d	+	−	+	−	d	+	+	−
Mannitol	−	−	−	−	−	−	−	−	−	−	−	−	d	−	−	−	−	−	−	−	−	−	−	+
Mannose	−	d	−	d	−	−	+	−	−	−	+	+	−	−	+	−	−	+	−	+	+	+	+	+
Melezitose	−	−	−	−	−	−	−	−	−	−	−	−	−	−	−	−	d	−	−	−	d	−	−	d
Melibiose	−	−	−	−	−	−	d	−	−	−	−	−	−	−	−	−	−	−	−	−	−	−	−	−
Raffinose	−	−	d	−	−	−	−	−	−	−	−	d	−	+	−	d	−	−	−	d	−	−	−	−
Rhamnose	−	−	−	−	−	−	d	−	−	−	−	−	d	−	−	−	−	−	−	+	−	−	−	−
Ribose	−	d	d	d	−	−	−	−	d	−	−	−	−	−	d	d	w	d	−	+	d	d	+	d
Salicin	−	−	−	−	−	d	d	−	−	−	−	+	−	+	−	d	−	+	−	d	d	−	d	d
Sorbitol	−	d	−	−	−	d	−	−	−	−	−	d	−	−	−	−	−	−	−	−	−	−	−	d
Sorbose	−	−	−	−	−	−	−	−	−	−	−	−	d	−	+	−	−	−	−	−	−	−	−	−
Starch	−	−	−	−	−	−	−	−	−	−	d	+	+	−	+	−	−	−	+	−	+	d	−	−
Sucrose	−	−	−	+	−	−	d	d	d	−	+	+	+	−	+	d	−	−	+	−	d	−	+	−
Trehalose	−	−	−	−	−	−	d	−	−	d	−	−	+	−	+	−	−	−	d	−	−	+	−	+
Xylose	−	d	d	−	−	−	−	−	−	d	+	−	+	−	−	−	+	−	−	+	−	−	−	+

TABLE 15.14

Biochemical reactions of species of genus **Clostridium** *Group II*

	12. C. ghoni	13. C. bifermentans	14. C. sordellii	15. C. lituseburense	16. C. limosum	17. C. subterminale	18. C. mangenoti	19. C. sporogenes	20. C. botulinum A, B, C, D, F	B, C, D, E, F	G	21. C. plagarum	22. C. acetobutylicum	23. C. histolyticum	24. C. aurantibutyricum	25. C. novyi, type A	Type B	Type C	26. C. perfringens	27. C. hæmolyticum	28. C. felsineum	29. C. chauvoei	30. C. septicum	31. C. difficile
H$_2$S produced	+	d	d	–	+	+	+	+	+		–	+	–	+	d	–			–	–	–	–	–	–
Urease	–	–	+	–	+	+	+	+	+		–	–	–	–	–	–			–	d	–	–	–	–
Gelatin liquefied	+	+	+	+	+	+	+	+	+	+	+	+	+	+	+	+	+	d	+	+	+	+	+	+
Casein hydrolyzed	+	+	+	+	+	+	+	+	+		–	+	+	d	+	–		–	d	–	d	–	–	–
Nitrates reduced	–	d	d	–	+	–	–	–	–		–	+	+	d	+	–	–	+	d	d	–	d	–	+
Acetylmethyl-carbinol produced	–	–	–	–	–	–	–	–	–		–	–	+	–	–	–			–	d	+	–	–	–
Toxin produced	–	–	+	–	+	–	–	–	+	+	+	–	–	+	+	–	+	+	–	+	–	–	–	–
Pathogenic for laboratory animals	–	–	+	–	v	–	–	–	+	+	+	–	–	+	–	+	+	–	+	+	–	w	+	w
Blood hemolyzed	–	v	v	+	+	+	–	+	+	+	+	–	–	+	–	+	+	+	w	d	+	+	+	–

meat broth with stringy sediment; better growth in glucose broth.

Fermentation products from peptone, yeast extract medium include acetic, isobutyric, isovaleric acids and smaller amounts of propionic and isocaproic acids.

Milk is digested.

Optimum temperature for growth is 30–37 C.

Has been found in soil.

Reference strain: ATCC 25761; NCIB 10639.

19. Clostridium sporogenes (Heller) Bergey *et al.* 1923, 329. (*Bacillus sporogenes* var. A Metchnikoff 1908, 944; *Metchnikovillus sporogenes* Heller 1922, 29; *Clostridium sporogenes* var A (Metchnikoff) Prévot 1938, 83.)

spo.ro′ge.nes. M.L. n. *spora* a spore; Gr. v. *gennaio* produce; M.L. part.adj. *sporogenes* spore-producing.

Straight rods, 0.3–0.4 by 1.4–6.6 μm. Motile with peritrichous flagella. Spores are oval, subterminal. Gram-positive. Cell wall contains DL-diaminopimelic acid; cell wall sugar is galactose.

Surface colonies on most solid media, 2–6 mm in diameter, raised whitish to yellowish center with gray rhizoids, "Medusa head" margin, semi-opaque, matt surface. On moist media, colonies are flat, thin, spreading.

Abundant growth in nutrient, cooked meat broth; grayish yellow sediment.

Acid production from carbohydrates often masked by ammonia from amino acid deamination. Fermentation products include large amounts of butyric acid with smaller amounts of acetic, iso-butyric, isovaleric and isocaproic acids and propyl, isobutyl and isoamyl alcohols.

Milk is digested.

Optimum temperature for growth is 30–40 C; grows at 25–45 C.

Has been found in soil, wounds, food, intestinal contents.

The G + C content of the DNA is 26 moles %.

Reference strain: ATCC 3584; NCIB 10696.

20. Clostridium botulinum (van Ermengem) Bergey *et al.* 1923, 328. (*Bacillus botulinus* van Ermengem 1896, 443; *Ermengemillus botulinus* (van Ermengem) Heller 1922, 28; *Botulobacillus botulinus* (van Ermengem) Orla-Jensen 1909, 343.)

bo.tu.li′num. L. n. *botulus* sausage; M.L. adj. *botulinum* pertaining to sausage.

Six toxin types, A, B, C, D, E, F, G, are differentiated on the serological specificity of the toxin, which is not necessarily correlated with the cultural properties of the organism.

Type A and Proteolytic Strains of Types B, C, D, and F. Straight to slightly curved rods 0.8–1.3 by 4.4–8.6 μm. Motile with peritrichous flagella. Spores are oval, subterminal with no exosporium. Gram-positive. Cell wall contains DL-diaminopimelic acid; cell wall sugar is glucose.

Colonies are circular, 3–8 mm in diameter, opaque center, rhizoid margin, translucent, gray, matt to semi-glossy surface.

Abundant growth in nutrient broth, cooked meat broth, with uniform turbidity, grayish sediment.

Acid production from carbohydrates often

masked by ammonia from amino acid deamination. Fermentation products include acetic, butyric acids with smaller amounts of propionic, isobutyric, isovaleric acids and propyl, isobutyl, butyl and isoamyl alcohols.

Milk is digested.

Toxin is formed; pathogenic for laboratory animals only through action of neurotoxin; no tissue invasion.

Optimum temperature for growth is 30–40 C.

Has been found in food, soil, feces, marine sediments.

The G + C content of the DNA is 26–28 moles %.

Reference strains: Type A—ATCC 25763; NCIB 10640; proteolytic Type B—ATCC 7949; NCIB 10657; proteolytic Type F—ATCC 25764; NCIB 10658.

Type E and Non-proteolytic Strains of Type B and Type F. Straight rods, 0.3–0.7 by 3.4–7.5 μm. Gram-positive, becoming Gram-negative in older cultures. Motile with peritrichous flagella. Spores are oval, subterminal. Spores of type E have characteristic appendages and exosporium; spores of type F have an exosporium but no appendages. Cell wall contains DL-diaminopimelic acid.

Surface colonies 1–3 mm in diameter, slightly irregular with lobate margin, translucent to semiopaque, mosaic structure, matt surface.

Poor to moderate growth in nutrient broth and cooked meat broth; abundant growth in broth with fermentable carbohydrate.

Fermentation products include acetic and butyric acids.

Gelatin is hydrolyzed; some strains of type E do not hydrolyze gelatin. Milk is coagulated with a soft curd, but not digested.

Toxin is formed; pathogenic for laboratory animals only through the action of neurotoxin; no tissue invasion.

Optimum temperature for growth is variable from 25–37 C.

Has been found in soil, marine sediments, food, fish, birds, mammals.

The G + C content of the DNA is 26–28 moles %.

Reference strains: Type B—ATCC 25765; NCIB 10642; Type E—ATCC 9564; NCIB 10660; Type F—ATCC 27321; NCIB 10641.

Non-proteolytic Strains of Types C and D. Straight rods, 0.5–0.7 by 3.4–7.9 μm. Motile with peritrichous flagella. Spores are oval, subterminal. Gram-positive. Cell wall contains DL-diaminopimelic acid.

Surface colonies are circular, slightly irregular, slightly lobate margin, slightly raised, translucent, grayish white, smooth, matt surface.

Moderate growth in nutrient broth or cooked meat broth with fermentable carbohydrate.

Fermentation products are acetic, propionic and butyric acids.

Gelatin is hydrolyzed; milk is not changed.

Toxin is formed; pathogenic for laboratory animals only through the action of the neurotoxin; no active tissue invasion before death.

Optimum temperature for growth is 30–37 C.

Has been found in feces and carcasses of animals and birds; in soil only in association with birds or animals.

The G + C content of the DNA is 26–28 moles %.

Reference strains: Type C—ATCC 25766; NCIB 10618; Type D—ATCC 25767; NCIB 10619.

Type G. Straight rods, 1.3–1.9 by 1.6–9.4 μm. Motile with peritrichous flagella. Spores are oval, subterminal. Gram-positive.

Surface colonies are circular, 0.5–1.5 mm in diameter, entire margin, raised, translucent, gray, smooth, glossy surface.

Moderate growth in nutrient broth or cooked meat medium.

Fermentation products from peptone, yeast extract medium include primarily acetic acid with smaller amounts of isobutyric, butyric, isovaleric and lactic acids and propyl and butyl alcohols.

Milk is digested slowly.

Toxin is formed; pathogenic for laboratory animals.

Optimum temperature for growth is 30–37 C.

Has been found in soil.

Reference strain: ATCC 27322; NCIB 10714.

21. **Clostridium plagarum** (Prévot) comb. nov. (*Bacillus* S Adamson 1919, 373; *Inflabilis plagarum* Prévot 1938, 77.)

pla.gar′um. L. n. *plaga* a blow; M.L. gen.pl.n. *plagarum* of wounds.

Straight to slightly curved rods, 1.3–1.6 by 4.4–5.5 μm. Not motile. Spores are oval, subterminal. Gram-positive.

Surface colonies are circular, 1–3 mm in diameter, slightly raised, entire margins, peaked center, white, translucent, glossy surface.

Moderate growth in broth media containing fermentable carbohydrate.

Fermentation products include acetic and butyric acids, with smaller amounts of propionic acid and butanol.

Milk undergoes stormy fermentation, then digestion.

Optimum temperature for growth is 37 C; grows 25–45 C.

Has been found in wounds, soil.

Reference strain: ATCC 25768; NCIB 10620.

22. **Clostridium acetobutylicum** McCoy, Fred, Peterson and Hastings 1926, 483. (*Clostridium acetonobutylicum* Prévot 1938, 80; *Clostridium acetobutyricum* Prévot 1940, 110.)

a.ce.to.bu.ty'li.cum. English n. *acetone;* M.L. adj. *butylicum* butylic; M.L. neut.adj. *acetobutylicum* referring to production of acetone and butyl alcohol.

Straight rod 0.6–0.9 by 2.4–4.7 μm; often granulose positive. Motile with peritrichous flagella. Spores are oval, subterminal; no exosporium, no appendages. Gram-positive. Cell wall contains DL-diaminopimelic acid.

Surface colonies are circular, 3–5 mm in diameter, raised, irregular margin, grayish white, translucent, glossy surface.

Little growth in nutrient broth. Abundant growth in nutrient broth containing fermentable carbohydrate.

Fermentation products include acetic and butyric acids and butanol. Acetone produced.

Glucose gelatin hydrolyzed; milk coagulated, stormy fermentation; milk digested slowly by some strains.

Fixes atmospheric nitrogen.

Optimum temperature for growth is 37 C.

Has been found in soil.

The G + C content of the DNA is 28–29 moles %.

Reference strain: ATCC 824; NCIB 8052.

23. Clostridium histolyticum (Weinberg and Séguin) Bergey *et al.* 1923, 328. (*Bacillus histolyticus* Weinberg and Séguin 1916, 449; *Weinbergillus histolyticus* (Weinberg and Séguin) Heller 1922, 31.)

his.to.ly'ti.cum. Gr. n. *histus* tissue; Gr. adj. *lyticus* dissolving; M.L. neut.adj. *histolyticum* tissue-dissolving.

Straight rods, 0.6–1.0 by 1.6–3.1 μm. Sluggishly motile with peritrichous flagella. Spores are oval, subterminal; no exosporium, no appendages. Gram-positive.

Surface colonies are irregularly circular, 0.5–2 mm in diameter, convex, entire margin, transparent to opaque depending upon the sporulation, gray, glossy surface.

Abundant growth in nutrient broth, cooked meat broth, uniform turbidity, grayish sediment.

Fermentation product from peptone, yeast extract medium is primarily acetic acid.

Collagen hydrolyzed. Milk is digested.

Optimum temperature for growth is 37 C.

Aerotolerant; does not sporulate aerobically.

Has been found in soil, wounds.

Reference strain: ATCC 19401; NCIB 503.

24. Clostridium aurantibutyricum Hellinger 1944, 46.

au.ran.ti.bu.ty'ri.cum. M.L. n. *aurantium* orange; M.L. n. *acidum butyricum* butyric acid; M.L. neut.adj. *aurantibutyricum* probably intended to mean the orange-colored organism producing butyric acid.

Straight rods 0.5–0.8 by 2.8–6.3 μm. Motile, with peritrichous flagella. Gram-positive, rapidly becoming Gram-negative; often granulose positive. Spores are oval, subterminal. Cell wall contains DL-diaminopimelic acid; cell wall sugars are glucose, galactose.

Surface colonies are 1–2 mm in diameter, irregular margin, slightly raised, translucent, gray to pink-orange, glossy surface.

Slight growth in broth with or without fermentable carbohydrate.

Fermentation products include acetic with smaller amounts of butyric acid and butanol.

Stormy fermentation in milk.

Optimum temperature for growth is 30–37 C.

Has been found in rotting hibiscus stumps, flax.

The G + C content of the DNA is 28 moles %.

Reference strain: ATCC 17777; NCIB 10659.

25. Clostridium novyi (Migula) Bergey *et al.* 1923, 326. (*Bacillus novyi* Migula 1900, 872.)

no'vy.i. M.L. gen. n. *novyi* pertaining to F. G. Novy, American bacteriologist.

Type A. Straight rods, 1.1–1.4 by 1.6–10.1 μm. Motile with peritrichous flagella. Spores are oval, subterminal and have no exosporium or appendages. Gram-positive, becoming Gram-negative. Cell wall contains DL-diaminopimelic acid.

Surface colonies 1–3 mm in diameter, flat, semitransparent, lobate margin, gray, glossy surface.

Poor growth in nutrient broth, cooked meat broth without fermentable carbohydrate. Gray, flocculent sediment.

Fermentation products include acetic, propionic, butyric acids. Propionic acid produced from lactic acid.

Milk is slowly coagulated.

Produces *C. novyi* alpha, gamma, delta and epsilon toxins.

Optimum temperature for growth is 40–45 C.

Has been found in soil, wounds, marine sediments, feces.

The G + C content of the DNA is 23 moles %.

Reference strain: ATCC 17861; NCIB 10661.

Type B. Straight to curved rods, 1.4–2.5 by 4.7–22.5 μm. Motile with peritrichous flagella. Spores are oval, subterminal, with no exosporium and no appendages. Gram-positive. Cell wall contains DL-diaminopimelic acid.

Surface colonies are irregular, 1–3 mm in diameter, slightly raised, gray, translucent, glossy surface.

Moderate growth in cooked meat broth; better growth in broth media with fermentable carbohydrate.

Fermentation products include major amounts

of acetic, isobutyric, isovaleric acids with smaller amounts of propionic and isocaproic acids, with ethyl, propyl and isobutyl alcohols. Propionic acid is formed from lactic acid.

Milk is slowly digested by some strains.

Nitrate reduction is variable depending upon the basal medium.

Produces alpha, beta, zeta and eta toxins.

Optimum temperature for growth is 37 C; grows 25–45 C.

Has been found in infections of man and animals.

Reference strain: ATCC 25758; NCIB 10626.

Type C. Straight to slightly curved rods, 0.8–1.3 by 2.0–15.4 μm. Not motile. Oval, subterminal spores; no exosporium, no appendages. Gram-positive.

Surface colonies are 1–2 mm in diameter, elongate, irregular, rhizoid, slightly raised margins, translucent, gray, glossy pebbled surface.

Slight growth in nutrient broth, cooked meat broth without fermentable carbohydrate. White flocculent sediment.

Weakly ferments glucose, inositol, ribose. Fermentation products include butyric and propionic acids with smaller amounts of acetic acid. Propionic acid formed from lactic acid.

Gelatin is sometimes hydrolyzed. Milk unchanged. Indole formed occasionally. Blood agar is not or only slightly hemolyzed.

Produces small amounts of gamma toxin.

Optimum temperature for growth is 45 C.

Isolated from water buffalo with osteomyelitis.

Reference strain: VPI 2383; NCIB 9747.

26. Clostridium perfringens (Veillon and Zuber) Hauduroy *et al.* 1937, 119. (*Bacillus perfringens* Veillon and Zuber 1898, 539; *Bacterium welchii* Migula 1900, 392.)

per.frin′gens. L. part.adj. *perfringens* breaking through.

Straight rods, 0.9–1.3 by 3.0–9.0 μm. Gram-positive. Not motile. Spores are oval, subterminal and rare on usual media; no exosporium, no appendages. Cell walls contain LL-diaminopimelic acid; cell wall sugars are glucose, galactose, rhamnose.

Surface colonies are 2–5 mm in diameter, circular with occasional spreading margin, raised, grayish yellow, translucent, glossy surface.

Moderate growth in nutrient broth; profuse and rapid growth in broth with fermentable carbohydrate.

Fermentation products include acetic and butyric acids and butanol. Acetylmethylcarbinol produced by a few strains.

Nitrate reduction is variable depending upon the basal medium. Stormy fermentation in milk.

Toxin is formed. Five types, A, B, C, D and E,

are distinguished on the basis of major lethal toxins. Pathogenic for man and animals.

Optimum temperature for growth is 45 C; grows at 20–50 C.

Has been found in soil, marine sediments, wounds, feces.

The G + C content of the DNA is 24–27 moles %.

Reference strains: Type A—ATCC 13124; NCIB 6125; Type B—ATCC 3626; NCIB 10691; Type C—ATCC 3628; NCIB 10662; Type D—ATCC 3629; NCIB 10663; Type E—ATCC 27324; NCIB 10748.

27. Clostridium haemolyticum (Hall) Scott, Turner and Vawter 1935, 172. (*Clostridium hemolyticus bovis* Vawter and Records 1927, 497; *Bacillus hemolyticus* Hall 1929, 156.)

hae.mo.ly′ti.cum. Gr. n. *haema* blood; Gr. adj. *lyticus* dissolving: M.L. neut.adj. *haemolyticum* blood dissolving, hemolytic.

Straight rods, 0.9–1.6 by 2.4–6.6 μm. Motile with peritrichous flagella. Spores are oval, subterminal. Gram-positive. Cell walls contain DL-diaminopimelic acid.

Surface colonies circular to slightly irregular, 1–3 mm in diameter, undulate margins, low convex, translucent, gray, rough glossy surface.

Slight growth in nutrient broth, cooked meat medium, without fermentable carbohydrate. Flocculent sediment.

Fermentation products include acetic, propionic, butyric acids with smaller amounts of valeric acid. Propionic acid is produced from lactic acid.

Gelatin is hydrolyzed, often quite slowly. Milk is slowly coagulated or unchanged.

Toxin is formed; *C. novyi* beta, zeta, eta toxins produced. Pathogenic for cattle, sheep, laboratory animals.

Optimum temperature for growth is 37 C; does not grow at 15 C.

Strict anaerobe.

Has been isolated from liver infections of cattle, sheep.

The G + C content of the DNA is 21 moles %.

Reference strain: ATCC 9650; NCIB 10664.

28. Clostridium felsineum (Carbone and Tombolato) Bergey *et al.* 1939, 766. (*Bacillus felsineus* Carbone and Tombolato 1917, 563; *Clostridium felsinus* (sic) (Carbone and Tombolato) Bergey *et al.* 1930, 453; *Clostridium felsinae* (Carbone and Tombolato) Bergey *et al.* 1934, 490.)

fel.si′ne.um. L. n. *Felsinea* Latin name for Bologna, Italy; M.L. neut.adj. *felsineum* pertaining to Bologna.

Straight rods, 0.5–0.9 by 3.1–6.1 μm. Motile

with peritrichous flagella. Spores are oval, subterminal; no exosporium, no appendages. Gram-positive. Cell wall contains DL-diaminopimelic acid; cell wall sugars are glucose, galactose, rhamnose.

Surface colonies are slightly irregular, 0.5–2 mm in diameter, entire margins, low convex, translucent, cream color to orange-brownish on aging, pitted glossy surface.

Slight growth in nutrient broth without fermentable carbohydrate; abundant growth with fermentable carbohydrate; ropy sediment.

Fermentation products include acetic and butyric acids with small amounts of propionic acid and propyl and butyl alcohols.

Stormy fermentation of milk. No growth on blood agar or egg yolk agar.

Optimum temperature for growth is 25–37 C.

The G + C content of the DNA is 26 moles %.

Reference strain: ATCC 17788; NCIB 10690.

29. Clostridium chauvoei (Arloing, Cornevin and Thomas) Scott 1928, 259. (*Bacterium chauvoei* Arloing, Cornevin and Thomas 1887, 82; *Bacillus chauvaei* (*sic*) (Arloing *et al.*) Trevisan 1889, 22; *Clostridium chauvaei* (*sic*) (Arloing *et al.*) Scott 1928, 259.)

chau.voe′i. M.L. gén. n. *chauvoei* pertaining to Professor J. A. B. Chauveau, French bacteriologist.

Straight rods, 0.5–0.8 by 1.6–3.4 μm. Motile with peritrichous flagella. Spores are oval, subterminal. Gram-positive, irregularly staining.

Surface colonies 1–3 mm, usually circular, entire margins, translucent, grayish white, matt to glossy surface.

Slight to moderate growth in nutrient broth or cooked meat broth with or without fermentable carbohydrate. Gas production continues after growth has ceased.

Fermentation products include acetic and butyric acids and butyl alcohol.

Milk is unchanged, or slowly coagulated.

Weak toxin is formed; pathogenic for mice, guinea pigs.

Optimum temperature for growth is 37 C; does not grow above 42 C.

Has been found in bovine infections and bovine, canine intestinal contents.

Reference strain: ATCC 10092; NCIB 10665.

30. Clostridium septicum (Macé) Ford 1927, 726. (*Bacillus septicus* Macé 1889, 445; *Vibrio septicus* (Macé) Rottgardt 1926, 553.)

sep′ti.cum. Gr. adj. *septicum* putrefactive.

Rods, 1.1–1.6 by 3.1–14.1 μm. Motile with peritrichous flagella. Spores are oval, subterminal; no exosporium, no appendages. Cell wall sugars are glucose, galactose, rhamnose, mannose. Cell wall does not contain diaminopimelic acid. Gram-positive.

Surface colonies irregularly circular, 1–5 mm in diameter, slightly raised, translucent, gray, glossy surface, often swarming.

Moderate growth in nutrient broth or cooked meat broth without fermentable carbohydrate; profuse growth with fermentable carbohydrate.

Fermentation products include acetic and butyric acids with smaller amounts of ethyl, isobutyl and butyl alcohols.

Milk is slowly coagulated. Nitrate reduced by some strains.

Has been found in soil, infections in man and animals, intestinal contents.

Reference strain: ATCC 12464; NCIB 947.

31. Clostridium difficile (Hall and O'Toole) Prévot 1938, 84. (*Bacillus difficilis* Hall and O'Toole 1935, 390.)

dif′fi.cile. L. neut.adj. *difficile* difficult.

Rods, 1.3–1.6 by 3.1–6.4 μm. Motile with peritrichous flagella. Spores are oval, subterminal, becoming terminal. Gram-positive.

Surface colonies circular, 3–5 mm in diameter, entire margin, low convex, translucent, whitish, matt surface.

Moderate growth in nutrient broth with or without fermentable carbohydrate; granular sediment.

Fermentation products include small amounts of acetic, isobutyric, isovaleric, valeric, butyric, isocaproic acids, isobutanol and hexanol.

Milk is unchanged. Blood agar is not hemolyzed by most strains.

Weak toxin is produced by some strains. Pathogenic for laboratory animals.

Optimum temperature for growth is 30–37 C; grows at 25–45 C.

Has been found in infections of man, intestinal contents.

The G + C content of the DNA is 28 moles %.

Reference strain: ATCC 9689; NCIB 10666.

Group III

32. Clostridium sphenoides (Douglas, Fleming and Colebrook) Bergey *et al.* 1923, 331. (*Bacillus sphenoides* Douglas, Fleming and Colebrook in Bulloch *et al.* 1919, 43; *Douglasillus sphenoides* Heller 1922, 14; *Plectridium sphenoides* Prevot 1938, 88.)

sphe.noi′des. Gr. adj. *sphenoides* wedge-shaped.

Rods, 0.3–0.5 by 1.6–6.7 μm. Motile with peritrichous flagella. Spores are spherical, terminal to subterminal; thick exosporium, no appendages. Gram-positive. Cell wall contains DL-diaminopimelic acid.

TABLE 15.15

Distinguishing characteristics of species of genus **Clostridium** *Group III*
Spores terminal, gelatin not liquefied

	Motility	Indole	Glucose	Mannitol	Lactose	Ribose	Trehalose	Maltose	Mannose	Aero-tolerance
32. *C. sphenoides*	+	+	+	+	V[a]	V	V	+	+	−
33. *C. indolis*	+	+	+	d	+	−	−	+	d	−
34. *C. scatologenes*	+	+	+	−	−	+	−	−	+	−
35. *C. malenominatum*	+	+	−	−	−	−	−	−	−	−
36. *C. tertium*	+	−	+	+	+	+	+	+	+	+
37. *C. sartagoformum*	+	−	+	+	+	V	+	+	+	−
38. *C. cellobioparum*	+	−	+	w	w	+	−	+	+	−
39. *C. thermosaccharolyti-cum*	+	−	+	−	+	d	d	+	+	−
40. *C. pseudotetanicum*	+	−	+	−	+	+	+	+	+	−
41. *C. carnis*	+	−	+	−	d	d	−	+	+	+
42. *C. paraputrificum*	+	−	+	−	+	V	−	+	+	−
43. *C. aminovalericum*	+	−	+	−	−	−	−	−	−	−
44. *C. glycolicum*	+	−	+	−	−	−	−	−	−	−
45. *C. sporosphaeroides*	−	−	−	−	−	−	−	−	−	−
46. *C. cochlearium*	+	−	−	−	−	−	−	−	−	−
47. *C. ramosum*	−	−	+	V	+	V	+	+	+	−
48. *C. innocuum*	−	−	+	+	−	d	+	−	+	−
49. *C. barkeri*	−	−	+	+	−	+	−	−	−	−
50. *C. perenne*	−	−	+	−	+	+	−	+	+	−

ᵃ V = variable; w = weak.

Surface colonies circular, 1–2 mm in diameter, low convex, entire margins, translucent, grayish, glossy surface.

Moderate growth in nutrient broth, cooked meat broth without fermentable carbohydrate; profuse growth with fermentable carbohydrate.

Fermentation products include acetic acid, ethyl and propyl alcohols.

Milk is slowly coagulated, or unchanged.

Optimum temperature for growth is 37 C.

Has been found in infected wounds of man.

The G + C content of the DNA is 41 moles %.

Reference strain: ATCC 19403; NCIB 10627.

33. Clostridium indolis McClung and McCoy 1957, 674. (*Terminosporus indologenes* Bezjak 1952, 101.)

in.do′lis. M.L. adj. *indolis* pertaining to indole.

Straight or slightly curved rods, 0.5–0.8 by 1.3–10.2 μm. Motile with peritrichous flagella. Spores are oval to spherical, terminal; no exosporium, no appendages. Gram-positive. Cell wall contains DL-diaminopimelic acid.

Surface colonies circular, 1–3 mm in diameter, low convex, slightly irregular margin, translucent, white, matt surface.

Moderate growth in nutrient broth without fermentable carbohydrate; profuse growth with fermentable carbohydrate.

Fermentation products include major amounts of acetic acid, with smaller amounts of propionic and butyric acids and ethyl and propyl alcohols.

Milk is slowly coagulated.

Optimum temperature for growth is 37 C; grows at 25 C but not at 45 C.

Has been found in infected wounds of man.

The G + C content of the DNA is 44 moles %. Reference strain: ATCC 25771; NCIB 9731.

34. Clostridium scatologenes (Weinberg and Ginsbourg) Prévot 1948, 191. (*Clostridium scatol* Fellers and Clough 1925, 128; *Bacillus scatologenes* Weinberg and Ginsbourg 1927, 54.)

sca.to.lo′gen.es. Gr. n. *skatos* dung; Gr. v. *gennaio* produce; M.L. part.adj. *scatologenes* presumably meaning an organism that produces a dung-like odor.

Straight rods, 1.5 by 10 μm. Motile with peritrichous flagella. Spores are oval, terminal. Gram-positive.

Surface colonies, circular, 0.5–1.0 mm in diameter, convex, irregular, translucent, gray, matt surface.

Moderate to profuse growth in nutrient broth with or without fermentable carbohydrate.

Fermentation products include acetic and butyric acids with smaller amounts of propyl, butyl, isocaproic and caproic acids and propyl alcohol.

Milk is unchanged.

Optimum temperature for growth is 37 C; grows at 25 C, not at 45 C.

Has been found in contaminated food.

Reference strain: ATCC 25775; NCIB 8855.

35. Clostridium malenominatum (Weinberg et al.) Spray 1948, 786. (Bacillus malenominatus Weinberg, Nativelle and Prévot 1937, 763.)

ma.le.nom.i.na'tum. L. pref. *mal* ill; L. inf. *nominare* to name; M.L. past.part. *malenominatum* poorly named.

Straight rods, 0.5–0.8 by 2.4–7.9 μm. Motile with peritrichous flagella. Spores are oval to spherical, terminal. Gram-positive.

Surface colonies circular, 0.5–2 mm in diameter, flat, slightly irregular margins, translucent, grayish white, glossy surface.

Moderate growth in nutrient broth with or without fermentable carbohydrate.

Fermentation products from peptone, yeast extract medium include acetic and butyric acids, with smaller amounts of propionic acid.

Milk is unchanged.

Optimum temperature for growth is 37 C.

Has been found in feces.

Reference strain: ATCC 25776; NCIB 10667.

36. Clostridium tertium (Henry) Bergey et al. 1923, 332. (Bacillus tertius Henry 1917, 347; Henrillus tertius (Henry) Heller 1922, 15; Plectridium tertium (Henry) Prévot 1938, 87.)

ter'ti.um. L. neut.adj. *tertium* third.

Straight rods 0.6–1.1 by 2.4–3.9 μm. Motile with peritrichous flagella. Spores are oval, terminal; exosporium present, no appendages. Gram-positive. Cell wall does not contain diaminopimelic acid. Cell wall sugars are glucose, mannose.

Surface colonies circular, 1–4 mm in diameter, low convex, slightly irregular margins, translucent to opaque, white to gray, matt surface.

Moderate growth in nutrient broth without carbohydrate; profuse growth with fermentable carbohydrate.

Fermentation products include acetic acid with smaller amounts of butyric acid.

Milk is coagulated. Nitrate is reduced by some strains. Blood agar is not hemolyzed by most strains.

Optimum temperature for growth is 37–45 C; grows at 25–45 C.

Aerotolerant.

Has been found in wounds, soil, feces.

The G + C content of the DNA is 24–26 moles %.

Reference strain: ATCC 14573; NCIB 10697.

37. Clostridium sartagoformum Partansky and Henry 1935, 564.

sar.ta.go.for'mum. L. n. *sartago* frying pan; L. n. *forma* shape; M.L. adj. *sartagoformum* probably intended to mean shaped like a frying pan.

Straight to slightly curved rods, 0.6–0.9 by 3.1–8.0 μm. Motile. Spores are oval, terminal. Gram-positive. Cell wall does not contain diaminopimelic acid.

Surface colonies circular, 1–3 mm in diameter, entire margins, flat, gray, translucent, matt surface.

Moderate growth in nutrient broth without fermentable carbohydrate; profuse growth with fermentable carbohydrate.

Fermentation products include major amounts of acetic and butyric acids.

Milk is unchanged.

Optimum temperature for growth is 37–45 C; grows at 25–45 C.

Has been found in soil, mud.

Reference strain: ATCC 25778; NCIB 10668.

38. Clostridium cellobioparum Hungate 1944, 503. (Clostridium cellobioparus (sic) Hungate 1944, 503.)

cel.lo.bi'o.par.um. M.L. n. *cellobiosum* cellobiose; L. verb.adj. suff. *parus* producing; M.L. neut.adj. *cellobioparum* cellobiose-producing.

Straight rods, 0.5–0.6 by 1.4–3.3 μm. Motile with peritrichous flagella. Spores oval to spherical, terminal. Gram-negative. Cell wall does not contain diaminopimelic acid.

Surface colonies circular, 0.5–1.5 mm in diameter, entire margins, convex, opaque, creamy white to yellowish, glossy surface.

Slight growth in nutrient broth without fermentable carbohydrate; profuse growth with fermentable carbohydrate. Ropy sediment.

Ferments cellulose. Fermentation products include acetic acid. Acetylmethylcarbinol not formed, or in small amount.

Milk unchanged.

Optimum temperature for growth is 30–37 C; does not grow at 25 or 45 C.

Has been found in rumen contents.

The G + C content of the DNA is 25 moles %.

Reference strain: ATCC 15832; NCIB 10669.

39. Clostridium thermosaccharolyticum McClung 1935, 200. (Terminosporus thermosaccharolyticus (McClung) Prévot 1938, 86.)

ther.mo.sac'cha.ro.ly'ti.cum. Gr. adj. *thermus* hot; Gr. n. *saccharum* sugar; Gr. adj. *lyticus* dissolving; M.L. neut.adj. *thermosaccharolyticum* pre-

TABLE 15.16
Carbohydrates fermented by species of genus Clostridium Group III

	32. C. sphenoides	33. C. indolis	34. C. scatologenes	35. C. malenominatum	36. C. tertium	37. C. sartagoformum	38. C. cellobioparum	39. C. thermosaccharolyticum	40. C. pseudotetanicum	41. C. carnis	42. C. paraputrificum	43. C. aminovalericum	44. C. glycolicum	45. C. sporosphaeroides	46. C. cochlearium	47. C. ramosum	48. C. innocuum	49. C. barkeri	50. C. perenne
Adonitol	-	-	-	-	+	-	-	+	-	d	d	-	-	-	-	+	-	-	-
Amygdalin	-	d	-	-	+	-	-	+	+	d	d	-	-	-	-	+	-	-	-
Arabinose	d	d	+	-	d	-	+	+	+	-	d	-	-	-	d	d	-	-	-
Cellobiose	+	+	-	-	+	+	+	+	+	+	d	w	-	-	-	+	+	-	+
Cellulose	-	-	-	-	-	+	-	+	-	-	-	-	-	-	-	-	-	-	+
Dulcitol	-	-	-	-	d	-	+	-	-	-	-	-	d	-	-	d	-	-	-
Esculin	d	+	-	-	d	-	-	d	-	-	d	-	-	-	-	+	-	-	-
Erythritol	-	-	-	-	-	-	-	+	-	-	-	-	-	-	-	-	-	-	-
Fructose	+	+	+	-	+	+	+	+	+	d	+	-	+	-	-	+	+	+	+
Galactose	d	+	d	-	+	+	+	+	+	+	+	+	-	-	-	+	+	-	+
Glucose	+	+	+	-	+	+	+	+	+	+	+	+	+	+	-	+	+	+	+
Glycerol	d	-	+	-	-	-	w	-	+	-	+	-	-	-	-	-	-	-	-
Glycogen	d	-	-	-	+	+	-	+	+	-	-	-	-	-	-	-	-	-	-
Inositol	d	-	-	-	-	-	-	+	-	-	-	-	-	-	-	-	-	-	-
Inulin	-	-	-	-	d	-	-	+	-	d	-	-	-	-	-	+	-	-	-
Lactose	d	+	-	-	+	+	w	+	+	d	+	-	-	-	-	+	-	-	+
Maltose	+	+	-	-	+	+	+	+	+	+	+	-	-	-	-	+	-	-	+
Mannitol	+	d	-	-	+	+	w	-	-	-	-	-	-	-	d	+	+	+	-
Mannose	+	d	+	-	+	+	+	+	+	+	+	-	-	-	-	+	+	-	+
Melezitose	d	d	-	-	d	d	-	-	-	-	-	-	-	-	-	-	-	-	-
Melibiose	-	-	-	-	+	+	+	-	+	-	-	-	-	-	-	+	-	-	-
Raffinose	+	+	-	-	d	-	+	d	+	d	-	-	-	-	-	+	-	-	-
Rhamnose	+	d	+	-	d	-	+	-	-	-	d	-	-	-	-	d	-	-	-
Ribose	d	d	+	-	+	d	+	d	+	d	d	-	-	-	-	d	d	+	+
Salicin	+	+	-	-	+	+	+	+	+	+	+	-	-	-	-	+	+	-	+
Sorbitol	-	-	-	-	d	-	+	-	-	-	d	-	-	+	-	-	-	+	-
Sorbose	d	-	-	-	-	-	d	-	-	d	-	-	-	-	-	d	-	-	-
Starch	-	-	-	-	+	+	+	+	+	d	-	-	-	-	-	-	-	-	-
Sucrose	+	+	-	-	+	+	-	+	+	+	+	-	-	-	-	+	d	-	+
Trehalose	d	+	-	-	+	+	-	d	+	-	d	-	-	-	-	+	+	-	-
Xylose	+	+	-	-	+	+	+	+	+	-	d	-	+	-	-	d	d	+	+

TABLE 15.17
Biochemical reactions of species of genus Clostridium Group III
Gelatin not liquefied, casein not hydrolyzed. Urease and lipase not produced.

	32. C. sphenoides	33. C. indolis	34. C. scatologenes	35. C. malenominatum	36. C. tertium	37. C. sartagoformum	38. C. cellobioparum	39. C. thermosaccharolyticum	40. C. pseudotetanicum	41. C. carnis	42. C. paraputrificum	43. C. aminovalericum	44. C. glycolicum	45. C. sporosphaeroides	46. C. cochlearium	47. C. ramosum	48. C. innocuum	49. C. barkeri	50. C. perenne
Lecithinase produced	-	-	-	-	-	-	-	-	-	-	-	-	-	-	-	-	-	-	d
H₂S produced	+	+	+	+	-	-	-	-	-	-	-	-	+	+	d	-	-	-	w
Nitrates reduced	d	+	-	+	d	-	-	-	-	-	-	d	d	-	-	-	-	-	+
Acetylmethylcarbinol produced	-	+	-	-	-	-	d	-	-	-	-	-	-	-	-	d	-	+	-
Toxin produced	-	-	-	-	-	-	-	-	-	+	-	-	-	-	-	-	-	-	-
Pathogenic for laboratory animals	-	-	w	-	-	-	-	-	-	+	-	-	-	-	-	-	-	-	-
Blood hemolyzed	-	-	w	-	-	+	-	-	-	+	-	w	-	-	-	-	-	+	-

sumably intended to mean thermophilic and sugar-fermenting.

Straight to slightly curved rods 0.5 by 3.1–5.5 μm. Motile with peritrichous flagella. Spores are oval to spherical, terminal; no exosporium, no appendages. Gram-negative.

Surface colonies circular, low flat with raised center, irregular lobate margin, translucent, grayish, glossy surface.

Slight growth in nutrient broth with or without fermentable carbohydrate.

Fermentation products include acetic and butyric acids.

Milk is coagulated. No growth on blood agar. No growth on egg yolk agar.

Optimum temperature for growth is 55 C; no growth below 45 C.

The G + C content of the DNA is 26 moles %. Has been found in soil.

Reference strain: ATCC 7956; NCIB 9385.

40. Clostridium pseudotetanicum (Prévot) *comb. nov.* (Pseudotetanusbacillus Tavel and Lanz 1893, 162; *Bacillus pseudotetani* Migula 1900, 598; *Plectridium pseudotetanicum* Prévot 1938, 87.)

pseu.do.te.tan'i.cum. Gr. adj. *pseudes* false; M.L. gen.n. *tetani* specific epithet; M.L. adj. ending *-icus* to intensify characteristic; M.L.adj. *pseudotetanicum* false (*Clostridium*) *tetani*.

Straight rods, 0.9–1.3 by 1.9–3.4 μm. Motile with peritrichous flagella. Spores are oval, terminal. Gram-positive.

Surface colonies, irregularly circular, 1–5 mm in diameter, diffuse flat margin, translucent, gray to white, matt surface.

Moderate growth in nutrient broth without fermentable carbohydrate; profuse growth with fermentable carbohydrate.

Fermentation products include acetic and butyric acids, with lesser amounts of propionic acid.

Milk is coagulated.

Optimum temperature for growth is 30–37 C; grows at 25–45 C.

Has been found in infections of man.

The G + C content of the DNA is 27 moles %. Reference strain: ATCC 25779; NCIB 10630.

41. Clostridium carnis (Klein) Spray 1939, 750. (*Bacillus carnis* Klein 1904, 459; *Plectridium carnis* Prévot 1938, 87.)

car'nis. L. gen. n. *carnis* of flesh.

Straight to slightly curved rods, 0.6–0.9 by 2.2–6.9 μm. Motile with peritrichous flagella. Spores are oval, terminal. Gram-positive.

Surface colonies circular, 1–2 mm in diameter, raised, entire margins, translucent, grayish white, glossy surface.

Slight growth in nutrient broth with or without fermentable carbohydrate.

Fermentation products include acetic and butyric acids.

Milk is unchanged.

Optimum temperature for growth is 37 C, no growth at 25 or 45 C.

Has been found in infections of man and animals, soil.

The G + C content of the DNA is 25 moles %. Reference strain: ATCC 25777; NCIB 10670.

42. Clostridium paraputrificum (Bienstock) Snyder 1936, 402. (*Bacillus diaphthirus* Trevisan 1889, 15; *Bacillus paraputrificus* Bienstock 1906, 413.)

pa.ra.pu.tri'fi.cum. Gr. pref. *para* beside; M.L. n. *putrificum*, a specific epithet; M.L. neut.adj. *paraputrificum* resembling (*Clostridium*) *putrificum*.

Straight to slightly curved rods 0.6–0.8 by 3.1–6.6 μm. Motile with peritrichous flagella. Spores are oval, terminal. Gram-positive becoming Gram-negative quickly. Cell wall does not contain diaminopimelic acid. Cell wall sugars are glucose, galactose, rhamnose.

Surface colonies irregularly circular, 5–7 mm in diameter, diffuse margin, flat, translucent, grayish white, glossy surface.

Moderate growth in nutrient broth without fermentable carbohydrate; profuse growth with fermentable carbohydrate.

Fermentation products include acetic and butyric acids with smaller amounts of propionic acid.

Milk is coagulated. Nitrate is not reduced by most strains.

Optimum temperature for growth is 30–37 C; grows at 25–45 C.

Has been found in soil, feces.

The G + C content of the DNA is 26–27 moles %.

Reference strain: ATCC 25780; NCIB 10671.

43. Clostridium aminovalericum Hardman and Stadtman 1960, 552.

a.mi'no.va.ler'i.cum. M.L. adj.suff. *-icum* intensifying a character; M.L. neut.adj. *aminovalericum* referring to ability to ferment aminovaleric acid strongly.

Straight rods 0.3–0.5 by 1.5–5.2 μm. Motile with peritrichous flagella. Spores are spherical, terminal. Gram-positive. Cell wall does not contain diaminopimelic acid.

Surface colonies irregularly circular, 1–2 mm in diameter, entire margins, flat, translucent, gray, matt surface.

Moderate growth in nutrient broth with or without fermentable carbohydrate.

Ferments glucose, cellobiose weakly, aminovaleric acid strongly.

Fermentation products include major amounts

of acetic acid. Acetylmethylcarbinol is not produced.

Milk is unchanged. Blood agar is weakly hemolyzed.

Optimum temperature for growth is 37 C; grows 25–45 C.

Has been found in sewage sludge.

Type strain: ATCC 13725; NCIB 10631.

44. Clostridium glycolicum Gaston and Stadtman 1963, 361.

gly.co′li.cum. M.L. adj.suff. -icum intensifying a character; M.L. neut.adj. glycolicum referring to ability to ferment ethylene glycol.

Straight to slightly curved rods, 0.2–0.5 by 1.8–7.3 μm. Motile with peritrichous flagella. Spores are oval, terminal. Gram-positive. Cell wall does not contain diaminopimelic acid.

Surface colonies circular, 3–4 mm in diameter, erose margin, low convex, opaque, white, matt surface.

Moderate growth in nutrient broth without fermentable carbohydrate; profuse growth with fermentable carbohydrate.

Ferments ethylene glycol.

Fermentation products include large amounts of acetic and isovaleric acids, with smaller amounts of propionic acid, propyl, isocaproic and caproic alcohols.

Milk is unchanged. Indole is not formed by most strains.

Optimum temperature for growth is 25–37 C; grows at 25–45 C.

Has been found in soil.

The G + C content of the DNA is 29 moles %.

Type strain: ATCC 14880; NCIB 10632.

45. Clostridium sporosphaeroides Soriano and Soriano 1948, 39.

spo.ro.sphae.roi′des. Gr. n. sporus seed; Gr. adj. sphaeroides globular; M.L. neut.adj. sporosphaeroides spherical spores.

Straight to slightly curving rods, 0.5–0.6 by 4.0–8.0 μm. Not motile. Spores are oval to spherical, terminal. Gram-positive.

Surface colonies, circular, 1–2 mm in diameter, slightly raised, slightly lobate margin, translucent, gray, glossy surface.

Poor to moderate growth in nutrient broth, cooked meat broth. Ropy sediment.

Carbohydrates not fermented. Fermentation products from peptone, yeast extract medium include acetic, propionic and butyric acids with smaller amounts of propyl, isobutyl and amyl alcohols.

Milk is unchanged.

Optimum temperature for growth is 37–45 C.

Has been found in canned food.

Reference strain: ATCC 25781; NCIB 10672.

46. Clostridium cochlearium (Douglas, Fleming and Colebrook) Bergey et al. 1923, 333. (Bacillus cochlearius Douglas, Fleming and Colebrook in Bulloch et al. 1919, 40; Clostridium cochlearum (sic) (Douglas et al.) Bergey et al. 1923, 333.)

coch.le.a′ri.um. L. n. cochlear spoon; M.L. neut. adj. cochlearium resembling a spoon.

Straight to curved rods, 0.8–1.3 by 2.8–14.1 μm. Motile with peritrichous flagella. Spores are oval, terminal. Gram-positive.

Surface colonies, irregular, 1–3 mm in diameter, lobate margin, flat, translucent, gray, glossy surface.

Moderate to profuse growth in nutrient and cooked meat broths.

Carbohydrates are not fermented.

Fermentation products from peptone, yeast extract medium include acetic, propionic and butyric acids.

Milk is unchanged. Hydrogen sulfide may or may not be produced.

Optimum temperature for growth is 37–45 C; grows at 25–45 C.

Has been found in soil, wounds, feces.

Reference strain: ATCC 17787; NCIB 10633.

47. Clostridium ramosum (Vuillemin) Holdeman, Cato and Moore 1971, 39. (Nocardia ramosa Vuillemin 1931, 32.)

ra.mo′sum. L. neut.adj. ramosum much branched.

Rods 0.3–0.5 by 1.9–4.9 μm, often in short chains. Not motile. Spores are oval to spherical, terminal. Gram-positive. Cell walls contain DL-diaminopimelic acid.

Surface colonies, circular, 1–2 mm in diameter, entire margin raised, peaked, translucent, white, glossy surface.

Slight growth in nutrient broth, cooked meat broth, without fermentable carbohydrate; profuse growth with fermentable carbohydrate.

Fermentation products include acetic and formic acids, with smaller amounts of lactic acid.

Milk is coagulated.

Optimum temperature for growth is 30–37 C. Grows at 25–45 C.

Has been found in human and animal infections and in feces.

The G + C content of the DNA is 27 moles %.

Neotype strain: Prevot 113-1; VPI 0427; ATCC 25582; NCIB 10673.

48. Clostridium innocuum Smith and King 1962, 939.

in.noc′u.um. L. neut.adj. innocuum harmless.

Rods, 0.8–1.4 by 1.6–3.5 μm. Sporing cells are larger than vegetative cells. Not motile. Spores are oval, terminal. Gram-positive. Cell walls do not contain diaminopimelic acid. Cell wall sugars are glucose, galactose.

Surface colonies, circular, 2–3 mm in diameter, low convex, entire margin, translucent, white, glossy surface.

Slight growth in nutrient broth, cooked meat broth, without fermentable carbohydrate; moderate growth with fermentable carbohydrate.

Fermentation products include butyric acid, with smaller amounts of acetic acid and isoamyl alcohol.

Milk is unchanged.

Optimum temperature for growth 37–45 C.

Has been found in human infections.

The G + C content of the DNA is 43 moles %.

Type strain: ATCC 14501; NCIB 10674.

49. Clostridium barkeri Stadtman, Stadtman, Paston and Smith 1972, 760.

bar'ker.i. M.L. gen.n. *barkeri* pertaining to Professor H. A. Barker, American biochemist.

Rods, 0.3–0.5 by 1.6–9.7 µm. Not motile. Spores are oval, terminal. Cell wall does not contain diaminopimelic acid. Gram-positive.

Surface colonies, circular, 0.5–2.5 mm in diameter, convex, entire margin, fine granular structure, translucent, white, glossy surface.

Moderate growth in nutrient broth, cooked meat broth with or without fermentable carbohydrate.

Ferments nicotinic acid.

Fermentation products include butyric and lactic acids and smaller amounts of propyl, isobutyl, butyl and isoamyl alcohols.

Milk is unchanged.

Optimum temperature for growth 37 C; grows at 25–45 C.

Has been found in Potomac river mud.

Type strain: ATCC 25849; NCIB 10623.

50. Clostridium perenne (Prévot) McClung and McCoy 1957, 673. (*Acuformis perennis* Prévot 1940, 576.)

pe.ren'ne. L. neut.adj. *perenne* perpetual.

Straight to slightly curved rods, 0.8–1.3 by 3.1–12.8 µm. Not motile. Spores are oval, terminal. Gram-positive. Cell wall contains DL-diaminopimelic acid.

Surface colonies, irregularly circular, 1–3 mm in diameter, irregular lobate margin, slightly raised, translucent to opaque, gray, mosaic structure, glossy surface.

Slight growth in nutrient broth without fermentable carbohydrate; moderate to profuse growth with fermentable carbohydrate.

Fermentation products include acetic and butyric acids and propyl alcohol.

Milk is coagulated with a soft curd. Hydrogen sulfide production weak. Lecithinase sometimes produced on egg yolk agar.

Optimum temperature for growth 30–37 C; grows at 25–45 C.

Has been found in wound infections of man and in feces.

The G + C content of the DNA is 25 moles %.

Reference strain: ATCC 25782; NCIB 10675.

Group IV

51. Clostridium cadaveris (Klein) McClung and McCoy 1957, 672. (*Bacillus cadaveris* Klein 1899, 280; *Plectridium cadaveris* (Klein) Prévot 1938, 88; Not *Clostridium cadaveris* Sternberg 1890, 213.)

ca.dav'er.is. L. n. *cadaver* dead body; L. gen. n. *cadaveris* of a corpse.

Straight rods, 0.5–0.8 by 2.4–4.7 µm. Motile with peritrichous flagella. Spores are oval, terminal. Gram-positive. Cell wall contains DL-diaminopimelic acid.

Surface colonies, circular, 1–2 mm in diameter, convex, translucent, white, glossy surface.

Slight growth in nutrient broth without fermentable carbohydrate; profuse growth with heavy sediment in glucose broth.

Ferments glucose. Variable on fructose. Fermentation products include acetic, butyric acids and smaller amounts of propionic, isobutyric and isovaleric acids and ethyl, propyl, isobutyl and butyl alcohols.

Blood agar is only slightly hemolyzed or not at all. Milk digested slowly.

Optimum temperature for growth 25–37 C; grows at 25–45 C.

Has been found in infections of animals and man and in feces.

The G + C content of the DNA is 27 moles %.

Reference strain: ATCC 25783; NCIB 10676.

52. Clostridium lentoputrescens Hartsell and Rettger 1934, 511. (*Plectridium lentoputrescens* (Hartsell and Rettger) Prévot 1938, 88.)

len.to.pu.tres'cens. L. adj. *lentus* slow; L. part. adj. *putrescens* decaying; M.L. neut.adj. *lentoputrescens* slowly decaying.

Rods, 0.3–0.7 by 1.9–6.9 µm. Motile with peritrichous flagella. Spores are spherical, terminal. Gram-positive. Cell wall contains DL-diaminopimelic acid.

Surface colonies circular, 1–3 mm in diameter, slightly irregular margins, low convex, translucent, grayish white, glossy surface.

Slight to moderate growth in nutrient broth; better growth in cooked meat medium.

Ferments glucose weakly if at all. Fermentation products include acetic, propionic, butyric acids and propyl and isobutyl alcohols.

Milk may or may not be digested.

Optimum temperature for growth is 37 C.

TABLE 15.18

Distinguishing and other characteristics of species of genus **Clostridium** *Group IV*

Spores terminal, gelatin liquefied

Lipases, urease and acetylmethylcarbinol not produced; nitrates not reduced.

	51. *C. cadaveris*	52. *C. lentoputrescens*	53. *C. putrificum*	54. *C. oceanicum*	55. *C. tetani*	56. *C. putrefaciens*
Milk digested	+	d	+	+	d	−
Casein hydrolyzed	+	+	+	+	−	−
Indole produced	+	+	−	−	d	−
Glucose fermented[a]	+	−	+	+	−	+
Maltose fermented	−	−	−	+	−	−
Spore shape[b]	O	S	OS	O	S	OS
Grows at 0 C	−	−	−	−	−	+
H₂S produced	+	+	+	+	d	−
Toxin formed	−	−	−	−	+	−
Pathogenic to laboratory animals	−	−	−	−	+	−
Blood hemolyzed	w	−	+	+	+	+
Lecithinase produced	−	−	−	+	−	−

[a] Other carbohydrates listed in Table 15.15 are not fermented unless mentioned in descriptions
[b] O = oval; S = spherical.

Has been found in soil, feces.

Reference strain: possible type or cotype strain, ATCC 17794; NCIB 10629.

53. Clostridium putrificum (Trevisan) Reddish and Rettger 1922, 9. (*Pacinia putrifica* Trevisan 1889, 23; *Bacillus putrificus* (Trevisan) Bienstock 1899, 861.)

pu.tri′fi.cum. L. neut.adj. *putrificum* making rotten.

Straight and slightly curved rods, 0.3–0.5 by 3.1–7.9 μm. Motile with peritrichous flagella. Spores are oval, terminal. Gram-positive. Cell wall contains DL-diaminopimelic acid.

Surface colonies, circular, 1–2 mm in diameter, irregular margins, raised, translucent, white, glossy surface.

Moderate growth in nutrient broth, cooked meat broth without fermentable carbohydrate; profuse growth in glucose broth.

Ferments glucose, with the production of acetic acid and smaller amounts of propionic, isobutyric, butyric, isovaleric, isocaproic acids, and ethyl, propyl, isobutyl and butyl alcohols.

Optimum temperature for growth 37–45 C.

Has been found in soil, feces and wounds.

The G + C content of the DNA is 22 moles %.

Reference strain: ATCC 25784; NCIB 10677.

54. Clostridium oceanicum Smith 1970, 811.

o.ce.an′i.cum. L. adj. *oceanicum* belonging to the sea.

Straight to curved rods, 0.5–1.0 by 2.0–11 μm. Motile with peritrichous flagella. Spores are oval to spherical, terminal. Gram-positive. Cell wall does not contain diaminopimelic acid.

Surface colonies, irregularly circular, 2–6 mm in diameter, undulate margin, raised, translucent, gray, glossy surface.

Slight growth in nutrient and cooked meat broths without fermentable carbohydrate; moderate growth with fermentable carbohydrate.

Ferments fructose, galactose, glucose, maltose and mannose. Fermentation products include acetic, isobutyric, butyric, isovaleric acids and propyl and butyl alcohols.

Optimum temperature for growth is 30–37 C; grows at 20 C, not at 45 C.

Has been found in marine sediments.

The G + C content of the DNA is 27–28 moles %.

Type strain: ATCC 25647; NCIB 10625.

55. Clostridium tetani (Flügge) Bergey *et al.* 1923, 330. (*Bacillus tetani* Flugge 1886, 274.)

te′ta.ni. Gr. n. *tetanus* tetanus; M.L. gen.n. *tetani* of tetanus.

Straight rods, 0.5–1.1 by 2.4–5.0 μm. Most strains

motile with peritrichous flagella. Spores are spherical, terminal; no exosporium, no appendages. Gram-positive, becoming Gram-negative. Cell wall contains DL-diaminopimelic acid; cell wall sugars are glucose, rhamnose with traces of galactose, mannose.

Surface colonies, circular, 4–6 mm diameter, irregular margins, flat, translucent, gray, matt surface, often with swarming on moist agar.

Slight growth in nutrient broth; better growth in cooked meat broth.

Very rarely, strains have been reported to ferment glucose.

Fermentation products from peptone, yeast extract medium include acetic and butyric acids with smaller amounts of propionic acid, ethyl and butyl alcohols.

Milk is slowly clotted, or precipitated, or left unchanged. Hydrogen sulfide production is variable, depending on the medium. Indole production is variable.

Optimum temperature for growth 37 C.; no growth at 45 C.

Has been found in wounds of man and animals, in soil and feces.

The G + C content of the DNA is 25 moles %. Reference strain: ATCC 8033; NCIB 10628.

56. Clostridium putrefaciens (McBryde) Sturges and Drake 1927, 175. (*Bacillus putrefaciens* McBryde 1911, 50; *Palmula putrefaciens* (McBryde) Prévot 1938, 89; *Acuformis putrefaciens* (McBryde) Prévot 1940, 168.)

pu.tre.fa′ci.ens. L. adj. *putrefaciens* putrefying.

Rods, 1.5–1.75 by 7.5 μm, often in long filaments. Not motile. Spores are oval or spherical, terminal. Gram-positive.

Surface colonies, irregularly shaped, 1–4 mm, low, flat, tangled rhizoids, opaque to translucent, gray, matt surface.

Slight to moderate growth in nutrient broth, better growth in cooked meat broth.

Ferments glucose. Variable weak fermentation of fructose, galactose, mannose, sorbose, xylose. Fermentation products include acetic, butyric and formic acids.

Optimum temperature for growth is 20–25 C; grows at 0–30 C.

Has been found in pork.

Reference strain: ATCC 25786; NCIB 9836.

Group V

57. Clostridium acidiurici (Liebert) Barker 1938, 323. (*Bacillus acidi-urici* Liebert 1909, 1001.)

a.ci.di.u′ri.ci. M.L. n. *acidum uricum* uric acid; M.L. gen.n. *acidiurici* of uric acid.

Rods, 0.5–0.7 by 2.5–4.0 μm. Motile. Gram-variable to Gram-negative. Spores oval, terminal.

Surface colonies on uric acid agar, 1–2 mm, raised, entire margins, opaque, white.

No growth in gelatin medium, milk or the usual laboratory media.

Does not ferment any carbohydrates.

Distinctive characteristics: Requires uric acid or other purines as a source of carbon and energy.

Isolated from soil.

Reference strain: NCIB 10678.

58. Clostridium brevifaciens Bucher 1961, 644.

bre.vi.fa′ci.ens. L. adj. *brevis* short; L. part. adj. *faciens* making; M.L. part.adj. *brevifaciens* referring to the shortening of larvae infected with the organism.

Rods, 0.9–1.3 by 3–10 μm. Motile. Gram-negative.

Surface colonies: 50–150 μm, elongated, irregular margin, only a few cells thick, colorless, transparent.

No growth in the usual laboratory media.

Requires alkaline environment (pH 8.5–10.2), a high concentration of potassium or sodium and a growth factor from apple leaves.

Isolated from intestinal tract of the larvae of *Malacosoma pluviale* suffering from brachytosis.

59. Clostridium kluyveri Barker and Taha 1942, 362. (*Terminosporus kluyveri* (Barker and Taha) Prévot 1948, 198.)

kluy′ve.ri. M.L. gen.n. *kluyveri* named for Professor A. J. Kluyver, Dutch microbiologist.

Rods, 0.9–1.1 by 3.0–11.0 μm. Gram-negative. Motile.

No growth in the usual laboratory media.

Requires a high concentration of yeast extract in the medium.

Ferments ethanol to yield caproic acid.

Isolated from black mud of fresh water and marine origin.

Reference strain: ATCC 8527; NCIB 10680.

60. Clostridium malacosomae Bucher 1961, 645. (*Paraplectrum malacosomase* (Bucher) Prévot 1966, 210.)

ma.la.co.so′ma.e. L. gen.n. *malacosomae* of *Malacosoma* (*pluviale*) whose larvae act as host to the bacterium.

Rods, 1 by 4–7 μm. Gram-negative. Not motile.

Surface colonies: 0.5 mm, circular, flat irregular margins, translucent, colorless, glossy.

No growth in the usual laboratory media.

Requires alkaline environment (pH 8.5–10.2), a high concentration of potassium or sodium, and a growth factor from apple leaves.

Isolated from intestinal tract of larvae of *Malacosoma pluviale*.

61. Clostridium thermocellum Viljoen, Fred and Peterson 1926, 7.

ther.mo.cel′lum. Gr. adj. *thermus* hot; M.L. n. *cellulosum* cellulose; M.L. adj. *thermocellum* (probably intended to mean) a thermophile that digests cellulose.

Rods, 0.6–0.7 by 2.5–3.5 μm. Gram-negative. Motile. Oval, terminal spores.

Surface colonies watery, slightly convex, frequently with an insoluble yellowish pigment.

No growth on most media unless they contain cellulose, cellobiose, xylose or one of the hemicelluloses. Fermentation products include formic, acetic, lactic and succinic acids and ethyl alcohol.

Isolated from horse manure, soil, marine mud, human feces.

Reference strain: ATCC 27405; NCIB 10682.

Species incertae sedis

Species for which strains were not available for study.

a. *Clostridium albuminolyticum* González 1956, 329.
b. *Clostridium aurantiacum* Suto, Kurashima, Namba, Matsuki and Furusaka 1960, 121.
c. *Endosporus azotophagus* Tchan and Pochon 1950, 418.
d. *Clostridium butanologenum* Asai and Haruda 1943, 872.
e. *Clostridium catenaforme* Zarma 1963, 675.
f. *Plectridium causophilum* Lebert 1949, 500.
g. *Clostridium ellipsosporogenes* Rotmistrov 1939, 66.
h. *Plectridium imitans* Novotny 1953, 101.
i. *Clostridium kainantoi* Takeda and Matsui 1955, 78.
j. *Inflabilis lacustris* Prévot, Thouvenot, Partrigalla and Sillioc 1956, 931.
k. *Bacillus linumus* Nakahama 1940, 41.
l. *Clostridium maebashi* Iseki, Furukawa and Yamamoto 1959, 509.

m. *Clostridium omnivorum* Prévot, Thouvenot and Kaiser 1959, 433.
n. *Clostridium saturni-rubrum* Prévot 1964, 1035.
o. *Clostridium sporoappendiculatum* Krasil'nikov, Duda and Sokolov 1964, 455.
p. *Clostridium sporocomosum* Krasil'nikov, Duda and Sokolov 1964, 457.
q. *Clostridium sporocristum* Krasil'nikov, Duda and Sokolov 1964, 456.
r. *Bacillus sporofasciens* Krasil'nikov, Duda and Sokolov 1964, 435.
s. *Clostridium sporofilum* Krasil'nikov, Duda and Sokolov 1964, 456.
t. *Bacillus sporopenatus* Krasil'nikov, Duda and Sokolov 1964, 434.
u. *Clostridium sporopenitum* Krasil'nikov and Duda 1963, 454.
v. *Clostridium sporopilosum* Krasil'nikov, Duda and Sokolov 1964, 457.
w. *Clostridium spororenalis* Krasil'nikov, Duda and Sokolov 1964, 458.
x. *Bacillus sporoscopolus* Krasil'nikov, Duda and Sokolov 1964, 435.
y. *Clostridium sporosetosum* Krasil'nikov, Duda and Sokolov 1964, 457.
z. *Clostridium sporospilum* Krasil'nikov, Duda and Sokolov 1964, 456.
aa. *Clostridium sporospinum* Krasil'nikov, Duda and Sokolov 1964, 457.
bb. *Clostridium sporostellatum* Krasil'nikov, Duda and Sokolov 1964, 457.
cc. *Clostridium sporotrichum* Krasil'nikov, Duda and Sokolov 1964, 456.
dd. *Clostridium sporoverrucosum* Krasil'nikov, Duda and Sokolov 1964, 457.
ee. *Plectridium thermocausophilum* Jacob 1961, 34.
ff. *Clostridium thermocellulolyticum* (Pochon) Ostertag 1952, 501. (*Terminosporum thermocellulolyticus* Pochon 1942, 354.)
gg. *Clostridium thermofermentans* Tidel'skaya 1939, 188.
hh. *Clostridium toanum* Baba 1943, 191.

Genus IV. **Desulfotomaculum** *Campbell and Postgate 1965, 361*

L. Leon Campbell

De.sul.fo.to.ma′cu.lum. L. pref. *de* from; L. n. *sulfur* sulfur; L. n. *tomaculum* sausage; M.L. neut.n. *Desulfotomaculum* a sausage (-shaped organism) that reduces sulfur compounds.

Straight or curved rods, 0.3–1.5 by 3–6 μm, with rounded ends, usually single but sometimes in chains. **Motile with peritrichous flagella. Spores oval to round, terminal to subterminal, causing slight swelling of the cells. Gram-negative.** Produce black colonies in lactate-sulfate agar containing ferrous salts.

Chemoorganotrophs; **metabolism respiratory.**

Sulfates, sulfites and reducible sulfur compounds act as electron acceptors and are reduced to H₂S. Lactate and pyruvate are universal electron donors; limited range of substrates utilized, rarely includes carbohydrates; acetate not oxidized. **Oxidation of organic substrates incomplete leading to formation of acetate or homologue and CO₂. Cells contain a cyto-**

chrome of the protoheme class. Specialized media containing a reducible sulfur compound and organic growth factors are required for growth (Adams and Postgate, 1959; Campbell *et al.*, 1957; Postgate and Campbell, 1963).

Strict anaerobes.

Temperature optimum 35–55 C; maximum 70 C. Some strains grow at 30 C.

Common inhibitants of soil, fresh water, geo-thermal regions, certain spoiled foods, intestines of insects, and in rumen content. None has been reported to be pathogenic to man, guinea pig, mouse, rat or rabbit.

The G + C content of the DNA ranges from 41.7–45.5 moles % (buoyant density).

Type species: *Desulfotomaculum nigrificans* (Werkman and Weaver) Campbell and Postgate 1965, 360.

Description of the species of genus **Desulfotomaculum**

1. **Desulfotomaculum nigrificans** (Werkman and Weaver) Campbell and Postgate 1965, 361. (*Clostridium nigrificans* Werkman and Weaver 1927, 63; *Sporovibrio desulfuricans* (Beijerinck) Starkey 1938, 300.)

nig.ri'fi.cans. L. part.adj. *nigrificans* blackening.

Straight to curved rods 0.3–0.5 by 3–6 μm with rounded ends. Sometimes lenticulate to swollen. "Twisting and tumbling" motility.

Utilizes lactate, pyruvate, glucose and ethanol as hydrogen donors for sulfate reduction. Acetate and formate are not utilized. Pyruvate usually supports growth in presence or absence of sulfate. H₂S formed from cystine.

Thermophilic, temperature range 45–70 C. Optimal temperature 55 C. Can be adapted to grow slowly at 30–37 C.

Growth is inhibited by <0.1 μg/ml of hibitane.

The G + C content of the DNA is 44.7–46.6 moles % (buoyant density).

Type strain not extant. Suggested neotype: Delft 74T; NCIB 8395; ATCC 19998.

2. **Desulfotomaculum ruminis** Campbell and Postgate 1965, 361.

ru.min'is. L. n. *rumen* throat, adopted for first stomach (rumen) of a ruminant; L. gen.n. *ruminis* of a rumen.

Straight to curved rods, 0.5 by 3–6 μm, with rounded ends. Sometimes paired. Slight tumbling motility. Grows in specialized media containing lactate, pyruvate or formate (but not acetate or ethanol) plus sulfate and produces H₂S. Pyruvate supports growth with or without sulfate.

Mesophilic, temperature range 30–48 C; optimal temperature 37 C.

Growth is inhibited by 1 μg/ml of hibitane.

The G + C content of the DNA is 45.5 moles % (buoyant density).

Type strain: Coleman DL; NCIB 8452; ATCC 23193 (Campbell and Postgate, 1969).

3. **Desulfotomaculum orientis** (Adams and Postgate) Campbell and Postgate 1965, 361. (*Desulfovibrio orientis* Adams and Postgate 1959, 256.)

or.i.en'tis. L. part.adj. *oriens* rising (sun), hence the orient; L. gen.n. *orientis* of the orient.

Fat curved rods, 1.5 by 5 μm, sometimes paired. "Tumbling and twisting" motility. Spore round, central or paracentral (on rare occasions terminal), slightly swelling the cells.

Grows in specialized media containing lactate or pyruvate (but not formate, acetate or ethanol), plus sulfate and thioglycolate and produces H₂S. Does not grow without sulfate, even with pyruvate.

Mesophilic, temperature range 30–42 C. Optimal temperature between 30 and 37 C.

Growth is inhibited by <0.1 μg/ml of hibitane.

The G + C content of the DNA is 41.7 moles % (buoyant density).

Type strain: Singapore I; NCIB 8382; ATCC 19365.

Genus V. **Sporosarcina** Kluyver and van Niel 1936, 401

T. Gibson†

(*Sporosarcina* Orla-Jensen 1909, 340. Illeg. Rule 12c.)

Spo.ro.sar.ci'na. M.L. n. *spora* a spore; M.L. fem.n. *Sarcina* generic name; M.L. fem.n. *Sporosarcina* spore-forming *Sarcina*.

Cells spherical, occurring in regular tetrads or packets. May be motile or non-motile. **Endospores formed.** Gram-positive.

Chemoorganotrophs; energy-yielding metabolism strictly respiratory, oxygen being the terminal electron acceptor. Growth reported only in complex media; minimal nutritional requirements unknown.

Strict aerobes.

Type species: *Sporosarcina ureae* (Beijerinck) Kluyver and van Niel 1936, 401.

---

† Deceased.

Further Comments

Orla-Jensen's (1909) proposal of the name *Sporosarcina* was equivocal and therefore illegitimate (Rule 12c). Kluyver and van Niel accepted this alternative of Orla-Jensen and circumscribed the genus.

The characteristics of *Sporosarcina* are partly those of the genus *Micrococcus* and partly those of the genus *Bacillus*. The vegetative cells are like those of *Micrococcus* in their constantly spherical shape, their method of division in two or three perpendicular planes and the composition of their cell walls. The production of endospores is a link with *Bacillus* as is motility and the base composition of the DNA. The briefly described *Urosarcina dimorpha* Beijerinck 1901, 53, which forms endospores but lacks motility, might well be considered for transfer to the genus *Sporosarcina*.

Description of the species of genus Sporosarcina

1. **Sporosarcina ureae** (Beijerinck) Kluyver and van Niel 1936, 401. (*Planosarcina ureae* Beijerinck 1901, 52; *Sporosarcina ureae* (Beijerinck) Orla-Jensen 1909, 340, Illeg.; *Sarcina ureae* (Beijerinck) Löhnis 1911, 138.)

u're.ae. Gr. n. *urum* urine; M.L. n. *urea* urea; M.L. gen.n. *ureae* of urea.

Cells spherical, 1.2–2.5 μm in diameter, forming tetrads, more rarely cubical bundles. Motile by several randomly spaced flagella on each cell. Endospores, 0.8–1.0 μm in diameter, formed centrally. On germination the spore swells and acquires the properties of a vegetative cell (Thompson and Leadbetter, 1963). The chief amino acids of the vegetative cell wall are alanine, glutamic acid, lysine and glycine; diaminopimelic acid is absent (Cummins and Harris, 1956). Spores contain dipicolinic acid, 4–7% of dry weight (Thompson and Leadbetter, 1963).

Colonies on agar are circular, gray, microscopically coarsely granular, becoming opaque and yellowish, brownish or orange in different strains and on different media. A suitable medium for vegetative growth contains peptone, meat extract, NH₄Cl, each 0.5%, at pH 8–8.5.

Broth becomes turbid; later an easily dispersed sediment; a granular growth may form on the walls of the tube.

No change in litmus milk or slowly becomes alkaline.

Carbon sources in a medium that permits some growth in their absence are acetate, lactate, succinate, malate and citrate.

No action on glucose, starch, gelatin (within 4 weeks), casein, uric acid or blood. Arginine dihydrolase negative.

Urea converted to ammonium carbonate moderately actively.

Nitrate reduced to nitrite by most strains; nitrite persists.

Tributyrin slowly hydrolyzed.

Active growth at 22–30 C; no growth at 37 C. Optimum for sporulation about 22 C. Vegetative cells show few survivors after 10 min at 65 C. Spores withstand 85 C for 10 min; at 99.5 C only a small fraction is viable at 10 min (MacDonald and MacDonald, 1962).

Optimum pH about 8.8. pH limits: 6.4 and 9.4. Optimum for sporulation pH 6.8–7.0.

Delayed or no growth in broth containing 7.5% NaCl (w/v).

The G + C content of the DNA has been reported to range from 40–43 moles % (T_m).

Type strain: ATCC 6473; NCIB 9251. This is a derivative of the strain deposited by Beijerinck in the Kral Collection. Kocur and Martinec (1963) found no essential difference between the nine strains studied, including ATCC 6473 (incorrectly designated 6474) and there was no need to propose a neotype.

GENUS OF UNCERTAIN AFFILIATION

Genus Oscillospira Chatton and Pérard 1913, 1159

T. GIBSON[†]

Os.cil.lo.spi'ra. L. n. *oscillum* a swing; L. n. *spira* a spiral; M.L. fem.n. *Oscillospira* the oscillating spiral.

Large rods or filaments 3–6 μm in diameter, divided by closely spaced cross walls into numerous disc-shaped cells. Reproduction by transverse fission. Motile by means of numerous lateral flagella. Endospores may be formed. Gram-negative.

Growth in pure culture has not been reported.

Exposure to air abolishes motility, thus suggesting the organisms are anaerobic.

Occur in the alimentary tract of herbivorous animals.

Type species (monotype): *Oscillospira guilliermondi* Chatton and Pérard 1913, 1159.

[†] Deceased.

Further Comments

Organisms of this genus have the multicellular structure shown by *Caryophanon* but have been distinguished from that genus by their usually larger size, the production of endospores (a property that is frequently not detected), the probability that they are anaerobes, their occurrence in the alimentary tract of animals and the fact that they have not been grown in pure culture. The reported difference in Gram reaction may deserve further study. The endospores have not been seen to germinate, and it is not known if they possess all the properties that characterize endospores of other genera.

Description of the species of genus **Oscillospira**

1. **Oscillospira guilliermondi** Chatton and Pérard 1913, 1159. (*Oscillaria caviae* Simons 1920, 367 according to Simons 1922, 502.)

guil.lier.mon′di. M.L. gen.n. *guilliermondi* of Guilliermond; named for A. Guilliermond, a French biologist.

Large, often curved organism, 3–6 by 10–40 μm in size. Larger or smaller forms may be produced. The rod has rounded ends, and may taper to one pole. Closely spaced cross walls, formed by diaphragm-like ingrowth from the outer wall, divide the rod into disc-shaped cells not more than 2 μm long.

An endospore, about 2.5 by 4 μm in size which lies longitudinally in the rod and occupies as much space as several disc-shaped cells may be formed. Rarely, there may be two spores in a single rod. The spore is refractile and resists cold stains; its other properties are unknown. The occurrence of sporulation is variable; the host's diet is thought to be a controlling factor.

The cells frequently contain much polysaccharide which gives a reddish to mauve color with iodine.

Originally described in cecal contents of the guinea pig. Organisms that have the internal structure of *Oscillospira guilliermondi* have been found in the alimentary tract, chiefly in the rumen or caecum, of several species of herbivorous animals. Some differ from the original description of that species in certain morphological features. Moir and Masson (1952) identified as *Oscillospira guilliermondi* an organism in the rumen of sheep which had a spherical, not an elliptical, spore.

Illustrations: Chatton and Pérard, 1913, 1161.

Important Notes

for

Users of this Edition

1. Always read both generic and species descriptions because characters listed in the generic description are not usually listed in the species descriptions.

2. In tables, characters common to all taxa are not shown but may be listed in footnotes.

3. Generally in tables (exceptions are clearly indicated in footnotes, q.v.) the meanings of symbols are as follows:

+   more than 90% strains positive
−   more than 90% strains negative
d   11–89% strains positive
( )   delayed reaction
w   weak reaction
D   Different reactions in different taxa (species of a genus or genera of a family)
v   strain instability (NOT differences between strains)

PART 16

GRAM-POSITIVE, ASPOROGENOUS, ROD-SHAPED BACTERIA*

FAMILY I. **LACTOBACILLACEAE** WINSLOW, BROADHURST, BUCHANAN, KRUMWIEDE, ROGERS AND SMITH 1917, 561

Lac.to.ba.cil.lac′e.ae. M.L. masc.n. *Lactobacillus* type genus of the family; -*aceae* ending to denote a family; M.L. fem.pl.n. *Lactobacillaceae* the *Lactobacillus* family.

Straight or curved rods usually occurring singly or in chains. Non-motile, rare strains motile. Gram-positive. Anaerobic or facultative. Complex organic nutritional requirements. Highly saccharoclastic. At least half the end product carbon from carbohydrate metabolism is lactate. Lactate is not attacked anaerobically. Catalase negative (no porphyrin in apoprotein). Benzidine reaction negative. Pathogenicity unusual. Found in fermenting animal and plant products where carbohydrates are available; also found in the mouth, vagina and intestinal tracts of various warmblooded animals, including man.

Genus I. **Lactobacillus** *Beijerinck 1901, 212. Nom. cons.* Opin. 38, Jud. Comm. 1971, 104

Morrison Rogosa

Lac.to.ba.cil′lus. L. n. *lac, lactis* milk; L. dim.n. *bacillus* a small rod; M.L. masc.n. *Lactobacillus* milk rodlet.

Rods, varying from long and slender to short coccobacilli. Chain formation common, particularly in later logarithmic phase of growth. Motility unusual; when present, by peritrichous flagella. Non-sporing. Gram-positive becoming Gramnegative with increasing age and acidity. Some strains exhibit bipolar bodies, internal granulations or a barred appearance with the Gram reaction or methylene blue stain.

Metabolism fermentative even though growth generally occurs in air; some are strict anaerobes on isolation.

Characteristically saccharoclastic; glucose fermented decreasing pH 1 or more units. At least half of end product carbon is lactate. Lactate is not fermented. Additional products may be acetate, formate, succinate, CO_2 or ethanol. Volatile acids with more than 2 carbon atoms are not produced.

Nitrate reduction highly unusual and then only where terminal pH is poised above 6.0. Gelatin not liquefied. Casein not digested but minute amounts of soluble nitrogen produced by some strains of some species. Indole and H_2S not produced.

Catalase and cytochrome negative (porphyrins absent); however, rare strains decompose peroxide by a pseudocatalase (porphyrin not present); benzidine reaction negative.

Pigment production rare; if present, yellow or orange to rust or brick red.

Complex nutritional requirements for amino acids, peptides, nucleic acid derivatives, vitamins, salts, fatty acids or fatty acid esters and fermentable carbohydrates. Nutritional requirements are generally characteristic for each species.

Surface growth on solid media often enhanced by anaerobiosis and 5–10% CO_2.

Temperature range 5–53 C; optimum generally 30–40 C.

Aciduric, optimal pH usually 5.5–5.8 or less, and

---

* **Editorial Note.** The division of genera between Part 16 and the first section of Part 17 is arbitrary and readers should consult both. Some genera have traditionally been associated with the actinomycetes and are so treated in this edition.

generally growing at 5.0 or less; at neutral or initial alkaline reactions lag phase may be lengthened or total growth yield reduced.

Found in dairy products and effluents, grain and meat products, water, sewage, beer, wine, fruits and fruit juices, pickled vegetables, sourdough and mash; also parasitic in the mouth, intestinal tract and vagina of many homothermic animals including man. Pathogenicity is highly unusual.

The G + C content of the DNA ranges from 34.7 ± 1.4–53.4 ± 0.5 moles % (buoyant density).

Type species: *Lactobacillus delbrueckii* (Leichmann) Beijerinck 1901, 229. Opin. 38, Jud. Comm. 1971, 104.

Further Comments

The following, included in THE MANUAL, 7th ed., are here excluded from *Lactobacillus:*

1. *L. caucasicus* Beijerinck 1901, the previous type species, is a rejected name (Opin. 38, Jud. Comm. 1971, 104).

2. *Lactobacillus bifidus* is transferred to *Bifidobacterium.*

3. Strains whose descriptions are consistent with the original description of *L. thermophilus* Ayers and Johnson 1924, 291 produce endospores, catalase and cytochromes and are probably very similar, if not identical, to *Bacillus coagulans* (Sharpe, 1962; Kitahara and Suzuki, 1963).

4. *L. pastorianus* (van Laer) Bergey *et al.* 1923 (see *Species incertae sedis*).

Various spore-bearing rods which produce lactic acid, are facultative or aerobic and catalase positive, have generally correctly been assigned to the genus *Bacillus*. Among these are: *Bacillus coagulans* Hammer 1915, 129; *Bacillus circulans* Jordan 1890, 831; *Bacillus lactis-termophilus* (sic) Gorini 1894, 16 (not validly published -Rule 14a (1)); *Bacillus thermo-acidurans* Berry 1933, 72; *Bacillus dextrolacticus* Werkman and Andersen 1940, 187; *Lactobacillus sporogenes* Horowitz-Wlassowa and Nowotelnow 1932, 333.

Other spore-forming rods which produce lactic acid homofermentatively, are catalase negative,

not strict anaerobes and otherwise closely resemble lactobacilli have been placed in the genus *Sporolactobacillus* (p. 550). Organisms similar to *Sporolactobacillus* isolated from the rhizosphere of various wild plants are *Bacillus laevolacticus* Nakayama and Yanoshi 1967, 149 and *Bacillus racemilacticus* Nakayama and Yanoshi 1967, 149.

Unnamed catalase negative rods from chicken meat, presumably related to both lactobacilli and aerobic sporeformers, were isolated by Thornley and Sharpe (1959). Glucose was converted homofermentatively to lactic acid but growth did not occur in media favoring the growth of lactobacilli, spores were not observed and some strains had coryneform morphology.

The microorganisms of the San Francisco sourdough bread process probably include heterofermentative lactobacilli having a high CO_2 and maltose requirement, an apparent inability to ferment any other carbohydrate and a need for fatty acids as satisfied by Tween 80. Kline and Sugihara (1971) did not name the organism.

Cillobacterium cellulosolvens Bryant *et al.* 1958, 533 produces >90% lactic acid from cellobiose. It has been suggested that the organism be included in *Lactobacillus*. However, similar rumen strains also produce butyrate and valerate and van Gylswyk and Hoffman (1970) suggest broadening the Bryant *et al.* (1958) definition of *Cillobacterium* to include these characteristics. It would, of course, be odd to include cellulose-digesting bacteria in *Lactobacillus*.

Strictly anaerobic *Lactobacillus* strains are known. Those from the rumen resemble or are *L. lactis* (Mann and Oxford, 1954; Jensen *et al.*, 1956), *L. leichmannii* or similar organisms. In addition, however, some anaerobic strains now assigned to other genera probably are lactobacilli and such organisms as *Eubacterium crispatum*, Brygoo and Aladame 1953, 641; *Eubacterium disciformans* (Massini) Prévot 1938, 295; and *Catenabacterium catenaforme* (Eggerth) Prévot 1938, 296, are here provisionally listed under *Species incertae sedis*.

Key to the species of genus **Lactobacillus**

I. Homofermentative. Lactic acid is the major product from glucose (generally 85% or more).

   A. No gas from glucose or gluconate; ribose not fermented, thiamine not required; aldolase activity; D- or L- or DL-lactic acid produced; G + C = 34.7–50.8%; generally grows at 45 C or higher, generally not at 20 C and not at 15 C. Colonies normally rough becoming smooth and compact in the presence of Tween 80 or sodium oleate (Rogosa and Mitchell, 1950).

      1. Produce D(−)-lactic acid.*

          1. *L. delbrueckii*

          2. *L. leichmannii*

          3. *L. jensenii*

---

* See also No. 27, *L. vitulinus*.

               4. *L. lactis*
               5. *L. bulgaricus*

2. Produce DL-lactic acid.

               6. *L. helveticus*
               7. *L. acidophilus*

3. Produce mainly L(+)- and very small amounts of D(−)-lactic acid.*

               8. *L. salivarius*

B. No gas from glucose. Gas from gluconate. Ribose, when fermented, yields lactic and acetic acids without gas. Thiamine not required. Aldolase activity. Glucose 6-phosphate dehydrogenase and inducible 6-phosphogluconate dehydrogenase activity. Variable growth at 45 C. Growth at 15 C. G + C = 45–46.4%.

   1. Produce chiefly L(+)-lactic acid. Ribose fermented.

               9a–9c. *L. casei* subspp. *casei, rhamnosus, alactosus*
               9d. *L. casei* subsp. *tolerans*
               10. *L. xylosus*

   2. Produce DL-lactic acid. Ribose fermented.

               9e. *L. casei* subsp. *pseudoplantarum*
               11. *L. plantarum*
               12. *L. curvatus*

   3. Produce DL-lactic acid. Ribose fermentation equivocal.

               13a. *L. coryniformis* subsp. *coryniformis*

   4. Produce D(−)-lactic acid. Ribose not fermented. Favorable pH about 5.

               13b. *L. coryniformis* subsp. *torquens*
               14. *L. homohiochii*

II. Heterofermentative, producing about 50% of end products from glucose as lactic acid, with considerable amounts of CO_2, acetic acid and ethanol; mannitol from fructose.

A. Gas from glucose and gluconate. Ribose fermented to lactic and acetic acids without gas. Thiamine required. DL-Lactic acid produced. Aldolase is not demonstrable. Glucose 6-phosphate dehydrogenase activity and equal or very much less 6-phosphogluconate dehydrogenase activity are demonstrable.

   1. Growth at 45 C; no growth at 15 C.

               15. *L. fermentum*

   2. Variable growth at 45 and 15 C; no growth at 48 C.

               16. *L. cellobiosus*

   3. Generally growth at 15 C; no growth at 45 C.

               17. *L. brevis*
               18. *L. buchneri*
               19. *L. viridescens*
               20. *L. coprophilus*

III. Heterofermentative. DL-Lactic acid, CO_2, acetate produced. Less well known organisms. Slower growing or "indifferent" toward most carbohydrates. Nutrition, enzymology, G + C, cell wall chemistry, gluconate and ribose fermentations generally not studied.

Malate, fructose, gluconate, ribose and xylose fermented; variable or less fermentation of glucose, galactose, sucrose and maltose; ethanol tolerance 15–18%; optimal initial pH range 4.5–5.5; optimal temperature 30–35 C.

               21. *L. hilgardii*

Glucose and fructose fermented; malate, and generally no other carbohydrate fermented; ethanol tolerance >20%; optimal initial pH range 4.5–5.5; optimal temperature 25–30 C.

               22. *L. trichodes*

Fructose fermented, weak malate activity and weak or negative glucose fermentation, gluconate fermented, ribose not fermented, no other carbohydrates fermented; ethanol tolerance 15%; optimal initial pH range 4.5–5.5; optimal temperature 25–30 C.

---

* See also No. 26, *L. ruminis.*

23. L. fructivorans

Arabinose fermented. Glucose, fructose and galactose may be fermented. Malate and other carbohydrates not fermented. Ethanol tolerance 15%; optimal initial pH range 5–7; optimal temperature <30 C and good growth at 10 C.

24. L. desidiosus

Hiochic acid (D-mevalonic acid) essential for growth. Favorable pH about 5.

25. L. heterohiochii

Description of the species of genus Lactobacillus

1. Lactobacillus delbrueckii (Leichmann) Beijerinck 1901, 229. *Nom. cons.* Opin. 38, Jud. Comm. 1971, 104. (*Bacillus Delbrücki* (sic) Leichmann 1896, 284; *Bacillus acidificans longissimus* Lafar 1896, 195; *Bacterium delbrücki* (sic) (Leichmann) Migula 1900, 406; *Lactobacillus delbrücki* (sic) (Leichmann) Beijerinck 1901, 230; *Thermobacterium cereale* Orla-Jensen 1919, 164; *Lactobacterium delbrücki* (sic) (Leichmann) van Steenberge 1920, 816; *Ulvina delbrücki* (sic) (Leichmann) Pribram 1933, 75; *Plocamobacterium delbrücki* (sic) (Leichmann) Pribram 1933, 77.)

del.bruec′ki.i. M.L. gen.n. *delbrueckii* of Delbrück; named for M. Delbrück, a German bacteriologist.

Rods, 0.5–0.8 by about 2–9 μm, with rounded ends occurring singly and in short chains. Internal granulations are revealed by methylene blue stains. Non-motile. Colony normally rough and non-pigmented.

Acid without gas from glucose and other carbohydrates (see Table 16.1). Maltose fermented (but negative variants occur and can be selected).

D(−)-Lactic acid produced homofermentatively. Ammonia generally produced from arginine. No acidity in milk.

Cell walls contain glycerol teichoic acid; the peptidoglycan is of the L-lysine-D-aspartate type. Serological group not detectable.

Calcium pantothenate, niacin and riboflavin required; neotype strain requires thymidine. Thiamine, pyridoxal, folic acid, or vitamin B$_{12}$ not required.

No growth at 15 C; growth at 45 C and frequently at 50–52 C; optimum 40–44 C.

The G + C content of the DNA is 50.0 moles % (buoyant density).

Isolated from distillery sour potato mash, grain and vegetable mashes fermenting at >41 C.

Neotype strain: ATCC 9649 (Hansen, 1968). Isolated from distillery sour grain mash incubated at 45 C.

Comments: Rogosa and Hansen (1971) discuss the rejection of *L. caucasicus* as the type species (Opin. 38, Jud. Comm. 1971, 104), its replacement by *L. delbrueckii* and some of the nomenclatural problems; the original descriptions of *L. fermentum* and *L. delbrücki* (sic) were confused by Beijerinck (1901) so that *L. fermentum* and *L. delbrücki*

were considered as two elements as follows. "Le nom correct de cette bactérie serait *Lactobacillus fermentum* var. *delbrücki*. Pour la simplicité j'écris seulement *L. delbrücki*." Thus, *L. delbrueckii* has probably never been validly published, although it or other orthographic variants, have been used as such since Beijerinck's time. Rogosa and Hansen (1971) have requested the Judicial Commission to recognize and clarify this situation by issuing an Opinion conserving the name *Lactobacillus fermentum* Beijerinck. This text, including the descriptions of *L. delbrueckii* and *L. fermentum*, has been written in anticipation of such action.

At the time the neotype strain was isolated, all known culture collection strains named *L. delbrückii* were incorrectly designated; they were typical strains of *L. casei* subsp. *rhamnosus* (all probably derived originally from one strain).

The present neotype is nearly identical with the description by Henneberg (1903) of a pure culture of *Bacillus delbrücki*. Henneberg observed a very weak fermentation of galactose; the neotype strain does not ferment galactose; however, early samples of "pure galactose" might easily contain 30 or more per cent of glucose. Otherwise, the neotype and Henneberg's strains are identical in 18 fermentations and all other characteristics.

2. Lactobacillus leichmannii (Henneberg) Bergey *et al.* 1923, 249. (*Bacillus Leichmanni* I (sic) Henneberg 1903, 330; *Bacillus Leichmanni* III (sic) Henneberg 1903, 331; *Lactobacillus leichmanni* (sic) (Henneberg) Bergey *et al.* 1923, 249; *Bacterium leichmanni* I (sic) (Henneberg) Henneberg 1926, 107; *Lactobacterium leichmannii* (Henneberg) Krasil'nikov 1949, 215.)

leich.man′ni.i. M.L. gen.n. *leichmannii* of Leichmann; named for G. Leichmann, a German bacteriologist.

Rods, with rounded ends, 0.6 by 2.0–4.0 μm, occurring singly and in short chains. Granules stain with methylene blue. Non-motile. Colony normally rough. Non-pigmented.

D(−)-Lactic acid produced homofermentatively. Ammonia generally produced from arginine. No acid in milk but lactose is fermented in other suitable media.

Cell walls contain glycerol teichoic acid; the peptidoglycan is of the L-lysine-D-aspartate type. Serological group not determinable.

TABLE 16.1—Carbohydrate reactions of the species of genus **Lactobacillus**

	1. L. delbrueckii	2. L. leichmannii	3. L. jensenii	4. L. lactis	5. L. bulgaricus	6. L. helveticus	7. L. acidophilus	8. L. salivarius	9a-c. L. casei	9d. L. casei subsp. tolerans	9e. L. casei subsp. pseudoplantarum	10. L. xylosus	11. L. plantarum	12. L. curvatus	13a. L. coryniformis subsp. coryniformis	13b. L. coryniformis subsp. torquens	14. L. homohiochii	15. L. fermentum	16. L. cellobiosus	17. L. brevis	18. L. buchneri	19. L. viridescens	20. L. coprophilus	21. L. hilgardii	22. L. trichodes	23. L. fructivorans	24. L. desidiosus	25. L. heterohiochii
Amygdalin	−[a]	+	+	−	−	−	+	−	+	−	+	+	+[*]	−	−	−	−	d	+	−	−	−	0	−	0	−	0	−
Arabinose	−	−	+	−	−	−	+	−	−	−	−	+	d[*]	−	−	−	−	d	+	+	+	−	+	−	−	−	+	−
Cellobiose	−	+	+	+	−	+	+	+	+	+	+	+	+	+	+	+	+	−	+	−	+	+	+	−	+	−	−	+
Fructose	+	+	+	+	+	+	+	+	+	+	+	+	+	+	+	+	+	+	+	+	+	+	+	+	+	+	+	+
Galactose	※	+	+	+	+	+	+	+	+	+	+	+	+	+	d	d	+	+	+	※	※	+	+	d	+	※	※	+
Glucose (acid)	+	+	+	+	+	+	+	+	+	+	+	+	+	+	+	+	+	+	+	+	+	+	+	+	+	+	+	+
Glucose (gas)	−	−	−	−	−	−	−	−	−	−	−	−	−	−	−	−	−	+	+	+	+	−	+	+	+	+	+	+
Gluconate	−	+	+	+	−	−	d	−	+	+	+	+	+	d	d	d	+	+	+	+	+	+	+	+	−	+	※	+
Lactose	+	+	−	+	+	+	+	+	+	+	+	+	+	−	−	−	+	+	+	−	−	−	−	−	−	−	−	−
Maltose	d	−	+	+	−	−	+	+	(d)	+	+	+	+	+	+	d	−	+	+	+	+	+	+	d	d	d	※	+
Mannitol	−	−	−	−	−	−	−	+	+	+	+	+	+	+	+	+	+	−	−	−	−	−	+	−	−	−	※	−
Mannose	+	+	+	+	+	+	+	+	*	+	+	+	+	+	+	+	+	+	+	+	+	+	+	+	−	+	※	+
Melezitose	−	−	−	−	−	−	−	+	+	+	+	+	+	d	+	+	+	−	−	−	−	−	−	d	−	−	−	−
Melibiose	+	−	+	+	−	−	d	+	+	+	+	+	+	−	d	−	−	+	+	+	+	−	+	+	−	−	+	−
Raffinose	−	−	+	−	−	−	d	+	+	+	+	+	+	−	+	−	−	+	+	+	+	−	+	+	−	−	−	−
Rhamnose	−	−	−	−	−	−	−	−	−	−	−	−	d	−	d	−	+	−	−	−	−	−	−	−	−	−	−	+
Ribose	−	−	−	+	−	+	d	+	*	+	+	+	+	−	+	−	−	+	+	+	+	d	−	d	−	−	−	+
Salicin	−	+	+	+	+	+	+	+	+	+	+	+	+	+	+	+	+	+	+	※	※	d	+	+	−	−	※	−
Sorbitol	−	−	+	+	+	−	+	*	+	+	+	+	+	+	−	−	−	+	+	+	※	d	+	+	−	−	−	−
Sucrose	+	+	+	+	+	+	+	+	+	+	+	+	+	+	+	+	−	+	+	+	+	−	+	+	−	−	+	+
Trehalose	−	+	+	+	+	*	+	+	+	+	+	+	+	+	+	+	+	d	+	d	d	d	+	d	−	−	−	−
Xylose	−	−	−	+	−	−	+	+	(d)	+	+	+	d[*]	−	d	−	−	d	+	d	d	d	+	d	−	−	−	−
Esculin	−	+	−	※	−	−	+	*	+	+	+	+	+	+	d	−	0	−	+	−	−	−	+	−	−	−	0	0

[a] The symbols used are: + = positive reaction by 90% or more strains; d = some strains +, others − (about 89-11% positive); − = negative reaction by most strains (90% or more); ( ) = delayed reaction; w = weak reaction; ※ = weak, slow or negative; * = see text; 0 = not tested. Combinations of symbols, e.g.: (−) = rarely positive, and then slowly; −w = negative or weak reaction.

Negative reactions by all strains from L-erythritol, D-erythrose, D-fucose, D-glucoheptose, α-methyl-D-xyloside and perseitol. All negative from α-methyl-D-glucoside except that Nos. 12, 17, 19 and 20 are variable; negative from α-methyl-D-mannoside except No. 12 variable. Except for some strains of L. casei, adonitol and sorbose are not fermented; glycerol, inositol, inulin and starch, dextrin, dulcitol are very rarely fermented. Tagatose (100%) by L. plantarum and variably by L. curvatus; these three strains of L. casei; turanose (100%) by L. plantarum and some strains of L. plantarum. ...atose (100%), turanose (100%) and D-lyxose (95%) fermented by L. salivarius subsp. salivarius and some strains of L. salivarius subsp. salivarius. D-Arabinitol is fermented by L. salivarius subsp. salivarius; these pentitols substrates are not fermented by any other species. ...xylitol is fermented by L. salivarius subsp. salivarius; these pentitols are not fermented by...

Complex nutrition. Calcium pantothenate, niacin, folic acid required; Vitamin B_{12} required or stimulatory; riboflavin, pyridoxal, thiamine not required.

No growth at 15 C; growth at 45 C; optimum 35–40 C.

The G + C content of DNA is 50.8 ± 0.5 moles % (buoyant density).

Isolated from compressed yeast, grain mash.

Neotype strain: ATCC 4797 (Rogosa and Hansen, 1971).

Comments: Previous editions of THE MANUAL considered *Bacillus Leichmanni* I (Henneberg, 1903) as the original synonym. However, Henneberg (1903) named two other organisms: *Bacillus Leichmanni* II (*sic*) and *Bacillus Leichmanni* III (*sic*). The proposed neotype strain appears closely related to *Bacillus Leichmanni* I (*sic*) and even more closely to *Bacillus Leichmanni* III (*sic*). *Bacillus Leichmanni* II (*sic*) appears to be *L. plantarum*. Descriptions of temperature relations are contradictory and the text is not understood by any of several competent native German-speaking scientists whom I have consulted (see Rogosa and Hansen (1971) for further details).

3. **Lactobacillus jensenii** Gasser, Mandel and Rogosa 1970, 221.

jen.sen′i.i. M.L. gen.n. *jensenii* of Jensen; named for S. Orla-Jensen, Danish microbiologist.

Seven strains are phenotypically indistinguishable from *L. leichmannii* except the lactic dehydrogenases have different electrophoretic mobilities in starch gels (Gasser, 1970). Whereas the G + C composition of *L. leichmannii* is 50.8 ± 0.5 moles %, that of *L. jensenii* is 36.1 ± 1.2.

Isolated from human vaginal discharge and blood clot.

Type strain: 62 G; CIPP 6917; ATCC 25258 (Gasser *et al.*, 1970).

4. **Lactobacillus lactis** (Orla-Jensen) Bergey *et al.* 1934, 303. (*Bacillus lactis acidi* Leichmann 1896, 283; *Bacterium lactis acidi* (Leichmann) Leichmann 1896, 778; *Thermobacterium lactis* Orla-Jensen 1919, 164; *Lactobacillus lactis-acidi* (Leichmann) Bergey *et al.* 1923, 248; *Lactobacterium caucasicum* var. *lactis* (Orla-Jensen) Krasil′nikov 1949, 211.)

lac′tis. L. n. *lac* milk; L. gen.n. *lactis* of milk.

Rods, less than 2 μm wide, often appearing as long forms with a tendency to grow into threads, often strongly curling, occurring singly or in pairs in young, vigorous cultures. Generally, granules are demonstrable with methylene blue stains. Non-motile. Colony normally rough, 1–3 mm in diameter and non-pigmented being white to light gray.

D(−)-Lactic acid produced by homofermenta-

tion. Negative, weak or variable reactions with esculin (Table 16.1).

Milk coagulated with a final acidity of about 1.6% lactic acid. Ammonia not produced from arginine.

Cell walls contain glycerol teichoic acid; the peptidoglycan is of the L-lysine-D-aspartate type. There is no distinguishing hexose or pentose. Serological group E (same as *L. buchneri* and *L. brevis*). Serological groups and methods for all members of *Lactobacillus* are those of Sharpe (1955) and Sharpe and Wheater (1957).

Requires calcium pantothenate, niacin, riboflavin; exceptional strains may require vitamin B_{12} (cyanocobalamines); thiamine, pyridoxal or pyridoxamine, folic acid and thymidine not required.

No growth at 15 C; growth at 45 C or even 50–52 C; optimal growth at 40–43 C.

The G + C content of the DNA of two strains was 50.3 ± 1.4 moles % (buoyant density). There is 86% reassociation of the DNA of *L. lactis* ATCC 12315 and the DNA of *L. bulgaricus* ATCC 11842 (Simonds *et al.*, 1971).

Isolated from milk, cheese and starter cultures used in the manufacture of cheese.

Type strain: ATCC 12315 (*Thermobacterium lactis* No. 10 Orla-Jensen 1919, 164; Rogosa and Hansen, 1971).

5. **Lactobacillus bulgaricus** (Orla-Jensen) Rogosa and Hansen 1971, 181. (*Thermobacterium bulgaricum* Orla-Jensen 1919, 164.)

Names of organisms whose relationship to *L. bulgaricus* is uncertain: *Lactobacillus longus* Beijerinck 1901, 217; Bacille A Grigoroff 1905, 716; *Bacillus bulgaricus* Luerssen and Kühn 1907, 241; *Bacterium bulgaricum* (Luerssen and Kühn) Buchanan and Hammer 1915, 250; *Acidobacterium bulgaricum* (Luerssen and Kühn) Schlirf 1925, 116; *Plocamobacterium bulgaricum* (Luerssen and Kühn) Lehmann and Neumann 1927, 511; *Bacterium giogurt* de′Rossi 1927, 743; *Lactobacterium bulgaricum* (Luerssen and Kühn) Krasil′nikov 1949, 212.

bul.ga′ri.cus. M.L. adj. *bulgaricus* Bulgarian.

This species is closely related to *Lactobacillus lactis*, being morphologically indistinguishable, producing the same amount of D(−)-lactic acid in milk, having the same general cell wall structure and group antigen, including glycerol teichoic acid and a peptidoglycan of the L-lysine-D-aspartate type, apparently identical lactic acid dehydrogenases and similar G + C moles % in the DNA (50.3). The only significant difference is that *L. bulgaricus* ferments fewer sugars than *L. lactis* (Table 16.1). The latter might be a mutant or variant of the former.

Type strain: ATCC 11842 (original strain No. 14

from Orla-Jensen (1919) (Rogosa and Hansen, 1971)).

Comments: Previous descriptions in THE MANUAL were composites of two organisms, one producing D(−)- and the other producing DL-lactic acid; sugar fermentations of this "organism" reported from various sources were in conflict; from study of the actual strains on which THE MANUAL'S descriptions were based, it is definite that the two organisms involved are L. bulgaricus and L. jugurti (the latter a maltose negative variant of L. helveticus). In DNA-DNA hybridization experiments there is no reassociation of L. bulgaricus DNA and L. jugurti DNA (Simonds et al., 1971). These, and sometimes L. lactis and other lactobacilli, often occur simultaneously in such sour milks as yogurt, etc. (see Rogosa and Hansen (1971) for documentation).

6. **Lactobacillus helveticus** (Orla-Jensen) Bergey et al. 1925, 184. (Bacillus ε von Freudenreich 1895, 173; Bacillus casei ε von Freudenreich and Thöni 1904, 532; Caseobacterium ε Orla-Jensen 1909, 337; Thermobacterium helveticum Orla-Jensen 1919, 164; Lactobacillus helveticum (sic) (Orla-Jensen) Bergey et al. 1925, 184; Plocamobacterium helveticum (Orla-Jensen) Pribram 1933, 78; Lactobacterium helveticum (Orla-Jensen) Krasil'nikov 1949, 212.)

hel.ve′ti.cus. L. adj. helveticus Swiss.

Rods 0.6–1.0 by 2.0–6.0 μm occurring singly and in chains. No metachromatic granules with methylene blue (different from L. delbrueckii, L. bulgaricus, L. leichmannii and L. lactis).

In agar pour plates, colonies are 2–3 mm in diameter or less, normally opaque, white to light gray and rough to rhizoid. Growth of agar streak cultures is greatly enhanced by anaerobiosis and 5% CO$_2$. In media with Tween 80 or sodium oleate, colonies tend to be larger and smoother.

DL-Lactic acid produced by homofermentation. Often weak, slow or no acidity from fructose and mannose. Trehalose occasionally fermented.

Ammonia not produced from arginine.

Cell walls contain glycerol teichoic acid; the peptidoglycan is of the L-lysine-D-aspartate type; arabinose, rhamnose, galactose and mannose residues not present. Serological group A (Sharpe and Wheater, 1957).

Requires complex media. Grows well in milk producing a high acidity of 2% or more of lactic acid. Media containing whey, tomato juice, liver or carrot digests, casein digests + yeast extract containing fermentable carbohydrate, support good growth. In nutritionally defined media, calcium pantothenate, niacin, riboflavin, pyridoxal or pyridoxamine and magnesium are re-

quired. Exogenous thiamine, folic acid, vitamin B$_{12}$ and thymidine are not required.

Temperature relations: Optimum, 40–42 C. No growth at 15 C. Maximum, 50–53 C.

The G + C content of the DNA clusters around 39.3 moles % (buoyant density).

Isolated from sour milk, cheese starter cultures and cheese, particularly Emmental and Gruyére cheese.

Neotype strain: ATCC 15009 (Thermobacterium helveticum No. 12 Orla-Jensen 1919, 164; Rogosa and Hansen, 1971).

Comments: Thermobacterium jugurt Orla-Jensen 1919, 164 (Lactobacillus jugurti (Orla-Jensen) Rogosa and Sharpe 1959, 333) is a biotype of this species differing only in not fermenting maltose. The cell wall peptidoglycan is of the L-lysine-D-aspartate type and there is also glycerol teichoic acid as in L. hevleticus. The range of G + C content of the DNA is the same as for L. helveticus, namely, 38–40 moles % (buoyant density).

7. **Lactobacillus acidophilus** (Moro) Hansen and Mocquot 1970, 326. (Bacillus acidophilus Moro 1900, 115; Thermobacterium intestinale Orla-Jensen, Orla-Jensen and Winther 1936, 331.)

a.ci.do′phi.lus. L. adj. acidus sour; M.L. neut.n. acidum acid; Gr. adj. philus loving; M.L. adj. acidophilus acid-loving.

Rods with rounded ends, generally 0.6–0.9 by 1.5–6 μm, occurring singly, in pairs and in short chains. Non-motile. Non-flagellated.

Colony usually rough. Microscopic examination generally reveals twisted or fuzzy filamentous projections with dark felt-like mass in the center. Deep colonies are irregularly shaped with radiate or ramified projections. No characteristic pigment.

Glycogen fermented by some strains and generally weakly. Some strains ferment melibiose, raffinose or both. Homofermentative, producing DL-lactic acid. Generally less than 10% of other carbohydrate fermentation products.

Ammonia not produced from arginine. Acidity and coagulation of milk variable; acidity varying from 0.3–1.9% lactic acid.

Cell wall peptidoglycan is of the L-lysine-D-aspartate type; teichoic acid generally absent; in some strains small amounts of glycerol teichoic acid are detectable. Cell walls do not contain any distinguishing hexoses or pentoses (Cummins and Harris, 1956). Strains appear serologically diverse and no group reactions have been demonstrated.

Acetate or mevalonic acid, riboflavin, calcium pantothenate, niacin and folic acid required. Exogenous thiamine, pyridoxal and thymidine not required. Vitamin B$_{12}$ (cyanocobalamines) generally not required. Mutant strains may require deoxyribosides.

No growth at 15 C and may not grow at 22 C; growth generally at 45 C and may grow at 48 C; optimum 35–38 C. Growth occurs at initial pH values of 5–7; optimum 5.5–6.0.

The G + C content of the DNA of six strains was 36.7 ± 0.7 moles % (buoyant density).

Originally isolated from feces of infants; also isolated from mouth and vagina of human young and adults; intestinal tract of turkeys and chickens; and mouth and intestinal tract of rats and hamsters.

Neotype strain: ATCC 4356 (Hansen and Mocquot, 1970).

Comment: "Doderlein's bacillus" is a vague term for aciduric, Gram-positive rods of the human vagina and may comprise mixtures of *L. acidophilus*, *L. casei*, *L. fermentum*, *L. cellobiosus* or even *Leuconostoc mesenteroides* (Rogosa and Sharpe, 1960). Similarly, in the older literature *L. acidophilus* was often used as a generic term for lactobacilli of dental origin comprising eight or more *Lactobacillus* species (Rogosa *et al.*, 1953).

8. Lactobacillus salivarius Rogosa, Wiseman, Mitchell and Disraely 1953, 691.

sal.i.var'i.us. L. adj. *salivarius* salivary.

Rods, 0.6–0.9 by 1.5–5 μm, with rounded ends, occurring singly, in pairs and in chains of varying length. Non-motile and non-flagellated. Gram-positive.

Colony normally rough, 1–3 mm in diameter, non-pigmented being white to light gray.

L(+)-Lactic acid with small and variable amounts of D(−)-lactic acid produced by homofermentation. No growth with, or gas from, gluconate.

Rogosa *et al.* (1953) recognized two subspecies, a division confirmed by Rogosa and Sharpe (1959). In addition to the properties already described, *Lactobacillus salivarius* subsp. *salivarius* Rogosa *et al.* 1953 fermented rhamnose, but not salicin and esculin; whereas *Lactobacillus salivarius* subsp. *salicinius* Rogosa *et al.* 1953 fermented salicin and esculin, but not rhamnose.

Ammonia not produced from arginine. Milk coagulated with a final accumulation of about 0.9% lactic acid.

Cell walls contain glycerol teichoic acid; peptidoglycan is of the L-lysine-D-aspartate type. Serological group G.

Nutritional requirements complex with exogenous requirements for calcium pantothenate, niacin, riboflavin and folic acid; pyridoxal or pyridoxamine, thiamine, vitamin B_{12} and thymidine not required.

No growth at 15 C; growth variable at 45 C; growth optimal from 35–40 C.

The G + C content of DNA of both subspecies is 34.7 ± 1.4 moles % (buoyant density).

Isolated from the mouth and intestinal tracts of the hamster, the mouth of man and the intestinal tract of the hen. This organism was the predominant homofermentative lactobacillus in hamsters sampled in the U. S. A. and England.

Type strains: *L. salivarius* subsp. *salivarius* ATCC 11741 (Hansen, 1968); *L. salivarius* subsp. *salicinius* ATCC 11742 (Hansen, 1968).

Comment: Report of motility by Gemmell and Hodgkiss (1964) was based on faulty methodology and confusion of culture collection numbers. Re-investigation by a number of laboratories confirmed lack of motility and flagellation (see electron micrograph; Kandler, 1967).

9. Lactobacillus casei (Orla-Jensen) Hansen and Lessel 1971, 71. (*Bacillus* α von Freudenreich 1890, 266; *Bacillus* α von Freudenreich 1891, 20; *Bacillus casei* α von Freudenreich and Thöni 1904, 532; *Caseobacterium vulgare* Orla-Jensen 1916, 35; *Bacterium casei* α Orla-Jensen 1916, 35; *Streptobacterium casei* Orla-Jensen 1919, 166; *Lactobacillus casei* Bergey *et al.* 1923, 253; *Lactobacterium casei* (Orla-Jensen) Krasil'nikov 1949, 215) (none of these names was validly published; see Hansen and Lessel, 1971).

ca'se.i. L. n. *caseus* cheese; L. gen.n. *casei* of cheese.

Short or long rods, generally less than 1.5 μm wide, often with square ends and tending to form chains. Flagella absent and non-motile.

Pour plate of deep colonies smooth, lens or diamond-shaped, white to very light yellow. Growth in broth even heavy turbidity.

Sorbitol and sorbose usually fermented. Maltose and sucrose often slowly fermented and negative variants may sometimes be selected from a positive population. Glycogen and starch not attacked.

L(+)-Lactic acid is produced in excess of D(−)-lactic acid; the resultant optical rotation is dextrorotatory. Fructose 1,6-diphosphate aldolase present. But at least part of the hexose monophosphate shunt is also present in which there is divergence from glycolysis at the glucose 6-phosphate level (triphosphopyridine nucleotide-linked glucose 6-phosphate dehydrogenase activity) and pentoses participate catalytically (Buyze, thesis, University of Utrecht, 1955; Buyze *et al.*, 1957; findings confirmed in this author's laboratory). Ribose is fermented to lactic and acetic acids without CO_2 production; inducible growth with 4% gluconate is rapid and abundant with CO_2 production.

Ammonia not produced from arginine.

Cell wall peptidoglycan is of the L-lysine-D-aspartate type. Cell walls contain rhamnose, galactose, glucose, mannose, glucosamine, galactosamine and an unknown hexosamine; diaminopimelic acid (present in *Lactobacillus plantarum*) is not found in *L. casei*. Teichoic acids have not been detected in cell walls but glycerol teichoic acid has been found in non-cell wall substance.

Riboflavin, folic acid, calcium pantothenate and niacin required. Pyridoxal, or pyridoxamine, required or stimulatory. Thiamine, vitamin B_{12} and thymidine not required.

Three subspecies were recognized by Rogosa *et al.* (1953), all of which grow at 15 and often as low as 6 C. They are differentiated as follows:

D-aspartate type, but unlike *L. casei* subsp. *casei*, the cell wall contains teichoic acid. The teichoic acid is of the glycerol type and is unusual since there is no detectable D-alanine, a common constituent of the regular teichoic acids.

Isolated only from pasteurized milk.

9e. *Lactobacillus casei* subsp. *pseudoplantarum* Abo-Elnaga and Kandler 1965, 26.

pseu'do.plan.ta'rum. Gr. adj. *pseudes* false; M.L. gen.pl.n. *plantarum* a specific epithet; M.L. gen.pl.n. *pseudoplantarum* the false (*L.*) *plantarum*.

Thirty-eight strains differ from remaining *L. casei* in producing DL-lactic acid. Fermentation

Subspecies	Growth at 45 C.	Lactose fermented	Rhamnose fermented	Group antigen	Type or Neotype, ATCC
a. *casei*	−	+	−	B or C	393 (Hansen and Lessel, 1971)
b. *alactosus*	−	−	−	B or C	27216 (Mills and Lessel, 1973)
c. *rhamnosus*	+	+	+	C	7469 (Hansen, 1968)

At least 90% of the strains of these three subspecies grow at the expense of and ferment pyruvate; all other lactobacilli are negative.

The G + C content of the DNA of seven strains (including all varieties of the species) was 46.4 ± 0.8 moles % (buoyant density).

Isolated from milk and cheese, dairy products and dairy environments, sour dough, cowdung, silage, and human mouth, human intestinal contents and stools and the human vagina. This species has never been isolated from oral or intestinal samplings from the rat, hamster, mouse or rabbit.

Comments: Abo-Elnaga and Kandler (1965) describe four strains of *L. casei* subsp. *rhamnosus* fermenting arabinose. This result is highly unusual and the author has never personally isolated such strains.

The neotype strain, ATCC 7469, of *L. casei* subsp. *rhamnosus* was often confused with *L. helveticus* because the outmoded and illegitimate usage of *L. casei* ε (originally intended to apply to organisms later named *L. helveticus*) was mistakenly applied to *L. casei* subsp. *rhamnosus* (see Sharpe and Wheater, 1957).

9d. *Lactobacillus casei* subsp. *tolerans* Abo-Elnaga and Kandler 1965, 26.

tol'er.ans. L. pres.part. *tolerans* tolerating, enduring.

Six strains had a G + C moles % of 46.9 and differed from the species modal character in surviving at 72 C for 40 sec and in fermenting only lactose. However, their strain M7/74, in our laboratory, ferments the substrates shown in Table 16.1. The cell wall peptidoglycan is of the L-lysine-

reactions of strain M40 (Abo-Elnaga and Kandler, 1965) are shown in Table 16.1. The DNA of strain M40 has a G + C moles % of 45.9 (buoyant density).

Isolated from Tilsit cheese.

10. Lactobacillus xylosus Kitahara 1938, 1449.

xy.los'us. M.L. adj. *xylosus* of xylose, pertaining to xylose.

Lactobacillus xylosus Kitahara 1938, 1449 and Kitahara (1940) was originally described as a homofermentative organism producing L(+)-lactic acid but otherwise resembling *L. plantarum*. *L. xylosus* fermented xylose but not arabinose and was thus differentiated from *L. arabinosus* Fred *et al.* 1921, 410 (fermenting arabinose but not xylose) and *L. pentosus* Fred *et al.* 1921, 410 (fermenting both arabinose and xylose). However, *L. xylosus*, ATCC 15577, deposited by Kitahara as the type strain, more closely resembles *L. casei* than *L. plantarum*: L(+)-lactic acid is the chief product from glucose. Dextrin, esculin, lactose, starch reactions are negative or equivocal; glycerol, glycogen are not fermented. As in *L. casei*, the cell wall peptidoglycan linkage is of the L-lysine-D-aspartate type. The L-lactic acid dehydrogenase (LDH) is similar to that of *L. casei* in being activated by fructose 1,6-diphosphate and manganese and by its action in the physiological direction only (pyruvate to lactic acid); however, electrophoretic mobility in disc gel electrophoresis at pH 7.9 is different from that of all the subspecies of *L. casei*. *L. xylosus* is also different from *L. casei* in not attacking esculin melezitose and sorbitol, in not growing on pyruvate and malate, and, of course, in fermenting xylose. There are only rather distant immunological similarities between the aldolases of *L. casei* and *L. xylosus*.

The G + C content of the DNA is 39.4 moles % (chemical analysis, Suzuki and Kitahara, 1964).

11. Lactobacillus plantarum (Orla-Jensen) Bergey *et al.* 1923, 250. (*Streptobacterium plantarum* Orla-Jensen 1919, 174; *Lactobacillus plantari* (*sic*) (Orla-Jensen) Bergey *et al.* 1923, 250; *Lactobacterium plantarum* (Orla-Jensen) Krasil'nikov 1949, 216; *Lactobacillus arabinosus* Fred, Peterson and Anderson 1921, 410; *Lactobacillus pentosus* Fred *et al.* 1921, 410; *Bacillus rudensis* Davis and Mattick 1929, 50; *Lactobacillus rudensis* (Davis and Mattick) Davis 1937, 374; *Lactobacillus plantarum* var. *rudensis* (Davis and Mattick) Breed and Pederson 1938, 667; *Lactobacillus plantarum* var. *mobilis* Harrison and Hansen 1950, 446.)

Names of organisms whose relationship to *L. plantarum* is uncertain: *Bakterium pabuli acidi* II (*sic*) Weiss 1899, 149; *Bacillus Beijerincki* (*sic*) Henneberg 1903, 318; *Bacillus Listeri* (*sic*) Henneberg 1903, 329; *Bacillus Wortmanni* (*sic*) Henneberg 1903, 330; *Bacillus Leichmanni* II (*sic*) Henneberg 1903, 331; *Bacillus Maerckeri* (*sic*) Henneberg 1903, 331; *Bacterium brassicae* Wehmer 1903, 628; *Lactobacterium listeri* (Henneberg) van Steenberge 1920, 814; *Lactobacillus brassicae* (Wehmer) Le Fevre 1922, 25; *Lactobacillus pabuli-acidi* (*sic*) Bergey *et al.* 1923, 247; *Lactobacillus beijerincki* (*sic*) (Henneberg) Bergey *et al.* 1923, 248; *Lactobacillus listeri* (Henneberg) Bergey *et al.* 1923, 248; *Lactobacillus cucumeris* (Henneberg) Bergey *et al.* 1923, 250; *Lactobacillus pabuliacidi* (Weiss) Bergey *et al.* 1925, 181; *Bacterium busae asiaticae* Tschekan 1929, 89; *Lactobacillus busaesiaticus* (Tschekan) Bergey *et al.* 1930, 288; *Lactobacillus wortmanni* (*sic*) (Henneberg) Bergey *et al.* 1930, 288.

plan.ta'rum. L. fem.n. *planta* a sprout; M.L. n. *planta* a plant; M.L. gen.pl.n. *plantarum* of plants.

Rods with rounded ends, straight, generally 0.9–1.2 μm wide by 3–8 μm long, occurring singly, in pairs, or in short chains. Motility, and flagellation ordinarily absent but motile, peritrichously flagellated strains have been described (Harrison and Hansen, 1950; Langston and Bouma, 1960).

Anaerobically, surface colonies are about 3 mm wide, raised, round, smooth, compact, white and occasionally light or dark yellow. Growth in broth results in an even heavy turbidity.

α-Methyl-D-glucoside and melezitose often fermented; some strains ferment α-methyl-D-mannoside. Some strains ferment arabinose and some ferment both arabinose and xylose and have been known as *L. arabinosus* and *L. pentosus*, respectively.

DL-Lactic acid produced. Fructose 1,6-diphosphate aldolase present and also hexose monophosphate shunt activity. Growth on gluconate with CO_2 production. Ribose is fermented to 1 mole of lactic and 1 mole of acetic acid. Where other pentoses are fermented, the products are the same as for ribose.

Nitrates generally not reduced but rare strains may reduce nitrate if the pH is poised at 6.0 or higher. Ammonia is not produced from arginine. Growth in media with 4% sodium taurocholate. Milk is acidified and may be coagulated; 0.3–1.2% titratable acid produced.

Cell wall peptidoglycan is of the directly cross-linked meso-diaminopimelic acid type; ribitol teichoic acid and glucose are present; galactosamine is not present. Serological group D. These characteristics are different from *L. casei*.

Growth at 15 C, generally not at 45 C, optimal usually 30–35 C.

Calcium pantothenate and niacin required; thiamine, pyridoxal *or* pyridoxamine, folic acid, vitamin B_{12}, thymidine *or* deoxyribosides not required; riboflavin generally not required.

The G + C of the DNA of six strains was 45 ± 1 moles % (buoyant density).

Isolated from dairy products and environments, fermenting plants, silage, sauerkraut, pickled vegetables, spoiled tomato products, sourdough, cowdung, and the human mouth, intestinal tract and stools. This species has never been found in oral and intestinal samplings of rats, mice, hamsters, rabbits and guinea pigs.

Type strain: ATCC 14917 (Orla-Jensen's *Streptobacterium plantarum* No. 39); (Hansen, 1968).

Comments: Keddie (1959) found about 50% atypical strains from herbage and silage. These strains had some resemblance to *L. plantarum* but did not ferment maltose and mannitol, occasionally did not ferment melibiose, were esculin negative and some were polarly flagellated (see also Cunningham and Smith, 1940). Keddie (1959) also described sucrose negative strains, some polarly flagellated or having curved rods or other heterogeneous properties. It is difficult to assign any of these atypical strains to present species. Some of these strains are very probably similar to some new species given names by Abo-Elnaga and Kandler, 1965 (see *L. curvatus*, *L. coryniformis* subsp. *coryniformis*, *L. casei* subsp. *tolerans* and *L. casei* subsp. *pseudoplantarum*).

Strains from the rat were named *L. plantarum* by Jordan *et al.* (1959) but many were so highly atypical that this species designation is doubtful.

12. Lactobacillus curvatus (Troili-Petersson) Abo-Elnaga and Kandler 1965, 19. (*Bacterium curvatum* Troili-Petersson 1903, 137.)

cur.va′tus. L. v. *curvo* curve; L. past part. *curvatus* curved.

Curved, bean-shaped rods, rounded ends, 0.7–0.9 by 1.0–1.2 μm, occurring in short chains or closed rings of generally four cells or horseshoe-shaped forms. Non-sporing. Some strains at first motile; motility lost on subculture. Flagellation not described.

Colonies generally somewhat smaller than those of *L. plantarum* but usually of same appearance.

No diaminopimelic acid (DAP) in cell wall (different from *L. plantarum*); the peptidoglycan is of the L-lysine-D-aspartate type.

Homofermentative. Acid and no gas from glucose; DL-lactic acid chief product.

Esculin split (98%). No growth in 4% taurocholate (unlike *L. plantarum*).

Lactose variable (41% positive); acidity in milk variable (59% negative); questionable or slight fermentations of starch and turanose.

Growth at 15 C; no growth at 45 C; optimal range 30–37 C. The G + C content of the DNA is 43.9 moles % (buoyant density).

Isolated from cowdung, dairy barn air, milk and silage, endocarditis.

13a. Lactobacillus coryniformis subsp. **coryniformis** Abo-Elnaga and Kandler 1965, 18.

co.ry′ni.form′is. Gr. n. *coryne* a club; L. adj. *formis* shaped; M.L. adj. *coryniformis* club-shaped.

Short often coccoid rods, occasionally somewhat pear-shaped, 0.8–1.1 by 1.0–3.0 μm, occurring singly, in pairs or short chains.

Colonies and growth indistinguishable from *L. plantarum* (but there is no diaminopimelic acid in cell wall. The peptidoglycan is of the L-lysine-D-asparate type).

Homofermentative. Acid and no gas from glucose; D(−)-lactic acid is the chief product with some L(+)-lactic acid. Gas from gluconate. Ribose fermented slowly or equivocally.

No growth in 4% taurocholate. Usually no acidity in milk (84%). Esculin variable.

Growth at 15 C; slight or no growth at 45 C; optimum 30–37 C.

The G + C content of the DNA is 45 moles % (T_m).

Found mainly in silage (75%), also from cowdung and dairy barn air. None has been isolated from milk, milking machines or cheese.

Type strain: Not designated; suggested reference strain ATCC 25602.

13b. Lactobacillus coryniformis subsp. *torquens* Abo-Elnaga and Kandler 1965, 19.

torqu′ens. L. pres.part. of *tourquere* to twist.

Five strains differ from the type subspecies of the species in producing D(−)-lactic acid. Milk is acidified. The carbohydrates fermented are the same as those fermented by *L. coryniformis* subsp. *coryniformis* except that the suggested reference strain of *L. coryniformis* subsp. *torquens* does not ferment rhamnose, salicin, sorbitol and sucrose (Table 16.1).

The peptidoglycan of the cell wall is of the L-lysine-D-asparate type.

Isolated from cowdung and dairy barn air. None has been isolated from milk or milk products.

The G + C content of the DNA is 45.4 moles %.

Type strain: Not designated; suggested reference strain ATCC 25600.

14. Lactobacillus homohiochii Kitahara, Kaneko and Goto 1957, 118.

ho′mo.hi.o′chi.i. Gr. adj. *homos* like, equal; Japanese n. *hiochi* spoiled saké; M.L. gen.n. *homohiochii* probably intended to mean homofermentative lactobacillus of hiochi.

Rods, with rounded ends, 0.7–0.8 by 2–4 μm or occasionally 6 μm in length. Non-motile.

Characteristics in solid and fluid media, as well as the nutritive requirement for hiochic acid (D-mevalonic acid) even in the presence of acetate, are very similar to *L. heterohiochii*. Resistant to 13–16% ethanol.

Homofermentative; hexoses are converted almost exclusively to chiefly D(−)- with some L(+)-lactic acid.

Microaerophilic.

Growth temperature; optimum about 30 C, limited or no growth at 40 C; no growth at 45 C.

Initial pH for growth: Acidophilic; optimum about 5, slight growth at 5.5, no growth at 6 or higher.

G + C content of the DNA 46 moles% (T_m).

Source: Spoiled saké.

Type strain: Not designated; suggested reference strain ATCC 15434.

15. Lactobacillus fermentum Beijerinck 1901, 233. (*Bacillus δ* von Freudenreich 1895, 175; *Bacillus casei δ* von Freudenreich and Thöni 1904, 532; *Lactobacterium fermentum* (Beijerinck) van Steenberge 1920, 816; *Lactobacillus fermenti* Beijerinck (*sic*) according to Bergey *et al.* 1923, 252; *Betabacterium Jensenii* (*sic*) Frank 1936 in Orla-Jensen 1943, 91; *Bacterium gayoni* Müller-Thurgau and Osterwalder 1917, 34; *Betabacterium longum* Orla-Jensen 1919, 175; *Lactobacillus gayoni* (Müller-Thurgau and Osterwalder) Pederson 1929, 23; *Lactobacillus longus* Orla-Jensen) Bergey *et al.* 1934, 312; *Lactobacterium longum* (Orla-Jensen) Krasil′nikov 1949, 220.)

fer.men′tum L. neut.n. *fermentum* ferment, yeast.

Rods, variable in size, usually short, 0.5–1.0 by 3.0 or more μm, sometimes in pairs or chains. Non-motile.

Colonies are generally flat, circular or irregular to rough, often translucent. Non-pigmented, but rare strains produce rusty orange pigment.

Heterofermentative. Acid and gas from glucose; growth at the expense of 4% gluconate with copious CO_2 production. D-Ribose fermented to lactic and acetic acids without gas. Mannitol produced from fructose.

Glucose probably metabolized by hexose monophosphate shunt (glucose 6-phosphate dehydrogenase and 6-phosphogluconate dehydrogenase activity). Fructose 1,6-phosphate aldolase absent.

Galactose, lactose, melibiose and raffinose usually fermented (96, 92, 98, 98% of strains, respectively). Arabinose, xylose, α-methyl-D-glucoside and trehalose usually not fermented (79, 80, 88, 88% of strains, respectively). Generally little or slow acidity in milk.

DL-Lactic acid produced (50% of total glucose carbon); other major products are acetate, ethanol and CO_2.

Nitrates not reduced (very rare exceptions). Ammonia produced from arginine.

Cell walls contain glycerol teichoic acid; the peptidoglycan is of the L-ornithine-D-aspartate type.

Calcium pantothenate, niacin, thiamine required. Riboflavin, pyridoxal, folic acid not required.

No growth at 15 C; growth at 45 C; freshly isolated strains may have optimum from 41–42 C. Historically, higher temperature gas-producing rods have been lumped in this species.

The G + C content of the DNA is 53.4 ± 0.5 moles % (buoyant density).

Isolated from yeast, milk products, sourdough, fermenting plants or products, wine, manure, silage, mouth and feces of humans and rats, and cecum of turkeys.

Neotype strain: ATCC 14931 (original strain Orla-Jensen (1919) *Betabacterium longum* No. 28).

Comments: Rogosa and Hansen (1971) have requested the Judicial Commission to conserve the name *Lactobacillus fermentum* Beijerinck 1901, 233 (see "Comments" under *L. delbrueckii*).

16. Lactobacillus cellobiosus Rogosa, Wiseman, Mitchell and Disraely 1953, 693.

cello.bi.o′sus. M.L. adj. *cellobiosus* pertaining to cellobiose, a sugar derived from cellulose.

Rods, variable in size, may be short, often 0.5–1.0 by 3–5 μm or more in length. Non-motile and non-flagellated.

Colonies vary from smooth, raised, butyrous, grayish white to rough or cauliflower or doughnut-like forms; mixtures of these forms are common in pure cultures.

Glucose fermented heterofermentatively; major products are lactate, acetate, CO_2 and ethanol. Inducible growth on gluconate with CO_2 production; D-ribose fermented without gas production yielding lactic and acetic acids.

Glucose 6-phosphate and 6-phosphogluconate dehydrogenases present; fructose 1,6-diphosphate aldolase absent.

Eighty-four per cent of strains ferment xylose; weak, slow or negative with mannose, lactose and salicin.

Ammonia produced from arginine. Little or no acidity in milk.

Cell wall peptidoglycan is of the L-ornithine-D-aspartate type; glycerol teichoic acid present; group antigen not determinable.

Complex nutritional requirements. Calcium pantothenate, niacin and thiamine are required; riboflavin, pyridoxal, folic acid and vitamin B_{12} not required.

Growth is variable at 15 C and negative at 45 C; optimum 30–35 C.

The G + C content of the DNA is 53.1 ± 0.8 moles % (buoyant density).

Isolated from the mouth of man.

Type strain: ATCC 11739 (Hansen, 1968).

Comments: This species and *L. fermentum* are very similar but *L. cellobiosus* ferments cellobiose and splits esculin. Otherwise phenotypically, and in the G + C content of the DNA (53.4 ± 0.5 for *L. fermentum* and 53.1 ± 0.8 for *L. cellobiosus*) there is great resemblance. The neotype strain of *L. fermentum* ATCC 14931 and the type strain of *L. cellobiosus* ATCC 11739 have cell wall peptidoglycans of the L-ornithine-D-asparate type. Other presumed strains of *L. fermenti* (*sic*) have peptidoglycans of the L-lysine-D-aspartate type (Cummins and Harris, 1956; Ikawa and Snell, 1960; Kandler, 1967; Kandler, 1970). Williams and Sadler (1971) criticized these cited reports as erroneous and stated that both *L. fermentum* and *L. cellobiosus* peptidoglycans contain ornithine and not lysine. However, Williams and Sadler (1971) suspected the authenticity of *L. cellobiosus*, NCIB 4037, and *L. buchneri*, NCIB 8837. Subsequent examination by Mandel showed that the purported *L. cellobiosus* NCIB 4037 has a G + C content of 45.5 moles % (consistent with *L. buchneri*) and the purported *L. buchneri*, NCIB 8837, has a G + C content of 51.3 moles % (consistent with *L. cellobiosus*). Historically, heterofermentative lactobacilli growing at high temperatures (45 C or higher) have been designated as *L. fermenti* (*sic*); a small number of differentiating phenotypic characteristics were used and none of these included modern or sophisticated biochemical, enzymological or genetic data. *L. fermenti* (*sic*) has been heterogeneous comprising strains generally not fermenting arabinose and xylose, but also including strains fermenting one or both of these pentoses. Thus, Reuter has suggested types of *L. fermenti* (*sic*) and his type II, ATCC 23272, although generally fitting older concepts of *L. fermentum*, has a G + C content of only 40 moles % (as compared to 53 % for the neotype strain). In addition, there are occasional strains of *L. brevis* and *L. buchneri* capable of growth at 45 C and these may have been wrongly designated as *L. fermenti* (*sic*) (particularly those presumably fermenting arabinose, xylose or both); these occasional thermoduric strains of *L. brevis* and *L. buchneri* generally grow at 15 C (usually not tested in earlier studies) whereas *L. fermentum* does not grow at this low temperature. With this confusion in clear criteria and mistakes in labeling within culture collections, it is not surprising that different compositions of *L. fermentum* peptidoglycans have been reported. Recently, Kandler (1972, personal communication) has re-examined the neotype strain of *L. fermentum* and a number of other strains with the same modal characteristics and their peptidoglycans are of the L-ornithine-D-aspartate type; other labeled strains of *L. fermenti* (*sic*) such as strain ATCC 9338 (used in most or perhaps all of the early studies) have peptidoglycans of the L-lysine-D-aspartate type; however, G + C and enzymology studies, etc. indicate that some at least, are *L. brevis* or hitherto undetected new species. Obviously, further studies of the latter heterogeneous group are required. Finally, the electrophoretic mobilities of the glucose 6-phosphate dehydrogenases of the neotype strain of *L. fermentum* and the type strain of *L. cellobiosus* are similar. If current DNA-DNA homology studies should indicate great homology, then *L. cellobiosus* might be considered as a subspecies of *L. fermentum*.

17. **Lactobacillus brevis** (Orla-Jensen) Bergey *et al.* 1934, 312. (*Bacillus* γ von Freudenreich 1891, 22; *Bacillus casei* γ von Freudenreich and Thöni 1904, 532; *Betabacterium breve* Orla-Jensen 1919, 175; *Lactobacterium breve* (Orla-Jensen) Krasil'nikov 1949, 217; *Saccharobacillus pastorianus* var. *berolinensis* Henneberg 1901, 383; *Bacillus lindneri* Henneberg 1901, 384; *Bacillus fasciformis* Schönfeld and Rommel 1902, 585; *Bacillus brassicae fermentatae* Henneberg 1903, 332; *Bacillus panis fermentati* Henneberg 1903, 341; *Bacterium soya* Saito 1907, 154; *Bacillus acidophil-aerogenes* Torrey and Rahe 1915, 437; *Lactobacillus pentoaceticus* Fred, Peterson and Davenport 1919, 357; *Lactobacillus lindneri* (Henneberg) Bergey *et al.* 1923, 245; *Lactobacillus berolinensis* (Henneberg) Bergey *et al.* 1923, 246; *Lactobacillus panis* (Henneberg) Bergey *et al.* 1923, 251; *Lactobacillus soya* (Saito) Bergey *et al.* 1923, 251; *Lactobacillus fermentatae*

Bergey *et al.* 1923, 252; *Bacterium lindneri* (Henneberg) Henneberg 1926, 123; *Saccharobacillus pastorianus* var. *berolinensis fasciformis* Henneberg 1926, 123; *Plocamobacterium pentoaceticum* (Fred *et al.*) Pribram 1933, 78; *Betabacterium arabinosaceum* Frank 1936 as cited by Orla-Jensen 1943, 91; *Lactobacillus rudensis* Davis 1937, 374; also see subjective synonymy of *L. plantarum*; *Lactobacillus brevis* var. *rudensis* (Davis and Mattick) Breed and Pederson 1938, 667; *Bacillus belorinensis* (*sic*) (Henneberg) Otani 1939, 149; *Lactobacillus pentoaceticus* var. *magnus* Iwasaki 1940, 148; *Lactobacterium fasciformis* (Schönfeld and Rommel) Krasil'nikov 1949, 219; *Lactobacterium lindneri* (Henneberg) Krasil'nikov 1949, 219; *Lactobacillus pastorianus* var. *brownii* Shimwell 1949, 26; *Lactobacillus brownii* (Shimwell) Shimwell 1949, 29; *Lactobacillus pastorianus* var. *diastaticus* Andrews and Gilliland 1952, 195.)

bre'vis. L. adj. *brevis* short.

Rods, generally short and straight, 0.7–1.0 by 2.0–4.0 μm, with rounded ends, occurring singly or in short chains. Gram or methylene blue stains may reveal bipolar or other granulations.

Colonies generally rough or intermediate, flat, may be nearly translucent. Generally non-pigmented; some strains pigmented orange to red.

Heterofermentative. Acid and gas from glucose autoclaved in media; separately sterilized glucose is catabolized aerobically but not anaerobically. Acid and gas from fructose anaerobically. Growth at the expense of 4% gluconate with copious CO_2 production. D-Ribose fermented to lactic and acetic acids without gas. Pyruvate dismutation anaerobically yields lactate + acetate + CO_2; aerobically, lactate slowly yields acetate + CO_2; acetoin is not formed.

Glucose has been reported to be metabolized by the hexose monophosphate shunt (Buyze, 1955; Buyze *et al.*, 1957). There is glucose 6-phosphate dehydrogenase activity but the extremely low 6-phosphogluconate dehydrogenase activity raises some doubt that the hexose monophosphate shunt is a primary pathway for the aerobic glucose dissimilation.

Fructose 1,6-diphosphate aldolase activity is absent.

Mannitol produced from fructose. α-Methyl-D-glucoside fermented by 80% of strains. Little or no acidity in milk.

DL-Lactic acid produced (50% of total glucose carbon); other major products are acetate, ethanol and CO_2.

Very rare strains decompose peroxide by a pseudocatalase (not containing porphyrin). Ammonia produced from arginine.

Cell walls contain glycerol teichoic acid; the peptidoglycan is of the L-lysine-D-aspartate type.

Serological group E (same as *L. bulgaricus* and *L. lactis*).

Complex nutrition. Calcium pantothenate, niacin, thiamine, folic acid required; riboflavin, pyridoxal, vitamin B_{12} not required.

Growth at 15 C; no growth at 45 C; optimum about 30 C.

The G + C content of the DNA is 42.7 ± 1.5–46.4 ± 1.0 moles % (buoyant density).

Isolated from milk, kefir, cheese, sauerkraut, spoiled tomato products, sourdough, certain soils, ensilage, cow manure, feces, and the mouth and intestinal tract of humans and rats.

Type strain: ATCC 14869 (Orla-Jensen's (1919) original strain *Betabacterium breve* No. 14).

Note. The somewhat variable DNA composition of *L. brevis* suggests that this species is internally heterogeneous.

18. **Lactobacillus buchneri** (Henneberg) Bergey *et al.* 1923, 251. (*Bacillus Buchneri* (*sic*) Henneberg 1903, 163; *Lactobacillus lycopersici* Mickle 1924, 404; *Bacterium buchneri* (Henneberg) Henneberg 1926, 111; *Ulvina buchneri* (Henneberg) Pribram 1933, 75; *Lactobacterium buchneri* (Henneberg) Krasil'nikov 1949, 218; *Bacillus Hayducki* (*sic*) Henneberg 1903, 330; *Bacillus Wehmeri* (*sic*) Henneberg 1903, 331; *Bacterium mannitopoeum* Müller-Thurgau 1908, 396; *Lactobacillus mannitopoeus* (Müller-Thurgau) Pederson 1929, 31; *Lactobacillus mannitopoeus* var. *fermentus* Iwasaki 1940, 148.)

buch'ner.i. M.L. gen.n. *buchneri* of Buchner; named for E. Buchner, a German bacteriologist.

Identical with *L. brevis* except *L. buchneri* (1) ferments melezitose; (2) does not require folic acid (Rogosa *et al.*, 1953; Rogosa and Sharpe, 1959; Abo-Elnaga and Kandler, 1965) and many strains require riboflavin. However, Franklin and Sharpe (1964) found many melezitose-fermenting strains which were atypical, requiring folic acid and pyridoxal; (3) has a G + C content of the DNA of 44.8 ± 1.1 (three strains).

Neotype strain: ATCC 4005 (Rogosa and Hansen, 1971).

Comment: The original description of the temperature relations of *Bacillus Buchneri* (*sic*) (Henneberg, 1903) is contradictory although one interpretation could be that the maximum temperature at which acid was produced in mash initially was 47 C and the optimum 39–40 C; later, however, greatest amounts of acid were produced between 23–30 C. Present strains grow optimally at about 30 C.

19. **Lactobacillus viridescens** Niven and Evans 1957, 758. (*Lactobacillus corynoides* subsp. *corynoides* Abo-Elnaga and Kandler 1965, 125;

Lactobacillus corynoides subsp. *minor* Abo-Elnaga and Kandler 1965, 128; *Lactobacillus viridescens* subsp. *minor* Kandler and Abo-Elnaga 1966, 754.)

vir.i.des'cens. M.L. *viridescens* pres.part. of *viridescere* to grow green, greening.

Small rods, may be coccoid, 0.8 by 2.0–4.0 μm, occurring singly or in pairs. Ends usually rounded but may be tapered. Non-motile.

Deep colonies generally smooth, compact, 0.5–1.0 mm. Non-pigmented.

Heterofermentative. DL-Lactic acid and CO_2 from glucose. Mannose, fructose and maltose also fermented. Sucrose usually fermented (75% of strains) generally forming large quantities of a mucoid polysaccharide; some strains (20%) ferment trehalose.

Hippurate not attacked.

No change in milk. No ammonia from arginine.

Cell wall peptidoglycan is of the L-lysine-L-alanine-L-serine type. The cell wall composition and general structure bears marked resemblances to some strains of *Leuconostoc* (Kandler *et al.*, 1967).

Growth may be slow but is stimulated by added 5–10% CO_2, Tween 80, citrate, Mn^{++} (4 mg/liter) and thiamine (1 mg/liter) in complex media. Pantothenate, niacin, thiamine, riboflavin, biotin are required; folic acid and pyridoxal may be stimulatory.

Growth at 5 and 15 C; negative or variable at 40 C; no growth at 45 C; optimum about 30 C.

The G + C of the DNA is 37.5 moles % for the type strain, 42.3 % (both buoyant density) for strain NCDO 403 and 41.8% for one strain of *L. corynoides* subsp. *corynoides*.

Isolated from discolored cured meat products such as sausage and bologna.

Type strain: S38A; ATCC 12706 (Lessel and Rogosa, 1971).

Comments: Abo-Elnaga and Kandler (1965) named what they thought to be a new species (48 strains from raw and pasteurized milk and milking machine slime) as *Lactobacillus corynoides* subsp. *corynoides* and *Lactobacillus corynoides* subsp. *minor*. These have the same general phenotypic characteristics (including cell wall composition and structure) as the type strain of *L. viridescens*. Kandler and Abo-Elnaga (1966) later recognized the synonomy of the two subspecies of *L. corynoides* and *L. viridescens*. *L. corynoides* subsp. *corynoides* and *L. corynoides* subsp. *minor* corresponded to their terminology of *L. viridescens* subsp. *viridescens* and a new subspecies *L. viridescens* subsp. *minor*, respectively.

The divergence of G + C content of two phenotypically similar strains of *L. viridescens* and their physiological and structural similarity to some *Leuconostoc* suggest further study.

20. Lactobacillus coprophilus Kandler and Abo-Elnaga 1966, 757.

cop.roph'il.us. Gr. n. *copros* dung; Gr. adj. *philus* loving; M.L. adj. *coprophilus* dung-loving.

Short rods with tendency to thicken at one end; 0.8–1.0 μm wide by 2–4 μm long; arranged singly or in short chains. Microaerophilic; poor growth aerobically.

Colonies on MRS agar incubated anaerobically reach 2 mm diameter, are white with smooth edges and surfaces. Abundant turbidity in MRS broth (De Man *et al.*, 1960) with later sediment.

Heterofermentative. Products from glucose are DL-lactic acid, acetic acid, ethanol and CO_2. Products from pentoses are lactic and acetic acids.

No growth in litmus milk. Growth in 6.5% NaCl. Ammonia produced from arginine.

Cell wall peptidoglycan is of the L-lysine-L-alanine-L-alanine type.

Optimum temperature about 35 C. Growth at 15 C but not at 45 C.

The G + C content of the DNA is 41 moles %.

Six strains isolated from cowdung.

Type strain: Not designated.

Comment: *L. coprophilus* subsp. *confusus* Holzapfel and Kandler 1969, 655 differs in not fermenting arabinose, in fermenting salicin consistently, in producing a dextran-like slime and having a peptidoglycan containing alanine, glutamic acid, lysine (3:1:1). Holzapfel and Kandler (1969) consider strain ATCC 10881, previously classified as *Leuconostoc mesenteroides* to be *L. coprophilus* subsp. *confusus*. Sharpe, Garvie and Tilbury (1972) propose re-naming these strains as *Lactobacillus confusus*.

21. Lactobacillus hilgardii Douglas and Cruess 1936, 115. (Lactobacillus Type II Fornachon 1943, 58.)

hil.gar'di.i. M.L. gen.n. *hilgardii* named for Hilgard.

Rods, with rounded ends occurring singly, in short chains and frequently in long filaments. Stained cells are 0.5–0.8 by 2.0–4.0 μm. Individual filaments may be 15 or more μm long. No motility. No flagella. Gram-positive becoming Gram-negative and granulated with age.

Growth moderate, raised, hyaline to white and cream-colored with entire edges on slants of tryptone yeast extract glucose agar, tryptone tomato juice glucose agar or yeast infusion glucose agar. In such broths, there is moderate turbidity within 3–4 days with subsequent clearing and sediment.

Heterofermentative. Products from glucose are DL-lactic acid (about 50%), acetic acid, CO_2, ethanol and glycerol. These products plus mannitol are formed from fructose. D-Xylose is fermented to DL-lactic acid and acetic acid.

Erratic utilization of substrates occurs at pH 6.8–7.0, but in the initial pH range of 4.5–5.5 significant acid (>0.5–1.0 pH unit decrease) is formed from glucose, galactose, fructose, lactose, maltose, raffinose and sucrose. Citric and L-malic acids are attacked. A wide range of other carbohydrates and polyols is not attacked.

Gelatin not liquefied. Litmus milk unchanged.

The cell wall peptidoglycan is of the L-lysine-D-aspartate type.

CO_2 (5% or more) facilitates initial isolation and cultivation. Grows in presence of 15–18% (v/v) ethanol in media or in wine containing 1% yeast autolysate.

Grows well from 28–34 C; minimal or no growth at 15 C; may not grow at 40–43 C; no growth at 45 C. Growth between pH 3.8–8.0; however, optimal initial pH range for growth is 4.5–5.5.

G + C content of the DNA is 40.3 moles %.

Isolated from California table wines.

Type strain not designated; suggested reference strain: ATCC 8290.

22. Lactobacillus trichodes Fornachon, Douglas and Vaughn 1949, 129. (*Lactobacillus* Type I Fornachon 1943, 57.)

tri.cho'des. Gr. adj. *trichodes* like a hair.

Rods, 0.4–0.6 by 2–4 μm, occurring singly, in pairs and in chains. Very long, thread-like chains and filaments frequently form a tangled mass. Non-motile.

Colonies in poured plates of autolyzed yeast glucose agar develop slowly and are small, subsurface, creamy white and rough or irregular in shape. Surface growth, if any, is scant. In broth, the turbidity after 3–5 days exhibits a very pronounced silky, wavy appearance when shaken gently. After 2–3 weeks cells settle forming a compact sediment. Sometimes growth appears as a flocculent deposit consisting of long, tangled filaments while the liquid above remains clear.

Heterofermentative. Malic acid, citric acid and tartaric acid are not attacked. From glucose chief products are lactic and acetic acids, CO_2 and ethanol; lactic and acetic acids, CO_2 and mannitol are formed from fructose.

Growth occurs with appropriate media containing 15% ethanol. Grows vigorously in wine containing 20% v/v ethanol; some strains grow with 21% ethanol. Vigorous growth has been obtained only in wine in and media containing yeast autolysate plus 20% ethanol. No growth in nutrient broth, glucose peptone broth, Bacto yeast extract broth, grape juice and litmus milk.

Optimum temperature is 25–30 C in ethanol-free media and 20–25 C in wine or other media of high ethanol content. Optimum initial pH range is from 4.5–5.5; usually no growth >5.8 or <3.5.

FAMILY I. LACTOBACILLACEAE 591

Isolated from dessert and appetizer wines containing 20% ethanol, and lees in California, Australia, France, Spain. In California the organism has been commonly referred to as the "hair bacillus," "cottony bacillus," "cottony mold," or "Fresno mold."

G + C: 42.7 moles % (T_m), one strain.

Type strain not designated; suggested reference strain ATCC 27394.

Note. Some of the heterofermentative species found in wines and fruit juices, such as *L. hilgardii* and *L. trichodes*, are probably only part of the total *Lactobacillus* population of very markedly acidophilic organisms often contributing to the flavor of wines through the malolactic fermentation (conversion of malic acid to lactic acid and CO_2) (see Fornachon (1943), Fornachon *et al.* (1949), Vaughn *et al.* (1949), Pilone *et al.* (1966) and Peynaud and Sapis-Domercq (1970)).

23. Lactobacillus fructivorans Charlton, Nelson and Werkman 1934, 1.

fruct.i.vor'ans. L. n. *fructus* fruit, L. inf. *vorare* to eat; M.L. pres.part. *fructivorans* fruit-eating, intended to mean fructose-devouring.

Rods, 0.5–0.8 by 1.5–4 μm, occurring singly, in pairs and often in chains with more or less long curved or coiled filaments.

Heterofermentative. Acid and gas from fructose; variable, slow or negative in glucose, sucrose, maltose, inulin and L-malate; negative in arabinose, xylose, ribose, citrate and a wide range of 45 other carbohydrates and polyols.

Chief products from fructose: DL-Lactic and acetic acids, CO_2 and mannitol.

Gelatin not liquefied. Litmus milk unchanged.

The cell wall peptidoglycan is of the L-lysine-D-aspartate type.

Growth is generally slow, often not visually apparent until 4–5 days or as long as 2 weeks. In fluid cultures, growth is sedimented with clear supernatant for some time. CO_2 facilitates initial isolation. Grows in media with 15% v/v ethanol.

Growth optimum 25–30 C, may not grow at 40–43 C, no growth at 45 C, minimal or no growth at 15 C. Optimum pH range for fructose and glucose fermentation is 4.5–5.5.

G + C content of the DNA 39.8–40.8 moles % (buoyant density).

Isolated from spoiled mayonnaise and salad dressings.

Type strain: Not designated; suggested reference strains ATCC 8288; NRRL strains B-3796 through B-4003.

Note. The specific epithet *fructivorans* has been incorrectly written as *fructovorans* in THE MANUAL (6th ed.) and in *Index Bergeyana*. *Lactobacillus fructosus* Kodama 1956, 705 has been reported to produce D(−)-lactic acid; otherwise it appears phenotypically identical with *L. fructivorans*.

24. Lactobacillus desidiosus Vaughn, Douglas and Fornachon 1949, 138. (*Betabacterium caucasicum* Orla-Jensen 1919, 175; *Bacterium caucasicum* Orla-Jensen 1919, 175; *Betabacterium pentoaceticum* Orla-Jensen *et al.* 1947, 112.)

des.id.i.o'sus. L. adj. *desidiosus* inactive, indolent.

Rods, 0.7–1.0 by 2.0–4.0 μm, occurring singly, with frequent tendency toward filamentation.

Heterofermentative. DL-Lactic acid, acetic acid and CO_2 produced from glucose if it is attacked. Arabinose may be the only carbohydrate fermented; some strains may also weakly ferment glucose, fructose and galactose. Dextrin, malate and citrate not fermented; gluconate and ribose reactions not tested; probably no other carbohydrates or polyols fermented.

Growth in pure culture is slow. Yeast extract stimulatory; growth in media with 15% v/v ethanol. Rods tend to clump together resembling miniature kefir grains. Optimum initial pH range 5–7.

Optimum temperature: <30 C; comparatively good growth at 10 C; no growth at 37–40 C.

G + C: Not determined.

Source: Kefir grains.

Type strain: Not designated.

25. Lactobacillus heterohiochii Kitahara, Kaneko and Goto 1957, 117.

het'er.o.hi.o'chi.i. Gr. adj. *heteros* different, other; Japanese n. *hiochi* spoiled saké; M.L. gen.n. *heterohiochii* probably intended to mean heterofermentative lactobacillus of hiochi.

Rods, with rounded ends, 0.7–0.8 by 2–6 μm, occasionally as long as 10–20 μm. No flagella. Non-motile.

Colonies white, glistening, pinhead or fist form. Stab culture filiform or papillate growth after 3–5 days. Streak culture very limited or no growth. Fluid culture turbid after 2–4 days with silky luster when shaken; firm sediment with clear supernatant fluid in older cultures.

Ferments glucose and fructose. Gluconate and ribose reactions slowly positive. Does not ferment other carbohydrates such as mannose, galactose, pentoses, oligosaccharides, polysaccharides, polyols and glucosides.

Glucose fermentation products: 50% DL-lactic acid, ethanol, acetic acid and CO_2.

Fructose fermentation products: Mannitol, DL-lactic acid, acetic acid, CO_2 and some ethanol.

No growth in milk or in media usually optimal for lactobacilli; growth in saké due to the essentiality of its hiochic acid (now known to be D-mevalonic acid); growth in otherwise optimal lactobacillus media supplemented with D-mevalonic acid. Resistant to 13–16% or more ethanol.

Microaerophilic.

Growth temperature: Optimum about 25–30 C, maximum <40 C.

Initial pH for growth: Very acidophilic, optimum about 5, slight growth at 6, no growth at 7.

G + C not determined.

Isolated from spoiled saké, known as "hiochi" in Japan.

Type strain: Not designated; suggested reference strain ATCC 15435.

Descriptions of the two recently described strictly anaerobic species, *L. ruminis* and *L. vitulinis*, were added in proof.

26. **Lactobacillus ruminis** Sharpe, Latham, Garvie, Zirngibl and Kandler 1973, 47.

ru.min'is. L. n. *rumen* the throat; M.L. n. *rumen* the rumen; M.L. gen.n. *ruminis* of the rumen.

Rods, non-sporing. Motile by peritrichous flagella; motility not always easy to demonstrate and often sluggish, best demonstrated in certain media containing low concentrations of glucose. Gram-positive.

Homofermentative, producing mostly L(+)- and about 5% D(−)-lactic acid. Fructose, galactose, glucose, mannose, maltose, sucrose, cellobiose, melibiose, raffinose, amygdalin, esculin and salicin fermented. Arabinose, ribose, xylose, rhamnose, melezitose, trehalose, inulin, glycogen, glycerol, inositol, mannitol and sorbitol not fermented. Lactose fermented by some strains.

Arginine not hydrolyzed. Catalase negative.

Cell wall contains *meso*-DAP but not D-aspartic acid or L-lysine. Galactose present, rhamnose in trace amounts.

Anaerobic. Grows at 45 C but not at 15 C.

Isolated from the rumen of a cow.

The G + C content of the DNA is 43.7 ± 0.1 moles %.

Type strain: RF1.

27. **Lactobacillus vitulinus** Sharpe, Latham, Zirngibl and Kandler 1973, 47.

vi.tu.lin'us. L. masc.adj. *vitulinus* of a calf.

Rods, non-sporing. Not motile. Gram-positive.

Homofermentative, producing D(−)-lactic acid. Fermentation pattern similar to that of *L. ruminis*, except lactose fermented, and inulin, trehalose and sorbitol may be fermented.

Arginine not hydrolyzed.

Cell wall contains *meso*-DAP; galactose and rhamnose absent.

Anaerobic. Grows at 45 C but not at 15 C.

Isolated from the rumen of a calf but has been found in the rumen contents of older animals.

The G + C content of the DNA is 35.7 ± 1.3 moles %.

Type strain: T185.

Species incertae sedis

a. *Lactobacillus batatas* Kitahara 1949, 23.

A coccus or very small rod which produces D(−)-lactic acid and CO_2 from glucose. Like a number of *Leuconostoc*, does not produce gas from gluconate, ammonia from arginine or ferment ribose. G + C ratio of the DNA is 57.6 moles %. Could be classified as a *Leuconostoc*.

b. *Lactobacillus bifermentans* Pette and van Beynum 1943, 339.

Has the general characteristics of a homofermentative lactobacillus producing DL-lactic acid from glucose. However, after lengthy induction, lactate is degraded anaerobically to acetate, ethanol, CO_2, H_2 and occasionally traces of propionate. This anaerobic attack on lactate is inconsistent with definitions of *Lactobacillus* and the taxonomic position of this organism is uncertain.

c. *Lactobacillus intermedium* (*sic*) (Müller-Thurgau and Osterwalder) Bergey *et al.* 1930, 295. (*Bacterium intermedium* Müller-Thurgau and Osterwalder 1917, 33.)

Strains reported to produce L(+)-lactic acid and to have a G + C content of 43.2–45.2 moles % (ATCC 25371 and 25372) have in our laboratory produced DL-lactic acid and their G + C content was determined as 54. They thus resemble *L. fermentum*.

d. *Lactobacillus malefermentans* Russell and Walker 1953, 162.

Strain ATCC 11306 is heterofermentative; glucose and gluconate is fermented with production of CO_2 and acid; ribose also fermented but 45 other carbohydrates are not. The cell wall peptidoglycan is of the L-lysine-D-aspartate type.

e. *Lactobacillus mali* Carr and Davies 1970, 774.

Described as a catalase positive, homofermentative rod. However, ATCC 27053, deposited as *L. mali*, is catalase negative. The organism is not clearly recognizable from its original description, although its relationship to *L. casei* was suggested.

f. *Lactobacillus parvus* Russell and Walker 1953, 312.

The type strain (NCIB 8516; ATCC 11305) is phenotypically and in G + C moles % identical with *L. buchneri*.

g. *Lactobacillus frigidus* Bhandari and Walker 1953, 332.

The type strain (NCIB 8518; ATCC 11307) is phenotypically and in G + C moles % identical with *L. buchneri*.

h. *Lactobacillus pastorianus* (van Laer) Bergey *et al.* 1923, 246. (*Saccharobacillus pastorianus* van Laer 1892, 43; *Bacillus pastorianus* (van Laer) Macé 1897, 957; *Lactobacterium pastorianum* (van Laer) van Steenberge 1920, 816.)

Has the characteristics of *L. brevis* and is considered a synonym of *L. brevis* (Min. 7, International Committee on Lactobacilli and Closely Related Organisms, Moscow meeting, 1966).

The name has been widely used almost as a generic, or even colloquial term for many strains from breweries.

i. *Lactobacillus sake* Katagiri, Kitahara and Fukami 1934, 157.

Except that it does not ferment mannitol, has the phenotypic characteristics and G + C content (42.2 moles %) similar to *L. plantarum*. The cell wall peptidoglycan is of the L-lysine-D-aspartate type.

Type strain: Kitahara T. S.; ATCC 15721. Isolated from saké starter. However, other strains recently received as *L. sake* appear quite dissimilar and very probably more work will show that strains have been confused in culture collections.

j. *Bacillus caucasicus* von Freudenreich 1897, 135.

A heterofermentative organism, probably a lactobacillus from kefir grains. Whether or not it was similar to *L. desidiosus* can not be determined.

k. *Betabacterium vermiforme* (Ward) Mayer 1938, 47. (*Bacterium vermiforme* Ward 1892, 149.)

Originally isolated from ginger beer fermentation. ATCC strain 13133 is indistinguishable from *L. buchneri*. Although reported to produce slime, other heterofermentative lactobacilli, such as *L. buchneri*, *L. brevis*, *L. viridescens*, produce dextran polysaccharides from sucrose and consequently may be confused with some *Leuconostoc* species.

l. *Lactobacillus catenaforme* (Eggerth) Moore and Holdeman 1970, 15. (*Bacteroides catenaformis* Eggerth 1935, 286; *Catenabacterium catenaforme* (Eggerth) Prévot 1938, 296.)

Anaerobic, non-motile rods, having general properties of *Lactobacillus*; D(−)-lactic acid is the main product from glucose. Twenty-one strains from human feces, intestinal and pleural infec-

tions have been studied. G + C moles % of the DNA of Pasteur Institute (Paris) strain 1871 is 32.7 (Sebald, Gasser and Werner, 1965); two other strains give values of 31–33. *Catenabacterium catenaforme*, strain E194e from Reuter is exceptional in being motile, but 97% of the product from glucose is D(−)-lactic acid; in general, these strains have a fermentation pattern (including little or no acid in milk) resembling *L. leichmannii*.

Editorial Note. The authorities for the new combinations *Lactobacillus catenaforme*, *L. crispatus* and *L. minutus* are confirmed in Moore and Holdeman, 1973.

m. *Lactobacillus crispatus* (Brygoo and Aladame) Moore and Holdeman 1970, 15. (*Eubacterium crispatum* Brygoo and Aladame 1953, 641.)

Original Brygoo and Aladame (1953) description is contradictory; pure acetic acid is produced but lactic acid is apparently also a major product! The Anaerobe Laboratory (1970), reports that a co-type (*sic*) strain, Prévot Collection No. II, produces >90% lactic acid from glucose (about 75% D(−)- and 25% L(+)-lactic acid). Growth is favored by anaerobiosis but heavy inocula grow aerobically. Isolated from "mucus of dental origin" (?).

n. *Lactobacillus disciformans* (Prévot) Moore and Holdeman 1970, 15. (*Eubacterium disciformans* Prévot 1938, 295.)

One strain from a lung abscess, 364B (Prévot's original collection), is an anaerobic non-motile rod producing >95% lactic acid from glucose. Twelve or more carbohydrates are fermented.

o. *Lactobacillus minutus* (Hauduroy *et al.*) Moore and Holdeman 1972, 63. (*Bacteroides minutum* Hauduroy Ehringer, Urbain, Guillot and Magrou 1937, 64; *Eubacterium minutum* (Hauduroy *et al.*) Prevot 1938, 295).

Anaerobic, small rods, producing chiefly D(−)-with some L(+)-lactic acid from glucose. Some acid generally also from fructose and mannose; most carbohydrates not fermented. Seven strains from "human clinical material."

GENERA OF UNCERTAIN AFFILIATION

Genus **Listeria** *Pirie 1940, 383 Nom. cons.* Opin. 12. Jud. Comm. 1954, 151

H. P. R. SEELIGER AND H. J. WELSHIMER

(*Listerella* Pirie 1927, 164 *Nom. rej.* Opin. 14. Jud. Comm. 1954, 157; Not *Listerella* Jahn 1906, 540; Not *Listerella* Cushman 1933, 36.)

Lis.te′.ri.a. M.L. fem.n. *Listeria* named for Lord Lister, an English surgeon and discoverer of antisepsis.

Small, coccoid, Gram-positive rods with a **tendency to produce chains** of three to five or more cells, and to produce, in the rough state, elongated to filamentous forms. Smears from 18–24 hr old colonies show typical diphtheroid palisade arrangement with a few V or Y forms. Do not produce spores or capsules, not acid-fast. Gram-positive but in older cultures may be Gram-nega-

TABLE 16.2
Characteristics differentiating the species of genus **Listeria** *and its differentiation*
from genus **Erysipelothrix**

	1. *L. mono-cytogenes*	2. *L. denitrificans*	3. *L. grayi*	4. *L. murrayi*	*Erysipelothrix*
Mannitol acid	−	−	+	+	−
$NO_3 \rightarrow NO_2$	−	+	−	+	−
Motile	+	+	+	+	−
Esculin hydrolyzed	+	+	+	+	−
Catalase	+	+[a]	+	+	−
Beta-Hemolysis	+	−	−	−	−
Glucose fermented	+	+	+	+	−
Glucose oxidized	+	+	+	+	+
Pathogenic to mice	+	+?	−	−	+
G + C ratio	38	56 ± 1	38 ± 1	38 ± 1	36[b]

[a] Based on ATCC strain, a subculture of what is presumably the only isolate.

[b] Reported by Stuart and Welshimer (1974); Flossmann and Erler (1972) reported a G + C ratio of 38–40 moles %.

tive. **Motile** by peritrichous flagella when grown at 20–25 C. At 37 C only a few flagella found, usually one polar flagellum, occasionally two to four, sometimes none.

Acid, but no gas, from glucose and several other carbohydrates (Tables 16.2, 16.3). **Esculin is hydrolyzed.** Indole not produced, urea not hydrolyzed. Gelatin, casein and milk are not hydrolyzed. **Usually catalase positive.**

Aerobic to microaerophilic (growth enhanced under reduced oxygen and 5–10% CO_2). Growth occurs between 4 and 38 C, particularly when the medium contains small amounts of glucose.

Several antigenic types can be distinguished by means of O- and H-antigens. These share some partial O-antigenic factors with other Gram-positive bacteria but not with *Erysipelothrix*. Sensitive to many antibiotics, but not to polymyxin B.

Found in the feces of animals and man, on vegetation and in silage. Parasitic on poikilothermic and warm-blooded animals, including man. The G + C content of the DNA is 38 moles %, except *L. denitrificans* which is 56 moles %.

Type species: *Listeria monocytogenes* (Murray, Webb and Swann) Pirie 1940, 383.

Description of the species of genus **Listeria**

1. **Listeria monocytogenes** (Murray *et al.*) Pirie 1940, 383. (*Bacterium monocytogenes* Murray, Webb and Swann 1926, 408; *Listerella hepatolytica* Pirie 1927, 164; *Bacterium monocytogenes hominis* Nyfeldt 1932, 112; *Corynebacterium parvulum* Schultz, Terry, Brice and Gebhardt 1934, 1023; *Erysipelothrix monocytogenes* (Murray *et al.*) Wilson and Miles 1946, 401; *Corynebacterium infantisepticum* Potel 1950, 490.)

mo.no.cy.to'.ge.nes. M.L. n. *monocytum* a blood cell, monocyte; Gr. v. *gennaio* produce; M.L. adj. *monocytogenes* monocyte-producing.

Small coccoid rods, 0.4–0.5 by 0.5–2.0 μm, with rounded ends, slightly curved in some culture media, occurring singly and in V-shaped or parallel pairs. In rough cultures elongated rods and long filaments, up to 50–100 μm, occur. Motile by means of four peritrichous flagella at 20–25 C; fewer flagella or one flagellum at 37 C.

Cell walls contain about 20% hexose (glucose and galactose), 5% hexosamine and 5% protein

with alanine, glutamatic acid, diaminopimelic acid, aspartic acid and leucine. They do not contain glycine or serine, thus distinguishing them from *Erysipelothrix rhusiopathiae* (see also Table 16.2).

In a 0.25% agar, 8.0% gelatin and 1.0% glucose semi-solid medium, growth along the stab in 24 hrs at 37 C is followed by irregular, cloudy extensions into the medium; growth spreads slowly through the entire medium. An umbrella-like zone of maximal growth occurs 3–5 mm below the surface.

Sheep liver extract agar colonies circular, smooth, butyrous, slightly flattened, transparent by transmitted and milky by reflected light. Smooth, intermediary and rough colonies may be distinguished.

On sheep liver extract agar slant growth confluent, flat, transparent, butyrous.

Growth on peptone agar is thinner than on liver extract agar.

Good growth on blood agar with a narrow zone

of hemolysis around colonies, varying with species of blood. A few strains are strongly, others poorly, hemolytic.

Carbohydrate reactions are shown in Table 16.3. End product of fermentation is mainly lactic acid.

All cultures have a penetrating, rather acid odor.

Hydrogen sulfide not produced, nitrates not reduced to nitrites, acetoin produced (when tested by Barritt's method).

After several months incubation some strains produce a yellowish or reddish pigment.

Require for growth, biotin, riboflavin, thiamine and thioctic acid; cysteine, glutamine, isoleucine, leucine, valine and other amino acids. Arginine, histidine, methionine and tryptophan have a growth-stimulating effect.

Tween 80 and Tween 20 hydrolyzed. A phospholipinase may cause varying degrees of opacity in egg yolk media; the strains with pronounced lipolytic action usually are strongly hemolytic.

Optimum temperature 37 C; grows at all temperatures down to 2.5 C. Thermal death point 58–59 C in 10 min.

Survives 8 weeks in 20% NaCl at 4 C.

Sensitive to tetracycline, chloramphenicol, erythromycin, ampicillin, neomycin and other antibiotics; less sensitive to penicillin and sulfonamides; resistant to polymyxin B.

Intravenous or intraperitoneal injection of cultures into rabbits results in a very marked increase in monocytes circulating in the blood. Monocytosis is also induced by a chloroform-soluble lipid extract which in addition to lipid substance contains hexose and hexosamine. Infection is characterized by necrotic or granulomatous foci in various organs. Causes conjunctivokeratitis when instilled into the conjunctiva of rabbits, hamsters and guinea pigs.

Several types and subtypes may be distinguished by flagellar and somatic antigens. They bear no relation to the host species or to the geographical area from which they were isolated. The various types exhibit a partial somatic antigenic relationship with other bacteria.

Strains belonging to serotype 5 differ in the following characteristics: pronounced beta-hemolysis on 5% sheep blood agar, failure to produce acid from trehalose, salicin, xylose, dextrin and maltose.

Strains from fecal matter and environmental sources are frequently non-hemolytic on 5% sheep blood agar. They usually do not belong to serotypes found in pathological material from listeric animals and man. They are apathogenic to white mice and chicken embryos, and fail to produce conjunctivitis in the eyes of rabbits.

Widely distributed in nature. Isolated from healthy ferrets, insects, sewage, decaying vegeta-

TABLE 16.3

Acid production from carbohydrates by species of genus **Listeria**[a]

	1. *L. monocytogenes*	2. *L. denitrificans*	3. *L. grayi*	4. *L. murrayi*
L-Arabinose	−[b]	+	−	−
D-Galactose	d	+	+	(+)
Glycogen	−	+	−	−
Lactose	(d)	+	+	+
Mannitol	−	−	+	+
Melezitose	d	−	−	−
Melibiose	−	(+)	−	−
Rhamnose	d	−	−	d
Sucrose	(d)	+	−	−
Xylose	−	+	−	−

[a] All species produce acid but no gas in 24–48 hrs from amygdalin, esculin, cellobiose, dextrin, fructose, glucose, maltose, mannose, salicin, starch and trehalose. No acid in 21 days from adonitol, dulcitol, erythritol, inositol, inulin or raffinose.

[b] Key: + = acid produced 24–48 hrs (90% or more strains); − = no acid produced 21 days (90% or more strains); d = some strains positive, some negative; (+) = acid produced slowly (3–7 days); (d) = some strains produce acid slowly (3–7 days), other strains negative.

tive matter, silage, soil and fertilizer, and fecal specimens of chinchillas, ruminants and man. Also isolated from lesions in organs, from meconium, from blood and cerebrospinal fluid of man, and over 50 species of warm-blooded and poikilothermic animals. Causes meningitis, encephalitis, septicemia, endocarditis, abortion, abscesses and local purulent lesions. Many cases have proved fatal (for details see Seeliger (1961)).

Type strain: Murray 53-XIII; ATCC 15313; NCTC 10357.

2. **Listeria denitrificans** Prévot 1961, 512. (*Listeria* L26 Sohier, Benazet and Piéchaud 1948, 55.)

de.ni.tri'.fi.cans. M.L. inf. *denitrificare* to denitrify; M.L. part.adj. *denitrificans* denitrifying.

Small, non-spore-forming, non-acid-fast rods with rounded ends, often forming small chains or diploforms. In smooth colonies coccoid or diphtheroid rods occur, in rough colonies long filaments, occasionally Y forms; actively motile at 22 and 37 C.

On nutrient agar, S forms produce grayish, convex colonies; R forms produce rough type colonies with a depressed center and a dented margin. In aged cultures a yellowish coloration becomes noticeable.

Grows aerobically. Zone of maximal growth below the surface of semi-solid agar stab.

Methyl red positive. Voges-Proskauer negative. Nitrate reduced to nitrite. H₂S not produced.

Serologically distinct, not related to *L. monocytogenes*, *L. grayi* or *E. rhusiopathiae*.

Isolated from cooked blood of beef. Natural habitat not known. Pathogenic to rats and mice when injected intraperitoneally; does not cause conjunctivitis when instilled into the eyes of rabbits or guinea pigs.

The G + C content of the DNA is 56 moles %.

Type strain: Prévot 55134; ATCC 14870.

Comment: In view of hybridization studies, the base composition and differences in biochemical properties, the classification of this species within the genus *Listeria* needs reconsideration (Stuart and Welshimer, 1973).

3. **Listeria grayi** Errebo Larsen and Seeliger 1966, 35.

gray'i M.L. gen. n. *grayi* of Gray; named for M. L. Gray, known for his work in the field of Listeriosis.

Small, non-spore-forming, non-acid-fast rods with rounded ends, occurring occasionally in palisades, also as diploforms or in short chains. Motile by four peritrichous flagella at 20–25 C, with fewer flagella or one polar flagellum at 37 C.

On nutrient agar small (0.75 mm in diameter), translucent, grayish colonies. No pigment at prolonged incubation. Grows aerobically at 20 and 37 C. In semi-solid agar stab culture, formation of an umbrella-like zone of maximum growth occurs 3–5 mm below the surface.

Small amounts of H₂S produced on appropriate media (demonstrable with lead acetate paper only). Acetoin produced (when tested by Barritt's method). Nitrate not reduced to nitrite. Oxidase negative. Tolerant to 6% NaCl. pH range of growth 5–9, no growth at pH 9.6.

Sensitive to penicillin.

Serologically related by partial somatic antigens to *L. monocytogenes*, H-antigen different from known H-antigens of *L. monocytogenes*.

Usually not pathogenic after intraperitoneal or intradural injection to white mice; 5 × 10⁸ cells may be toxic to mice, strain C 57/Leaden. Not pathogenic after intravenous injection into pregnant rabbits. Does not cause conjunctivitis when instilled into the eyes of rabbits or guinea pigs.

Isolated from feces of chinchillas. Natural habitat not known.

Type strain: Seeliger 332/64; ATCC 19120.

Editorial Note. Stuart and Welshimer (1974) suggest a new monospecific genus *Murraya* to include *Murraya grayi* subsp. *grayi* (here *Listeria grayi*) and *M. grayi* subsp. *murrayi* (here *L. murrayi*).

4. **Listeria murrayi** Welshimer and Meredith 1971, 7.

mur'ray.i. M.L. gen.n. *murrayi* of Murray; named for E.G.D. Murray, co-discoverer of *L. monocytogenes*.

Straight rods 0.6–0.8 by 0.7–2.5 μm with rounded ends; occurring singly, occasionally in pairs; no metachromatic granules with Albert stain. Actively motile at 20–25 C with three to five peritrichous flagella; motile organisms absent or rare at 37 C, with an occasional organism having two to four flagella.

Grows best on tryptose blood agar base with 1% glucose added; colonies round, smooth, butyrous, slightly raised, 1–1.5 mm in diameter in 24 hrs at 37 C. By reflected light colonies gray with tinge of blue; by transmitted daylight colonies blue; become tinged with yellow as they enlarge. Yellow pigmentation after 48–72 hrs incubation at 20–25 C on tryptose agar without added glucose. Growth is thinner on nutrient agar than on tryptose-glucose agar.

On sheep blood agar colonies white and non-hemolytic, growth poorer than on tryptose-glucose agar. Slight darkening of blood around crowded colonies.

In brain heart infusion broth heavy growth with accumulation of mucoid deposit which rises in a corkscrew fashion in a confluent mass on swirling tube. In nutrient broth growth poor with little of the tenacious mucoid sediment.

Gelatin, inspissated serum and coagulated egg are not attacked. Acid odor prominent when grown on agar medium containing glucose. All cultures methyl red positive, Voges-Proskauer positive, indole negative. Nitrates reduced to nitrites. Catalase positive.

Aerobic; facultatively anaerobic in presence of carbon dioxide but growth thin.

Optimum temperature 37 C; grows at 4 and 45 C.

Does not produce conjunctivitis when instilled into the eyes of rabbits. Not pathogenic on intravenous injection into rabbits. Intraperitoneal injection of 10⁸ cells into 20-g Rockland Farm SW mice not pathogenic.

Serologically distinct from *L. monocytogenes* and *L. denitrificans*. Some H- and O-antigens shared with *L. grayi*.

Isolated from leaves of corn (maize) at end of growing season; natural habitat probably vegetation and soil.

Type strain: F9; ATCC 25401.

Genus **Erysipelothrix** Rosenbach 1909, 367

H. P. R. Seeliger

E.ry.si.pe'.lo.thrix. Gr. neut.n. *erysipelas* erysipelas; Gr. fem.n. *thrix* hair; M.L. fem.n. *Erysipelothrix* erysipelas thread.

Rod-shaped organisms with a **tendency to form long filaments.** The filaments may also thicken and show characteristic granules. **Non-motile.** Neither capsules nor spores produced. **Gram-positive,** older cultures having a tendency to become Gram-negative.

Acid but no gas from glucose and from certain other carbohydrates. **Catalase negative. Esculin not hydrolyzed.** Produce alpha-hemolysis but **no beta-hemolysis** on blood agar.

Aerobic; grows better in an atmosphere with reduced oxygen and containing 5–10% CO_2.

Temperature range 16–41 C. Most rapid growth at 37 C, maximum cell crop at 33 C.

Parasitic on mammals, birds and fish.

Type species: *Erysipelothrix rhusiopathiae* (Migula) Buchanan 1918, 55.

Description of the species of genus **Erysipelothrix**

1. Erysipelothrix rhusiopathiae (Migula) Buchanan 1918, 55. *Epit. spec. cons.* Opin. 32, Jud. Comm. 1970, 9. (*Bacillus insidiosus* Trevisan 1885, 100; *Bacillus rhusiopathiae suis* Kitt 1893, 284; *Bacterium rhusiopathiae* Migula 1900, 431; *Erysipelothrix porci* Rosenbach 1909, 367; *Erysipelothrix erysipeloides* (*sic*) Rosenbach 1909, 367; *Erysipelothrix murisepticus* (*sic*) Rosenbach 1909, 367; *Erysipelothrix insidiosa* (Trevisan) Langford and Hansen 1953, 18; Bacillus des Schweinerotlaufs Loeffler 1886, 46.)

rhu.si.o.pa'.thi.ae. Gr. adj. *rhusios* red; Gr. n. *pathos* disease; M.L. gen.n. *rhusiopathiae* of red disease (swine erysipelas).

Cells in smooth colonies are short, slender, straight or slightly curved rods, 0.2–0.4 by 0.5–2.5 μm. Cells in rough and in some smooth colonies vary from short forms to long filamentous structures. Thick rods may be present singly, in chains or in entangled masses.

The cell walls contain lysine, glycine, serine, glutamic acid and alanine; they do not contain DL-diaminopimelic acid, which distinguishes them from those of *Listeria monocytogenes* (Mann, 1969).

In gelatin stabs most strains develop, in less than 48 hrs, lateral radiating projections resulting in the typical "test-tube brush" appearance; no spreading on the surface.

Fully developed agar colonies, 1.0–1.5 mm in diameter, are circular and entire, transparent with a bluish sheen by reflected light. Tellurite agar colonies are grayish and pinpoint in 24 hrs, later increasing in size and becoming jet black.

On prolonged incubation on blood agar there is first a greening, then a slight but definite clearing around the colonies.

Acid but no gas from glucose, galactose, fructose and lactose in 1% peptone water with 5% horse serum. Acid may be produced from xylose and melibiose. Usually no acid from glycerol, sorbitol, mannitol, inositol, rhamnose, sucrose, trehalose, melezitose, raffinose, starch, inulin or salicin. Acid production is poor or inconsistent when basal medium contains only 1% peptone in water. The addition of yeast autolysate to media for fermentation studies is recommended.

Gelatin not liquefied. Hydrogen sulfide produced. Indole and acetoin not produced, urea not hydrolyzed, nitrate usually not reduced to nitrite, Tween 80 or Tween 20 not split. Litmus milk unchanged.

Requires riboflavin and small amounts of oleic acid for growth.

Resistant to polymyxin B and sulfonamides; sensitive to penicillin, streptomycin, neomycin and other antibiotics; tolerates up to 0.2% phenol; tolerates 0.5% potassium tellurite.

Serologically distinct from all species of *Listeria.* On the basis of somatic antigens several serotypes can be distinguished. They bear no relation to the host species. Some strains do not possess a type-specific antigen and do not induce antibody formation (Dédié, 1949; Kucsera, 1973).

Causes swine erysipelas but susceptibility of swine is variable. Occasionally pathogenic to man, causing erysipeloid. Mice are very susceptible; septicemia is produced. Rabbits are less susceptible and guinea pigs quite resistant. Fish handlers are especially subject to erysipeloid infections derived from fish.

Widely distributed in nature.

Working strain: NCTC 8163.

Genus **Caryophanon** Peshkoff 1939, 241

T. GIBSON†

Ca.ry.oph'a.non. Gr. n. *caryum* nut, kernel, nucleus; Gr. adj. *phanus* bright, conspicuous; M.L. neut.n. *Caryophanon* that which has a conspicuous nucleus.

Large rods or filaments up to 3 μm in diameter, divided by closely spaced cross walls into numerous disc-shaped cells which are less than 1 μm long when growth is active. In living organisms cross walls show as dark lines, partly complete, partly developing by ingrowth from the external wall. Motile by lateral flagella. Spores not produced. Gram-positive.

Chemoorganotrophs; energy-yielding metabolism strictly respiratory, using oxygen as the terminal electron acceptor.

Strict aerobes.

Type species: *Caryophanon latum* Peshkoff 1939, 241 (Peshkoff 1948, 1004).

Description of the species of genus **Caryophanon**

1. **Caryophanon latum** Peshkoff 1939, 241.
la'tum. L. neut. adj. *latum* broad.

Description supplemented from Pringsheim and Robinow, 1947 and Provost and Doetsch, 1962.

Freshly isolated cultures in the early stages of growth contain rods or filaments 3 by 6–30 μm or longer. Rods have rounded ends and may occur in chains. Cross walls, formed at short intervals by diaphragm-like ingrowth from the external wall, divide the rod into numerous disc-shaped cells. As growth proceeds transverse fission may give rise to shorter, even spherical, forms. The size and shape of the organism vary with the conditions of culture. Size frequently diminishes during laboratory cultivation; finally a proportion of the rods in a culture may not exceed 1 μm in diameter. Motile; flagella lateral, numerous. Gram-positive.

The chief amino acids in the cell wall are alanine, glutamic acid and lysine; diaminopimelic acid is absent (Becker *et al.*, 1967). Cell walls are readily dissolved by lysozyme (Tyeryar and Doetsch, 1962).

On dung extract agar or peptone-yeast extract-acetate agar colonies are circular and glossy, become moderately opaque and up to 2 mm in diameter. At low magnifications colonies show the structure and rough outline produced by large rods. On peptone-meat extract agar colonies are much smaller. Growth in liquid media much restricted if not vigorously aerated. Little or no growth occurs in various liquids that support growth if solidified with agar.

Acetate, butyrate and poly-β-hydroxybutyrate are utilized. Glucose and other sugars, starch and cellulose are not utilized.

No action on gelatin, casein, uric acid or urea. Nitrates not reduced to nitrites. Tributyrin very weakly hydrolyzed. Indole not formed. Catalase produced.

Cells accumulate poly-β-hydroxybutyrate in large amounts in media containing acetate and butyrate, each 0.5%.

A liquid medium which has supported growth (not of all the strains examined) contains acetate, glutamate and NH_4^+ as C and N sources, also thiamine and biotin (Kele and McCoy, 1970).

Strict aerobe.

Grows at 10 and 37 C, not at 45 C. Growth occurs at pH 6 to 8, optimally at pH 7.6–8 (Weeks and Kelley, 1958).

Originally isolated from cowdung. Has been detected on cow feces less than 1 day to a few days after being voided. Isolated once from decaying *Pleurotus* on the stump of a tree (R. E. Buchanan, unpublished).

Type strain: NCIB 9533.

Comment: Peshkoff (1939, 241) recognized a second species, *Caryophanon tenue*, which differs from *C. latum* in its smaller diameter (1.5 μm). This distinction may disappear during cultivation on artificial media owing to a decrease in size of *C. latum*.

---

† Deceased.

PART 17

ACTINOMYCETES AND RELATED ORGANISMS*

CORYNEFORM GROUP OF BACTERIA

MORRISON ROGOSA, C. S. CUMMINS, R. A. LELLIOTT AND R. M. KEDDIE

The family Corynebacteriaceae Lehmann and Neumann 1907, 500 was recognized in the 6th edition of THE MANUAL and included the genera *Corynebacterium, Listeria* and *Erysipelothrix;* in the 7th Edition *Microbacterium, Cellulomonas* and *Arthrobacter* were added.

The family Brevibacteriaceae Breed 1953, 13, with the genera *Brevibacterium* and *Kurthia*, was first included in the 7th edition of THE MANUAL; since the type genus, *Brevibacterium*, is here considered a *genus incertae sedis*, the family has no standing in nomenclature and leaves *Kurthia* unplaced.

The coryneform bacteria present a number of unresolved problems in taxonomy and classification.

The rather widespread use of the terms "coryneform" or "diphtheroid" to describe any non-sporing Gram-positive rod of irregular outline has given rise to considerable confusion as to which organism or groups of organisms should properly constitute the genus *Corynebacterium*. Davis and Newton (1969) examined 70 strains of coryneform bacteria bearing the generic names *Corynebacterium, Arthrobacter, Brevibacterium, Microbacterium, Cellulomonas, Kurthia, Mycobacterium, Nocardia, Jensenia, Listeria* and *Erysipelothrix*. If one considers organisms which, at least at some stage in their growth cycle on artificial media, produce "straight to slightly curved rods with irregularly staining segments, and sometimes granules" (see generic description of *Corynebacterium*), one may add to the list *Bifidobacterium, Actinomyces, Propionibacterium* and the more recently described genera such as *Rothia, Arachnia* and *Bacterionema*.

Other evidence suggests *Mycobacteriaceae* and *Nocardiaceae* might be grouped with Corynebacteriaceae and *Propionibacteriaceae;* however, experts are unable to reconcile certain differences of opinion.

Some of the above mentioned families are recognized in this part of THE MANUAL. However, it seems unwise to recognize Corynebacteriaceae and instead a working concept is presented of a "Coryneform Group of Bacteria" to include *Corynebacterium, Arthrobacter* (with the related genera *Brevibacterium* and *Microbacterium* as *genera incertae sedis*), *Cellulomonas* and tentatively *Kurthia*.

The taxonomic position of the plant pathogenic corynebacteria poses a considerable problem. In purely morphological terms these organisms fit the generic description and, in at least one instance (*C. fascians*), are serologically related to the human and animal pathogens. However, some are motile and all of those so far examined (except *C. fascians*) show differences in cell wall composition of a kind usually found between genera in other groups (see Table 17.1). Moreover, the DNA base ratios of the plant pathogens appear to be distinctly higher than those of the animal and human pathogens. To meet these difficulties the species have been described in three separate sections:

I. Human and Animal Parasites and Pathogens.

II. Plant Pathogenic Corynebacteria.

III. Non-pathogenic Corynebacteria.

Characteristics common to the species in each section are listed at the beginning of each section.

* **Editorial Note.** The division of genera between Part 16 and the first section of Part 17 is arbitrary and readers should consult both. Some genera have traditionally been associated with the actinomycetes and are so treated in this edition.

The anaerobic diphtheroids, represented by strains previously described as *C. acnes*, *C. avidum* and *C. granulosum*, are excluded from the genus because they are anaerobic organisms producing propionic acid as the major end-product of fermentation and because their cell wall composition differs markedly from that of the type species (Table 17.1, group 4). These strains are now included in the genus *Propionibacterium*.

Further Comments

The position of *C. pyogenes* and *C. haemolyticum* is discussed in Addendum I of Section I. These strains represent a well defined group and are a common and well recognized cause of pyogenic infections in domestic animals. However, they bear little similarity to other animal pathogenic corynebacteria; they are generally catalase negative, produce a soluble hemolysin, have a very different cell wall composition (Table 17.1, group 3) and share a cell wall polysaccharide antigen with streptococci of Lancefield group G. Their final classification must await further investigation.

A number of species of human and animal origin have been listed in Addendum II—*Species incertae sedis*. Descriptions of these are generally inadequate.

The reported G + C contents of the DNA of corynebacteria cover a wide range—48 to about 70%. However, except for a figure of 67.5% for *C. pseudotuberculosis* (Sukapure *et al.*, 1970), determinations by Tm and buoyant density would give values for the human and animal pathogens in a much narrower range—57–60%. Since the DNA from both *C. pyogenes* and *P. acnes* have G + C contents of 58–59%, it would appear that determinations of DNA base ratios are of limited value in the taxonomy of coryneform organisms.

No diagnostic key has been attempted, but a table (Table 17.2) of some differential features has been included. Descriptions of the species include information on the cell wall composition for the principal sugars identified in cell wall hydrolysates, and for the diamino acid of the peptidoglycan, since these are the features most readily used in taxonomy. The detailed structure

TABLE 17.1

Distinctive cell wall components of corynebacteria

Groups of strains	Sugars	Cell wall components (amino acids in peptidoglycan)	References
1. *C. diphtheriae* and most other human and animal pathogenic corynebacteria	Arabinose, galactose (mannose, glucose may also be present)	ala, glu, *meso*-DAP[a]	Cummins and Harris, 1958; Cummins, 1962
2. Plant pathogenic corynebacteria:			
(i) *C. fascians*	Arabinose, galactose	ala, glu, *meso*-DAP	Cummins, 1962
(ii) *C. poinsettiae*[b]	Rhamnose, mannose, galactose glucuronic acid	gly, glu, homoser, ala, orn	Perkins, 1967 Perkins, 1970
C. betae	Rhamnose, mannose, fucose glucuronic acid		
(iii) *C. tritici*	Xylose, glucose, mannose	gly, glu, DAB[c], ala	Perkins, 1968
C. insidiosum[b]	Rhamnose, mannose, galactose and fucose		Perkins, 1971[d]
C. sepedonicum[b]	Rhamnose, mannose, galactose		
3. *C. pyogenes*	Rhamnose, glucose	ala, glu, lys	Barksdale *et al.*, 1957
C. haemolyticum	Rhamnose		
4. *Propionibacterium acnes* (*C. acnes*) and "anaerobic diphtheroids"	Galactose and/or glucose and/or mannose (not all strains have all three sugars)	ala, glu, gly, L-DAP (occasional strain has *meso*-DAP)	Johnson and Cummins, 1972

[a] DAP = α, ϵ, diaminopimelic acid.

[b] Results of cell wall analysis of other strains do not always agree, see, e.g. Yamada and Komagata 1970.

[c] DAB = diaminobutyric acid.

[d] Personal communication.

of the peptidoglycan has been determined for a number of coryneform organisms (Fiedler *et al.*, 1970) and it appears to be of the same type for all human and animal pathogens described in Section I.

Similarities between *Corynebacterium* (Section I), *Mycobacterium* and *Nocardia*. The human and animal parasites in the genus *Corynebacterium* as here defined show a number of features which indicate a rather close relationship to the mycobacteria and nocardias. Prominent among these features are the possession of arabinose and galactose as common cell wall sugars, and the presence of a common cell wall antigen (Cummins, 1962). More recent work on the lipids of mycobacteria, corynebacteria and nocardias has reemphasized this relationship, and it now appears probable that these three groups of organisms are characterized by the presence of α-branched, β-hydroxy acids (mycolic acids) of the general type

$$\overset{\beta}{R_1-CH}-\overset{\alpha}{CH}\ COOH.$$
$$\underset{OH}{\vert}\qquad\underset{R_2}{\vert}$$

In mycobacteria the principal mycolic acid found in human strains of *M. tuberculosis*, for example, has the crude formula $C_{78}H_{152}O_3$, while others have similar structures, with numbers of carbon atoms between C_{79} and C_{85} (Etemadi and Lederer, 1965). In strains of *Nocardia*, similar acids are found, except that the number of carbon atoms varies from C_{48}–C_{58} (Bordet *et al.*, 1965).

In corynebacteria, corynemycolic acids occur, with a still smaller number of carbon atoms, C_{32}–C_{36}; these acids have been isolated from *C. diphtheriae*, *C. pseudotuberculosis* (*C. ovis*) and *C. pseudodiphtheriticum* (*C. hofmannii*) (Senn *et al.*, 1967; Lacave, Asselineau and Toubiana, 1967; Welby-Guisse, Laneelle and Asselineau, 1970). Furthermore *C. diphtheriae* contains a toxic glycolipid which is a 6-6' diester of trehalose, containing corynemycolic acid ($C_{32}H_{64}O_3$) and corynemycolenic acid ($C_{32}H_{62}O_3$) in equimolar ratios. This glycolipid is therefore a lower mycolic acid analogue of the cord factor of *M. tuberculosis*, which is trehalose-6-6' dimycolate (see Senn *et al.*, 1967; Kato, 1970).

Mycolic acids have not been reported from any other coryneforms. In contrast, for example, in strains of *Propionibacterium* (including *P. acnes*) the principal fatty acids are C_{15} branched chain

TABLE 17.2

Some distinctive features of species of genus **Corynebacterium** *parasitic on man and animals*

Species	Hemo-lysis	Fermen-tation of sucrose	Reduc-tion of nitrate	Gela-tinase	Urease	Other features
1. *C. diphtheriae*	+	–	+	–	–	Pathogenic: specific exotoxin: specific bacteriophages
2. *C. pseudotuberculosis*	+	d[a]	d[a]	d[a]	d[a]	Pathogenic: specific exotoxin
3. *C. xerosis*	–	+	+	–	–	Not pathogenic: barred morphology often predominant
4. *C. renale*	–	–	–	–	+	Pathogenic: causes pyelitis and cystitis in exptl. animals. Zone of clearing on milk agar: alkaline reaction and burgundy-color in litmus milk
5. *C. kutscheri*	–	+	–	–	+	Specific parasite of rats and mice
6. *C. pseudodiphtheriti-cum*	–	–	+	–	+	Short, regular rods: strongly Gram-positive; no acid produced from any carbohydrate
7. *C. equi*	–	–	+	–	–	Oval or coccoid forms on solid media: capsulated: salmon-pink pigment: no acid produced from any carbohydrate
8. *C. bovis*	–	–	–	–	+	Lipolytic for butterfat: will grow in media containing 9% NaCl: requires long chain fatty acids

[a] Variable results reported in the literature.

fatty acids (Moss and Cherry, 1968; Dowell *et al.*, 1967). Etemadi (1963) specifically noted that C_{32} acids were not found in strain 936-B of *C. parvum*, here considered a synonym of *P. acnes*.

Genus I. **Corynebacterium** *Lehmann and Neumann 1896, 350*

C. S. CUMMINS, R. A. LELLIOTT AND M. ROGOSA†

Co.ry.ne.bac.ter′i.um. Gr. n. *coryne* a club; Gr. n. *bakterion* a small rod; M.L. neut.n. *Corynebacterium* a club bacterium.

Section I. Human and Animal Parasites and Pathogens

Straight to slightly curved rods with irregularly stained segments, and sometimes granules. Frequently show club-shaped swellings. Snapping division produces angular and palisade (picket fence) arrangements of cells. Generally non-motile. Gram-positive, although some species (e.g. *C. diphtheriae*) lose the stain easily, especially in old cultures, while others are more tenacious; granules, however, are strongly Gram-positive. Not acid-fast.

Cell wall composition is characterized by the presence of *meso*-diaminopimelic acid as the diamino acid of the peptidoglycan, and by a polysaccharide containing arabinose, galactose and often mannose (Table 17.1).

Chemoorganotrophs: Carbohydrate metabolism mixed fermentative and respiratory.

Do not produce soluble hemolysins but on solid media containing blood lysis may occur of the red cells in contact with the colonies. Some pathogenic species produce exotoxins.

Aerobic and facultatively anaerobic; grow best aerobically often with a surface pellicle.

Catalase positive.

Reported G + C content of the DNA varies from 52–68%, but, if determinations by Tm and buoyant density only are considered, most species are probably in the range 57–60%.

Probably widely distributed in nature.

Type species: *Corynebacterium diphtheriae* (Kruse) Lehmann and Neumann 1896, 350.

Description of the species of genus **Corynebacterium** (human and animal parasites and pathogens)

1. **Corynebacterium diphtheriae** (Kruse) Lehmann and Neumann 1896, 350. (*Microsporon diphthericum* (*sic*) Klebs 1875, 221; *Microsporon diphtheriticum* Klebs 1883, 143; *Bacillus diphtheriae* Kruse *in* Flugge 1886, 460; *Pacinia loeffleri* Trevisan 1889, 23; *Bacterium diphtheriae* (Kruse) Migula 1900, 499; *Mycobacterium diphtheriae* (Kruse) Krasil′nikov 1941, 98.)

Vernacular name: Klebs-Loeffler bacillus.

diph.the′ri.ae. Gr. n. *diphtheria* leather, skin; M.L. fem.n. *diphtheria* a disease in which a leathery membrane forms in the throat; M.L. gen.n. *diphtheriae* of diphtheria.

Straight or slightly curved rods, frequently swollen at one or both ends, 0.3–0.8 by 1.0–8.0 μm: usually stain unevenly and often contain metachromatic granules (polymetaphosphate) which stain bluish purple with methylene blue. Gram-positive, but rather easily decolorized, especially in old cultures.

On blood agar colonies vary considerably in size and appearance depending on type (see below) but usually 1–3 mm at 24 hr; may show a narrow band of hemolysis around and/or under colony but soluble hemolysin not produced. Colonies on blood tellurite agar (0.04% potassium tellurite)

gray to black, appearance depends on type (see below). On Loeffler's medium growth good, grayish to cream colored; no liquefaction.

Three distinct cultural types recognized (see McLeod, 1943) called *gravis*, *intermedius* and *mitis*, in accordance with the clinical severity of the cases from which the different strains were most frequently isolated: distinguishing features of the three types are shown in Table 17.3; some strains have mixtures of characters which make it difficult to place them in any one type; so-called *minimus* strains (Frobisher, Adams and Kuhns, 1945) appear to be identical with *intermedius* (Johnstone and McLeod, 1949; Freeman and Minzel, 1950). Toxigenic strains of all three cultural types produce an identical toxin, most non-toxigenic strains correspond in description to *mitis;* it is possible that *gravis* strains produce an additional toxic substance (for discussion, see McLeod, 1943).

Typical strains produce acid from glucose and maltose, but not from sucrose; however, a few toxigenic strains have been shown to produce acid from sucrose (Frobisher, Adams and Kuhns, 1945; Johnstone and McLeod, 1949); aerobic and facul-

† C. S. Cummins prepared Section I—Human and Animal Parasites and Pathogens; R. A. Lelliott, Section II—Plant Pathogenic Corynebacteria; M. Rogosa, Section III—Non-pathogenic Corynebacteria.

TABLE 17.3

Distinguishing features of **Gravis, Intermedius** *and* **Mitis** *strains of* **C. diphtheriae**

Character	Gravis	Intermedius	Mitis
Morphology	Short, irregular rods, few metachromatic granules	Long, barred forms, few granules	Long, curved pleomorphic rods, many granules
Colonies on blood-tellurite medium	Gray or black center, pale periphery; crenated edge with radial striations; 2–3 mm at 24 hr; colony brittle to touch	Small, grayish black, rough or smooth, less than 1 mm at 24 hr	Dark gray to black, smooth shiny, entire edge. 1–2 mm at 24 hr. Colony soft and smears easily.
Type of growth in broth	Surface pellicle, clear fluid, coarse deposit	Fine granular turbidity which settles later	Diffuse turbidity with deposit
Reversal of pH in broth	Early	Late	pH does not revert
Fermentation of starch	+	–	–

tative, mixed fermentative and respiratory metabolism (Tasman and Brandwijk, 1938, 1940); does not produce indole, urease or gelatinase; optimum temperature 37, range 15–40 C.

Characteristic cell wall sugars are arabinose, galactose and mannose: diamino acid of peptidoglycan is *meso*-DAP (Cummins and Harris, 1956; four strains).

A toxic glycolipid is present which is a 6-6′ diester of trehalose, containing corynemycolic acid ($C_{32}H_{64}O_3$) and corynemycolenic acid ($C_{32}H_{62}O_3$) in equimolar proportions (Senn *et al.*, 1967; Kato, 1970).

Pimelic acid, nicotinic acid and β-alanine required by almost all strains; β-alanine may be replaced by pantothenate: pimelic acid may be replaced by biotin: strains which require *p*-aminobenzoic acid or purines, in addition, have been described (for review, see Koser, 1968, 369–374).

Most strains produce a highly lethal exotoxin; however non-toxigenic strains occur which are typical in all other respects; many of these are of the *mitis* variety: ability to produce toxin is determined by the presence of prophage carrying a specific determinant called *tox*+, and toxigenicity is induced in non-toxigenic strains by making them lysogenic for phages of this type (see, for example, Holmes and Barksdale, 1969).

Agglutination tests using whole cell suspensions show a large number of serological types, but little cross-reaction between strains of the three cultural varieties, each of which form a more or less separate group of agglutinating serotypes (see, e.g. Hewitt, 1947): however, all strains examined show a common cell wall antigen shared with other corynebacteria which have arabinose and galac-

tose as major cell wall sugars (Cummins, 1954, 1962).

Many strains carry bacteriophage lytic for other strains, and at least 19 lysotypes have been distinguished (Saragea and Maximescu, 1966): conversion of non-toxigenic strains to toxigenicity is mediated by phage: some *C. diphtheriae* phages will lyse strains of *C. pseudotuberculosis* (Carne, 1968) and also some of the strains described as *C. ulcerans* (see below).

The G + C content of the DNA varies from 51.8–60% (Shapiro, 1968; Hill, 1966); it seems probable that the correct figure is near 60%.

Isolated from cases of diphtheria and from the nasopharynx of healthy carriers.

Reference strain: McLeod 467E; NCTC 3984; ATCC 19409.

Further Comments

Strains intermediate between *C. diphtheriae* and *C. pseudotuberculosis*, called *Corynebacterium ulcerans* by Gilbert and Stewart (1926–1927), resemble *C. diphtheriae*, gravis type, but differ in that they liquefy gelatin, ferment trehalose slowly and do not reduce nitrate to nitrite. They have been isolated mainly from the nasopharynx of healthy persons, or of patients with a diphtheria-like disease, but one such strain has been isolated from acute mastitis in a cow (Jebb, 1948). Many of these strains produce diphtheria toxin, and toxigenicity can be induced in non-toxigenic strains by phage from PW8 strains of *C. diphtheriae* (Maximescu, 1968). However, these strains also appear to produce a toxin resembling that of *C. pseudotuberculosis* (Petrie and McClean, 1934), and the reaction they produce on intradermal injection into guinea pig skin is not completely neutralized by diphtheria antitoxin.

A discussion of this and many other aspects of the biology of *C. diphtheriae* will be found in Barksdale (1970).

2. Corynebacterium pseudotuberculosis
(Buchanan) Eberson 1918, 294. (*Bacillus pseudotuberculosis-ovis* Lehmann and Neumann 1896, 362; *Bacillus pseudotuberculosis* Buchanan 1911, 238; *Corynebacterium ovis* Bergey *et al.* 1923, 388; *Corynebacterium pseudotuberculosis-ovis* (Lehmann and Neumann) Hauduroy *et al.* 1937, 159; *Corynebacterium preisz-nocardi* Hauduroy *et al.* 1937, 159; *Mycobacterium tuberculosis-ovis* Krasil'nikov 1941, 108.)

Vernacular name: Preisz-Nocard bacillus.

pseu.do.tu.ber.cu.lo' sis. Gr. adj. *pseudes* false; M.L. fem.n. *tuberculosis* tuberculosis; M.L. gen.n. *pseudotuberculosis* of false tuberculosis.

Appearance very similar to that of *C. diphtheriae*, especially the *gravis* type: small irregular rods, 0.5–0.6 by 1.0–3.0 μm, staining irregularly, with club forms and metachromatic granules: in electron micrographs a floccular electron-dense layer in found external to the cell wall (Hard, 1969) which probably represents the toxic surface lipid described by Carne, Wickham and Kater (1956) and is analogous to the "cord-factor" of *M. tuberculosis* (Carne, Wickham and Kater, 1956; LaCave, Asselineau and Toubiana, 1967).

On blood agar yellowish white, opaque convex colonies with matt surface, about 1 mm at 24 hr, often with a narrow zone of hemolysis around the colony: Lovell and Zaki (1966) demonstrated a cell-free hemolysin in 9 out of 11 strains; hemolytic activity much enhanced by diffusible products from *C. equi* and *C. renale* (Fraser, 1964, 50). Blood tellurite colonies small, uniformly blackish, low convex with matt surface, more uniform in color than *C. diphtheriae*. Yellowish, friable growth on Loeffler serum slopes after 24 hr, no liquefaction. Scanty growth in *broth* with slight pellicle and sediment: no general turbidity.

Aerobic and facultative; glucose fermented, not oxidized (Davis and Newton, 1969). All, or almost all, strains produce acid from glucose, galactose, maltose and mannose; variable results reported for lactose, sucrose, xylose, dextrin, arabinose, mannitol, glycerol; starch and trehalose not fermented (Jebb, 1948, four strains); however, NCTC 3450 is reported to hydrolyze starch (Davis and Newton, 1969). Nitrate reduction variable; Knight (1969) reported that equine strains reduced nitrate to nitrite, while ovine strains could not. Some strains reported as hydrolyzing urea, e.g. NCTC 3450 (Davis and Newton, 1969) and 3 out of 8 strains from chronic abscesses in horses (Hughes

and Biberstein, 1959). Gelatin liquefaction variable, e.g. one strain out of four positive (Jebb, 1948); eight strains out of eight positive (Hughes and Biberstein, 1959).

The results reported in the literature for fermentation and other tests on this organism seem to be unusually variable, but it is difficult to tell whether this is due to differences in technique or to strain variation.

Cell wall sugars are arabinose, galactose, glucose and mannose: diamino acid of peptidoglycan is *meso*-DAP (Cummins and Harris, 1956; NCTC 3450).

Nutritional requirements not known in detail; growth much improved by the addition of blood or serum to medium; increased CO_2 concentration may facilitate primary isolation (Knight, 1969).

Cell-free filtrates of most strains are lethal to guinea pigs, mice, rabbits and sheep: the toxic material can be toxoided with formalin, and antitoxic sera produced; crude toxin causes intense cellular and fluid exudation (Jolly, 1965); toxin is probably identical with soluble hemolysin (Lovell and Zaki, 1966). All strains produce antigenically similar toxin (Doty *et al.*, 1964).

Agglutination tests with suspensions of intact cells generally impossible because of autoagglutination, probably due to surface lipids (see above): cell walls contain antigenic determinant common to type species and to other corynebacteria having arabinose and galactose as major cell wall sugars.

Strains of *C. pseudotuberculosis* are sensitive to some of the bacteriophages used in typing *C. diphtheriae*, although none of the strains tested falls exactly into any of the *C. diphtheriae* lysotypes (Carne, 1968): strains of *C. diphtheriae* are sensitive to bacteriophages from *C. pseudotuberculosis* (Saragea, Meitert and Bica-Popii, 1966).

Originally isolated from necrotic areas in the kidney of a sheep.

Causes ulcerative lymphangitis, abscesses and other chronic purulent infections in sheep, goats, horses and other warm-blooded animals: occasional infections in man.

The G + C content of the DNA reported as 52.5 moles % for one strain (Bouisset, Breuillard and Michel, 1963, quoted by Hill, 1966). However, more recently Sukapure *et al.* (1970) have reported the G + C content of strain 1076 as 67.5%.

Reference strain: NCTC 3450; ATCC 19410.

3. Corynebacterium xerosis Lehmann and
Neumann 1899, 385. (*Pacinia neisseri* Trevisan 1889, 23; *Bacillus xerosis* Lehmann and Neumann 1896, 361; *Corynebacterium xerosis* (Lehmann and Neumann) Lehmann and Neumann 1899, 385; *Bacterium xerosis* (Lehmann and Neumann) Migula 1900, 501; *Bacterium colomatti* Chester 1901,

186; *Corynebacterium conjunctivae* Lewandowsky 1904, 351; *Mycobacterium xerosis* (Lehmann and Neumann) Krasil'nikov 1941, 99.)

xe.ro'sis. Gr. fem.n. *xerosis* a parched skin. M.L. gen.n. *xerosis* of xerosis.

Irregularly staining, often barred rods, with occasional granules and club forms.

Small circular colonies (1 mm or less at 24 hr) on plain agar may be rough or smooth. On blood agar colonies larger than on media without blood, may be pale yellow to tan color, no hemolysis, colonies rough or smooth. Granular deposit in broth with clear supernatant.

Aerobic and facultative: glucose and gluconate largely broken down by pentose phosphate pathway (Zagallo and Wang, 1967, ATCC 7084), does not liquefy serum, indole not produced, ferments glucose, maltose, galactose and sucrose, producing acid but no gas: good growth at 22 and 37 C.

Cell wall sugars are arabinose, galactose and mannose: diaminoacid of peptidoglycan is *meso*-DAP (three strains, including NCTC 9735: Cummins and Harris, 1956; Cummins, 1971). Cell walls contain antigen common to other corynebacteria which have arabinose and galactose as distinctive cell wall sugars (Cummins, 1962).

Requires amino acids but not vitamins (Smith, 1969, three strains, including ATCC 373 and ATCC 7064).

Not pathogenic.

Originally isolated from conjunctival sac in man. Presumably inhabits skin and mucous membranes of man.

The G + C content of the DNA ranges from 55–59 moles % (Bouisset, Breuillard and Michel, 1963; Marmur and Doty, 1962).

Reference strain: ATCC 373.

Organisms Possibly Related to C. xerosis

a. *Lipophilic diphtheroids from skin.* A large collection of lipophilic diphtheroids from skin was investigated by Smith (1969) to determine their relationship, if any, to *C. xerosis.* All of these strains required lipid for growth; in most cases, this could be supplied by Tween 80. The strains were divided tentatively into seven groups on the basis of fermentation reactions, but Smith did not consider any of the groups identical with *C. xerosis,* quite apart from the fact that *C. xerosis* itself is not lipophilic.

Two of these strains of lipophilic diphtheroids have been examined for cell wall composition and found to have the "arabinose-galactose-*meso*-DAP" pattern of components (Cummins, 1971). There seems little doubt that these strains belong in *Corynebacterium,* but they have not been named,

and do not fit exactly into any existing species. They appear to be most like *C. xerosis.*

b. *Corynebacterium minutissimum.* This is the name given to an organism or a group of organisms isolated from the lesions of erythrasma, a skin disease characterized by scaly plaques which fluoresce coral red under Wood's light (peak transmission—365 nm). The strains are nutritionally exacting, but can be grown aerobically on a solid medium containing tissue culture medium and 20% bovine fetal serum: the colonies then show a coral-red to orange fluorescence like that of the skin lesions. They will grow on blood agar, but the characteristic fluorescence may not be apparent.

The relationship of these strains to the lipophilic diphtheroids described by Smith (1969) is not certain. (For *C. minutissimum,* see Castellani and Chalmers, 1913; Sarkany, Taplin and Blank, 1961, 1962, McBride *et al.,* 1970.)

4. **Corynebacterium renale** (Migula) Ernst 1906, 89. (*Bacillus pyelonephritidis boum* Höflich 1891, 356; *Bacillus pyelonephritidis bovis* Lucet 1893, 329; *Bacterium renale* Migula 1900, 504; *Bacillus pyelonephritis bovis* (*sic*) Künnemann 1903, 129; *Bacillus renalis* (Migula) Künnemann 1903, 150; *Bacillus pyelonephritidis* Glage 1903, 166; *Corynebacterium ovale* Welsch and Thibaut 1948, 207; *Mycobacterium renale* (Migula) Krasil'-nikov 1949, 165.)

re.na'le. L. neut.adj. *renale* pertaining to the kidneys.

A rather large irregularly staining bacillus, 0.7 by 3.0 μm or more, often with pointed ends; metachromatic granules present; fimbriae (pili) present (Yanagawa, Otsuki and Tokui, 1968; Yanagawa and Otsuki, 1970).

Dry, granular, cream to pale yellow colonies on nutrient or blood agar not hemolytic, growth not improved by blood, but yellow pigment may be more marked. On milk agar (nutrient agar + 10% sterilized milk) pale yellow colonies, surrounded at 48 hr by halo or zone of clearing; reaction not shown on media containing either casein or butter fat (Lovell, 1946). Litmus milk at 7–10 days, medium alkaline with clear burgundy-colored supernatant and granular deposit.

Aerobic, facultative, glucose fermented, not oxidized, with production of acid. Maltose, mannitol, lactose, sucrose, salicin not fermented.

Serum not liquefied. Acetoin and uricase produced.

Arabinose, galactose, glucose and mannose found as cell wall sugars; diamino acid of peptidoglycan is *meso*-DAP (Cummins and Harris, 1956, one strain).

Acid extracts (N/20 HCl at 100 C for 15 min) of 22 out of 26 strains reacted strongly in precipitin reactions with antisera to three of the strains (Lovell, 1946): at least three serological types have been distinguished (Yanagawa, Basri and Otsuki, 1967): cell walls contain antigenic determinant common to other corynebacteria with arabinose and galactose as major cell wall sugars (Cummins, 1962, one strain).

G + C content of the DNA reported as 53 moles % (one strain, Bouisset, Breuillard and Michel, 1963).

Isolated from cases of cystitis and pyelonephritis in cattle.

Reference strain: NCTC 7448; ATCC 19412.

5. Corynebacterium kutscheri (Migula) Bergey et al. 1925, 395. (Bacillus pseudotuberculosis murium Kutscher 1894, 338; Bacterium kutscheri Migula 1900, 372; Mycobacterium pseudotuberculosis Chester 1901, 355; Corynebacterium murium Bergey et al. 1923, 386.)

kut′scher.i. M.L. gen.n. kutscheri of Kutscher, the bacteriologist who first isolated the species.

Irregularly staining slender rods, often clubbed, sometimes with pointed ends; metachromatic granules present.

Small, thin, yellowish or grayish white serrate colonies on nutrient agar. Abundant growth on Loeffler's medium. Broth turbid with sediment, no pellicle. Litmus milk no change.

Aerobic, facultative. Acid from glucose, fructose, maltose, mannose, salicin, sucrose. Reduces potassium tellurite.

All of five strains agglutinated to titer with a serum prepared against one of them (Pierce-Chase, Fauve and Dubos, 1964).

Cell wall sugars are arabinose, galactose, mannose and rhamnose: diamino acid of peptidoglycan is meso-DAP (Cummins and Harris, 1956, one strain).

This organism appears to be a frequent, if not invariable, parasite of mice and rats. Its pathogenicity for normal animals is low, but if their resistance to infection is altered, for example by cortisone or dietary deficiency, extensive pseudo-tuberculous lesions develop from which C. kutscheri can readily be isolated. However, the organism has not so far been isolated from the tissue of normal untreated animals.

For a thorough discussion of the question of latent infection with this organism, see Pierce-Chase, Fauve and Dubos, 1964; Fauve, Pierce-Chase and Dubos (1964).

Reference strain: NCTC 949.

6. Corynebacterium pseudodiphtheriticum Lehmann and Neumann 1896, 361. (Bacillus pseudodiphtheriticus (Lehmann and Neumann) Kruse 1896, 476; Bacterium pseudodiphtheriticum (Lehmann and Neumann) Migula 1900, 503; Mycobacterium pseudodiphthericum (sic) (Lehmann and Neumann) Chester 1901, 355; Corynebacterium pseudodiphthericum (sic) (Lehmann and Neumann) Bergey et al. 1925, 393.)

Vernacular name: Hofmann's bacillus. Often referred to as C. hofmannii Holland 1920, 22, a name not validly published.

pseu.do.diph.the.ri′ti.cum. Gr. adj. pseudes false M.L. fem.n. diphtheria diphtheria; M.L. adj. diphtheriticus diphtheritic; M.L. neut. adj. pseudodiphtheriticum relating to false diphtheria.

Short rather regular rods, 0.5–2.0 μm by 0.3–0.5 μm, which stain evenly except for a transverse medial unstained septum: club forms and metachromatic granules minimal or absent: in stained smears the organisms often lie in rows with the long axes parallel. Gram-positive, less readily decolorized than most other corynebacteria.

Good growth on all media, whether or not they contain blood or serum. On blood agar growth white to cream colored, colonies regular and smooth, butyrous consistency.

Aerobic and facultatively anaerobic. Acid not produced from any carbohydrate tested. Optimum temperature 37 C.

Major cell wall sugars are arabinose, galactose and glucose: diamino acid of peptidoglycan is meso-DAP (Cummins and Harris, 1956, two strains).

Cell walls contain antigenic determinant common to other corynebacteria which have arabinose and galactose as major cell wall sugars (Cummins, 1962).

Cells contain corynemycolic acids, mainly a mixture of $C_{32}H_{64}O_3$, $C_{34}H_{66}O_3$ and $C_{36}H_{68}O_3$ (Welby-Guise, Laneelle and Assenlineau, 1970).

Found in nasopharyngeal mucosa of man. Not pathogenic: No toxins produced.

Suggested reference strain: ATCC 10700.

7. Corynebacterium equi Magnusson 1923, 36. (Corynebacterium pyogenes (equi) Miessner and Wetzel 1923, 454; Corynebacterium (pyogenes) equi roseum Lutje 1923, 561; Mycobacterium equi (Magnusson) Jensen 1934, 33; Corynebacterium magnusson-holth Plum 1940, 20; Corynebacterium purulentus Holtman 1945, 161.)

e′qui. L. n. equus horse; L. gen.n. equi of the horse.

Rods variable in length between almost coccoid and rather long curved clubbed forms: oval or coccoid forms common in material from the animal body or cultures on solid media; longer bacillary form found in broth cultures especially. Capsulated (India ink method; Bruner and Edwards, 1941): has been reported to be acid-fast on occasions (Bendixen and Jepsen, 1940, quoted by Jensen, 1953; Jensen 1934).

Grows well on all ordinary media, the growth being mucoid and pinkish in color. No liquefaction of serum. Black moist colonies on agar containing 1:500 potassium tellurite.

Aerobic and facultatively anaerobic (Cotchin, 1943). Glucose fermented, not oxidized (NCTC 1621, Davis and Newton, 1969). Acid not produced from any carbohydrate in usual tests. However, glucose stimulates growth and is slowly utilized: stimulatory effect may be due to acetate produced from the glucose (Pradip, Larson and McClesky, 1966; Bruner and Edwards, 1941). Catalase produced; nitrate vigorously reduced to nitrite, NH_3 not produced (Bruner and Edwards, 1941).

Optimum temperature 37 C, but grows almost as well at 20 C (Magnusson, 1938; Cotchin, 1943).

Nutritional requirements not known in detail. In synthetic medium with $(NH_4)_2SO_4$ and acetate, it appeared that thiamine was stimulatory, but not essential (Pradip, Lawson and McClesky, 1966).

Unusually resistant to oxalic acid and 15% sulfuric acid. Will stand 2.5% oxalic acid for 60 min; may withstand 15% sulfuric acid for 45 min. 2.5% oxalic acid is recommended for use in isolation (Karlson et al., 1940; Cotchin, 1943).

Cell walls contain the antigen common to other corynebacteria with arabinose and galactose as major cell wall sugars. (NCTC 1621; Cummins 1962). A species specific antigen is present (Bruner and Edwards, 1941), and there are a number of different serological types as judged by agglutination tests with bacterial suspensions or precipitin tests with acid extracts (see also Cotchin 1943).

Major cell wall sugars are arabinose and galactose: the diamino acid of peptidoglycan is meso-DAP (NCTC 1621: Cummins and Harris, 1956).

Originally isolated from pneumonia in foals; apart from cases of broncho-pneumonia in foals, it has been isolated from the genital tract of mares, from a buffalo cow following abortion, from aborted equine fetuses, and from the submaxillary lymph glands of swine (see Bruner and Edwards, 1941; Cotchin 1943). Will survive in soil for at least a year (Wilson, 1955). Low pathogenicity for laboratory animals; peritonitis or local abscesses may occur if large doses are injected by appropriate route; intravenous injections of living organisms usually without effect.

Reference strain NCTC 1621; ATCC 6939.

8. **Corynebacterium bovis** Bergey *et al.* 1923, 388.

bo'vis. L. n. *bos* the ox; L. gen.n. *bovis* of the ox.

Irregular rods 0.5–0.7 by 2.5–3.0 μm, usually barred and clubbed; coccobacillary forms may occur.

On nutrient agar + 0.1% Tween 80 colonies white to cream colored, circular with regular edge, slightly shiny, about 1 mm at 24–28 hr. Calf blood agar: no hemolysis.

Aerobic and facultative, most strains produce acid from glucose anaerobically by the Hugh and Leifson test (Hugh and Leifson, 1953). All or almost all strains ferment glucose, fructose, maltose and glycerol; results variable with arabinose, ribose, lactose, trehalose, melibiose and dextrin. Produces oxidase. Lipolytic for butter fat. Acetoin produced. Does not hydrolyze starch or casein; coagulated serum not liquefied. Will grow in broth containing 9% NaCl.

Can use ammonium salts as sole nitrogen source, but casein hydrolysate accelerates growth; 11 out of 42 strains required nicotinic acid if casein hydrolysate was absent, 6 out of 42 strains had an absolute requirement for nicotinic acid. Requires unsaturated long chain fatty acids (palmitoleic, ricinoleic, oleic) (Smith, 1969; Skerman and Jayne-Williams, 1966).

Cell wall sugars arabinose, galactose and mannose; diamino acid of peptidoglycan is *meso*-DAP (Cummins, 1971: ATCC 7715; NCTC 3224; NIRD 61).

Found in aseptically drawn milk. Probably commensal on the cow's udder. May be a cause of bovine mastitis (Cobb and Walley, 1962).

Reference strain: ATCC 7715; NCTC 3224; NIRD 61.

Further Comments

The species description given here is summarized from the results of Jayne-Williams and Skerman (1966) on 42 strains, and corresponds to their group I. They noted the technical difficulties of using classical fermentation tests with these organisms, and eventually adopted a modification of the method of Clark and Cowan (1952), using concentrated washed suspensions and pure substrate in distilled water + indicator.

9. **Corynebacterium paurometabolum** Steinhaus 1941, 783.

pau.ro.me.ta'bo.lum. Gr. adj. *paurus* little; Gr. adj. *metabolus* changeable; M.L. neut.adj. *paurometabolum* little changeable, probably meaning producing little change.

Rods 0.5–0.8 by 1.0–2.5 μm, occurring singly, in pairs and in masses: metachromatic granules present.

White to gray, entire, circular, small, dry, granular colonies on nutrient agar. Slight α-hemolysis on blood agar. Pellicle on broth with granular sediment but no general turbidity. Litmus milk alkaline.

Aerobic. No fermentation of any carbohydrate

tested. Nitrates not reduced. Indole not formed. No liquefaction of gelatin.

Principal cell wall sugars are arabinose and galactose: diamino acid of peptidoglycan is *meso*-DAP (Cummins, 1971, ATCC 8368).

Not pathogenic for guinea pigs.

Isolated from the mycetome and ovaries of the bed bug, *Cimex lectularius* L.

Type strain: ATCC 8638.

Addendum I

a. *Corynebacterium pyogenes* (Glage) Eberson 1918, 23. (*Bacillus pyogenes* Glage 1903, 173.)

py.o′ge.nes. Gr. n. *pyum* pus. Gr. v. *gennaio* produce. M. L. adj. *pyogenes* pus-producing.

Rods 0.2 by 0.3–2 μm; smallest forms appear as scarcely visible dots (common in old abscesses): chains of coccal forms occur resembling streptococci but short diphtheroid forms with clubs usually also present: non-motile: Gram-positive, although rather easily decolorized.

At 24 hr, tiny colonies on blood agar surrounded by zone of β-hemolysis 2–3 times the diameter of the colony (large colony types may occur, see below). On serum slopes pits of liquefaction around colonies, sometimes whole slope liquefied. Litmus milk acid, clot digested.

Aerobic and facultative. Type of metabolism not known for certain, but probably strictly fermentative. Acid from glucose, fructose, galactose, xylose, maltose, lactose, dextrin and starch; no fermentation of mannitol, sorbitol, dulcitol, glycerol, inositol, arabinose, rhamnose, raffinose, sucrose and trehalose. Catalase negative (but see comments below). Nitrates not reduced; indole not produced. Optimum temperature 37, range 20–40 C.

Nutritional requirements not known in detail: growth of all strains much improved by blood or serum: in some cases, growth enhanced by 5–10% CO_2.

Characteristic cell wall sugar components are rhamnose and glucose; diamino acid of peptidoglycan is lysine.

Cell wall polysaccharides in formamide extracts of 10 strains gave sharp precipitin reactions with antisera to NCTC 5224. These extracts crossreacted also with antisera to group G streptococci, but not with antisera to other Lancefield groups.

Culture filtrates fatal to mice and rabbits on i.v. injection: soluble hemolysin produced, active against human, guinea pig, horse and rabbit erythrocytes: both toxic and hemolytic activities of crude cell-free extracts are neutralized by antitoxin (Lovell, 1937, 1941, 1944; Barksdale *et al.*, 1957).

G + C content of the DNA 58 ± 1% moles % (Cummins, 1971).

Originally isolated from bovine pus (Lucet, 1893); frequently isolated from acute pyogenic lesions in cattle, sheep and pigs, sometimes specifically associated with *Fusiformis necrophorus* (Roberts, 1967, 1968). Also occasionally from acute pharyngitis, urethral exudates and cutaneous lesions in man (Barksdale *et al.*, 1957). Presumably a parasite on mucous surfaces of warm-blooded animals.

Further Comments

1. *C. pyogenes* and *C. haemolyticum*. The description of *C. pyogenes* given above is based largely on the results of Barksdale *et al.* (1957) and applies essentially to the small colony strains of *C. pyogenes*. Large colony mutants of *C. pyogenes* can occur, however, and these resemble closely a hemolytic coryneform organism originally isolated by MacLean, Liebow and Rosenberg (1946), and called by them *C. haemolyticum*. The main points of difference between *C. haemolyticum*, the large colony variant of *C. pyogenes* (= *C. pyogenes* L) and the usual small colony type of *C. pyogenes* (= *C. pyogenes* S) are given in Table 17.4.

The interrelationships of these various organisms are not yet absolutely clear, and would probably be clarified further by studies of DNA base ratios and DNA/DNA homologies. The relationship of *C. pyogenes* to group G streptococci also needs exploration.

2. *Catalase production by C. pyogenes. C. pyogenes* is described as catalase negative and this undoubtedly is true of the great majority of strains (e.g. Roberts, 1967; Jayne-Williams and Skerman, 1966). However, *C. pyogenes* ATCC 8104 has shown weak catalase activity (Cummins, 1971; Reid and Joya, 1969). Since the large colony types of *C. pyogenes*, and also strains of *C. haemolyticum*, usually show weakly positive catalase activity, the position of catalase production in this group of organisms resembles that described by Whittenbury (1964) in the lactic acid bacteria, where some strains of *Streptococcus*, *Pediococcus* and *Lactobacillus* showed catalase activity which was inhibited by 0.1 M sodium azide, but where it is fair to describe the genera concerned as being broadly catalase negative.

Addendum II. Species incertae sedis

b. *Corynebacterium enzymicum* (Mellon) Eberson 1918, 29. (*Bacillus enzymicus* Mellon 1917, 297; *Mycobacterium flavum* var. *enzymicum* Krasil′nikov 1941, 116.)

en.zy′mi.cum. Gr. n. *zyme* leaven; M.L. n.

TABLE 17.4
Distinguishing features of **C. pyogenes** *and* **C. haemolyticum**[a]

Character	C. pyogenes S	C. pyogenes L	C. haemolyticum
Cell size (approx.), μm	0.2 by 2.0	0.2 by 2.0–10.0	0.2 by 2.0–10.0
Colony size (24 hr blood agar), mm	<1	1–2	1–2
Fermentation of:			
Xylose	+5/5	−8/8	−2/2
Inositol	−5/5	+7/8	−2/2
Starch	+4/5	−8/8	−2/2
Gelatin liquefaction	+5/5	−8/8	−2/2
Soluble hemolysin	+	−	−
Catalase production[b]	−	w	w
Cell wall sugars	Rhamnose,[c] glucose	Rhamnose	Rhamnose
Reaction of formamide extracts with antisera to:			
C. pyogenes S	++	−	−
C. haemolyticum	−	++	++

[a] Data from Barksdale *et al.*, 1957.

[b] For comments on catalase production in these organisms, see text: w = weak positive.

[c] Small amounts of mannose have also been found in the cell walls of some strains of *C. pyogenes* S.

enzymicus related to enzyme; name given by Mellon because of the wide range of sugars fermented.

Pleomorphic, beaded and club-shaped rods, sometimes coccoid forms, Gram-positive, non-motile; originally described as being capsulated in the animal body (Mellon, 1917).

On blood or glucose agar bacillary form produces very small colorless colonies; coccoid form, heavy yellowish white moist growth. Fine moist confluent growth on Loeffler's medium. In glucose broth bacillary forms give granular sediment; coccoid form, diffuse growth. Litmus milk acid, clot in 48–72 hr.

Aerobic, facultative. Acid from glucose, lactose, maltose, sucrose, dextrin, inulin, salicin and glycerol. Indole not produced. Slight reduction of nitrate. Optimum temperature 37 C.

Pathogenic for rabbits, guinea pigs and mice; the original strain, on i.v. injection of living cultures, produced lesions in the gall bladder, joints, kidney and muscles of rabbits.

Isolated from lung puncture fluid in a fatal case of bronchopneumonia in man.

One of the features of this organism as originally described by Mellon was the development of a more or less permanent coccoid or diploccoccal phase after allowing cultures of the diphtheroid forms to stand for several days at 25–30 C, and then subculturing (Mellon, 1917).

c. *Corynebacterium hoagii* (Morse) Eberson 1918, 11. (*Bacillus hoagii* Morse 1912, 281.)

hoa'gi.i. M.L. gen.n. *hoagii* of Hoag; named for Dr. Louis Hoag, who first isolated this species.

Rods 0.8–1.0 by 1.0–3.0 μm, occurring singly. Short forms show polar staining, larger forms are barred and slightly club shaped; Gram-positive, non-motile.

Small, pale pink, dull, entire colonies on nutrient agar. Broth turbid with slight pink sediment. Litmus milk slightly alkaline with pink sediment.

Aerobic. Acid from glucose and sucrose but not from maltose. Nitrites not produced from nitrates; indole not produced; gelatin not hydrolyzed. Optimum temperature 30 C.

Isolated from human throat: also found as aerial contaminant of cultures.

d. *Corynebacterium striatum* (Chester) Eberson 1918, 22. (*Bacterium striatum* Chester 1901, 171.)

stri.a'tum. L. part.adj. *striatus* grooved.

Pleomorphic Gram-positive rods, often club shaped, 0.25–0.5 by 2.0–3.0 μm; coccoid forms and long filaments found in old cultures: metachromatic granules present, often regularly arranged to produce a segmented effect: non-motile; not acid fast.

Rather slow growth on agar, white smooth entire colonies about 1 mm diameter at 48 hr; some strains produce a yellowish green pigment soluble in the medium. On blood agar slight hemolysis around deep colonies. Clear supernatant in broth, no pellicle, finely granular white sediment. Moderate growth on Loeffler's medium, no liquefaction.

Aerobic and facultative. Ferments glucose, fructose, mannose, trehalose, dextrin, glycogen and usually also lactose, maltose and starch;

about 50% of strains ferment galactose, occasional strain ferments sucrose. Nitrates not reduced to nitrites; no production of acetoin or indole. Catalase produced. Gelatin liquefaction variable, about 50% strains positive.

Originally isolated from human nasopharynx: also from milk of cows with mastitis. Twenty-four-hour cultures fatal to guinea pigs and mice on intramuscular injection.

Further Comments

As in the 7th edition of THE MANUAL, this description is based in part on that of Munch-Petersen (1954), who studied strains isolated from cattle, and in part on the descriptions of *Bacterium striatum* (Chester, 1901) and *Bacillus flavidus* (Morse, 1912). There do not appear to have been any detailed comparisons made between strains from human and bovine sources, and until this is done there must remain some doubt as to whether more than one group of strains is being described under the name *C. striatum*.

Eberson (1918) described *C. striatum* as an organism found in normal nasal mucus, with the following main features: (1) a thick organism, with clear-cut barring and large and irregular granules; (2) heavy growth on serum with white to yellow pigment; and (3) fermentation of glucose and maltose but not sucrose.

e. *Corynebacterium murisepticum* von Holzhausen 1927, 98. (*Mycobacterium murisepticum* (von Holzhausen) Krasil'nikov 1949, 165.)

mu.ri.sep'ti.cum. L. n. *mus, muris*, a mouse; Gr. adj. *septicus* septic; L. neut.adj. *murisepticum* mouse-poisoning or mouse-infecting.

Slender rods 1.2–1.5 µm long, with polar granules; rods grow out into long filaments: Gram-positive: non-motile.

Growth good on Loeffler's medium, on potato and in egg glycerol broth; growth poor in gelatin. Litmus milk acid, not clot.

Aerobic, facultative. Produces acid from glucose, fructose, galactose, maltose, lactose, sucrose, inulin and mannitol. H₂S produced; indole not produced. Optimum temperature, 37 C.

Isolated from septicemia in mice. Pathogenic for mice.

f. *Corynebacterium nephridii* Büsing, Döll and Freytag 1953, 78.

This organism was isolated from the kidney of the medicinal leech (Büsing, Döll and Freytag, 1953; Büsing and Freytag, 1953–1954). However, other isolations have not been reported, and Hart, Larson and McClesky (1965) who reexamined the original strain, found it to be Gram-negative and proposed that it be transferred to the genus *Achro-*

mobacter. The status of this organism must remain in doubt until more strains have been examined. In the meantime, it is mentioned in this section for purposes of cross-reference.

g. *Corynebacterium phocae* Svenkerud, Rosted and Thorsaug 1951, 168.

pho'cae. M.L. n. *phoca* name of genus of seals; M.L. gen.n. *phocae* of seals.

Rods 0.4–0.6 by 0.7–2.0 µm, occurring in chains of three to seven or more cells, or in pairs; occasional single cells or very long forms (10–15 µm) the latter sometimes curved. Gram-positive, non-motile.

Very small (0.1 mm) circular smooth colonies on agar, transparent and colorless at 24 hr, opalescent, whitish and larger (0.5–1.0 mm) at 48 hr. Colonies on blood agar similar but slightly larger; no hemolysis. No growth on tellurite medium.

Aerobic, facultative. Acid from glucose, fructose, sucrose, maltose, trehalose and salicin; starch, esculin and α-methylglycoside hydrolyzed; acetoin produced. No liquefaction of gelatin or coagulated serum; indole not produced; nitrates not reduced to nitrites; no production of catalase or urease; citrate not utilized.

Isolated from an erysipelas occurring in the transition between the corium and the blubber of seals.

Further Comments

This organism is catalase negative and does not grow on media containing potassium tellurite, which suggests that despite its coryneform morphology, it does not belong in *Corynebacterium*.

h. *Corynebacterium vaginalis* Zinnemann and Turner 1963, 214. (See *Haemophilus vaginalis*, page 368.)

i. *Microbacterium flavum* Orla-Jensen 1919, 181.

Robinson (1966) showed that *Microbacterium flavum* has a cell wall structure of the "arabinose-galactose-*meso*-DAP" type, and is thus similar to *C. diphtheriae* but quite different from strains of *M. lacticum* and *M. liquefaciens*. There were also differences in enzyme patterns (catalase and esterase), absorption spectra of pigments, and heat resistance. Further, it has been shown by Schleifer (1970) that *M. flavum* has a peptidoglycan composition essentially identical to that in *C. diphtheriae*, and its G + C content is 59% as compared with 64% for *M. lacticum* and 36% for *M. thermosphactum* (Collins-Thompson, Sporhaug and Witter, 1971).

It appears that *M. flavum* should ultimately be transferred to *Corynebacterium*, provided that examination of a larger number of strains substantiates the above results.

Section II. Plant Pathogenic Corynebacteria

Morphologically similar to the human and animal pathogens but less pleomorphic and there is generally less evidence of snapping division. Stain unevenly (granular) and on glycerol media have fatty inclusions which stain with Sudan Black B. Intracellular starch is not accumulated. Some species motile.

Starch hydrolysis is weak or negative.

Nitrate reductase negative; coagulase negative; indole, M.R. and V.P. negative; H₂S or urease rarely produced. Strains examined are sensitive to triphenyltetrazolium chloride (Lovrekovitch and Klement, 1960).

Require organic growth factors.

Strict aerobes.

The G + C content of the DNA is probably in the range 65–75 moles %.

Further Comments

Although there are detailed descriptions of most of the plant pathogenic corynebacteria, many of them are based on methods no longer in general use and are misleading in today's context. Two comparative studies (Ramamurthi, 1959; Carrier, 1963) using more modern methods are available but because the numbers of strains used were comparatively small—one used 33 isolates of six species and the other 27 isolates of nine species—they can give little indication of the intraspecific variation to be expected; nor can the authenticity of apparent interspecific similarities and differences be judged. The results obtained by them often differ for the same character, probably because different methods were used.

For these reasons it is felt that an attempt to amend the original authors' description by including the results of later studies would still be misleading. Instead the specific descriptions given here include only characters on the state of which different sources agree, or which are not expected to be subject to major intraspecific variation.

An attempt is made to obtain some grouping of similar species by using not only the characters given in Table 17.5 but more particularly their serological relationships (Rosenthal and Cox 1953, 1954; Mushin *et al.*, 1959; Lazar, 1968; Lazar and Graham, 1970), the amino acid composition of their cell walls, and their minimal nutritional requirements. Thus *C. sepedonicum* is not grouped with *C. insidiosum* and *C. michiganense* which it otherwise resembles, because it appears to share few nucleoprotein or polysaccharide antigens with them. Those species for which no comparative studies are available are in an Addendum; there is at present no reason to suppose that they are less valid members of the genus than those in groups 1–5.

Description of the species of genus **Corynebacterium** (*plant pathogens*)

Group 1.

1. **Corynebacterium fascians** (Tilford) Dowson 1942, 313. (*Phytomonas fascians* Tilford 1936, 394; *Bacterium fascians* (Tilford) Lacey 1939, 262; *Pseudobacterium fasciens* (*sic*) (Tilford) Krasil'nikov 1949, 225.)

fas'ci.ans. L. part.adj. *fascians* binding together, bundling.

Cells after 2–3 days on nutrient agar are 0.5–0.9 by 1.5–4.0 μm, slightly curved rods arranged singly, angularly and as palisades. Bending division was at first reported (Komagata *et al.*, 1969) but later snapping division (Yamada and Komagata, 1970). Not markedly pleomorphic. On mannitol agar most strains show elongated forms and some chains (Mohanty, 1951). The development of L forms (Lacey, 1961) and (on a synthetic low-nitrogen medium) branching acid-fast, nocardia-like cells (Lacey, 1955) have been reported. Non-motile. Gram-positive, usually with strongly stained granules.

Colonies of freshly isolated cultures are round, entire, convex, opaque, moist, semi-fluidal and cream to yellow on nutrient agar; growth is slow and discrete colonies appear in about 3 days. Growth is heavier and yellow to orange on nutrient glucose agar. Stable, rough, matt variants, not associated with loss of virulence, occur.

Thirteen carotenoid pigments have been identified (Prebble, 1968).

Breaks down the plant hormone β-indole acetic acid rapidly.

Thiamine is an essential growth factor (Starr, 1949) and biotin, arginine, histidine, purine and pyrimidines are stimulatory (Ramamurthi, 1959). A strain with biotin as an essential growth factor has been reported (Mohanty, 1951). But reported to grow in Koser's citrate medium through four transfers (Carrier, 1963).

Optimum growth temperature 24–27 C; maximum 37 C.

Causes a fasciation disease of many plants.

The G + C content of the DNA of two strains is 62.9 and 67.3 moles % (T_m) (Yamada and Komagata, 1970).

TABLE 17.5

Differential characters for plant pathogenic species of genus **Corynebacterium**[a]

	Group											
	1	2				3	4		5			
	1. C. fascians	2. C. rathayi	3. C. agropyri	4. C. tritici	5. C. iranicum	6. C. sepedonicum	7. C. insidiosum	8. C. michiganense	9. C. flaccumfaciens	10. C. poinsettiae	11. C. betae	12. C. oortii
Motility	−	−	−	+	·	−	−	−[b]	+	+	+	+
Growth slow[c]	+	+	+	+	+	+	+	+	−	−	−	−
Growth opaque	+	+	+	+	+	+	+	±	±	±	−	+
Growth at 37 C	+	−	−	−	−	−	−	(+)	(+)	(+)	(+)	(+)
Urease	(+)	−	·	−	·	−	−	−				
Esculin hydrolyzed[d]		+	·	+	·	+	+	+	+	+	+	−
Medium yellow pigment	+	+	+	+	+	−	−	(+)	d	−	−	−
Pink pigment	−	−	−	−	−	−	−	(−)	(−)	+	−	−
Indigoidine produced	−	−	−	−	−	−	+	−	−	−	−	−
Causes hypertrophies of stems	+	−	−	−	−	−	−	−	−	−	−	−
Causes gumming of inflorescences	−	+	+	+	+	−	−	−	−	−	−	−
Causes wilts and/or leaf spots	−	−	−	−	+	+	+	+	+	+	+	+

[a] + All of the few strains examined reported to be positive; ± translucent at 3 days, becoming more opaque later; − all of the few strains examined reported to be negative; (+) most strains positive; (−) most strains negative; d strains differ.

[b] Some authors claim to have detected motility.

[c] Colonies on nutrient dextrose agar <1 mm diameter after 3 days.

[d] Carrier, 1963.

Further Comments

Serologically more closely related to *C. diphtheriae* and to group 5 than to other plant or animal corynebacteria (Lazar, 1968).

Lacey (1955), Ramamurthi (1959) and Carrier (1963) have commented on the similarity of *C. fascians*, and particularly the rough forms, to *Nocardia* spp. In cell wall components, mode of division and serological relatedness, however, it resembles *C. diphtheriae* more.

Group 2.

C. rathayi, C. agropyri, C. tritici and *C. iranicum* cause a similar disease of some cereals and grasses and are probably identical. However, until further comparative studies have been made they are treated as separate species.

They were described separately for different hosts but it has been shown that both *C. rathayi* and *C. tritici* will affect wheat (*Triticum* spp).

(Sabet, 1954) and cocksfoot grass (*Dactylis glomerata*) (Williams, 1964). *C. agropyri* does not appear to have been reported since its discovery by O'Gara (1916). The separation of *C. iranicum* from *C. tritici*, both of which affect wheat, does not seem justified by the evidence. *C. rathayi* and *C. tritici* are known to be transmitted by eelworms (*Anguina* spp.).

2. **Corynebacterium rathayi** (Smith) Dowson 1942, 313. (*Aplanobacter rathayi* Smith 1913, 926; *Bacterium rathayi* (Smith) Aujeszky 1914, 93; *Phytomonas rathayi* (Smith) Bergey, Harrison, Breed, Hammer and Huntoon 1923, 192; *Erwinia rathayi* (Smith) Gram, Jorgensen and Rostrup 1929, 464; *Agrobacterium rathayi* (Smith) Savulescu 1947, 10; *Pseudobacterium rathayi* (Smith) Krasil'-nikov 1949, 225.)

rath'ay.i. M.L. gen.n. *rathayi* of Rathay, named for E. Rathay, an Austrian plant pathologist, who first isolated the organism.

Cells 0.5–0.75 by 0.95–1.3 μm, short rods with bluntly rounded ends, usually singly or in pairs joined end to end. Not pleomorphic and no evidence of snapping division. Non-motile.

Grows very poorly on nutrient agar; growth heavier on nutrient glucose agar. Often, and particularly with freshly isolated cultures, discrete colonies cannot be obtained and it grows only when densely seeded. Growth is better and faster on cooked potato. Colonies differ markedly in size, are bright yellow, entire and convex.

Growth reported only on complex media; minimal nutritional requirements unknown. No growth in Koser's citrate.

Optimum temperature for growth about 25 C; maximum between 30 and 37 C.

Causes a gumming disease of cocksfoot grass (*Dactylis glomerata*).

3. Corynebacterium agropyri (O'Gara) Burkholder 1948, 395. (*Aplanobacter agropyri* O'Gara 1916, 343; *Phytomonas agropyri* (O'Gara) Bergey, Harrison, Breed, Hammer and Huntoon 1923, 190; *Bacterium agropyri* (O'Gara) Stapp 1928, 37; *Agrobacterium agropyri* (O'Gara) Savulescu 1947, 10; *Pseudobacterium agropyri* (O'Gara) Krasil'nikov 1949, 240; *Empedobacter agropyri* (O'Gara) Brisou 1958, 182.)

ag.ro.py'ri. M.L. gen.n. *agropyri* of *Agropyron*, generic name of a grass.

Cannot be differentiated by its description (O'Gara, 1916) from *Corynebacterium rathayi*.

Causes a gumming disease of western wheat grass (*Agropyron smithii*).

4. Corynebacterium tritici (Hutchinson) Burkholder 1948, 400. (*Pseudomonas tritici* Hutchinson 1917, 174; *Phytomonas tritici* (Hutchinson) Bergey, Harrison, Breed, Hammer and Huntoon 1930, 248; *Bacterium tritici* (Hutchinson) Elliott 1930, 234; *Agrobacterium tritici* (Hutchinson) Savulescu 1947, 10.)

tri'ti.ci. M.L. gen.n. *tritici* of *Triticum*, generic name of wheat.

Said to be motile with one polar flagellum. Otherwise not significantly different from *Corynebacterium rathayi*.

Causes a gumming disease of wheat (*Triticum aestivum*).

5. Corynebacterium iranicum Scharif 1961, 21.

i.ran'i.cum. M.L. adj. *iranicum* pertaining to Iran.

Its description (Scharif, 1961) is not significantly different from *Corynebacterium rathayi*.

Causes a gumming disease of wheat (*Triticum aestivum*).

Group 3.

6. Corynebacterium sepedonicum (Spieckermann and Kotthoff) Skaptason and Burkholder 1942, 441. (*Bacterium sepedonicum* Spieckermann and Kotthoff 1914, 674; *Aplanobacter sepedonicum* (*sic*) (Spieckermann and Kotthoff) Smith 1920, 207; *Phytomonas sepedonica* (Spieckermann and Kotthoff) Magrou 1937, 411; *Mycobacterium sepedonicum* (Spieckermann and Kotthoff) Krasil'nikov 1949, 190; *Pseudobacterium sepedonicum* (Spieckermann and Kotthoff) Krasil'nikov 1949, 225.)

se.pe.don'i.cum. Gr. n. *sepedon* rottenness, decay; M. L. neut.adj. *sepedonicum* leading to decay.

Cells 0.4–0.6 by 0.8–1.2 μm, predominantly wedge-shaped but coccoid and curved and straight, rod-shaped cells usually also present. Predominantly single cells but some V, Y and palisade arrangements usually present. Bending (as opposed to snapping) division is reported (Yamada and Komagata, 1970). Non-motile.

Growth on all media is very slow and often erratic unless a heavy inoculum is used. On nutrient glucose agar, on which growth is optimal, colonies become visible after about 5 days and rarely exceed 1 mm diameter; they are smooth, entire, low convex, semi-fluidal when freshly isolated but become butyrous with prolonged subculturing, white or cream to pale yellow, opaque and glistening.

Cell walls are reported by Yamada and Komagata (1970) to contain LL-diaminopimelic acid and lysine. See also Table 17.1.

Thiamine, biotin and nicotinic acid (Starr, 1949) and histidine, and purines and pyrimidines (Ramamurthi, 1959) are essential growth factors. Lachance (1962) reported that asparagine and methionine are essential and histidine and leucine stimulatory. Paquin and Lachance (1970) found that cystine was strongly inhibitory and some other amino acids moderately inhibitory.

Aerobe. Probably a microaerophile; shaken, shallow broth cultures often fail to grow when equivalent still cultures grow adequately.

Optimum temperature for growth 20–23 C; minimum 3–4 C, maximum 30–31 C.

Causes a vascular wilt and tuber rot of potato (*Solanum tuberosum*). Can also affect some other *Solanum* spp.

The G + C content of the DNA of one strain is 69.8 moles % (T_m) (Yamada and Komagata, 1970).

Group 4.

The AB serotypes of both species are identical in agglutinogen complement (Rosenthal and Cox, 1953).

7. **Corynebacterium insidiosum** (McCulloch) Jensen 1934, 41. (*Aplanobacter insidiosum* McCulloch 1925, 497; *Bacterium insidiosum* (McCulloch) Stapp 1928, 178; *Phytomonas insidiosa* (McCulloch) Bergey, Harrison, Breed, Hammer and Huntoon 1930, 278; *Erwinia insidiosa* (McCulloch) Jensen 1934, 41; *Mycobacterium insidiosum* (McCulloch) Krasil'nikov 1941, 102; *Burkholderiella insidiosa* (McCulloch) Săvulescu 1947, 21.)

in.si.di.o'sum. L. neut.adj. *insidiosum* deceitful, insidious.

Cells 0.4–0.5 by 0.7–1.0 μm, predominantly wedge-shaped but coccoid, and curved and straight, rod-shaped cells usually also present. Bending (as opposed to snapping) division is reported by Yamada and Komagata (1970). Predominantly single cells but some V, Y and palisade arrangements are usually present. Not motile.

Growth on all media is slow. On nutrient glucose agar, on which growth is optimal, colonies are 0.5–1.0 mm after 5 days and 5–6 mm after 9 days. They are circular, smooth, entire or irregular, flat to low convex, opaque, white at first but later becoming pale yellow. Freshly isolated cultures are semi-fluidal and non-viscid but often become butyrous after prolonged subculturing. Dark blue to violet granules of indigoidine develop, usually first at the edges of confluent growth but later in the center of colonies, after 6–7 days at 15–20 C; higher temperatures inhibit or reduce pigment formation.

Cell walls are reported by Yamada and Komagata (1970) to contain LL-diaminopimelic acid. See also Table 17.1.

Thiamine, biotin and nicotinic acid (Starr, 1949) and histidine and purine and pyrimidines (Ramamurthi 1959) are essential growth factors.

Optimum temperature for growth 21–24 C; minimum<1 C, maximum 28–31 C varying with strain.

Causes a vascular wilt and stunting of lucerne (alfalfa) (*Medicago sativa*), some other *Medicago* spp. and *Melilotus alba*.

The G + C content of the DNA of one strain is reported as 78.1 moles % (T_m) (Yamada and Komagata, 1970).

8. **Corynebacterium michiganense** (Smith) Jensen 1934, 47. (*Bacterium michiganense* Smith 1910, 794; *Pseudomonas michiganensis* (Smith) Stevens 1913, 30; *Aplanobacter michiganense* (Smith) Smith 1914, 165; *Phytomonas michiganensis* (Smith) Bergey, Harrison, Breed, Hammer and Huntoon 1923, 91; *Mycobacterium flavum* subsp. *michiganense* (Smith) Krasil'nikov 1941, 119.)

mi.chi.ga.nen'se. M.L. adj. *michiganense* pertaining to Michigan (State, U.S.A.).

Cells 0.6–0.7 by 0.7–1.2 μm, predominantly

wedge-shaped but coccoid, and curved and straight, rod-shaped cells are usually also present. Bending (as opposed to snapping) division is reported by Yamada and Komagata (1970). Predominantly single cells but some V, Y and palisade arrangements are usually present. Authors differ (see review by Carrier, 1963) but there is no unequivocal evidence for motility.

Growth is slow on all media. On nutrient glucose agar, on which growth is optimal, colonies are 1 mm in 5 days and 2–3 mm in 7–8 days; they are smooth, entire, convex, semi-fluidal when freshly isolated but become butyrous with prolonged subculturing, pale yellow becoming deeper yellow, opaque and glistening. Pink, red, orange and non-pigmented variants occur uncommonly.

LL-Diaminopimelic acid, alanine, glutamic acid and glycine are reported by Yamada and Komagata (1970) in the cell walls and valine, aspartic acid, threonine, leucine, iso-leucine and small amounts of methionine and lysine by Cabezas de Herrera and Moreno (1969) who also reported the sugars arabinose, glucose, galactose, mannose, ribose and rhamnose.

The major pigments are carotenoids; lycopene, spirilloxanthin, cryptoxanthin, β-carotene and canthaxanthin occur in different combinations in the wild type and in colored mutants (Saperstein *et al.*, 1954).

Thiamine, biotin, nicotinic acid, and in one culture tryptophan (Starr, 1949) and histidine, and purines and pyrimidines (Ramamurthi, 1959) are essential growth factors. Unspecified amino acids are stimulatory.

Optimum temperature for growth 24–27 C; minimum 1 C, maximum 36–37 C.

Causes a vascular wilt, canker and leaf and fruit spot of tomato (*Lycopersicon esculentum*) and some other solanaceous plants.

The G + C content of the DNA of six strains ranged from 67.3–70.7 moles % (T_m).

Group 5.

C. flaccumfasciens and *C. poinsettiae* are identical in agglutinin complement (Rosenthal and Cox, 1954); the nucleoprotein antigens of a pink variant of *C. flaccumfaciens* and of *C. poinsettiae* are apparently identical and the two species could not be distinguished by other methods (Lazar and Graham, 1970).

9. **Corynebacterium flaccumfaciens** (Hedges) Dowson 1942, 313. (*Bacterium flaccumfaciens* Hedges 1922, 433; *Phytomonas flaccumfaciens* (Hedges) Bergey, Harrison, Breed Hammer and Huntoon 1923, 178; *Pseudomonas flaccumfaciens* (Hedges) Stevens, 1925, 27.)

flac.cum.fa'ci.ens. L. adj. *flaccus* flabby; L. part.-

adj. *faciens* making; M.L. part.adj. *flaccumfaciens* wilt-making.

Cells 0.3–0.5 by 0.6–3.0 μm (most cells about 1.5 μm long), predominantly slightly curved rods with some straight rods and some wedge-shaped forms. Predominantly single cells but some V, Y and palisade arrangements are usually present. Bending (as opposed to snapping) division is reported by Yamada and Komagata (1970). Motile by one or rarely two or three polar or subpolar flagella; reported to be polar at 24–26 C and degenerate peritrichous at 20 C (Schuster *et al.*, 1968).

At 3 days colonies of freshly isolated strains on nutrient glucose agar are 2–4 mm diameter, smooth, entire, low convex, non-viscid, semi-fluidal, translucent to semi-opaque, cream to pale yellow, shining; with further incubation they become more yellow, more opaque and 4–7 mm in diameter. A variant that produces orange or salmon pink growth depending on temperature, and a yellow variant that produces a blue to purple water-soluble pigment have been described as *C. f.* var. *aurantiacum* Schuster and Christianson 1957, 52 and *C. f.* var. *violaceum* Schuster, Vidaver and Mandel 1968, 427 respectively; otherwise they were not distinguishable from the type (Schuster *et al.*, 1968). After prolonged subculture colonies are often smaller and more butyrous.

Cell walls are reported by Yamada and Komagata (1970) to contain lysine, see also Table 17.1.

Thiamine, biotin and pantothenate (Starr, 1949) and purines and pyrimidines (Ramamurthi, 1959) are essential growth factors; most amino acids are stimulatory.

Optimum temperature for growth 20–30 C; minimum less than 1.5 C, maximum 37–40 C.

The G + C content of the DNA of three strains determined by T_m is 68.3 and 68.5 (Yamada and Komagata, 1970) and 74.5 (Sukapure *et al.*, 1970) and of nine strains by buoyant density is 71 moles % (Schuster *et al.*, 1968).

Causes a vascular wilt of bean (*Phaseolus vulgaris*).

10. **Corynebacterium poinsettiae** (Starr and Pirone) Burkholder 1948, 399. (*Phytomonas poinsettiae* Starr and Pirone 1942, 1080; *Bacterium poinsettiae* (Starr and Pirone) Hauduroy, Ehringer, Guillot, Magrou, Prévot, Rosset and Urbain 1953, 95.)

poin.set'ti.ae. M.L. gen.n. *poinsettiae* of *Poinsettia*, a genus of flowering plants.

Cells 0.2–0.8 by 0.5–8.5 μm (most cells 0.3–0.6 by 1.0–3.0 μm), predominantly slightly curved or straight rods but some wedge-shaped and coccoid forms. Predominantly single cells but some V, Y and palisade arrangements are usually present.

Bending (as opposed to snapping) division reported by Yamada and Komagata (1970). Motile with one, or rarely two, polar or lateral flagella (*cf. Corynebacterium flaccumfaciens*). Gram-positive; often Gram-negative with positive granules in old cultures. Kuhn and Starr (1962) have observed a change from Gram-positive to Gram-negative reaction and an arthrobacter-like cyclic alteration; large cocci germinate bipolarly to yield "spindles" or unipolarly to yield "grape seeds" and rods which may develop lateral flagella, sometimes become branched, but ultimately fragment to form cocci.

At 3-4 days colonies of freshly isolated strains on nutrient glucose agar are 2–4 mm in diameter, smooth, entire, raised, non-viscid, semi-fluidal, translucent, pale salmon pink. Growth is red-pink on media low, and orange-yellow on media high, in thiamine (Starr and Saperstein, 1953).

Cell walls reported by Yamada and Komagata (1970) to contain lysine. See also Table 17.1.

The carotenoid pigments lycoxanthin, cryptoxanthin and spirilloxanthin are produced (Starr and Saperstein, 1953).

Thiamine, biotin and pantothenate (Starr, 1949) and histidine, purines and pyrimidines (Ramamurthi, 1959) are essential growth factors. Most amino acids are stimulatory.

Optimum temperature for growth 24–27 C; minimum about 3 C, maximum 37–40 C.

The G + C content of the DNA of three strains is 70.0 and 72.8 (T_m) and 71 (buoyant density) moles %.

Causes a stem canker and leaf spot of the poinsettia (*Euphorbia pulcherrima*).

11. **Corynebacterium betae** Keyworth, Howell and Dowson 1956, 89.

bet'ae. M.L. gen.n. *betae* of *Beta* the generic name of beet.

Cells 0.3–0.5 μm by 0.8–1.6 μm, predominantly curved and straight rods but some wedge-shaped and coccoid forms. Predominantly single cells but some V, Y and palisade arrangements are usually present. Motile by one to three polar or subpolar flagella at 25 C; original description—3 peritrichous flagella (temperature not stated) *cf. C. flaccumfaciens*.

At 3 days colonies on nutrient glucose agar are 2–3 mm in diameter, smooth, entire, raised, non-viscid, fluidal, translucent, cream to pale yellow.

Optimum temperature for growth about 25 C; minimum 4 C or less, maximum 37–39 C.

Causes a vascular wilt and leaf spot of red beet (cv. of *Beta vulgaris*)

12. **Corynebacterium oortii** Saaltink and Maas Geesteranus 1969, 126.

oor.ti′i. M.L. gen.n. *oortii* of Oort, named for Dr. A. J. P. Oort, a Dutch professor of plant pathology.

Cells 0.5–1.1 by 1.3–2.6 μm (average 0.8–1.9 μm), mostly curved rods with rounded ends, some club- and wedge-shaped. Motile by one or rarely two polar or subpolar flagella.

After 3 days colonies on nutrient glucose agar are 2–3 mm diameter, smooth, raised, entire, non-viscid, semi-fluidal, semi-opaque to opaque, shining, cream colored becoming pale yellow.

Nitrates not reduced to nitrites in 2 or 5 days; reported to be reduced at 10 days (Saaltink and Maas Geesteranus, 1969).

Optimum temperature for growth 25–30 C; minimum 5 C and maximum 37 C.

Causes a vascular, and leaf and bulb spot, disease of tulips (*Tulipa* spp.).

Addendum I

No comparative studies appear to have been made between the following species and those in groups 1–5. Cultures of *C. humiferum*, *C. humuli* and *C. hypertrophicans* are probably not obtainable. The descriptions given here are from those of their authors.

a. *Corynebacterium beticola* Abdou 1969, 163.

Cells 0.5–0.7 by 0.9–1.8 μm, rods motile by one polar flagellum. Non-sporing. Gram-variable with Gram-positive granules. Not acid fast.

Colonies on nutrient dextrose agar are about 1 mm diameter in 2 days, smooth, convex, non-viscid, translucent and cream colored.

Ferments many carbohydrates usually with gas production.

Catalase positive, nitrates reduced to nitrites, urease negative, produces levan.

Facultative anaerobe.

Optimum temperature for growth 20–25 C; minimum 0 C, maximum 37 C.

Tolerates 0.1% triphenyltetrazolium chloride.

Causes a rot of sugar beet (cv. of *Beta vulgaris*) roots.

b. *Corynebacterium ilicis* Mandel, Guba and Litsky 1961, 61.

Cells rods, motile by one to several subpolar flagella. Gram-positive.

Growth slow; colonies yellow, smooth to mucoid on agar media. Grows in ammonium-glucose minimal medium.

Slight acid from glucose, sucrose and mannitol.

Catalase, cytochrome oxidase and urease positive. Tyrosinase, Voges-Proskauer and nitrate reductase negative. Casein hydrolyzed but not starch.

Yellow, carotenoid pigment.

No anaerobic growth with glucose and nitrate.

Optimum growth temperature 25 C; grows slowly between 5 and 35 C.

Causes a shoot and branch blight of American holly (*Ilex opaca*).

The G + C content of the DNA of one strain is 60.0 moles % (buoyant density, Schuster *et al.*, 1968).

c. *Corynebacterium humiferum* Seliskar 1952, 9.

Cells 0.4–0.7 by 0.7–2.8 μm; pleomorphic rod, commonly curved or club shaped. Predominantly single with occasional palisade and angular arrangements. Non-motile. Gram-positive.

Colonies on nutrient glucose agar are 1–2 mm diameter, circular, smooth, entire, convex, translucent and shining; growth is slow.

No growth in Koser's citrate. Nitrates not reduced.

Strict aerobe.

Optimum temperature for growth 24–28 C; minimum 6 C, maximum 34–36 C.

Associated with a wet wood condition of the heart wood of Lombardy poplar (*Populus nigra* var. *italica*) and aspen (*Populus tremuloides*).

d. *Corynebacterium humuli* Stow and Ihara 1955, 298.

Cells 0.5–0.7 by 0.7–2.0 μm. Short, straight rods in young cultures; some slightly curved longer rods in older cultures. Motile with five to eight polar flagella. Gram-positive to Gram-variable; granular staining in cells of old cultures.

Growth on nutrient agar rapid, grayish white, glistening, circular, smooth, convex or flat, entire or sometimes undulate margin, amorphous, opaque.

Acid but no gas from a few carbohydrates.

Strict aerobe.

Optimum temperature for growth 25 C. Does not grow at 37 C.

Occurs in the intercellular spaces and vascular tissues of healthy and fasciated hop plants (*Humulus* spp.). Probably not the cause of the hyperplasias (Stow and Ihara, 1955).

e. *Corynebacterium hypertrophicans* (Stahel) Burkholder 1948, 398. (*Pseudomonas hypertrophicans* Stahel 1933, 447.)

Cells 0.6–0.8 by 1.2–2.8 μm, rods predominantly single but chains 10–20 μm long can be present. In freshly isolated strains cells conglomerate in packets of 50–100 cells arranged in parallel. Motile by one polar flagellum. Bipolar, and also sometimes one to three central, bodies that are visible without staining are present. Non-sporing. Gram-positive, not acid fast.

Colonies on nutrient sucrose agar are about 2 mm diameter in 3 days, circular, slightly raised, whitish and translucent.

Strict aerobe.

Causes a witches broom disease of *Eugenia latifolia*.

Section III. Non-pathogenic Corynebacteria

Organisms of Section III have been most commonly isolated from soils, water, air and contaminated bacteriological media or blood *etc*. They are often poorly described and some have been at first considered *Brevibacterium* or other species and later renamed as *Corynebacterium* species. Because of the heavy or sole reliance on morphological criteria, and because at some stage of growth there may be morphological similarities, the organisms of this group could be confused with members of at least 10 other genera (see *Brevibacterium* (page 625) and the introduction to the Coryneform Group of Bacteria (page 599). In most cases there are insufficient data to permit differentiation of species. Therefore, the following names are listed with minimal comment; most have been given in Addendum I (Hatt and Zvirbulis, 1967) and Addendum III (Zvirbulis and Hatt, 1969) to *Index Bergeyana*.

1. **Corynebacterium acetoacidophilum** Shiio, Mitsugi, Otsuga and Tsunoda 1964, 443 (U.S. Patent 3,117,915; ATCC 13870).

2. **Corynebacterium acetophilum** Harada, Seto and Murooka 1968, 169. Source: Soil.

3. **Corynebacterium aurantiacum** Iizuka, Tanabe, Fukumura and Kato 1967, 140 (*Not Corynebacterium aurantiacum* Eberson 1918, 21). Source: Polluted water lines in a nylon manufacturing plant, Nagoya, Japan.

4. **Corynebacterium callunae** Lee and Good 1963, 1349. (U.S. Patent 3,087,863). Not val. pub., Rule 11. Reference strain: NRRL B-2244; ATCC 15991.

5. **Corynebacterium citreum-mobilis** (*sic*) Köhler 1955, 275. Not Val. Pub. Rule 14a (1). Source: Contaminant on a blood plate:

6. **Corynebacterium ethanolaminophilum** Harada, Murooka and Izumi 1967, 485. A non-spore-forming bacterium E 17, Harada and Murooka 1966, 192. Source: Soil.

7. **Corynebacterium flaccumfaciens** var. **non-myxogenes** Sasaki and Shiio 1960, 31. Not Val. Pub. Rule 14a (2).

8. **Corynebacterium glutamicum** (Kinoshita, Nakayama and Akita) Abe, Takayama and Kinoshita 1967, 299. (*Micrococcus glutamicus* Kinoshita, Nakayama and Akita 1958, 176. *Index Bergeyana*, 684.)

9. **Corynebacterium herculis** Dunn, Fuld, Kusmierek, Lim and Wang 1964, 147 (U.S. Patent 3,120,472). Type strain: ATCC 13868. Produces L-glutamic acid. Source: Soil.

10. **Corynebacterium hydrocarboclastus** (*sic*) Iizuka and Komagata 1964, 211. (*Corynebacterium hydrocarboclastus* (*sic*) Iizuka and Komagata in Shiio, Otsuka, Ishii, Kaysuya and Iizuka 1963, 23. Not Val. Pub. Rule 14a (2); *Corynebacterium oleophilus* (*sic*) Iizuka and Komagata in Shiio *et al.* 1963, 23 Not Val. Pub. Rule 14a (2).)
Produces L-glutamic acid (U.S. Patent 3,406,095). From petroleum zone soil, Japan. Type strain: IAM 1484; ATCC 15961.

11. **Corynebacterium lilium** Lee and Good 1963, 1349 (U.S. Patent 3,087,863). Not Val. Pub., Rule 11.

12. **Corynebacterium luteum** (Söhngen) Krasil'nikov 1941, 113. Not Val. Pub. Rule 12d. (*Mycobacterium luteum* Söhngen 1913, 599. *Ind. Berg.* 729, *q.v.*)
Comment: Krasil'nikov (1941, 113) incorrectly ascribes *Corynebacterium luteum* to Kisskalt 1918.

13. **Corynebacterium mediolanum** (*sic*) Mamoli 1939, 1863. Val. Pub. Leg. (Von uns isolierte reine dehydrierende, Stamm, Mamoli and Vercellone 1939, 319.) Source: Yeast (Mailänder Hefe).

14. **Corynebacterium melassecola** Goto, Nishio, Kojima, Hayakawa and Araki 1967. U.S. Patent 3,355,359. Not Val. Pub. Rule 11.

15. **Corynebacterium mycetoides** (Castellani) Ortali and Capocacci 1956, 490. (*Micrococcus mycetoides* Castellani 1942, 553.)

16. **Corynebacterium nubilum** var. **nanum** Jensen 1934, 44. Source: Garden soil.

17. **Corynebacterium roseum** Iizuka, Tanabe, Fukumura and Kato 1967, 132. Source: Polluted water lines in a nylon manufacturing plant, Nagaya, Japan.

18. **Corynebacterium sanguinis** Getzel 1941, 179. Source: Human and animal blood.

Genus II. **Arthrobacter** Conn and Dimmick 1947, 300

R. M. KEDDIE

Ar.thro.bac'ter. Gr. n. *arthrus* a joint; M.L. masc.n. *bacter* the masculine equivalent of the Gr. neut.n. *bactrum* a rod; M.L. masc.n. *Arthrobacter* a jointed rod.

Cells which in complex media undergo a marked change in form during the growth cycle. Older cultures (generally 2–7 days) are composed entirely or largely of coccoid cells. In some strains the coccoid cells are uniform in size and spherical, and resemble micrococci; in others they are spherical to ovoid or slightly elongate. In some cultures larger coccoid cells some 2–4 times the size of the remainder may occur; these may predominate under some cultural conditions. On transfer to fresh complex medium growth occurs by enlargement (swelling) of the coccoid cells followed by elongation from one or occasionally two parts of the cell to give rods which usually have a diameter less than that of the enlarged coccoid cell (Plate 17.1, Figs. 1–4). In the larger coccoid cells outgrowth may occur at two, three or rarely four parts of the cell (Plate 17.1, Figs. 5–6). In both cases subsequent growth and division give rise to irregular rods which vary considerably in size and shape and include straight, bent and curved, wedge-shaped and club-shaped forms (Plate 17.1, Fig. 7). A proportion of the rods are arranged at an angle to each other giving V formations but more complex angular arrangements often occur. Post-fission outgrowths, usually from the proximal ends of one or both cells of a pair of rods (Plate 17.1, Fig. 8), and bud-like outgrowths from segments of septate rods (Fig. 9) especially in richer media, may give the appearance of rudimentary branching but true mycelia are not formed. As the exponential phase proceeds the rods become shorter and are eventually replaced by the coccoid cells characteristic of stationary phase cultures (Plate 17.1, Fig. 10). The coccoid cells are formed either by a gradual shortening of the rods at each successive division or, especially in richer media, by multiple fragmentation of larger rods (Plate 17.1, Fig. 11). The rods are non-motile or motile by one subpolar or a few lateral flagella. Do not form endospores.

Gram-positive. However, **the rods may be readily decolorized and may show only Gram-positive granules in otherwise Gram-negative cells.** Coccoid cells are Gram-positive but may be weakly so. **Not acid fast.**

The cell walls do not contain both meso-diaminopimelic acid and arabinose.

Chemoorganotrophs: **Metabolism respiratory, never fermentative.** Molecular oxygen is the terminal electron acceptor.

Little or no acid is produced from glucose in peptone medium. **Do not attack cellulose.** Catalase positive.

All species grow in a medium containing soil extract and yeast extract. Nutrient agar (peptone + meat extract) is suitable for laboratory strains which do not require *terregens* factor or vitamin B_{12} but may be inhibitory to newly isolated strains (Topping, 1937).

Strict aerobes.

Temperature optimum 20–30 C; most strains grow at 10 C but usually not at 37 C. **Do not survive heating at 63 C for 30 min in skim milk.** Grow best at a neutral to slightly alkaline pH.

The **G + C content** of the DNA of the species examined ranges from *ca.* **60–72 moles** % (T_m).

Type species: *Arthrobacter globiformis* (Conn) Conn and Dimmick 1947, 301.

Further Comments

Members of the genus *Arthrobacter* are among the dominant bacteria in soil, at least as assessed by the usual plating methods. However, present evidence suggests that the species already described do not represent the diversity of the soil arthrobacters (Keddie *et al.*, 1966; Skyring and Quadling, 1969). The heavy dependence on morphological features in the generic definition also generates problems: it is not uncommon to isolate saprophytic coryneform bacteria from soil and other habitats in which the transformation of rods to coccoid forms is limited or incomplete (see Keddie *et al.*, 1966; Skyring and Quadling, 1968). The existence of such forms presents a taxonomic and determinative problem which has not yet been satisfactorily resolved. Also, in recent years it has become clear that organisms which show the sequence of morphological changes and staining properties of the genus *Arthrobacter* occur in a wide variety of habitats (see e.g. Keddie *et al.*, 1966; Mulder *et al.*, 1966; Schefferle, 1966). Although limited comparative studies have been made, much further work is needed to determine which if any of the strains from other habitats should be classified in the genus *Arthrobacter*. (See also *Brevibacterium* p. 625.)

The sequence of morphological changes that occurs during the growth cycle is a distinguishing characteristic of the genus (for detailed accounts see Sacks, 1954; Sguros, 1955, 1957; Sundman, 1958; Stevenson, 1961; Starr and Kuhn, 1962;

Ensign and Wolfe, 1963), but the degree of cellular morphogenesis is dependent on environmental conditions, particularly the nutritional status of the medium. Coccoid cells of *A. globiformis* inoculated into a mineral salts, glucose, biotin medium, multiply with up to only a 2-fold increase in cell size and rapid reversion to the coccoid form occurs. Addition of increasing concentrations of complex constituents (yeast extract *etc.*) to the medium results in progressively greater elongation and irregularity of the rods and with higher concentrations the final transformation into coccoid cells may be considerably delayed (Stevenson, 1961; Veldkamp, van den Berg and Zevenhuizen, 1963). Ensign and Wolfe (1963) have also shown that a strain they named *A. crystallopoietes* (see *A. globiformis, Further Comments*) remained in the coccal stage throughout the growth cycle in a mineral salts + glucose medium. In similar media containing certain single amino acids or organic acids, and in complex media, the more usual morphological changes occurred. Care must therefore be exercised in the choice of medium for morphological studies: nutritionally adequate media with relatively low concentrations of complex organic constituents are probably the most generally satisfactory for this purpose e.g. agar media similar to the YS medium of Lochhead and Burton (1955).

The terms "arthrospore" applied to the more usual coccoid cells, and "cystite" are not used in the present descriptions. The large coccoid cells or "cystites" have frequently been considered to be specialized spore structures because initial outgrowth of slender rods from two or three parts of the cell gives the appearance of germ tubes emerging from a spore. Large coccoid cells normally comprise at most only a small proportion of the population of older cultures but Stevenson (1963) noted that the entire population of a culture of *A. globiformis* was composed of "cystites" when inoculum from a rich medium was introduced into a minimal medium containing a relatively high concentration (1%) of glucose. Stevenson (1963) suggested that "cystites" are really morphologically aberrant forms resulting from unbalanced growth in conditions of nutritional stress and found no evidence to support theories that they are specialized spore structures. Mulder *et al.* (1966) reported that growth of soil arthrobacters in a glucose-containing medium with a high carbon/nitrogen ratio resulted in populations composed mainly of large coccoid cells which they considered to be "cystites." The accumulation of large quantities of intracellular polysaccharide was thought to be responsible for "cystite" formation (Mulder and Zevenhuizen, 1967). The terms "arthrospore" and "cystite"

therefore have a somewhat uncertain meaning and imply a structure and function not known to exist.

The characteristic angular arrangement of cells giving the appearance of V, Y and similar formations may arise in different ways: by snapping post-fission movements, by "angular growth" (Starr and Kuhn, 1962), or by outgrowth of rods at an acute angle from pairs of adjacent coccoid cells (Sguros, 1957; Starr and Kuhn, 1962).

Previous descriptions of the Gram-reaction of *Arthrobacter* spp. are confusing and often contradictory. The most common term used is "Gram-variable" although in the 7th edition of THE MANUAL some species were described as Gram-negative both in the rod and coccoid forms. However, the cell wall composition is typical of Gram-positive bacteria (Cummins and Harris, 1959; Keddie *et al.*, 1966) and Keddie *et al.* (1966), in a study of a number of named strains and many new isolates, invariably found some degree of Gram-positivity at all stages of growth: some strains were quite strongly Gram-positive. They are therefore described as Gram-positive although it is recognized that they may be readily decolorized.

More exacting techniques have led to a revision of the nutritional requirements originally described for certain species. In particular it has been shown that the type species *A. globiformis* comprises both non-exacting strains and strains which require biotin (see Morris, 1960; Chan and Stevenson, 1962; Keddie *et al.*, 1966). The type strain ATCC 8010, NCIB 8907 does not require biotin (Keddie, unpublished) whereas a second, widely used Conn strain, ATCC 4336, NCIB 8602, does (Morris, 1960; Chan and Stevenson, 1962).

It is now clear that many of the classical differential features used in species descriptions in the 7th edition of THE MANUAL have a limited value in the internal taxonomy of this genus. Revision of the nutritional requirements of some species has also blurred distinctions formerly made between species. Cell wall composition and detailed nutritional requirements are much more useful but unfortunately these techniques are on the whole rather laborious for routine determinative purposes. However certain tests for simple nutritional requirements are not unduly onerous although an adequate technique must be used (see Owens and Keddie, 1969) and simple rapid methods have been devised for the detection of some of the determinatively useful cell wall components (Becker *et al.*, 1964; Murray and Proctor, 1965; Boone and Pine, 1967).

More recently Owens and Keddie (unpublished)

Bill –

have applied to a collection of soil and herbage coryneform bacteria, including named strains and new isolates of *Arthrobacter* spp., the methods developed by Stanier *et al.* (1966) for the aerobic pseudomonads, whereby a large number of organic compounds is tested for the ability to serve as carbon + energy sources. The results obtained are sufficiently promising to justify their inclusion for those species studied so far. In particular they allow a much more definite circumscription of the type species and illustrate well the close relationship to it of certain strains formerly considered to be distinct species. These studies also reveal that strains of *A. globiformis* show a nutritional versatility similar in extent to that found in some members of the genus *Pseudomonas*. The method used by Owens and Keddie was similar to that used by Stanier *et al.* except that the mineral base used was that described by Owens and Keddie (1969) with yeast extract (0.02%, w/v) and vitamin B_{12} (2 μg/l) incorporated to allow growth of all nutritionally exacting species except those re-

quiring the *terregens* factor. A multiloop inoculator (Tarr, 1958) was used in preference to the replica plating technique using velvet pads; the bacterial suspensions used for inoculation were prepared in a suitable diluent (Owens and Keddie, 1969). Since the basal medium allows some growth it is essential to use appropriate controls.

A key for the differentiation of species has not been constructed as this would have little value at the present time for reasons already given. Lists of characters based on "ideal phenotypes" (Stanier *et al.*, 1966) have been prepared for two species which are better characterized and of common occurrence; they should be used in conjunction with data on cell wall composition, in particular the diamino acid present (Table 17.6). The remaining species are characterized mainly on the basis of their minimal nutritional requirements and cell wall composition. Few strains have been isolated and little is known about their distribution in nature. Some of their characteristics are given in Table 17.6.

Description of the species of the genus **Arthrobacter**

1. **Arthrobacter globiformis** (Conn) Conn and Dimmick 1947, 301. (*Bacterium globiforme* Conn 1928, 3; *Achromobacter globiformis* (Conn) Bergey *et al.* 1930, 226; *Mycobacterium globiforme*

(Conn) Krasil'nikov 1941, 97; *Corynebacterium globiforme* (Conn) Wood 1950, 132. Subj. synonyms: *Bacterium cocciforme* Migula 1900, 439; *Arthrobacter globiforme* var. *aurescens* Clark 1951, 180;

TABLE 17.6
Some characteristics of species of genus **Arthrobacter**[a]

	A. globiformis	A. simplex	A. tumescens	A. citreus	A. terregens	A. flavescens	A. duodocadis
Cell walls contain[b]							
LL-DAP	·	+	+	·	·	·	+
Lysine	+	·	·	+	+	+	+
Glycine	·	+	+	·	+	+	+
Aspartic acid	·	·	·	·	+	+	+
Serine	·	·	·	·	·	·	+
Growth factor requirements[c]							
Biotin	d	−	−	+	+	+	−
Thiamine	−	−	+	+	+	+	+
Pantothenic acid	−	−	−	−	+	−	−
Nicotinic acid	−	−	−	+	−	−	−
B_{12}	−	−	−	−	−	−	+
Terregens factor[d]	−	−	−	+	+	+	−
Gelatin hydrolyzed	+	+	+	+	−	+	+
Starch hydrolyzed	d	−	d	−	−	+	−
$NO_3 \rightarrow NO_2$	d	+	d	+	+	+	+

[a] Characteristics of *A. globiformis* based on some 22 strains, of *A. tumescens* some 6 strains, others usually 1 strain, the type.

[b] +, present as a major component; '·', absent or only small amounts detected; d, some strains positive, some negative. Alanine and glutamic acid are major components of all species.

[c] Some species require amino acids in addition.

[d] Or other sideramines, or certain chelators.

Arthrobacter aurescens Phillips 1953, 241; *Arthrobacter pascens* Lochhead and Burton 1953, 18; *Arthrobacter ramosus* Jensen 1960, 131. Probable subj. synonyms: *Corynebacterium ureafaciens* Krebs and Eggleston 1939, 320; *Arthrobacter oxydans* Sguros 1954, 22; *Arthrobacter ureafaciens* (Krebs and Eggleston) Clark 1955, 111; *Arthrobacter nicotianae* Giovannozzi-Sermanni 1959, 85; *Arthrobacter crystallopoietes* Ensign and Rittenberg 1963, 149; *Arthrobacter atrocyaneus* Kuhn and Starr 1960, 179.)

glo.bi.for'mis. L. n. *globus* ball, globe; L. n. *forma* shape; M.L. adj. *globiformis* spherical.

Morphology generally as for genus. Coccoid cells *ca.* 0.6–0.8 μm in diameter; rods *ca.* 0.5–0.8 by 1–4 or more μm.

Generally non-motile; occasional motile strains occur (Grainger and Keddie, unpublished).

Galactose is the only sugar which occurs invariably in the cell wall (Cummins and Harris, 1959; Gillespie, 1963; Keddie *et al.*, 1966).

A very wide range of organic compounds is utilized as sole or principal source of carbon + energy for growth; in a study of 22 strains Owens and Keddie (unpublished) found that on the average individual strains utilized 85 out of 180 compounds tested. The ability to utilize uric acid, glycine, creatine, betaine and sarcosine in this way was considered to be a constellation of features characteristic of this species.

Other characteristics are given in Tables 17.6 and 17.7.

Growth on nutrient agar and similar media generally shows no distinctive pigmentation; occasional strains have a lemon yellow non-diffusible pigment.

Some strains are nutritionally non-exacting and grow in a suitable mineral salts medium with an ammonium salt or nitrate as sole nitrogen source with glucose as carbon + energy source; many strains require biotin in addition.

Growth occurs at 10 C but usually not at 37 C; optimum *ca.* 25–30 C. The usual source is soil.

The G + C content of the DNA is 60–64.4 moles % (T_m, 10 strains, Marmur *et al.* 1963; Jones and Bradley, 1964; Yamada and Komagata, 1970).

Type strain: ATCC 8010; NCIB 8907 does not require biotin. The widely used Conn strain, ATCC 4336, NCIB 8602, requires biotin.

Further Comments

A salient feature of this species is its nutritional versatility coupled with the ability to grow in simple media without exogenous organic growth factors or with biotin as the sole vitamin supplement.

TABLE 17.7

Characteristics of ideal phenotype of **Arthrobacter globiformis**[a]

These characteristics, when used with cell wall composition (Table 17.6), are of greatest value in recognition of strains of *A. globiformis*.

1. Vitamin requirement: None or biotin only	22
2. Inorganic nitrogen utilized as sole nitrogen source[b]	22
Utilizes as carbon + energy source:	
3. Glycine	22
4. Tyramine	22
5. Uric acid	22
6. p-Hydroxybenzoate and/or 3:4-dihydroxybenzoate	22
7. m-Hydroxybenzoate	20
8. D-Xylose	20
9. Melezitose	20
10. D-Glucuronate	19
11. α-D-Glucosamine and/or N-Acetylglucosamine	19
12. Meso-inositol	18
13. 2-Oxoglutarate	17

[a] The ideal phenotype is positive for all of these characteristics. The figures indicate the number of strains positive for each characteristic out of 22 tested (Owens and Keddie, unpublished).

Of the 22 strains tested, 3 deviated from the ideal phenotype in three characters, the remainder conformed exactly or at most deviated in two characters. Of more than 90 coryneform strains, other than *A. globiformis*, tested by Owens and Keddie more than 90% deviated in seven or more characters and none deviated in less than five.

[b] See Owens and Keddie, 1969, for a suitable method.

In a study of 22 strains (including one Conn strain) considered to belong to this species, Owens and Keddie (unpublished) found that, of some 180 organic compounds tested, individual strains utilized, on average, 85 as sole or principal sources of carbon + energy. Forty-eight substrates were utilized by 20 or more strains; they included representatives of the following groups of compounds: carbohydrates and sugar derivatives, fatty acids, dicarboxylic acids, hydroxy-acids, oxo-acids, glycols and polyalcohols, non-nitrogenous aromatic compounds, amino acids, amines and heterocyclic compounds. The 13 characteristics, which were considered most useful in recognizing this species from an assemblage of more than 100 strains of coryneform bacteria, which included named strains of *Arthrobacter* and *Cellulomonas* and isolates from soil and herbage, are given in Table 17.7.

A. pascens, A. aurescens, A. ramosus and *A. atrocyaneus* conform with the revised description of *A. globiformis* in minimal nutritional requirements (Keddie *et al.*, 1966; Owens and Keddie, 1969), cell wall composition (Cummins and Harris, 1959; Keddie *et al.*, 1966) and in the range and constellation of organic substances utilized as sole or principal carbon + energy sources (Owens and Keddie, unpublished). With the exception of *A. atrocyaneus*, the G + C values are similar to those reported for *A. globiformis* (Yamada and Komagata, 1970). *A. pascens* and *A. aurescens* were wrongly described as facultatively anaerobic in the 7th edition of THE MANUAL; they are obligately aerobic. *A. pascens, A. aurescens* and *A. ramosus* are therefore considered subjective synonyms of *A. globiformis*. However, although the single strain of *A. atrocyaneus* resembles *A. globiformis* in a large number of phenotypic traits, Yamada and Komagata (1970) reported that the G + C content of the DNA is 69.5 moles % and, in conflict with the findings of Keddie *et al.*, that the cell walls contained moderate amounts of LL-DAP and not lysine (Yamada and Komagata, 1970). Despite this conflicting evidence, for the present, *A. atrocyaneus* is perhaps best considered a possible synonym of *A. globiformis*. A distinctive feature of this strain is the production of an intense non-diffusible blue pigment, indigoidine (Kuhn *et al.*, 1965), in peptone media containing certain sugars and sugar alcohols (Kuhn and Starr, 1960).

A. ureafaciens, A. oxydans, A. nicotianae and *A. crystallopoietes* have been studied less extensively by the methods described above but all have similar minimal nutritional requirements and show a high degree of conformity in the characters described for the ideal phenotype of *A. globiformis*. The names *A. ureafaciens, A. oxydans, A. nicotianae* and *A. crystallopoietes* are therefore regarded as probable subjective synonyms of *A. globiformis*. Distinctive features recorded for *A. ureafaciens* in the 7th edition of THE MANUAL were yellow chromogenesis and production of urea from creatine, creatinine and uric acid (see also Krebs and Eggleston, 1939). The description was based on a single strain (culture NC) isolated from a creatinine enrichment by Dubos and Miller (1937) and named by them *Corynebacterium creatinovorans*, but the name was not validly published (see Clark, 1955). The culture was further studied by Krebs and Eggleston (1939) who named it *C. ureafaciens*. These earlier descriptions clearly indicated that creatine (Dubos and Miller, 1937) and uric acid (Krebs and Eggleston, 1939) were utilized as sole sources of carbon +

energy: both of these substances are characteristic and universal or near universal substrates of *A. globiformis* (Owens and Keddie, unpublished).

The main feature which distinguished *A. oxydans* from *A. globiformis* is its ability to utilize nicotine as sole carbon + energy source with the consequent production of a diffusible blue pigment (Sguros, 1954, 1955), a trait also possessed by *A. nicotianae* (Giovannozzi-Sermanni, 1959) which differs from *A. oxydans* in minor respects. *A. ureafaciens, A. oxydans* and *A. nicotianae* have cell wall compositions (Cummins and Harris, 1959) and G + C values (Yamada and Komagata, 1970) similar to those of *A. globiformis*. Similarly the ability to utilize 2-hydroxypyridine as sole carbon + energy source giving a crystalline blue pigment as an oxidation product was a salient distinguishing feature of *A. crystallopoietes* (Ensign and Rittenburg, 1963). It is clear from the data of Krulwich *et al.* (1967), that the cell wall composition of *A. crystallopoietes* is similar to that of *A. globiformis*. This strain has been the subject of detailed studies of cellular morphogenesis in *Arthrobacter* (Ensign and Wolfe, 1963; Krulwich *et al.*, 1967; Krulwich and Ensign, 1968, 1969). Since it is now clear that different strains of *A. globiformis* characteristically utilize an extremely wide and diverse range of organic substances, the utilization of unusual substrates by individual strains can no longer be considered in any way exceptional or worthy *per se* of separate specific status for these strains.

It should be noted that although a number of former species, now considered to represent strains of one species, bear a considerable degree of phenotypic similarity to each other and to strains of *A. globiformis*, Zagallo and Wang (1962) reported that the major pathways of carbohydrate catabolism in *A. globiformis* and *A. ureafaciens* differed from those in *A. pascens* and *A. atrocyaneus*. Further biochemical studies may thus reveal the existence of subgroups within what is here considered as one species.

2. **Arthrobacter simplex** (Jensen) Lochhead 1957, 608. (*Corynebacterium simplex* Jensen 1934, 43.)

sim'plex. L. adj. *simplex* simple.

Morphology generally as for genus. Older cultures are composed of coccoid cells and very short rods *ca.* 0.4–0.5 by 0.5–0.8 μm; large coccoid forms are uncommon. The irregular rods occurring in late lag and exponential phase cultures are *ca.* 0.4–0.5 by 1.0–3.0 or more μm. The rods are stated to be less variable in form than in other species of the genus. Motile and non-motile strains occur. When motile, from one to four flagella occur in polar (*sic*) or lateral positions (Clark and Carr, 1951).

Owens and Keddie (unpublished) found that the type strain utilized about 60 of 180 compounds tested, as sole or principal sources of carbon + energy. They included a very narrow range of carbohydrates and sugar derivatives, a wide range of fatty acids, simple alcohols and amino acids, together with some hydroxy-acids, oxo-acids, amines, pyrimidines and phenol.

Other characteristics are given in Table 17.6. Growth on nutrient agar shows no distinctive pigmentation.

Nutritionally non-exacting; growth occurs in a suitable mineral salts medium with an ammonium salt or nitrate as sole nitrogen source and glucose as carbon + energy source.

Growth occurs at 10 and 37 C; optimum 26–37 C. The only known source is soil.

The G + C content of the DNA is 71.7 moles % (Yamada and Komagata, 1970, type strain).

Type strain: ATCC 6946; NCIB 8929 (NCIB 8913).

3. **Arthrobacter tumescens** (Jensen) Conn and Dimmick 1947, 302. (*Corynebacterium tumescens* Jensen 1934, 45.)

tumes'cens. L. part.adj. *tumescens* swelling up.

Morphology generally as for genus. Coccoid cells *ca.* 0.5–0.6 μm in diameter; rods *ca.* 0.5–0.8 by 2.0–6.0 μm. Generally non-motile; occasional motile strains occur (Grainger and Keddie, unpublished).

A relatively wide range of organic compounds is utilized as sole or principal source of carbon + energy for growth: in a study of six strains, including the type strain, Owens and Keddie (unpublished) found that, on average, individual strains utilized 58 out of 180 compounds tested.

Other characteristics are given in Tables 17.6 and 17.8.

Growth on nutrient agar and similar media has no distinctive pigmentation.

Thiamine is the only organic growth factor required: when so provided, utilizes an ammonium salt or nitrate as sole nitrogen source in a suitable mineral medium with glucose as carbon + energy source (Keddie *et al.*, 1966; Owens and Keddie, 1969).

Growth occurs at 10 C and may or may not occur at 37 C; optimum *ca.* 30 C.

The only known source is soil.

The G + C content of the DNA is 69.8 moles % (Yamada and Komagata, 1970, type strain).

Type strain: ATCC 6947: NCIB 8914.

Further Comments

In a study of six strains (including the type strain) considered to belong to this species, Owens

TABLE 17.8
Characteristics of ideal phenotype of
Arthrobacter tumescens[a]

These characteristics, when used with cell wall composition (Table 17.6), are of greatest value in the recognition of strains of *A. tumescens*.

1. Vitamin requirement: Thiamine only	6
2. Inorganic nitrogen utilized as sole nitrogen source[b]	6
Utilizes as carbon + energy source:	
3. D-Mannose	6
4. Raffinose	6
5. α-D-Glucosamine	6
6. Acetate	6
7. Crotonate	6
8. Citrate	6
9. Meso-inositol	6
10. D-Alanine	6
11. Thymine and/or uracil	6
12. Suberate and/or azelate	5
13. D-Glucuronate	0
14. p-Dihydroxybenzoate and/or 3:4-dihydroxybenzoate	0
15. L-Ornithine	0
16. D-Phenylalanine	0
17. Uric acid	0

[a] The ideal phenotype is positive for characteristics 1–12 and negative for 13–17. The figures indicate the number of strains positive for each characteristic out of six tested (Owens and Keddie, unpublished).

Of six strains tested, five showed perfect conformity with the ideal phenotype and one strain deviated in one character. Of strains other than *A. tumescens* tested, all deviated in six or more characters.

[b] See Owens and Keddie (1969) for method.

and Keddie (unpublished) found that individual strains generally utilized as sole or principal sources of carbon + energy, between 47 and 60 of some 180 organic compounds tested and one strain utilized 70 compounds. All six strains utilized a common core of 36 substrates which included a wide range of carbohydrates and sugar derivatives and representatives of: fatty acids, dicarboxylic acids, hydroxy-acids, oxo-acids, polyalcohols, amino acids, amines and pyrimidines. The 17 characteristics which were found most useful in recognizing this species from an assemblage of more than 100 strains of coryneform bacteria, which included named strains of *Arthrobacter* and *Cellulomonas* and isolates from soil and herbage, are given in Table 17.8.

In the 7th edition of THE MANUAL *A. tumescens* was wrongly described as catalase negative. Further differences from the previous description are in nutritional requirements and temperature relationships. The previous description also includes the statement that citrates are not utilized as sole carbon source whereas Owens and Keddie (unpublished) found that citrate was utilized as sole or major carbon + energy source in otherwise nutritionally adequate media.

4. Arthrobacter citreus Sacks 1954, 342.

cit're.us. M.L. adj. *citreus* lemon colored.

Description based on type strain only.

Morphology generally as for genus. Coccoid cells *ca.* 0.7 μm in diameter; rods *ca.* 0.8 by 2.0–5.0 μm. The rods are feebly motile.

Galactose is the only sugar which occurs in the cell wall (Cummins and Harris, 1959).

Owens and Keddie (unpublished) found that the type strain utilized about 40 of 180 compounds tested as sole or principal sources of carbon + energy. They included a relatively wide range of carbohydrates, sugar derivatives and amino acids, together with some fatty acids, dicarboxylic acids, hydroxy-acids, oxo-acids, non-nitrogenous aromatic compounds, amines and heterocyclic compounds; of those tested no simple alcohols, polyalcohols or glycols were utilized.

Urease is not produced.

Other characteristics are given in Table 17.6.

Growth on nutrient agar and similar media, and on soil extract agar has a lemon yellow non-diffusible pigment which is insoluble in ether and acetone.

Biotin, thiamine, nicotinic acid, tyrosine, methionine, cystine and sideramines such as ferrichrome and mycobactin are required for growth; glutamic acid, while not essential, is required for maximum growth. Sideramines can be replaced by certain synthetic metal chelators (Seidman and Chan, 1969, 1970).

Growth occurs at 10 C but not at 37 C; optimum 25–30 C.

Isolated from chicken feces but probably a dust or soil contaminant.

The G + C content of the DNA is 62.9 moles % (T_m, type strain, Yamada and Komagata, 1970).

Type culture: ATCC 11624; NCIB 8908.

5. Arthrobacter terregens Lochhead and Burton 1953, 17.

ter're.gens L. n. *terra* soil; L. part.adj. *egens* requiring; M.L. part.adj. *terregens* soil-requiring.

Description based on type strain only.

Morphology generally as for genus. Coccoid cells *ca.* 0.6–0.9 μm in diameter; rods *ca.* 0.6–0.8 by 1.0–5.0 μm. Motility not observed.

Urease is not produced.

Other characteristics are given in Table 17.6.

Growth on nutrient agar supplemented with soil extract and on soil extract agar is yellowish brown.

Terregens factor, biotin, thiamine, pantothenic acid (Burton and Lochhead, 1953; Keddie *et al.*, 1966) and L-methionine (Owens and Keddie, 1969) are required for growth. Most of the usual laboratory media will not support growth unless supplemented with a source of *terregens* factor or similar factors and preferably also yeast extract. Suitable sources of *terregens* factor are soil extract or culture filtrate of strain ATCC 13346 (*A. pascens* Lochhead and Burton 1953). The *terregens* factor is replaceable by other sideramines such as coprogen, ferrichrome (Burton, Sowden and Lochhead, 1954) and mycobactin (Reich and Hanks, 1961); it can also be replaced by suitable concentrations of certain synthetic metal chelators e.g. 8-hydroxyquinoline (Morrison, Antoine and Dewbrey, 1965). When provided with essential growth factors and glucose as carbon + energy source, ammonium salts are utilized as major nitrogen source (Owens and Keddie, 1969).

Growth occurs at 10 C but not at 37 C; optimum *ca.* 20–26 C.

Isolated from soil.

Type strain: ATCC 13345; NCIB 8908.

Further Comments

Burton and Lochhead (1953) reported that in an otherwise adequate medium containing an ammonium salt, the nitrogen requirements of *A. terregens* could be satisfied by L-glutamic acid although this amino acid could be replaced by combinations of other amino acids. Owens and Keddie (1969) on the other hand reported that the requirements for amino acids could be satisfied by L-methionine in otherwise suitable media containing an ammonium salt; L-methionine was not replaceable by L-glutamic acid (Keddie, unpublished). The most likely explanation for this apparent contradiction is that many commercial samples of L-glutamic acid and other amino acids are contaminated by L-methionine (Guirard and Snell, 1964; Demain, 1965).

6. Arthrobacter flavescens Lochhead 1958, 170.

fla.ves'cens. L. v. *flavesco* become golden yellow; L. part.adj. *flavescens* becoming yellow.

Description based on type strain only.

Morphology generally as for genus. Coccoid cells *ca.* 0.5 μm; rods *ca.* 0.4–0.6 by 1.0–6.0 μm. Motility not observed.

Growth on suitably supplemented nutrient agar has a yellow non-diffusible pigment: on suitable media containing "Casamino acids" the pigment is lemon yellow.

Other characteristics are given in Table 17.6.

Terregens factor (see *A. terregens* for sources), biotin and thiamine are required for growth; the *terregens* factor is replaceable by other sideramines such as coprogen, ferrichrome (Lochhead, 1958) and mycobactin (Morrison and Dewbrey, 1966). When supplied with *terregens* factor, vitamins and glucose as carbon + energy source, ammonium salts or nitrate are utilized as nitrogen source (Owens and Keddie, 1969).

Temperature range for growth *ca.* 8–40 C; good growth at 20–32 C.

Isolated from soil.

Type strain: ATCC 13348; NCIB 9221.

7. Arthrobacter duodecadis Lochhead 1958, 170.

du.o.dec′a.dis. Gr. n. *duodecado* twelve; M.L. gen.n. *duodecadis* of twelve, referring to the requirement of the organism for vitamin B_{12}.

Description based on type strain only.

Morphology generally as for genus. Coccoid cells *ca.* 0.4–0.6 μm in diameter; rods *ca.* 0.5–0.6 by 1.5–4.0 μm. Motility not observed.

Owens and Keddie (unpublished) found that the type strain utilized only about 40 of 180 compounds tested as sole or principal sources of carbon + energy. They included a wide range of carbohydrates, sugar derivatives and fatty acids, a few amino acids together with some dicarboxylic acids, hydroxy-acids, oxo-acids, polyalcohols, non-nitrogenous aromatic compounds and amines; of those tested no simple alcohols or heterocyclic compounds were utilized.

Growth on suitably supplemented nutrient agar or similar media shows no distinctive pigmentation; inclusion of soil extract results in a brownish pigmentation.

Other characteristics are given in Table 17.6.

Vitamin B_{12} and thiamine are required for growth; vitamin B_{12} is not replaceable by thymidine, methionine or factor B. Many of the usual laboratory media will not support growth unless supplemented with vitamin B_{12} and preferably also yeast extract, or with soil extract. When provided with essential vitamins and glucose as carbon + energy source, ammonium salts but not nitrate are utilized as nitrogen source (Owens and Keddie, 1969).

Growth occurs at 10 C but not at 37 C; good growth at 20–30 C.

Isolated from soil.

Type strain: ATCC 13347; NCIB 9222.

Species incertae sedis

The following putative species have not been examined for the characteristics used in species descriptions in the present revision of the genus and therefore their validity cannot yet be assessed.

a. *Arthrobacter luteus* Kaneko, Kitamura and Yamamoto 1969, 322. This species was stated to be facultatively anaerobic and therefore does not conform with the revised description of *Arthrobacter*.

b. *Arthrobacter marinus* Cobet, Wirsen and Jones 1970, 165.

c. *Arthrobacter variabilis* Müller 1961, 524.

d. *Arthrobacter viscosus* Gasdorf, Benedict, Cadmus, Anderson and Jackson 1965, 150.

e. *Arthrobacter polychromogenes* Schippers-Lammertse, Muijsers and Klatser-Oedekerk 1963, 13. This organism is very similar to *A. atrocyaneus*, here considered a possible subjective synonym of *A. globiformis*.

f. *Arthrobacter consociatus* Rivière 1961, 799.

g. *Arthrobacter nicotinovorus* (Bucherer) Niemer, Bucherer, Zeitler and Stadler 1964, 282. (*Bacterium nicotinovorum* Bucherer 1942, 169.)

Genera incertae sedis

Genus A. **Brevibacterium** *Breed 1953, 13*

MORRISON ROGOSA AND R. M. KEDDIE

In the 7th edition of THE MANUAL this genus was used as a repository for a number of little known or ill defined bacteria formerly in *Bacterium*, a genus recognized in the 6th edition but not in the 7th.

Brevibacterium and its type species *B. linens* were described as typically short, unbranched rods, reproducing by simple cell division, and placed in the family Brevibacteriaceae proposed by Breed in 1953. There was no mention of coryneform morphology.

Breed's description of *B. linens* must have been erroneous because present strains have the morphological diversity and certain other properties of the genus *Arthrobacter*. Mulder *et al.* (1966), Mulder and Antheunisse (1963), Mulder (1964), Schefferle (1957, 1966) and da Silva and Holt (1965) all indicate a close relationship of *B. linens* to *A. globiformis*. Komagata *et al.* (1969) observed that the mode of cell division in *B. linens*, ATCC 8377, was the "bending" type like that found in six species of *Arthrobacter* studied and da Silva and Holt (1965) even proposed the transfer of *B. linens* to the genus *Arthrobacter* as *Arthrobacter linens* (Wolff) da Silva and Holt 1965, 925.

Fiedler *et al.* (1970) studied the cell wall peptidoglycan of 147 strains of *Microbacterium*, *Coryne-*

bacterium, Arthrobacter (58), Brevibacterium (34) and unidentified coryneform organisms. They state "The strains which are presently ascribed to the genus Brevibacterium should probably be transferred to either Corynebacterium or Arthrobacter." The obvious coryneform morphology of most of these organisms was overlooked by Breed when creating this genus. The variation of the cell wall murein types found in the strains of Brevibacterium match very well the variation found in corynebacteria and arthrobacter.

Yamada and Komagata (1970) found that the G + C moles % in named strains of Brevibacterium varied from 46.6–70.5. However, glutamic acid producing bacteria, whether they were previously named Brevibacterium, Corynebacterium, or Micrococcus species, had a narrow range of G + C values with a mean value of 53.1 ± 1.3. All had a "snapping" mode of cell division and had DAP in their cell walls like C. diphtheriae (Yamada and Komagata, 1970). Among these were the following strains: B. divaricatum, NRRL B-2312; B. flavum, ATCC 14067; B. lactofermentum ATCC 13869; Corynebacterium callunae NRRL B-2244; C. herculis ATCC 13868; C. lilium NRRL B-2243; and Micrococcus glutamicus ATCC 13032. M. glutamicus had the typical "snapping" division without any suggestion of a Micrococcus arrangement. A large number of patent strains listed in the ATCC catalogue as Corynebacterium glutamicum are probably very similar or identical to M. glutamicus.

The family Brevibacteriaceae, the genus Brevibacterium and the 23 species listed in the 7th edition of THE MANUAL are here considered incertae sedis. All are listed together with an indication, where possible, of the genus to which they seem related. A large number of species have been ascribed to this genus since the last edition. A partial list is given and many are of industrial importance. A number which exist only in the patent literature have not been included.

Species incertae sedis

Species a to v were listed in the 7th edition of THE MANUAL.

a. Brevibacterium linens (Wolff) Breed 1953, 13. (Bacterium linens Wolff 1910, 422.)

Related to Arthrobacter, particularly A. globiformis.

Originally isolated from dairy products.

b. Brevibacterium acetylicum (Levine and Soppeland) Breed 1957, 502. (Flavobacterium acetylicum Levine and Soppeland 1926, 46.)

A motile rod with peritrichous flagella.

c. Brevibacterium erythrogenes (Lehmann and Neumann) Breed 1953, 13. (Bacterium erythrogenes Lehmann and Neumann 1896, 253.)

Very probably a variant of B. linens.

d. Brevibacterium healii (Buchanan and Hammer) Breed 1953, 14. (Bacterium healii Buchanan and Hammer 1915, 249.)

Morphology and growth in gelatin like Kurthia zopfii.

e. Brevibacterium lipolyticum (Huss) Breed 1953, 14. (Bactridium lipolyticum Huss 1908, 474.)

Described as a motile coccoid organism forming chains like streptococci. It could have been a streptococcus capable of splitting milk fats. Apparently it was found rarely and was stated to be different from the commonly found lipolytic diphtheroid organisms from the cow's udder such as Bacillus pseudodiphtheria Bergey 1904, 11; Bacterium lipolyticum Evans 1918, 576; Corynebacterium bovis Bergey et al. 1923, 388; Corynebacterium lipolyticum (Evans) Jayne-Williams and Skerman 1966 (see Jayne-Williams and Skerman, 1966, for an historical discussion).

Originally isolated from water.

f. Brevibacterium brunneum (Copeland) Breed 1957, 494. (Bacillus brunneus Copeland 1899, 348.)

Gram-stain not recorded but THE MANUAL states it may be very similar or identical with Corynebacterium bruneum Lehmann and Neumann 1927, 708.

g. Brevibacterium fulvum (Zimmermann) Breed 1953, 14. (Bacillus fulvus Zimmermann 1890, 92.)

The description is incomplete and trivial.

h. Brevibacterium fuscum (Zimmermann) Breed 1953, 14. (Bacillus fuscus Zimmermann 1890, 118.)

An aerobic organism forming brownish, yellowish or orange pigments.

i. Brevibacterium helvolum (Zimmermann) Lochhead 1955, 115. (Bacillus helvolus Zimmermann 1890, 100.)

The various descriptions of this organism are confused and contradictory; the original is too poor to permit subsequent identification. Two strains listed as Brevibacterium helvolum, ATCC 11822 and Arthrobacter globiformis, NCIB 8717, are derived from Jensen's (1934) strain Ca3 named Corynebacterium helvolum by him. These strains conform closely to the description of A. globiformis in the current MANUAL. Therefore, the Corynebacterium helvolum of Jensen (1934) is a subjective synonym of A. globiformis (see Lochhead (1955) for a full discussion).

Originally isolated from marine sources.

j. Brevibacterium immotum (ZoBell and Upham) Breed 1953, 14. (Bacterium immotum ZoBell and Upham 1944, 271.)

Morphologically like Kurthia.

k. *Brevibacterium marinopiscum* (ZoBell and Upham) Breed 1953, 14. (*Bacterium marinopiscus* (sic) ZoBell and Upham 1944, 258.)

Morphologically like *Kurthia*.

l. *Brevibacterium sociovivum* (ZoBell and Upham) Breed 1953, 14. (*Bacterium sociovivum* ZoBell and Upham 1944, 269.)

Morphologically like *Kurthia; B. immotum, B. marinopiscum* and *B. sociovivum* are minor variants of each other.

m. *Brevibacterium stationis* (ZoBell and Upham) Breed 1953, 14. (*Achromobacter stationis* ZoBell and Upham 1944, 273.)

An ovoid rod, 0.4 by 0.5–0.6 μm, and could even be a coccus. Labeled strains and *B. ammoniagenes* are 90% similar by numerical taxonomy (Chatelain and Second, 1966).

n. *Brevibacterium maris* (Harrison) Breed 1953, 14. (*Flavobacterium maris* Harrison 1929, 229.)

A coccoid rod isolated from the skin of fish. It may be an Arthrobacter forming orange-yellow or red-orange pigments.

Originally isolated from insects.

o. *Brevibacterium imperiale* (Steinhaus) Breed 1953, 13. (*Bacterium imperiale* Steinhaus 1941, 777.)

A small motile rod isolated from the alimentary tract of the imperial moth, *Eacles imperialis* Dr.

p. *Brevibacterium incertum* (Steinhaus) Breed 1953, 13. (*Bacterium incertum* Steinhaus 1941, 776.)

A motile monotrichous short rod, with ambiguous Gram-staining reaction, isolated from the ovaries of the lyreman cicada, *Tibiean linnei* Smith and Grossbeck.

q. *Brevibacterium insectiphilium* (Steinhaus) Breed 1953, 13. (*Bacterium insectiphilium* Steinhaus 1941, 777.)

A coccoid to rod-like organism isolated from the body wall of the bag worm, *Thyridopteryx ephemeraeformis* Haw.

r. *Brevibacterium minutiferula* (Steinhaus) Breed 1953, 13. (*Bacterium minutiferula* Steinhaus 1941, 778.)

A very small rod, possibly coccoid, 0.4–0.9 by 0.7–1.0 μm. Isolated once from a triturated specimen of the mud-dauber wasp, *Sceliphron cementarium* Dru.

s. *Brevibacterium quale* (Steinhaus) Breed 1953, 13. (*Bacterium qualis* (sic) Steinhaus 1941, 774.)

The specific epithet, *qualis* of what kind, is appropriate to the limited description of an ellipsoidal organism, perhaps a rod, isolated from the alimentary tract of the tarnished plant bug, *Lygus pratensis* L.

t. *Brevibacterium tegumenticola* (Steinhaus) Breed 1953, 13. (*Bacterium tegumenticola* Steinhaus 1941, 775.)

A coccoid rod isolated once from the integument of the bed bug, *Cimex lectularius* L.

Originally isolated from various sources.

u. *Brevibacterium ammoniagenes* (Cooke and Keith) Breed 1953, 14. (*Bacterium ammoniagenes* Cooke and Keith 1927, 318.)

The morphological and biochemical characteristics are those of coryneform bacteria abundant in poultry litter (Schefferle, 1966).

v. *Brevibacterium sulfureum* (Bergey et al.) Breed 1957, 502. (*Flavobacterium sulfureum* Bergey, Harrison, Breed, Hammer and Huntoon 1923, 103.)

A motile, peritrichously flagellated, Gram-positive, yellow-pigmented rod.

The following species were not listed in the 7th edition of THE MANUAL or in *Index Bergeyana*. They are cited in either Hatt and Zvirbulis (1967) or Zvirbulis and Hatt (1969).

Isolated from insects

w. *Brevibacterium protophormiae* Lysenko 1959, 41.

Isolated from a dipteran, *Protophormia terraenovae*, in sewage disposal plant.

Type strain: M576; CCEB 282; ATCC 19271.

x. *Brevibacterium saperdae* Lysenko 1959, 41.

Isolated from body cavity of dead larva, *Saperda carcharias* L. Type strain: 48-1-4; CCEB 366; ATCC 19272.

Isolated from various sources such as soil, rice paddies, sewage, vegetables, fruits, marine products, human pathological material (sic) and "natural sources." Many are of industrial interest; patented cultures are asterisked; not validly published names are not included.

y. *Brevibacterium flavum** Okumura, Tsugawa, Tsunoda, Kono, Matsui and Miyachi 1962, 147.

Produces L-glutamic acid.

z. *Brevibacterium immariophilum** Okumura et al. 1962, 147.

aa. *Brevibacterium lactofermentum* Okumura et al. 1962, 147.

bb. *Brevibacterium roseum** Okumura et al. 1962, 147.

cc. *Brevibacterium saccharolyticum** Okumura et al. 1962, 147.

dd. *Brevibacterium divaricatum* Su and Yamada 1960, 74.

Produces L-glutamic acid.

ee. *Brevibacterium leucinophagum* Kinney and Werkman 1960, 216.

Type strain: 14A2; ATCC 13809.

ff. *Brevibacterium liquefaciens* Okabayashi and Masuo 1960, 1087.

Type strain: ATCC 14929.

gg. *Brevibacterium pentoso-alanicum* Yamada and Hirose 1960, 626.

hh. *Brevibacterium pentoso-aminoacidicum* Yamada and Hirose 1961, 626.

ii. *Brevibacterium lyticum* Takayama, Udagawa and Abe 1960, 653.

Type strain: ATCC 15921

jj. *Brevibacterium albidum* Komagata and Iizuka 1964, 500.

Type strain: Y-3-2; IAM 1631; ATCC 15831.

kk. *Brevibacterium citreum* Komagata and Iizuka 1964, 498.

Type strain: 2Y-10; IAM 1514; ATCC 15828.

ll. *Brevibacterium luteum* Komagata and Iizuka 1964, 499.

Type strain: 2Y-12; IAM 1632; ATCC 15830.

mm. *Brevibacterium testaceum* Komagata and Iizuka 1964, 497.

Type strain: Rp-3; IAM 1561; ATCC 15829.

nn. *Brevibacterium pusillum* Iizuka and Komagata 1965, 2.

Type strain: 100; IAM 1478; ATCC 19096.

oo. *Brevibacterium alanicum* Ogawa *et al.* 1959, 46.

pp. *Brevibacterium aminogenes* Ota and Tanaka 1959, 51.

qq. *Brevibacterium chromogenes* Inoue and Niida 1962, 4.

rr. *Brevibacterium frigoritolerans* Delaporte and Sasson 1967, 2260.

ss. *Brevibacterium halotolerans* Delaporte and Sasson 1967, 2259.

tt. *Brevibacterium fermentans* Chatelain and Second 1966, 641.

uu. *Brevibacterium oxydans* Chatelain and Second 1966, 642.

Genus B. **Microbacterium** *Orla-Jensen 1919, 179*

MORRISON ROGOSA AND R. M. KEDDIE

Small diphtheroid rods with rounded ends; angular and palisade arrangements of cell masses are typical. Non-motile. Gram-positive. Good growth on surface of solid media in air. Weak acid production (chiefly L(+)-lactic acid) from fermented carbohydrates. Catalase positive. Optimal growth temperature about 30 C; limited or variable growth at 15 and 35 C; no growth at 10 and 40 C; survives 72 C for 15 min or more in skim milk. Found chiefly in dairy products and on dairy utensils.

Further Comments

This genus has long been a subject of controversy; should it be recognized and if so where does it belong? While recognized in previous editions of THE MANUAL it has been placed in *Bacteriaceae* (5th ed.), *Lactobacteriaceae* (6th ed.) and *Corynebacteriaceae* (7th ed.). Jensen (1934) suggested transferring *M. lacticum* and *M. liquefaciens* to *Corynebacterium* whereas *M. flavum* was tentatively allotted to *Mycobacterium*.

Originally the genus contained four species: *M. flavum, M. lacticum, M. liquefaciens* and *M. mesentericum*. Wittern (1933) suggested that *M. mesentericum* should be assigned to the genus *Mycobacterium* and Orla-Jensen (1943) agreed that it belonged either in *Mycobacterium* or *Corynebacterium;* in the 7th and present edition it has been assigned to the genus *Nocardia*. (See *Nocardia mesenterica.*) The above description applies to the three remaining original species.

A fifth organism, *Microbacterium thermosphactum* McLean and Sulzbacher 1953, 428 was reported to differ from the foregoing in having a lower optimal growth temperature (20–22 C), even growing at

0–1 C, in having a low heat resistance (no survival at 63 C for 5 min), and in its source (meat products). There are internal contradictions, mistakes or oversights in McLean and Sulzbacher's (1953) original report. They stated that *M. thermosphactum* did not have morphological characteristics of *Mycobacterium* or *Corynebacterium* but they overlooked the remarkable morphological resemblance of *M. thermosphactum* to *Kurthia;* their Figure 3 (and less so Fig. 2) shows the typical *Kurthia*-like exponential phase long chains of regular rods, their Figure 4 shows the coccoid forms typical of most stationary phase *Kurthia* cultures, and their Figure 1 shows aberrant forms sometimes seen in *Mycobacterium, Corynebacterium, Arthrobacter* or other genera. Also, they stated that *M. thermosphactum* produced L(+)-lactic acid homofermentatively and it was thus differentiated from other members of *Microbacterium* which produced dextro lactic acid; however, Orla-Jensen (1919) clearly meant dextrorotation, not molecular configuration, and thus all microbacteria produce L(+)-lactic acid.

Robinson (1966) concurred with earlier proposals that *M. flavum* (based on cell wall composition and electrophoretic mobilities of certain enzymes) be transferred to *Corynebacterium* because of the great similarities of *M. flavum* to human and animal corynebacteria. Robinson (1966) would, however, retain *Microbacterium* to accommodate the heat resistant *M. lacticum* and *M. liquefaciens*. However, this heat resistance may be a tenuous differentiating characteristic for the simple reason that some strains isolated from sources not heat treated are essentially indistinguishable from the vast cluster of strains which were isolated from heated sources (all of which strains must then logically be heat resistant).

Robinson (1966) did not study *M. thermosphactum*. Davis and Newton's (1969) numerical taxonomy study of phenotypic characteristics indicated that the inclusion of *M. lacticum* in the cluster of arthrobacters containing *Arthrobacter terregens* and *A. flavescens* is not unreasonable (one G + C value cited below supports this relationship).

Davidson and Hartree (1968) found catalase activity, positive benzidine reactions and cytochrome bands a, b₁ and a₃ to be dependent on media and temperature; cells incubated at 20 C were positive, but cells of *M. thermosphactum* incubated at 30 C were negative or weak (there have been previously unexplained conflicting reports on catalase, benzidine reactions, and presence of cytochromes). *M. lacticum* and *M. flavum* had classical cytochrome bands like *Saccharomyces cerevisiae*. Davidson *et al.* (1968) found that *M. thermosphactum* strains were pleomorphic with rod and coccoid stages. The long rod forms resembled kurthiae. However, kurthiae are motile and obligately aerobic, and are methyl red negative, whereas *M. thermosphactum* is non-motile, facultatively anaerobic and usually methyl red positive.

The amino acids of the cell-wall murein (peptidoglycan) of *M. flavum* and *M. thermosphactum* have a primary structure similar to that of the murein of *Corynebacterium diphtheriae* and are thus closely related to the human and animal pathogenic corynebacteria; Schleifer (1970) stated that they ". . . should be removed from the genus *Microbacterium*." *M. lacticum* and *M. liquefaciens*, on the other hand, have mureins significantly different from those of human and animal corynebacteria and show greatest similarity to certain plant pathogenic corynebacteria (Schleifer, 1970; Robinson, 1966).

Collins-Thompson *et al.* (1972) have studied some enzyme reactions and the G + C moles % of the DNA in *Microbacterium*. The G + C moles % of *M. flavum* (58) and of *M. lacticum* (63) correspond with the corynebacteria (48–59) and the *Arthrobacter globiformis* cluster of arthrobacters (60–64), respectively; Yamada and Komagata (1970) found a G + C value of 69.3 moles % for the same strain of *M. lacticum*, ATCC 8180, studied by Collins-Thompson *et al.*; however, Komagata *et al.* (1969) observed that this strain has a "bending" mode of cell division like seven other studied strains representing six species of *Arthrobacter*. Both *M. lacticum* and *M. flavum* have relatively low activity of glycolytic enzymes but a relatively high activity of enzymes associated with the tricarboxylic acid cycle; *M. thermosphactum* behaves oppositely and in addition has a G + C value of 36 moles % (like many streptococci).

In summary:

a. *Microbacterium lacticum* Orla-Jensen 1919, 179.

b. *Microbacterium liquefaciens* Orla-Jensen 1919, 182.

The above species show some properties in common with both *Arthrobacter* and plant pathogenic corynebacteria and further comparative studies are required.

c. *Microbacterium flavum* Orla-Jensen 1919, 181.

Three strains have a cell wall structure and G + C moles % resembling *Corynebacterium diphtheriae*. *M. flavum* should ultimately be transferred to *Corynebacterium* if examination of more strains confirms these results.

d. *Microbacterium thermosphactum* McLean and Sulzbacher 1953, 428.

An anomalous organism having morphological features resembling *Kurthia* but having a metabolism and a DNA composition of some lactic acid bacteria, i.e. *Streptococcus*. Its position is still uncertain.

Genus III. Cellulomonas Bergey et al. 1923, 154, emend. mut. char. Clark 1952, 50

R. M. KEDDIE

Cel.lu.lo.mo'nas. M.L. n. *cellulosa* cellulose; Gr. n. *monas* a unit, monad; M.L. fem.n. *Cellulomonas* cellulose monad.

In young cultures, irregular rods *ca.* 0.5 μm in diameter by 0.7–2.0 μm or more in length, which may be straight, angular or slightly curved and occasionally club-shaped or beaded; a proportion of the rods are arranged at an angle to each other giving V formations (Plate 17.1, Fig. 12). Occasionally cells may show rudimentary branching but true mycelia are not formed. As cultures age the rods become shorter but only a small propor- tion of coccoid cells is seen in cultures a week or more old. Motile by one (usually polar or subpolar) or a few lateral flagella, or non-motile. Do not form endospores. Gram-positive: the cells are readily decolorized and often a mixture of Gram-positive and Gram-negative rods is seen, sometimes with the latter predominating; often one or more Gram- positive granules are seen in otherwise Gram- negative cells. Not acid-fast.

Chemoorganotrophs: Metabolism primarily respiratory using molecular oxygen as terminal electron acceptor; most strains produce acid from glucose under both aerobic and anaerobic conditions (see *Further Comments*).

Catalase positive. **Cellulose is attacked by all strains recognized at present.**

The cell walls contain neither meso-diaminopimelic acid nor arabinose. The walls are typical of Gram-positive organisms; they contain alanine, glutamic acid and lysine (Keddie et al., 1966; Yamada and Komagata, 1970), one or more sugars may be present but galactose is absent (Keddie *et al.*, 1966). Recent work (Lamey and Keddie, unpublished) has shown that the diamino acid in several strains tested is ornithine and not lysine. The methods used by Keddie *et al.* (1966) and Yamada and Komagata (1970) did not allow these two amino acids to be distinguished.

Biotin and thiamine are the only exogenous organic growth factors required by strains now recognized (Keddie *et al.*, 1966) and when provided with these vitamins growth occurs in suitable mineral media with glucose as carbon + energy source and an ammonium salt (or nitrate for most strains) as nitrogen source (Owens and Keddie, 1969).

Aerobic; most strains also capable of anaerobic growth (see *Further Comments*).

Temperature optimum ca. 30 C. Do not survive heating at 63 C for 30 min in skim milk. Grow well at neutral pH.

The G + C content of the DNA ranges from 71.7–72.7 moles % (T_m) (Yamada and Komagata, 1970).

Type species: *Cellulomonas flavigena* (Kellerman and McBeth) Bergey *et al.* 1923, 165.

Further Comments

Members of the genus *Cellulomonas* are coryneform bacteria whose salient feature is the ability to attack cellulose. This property is universal among authentic strains because it was used in their original isolation (see e.g. Kellerman *et al.*, 1913) and is extremely stable (Clark, 1951). However, the possibility that strains occur which do not attack cellulose has not been excluded in the generic definition.

Conflicting opinions have been expressed about the Gram-reaction of *Cellulomonas* spp. In previous editions of THE MANUAL they have been described as Gram-negative and Gram-variable. Mulder and Antheunisse (1963) considered two named strains which they examined to be Gram-negative. Keddie *et al.* (1966) noted that the six named strains they examined were weakly Gram-positive and confirmed that the cell wall composition was typical of Gram-positive organisms. They are therefore described as Gram-positive although it is recognized that they may be readily decolorized.

The relationship to oxygen of members of the genus is also problematical. Most authentic named strains tested give a significant but much reduced amount of growth in anaerobic conditions in a variety of glucose-supplemented media, a few give equivocal results (Keddie, unpublished). On the other hand using a variety of media and a method similar to that described by Hugh and Leifson (1953) all the strains tested gave acid from glucose under both aerobic and anaerobic conditions.

Ten species were recognized in the 7th edition of THE MANUAL. The descriptions of six were based on studies of authentic strains (Clark, 1953) but the species were distinguished by relatively minor differences in a few features. However, recent studies have emphasized the close resemblance of *C. biazotea*, *C. cellasea*, *C. flavigena*, *C. gelida* and *C. uda* to each other in cell wall composition and in requirements for vitamins (Keddie *et al.*, 1966) and substrate nitrogen (Owens and Keddie, 1969). Owens and Keddie (unpublished) found that of 180 carbon and energy sources tested, the same strains utilized on the average only 22, and in addition to attacking cellulose, utilized a common core of 11 substrates which included acetate and 10 carbohydrates. A number of other carbohydrates and related substances, simple organic acids and a few amino acids were the only other substrates utilized.

For these reasons in the present revision of the genus only one species is recognized in which minor variations in chromogenesis, motility and nitrate reduction may occur.

Since the specific epithet *flavigena* (*Bacillus flavigena* Kellerman and McBeth 1912, 488) antedates *Bacillus biazoteus* Kellerman, McBeth, Scales and Smith 1913, 506, the type species is therefore *Cellulomonas flavigena*. *C. biazotea*, *C. cellasea* (Kellerman *et al.* 1913, 508) Bergey *et al.* 1923, 158, *C. gelida* (Kellerman *et al.* 1913, 510) Bergey *et al.* 1923, 162 and *C. uda* (Kellerman *et al.* 1913, 514) Bergey *et al.* 1923, 166, are considered subjective synonyms of *C. flavigena*. *C. fimi* (McBeth and Scales 1913, 30) Bergey 1923, 166 has not been examined by the methods mentioned above, but was distinguished from *C. biazotea* by its ability to ferment xylose and arabinose. Owens and Keddie (unpublished) found xylose and arabinose were used by all members of the genus examined, so this distinction is of doubtful validity. *C. fimi* is therefore considered a possible subjective synonym of *C. flavigena*.

Description of the species of genus **Cellulomonas**

1. **Cellulomonas flavigena** (Kellerman and McBeth) Bergey *et al.* 1923, 165. (*Bacillus flavigena* (*sic*) Kellerman and McBeth 1912, 488; *Bacterium flavigena* (*sic*) (Kellerman and McBeth) McBeth and Scales 1913, 22; *Aplanobacter flavigenum* (Kellerman and McBeth) Magrou and Prévot 1948, 104; *Pseudobacterium flavigenum* (Kellerman and McBeth) Krasil'nikov 1949, 235; *Empedobacter flavigena* (Kellerman and McBeth) Brisou 1958, 182.)

fla.vi′ge.na. L. adj. *flavus* yellow; L. v. *gigno* produce; M.L. adj. *flavigena* yellow producing.

Morphology and physiology as for genus.

Grows well or moderately well on peptone meat extract media at near neutral pH and 30 C. Most strains produce a yellow non-diffusible pigment on nutrient agar and similar media.

Filter paper strip in 0.5% peptone solution is reduced to a pulpy mass or weakened sufficiently so that the fibers separate on slight agitation. Starch hydrolyzed. Acetylmethylcarbinol not produced.

Of 180 organic compounds, only between 16 and 28 were utilized as carbon and energy sources by individual strains (Owens and Keddie, unpublished). These comprised mainly carbohydrates and sugar derivatives but a few organic acids and amino acids were utilized. Eleven characteristics were found of greatest value in distinguishing the species and these also constitute what is considered to be "the ideal phenotype" in the sense of Stanier *et al.* (1966) (Table 17.9).

Gelatin slowly hydrolyzed. Most strains reduce nitrate to nitrite.

Aerobes; most strains are facultative anaerobes: some strains give equivocal results for anaerobic growth.

Usually found in soil.

Reference strain: ATCC 482; NCIB 8073.

Species incertae sedis

Cultures are not available and from the original descriptions of Gram-reaction and morphology they cannot be placed unequivocally in the genus. In other features they differ in minor respects from *C. flavigena*.

TABLE 17.9
Characters of greatest value in the recognition of strains of **Cellulomonas flavigena**[a]

	Ideal phenotype[b]
1. Biotin and thiamine required	+
2. Inorganic nitrogen utilized as sole nitrogen source[c]	+
3. Cellulose attacked	+
Utilization of:[d]	
4. D-Ribose	−
5. Acetate	+
6. Propionate and/or butyrate and/or pentoate	−
7. One or more of: malonate, succinate, glutarate, adipate, pimelate, suberate, azelate	−
8. One or more of: ethanol, propanol, butanol, pentanol, hexanol, heptanol, octanol, nonanol	−
9. *p*-Hydroxybenzoate and/or *m*-hydroxybenzoate and/or 3:4-dihydroxybenzoate	−
10. L-Asparagine	−
11. 1:4-Butanediamine	−

[a] Of more than 100 strains of coryneform bacteria tested by Owens and Keddie other than *C. flavigena*, including named strains of *Arthrobacter* and soil and herbage isolates, more than 95% deviated in six or more characters, and none deviated in less than three characters.

[b] + Indicates 7/7 strains positive; − indicates 7/7 negative (Owens and Keddie, unpublished).

[c] See Owens and Keddie (1969) for method.

[d] See page 620 for method.

a. *Cellulomonas acidula* (Kellerman *et al.*) Bergey *et al.* 1923, 167.

b. *Cellulomonas aurogena* (Kellerman *et al.*) Bergey *et al.*, 1923, 157.

c. *Cellulomonas galba* (Kellerman *et al.*) Bergey *et al.* 1923, 157.

d. *Cellulomonas pusilla* (Kellerman *et al.*) Bergey *et al.* 1923, 161.

Genus IV. **Kurthia** *Trevisan 1885, 92. Nom cons.* Opin. 13, Jud. Comm. 1954, 152

R. M. KEDDIE AND M. ROGOSA

(*Zopfius* Wenner and Rettger 1919, 351.)

Kurth′i.a. M.L. fem.n. *Kurthia* named for H. Kurth, the German bacteriologist who described the type species.

In young cultures (18–24 hr): **regular, unbranched rods** with rounded ends **occurring in chains** which are often parallel (Plate 17.1, Figs. 13–14); the rods are *ca.* 0.8 μm in diameter and vary in length according to the stage of growth but are generally *ca.* 2–8 μm long. **Older cultures** (3–7 days) are **usually composed of coccoid cells** formed by fragmentation of the rods; in some strains short rods are the dominant forms in such cultures. The rods are motile by peritrichous flagella. Do not form endospores.

Gram-positive. Not acid fast.

Chemoorganotrophs: **Metabolism respiratory,** never fermentative.

Molecular oxygen is the terminal electron acceptor.

Catalase positive. Oxidase negative. Do not reduce nitrate.

Do not produce acid from carbohydrates or polyols; media become alkaline presumably from peptones.

Strict aerobes.

Temperature optimum 25–30 C; range *ca.* 5–37 C. Some strains grow at 45 C, none at 50 C. Grow well in presence of 4–6% NaCl.

Type species: *Kurthia zopfii* (Kurth) Trevisan 1885, 92.

Description of the species of genus **Kurthia**

1. **Kurthia zopfii** (Kurth) Trevisan 1885, 92. *Nom. cons.* Opin. 13, Jud. Comm. 1954, 152. (*Bacterium zopfii* Kurth 1883, 98; *Helikobacterium zopfii* (Kurth) Escherich 1886, 6; *Bacillus zopfii* (Kurth) Macé 1889, 507; *Zopfius zopfii* (Kurth) Wenner and Rettger 1919, 351.)

zop'fi.i. M.L. gen.n. *zopfii* of Zopf; named for W. Zopf, a German botanist.

Description based on the original descriptions (Kurth, 1883) and those of Wenner and Rettger (1919) and Gardner (1969).

Morphology as for genus. The regular rods characteristic of exponential phase cultures are *ca.* 0.8 μm by 3–4 μm. Motile by numerous peritrichous flagella. Non-motile strains occur.

Surface colonies on nutrient agar are rhizoid; the edges of young colonies show a characteristic "Medusa-head" appearance under low magnification. Growth is not pigmented.

Produce acid from ethanol but not glycerol.

Gelatin, starch and esculin are not hydrolyzed. However, growth on gelatin is a useful diagnostic feature; on gelatin slopes inoculated with a single central streak outgrowths are produced which apparently follow the lines of stress in the medium; the resultant appearance, resembling a feather, is characteristic. This feature received much attention in the earlier literature (see Kufferath, 1911; Gardner, 1969). In gelatin stabs arborescent growth is usually produced.

Urease, lecithinase, H₂S and indole are not produced.

Growth reported only in complex media; minimal nutritional requirements unknown.

Originally isolated from intestinal contents of chickens; has also been isolated from hen manure, stagnant fresh water, fresh and putrefying meats, a variety of meat products, working surfaces in meat plants, and milk. Not known to be pathogenic.

Suggested working strain: ATCC 10538; NCTC 4043, Sneath and Skerman 1966, 62.

Further Comments

In the 7th edition of THE MANUAL the type species was stated to be facultatively anaerobic, presumably quoting Wenner and Rettger (1919), and this statement also appeared in the generic description. However, more recent studies have shown that the type strain and all new isolates of *K. zopfii* are obligately aerobic (Gardner, 1969; Keddie, unpublished).

A number of strains were described by Gardner (1969) which have the general characteristics of *K. zopfii* but which differ in the following respects: growth occurs at 44 C, acid is produced from glycerol but not from ethanol, growth on nutrient agar has a yellow or cream pigmentation.

Kurthia zenkeri (Hauser) Bergey *et al.* 1925, 215 is usually considered a synonym of *K. zopfii*; its differentiation from the latter species was based on its inability to form arborescent growth in a gelatin stab.

Species incertae sedis

a. *Kurthia variabilis* Severi 1946, 108.

b. *Kurthia bessonii* (Hauduroy *et al.*) Severi 1946, 108. (*Listerella bessoni* (*sic*) Hauduroy *et al.* 1937, 271.)

The above two species were originally isolated from human feces in mild cases of food poisoning; *K. bessonii* was also isolated from normal feces. Their rather limited descriptions differ from that of the type species in relatively minor respects; *K. bessonii* in being proteolytic and *K. variabilis* in having higher maximum and optimum growth temperatures and in producing small amounts of H₂S. Both species were stated to be facultatively anaerobic.

c. *Microbacterium thermosphactum* McLean and Sulzbacher 1953, 428.

Has some morphological characteristics in common with *Kurthia*, but is non-motile and ferments carbohydrates producing L(+)-lactic acid as the principal product. Its taxonomic position is uncertain. (See *Microbacterium*, page 628.)

FAMILY I. **PROPIONIBACTERIACEAE** DELWICHE 1957, 569

Pro.pi.on.i.bac.te.ri.a'ce.ae. M.L. neut.n. *Propionibacterium* the type genus of the family; *-aceae* ending to denote a family; M.L. fem.pl.n. *Propionibacteriaceae* the *Propionibacterium* family.

Gram-positive, non-spore-forming, anaerobic to aerotolerant, pleomorphic, branching or regular rods or filaments. Where saccharolytic, **major products include CO₂, propionic and acetic acids or mixtures of organic acids often including butyric, formic, lactic or other monocarboxylic acids. Some species are** not saccharolytic. Lactic acid alone or lactic with an equal or greater amount of acetic acid does not accumulate as the only major products. Growth often enhanced by CO_2. Inhabitants of the skin, respiratory and intestinal tracts of most animals, and of animal products. Several species are found in soft tissue infections.

Key to the genera of family **Propionibacteriaceae**

I. Not motile.
Anaerobic to aerotolerant.
Produce propionic acid and acetic acids and lesser amounts of isovaleric, formic, succinic and lactic acids.
Genus I. *Propionibacterium.*
II. Motile or non-motile.
Obligately anaerobic. Fermentative or non-fermentative.
Fermentative species produce mixtures of organic acids, often including large amounts of butyric, acetic, formic and lactic or other mono-carboxylic acids.
Genus II. *Eubacterium.*

Genus I. **Propionibacterium** Orla-Jensen 1909, *337*

W. E. C. MOORE AND LILLIAN V. HOLDEMAN

Pro.pi.on.i.bac.te'ri.um. M.L. n. *acidum propionicum* propionic acid; Gr. dim.n. *bakterion* a small rod; M.L. neut.n. *Propionibacterium* propionic (acid) bacterium.
Microaerophilic to anaerobic species formerly in the genus *Corynebacterium* have been transferred to this genus.

Gram-positive, non-spore-forming, non-motile rods. Usually pleomorphic, diphtheroid or club-shaped with one end rounded and the other end tapered or pointed and stained less intensely. Cells of some cultures may be coccoid, elongate, bifid or even branched. Cells usually arranged in singles, pairs or V and Y configurations, short chains or clumps in "Chinese character" arrangement.

Chemoorganotrophs: Metabolize carbohydrates, peptone, pyruvate or lactate. **Fermentation products include combinations of propionic and acetic acids and frequently lesser amounts of isovaleric, formic, succinic or lactic acids and carbon dioxide. All species produce acid from glucose.**

Anaerobic to aerotolerant. Although most strains in this genus grow most rapidly under strictly anaerobic conditions, many strains grow well in deep tubes of peptone-yeast extract-glucose broth exposed to air when a large inoculum is used.

Tween 80 stimulates growth of most strains.

Growth most rapid at 30–37 C and pH near 7. Some strains grow at 25 and 45 C. Most strains grow in glucose broth with 20% bile or 6.5% NaCl.

Ammonia produced from proteinaceous material by most strains. Hippurate usually not hydrolyzed; neutral red usually not reduced.

Growth may be white, gray, pink, red, yellow or orange.

The G + C content of the DNA of the described species ranges from 59–66 moles % (T_m).

Found in dairy products or from the skin of man and intestinal tract of man and animals. Some species may be pathogenic.

Characteristics by which species in the genus can be differentiated are given in the key and in Tables 17.10, and 17.11. Other characteristics of the species in the genus are given in the text concerning each species. The drawings (Fig. 17.A) are composites from several cultures. Individual cultures usually show somewhat less variation in cellular morphology than is depicted. (Scale for drawings: $\%_6$ inch = 10 μm.)

Type species: *Propionibacterium freudenreichii* van Niel 1928, 162.

Further Comments

Genetic relationships among previously described species of this genus have been studied by DNA/DNA competition experiments (Johnson and Cummins, 1972). The results demonstrate that there are eight groups of closely related strains, here recognized as species. Several species include a continuum of variant strains, each differing from the next by only one or a few characteristics. An especially large amount of phenotypic variation can occur among strains of *P. acnes* that show high DNA nucleotide sequence similarities.

Key for species of genus **Propionibacterium**

I. Esculin hydrolyzed.
   A. Gelatin not completely hydrolyzed.
      1. Mannitol not acid.
         a. Nitrate reduced.
            i. Lactose not acid.

                         1a. *P. freudenreichii* subsp. *freudenreichii*

            ii. Lactose acid.

                         1b. *P. freudenreichii* subsp. *globosum*

         b. Nitrate not reduced.
            i. Trehalose not acid.

                         1c. *P. freudenreichii* subsp. *shermani:*

            ii. Trehalose acid.

                2. *P. theonii*

      2. Mannitol acid.
         a. Nitrate reduced.

                3. *P. acidi-propionici*

         b. Nitrate not reduced.

                4. *P. jensenii*

   B. Gelatin completely hydrolyzed.

                5. *P. avidum*

II. Esculin not hydrolyzed.
   A. Indole produced.

                6. *P. acnes*

   B. Indole not produced.
      1. Adonitol, erythritol, maltose and ribose acid.
                7. *P. lymphophilum*
      2. Adonitol, erythritol, maltose or ribose not acid.
         a. Nitrate reduced.

                6. *P. acnes**

         b. Nitrate not reduced.

                8. *P. granulosum*

Description of the species of genus **Propionibacterium**

1. **Propionibacterium freudenreichii** van Niel 1928, 162. (*Bacterium acidi propionici a* von Freudenreich and Orla-Jensen 1907, 532; *Bacterium acidi propionici d* Sherman 1921, 387; *Propionibacterium shermanii* van Niel 1928, 163. Probable synonyms (labeled strains not known to be extant): *Propionibacterium globosum* Sakaguchi, Iwasaki and Yamada 1941, 131; *Propionibacterium orientum* Sakaguchi, Iwasaki and Yamada 1941, 133; *Propionibacterium coloratum* Sakaguchi, Iwasaki and Yamada 1941, 133; *Propionibacterium casei* Janoschek 1944, 333.)

freu.den.reich'i.i. M.L. gen.n. *freudenreichii* of Freudenreich; named for Edouard von Freudenreich, the Swiss bacteriologist who isolated this species.

---

* For strains that are catalase negative and produce acid in sucrose also see *Arachnia propionica.*

Description based on literature descriptions, including those of Sakaguchi *et al.* (1941), Janoschek (1944) and Werkman and Brown (1933) and on study of two strains of *P. freudenreichii* (Williams E1.51 and ATCC 6207) and seven strains of *P. shermanii* (van Niel 1.11, IAM 1714, and ATCC 8262, 9615, 9616, 9617, 13673) (Tables 17.10, 17.11 and Fig. 17.A).

Surface colonies on horse blood agar (2 days) are 0.2–0.5 mm, circular, entire, convex to pulvinate, semi-opaque, glistening, gray to white (may become cream, tan or pink). Colonies in deep agar are lenticular, to 4 mm, white, tan or pink.

Glucose broth cultures are turbid with a smooth or granular sediment, or clear with a granular sediment and terminal pH of 4.5–4.9.

Anaerobic to aerotolerant. Rarely grow on the surface of agar incubated aerobically, grow in deep broth incubated aerobically, but more slowly than anaerobically.

Strains may require pantothenic acid, biotin or thiamine, but do not require para-aminobenzoic acid (Delwiche, 1949).

The major long chain fatty acids produced in thioglycollate cultures (Moss *et al.*, 1969) are 12-methyltetradecanoic (about 43%) and a 17-carbon branched-chain acid (about 12%).

Cell walls contain *meso*-diaminopimelic acid and major amounts of galactose and moderate amounts of mannose and rhamnose; no glucose. The G + C content of the DNA ranges from 64–67 moles % (T_m, Johnson and Cummins, 1972).

Isolated from raw milk, Swiss cheese and other dairy products.

Further Comments

Strains in this species formerly were differentiated into separate species on the basis of reduction of nitrate and fermentation of lactose as follows:

	NO$_3$ reduced	Acid from lactose and acid curd in milk
1a. *P. freudenreichii*	+	−
1b. *P. globosum*	+	+
1c. *P. shermanii*	−	+

P. shermanii also ferments maltose, melezitose and melibiose, which *P. freudenreichii* does not.

Although these organisms are very similar, recognition of *P. freudenreichii* subsp. *freudenreichii*, *P. freudenreichii* subsp. *shermanii* and *P. freudenreichii* subsp. *globosum* (synonym of *P. casei*) may be warranted because of their significance in cheese manufacture.

2. **Propionibacterium thoenii** van Niel 1928, 164. (*Bacterium acidi propionici* var. *rubrum* Thöni

and Allemann 1910, 29; *Propionibacterium rubrum* van Niel 1928, 164. Probable synonym (labeled strains not known to be extant) *Propionibacterium sanguineum* Janoschek 1944, 336.)

thoe'ni.i. M.L. gen.n. *thoenii* of Thöni; named for J. Thöni, the Swedish bacteriologist who isolated this organism.

Description based on study of the type strain and ATCC 4871 (*P. rubrum* van Niel 23) and ATCC 4872 (*P. rubrum*, van Niel 19) (Tables 17.10, 17.11 and Fig. 17.A).

Surface colonies on horse blood agar (2 days) are punctiform to 0.5 mm, circular, entire to pulvinate, glistening and white, pink or orange. Colonies in deep agar are lenticular and white, orange or red.

Glucose broth cultures are clear or turbid with ropy sediment and terminal pH of 4.7–4.9.

Anaerobic to aerotolerant or facultative. May grow as well in aerobic as in anaerobic conditions.

Pantothenic acid, biotin, thiamine, or para-aminobenzoic acid may be required for growth (Delwiche, 1949).

The major long chain fatty acid produced in thioglycollate cultures (Moss *et al.*, 1969) is 13-methyltetradecanoic (34%); one of two strains produced a 17-carbon branched-chain acid (20%).

Cell walls contain L-diaminopimelic acid and small to moderate amounts of glucose and galactose and none to traces of mannose.

The G + C content of the DNA is 66–67 moles % (T_m, Johnson and Cummins, 1972).

Isolated from dairy products.

Type strain: ATCC 4874; van Niel 15.

3. **Propionibacterium acidi-propionici** Orla-Jensen 1909, 337. Jud. Comm. 1958, 162 Opinion 20(4). (*Bacillus acidi propionici* von Freudenreich and Orla-Jensen 1907, 532; *Propionibacterium pentosaceum* van Niel 1928, 163; *Propionibacterium arabinosum* Hitchner 1932, 41.)

a.ci'di.-pro.pi.on'i.ci. M.L. n. *acidum propionicum* propionic acid; M.L. gen.n. *acidi-propionici* of propionic acid.

Description based on study of the type strain, ATCC 4875 (*P. pentosaceum*, van Niel 4), ATCC 4965 (*P. arabinosum*, Hitchner) and IAM 1725. Phenotypic characteristics of these strains are similar to original descriptions (Tables 17.10, 17.11 and Fig. 17.A).

Surface colonies on horse blood agar (2 days anaerobic incubation) are punctiform to 1 mm, circular to slightly irregular, convex to pulvinate, entire or slightly scalloped, gray or white, semi-opaque; usually non-hemolytic, but may show slight β-hemolysis under area of confluent growth. Colonies in deep agar are white, becoming pink after continued incubation.

TABLE 17.10

Fermentation reactions of the species of genus **Propionibacterium**[a]

	1a. *P. freudenreichii* ss. *freudenreichii*	1b. *P. freudenreichii* ss. *globosum*	1c. *P. freudenreichii* ss. *shermanii*	2. *P. thoenii*	3. *P. acidipropionici*	4. *P. jensenii*	5. *P. avidum*	6. *P. acnes*	7. *P. lymphophilum*	8. *P. granulosum*
Products from peptone-yeast extract-glucose broth	PASfl	PAsl	PAS (fl)	PA(sl *fiv*)	PAs (*lfiv*)	PA(*sfiv*)	PA(*sfiv*)	PA (*Lsfiv*)	PASlf	PA(*siv*)
Products from lactate	P^-	PA	PA	PA^-	PA	PA	PA^s	P^-	PA^-	PA^s
Acid produced from:										
Adonitol	v	·	$+^w$	$+^w$	w^+	$+^w$	v	$-^+$	$+$	$-$
Amygdalin	$-^w$	$-$	$-$	$+^w$	w^-	w^-	$-^w$	$-$	w^-	$-^w$
Arabinose	$+^w$	$+$	$+^w$	$-^+$	$+$	v	v	$-^+$	$-$	$-^w$
Cellobiose	w^-	w	$-^+$	w^-	$+^w$	d	$-^w$	$-^+$	$-$	$-$
Dulcitol	$-$	$-$	$-$	$-$	$-$	$-$	$-$	$-$	$-$	$-$
Erythritol	$+^-$	$+$	$+^w$	$+$	$+^w$	$+^w$	$+^w$	$-^+$	$+$	$-$
Esculin	$-$	$-$	$-$	$+$	w^-	$-^w$	$-$	$-$	$-$	$-$
Esculin hydrol.	$+$	$+$	$+$	$+$	$+$	$+$	$+$	$-$	$-$	$-$
Fructose	$+$	w	$+^w$	$+$	$+$	$+$	$+$	$+^-$	$+^-$	$+$
Galactose	$+^w$	$+$	$+$	$+$	$+$	$+$	$+$	$+^-$	$-^w$	w^-
Glycerol	$+^w$	$+$	$+$	$+$	$+$	$+$	$+$	$+^-$	$-$	$+$
Glycogen	w^-	$-$	$-$	$-^+$	v	$-^w$	$-$	$-$	$-$	$-$
Inositol	v	w	$+^-$	$+^-$	$+$	$+^-$	$-^+$	$-^w$	v	$-$
Inulin	$-$	$-$	$-$	$-$	$-^w$	$-^w$	$-$	$-$	$-$	$-$
Lactose	$-$	$+$	$+$	d	$+$	v	v	$-$	$-$	$-$
Maltose	w^-	w	w	$+^w$	$+$	$+^-$	$+$	$-^w$	$+$	$+^-$
Mannitol	$-$	$-$	$-$	$-$	$+$	$+$	$-^+$	d	$-$	$-^+$
Mannose	$+^w$	$+$	$+$	$+$	$+$	$+$	$+$	$+^-$	$-$	$+^w$
Melezitose	$-$	w	w^-	v	$+^w$	$+^w$	$+^-$	$-$	$-$	v
Melibiose	$-$	w	d	$+^-$	v	$+^w$	$-^+$	$-$	$-$	v
Raffinose	$-$	$-$	$-$	$-^+$	d	$+^w$	$-^+$	$-$	$-$	d
Rhamnose	$-$	$-$	$-$	$-$	$+^w$	$-$	$-$	$-$	$-$	$-$
Ribose	$+^w$	$+$	v	$+$	$+^w$	$+^w$	$+^w$	$-^+$	v	v
Salicin	$+^w$	$-$	$-^w$	$+^w$	$+$	$+^-$	v	$-$	$-$	$-^w$
Sorbitol	$-$	$-$	$-^+$	v	$+$	$-^+$	$-^+$	$-^+$	$-$	$-$
Sorbose	$-$	w	w^-	v	$+^-$	$-^w$	$-^w$	$-^w$	$-$	$-^w$
Starch	$-$	w	w^-	$+^w$	$+^w$	v	$-^+$	$-^w$	w^-	$-^w$
Starch hydrol.	$-$	·	$-$	$-^+$	$-$	$-$	$-$	$-$	d	$-$
Sucrose	w^-	$-$	$-^+$	$+^w$	$+$	$+^w$	$+$	$-^+$	d	$+^w$
Trehalose	$-$	$-$	$-$	$+$	$+^w$	$+^w$	$+^w$	$-^+$	$-$	w^-
Xylose	$-^w$	w	w^-	$-^w$	$+^w$	v	$-^+$	$-^w$	$-$	$-^w$

[a] Products from PYG broth: Capital letters indicate >1 meq acid/100 ml broth; small letters <1 meq/100 ml. A, acetic; F, formic; *iv*, isovaleric; L, lactic; P, propionic; S, succinic; ( ) products of occasional strains.

+, strong acid, i.e. pH 5.5 or below in 90–100% of strains; w, weak acid, i.e. pH 5.5–6.0 in 90–100% of strains. (Uninoculated xylose and arabinose, under CO_2 (prereduced media), are often pH 5.9. For cultures in these media, we consider pH 5.4–5.7 weak acid and pH below 5.4 strong acid); −, acid not produced, i.e. pH 6.0 in 90–100% of strains; d, reaction positive 40–60% strains; v, reaction variable within a strain. When any of the above used as superscripts, reaction of 10–40% of strains; ·, not tested.

Acid products determined by procedures given in Holdeman and Moore, 1972.

Glucose broth cultures are turbid with smooth or ropy sediment and terminal pH of 4.1–4.9.

Anaerobic to aerotolerant or facultative. May grow as well in aerobic as in anaerobic conditions.

Pantothenic acid and biotin required for growth; thiamine stimulates growth (Delwiche, 1949).

The major long chain fatty acids produced in thioglycollate cultures (Moss *et al.*, 1969) are 13-methyltetradecanoic (17–40%) and 12-methyltetradecanoic (12–23%).

Cell walls contain L-diaminopimelic acid and moderate amounts of glucose and trace to moderate amounts of galactose and mannose.

The G + C content of the DNA is 66–68 moles % (T_m, Johnson and Cummins, 1972).

Isolated from dairy products.

Type strain: ATCC 25562; V.P.I. 0339; Prévot 14 X; Orla-Jensen.

4. Propionibacterium jensenii van Niel 1928,

TABLE 17.11
Biochemical reactions of the species of genus **Propionibacterium**[a]

	1a. P. freudenreichii subsp. freudenreichii	1b. P. freudenreichii subsp. globosum	1c. P. freudenreichii subsp. shermanii	2. P. thoenii	3. P. acidipropionici	4. P. jensenii	5. P. avidum	6. P. acnes	7. P. lymphophilum	8. P. granulosum
Gelatin	−	−	−^{w}	−	−^{w}	−	d	d$^-$	w	w$^-$
Milk	−	c	c$^-$	−^{c}	c$^-$	−^{c}	dc$^-$	dc$^-$	−	c$^-$
Indole produced	−	−	−	−	−	−	−	+$^-$	−	−
Nitrate reduced	+	+	−	−	+	−	−	+$^-$	d	−
Catalase	w	+	+	+	v	+$^-$	+	+$^-$	v	+
Gas	2+	−	+$^-$	+$^-$	+^{2}	−$^+$	−^{4}	−^{2}	+$^-$	v
Acetoin	−	·	−	−$^+$	−^{w}	−^{w}	−	−	−	−
Growth in 20% bile	−	·	+^{2}	+$^-$	4$^-$	v	4$^+$	−^{2}	+	v
Lecithinase (EYA)	−	·	−	−	−	−	−	−	−	−
Lipase (EYA)	−	−	−	−	−	−	−$^+$	−	−	−
Hemolysis	b^w	·	−^{b}	b^w	−^{w}	−^{b}	b$^-$	−^{b}	−	−^{b}
Propionate ← threonine	−	−	−	−	−	−	−	−$^+$	−	−

[a] +, positive reaction—90–100% of strains; −, negative reaction—90–100% of strains; v, reaction variable within strains; hemolysis—b = beta; w = weak; gas and growth—relative amounts indicated by +, 2, 3, 4; s = stimulated. Milk—c, curd, d, digested; gelatin—w, liquefaction of chilled cultures in less than half the time of the uninoculated controls.

Any of the above used as superscripts indicates reaction of 10–40% of strains; · not tested.

Reactions determined by procedures given in Holdeman and Moore (1972).

163. (*Bacterium acidi propionici b* von Freudenreich and Orla-Jensen 1907, 532; *Propionibacterium jensenii* var. *raffinosaceum* van Neil 1928, 162; *Propionibacterium peterssonii* van Niel 1928, 163; *Propionibacterium technicum* van Niel 1928, 164; *Propionibacterium raffinosaceum* Werkman and Kendall 1931, 30; *Propionibacterium zeae* Hitchner 1932, 41. Probable synonyms (labeled strains not known to be extant): *Propionibacterium amylaceum* Sakaguchi, Iwasaki and Yamada 1941, 131; *Propionibacterium amylaceum* var. *auranticum* Sakaguchi, Iwasaki and Yamada 1941, 133; *Propionibacterium japonicum* Sakaguchi, Iwasaki and Yamada 1941, 132; *Propionibacterium pituitosum* Janoschek 1944, 336.)

jen.se'ni.i. M.L. gen.n. *jensenii* of Jensen; named for S. Orla-Jensen, the Danish bacteriologist who isolated this organism.

Description based on literature descriptions, including those of Werkman and Brown (1933), Sakaguchi *et al.* (1941) and Janoschek (1944), and on study of 13 strains including 3 of *P. jensenii* (ATCC 4867 (van Niel 24), ATCC 4868 (van Niel 29), ATCC 4869 (van Niel 1)), 2 of *P. technicum* (ATCC 14073 (van Niel E.6.1) and ISL 106 (van Niel 22)) 1 of *P. raffinosaceum* (ISL 103 (van Niel 29)), 1 of *P. peterssonii* (ATCC 4870 (van Niel 20)),

1 of *P. zeae* (ATCC 4964 (Hitchner)), and ATCC 4871. Phenotypic characteristics of these strains are similar to original descriptions of *P. jensenii*, *P. raffinosaceum*, *P. technicum*, *P. peterssonii* or *P. zeae* (Tables 17.10, 17.11 and Fig. 17.A).

Surface colonies on horse blood agar (2 days) are punctiform, circular, entire, convex to pulvinate, glistening, semi-opaque and white, cream, or pink; 1 out of 13 strains β-hemolytic. Colonies in deep agar are minute to 4 mm, lenticular and white, pink or red-brown.

Glucose broth cultures are turbid or clear with smooth, granular or ropy sediment and terminal pH of 4.4–4.9.

Anaerobic to aerotolerant or facultative. Some strains grow as well aerobically as anaerobically.

Pantothenic acid, biotin or *p*-aminobenzoic acid may be required for growth (Delwiche, 1949).

The major long chain fatty acids produced in thioglycollate cultures (Moss *et al.*, 1969) are 13-methyltetradecanoic acid (about 31%) and 12-methyltetradecanoic acid (about 17%).

Cell walls contain L-diaminopimelic acid and moderate amounts of glucose and trace to small amounts of galactose and mannose.

The G + C content of the DNA ranges from 65–68 moles % (T_m, Johnson and Cummins, 1972).

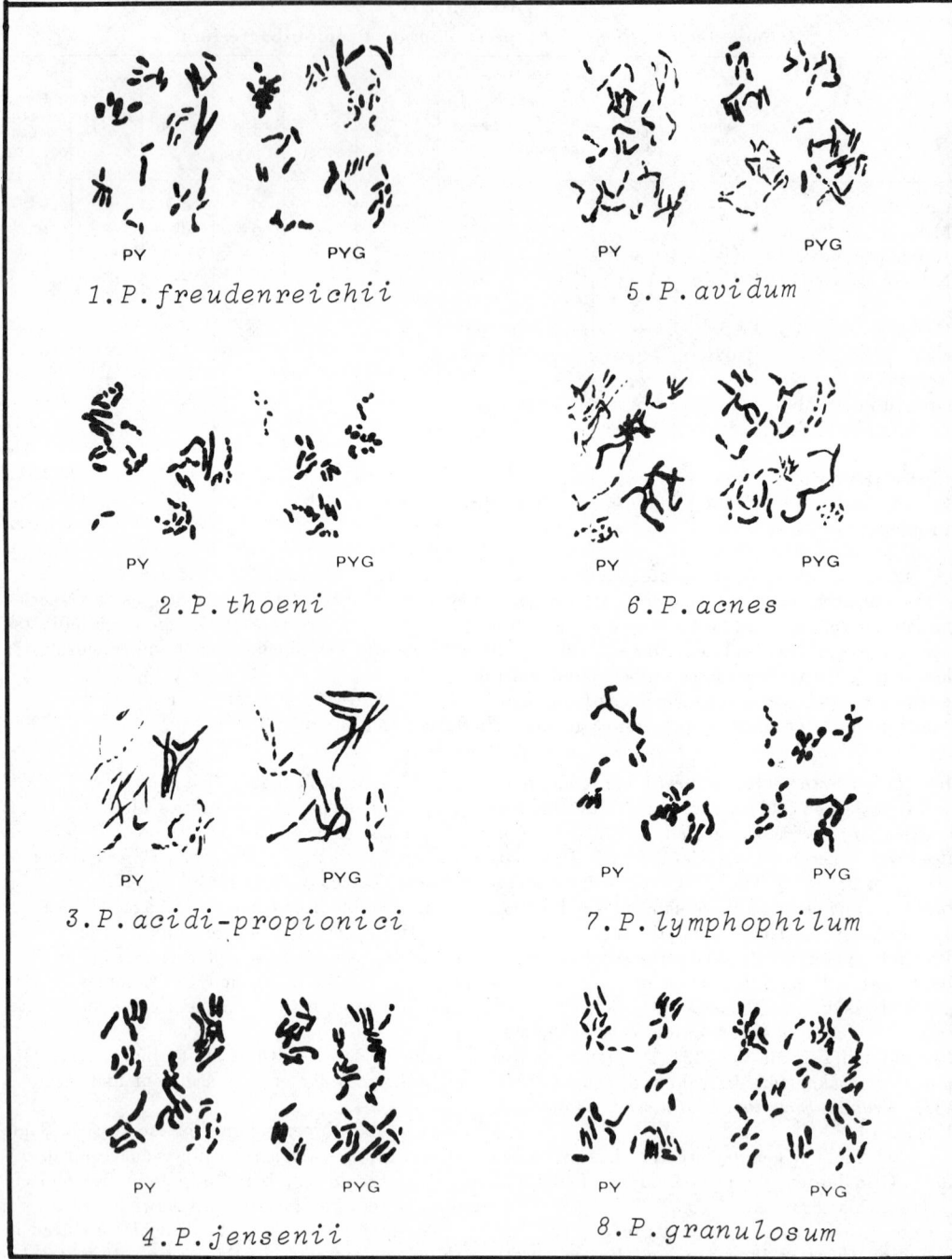

Fig. 17.A. Morphology of the species of genus *Propionibacterium*. Drawings are composites of several cultures; individual cultures usually show somewhat less variation. Scale: 9/16 inch = 10 μm. PY, peptone-yeast extract broth cultures; PYG, PY-glucose broth cultures.

Isolated from dairy products, silage and occasionally from infections.

Further Comments

All available strains in this species showed 78–100% similarity to *P. jensenii* van Niel 29, ATCC 4868 in nucleotide sequence. Strain ATCC 4964, isolated by Hitchner from silage, showed the lowest (78%) similarity. Its cultural and biochemical characteristics are similar to the other strains. Hitchner (1934) found it to be serologically distinct from other strains of *P. jensenii*.

5. **Propionibacterium avidum** (Eggerth) Moore and Holdeman 1969, 7. (*Bacteroides avidus* Eggerth 1935, 289; *Corynebacterium avidum* (Eggerth) Prévot 1938, 304; *Mycobacterium avidum* (Eggerth) Krasil'nikov 1949, 167.)

a'vi.dum. L. neut.adj. *avidum* greedy, voracious.

Description based on study of 20 strains including ATCC 25577 (Tables 17.10, 17.11 and Fig. 17.A).

Surface colonies on horse blood agar (2 days) are punctiform to 0.5 mm, circular, entire, convex to pulvinate, semi-opaque, glistening, white.

Glucose broth is clear or turbid with smooth or ropy sediment and a terminal pH of 4.1–5.1.

Anaerobic to aerotolerant. Upon initial isolation, may grow on the surface of agar incubated aerobically and usually grow in deep broth medium incubated aerobically. After several transfers, strains usually grow exceedingly well on the surface of aerobic agar plates.

Cell walls of four strains tested contain L-diaminopimelic acid and equal amounts of galactose and glucose, and equal or lesser amounts of mannose. Cell walls of two strains contain *meso*-diaminopimelic acid, no galactose and equal amounts of glucose and mannose.

The G + C content of the DNA is 62–63 moles % (T_m, Johnson and Cummins, 1972).

Isolated from blood, pus, infected wounds, brain, submaxillary abscess, other types of tissue abscesses, and feces.

Reference strain: ATCC 25577; V.P.I. 0179; CIPP (Prévot) 1689B.

6. **Propionibacterium acnes** (Gilchrist) Douglas and Gunter 1946, 22. (*Bacillus acnes* Gilchrist 1901, 425; *Bacillus parvus liquefaciens* Jungano 1908, 618; *Bacillus anaerobius diphtheroides* Massini 1914, 125; *Corynebacterium acnes* (Gilchrist) Eberson 1918, 10; *Corynebacterium parvum infectiosum* Mayer 1926, 370; *Corynebacterium adamsoni* (*sic*) Prévot 1938, 304; *Corynebacterium anaerobium* Prévot 1938, 304; *Corynebac-*

terium liquefaciens Prévot 1938, 304; NOT *C. liquefaciens* Andrewes *et al.* 1923, 408 or *C. liquefaciens* (Orla-Jensen) Jensen 1934, 49 (Murray in THE MANUAL, 7th ed.); *Corynebacterium parvum* Prévot 1940, 202; *Actinobacterium liquefaciens* Beerens and Goudaert 1952–53, 119. Probable synonyms: *Bacillus hepatodystrophicans* Kuczynski 1929, 37; *Corynebacterium hepatodystrophicans* (Kuczynski) Prévot 1938, 304; *Corynebacterium renale cuniculi* Prévot 1938, 304.)

ac'nes. Gr. n. *acme* a point; incorrectly transliterated as M.L. n. *acne* acne; M.L. gen.n. *acnes* of acne.

Description based on study of the neotype strain and 323 other similar strains (Tables 17.10, 17.11 and Fig. 17.A).

Colonies in deep agar are lenticular, minute to 4 mm, white; colonies of some strains become tan, pink or orange in 3 weeks. Surface colonies on blood (horse or rabbit) agar (2–3 days) are punctiform to 0.5 mm, circular, entire to pulvinate, translucent to opaque, white to gray, glistening. Reaction produced on egg yolk medium after incubation for 6–8 days (Werner, 1967).

Glucose broth cultures are turbid or clear with granular, flocculent or ropy sediment and terminal pH of 4.7–6.2. It is necessary to have rapid growth (heavy inoculum or Tween 80 in the medium) to obtain the lower pH values.

Anaerobic to aerotolerant. Two per cent of the strains tested grew on the surface of blood agar plates incubated aerobically; 30% grew slightly to well on the surface of blood agar plates incubated in a candle jar. Many strains grew in deep glucose broth without special provisions for anaerobiosis.

The total quantity of acid (especially the proportion of lactic acid) produced from fermentable carbohydrates is highly variable, apparently depending on both the rate and the amount of growth. Lactate is converted to propionate by most strains, but often only if the initial oxidation-reduction potential of the medium is sufficiently low or if the initial growth rate is rapid. The major long chain fatty acid produced in thioglycollate cultures (Moss *et al.*, 1969) is 13-methyltetradecanoic acid (32 to 62%).

When agar surface cultures are exposed to air for 30 min or more before testing, about one-third of the strains are catalase negative; two-thirds show slight catalase activity; a few show moderate catalase activity. Even in positive strains, catalase activity may not be detected unless anaerobic cultures are exposed to air before testing.

On the basis of cell wall composition and cell wall antigens of 62 strains of *P. acnes*, there are

Editorial Note. The authorities for the new combinations *Propionibacterium avidum* and *P. granulosum* (species 8) are confirmed in Moore and Holdeman, 1973.

TABLE 17.12

Characteristics of groups of **P. acnes** *(adapted in part from Johnson and Cummins, 1972)*

Characteristic	Group I (49 strains[a])	Group II (19 strains[b])
Agglutination		
Group I cell wall antiserum absorbed with Group II	+	−
Group II cell wall antiserum absorbed with Group I	−	+
Cell wall composition		
Galactose	+	−
Glucose	+	+
Mannose	±	±
DAP-LL	49 strains	14 strains
DAP-meso	0 strains	2 strains
Relative similarity value (DNA competition experiments)		
Group I	90–100%	90–100%
Group II	90–100%	90–100%
Acid from:		
Sorbitol	86%	0%
Glycerol	72%	60%
Galactose	65%	53%
Indole produced	98%	75%
Definite β-hemolysis[c]	18%	0%

[a] Including ATCC 6919 (neotype strain), ATCC 11827 and NCDC 554.

[b] Including ATCC 11828, ATCC 11829 and NCDC 605.

[c] Many strains produce slight β-hemolysis on extended incubation.

two groups in the species (Johnson and Cummins, 1972). Some of the morphological and cultural reactions of the strains are somewhat correlated with the cell wall groups (Table 17.12). Strains of group II are often more coccoid than those of group I and are most similar to previous descriptions of *C. parvum* and *C. adamsoni*, which are synonyms of *P. acnes*.

Pathogenic for mice (Eberson, 1918) and for guinea pigs (Mandin, 1955 in Prévot *et al.*, 1967). For a discussion of pathogenic properties of *P. acnes*, see Prévot *et al.*, 1967.

The G + C content of the DNA of both serological groups is 57–60 moles % (T_m, Johnson and Cummins, 1972).

Isolated from the normal skin (approximately 10^4/sq. in.), intestinal contents, wounds, blood, pus, soft tissue abscesses.

Neotype strain: Ponsonby; NCTC 737; ATCC 6919 (Zierdt *et al.*, 1968).

Further Comments

Strains of *P. acnes* are the most common contaminants in anaerobic cultures. This creates a special problem when studying known strains of the species, since it is usually impossible to distinguish the contaminant with certainty. The organism is also a common contaminant of clinical specimens, presumably from the skin of the patient or attendant. However, in several well documented clinical infections, *P. acnes* has been isolated repeatedly from blood and bone marrow in chronic infections. In some abscess material *P. acnes* may be the only organism isolated or may be present, along with other organisms, as one of the predominant organisms present. A high percentage of the clinically normal adult human population carries antibody activity against *P. acnes* (Pasteur Institute (Paris), unpublished data).

7. **Propionibacterium lymphophilum** (Torrey) Johnson and Cummins 1972, 1057. (*Bacillus lymphophilus* Torrey 1916, 79; *Corynebacterium lymphophilum* (Torrey) Eberson 1918, 23; *Mycobacterium lymphophilum* (Torrey) Krasil'nikov 1949, 166.)

lym.pho'phi.lum. Gr. n. *lympho* lymph; Gr. neut.adj. *philum* loving. M.L. adj. *lymphophilum* lymph-loving.

Description based on study of two strains including CIPP (Prévot) 1519F (V.P.I. 0202) (Tables 17.10, 17.11 and Fig. 17.A).

Surface colonies on horse blood in 4 days are punctiform to 0.5 mm, circular, entire, convex to pulvinate, white, glistening, smooth.

Glucose broth cultures (24 hr) are turbid, becoming clear, with a ropy sediment and terminal pH of 5.4–5.7.

Anaerobic, producing no growth on agar surface incubated aerobically but growth develops in deep broth incubated aerobically.

Cell walls contain lysine (but no diaminopimelic acid) and galactose, glucose and mannose. Cell wall material is antigenically distinct from the other species in the genus.

The G + C content of the DNA is 53–54 % (T_m, Johnson and Cummins, 1972).

Isolated from urinary tract infections (CIPP (Prévot) 1519F, V.P.I. 0202) and mesenteric ganglion of a monkey inoculated with *Actinobacterium* (CIPP (Prévot) SB, V.P.I. 0383). Strains originally described were from lymph glands in Hodgkins disease.

8. **Propionibacterium granulosum** (Prévot) Moore and Holdeman 1970, 15. (*Bacille granuleux*

Jungano 1909, 122; *Corynebacterium granulosum* Prévot 1938, 304; *Corynebacterium pyogenes bovis* Prévot 1940, 204. According to the description of *C. pyogenes bovis* (Prévot *et al.* 1967, 1749), this organism could not be distinguished from some strains of *P. granulosum* as described in this Manual.)

gra.nu.lo'sum. L. neut.adj. *granulosum* full of granules.

Description based on study of the reference strain and 36 other strains with similar characteristics (Tables 17.10, 17.11 and Fig. 17.A).

Surface colonies on horse blood agar (2–3 days) are minute to 1 mm, circular, entire, convex to pulvinate, white or gray-white, translucent, glistening.

Glucose broth cultures are turbid or clear with a flocculent, granular or ropy sediment and terminal pH of 5.0–5.3.

Anaerobic to aerotolerant. Available strains did not grow on the surface of blood agar plates incubated aerobically; however, some strains grew in deep broth cultures incubated aerobically.

Cell walls contain L-diaminopimelic acid and moderate amounts of galactose, small amounts of mannose and none to traces of glucose.

The G + C content of the DNA is 61–63 moles % (T_m, Johnson and Cummins, 1972).

Isolated from the intestinal tract and from abscesses.

Reference strain: ATCC 25564; V.P.I. 0507 (isolated as a contaminant from a culture labeled *Staphylococcus aerogenes*).

Nomen dubium

The original strain of this species has been lost.

a. *Corynebacterium diphtheroides* (Jungano) Prévot 1938, 304. (*Bacille diphthéroïde* Jungano 1909, 113.)

The original description by Jungano (1909, 114) is of an organism isolated from the rat intestinal tract. The organism is morphologically similar to propionibacteria or to *Eubacterium alactolyticum*. According to Jungano, it produced much gas, did not liquefy gelatin or coagulate milk or digest egg white. It did not ferment lactose, sucrose or dextrin. The pH in glucose cultures after 14 days was 6.17 ("evaluated with H_2SO_4 as pH 1" (transl.)). It produced indole and was not pathogenic for guinea pigs or rabbits.

Many strains of *P. acnes* fit this description except that they produce only moderate amounts of gas (which, however, does break deep agar columns). *P. acnes* is part of the normal flora of intestinal tracts.

CIPP (Prévot) strains 379 and 1626, specifically mentioned in the description of *C. diphtheroides* by Prévot *et al.* (1967, 766), are *Eubacterium alactolyticum* and are indole negative. None of the 34 strains of *E. alactolyticum* available to us was isolated from intestinal contents.

Thus, the identity of Jungano's organism with *P. acnes* depends on his meaning of "much gas." If Jungano's test for indole was not reliable (gave false positives), the organism could be *E. alactolyticum*; however, it is unlikely that he would have found *E. alactolyticum* in intestinal contents.

Genus II. **Eubacterium** *Prévot 1938, 294*

LILLIAN V. HOLDEMAN AND W. E. C. MOORE

Eu.bac.te'ri.um. Gr. pref. *eu-* good-, well-, beneficial (*not* as opposed to pseudo-); Gr. neut.dim.n. *bakterion* a small rod; M.L. neut.n. *Eubacterium* beneficial bacterium.

Gram-positive, obligately anaerobic, non-spore-forming rods, uniform or pleomorphic, non-motile or motile. Chemoorganotrophs, saccharoclastic or non-saccharoclastic. Usually produce mixtures of organic acids from carbohydrates or peptone, often including large amounts of butyric, acetic, or formic acids. Do not produce:

1. Propionic as a major acid product (see *Propionibacterium*).

2. Lactic as the sole major acid product (see *Lactobacillus*).

3. Succinic (in the presence of CO_2) and lactic with small amounts of acetic or formic acids (see *Actinomyces*).

4. Acetic and lactic (acetic $\geq$ lactic), with or without formic, as the sole major acid products (see *Bifidobacterium*).

The species and strains within species vary in sensitivity to oxygen; some can be cultured only in pre-reduced media.

Catalase usually not produced (trace amounts detected in some strains); hippurate usually not hydrolyzed.

Growth usually most rapid at 37 C and pH near 7.

Found in cavities of man and other animals, animal and plant products, infections of soft tissue, and soil. Some species may be pathogenic.

Characteristics by which species in the genus can be differentiated are given in Tables 17.13–17.17.

The information on the volatile acids produced by many of the species in Table 17.17 was obtained by Duclaux distillation, a method not as reliable as the chromatographic method used for the species in Tables 17.13 and 17.15. Other characteristics are given in the text. Cellular morphology is illustrated in Figures 17.B and 17.C. Methods presented in Holdeman and Moore (1972) were used in the cultural, biochemical and chromatographic characterization of all strains listed by us.

All descriptions are based on strains in which no spores have been detected. Heat-resistant spores, often difficult to detect, have been found in organisms similar to some described species, especially motile or filamentous species; these strains are members of the genus *Clostridium*.

Type species: *Eubacterium foedans* (Klein) Prévot 1938, 294. No strains conforming to the original description have been isolated since Klein's report in 1908.

Further Comments

We are indebted to Dr. A. R. Prévot, Pasteur Institute, Paris for strains of many species described in this genus. The Prévot Collection of anaerobes was obtained with the assistance of The American Type Culture Collection.

Key to species 1–20 of the genus **Eubacterium**

I. No growth at 37 C; growth at 20 C
  1. *E. foedans*
II. Growth at 37 C.
  A. Butyric acid produced.
    1. Caproic acid produced.
      2. *E. alactolyticum*
    2. Caproic acid not produced.
      a. Mannitol acid.
        b. Raffinose acid.
          3. *E. rectale*
        bb. Raffinose not acid.
          c. Erythritol acid.
            4. *E. limosum*
          cc. Erythritol not acid.
            5. *E. ruminantium*
      aa. Mannitol not acid.
        b. Indole produced.
          6. *E. saburreum*
        bb. Indole not produced.
          c. Fructose acid.
            d. Mannose acid.
              e. Esculin hydrolyzed.
                f. Maltose acid.
                  g. Nitrate reduced.
                    h. Glycogen acid.
                      7. *E. budayi*
                    hh. Glycogen not acid.
                      8. *E. nitritogenes*
                  gg. Nitrate not reduced.
                    h. Starch hydrolyzed.
                      3. *E. rectale*
                    hh. Starch not hydrolyzed.
                      9. *E. ventriosum*
                ff. Maltose not acid.
                  g. Lactose acid.
                    10. *E. multiforme*
                  gg. Lactose not acid.
                    11. *E. cylindroides*

ee. Esculin not hydrolyzed.

> 12. *E. moniliforme*

dd. Mannose not acid.

e. Starch acid.

f. Starch hydrolyzed.

> 3. *E. rectale*

ff. Starch not hydrolyzed.

> 5. *E. ruminantium*

ee. Starch not acid.

> 13. *E. tortuosum*

cc. Fructose not acid.

d. Sucrose acid.

> 14. *E. cellulosolvens*

dd. Sucrose not acid.

> 15. *E. combesii*

B. Butyric acid not produced.

1. Indole produced.

> 16. *E. tenue*

2. Indole not produced.

a. Fructose acid.

b. Maltose acid.

c. Inositol acid.

> 17. *E. fissicatena*

cc. Inositol not acid.

d. Amygdalin acid.

> See *Clostridium ramosum*

dd. Amygdalin not acid.

e. Rhamnose acid.

> 18. *E. contortum*

ee. Rhamnose not acid.

> 19. *E. aerofaciens*

bb. Maltose not acid.

> See *Lachnospira multiparis*

aa. Fructose not acid.

> 20. *E. lentum*

See Table 17.17 for key-table of other species in the genus.

Clostridium ramosum includes organisms formerly recognized as *Eubacterium filamentosum* (*Catenabacterium filamentosum*); most of these strains do not sporulate readily and are not readily recognized as clostridia.

Description of the species of genus **Eubacterium**

1. Eubacterium foedans (Klein) Prévot 1938, 294. (*Bacillus foedans* Klein 1908, 1834.)

foe′dans. L. part.adj. *foedans* making foul or filthy.

No known strains are extant. Description from Klein, 1908. No organisms matching this description have been reported since Klein's original report. Prévot *et al.* (1967) list characteristics of a strain from African soil that appears to differ in several respects. We were unable to isolate strains with characteristics of *E. foedans* from spoiled hams.

Cells usually 0.4 by 3–5 μm (length varies from 1.6–14 μm), straight or curved with rounded ends, often in long chains (Fig. 17B); non-motile.

Optimum temperature 20 C; no growth at 10 C or 37 C.

In glucose broth gas, fetid odor, alkaline reaction and cottony sediment. Gas and strong fetid odor produced in pork glucose or gelatin glucose media.

Gelatin glucose liquefied in 8 weeks at 20 C. Milk not coagulated but strongly fetid.

Not pathogenic for guinea pigs (subcutaneous inoculation).

Isolated from putrefied ham.

2. Eubacterium alactolyticum (Prévot and Taffanel) Holdeman and Moore 1970, 23. (*Ramibacterium alactolyticum* Prévot and Taffanel 1942, 261; *Ramibacterium dentium* Vinzent and Reynes 1947, 595; *Ramibacterium pleuriticum* Prévot, Raynaud and Digeon 1947, 483.)

a.lac.to.ly'ti.cum. Gr. pref. *a* not; L. n. *lac, lactis* milk; Gr. adj. *lyticus* dissolving; M.L. neut. adj. *alactolyticum* not milk digesting.

Description from Holdeman *et al.* (1967, 323) and Holdeman and Moore (1972, 39) who studied neotype strain and 37 other strains with similar characteristics (Tables 17.13, 17.14 and Fig. 17.B).

Surface colonies on horse blood agar (2–3 days) punctate to 0.5 mm, circular, entire, convex to pulvinate, smooth, shiny. Smaller colonies may be translucent, larger colonies usually are opaque.

Glucose broth cultures are usually turbid with granular or smooth (occasionally ropy or flocculent) sediment; a few strains produce sediment without turbidity in broth cultures. Terminal pH in glucose cultures usually 5.0–5.6; that of a few strains is 5.8–6.0.

Most strains grow at 30 C; some strains grow at 25 C and 45 C.

Cell walls contain *meso*-diaminopimelic acid.

Isolated from dental tartar; various kinds of infections including purulent pleurisy, jugal cellulitis, postoperative wounds; abscesses of the brain, lung, intestinal tract, and mouth.

Neotype strain: Prévot collection DO-4; ATCC 23263 (Holdeman *et al.*, 1967, 323).

3. Eubacterium rectale (Hauduroy *et al.*) Prévot 1938, 294. (*Bacteroides rectalis* Hauduroy, Ehringer, Urbain, Guillot and Magrou 1937, 72; *Pseudobacterium rectale* (Hauduroy *et al.*) Krasil'nikov 1949, 230.)

rec.ta'le. M.L. n. *rectum* the straight bowel, rectum; M.L. neut.adj. *rectale* rectal.

Description from Prévot *et al.* (1967) and Holdeman and Moore (1972) who studied 23 strains (Tables 17.13, 17.14 and Fig. 17.B).

Surface colonies on horse blood agar (2 days) or agar in roll tubes 0.5–2 mm, circular-irregular, entire-scalloped, convex, translucent, may be mottled when viewed by obliquely transmitted light; non-hemolytic (hemolytic, Grootten, 1929). Some strains will not grow on the surface of blood agar plates incubated in an anaerobe jar.

Glucose broth cultures turbid with smooth or flocculent sediment and terminal pH of 4.7–5.5, usually around 5.0.

Most strains grow at 25–45 C.

Meso-diaminopimelic acid detected in three strains tested.

Isolated from rectal abscess (Grootten) and from feces.

Reference strain: ATCC 25578.

4. Eubacterium limosum (Eggerth) Prévot 1938, 295. (*Bacteroides limosus* Eggerth 1935, 290; *Butyribacterium rettgeri* Barker and Haas 1944, 303; *Mycobacterium limosum* (Eggerth) Krasil'nikov 1949, 166; *Butyribacterium limosum* (Eggerth) Moore and Cato 1965, 79.)

li.mo'sum. L. neut.adj. *limosum* full of slime, slimy.

Description from Holdeman and Moore (1972) who studied type strain and 49 similar strains (Tables 17.13, 17.14 and Fig. 17.B).

Surface colonies on horse blood agar (2 days) punctiform (2 mm), circular, entire, convex, translucent to slightly opaque, sometimes with mottled appearance when viewed by obliquely transmitted light.

Glucose broth cultures turbid with stringy or smooth sediment and terminal pH of 4.5–5.2.

Carbon dioxide produced in peptone-yeast extract-glucose (PYG) culture in addition to acids listed. Products in peptone-yeast extract (PY) cultures are acetate, isobutyrate, butyrate (small amounts) and isovalerate, sometimes with traces of propionate.

Synthesizes vitamin B_{12}.

Some strains grow equally well at 25, 37 and 45 C.

Isolated from feces of man, intestinal contents of rats, poultry and fish, various infections (rectal abscesses, blood, wounds), and mud.

Type strain: Eggerth; ATCC 8486.

5. Eubacterium ruminantium Bryant 1959, 140.

ru.mi.nan'ti.um. M.L. pl.n. *ruminantia* ruminants; M.L. gen.pl.n. *ruminantium* of ruminants.

Description from Bryant (1959) who studied 20 strains and from Holdeman and Moore (1972) who studied 2 strains (Tables 17.13, 17.14 and Fig. 17.B).

Cells are Gram-positive in 16-hr cultures, becoming Gram-negative; not encapsulated.

Deep agar colonies are lenticular and do not produce gas. Surface colonies on rumen fluid-glucose-cellobiose agar are entire, low convex, smooth, translucent to opaque, light buff colored.

Glucose broth cultures turbid in 18 hr with

Editorial Note. The authorities for the new combinations *Eubacterium alactolyticum*, *E. saburreum* (species 6), *E. budayi* (species 7), *E. multiforme* (species 10), *E. cylindroides* (species 11), *E. moniliforme* (species 12), *E. cellulosolvens* (species 14), *E. combesii* (species 15) and *E. tenue* (species 16) are confirmed in Moore and Holdeman, 1973.

TABLE 17.13

Fermentation reactions of species of genus **Eubacterium,** *species 1–10[a]*

	1. E. foedans	2. E. alacto-lyticum	3. E. rectale	4. E. limosum	5. E. rumi-nantium	6. E. saburreum	7. E. budayi	8. E. nitrit-ogenes	9. E. ventri-osum	10. E. multi-forme
Products from peptone-yeast extract-glucose	·	Abc(F c'sl)	LB(sfa)	LAb(sp)	LFb(ap)	albs24	Lba	LBAs (fp)	Fbla(s)	LAB (fsp)
Products from lactate	·	Bᶜ⁻	—	AB⁻	—	—	—	⁻B	—	APB
Products from pyruvate	·	AFBC⁻	AB⁻	ABiv⁻	—	ABL	AF⁻	ABᶠ	AF⁻	AB
Acid produced from:										
Adonitol	·	d	—⁺	+⁻	—	—	—	—	—	—
Amygdalin	·	—⁺	d	—	—	—	—ʷ	—ʷ	—⁺	—
Arabinose	·	—ʷ	+ʷ	w⁻	+	w	w	—	w⁻	—
Cellobiose	·	—	+ʷ	—ʷ	d	—	+	+⁻	d	—ʷ
Dulcitol	·	—	—	—	—	—	—	—	—	—
Erythritol	·	—ʷ	—ʷ	+ʷ	—	w	—	—	—ʷ	—
Esculin	·	—ʷ	—⁺	—ʷ	—	—	—ʷ	—ʷ	d	—
Esculin hydrol.	·	—	+	+	+	+	+	+	+	+
Fructose	·	+	+	+	w	w	+	+ʷ	+ʷ	+ʷ
Galactose	·	—ʷ	+	—⁺	—	w	w	d	+	—⁺
Glucose	—	+ʷ	+	+	+	+	+	+	+	+
Glycerol	·	—	—	—	—	—	—	d	—ʷ	—
Glycogen	·	—	+⁻	—	—	—	w	—	—ʷ	—
Inositol	·	—	—	—	—	—	—	—	—ʷ	—
Inulin	·	—	+⁻	—	—	—	—	—	—	—
Lactose	·	—	+ʷ	—ʷ	d	w	w⁺	—ʷ	+⁻	w⁺
Maltose	·	—	+ʷ	—⁺	d	w	w	+ʷ	+	—
Mannitol	·	+⁻	—⁺	+ʷ	—ʷ	—	—	—	—	—
Mannose	·	—ʷ	d	—⁺	—	w	+	+	+ʷ	+
Melezitose	·	—ʷ	+⁻	—	—	—	—	—	—ʷ	—
Melibiose	·	—	+⁻	—	—	—	—ʷ	—	d	—
Raffinose	·	—	+ʷ	—	—	w	—	—	d	—
Rhamnose	·	—	—⁺	—ʷ	—	—	—	—	d	—
Ribose	·	—⁺	d	+ʷ	·	—	w⁺	—⁺	d	—
Salicin	·	—	+⁻	—ʷ	d	—	w⁻	d	—ʷ	—
Sorbitol	·	—⁺	+⁻	—ʷ	—	—	—	—	—	—
Sorbose	·	—	—⁺	—ʷ	—	—	—⁺	w⁻	—	—
Starch	·	—	+ʷ	—ʷ	+	w	w⁺	—ʷ	w⁺	—
Starch hydrol.	·	—	+	—⁺	—	—	+⁻	—	—	—
Sucrose	·	—	+ʷ	—ʷ	d	+	—	—ʷ	d	—
Trehalose	·	—	—⁺	—	—	w	—	—⁺	—	—ʷ
Xylose	·	—ʷ	+	d	d	w	—ʷ	—	w⁺	—

[a] Products from PYG broth, lactate and pyruvate: Capital letters indicate $\geq$ 1 meq acid/100 ml broth; small letters < 1 meq/100 ml. A, acetic; B, butyric; C, caproic; C′, caprylic; F, formic; iB, isobutyric; iv, isovaleric; L, lactic; P, propionic; S, succinic; 2, ethanol; 4, butanol; ( ) products of occasional strains. +, strong acid—pH 5.5 or below—for 90–100% of strains; —, negative reaction for 90–100% of strains; w, weak acid—pH 5.5–6.0—for 90–100% of strains. (Uninoculated xylose and arabinose, under CO_2, are often pH 5.9; for cultures in these media pH 5.4–5.7 are considered weak acid and pH below 5.4 strong acid.); d, reaction positive for 40–60% of strains.
When any of the above used as superscripts, reaction of 10–40% of strains.

smooth or ropy sediment and terminal pH of 5.0–5.5.

Nine of 20 strains fermented xylan; none fermented gum arabic.

Optimum growth temperature is 30–37 C, no growth at 22 or 50 C. Less growth produced when 0.5% yeast extract and 1.5% trypticase replaced rumen fluid in the medium.

Isolated from bovine rumen contents where it represents up to 7.3% of the total isolates.

Type strain: Bryant, GA195.

Further Comments

Bryant described two biotypes, with eight strains in biotype 1 and five in biotype 2. The other seven strains seemed to be intermediates between the two biotypes.

TABLE 17.14

Biochemical reactions of species of genus **Eubacterium**, *species 1–10[a]*

	1. E. foedans	2. E. alactolyticum	3. E. rectale	4. E. limosum	5. E. ruminantium	6. E. saburreum	7. E. budayi	8. E. nitritogenes	9. E. ventriosum	10. E. multiforme
Gelatin digested	+	−	−w	−+	w	−	−	−w	−	+
Milk	−	−c	c−	−	a	c	−	−	c−	cd
Meat	·	−	−	−	−	−	−	−	−	−
Indole produced	·	−	−	−	−	+	−	−	−	−
Nitrate reduced	·	−	−	−	d	−	+	+	−	+
Neut. red reduced	·	+	+	+	−	+	+	d	+	+
Gas	4	3−	+3	+4	+−	3	4	42	2−	4
NH₃	·	−+	−w	+−	−	+	+	+w	−w	+
Acetoin	·	−	−+	+−	w−	−	−	d	d	−w
Growth in 20% bile	·	+−	+3	4+	+2	−	s4	2+	24	2
Growth in 6.5% NaCl	·	d	−	−w	−	−	−	−	−+	−
Growth at pH 8.5	·	d	d	+−	w	·	+	+−	−+	+
Lecithinase (EYA)	·	−	−	−	−	−	+	−w	−	−
Lipase (EYA)	·	−	−	−	·	−	−	−	−	−
Hemolysis	·	−	−	−b	·	a	−b	−b	−	b−
Motility	−	−	d	−	−	−	−	−	−	+
Threonine ← propionate	·	−w	−	−+	−	−	+−	+	−+	+

[a] +, positive reaction—90–100% of strains; −, negative reaction—90–100% of strains; d, reaction + for 40–60% of strains; hemolysis—a = alpha; b = beta; w = weak; gas and growth—relative amounts indicated, by +, 2, 3, 4; s = stimulated; EYA—McClung-Toabe egg yolk agar. Milk—c, curd, d, digested; gelatin—w, liquefaction of chilled cultures in less than half the time of the uninoculated controls; meat—d, digested.

Any of the above used as superscripts indicates reaction of 10–40% of strains.

Reactions determined by procedures given in Holdeman and Moore (1972).

6. **Eubacterium saburreum** (Prévot) Holdeman and Moore 1970, 23. (*Catenabacterium saburreum* Prévot 1966, 171; *Leptotrichia aerogenes* Hofstad 1967, 548.)

sa.bur′re.um. L. n. *saburra* sand; M.L. neut.adj. *saburreum* sandy.

Nineteen strains were isolated and described but not named by Theilade and Gilmour. Description from Theilade and Gilmour (1961), Hofstad (1967) and study of one strain (Tables 17.13, 17.14 and Fig. 17.B).

Surface colonies on blood agar are 1–4 mm, flat, with interlaced filamentous rhizoid edges and small slightly raised granular centers which penetrate the agar. Isolated strains do not grow in 5% serum-infusion medium broth until the 10th or 12th transfer on agar. Blood agar is required for continued subculture.

Cultures in broth produce large, light, hairy granules that adhere firmly to the sides of the tube. Terminal pH in culture medium with a fermentable carbohydrate is 4.5–5.5.

α-Methyl glucoside is fermented, sometimes only weakly.

Produces moderate amounts of CO_2 (personal communication, E. Theilade).

Cell walls contain glucose and rhamnose, no galactose.

Isolated from the mouth (*materia alba*).

7. **Eubacterium budayi** (Le Blaye and Guggenheim) Holdeman and Moore 1970, 23. (*Bacillus cadaveris butyricus* Buday 1898, 374; *Bacterium budayi* Le Blaye and Guggenheim 1914, 402; *Eubacterium cadaveris* Prévot 1938, 295; *Pseudobacterium cadaveris* (Prévot) Krasil'nikov 1949, 230.)

bu′day′i. M.L. gen.n. *budayi* of Buday; named for the bacteriologist who first isolated the organism.

Description from Prévot *et al.* (1967) and study of Prévot collection ECI isolated from pond mud (Tables 17.13, 17.14 and Fig. 17.B).

Surface colonies on McClung-Toabe egg yolk agar (2 days) are 4–5 mm, irregular, low convex, scalloped, yellowish and are surrounded by a small opaque zone in the agar. Colonies in deep agar look like snowflakes, often with rhizoids.

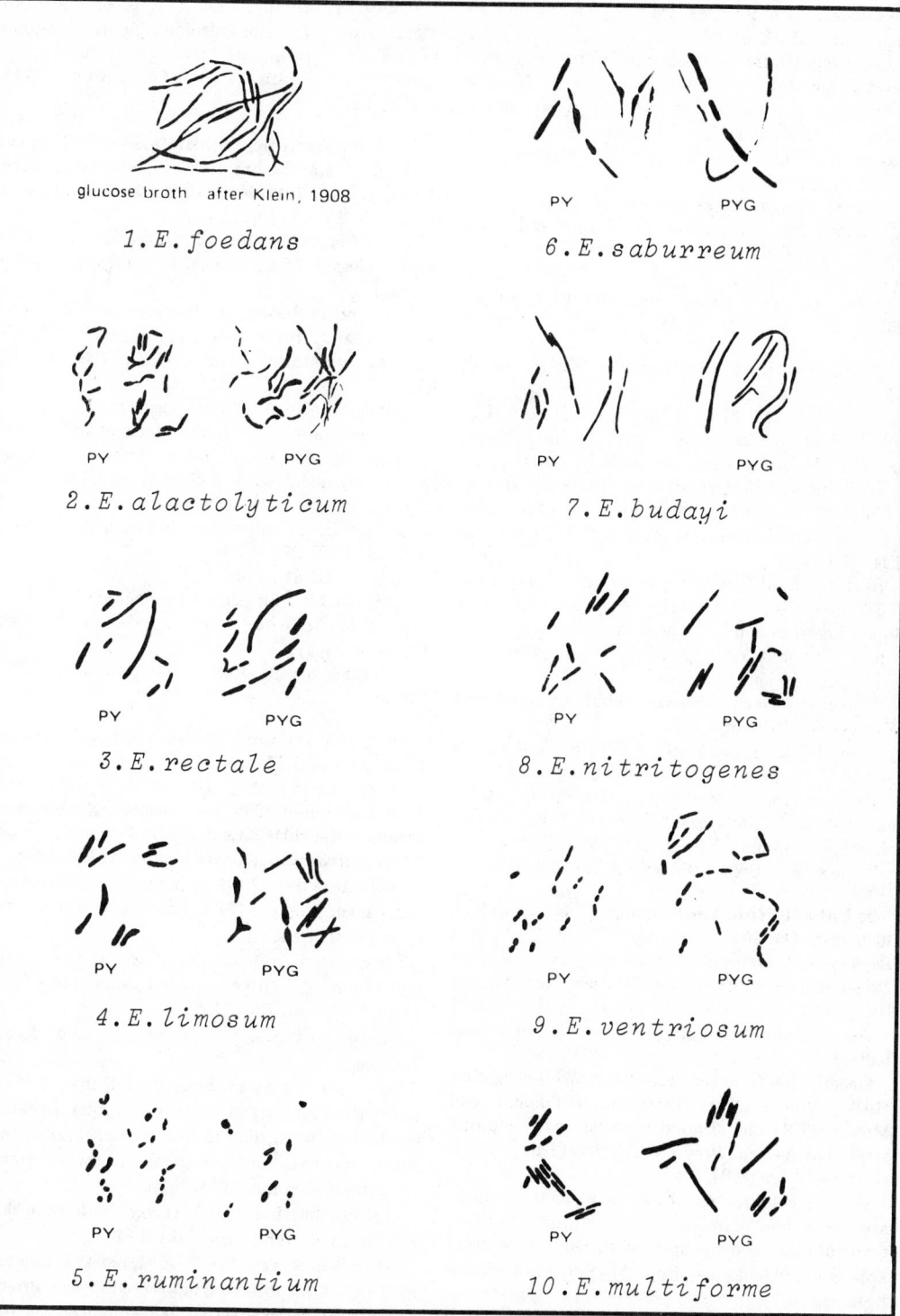

Fig. 17.B. Morphology of species 1 to 10 of the genus *Eubacterium*. Drawings are composites of several cultures; individual cultures usually show less variation. Scale: 9/16 inch = 10 μm. PY, peptone-yeast extract broth cultures; PYG, PY-glucose broth cultures.

Glucose broth cultures are turbid with smooth sediment and pH of 5.

Gelatin partially liquefied, slowly (Prévot *et al*.); gelatin not liquefied (our laboratory). Nitrate reduced to nitrite in the presence of lactose or galactose (Prévot *et al*.). Milk not changed (our laboratory); milk curdled with slight retraction of the curd (Prévot *et al*.)

Growth most rapid at 37–45 C.

Isolated from a cadaver by Buday and from poorly sterilized catgut, mud and soil by Prévot *et al*.

Reference strain: Prévot collection ECI; ATCC 25541.

8. Eubacterium nitritogenes Prévot 1940, 355.

ni.tri.to'ge.nes. M.L. n. *nitritum* nitrite; M.L. verbal suff. *-genes* from Gr. v. *gennaio* beget, produce; M.L. adj. *nitritogenes* nitrite-producing.

Description from Prévot *et al*. (1967) and Holdeman and Moore (1972) who studied the type strain and three similar strains (Tables 17.13, 17.14 and Fig. 17.B).

Surface colonies on horse blood agar (2 days) are 0.5–2 mm, circular–slightly irregular with scalloped edge, low convex, translucent–opaque, sometimes with mottled appearance when viewed by obliquely transmitted light.

Glucose broth cultures are turbid with sediment and terminal pH of 5.1–5.3.

Growth most rapid at pH of 6.5–7.8. Grows at 30 and 45 C; three out of four strains grow at 25 C.

Isolated from peptic digest of meat (pH 2.5), antarctic soil, intestinal contents of carp, canned cheese, and human infections.

Type strain: Prévot TA20A; ATCC 25547.

9. Eubacterium ventriosum (Tissier) Prévot 1938, 295. (*Bacillus ventriosus* Tissier 1908, 204; *Bacteroides ventriosus* (Tissier) Eggerth 1935, 281; *Pseudobacterium ventriosum* (Tissier) Krasil'nikov 1949, 232.)

ven.tri.o'sum. L. neut. adj. *ventriosum*, pot-bellied.

Description from Eggerth (1935), Weinberg *et al*. (1937), Prévot *et al*. (1967) and Holdeman and Moore (1972) who studied nine strains, including strain 1736A of the Prévot collection (Tables 17.13, 17.14 and Fig. 17.B).

Surface colonies on horse blood agar (2 days) are 0.5–3 mm, circular, entire-diffuse, convex, translucent, smooth, shiny, with slightly mottled appearance when viewed by obliquely transmitted light.

Glucose broth cultures turbid with smooth, ropy or granular sediment and terminal pH of 4.6–5.4.

Grows at 30 and 45 C, most strains grow at 25 C.

Isolated from human and dog feces, mouth abscess, neck infection, purulent pleurisy, pulmonary abscess and material from a bronchiectasis.

Reference strain: Prévot collection 1736A; ATCC 25579.

10. Eubacterium multiforme (Distaso) Holdeman and Moore 1970, 23. (*Bacillus multiformis* Distaso 1911, 101; *Cillobacterium multiforme* (Distaso) Prévot 1938, 297.)

mul.ti.for'me. L. adj. *multus* much, many; L. n. *forma* shape; M.L. neut.adj. *multiforme* many-shaped.

Description based on Distaso (1911), Prévot *et al*. (1967), Holdeman and Moore (1972) who studied two strains (Tables 17.13, 17.14 and Fig. 17.B).

Cells motile with peritrichous flagella.

Surface colonies on horse blood agar (2 days) 1–2 mm, circular, erose, convex, translucent, gray-white, smooth, with mosaic appearance when viewed by obliquely transmitted light.

Glucose broth cultures turbid with smooth sediment and pH of 5.3–5.6.

Grows well at 25 and 45 C.

Meso-diaminopimelic acid detected.

Isolated from dog feces and soil of the Ivory Coast (Africa).

Reference strain: Prévot collection 06A; ATCC 25552.

11. Eubacterium cylindroides (Rocchi) Holdeman and Moore 1970, 23. (*Bacterium cylindroides* Rocchi 1908, 479; *Ristella cylindroides* (Rocchi) Prévot 1938, 292; *Pseudobacterium cylindroides* (Rocchi) Krasil'nikov 1949, 243; *Bacteroides cylindroides* (Rocchi) Kelly 1957, 433.)

cy.lin.dro.i'des. Gr. n. *cylindrus* a cylinder; Gr. n. *idus* form, shape; M.L. neut.adj. *cylindroides* cylinder-shaped.

Description from Rocchi (1908) and Holdeman and Moore (1972) who studied 23 strains (Tables 17.15, 17.16 and Fig. 17.C).

Gram-positive only in young cultures; destain very easily.

Surface colonies on horse blood agar (2 days) punctate (2 mm), circular to slightly irregular, entire to diffuse, flat to low convex, translucent, sometimes with mottled appearance when viewed by obliquely transmitted light.

Glucose broth cultures turbid with smooth or ropy sediment and terminal pH of 4.8–5.5.

Growth most rapid at 37–45 C; most strains grow at 25 C. Glucose or fructose enhances growth. Growth of some strains also enhanced by 0.1% Tween 80 or 5% rumen fluid.

Isolated from human feces.

Further Comments

This species was isolated from feces and described by Rocchi in 1908 as an anaerobic, Gramnegative bacillus that stains unevenly and grows at 38 C. Rocchi's strains are not known to be extant. Until recently, isolation had not been reported since Rocchi's original work possibly because Rocchi's original report of optimum temperature had been erroneously quoted as 18 C. Since the cells decolorize easily, we believe that the organisms described here probably are representative of *B. cylindroides* Rocchi; their characteristics are very similar to those originally described for the species.

12. **Eubacterium**   **moniliforme**   (Repaci) Holdeman and Moore 1970, 23. (*Bacillus moniliforme* Repaci 1910, 412; *Bacillus repazii* Herter 1917, 750; *Cillobacterium moniliforme* (Repaci) Prévot 1938, 296.)

mo.ni.li.for'me. L. n. *monile, monilis* a necklace; L. n. *forma* shape; M.L. neut.adj. *moniliforme* necklace-shaped.

Description from Weinberg *et al.* (1937), Prévot *et al.* (1967) and Holdeman and Moore (1972, 39) who studied 11 strains, including Prévot collection 2055 (Tables 17.15, 17.16 and Fig. 17.C).

Cells of young cultures motile. With Leifson's flagella stain, six strains had peritrichous flagella;

TABLE 17.15

Fermentation reaction of species of genus **Eubacterium**, *species 11–20*[a]

	11. E. cylindroides	12. E. moniliforme	13. E. tortuosum	14. E. cellulosolvens	15. E. combesii	16. E. tenue	17. E. fissicatena	18. E. contortum	19. E. aerofaciens	20. E. lentum
Products from peptone-yeast extract-glucose	Lb(asf)	LBa (fsp2)	Lbs(af)	Lfsa (bp)	AB *ibiv*p	Af(lsp *ibiv*)	AF2(s)	Af(sl2)	LAF (s2)	(afsl)
Products from lactate	–	_A	–	–	_AB		–	–		–
Products from pyruvate	AB^F–	AB^Fl	ABFL–	–	A^B	AF–	AF^L2	AF^2	A^FL–	_AF
Acid produced from:										
Adonitol	–	_w	–	–	–	–	–	–	–	–
Amygdalin	–	–	w–	w	–	–	–	–	–	–
Arabinose	–	–	_w	–	–	–	_+	+^w	_+	–
Cellobiose	_w	–	d	+^w	–	–	–	+–	+–	–
Dulcitol	–	–	–	–	–	–	–	w–	–	–
Erythritol	–	–	–	–	–	–	–	–	–	–
Esculin	_w	–	–	+^w	–	–	–	+–	–	–
Esculin hydrol.	+	–	+	+	+	–	+	+	+–	–
Fructose	+^w	+	w+	w–	–	_w	+^w	+^w	+	–
Galactose	_+	d	w+	+–	–	–	+^w	+^w	+^w	–
Glucose	+	+^w	+	+	+^w	w	+	+	+	–
Glycerol	–	–	–	–	–	–	–	w	–	–
Glycogen	–	–	–	–	–	–	–	+	–	–
Inositol	–	–	–	–	–	–	+	–	–	–
Inulin	d	–	_+	+–	–	–	_+	d	–	–
Lactose	–	–	d	w–	–	–	_+	d	+–	–
Maltose	–	d	w+	+–	_w	_w	w+	+	+^w	–
Mannitol	–	–	–	–	–	–	_w	–	–	–
Mannose	+^w	+	–	–	–	–	w–	d	+	–
Melezitose	–	–	–	–	–	–	–	–	–	–
Melibiose	–	–	w–	–	–	–	–	+^w	_w	–
Raffinose	_w	–	w–	+–	–	–	_w	+^w	_+	–
Rhamnose	–	–	–	–	–	–	+	+^w	–	–
Ribose	_w	_w	–	–	–	–	d	+^w	–	–
Salicin	d	_w	w–	+^w	–	–	d	+	+^w	–
Sorbitol	–	–	–	–	–	–	–	–	–	–
Sorbose	–	_w	_w	·	–	–	_w	+^w	–	–
Starch	–	_w	–	_	–	_w	–	d	_	–
Starch hydrol.	+–	_+	–	–	–	_+	–	–	–	–
Sucrose	d	–	+^w	+^w	–	–	+	+	+–	–
Trehalose	_+	–	–	–	–	–	w+	_w	_+	–
Xylose	_+	–	_w	–	_	–	+–	d	–	–

[a] For symbols see Table 17.13.

TABLE 17.16

Biochemical reactions of species of genus **Eubacterium**, *species 11–20*

	11. E. cylindroides	12. E. moniliforme	13. E. tortuosum	14. E. cellulosolvens	15. E. combesii	16. E. tenue	17. E. fissicatena	18. E. contortum	19. E. aerofaciens	20. E. lentum
Gelatin digested	−w	−	−	−w	+	+	−	−	−w	−w
Milk	−	−	a−	c	cd	cd	a−	c	ac−	−
Meat	−	−	−	−	−d	−d	−	−	−	−
Indole produced	−	−	−	−	−	+	−	−	−	−
Nitrate reduced	−	d	d	−	−	−	−	−	−	+−
Neutral red reduced	+	+	+	+	d	+	+	+	+	+−
Gas	−3	4	−2	−3	24	24	43	24	4−	−+
NH3	w−	+	w−	−	+	+4	+−	−+	+−	+
Acetoin	−w	−+	−w	−w	−	d	−	d	−w	−
Growth in 20% bile	4−	4+	−4	+	42	+2	4	24	4−	+3
Growth in 6.5% NaCl	d	−	−	−	−	−	−	−	+−	d
Growth at pH 8.5	+−	+−	+−	+	−	−+	+	d	+−	+−
Lecithinase (EYA)	−	−	−	·	−	+	−	−	−	−
Lipase (EYA)	−	−	−	·	−+	−	−	−	−	−
Hemolysis	−w	−w	−a	·	bw	w−	b−	w	−+	−
Motility	−	+−	−	+	+−	d	−+	−	−	−
Threonine ← propionate	−	−+	−	−	−+	+	−	−	−	−

a For symbols see Table 17.14.

no flagella could be found in flagella stains of two other motile strains.

Surface colonies on horse blood agar (2 days) usually are large (2–8 mm), circular with irregular edges, pulvinate or umbonate, opaque.

Glucose broth cultures are turbid with smooth sediment and terminal pH of 5–5.6.

All strains grew at 45 C, most grew at 30 C and some at 25 C.

Meso-diaminopimelic acid detected in two out of two strains tested.

Isolated from blood and various kinds of infections.

Reference strain: Prévot collection 2055; ATCC 25546.

Further Comments

Although no spores have been detected in the strains on which this description is based, some strains, otherwise like *E. moniliforme*, produce spores and therefore belong to the genus *Clostridium*.

13. **Eubacterium tortuosum** (Debono) Prévot 1938, 295. (*Bacillus tortuosus* Debono 1912, 233; *Bacteroides tortuosus* (Debono) Bergey *et al.* 1923, 259; *Mycobacterium flavum* var. *tortuosum* Krasil'nikov 1949, 189.)

tor.tu.o'sum. L. neut.adj. *tortuosum* full of windings.

Description from Debono (1912), Prévot *et al.*

(1967) and Holdeman and Moore (1972) who studied nine strains (Tables 17.15, 17.16 and Fig. 17.C).

Surface colonies on horse blood agar (2 days) 0.5–4 mm, circular, entire-erose-diffuse, convex-umbonate, translucent, gray-white, dull.

Glucose broth cultures have flocculent-gelatinous (occasionally granular) sediment, no turbidity and terminal pH of 5.3–5.6.

Most strains grow at 30 and 45 C, some grow at 25 C.

Isolated from human feces, intestinal contents of mice, human infection, turkey liver lesions, turkey enteritis, soil, and fresh water.

Reference strain: VPI 1084B; ATCC 25548.

14. **Eubacterium cellulosolvens** (Bryant *et al.*) Holdeman and Moore 1972, 39. (*Cillobacterium cellulosolvens* Bryant, Small, Bouma and Robinson 1958, 529.)

cel.lu.lo.sol'vens. M.L. n. *cellulosum* cellulose; L. part.adj. *solvens* dissolving; M.L. part.adj. *cellulosolvens* cellulose-dissolving.

Description from Bryant *et al.* (1958), van Gylswyk and Hoffman (1970) and from study of one strain from van Gylswyk and Hoffman (Tables 17.15, 17.16 and Fig. 17.C).

Cells motile with peritrichous flagella.

Surface colonies 3–5 mm, circular, entire, flat to slightly convex, translucent, light tan. Deep colonies are lenticular.

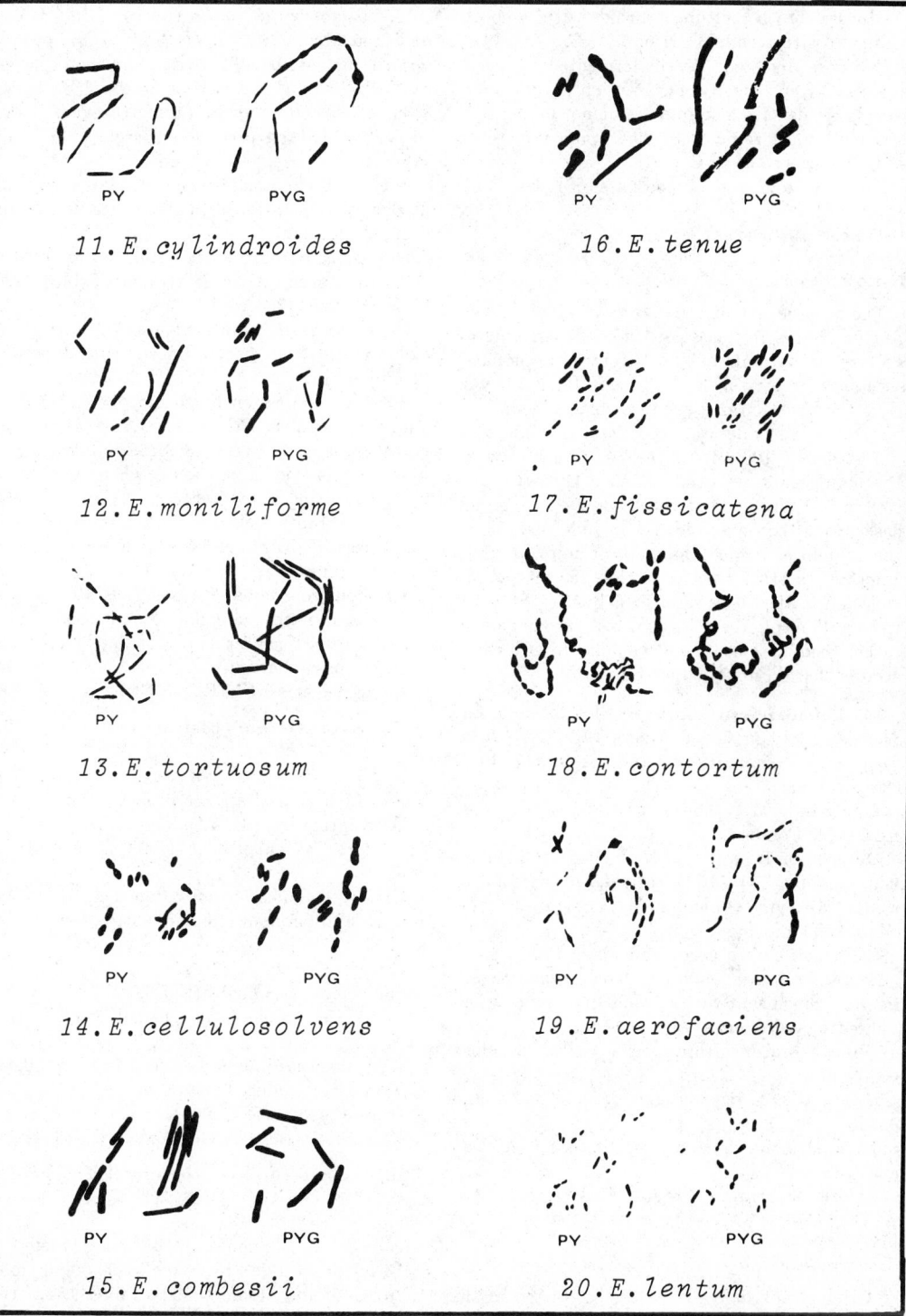

Fig. 17.C. Morphology of species 11 to 20 of the genus *Eubacterium*. Drawings are composites of several cultures; individual cultures usually show less variation. Scale: 9/16 inch = 10 μm. PY, peptone-yeast extract broth cultures; PYG, PY-glucose broth cultures.

Glucose broth cultures turbid with smooth sediment with terminal pH of 5.4–5.8.

Cellulose digested. Pectin fermented by some strains; xylan usually not fermented, or only weakly so; dextrin and gum arabic not fermented.

Growth most rapid at 37–39 C, poor growth at 45 C, no growth at 22 C.

Isolated from rumen contents of sheep and cows.

Type strain: Bryant B348.

Further Comments

The strains studied by Bryant *et al.* (1958) produce primarily lactate with small amounts of acetate and formate but no butyrate, propionate, succinate or hydrogen. Strains isolated by van Gylswyk and Hoffman produce small amounts of butyrate and valerate. We detected butyrate (0.25–0.4 meq/100 ml) in the one van Gylswyk-Hoffman strain we studied. Van Gylswyk and Hoffman suggested that the Bryant *et al.* species description be amended to include organisms that produce a small amount of butyrate and valerate. In other characteristics, the strains are very much like those described by Bryant *et al.* Recognition of these organisms as eubacteria rather than lactobacilli will be difficult with those strains that do not produce butyrate.

15. Eubacterium combesii (Prévot and Laplanche) Holdeman and Moore 1970, 23. (*Cillobacterium combesi* Prévot and Laplanche 1947, 688.)

com.be'si.i. M.L. gen.n. *combesii* of Combes; named for Combes.

Description from Prévot *et al.* (1967) and Holdeman and Moore (1972) who studied Prévot Collection A13D (syntype) and one other strain (Tables 17.15, 17.16 and Fig. 17.C).

Cells motile with peritrichous flagella.

Surface colonies on horse blood agar convex, lobate, translucent with mosaic appearance when viewed by obliquely transmitted light.

Glucose broth cultures turbid with flocculent sediment.

Grows at 30 C but not well, if at all, at 25 and 45 C.

Isolated from human infection and African soil.

Reference strain: Prévot Collection A13D; ATCC 25545.

Further Comments

Although no spores have been detected in the strains on which this description is based, some strains, otherwise like *E. combesii*, produce spores and have been identified by us as *Clostridium hastiforme* (Holdeman and Moore 1972, 68).

16. Eubacterium tenue (Bergey *et al.*) Holdeman and Moore 1970, 23.(*Bacillus tenuis spatuliformis* Distaso 1911, 101; *Bacteroides tenuis* Bergey *et al.* 1923, 623; *Cillobacterium spatuliforme* Prévot 1938, 297; *Bacillus spatuliformis* (Prévot) Prévot 1940, 79; *Cillobacterium tenue* (Bergey *et al.*) Clise 1957, 567.)

te'nu.e. L. neut.adj. *tenue* slender (originally used with *spatuliformis* to indicate forms like slender spatulas).

Description from Distaso (1911), Prévot *et al.* (1967) and study of three strains (Tables 17.15, 17.16 and Fig. 17.C).

Cells motile in young cultures by microscopic examination; no flagella seen with Leifson's flagella stain.

Surface colonies on horse blood agar (2 days) slightly irregular with lobate-diffuse edges, flat, translucent, gray-white, smooth, dull, with granular or mottled appearance when viewed by obliquely transmitted light.

Glucose broth cultures turbid with sediment and terminal pH of 5.9–6.0. Putrid odor.

Growth most rapid at 24–37 C.

Isolated from feces of a dog (Distaso), pleural fluid (man with malignant lymphogranuloma), case of breast cancer (Prévot) and deep-seated abscess (post-abortion).

Reference strain: ATCC 25553.

17. Eubacterium fissicatena Taylor 1972, 462.

fiss.i.ca.te'na. L. n. *fissi* a cleft; L. n. *catena* a chain; M.L. n. *fissicatena* a broken chain.

Description from study of eight strains received from Dr. Taylor (Tables 17.15, 17.16 and Fig. 17.C).

Surface colonies on horse blood agar (2 days) pinpoint to 0.5 mm, circular, entire, low convex, smooth.

Glucose broth cultures turbid with smooth sediment and terminal pH of 5.0–5.5.

All strains grew at 25 C; usually no growth at 45 C.

Hydroxyethylflavine produced from riboflavin. CO_2 and H_2 produced.

Isolated from the alimentary tract of goats.

Reference strain: Taylor A2A; NCIB 10446.

18. Eubacterium contortum (Prévot) Holdeman, Cato and Moore 1971, 306. (*Catenabacterium contortum* Prévot 1947, 414.)

con.tor'tum. L. neut.adj. *contortum* twisted.

Description from Prévot *et al.* (1967) and Holdeman, Cato and Moore (1971) (Tables 17.15, 17.16 and Fig. 17.C).

Surface colonies on horse blood agar (2 days) 0.5 mm, circular, entire to erose, low convex, translucent.

Glucose broth culture usually turbid with heavy sediment (gelatinous or granular) and terminal pH of 4.8–5.0.

Urease produced (Huet and Aladame, 1952); urease not produced (our laboratory).

Isolated from cases of appendicitis, salpingitis, rectocolitis, purulent pleurisy, and human feces.

Type strain: Prévot 113 VI; VPI 0119; ATCC 25540 (Holdeman et al., 1971, 304).

19. **Eubacterium aerofaciens** (Eggerth) Prévot 1938, 295. (*Bacteroides aerofaciens* Eggerth 1935, 282; *Pseudobacterium aerofaciens* (Eggerth) Krasil'nikov 1949, 232.)

ae.ro.fa′ci.ens. Gr. n. *aer* air, gas; L. v. *facio* make, produce; M.L. part.adj. *aerofaciens* gas-producing.

Description from Eggerth (1935), Prévot et al. (1967) and Moore, Cato and Holdeman (1971) who studied 53 strains (Tables 17.15, 17.16 and Fig. 17.C).

Surface colonies on horse blood agar (2 days) 1–3 mm, circular, entire to erose, convex, translucent or slightly opaque, smooth, glistening.

Glucose broth cultures turbid with smooth sediment and terminal pH of 4.6–5.1.

Most strains grow at 25–45 C.

Isolated from human feces, cecal contents of chickens, contents of dog and hog large intestine, and various kinds of infections (purulent pleurisy, peritonitis, postoperative wound and chronic furuncles).

Neotype strain: VPI 1003; ATCC 25986 (Moore et al., 1971, 307).

20. **Eubacterium lentum** (Eggerth) Prévot 1938, 295. (*Bacteroides lentus* Eggerth 1935, 280; *Pseudobacterium lentum* Krasil'nikov 1949, 232. Probable synonym, *Coccobacillus oviformis* Tissier 1908, 203; *Bacteroides oviformis* (Tissier) Levine and Soppeland 1926, 35.)

len′tum. L. neut.adj. *lentum* slow.

Description based on Moore, Cato and Holdeman (1971) who studied 53 strains (Tables 17.15, 17.16 and Fig. 17.C).

Surface colonies on horse blood agar 0.5–2 mm, circular, entire to erose, raised to low convex, translucent to semi-opaque, dull to shiny, smooth, sometimes with mottled appearance when viewed by obliquely transmitted light.

Glucose broth cultures moderately turbid with small smooth (occasionally stringy) sediment; pH not changed.

Most strains grow at 30 and 45 C; some grow at 25 C.

Isolated from human feces and from blood, postoperative wounds and various kinds of abscesses (brain, rectal, scrotal, pelvic, etc.).

Neotype strain: Prévot 1899 B; VPI 0255; ATCC 25559 (Moore et al., 1971, 299).

Further Comments

There are anaerobic Gram-positive non-sporing rods that ferment carbohydrates in media with 0.1% Tween 80 but not in media without Tween 80. These strains are not *E. lentum*, but could be confused with *E. lentum* if not tested in media with 0.1% Tween 80.

(Strains of species 21–28 were not available for study, but their descriptions are sufficient to recognize them as species of *Eubacterium*, should they be reisolated.)

21. **Eubacterium endocarditidis** (Prévot) comb. nov. (*Cillobacterium endocarditis* (sic) Prevot 1938, 296.)

en.do.car.di′ti.dis. M.L. n. *endocardium* heart lining; Gr. suff. *-itis* disease of; M.L. gen.n. *endocarditidis* of endocarditis.

Description from Prévot et al. (1967) and Weinberg et al. (1937) (Table 17.17).

Cells are pleomorphic and actively motile. Colonies in deep agar are lenticular with lateral buds.

Glucose broth cultures (24 hr) are turbid with sediment and abundant gas production. Poor growth in peptone broth.

Produces butyric acid and acetic, formic, propionic or lactic acid, NH_3, H_2S, amines, sometimes aldehydes and ketones and gas.

Coagulated proteins are not digested.

Neutral red reduced (Prévot); not reduced (Weinberg et al.). Sulfates not reduced.

Intramuscular injection of 2–5 ml (of broth cultures) produces death of guinea pigs in 2–5 days with extensive hemorrhagic edema, peritoneal hemorrhage and visceral congestion.

Isolated from blood in a case of febrile endocarditis, soil and river water.

22. **Eubacterium helminthoides** (Lewkowicz) comb. nov. (*Catenabacterium helminthoides* (Lewkowicz) Prévot 1938, 295; *Bacillus helminthoides* Lewkowicz 1901, 651.)

hel.min.tho.i′des or hel.min.thoi′des. Gr. adj. *helminthoides* worm-like.

Description based on Weinberg et al. (1937, 605) and Prévot et al. (1967) (Table 17.17).

Cells are 0.7–1 by 3–20 µm, straight or curved with rounded ends, Gram-positive in young cultures, Gram-variable in older cultures. Filaments are curved or curled, variable in width, with central or terminal swellings. Non-motile.

Colonies in deep agar after 24 hr are 3–5 mm, round, grayish, rhizoid (by microscopic examination). Gas splits the agar. Butyric odor. Surface

TABLE 17.17

Characteristics of species 21–28 of genus **Eubacterium**[a]

Species	Motile	Optimum temp.	Gelatin digested	Milk	NO₃ reduction	Indole produced	Gas[b]	Fructose	Glucose	Lactose	Maltose	Mannitol	Starch	Sucrose	Additional characteristics
21. *E. endocarditidis*	+	37	+	–	–		4	–	+		+		–	+	Acid: arabinose; no acid: dextrin, inulin, salicin
22. *E. helminthoides*	–	37	**+**	–°	**–**		4		+		**+**	+		+	Acid: glycerol, sorbose
23. *E. pseudotortuosum*	–	37	**+**	–	**+**		4	+	+	+	+		+	+	Acid: galactose, sorbitol
24. *E. obsti*	–	37	**+**	–	–	–	4		d		**–**				"May ferment glucose; other carbohydrates are not fermented"
25. *E. ethylicum*	–	37	–	C	–	–	4	+	+	+		+	+	+	Acid: glycerol, no acid: amygdalin; hydrolyzes starch
26. *E. helwigiae*	–	55	–	a	+	–	+	+	+	+	+			+	Acid: mannose, galactose; hydrolyzes starch.
27. *E. ureolyticum*	–	†	–	C	+ᵂ		**–**	+	+	+	+				Acid: galactose; **hydrolyzes urea**
28. *E. parvum*	–	37	–	C	–		–	+	+	+	+		+		Acid: galactose

[a] + = positive reaction or acid produced; – = negative reaction or no acid produced; d = reaction may be positive or negative; ° = see text; C = coagulated; a = acid (milk). ᵂ nitrate reduced in the presence of galactose and lactose. Useful distinguishing characteristics are given in **boldface.**

[b] 4 = abundant gas; + = gas produced. † optimum temperature not given, presumably is 37 C.

colonies are 1.5–2.5 mm in 24 hr, thin, transparent, grayish, rhizoid with less dense centers.

No growth in nutrient broth. Glucose broth cultures are turbid with gray flocculent sediment. The broth clears in 12 days; gas and rancid odor are produced.

Produces NH_3, H_2S, acetoin, formic, butyric and lactic acids, and alcohol.

Milk is not coagulated (coagulated by a variety of this species isolated from mollusks from rice paddies), coagulated proteins not digested. Neutral red and phenosafranin not reduced (reduced by the mollusk strains).

Produces small abscesses experimentally in rabbits.

Isolated from the mouth of nursing infants, intestinal contents of mollusks, and pond mud.

23. Eubacterium pseudotortuosum Prévot 1947, 413.

pseu.do.tor.tu.o′sum. Gr. adj. *pseudes* false; L. adj. *tortuosum* a specific epithet; M.L. neut.adj. *pseudotortuosum* not the true (*Eubacterium*) *tortuosum*.

Description from Prévot *et al.* (1967) (Table 17.17).

Cells are 0.4–0.5 by 3–4 μm, straight or curved, in tortuous chains or in curving filaments. Cells are easily decolorized.

Deep agar colonies are lenticular, producing gas.

Cultures in peptone broth are turbid; in glucose broth, densely turbid with abundant gas.

Optimum pH is 7.4.

Produces NH_3, H_2S, aldehydes, ketones, acetoin, formic, butyric and lactic acids, and alcohols.

Coagulated proteins are not digested. Reduces phenosafranin.

Not pathogenic for mice or guinea pigs.

Isolated from a bronchial infection and a case of acute appendicitis.

24. Eubacterium obstii Prévot 1938, 294.

(Bacillus B Obst 1919, 168; *Pseudobacterium obsti* (Prévot) Krasil′nikov 1949, 230.)

ob′sti.i. M.L. gen.n. *obstii* of Obst; named for M. Obst, who first isolated the organism.

Description from Prévot *et al.* (1967) and Weinberg *et al.* (1937) (Table 17.17).

Cells are short, straight, non-motile.

Deep agar colonies are discoid, sometimes rhizoid, transparent, grow below the top 2 cm of the exposed agar surface and produce abundant gas.

Surface colonies on blood agar in 48 hr are small, discoid, hemolytic.

Peptone broth cultures are slightly turbid. Glucose broth cultures are turbid with gas.

Produces NH_3, H_2S, amines, aldehydes, ketones, and acetic and butyric acids.

Coagulated proteins are not digested. Sulfates are not reduced, safranin is reduced.

Intraperitoneal injection of 1 ml of culture caused death of guinea pigs in 18 hr with fluid and gas in the peritoneal cavity. The same symptoms were observed in mice injected with 2 drops of culture.

Isolated from the stomachs of sardines and herrings swollen by gas, and from cheese.

25. Eubacterium ethylicum Prévot 1938, 295.

(*Bacillus gracilis ethylicus* Achalme and Rosenthal 1906, 1925; *Pseudobacterium ethylicum* (Prévot) Krasil′nikov 1949, 230.)

e.thy′li.cum. M.L. n. *ethyl* the ethyl radical; M.L. neut.adj. *ethylicum* pertaining to ethyl.

Description from Prévot *et al.* (1967) and Weinberg *et al.* (1937) (Table 17.17).

Cells are 0.6–0.7 by 2–3 μm, straight or slightly curved, granular, occur singly or in chains of two to four cells; non-motile.

Colonies in deep agar are 1–2 mm, lenticular or irregular.

Glucose or peptone broth cultures produce faint turbidity and a flocculent sediment which leaves the broth clear.

Produces acetic and butyric acids and traces of lactic acid, ethanol, NH_3 and abundant gas but no odor.

Coagulated proteins not attacked.

Intravenous or subcutaneous injection of 2–5 ml (of broth cultures) killed rabbits or guinea pigs in several days with visceral degeneration. Several smaller doses produced cachexia and death in 3–20 weeks without depressed appetite.

Isolated from a case of gastritis and from subacute appendicitis.

26. Eubacterium helwigiae Holdeman and Moore 1972, 46. (*Catenabacterium ruminantium* Stellmach-Helwig 1961, 46. NOT *Eubacterium ruminantium* Bryant 1959, 140.)

hel.wig′i.ae. L. gen.n. *helwigiae* of Helwig; named for Ruth Stellmach-Helwig, the bacteriologist who first described this organism.

Description from Stellmach-Helwig (1961) (Table 17.17).

Cells vary in size, depending upon the culture medium. In 0.5% basal broth medium, cells are 0.4 by 2.3–3.3 μm, uniform with rounded ends; in 2% basal broth medium, 0.5 by 1.8–2.6 μm with vacuoles and granules; in semi-solid agar, 0.3 by 3.3–7.5 μm; and on solid agar, 0.4 by 3–4.8 μm, with slight swellings. Long chains are not formed.

Growth is stimulated by CO_2 and by rumen fluid.

Does not hydrolyze citrate or casein (*sic*). Produces H_2S but not acetoin.

Isolated from the rumen contents of a sheep.

27. Eubacterium ureolyticum Huet and de Cadore 1954, 242.

u.re.o.ly'ti.cum. M.L. n. *urea* urea; Gr. adj. *lyticus* dissolving; M.L. neut.adj. *ureolyticum* urea dissolving.

Description from Huet and de Cadore 1954 (Table 17.17).

Cells are 0.8 by 3 μm, non-motile, granular.

Deep agar colonies are irregular, like snowflakes. Poor growth in peptone broth.

Produces an extremely active urease. Produces H₂S, butyric, acetic and lactic acids, amines and aldehydes.

Proteins not digested. Neutral red reduced.

Isolated from sheep feces on VF urea agar.

28. Eubacterium parvum Prévot 1938, 295.

par'vum. L. neut.adj. *parvum* small.

Description from Prévot *et al.* (1967) and Weinberg *et al.* (1937) (Table 17.17).

Cells are 0.5 by 1–1.5 μm in pairs or chains. In older cultures cells stain unevenly and appear as chains of cocci. Filaments may form. Non-motile.

Deep agar colonies are small, gray-white and lenticular. No gas is produced.

Glucose broth cultures have dense turbidity with heavy sediment.

Produces NH₃, aldehydes, ketones, acetylmethylcarbinol, formic, butyric and lactic acids, and alcohol.

Coagulated proteins are not digested. Reduces phenosafranin.

Not pathogenic for rabbits or guinea pigs.

Isolated from feces of foals, human appendicitis, purulent pleurisy, brain abscesses, tonsilitis and purulent bronchitis.

Species incertae sedis

Strains of species a–f not available for study.

a. *Catenabacterium hemicellulolyticum* Prévot 1966, 173.

b. *Catenabacterium leptotrichoides* Prévot 1938, 296.

c. *Catenabacterium rotans* Patočka and Šebek 1951, 309.

d. *Cillobacterium silvestris* Lanthiez *in* Prévot 1948, 167.

If the products reported (Duclaux distillation) are correct, this species is, by present definition, a member of the genus *Propionibacterium*. Because Duclaux analyses have proved unreliable, we have listed the species as *incertae sedis* in the genus *Eubacterium*. Only two isolations of this species have been reported.

e. *Eubacterium niosii* (Hauduroy *et al.*) Prévot 1938, 294. (Anaerober Bacillus, Niosi 1911, 193; *Bacteroides niosii* Hauduroy *et al.* 1937, 65; *Pseudo-*

bacterium niosii (Hauduroy *et al.*) Krasil'nikov 1949, 230.)

f. *Eubacterium sarcosinogenum* Szulmajster and Kaiser 1960, 776.

g. *Eubacterium typhi-exanthematici* (Plotz) Prévot 1938, 295. (*Bacillus typhi-exanthematici* Plotz 1915, 1; *Corynebacterium typhi* Topley and Wilson 1936, 349.)

Cultures in PYG with 0.1% Tween 80 produced 2.1 meq of acetic and 0.9 meq of lactic acids/100 ml of culture, with small amounts of formic and succinic acids and alcohols, similar to members of the genus *Bifidobacterium*. However, in the same medium without Tween 80 where there was less growth, only 0.8 meq of lactic acid/100 ml of culture with traces of propionic acid and alcohols were produced. This suggests that the organism might properly belong to the genus *Propionibacterium*. Additional characteristics must be determined to establish the taxonomic position of this species.

Nomina dubia

Strains of the following are not extant. Descriptions may be found in the original references and in Prévot *et al.* (1967).

Eubacterium biforme (Eggerth) Prévot 1938, 295. (*Bacteroides biformis* Eggerth 1935, 283; *Pseudobacterium biforme* (Eggerth) Krasil'nikov 1949, 231.)

Authors' Note. Since the preparation of this manuscript, many strains conforming to the original description have been isolated from human feces, and an emended description of *E. biforme* is to be published.

Eubacterium minutum (Hauduroy *et al.*) Prévot 1938, 295. (*Bacteroides minutus* Hauduroy *et al.* 1937, 64; indistinguishable from *E. lentum.*)

Eubacterium poeciloides (Roger and Garnier) Prévot 1938, 295. (*Bacillus poeciloides* Roger and Garnier 1906, 873; *Bacteroides poeciloides* (Roger and Garnier) Hauduroy *et al.* 1953, 251.)

Catenabacterium lottii Prévot 1938, 296 [Weinberg *et al.* 1937 (Bacillus β of Lotti).]

Catenabacterium nigrum Prévot 1938, 296.

Cillobacterium guinaeensis Digeon in Prévot 1948, 166.

Cillobacterium meningitis (sic) Prévot 1938, 297 [Weinberg *et al.* 1937 (Bacillus of Ghon, Mucha, and Müller).]

Ramibacterium pseudoramosum (Distaso) Prévot 1938, 296. (*Bacillus pseudoramosus* Distaso 1912, 441.)

Nomen confusum

Ramibacterium ramosoides (Runeberg) Prévot 1938, 296. (*Bacillus ramosoides* Runeberg 1908, 449.)

The original description of Runeberg was based

on 11 strains and at least three distinctly different organisms. Presumably one was *Propionibacterium acnes* from normal skin flora, an organism which accounts for more than half of the laboratory contamination in laboratories working with anaerobes.

Addendum I

The following species, previously assigned to the genera *Eubacterium, Catenabacterium* or *Cillobacterium* do not possess the characteristics of the genus.

The first three produce lactic acid as the sole major product; they are discussed further under *Lactobacillus*.

Catenabacterium catenaforme = *Lactobacillus catenaforme* (Eggerth) Moore and Holdeman 1970, 15.

E. disciformans = *Lactobacillus disciformans* (Prévot) Moore and Holdeman 1970, 15.

E. crispatum = *Lactobacillus crispatus* (Brygoo and Aladame) Moore and Holdeman 1970, 15.

E. asteracearum = *Erwinia asteracearum*, see page 339.

E. salmonis = *Aeromonas salmonicida*, see page 348.

E. quartum Prévot 1938, 294 (Probably *Clostridium*) Two of three labeled strains, which otherwise conformed to the description, formed spores.

E. quintum Prévot 1940, 65 (*Clostridium*) Only available labeled strain, which otherwise conformed to the description, formed spores.

Catenabacterium filamentosum Prévot 1938, 295 (*Eubacterium filamentosum* (Prévot) Holdeman and Moore 1970, 23) = *Clostridium ramosum*.

Cillobacterium thermophilum Prévot, Thouvenot, Pitre, and Bressou 1954, 778 = *Bifidobacterium thermophilum* (Prévot *et al.*) Holdeman and Moore 1972, 54.

ORDER I. ACTINOMYCETALES BUCHANAN 1917, 162

DAVID GOTTLIEB

Ac.ti.no.my.ce.ta′les. M.L. masc.n. *Actinomyces* type genus of the order; -*ales* ending to denote an order; M.L. pl.fem.n. *Actinomycetales* the *Actinomyces* order.

Bacteria that tend to form branching filaments which in some families develop into a mycelium. The filaments may be extremely short as in the *Mycobacteriaceae* and *Actinomycetaceae* or well developed as in *Streptomycetaceae*. The diameter varies from 0.5–2.0 μm, generally less than 1.0 μm. Filaments are not always observed because in certain families the filaments tend to fragment and can be seen only in some cultural stages of development or in host tissue. They are sometimes rare as in *Mycobacterium*. Fragmentation of filaments leads to the formation of coccoid, elongate or diphtheroid elements. In some families true spores are formed on aerial and/or substrate hyphae. Spores may be produced singly on the hypha, as a pair or as chains of various numbers of spores. If sufficient numbers of spores are present the chains can be straight, looped or spiral. Such chains arise singly from the hyphae or in a verticillate manner. In the *Actinoplanaceae*, spores are borne in a sporangium and are motile or non-motile, depending on the genus.

Gram-positive, though this reaction may vary with the age of the culture. The mycobacteria are acid-alcohol-fast and some members of the *Nocardiaceae* are weakly acid-fast. All members of the order are aerobic except for genera in the *Actino-*mycetaceae, which may be anaerobic, facultatively anaerobic or aerobic.

The *Actinomycetales* are commonly isolated from soil and less commonly from fresh water. Forms pathogenic to man and animals exist in at least four families. Spores can also be allergenic to man. Some species are pathogenic to plants; others are obligate symbionts in plant root nodules and fix nitrogen.

Further Comments

New criteria for the identification of taxa in the *Actinomycetales* are now being generally adopted. In *Streptomyces* and *Streptoverticillium*, standard procedures of culture and observation are now used to characterize species (Shirling and Gottlieb, 1966).

Cell wall composition also has become widely accepted as an aid in the identification of genera. Four cell wall types are now widely accepted (Becker *et al.*, 1965; Lechevalier *et al.*, 1966; Lechevalier and Lechevalier, 1970) (Table 17.18). Other cell wall patterns have been proposed (Lechevalier and Lechevalier, 1967; Sukapure *et al.*, 1970) but more species and strains must be studied to ascertain whether these types are indeed characteristic of the genera in which they have been found. Sugar patterns of whole cells have also been suggested and combinations of cell wall and sugar patterns seem promising, although at present there are some anomalies with present classifications (Tables 17.19 and 17.20).

TABLE 17.18
Major constituents of cell wall types of actinomycetes[a]

Cell wall type	DAP meso-	DAP LL-	Gly-cine	Arabi-nose	Galac-tose
I	–	+	+	–	–
II	+	–	+	–	–
III	+	–	–	–	–
IV	+	–	–	+	+

[a] References: Becker *et al.*, 1965; Yamaguchi, 1965; Lechevalier and Lechevalier, 1970.

All preparations also contain major amounts of glucosamine, muramic acid, alanine and glutamic acid.

The chromatographic method used does not differentiate between meso- and D-forms of 2–6 diaminopimelic acid (DAP); since D-DAP is relatively rare it is assumed it is a minor constituent, if present.

TABLE 17.19
Whole cell sugar patterns of aerobic actinomycetes

Pattern	Arabinose	Galactose	Xylose	Madurose
A	+	+	–	–
B	–	–	–	+
C	–	–	–	–
D	+	–	+	–

TABLE 17.20
*Distribution of cell wall types and whole cell sugar patterns in genera of **Actinomycetales***

Cell wall type	Sugar pattern	Genera
I	NC[a]	*Streptomyces, Streptoverticillium, Microellobosporia, Sporichthya*
II	D	*Micromonospora, Actinoplanes Ampullariella, Amorphosporangium, Dactylosporangium*
III	B	*Microbispora, Streptosporangium Spirillospora, Planomonospora, Dermatophilus, Nocardia-madurae* type (*Actinomadura*)
III	C	*Actinobifida, Thermoactinomyces Geodermatophilus*
IV	A	*Mycobacterium, Nocardia, Pseudonocardia Thermomonospora, Micropolyspora*

[a] NC, no characteristic sugar pattern.

Base ratios in DNA may be helpful in placing some atypical organisms into the correct genera. Similarly, the homology of DNA from an unidentified species can be determined with the DNA from a known species, thus aiding in the identification of the actinomycete (Tewfik and Bradley, 1967).

Members of the *Actinomycetales*, despite the filamentous growth of some members, are now generally accepted to be bacteria and not fungi. The properties relating them to bacteria are: the absence of a nuclear membrane, small hyphal diameters, sensitivity to lysozyme, chemical nature of the cell wall, sensitivity to antibacterial agents, and flagella, when produced, of the bacterial type.

Key to the families of order **Actinomycetales**

I. Mycelium not formed; branching filaments may be produced; cells may be rod, diphtheroid or coccoid; no spores formed.
  A. Not acid-alcohol-fast; usually facultatively anaerobic; some anaerobic or aerobic; most do not contain 2,6-diaminopimelic acid in the cell walls.
                Family I. *Actinomycetaceae*, p. 659
  B. Acid-alcohol-fast, at least in some stages of growth; cell wall type IV.
                Family II. *Mycobacteriaceae*, p. 681
II. True mycelium produced.
  A. Symbionts in plant nodules with a free stage in soil.
                Family III. *Frankiaceae*, p. 701
  B. Saprophytes or facultative parasites.
    1. Spores borne inside sporangia.
                Family IV. *Actinoplanaceae*, p. 706
    2. Spores not borne inside sporangia.

a. Mycelial filaments divide transversely and in at least two longitudinal planes to form masses of motile, coccoid elements. Aerial mycelium usually absent; cell wall type III.
Family V. *Dermatophilaceae*, p. 723

b. Mycelial filaments commonly fragment to give coccoid or elongate elements that are usually non-motile, though a few species are reported to be motile. Aerial spores are occasionally produced but usually are absent; cell wall type IV. Sometimes acid-fast.
Family VI. *Nocardiaceae*, p. 726

c. Mycelial filaments tend to remain intact and not fragment. Usually abundant aerial mycelium and long spore chains (5–50 or more); cell wall type I.
Family VII. *Streptomycetaceae*, p. 747

d. Mycelial filaments remain intact; spores formed singly, in pairs or short chains on either or both aerial or substrate mycelium. Cell wall type (II, III, or IV) varies with the genus.
Family VIII. *Micromonosporaceae*, p. 846

FAMILY I. **ACTINOMYCETACEAE** BUCHANAN 1918, 403

JOHN M. SLACK*

Ac.ti.no.my.ce.ta'ce.ae. M.L. masc.n. *Actinomyces* type genus of the family, *-aceae* ending to denote a family; M.L. pl.n. *Actinomycetaceae* the *Actinomyces* family.

Gram-positive bacteria, predominantly diphtheroid in shape, which tend to form **branched filaments** in tissue or in some stages of cultural development. **Fragmentation of the filaments readily occurs** producing diphtheroid or coccoid forms. **Aerial mycelium and spores are not formed. Non-acid-fast and non-motile.** Usually grow as **facultative anaerobes** but **some are anaerobic** and **some are aerobic;** carbon dioxide usually stimulates growth. **Carbohydrates are fermented. Catalase may or may not be formed. Chemoorganotrophs.**

Key to the genera of family **Actinomycetaceae**

I. Anaerobic to facultatively anaerobic.
  A. Catalase negative or positive.
    1. Most species form a filamentous microcolony; filaments transitory, diphtheroid cells are predominant; fermentative, glucose fermentation products include acetic, formic, lactic and succinic acids but not propionic acid; cell wall contains neither diaminopimelic acid nor arabinose.
      Genus I. **Actinomyces,** p. 660
  B. Catalase negative.
    1. Filamentous microcolony; filaments transitory, diphtheroid cells and spheroplasts common, fermentative, glucose fermentation products are primarily propionic and acetic acids; cell wall contains diaminopimelic acid but not arabinose.
      Genus II. **Arachnia,** p. 668
    2. Smooth microcolony; filaments usually not formed, diphtheroid cells and bifid forms are common; fermentative, glucose fermentation products are primarily acetic and lactic acids; cell wall contains neither diaminopimelic acid nor arabinose.
      Genus III. **Bifidobacterium,** p. 669
II. Aerobic to facultatively anaerobic.
  A. Catalase positive or negative.

---

* Much of the information in this section has been from the work and reports of the "Subgroup on Taxonomy of Microaerophilic Actinomycetes," an authorized Subgroup of the Subcommittee on Taxonomy of the *Actinomycetales* of the International Committee on Systematic Bacteriology, International Association of Microbiological Societies. For membership and other information see Slack, 1968.

1. Filamentous microcolony; cell types include rods, filaments and characteristically filaments with a bacillus-body at one end; some strict anaerobes; fermentative, products from glucose grown cultures include CO_2, formic, acetic, propionic and lactic acids; cell wall contains both diaminopimelic acid and arabinose.

Genus IV. **Bacterionema**, p. 676

B. Catalase positive.

1. Smooth microcolony; growth at any given time may yield exclusively coccoid, diphtheroid or filamentous forms or a mixture of any of these; grows best aerobically; fermentative, glucose fermentation product is primarily lactic acid, propionic acid is not produced; cell wall contains neither diaminopimelic acid nor arabinose.

Genus V. **Rothia**, p. 679

Genus I. **Actinomyces** *Harz 1877, 485*

JOHN M. SLACK

(Not *Actinomyce* Meyen 1827, 442; *Discomyces* Rivolta 1878, 208; *Actinocladothrix* Affanassieff and Schulz 1889, 684.)

Ac.ti.no.my'ces Gr. n. *actis, actinis* ray; Gr. n. *myces* fungus; M.L. masc.n. *Actinomyces* ray fungus.

Gram-positive, irregularly staining bacteria; non-acid-fast, non-spore-forming and non-motile. Filaments with true branching may predominate and are particularly evident in 18–48 hr microcolonies. **Diphtheroid cells or branched rods are common;** V, Y and T forms occur (Plate 17.2, Figs. 1, 2). **Filaments,** 1 μm or less in diameter, **varying in length and degree of branching** occur in most strains.

Chemoorganotrophs. Carbohydrates are fermented with the production of **acid but no gas. End-products from glucose fermentation** may include **acetic, formic, lactic** and **succinic acids** but **not propionic acid.**

Proteolytic activity rare, and, if present, is weak. **Indole not produced; urease negative.** Some species may show greening or complete lysis of rabbit red blood cells.

Cell wall peptidoglycan contains alanine, glutamic acid and lysine with either (1) **aspartic acid or** (2) **ornithine;** diaminopimelic acid does not occur in any currently recognized species. Cell wall carbohydrates may include glucose, galactose, rhamnose and other deoxy sugars; **arabinose does not occur.**

Organic nitrogen is required for growth.

Facultative anaerobes; most are preferentially anaerobic, **one species grows well aerobically.** CO_2 is required for maximum growth. **Catalase negative or positive.**

Certain species are pathogenic for man and/or animals.

Type species: *Actinomyces bovis* Harz 1877, 485.

Further Comments

Isolation. Clinical material should be inoculated into freshly boiled thioglycollate medium (pref-

erably with addition of serum or other enrichment) and also streaked onto a brain heart infusion agar (BHIA) and two heart infusion blood agar plates. Incubate the BHIA and one blood plate anaerobically in an atmosphere of $N_2:H_2:$-$CO_2(80:10:10)$ with a catalyst, otherwise in $N_2:$-$CO_2(95:5)$. Incubate the second blood plate aerobically with addition of 5% CO_2. Observe for microcolonies in 18–24 hr and mature colonies in 7–14 days.

Genera with similar morphology. Cell and colony morphology provides the initial basis for considering an organism to be an *Actinomyces* but is not sufficient to differentiate *Actinomyces* from certain species of *Corynebacterium, Bifidobacterium, Propionibacterium, Bacterionema, Rothia* or *Arachnia.* Confirmation of an organism as an *Actinomyces* and species identification is dependent upon physiological, cell wall, and antigenic characteristics.

Microcolony morphology, particularly on BHIA (Plate 17.2, Figs. 3–4; Plate 17.3, Fig. 2) is frequently helpful in identification because a "spider" colony with multiple branching filaments or a granular-centered colony with peripherally extended branched filaments is highly suggestive of an *Actinomyces* species. However, *A. bovis, A. odontolyticus* and occasional strains of *A. israelii A. naeslundii* and *A. viscosus* form smooth microcolonies without branched filaments. Thus, colonial morphology is not always of differential value.

Physiological characteristics. Actinomyces species vary in tolerance to oxygen and in CO_2 requirements (Gerencser and Slack, 1967).

In order to obtain reproducible results in various biochemical tests it is to be emphasized that an adequate basal medium, sufficient inoculum and proper oxygen concentrations are essential. In addition, best results are obtained by the use of standardized techniques (Howell *et al.*, 1959; Slack, 1968; Georg, 1970). Each species produces a fermentation pattern which is distinctive (Table 17.22) and when considered with other characteristics (Tables 17.21) is of diagnostic value.

Cell wall. In the two possible peptidoglycan patterns the amount of aspartic acid may be minimal in pattern 1 and as an additional component in pattern 2. No species currently recognized contains diaminopimelic acid or arabinose. These results are based on the study of 80 strains including *A. bovis*, *A. israelii*, *A. naeslundii* and *A. odontolyticus* (Cummins, 1962; Pine and Boone, 1967; DeWeese *et al.*, 1968).

Antigenic characteristics. Any species can be identified through the combined use of morphology and serology. Fluorescent antibody techniques provide rapid and specific results (Slack and Gerencser, 1966; Lambert *et al.*, 1967). Gel diffusion techniques using culture filtrates as antigens are useful but are presently less specific for species identification (King and Meyer, 1963; Georg *et al.*, 1964).

Confusion in nomenclature. The misuse of two generic names in the current literature causes nomenclatural misunderstandings. *Actinobacterium* is often considered an objective synonym of *Actinomyces* (see *A. israelii*), but is not valid when used as a generic name of any species of *Actinomyces*. The use of *Actinomyces* as a generic name for aerobic streptomycetes does not conform to the current definition of *Actinomyces*, and these filamentous organisms should be classified as species of *Streptomyces*.

Description of the species of genus **Actinomyces**

1. **Actinomyces bovis** Harz 1877, 485. (*Discomyces bovis* (Harz) Rivolta 1878, 208; *Sarcomyces bovis* (Harz) Rivolta 1879, 146; *Oospora bovis* (Harz) Sauvageau and Radais 1892, 271; *Actinocladothrix bovis* (Harz) Gasperini 1892, 183; *Nocardia bovis* (Harz) Blanchard 1896, 857; *Streptothrix bovis* (Harz) Chester 1901, 361; *Cladothrix bovis* (Harz) Macé 1901, 1082; *Sphaerotilus bovis* (Harz) Engler 1907, 5; *Proactinomyces bovis* (Harz) Henrici 1939, 409.)

bo'vis. L. n. *bos* the ox; L. gen.n. *bovis* of the ox.

Description is based in part on the work of Erikson, 1940; Thompson, 1950; King and Meyer, 1957; Pine *et al.*, 1960; Waksman, 1961; Slack 1968; Georg *et al.*, 1964; Georg, 1970.

Microcolonies in 18–24 hr on brain heart infusion agar (observed at 100 X) are circular, entire, flat, granular to smooth, soft and may show a granular dense central core (Plate 17.2, Fig. 3). Occasional strains may form filamentous colonies (resembling *A. israelii*) and some strains may produce both smooth and filamentous colonies. Macrocolonies in 7–14 days on brain heart infusion agar or blood agar are 0.5–1.0 mm in diameter, low convex to convex, circular, entire, opaque, white, butyrous with smooth or granular surface. Some strains may produce irregular, heaped up colonies resembling those of *A. israelii*. Aerial hyphae have not been reported. Pigments are not produced. No hemolysis on blood agar.

In liquid media a diffuse growth is usually produced which settles to the bottom and has a flaky appearance when shaken. Some isolates produce a more viscous growth; others a granular growth like *A. israelii* except the granules are usually soft and easily broken.

Fermentation reactions are given in Table 17.22 and other biochemical reactions in Table 17.23.

End-products from glucose are as for genus.

TABLE 17.21

Characteristics differentiating the species of genus
Actinomyces[a]

	Species				
	1. *Actinomyces bovis*	2. *Actinomyces odontolyticus*	3. *Actinomyces israelii*	4. *Actinomyces naeslundii*	5. *Actinomyces viscosus*
Catalase	−	−	−	−	+
Microcolony, smooth or filamentous	S	S	F	F	F
Red mature colony on blood agar	−	+	−	−	−
Nitrate → Nitrite	d	+	d	+	+
Starch hydrolysis— wide zone	+	−	−	−	−
Fermentation (acid only)					
Arabinose	−	d	d	−	−
Ribose	−	−	+	−	−
Xylose	d	d	+	−	−
Soluble starch	+	+[s]	d	d	d
Serological group (fluorescent antibody)	B	E	D	A	F
Serotype	1, 2	1, 2	1, 2	1, 2	1, 2

[a] + = 90% of strains positive; − = 90% of strains negative; d = different reactions, positive or negative; +[s] = positive, slow.

No specific growth requirements reported, except some strains are serophilic.

Facultative anaerobe, grows best in presence of CO_2. Aerobic growth sparse, if any. In tubes of deep agar there is a dense zone of growth 1-2 cm below the surface with isolated soft lobate colonies deeper in the agar. Optimum temperature 35-37 C.

The major amino acids in the cell wall are alanine, aspartic acid, glutamic acid and lysine; diaminopimelic acid and ornithine are not present. Aspartic acid may be absent in filamentous strains (Pine and Boone, 1967). The sugars include: rhamnose, glucose, mannose, fucose and deoxytalose (variable); arabinose is not present.

Sensitive to penicillin, streptomycin, tetracyclines, cephalosporin and lincomycin (Boand and Novak, 1949; Howell, 1953; Lerner, 1967).

Serologically designated as group B with serotypes 1 and 2; filamentous strains are predominantly serotype 2. With the fluorescent antibody technique there is no cross-reaction between the two serotypes or with other species of *Actinomyces* (Slack and Gerencser, 1970). Gel diffusion and cell wall agglutination techniques may also be used (Cummins, 1962; King and Meyer, 1963; Georg *et al.*, 1964; Slack and Gerencser, 1966).

Causes actinomycosis in cattle (presumably the principal habitat). Similar infections have been described in swine, horses and other animals but the species involved cannot be determined because of inadequate description. Infections in man have not been established. Experimental infections have been produced in hamsters and mice (Meyer and Veges, 1950; Hazen and Little, 1958; Pine *et al.*, 1960; Georg and Coleman, 1968). "Sulfur" granules are produced in animal infections which contain a polysaccharide-protein complex as well as CaO, P_2O_5 and acid phosphate probably occurring as a poorly crystallized apatite (Pine and Overman, 1966; Frazier and Fowler, 1967).

The G + C content of the DNA of the suggested neotype strain is 63 moles % (T_m).

Suggested neotype strain: ATCC 13683.

2. Actinomyces odontolyticus Batty 1958, 455.

o.dont.o.lyt′i.cus. Gr. n. *odous, odontos* tooth; Gr. adj. *lyticus* dissolving M.L. adj. *odontolyticus* tooth dissolving.

Microcolonies in 24-48 hr on brain heart infusion agar (observed at 100 ×) are usually smooth, entire and may have dense centers. A few strains form filamentous microcolonies resembling the "spider colony" of *A. israelii*. Small colonies (up to 1 mm) on blood agar are circular to irregular, low convex, smooth to slightly granular and entire. May produce an area of greening around the colonies and resemble alpha-streptococci. Macrocolonies on BHIA after 7-14 days incubation are 1-2 mm in diameter, circular to irregular, raised, convex or umbonate, smooth or granular, entire or irregular, opaque, white and soft. Strains which produce filamentous microcolonies may produce rough, mature colonies resembling the "molartooth" colonies of *A. israelii*. On rabbit, sheep or horse blood agar mature colonies resemble those on BHIA except that the colony is dark red by transmitted light. This pigment may appear in 2-14 days and may develop best when an anaerobically grown culture is allowed to stand aerobically at room temperature. Pigmented colonies may be difficult to subculture.

Growth in liquid media is usually even and turbid but sometimes flocculent.

Fermentation reactions are given in Table 17.22 and other biochemical reactions in Table 17.23.

End-products from glucose grown cultures include acetic, formic, succinic and lactic acids but only one strain was tested.

Most strains require serum for good growth.

Some strains may grow aerobically on blood agar. Optimum temperature is 35-37 C.

The cell wall contains ornithine. Sugars include deoxytalose, fucose, galactose, glucose, mannose and rhamnose; arabinose is not present. Results based on five strains.

Antibiotic sensitivities have not been reported.

Serologically designated as group E with serotypes 1 and 2. Identifiable using the fluorescent antibody technique with low titer cross-reaction only with *A. viscosus*. Low titered cross-reactions have been reported with *A. naeslundii* and *A. israelii* using gel diffusion or cell wall agglutination techniques (Gerencser and Slack, 1967; Lambert *et al.*, 1967; Snyder *et al.*, 1967; Brock and Georg, 1969; Slack *et al.*, 1969 and 1970).

Isolated from deep dentinal caries in man (Batty, 1958) but the relationship of this organism to the disease has not been established. Minimal lesions have been produced in mice (Georg and Coleman, 1968). Normal habitat is the oral cavity of man.

The G + C content of the DNA of the cotype strain is 62 moles % (T_m).

Cotype strain: ATCC 17929; NCTC 9935 (Batty, 1958).

3. Actinomyces israelii (Kruse) Lachner-Sandoval 1898, 64. (*Streptothrix israeli* (sic) Kruse 1896, 56; *Discomyces israeli* (Kruse) Gedoelst 1902, 163; *Actinobacterium israeli* (Kruse) Sampietro 1908, 408; *Cohnistreptothrix israeli* (Kruse) Pinoy, 1913, 931; *Nocardia israeli* (Kruse) Castellani and Chalmers 1913, 814; *Oospora israeli* (Kruse) Sartory 1920, 773; *Brevistreptothrix israeli* (Kruse)

TABLE 17.22

Acid production[a] from carbon compounds by species of genus **Actinomyces**

	1. A. bovis	2. A. odonto- lyticus[b]	3. A. israelii	4. A. naeslundii	5. A. viscosus	a. A. eriksonii	b. A. humiferus	c. A. suis
Acetate	0/1[c]	0/0	0/4	0/1	0/2	0/12	0/27	0/0
Adonitol	0/1	6/30	0/4	0/12	0/2	0/12	0/27	7/10
Arabinose	2/26	15/30	57/223	2/54	0/56	12/12	25/27	9/10
Cellobiose	0/1	0/3	53/65	1/6	4/22	12/12	18/27	0/0
Dextrin	1/1	0/0	2/4	1/1	0/2	1/1	27/27	0/0
Dulcitol	0/1	0/30	0/4	0/12	0/2	0/12	9/27	1/10
Erythritol	0/1	0/0	0/4	0/1	0/2	0/12	0/0	0/0
Fructose	12/12	0/0	4/4	1/1	29/34	5/5	27/27	10/10
Galactose	2/2	0/0	4/4	1/1	24/34	5/5	25/27	10/10
Glucose	37/37	30/30	223/223	66/66	56/56	12/12	27/27	10/10
α-Methyl-D-glucoside	0/1	0/0	0/68	0/35	0/56	8/12	0/0	0/0
Glycerol	8/26	26/30	0/223	11/12[d]	20/23[d]	0/12	20/27	7/10
Glycogen	1/1	0/3	0/65	0/6	2/22	5/5	0/0	0/0
Inositol	1/1	14/30	137/223	40/52	48/56	0/12	2/27	0/0
Inulin	0/12	0/30	61/158	25/53	22/56	4/12	1/27	9/10
Lactate	0/1	0/0	0/4	0/1	0/2	0/12	0/27	0/0
Lactose	33/33	23/30	197/223	33/42	48/56	12/12	15/27	10/10
Maltose	29/33	29/30	158/158	32/42	32/34	12/12	25/27	10/10
Mannitol	4/37	0/30	149/223	0/56	0/56	12/12	24/27	5/10
Mannose	8/12	0/0	44/65	3/3	31/34	1/1	25/27	0/0
α-Methyl-D-manno- side	0/1	0/0	0/68	0/35	0/0	0/1	0/0	0/0
Melezitose	0/1	0/30	3/68	0/35	0/0	12/12	25/27	0/0
Melibiose	0/1	0/3	4/6	6/6	16/22	12/12	24/27	0/0
Pyruvate	0/1	0/0	0/4	0/1	0/2	4/12	27/27	0/0
Raffinose	0/16	15/30	118/223	50/53	53/56	12/12	24/27	10/10
Rhamnose	2/12	13/30	4/56	2/12	0/56	2/12	25/27	5/10
Ribose	0/12	0/0	123/158	8/42	0/34	1/1	3/27	0/0
Salicin	7/33	21/30	134/223	24/54	45/56	12/12	8/27	0/0
Sorbitol	0/1	1/30	2/61	0/1	0/34	12/12	1/27	2/10
Potato starch	13/14	0/0	0/4	0/1	0/2	1/1	0/0	0/0
Soluble starch	13/14	30/30	21/64	9/12	21/22	10/10	27/27	10/10
Sucrose	28/33	27/30	158/158	37/42	54/56	12/12	27/27	9/10
Trehalose	0/1	20/30	4/6	8/11	15/22	12/12	19/27	10/10
Xylose	11/32	16/30	206/223	1/56	0/56	12/12	24/27	10/10

[a] Acid but no gas produced.

[b] Medium thioglycollate fermentation base + 2% rabbit serum.

[c] Number of strains positive/number of strains tested.

[d] See species description.

Lignières 1924, 19; *Corynebacterium israeli* (Kruse) Haupt and Zeki 1933, 95; *Proactinomyces israeli* (Kruse) Negroni 1934, 1240.)

is.ra.e′li.i. M.L. gen.n. *of Israel;* named for Professor James Israel, one of the original isolators of this organism.

Description is based in part on the work of Negroni and Bonfiglioli, 1937; Erikson, 1940; Rosebury, 1944; Howell *et al.*, 1959; Waksman,

1961; Buchanan and Pine, 1962; Slack, 1968; Brock and Georg, 1969; Slack *et al.* 1969; Georg *et al.*, 1964; Georg, 1970.

Microcolonies in 18–24 hr on brain heart infusion agar (observed at 100 ×) are composed of branching filaments of varying lengths originating from a single point and without a distinct dense center (Plate 17.2, Fig. 4). Some cultures (20/64) may produce colonies made up of short angular branch-

TABLE 17.23
Biochemical reactions of species of genus **Actinomyces**

	1. *A. bovis*	2. *A. odontolyticus*	3. *A. israelii*	4. *A. naeslundii*	5. *A. viscosus*	a. *A. eriksonii*	b. *A. humiferus*
Catalase	0/30[a]	0/30	0/192	0/54	56/56[b]	0/12	0/27
Indole	0/14	0/230	0/223	0/49	0/47	0/12	0/27
$NO_3 \rightarrow NO_2$	3/12	230/230	109/220	49/55	21/22	0/12	0/27
Methyl red	12/12	202/203	57/63	12/12	21/22	12/12	27/27
Voges-Proskauer	0/8	0/203	0/192	0/37	0/48	0/12	4/27
H_2S (lead acetate paper)							
Triple sugar iron	10/10	26/30	26/26	6/14	22/22	12/12	0/0
BHIA	6/10	0/30	2/89	4/12	0/22	9/12	18/27
Urease production	0/10	0/30	0/26	0/12	0/22	0/12	0/27
Esculin hydrolysis	13/13	19/30	24/26	9/12	21/22	7/12	24/27
Starch hydrolysis							
Wide zone	30/30	0/203	0/63	0/23	0/48	0/12	0/27
Narrow zone	0/30	0/203	26/63	9/23	41/48	12/12	27/27
Gelatin hydrolysis							
Tube	1/24	0/230	0/89	0/59	0/48	0/12	12/27
Plate	0/0	0/0	34/89	0/0	0/0	0/0	0/0
Litmus milk							
Acid only	4/15	14/30	18/26	23/52	13/23	5/12	22/27
Acid and clot	8/15	10/30	8/26	24/52	10/23	7/12	0/0
No change	3/15	6/30	0/26	5/52	0/23	0/12	5/27

[a] Number of strains positive/number of strains tested.

[b] Aerobic + CO_2 cultures used, if anaerobic cultures are used expose to air 30 min before testing.

ng filaments or small rough compact colonies with no filaments at the edge. Macrocolonies on BHIA after 7–14 days incubation are rough and 0.5–3 mm in diameter (Plate 17.3, Fig. 1). These colonies are circular to irregular; convex, pulvinate or heaped, undulate, erose or lobate; white to cream-white and friable to hard. The surface may be granular or convoluted producing colonies described as "molar-tooth," "raspberry-like" or "bread crumb-like." Some strains produce smooth colonies which are circular to irregular; convex, umbonate or pulvinate; entire, opaque; grayish white and soft. The colonies have a similar appearance on blood agar and are non-hemolytic. No aerial hyphae have been observed. Pigments are not produced.

In liquid media growth is usually as discrete compact masses or "granules" of variable size in a clear medium. Some strains (0/53 serotype 1; 9/11 serotype 2) produce a diffuse, viscous growth.

Fermentation reactions are given in Table 17.22 and other biochemical reactions in Table 17.23.

End-products from glucose-grown cultures are same as for genus with the ratio of lactic to succinic acid approximating 2:1 (Buchanan and Pine, 1962; Li and Georg, 1968).

Chemically defined media have been reported (Christie and Porteus, 1962).

In tubes of deep agar there is no growth in the upper 1 cm, then a zone of dense growth. Lobate colonies deeper in the agar may become filamentous. Optimum temperature 35–37 C.

The cell wall contains ornithine; some strains contain appreciable amounts of aspartic acid, glycine and leucine. The sugars include galactose with only small amounts of other sugars (Cummins and Harris, 1958; Cummins, 1962; Boone and Pine, 1968; DeWeese *et al.*, 1968).

Sensitive to penicillin, tetracyclines, chloramphenicol, cephalosporin and lincomycin (Holm, 1948; Garrod, 1952; Howell, 1953; Suter and Vaughan, 1955; Lerner, 1967).

Serologically designated as group D with serotypes 1 and 2. Identifiable using the fluorescent antibody technique. Cross-reactions do not occur with other species except at low titer with *A. naeslundii* and *A. viscosus;* some strains cross react at low titer with *Propionibacterium acnes.* Serological identification is also possible using gel diffusion or cell wall agglutination techniques (Cummins, 1962; King and Meyer, 1963; Slack and Gerencser, 1966; Snyder *et al.*, 1967; Lambert *et al.*, 1967; Brock and Georg, 1969; Slack *et al.*, 1969 and 1970).

Causes human actinomycosis and occasionally infections in cattle. Experimental infections have been produced in hamsters, mice and rabbits (Meyer and Verges, 1950; Hazen and Little, 1958; Coleman and Georg, 1969). "Sulfur" granules are produced in human and animal infections (Widra, 1963). The source of infection is endogenous as the normal habitat is the oral cavity of man including tonsillar crypts and dental calculus (Emmons, 1938; Slack, 1942; Howell et al., 1962; Blank and Georg, 1968).

The G + C content of the DNA of the suggested neotype strain is 60 moles % (T_m).

Suggested neotype strain: ATCC 12102.

4. Actinomyces naeslundii Thompson and Lovestedt 1951, 175.

naes.lund'i.i. M.L. gen.n. of Naeslund; named for Carl Naeslund who first described this organism as "group C Actinomyces" but did not give it a specific epithet.

Description is based in part on the work of Naeslund, 1925; Thompson and Lovestedt, 1951; Garrod, 1952; Howell et al., 1959 and 1962; Georg et al., 1964; Georg, 1970; Slack, 1968.

Microcolonies in 18–24 hr on brain heart infusion agar (observed at 100 X) have a dense mass of diphtheroid cells and tangled filaments at their centers surrounded by a periphery of radiating, curved and branched filaments (Plate 17.3, Fig. 2). Occasionally the colonies are not filamentous but are flat and round with entire edges. Macrocolonies on BHIA after 7–14 days incubation are 1–3 mm in diameter, convex, round with entire or fluted margin, smooth or finely granular surface, white and soft. Less commonly, the colonies are heaped or irregularly lobate and may resemble the "molar-tooth" colony of A. israelii. A single strain may produce both types of colonies but on repeated subculture usually produces only the smoother colony.

In liquid media growth is usually as a flocculent mass toward the top with some discreet soft granules but also.

Fermentation reactions are given in Table 17.22; those reported for glycerol are in thioglycollate broth, in Actinomyces fermentation broth the result was 0/40. Other biochemical reactions are given in Table 17.23.

Most strains will grow in air on solid media. Optimum temperature is 35–37 C, no growth at 24 C.

The cell wall contains ornithine; glycine, leucine and aspartic acid are present in small amounts. Sugars include glucose, mannose, rhamnose, 2-deoxytalose and fucose (Cummins and Harris, 1958; Boone and Pine, 1968; DeWeese et al., 1968).

Antibiotic sensitivities have not been reported. Serologically designated as group A, with serotypes 1 and 2. Identifiable using the fluorescent antibody technique with low titer cross-reactions

with A. israelii and A. viscosus. Additional cross-reactions with A. odontolyticus and Arachnia propionica using gel diffusion or cell wall agglutination techniques (Cummins, 1962; King and Meyer, 1963; Georg et al., 1964; Slack and Gerencser, 1966 and 1970; Lambert et al., 1967).

Human infections have been reported. Experimental infections produced in mice (Howell et al., 1959; Coleman et al., 1969; Coleman and Georg, 1969). Periodontal destruction has been produced in rats (Socransky et al., 1970). Habitat is oral cavity of man including tonsillar crypts and dental calculus (Thompson and Lovestedt, 1951; Howell et al., 1962; Blank and Georg, 1968).

Suggested neotype strain: ATCC 12104.

5. Actinomyces viscosus (Howell, Jordan, Georg and Pine) Georg, Pine and Gerencser 1969, 292. (Odontomyces viscosus Howell, Jordan, Georg and Pine 1965, 65.)

vis.co'sus. L. adj. viscosus sticky.

Description is based in part on the work of: Howell, 1963; Howell and Jordan, 1963; Gerencser and Slack, 1969.

Microcolonies in 18–24 hr on brain heart infusion agar (incubated aerobically with CO_2) have a dense center with a filamentous fringe (observed at 100 X). After being transferred several times some strains will produce smooth microcolonies which are circular with a highly granular surface, entire edge and some have a small optically dark center. Macrocolonies on BHIA after 7 days incubation are circular, convex to heaped, smooth, entire, cream to white, glistening and opaque; soft or mucoid in consistency. Aerial hyphae have been reported (Howell, 1963). Hamster strains often have an eccentric pit near the apex and a granular or frosty appearance. Human strains show various radial or concentric striations and sometimes shallow central depressions. "Smooth" mature colonies are round, entire, low convex, smooth and transparent with an optically dark central "core."

In liquid media growth is diffuse with a viscous (strain variable) sediment which gives a mucoid or viscous rope when swirled.

Fermentation reactions are given in Table 17.22; glycerol was tested in thioglycollate broth, with Actinomyces fermentation broth the result was 0/56. Other biochemical reactions are given in Table 17.23.

End-products from glucose grown cultures include lactic acid as a major product with small amounts of formic, acetic and succinic acids (Howell et al., 1965).

Facultative anaerobe, grows better with CO_2; best growth obtained aerobically with CO_2; fixes CO_2 to form succinic acid.

The cell wall contains ornithine. Sugars include galactose, glucose, mannose and rhamnose (Howell et al., 1965; Gerencser and Slack, 1969).

Antibiotic sensitivities have not been reported. Serologically designated as group F with serotypes 1 and 2. Identifiable using the fluorescent antibody technique with low titer cross-reactions with A. naeslundii and A. odontolyticus. Gel diffusion tests show cross-reactions with A. viscosus, A. israelii and A. naeslundii (Lambert et al., 1967; Snyder et al., 1967; Gerencser and Slack, 1969 and 1970).

Periodontal disease with subgingival plaque occurs spontaneously in hamsters and may be produced experimentally with hamster isolates. Experimental infections have been produced in mice with hamster strains (Jordan and Keyes, 1964; Howell and Jordan, 1963; Georg and Coleman, 1968). Pathogenicity for man has not been established. This species has been isolated from the oral cavity of hamsters, rats and man (Howell et al., 1965, Gerencser and Slack, 1969).

The G + C content of the DNA of the type strain is 63 moles % (T_m).

Type strain: ATCC 15987 (Howell et al., 1965).

Species incertae sedis

Actinomyces eriksonii Georg, Robertstad, Brinkman and Hicklin 1965, 88.

er.ik.son'i.i. M.L. gen.n. eriksonii of Erikson; named for Dagny Erikson, in recognition of her classical studies on Actinomyces.

Description is based primarily on the study of 12 strains by Georg et al., 1965.

Diphtheroid or filamentous cells, generally 1 μm or less in diameter. Some cells are highly branched and occasionally clubbed. Bifurcated ends are common. Gram-positive.

Microcolonies in 48 hr on BHIA (observed at 100 ×) are flat, granular with a dense central core and no projecting filaments. Macrocolonies on BHIA after 7–14 days are dull white, soft, slightly convex to conical with a smooth surface and entire edge; the older colonies are more irregular. In liquid media diffuse growth with lobular soft colonies.

Fermentation reactions are given in Table 17.22 and other biochemical reactions in Table 17.23.

Strict anaerobe; CO_2 does not improve anaerobic growth or allow aerobic growth. Optimum temperature 37 C.

End-products from glucose-grown cultures include acetic, lactic and formic acids but not propionic; the molar ratio of acetic to lactic acid approximates 2:1.

The major amino acids in the cell wall are alanine, glutamic acid and lysine; galactose and small amounts of other sugars (Georg et al., 1964; Boone and Pine, 1968; DeWeese et al., 1968).

Antibiotic sensitivities have not been reported. Serologically identifiable by the fluorescent antibody technique and there are no cross-reactions with other species of Actinomyces. A one-line cross with A. israelii is demonstrable with the gel diffusion technique.

Has caused pulmonary and subcutaneous abscesses in man; isolated from sputum and tonsillar crypts. Minimal lesions have been produced experimentally in mice.

The G + C content of the DNA of the type strain is 62 moles % (T_m) (Georg et al., 1965).

Type strain: ATCC 15423.

Further Comments

On the basis of similarity in end-products from glucose fermentation and cell wall analysis coupled with some evidence of serological cross-reactions, A. eriksonii could be considered a species of Bifidobacterium (also see Further Comments under Bifidobacterium). On the basis of its filamentous morphology and pathogenicity it could continue to be considered a species of Actinomyces. This question can only be resolved by a comprehensive study of a large number of strains.

b. Actinomyces humiferus Gledhill and Casida 1969, 118.

hu.mif'er.us. L. n. humus soil; L. v. fero bear; M.L. adj. humiferus soil-borne.

Description is primarily from Gledhill and Casida, 1969.

Branched, filamentous cells, often with swollen terminal regions which fragment into diphtheroid or coccoid cells of varied size and shape. Gram-positive, not acid fast, non-motile and no spores.

Microcolonies in 18–24 hr on heart infusion agar are branching and filamentous or "spider-like." Macrocolonies in 7–10 days are small (less than 1 mm) opaque, entire, convex with a dark central area. Rough colony variation occasionally observed. Pigmentation not evident; no aerial hyphae.

In liquid media growth is granular or flocculant forming a white sediment without turbidity.

Production of acids from carbon compounds is given in Table 17.22. Other compounds tested: for utilization-fumarate 27/27, α-ketoglutarate 25/27; gluconate 6/27, citrate, propionate, succinate, oxalate all 0/27; for fermentation-gentiobiose 23/27, turanose 27/27.

Other biochemical reactions are given in Table 17.23; further tests: hydrolysis of casein 23/27, tributyrin 0/27; decomposition of tyrosine and xanthine 0/27; benzidine, oxidase, ammonia from peptone or arginine all 0/27.

Organic nitrogen required for growth.

Aerobic but will grow under reduced oxygen; poor or no growth anaerobically. Growth is not stimulated and may be inhibited by increased CO_2. In tubes of deep agar maximum growth is in the upper 1 cm. Optimum temperature 30 C; initial isolates unable to grow at 37 C.

End-products from glucose include lactic, succinic and acetic acids but not propionic.

The major amino acids in the cell wall are alanine, glutamic acid, lysine, ornithine and aspartic acid; the principal sugar is rhamnose with traces of glucose and fucose.

Cells sensitive to lysozyme.

Serologically there is no cross-reaction demonstrable by the fluorescent antibody technique using antisera specific for species of *Actinomyces* or *Rothia*.

Isolated from the soil; experimental infections could not be induced in mice.

The G + C content of the DNA averaged for 12 strains is 73 moles % (density gradient).

Type strain: *Actinomyces humiferus* ATCC 25174 (Gledhill and Casida, 1969).

Further Comments

This organism resembles *Actinomyces* on the basis of morphology, cell wall composition and end-products of glucose fermentation but differs in growing at 30 C, poorly or not at all at 37 C, being lysozyme sensitive and in its source. At present there is not sufficient information to judge the relationship of this organism to *Actinomyces*.

c. *Actinomyces suis* Grässer 1957, 148. (Not *Actinomyces suis* Gasperini 1892, 183.)

su'is. L. n. *sus* the hog, swine; L. gen.n. *suis* of the hog.

Grässer (1957) isolated two groups of *Actinomyces* from granules in udder actinomycosis of swine. One group was identical with *A. israelii* but the other group differed biochemically and serologically from *A. israelii* and *A. bovis* and was designated as *A. suis*. The following brief description is taken from Grässer 1962, 1963.

Three distinct colony types have been observed: smooth, transitional smooth-rough and rough colonies. Smooth forms are circular, completely smooth surface, fine granular or very rarely coarse granular. Colony consistency may vary from soft, buttery to tough, solid adhering tenaciously to the agar surface. The size of the colonies varies between 1–2 mm. Colonies on serum agar appear white. Brown to reddish brown colonies predominate on blood agar. Short bacterial forms are present on initial isolation giving rise to branched filaments on subculture. Filaments are observed at the periphery of smooth colonies more frequently than is the case with *A. bovis*. Transitional forms may appear upon subculture. Continued cultivation may produce convoluted, molar type colonies.

Microscopic examination of Gram-stained smears prepared from smooth colonies reveals rods of medium length and uniform thickness with variable staining characteristics. Branching, Y- and V-shaped forms are rarely observed. Gram-stained preparations of the rough colonies are dominated by long, curved, swollen, V- and Y-shaped forms with frequent branching. Pleomorphism increases with age as evidenced by the presence of irregularly stained short, coccoid and fragmented rods.

Type of growth in serum and glucose broth varies with colony type from uniformly turbid, to discrete compact masses or slimy-firm deposits.

All strains described are facultative. Aerobic growth is poor after 4 days, but visible colonies are evident anaerobically after 2 days.

Carbohydrate fermentation are shown in Table 17.22; acid is produced from esculin 6/6.

Gelatin is not hydrolyzed and no or poor growth in litmus milk.

Two serological types. Cross-reactions occur with *A. bovis* and *A. israelii*.

Isolates have been made from swine mastitis. Significance of this organism in the etiology of this pathological condition is still obscure. Efforts to produce experimental infections in hamsters have been unsuccessful (0/7).

Further Comments

A culture has not been available from Grässer but similar organisms have been isolated from swine (R. Robertstad, personal communication).

Nomina dubia

Actinomyces baudeti Brion 1942, 157. Inadequate description, no existing culture and illegitimate name.

Actinomyces discofoliatus Grüter 1933, 510. Inadequate description and no existing culture.

Actinomyces silberschmidti (Chalmers and Christopherson) Dodge 1935, 711. Inadequate description and no existing culture.

Actinomyces suis Gasperini 1892, 183. (not *A. suis* Grässer 1957, 148.) Not validly published, name listed only as synonym of *A. bovis* with no description.

Genus II. Arachnia *Pine and Georg 1969, 269*

LEO PINE AND LUCILLE K. GEORG

A.rach'nia. Gr. n. *arachnion*, a cob web; M.L. fem.n. *arachnia* referring to filamentous microcolonies.

Branched diphtheroid rods 0.2–0.3 μm by 3–5 μm, and **branched filaments** 5–20 μm or longer. Cells occasionally of uneven diameter, frequently with swollen or clubbed ends. **Swollen coccoid cells** and/or spheroplasts, of diameter to 5 μm, formed during stationary growth phase. **Gram-positive, non-acid fast, non-spore-forming** and **non-motile.**
Microcolonies definitely **filamentous,** composed of branched non-septate and septate filamentous elements often originating from a common center. **No aerial filaments.**
Chemoorganotrophs: Metabolism fermentative, carbohydrates being the fermentable substrates; **fermentation of glucose anaerobically yields carbon dioxide, acetic acid, propionic acid** and traces of DL-lactic and succinic acids; glucose fermented in air to carbon dioxide and acetic acid, but **fermentative products are not oxidized beyond acetate. Does not ferment lactate or pyruvate. Does not contain catalase.**
Facultative anaerobe, CO_2 is not required for aerobic or anaerobic growth.
Cell wall contains LL-diaminopimelic acid.
Pathogenic for man and laboratory mice.
Type species: *Arachnia propionica* Pine and Georg 1969, 269.

Further Comments

The genus *Arachnia* (Pine and Georg, 1969) was created for those organisms that have a definite filamentous phase either in the microcolony *in vitro* or in clinical material, but that differ markedly from *Actinomyces* species by having diaminopimelic acid in their cell walls and by fermenting glucose with the production of propionic acid as a major product. In these two latter respects the genus *Arachnia* is similar to *Propionibacterium.*

The genus *Arachnia* differs from the genus *Propionibacterium* by the consistent absence of catalase under a variety of experimental conditions, its limited and late fermentation of glycerol, its formation of filamentous colonies on agar medium, its pathogenicity in man and animals causing typical actinomycosis, its production of filamentous elements in pathological material, and finally its limited utilization of glucose (Pine and Hardin, 1959; Buchanan and Pine, 1962). Its DNA shows no homology with that of the two major groups of propionic acid-producing bacteria (Johnson and Cummins, 1972).
Thus, the genus *Arachnia* is closely related to *Propionibacterium* by its glucose fermentation and cell wall composition, but is also closely related to the genus *Actinomyces* not only by its morphological aspects but in its medical aspects, producing lacrimal infections and true cases of disseminated actinomycosis with the production of "sulfur granules" with or without clubs (Gerencser and Slack, 1967).
A consideration of the overall characteristics of the genus *Arachnia* emphasizes that the genus can only be distinguished from *A. israelii* or *A. naeslundii* by the demonstration of diaminopimelic acid in its cell wall and by the fermentation of glucose with the high production of propionic acid or by antigenic analyses (Gerencser and Slack, 1967). Demonstration of DAP and other cell wall components may be accomplished using the cell wall preparation and procedure described by Boone and Pine (1968).
For general procedures relating to the isolation and culture of the organisms, see *Further Comments* on the genus *Actinomyces* this volume and the publications: Pine (1963), Georg *et al.* (1964), Slack (1968), Georg (1970).

Description of the species of genus **Arachnia**

1. **Arachnia propionica** (Buchanan and Pine) Pine and Georg 1969, 269. (*Actinomyces propionicus* Buchanan and Pine 1962, 305.)
pro.pi.on'i.ca. M.L. fem.adj. *propionica* pertaining to propionic acid.
Microcolonies in 18–24 hr on brain heart infusion agar (observed at 100 ✕) are composed of branching filaments originating from a single point, indistinguishable from those of *A. israelii.* In 7–14 days on BHIA macrocolonies are circular to irregular; convex, umbonate or heaped;

entire edge; smooth or convoluted surface; white; opaque and soft. They may resemble the "molar-tooth" colonies produced by *A. israelii* and *A. naeslundii.*
In liquid media cells usually grow in compact masses of variable size which dissociate with age.
Acid but no gas is produced from various carbon sources (Table 17.24). Biochemical reactions are listed in Table 17.25.

TABLE 17.24

Acid production from carbon sources by strains of **Arachnia propionica**

Substrate	Strains positive[a]	Substrate	Strains positive[a]	Substrate	Strains positive[a]
Acetate	0/1	Glycerol	4/14	Melibiose	0/1
Adonitol	10/10	Glycogen	0/10	Pyruvate	0/10
Arabinose	0/14	Inositol	4/13	Raffinose	8/8
Cellobiose	0/8	Inulin	0/10	Rhamnose	0/14
Dextrin	0/1	Lactate	0/10	Ribose	0/1
Dulcitol	0/10	Lactose	14/14	Salicin	2/14
Erythritol	0/1	Maltose	14/14	Sorbitol	10/10
Fructose	4/4	Mannitol	14/14	Potato starch	1/1
Galactose	14/14	Mannose	4/4	Soluble starch	11/14
Glucose	14/14	α-Methyl-D-		Sucrose	14/14
α-Methyl-D-		mannoside	0/1	Trehalose	14/14
glucoside	0/1	Melezitose	0/4	Xylose	0/14

[a] Number of strains positive/number of strains tested.

TABLE 17.25

Biochemical reactions of strains of
Arachnia propionica

Test	Strains positive[a]	Test	Strains positive[a]
Catalase	0/14	Starch hydrolysis	
Indole	0/13	Wide zone	0/8
$NO_3 \rightarrow NO_2$	14/14	Narrow zone	8/8
Methyl red	6/10	Gelatin hydroly-	
Voges-Proskauer	0/10	sis	
H_2S (lead ace-		Tube	0/8
tate paper)		Plate	5/8
BHIA	0/8	Litmus milk	
Triple sugar	6/8	Acid only	2/8
iron		Acid and clot	4/8
Urease	0/13	No change	2/8

[a] Number of strains positive/number of strains tested.

Growth reported on complex media only; minimal nutritional requirements unknown.

The major amino acids in the cell wall are alanine, glutamic acid, glycine and LL-diaminopimelic acid. Sugars are galactose and small amounts of glucose (Buchanan and Pine, 1962; DeWeese *et al.*, 1968).

Serologically two serotypes by fluorescent antibody technique without cross-reactions with *Actinomyces* species (Gerenscer and Slack, 1968).

The G + C content of the DNA is 63–65 moles % (T_m) with five strains; there are no reports pertinent to bacteriophage, habitat or antibiotic sensitivities.

Causes human actinomycosis and lacrimal canaliculitis; pathogenic for laboratory mice (Buchanan and Pine, 1962; Gerencser and Slack, 1968).

Type strain: ATCC 14157.

Genus III. **Bifidobacterium** *Orla-Jensen 1924, 472*

Morrison Rogosa*

(*Tissieria* Pribram 1929, 376; *Bifidibacterium* Prévot 1938, 301.)

Bif′i.do.bact.er.i.um. L. adj. *bifidus* cleft, divided; Gr. dim.n. *bakterion* a small rod; M.L. neut.n. *Bifidobacterium* a cleft rodlet.

Rods highly variable in appearance; uniform or branched, bifurcated **Y** and **V** forms and **club** or **spatulate** forms are generally characteristic of freshly isolated strains. **Morphology** may be influenced by nutritional conditions and on subculture straight or curved rods, of varying width within a rod, may possess breaks resembling branching. Gram-positive, non-acid-

* This section was prepared in cooperation with the "Subgroup on Bifid Bacteria" of the International Committee on Systematic Bacteriology, International Association of Microbiological Societies and particularly with the members H. Beerens, P. A. Hansen, O. Kandler, G. Reuter, T. Mitsuoka, and F. Gasser.

fast, non-spore-forming, non-motile bacteria. **Often stain irregularly,** occasionally two or more granules may stain with methylene blue while the remainder of the cell may be unstained.

Saccharoclastic (Table 17.26); **gas is not produced. Glucose is fermented primarily to acetic and L(+) lactic acid,** generally in a molar ratio of 3:2. Small amounts of formic and succinic acids produced. CO_2, **butyric and propionic acids are not produced.** Fructose-1,6-diphosphate aldolase is absent; glucose-6-phosphate dehydrogenase activity is generally absent but is demonstrable in strains from bees and in *B. adolescentis.* Glucose is degraded by the fructose-6-phosphate shunt in which phosphoketolase cleaves fructose-6-phosphate into acetylphosphate and erythrose-4-phosphate; pentosephosphates are then formed by transaldolase and transketolase; xylulose-5-phosphate phosphoketolase splits substrate into acetylphosphate and glyceraldehyde-3-phosphate; through a reaction sequence also occurring in glycolysis, lactate is formed from glyceraldehyde-3-phosphate.

Catalase negative (*B. indicum* positive when grown in presence of hemin); benzidine reaction negative; indole negative; nitrates not reduced. Anaerobic, although there may be infrequent oxygen tolerance in the presence of CO_2.

Optimal temperature 36–38 C.

The G + C content (cesium chloride density or T_m) varies from 57.2–64.5 moles % (211 strains).

Type species: *Bifidobacterium bifidum* (Tissier) Orla-Jensen 1924, 472.

Further Comments

The first description of *Lactobacillus bifidus* in THE MANUAL (5th edition) was a composite one of *Bifidobacterium bifidum* and *Lactobacillus acidophilus.* Weiss and Rettger (1934) were also unable to define them clearly. The 6th edition made some attempt to differentiate between the two Weiss and Rettger organisms, but the 7th edition blurred any distinction.

Orla-Jensen (1919) had already recognized the separate generic status of bifid bacteria; the bifid bacteria are heterofermentative, producing acetic acid as the major product and L(+)-lactic acid but no CO_2 from glucose; with *L. acidophilus* (*Thermobacterium intestinale*) DL-lactic acid is the major product. Weiss and Rettger (1934) were probably unaware of the earlier Orla-Jensen (1919) work because they did not cite it. Orla-Jensen *et al.* (1936) again demonstrated the distinction between the two organisms. Weiss and Rettger (1938) cited a personal communication from Orla-Jensen objecting to their earlier conclusions that "The bulk of the acids produced by both types of organisms (*L. bifidus* and *L. acidophilus*) is inactive lactic acid" and that "... *L. bifidus* should be regarded as a variant of the species in which *L. acidophilus* is the central type." Weiss and Rettger (1938) corrected their previous descriptions and used the nomenclature *L. bifidus* type I and *L. bifidus* type II or *L. parabifidus.* The descriptions of these organisms are consistent with present concepts of *L. acidophilus* and *Bifidobacterium bifidum,* respectively.

Tissier (1900) believed that his organism was confined to nursing infants. This view was widely accepted and therefore isolates from adult feces were considered as different organisms. It is reflected in a strongly worded statement by Weiss and Rettger (1938) who insisted that their *Lactobacillus bifidus* type I "is the one originally isolated by Tissier from the stools of breast-fed infants, namely, his '*B. bifidus*'." These notions were mistaken because bifid organisms have been isolated from infants, adults and animals. An original Tissier (1900) strain has since been found extant at the Pasteur Institute in Paris and was studied by Werner and Seeliger (1964) and Dehnert (1965) and has the attributes of group 1 and 2. Furthermore, extant Weiss and Rettger strains which they named *Lactobacillus bifidus* have the characteristics of *Lactobacillus acidophilus.* Thus, *Lactobacillus bifidus* type I is not a synonym of *Bacillus bifidus* Tissier as they stated. Therefore, *Lactobacillus bifidus* type II, *Bacteroides bifidus* group 2 (Eggerth, 1935), *Bacterium bifidum* Orla-Jensen *et al.* (1936), *Bacillus bifidus* Tissier (1900), *Bifidobacterium bifidum* Orla-Jensen (1924) and *Actinomyces parabifidus* Pine and Georg (1965) are here considered as synonyms.

Reuter (1963) has given names to what had been considered biotypes, namely, *Bifidobacterium adolescentis, B. longum, B. breve, B. parvulorum, B. infantis, B. liberorum* and *B. lactentis.* The cell wall characteristics (reported by O. Kandler to the Subgroup on Bifid Bacteria) are different from those of *B. bifidum. B. longum* also has ornithine in the peptidoglycan tetrapeptide but the cross-linking peptide is L-serine-L-alanine-L-threonine-L-alanine. *B. infantis, B. parvulorum* and *B. breve* contain lysine in the tetrapeptide and are cross-linked by a single glycine residue. *B. aldolescentis* has a lysine type of tetrapeptide and is cross-linked by a single D-aspartic acid or a tripeptide (L-serine-L-alanine-L-alanine). Six bovine rumen strains had an ornithine type of tetrapeptide cross-linked by an L-alanine-L-alanine dipeptide.

The previous groupings of Dehnert (1957, 1960) Lerche and Reuter (1961), Seeliger and Werner

(1963) and Werner (1966) would have placed *B. bifidum* in groups 1 or 2; *B. infantis* in groups 3 or 4; *B. breve* in groups 3 or 6 and *B. longum* in Group 5, based on some carbohydrate fermentation differences and sources of isolation. Bifids have been isolated from the intestinal tract or feces of human infants and adults, pigs, fowl, mice, rats, guinea pigs and bees, from ovine and bovine rumens, and from the human vagina.

Subsequently, DNA-DNA homology studies (Scardovi et al., 1971) have shown that *B. bifidum* is genetically distinct; *B. adolescentis* also appears genetically distinct; *B. infantis*, *B. liberorum* and *B. lactentis* have homologous DNA and are therefore here merged under *B. infantis* as suggested by Scardovi *et al.* (1971); *B. breve* and *B. parvulorum* are genetically homologous and are

here merged under *B. breve*; *B. ruminale* Scardovi *et al.* and *B. thermophilum* Mitsuoka are genetically homogeneous and are here merged under *B. thermophilum*.

Additional references on which description is based are: de Vries *et al.* (1967); de Vries and Stouthamer (1967, 1968); Sebald, Gasser and Werner (1965); Pine and Howell (1956); Lerche and Reuter (1961); Reuter (1963, 1971); Kandler (1967, 1970); Kandler *et al.* (1968); Sundman *et al.* (1959); Zillikin *et al.* (1954) (including pertinent references cited therein); Kojima *et al.* (1968); Matteuzzi *et al.* (1971) Mitsuoka (1969); Scardovi (1964); Scardovi and Trovatelli (1965, 1969) and Scardovi *et al.* (1969, 1970, 1971).

Substrates fermented by the various species named in the literature are shown in Table 17.26.

Description of the species of genus **Bifidobacterium**

1. Bifidobacterium bifidum (Tissier) Orla-Jensen 1924, 472. (*Bacillus bifidus communis* Tissier 1900, 85; *Bacillus bifidus* Tissier 1900, 86; *Bacteroides bifidus* (Tissier) Castellani and Chalmers 1919, 960; *Bacterium bifidum* (Tissier) Lehman and Neumann 1927, 513; *Tissieria bifida* (Tissier) Pribram 1929, 376; *Nocardia bifida* (Tissier) Vuillemin 1931, 132; *Actinomyces bifidus* (Tissier) Nannizzi 1934, 13; *Actinobacterium bifidum* (Tissier) Puntoni 1937, 167; *Bifidibacterium bifidum* (Tissier) Prévot 1938, 303; *Lactobacillus bifidus* type II Weiss and Rettger 1938, 18; *Lactobacillus parabifidus* Weiss and Rettger 1938, 18; *Cohnistreptothrix bifidus* (Tissier) Negroni and Fischer 1944, 327; *Actinomyces parabifidus* (Weiss and Rettger) Pine and Georg 1965, 154.)

bi'fi.dum. L. neut.adj. *bifidum* cleft, divided one.

Gram-positive but may stain irregularly as the culture ages; methylene blue may stain internal granules but not the entire cell; rods highly variable in appearance.

Surface colonies on agar plates incubated anaerobically are usually circular, convex or lens shaped; whitish, opaque, with smooth to mucoid soft surface. Mycelium not formed. In agar deeps colonies are variable in shape and no growth near the surface.

In glucose broth incubated anaerobically good growth with turbidity and eventual clearing except for flocculent precipitate. Final pH is 4.0–4.8.

Freshly prepared milk is acidified and often coagulated, particularly with cysteine added; 1.3–1.4% titratable acidity may sometimes be detected but range of acidity is wide and may be as little as 0.1% particularly with group 1 and 2 strains (see *Further Comments*).

Gelatin not hydrolyzed. H_2S negative. Ammonia not produced from arginine. Esculin hydrolyzed (loss of long wave ultraviolet fluorescence).

Organic nitrogen required for growth; growth dependent on the presence of fermentable carbohydrate.

The cell wall murein contains muramic acid and glucosamine, a tetrapeptide consisting of L-alanine, D-glutamic acid amide, L-ornithine and D-alanine; the cross-linking dipeptide bridge between the ε-amino group of ornithine and the carboxyl group of D-alanine in the adjacent tetrapeptide consists of L-serine and L-aspartic acid.

Optimum temperature 36–38 C; good growth from 32–38 C; variably limited growth 23–25 C; no growth at 20 C or below or at 45 C.

Optimum initial pH 6–7; little or no growth at pH 5.5 or less.

Anaerobic; dies rapidly in aerobic subcultures. Where investigated, CO_2 required for growth.

Bacteriophage has been reported (Youssef *et al.* 1970).

Pathogenicity not reported for man or animals.

Originally isolated from stools of breast-fed infants. Found in alimentary tract and stools of breast and bottle-fed infants and human adults.

Type strain: Ti (Tissier) Anaerobe Collection, Institute Pasteur, Lille, France.

2. Bifidobacterium adolescentis Reuter 1963, 502.

ad.o.les.cent'is. L. n. *adolescens* adolescent; M.L. gen.n. *adolescentis* of an adolescent.

Reuter (1963) described a group of short, curved, occasionally bifurcated and anaerobic rods; he proposed four biotypes (a, b, c, d) based on sero-

TABLE 17.26

Fermentation reactions of named species of genus **Bifidobacterium**[a]

Previously named species	Arabi-nose	Xy-lose	Ri-bose	Gluco-nate	Cello-biose	Lac-tose	Man-nitol	Melez-itose	Sali-cin	Starch	Tre-halose
1. *B. bifidum* Orla-Jensen	–	–	–	–	+	+	–	–	–	–	–
2. *B. adolescentis* Reuter	+	+	+	+	+	+	v	v	+	v	v
3. *B. infantis* Reuter	–	–	+	–	v	+	–	–	v	v	v
B. liberorum Reuter	–	+	+	–	v	+	–	v	v	v	v
B. lactentis Reuter	–	+	+	–	v	+	+	–	–	v	v
4. *B. breve* Reuter	–	–	+	–	+	+	+	v	v	v	v
B. parvulorum Reuter	–	–	+	–	v	+	+	–	–	+	–
5. *B. longum* Reuter	+	+	+	–	–	+	–	+	–	v	v
B. longum subsp. *animalis* Mit-suoka	+	+	+	–	–	+	–	–	v	v	v
6. *B. pseudolongum* Mitsuoka	+	+	+	–	v	v	–	v	v	+	–
7. *B. thermophilum* Mitsuoka	–	–	–	–	v	v	–	v	v	+	v
8. *B. suis* Matteuzzi et al.	+	+	–	–	–	+	–	–	–	–	–
9. *B. asteroides* Scardovi et al.	+	+	+	+	+	–	–	–	+	–	–
10. *B. indicum* Scardovi et al.	–	–	+	+	+	–	–	–	+	–	–
11. *B. coryneforme* Scardovi et al.	+	+	+	+	+	–	–	–	+	–	–

[a] + = positive; – = negative; v = variable, slight, delayed or erratic reactions by the same strain or related strains (so-called biotypes).

All bifidobacteria ferment glucose, galactose and fructose (sometimes slowly). *B. bifidum* biotype a (from adults) slowly ferments sucrose and melibiose; *B. bifidum* biotype b (from infants) does not ferment sucrose but vigorously ferments melibiose; neither ferments maltose or raffinose. Other bifidobacteria (nearly 100%) ferment maltose, melibiose, raffinose and sucrose. Inositol is fermented only by some strains of *B. liberorum* and *B. lactentis*. Inulin is fermented only by some strains of *B. adolescentis*, *B. infantis* and *B. liberorum*. Esculin and amygdalin are fermented by *B. adolescentis* and *B. breve*; *B. infantis*, *B. liberorum*, *B. longum* are negative and *B. thermophilum* is equivocal; remaining have not been tested. Sorbitol is fermented only by certain strains of *B. breve* and *B. adolescentis*. Mannose is fermented only by certain strains of *B. breve*, *B. longum*, *B. suis* and generally slowly, variably, or not at all by the others. Bifidobacteria do not ferment adonitol (ribitol), dulcitol, erythritol, glycerol, rhamnose and α-methyl-D-mannoside.

logical reactions and differences mainly in the fermentations of mannitol and sorbitol.

End-products from growth on glucose: acetic and L(+)-lactic acids by pathway as described for the genus.

However, gluconate fermentation is inducible with acid and gas (CO_2) production, indicating that at least the oxidative parts of the hexose monophosphate shunt, i.e. glucose-6-phosphate and 6-phosphogluconate dehydrogenases may be present but are not significantly expressed during growth with glucose; the only other known gluconate fermenting bifids are from bees, i.e. *B. asteroides*, *B. indicum* and *B. coryneforme*; there is no DNA homology between these and strains otherwise phenotypically identifiable as *B. adolescentis* (Scardovi et al., 1971).

Optimum growth at about 35–37 C; no growth at 46.5 C and 20 C.

Isolated from human adult and infant feces, the appendix, dental caries, and the vagina.

Type strain: E 194a; ATCC 15703 biotype a (Reuter 1971), isolated from human adult feces.

Note. DNA-DNA homology studies (Scardovi et al., 1971) indicate that there is often little genetic relatedness between strains having the biochemical phenotypic characteristics of Reuter's (1963) biotype groups. For example, some biotype a strains have DNA completely homologous with some group b or c reference DNA; generally only DNA from group b strains show complete homology with group b reference DNA; generally biotype c DNA-DNA homology is only strain specific and this may be frequently so with biotype d strains. Because of the high frequency of biotype b strains from dental caries and their relatively high DNA homogeneity, Scardovi et al., (1971) suggest these may be genetically different from other strains phenotypically identifiable as *B. adolescentis* and would tentatively consider them as group "dentium" distinct from *B. aldolescentis*. An unassigned group from waste waters whose cells are arranged in irregular, sometimes long chains and which is phenotypically like biotype c but genetically distinct, Scardovi et al. (1971) consider as unassigned group "catenula-tum"; certain other genetically distinct strains from waste waters have an angular cell arrangement, are not completely phenotypically identifiable as *B. adolescentis* because of a failure to ferment gluconate, but are tentatively referred to as unassigned group "angulatum" (Scardovi *et al.*, 1971). Obviously, phenotypy and genotypy do not necessarily correspond in the bifidobacteria resembling *B. adolescentis*.

3. Bifidobacterium infantis Reuter 1963, 502. (*Bifidobacterium liberorum* Reuter 1963, 504; *Bifidobacterium lactentis* Reuter 1963, 504.)

in.fant′is. L. n. *infans* an infant; M.L. gen.n. *infantis* of an infant.

Cells are small, thin, often spherical or bubble shaped, often containing central granules thus resembling an eye, with no branching tendency. Strains previously designated as *B. liberorum* and *B. lactentis* ferment xylose; inositol is fermented by *B. infantis* biotype b and *B. liberorum* and more slowly by *B. lactentis*; cellobiose and trehalose are fermented by all except *B. infantis* b; salicin is fermented by *B. infantis* a and *B. liberorum*; and mannitol is fermented by *B. lactentis*; some of these minor differences are strain differences and may vary with media and other test conditions.

Anaerobic. No growth at 46.5 C and 20 C.

Most pertinent literature indicates that these designated strains are the predominant bifidobacteria in the feces of breast-fed infants; they would previously have been placed in Dehnert's (1957, 1960) group III or IV.

Scardovi *et al.* (1971), studying DNA homology of type strains, found a close genetic relatedness among strains designated as *B. infantis*, *B. liberorum* and *B. lactentis* and proposed merging them into a single species for which the name *B. infantis* has page priority. The author has accepted his proposal.

Type strain; S12; ATCC 15697 (Reuter, 1971).

4. Bifidobacterium breve Reuter 1963, 502. (*Bifidobacterium parvulorum* Reuter 1963, 502.)

bre′ve. L. neut.adj. *breve* short.

Cells short, slender or thick, often club-shaped rods, with or without bifurcations. Granules sometimes demonstrable by Gram-stain. Cells may autoagglutinate in saline.

Colonies convex to pulvinate, smooth or undulating surface, entire, 2-3 mm in diameter, soft consistency.

Positive reactions in amygdalin, esculin and cellobiose may be slow. Strains previously designated as *B. breve* have been described as fermenting sorbitol and mannitol, whereas *B. parvulorum* was negative.

End-products from glucose: acetic and L(+)-lactic acids; no gas.

Anaerobic. No growth at 46.5 C and 20 C.

Isolated from infant feces and from the vagina.

DNA homology: type strains and other strains previously designated as *B. breve* and *B. parvulorum* demonstrate DNA homology of 88–94% and in reciprocal reactions 98–106%. Certain strains of *B. infantis*, *B. liberorum* and *B. lactentis* show 45–50% DNA homology with *B. breve*. The DNA from other species has little or no homology with that of *B. breve* (Scardovi *et al.*, 1971). *B. breve* would have been placed previously in Dehnert's (1957, 1960) group III or Werner and Seeliger's (1963) group VI.

Type strain: S1; ATCC 15700 (Reuter, 1971).

5. Bifidobacterium longum Reuter 1963, 502.

long′um. L. neut.adj. *longum* long.

Cells long, curved, club-shaped, swollen or dumb-bell shaped rods which may be bifurcated. Gram-variable.

Colonies convex to pulvinate, entire, 2–5 mm in diameter, soft, moist, shining or slimy.

End-products from glucose: acetic and L(+)-lactic acids; no gas.

Bifidobacteria which ferment pentoses but not gluconate have been customarily assigned to this species and would have previously been placed in Dehnert's (1957, 1960) group V.

Anaerobic.

No growth at 46.5 C and 20 C.

Isolated from feces of infants and adults, and from the intestine of rats, guinea pigs and calves.

DNA homology values vary (generally between 60–92%) and depend on reference DNA used (further work is required—see Scardovi *et al.*, 1971).

Type strain: E194b; ATCC 15707 (Reuter, 1971).

Further Comments

Further work is required to show whether *Bifidobacterium longum* Reuter subsp. *animalis* Mitsuoka biotype a 1969, 60 is a synonym of *B. longum* Reuter 1963, 502 or an independent species.

6. Bifidobacterium pseudolongum Mitsuoka 1969, 60. (*Bifidobacterium globosum* Scardovi, Trovatelli, Crociani and Sgorbati 1969, 290.)

pseu.do.long′um. Gr. adj. *pseudés* false; L. adj. *longum* specific epithet; M.L. neut.adj. *pseudolongum* false (*B.*) *longum*.

Young cells have coryneform appearance with short, coccoidal to curved or tapered cells, often arranged angularly, singly, doubly or rarely in short chains. In 90% air + 10% CO_2 cells on slopes have bifurcations or short cross-branchings often with enlarged ends or longer elements may be curved like vibrios.

Colonies smooth, convex, entire edges, cream to white, glistening, soft, easily emulsified.

Fluid cultures easily dispersible compact sediment; supernatant often clear.

End-products from glucose: acetic and L(+)-lactic acids (ca. 3:1); no gas. Catabolic pathway during growth on glucose as described for the genus; however, low specific activities of fructose-1,6-diphosphate and 6-phosphogluconate dehydrogenases in some cell extracts make mechanism of glucose utilization unclear.

Ammonia not produced from arginine. CO_2 not evolved from malate or citrate.

Growth factors required pantothenate, riboflavin, thiamine and folic acid; adenine, guanine, xanthine, uracil not required; ammonia satisfies nitrogen requirements.

Growth in stabs incubated in air begins a few millimeters below the surface; very few strains require CO_2 in stabs; none or limited delayed growth on slopes incubated in air; some growth on slopes incubated in 90% air + 10% CO_2. Thus, growth in high oxygen tensions occurs only if CO_2 is present, whereas, CO_2 for anaerobic growth is stimulatory but dispensable.

Temperature optimum 39–40 C; usually grows at 45 C; no growth at 20 C.

pH optimum 6.5–7.0; growth retarded at pH 6.0. No growth at pH 8.0.

The G + C content of the DNA is 64.5 ± 1.4 moles % (T_m).

DNA-DNA homology with *B. globosum* Scardovi *et al.* 1969, 290 is 69–73%; (see Scardovi *et al.*, 1971); homology with the DNA of other groups or species of bifidobacteria is negative or insignificant.

Isolated from feces of swine, chickens, rat and mouse, ovine and bovine rumens.

Type strain: PNC-2-9G, type a from feces of swine (Mitsuoka, 1969); ATCC 25526.

Further Comments

Mannitol-fermenting strains resembling *B. pseudolongum* Mitsuoka (1969) as reported by Scardovi *et al.* (1969) have been determined to be *B. adolescentis.*

7. **Bifidobacterium thermophilum** Mitsuoka 1969, 59. (*Bifidobacterium ruminale* Scardovi, Trovatelli, Crociana and Sgorbati 1969, 291.)

ther.mo'phil.um. Gr. n. *therme* heat; Gr. adj. *philus* loving; M.L. adj. *thermophilum* heat-loving.

Slender rods 3–8 μm long, slightly curved often with tapered ends, protuberances or irregularities near the junction of paired cells, branchings rare, arranged singly or in pairs, never in clumps or in angular disposition. Aged cells often banded and take Gram-stain irregularly.

Ferments the same sugars as *B. pseudolongum* except that arabinose, xylose and ribose are not fermented; end-products from glucose acetic and L(+)-lactic acids; no gas.

Serologically distinct from *B. pseudolongum* but cross-reacts with group II rumen strains of Scardovi *et al.* (1969).

pH optimum 6.5–7.0; no growth at <5.0 (more acid tolerant than *B. globosum*); no growth at 8.0.

Usually grows at 46.5 C.

The G + C content of the DNA 60 ± 1.5 moles % (T_m). Complete homology with *B. ruminale*; 18–54% with *B. pseudolongum.*

Isolated from the bovine rumen, feces of swine and chickens.

Type strain: P2-91, type a from feces of swine (Mitsuoka, 1969); ATCC 25525.

8. **Bifidobacterium suis** Matteuzzi, Crociani, Zani and Trovatelli 1971, 393.

su'is. L. n. *sus* the hog, swine; L. gen.n. *suis* of the hog.

Cells slender, elongated, 2–6 μm long, with rare terminal bifurcations of clubs.

Colonies circular, soft, smooth, white, with entire margins.

Fluid cultures turbid, clearing after 1–2 days, with dispersible sediment.

End-products from glucose: lactic and acetic acid mole ratios vary from 1:1.7–1:2.0. Glucose catabolic pathway similar to *B. bifidum* although *B. suis* cell-free extracts may have fructose-1,6-diphosphate aldolase activity and dehydrogenase activities with glucose-1,6-phosphate and 6-phosphogluconate; however, gluconate does not appear to be normally fermented by growing cells.

Acetylmethylcarbinol not produced. Skim milk acidified and coagulated in 1–2 days.

No growth in carbohydrate-free medium. Riboflavin is the only demonstrable vitamin required for growth.

Anaerobic, but weak growth on slopes in air + CO_2; no growth on slopes in air.

Optimum temperature 38–39 C; minimum, 19–20 C; maximum, 44.5–45 C; does not survive at 60 C for 30 min.

Optimum pH 7–8; growth between 5.3–9.4

The G + C content of the DNA is 62 moles % (T_m). Genetically not related to other species of bifidobacteria from man and animals other than the pig.

Type strain: SU859 from the Collection of Scardovi, Istituto di Microbiologia Agraria Universita di Bologna, Italia.

9. **Bifidobacterium asteroides** Scardovi and Trovatelli 1969, 83. (*Bacillus constellatus* White 1921, 69.)

as.ter.oi'des. Gr. adj. *asteroides* star-like.

Cells grown anaerobically, under 5% or more CO_2, in fresh rich media are 2–2.5 μm long, pear shaped or slightly curved and tend to have pointed ends; frequently arranged radially in star-like groups around a mass of common hold-fast material (phase-contrast microscopy). Nutritional or CO_2 deficiencies, or growth with certain sugars, may induce clavate or spatulate cells with occasional central swellings, irregular or cross-like branchings in the central part of the cell body.

Colonies are circular, smooth, convex, with entire edge and glistening surface; consistency is such that entire colony is removed by needle and can hardly be emulsified in water. Deep colonies are lens shaped with entire margins.

Growth in static fluid culture tends to adhere to the glass walls and to leave the liquid clear.

CO_2 required for growth in all media including stabs; aerobic growth on agar slopes is limited or restricted and occurs only if air is enriched with CO_2 (350 strains).

Trehalose (6%) and melezitose (4%) occasionally fermented; some strains may not ferment maltose (24%), cellobiose (12%) or pentoses (6%).

Acids and CO_2 from gluconate with intermediate pentose production; glucose-6-phosphate and 6-phosphogluconate dehydrogenase activities in cell extracts but apparently not in growing cultures where CO_2 is not formed from glucose. Acetate and L(+)-lactate are chief products from glucose and glucose catabolic pathway is similar to that of *B. bifidum*.

H_2O_2 decomposed when cells are grown in the presence of O_2.

Litmus milk unchanged. Ammonia not produced from arginine. CO_2 not produced from citrate and malate.

Biotin, pyridoxine, nicotinic acid, riboflavin, thiamine and pantothenate may be required; folic acid, p-aminobenzoic acid, adenine, guanine, xanthine and uracil are not required for growth.

Optimum temperature about 35 C; no growth at about 20 C or at 42 C.

The tetrapeptide in the cell wall peptidoglycan consists of L-alanine, D-glutamic acid amide, L-lysine and D-alanine; one glycine residue crosslinks adjacent tetrapeptides between the ε-amino group of lysine and the carboxyl group of a C-terminal D-alanine.

The DNA is about 30% homologous with DNA from *B. indicum*, but there is no significant homology with DNA from bifids isolated from animals other than bees.

Pathogenicity not reported for man or animals.

Isolated from the hind gut of live honeybees *Apis mellifica* L. and its varieties *A. m. ligustica*

Spin. and *A. m. caucasica* G; not found in *Apis indica* F.

Type strain: C51; ATCC 25910; from the collection of Scardovi and Trovatelli, Istituto di Microbiologia Agraria, Universita di Bologna, Italia.

10. Bifidobacterium indicum Scardovi and Trovatelli 1969, 84.

in'di.cum. M.L. neut.adj. *indicum* from specific epithet of bee, *Apis indica*.

Cells from solid media are 2–2.5 μm long, generally in pairs, with slightly bifurcated ends to give bone-like appearance; older cells are characteristically staghorn-shaped; star-like clusters not observed.

Colonies indistinguishable from those of *B. asteroides*.

In liquid media there is even turbidity and dispersible sediment.

Acids and CO_2 from gluconate; ribose fermented.

Catalase positive only when grown in presence of hemin.

Oxygen tolerance similar to *B. asteroides*; CO_2 required aerobically and anaerobic CO_2 requirement may be equivocal.

Remaining characteristics are very similar to those of *B. asteroides*.

DNA-DNA reciprocal homology with *B. asteroides* is about 30%. There is no homology with DNA of bifids from any animal species other than bees.

Isolated from the intestine of *Apis indica* F., a bee from Malaysia.

Type strain: C410; ATCC 25912; from the collection of Scardovi and Trovatelli; Istituto di Microbiologia Agraria, Universita di Bologna, Italia.

11. Bifidobacterium coryneforme Scardovi and Trovatelli 1969, 85.

cor.yn'e.form.e. Gr. n. *coryne* a club; L.n. *forma* shape: M.L. neut.adj. *coryneforme* club-shaped.

Cell morphology is suggestive of corynebacteria; short cells 1–1.5 μm long, often lanceolate or of irregular shape with knobs or rudimentary branching, mostly in pairs with angular disposition. Star-like groups generally absent.

Colonies indistinguishable from those of *B. asteroides* and *B. indicum*.

Uniform turbidity in liquid media; dispersible sediment.

Growth scanty in Brewer thioglycollate medium suitable for *B. asteroides* and *B. indicum*; good growth in MRS medium; no growth on slopes incubated in air + CO_2; strict anaerobe; CO_2 apparently unnecessary for growth initiation.

Catalase negative under all conditions.

Acids and CO_2 from gluconate; ribose fermented.

Remaining characteristics are very similar to those of B. asteroides and B. indicum.

DNA-DNA reciprocal homology with B. asteroides and B. indicum ranges from 60–86%. There

is no homology with DNA of bifids from any other animal species other than bees.

Isolated occasionally (11 strains) from the intestine of Apis mellifica L.

Type strain: C215; ATCC 25911; from the collection of Scardovi and Trovatelli, Istituto di Microbiologia Agraria, Universita di Bologna, Italia.

Genus IV. **Bacterionema** Gilmour, Howell and Bibby 1961, 139

MARION N. GILMOUR

Bac.ter.i.o.ne′ma. M.L. n. bacter, masc.equiv. of Gr. neut.n. bactrum, a staff or rod; Gr. n. nema a thread; M.L. n. Bacterionema a thread-shaped long rod.

Cells are **pleomorphic**, comprising **non-septate** and **septate filaments** 1–1.5 μm by 20–200 μm and **bacilli** 1.5–2.5 μm by 3–10 μm. **Characteristic morphology is a bacillus attached to a filament** ("whip-handle," Plate 17.4, Fig. 1). **Reproduction by filament septation and fragmentation** to form rods which germinate to yield one to four filaments. Branching is frequent with aerobic and/or acid conditions. **Gram-positive, non-acidfast, no endospores** and **non-motile.** Metachromatic granules formed.

Microcolonies are spider-like (Plate 17.4, Fig. 2).

Chemoorganotroph; carbohydrates are usually **fermented to yield acid and some gas.** Products from glucose-grown cultures include acetylmethylcarbinol and: with aerobiosis, CO_2 and propionic acid are major, formic, acetic and lactic acids are minor and/or variable; with reduced oxygen, CO_2 and lactic acid are major,

formic, acetic, propionic and succinic acids are minor.

Facultative anaerobe; depending upon cultivation conditions, some isolates may appear to be strict aerobes or anaerobes, but later adapt to facultative growth.

Type species: Bacterionema matruchotii (Mendel) Gilmour, Howell and Bibby 1961, 139.

Further Comments

The genus Bacterionema was originated by Gilmour, Howell and Bibby (1961) to separate these branching filamentous organisms from the physiologically different, non-branching filament, Leptotrichia.

The genus is currently placed in the family Actinomycetaceae because of its filamentous branching morphology, but is excluded from the Actinomyces and Nocardia because of cell size, the whip-handle cells and physiological char-

TABLE 17.27
Acid production from carbon sources by strains of **Bacterionema matruchotii**[a]

Substrate	Strains positive	Substrate	Strains positive	Substrate	Strains positive
Arabinose	0/151	Glycerol	0/150	Raffinose	78/159
Cellobiose	0/90	Inositol	0/33	Rhamnose	0/152
Dextrin	154/155	Inulin	0/221	Salicin	118/129
Dulcitol	0/145	Lactose	8/262	Sorbitol	0/144
Fructose	173/173	Maltose	234/249	Soluble starch	1/22
Galactose	0/170	Mannitol	8/182	Sucrose	269/270
Glucose	317/317	Mannose	122/123	Trehalose	3/134
α-Methyl-D-glucoside	0/118	Melezitose	0/12	Xylose	0/151
		Melibiose	0/102		

[a] Number of strains producing acid in stationary aerobic culture/number of strains tested. The ratios were similar for both aerobically and anaerobically derived and maintained strains, incubated aerobically or anaerobically. Exceptions: aerobic shake cultures, maltose 0/21; anaerobic jar with heated catalyst, maltose 29/42, mannose, 27/42, raffinose 17/56, salicin 0/42.

Acetate, adonitol, erythritol, glycogen, α-methyl-D-mannoside, pyruvate, ribose and potato starch not tested.

TABLE 17.28

Biochemical reactions[a] of strains of **Bacterionema matruchotii**

Test	Strains positive[b]	Test	Strains positive[b]
Benzidine[c]	25/25	Ammonia production	0/134
Catalase[d]	195/195	Hippurate hydrolysis	96/99
Indole	0/281	Starch hydrolysis[f]	67/67
Nitrate → nitrite	305/307	Gelatin hydrolysis	0/264
Methyl red	0/116	Litmus milk	0/117
Voges-Proskauer[e]	171/172	Lipase production	0/44
H₂S	0/176	Acid phosphatase	0/44
Urease production	25/56	Growth in 10% bile	0/36
Esculin hydrolysis	89/99	Growth in 6.5% NaCl	0/36

[a] From stationary aerobic cultures. Unless otherwise specified, ratio values are similar with aerobically and anaerobic jar grown cultures from either aerobically or anaerobically derived strains. Anaerobic jars were with heated catalyst.

[b] Number of strains positive/number of strains tested.

[c] Aerobically derived stains grown aerobically.

[d] Aerobically derived strains grown aerobically and when grown in an anaerobic jar 22/60 with 32/60±. Anaerobically derived strains when grown aerobically 29/38 and when grown in an anaerobic jar 6/38, with 23/38±.

[e] With aerobic shaking 33/37; with anaerobic jar stationary 5/35.

[f] Anaerobic jar 17/71.

cteristics (Bibby, 1935; Bibby and Berry, 1939; Davis and Baird-Parker, 1959; Gilmour and Beck, 1961; Howell and Pine, 1961; Melville, 1965; Snyder *et al.*, 1967). However, from studies on a small number of aerobically isolated and maintained strains, it has been suggested that there is a closer relationship to *Corynebacterium* and *Mycobacterium* (Davis and Baird-Parker, 1959; Schmidt and Richardson, 1962; Melville, 1965; Pine and Georg, 1965).

The genus is presented with but one species because the majority of isolates are very similar by the criteria used to date. However, Bulleid (1925) and Morris (1954) described 27 anaerobically derived isolates which were morphologically similar but which appeared to be physiologically different. The Morris strains were not cultivable in air, and were lost before complete characterization. Others have reported the loss of similar isolates. Successful serial strict anaerobic growth of many *Bacterionema* strains requires the use of minimal air exposure and of media which are either freshly prepared or to which hemin is added (e.g. 0.2 μg hemin/ml to one-half strength brain heart infusion supplemented with 0.1–0.2% yeast extract). Because these conditions have not yet been used routinely during isolation, the above culture losses are explicable. With the appropriate cultivation conditions, successful culture of such isolates probably will allow other species to be established.

However, it is also possible that with additional information isolates which are currently placed in the single species will be separated. Although they are similar in sugar fermentation patterns and other biochemical tests, the strain differences in catalase reactions and gaseous growth responses of initial isolates may become sufficiently amplified to warrant separate speciation.

Description of the species of genus **Bacterionema**

1. **Bacterionema matruchotii** (Mendel) Gilmour, Howell and Bibby 1961, 139. (*Cladothrix matruchoti* Mendel 1919, 584; *Oospora matruchoti* (Mendel) Sartory 1920, 813; *Actinomyces matruchoti* (Mendel) Nannizzi 1934, 51.)

ma.tru.cho′ti.i. M.L. gen.n. *matruchotii*, of Matruchot; named for Professor Matruchot, a French mycologist.

Pertinent current or review articles relating to morphology and physiology are: Ludwig, 1955; Howell and Pine, 1961; Gilmour, 1961, 1962; Gilmour and Beck, 1961; Kroeger and Sibal, 1961; Takazoe *et al.*, 1963; Takazoe and Nakamura, 1965; Kasai, 1965; Gilmour and Bibby, 1966; Winford and Haberman, 1966; Takazoe and Ennever, 1969.

A bacillus body attached to a filament is characteristic (Plate 17.4, Fig. 1). Adherent spherical elements less than 0.4 μm, occur primarily during stationary phase, but have no known function. Mesosomes formed. The quantity of polymetaphosphate metachromatic granules correlates inversely with the intracellular deposition of hydroxyapatite crystals in decline phase cultures. Lipid inclusions formed. Cell wall, Grampositive type by electron microscopy, having an additional constant outer thin dense-light-dense layer which gives rise to minute spherical elements adjoining the filaments. Filament wall thickness 8–10 nm, bacillus wall thickness 45–50 nm. Thin slime layer. Cells grown anaerobically are larger and more filamentous than those grown aerobically, but are otherwise similar.

Gram-positive, but Gram-negative filaments containing Gram-positive granules may occur and these correlate with the loss of culture transferability.

Morphology of young microcolonies is not affected by aerobic or anaerobic incubation conditions, and is similar for both aerobically and anaerobically isolated and maintained strains. Microcolonies observed at 20 $\times$ are flat, filamentous, spider-like (Plate 17.4, Fig. 2), may have dense centers with aerobic incubation, and are composed of non-septate, septate and fragmenting filaments of varying lengths, and germinating bacilli (with one-half strength brain heart infusion supplemented with 0.2% yeast extract, observed after 10 hr aerobic incubation from a 24 hr aerobic inoculum, and after 16 hr anaerobic incubation from a 48 hr anaerobically incubated inoculum). They should be used in conjunction with a broth dilution technique to purify cultures. The direct observation of coverslip covered plate cultures at 980 $\times$ magnification is more reliable than stained smears for assessing culture purity.

Macrocolony appearance is variable. Aerobically incubated surface colonies are 0.5–1.5 mm; and can be circular, convex, rough, with an almost entire or filamentous margin; or, irregular, molar toothed, rough, with an entire to filamentous margin at the base; or, irregular, with a low convex rough center and raised curled up lobate margin. The three colony types are opaque, tough and adherent to the medium. After R → S transition, which frequently occurs with aerobic incubation, the colonies are circular to irregular, convex or umbonate, rough, with entire or lobate non-filamentous margins, opaque, soft, non-adherent and the cells are shorter, more bacillary and frequently branched as Y forms.

With anaerobic jar incubation, surface R colonies and initial isolates are 1–2 mm, filamentous, flat with filamentous edges, opaque at the center to translucent at the edge, tough and adherent to the medium. This form is usually maintained indefinitely with serial anaerobic transfer and is found with both aerobically and anaerobically derived strains. When R → S transition occurs, mature colonies are irregular, low convex, rough, with or without a narrow filamentous edge, or, irregular, molar toothed with filamentous edge at the base, and both types are tough to soft and non-adherent to the medium.

Subsurface colonies in agar deeps or pour plates incubated in air, are fluffy and resemble a ball of hair. In liquid media, usually discrete, compact masses or "granules" of variable size are formed. Smooth aerobic variants tend to produce a diffuse growth with incubation in air or in an anaerobic jar.

Chemoorganotroph; fermentative and respiratory. Carbohydrates fermented to acid and some CO_2 (not demonstrable by Durham tube). Respiration of glucose, which occurs in some media, and of acetate, pyruvate and lactate is energetically useful. T.C.A. cycle acids can be oxidized. Lactate is the preferred respiratory substrate; CO_2 and acetic acid are produced.

Riboflavin, thiamine, nicotinic acid, pantothenic acid and cysteine are required for growth in a synthetic medium containing 18 amino acids and glucose. Hemin (0.2 μg/ml of medium) stimulates growth under anaerobic conditions, but can be inhibitory with aerobic incubation. CO_2 is stimulatory under all growth conditions.

Optimal pH for growth initiation is 6.5–7.5 limiting range is 5.5–8.5. Optimal temperature is 37 C with little or no growth at room temperature; no reports at higher temperatures. Killed at 60–65 C for 10 min (five strains).

Oxygen requirements are not strict; of 130 strains derived and maintained anaerobically, all grew both aerobically and anaerobically; of 183 aerobically derived strains, 147 grew both aerobically and anaerobically and 36 were aerobes. However, anaerobic cultivation conditions were not necessarily optimal in all studies.

Relatively large electrical potentials are produced by both resting and glucose utilizing cultures (Parker, 1967).

Bacteriophages have not been isolated.

Diaminopimelic acid is present in cell walls. Reports of other major components vary and include: glutamic acid, alanine, arabinose and glucose (Davis and Baird-Parker, 1959), glutamic acid, alanine and glucose (Pine and Georg, 1965) glutamic acid, alanine, leucine, isoleucine, phenylalanine, methionine, glucose, galactose and mannose (Boone and Pine, 1968). Baboolal (1969) found glucose, arabinose, galactose, muramic acid, DL-DAP, alanine, glycine, glutamic and aspartic acids in six strains.

Serologically, the aerobically derived isolates studied comprise one group; serotypes have not been reported. Cross-reactions with *Mycobacterium*, *Nocardia*, *Actinomyces* and *Corynebacterium* have been reported. Agglutination, complement fixation, precipitin, gel diffusion and fluorescent antibody techniques have been used (Schmidt and Richardson, 1962; Sibal *et al.*, 1962; Snyder *et al.*, 1967).

The antibiotic sensitivities given are the ratios of numbers of strains sensitive/strains tested, and are the same with either aerobic or anaerobic conditions unless otherwise specified: 13/14 to bacitracin, carbomycin, dihydrostreptomycin, erythromycin, neomycin, oxytetracycline, penicillin, and tetracycline; 8/14 to chlorotetracycline; 4/14 to triple sulfa; 11/14 aerobically, 5/5 anaerobically to chloramphenicol; 4/14 aerobically, 0/5 anaerobically to polymyxin (Gilmour and Beck, 1961).

Subcutaneous or intradermal injections of live suspensions into mice produce nodules or abscesses 1–1.5 cm in diameter. Similar lesions are produced following intravenous or intraperitoneal injections (Howell and Pine, 1961).

Found in the oral cavity of man and primates, particularly in calculus and plaque deposits on the teeth (Howell *et al.*, 1962; Cock and Bowen, 1967).

The G + G content of the DNA is 55–57 moles % (T_m) (Page and Krywolap, Abstract G101, *Bacteriol. Proc.* 1973). No homology studies have been reported.

Cultures can be maintained by; lyophilization in skimmed milk (viability 1–2 years); storing glycerol broth suspensions at −75 C (viability >1 year); or two sequential transfers, each with 2–3 day incubation, followed by storage at 3 C for 1–2 months.

Neotype strain: ATCC 14266; NCTC 10254 (Gilmour *et al.*, 1961).

Genus V. **Rothia** *Georg and Brown 1967, 86*

Lucille K. Georg

Roth'ia. M.L. fem.n. *Rothia;* named for Dr. Genevieve D. Roth, who performed basic studies with these organisms.

Coccoid, diphtheroid or filamentous cells (Plate 17.4, Fig. 3). **Filamentous forms branched**, usually 1 μm in **diameter**; however, diameter may be considerably greater, up to 5 μm, due to irregular swellings and clubbed ends. Growth at any given time may consist exclusively of coccoid, diphtheroid or filamentous forms or may be a mixture of any of these. **Gram-positive, non-acid fast, non-spore-forming, non-motile.**

Mature colonies are creamy white. They may be **smooth or rough** surfaced (Plate 17.4, Fig. 4). Most are of **soft texture**; extremely rough colonies may be dry and crumbly. **No aerial hyphae** produced.

Catalase positive.

Chemoorganotroph. Ferments carbohydrates with production of acid but no gas. Major product of glucose fermentation is **lactic acid**, other products may include acetic, formic and succinic acids **but not propionic acid**. May be weakly proteolytic.

Cell wall constituents include alanine, glutamic acid, lysine and large amounts of galactose, but **no arabinose or diaminopimelic acid** (Davis and Baird-Parker, 1959; Cummins, personal communication).

Aerobic; some strains able to grow slowly at reduced oxygen pressures; CO_2 does not stimulate growth.

Optimal temperature 35–37 C.

Type species: *Rothia dentocariosa* (Onishi) Georg and Brown 1967, 86.

Further Comments

The genus *Rothia* was created to accommodate organisms which are common inhabitants of the normal mouth and throat and resemble *Actinomyces* species morphologically, but grow better aerobically, and differ significantly from members of the genera *Actinomyces* and *Nocardia* in their physiology and cell wall constituents.

For primary isolation and maintenance, enriched media such as brain heart infusion or trypticase soy, are recommended with aerobic incubation at 37 C. Cultures in deep agar butts of these media remain viable 3–4 months at room temperature. For long term storage, lyophilization of suspensions in milk is satisfactory. Little or no growth is obtained on media with inorganic nitrogen or simple peptones, such as Czapek or Sabouraud dextrose agar. A satisfactory fermentation medium is meat extract peptone base with Andrade's indicator at pH 7.4 (meat extract 3.0 g, peptone 10.0 g, NaCl 5.0 g, Andrade's indicator 10.0 ml, distilled water to 1000 ml).

Catalase test is best performed by flooding H_2O_2 over aerobic growth on slants and observing stream of bubbles. Anaerobic cultures must be exposed to air 15–20 min before testing. The Hugh and Leifson test (1953) is useful in distinguishing *Rothia* from *Nocardia*; *Rothia* ferments glucose, whereas *Nocardia* oxidizes or does not attack glucose in this medium.

Description of the species of genus **Rothia**

1. Rothia dentocariosa (Onishi) Georg and Brown 1967, 86. (*Actinomyces dentocariosus* Onishi 1949, 282; *Nocardia dentocariosus* (Onishi) Roth 1957, 1115; *Nocardia salivae* Davis and Freer 1960, 165.)

den.to.car.i.o'sa. M.L. n. *dens, dentis* tooth; M.L. adj. *cariosus* decayed or decaying; M.L. n. *dentocariosa* decayed tooth.

Description is based on the study of 50 isolates (Brown *et al.*, 1969).

Cultures may consist of coccoid, diphtheroid or filamentous forms or a mixture of any of these (Plate 17.4, Fig. 3). Cells in 2–3-day-old broth cultures are frequently entirely coccoid. Diphtheroidal or filamentous forms are common in old broth cultures or on solid media.

Colonies, grown aerobically on brain heart infusion (BHI) agar are approximately 1 mm at 24 hr. They are smooth with entire or fringed borders. However, 24-hr colonies grown under

TABLE 17.29
Acid production[a] from various carbon sources by strains of **Rothia dentocariosa**

Substrate	Strains positive[b]	Substrate	Strains positive[b]	Substrate	Strains positive[b]
Acetate	0/16	Glycerol	36/50[c]	Melibiose	0/16
Adonitol	0/16	Glycogen	0/16	Pyruvate	0/16
Arabinose	0/50	Inositol	0/50	Raffinose	0/16
Cellobiose	0/16	Inulin	0/16	Rhamnose	4/16
Dextrin	4/16[c]	Lactate	0/16	Ribose	0/16
Dulcitol	0/16	Lactose	0/50	Salicin	50/50
Erythritol	0/16	Maltose	50/50	Sorbitol	0/16
Fructose	16/16	Mannitol	0/50	Soluble starch	0/50
Galactose	11/16	Mannose	14/16	Sucrose	50/50
Glucose	50/50	α-Methyl-D-		Trehalose	16/16
α-Methyl-D-		mannoside	0/0	Xylose	0/50
glucoside	8/16	Melezitose	16/16		

[a] Acid but no gas produced, after 7 days.
[b] Number of strains positive/number of strains tested.
[c] Of these isolates, 100%+ after 21 days.

TABLE 17.30
Biochemical reactions of strains of **Rothia dentocariosa**

Test	Strains positive[a]	Test	Strains positive[a]
Catalase	50/50[b]	Urease production	0/50
Indole	0/50	Esculin hydrolysis	50/50
$NO_3 \rightarrow NO_2$	50/50	Methyl red	variable
NO_2 reduction	50/50[c]	Voges-Proskauer	variable
H_2S (lead acetate paper)		Gelatin hydrolysis (tube)	0/50[d]
Triple sugar iron	48/50	Litmus milk	
HIA	1/50	No change	50/50

[a] Number of strains positive/number of strains tested, after 7 days.
[b] Catalase positive in aerobic cultures. Anaerobic cultures must be exposed to air (15–20 min) befor testing.
[c] Medium used, trypticase soy broth with 0.01% KNO_2.
[d] Thiogel stabs negative at 21 days. More sensitive methods yield variable results.

anaerobic conditions are microscopic and highly filamentous resembling the "spider" colonies of *Actinomyces* species.

Macrocolonies grown aerobically on BHI agar or trypticase soy (TS) agar average 4-6 mm in 7 days. They may be convex and smooth reflecting a bacillary or coccoid micromorphology or raised with highly convoluted surfaces reflecting a filamentous morphology. Very smooth and extremely rough colonies may appear on the same plate (Plate 17.4, Fig. 4). All mature colonies are creamy white and most are of soft texture; extremely rough colonies develop a dry, crumbly texture. No aerial hyphae are produced. No hemolysis observed on blood agar.

Acid but no gas is produced from various carbon sources in 7 days (Table 17.29). Biochemical reactions are given in Table 17.30.

Good growth on complex media only; minimal nutritional requirements unknown.

Aerobic; half of the strains show moderate growth under reduced oxygen conditions and about 40% produce slight growth anaerobically. CO_2 does not stimulate growth.

Optimum temperature 35-37 C. Does not survive 1-hr exposure at 57 C or long term storage at 5 C. Growth range extends from pH 5.4-8.8. Maximum growth occurs at pH 7.0 (Roth and Thurn, 1962).

Cell wall constituents include alanine, glutamic acid, lysine and large amounts of galactose; but no arabinose or diaminopimelic acid (Davis and Baird-Parker, 1959; Cummins, C. S., personal communication).

According to Hammond (1970), fructose appears to be a determinant of serological specificity.

The G + C content of the DNA for 11 strains was 65.4-69.7 moles % (T_m), (Hammond, 1970).

Abcess formation has been demonstrated experimentally in mice (Roth and Flanagan, 1969). No natural infections reported in man or animals.

Neotype strain: CDC X599; ATCC 17931.

FAMILY II. **MYCOBACTERIACEAE** CHESTER 1897, 63

ERNEST H. RUNYON, LAWRENCE G. WAYNE AND GEORGE P. KUBICA*

(*Proactinomycetaceae* Lehmann and Haag in Lehmann and Neumann 1927, 674.) My.co.bac.te.ri.a′ce.ae. M.L. neut.n. *Mycobacterium* type genus of the family; -aceae ending to denote a family; M.L. pl.fem.n. *Mycobacteriaceae* the *Mycobacterium* family.

Description as for genus *Mycobacterium*.

Further Comments

The genus *Mycococcus* Krasil'nikov 1938, 335 and 1949, 198 was included in the *Mycobacteriaceae* in the 7th edition of THE MANUAL 1957, 707, but is here excluded on the basis of insufficient characterization. Very few specifically named strains have been available for study; these have been heterogeneous and apparently more closely allied with other families than with *Mycobacteriaceae*. Also included in the last edition of THE MANUAL with the family *Mycobacteriaceae* but designated as of doubtful relationship were bacteria given the specific epithet *rhodochrous*. Strains carrying this label have been ascribed by different investigators to many different genera, including most prominently *Mycobacterium* and *Nocardia*. The current description excludes these organisms from *Mycobacterium* on the basis of complete or almost complete lack of acid-fastness, and on distinctive immunological, phage susceptibility and biochemical properties. The cultures which formerly carried this epithet are treated as members of species within the genus *Nocardia*. See *Nocardia rubra* and others (Nocardia species 21-26, 28, 29). The taxonomic status of these organisms is currently unsettled. Two generic names have been proposed, *Jensenia* Bisset and Moore 1950, 280 and *Gordona* Tsukamura 1971, 19, but more information is needed. See Goodfellow *et al.*, 1972 and Ridell and Norlin, 1973.

---

* Dr. Runyon prepared the descriptions of the family, genus and type species, Dr. Wayne of the slowly growing species and Dr. Kubica of the rapidly growing species.

Genus I. Mycobacterium *Lehmann and Neumann 1896, 363*

(*Coccothrix* Lutz 1886, 98 (not validly published); *Sclerothrix* Metchnikoff 1888, 70 illeg.; Not *Sclerothrix* Kützing 1843, 299 and 1849, 319 (a genus of algae); *Mycomonas* Orla-Jensen 1909, 329 illeg.)

My.co.bac.te'.ri.um. Gr. n. *myces* a fungus; Gr. neut.dim.n. *bakterion* a small rod; M.L. neut.n. *Mycobacterium* a fungus rodlet.

Slightly curved or straight **rods,** 0.2–0.6 by 1.0–10 μm, sometimes branching; **filamentous or mycelium-like growth** may occur but **on slight disturbance usually becomes fragmented into rods or coccoid elements. Acid-alcohol-fast** at some stage of growth. Not readily stainable by Gram's method but usually considered Gram-positive. Non-motile. No endospores, conidia or capsules, no grossly visible aerial hyphae.

Genus includes obligate parasites, saprophytes and intermediate forms differing in nutritional requirements. Saprophytic strains grow on very simple substrates; others require more complex media or supplements for growth (mycobactin); and others have not been cultivated outside living cells. Species differ in amidase, catalase and other enzyme activity; all are **aerobic** although from dispersed seeding in tubed agar medium, growth of some species occurs only in the depths of the medium.

Lipid content of cells and especially cell walls high; included are waxes having component chloroform-soluble mycolic (fatty) acids with long, branched, about 80-carbon-atom chains. **Diffusible pigment rare.** Colonies of some species are regularly or variably yellow or orange, usually due to carotenoid pigments, the formation of which may or may not require exposure to light. The cell wall peptidoglycolipid contains *meso*-diaminopimelic acid, alanine, glutamic acid, glucosamine, muramic acid, arabinose and galactose (wall type IV). Mycobacteria hydrolyze phenolphthalein disulfate although some species do so only slowly.

Growth slow or very slow; easily visible colonies are produced from dilute inoculum after 2 days to 8 weeks incubation at optimum temperature, which may be near 40 C for some species but closer to 30 C for others. Although no taxonomic division of the genus on growth rate is recognized, the terms "rapid" and "slow" growers are used for convenience in identification.

Diseases produced include tuberculosis, leprosy and other usually chronic more or less necrotizing, limited or extensive granulomas.

Found in soil, water, warm-blooded and cold-blooded animals.

The G + C content of the DNA ranges from 62–70 moles % (T_m; buoyant density) (Wayne and Gross, 1968; Tewfik and Bradley, 1967).

Type species: *Mycobacterium tuberculosis* (Zopf) Lehmann and Neumann 1896, 363.

Further Comments

For convenience the genus may be divided into three groups:

1. *Slow growers.* Grossly visible colonies apparent only after 7 or more days on media seeded with fresh inocula, diluted sufficiently to yield well isolated colonies. Microcolonial test positive.

Species 1–18. See Tables 17.31 and 17.33.

2. *Rapid growers.* Grossly visible colonies in less than 7 days under similar conditions and at 25 and 37 C. Microcolonial test negative.

Species 19–28. See Tables 17.32 and 17.33.

3. Organisms which have special growth requirements or have not been cultivated *in vitro.*

Species 29–31.

In many cases it is unnecessary to carry an identification to the level of an individual species. For example, in clinical diagnosis, it may suffice to identify an organism as belonging to a complex of similar organisms, with similar significance. In these situations, the complexities of the tests required for speciation may outweigh the value received from the identification. For this reason Table 17.33 is offered as a guide to a minimal test protocol which will permit identification of some organisms at a level usually considered to be sufficient for routine practical purposes. Identification beyond these levels may require application of a larger series of tests presented in the master lists (Tables 17.31 and 17.32), as well as consideration of properties described in the narrative portions of the text. It should be recognized that intermediate growth rates are frequently seen, especially in old laboratory strains, and the groupings used are relative.

Like other organisms, a given strain of a *Mycobacterium* species may lose its capacity to exhibit a key property (e.g. pigment or a particular enzyme) and still be considered to belong to that species, on the basis of a large number of other properties. Mycobacteria are not well characterized by bacteriologic testing techniques useful for speciating members of other genera. For this reason, it is essential that tests be performed precisely as described in the literature cited (see Table 17.34).

Acid-alcohol-fastness is partially or completely lost at some stage of growth by a variable proportion of the cells of some species. Cells of rapid growers may be less than 10% acid-fast. Beading with metachromatic granules and banding with non-stained areas are common.

Most strains of mycobacteria form more than one kind of colony, but colonies of some species, as *M. tuberculosis*, are regularly rough; some, as of *M. intracellulare* on primary culture from clinical specimens, are more commonly smooth. Primary culture smooth colonies may often be very thin, *in vitro* adapted colonies in contrast are high domed. Cells of rough strains are usually compacted in curving strands; cells of smooth strains are not visibly oriented in any pattern. Some colonies, as of *M. fortuitum* and *M. xenopi*, in early growth may be mycelial, older ones exhibiting branching filamentous extensions on and into some media such as cornmeal glycerol agar;

fragmentation to bacilli usually occurs in smear preparation. Aerial filamentous extensions rare, never visible without magnification (× 30–100).

According to currently available data, the mycolic acids of mycobacteria are distinctive. Corresponding acids of *Corynebacterium* and *Nocardia* species tested have smaller molecules, insoluble in chloroform.

May be isolated from sputum, soil or other contaminated sources by primary digestion and decontamination with alkali, acid or hypochlorite solution, followed by seeding on egg medium containing malachite green. Other media may be employed. Alternatively the digest, neutralized, may be injected into mice, and cultures obtained about 2 weeks later from internal organs. Isolation of mycobacteria from soil also has been attained by use of nutrient medium in which paraffin is the only added carbon source.

Description of the species of genus Mycobacterium

1. **Mycobacterium tuberculosis** (Zopf) Lehmann and Neumann 1896, 363. (*Bacterium tuberculosis* Zopf 1883, 67; *Bacillus tuberculosis* (Zopf) Klein 1884, 34; *Mycobacterium tuberculosis typus humanus* Lehmann and Neumann 1907, 550; *Mycobacterium tuberculosis* var. *hominis* Bergey *et al.* 1934, 536.)

tu.ber.cu.lo'sis. L. dim.n. *tuberculum* a small swelling, tubercle; Gr. suff. *-osis* characterized by; M.L. gen.n. *tuberculosis* of tuberculosis.

The following description is based on strain H37Rv.

Rods, ranging in size from 0.3–0.6 by 1–4 μm, straight or slightly curved, occurring singly and in occasional threads. Stain uniformly or irregularly, often showing banded or beaded forms. Strongly acid-fast and acid-alcohol-fast as demonstrated by Ziehl Neelsen or fluorochrome procedures. Growth tends to be in serpentine, cord-like masses in which the bacilli show a parallel orientation. Colonies of avirulent forms are less compact. Growth in all media is slow, requiring several days or weeks for development, depending upon the medium and the size of inoculum; generation time *in vitro* under optimal conditions 14–15 hr.

On most solid media, colonies are rough, raised, thick, with a nodular or wrinkled surface and an irregular thin margin; may become somewhat pigmented (off white to faint buff or even yellow). Colonies on oleic acid albumin agar are flat, rough, corded, dry and usually non-pigmented.

In liquid media lacking a dispersing agent, growth begins on the bottom of the tube as a film which extends up the sides of the container; this

will eventually form a pellicle which will extend up the sides of the tube above the medium. When the tube is shaken, the bottom growth swirls through the medium as typical small, loose floccules. A spreading pellicle without growth on the bottom may result from surface inoculation. Pellicles, with age, become thick and wrinkled. In Dubos' Tween-albumin medium growth is diffuse, settling if undisturbed, but readily dispersed. From dispersed seeding in tubed agar medium, evident growth is confined mainly to the surface.

Optimum temperature: 37 C, some growth at 30–34 C.

Optimum pH in range 6.4–7.0. Growth at 37 C is stimulated by incubation in air with 5–10% added CO_2 and by inclusion of glycerol to 0.5% in the medium.

Nitrate reduction vigorous. Niacin positive. Catalase activity relatively weak and lost after heating at 68 C.

Strain to strain differences in tubercle bacilli have been demonstrated by their different response patterns to a number of phages.

M. tuberculosis produces tuberculosis in man, other primates, dogs and some other animals which have contact with man. Experimentally, from inoculum of 0.01 mg, it is highly pathogenic for guinea pigs and hamsters, but relatively non-pathogenic for rabbits, cats, goats, bovine animals or domestic fowls. Inocula of 0.001–1 mg are used for experimental disease production in mice. Strains of lower virulence for experimental animals have been isolated from cases of lupus, scrofuloderma and urogenital tuberculosis (Griffith, 1957; Lind and Obrant, 1962). Attenuation of virulence may occur spontaneously upon subculture in artificial media. Virulence can be maintained by selection of appropriate portions of growth on suitable media or by animal passage.

Many strains of *M. tuberculosis* isolated from patients from southern India cause only localized lesions in guinea pigs and the disease tends to regress. These strains are catalase active, but are susceptible to peroxide and to isoniazid (Mitchison *et al.*, 1963).

Streptomycin, para-aminosalicylic acid, isoniazid and at least nine other drugs are in use because of their inhibitory effect on *M. tuberculosis*, and susceptibility to their action is useful in recognition of the species. Spontaneous mutants resistant to one of these drugs may replace the parent strain if treatment is improper. Resistance to isoniazid is regularly accompanied by changes in other properties, such as loss of peroxidase and catalase activity and attenuation of virulence for guinea pigs. Strains still catalase positive may or may not be virulent for guinea pigs, whereas catalase negative strains almost uniformly lack the ability to produce progressive disease in guinea pigs.

Antigenic character: Infected animals including man exhibit delayed hypersensitivity to crude or purified *M. tuberculosis* culture filtrates (tuberculins) and less sensitivity to tuberculin-like preparations from other mycobacteria. Disease caused by *M. bovis* is not distinguishable from that due to *M. tuberculosis* by use of commonly available tuberculins. Infections by other species of mycobacteria result in much less sensitivity to tuberculin, although if "second strength" (250 tuberculin units) PPD tuberculin is used, skin tests may erroneously be interpreted as indicating infection with *M. tuberculosis*.

By agglutination and agglutinin-absorption techniques, *M. tuberculosis* is antigenically homogeneous, is very similar to *M. bovis* and *M. microti* but distinct from other species. Immunoelectrophoresis, double diffusion precipitin tests in gel, hemagglutination and other serologic techniques in general confirm these findings. Several antigens are shared by *M. tuberculosis*, *M. bovis* and *M. kansasii*, but each of these other species probably has one or more distinctive antigens (Castelnuovo and Morellini, 1965). Relationship to *M. leprae* is seen in the presence of strong *M. tuberculosis*-antigen-precipitating antibodies in sera of lepromatous leprosy patients (Rees *et al.*, 1965).

Neotype strain: H37Rv (Kubica *et al.*, 1972); a virulent strain isolated from a patient in Saranac Lake Sanatorium (Steenken *et al.*, 1934) and obtainable from Trudeau Institute, Inc., Saranac Lake, New York 12983.

2. **Mycobacterium microti** Reed in Breed *et al.* 1957, 703. (*Mycobacterium tuberculosis* var. *muris* Brooke 1941, 816; *Mycobacterium muris* Smith and Conant in Smith *et al.* 1948, 413; Not *Mycobacterium muris* Simmons 1927, 15.)

mic.ro′ti. M.L. masc.n. *Microtus* a genus that includes the vole; M.L. gen.n. *microti* of *Microtus*.

Common name: Vole bacillus (Wells, 1937).

Source of description: Runyon *et al.* (1967) and Käppler (1968).

Rods. Primary growth on glycerol-free egg media in 28–60 days. May adapt to tolerance to glycerol. Colony morphology variable. Optimum temperature 37 C.

Cause of naturally acquired generalized tuberculosis in the vole. Local lesions produced in guinea pigs, rabbits and calves. Lose pathogenicity on repeated subculture.

Immunologically closely related to *M. tuberculosis* and *M. bovis*.

Further Comments

An intermediate form between *M. tuberculosis* and *M. bovis*.

3. **Mycobacterium bovis** Karlson and Lessel 1970, 280. (*Mycobacterium tuberculosis typus bovinus* Lehmann and Neumann 1907, 550; *Mycobacterium tuberculosis* var. *bovis* Bergey *et al.* 1934, 537.)

bo′vis. L. n. *bos* the ox; L. gen.n. *bovis* of the ox.

Common name: bovine tubercle bacillus (Th. Smith, 1896).

Source of description: Runyon *et al.*, 1967; Karlson and Lessel, 1970.

Short to moderately long rods. On primary isolation growth is very poor on glycerol-containing media, although repeated subculture permits adaptation to growth on such media. Furthermore, freshly isolated cultures of *M. bovis* are microaerophilic; inocula dispersed into liquid, semisolid or solid agar media grow in the medium but not on the surface, as distinguished from *M. tuberculosis* which is highly aerobic. On repeated subculture, *M. bovis* will adapt to aerobic growth. Dilute inocula on egg media yield small rounded white colonies, with irregular edges and a granular surface after 21 days or more of incubation at 37 C. Colonies on transparent oleic acid albumin agar thin, flat, generally corded; not easily emulsified in absence of a detergent.

Strains usually lose catalase on acquiring resistance to isoniazid. Some strains resistant to para-amino salicylic acid on first isolation.

Originally isolated from tubercles in cattle; generally more pathogenic for animals than is *M. tuberculosis*. Produces tuberculosis in cattle, both domestic and wild ruminants, man and other primates, carnivores including dogs and cats,

*Properties of slowly growing species of genus **Mycobacterium**[a]*

No.	Property	1. M. tuberculosis	2. M. microti*	3. M. bovis	4. M. africanum*	5. M. kansasii*	6. M. marinum	7. M. simiae	8. M. gastri	9. M. nonchromogenicum*	10. M. terrae*	11. M. triviale	12. M. gordonae	13. M. scrofulaceum	14. M. paraffinicum*	15. M. intracellulare	16. M. avium	17. M. xenopi	18. M. ulcerans
3	Urease	+	+	+	+	+	+		+		-	-	∓	+	-	-	-	-	-
5	Nicotinamidase	+	+	-	+	+	+		+	+	-	-	-	+	-	+	+	+	
6	Pyrazinamidase	+	+	±	+	±	+		-	+	-	-	-	+	-	±	±	+	
31	Arylsulfatase (2 week)	-	-	∓		∓	+		+	+	-	-	+	∓	-	-	∓	∓	+
32	Growth on 5% NaCl	-	-	∓		+	+		-	-	-	+	+	-	-	-	-	-	-
33	Tellurite red'n (3 day)					-	-	see	-		-	-	-	-		±	±	-	
	(9 day)	±		-	-	+	+	text	±		+	+	±	±	+	+	+	±	±
34	Nitrate reduction	+	+	±	-	+	-		-	-	+	+	-	-	-	-	-	-	+
35	Catalase 68 C	-	-	±	-	+	+		±	+	+	+	±	+	-	±	+	+	∓
36	Tween agar opac. (4 week)																		
37	Nitrite red'n (7 day)	-	-	-	-	+	-		-	+	+	-	+	∓	+	-	-	∓	-
39	Niacin	+	+	-	-	∓	-		-	-	-	-	-	-	-	-	-	-	-
40	Tween 80 hydrol. (5 day)	±	-	-	-	+	+		-	+	+	+	+	∓	-	-	-	-	∓
41	Catalase >45 mm foam	∓	-	-	-	+	-		+	+	+	+	+	±	+	-	-	∓	±
42	Pigment	N	N	N	N	P	P		N	N	N	N	S	S	S	N	N	N[b]	N[b]
43	Resists 10 µg/ml T2H	+	-	+	+	+	+		∓	+	+	+	+	+	+	+	+	+	+
44	Resists 1 µg/ml isoniazid	-	-	+	-	∓	∓		+	+	+	+	+	∓	+	+	±	∓	∓
45	Microcolonial test	+	+	+	+	+	+		+	+	+	+	+	+	-	+	+	+	-
46	Growth at 25 C	-	-	∓	+	+	+		±	+	+	+	∓	±	-	±	±	-	-
	37 C	+	+	+		+	-		-	+	+	+	-	+	-	+	+	+	-
	40 C	+		+		+	-		-	+	+	-	-	∓	-	+	+	+	-
	45 C	-	-	-		-	-		-	-	-	-	-	-	-	-	-	+	-
	52 C	-	-	-		-	-		-	-	-	-	-	-	-	-	-	-	-

[a] Data were assembled from cited references as well as from unpublished data compiled by members of the Advisory Committee for the Family. Except for species marked with an asterisk results are expressed as follows: + = more than 84% of strains positive; ± = between 50–84% of strains positive; ∓ = between 16–49% of strains positive; − = less than 16% of strains positive. Data of strains marked with an asterisk are based on a single strain. P = photochromogenic, i.e. pigment is produced only on exposure to light; S = scotochromogenic, i.e. pigment is produced even when cultures are grown or maintained in the dark; N = no pigment, at least in young cultures. Details of techniques and interpretation of the properties cited may be found in references cited in Table 17.34. All species tested are negative for acetamidase (1), benzamidase (2) and iron uptake (38).

[b] Characteristic yellow color may be slow to develop.

swine, parrots and possibly some birds of prey. Experimentally highly pathogenic for rabbits, guinea pigs and calves; at least moderately pathogenic for hamsters and mice; slightly pathogenic for dogs, cats, horses and rats; not pathogenic for most fowl. Loss of virulence for guinea pigs and rabbits and loss of catalase activity accompany a loss of sensitivity to isoniazid as for *M. tuberculosis*. Certain strains isolated from cases of lupus and scrofuloderma in man have low pathogenicity for animals (Griffith, 1957).

Antigenic structure: Tuberculins prepared from *M. tuberculosis* and *M. bovis* are ordinarily indistinguishable in their action. Worthington and Kleeberg (1967) demonstrated that desensitization was more specific than sensitization for distinguishing among mycobacterial species, and *M. bovis* was sharply distinguished from three other species; *M. tuberculosis* was not included in this study.

Immunodiffusion and immunoelectrophoretic analysis of culture filtrates and extracts indicate that *M. bovis* and *M. tuberculosis* share the demonstrable antigens. Any differences observed were considered to represent absence of individual antigens from particular strains, rather than being species related (Castelnuovo *et al.*, 1958; Lind, 1959; Tuboly, 1965). However, use of adsorbed sera appeared to permit distinction between these two species (Jensen *et al.*, 1968).

The bacillus Calmette-Guérin (BCG) conforms to the properties described for *M. bovis* except that it is much attenuated in pathogenicity and grows well on glycerinated media (Runyon *et al.*, 1967).

Proposed neotype strain: Yoder 18802–887; ATCC 19210 (Karlson and Lessel 1970, 280).

4. Mycobacterium africanum Castets, Rist and Boisvert 1969, 321.

a.fri.ca'num. English n. *African* a native of Africa; M. L. gen.pl.n. *africanum* of Africans.

Source of descriptions: Castets *et al.* (1969) and personal communications.

Rods average 3 μm. When grown on egg medium at 37 C, colonies are flat, dull and rough. Sodium pyruvate stimulates growth in egg medium. Growth homogeneous in Dubos' medium with Tween 80, and granular in Youman's medium with bovine serum. Growth in Lebek agar deeps 15 mm below surface.

Resistant to 2 μg/ml thiacetazone.

Isolated from sputum of a tuberculosis patient in Senegal and a cause of human tuberculosis in tropical Africa. In guinea pigs 0.01 and 1 mg injected subcutaneously exhibit irregular pathogenesis, of lower order than *M. tuberculosis* of normal virulence. Generalized lesions seen by

the 3rd month. Limited virulence on intravenous injection of 0.01 mg to rabbits. In mice, 0.5 mg injected intravenously causes pulmonary lesions and considerable granulation, with 50% dead by 65 days.

Type strain: ATCC 25420.

Further Comment

Intermediate in properties between *M. bovis* and *M. tuberculosis*. Status as distinct species to be evaluated.

5. Mycobacterium kansasii Hauduroy 1955, 73. (Subj. syn. *M. luciflavum* Manten 1957, 363.)

kan.sas'i.i. Kansas, a geographic place name. M.L. gen.n. *kansasii* of Kansas.

Common name: Group I photochromogen; yellow bacillus.

Sources of description: Wayne, 1966; Wayne and Doubek, 1968; Buhler and Pollak, 1953; Kestle *et al.*, 1967; Subcomm. on Mycobacteria, Am. Soc. Microbiol. 1962.

Moderately long to long rods; broaden and exhibit marked cross-barring on incubation in the presence of sources of fatty acids. Dilute inocula on inspissated egg media yield smooth to rough colonies after 7 or more days of incubation at 37 C. On oleic acid albumin agar, flat, smooth or somewhat granular surface, regular or slightly undulating margins, dense central spot, exhibit same pigment pattern (photochromogenicity) as on egg media. Most strains appear somewhat rough microscopically, but are readily emulsified in water. Some strains are so rough as to resist emulsification. Colonies grown in dark are nonpigmented; when grown in light or when exposed briefly to light when colonies are young, become brilliant yellow. Rarely, strains produce no pigment, even after exposure to light, or produce deep orange pigment, even when grown in dark. The usual photochromogenic strains and, most prominently, the occasional scotochromogenic strains, if grown in a lighted incubator, form dark red crystals of β-carotene on the surface and inside of colony. The crystals are useful in identification (Runyon, 1965).

Most strains are strongly catalase positive; less commonly, weakly positive and inactivated at 68 C for 20 min, and these types appear less pathogenic for man.

Isolated from human pulmonary lesion. Causes chronic human pulmonary disease resembling tuberculosis, although some uncommon forms with unusually weak catalase activity, have been found in clinical specimens without being implicated in disease (Buhler and Pollak, 1953; Wayne 1962; Hobby *et al.*, 1967).

Subcutaneous inoculation may cause local

lesions but no gross visceral lesions or death in guinea pigs. Intraperitoneal or intravenous inoculation produce self-limiting visceral lesions. Pathogenicity for guinea pigs not markedly enhanced by cortisone administration, but pretreating animals' lungs with coal dust may intensify extent of disease (Pollak and Buhler, 1955; Tacquet et al., 1967). The usual forms cause self-limiting ulceration on intradermal inoculation of guinea pigs with 10^{-4} mg of bacilli, whereas the low catalase forms require about 100 times larger inoculum. Intraperitoneal inoculation of hamsters usually causes death, with lesions of lymph nodes, spleen, liver and occasional invasion of capsule of the kidney; virulence enhanced by cortisone. Some deaths occur in mice inoculated intraperitoneally, but most exhibit self-limiting granulomas of liver, spleen and lymph nodes. Rats exhibit minimal lesions and chickens none (Pollak and Buhler, 1955). Rabbits, inoculated intravenously with 5 mg, develop macroscopic lesions of joints and tendon sheaths, and some gross lesions of liver and lung; rarely fatal (Engbaek et al., 1964). Intratracheal inoculation of rhesus monkey causes self-limited disease, which regresses with time (Grover et al., 1957).

Most frequently isolated from pulmonary secretions or actual tubercles of man. Occasionally associated with lesions of lungs or lymph nodes of deer, swine and cattle (Tacquet et al., 1964; Worthington and Kleeberg, 1964; Pattyn et al., 1967). Natural sources of infection unclear. Extensive soil sampling fails to yield isolates of M. kansasii (Wolinsky and Rynearson, 1968). Low catalase strains have been isolated from water. Some high catalase strains have also been isolated from tap water (Bailey et al., 1970).

Antigenic structure: Magnusson (1967) distinguished M. kansasii from 10 other mycobacterial species, including M. marinum, by means of dermal hypersensitivity. Although some degree of cross-reactivity does occur, dermal desensitization is effected only by homologous antigen (Worthington and Kleeberg, 1967). The use of sensitized guinea pig intestine by the Schultz-Dale technique permitted differentiation into two or possibly three antigenic patterns (Jensen et al., 1966). Only one agglutinating serotype has been established, accounting for 154 of 155 tested smooth strains whose identity has been established by biochemical methods (Hobby et al., 1967). Rough strains could not be typed because of spontaneous agglutination. Both high and low catalase varieties exhibit the same agglutinating serotype (Wayne, 1966) and this appears to be associated with a specific phenol-soluble antigen detectable by immunodiffusion (Wayne, 1971). This species is also of homogeneous serotype according to immunofluorescence tests (Jones et al., 1965). Immunodiffusion and immunoelectrophoresis studies on culture filtrates or lysates demonstrate as many as 19 lines of precipitate against homologous serum. The pattern is distinct for this species, although a number of antigens are shared with M. tuberculosis (Castelnuovo and Morellini, 1962). Non-pigmented and scotochromogenic variants exhibit the same immunoelectrophoretic pattern as do the photochromogenic strains (Gimpl and Vandor, 1967).

Type strain: ATCC 12478.

Further Comments

Because this organism may resemble M. marinum Aronson (1926) which is also photochromogenic and which may adapt in the laboratory to growth at 37 C, it is best to confirm identity with a battery of biochemical tests or specific agglutination of smooth strains. M. marinum grows rapidly at 25 C, whereas M. kansasii is slow to develop at this temperature. Differences are also observed by the thin layer chromatography of bacillary lipids (Szulga et al., 1966).

6. **Mycobacterium marinum** Aronson 1926, 320. (*Mycobacterium platypoecilus* (sic) Baker and Hagan 1942, 252; *Mycobacterium balnei* Linell and Norden 1952, 890; subjective synonymy proposed by Bojalil 1959, 169.)

ma.ri'num. L. adj. *marinus* of the sea, marine.

Sources of description: Schaefer and Davis, 1961; Wayne and Doubek, 1968; Clark and Shepard 1963.

Moderately long to long rods with frequent cross-barring. Dilute inocula on inspissated egg media yield smooth to rough colonies after 7 or more days of incubation at 30 C. Colonies grown in the dark are non-pigmented; when grown in light or when exposed briefly to light when colonies are young, become brilliant yellow. On oleic acid albumin agar, smooth colonies; interconvertible between domed, with entire margin, and domed center, with flattening or irregularity at periphery; exhibit same pigment pattern (photochromogenicity) as on egg media. Growth in temperature range of 25-35 C but usually not at 37 C; may adapt to growth at 37 C.

Isolated from diseased fish, and aquariums (Aronson, 1926). In man, frequently seen in epidemic form as skin lesions resulting from abrasions occurring in swimming pools harboring the organism (Linell and Nordén, 1954). Causes cutaneous granulomas ("swimming pool granuloma") in man, usually on elbow, but also found on knee, foot, finger and toe; papules or nodules, sometimes ulcerating; usually heal spontaneously over a period of months (Nordén and Linell, 1951;

Schaefer and Davis, 1961). Mice receiving a large inoculum intraperitoneally develop ulcerations on tail, paws and scrotum; visceral lesions and death sometimes occur; after intravenous inoculation lesions limited to tail; footpad inoculation leads to local swelling and some ulceration (Fenner, 1956). Guinea pigs inoculated subcutaneously or by inhalation develop no disease; intraperitoneal inoculation occasionally leads to scrotal lesions. Rats inoculated intraperitoneally develop no disease, but nodules may occur in omentum and hilar lymph nodes. Chickens develop no lesions when inoculated intraperitoneally or intravenously, but chick embryos maintained at 33 C (but not 37 C) acquire fatal infection. Rabbits develop local lesions when inoculum is applied to abraded skin sites, and may develop granuloma with caseous necrosis in scrotum after intraperitoneal or intravenous inoculation. Representatives of 50 poikilothermic species (reptiles, amphibians and fish) have been found susceptible to fatal systemic infection when maintained at 30 C (Clark and Shepard, 1963).

Antigenic structure: Magnusson (1967) distinguished *M. marinum* from 10 other mycobacterial species, including *M. kansasii*, by dermal hypersensitivity. The use of guinea pig intestine by the Schulz-Dale technique demonstrated a distinct antigenic pattern (Jensen *et al.*, 1966). One agglutinating serotype has been established, which reacted to all of 21 strains tested (Schaefer, 1965). No cross-reaction seen with *M. kansasii* by this technique or by immunodiffusion with phenol soluble antigen (Wayne, 1971). Similarly, only one serotype established by immunofluorescence (Jones and Kubica, 1968). Castelnuovo and Morellini (1962), employing immunoelectrophoretic analysis, reported numerous precipitate lines of identity between *M. balnei*, *M. platypoecilus* and *M. marinum* and concluded these represented a single species.

Infection may cause numerous low grade conversions of tuberculin reaction, suggestive of a tuberculosis epidemic (Mullohan and Romer, 1961).

See also discussion of *M. kansasii*.

Type strain: ATCC 927.

7. Mycobacterium simiae Karassova, Weissfeiler and Krasznay 1965, 282.

si'mi.ae. L. n. *simia* the ape; L. gen.n. *simiae* of the ape.

Sources of description: Karassova *et al.*, 1965; Weiszfeiler *et al.*, 1968; Käppler, 1968; Magnusson, personal communication.

The original publication of the species described photochromogenic mycobacteria which were niacin negative and positive for catalase and peroxi-

dase. Variable results were reported for nitrate reduction, lipase and amidases. No type culture was specified in the original publication. Käppler provided a detailed biochemical description of one of the original strains (No. 61), reporting it as negative for nitrate reduction but positive for phosphatases, Tween 80 hydrolysis, and nicotinamidase and pyrazinamidase. This differs from the original report of Karassova *et al.*, who found this strain negative for all amides tested. Magnusson subdivided cultures received as *M. simiae* into two groups, on the basis of delayed hypersensitivity reactions, with strain No. 61 falling into one group, and the original authors' strain No. 29 typical of the second group. Karassova *et al.* describe strain No. 29 as positive for nitrate reduction, Tween 80 hydrolysis, acetamidase and urease. Weiszfeiler *et al.* report some similarity in antigenic structure between No. 61 and *M. kansasii*, *M. avium*, B.C.G. and *M. balnei* (*marinum*), but note that this strain may not yet be considered typical of *M. simiae*.

The taxonomic status of this species remains unclear.

Type strain: Strain 29; ATCC 25275, designated by Weiszfeiler, personal communication.

8. Mycobacterium gastri Wayne 1966, 923.

gas'tri. Gr. n. *gaster* the stomach; M.L. gen.n *gastri* of the stomach.

Common name: "J" group.

Sources of description: Wayne, 1966; Kestle *et al.*, 1967; Wayne and Doubek, 1968.

Moderately long to long rods with cross-barring frequently seen. Dilute inocula on inspissated egg media yield smooth to rough white colonies after 7 or more days of incubation at 37 C. On oleic acid albumin agar flat, smooth or somewhat granular surface, regular or slightly undulating margins, dense central spot; some strains rough. Capable of growth in temperature range of 25–40 C.

Strains usually lose catalase activity on acquiring resistance to isoniazid.

Fails to produce progressive disease in the guinea pig, but usually capable of producing local ulceration at site of intradermal inoculation of 10^{-2}–10^{-3} mg of bacilli (Wayne, 1966).

Isolated from human gastric lavage specimens. Found in human gastric lavage or sputum specimens as casual residents, not considered etiologic agent of disease (Wayne, 1966; Kestle *et al.*, 1967). Also found in soil (Wolinsky and Rynearson, 1968).

Antigenic analysis: Although closely related to *M. kansasii* in biochemical terms, this species is not agglutinated by *M. kansasii*-typing serum (Wayne, 1966) and the phenol soluble antigen does not cross react with this serum (Wayne, 1971).

Norlin *et al.* (1969), found a very high similarity between *M. gastri* and *M. kansasii* by immunodiffusion techniques; both *M. gastri* and low catalase strains of *M. kansasii* lacked an antigen found in the more common high catalase strains of the latter species.

Type strain: ATCC 15754.

Further Comment

Similar to non-pigmented strains of *M. kansasii*, but may be differentiated on basis of nitrate reduction, catalase, isoniazid susceptibility and serotype.

9. Mycobacterium nonchromogenicum Tsukamura 1965, 110.

Subsequent to the original publication of this name, Tsukamura (1966) proposed that the name of these organisms be changed to *M. terrae;* such a proposal probably does not constitute valid publication of the new name under Rule 12d, International Code of Nomenclature of Bacteria (1966). Under these circumstances, the name *M. terrae*, applied independently by Wayne (1966) to another group of organisms, would appear to be validly published. *M. nonchromogenicum* Tsukamura is similar to *M. terrae* Wayne (Wayne, 1967) but differs in some properties. By a number of diagnostic criteria employed, they may be grouped together for practical purposes as an *M. terrae* complex (Wayne and Doubek, 1968), inasmuch as neither appears to be associated with disease. See *M. terrae* below.

Reactions in Table 17.31 based on type culture (Käppler, 1968, and personal observations).

Lacks α- and β-esterase activity. Occasional strains develop a pink pigment.

Bacilli persist in tissues of mice without evidence of pathogenicity (Tsukamura, 1967).

Isolated from mice injected with soil.

Type strain: ATCC 19530; NCTC 10424.

10. Mycobacterium terrae Wayne 1966, 922.

ter'rae. L. n. *terra* earth. M.L. gen.n. *terrae* of the earth.

Common name: Radish bacillus (Richmond and Cummings, 1950).

Sources of description: Wayne, 1966; Kestle *et al.* 1967; Wayne and Doubek, 1968; Käppler, 1968.

Moderately long to long rods. Dilute inocula on inspissated egg media or oleic acid albumin agar yield smooth to rough, white to buff colonies after 7 or more days of incubation at 37 C.

Type strain exhibits α- and β-esterase activity.

Isolated from sputum and gastric lavage specimens from humans; considered casual residents rather than pathogens. Have also been isolated from soil (Wolinsky and Rynearson, 1968).

Produces neither local nor systemic lesions after inoculation of 10^{-1} mg, intradermally, to guinea pigs.

Type strain: ATCC 15755.

Further Comment

Some differences in arylsulfatase activity, growth temperatures, drug susceptibilities, colonial morphology and amidase activities appear to exist between *M. terrae* and *M. nonchromogenicum* (see *M. nonchromogenicum* above). Nevertheless, a question exists as to whether these should be maintained as separate species; a decision awaits further studies. *M. novum* Tsukamura (1967) appears intermediate between these two species and may, for the present, be treated as a synonym of *M. terrae*.

11. Mycobacterium triviale Kubica in Kubica, Silcox, Kilburn, Smithwick, Beam, Jones and Stottmeier 1970, 162.

tri.vi.a'le. L. n. *trivialis* that which belongs to the cross-roads i.e. common, of little importance; L. neut.adj. *triviale* of little importance.

Common name: V subgroup bacillus.

Source of description: Kubica *et al.*, 1970; Kestle *et al.*, 1967; also unpublished data.

From dilute inocula mature colonies do not appear on solid media for over a week. Colonies on egg medium are rough, dry, heaped up and nonchromogenic. On oleic acid agar characteristic rough R colonies are seen, which are easily confused with those of *M. tuberculosis*. Grows poorly, if at all, on corn meal agar.

Resistant to thiacetazone.

Isolated from sputum, but not considered a pathogen.

Antigenic structure: Specific bands found by immunodiffusion analysis of protoplasmic extract, and specific reaction by fluorescent antibody and agglutination tests on whole cells (Kubica *et al.*, 1970).

Type strain: ATCC 23292.

12. Mycobacterium gordonae Bojalil, Cerbón and Trujillo 1962, 344.

gor.do'nae. L. gen.n. *gordonae* of Gordon; named after American bacteriologist Ruth E. Gordon.

Common name: Tap water scotochromogen.

Sources of descriptions: Bönicke, 1962; Wayne *et al.*, 1967; Kestle *et al.*, 1967; Wayne and Doubek, 1968.

Moderate to long rods. Colonies on inspissated egg medium usually smooth and yellow or orange in 7 or more days' incubation at 37 C. Although pigment is produced when cultures are grown in the dark, the color is often intensified by growing in continuous light. On oleic acid albumin agar, smooth yellow to orange colonies; may be domed, with entire margin or have domed center, with flattening and irregularity at periphery; occasionally rough. Growth optimal at 35 C.

Frequently encountered as casual resident in human sputum and gastric lavage specimens. Also found in water taps and soil. Rarely if ever implicated in disease processes (Wolinsky and Rynearson, 1968).

Antigenic structure: These organisms have been subjected to very little immunologic analysis. They can be distinguished from *M. scrofulaceum* by delayed hypersensitivity skin tests (Runyon and Dietz, 1971). The methanol- and acetone-soluble cell fraction does not produce a precipitate line in immunodiffusion studies against antiserum specific for similar extracts of *M. scrofulaceum* (Wayne *et al.*, 1967). These bacilli are not agglutinated by sera produced against *M. scrofulaceum* (*marianum*), *M. avium*, *M. intracellulare*, *M. kansasii* or *M. marinum* (Hobby *et al.*, 1967).

Type strain: ATCC 14470.

Further Comments

The organisms Bönicke (1961) referred to as *M. aquae* Galli-Valerio appear to correspond to the type strain of *M. gordonae* (Wayne, 1970); *M. aquae* is a subjective synonym of *M. smegmatis* (Gordon and Smith, 1955).

13. **Mycobacterium scrofulaceum** Prissick and Masson 1956, 802. (*M. marianum* Suzanne and Penso 1953, 383; see discussion of synonymy and nomenclature, Wayne and Lessel, 1969.)

scro.fu.la′ce.um. L. n. *scrofula* tuberculous lymphadenitis, L. gen.n. *scrofulaceum* of scrofula.

Common name: Scrofula scotochromogen.

Sources of description: Wayne *et al.*, 1967; Penso *et al.*, 1957; Kestle *et al.*, 1967; Wayne and Doubek, 1968.

Short to long rods or filaments. Colonies on inspissated egg medium usually smooth and yellow to orange in 7 or more days incubation at 37 C. On oleic acid albumin agar, smooth yellow colonies, some domed with entire margin and others domed center and flattening and irregularity at periphery. Occasional strains rough. Growth optimal at 35 C.

Isolated from closed lesion of cervical lymphadenitis in a child (Prissick and Masson, 1956). Most commonly encountered in human secretions. Found in pus from suppurating cervical lymph nodes (especially of children) and considered etiologic agent of the lesions (Prissick and Masson, 1957). Also found in human sputum and gastric lavage specimens, usually as a casual resident, but occasionally associated with pulmonary disease. Occasionally found in soil (Kestle *et al.*,

1967; Wolinsky and Rynearson, 1968). Serotypes of this species have been found in swine (Schaefer, 1968).

In human disease most commonly implicated as etiologic agent of cervical lymphadenitis in children. Of limited pathogenicity in experimental animals. Does not cause extensive generalized disease or death in rats, hamsters or chickens; occasional lymph node involvement, and rarely localized lesions in liver or spleen. In guinea pigs inoculated subcutaneously produces abscesses at site of inoculation and enlargement of regional lymph nodes; intraperitoneal inoculation causes enlargement and suppuration of regional lymph nodes and consistently a peritonitis of various degrees and rarely lesions of liver or spleen, but not generalized disease or death (Penso *et al.*, 1957; Prissick and Masson, 1957).

Antigenic structure: Magnusson (1962) distinguished *M. scrofulaceum* from nine other species of mycobacteria and from a number of nocardias on the basis of dermal hypersensitivity; he (Magnusson, 1967) equated *M. scrofulaceum* with *M. marianum*. Runyon and Dietz (1971) were able to distinguish these from *M. gordonae* by this technique. Schaefer (1965, 1968) demonstrated the existence of three serotypes identified as "scrofulaceum," "Lunning" and "Gause" by bacillary agglutination, with some cross-reactivity with cultures which had in addition properties of other species. In immunoelectrophoretic studies of bacillary extracts, this species gives a distinct and well defined reaction pattern (Castelnuovo and Morellini, 1962). A methanol- and acetone-soluble extract of bacilli produced a specific immunodiffusion precipitin band, common to those bacteria fitting the biochemical pattern of *M. scrofulaceum;* included were cultures identified as *M. marianum*, *M. scrofulaceum* and the "Gause" strain (Wayne *et al.*, 1967).

Type strain: ATCC 19981.

Further Comments

M. scrofulaceum and *M. marianum* are considered subjective synonyms. Under rules of priority, *M. marianum* would be the legitimate name; however, "*M. marianum*" is judged an orthographic (although not an etymologic) variant of "*M. marinum*" Aronson, 1926, and because of the ease of confusion between these names, is considered illegitimate. For this reason, *M. scrofulaceum* is accepted as the legitimate name for this species. For detailed discussion, see Wayne and Lessel (1969).

14. **Mycobacterium paraffinicum** Davis, Chase and Raymond 1956, 314.

pa.raf.fin′i.cum. L. n. *paraffin* little affinity;

paraffins, a class of chemical compounds; M.L. gen.n. *paraffinicum* of paraffin.

Sources of description: Davis *et al.*, 1956; Käppler, 1968. Also unpublished data of present authors.

Slender rods, 0.5–0.7 by 3–7 μm. Pellicle growth from liquid mineral salts medium shows tendency to cording. Grows on paraffins of C_2 to C_{10}, and possibly higher, as sole carbon source. Yellow waxy wrinkled colonies on mineral salts medium. Isolated from soil.

Type strain: ATCC 12670.

Further Comments

This organism is very similar to *M. scrofulaceum*. Further study is required before a decision can be made to reduce it to subjective synonymy with the latter species. If such a decision is made, the name *"scrofulaceum"* would have priority, as it was published effectively a few weeks before *"M. paraffinicum."*

15. **Mycobacterium intracellulare** (Cuttino and McCabe) Runyon 1965, 258

(*Nocardia intracellularis* Cuttino and McCabe 1949, 16; *Mycobacterium intracellularis* (*sic*) Runyon 1965, 258. Subj. syn. *M. brunense* Kazda 1967, 207.)

in'tra.cel.lu.lar'e. L. prep. *intra* within; L. n. *cella* small room; L. adj. *intracellulare* within cell.

Common name: Battey bacillus.

Sources of description: Wayne, 1966; Kestle *et al.*, 1967; Engbaek *et al.*, 1968; Pattyn, 1967.

Rods short to long. In new growth transiently filamentous, eventually becoming coccobacillary. Dilute inocula on inspissated egg media yield usually smooth non-pigmented colonies after 7 or more days of incubation at 37 C; on aging colonies may become yellow. On oleic acid albumin agar, smooth, thin, transparent, lobed, non-pigmented colonies, often becoming thicker (domed) in center; rarely rough colonies occur.

May cause a severe chronic pulmonary disease in man. Limited lesions in swine.

Experimentally causes limited disease in chickens and mice, much less severe than *M. avium*, and rarely fatal. In general chicken pathogenicity corresponds to serotype, with *M. avium* serotypes I and II more consistently pathogenic than the *M. intracellulare* serotypes. Forms intermediate between *M. avium* and *M. intracellulare* in terms of colonial morphology, cultural behavior and chicken pathogenicity occur. Growth at 42 C has been reported to increase pathogenicity of inoculum (Scammon *et al.*, 1964). Lesions in guinea pigs usually limited to site of inoculation. Hamsters inoculated intratesticularly show extensive local lesions and, frequently, secondary focal lesions in

liver, spleen and abdominal lymph nodes. Rabbits receiving 10^{-3} mg intravenously develop no demonstrable lesions (Feldman and Ritts, 1963) but 5 mg, although not causing death, produce rare macroscopic lesions in visceral organs, and usually moderate to severe lesions of joints and tendon sheaths (Engbaek *et al.*, 1964).

Isolated from fatal systemic disease in a child. Most frequently encountered in pulmonary secretions from people suffering with tuberculosis-like disease, and from surgical specimens from such patients. When isolated from human secretions, usually considered etiologic agent of pulmonary disease, although occasionally isolated as apparent casual resident. Also isolated from disease processes in cattle and swine. Occasionally found in soil (Wolinsky and Rynearson, 1968).

Antigenic structure: Magnusson (1962, 1967) distinguished *M. intracellulare* from 10 other mycobacterial species, including *M. avium* by dermal hypersensitivity, which did not permit division of this species into serotypes as outlined below. The use of sensitized guinea pig intestine by the Schultz-Dale technique separated this species from seven others including *M. avium* (Jensen *et al.*, 1966). At least 15 agglutinating serotypes or serogroups established among cultures identifiable as *M. intracellulare* (Schaefer, 1968; Hobby *et al.*, 1967; Saito and Kubica, 1968). Members of these serotypes are predominantly non-pathogenic for chickens and were isolated mainly from man, cattle and swine, as opposed to two or possibly three distinct serotypes of *M. avium*, which are virulent for chickens, and less frequently associated with mammalian disease. Jones *et al.* (1965) were able to make some divisions along comparable lines by immunofluorescence techniques; Bennedsen (1968), employing an indirect test with sera obtained early in experimental infections of rabbits, distinguished seven immunofluorescence serotypes which corresponded to the agglutinating patterns.

Type strain: ATCC 13950.

Further Comments

There is debate as to whether *M. intracellulare* is most appropriately treated as a variety of *M. avium* or as a distinct and separate species (Wayne, 1966; Runyon, 1967). A distinction has been made between the species on the basis of correlation of pathogenicity with serotype (Schaefer, 1968) but a number of strains have been described with biochemical and pathogenic behavior intermediate between these two species (Scammon *et al.*, 1964; Pattyn, 1967). There is a feeling that serotype Davis strains, which include the so-called *M. brunense* (Kubin *et al.*, 1969), resemble *M. avium* more closely than *M. intracellulare*.

16. **Mycobacterium avium** Chester 1901, 356 *nom. cons.* Opin. 47, Jud. Comm. 1973, 472. (Tuberculose des oiseaux, Strauss and Gamaléia 1891, 466; *Bacillus tuberculosis gallinarum* Sternberg, 1892, 392; *Mycobacterium tuberculosis avium* Lehmann and Neumann, 1896, 370; *Mycobacterium tuberculosis typus gallinaceus* Lehmann and Neumann, 1907, 553.)

a'vi.um. L. n. *avis* a bird; L. gen.pl.n. *avium* of birds.

Common name: Avian tubercle bacillus.

Sources of description: Wayne, 1966; Kestle *et al.*; 1967; Engbaek *et al.*, 1968; Pattyn, 1967.

Short to long rods, some filaments. Dilute inocula on inspissated egg media yield usually smooth non-pigmented colonies after 7 or more days of incubation at 37 C; on aging colonies may become yellow. On oleic acid albumin agar, smooth, thin, transparent, lobed non-pigmented colonies. Occasionally rough strains are encountered.

Produces tuberculosis in domestic fowls and other birds; in pigs a localized disease. Experimentally in the rabbit and mouse it usually proliferates without macroscopic tubercles, producing disease of the Yersin type. Not pathogenic for guinea pig and rat. Strains of *M. avium* serotypes occasionally have been implicated in human pulmonary disease, although *M. avium*-like organisms causing human disease are usually more similar to *M. intracellulare* and fit one of that species' serotypes (Hobby *et al.*, 1967). Lesions in cattle may be caused by either *M. avium* or *M. intracellulare* serotypes (Schaefer, 1968). *M. avium* serotypes are highly virulent, type II being most consistently so, and representing the more common bird pathogen in nature (Schaefer, 1968); inocula of .01 mg will kill chickens. Chickens infected with 5 mg of moist bacilli, intravenously, die within 2 months, with gross lesions in spleen and microscopic lesions in lung and spleen (Engbaek *et al.*, 1968). Resistance to 1–2 mg of isoniazid/ml is accompanied by loss of virulence for chickens (Tacquet *et al.*, 1958). For more detailed description see Feldman (1938). An inoculum of 0.01 mg will kill rabbits. Rabbits inoculated with 5 mg moist bacilli, intravenously, usually die within 40 days with macroscopic lesions of spleen, and occasionally lungs, and microscopic lesions of spleen and lungs. Animals surviving 3 months show lesions of joints and tendon sheaths (Engbaek *et al.*, 1964, 1968). Guinea pigs generally exhibit only local lesions although 1 mg of bacilli intravenously will cause some deaths. Mice inoculated intravenously with 10 mg are variously affected, with some deaths.

Antigenic structure: Magnusson (1962, 1967) distinguished *M. avium* from 10 other mycobacterial species, including *M. intracellulare*, by dermal hypersensitivity; this technique did not distinguish between the serotypes I and II described below. Although some degree of cross-reactivity occurs, dermal desensitization is effected only by homologous antigen (Worthington and Kleeberg, 1967). The use of sensitized guinea pig intestine by Schultz-Dale technique differentiated *M. avium* from at least five other species, including *M. intracellulare* (Jensen *et al.*, 1966). Two agglutinating serotypes (I and II) were established for *M. avium* (Schaefer, 1965; Saito and Kubica, 1968); bacilli isolated from diseased birds almost always fall into one of these types, with type II the most frequent cause of natural bird infection. Subsequently Marks *et al.* (1969) showed that type II could be divided into two subtypes, with some cross-reaction. They proposed that the serotype designations I and II be replaced by types 1, 2 and 3. Type II is also the most consistently pathogenic in experimental infection of chickens. Occasionally a member of one of these serotypes is isolated from human infection. Bennedsen (1968), employing an indirect immunofluorescence test, found reactions specific for each of the two original serotypes; two cross-reactions with *M. intracellulare* strains were seen. Immunodiffusion and immunoelectrophoresis studies on culture filtrates or lysates demonstrate as many as nine lines of precipitate against homologous serum, but some represent antigens common to a number of myocobacteria, e.g. four each common to *M. tuberculosis* and *M. bovis*, six common to *M. paratuberculosis*, and only two each common to *M. phlei*, *M. smegmatis* and *M. marinum* (Lind and Norlin, 1963; Tuboly, 1965). One precipitin line when *M. avium* antigen is tested against *M. tuberculosis* antiserum does not coincide with any line produced with *M. tuberculosis* antigen against this serum (Castelnuovo *et al.*, 1959; Lind, 1959).

Isolated from tubercles in fowls. Widely distributed as the causal agent of tuberculosis in birds and less frequently in lesions or lymph nodes of cattle, swine and other animals. Rarely found in soil or as etiologic agent of human disease. Avian serotypes 1 and 2 are widely distributed through at least three continents (Wolinsky and Rynearson, 1968), although the new type 3 is seen only in Europe (Marks *et al.*, 1969).

Proposed neotype: SSC 1336; ATCC 25291 (Engbaek *et al.* 1971, 192).

Further Comments

See *Further Comments* under *M. intracellulare*.

17. **Mycobacterium xenopi** Schwabacher 1959, 59. (*Mycobacterium xenopei* (*sic*) Schwabacher 1959, 59; (subj. syn. *M. littorale* Marks 1964, 479.)

xe.no'pi. *Xenopus* a genus of toad; M.L. gen.n. *xenopi* of *Xenopus*.

Sources of description: Runyon (1968); Engbaek *et al.*, 1967; Boisvert, 1965; Kestle *et al.*, 1967.

Long to filamentous rods. Dilute inocula on inspissated egg media yield smooth, non-pigmented colonies after 14 or more days of incubation at 37 C; on aging, most colonies become yellow. On Middlebrook 7H10 agar, colonies have compact centers, surrounded by a fringe of microscopically evident branching filaments on the agar surface. Colonies become adherent to medium by a button-like growth into the agar. Optimal growth at 40–45 C.

Occasionally associated with chronic pulmonary disease, but more frequently isolated from human secretions without associated disease, infrequently in disease of genitourinary tract (Marks and Schwabacher, 1965; Engbaek *et al.*, 1967). Isolated from skin granulomas of the toad *Xenopus laevis* (Marks and Schwabacher, 1965). Variable response in different strains of mice inoculated intraperitoneally with 0.2–10 mg, but few animals die, and limited numbers of macroscopic lesions appear in liver, spleen, kidney or lung. Guinea pigs receiving 4 mg intramuscularly develop caseous abscess at site of inoculation; 1 mg intravenously causes no macroscopic lesions and intraperitoneally causes macroscopic nodules in omentum of some animals; 0.1 mg intracutaneously causes swelling and ulceration at site. Hens receiving 4 mg intramuscularly develop no gross lesions; 5 mg intravenously usually kill with lesions in liver and/or spleen, but 1 mg by this route induces no gross lesions. Rabbits receiving 4 mg intramuscularly develop caseous abscesses at site of inoculations; 5 mg intravenously cause few macroscopic lesions, or lesions of joints or tendon sheaths (Schwabacher, 1959; Engbaek *et al.*, 1967).

Antigenic structure: Distinguished from other mycobacteria by comparative dermal hypersensitivity reactions (Magnusson, 1967). Exhibits four species-specific antigens by immunodiffusion tests (Beck and Stanford, 1968).

Type strain: ATCC 19250; NCTC 10042.

18. **Mycobacterium ulcerans** MacCallum, Tolhurst and Buckle in Fenner 1950, 817.

ul'ce.rans. L. part.adj. *ulcerans* making sore, causing to ulcerate.

Sources of description: MacCallum *et al.*, 1948; Pattyn *et al.*, 1964–66; also unpublished data based on authors' experience.

Moderately long rods. Growth on inspissated egg medium evident after 4 weeks incubation at 30–33 C as minute, transparent domed colonies. On aging colonies become low convex to flat with

irregular outline and rough surface, yellow. Rough corded colonies on oleic acid albumin agar. Capable of growth between 30 and 33 C, but little growth at 25 C and usually none at 37 C.

Causes skin ulcers in man characterized by indolent extension from areas of inconspicuous induration to involve large areas with undermining of edges. Rats and mice are infected experimentally; guinea pigs, rabbits, fowls and lizards are resistant. Experimentally inoculated rats develop hemorrhagic necrotic lesions surrounded by zones of cellular accumulations consisting of leucocytes, lymphocytes and macrophages. There are no giant cells. The necrotic and cellular zones show large clumps of acid-fast bacilli in the extracellular spaces and in macrophages. Human lesions do not show inflammatory responses but consist of areas of lipid necrosis and tissue breakdown. Lesions develop only in cooler parts of experimental animals' bodies. Thus, inoculation of mouse footpad consistently causes local lesions; inoculation by intranasal, intraperitoneal or intravenous route causes no visceral lesions, but after long incubation results in ulcerating lesions in hairless peripheral parts of the body and on the scrotum (Fenner, 1956).

Originally isolated from human skin lesion in Australia; has been isolated from ulcerative skin infections of man in Mexico, New Guinea, Malaya and Africa.

Antigenic structure: In complement fixation tests with sera of rabbits immunized with human, bovine and murine types of tubercle bacilli, *M. ranae* and *M. phlei*, the heat-killed, washed bacilli serving as antigens, *M. ulcerans* was found to be antigenically distinct from the other pathogenic species of *Mycobacterium* tested. This conclusion was supported by skin-sensitivity reactions in guinea pigs (Fenner and Leach, 1952).

Reference strain: ATCC 19423.

19. **Mycobacterium phlei** Lehmann and Neumann 1899, 411 *proparte*. (*Bacterium phlei* (Lehmann and Neumann) Grimme 1902, 81; *Sclerothrix phlei* (Lehmann and Neumann) Vuillemin 1931, 160; subj. syn. *Mycobacterium moelleri* Chester 1901, 358; Timotheebacillus or Grasbacillus I Moeller 1898, 376.)

phle'i. M.L. neut.n. *Phleum* a grass genus, timothy; M.L. gen.n. *phlei* of timothy.

Common names: Timothy bacillus, Hay bacillus.

Sources of description: Gordon and Smith, 1953; Gordon and Mihm 1959; Tsukamura *et al.*, 1968.

Short rods, 1.0–2.0 μm in length, rarely longer. Acid-fast staining, particularly after prolonged incubation (5–7 days), may be very irregular (5–100% of cells acid-fast). Dilute inocula on inspissated egg media usually produce rough, coarsely wrinkled, deep yellow to orange colonies after 2–5 days incubation; a few cultures smooth, soft and butyrous. On oleic acid albumin agars growth not as abundant; colonies may be smooth with a domed center surrounded by a flat, translucent skirt with either entire or irregular edges and dark granules near the center of the colony; rough colonies are flat and granular to loosely corded in appearance, with granules near the center and irregular edges (Jones and Kubica, 1965). Capable of growth from 22–52 C.

Not pathogenic for mouse, rat, guinea pig, rabbit, chicken, frog or carp (Penso *et al.*, 1951; Durr *et al.*, 1959).

Originally isolated from hay and grass. Reportedly widely distributed in nature, although recent soil surveys have revealed few strains (Wolinsky and Rynearson, 1968).

Antigenic structure: One homogeneous group demonstrated by immunofluorescence (Jones and Kubica, 1968), species-specific sensitins (Magnusson, 1962) and immunodiffusion and immunoelectrophoretic techniques (Castelnuovo *et al.*, 1960; Gimpl and Lanyi, 1965; Lind, 1960; Norlin, 1965).

Suggested working type: ATCC 11758; NCTC 8151.

20. Mycobacterium vaccae Bönicke and Juhasz 1964, 133.

vac'cae. L. n. *vacca* a cow; M.L. gen.n. *vaccae* of the cow.

Sources of description: Bönicke and Juhasz, 1964; Tsukamura *et al.*, 1968.

Short, plump rods 1.0–4.0 μm in length, slightly curved with rounded or occasionally thickened ends; branching rare but occasional Y-shaped cells observed. Acid-fast staining may be irregular in older cultures. Dilute inocula on inspissated egg media usually produce, after 2–3 days incubation, smooth, moist, shiny, butyrous, dome-shaped colonies with deep yellow to orange pigmentation. Most strains extremely light sensitive being non-chromogenic if grown 1–2 days in total darkness, but acquiring yellow pigment when exposed only briefly to light. Rough colonies as well as non-pigmented ones are occasionally encountered. Growth good from 22–40 C; at 17 and 42 C growth is restricted and pigment production inhibited.

Isolated from lacteal glands of cattle. Apparently widely distributed in nature; found in meadows, pastures, watering ponds, wells and even occasionally in skin lesions of cows.

Type strain: ATCC 15483.

Further Comment

According to Tsukamura *et al.* (1968) this is a heterogeneous species. Our observations would support this conclusion. Undoubtedly many of the strains described by Jones and Kubica (1963, 1965) as Rapid Grower C are members of this species.

21. Mycobacterium diernhoferi Bönicke and Juhasz 1965, 292.

diern.hof'er.i. M.L. gen.n. *diernhoferi* of Diernhofer; named after Austrian professor Dr. Karl Diernhofer, Veterinary School, Vienna.

Source of description: Bönicke and Juhasz 1965; personal communications.

Short, plump rods 0.5–0.8 by 1.0–3.0 μm, often with thick, rounded ends. Primary branching never observed. Organisms from young cultures acid fast, but staining in older cultures may be irregular. Dilute inocula on inspissated egg media usually yield gray to dirty yellow pigmented colonies after 3 days incubation. Prolonged exposure to light causes the pigment to change to dark brown or gray-black. Colonies generally smooth, hemispheric and glistening, with entire edge; dissociation from smooth to rough colony type seldom occurs. Good growth from 22–37 C; growth somewhat inhibited at 17 and 40 C and completely inhibited at 42 C.

Does not produce disease in mice. Isolated from a drinking trough for cattle. Found in environment of domestic cattle, but not yet isolated from milk.

Antigenic structure: By species-specific sensitins it is possible to distinguish *M. diernhoferi* from other rapidly growing mycobacteria (Takeya personal communication).

Type strain: ATCC 19340.

22. Mycobacterium smegmatis (Trevisan) Lehmann and Neumann 1899, 403. (Smegma bacillus Alvarez and Tavel 1885, 303; *Bacillus smegmatis* Trevisan 1889, 14; *Bacterium smegmatis* (Trevisan) Migula 1900, 497; *Mycobacterium paratuberculosis smegmatis* Bustinza 1951, 151.)

smeg.ma'tis. Gr. n. *smegma* an ungent or ointment, a detergent, in M.L. sebaceous humor; M.L. gen.n. *smegmatis* of smegma.

Sources of description: Gordon and Smith, 1953; Gordon and Mihm, 1959; Tsukamura *et al.*, 1968.

Slender rods 3.0–5.0 μm in length, sometimes curved, with branching or Y-shaped cells observed; cells occasionally swollen, with beaded or ovoid, deeper staining bodies. Acid-fast staining after incubation for 5 days may be irregular (10–80% of cells acid-fast). Dilute inocula on inspissated egg media usually produce abundant

TABLE 17.32

Properties of rapidly growing species of genus **Mycobacterium**[a]

No.	Property	19. M. phlei	20. M. vaccae	21. M. diern-hoferi	22. M. smeg-matis	23. M. tham-nopheos	24. M. flaves-cens	25. M. fortu-itum	26. M. peregri-num	27a. M. chelonei subsp. chelonei	27b. M. chelonei subsp. ab-scessus
	Amidases										
1*	Acetamidase	−	+	+	+	+	−	+	+	+	+
2*	Benzamidase	−	+	−	V	+	−	−	−	−	−
4*	Isonicotinamidase	−	+	−	+	−	−	−	−	−	−
5	Nicotinamidase	+	+	+	+	+	+	−	−	+	+
6	Pyrazinamidase	+	+	+	+	+	+	−	−	+	+
7	Allantoinase	−	+	−	−	+	−	+	+	−	−
8	Succinamidase	−	−	−	+	−	−	−	−	−	−
9	Decompose phenylalanine	+	+	−	−	−	−	−	−	−	−
	Carbon source										
10	Benzoate	−	V	−	+	−	−	−	−	−	−
11	Citrate	+	+	−	+	+	−	+	+	+	−
12	Malate	+	+	V	+	+	−	+	+	+	+
13	Oxalate	−	−	−	+	−	−	−	−	−	−
14	Succinate	+	+	+	+	+	V	+	+	+	+
	Acid production from carbo-hydrates										
15	Arabinose	+	±	+	+	−	−	−	−	−	−
16	Dulcitol	−	−	−	+	−	−	−	−	−	−
17	Fructose	+	+	+	+	V	±	+	+	−	−
18	Galactose	+	−	−	+	±	−	−	−	−	−
19	Inositol	−	+	+	+	−	−	−	−	−	−
20	Mannitol	+	+	+	+	+	V	−	+	−	−
21	Mannose	+	+	+	+	+	+	+	+	+	+
22	Rhamnose	−	+	−	+	−	−	−	−	−	−
23	Sorbitol	+	V	−	+	+	V	−	−	−	−
24	Trehalose	+	+	+	+	+	V	+	+	+	+
25	Xylose	+	+	+	+	−	−	−	−	−	−
26	Survives 60 C/4 hr	+	−	−	−	−	−	−	−	−	−
27	MacConkey agar	−	−	−	−	−	−	+	+	d	+
28	Malachite green, 0.01%	V	−	−	V	−	−	+	+	+	+
29	Methyl violet, 0.01%	V	−	−	V	−	−	+	+	+	+
30	Pyronin B, 0.01%	V	−	−	+	−	−	+	+	+	+
31	Arylsulfatase (3-day)	−	−	−	−	−	−	+	+	+	+
	(4 week)	+	+	+	+	+	+	+	+	+	+
32	Growth on 5% NaCl	+	+	−	+	+	+	+	+	−	+
33	Tellurite reduction (3 day)	−	+	+	+	−	∓	±	±	−	±
34	Nitrate reduction	+	V	∓	+	−	+	+	+	−	−
35	Catalase, 68 C	+	+	−	∓	−	+	+	+	+	+
38	Iron uptake	+	+	+	+	+	−	+	+	+	+
39	Niacin	−	−	−	−	−	−	−	−	−[b]	−
40	Tween hydrolysis (5 day)	+	+	−	+	+	+	±	±	−	−
41	Catalase >45 mm foam	+	+	−	+	−	+	+	+	+	+
42	Pigment	S	P/S	N/P	N	N	S	N	N	N	N
46	Growth at 25–32 C	+	+	+	+	+	+	+	+	+	+
	37 C	+	+	+	+	−	+	+	+	+	+
	40 C	+	+	+	+	−	+	+	−	V	+
	45 C	+	−	−	+	−	−	−	−	−	−
	52 C	+	−	−	−	−	−	−	−	−	−

[a] Data assembled and results expressed as for Table 17.31, except the notation V which indicates character is inconstant, i.e. the same strain may sometimes test positive, sometimes negative.

All species tested give a negative microcolonial test and are negative for salicylamidase and malonamidase, positive for urease (4 hr test). All species positive on glucose, negative on lactose and raffinose.

[b] Some strains of *M. chelonei* may be niacin positive (the former *M. borstelense* var. *niacinogenes*).

TABLE 17.33

Selected properties for routine identification of species and/or complexes of mycobacterial species[a]

No.	Property	1. M. tuberculosis	3. M. bovis	5. M. kansasii	6. M. marinum	8. M. gastri	9–11. M. nonchromogenicum-terrae-triviale complex	12. M. gordonae	13. M. scrofulaceum	15–17. M. avium-intracellulare-xenopi complex[b]	18. M. ulcerans	19. M. phlei	20. M. vaccae	21. M. diernhoferi	22. M. smegmatis	23. M. thamnopheos	24. M. flavescens	25–26. M. fortuitum-peregrinum complex	27. M. chelonei complex
3	Urease	+	+	+	+	+	−	−	+	−		+	+	+	+	+	+	+	+
27	MacConkey agar																	+	+
31	Arylsulfatase (3 day)	−	−	−	−	−		−	−			−	−	−	−	−	−	+	d
32	Growth on 5% NaCl						d					+	+	−	+	+	−	+	+
33	Tellurite red'n (3 day)	+	−	−	−	−	+	−	−	d		−	−	−	+	−	−	+	d
34	Nitrate reduction	+	−	+	−	−	+	−	−	−	−	+	d	+	+	−	d	d	d
35	Catalase 68 C	−	−	+	−	−	+	+	+	+	+	+	+	d	d	−	+	+	+
38	Iron uptake	−	−	−	−	−	−	−	−	−	−	+	+	−	+	+	+	+	+
39	Niacin	+	−	−	−	−	−	−	−	−	∓	−	−	+	−	+	−	+	−
40	Tween 80 hydrol. (5 day)	d	−	+	+	+	+	+	−	−	−	+	+	−	+	+	+	+	−
41	Catalase: >45 mm foam	−	−	+	−	−	+	+	+	−	−	+	+	−	+	+	+	+	+
42	Pigment	N	N	P	P	N	N	S	S	N	N	S	P/S	N/P	N	N	S	N	N
43	Resists 10 µg/ml T2H	+	−	+	+	+	+	+	+	+	+	+	+	+	+	−	+	+	+
46	Growth at 37 C	+	+	+	d	+	+	+	+	+	−	+	+		+	−	+	+	+
	45 C	−	−	−	−	−	−	−	−	d	−	+	−	+	+	−	−	−	−
	52 C	−	−	−	−	−	−	−	−		−	+	−	−	−	−	−	−	−

[a] Identifications based on this table will serve most practical needs; more definitive speciation may be obtained by use of additional properties given in Tables 17.31 and 17.32. See footnotes to Table 17.31 for meaning of symbols +, −, N, S and P; d = strain-to-strain variation, some strains positive, others negative.

[b] *M. xenopi* is distinguished on cellular and colonial morphology, drug susceptibility and late pigment production.

growth of finely wrinkled to coarsely folded colonies, creamy white in color after 2–4 days incubation. Smooth glistening, butyrous colonies commonly seen. Pigmented colonies rare, but may be observed in older cultures and more commonly on non-egg-containing media. On oleic acid albumin agar rough colony forms produce slightly domed, smooth-textured, non-corded but granular colonies. The smooth form is domed, has a narrow, translucent apron with entire edge, and dark granules are distributed throughout the central portion of the colony (Jones and Kubica, 1965). Capable of growth from 25–45 C.

Small inocula not pathogenic for guinea pigs, mice, hamsters or chickens, although positive cultures sometimes obtained from spleens of mice and/or guinea pigs (Durr *et al.*, 1959). Isolated from smegma. Once commonly found in soil and water, though recent isolations from these sources have been infrequent.

Antigenic structure: One homogeneous group identifiable by species-specific sensitins (Magnusson, 1962) and immunodiffusion and immunologic techniques (Castelnuovo *et al.*, 1960; Gimpl and Lanyi, 1965; Lind, 1960; Norlin, 1965).

Suggested neotype: ATCC 14468.

23. Mycobacterium thamnopheos Aronson 1929, 215.

tham.no'phe.os. M.L. masc.n. *Thamnophis* garter snake, a genus of snakes; M.L. gen.n. *thamnopheos* of the garter snake.

Sources of descriptions: Original description supplemented by thesis of Bynoe, McGill University, Montreal, 1931, and by Tsukamura *et al.*, 1968.

Long, slender rods 4.0–7.0 μm in length, slightly curved; beaded and barred forms often observed. Alcohol-fast; not strongly acid-fast, resisting only momentary decolorization with acid alcohol. On inspissated egg media small raised, moist, convex colonies, non-pigmented to pink or salmon colored appear usually within 5–7 days. Capable of growth from 10–35 C (Bynoe); no growth at 37 C (Aronson).

Experimentally produces generalized disease in snakes, frogs, lizards and fish, but not pathogenic for guinea pigs, rabbits or fowl. Isolated from lungs and livers of garter snakes (*Thamnophis sirtalis*) and possibly a parasite in other cold-blooded vertebrates.

Type strain: ATCC 4445.

Further Comments

Lechevalier *et al.* (1971) suggest that on the basis of lipid composition, *M. thamnopheos* ATCC 4445 should be placed in the genus *Nocardia*. More

strains must be studied before this proposal can be assessed.

24. Mycobacterium flavescens Bojalil, Cerbón and Trujillo 1962, 344.

fla.ves'cens. L. v. *flavesco* become golden yellow; L. pres.part. *flavescens* becoming yellow.

Sources of description: Bojalil *et al.*, 1962; Wayne *et al.*, 1967.

Rod-shaped organisms. Dilute inocula on inspissated egg media usually produce after 7–10 days incubation soft, butyrous, intensely orange-colored colonies which adhere strongly to the medium and are difficult to remove. Capable of growth from 25–42 C. Although this organism is intermediate in growth rate, its metabolic and physiologic activities are more similar to those of the rapidly growing species (Wayne, 1967; Wayne *et al.*, 1967).

Isolated from drug-treated tuberculous guinea pigs.

Type strain: ATCC 14474; NCTC 10271.

Further Comments

Not believed to be pathogenic for lower animals or man (Wayne, 1959; Wayne *et al.*, 1967).

25. Mycobacterium fortuitum da Costa Cruz 1938, 299. (*Mycobacterium giae* Dārziņš 1950, 32; *Mycobacterium minetti* Penso, Castelnuovo, Guadiano, Princivalle, Vella and Zampieri 1952, 493; strains of *Mycobacterium* from cows 18, 19, 70 and 75 Minett 1932, 317.

for.tu'i.tum. L. neut.adj. *fortuitum* casual, accidental.

Stanford and Gunthorpe (1969) have presented evidence that *M. ranae* (Küster) Bergey *et al.* 1923 is the prior and legitimate name. Because strains of *M. smegmatis* also have been widely distributed under the label *M. ranae* and because *M. fortuitum* has for decades been universally used for the taxon of da Costa Cruz, request for conservation of this name has been made to the Judicial Commission (Runyon, 1971).

Sources of description: da Costa Cruz, 1938; Penso *et al.*, 1952; Gordon and Smith, 1955; Gordon and Mihm, 1959; Tsukamura *et al.*, 1968.

Rods 1.0 to 3.0 μm in length. Coccoid and short forms, to long, slender rods, occasionally beaded or swollen with an ovoid, non-acid-fast body at one end. In pus, long and filamentous forms with definite branching. Ten to 100% of cells acid-fast after 5 days incubation at 28 C. Dilute inocula on inspissated egg media usually produce abundant growth within 2–4 days. Colonies may be smooth, soft, butyrous, hemispheric or waxy and multilobate or rosette clustered; also common are dull, waxy, rough colonies with heaped up centers. Colonies generally are off white to cream colored, however, when grown on malchite green-containing media, the dye is commonly absorbed by the colonies, imparting to them a green color (Hartwig et al., 1962). On oleic acid albumin agar the smooth colony types yield domed colonies with entire edges and dark central areas. On corn meal agar, smooth M. fortuitum colonies produce an extensive network of peripheral filaments. Rough colony variants on both oleic acid albumin and corn meal agars are dense, and corded; filamentous extensions seen in early growth, although inconspicuous in mature colonies. Capable of growth from 25–37 C. Most strains grow at 40 and 22 C, but growth at 45 or 17 C is usually inhibited.

Generalized disease in guinea pigs, rabbits and mice rarely seen, even with massive doses of young cultures. Local lesions commonly observed in kidneys of mice, guinea pigs, rabbits, monkeys and calves; lesions of ear lead to characteristic spinning behaviour in mice (Penso et al., 1952; Wells et al., 1955). Isolated from a cold abscess of man. Several strains isolated from lymph glands of cattle, from pulmonary disease and local abscesses of man, and from systemic, nodular infection of the frog Gia. Found in soil and infections of man, cattle and cold-blooded animals.

Antigenic structure: Schaefer (1967) described two serotypes of M. fortuitum; Magnusson (1962) reported that this organism could be identified by dermal hypersensitivity techniques. Immunodiffusion and immunoelectrophoretic studies support the uniqueness of this species (Castelnuovo et al., 1960; Gimpl and Lanyi, 1965; Norlin, 1965).

Suggested neotype: ATCC 6841.

Further Comments

Although M. peregrinum is dealt with separately in this chapter, it may well be a subjective synonym of M. fortuitum. A numerical taxonomic analysis of several hundred in vitro tests has shown these two species to differ in only six properties: M. fortuitum grows at 42 C, grows on 0.05 M nitrite, does not produce acid from mannitol, lacks pyrazinamidase and nicotinamidase activity, but is positive in propionamidase; M. peregrinum gives exactly the opposite reactions. Although none of these reactions was consistently positive or negative for many strains of each species, the separations have been corroborated by the chemical and serologic studies of Jenkins et al. (1971).

26. **Mycobacterium peregrinum** Bojalil, Cerbón and Trujillo, 1962, 343.

per.e.grin'um. L. neut.adj. *peregrinum* strange, foreign.

Source of description: Bojalil et al., 1962; and personal communications.

Rod-shaped organisms. Dilute inocula on most mycobacterial culture media usually produce visible growth after 2–4 days. On inspissated egg media colonies are commonly smooth, buff or pale straw colored. On oleic acid albumin agar both smooth and rough, non-pigmented colony forms have been observed. Capable of growth from 22–37 C; usually does not grow at 40 C from dilute inocula.

The reactions for production of acid from carbohydrates differ from those reported for the type culture in the following: type culture reported as producing acid from arabinose, galactose, inositol, rhamnose, sorbitol and xylose. Investigation of larger numbers of strains classified as M. peregrinum yielded the sugar reactions shown in Table 17.32.

Neither localized disease nor the spinning phenomenon observed with M. fortuitum can be consistently demonstrated when mice are inoculated with M. peregrinum. Inability to recover organisms from the tissues further indicates that this species has very little virulence for the healthy mouse.

One strain isolated from bronchial aspirations of a child with respiratory symptoms, another from nasal exudate of a cow. Isolated from the sputum of man, though not always under conditions indicating complicity in human disease; also found in the soil.

Antigenic structure: Jenkins et al. (1971) have shown this species to be serologically and chemically distinct from other rapid growers (see *Further Comments* under M. fortuitum).

Type strain: ATCC 14467.

27. **Mycobacterium chelonei** Bergey et al. 1923, 376. (*Mycobacterium abscessus* Moore and Frerichs, 1953, 163; *Mycobacterium runyonii* Bojalil, Cerbón and Trujillo, 1962, 343; *Mycobacterium borstelense* Bönicke, 1965, 535.)

che.lon'e.i. Gr.n. *chelone* a tortoise; M.L. gen.n. *chelonei* of a tortoise.

Sources of description: Tsukamura et al., 1968; Saito et al., 1968; Stanford and Beck, 1969; Stanford et al., 1972; Kubica et al., 1972.

Organisms pleomorphic, ranging from long, narrow to short, thick rods (0.2–0.5 by 1–6 μm), with coccoid forms (0.5 μm diameter) also reported. Young cultures (less than 5 days) are strongly acid-fast, however non-acid-fast forms begin to develop from the 5th day onward. After 3–4 days incubation on most media, dilute inocula yield colonies which are smooth, moist, shiny and non-chromogenic or creamy buff in color. Rough colonies occasionally observed, often after prolonged incubation (3 weeks). M. chelonei grown on corn meal agar does not exhibit the extensive network of filaments observed in *M. fortuitum*; rather they yield semi-transparent, finely granular colonies with either entire or multilobate edges. The rough colony form on oleic acid albumin agar has an overall smooth glistening appearance over an underlying rugulose structure. Capable of growth from 22–40 C. No growth at 42 C.

As originally described the organism produced only transient lesions in mice, hamsters, guinea pigs and rabbits. The original report described only limited pathology in mice following intraperitoneal infection. More recently reports of cooperative studies circulated through these authors' hands have revealed that the intravenous infection of mice with *M. chelonei* causes gross lesions visible in spleen, liver, lung and kidney of the infected animal. Has produced pathologic changes in the synovial tissue of knee and abscesslike lesions of gluteal region of man. Occasionally isolated from sputum (with and without related disease). Also found in soil.

Antigenic structure: Using absorbed sera, Jenkins *et al.* (1971) have shown this species to be serologically and chemically distinct from other closely related rapid growers (see *Further Comments* under *M. fortuitum*).

Stanford and Beck (1969), Stanford *et al.* (1972) and Kubica *et al.* (1972) have presented evidence that *M. abscessus* and *M. borstelense* are very similar and should be considered as synonyms of *M. chelonei*. Although the two species could not be clearly separated by Stanford and his colleagues (1969, 1972) or by serologic data, Jenkins *et al.* (1971) showed they had different lipid composition. Numerical taxonomic analysis of several hundred *in vitro* characteristics have shown *M. abscessus* strains to differ in some characteristics from those labeled *M. chelonei* and *M. borstelense*, and it has been recommended that two subspecies be recognized (Kubica *et al.*, 1972).

27a. *Mycobacterium chelonei* subsp. *chelonei* Bergey *et al.* 1923, 376.

As well as the properties listed in Table 17.32, this subspecies loses acid phosphatase at 100 C, does not grow in the presence of 0.2% picric acid, 1% deoxycholate or M/40 nitrite, cannot grow with nicotinamide or nitrite as the sole source of nitrogen.

Type strain: NCTC 947.

27b. *Mycobacterium chelonei* subsp. *abscessus* (Moore and Frerichs) Kubica, Baess, Gordon, Jenkins, Kwapinski, McDurmont, Pattyn, Saito, Silcox, Stanford, Takeya and Tsukamura 1972, 68.

As well as the differences listed in Table 17.32, does not lose acid phosphatase at 100 C, grows in the presence of 0.2% picric acid, 1% deoxy-

TABLE 17.34

References to specific techniques and interpretation of properties cited in Tables 17.31, 17.32 and 17.33

Property numbers	Reference
1, 2, 3, 4, 5, 6, 7, 8	Bönicke, 1961, 7
9	Gordon, 1966, 329
10, 11, 12, 13, 14, 15, 16, 17, 18, 19, 20, 21, 22, 23, 24, 25, 26	Gordon and Mihm, 1959, 736
27, 28, 29, 30	Jones and Kubica, 1963, 355
31	Jones *et al.*, 1966, 790
32	Kestle *et al.*, 1967, 1041
33	Kilburn *et al.*, 1969, 94
34[a] (rapid growers)	Kubica and Dye, 1967, 44, 70
35	Kubica and Pool, 1960, 387
36	Wayne, 1966, 919
37	Wayne and Doubek, 1965, 738
34[a] (slow growers), 38, 39, 40, 41, 42, 43, 44	Wayne and Doubek, 1968, 925
45	Wayne *et al.*, 1957, 451
46	Wayne *et al.*, 1967, 88

[a] Note that different techniques are cited for nitrate reduction (property 34) among slowly and rapidly growing mycobacteria.

cholate, or M/40 nitrite, and with nicotinamide or nitrite as the sole source of nitrogen.

Type strain: ATCC 19977.

28. **Mycobacterium paratuberculosis** Bergey *et al.* 1923, 374. (*Mycobacterium enteritidis* Lehmann in Lehmann and Neumann 1927, 755; *Bacterium paratuberculosis* (Bergey *et al.*) Miessner and Berge *in* Kolle and Wasserman 1929, 784; *Bacillus paratuberculosis* (Bergey *et al.*) Krasil'nikov 1941, 109; *Mycobacterium johnei* Francis 1943, 140; Darmtuberculose Johne and Frothingham 1895, 438).

pa.ra.tu.ber.cu.lo′sis. Gr. pref. *para* beside, related; M.L. n. *tuberculosis* tuberculosis; M.L. fem.n. *paratuberculosis* tuberculosis-like, paratuberculosis.

Common name: Johne's bacillus.

Description taken from M'Fadyean (1907) and Twort and Ingram (1913).

Plump rods, 1–2 μm in length, staining uniformly, but occasionally the longer forms show alternately stained and unstained segments.

This organism is difficult to cultivate; in primary cultures, it had originally been grown only in media containing dead tubercle bacilli or other dead acid-fast bacteria (Boquet, 1928). In a few instances cultures have been acclimatized to a synthetic medium free from added dead bacteria (Dunkin, 1933; Watson, 1935; Morrison, 1965).

Incorporation of purified extracts of myco-bacteria (mycobactins) eliminates requirement of addition of dead acid-fast bacilli into the medium to permit growth. The mycobactins represent a family of iron-chelating growth factors derived from different species of mycobacteria (White and Snow, 1968; Snow, 1965).

Colonies on glycerol agar containing heat-killed *M. phlei* are just discernible after 4–6 weeks and are dull white, raised and circular. On Dorset's glycerol egg medium containing heat-killed *M. phlei*, colonies appear after 4–6 weeks, and are minute, dull white, raised and circular, with a thin, slightly irregular margin. Older colonies become more raised, radially striated or irregularly folded and dull yellowish white. On Dorset's glycerol egg medium containing sheep's brain and heat-killed *M. phlei* growth is slightly more luxuriant than that described above.

Isolated from the intestinal mucosa of cattle suffering from Johne's disease, a chronic diarrhea. Apparently an obligate parasite in nature. The organisms isolated from sheep are reported to be more difficult to cultivate than are those from cattle (Dunkin and Balfour-Jones, 1935).

Produces Johne's disease in cattle and sheep. Experimentally produces a similar disease in goats also. Guinea pigs, rabbits, rats and mice are not affected. Very large doses in laboratory animals produce slight nodular local lesions comparable with those produced by *M. phlei*.

Antigenic structure: Johnin, prepared as tuberculin, gives positive reactions in cattle with Johne's disease. According to M'Fadyean and Sheather (1916), tuberculous animals may also give a reaction. Plum (1925) showed that animals sensitized to avian tuberculin react to johnin and that avian tuberculin causes a reaction in some animals infected with Johne's bacillus.

In immunoelectrophoretic studies, Tuboly (1965) demonstrated the presence of four antigens in common with *M. avium* and five in common with *M. tuberculosis*.

29. **Mycobacterium leprae** (Hansen) Lehmann and Neumann 1896, 372. (*Bacillus leprae* Hansen 1880, 32.)

lep'rae. Gr. n. *lepra* leprosy; M.L. gen.n. *leprae* of leprosy.

Common name: Leprosy bacillus or Hansen's bacillus.

Description of organisms seen in leprosy tissue from Hansen (1874) and Topley and Wilson (1936).

Rods, 0.3–0.5 by 1.0–8.0 μm, with parallel sides and rounded ends, staining evenly or at times beaded. When numerous, as from lepromatous cases, generally arranged in clumps, rounded masses or in groups of bacilli side by side. Strongly acid-fast.

Leprosy bacilli concentrated directly from human tissues have been demonstrated to possess cytochrome oxidase and alkaline phosphatase (Chatterjee *et al.*, 1956) and phenol oxidase (Prabhakaran *et al.*, 1968) activity. Fisher and Barksdale (1971) report finding bacilli of leprosy lesions to be distinctive in losing acid-fastness on extraction with pyridine.

Non-cultivable acid-fast bacilli from human leprous tissue multiply, with a generation time of 20–30 days, when inoculated into footpads of mice (Shepard, 1960; Shepard and Chang, 1962). Under these conditions of *in vivo* cultivation, the leprosy bacilli do not invade deep tissues, and their multiplication can be inhibited by diaminodiphenyl sulfone, isoniazid, para-aminosalicylic acid and cycloserine. Experimental transmission of leprosy to man has not been successful. Experimental transmission to immunosuppressed mice has produced a lepromatous-like model infection (Rees *et al.*, 1967).

Causes leprosy in man. In the lepromatous form of the disease, bacilli are so abundant in the tissue as to produce stuffed-cell granulomas; in the tuberculoid and neural lesions organisms are rare. Obligate intracellular parasite in man. Confined largely to the skin (expecially to convex and exposed surfaces), testes and to peripheral nerves. Probably do not grow in the internal organs.

Further Comments

Though not yet cultivated *in vitro*, these bacilli were the first to be recognized as a cause of human disease (Hansen, 1874). Bacteriologic identification depends on: (a) acid-fast staining; (b) failure of the organism to multiply in bacteriological media; and (c) ability to undergo limited multiplication in the mouse footpad (Shepard, 1960). Heated suspensions of the bacilli (obtained from nodules) produce a positive lepromin reaction in 75–97% of normal persons and of tuberculoid cases of leprosy but usually produce no reaction in lepromatous individuals (Mitsuda: see Hayashi, 1932).

Many organisms have been isolated from leprous tissues, some of which are acid-fast and which have been styled *M. leprae*. None of these strains, after adequate study, has been accepted as *M. leprae*. Hanks (1941) found that acid-fast cultures of this type were recoverable only from lesions located proximal to open ulcers in the skin, and presumably are exogenous contaminants.

30. **Mycobacterium lepraemurium** Marchoux and Sorel 1912, 700. (*Mycobacterium leprae murium*

(sic) Marchoux and Sorel 1912, 700; Bacillus der Rattenlepra, Stefansky 1903, 481.

lep.rae.mu′ri.um. Gr. n. *lepra* leprosy; L. n. *mus*, the mouse; M.L. gen.pl.n. *lepraemurium* of leprosy of mice.

Common name: Rat leprosy bacillus.

Rods, 3–5 μm in length, with slightly rounded ends. When stained, the cells often show an irregular appearance. The densely and uniformly stained forms appear to be infective for animals, in contrast to the "degenerate" unevenly stained forms (Rees *et al.*, 1960). Strongly acid-fast.

Like the human leprosy bacillus, this organism has not been cultivated *in vitro*, but it can be passed experimentally through rats, mice and hamsters. Some elongation of cells with synthesis of macromolecular substances has been observed in culture medium, but multiplication of cells did not occur (Draper and Hart, 1968).

The bacilli from lesions are not bound together in clumps, rounded masses and palisades as in human leprosy. For further details, see review by Lowe (1937).

Hydrogen transfer capacity of bacilli from tissues can be demonstrated by tetrazolium reduction and some correlation can be made between this property and infectiousness of bacilli (Hanks, 1954). Multiplication of *M. lepraemurium* in mice is inhibited by isoniazid, streptomycin, iproniazid and viomycin, and, to a lesser degree, Promin and diaminodiphenyl sulfone (Hobby *et al.*, 1954).

Cell walls of *M. lepraemurium* contain arabinose and galactose as principal cell wall sugars, and alanine, glutamic acid and diaminopimelic acid as mucopeptide amino acids, and a high proportion of lipid, as is typical of other, cultivable mycobacteria (Cummins *et al.*, 1967).

A cause of endemic disease of rats in various parts of the world, having been found in Odessa, Berlin, London, New South Wales, Hawaii, San Francisco and elsewhere. The natural disease occurs chiefly in the skin and lymph nodes, causing induration, alopecia and eventual ulceration.

Nodular diseases of the skin of other animals have been described, e.g. disease of buffalo in India, of a frog in South America, and of cats in Australia, has been associated with acid-fast bacilli that have not yet been cultivated on artificial media.

Species incertae sedis

The species have been described but until further information is available they are considered as *incertae sedis*.

a. *Mycobacterium acapulcensis* Bojalil, Cerbón and Trujillo 1962, 343; ATCC 14473.

b. *Mycobacterium aurum* Tsukamura 1966, 266; ATCC 23366; NCTC 10437.

c. *Mycobacterium chitae* Tsukamura 1966, 203; ATCC 19627.

d. *Mycobacterium parafortuitum* Tsukamura 1966, 12; ATCC 19686; NCTC 10411.

e. *Mycobacterium thermoresistibile* Tsukamura 1966, 266; NCTC 10409.

f. *Mycobacterium album* Söhngen 1913, 599.

g. *Mycobacterium methanicum* Nechaeva 1949, 316.

h. *Mycobacterium azot-absorptum* L'vov and Lyubimov 1965, 251.

i. *Mycobacterium sarni* Riccardo, Formisano and Coppola 1966, 286.

j. *Mycobacterium butanitrificans* Coty 1967, 29.

k. *Mycobacterium brunense* Kazda 1967, 207.

FAMILY III. **FRANKIACEAE** BECKING 1970, 201

J. H. BECKING

Frank.i.a′ce.ae. M.L. n. *Frankia* type genus of family; -aceae M.L. pl. ending to denote family; M.L. pl.n. *Frankiaceae* the Frankia family.

Symbiotic, filamentous, mycelium-forming bacteria which induce and live in root nodules on a wide variety of non-leguminous dicotyledonous plants belonging to 6 orders, 7 families and 14 genera (Table 17.35). The nodules are capable of fixing molecular nitrogen. These bacteria have also a free stage in the soil.

Further Comments

The taxonomic position of these bacteria is not clear. The segmentation of the vegetative mycelium into elongate or coccoid elements is a feature of the *Nocardiaceae*; on the other hand, the terminal swellings on the mycelial hyphae may be regarded as degenerate sporangia, a feature of the *Actinoplanaceae*. In contrast to both of these families, these organisms have not yet been grown on artificial media. A study of the cell wall components (Becking, unpublished) showed they contain *meso*-DAP (but no LL-DAP), arabinose, galactose and glycine. They are therefore related

TABLE 17.35

Non-leguminous nodule-bearing plants with **Frankia** *symbiosis*

Plant order	Plant family	Plant genus	*Frankia* species
Fagales	Betulaceae	*Alnus*	1. *F. alni*
		Elaeagnus	
Rhamnales	⎰ Elaeagnaceae	*Hippophaë*	2. *F. elaeagni*
		Shepherdia	
	⎱ Rhamnaceae	*Discaria*	3. *F. discariae*
		Ceanothus	4. *F. ceanothi*
Coriariales	Coriariaceae	*Coriaria*	5. *F. coriariae*
Rosales	Rosaceae	*Dryas*	6. *F. dryadis*
		Purshia	7. *F. purshiae*
		Cercocarpus	8. *F. cercocarpi*
Myricales	Myricaceae	*Myrica*	
		Gale	9. *F. brunchorstii*
		Comptonia	
Casuarinales	Casuarinaceae	*Casuarina*	10. *F. casuarinae*

to *Actinomyces* cell wall types II and IV (Table 17.18). It is here accepted as a separate mono-generic family because of its distinctive morphologic and physiologic characteristics.

Genus I. **Frankia** *Brunchorst 1886, 174*

(*Frankiella* Maire and Tison 1909, 242.)

Frank′i.a. M.L. fem.n. *Frankia* generic name, dedicated to B. Frank, a Swiss microbiologist.

A true mycelium is produced; the mycelium is septate and branched, but branching is not necessarily correlated with cross-wall formation. The hyphae of most species are 0.3–0.5 μm in diameter, but may be larger in some species (Table 17.36). In most cases the slender hyphae in the host tissue lack the refractivity necessary for study by light microscopy. Electron micrographs show that the hyphae contain fine strands of DNA not enclosed by a nuclear membrane (Plate 17.5); they are thus procaryotic. Distinct membranous bodies, most probably identical with the so-called plasmalemmosomes (Edwards and Stevens, 1963) or mesomes (Fitz-James, 1960; Chen, 1964) of some actinomycetes and Gram-positive bacteria can be observed.

In an active, nitrogen-fixing nodule, the center of the host cell is filled by a hyphal mass; **near the periphery, spherical or club-shaped terminal swellings** (Plate 17.6, Fig. 1.) **are formed on radially arranged hyphae close to the plant cell wall. These spherical bodies are often called "vesicles."** Both vesicles and club-shaped structures have been considered as a kind of sporangium but there is little evidence for this view; the vesicles lack complete cell walls and in older stages are disintegrated by the living host-cell cytoplasm surrounding them. **The vesicles and club-shaped structures are prob-** **ably associated with nitrogen fixation** as they are abundant in tissues actively fixing nitrogen. In acetylene reduction tests, tissue slices of alder root nodules containing the endophyte predominantly in the vesicular form showed much more nitrogen fixation that similar slices containing the endophyte but poor in vesicles.

In host cells which have not survived invasion by *Frankia* species, vesicles or club-shaped structures are not found; instead the **mycelium fragments into small particles or bacteria-like cells** (Plate 17.6, Fig. 2) sometimes termed "bacteroids," which fill the host cell completely (Table 17.36). When the host cell ruptures these particles are released into adjacent host cells or into the soil. They may represent a resting stage, capable of surviving in soil and of propagating the endophyte. Species in which bacteria-like cells are commonly found nodulate easily in water cultures when crushed nodules are applied to the root system. In species where these cells have not been observed, nodulation has been found only when planted in habitat soil.

Hyphae, vesicles and club-shaped structures are Gram-variable. The rod-shaped structures or bacteria-like cells are Gram-positive.

Molybdenum is required for the nitrogen-fixation process; non-leguminous nodulated plants deficient in this element show restricted nitrogen fixation and when grown under nitrogen-deficient conditions in soil or water culture produce only poor growth (Hewitt and Bond, 1961; Bond and Hewitt, 1961; Becking, 1961). **Cobalt is also essential** for nitrogen fixation (Bond and Hewitt, 1962; Hewitt and Bond, 1966). Vitamin B_{12} analogues, containing cobalt, have been isolated from non-leguminous root nodules (Kliewer and Evans, 1962, Bond *et al.*, 1965).

No species has as yet been grown on artificial media; in previous claims of isolation Koch's postulates were not fulfilled. The organisms have been grown in root nodule tissue culture, but are often lost in subsequent transplants of the tissue because of the difficulty of infecting newly formed callus tissue (Becking, 1965, 1966, 1970). So far the endophyte-containing callus tissue does not show nitrogen fixation, probably because the callus tissue can grow only on a complex organic medium containing much combined nitrogen (Becking, unpublished).

Probably micro-aerophilic. In *Alnus* and *Hippophaë* optimum nitrogen fixation took place in the presence of 12% oxygen, in *Myrica* and *Casuarina* in about 20% (Bond, 1959, 1961, 1964).

The organisms inhabit cortical parenchyma of root nodule tissues as symbionts, but are also present in a free state in soil, presumably as a viable, resting stage, since sterile non-leguminous seedlings nodulate when planted in suitable soil.

The endophyte enters the plant through a root hair, which has undergone strong curvature as a result of auxin formation by the endophyte (Beck-

ing, 1966, 1968, 1970). It reaches the cortex of the root by means of an infection thread and by proliferation of plant cells of a side root primordium a nodule is formed. A root nodule is thus a deformed side root of inhibited or arrested growth. The infected cells are enlarged and have a large and misshapen nucleus, which is probably polyploid (Becking, 1966, 1968, 1970). Two types of nodules are produced; in most plants a coralloid nodule is produced by continuous branching of the deformed side root with reduced apical meristematic activity; in *Casuarina* and representatives of the *Myricaceae*, the apex of each nodule lobe gives rise to a normal but negatively geotropic root and the nodules in these species become covered with upward growing rootlets. Such a root nodule may be called a rhizothamnion. Upward growing rootlets are also formed on the nodular lobes in some ineffective cross-inoculation associations of alder species, such as the root nodule endophyte of *Alnus glutinosa* on *A. rubra* and on *A. jorullensis* (pseudo-rhizothamnia). Moreover, it is not certain that *Comptonia peregrina* (*Myricaceae*) produces rhizothamnia (see below).

The endophyte is specific for a certain plant genus or a group of related genera. In contrast to the *Rhizobiaceae* no exceptions are known; the barriers of incompatibility are strict. It therefore seems appropriate to give species status to these specific cross-inoculations groups.

Type species: *Frankia alni* (Woronin) Von Tubeuf 1895, 118.

Description of the species of genus **Frankia**

1. **Frankia alni** (Woronin) Von Tubeuf 1895, 118. (*Schinzia alni* Woronin 1866, 6; *Plasmodiophora alni* (Woronin) Möller 1885, 105; *Frankia subtilis* Brunchorst 1886, 174; *Frankiella alni* (Woronin) Maire and Tison 1909, 242; *Aktinomyces alni* (*sic*) (Woronin) Peklo 1910, 505; *Actinomyces alni* (Woronin) Roberg 1934, 482; *Proactinomyces alni* (Woronin) Krasil'nikov 1941, 74; *Streptomyces alni* (Woronin) Fiuczek 1959, 285; *Nocardia alni* (Woronin) Waksman 1961, 35.)

al'ni. L. gen.fem.n. *alni* of *Alnus* the alder.

Hyphae are filamentous, branched and septate. Membranous bodies (mesosomes) observed by electron microscopy are very conspicuous and associated with cell wall formation. Hyphae are embedded in a thick layer of polysaccharide in the host cell and enclosed in a double membrane, the invaginated cytoplasmic membrane of the host cell.

Spherical vesicles, normally 3–5 μm in diameter but occasionally, as in *Alnus rugosa* (= *A.*

undulata), may be up to 6–8 μm in diameter according to Brunchorst (1886). Within the vesicles cell division is uncoordinated and anomalous, so that cell walls are mostly incomplete. In a later stage the vesicles are resorbed by the host, losing their contents and leaving behind only their cell walls and membranes.

In contrast the "bacteroids" are the result of complete cell division; they are polyhedral shaped, arranged in groups, each of which is separated from its neighbors by a thick polysaccharide layer. These cells are never absorbed by the host as they occur only in dead cells. Initially cell walls are thin but thicken as the cell matures and there are inclusions of lipid reserve material. The polyhedral-shaped cells are released by rupture of the host cell wall and probably function in the endophyte's survival and dispersal.

Root nodules are coralloid and up to 8 cm in diameter or the size of a tennis ball. In general they are present only in the top layers of the soil.

The nitrogen-fixing ability of these nodules has been proven in water culture experiments and by ^{15}N-tests (Bond, 1955, 1959; Becking, 1966, 1968, 1970).

Symbiont of all species of *Alnus* so far investigated. However, *Alnus* is the only genus of the family *Betulaceae* which bears nodules. The endophyte of *Alnus* is the most studied. Well known nodulated species are: *Alnus glutinosa* (common alder) and *A. incana* (gray alder) (Europe), *A. rubra* (red alder) (North America), *A. jorullensis* (South America), *A. japonica* and *A. sieboldiana* (Japan). There is incomplete compatibility, resulting in poor nodulation and nitrogen fixation, between the endophytes of *A. glutinosa* and *A. rubra* (Becking, 1966, 1968) and between *A. glutinosa* and *A. jorullensis* (Rodriquez-Barrueco, 1966).

2. **Frankia elaeagni** (Schröter) Becking 1970, 208. (*Plasmodiophora elaeagni* Schröter 1886, 134; *Frankiella elaeagni* (Schröter) Maire and Tison 1909, 243; *Tetramyxa elaeagni* (Schröter) Yendo in Yendo and Takase 1932, 124; *Actinomyces elaeagni* (Schröter) Roberg 1934, 482; *Proactinomyces elaeagnii* (Schröter) Krasil'nikov 1941, 75.)

el.ae.ag'ni. L. gen.fem.n. *elaeagni* of *Elaeagnus.* Morphology as in Table 17.36.

Root nodules are coralloid and up to 5 cm in diameter; usually present only in the top layers of soil. Their nitrogen-fixing ability has been proven by water culture experiments and ^{15}N-tests (Bond *et al.*, 1956, 1957, 1958; Gardner and Bond, 1957; Gardner, 1958).

Symbiont of root nodules of all species of *Elaeagnus* (oleaster), *Hippophaë* (sea buckthorn) and *Shepherdia* (buffalo berry). There is complete compatibility between the endophyte and all hosts of these genera and species. Also found in soil, although its form there is unknown.

3. **Frankia discariae** Becking 1970, 210. dis.car'i.ae. L. gen.fem.n. *discariae* of *Discaria.* Morphology as in Table 17.36.

Root nodules are coralloid and up to 3 cm in diameter; usually present in the top layers of soil only. Their nitrogen-fixing ability has been proven by ^{15}N-experiments (Morrison, 1961).

Symbionts of root nodules on *Discaria toumatou.*

TABLE 17.36

Characteristics differentiating the species of genus **Frankia**[a]

Species	Hyphae				Bacteria-like cells or "bacteroids"
	Diameter	Terminal swellings			
		Spherical(S) or club-shaped (C)	Diameter[b]	Length	
	μm		μm	μm	μm
A. Nodules coralloid					
1. *F. alni*	0.3–0.5	S	3.0–5.0 6.0–8.0[c]		0.5–1.5
2. *F. elaeagni*	0.3–0.5	S	2.5–4.0 1.8–2.2[c]		0.3–0.9
3. *F. discariae*	0.3–0.4	S	*ca.* 4.0		Not known
4. *F. ceanothi*	0.3–0.4	SC	1.5–3.0[d] aver. 2.0–2.5		Not known
5. *F. coriariae*	0.4–0.7	C	1.2–1.3	9.0–12.0	Not known
6. *F. dryadis*	0.5–0.8	C	1.5–2.0	1.5–5.0	1.0–2.0
7. *F. purshiae*	0.3–0.5	SC	2.2–4.4	2.2–5.5	Not known
8. *F. cercocarpi*	0.3–0.5	C	3.0	4.0	Not known
B. Rhizothamnia produced					
9. *F. brunchorstii*	1.2–2.8	C	1.6–2.4	7.5–12.5	1.5–2.5
10. *F. casuarinae*	0.3–0.5	C	0.6–1.5	3.0–4.0	0.4–1.0

[a] Species 2, 3, 7, 8 have been studied by light microscopy only; all others by both light and electron microscopy. Data on species 4, 5, 6, 10 are unpublished results of author.

[b] Diameter of spherical cells or vesicles; diameter of club-shaped structures at widest point.

[c] Rarely these dimensions.

[d] Club-shaped in younger stages with gradually tapering base.

a native of New Zealand. Also present in soil although its form is unknown.

4. Frankia ceanothi Atkinson 1892, 177.

ce.an.oth'i. L. gen.masc.n. *ceanothi* of *Ceanothus* (California lilac or tea bush).

Morphology as in Table 17.36. The vesicles are smaller than in other species and in their younger stages have a gradually tapering base and are more or less club shaped.

Root nodules coralloid, 5–6 cm in diameter and usually in the top layers of the soil. Their nitrogen-fixing capacity has been proven by pot experiments (Quick, 1944; Hellmers and Kelleher, 1959; Russell and Evans, 1966) and by ¹⁵N-tests (Delwiche *et al.*, 1965; Bond, 1967).

Symbionts of root nodules of *Ceanothus;* of about 50 described species only about half have been investigated for nodulation; well known nodulated species are: *C. americanus, C. leucodermis* (chaparral whitethorn), *C. sanguineus* and *C. velutinus* (deer brush). Nodulation has been reported for *C. azureus, C. cordulatus* (snow brush), *C. cuneatus, C. delilianus, C. divaricatus, C. diversifolius* (pine mat), *C. fendleri, C. foliosus, C. greggii, C. griseus, C. impressus, C. incanus, C. integerrimus, C. jepsonii, C. microphyllus, C. oliganthus, C. ovatus, C. parvifolius, C. prostatus* (squaw carpet), *C. rigidus, C. sorediatus* and *C. thyrsiflorus.* The genus is limited to North America.

5. Frankia coriariae Becking 1970, 212.

cor.i.ar'i.a. L. gen.fem.n. *coriariae* of *Coriaria*.

Morphology as in Table 17.36. The club-shaped hyphae have rounded ends. Bacteroids have not been found.

Root nodules are coralloid, up to 5 cm in diameter; they occur mostly in the top layers of the soil. Their nitrogen-fixing capacity has been proven in pot experiments (Kataoka, 1930) and in vermiculite cultures and ¹⁵N-tests (Harris and Morrison, 1958; Bond, 1962).

Symbionts of root nodules of *Coriaria;* nodulation has been reported on *C. myrtifolia* in southern Europe (Bond and Montserrat, 1958), *C. japonica* in Japan (Kataoka, 1930) and on eight New Zealand species and their hybrids: *C. angustissima, C. arborea, C. kingiana, C. lurida, C. plumosa, C. pottsiana, C. pteridoides* and *C. sarmentosa.* The endophyte of *C. myrtifolia* is completely incompatible with that of *C. japonica*, suggesting a strong adaptation to a particular host species (Bond, 1962); until more information is available separation of these endophytes into two species is not recommended.

6. Frankia dryadis Becking 1970, 212.

dry.a'dis. M.L. gen.n. *dryadis* of *Dryas* (dryad or mountain avens).

Morphology as in Table 17.36. The club-shaped extremities of the hyphae have rounded ends; the smaller ones are more or less round and look like vesicles. Sometimes intermediates between hyphae and club-shaped structures are observed, apparently not full grown stages of the latter.

Root nodules are coralloid, up to 3 cm in diameter; usually found 15 cm or more beneath the surface and usually 30–50 cm away from the base of the stem, near the ends of tiny feeder rootlets. They are not found on recently germinated seedlings, but begin to appear in the 2nd or 3rd year on most plants. Root nodules never occur in silt soil but only in sandy soils or gravel beds. Their nitrogen-fixing capacity was suspected from ecologic studies (Crocker and Major, 1955; Crocker and Dickson, 1957; Schoenike, 1957–58) but was proven by ¹⁵N-tests by Lawrence *et al.*, 1967.

Symbionts on root nodules of *Dryas drummondii* in Alaska (Allen *et al.*, 1964; Lawrence *et al.*, 1967); also reported in *Dryas octopetala* and *D. integrifolia* in Glacier Bay, Alaska, according to Sprague, cited by Lawrence *et al.*, 1967. However, in Scotland and continental Europe the common species, *D. octopetala*, is not nodulated.

7. Frankia purshiae Becking 1970, 213.

pur.shi'ae. L. gen.fem.n. *purshiae* of *Purshia* (bitterbrush or antelope bush).

Hyphae, 0.3–0.5 μm in diameter, filamentous, branched, septate. The presence of other structures and forms has not yet been determined.

Root nodules coralloid, up to 4 cm in diameter and generally more frequent in the top layers of soil. Their nitrogen-fixing ability was suggested by pot experiments (Wagle and Vlamis, 1961) and confirmed by ¹⁵N-tests (Webster *et al.*, 1967).

Symbionts of root nodules of *Purshia tridentata* and *P. glandulosa*, which range along the west coast of North America from southern British Colombia to Mexico.

8. Frankia cercocarpi Becking 1970, 213.

cer.co.car'pi. L. gen.masc.n. *cercocarpi* of *Cercocarpus* (mountain mahogany).

Hyphae, 0.3–0.5 μm in diameter, filamentous, branched and septate. The presence of other structures and forms has not yet been determined.

Root nodules coralloid, up to 5 cm in diameter and often equally distributed over the root system. Their nitrogen-fixing capacity has been demonstrated by pot experiments (Vlamis *et al.*, 1964); ¹⁵N-tests have not been performed.

Symbiont of root nodules on *Cercocarpus betuloides* in the western United States.

9. Frankia brunchorstii Möller 1890, 224. (*Aktinomyces myricae* Peklo 1910, 505; *Proactinomyces myricae* (Peklo) Krasil'nikov 1941, 75.)

brun.chor'sti.i. M.L. gen.masc.n. *brunchorstii* of Brunchorst; named for J. Brunchorst, a German botanist, who first studied these organisms.

Morphology as in Table 17.36. The thickest hyphae may be composed of several hyphae, although electron microscopy reveals that the species has wider hyphae than any other described species of *Frankia*. The clubs at the end of the hyphae have rounded ends and are divided by septa into four to six compartments.

Rhizothamnia are produced. The nodules are up to 6 cm in diameter and usually occur in the uppermost layers of the soil. Their nitrogen-fixing capacity has been confirmed by pot experiments and ^{15}N-tests (Bond, 1951; Leaf *et al.*, 1959; Sloger and Silver, 1965). *Comptonia peregrina* may produce coralloid nodules (Ziegler, 1959–60).

Symbionts of root nodules of species of *Gale, Myrica* and *Comptonia*. Well known nodulated species are: *Gale palustris* (= *Myrica gale*) (sweet gale) in Europe and Asia; *Myrica cerifera* (wax myrtle) (southern North America), *M. cordifolia* (southern Africa), *M. javanica* (Asia) and *Comptonia peregrina* (= *Myrica asplenifolia*) in eastern North America. The endophytes of *Gale palustris, Myrica cerefera* and *Myrica cordifolia* are completely incompatible (Gardner and Bond, 1966),

but more information is needed before the status of a species is given to each of these endophytes.

10. **Frankia casuarinae** Becking 1970, 211.

ca.su.ar.in'ae. L. gen.fem.n. *casuarinae* of *Casuarina* (ru, forest oak, beefwood).

Morphology as in Table 17.36.

Rhizothamnia are produced. The root nodules are up to 7 cm in diameter or the size of a tennis ball; they occur mainly in the top layers of the soil. The branching of the individual lobes of the nodules is not always dichotomous, three to four branches may arise from the same point. The nitrogen-fixing capacity of the nodules has been confirmed by water culture experiments and ^{15}N-tests (Bond, 1957, 1958, 1959).

Symbionts of root nodules of *Casuarina;* all species investigated in their native tropical and subtropical habitat are nodulated. Well known nodulated species are *Casuarina cunninghamiana, C. equisetifolia* (common ru), *C. glauca, C. junghuhniana, C. lepidophloia, C. montana, C. muricata, C. stricta, C. sumatrana, C. tenuissima* and *C. triangularis. C. equisetifolia* is a beach-growing, extremely salt-tolerant species. Complete compatibility exists in reciprocal inoculations with the endophyte of all species of *Casuarina.*

FAMILY IV. **ACTINOPLANACEAE** COUCH 1955, 269

J. N. COUCH AND C. E. BLAND

(*Actinosporangiaceae* Couch 1955, 149; *Streptosporangiaceae* Krasil'nikov 1964, 6.) Ac.ti.no.pla.na'ce.ae. M.L. n. *Actinoplanes* type genus of family; -*aceae* ending to denote family; M.L. pl.n. *Actinoplanaceae* the *Actinoplanes* family.

Bacteria that form a distinct mycelium, intramatrical but in some forms also aerial. The diameter of the filaments 0.2–2.6 μm. Sporangia characteristic of the family, spherical, sub-spherical, lobed, cylindrical, digitate, club-shaped, to very irregular in shape; formed always in the air above the surface of the culture media on branched or unbranched, septate, palisade hyphae or on undifferentiated hyphae (Plate 17.7). Palisade hyphae up to 2 μm in diameter. Spores arranged in the sporangia in one or more spirals, or in parallel rows, or irregularly, few to many in a sporangium or in single file with one to several spores in one sporangium. Sporangial dehiscence variable. Spores spherical, sub-spherical, oval, rod-shaped or spiral, with flagella and motile or without flagella and non-motile; contain one or several small spherical lipoid globules. Conidia formed in some species on penicillate conidiophores or in moniliform series. Gram-positive; not acid fast.

Cell wall composition of some genera and species of the family is given in Table 17.37.

Growth on certain agars brilliantly colored to white; flocculent, pasty to compact and roughish.

Chemoorganotrophs: Metabolism respiratory. Molecular oxygen is the universal electron acceptor.

Aerobic.

Temperature range: 15–45 C.

Abundant in all soils which contain humus and less common in fresh water habitats, on dead plant parts as pollen, leaves, etc. and animal material as hair and other keratin and chitin substrates.

Further Comments

Since the family *Actinoplanaceae* was established (Couch, 1955) it has grown to include 10 genera which fall into two distinct groups: those with large, spherical to irregular multisporous sporangia and those with small filiform or club-

TABLE 17.37

Cell wall composition[a] of some species in the family **Actinoplanaceae**

Organism (morphologic designation)	Strain no.	DAP	HDAP	Galactose	Glucose	Mannose	Arabinose	Xylose	Deoxyhexose
Actinoplanes philippinensis	2	+++		+++	+++	++	++	++	++
A. utahensis	260		+++	+++		++		TR	TR
A. missouriensis	431		+++	+++	++	TR	TR	TR	
Amorphosporangium auranticola	253	++	+	+++	+++	++	++	++	++
Ampullariella regularis	79		+++	+++	+++	++	++	++	++
A. regularis	168		+++	+++	+++	++	++	++	++
A. regularis	28		+++					+++	++
A. digitata	33		+++	+++			++		
A. lobata	72		+++	+++		++		TR	
A. campanulata	65	++	+	+++		++			
Pilimelia anulata	1	+++		+	+++	TR	++	++	
P. terevasa	1	+++		+	+++	+++	++	+	
Spirillospora albida (y)	761	+++	+	+					
Streptosporangium album (y)	S-16-1	++	TR	+++	+	++			
S. amethystogenes (y)	S-5	++	TR						
S. roseum (y)	276-1	++	TR	TR	TR	TR			
S. viridialbum (y)	S-20	++	TR	TR	TR	+			
S. vulgare (y)	S-1	++		TR	TR	+			

[a] With the exception of *Spirillospora* and *Streptosporangium* which lack glycine as a major cell wall component, all members of the Actinoplanaceae studied thus far have major amounts of glucosamine, muramic acid, glutamic acid, glycine and alanine as cell wall components. *Ampullariella lobata* and *A. campanulata* also contain heptose and traces of galactosamine. Determinations by Szaniszlo and Gooder (1967) except those marked (y) which are by Yamaguchi (1967).

shaped sporangia containing one to four to several spores. The latter suggests a condition described by Lachner-Sandoval (1898), Drechsler (1919) and more recent workers in which the mature spores of certain species of *Streptomyces* remain in chains, held together by the enveloping hyphal wall (Vernon, 1955). Thus it appears that in certain species of *Streptomyces* the so-called conidial chain could be interpreted as a sporangium. This is not to imply that the two families should at present be merged but only to indicate that they are closely related.

Genera and species incertae sedis

Actinosporangium violaceus Krasil'nikov and Yuan (1961) and *Streptosporangium indianesis* (*sic*) Gupta (1965) were described as producing unwalled sporangia by the fusion of several sporangia. However, the lack of a sporangial wall indicates the absence of true sporangia (i.e. a spore-containing structure) and therefore a highly questionable placement of these organisms as members of the *Actinoplanaceae*.

Intrasporangium calvum Kalakoutskii, Kirillova, and Krasil'nikov (1967) was described as producing "sporangium-like" vesicles in its hyphae. Lechevalier and Lechevalier (1969) found that the vesicles contained no spores, but were likely a reaction to the culture media. Similar vesicles were described and drawn by Couch (1963, Fig. 39) for *Ampullariella regularis*.

Elytrosporangium brasiliensis Falcão de Morais, Batista and Massa (1966) has been studied by Thiemann (1968) and found to have no taxonomic standing since it behaves like a typical streptomycete. Further study of this organism is needed to establish its taxonomic position.

Key to the genera of family Actinoplanaceae

I. Sporangia spherical, cylindrical or highly irregular and containing a few to several thousand spores per sporangium. Spores in coiled or in parallel chains within the sporangia.

A. Sporangia spherical, subspherical to very irregular; spores arranged in coiled chains within the sporangia.

1. Spores motile.
   a. Zoospores spherical to subspherical with a tuft of polar flagella; agar colonies usually orange or reddish orange.

             Genus I. *Actinoplanes*, p. 708

   b. Zoospores rod-shaped, curved or spiral with one to three sub-polar flagella; agar colonies white, pale yellow, light gray or bright blue.

             Genus II. *Spirillospora*, p. 711

2. Spores non-motile, spherical or slightly elongated.

             Genus III. *Streptosporangium*, p. 711

B. Sporangia exceedingly irregular, spores short rods.

             Genus IV. *Amorphosporangium*, p. 715

C. Sporangia cylindrical, ovoid or irregular; spores rod-shaped, arranged in parallel chains within the sporangia.
   1. Zoospores rod-shaped, 0.5–1.0 by 2.0–4.0 μm, with a polar tuft of flagella; agar colonies orange, brown, green-brown or black.

             Genus V. *Ampullariella*, p. 716

   2. Zoospores rod-shaped, 0.3–0.7 by 0.8–1.5 μm, with a single polar flagellum or one to four lateral flagella; agar colonies yellow-brown, yellow-gray, yellow or bright lemon yellow; organism usually occurring on hair and perhaps other keratin and chitin material.

             Genus VI. *Pilimelia*, p. 718

II. Sporangia club-shaped, finger-like or pyriform containing one to six spores each occurring singly or arranged in double parallel rows on the aerial hyphae.

A. Sporangia containing one spore each, arranged in double parallel rows along the aerial hyphae.

             Genus VII. *Planomonospora*, p. 719

B. Sporangia finger-like or club-shaped, two to several spores arranged linearly within each sporangium.
   1. Sporangia finger-like or linear; zoospores exhibiting lophotrichous or peritrichous flagella.
      a. Zoospores rod-shaped, formed in longitudinal *pairs* on the aerial mycelium.

             Genus VIII. *Planobispora*, p. 720

      b. Zoospores oval to pyriform, formed in a single row of three to four in sporangia grown directly from the vegetative mycelium.

             Genus IX. *Dactylosporangium*, p. 721

   2. Sporangia club-shaped; zoospores uniflagellate.

             Genus X. *Kitasatoa*, p. 722

Genus I. **Actinoplanes** *Couch 1950, 89 emend. mut. char. 1955, 153**

J. N. COUCH AND C. E. BLAND

Ac.ti.no.pla′nes. Gr. n. *actis, actinis* a ray, beam; Gr. n. *planes* a wanderer; M.L. masc.n. *Actinoplanes* literally, a ray wanderer; intended to signify an actinomycete with swimming spores.

Sporangia, 3–20 by 6–30 μm, spherical, subspherical, cylindrical with rounded ends, to very irregular. **Spores, globose to subglobose, 1–1.5** μm in diameter, in coils, nearly straight chains, or irregularly arranged in sporangia; **motile** by a tuft of polar flagella 2–6 μm in length (see Plate 17.7).

Hyphae, 0.2–2.6 μm in diameter, branched, irregularly coiled, twisted or straight with few septa.

Vertical pallisade hyphae formed on certain agars; **aerial mycelium scanty,** except in *A. armeniacus.*

Most species brilliantly colored on peptone Czapek and certain other agars; colors: orange, red, yellow, violet and purple. Some form diffusible pigments which color the agar blue, red, yellow, brownish or greenish. No organic growth factors are required.

---

* Colors mentioned in genera described by Couch and Bland are after Ridgeway.

H₂S is produced by some species. Cell wall composition is given in Table 17.37.

Strict aerobes.

Temperature range 18–35 C.

Occur on a wide variety of plant material, less often on parts of dead animals such as hair, hoofs, snake skin.

The G + C content of the DNA (of two species studied) ranges from 72.1–72.6 moles %.

Type species: *Actinoplanes philippinensis* Couch 1950, 89.

Further Comments

This is the commonest genus of the family, making up three quarters of the isolates in the University of North Carolina collection. It is closely related to *Ampullariella* and *Amorphosporangium* but can be distinguished from them by sporangial shape and spore structure.

Key to the species of genus **Actinoplanes**

I. Sporangia globose to subglobose.
  1. Sporangia usually more than 15 μm in diameter.
    a. Sporangia 8–25 μm in diameter.
                1. *A. philippinensis*
    b. Sporangia up to 50 μm in diameter.
                2. *A. armeniacus*
  2. Sporangia 6–14 μm in diameter.
                3. *A. missouriensis*
II. Sporangia irregular in shape, 5–18 μm in diameter.
                4. *A. utahensis*

Description of the species of genus **Actinoplanes**

1. **Actinoplanes philippinensis** Couch 1950, 89.

phil.ip.pi.nen′sis. M.L. adj. *philippinensis* pertaining to the Philippines.

Sporangia 8.4–25.0 μm; mostly spherical when mature. Spores arranged in coils in the sporangium, 1.0–1.2 μm in diameter, motile by polytrichous polar flagella, 2–2.8 μm in length. Hyphae 0.5–1.5 μm wide, branched and sparingly septate.

On Czapek agar growth poor to good, rarely very good, colonies from point inocula 3–10 mm in diameter, rarely up to 25 mm in diameter after 2 months; flat or slightly elevated; margin smooth or scalloped; color light buff to tawny, changing in some old cultures to purplish brown. Sectoring frequent. In vertical section, the growth consists of a compact upper layer, made up mostly of characteristic long conspicuous vertical palisade hyphae and a lower region of loosely arranged hyphae. Sporangia abundant in some cultures and not formed in others; spherical to irregular, at times occurring beneath the surface in old cultures owing to overgrowth by palisade hyphae; some times a new layer of sporangia is formed over the first layer. Odor slightly fragrant. Agar usually pale yellow.

Growth on peptone Czapek agar good to very good, colonies from point inocula 10–18 mm in diameter after 2 months; consisting of heaped convolutions in the center, becoming concentric rings of narrow ridges with narrow radial grooves towards the margin, usually with an elevated or radially ridged and grooved margin. Surface shiny. Color brilliant, near apricot orange or orange chrome. Sporangia absent or very rare. Palisades not formed. Smaller hyphae form vast numbers of bacterioid spheres and rods which, when the material is crushed, break off and resemble *Nocardia*.

On casein agar growth good, colonies from point inocula 7–10 mm after 2 months; surface raised with narrow anastomozing ridges, margin scalloped, no sporangia formed; colony color zinc orange, agar cleared below colony.

On tyrosine agar growth poor, colonies from point inocula 3 mm in diameter after 2 months; surface smooth, margin undulate, color near apricot buff; no sporangia; tyrosine crystals cleared beneath colonies.

Isolated from soil from a rice paddy, Philippine Islands, 1945.

Type strain: UNCC P-15; ATCC 12427.

2. **Actinoplanes armeniacus** Kalakutskii and Kuznetsov 1964, 620.

ar.men.i.a′cus M.L. adj. *armeniacus* pertaining to Armenia, a Soviet Republic.

Sporangia spherical, 20–50 μm in diameter, formed on stalks; spores spherical to ovoid, 1.0–1.5 μm in diameter, with peritrichous flagella;

aerial mycelium whitish, arranged frequently in several concentric rings (up to six or seven) in each colony; conidiophores formed in chains on the aerial mycelium; substrate mycelium sparingly septate.

On Czapek agar with glucose aerial mycelium white, scarce; substrate mycelium sandy. Culture medium not discolored. On Czapek agar with starch no aerial mycelium, substrate mycelium light orange. On Czapek agar without carbon source aerial mycelium white, very sparse; substrate mycelium colorless.

On peptone-corn steep agar no aerial mycelium; substrate mycelium yellowish brown.

On meat extract agar no aerial mycelium; substrate mycelium grayish.

Nitrates are not reduced to nitrites; starch is hydrolyzed, gelatin liquefied, no growth on cellulose.

Type strain: 26A-32; RIA807; ATCC 15676.

Further Comments

The spherical, exceedingly large sporangia formed on stalks and the whitish aerial mycelium frequently arranged in six to seven concentric zones are distinguishing characteristics which separate this from the other species of *Actinoplanes*. This may represent a new genus.

3. Actinoplanes missouriensis Couch 1963, 69.

miss.ou.ri.en'sis. M.L. adj. *missouriensis* pertaining to the state of Missouri.

Sporangia usually globose to sub-globose, at times irregular, rarely lobed or digitate, 6–14 μm in greatest diameter. Spores arranged in irregular coils within the sporangium, 1.0–1.2 μm, weakly motile by polytrichous (6–20) polar flagella. Sporangial dehiscence by the swelling of an intersporal substance which pushes the spores apart, eventually causing the thick, fragile wall to break, allowing the spores to escape. Hyphae usually less than 1 μm wide, with many short branches; palisade hyphae rare.

Growth on Czapek agar very good, colonies from point inocula 13–14 mm in diameter after 5 weeks; the center elevated with several mounds with shallow radial grooves between; border flat; color mostly ochraceous salmon, but with pale lavender areas, also with whitish areas made powdery by sporangia; sporangia abundant. Agar uncolored to pale lavender.

On peptone Czapek agar growth very good to excellent; colonies from point inocula 13–18 mm in diameter after 5 weeks, heaped up with coarse convolutions; color brilliant zinc orange to ochraceous orange; sporangia none to many in places. Agar uncolored to pale lavender.

On casein agar growth good; colonies from point inocula 9–10 mm in diameter after 5 weeks, elevated, convoluted; color near zinc orange, ochraceous salmon, to light ochraceous salmon; sporangia none; agar pale yellow. Nearly all of the casein hydrolyzed.

Most of tyrosine in tyrosine agar used after 5 weeks; growth barely good, colonies from point inocula 6–9 mm in diameter, center elevated, border flat; smear confluent, minutely bumpy; color near zinc orange; agar not discolored; sporangia none.

Type strain: UNCC 431 (Couch 1963, 53).

4. Actinoplanes utahensis Couch 1963, 67.

u.tah.en'sis. M.L. adj. *utahensis* pertaining to the state of Utah.

Sporangia very irregular in shape and size; usually rounded but with a wavy wall, lobed, pyriform, club-shaped or digitate, 5–18 μm in greatest diameter. Spores arranged in irregular coils and rounded within the sporangium, subglobose, 1.0–2.0 μm in diameter, motile by polytrichous polar flagella. Vegetative hyphae 0.2–1.2 μm wide; sporangiophores 1–2 μm wide and usually unbranched.

Growth on Czapek agar good, colonies from point inocula 15–17 mm in diameter after 5 weeks, center with a few convolutions, border flat; smear confluent, surface low, flattish, with many minute bumps; color a brilliant apricot orange to salmon orange; sporangia sparse, shape as on pollen or filiform, 6–9 μm in diameter. A few microspores; agar not discolored.

Growth on peptone Czapek very good, colonies from point inocula 11–12 mm in diameter with elevated convolutions after 5 weeks; smear confluent with elevated convolutions along the margin and a few scattered ones elsewhere, surface, flattish, minutely bumpy; color apricot orange to ferruginous in the center of smear; sporangia none; microspores abundant, formed singly, in clusters or in chains; agar slightly darkened.

On casein agar growth good, colonies from point inocula 9–12 mm in diameter, flattish, bumpy, or convoluted; smear confluent, surface mostly flat with conspicuous bumps, 1–2 mm in diameter; color zinc orange to tawny or fading to almost hyaline; sporangia none; agar amber colored. Nearly all casein hydrolyzed after 4–5 weeks.

Growth on tyrosine agar good, colonies from point inocula 5–7 mm in diameter, flat but with a few bumps in center; smear continuous, very thin, flat; color near zinc orange, sporangia none; agar not discolored. Tyrosine cleared under smear and partly under point inocula.

Type strain: UNCC 260 (Couch 1963, 53).

Genus II. **Spirillospora** Couch 1963, 61

J. N. COUCH AND C. E. BLAND

Spi.ril.lo.spo'ra. G. n. *speira* a coil: G. n. *spora* a seed; M.L. fem.n. *Spirillospora* spiral spores.

Hyphae 0.2–1 μm in diameter. Sporangia 5–24 μm in diameter, spherical to vermiform. **Spores** developed from one or more coils in the sporangium, **short to long rods to spiral** in shape, motile by one to three sub-polar flagella (Plate 17.7).

May be grown on a wide variety of culture media.

For cell wall composition see Table 17.37.

Temperature range 18–35 C optimum 25 C.

The G + C content of the DNA of the single species examined is 72.9 moles %.

Type species: *Spirillospora albida* Couch 1963, 65.

Description of the species of genus **Spirillospora**

1. **Spirillospora albida** Couch 1963, 65.

al'bi.da. L. fem.adj. *albida* white.

Sporangia usually spherical, 5–24 μm in diameter, average 10 μm. Spores developed from one or more coils in the sporangium, short to long rods, bent rods to spiral in shape; weakly but definitely motile with one to three subpolar flagella. Mycelium white to pale yellowish; hyphae 0.2–1 μm in diameter. About one third of the isolates to date produce in old cultures (4 weeks) a blue diffusible pigment which colors the agar and mycelium a deep blue; the pigment has been studied and named spirillimycin (Domnas, 1968).

Growth fair on Czapek agar; colonies from point inocula 10–13 mm in diameter after 2 months; smear more or less confluent, mycelium within the agar thin and hard to see, usually white; aerial mycelium sparse, white; conidia in moniliform series and in coils fairly abundant; sporangia frequently in tufts on aerial hyphae, formed singly or on branched sporangiophores, at times in concentric zones, sizes and shapes of sporangia and spores as on pollen. No diffusible pigment.

Growth good on peptone Czapek, diameter of colonies from point inocula as on Czapek but growth much thicker and more compact, center slightly elevated and covered with a white tomentum; mycelium within agar pale yellowish; coils and conidia as on Czapek agar; sporangia few, 5–10 μm in diameter; odor slight; no diffusible pigments in 14 isolates but deep blue diffusible pigments in 8 isolates.

On casein agar growth fairly good, colonies from point inocula 5 mm in diameter after 2 months, most of growth within agar, compact; center with closely spaced bumps, each about 0.7 mm broad; smear consisting of many separate mounds 0.2–2 mm in diameter; color pale buffy pink; surface slick; no aerial hyphae and no sporangia; agar pale yellow. Nearly all of the casein hydrolyzed.

Colonies on tyrosine agar from point inocula 0.8–1 cm in diameter after 6 weeks, extending into agar 2–3 mm, forming at the surface a dense, thick, cottony pile of aerial hyphae on which are formed a considerable number of oval to spherical conidia but no sporangia; agar near clay color. Crystals about half cleared beneath cultures.

Type strain: UNCC 1030; ATCC 15331 (Couch 1963, 53).

Genus III. **Streptosporangium** Couch 1955, 148

J. N. COUCH AND C. E. BLAND

Strep'to.spo.ran.gi.um. Gr. adj. *streptus* twisted; Gr. n. *spora* a seed; Gr. n. *angium* a vessel; M.L. neut.n. *Streptosporangium* spores coiled within a sporangium.

Sporangia spherical to ovoid; 7–48 μm (usually 7–20 μm). **Sporangiospores arranged in a single coil within the sporangium,** spherical to ovoid, 1.0–1.3 μm by 1.5–3.5 μm, **non-motile** (Plate 17.7). Mycelium as in *Streptomyces;* hyphae much branched, sparingly septate; 0.5–1.2 μm in diameter, palisade hyphae absent. Growth less vigorous than *Actinoplanes* on various agars but usually with greater development of aerial hyphae.

Organic growth factors required by some species.

For cell wall composition see Table 17.37.

Strict aerobes.

Temperature range 18–35 C (55 C in one species).

The G + C content of the DNA of the two species studied is 69.5–70.6 moles %.

Type species: *Streptosporangium roseum* Couch 1955, 151.

Key to the species of genus **Streptosporangium**

I. Aerial mycelium on oatmeal agar pale pink.
  A. Spores 1.0–1.3 by 1.5–1.9 µm.
    1. No growth at 42 C.
      a. Violet crystals not produced.
              1. *Streptosporangium roseum*
              2. *Streptosporangium vulgare*
      b. Violet crystals produced.
              3. *Streptosporangium amethystogenes*
    2. Growth at 42 C; violet crystals not produced.
      a. Starch hydrolyzed.
              4. *Streptosporangium pseudovulgare*
      b. Starch not hydrolyzed.
              5. *Streptosporangium nondiastaticum*
  B. Spores 0.6–0.8 by 1.5–3.5 µm, three times longer than wide.
              6. *Streptosporangium longisporum*
II. Aerial mycelium on oatmeal agar other than pink.
  A. Aerial mycelium white.
    1. Spores spherical 0.2 by 0.4 µm.
              7. *Streptosporangium viridogriseum*
      a. Nitrate reduced to nitrite.
              7a. *Streptosporangium viridogriseum* subsp. *viridogriseum*
      b. Nitrate not reduced.
              7b. *Streptosporangium viridogriseum* subsp. *kofuense*
    2. Spores spherical to ovoid 0.8–2 by 1.5–1.9 µm.
      a. Sporangia 6–8 µm.
              8. *Streptosporangium album*
      b. Sporangia 10–30 µm.
              9. *Streptosporangium albidum*
  B. Aerial mycelium gray, yellowish or greenish.
    1. Sporangia 6–14 µm; aerial mycelium light gray to yellowish.
              10. *Streptosporangium viridialbum*
      a. Nitrates not reduced.
              10a. *Streptosporangium viridialbum* subsp. *viridialbum*
      b. Nitrates reduced to nitrites.
              10b. *Streptosporangium viridialbum* subsp. *reducens*
    2. Sporangia 8–11 µm; aerial mycelium dark gray.
              11. *Streptosporangium rubrum*

Description of the species of genus **Streptosporangium**

1. Streptosporangium roseum Couch 1955, 151. (*Angiococcus moliroseus* Peterson 1959, 169.) ro′se.um L. neut.adj. *roseum* rose colored.

Mycelium, well developed on boiled grass leaves or on pieces of filter paper in sterile soil water; white at first, changing to pink with production of sporangia. The aerial mycelium appears as minute tufts of hyphae which grow to form mounds, up to 2 mm in diameter, arranged more or less in concentric circles and usually minutely pock marked. Sporangia are spheres 7–19 µm, usually 8–9 µm, in diameter, white in small groups, pink in large masses; a few to many sporangia appear first on scattered single hyphae, apical on the main thread or on short lateral branches; eventually mounds may be solid masses of sporangia. Immersion of the mature sporangium in water causes an intersporal substance to swell pushing the wall out into a cone shaped projection about half as long as the diameter of the sporangium; the spores are forcibly ejected through an opening in the cone; the sporangial wall persists several hours after spore discharge. Spores are spherical, sub-spherical, short rods or bent rods, 1.8–2 µm in diameter, each containing a small shiny lipid globule; non-motile.

On Czapek agar growth fair, cultures from point inocula 7–12 mm in diameter after 6 weeks; usually flat, level with agar surface; concentric zonation distinct or absent; surface glossy or powdery. Color usually white, sometimes pinkish buff or cream-buff. Sporangia absent to fairly abundant, formed some distance above the surface of the agar. In some cultures coils are formed which break up into conidia as in *Streptomyces*.

On peptone Czapek agar growth good, cultures from point inocula 15–20 mm in diameter after 6 weeks; flat or with a few low radial or irregular ridges and grooves. Margin fringed or entire. Aerial hyphae at times in white concentric rings and giving a powdery appearance instead of a glossy surface. Color usually olive-buff to deep olive-buff. Sporangia rare.

On potato glucose agar growth good, cultures from point inocula 10–18 mm in diameter after 2 months; center elevated with irregular bumps and ridges. Margin flat, color at first creamy, becoming tawny and then carob brown or Kaiser brown, after which white floccose spots of hyphae appear, spreading usually to cover the entire surface. Sporangia usually formed in vast numbers, the white areas becoming rosy pink as the sporangia mature; the pinkish areas frequently minutely pocked. Surface moist at first, appearing dry and floccose as aerial hyphae and sporangia are formed. Agar colored reddish brown with a vinaceous tinge.

On Emerson's agar (YpSs) growth good, cultures from point inocula about 20 mm in diameter after 6 weeks, composed of a whitish central area, 4–6 mm wide, made up of elevated, irregular bumps and ridges, which abruptly change into radial ridges and grooves, sloping down to a flat, white border, 1–2 mm wide, and composed of minute concentric circles of white hyphae. Ridges and grooves vinaceous brown, sometimes covered with a whitish down. Margin smooth or scalloped, ending abruptly. Surface dry. Sporangia formed abundantly first in the center as the white changes to pink. Agar colored pale vinaceous brown.

Type strain: UNCC 27; ATCC 12428.

2. **Streptosporangium vulgare** Nonomura and Ohara 1960, 407.

vul.ga're L. neut.adj. *vulgare* common.

Sporangia 6–8 μm, spores 1.0–1.2 by 1.5–1.9 μm. Aerial hyphae 0.7–1.0 μm wide.

On oatmeal agar-Y aerial mycelium pale pink to pink; soluble pigment pale yellow to yellow; no crystals produced after 1 month at 30 C.

On starch agar aerial mycelium lacking or scanty; substrate mycelium yellow with orange tinge or pale orange after 15 days.

Substrate mycelium on glycerol asparagine agar pale orange to pale rose.

On potato agar growth pale yellow with orange tinge or pale orange. Soluble pigment may be pale orange.

Utilizes inositol and rhamnose.

Requires thiamine for growth.

Nitrates not reduced to nitrites.

Melanin not produced.

Temperature range 18–35 C.

Type strain: S-1; NRRL B-2633.

3. **Streptosporangium amethystogenes** Nonomura and Ohara 1960, 407.

a.me.thy.sto'ge.nes. L. adj. *amethystinus* of the color amethyst; Gr. v. *gennaio* produce; M.L. adj. *amethystogenes* amethyst color producing.

Sporangia 6–8 μm, spores 1.0–1.3 by 1.5–1.9 μm. Aerial mycelia 0.7–1 μm in width.

On oatmeal agar-Y aerial mycelium scanty, pale pink; growth pale pink; violet crystals produced after 1 month at 30 C.

No aerial mycelium on starch agar. Growth pale yellow after 15 days.

On glycerol asparagine agar substrate mycelium pale yellow to brownish white.

Growth pale yellow, brown to grayish brown on potato agar.

Requires thiamine and biotin for growth.

Nitrates reduced to nitrites.

Melanin not produced.

Type strain: S-5; NRRL B-2639.

4. **Streptosporangium pseudovulgare** Nonomura and Ohara 1969, 708.

pseu.do.vul.gar'e. Gr. adj. *pseudes* false; specific epithet *vulgare;* M.L. adj. *pseudovulgare* false *S. vulgare.*

Sporangia 7–10 μm, globose, formed on the aerial mycelium, sessile or on short stalks. Aerial mycelium pale pink. Spores 1.2 by 1.5 μm.

On oatmeal agar with yeast extract and yeast starch agar aerial mass color pink, reverse side of colony pale brown, soluble pigment yellowish brown.

Starch hydrolyzed. Utilizes rhamnose or inositol.

Gelatin liquefied. Nitrates reduced to nitrites. No melanoid pigment produced.

Growth to 55 C.

Type strain: S₂-32; CBS.

5. **Streptosporangium nondiastaticum** Nonomura and Ohara 1969, 708.

non.di.as.ta'ti.cum M.L. pref. *non* not; M.L.

adj. *diastaticus* diastatic; M.L. neut.adj. *non-diastaticum* not starch digesting.

Sporangia 10–15 μm in diameter, globose, formed on the aerial mycelium, sessile or on short stalks. Aerial mycelium pink. Spores 1.3 by 1.5 μm.

On oatmeal agar with yeast extract aerial mass color pink reverse pale yellow-brown; soluble pigment pale yellow-brown.

On yeast starch agar aerial mass pink, reverse orange, soluble pigment yellow-brown.

Starch not hydrolyzed.

Gelatin liquefied. Nitrates reduced to nitrites. Melanoid pigments not produced.

Growth to 42 C.

Type strain: S₂-31.

6. Streptosporangium longisporum Schäfer 1969, 368.

lon.gi.spor'um. L.adj. *longus* long; M.L. n. *spora* spore; M.L. neut.adj. *longisporum* long spored.

Sporangia spherical, usually 7–13 μm in diameter but occasionally up to 18 μm, colorless to reddish with age; sporangial wall relatively thick and persistent; substrate hyphae 0.4–1.0 μm wide, septate; aerial hyphae 0.6–1.0 μm wide, sparingly septate; spores rod shaped or allantoid, 0.6–0.9 by 1.5–3.5 μm.

Growth on Czapek agar moderate; substrate mycelium red; aerial mycelium white.

On peptone yeast extract agar growth good; substrate mycelium red; no aerial mycelium.

Growth on tyrosine agar good; substrate mycelium reddish brown; colonies raised and covered with aerial mycelium; melanin not produced.

Nitrate reduction poor; casein completely hydrolyzed. Soluble pigments not produced.

Type strain: S₆₆; ATCC 25212; CBS 184.69.

7. Streptosporangium viridogriseum Okuda, Furumai, Watanabe, Okugawa and Kimura 1966, 126.

vi.ri.do gri'se.um. L. adj. *viridis* green; L. adj. *griseus* gray; M.L. neut. adj. *viridogriseum* greenish gray.

Sporangia 20–48 μm in diameter (29 μm average); spores spherical, 0.2–0.4 μm in diameter, non-flagellated; aerial mycelium non-septate, 0.4–0.6 μm wide, dividing into short rods or conidia, 0.4–0.6 by 1.0–2.4 μm; sporangiophore 20–48 μm long (38 μm average) and 0.4–0.6 μm in diameter.

Growth on 21 media pale yellowish to yellowish brown; aerial mycelium white and cottony, later turning greenish gray. Soluble pigments not produced.

A number of carbon sources give good growth when tested by the method of Pridham and Gottlieb (1948). Arabinose, inositol, lactose, rhamnose and xylose used to some extent; dulcitol, inulin and salicin not utilized.

Growth enhanced by calcium and magnesium ions.

Starch hydrolyzed. Melanin not produced. Tyrosine crystals not utilized.

Produces sporoviridin (Okuda *et al.*, 1966), a glycosidic antibiotic active against Gram-positive bacteria, yeasts and *Trichophyton*.

Temperature range 27–37 C at pH 4.0–9.0; growth to 55 C at pH 6.

Type strain: TA 597; NIHJ-AT523.

7a. *Streptosporangium viridogriseum* subsp. *viridogriseum* subsp. nov.

Morphology as for species.

Gelatin liquefied, nitrates reduced to nitrites.

7b. *Streptosporangium viridogriseum* subsp. *kofuense* Nonomura and Ohara 1969, 708.

Sporangia 12–20 μm in diameter; spores 0.4–1.0 μm.

Gelatin not liquefied. Nitrates not reduced to nitrites.

Type strain: S₂-28.

8. Streptosporangium album Nonomura and Ohara 1960, 407.

al'bum. L. neut.adj. *album* white.

Sporangia 6–8 μm. Aerial mycelia 0.7–1.0 μm wide, usually white. Spores 1.0–1.3 by 1.5–1.9 μm.

On oatmeal agar-Y aerial mycelium white, no soluble or crystalline pigment produced after 1 month at 30 C. Thin growth on starch agar after 15 days.

On glycerol asparagine agar substrate mycelium pale yellow.

On potato agar, growth flat and yellow-brown; soluble pigment pale yellow.

Requires both thiamine and biotin for growth.

Gelatin liquefied. Nitrates not reduced to nitrites. Melanin not produced.

Produces a sporoviridin-like antibiotic.

Type strain: S-16; NRRL B-2635.

9. Streptosporangium albidum Furumai, Ogawa and Okuda 1968, 179.

al'bi.dum. L. neut. adj. *albidum* white.

Sporangia spherical, 10–30 μm (25 μm average); sporangial wall thick and elastic. Aerial mycelium thin and white after several days growth, becoming cottony or floccose and brownish white with age. Spores elliptical, 1.0–1.4 by 1.4–1.6 μm.

Growth on glucose nitrate agar wrinkled and colorless to pale yellowish brown; aerial mycelium white and cottony; good sporangial development.

Growth wrinkled and yellowish brown on glucose peptone agar; aerial mycelium scant, white.

On tyrosine agar growth colorless and transparent; aerial mycelium cottony, brownish white; sporangia sparse.

Soluble pigments not formed.

Does not require biotin or thiamine for growth.

Does not utilize dextrin or salicin but exhibits moderate to fair utilization of other carbon sources recommended by Pridham and Gottlieb (1948). Cellulose and starch not hydrolyzed.

Gelatin not liquefied. Nitrates reduced to nitrites.

Type strain: MCRL-048.

10. Streptosporangium viridialbum Nonomura and Ohara 1960, 407.

vi.ri.di.al′bum. L. adj. *viridis* green; L. adj. *albus* white; M.L. neut.adj. *viridialbum* greenish white.

Sporangia 6–8 μm. Spores 1.0–1.3 by 0.5–1.9 μm. Aerial mycelia 0.7–1.0 μm wide, greenish white.

On oatmeal agar-Y aerial mycelium yellowish gray to greenish white at maturity; soluble or crystalline pigments not produced after 1 month at 30 C.

Growth on starch agar pale yellow and without aerial mycelium after 15 days. Substrate mycelium pale yellow on glycerol asparagine agar.

Growth flat and yellowish brown on potato agar; soluble pigment pale yellow.

Requires thiamine and biotin for growth.

Inositol not utilized. Nitrates not produced from nitrates.

Melanin not produced.

Type strain: S-20.

10a. *Streptosporangium viridialbum* subsp. *viridialbum subsp. nov.*

Description as for species.

10b. *Streptosporangium viridialbum* subsp. *reducens* Nonomura and Ohara 1969, 708.

Inositol utilized. Nitrates reduced to nitrites. Type strain: S₂-24.

11. Streptosporangium rubrum Potekhina 1965, 292.

ru′brum. L. neut.adj. *rubrum* red.

Sporangia spherical, 6.0–14.0 μm in diameter; sporangial wall thin, resistant to stains, and with slight net-like thickenings; sporangiophores branched, 0.5–1.2 by 1.6–27 μm long. Spores spherical 0.8–1.4 μm, rarely ovoid or rod shaped, the latter 1.2–1.8 μm long.

Growth on Czapek agar sparse; aerial mycelia snow white to pale pink; substrate mycelia poorly developed, pale pink to white; sporangia numerous.

On meat peptone agar with glucose growth is slow but good; aerial mycelium sparse, occurring along edge of colony, white or dirty pink; substrate mycelium cream pink to reddish and finally dirty carmine red with age; sporangia lacking.

Starch hydrolyzed. Cellulose decomposed after inoculations with soil by Pushkinskaya's method (1954).

No liquefaction of gelatin. Nitrates weakly reduced to nitrites.

Type: not designated.

Further Comments

Couch states that the spores of *S. roseum* are 1.8–2.0 μm in diameter, almost twice as large as the characteristic spherical spores of *S. rubrum*. Nonomura and Ohara (1960) say that the spores of the related species *S. roseum* and *S. vulgare* are rod shaped.

Genus IV. **Amorphosporangium** *Couch 1963, 65*

J. N. COUCH AND C. E. BLAND

A.mor.pho.spo.ran′gi.um. Gr. adj. *amorphos* without form; M.L. n. *spora* spore; Gr. n. *angium* a vessel; M.L. n. *Amorphosporangium* irregularly shaped sporangium.

Sporangia very irregular in shape, much lobed, 6–25 μm wide by 8–15 μm high, width usually greater than height. Spores rod-shaped, 0.5–0.7 by 1.0–1.5 μm; motile by two to three polar flagella (see Plate 17.7).

Mycelium brilliant shades of orange and in this and cell wall composition (see Table 17.37) resembles *Actinoplanes*.

No organic growth factors required.

Type species: *Amorphosporangium auranticolor* Couch 1963, 65.

Further Comments

Amorphosporangium globisporus Thiemann 1967, 233 is not included here as its correct placement is in doubt; it apparently belongs in the genus *Actinoplanes* (Thiemann, personal communication).

Description of the species of genus **Amorphosporangium**

1. Amorphosporangium auranticolor Couch 1963, 65.

au.ran.′ti.co.lor M.L. n. *aurantium* an orange; L. n. *color* tint, hue; M.L. adj. *auranticolor* orange colored.

Morphology as for genus.

Growth on Czapek agar very good, point inocula 10–13 mm in diameter after 6 weeks, center elevated, margin flat or sometimes entire growth elevated and convoluted; smear confluent, with conspicuous grooves which extend deep into the agar, forming irregular or somewhat rectangular areas; surface slick, wet; color very brilliant apricot orange or orange chrome; sporangia formed abundantly during the first year of culture, formed in clusters, very irregular in shape as on pollen but much smaller, 6–12 μm broad by 5–7 μm high; agar uncolored or yellowish; odor slight.

On peptone Czapek agar growth very good, point inocula 10–17 mm in diameter after 6 weeks outline of margin scalloped, very irregular; surface elevated, intestiniform or convoluted; texture toughish to pasty or waxy; smear confluent; surface of smear convoluted on the margin adjacent to the point inocula, rest of smear with closely arranged grooves and many mounds, surface slick and shiny; color scarlet to flame scarlet, sporangia none. Agar

uncolored to yellowish greenish; odor slightly fragrant, pleasant.

Growth on casein agar good, all or nearly all cleared after 2 months; point inocula about 1 cm in diameter, much elevated, convoluted, pasty to toughish; smear bumpy, confluent; color brilliant apricot orange; surface slick and wet; sporangia none; agar yellow to amber; odor slight.

On tyrosine agar growth fair to barely good; tyrosine mostly used under point inocula and partly under smear, growth of smear thin and flat; color buffy orange; sporangia none; microspores fairly abundant, formed singly or in groups; agar very dark, near chestnut.

Collected twice from soil near Dunphy, Nevada. Type strain: UNCC 253; ATCC 15330.

Genus V. **Ampullariella** *Couch 1964, 29*

J. N. COUCH AND C. E. BLAND

(*Ampullaria* Couch 1963, 55.)
Am.pul.la.ri.el′la. L. n. *ampulla* flask, bottle; M.L. dim. ending *-ella*; M.L. fem.n. *Ampullariella* a small bottle, to indicate bottle-shaped sporangia.

Sporangia bottle- or flask-shaped, digitate or lobate; 5–20 by 8–30 μm. **Spores arranged in parallel chains within the sporangium, rodshaped,** 0.5–1 by 2–4 μm, **motile** by polytrichous, polar flagella of 3.5–6 μm in length (see Plate 17.7). Hyphae much as in *Actinoplanes*.

For cell wall composition of various species see Table 17.37.
Temperature range 18–35 C; optimum 25 C.
The G + C content of the DNA, of the single species examined is 72.3 moles %.
Type species: *Ampullariella regularis* (Couch) Couch 1964, 29.

Key to the species of genus **Ampullariella**

I. Sporangia bottle-shaped, cylindrical, 5–14 by 8–30 μm.
                1. *Ampullariella regularis*
II. Sporangia irregular, rarely cylindrical.
    A. Sporangia mostly bell-shaped, 5–15 by 6–12 μm.
                2. *Ampullariella campanulata*
    B. Sporangia mostly irregular, lobed or fan-shaped in vertical section, 4–20 by 12–26 μm.
                3. *Ampullariella lobata*
    C. Sporangia mostly digitate, 3–9 by 6–15 μm.
                4. *Ampullariella digitata*

Description of the species of genus **Ampullariella**

1. Ampullariella regularis (Couch) Couch 1964, 29. (*Ampullaria regularis* Couch, 1963, 57.)
reg.u.lar′is. L. adj. *regularis* regular.
Sporangia cylindrical, 5–14 by 8–30 μm; the distal end flattened or slightly rounded, the stalk end mound-shaped with a short part of the stalk remaining attached to the sporangium and thus resembling a bottle with a cork. Spores arranged in vertical parallel rows in the sporangia; 0.5–1 by 2–4 μm. Sporangial dehiscence caused by the swelling of an intersporal substance which distorts the

shape to subglobose or club-shaped or to form a lateral bulge, the wall eventually opening by a lateral split from which the spores ooze and slowly become motile; the wall persistent. In one isolate the entire spore mass is pushed out through an opening in the top of the sporangium. Hyphae well developed on the pollen of the tree *Liquidambar styraciflua*, in places dichotomously branched, circinate.

Growth on Czapek agar good, colonies from point inocula 7–10 mm in diameter after 6 weeks;

surface of colonies rather flat; color coral red, salmon orange with albino areas; sporangia none to many; agar uncolored to greenish yellow; some cultures with brush-like conidiophores; palisade hyphae fairly distinct.

On peptone Czapek agar growth very good to excellent, colonies from point inocula 15–20 mm in diameter after 6 weeks; convoluted in center, with descending radial ridges toward the flat margin, color zinc orange, orange chrome, orange rufous to apricot orange, or rufous with vinaceous gray sectors; sporangia none, except in isolate 168; palisades not formed; agar slightly yellowish.

All casein of casein agar hydrolyzed, growth good, colonies from point inocula 7–10 mm in diameter after 6 weeks; surface bumpy, convoluted or ridged; smear thin and flattish, smooth or with a few bumps and convolutions; color ochraceous salmon, flesh ochre or ferruginous; sporangia none or rarely a few; agar pale yellowish to yellowish brown to deep brown.

In tyrosine agar slight to no clearing of crystals; growth fair, point inocula 3–10 mm in diameter but very thin after 6 weeks; smear continuous but thin, flat; color salmon-buff to coral-pink; no sporangia; agar uncolored to dark auburn.

Type strain: UNCC 79.

Three biotypes are recognized:

a. On casein agar, the casein is hydrolyzed and the agar becomes a dark brown. Tyrosine is hydrolyzed little if at all and the agar not darkened or very slightly. Type strain: 79.

b. Casein is hydrolyzed but the agar is not darkened. Reaction on tyrosine as in a. Type strain: 28.

c. Represented so far by a single isolate; darkens casein but not tyrosine agar. Also produces large numbers of sporangia on peptone Czapek agar and many on Czapek agar, a characteristic retained after 12 years in the laboratory and one retained by few other strains. Type strain: 168.

2. **Ampullariella campanulata** (Couch) Couch 1964, 29. (*Ampullaria campanulata* Couch 1963, 59.)

cam.pan.u.la'ta. L. dim.n. *campanula* bell; M.L. adj. *campanulata* bell-like.

Sporangia 5–15 by 6–12 μm; bell-shaped or in vertical section fan-shaped, frequently papillate at the top because of the unequal length of the rows of spores. Spore arrangement, size and shape, as well as mycelium on pollen, much as in *A. regularis*.

Growth on Czapek agar good to very good, colonies from point inocula 10–18 mm in diameter after 6 weeks, flattish to elevated; color variable,

ferruginous, Kaiser brown (colors after Ridgway) coral-red, mahogany to dull violet-black or black; in some isolates all of these colors produced in one smear; sporangia none to many; agar uncolored or yellowish, pale greenish yellow or purplish brown or slightly greenish brown.

Growth on peptone Czapek agar very good, colonies from point inocula up to 18 mm in diameter after 6 weeks; elevated, convoluted and minutely bumpy in the center, marginal area ridged or flat, smear convoluted or consisting of numerous small or large mounds (up to 5 mm in diameter), with a smooth, rough, or cracked surface; color flame scarlet, English red, orange rufous, grenadine red, to Sanford's brown or darker; sporangia none; agar uncolored to brown. Cultures very variable on peptone Czapek agar.

On casein agar all casein usually used after 2 months; growth good to very good, colonies from point inocula 10–18 mm in diameter after 2 months, flattish, except for a few convolutions and bumps; color, coral-red, bittersweet orange, flame scarlet to scarlet; sporangia none; agar yellowish.

In tyrosine agar no utilization to complete clearance beneath and close to culture; growth fair to good, colonies from point inocula 4–10 mm in diameter after 6 weeks, mostly flat but in places convoluted or with scattered, conspicuous bumps; color mostly coral-pink to Jasper pink; sporangia none; agar not discolored.

Isolations have been made from soils from North Carolina, Georgia, Kansas, Ohio and from Egypt, Argentina and Tahiti. Next in abundance to *A. regularis*.

Type strain: UNCC 65; ATCC 15348.

3. **Ampullariella lobata** (Couch) Couch 1964, 29. (*Ampullaria lobata* Couch 1963, 59.)

lo.ba'ta. M.L. fem.adj. *lobata* lobed.

Sporangia 4–20 by 12–23 μm; variable in size and shape rarely cylindrical but typically irregular and lobed; distinct lobes may give the appearance of two to three sporangia fused together, or one to several long finger-like branches may protrude from near the base of the larger sporangium, at times irregularly fan-shaped in section. Spores 0.4–1.2 μm, motile by several polar flagella, 3.5 μm in length. Hyphae as in *A. regularis*.

Growth on Czapek agar poor to excellent, colonies from point inocula, up to 25 mm after 6 weeks; flattish; color albino to coral-red to dragon's blood red; palisade hyphae abundant, septate and sparingly branched; sporangia absent on albino areas, abundant in scattered patches on pigmented areas; agar uncolored to pale yellowish green.

Growth on peptone Czapek agar very good, colonies from point inocula 15–22 mm in diameter after 6 weeks; surface bumpy to convoluted; colors of growth and agar much as on Czapek agar; sporangia none.

All casein in casein agar used after 2 months; growth good, colonies from point inocula up to 10 mm diameter after 6 weeks, surface elevated with folds, convolutions and ridges; color light coral-red to old rose to ferruginous; agar light brown; sporangia none.

In tyrosine agar slight to fairly complete clearing of crystals beneath colonies; growth poor to fair, colony consisting of many minute flat patches 2–10 mm in diameter after 6 weeks; pale buff to coral-pink with a few brilliant coral-red specks; agar slightly darkened; sporangia none.

Type strain: UNCC 72; ATCC 15350.

4. Ampullariella digitata (Couch) Couch 1964, 29. (*Ampullaria digitata* Couch 1963, 61.)

di.gi.ta'ta. L. fem. adj. *digitata* having fingers.

Sporangia 3–9 by 4–14 μm; very irregular, sub-cylindrical, lobed and frequently digitate; rarely bottle-shaped. Spores as in *A. regularis*.

Growth on Czapek agar fair, colonies from point inocula about 5 mm in diameter, flat with an entire or fimbriate margin; color pinkish cinnamon or cinnamon rufous to mahogany, margin usually lighter, near cinnamon to albino, gray- to slate-colored sectors abundant; sporangia abundant on the gray sectors; agar distinctly greenish yellow.

Growth on peptone Czapek agar good to excellent, from point inocula 10–28 mm in diameter, convoluted and ridged; the streaked area convoluted and ridged where crowded; color burnt sienna to mahogany to blackish brown, occasionally with lavender sectors; sporangia none; agar greenish brown to very dark brown.

Casein in casein agar usually completely utilized after 2–3 months; growth of point inocula fair to good, elevated, wrinkled or convoluted; smear confluent or consisting of minute conical patches; color brick red, sometimes with black sectors or very dark with brick red sectors; sporangia none; agar very dark brown.

In tyrosine agar crystals very slightly to mostly cleared under growth; growth fair to good, from point inocula 4–10 mm in diameter, elevated, bumpy or slightly convoluted in center; smear consisting of many tiny mounds, mostly less than 1 mm in diameter; flesh color to dirty buff; sporangia none; agar darkened throughout but particularly where the crystals of tyrosine have been all or partly dissolved.

The G + C content of the DNA is 72.1 moles %.

Type strain: UNCC 33; ATCC 15349.

Genus VI. Pilimelia Kane 1966, 225

WILMA KANE HANTON

Pi.li.mel'i.a. L. n. *pilus* a hair; Gr. fem.n. *Melia* a nymph beloved by the river god Inachus; M.L. fem.n. *Pilimelia* an aquatic organism growing on hair substrate.

Sporangia large, globose or cylindrical. **Rod-shaped spores produced end to end in parallel chains, approximately 1,000 per sporangium;** motile by means of flagella (Plate 17.7). Loses ability to produce sporangia on repeated subculturing.

Hyphae up to 2.2 μm wide and up to 300 μm long, irregularly septate. **Requires complex organic substrate** such as peptone for growth. Growth on peptone-containing agar **usually yellow** yellow-brown or gray; colonies compact 1–2 mm, soft pasty, **slight or no penetration of mycelium into agar.**

For cell wall composition see Table 17.37.

Obligate aerobes requiring Petri dish cultivation for maintenance; poor growth on primary test tube culture and no growth upon subculture.

World-wide distribution in soil.

Type species: *Pilimelia terevasa* Kane 1966, 225.

Further Comments

Both of the described species have a single polar flagellum. Other undescribed species have one to four sub-polar or lateral flagella (Hanton, unpublished data; Bland, 1968).

Although Pilimelia has been isolated only from keratinic substrates, after less than a year in culture it loses the ability to grow on natural keratin.

Description of the species of genus Pilimelia

1. Pilimelia terevasa Kane 1966, 225.

ter.e.vas'a. L. adj. *teres* rounded; L. pl.n. *vasa* vessels; M.L. pl.n. *terevasa* signifying rounded sporangia.

On soil-water on hair substrate: Sporangia globose, up to 24 μm in diameter. Spores rod-shaped, 0.3–0.6 by 1.2–1.5 μm, formed in parallel chains within the sporangium. Hyphae 1.2–2.2 by up to 56 μm long.

On peptone-containing agar, colonies soft or

pasty, 1–2 mm in diameter, yellow or yellow-brown, occasionally producing a diffusible melanin-like pigment.

Carbohydrates not utilized in broth shake cultures (Leonard, 1968). No individual or combination of purified amino acids utilized without presence of peptone. Growth optimal on 2–5% peptone, inhibited by 6% peptone; no other growth requirements found. A melanin-like pigment is formed when L-tyrosine is added to peptone media. Growth is slow the first 3 days, rapid from the 4th–12th day; after the 12th day autolysis begins. A yield of 100–200 mg dry weight of mycelia/100 ml of broth media is obtainable. Ammonia is released during growth. Struvite crystals ($MgNH_4PO_4 \cdot 6H_2O$) are formed in medium when $MgSO_4$ and K_2HPO_4 are present in concentrations greater than 0.25 g/liter and 0.5 g/liter respectively, and when ammonium accumulation of 0.2 mg/ml or more occurs.

Intracellular extracts contain major amounts of *meso*-diaminopimelic acid, glutamic and aspartic acids, glycine and alanine. Minor amounts of valine, threonine and isoleucine are also present as well as a fourth unidentified amino acid. Carbohydrates not present. Cytochrome a, b and c are present.

Cell walls contain 2,6-diaminopimelic acid, glucose, mannose and arabinose and minor amounts of galactose and xylose (Szaniszlo, 1967).

The intracellular pigment has an absorption spectrum of a maximum peak at 452, two lesser peaks at 425 and 479, and a shoulder peak at 400 (Szaniszlo, 1967).

Obligate aerobe requiring shake culture in broth.

Temperature range 15–35 C; 30 C optimum. pH range 6.5–7.6; initial optimum in broth pH 6.9.

Type strain: ATCC 25603.

2. Pilimelia anulata Kane 1966, 225.

an'u.lat.a. L. fem.adj. *anulata* ornamented with a ring.

On soil-water on hair substrate: Sporangia cylindrical, up to 11.2 μm wide and up to 35 μm long. Spores rod-shaped, 0.3–0.7 by 0.8–1.3 μm, in parallel chains up to 35 spores long. Hyphae up to 1.8 μm wide and up to 300 μm long.

Conidia occasionally produced below the liquid surface.

On peptone-containing agar yellow or yellow-gray, occasionally producing a diffusible melanin-like pigment, soft or pasty colonies.

Cell wall analysis shows 2,6-diaminopimelic acid, glucose, arabinose and xylose to be present with a minor amount of galactose and a trace amount of mannose present (Szaniszlo, 1967).

Obligate aerobe requiring Petri plate culture for maintenance.

Type strain: ATCC 25604.

Genus VII. **Planomonospora** *Thiemann, Pagani and Beretta 1967, 28*

JOSEF E. THIEMANN

Pla.no.mo.no'spo.ra. Gr. adj. *planos* wanderer; Gr. adj. *monos* single; Gr. fem.n. *spora* a seed; M.L. fem.n. *Planomonospora* a motile single-spored (organism).

Substrate and aerial mycelia formed, the former growing profusely into the medium and forming a compact layer on surface of agar.

Sporangia formed on aerial mycelia only, each containing a single large spore motile by peritrichous flagella. Not acid-fast. Gram-positive.

Grows on a variety of substrates; best growth on complex media. Pigments may give characteristic colors to the aerial and/or vegetative mycelium. Cell wall contains *meso*-diaminopimelic acid.

Starch hydrolyzed. Nitrates reduced to nitrites. Aerobic.

Temperature optimum 28–37 C; no growth at 45 C.

Found in soils, relatively rare occurring in 10 of 453 samples of temperate and tropical soils.

Type species: *Planomonospora parontospora* Thiemann, Pagani and Beretta 1967, 29.

Description of the species of genus **Planomonospora**

1. Planomonospora parontospora Thiemann, Pagani and Beretta 1967, 29.

pa.ron.to'spo.ra. Gr. v. *pareimi* to be side by side; Gr. n. *spora* a seed; M.L. n. *parontospora* spores side by side.

Vegetative mycelium 0.6–0.8 μm in diameter, not fragmented, rose colored. No soluble pigments.

Aerial mycelium 1 μm in diameter, white with faint rose tinge.

Sporangiospore fusiform, 1.0–1.5 by 3.5–4.5 μm. Sporangia arranged in a double parallel row (Plate 17.8, Figs. 1, 2); the mature fertile hyphae become characteristically bent. Spores liberated through an operculum at upper part of the sporangium.

Grows well on various organic media; poor or no growth on various synthetic media. Oatmeal agar gives abundant, rose-colored growth and sporangia formation. Similar growth on Hickey and Tresner agar. Good growth, smooth, flat, light rose in color with moderate development of aerial hyphae on glucose asparagine agar. Poor growth on glycerol asparagine agar. Moderate hyaline growth on Czapek glucose agar, aerial mycelia light rose. Poor growth on starch, tyrosine and skim milk agars. No growth on cellulose agar, peptone iron agar or calcium malate agar.

Arabinose, dextrin, fructose, galactose, glucose, inulin, lactose, mannitol, mannose, rhamnose, starch and xylose are utilized. Dulcitol, glycerol inositol, maltose, raffinose, ribose, sorbitol, sorbose and sucrose not utilized. Starch hydrolyzed.

Casein in skim milk agar slowly hydrolyzed; no coagulation but complete peptonization of litmus milk. Tyrosine not hydrolyzed. Gelatin not liquefied.

Optimum pH 7–8; no growth at pH 5 or lower.

Originally isolated from soil sample from Chile; has also been isolated from soils from Argentina, Venezuela, Peru, Italy and India.

Type strain: B677; ATCC 23863.

1a. *Planomonospora parontospora* subsp. *parontospora*
Description as for species.

1b. *Planomonospora parontospora* subsp. *antibiotica* Thiemann, Coronelli, Pagani, Beretta, Tamoni and Arioli 1968, 528.
Description as for species but gelatin hydrolyzed, litmus milk not peptonized; tyrosine digested with formation of a light brown diffusible pigment.

An antibiotic, sporangiomycin, formed.

Isolated from soils from Argentina and Venezuela.

Type strain: B-987; ATCC 23864.

2. **Planomonospora venezuelensis** Thiemann 1970, 247.

ve.ne.zuel.en'sis. M.L. adj. *venezuelensis* pertaining to Venezuela.

Vegetative mycelium 1.0 μm in diameter, twisted, highly branched. Not fragmented, occasionally septate. Violet-brown. Soluble, amber-brown to violet diffusible pigment formed on some media.

Aerial mycelium slender, 0.5–0.6 μm in diameter, long, wavy, white to grayish white.

Sporangia 4.5–5.5 by 1.0 μm, attached to aerial mycelium by a short sporangiophore. Sporangia formed singly along the aerial hyphae, or in groups, giving a characteristic palm-leaf pattern (Plate 17.8, Fig. 3). Spores fusiform, 1.0 by 3.0–3.5 μm; always liberated through an operculum at upper part of sporangium.

Moderate violet-colored growth on oatmeal agar; aerial mycelia light gray; traces of brown-violet soluble pigment. Good growth on Hickey and Tresner agar, wrinkled surface, traces of white aerial mycelia. Slight growth on glucose asparagine agar, smooth surface, dark violet-brown; traces of white aerial mycelium. Moderate growth on glucose asparagine agar. No growth on Czapek glucose agar. Moderate growth on glycerol asparagine agar, tyrosine agar and skim milk agar.

Arabinose, fructose, glucose, mannitol, rhamnose and xylose support good growth. Inositol, raffinose and sucrose support only moderate growth. Cellulose not metabolized. Some strains fail to grow on fructose, inositol and raffinose.

Gelatin liquefied; casein not hydrolyzed. Tyrosine hydrolyzed. H_2S produced.

Isolated from soil sample from Venezuela.

Type strain: B-1072; ATCC 23865.

Genus VIII. **Planobispora** *Thiemann and Beretta 1968, 157*

JOSEF E. THIEMANN

Pla.no.bi'spo.ra. Gr. adj. *planos* wanderer; L. adj. *bis* twice, double; Gr. n. *spora* a seed; M.L. fem.n. *Planobispora* a motile double-spored (organism).

Substrate mycelium 1 μm in diameter, well developed, highly branched. Gram-positive. Not acid fast.

Aerial mycelium 1.0 μm, long and sparsely branched; hyphae usually parallel to agar surface.

Sporangia containing a longitudinal pair of large spores formed only on aerial mycelia; 1.0–1.2 by 6–8.0 μm (Plate 17.8, Fig. 4). Spores motile by peritrichous flagella.

Cell wall contains *meso*-diaminopimelic acid.

Arabinose, glucose and xylose support good growth; raffinose, sucrose and cellulose not metabolized. Starch hydrolyzed. Nitrates reduced to nitrites.

Aerobic.

Optimum temperature 28–37 C; no growth at 45 C.

Isolated from soil collected from river bank in Venezuela. Rare, found in only 2 of 454 samples.

Type species: *Planobispora longispora* Thiemann and Beretta 1968, 157.

Description of the species of genus **Planobispora**

1. Planobispora longispora Thiemann and Beretta 1968, 157.

lon.gi.spo'ra. L. adj. *longus* long; M.L. n. *spora* spore; M.L. adj. *longispora* long spored.

Vegetative mycelium hyaline to cream colored. No soluble pigments formed.

Aerial mycelium white.

Sporangia formed singly along the aerial mycelium or in groups on short side branches. Sporangiophores 1.0–3.0 μm long. Sporangia smooth walled. Spores straight to slightly curved with rounded ends, 1.0–1.2 by 2.6–4.0 μm.

Good growth on oatmeal agar, powdery surface; abundant formation of white aerial mycelium and sporangia. Good growth on Hickey and Tresner and nutrient agar; no aerial mycelium. Poor growth on glucose asparagine agar; no aerial mycelium. No growth on Czapek-Dox medium.

Good growth on mannitol, moderate growth on fructose, inositol and rhamnose.

Gelatin partially hydrolyzed; casein hydrolyzed. Milk coagulated and peptonized. H₂S produced. Tyrosine not hydrolyzed.

pH optimum 6–8; no growth at pH 5 or lower.

Type strain: Pb-1075; ATCC 23867.

2. Planobispora rosea Thiemann 1970, 251.

ro'se.a. L. fem.adj. *rosea* rose colored.

Vegetative mycelia in most media rose colored, septate, non-fragmenting, highly irregular with numerous short ramifications.

Aerial mycelium always a light rose tinge.

Sporangia attached to aerial mycelia by a sporangiophore 1.0–2.0 μm long. Spores straight, fusiform, always with rounded ends, 1.0–1.2 by 3.0–3.5 μm.

Good growth on oatmeal agar, smooth, moderate formation of aerial mycelium and sporangia. Good growth on Hickey and Tresner agar, slightly wrinkled, yellow-amber; abundant aerial mycelium and sporangia formation. Moderate, smooth, flat growth on glucose asparagine agar; aerial mycelium scarce. Practically no growth on Czapek-Dox medium. No growth on peptone-yeast extract-iron agar.

Fructose and inositol support good growth; mannitol and rhamnose are not used.

Gelatin slightly liquefied. Litmus milk not coagulated or peptonized. Tyrosine hydrolyzed.

Type strain: Pb-1435; ATCC 23866.

Genus IX. **Dactylosporangium** Thiemann, Pagani and Beretta 1967, 43

JOSEF E. THIEMANN

Dac.ty.lo.spo.ran'gi.um. Gr. n. *dactylos* finger, finger-shaped; M.L. n. *sporangium* spore case; M.L. neut.n. *Dactylosporangium* finger-shaped sporangium.

Substrate mycelium 0.5–1.0 μm in diameter; hyphae long, irregularly branched, twisted and occasionally coiled. Septation occurs but is not frequent. Do not fragment. Produce tough leathery growth on surface of agar. Gram-positive. Not acid-fast.

Aerial mycelium not formed or is in a very rudimentary form as short and usually unbranched hyphae.

Sporangia, 1.0–1.2 by 4.0–6.0 μm, **formed in clusters on surface of solid media,** emerging from vegetative mycelium on short sporangiophores (Plate 17.8, Figs. 5–9). Usually straight but in some strains long, wavy and branched sporangia can be found. Each sporangium contains a **single row of three to five spores. Large globose spores** or "sporangioles" **formed on the mycelium embedded in the agar.**

Spores motile usually by polytrichous polar flagella, in some strains peritrichous.

Cell wall contains *meso*-diaminopimelic acid, pentoses and hexoses.

Utilizes arabinose, dextrin, fructose, galactose, glucose, inulin, lactose, maltose, mannitol, mannose, raffinose, rhamnose, starch, sucrose and xylose. Does not utilize dulcitol, glycerol, inositol or sorbitol.

Casein hydrolyzed. H₂S produced. Litmus milk not coagulated but completely peptonized.

Grows on a variety of substrates.

Aerobic.

Optimum temperature 37 C; no growth at 45 C. pH optimum 6–7; very slight growth at pH 5.0.

Isolated from cultivated soil; widely distributed.

Type species: *Dactylosporangium aurantiacum* Thiemann, Pagani and Beretta 1967, 43.

Description of the species of genus **Dactylosporangium**

1. Dactylosporangium aurantiacum Thiemann, Pagani and Beretta 1967, 43.

au.ran.ti'ac.um. M.L. neut.adj. *aurantiacum* orange colored.

Morphology as for genus.

Abundant growth on oatmeal agar, pale orange, smooth surface. Moderate growth on Hickey and Tresner agar, hyaline to very pale orange. On glucose asparagine agar moderate growth, smooth and flat, whitish cream in color. Similar on glycerol asparagine agar but hyaline. Good growth on nutrient agar, orange.

Sorbose and ribose not utilized.

Tyrosine and gelatine not hydrolyzed. Nitrates reduced to nitrites.

Little growth at 22 C.

Type strain: D-748; ATCC 23491.

2. **Dactylosporangium thailandense** Thiemann, Pagani and Beretta 1967, 43.

thai.lan.den′se. M.L. adj. *thailandense* pertaining to Thailand.

Morphology as for genus.

Good growth on oatmeal agar, slightly wrinkled, light orange-brown. Traces of light amber soluble pigment. On Hickey and Tresner agar, good growth, wrinkled, brown with rose tinge; soluble pigment brown with reddish tinge. Moderate, light orange growth on glucose asparagine agar. Similar on glycerol asparagine agar with traces of rudimentary aerial mycelium. Good growth, pale orange on nutrient agar. No growth on Czapek-Dox agar.

Ribose utilized.

Gelatin and tyrosine hydrolyzed. Nitrates not reduced.

Type strain: D-449; ATCC 23490.

Genus X. **Kitasatoa** *Matsumae and Hata in Matsumae, Ohtani, Takeshima and Hata, 1968, 616*

J. N. COUCH AND C. E. BLAND

Ki.ta.sa.to′a. M.L. fem.n. *Kitasatoa* named for Kitasato, a Japanese bacteriologist.

Sporangia club-shaped, 2–2.5 by 5 μm. Spores diplococcus-like, spherical, ellipsoidal or cylindrical, arranged in a single chain within the sporangium and motile by a single polar flagellum. Aerial hyphae abundant and of two types: one type, 1.0–1.2 μm wide, produces long chains of

TABLE 17.38

Characteristics of species of genus **Kitasatoa**

		1. *K. purpurea* Matsumae and Hata in Matsumae et al., 1968, 617	2. *K. diplospora* Matsumae, Ohtani and Hata in Matsumae et al., 1968, 621	3. *K. kauaiensis* Matsumae, Ohtani and Hata in Matsumae et al., 1968, 621	4. *K. nagasakiensis* Matsumae and Hata in Matsumae et al., 1968, 622
Zoospore		Globose single or in pairs 2.0–2.8 by 1.6–2.3 μm	Ellipsoidal single or in pairs 2.5–3.0 by 1.5 μm	Cylindrical single or in pairs 2.4–5.0 by 1.1–1.35 μm	Rod single or in parallel 2.9 by 0.8 μm
Conidia		Ellipsoidal, smooth 1.5–1.8 by 0.8–1.0 μm	Cylindrical, smooth 1.5–1.3 by 0.7 μm	Cylindrical, warty 1.2–1.5 by 0.8 μm	Ellipsoidal, smooth 1.2–1.6 by 0.8–1.0 μm
Glycerol-Czapek's agar	G.[a] A.M. S.P.	Dark brownish purple White Dark brownish purple	Pale yellowish brown White Light brown	Reddish brown White Reddish brown	Grayish brown White Pale brown
Sucrose-Czapek's agar	G. A.M. S.P.	Grayish yellow brown Light pinkish gray None	Pale yellow Pinkish orange None	Colorless Brownish white None	Brownish white Dark brownish gray None
Oatmeal agar	G. A.M. S.P.	Yellowish brown Light brownish gray Yellowish brown	Yellowish gray Brownish white Grayish yellow brown	Pale olive White Pale olive gray	Light brownish gray Gray white Brownish black
Milk coagulation		−	−	±	+
Milk peptonization		−	−	−	+
Hemolysis		−	+	±	−
Serum liquefaction		−	+	−	−
Collection site		Kauai Island, Hawaii	Kauai Island, Hawaii	Kauai Island, Hawaii	Nagasaki Prefecture, Japan
Type strain		KA-279	KA-280	KA-281	KA-282

[a] G., growth; A.M., aerial mycelium; S.P., soluble pigment.

cylindrical conidiospores; the other 1.2–1.5 μm wide produces either terminal sporangia or rounded vesicular bodies. Vegetative mycelium on liquid or solid media produces club-shaped sporangia at the hyphal tips; vegetative hyphae 0.8–1.2 μm wide, highly branched with few septa.

No organic growth factors required. Produce the antibiotic chloramphenicol. Temperature range 10–37 C; pH range 5.0–9.0.

Type species: *Kitasatoa purpurea* Matsumae and Hata in Matsumae, Ohtani, Takeshima and Hata 1968, 617.

Description of the species of genus **Kitasatoa**

Morphologically, all four members of the genus are very similar. Differentiation into species is based primarily on differences in cultural findings.

Their distinguishing features are given in Table 17.38.

FAMILY V. **DERMATOPHILACEAE** AUSTWICK 1958, 42, *emend. mut. char.* Gordon 1964, 521

MORRIS A. GORDON

Der.ma.to.phi.la′ce.ae. M.L. n. *Dermatophilus* type genus of family; -*aceae* ending to denote family; M.L. fem.pl.n. *Dermatophilaceae* the *Dermatophilus* family.

Mycelial filaments or muriform thalli which divide transversely and in at least two longitudinal planes to form masses of coccoid or cuboid cells which characteristically become motile. Aerial mycelium ordinarily absent but sometimes inducible by 10% CO_2. Gram-positive. Not acid fast.

Cell walls of type III (Becker *et al.*, 1965) (see Table 17.18). Pigment production common. Aerobic.

Includes pathogenic organisms causing skin lesions in mammals, including man.

Type genus: *Dermatophilus* Van Saceghem 1915, 357.

Key to the genera of family **Dermatophilaceae**

I. Mycelium of narrow tapering filaments with lateral branching at right angles; septa formed in transverse and in horizontal and vertical longitudinal planes. Weakly fermentative. Madurose present.

Genus I. *Dermatophilus*

II. Mycelium rudimentary. Produce a muriform, tuber-shaped, non-capsulated, multilocular thallus containing masses of cuboid cells 0.5–2.0 μm in diameter. Madurose absent.

Genus II. *Geodermatophilus*

Genus I. **Dermatophilus** *Van Saceghem 1915, 357, emend. mut. char. Gordon 1964, 521*

MORRIS A. GORDON

(*Dermatophylus* (*sic*) Van Saceghem 1915, 568; *Polysepta* Thompson and Bisset 1957, 590.)

Der.ma.toph′il.us. Gr. n. *derma* skin; Gr. adj. *philus* loving; M.L. masc.n. *Dermatophilus* skin loving.

Aerial mycelium ordinarily absent. Substrate mycelium consists of **long tapering filaments, branching laterally** at **right angles; septa formed in transverse and in horizontal and vertical longitudinal planes, giving rise to up to eight parallel rows of coccoid cells (spores), each of which becomes motile by a tuft of flagella.** Gram-positive. Cell walls of type III (Becker *et al.*, 1965) (see Table 17.18). Whole cell hydrolysates contain madurose (Lechevalier *et al.*, 1971).

Chemoorganotrophs. **Non-fermentative** but **acid is produced** from certain carbohydrates. **Catalase positive.** Not acid fast.

Growth reported only on complex media; minimum nutritional requirements unknown.

Aerobic and facultatively anaerobic.

Temperature optimum *ca.* 37 C.

Pathogenic for mammals, invading only the uncornified epidermis.

Type species: *Dermatophilus congolensis* Van Saceghem 1915, 357.

Description of the species of genus **Dermatophilus**

1. **Dermatophilus congolensis** Van Saceghem 1915, 357. (*Dermatophylus congolense (sic)* Van Saceghem 1915, 568; *Actinomyces dermatonomus* Bull 1929, 313; *Tetragenus congolensis* Van Saceghem 1934, 597 (growth form acc. to Lessel in Ind. Berg. 1143); *Actinomyces congolensis* (Van Saceghem) Hudson 1937, 1460; *Nocardia dermatonomus* (Bull) Henry 1952, 1; *Streptothrix bovis* Snijders and Jansen 1955, 242; *Dermatophilus dermatonomus* (Bull) Austwick 1958, 5; *Dermatophilus pedis* (Thompson and Bisset) Austwick 1958, 43.)

con.go.len'sis. M.L. adj. *congolensis* pertaining to the Congo; named for the Belgian Congo.

Hyphae, 0.5–1.5 μm in diameter, develop from germ tubes. After several transverse and longitudinal divisions, hyphae may be up to 5 μm diameter, with branches at right angles tapering to non-septate apices (Plate 17.9, Fig. 1). Become converted entirely into eight-ranked packets of isodiametric segments encased in a gelatinous sheath (Plate 17.9, Fig. 2.3). Each segment is released as a motile spore, bearing a tuft of five to many flagella (Plate 17.9, Fig. 4). The spores subsequently lose motility and germinate.

On meat infusion-horse blood agar colonies rough, often becoming viscous; adherent through invasion of substrate by hyphae; often white to gray at first, usually becoming orange to yellow; β-hemolytic. Good growth on Loeffler's medium, light yellow; medium liquefied by most strains. No growth on Sabouraud dextrose agar, Czapek agar or tomato paste-oatmeal agar. Broth is clear with a flocculent or ropy sediment; sometimes with a surface ring of growth.

Acid from glucose and fructose; transient acid (within 48 hr) from galactose; often late production from maltose. Acid not produced from lactose, sucrose, xylose, dulcitol, mannitol, sorbitol or salicin. Starch hydrolyzed.

Gelatin stab liquefied rapidly by most strains, slowly or not at all by others. Bromcresol purple milk peptonized by most strains. Casein hydrolyzed. Tyrosine and xanthine not hydrolyzed. Urease and catalase produced. Indole not formed; methyl red and Voges Proskauer tests negative. Nitrates not reduced.

Aerobic, facultatively anaerobic; in an atmosphere of 10% carbon dioxide at 37 C growth is accelerated, aerial hyphae may be formed, while septation and spore formation are delayed.

Etiological agent of naturally occurring streptotrichosis in cattle, sheep, horse, goat, deer and other herbivora, and rarely in man; similar infection inducible in guinea pigs; mice and rabbits also susceptible.

Reference strain: ATCC 14637.

Genus II. **Geodermatophilus** Luedemann 1968, 1857

George M. Luedemann

Ge.o.der.ma.to.phi'lus. Gr. n. *ge* earth; M.L. masc.n. *Dermatophilus* a genus of the *Actinomycetales;* M.L. masc.n. *Geodermatophilus* soil or earth bound dermatophilus-like organisms.

Organisms producing a **muriform, tuber-shaped, non-capsulated, holocarpic multilocular thallus** (see Cross, 1970 and glossary) containing masses of cuboid cells 0.5–2.0 μm in diameter. The thallus under favorable environmental conditions breaks up releasing cuboid and coccoid non-motile cells. Some of these cells may develop into **elliptical to lanceolate zoospores** which appear to be **propelled by a terminal tuft of long flagella.** Germinating spores or resting zoospores may divide directly to produce a new thallus or may produce a germ tube and an irregularly constricted and branched filament. The contents of these tubes first divide transversely by septa that do not appear to involve the outer layer of the cell wall, giving rise to a tube of longitudinally compressed cells. Septa are formed later in horizontal and vertical longitudinal planes and give rise to rows of cuboidal cells. **Mycelium rudimentary, aerial mycelium not produced.** Gram-positive.

Cell wall type III (see Table 17.18). Whole cell hydrolysates do not contain madurose (Lechevalier *et al.*, 1971).

Aerobic.

Type species: *Geodermatophilus obscurus* Luedemann 1968, 1857.

Further Comments

The organisms in this genus appear related to *Dermatophilus* through an obscure filamentous phase. This filamentous condition may be difficult to discover and sometimes may be induced by using a water culture or lean broth medium. The filamentous phase does not appear as a true mycelium but rather like the pseudomycelium seen in certain yeasts as loosely united filaments representing elongations of buds, characteristically pinched and constricted at irregular intervals. These filaments or tubes resemble the holocarpic sporangia of fungi, producing within the tubes spores, which, when liberated, may germinate directly or become zoospores. Occasionally cells of the thallus may germinate *in situ*, emerging as lanceolate zoospores or may produce a germ tube and small filament. The zoospores appear unique in that additional zoospores may develop as terminal buds often forming sluggishly motile chains of lanceolate or elliptical cells. The filamentous phase and zoospore stage may be difficult to observe in some isolates. In other isolates the thallus is reduced to a few cells and may appear reminiscent of miniature fungal dictyospores or sarcinaform packets. The muriform, multilocular thallus and the large number of morphological forms capable of being produced under different conditions of culture and by various strains are the most distinctive and constant features of members of this genus.

Description of the species of genus **Geodermatophilus**

1. Geodermatophilus obscurus Luedemann 1968, 1857.

ob.scur'us L. adj. *obscurus* dark, obscure, indistinct.

Thalli appear greenish black by transmitted light and vary in size from a few cuboidal cells to many cells arranged in cushion or tuber-shaped aggregates. Presence and abundance of zoospores, germ tubes and filaments vary with the strain.

Colonies on agar appear dark brown to black after 30 days incubation (26–28 C), are flat to plicate, granular in texture, dry, odor dank. Growth good on yeast extract-starch-sucrose-malt extract agar; poor on brain heart infusion agar.

Carbohydrate utilization and acid production varies with the subspecies (Table 17.39). Starch is hydrolyzed.

Casein not hydrolyzed. Gelatin hydrolysis and nitrate reduction vary with the subspecies. Blood not hemolyzed.

Growth optimal at 24–28 C, reduced at 37 C, none at 50 C. In aqueous suspensions, few cells survive 30 min at 60 C.

1a. Geodermatophilus obscurus subsp. **obscurus** Luedemann 1968, 1857.

Thalli varying in size from a few cells to many cells arranged in cushion or tuber-shaped aggregates. Thalli disintegrate giving rise to cuboidal and coccoidal cells and elliptical to lanceolate zoospores. Zoospores often budding while motile, giving rise to two- or three-celled motile units. Post-motile spores may produce germ tubes and filaments or may undergo enlargement, septation and thallus formation without an intervening germ-tube-filament stage.

Colony growth on yeast extract-casein hydrolysate-starch-dextrose agar or yeast extract-glycerol agar good; colony plicate, texture characteristically granular-friable, black.

Carbohydrate utilization and acid production given in Table 17.39.

Gelatin is not hydrolyzed. Nitrate reduction weak or not at all.

Isolated from a soil sample from the Amargosa Desert of Nevada, U.S.A.

Type strain: ATCC 25078.

1b. Geodermatophilus obscurus subsp. **amargosae** Luedemann 1968, 1857.

a.mar.go'sae. M.L. gen.n. *amargosae* of the Amargosa Desert.

Macromorphology similar to *G. obscurus*; micromorphology differs in that germinating cells often give rise to long, slender, unbranched filaments which enlarge and undergo septation to produce relatively long, broad, blunt-ended thalli (Plate 17.10) or the germinating cells may enlarge, sep-

TABLE 17.39

Carbohydrate utilization and acid production in **Geodermatophilus obscurus** *subspecies[a]*

Carbohydrate	Subspecies			
	a. *obscurus*	b. *amargosae*	c. *utahensis*	d. *dictyosporus*
D-Arabinose	0a	0	2a	0a
L-Arabinose	3A	3A	1	3A
D-Galactose	3	3A	3	3
D-Glucose	3a	3A	3a	3a
Glycerol	3a	3A	1	3A
Inositol	3	2	2	0
β-Lactose	0	1	0	3
D-Fructose	3a	3A	2	3A
D-Mannitol	3	3	3a	3
Melezitose	0	1	1	3
L-Rhamnose	1a	2A	1	1A
D-Ribose	0A	1A	1a	1a
Sucrose	3	3	3	3
D-Xylose	3a	3A	3a	3A

[a] 0, no growth; 1, poor growth; 2, fair growth; 3, good growth. A, acid production relatively consistent; a, sporadic or transient production of acid. Dulcitol, α-melibiose and raffinose are not utilized.

tate and form new thalli directly. Zoospores have rarely been observed in this isolate.

Slight differences in carbohydrate utilization (Table 17.39).

Gelatin is weakly hydrolyzed; nitrate is not reduced.

Isolated from a soil sample from the Amargosa Desert in Nevada, U.S.A.

Type strain: ATCC 25081.

1c. *Geodermatophilus obscurus* subsp. *utahensis* Luedemann 1968, 1857.

u.tah.en'sis. M.L. adj. *utahensis* belonging to Utah, one of the United States.

This subspecies is distinct in several respects. It is a prolific producer of zoospores, which bud to produce motile chains of two to three cells; germinating zoospores often produce a well branched pseudomycelium (Plate 17.10).

Growth of colonies on yeast extract-casein hydrolysate-starch-dextrose agar fair, colony flat, texture granular. Growth on agar containing yeast extract and 5–10% glycerol is good but when the glycerol content is reduced to 2%, as in carbohydrate utilization tests, growth is poor.

Carbohydrate utilization and acid production often poor (Table 17.39).

Gelatin is not hydrolyzed; nitrate is reduced.

Isolated from a soil sample from Zion National Park, Utah, U.S.A.

Type strain: ATCC 25079.

1d. *Geodermatophilus obscurus* subsp. *dictyosporus* Luedemann 1968, 1858.

dic.ty.o.spo'rus. Gr. n. *dictyon* a net; Gr. fem.n. *spora* a seed; M.L. n. *dictyosporus* netted spore, referring to longitudinal and vertical septations.

Multilocular thalli vary in size, often small, superficially resembling dictyospores of fungi. Thalli formed directly from resting zoospores are small and tend to characterize this isolate (Plate 17.10). Often a zoospore appears to arise directly from a locule of a small thallus, indicating *in situ* germination of some spores. Thalli appear olive green to greenish black in transmitted light.

Growth on yeast extract-casein hydrolysate-starch-dextrose agar good; colony plicate, dark grayish brown. Growth on yeast extract-glycerol agar good; colony granular, plicate, strong brown. Colonies on some agar media in early stages of development have a salmon pink peripheral border which later turns dark brown or black.

Lactose and melezitose utilized (Table 17.39).

Gelatin hydrolyzed; nitrate reduction weak or negative.

Isolated from a soil sample from Westgard Pass, California, U.S.A.

Type strain: ATCC 25080.

FAMILY VI. **NOCARDIACEAE** CASTELLANI AND CHALMERS 1919, 1040

Norvel M. McClung

No.car'di.a'ce.ae. M.L. fem.n. *Nocardia* type genus of the family; -*aceae* ending to denote family; M.L. fem.pl.n. *Nocardiaceae* the Nocardia family.

Aerobic actinomycetes having cell walls of type IV *sensu* Lechevalier and Lechevalier (1965) consisting of *meso*-diaminopimelic acid, arabinose and galactose as distinguishing components. Gram-positive. Mycelium production may be rudimentary or extensive. It can be limited to the substrate or both aerial and substrate hyphae may be common. Spore production varies with the genus.

Key to the genera of family **Nocardiaceae**

I. Spores not produced on differentiated hyphae. Reproductive bodies are mycelial fragments, formed irregularly in substrate or aerial hyphae.

Genus I. *Nocardia*, p. 726

II. Spores produced on differentiated hyphae. Spores borne in chains. Cylindrical spores formed acropetally.

Genus II. *Pseudonocardia*, p. 746

Genus I. **Nocardia** *Trevisan 1889, 9; Nom. cons.* Opin. 13, Jud. Comm. 1954, 153: see Int. Code of Nom. 1958, 166 and note in Index Bergeyana p. 753

No.car'di.a. M.L. fem.n. *Nocardia* named for Professor Edmond Nocard, who first described the type species.

Microscopic morphology varies from forms producing sparse mycelium due to early division into coccoid and/or bacillary fragments, to those in which fragmentation is delayed, allowing abundant mycelial production (Plate 17.11, Figs. 1–3). Branched filaments may be found in most cultures but are not common in older cultures of some species. Aerial hyphae are formed by some organisms which may fragment irregularly into coccoid and bacillary cells similar to those produced by substrate mycelia. Members of the genus may be divided into **three morphological groups according to the degree of mycelial development.**

Group I. Members of this group have a limited amount of mycelial development due to the onset of fragmentation in the center of the colony after 12–14 hr incubation (Plate 17.11, Fig. 1). The colonial texture is soft, butyrous or mucoid.

Group II. In members of this group, fragmentation is delayed for 18–20 hr so that an extensively branched mycelium develops (Plate 17.11, Fig. 2). Aerial hyphae are not produced. Colonial texture tends to be flaky, pasty or doughy with a matt surface.

Group III. In members of this group, fragmentation is often delayed for several days allowing abundant mycelial development (Plate 17.11, Fig. 3). Aerial hyphae are commonly produced by members of this group, which may fragment into unequal coccoid and bacillary elements. The wall of the hypha remains as the wall of the fragments in the manner of arthrospore production in fungi; hence they are not considered spores *sensu stricto*, but mycelial fragments. Colonial texture of members of this group may be crusty, cartilaginous or leathery.

Characteristically, a colony starts by elongation of a coccoid or bacillary mycelial fragment. A branch begins to form after several hours incubation with secondary and tertiary branches following, though fragmentation usually precludes branching in Group I nocardias.

Pre- and post-fission bending can be characterized and related to the morphological groups (McClung, 1949). Fragmentation begins in the center of the young colony producing bacillary elements which continue to elongate and again fragment. The outermost bacillary elements in the center of the microcolony are surrounded by branching peripheral mycelium.

Coccoid cells which have been called **microcysts or chlamydospores** are produced by some members of Groups I and II. These structures are more heat resistant than vegetative cells, surviving 80 C for several hours.

Gram-positive. Some species **acid-fast to partially acid-fast.**

Obligate aerobes. Non-motile.

Pigments are produced by several species of *Nocardia;* carotenoid pigments are the most common and best known.

Generally resistant to lysozyme. Virulent phages are known for a few species; evidence for a lysogenic system has been reported for at least one species.

The G + C content of nocardial DNA varies from *ca* 60–72 moles %. Distribution appears discontinuous and can be roughly correlated with the morphological groups: Group I, 61–63% G + C; Group II, 66–68% G + C and Group III, 68–72% G + C.

Type species: *Nocardia farcinica* Trevisan 1889, 9.

Further Comments

Separation of Group I *Nocardia* from certain mycobacteria and corynebacteria presents problems as it cannot be done on the basis of cell wall composition. Generally the microscopic morphology of young colonies and cultural characteristics may be used for differentiation, but some cultures may be assigned to the genera *Corynebacterium*, *Mycobacterium* or *Nocardia* with equal validity according to currently accepted criteria. The determination of the type of mycolic acid present may be useful. The mycolic acids of the mycobacteria contain some 80 carbon atoms, those of the nocardias some 50 and those of the corynebacteria some 32 carbon atoms (Asselineau, 1966; Laneele *et al.*, 1969). Determination of the pyrolytic products of these α-branched, β-hydroxylated fatty acids by gas chromatography has been suggested as a means of distinguishing representatives of these three genera (Lechevalier *et al.*, 1971).

Key to the species of genus **Nocardia**

I. Extensive mycelium produced because fragmentation does not begin until after 5 days incubation. Colonial texture, leathery, dry, crusty or flaky. Thin aerial mycelium commonly present. Long, branched filaments persist in old cultures. Morphological Group III.

    A. Acid-fast or partially acid-fast.

        1. Substrate mycelium white, gray or buff.

                1. *Nocardia farcinica*

        2. Substrate mycelium some shade of yellow, orange or red.

            a. Pathogenic.

            b. Nitrates reduced to nitrites.

   c. Xanthine hydrolyzed.
     2. *Nocardia otitidis-caviarum*
  cc. Xanthine not hydrolyzed.
   d. Gelatin hydrolyzed.
     3. *Nocardia brasiliensis*
   dd. Gelatin not hydrolyzed.
     4. *Nocardia asteroides*
  bb. Nitrates not reduced to nitrites.
    5. *Nocardia transvalensis*
 aa. Non-pathogenic.
  b. Gelatin hydrolyzed.
   c. Nitrates reduced to nitrites.
    6. *Nocardia formicae*
  cc. Nitrates not reduced to nitrites.
    7. *Nocardia coeliaca*
  bb. Gelatin not hydrolyzed.
   c. Brownish exopigment on organic media.
    8. *Nocardia polychromogenes*
  cc. Exopigment not produced.
    9. *Nocardia paraffinae*
AA. Not acid-fast.
 1. Chemoautrophic.
  a. Substrate mycelium yellow, citron or ochre.
    10. *Nocardia petroleophila*
 aa. Substrate mycelium not as above.
  b. Colonies with raised center surrounded by a wide margin, suggesting Saturn and her rings.
    11. *Nocardia saturnea*
 2. Heterotrophic.
    12. *Nocardia kuroishii*
II. Mycelial development limited due to fragmentation beginning after 20 hr of incubation. Colonial texture soft, pasty or butyrous. Hyphal fragments up to 15–20 μm long persist in mature cultures. Microcysts 0.8–1.5 μm diameter common in older cultures. Morphological Group II.
 A. Colonies white, cream or buff.
    13. *Nocardia rugosa*
 AA. Colonies some shade of yellow, pink or red.
  1. Gelatin hydrolyzed.
    14. *Nocardia rhodnii*
  2. Gelatin not hydrolyzed.
   a. Lysozyme resistant.
    b. Penicillin sensitive.
     c. Acid from L-arabinose.
      15. *Nocardia vaccinii*
     cc. No acid from L-arabinose.
      16. *Nocardia minima*
    bb. Penicillin resistant.
     c. Urea hydrolyzed.
      17. *Nocardia blackwellii*
     cc. Urea not hydrolyzed.
      18. *Nocardia convoluta*
   aa. Lysozyme sensitive.
    b. DNA hydrolyzed.

    c. Casein hydrolyzed.
        19. *Nocardia cellulans*
   cc. Casein not hydrolyzed.
      d. Nitrates reduced to nitrites.
        20. *Nocardia lutea*
     dd. Nitrates not reduced to nitrites.
        e. Blue crystal of indigotin from indole agar.
          21. *Nocardia globerula*
        ee. Not as above.
          22. *Nocardia rubropertincta*
  bb. DNA not hydrolyzed.
    c. Dextrin utilized as carbon and energy source.
      d. Nitrates reduced to nitrites.
        e. Phenol utilized as carbon and energy source.
          23. *Nocardia corallina*
        ee. Phenol not utilized as carbon and energy source.
          24. *Nocardia salmonicolor*
     dd. Nitrates not reduced to nitrites.
        25. *Nocardia rubra*
   cc. Dextrin not utilized as carbon and energy source.
        26. *Nocardia opaca*

III. Mycelium production extremely limited due to the early (12–14 hr) initiation of fragmentation. Branching is also sparse causing an ephemeral mycelium which soon fragments into unbranched hyphal elements. Aerial mycelium lacking. Colonial texture soft, pasty and sometimes mucoid. Short bacillary and coccoid cells predominate in older cultures. Morphological Group I.
  A. Partially acid-fast.
    1. Casein hydrolyzed.
        27. *Nocardia calcarea*
    2. Casein not hydrolyzed.
      a. Nitrates reduced to nitrites.
        28. *Nocardia restricta*
     aa. Nitrates not reduced to nitrites.
        29. *Nocardia erythropolis*
  AA. Not acid-fast.
    1. Isolated from marine environments.
      a. Produce acid from lactose. Lemon-yellow colonies.
        30. *Nocardia marina*
     aa. Do not produce acid from lactose. Orange-yellow colonies.
        31. *Nocardia atlantica*

Description of the species of genus **Nocardia**

1. Nocardia farcinica Trevisan 1889, 9. *Nom. cons.* Opin. 13 Jud. Comm. 1954, 153: see Int. Code of Nom. 1958, 166 and note in *Index Bergeyana* p. 753. (*Streptotrix farcinica* (*sic*) (Trevisan) Rossi-Doria 1891, 424; *Bacillus farcinicus* (Trevisan) Gasperini 1892, 183; *Actinomyces farcinicus* (Trevisan) Gasperini 1892, 222; *Oospora farcinica* (Trevisan) Sauvageau and Radais 1892, 248; *Actinomyces bovis farcinicus* Gasperini 1894, 684; *Streptothrix farcini bovis* Kitt 1899, 511; *Bacterium nocardi* Migula 1900, 345; *Streptothrix nocardii* (Migula) Foulerton and Jones 1901, 55; *Clado-*

thrix farcinica (Trevisan) Macé 1901. 1092; *Bacillus nocardi* (Migula) Matzuschita 1902, 524; *Discomyces farcinicus* (Trevisan) Krasil'nikov 1941, 88.)

    far.ci'ni.ca. L. v. *farcio* stuff; L. n. *farcinimum* a disease of horses; Fr. n. *farcin* farcy or glanders; M.L. fem.adj. *farcinica* relating to farcy.

    (This description is based on ATCC strain 3318.)

    Microcolonies formed from bacillary and coccoid cells by elongation and germ-tube formation. Primary branching begins after some 24 hr of incubation. An extensive mycelium is formed as fragmentation does not start until cultures are 3–5 days old. Sparse, grayish, aerial mycelium appears on most media as cultures age. In older cultures the products of fragmentation consist of hyphal elements, bacilli and coccoids. Long, branched, unfragmented hyphae persist. Morphological Group III. Acid-fast.

Slow growing, flat, crusty, grayish colonies on nutrient agar covered by a thin, whitish, aerial mycelium.

On nutrient gelatin small, flaky, irregular colonies having a granular appearance. No liquefaction.

Irregular, grayish colonies on glucose mineral salts medium; few aerial hyphae at borders.

Grayish yellow, dry, crumbly leathery growth on potato plug. Aerial hyphae at top of slant.

Acid produced from D-fructose, D-glucose and mannose. Esculin, Tweens 20, 40 and urea are hydrolyzed. The following compounds serve as sole energy and carbon sources: D-fructose, D-glucose, maltose, mannose, paraffin, sodium acetate, sodium butyrate, sodium H-malate, sodium propionate, sodium pyruvate, sodium succinate and testosterone.

Nitrate reduction variable.

Litmus milk alkaline after 5 days; no further change.

Temperature range: 20–40 C. Optimum: 35 C. pH range: 6–10. Optimum: 7.5.

Not inhibited by 5% NaCl. Not sensitive to penicillin (5 i.u. discs). Lysozyme resistant.

Source: Presumably from a case of bovine farcy.

The G + C content of the DNA is 71.0 moles %.

Reference strain: ATCC 3318 (Sneath and Skerman 1966, 86).

Further Comments

Unfortunately, the epithet *farcinica* remains as that of the type species of the genus. According to Gordon and Mihm (1957, 21), this strain (ATCC 3318) is among those labeled *N. farcinica*, which may not be typical of the species due to a misinterpretation of the description of the original *N. farcinica* culture. They recommended the reduction of this epithet to synonymy with *N. asteroides*, a better known and more commonly isolated species. We support this recommendation. Tsukamura (1969) placed *N. asteroides* and *N. farcinica* in different groups. Members of the latter group were positive for: growth at 45 C, acetamidase, benzamidase, urease, 1,3 butylene glycol as carbon source and acid production from rhamnose. Members of the former were negative in these reactions. Goodfellow's results do not agree. He placed ATCC 3318 along with three other isolates la-

beled *N. farcinica* in cluster 1, the *N. asteroides* cluster, subgroup 1C (1971, 41).

Editorial Note. The status of *Nocardia farcinica* is confused. A type strain probably does not exist. Some strains labeled *N. farcinica* contain nocardomycolic acids (ATCC 3318), others mycolic acids (NCTC 4524; CIP 378; ATCC 13781) and therefore belong in the genus *Mycobacterium* (Lechevalier *et al.*, 1971; Chamoiseau and Asselineau, 1970). NCTC 4524 is supposedly related to ATCC 3318 which is considered by some to be *N. asteroides*. *N. farcinica* has therefore been considered a *nomen dubium* by Lechevalier *et al.* (1971) and Ridell and Norlin (1973).

However, it must be remembered that *Nocardia* is a conserved generic name with *N. farcinica* as the type species (Opin. 13, Jud. Comm. 1954, 153), and that *Nocardia* Trevisan 1889 has precedence over *Mycobacterium* Lehmann and Neumann 1896. The problem can be settled only by action of the Judicial Commission and until this is done *N. farcinica* must remain as the type species of the genus.

2. **Nocardia otitidis-caviarum** Snijders 1924, LXXXVII. (*Actinomyces caviae* Erikson 1935, 10; *Nocardia caviae* (Erikson) Erikson 1935, 10.)

o.ti'ti.dis-cav.i.ar'um. M.L. n. *otitis* inflammation of the ear; M.L. n. *Cavia* (gen. pl. *caviarum*) generic name of the cavy or guinea pig; M.L. gen.n. *otitidis-caviarum* of ear disease of guinea pigs.

Coccoid and bacillary mycelial fragments produce extensively branched microcolonies. Fragmentation is delayed for 5 days, allowing the development of abundant substrate and aerial mycelium. The latter is branched and sometimes coiled. Bacillary fragments are produced in the center of the microcolony and cell division proceeds at an accelerated rate as the colony ages. Both aerial and substrate mycelium fragment into smaller and smaller hyphal segments, some of which become coccoids in older cultures. Morphological Group III.

Acid-fast on certain media such as Dubos oleic acid agar.

Colonies on nutrient agar are raised, convoluted, irregular and crusty; dull buff-pink in color. Thin, whitish, aerial mycelium in older cultures.

Scant growth on nutrient gelatin; no liquefaction.

Raised, convoluted, granular, irregular growth on potato plug. Grayed substrate growth covered with whitish, aerial hyphae.

Acid produced from arbutin, D-fructose, D-glucose, glycerol, inositol, maltose, mannitol and trehalose.

The following serve as sole carbon sources:

D-fructose, D-glucose, glycerol, maltose, mannitol, paraffin, sodium acetate, sodium butyrate, sodium propionate, sodium pyruvate, testosterone and trehalose.

Hypoxanthine, xanthine and urea are hydrolyzed.

Litmus milk alkaline, coagulated. Nitrate reduced.

Temperature range: 20–45 C. Optimum: 30 C. Completely inhibited at 50 C. pH range: 7–10. Optimum: 7.5.

Not inhibited by 7% NaCl. Not sensitive to penicillin (5 i.u. discs). Lysozyme resistant.

Produced 100% mortality in white mice within 7 days, when gastric mucin was used as an adjuvant. Rabbits and guinea pigs usually survive, but develop lesions according to the route of inoculation.

Isolated from ear infection of a guinea pig in Sumatra and from soils in the United States and India.

The G + C content of the DNA is 65.4 moles %.

Reference strain: ATCC 14629 (Sneath and Skerman 1966, 84).

Further Comments

This organism is related to *N. asteroides* and has been assigned to cluster 2, subgroup B by Goodfellow (1971, 47). The degree of relatedness shown by DNA hybridization experiments between a strain of *N. otitidis-caviarum* and a strain of *N. asteroides* was 73%; while under the same conditions, two strains of *N. asteroides* were 88% related (unpublished data; Franklin and McClung, 1972).

3. **Nocardia brasiliensis** (Lindenberg) Pinoy 1913, 936. (*Discomyces brasiliensis* Lindenberg 1909, 279; *Streptothrix brasiliensis* (Lindenberg) Greco 1916, 724; *Oospora brasiliensis* (Lindenberg) Sartory 1920, 786; *Actinomyces brasiliensis* (Lindenberg) Gomes 1923, 154; *Actinomyces violaceus* subsp. *brasiliensis* Krasil'nikov 1941, 16.)

bra.si.li.en'sis. M.L. adj. *brasiliensis* pertaining to Brazil, South America.

Branched filaments forming extensive mycelium on most media; about 1 μm in diameter. Fragmentation begins in center of colony after about 4 days incubation, and produces irregular bacillary and coccoid cells. Substrate mycelium yellow-orange to brown with thin, white aerial hyphae on the surface. Colonies granular, heaped, convoluted. Yellow-brown exopigment on most organic media. Morphological Group III. Acid-fast.

Colonies on nutrient agar orange-buff, granular, heaped with scant white aerial hyphae at edges.

On glycerol nitrate agar heaped, coral colonies with scant aerial mycelium at margins.

On yeast extract-glucose agar colonies yellow to orange to tan, raised, wrinkled. White, aerial hyphal production variable from none to abundant, depending on strain. Yellow soluble pigment produced in old cultures.

Orange-tan, dry, granulated growth with scant white aerial mycelium on potato plug.

Acid produced from arbutin, D-arabinose, D-fructose, D-galactose, DL-inositol, glucose, glycerol, mannitol, mannose, sorbitol and trehalose.

Esculin, casein, gelatin, guanine, hypoxanthine, keratin, tyrosine and urea are hydrolyzed.

The following serve as sole carbon and energy sources: acetate, butyrate, citrate, D-fructose, D-galactose, glucose, inositol, malate, maltose, mannitol, mannose, paraffin, L-proline, propionate, pyruvate, sebacic acid, sorbitol and succinate.

Slow coagulation of litmus milk followed by peptonization; alkaline; yellow-orange pellicle.

Nitrate reduced.

Temperature range: 10–45 C. No growth at 50 C. pH range: 6–9. Grows well in 5% NaCl. Completely inhibited in 7% NaCl.

Not sensitive to penicillin (5 i.u. discs). Lysozyme resistant.

Produces multilobed granules in white mice, guinea pigs and rabbits.

Isolated from various kinds of nocardiosis including mycetoma. We find this organism more commonly implicated in nocardiosis in the tropics than in temperate regions.

The G + C content of the DNA ranges from 67–68 moles %.

Reference strain: ATCC 19296 (Sneath and Skerman 1966, 84).

Further Comments

N. brasiliensis resembles *N. asteroides* in many ways. These species can be distinguished in that the latter does not decompose casein or tyrosine, or produce acid from inositol or mannitol while the former does (Gordon and Mihm, 1959). Bojalil and Cerbón (1959) separated the two species by showing that *N. brasiliensis* produces spherical colonies on the sides of tubes of 0.4% gelatin medium and a positive ninhydrin reaction while *N. asteroides* does not. Cluster 5 of Goodfellow (1971, 48).

4. **Nocardia asteroides** (Eppinger) Blanchard 1896, 856. (*Cladothrix asteroides* Eppinger 1891, 309; *Streptotrix eppingerii* (sic) Rossi-Doria 1891, 423; *Streptotrix asteroides* (sic) (Eppinger) Gasperini 1892, 183; *Oospora asteroides* (Eppinger) Sauvageau and Radais 1892, 252; *Actinomyces asteroides* (Eppinger) Gasperini 1894, 86; *Actinomyces eppingeri* (Rossi-Doria) Berestnev 1897, 142; *Discomyces asteroides* (Eppinger) Gedoelst 1902, 173; *Actinomyces eppinger* (Rossi-Doria) Namyslowski 1912, 566; *Asteroides asteroides* (Eppinger) Puntoni and

Leonardi 1935, 92; *Proactinomyces asteroides* (Eppinger) Baldacci 1937, 141.)

as.ter.o.i'des. Gr. adj. *asteroides* starlike.

Growth begins after about 10 hr of incubation by elongation and germ tube formation from bacillary and coccoid elements. Growth is slow and primary branching begins after about 24 hr incubation. There is extensive mycelial development because fragmentation does not begin until after 4 days of incubation. Filaments begin to break up in the center of microcolonies while peripheral hyphae continue to grow and branch. Aerial mycelium production varies according to strain of organism and cultural conditions. In older cultures, bacilli and coccoid cells are mixed with hyphal elements. Long branched hyphae persist. Morphological Group III. Acid-fast.

On nutrient agar raised, heaped, folded, granular colonies with irregular borders. Most strains some shade of yellow-orange. A thin aerial mycelium usually produced along the margins of colonies.

Thin, flaky colonies on nutrient gelatin; no liquefaction.

Thin, flaky, irregular, yellow-orange colonies on glucose mineral salts.

On glucose-yeast extract agar heaped, folded, irregular colonies becoming deep orange-red in color and covered with white aerial mycelium.

Acid produced from adonitol, arbutin, dextrin, D-fructose, D-glucose and mannose.

Esculin, allantoin, benzidine, Tweens 20, 40, 60 and urea are hydrolyzed.

The following serve as sole energy and carbon sources: adipic acid, D-fructose, D-glucose, glycerol, maltose, mannitol, mannose, paraffin, sebacic acid, sodium acetate, sodium butyrate, sodium H-malate, sodium propionate, sodium pyruvate, sodium succinate and testosterone.

Litmus milk alkaline in 1 week; no further change.

Nitrate reduced.

Temperature range: 10–50 C. Optimum: 28–30 C. pH range: 6–10. Optimum: 7.5.

Not inhibited by 7% salt. Not sensitive to penicillin (5 i.u. discs). Lysozyme resistant.

Isolates of this organism differ markedly in pathogenicity for mice. Most strains require an adjuvant before tissue invasion occurs. However, some strains kill mice by acute intoxication in 1 or 2 days. Rabbits and guinea pigs are more susceptible than mice.

Isolated from various lesions in humans and other animals; also from soils.

The G + C content of the DNA ranges from 67.0–69.4 moles %.

Reference strain: ATCC 19247 (Sneath and Skerman 1966, 86).

Further Comments

This taxon is heterogeneous and has been shown to include members differing in microscopic colonial morphology (McClung, 1954), carbon compound utilization (McClung, 1954) and pathogenicity for mice (Uesaka *et al.*, 1941). Tsukamura (1969) divided strains received as *N. asteroides* into two groups, and Goodfellow (1971, 46) divided 42 strains into five subgroups. In a study of nine strains of *N. asteroides* differing in physiological characteristics and pathogenicity for mice, Franklin and McClung (unpublished data) found a high degree of relatedness in DNA hybridization experiments (85–96%).

5. **Nocardia transvalensis** Pijper and Pullinger 1927, 155. (*Actinomyces transvalensis* (Pijper and Pullinger) Nannizzi 1934, 36; *Proactinomyces transvalensis* (Pijper and Pullinger) Krasil'nikov 1941, 67.)

trans.va.len'sis. M.L. adj. *transvalensis* pertaining to the Transvaal, South Africa.

Extensive substrate mycelium and aerial mycelium produced from germinating bacillary elements. Fragmentation is delayed for 5 or more days, thus microcolonies consisting of elaborate widespread mycelium result. The substrate stroma is some shade of pink to orange on most media and the aerial mycelium is white. The substrate hyphae fragment into shorter and shorter rods as does the aerial mycelium. Colorless exudate droplets formed on most media. Morphological Group III. Acid-fast.

Poor growth on nutrient agar; small pinkish colonies become covered with white aerial hyphae with age.

Scant growth on nutrient gelatin; no liquefaction.

Small, convoluted, colonies on glycerol nitrate agar; coral with scant white aerial hyphae.

On glucose peptone agar small, raised salmon colonies with a thin covering of white aerial hyphae.

Poor growth on starch agar; no hydrolysis.

On potato plug raised, heaped, convoluted, salmon-colored growth, covered with white aerial mycelium.

Litmus milk alkaline; no coagulation or peptonization.

Nitrate not reduced.

Optimum temperature: 28–30 C. Optimum pH: 7.5.

Isolated from mycetoma of the foot in South Africa.

The G + C content of the DNA is 69.0 moles %.

Reference strain: None designated.

Further Comments

This organism was considered a synonym of *N. brasiliensis* by Gonzales-Ochoa and Sandoval (1956, 17). It did not fit any of the species groups of Tsukamura (1969, 275), and could not be assigned to any cluster group by Goodfellow (1971, 54).

6. **Nocardia formicae** Harris and Woodruff 1953, 609. (*Streptomyces griseus* subsp. *formicus* Pridham and Lyon 1969, 219.)

for.mi'cae. M.L. fem.n. *Formica* generic name of the ant; M.L. gen.n. *formicae* of *Formica*.

Extensive mycelial development with fragmentation delayed for more than 5 days incubation. Filaments long and widely separated with cytoplasmic condensations irregularly produced. In submerged culture, branched rods up to 10 μm long. Morphological Group III. Not acid-fast.

Rugose, flat, flaky colonies on nutrient agar; no aerial mycelium.

Scant to no growth on sucrose nitrate agar.

Colonies tan, flat on glucose asparagine agar covered with a grayish white aerial mycelium.

Grayish tan colonies on glucose peptone agar gradually covered with thin, white aerial mycelium. Reverse brown; produces a brownish soluble pigment on this medium.

Scant to no growth on potato plug.

Acid produced from glucose, glycerol, lactose and maltose. Starch hydrolyzed.

Paraffin: not utilized.

Gelatin rapidly liquefied; no soluble pigment.

Nitrate reduced.

Litmus milk alkaline reaction; peptonization.

Produces noformicin, an anti-*Trichomonas* substance which is also active against swine influenza virus.

Optimum temperature: 28 C. Good growth at 40 C. pH range: 6.0-8.0.

Isolated from abandoned African ant nest in a log of imported mahogany.

Type strain: NRRL 2470.

Further Comments

This organism probably represents an asporogenous strain of *Streptomyces*. Proper designation depends on further study, especially determination of cell wall composition.

7. **Nocardia coeliaca** (Gray and Thornton) Waksman and Henrici 1948, 906. (*Mycobacterium coeliacum* Gray and Thornton 1928, 88; *Flavobacterium coeliacum* (Gray and Thornton) Bergey *et al.* 1930, 156; *Proactinomyces coeliacus* (Gray and Thornton) Reed 1939, 836; *Mycobacterium flavum* var. *coeliacum* Krasil'nikov 1941, 115.)

coe.li'a.ca. Gr. adj. *coeliacus* suffering in the bowels; L. fem.adj. *coeliaca* relating to the bowels.

Bacilli and coccoids germinate by elongation and germ tube formation. An extensive substrate mycelium is formed. A white aerial mycelium covers the stroma on most media. Fragmentation begins after about 5 days incubation in both subsurface and aerial mycelium producing rods which become shorter and shorter. Coccoid microcysts produced in older cultures. Morphological Group III. Acid-fast.

Colonies on nutrient agar small, raised, rugose with irregular margins; white to buff in color.

Raised, irregular growth on nutrient gelatin with a matt surface; buff-white to yellowish in color; liquefaction.

Raised, rugose, convoluted, crumbly growth on potato glycerol agar, cream to buff to orange to brown as culture ages.

Acid produced from D-fructose, D-galactose, ethanol, cellobiose, glycerol, L-arabinose, L-rhamnose, mannitol, mannose, sucrose and xylose.

The following are utilized as sole carbon and energy sources: D-glucose, glycerol, maltose, mannose, paraffin, phenol, sebacic acid, sodium acetate, sodium butyrate, sodium H-malate, sodium malonate, sodium octoate, sodium propionate, sodium pyruvate and testosterone.

Adenine, allantoin, casein, gelatin, guanine, hypoxanthine, tyrosine, xanthine and urea are hydrolyzed.

Litmus milk alkaline, no coagulation.

Nitrate not reduced.

Optimum temperature: 25-28 C. Optimum pH range: 7.6-8.0.

Penicillin sensitive (5 i.u. discs). Lysozyme sensitive.

Isolated from soils in Great Britain and Australia.

The G + C content of the DNA is 63 moles %.

Reference strain: ATCC 13181 (Sneath and Skerman 1966, 86).

Further Comments

This organism was assigned to minor cluster 12 by Goodfellow (1971, 52). It exhibits KDPG-aldolase activity according to Kersters (1970, 304).

8. **Nocardia polychromogenes** (Vallée) Waksman and Henrici 1948, 897. (*Streptothrix polychromogene* (*sic*) Vallée 1903, 288; *Streptothrix polychromogenes* Vallée 1903, 292; *Streptothrix pluricromogena* (*sic*) Caminita 1907, 198; *Oospora polychromogenes* (Vallée) Sartory 1920, 822; *Actinomyces polychromogenes* (Vallée) Lieske 1921, 32; *Proactinomyces polychromogenes* (Vallée) Jensen 1931, 363; *Actinomyces plurichromogenus* (*sic*) Dodge 1935, 737; *Actinomyces madurae* subsp. *plurichromogenus* Krasil'nikov 1941, 20.)

po.ly.chro.mo'ge.nes. Gr. adj. *poly* many; Gr. n.

chromus color; Gr. v. *gennaio* produce; M.L. adj. *polychromogenes* producing many colors.

Growth begins after about 14 hr of incubation by elongation of rods 5–15 by 0.8 µm. Growth is slow and primary branches do not form until 24 or more hr incubation. Secondary branching is also slow and occurs some 14 hr after primary branches are produced. After 40 hr incubation, autolysis destroys part of the original hypha. Branches developed from it continue to grow and proliferate. Growth is slow and branches tend to be long and widely spaced. Fragmentation begins after 5 days of incubation resulting in bacillary and coccoid cells. Scant aerial mycelium covers stroma on some media. An intracellular pigment (coral pink, Ridgway Plate XIII d) and a brownish exopigment produced. Morphological Group III. Partially acid-fast.

Colonies appear on nutrient agar after 3 days as small, raised, rough, dark orange colored structures; they grow slowly and after 1 week are raised, rough, crumbly, heaped and a brown-orange color.

Small, raised, crumbly colonies on nutrient gelatin; no liquefaction.

Growth slow on glucose mineral salts agar; after 5 days raised, rough, folded, brown-orange in color.

Growth slow on potato plug; heaped, warty, crumbly, red-brown after 2 weeks incubation.

Acid produced from adonitol, arbutin, dextrin, D-fructose, D-glucose, glycerol and mannose.

Esculin, benzidine, Tweens 20, 40, 60 and urea are hydrolyzed.

The following compounds serve as sole carbon and energy sources: adipic acid, D-fructose, D-glucose, glycerol, maltose, mannose, mannitol, paraffin, sebacic acid, sodium acetate, sodium butyrate, sodium H-malate, sodium succinate, sodium pyruvate and testosterone.

Litmus milk slowly becomes alkaline; no coagulation; slow peptonization (after 2 weeks).

Nitrate reduced.

Temperature range: 20–45 C. Optimum: 25 C. pH range: 7–10. Optimum: 7.5.

Not inhibited by 7% NaCl, 0.01% sodium azide, 0.1% potassium tellurite or 0.01% tetrazolium. Not sensitive to penicillin (5 i.u. discs). Lysozyme resistant.

Isolated from the blood of a horse, and soils in France and Australia.

Reference strain: None designated.

Further Comments

This organism was assigned to subgroup 1A of the *N. asteroides* cluster by Goodfellow (1971, 46).

9. **Nocardia paraffinae** (Jensen) Waksman and Henrici 1948, 901. (*Proactinomyces paraffinae* Jensen 1931, 362.)

pa.raf.fi′nae. M.L. n. *paraffina* paraffin; M.L. gen.n. *paraffinae* of paraffin.

Microcolonies develop from coccoids and bacillary elements by elongation and germ tube formation. Branching is extensive as the young colony develops. Fragmentation begins in the center of the young colony after 36 hr incubation, at first producing long branched hyphae, which become shorter and shorter as the culture ages. After 5–6 days peripheral vegetative hyphae may produce chains of spore-like bodies 0.8–1.0 by 1.2–1.5 µm. Scant aerial hyphae produced on some media. Morphological Group III. Partially acid-fast.

On nutrient agar raised, irregular, crusty colonies having a brownish yellow color. Aerial mycelium scant and whitish.

Colonies on glucose mineral salts agar granular, convoluted, irregular of a yellowish orange color; crumbly texture.

Granular, crumbly raised growth on potato plug with scant white aerial mycelium produced.

Acid produced from arbutin, dextrin, D-fructose, D-glucose, glycerol, maltose and mannose.

The following serve as sole carbon sources: D-fructose, D-glucose, glycerol, mannose, paraffin, sodium acetate, sodium benzoate, sodium butyrate, sodium propionate, sodium pyruvate and sodium succinate.

Esculin, Tweens 20, 40, 60 and urea hydrolyzed.

Litmus milk alkaline, no coagulation or peptonization.

Nitrate reduced.

Temperature range: 10–40 C. Optimum pH: 7.4.

Not inhibited by 6% NaCl. Not sensitive to penicillin (5 i.u. discs). Lysozyme resistant.

Found in soils.

The G + C content of the DNA is 68.5 moles %.

Reference strain: Not designated.

Further Comments

This organism was assigned to cluster 1, subgroup D of the *N. asteroides* cluster by Goodfellow (1971, 47).

10. **Nocardia petroleophila** Hirsch and Engel 1956, 445.

pe.tro′le.oph′ila. Gr. n. *petro* stone; L. n. *oleum* oil; Gr. adj. *philus* loving; M.L. adj. *petroleophila* loving stone oil, petroleum.

Description from Hirsch and Engel, 1956, and Hirsch, 1958.

Growth slow but abundant on all mineral salts media without a carbon source. Growth rate increases with a petroleum atmosphere. The organism will also grow on certain organic media, but without aerial mycelium production and with rapid fragmentation of substrate mycelium into bacillary elements. Filaments long and monopodically branched from 0.6–1.2 μm in diameter. Substrate mycelium yellowish, citron or ochre-yellow depending on the carbon or nitrogen source. Substrate mycelium fragments into bacillary elements under both autotrophic and heterotrophic growth conditions, but much more rapidly under the latter. Snow-white aerial mycelium is abundantly produced on mineral media, which breaks up into spore-like fragments with age. Morphological Group II on organic media. Morphological Group III on mineral media. Not acid-fast.

Slow but abundant growth on mineral salts agar; colonies up to 5 mm diameter, covered with snow-white aerial hyphae; under surface of colony butter yellow. Mineral salts agar with 0.1% formate, oxalate or acetate: growth sparse on formate, slow on acetate, and at the same rate as with mineral medium on acetate. Colonies snow-white.

Poor growth on nutrient gelatin, no liquefaction.

Poor growth on nutrient agar, whitish yellow colonies up to 0.5 mm diameter, reverse of colonies yellow, white aerial mycelium.

Sparse growth on glucose asparagine agar, colonies up to 0.3 mm diameter, reverse of colonies yellow, white aerial hyphae.

No growth on potato plug.

Growth slight on starch nitrate agar; colonies up to 0.5 mm diameter with snow-white aerial mycelium and pale yellow underside. Starch not hydrolyzed.

Good growth with nitrate; nitrate not reduced. Only slight growth with nitrite and ammonia.

Litmus milk not coagulated or peptonized; pH unchanged.

Not pathogenic for white mice. No effect after an intraperitoneal injection of 1 ml of a suspension in 3 months.

Optimum temperature between 25–28 C. Slight growth at 37 C, no growth at 50 C. pH range: 6.8–8.3.

Not inhibited by 10% NaCl.

Isolated from soils from Germany.

Type culture: ATCC 15777 (Hirsch and Engel 1956, 450).

Further Comments

This organism is one of two chemoautotrophs described for the genus. Their ability to utilize carbon dioxide from the air as both a carbon and energy source was demonstrated by Hirsch (1958, 373) using $^{14}CO_2$.

11. **Nocardia saturnea** Hirsch 1960, 401.

sa.turn′e.a. L. n. *Saturnus* lit. the sower, the planet Saturn; M.L. adj. *saturnea* pertaining to Saturn.

Description from Hirsch, 1960.

Richly branched substrate mycelium produced. Secondary branches appear to be associated with special cytoplasmic granules. The filaments fragment into rods, especially on organic media. The surface of the colony is covered with monodial, branched, snow-white mycelium which later becomes yellowish. Chains of 5–10 "segmentation spores" are produced in the aerial hyphae. Morphological Group III. Not acid-fast.

Slow, but good growth on silica gel plates. Colonies up to 4 mm in diameter with a raised central cushion surrounded by an extended margin in the form of Saturn and its rings. The ring can become higher and wider than the middle part of the colony.

Sparse to poor growth on agars.

On starch nitrate agar small colonies with underside yellowish and covered with white aerial mycelium; starch not hydrolyzed.

Small colonies on beef extract-peptone agar, bright yellow to brownish without aerial hyphae.

On Kligler-H_2S agar substrate mycelium yellowish; no aerial mycelium; H_2S not produced.

No growth on gelatin, nutrient gelatin, litmus milk, malt extract agar, potato or carrot plug.

Nitrate not reduced.

Optimum temperature between 28–30 C.

Isolated from air and compost in Germany.

Type culture: ATCC 15809 (Hirsch 1960, 404).

Further Comments

This chemoautotroph is closely related to *N. petroleophila* from which it is separated by its Saturn-shaped colonies on mineral media, its hyphal morphology and its differences in cultural behavior. This organism could not be assigned to any cluster by Goodfellow (1971, 54).

12. **Nocardia kuroishii** Uesaka 1952, 76.

ku.ro.i′shi.i. Japanese adj. *kuro* black; Japanese n. *ishi* stone; Kuroishi name of village in Kōchi prefecture, Japan; M.L. gen.n. *kuroishii* of Kuroishi.

Rather extensive substrate mycelium produced which soon becomes covered with an abundant aerial hyphae. After 5 days of incubation, the substrate mycelium fragments into bacillary elements followed by a similar process in the aerial hyphae. A brownish stroma produced on most media. Morphological Group III. Partially acid-fast.

Good growth on nutrient agar; yellowish at first then becoming red-brown with age; colonies flat, thin, leathery with thin, white, aerial hyphae as margins.

On nutrient gelatin yellowish brown colonies,

flat, flaky with no aerial mycelium; no liquefaction; a yellowish soluble pigment diffuses throughout medium.

Colonies on glycerol nitrate agar thin, pale yellow, flat, crusty; white aerial mycelium in irregular groups; yellowish soluble pigment produced.

Good growth on glucose peptone agar; yellowish, becoming reddish brown; colonies flat, irregular and crusty with thin, white, aerial hyphae at margins; brown-red soluble pigment diffuses into agar.

Fair growth on potato plug; reddish brown becoming dark brown; brown soluble pigment; white aerial mycelium covers stroma.

Good growth on starch agar; colonies reddish brown with white aerial mycelium produced at margins.

The following serve as sole carbon and energy sources: fructose, glucose, glycerol, lactose, mannitol, mannose and sucrose.

Litmus milk alkaline, brownish pigment; slow peptonization.

Nitrate not reduced.

Optimum temperature: 25–30 C. Optimum pH: 6.8.

Produces an antibiotic substance, neonocardin. Isolated from soil.

Reference strain: None designated.

13. **Nocardia rugosa** DiMarco and Spalla 1957, 28.

ru.go′sa. L. adj. *rugosa* full of wrinkles.

Microcolonies consist of a rather extensively branched mycelium which begins to fragment beginning in the center of the colony into short branched hyphae after 24 hr incubation. The fragmenting central hyphae become shorter and shorter as the culture ages. Microcysts are found in older cultures. Aerial mycellum not produced. Morphological Group II. Not acid-fast.

Raised cream-colored colonies on nutrient agar which are rough, convoluted and have a doughy consistency; brownish soluble pigment after 2 weeks incubation.

Tannish, heaped, folded colonies with irregular margins on nutrient gelatin; gelatin is liquefied.

Colonies on glucose asparagine agar creamy, raised, wrinkled with a moist surface and irregular margins.

Heavily convoluted, irregular, heaped, cream-colored colonies on glycerol peptone agar; a brown soluble pigment produced in older cultures.

Smooth, folded tannish cream colonies on potato agar with pasty consistency and irregular margins.

Acid produced from adonitol, L-arabinose, galactose, glucose, glycerol, D-mannitol, rhamnose and ribose.

Litmus milk alkaline; coagulation; no peptonization.

Optimum temperature: 34 C. Survives a temperature of 60 C for 1½ hr, but not for 3 hr. Optimum pH: 7.4.

Isolated from cattle rumen.

Reference strain: None designated.

Further Comments

This organism is said to produce vitamin B_{12} (DiMarco and Spalla, 1957). Goodfellow (1971, 54) used the name for strains which could not be assigned to any of his clusters.

14. **Nocardia rhodnii** (Erikson) Waksman and Henrici 1948, 914. (*Actinomyces rhodnii* Erikson 1935, 37, *Streptomyces rhodnii* (Erikson) Pridham *et al.* 1958, 58.)

rhod′ni.i. M.L. masc.n. *Rhodnius* generic name of the reduvid bug; M.L. gen.n. *rhodnii* of *Rhodnius*.

Microcolonies consisting of an extensively branched mycelium. Fragmentation begins after 24 hr incubation in the central part of a small colony. Shorter and shorter branched bacillary elements are formed as the culture ages. Scant, short, aerial hyphae produced. The center of microcolonies becomes filled with bacillary and coccoid elements while the peripheral hyphae continue to elongate and branch. Microcysts are produced in older cultures. Morphological Group II. Not acid-fast.

Colonies on nutrient agar convoluted, salmon colored having a granular texture and irregular margins; scant, whitish, aerial mycelium is present.

Raised, round, pale pink colonies on nutrient gelatin; rapid liquefaction.

Abundant, coral-pink, heaped, convoluted growth on glucose peptone agar.

Heaped, irregular, salmon-colored mass produced on potato plug with scant, whitish, aerial mycelium as drying occurs.

Acid produced from ethanol, D-fructose, maltose, mannitol, mannose, sorbitol, sucrose and trehalose.

The following are used as sole carbon and energy sources: adipic acid, H-malate, *n*-butyrate, *n*-octoate, propionate, sebacic acid and tartrate.

Orange pellicle on litmus milk, becomes alkaline with age, no coagulation or peptonization.

Nitrate reduced.

Optimum temperature: 28 C. Penicillin sensitive. Lysozyme sensitive.

Isolated from the reduvid bug, *Rhodnius prolixus*.

Reference strain: None designated.

Further Comments

This organism is similar to some strains of *N. asteroides*. It was assigned to cluster 3 by Goodfellow (1971, 47).

15. Nocardia vaccinii Demaree and Smith 1952, 251.

vac.cin′i.i. M.L. n. *Vaccinium* generic name of the blueberry; M.L. gen.n. *vaccinii* of *Vaccinium*.

Branching mycelium, which results from the germination of bacilli and coccoids, produces microcolonies in which fragmentation begins after 24 hr of incubation. The central branched hyphal elements become shorter as fragmentation proceeds while border hyphae continue to elongate and branch. Microcysts found in older cultures. White aerial hyphae produced. Morphological Group II. Acid-fast.

Colonies on nutrient agar small, flat, reddish with irregular borders; a thin white aerial mycelium produced; brownish soluble pigment formed in older cultures.

Colonies on nutrient gelatin irregular, granular, gray; substrate becomes reddish; no liquefaction.

Scant, small, gray colonies on glucose nitrate agar.

Raised, granular, rough growth having a gray aerial mycelium over a dark red substrate on potato plug.

Small pinkish to orange to salmon colonies having a crumbly texture on starch agar.

Acid produced from L-arabinose, cellobiose, D-fructose, D-glucose, glycerol, mannitol, mannose, sucrose, trehalose and xylose.

The following used as sole carbon and energy sources: citrate, glucose, glycerol, H-malate, malonate, mannitol, mannose, octoate, paraffin and sodium lactate.

Cellulose, chitin, DNA, starch, Tweens 20, 40, 60 and urea are hydrolyzed.

Litmus milk slowly becomes alkaline; no coagulation or peptonization.

Nitrate reduced.

Temperature range: 10–35 C. Optimum: 25–28 C. No growth at 37 C. pH range: 6–9. Optimum: 7.4.

Penicillin sensitive (5 i.u. discs). Lysozyme resistant.

Isolated from galls on blueberry plants.

Reference strain: ATCC 11092 (Sneath and Skerman 1966, 86).

Further Comments

This organism resembles *N. minima* according to its authors. It was assigned to cluster 11, one of the minor clusters of Goodfellow (1971, 52).

16. Nocardia minima (Jensen) Waksman and Henrici 1948, 902. (*Proactinomyces minimus* Jensen 1931, 365).

mi′ni.ma. L. sup.fem.adj. *minima* least.

Coccoid and bacillary cells germinate by elongation and germ tube production. Fragmentation does not begin until after 96 hr incubation, thus an extensively branched microcolony is formed. Fragmentation proceeds from the center of the colony outward, resulting in the production of bacillary elements 0.5–0.8 by 10–20 μm. Smaller and smaller fragments are produced as culture ages. In older cultures, mostly cocci and coccobacilli are present which stain irregularly. Morphological Group II. Partially acid-fast.

Colonies on nutrient agar raised, wrinkled, rough surfaced with irregular margins; pinkish yellow to orange becoming brownish with age; scant whitish aerial mycelium produced, especially at edges of colonies.

Irregular, heaped, warty growth on potato plug; coral to orange in color; sparse, white, aerial hyphae are present.

Acid produced from arbutin, dextrin, D-fructose, D-glucose, glycerol, maltose, mannose, paraffin, sodium butyrate, sodium propionate, sodium pyruvate and sodium succinate.

Esculin, Tweens 20, 40, 60 and urea are hydrolyzed.

Nitrate reduced.

Gelatin not liquefied.

Temperature range: 10–45 C. Optimum 28 C. pH range: 6.5–9. Optimum: 7.5.

Penicillin sensitive (5 i.u. discs). Lysozyme resistant.

Resistant to 5% NaCl and 0.001% sodium azide.

Reference strain: CBS Strain; PSA 28 (Sneath and Skerman 1966, 86).

Further Comments

This organism was assigned to cluster 1, subgroup B, the *N. asteroides* cluster by Goodfellow (1971, 46).

17. Nocardia blackwellii (Erikson) Waksman and Henrici 1948, 910. (*Actinomyces blackwellii* Erikson 1935, 37; *Streptomyces blackwellii* (Erikson) Pridham *et al.* 1958, 58.)

black.wel′li.i. M.L. gen.n. *blackwellii* of Blackwell; named for Blackwell.

Coccobacilli elongate and produce a branched mycelium composed of hyphae having a diameter of about 1 μm. Short aerial hyphae produced; substrate mycelium pink to orange on most media. Fragmentation begins in the center of the colony producing bacilli and coccobacilli, some of which develop into microcysts. Morphological Group II. Acid-fast.

Grows moderately well on nutrient agar, forming thin, dry colonies with white aerial hyphae.

Pinkish yellow colonies on gelatin with white aerial hyphae; no liquefaction.

Well developed, irregular, convoluted, pinkish orange growth with no aerial hyphae on glycerol nitrate agar.

Colonies on potato plug become dull coral with aerial mycelium.

Convoluted, bright yellow, surface growth on litmus milk; no coagulation or peptonization; acid not produced.

Flakes produced on broth, forming ring with white aerial hyphae at surface.

Acid produced from adonitol, arbutin, dextrin, D-fructose, D-glucose and mannose.

Esculin, allantoin, benzidine, Tweens 20, 40, 60 and urea are hydrolyzed.

The following compounds serve as sole energy and carbon sources: D-fructose, D-glucose, glycerol, mannitol, mannose, paraffin, sebacic acid, sodium acetate, sodium butyrate, sodium H-malate, sodium propionate, sodium pyruvate, sorbitol and testosterone.

Temperature range: 20–45 C. Optimum: 30 C. pH range: 6.5–9. Optimum: 7.2.

Not inhibited by 7% NaCl. Penicillin resistant (5 i.u. discs). Lysozyme resistant.

Isolated from hock joint of a foal.

Reference strain: None designated.

Further Comments

This organism is probably synonymous with *N. asteroides* to which it was assigned by Gordon and Mihm, 1957. It falls into cluster 1, subgroup A of Goodfellow (1971, 46).

18. **Nocardia convoluta** Chalmers and Christopherson 1916, 257. (*Discomyces convolutus* (Chalmers and Christopherson) Neveu-Lemaire 1921, 44; *Oospora convoluta* (Chalmers and Christopherson) Sartory 1920, 769; *Actinomyces convolutus* (Chalmers and Christopherson) Brumpt 1927, 1195; *Actinomyces hominis* subsp. *convolutus* Krasil'nikov 1941, 29.)

con.vo.lu′ta. L. fem.part. adj. *convoluta* wound together.

Microcolonies develop from coccoid and bacillary elements. Branching fairly extensive. Fragmentation begins in the center of the young colony after 18 hr of incubation. The ends of hyphal fragments become pointed and bend into a scimitar shape, producing a swirling cell arrangement. Scant aerial mycelium develops which fragments into bacillary elements as the culture matures. Coccoid and bacillary cells predominate in older cultures. Microcysts produced on certain media.

Morphological Group II. Partially acid-fast on Dubos oleic acid agar.

Colonies on nutrient agar raised, convoluted, irregular, gray-orange in color, and covered by a scant white aerial mycelium.

Heaped, convoluted colonies on nutrient gelatin; no liquefaction.

Heavy, heaped, convoluted, salmon-colored growth on potato plug; sparse, white, aerial mycelium.

Irregular, folded colonies on glucose nitrate agar, thin white aerial mycelium at borders.

Acid produced from D-fructose, D-galactose, glycerol, mannitol and mannose.

The following serve as sole carbon and energy sources: hydroxybenzoic acid, pimelic acid, sodium acetate, sodium butyrate, sodium propionate, sodium succinate and testosterone.

Tweens 20, 40 and 60 are hydrolyzed.

Nitrate reduced.

Temperature range: 10–40 C. Optimum: 30 C. pH range: 6–10. Optimum: 7.4.

Resistant to penicillin. Lysozyme resistant.

Isolated from yellow grain mycetoma in Sudan.

The G + C content of the DNA is 66.0 moles %.

Reference strain: None designated.

Further Comments

The curved bacilli with pointed ends arranged in palisades produce a distinctive microscopic morphology. This organism was assigned to a minor cluster by Goodfellow (1971, 64).

19. **Nocardia cellulans** Metcalf and Brown 1957, 569.

cel.lu′lans. L. part.adj. *cellulans* cell-making.

Coccoid and bacillary cells germinate by elongation producing a branched mycelium about 1 μm in diameter. Fragmentation begins at about 96 hr in the center of microcolonies, while peripheral hyphae continue to elongate and branch. Microcyst production as terminal spherical hyphal swellings begins after 7 days incubation.

Morphological Group II. Partially acid-fast.

Raised, cream-colored, soft colonies having a matt surface on agar media; no aerial hyphae. Growth cream colored on most media; bright yellow on yeast extract peptone agar.

On glucose agar long hyphae produced even after 28 days incubation, with numerous branched bacillary cells.

Cellulose hydrolyzed; long branched and unbranched filaments predominate during cellulose decomposition. Spherical microcysts very common in older cultures.

Acid from glucose, maltose and sucrose. Starch not hydrolyzed.

Paraffin not used as sole carbon source without a yeast extract supplement.

Fix atmospheric nitrogen in amounts of 12 mg N/g cellulose decomposed. Also fixes nitrogen using glucose, mannitol and sucrose as carbon sources.

Acid and curd produced in litmus milk.

Nitrate reduced.

Esculin, casein, gelatin, keratin, chitin, DNA and urea are hydrolyzed.

Optimum temperature: 28 C. pH range: 6–9.

Penicillin sensitive. Lysozyme sensitive.

Reference strain: ATCC 12830 (Sneath and Skerman 1966, 86).

Further Comments

This is one of two nitrogen-fixing *Nocardia* isolated by Metcalf and Brown (1957, 567) from soil of a chalk grassland plant community in Great Britain. It was assigned to cluster 10, subgroup A, the *N. turbata* cluster by Goodfellow (1971, 49).

20. **Nocardia lutea** Castellani and Chalmers 1919, 1062. (*Actinomyces luteus* (Castellani and Chalmers) Brumpt 1927, 1206.)

lu′te.a. L. fem.adj. *lutea* yellow.

Extremely pleomorphic bacilli, cocci, coccobacilli, X- and Y-shaped cells produce microcolonies by elongating and branching. Angular bends are common, resulting in irregular fragmentation beginning in the center of the small colony. Fragmentation begins after 20 hr of incubation. Scant, whitish, aerial mycelium produced on some media. Morphological Group II. Partially acid-fast on Dubos oleic acid medium.

Colonies on nutrient agar round, raised, mucoid with entire margins, pinkish yellow in color.

Colonies on nutrient gelatin light coral in color irregular having a convoluted surface and filamentous margins, becoming mucoid with age; no liquefaction.

Wrinkled, membranous, pinkish colonies on glycerol nitrate agar.

Acid produced from dextrin, D-fructose, D-glucose, glycerol, maltose, mannitol, mannose, sorbitol and trehalose.

The following serve as sole carbon sources: D-fructose, D-glucose, glycerol, mannitol, mannose, paraffin, sodium butyrate, sodium malate, sodium propionate, sodium pyruvate, sorbitol and trehalose.

Esculin, DNA, Tween 60 and urea hydrolyzed.

Litmus milk alkaline, no peptonization.

Nitrate reduced.

Temperature range: 10–40 C. Optimum: 30 C. pH range: 6–10. Optimum: pH 7.4.

Penicillin sensitive (5 i.u. discs). Lysozyme sensitive. Not inhibited by 7% NaCl or 0.02% sodium azide.

Isolated from a case of actinomycosis of the lachrymal gland of man.

The G + C content of the DNA is 69.8 moles %.

Reference strain: Not designated.

Further Comments

According to Erikson (1935, 30) this organism is closely related to saprophytes such as *N. rubra* and *N. polychromogenes*. Its qualifications as a pathogen have never been determined. Goodfellow (1971, 51) places one strain in subgroup C and another on subgroup D of cluster 14 which makes it a member of the "rhodochrous-complex." However, *N. lutea* differs from *N. corallina*, *N. rubropertincta* and *N. salmonicolor* of the same subgroup of Goodfellow in that it is positive for KDPG aldolase and the others are not (Kersters, 1970, 304).

21. **Nocardia globerula** (Gray) Waksman and Henrici 1948, 903. (*Mycobacterium globerulum* Gray 1928, 265; *Proactinomyces globerulus* (Gray) Reed 1939, 838.)

glo.be′ru.la. M.L. fem.dim.adj. *globerula* globular.

Irregular bacilli and coccoids germinate by elongation and branching. Fragmentation begins after 20 hr of incubation resulting in the formation of bacillary fragments in the center of the colony. Coccoid microcysts prevalent as culture ages. Capsules produced on some media. Morphological Group II. Partially acid-fast on Dubos oleic acid medium.

Colonies on nutrient agar circular with undulate margins, pasty texture, have a matt surface when young, but become mucoid with age; white at first becoming pinkish coral.

Convex, irregular, round light pink colonies with a mucoid surface on nutrient gelatin; no hydrolysis.

Smooth, mucoid light salmon-pink colonies with irregular margins on potato glycerol agar.

Acid produced from dextrin, D-fructose, D-glucose, glycerol, maltose, mannitol, mannose, sorbitol and trehalose.

The following serve as sole carbon sources: D-

Editorial Note. The taxonomic status of the "rhodochrous complex" (*Nocardia* species 21–26, 28, 29) is not settled. Two generic names have been proposed, *Jensenia* Bisset and Moore 1950, 280 and *Gordona* Tsukamura 1971, 19, but more information is needed. See Goodfellow *et al.*, 1972, Ridell and Norlin, 1973, and "Further Comments" under these species.

fructose, D-glucose, glycerol, mannitol, mannose, paraffin, sodium butyrate, sodium malate, sodium propionate, sodium pyruvate, sorbitol and trehalose.

Esculin, DNA, Tween 60 and urea hydrolyzed.

Litmus milk alkaline, no peptonization.

Indole not produced from tryptophan.

Indigotin produced as blue crystals on indole agar.

Nitrate not reduced.

Temperature range: 10–40 C. Optimum: 28 C. pH range: 6–10. Optimum: pH 7.4.

Penicillin sensitive (5 i.u. discs). Lysozyme sensitive.

Not inhibited by 7% NaCl, or 0.02% sodium azide.

Isolated from soil in Great Britain.

The G + C content of the DNA is 67.0 moles %.

Reference strain: ATCC 19370 (Sneath and Skerman 1966, 84).

Further Comments

This organism belongs to subgroup D of cluster 14 of Goodfellow (1971, 51), hence, is one of the "rhodochrous-complex" whose taxonomic position is unsettled. It is retained in *Nocardia* until definitive studies may place it elsewhere.

22. **Nocardia   rubropertincta**   (Hefferan) Waksman and Henrici 1948, 904. (*Bacillus rubropertinctus* Hefferan 1904, 460; *Serratia rubropertincta* (Hefferan) Bergey *et al.* 1923, 96; *Mycobacterium rubropertinctum* (Hefferan) Ford 1927, 255; *Proactinomyces rubropertinctus* (Hefferan) Reed 1939, 835.)

rub.ro.per.tinc'ta. L. adj. *ruber* red; L. pref. *per* very; L. part.adj. *tinctus* dyed, colored; M.L. fem.adj. *rubropertincta* heavily dyed red.

Germination of bacillary and coccoid cells begins after 10 hr incubation followed by rapid branching in 14 hr. Fragmentation is delayed until after 24 hr, allowing a much branched microcolony to develop. Fragmentation begins with the central hyphae of the microcolony producing long branched filaments, the ends of which continue to grow. The breaking up of these hyphae becomes rapid with continued incubation. Finally most of the hyphae become fragmented to bacillary elements 1 by 5–8 μm, some of which become shorter and some "round up" to become microcysts 1–1.5 μm in diameter. Morphological Group II. Partially acid-fast.

Colonies on nutrient agar elevated, rough, folded, dull with irregular margins, and a coral red color.

Granular, crumbly colonies with undulate margins on nutrient gelatin; no liquefaction.

Irregular, granular, growth in glucose mineral salts medium; pigmentation less intense than on organic media.

Heavy, heaped, folded, intense orange-red growth develops slowly on potato plug.

Acid is produced from dextrin, D-fructose, D-glucose, ethanol, glycerol, maltose, mannitol, mannose, sorbitol and trehalose.

Esculin, adenine, benzidine, DNA, Tweens 20, 40, 60 and urea are hydrolyzed.

These compounds serve as sole carbon and energy sources: acetamide, D-fructose, D-glucose, glycerol, L-alanine, L-arabinose, mannitol, mannose, paraffin, sodium acetate, sodium butyrate, sodium citrate, sodium H-malate, sodium propionate, sodium pyruvate and testosterone.

Nitrate reduced by most strains.

Litmus milk slowly becomes alkaline; no coagulation, no peptonization.

Temperature range: 10–40 C. Optimum: 28–30 C. pH range: 6–10. Optimum: 7.0.

Not inhibited by 7% NaCl. Penicillin sensitive (5 i.u. discs). Lysozyme sensitive.

Has been isolated from various soils, butter and air.

Reference strain: None designated.

Further Comments

This organism is qualitatively indistinguishable from strains of *N. corallina*, *N. rubra*, *N. salmonicolor*, *N. lutea*, *N. globerula* and *N. opaca*. They can, however, be distinguished quantitatively as shown by Goodfellow (1971, 50, 51) who assigns them all (along with others) to cluster 14, the "rhodochrous group," but to different subgroups. *N. rubropertincta* was listed in subgroup D.

23. **Nocardia corallina** (Bergey *et al.*) Waksman and Henrici 1948, 902. (*Bacillus mycoides corallinus* Hefferan 1904, 459; *Serratia corallina* Bergey *et al.* 1923, 93; *Streptothrix corallinus* (Bergey *et al.*) Reader 1926, 1; *Proactinomyces corallinus* (Bergey *et al.*) Jensen 1932, 364; *Anthracillus corallinus* (Bergey *et al.*) Pribram 1933, 89.)

co.ral.li'na. L. fem.adj. *corallina* coral-red.

Rather extensive mycelium develops from coccus and bacillary elements. Fragmentation begins in the center of microcolonies after about 20 hr incubation. Division proceeds to produce shorter and shorter fragments while the marginal hyphae continue to elongate and branch. Microcysts prevalent in older cultures. Striking coral-red intracellular pigment produced on all media.

Morphological Group II. Partially acid-fast on Dubos oleic acid medium.

Colonies on nutrient agar glistening, raised brilliant coral-red with filamentous or arborescent margins; no aerial hyphae produced.

Round, smooth, glistening colonies with a filamentous border on nutrient gelatin; no liquefaction.

Colonies on potato glycerol agar raised, granular, coral-red becoming brown-orange with age.

Acid from D-fructose, D-glucose, glycerol, maltose, mannitol, mannose and sorbitol.

Esculin and tyrosine hydrolyzed.

Carbon sources: dextrin, D-glucose, glycerol, mannitol, mannose, naphthalene, paraffin, phenol, m-cresol, sebacic acid, sodium acetate, sodium pyruvate and sorbitol.

Litmus milk alkaline, no coagulation or peptonization.

Nitrate reduced.

Optimum temperature: 25–28 C.

Optimum pH: 6.8–8.0.

Not inhibited by 7% NaCl.

Penicillin sensitive (5 i.u. discs). Lysozyme sensitive.

Isolated from soils from Great Britain, Australia and the United States.

The G + C content of the DNA ranges from 62.3–70.95 moles %.

Reference strain: None designated.

Further Comments

This species is typical of morphological Group II *Nocardia*. It is closely related to *N. opaca*, *N. rubra*, *N. salmonicolor* and *N. lutea*, as well as *N. rubropertincta*, all of which fall into Goodfellow's cluster 14, the "rhodochrous group." However, these organisms have all of the distinctive characteristics of nocardiae including the type of mycolic acid (Lechevalier *et al.*, 1971, 313), hence are retained in the genus. Further studies are necessary and possibly new criteria found before these organisms can be confidently placed in their proper systematic niche.

24. **Nocardia salmonicolor** (den Dooren de Jong) Waksman and Henrici 1948, 904. (*Mycobacterium salmonicolor* den Dooren de Jong 1927, 216; *Flavobacterium salmonicolor* (den Dooren de Jong) Bergey *et al.* 1930, 157; *Proactinomyces salmonicolor* (den Dooren de Jong) Jensen 1943, 368.

sal.mo.ni'co.lor. L. n. *salmo, salmonis* salmon; M.L. adj. *salmonicolor* salmon-colored.

Microcolonies initiated by elongation of bacillary or coccoid elements. Branching begins after about 12 hr incubation. Rapid mycelial formation follows, producing a well developed microcolony before the beginning of fragmentation at about 22 hr of incubation. The rate of cell division increases rapidly in the center of the young colony while border hyphae continue growing and branching. After 72 hr incubation hyphal fragments,

bacilli and coccoids are found with little or no evidence of branching. Microcysts produced on some media in older cultures.

Morphological Group II. Partially acid-fast.

Raised, rough, folded salmon-pink colonies on nutrient agar.

Colonies on nutrient gelatin small, raised, irregular, orange in color; no liquefaction.

Well developed red-orange colonies on glycerol asparagine agar; granular with crenate margins; faint white aerial hyphae at edges of some colonies.

Heavy irregular growth, tending to heap and fold on glucose peptone agar; yellow pigment becoming red-orange with age.

Heavy, heaped, folded, deep red-orange growth on potato plug.

Acid produced from dextrin, D-fructose, D-glucose, ethanol, glycerol, maltose, mannitol, mannose, sorbitol and trehalose.

Esculin, benzidine, Tweens 20, 40 and 60 and tyrosine are hydrolyzed.

The following compounds can serve as sole carbon and energy sources: acetamide, adipic acid, citrate, dextrin, D-fructose, D-glucose, DL-norleucine, glycerol, lactate, H-malate, L-tyrosine, mannitol, mannose, paraffin, *p*-cresol, pimelic acid, sebacic acid, sodium acetate, sodium butyrate, sodium octoate, sodium propionate, sodium pyruvate, sodium succinate and testosterone.

Litmus milk alkaline; slight clearing with age.

Nitrate reduced.

Temperature range: 10–40 C. Optimum: 28 C.

pH range: 5–10. Optimum: 7–7.5.

Not inhibited by 5% NaCl. Penicillin sensitive (5 i.u. discs). Lysozyme sensitive.

Isolated from soil from Rothamsted, England by ethylamine enrichment.

Reference: None designated.

Further Comments

This organism assigned to subgroup 14 C, one of the cluster groups for the "rhodochrous-complex" by Goodfellow (1971, 50).

25. **Nocardia rubra** (Krasil'nikov) Waksman and Henrici 1948, 905. (Not *Nocardia rubra* (Kruse) Chalmers and Christopherson 1916, 265; *Proactinomyces ruber* Krasil'nikov 1938, 143.)

rub'ra. L. fem.adj. *rubra* red.

Growth begins after about 10 hr incubation by elongation of rods 3–5 by 0.7 μm. Branching begins after 14 hr with secondary and tertiary branching quickly following. Initial fragmentation which divides the mycelium into long branched hyphae occurs after 20 hr of incubation. Fragmentation proceeds at an accelerated rate in the center of the microcolony while the peripheral hyphae continue to grow, branch and fragment. Microcysts common in old cultures. An intracellular pigment (flame scarlet Ridgway Plate II 9) is produced by most strains. Aerial mycelium lacking. Morphological Group II. Partially acid-fast on Dubos oleic acid agar.

Colonies on nutrient agar flat, rough, bright red with a pasty consistency and undulate margins.

Colonies on nutrient gelatin somewhat raised, heaped, irregular; no liquefaction.

On glucose mineral salts medium flat, granular colonies with irregular margins; pigment paler than on organic media.

Heaped, mucoid growth with brilliant red pigment on potato plug.

Acid produced from D-fructose, dextrin, D-glucose, maltose, mannitol, mannose, sorbitol, sucrose and trehalose.

Esculin, benzidine, Tweens 20, 40 and 60 and tyrosine are hydrolyzed.

The following serve as sole carbon and energy sources: acetamide, adipic acid, D-fructose, D-glucose, glycerol, mannitol, mannose, sorbitol, sucrose and trehalose.

Litmus: alkaline reaction; no coagulation or peptonization.

Nitrate not reduced.

Temperature range: 10–40 C. Optimum: 25–28 C. pH range: 6–10. Optimum: 7–7.5.

Not inhibited by 7% NaCl. Resistant to 0.01% sodium azide, 0.1% potassium tellurite and 0.01% tetrazolium.

Penicillin sensitive (5 i.u. discs). Lysozyme sensitive.

Isolated from various kinds of soils.

The G + C content of the DNA is 67–71.7 moles %.

Reference strain: None designated.

Further Comments

This organism is very typical of McClung's (1949, 154) Group II. It has been assigned to the "rhodochrous group" by Goodfellow (1971, 50), along with *N. corallina*, *N. salmonicolor*, *N. globerula* and several other carotenoid pigment producing strains. He assigned it to subgroup 14 C.

26. **Nocardia opaca** (den Dooren de Jong) Waksman and Henrici 1948, 897. (*Mycobacterium opacum* den Dooren de Jong 1927, 216; *Proactinomyces opacus* (den Dooren de Jong) Jensen 1932, 369.)

o.pa′ca. L. fem.adj. *opaca* shaded, dark, opaque.

Coccoid and bacillary elements germinate by elongation and germ tube formation. An extensively branched microcolony produced. Central filaments fragment into 0.8–1.0 by 5–15 μm hyphae, beginning after 24 hr incubation. The central hyphal elements become shorter and shorter as fragmentation continues, while growth continues at the margin. Coccoid microcysts, bacillary and hyphal fragments characterize older cultures. Morphological Group II. Partially acid-fast.

Colonies on nutrient agar raised, round pale buff-pink with a matt surface and filamentous to entire margins.

Colonies on gelatin convex, round pale pink with a shiny surface; no liquefaction.

Dry, rough, heaped, grayish pink colonies on potato plug.

Acid produced from D-fructose, D-glucose, glycerol, maltose, mannitol, mannose, sorbitol and sucrose.

The following serve as sole carbon sources: D-glucose, glycerol, maltose, mannitol, mannose, paraffin, sorbitol, sucrose and trehalose.

Esculin, Tweens 20, 40 and 60 and urea are hydrolyzed.

Litmus milk alkaline; no coagulation or peptonization.

Nitrate reduced.

Saturated, long chain aliphatic hydrocarbons are used as energy sources (Webley, 1954, 420).

Temperature range: 10–40 C. Optimum: 30 C. pH range 6–9. Optimum: 7.2.

Not inhibited by 7% NaCl or 0.02% sodium azide.

Penicillin sensitive (5 i.u. discs). Lysozyme sensitive.

Isolated from soils.

The G + C content of the DNA ranges from 66–68 moles %.

Reference strain: Not designated.

Further Comments

This organism is similar to *N. corallina* and *N. lutea*. It has been assigned to cluster 14, subgroup E, the "rhodochrous complex" by Goodfellow (1971, 51). *N. opaca* has KDPG-aldolase activity according to Kersters (1970, 304).

27. **Nocardia calcarea** Metcalf and Brown 1957, 568.

cal.car′e.a. L. fem.adj. *calcarea* chalky.

Coccoid and bacillary cells produce a sparsely branched mycelium, which fragments into branched hyphae up to 10 μm in length in about 12 hr. Fragmentation begins in the center of young colonies, peripheral hyphae continue to elongate and branch. Pre-fission bending and post-fission snapping movements common. Microcysts common as cultures age. Morphological Group I. Partially acid-fast.

Circular raised colonies without aerial hyphae on agar media; cream to pinkish tan in color, matt surface; soft and easily removed from surface of medium.

Short rods and cocci in 24 hr on sucrose agar with a few branched hyphae up to 20 μm long.

On glucose and mannitol agars hyphae longer and more persistent than on sucrose agar with occasional chains of spherical microcysts.

Branched filaments very rare after 3 days on yeast extract peptone agar; fragmentation begins in 10 hr; coccoid microcysts common.

Fixes atmospheric nitrogen in amounts of 2–4.5 mg N/g glucose, sucrose or mannitol in the medium.

Acid from glucose and sucrose.

Uses glucose, maltose, sucrose and paraffin as sole carbon sources.

Slight alkaline reaction in litmus milk; no peptonization.

Gelatin not liquefied. Nitrate reduced.

Casein, chitin, DNA and keratin are hydrolyzed.

Optimum temperature: 28 C. pH range 6–8.

Isolated from chalky grasslands of Great Britain.

Reference strain: ATCC 19369 (Sneath and Skerman 1966, 84).

Further Comments

This organism was assigned to cluster 10 subgroup B, the *N. turbata* cluster by Goodfellow (1971, 49).

28. Nocardia restricta (Turfitt) *comb. nov.* (*Proactinomyces restrictus* Turfitt 1944, 491; *Mycobacterium restrictum* (Turfitt) Krasil'nikov 1949, 178.)

re.stric'ta. L. fem.part.adj. *restricta* restricted or limited.

Short rods and cocci germinate by elongation and germ tube formation. Mycelium production is limited due to onset of fragmentation after 12–14 hr incubation. Microcolonies consist of short hyphal fragments, sometimes with branches which tend to be oriented parallel to each other, due to pre-fission movements. The central bacilli become shorter and shorter with age. Coccoid microcysts abundantly produced in older cultures. Morphological Group I. Partially acid-fast on Dubos oleic acid agar.

Colonies on nutrient agar, small, up to 2 mm diameter, irregularly round, smooth, cream colored, with entire edges.

Growth restricted on nutrient gelatin; no liquefaction.

Heavy, mucoid, gray-orange growth on potato plug.

Acid produced from D-fructose, D-glucose, glycerol, maltose, mannitol, mannose, sucrose and trehalose.

The following serve as sole carbon sources: D-alanine, L-alanine, D-fructose, D-glucose, glyc-

erol, maltose, mannitol, mannose, paraffin, sodium acetate, sodium butyrate, sodium citrate, sodium gluconate, sodium propionate, sodium pyruvate, sorbitol and trehalose.

Acid slowly produced in litmus milk; no coagulation or peptonization.

Nitrate reduced rapidly to nitrite.

Esculin, Tweens 20, 40 and 60 and urea hydrolyzed.

Temperature range: 10–40 C. Optimum: 25 C. pH range: 6–9. Optimum: 7.5.

Not inhibited by 7% NaCl. Sensitive to penicillin (5 i.u. discs). Lysozyme sensitive.

Isolated from soils in Great Britain.

Reference strain: Not designated.

Further Comments

This organism is closely related to *N. erythropolis* and was first studied because of its ability to decompose cholesterol. Mutant auxotrophs have been produced (Beaudoin, 1966, 16), sensitivity to a phage designated R-1 being one marker. It was assigned to cluster 14, subgroup E, the "rhodochrous group" by Goodfellow (1971, 51).

29. Nocardia erythropolis (Gray and Thornton) Waksman and Henrici 1948, 898. (*Mycobacterium erythropolis* Gray and Thornton 1928, 87; *Actinomyces erythropolis* (Gray and Thornton) Bergey *et al.* 1930, 472; *Proactinomyces erythropolis* (Gray and Thornton) Jensen 1932, 371.)

e.ry.thro'po.lis. Gr. adj. *erythrus* red; Gr. n. *polis* a city; M.L. n. *erythropolis* red city.

Growth begins by elongation of rods 5–15 by 0.8 μm. Initial branched mycelium not always produced as fragmentation may precede branching. Sometimes as many as three short branches appear before fragmentation begins. Fragmentation usually occurs after 14 hr of incubation in the center of the microcolony and may be multiple. Most of the divisions are at the apex of acutely bent hyphae-producing ends which grow parallel to each other. Fragmentation continues in the center of the colony while border hyphae grow and occasionally branch. Coccoid cells not produced. Colonial texture is pasty and intracellular pigment (light pinkish cinnamon, Ridgway Plate XXIX 15 d) is produced. Morphological Group I. Partially acid-fast on Dubos oleic acid medium.

Colonies on nutrient agar smooth, raised, granular, pinkish gray with entire margins.

Raised, circular colonies, becoming mucoid on nutrient gelatin; no liquefaction.

Dry, heaped, granular colonies becoming grayish coral-tan in color on potato glycerol agar.

Acid produced from arbutin, dextrin, D-fructose, D-glucose, glycerol, maltose, mannitol, mannose, sucrose and trehalose.

The following serve as carbon and energy sources: acetate, citrate, glucose, inositol, mannitol, mannose, paraffin, phenol, pyruvate, trehalose, sorbitol and succinate.

The following are hydrolyzed: adenine, esculin, Tweens 20, 40, 60, tyrosine and urea.

Litmus milk pink pellicle, alkaline. No coagulation.

Nitrites not produced from nitrates.

Temperature range: 10–40 C. Optimum: 28 C. pH range 6–10; optimum 7.2.

Not inhibited by 7% NaCl. Sensitive to penicillin (5 i.u. discs). Lysozyme sensitive.

Isolated from soils from Great Britain and the United States.

The G + C content of the DNA is 62.5 moles %.

N. erythropolis auxotrophs have been shown to recombine with auxotrophs of *Jensenia canicruria*. Thus the latter represents a specific mating type of *N. erythropolis*. Both organisms are also sensitive to nocardiophages MNP1, MNP7, φEC and φC of Adams *et al.* (1967, 247).

Reference strain: ATCC 25544 (Sneath and Skerman 1966, 86).

Further Comments

Both *N. erythropolis* and *Nocardia (Jensenia) canicruria* fall into cluster 14, subgroup 14 D of Goodfellow (1971, 50). Various authors have suggested that both organisms are synonymous with *Mycobacterium rhodochrous*. We agree with Adams, Adams, and Brownell (1970, 144) and retain this taxon in the genus *Nocardia*.

30. **Nocardia marina** Waksman 1957, 740. (*Proactinomyces flavus* Humm and Shepard 1946, 77; *Proactinomyces citreus* subsp. *marinae* Krasil'-nikov 1949, 141.)

Comment. The epithet *marina* was used by Waksman in lieu of *flava* to avoid the creation of a later homonym of *Nocardia flava* (Krasil'nikov) Waksman and Henrici."

ma.ri'na. L. fem.adj. *marina* of the sea, marine.

Microcolonies consisting of peripheral branched hyphae with fragmentation producing bacilli 0.8–1.0 by 5–10 μm in the center. Branching sparse. Coccobacilli and coccoid microcysts common in older cultures. Morphological Group I. Not acid-fast.

Bright lemon-yellow colonies on nutrient agar which are round, flat with a raised center and undulate margins; surface smooth with a matt finish.

Agar slowly digested. Alginic acid decomposed. Cellulose and starch hydrolyzed.

Acid from arabinose, cellobiose, fructose, galactose, glucose, glycerol, lactose, mannitol, maltose, mannose, rhamnose, salicin, sucrose and xylose.

No acid from dulcitol, inositol, inulin or sorbitol. Acetic, butyric and lactic acids are used as carbon sources. Citric, gluconic, maleic, malic, malonic, oxalic, propionic, succinic, tartaric and iso-valeric acids are not used as carbon sources.

D-Arginine used as both carbon and nitrogen sources. Aspartic acid, cystine, glycine, glutamic acid, L-leucine and tyrosine used only as nitrogen sources.

Acetylmethylcarbinol not produced.

Ammonia, nitrite and nitrate serve as nitrogen sources. Ammonia produced from nitrite, nitrate, asparagine, peptone and urea. Urea used as a nitrogen source. Catalase positive.

Crateriform liquefaction of gelatin, becoming infundibuliform and after several days, stratiform.

Milk alkaline followed by peptonization; no coagulation.

Nitrate not reduced. Indole not produced.

Indigotin not produced from indole.

Hydrogen sulfide not produced.

Chitin hydrolyzed.

Optimum temperature between 25–30 C.

Grows well on media prepared with distilled water and in salinities through 6% sea salt; slightly less growth at salinities above 5%. Morphology and pigmentation apparently unaffected by salinity.

Isolated from marine sediments of South Atlantic coast of the United States.

Reference strain: None designated.

Further Comments

This organism is morphologically and physiologically similar to *N. atlantica*. It was assigned to cluster 10, subgroup B, the *N. turbata* group by Goodfellow (1971, 49).

31. **Nocardia atlantica** (Humm and Shepard) Waksman 1957, 741. (*Proactinomyces atlantica* Humm and Shepard 1946, 78.)

at.lan'ti.ca. M.L. fem.adj. *atlantica* pertaining to the Atlantic ocean.

Branched filaments 0.4–0.6 μm in diameter, which break up after 10–12 hr into coccobacilli and cocci. Older cultures entirely coccoid cells and microcysts. Morphological Group I. Not acid-fast.

Colonies bright yellow to orange-yellow on most media, having a doughy consistency and smooth surface with a slightly raised center.

Agar slowly digested. Alginic acid decomposed. Cellulose hydrolyzed. Starch not hydrolyzed.

Acid from arabinose, cellobiose, fructose, glucose, maltose, mannose, raffinose, rhamnose, salicin and sucrose. No acid from dulcitol, lactose, mannitol or sorbitol. Gluconic, lactic and malonic acids utilized. Acetic, butyric, citric, maleic, malic, oxalic, propionic and iso-valeric acids not utilized.

D-Arginine and glutamic acid serve as both nitrogen and carbon sources, DL-alanine, aspartic acid, cystine, glycine, L-leucine, DL-phenylalanine, L-proline and tyrosine used only as nitrogen sources.

Acetylmethylcarbinol not produced. Catalase positive.

Ammonia, nitrite and nitrate serve as poor nitrogen sources. Ammonia slowly produced from nitrite, nitrate, asparagine and peptone. Urea used as a nitrogen source; ammonia does not accumulate.

Crateriform liquefaction of gelatin becoming stratiform with age.

Acid coagulation of milk followed by slow peptonization.

Nitrate reduced.

Indigotin not produced from indole.

Hydrogen sulfide and indole not produced.

Chitin hydrolyzed.

Optimum temperature between 28–30 C.

Grows well in media prepared with distilled water and in all salinities through 6% sea salt, with no apparent affect on morphology or pigmentation.

Isolated from seaweed and marine sediments of the South Atlantic coast of the United States.

Type strain: Not extant.

Genera incertae sedis

1. *Jensenia* Bisset and Moore 1950, 280.
2. *Actinomadura* Lechevalier and Lechevalier 1968, 400.
3. *Oerskovia* Prauser, Lechevalier and Lechevalier 1970, 534, *emend. mut. char.* Lechevalier 1972, 263.
4. *Gordona* Tsukamura 1971, 15.

Species incertae sedis

(Asterisks indicate cultures not available or unknown.)

1. *Nocardia aerocolonigenes* (Shinobu and Kawato) Pridham 1970, 32 has been reduced to synonymy with *Streptomyces aerocolonigenes* Shinobu and Kawato 1960, 215.
2. * *Nocardia actinomorpha* (Gray and Thornton) Waksman and Henrici 1948, 912.
3. * *Nocardia alba* (Krasil'nikov) Waksman 1953, 153.
4. * *Nocardia alni* (Woronin) Waksman 1961, 35. Editor: A synonym of *Frankia alni* q.v.
5. *Nocardia aurantia* SC2318 E. R. Squibb & Sons (Helen Croll); ATCC 12674. (*Nocardia aur-*

entia (*sic*) Sih and Weisenborn (United States Patent 3,065,146, Nov. 20, 1962.) Not validly published.

6. *Nocardia butanica* ATCC 21197 Kyowa Ferm. Not validly published, no author.

7. * *Nocardia cuniculi* Erikson 1935, 32. Invalidation of strain A 6864 by Erikson, 1935. See Gordon, 1957.

8. * *Nocardia dicksonii* (Erikson) Waksman 1961, 41; not adequately described.

9. *Nocardia dassonvillei* (Brocq-Rousseau) Liégard and Landrieu 1911, 426. (*Actinomadura* sp. Lechevalier and Lechevalier 1968, 400.) Type III wall. IMRU 509.

10. * *Nocardia fastidiosa* Suter 1951, 675.

11. *Nocardia histidans* Stapley, Demney, Miller and Woodruff 1967, 596; ATCC 21021.

12. * *Nocardia ivorensis* Combes, Kauffmann and Vazart 1957.

13. *Nocardia leishmani* Chalmers and Christopherson 1916, 255. Syn. of *Nocardia asteroides*.

14. * *Nocardia listeri* (Erikson) Waksman 1961, 47; inadequate description.

15. *Nocardia madurae* (Vincent) Blanchard 1896, 868. (*Actinomadura* sp. Lechevalier and Lechevalier 1968, 400.) Type III wall.

16. * *Nocardia mesenterica* (Orla-Jensen) Waksman and Henrici 1948, 907. Could be *Lactobacillus* . . .?

17. *Nocardia neoopaca* ATCC 21499; not validly published. "*rhodochrous* complex."

18. * *Nocardia nigra* (Krasil'nikov) Waksman 1953, 149; inadequate description.

19. * *Nocardia panjae* (Erikson) Waksman 1961, 51; poor description.

20. *Nocardia pellegrino* ATCC 15998; not validly published, no authors.

21. *Nocardia pelletieri* (Laveron) Pinoy *in* Thiroux and Pelletier 1912, 589; ATCC 14816; NCTC 4162.

22. * *Nocardia pulmonalis* (Burnett) Waksman and Henrici 1948, 901; inadequate description.

23. * *Nocardia serophila* (Sartory and Baily) Waksman 1961, 57; inadequate description.

24. * *Nocardia sumatrae* (Erikson) Waksman 1961, 57; *Nocardia cuniculi* de Mello?

25. *Nocardia sylvodorifera* ATCC catalogue 1934, 51, No. 7372. Syn. *Nocardia asteroides* q.v. Gordon and Mihm 1957, 21.

26. *Nocardia turbata* Erikson 1954, 208. *Oerskovia* sp. Lechevalier and Lechevalier.

27. *Nocardia tenuis* (Castellani) Castellani 1913, 820; ATCC 15907; description lacking.

28. * *Nocardia uniformis* Marton and Szabó 1959, 131; inadequate description.

29. * *Nocardia upcottii* (Erikson) Waksman 1959, 1961, 59; inadequate description.

30. *Nocardia variabilis* (Cohn) Waksman 1961, 60; related to *Actinomyces carneus* Gasperini 1894, *A. ochroleucus* and *A. ochraceus* Neukirch 1902, according to Waksman.

31. * *Nocardia viridis* (Krasil'nikov) Waksman and Henrici 1948, 908; could be synonym of *Streptomyces viridis*, *S. viridans* or several others.

32. * *Nocardia fordii* (Erikson) Waksman 1953, 159; description poor.

33. * *Nocardia maculata* (Millard and Burr) Waksman and Henrici 1948, 913; description poor.

34. * *Nocardia africana* Pijper and Pullinger 1927, 155.

35. * *Nocardia fructifera* (Krasil'nikov) Waksman 1953, 155; description inadequate.

36. * *Nocardia flavescens* (Jensen) Waksman and Henrici 1948, 913.

37. * *Nocardia citrea* (Krasil'nikov) Waksman and Henrici 1948, 908; could be *Nocardia lutea*, poor description.

38. * *Nocardia flava* (Krasil'nikov) Waksman and Henrici 1948, 908, description lacking.

39. * *Nocardia albicans* (Krasil'nikov) Waksman 1953, 146. Description inadequate.

40. * *Nocardia pretoriana* Pijper and Pullinger 1927, 154 . . . probably *Actinomadura* sp (personal opinion).

Genus II. **Pseudonocardia** *Henssen 1957, 408*

A. Henssen

Pseu.do.no.car'di.a. Gr. adj. *pseudes* false; M.L. n. *Nocardia* a genus of actinomycetes; M.L. fem.n. *Pseudonocardia* the false *Nocardia*.

Substrate hyphae 0.4–2.6 μm in diameter; aerial hyphae 0.4–1.8 μm in diameter. A characteristic feature is the growth of the hyphae by budding; a constriction is produced just behind the tip of the terminal segment, the tip then elongates to form a new segment, another constriction is formed near the tip and the process repeated (see Figs. 13 and 15, IJSB 1971, *21:* 32).

Spores may be formed on aerial or substrate mycelium in three ways:

1. The usual way is by successive acropetal formation, designated as Pseudonocardia-type (Henssen and Schnepf, 1967), the spore chains arise terminally or laterally.

2. Fragmentation spores produced irregularly along the hyphae.

3. Two or three blastospores develop per segment in a basipetal direction (see Figs. 4 to 7, IJSB 1971, *21:* 30). Spores are smooth walled or spiny and vary greatly in size, usually 0.5–1.0 μm wide by 1.5–3.0 μm long (length may vary from 1.0–4.5 μm).

Gram-positive, not acid-fast. Cell wall type IV (Table 17.18).

Colonies colorless or yellow to orange. Soluble pigment rarely formed. Aerial mycelium white, powdery or forming a thick cover.

Starch not hydrolyzed, milk unchanged.

Grows well on a variety of organic and synthetic media. Minimum growth requirements not known.

Aerobic.

Type species: *Pseudonocardia thermophila* Henssen 1957, 408.

Description of the species of genus **Pseudonocardia**

1. **Pseudonocardia thermophila** Henssen 1957, 408.

ther.mo'phi.la. Gr. n. *thermus* heat; Gr. adj. *philus* loving; M.L. fem.adj. *thermophila* heat loving.

Segments of the substrate mycelium limited by septa, often zig-zag shaped. May produce fragmentation spores, 1.5–1.8 by 2.5 μm.

Aerial hyphae may or may not be limited by septa. Hyphae often zig-zag shaped and bear straight chains of blastospores. Chains of blastospores also arise directly from the substrate mycelium, usually unbranched. Under certain conditions fragmentation spores produced in the aerial hyphae in a basipetal direction starting at the terminal segment or, more rarely, intercalary in the hyphae.

Good growth on Casamino-peptone-Czapek agar, colonies yellow to orange, thick cover of white aerial mycelium. Similar on yeast starch agar and limited amount of an orange soluble pigment produced. On nutrient agar and yeast agar, good growth, colonies yellow, thick cover of white aerial mycelium, no soluble pigment. On asparagine-glycerol and yeast-glucose agar, good growth, colonies yellowish, aerial mycelium limited. No growth on oatmeal agar.

Gelatin is not liquefied. Nitrate reduced slowly to nitrite.

Optimum temperature 40–50 C, growth slight to 28 C; at 60 C grows better anaerobically.

Isolated from fresh and rotten manure.

Type strain: MB A 18; ATCC 19285; CBS 277.66.

2. Pseudonocardia spinosa Schäfer in Henssen and Schäfer 1971, 31.

spi.no'sa. L. fem.adj. *spinosa* spiny.

Substrate mycelium forms a compact mass, composed of non-septate hyphae 0.4–2.6 μm in diameter. Hyphae constricted at intervals, segments irregular, globose or ovoid in shape.

Aerial hyphae 0.4–1.0 μm in diameter, branched and articulate, without cross walls. The segments act as spores.

Spores 2.5–4.5 μm long, often thickened at the apex, irregularly covered with tiny spines.

Grows slowly on a variety of media; the first sign of aerial mycelium is visible usually after 2 weeks incubation.

Good growth on Casamino-peptone-Czapek agar, substrate mycelium a compact mass on the agar surface, yellow, white aerial mycelium forms a thick crust or may develop only scantily. Good growth on yeast-starch agar, colonies yellow, aerial mycelium white. Moderate to good growth on asparagine-glycerol agar, substrate mycelium yellow, abundant white aerial mycelium. Moderate to good growth on oatmeal agar.

Optimum temperature 20–30 C, no growth at 37 C.

Isolated from soil.

Type strain: MB SF 1; ATCC 25924; CBS 818.70.

FAMILY VII. **STREPTOMYCETACEAE** WAKSMAN AND HENRICI 1943, 339

Thomas G. Pridham and Homer D. Tresner*

Strep.to.my.ce.ta'ce.ae. M.L. masc.n. *Streptomyces* type genus of the family; *-aceae* ending to denote a family; M.L. fem.pl.n. *Streptomycetaceae* the *Streptomyces* family.

Vegetative hyphae (0.5–2.0 μm in diameter) produce a well developed branched mycelium that does not fragment readily. Reproduction by germination of the aerial spores; sometimes by growth of fragments of the vegetative mycelium.

Gram-positive. Aerobic. *Sporichthya* reported as Gram-variable and facultatively anaerobic.

Cell walls contain L-diaminopimelic acid, glycine and no major amounts of arabinose (cell wall Type I of Lechevalier *et al.* 1966, 153).

Primarily soil forms, sensitive to anti-bacterial agents including antibiotics.

G + C content of the DNA, in two genera examined (*Streptomyces* and *Streptoverticillium*) ranged from 69–73 moles %.

Genera incertae sedis

Seven genera and subgenera may be tentatively included in the family *Streptomycetaceae*, as the cell wall of most strains studied is Type I, but are placed under *genera incertae sedis* because of inadequate descriptions.

1. *Actinopycnidium* Krasil'nikov 1962, 250.

2. *Actinosclerotium* Tešić 1966, 91.

3. *Actinosporangium* Krasil'nikov and Yuan 1961, 113.

4. *Elytrosporangium* Falcão de Morais, Batista and Massa 1966, 162.

5. *Intrasporangium* Kalakoutskii, Kirillova and Krasil'nikov 1967, 80.

6. *Microtetraspora* Thiemann, Pagani and Beretta 1968, 295.

7. *Streptopycnidium* Tešić 1966, 91.

8. *Thermostreptomyces* Craveri and Pagani 1962, 118.

Key to the genera of family **Streptomycetaceae**

I. Sporangia-like vesicles not formed.
  A. Non-motile aerial spores produced.
    1. Spores not borne on verticillate sporophores.
Genus   I. Streptomyces, p. 748

---

* The authors are indebted to many specialists, too numerous to name here, for providing necessary materials, translations, information and suggestions, and, in some cases, cultures. We are particularly indebted to Mr. William H. Trejo for assistance in preparing the section on *species incertae sedis* for the genus *Streptomyces;* and to Dr. Elwood B. Shirling with whom liaison has been maintained with reference to type strain designations and other critically necessary information.

2. Spores borne on verticillate sporophores.

    Genus II. *Streptoverticillium*, p. 829

B. Motile aerial spores produced.

    Genus III. *Sporichthya*, p. 842

II. Sporangia-like vesicles formed.

    Genus IV. *Microellobosporia*, p. 843

Genus I. Streptomyces *Waksman and Henrici 1943, 339*

Thomas G. Pridham and Homer D. Tresner

(*Streptothrix* Cohn 1875, 204; Not *Streptothrix* Corda 1839, 27.)
Strep.to.my′ces. Gr. adj. *streptos* pliant, bent; Gr. n. *myces* fungus; M.L. masc.n. *Streptomyces* pliant, or bent fungus.

Slender, **coenocytic** hyphae, 0.5–2.0 μm diameter. **The aerial mycelium at maturity forms chains of three to many spores 0.5–2.0 μm in diameter.** The cell wall contains L-diaminopimelic acid and no readily detectable arabinose. Gram-positive.

On isolation, **colonies are small (1–10 mm** diameter) discrete and lichenoid, leathery or butyrous; **initially relatively smooth surfaced** but **later develop a weft of aerial mycelium** that may appear granular, powdery, velvety or floccose. **Produce a wide variety of pigments** responsible for colors of vegetative mycelium, aerial mycelium and substrate. Many strains produce one or more anti-bacterial, anti-fungal, anti-algal, anti-viral, anti-protozoal or anti-tumor antibiotics. Many strains are sensitive to anti-bacterial agents (antibiotics and synthetic compounds) and to actinophages.

Heterotrophs: **Highly oxidative,** acids not generally detectable in fermentation tests.

Glucose is utilized for growth by all species tested.

Generally hydrolyze one or more of the following: gelatin, casein and starch. Generally reduce nitrates to nitrites.

Aerobes.

Temperature optimum 25–35 C; some species grow at temperatures within the thermophilic range.

Optimal pH range for growth 6.5–8.0.

Species of *Streptomyces* may form synemmata-like, sclerotial or pycnidial structures. There are isolated reports on the occurrence of a definite sexual cycle comprising formation and fusion of motile gametes. Hybridization of wild-type strains has not been reported, but certain auxotrophic mutant strains may be induced to hybridize. Heterokaryosis occurs.

Type species: *Streptomyces albus* (Rossi Doria) Waksman and Henrici 1943, 339.

Further Comments

Beyond these general characteristics, groups within the genus *Streptomyces* may be recognized, based on the following:

A. Color of mature sporulated aerial mycelium (spores *en masse*): White; Gray (gray to brownish); Yellow (yellowish to greenish yellow); Red (tan, pink and rose shades); Blue (bluish to grayish blue shades); Green (greenish to grayish green shades); Violet.

B. Spore chain morphology: (See Plate 17.12).
  1. *Rectus Flexibilis* (RF)—spores in straight or flexuous chains.
  2. *Spira* (S):
     a. Spore chains in the form of hooks, open loops or greatly extended coils of large diameter (5–10 μm) (*Retinaculum-Apertum*).
     b. Spore chains in the form of short, gnarled or compact coils or extended, long and open coils (*Spira*).

C. Melanoid pigments: Ability to produce brown, dark brown or black melanin-like pigments (commonly but usually incorrectly referred to as "diffusible or soluble pigments") which color the media in or on which the organisms are grown.

D. Spore wall ornamentation: (See Plate 17.13). Glabrous (smooth), Verrucose (warty), Echinulate (spiny) and Hirsute (hairy).

E. Ability to utilize, under standardized conditions, particular carbon-containing compounds for growth. Those commonly used are: D-glucose, D-xylose, L-arabinose, L-rhamnose, D-fructose, D-galactose, raffinose, D-mannitol, *i*-inositol and sucrose.

F. Ability or inability to produce particular antibiotic factors: groupings depend upon precise identification of particular antibiotics or families of antibiotics; cross-antag-

TABLE 17.40

Primary key to tables of species[a]

Melanoid pigment	Spore surface	White (W) RF	White (W) S	Gray (GY) RF	Gray (GY) S	Yellow (Y) RF	Yellow (Y) S	Red (R) RF	Red (R) S	Blue (B) RF	Blue (B) S	Green (G) RF	Green (G) S	Violet (V) RF	Violet (V) S
C+	SM	17.41a (751)	17.41c (751)	17.42a (758)	17.42c (764)	17.43a (793)		17.44a (804)	17.44c (808)	17.45a (820)	14.45b (820)				
	WTY								17.44d (812)		17.45c (820)				
	SPY		17.41d (751)		17.42d (768)		17.43c (802)		17.44e (812)		17.45d (820)				17.47b (826)
	H		17.41e (751)		17.42e (768)						17.45e (820)				
C−	SM	17.41b (751)	17.41f (753)	17.42b (762)	17.42f (771)	17.43b (795)	17.43d (802)	17.44b (808)	17.44f (814)				17.46a (825)	17.47a (826)	
	WTY				17.42g (782)				17.44h (818)						
	SPY				17.42h (782)				17.44g (818)		17.45f (820)		17.46b (825)		
	H				17.42i (786)								17.46c (825)		
?	?	17.41g (756)		17.42j (788)		17.43e (802)		17.44i (818)				17.46d (825)			

[a] Melanoid pigments produced C+, not produced C−. Spore wall ornamentation: SM = smooth, WTY = warty, SPY = spiny, H = hairy (see Plate 17.13). Spore color *en masse* indicated as White (W), Gray (GY) *etc.* RF = Rectus flexibilis; spore chains straight or flexuous: S = Spira; Spore chains in form of hooks, open loops and coils (see Plate 17.12).

Figures in parentheses indicate page on which tables on utilization of carbon compounds may be found for each particular group or groups; descriptions of species covered immediately follow the table or begin on a nearby page.

onism spectra; antibiotic spectra and the like.

G. Color of vegetative mycelium: colorless, yellow, brown to black, orange, pink, red and red-blue violet.

The list of described species of *Streptomyces* has grown to many hundreds and proposals of new species continue. To give adequate systematic consideration to this large assemblage, it was necessary to depart from the format and taxonomic treatment employed in previous editions of THE MANUAL. Certain ground rules were established and followed in selecting strains and taxa for inclusion:

1. Only names of species and subspecies published prior to July 1, 1967 were considered, with one exception: new combinations for streptomycetes, first published under other generic names prior to July 1, 1967, e.g. *Actinomyces*, *Chainia* and *Nocardia*, also were considered.

2. If the original strain is no longer extant, only an officially designated neotype strain was accepted. *Streptomyces albus* is the only species in this category.

3. When one or more strains are currently available and precisely designated, but none has been designated as type or neotype, a suitable strain was designated as a "reference strain" for inclusion in THE MANUAL.

4. Unless a reasonably clear definition of the color of aerial mycelium (spore color *en masse*)

was apparent from the available published characterizations of the type or reference strain, or from actual observations, that species is included in the list of *species incertae sedis* (p. 827).

5. Names of valid and legitimate species for which there are no strain designations in the original descriptions and efforts to elicit strain designations were unsuccessful, and those for which strains are not known to be extant are also included as *species incertae sedis*.

Data used for characterizing *Streptomyces* were derived from three main sources: (1) from the original descriptions of the taxa; (2) from publications resulting from the International *Streptomyces* Project (Shirling and Gottlieb, 1968, 1969); and (3) from the personal observations of the authors. Wherever possible, descriptions are based on published data on type strains or on observation of cultures of type strains. Where two ATCC numbers are given in descriptions, the second is that of the International *Streptomyces* Project (ISP) strain.

The large number of species recognized necessitated conservation of space. Table 17.40 is a primary key to the groups based on spore color, spore chain morphology, spore wall ornamentation and melanin pigment production: the page is given on which the Table appears in which carbon utilization is consolidated and on or near which the descriptions of the species in that group begin.

Description of the species of genus **Streptomyces**

Editorial Note. The presentation of the species of *Streptomyces* is different from that of other genera. As a result the type species is no. 15 instead of no. 1. Because of the application of criteria regarded as newer and more precise than those used previously, subspecies are not always listed after the primary species. Also, subspecies are assigned numbers rather than lettered as in other genera; there are thus actually 416 species and 47 subspecies listed.

Active programs are being conducted on *Streptomyces* and related genera by the Subcommittee on Actinomycetes of the International Committee on Systematic Bacteriology and by the Subcommittee's International *Streptomyces* Project (ISP). Although all of the presentations in THE MANUAL represent the current "state of the art," this term is probably more applicable to this genus. This is indicated by the large number of species recognized here, as of July, 1967, and the additional

species described by Krasil'nikov (1970) and Shirling and Gottlieb (1972), by the fact that the primary species of four subspecies recognized are considered *incertae sedis* and by the need for the more historical treatment used.

The histories of strains were traced as accurately as possible to determine source, type, neotype or reference strains, and to provide more complete characterizations. Where no reference is given the type was designated in the original description; other designations are mentioned where known. Some strain references are of historical value only; for example, the CBS identified early strains by the name of the depositor, even though he may have deposited several different species; these are indicated in quotation marks and should not be used for ordering strains. Many patents are mentioned; although names given in patents are not legitimate, they are often of historical and practical interest.

TABLE 17.41a–e

White series[a]

Names of species and subspecies	Strain desig- nation	Utilization of carbon compounds (P & G basal)											
		No carbon control	d-Glucose	d-Xylose	l-Arabinose	l-Rhamnose	d-Fructose	d-Galactose	Raffinose	d-Mannitol	i-Inositol	Salicin	Sucrose
17.41a W; RF; C+; SM													
1. Streptomyces albolongus**[b]	T		+	+	+	−	−	+	+	−	−	+	−
2. S. viridaris	T		+		+			+		−	−		+
17.41b W; RF; C−; SM													
3. S. alboniger	T	−	+	+	+	−	+	+	−	+	+	+	−
4. S. albosporeus subsp. labilomyceticus (see no. 392)***[c]	T		+	+	−	−	−	+	+	−		+	+
5. S. albovinaceus	T	−	+	+	+	+	+	+	−	+	−	+	−
6. S. aureocirculatus	T		+	−	−	+	+	−	+	+			−
7. S. baarnensis	T	−	+	+	+	+	+	+	−	+	+		
8. S. clavifer	T	−	+	+	−	+	+	−	+	−			−
9. S. galtieri	T	−	+	−	−	−	+	+	−	−	−		+
17.41c W; S; C+; SM													
10. S. bobili	T	−	+	+	+	+	+	+	+	−	+	−	+
11. S. longispororuber	R												
17.41d W; S; C+; SPY													
12. S. longisporus	R		+	+	+	+	+		+	+	+		+
17.41e W; S; C+; H													
13. S. herbeus	T		+	−	−	+		+	−	+	−		−

[a] In all tables the reactions for utilization of carbon compounds are for the type (T), neotype (N) or reference (R) strain listed and were obtained on P & G basal medium (Pridham and Gottlieb, 1948) unless noted otherwise. + indicates carbon compound utilized; − not utilized; ± very slight utilization.

[b] ** (Double asterisk) indicates carbon utilization basal medium not specified.

[c] *** (Triple asterisk) indicates carbon utilization basal medium, Czapek's solution.

1. **Streptomyces albolongus** Tsukiura, Okanishi, Koshiyama, Ohmori, Miyaki and Kawaguchi 1964, 225.

al.bo.lon′gus. L. adj. *albus* white; L. adj. *longus* long; M.L. adj. *albolongus* white and long.

Produces proceomycin, an anti-bacterial antibiotic; inhibited by streptomycin.

Type strain: Bristol-Banyu 304R7 (single isolate).

2. **Streptomyces viridaris** (*sic*) (Krasil'nikov and Egorova) Pridham 1970, 31. (*Actinomyces viridaris* Krasil'nikov and Egorova in Krasil'nikov 1965, 175.)

vi.ri.da′ris. L. v. *virido* become green. M.L. adj. *viridaris* producing green.

Exhibits anti-bacterial activity; forms green vegetative mycelium on some media.

Type strain: INMI 1876 (Pridham, 1965, 31) listed as the typical representative of the taxon in the original description.

3. Streptomyces albo-niger (sic) Porter, Hewitt, Hesseltine, Krupka, Lowery, Wallace, Bohonos and Williams 1952, 409.

al.bo-ni'ger. L. adj. *albus* white, L. adj. *niger* black, M.L. adj. *albo-niger* whitish black.

Produces puromycin; forms black to greenish black diffusible pigment when grown on starch agar, glucose agar, potato plugs and certain other media; spore chains typically flexuous; NaCl tolerance $\geq 10\%$ but $<13\%$.

Type strain: P-638; ATCC 12461; ATCC 19722 (Waksman, 1961, 170).

4. Streptomyces albosporeus subsp. **labilomyceticus** Okami, Suzuki and Umezawa 1963, 154. (*Streptomyces albosporeus* var. *labiolomyceticus, lapsus calami*) Okami, Suzuki and Umezawa 1963, 154.)

subsp. la'bil.o.my.ce'ti.cus. L. adj. *labilis* perishable, unstable; English n. *labilomycin* name of an unstable antibiotic; L. n.suff. -*icus* belonging to; M.L. adj. *labilomyceticus* belonging to labilomycin.

Produces labilomycin; vegetative mycelium red, yellow, violet, brown or orange.

Type strain: A955-Y3; NIHJ 425 (single isolate).

5. Streptomyces albovinaceus (Kudrina) Pridham, Hesseltine and Benedict 1958, 57. (*Actinomyces albovinaceus* Kudrina in Gauze *et al.* 1957, 118.)

al.bo.vi.na'ce.us. L. adj. *albus* white; L. adj. *vinaceus* of or belonging to wine or the grape; M.L. *albovinaceus* white wine or white grape, but probably refers to pink or red-tinged aerial mycelium, vegetative mycelium and diffusible pigments.

Reportedly exhibits anti-bacterial activity; aerial mycelium, vegetative mycelium and diffusible pigments exhibit tinges of pink or red; spore chains typically flexuous; NaCl tolerance $\geq 4\%$, but $<7\%$.

Type strain: INA 273/53; ATCC 15823; ATCC 19723 (Gauze in Gottlieb 1968, 20).

6. Streptomyces aureocirculatus (Krasil'nikov and Yuan) Pridham 1970, 8. (*Actinomyces aureocirculatus* Yuan in Konova 1962, 188 (Not Val. Pub.); *Actinomyces aureocirculatus* Krasil'nikov and Yuan 1965, 33.)

au.re.o.cir.cu.la'tus. L. adj. *aureus* golden; L. part.adj. *circulatus* curled; M.L. adj. *aureocirculatus* golden-curled.

Exhibits anti-bacterial activity consisting of at least two components; good growth on Czapek's solution agar.

Type strain: INMI 735; RIA 682; ATCC 15851; ATCC 19823 (single isolate).

7. Streptomyces baarnensis Pridham, Hesseltine and Benedict 1958, 74.

baarn.en'sis. Dutch *Baarn*, a community in the Netherlands province of Utrecht; M.L. adj. *baarnensis* relating to Baarn.

Excellent growth on Czapek's solution agar; inhibited by streptomycin.

Type strain: CBS (Dreyfus 472); CBS 306.55; CBS 665.68; ATCC 23885 (single isolate).

8. Streptomyces clavifer (Millard and Burr) Waksman in Waksman and Lechevalier 1953, 103. (*Actinomyces clavifer* Millard and Burr 1926, 630.)

cla'vi.fer. L. adj. *clavifer* club-bearing.

Poor to fair growth on Czapek's solution agar; NaCl tolerance $\geq 10\%$, but $<13\%$.

Type strain: CBS 101.27 (by deposit by Millard in 1927).

9. Streptomyces galtieri Goret and Joubert 1951, 126.

gal.ti.er'i. M.L. gen.n. *galtieri* of Galtier; named for Professor Galtier of the Veterinary School, Lyons, France.

Poor growth on Czapek's solution agar; isolated from cases of canine septicemia (thoracic, abdominal and brain lesions); NaCl tolerance $\geq 7\%$, but $<10\%$, said to dissociate into two biotypes designated A and B. Possibly a *Nocardia*.

Type strain: Goret and Joubert AB; CBS (Goret); CBS 368.49.

10. Streptomyces bobili (Waksman and Curtis) Waksman and Henrici 1948, 937. (*Actinomyces bobili* Waksman and Curtis 1916, 121; *Streptomyces bobiliae* (sic) (Waksman and Curtis) Waksman and Henrici 1948, 937.)

bo.bi'li. M.L. n. *bobili* named for Bobili; the nickname of an individual.

Produces a cinerubin-like antibiotic; inhibited by streptomycin; excellent growth on Czapek's solution agar; forms reddish to violet-reddish vegetative mycelium on some media; does not always exhibit melanin-like chromogenicity; NaCl tolerance $\geq 4\%$, but $<7\%$.

Type strain: IMRU 3310 (single isolate).

11. Streptomyces longispororuber Waksman in Waksman and Lechevalier 1953, 99. (*Actinomyces longisporus ruber* Krasil'nikov 1941, 22 (Not Val. Pub.); *Streptomyces longisporus ruber* (Krasil'nikov) Pridham *et al.* 1958, 64 (Not Val. Pub.).)

lon.gi.spo.ro.ru'ber. L. adj. *longus* long; Gr. n. *spora* a seed; M.L. n. *spora* a spore; L. adj. *ruber* red; M.L. adj. *longispororuber* long-spored, red.

Spore chains of atypical *Retinaculum-Apertum*

type. Characterization taken from Hütter 1967, 23 and 289–290.

Reference strain: INA 11668/54. None of Krasil'nikov's original isolates is extant.

12. **Streptomyces longisporus** (Krasil'nikov) Waksman in Waksman and Lechevalier 1953, 39. (*Actinomyces longisporus* Krasil'nikov 1941, 47.)

lon.gi'spo.rus. L. adj. *longus* long; Gr. n. *spora* a seed; M.L. n. *spora* a spore; M.L. adj. *longisporus* long-spored.

Reference strain: INA 4417/56; ATCC 23931 (suggested as a possible neotype strain by Krasil'nikov and Preobrazhenskaya in Shirling

and Gottlieb 1968, 432). None of Krasil'nikov's original isolates is extant.

13. **Streptomyces herbeus** (Krasil'nikov and Egorova) Pridham 1970, 18. (*Actinomyces herbeus* Krasil'nikov and Egorova in Krasil'nikov 1965, 201.)

her'be.us. L. adj. *herbeus* grass-green.

Exhibits anti-bacterial activity; excellent growth on Czapek's solution agar; forms green-colored vegetative mycelium on some media; hairs on spores appear to have knob-like tips. This taxon might also be placed in the Gray color series. Type strain: INMI 2389 (single isolate).

TABLE 17.41f
White series[a]

Names of species and subspecies	Strain desig-nation	Utilization of carbon compounds (P & G basal)										
		No carbon control	D-Glucose	D-Xylose	L-Arabinose	L-Rhamnose	D-Fructose	D-Galactose	Raffinose	D-Mannitol	i-Inositol	Salicin	Sucrose
17.41f W; S; C−; SM													
14. S. albofaciens	T		+	±	+	−	+		+	+	+		±
15. S. albus (see nos. 16, 17 and 35 and *incertae sedis* nos. D 2 and 3).	N	−	+	+	−	−	±	+	−	+	−	+	
16. S. albus subsp. bruneomycini (see nos. 15, 17 and 35 and *incertae sedis* nos. D 2 and 3).	T		−	−	−	+	−	+					
17. S. albus subsp. pathocidicus (see nos. 15, 16 and 35 and *incertae sedis* nos. D 2 and 3).	T	−	+	+	+	+	−	+	−	−	+	−	
18. S. almquisti	T	−	+	+	−	−	+	−	+	−	+		−
19. S. aminophilus	T	−	+	+	−	−	+	+	−	+	−	+	
20. S. cacaoi	T	−	+	+	+		−		±	+	−		±
21. S. chrestomyceticus	T	−	+	−	−	−	+	+	−	−	−		
22. S. flocculus	T	−	+	+	+	−	+	+	+	+	+	+	
23. S. gibsonii	T	−	+	+	+	−		−	−	+	−	+	
24. S. herbescens	T		+	+	+	+	−	+	+	+			+
25. S. iodoformicus	T		+	−	+	−	+	+	−	−			+
26. S. ochraceiscleroticus	T		+	+	+	−	+	+	+	+	+	+	
27. S. rangoon	T		+	+	±	−	+		+	−			−
28. S. rimosus (see nos. 29 and 30).	T	−	+		−	+	+	+	+			
29. S. rimosus subsp. paromomycinus (see nos. 28 and 30).	T		+	−	−	−	+	+	+	+	+		−
30. S. rimosus subsp. pseudoverticillatus (see nos. 28 and 29).	T	−	+	−	+	−	+	+	+	+	+	+	
31. S. spiroverticillatus	T	−	+	+	+	−	+	−	−	−			
32. S. subflavus	T		+	−	+	−	+	−	+	+			−
33. S. varsoviensis	T	−	+	−	−	−	+	+	−	+	+	+	−
34. S. xantholiticus	T		+	−	−	−	+						−

[a] For explanation of meanings of symbols see footnotes to Tables 17.40 and 17.41a–e.

14. **Streptomyces albofaciens** Thirumalachar and Bhatt 1960, 63.

al.bo.fa′ci.ens. L. adj. *albus* white; L. v. *facio* make; M.L. part. adj. *albofaciens* making white.

Produces oxytetracycline. Characterization, in part, in Shirling and Gottlieb 1968, 288.

Type strain: 27-A; ATCC 23873 (single isolate).

15. **Streptomyces albus** (Rossi Doria) Waksman and Henrici 1943, 339. (*Streptotrix* (*sic*) *alba* Rossi Doria 1891, 421.)

al′bus. L. adj. *albus* white.

Exhibits only slight anti-bacterial activity; inhibited by streptomycin; poor growth on Czapek's solution agar; gelatin is liquefied; casein is hydrolyzed; starch is not hydrolyzed; tyrosine and xanthine are solubilized; non-acid-fast; optimum temperature for growth 25–44 C; whole cell hydrolysates contain L-diaminopimelic acid and no major amounts of arabinose or galactose; NaCl tolerance $\geq$ 13%, but <15%; neotype strain isolated from straw.

Neotype strain: IMRU 3004; ATCC 3004. Opinion 29, Jud. Comm. 1963, 123.

16. **Streptomyces albus** subsp. **bruneomycini** (Kudrina, Ol'khovatova, Murav'eva and Gauze) Pridham 1970, 6. (*Actinomyces albus* subsp. *bruneomycini* Kudrina, Ol'khovatova Murav'eva and Gauze 1966, 403.)

subsp. bru′ne.o.my.ci.ni. M. L. adj. *bruneus* dark brown; M.L. gen.suffix *-mycini* for names of antibiotics; M.L. gen.n. *bruneomycini* of bruneomycin.

Produces the anti-bacterial and anti-tumor antibiotic bruneomycin.

Type strain: INA 471/63 (single isolate).

17. **Streptomyces albus** subsp. **pathocidicus** Nagatsu, Anzai and Suzuki 1962, 103.

subsp. pa.tho.ci′di.cus. Gr. comb. form *patho-* from Gr. n. *pathos* disease; L. v. suff. *-cida* from L. v. *caedo* to cut, to kill; L. adj.n.suff. *-icus* belonging to; M.L. adj. *pathocidicus* belonging to pathocide, but obviously referring to the antibiotic pathocidin.

Produces blasticidin S and pathocidin (8-azaguanine); NaCl tolerance $\geq$4%, but <7%.

Type strain: IPCR B-28; ATCC 14510 (single isolate).

18. **Streptomyces almquistii** (Duché) Pridham, Hesseltine and Benedict 1958, 74. (*Actinomyces almquisti* Duché 1934, 278.)

alm′quis.ti.i. M.L. gen.n. *almquistii* of Almquist; named for an early investigator of *Actinomycetales*.

Exhibits slight anti-bacterial activity; inhibited by streptomycin; poor growth on Czapek's solution agar; forms yellowish vegetative mycelium on some media.

Type strain: CBS (*A. albus* Krainsky (Waksman and Curtis); ATCC 618 (single isolate obtained by Duché from CBS as *A. albus* Krainsky (Waksman and Curtis) in 1932 and renamed as *A. almquisti*).

19. **Streptomyces aminophilus** Foster in Oswald *et al.* in Hütter 1961, 370. (*Streptomyces aminophilus* Foster in Oswald, Reedy and Randall 1956, 236 (Not Val. Pub.), *Streptomyces aminophilus* in Campbell *et al.* 1956, 240 (Not Val. pub.); *Streptomyces aminophilus* in W. E. Wooldridge, German Patent 1,000,966, July 4, 1957 (Not. Val. Pub.); *Streptomyces aminophilus* in Pridham *et al.* 1958, 76 (Not Val. Pub.); *Streptomyces aminophilus* in Waksman 1961, 324 (Not Val. Pub.)

a.mi.no′phi.lus. M.L. n. *aminum* amine; Gr. adj. *philus* loving; M.L. adj. *aminophilus* amine- nitrogen loving.

Exhibits anti-bacterial activity; produces perimycin (Nepera 1968, antibiotic 1968, aminomycin, fungimycin), a heptaenic anti-fungal antibiotic; inhibited by streptomycin; poor growth on Czapek's solution agar; NaCl $\geq$13 but <15%.

Type strain: Nepera 1968; NRRL 2390; ATCC 13558; ATCC 14961; ATCC 23878 (single isolate).

20. **Streptomyces cacaoi** (Waksman) Waksman and Henrici 1948, 951. (*Actinomyces cacaoi* Waksman in Bunting 1932, 516.)

ca.ca′o.i. Mexican Spanish *cacao* the cacao; M.L. gen.n. *cacaoi* of cacao.

Originally isolated from musty cacao beans from Nigeria; poor growth on Czapek's solution agar; NaCl tolerance $\geq$13%, but <15%.

Type strain: Bunting III (203H); IMRU 3082 (Waksman, 1961, 183).

21. **Streptomyces chrestomyceticus** Canevazzi and Scotti 1959, 248.

chres′to.my.ce′ti.cus. Gr. adj. *chrestos* useful; English n. *chrestomycin* name of an antibiotic; L. adj.suff. *-icus* belonging to; M.L. adj. *chrestomyceticus* belonging to chrestomycin.

Produces neomycins E and F (paromomycins I and II) exhibits anti-fungal activity; slight to no inhibition by streptomycin; moderate growth on Czapek's solution agar.

Type strain: NCIB 8995; CMI 79589; IMRU 3835; ATCC 14957 (single isolate).

22. **Streptomyces flocculus** (Duché) Waksman and Henrici 1948, 955. (*Actinomyces flocculus* Duché 1934, 300.)

floc′cu.lus. L. n. *floccus* a flock of wool; Med.L. dim.adj. *flocculus* like a small flock of wool.

Produces ferrioxamine E; relatively poor growth on Czapek's solution agar.

Type strain: MNHN-373; LC-373; NRRL B-2843 (single isolate).

23. Streptomyces gibsonii (Erikson) Waksman and Henrici 1948, 963. (*Streptothrix* sp. in Gibson 1920, 357; *Oospora* de A. G. Gibson in Sartory 1923, 776; *Actinomyces Gibsoni* (*sic*) Dodge 1935 (Aug. 1) 722; *Actinomyces Gibsonii* Erikson 1935, (Sept. 25), 36; *Nocardia gibsonii* (Erikson) Waksman in Waksman and Lechevalier 1953, 155 (Illeg.).)

gib.so′ni.i. M.L. gen.n. *gibsonii* of Gibson, named for A. G. Gibson who first isolated the organism.

Exhibits slight anti-bacterial activity; poor growth on Czapek's solution agar; isolated from a monkey injected with strain NCTC 450 subsequently named *Actinomyces upcottii* Erikson 1935, 36 originally obtained from the spleen in a case of acholuric jaundice by Dr. A. G. Gibson in 1920.

Type strain: Gibson; 200; NCTC 4575; IMRU 3420; ATCC 6852 (single isolate).

24. Streptomyces herbescens (Krasil'nikov and Egorova) Pridham 1970, 18. (*Actinomyces herbescens* Krasil'nikov and Egorova in Krasil'nikov 1965, 166.)

her.bes′cens. L. v. *herbesco* become green; L. part.adj. *herbescens* becoming green.

Exhibits anti-bacterial and anti-fungal activity; forms green-colored vegetative mycelium on some media.

Type strain: INMI 1252.

25. Streptomyces iodoformicus (Kirillova and El-Registan) Pridham 1970, 19. (*Actinomyces iodoformicus* Kirillova and El-Registan in Krasil'nikov 1965, 314.)

i.o.do.for.mi′cus. M.L. n. *iodoformum* iodoform; M.L. adj. *iodoformicus* related to iodoform.

Exudes characteristic odor resembling that of iodoform; may form a brown pigment on some media; said to exhibit no anti-microbial activity. Possibly a *Nocardia* because of lack of proteolytic and amylolytic activity.

Type strain: INMI 18-18 (Pridham, 1970, 19).

26. Streptomyces ochraceiscleroticus Pridham 1970, 22. (*Chainia ochracea* Kuznetsov 1962, 539.)

och.ra′ce.i.scle.ro′ti.cus. Gr. n. *ochra* ochre; M.L. adj. *ochraceus* like-ochre, rust colored; L. n. *sclerotium* sclerotium; M.L. adj. *ochraceiscleroticus* sclerotium with rust color.

Exhibits anti-bacterial activity and slight anti-fungal activity; excellent growth on Czapek's solution agar; forms sclerotia on some media.

Type strain: 10A-30; RIA 710; ATCC 15814 (single isolate).

27. Streptomyces rangoon (Erikson) Pridham Hesseltine and Benedict 1958, 61. (*Actinomyces rangoon* (*sic*) Erikson 1935, 37; *Nocardia rangoonensis* (*sic*) Waksman and Henrici 1948, 911.)

ran.goon. M.L. n. *rangoon* referring to Rangoon, Burma.

Isolated from fatal case of human pulmonary streptothricosis. Characterization, in part, in Hütter, 1961, 370 and in Shirling and Gottlieb, 1969, 472.

Type strain: NCTC 1678; ATCC 6860 (single isolate).

28. Streptomyces rimosus Sobin, Finlay and Kane in Waksman in Waksman and Lechevalier 1953, 47. (*Streptomyces rimosus* in Finlay *et al.* 1950, 85 (Not Val. Pub.); *Streptomyces rimosus* in Sobin, Finlay and Kane, U.S. Patent 2,516,080, July 18, 1950 (Not Val. Pub.).)

ri.mo′sus. L. adj. *rimosus* full of fissures.

Produces oxytetracycline and rimocidin; not inhibited by streptomycin; converts alkaloids; poor growth on Czapek's solution agar; NaCl tolerance $\geq 13\%$, some strains tolerate 15%.

Type strain: S3279; Pfizer FD 10326; NRRL 2234 ATCC 10970; IMRU 3558; ATCC 23955 (single isolate).

29. Streptomyces rimosus. subsp. **paromomycinus** Coffey, Anderson, Fisher, Galbraith, Hillegas, Kohberger, Thompson, Weston and Ehrlich 1959, 730. (*Streptomyces rimosus* forma *paromomycinus* Coffey *et al.* 1959, 730; *Streptomyces rimosus* forma *paromomycinus* in Parke, Davis & Company, Belgian Patent 547,976, June 15, 1956 (Not Val. Pub.); *Streptomyces rimosus* forma *paromomycinus* in Frohardt *et al.*, U.S. Patent 2,916,485, December 8, 1959 (Not Val. Pub.).)

subsp. par.o.mo.my.ci′nus. Gr. adj. *paromoios* closely resembling; M.L. suff. -*mycin* for antibiotic names; M.L. adj. *paromomycinus* closely resembling mycin (presumably taken from the name of an antibiotic it produces).

Produces the aminocyclitol anti-bacterial antibiotics neomycins E and F (paromomycins I and II); produces the glutarimide anti-fungal antibiotic (streptimidone); inhibited by streptomycin; fair growth on Czapek's solution agar.

Type strain: P-D 04998; NRRL 2455; ATCC 14827 (single isolate).

30. Streptomyces rimosus subsp. **pseudoverticillatus** Cross 1962, 833. (*Streptomyces rimosus* forma *pseudoverticillatus* Cross, 1962, 833.)

pseu.do.ver.ti.cil.la′tus. Gr. adj. *pseudes* false; L. n. *verticillus* a whorl; M.L. adj. *pseudoverticillatus* false whorled.

Produces oxytetracycline; exhibits anti-fungal activity; not inhibited by streptomycin; poor growth on Czapek's solution agar; not a typical verticillate form. Characterization, in part, by T. G. Pridham and A. J. Lyons, Jr.

Type strain: A5279; CUB 125 (personal communication T. Cross to T. G. Pridham, 1969).

31. Streptomyces spiroverticillatus Shinobu 1958, 93.

spiro.ver.ti.cil.la'tus. Gr. n. *spira* a coil, spiral; L. n. *verticillus* a whorl; M.L. adj. *verticillatus* whorled; M.L. adj. *spiroverticillatus* coiled and whorled.

Exhibits anti-bacterial and anti-fungal activity; inhibited by streptomycin; poor growth on Czapek's solution agar; forms unusually long spore chains that terminate in the typical *Retinaculum-Apertum* type structure; not a typical verticillate form. Characterization, in part, in Shirling and Gottlieb, 1968, 172.

Type strain: OEU 508; ATCC 19811 (Shinobu in Shirling and Gottlieb 1968, 172).

32. Streptomyces subflavus (Krasil'nikov, Korenyako and Nikitina) Pridham 1970, 26. (*Actinomyces subflavus* Krasil'nikov, Korenyako and Nikitina in Krasil'nikov 1965, 227.)

sub.fla'vus. L. pref. *sub-* less than, somewhat; L. adj. *flavus* yellow; L. adj. *subflavus* somewhat yellowish.

Exhibits slight anti-bacterial activity; good growth on Czapek's solution agar.

Type strain: INMI 434; ATCC 19846 (single isolate).

33. Streptomyces varsoviensis Kurylowicz and Woznicka 1967, 1. (*Actinomyces varsoviensis* in Kurylowicz and Ulak, Polish Patent 43,565, July 5, 1963 (Not Val. Pub.); *Actinomyces varsoviensis* in Kurylowicz and Ulak, British Patent 963,886, July 15, 1964 (Not. Val. Pub.); *Actinomyces* (*Streptomyces*) (*sic*) *varsoviensis* Kurylowicz and Woznicka 1967, 1.)

var.so'vi.en'sis. M.L. *Varsovia* Warsaw; M.L. adj. *varsoviensis* pertaining to Warsaw; named for Warsaw, Poland.

Produces oxytetracycline; exhibits anti-fungal activity: NaCl tolerance $\geq 13\%$, but $< 15\%$.

Characterization, in part, in Shirling and Gottlieb, 1969, 488.

Type strain: 13-1; ATCC 14631c; NCIB 9522; CBS 357.64 (Kurylowicz and Woznicka in Shirling and Gottlieb, 1969, 488).

34. Streptomyces xantholiticus (*sic*) (Konev and Tsyganov) Pridham 1970, 31. (*Actinomyces xantholiticus* (*sic*) Konev and Tsyganov 1962, 1026.)

xan.tho.li'ti.cus. Gr. adj. *xanthus* yellow; Gr. adj. *lytos* soluble; M.L. adj. *xantholiticus* yellow and soluble (referring to the yellow color of the vegetative mycelium and the tendency of the organism to lyse when maintained on some solid media).

Exhibits anti-bacterial activity; produces antibiotics 1130/12 (xanthalycins A and B), pentaenic anti-fungals; also produces tetraenic anti-fungal activity; spore chains atypical *Retinaculum-Apertum* type.

Type strain: LIA 1130/12 (single isolate).

35. Streptomyces albus subsp. **fungatus** (Solov'eva and Rudaya) Pridham 1970, 6. (*Actinomyces albus* (Rossi Doria, 1891) Gasperini 1894, strain VNIIA 604-36 in Solov'eva and Rudaya 1959, 5–10; *Actinomyces albus* subsp. *fungatus* Solov'eva and Rudaya 1959, 8; *Streptomyces albus* var. *fungistaticus* in Khokhlov and Liberman 1960, 81 (*lapsus calami*).)

subsp. fun.ga'tus. L. n. *fungus* fungus; L. suff. *-atus* provided with; M.L. *fungatus* (probably referring to the anti-fungal activity of the organism).

Forms coiled chains of spores (section *Spira*); does not form melanoid pigments; produces the anti-bacterial and anti-fungal albofungin mixture (*ca.* 10 components, Rosenfeld *et al.* 1963, 320) and albonoursin; poor growth on Czapek's solution agar; no information on ornamentation of spore walls.

TABLE 17.41g

White series[a]

Names of species and subspecies	Strain designation	Utilization of carbon compounds (P & G basal)											
		No carbon control	D-Glucose	D-Xylose	L-Arabinose	L-Rhamnose	D-Fructose	D-Galactose	Raffinose	D-Mannitol	i-Inositol	Salicin	Sucrose
17.41g W; ?; ?; ?													
35. S. albus subsp. fungatus	T												
36. S. hydrogenans	T												
37. S. vendargus	T		+	−	−	−	+		+	+	+		−

[a] For explanation of meanings of symbols see footnotes to Tables 17.40 and 17.41a–e.

Type strain: VNIIA 604-36 (single isolate).

36. Streptomyces hydrogenans Lindner, Junk, Nesemann and Schmidt-Thomé 1958, 117. (*Streptomyces hydrogenans* in Lindner *et al.*, German Ausgelegeschrift 1,016,263, Sept. 26, 1957 (Not Val. Pub.).)

hy.dro′gen.ans. Gr. n. *hydro* water; L. v. *geno* produce; pres.part. *genans* producing; M.L. part.-adj. *hydrogenans* water producing.

Produces 20 β-keto reductase; no information on morphology of chains of spores, formation of melanoid pigment or spore wall ornamentation.

Type strain: FHP 678; ATCC 19631 (single isolate).

37. Streptomyces vendargus Pridham, Hesseltine and Benedict 1958, 58. (*Streptomyces Vendargus* (*sic*) in N. V. Koninklijke Nederlandsche Gist -en Spiritusfabriek, Australian Patent Application 3985/54, April 21, 1955 (Not Val. Pub.); *Streptomyces vendargus* in A. O. Struyk and A. A. Steethman, Canadian Patent 514,164, June 28, 1955 (Not Val. Pub.); *Streptomyces vendargensis* (*sic*) in N. V. Koninklijke Nederlandsche Gist -en Spiritusfabriek, British Patent 764,198, December 19, 1956 (Not Val. Pub.); *Streptomyces vendargus* in A. P. Struyk and A. A. Steethman, Australian Patent 206,799, April 21, 1957 (Not Val. Pub.).)

vend′arg.us. M.L. adj. *vendargus* (of unknown derivation; possibly pertaining to a place in France).

Produces oxytetracycline and Vengicide (an anti-fungal purine antibiotic). Questionable information on morphology of spore chains and none on ornamentation of spore walls. Characterization, in part, in Shirling and Gottlieb, 1969, 490.

Type strain: CBS 154.57 (single isolate).

See Table 17.42a.

38. Streptomyces achromogenes (*sic*) Okami and Umezawa in Umezawa, Takeuchi, Okami and Tazaki 1953, 268.

a.chro.mo′ge.nes. Gr. adj. *achromous* colorless; Gr. v. *gennaio* produce; M.L. adj. *achromogenes* colorless producing (probably intended for *achromatogenes* not producing color).

Produces achromoviromycin and sarcidin; inhibited by streptomycin; poor growth on Czapek's solution agar; some spore chains of atypical *Retinaculum-Apertum* type; NaCl tolerance ≧7%, but <10%.

Type strain: Z-4-1; NIHJ 213; ATCC 12767 (single isolate).

39. Streptomyces antibioticus (Waksman and Woodruff, 1941) Waksman and Henrici 1948, 942. (*Actinomyces* sp. in Waksman and Woodruff 1940, 586; *Actinomyces antibioticus* Waksman and Woodruff 1941, 246.)

an.ti.bi.o′ti.cus. Gr. pref. *anti-* against; Gr. n. *bius* life; M.L. adj. *antibioticus* against life, antibiotic.

Produces the actinomycin X(B) complex; inhibited by streptomycin; poor growth on Czapek's solution agar; spore chains typically flexuous; NaCl tolerance ≧7%, but <10%.

Type strain: IMRU 3435; ATCC 8663 (single isolate).

40. Streptomyces bikiniensis Johnstone and Waksman 1947, 294. (*Actinomyces bikiniensis* (Johnstone and Waksman, 1947) Krasil'nikov 1949, 100.)

bi.ki.ni.en′sis. M.L. adj. *bikiniensis* pertaining to Bikini atoll.

Produces streptomycin II; very slight inhibition by streptomycin; poor growth on Czapek's solution agar; vegetative growth on most media is gray, black or deep brown; spore chains typically flexuous; NaCl tolerance ≧7% but <10%.

Type strain: IMRU 3514; ATCC 11062 (single isolate).

41. Streptomyces cacaoi subsp. **asoensis** Isono, Nagatsu, Kawashima and Suzuki 1965, 850. (*Streptomyces cacaoi* in Suzuki *et al.* 1965, 131).

subsp. aso.en′sis. M.L. adj. *asoensis* probably pertaining to Aso, a geographical area in Japan.

Produces the polyoxin complex of selectively specific anti-fungal antibiotics comprised of polyoxins A through L. Polyoxin C apparently exhibits no biological activity.

Type strain; IPCR 20-52 (designated herein).

42. Streptomyces cinereoruber Corbaz, Ettlinger, Keller-Schierlein and Zähner 1957, 330.

ci.ner.e.o.ru′ber. L. adj. *cinereus* ashy; L. adj. *ruber* red; M.L. adj. *cinereoruber* ashy red.

Produces rhodomycin A and rhodomycin B; inhibited by streptomycin; poor growth on Czapek's solution agar; vegetative growth and diffusible pigment in tints and shades of red on some media.

Type strain: ETH 7451 (Hütter 1967, 269).

43. Streptomyces cinereoruber subsp. **fructofermentans** Corbaz, Ettlinger, Keller-Schierlein and Zähner 1957, 331.

subsp. fruc.to.fer.men′tans. L. n. *fructus* fruit; L. part.adj. *fermentans* fermenting; M.L. part.adj. *fructofermentans* fruit fermenting (but pertaining to ability of the organism to utilize L-rhamnose, D-fructose and D-sorbitol).

Produces cinerubin A and cinerubin B; inhibited by streptomycin; moderate growth on Czapek's

TABLE 17.42a

Gray series[a]

Names of species and subspecies	Strain designation	Utilization of carbon compounds (P & G basal)											
		No carbon control	D-Glucose	D-Xylose	L-Arabinose	L-Rhamnose	D-Fructose	D-Galactose	Raffinose	D-Mannitol	i-Inositol	Salicin	Sucrose
17.42a GY; RF; C+; SM													
38. S. achromogenes (see no. 80).	T	−	+	+	+	+	+	+	−	+	+	+	−
39. S. antibioticus	T	−	+	+	+	+	+	+	−	+	+	−	−
40. S. bikiniensis	T	−	+	+	−	−	−	+	−	−	−	±	±
41. S. cacaoi subsp. asoensis (see no. 20).	T		+	+	+		+	+	+		+		+
42. S. cinereoruber (see no. 43).	T		+	+	+	−	−	+	−	−	−	+	−
43. S. cinereoruber subsp. fructofermentans (see no. 42).	T	−	+	+	+	+	+	+	−	−	−	+	
44. S. cylindrosporus subsp. piceus (see *incertae sedis* no. A3).	T												
45. S. ederensis	T												
46. S. fulvoviolaceus	T	−	+	+	+	+	+	+	+	+	+	+	+
47. S. fulvoviridis	T	−	+	+	+	+	+	−	+	−	−	−	
48. S. gardneri	T	−	+	+	+	+	+	+	−	−	+		
49. S. globosus	R	−	+	+	+	−							
50. S. griseorubiginosus	R		+	+	+	+	+		+	+	+		+
51. S. herbaricolor	T	−	+	+	−	+	+	−	−			−	+
52. S. indigoferus	T	−	+	+	−	−	−	+	−	−			−
53. S. litmocidini	T		+	±	+	−	±	−	−				−
54. S. narbonensis (see no. 73).	T	−	+	+	−	+	+	−	−				+
55. S. nashvillensis	T	−	+	+	−	−	−	+	−			+	±
56. S. noboritoensis	T		+	+	−	+	+	+	−	+	+	+	±
57. S. phaeopurpureus	T	−	+	+	+	+	+	+	+	+	+	−	
58. S. purpeofuscus	T	−	+	+	−	−	+	−	−	−			−
59. S. showdoensis	T		+	+	±	−	+	+	−	−		+	±
60. S. tanashiensis (see no. 265).	T	−	+	+	−	−	+	−	−		+		
61. S. violaceorectus	T		+	+	−	+	−	−	+				+
62. S. zaomyceticus	T	−	+	+	+	−	+	−	−	−	+	+	+

[a] For explanation of meanings of symbols see footnotes to Tables 17.40 and 17.41a–e.

solution agar; vegetative growth and diffusible pigment in tints and shades of red on some media.

Type strain: ETH 6143; NRRL 2588 (single isolate).

44. Streptomyces cylindrosporus subsp. piceus Pridham 1970, 35. (*Actinomyces cylindrosporus* subsp. *atratus* (*sic*) Yen and Chou 1964, 432.)

subsp. pi′c.e.us. L. adj. *piceus* pitch black.

Exhibits no anti-microbial activity; forms nearly black vegetative mycelium and diffusible pigment; isolated from forest soil.

Type strain IMASP B8-35 (single isolate).

45. Streptomyces ederensis Wallhäusser, Nesemann, Präve and Steigler 1966, 734. (*Strepto-* *myces ederensis* Lindner, Wallhäusser and Huber in Canadian Patent 672,917, October 22, 1963 (Not Val. Pub.).

eder.en′sis. M. L. adj. *ederensis* pertaining to Eder; named for the Eder valley in Germany.

Produces the moenomycin complex of antibacterial antibiotics comprised of moenomycins A, B_1, B_2 and C. Characterization taken from Wallhäusser *et al.* 1966 and Lindner and Huber in Canadian Patent 672,917, Oct. 22, 1963.

Type strain: 2861; ATCC 15304 (single isolate).

46. Streptomyces fulvoviolaceus (Artamonova and Krasil'nikov) Pridham 1970, 15. (*Actinomyces fulvoviolaceus* Artamonova and Krasil'nikov in Rautenshtein 1960, 335.)

ful'vo.vi.o.la'ce.us. L. adj. *fulvus* reddish yellow; L. adj. *violaceus* violet colored; M.L. adj. *fulvoviolaceus* reddish yellow, violet colored.

Produces the anti-bacterial and anti-viral antibiotic 9700 complex consisting of at least two components (α-rubromycin and β-rubromycin); inhibited by streptomycin; excellent growth on Czapek's solution agar; forms brown to gray-brown vegetative growth tinged with violet; forms red to violet diffusible pigment on some media.

Type strain: INMI 9700; ATCC 15862 (single isolate).

47. Streptomyces fulvoviridis (Kuchaeva, Krasil'nikov, Skryabin and Taptykova) Pridham 1970, 16. (*Actinomyces fulvoviridis* Kuchaeva, Krasil'nikov, Skryabin and Taptykova in Rautenshtein 1960, 251; *Actinomyces oleaceus* nov. spec. Krassil'nikov, Kutchayeva and Skriabin in Suppl. to Abstr. Commun., Symp. Antibiot., Prague, Czech., May 18-23, 1959 (no page number) (Not Val. Pub.); *Actinomyces fulvoviridis* n. sp. Kutchayeva, Krasil'nikov and Skriabin 1960, 58 (Illeg.).

ful.vo.vi'ri.dis. L. adj. *fulvus* reddish yellow; L. adj. *viridis* green; M.L. adj. *fulvoviridis* reddish yellow, green.

Exhibits selective anti-bacterial and anti-fungal activity; inhibited by streptomycin; produces vitamin B_{12}; poor growth on Czapek's solution agar; forms vegetative mycelium colored with tints and shades of green and brown on some media; forms bright yellow diffusible pigment on some media. Characterization, in part, in Kuchaeva *et al.* in Rautenshtein (ed.), 1960, 226 and in Shirling and Gottlieb, 1968, 321.

Type strain: VI-16-3 CSAV; INMI 16-3; RIA 660; ATCC 15863; ATCC 23909 (Krasil'nikov in Shirling and Gottlieb, 1968, 321).

48. Streptomyces gardneri (Waksman) Waksman 1961, 215. (*Proactinomyces* sp in. Gardner and Chain 1942, 123; *Proactinomyces gardneri* Waksman in Waksman, Horning, Welsch and Woodruff 1942, 289; *Nocardia gardneri* (Waksman) Waksman and Henrici 1948, 914.)

gard'ne.ri. M.L. gen.n. *gardneri* of Gardner; named for Professor A. D. Gardner, one of the two who first isolated the organism.

Produces proactinomycins A, B and C; inhibited by streptomycin; excellent growth on Czapek's solution agar; spore chains typically straight; whole cell hydrolysates contain L-diaminopimelic acid.

Type strain: NCTC 6531; ATCC 9604; IMRU 3834; ATCC 23911 (single isolate).

49. Streptomyces globosus (Krasil'nikov) Waksman in Waksman and Lechevalier 1953, 68. (*Actinomyces globosus* Krasil'nikov 1941, 58.)

glo.bo'sus. L. adj. *globosus* spherical (referring to the shape of the spores when examined with the light microscope).

Exhibits anti-bacterial and anti-fungal activity; inhibited by streptomycin; excellent growth on Czapek's solution agar; typically forms straight and short chains of spores. Melanin-like chromogenicity is not expressed to great degree with strain IMRU 3763.

Reference strain: Todorovic; IMRU 3763; NRRL B-2292. None of Krasil'nikov's original isolates is extant.

Note. Waksman 1961, 218-219 designated strain Sanchez-Marroquin DI-15 = IMRU 3736 = ATCC 14979 as the type strain. Strain Sanchez-Marroquin DI-15 would be placed in section *Retinaculum-Apertum* color series Red, and clearly has no relationship to the taxon originally described by Krasil'nikov.

50. Streptomyces griseorubiginosus (Ryabova and Preobrazhenskaya) Pridham, Hesseltine and Benedict 1958, 62. (*Actinomyces griseorubiginosus* Ryabova and Preobrazhenskaya in Gauze *et al.* 1957, 193.)

gri.se.o.ru.bi.gin.o'sus. M.L. adj. *griseus* gray; o.var. of L. adj. *robiginosus* (sic) rusty; M.L. adj. *griseorubiginosus* gray, rusty (referring to the gray aerial mycelium and rosy reddish vegetative mycelium and diffusible pigment on a chemically defined medium).

May exhibit anti-bacterial and anti-yeast activity; forms pH-sensitive red, brown or brownish red vegetative mycelium and diffusible pigment on some media (pink or red under alkaline conditions, yellow-brown or gray under acidic conditions); some spore chains of atypical *Retinaculum-Apertum* type may form; sclerotia-like bodies may form; some discrepancies in spore wall ornamentation have been reported, i.e. spiny spores noted in one instance. Characterization, in part, in Shirling and Gottlieb, 1969, 438.

Reference strain: INA-7712 (designated by Gauze in Gottlieb, 1968, 20, but later stated by Gauze to be "a neotype strain" in Shirling and Gottlieb, 1969, 438).

51. Streptomyces herbaricolor Kawato and Shinobu 1959, 114. (*Streptomyces viridofaciens* (sic) Kawato and Shinobu 1959, 53 (Not Val. Pub.).

her.bar.i'co.lor. L. adj. *herbarius* botanical; L. n. *color* color; M.L. adj. *herbaricolor* probably a *lapsus linguae* for *herbicolor* grass colored, green (referring to the grass green diffusible pigment produced by the organism on chemically defined media).

Inhibited by streptomycin; excellent growth on

Czapek's solution agar; forms green vegetative growth and diffusible pigment on some media; spore chains typically straight; NaCl tolerance <4%.

Type strain: OEU 608; NRRL B-3299; ATCC 23922 (single isolate).

52. Streptomyces indigoferus Shinobu and Kawato 1960, 49.

in.di.go'fer.us. French *indigo* the dye indigo (from India); L. suff. *-fer* (*-ferus*) from L. v. *fero* bear; M.L. adj. *indigoferus* bearing (producing) indigo and referring to production of blue to green diffusible pigments on chemically defined media.

Inhibited by streptomycin; excellent growth on Czapek's solution agar; forms green or blackish blue vegetative growth and diffusible pigment on some media; spore chains typically straight; NaCl tolerance <4%; sparse formation of aerial mycelium; whole cell hydrolysates contain L-diaminopimelic acid.

Type strain: OEU 709; IFO 3868; ATCC 23924 (Shinobu in Shirling and Gottlieb, 1968, 336).

53. Streptomyces litmocidini (Ryabova and Preobrazhenskaya) Pridham, Hesseltine and Benedict 1958, 65. (*Actinomyces litmocidini* Ryabova and Preobrazhenskaya in Gauze *et al.* 1957, 187.)

lit.mo.ci.di'ni. M.L. gen.n. *litmocidini* of litmocidin.

Produces litmocidin (a mixture of granaticin and a granaticin-like antibiotic); inhibited by streptomycin; poor growth on Czapek's solution agar; forms purple vegetative growth and diffusible pigment on some media; NaCl tolerance ≧7%, but <10%.

Type strain INA 1823/55; ATCC 19780 (Gauze in Gottlieb, 1968, 20).

54. Streptomyces narbonensis Corbaz, Ettlinger, Gäumann, Keller, Kradolfer, Kyburz, Neipp, Preolog, Reusser and Zähner 1955, 935. (*Streptomyces narboensis* (*sic*) and *Streptomyces narbonesis* (*sic*) Corbaz *et al.* 1955 in Okami *et al.* 1962, 148 (*lapsi calami*).)

nar.bo.nen'sis. M.L. adj. *narbonensis* belonging to Narbonne, a small community near Cannes on the Côte d'Azur, France, the source of the soil from which the organism was isolated.

Produces narbomycin; inhibited by streptomycin; fair growth on Czapek's solution agar; NaCl tolerance ≧4%, but <7%.

Type strain: ETH 7346; ATCC 19790 (single isolate).

55. Streptomyces nashvillensis McVeigh and Reyes 1961, 312.

nash.vil.len'sis. M.L. adj. *nashvillensis* be-

longing to Nashville, a city in Tennessee, the source of the soil from which the organism was isolated.

Exhibits anti-microbial activity; inhibited by streptomycin; poor growth on Czapek's solution agar; spore chains typically straight.

Type strain: McVeigh V-8; NRRL B-2606 (single isolate).

56. Streptomyces noboritoensis Isono, Yamashita, Tomiyama, Suzuki and Sakai 1957, 21. (*Streptomyces noboritoensis* in Sumiki, Nakamura, Kawasaki, Yamashita, Anzai, Isono, Serizawa, Tomiyama and Suzuki 1955. 170 (Not Val. Pub.).)

no.bo.ri.to.en'sis. M.L. adj. *noboritoensis* belonging to noborito, referring to Inada-noborito, Kawasaki City, Kanagawa Prefecture, Japan, the source of the soil from which the organism was isolated.

Produces hygromycin (homomycin), blastomycin (an anti-fungal antibiotic) and one other antibacterial antibiotic; grows on Czapek's solution agar. Characterization, in part, in Shirling and Gottlieb 1969, 457.

Type strain: 97; IPCR(97); KCC 65; NIHJ 163; OEU (A-3) (Suzuki and Tagisawa in Shirling and Gottlieb, 1969, 457).

57. Streptomyces phaeopurpureus Shinobu 1957, 63.

phae'o.pur.pur'e.us. Gr. adj. *phaeus* brown; L. adj. *purpureus* purple colored; M.L. adj. *phaeopurpureus* brown, purple colored.

Inhibited by streptomycin; excellent growth on Czapek's solution agar; forms red-brown and purple-colored vegetative mycelium on some media.

Type strain: OEU 146; ATCC 23946 (single isolate).

58. Streptomyces purpeofuscus (*sic*) Yamaguchi and Saburi 1955, 207.

pur.pe.o.fus'cus. M.L. variant *purpe-* of L. adj. *purpureus* purple; L. adj. *fuscus* dark, tawny; M.L. adj. *purpeofuscus* dark purple, referring to color of vegetative mycelium.

Exhibits anti-bacterial, anti-fungal and antitrichomonal activity; inhibited by streptomycin; poor growth on Czapek's solution agar; forms gray-black, violet-black and violet-brown vegetative growth on some media; NaCl tolerance <4%.

Produces negamycin.

Type strain: H-5080; NRRL B-1817; ATCC 23952 (single isolate).

59. Streptomyces showdoensis Nishimura, Mayama, Komatsu, Kato, Shimaoka and Tanaka 1964, 150.

show.do.en'sis. M.L. adj. *showdoensis* belonging to Shodo, an island in Kagawa Prefecture, Japan, the source of the soil from which the organism was isolated.

Produces four antibiotics (showdomycin, an actinomycin, a macrolide antibiotic, and one other antibiotic); NaCl tolerance ≧4%, but <7%.

Type strain: SRL Z-452; ATCC 15105 (single isolate).

60. Streptomyces tanashiensis Hata, Ohki and Higuchi 1952, 529. (*Streptomyces* No. 144, a variant strain to *Streptomyces aureus* (*sic*) in Hata, Higuchi, Sano and Sawachika 1949, 229; *Streptomyces* No. 144 resembling or similar to *Streptomyces aureus* in Hata, Higuchi, Sano and Sawachika 1949–1950, 313.)

ta.na.shi.en'sis. M.L. adj. *tanashiensis* belonging to Tanashi-machi, a town near Tokyo, Japan, the source of the soil from which the organism was isolated.

Produces luteomycin and exhibits anti-fungal activity; inhibited by streptomycin; poor growth on Czapek's solution agar; spore chains typically flexuous; NaCl tolerance ≧4%, but <7%.

Type strain: No. 144; KITA 144; ATCC 23967 (single isolate).

61. Streptomyces violaceorectus (Ryabova and Preobrazhenskaya) Pridham, Hesseltine and Benedict 1958, 63. (*Actinomyces violaceorectus* Ryabova and Preobrazhenskaya in Gauze *et al.* 1957, 182.)

vi.o.la.ce.o.rec'tus. L. adj. *violaceus* violet colored; L. adj. *rectus* straight; M.L. adj. *violaceorectus* violet colored, straight (referring to the color of the vegetative mycelium and diffusible pigment on some media and to the structure of sporophores).

Exhibits anti-bacterial and anti-fungal activity; forms vegetative mycelium and diffusible pigment colored in tints and shades of red or purple; pigment is pH sensitive (violet or purple under alkaline conditions, pink or orange under acidic conditions), color of aerial mycelium might also be interpreted as being red. Characterization, in part, in Shirling and Gottlieb, 1969, 497.

Type strain: INA 506 (Gauze in Gottlieb, 1968, 20).

62. Streptomyces zaomyceticus Hinuma 1954, 134.

za.o.my.ce'ti.cus. English n. *zaomycin* an antibiotic named after the Japanese Mt. Zao, the source of the soil from which the organism was isolated; L. n.suff. *-icus* belonging to; M.L. adj. *zaomyceticus* belonging to zaomycin.

Produces zaomycin; inhibited by streptomycin; poor growth on Czapek's solution agar; spore chains typically flexuous.

Note. Hinuma (1954) stated that a complete

description of the species would be published elsewhere. No record of this publication has been found.

Type strain: No. N-187; MTHU N-187; NRRL B-2038 (single isolate).

See Table 17.42b.

63. Streptomyces aburaviensis. Nishimura, Kimura, Tawara, Sasaki, Nakajima, Shimaoka, Okamoto, Shimohira and Isono 1957, 206. (*Streptomyces* S-66 in Nishimura *et al.* 1957, 205.)

a.bu.ra.vi.en'sis. M.L. adj. *aburaviensis* of Aburabi, Shiga Prefecture, Japan, the source of the soil from which the organism was isolated.

Produces aburamycin and exhibits anti-fungal activity; inhibited by streptomycin; poor growth on Czapek's solution agar; spore chains typically straight; spores may be rough walled when viewed by electron microscopy.

Type strain: SRL S-66; ISM 1083; ATCC 23869 (single isolate).

64. Streptomyces caeruleus (Baldacci) Pridham, Hesseltine and Benedict 1958, 60. (*Actinomyces caeruleus* Baldacci 1944, 180.)

cae.ru'le.us. L. adj. *caeruleus* dark blue, azure.

Produces caerulomycin; poor growth on Czapek's solution agar and Pridham and Gottlieb carbon-utilization media; forms bluish black to greenish black diffusible pigment on some media; spore chains typically flexuous; isolated from rice straw; requires alkaline environment for optimal growth.

Type strain: IPV 930 (Baldacci in Shirling and Gottlieb 1972, 279).

65. Streptomyces catenulae Davisson and Finlay in Waksman 1961, 190. (*Streptomyces catenulae* in Davisson and Finlay, United States Patent 2,895,876, July 21, 1959 (Not Val. Pub.).)

ca.ten'u.lae. L. dim.n. *catenula* a small chain; L. gen.dim.n. *catenulae* of a small chain.

Produces neomycins E and F (catenulin = paromomycin I, II); exhibits anti-fungal activity; inhibited by streptomycin; poor growth on Czapek's solution agar; NaCl tolerance ≧10%, but <13%; whole cell hydrolysates contain L-diaminopimelic acid; typically forms short dichotomously arranged chains of spores.

Type strain: Pfizer 5541-6A; Pfizer 6563; ATCC 12476; ATCC 23893 (single isolate).

66. Streptomyces chrysomallus subsp. **fumigatus** Frommer 1959, 202.

subsp. fu.mi.ga'tus. L. part.adj. *fumigatus* smoked.

Produces the actinomycin C complex; inhibited by streptomycin; poor growth on Czapek's solution agar.

Type strain: DOA 1196; NRRL B-2289 (the

TABLE 17.42b

Gray series[a]

Names of species and subspecies	Strain designation	Utilization of carbon compounds (P & G basal)											
		No carbon control	D-Glucose	D-Xylose	L-Arabinose	L-Rhamnose	D-Fructose	D-Galactose	Raffinose	D-Mannitol	i-Inositol	Salicin	Sucrose
17.42b GY; RF; C−; SM													
63. S. aburaviensis	T	−	+	±	−	−	±	−	−	−	−	−	−
64. S. caeruleus	T	−	+	−	−	−							
65. S. catenulae	T	−	+	−	−	−	+	+	−	+	−	−	
66. S. chrysomallus subsp. fumigatus (see no. 288).	T	−	+	+	+	−	−	+	−	−	−	−	
67. S. flavogriseus	T	−	+	+	+	+							
68. S. flavovirens	T	−	+	+	+	+	+	+	+	+	−	−	
69. S. gelaticus	T	−	+	+	−	+	−	+	+	−	−	+	+
70. S. griseolus	T	−	+	+	+	−	+	+	+	−	−	+	
71. S. halstedii	T	−	+	+	+	−	+	+	−	−	+	−	
72. S. misakiensis	T		+	−	−	−	+	−	+	−	−	−	+
73. S. narbonensis subsp. josamyceticus (see no. 54).	T		+	+	+	−	+	+	−	−	−	−	+
74. S. nigrifaciens	T	−	+	+	+	+	+	+	+	+	+	−	
75. S. nitrosporeus	T	−	+	+	+	+	−	+	−	−	−	−	
76. S. olivarius	T												
77. S. omiyaensis	T	−	+	+	−	+	−	+	−	−	−		
78. S. ramulosus	T	−	+	−	−	−	±	+	+	+	−		
79. S. xanthocidicus	T		+	+	+	−	+	+	−	−	−	−	+

[a] For explanation of meanings of symbols see footnotes to Tables 17.40 and 17.41a–e.

only one of three strains isolated that allowed adequate characterization; designated herein).

67. Streptomyces flavogriseus (Duché) Waksman in Waksman and Lechevalier 1953, 55. (*Actinomyces flavogriseus* Duché 1934, 341.)

fla.vo.gri′se.us. L. adj. *flavus* yellow; M.L. adj. *griseus* gray; M.L. adj. *flavogriseus* yellowish gray.

Exhibits slight, if any, anti-microbial activity; inhibited by streptomycin; poor growth on Czapek's solution agar; NaCl tolerance ≧7%, but <10%.

Type strain: Heim; "CBS (Duché)," CBS 101.34 (single isolate).

68. Streptomyces flavovirens (Waksman) Waksman and Henrici 1948, 940.(*Actinomyces 128* in Waksman 1919, 117; *Actinomyces flavovirens* Waksman 1923, 352.)

fla.vo.vi′rens. L. adj. *flavus* yellow; L. part.adj. *virens* being green; M.L. part.adj. *flavovirens* yellow-green.

Produces an actinomycin complex; inhibited

by streptomycin; poor growth on Czapek's solution agar; spore chains typically flexuous; NaCl tolerance ≧7%, but <10%.

Type strain: 128; IMRU 3320; ATCC 3320; ATCC 19758 (single isolate).

69. Streptomyces gelaticus (Waksman) Waksman and Henrici 1948, 952. (*Actinomyces 104* in Waksman 1919, 165; *Actinomyces gelaticus* Waksman 1923, 356.)

ge.la′ti.cus. L. part.adj. *gelatus* congealed, jellied; M.L. adj. *gelaticus* resembling hardened gelatin.

Slight, if any, anti-microbial activity exhibited; inhibited by streptomycin; poor growth on Czapek's solution agar; sparse formation of aerial mycelium; whole cell hydrolysates contain L-diaminopimelic acid; forms green to black vegetative mycelium on some media; NaCl tolerance ≧7%, but <10%.

Type strain: 104; IMRU 3323; ATCC 3323; ATCC 23912 (single isolate).

70. Streptomyces griseolus (Waksman) Waksman and Henrici 1948, 938. (*Actinomyces* 96 in Waksman 1919, 121; *Actinomyces griseolus* Waksman 1923, 369.)

gri.se′o.lus. Med.L. adj. *griseus* gray; M.L. dim.-adj. *griseolus* somewhat gray.

Exhibits slight anti-microbial activity; inhibited by streptomycin; poor growth on Czapek's solution agar; forms gray to black vegetative mycelium on some media; spore chains typically flexuous; NaCl tolerance ≧7%, but <10%.

Type strain: 96; IMRU 3325; ATCC 3325; ATCC 19764 (single isolate).

71. Streptomyces halstedii (Waksman and Curtis) Waksman and Henrici 1948, 953. (*Actinomyces halstedii* Waksman and Curtis 1916, 124.)

hal.ste′di.i. M.L. gen.n. *halstedii* of Halsted; named for Professor Halsted of Rutgers University.

Exhibits little or no anti-microbial activity; inhibited by streptomycin; poor growth on Czapeks' solution agar; forms gray to black vegetative mycelium on some media; spore chains typically flexuous; NaCl tolerance ≧7%, but <10%.

Type strain: IMRU 3328; ATCC 10897; ATCC 19770 (single isolate).

72. Streptomyces misakiensis Nakamura 1961, 86. (*Streptomyces misakiensis* in Isono, Anzai and Suzuki 1958, 264 (Not Val. Pub.).)

mi.sa.ki.en′sis. M.L. adj. *misakiensis* belonging to misaki, referring to Misakicho, Kanagawa Prefecture, Japan, the source of the soil from which the organism was isolated.

Produces tubermycin A and α-phenazine carboxylic acid (tubermycin B), anti-bacterial antibiotics; forms orange or red vegetative mycelium on some media. Characterization, in part, in Shirling and Gottlieb, 1968, 348.

Type strain: 1755; IPCR (7617); ATCC 23938 (single isolate).

73. Streptomyces narbonensis subsp. **josamyceticus** Osono, Oka, Watanabe, Numazaki, Moriyama, Ishida, Suzaki, Okami and Umezawa 1967, 174.

subsp. jo.sa.my.ce′ti.cus. English n. *josamycin* name of an antibiotic; L. n.suff. *-icus* belonging to; M.L. adj. *josamyceticus* belonging to josamycin.

Produces the macrolide antibiotic josamycin.

Type strain: A204-P2 (single isolate).

74. Streptomyces nigrifaciens Waksman 1961, 247. (Actinomyces 145 in Waksman 1919, 167.)

ni′gri.fa′ci.ens. L. adj. *niger* black; L. part.adj. *faciens* producing; M.L. part.adj. *nigrifaciens* producing black pigment.

Exhibits very slight anti-microbial activity; inhibited by streptomycin; poor growth on Czapek's solution agar; spore chains typically flexuous; NaCl tolerance ≧7%, but <10%.

Type strain: 145; IMRU 3067 (single isolate).

75. Streptomyces nitrosporeus Okami 1952, 477.

ni.tro.spo′re.us. Gr. n. *nitrum* nitre; M.L. n. *nitras* nitrate; Gr. n. *spora* a seed; M.L. n. *spora* a spore; M.L. adj. *nitrosporeus* nitrate spored, name is based on rapid spore formation accompanied by vigorous reduction of nitrate by the organism according to Okami 1952, 479.

Produces nitrosporin; inhibited by streptomycin; poor growth on Czapek's solution agar; spore chains typically flexuous; forms green to black vegetative growth on some media; NaCl tolerance ≧7%, but <10%.

Type strain: 0-20; NIHJ 21; ATCC 12769; ATCC 19792 (single isolate).

76. Streptomyces olivarius (Kuchaeva, Krasil'nikov and Skryabin) Pridham 1970, 23. (*Actinomyces olivarius* Kuchaeva, Krasil'nikov and Skryabin 1960, 58.)

o.li.va′ri.us. L. n. *oliva* olive; L. n. stem suff. *-arius* belonging to; M.L. adj. *olivarius* pertaining to olive.

Exhibits anti-bacterial and anti-fungal activity; produces bright green-colored diffusible pigment on chemically defined iron citrate media.

Type strain: INMI 1300a (Pridham 1970, 23).

77. Streptomyces omiyaensis Umezawa and Okami 1950, 293.

o.mi.ya.en′sis. M.L. adj. *omiyaensis* of Omiya City near Tokyo, Japan, the source of the soil from which the organism was isolated.

Produces chloramphenicol; inhibited by streptomycin; poor growth on Czapek's solution agar; spore chains typically flexuous; spores are rough walled when viewed by electron microscopy. Characterization taken from Pridham and Lyons 1965, 338.

Type strain: 102, NRRL B-1587 (single isolate).

78. Streptomyces ramulosus Ettlinger, Gäumann, Hütter, Keller-Schierlein, Kradolfer, Neipp, Prelog and Zähner 1958, 217.

ram.u.lo′sus. L. adj. *ramulosus* much branched.

Produces acetomycin and exhibits anti-fungal activity; inhibited by streptomycin; poor growth on Czapek's solution agar; whole cell hydrolysates contain L-diaminopimelic acid; typically forms short dichotomously arranged chains of spores; forms violet to red-brown or green vegetative mycelium on some media. Characterization taken from Pridham and Lyons, 1965, 341.

Type strain: ETH 17653; ATCC 19802 (single isolate).

79. **Streptomyces xanthocidicus** Nagatsu, Asahi and Suzuki in Asahi, Nagatsu and Suzuki 1966, 196.

xan.tho.ci'di.cus. Gr. adj. *xanthus* yellow; L. root *cid-* of v. *caedo* to cut; M.L. part.adj. *xanthocidicus* pertaining to yellow and to cut, probably referring to the name given the antibiotic produced which, in turn, was probably derived from its activity against *Xanthomonas oryzae*.

Produces xanthocidin; probably good growth on Czapek's solution agar (sucrose nitrate agar).

Type strain: IPCR 51-4 (single isolate).

TABLE 17.42c

Gray series[a]

Names of species and subspecies	Strain designation	No carbon control	D-Glucose	D-Xylose	L-Arabinose	L-Rhamnose	D-Fructose	D-Galactose	Raffinose	D-Mannitol	i-Inositol	Salicin	Sucrose
17.42c			GY; S; C+; SM										
80. S. achromogenes subsp. rubradiris (see also no. 38).	T	±	+	+	+	+	+	+	+	+	±	±	+
81. S. anandii	T		+	+	+	−	+	+	+	+	+	−	
82. S. aurantiogriseus	R		+	+	+	+	+		+	+	+		+
83. S. bobili subsp. sporificans (see also no. 10).	T												
84. S. cinerochromogenes	T		−	+	+			−	−	−		+	+
85. S. cirratus	T	−	+	+	+	−	+	+	−	−	−	−	+
86. S. collinus	T		+	+	+	+	+		+	+	+		+
87. S. eurythermus	T	−	+	+	+	−	+		+	+	+	−	+
88. S. galbus	T	−	+	+	+	−	+		−	+	+		−
89. S. galilaeus	T	−	+	+	+	+							
90. S. griseoruber	T	−	+	+	+	+	+		−		+	+	−
91. S. griseosporeus	T		+	+	+	+	+	+	+	+	+	+	+
92. S. hygroscopicus subsp. ossamyceticus (see also nos. 152, 153, 154, 155 and *Incertae sedis* no. B17).	T	−	+	+	+	+		+	+	+	+	−	+
93. S. kurssanovii	T	−	+	+	+	−	+	+	+	−	−	−	+
94. S. luteogriseus	T	−	+	+	+	+	+	+	+	+	+	+	+
95. S. massasporeus	T	−	+	+	+	+	+	+	+	+	+	+	+
96. S. mirabilis	R	−	+	+	+	+							
97. S. multispiralis	T		+	+	+	−	+	+	+	+	+	+	+
98. S. naganishii	T		+	+	+	+	+	+	+	+	+	+	
99. S. neyagawaensis	T	−	+	+	+	+	+	+	+	+	+	+	+
100. S. nojiriensis	T		+	−	−	−	−	−	−	−	−	+	
101. S. olivochromogenes	T	−	+	+	+	+	+	+	+	+	+		
102. S. phaeofaciens	T	−	+	+	+	+							
103. S. pulveraceus	T	−	+	−	+	+	+	+		−		+	−
104. S. rameus	T	−	+	+	−	+	+	+	+	−		+	+
105. S. resistomycificus	T		+	+	+	+	+		+	+	+		+
106. S. rishiriensis	T	−	+	+	+	+	+	+	+		+	+	+
107. S. thermoviolaceus (see also no. 267).	T												
108. S. violaceochromogenes	R		+	+	+	+	+		+	+	+		+

[a] For explanation of meanings of symbols see footnotes to Tables 17.40 and 17.41a–e.

80. Streptomyces achromogenes subsp. **rubradiris** (*sic*) Bhuyan, Owen and Dietz 1965, 93.

subsp. rub'ra.dir'is. L. adj. *rubidus* reddish; L. n. *iris* the rainbow; M.L. adj. *rubradiris* reddish rainbow.

Produces rubradirin (antibiotic U-11,092); inhibited by streptomycin; poor growth on Czapek's solution agar.

Type strain: UC 2630; NRRL 3061 (single isolate).

81. Streptomyces anandii Batra and Bajaj 1965, 242.

a.nan.di'i. M.L. gen.n. *anandii* of Anand, Gujarat, India, the source of the soil from which the organism was isolated.

Exhibits anti-bacterial activity; produces a pentaenic anti-fungal antibiotic, pentaene G_8; inhibited by streptomycin; excellent growth on Czapek's solution agar.

Type strain: G8; ITCC 1233; ATCC 19388 (designated by Batra for the ISP).

82. Streptomyces aurantiogriseus (Preobrazhenskaya) Pridham, Hesseltine and Benedict 1958, 67. (*Actinomyces aurantiogriseus* Preobrazhenskaya in Gauze *et al.* 1957, 74.)

au.ran. ti.o.gri'se.us. L. n. *aurum* gold; M.L. n. *Aurantium* generic name of the orange; M.L. adj. *griseus* gray; M.L. adj. *aurantiogriseus* orange, gray.

Exhibits anti-bacterial activity; aerial mycelium initially colored orange or red, later becoming gray. Characterization in part, in Shirling and Gottlieb, 1968, 297. Hütter, 1964, 642 comments that two kinds of single spore cultures can be isolated from the type strain: colonies with gray-colored aerial mycelium and coiled chains of spores, and colonies with orange-red aerial mycelium and straight, sterile aerial hyphae.

Reference strain: INA 10369/58; ATCC 23883 (incorrectly designated as the type strain by Gauze in Gottlieb 1968, 20).

83. Streptomyces bobili subsp. **sporificans** Pridham 1970, 34. (*Streptomyces* sp., strain L.A. 5937 in Sensi and Timbal 1959, 160; *Streptomyces bobiliae* subsp. *sporificans* in Sensi and Pagani, British Patent 920,799, March 13, 1963 (Not Val. Pub.).)

subsp. spo.ri'fi.cans. Gr. n. *spora* seed; M.L. n. *spora* a spore; L. v. *facio* to make; L. comb.adj. ending *-ficans* producing; M.L. part.adj. *sporificans* making spores.

Produces the antibiotic L.A. 5937 complex, a sideromycin group anti-bacterial that does not contain iron; spore surfaces reportedly smooth to warty.

Type strain: L.A. 5937; PV 18496 (single isolate).

84. Streptomyces cinerochromogenes (*sic*) Miyairi, Takashima, Shimizu and Sakai 1966, 58.

cin'er.o.chro.mo'ge.nes. L. adj. *cinereus* ashy; Gr. n. *chroma* color; Gr. v.suff. *-genes* producing; M.L. adj. *cinerochromogenes* producing ashy color.

Produces the anti-bacterial antibiotics cineromycin A and cineromycin B.

Type strain: FUJI 50 (single isolate).

85. Streptomyces cirratus Koshiyama, Okanishi, Ohmori, Miyake, Tsukiura, Matsuzaki and Kawaguchi 1963, 61.

cir.ra'tus. L. adj. *cirratus* curled, in ringlets.

Produces the anti-bacterial antibiotics cirramycin A and cirramycin B; inhibited by streptomycin; poor growth on Czapek's solution agar; chains of spores may be of *Retinaculum-Apertum* type; melanin-like chromogenicity may not be very positively expressed; NaCl tolerance $\geqq 4\%$, but $<7\%$.

Type strain: Okanishi A9745; Bristol-Banyu 12090; ATCC 14699 (single isolate).

86. Streptomyces collinus Lindenbein 1952, 380.

col.li'nus. L. adj. *collinus* hilly, mounded.

Produces collinomycin and rubromycin; forms red vegetative mycelium on some media; NaCl tolerance $\geqq 4\%$, but $<7\%$.

Type strain: Ist 301; ATCC 19743 (single isolate).

87. Streptomyces eurythermus Corbaz, Ettlinger, Gäumann, Keller-Schierlein, Neipp, Prelog, Reusser and Zähner 1955, 1202.

eur.y.ther'mus. Gr. pref. *eury* wide; Gr. n. *thermus* heat; M.L. *eurythermus* wide, heat.

Produces angolamycin; inhibited by streptomycin; excellent growth on Czapek's solution agar; forms red-brown to black vegetative mycelium on some media; spore chains of atypical *Retinaculum-Apertum* type; NaCl tolerance $\geqq 7\%$, but $<10\%$.

Type strain: ETH 6677; Ciba A6677; ATCC 14975; ATCC 19749 (Hütter in Shirling and Gottlieb, 1968, 106).

88. Streptomyces galbus Frommer 1959, 195.

gal'bus. L. adj. *galbus* greenish yellow.

Produces the actinomycin X complex; inhibited by streptomycin; poor growth on Czapek's solution agar; forms bright orange or bright yellow vegetative mycelium and diffusible pigment on some media; spore chains of atypical *Retinaculum-Apertum* type: spores may be warty.

Type strain: Wind 731; NRRL B-2283; ATCC 23910 (Frommer in Shirling and Gottlieb, 1968, 321).

89. Streptomyces galilaeus Ettlinger, Corbaz and Hütter 1958, 356.

ga.li.lae'us. L. n. *Galilaea* a province in Palestine, apparently the source of the soil (Newi Yusha, Israel) from which the organism was isolated.

Produces ferrimycins A_1 and A_2 (pilosomycins A and B) and cinerubins A and B; inhibited by streptomycin; fair growth on Czapek's solution agar; forms red, red-brown or yellow-red vegetative mycelium on some media; NaCl tolerance $\geqq 4\%$, but $<7\%$.

Type strain: ETH 18822; Ciba A18822; NRRL 2722; ATCC 14969 (single isolate).

90. Streptomyces griseoruber Yamaguchi and Saburi 1955, 220.

gri.se.o.ru'ber. M.L. adj. *griseus* gray; L. adj. *ruber* red; M.L. adj. *griseo-ruber* grayish red.

Exhibits anti-trichomonal activity; poor growth on Czapek's solution agar; forms red to purple vegetative mycelium on some media; spores phalangiform; NaCl tolerance $\geqq 4\%$, but $<7\%$.

Type strain: H-4650; ATCC 23919 (single isolate).

91. Streptomyces griseosporeus Niida and Ogasawara 1960, 23.

gri.se.o.spo're.us. M.L. adj. *griseus* gray; M.L. n. *spora* a spore; M.L. adj. *griseosporeus* gray spored.

Produces the anti-bacterial antibiotic taitomycin.

Type strain: Meiji Seika B-793; 1104 (single isolate).

92. Streptomyces hygroscopicus subsp. **ossamyceticus** Schmitz, Jubinski, Hooper, Crook, Price and Lein 1965, 82.

subsp. os.sa.my.ce'ti.cus. English n. *ossamycin* name of an antibiotic named for Mount Ossa of Greek mythology; L. adj.suff. *-icus* belonging to; M.L. adj. *ossamyceticus* belonging to ossamycin.

Produces ossamycin (an anti-tumor, anti-protozoal and anti-fungal antibiotic); moderate growth on Czapek's solution agar; hygroscopic.

Type strain: C8158; ATCC 15420 (single isolate).

93. Streptomyces kurssanovii (Preobrazhenskaya, Kudrina, Ryabova and Blinov) Pridham, Hesseltine and Benedict 1958, 69. (*Actinomyces kurssanovii* Preobrazhenskaya, Kudrina, Ryabova and Blinov in Gauze *et al.* 1957, 156.)

kurs.sanov'i.i. M.L. gen.n. *kurssanovii* of Kursanov, possibly named after L. I. Kursanov, a Russian microbiologist.

Exhibits anti-bacterial activity; inhibited by streptomycin, poor growth on Czapek's solution agar; spore chains of atypical *Retinaculum-Apertum* type; NaCl tolerance $\geqq 4\%$, but $<7\%$.

Note. Hütter (1967, 321) comments that three strains originally assigned to this species are quite different from each other.

Type strain: INA 10294; ATCC 15824; ATCC 19774 (Gauze in Gottlieb, 1968, 20).

94. Streptomyces luteogriseus Schmitz, Deak, Crook and Hooper 1964, 89.

lute.o.gri'se.us. L. adj. *luteus* golden yellow; M.L. adj. *griseus* gray; M.L. adj. *luteo-griseus* grayish, golden-yellow, referring to the yellowish gray color of sporulating aerial mycelium on certain media.

Produces peliomycin, an anti-neoplastic antibiotic; very poor growth on Czapek's solution (sucrose nitrate) agar; NaCl tolerance $\geqq 7\%$, but $<10\%$.

Type strain: C-4657; ATCC 15072 (single isolate).

95. Streptomyces massasporeus Shinobu and Kawato 1959, 283.

mas.sa.spo're.us. L. n. *massa* mass, lump; M.L. n. *spora* a spore; M.L. adj. *massasporeus* mass, spore, referring to the coalescence of spores into moist masses.

Exhibits anti-microbial activity; inhibited by streptomycin; poor growth on Czapek's solution agar; pigment of vegetative mycelium is red with acid, blue or violet with alkali; aerial mycelium color also might be interpreted as red; spores coalesce into masses; NaCl tolerance $\geqq 7\%$, but $<10\%$.

Type strain: OEU 602; ATCC 19785 (single isolate).

96. Streptomyces mirabilis Ruschmann 1952, 543. (*Actinomyces mirabilis* (Ruschmann) Kudrina in Gauze *et al.* 1957, 107.)

mi.ra'bi.lis. L. adj. *mirabilis* marvelous.

Produces miramycin; inhibited by streptomycin; poor growth on Czapek's solution agar; NaCl tolerance $<4\%$.

Reference strain: AC 680; NRRL B-2400.

97. Streptomyces multispiralis Yamamoto, Iwasa, Shibata, Mizuno and Miyake 1965, 360.

mul.ti.spi.ra'lis. L. adj. *multus* much, many; Gr. n. *spira* a spiral; M.L. adj. *multispiralis* many spirals.

Produces neohumidin, an anti-bacterial and anti-fungal antibiotic.

Type strain: No. 70794 (single isolate).

98. Streptomyces naganishii Yamaguchi and Saburi 1955, 219.

na.ga.nish'i.i. M.L. gen.n. *naganishii* of Naganishi, named for Professor H. Naganishi of the University of Hiroshima, Japan.

Exhibits anti-bacterial, anti-fungal and anti-trichomonal activity; inhibited by streptomycin; poor growth on Czapek's solution agar; NaCl tolerance ≧7%, but <10%.

Type strain: H-4871; ATCC 23939.

99. Streptomyces neyagawaensis Yamamoto, Nakazawa, Horii and Miyake 1960, 268. (*Streptomyces neyagawaensis* in Horii, Miyake, Yamamoto and Nakazawa 1959, 264 (Not Val. Pub.); *Streptomyces neyagawaesis* (*lapsus calami*) in Horii *et al.* 1959, 342.)

ne.ya.ga.wa.en'sis. M.L. adj. *neyagawaensis* belonging to Neyagawa City, Japan, near which the soil was obtained from which the organism was isolated.

Produces the anti-fungal antibiotic folimycin; inhibited by streptomycin; excellent growth on Czapek's solution agar; spores may have rough surfaces.

Type strain: 41895; IFO 3784; NRRL B-3092 (single isolate).

100. Streptomyces nojiriensis Ishida, Kumagai, Niida, Hamamoto and Shomura 1967, 64.

no.jir.i.en'sis. M.L. adj. *nojiriensis* belonging to Nojiri, named for Lake Nojiri at Nagano, Japan, the source of the soil from which the organism was isolated.

Produces nojirimycin, an anti-bacterial antibiotic; poor growth on Czapek's solution agar.

Note. Two additional strains producing nojirimycin, but not included in this taxon, were reported by Ishida *et al.* 1967, 62.

Type strain: SF-426 (single isolate).

101. Streptomyces olivochromogenes (Waksman) Waksman and Henrici 1948, 941. (*Actinomyces* 205 in Waksman 1919, 106; *Actinomyces olivochromogenus* (*sic*) Waksman 1923, 360; *Streptomyces olivochromogenus* (*sic*) (Waksman) Waksman and Henrici 1948, 941.)

o.li.vo.chro.mo'ge.nes. L. n. *oliva* olive; Gr. n. *chroma* color; Gr. v.suff. *-genes* producing; M.L. adj. *olivochromogenes* producing an olive color.

Comment: Krasil'nikov 1949, 107 and Waksman 1957, 772 changed the spelling of the epithet "*olivochromogenus*" to "*olivochromogenes.*"

Exhibits anti-bacterial activity; inhibited by streptomycin; poor to fair growth on Czapek's solution agar. Characterization taken from Pridham, Lyons and Seckinger, 1965, 206 and 227.

Type strain: 205; Waksman 3336; ATCC 3336 (single isolate).

102. Streptomyces phaeofaciens Maeda, Okami, Taya and Umezawa 1952, 327.

phae.o.fa'ci.ens. Gr. adj. *phaeus* brown; L. part.adj. *faciens* producing: M.L. part.adj. *phaeofaciens* producing brown, referring to production of brown diffusible pigment.

Produces the anti-fungal antibiotic, phaeofacin; poor growth on Czapek's solution agar; spores phalangiform; NaCl tolerance ≧7%, but <10%.

Type strain: T-23; NIHJ 226; ATCC 15034 (single isolate).

103. Streptomyces pulveraceus Shibata, Higashide, Kanzaki, Yamamoto and Nakazawa 1961, 172.

pul.ver.ac'e.us. L. n. *pulvis, pulveris* powder; M.L. adj. *pulveraceus* powdery.

Produces neomycins E and F (paromomycin and paromomycin II), zygomycin B, cycloheximide and naramycin B; probably grows poorly on Czapek's solution agar.

Type strain: No. 45449; IFO 3855 (single isolate).

104. Streptomyces rameus Shibata 1959, 398.

ra.me'us. L. n. *ramus* a branch; M.L. adj. *rameus* pertaining to branches.

Produces streptomycin.

Type strain: No. 43797 (single isolate).

105. Streptomyces resistomycificus Lindenbein 1952, 376. (*Streptomyces resistomycificus* Lindenbein and Olfermann in Brockmann and Schmidt-Kastner 1951, 479 (Not Val. Pub.).)

re.sis.to.my.ci'fi.cus. L. v. *resisto* resist; Gr. n. *myces* fungus; L. v. *facio* to make; L. v.suff. *-ficus* producing; M.L. adj. *resistomycificus* making resistant to a fungus; producing resistomycin.

Produces resistomycin; inhibited by streptomycin; excellent growth on Czapek's solution agar; NaCl tolerance ≧4%, but <7%. Characterization in part in Shirling and Gottlieb, 1968, 165.

Type strain: Pürk 262; IMRU 3658; ATCC 19804 (Waksman, 1961, 266).

106. Streptomyces rishiriensis Kawaguchi, Tsukiura, Okanishi, Miyaki, Ohmori, Fujisawa and Koshiyama 1965, 3.

ri.shir.i.en'sis. M.L. adj. *rishiriensis* belonging to Rishiri; named for Rishiri Island, Hokkaido, Japan, the source of the soil from which the organism was isolated.

Produces the coumermycin complex (coumermycins A_1, A_2, B, C and D); inhibited by streptomycin; moderate growth on Czapek's solution agar; spore chains may be of *Retinaculum-Apertum* type.

Type strain: 404Y3; ATCC 14812 (single isolate).

107. Streptomyces thermoviolaceus Henssen 1957, 388. (*Streptomyces thermoviolaceus* subsp. *pingens* in Henssen 1957, 388; *Actinomyces thermoviolaceus* in Agre and Kudryavtseva in Krasil'nikov 1965, 315.)

ther.mo.vi.o.la'ce.us. Gr. n. *thermus* heat; L. adj. *violaceus* violet colored; M.L. adj. *thermoviolaceus* heat, violet colored (probably referring to the thermophilic nature of the species and the violet color of the aerial mycelium).

A eurythermal thermophile and facultative anaerobe; grows well at 28–50 C with formation of aerial mycelium best at higher temperatures; grows only slightly at 60 C and then only under anaerobic conditions; forms yellow, yellow-brown, orange or orange-yellow vegetative mycelium on some media; forms violet, or blackish violet diffusible pigment on some media; spore chains of atypical *Retinaculum-Apertum* type; spores may appear somewhat warty when viewed with the electron microscope; originally isolated from mixed fresh horse and swine manure.

Type strain: R 77; ATCC 19283; CBS 278.66 (Henssen and Schnepf, 1967, 217).

108. **Streptomyces violaceochromogenes** (Ryabova and Preobrazhenskaya) Pridham 1970, 28. (*Actinomyces violaceus chromogenes* n. subsp.

(*sic*) Krasil'nikov 1949, 55; *Actinomyces violaceochromogenes* (Krasil'nikov) Ryabova and Preobrazhenskaya in Gauze *et al.* 1957, 183); *Actinomyces violochromogenes* (*sic*) Artamonova and Krasil'nikov in Rautenshtein 1960, 334.)

vi.o.la.ce.o.chro.mo'ge.nes. L. adj. *violaceus* violet; Gr. n. *chroma* color; Gr. v.suff. *-genes* producing; M.L. adj. *violaceochromogenes* producing violet color.

Probably exhibits anti-microbial activity; aerial mycelium color also may be in the Red series; forms pH-dependent red- to violet-colored vegetative mycelium on some media. Characterization, in part, in Shirling and Gottlieb, 1969, 494.

Reference strain: INA 425 (selected by Preobrazhenskaya and approved by Krasil'nikov, as the strain to be used in the ISP, i.e. a potential neotype strain for the species). Strain INMI 2929; ATCC 15893 also is a potential neotype strain.

Note. See Shirling and Gottlieb (1969, 494–496 and 500–501) for further discussion of the confusing nomenclatural and neotype strain problem connected with the name of the species.

TABLE 17.42d, e

Gray series[a]

Names of species and subspecies	Strain designation	Utilization of carbon compounds (P & G basal)											
		No carbon control	D-Glucose	D-Xylose	L-Arabinose	L-Rhamnose	D-Fructose	D-Galactose	Raffinose	D-Mannitol	*i*-Inositol	Salicin	Sucrose
17.42d GY; S; C+; SPY													
109. S. afghaniensis***	T		+	+	+	+	+	+	+	±	±		+
110. S. arenae	T	−	+	+	+	+	+	+	+	+	+		+
111. S. atrocyaneus	T												
112. S. chromofuscus	R		+	+	+	+		+	+	+			−
113. S. durhamensis	T	−	+	+	+	−	+	+	+	+	+	−	
114. S. echinatus	T	−	+	+	+	+	+	+	+	+	−	−	
115. S. filipinensis	T	−	+	+	+	+	+	+	+	+	−		+
116. S. fimbriatus	R		+	+	+	+	+	+	+	+			
117. S. griseochromogenes	T		+	+	+	−	+	+	+			−	
118. S. iakyrus	T	−	+	+	+	+	+	+	±	+	+	+	
119. S. lucensis	T		+	+	+	−	+	−		+	−		+
120. S. malachitofuscus	T		+	+	+	+	+	−		+	+		+
121. S. malachitorectus	T		+	+	+	+	+	−		+	+		+
17.42e GY; S; C+; H													
122. S. pilosus	T		+	+	+	+	+	−		+	+		−

[a] For explanation of meanings of symbols and asterisks, see footnotes to Tables 17.40 and 17.41a–e.

109. Streptomyces afghaniensis Shimo, Shiga, Tomosugi and Kamoi 1959, 1.

af.ghan.i.en'sis. M.L. adj. *afghaniensis* relating to Afghanistan, the source of the soil from which the organism was isolated.

Produces the anti-bacterial antibiotic, taitomycin; does not always form melanin-like pigments.

Type strain: 772; ATCC 23871 (single isolate).

110. Streptomyces arenae Pridham, Hesseltine and Benedict 1958, 67.

a.ren'ae. L. n. *arena* sand, L. gen.n. *arenae* of sand, referring to a sandy area near Zion, Illinois, U.S.A., from which the organism was isolated.

Comment: The origin of this name is in Abbott Laboratories, British Patent 719,230, December 1, 1954.

Produces mycomycetin, an antibiotic effective against mycobacteria; inhibited by streptomycin; excellent growth on Czapek's solution agar.

Type strain: NA269-M2; NRRL 2377 (single isolate).

111. Streptomyces atrocyaneus (Yen and Chou) Pridham 1970, 33. (*Actinomyces atrocyaneus* Yen and Chou 1964, 428.)

at.ro.cy.an'e.us. L. adj. *ater* black; Gr. adj. *cyaneus* dark blue; M.L. adj. *atrocyaneus* black, dark blue.

Does not exhibit anti-microbial activity; forms blackish blue vegetative mycelium on some media; isolated from forest soil.

Type strain: IMASP B15-27 (single isolate).

112. Streptomyces chromofuscus (*sic*) (Preobrazhenskaya, Blinov and Ryabova) Pridham, Hesseltine and Benedict 1958, 68. (*Actinomyces chromofuscus* Preobrazhenskaya, Blinov and Ryabova in Gauze *et al.*, 1957, 176.)

chro.mo.fus'cus. Gr. n. *chroma* color; L. adj. *fuscus* dark, tawny; M.L. adj. *chromofuscus* dark or tawny colored.

Exhibits anti-microbial activity. Characterization, in part, in Shirling and Gottlieb, 1968, 307.

Reference strain: INA 13638/58; ATCC 23896 (incorrectly designated as the type strain by Gauze in Gottlieb, 1968, 20).

113. Streptomyces durhamensis M.A.Gordon and Lapa 1966, 754.

dur.ham.en'sis. M.L. adj. *durhamensis* belonging to Durham; named for Durham, North Carolina, U.S.A.

Produces durhamycin, an anti-fungal pentaenic antibiotic; inhibited by streptomycin; excellent growth on Czapek's solution agar; spore chains of atypical *Retinaculum-Apertum* type; isolated from soil of potted tomato plant sprayed with 2,4-trichlorophenoxyacetic acid.

Type strain: Warren; M. A. Gordon 59123; ATCC 23194 (single isolate).

114. Streptomyces echinatus Corbaz, Ettlinger, Gäumann, Keller-Schierlein, Kradolfer, Neipp, Prelog, Reusser and Zähner 1957, 199.

ech.in.at'us. Gr. n. *echinus* a hedgehog; M.L. adj. *echinatus* like a hedgehog, bristly.

Produces the anti-bacterial antibiotic, levomycin; inhibited by streptomycin; fair growth on Czapek's solution agar; NaCl tolerance ≧7%, but <10%.

Type strain: ETH 8331; NRRL 2587; ATCC 19748 (single isolate).

115. Streptomyces filipinensis Ammann, Gottlieb, Brock, Carter and Whitfield 1955, 559. (Hitherto undescribed streptomycete in Gottlieb, Ammann and Carter, 1955, 219.)

fi.li.pi.nen'sis. M.L. adj. *filipinensis* belonging to the Philippines, the source of the soil from which the organism was isolated.

Produces filipin, an anti-fungal pentaenic antibiotic complex; inhibited by streptomycin; excellent growth on Czapek's solution agar; NaCl tolerance ≧4%, but <7%.

Type strain: 114-8; NRRL 2437; ATCC 23905 (single isolate).

116. Streptomyces fimbriatus (Millard and Burr) Waksman in Waksman and Lechevalier 1953, 104. (*Actinomyces fimbriatus* Millard and Burr 1926, 639.)

fim.bri.a'tus. L. adj. *fimbriatus* fibrous, fringed.

Produces septacidin, an anti-tumor and anti-fungal purine antibiotic; not inhibited by streptomycin; excellent growth on Czapek's solution agar.

Note. Millard and Burr's original single isolate (no longer extant) was obtained from a case of common potato scab.

The reference strain is discussed in Dutcher *et al.*, 1963, 40; 1964, 83; United States Patent 3,155,647, November 3, 1964; and von Saltza *et al.*, 1964, 15Q.

Reference strain: SC-3683; ATCC 15051 (single isolate, a potential neotype strain). The original single isolate of Millard and Burr is no longer extant.

117. Streptomyces griseochromogenes Fukunaga in Fukunaga, Misato, Ishii and Asakawa 1955, 181.

gri.se.o.chro.mo'ge.nes. M.L. adj. *griseus* gray; Gr. n. *chroma* color; Gr. v.suff. *-genes* producing; M.L. adj. *griseochromogenes* producing gray color.

Produces blasticidins A, B, C and S and cytomycin; excellent growth on Czapek's solution agar; some spore chains of atypical *Retinaculum-Apertum* type; NaCl tolerance $\geq 4\%$, but $<7\%$.

Type strain: 2A-327; ATCC 14511 (single isolate).

118. **Streptomyces iakyrus** de Queiroz and Albert 1962, 33.

iaky.rus. Amazonian oral aboriginal language (Nheêngatû) adj. *iakryus* green, referring to the color of the vegetative mycelium and diffusible pigment on some media.

Produces the pigmented peptidic antibacterial antibiotics iaquirina I, iaquirina II and iaquirina III; inhibited by streptomycin; excellent growth on Czapek's solution agar; forms green or yellowish green vegetative mycelium and diffusible pigment on some media; aerial mycelium color might also be interpreted as red; NaCl tolerance $\geq 7\%$, but $<10\%$.

Type strain: IAUR 3923; 3119; ATCC 15375 (single isolate).

119. **Streptomyces lucensis** Arcamone, Bertazzoli, Canevazzi, DiMarco, Ghione and Grein 1957, 119.

lu.cen'sis. M.L. adj. *lucensis* belonging to Lucca, a city in Italy, the source of the soil from which the organism was isolated.

Produces the tetraenic anti-fungal antibiotic, lucensomycin (etruscomycin); NaCl tolerance $\geq 7\%$, but $<10\%$. Characterization, in part, taken from Waksman, 1961, 237 and Shirling and Gottlieb, 1969, 447.

Type strain: FI 1163; IMRU 3783 (single isolate).

120. **Streptomyces malachitofuscus** (Preobrazhenskaya, Maksimova and Blinov) Pridham 1970, 22. (*Actinomyces malachitofuscus* Preobrazhenskaya, Maksimova and Blinov 1964, 963.)

mal a.chit.o.fus'cus. Gr. n. *malache* the mallow; L. adj. *fuscus* dark or tawny. M.L. adj. *malachitofuscus* mallow, dark (dark green).

Exhibits anti-bacterial activity; forms green vegetative mycelium on some media; spore wall ornamentation may be somewhat hairy.

Comment: Strain INA 7171 is characterized in detail in the original description, whereas strain INA 739 subsequently was selected as the type strain. Characterization, in part, in Shirling and Gottlieb, 1969, 450.

Type strain: INA 739 (Gauze in Shirling and Gottlieb, 1969, 450).

121. **Streptomyces malachitorectus** (Preobrazhenskaya, Maksimova and Blinov) Pridham 1970, 22. (*Actinomyces malachitorectus* Preobrazhenskaya, Maksimova and Blinov 1964, 963.)

mal'a.chit.o.rec'tus. Gr. n. *malache* the mallow; L. adj. *rectus* straight; M.L. adj. *malachitorectus* mallow, straight (straight and mallow green).

Exhibits anti-bacterial activity: forms green vegetative mycelium and diffusible pigment on some media. Characterization, in part, in Shirling and Gottlieb, 1969, 450.

Type strain: INA 8954; ATCC 25472 (Gauze in Shirling and Gottlieb, 1969, 145).

122. **Streptomyces pilosus** Ettlinger, Corbaz and Hütter 1958, 347.

pi.lo'sus. L. adj. *pilosus* hairy, shaggy.

Exhibits anti-microbial activity; NaCl tolerance $\geq 7\%$, but $<10\%$. Characterization, in part, in Shirling and Gottlieb, 1968, 157.

Type strain: ETH 11686; ATCC 19797.

See Table 17.42f.

123. **Streptomyces albidofuscus** Maeda, Kosaka, Okami and Umezawa 1953, 140.

al.bi.do.fus'cus. L. adj. *albidus* white; L. adj. *fuscus* dark, tawny; M.L. adj. *albidofuscus* whitish tawny.

Produces pyridomycin; inhibited by streptomycin; poor growth on Czapek's solution agar; sparse formation of aerial mycelium; Maeda *et al.*, 1953, 140, reported formation of loose spirals; the ISP work (Shirling and Gottlieb, 1968, 364) reported possibly coiled or straight to flexuous chains of spores; Pridham and Lyons, 1969, 199, reported straight chains of spores.

The type strain has also been given the epithet "*pyridomyceticus*" by Yagashita 1955, 201 and Umezawa and Okami 1957, 172.

Type strain: 451-A8; NIHJ 284; NRRL B-2517 (single isolate).

124. **Streptomyces albogriseolus** Benedict, Shotwell, Pridham, Lindenfelser and Haynes 1954, 653.

al.bo.gri.se'o.lus. L. adj. *albus* white; M.L. adj. *griseus* gray; M.L. dim.adj. *albogriseolus* whitish gray.

Produces the neomycin complex; inhibited by streptomycin; excellent growth on Czapek's solution agar.

Type strain: 7-A; NRRL B-1305 (single isolate).

125. **Streptomyces ambofaciens** Pinnert-Sindico 1954, 702. (*Streptomyces* 3486 in Pinnert-Sindico, 1954, 702.)

am.bo.fa'ci.ens. L. adj. *ambo* both; L. part.adj. *faciens* producing, M. L. part.adj. *ambofaciens* producing both, referring to the production of two different antibiotics by the organism.

Produces spiramycins I, II and III and netropsin; inhibited by streptomycin; fair growth on Czapek's solution agar; spore chains of atypical

TABLE 17.42f

Gray series[a]

Names of species and subspecies	Strain designation	Utilization of carbon compounds (P & G basal)											
		No carbon control	D-Glucose	D-Xylose	L-Arabinose	L-Rhamnose	D-Fructose	D-Galactose	Raffinose	D-Mannitol	i-Inositol	Salicin	Sucrose

17.42f GY; S; C−; SM

Names of species and subspecies	Strain designation	No carbon control	D-Glucose	D-Xylose	L-Arabinose	L-Rhamnose	D-Fructose	D-Galactose	Raffinose	D-Mannitol	i-Inositol	Salicin	Sucrose
123. S. albidofuscus	T	−	+	−	−	−	±	±	−	−	−	−	−
124. S. albogriseolus	T	−	+	+	+	+	+	+	+	+	+	+	+
125. S. ambofaciens	T	−	+	+	+	+	+		−	+	+		+
126. S. anthocyanicus	T		+		+	+	+	+		−	−	+	
127. S. antimycoticus	T	−	+	+	+	+	+		−	+	+	+	+
128. S. argenteolus	T	−	+	+	+	+	+	+		+	−	+	−
129. S. atratus	T	−	+	+	−	+	+	+	+	−		+	
130. S. aureofaciens	T	−	+	±	+	−	+	+	−	−	−	−	+
131. S. avellaneus	T		+	±	−		+		−		−		+
132. S. caesius	T		+		−	+	+	+	−		+	+	−
133. S. carnosus	T		+	+	+	+							
134. S. chibaensis	T	−	+	+	+	+	+	+	+		+		+
135. S. coelescens	T		+		+				−				
136. S. coelicolor subsp. achrous (see also nos. 137, 138, 139 and 290).	R	−	+	+	+	+	+	+	+	+	+		
137. S. coelicolor subsp. coelicoferus (see also nos. 136, 138, 139 and 290).	T		+		+	+	+	+	−	+	+		+
138. S. coelicolor subsp. coelicolatus (see also nos. 136, 137, 139 and 290).	T		+		+	+	+	+	+	−	+		−
139. S. coelicolor subsp. coelicovarians (see nos. 136, 137, 138 and 290).	T		+		+	+	+	+	−	−	+		−
140. S. corchorusii	T	−	+	+	+	+	+		+	+	+		+
141. S. cyanogenus	T		+	+	+	+	+		+	+	+		+
142. S. diastaticus subsp. ardesiacus (see also *incertae sedis* no. A4).	T	−	+	+	+	+							
143. S. diastatochromogenes subsp. bracus (see also *incertae sedis* no. A5).	T		+	−	−	+	+	+	+	+	+	+	+
144. S. endus	T	−	+	+	+	+	+	+		+	−	+	−
145. S. erumpens	T	−	+		+	−	+	+	+	+	+		−
146. S. griseoaurantiacus	T		+		+	+	+	+	+		−		−
147. S. griseofuscus	T		+	+	+		+			+	−		−
148. S. griseolosuffuscus	T	−	+	+	+		+	+		−	−		−
149. S. griseoluteus	T	−	+	+	−	−	+	+	−		−		−
150. S. griseus subsp. difficilis (see also nos. 302, 303, 304, 305 and *incertae sedis* nos. B15 and B16).	T												
151. S. humidus	T		+	+	+	+	+	+	−	+	+	+	−
152. S. hygroscopicus (see also nos. 92, 153, 154, 155 and *incertae sedis* no. B17).	R	−	+	+	+	+	+		−		−	+	
153. S. hygroscopicus subsp. angustmyceticus (see also nos. 92, 152, 154, 155 and *incertae sedis* no. B17).	T	−	+	−	−	−		±	±	+	−	−	+

TABLE 17.42f—*Continued*

Names of species and subspecies	Strain designation	Utilization of carbon compounds (P & G basal)											
		No carbon control	D-Glucose	D-Xylose	L-Arabinose	L-Rhamnose	D-Fructose	D-Galactose	Raffinose	D-Mannitol	i-Inositol	Salicin	Sucrose
154. S. hygroscopicus subsp. decoyicus (see also nos. 92, 152, 153, 155 and *incertae sedis* no. B17).	T	−	+	+	−	−	+	+	−	+	+	−	
155. S. hygroscopicus subsp. globosus (see also nos. 92, 152, 153, 154 and *incertae sedis* no. B17).	T	−	+	+	−	−	+	+	+	+	+	−	+
156. S. libani (see also no. 157).	T		+	+	−	−	+		+		+		+
157. S. libani subsp. rufus (see also no. 156).	T		+	+	+	−	+		+		+		+
158. S. lividans	T				−		+	+		+	+		−
159. S. lusitanus	T	−	+	−	±	−	+		−	−	±	−	+
160. S. lydicus	T	−	+	+	+	−	+	+		+	+	−	+
161. S. melanosporofaciens	T		+	+	+	+	+		+	+	+		
162. S. misionensis	T		+	+	+	+	+	+	+	+	+		
163. S. murinus	T	−	+	+	−	−	+		+		+		
164. S. mutabilis	T	−	+	+	+	+	+	+	+	+	+	−	±
165. S. nigrescens	T		+	+	−		+		+	+	+		+
166. S. nodosus	T	−	+			+	+		+		+		
167. S. nogalater	T	−	+	+	+	+	+	+	+	+	+	−	
168. S. olivaceiscleroticus	T	−	+	+	+	+							
169. S. olivaceoviridis	T		+	+	+	+	+		+	+	+		+
170. S. olivaceus	T	−	+	+	+	+	+	+	+	+	+	−	−
171. S. parvullus	T	−	+	+	+	+	+		−	+	+	+	+
172. S. platensis	T	−	+	−	−	−	+		+	+	+		
173. S. plicatus	T	−	+	+	+	+	+	+		+	+		−
174. S. poonensis	T	−	+	+	+	+	+	+		+	+	+	
175. S. psammoticus	T	−	+	−	−	−	+		−	−	−	−	+
176. S. purpurogeniscleroticus	T	−	+	+	+	+	+	+	+	+	+		
177. S. recifensis	T		+	+	+	−	+	+	+	+	−	+	+
178. S. rochei	T	−	+	+	+	+	+	+		+	+	+	−
179. S. rokugoensis	T	−	+	+	+	+	+			+	+	+	
180. S. roseodiastaticus	T	−	+	+	+	+							
181. S. rutgersensis subsp. castelarense (see also no. 319).	T		+	+	+	+	+	+	+	+	−	+	−
182. S. sayamaensis	T		−	−	−		+	+	−	−	−	−	+
183. S. sendaiensis	T		−	−	+	−	+	+		+	+	+	
184. S. sioyaensis	T		+	+		−	+	+	+	+	−	+	+
185. S. tendae	T	−	+	+	+	+	+		−	+	+		+
186. S. thermovulgaris	T		+	+	+	+	+		+	+	+		+
187. S. tricolor	T												
188. S. tubercidicus	T	−	+	−	−	−	+		+	+	+		+
189. S. tumemacerans	T	−	+	−	−	+	+	+	−		+		−
190. S. vastus	T		+	+	+	+	+	+	+	+	+	+	+
191. S. violaceolatus	T		+	+	+	+	+		+	+	+		+
192. S. violaceus-niger (see also *incertae sedis* no. B33).	R	−	+	+	+	+	+		+	+	+	+	−
193. S. violaceus-ruber	T	−	+	+	+	+	+	+	−	+	+	+	−
194. S. viridifaciens	T	−	+	−	−	−	+	+	−	−	−	−	+

a For explanation of meanings of symbols see footnotes to Tables 17.40 and 17.41a–e.

Retinaculum-Apertum type; forms dark brown, dark blue or almost black diffusible pigment when grown on inorganic salts-starch agar; NaCl tolerance ≥10%, but <13%.

Type strain: 3486; ATCC 23877 (single isolate).

126. Streptomyces anthocyanicus (Krasil'nikov, Sorokina, Alferova and Bezzubenkova) Pridham 1970, 7. (*Actinomyces anthocyanicus* Krasil'nikov, Sorokina, Alferova and Bezzubenkova in Krasil'nikov 1965, 118.)

an.tho.cy.an'i.cus. Gr. n. *anthos* a flower; *cyanicus* presumably based on Gr. adj. *cyaneus* dark blue, M.L. adj. *anthocyanicus* presumably referring to dark blue color or anthocyanin.

Exhibits slight anti-bacterial activity; spore chains of typical *Retinaculum-Apertum* type; forms blue-colored vegetative mycelium and diffusible pigment on some media.

Type strain: INMI 69; ATCC 19821 (designated by Krasil'nikov for the ISP).

127. Streptomyces antimycoticus Waksman 1957, 799. (*Streptomyces* sp. in Leben, Stessel and Keitt 1952, 159.)

an.ti.my.co'ti.cus. Gr. pref. *anti-* against; Gr. n. *myces* fungus; M.L. adj. *antimycoticus* anti-fungal.

Produces the endomycin complex (a mixture of polyenic and non-polyenic anti-fungal antibiotics designated as endomycins A, B, C and D); exhibits anti-bacterial activity; inhibited by streptomycin; moderate growth on Czapek's solution agar; spore walls appear somewhat rough; hygroscopic.

Type strain: A158; NRRL 2421; ATCC 23880 (single isolate).

128. Streptomyces argenteolus Tresner, Davies and Backus 1961, 74. (Actinomycete species in Perlman, Fried, Titus and Langlykke, U.S. Patent 2,709,705, May 31, 1955; *Streptomyces argenteolus* in Fried, Perlman, Langlykke and Titus, U.S. Patent 2,855,343, October 7, 1958 (Not Val. Pub).)

ar.gen.te.o'lus. L. adj. *argenteolus* of silver.

Produces anti-bacterial antibiotics soluble in *n*-butanol; inhibited by streptomycin; produces vitamin B_{12}; carries out 16 α-hydroxylation of steroids; poor growth on Czapek's solution agar; NaCl tolerance ≥4%, but <7%.

Type strain: MD 2428; ATCC 11009; ATCC 23882 (single isolate).

129. Streptomyces atratus Shibata, Higashide Yamamoto and Nakazawa 1962, 230. (See Japanese Patent 17,398, September 26, 1961.)

a.tra'tus. L. adj. *atratus* clothed in black.

Produces rufomycin A, rufomycin B and other anti-bacterial activity; probably grows poorly on Czapek's solution agar; spore chains of atypical *Retinaculum-Apertum* type; forms gray to black vegetative mycelium on some media.

Type strain: 46408; IFO 3897; ATCC 14046 (single isolate).

130. Streptomyces aureofaciens Duggar 1948, 177.

au.re.o.fa'ci.ens. L. adj. *aureus* golden; L. part.adj. *faciens* producing; M.L. part.adj. *aureofaciens* producing golden, referring to the production of a golden yellow pigment in the vegetative mycelium of the organism.

Produces 7-chlortetracycline, other antibiotics in the tetracycline family and ayfactins A and B (antifungal heptaene antibiotics); inhibited by streptomycin; excellent growth on Czapek's solution agar; both straight and atypical *Retinaculum-Apertum* type spore chains may be formed; spores phalangiform; NaCl tolerance ≥2%, but <4%. Numerous natural variants and derived mutant strains of this species have been reported in the literature.

Type strain: A-377; NRRL 2209; ATCC 10762; ATCC 23884 (Ettlinger, Corbaz and Hütter, 1958, 355, listed as a "typical strain" in Duggar, 1948, 179).

131. Streptomyces avellaneus Baldacci and Grein 1966, 195.

a'vel.la'ne.us. L. adj. *avellaneus* hazel colored, referring to color of the aerial mycelium of the organism.

Produces tetracycline; poor growth on Czapek's solution agar; both straight and *Retinaculum-Apertum* type spore chains may be formed; spores phalangiform; NaCl tolerance ≥2%, but <4%.

Type strain: 2758 FI; IMRU 3911; NRRL B-3447 (single isolate).

132. Streptomyces caesius (Krasil'nikov, Sorokina, Alferova and Bezzubenkova) Pridham 1970, 10. (*Actinomyces caesius* Krasil'nikov, Sorokina, Alferova and Bezzubenkova in Krasil'nikov 1965, 107.)

cae'si.us. L. adj. *caesius* bluish gray, probably referring to the color of the aerial mycelium of the organism.

Exhibits anti-bacterial and anti-fungal activity; forms blue-colored vegetative mycelium and diffusible pigment on some media.

Type strain: INMI 118 (Krasil'nikov *et al.*, in their original description, state strain INMI 118 "might be" the type strain of the species.)

133. Streptomyces carnosus (Millard and Burr) Waksman in Waksman and Lechevalier 1953, 105. (*Actinomyces carnosus* Millard and Burr 1926, 602.)

car.no'sus. L. adj. *carnosus* pertaining to flesh.

Isolated from small, unruptured potato scab; inhibited by streptomycin; moderate growth on Czapek's solution agar; NaCl tolerance ≧10%, but <13%.

Type strain: 1; "CBS (Millard)"; CBS 100.27; NRRL B-1581 (single isolate).

134. Streptomyces chibaensis Suzuki, Nakamura, Okuma and Tomiyama 1958, 81.

chi.ba.en'sis. M.L. adj. *chibaensis* belonging to Chiba City, Japan, the source of the soil from which the organism was isolated.

Produces acetylene dicarboxamide (aquamycin) an anti-bacterial and anti-tumor antibiotic; exhibits anti-fungal activity; excellent growth on Czapek's solution agar; forms bright yellow vegetative mycelium and diffusible pigment on some media.

Type strain: (a single isolate carrying no strain designation in the original publication); 77-SN-2; ATCC 23895 (Suzuki in Shirling and Gottlieb, 1968, 307).

135. Streptomyces coelescens (Krasil'nikov, Sorokina, Alferova and Bezzubenkova) Pridham 1970, 21. (*Actinomyces coelecsens* Krasil'nikov, Sorokina, Alferova and Bezzubenkova in Krasil'nikov 1965, 110.)

coel.es'cens. L. n. *coelum* heaven, sky; L. adj.v. termination *escens* beginning, slightly; M.L. adj. *coelescens* slightly blue.

Exhibits anti-bacterial and selective anti-fungal activity; good growth on Czapek's solution agar; forms blue-colored vegetative mycelium and diffusible pigment on some media.

Type strain: INMI 20-41; ATCC 19830 (Krasil'nikov in Shirling and Gottlieb 1972, 286).

136. Streptomyces coelicolor subsp. **achrous** (Ryabova and Preobrazhenskaya) Pridham, Hesseltine and Benedict 1958, 75. (*Actinomyces coelicolor* var. *achrous* Ryabova and Preobrazhenskaya in Gauze *et al.* 1957, 191.)

subsp. a'chro.us. Gr. adj. *achrous* colorless.

Exhibits slight anti-bacterial activity; inhibited by streptomycin; excellent growth on Czapek's solution agar; forms deep blue vegetative mycelium and diffusible pigment on some media; NaCl tolerance ≧7%, but <10%.

Reference strain: Waksman; ATCC 3355.

Note. Strain ATCC 3355 was identified originally by Waksman as *Actinomyces violaceus-ruber.*

137. Streptomyces coelicolor subsp. **coelicoferus** (Krasil'nikov, Sorokina, Alferova and Bezzubenkova) Pridham 1970, 11. (*Actinomyces coelicoferus* n. subsp. (*sic*) Krasil'nikov *et al.* in Krasil'nikov 1965, 105.)

subsp. coe.li.co.fe'rus. L. n. *coelum* heaven, sky; L. n. *color* color; L. suff. *fer* (*ferus*) from v. *fero* bear; M.L. adj. *coelicoferus* bearing sky blue color.

Exhibits anti-bacterial activity; forms blue-colored vegetative mycelium and diffusible pigment on some media.

Type strain: INMI 1250 (single isolate).

138. Streptomyces coelicolor subsp. **coelicolatus** (Krasil'nikov, Sorokina, Alferova and Bezzubenkova) Pridham 1970, 11. (*Actinomyces coelicolatus* n. subsp. (*sic*) Krasil'nikov *et al.* in Krasil'nikov 1965, 107.)

subsp. coe.li.co.la'tus. L. n. *coelum* heaven, sky; L. n. *color* color; L. adj.suff. -*atus* provided with; M.L. adj. *coelicolatus* provided with sky (blue) color.

Antagonistic only to some other streptomycetes; forms blue-colored vegetative mycelium and diffusible pigment on some media.

Type strain: INMI 464 (single isolate).

139. Streptomyces coelicolor subsp. **coelicovarians** (Krasil'nikov, Sorokina, Alferova and Bezzubenkova) Pridham 1970, 12. (*Actinomyces coelicovarians* n. subsp. (*sic*) Krasil'nikov 1965, 105.)

subsp. coe.li.co.va'ri.ans. L. n. *coelum* heaven, sky; L. n. *color* color; L. pref. *con-* with; L. part.adj. *varians* varying; M.L. part.adj. *coelicovarians* with varying sky (blue) color.

Exhibits anti-bacterial activity; forms blue-colored vegetative mycelium and diffusible pigment on some media.

Type strain: INMI 62 (Pridham, 1970, 12).

140. Streptomyces corchorusii (*sic*) Ahmad and Bhuiyan 1958, 143.

cor'cho.rus.i.i. M.L. n. *Corchorus* generic name; M.L. gen.n. *corchorusii* (*sic*) of *Corchorus;* the organism was isolated from soil of a field of jute (*Corchorus capsulatus*). The genitive of *Corchorus* is *Corchori.*

Exhibits anti-fungal activity; apparently grows on Czapek's solution agar; NaCl tolerance ≧4%, but <7%. Characterization, in part, in Shirling and Gottlieb, 1969, 420.

Type strain: NCIB 9476 (single isolate).

141. Streptomyces cyanogenus (Krasil'nikov, Sorokina, Alferova and Bezzubenkova) Pridham 1970, 14. (*Actinomyces cyanogenus* Krasil'nikov, Sorokina, Alferova and Bezzubenkova in Krasil'nikov 1965, 114.)

cy.an.o. ge'nus. Gr. adj. *cyaneus* dark blue; Gr. v.suff. *genes* producing; M.L. adj. *cyanogenus* producing dark blue.

Inhibits some other streptomycetes; forms red

or blue vegetative mycelium on some media; vegetative mycelium pigment is pH sensitive (blue under alkaline conditions, red under acidic conditions); aerial mycelium color might also be interpreted as red. Characterization, in part, in Shirling and Gottlieb, 1969, 423.

Type strain: INMI 1112-7; ATCC 19836 (Krasil'nikov in Shirling and Gottlieb 1969, 423).

142. Streptomyces diastaticus subsp. **ardesiacus** (Baldacci, Grein and Spalla) Pridham, Hesseltine and Benedict 1958, 67. (*Actinomyces diastaticus* var. *ardesiacus* Baldacci, Grein, and Spalla 1955, 136.)

subsp. ar.de'si.a.cus. Italian n. *ardesia* slate; L. suff. *-acus* belonging to; M.L. adj. *ardesiacus* slate colored.

Inhibited by streptomycin; excellent growth on Czapek's solution agar.

Type strain: IPV 755; NRRL B-1773 (single isolate).

143. Streptomyces diastatochromogenes subsp. **bracus** Sakagami, Yamabayashi and Sekine 1966, 108. (*Streptomyces* No. 658 strain in Sakagami, Sekine, Yamabayashi, Kitaura, Ueda and Kosaka 1966, 99.)

subsp. bra'cus. L. n. *bractea*, hence *bracus* thin-plated (referring to the hexagonal plate-like crystals of bramycin produced by the organism).

Produces the anti-fungal antibiotic, bramycin; moderate growth on Czapek's solution agar (sucrose nitrate agar); forms pale chartreuse diffusible pigment on some media.

Type strain: 658 (single isolate).

144. Streptomyces endus Anderson and Gottlieb 1952, 302. (*Streptomyces* 9-20 in Gottlieb, Bhattacharyya. Carter and Anderson 1951, 393.)

en.dus. Gr. pref. *endo* inside, within; M.L. pref. *endon* inside, within, hence M.L. adj. *endus* referring to the site (inside the hyphae) of formation of the antibiotic endomycin.

Produces a polyenic anti-fungal (endomycins A and B) and non-polyenic anti-fungal mixture including enhygrofungin exhibits anti-bacterial activity; inhibited by streptomycin; moderate growth on Czapek's solution agar; cultures are hygroscopic with surfaces becoming covered with moist brown to black areas; spores phalangiform, sometimes appearing roughened; NaCl tolerance ≧7%, but <10%.

Type strain: 9-20; NRRL 2339; ATCC 23904 (single isolate).

145. Streptomyces erumpens Calot and Cercos 1963, 159.

er.um'pens. L. part.adj. *erumpens* bursting forth.

Exhibits anti-bacterial activity; inhibited by streptomycin; produces the tetraenic anti-fungal, antibiotic 17732 (tetrins A and B and another polyene) poor growth on Czapek's solution agar. *Note.* See also Calot and Cercos, 1963, 303.

Type strain: INTA 17732; NRRL B-3163 (single isolate).

146. Streptomyces griseoaurantiacus (Krasil'nikov and Yuan) Pridham 1970, 17. (*Actinomyces griseoaurantiacus* Krasil'nikov and Yuan in Krasil'nikov 1965, 52.)

gri'se.o.au.ran.ti'a.cus. M.L. adj. *griseus* gray; L. n. *aurum* gold; M.L. n. *Aurantium* generic name of the orange; M.L. adj. *griseoaurantiacus* orange colored with gray.

Exhibits anti-microbial activity consisting of at least three factors; selectively inhibits *Debaryomyces* species; good growth on Czapek's solution agar; forms orange, red, violet-red or red-orange vegetative mycelium on some media.

Type strain: INMI AK-5; ATCC 19840 (single isolate).

147. Streptomyces griseofuscus Sakamoto, Kondo, Yumoto and Arishima 1962, 98. (*Streptomyces* No. 1068 in Sakamoto 1959, 169; *Streptomyces* No. 1068 in Ogawa, Itō, Inoue and Nishio 1960, 353; *Streptomyces* No. 1068 in Arishima and Sakamoto, Japanese Patent 1,148, March 7, 1961. *Streptomyces* No. 1068 in Arishima and Sakamoto, Japanese Patent 15,945, September 11, 1961.)

gri.se.o.fus'cus. M.L. adj. *griseus* gray; L. adj. *fuscus* dark, tawny; M.L. adj. *griseofuscus* gray, tawny.

Produces bundlins A and B (anti-bacterial antibiotics) and moldcidin A and pentamycin (pentaenic anti-fungal antibiotics); poor growth on Czapek's solution agar; aerial mycelium color might also be interpreted as red. Characterization, in part, in Shirling and Gottlieb, 1968, 326.

Type strain: Meiji Seika 1068; ATCC 23916 (single isolate).

148. Streptomyces griseolosuffuscus Fügner and Bradler 1963, 179.

gri.se'o.lo.suf.fus'cus. M.L. dim.adj. *griseolus* somewhat gray; L. adj. *suffuscus* darkish; M.L. adj. *griseolosuffuscus* somewhat gray, darkish.

Exhibits anti-mycobacterial activity; produces the pentaenic anti-fungal antibiotic fungichromin.

Type strain: IMET JA 3708 (single isolate).

149. Streptomyces griseoluteus Umezawa, Hayano, Maeda, Ogata and Okami 1950, 112.

gri'se.o.lu'te.us. M.L. adj. *griseus* gray; M.L. adj. *luteus* yellow; M.L. adj. *griseoluteus* grayish yellow.

Produces griseoluteins A and B; exhibits slight

anti-fungal activity; inhibited by streptomycin; poor growth on Czapek's solution agar; spore chains of atypical *Retinaculum-Apertum* type; NaCl tolerance ≥10%, but <13%.

Type strain: P-37; NIHJ 22; IMRU 3729; ATCC 12768 (single isolate).

150. Streptomyces griseus subsp. **difficilis** (Yen and Chou) Pridham 1970, 38. (*Actinomyces griseus* var. *difficilis* Yen and Chou 1964, 434.)

subsp. dif.fi′ci.lis. L. adj. *difficilis* difficult, troublesome.

Exhibits anti-fungal activity; isolated from forest soil.

Type strain: IMASP Y1-11 (Pridham, 1970, 38).

151. Streptomyces humidus Nakazawa and Shibata in Imamura, Hori, Nakazawa, Shibata, Tatsuoka and Miyake 1956, 648.

hu′mi.dus. L. adj. *umidus* (less correctly *humidus*) wet, damp, moist.

Produces dihydrostreptomycin, humidin and cobalamines; inhibited by streptomycin; poor growth on Czapek's solution agar; spore chains of atypical *Retinaculum-Apertum* type; spores phalangiform; NaCl tolerance ≥4%, but <7%.

Note. This taxon also is characterized in Nakazawa, Shibata, Tanabe and Yamamoto 1958, 321.

Type strain: 23572; IFO 3520; ATCC 12760; ATCC 23923 (single isolate).

152. Streptomyces hygroscopicus (Jensen) Waksman and Henrici 1948, 953. (*Actinomyces hygroscopicus* Jensen 1931, 357.)

hy.gro.scop′i.cus. Gr. adj. *hygrus* moist; Gr. n. *scopus* watcher; M.L. adj. *hygroscopicus* detecting moisture, covered with moisture, hygroscopic.

Produces the hygromycin complex; inhibited by streptomycin; moderate growth on Czapek's solution agar; spores phalangiform and may appear roughened; NaCl tolerance ≥7%, but <10%.

Reference strain: NRRL 2387 (a potential neotype strain). None of Jensen's original isolates is extant.

153. Streptomyces hygroscopicus subsp. **angustmyceticus** Yüntsen, Ohkuma, Ishii and Yonehara 1956, 200. (New strain 6A-704 in Yüntsen, Yonehara and Ui 1954, 113; *Streptomyces hygroscopicus* (Jensen) Waksman and Henrici 1948, 953 in Sakai, Yüntsen and Ishikawa 1954, 117.)

subsp. an.gust.my.cet′i.cus. L. adj. *angustus* narrow; Gr. n. *myces* fungus; M.L. adj. *myceticus* fungus-like; M.L. adj. *angustmyceticus* like a narrow fungus, but referring to the narrow spectrum of antibacterial activity of the organism, hence angustmycin.

Produces angustmycin A, angustmycin B (adenine) and angustmycin C—anti-mycobacterial

antibiotics; exhibits anti-fungal activity; inhibited by streptomycin; excellent growth on Czapek's solution agar; hygroscopic; NaCl tolerance ≥10%, but <13%.

Type strain: 6A-704; IFO 3934; ATCC 15484 (single isolate).

154. Streptomyces hygroscopicus subsp. **decoyicus** Vavra, Dietz, Churchill, Siminoff and Koepsell 1959, 427. (*Streptomyces hygroscopicus* var. *decoyinine* in Schroeder and Hoeksema 1959, 419 (Not Val. Pub.); *Streptomyces hygroscopicus* var. *decoyicus* in Eble, Hoeksema, Boyack and Savage 1959, 419 (Not Val. Pub.); *Streptomyces hygroscopicus* var. *decoyicus* in Lewis, Reames and Rhuland 1959, 421 (Not Val. Pub.); *Streptomyces hygroscopicus* var. *decoyinine* in DeBoer *et al.*, U.S. Patent 3,094,460, June 18, 1963 (Not Val. Pub.).)

subsp. de.coy′i.cus English n. *decoyinine* name of an antibiotic; L. n.stem suff. -*cus* denoting possession; M.L. adj. *decoyicus* possessing decoyinine.

Produces angustmycins A and C; exhibits anti-fungal activity; inhibited by streptomycin; excellent growth on Czapek's solution agar; hygroscopic; NaCl tolerance ≥10%, but <13%.

Type strain: NRRL 2666 (single isolate).

155. Streptomyces hygroscopicus subsp. **glebosus** Ohmori, Okanishi and Kawaguchi 1962, 26. (*Streptomyces* strain No. 12096 in Okanishi, Koshiyama, Ohmori, Matsuzaki, Ohashi and Kawaguchi 1962, 13; Streptomyces strain No. 12096 in Mayaki, Tsukiura, Wakae and Kawaguchi 1962, 25.)

subsp. gle.bo′sus. L. n. *glaeba* sod, soil, lump, M.L. adj. *glebosus* lumpy.

Produces glebomycin; inhibited by streptomycin; excellent growth on Czapek's solution agar; ridged spores by carbon repligraphy according to Dietz and Matthews, 1968, 935; hygroscopic; NaCl tolerance ≥10%, but <13%.

Type strain: 12096; Bristol A-9634; ATCC 14607 (single isolate).

156. Streptomyces libani Baldacci and Grein 1966, 196.

li′ba.ni. L. gen.n. *libani* of Lebanon, the source of the soil from which the organism was isolated.

Produces libanomycin; exhibits anti-fungal activity; moderate growth on Czapek's solution agar; hygroscopic; aerial mycelium typically colored in shades of hazel-nut brown (*avellaneus*).

Type strain: 2343 FI; IPV 1945; NRRL B-3446.

157. Streptomyces libani subsp. **rufus** Baldacci and Grein 1966, 196.

subsp. ru′fus. L. adj. *rufus* red, reddish.

Produces libanomycin; exhibits anti-fungal

activity; inhibited by streptomycin; moderate growth on Czapek's solution agar; aerial mycelium typically colored in shades of hazel-nut brown (*avellaneus*); hygroscopic; forms reddish violet diffusible pigment with some media.

Type strain: 2501 FI; IPV 1942; NRRL B-3445 (single isolate).

158. Streptomyces lividans (Krasil'nikov, Sorokina, Alferova and Bezzubenkova) Pridham, 1970, 21. (*Actinomyces lividans* Krasil'nikov, Sorokina, Alferova and Bezzubenkova in Krasil'nikov 1965, 109.)

li'vi.dans. L. part.adj. *lividans* becoming bluish, black and blue.

Exhibits anti-bacterial activity; forms blue-colored vegetative mycelium and diffusible pigment on some media.

Type strain: INMI 32-13; ATCC 19844 (Pridham 1970, 21).

159. Streptomyces lusitanus Villax 1963, 661.

lu.si.ta'nus. L. n. *Lusitania* name of ancient Portugal; M.L. adj. *lusitanus* of Portugal.

Produces tetracycline and 7-chlortetracycline; no growth on Czapek's solution agar; spore chains of atypical *Retinaculum-Apertum* type; spores phalangiform; NaCl tolerance ≧2%, but <4%.

Type strain: CBS 101-A (single isolate).

160. Streptomyces lydicus DeBoer, Dietz, Silver and Savage 1956, 886.

ly'di.cus. L. n. *Lydia* an ancient state in Asia Minor; M.L. adj. *lydicus* of Lydia.

Produces streptolydigin (anti-bacterial), lydimycin (anti-fungal), actithiazic acid (anti-bacterial) and factor c (anti-bacterial); excellent growth on Czapek's solution agar; hygroscopic; NaCl tolerance ≧10%, but <13%.

Type strain: NRRL 2433 (single isolate, subsequently listed in DeBoer *et al.*, U.S. Patent 3,160,560, December 8, 1964).

161. Streptomyces melanosporofaciens Arcamone, Bertazzoli, Ghione and Scotti 1959, 215. (*Streptomyces melanosporus* Arcamone 1960, 96 (Not Val. Pub.); *Streptomyces melanosporus* var. *melanosporofaciens* (*sic*) in Umezawa 1967, 266.)

me.la.no.spo.ro.fa'ci.ens. Gr. adj. *melas* black; M.L. n. *spora* a spore; L. part.adj. *faciens* producing; M.L. part.adj. *melanosporofaciens* black spore producing.

Produces the anti-bacterial and anti-fungal antibiotics melanosporin and elaiophylin; poor growth on Czapek's solution agar; spores might be interpreted as being warty; hygroscopic. Characterization, in part, in Shirling and Gottlieb, 1969, 452.

Type strain: 1573; Farmitalia No. 1573 (single isolate).

162. Streptomyces misionensis Cercos, Eilberg, Goyena, Souto, Vautier and Widuczynski 1962, 22.

mi.si.on.en'sis. M.L. adj. *misionensis* belonging to Misiones, a province in Argentina, South America, the source of the soil from which the organism was isolated.

Produces the pentaenic anti-fungal antibiotic, misionin; inhibited by streptomycin; poor growth on Czapek's solution agar; hygroscopic; NaCl tolerance ≧7%, but <10%.

Type strain: INTA 3944; ATCC 14991 (single isolate).

163. Streptomyces murinus Frommer 1959, 198.

mu.ri'nus. L. adj. *murinus* of mice; M.L. adj. *murinus* mouse-gray; referred to as reddish-gray in original description.

Produces the actinomycin X complex; exhibits anti-fungal activity; inhibited by streptomycin; poor growth on Czapek's solution agar; spores phalangiform; hygroscopic; NaCl tolerance ≧7%, but <10%.

Type strain: Ital 1131; ATCC 19788.

164. Streptomyces mutabilis (Preobrazhenskaya and Ryabova) Pridham, Hesseltine and Benedict 1958, 69. (*Actinomyces mutabilis* Preobrazhenskaya and Ryabova in Gauze *et al.* 1957, 166.)

mu.ta'bi.lis. L. adj. *mutabilis* changeable, so named because the organism could not be assigned to any of the species known in the literature.

Exhibits anti-bacterial and anti-fungal activity; poor growth on Czapek's solution agar; spore chains of atypical *Retinaculum-Apertum* type; NaCl tolerance ≧10%, but <13%.

Type strain: INA B-472; ATCC 19789 (Gauze in Gottlieb, 1968, 20).

165. Streptomyces nigrescens (Sveshnikova) Pridham, Hesseltine and Benedict 1958, 70. (*Actinomyces nigrescens* Sveshnikova in Gauze *et al.* 1957, 146.)

ni.gres'cens. L. part.adj. *nigrescens* becoming black.

Exhibits anti-bacterial and anti-fungal activity; hygroscopic. Characterization, in part, in Shirling and Gottlieb, 1968, 353.

Type strain: INA 1800/54; ATCC 23941 (Gauze in Gottlieb, 1968, 20).

166. Streptomyces nodosus Trejo in Waksman 1961, 250. (Streptomycete culture M 4575 in Gold, Stout, Pagano and Donovick 1956, 579; *Streptomyces* species (M4575) in Vandeputte, Wachtel and Stiller 1956, 587; species of *Streptomyces* in Bartner, Zinnes, Moe and Kulesza 1958, 53 *inter alia*.)

no.do′sus. L. adj. *nodosus* knotty.

Produces amphotericin A (tetraenic anti-fungal) and amphotericin B (heptaenic anti-fungal); poor growth on Czapek's solution agar; reddish brown to black vegetative mycelium on some media; spores phalangiform; NaCl tolerance ≧4%, but <7%.

Type strain: M 4575; SC 2388; IMRU 3694; ATCC 14899; ATCC 23942 (single isolate).

167. Streptomyces nogalater Bhuyan and Dietz 1966, 838. (*Streptomyces nogalater* var. *nogalater* in Bhuyan, Kelly and Smith, U.S. Patent 3,183,157, May 11, 1965 (Not. Val. Pub.) *inter alia*; *Steptomyces nogalater* var. *nogalater* Bhuyan and Dietz 1966, 838.)

Comment: No other variety or subspecies of *Streptomyces nogalater* has been proposed in the literature.

no.gal.at′er. Spanish n. *nogal* walnut; L. adj. *ater* black; M.L. adj. *nogalater* black walnut, referring to the production (by the organism) on most media of an odor like that of black walnuts.

Produces nogalamycin, an anti-bacterial and anti-tumor antibiotic; exhibits anti-fungal activity; inhibited by streptomycin; spore chains of atypical *Retinaculum-Apertum* type; forms bright orange vegetative mycelium on some media; NaCl tolerance ≧7%, but <10%.

Type strain: NRRL 3035 (single isolate).

168. Streptomyces olivaceiscleroticus Pridham 1970, 41. (*Chainia olivaceus* (*sic*) Thirumalachar in Kalakoutskii and Krasil'nikov in Rautenshtein 1960, 45.)

o.li.va′ce.i.scler.o′ti.cus. M.L. adj. *olivaceus* olive colored; L. n. *sclerotium* sclerotium; M.L. adj. *olivaceiscleroticus* sclerotium with olive color.

Exhibits anti-fungal activity; inhibited by streptomycin; excellent growth on Czapek's solution agar; forms sclerotia on some media.

Type strain: IMRU 3751; ATCC 15722 (Pridham, 1970, 41).

169. Streptomyces olivaceoviridis (Preobrazhenskaya and Ryabova) Pridham, Hesseltine and Benedict 1958, 65. (*Actinomyces olivaceoviridis* Preobrazhenskaya and Ryabova in Gauze *et al.* 1957, 163.)

o.li.va′ce.o.vi′ri.dis. M.L. adj. *olivaceus* olive colored; L. adj. *viridis* green; M.L. adj. *olivaceoviridis* olive-green colored, referring to the gray-olive colored aerial mycelium and greenish gray-brown vegetative mycelium on a chemically defined medium.

Exhibits marked activity against Gram-positive bacteria and lesser activity against Gram-negative bacteria; forms some spore chains of atypical *Retinaculum-Apertum* type; color of aerial mycelium might also be interpreted as yellow. Characterization, in part, in Shirling and Gottlieb, 1969, 458.

Type strain: INA 11584 (Gauze in Gottlieb 1968, 20.)

170. Streptomyces olivaceus (Waksman) Waksman and Henrici 1948, 950. (*Actinomyces 206* in Waksman 1919, 168; *Actinomyces olivaceus* Waksman 1923, 354.)

o.li.va′ce.us. M.L. adj. *olivaceus* olive colored, apparently referring to the color of vegetative mycelium.

Inhibited by streptomycin; moderate growth on Czapek's solution agar; NaCl tolerance ≧10%, but <13%.

Type strain: 206; IMRU 3335; ATCC 3335; ATCC 19794 (single isolate).

171. Streptomyces parvullus (*sic*) Waksman and Gregory 1954, 1055.

par.vu′l.lus. L. adj. *parvulus* very small.

Produces actinomycin D (=actinomycin D$_{IV}$, C$_1$, I$_1$, X$_1$, IV, A$_{IV}$, or B$_{IV}$); inhibited by streptomycin; moderate growth on Czapek's solution agar; NaCl tolerance ≧10%, but <13%.

Type strain: G-375; IMRU 3677; ATCC 12434; ATCC 19796 (single isolate).

172. Streptomyces platensis Tresner and Backus 1956, 244. (*Streptomyces platensis* in Eli Lilly and Company, British Patent 713,795, August 18, 1954 (Not Val. Pub.); *Streptomyces platensis* in J. M. McGuire, Canadian Patent 520,836, January 17, 1956 (Not Val. Pub.).)

plat.en′sis. Gr. adj. *platys* = *platos* flat, broad, wide; M.L. adj. *platensis* belonging to flat.

Produces oxytetracycline and 7-chlortetracycline; inhibited by streptomycin; hydroxylates alkaloids; moderate growth on Czapek's solution agar; forms orange-brown to red-brown vegetative mycelium on some media; forms a wine-colored diffusible pigment on some media; hygroscopic.

Type strain: MS-5353; NRRL 2364; ATCC 13865; ATCC 23948; CBS 310.56 (single isolate).

173. Streptomyces plicatus Pridham, Hesseltine and Benedict 1958, 65. (*Streptomyces plicatus* in Parke, Davis & Company, British Patent 707,332, April 14, 1954 (Not Val. Pub.). *Streptomyces plicatus* in Haskell *et al.* 1958, 743 (Not Val. Pub.); *Streptomyces plicatus* in Haskell 1958, 747 (Not Val. Pub.).)

pli′c.a′tus. L. past.part.adj. *plicatus* folded, coiled.

Produces amicetin, bamicetin, plicacetin and two other antibacterial antibiotics; inhibited by streptomycin; excellent growth on Czapek's solution agar; some spore chains of atypical *Retinaculum-Apertum* type.

Type strain: PD 04918; NRRL 2428 (single isolate).

174. Streptomyces poonensis (Thirumalachar) Pridham 1970, 42. (*Chainia poonensis* Thirumalachar in Kalakoutskii and Krasil'nikov in Rautenshtein 1960, 45.)

poon.en'sis. M.L. adj. *poonensis* belonging to Poona, India, the source of the soil from which the organism was isolated.

Exhibits slight anti-bacterial activity; exhibits anti-fungal activity; inhibited by streptomycin; excellent growth on Czapek's solution agar; sclerotia formed on some media; forms dull violet-colored vegetative mycelium and diffusible pigment on some media; whole cell hydrolysates contain L-diaminopimelic acid. Characterization in Pridham and Lyons, 1969, 211.

Type strain: IMRU 3752; ATCC 15723 (Pridham 1970, 42).

175. Streptomyces psammoticus Virgilio and Hengeller 1960, 167. (*Streptomyces feofaciens* in Bellenghi, W. German Patent Application L 21111, March 8, 1956 (Not Val. Pub.); *Streptomyces feofaciens* in Lepetit S.p.A., British Patent 755,139, May 22, 1957 (Not Val. Pub.).)

psam.mo'ti.cus. Gr. n. *psamma* sand; M.L. adj. *psammoticus* sandy.

Produces tetracycline; spore chains of atypical *Retinaculum-Apertum* type; spores phalangiform; NaCl tolerance ≥2%, but <4%.

Type strain: Univ. of Pavia S4623/33; P-19; CBS 175.61; CBS 266.65 (single isolate).

176. Streptomyces purpurogeneiscleroticus Pridham 1970, 43. (*Chainia purpurogena* Thirumalachar in Thirumalachar and Sukapure 1964, 165.)

pur.pur.o.ge.ni.scler.o'ti.cus. L. adj. *purpureus* purple colored; Gr. v.suff. *-genes* producing; L. n. *sclerotium* sclerotium; M.L. adj. *purpurogeniscleroticus* sclerotium along with producing purple color.

Exhibits anti-bacterial activity; exhibits anti-fungal activity; excellent growth on Czapek's solution agar; forms sclerotia on some media.

Type strain: Thirumalachar C-3; NRRL B-2952 (Pridham, 1970, 43).

177. Streptomyces recifensis (Gonçalves de Lima, Machado, Araujo, Falcão de Morais and Biermann) Falcão de Morais, Gonçalves de Lima and Maia 1957, 249. (*Nocardia recifei* Gonçalves de Lima, Machado, Araujo, Falcão de Morais and Biermann 1955, 26.)

re.cif.en'sis. M.L. adj. *recifensis* belonging to Recife, Brazil, the source of the soil from which the organism was isolated.

Exhibits anti-bacterial activity; some spore chains of atypical *Retinaculum-Apertum* type;

spores phalangiform; NaCl tolerance ≥7%, but <10%.

Type strain AX-18; IAUR 3054; ATCC 19803 (single isolate).

178. Streptomyces rochei Berger, Jampolsky and Goldberg in Waksman in Waksman and Lechevalier 1953, 40. (*Streptomyces rochei* in Berger, Jampolsky and Goldberg 1949. 476 (Not Val. Pub.); *Actinomyces rochei* (Berger, Jampolsky and Goldberg) Preobrazhenskaya, Blinov and Ryabova in Gauze *et al.* 1957, 133.)

ro'che.i. M.L. gen.n. *rochei* of Roche.

Produces borrelidin, an anti-bacterial and anti-spirochetal antibiotic; exhibits anti-fungal activity; inhibited by streptomycin; moderate growth on Czapek's solution agar. Characterization, in part, in Waksman, 1961, 267, and in Shirling and Gottlieb, 1968, 368.

Type strain: X-15; IMRU 3602; NRRL B-1559; ATCC 10739; CBS 224.46 (single isolate).

179. Streptomyces rokugoensis Funaki and Tsuchiya 1958, 1097.

ro.ku.go.en'sis. M.L. adj. *rokugoensis* belonging to Rokuyo, a place in Japan, the source of the soil from which the organism was isolated.

Produces the neomycin complex; inhibited by streptomycin; forms some spore chains of atypical *Retinaculum-Apertum* type.

Type strain: Meiji Myogyo (Funaki); NRRL B-2016 (single isolate).

180. Streptomyces roseodiastaticus (Duché) Waksman in Waksman and Lechevalier 1953, 27. (*Actinomyces roseodiastaticus* Duché 1934, 329.)

ro.se.o.di.a.sta'ti.cus. L. adj. *roseus* rosy; M.L. adj. *diastaticus* diastatic, starch digesting; M.L. adj. *roseodiastaticus* rosy, diastatic.

Inhibted by streptomycin; excellent growth on Czapek's solution agar.

Type strain: "CBS (Duché)," CBS 102.34 (single isolate obtained by Duché from the CBS under the name *Actinomyces diastaticus* and renamed by him in 1934).

181. Streptomyces rutgersensis subsp. **castelarense** (*sic*) Cercos 1954, 263. (*Streptomyces* strain D.I.N.R. 41 in Cercos and Rosemblit 1950, 98; *Streptomyces* D.I.N.R. 41 in Cercos 1953, 53.)

subsp. cas.tel.ar.en'se. M.L. neut.adj. *castelarense* (*sic*) correctly, *castelarensis* from Castelar, Argentina, South America, the source of the organism (from dust).

Produces camphomycin (two components); probably grows well on Czapek's solution agar; hygroscopic: spores short, cylindrical, phalangiform and may appear roughened; NaCl tolerance ≥4%, but <7%.

Type strain: IMA B-23-7; DINR 41; INTA 41; CBS 309.55; ATCC 15191 (single isolate).

182. Streptomyces sayamaensis Arishima, Sekizawa, Sato and Miwa 1955, 816.

sa.ya.ma.en'sis. M.L. adj. *sayamaensis* belonging to Sayama hill in Japan, the source of the soil from which the organism was isolated.

Produces tetracycline and 7-chlortetracycline; spore chains straight and of atypical *Retinaculum-Apertum* type; spores phalangiform; NaCl tolerance ≥4%, but <7%.

Type strain: 310 (single isolate).

183. Streptomyces sendaiensis Katô 1959, 79.

sen.dai.en'sis. M.L. adj. *sendaiensis* of Sendai, Japan, the source of the soil from which the organism was isolated.

Produces chloramphenicol.

Type strain: MTHU C-121 (single isolate).

184. Streptomyces sioyaensis Nishimura, Okamoto, Mayama, Ohtsuka, Nakajima, Tawara, Shimohira and Shimaoka 1961, 257.

si.o.ya.en'sis. M.L. adj. *sioyaensis* of Sioya (*sic*) referring to Shioya, Kobe, Japan, the source of the soil from which the organism was isolated.

Produces the thiostrepton-like anti-bacterial antibiotic siomycin. Characterization, in part, in Shirling and Gottlieb, 1968, 170.

Type strain: H-690; SRL(H-690); ATCC 13989; ATCC 19810 (single isolate).

185. Streptomyces tendae Ettlinger, Corbaz and Hütter 1958, 351.

ten'dae. M.L. gen.n. *tendae* of Tende, Germany, the source of the soil from which the organism was isolated.

Produces carbomycin; inhibited by streptomycin; excellent growth on Czapek's solution agar; some spore chains of a atypical *Retinaculum-Apertum* type.

Type strain: ETH 11313; ATCC 19812.

186. Streptomyces thermovulgaris Henssen 1957, 391.

ther.mo.vul.ga'ris. Gr. n. *therme* heat; L. adj. *vulgaris* common; M.L. adj. *thermovulgaris* heat, common.

A eurythermal thermophile; excellent growth at 60 C, good growth at 40–50 C, slight growth with sparse formation of aerial mycelium at 28 C; reportedly grows better under anaerobic conditions at the higher temperatures; isolated principally from a variety of fresh and rotted animal manures, the type strain from fresh cow manure. Characterization, in part, in Shirling and Gottlieb, 1969, 485.

Type strain: R₁₀; ATCC 19284; CBS 276.66 (Henssen and Schnepf, 1967, 218).

187. Streptomyces tricolor (Wollenweber) Waksman 1960, 158. (*Actinomyces tricolor* Wollenweber 1920, 13.)

tri'col.or. L. pref. *tri-* three; L. n. *color* color; M.L. *tricolor* of three colors.

Forms yellow-, red- or blue-colored vegetative mycelium on some media; forms blue diffusible pigment on some media; isolated from flat scab of potato. Characterization based, in part, in Hütter, 1967, 312.

Type strain: CBS (Wollenweber) CBS 103.21 (single isolate).

188. Streptomyces tubercidicus Nakamura 1961, 90. (Unidentified actinomyces strain (*sic*) or streptomycete (*sic*) in Anzai, Nakamura and Suzuki 1957, 201; *Streptomyces tubercidicus* in Suzuki and Marumo 1960, 360 (Not Val. Pub.).)

tu.ber.ci'di.cus. L. n. *tuber* nodule; L. combining form *-cid-* from L. v. *caedo* to kill; L. n.stem suff. *-icus* to denote possession; M.L. adj. *tubercidicus* nodule destroying, referring to antitumor activity of the antibiotic produced by the organism.

Produces tubercidin, an antimycobacterial and antitumor antibiotic; forms pH-dependent pigment (yellowish with acid, pinkish with alkali); hygroscopic. Characterization, in part, in Shirling and Gottlieb, 1969, 485.

Type strain: IPCR 585 (single isolate).

189. Streptomyces tumemacerans (Krasil'nikov and Koveshnikov) Pridham 1970, 27. (*Actinomyces tumemacerans* Krasil'nikov and Koveshnikov 1962, 589.)

tu.me.ma'ce.rans. L. v. *tumeo* swell, form a tumor; L. part.adj. *macerans* to soften by steeping, to ret; M.L. part.adj. *tumemacerans* softening a tumor.

Produces four antibiotics; P-42A (cycloheximide-like), P-42B (nystatin-like), P-42E (rimocidin-like) and P-42S (streptomycin-like or kanamycin-like) according to Tokhtamuratov *et al.* (1964, 205; 1965, 30). See also Umezawa 1967, 933; reportedly dissolves plant tumors caused by *Agrobacterium tumefaciens* (Smith and Townsend) Conn 1942 without killing the bacteria.

Type strain: INMI P-42 (single isolate).

190. Streptomyces vastus Szabó and Marton 1958, 245.

vast'us. L. adj. empty, vast—referring to the occurrence of the organism in the Hortobagy Puszta (eastern Hungary).

Exhibits slight anti-microbial activity; forms blue-colored vegetative mycelium on some media; NaCl tolerance ≥5%.

Type strain: CBS (Szabó A-10); CBS 290.60 (deVries in Shirling and Gottlieb, 1969, 490).

191. Streptomyces violaceolatus (Krasil'nikov, Sorokina, Alferova and Bezzubenkova) Pridham 1970, 28. (*Actinomyces violaceolatus* Krasil'nikov, Sorokina, Alferova and Bezzubenkova in Krasil'nikov 1965, 113.)

vi.o.la.ce.o'la'tus. L. adj. *violet*; L. adj. *latus* broad; M.L. adj. *violaceolatus* violet, broad.

Forms blue-, purple- or red-colored vegetative mycelium and diffusible pigment depending upon pH. Characterization, in part, in Shirling and Gottlieb, 1969, 496.

Type strain: INMI 4; ATCC 19847 (Krasil'nikov in Shirling and Gottlieb, 1969, 496).

192. **Streptomyces violaceus-niger** (Waksman and Curtis) Pridham, Hesseltine and Benedict 1958, 63. (*Actinomyces violaceus-niger* Waksman and Curtis 1916, 111; *Streptomyces violaceoniger* (*sic*) (Waksman and Curtis) Waksman and Henrici 1948, 947.)

vi.o.la'ce.us-ni'ger. L. adj. *violaceus* violet; L. adj. *niger* black; M.L. adj. *violaceus-niger* violet-black.

Exhibits anti-bacterial and anti-fungal activity (nigericin complex?); excellent growth on Czapek's solution agar; gray to black vegetative mycelium on some media; hygroscopic; spore walls may appear roughened; NaCl tolerance $\geq 4\%$, but $<7\%$.

Reference strain: R. G. Benedict A-975; NRRL B-1476 (original single isolate of Waksman and Curtis no longer extant).

193. **Streptomyces violaceus-ruber** (Waksman and Curtis) Pridham 1970, 44. (*Actinomyces violaceus-ruber* Waksman and Curtis 1916, 127; *Actinomyces violaceons* (*sic*) in Waksman and Curtis 1916, 110; *Actinomyces violaceus* (*sic*) in Waksman and Curtis 1916, 178; *Streptomyces violaceoruber* Waksman in Kutzner and Waksman 1959, 535.)

vi.o.la'ce.us-ru'ber. L. adj. *violaceus* violet; L. adj. *ruber* red; M.L. adj. *violaceus-ruber* violet-red.

Exhibits slight anti-bacterial activity; inhibited by streptomycin; excellent growth on Czapek's solution agar; forms pH-dependent blue to red vegetative mycelium and diffusible pigment on some media; NaCl tolerance $\geq 7\%$, but $<10\%$.

Comment: The antibiotic activity may be due to celicomycin (actinomycin K). Much genetic work on streptomycetes has been conducted with strains identified as *Streptomyces coelicolor* (Sermonti and Casciano, 1963). These strains and derived mutants probably are more closely related to *Streptomyces violaceus-ruber* than to *S. coelicolor* (Pridham et al., 1965).

Type strain: IMRU 3030; ATCC 14980; ATCC 19816 (single isolate).

194. **Streptomyces viridifaciens** Pridham, Hesseltine and Benedict 1958, 65. (*Streptomyces*

viridifaciens in Gourevitch and Lein, U.S. Patent 2,712,517, July 5, 1955 (Not Val. Pub.).)

vi.ri.di.fa'ci.ens. L. adj. *viridis* green; L. part.-adj. *faciens* producing; M.L. part.adj. *viridifaciens* green-producing.

Produces tetracycline, 7-chlortetracycline and ayfactins A and B (heptaenic anti-fungal antibiotics); inhibited by streptomycin; moderate growth on Czapek's solution agar; spore chains of atypical *Retinaculum-Apertum* type; spores phalangiform; NaCl tolerance $\geq 4\%$, but $<7\%$.

Type strain: Bristol 567201; ATCC 11989 (single isolate).

See Table 17.42g–h.

195. **Streptomyces atroolivaceus** (Preobrazhenskaya, Blinov and Ryabova) Pridham, Hesseltine and Benedict 1958, 68. (*Actinomyces atroolivaceus* Preobrazhenskaya, Blinov and Ryabova in Gauze et al. 1957, 143.)

at.ro.o.li.va'ce.us. L. adj. *ater* black; M.L. adj. *olivaceus* olive colored; M.L. adj. *atroolivaceus* of a dark olive color, referring to the pigment on an organic medium.

Exhibits anti-microbial activity; NaCl tolerance $\geq 7\%$, but $<10\%$.

Comment: Non-coiled chains of spores observed with the type strain (Shirling and Gottlieb, 1968, 84).

Type strain: INA 4776/54; ATCC 19725 (Gauze in Gottlieb, 1968, 20).

196. **Streptomyces cyanocolor** (Krasil'nikov, Sorokina, Alferova and Bezzubenkova) Pridham 1970, 14. (*Actinomyces cyanocolor* Krasil'nikov, Sorokina, Alferova and Bezzubenkova in Krasil'nikov 1965, 114.)

cy'an.o.co.lor. Gr. adj. *cyaneus* dark blue; L. n. *color* color; M.L. adj. *cyanocolor* dark blue colored.

Exhibits no anti-microbial activity except against some other streptomycetes; produces anthocyanidin pigments; forms pH-dependent red or blue-colored vegetative mycelium and diffusible pigments on some media.

Type strain: INMI 31-23; ATCC 19835 (Krasil'nikov in Shirling and Gottlieb, 1969, 421).

197. **Streptomyces graminofaciens** Charney, Fisher, Curran, Machlowitz and Tytell 1953, 1283.

gra.mi.no.fa'ci.ens. L. n. *gramen* grass; L. part.adj. *faciens* producing; M.L. part.adj. *graminofaciens* grass producing, probably refers to production of (strepto)gramin.

Produces streptogramins A and B which act synergistically; inhibited by streptomycin; excellent growth on Czapek's solution agar; brown to black vegetative mycelium on some media; NaCl tolerance $\geq 7\%$, but $<10\%$.

Type strain: MA-317 = ATCC 12705 (single isolate).

TABLE 17.42g-h
Gray series[a]

Names of species and subspecies	Strain designation	Utilization of carbon compounds (P & G basal)											
		No carbon control	D-Glucose	D-Xylose	L-Arabinose	L-Rhamnose	D-Fructose	D-Galactose	Raffinose	D-Mannitol	i-Inositol	Salicin	Sucrose
17.42g GY; S; C−; WTY													
195. S. atroolivaceus (see also no. 204).	T		+	+	+	+	+						
196. S. cyanocolor	T		+	−		+	+	+	−		+		−
197. S. graminofaciens	T	−	+	+	+	+							
198. S. griseoplanus	T	−	+	+	+	−	+	+	+	−	−	−	−
17.42h GY; S; C−; SPY													
199. S. albaduncus	T		+	+	+	+	+	+	+	+	+	+	±
200. S. albospinus	T		+	±	−	−	+	+	+	+	+	+	−
201. S. albulus	T	−	+	−	−	−	+	+	−	+	+	+	
202. S. althioticus	T		+	+	+	+	+	+	−	+	+		±
203. S. arabicus	T	−	+	+	+	+	+		−	+	+		
204. S. atroolivaceus subsp. mutomycini (see also no. 195).	T												
205. S. canus	T	−	+	+	+	+	+	+	±	+	+		+
206. S. chattanoogensis	T	−	+	−	−	−	+		+	+	+		+
207. S. chlorobiens	T		+	+	+	+		+	+	+	+		+
208. S. cuspidosporus	T	±	+	+	+	+	+	+	+	+	+	+	+
209. S. gancidicus	T		+	+	+	+	+	+	−	+	+	−	+
210. S. griseoflavus	R		+	+	+	+	+		−	+	+		+
211. S. griseoincarnatus	T		+	+	+	+	+		−	+	±		+
212. S. griseorubens	T		+	+	±	+	+		−	+	±		
213. S. macrosporeus	T	−	+	+	+	+			−	+	+		−
214. S. malachiticus	T	−	+	+	+	+	+	+	+	+		−	+
215. S. matensis	T	−	+	+	+	+	+	+	−	+	+		+
216. S. noursei	T	−	+	−	−	−	+	+	−	+	+	−	+
217. S. olivoviridis	T	−	+	+	+	+	+	+	−	+	−		+
218. S. pseudogriseolus	T	−	+	+	+	+	+	+	−	+	+	+	
219. S. rubiginosus	T		+	+	±	+	+		−	+	+		+
220. S. sparsogenes	T	−	+	+	+	+	+	±	+	+	±	−	+
221. S. viridiviolaceus	T												
222. S. virido-diastaticus	T		+	+	+	+	+			+	+		

[a] For explanation of meanings of symbols see footnotes to Tables 17.40 and 17.41a–e.

198. Streptomyces griseoplanus Backus, Tresner and Campbell 1957, 536. (*Streptomyces griseoplanus* in DeVoe *et al.* 1957, 730 (Not Val. Pub.).)

gri.se.o.pla'nus. M.L. adj. *griseus* gray; M.L. adj. *planus* flat, level; M.L. adj. *griseoplanus* flat, gray (referring to the restricted, flat, plane growth and grayish spore color *en masse* of the organism).

Produces alazopeptin, an antitumor antibiotic and other antibiotics (components X, Y_1 and Z); poor growth on Czapek's solution agar; NaCl tolerance <4%.

Type strain: Lederle AA-223; ATCC 19766 (Tresner in Shirling and Gottlieb, 1968, 124).

199. Streptomyces albaduncus Tsukiura, Okanishi, Ohmori, Koshiyama, Miyaki, Kitazima and Kawaguchi 1964, 41. (*Streptomyces albaduncus* in Kawaguchi *et al.*, Belgian Patent 634,041, December 24, 1963 (Not Val. Pub.).)

al.ba.dun'cus. L. adj. *albus* white; L. adj. *uncus* hooked, crooked; M.L. adj. *alba(d)uncus* white, hooked, probably referring to color of aerial mycelium and nature of spore chains of the organism.

Produces danomycin, an anti-bacterial antibiotic; exhibits anti-fungal activity, inhibited by streptomycin; moderate growth on Czapek's solution agar; some spore chains of atypical *Retinaculum-Apertum* type; originally described as having white aerial mycelium with gray mutants.

Type strain: Bristol-Banyu 13246; ATCC 14698 (single isolate).

200. Streptomyces albospinus (*sic*) Wang, Hamada, Okami and Umezawa 1966, 217.

al.bo.spin'us. L. adj. *albus* white; L. adj. *spineus* spiny; M.L. adj. *albospinus* white, spiny, referring to the color of the aerial mycelium and nature of spore wall ornamentation.

Produces spinamycin, a non-polyenic antifungal antibiotic.

Type strain: M750-G1 (single isolate).

201. Streptomyces albulus Routien in Pridham and Lyons 1969, 194. (*Streptomyces albulus* in Rao and Cullen, Abstrs. Papers 134th Meet. Amer. Chem. Soc., Sept. 7–12, Chicago, Illinois, 1958 (Not Val. Pub.); *Streptomyces albulus* in Rao *et al.* German Patent 1,052,065, March 5, 1959 (Not Val. Pub.); *Streptomyces albulus* in Rao and Cullen 1960, 1127 (Not Val. Pub.); *Streptomyces albulus* in Rao 1960, 1129 (Not Val. Pub.); *Streptomyces albulus* in Charles Pfizer and Company, Inc., British Patent 866,600, April 26, 1961 (Not Val. Pub.); *Streptomyces albulus* in Rao 1962, 123 (Not Val. Pub.): *Streptomyces* sp. in Rao, U.S. Patent 3,095,418, June 25, 1963.)

al.bu'lus. L. adj. *albulus* whitish.

Exhibits anti-bacterial activity; produces cycloheximide, acetoxycycloheximide, desacetyl-dehydroacetoxycycloheximide (actiphenol), two cycloheximide diastereoisomers, nystatin and albonoursin; poor growth on Czapek's solution agar.

Type strain: Pfizer BA 4105; ATCC 12757 (single isolate).

202. Streptomyces althioticus Yamaguchi, Nakayama, Takeda, Tawara, Maeda, Takeuchi and Umezawa 1957, 196.

al.thi.o'ti.cus. English n. *althiomycin* name of a sulfur-containing antibiotic; L. n.suff. *-icus* belonging to; M.L. adj. *althioticus* belonging to althiomycin.

Produces althiomycin, an anti-bacterial antibiotic; growth on Czapek's solution agar with

formation of purplish vegetative mycelium and diffusible pigment; some spore chains of atypical *Retinaculum-Apertum* type; NaCl tolerance ≧10%, but <13%.

Type strain: Nippon Kayaku 245-Z2; NIHJ 75; ATCC 19724 (single isolate).

203. Streptomyces arabicus Shibata, Nakazawa, Miyake, Inoue and Okabori 1957, 32. (*Streptomyces arabicus* in Yamaguchi and Saburi 1955, 204 (Not Val. Pub.).)

a.ra'bi.cus. L. adj. *arabicus* of Arabia, the source of the soil from which the organism was isolated.

Produces resistomycin (croceomycin); exhibits non-polyenic anti-fungal activity; exhibits anti-trichomonal activity; inhibited by streptomycin; excellent growth on Czapek's solution agar; some spore chains of atypical *Retinaculum-Apertum* type; NaCl tolerance ≧10%, but <13%.

Type strain: 6762; IFO 3406; ATCC 23881 (single isolate).

204. Streptomyces atroolivaceus subsp. **mutomycini** (Gauze, Maksimova, Popova, Brazhnikova, Uspenskaya, and Rossolimo) Pridham 1970, 8. (*Actinomyces atroolivaceus* subsp. *mutomycini* Gauze, Maksimova, Popova, Brazhnikova, Uspenskaya and Rossolimo 1959, 21.)

subsp. mu.to.mycin.i. L. v. *muto* shift, change, alter; M.L. suff. *-mycin* for antibiotic names; M.L. gen.n. *mutomycini* of mutomycin.

Produces mutomycin, an antibiotic effective against respiratory-deficient staphylococci, tumors and viruses.

Type strain: INA 4305 (single isolate).

205. Streptomyces canus Heinemann, Kaplan, Muir and Hooper 1953, 1239.

ca'nus. L. adj. *canus* white, gray.

Produces amphomycin, an anti-bacterial antibiotic; inhibited by streptomycin; excellent growth on Czapek's solution agar; some spore chains of atypical *Retinaculum-Apertum* type; NaCl tolerance ≧7%, but <10%.

Type strain: Bristol Labs. 456786 (A6786); ATCC 12237; ATCC 19737 (Waksman, 1961, 190).

206. Streptomyces chattanoogensis Burns and Holtman 1959, 398.

chat.ta.noo.gen'sis. M.L. adj. *chattanoogensis* belonging to Chattanooga, Tennessee, the source of the soil from which the organism was isolated.

Exhibits slight anti-bacterial activity; produces pimaricin, a tetraenic anti-fungal antibiotic; inhibited by streptomycin; excellent growth on Czapek's solution agar; NaCl tolerance ≧7%, but <10% (original description states inhibition of growth with 3% NaCl).

Type strain: J-23; ATCC 13358; ATCC 19739 (single isolate).

207. Streptomyces chlorobiens (*sic*) (Krasil'-nikov and Egorova) Pridham 1970, 10. (*Actinomyces chlorobiens* Krasil'nikov and Egorova in Krasil'nikov 1965, 194.)

chlo.ro'bi.ens. Gr. adj. *chlorus* greenish yellow, green; Gr. n. *bios* life; M.L. part.adj. *chlorobiens* living, greenish yellow.

Exhibits anti-bacterial activity; forms green-colored vegetative mycelium on some media.

Type strain: INMI 6166.

208. Streptomyces cuspidosporus Higashide, Hasegawa, Shibata, Mizuno and Akaike 1966, 2.

cu′spi.do.spo′rus. L. n. *cuspis -idis* point; M.L. n. *spora* a spore; M.L. adj. *cuspidosporus* spore with points or spines.

Produces sparsomycin and tubercidin and several other antibiotics: excellent growth on Czapek's solution agar; forms green to blue to yellowish green diffusible pigment in some media.

Type strain: B-79; RTCI B-79; IFO 12378 (single isolate).

209. Streptomyces gancidicus Suzuki 1957, 538. (*Streptomyces* sp. strain AAK-84 in Suzuki 1955, 86; *Streptomyces* No. AAK-84 in Aiso, Arai, Suzuki and Takamizawa 1956, 97.)

gan.ci′di.cus. English n. *gancidin* name of an antibiotic; L. n.suff. *-icus* belonging to; M.L. adj. *gancidicus* belonging to gancidin.

Produces the gancidin complex (components A and W) effective against Gram-positive bacteria and tumors; inhibited by streptomycin; moderate growth on Czapek's solution agar.

Type strain: AAK-84; IFM 1024 (AAK-84); NRRL B-1872 (single isolate).

210. Streptomyces griseoflavus (Krainsky) Waksman and Henrici 1948, 948. (*Actinomyces griseoflavus* Krainsky 1914, 694.)

gri.se.o.fla′vus. M.L. adj. *griseus* gray; L. adj. *flavus* yellow; M.L. adj. *griseoflavus* grayish yellow.

Forms yellow vegetative mycelium and diffusible pigment on some media; NaCl tolerance ≧7%, but <10%. Characterization, in part, in Shirling and Gottlieb, 1969, 433.

Reference strain: Ciferri A 28 Nr. 1118; CBS (Ciferri); CBS 409.52 (suggested as the neotype strain by Hütter, 1967, 20). None of Krainsky's original isolates is extant.

211. Streptomyces griseoincarnatus (Preobrazhenskaya, Ryabova and Blinov) Pridham, Hesseltine and Benedict 1958, 69. (*Actinomyces griseoincarnatus* Preobrazhenskaya, Ryabova and Blinov in Gauze *et al.* 1957, 169.)

gri.se.o.in.car.na′tus. M.L. adj. *griseus* gray; L. adj. *incarnatus* flesh colored; M.L. adj. *griseoincarnatus* grayish flesh-colored, referring to changes in color of the aerial mycelium.

Exhibits anti-bacterial activity; forms red or orange diffusible pigment on some media. Characterization, in part, in Shirling and Gottlieb, 1968, 328.

Type strain: INA 9673/55; ATCC 23917 (Gauze in Gottlieb, 1968, 20).

212. Streptomyces griseorubens (Preobrazhenskaya, Blinov and Ryabova) Pridham, Hesseltine and Benedict 1958, 65. (*Actinomyces griseorubens* Preobrazhenskaya, Blinov and Ryabova in Gauze *et al.* 1957, 144.)

gri.se.o.ru′bens. M.L. adj. *griseus* gray; L. part.adj. *rubens* blushing, reddening: M.L. part. adj. *griseorubens* gray-reddening, referring to color of the aerial mycelium, vegetative mycelium and diffusible pigment.

Exhibits anti-bacterial and anti-fungal activity; some spore chains of atypical *Retinaculum-Apertum* type; NaCl tolerance ≧7%, but <10%.

Characterization, in part, in Shirling and Gottlieb, 1968, 126.

Type strain: INA 6124/54; ATCC 19909; ATCC 19767 (Gauze in Gottlieb, 1968, 20).

213. Streptomyces macrosporeus Ettlinger, Corbaz and Hütter 1958, 346.

mac.ro.spo′re.us. Gr. adj. *macrus* large; Gr. n. *spora* a seed; M.L. n. *spora* a spore; M.L. adj. *macrosporeus* large spored.

Produces carbomycin A; inhibited by streptomycin; fair growth on Czapek's solution agar; some spore chains of atypical *Retinaculum-Apertum* type; spores reportedly (1.7–2.0 by 1.5–2.0 μm) about twice the usual size of streptomycete spores; NaCl tolerance ≧4%, but <7%.

Type strain: ETH 7534; ATCC 19783.

214. Streptomyces malachiticus (Kudrina, Preobrazhenskaya and Ryabova) Pridham 1970, 22. (*Actinomyces malachiticus* Kudrina, Preobrazhenskaya and Ryabova in Gauze *et al.* 1957, 162; *Streptomyces malachitus* (Kudrina, Preobrazhenskaya and Ryabova in Gauze *et al.*, 1957) Pridham, Hesseltine and Benedict 1958, 69 (*lapsus calami.*)

mal′a.chit.icus. Gr. n. *malache* a mallow (referring to green color of leaves of mallow); M.L. adj. *malachiticus* belonging to mallow, referring to the color of the vegetative mycelium.

Exhibits slight anti-bacterial activity; moderate growth on Czapek's solution agar; some spore chains of atypical *Retinaculum-Apertum* type.

Type strain: INA 399/54; ATCC 19918; ATCC 19784 (Gauze in Gottlieb, 1968, 20).

215. Streptomyces matensis Margalith, Beretta and Timbal 1959, 71.

mat.en′sis. M.L. adj. *matensis* belonging to mat (of uncertain derivation).

Produces the althiomycin-like antibacterial

antibiotic matamycin; inhibited by streptomycin; fair growth on Czapek's solution agar; some spore chains of atypical *Retinaculum-Apertum* type.

Type strain: Lepetit ME/17; NRRL B-2576 (single isolate).

216. Streptomyces noursei Brown, Hazen and Mason 1953, 609. (Actinomycete 48240 in Brown and Hazen 1949, 454; a soil actinomycete in Hazen and Brown 1950, 423; Actinomycete No. 48240 in Hazen and Brown 1951, 93.)

nour'se.i. M.L. gen.n. *noursei* of Nourse, referring to the owner of the farm where soil was obtained from which the organism was isolated.

Exhibits anti-bacterial activity; produces nystatin and cycloheximide; poor growth on Czapek's solution agar; NaCl tolerance $\geq 7\%$, but $<10\%$.

Type strain: 48240; IMRU 3771; ATCC 11455 (single isolate).

217. Streptomyces olivoviridis (Kuchaeva, Krasil'nikov, Skryabin and Taptykova) Pridham 1970, 23. (*Actinomyces olivoviridis* Kuchaeva, Krasil'nikov, Skryabin and Taptykova in Rautenshtein 1960, 251; *Actinomyces olivoviridis* n. sp. (*sic*) in Krasil'nikov, Kutchayeva and Skryabin, Suppl. to Program, Symposium on Antibiotics, Prague, Czechoslovakia May 18–23, 1959 (Not Val. Pub.); *Actinomyces olivoviridis* n. sp. (*sic*) Kutchayeva, Krasil'nikov and Skryabin 1960, 58 (Not Val. Pub.); *Actinomyces olivovirilis* (*sic*) Kuchaeva, Krasil'nikov, Skryabin and Taptykova in Rautenshtein 1960, 248.)

o.li.vo.vi'ri.dis. L. n. *oliva* olive; L. adj. *viridis* green; M.L. adj. *olivoviridis* olive-green.

Exhibits anti-bacterial and anti-fungal activity; inhibited by streptomycin; poor growth on Czapek's solution agar; reportedly forms bright green vegetative mycelium and diffusible pigment with chemically defined iron citrate media and green or brown shades with other media; forms spore chains of atypical *Retinaculum-Apertum* type.

Type strain: INMI 1475; RIA 661; ATCC 15882; ATCC 23944 (Krasil'nikov in Shirling and Gottlieb, 1968, 355).

218. Streptomyces pseudogriseolus Okami and Umezawa in Okami, Utahara, Ōyagi, Nakamura, Umezawa, Yanagisawa and Tunematsu 1955, 128.

pseu.do.gri.se'o.lus. Gr. adj. *pseudes* false; M.L. dim.adj. *griseolus* specific, epithet; M.L. dim.adj. *pseudogriseolus* the false *griseolus* referring to resemblance to *S. griseolus*.

Produces xanthomycin; inhibited by streptomycin; excellent growth on Czapek's solution agar; NaCl tolerance $\geq 10\%$, but $<13\%$.

Type strain: 534; H-16-C; NIHJ 224; ATCC 12770; ATCC 23949 (single isolate).

219. Streptomyces rubiginosus (*sic*) (Preobrazhenskaya, Blinov and Ryabova) Pridham, Hesseltine and Benedict 1958, 70. (*Actinomyces rubiginosus* Preobrazhenskaya, Blinov and Ryabova in Gauze *et al.* 1957, 134.)

ru.bi.gin.o'sus. o.var. of L. adj. *robiginosus* (*sic*) rusty, referring to the red-gray-brown color of vegetative mycelium.

Exhibits anti-bacterial and anti-fungal activity. Characterization, in part, in Shirling and Gottlieb, 1968, 374.

Type strain: INA 11852; ATCC 19927; ATCC 23961 (Gauze in Gottlieb, 1968, 20).

220. Streptomyces sparsogenes Owen, Dietz and Camiener 1963, 772. (*Streptomyces sparsogenes* var. *sparsogenes* Owen, Dietz and Camiener 1963, 772.)

spar.so'gen.es. L. part.adj. *sparsus* scattered; Gr. v.suff. *-genes* producing; M.L. adj. *sparsogenes* scattered production, but probably referring to sparse formation of aerial mycelium.

Produces sparsomycin, tubercidin and several other antibiotics; excellent growth on Czapek's solution agar; hygroscopic; NaCl tolerance $\geq 4\%$, but $<7\%$.

Comment: No other varieties or subspecies of this taxon have been proposed. Characterization, in part, in Shirling and Gottlieb, 1969, 481.

Type strain: UC 2474; NRRL 2940; NCIB 9449 (single isolate, designated in Pike, Wiley and Slechta, U.S. Patent 3,167,540, January 26, 1965).

221. Streptomyces viridiviolaceus (Ryabova and Preobrazhenskaya) Pridham, Hesseltine and Benedict 1958, 70. (*Actinomyces viridiviolaceus* Ryabova and Preobrazhenskaya in Gauze *et al.* 1957, 188.)

vi.ri.di.vi.o.la'ce.us. L. adj. *viridis* green; L. adj. *violaceus* violet; M.L. adj. *viridiviolaceus* green, violet, referring to the greenish color of the aerial mycelium and the violet color of diffusible pigment.

Exhibits anti-bacterial and anti-fungal activity; forms violet-colored vegetative mycelium and diffusible pigment on some media.

Type strain: INA 5726/56 (Gauze in Gottlieb, 1968, 20).

222. Streptomyces virido-diastaticus (*sic*) (Baldacci, Grein and Spalla) Pridham, Hesseltine and Benedict 1958, 67. (*Actinomyces virido-diastaticus* (*sic*) Baldacci, Grein and Spalla 1955, 133; *Actinomyces viridodiastaticus* in Baldacci, Grein and Spalla 1955, 140.)

vi.ri.do-di.a.sta'ti.cus. L. adj. *viridis* green; M.L. n. *diastasum* the enzyme diastase, hence M.L. adj. *diastaticus;* diastatic; M.L. adj. *virido-diastaticus* green-diastatic.

TABLE 17.42i

Gray series[a]

Names of species and subspecies	Strain designation	No carbon control	D-Glucose	D-Xylose	L-Arabinose	L-Rhamnose	D-Fructose	D-Galactose	Raffinose	D-Mannitol	i-Inositol	Salicin	Sucrose
						Utilization of carbon compounds (P & G basal)							
17.42i GY; S; C−; H.													
223. S. calvus	T	−	+	+	+	+	+	+	+	+	+	+	+
224. S. cyanoalbus	T	−	+	+	+	+	+	+	+	+	−	−	+
225. S. finlayi	T		+	+	+	+	−	−	−	−	−	−	±
226. S. flaveolus	T	−	+	+	+	+	+	+	+	+	+	+	+
227. S. geysiriensis	T												
228. S. herbiferis	T		+	+	+	+		+		+			+
229. S. pactum	T	−	+	−	−	−	−	+	−	−	−	−	−

[a] For explanation of meanings of symbols see footnotes to Tables 17.40 and 17.41a–e.

Some spore chains may be of atypical *Retinaculum-Apertum* type. Characterization, in part, in Shirling and Gottlieb, 1969, 500.

Type strain: IPV 334a; ISP 5249 (Baldacci in Shirling and Gottlieb, 1969, 500).

223. Streptomyces calvus Backus, Tresner and Campbell 1957, 533. (*Streptomyces calvus* (n. sp.), Lederle strain T3018 in Thomas, Singleton, Lowery, Sharpe, Pruess, Porter, Mowat and Bohonos 1957, 716 (Not Val. Pub.); *Streptomyces calvus* in Hewitt, Gumble, Taylor and Wallace 1957, 722 (Not Val. Pub.).)

cal′vus. L. adj. *calvus* bald, referring to the sparse formation of aerial mycelium of the organism.

Produces nucleocidin, an anti-bacterial, anti-protozoal and anti-trypanosomal purine; inhibited by streptomycin; excellent growth on Czapek's solution agar; sparse formation of aerial mycelium on most media; optimal temperature reportedly 32–37 C; NaCl tolerance ≧7%, but <10%.

Type strain: Lederle T-3018; ATCC 13382; ATCC 23890 (single isolate).

224. Streptomyces cyanoalbus (Krasil'nikov and Agre) Pridham 1970, 13. (*Actinomyces cyanoalbus* Krasil'nikov and Agre in Rautenshtein 1960, 273.)

cy.an.o.al′bus. Gr. adj. *cyaneus* dark blue; L. adj. *albus* white; M.L. adj. *cyanoalbus* blue, white, referring to formation of colorless (white) or blue vegetative mycelium.

Reportedly exhibits anti-fungal activity; excellent growth on Czapek's solution agar; forms blue or green vegetative mycelium on some media. Comment: Some investigators have reported

spiny spores for the type strain *cf.* Shirling and Gottlieb, 1968, 314.

Type strain: INMI 414; ATCC 15859; ATCC 23902 (Krasil'nikov in Shirling and Gottlieb, 1968, 314).

225. Streptomyces finlayi (Szabó, Marton, Buti and Pártai) Pridham 1970, 35. (*Actinomyces finlayi* Szabó *et al.* 1963, 209.)

fin.lay′i. M.L. gen.n. *finlayi* of Finlay, named for A. C. Finlay, discoverer of oxytetracycline.

Exhibits slight activity against *Micrococcus luteus* (*Sarcina lutea*) and some streptomycetes; forms green to yellowish green vegetative mycelium on some media; some spore chains of atypical *Retinaculum-Apertum* type; isolated from soil and also reported to represent a considerable proportion of the intestinal microflora of larvae of St. Mark's fly (*Bibio marci* L.) according to Szabó, Marton, Ferenczy and Buti 1967, 239.

Type strain: R-1-30; FBUA 1869; ATCC 23906.

226. Streptomyces flaveolus (Waksman) Waksman and Henrici 1948, 936. (*Actinomyces* 168 in Waksman 1919, 134; *Actinomyces flaveolus* Waksman 1923, 368; *Streptomyces flaveolus* subsp. *flaveolus* Waksman in Pridham, Lyons and Seckinger 1965, 220.)

fla.ve′o.lus. L. adj. *flavus* yellow; L. dim.adj. *flaveolus* somewhat yellow.

Exhibits slight anti-microbial activity; inhibited by streptomycin; excellent growth on Czapek's solution agar; NaCl tolerance ≧7%, but <10%.

Comment: See Pridham, Lyons and Seckinger, 1965, 220, for further discussion of this taxon.

Type strain: 168; IMRU 3319; ATCC 3319; ATCC 19754 (single isolate).

227. Streptomyces geysiriensis Wallhäusser, Nesemann, Präve and Steigler 1966, 734. (*Streptomyces geysiriensis* n. sp. 3888 in Lindner, Wallhäusser and Huber, German Auslegeschrift 1,113,791, September 14, 1961 (Not Val. Pub.); *Streptomyces geysiriensis* n. sp. 3888 in Lindner, Wallhäusser and Huber, Canadian Patent 672,917, October 22, 1963 (Not Val. Pub.).)

gey'sir.i.en'sis. Icel. *geysir* a geyser; M.L. adj. *geysiriensis* belonging to a geyser, referring to the source of the organism (an Iceland geyser).

Produces the moenomycin complex of antibacterial antibiotics (moenomycins A, B₁, B₂ and C); moderate growth on Czapek's solution (synthetic) agar.

Type strain: 3888; F.H. 3888; ATCC 15303 (single isolate).

228. Streptomyces herbiferis (*sic*) (Krasil'nikov and Egorova) Pridham 1970, 18. (*Actinomyces herbiferis* (*sic*) Krasil'nikov and Egorova in Krasil'nikov 1965, 198.)

her'bi.fer.is. L. adj. *herbifer* producing grass, grassy; L. gen.adj. *herbiferi* of grassy (green).

Forms grass-green vegetative mycelium on some media.

Type strain: INMI 10 (single isolate).

229. Streptomyces pactum (*sic*) Bhuyan, Dietz and Smith 1962, 185. (*Streptomyces pactum* var. *pactum* (*sic*) Bhuyan, Dietz and Smith 1962, 185; *Actinomyces pactum* var. *pactum* (*sic*) in Preobrazhenskaya 1966, 858.)

pac'tum. Gr. adj. *pactos* solid, firm, coagulated, referring to the compactness of the coiled chains of spores; L. adj. *pactus* settled.

Produces pactamycin, an anti-bacterial and anti-tumor antibiotic; inhibited by streptomycin; poor growth on Czapek's solution agar; NaCl tolerance ≧4%, but <7%.

Comment: No other varieties of this taxon have been proposed in the literature.

Type strain: UC 2432; NRRL 2939 (single isolate).

See Table 17.42j.

230. Streptomyces akitaensis Soeda and Fujita 1959, 296. (*Streptomyces akitaensis* in Soeda and Fujita 1959, 293 (Not Val. Pub.).)

a.ki.ta.en'sis. M.L. adj. *akitaensis* belonging to akita (epithet of uncertain derivation).

Produces akitamycin, an anti-fungal tetraene; poor growth on Czapek's solution agar.

Type strain: H-5504 (single isolate).

231. Streptomyces akiyoshiensis Tatsuoka, Miyake, Hitomi, Ueyanagi, Iwasaki, Yamaguchi,

Kanazawa, Araki, Tsuchiya, Hiraiwa, Nakazawa and Shibata 1961, 39. (Streptomycete strain in Kanazawa, Tsuchiya and Araki 1960, 924.)

a.ki.yo.shi.en'sis. M.L. adj. *akiyoshiensis* belonging to akiyoshi (epithet of uncertain derivation).

Produces HON (δ-hydroxy-α-oxo-L-norvaline), an anti-mycobacterial compound; probably grows poorly on Czapek's solution agar; forms violet to violet-brown to black vegetative growth on some media.

Type strain: H-8998; IFO 3810; ATCC 13479 (single isolate).

232. Streptomyces alanosinicus Thiemann and Beretta 1966, 158. (*Streptomyces alanosinicus* in Thiemann, Murthy and Coronelli, South African Patent 65/3609, January 1, 1966 (Not Val. Pub.).)

al'an.o.si'ni.cus. English n. *alanosine* name of an antibiotic; L. n.stem suff. *-icus* belonging to; M.L. adj. *alanosinicus* belonging to alanosine.

Produces alanosine, an anti-tumor, anti-viral and anti-fungal antibiotic; excellent growth on Czapek's solution agar; spores may be echinulate.

Type strain: V/119; ATCC 15710 (single isolate).

233. Streptomyces albidus subsp. **invertens** (Kudrina) Pridham, Hesseltine and Benedict 1958, 66. (*Actinomyces albidus* var. *invertens* Kudrina in Gauze *et al.* 1957, 115.)

subsp. in.ver'tens. L. part.adj. *invertens* inverting; referring to the strong sucrose inverting activity of the organism.

Exhibits anti-bacterial activity.

Reference strain: INA 5242/54.

234. Streptomyces albochromogenes Tanaka, Karasawa, Miyairi, Shinjo, Nishimura and Umezawa 1958, 270. (Actinomycete strain no. 314 C₁ in Tanaka, Yamazaki, Okabe and Umezawa 1957, 189.)

al.bo.chro.mo'ge.nes. L. adj. *albus* white; Gr. n. *chroma* color; Gr. v.suff.*-genes* producing; M.L. adj. *albochromogenes* producing white color.

Produces raromycin, an anti-bacterial and anti-tumor antibiotic.

Type strain: IAM 314 C₁ (single isolate).

235. Streptomyces ansochromogenes (Yen and Zhang) Pridham 1970, 32. (*Actinomyces ansochromogenes* Yen and Zhang 1964, 264.)

an.so.chro.mo'ge.nes. L. n. *ansa* handle; Gr. n. *chroma* color; Gr. v.suff.*-genes* producing, M.L. adj. *ansochromogenes* handle, producing color, possibly referring to the morphology of the organism and color production.

Exhibits anti-fungal activity and slight anti-bacterial activity; forms brown to black vegeta-

TABLE 17.42j
Gray series[a]

Inadequately described species in the gray series; names of species and subspecies	Strain designation	Morphological section	Melanoid pigment formation	Spore surface configuration	No carbon control	D-Glucose	D-Xylose	L-Arabinose	L-Rhamnose	D-Fructose	D-Galactose	Raffinose	D-Mannitol	i-Inositol	Salicin	Sucrose
						17.42j GY; ?; ?; ?										
230. S. akitaensis	T	S	−?			+	−	−	+	−	+	±	+	−	±	−
231. S. akiyoshiensis**	T	S	+		−		+	+	+	+	+	+	+	+	+	−
232. S. alanosinicus	T	S	+			+	+	+	−	+	+	+	+	+	+	
233. S. albidus subsp. invertens	R	S	+?			+	+	+	−	+	+	+	+	+	+	
234. S. albochromogenes**	T	S	+?			+	+	+	+	+	+	+	+	−		+
235. S. ansochromogenes (see also no. 236).	T	S	+		−	+	+	−	−	+	+	−	+	−		
236. S. ansochromogenes subsp. pallens (see also no. 235)	T	S	+		−	+	+	−	−	+	+	−	+	−		
237. S. avidinii	T	S	+			−	−				−					
238. S. carcinomycicus	T	S	−			−	−		−			+			+	
239. S. castaneoglobisporus	T	RF	+			+	+	−	+	+	+	+	+	+	−	
240. S. castaneus	T		+			+	+	−	+	−	+	−	−	−		+
241. S. cyanoflavus	T	RF	+		−	+	−	−	−	+		+	−	+	−	+
242. S. djakartensis	T	S	+													
243. S. erythrochromogenes subsp. narutoensis (see also incertae sedis no. A6).	T															
244. S. glomerochromogenes	T	S	−		−	+	−		−	−	−	−	+	−		
245. S. grisinus	T	S	−	Hairy?												
246. S. haranomachiensis	T	RF	+			+	+	−	−	+	+	+	+	+	+	+
247. S. hygrostaticus	R	S	−			+	+	+			+	+	+	+	+	+
248. S. insulatus	T	S		Smooth												
249. S. inversochromogenes	T	S		Smooth	−	+	+	+	−	±	+	+	+	+	−	−
250. S. kitazawaensis	T	RF	+		−	+	+	+	−	±	+	+	+	+	+	−
251. S. mariensis**	T	S	−				−	+		+	+	−	+	+	+	+
252. S. minutiscleroticus	T	S	−			+		+	+	+	+	+	+		+	−
253. S. mitakaensis	T	S	−			+		+	+	+	+	+	+	+	+	
254. S. nigrogriseolus	T	RF		Smooth												
255. S. ogaensis	T	S	+			+	−			−	−	+		+	−	
256. S. piedadensis	T		−		−	+	+	+	+	+	+	+	−	−	+	
257. S. regensis	T	S	+			+	+	−	−	+	−		+	+		+
258. S. robefuscus	T	S	+			+	+	+	+	+	+		+			+
259. S. robeus	T	S	+			+	+	+	+	+	+		+			+
260. S. robustrus	T	S	+			+	+	+	+	+	+		+			+
261. S. roseogriseolus	T	RF		Smooth												
262. S. roseogriseus	T	S		Smooth												
263. S. sahachiroi	T	S	−			+	+	−		+	+	−		+	−	+	+
264. S. senoensis	T	RF	+			+	+	−		+	−	−	−	−		+
265. S. tanashiensis subsp. cephalomyceticus*** (see also no. 60).	T	RF	+?			+	+	+	−		+	−	−	−	+	+
266. S. thermonitrificans	T	RF	+			+	−	−	−	−	+	−	+	+	−	−
267. S. thermoviolaceus subsp. apingens (see also no. 107).	T	S		Warty?												
268. S. viridoniger	T	S	−			+	+	−		+	+	+	+	+	+	+
269. S. werraensis	T	S	−													

[a] For explanation of meanings of symbols and asterisks, see footnotes to Tables 17.40 and 17.41a–e.

tive mycelium and red-brown diffusible pigment on some media.

Type strain: IMASP 9-252 (Pridham 1970, 32).

236. Streptomyces ansochromogenes subsp. **pallens** (Yen and Zhang) Pridham 1970, 33. (*Actinomyces ansochromogenes* var. *pallens* Yen and Zhang 1964, 264.)

subsp. pal'lens. L. adj. *pallens* pale, greenish.

Exhibits antibacterial and antifungal activity; forms brown to black vegetative mycelium and yellow-brown diffusible pigment on some media.

Type strain: IMASP 9-12 (Pridham 1970, 33).

237. Streptomyces avidinii Stapley, Mata, Miller, Demny and Woodruff 1964, 20.

a.vi.di′ni.i. English n. *avidin* the name of a biotin-binding protein; M.L. gen.n. *avidinii* of avidin.

Produces the antibiotic MSD-235 complex, streptavidin and antibiotic MSD 235S, a synergistic anti-bacterial complex; poor growth on Czapek's solution agar.

Type strain: MA-833 (single isolate).

238. Streptomyces carcinomycicus Hosoya and Soeda in Hosoya 1955, 128. (*Streptomyces gannmycicus* (*sic*) Hosoya and Soeda in Hosoya 1955, 130; *Streptomyces ganmycicus* (*sic*) in Harada and Tanaka 1956, 113; *Streptomyces carzinomycicus* (*sic*) in Sugawara and Onuma 1957, 140; *Streptomyces carcinomicicus* (*sic*) in Soeda 1959, 300; *Streptomyces carcinomyceticus* (*sic*) in Waksman 1961, 324.)

car.cin.o.my′ci.cus. English n. *carcinomycin* name of an antibiotic; L. suff. *-icus* belong to; M.L. adj. *carcinomycicus* belonging to carcinomycin.

Produces carcinomycin, an anti-tumor antibiotic; reportedly forms pale purple diffusible pigment on calcium malate agar.

Type strain: H-5342; IID (H-5342); JDA (H-5342) (single isolate).

239. Streptomyces castaneoglobisporus (Yen) Pridham 1970, 34. (*Actinomyces castaneoglobisporus* Yen 1957, 208); *Actinomyces castaneoglobosus* (*sic*) in Yen and Zhang 1964, 267 (illeg.).)

ca.sta′ne.o.glo.bi′spo.rus. L. n. *castanea* chestnut tree, chestnut; L. n. *globus* a round body; M.L. n. *spora* a spore; M.L. adj. *castaneoglobisporus* chestnut-brown, round spored (referring to color of vegetative mycelium and diffusible pigment and to morphology of spores of the organism).

Exhibits slight anti-bacterial and anti-fungal activity; forms red-brown to black vegetative mycelium and diffusible pigment on some media.

Type strain: IMASP A.S. 4.149 (single isolate).

240. Streptomyces castaneus (Yen) Pridham 1970, 35. (*Actinomyces castaneus* Yen 1957, 474.)

ca.sta′ne.us. L. n. *castanea* chestnut tree, chestnut (referring to color of the vegetative mycelium and diffusible pigment of the organism).

Exhibits anti-bacterial activity and slight anti-fungal activity; forms reddish black vegetative mycelium and diffusible pigment on some media. Characterization taken from Yen and Zhang, 1964, 268.

Type strain: IMASP A.S. 4.174 (single isolate).

241. Streptomyces cyanoflavus (*sic*) Funaki and Tsuchiya in Funaki, Tsuchiya, Maeda and Kamiya 1958, 144. (*Streptomyces cyaneoflavus* Funaki *et al.* 1962, 15 (*lapsus calami*).)

cy.an.o.fla′vus. Gr. adj. *cyaneus* dark blue; L.

adj. *flavus* yellow; M.L. adj. *cyanoflavus* dark blue, yellow, referring to colors of diffusible pigments formed by the organism.

Produces cyanomycin, an anti-bacterial antibiotic with pH-indicator properties and antibiotic 4738-A, a pyrrothine-type antibiotic similar to aureothricin; forms blue or greenish blue diffusible pigment on some media.

Type strain: 4738; Meiji 4738; OEU A-8 (single isolate).

242. Streptomyces djakartensis Huber, Wallhäusser, Fries, Steigler and Weidenmüller 1962, 1191. (*Streptomyces djakartensis* in Lindner, Huber, and Wallhäusser, German Patent 1,077,381, March 10, 1960 (Not Val. Pub.).)

dja.kart.en′sis. M.L. adj. *djakartensis* belonging to Djakarta, Indonesia, the source of the soil from which the organism was isolated.

Produces niddamycin (3-desacetyl carbomycin B), an antibacterial macrolide; good growth on Czapek's solution agar.

Type strain: 3463; F.H. 3463; ATCC 13441 (single isolate).

243. Streptomyces erythrochromogenes subsp. **narutoensis** Kondo, Yumoto, Miyakawa, Hamamoto, Sezaki, Sato and Niida 1962, 9.

subsp. na.ru′to.en′sis. M.L. adj. *narutoensis* belonging to Naruto, Japan.

Produces streptomycin; forms violet to red-violet vegetative mycelium on some media.

Type strain: Meiji Seika E-123 (single isolate).

244. Streptomyces glomerochromogenes (Yen and Zhang) Pridham 1970, 36. (*Actinomyces glomerochromogenes* Yen and Zhang 1964, 263.)

glo′mer.o.chro.mo′ge.nes. L. v. *glomero* form into a ball; Gr. n. *chroma* color; Gr. v.suff. *-genes* producing; M.L. adj. *glomerochromogenes* producing a ball-like color (probably intended to refer to formation of balls or masses of spores and production of dark pigment).

Exhibits anti-bacterial and anti-fungal activity; forms brown to black vegetative mycelium.

Type strain: IMASP 9-90 (single isolate).

245. Streptomyces grisinus (Krasil'nikov) Pridham 1970, 17. (*Actinomyces griseus* (*sic*) in Krasil'nikov, Belozerskii, Rautenshtein, Nikitina, Sokolova and Uryson 1956, 1117; *Actinomyces* sp., strain No. 15 (and others) in Krasil'nikov *et al.* 1957, 418; *Actinomyces grisinus* Krasil'nikov 1958, 263.)

gri.si′nus. M.L. adj. *griseus* gray; L. adj.suff. *-inus* belonging to; M.L. adj. *grisinus* belonging to gray.

Produces the peptidic anti-bacterial and anti-fungal antibiotic grizein (grizin, grisine, grisemin, grisemine, grizemin or grisin).

Comment: There is a possibility that hirsute spores are formed by strains of this species based on comments by Preobrazhenskaya *et al.* 1960, 55 with regard to spore wall ornamentation of strains of *Actinomyces griseus* Krainsky 1914, 682 *sensu* Krasil'nikov.

Type strain: 15 = INMI 15 (Krasil'nikov *et al.*, 1957, 418).

246. Streptomyces haranomachiensis Matsumoto 1961, 143.

ha.ra.no.mach.i.en′sis. M.L. adj. *haranomachiensis* belonging to Haranomachi (probably a Japanese place name).

Produces vancomycin; excellent growth on Czapek's solution agar; forms pink diffusible pigment on some media; may be a *Nocardia*.

Type strain: K-288; MTHU K-288 (single isolate).

247. Streptomyces hygrostaticus Furushiro, Shimizu, Sakai, Minogato and Fujisawa 1958, 24.

hy.gro.sta′ti.cus. Gr. adj. *hygrus* moist; L. n. *statio* standing still; L. adj.suff. *-icus* belonging to; M.L. adj. *hygrostaticus* belonging to moist, standing still (probably intended to refer to hygroscopic nature of aerial mycelium and biostatic activity).

Produces the musarin-like anti-bacterial and anti-fungal antibiotic, hygrostatin; produces leucocidin, an anti-bacterial antibiotic; hygroscopic.

Reference strain: FUJI (Sakai 662). Confusion exists as to type strain as more than one strain is referred to in the literature.

248. Streptomyces insulatus Okami and Umezawa in Takita, Ohi, Okami, Maeda and Umezawa 1962, 46. (*Streptomyces islandicus* Okami and Umezawa in Takita *et al.* 1962, 46 (illeg.).)

in.su.la′tus. L. n. *insula* island; L. n.suff. *-atus* provided with; M.L. adj. *insulatus* provided with island, referring to Oshima Island, Japan, the source of the soil from which the organism was isolated.

Produces rufomycins A and B; and ilamycins A₂, B₁, B₂, C₁ and C₂, all primarily anti-mycobacterial antibiotics.

Comment: The original strain isolated (A-165-Z1) gave rise to two types (A and B). Type B was stated to be degenerate.

Type strain: A-165-Z1-A; NIHJ 364; ATCC 14093 (single isolate).

249. Streptomyces inversochromogenes (Yen and Zhang) Pridham 1970, 40. (*Actinomyces inversochromogenes* Yen and Zhang 1964, 262.)

in.ver′so.chro.mo′ge.nes. L. v. *inverto* invert; change; Gr. n. *chroma* color; Gr. v.suff. *-genes* producing; M.L. adj. *inversochromogenes* invert, producing color.

Exhibits anti-bacterial activity and slight antifungal activity; reportedly forms dextrorsely coiled chains of spores with one to three turns; forms brown to black vegetative mycelium.

Type strain: IMASP 9–17 (single isolate).

250. Streptomyces kitazawaensis Harada and Tanaka 1956, 117. (*Streptomyces* sp., strain No. 48-B-3 in Harada, Nara and Okamoto 1956, 6; *Streptomyces kitazawaensis* in Harada, Kubo and Itagaki 1956, 9 (Not Val. Pub.); *Streptomyces kitazawaensis* in Nakayama, Okamoto and Harada 1956, 63 (Not val. pub.).)

ki.ta.za.wa.en′sis. M.L. adj. *kitazawaensis* belonging to Kitazawa, a Japanese place name.

Produces carzinocidin and the antimycin A complex; grows poorly on Czapek's solution agar.

Type strain: 48-B-3 (single isolate).

Comment: A second strain, 21-A-2, isolated at a later date also is associated with the name of the taxon. See Nakayama, Okamoto and Harada, 1956, 63.

251. Streptomyces mariensis Soeda 1959, 300.

ma.ri.en′sis. M.L. adj. *mariensis* belonging to mari (derivation of epithet unknown).

Produces marinamycin, an anti-tumor antibiotic; poor growth on Czapek's solution agar.

Type strain: S-34; JDA S-34 (single isolate).

252. Streptomyces minutiscleroticus (Thirumalachar) Pridham 1970, 41. (*Chainia minutisclerotica* Thirumalachar in Thirumalachar, Rahalkar, Desmukh and Sukapure 1965, 7.)

mi.nu.ti.scl′er.oti.cus. L. part.adj. *minutus* small (literally diminished); L. n. *sclerotium* sclerotium; M.L. adj. *minutiscleroticus* small sclerotium.

Produces antibiotic M5-18903, an optical antipode of aburamycin; exhibits antifungal activity; said to form small sclerotia on some media.

Type strain: HACC 147; ATCC 17757; ATCC 19346 (Pridham, 1970, 41).

253. Streptomyces mitakaensis Arai, Nakamura, Sakagami, Fukuhara and Yonehara 1956, 193.

mi.ta.ka.en′sis. M.L. adj. *mitakaensis* belonging to Mitaka City, Japan, the source of the soil from which the organism was isolated.

Produces mikamycins A and B, macrolide antibacterial antibiotics; good growth on Czapek's solution agar.

Comment: The taxon is more fully characterized in Arai, Karasawa, Nakamura, Yonehara and and Umezawa, 1958, 14.

Type strain: 74-4; IAM (74-4); NIHJ 77; ATCC 15295 (single isolate).

254. Streptomyces nigrogriseolus (Yen and

Chou) Pridham 1970, 41. (*Actinomyces nigro-griseolus* Yen and Chou 1964, 427.)

ni.gro.gri.se′o.lus. L. adj. *niger* black; M.L. dim.adj. *griseolus* somewhat gray; M.L. dim.adj. *nigrogriseolus* somewhat gray (spotted with) shiny black.

Exhibits anti-bacterial and anti-fungal activity; forms black vegetative mycelium on some media.

Type strain: IMASP B1–12 (Pridham, 1970, 41).

255. Streptomyces ogaensis Nagatsu and Suzuki 1963, 205. (*Streptomyces ogaensis* in Ohkuma, Nagatsu, Itakura, Suzuki and Sumiki 1962, 152 (Not Val. Pub.).)

o.ga.en′sis. M.L. adj. *ogaensis* belonging to Oga Peninsula, Japan, the source of the soil from which the organism was isolated.

Produces cervicarcin, an anti-tumor antibiotic; reportedly exhibits no activity against bacteria, molds or yeasts.

Type strain: 13-90; IPCR 13-90 (single isolate).

256. Streptomyces piedadensis (Castellani) Pridham 1970, 42. (*Nocardia piedadensis* Castellani 1964, 334.)

pie′d.a.den sis. Etymology uncertain. M.L. adj. *piedadensis* belonging to piedad.

Exhibits no anti-microbial activity; inhibited by streptomycin; excellent growth on Czapek's solution agar; isolated from a case of "macroulcus perstans" of the leg. The type strain contains L-diaminopimelic acid in whole cell hydrolysates. Characterization taken from Pridham and Lyons, 1969, 210.

Type strain: CBS 459.65 (Pridham, 1970, 42).

257. Streptomyces regensis Gupta, Sobti and Chopra 1963, 15.

reg.en′sis. M.L. adj. *regensis* belonging to reg (unknown derivation, possibly referring to a place in India).

Produces an actinomycin complex; grows on Czapek's solution agar; forms green vegetative mycelium and diffusible pigment on some media.

Type strain: X-5263 (single isolate).

258. Streptomyces robefuscus (Krasil'nikov and Vinogradova) Pridham 1970, 25. (*Actinomyces robefuscus* Krasil'nikov and Vinogradova in Rautenshtein 1960, 223.)

ro be.fus′cus. L. n. *robur* strength, oak; L. adj. *fuscus* dark, tawny; M.L. adj. *robefuscus* similar to dark oak (referring to dark oak color of vegetative mycelium).

Exhibits anti-bacterial and anti-fungal activity; reportedly produces autoinhibitory "necrohormones" on Czapek's medium only; vegetative mycelium typically some tint or shade of brown on most media.

Type strain: INMI 3 (Pridham, 1970, 25).

259. Streptomyces robeus (Krasil'nikov and Vinogradova) Pridham 1970, 25. (*Actinomyces robeus* Krasil'nikov and Vinogradova in Rautenshtein 1960, 222.)

ro′be.us. L. n. *robur* strength, oak; M.L. adj. *robeus* similar to oak (referring to the color of the vegetative mycelium and diffusible pigment of the organism).

Exhibits slight anti-bacterial activity and activity against molds, but not yeasts; reportedly produces autoinhibitory "necrohormones" on Czapek's medium only; vegetative mycelium typically some tint or shade of brown on most media.

Type strain: INMI 8 (Pridham, 1970, 25).

260. Streptomyces robustrus (*sic*) (Krasil'-nikov and Vinogradova) Pridham 1970, 25. (*Actinomyces robustrus* (*sic*) Krasil'nikov and Vinogradova in Rautenshtein 1960, 223.)

ro′b.us′trus. L. n. *robur* strength, oak; M.L. adj. *robusteus* similarity to oak (referring to color of fumed oak as color of the vegetative mycelium).

Note. Probably *robustrus* is a *lapsus calami* for *robusteus*.

Exhibits anti-bacterial and anti-fungal activity; reportedly produces autoinhibitory "necrohormones" on Czapek's medium only; vegetative mycelium typically some tint or shade of brown on most media.

Type strain: INMI 5 (Pridham, 1970, 25).

261. Streptomyces roseogriseolus (Yen and Chou) Pridham 1970, 43. (*Actinomyces roseogriseolus* Yen and Chou 1964, 426.)

ro′se.o.gri.se′o.lus. L. adj. *roseus* rose colored; M.L. dim.adj. *griseolus* somewhat gray; M.L. dim.adj. *roseogriseolus* rose colored, somewhat gray.

Exhibits anti-bacterial and anti-fungal activity; reportedly inhibits only *Sporobolomyces* of nine yeasts tested; forms pale rose-colored vegetative mycelium on some media.

Type strain: IMASP Y3-10 (single isolate).

262. Streptomyces roseogriseus (Yen and Chou) Pridham 1970, 43. (*Actinomyces roseogriseus* Yen and Chou 1964, 425.)

ro′se.o.gri′se.us. L. adj. *roseus* rose colored; M.L. adj. *griseus* gray; M.L. adj. *roseogriseus* rose colored, gray.

Exhibits anti-bacterial activity; forms red to yellow vegetative mycelium on some media.

Type strain: IMASP Y18-13 (Pridham, 1970, 43).

263. Streptomyces sahachiroi Hata, Koga, Sano, Kanamori, Matsumae, Sugawara, Hoshi, Shima, Ito and Tomizawa 1954, 107.

sa.ha′chi.roi. M.L. gen.n. *sahachiroi* of sahachiro

(unknown derivation, possibly a Japanese place name).

Produces carzinophilin A, an anti-tumor antibiotic; exhibits anti-bacterial activity; grows on Czapek's solution agar.

Type strain: K-534, KITA (K534) (single isolate).

264. Streptomyces senoensis Kanda, Asano and Shinobu 1962, 223. (*Streptomyces senoensis* in Kawato and Shinobu 1960, 59 (Not Val. Pub.); *Streptomyces senoensis* in Kawato and Shinobu 1962, 215 (Not Val. Pub.).)

se.no.en'sis. M.L. adj. *senoensis* of Seno Station, Hiroshima, Japan, the source of the soil from which the organism was isolated.

Produces senomycin, an antibacterial antibiotic; poor to moderate growth on Czapek's solution agar.

Type strain: 2750E (single isolate).

265. Streptomyces tanashiensis subsp. **cephalomyceticus** Matsumae 1960, 143.

subsp. ceph'a.lo.my.ce'ti.cus. Gr. n. *enkephalos* the brain; English n. *cephalomycin* name of an anti-viral antibiotic; L. n.suff. *-icus;* M.L. adj. *cephalomyceticus* belonging to cephalomycin.

Produces cephalomycin, an antibiotic especially active against Japanese encephalitis virus; also produces a luteomycin-like antibiotic; poor growth on Czapek's solution agar; forms a red pigment with cytochrome-like absorption spectrum.

Type strain: Z-1120, KITA (Z-1120) (single isolate).

266. Streptomyces thermonitrificans Desai and Dhala 1967, 137.

ther.mo.ni.tri'fi.cans. Gr. n. *therme* heat; M.L. part.adj. *nitrificans* nitrifying; M.L. part.adj. *thermonitrificans* heat, nitrifying, referring to thermophily and vigorous nitrate reduction of the organism.

Exhibits no anti-microbial activity, but displays weak lytic activity on heat-killed bacterial cells; an obligate thermophile, growing at 37–50 C, with optimum at 45–50 C; no growth at 27 C or below.

Type strain: NCIM-2007; ATCC 23385 (single isolate).

267. Streptomyces thermoviolaceus subsp. **apingens** Henssen 1957, 390.

subsp. a.pin'gens. Gr. pref. *a-* denoting negation; L. part.adj. *pingens* coloring; M.L. part.adj. *apingens* not coloring.

Excellent growth on Czapek's solution agar; a eurythermal thermophile and facultative aerobe; grows best at 50 C under anaerobic conditions; slight growth at 28 C and 60 C; forms yellow, orange or brown vegetative mycelium on some

media; originally isolated from mixed fresh horse and swine manure.

Type strain: R_{89}; A_{34}; CBS 140.67; ATCC 19994 (Henssen and Schnepf, 1967, 217).

268. Streptomyces viridoniger Szabó and Marton 1958, 250.

vi.ri.do.ni'ger. L. adj. *viridis* green; L. adj. *niger* black; M.L. adj. *viridoniger* green, black.

Exhibits slight anti-bacterial activity; forms green to blackish green vegetative mycelium and diffusible pigment on some media; reportedly characterized by sparse formation of aerial mycelium and frequent occurrence of chlamydospores; NaCl tolerance $\leq$ 10–12%.

Type strain: B-1-1/a (designated herein).

269. Streptomyces werraensis Wallhäusser, Huber, Nesemann, Präve and Zepf 1964, 357. (*Streptomyces werraensis* in Wallhäusser and Huber, German Patent 1,142,989, January 31, 1963 (Not Val. Pub.).)

wer.ra.en'sis. M.L. adj. *werraensis* belonging to River Werra, Germany (referring to "werramycin," the name originally assigned to the antibiotics produced).

Produces the anti-bacterial antibiotics monactin, dinactin and trinactin; produces the inactive compound, nonactin.

Type strain: FH 3582; ATCC 14424 (single isolate).

See Table 17.43a.

270. Streptomyces alboflavus (Waksman and Curtis) Waksman and Henrici 1948, 954. (*Actinomyces alboflavus* Waksman and Curtis 1916, 120.)

al.bo.fla'vus. L. adj. *albus* white; L. adj. *flavus* yellow; M.L. adj. *alboflavus* whitish yellow.

Exhibits marked anti-bacterial and anti-fungal activity; does not always exhibit melanin-like chromogenicity; inhibited by streptomycin; moderate growth on Czapek's solution agar; not considered a typical representative of the *Rectus-Flexibilis* spore chain type, i.e. forms apparently straight or flexuous chains of spores in arrangements other than typical flexuous, straight, cobweb (randomly arranged) or dichotomous.

Type strain: IMRU 3008; ATCC 12626; ATCC 23874 (single isolate).

271. Streptomyces bacillaris (Krasil'nikov) Pridham 1970, 9. (*Actinomyces bacillaris* Krasil'nikov 1958, 258.)

ba.cil.l'aris. L. dim.n. *bacillum* a small rod; M.L. n. *bacillus* a rodlet; L. adj.suff. *-aris* pertaining to; M.L. adj. *bacillaris* pertaining to a rodlet.

Exhibits slight anti-bacterial and anti-fungal activity; inhibited by streptomycin; excellent growth on Czapek's solution agar; spore chains typically flexuous.

TABLE 17.43a
Yellow series[a]

Names of species and subspecies	Strain desig-nation	Utilization of carbon compounds (P & G basal)											
		No carbon control	D-Glucose	D-Xylose	L-Arabinose	L-Rhamnose	D-Fructose	D-Galactose	Raffinose	D-Mannitol	i-Inositol	Salicin	Sucrose
17.43a Y; RF; C+; SM													
270. S. alboflavus	T	−	+	+	+	−	+	+	+	+	+	−	+
271. S. bacillaris	T	−	+	+	−	−	+	+	+	+	+	+	
272. S. cavourensis	T	−	+	+	−	−	+		−	+	−		
273. S. cyaneofuscatus	T	−	+	+	−	+	+	+	−	+	−	+	+
274. S. fulvissimus	R	−	+	+	+	−	+		−	+	+	+	
275. S. griseobrunneus	T	−	+	+	−	−	+	+	+	+	−	+	+
276. S. michiganensis	T	−	+	+	−	−	+	±	+	+	−		−
277. S. tsusimaensis	T		+	+	+	−	+	+	−	+	−	+	+
278. S. xanthochromogenus (*sic*)	T		+	+	±	±	+		±	+	±		±

[a] For explanation of meanings of symbols see footnotes to Tables 17.40 and 17.41a–e.

Type strain: INMI 445; ARI 445; ATCC 15855 (Pridham, 1970, 9).

272. Streptomyces cavourensis Giolitti 1958 in Waksman 1961, 191 (Strain of the *S. tanaschiensis* (*sic*) type in Craveri and Giolitti 1957, 1307; *Streptomyces cavourensis* in Giolitti and Craveri, Belgian Patent 560,930 March 18, 1958 (Not Val. Pub.); streptomyces no. 829 related to *Streptomyces tanashiensis* (*sic*) in Giolitti, Abstrs. VII Int. Cong. Microbiol. 1958, 38 Abstract 22k.)

ca.vour.en'sis. M.L. adj. *cavourensis* of Cavour, named for Conte di Cavour, an Italian statesman and hero, 1810–1861.

Exhibits slight anti-bacterial activity; produces flavensomycin, an anti-fungal and anti-insecticidal antibiotic; inhibited by streptomycin; moderate growth on Czapek's solution agar; NaCl tolerance ≧7%, but <10%.

Type strain: 829, NRRL 2740; IMRU 3758; CMI 70852 (single isolate).

273. Streptomyces cyaneofuscatus (Kudrina) Pridham, Hesseltine and Benedict 1958, 58. (*Actinomyces cyaneofuscatus* Kudrina in Gauze *et al.* 1957, 85.)

cy.an.e.o.fus.ca'tus. Gr. adj. *cyaneus* dark blue; L. adj. *fuscus* dark, tawny; L. n.stem suff. *-atus* provided with; M.L. adj. *cyaneofuscatus* provided with dark blue, tawny, referring to different pigments formed.

Exhibits anti-bacterial activity; inhibited by streptomycin; excellent growth on Czapek's solution agar; forms a stable (non-pH-sensitive) blue, diffusible pigment on some chemically defined media; NaCl tolerance ≧7%, but <10%.

Type strain: INA 99/54; ATCC 19746 (Gauze in Gottlieb, 1968, 20).

274. Streptomyces fulvissimus (Jensen) Waksman and Henrici 1948, 946. (*Actinomyces fulvissimus* Jensen 1930, 66.)

ful.vis'si.mus. L. sup.adj. *fulvissimus* very yellow.

Exhibits anti-bacterial and anti-fungal activity; excellent growth on Czapek's solution agar; forms orange, red or bright yellow vegetative mycelium on some media; NaCl tolerance ≧7%, but <10%.

Reference strain: NRRL B-1453; IMRU 3665; NCIB 9609. None of Jensen's original strains is extant.

275. Streptomyces griseobrunneus Waksman 1961, 220 (*Actinomyces* 218 in Waksman 1919, 125.)

gri.se.o.brun'ne.us. M.L. adj. *griseus* gray; M.L. adj. *brunneus* dark brown; M.L. adj. *griseobrunneus* grayish dark-brown colored.

Exhibits slight anti-bacterial and anti-fungal activity; inhibited by streptomycin; excellent growth on Czapek's solution agar; NaCl tolerance ≧7%, but <10%.

Type strain: 218; IMRU 3068 (single isolate).

276. Streptomyces michiganensis Corbaz, Ettlinger, Keller-Schierlein and Zähner 1957, 205.

mi'chi.gan.en'sis. M.L. adj. *michiganensis* belonging to Michigan, the source of the soil from which the organism was isolated.

Produces the actinomycin X complex; also reported to produce the actinomycin F or Z complex; poor growth on Czapek's solution agar; NaCl tolerance ≧ 7%, but <10%.

Type strain: ETH 9001; ATCC 14970; ATCC 19786 (single isolate).

277. Streptomyces tsusimaensis Nishimura, Mayama, T. Kimura, A. Kimura, Kawamura, Tawara, Tanaka, Okamoto and Kyotani 1964, 13.

tsu.si ma.en'sis. M.L. adj. *tsusimaensis* belonging to Tsushima, an island in Japan, the source of the soil from which the organism was isolated.

Produces nonactin (a biologically inactive compound) and vallinomycin, an anti-bacterial and anti-fungal antibiotic.

Type strain: N-329; SRL N-329 (single isolate).

278. Streptomyces xanthochromogenus (*sic*) Arishima, Sakamoto and Sato 1956, 469.

xan'tho.chro.mo'ge.nus. Gr. adj. *xanthus* yellow; Gr. n. *chroma* color; Gr. v.suff. *-genes* producing; M.L. adj. *xanthochromogenus* (*sic*.) producing yellow color.

Produces xanthicin, an antifungal antibiotic; NaCl tolerance ≥ 7%, but <10%.

Type strain: 689; ATCC 19818 (single isolate).

See Table 17.43b.

279. Streptomyces albidoflavus (Rossi Doria) Waksman and Henrici 1948, 949. (*Streptotrix albido flava* (*sic*) Rossi Doria 1891, 407 (Illeg.); *Actinomyces albidoflavus* (Rossi Doria) Gasperini 1894, 87; *Streptothrix albidoflava* (*sic*) Rossi Doria 1891, 407 (Illeg.); *Streptothrix albido* (Rossi Doria) Chester 1901, 365 (Illeg.); *Cladothrix albido-flava* (Rossi Doria) Macé 1901, 1097 (Illeg.); *Nocardia albida* (Rossi Doria) Chalmers and Christopherson 1916, 271 (Illeg.).)

al.bi.do.fla'vus. L. adj. *albidus* white; L. adj. *flavus* yellow; M.L. adj. *albidoflavus* whitish yellow.

Exhibits anti-bacterial and anti-fungal activity; poor growth on Czapek's solution agar; NaCl tolerance ≥ 10%, but <13%.

Reference strain: "CBS (Duché)"; CBS 416.34; none of Rossi Doria's strains is extant.

280. Streptomyces alboviridis (Duché) Pridham, Hesseltine and Benedict 1958, 74. (*Actinomyces alboviridis* Duché 1934, 317.)

al.bo.vi'ri.dis. L. adj. *albus* white; L. adj. *viridis* green; M.L. adj. *alboviridis* whitish green.

Exhibits anti-bacterial and anti-fungal activity; inhibited by streptomycin; moderate growth on Czapek's solution agar; spore chains typically flexuous.

Reference strain: Nicot 68; LCP 68; NRRL B-3633 (isolated by Nicot in 1951 and identified by Labourer and Duché; original strain of Duché is no longer extant).

281. Streptomyces anulatus (Beijerinck) Waksman in Breed *et al.* 1957, 755. (*Actinomyces Streptothrix annulatus* (*sic*) Beijerinck 1912, 7

(Illeg.); *Actinomyces annulatus* (*sic*) Beijerinck 1912, 4; *Streptomyces annulatus* (*sic*) (Beijerinck) Waksman in Waksman and Lechevalier 1953, 40.)

a.nu.la'tus. L. adj. *anulatus* furnished with a ring.

Exhibits anti-bacterial and anti-fungal activity; inhibited by streptomycin; poor growth on Czapek's solution agar; spore chains typically flexuous. This species originally was proposed because of its formation of characteristic annular rings of growth; many other species of streptomycetes also exhibit this phenomenon.

Type strain: "CBS (Beijerinck)"; CBS 100.18 (Beijerinck deposited this strain in CBS in 1918).

282. Streptomyces badius (Kudrina) Pridham, Hesseltine and Benedict 1958, 58. (*Actinomyces badius* Kudrina in Gauze *et al.* 1957, 87.)

ba'di.us. L. adj. *badius* brown.

Exhibits anti-bacterial and anti-fungal activity; inhibited by streptomycin; poor growth on Czapek's solution agar; forms deep brown vegetative mycelium and diffusible pigment on some media; spore chains typically flexuous; NaCl tolerance ≥ 7%, but <10%.

Type strain: INA 1203/53; ATCC 19729; ATCC 19888 (Gauze in Gottlieb 1968, 20).

283. Streptomyces californicus (Waksman and Curtis) Waksman and Henrici 1948, 936. (*Actinomyces californicus* Waksman and Curtis 1916, 122.)

ca.li.for'ni.cus. M.L. adj. *californicus* belonging to California, the source of the soil from which the organism was isolated.

Produces the viomycin complex (components A, B and C), the griseorhodin complex, and the antibiotic of Thrum *et al.* (1967); inhibited by streptomycin; moderate growth on Czapek's solution agar; forms red- to violet-colored vegetative mycelium and diffusible pigment on some media; pigment is pH-sensitive (red with acid, blue with alkali); spore chains typically flexuous; NaCl tolerance ≥ 7%, but <10%. See Pridham, Lyons and Seckinger (1965, 217) and Lyons and Pridham (1966, 1) for further discussion of this and related taxa.

Type strain: IMRU 3312; ATCC 3312; ATCC 19734 (single isolate).

284. Streptomyces canescens Waksman in Breed *et al.* 1957, 768. (*Streptomyces canescus* (*sic*) Hickey, Corum, Hidy, Cohen, Nager and Kropp 1952, 473.)

ca.nes'cens. L. part.adj. *canescens* becoming hoary or white.

Exhibits anti-bacterial activity; produces ascosins A and B, heptaenic anti-fungal antibiotics; poor growth on Czapek's solution agar; inhibited

TABLE 17.43b
Yellow series[a]

Names of species and subspecies	Strain designation	No control carbon	D-Glucose	D-Xylose	D-Arabinose	L-Rhamnose	D-Fructose	D-Galactose	Raffinose	D-Mannitol	i-Inositol	Salicin	Sucrose
17.43b Y; RF; C−; SM													
279. S. albidoflavus	R	−	+	+	+	−	+		−	+	−		−
280. S. alboviridis	R	−	+	+	−	+	+		−	+	−		−
281. S. anulatus	T	−	+	+	+	+	+		−	+	−		−
282. S. badius	T	−	+	+	+		+		−	+	−		−
283. S. californicus	T	−	+	+	−		+	+		+	−		−
284. S. canescens	T	−	+	−	+	−	+			−	−		−
285. S. celluloflavus	T	−	+	−	−	−							
286. S. cellulosae	T	−	+	+	+	+	+				+	+	
287. S. champavatii	T	−	+	+	+		+	+		+	−		−
288. S. chrysomallus (see no. 66).	T	−	+	+	+	+	+	+		+	−	+	−
289. S. citreofluorescens	T	−	+	+	+	+	+	+		+	−		−
290. S. coelicolor (see no. 136, 137, 138 and 139).	T		+	+	+		+			+			
291. S. felleus	R	−	+	+	+	+				+	−	+	
292. S. fimicarius	T	−	+	+	+	+	+	+		+	−		
293. S. floridae	T		+	+	−	−	+	+		+	−		
294. S. fluorescens	T	−	+	+	+	−	+	+		+	−	+	−
295. S. globisporus (see nos. 296, 297, 298 and incertae sedis no. B14).	R	−	+	+	+	+	+			+			−
296. S. globisporus subsp. caucasicus (see nos. 295, 297, 298 and incertae sedis no. B14).	T	−	+	+	+	−							
297. S. globisporus subsp. flavofuscus (see nos. 295, 296, 298 and incertae sedis no. B14).	T	−	+	+	+	+							
298. S. globisporus subsp. vulgaris (see nos. 295, 296, 297 and incertae sedis no. B14).	R		+	+	+	+	+	+		+			−
299. S. gougeroti	T	−	+	−	+	−							
300. S. griseinus	T		+	+	+	+	+	+		+	−	−	
301. S. griseoloalbus	T		+	+	+	+	+			+	+		+
302. S. griseus (see nos. 150, 303, 304, 305 and incertae sedis nos. B15 and B16).	R	−	+	+	−	−	+	+	−	+	−	+	−
303. S. griseus subsp. alpha (see nos. 150, 302, 304, 305 and incertae sedis nos. B15 and B16).	T	−	+	+	−	+							
304. S. griseus subsp. cretosus (see nos. 150, 302, 303, 305 and incertae sedis nos. B15 and B16).	R	−	+	+	−	+	+		−		−	+	
305. S. griseus subsp. solvifaciens (see nos. 150, 302, 303, 304 and incertae sedis nos. B15 and B16).	T	−	+	+	+	−	+		−	+	−	+	
306. S. intermedius	R	−	+	+	+	−	+	+	+	+	−	+	+
307. S. kanamyceticus	T	−	+	+	+	−	+	+	+	+	−	+	
308. S. levoris	T	−	+	+	+	−	+	+		+	−	+	−

795

TABLE 17.43b—*Continued*

Names of species and subspecies	Strain desig-nation	Utilization of carbon compounds (P & G basal)											
		No carbon control	D-Glucose	D-Xylose	L-Arabinose	L-Rhamnose	D-Fructose	D-Galactose	Raffinose	D-Mannitol	i-Inositol	Salicin	Sucrose
17.43b Y; RF; C—; SM—(*Continued*)													
309. S. limosus	T		+	+	+	−	+		−	+	−		−
310. S. lipmanii	T	−	+	+	−	+	+	+	+	+	−	+	±
311. S. microflavus	T	−	+	+	−	+	+	+	−	+	−	+	
312. S. odorifer	R	−	+	+	+	−	+		−	+	+	+	
313. S. parvus	R	−	+	+	+	+	+	+	−	−		−	−
314. S. pluricolorescens	T	−	+	+	−	+	+		−	−			−
315. S. pneumonicus	T		+	+	−	−	+		+	−	+		
316. S. praecox	T	−	+	+	+	+	+	+	+	+		+	
317. S. puniceus	T	−	+	+	+	−	+	+	−	+		+	±
318. S. raffinosus	T	−	+	+	+	−	+	+	+	+		+	+
319. S. rutgersensis	T	−	+	+	+	−	+	+	+	+		+	
320. S. sampsonii	T	−	+	+	+	−	+		−	+		−	−
321. S. setonii	T	−	+	+	+	+	+		−	+		−	−
322. S. sindenensis	T	−	+	+	+	−	+	+	−	+		+	
323. S. sulphureus	R	−	+	+	+	−	+		+		+	−	+
324. S. willmorei	T	−	+	+	−	+							

a For explanation of meanings of symbols see footnotes to Tables 17.40 and 17.41a–e.

by streptomycin; NaCl tolerance ≥ 10%, but <13%.

Type strain: NRRL 2419; IMRU 3847; ATCC 15731; ATCC 19736 (Cohen listed the designation of the original single isolate in U.S. Patent 2,723,216, November 8, 1955).

285. Steptomyces celluloflavus Nishimura, Kimura and Kuroya 1953, 64.

cel.lu.lo.fla'vus. M.L. n. *cellulosum* cellulose; L. adj. *flavus* yellow; M.L. adj. *celluloflavus* cellulose, yellow (intended to refer to the yellow streptomycete that attacks cellulose).

Produces aureothricin; inhibited by streptomycin; poor growth on Czapek's solution agar; sparse formation of aerial mycelium.

Type strain: 39a; SRL 39a; NRRL B-2493 (single isolate).

286. Streptomyces cellulosae (*sic*) (Krainsky) Waksman and Henrici 1948, 938. (*Actinomyces cellulosae* (*sic*) Krainsky 1914, 683.)

cel.lu.lo'sae. M.L. n. *cellulosum* cellulose; M.L. gen.n. *cellulosi* of cellulose (probably intended to mean the species that degrades cellulose).

Exhibits slight anti-bacterial activity; excellent growth on Czapek's solution agar; spore chains typically flexuous.

Type strain: "CBS (Krainsky)"; CBS 122.18 (deposited in CBS in 19[...]

287. Streptomyces champavatii Uma and Narasimha Rao 1959, 133. (*Streptomyces* sp. in Narasimha Rao and Uma 1958, 115.)

cham.pa.va'ti.i. M.L. gen.n. *champavatii* of Champavathi, named after the Champavathi River in Andrha Pradesh, India.

Produces champamycins A and B (heptaenic anti-fungal antibiotics) and champavatin, a non-polyenic anti-fungal antibiotic; produces vitamin B₁₂; poor growth on Czapek's solution agar; forms green vegetative mycelium and diffusible pigment on some media. Characterization taken from Narasimha Rao and Uma (1958), Uma and Narasimha Rao (1959) and Rao and Narasimha Rao (1967).

Type strain: 1033 (designated herein).

288. Streptomyces chrysomallus Lindenbein 1952, 369. (Actinomyces (*sic*)-strain isolated by von Plotho in Brockmann and Grubhofer 1949, 376; *Streptomyces*-strain in Brockmann and Grubhofer 1950, 494; *Streptomyces chrysomallus* in Brockmann, Grubhofer, Kass and Kalbe 1951, 260 (Not Val. Pub.).)

chry.so'mal.lus. Gr. adj. *chrysomallus* with golden wool.

Produces the actinomycin C complex; exhibits anti-fungal activity; inhibited by streptomycin; excellent growth on Czapek's solution agar; forms bright yellow vegetative mycelium and diffusible pigment on some media; spore chains typically flexuous; NaCl tolerance $\geq$ 7%, but <10%.

Type strain: Lindenbein 1a; NRRL 2250; IMRU 3657; ATCC 11523 (stated to be the type strain in Ettlinger, Corbaz and Hütter 1958, 349; Waksman 1961, 194 and Hütter 1963, 212.

Comment: It should be noted that P. Wilde incorrectly selected yet another strain, Schön. 192 = ATCC 23209, as the type strain for the ISP in Shirling and Gottlieb 1968, 97.

289. Streptomyces citreofluorescens (Korenyako, Krasil'nikov, Nikitina and Sokolova) Pridham 1970, 10. (*Actinomyces citreofluorescens* Korenyako, Krasil'nikov, Nikitina and Sokolova in Rautenshtein 1960, 156.)

cit're.o.flu.o.res'cens. L. n. *Citrus* name of genus which includes the lemon. M.L. v. *fluoresco* fluoresce; M.L. part.adj. *citreofluorescens* with a yellow fluorescence.

Produces three antibiotics: the actinomycin C complex, a colorless anti-bacterial antibiotic, and a water-soluble anti-fungal antibiotic; inhibited by streptomycin; moderate growth on Czapek's solution agar; forms bright yellow vegetative mycelium and diffusible pigment on some media; spore chains flexuous.

Type strain: INMI 2292; RIA 648; ATCC 15858; ATCC 23898 (Krasil'nikov in Shirling and Gottlieb, 1968, 311).

290. Streptomyces coelicolor (Müller) Waksman and Henrici 1948, 935. (*Streptothrix coelicolor* Müller 1908, 197 (Illeg.); *Cladothrix coelicolor* (Müller) Macé 1913, 758; *Nocardia coelicolor* (Müller) Chalmers and Christopherson 1916, 271; *Actinomyces coelicolor* (Müller) Lieske 1921, 28; *Corynebacterium coelicolor* (Müller) Müller 1950, 274 (Not Val. Pub.).)

coe.li'co.lor. L. n. *coelum* heaven, sky (blue); L. n. *color* color; M.L. n. *coelicolor* sky (blue) color.

Produces heptaenic anti-fungal activity; inhibited by streptomycin; poor growth on Czapek's solution agar; may form sclerotia and substrate spores; potato plugs are pigmented dark blue in 4–7 days; forms reddish, yellowish, brown or green diffusible pigments, the brown and yellowish pigments changing to green at alkaline pH, spore chains typically flexuous; NaCl tolerance $\geq$ 13%, but <15%.

Note. A number of strains of streptomycetes bearing this name and used in genetic studies have been incorrectly identified. See Kutzner and Waksman (1959, 528) for a discussion of this taxon.

Type strain: "CBS (Müller)"; CBS 210.27; ATCC 23899 (deposited by Müller in CBS in 1927).

291. Streptomyces felleus (*sic*) Lindenbein

1952, 374. (*Actinomyces* sp. in Benfey, B., Dissertation, Göttingen; *Actinomyces* sp. in Brockmann and Henkel 1950, 138; *Streptomyces* sp. strain 326 in Brockmann and Henkel 1951, 284; *Actinomyces* Bo-105 = *Streptomyces felleus* Lindenbein in Brockmann, Genth and Strufe 1952, 426 (Not Val. Pub.).)

fel'le.us. Variant of L. gen.n. *fellis* of gall or bile (pertaining to the bitter taste of proactinomycin A).

Produces proactinomycin A; inhibited by streptomycin; poor growth on Czapek's solution agar; spore chains typically flexuous; NaCl tolerance $\geq$ 4%, but <7%.

Reference strain: Bo-105; NRRL 2251; IMRU 3659.

Comment: The earliest strain designation on record is that in Brockmann and Henkel 1951, 284 (strain 326); in 1952 Lindenbein listed three strains (W26, 326 and Gütt 467) in his proposal of the new species *S. felleus*; in 1952, the name *S. felleus*, although not validly published because it appeared in British Patent 682,045, was associated with only one strain designation (Bo-105) and the notation that it had been deposited at NRRL. Subsequently Waksman (1961, 206) designated strain IMRU 3659 which he had obtained as NRRL 2251 as the "Type culture." Still later, Wilde in Shirling and Gottlieb (1968, 110) selected strain Söt 26 as the type strain for the ISP.

292. Streptomyces fimicarius (Duché) Waksman and Henrici 1948, 940. (*Actinomyces fimicarius* Duché 1934, 346.)

fi.mi.ca'ri.us. L. n. *fimus* dung, manure; L. adj. *carus* dear, loving; M.L. adj. *fimicarius* dung-loving.

Exhibits slight anti-microbial activity; inhibited by streptomycin; excellent growth on Czapek's solution agar; spore chains typically flexuous; originally isolated from manure.

Type strain: "CBS (Duché)"; CBS 420.34 (deposited by Duché in CBS in 1934).

293. Streptomyces floridae Bartz, Ehrlich, Mold, Penner and Smith 1951, 4.

flo.ri'dae. M.L. gen.n. *floridae* of Florida, the source of the soil from which the organism was isolated.

Produces the viomycin complex; inhibited by streptomycin; poor growth on Czapek's solution agar; forms dull violet to red-brown vegetative mycelium and diffusible pigment on some media; spore chains typically flexuous.

Type strain: PD 04833; PD A5014a; NRRL 2423 (Ettlinger, Corbaz and Hütter, 1958, 349).

294. Streptomyces fluorescens (Krasil'nikov) Pridham 1970, 15. (*Actinomyces fluorescens* Krasil'nikov 1958, 258.)

flu.o.res'cens. M.L. v. *fluoresco* fluoresce; M.L.

pres.part. *fluorescens* fluorescing, referring to the yellow fluorescent pigment produced by the organism.

Produces the actinomycin X complex; exhibits anti-fungal activity; inhibited by streptomycin; poor growth on Czapek's solution agar; forms bright yellow vegetative mycelium and diffusible pigment on some media.

Note. See Korenyako, Sokolova and Nikitina (Suppl. to Abstr. Symp. Antibiot., Prague, Czech. 1959), Korenyako, Sokolova and Nikitina (1960, 59) and Korenyako, Krasil'nikov, Nikitina and Sokolova in Rautenshtein (1960, 133) for further discussion of this taxon.

Type strain: INMI 592; RIA 647; ATCC 15860; ATCC 23907 (Krasil'nikov in Shirling and Gottlieb, 1968, 318).

295. Streptomyces globisporus (Krasil'nikov) Waksman in Waksman and Lechevalier 1953, 39. (*Actinomyces globisporus* Krasil'nikov 1941, 48.)

glo.bi'spo.rus. L. n. *globus* a round body; M.L. n. *spora* a spore; M.L. adj. *globisporus* round spored (as determined by light microscopy).

Exhibits anti-bacterial and anti-fungal activity; inhibited by streptomycin; excellent growth on Czapek's solution agar. Characterization, in part, in Shirling and Gottlieb, 1968, 324.

Reference strain: INMI 2302; RIA 335; ATCC 15864; ATCC 23913 (Krasil'nikov selected this strain as the "type strain" in Shirling and Gottlieb, 1968, 324.) Strain INMI 2302 is not a descendant of one of the original isolates of this taxon.

296. Streptomyces globisporus subsp. **caucasicus** (Kudrina) Pridham, Hesseltine and Benedict 1958, 59. (*Actinomyces globisporus* var. *caucasicus* Kudrina in Gauze *et al.* 1957, 79.)

subsp. cau.ca'si.cus. Gr. n. *Caucasia* region of the Caucasus; M.L. adj. *caucasicus* belonging to the Caucasus.

Exhibits anti-bacterial and anti-fungal activity; inhibited by streptomycin; poor growth on Czapek's solution agar; spore chains typically flexuous; NaCl tolerance $\geq$ 10%, but <13%.

Type strain: INA 13195/54; CBS 120.60; ATCC 19907 (designated herein).

297. Streptomyces globisporus subsp. **flavofuscus** (Kudrina) Pridham, Hesseltine and Benedict 1958, 59. (*Actinomyces globisporus* var. *flavofuscus* Kudrina in Gauze *et al.* 1957, 81.)

subsp. fla.vo.fus'cus. L. adj. *flavus* yellow; L. adj. *fuscus* dark, tawny; M.L. adj. *flavofuscus* dark yellow.

Exhibits anti-bacterial and anti-fungal activity; inhibited by streptomycin; poor growth on Czapek's solution agar; spore chains typically flexuous; NaCl tolerance $\geq$ 10%, but <13%.

Type strain: (none by original designation) is

INA 1565/53; CBS 121.60; ATCC 19908 (designated herein).

298. Streptomyces globisporus subsp. **vulgaris** (Krasil'nikov) Pridham, Hesseltine and Benedict 1958, 58. (*Actinomyces globisporus vulgaris* n. subsp. (*sic*) Krasil'nikov 1941, 49; *Streptomyces globisporus vulgaris* (Krasil'nikov) Pridham, Hesseltine and Benedict 1958, 58.)

subsp. vul.ga'ris. L. adj. *vulgaris* common.

Exhibits anti-bacterial and slight anti-fungal activity; may possibly produce pneumocin, an anti-pneumococcal antibiotic; inhibited by streptomycin: poor growth on Czapek's solution agar. Characterization taken from Nikitina *et al.* in Rautenshtein (1960, 99), Hütter (1963, 191) and from Shirling and Gottlieb (1969, 502).

Reference strain: INMI 1034; RIA 334; ATCC 15895; selected by Krasil'nikov as the "type" strain in Shirling and Gottlieb, 1969, 502, although none of his original isolates is extant.

299. Streptomyces gougerotii (Duché) Waksman and Henrici 1948, 947. (*Actinomyces gougeroti* (*sic*) Duché 1934, 272; *Actinomyces gougerotii* Duché in Hütter 1967, 75; *Streptomyces gougerotii* Waksman and Henrici in Hütter 1967, 75.)

gou.ge.ro'ti.i. M.L. gen.n. *gougerotii* of Gougerot; named for Professor Gougerot from whom the original culture was obtained.

Exhibits slight anti-fungal activity; inhibited by streptomycin; poor growth on Czapek's solution agar; spore chains typically flexuous; NaCl tolerance $\geq$ 7%, but <10%.

Type strain: "CBS (Duché)"; CBS 422.34; IMRU 3590; ATCC 10975 (single isolate).

300. Streptomyces griseinus Waksman 1959, 1045. (Grisein-producing strain 3478 of *Streptomyces griseus* in Reynolds and Waksman 1948, 739; Waksman 1950, 66, Waksman in Waksman and Lechevalier 1953, 26 *inter allii*.)

gri.se.in'us. M.L. adj. *griseus* gray; English n. *grisein* name of an antibiotic; L. adj.suff. *-inus* belonging to; M.L. adj. *griseinus* belonging to grisein, the antibiotic produced by the organism.

Produces the iron-containing grisein complex (components A, B, C and D) and streptocin; exhibits non-polyenic anti-fungal activity; produces vitamin B_{12}; inhibited by streptomycin; poor growth on Czapek's solution agar; spore chains typically flexuous; NaCl tolerance $\geq$ 7%, but <10%.

Type strain: 25G; IMRU 3478; ATCC 3478; ATCC 23915 (single isolate).

301. Streptomyces griseoloalbus (Kudrina) Pridham, Hesseltine and Benedict 1958, 58. (*Actinomyces griseoloalbus* Kudrina in Gauze *et al.* 1957, 112.)

gri.se′o.lo.al′bus. M.L. adj. *griseolus* somewhat gray; L. adj. *albus* white; M.L. adj. *griseoloalbus* somewhat grayish white.

Produces the grisein (albomycin) complex, exhibits anti-fungal activity. Characterization taken from Kudrina in Gauze *et al.* 157, 112; Hütter 1963, 191; and Shirling and Gottlieb 1969, 436.

Type strain: INA 1875/54; (Gauze in Gottlieb 1968, 20).

302. Streptomyces griseus (Krainsky) Waksman and Henrici 1948, 948. (*Actinomyces griseus* Krainsky 1914, 662.)

gri′se.us. M.L. adj. *griseus* gray.

Produces streptomycin, mannosidostreptomycin and streptocin; produces cycloheximide, an antifungal antibiotic; not inhibited by streptomycin; poor growth on Czapek's solution agar; spore chains typically flexuous.

Reference strain: 18–16; IMRU 3463; ATCC 23921; ATCC 23345; none of Krainsky's original isolates is extant (Waksman, 1961, 226, designated strain IMRU 3463 as the "type culture").

303. Streptomyces griseus subsp. **alpha** (Ciferri) Pridham 1970, 37. (*Actinomyces albus* subsp. α (*sic*) Ciferri 1927, 83.)

subsp. al′pha. Gr. n. *alpha* first letter of the Greek alphabet.

Exhibits slight anti-bacterial activity; inhibited by streptomycin; poor growth on Czapek's solution agar; spore chains typically flexuous. Reported as a cause of musty odor in cacao beans (*Theobroma cacao* L.)

Type strain: "CBS (Ciferri)"; CBS 219.25; NRRL B-2249 (Pridham, 1970, 37).

304. Streptomyces griseus subsp. **cretosus** Pridham 1970, 37. (*Oospora cretacea* Krüger 1905, 286 (Illeg.); *Actinomyces cretaceus* (Krüger) Krasil'nikov 1941, 34; *Streptomyces cretaceus* (Krüger) Waksman 1950, 143.)

cre.tos′us. L. adj. *cretosus*, chalky.

Exhibits slight anti-microbial activity; inhibited by streptomycin; moderate growth on Czapek's solution agar. Krüger's original isolations were from zonate scab of beets; Wollenweber's isolations (reference strain) from potatoes.

Reference strain: "CBS (Wollenweber)"; CBS 137.21; IMRU 3005; ATCC 3005; NRRL B-2252; none of Krüger's original isolates is extant.

305. Streptomyces griseus subsp. **solvifaciens** Pridham 1970, 38. (*Actinomyces* sp. in Welsch 1941, 801 and earlier papers; *Actinomyces* G in Welsch 1942, 572; *Streptomyces albus* G in Welsch 1947, 35 (thesis).).

subsp. solv.i.fa′ci.ens. L. v. *solvo* to loosen; L. v. *facio* to make; M.L. part.adj. *solvifaciens*

making loose, dissolving, referring to the lytic activity of actinomycetin.

Exhibits anti-bacterial and anti-fungal activity; produces actinomycetin, now considered a general term for a number of different lytic enzymic antibacterial and anti-viral factors; inhibited by streptomycin; poor growth on Czapek's solution agar; spore chains typically flexuous.

Type strain: "G" of Welsch; NRRL B-1561 (Pridham, 1970, 38).

306. Streptomyces intermedius (Krüger). Waksman in Waksman and Lechevalier 1953, 116 (*Oospora intermedia* Krüger 1905, 289 (Illeg.); *Actinomyces intermedius* (Krüger) Wollenweber 1920, 13.)

in.ter.me′di.us. L. adj. *intermedius* intermediate.

Exhibits slight anti-bacterial activity and heptaenic anti-fungal activity; inhibited by streptomycin; moderate growth on Czapek's solution agar; spore chains typically flexuous; NaCl tolerance $\geq$ 10%, but <13%.

Note. Krüger's original isolates were obtained from sugar beet scab; Wollenweber's from soil producing scab disease, probably of potatoes.

Reference strain: "CBS (Wollenweber)"; CBS 101.21; IMRU 3329; ATCC 3329; none of Krüger's original isolates is extant.

307. Streptomyces kanamyceticus Okami and Umezawa in Umezawa, Ueda, Maeda, Yagishita, Kondō, Okami, Utahara, Ōsato, Nitta and Takeuchi 1957, 183. (Kanamycin-producing strain in Takeuchi, Hikiji, Nitta, Yamazaki, Abe, Takayama and Umezawa 1957, 107.)

kan.a.my.ce′ti.cus. English n. *kanamycin* name of an antibiotic; L. adj.suff. -*icus* belonging to; M.L. adj. *kanamyceticus* of kanamycin.

Produces kanamycins A, B and C; exhibits antifungal activity; inhibited by streptomycin; excellent growth on Czapek's solution agar; sparse formation of aerial mycelium; NaCl tolerance $\geq$ 7%, but <10%; whole cell hydrolysates contain L-diaminopimelic acid. Considered as an "odd" member of Section *Rectus-Flexibilis* by Pridham and Lyons (1965, 341).

Type strain: NIHJ K2J; NIHJ 235; ATCC 12853.

308. Streptomyces levoris (Krasil'nikov) Pridham 1970, 21. (*Actinomyces levoris* Krasil'nikov 1958, 258.)

le.vo′ris. L. gen.n. *levoris* of smoothness.

Exhibits anti-bacterial and anti-fungal activity; inhibited by streptomycin; poor growth on Czapek's solution agar.

Note. It is not clear whether the type strain for this taxon produces either or both of the antifungal antibiotics, levorin (candicidins?) or levorostatin (see Umezawa, 1967, 822). For further

discussion of this taxon see Korenyako, Krasil'-
nikov and Nikitina in Rautenshtein, 1960, 116.

Type strain: INMI 2725; RIA 338; ATCC 15876;
ATCC 23929; (Krasil'nikov in Shirling and
Gottlieb, 1968, 341).

309. Streptomyces limosus Lindenbein 1952,
379.

li.mo'sus. L. adj. *limosus* slimy, referring to the
river bank slime from which the organism was
isolated.

Produces limocrocin, a yellow polyenic anti-
bacterial compound which turns deep blue in the
presence of concentrated H_2SO_4; NaCl tolerance
$\geq$ 10%, but <13%. Characterization, in part, in
Shirling and Gottlieb, 1968, 138.

Type strain: BöBr 136; ATCC 19778; ETH
23898 (probably a single isolate, but also stated to
be the type by Hütter, 1967, 295).

310. Streptomyces lipmanii (Waksman and
Curtis) Waksman and Henrici 1948, 952. (*Actinomyces lipmanii* Waksman and Curtis 1916, 123.)

lip.man'i.i. M.L. gen.n. *lipmanii* of Lipman;
named for Professor J. G. Lipman of the New Jersey Agricultural Experiment Station.

Exhibits slight anti-microbial activity; inhibited by streptomycin; poor growth on Czapek's
solution agar; spore chains typically flexuous;
NaCl tolerance $\geq$ 7%, but <10%.

Type strain: IMRU 3331; ATCC 3331; ATCC
19779 (single isolate).

311. Streptomyces microflavus (Krainsky)
Waksman and Henrici 1948, 950. (*Actinomyces
microflavus* Krainsky 1914, 686; *Micromonospora
microflava* (Krainsky) Duché 1934, 29.)

mic.ro.fla'vus. Gr. adj. *micrus* small; L. adj.
flavus yellow; M.L. adj. *microflavus* small, yellow.

Exhibits slight anti-microbial activity; inhibited
by streptomycin; moderate growth on Czapek's
solution agar; hydroxylates steroids; spore chains
typically flexuous; NaCl tolerance $\geq$ 7%, but
<10%.

Type strain: "CBS (Krainsky)"; CBS 124.18;
IMRU 3332; ATCC 3332; ATCC 13231 (single
isolate).

312. Streptomyces odorifer (Rullmann) Waks-
man in Waksman and Lechevalier 1953, 79. (*Clado-
thrix odorifera* Rullmann 1895, 44; *Oospora
odorifera* (Rullmann) Lehmann and Neumann
1896, 392 (Illeg.); *Actinomyces odoriferus rullmanni
I* Berestnev 1897, 167 (Not Val. Pub.); *Cladothrix
odoriferus rullmanni I* Berestnev 1897, 167 (Not
Val. Pub.); *Actinomyces odorifer* (Rullmann)
Lachner-Sandoval 1898, 65; *Streptothrix odorifera*
(Rullmann) Foulerton and Jones 1902, 112 (Illeg.);
Nocardia odorifera (Rullmann) Castellani and
Chalmers 1913, 818.)

o.do'ri.fer. L. adj. *odorifer* fragrant.

Exhibits anti-bacterial and anti-fungal activity;
excellent growth on Czapek's solution agar; NaCl
tolerance $\geq$ 10%, but <13%.

Reference strain: Král Collection culture; IMRU
3334; ATCC 6246.

313. Streptomyces parvus (Krainsky) Waks-
man and Henrici 1948, 939. (*Actinomyces parvus*
Krainsky 1914, 685; *Nocardia parva* (Krainsky)
Chalmers and Christopherson 1916, 268.)

par'vus. L. adj. *parvus* small.

Produces the actinomycin C complex; exhibits
slight anti-fungal activity; inhibited by streptomycin; moderate growth on Czapek's solution
agar forms bright yellow vegetative mycelium and
diffusible pigment on some media; spore chains
typically flexuous; NaCl tolerance $\geq$ 10%, but
<13%.

Reference strain: NRRL B-1455; IMRU 3686;
ATCC 12433. None of Krainsky's original isolates is extant.

314. Streptomyces pluricolorescens Okami
and Umezawa in Waksman 1961, 259. (Actino-
mycete strain in Takeuchi, Nitta and Umezawa
1956, 22; *Streptomyces pluricolorescens* in Maeda,
Takeuchi, Nitta, Yagishita, Utahara, Osato,
Ueda, Kondo, Okami and Umezawa 1956, 75 (Not
Val. Pub.); *S. pluricolorescens* n. sp. in Takeuchi,
Hikiji, Nitta and Umezawa 1957, 143 (Not Val.
Pub.); *Streptomyces pluricolorescens* in Pridham,
Hesseltine and Benedict 1958, 77 (Not Val. Pub.);
Streptomyces pluricolorescens in Umezawa, Take-
uchi, Okami, Nitta and Maeda, Japanese Patent
7,598 August 29, 1959 (Not Val. Pub.).)

plu.ri.co.lor.es'cens. L. comp.adj. *plus, pluris*
more, many; L. n. *color* color; M.L. part.adj.
pluricolorescens becoming many-colored or varie-
gated.

Produces pluramycins A and B, anti-bacterial
and anti-tumor antibiotics; inhibited by streptomycin; poor growth on Czapek's solution agar;
spore chains typically flexuous; reportedly forms
wine-red diffusible pigment on some chemically
defined media; NaCl tolerance $\geq$ 7%, but <10%.
Characterization, in part, in Shirling and Gottlieb,
1968, 159.

Type strain: 91-T1; NIHJ 238; ATCC 19798
(single isolate).

315. Streptomyces pneumonicus (Krasil'-
nikov, Nikitina and Kondrat'eva) Pridham 1970,
24. (*Actinomyces pneumonicus* Krasil'nikov, Nikitina and Kondrat'eva in Rautenshtein 1960, 160.)

pneu.mo'ni.cus. M.L. n. *pneumonia* pneumonia;
M.L. adj.suff. *-icus* pertaining to; M.L. adj.
pneumonicus related to pneumonia, referring to
the activity of the organism against pneumococci.

Produces pneumocin, an anti-bacterial antibiotic reportedly quite effective against penumococci; exhibits slight anti-fungal activity; not inhibited by streptomycin; very good growth on Czapek's solution agar; forms red to red-brown vegetative mycelium on some media.

Type strain: INMI 367 (single isolate).

316. **Streptomyces praecox** (Millard and Burr) Waksman in Waksman and Lechevalier 1953, 107. (*Actinomyces praecox* Millard and Burr 1926, 633.)

prae′cox. L. adj. *praecox* premature, precocious.

Exhibits slight anti-microbial activity; inhibited by streptomycin; poor growth on Czapek's solution agar; spore chains typically flexuous; NaCl tolerance ≥ 7%, but <10%; isolated from unruptured knob-like scab of potatoes.

Type strain: 20; "CBS (Millard)"; CBS 104.27; IMRU 3374; ATCC 3374 (single isolate).

317. **Streptomyces puniceus** Patelski in Routien and Hofmann 1951, 387.

pu.ni′ce.us. L. adj. *puniceus* reddish, purple.

Produces viomycins A, B and C (anti-bacterial antibiotics); exhibits slight anti-fungal activity; inhibited by streptomycin; poor growth on Czapek's solution agar; forms violet-colored vegetative mycelium on some media; spore chains typically flexuous; NaCl tolerance ≥ 7%, but <10%.

Type strain: Pfizer 1314-5; FD 3568; ATCC 19801 (single isolate).

318. **Streptomyces raffinosus** (Krasil'nikov) Pridham 1970, 24. (*Actinomyces raffinosus* Krasil'-nikov 1958, 258.)

raf′fi.no.sus. M.L. adj. *raffinosus* of raffinose, referring to ability to utilize raffinose.

Exhibits slight anti-microbial activity; inhibited by streptomycin; excellent growth on Czapek's solution agar.

Type strain: INMI 058; RIA 337; ATCC 15883 (Pridham, 1970, 24).

319. **Streptomyces rutgersensis** (Waksman and Curtis) Waksman and Henrici 1948, 952. (*Actinomyces rutgersensis* Waksman and Curtis 1916, 123.)

rut.ger.sen′sis. M.L. adj. *rutgersensis* belonging to Rutgers; named for Rutgers University, New Brunswick, New Jersey, U.S.A.

Exhibits anti-bacterial and anti-fungal activity; inhibited by streptomycin; poor growth on Czapek's solution agar; hooked or coiled chains of spores reportedly occasionally occur; NaCl tolerance ≥ 10%, but <13%.

Type strain: IMRU 3350; ATCC 3350; ATCC 19809 (single isolate).

320. **Streptomyces sampsonii** (Millard and Burr) Waksman in Waksman and Lechevalier 1953, 155. (*Actinomyces sampsonii* Millard and Burr 1926, 614.)

samp.so′ni.i. M.L. gen.n. *sampsonii* of Sampson, a patronymic.

Isolated from medium ruptured common potato scab. Characterization, in part, in Shirling and Gottlieb, 1969, 480.

Type strain: 8; "CBS (Millard)"; IMRU 3371 (single isolate).

321. **Streptomyces setonii** (Millard and Burr) Waksman in Waksman and Lechevalier 1953, 107. (*Actinomyces setonii* Millard and Burr 1926, 604.)

se.to′ni.i. M.L. gen.n. *setonii* of Seton; named for a person, Seton.

Exhibits slight anti-bacterial activity; exhibits anti-fungal activity; poor growth on Czapek's solution agar; aerial mycelium may become dull red on some media; isolated from unruptured potato scab. Characterization, in part, in Shirling and Gottlieb, 1969, 481.

Type strain: 2; "CBS (Millard)"; CBS 105.27; IMRU 3375 (single isolate).

322. **Streptomyces sindenensis** Nakazawa and Fujii 1957, 109. (*Streptomyces sindenensis* in Nakazawa, Fujii, Inoue, Hitomi, Miyake and Kaneko 1954, 168 (Not. Val. Pub.).)

sin.den.en′sis. M.L. adj. *sindenensis* belonging to Sinda Village, Osaka Prefecture, Japan, the source of the soil from which the organism was isolated.

Produces the amicetin complex; inhibited by streptomycin; moderate growth on Czapek's solution agar; spore chains typically flexuous; NaCl tolerance ≥ 10%, but <13%.

Note. Strain Nakazawa 5866, identified as a strain of *S. sindenensis* by Nakazawa 1964, 69, previously was named *Actinopycnidium elongatum* by Krasil'nikov 1962, 252. Strain Nakazawa 5866 is reported to produce the amicetin complex and to have a sexual phase in its life cycle (Nakazawa 1968, 278). Pink fruit bodies form on glucose asparagine agar and contain motile isogametes which fuse to yield zygotes. Crystalline bodies that are viable also form.

Type strain: 1071; IFO 3399; ATCC 23963 (single isolate).

323. **Streptomyces sulphureus** (Gasperini) Waksman 1961, 278. (*Actinomyces sulphureus* Gasperini 1894, 78; *Streptothrix sulphurea* (Gasperini) Caminiti 1907, 197 (Illeg.); *Nocardia sulphurea* (Gasperini) Vuillemin 1931, 129.)

sul.phu′re.us. L. adj. *sulfureus* of sulfur, referring to the bright sulfur-yellow color of the aerial mycelium.

Exhibits slight anti-bacterial activity; exhibits heptaenic anti-fungal activity; excellent growth on Czapek's solution agar; spore chains typically flexuous; NaCl tolerance $\geq$ 10%, but <13%.

Reference strain: IMRU 3007; ATCC 3007; NRRL B-1331; any strains Gasperini might have studied are no longer extant.

Comment: Yet another strain (Gasperini?; IPV 510; ISP 5104) was used in the ISP study as a suggested neotype strain. Its characteristics differ from those cited here.

324. Streptomyces willmorei (Erikson) Waksman and Henrici 1948, 966. (*Actinomyces willmorei* Erikson 1935, 36.)

will.mo're.i. M.L. gen.n. *willmorei* of Willmore; named for J. G. Willmore, the surgeon who first isolated the organism.

Inhibited by streptomycin; moderate growth on Czapek's solution agar; spore chains typically flexuous; NaCl tolerance $\geq$ 7%, but <10%; isolated from a case of streptotrichosis of the liver.

Type strain: NOTC 1856; ATCC 6867; IMRU 3332; IMRU A6867 (single isolate).

325. Streptomyces hawaiiensis Cron, Whitehead, Hooper, Heinemann and Lein 1956, 63.

ha.wai'i.en.sis. M.L. adj. *hawaiiensis* belonging to Hawaii, the source of the soil from which the organism was isolated.

Produces thiostrepton, an anti-bacterial antibiotic; exhibits slight anti-fungal activity; inhibited by streptomycin; fair growth on Czapek's solution agar; NaCl tolerance $\geq$ 7%, but <10%. Characterization, in part, in Shirling and Gottlieb, 1968, 130.

Type strain: 678506; Bristol A8506; ATCC 12236; ATCC 19771 (single isolate).

326. Streptomyces albohelvatus (Krasil'nikov, Korenyako and Nikitina) Pridham 1970, 5. (*Actinomyces albohelvatus* Krasil'nikov, Korenyako and Nikitina in Krasil'nikov 1965, 224.)

al.bo'hel.va'tus. L. adj. *albus* white; L. adj. *helvatus* honey-yellow; M.L. adj. *albohelvatus* whitish honey-yellow.

Exhibits slight anti-bacterial activity; inhibited by streptomycin; poor growth on Czapek's solution agar.

Type strain: INMI 1349; ATCC 19820 (Krasil'nikov in Shirling and Gottlieb, 1969, 399).

327. Streptomyces aurigineus (Krasil'nikov,

TABLE 17.43c–e

Yellow series[a]

Names of species and subspecies	Strain designation	Utilization of carbon compounds (P & G basal)											
		No carbon control	D-Glucose	D-Xylose	L-Arabinose	L-Rhamnose	D-Fructose	D-Galactose	Raffinose	D-Mannitol	i-Inositol	Salicin	Sucrose
17.43c Y; S; C+; SPY													
325. S. hawaiiensis	T	−	+	+	+	+	+	+	+	+	+		+
17.43d Y; S; C−; SM													
326. S. albohelvatus	T		+		+	−		+	−	−	−		−
327. S. aurigineus	T		+	−	+	−	−	+	−	−	−		−
328. S. canarius	T	−	+	+	+	+	+	+	+	+	+	−	+
329. S. chryseus	T		+		+	−		+	−	−	−		
330. S. flavidovirens	T	−	+	+	+	+		−	−	−			+
331. S. helvaticus	T		+		+	−		+	−	−	−		−
332. S. longisporoflavus	R	−	+	+	+	+	+	+	−	+	−	+	−
333. S. niveus	T	−	+	+	+	−		−	−	+	−	−	
334. S. paucidiastaticus	T		+	−	+		+	+	+		+	−	−
335. S. spheroides	T	−	+	+	−	+	+		−	+	−		
17.43e Y:?:C−:?													
336. S. pimprina	R		+		+	+	+	+	+	+		+	+

[a] For explanation of meanings of symbols see footnotes to Tables 17.40 and 17.41a–e.

Korenyako and Nikitina) Pridham 1970, 9. (*Actinomyces aurigineus* Krasil'nikov, Korenyako and Nikitina in Krasil'nikov 1965, 220.)

au.ri.gi.ne′us. L. adj. *aurigineus* yellowish.

Produces the actinomycin B complex.

Type strain: INMI 2375; ATCC 19827 (Krasil'-nikov in Shirling and Gottlieb, 1969, 409).

328. Streptomyces canarius Vavra and Dietz

1965, 76. (*Streptomyces canarius* var. *canarius* in Vavra and Dietz, Abstrs. Papers 4th Intersci. Conf. Antimicrob. Agents Chemother. 1964, 21 (Not Val. Pub.); *Streptomyces canarius* var. *canarius* in Vavra and Bergy, German Patent 1,187,767, February 25, 1965 (Not Val. Pub.); *Streptomyces canarius* var. *canarius* in Vavra and Bergy, U.S. Patent 3,183,156, May 11, 1965 (Not Val. Pub.); *Streptomyces canarius* var. *canarius* Vavra and Dietz 1965, 76.)

ca.na′ri.us. L. adj. *canarius* relating to the Canary Islands, referring to production of a bright canary yellow pigment on a variety of media.

Exhibits slight anti-bacterial activity; produces canarius (*sic*), an anti-viral and anti-yeast antibiotic; inhibited by streptomycin; excellent growth on Czapek's solution agar; forms orange to orange-yellow vegetative mycelium and diffusible pigment on some media; NaCl tolerance $\geq$ 4%, but <7%.

Note. No other subspecies of this taxon have been described.

Type strain: UC 2591; NRRL 2976 (single isolate).

329. Streptomyces chryseus (Krasil'nikov,

Korenyako and Nikitina) Pridham 1970, 10. (*Actinomyces chryseus* Krasil'nikov, Korenyako and Nikitina in Krasil'nikov 1965, 224.)

chry′se.us. M.L. adj. *chryseus* golden.

Exhibits anti-bacterial activity; good growth on Czapek's solution agar.

Type strain: INMI 1007B; ATCC 19829 (single isolate).

330. Streptomyces flavidovirens (Kudrina)

Pridham, Hesseltine and Benedict 1958, 66. (*Actinomyces flavidovirens* Kudrina in Gauze *et al.*, 1957, 90.)

fla.vi.do.vi′rens. L. adj. *flavidus* yellowish; L. part.adj. *virens* being green; M.L. part.adj. *flavidovirens* being yellowish green.

Exhibits anti-bacterial activity; exhibits slight anti-fungal activity; excellent growth on Czapek's solution agar; spore chains of atypical *Retinaculum-Apertum* type; NaCl tolerance $\geq$ 4%, but <7%.

Type strain: INA 12287; ATCC 19900 (Gauze in Gottlieb, 1968, 20.)

331. Streptomyces helvaticus (*sic*) (Krasil'nikov, Korenyako and Nikitina) Pridham 1970, 18.

(*Actinomyces helvaticus* (*sic*) Krasil'nikov, Korenyako and Nikitina in Krasil'nikov 1965, 224.)

helv.va′ti.cus. M.L. n. *Helvetia* Switzerland: M.L. adj.orth.var. *helvaticus* belonging to Switzerland.

Exhibits anti-bacterial activity; good growth on Czapek's solution agar.

Type strain: INMI 1013B; ATCC 19841 (single isolate).

332. Streptomyces longisporoflavus Waksman in Waksman and Lechevalier 1953, 94. (*Actinomyces longisporus flavus* Krasil'nikov 1941, 30

(Not Val. Pub.); *Actinomyces longisporoflavus* Krasil'nikov 1941 in Kudrina in Gauze *et al.* 1957, 95; *Streptomyces longisporus flavus* (Krasil'nikov) Pridham, Hesseltine and Benedict 1958, 66 (Not Val. Pub.).)

lon.gi.spo.ro.fla′vus. L. adj. *longus* long; M.L. n. *spora* a spore; L. adj. *flavus* yellow; M.L. adj. *longisporoflavus* long-spored, yellow.

Exhibits anti-bacterial activity; excellent growth on Czapek's solution agar.

Reference strain: INA 81/53; ATCC 23932; none of Krasil'nikov's original isolates is extant.

333. Streptomyces niveus Smith, Dietz, Sokolski and Savage 1956, 135. (*Streptomyces niveus* in

F.-k. Lin and Coriell, Abstrs. 3rd Ann. Symp. on Antibiot., November 2–4, Washington, D.C., 1955 (Not Val. Pub.); *Streptomyces niveus* in Hoeksema, Johnson and Hinman 1955, 6710 (Not Val. Pub.); *Streptomyces niveus* in Lin and Coriell in Welch and Wright 1955, 670 (Not Val. Pub.); *Actinomyces niveus* (Smith *et al.*) Kuznetsov *et al.* 1966, 841 (Illeg.); Not *Actinomyces nivea* Krainsky in Chalmers and Christopherson 1916, 270 (a *lapsus calami*); Not *Actinomyces niveus* Oda 1935, 1216.)

ni′ve.us. L. adj. *niveus* snow-white, referring to the color of the aerial mycelium of the organism.

Produces novobiocin and isonovobiocin, anti-bacterial antibiotics; inhibited by streptomycin; moderate growth on Czapek's solution agar; coils of spore chains are long, and small in diameter; NaCl tolerance $\geq$ 7%, but <10%.

Type strain: NRRL 2466; ATCC 19793 (single isolate).

334. Streptomyces paucidiastaticus Fügner and Bradler 1963, 186.

pau′ci.di.a.sta′ti.cus. L. adj. *paucus* few, little; M.L. adj. *diastaticus* diastatic, starch-digesting; M.L. adj. *paucidiastaticus* little or slightly diastatic.

Exhibits anti-bacterial activity; produces anti-fungal pentaene JA 4015; poor growth on Czapek's solution agar.

Type strain: IMET JA 4015 (single isolate).

335. Streptomyces spheroides Wallick, Harris, Reagan, Ruger and Woodruff 1956, 911. (*Streptomyces spheroides* in Wallick, Harris, Reagan, Ruger and Woodruff, Abstrs. 3rd Ann. Symp. on Antibiot., November 2–4, Washington D.C., 1955 (Not Val. Pub.); *Streptomyces spheroides* in Kaczka, Wolf, Rathe and Folkers 1955, 6404 (Not Val. Pub.); *Streptomyces spheroides* in Wallick *et al.* in Welch and Wright 1955, 670 (Not Val. Pub.); *Actinomyces spheroides* in Bitteeva 1962, 601 (Illeg.).)

spher.oi′des. Gr. n. *sphaira* a ball, a sphere; Gr. adj.suff. *-oideos* form of, type of; M.L. adj. *spheroides* form of a ball referring to the characteristic compact coils of spores, resembling spheres.

Produces novobiocin and other analogues; moderate growth on Czapek's solution agar; NaCl tolerance ≥ 4%, but <7%.

Type strain: MA-319; NRRL 2449; ATCC 23965 (single isolate).

336. Streptomyces pimprina Thirumalachar in Thirumalachar, Menon and Bhatt 1961, 136.

pim.pri′na. M.L. adj. *pimprina* like or belonging to Pimpri near Poona, India, referring to the source of the soil from which the organism was isolated.

Spore wall ornamentation category unknown; spore chains possibly may be coiled; no melanoid pigment formed; produces hamycin (a major and a minor component) and hamycin X (heptaenic anti-fungal antibiotics), acetopyrrothine, isobutyropyrrothine, aureothricin, antibiotic A and compound D1 (inactive); said to grow poorly on Czapek's solution agar; forms yellow to purple vegetative mycelium on some media.

Reference strain: HACC 2510 (listed as the sole strain in U.S. Patent 3,261,751, July 19, 1966).

337. Streptomyces capoamus Goncalves de Lima, Albert and Gonçalves de Lima, 1964, 317.

ca.po.am′us. Nheêngatû Amazonian dialect *capoama* island; M.L. adj. *capoamus* island, referring to Ascension Island, the source of the soil from which the organism was isolated.

Produces ciclacidins A and B and ciclamicin (anti-bacterial and anti-tumor antibiotics); poor growth on Czapek's solution agar; forms pink, red

TABLE 17.44a

Red series[a]

Names of species and subspecies	Strain designation	No carbon control	D-Glucose	D-Xylose	L-Arabinose	L-Rhamnose	D-Fructose	D-Galactose	Raffinose	D-Mannitol	i-Inositol	Salicin	Sucrose
17.44a R; RF; C+; SM													
337. S. capoamus	T		+	+	+	−	+	+		+	+	+	+
338. S. cinnabarinus	T		+	+	+	+	+		+	+	+		+
339. S. crystallinus	R		+										−
340. S. flavotricini	T		+	−	−	−	±						−
341. S. gobitricini	T	−	+	+	+	+	+	+	−		+	−	
342. S. lincolnensis	T	−	+	+	+	+	+	+	+	+	+	+	+
343. S. melanogenes	T	−	+	+	−		+	+		+	+		±
344. S. phaeochromogenes (see no. 345).	R	−	+	+	+	+	+	+	+	+	+	+	
345. S. phaeochromogenes subsp. chloromyceticus (see no. 344).	T	−	+	+	+		+	+	−	−	−	+	
346. S. pseudovenezuelae	T	+	+	−	+	−	+	−	−	+			−
347. S. roseoviridis	T	−	+	+	+	−	−	+	−	−	−	−	
348. S. spectabilis	T	−	+	+	−	−	+	+	+	+	−	−	
349. S. subrutilus	T	+	−	−	−		+	+	−	−	−		+
350. S. umbrinus	T	−	+	+	+	+	+	+	+	+	+	−	
351. S. venezuelae	T	−	+	+	+	+	+	+	−	−	−	+	
352. S. xanthophaeus	T		+	−	−	−	−	−	−	−	−		−

[a] For explanation of meanings of symbols see footnotes to Tables 17.40 and 17.41a–e.

or reddish violet vegetative mycelium and diffusible pigment on some media; spore chains possibly may be in aberrant coils.

Note. One of the ciclacidins is identical with *n*-pyrromycinone (*bis*-anhydro-rutilantinone) according to Gonçalves de Lima *et al.* 1968, 471.

Type strain: IA-37; AIC 3122; IAUR 3122; IAUR 4670; IAUR 4670-IA-37; ATCC 19006 (single isolate).

338. Streptomyces cinnabarinus (Ryabova and Preobrazhenskaya) Pridham, Hesseltine and Benedict 1958, 62. (*Actinomyces cinnabarinus* Ryabova and Preobrazhenskaya in Gauze *et al.* 1957, 196.)

cin.na.bar'i.nus. M.L. adj. *cinnabarinus* of cinnabar, referring to the vermilion color of vegetative mycelium and diffusible pigment.

Reportedly exhibits no anti-microbial activity except against *Kloeckera brevis* Lodder; forms reddish-colored vegetative mycelium and diffusible pigment on some media. The original description for this taxon would place the organism in the Gray color series. Characterization, in part, in Shirling and Gottlieb, 1969, 416.

Type strain: INA 1242; (Gauze in Gottlieb, 1968, 20).

339. Streptomyces crystallinus Tresner, Davies and Backus 1961, 74. (*Streptomyces cristallinus* (*sic*) in Jacques Loewe Research Foundation, Inc., German Patent 934,429, October 20, 1955 (Not. Val. Pub.); *Streptomyces crystallinus* in Jacques Loewe Research Foundation, Inc., British Patent 758,276, October 3, 1956 (Not Val. Pub.); *Streptomyces crystallinus* in Jacques Loewe Research Foundation, Inc., Japanese Patent 5897, August 3, 1957 (Not Val. Pub.).)

crys.tal.lin'us. L. adj. *crystallinus* of crystal, referring to crystals formed by the organism in some media.

Produces hygromycin A and other antibiotics; exhibits anti-fungal activity; inhibited by streptomycin; poor growth on Czapek's solution agar; spore chains straight to long and flexuous; forms light to dark brown vegetative mycelium and diffusible pigment on many media.

Reference strain: Lederle T-1384; NRRL B-3629.

340. Streptomyces flavotricini (Preobrazhenskaya and Sveshnikova) Pridham, Hesseltine and Benedict 1958, 60. (*Actinomyces flavotricini* Preobrazhenskaya and Sveshnikova in Gauze *et al.* 1957, 49).

fla.vo.tri.ci'ni. L. adj. *flavus* yellow; o.v. *tricini* from Gr. n. *thrix* the hair; M.L. adj. *flavotricini* of yellow hair, probably referring to yellow diffusible pigment and formation of a streptothricin-like antibiotic.

Reportedly exhibits marked anti-bacterial and anti-fungal activity; NaCl tolerance $\geq$ 4%, but <7%. Characterization, in part, in Shirling and Gottlieb, 1968, 114.

Reference strain: INA 11669/58; ATCC 19757; Gauze designated this as the "type strain" in Gottlieb, 1968, 20.

341. Streptomyces gobitricini (Preobrazhenskaya and Sveshnikova) Pridham, Hesseltine and Benedict 1958, 67. (*Actinomyces gobitricini* Preobrazhenskaya and Sveshnikova in Gauze *et al.* 1957, 34; *Actinomyces globitricini* (*sic*) Gauze *et al.* 1957 in Hütter 1967, 181.)

go.bi.tri.ci'ni. English n. *Gobi* the Gobi Desert in Mongolia; o.v. *tricini* from Gr. n. *thrix* the hair; M.L. adj. *gobitricini* of gobi hair, referring to the Gobi desert, the first source of the soil from which the organism was isolated, and probably to formation of a streptothricin-like antibiotic.

Exhibits anti-bacterial and anti-fungal activity; inhibited by streptomycin; moderate growth on Czapek's solution agar; some spore chains of atypical *Retinaculum-Apertum* type; NaCl tolerance $\geq$ 7%, but <10%.

Type strain: INA 5618; CBS (USSR, ARI); CBS 123.60 (designated herein).

342. Streptomyces lincolnensis Mason, Dietz and DeBoer 1963, 555. (*Streptomyces lincolnensis* var. *lincolnensis* in Mason, Dietz and DeBoer, Abstrs. Papers, 2nd Intersci. Conf. Antimicrob. Agent Chemother., 1962 (Not Val. Pub.); *Streptomyces lincolnensis* var. *lincolnensis* in Bergy, Herr and Mason, Belgian Patent 619,645, January 2, 1963 (Not Val. Pub.); *Streptomyces lincolnensis* var. *lincolnensis* in Bergy, Herr and Mason, U.S. Patent 3,086,912, April 23, 1963 (Not Val. Pub.); *Streptomyces lincolnensis* var. *lincolnensis* Mason, Dietz and DeBoer 1963, 555.)

lin.coln.en'sis. M.L. adj. *lincolnensis* belonging to Lincoln, referring to the source of the soil from which the organism was isolated, *viz.* Gehring, near Lincoln, Nebraska, U.S.A.

Produces lincomycin and related anti-bacterial antibiotics (lincomycins A, B, C and D) by directed fermentations; excellent growth on Czapek's solution agar; NaCl tolerance $\geq$ 7%, but <10%.

Note. No other subspecies of this taxon has been described.

Type strain: UC 2376; NRRL 2936; NCIB 9143 (a single isolate designated in patents listed above).

343. Streptomyces melanogenes Sugawara and Onuma 1957, 141. (*Streptomyces melanogenes* in Sugawara, Matsumae and Hata 1957, 133 (Not Val. Pub.); Not *Actinomyces melanogenes* Rubentschik 1928, 312.)

me.la.no'ge.nes. Gr. adj. *melas* black: Gr. v.suff. -*genes* producing; M.L. adj. *melanogenes* producing black.

Produces the peptidic, anti-bacterial, anti-tumor and anti-ascaris antibiotic, melanomycin; spore chains typically straight; reportedly forms red, bluish black, or bluish green vegetative mycelium on some media. Characterization, in part, in Shirling and Gottlieb, 1968, 348. This species might also be placed in the Gray color series.

Type strain: KITA V-1179; ATCC 23937 (single isolate not listed in original description but subsequently designated as the type strain by Hata in Shirling and Gottlieb, 1968, 348).

344. Streptomyces phaeochromogenes (Conn) Waksman in Breed, Murray and Smith 1957, 778. (*Actinomyces pheochromogenus* (*sic*) Conn 1917, 16; *Streptomyces pheochromogenus* (Conn) Pridham, Hesseltine and Benedict 1958; 74; *Streptomyces phaeochromogenus* (*sic*) (Conn) Waksman and Henrici 1948, 943.)

phae.o.chro.mo'ge.nes. Gr. adj. *phaeus* brown; Gr. n. *chroma* color; Gr. v.suff. -*genes* producing; M.L. adj. *phaeochromogenes* producing brown color.

Exhibits anti-bacterial activity; inhibited by streptomycin; excellent growth on Czapek's solution agar; spore chains typically straight; NaCl tolerance ≥ 7%, but <10%.

Reference strain: IMRU 3338; ATCC 3338; ATCC 23945; "CBS (Waksman)"; CBS 282.30; CBS 288.60. None of Conn's original 20 strains is extant.

345. Streptomyces phaeochromogenes subsp. **chloromyceticus** Okami 1948, 503. (*Streptomyces* No. 163 strain in Umezawa, Tazaki, Kanari, Okami and Fukuyama 1948, 358; chloromycetin-producing strain in Umezawa, Hayano and Ogata 1948, 339; *Actinomyces pheochromogenus* (*sic*) var. *chloromycetiens* (*sic*) in Okami 1949, 595; *Streptomyces pheochromogenus* (*sic*) var. *chloromyceticus* Okami 1948, 503.)

subsp. chlo'ro.my.ce'ti.cus. English n. *Chloromycetin* trade name of an antibiotic; L. n.suff. -*icus* belonging to; M.L. adj. *chloromyceticus* belonging to Chloromycetin.

Produces chloramphenicol; inhibited by streptomycin; poor growth on Czapek's solution agar.

Type strain: No. 163; 0-163; NIHJ 38 (0-163); NRRL B-2119 (single isolate).

346. Streptomyces pseudovenezuelae (Kuchaeva, Krasil'nikov, Taptykova and Gesheva) Pridham 1970, 24. (*Streptomyces* sp. 3774 in Murat, Stinebring, Schaffner and Lechevalier 1959, 109; *Actinomyces pseudovenezuelae* Kuchaeva, Krasil'nikov, Taptykova and Gesheva 1961, 114.)

pseu.do.ve.ne.zu.e'lae. Gr. adj. *pseudes* false;

M.L. gen.n. *venezuelae* a specific epithet; M.L. gen.n. *pseudovenezuelae* the false venezuelae.

Produces chloramphenicol and exhibits activity against brucellae in guinea pig monocytes.

Type strain: IMRU 3774; RIA 742; ATCC 21951 (Krasil'nikov in Shirling and Gottlieb, 1968, 362).

347. Streptomyces roseoviridis (Preobrazhenskaya) Pridham, Hesseltine and Benedict 1958, 61. (*Actinomyces roseoviridis* Preobrazhenskaya in Gauze et al. 1957, 57.)

ro.se.o.vi'ri.dis. L. adj. *roseus* rosy; L. adj. *viridis* green; M.L. adj. *roseoviridis* rosy green, referring to the rosy aerial mycelium and green vegetative mycelium and diffusible pigment.

Exhibits anti-bacterial and anti-fungal activity; poor growth on Czapek's solution agar; spore chains typically flexuous.

Type strain: INA 3617; ATCC 23959 (Gauze in Gottlieb, 1968, 20).

348. Streptomyces spectabilis Mason, Dietz and Smith 1961, 118. (*Streptomyces spectabilis* in Siminoff, Smith, Sokolski and Savage 1957, 576 (Not Val. Pub.); *Streptomyces spectabilis* in Folkertsma, Sokolski and Snyder 1958, 114 (Not Val. Pub.); *Streptomyces spectabilis* in The Upjohn Company, British Patent 811,757, April 8, 1959 (Not Val. Pub.), *inter alia*.)

spec.ta'bi.lis. L. adj. *spectabilis* visible, notable, remarkable.

Produces the streptovaricin complex (components A, B, C, D and E), actinospectacin and prodigiosin; inhibited by streptomycin; poor growth on Czapek's solution agar; spore chains sometimes of a type resembling section *Biverticillus*; forms vegetative mycelium in tints and shades of orange to red; NaCl tolerance ≥ 7%, but <10%.

Type strain: NRRL 2494 (first single isolate referred to in publication and patents).

349. Streptomyces subrutilus Arai, Kuroda, Yamagishi and Katoh 1964, 25.

sub.ru'ti.lus. M.L. adj. *subrutilus* reddish.

Produces hydroxystreptomycin; spore chains typically straight, serologically different from streptoverticillia that produce hydroxystreptomycin.

Type strain: No. 713; IFM 713 (single isolate).

350. Streptomyces umbrinus (Sveshnikova) Pridham, Hesseltine and Benedict 1958, 61. (*Actinomyces umbrinus* Sveshnikova in Gauze et al. 1957, 62.)

um.bri'nus. M.L. adj. *umbrinus* wood brown, the color of the aerial mycelium.

Exhibits anti-bacterial activity; inhibited by streptomycin; excellent growth on Czapek's solution agar; forms purple-brown vegetative mycelium on some media; spore chains typically flexuous; NaCl tolerance ≧ 7%, but <10%.

Type strain: INA 1703/53; ATCC 19929 (Gauze in Gottlieb, 1968, 20).

351. **Streptomyces venezuelae** Ehrlich, Gottlieb, Burkholder, Anderson and Pridham 1948, 467. (*Streptomyces* sp. in Ehrlich, Bartz, Smith and Joslyn 1947, 417; actinomycete very similar to, if not identical with the streptothricin-producing *Streptomyces lavendulae* in Carter, Gottlieb and Anderson 1948, 113; *Streptomyces* sp. in Bartz 1948, 445; *Streptomyces* isolate 8-44 in Gottlieb, Bhattacharyya, Anderson and Carter 1948, 409; soil actinomycete (*Streptomyces*) in Smith, Joslyn, Gruhzit, McLean, Penner and Ehrlich 1948, 425; isolate 8-44, identified as *S. lavendulae* or *S. reticuli* in Pridham and Gottlieb 1948, 107; *Actinomyces venezuelae* (Ehrlich *et al.*) Krasil'nikov 1949, 66 (Illeg.).)

ve.ne.zu.e'lae. M.L. n. *venezuela* Venezuela; M.L. gen.n. *venezuelae* of Venezuela.

Produces chloramphenicol; inhibited by streptomycin; poor growth on Czapek's solution agar; spore chains typically straight; NaCl tolerance ≧ 4%, but <7%. This taxon might also be placed in the Gray color series.

Type strain: Burkholder No. A65; P.D. 04745; NRRL 2277; ATCC 10712.

352. **Streptomyces xanthophaeus** Lindenbein 1952, 378.

xan.tho.phae'us. Gr. adj. *xanthophaes* golden, gleaming; M.L. adj. *xanthophaeus* shining like gold.

Produces geomycin, a streptothricin-group antibiotic; spore chains of the *Retinaculum-Apertum* type occasionally form, otherwise spore chains typically are straight; spores cylindrical and phalangiform; NaCl tolerance ≧ 4%, but <7%. This taxon might also be placed in the Gray color series. Characterization, in part, in Shirling and Gottlieb 1968, 180.

Type strain: Wüst 70; ATCC 19819 (single isolate).

See Table 17.44b, c.

353. **Streptomyces aureomonopodiales** (*sic*) (Krasil'nikov and Yuan) Pridham 1970, 8. (*Actinomyces aureomonopodiales* (*sic*) Krasil'nikov and Yuan in Krasil'nikov 1965, 43.)

au.re.o.mo.no.po.di.a'les. L. n. *aurum* gold; Gr. n. *monopodios* a table with one foot (Bot., successive lateral branches from a primary axial stem); L. adj.stem. suff. -*alis* pertaining to; M.L. adj. *aureomonopodialis* pertaining to golden, monopodial.

Exhibits anti-bacterial activity (at least two factors) and anti-fungal activity; moderate growth

on Czapek's solution agar; forms crimson to reddish orange vegetative mycelium on some media; peptone iron agar is blackened; whole cell hydrolysates contain L-diaminopimelic acid.

Type strain: INMI 5008; RIA 680; ATCC 15879.

Note. Despite the fact that Krasil'nikov and Yuan clearly designated strain INMI 5008 as the type strain, Krasil'nikov selected INMI 1510 as the "type" in Shirling and Gottlieb (1969, 407).

354. **Streptomyces exfoliatus** (Waksman and Curtis) Waksman and Henrici 1948, 951. (*Actinomyces exfoliatus* Waksman and Curtis 1916, 116.)

ex.fo.li.a'tus. L. part.adj. *exfoliatus* stripped of leaves.

Exhibits slight anti-bacterial activity; inhibited by streptomycin; excellent growth on Czapek's solution agar; spore chains typically flexuous; NaCl tolerance ≧ 7%, but <10%.

Type strain: IMRU 3316; ATCC 12627; ATCC 19750 (single isolate).

355. **Streptomyces filamentosus** Okami and Umezawa in Okami, Okuda, Takeuchi, Nitta and Umezawa 1953, 153.

fi.la.men.to'sus. L. v. *filare* to spin; M.L. adj. *filamentosus* full of threads or filaments.

Produces caryomycin, an anti-bacterial antibiotic; inhibited by streptomycin; excellent growth on Czapek's solution agar; spore chains typically flexuous.

Type strain: 1-C-9; NIHJ 256; ATCC 19753 (single isolate).

356. **Streptomyces prunicolor** (Ryabova and Preobrazhenskaya) Pridham, Hesseltine and Benedict 1958, 63. (*Actinomyces prunicolor* Ryabova and Preobrazhenskaya in Gauze *et al.* 1957, 184.)

pru'ni.co.lor. L. n. *prunus* plum; L. n. *color* color; M.L. adj. *prunicolor* plum colored, referring to the color of the vegetative mycelium of the organism.

Presumably exhibits anti-bacterial activity; forms reddish brown or reddish purple vegetative mycelium on some media. Characterization, in part, in Shirling and Gottlieb, 1969, 468.

Reference strain: INA 8805/64; Gauze designated strain INA 8805/54 as the "type" in Gottlieb 1968, 20; subsequently, the designation was corrected to 8805/64 when the strain was made available for the ISP. None of the original isolates is extant.

357. **Streptomyces roseofulvus** (Preobrazhenskaya) Pridham, Hesseltine and Benedict 1958, 61. (*Actinomyces roseofulvus* Preobrazhenskaya in Gauze *et al.* 1957, 55.)

ro.se.o.ful'vus. L. adj. *roseus* rosy; L. adj.

TABLE 17.44b, c

Red series[a]

Names of species and subspecies	Strain desig-nation	Utilization of carbon compounds (P & G basal)											
		No carbon control	D-Glucose	D-Xylose	L-Arabinose	L-Rhamnose	D-Fructose	D-Galactose	Raffinose	D-Mannitol	i-Inositol	Salicin	Sucrose
17.44b　R; RF; C−; SM													
353. S. aureomonopodiales	T	−	+	+	+	−	+	+	−	+	+	+	
354. S. exfoliatus	T	−	+	+	+	+	+	+	+	−	−	+	+
355. S. filamentosus	T	−	+	+	+	−	−	+	−	−	−	−	+
356. S. prunicolor	R		+	+	+	+	+		+	+	+		
357. S. roseofulvus	T	−	+	+	+	+	+	+	+	−	−	+	+
358. S. roseolus	T		+	+	+	+	±	−	−	−	−	−	−
359. S. roseosporus	T		+	+	+	+	−	−	−	−	−	+	−
360. S. rubiginosohelvolus	T		+	+	+	+	+	−	+	−	−		
361. S. termitum	T		+	+	−	−	−	−	−	−			
17.44c　R; S; C+; SM													
362. S. cinnamonensis (see no. 420)	T	−	+	−	−	−	+	−	−	−	−	−	+
363. S. colombiensis	T	−	+										
364. S. goshikiensis	T	−	+	−	−	−	+	−	−	−	−		
365. S. katrae	T	−	+	−	−	−	+	+	+	−	−	−	
366. S. lavendofoliae	T	−	+	+	+	−	−	+	−	+	−	−	
367. S. lavendulae (see nos. 368, 369 and 370).	R	−	+	−	−	−	+	+	−	−	−	+	−
368. S. lavendulae subsp. avireus (see nos. 367, 369 and 370).	T	−	+	−	−	−	+	−	−	−	−	+	−
369. S. lavendulae subsp. brasilicus (see nos. 367, 368 and 370).	T	−	+	−	−	−	+	−	−	−	−	−	−
370. S. lavendulae subsp. grasserius (see nos. 367, 368 and 369).	T	−	+	−	−	−	+	−	−	−	−	+	−
371. S. lavendulocolor	T	−	+	+	+	−	−	+	−	−	+	−	−
372. S. luridus	T		+	+	+	−	±	−	−	+	−	−	
373. S. orchidaceus	T												
374. S. racemochromogenes	T		+	−	+	−	−	−	−	−	−		+
375. S. syringae	T		+	−	−	−	+	−	−	−	−	−	
376. S. toxytricini	T		+	−	−	−	±	−	−	±	−	−	
377. S. tuirus	T		+	+	+	+	+	+	+	+	+	−	+
378. S. vinaceus	T	−	+	−	−	−							
379. S. virginiae	T	−	+	−	−	−	+	−	−	−	−	+	+

[a] For explanation of meanings of symbols see footnotes to Tables 17.40 and 17.41a–e.

fulvus deep yellow; M.L. adj. *roseofulvus* rosy, deep yellow, referring to color of aerial mycelium and vegetative mycelium, respectively.

Exhibits anti-bacterial activity; inhibited by streptomycin; moderate growth on Czapek's solution agar; spore chains typically straight.

Type strain: INA 14535; ATCC 19805; ATCC 19921; (Gauze in Gottlieb, 1968, 20).

358. **Streptomyces roseolus** (Preobrazhenskaya and Sveshnikova) Pridham, Hesseltine and Benedict 1958, 61. (*Actinomyces roseolus* Preobrazhenskaya and Sveshnikova in Gauze *et al.* 1957, 37.)

ro.se.o'lus. L. dim.adj. *roseolus* somewhat rosy, referring to the color of the aerial mycelium.

Exhibits anti-bacterial activity. Characterization, in part, in Shirling and Gottlieb, 1968, 167.

Type strain: INA 5449/54; ATCC 23210; (Gauze in Gottlieb, 1968, 20).

359. **Streptomyces roseosporus** Falcão de Morais and Dália Maia 1961, 41. (*Streptomyces venezuelae* subsp. *roseospori* (*sic*) Falcão de Morais, Dália Maia and Souto Maior Genn 1958, 102.)

ro.se.o.spo'rus. L. adj. *roseus* rosy; M.L. n. *spora* a spore; M.L. adj. *roseosporus* rosy-spored.

Probably exhibits anti-bacterial activity; probably inhibited by streptomycin; probably does not grow on Czapek's solution agar; chains typically straight. Characterization, in part, in Shirling and Gottlieb, 1968, 370.

Type strain: IAUR 4192; ATCC 23958 (implied as a single isolate in Falcão de Morais and Dália Maia 1961, 41 and designated by Falcão de Morais as the type strain in Shirling and Gottlieb, 1968, 370.)

360. **Streptomyces rubiginosohelvolus** (*sic*) (Kudrina) Pridham, Hesseltine and Benedict 1958, 59. (*Actinomyces rubiginosohelvolus* (*sic*) Kudrina in Gauze *et al.* 1957, 89.)

ru.bi.gin.o.so.hel'vo.lus. o.v. of L. adj. *robiginosus* (*sic*) rusty; L. adj. *helvolus* somewhat yellow; M.L. adj. *rubiginosohelvolus* rusty, somewhat yellow.

Exhibits anti-bacterial and anti-fungal activity; yellow pigment of vegetative mycelium changes to pink on addition of alkali. Characterization, in part, in Shirling and Gottlieb, 1968, 372.

Type strain: INA 10/53; ATCC 19926; ATCC 23960 (Gauze in Gottlieb, 1968, 20).

361. **Streptomyces termitum** Duché, Heim and Labourer in Heim 1951, 359. (*Streptomyces termitarum* (*sic*) Duché et Heim (*sic*) in Catalogues des Collections Vivantes, Herbiers et Documents. VI. La Mycotheque, 3rd Suppl., 1953, Mus. Nat. d'Hist. Natur., Lab. de Cryptogamie, Paris, France, 15 (Not Val. Pub.).)

ter.mit'um. L. n. *termes* a woodworm, termite; L. gen.pl.n. *termitum* of termites, referring to the source of the organism.

Poor growth on Czapek's solution agar; spore chains typically flexuous; reportedly isolated regularly, or at different times, from the mucus produced by *Antennopsis*, a genus of termites. Characterization, in part, in Shirling and Gottlieb, 1969, 483.

Type strain: L.C. 620 (Nicot in Shirling and Gottlieb, 1969, 483).

362. **Streptomyces cinnamonensis** Okami in Maeda, Okami, Kosaka, Taya and Umezawa 1952, 572.

cin.na.mo.nen'sis. Gr. n. *cinnamum* cinnamon; M.L. adj. *cinnamonensis* belonging to cinnamon, referring to the color of the aerial mycelium.

Produces actithiazic acid, a biotin antagonist and anti-mycobacterial antibiotic; exhibits anti-fungal activity; spore chains of typical *Retinaculum-Apertum* type; spores phalangiform; NaCl tolerance $\geq$ 4%, but <7%.

Type strain: 154-T3; NIHJ 35; ATCC 12308 (single isolate).

363. **Streptomyces colombiensis** Pridham, Hesseltine and Benedict 1958, 76. (*Streptomyces colombiensis* in Merck and Company, British Patent Application 18,236, 1949 (Not Val. Pub.); *Streptomyces colombiensis* in Wood and Hendlin, U.S. Patent 2,595,499, May 6, 1952 (Not Val. Pub.).)

co.lom.bi.en'sis. M.L. adj. *colombiensis* belonging to Colombia.

Exhibits anti-fungal activity; produces vitamin B_{12}; inhibited by streptomycin; poor growth on Czapek's solution agar.

Type strain: Merck MA-52; NRRL B-1990 (single isolate).

364. **Streptomyces goshikiensis** Niida in Shirling and Gottlieb 1968, 324. (*Streptomyces goshikiensis* Niida in Kondo, Sakamoto and Yumoto 1961, 365 (Not Val. Pub.); *Streptomyces goshikiensis* n. sp. in Kondo, Miyakawa, Yumoto, Sezaki, Shimura, Sato and Hara 1962, 157 (Not Val. Pub.); *Streptomyces goshikiensis* in Kondo *et al.*, Japanese Patent 26,948, December 28, 1963 (Not Val. Pub.); *Streptomyces goshikiensis* in Kondo *et al.*, Japanese Patent 10,244, June 11, 1964 (Not Val. Pub.).)

go.shi.ki.en'sis. M.L. adj. *goshikiensis* belonging to goshiki (of unknown derivation).

Produces the macrolide anti-bacterial antibiotics, bandamycins A and B; produces a pentaenic anti-fungal antibiotic; spore chains of typical *Retinaculum-Apertum* type. Characterization, in part, in Shirling and Gottlieb, 1968, 324.

Type strain: Meiji (Oda); ATCC 23914 (Oda in Shirling and Gottlieb, 1968, 324; probably a single isolate).

365. **Streptomyces katrae** Gupta and Chopra 1963, 1.

kat'rae. M.L. gen.n. *katrae* of Katra; named for Katra, Jammu Province, India, the source of the soil from which the organism was isolated.

Exhibits anti-bacterial and anti-fungal activity; inhibited by streptomycin: fair growth on Czapek's solution agar; forms green vegetative mycelium on some media; spore chains of typical *Retinaculum-Apertum* type.

Type strain: RRL 5036; NRRL B-3093 (single isolate).

366. **Streptomyces lavendofoliae** (Kuchaeva, Krasil'nikov, Taptykova and Gesheva) Pridham 1970, 19. (*Actinomyces lavendofoliae* Kuchaeva,

Krasil'nikov, Taptykova and Gesheva 1961, 120.)
la.ven'do.fo'li.ae. Med.L. n. *lavendula* lavender;
L. n. *folium* a leaf; M.L. gen.n. *lavendofoliae* of
lavender leaf, referring to the color of the aerial
mycelium.

Reportedly produces the streptothricin complex; inhibited by streptomycin; poor growth on
Czapek's solution agar; spore chains of typical
Retinaculum-Apertum type.

Type strain: INA 3613; RIA 750; ATCC 15872;
ATCC 23928 (single isolate).

367. **Streptomyces lavendulae** (Waksman and
Curtis) Waksman and Henrici 1948, 944. (*Actinomyces lavendulae* Waksman and Curtis 1916, 126.)
la.ven'du.lae. Med.L. n. *lavendula* lavender;
Med.L. gen.n. *lavendulae* of lavender color.

Produces the streptothricin complex; inhibited
by streptomycin; poor growth on Czapek's solution agar; spore chains of typical *Retinaculum-Apertum* type; spores phalangiform; NaCl tolerance ≥ 4%, but <7%.

Reference strain: IMRU 3440; ATCC 8664.

Note. The original strain isolated by Waksman
and Curtis is available only as a non-viable dried
herbarium specimen. The determinable characters
of this specimen do not agree with those of viable
strains currently recognized as being members of
the taxon (Pridham, Lyons and Seckinger, 1965,
225).

368. **Streptomyces lavendulae** subsp. **avireus**
(Kuchaeva,   Krasil'nikov,   Taptykova   and
Gesheva) Pridham 1970, 19. (*Actinomyces lavendulae avireus* n. subsp. (*sic*) Kuchaeva, Krasil'-
nikov, Taptykova and Gesheva 1961, 119.)
subsp. a.vi're.us. Gr. pref. *a*- signifying a negation or absence of; L. n. *virus* poison, virus; M.L.
n. *avireus* absence of virus (referring to the lack of
anti-viral activity).

Produces the streptothricin VI complex; reportedly has no effect on tobacco mosaic virus;
poor growth on Czapek's solution agar; spore
chains of typical *Retinaculum-Apertum* type.

Type strain: IMRU 3516 (single isolate).

369. **Streptomyces lavendulae** subsp. **brasilicus** Falcão de Morais, Maia and Genn 1958, 76.
subsp. bra.sil'i.cus. M.L. adj. *brasilicus* from
Brazil.

Exhibits streptothricin-like anti-microbial activity; poor growth on Czapek's solution agar;
spore chains of typical *Retinaculum-Apertum* type.

Type strain: FC 275; IAUR 4142; NRRL B-2937
(designated herein).

370. **Streptomyces lavendulae** subsp. **grasserius** (Kuchaeva, Krasil'nikov, Taptykova and
Gesheva) Pridham 1970, 20. (*Streptomyces griseo-*

lavendus in Sumiki *et al.*, Japanese Patent 6,296,
August 15, 1957 (Not Val. Pub.); *Actinomyces
lavendulae grasserius* n. subsp. (*sic*) Kuchaeva,
Krasil'nikov, Taptykova and Gesheva 1961, 119.)
subsp. gras'se.ri.us. French n. *grasserié* a disease
of silkworms; M.L. adj. *grasserius* grasserial
(referring to the activity of the organism against
silkworm jaundice virus, grasserié).

Produces the grasseriomycin complex (a streptothricin complex); inhibited by streptomycin;
poor growth on Czapek's solution agar; spore
chains of typical *Retinaculum-Apertum* type.

Type strain: IAM 2A-458; ATCC 15875 (single
isolate).

371. **Streptomyces lavendulocolor** (Kuchaeva, Krasil'nikov, Taptykova and Gesheva)
Pridham 1970, 20. (*Actinomyces lavendulocolor*
Kuchaeva, Krasil'nikov, Taptykova and Gesheva
1961, 120; *Actinomyces lavendolocolor* in Kuchaeva
et al. 1961, 121 (*lapsus calami*); *Actinomyces
lavendocolor* in the American Type Culture Collection, Catalogue of Strains, 8th ed. 1968, 6 and
Shirling and Gottlieb 1968, 339 (*lapsus calami*).)
la.ven'du.lo.co.lor. Med.L. n. *lavendula* the
lavender; L. n. *color* color; M.L. adj. *lavendulocolor*
lavender colored.

Said to produce the streptothricin complex;
inhibited by streptomycin; said to inhibit some
other streptothricin complex-producing strains;
poor growth on Czapek's solution agar; spore
chains of typical *Retinaculum-Apertum* type.

Type strain: INA 4518; ATCC 15871; ATCC
23925 (single isolate).

372. **Streptomyces luridus** (Krasil'nikov,
Korenyako, Meksina, Valedinskaya and Veselov)
Waksman 1961, 237. (*Actinomyces luridus* Krasil'-
nikov, Korenyako, Meksina, Valedinskaya and
Veselov 1957, 563.)
lu'ri.dus. L. adj. *luridus* pale yellow, ghostly
pallid (probably referring to the color of the vegetative mycelium).

Exhibits anti-bacterial activity; exhibits antifungal activity; produces luridin, an anti-viral
antibiotic; inhibited by streptomycin; good
growth on Czapek's solution agar; spore chains of
atypical *Retinaculum-Apertum* type; NaCl tolerance ≥ 4%, but <7%. Characterization, in part,
in Shirling and Gottlieb, 1968, 142.

Note. The original description records this taxon
as being non-chromogenic i.e. no substrate pigments formed.

Type strain: INMI 111; ATCC 19782 (Krasil'-
nikov in Shirling and Gottlieb, 1968, 142).

373. **Streptomyces orchidaceus** Hütter 1964,
626. (*Streptomyces orchidaceus* in Welch, Putnam
and Randall 1955, 72 (Not Val. Pub.); *Strepto-*

myces orchidaceus in Harned, Hidy and LaBaw 1955, 204 (Not Val. Pub.); *Streptomyces orchidaceus* in Hidy *et al.* 1955, 2345 (Not Val. Pub.); *Streptomyces orchidaceus* or *S. orchidaceous* (*sic*) in Commercial Solvents Corporation, British Patent 768,007, February 13, 1957 (Not Val. Pub.); *Streptomyces orchidaceus* in Harned and LaBaw, German Patent 958,242, February 14, 1957 (Not Val. Pub.); *Streptomyces orchidaceus* in Runge, U.S. Patent 2,815,348, December 3, 1957 (Not Val. Pub.); *Streptomyces orchidaceus* in Pridham, Hesseltine and Benedict 1958, 77 (Not Val. Pub.); *Streptomyces orchidaceous* (*sic*) in Harned, U.S. Patent 3,090,730, May 21, 1963 (Not Val. Pub.); *Streptomyces orchidaceous* (*sic*) in Hodge, U.S. Patent 3,428,525, February 18, 1969 (Not Val. Pub.).)

or.chi.da′ce.us. M.L. n. *Orchis, Orchidis* generic name of an orchid; L. adj.suff. *-aceus* pertaining to; M.L. adj. *orchidaceus* pertaining to orchid; referring to a blue-red hue.

Produces D-4-amino-3-isoxazolidone (cycloserine) and *O*-carbamyl-D-serine; spore chains of typical *Retinaculum-Apertum* type.

Type strain: JN-21; NRRL 2454 (single isolate).

374. **Streptomyces racemochromogenes** Sugai 1956, 171. (*Streptomyces phaeochromogenus* (*sic*) 2-229 in Otani and Sugai 1953, 257; *Streptomyces phaeochromogenus* (*sic*) 2-229 in Otani and Sugai 1953, 372; *Streptomyces* No. 229 in *S. lavendulae* group in Otani *et al.* 1955, 400; *Streptomyces racemochromogenus* (*sic*) in Sugai 1956, 171.)

Note. Sugai (1956, 176) refers to work with this taxon as early as 1951. No references are given for this work.

ra.ce′mo.chro. mo′gen.es. L. n. *racemus* a raceme or cluster of berries; M.L. adj. *chromogenes* producing color; M.L. adj. *racemochromogenes* raceme, producing color (probably referring to morphology of spore chains and to chromogenicity).

Produces the racemomycin complex (components A, B, C and O), a streptothricin complex; sometimes forms blue to green vegetative mycelium on some media; spore chains of typical *Retinaculum-Apertum* type. Characterization, in part, in Shirling and Gottlieb, 1968, 366.

Type strain: 2-229; 229; ATCC 23954 (single isolate).

375. **Streptomyces syringae** (Kuchaeva, Krasil′nikov, Taptykova and Gesheva) Pridham 1970, 26. (*Actinomyces syringae* Kuchaeva, Krasil′-nikov, Taptykova and Gesheva 1961, 113; *Ast. cyringae* (*sic*) in Kuchaeva *et al.* 1961, 124; Not *A. syringi* Gauze *et al.* 1957 in Hütter 1964, 649 (*lapsus calami*).)

sy.rin′gae. Gr. n. *syrinx, syringis* a pipe or tube;

M.L. n. *Syringa* generic name of the lilac; M.L. gen.n. *syringae* of the lilac, probably referring to the color of aerial mycelium.

Produces streptin; exhibits anti-fungal activity; not inhibited by streptomycin; poor growth on Czapek′s solution agar; spore chains of typical *Retinaculum-Apertum* type.

Note. Formerly named *Streptomyces lavendulae* or *Streptomyces reticulus-ruber*.

Type strain: Merck 3R14 (M1); IMRU 3435 (Pridham, 1970, 26).

376. **Streptomyces toxytricini** (Preobrazhenskaya and Sveshnikova) Pridham, Hesseltine and Benedict 1958, 68. (*Actinomyces toxytricini* Preobrazhenskaya and Sveshnikova in Gauze *et al.* 1957, 47.)

tox.y.tri.ci′ni. Probably from English n. *toxythricin* probably a suggested, but never used, antibiotic name; hence, M.L. gen.n. *toxytricini* of toxythricin.

Exhibits anti-bacterial and anti-fungal activity; spore chains of typical *Retinaculum-Apertum* type; NaCl tolerance $\geq 4\%$, but $<7\%$. Characterization, in part, in Shirling and Gottlieb, 1968, 174.

Type strain: INA 13887/54; ATCC 19813 (Gauze in Gottlieb, 1968, 20).

377. **Streptomyces tuirus** Albert and Malaquias de Queiroz 1963, 43.

tu′ir.us. Nheêngatû Amazonian dialect *tuira* violet, violet-blue; M.L. adj. *tuirus* violet, violet-blue (referring to the color of vegetative mycelium and diffusible pigment).

Produces tuoromycin, a pigmented anti-bacterial antibiotic; inhibited by streptomycin; moderate growth on Czapek′s solution agar; forms red to violet vegetative mycelium on some media; forms red to red-brown diffusible pigment on some media.

Type strain: 3121; IAUR 4564; ATCC 19007 (single isolate).

378. **Streptomyces vinaceus** Jones 1952, 47. (Not *Actinomyces vinaceus* or *Actinomyces vinaceous* in Ciba Limited, British Patent 651,269, March 14, 1951 (Not Val. Pub.); Not *Actinomyces vinaceus* Mayer, Crane, DeBoer, Konopka, Marsh and Eisman 1951, 282–284 (Not Val. Pub.); Not *Streptomyces vinaceus* Mayer *et al.* (*sic*) in Waksman in Waksman and Lechevalier 1953, 42 (Illeg.).)

vi.na′ce.us. L. adj. *vinaceus* of or belonging to wine or the grape, referring to the color of the aerial mycelium.

Produces vitamin B_{12}; inhibited by streptomycin; poor growth on Czapek′s solution agar; spore chains of typical *Retinaculum-Apertum* type.

Type strain: Jones 8542-1; NRRL 2382 (a single

isolate designated in Bennett, U.S. Patent 2,681,881, June 22, 1954).

379. Streptomyces virginiae Grundy, Whitman, Rdzok, Rdzok, Hanes and Sylvester 1952, 399.

vir.gi′ni.ae. M.L. gen.n. *virginiae* of Virginia, referring to the source of the soil (near Roanoke, Virginia) from which the organism was isolated.

Produces actithiazic acid, a biotin antagonist and anti-mycobacterial antibiotic; exhibits slight anti-fungal activity; poor growth on Czapek's solution agar; spore chains of typical *Retinaculum-Apertum* type; NaCl tolerance ≧ 4%, but <7%.

Type strain: NA 255-B8; NRRL B-1446; ATCC 19817.

380. Streptomyces lateritius (Sveshnikova) Pridham, Hesseltine and Benedict 1958, 67. (*Actinomyces lateritius* Sveshnikova in Gauze *et al.* 1957, 70).

la.ter.it′i.us. L. n. *later* brick; M.L. adj. *lateritius* o.v. of *latericius* of bricks, brick red (referring to the color of the aerial mycelium).

Exhibits antibacterial activity; forms blue vegetative mycelium under alkaline conditions (red under acid conditions) on some media; forms blue or violet diffusible pigment on some media; spore chains of atypical *Retinaculum-Apertum* type; spores phalangiform; NaCl tolerance ≧ 4%, but <7%. Characterization, in part, in Shirling and Gottlieb, 1968, 136.

Type strain: INA 6993; ATCC 19776; ATCC 19913 (Gauze in Gottlieb, 1968, 20).

381. Streptomyces flavovariabilis (Korenyako and Nikitina) Pridham 1970, 15. (*Actinomyces flavovariabilis* Korenyako and Nikitina in Krasil'nikov 1965, 304.)

fla.vo.va.ri.a′bi.lis. L. adj. *flavus* yellow; L. adj. *variabilis* variable; M.L. adj. *flavovariabilis* yellow, variable.

Exhibits anti-bacterial and anti-tumor activity; grows on Czapek's solution agar.

Type strain: INMI 702 (single isolate).

382. Streptomyces janthinus (Artamonova and Krasil'nikov) Pridham 1970, 19. (*Actinomyces janthinus* Artamonova and Krasil'nikov in Rautenshtein 1960, 334.)

jan′thi.nus. L. adj. *ianthinus* violet-colored,

TABLE 17.44d, e
Red series[a]

Names of species and subspecies	Strain desig-nation	Utilization of carbon compounds (P & G basal)											
		No carbon control	D-Glucose	D-Xylose	L-Arabinose	L-Rhamnose	D-Fructose	D-Galactose	Raffinose	D-Mannitol	i-Inositol	Salicin	Sucrose
17.44d R; S; C+; WTY													
380. S. lateritius	T		+	+	+	+	+		−	−	±		−
17.44e R; S; C+; SPY													
381. S. flavovariabilis	T		+	+	+	+	+	+	+	+			+
382. S. janthinus	T	−	+	+	+	+	+	+	+	+	+	+	+
383. S. purpurascens	T		+	+	+	+	+		+	+	+	+	+
384. S. roseospinus	T		+	+	+	−	−	+	−	+			+
385. S. roseoviolaceus	T		+	+	+	+	+		+	+	+		+
386. S. violaceus (see Nos. 387 and 388)	R	−		+	+	+	+	+	+	+		+	+
387. S. violaceus subsp. confinus (see Nos. 386 and 388)	T	−	+		+	+	+	+	+			+	+
388. S. violaceus subsp. vicinus (see Nos. 386 and 387)	T	−		+		+	+	+	+			+	+
389. S. violarus	T	−		+		+	+	+	+			+	+
390. S. violatus	T	−		+	+	+	+	+	+			+	+
391. S. yokosukanensis	T	−		+	+	+	+	+	+	+	+	+	+

[a] For explanation of meanings of symbols see footnotes to Tables 17.40 and 17.41a–e.

referring to color of vegetative mycelium and diffusible pigment.

Exhibits anti-bacterial and slight anti-fungal activity; inhibited by streptomycin; excellent growth on Czapek's solution agar; forms violet- to orange-colored vegetative mycelium on some media.

Type strain: INMI 117; RIA 659; ATCC 15870; ATCC 23925 (Krasil'nikov in Shirling and Gottlieb, 1968, 337).

383. Streptomyces purpurascens Lindenbein 1952, 371. (*Streptomyces* sp. in Brockmann and Bauer 1950, 492; *Streptomyces purpurascens* nov. spec. in Brockmann, Bauer and Borchers 1951, 700 (Not Val. Pub.).)

pur.pur.as′cens. L. part.adj. *purpurascens* making purple.

Produces rhodomycin A, isorhodomycin A and rhodomycin B; sometimes forms red to violet, pH-sensitive, vegetative mycelium and diffusible pigment (blue under alkaline conditions, red or orange under acid conditions). Characterization, in part, in Shirling and Gottlieb, 1969, 470.

Type strain: Maria 515 (Wilde in Shirling and Gottlieb, 1969, 470).

384. Streptomyces roseospinus Suzuki, Ozawa and Tanabe 1964, 335. (*Streptomyces* No. 49 in Suzuki and Tanabe 1963, 623; Actinomycete strain No. 49 in Suzuki, Ozawa and Tanabe 1964, 81.)

ro.se.o.spin′us. L. adj. *roseus* rosy; L. n. *spina* thorn, spine; M.L. adj. *roseospinus* rosy, spine, referring to color of aerial mycelium and nature of spore wall ornamentation.

Produces α-galactosidase. Characterization taken from Suzuki *et al.* 1965.

Type strain: No. 49 (single isolate).

385. Streptomyces roseoviolaceus (Sveshnikova) Pridham, Hesseltine and Benedict 1958, 68. (*Actinomyces roseoviolaceus* Sveshnikova in Gauze *et al.* 1957, 67.)

ro.se.o.vi.o.la′ce.us. L. adj. *roseus* rosy; L. adj. *violaceus* violet colored; M.L. adj. *roseoviolaceus* rosy, violet colored, referring to color of aerial mycelium, vegetative mycelium and diffusible pigment.

Exhibits anti-bacterial and anti-fungal activity; forms pH-sensitive red to blue vegetative mycelium and diffusible pigment (violet or blue-violet under alkaline condition; violet, red or pinkish orange under acid conditions). This taxon might also be placed in the Violet color series. Characterization, in part, in Shirling and Gottlieb, 1969, 474.

Type strain: INA 1020/54 (Gauze in Gottlieb, 1968, 20).

386. Streptomyces violaceus (Rossi Doria) Waksman in Waksman and Lechevalier 1953, 43. (*Streptotrix* (*sic*) *violacea* Rossi Doria 1891, 411 (Illeg.); *Oospora violacea* (Rossi Doria) Sauvageau and Radais 1892, 252 (Illeg.); *Actinomyces violaceus* (Rossi Doria) Gasperini 1894, 84; *Cladothrix violacea* (Rossi Doria) Macé 1897, 1032 (Illeg.); *Nocardia violacea* (Rossi Doria) Chalmers and Christopherson 1916, 270; *Discomyces violaceus* (Rossi Doria) Brumpt 1922, 995 (Illeg.).)

vi.o.la′ce.us. L. adj. *violaceus* violet colored, referring to the color of the vegetative mycelium and diffusible pigment of the organism.

Produces the anthracycline anti-bacterial mycetin complex (components A, B_1, B_2 and C); exhibits anti-fungal and anti-viral activity; excellent growth on Czapek's solution agar; vegetative mycelium and diffusible pigments are pH sensitive (violet under alkaline conditions; red under acid conditions). For more detailed discussion see Artamonova and Krasil'nikov in Rautenshtein (1960, 275).

Reference strain: INMI 1; RIA 656; ATCC 15888; none of Rossi Doria's original isolates is extant.

387. Streptomyces violaceus subsp. **confinus** (Artamonova and Krasil'nikov) Pridham 1970, 28. (*Actinomyces violaceus confinus* n. subsp. (*sic*) Artamonova and Krasil'nikov in Rautenshtein 1960, 329.)

subsp. con.fi′nus. L. adj. *confinis* related.

Exhibits anti-bacterial and anti-fungal activity; may exhibit anti-viral activity; good growth on Czapek's solution agar; forms violet-colored vegetative mycelium and diffusible pigment on some media; utilizes sorbitol with Pridham and Gottlieb basal agar.

Type strain: INMI 829 (Pridham, 1970, 28).

388. Streptomyces violaceus subsp. **vicinus** (Artamonova and Krasil'nikov) Pridham 1970, 29. (*Actinomyces violaceus vicinus* n. subsp. (*sic*) Artamonova and Krasil'nikov in Rautenshtein 1960, 329.)

subsp. vi.ci′nus. L. adj. *vicinus* similar.

Exhibits anti-bacterial, anti-fungal and anti-viral activity; forms violet-colored vegetative mycelium and diffusible pigment on some media.

Type strain: INMI 1022 (Pridham, 1970, 29).

389. Streptomyces violarus (*sic*) (Artamonova and Krasil'nikov) Pridham 1970, 30. (*Actinomyces violarus* Artamonova and Krasil'nikov in Rautenshtein 1960, 334.)

vi.o.la′rus. L. adj. *violaris* of or belonging to violets, violet, referring to the color of the vegetative mycelium.

Produces the anti-bacterial violarin complex including violarin B (mycetin A, rhodomycin, antibiotic 12-12, antibiotic 452-7); exhibits anti-fungal and anti-viral activity; good growth on Czapek's solution agar; forms pH-sensitive violet-colored vegetative mycelium and diffusible pigment on some media (blue under alkaline conditions, red under acid conditions); does not utilize sorbitol with Pridham and Gottlieb basal agar.

Note. The strain designation INMI 1212 also is associated with *Streptomyces violens*, p. 828.

Type strain: INMI 1212; RIA 157; ATCC 15891 (Krasil'nikov in Shirling and Gottlieb, 1969, 499).

390. Streptomyces violatus (Artamonova and Krasil'nikov) Pridham 1970, 30. (*Actinomyces violatus* Artamonova and Krasil'nikov in Rautenshtein 1960, 334.)

vi.o.la'tus. L. n. *viola* the violet; L. adj. *violatus* flavored with violet, referring to the color of the vegetative mycelium and diffusible pigment of the organism.

Exhibits anti-bacterial and anti-fungal activity; excellent growth on Czapek's solution agar; forms violet to red vegetative mycelium and diffusible pigment on some media; does not utilize sorbitol with Pridham and Gottlieb basal agar.

Type strain: INMI 1205; RIA 708; ATCC 15892 (single isolate).

391. Streptomyces yokosukanensis (*sic*) Nakamura 1961, 97. (*Streptomyces* sp. in Isono and Suzuki 1960, 270.)

yo'ko.su'ka.nen'sis. M.L. adj. *yokosukanensis* (*sic*) belonging to Yokosuka City, Kanagawa Prefecture, Japan, the source of the soil from which the organism was isolated.

Produces 9-β-D-ribofuranosylpurine (nebularine), an anti-mycobacterial, anti-candidal, anti-tumor antibiotic; exhibits anti-fungal activity; poor growth on Czapek's solution agar; may form sclerotia.

Type strain: IPCR B-34; CBS 277.65 (single isolate).

TABLE 17.44f

Red series[a]

Names of species and subspecies	Strain desig-nation	Utilization of carbon compounds (P & G basal)											
		No carbon control	D-Glucose	D-Xylose	L-Arabinose	L-Rhamnose	D-Fructose	D-Galactose	Raffinose	D-Mannitol	i-Inositol	Salicin	Sucrose
17.44f R; S; C−; SM													
392. S. albosporeus	R	−	+	+	+	+	+	+	+	+	+	−	
393. S. aurantiacus	R		+	±	+	+	+	+	±	+	+	−	±
394. S. aureoverticillatus	T	−	+	+	+	−	+	+	+	+	+	+	
395. S. aurini	R												
396. S. cremeus	T	−	+	+	+	−	+	+	−	−	−	−	
397. S. daghestanicus	T		+	+	+	+	+	−	+	−	−		
398. S. fradiae	T	−	+	+	+	−	+	+	−	−	−		
399. S. fragilis	T	−	+	+	+	−	+	−	−	−			±
400. S. fumanus	T		+	+	+	+	+	+	+	−	−		
401. S. glomeroaurantiacus	T	−	+	+	−	−	+	+	−	+	+	−	−
402. S. griseoviridis	T	−	+	+	+	+	+	+	−	+	−		
403. S. niveoruber	T	−	+	+	+	+							
404. S. peucetius	T**		+	+	−	−	+		+	+			+
405. S. phaeoviridis	T**	−	+	+	+	+	+		+	+			+
406. S. roseiscleroticus	T	−	+	+	+	+	+	+	−	+	−		
407. S. roseoflavus	T	−	+	+	+	−	−	−	−	−			
408. S. roseolilacinus	T	−	+	−	+	±	−	−	−	−			
409. S. rubro-cyaneus	T		+	+	+	+	+	+	+	+	+		
410. S. tauricus	T		+	+	+	+	+		+	−			
411. S. vinaceus-drappus	T		+	+	+	+	+		+	+	+	+	
412. S. virocidus	T		+	+	+	+	+	−	+	+	+		+

[a] For explanation of meanings of symbols and asterisks, see footnotes to Tables 17.40 and 17.41a–e.

392. Streptomyces albosporeus (Krainsky) Waksman and Henrici 1948, 954. (*Actinomyces albosporeus* Krainsky 1914, 649; *Nocardia albosporea* (Krainsky) Chalmers and Christopherson 1916, 268.)

al.bo.spo're.us. L. adj. *albus* white; M.L. n. *spora* a spore; M.L. adj. *albosporeus* white spored.

No antimicrobial activity detected; inhibited by streptomycin; excellent growth on Czapek's solution agar; very sparse formation of aerial mycelium; forms yellow, red, red-brown, violet- or orange-colored vegetative mycelium on some media; NaCl tolerance $\geq$ 10%, but <13%.

Reference strain: Waksman and Curtis 367; IMRU 3003; none of Krainsky's original isolates is extant.

393. Streptomyces aurantiacus (Rossi Doria) Waksman in Waksman and Lechevalier 1953, 53. (*Streptotrix* (sic) *aurantiaca* Rossi Doria 1891, 417 (Illeg.); *actinomyces* (sic) *aurantiacus* (Rossi Doria) Gasperini 1892, 222; *Actinomyces aurantiacus* (Rossi Doria) Gasperini 1894, 84; *Cladothrix aurantiaca* (Rossi Doria) Macé 1897, 1033; *Nocardia aurantiaca* (Rossi Doria) Chalmers and Christopherson 1916, 268.)

au.ran.ti'a.cus. M.L. n. *Aurantium* generic name of the orange; M.L. adj. *aurantiacus* orange colored.

Exhibits anti-bacterial and anti-fungal activity; forms red-brown; yellow-pink or red-orange vegetative mycelium on some media; pigments formed are insoluble in water, but soluble in organic solvents and are not pH sensitive. Characterization, in part, in Shirling and Gottlieb, 1969, 404.

Reference strain: INMI 1373; ATCC 19822; none of Rossi Doria's original isolates is extant.

394. Streptomyces aureoverticillatus (Krasil'nikov and Yuan) Pridham 1970, 9. (*Actinomyces aureoverticillatus* Krasil'nikov and Yuan 1960, 487.)

au.re.o.ver.ti.cil.la'tus. L. adj. *aureus* golden; M.L. adj. *verticillatus* whorled; M.L. adj. *aureoverticillatus* golden, whorled (referring to color of the vegetative mycelium of the organism and nature of its morphology).

Exhibits anti-bacterial and anti-fungal activity; moderate growth on Czapek's solution agar; forms purplish orange, orange-red, red, yellow and orange-brown vegetative mycelium on some media. The pigment of the vegetative mycelium is insoluble in water and pH sensitive (yellow under alkaline conditions, red under acid conditions); acetone:ethanol (3:1) extracts exhibit one maximum at 530 nm in hexane; spore chains of atypical *Retinaculum-Apertum* type.

Comment: This taxon is not whorled in the sense of the genus *Streptoverticillium* Baldacci 1958, 15.

Type strain: INMI 1077; RIA 679; ATCC 15854; ATCC 19726 (single isolate).

395. Streptomyces aurini (Preobrazhenskaya) Pridham, Hesseltine and Benedict 1958, 67. (*Actinomyces aurini* Preobrazhenskaya in Gauze *et al.* 1957, 54.)

au.ri'ni. L. adj. *aureus* golden; possibly o.v. of L. suff. *-inus* pertaining to, like; M.L. adj. *aurini* golden like, referring to the golden yellow diffusible pigment formed by the organism.

Exhibits anti-bacterial and anti-fungal activity. Characterization, in part, taken from Hütter (1964, 617).

Reference strain: INA 6001; ETH 24183.

396. Streptomyces cremeus (Kudrina) Pridham, Hesseltine and Benedict 1958, 66. (*Actinomyces cremeus* Kudrina in Gauze *et al.* 1957, 93; *A. creneus* in Gauze in Gottlieb 1968, 20 (*lapsus calami*).)

cre'me.us. L. adj. *cremeus* cream-white.

Produces cremomycin, an anti-bacterial and anti-fungal antibiotic; inhibited by streptomycin; poor growth on Czapek's solution agar; spore chains may be of *Rectus-Flexibilis* or atypical *Retinaculum-Apertum* types; NaCl tolerance $\geq$ 7%, but <10%.

Comment: This taxon might also be placed in the Yellow color series. Characterization, in part, taken from M. E. Bergy and T. R. Pyke, U.S. Patent 3,350,269, October 31, 1967.

Type strain: INA 815/54; NRRL 3241; ATCC 19897 (Gauze in Gottlieb, 1968, 20).

397. Streptomyces daghestanicus (Sveshnikova) Pridham, Hesseltine and Benedict 1958, 67. (*Actinomyces daghestanicus* Sveshnikova in Gauze *et al.* 1957, 59.)

da.ghe.stan'i.cus. M.L. adj. *daghestanicus* belonging to Daghestan, A.S.S.R., the source of the soil from which the organism was isolated.

Exhibits marked anti-bacterial activity; some spore chains of *Retinaculum-Apertum* type; NaCl tolerance $\geq$ 7%, but <10%. Characterization, in part, in Shirling and Gottlieb, 1968, 104.

Type strain: INA 2656/55; ATCC 19747; (Gauze in Gottlieb, 1968, 20).

398. Streptomyces fradiae (Waksman and Curtis) Waksman and Henrici 1948, 954. (*Actinomyces fradii* (sic) Waksman and Curtis 1916, 125; *Streptomyces fradii* (sic) Waksman and Henrici 1948, 954 according to Pridham, Hesseltine and Benedict 1958, 60; *Streptomyces fradiae* (Waksman and Curtis) subsp. *fradiae* Waksman and Curtis in Pridham, Lyons and Seckinger 1965, 222.)

fra'di.ae. M.L. gen.n. *fradiae* of Fradia; a patronymic.

Produces the neomycin complex, (components A through F), fradicin (an anti-fungal hexaene) and an anti-*Bacillus subtilis* factor; inhibited by streptomycin; poor to fair growth on Czapek's solution agar; spore chains of atypical *Retinaculum-Apertum* type; NaCl tolerance $\geqq$ 7%, but <10%.

Type strain: IMRU 3535, ATCC 10745; ATCC 19760 (single isolate).

399. Streptomyces fragilis Anderson, Ehrlich, Sun and Burkholder 1956, 105. (*Streptomyces* sp. in Stock *et al.* 1954, 71; *Streptomyces* (Parke, Davis No. 04926) in Ehrlich *et al.* 1954, 72; *Streptomyces* sp. in Bartz *et al.* 1954, 72; *Streptomyces*, PD 04926 in Coffey *et al.* 1954, 775; *Streptomyces* sp. in Fusari *et al.* 1954, 2878; *Streptomyces* sp. in Fusari *et al.* 1954, 2881; *Streptomyces* sp. in Moore *et al.* 1954, 2884; *Streptomyces* sp. in Sugiura and Stock 1955, 127; *Streptomyces* sp. in Norman 1955, 213; *Streptomyces fragilis* Anderson *et al.* in Coffey *et al.* 1954, 791 (Not Val. Pub.); *Streptomyces* C1437 in Anderson *et al.* 1956, 101; *Streptomyces* C10076 in Anderson *et al.* 1956, 101.)

fra'gi.lis. L. adj. *fragilis* fragile.

Produces *O*-diazoacetyl-L-serine (azaserine), an anti-bacterial, anti-fungal, anti-protozoal and anti-tumor antibiotic; inhibited by streptomycin: poor growth on Czapek's solution agar; spore chains of atypical *Retinaculum-Apertum* type; NaCl tolerance $\geqq$ 4, but <7%.

Type strain: C1437; PD 04926; NRRL 2424; IMRU 3732; ATCC 23908.

400. Streptomyces fumanus (Sveshnikova) Pridham, Hesseltine and Benedict 1958, 67. (*Actinomyces fumanus* Sveshnikova in Gauze *et al.* 1957, 61; *Actinomyces fumarius* (*sic*) in Danga and Gottlieb 1959, 43.)

fu.man'us. M.L. adj. *fumanus* smoky, probably referring to the color of the vegetative mycelium of the organism.

Exhibits anti-bacterial and anti-fungal activity. Characterization, in part, in Shirling and Gottlieb, 1969, 431.

Type strain: INA 10256/54; ATCC 19904 (Preobrazhenskaya in Shirling and Gottlieb, 1969, 431).

401. Streptomyces   glomeroaurantiacus (Krasil'nikov and Yuan) Pridham 1970, 17. (*Actinomyces glomeroaurantiacus* Krasil'nikov and Yuan in Krasil'nikov 1965, 50.)

glo'mer.o.au.ran.ti'a.cus. L. v. *glomero* form into a ball; M.L. adj. *aurantiacus* orange colored; M.L. adj. *glomeroaurantiacus* orange-colored ball.

Exhibits anti-bacterial activity (at least three factors); exhibits anti-fungal activity; inhibited

by streptomycin; poor growth on Czapek's solution agar; forms red to orange vegetative mycelium on some media; coiling of chains of spores may be limited and aerial mycelium color might be of the White series.

Type strain: INMI 1464; RIA 683; ATCC 15866.

402. Streptomyces griseoviridis Anderson, Ehrlich, Sun and Burkholder 1956, 114. (A streptomyces (*sic*) in Bartz *et al.* 1955, 777; an actinomycete in Haskell *et al.* 1955, 784; *Streptomyces* P-D 04955 in Ehrlich *et al.* 1955, 790; *Streptomyces* strain in Ames *et al.* 1955, 4260; *Streptomyces* culture A9071 in Anderson *et al.* 1956, 109; *Streptomyces griseoviridus* (*sic*) Anderson *et al.* 1956, 114; Not *Actinomyces griseoviridis* or *griseo-viridis* Nicolaieva 1915, 239.)

Note. See Waksman (1961, 225) for change in spelling of epithet.

gri.se.o.vi'ri.dis. M.L. adj. *griseus* gray; L. adj. *viridis* green; M.L. adj. *griseoviridis* gray-green.

Produces griseoviridin and etamycin (viridogrisein); inhibited by streptomycin; moderate growth on Czapek's solution agar; some spore chains of atypical *Retinaculum-Apertum* type; yellow aerial mycelium might also be noted; NaCl tolerance $\geqq$ 7%, but <10%.

Type strain: A9071; P-D 04955; NRRL 2427; IMRU 3735; ATCC 23920.

403. Streptomyces niveoruber Ettlinger, Corbaz and Hütter 1958, 350.

ni've.o.ru'ber. L. adj. *niveus* snow-white; L. adj. *ruber* red; M.L. adj. *niveoruber* snow-white, red (referring to the white color of the aerial mycelium and red color of the vegetative mycelium).

Produces cinerubins A and B; excellent growth on Czapek's solution agar; forms red vegetative mycelium on some media. This taxon might also be placed in the White color series.

Type strain: ETH 17860; ATCC 14971.

404. Streptomyces peucetius Grein, Spalla, DiMarco and Canevazzi 1963, 109. (*Streptomyces* sp. in DiMarco *et al.* 1963; a streptomyces in Dubost *et al.* 1963, 1813; *Streptomyces peucetius* in Grein and Spalla 1962, 175 (Not Val. Pub.); *Streptomyces peucetius* in Societa Farmaceutici Italia, Italian Patent 22651/62, 1962 (Not Val. Pub.); *Streptomyces peuceticus* (*sic*) in Umezawa 1967, 159 and Hütter 1967, 322 (*lapsus calami*); *Streptomyces peucieticus* (*sic*) in Umezawa 1967, 264 (*lapsus calami*).)

peu.ce'ti.us. M.L. adj. *peucetius* pertaining to Peucetia, an ancient name for Central Puglia in Italy, the source of the soil from which the organism was isolated.

Produces daunomycin, an anti-bacterial and anti-tumor antibiotic; produces polyenic anti-fungal activity; excellent growth on Czapek's solution agar; reported to form coremia; forms pinkish vegetative mycelium on some media; spore chains of atypical *Retinaculum-Apertum* type.

Type strain: F.I. 1762; 1762; IMI 10135; NCIB 9475 (an apparent single isolate designated in various patents).

405. Streptomyces phaeoviridis Shinobu 1957, 63.

phae.o.vi'ri.dis. Gr. adj. *phaeus* brown; L. adj. *viridis* green; M.L. adj. *phaeoviridis* brown-green.

Does not exhibit anti-microbial activity; inhibited by streptomycin; excellent growth on Czapek's solution agar; forms red-brown vegetative mycelium and diffusible pigment on some media; spore chains of atypical *Retinaculum-Apertum* type.

Type strain: OEU 503; NRRL B-2258 (single isolate).

406. Streptomyces roseiscleroticus Pridham 1970, 43. (*Chainia rosea* Thirumalachar in Thirumalachar, Sukapure, Rahalkar and Gopalkrishnan 1966, 10.)

ro.se.i.scle'r.oti.cus. L. adj. *roseus* rosy; L. n. *sclerotium* a sclerotium; M.L. adj. *roseiscleroticus* rosy, belonging to sclerotium.

Exhibits anti-bacterial and anti-fungal activity; inhibited by streptomycin; excellent growth on Czapek's solution agar; forms red to red-orange vegetative mycelium on some media; forms sclerotia; whole cell hydrolysates contain L-diaminopimelic acid (Pridham and Lyons, 1969, 211).

Type strain: HACC 144; ATCC 17755; CBS 226.65 (single isolate).

407. Streptomyces roseoflavus Arai 1951, 218. (*Streptomyces* strain No. 320 in Aiso *et al.* 1950, 87; *Actinomyces roseoflavus* (Arai) 1951 (*sic*) in Preobrazhenskaya in Gauze *et al.* 1957, 52 (Illeg.).)

ro.se.o.fla'vus. L. adj. *roseus* rosy; L. adj. *flavus* yellow; M.L. adj. *roseoflavus* rose, yellow.

Produces flavomycin, an anti-bacterial antibiotic and mycelin, an anti-fungal antibiotic; poor growth on Czapek's solution agar; forms spore chains of atypical *Retinaculum-Apertum* type; NaCl tolerance <4%.

Type strain: 320; Arai 320; IMRU 3572; NRRL B-2789 (single isolate).

Note. Waksman (1961, 270) incorrectly designated strain ATCC 13,167 = IMRU 3672 as the type strain.

408. Streptomyces roseolilacinus (Preobrazhenskaya and Sveshnikova) Pridham, Hesseltine and Benedict 1958, 68. (*Actinomyces roseolilacinus* Preobrazhenskaya and Sveshnikova in Gauze *et al.* 1957, 35.)

ro.se.o.li.la.ci'nus. L. adj. *roseus* rosy; L. adj. *lilacinus* lilac colored; M.L. adj. *roseolilacinus* rose, lilac colored, referring to color of the aerial mycelium of the organism.

Exhibits anti-bacterial and anti-fungal activity; poor growth on Czapek's solution agar; NaCl tolerance ≥ 4%, but <7%. Characterization, in part, in Shirling and Gottlieb, 1968, 167.

Type strain: INA 14250; CBS 264.66; ATCC 19806; ATCC 19922 (Gauze in Gottlieb 1968, 20).

409. Streptomyces rubro-cyaneus (*sic*) (Krasil'nikov and Asem Khusein) Pridham 1970, 26. (*Actinomyces rubro-cyaneus* (*sic*) Krasil'nikov and Asem Khusein in Krasil'nikov 1965, 158.)

rub.ro-cy.an'e.us. L. adj. *ruber* red; Gr. adj. *cyaneus* dark blue; M.L. adj. *rubrocyaneus* red-dark blue.

Exhibits no anti-microbial activity; inhibited by streptomycin; forms blue-green vegetative mycelium on some media.

Comment: Krasil'nikov and Asem Khusein (1965, 158) have extensively characterized some blue-green and red pigments from strains of this taxon.

Type strain: INMI 21 (Pridham, 1970, 26).

410. Streptomyces tauricus (Ivanitskaya, Upiter, Sveshnikova and Gauze) Pridham 1970, 27. (*Actinomyces tauricus* Ivanitskaya, Upiter, Sveshnikova and Gauze 1966, 974.)

tau'ri.cus. L. adj. *tauricus* Taurian, Thracian of the Crimea, Albanian.

Produces tauromycetin, an anti-tumor antibiotic; forms red, violet or reddish violet vegetative mycelium on some media; said to blacken peptone iron agar.

Type strain: INA 8173 Pridham 1970, 27; also selected by Gauze for the ISP.

Comment: The only strain designation given in the original description by Ivanitskaya *et al.* 1966, 973 is 13170. The relationship of strain INA 8173 and 13170 is unknown.

411. Streptomyces vinaceus-drappus Pridham, Hesseltine and Benedict 1958, 68. (*Streptomyces* sp. in DeBoer, Caron and Hinman 1953, 499; *Streptomyces vinaceus-drappus* in Hinman, Caron and DeBoer 1953, 4L (Not Val. Pub.); *Streptomyces vinaceus-drappus* in Flynn and Woolf 1953, 4L (Not Val. Pub.); *Streptomyces vinaceus-drappus* in Hinman, Caron and DeBoer 1953, 5864 (Not Val. Pub.); *Streptomyces viaceusdrappus* (*sic*) in Flynn, Hinman, Caron and Woolf 1953, 5867 (Not Val. Pub); *Streptomyces vinaceus-drappus* in The Upjohn Company, British Patent 708,686, May 5, 1954 (Not Val. Pub.).)

vi.na'ce.us-drapp.us. L. adj. *vinaceus* of or belonging to wine or the grape; Low L. n. *drappus* of unknown origin, now referring to the color "drab;"

M.L. adj. *vinaceus-drappus* of wine-drab, referring to the drab wine color of the aerial mycelium and spores of the organism.

Produces the anti-bacterial antibiotics amicetin, amicetin A, amicetin B (plicacetin), amicetin C and streptolin; excellent growth on Czapek's solution agar. This taxon might also be placed in the Gray color series.

Type strain: D-13; UC 2007; NRRL 2363 (Dietz in Shirling and Gottlieb, 1969, 494).

412. Streptomyces virocidus (Kuchaeva, Krasil'nikov, Taptykova and Gesheva) Pridham 1970, 31. (*Actinomyces virocidus* Kuchaeva, Krasil'nikov, Taptykova and Gesheva 1961, 120.)

vir'o.ci.dus. L. n. *virus* slime, poison; now a virus; M.L. adj. *virocidus* virucidal.

Produces virusin, an anti-bacterial, anti-fungal and anti-viral antibiotic.

Type strain: INMI 1609 (Pridham 1970, 31; the "basic representative" in Kuchaeva *et al.* 1961).

413. Streptomyces erythraeus (Waksman) Waksman and Henrici 1948, 938. (*Actinomyces* 161 in Waksman 1919, 112; Actinomyces erythreus (*sic*) Waksman in THE MANUAL 1923, 370; *Streptomyces erythreus* (*sic*) (Waksman) (Waksman and Henrici 1948, 938.)

Comment: See Waksman in Breed *et al.*, 1957, 766, for correction in spelling of *erythreus* to *erythraeus*.

e.ryth'rae.us. Gr. adj. *erythraeus* red.

Produces the erythromycin complex (components A, B and C); inhibited by streptomycin; excellent growth on Czapek's solution agar; forms pinkish diffusible pigment on some media; spore chains of atypical *Retinaculum-Apertum* type: NaCl tolerance $\geq 7\%$, but $<10\%$.

Reference strain: Lilly M5-12559; NRRL 2338; IMRU 3737; ATCC 11635.

Note. Progeny of the original isolate of Waksman and Curtis is available only as two nonviable herbarium specimens (see Pridham *et al.*, 1965, 219).

414. Streptomyces luteofluorescens Shinobu 1962, 115.

lu.te.o.flu.o.res'cens. L. adj. *luteus* yellow; M.L. v. *fluoresco* fluoresce; M.L. part. adj. *luteofluorescens* yellow, fluorescing (referring to formation of fluorescent yellow diffusible pigment).

Spore chains short and tend to form loops, hooks or partial coils of atypical *Retinaculum-Apertum* type; forms yellow, green-yellow, red-orange or yellow-pink vegetative mycelium on some media;

TABLE 17.44g, h, i

Red series[a]

Names of species and subspecies	Strain desig-nation	Utilization of carbon compounds (P & G basal)											
		No carbon control	D-Glucose	D-Xylose	L-Arabinose	L-Rhamnose	D-Fructose	D-Galactose	Raffinose	D-Mannitol	i-Inositol	Salicin	Sucrose
17.44g R; S; C−; SPY													
413. S. erythraeus	R	−	+	+	+	+	+	+	+	+	+	+	
17.44h R; S; C−; WTY													
414. S. luteofluorescens	T		+	+	+	+	+		−	+	−		+
17.44i R; ?; ?; ?													
415. S. erythrogriseus	T			+		+			−	+		+	+
416. S. garyphalus	T												
417. S. lavendularectus	T	−	+	−	−	−	−	+	−	−	−	−	−
418. S. nagasakiensis	T		+	+	+	+	+	+	−				+
419. S. rubrolavendulae	T	−	+	+	+	+	+	+	−	+	+	−	−
420. S. cinnamonensis subsp. proteo-lyticus (see no. 362).	T												

[a] For explanation of meanings of symbols see footnotes to Tables 17.40 and 17.41a–e.

reportedly forms yellow fluorescent diffusible pigment. This taxon might also be placed in the Yellow color series. Characterization, in part, in Shirling and Gottlieb, 1969, 448.

Type strain: OEU 719 (single isolate).

415. Streptomyces erythrogriseus Falcão de Morais and Dália Maia 1959, 64.

e.ry.thro.gri′ise.us. Gr. adj. *erythraeus* red; M.L. adj. *griseus* gray; M.L. adj. *erythrogriseus* red, gray (referring to change in color of aerial mycelium from gray to red).

Spore wall ornamentation category unknown; spore chains in coils (section *Spira*); no melanoid pigment formed; produces the anti-bacterial indicator antibiotic erygrisin; exhibits anti-yeast activity; moderate to excellent growth on Czapek's solution agar; forms red, red-brown or red-violet vegetative mycelium and diffusible pigment on some media. This taxon might also be placed in the Gray color series.

Type strain: 4165; IAUR 1173; designated herein and selected by Falcão de Morais for the ISP.

416. Streptomyces garyphalus Harris, Ruger, Reagan, Wolf, Peck, Wallick and Woodruff 1955, 185. (*Streptomyces* sp. in Kuehl *et al.* 1955, 2344.)

gar′y.phal′us. Aztec (Nahuatl dialect) *garyphalus* soil or region.

Spore wall ornamentation category unknown; spore chains straight (section *Rectus-Flexibilis*); possibly forms melanoid pigment; produces D-4-amino-3-isoxazolidone (cycloserine).

Type strain: Merck 106–7 (the only strain definitively characterized in the original publication, and designated herein).

417. Streptomyces lavendularectus (Krasil'nikov) Pridham 1970, 40. (*Actinomyces lavendularectus* Krasil'nikov and Kuchaeva 1960 in Yen and Lu 1964, 238.)

la.ven.du.la.rec′tus. Med.L. gen.n. *lavendulae* of lavender color; L. adj. *rectus* straight; M.L. adj. *lavendularectus* lavender colored, straight.

Spore wall ornamentation category unknown; spore chains straight (section *Rectus-Flexibilis*); ability to form melanoid pigment unknown; exhibits anti-bacterial activity; exhibits anti-fungal activity.

Note. The antibiotic activities reported for the type strain are characterized by paper chromatography in the paper of Yen and Lu.

Type strain: INMI 1793 (Pridham, 1970, 40). Comment: Yen and Lu ascribe the name *A. lavendularectus* to Krasil'nikov and Kuchaeva (1960) on the basis of a personal communication. The publication referred to has not been located.

418. Streptomyces nagasakiensis Aburatani 1959, 176.

na.ga.sa.ki.en′sis. M.L. adj. *nagasakiensis* belonging to Nagasaki City, Japan, the source of the soil from which the organism was isolated.

Spore wall ornamentation category unknown; spore chains straight (section *Rectus-Flexibilis*); no melanoid pigment formed; produces D-4-amino-3-isoxazolidone (cycloserine) and two other peptides; exhibits anti-yeast activity.

Type strain: 5915 (single isolate).

419. Streptomyces rubrolavendulae (Yen) Pridham 1970, 44. (*Actinomyces rubrolavendulae* Yen 1957, 209.)

rub.ro.la.ven′du.lae. L. adj. *ruber* red; Med.L. gen.n. *lavendulae* of lavender color: M.L. gen.n. *rubrolavendulae* of red lavender color.

Spore wall ornamentation category unknown; spore chains of *Retinaculum-Apertum* type; no melanoid pigment formed; exhibits anti-bacterial activity; exhibits anti-mold activity; inhibits only *Sporobolomyces* of four yeasts tested; forms orange-yellow to red vegetative mycelium on some media; forms orange-violet to violet-red diffusible pigment in some media.

Characterization, in part, taken from Yen and Lu, 1964, 239, wherein some of the antibiotic activities are more fully characterized by paper chromatography.

Type strain: IMASP 2737 (Pridham, 1970, 44).

420. Streptomyces cinnamonensis subsp. **proteolyticus** (Sveshnikova) Pridham, Hesseltine and Benedict 1958, 60. (*Actinomyces cinnamonensis* var. *proteolyticus* Sveshnikova in Gauze *et al.* 1957, 63.)

subsp. pro′te.o.lyt′i.cus. M.L. adj. *proteolyticus* proteolytic.

Spore wall ornamentation category unknown; spore chains straight (section *Rectus-Flexibilis*); ability to form melanoid pigment unknown; may or may not exhibit anti-microbial activity.

Type strain: INA 14013; ATCC 19893 (designated herein).

See Table 17.45a-f.

421. Streptomyces ashchabadicus (Preobrazhenskaya) Pridham 1970, 8. (*Actinomyces ashchabadicus* Preobrazhenskaya 1966, 857.)

ash.cha.bad′i.cus. M.L. adj. *ashchabadicus* belonging to Ashkhabad, a city of the Turkomen Republic of Central Asia.

Exhibits slight antibiotic activity against Gram-positive bacteria; forms yellow to brownish yellow vegetative growth on some media.

Type strain: INA 13496.

422. Streptomyces polychromogenes Hagemann, Pénasse and Teillon in Hütter 1964, 615. (*Streptomyces* sp., souche T.4473 in Janot *et al.* 1954, 440; *Streptomyces polychromogenus* (*sic*) in

TABLE 17.45a–f
Blue series

Names of species and subspecies	Strain desig-nation	No carbon control	D-Glucose	D-Xylose	L-Arabinose	L-Rhamnose	D-Fructose	D-Galactose	Raffinose	D-Mannitol	i-Inositol	Salicin	Sucrose
17.45a B; RF; C+; SM													
421. S. ashchabadicus	T												
422. S. polychromogenes	T	−	+	+	+	−	+	+	−	−	−		+
17.45b B; S; C+; SM													
423. S. amakusaensis	T	−	+	−	±	−	−	−	−	−	−	−	±
424. S. caelestis	T	−	+	+	+	+	+	+	+	−	+	−	+
17.45c B; S; C+; WTY													
425. S. azureus	T	−	+	+	+	+	+	+	+	+	+		+
17.45d B; S; C+; SPY													
426. S. bellus	T	−	+	+	+	+	+		+	+	+		+
427. S. chartreusis	T	−	+	+	+	+	+		+	+	+	+	+
428. S. coeliatus	T		+	−	+	+	+	+	+	+			+
429. S. coerulatus (see Nos. 430 and 444).	T		+		−	+	+	+	−	+	+		+
430. S. coerulatus subsp. amylolyticus (see nos. 429 and 444).	T		+		−	+	+	+	+	+	+		+
431. S. coeruleofuscus (see *incertae sedis* no. B10).	R		+	+	+	+	+		+	+	+		+
432. S. coeruleorubidus	T	−	+	+	+	+	+	+	+	+	+	+	+
433. S. coerulescens	T	−	+	+	+	+	+		+	+	+		+
434. S. curacoi	T	−	+	+	+	+	+	+	+	+	+	+	+
435. S. cyaneus	R	−	+	+	+	+							
436. S. cyanoglomerus	T		+		+	+	+	+	+	+			−
437. S. indigocolor	T		+	−	+	+	+	+	+	+			−
438. S. lanatus	T	−	+	+	+	+	+		+	+	+		+
439. S. lazureus	T		+	−	+	+	+	−	+	+			−
440. S. valynus	T												
441. S. viridochromogenes	T	−	+	+	+	+	+	+	+	+	+		
17.45e B; S; C+; H													
442. S. glaucescens	T												
17.45f B; S; C−; SPY													
443. S. bluensis	T		+	+	+	+	+	+	+	+	+	+	+
444. S. coerulatus subsp. anaseuli (see nos. 429 and 430).	T		+			+	+	+		+	+		
445. S. coeruleoroseus	T												
446. S. ipomoeae	R		+	+	+	+	+		+	+	+		+
447. S. spinosus	T												

[a] For explanation of meanings of symbols see footnotes to Tables 17.40 and 17.41a–e.

Hagemann, Pénasse and Teillon 1955, 240 (Not Val. Pub.); *Actinomyces polychromogenes* Hagemann *et al.* 1955 in Preobrazhenskaya 1966, 854 (Illeg.).)

po.ly.chro.mo'ge.nes. Gr. adj. *poly* many; Gr. n. *chroma* color; Gr. v.suff. *-genes* producing; M.L. adj. *polychromogenes* producing many colors, referring to characteristic variation of pigmentation.

Produces antibiotic T.4473 and *O*-carbamyl-D-serine (slightly active); exhibits anti-fungal activity; inhibited by streptomycin; excellent growth on Czapek's solution agar; spore chains typically straight; spores phalangiform; forms green or yellow-green vegetative growth on some media; NaCl tolerance $\geq 4\%$, but $<7\%$. The type strain also has been placed in the Red color series (Shirling and Gottlieb, 1969, 464).

Type strain: Roussel UCLAF T.4473; ATCC 12595; CBS 311.56 (single isolate).

423. Streptomyces amakusaensis Nagatsu, Anzai, Ohkuma and Suzuki 1963, 209.
(*Streptomyces amakusaensis* in Anzai, Ohkuma, Nagatsu and Suzuki 1962, 110 (Not Val. Pub.); *Streptomyces amakusaensis* in Ohkuma, Anzai and Suzuki 1962, 115 (Not Val. Pub.); *Streptomyces amakusaensis* in Anzai 1962, 117 (Not Val. Pub.); *Actinomyces amacusaensis* (*sic*) Nagatsu *et al.* in Preobrazhenskaya 1966, 853.)

am.a.ku.sa.en'sis. M.L. adj. *amakusaensis* belonging to Amakusa Island, Japan, the source of the soil from which the organism was isolated.

Produces *N*-formyl-*trans-p*-methoxy styryl amine (tuberin), an anti-mycobacterial antibiotic; exhibits other anti-bacterial and anti-fungal activities; slight inhibition by streptomycin; poor growth on Czapek's solution agar; does not always exhibit melanin-like chromogenicity.

Type strain: IPCR 10-101; CBS 280.65; ATCC 23876 (single isolate).

424. Streptomyces caelestis DeBoer, Dietz, Wilkins, Lewis and Savage 1955, 831.
(*Streptomyces celestis* (*sic*) NRRL 2418 in Hinman *et al.*, U.S. Patent 2,851,463, September 9, 1958 (*lapsus calami*); *Actinomyces caelestis* n. sp. (*sic*) and *Actinomyces caelestis* (DeBoer *et al.*) in Krasil'nikov 1965; 120; *Actinomyces coelestis* (*sic*) DeBoer *et al.*, in Preobrazhenskaya 1966, 852.)

cae.les'tis. L. adj. *caelestis* of the sky, heavenly (referring to the blue color of the aerial mycelium and spores).

Produces celesticetin; exhibits slight anti-fungal activity; inhibited by streptomycin; excellent growth on Czapek's solution agar; spore chains of atypical *Retinaculum-Apertum* type; NaCl tolerance $\geq 7\%$, but $<10\%$.

Type strain: D-52; UC 2011; NRRL 2418; ATCC 14924; ATCC 15804; ATCC 19733 (single isolate).

425. Streptomyces azureus Kelly, Kutscher and Tuoti 1959, 1334.
(*Streptomyces* sp. in Pagano, Weinstein, Stout and Donovick 1956, 554; *Streptomyces* culture in Steinberg, Jambor, Suydam and Soriano 1956, 562; *Streptomyces* sp. in Donovick, Pagano and Vandeputt, German Patent 1,007,955 May 9, 1957; *Streptomyces* sp. 3705 in Olin Mathieson Chemical Company, British Patent 795,570 May 28, 1958; *Actinomyces azureus* (Kelly *et al.*) Preobrazhenskaya 1966, 853.)

Note. The earliest published use of the name *Streptomyces azureus* is in Kelly *et al.* 1959, 1334 who refer to the description in Pagano *et al.* 1956, 554.

az'ur.e.us. L. adj. *azureus* azure-blue, referring to the color of the aerial mycelium of the organism.

Produce thiostrepton, a polypeptidic antibacterial antibiotic: NaCl tolerance $\geq 7\%$, but $<10\%$. Spores have been characterized from smooth to rough to warty in reports on electron microscopy of this taxon. Characterization, in part, taken from Pagano *et al.* 1956, 554; Hütter 1962, 27 and Trejo and Bennett 1963, 676.

Type strain: SC-2364; IMRU 3705; ATCC 14921; ATCC 19728 (single isolate).

426. Streptomyces bellus Margalith and Beretta 1960, 193.
(*Actinomyces bellus* (Margalith *et al.*) Preobrazhenskaya 1966, 855.)

bell'us. L. adj. *bellus* pretty, handsome.

Produces althiomycin (matamycin), an antibacterial antibiotic; inhibited by streptomycin; excellent growth on Czapek's solution agar; forms pink to orange-pink to orange vegetative mycelium on some media.

Type strain: A/870; ATCC 14925; ATCC 23886 (single isolate).

427. Streptomyces chartreusis Leach, Calhoun, Johnson, Teeters and Jackson 1953, 4011.
(*Actinomyces chartreusis* (Calhoun *et al.*) Preobrazhenskaya 1966, 852.)

char.treu'sis. French n. *chartreuse* a Carthusian monastery famed for a sweet yellow liqueur, hence the color "chartreuse;" M.L. adj. *chartreusis* yellow, referring to color of the diffusible pigment formed by the organism.

Produces chartreusin, an anti-bacterial antibiotic; exhibits slight anti-fungal activity; inhibited by streptomycin; excellent growth on Czapek's solution agar; NaCl tolerance $\geq 7\%$, but $<10\%$.

Note: This taxon is more fully characterized in Calhoun and Johnson, 1956, 294.

Type strain: K-180; UC 2012; NRRL 2287; ATCC 14922; ATCC 19738 (single isolate).

428. Streptomyces coeliatus (Krasil'nikov, Sorokina, Alferova and Bezzubenkova) Pridham 1970, 11. (*Actinomyces coeliatus* Krasil'nikov, Sorokina, Alferova and Bezzubenkova in Krasil'nikov 1965, 89.)

coel.li.a'tus. L. n. *coelum* sky; L. adj.suff. *-atus* provided with; M.L. adj. *coeliatus* provided with blue.

Exhibits anti-bacterial and slight anti-fungal activity; forms blue-colored vegetative mycelium and diffusible pigment on some media.

Type strain: INMI 37-я; ATCC 19833 (Krasil'nikov in Shirling and Gottlieb 1969, 418).

429. Streptomyces coerulatus (Krasil'nikov, Sorokina, Alferova and Bezzubenkova) Pridham 1970, 13. (*Actinomyces coerulatus* Krasil'nikov, Sorokina, Alferova and Bezzubenkova in Krasil'nikov 1965, 88.)

coe.ru.la'tus. L. adj. *coeruleus* dark blue, azur; L. adj.suff. *-atus* provided with; M.L. adj. *coerulatus* provided with dark blue.

Exhibits anti-bacterial and anti-fungal activity; forms blue-colored vegetative mycelium and diffusible pigment on some media.

Type strain: INMI 1057.

430. Streptomyces coerulatus subsp. **amylolyticus** (Krasil'nikov, Sorokina, Alferova and Bezzubenkova) Pridham 1970, 12. (*Actinomyces coerulatus amylolyticus* (*sic*) Krasil'nikov, Sorokina, Alferova and Bezzubenkova in Krasil'nikov 1965, 88.)

subsp. a.my.lo.ly'ti.cus. Gr. n.*amylum* fine meal, starch; Gr. adj. *lyticus* loosening, dissolving; M.L. adj. *amylolyticus* starch-dissolving.

Exhibits anti-bacterial and anti-fungal activity; forms blue-colored vegetative mycelium and diffusible pigment on some media.

Type strain: INMI 1031-4 (single isolate).

431. Streptomyces coeruleofuscus (Preobrazhenskaya) Pridham, Hesseltine and Benedict 1958, 67. (*Actinomyces coeruleofuscus* Preobrazhenskaya in Gauze *et al.* 1957, 128.)

coe.ru'le.o.fus'cus. L. adj. *coeruleus* dark blue, azure; L. adj. *fuscus* dark, tawny; M.L. adj. *coeruleofuscus* dark blue, tawny (referring to the bluish aerial mycelium and brownish vegetative mycelium of the organism).

Exhibits anti-bacterial and anti-fungal activity; NaCl tolerance $\geq$ 7%, but <10%.

Reference strain: INA 2922/57; ATCC 19741 (designated by Gauze as the "type strain" in Gottlieb, 1968, 20).

Note. Preobrazhenskaya (1966) designated strains INA 6920 and INA 12145 as typical. None of the three strains listed here is mentioned in the original description.

432. Streptomyces coeruleorubidus (Preobrazhenskaya) Pridham, Hesseltine and Benedict 1958, 67. (*Actinomyces coeruleorubidus* Preobrazhenskaya in Gauze *et al.* 1957, 125.)

coe.ru'le.o.ru.bi'dus. L. adj. *coeruleus* dark blue, azure; L. adj. *rubidus* dark red; M.L. adj. *coeruleorubidus* dark blue, dark red (referring to the bluish aerial mycelium and red vegetative mycelium of the organism on chemically defined media).

Exhibits anti-bacterial and anti-fungal activity; inhibited by streptomycin; excellent growth on Czapek's solution agar; forms red or orange vegetative mycelium on some media; NaCl tolerance $\geq$ 7%, but <10%.

Type strain: INA 12531/54; ATCC 13740; ATCC 23900 (Preobrazhenskaya, 1966, 850).

433. Streptomyces coerulescens (Preobrazhenskaya) Pridham, Hesseltine and Benedict 1958, 67. (*Actinomyces coerulescens* Preobrazhenskaya in Gauze *et al.* 1957, 120.)

coe.ru.les'cens. L. adj. *coeruleus* dark blue, azure; M.L. part.adj. *coerulescens* becoming blue, slightly blue (referring to the color of the aerial mycelium on a chemically defined medium).

May possibly produce coerulomycin (similar to chartreusin); excellent growth on Czapek's solution agar; forms yellow, yellow-orange, yellow-brown or gray-brown vegetative mycelium; edges of vegetative growth may be colored red-brown; NaCl tolerance $\geq$ 7%, but <10%.

Type strain: INA 4562; ATCC 19896; ATCC 19742 (Preobrazhenskaya, 1966, 850).

434. Streptomyces curacoi Cataldi in Trejo and Bennett 1963, 683 (*Streptomyces cura-coi* (*sic*) and *Streptomyces curacoi* in Galmarini and Doulofeu 1961, 76 (Not Val. Pub.); *Streptomyces cura-coi* (*sic*) in Cataldi *et al.* U.S. Patent 3,015,607, January 2, 1962 (Not Val. Pub.); *Actinomyces curacoi* Cataldi *et al.* 1962 in Preobrazhenskaya 1966, 855.)

cu.ra.co'i. M.L. gen.n. *curacoi* of Cura-Co in the province of La Pampa, Argentina.

Produces curamycin, an anti-bacterial and anti-viral antibiotic; moderate growth on Czapek's solution agar; NaCl tolerance $\geq$ 4%, but <7%.

Type strain: SC 3064; ATCC 13385; ATCC 19745 (single isolate).

435. Streptomyces cyaneus (Krasil'nikov) Waksman in Waksman and Lechevalier 1953, 42. (*Actinomyces cyaneus* Krasil'nikov 1941, 14.)

cy.an'e.us. Gr. adj. *cyaneus* dark blue.

Exhibits slight anti-bacterial and anti-fungal activity; inhibited by streptomycin; excellent growth on Czapek's solution agar; forms violet to reddish blue vegetative mycelium and diffusible pigment on some media, pigment not changing with pH; NaCl tolerance $\geq 4\%$, but $<7\%$.

Reference strain: Todorović; Kutzner H-112; IMRU 3761; ATCC 14923; the original single isolate of Krasil'nikov is no longer extant.

436. Streptomyces cyanoglomerus (Krasil'nikov, Sorokina, Alferova and Bezzubenkova) Pridham 1970, 14. (*Actinomyces cyanoglomerus* Krasil'nikov, Sorokina, Alferova and Bezzubenkova in Krasil'nikov 1965, 86.)

cy.an.o.glo'mer.us. Gr. adj. *cyaneus* dark blue; L. v. *glomero* form into a ball; M.L. adj. *cyanoglomerus* formed into a dark blue ball.

Exhibits anti-bacterial activity; inhibits only *Candida* of a number of molds and yeasts tested; forms blue-colored vegetative mycelium and diffusible pigment on some media.

Comment: Three subspecies were proposed by Krasil'nikov *et al.* in Krasil'nikov 1965, 86; the names are considered illegitimate and are not treated here.

Type strain: INMI 31-M.

437. Streptomyces indigocolor (Krasil'nikov, Sorokina, Alferova and Bezzubenkova) Pridham 1970, 18. (*Actinomyces indigocolor* Krasil'nikov, Sorokina, Alferova and Bezzubenkova in Krasil'nikov 1965, 87.)

in'di.go.co.lor. Gr. n. *indikon* indigo; L. n. *indicum* indigo; Spanish n. *indigo* indigo; L. n. *color* color; M.L. adj. *indigocolor* indigo blue colored.

Inhibits some streptomycetes but not other bacteria or fungi; forms blue-colored vegetative mycelium and diffusible pigment on some media.

Type strain: INMI 206; ATCC 19842 Pridham 1970, 18; Krasil'nikov *et al.* stated that strain INMI 206 "might be" the type strain of two described.

438. Streptomyces lanatus Frommer 1959, 204.

la.na'tus. L. adj. *lanatus* wooly, referring to the nature of the aerial mycelium of the organism.

Produces the actinomycin X complex; exhibits anti-fungal activity; inhibited by streptomycin; excellent growth on Czapek's solution agar; forms brown to reddish brown vegetative mycelium and diffusible pigment on some media.

Type strain: SV 1944; ATCC 19775 (single isolate).

439. Streptomyces lazureus (*sic*) (Krasil'nikov, Sorokina, Alferova and Bezzubenkova) Pridham 1970, 21. (*Actinomyces lazureus* (*sic*) Krasil'nikov, Sorokina, Alferova and Bezzubenkova in Krasil'nikov 1965, 88.)

la.zur'e.us. M.L. adj. *lazurius* blue, referring to the blue color of the vegetative mycelium and diffusible pigment.

Inhibits some streptomycetes, but not other bacteria or fungi; forms blue-colored vegetative mycelium and diffusible pigment on some media.

Type strain: INMI 383-K; ATCC 19843 (Pridham, 1970, 21).

440. Streptomyces valynus (Preobrazhenskaya) Pridham 1970, 27. (*Actinomyces valynus* Preobrazhenskaya 1966, 856; *Actinomyces valinus* (*sic*) in Preobrazhenskaya 1966, 850.)

val.y'nus. Of unknown derivation; possibly referring to a colleague of T. P. Preobrazhenskaya, or to a college in Russia.

Exhibits weak anti-microbial activity, if any; forms pH-dependent deep blue vegetative mycelium and diffusible pigment on some media; forms brown vegetative mycelium on some media.

Type strain: INA 612.

441. Streptomyces viridochromogenes (Krainsky) Waksman and Henrici 1948, 942. (*Actinomyces viridochromogenes* Krainsky 1914, 684; *Actinomyces viridochromogenus* in Waksman and Curtis 1916, 114 (*lapsus calami*); *Actinomyces viridichromogenes* (*sic*) Krasil'nikov 1941, 32 (*lapsus calami*).)

vi.ri.do.chro.mo'ge.nes. L. adj. *viridis* green; Gr. n. *chroma* color; Gr. v.suff. *-genes* producing; M.L. adj. *viridochromogenes* producing green color.

Exhibits very slight anti-microbial activity; inhibited by streptomycin; fair to moderate growth on Czapek's solution agar; forms green to black vegetative mycelium on some media.

Reference strain: NRRL B-1511; ATCC 14920; none of Krainsky's original isolates is extant.

442. Streptomyces glaucescens (Preobrazhenskaya) Pridham, Hesseltine and Benedict 1958, 67. (*Actinomyces glaucescens* Preobrazhenskaya in Gauze *et al.* 1957, 122.)

glau.ces'cens. L. adj. *glaucus* bluish gray; M.L. adj. *glaucescens* slightly bluish gray, referring to the bluish green color of the aerial mycelium on a chemically defined medium.

Exhibits anti-bacterial and anti-fungal activity; moderate growth on Czapek's solution agar; forms red- or orange-colored vegetative mycelium on some media; spore wall ornamentation might also be interpreted as spiny; taxon might also be placed in the Green color series; NaCl tolerance $\geq 4\%$, but $<7\%$. Characterization, in part, in Shirling and Gottlieb, 1968, 120.

Type strain: INA 8731; ATCC 19761; ATCC 23622 (Hütter, 1962, 32; Gauze in Gottlieb, 1968, 20).

443. Streptomyces bluensis Mason, Dietz and

Hanka 1963, 608. (*Streptomyces bluensis* var. *bluensis* Mason, Dietz and Hanka 1963, 608; *Actinomyces bluensis* var. *bluensis* (Mason *et al.*) Preobrazhenskaya 1966, 858.)

blu.en'sis. Old French adj. *bleu* blue; M.L. adj. *bluensis* belonging to blue, referring to the blue color of the aerial mycelium.

Produces glebomycin (bluensomycin); excellent growth on Czapek's solution agar.

Comment: No other varieties of this species have been proposed.

Type strain: U-12898; NRRL 2876 (a single isolate).

444. Streptomyces coerulatus subsp. anaseuli (*sic*) (Krasil'nikov, Sorokina, Alferova and Bezzubenkova) Pridham 1970, 12. (*Actinomyces coerulatus anaseuli* (*sic*) Krasil'nikov, Sorokina, Alferova and Bezzubenkova in Krasil'nikov 1965, 88.)

subsp. an'a.seul.i. Subspecific epithet of uncertain origin.

Exhibits anti-bacterial activity; forms blue-colored vegetative mycelium and diffusible pigment on some media.

Type strain: INMI 243-13 (single isolate).

445. Streptomyces coeruleoroseus (Preobrazbenskaya) Pridham 1970, 13. (*Actinomyces coeruleoroseus* Preobrazhenskaya 1966, 857.)

coe.ru'le.o.ro'se.us. L. adj. *coeruleus* dark blue, azure; L. adj. *roseus* rose colored; M.L. adj. *coeruleoroseus* dark blue, rose colored.

Exhibits weak antibiotic activity against *Chlorella*; forms pink, brownish pink and pinkish yellow vegetative mycelium and diffusible pigment on some media.

Type strain: INA 9106.

446. Streptomyces ipomoeae (Person and Martin) Waksman and Henrici 1948, 958. (*Actinomyces ipomoea* (*sic*) Person and Martin, 1940, 923; *Streptomyces ipomoea* (*sic*) (Person and Martin) Waksman and Henrici 1948, 958.)

i.po.moe'ae. M.L. n. *Ipomoea* generic name of the sweet potato; M.L. gen.n. *ipomoeae* of *Ipomoea*, referring to the source of the organism.

Causes soft rot of sweet potatoes (*Ipomoea* sp.). Characterization, in part, in Shirling and Gottlieb, 1969, 442.

Reference strain: Martin 9820; none of Person's and Martin's original isolates is extant.

447. Streptomyces spinosus (Preobrazhenskaya) Pridham 1970, 26. (*Actinomyces spinosus* Preobrazhenskaya 1966, 856.)

spi.no'sus. L. adj. *spinosus* thorny, prickly.

Exhibits weak anti-microbial activity, if any; forms yellowish vegetative mycelium on some media; spines reportedly longer than the length of the spores.

Type strain: INA 3763.

See Table 17.46a–d.

448. Streptomyces griseomycini (Preobrazhenskaya, Blinov and Ryabova) Pridham, Hesseltine and Benedict 1958, 69. (*Actinomyces griseomycini* Preobrazhenskaya, Blinov, and Ryabova in Gauze *et al.* 1957, 136.)

gri.se.o.my.ci'ni. M.L. adj. *griseus* gray; M.L. suff. *-mycin* for antibiotic names; M.L. gen.adj. *griseomycini* of gray, antibiotic (referring to gray aerial mycelium and antibiotic activity).

Exhibits anti-bacterial activity; NaCl tolerance ≧7%, but <10%; some spore chains of atypical *Retinaculum-Apertum* type; taxon might also be placed in the Gray color series. Characterization, in part, in Shirling and Gottlieb, 1968, 124.

Type strain: INA 13984; ATCC 19765 (Gauze in Gottlieb, 1968, 20).

449. Streptomyces griseostramineus (Preobrazhenskaya, Kudrina, Blinov and Ryabova) Pridham, Hesseltine and Benedict 1958, 65. (*Actinomyces griseostramineus* Preobrazhenskaya, Kudrina, Blinov and Ryabova in Gauze *et al.* 1957, 155.)

gri.se.o.stra.mi'ne.us. M.L. adj. *griseus* gray; L. adj. *stramineus* straw-colored; M.L. adj. *griseostramineus* gray, straw-colored (referring to the gray aerial mycelium and straw-yellow vegetative mycelium on a chemically defined medium).

Exhibits anti-bacterial and anti-fungal activity; NaCl tolerance ≧7%, but <10%; spore wall ornamentation might also be considered spiny; taxon might also be placed in the Gray color series. Characterization, in part, in Shirling and Gottlieb, 1968, 126.

Type strain: INA 10381; ATCC 19768 (Gauze in Gottlieb, 1968, 20).

450. Streptomyces prasinosporus Tresner, Hayes and Backus 1966, 162.

pra.si.no'spo.rus. L. adj. *prasinus* green; M.L. n. *spora* a spore; M.L. adj. *prasinosporus* green-spored.

Forms reddish vegetative mycelium on some media; NaCl tolerance ≧7%, but <10%.

Type strain: BD-278; ATCC 17918.

451. Streptomyces ghanaensis Wallhäusser, Nesemann, Präve and Steigler 1966, 734. (*Streptomyces ghanaensis* in F. Lindner *et al.*, German Patent 1,113,791, September 14, 1961 (Not Val. Pub.); *Streptomyces ghanaensis* in F. Lindner *et al.*, Canadian Patent 672,917, October 22, 1963 (Not Val. Pub.); *Streptomyces ghanaensis* in Farbwerke Hoechst A.-G., British Patent 977,327, December

TABLE 17.46a–d

Green series[a]

Names of species and subspecies	Strain desig-nation	Utilization of carbon compounds (P & G basal)											
		No carbon control	D-Glucose	D-Xylose	L-Arabinose	L-Rhamnose	D-Fructose	D-Galactose	Raffinose	D-Mannitol	i-Inositol	Salicin	Sucrose
17.46a GN; S; C+; H.													
448. S. griseomycini	T		+	+	+	+	+		−	+	+			−
449. S. griseostramineus	T		+	+	+	+	+			+	+			−
450. S. prasinosporus	T		+	+	+	+	+			+	+	+	−
17.46b GN; S; C−; SPY													
451. S. ghanaensis	T												
452. S. hirsutus	T	−	+	+	+	+	+		+	+	+		+
453. S. prasinus	T	−	+	+	+	+	+	+	−	+	+	−	+
454. S. viridosporus	T	−	+	+	+	+	+	+	−	+	+	−	±
17.46c GN; S; C−; H													
455. S. acrimycini	T	−	+	+	−	+	+		−	+	+		−
456. S. bambergiensis	T												
457. S. prasinopilosus	T	−	+	+	+	+	+	+	−	+	+	−	
17.46d GN; ?; ?; ?													
458. S. horton	T	−	+	+	+	−	+	+		+	−		−

[a] For explanation of meanings of symbols see footnotes to Tables 17.40 and 17.41a–e.

9, 1964 (Not Val. Pub.); *Actinomyces ghanaensis,* 1964 (*sic*) in Preobrazhenskaya 1966, 858.)

ghan.a.en'sis. M.L. adj. *ghanaensis* belonging to Ghana, the source of the soil from which the organism was isolated.

Produces the moenomycin complex of antibacterial antibiotics (components A, B, B_1 and C); moderate growth on synthetic agar.

Type strain: 4092; F.H. 4092; ATCC 14672 (single isolate).

452. **Streptomyces hirsutus** Ettlinger, Corbaz and Hütter 1958, 344.

hir.sut'us. L. adj. *hirsutus* shaggy, bristly, with stiff hairs.

Non-antagonistic; fair growth on Czapek's solution agar; some spore chains of atypical *Retinaculum-Apertum* type; NaCl tolerance ≧7%, but <10%. Characterization, in part, in Shirling and Gottlieb, 1968, 134.

Type strain: ETH 16660; ATCC 19773.

453. **Streptomyces prasinus** Ettlinger, Corbaz and Hütter 1958, 343.

pra.sin'us. L. adj. *prasinus* green.

Non-antagonistic; excellent growth on Czapek's solution agar; some spore chains of atypical *Retinaculum-Apertum* type; NaCl tolerance ≧7%, but <10%. Characterization, in part, in Shirling and Gottlieb, 1968, 159.

Type strain: ETH 13815; ATCC 19800.

454. **Streptomyces viridosporus** Pridham, Hesseltine and Benedict 1958, 67. (*Streptomyces viridosporus* in Parke, Davis & Company, British Patent 712,547, July 28, 1954 (Not Val. Pub.); (*Streptomyces viriodosporus* (*sic*) in Pridham, Hesseltine and Benedict 1958, 67).)

vi.ri.do.spo'rus. L. adj. *viridis* green; M.L. n. *spora* a spore; M.L. adj. *viridosporus* green-spored.

Exhibits slight anti-bacterial activity; produces the tetraenic anti-fungal antibiotic, sistomycosin; inhibited by streptomycin; excellent growth on Czapek's solution agar; some spore chains of atypical *Retinaculum-Apertum* type.

Type strain: P-D 04889, NRRL 2414 (single isolate).

455. **Streptomyces acrimycini** (Preobraz-

henskaya, Blinov and Ryabova) Pridham, Hesseltine and Benedict 1958, 65. (*Actinomyces acrimycini* Preobrazhenskaya, Blinov and Ryabova in Gauze *et al.* 1957, 140.)

a.cri.my.ci′ni. L. adj. *acer* sharp, keen, pungent; M.L. suff. *-mycin* for antibiotic names; M.L. gen.adj. *acrimycini* of the sharp antibiotic.

Exhibits anti-bacterial and anti-fungal activity; inhibited by streptomycin; fair growth on Czapek's solution agar; some spore chains of atypical *Retinaculum-Apertum* type; NaCl tolerance ≧7%, but <10%. Characterization, in part, in Shirling and Gottlieb, 1968, 80.

Type strain: INA 7699; ATCC 19720; ATCC 19885 (Gauze in Gottlieb, 1968, 20).

456. Streptomyces bambergiensis Wallhäusser, Nesemann, Präve and Steigler 1966, 734. (*Streptomyces bambergiensis* in F. Lindner *et al.*, German Patent 1,113,791, September 14, 1961 (Not Val. Pub.); *Streptomyces bambergiensis* in F. Lindner *et al.*, Canadian Patent 672,917, October 22, 1963 (Not Val. Pub.); *Streptomyces bambergiensis* in Farbwerke Hoechst A.-G., British Patent 977,327, December 9, 1964 (Not Val. Pub.); *Actinomyces bambergiensis* 1964 (*sic*) in Preobrazhenskaya 1966, 858.)

bam.ber.gi.en′sis. M.L. adj. *bambergiensis* belonging to Bamberg, Germany; the source of the soil from which the organism was isolated.

Produces the moenomycin complex of antibacterial antibiotics (components A, B, B₁ and C); moderate growth on synthetic agar.

Type strain: 3263; F.H. 3263; ATCC 13879 (single isolate).

457. Streptomyces prasinopilosus Ettlinger, Corbaz and Hütter 1958, 345.

pra.si.no′pi.lo.sus. L. adj. *prasinus* green; L. adj. *pilosus* hairy; M.L. adj. *prasinopilosus* green-hairy.

Exhibits anti-bacterial activity; moderate growth on Czapek's solution agar; forms red to orange vegetative mycelium on some media; some spore chains of atypical *Retinaculum-Apertum* type; NaCl tolerance ≧4%, but <7%.

Type strain: ETH 13765; ATCC 19799 (single isolate).

458. Streptomyces horton (Erikson) Pridham, Hesseltine and Benedict 1958, 60. (microorganism in St. John-Brooks 1931, M.R.C. System of Bacteriology, Vol. 8; *Actinomyces horton* Erikson 1935, 36; *Streptomyces hortonensis* (*sic*) (Erikson) Waksman and Henrici 1948, 962; *Streptomyces hortensis* (*sic*) in M.R.C. Memorandum No. 21, p. 2, 1951 (Not Val. Pub.).)

hor′ton. English n. *Horton*; named for Horton War Hospital, Epsom, England, the source of the sample from which the organism was isolated.

Spore wall ornamentation category unknown; spore chains possibly straight to flexuous (section *Rectus-Flexibilis*); no melanoid pigment formation; non-antagonistic, although there is one report that the organism is bacteriostatic; poor growth on Czapek's solution agar; forms brown to gray-brown vegetative mycelium on some media; very sparse formation of aerial mycelium; whole cell hydrolysates contain L-diaminopimelic acid; originally isolated from pus containing typical actinomycotic granules from a parotid abscess. Characterization, in part, taken from Pridham and Lyons, 1969, 197.

Type strain: St. John-Brooks (Horton); NCTC 600 (single isolate).

TABLE 17.47a, b
Violet series[a]

Names of species and subspecies	Strain designation	Utilization of carbon compounds (P & G basal)											
		No carbon control	D-Glucose	D-Xylose	L-Arabinose	L-Rhamnose	D-Fructose	D-Galactose	Raffinose	D-Mannitol	i-Inositol	Salicin	Sucrose
17.47a V; RF; C−; SM													
459. S. rectiviolaceus	T	−	+	+	+	+	+	+	+	+	+		+
17.47b V; S; C+; SPY													
460. S. lilacinofulvus	T												
461. S. mauvecolor***	T		+	−	+	−	−	+	+	−	−	+	−
462. S. violans	T	−	+		+	+	+	+	+		+		+
463. S. violascens	T	−	+	+	+	−	+		+	−	±		±

[a] For explanation of meanings of symbols and asterisks, see footnotes to Tables 17.40 and 17.41a–e.

459. Streptomyces rectiviolaceus (Artamonova) Pridham 1970, 25. (*Actinomyces violaceus* var. *rectus* in Krasil'nikov and Asem Khusein in Krasil'nikov 1965, 140 (Not Val. Pub.); *Actinomyces rectiviolaceus* Artamonova in Krasil'nikov 1965, 234.)

rec'ti.vio.la'ce.us. L. adj. *rectus* straight; L. adj. *violaceus* violet colored; M.L. adj. *rectiviolaceus* straight, violet colored.

Exhibits anti-bacterial and anti-fungal activity; forms violet or red-colored vegetative mycelium on some media.

Type strain: INMI 563 (Pridham, 1970, 25).

460. Streptomyces lilacinofulvus (Yen and Chou) Pridham 1970, 40. (*Actinomyces lilacinofulvus* Yen and Chou 1964, 424.)

li.la'cin.o.ful'vus. L. adj. *lilacinus* lilac colored; L. adj. *fulvus* deep yellow; M.L. adj. *lilacinofulvus* lilac colored, deep yellow.

Exhibits anti-bacterial activity and anti-fungal activity; forms dark yellow vegetative mycelium on some media; isolated from forest soil in China.

Type strain: IMASP Y1-1 (Pridham, 1970, 40).

461. Streptomyces mauvecolor Okami and Umezawa in Murase, Hikiji, Nitta, Okami, Takeuchi and Umezawa 1960, 114.

mau've.co.lor. L. n. *malva* mallow, a plant with violet-colored petals, whence French *mauve*; L. n. *color* color; M.L. adj. *mauvecolor* mauve colored.

Produces peptimycin, a peptidic anti-tumor antibiotic; probably grows poorly on Czapek's solution agar.

Type strain: 1112-A3 (single isolate).

462. Streptomyces violans (*sic*) (Artamonova and Krasil'nikov) Pridham 1970, 29. (*Actinomyces violans* (*sic*) Artamonova and Krasil'nikov in Rautenshtein 1960, 336.)

vi.o.lans'. L. part. *violans* violating, but probably intended from L. n. *viola* the violet referring to the pink to violet color of the vegetative and aerial mycelium of the organism.

Exhibits anti-bacterial and anti-fungal activity: good growth on Czapek's solution agar; forms violet-colored vegetative mycelium and diffusible pigment on some media; utilizes sorbitol with Pridham and Gottlieb basal agar.

Type strain: INMI 167 (single isolate).

463. Streptomyces violascens (Preobrazhenskaya and Sveshnikova) Pridham, Hesseltine and Benedict 1958, 68. (*Actinomyces violascens* Preobrazhenskaya and Sveshnikova in Gauze *et al.* 1957, 41.)

vi.o.la'scens. M.L. part.adj. from assumed *violaso. violascens* becoming violet.

Exhibits slight anti-microbial activity; poor growth on Czapek's solution agar; NaCl tolerance $\geqq 7\%$, but $<10\%$. Characterization, in part, in Shirling and Gottlieb, 1968, 380.

Type strain: INA 3959/54; ATCC 23968; (Gauze in Gottlieb, 1968, 20).

Species incertae sedis

A. Type strain not extant

1. *Streptomyces acidophilus* (Jensen) Waksman and Henrici 1948, 956. No reference strains known.

2. *Streptomyces candidus* (Krasil'nikov) Waksman in Waksman and Lechevalier 1953, 94.

3. *Streptomyces cylindrosporus* (Krasil'nikov) Waksman in Waksman and Lechevalier 1953, 68.

4. *Streptomyces diastaticus* (Krainsky) Waksman and Henrici 1948, 939.

5. *Streptomyces diastatochromogenes* (Krainsky) Waksman and Henrici 1948, 941.

6. *Streptomyces erythrochromogenes* (Krainsky) Waksman and Henrici 1948, 944.

7. *Streptomyces gracilis* (Millard and Burr) Waksman in Waksman and Lechevalier 1953, 106. *Nomen dubium.*

8. *Streptomyces roseochromogenus* (*sic*) (Jensen) Waksman and Henrici 1948, 937.

9. *Streptomyces roseocitreus* Kato 1953, 209.

10. *Streptomyces rubescens* (Jarach) Waksman and Henrici 1948, 956.

11. *Streptomyces scabies* (Thaxter) Waksman and Henrici 1948, 957. Many taxonomically different reference strains available.

12. *Streptomyces somaliensis* (Brumpt) Waksman and Henrici 1948, 965.

13. *Streptomyces thermodiastaticus* (Bergey) Waksman in Waksman and Lechevalier 1953, 102.

14. *Streptomyces thermofuscus* (Waksman, Umbreit and Cordon) Waksman and Henrici 1948, 957.

15. *Streptomyces thermophilus* (Gilbert) Waksman and Henrici 1948, 956.

16. *Streptomyces viridis* (Lombardo-Pellegrino) Waksman in Waksman and Lechevalier 1953, 101. *Nomen dubium.*

B. Type strain not designated

1. *Streptomyces agglomeratus* (Yen) Pridham 1970, 32.

2. *Streptomyces ahygroscopicus* (Chiu and Wu) Pridham 1970, 32.

3. *Streptomyces alma-ataensis* (Novogrudsky) Pridham 1970, 7. Availability of original strains doubtful.

4. *Streptomyces armillatus* Mancy-Courtillet and Pinnert-Sindico 1954, 580.

5. *Streptomyces atrolaccus* (Yen) Pridham 1970, 33.

6. *Streptomyces aurantiacogriseus* (Yen) Pridham 1970, 33.

7. *Streptomyces bottropensis* Waksman 1961, 182.

8. *Streptomyces caiusiae* Dhala, Poonawalla and Bhatnagar 1957, 76.

9. *Streptomyces casei* (Bernstein and Morton) Waksman in Waksman and Lechevalier 1953, 103.

10. *Streptomyces coeruleofuscus* subsp. *actinomycini* (Maksimova and Kovsharova) Pridham 1970, 13. Type strain designation in question.

11. *Streptomyces erythreus* subsp. *speleomycini* (Sabo and Preobrazhenskaya) Pridham 1970, 15.

12. *Streptomyces flavomacrosporus* (*sic*) (Yen) Pridham 1970, 36.

13. *Streptomyces fumigatiscleroticus* Pridham 1970, 36. Color of aerial mycelium indeterminate.

14. *Streptomyces globisporus* subsp. *tundromycini* (Kovalenkova) Pridham 1970, 16.

15. *Streptomyces griseus* subsp. *macrosporus* (Yen) Pridham 1970, 38.

16. *Streptomyces griseus* subsp. *segmentosus* (Yen) Pridham 1970, 38.

17. *Streptomyces hygroscopicus* subsp. *indica* Thirumalachar, Bringi, Deshmukh and Rahalkar 1964, 25.

18. *Streptomyces iverini* (Preobrazhenskaya, Blinov and Ryabova) Pridham, Hesseltine and Benedict 1958, 69.

19. *Streptomyces lilaceus* Pridham 1970, 40.

20. *Streptomyces litmogenes* Soong and Au 1962, 40.

21. *Streptomyces longissimus* (Krasil'nikov) Waksman in Waksman and Lechevalier 1953, 87.

22. *Streptomyces luteolutescens* (Yen) Pridham 1970, 41.

23. *Streptomyces mellinus* (Maksimova, Kovsharova and Proshlyakova) Pridham 1970, 22.

24. *Streptomyces primycini* Vályi-Nagy, Uri and Szilágyi 1956, 305.

25. *Streptomyces ramnaii* Ahmad, De and Rahman 1955, 179.

26. *Streptomyces syringini* (Preobrazhenskaya and Sveshnikova) Pridham, Hesseltine and Benedict 1958, 61.

27. *Streptomyces thermoflavus* (Kudrina and Maksimova) Pridham 1970, 27.

28. *Streptomyces tian-schanicus* (Novogrudsky) Pridham 1970, 27.

29. *Streptomyces toxicus* (Krasil'nikov) Pridham 1970, 27.

30. *Streptomyces utilis* Thirumalachar and Bhatt 1960, 61.

31. *Streptomyces variabilis* (Preobrazhenskaya, Ryabova and Blinov) Pridham, Hesseltine and Benedict 1958, 70. *Nomen ambiguum.*

32. *Streptomyces variabilis* var. *roseolus* (Preobrazhenskaya, Ryabova and Blinov) Pridham, Hesseltine and Benedict 1958, 70. *Nomen ambiguum.*

33. *Streptomyces violaceus-niger* subsp. *crystallomycini* (Gauze *et al.*) Pridham 1970, 29.

34. *Streptomyces viridogriseus* Thirumalachar in Thirumalachar and Menon 1962, 108.

C. Color of aerial mycelium indeterminate

1. *Streptomyces anthocyaneus* (Vetlugina and Shigayeva) Pridham 1970, 7.

2. *Streptomyces cyaneogriseus* (Yen) Pridham 1970, 35.

3. *Streptomyces flaviscleroticus* Pridham 1970, 36.

4. *Streptomyces fradiae* subsp. *italicus* Grein, Spalla and Cotta 1965, 304.

5. *Streptomyces humifer* Pridham 1970, 39.

6. *Streptomyces mayaensis* Taguchi and Yoshikawa 1961, 44.

7. *Streptomyces microsporus* (Yen) Pridham 1970, 41.

8. *Streptomyces novaecaesareae* Waksman and Henrici 1948, 951.

9. *Streptomyces paraguayensis* (deAlmeida) Waksman 1961, 255. Morphology indeterminate.

10. *Streptomyces sclerotialus* Pridham 1970, 44.

11. *Streptomyces verne* (Waksman and Curtis) Waksman and Henrici 1948, 936.

12. *Streptomyces violens* (Kalakoutskii and Krasil'nikov) Pridham 1970, 30.

D. Nomina dubia

1. *Streptomyces alborubidus* (Kudrina) Pridham, Hesseltine and Benedict 1958, 66.

2. *Streptomyces albus* subsp. *aromaticus* (Krasil'nikov) Pridham 1970, 5. Type strain not designated.

3. *Streptomyces albus* subsp. *odoratus* (Krasil'nikov) Pridham 1970, 6. Type strain not extant.

4. *Streptomyces aureus* Waksman and Henrici 1948, 943.

5. *Streptomyces auriscleroticus* Pridham 1970, 33.

6. *Streptomyces foersteri* (Cohn) Müller 1950, 291.

7. *Streptomyces oligocarbophilus* (Beyerinck and van Delden) Foster in Werkman and Wilson 1951, 384.

E. Nomina ambigua

1. *Streptomyces bicolor* (Preobrazhenskaya) Pridham, Hesseltine and Benedict 1958, 66. Type strain not designated.

2. *Streptomyces citreus* (Krainsky) Waksman and Henrici 1948, 946.

3. *Streptomyces glaucus* (Lehmann and Schütze in Lehmann and Neumann 1907, emend. Krasil'nikov 1941) Waksman in Waksman and Lechevalier 1953, 91.

F. Miscellaneous

1. *Streptomyces capuensis* Baldacci, Farina, Locci and Ragni 1965, 59. Derived mutant strain from insufficiently characterized wild type strain.

2. *Streptomyces cellostaticus* Hamada 1958, 174. Original description not seen.

3. *Streptomyces craterifer* (Millard and Burr) Waksman in Waksman and Lechevalier 1953, 105. Not a streptomycete.

4. *Streptomyces griseoloviolaceus* (Yen) Pridham 1970, 37. May be a *nomen nudum*.

5. *Streptomyces leidynematis* Hoffman 1953, 376. Color of aerial mycelium indeterminate; never cultivated; a *nomen dubium*.

6. *Streptomyces mekemicus* Ito, Noguchi and Yasumura 1963, 1–8. Original description not seen.

7. *Streptomyces melanochromogenes* Tsai, Su, Pao, Liang, Wu, Wu, Liu, Ch'u, Hsu and Kurylowicz 1957, 717. Original description not seen.

8. *Streptomyces melanocyclus* (Merker) Waksman and Henrici 1948, 956. Type strain not designated. *Nomen dubium*.

9. *Streptomyces melanosporeus* (Krainsky) Waksman and Henrici 1948, 955. Not a streptomycete.

10. *Streptomyces rubrocyano-diastaticus* (*sic*) subsp. *atrodiastaticus* (Baldacci, Grein and Spalla) Pridham, Hesseltine and Benedict 1958, 68. No validly published species established.

11. *Streptomyces rubrocyano-diastaticus* (*sic*) subsp. *impiger* (Baldacci, Grein and Spalla) Pridham, Hesseltine and Benedict 1958, 70. No validly published species established.

12. *Streptomyces rubrocyano-diastaticus* (*sic*) subsp. *piger* (Baldacci, Grein and Spalla) Pridham, Hesseltine and Benedict 1958, 70. No validly published species established.

13. *Streptomyces shiodaensis* Katagiri and Shoji 1964, 865. Original description not seen.

14. *Streptomyces terrestris* Thirumalachar 1962, 28. Original description not seen.

15. *Streptomyces virusinus* (Kuchaeva) Pridham 1970, 31. Original description not located.

Genus II. **Streptoverticillium** *Baldacci 1958, 15, emend. mut. char. Baldacci, Farina and Locci 1966, 168*

E. Baldacci and R. Locci

(*Verticillomyces* Shinobu 1965, 92.)

Strep.to.ver.ti.cil'li.um. Gr. adj. *streptus* pliant, easily twisted; L. n. *verticillus* whorl, whirl of a spindle; M.L. neut.n. *Streptoverticillium* a whorled actinomycete.

Substrate mycelium, 1–2 μm in diameter, branching. **The aerial mycelium is characteristic; three to six short branches (1–10 μm long, average 3–5 μm) are produced in a whorl (verticil) at more or less regular intervals along the aerial mycelium giving a "barbed wire" appearance at** *ca.* **100 × magnification. These branch in turn to produce 2–12 or more secondary branches in an umbel-like manner; chains of rounded to oblong spores form at the ends of the secondary branches.** A chain may be straight, flexuous or terminate in hooks.

Reproduction occurs either from particles of substrate or aerial mycelium or from germination of spores.

Gram-positive. Cell wall Type I (Table 17.18).

On primary isolation **colonies are small and discrete, initially smooth, later developing a weft of aerial mycelium** that may appear velvety or floccose. **Substrates containing starch or other polysaccharides are especially good for production of spores;** on less suitable media spores are not formed. **Produce a wide variety of pigments** and substrate and aerial mycelium usually colored (Table 17.48).

Hydrogen sulfide *per se* seldom produced on peptone-yeast extract-iron agar (peptone-iron agar), or in traces only, when lead acetate paper is used as a detecting agent; cultivation of streptoverticillia on this medium, however, often results in the formation of bluish black to black diffusible pigment which correlates with melanin production and the production of brown, deep brown or black diffusible pigments in other organic media. Starch and gelatin usually hydrolyzed. Tyrosine is utilized by all species; xanthine is utilized by only two species. Produce a variety of compounds (antibiotics) which exhibit anti-bacterial, anti-fungal, anti-protozoal and anti-tumor activity. Sensitive to anti-bacterial agents and to actinophages.

Aerobic. Optimum growth range about 25 C to about 35 C at pH 6.5–8.0. Saprophytic soil forms.

The G + C content of the DNA of the species examined ranges from 69–73 moles %.

Type species: *Streptoverticillium baldaccii* Farina and Locci 1966, 48.

Description of the species of genus **Streptoverticillium**

The species of *Streptoverticillium* may be grouped into 12 series based on the color of the substrate and aerial mycelium (Table 17.48). Morphological properties show such range that they are of minor value in differentiation; detailed measurements are given in Table III of Locci *et al.* 1969; photomicrographs of many of the species are given in the same publication.

Biochemical characteristics are summarized in Table 17.49.

The species descriptions deal mainly with cultural properties. All characteristics are based on the study of the type strain or representative strains of each species listed. The nomen species of each series is described first and most fully; other species in that series are compared with it.

TABLE 17.48

Color characteristics of series of genus **Streptoverticillium**

Series	Color of substrate mycelium	Color of aerial mycelium	Species
I. **Baldaccii**	pink-red to orange-red	pink, gray-pink and violet-pink	1. *Stv. baldaccii* 2. *Stv. fervens* 2b. *Stv. fervens* subsp. *melrosporus* 3. *Stv. rubrochlorinum*
II. **Biverticillatum**	colorless, reddish and orange, yellow to brick red	pinkish white	4. *Stv. biverticillatum* 5. *Stv. aureoversales* 6b. *Stv. pentaticum* subsp. *jenense* 7. *Stv. roseoverticillatum* 7b. *Stv. roseoverticillatum* subsp. *albosporum* 8. *Stv. rubroverticillatum*
III. **Hiroshimense**	brick red	beige to pink-beige	9. *Stv. hiroshimense*
IV. **Salmonis**	brick-red to orange	white with pink and yellow shades	10. *Stv. salmonis*
V. **Luteoverticillatum**	yellow, yellowish to brown	light yellow, yellowish to beige	11. *Stv. luteoverticillatum* 12. *Stv. olivoreticuli* 13. *Stv. waksmanii*
VI. **Griseocarneum**	brownish yellow	pinkish beige with lilac shades	14. *Stv. griseocarneum*
VII. **Cinnamoneum**	yellow to greenish yellow and brown-yellow	pinkish with beige and lilac shades	15. *Stv. cinnamoneum* 15b. *Stv. cinnamoneum* forma *azacoluta* 15c. *Stv. cinnamoneum* subsp. *albosporum* 15d. *Stv. cinnamoneum* subsp. *lanosum* 15e. *Stv. cinnamoneum* subsp. *sparsum* 16. *Stv. hachijoense*
VIII. **Ardum**	light yellow to yellowish to pinkish yellow	basically white with yellow, pink and gray	17. *Stv. ardum* 18. *Stv. abikoense* 19. *Stv. albireticuli* 20. *Stv. eurocidicum* 21. *Stv. kishiwadense* 22. *Stv. mashuense* 23. *Stv. olivoverticillatum* 24. *Stv. orinoci* 25. *Stv. parvisporogenes*

TABLE 17.48—*Continued*

Series	Color of substrate mycelium	Color of aerial mycelium	Species
IX. **Kentuckense**	yellow, yellowish to hazel-nut yellow	beige with shades toward yellow, pink and cinnamon	26. *Stv. kentuckense* 27. *Stv. album* 28. *Stv. distallicum* 29. *Stv. ehimense* 30. *Stv. flavopersicum* 31. *Stv. griseoverticillatum* 32. *Stv. netropsis* 33. *Stv. rectiverticillatum* 34. *Stv. septatum*
X. **Mobaraense**	yellow to greenish yellow	grayish green	35. *Stv. mobaraense* 36. *Stv. blastmyceticum* 37. *Stv. lavenduligriseum*
XI. **Lilacinum**	brown	pinkish white	38. *Stv. lilacinum* 39. *Stv. kashmirense*
XII. **Thioluteum**	brown-yellow to greenish	light yellow	40. *Stv. thioluteum*

Colors are from Ostwald's Manual (1939); specific tab numbers are given in Locci *et al.* (1969).

Only strains studied are described. Other species and subspecies, not available for the study, are considered *incertae sedis* until studied; these are listed in the Addendum.

Note. PA: potato dextrose agar (Baldacci *et al.* 1954). CN: Bacto Czapek agar. CC: Casamino acids Czapek agar (1 g/l Difco vitamin-free Casamino acids, replacing sodium nitrate. GA: glucose asparagine agar (ISP medium 5 with 1% glucose replacing glycerol). GY: glycerol asparagine agar (ISP medium 5). SA: inorganic salts-starch agar (ISP medium 4). MA: yeast extract malt extract agar (ISP medium 2). EA: Bacto Emerson agar. BA: Bennett agar (1% glucose, 0.1% Bacto beef extract, 0.1% yeast extract, 0.2% peptone, 1.5% agar). NA: Oxoid nutrient agar. ISP: International Streptomyces Project, see Shirling and Gottlieb, 1966.

Series I: Baldaccii

1. Streptoverticillium baldaccii Farina and Locci 1966, 48.

PA: abundant growth; reverse pink to salmon pink; aerial mycelium pale pink to salmon pink with whitish tufts.

CN: limited growth, reverse colorless to pinkish; serial mycelium traces of pink.

CC: slightly better than on CN; colors similar.

GA: good growth; reverse red; aerial mycelium pale pink with whitish overgrowth.

GY: good growth; reverse red; aerial mycelium pale pink.

SA: good growth; reverse red to cherry red; aerial mycelium pale pink.

MA: good growth; reverse orange-red; aerial mycelium pink with whitish patches.

EA and BA: good growth; reverse orange-red; aerial mycelium pink.

NA: good growth; reverse initially orange-red turning to brown-red; aerial mycelium poor, pink with violet shades; brown soluble pigment.

Optimal growth at 27 C; only substrate mycelium formed at 37 C; no growth at 45 C.

Type strain exhibits anti-bacterial and anti-fungal activity.

Type strain: IPV 1339; ATCC 23654.

2. Streptoverticillium fervens (DeBoer, Dietz, Evans and Michaels) Locci, Baldacci and Petrolini Baldan 1969, 23. (*Streptomyces fervens* DeBoer *et al.* 1960, 220.)

Sporulation limited, sporulated umbels very rare.

On PA, GA, GY, SA, MA, and BA aerial mycelium less cottony, tending toward pink; reverse colors darker, tending toward brownish red.

On CN and CC practically no growth.

On EA only traces of aerial mycelium. On NA no aerial mycelium. Good pigmentation.

2a. Streptoverticillium fervens subsp. *fervens comb. nov.*

Description as for species.

TABLE 17.49

Biochemical properties of species and subspecies of genus Streptoverticillium[a]

	Melanin production	H₂S production	Starch hydrolysis	Casein hydrolysis	Gelatin lique-faction	Growth on carbohydrates, ISP medium 9										
						Control (no sugar)	Arabi-nose	Cellu-lose	Fruc-tose	Glucose	Inositol	Manni-tol	Raffi-nose	Rham-nose	Sucrose	Xylose
Series I. Baldaccii																
1. *Stv. baldaccii*	++	tr	++	−	++	−	−	−	−	++	(+)	tr	tr	−	++	−
2. *Stv. fervens*	++	+	++	+	++	−	−	−	tr	++	tr	tr	tr	tr	++	tr
2b. *Stv. fervens* subsp. *melrosporus*	++	−	++	(+)	++	tr	−	−	(+)	++	+	(+)	(+)	(+)	(+)	+
3. *Stv. rubrochlorinum*	++	tr	++	−	++	−	−	−	−	++	tr	tr	−	−	(+)	−
Series II. Biverticillatum																
4. *Stv. biverticillatum*	++	tr	++	(+)	(+)	tr	−	−	(+)	++	tr	−	(+)	tr	(+)	(+)
5. *Stv. aureoversales*	++	−	++	−	++	−	−	−	−	++	+	−	tr	tr	++	−
6b. *Stv. pentaticum* subsp. *jenense*	++	−	+	(+)	++	tr	−	−	+	++	tr	−	−	−	++	−
7. *Stv. roseoverticillatum*	++	−	+	−	++	tr	−	−	−	++	+	−	−	−	++	−
7b. *Stv. roseoverticillatum* subsp. *albosporeum*	++	−	+++	++	++	−	−	−	tr	++	tr	−	tr	tr	−	−
8. *Stv. rubroverticillatum*	+	−	+	−	+	−	−	−	tr	+	+	+	tr	tr	tr	tr
Series III. Hiroshimense																
9. *Stv. hiroshimense*	+	tr	+	−	+	−	−	−	−	+++	+	−	(+)	−	tr	−
Series IV. Salmonis																
10. *Stv. salmonis*	+++	−	+	−	+++	−	−	−	−	+	+	+	tr	−	tr	−
Series V. Luteoverticillatum																
11. *Stv. luteoverticillatum*	+	tr	+	(+)	+	tr	−	tr	+	+	−	+	tr	−	++	−
12. *Stv. olivoreticuli*	−	−	+++	++	+	−	tr	−	+	+++	(+)	+	tr	−	(+)	tr
13. *Stv. waksmanii*	+	−	+	−	+	tr	+	tr	+	+	(+)	+	(+)	−	+	−
Series VI. Griseocarneum																
14. *Stv. griseocarneum*	−	tr	+	−	+	−	−	−	tr	+	+	−	tr	−	tr	−
Series VII. Cinnamoneum																
15. *Stv. cinnamoneum*	−	tr	+	−	+	tr	tr	tr	tr	+	+	tr	tr	tr	+	tr
15b. *Stv. cinnamoneum* forma *azacoluta*	−	(+)	+	(+)	+	tr	−	−	(+)	+	−	−	tr	tr	(+)	tr
15c. *Stv. cinnamoneum* subsp. *albosporum*	−	tr	+	+	+	−	+	−	−	−	(+)	−	−	−	−	−
15d. *Stv. cinnamoneum* subsp. *lanosum*	−	−	+	−	+	tr	−	(+)	(+)	+	+	(+)	(+)	tr	(+)	tr
15e. *Stv. cinnamoneum* subsp. *sparsum*	−	tr	+	−	+	−	−	−	tr	+	+	−	tr	−	−	−

16. *Stv. hachijoense*
Series VIII. Ardum
17. *Stv. ardum*
18. *Stv. abikoense*
19. *Stv. albireticuli*
20. *Stv. eurocidicum*
21. *Stv. kishiwadense*
22. *Stv. mashuense*
23. *Stv. olivoverticillatum*
24. *Stv. orinoci*
25. *Stv. parvisporogenes*
Series IX. Kentuckense
26. *Stv. kentuckense*
27. *Stv. album*
28. *Stv. distallicum*
29. *Stv. ehimense*
30. *Stv. flavopersicum*
31. *Stv. griseoverticillatum*
32. *Stv. netropsis*
33. *Stv. rectiverticillatum*
34. *Stv. septatum*
Series X. Mobaraense
35. *Stv. mobaraense*
36. *Stv. blastmyceticum*
37. *Stv. lavenduligriseum*
Series XI. Lilacinum
38. *Stv. lilacinum*
39. *Stv. kashmirense*
Series XII. Thioluteum
40. *Stv. thioluteum*

[a] Degree of activity is indicated in decreasing order +, (+), tr, −. Tyrosine crystals disappear from the medium with all species and varieties. Xanthine crystals disappear from the medium only with *Stv. kentuckense* (+) and *Stv. flavopersicum* +.

The type strain produces fervenulin and exhibits anti-fungal activity.

Type strain: UC 2293; NRRL 2755; IPV 2021.

2b. *Streptoverticillium fervens* subsp. *melrosporus* Mason, Lummis and Dietz 1965, 111.

Poor sporulation. Reverse colors dark with shades of red and brown red. On GA, GY, SA and MA aerial mycelium tending toward gray. On EA aerial mycelium tends toward lilac. Soluble pigment formation strong in nutrient broth.

Starch, casein and gelatin are hydrolyzed more strongly than the primary subspecies.

Good growth at 37 C, only traces of aerial mycelium formed; no growth at 45 C. Soluble pigment production stronger at 37 C.

Type strain produces melrosporus and exhibits anti-fungal activity.

Type strain: UC 2459; NRRL 3117; IPV 2022.

3. **Streptoverticillium rubrochlorinum** Locci, Baldacci and Petrolini Baldan 1969, 22. (*Actinomyces* 51-10, Konev 1964, 628; *Streptoverticillium* (*Actinomyces*) *rubrochlorinum* 51 Severinets and Kotenko 1965, 154.)

Sporulation less abundant and aerial mycelium less cottony and abundant than in *S. baldaccii*; colors tend toward violet-pink. Reverse colors darker, tending toward brownish red. These differences most noticeable on PA, EA, BA and NA. Soluble pigment stronger on NA. Practically no growth on CA and CC.

Traces of aerial growth produced at 37 C; no growth at 45 C.

The type strain produces tetraene 51-10, pentaene 51-10 and heptaene 51.10, and exhibits anti-bacterial activity.

Type strain: 51-10; LIA 0084; IPV 2007.

Series II. Biverticillatum

4. **Streptoverticillium biverticillatum** (Preobrazhenskaya) Farina and Locci 1966, 49. (*Actinomyces biverticillatus* Preobrazhenskaya in Gauze *et al.* 1957, 75; *Streptomyces biverticillatus* (Preobrazhenskaya) Pridham, Hesseltine and Benedict 1958, 72; *Streptoverticillium biverticillatus* (sic) (Preobrazhenskaya) Baldacci 1958, 25.)

PA: medium growth, initially colorless then reddish brown; aerial mycelium whitish with pink shades.

CN: traces of growth.

CC: poor growth; reverse colorless to dark red; traces of pinkish white aerial mycelium.

GA: good growth; reverse yellowish to reddish; aerial mycelium yellowish pink.

GY: good growth; reverse yellowish then darkening; aerial mycelium yellowish pink.

SA: good growth; reverse yellowish; aerial mycelium yellowish pale pink.

EA: good growth; reverse brownish red; aerial mycelium pink; soluble pigment brown.

MA: good growth; reverse reddish brown; aerial mycelium pinkish white.

BA: good growth; reverse brick red; aerial mycelium pink.

NA: medium growth; reverse orange-yellow; aerial mycelium pinkish white with raspberry pink patches; soluble pigment brown.

Good growth at 27 and 37 C, pinkish shades being stronger at the higher temperature; no growth at 45 C.

Reference strain: Fabian; CBS 211.62; IPV 1594.

5. **Streptoverticillium aureoversales** Locci, Baldacci and Petrolini Baldan 1969, 24. (The name *Actinomyces aureoversales* was used in a thesis (Yuan Chi-Shen, 1962, 188), which was abstracted but has not been validly published.)

On PA and NA reverse colors brown; traces of whitish aerial mycelium on PA, more abundant on NA.

Growth fair on CN and CC, the reverse being dirty pinkish.

On CC, GA, GY, SA and MA aerial mycelium is pinkish. On SA and GA reverse color is red.

Grows equally well at 27 and 37 C, aerial mycelium production slightly less at 37 C and slow in appearing.

The reference strain produces tetraene 380 and pentaene 380 and exhibits anti-bacterial activity.

Reference strain: INMI 380; IPV 2035; ATCC 15853.

6. **Streptoverticillium pentaticum** (see *Species incertae sedis*).

6b. *Streptoverticillium pentaticum* subsp. *jenense* (Fügner and Bradler) Locci, Baldacci and Petrolini Baldan 1969, 25. (*Streptomyces pentaticus* var. *jenensis* Fügner and Bradler 1963, 184.)

Reverse color definite red, particularly on CA, GY, SA and NA.

Aerial mycelium pale pink on EA, slightly deeper on SA and more accentuated with violet shades on NA; very light on GY. Traces of soluble pigment on PA, MA and BA.

Growth less abundant at 37 than at 27 C but aerial mycelium present.

Type strain produces fervenulin, fungichromin and prodigiosin.

Type strain: IMET JA4495; IPV 2002.

7. **Streptoverticillium roseoverticillatum** (Shinobu) Farina and Locci 1966, 49. (*Streptomyces roseoverticillatus* Shinobu 1956, 92; *Verti-*

cillomyces roseoverticillatus (Shinobu) Shinobu 1965, 152.)

Sporulation delayed and less abundant. Red color of reverse predominates over yellow on PA, GA, GY and SA. No growth present after 16 days at 37 C.

The type strain exhibits anti-bacterial and anti-fungal activity.

Type strain: Shinobu (OEU) 462; IPV 2003; ATCC 19807.

7a. *Streptoverticillium roseoverticillatum* subsp. *roseoverticillatum comb. nov.*

Description as for species.

7b. *Streptoverticillium roseoverticillatum* subsp. *albosporum* (Thirumalachar) Locci, Baldacci and Petrolini Baldan 1969, 25. (*Streptomyces roseoverticillatus* subsp. *albospora* Thirumalachar in Thirumalachar, Bringi, Deshmukh, Rahalkar, Indira and Gopalkrishnan 1964, 20.)

Characteristic red color of reverse is stronger, turning from orange-red to violet-red, dark red and brown. Aerial mycelium with stronger pink shades on MA and PA, flat and sparse (orange-red) on EA and NA. Growth poorer and devoid of aerial mycelium on BA and CB.

Type strain produces streptorubin B.

Type strain: 134; HACC 227; IPV 2069; ATCC 25189.

8. **Streptoverticillium rubroverticillatum** (Yen) Locci, Baldacci and Petrolini Baldan 1969, 26. (*Actinomyces rubroverticillatus* Yen 1956, 78.)

Spore chains straight, rarely slightly flexuous, single lateral chains and umbels also present. Reverse colors deeper on MA and NA; deeper pink color of aerial mycelium on PA. GA, SA and MA and paler on EA. Growth colorless in NB and CB. Soluble pigment formation much stronger on EA and NA. Growth and aerial mycelium poor at 37 C.

Reference strain: INA 3517; IPV 2052.

Series III. Hiroshimense

9. **Streptoverticillium hiroshimense** (Shinobu) Farina and Locci 1966, 49. (*Streptomyces hiroshimensis* Shinobu 1955, 46; *Verticillomyces hiroshimensis* (Shinobu) Shinobu 1965, 118.)

Growth on PA moderate; reverse brownish; aerial mycelium beige pinkish.

CN: traces of off-white growth.

CC: poor growth; reverse colorless; aerial mycelium sparse and pink; slight pigmentation of the medium.

GA: good growth; reverse initially yellowish then brown red, pinkish beige aerial mycelium.

GY: good growth; reverse dark red; aerial mycelium from white to pale pink.

SA: good growth; reverse raspberry pink; good aerial mycelium formation; beige-pink.

MA: good growth; reverse from dirty red to brown-red; aerial mycelium beige-pink.

EA: good growth; reverse from brick red to brown; beige pink aerial mycelium; traces of soluble pigment.

BA: good growth; reverse dirty red; aerial mycelium pink.

NA: poor growth; brown reverse; aerial mycelium light beige; traces of soluble pigment.

NB: yellowish and CB colorless growth; no aerial mycelium.

At 37 C growth and production of aerial mycelium are good, although less than at 27 C; no growth at 45 C.

Some strains do not produce H₂S, but utilize casein and fail to form aerial mycelium at 37 C.

The type strain exhibits anti-bacterial and anti-fungal activity.

Type strain: OEU 201; ATCC 19772; IPV 2015.

Series IV. Salmonis

10. **Streptoverticillium salmonis** (*sic*) (Baldacci, Farina and Locci) Locci, Baldacci and Petrolini Baldan 1969, 27. (Not *Streptomyces salmonicida* Rucker 1949, 661; *Verticillomyces salmonicida* (Rucker) Shinobu 1965, 157; *Streptoverticillium salmonicida* (Rucker) Baldacci, Farina and Locci 1966, 164.)

Good growth on PA, reverse brick red; aerial mycelium yellowish white; traces of soluble pigment.

CN: traces of growth only.

CC: very poor, colorless growth.

GA: good growth; orange-yellow reverse; aerial mycelium white with traces of pink and yellow.

GY: good growth; reverse orange-yellow; aerial mycelium whitish with shades of yellow.

MA: good growth; reverse brick red; aerial mycelium yellowish white.

EA: good growth; reverse brick red; aerial mycelium dirty white; brown soluble pigment.

BA: good growth; reverse dirty red; aerial mycelium dirty white.

NA: medium growth; reverse brown; aerial mycelium poor pink; brown soluble pigment.

NB: yellowish growth; brown soluble pigment.

CB: colorless growth.

The strain grows also at 37 C, very poor or no aerial mycelium being formed. No growth at 45 C.

The type strain exhibits anti-bacterial and anti-fungal activity.

Type strain: Lederle A-7604E; NRRL B-1472; IPV 2019.

Series V. Luteoverticillatum

11. Streptoverticillium luteoverticillatum (Shinobu) Locci, Baldacci and Petrolini Baldan 1969, 28. (*Streptomyces luteoverticillatus* Shinobu 1956, 93; *Verticillomyces luteoverticillatus* (Shinobu) Shinobu 1965, 133.)

Good growth on PA; reverse from yellowish to dirty yellowish brown; aerial mycelium dirty beige.

CN: traces of colorless growth.

CC: good growth; reverse from light yellow to dirty yellow; aerial mycelium whitish.

GA: good growth; reverse yellowish; aerial mycelium beige.

GY: good growth; reverse brown-yellow; aerial mycelium gray-white with beige shades.

SA: good growth; reverse yellowish; aerial mycelium beige to beige-pink.

MA: good growth; brown reverse; aerial mycelium white to yellowish beige; soluble pigment brown.

EA: good growth; reverse brown; aerial mycelium light yellowish to dirty beige.

BA: good growth; brown reverse; aerial mycelium light yellowish.

NA: poor growth; reverse brown; aerial mycelium poor, yellowish.

NB: dirty yellowish growth.

CB: whitish growth.

The species is able to grow at 37 C, although less abundantly than at 27 C, no aerial mycelium is produced. No growth at 45 C.

The type strain produces neutramycin and exhibits anti-fungal activity.

Type strain: OEU 486; ATCC 23933; IPV 2001; IPV 2025.

12. Streptoverticillium olivoreticuli (Arai, Nakada and Suzuki) Baldacci, Farina and Locci 1966, 162. (*Streptomyces olivoreticuli* Arai, Nakada and Suzuki 1957, 441.)

Reverse colors much lighter.

Aerial mycelium reduced, absent or in traces on CN, CC, EA and NA. Tends towards whitish gray on other media. On broths only suspended colorless to yellowish growth present. Soluble pigment formed on NA. No growth at 37 C after 15 days.

The type strain produces the viomycin complex and heptaene 100.

Type strain: 100; IFM 1018; ATCC 23943; IPV 2056.

13. Streptoverticillium waksmanii (Waksman) Baldacci, Farina and Locci 1966, 166. (*Actinomyces reticulus-ruber* Waksman 1919, 146; *Streptomyces rubrireticuli* (Waksman) Waksman and Henrici 1948, 945; *Streptoverticillium rubri-*

reticuli (Waksman and Henrici) Baldacci 1958, 25; *Verticillomyces rubrireticuli* (Waksman and Henrici) Shinobu 1965, 155.)

In addition to typical umbellate forms, irregular structures and transition forms also observed. Spore chains straight. Reverse colors slightly less strong. Aerial mycelium development generally less abundant (except on SA), tending toward beige shades on MA, EA, BA and NA. Growth better on CN; surface pellicle formed on CB. Soluble pigment formation on BA and NA. Poor growth at 37 C, no aerial mycelium.

The type strain produces antibiotic F-20.

Type strain: F-20; IMRU 3631; ATCC 12629; CBS 373.58; IPV 1725; IPV 2012.

Series VI. Griseocarneum

14. Streptoverticillium griseocarneum (Benedict, Stodola, Shotwell, Borud and Lindenfelser) Baldacci, Farina and Locci 1966, 158. (*Streptomyces griseocarneus* Benedict, Stodola, Shotwell, Borud and Lindenfelser 1950, 77; *Streptoverticillium griseocarneus* (sic) (Benedict, Stodola, Shotwell, Borud and Lindenfelser) Baldacci 1958, 25; *Verticillomyces griseocarneus* (Benedict, Stodola, Shotwell, Borud and Lindenfelser) Shinobu 1965, 113.)

Good growth on PA, reverse yellowish; aerial mycelium beige-pink with lilac shades; traces of soluble pigment.

CN and CC: very poor growth; isolated colorless colonies.

GA: poor growth; reverse brown yellowish; aerial mycelium poor, beige; brown soluble pigment.

GY: reverse brownish; aerial mycelium beige; traces of soluble pigment.

SA: good growth; reverse yellowish to light brown; aerial mycelium pinkish beige.

MA: good growth; reverse yellowish to brownish; aerial mycelium poor, pink with slight lilac shades.

EA: good growth; reverse dirty yellow; aerial mycelium poor, whitish.

BA: good growth; reverse brownish; beige aerial mycelium.

NA: medium growth; reverse brownish; poor, whitish aerial mycelium; brown soluble pigment.

CB: colorless growth, white pellicle.

The strain grows very poorly at 37 C. No growth at 45 C.

Some strains do not produce H_2S.

The type strain produces hydroxystreptomycin and rotaventin.

Type strain: A-637; NRRL B-1068; IPV 1959; ATCC 12628.

Series VII. Cinnamoneum

15. Streptoverticillium cinnamoneum (Benedict, Dvonch, Shotwell, Pridham and Lindenfelser) Baldacci, Farina and Locci 1966, 158. (*Streptomyces cinnamoneus* Benedict, Dvonch, Shotwell, Pridham and Lindenfelser 1952, 591; *Streptomyces cinnamomeus* forma *cinnamomeus* Pridham, Shotwell, Stodola, Lindenfelser, Benedict and Jackson 1956, 576; *Streptoverticillium cinnamomeus* forma *cinnamomeus* (Pridham *et al.*) Baldacci 1958, 25; *Verticillomyces cinnamomeus* forma *cinnamomeus* (Benedict *et al.*) Shinobu 1965, 104.)

Abundant growth on PA, reverse brown; aerial mycelium lilac-pink.

CN: limited growth; reverse colorless; aerial mycelium pale pink.

CC: limited growth; reverse colorless; aerial mycelium pink.

GA: good growth; reverse dirty yellow; aerial mycelium light pink and then hazel-nut brown; abundant exudate.

GY: good growth; reverse yellowish; aerial mycelium pinkish beige with light pink tufts.

SA: good growth; reverse yellowish to brown-yellow with greenish shades; aerial mycelium beige.

MA: good growth; reverse yellowish; aerial mycelium pinkish beige; exudate present.

EA: good growth; reverse yellow; poor aerial mycelium, whitish.

BA: good growth; reverse greenish yellow; aerial mycelium pinkish.

NA: poor growth; reverse colorless; aerial mycelium poor, beige.

CB: colorless growth; pink white pellicle.

NB: yellowish growth.

Grows at 37 C, less abundantly than at 27 C, and the aerial mycelium is whitish. No growth at 45 C.

15a. *Streptoverticillium cinnamoneum* subsp. *cinnamoneum* comb. nov.

Description as for species.

The type strain produces cinnamycin, a non-mobile factor and fungichromin.

Type strain: A-725; NRRL B-1285; ATCC 11874; CBS 293.64; IPV 2013.

15b. *Streptoverticillium cinnamoneum* forma *azacoluta* (Pridham, Shotwell, Stodola, Lindenfelser, Benedict and Jackson) Locci, Baldacci and Petrolini Baldan 1969, 32. (*Streptomyces cinnamomeus* forma *azacoluta* Pridham, Shotwell, Stodola, Lindenfelser, Benedict and Jackson 1956, 577; *Streptoverticillium cinnamomeus* forma *azacoluta* (Pridham, Shotwell, Stodola, Lindenfelser, Benedict and Jackson) Baldacci 1958, 25; *Verticillo-myces cinnamomeus* forma *azacoluta* (Pridham, Shotwell, Stodola, Lindenfelser, Benedict and Jackson) Shinobu 1965, 102.)

Reverse and aerial mycelium colors weaker (PA, CN and CC) in strain IPV 1978.

Some strains do not form H_2S.

The type strain produces duramycin, azacolutin A, azacolutin B and possibly one or more components of the phleomycin complex.

Type strain: S-205; NRRL B-1699; IPV 1978; ATCC 12686; CBS 369.58.

15c. *Streptoverticillium cinnamoneum* subsp. *albosporum* Thirumalachar in Rahalkar and Thirumalachar 1968, 96.

Culturally the subspecies shows a weakening of the reverse (except on PA) and aerial mycelium colors. Growth absent or very poor on CN and BA. Aerial mycelium scarce on CC, good on EA.

Carbohydrate utilization very limited: only inositol utilized. Growth good at 37 C and 27 C; aerial mycelium slightly inferior and lighter in color at 37 C.

The type strain produces pentaene HA-145.

Type strain HA-145; HACC 204; IPV 2066; ATCC 25186.

15d. *Streptoverticillium cinnamoneum* subsp. *lanosum* Thirumalachar in Radalkar and Thirumalachar 1968, 96.

Color of aerial mycelium weaker (e.g. on SA, PA, GA). Growth good on CN with aerial mycelium present. Aerial mycelium more abundant on EA and NA. Growth at 37 C and 27 C equally good; aerial mycelium poorer and lighter in color.

The type strain produces pentaene HA-176.

Type strain: HA-176; HACC 205; IPV 2067; ATCC 25187.

15e. *Streptoverticillium cinnamoneum* subsp. *sparsum* Thirumalachar in Rahalkar and Thirumalachar 1968, 96.

No growth on CC and traces only on CN and on BA. Substrate mycelium lighter on PA, with pinkish beige aerial mycelium. Growth slightly more abundant on NA and NB (surface pellicle). Aerial mycelium poorer on SA. Sucrose, xylose, mannitol, rhamnose, raffinose and cellulose utilized. Growth slightly better at 37 C than in strain IPV 2013.

The type strain produces pentaene HA-135.

Type strain: HA-106; HACC 203; IPV 2068; ATCC 25185.

16. Streptoverticillium hachijoense (Hosoya, Komatsu, Soeda and Sonoda) Locci, Baldacci and Petrolini Baldan 1969, 34. (*Streptomyces hachijoensis* Hosoya, Komatsu, Soeda and Sonoda 1952, 505.)

Slightly flexuous spore chains also present. True monoverticillate forms also present. Reverse colors light yellow. On GA pinkish white aerial mycelium. White surface growth present on NB. Aerial mycelium present at 37 C, but lighter in color than at 27 C.

The type strain exhibits anti-bacterial activity and produces trichomycins A, B and C.

Type strain: IAM H-2609; IPV 2014; ATCC 19769.

Series VIII. Ardum

17. Streptoverticillium ardum (DeBoer, Dietz, Lummis and Savage) Locci, Baldacci and Petrolini Baldan 1969, 34. (*Streptomyces ardus* DeBoer, Dietz, Lummis and Savage 1961, 17.)

Good growth on PA; reverse yellow with greenish shades to dirty yellow; aerial mycelium white to gray-white.

CN: traces of colorless growth.

CC: poor growth; colorless; aerial mycelium traces.

GA: medium growth; yellowish; poor aerial mycelium white.

GY: medium growth; reverse almost colorless with beige to yellowish patches; aerial mycelium whitish.

SA: medium growth; colorless with some tendency toward yellowish; poor, white-gray aerial mycelium.

MA: good growth, reverse from colorless to yellowish brown; traces of aerial mycelium after 30 days.

EA: good growth; reverse colorless to hazel-nut yellowish; aerial mycelium white.

BA: good growth; reverse from colorless to yellowish; aerial mycelium poor off-white.

NA: poor growth; reverse from colorless to yellowish; aerial mycelium poor, white.

On PI good white aerial mycelium is formed, differentially from other species.

NB: yellow growth; traces of soluble pigment.

CB: colorless growth.

Only substrate mycelium is formed at 37 C and growth is also inferior. No growth at 45 C.

The type strain produces porfiromycin and exhibits anti-fungal activity.

Type strain: UC 2500; NRRL 2817; IPV 2020.

18. Streptoverticillium abikoense (Umezawa, Tazaki and Fukuyama) Locci, Baldacci and Petrolini Baldan 1969, 35. (*Streptomyces abikoensis* Umezawa, Tazaki and Fukuyama 1951, 333; *Verticillomyces abikoensis* (Umezawa, Tazaki and Fukuyama) Shinobu 1965, 96.)

Spore chains straight. Reverse colors tend to appear darker. Aerial mycelium can be beige, pale pink to pinkish white (SA and MA); more

abundant on PA, SA, MA and GY. Soluble pigments present on EA. It utilizes starch, casein and more readily gelatine. Growth at 37 C equals that at 27 C, however aerial mycelium is less abundant and lighter in color. Brown soluble pigments at 37 C on PA.

The type strain produces abikoviromycin and exhibits polyenic anti-fungal activity.

Type strain: Z-1-6; ATCC 12766; CBS 487.62; IPV 2027.

19. Streptoverticillium albireticuli (Nakazawa) Locci, Baldacci and Petrolini Baldan 1969, 37. (*Streptomyces albireticuli* Nakazawa 1955, 649; *Verticillomyces albireticuli* (Nakazawa) Shinobu 1965, 98.)

True biverticillate structures also observed. Reverse darkening on GA, GY, BA, MA, EA and NA. Aerial mycelium absent on EA, NA and PI; in general aerial mycelium formation more abundant. Colors of aerial mycelium tend toward beige and pink-beige on MA, BA and SA. Soluble pigments on MA, BA and NA, traces on PA and EA. Pigmentation on NB stronger than in *S. ardum*.

The type strain produces enteromycin, carbomycin and eurocidin.

Type strain: Nakazawa 3724; IPV 2055; ATCC 19721.

20. Streptoverticillium eurocidicum (Okami, Utahara, Nakamura and Umezawa) Locci, Baldacci and Petrolini Baldan 1969, 36. (*Streptomyces eurocidicus* Okami, Utahara, Nakamura and Umezawa 1954, 102; *Verticillomyces eurocidicus* (Okami, Utahara, Nakamura, and Umezawa) Shinobu 1965, 111.)

Typical morphological structures, however true single monoverticillate forms also present. Reverse colors show stronger shades, definitely greenish on some media (SA). Aerial mycelium colors generally white to yellowish (SA) or to beige (PA and GA). Aerial mycelium abundant on MA and BA, and on broths, but absent on EA, NA and PI after 15 days. Soluble pigments on MA, BA, EA, NA and PA (traces only). It utilizes starch and casein; more poorly inositol, fructose and raffinose. No growth at 37 C after 15 days.

The type strain produces azomycin, tertiomycin A, tertiomycin B and eurocidin.

Type strain: 549-A-1; NRRL B-1676; IPV 1996.

21. Streptoverticillium kishiwadense (Shinobu and Kayamura) Locci, Baldacci and Petrolini Baldan 1969, 35. (*Streptomyces kishiwadensis* Shinobu and Kayamura 1964, 176; *Verticillomyces kishiwadensis* (Shinobu and Kayamura) Shinobu 1965, 122.)

Typical morphology but poorer sporulation.

Length of primary branches shorter (3–6 μm), spore chains straight and shorter. Stronger reverse colors on all media (particularly on CN, GY and MA). Aerial mycelium colors whitish with beige shades on MA and pinkish beige tones on GY, SA, MA and PA. Aerial mycelium absent on EA, NA and PI. It utilizes starch and casein and less abundantly inositol and raffinose. Arabinose scarcely utilized. Growth and aerial mycelium formation as good at 37 C as at 27 C, with slight color differences.

There are no reports of antibiotic activity for the type strain.

Type strain: OEU 738; IPV 2026; ATCC 25464.

22. Streptoverticillium mashuense (Sawazaki, Suzuki, Nakamura, Kawasaki, Yamashita, Isono, Anzai, Serizawa and Sekiyama) Locci, Baldacci and Petrolini Baldan 1969, 36. (*Streptomyces mashuensis* Sawazaki, Suzuki, Nakamura, Kawasaki, Yamashita, Isono, Anzai, Serizawa and Sekiyama 1955, 44; *Verticillomyces mashuensis* (Sawazaki, Suzuki, Nakamura, Kawasaki, Yamashita, Isono, Anzai, Serizawa and Sekiyama) Shinobu 1965, 137.)

Typical umbellate monoverticillate forms, compact structures with more crowded verticils along the main axis. Spore chains straight. Strain characterized by a better sporulation and growth on Czapek media. Aerial mycelium not present on NA and PI after 15 days. Reverse colors tending toward yellow (PA, CN and GA) and greenish yellow (CC and GY). Aerial mycelium generally more abundant than in *S. abikoense*. Aerial mycelium on EA. Inositol utilization poor. Growth as good at 37 C as at 27 C. Aerial mycelium color stronger at 37 C. Grows also at 45 C, with only traces of aerial mycelium.

The type strain produces streptomycin and exhibts polyenic anti-fungal activity.

Type strain: IPCR 449; ATCC 23934; CBS 279.65; IPV 1986.

23. Streptoverticillium olivoverticillatum (Shinobu) Baldacci, Farina and Locci 1966, 163. (*Streptomyces olivoverticillatus* Shinobu 1956, 91; *Verticillomyces olivoverticillatus* (Shinobu) Shinobu 1965, 146.)

Primary branches shorter and spore chains usually straight. Reverse colors tend to yellow and brown (SA, MA, BA and EA). Aerial mycelium not present on NA, EA and PI after 15 days. No soluble pigment on NB (15 days). Inositol and raffinose utilization poor, negative for fructose.

The type strain exhibits anti-bacterial and anti-fungal activity.

Type strain: OEU 383; IPV 2009; ATCC 25480.

24. Streptoverticillium orinoci Cassinelli, Grein, Orezzi, Pennella and Sanfilippo 1967, 358.

Typical umbellate monoverticillate forms; spore chains usually straight. Reverse color strong yellow on PA, GY, MA, EA and BA. Poor growth on NA and PI. Aerial mycelium whitish to gray-white on BA. Growth good also at 37 C with poorer aerial mycelium production than at 27 C.

The type strain produces ochramycin, neoantimycin and neoaureothin.

Type strain: FI 1882; IPV 1901; ATCC 23202.

25. Streptoverticillium parvisporogenes (ignotus 1960) Locci, Baldacci and Petrolini Baldan 1969, 37. (*Streptomyces parvisporogenes* ignotus 1960, Brit. Pat. 832,391.)

Number of spore chains per umbel higher, true biverticillate structures present. A swelling of the main axis at the insertion point of the primary branches can be observed. Reverse colors tend to darken on all media. Aerial mycelium very poor or absent on EA, NA and PI; white tending toward gray-white and yellowish white, sometimes with pink-yellow shades (SA). EA, MA and PA (traces only) pigmented. The strain hydrolyzes starch and utilizes tyrosine. Inositol utilization poor.

The type strain produces antibiotic PA-150 (compound 616) and exhibits anti-bacterial activity.

Type strain: BA-3572; ATCC 12568; IPV 1972.

Series IX. Kentuckense

26. Streptoverticillium kentuckense (Barr and Carman) Baldacci, Farina and Locci 1966, 160. (*Streptomyces kentuckensis* Barr and Carman 1956, 286; *Verticillomyces kentuckensis* (Barr and Carman) Shinobu 1965, 120.)

Good growth on PA, reverse yellow to yellowish brown; aerial mycelium pinkish beige; traces of soluble pigment.

CN and CC: poor, colorless growth; pinkish aerial mycelium.

GA: good growth; from yellowish to dirty yellow; aerial mycelium pinkish beige; traces of soluble pigment, brown.

GY: medium growth; reverse from ivory to dark brown; aerial mycelium medium; initially white then light beige.

SA: good growth; reverse from yellowish to dirty yellow; aerial mycelium initially white then pinkish beige.

MA: good growth, reverse brownish yellow; aerial mycelium initially white than pinkish beige; traces of soluble pigment.

EA: good growth; reverse brownish yellow; aerial mycelium white to pinkish beige.

BA: good growth; reverse brown-yellow; aerial mycelium initially white-beige than pinkish beige; brown soluble pigment.

NA: poor growth; reverse initially yellow with greenish shades then yellowish; aerial mycelium poor, off-white; brown soluble pigment.

CB: colorless growth; whitish aerial mycelium.

NB: yellowish growth with whitish aerial mycelium.

Growth at 37 C; no aerial mycelium, brownish red soluble pigment. No growth at 45 C.

Some cultures do not utilize casein or raffinose. The type strain produces raisnomycin.

Type strain: 5X1 RCV; ATCC 12691; IPV 1958.

27. Streptoverticillium album Locci, Baldacci and Petrolini Baldan 1969, 40.

Spore chains usually shorter. Reverse colors varying from yellow to hazel-nut yellow. Aerial mycelium colors lighter and tending toward pinkish beige on GA, GY, SA and MA. No soluble pigment. Growth as good at 37 C as at 27 C. Aerial mycelium better at 37 C.

The type strain produces acetopyrrothine (thiolutin).

Type strain: NRRL 2401; IPV 1993.

28. Streptoverticillium distallicum Locci, Baldacci and Petrolini Balden 1969, 42. (*Streptomyces distallicus* Arcamone, Bizioli, Canevazzi and Grein 1959 Pat. 1,039,198; Not Val. Pub.)

Spore chains straight to flexuous and ending in hooks. Reverse colors darker, after 15 days, and then tend to equal those of *S. kentuckense*. No great differences in aerial mycelium colors. Aerial mycelium poorer on EA, absent on liquid media. Growth at 37 C inferior to that at 27 C.

Type strain produces distamycins A, B and C and mycolutein.

Type strain: FI 1096/13; NRRL 2886; IPV 1983.

29. Streptoverticillium ehimense (Shibata, Honso, Tokui and Nakazawa) Locci, Baldacci and Petrolini Baldan 1969, 40. (*Streptomyces ehimensis* Shibata, Honso, Tokui and Nakazawa 1954, 168; *Verticillomyces ehimensis* (Shibata, Honso, Tokui and Nakazawa) Shinobu 1965, 109.)

Species characterized by longer primary branches (up to 15 μm); true biverticulate forms can also be observed. Good growth on Czapek media. Aerial mycelium slightly lighter in color with pinkish (PA, SA and BA) and beige (EA) shades. Soluble pigments present on EA, NA, traces on PA and BA. Raffinose utilized by some cultures. Poor growth at 37 C; aerial mycelium less abundant and lighter in color.

The type strain produces candimycin and exhibits anti-bacterial activity.

Type strain: 138; IPV 1995; ATCC 23903.

30. Streptoverticillium flavopersicum (Oliver, Goldstein, Bower, Holper and Otto) Locci, Baldacci and Petrolini Baldan 1969, 41. (*Streptomyces flavopersicus* Oliver, Goldstein, Bower, Holper and Otto 1961, 495.)

A few pseudoverticillate structures, consisting of long spore chains ending in hooks, also present. Reverse color lighter on BA. Aerial mycelium pinkish beige with various shades. Traces of yellowish soluble pigments present on GY and EA. Poor growth at 37 C.

The type strain produces actinospectacin and exhibits anti-fungal activity.

Type strain: M-141; NRRL 2820; IPV 2010; ATCC 19756.

31. Streptoverticillium griseoverticillatum (Shinobu and Shimada) Locci, Baldacci and Petrolini Baldan 1969, 39. (*Streptomyces griseoverticillatus* Shinobu and Shimada 1962, 170; *Verticillomyces griseoverticillatus* (Shinobu and Shimada) Shinobu 1965, 115.)

Shorter spore chains. Reverse usually lighter in color (PA, GA, MA, BA and NA) dirty yellow on PA. Aerial mycelium tending toward dirty pink (PA and EA) and beige-pink (MA). No soluble pigments. Poor aerial mycelium formation at 37 C.

The type strain produces takacidin.

Type strain: OEU 722; IPV 1976.

32. Streptoverticillium netropsis (Finlay, Hochstein, Sobin and Murphy) Baldacci, Farina and Locci 1966, 161. (*Streptomyces netropsis* Finlay, Hochstein, Sobin and Murphy 1951, 341; *Verticillomyces netropsis* (Finlay, Hochstein, Sobin and Murphy) Shinobu 1965, 144.)

Spore chains usually terminate in hooks. On PA, SA, BA, GA, GY and MA the reverse show reddish shades. Aerial mycelium tends toward pinkish on PA, GY, MA and SA. Soluble pigments also present on EA and GY (poor), reddish brown on GA and BA. Poor growth at 37 C.

The type strain produces netropsin and exhibits anti-fungal activity.

Type strain: 2937-6; FD 4779; NRRL 2268; ATCC 23940; IPV 1720.

33. Streptoverticillium rectiverticillatum (Krasil'nikov and Yuan) Locci, Baldacci and Petrolini Baldan 1969, 41. (*Actinomyces rectiverticillatus* Krasil'nikov and Yuan 1965, 49.)

Sporulation poorer, spore chains shorter. Some simple monoverticillate structures present. Growth good on Czapek media. Reverse colors usually lighter. Aerial mycelium from off-white with pinkish shades (PA, BA, GY and EA) to light pink (CC). Soluble pigments on nutrient media. Growth at 37 C poorer than at 27 C, traces of aerial mycelium.

The type strain exhibits anti-bacterial and anti-fungal activity.

Type strain: INMI 380; ATCC 19845; IPV 1852.

34. Streptoverticillium septatum Prokop 1964, 434.

Reverse colors lighter, from yellowish (GY) to hazel-nut yellow (PA, GA, MA, EA and BA). Aerial mycelium lighter in color on GA, poorer on BA and EA. Soluble pigment present on NB. Poorer growth at 37 C with traces only of aerial mycelium.

The type strain produces antibiotic M-741.

Type strain: M-741; NRRL 2974; IPV 2047.

Series X. Mobaraense

35. Streptoverticillium mobaraense (Kubo, Suzuki and Tamura) Locci, Baldacci and Petrolini Baldan 1969, 42. (Streptomyces mobaraensis Kubo, Suzuki and Tamura 1964, 47.)

Good growth on PA; reverse light beige to greenish yellow; aerial mycelium dirty white with greenish shades to greenish gray.

CN and CC: very poor growth; colorless; traces only of off-white aerial mycelium.

GA: good growth; reverse yellowish; aerial mycelium whitish.

GY: good growth; reverse beige to light brown to yellowish brown; aerial mycelium white to off-white beige.

SA: good growth; reverse brown-yellowish to dirty greenish yellow; aerial mycelium beige to dirty greenish beige.

MA: good growth; reverse brown-yellow; aerial mycelium white to dirty beige.

EA: good growth; reverse yellowish; aerial mycelium off-white.

BA: good growth; reverse yellow; aerial mycelium white with pale pink shades in patches.

NA: good growth; reverse yellowish; poor, white aerial mycelium.

CB: colorless growth.

NB: yellowish growth with surface pellicle.

The strain grows at 27 C as well as at 37 C. There are also no differences in the amount of aerial mycelium, which is greener in color at 37 C. Greenish shades of the reverse are also more accentuated. No growth at 45 C.

The type strain produces piericidin A, piericidin B and detoxin and exhibits anti-bacterial activity.

Type strain: IPCR 16-22; IPV 2058.

36. Streptoverticillium blastmyceticum (Watanabe, Tanaka, Fukuhara, Miyairi, Yonehara and Umezawa) Locci, Baldacci and Petrolini Baldan 1969, 43. (Streptomyces blastmyceticus Watanabe, Tanaka, Fukuhara, Miyairi, Yonehara and Umezawa 1957, 39.)

A few monoverticils also present, spore chains slightly longer. The species is melanin positive thus showing a darkening of the reverse also on GA, MA, BA and EA. Typical green-gray color of the aerial mycelium on PA and SA. Gray tufts also present on GA and GY. When aerial mycelium is not well developed it appears flat and off-white in color. Poor aerial mycelium on PI. At 37 C poor growth; devoid of aerial mycelium.

The type strain produces blastmycin (antimycin A₃) and exhibits anti-bacterial activity.

Type strain: IAM 455-D1; IPV 1994; ATCC 19731.

37. Streptoverticillium lavenduligriseum Locci, Baldacci and Petrolini Baldan 1969, 43. (Streptomyces lavenduligriseus Rao, Marsh and Brooks 1964, 267; U.S. Patent 3,155,583; Not Val. Pub.)

Straight-flexuous forms more frequent, with some spore chains ending in hooks. Some simple monoverticils present on xylose. Reverse sides of cultures generally darker in color, from dirty yellow to brown (PA, MA and EA). Aerial mycelium better developed on GA and NA; also present (whitish) on PI. Green-gray shades particularly evident on SA and GY (some pinkish). Poorer growth and aerial mycelium development at 37 C.

The type strain produces narangomycin.

Type strain: BA-6903; IPV 2048; ATCC 13306.

Series XI. Lilacinum

38. Streptoverticillium lilacinum (Nakazawa, Tanabe, Shibata, Miyabe and Takewaka) Locci, Baldacci and Petrolini Baldan 1969, 44. (Streptomyces lilacinus Nakazawa, Tanabe, Shibata, Miyabe and Takewaka 1956, 81; Verticillomyces lilacinus (Nakazawa, Tanabe, Shibata, Miyabe and Takewaka) Shinobu 1965, 126.)

Good growth on PA; reverse dark brown; aerial mycelium white with pinkish shades; brown soluble pigment.

CN: traces of growth only; colorless; no aerial mycelium.

CC: reverse colorless to brown yellowish with pink shades; aerial mycelium white with pink-beige shades; traces of soluble pigment, brown.

GA: poor growth; reverse colorless, yellowish to brown; aerial mycelium whitish with pink shades.

GY: medium growth; reverse colorless, yellowish to brown; aerial mycelium white with pinkish shades; traces of soluble pigment, reddish.

SA: good growth; reverse yellowish to brown; aerial mycelium pinkish.

MA: good growth; reverse brown; aerial mycelium pinkish white with lilac shades; brown soluble pigment.

EA: good growth; reverse brown; aerial mycelium dirty white; brown soluble pigment.

BA: good growth; reverse brown; aerial mycelium pink with lilac shades; reddish brown soluble pigment.

NA: very poor growth; reverse brown; traces of pinkish white aerial mycelium; brown soluble pigment.

CB: colorless growth; traces of pink pellicle.

NB: yellowish growth; brown soluble pigment.

Grows at 37 C, aerial mycelium is less pink in color. No growth at 45 C.

The type strain produces cladomycin and exhibits anti-fungal activity.

Type strain: 2305; NIHJ 71; IPV 1999; ATCC 23930.

39. Streptoverticillium kashmirense (Gupta and Chopra) Locci, Baldacci and Petrolini Baldan 1969, 45. (*Streptomyces kashmirensis* Gupta and Chopra 1963, 112.)

Some wavy chains of spores can be observed. Single side chains also present. Reverse colors lighter on MA and NA; aerial mycelium colors also lighter. However on MA there is a tendency toward whitish pink with lilac shades. Soluble pigments present on EA and BA. Growth as good at 27 C as at 37 C. No aerial mycelium at 37 C.

The type strain exhibits anti-bacterial and anti-fungal activity.

Type strain: RRL 37 A/9; NRRL B-3103; IPV 2023.

Series XII. Thioluteum

40. Streptoverticillium thioluteum (Okami) Baldacci, Farina and Locci 1966, 165. (*Streptomyces thioluteus* Okami 1952, 30; *Verticillomyces thioluteus* (Okami) Shinobu 1965, 161.)

Good growth on PA, reverse orangish yellow; aerial mycelium yellowish.

CN: very poor growth; colorless; whitish tufts of aerial mycelium.

CC: limited growth; reverse colorless to yellowish; aerial mycelium whitish yellow.

GA: good growth; reverse yellow to brown yellow; poor aerial mycelium, yellowish.

GY: good growth; reverse yellowish brown; aerial mycelium yellowish.

SA: good growth; reverse yellowish; aerial mycelium yellowish.

MA: good growth; reverse brown; aerial mycelium yellowish; soluble pigments present.

EA: good growth; brown reverse; aerial mycelium greenish yellow; traces of soluble pigment.

BA: good growth; reverse yellow to brown-yellow; aerial mycelium yellowish.

NA: medium growth; aerial mycelium greenish yellow; reverse brownish yellow; traces of aerial mycelium and pigment.

CB: colorless growth.

NB: yellowish growth.

Grows at 37 C, however no aerial mycelium is produced in 15 days. No growth at 45 C.

Some cultures form traces of H_2S and utilize rhamnose and casein.

The type strain produces propiopyrrothine (aureothricin) and aureothin.

Type strain 26A; ATCC 12310; IPV 2050.

Species incertae sedis

I. The following named species and subspecies have not been examined and are listed as *incertae sedis* (Locci *et al.*, 1969).

Note. A. = Actinomyces, S. = Streptomyces, Stv. = Streptoverticillium.

a) *S. abikoensum* var. *spiralis;* b) *S. alboverticillatus;* c) *A. aureofasciculus;* d) *S. biverticillatopsis;* e) *S. caespitosus;* f) *S. cendrugii;* g) *A. cinereoverticillatus;* h) *S. cinnamoneus* subsp. *monicae* and subsp. *terricola;* i) *A. circulatus* subsp. *monomycini* and subsp. *roseus;* j) *S. fervens* subsp. *phenomyceticus;* k) *S. flavoreticuli;* l) *A. fradioverticillatus;* m) *A. griseoverticillatus;* n) *Stv. griseoviridum;* o) *A. hachijoensis* subsp. *fuscatus;* p) *S. jammensis;* q) *S. kitasatoensis;* r) *A. lilacinoverticillatus;* s) *S. multifidus;* t) *Stv. mycoheptinicum;* u) *A. ochraceoverticillatus;* v) *S. pentaticus;* w) *S. phaeoverticillatus;* x) *S. reticuli* subsp. *aquamyceticus*, subsp. *latumcidicus*, subsp. *protomycicus* and subsp. *shimofusaensis;* y) *Stv. rutilum;* z) *S. thioluteus sterilus;* aa) *S. verticillus, S. verticillatus* subsp. *viridans.*

II. The following species names are not accepted herein as belonging to the genus *Streptoverticillium* (Locci *et al.* 1969).

i) *S. luteochromogenes;* ii) *S. luteoreticuli;* iii) *S. mediocidicus;* iv) *S. reticuli;* v) *S. thermotolerans;* vi) *S. verticillatus;* vii) *S.viridoflavus.*

Genus III. **Sporichthya** *Lechevalier, Lechevalier and Holbert 1968, 279*

Thomas G. Pridham and Homer D. Tresner

Spor.ich'thy.a. Gr. masc.n. *spora* seed; Gr. masc.n. *ichthyus* fish; M.L. fem.n. *Sporichthya* fish-like seed, referring to motile spores.

Hyphae, 0.5–1.2 μm in diameter, short (10–25 μm) branched, **form only aerial mycelium that is attached to the surface of solid media by means of holdfasts** which originate from the wall of the hyphal base. **The aerial hyphae divide into smooth-walled spores** which may be spherical (0.5–1.2 μm), rod-shaped (0.5–1.2 by 0.75–1.5 μm), balloon-shaped or pisciform (up to 6 μm long). Spores exhibit a collar-like scar and a flagellum originates at this end of the spore; up to three flagella arranged in a tuft are sometimes seen. **Spores become motile** when exposed to water. The **cell wall contains L-diaminopimelic acid. Gram-variable.**

On poor media, growth is mold-like; on rich media, bacteria-like.

Heterotrophic. Acids not generally detectable. Starch is hydrolyzed.

Casein is hydrolyzed; gelatin is not liquefied; hypoxanthine and tyrosine are not attacked.

Facultatively anaerobic.

Temperature optimum 28–37 C; growth range 22–42 C, no growth at 55 C.

Optimum pH for growth *ca.* 7.0.

Type species (monotype): *Sporichthya polymorpha* Lechevalier *et al.* 1968, 279.

Description of the species of genus **Sporichthya**

1. **Sporichthya polymorpha** Lechevalier, Lechevalier and Holbert 1968, 279.

pol.y.mor'pha. Gr. pref. *poly* many; Gr. fem.n. *morphe* shape; M.L. n. *polymorpha* many shapes, referring to diverse shapes of spores.

Morphology as for genus.

Old cultures, particularly on rich media, are mostly Gram-positive; younger cultures contain varying numbers of Gram-negative elements. Stains easily with methylene blue, Ziehl's carbolfuchsin and safranin. Sudan black stains for lipid are negative. Capsules not present.

When viewed under a high dry objective (60 ×) cells appear black, blue, red, brown or colorless due to light refraction. When black, the filaments are hydrophobic; when colorless, hydrophilic. Other colors probably represent stages in development of the substance responsible for hydrophobicity.

Growth has a chalky appearance and on poor media is definitely white, on rich media dirty white. Grows rapidly (growth visible in 24 hr at

28 C) on Czapek's solution agar, Bennett's agar, yeast extract glucose agar and yeast extract glucose agar supplemented with CaCO₃. It grows less rapidly on other media.

The organism grows very poorly in agitated liquid culture. In static liquid culture, a whitish pellicle forms.

Acid is produced from fructose but not from arabinose, galactose, glucose, glycogen, inulin, lactose, maltose, mannitol, mannose, α-methyl-D-mannoside, melezitose, melibiose, rhamnose, trehalose or xylose.

Facultatively anaerobic. Capable of reduced growth in an atmosphere of hydrogen and carbon dioxide; such growth capable of initiating further aerobic growth but not anaerobic growth.

Isolated from greenhouse soil at Rutgers, The State University, New Brunswick, New Jersey, U.S.A.

Type strain: M.P.L.P.2; IMRU 3913; ATCC 23823.

Genus IV. **Microellobosporia** *Cross, Lechevalier and Lechevalier 1963, 422*

Thomas G. Pridham

(*Macrospora* Tsyganov, Zhukova and Timofeeva 1964, 868 (English translation 773), Illeg. Rule 24b; *Microechinospora* Konev, Tsyganov, Minbaev and Morozov 1967, 309 (English translation 254).)

Mi.cro.ello.bos'por.i.a. Gr. adj. *micros* small; Gr. adj. *ellobos* enclosed in a pod; Gr. n. *spora* seed; M.L. fem.n. *Microellobosporia* small seeds (spores) in a pod.

Note. The genus *Microechinospora* Konev *et al.* 1967, 309 is considered an objective synonym of the genus *Microellobosporia* Cross *et al.* 1963, 422, based on personal communication from H. A. Lechevalier, 1968. In the English translation, the title of the Konev *et al.* paper cites *Echinospora gen. nov.*, a *lapsus calami.*

Slender hyphae, about 1 μm in diameter, developing a substrate mycelium which grows into and forms a compact layer on the top of the medium and **an aerial mycelium. Both substrate and aerial mycelia bear sporangia** on short sporangiophores; the **sporangia contain a single longitudinal row of non-motile sporangiospores,** usually one to five. Spores are formed by simultaneous division of an intrasporangial hypha.

Arthrospores are not produced. Cell wall Type I.

Heterotrophs. The type species hydrolyzes gelatin and starch. Carbon utilization of the various species is given in Table 17.50.

Aerobic.

Temperature optimum 30 C; growth occurs at 25–30 C. Optimal pH range for growth at 30 C 6.0–8.5.

Antibiotic production by some strains.

Type species: *Microellobosporia cinerea* Cross, Lechevalier and Lechevalier 1963, 422.

Key to the species of genus **Microellobosporia**

I. Surfaces of sporangia free of crystals or spines.
  A. Vegetative mycelium colored in tints and shades of violet.
    1. In Pridham and Gottlieb basal medium utilizes sodium acetate, lactose, maltose, D(+)-sorbitol and D-xylose.

    1. *Microellobosporia cinerea*
    2. Does not utilize sodium acetate, lactose or maltose and utilizes D(+)-sorbitol and D-xylose poorly.

    2. *Microellobosporia violacea*
  B. Vegetative mycelium not colored violet.

    3. *Microellobosporia flavea*
II. Surfaces of sporangia covered with crystals forming serrate spines.

    4. *Microellobosporia grisea*

Description of the species of genus **Microellobosporia**

1. Microellobosporia cinerea Cross, Lechevalier and Lechevalier 1963, 422.

ci.ne're.a. L. fem.adj. *cinerea* gray.

Long branching hyphae, about 1 μm in diameter, penetrate the medium and form compact colonies which are at first colorless but become pink to red-purple depending upon the medium and its pH. Similar hyphae produce an aerial mycelium.

Sporangia appear singly and laterally on the aerial hyphae after 3–4 days' incubation; later they are borne in greater numbers and also occur terminally. When mature, sporangia are 1.5–3.6 μm in diameter and 2–9 μm long, depending on the number and size of contained spores. Usually sporangia contain two to five spores; occasionally long sporangia containing six to seven spores can be seen; infrequently single-spored sporangia occur, most often on the short aerial hyphae growing from the substrate mycelium encircling the colony. The sporangial wall is thin and wrinkled and does not appear to bear any appendages; it originates from the apex of the sporangiophores, which is swollen. The sporangium grows from a lateral bud on a hypha to form a short club-shaped initial structure; the contents divide into spores which swell slightly to give the sporangium a beaded appearance.

Spores vary in size, even within a single sporangium, from 1.5–3.5 μm (average 2.5 μm) and are round to oval.

Good growth on potato agar; aerial mycelium scanty, white to light gray; colonies flesh pink; no diffusible pigment. Excellent growth on Cza-

pek's solution agar with glucose; aerial mycelium scanty, white to off-white; reverse light brown to purple-brown; purple-brown diffusible pigment. Moderate growth on soya glucose agar; aerial mycelium scanty, white to light gray; colonies pale brown to light purple brown to dirty pink; no diffusible pigment; abundant formation of sporangia on colonies lacking aerial mycelium. On Lindenbein's synthetic agar good growth; aerial mycelium scanty, white; reverse bright red to dirty pink; both colonies and diffusible pigment show indicator properties, i.e. with dilute alkali they become purplish pink, with dilute acid yellow to orange-pink. Excellent growth on glycerol asparagine agar; aerial mycelium none to moderate, light gray; colonies pink to red or pale yellow; may or may not produce pink diffusible pigment; abundant formation of sporangia on substrate mycelium.

For utilization of carbon sources see Table 17.50.

At least one antibiotic produced, effective against Gram-positive bacteria and mycobacteria.

Type strain: IMRU 3855 (monotype).

2. Microellobosporia violacea (Tsyganov, Zhukova and Timofeeva) *comb. nov.* (*Macrospora violaceus* (*sic*) Tsyganov, Zhukova and Timofeeva 1964, 868.)

vi.o.la'ce.a. L. fem.adj. *violacea* violet.

Morphology as for *M. cinerea*.

A violet diffusible pigment produced on the following media. Excellent growth on potato agar; aerial mycelium abundant, light mouse gray;

TABLE 17.50

Carbon utilization patterns of the species of genus **Microellobosporia**

Carbon source	Utilization[a]			
	1. *M. cinerea*	2. *M. violaceus*	3. *M. flavea*	4. *M. grisea*
No carbon control	−	−	−	0
D-(+) Glucose control	+++	+++	++	+++
D-(+) Xylose	++	‡	++	+++
L-(+) Arabinose	++	+	++	+++
L-(+) Rhamnose	+++	++	++	+++
Fructose	0	0	0	+++
D-(+) Galactose	++	++	++	+++
Maltose	++	−	+	++
Lactose	++	−	+	++
DL-Sucrose	++	0	+++	++
Raffinose	+++	+++	++	++
Starch	+++	++	++	++
Inulin	0	0	0	+++
Cellulose	0	0	0	‡
D-(−) Mannitol	+++	++	+++	++
D-(+) Sorbitol	+++	‡	++	‡
Sodium acetate	++	−	++	0
Sodium citrate	++	‡	+	++
Sodium malate	++	0	+	++

[a] Basal medium: Pridham and Gottlieb (1948, 107); − = growth absent; ‡ = very weak, nil, or doubtful growth; + = weak growth; ++ = satisfactory or moderate growth; +++ = good growth or grows well; 0 = no data.

growth dark violet. On Lindenbein's agar excellent growth, aerial mycelium grayish violet; colonies dark violet. Good growth on glycerol asparagine agar; no aerial mycelium, colonies pinkish violet; diffusible pigment pinkish violet.

Produces violacin, a pigmented antibiotic of the rhodomycin-cinerubin-mycetin type.

Type strain: LIA 2732/3 (monotype).

3. **Microellobosporia flavea** (*sic*) Cross, Lechevalier and Lechevalier 1963, 426.

fla′ve′a. M.L. fem.adj. *flavea* presumably from *flavus* yellow.

Morphology similar to *M. cinerea*.

Good growth on potato agar; aerial mycelium scanty, white; colonies white; no diffusible pigment. Excellent growth on Czapek's solution agar with glucose; aerial mycelium scanty, white to light gray; colonies straw yellow; no diffusible pigment. Good growth on soya glucose agar; aerial mycelium scanty, white; colonies yellow ochre; no diffusible pigment. Excellent growth on Lindenbein's synthetic agar; mycelium white; colonies yellowish to orange; no diffusible pigment. Good growth on glycerol asparagine agar; aerial mycelium scanty, white; colonies straw yellow; no diffusible pigment.

The type strain is antibiotically inactive; strain IMRU 3858 slightly active against Gram-positive bacteria.

Type strain: IMRU 3857 (lectotype).

4. **Microellobosporia grisea** (Konev, Tsyganov, Minbaev and Morozov) Pridham 1970, 17. (*Microechinospora grisea* Konev, Tsyganov, Minbaev and Morozov 1967, 309.)

gri′se.a. M.L. fem.adj. *grisea* gray.

Sporangia pyriform to clavate with surfaces covered with crystals forming serrate spines; each contains one to three (most commonly 1) sporangiospores.

Good growth on Czapek's solution agar with glucose; aerial mycelium short, white to glaucus gray; vegetative mycelium pink to light violet-pink; no diffusible pigment. Good growth on glycerol asparagine agar; aerial mycelium short, whitish to straw colored, sometimes pink tinge; vegetative mycelium yellowish to colorless; color of diffusible pigment not stated. No information on growth on potato agar, soya glucose agar or Lindenbein's agar.

No anti-bacterial activity but does exhibit mycelial anti-fungal activity.

Type strain: LIA P-147 (monotype).

FAMILY VIII. **MICROMONOSPORACEAE** KRASIL'NIKOV 1938, 272

E. KÜSTER

Mic.ro.mo.no.spo.ra'ce.ae. M.L. n. *Micromonospora* type genus of the family; *-aceae* ending to denote a family; M.L. fem.pl.n. *Micromonosporaceae* the *Micromonospora* family.

Aerial mycelium present, except in *Micromonospora*. Spores are formed singly, in pairs or short chains on either or both the aerial and substrate mycelium. Sporophores are lacking or very short, in some cases showing dichotomous branching. Generally aerobic, a few anaerobic species are known. Mostly mesophilic, some thermophilic. Primarily saprophytic soil forms.

Further Comment

The family *Micromonosporaceae*, as originally proposed by Krasil'nikov (1938), was monogeneric

(*Micromonospora*). Later Krasil'nikov (1964) added other genera. Six genera are included here, although they have few features in common, and further study may necessitate a change in the taxonomic position of one or more genera and several species.

Editorial Note. Some of our authors have suggested that the genera *Thermomonospora* and *Micropolyspora* (cell wall Type IV) should be transferred to the family *Nocardiaceae*. We have followed the suggestion of others that they remain in *Micromonosporaceae* for the present.

Key to the genera of family **Micromonosporaceae**

I. Aerial mycelium absent.
  A. Cell wall Type II.
    1. Single spores on substrate mycelium only.
             Genus I. *Micromonospora*, p. 846
II. Aerial mycelium present.
  A. Cell wall usually Type III. (One species in *Actinobifida* is Type II and one in *Thermomonospora* Type IV.)
    1. Single spores on both the aerial and substrate mycelium.
      a. Sporophores lacking or very short.
             Genus II. *Thermoactinomyces*, p. 855
      aa. Sporophores dichotomously branched.
             Genus III. *Actinobifida*, p. 856
    2. Single spores on aerial mycelium only.
             Genus IV. *Thermomonospora*, p. 858
    3. Longitudinal pairs of spores on aerial mycelium only.
             Genus V. *Microbispora*, p. 859
  B. Cell wall Type IV.
    1. Short chains of spores on both the aerial and substrate mycelium.
             Genus VI. *Micropolyspora*, p. 861

Genus I. **Micromonospora** *Ørskov 1923, 147*

GEORGE M. LUEDEMANN

Mic.ro.mo.no'spo.ra. Gr. adj. *micrus* small; Gr. adj. *monus* single, solitary; Gr. n. *spora* a seed; M.L. fem.n. *Micromonospora* small, single-spored (organism).

Well developed, branched, septate mycelium averaging 0.5 μm in diameter. **Spores borne singly, sessile or on short or long sporophores** which often occur in branched clusters. Sporophore development monopodial or in some cases sympodial.

Aerial mycelium absent or in some cultures

appearing irregularly as a restricted white or grayish bloom. Gram-positive. Not acid-fast. **Cell wall Type II.**

Colonies on agar limited in peripheral growth, rising steeply from the surface of the medium giving a convolute walnut-kernel-like appearance.

Most isolates are strongly proteolytic, cellulolytic and diastatic.

Most species aerobic, two anaerobic. Sensitive to pH environs below 6.0. Good growth occurs normally between 20–40 C but not above 50 C. Spores generally resistant to water temperatures of 60 C for 90 min but killed at 90 C in 15 min.

Usually saprophytic occurring as normal populations in soils and aquatic environments.

Type species: *M. chalcea* (Foulerton) Ørskov 1923, 156.

Further Comments

Immature or non-sporulating colonies of *Micromonospora* may often resemble the glabrous colonies of orange-colored species of *Nocardia*, *Actinoplanes* or *Amorphosporangium*. Sporulating colonies of micromonosporae may be distinguished from these genera by the development of a characteristic dark brown to black, waxy or moist layer composed of dark pigmented spores. Mature cultures of *Micromonospora* may often be composed of coccoid spores and mycelial fragments which resemble in microscopic appearance species of *Nocardia*. Despite the superficial resemblance to other genera, the manner by which micromonosporae produce their spores is distinctive and if the origin of the spore is sought, misidentifications can be avoided. Micromonosporae produce their spores as single, well defined, terminal or lateral structures which occur on the surface and throughout the colony but not in a powdery aerial mycelium characteristic of the streptomycetes, the genus *Actinobifida*, or *Thermoactinomyces*. The known mesophilic, aerobic micromonosporae are a relatively homogenous group. The position of the anaerobic species of *Micromonospora* within this group requires additional investigation. The paucity of species descriptions based upon clearly recognizable characteristics has hindered the development of a firm species concept in *Micromonospora*. The recognition of reliable morphological differences among the species has been slow in development. Echinulate spores, sympodially branched sporophores and mycelial pattern appear to be useful morphological characteristics (Luedemann, 1971). Among the physiological criteria, carbohydrate utilization patterns appear as helpful diagnostic aids for certain species.

Key to the species of genus **Micromonospora**

I. Aerobic.
  A. α-Melibiose utilized (for species in which carbohydrate data are lacking see IC or ID).
    1. Raffinose utilized.
      a. Sporophores arranged monopodially.
        i. Growth poor on Czapek's sucrose agar.
            1. *Micromonospora chalcea*
        ii. Growth fair to good on Czapek's sucrose agar.
            2. *Micromonospora halophytica*
    2. Raffinose not utilized.
      a. Sporophores arranged sympodially.
            3. *Micromonospora carbonacea*
      b. Spores often produced on monopodially arranged, short lateral spikes. In broth culture mycelium often formed into cluster colonies which fragment readily.
        i. D-Mannitol not utilized.
            4. *Micromonospora narashinoensis*
        ii. D-Mannitol utilized.
            See *Species incertae sedis* (p. 854).
      c. Spores produced monopodially either on long sporophores or in flower-like clusters.
            5. *Micromonospora melanosporea*
  B. α-Melibiose not utilized.
    1. L-Rhamnose utilized. Spores echinulate.
            6. *Micromonospora echinospora*
    2. L-Rhamnose not utilized. Spores aberrant.
            7. *Micromonospora purpurea*
  C. Produce characteristic pigments.
    1. Produce extensive dark brown to black pigment diffusing into agar.
            8. *Micromonospora purpureochromogenes*

　　2. Do not produce brown or black diffusible pigment.
　　　　a. Mycelial pigment green or blue-green.
　　　　　　i. Cellulose hydrolyzed.
　　　　　　　　　　9. *Micromonospora bicolor*
　　　　　　ii. Cellulose not hydrolyzed.
　　　　　　　　　　10. *Micromonospora coerulea*
　D. Pigments non-characteristic; colonies in shades of pale yellow, orange, brown or black.
　　1. Growth vigorous on protein-containing media.
　　　　　　　　11. *Micromonospora globosa*
　　2. Growth not vigorous on protein-containing media.
　　　　a. Does not grow on potato. Cellulose hydrolyzed.
　　　　　　　　12. *Micromonospora elongata*
　　　　b. Grows on potato. Cellulose not hydrolyzed.
　　　　　　i. Isolated from soil.
　　　　　　　　　　13. *Micromonospora parva*
　　　　　　ii. Clinical isolate.
　　　　　　　　　　14. *Micromonospora gallica*
II. Anaerobic.
　A. Cellulose not utilized; acetic and formic acids produced.
　　　　　　　　15. *Micromonospora acetoformici*
　B. Cellulose utilized; acetic and propionic acids produced.
　　　　　　　　16. *Micromonospora propionici*

Description of the species of genus Micromonospora

1. Micromonospora chalcea (Foulerton 1905) Ørskov 1923, 156. (*Streptothrix chalcea* Foulerton 1905, 1200; *Nocardia chalcea* (Foulerton) Chalmers and Christopherson 1916, 268: *Actinomyces chalceus* (Foulerton) Pribram 1919, 17; *Micromonospora chalcease* (*sic*) (Foulerton) Ørskov 1923, 156.)

chal'ce.a. Gr. fem.adj. *chalcea* copper, bronze.

Description based on study of strain ATCC 12452 (Luedemann and Brodsky, 1964).

Spores oval to spherical, 0.7–1.0 μm in diameter, brown to dark brown by transmitted light and smooth walled when viewed by phase-contrast microscopy; by electron microscopy the spores appear smooth walled or with minor irregularities (warts). Spores sessile or on short or long sporophores, produced randomly throughout the mycelium usually within 3–5 days after transfer onto nutrient media. Spores separate early from their sporophores. Mycelium does not appear to degenerate into polymorphic elements.

Young colonies typically raised and folded, reddish orange, turning brown, olive brown, dark brown and eventually black due to sporulation. Spore layer moist to dry, usually not viscid. Pale yellow fluorescent diffusible pigment often produced on agar media, particularly yeast starch agar.

The diagnostic carbohydrate utilization pattern is good growth on α-melibiose and raffinose, and poor growth on D-mannitol and L-rhamnose.

Good growth occurs on L-arabinose, D(+)-cellobiose, D-glucose, D-galactose, β-lactose, D-fructose, D(+)-mannose, soluble starch, sucrose, α,α-trehalose and D-xylose. Poor growth occurs on glycerol, dulcitol, inositol, D-ribose, salicin, D-sorbitol and L-sorbose. Starch hydrolyzed. Decomposes cellulose.

Gelatin liquefied. Milk digested. No melanoid-like pigments are produced. Nitrate reduction variable. Growth is poor on Czapek sucrose agar. Good growth occurs on autoclaved potato slice without the addition of $CaCO_3$. Maximum sodium chloride tolerance 4–5%.

Good growth occurs between 27–37 C, but no growth occurs at 45 C.

Organism has been used for the microbiological transformation of steroids (Shull, 1959, U. S. Patent 2,890,153).

Isolated from air, soil and aquatic environments. Jensen (1932) believed this species formed a large species group with a rather wide range of variation and probably represented the most frequently isolated species.

Neotype strain: ATCC 12452 (Luedemann, 1971, 252).

2. Micromonospora halophytica Weinstein, Luedemann, Oden and Wagman 1968, 436.

hal'o.phyt'i.ca. M.L. fem.adj. *halophytica* a plant which grows within the influence of salt water.

Spores randomly produced throughout a long branching mycelium on short or long sporophores or spores occasionally sessile. Spores ellipsoidal to spherical, up to 1.2 μm in diameter. Spores smooth walled appearing dark colored in older cultures. Vegetative mycelium does not normally break up into polymorphic elements.

Growth fair on Czapek's agar. No reaction on tyrosine agar or peptone iron agar. Growth poor on potato slice but good growth occurs if CaCO₃ is added.

Utilizes for growth L-arabinose, D-galactose, D-glucose, β-lactose, D-fructose, D(+)-mannose, α-melibiose, raffinose, starch, sucrose, α,α-trehalose and D-xylose. Poor growth occurs on adonitol, cellulose, dulcitol, inositol, D-mannitol, L-rhamnose, D-ribose and D-sorbitol. Hydrolyzes starch. Decomposes cellulose. Reduces nitrate.

Strains produce the halomicin antibiotic complex.

Grows aerobically in a temperature range of 18–40 C but not at 50 C.

2a. *Micromonospora halophytica* subsp. *halophytica* Weinstein *et al.* 1968, 437.

Growth good, folded, color orange to orange-brown. Sporulation abundant in older, brown-colored colonies. A light reddish brown diffusible pigment often produced on agar media.

Gelatin liquefied; milk digested.

Isolated from a salt pool in Syracuse, N.Y. by A. Woyciesjes.

Type strain: NRRL 2998.

2b. *Micromonospora halophytica* subsp. *nigra* Weinstein *et al.* 1968, 437.

ni'gra. L. fem.adj. *nigra* black.

Reddish brown diffusible pigment not produced. Produces a spore layer which turns the initially orange colony olive brown and later black.

Gelatin liquefaction and milk digestion variable.

Type strain: NRRL 3097.

3. **Micromonospora carbonacea** Luedemann and Brodsky 1965, 47.

car'bo.na'cea. L. n. *carbo* coal, charcoal; L. suff. *aceus* quality or nature of; M.L. adj. *carbonacea* charcoal-like, referring to color.

Spores at maturity are oval to spherical, 0.7–1.0 μm in diameter, brown by transmitted light and smooth walled when viewed by light microscopy; by electron microscopy the spores appear smooth walled or with only minor irregularities. In broth, sporulating areas are usually sparsely dispersed in definite clumps throughout the mycelial web which consists mostly of unbranched, long, loosely woven, fine mycelial strands. The spores remain firmly attached to the sympodial type of

sporophore and only in older cultures are found free. The mycelium does not degenerate into polymorphic bodies.

Fair to good growth on Czapek sucrose agar plus 0.1% CaCO₃. Only fair growth on autoclaved potato slice but good growth occurs on this substrate if CaCO₃ is added prior to autoclaving.

The diagnostic carbohydrate utilization pattern is good growth on α-melibiose and no growth on D-mannitol or L-rhamnose. Good growth on L-arabinose, D(+)-cellobiose, D-glucose, D-galactose, β-lactose, D-fructose, D(+)-mannose, soluble starch, sucrose, α,α-trehalose and D-xylose. Poor growth on dulcitol, glycerol, inositol, raffinose, D-ribose, salicin, D-sorbitol and L-sorbose.

Gelatin is liquefied. Milk is digested. No melanoid-like pigments are produced.

Good growth between 27–37 C, but no growth at 45 C.

Produces the everninomicin antibiotic complex.

3a. *M. carbonacea* subsp. *carbonacea* Luedemann and Brodsky 1965, 51.

On agar, spores are generally abundantly produced. Colonies raised, folded, initially orange but turning brown to black as spore layer develops. Characteristic blackish sporulating peripheral sectors similar to alluvial fans often a macroscopic diagnostic aid. Spore layer moist or dry but not viscid.

Nitrate is reduced.

Isolated from a soil sample from Olean, N.Y. by A. Woyciesjes.

Type strain: NRRL 2972.

3b. *M. carbonacea* subsp. *aurantiaca* Luedemann and Brodsky, 1965, 51.

au.ran'ti.a'ca. L. fem.adj. *aurantiaca* orange colored.

The orange to orange-red colonies of this subspecies produce few spores and are only occasionally found producing small carbon-like specks (spores) on the colony surface.

A pale yellowish diffusible pigment occasionally observed surrounding colonies of this strain on mannose and xylose agars.

Nitrate not reduced to nitrite.

Type strain: NRRL 2997.

4. **Micromonospora narashinoensis** (Shidara) Arai and Kuroda 1965, 33. (*Nocardia narashinoensis* Shidara 1955, 553.)

na.ra.shi.no'en.sis. English n. *Narashino*, town in Japan; L. suff. *-ensis* indicating place or origin; M.L. adj. *narashinoensis* of Narashino.

Description based on Arai and Kuroda (1965).

Spores spherical or elongate, 0.6–1.0 by 0.9–1.8 μm, either sessile or often on short sporophores. Sporulating cultures develop a branched mycelium within 48 hr and mature spores which may be released into the medium within 72 hr.

Colonies on yeast extract-glucose agar producing abundant, folded growth of a deep orange color, later covered with a sparse dark brown spore layer. No soluble pigment produced. Colonies on nutrient agar minute, orange to tan colored, spore layer developing at periphery of colony as a moist, drab, dark brown to black layer. Colonies on tyrosine agar minute, orange in color, sparse development of a brown spore layer, purplish diffusible pigment produced. Growth on potato plug folded, light orange in color becoming covered with a brown-colored spore layer. No growth in cellulose solution. Starch hydrolyzed.

Gelatin liquefied. Milk digested. Blood hemolyzed (β-form). Nitrates not reduced. Ammonium sulfate and glutamate utilized moderately well but growth much better with yeast extract.

Strain No. 76-N₃-5 produces the antibiotic nocardorubin (rufinosporin) active against Gram-positive bacteria.

Isolated from soil sample collected in Narashino, Chiba, Japan.

Further Comment

In *Index Bergeyana* (1966, 764) *Nocardia narashinoensis* is ascribed to Endo (1956, 228). Although Endo mentions *N. narashinoensis* n. sp. (1956, 228) and in an earlier publication (1955, 168), neither of these papers serves as a legitimate description for the species. See also comment under *Species incertae sedis* b. *Micromonospora sp.* ATCC 10026.

5. **Micromonospora melanosporea** (Krainsky) Baldacci and Locci 1961, 28. (*Actinomyces melanosporeus* Krainsky 1914, 687. *Nocardia melanosporea* (Krainsky) Chalmers and Christopherson 1916, 268; *Streptomyces melanosporeus* (Krainsky) Waksman and Henrici 1948, 955.)

mel'a.no.spor.ea. Gr. adj. *melas* black; Gr. n. *spora* seed. M.L. adj. *melanosporea* black spored.

Description based on Baldacci and Locci (1961).

Mycelium composed of short, branched or clustered sporophores, 1–4 μm in length. Spores borne singly, 0.9–1.1 μm in diameter.

Colonies grow vigorously on most substrates. Young vegetative colonies salmon pink or red in color, later covered with a black layer of spores. On potato agar growth smooth or folded, at first salmon pink then turning black. Some strains retarded in growth upon this medium. On Czapek's agar color is salmon pink to reddish, sporulation very slow. On nutrient agar growth fair, sporulation occurring late, color salmon pink to reddish. On glycerol asparagine agar growth is very

limited, sporulation occurring late. On glucose asparagine agar color of colonies initially salmon pink, later becoming black. On starch agar growth limited. On V-8 agar vegetative growth is good, salmon pink in color. The color of the vegetative mycelium varies among the strains in relation to the substrate used and cannot be used as a reliable diagnostic aid.

Most strains exhibit diastatic activity. All strains grow in milk and coagulation gradually occurs. Cellulose decomposed. Nitrates reduced. Growth occurs at 28 and 37 C but not at 50 C.

5a. *Micromonospora melanosporea* subsp. *melanosporea* (Krainsky) Baldacci and Locci 1961, 28.

Mycelium composed of short, branched sporophores.

Isolated from water, soil, and from a blood specimen from a woman with a mycetoma of the knee.

5b. *Micromonospora melanosporea* subsp. *corymbica* Baldacci and Locci 1961, 29.

cor'ymb.i.ca. L. n. *corymbus*, a flower cluster. Sporophores in clusters rather than branched.

Strain 562 exhibits moderate antibiotic activity against *Mycobacterium tuberculosis*, other strains are inactive.

Isolated from water and soil.

6. **Micromonospora echinospora** Luedemann and Brodsky 1964, 116.

e.chi'no.spo.ra. Gr. adj. *echinos* spiny appearance; Gr. n. *spora* seed. M.L. fem. n. *echinospora* spiny spore.

Spores at maturity spherical 1.0–1.5 μm in diameter, dark brown to black, appearing rough walled under a phase-contrast microscope; under an electron microscope the roughness appears due to blunt spines 0.1–0.2 μm long. Sporophores mostly observed as solitary structures although they may occur occasionally in small clusters on the same hypha. Spores adhere firmly to the sporophore until mature.

Most strains grow on Czapek sucrose agar. No growth occurs on autoclaved potato slice but good growth occurs on this medium if CaCO₃ is added prior to autoclaving.

The diagnostic carbohydrate utilization pattern appears as good growth on L-rhamnose and poor growth on α-melibiose. The organism utilizes for growth L-arabinose, D(+)-cellobiose, D-glucose, D(+)-mannose, sucrose, α,α-trehalose, D-xylose and soluble starch. D-galactose, β-lactose and D-fructose yield only poor to fair growth, however, some differences exist among the subspecies in regard to their ability to utilize these carbohydrates. Poor growth on dulcitol, glycerol, inositol, D-mannitol, raffinose, salicin, D-sorbitol and L-sorbose. Very slow growth on and decomposition of cellulose.

Gelatin liquefied. Milk digested. Melanoid-like pigments not produced. Nitrate reduced by some strains but the reaction is not always reproducible.

Most strains produce antibiotics of the gentamicin complex.

Good growth between 27–37 C; no growth at 45 C.

6a. *Micromonospora echinospora* subsp. *echinospora* Luedemann and Brodsky 1964, 116.

Colonies usually folded, orange-brown, maroon or characteristically dark purple. Aerial mycelium absent but occasionally a very short purplish gray bloom devoid of spores occurs on some colonies. Sporulation moderate, often found in cultures only after 2 or 3 weeks of incubation. Spore layer when present purplish black, waxy to dry, not moist or viscid.

Poor growth on D-ribose and poor to fair growth on D-galactose, β-lactose and D-fructose.

The organism was isolated from a soil sample from Jamesville, N.Y. by A. Woyciesjes.

Type strain: ATCC 15837; NRRL 2985.

6b. *Micromonospora echinospora* subsp. *ferruginea* Luedemann and Brodsky 1964, 116.

fer'ru.gin'e.a. L. fem.adj. *ferruginea* of the color of iron rust.

Colony color on a number of carbohydrate sources which sustain good growth in shades of orange and maroon; similar in appearance to iron rust.

Growth is fair to good on D-ribose.

Nitrate reduction negative.

Isolated from a soil sample from N.Y. state by A. Woyciesjes.

Type strain: ATCC 15836; NRRL 2995.

6c. *Micromonospora echinospora* subsp. *pallida* Luedemann and Brodsky 1964, 116.

pal'lid.a. L. fem.adj. *pallida* pale.

Purple mycelial pigments not present. Colony color pale, ranging from light ivory to light melon yellow except where abundant sporulation occurs and imparts a dark brown to black coloration. Luxuriant growth occurs on D-fructose, growth on D-ribose is slight but often abundant sporulation on this carbohydrate imparts a black color to the colony. Nitrate reduced.

Isolated from a soil sample from Jamesville, N.Y. by A. Woyciesjes.

Type strain: ATCC 15838; NRRL 2996.

7. **Micromonospora purpurea** Luedemann and Brodsky 1964, 116.

pur'pur.ea. L. fem.adj. *purpurea* purple colored, dull red with a slight dash of blue.

Structures resembling spores or sporophores rarely found, sporulation atypical or abortive. Terminal and intercalary chlamydospores often present.

Colonies raised, folded, characteristically purple or maroon colored in older cultures although all shades of orange and orange-brown may be encountered. Colonies dry or waxy, rarely moist, never viscid.

The organism grows on Czapek sucrose agar. No growth on autoclaved potato slice (pH of raw potato 5.8–6.2) but good growth if $CaCO_3$ is added.

The diagnostic carbohydrate utilization pattern appears as poor growth on α-melibiose, D-mannitol and L-rhamnose; fair to good growth occurs on D-ribose. The organism utilizes for growth L-arabinose, D(+)-cellobiose, D-glucose, D(+)-mannose, sucrose, α,α-trehalose, D-xylose and soluble starch. D-galactose, β-lactose and D-fructose yield only poor to fair growth. Growth poor on dulcitol, glycerol, inositol, raffinose, salicin, D-sorbitol and L-sorbose. Very slow growth and decomposition occur on cellulose.

Gelatin weakly liquefied. Milk digested. Melanoid-like pigments not produced. Nitrate reduced to nitrite.

Good growth between 27–37 C; no growth at 45 C.

Produces antibiotics of the gentamicin complex.

Isolated from a soil sample from Syracuse, N.Y. by A. Woyciesjes.

Type strain: ATCC 15835; NRRL 2953.

Further Comment

This strain probably represents an atypical form of *Micromonospora cchinospora*.

8. **Micromonospora purpureochromogenes** (Waksman and Curtis) Luedemann 1971, 244. (*Actinomyces purpeo-chromogenus* (sic) Waksman and Curtis 1916, 113; *Micromonospora fusca* Jensen 1932, 177; *Streptomyces purpeochromogenus* (sic) (Waksman and Curtis) Waksman and Henrici 1948, 943: *Streptomyces purpureochromogenes* (Waksman and Curtis) Waksman and Henrici 1957, 777; *Actinobifida fusca* (Jensen) Krasil'nikov and Agre 1965, 291; Not *Streptomyces purpureochromogenes* ATCC 3343; Not *Streptomyces purpureochromogenus* (sic) CBS.)

pur.pur.e.o.chro.mo.ge.nes. L. adj. *purpureus* purple colored; Gr. n. *chroma* color; Gr. v.suff. *-genes* producing; M.L. part.adj. *purpureochromogenes* producing purple color. (Note this color relates to the hue of the diffusible pigment when viewed by transmitted daylight.)

In broth, colonies have a tendency to disintegrate (lyse) into mycelial fragments and chlamydospores. Mycelium 0.4–0.6 μm diameter, spores are 0.8–1.2 μm in diameter and may be found singly or in clusters within the mycelial web. A predominantly monopodial system of branching occurs in the sporulating hyphae and may often be observed at the periphery of the colony.

Colonies on yeast extract-glucose agar at 21 days are dark brown in color; no aerial mycelium. A dark brown diffusible pigment is produced in most media upon which growth occurs. The diffusible pigment is water soluble and is precipitated upon acidification but soluble upon the addition of excess base. Growth on potato plugs is slow, colonies usually are orange to reddish orange in 8–10 days and dark brown to black at 21 days at which time a dark brownish black diffusible pigment may also be observed.

Carbohydrates utilized for growth are D-galactose, D-glucose, glycerol, β-lactose, D-fructose, α-melibiose, raffinose, sucrose and D-xylose. Poor growth on D-arabinose, L-arabinose, dulcitol, i-inositol, D-mannitol, melezitose, L-rhamnose and D-ribose.

On milk agar plates peptonization is poor and very little brown diffusible pigment is produced. In litmus milk peptonization is rarely more than 10%. Nitrate reduction is negative; some strains related to this strain give a variable reduction reaction.

Growth is slow and relatively poor on Czapek's agar, a small amount of brown diffusible pigment is produced. Growth and diffusible pigment production are reduced at NaCl concentrations of 1.5% and do not occur at 3%.

Isolated once from the California adobe soil.

Type strain: IMRU 3343; ATCC 27007.

9. Micromonospora bicolor Krasil'nikov 1941, 131.

bi.color. L. fem.adj. bicolor of two colors.

Description based on Krasil'nikov (1941).

Conidia oval, 1.0–1.2 by 0.8 μm; conidiophores long, 10.0–25.0 μm, arborescent.

On nutrient media colonies are green colored, smooth or slightly rough, closely adhering to the substrate. Colony surface covered with a dark brown or black layer of conidia. Pigment not diffusing into the medium. The culture grows poorly on common nutrient media and not at all on protein media, potato and wort agar. The organism grows well on synthetic media utilizing ammonium salts and nitrates as a source of nitrogen nutrition.

Glucose, fructose, sucrose, acetic and citric acids are used as sources of carbon.

Starch not hydrolyzed. Sucrose inverted. Grows well on cellulose and decomposes this substrate to a homogeneous, flocculent mass.

No growth occurs on gelatin. Milk unchanged.

Isolated from soil, sparsely distributed.

10. Micromonospora coerulea Jensen 1932, 177.

coe.ru'le.a. L. adj. coerulea dark colored, dark blue, dark green.

Colonies in shaken broth culture usually composed of long loosely woven hyphae. Spores borne singly, often on short or long lateral sporophores. Mycelium 0.4–0.6 μm in diameter. Spores spherical, 0.8–1.5 μm in diameter, smooth walled by phase-contrast microscopy. Mycelium rarely fragmenting except in old cultures.

Growth on glucose asparagine agar or starch casein agar fair to poor. Colonies of a granular texture, small and often glossy. Color dark bluish green. No diffusible pigment or aerial mycelium produced. Growth on 1% yeast extract-1% glucose agar good, texture granular, color of colony periphery deep orange yellow, colony center dark green. Growth on agar media often slow, requiring 3–5 weeks to develop. Growth initially a pale yellow-orange, later turning a yellow-green and finally a dark blue-green or greenish black. The blue-green mycelial pigment is water soluble and is an acid base indicator being green in the basic range and a blue-gray and precipitated in the acid range.

Good growth on D-galactose, D-glucose, D-fructose, D-mannitol, α-melibiose, raffinose, sucrose and D-xylose. Fair but often sporadic growth on β-lactose. Poor growth on L-arabinose, D-arabinose, dulcitol, glycerol, i-inositol, melezitose, D-ribose and L-rhamnose. Hydrolyzes starch. Cellulose poorly decomposed.

Maximum sodium chloride tolerance 1.5%. Grows well on potato plug without the addition of CaCO₃.

Hydrolysis of gelatin and milk poor. Nitrate not reduced to nitrite.

Grows well between 24–37 C but not at 45 C.

Isolated from a soil sample obtained from Mt. Haleakala, Island of Maui, Hawaiian Islands.

Neotype strain: 36; ATCC 27008 (Luedemann 1971, 248).

11. Micromonospora globosa Kriss 1939, 178.

glo.bo'sa. L. fem.adj. globosa spherical, globose.

Description based on Kriss, 1939.

Note. For discussion of the authority for the name see Luedemann (1969).

Four to 5 days after inoculation a well developed monopodially branched mycelium appears composed of fine hyphae 0.5–0.8 μm in diameter which do not disintegrate into fragments and in which cross-walls are not observed. Young colonies are tough and leathery but after 10–15 days mycelial decomposition takes place and the colonies become of a consistency similar to dough. In some cultures after 2–3 months irregular swellings and bulbs are found in the filaments which may be

2–3 times the size of spores and are thought to be involutionary forms similar to those seen in old cultures of other actinomycete species. Conidia are formed by a thickening of the ends of branches which become round and transformed into ball-shaped conidia which are separated by a transverse wall from the parent hypha. The conidia are 1.0–1.3 μm in diameter and occur singly at the ends of short branches. In good microscopic preparations single branches with conidia remind one of clusters of grapes. Optimum temperature for sporulation 35–37 C. Colonies on agar media are raised and plicate, often only weakly adhering to the substrate. Color of young colonies varying from pale yellow to red-orange depending upon composition of medium, pH and age. Later, often 10–15 days, a brownish black sporulation layer develops covering parts or the entire colony. Soluble pigments not formed.

Acid often detected on Czapek's glucose medium with litmus. Gas not formed. Sucrose inverted in Czapek's medium by all strains. Decomposition of starch occurs only under colonies. Growth and decomposition of cellulose not observed on Winogradsky's medium.

Proteolytic capacity differs in degree among strains studied. NH₃ is formed rapidly on beef peptone broth by some strains, more slowly by others. The degree of gelatin liquefaction varies among the strains. H₂S not produced in meat peptone broth. Milk is coagulated usually on the 3rd or 4th day and slowly peptonized. Acid reaction in litmus milk. Nitrates reduced to nitrites on Gilkey's medium.

Isolated from chestnut soil and rhizosphere of wheat from the Saratov Region of Russia. Eight similar strains of *Micromonospora* were isolated and formed the basis for this species description.

12. Micromonospora elongata Krasil'nikov 1941, 131.

e.lon′ga.ta. L. fem.part.adj. *elongata* elongated, stretched out.

Description based on Krasil'nikov, 1941.

Conidia oval 1.0–1.3 by 0.8 μm, produced on short conidiophores 2.0–3.0 μm; conidiophores sparsely branched.

Colonies on nutrient media gray-brown or straw colored, poorly developed, smooth or slightly wrinkled, closely adhering to the substrate. The surface of the colonies covered with a dark brown layer of conidia.

Grows poorly on common nutrient media and does not grow at all on gelatin, potato and in milk. Sucrose inverted. Grows readily on cellulose, which is decomposed rapidly. Nitrate reduced.

Isolated from Ashkhabad soils.

13. Micromonospora parva Jensen 1932, 177.

par′va. L. fem.adj. *parva* small, minor.

Description based on Jensen (1930, 1932).

In dextrose asparagine solution hyphae 0.3–0.5 μm in diameter, spores very scarce. On filter paper in a mineral nutrient solution after 16 days the inoculum is covered by a crust of spores which are spherical to oval, 1.0–1.3 μm in diameter. On dextrose asparagine agar after 16 days, growth scanty; after 30 days, somewhat better. Vegetative mycelium flat, smooth with myceloid edges spreading into medium, colorless to pale pink or pale orange. Sporulation scanty, giving rise to thin, grayish, moist crusts on the surface. No soluble pigment produced. On potato after 16 days, growth scanty; colonies appearing as small orange granules surrounded by white haloes; after 30 days, growth becomes better; vegetative mycelium raised, lichenoid, dark orange, no sporulation or soluble pigment.

Starch weakly hydrolyzed. Sucrose not inverted. Gelatin slowly liquefied. Milk unchanged or coagulated and slowly redissolved with faintly acid reaction. Cellulose not decomposed. Nitrate not reduced.

Encountered rarely in soil. Three strains were isolated from the soil by Jensen 125-I, 279-S2 and 279-S4.

14. Micromonospora gallica (Erikson) Waksman 1961, 296. (*Actinomyces gallicus* Erikson 1935, 36; *Streptomyces gallicus* (Erikson) Waksman and Henrici 1948, 959.)

gal′li.cus. L. fem.adj. *gallica* of or belonging to the Gauls.

Description based on Erikson (1935).

Sporulation similar to Ørskov's figures for *Micromonospora chalcea* but in *M. gallica* a definite, though limited segmentation of the substrate mycelium may be observed and also a few aerial hyphae are produced. Growth on potato plug at 16 days slow, a few minute translucent pink colonies produced; at 21 days the number of small colonies has increased and by 60 days colonies are 1–2 mm in diameter, tend to be heaped up and umbilicate, and are a bright coral (pinkish red) color. On Dorset egg medium at 7 days colonies are minute, becoming confluent and are tangerine colored. On inspissated serum at 3 days growth is abundant, membranous in texture and pink colored but becoming reddish brown after 5 days, colony reverse clear, no liquefaction at 20 days. On blood agar growth is abundant, colonies are small, pink, no hemolysis in 7 days. No growth occurs on glucose agar, glycerin agar or Czapek's agar.

On gelatin growth is scant, irregular and pink in color, liquefaction is slow and at 20 days has progressed only to a slight degree. In milk a clot is formed in 10 days and peptonization occurs in

15–17 days. In litmus milk no coagulation, peptonization or color change occurs within 18 days. Reaction on tyrosine agar is negative.

Isolated from a blood culture from a case of Banti's disease by Gibson.

Type strain: NCTC 4582.

15. Micromonospora acetoformici (*sic*) Sebald and Prévot 1962, 199.

a.ce'to.for'mi.ci. L. n. *acetum* vinegar; M.L. n. *acidum formicum* formic acid; M.L. gen.n. *acetoformici* of acetic and formic (acid production).

Description based on Sebald and Prévot (1962).

Mycelium 0.4–0.6 μm in diameter, non-septate, branched, 10–100 μm in length. Conidia spherical, 0.8–1.2 μm in diameter, sessile or borne on short conidiophores, abundantly produced on less nutritious media 2–5 days after inoculation. Mature conidia are released into the media.

On a medium such as Bryant's agar with cellulose and Na_2CO_3 or Pochon's medium (Sebald and Prévot, 1962) colonies develop rapidly, are white to ochre in color, floccose in texture. Mycelium compact toward the center of the colony, loosely woven at the periphery. Conidia randomly dispersed throughout the colony.

Glucose, fructose, maltose and starch fermented. Acetic, formic and lactic acid produced but not succinic acid. Galactose, glycerol, rhamnose, sucrose and xylose are not fermented. Cellulose, chitin and pectin not hydrolyzed.

Calcium carbonate appears necessary for growth and may constitute an essential carbon source. Although nitrates are reduced to nitrites, inorganic nitrogen sources do not support growth and organic nitrogen is indispensable. Milk peptonized, coagulation variable; gelatin liquefied slightly. Indol produced. Urease and catalase absent.

Strict anaerobe.

Good growth at 28 C, optimum 37 C. Spores resistant to 100 C for 15 min.

Isolated from the posterior region of the intestinal tract of the termite *Reticulitermes lucifugus* var. *saintonnensis*, using anaerobic conditions and a saline medium containing cellulose.

Further Comment

The thermal resistance of these spores indicates that they are probably endospores similar to those found in the genus *Thermoactinomyces* and casts some doubt as to whether this species belongs in the genus *Micromonospora*.

16. Micromonospora propionici Hungate 1946, 51.

pro'pi.on.i.ci. M.L. gen.n. *propionici* of propionic (acid).

Description based on Hungate, 1946.

Spores abundant, white, spherical, 0.8 μm in diameter, borne singly on short side branches. On cellulose-proteose peptone agar growth and spore formation slow, requiring 3–4 weeks at 30–40 C, growth better upon the addition of liver extract or powdered dry grass; colonies first visible as clear areas of digestion in the cellulose. As the size of the clear area increases, a white colony appears in the center. This colony consists of a gradually expanding hollow shell the outer surface composed of vegetative filaments, the adjacent inner portion containing numerous spores and the center relatively devoid of protoplasm. Colonies appear white, no colored pigments are formed.

Characteristic fermentation products of glucose or cellulose are CO_2, propionic and acetic acid. It is believed that glucose is the chief product of cellulose digestion.

Obligate anaerobe.

Isolated from the crushed alimentary tract of a worker termite, *Amitermes minimus*, using anaerobic shake tubes containing cellulose, proteose-peptone and agar. Also believed to have been isolated from a culture of protozoa from the rumen of cattle.

Further Comments

Index Bergeyana cites Hungate 1944, 381 as the valid publication date of this species. Question could be raised in respect to this date in regard to Rule 12a (2); it is undoubtedly one of the shortest, though by no means the most ambiguous, descriptions in the literature. Hungate's intended description appears to be as cited.

Species incertae sedis

a. *Micromonospora caballi* Morquer and Comby 1943, 27.

ca.bal'li. M.L. gen.n. *caballi* of the horse.

Description based on Morquer and Comby (1943).

On Sabouraud's agar the vegetative mycelium is long, abundantly branched, non-septate, 0.2–0.6 μm in diameter. Aerial hyphae produced which terminate in short chains of microarthrospores. Lateral spores produced which are considered microconidia and are 0.6–0.8 μm in diameter and are released by a gelatinization (lysis?) of the hyphae. Thick-walled resistant cells (chlamydospores) are produced and less commonly endospores are found. In old cultures the thallus breaks up into resistant arthrospores. Colonies on Sabouraud's agar sedum-like, on potato lichenoid. Upon peptone or amino nitrogen media the colony is covered with a whitish gray efflorescence.

Starch hydrolysis variable. Lipase, maltase and lactase produced.

Gelatin liquefied. Serum coagulated. Milk coagulated and peptonized.

Parasitic on the horse but not on the rabbit, rat or mouse. Isolated from a cutaneous actinomycosis of a horse.

Further Comments

The production of aerial hyphae which terminate in short chains of microarthrospores suggests that cell wall analysis might help clarify the generic position of this organism. Plate I of the original paper (1943) shows photomicrographs labeled *Actinomyces caballi*, perhaps an earlier name used by Morquer and Comby (1943).

b. *Micromonospora* sp. ATCC 10026.

This is an important but often confused strain

(see Luedemann, 1970). It was not given a specific epithet when the antibiotic, micromonosporin, was described (Waksman *et al.*, 1947), but Pridham and Gottlieb (1948) thought it to be *M. fusca* (*M. purpureochromogenes*). The CBS culture now designated *M. fusca* and deposited by Waksman in 1948 (personal communication, G. A. deVries, CBS) is strain 3450, the same as ATCC 10026. In 1953 the CBS strain 3450 was changed from *M.* sp. 3450 to *M. fusca* Jensen, probably on the authority of Pridham and Gottlieb, 1948. The ATCC culture and the CBS culture are in most respects similar and produce antibiotics which are chromatographically similar. The CBS strain 3450 bears a great deal of resemblance to *Micromonospora narashinoensis* and the strain ATCC 10026 might best be considered a strain or subspecies of *M. narashinoensis*.

Genus II. **Thermoactinomyces** *Tsiklinsky 1899, 501*

E. Küster

Ther.mo.ac.ti.no.my′ces. Gr. n. *thermos* heat; Gr. n. *actis, actinis* a ray; Gr. n. *myces* fungus; M.L. masc.n. *Thermoactinomyces* heat (-loving) ray fungus.

The genus *Thermoactinomyces* is **characterized by the formation of single spores on both the aerial and substrate mycelium.** Occasionally longitudinal pairs of spores are formed on the aerial mycelium. Simple or branching sporophores may be so short that the spores often appear to be produced sessily or directly on the mycelium. The genus comprises, so far, **only forms capable of growing at temperatures between 45–60 C.** Gram-variable.

Cell wall may be of Type III.

Aerobic.

Type species: *Thermoactinomyces vulgaris* Tsiklinsky 1899, 501.

Further Comment

In some respects this genus is similar to *Micromonospora* but clearly distinguished from it by the formation of aerial mycelium and by its thermophilic nature.

Description of the species of genus **Thermoactinomyces**

1. **Thermoactinomyces vulgaris** Tsiklinsky 1899, 501. (*Micromonospora vulgaris* (Tsiklinsky) Waksman, Umbreit and Cordon 1939, 51; *Thermoactinomyces thalpophilus* Waksman and Corke 1953, 378; *Micromonospora thalpophilus* (Waksman and Corke) Kosmachev 1962, 52; *Thermoactinomyces antibioticus* Craveri, Coronelli, Pagani and Sensi 1964, 515.)

vul.ga′ris. L. adj. *vulgaris* common.

Substrate mycelium composed of long, branching hyphae, 0.5 μm in diameter. Colonies are colorless.

Aerial mycelium, white to cream, powdery, composed of long hyphae and short branches, 1–2 μm in length.

Spores single, round to polygonal, smooth surface. Spores formed at the tips of very short sporophores or directly on the mycelium.

Cell wall Type III; type C sugar pattern.

Nutrient agar, malt extract and yeast extract-glucose agars are most suitable for cultivation. Rapid and excellent growth, aerial mycelium white to pale yellow, colonies yellow brown. No soluble pigment formed.

Moderate or poor growth on oatmeal, asparagine glucose and potato dextrose agars.

Gelatin liquefied, milk coagulated, starch hydrolysed, nitrate not reduced.

Optimum growth temperature 60 C, optimum sporulation at 55 C, good growth at 37 C, no growth at 27 and 70 C.

Isolated from soils, compost, manure, self-heated hay.

2. **Thermoactinomyces sacchari** Lacey 1971, 327.

sac′cha.ri. M.L. n. *Saccharum* generic name of sugar cane; M.L. gen.n. *sacchari* of sugar cane.

Substrate mycelium composed of long, branching septate hyphae, 0.6–0.8 μm in diameter. Colonies are colorless to cartridge-buff, becoming slimy and bacterial in appearance.

Aerial mycelium white, sparse, composed of short hyphae, short tufted. Spores single, angular, formed at the tips of short sporophores (up to 3 μm long).

Type C sugar pattern.

Good growth on nutrient agar with glucose, yeast extract, yeast extract-glucose and potato glucose agars. Poor growth on nutrient agar, chitin agar and basal mineral salts agar containing various carbohydrates sources tested.

Usually no soluble pigment formed, very occasionally a yellow-brown pigment is formed.

Melanin not formed, casein digested.

Optimum growth temperature 55–60 C, growth temperature range between 35–65 C. No growth below 33 and above 67 C resp.

Isolated from sugar cane, self-heated bagasse.

Nomina dubia

a. *Thermoactinomyces glaucus* Henssen 1957, 406. Was never obtained in pure culture and is not extant.

b. *Thermoactinomyces thermophilus* (Berestnev) Waksman 1961, 308. Was incompletely described and is not extant.

c. *Thermoactinomyces mesophilica*. Has been mentioned (Kalakoutskii *et al.*, 1968; Cross, 1969) but a valid description has not been found.

Genus III. **Actinobifida** *Krasil'nikov and Agre 1964, 939*

T. Cross

Ac.ti.no.bi′fi.da. Gr. n. *actinis* ray; L. adj. *bifidus* cleft, divided; M.L. fem.n. *Actinobifida* ray (fungus) with bifurcations.

Fine **Gram-positive mycelium** (0.7–1.0 μm) differentiated into **substrate mycelium and aerial mycelium; hyphae show characteristic dichotomous branching particularly evident in the sporophores.** The mycelium does not break up into regular rod-shaped or coccoid cells. **Single spores on substrate and aerial mycelium,** repeated dichotomous branching of the sporophores can lead to the formation of dense spore clusters. Young apical regions of the sporophore exhibit conspicuous **heart-shaped terminal swellings** which progressively develop into two, or occasionally as a result of further division, four spores. The **spores** of two species investigated in detail have a **structure similar to bacterial endospores,** contain **dipicolinic acid** and are **resistant to heat and desiccation.**

Characteristics differentiating the species are given in Table 17.51.

Chemoorganotrophs.

Strict aerobes.

Type species: *Actinobifida dichotomica* Krasil'nikov and Agre 1964, 939.

Further Comments

The spores of the actinomycete species *Thermoactinomyces vulgaris* Tsiklinsky 1899, 501, *Actinobifida dichotomica* Krasil'nikov and Agre 1964, 939,

and *Actinobifida chromogena* Krasil'nikov and Agre 1965, 287, are similar in that they have the structure and properties of bacterial endospores and so differ from all other actinomycete spores. These species form single spores on the aerial and substrate mycelia, have a Type III cell wall composition and on whole cell hydrolysis give a type C sugar pattern (according to the scheme of Lechevalier, 1968). There thus appears to be a close relationship between two species of the genus *Actinobifida* and the type species of the genus *Thermoactinomyces* and new evidence supporting the inclusion of the two genera in the separate family Thermoactinomycetaceae Baldacci and Locci 1966, 137.

Actinobifida alba has been shown to have a Type II cell wall but a strain exhibiting very similar features (CUB 357, Cross and Lacey, 1970, 212) has a Type III cell wall (Lechevalier, H. A., personal communication, October, 1968). *Thermomonospora curvata* Henssen 1957, 401 has dichotomously branched sporophores on the aerial mycelium only (Henssen and Schnepf, 1967) and a Type III cell wall composition. The generic boundaries between species placed in the genera *Thermoactinomyces, Actinobifida* and *Thermomonospora* require further clarification (for discussion see Cross and Lacey, 1970).

Description of the species of genus **Actinobifida**

1. **Actinobifida dichotomica** Krasil'nikov and Agre 1964, 939. (*Thermomonospora citrina* Manachini, Craveri and Craveri 1966, 86.)

di.chot′o.mi.ca. Gr. adj. *dichotomus* divided,

forked; -*ica* M.L. adj. ending of intensification; M.L. fem.adj. *dichotomica* dichotomous (branches).

Substrate hyphae long (about 0.7–0.8 μm diam-

TABLE 17.51

Differential characteristics of species of genus **Actinobifida**

	Aerial mycelium color	Reverse colony color	Optimum temperature
1. *A. dichotomica*	Pale yellow	Yellow	50–58 C
2. *A. alba*	White	Colorless	37–45 C
3. *A. chromogena*	Cinnamon	Brown	55–58 C

eter), penetrating agar media, showing dichotomous branching. Compact yellow colonies on most media. Single spores on short (1.0–10 μm) sporophores showing repeated dichotomous branching to give spore clusters (Locci, Baldacci and Petrolini, 1967; Cross, 1968).

Aerial hyphae pale yellow, short, dichotomously branched (about 0.7–1.0 μm in diameter).

Immature spores appear rounded, stain strongly with simple stains. Mature spores appear pentagonal, are phase bright, stain with difficulty, resistant to desiccation and heat ($D_{100°}$ = 77 min), contain dipicolinic acid. Electron micrographs show them to be endospores with well defined coat layers and cortex, and formed within a sporangium (Cross, Walker and Gould, 1968). Germination enhanced by addition of yeast extract and corn steep liquor to agar media.

Good growth on glucose asparagine agar, peptone corn agar, potato agar + yeast extract, meat extract-peptone-yeast extract agar and Emerson's agar. Aerial mycelium yellow to pale yellow, colonies lemon yellow to orange-yellow, exhibiting intense yellow fluorescence under ultraviolet light. Yellow pigment intracellular and carotenoid (Krasil'nikov, Eli-Registan and Agre, 1967).

Nitrate not reduced. Gelatin liquefied. H₂S not produced. Proteolytic. Xanthine and tyrosine not utilized. Starch hydrolyzed.

Temperature for growth: optimum 50–58 C, maximum 65 C, minimum 35 C.

Exhibits resistance to the antibiotic novobiocin, tolerating 100 μg/ml in agar media.

Not common in soil; can be isolated from air-dried, stored soils using selective agar containing novobiocin and cycloheximide (Cross, 1968).

Type strain: No. 114 (Dept. of Soil Biology, Lomonosov State University, Moscow).

2. Actinobifida alba Locci, Baldacci and Petrolini 1967, 88.

al′ba. L. fem.adj. *alba* white.

Substrate mycelium colorless to slightly yellowish, extensive and branching, single spores on short simple or, more commonly, on dichotomously branched sporophores. No soluble pigment.

Aerial mycelium white, short, regularly bifurcate sporophores.

Spores at first oval, then roundish, smooth (1.3–

1.5 by 1.6–2.0 μm), heat sensitive, killed within 5 min at 100 C when heated in dilute phosphate buffer.

Very good growth on Czapek + yeast extract agar, half strength nutrient agar (Oxoid) and Emerson's agar. No growth on Czapek's and glucose asparagine agar.

Slight reduction of nitrate in Czapek broth + yeast extract, negative in organic broth. Gelatin liquefaction, negative in tubes, slight in plates. Proteolytic. Xanthine and tyrosine not utilized. Starch hydrolyzed.

Optimum temperature for growth 45 C, for sporulation 37–45 C. No growth at 60 C, poor growth at 24 and 27 C.

Found in soil.

Type strain: IPV 1900.

3. Actinobifida chromogena Krasil'nikov and Agre 1965, 287.

chro.mo.ge′na. Gr. n. *chroma* color; Gr. v. *gennaio* produce; M.L. adj. *chromogenes* color producing.

Substrate mycelium cinnamon to dark brown. Single spores observed with difficulty, hyphae bear numerous chlamydospores.

Aerial mycelium cinnamon, sandy to brown; exhibiting characteristic dichotomous branching as well as irregular bifurcate branches. Spores spherical to oval with 14–18 short spines, heat resistant and containing dipicolinic acid. A cortex can be seen in thin sections, typical coat layers of bacterial endospores appear to be missing (Mach and Agre, 1968).

Good growth on peptone corn agar, weak growth on glucose asparagine and Czapek agars.

No assimilation of nitrate or ammonium salts. Gelatin liquefaction slow. Starch hydrolysis weak. Brown soluble pigment in liquid and agar media.

Optimum temperature for growth 55–58 C, maximum 68 C, minimum 35–37 C.

Type strain: No. 577 (Dept. of Soil Biology, Lomonosov State University, Moscow).

Addendum: The specific names *Actinobifida globosa* Krasil'nikov and Agre 1965, 291 and *Actinobifida fusca* Krasil'nikov and Agre 1965, 291 are considered as synonyms of *Micromonospora globosa* and *Micromonospora purpureochromogenes* q.v.

Genus IV. **Thermomonospora** *Henssen 1957, 398*

E. Küster

Ther.mo.mon'o.spo.ra. Gr. n. *thermos* heat; Gr. adj. *monos* single, solitary; Gr. fem.n. *spora* seed; M.L. fem.n. *spora* a spore; M.L. fem.n. *Thermomonospora* the heat (-loving) single-spored (organism).

Single spores formed only on aerial mycelium.

Vegetative hyphae flexuous and non-septate. Aerial mycelium white; aerial hyphae simple or branched, arising from side or terminal branches of vegetative hyphae. Single spores are usually formed acropetalously at the tip of the sporophores; sporophores occasionally branched di-chotomously. Gram-variable. Not acid-fast. Colonies colorless or yellow.

Cell wall may be of Type III or IV.

Facultative aerobe.

Thermophilic; optimal growth temperature 45–55 C.

Occurs in soils, manure and hay.

Type species: *Thermomonospora curvata* Henssen 1957, 401 (Henssen and Schnepf 1967, 227).

Description of the species of genus **Thermomonospora**

1. **Thermomonospora curvata** Henssen 1957, 401.

cur.va'ta. L. fem.adj. *curvata* curved.

Substrate mycelium composed of long, flexuous, branching hyphae, mainly curved in one direction.

Aerial mycelium chalk white. The hyphae are long (30–50 μm) and branching; small, sometimes dichotomous, branches act as sporophores; spores form at the tips.

Spores single, round to pyriform, 0.6–1.5 by 0.3–0.9 μm in size, often arranged in clusters. Spore surface is covered with short spines.

Cell wall Type III; type C sugar pattern (Lechevalier, 1968).

No soluble pigment formed.

Cellulose, yeast extract, yeast extract-starch, oatmeal and manure agars are most suitable for cultivation and spore formation. Good growth, aerial mycelium white, colonies yellow to orange.

Limited growth on asparagine-glucose, casein-glucose and meat extract agars.

Gelatin not liquefied; milk not changed after 16 days.

Starch not hydrolyzed. Slight reduction of nitrate to nitrite.

Optimum growth temperature 50 C, limited growth at 40 C and 65 C, no growth at 28 C.

Isolated from fresh manure, manure compost, straw.

2. **Thermomonospora viridis** (Schuurmans *et al.*) Küster and Locci 1963, 193. (*Actinomyces monosporus* Lehmann and Schütze in Schütze 1908, 50; *Micromonospora monospora* (Lehmann and Schütze) Krasil'nikov 1941, 131; *Thermoactinomyces monosporus* (Lehmann and Schütze) Waksman and Corke 1953, 377; *Thermoactinomyces monospora* (Lehmann and Schütze) Waksman and Corke 1953, 378; *Thermoactinomyces viridis* Schuurmans, Olson and San Clemente 1956, 61; *Thermopolyspora glauca* Corbaz, Gregory and Lacey 1963, 450.)

vir'i.dis. L. adj. *viridis* green.

Substrate mycelium made up of long, flexuous, branching hyphae, occasionally with curved endings.

Aerial mycelium, if present, usually grayish green in color. Short sporophores may appear which sometimes show a spike-like structure. Single spores on the aerial mycelium sessile or on short sporophores, two spores occasionally occur. Ovoid, 0.9–1.1 by 1.2–1.4 μm in size.

Cell wall Type IV; type A sugar pattern.

On nutrient agar (Oxoid) and Lab Lemco agar (Oxoid) growth good; vegetative mycelium colorless with dark green reverse, aerial mycelium grayish green, dark green soluble pigment.

Moderate to good growth on potato dextrose agar, aerial mycelium greenish.

Poor growth with little or no aerial mycelium on oatmeal agar and glycerol asparagine agar.

No growth on Czapek agar, V-8 Juice agar or potato slices.

Gelatin liquefied after 4 days. Milk coagulated, H₂S produced, strong hydrolysis of starch. Slight amount of nitrite from nitrates.

Melanin not formed.

Optimum growth temperature 55 C, optimum sporulation at 45 C, good growth at 37 C, no growth at 27 C and 65 C.

Isolated from hay, peat and various soils.

Further Comment

Some authors (Lechevalier and Lechevalier, 1967; Kalakoutskii *et al.* 1968; Cross 1969) suggest placing *T. viridis* in the genus *Micropolyspora* because of its cell wall composition (Type IV) and sugar pattern (type A); however, a final placement requires further examination.

a. *Thermomonospora fusca* Henssen 1957, 399.

b. *Thermomonospora lineata* Henssen 1957, 400.

Neither has been obtained in pure culture and strains are not extant.

Genus V. **Microbispora** Nonomura and Ohara 1957, 307

T. CROSS

(*Thermopolyspora* Henssen 1957, 394; *Waksmania* Lechevalier and Lechevalier 1957, 107.)

Mic.ro.bi'spo.ra. Gr. adj. *micrus* small; Gr. adj. *bis* two; Gr. n. *spora* a spore; M.L. fem.n. *Microbispora* the small two-spored (organism).

Spores in characteristic longitudinal pairs are formed on aerial mycelium; (Plate 17.8, Fig. 12) as a rule **no spores formed on substrate mycelium** (occasional production on substrate mycelium reported for one unidentified strain, Lechevalier and Lechevalier, 1967). Spores (1.2–1.8 μm diameter) may be sessile or on short sporophores; paired spores may be closely arranged along the aerial hyphae, giving the appearance of a catkin or borne at longer intervals (Plate 17.8, Figs. 10, 11). In most species aerial mycelium is pink.

Gram-positive. **Purified cell wall Type III.**

Several species form bronze-violet needle-like crystals of iodinin on certain solid media (e.g. oatmeal agar) and release this compound in submerged culture (Gerber and Lechevalier 1964) (Table 17.52). B vitamins, particularly thiamine essential for growth.

Aerobic to facultatively anaerobic. Mesophiles and facultative thermophiles.

Normal habitat appears to be soil and lake muds; one species, *M. rosea*, has been implicated in a case of pericarditis and pleuritis in man (Louria and Gordon, 1960).

Type species: *Microbispora rosea* Nonomura and Ohara 1957, 307.

Further Comments

For discussion on the priority of the generic name *Microbispora* over *Waksmania* and *Thermopolyspora* see Lechevalier, 1965.

Description of the species of genus **Microbispora**

For characteristics differentiating the species of the genus see Table 17.52.

1. Microbispora rosea Nonomura and Ohara 1957, 307. (*Waksmania rosea* Lechevalier and Lechevalier 1957, 107.)

ro'se.a. L. fem.adj. *rosea* rose colored.

Substrate mycelium penetrating the agar medium and forming tough compact colonies, colorless to orange-pink or red-brown.

Aerial mycelium white at first becoming pale pink with the formation of spores.

Good growth on oatmeal and starch agars with added yeast extract.

Type strain: ATCC 12950; IMRU 3757.

1a. *Microbispora rosea* subsp. *rosea* Nonomura and Ohara 1960, 404.

Reduces nitrate to nitrite.

1b. *Microbispora rosea* subsp. *nonnitrogenes* Nonomura and Ohara 1960, 404.

Unable to reduce nitrate to nitrite.

2. Microbispora aerata (Gerber and Lechevalier) *comb. nov.* (*Waksmania aerata* Gerber and Lechevalier 1964, 598.)

aer.a'ta. L. fem.adj. *aerata* covered with bronze.

Aerial mycelium off white to pale pink.

Abundant production of bronze iodinin crystals on oatmeal and "Pablum" agar (Lechevalier and Lechevalier, 1957).

Growth temperature range 28–55 C.

Type strain: ATCC 15448.

3. Microbispora amethystogenes Nonomura and Ohara 1960, 403.

am.e.thys.to'genes. L. adj. *amethysteus* amethyst colored; Gr. v. *gennaio* produce; M.L. adj. *amethystogenes* producing violet-colored (crystals).

On oatmeal-yeast extract agar forms a pale yellowish brown soluble pigment.

Growth temperature range 20–45 C.

Type strain: None specified.

3a. *Microbispora amethystogenes* subsp. *amethystogenes* Nonomura and Ohara 1960, 403.

Reduces nitrate to nitrite.

TABLE 17.52

Characteristics differentiating the species of genus **Microbispora**[a]

	1. M. rosea	2. M. aerata	3. M. amethystogenes	4. M. bispora	5. M. chromogenes	6. M. diastatica	7. M. parva	8. M. thermodiastatica	9. M. thermorosea
Aerial mycelium color: white	−	−	−	+	−	−	−	−	−
pink	+	+	+	−	+	+	+	+	+
Carbon sources:									
glucose	3	3	d	3	3	3	3	3	2
glycerol	3	3	3	0	3	3	3	2	3
arabinose	3	3	3	0	3	3	3	3	3
rhamnose	3	1	0	1	0	1	1	0	0
inositol	0	0	1	2	3	0	0	0	0
Starch hydrolysis	0	3	0	0	d	3	d	3	0
$NO_3 \rightarrow NO_2$	+	+	+	−	+	−	−	−	−
Gelatin liquefaction	d	1	d	0	d	d	1	1	1
Milk peptonization	d	3	2	0	d	1	1	3	1
Melanoid pigment	0	1	0	0	d	0	0	0	0
Soluble pigment	0	0	2	0	2	d	0	0	0
Iodinin production	−	+	+	−	−	−	d	−	−
Growth at: 25 C	+	−	+	−	+	+	+	−	−
50 C	−	+	−	+	−	−	+	+	+
55 C	−	+	−	+	−	−	−	+	+

[a] Extent of growth, utilization or pigment: 0 = none; 1 = slight; 2 = good; 3 = excellent. + = most strains positive; − = most strains negative; d = some (less than 90%) strains positive, some negative.

3b. *Microbispora amethystogenes* subsp. *non-reducans* Nonomura and Ohara 1960, 403.

Unable to reduce nitrate to nitrite.

4. Microbispora bispora (Henssen) Lechevalier 1965, 141. (*Thermopolyspora bispora* Henssen 1957, 395.)

bi.spo′ra. Gr. adj. *bis* two, paired; Gr. n. *spora* a spora; M.L. n. *bispora* two spores.

Aerial mycelium chalk-white, many lateral paired spores may be formed giving a catkin-like appearance.

Growth temperature range 50–60 C.

Reference strains: Two "type" strains have been listed, ATCC 15737 (IMRU 3759) and ATCC 19993 (R51, CBS 139.67), see Henssen and Schnepf 1967 for discussion.

5. Microbispora chromogenes Nonomura and Ohara 1960, 404.

chro.mo′ge.nes. Gr. n. *chroma* color; Gr. v. *gennaio* produce; M.L. adj. *chromogenes* producing color.

Oatmeal yeast extract agar, aerial mycelium distinct pink, soluble pigment dark purple-gray. Soluble pigment dark yellowish to olive green in yeast extract starch agar.

Type strain: None specified.

6. Microbispora diastatica Nonomura and Ohara 1960, 404.

di.a.sta′ti.ca. M.L. fem.adj. *diastatica* starch hydrolyzing.

On oatmeal yeast extract agar, aerial mycelium distinct pink, some strains produce pale yellow soluble pigment.

Type strain: None specified.

7. Microbispora parva Nonomura and Ohara 1960, 403.

par′va. L. fem.adj. *parva* small (feeble growth).

On oatmeal yeast extract agar, aerial mycelium white to pale pink, some strains produce few iodinin crystals after 1 month at 30 C.

Type strain: None specified.

8. Microbispora thermodiastatica Nonomura and Ohara 1969, 706.

ther.mo.di.a.sta′ti.ca. Gr. adj. *thermus* hot; M.L. adj. *diastaticus* starch hydrolyzing; M.L. fem.adj. *thermodiastatica* starch hydrolyzing, heat (loving organism).

On glucose agar, aerial mycelium pale pink. On oatmeal yeast extract agar, growth moderate to poor after 20 days at 50 C, aerial mycelium white to yellow-brown.

Type strain: M_2-59.

9. **Microbispora thermorosea** Nonomura and Ohara 1969, 707.

ther.mo.ros'e.a. Gr. adj. *thermus* hot; L. adj. *roseus* rose colored; M.L. fem.adj. *thermorosea* heat (loving) rose colored.

On glycerol agar, aerial mycelium pale pink. Growth on oatmeal yeast extract and starch agars poor after 20 days at 50 C, no aerial mycelium. Type strain: M₂-64.

Genus VI. **Micropolyspora** Lechevalier, Solotorovsky and McDurmont 1961, 13

T. CROSS

Mic.ro.po'ly.spo.ra. Gr. adj. *micrus* small; Gr. adj. *poly* many; Gr. n. *spora* a seed; M.L. fem.n. *Micropolyspora* the small many-spored (organism).

Substrate and aerial mycelium formed (about 1 μm in diameter) **both bearing short chains of 1–20 spores.** Spores (1.2–1.5 μm in diameter) are formed basipetally on lateral branches or terminally. Gram-positive.

Purified cell walls Type IV, contain *meso*-diaminopimelic acid, arabinose and galactose.

Mesophiles and facultative thermophiles.

Type species: *Micropolyspora brevicatena* Lechevalier, Solotorovsky and McDurmont 1961, 13.

Further Comments

The several species described in recent years have received little detailed study and future comparisons may show some to be synonymous. Recent work on the fine structure of developing *M. rectivirgula* spores (Dorokhova, Agre, Kalakoutski and Krasil'nikov, 1969, 1970) illustrates structures unique in the actinomycetes, the spores have very thick walls particularly at sites of contact between adjacent spores. Similar studies on other species may give an additional useful criterion for delimiting the genus.

The various species have only rarely been iso-lated from soil, they appear to suffer from the competition of other actinomycetes and bacteria when rich media and conventional dilution plates are used. The isolation of *M. faeni* from many samples of compost, hay and grain (by diluting in air and using an Andersen sampler (Andersen, 1958)) would indicate that at least one species of this genus is common in soil. Improvements in isolation techniques could reveal further species and provide information on their frequency and distribution in natural environments.

Though earlier, a genus name *Micropolispora* had been published (Shchepkina, 1940, 643) the name *Micropolyspora* has been retained (Lechevalier and Lechevalier, 1967).

Strains placed in the genus *Thermoactinopolyspora* Craveri and Pagani 1962, 118 are considered to belong to the genus *Micropolyspora* by Lechevalier and Lechevalier 1967, and Kalakoutski, Agre and Krasil'nikov 1968.

See Comment on *Thermomonospora viridis* (page 858) regarding the possible relationship of this species to the genus *Micropolyspora*.

For characteristics differentiating the species of the genus see Table 17.53.

Description of the species of genus **Micropolyspora**

1. **Micropolyspora brevicatena** Lechevalier, Solotorovsky and McDurmont 1961, 13.

brev.i.cat.e'na. L. adj. *brevis* short; L. n. *catena* chain; M.L. n. *brevicatena* short chain.

Substrate mycelium about 1 μm in diameter, long, branching, penetrating the agar medium and forming compact colonies which are at first whitish, becoming yellow orange. Spores formed within and at the surface of the agar. Aerial mycelium is not abundant on most media, hyphae long, branching, about 1 μm in diameter. Short chains of spores along as well as at the tip of the main hyphae, chains often curved or flexuous. Spores spherical to oblong sometimes pyriform, about 1.5 μm in diameter with a slightly warty surface. Spores in short chains (2–10), sometimes single, on short sporophores or sessile on the side of the hyphae.

One strain (IMRU 1084) forms a light brown soluble pigment after 2 weeks incubation; other strains formed no pigment.

Abundant slightly wrinkled pale orange growth on casein hydrolysate glycerol agar. Growth on yeast extract agar thin, pale orange with white powdery aerial mycelium. Growth on nutrient agar with glucose thin, powdery, pale orange with powdery white surface. No growth on starch agars or in nitrate broth.

L-Glutamic acid good source of nitrogen for growth, sporulation stimulated by the addition of L-arginine, MgSO₄, K₂HPO₄ and CaCl₂.

Optimum growth temperature 28–37 C; no growth at 45 C.

Four strains have been isolated, two from sputum of patients who had been treated for

TABLE 17.53

Characteristics differentiating the species of genus **Micropolyspora**

	1. M. brevicatena	2. M. angiospora	3. M. caesia	4. M. faeni	5. M. rectivergula	6. M. rubrobrunea	7. M. thermovirida	8. M. viridinigra
Color of aerial mycelium								
white/cream/yellow	+	+	−	+	+	−	−	−
blue/green/gray	−	−	+	−	−	+	+	+
Color of colony								
yellow/orange	+	+	−	+	+	−	−	−
green/gray	−	−	+	−	−	−	+	+
brown	−	−	−	−	−	+	−	−
Average no. spores in chain								
1–5	−	−	+	−	−	−	−	−
5–10	+	−	−	+	+	−	+	−
5–20	−	+	−	−	−	+	−	+
Spore surface								
smooth	−	−	+	+	+	−	−	−
warty	+	−	−	−	−	−	−	−
fine teeth/spines	−	+	−	−	−	+	+	+
Growth temperature C								
25	−	+	+	−	−	−	+	−
28	+	+	+	−	−	−	+	−
37	+	+	+	+	+	+	+	+
45–maximum	−	−	+	+	+	+	+	+

tuberculosis, one from an individual without active tuberculosis, and one from a pneumonic calf lung.

Reference strains: ATCC 15333; IMRU 1086W; ATCC 15725; IMRU 1084.

2. **Micropolyspora angiospora** Zhukova, Tsyganov and Morozov 1968, 728.

ang.io.spor′a. Gr. n. *angium* vessel; Gr. n. *spora* a spore; M.L. n. *angiospora* spores enclosed (in capsules).

White aerial mycelium on soya flour and oatmeal agars, substrate mycelium and colonies white to deep flesh or ochre colored. Spore chains (2–15 spores) are curved or irregularly coiled.

Spores spherical to oval (1.5–2.4 by 1.2–2.2 μm), surface spines enveloped in a translucent capsule.

Optimum growth temperature 27 C, minimum 21 C, maximum 37 C.

Type strain: LIA 3479/30.

3. **Micropolyspora caesia** Kalakoutskii 1964, 858.

cae′s.ia. L. fem.adj. *caesia* bluish gray.

Substrate mycelium white becoming gray-green on some media. Aerial mycelium white becoming grayish blue with abundant sporulation. The majority of spores on both substrate and aerial mycelia are borne singly either directly on the hyphae or on short sporophores. Paired spores and chains of two to four spores occasionally observed.

Spores round to oval, smooth, 0.6–1.0 μm in diameter.

Good growth on meat peptone, peptone corn and potato extract agars.

Produces antibiotics Micropolysporins A and B active mainly against Gram-positive bacteria.

No soluble pigments reported.

Optimum growth temperatures 28–45 C minimum 23 C, maximum 55 C.

Type strain: None specified.

4. **Micropolyspora faeni** Cross, Maciver and Lacey 1968, 354. (*Thermopolyspora polyspora* Henssen 1957, 396.)

fae′ni. L. n. *faenum* hay; L. gen.n. *faeni* of hay.

Substrate mycelium about 0.5–0.8 μm in diameter, branching, penetrating the agar medium and forming colonies which are at first colorless, later orange-yellow to yellow-brown. Short chains of spores abundant. Aerial mycelium about 1.0 μm in diameter, white and abundant on certain media. Short straight chains of spores borne both laterally and terminally (5–10 spores).

Spores globose to oval, 0.7–1.3 μm long, smooth; Gram-stain shows conspicuous unstained areas between spores in chain.

Good growth and aerial mycelium on yeast extract and nutrient agars, no soluble pigments. Excellent orange-yellow growth on skim milk agar, no aerial mycelium, no casein digestion.

Optimum growth temperature 50 C, minimum

and maximum growth temperatures (32–60 C) influenced by nature of supporting medium.

The principle cause of farmer's lung, a hypersensitivity disease resulting from the inhalation of high numbers of spores. Such high numbers can be encountered in overheated hay, grain and mushroom compost.

Type strain: A94; NCIB 9984.

5. Micropolyspora rectivirgula (Krasil'nikov and Agre) Prauser and Momirova 1970, 220.
(*Thermopolyspora rectivirgula* Krasil'nikov and Agre 1964, 106.)

rec'ti.vir.gu.la. L. adj. *rectus* straight; L. dim.n. *virgula* twig; M.L. n. *rectivirgula* straight branch.

Substrate mycelium 0.7–0.8 μm in diameter, colorless, yellowish or brownish yellow. Spore chains present. Aerial mycelium well developed on certain media; pale yellow, dark cream or sand colored. Straight chains of spores present. Spore chains on short sporophores, 3–10 spores in a chain, often 4–5. Spores ovate to spherical, 1.2–1.5 μm in diameter with a slightly tuberculate or smooth surface. Thick spore wall (70–100 nm) enclosed initially in a multilayered outer sheath. Good growth and aerial mycelium on peptone corn steep and glucose asparagine agars.

Optimum growth temperatures 45–55 C, minimum 30 C, maximum 65 C.

Type strain: INMI 683.

Further Comment

Recent comparative studies by Kalakoutskii, Agre and Krasil'nikov (1968) and phage sensitivity studies by Prauser and Momirova (1970) indicate a close similarity between *M. faeni* and *M. rectivirgula*.

6. Micropolyspora rubrobrunea Krasil'nikov, Agre and El-Registan 1968, 1072.

ru.bro.brun'ea. L. adj. *rubrus* red; L. adj. *bruneus* brown; M.L. fem.adj. *rubrobrunea* reddish brown.

Substrate mycelium forming red-brown, cinnamon or yellowish colonies which appear bright red under ultraviolet light. Aerial mycelium white at first, later (5–6 days) bluish green. Chains of 2–20 spores (occasional single spores) on aerial and substrate mycelia, chains curved often spirally. Spores irregular spherical or oval with fine spines.

Good growth and aerial mycelium on media with added yeast autolyzate. On Czapek's agar the scant aerial mycelium soon autolyzes. No soluble pigments.

Optimum growth temperatures 45–55 C, minimum 37 C, maximum 65 C.

Type strain: None specified.

7. Micropolyspora thermovirida Kosmachev 1964, 269.

ther.mo.vi'ri.da. Gr. n. *thermos* heat; L. adj. *viridis* green; M.L. fem.adj. *thermovirida* heat (loving) green organism.

Substrate mycelium dark gray (the presence of spores is not specifically mentioned by Kosmachev). Aerial mycelium pale green, grayish green or dark green, bearing chains of 1–15 occasionally 20 spores. Spores irregular, oval or round, 1.0–1.5 by 1.4–2.0 μm, with undulate outline and rounded short spines.

Good growth and aerial mycelium on starch-corn extract-yeast extract (Umezawa's) agar. Weak colorless growth on Czapek glucose agar, no aerial mycelium. No soluble pigments.

Optimum growth temperatures 40–50 C, minimum 24 C, maximum 57 C.

Type strain: None specified.

8. Micropolyspora viridinigra Krasil'nikov, Agre and El-Registan 1968, 1071.

vi.ri.di.ni'gra. L. adj. *viridis* green; L. adj. *niger* black; M.L. fem.adj. *viridinigra* green-black.

Substrate mycelium forming dirty brown, greenish brown or black colonies, which appear dark green or black under ultraviolet light. Aerial mycelium white at first, later (5–6 days) bluish green. Chains of 2–20 spores (occasional single spores) on aerial and substrate mycelia, chains curved often spirally. Spores irregular spherical or oval with fine spines.

Good growth and aerial mycelium on media with added yeast autolyzate. On Czapek's agar the scant aerial mycelium soon autolyzes. No soluble pigments.

Optimum growth temperatures 45–55 C, minimum 37 C, maximum 65 C.

Type strain: None specified.

ADDENDUM to *MICROMONOSPORACEAE*

Genera and species incertae sedis

George M. Luedemann

A. *Promicromonospora* Krasil'nikov, Kalakoutskii and Kirillova 1961, 107.

Pro.mic.ro.mo.no'spo.ra. L. pref. *pro-* before; M.L. fem.n. *Micromonospora* a genus of the *Actinomycetales*; M.L. fem.n. *Promicromonospora* a group of organisms thought to precede phylogenetically the genus *Micromonospora*.

Characteristic of this genus is the development

of single spores either on short sporophores or directly on the branches of the substrate mycelium (sessile). The spores at first are elongated but become round or oval, their diameter is usually larger than the mycelium, the spores ranging from 0.7–1.2 μm in diameter. Segmentation of the substrate mycelium occurs within 24–48 hr (above 20 C) resulting in the formation of diphtheroid- and coccoid-shaped cells resembling elements seen in *Nocardia* (*Proactinomyces*) and *Mycobacterium*. Aerial mycelium occurs at the periphery of some colonies but is scant or may not develop at all. The organisms resemble in some respects representatives of the genus *Micromonospora* and in other respects the genus *Nocardia* (*Proactinomyces*).

Type species: *Promicromonospora citrea* Krasil'nikov, Kalakoutskii and Kirillova 1961, 108.

A1. *Promicromonospora citrea* Krasil'nikov, Kalakoutskii and Kirillova 1961, 108.

cit′re.a. M.L. fem.adj. *citrea* lemon yellow colored.

Mycelium monopodially branched, 0.5–1.0 μm in diameter. Spores formed singly on substrate mycelium either on short sporophores or directly on the hyphae of the substrate mycelium (sessile). Spores not observed in chains or pairs and only rarely in clusters.

Colonies glabrous or with a slight aerial mycelium. Colony color yellow. Diffusible pigments not produced. Organism grows well on synthetic medium containing ammonium salts and organic media containing starch or potato. On starch yeast medium colonies rough or smooth, matt (dull) yellow, aerial mycelium grayish white on periphery of colony, seldom covering entire colony.

Carbohydrate assimilation using the Pridham and Gottlieb (1948) basal medium with 20 carbohydrates poor. By incorporating the carbohydrates into a dilute (1:20) beef peptone agar somewhat better growth was obtained with glucose, maltose, mannitol, starch, sucrose and xylose. Starch hydrolyzed. Cellulose not decomposed.

Gelatin not liquefied. Digestion and growth on milk poor. Reduction of nitrate to nitrite poor. Antibiotically inactive.

Growth of culture excellent on beef peptone agar, potato agar, starch yeast agar and potato slice. Poor growth on glucose asparagine agar, meat peptone gelatin, milk and carrot slice.

Aerobic.

Grows between 10–48 C, although poor growth occurs at extremes; no growth at 50 C; at 37 C growth is rapid and fragmentation of substrate mycelium is accentuated, aerial mycelium is inhibited.

Two strains (no. 18 and 25) were isolated from garden soil.

Further Comment

Cultures of this organism appear macroscopically similar to a streptomycete (white aerial mycelium) contaminated with a bacterium (yellow viscid sarcina-like colony). Repeated single colony isolation fails to separate the two colony forms and microscopic observation confirms the pattern of development described by the authors of the genus. Cell wall analysis (Group VI) and whole cell analysis (no diagnostic sugars) indicate few affinities with known actinomycete genera (M. P. Lechevalier unpublished data). The organism described appears distinct and difficult at this time to relate to a known family of the *Actinomycetales*.

B1. *Actinomonospora lusitanica* Castellani, De Brito and Pinto 1959, 35.

lu.si.ta′ni.a. L. n. *Lusitania* ancient name of the region roughly corresponding to modern Portugal; M.L. adj. *lusitanica* pertaining to Portugal.

Conidia produced singly at the apices of morphologically undifferentiated mycelial branches or laterally on short sporophores (pedicels), rarely sessile, never in clusters or chains. Conidia globular, subglobular or pyriform 2.5–3.0 μm in diameter. Terminal conidia (chlamydospores?) may give rise *in situ* to additional conidia (indeterminate development). Mycelium appearing septate under electron microscope. Arthrospores and endoconidia produced. Gram-positive, not acidfast.

Culture grows well on most common laboratory media. Aerial mycelium absent. Earthy or musty odor absent. On 2% glucose agar after 3–4 weeks growth is abundant, nodular or irregularly raised, occasionally membranous and rugose, firm and adherent to substrate. Color of colony red or purplish or red mixed with white zones. Occasionally culture entirely white. Pigment not diffusible. On Czapek's agar growth is only slightly less abundant but red or purplish colony pigment appears more intense. Growth is poor on glycerine agar, trypsin agar, calcium malate agar and MacConkey's medium. Growth on potato good, granular, bright red later becoming dark red or purplish. Similar growth on carrot.

Starch not hydrolyzed, sugars not fermented. Carbohydrates utilized: D-arabinose, D-galactose, D-glucose, glycerol, inositol, inulin, lactose, D-mannitol, D-sorbitol, soluble starch, sucrose and D-xylose. Maltose and raffinose not utilized. Arbutin not split.

Gelatin slowly liquefied. In milk coagulation and complete peptonization rarely occur. No production of indol or hydrogen sulfide. Voges-Proskauer negative. Nitrate not reduced to nitrite.

Growth poor on blood agar, no hemolysis.

Aerobic or facultatively microaerophilic.

Grows well between 29–37 C, not at 55 C, poorly at 4 C.

Not pathogenic for mice, rats or guinea pigs.

Isolated from an autochthonous case of yellow-grained mycetoma in Portugal.

Further Comments

It is difficult to place this organism generically. The original description and illustrations of the morphology of the spores and sporophores do not exclude it from the genus *Micromonospora*. According to the authors the terminal spores in *Actinomonospora* are large, 2–3 µm, and even though lateral spores are found they are not produced in grape-like clusters. These characteristics are not inconsistent with the morphological features found in species of *Micromonospora*. Several cultures of *Actinomonospora*, obtained from the CBS at different times and studied by the writer, were found to produce only the white waxy colonies noted also by Castellani *et al.* (1959). No spores were found and the culture was never observed to produce the characteristic red or purplish pigments noted by the original authors. Cell wall analysis of this culture is Type III (*Micromonospora* species belong to cell wall Type II). Whole cell analysis is reported as type B (M. P. Lechevalier unpublished data). (References to methods: Becker *et al.*, 1964; Lechevalier, 1968). G. Mungelluzzi (1966) reported obtaining cultures of *Actinomonospora lusitanica* from Castellani reconfirming the original observations.

Important Notes

for

Users of this Edition

1. Always read both generic and species descriptions because characters listed in the generic description are not usually listed in the species descriptions.

2. In tables, characters common to all taxa are not shown but may be listed in footnotes.

3. Generally in tables (exceptions are clearly indicated in footnotes, q.v.) the meanings of symbols are as follows:

+   more than 90 % strains positive
−   more than 90 % strains negative
d   11–89 % strains positive
( )   delayed reaction
w   weak reaction
D   Different reactions in different taxa (species of a genus or genera of a family)
v   strain instability (NOT differences between strains)

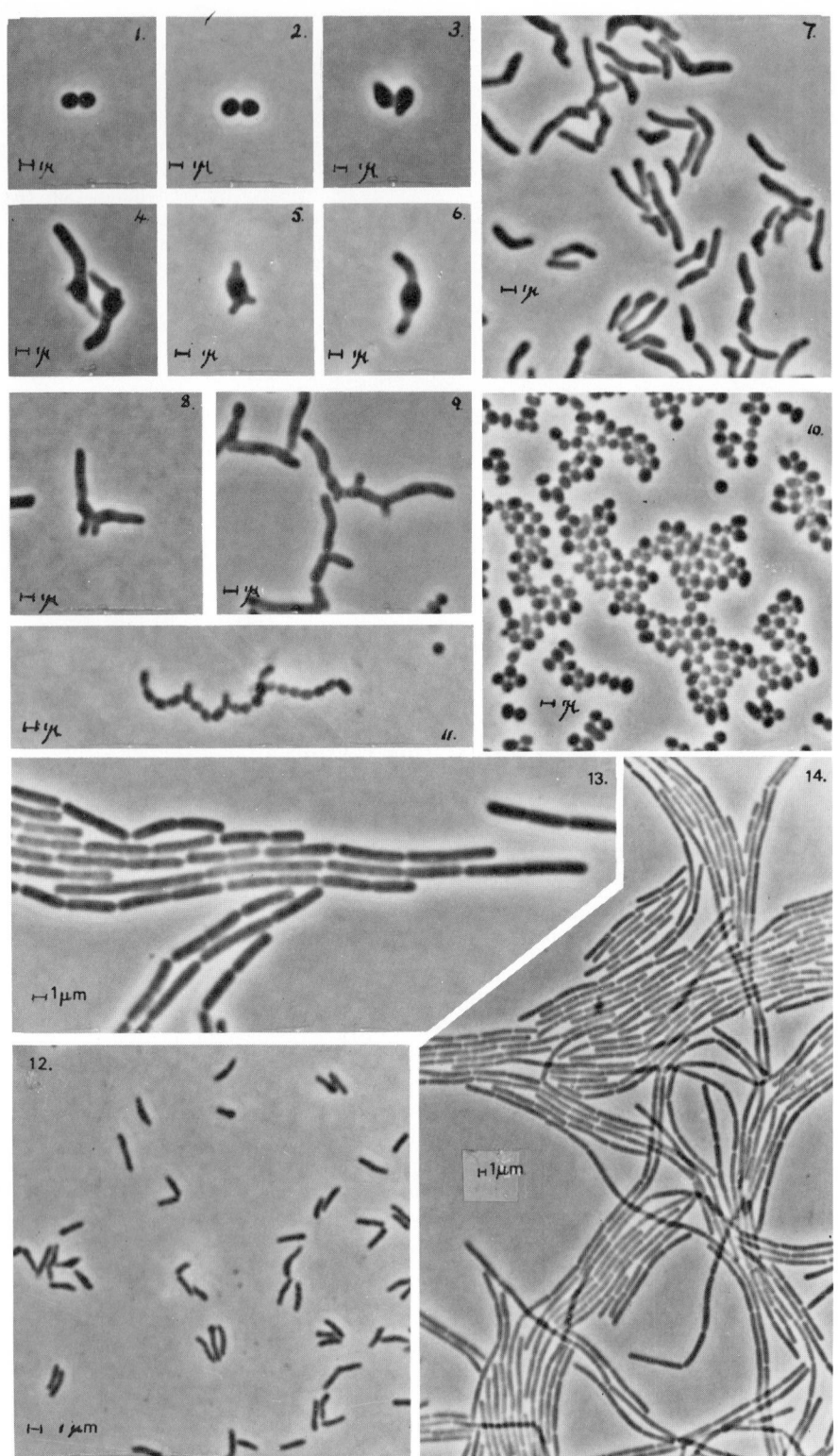

Plate 17.1. Morphology of Species of *Arthrobacter, Cellulomonas* and *Kurthia*.

A. globiformis (ATCC 8010, Figs. 1–7, 10; ATCC 4336, Figs. 8, 9, 11). Cultures were at 25 C.

On YSXA medium (0.1% w/v yeast extract, soil extract agar): Figures 1–4 slide cultures after 1.5, 4.25, 6 and 9.75 hr, respectively; Figures 7–8 after 24 hr.

On MYA medium (0.1% w/v yeast extract, mineral salts agar): Figures 5–6 after 5 hr; Figure 11 after 3 days.

On TSXA medium (0.25% w/v yeast extract and peptone, 25% v/v soil extract agar): Figure 9 after 11.5 hr; Figure 10 after 3 days.

Figure 12. *Cellulomonas flavigena* (ATCC 482) on TSXA medium after 45 hr at 30 C.

Figures 13–14. *Kurthia zopfii.* edge of colony on nutrient agar plate incubated at 30 C for 24 hr. All phase contrast by R. M. Keddie. Figure 14 × 940; all others × 1875.

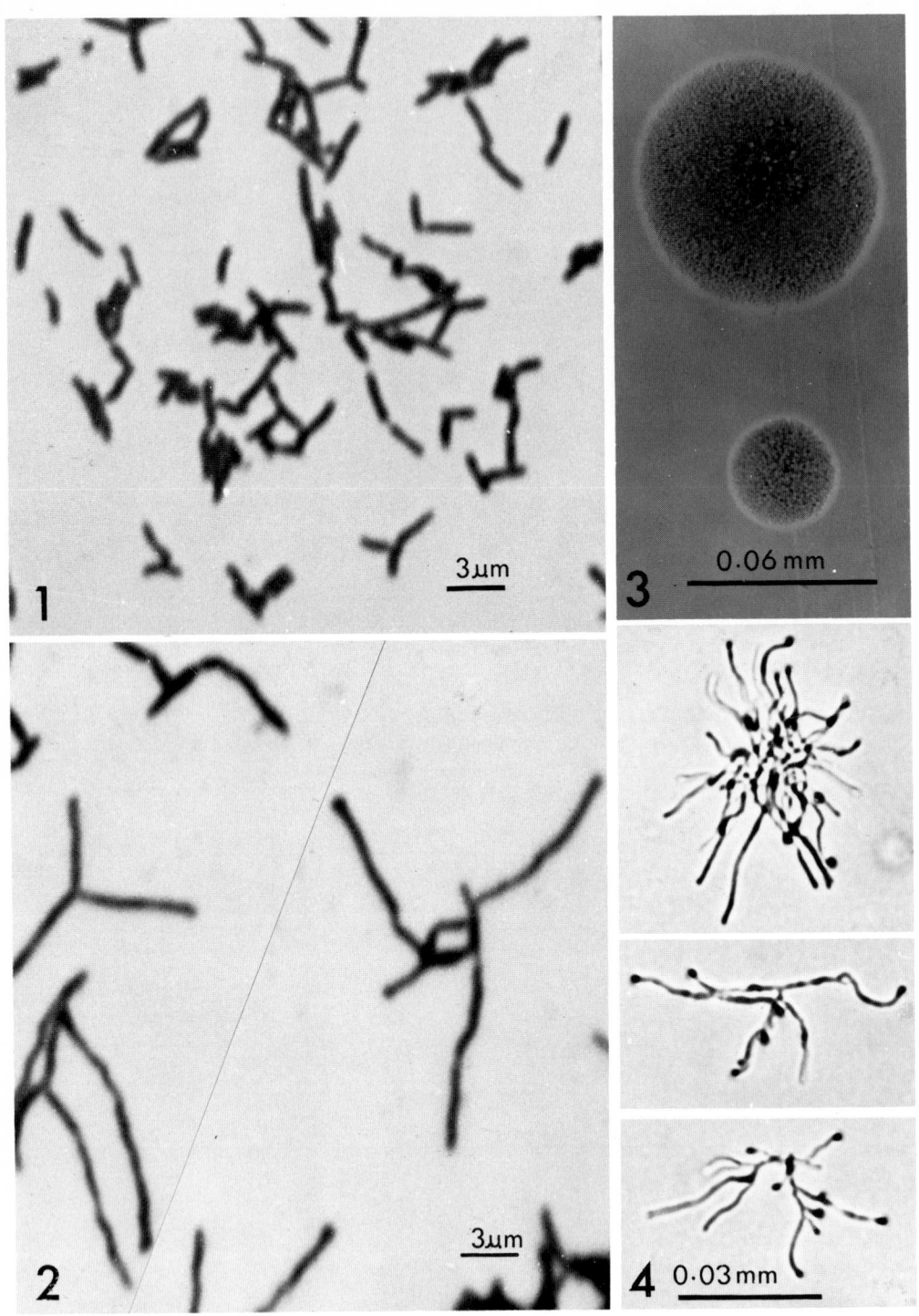

Plate 17.2. *Actinomyces*

Fig. 1. *Actinomyces israelii*, Gram-stain showing diphtheroid morphology.

Fig. 2. *A. israelii*, Gram-stain showing branching morphology.

Fig. 3. *A. bovis*, microcolony, 24 hour growth on brain heart infusion (BHI).

Fig. 4. *A. israelii*, microcolony, 24 hour growth on BHI.

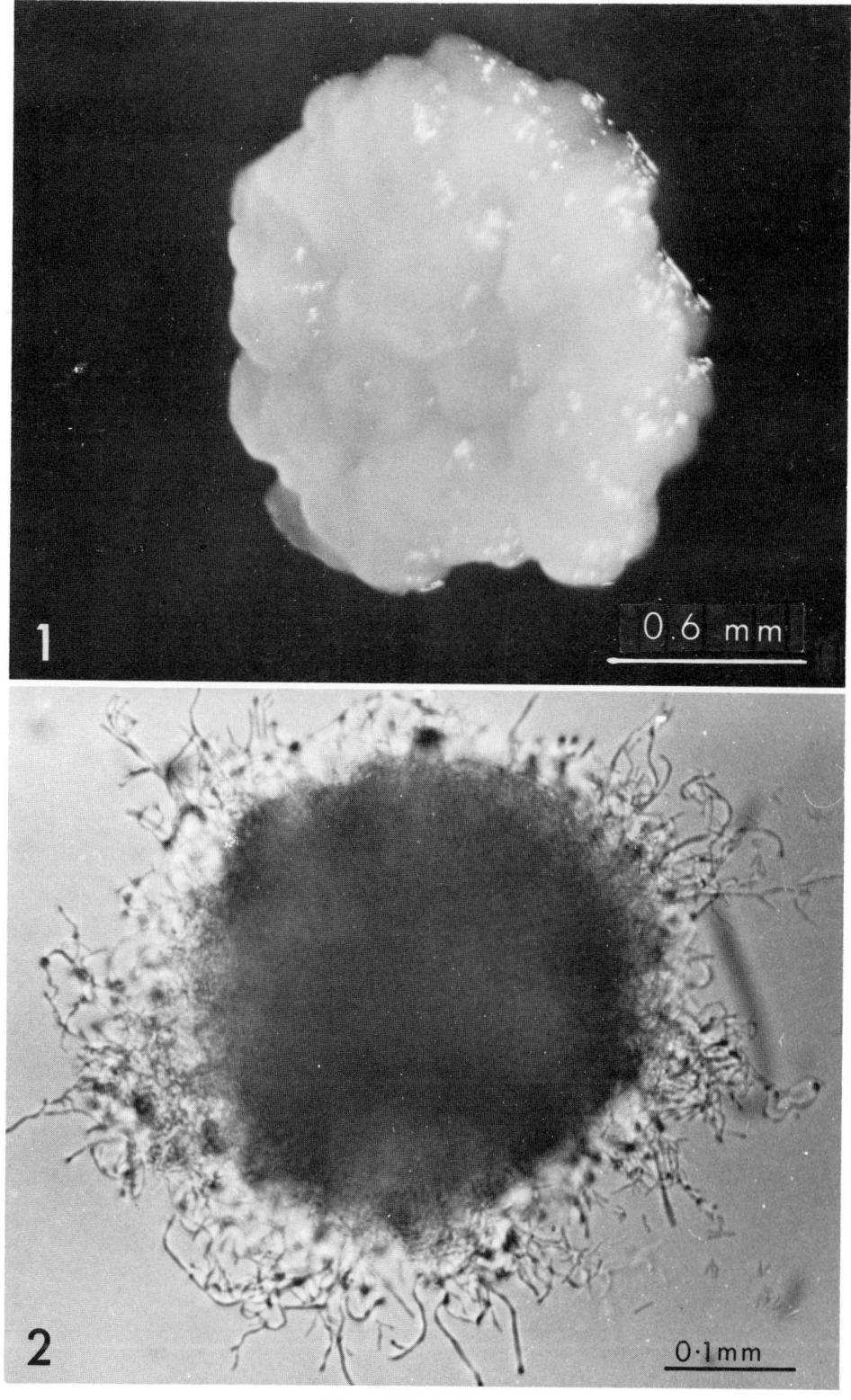

Plate 17.3. *Actinomyces*

Fig. 1. *A. israelii*, mature colony, 14 days growth on BHI.

Fig. 2. *A. naeslandii*, microcolony, 48 hour growth on BHI.

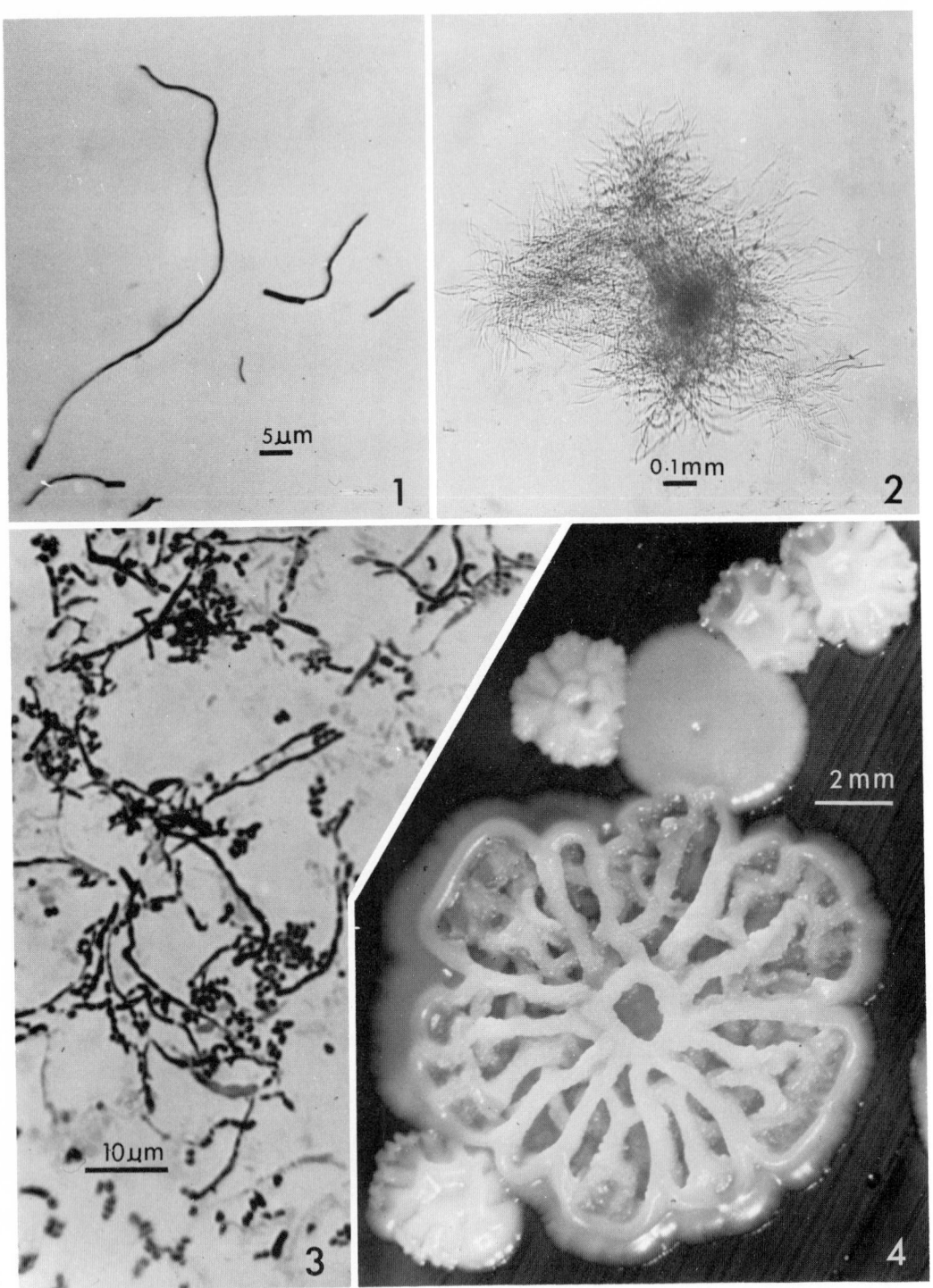

Plate 17.4. *Bacterionema* and *Rothia*

Figs. 1–2. *Bacterionema matruchotii*: Fig. 1. Stained by Gram-reaction from BHI plus yeast extract; Fig. 2. Typical microcolony.

Figs. 3–4. *Rothia dentocariosa*: Fig. 3. Stained by Gram-reaction showing coccoid and filamentous forms; Fig. 4. Rough and smooth colonies of a single isolate; both figures from 7-day-old culture on trypticase soy agar.

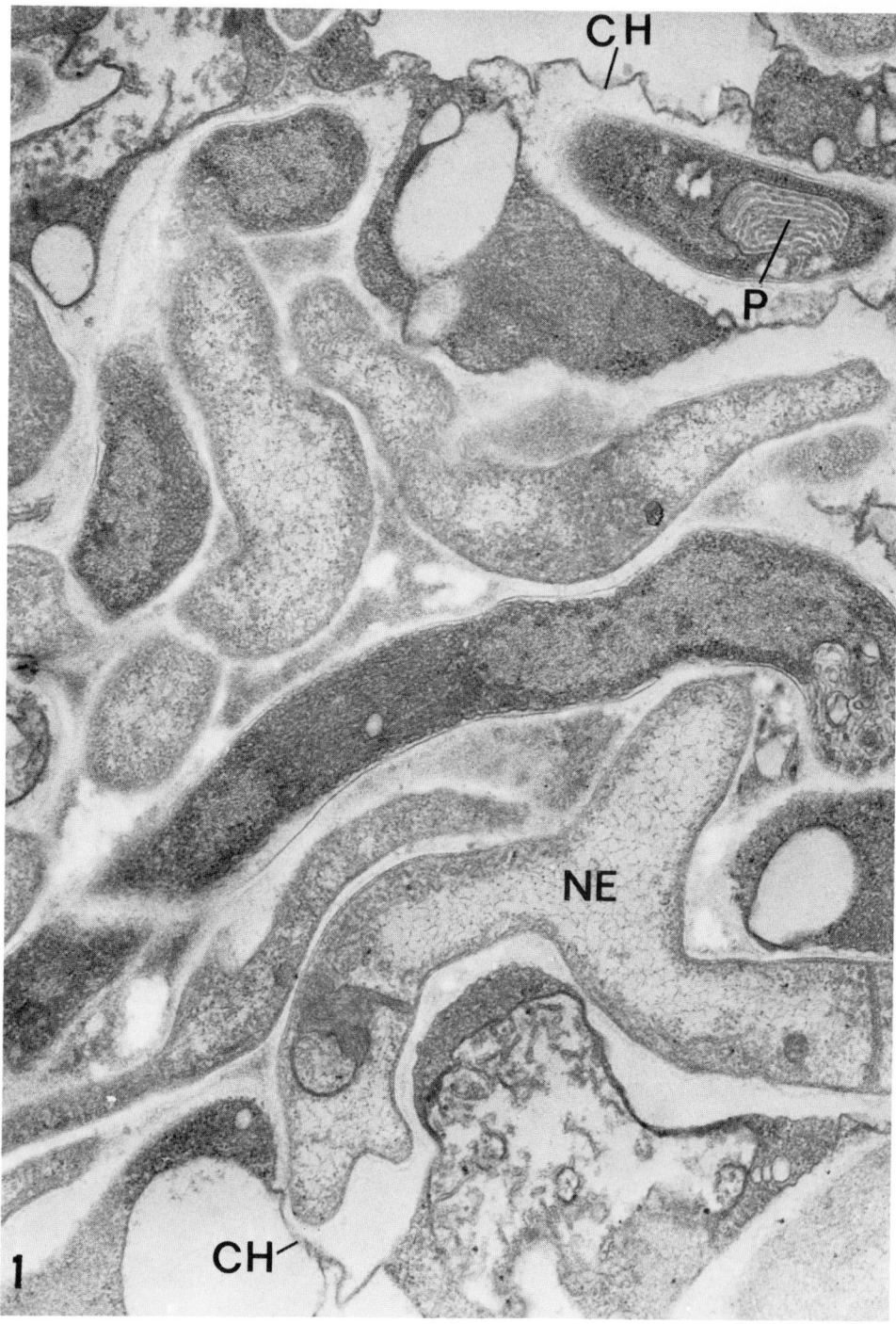

Plate 17.5. *Frankia*

Fig. 1. *Frankia alni.* Section through nodule of *Alnus glutinosa* showing hyphae in living host cell: NE, nuclear material; P, plasmalemmosome; CH, cytoplasmic membrane of host. Electron photomicrograph. × 47,000. From Becking, J. H., de Boer, W. E. and Houwink, A. L. 1964 Electron microscopy of the endophyte of *Alnus glutinoas.* Antonie van Leeuwenhoek 30: 343–376.

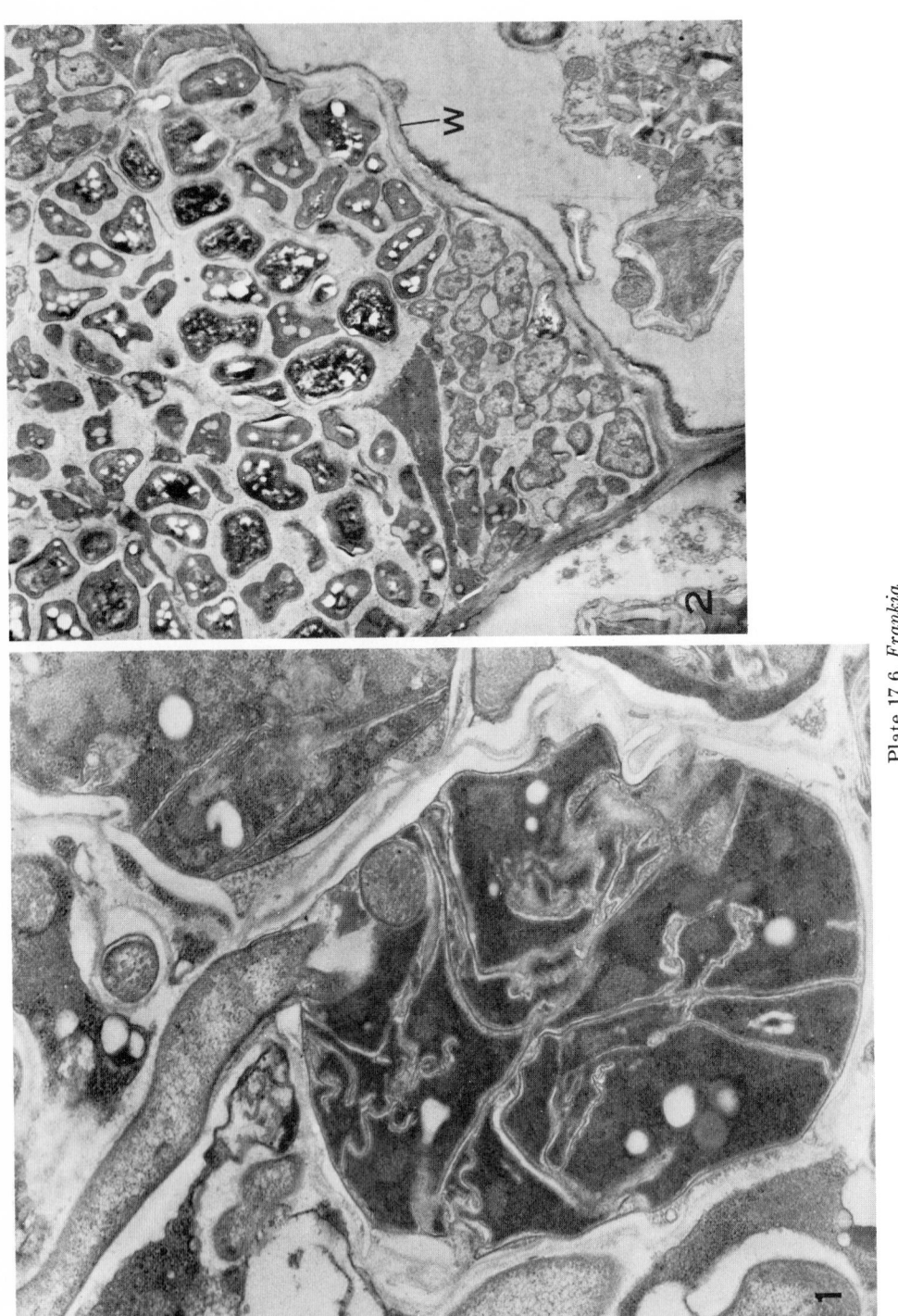

Plate 17.6. *Frankia*

Fig. 1. *Frankia alni.* Longitudinal section of vesicle at end of hypha in *Alnus glutinosa* host cell, showing complicated internal structure with many incomplete cell walls and the associated membranes. Electron photomicrograph. × 34,500. J. H. Becking *et al.*, 1964.

Fig. 2. Bacteria-like cells or "bacteroids" in dead host cell (*Alnus glutinosa*), arranged in groups in different stages of development. The irregular mature cells contain reserve material, probably lipid, resistant to sectioning. W, host-cell wall. Electron photomicrograph. × 8,500. J. H. Becking

872

SOME GENERIC TYPES IN THE ACTINOPLANACEAE

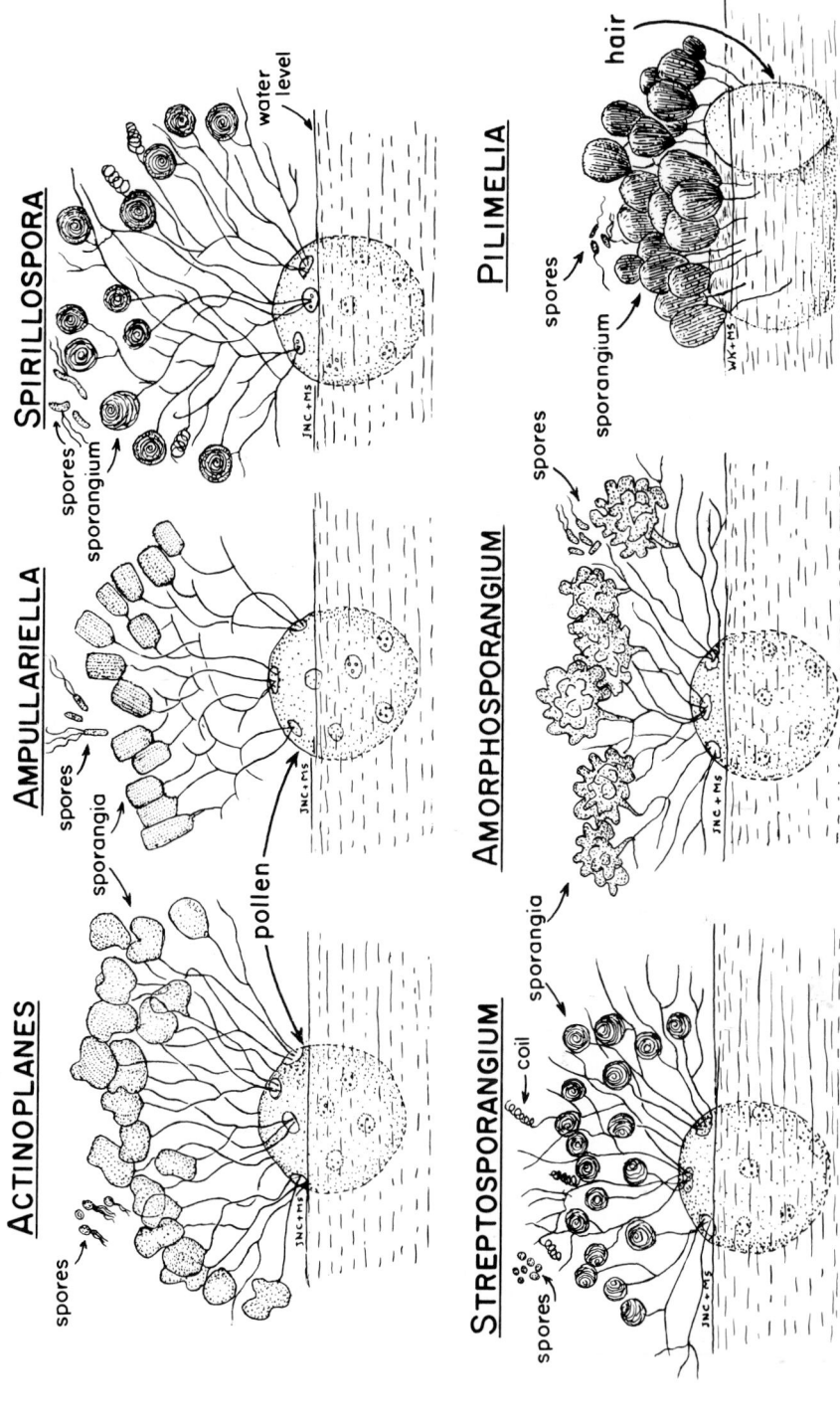

ACTINOPLANES
spores

AMPULLARIELLA
spores
sporangia
pollen

SPIRILLOSPORA
spores
sporangium
water level

STREPTOSPORANGIUM
spores
coil
sporangia

AMORPHOSPORANGIUM
spores
sporangia

PILIMELIA
hair
spores
sporangium

Plate 17.7. *Actinoplanaceae*

Semi-diagrammatic drawings of six of the commoner genera of the *Actinoplanaceae*, five growing on the pollen of *Liquidambar styraciflua* and one on a piece of human hair. The pollen floats at the surface of the water, the hyphae and sporangiophores emerging into the air through the small circular pits in the wall of the pollen grains. × 500.

From *McGraw-Hill Encyclopedia of Science and Technology*, volume 1. Copyright 1968. Used with permission of McGraw-Hill Book Company.

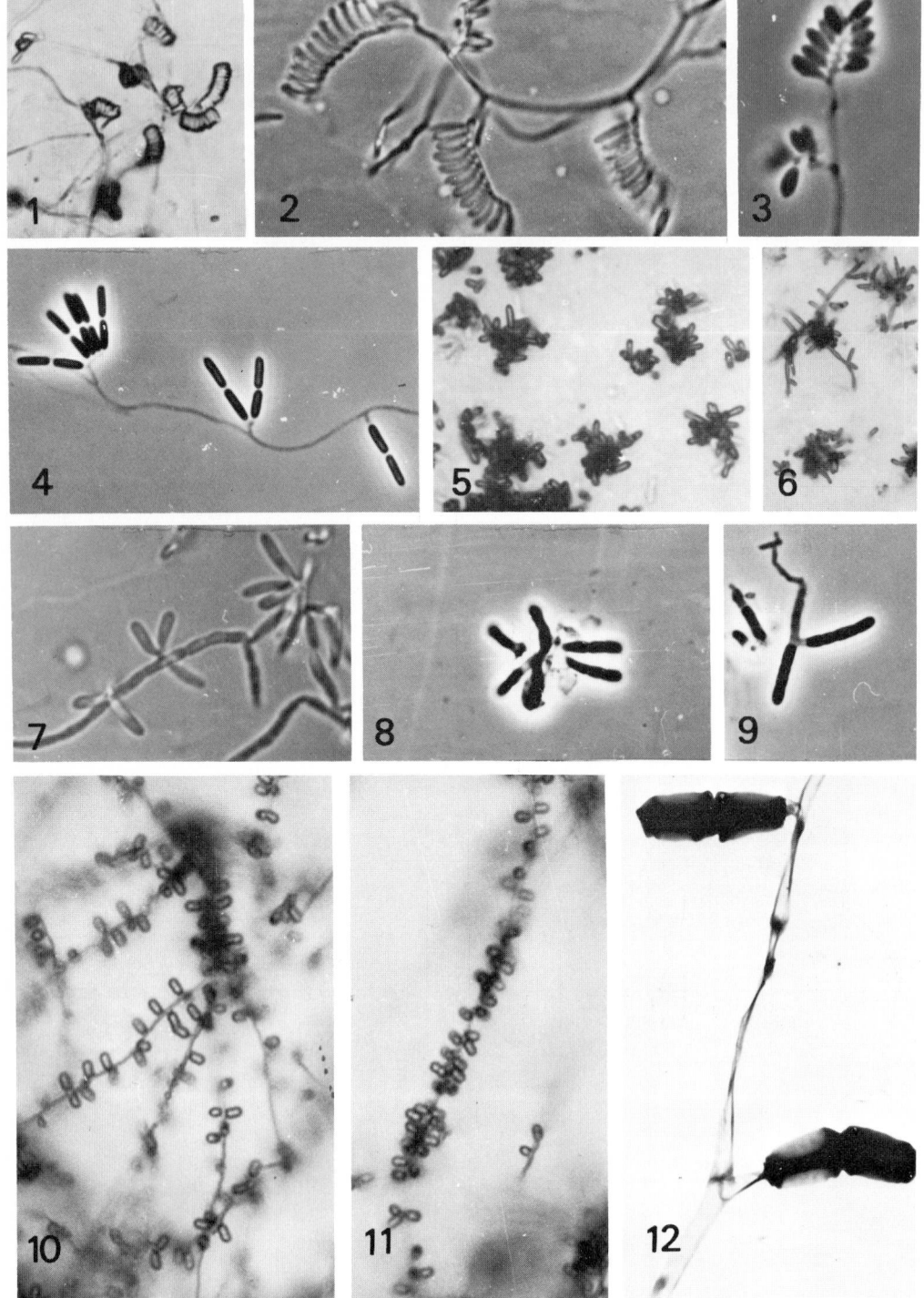

Plate 17.8. *Planomonospora, Dactylosporangium* and *Microbispora*

Figs. 1, 2. *Planomonospora parontospora* sporogenous hyphae. Fig. 1. Series of sporangia (× 534). Fig. 2. Developing on agar media (× 1335).

Fig. 3. *Pl. venezuelensis* wet mount showing "palm-leaf" pattern of the sporangia (× 1335).

Fig. 4. *Planobispora longispora* typical sporangia with longitudinal pair of sporangiospores attached to aerial mycelium (× 1335).

Figs. 5–9. *Dactylosporangium.*

Fig. 5. *D. aurantiacum* typical unbranched, finger-shaped sporangia formed at surface of culture media (× 534).

Fig. 6. *D. sp.* branched sporangia developing on surface of culture media (× 356).

Fig. 7. as Fig. 6 (× 1669).

Fig. 8. *D. aurantiacum* wet mount of sporangia (× 1335).

Fig. 9 as Fig. 8. *D. sp.* (× 1335).

Figs. 10–12. *Microbispora.*

Figs. 10–11. *M. rosea*, aerial mycelium with spores (× 712).

Fig. 12. Electron micrograph showing paired spores of *M. bispora* (× 1068).

Figs. 1–9 by J. E. Thiemann. Figs. 10–11 by T. Cross. Fig. 12 by A. Henssen.

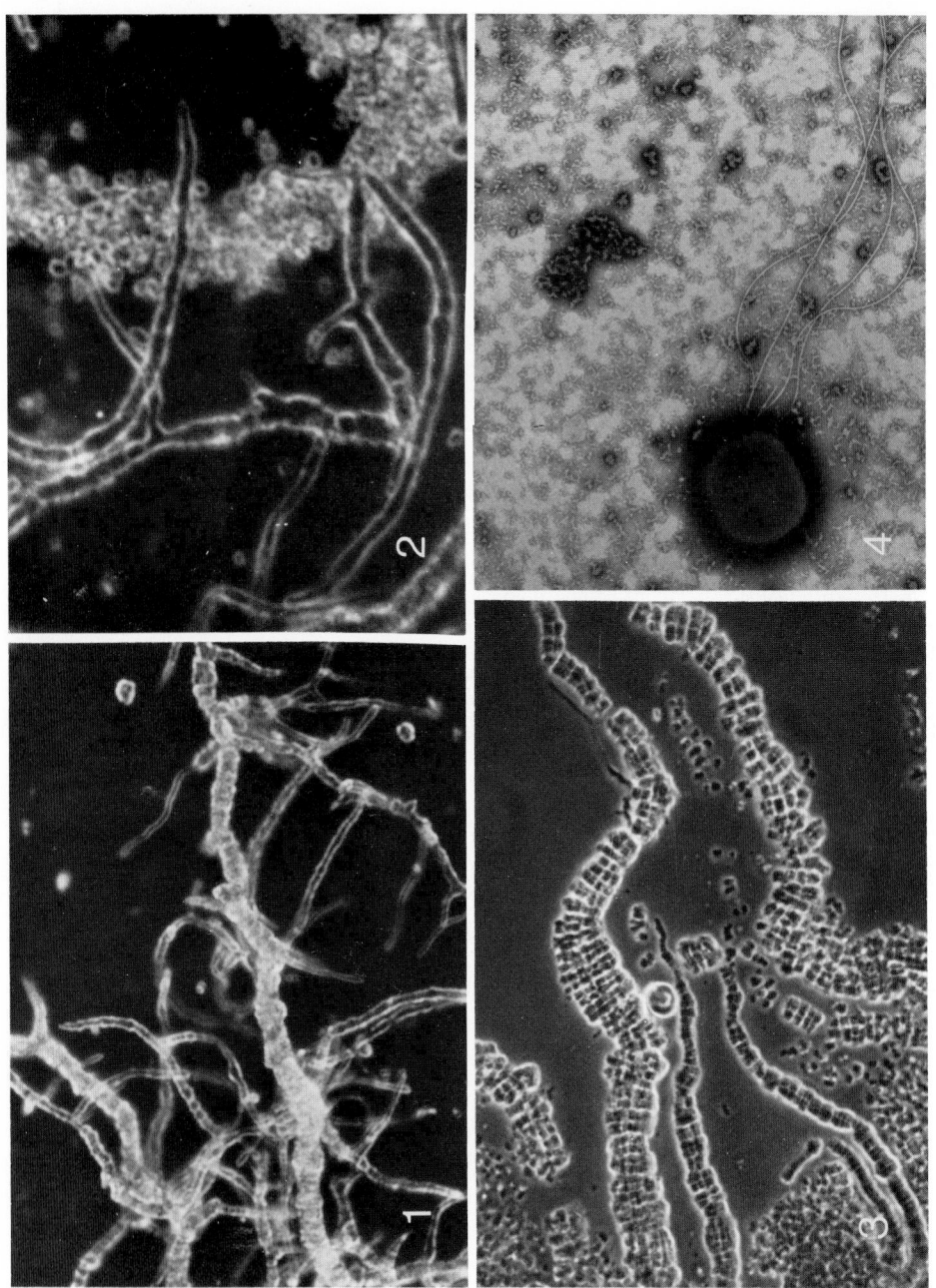

Plate 17.9. *Dermatophilus congolensis*

Fig. 1. *Dermatophilus congolensis*, wet mount of beef infusion peptone broth culture, showing various stages in development, from fine hyphae to multiseptate coarse filaments transforming into chains of coccal packets. Dark-field microscopy. $\times$ 945.

Fig. 2. Same preparation as in Fig. 1; final transformation of mature hypha into an agglomeration of motile spores, in process of dispersion. Dark-field microscopy. $\times$ 2100.

Figs. 1 and 2 from Streptothricosis: a new zoonotic disease. New York State Journal of Medicine *61*: 1283–1287, 1961; Fig. 3, A and B by permission of The Medical Society of the State of New York.

Fig. 3. Fully segmented hyphae forming cubical packets of coccoid spores; wet mount of broth culture. Dark-phase contrast. $\times$ 1750.

Fig. 4. Encapsulated motile spore of *D. congolensis* with tuft of flagella; negative stain (phosphotungstic acid). X 17,500.

Figs. 3 and 4 from The Genus Dermatophilus, J. Bacteriol., 88: 509–522, 1964, Figs. 33 and 25, respectively. By permission of the American Society for Microbiology.

877

1 _____ **2** _____ **3** _____

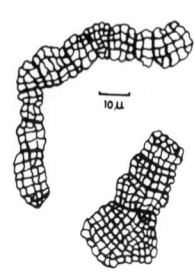 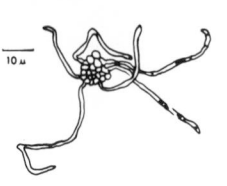 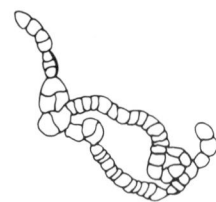 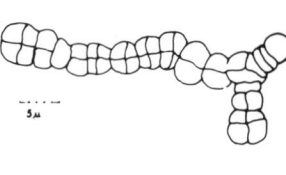

G. obscurus subsp. amargosae

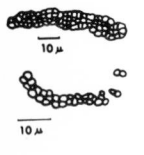 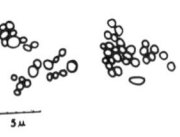 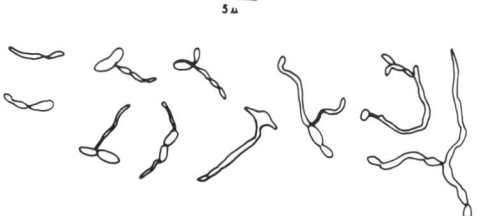 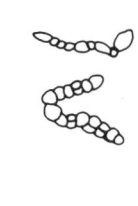

G. obscurus subsp. utahensis

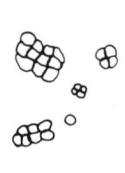 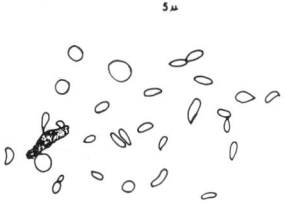 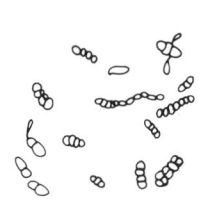

G. obscurus subsp. dictyosporus

Plate 17.10

Line drawings illustrating the variation in the morphological development of three subspecies of *Geodermatophilus obscurus* and the complexity of the life cycle.

Left to right: (*1*) variation in size and cell number of multiloculate thalli; (*2*) breakdown of thalli into component cells which may germinate (*top*) directly into filaments, (*center*) to form motile zoospores which then come to rest and form filaments, or (*bottom*) to form motile or non-motile cells from which small thalli develop directly rather than first developing a filamentous stage; (*3*) early stages in the formation of thalli illustrating muriform septation.

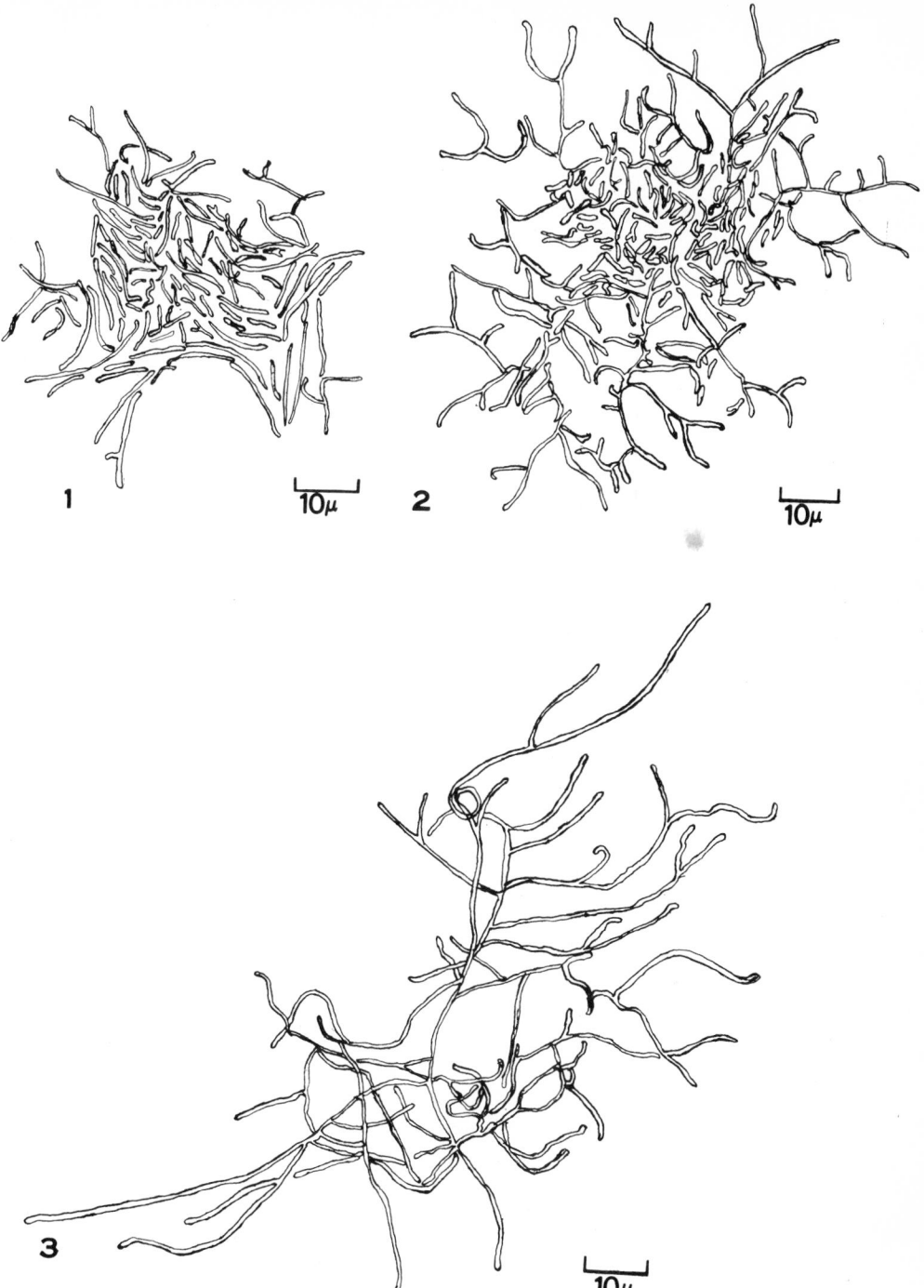

Plate 17.11. Mycelial development of three groups of *Nocardia*
Fig. 1. *Nocardia erythropolis* 27 hr at 28 C. Group I. Note sparse branching.
Fig. 2. *Nocardia rubra* 27 hr at 28 C. Group II.
Fig. 3. *Nocardia* sp. 27 hr at 28 C. Group III. Note extensive mycelial development and lack of fragmentation.
From *Lloydia* 12: 137–177. By permission of the editor.

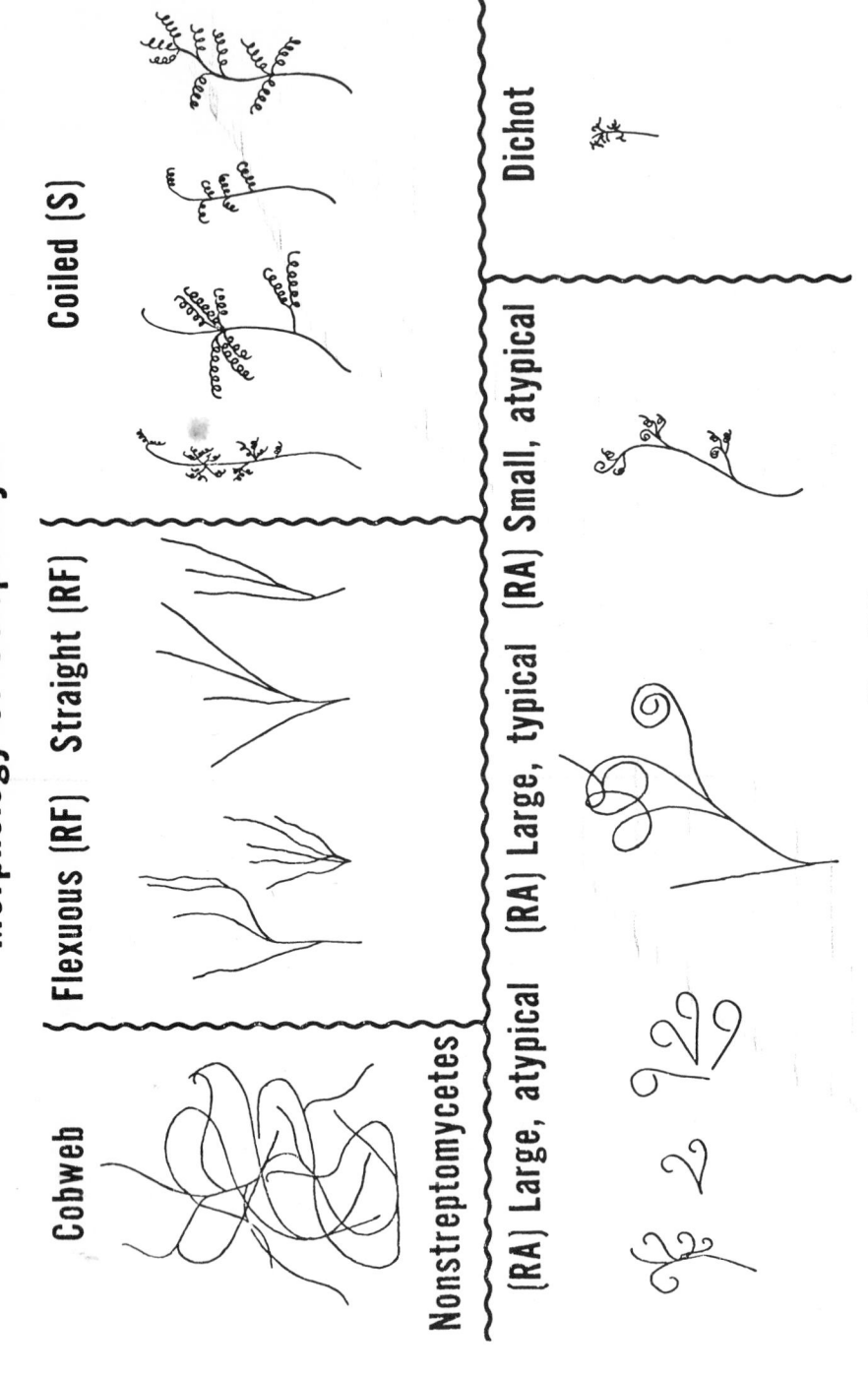

Plate 17.12

Morphology of spore chains of streptomycetes.

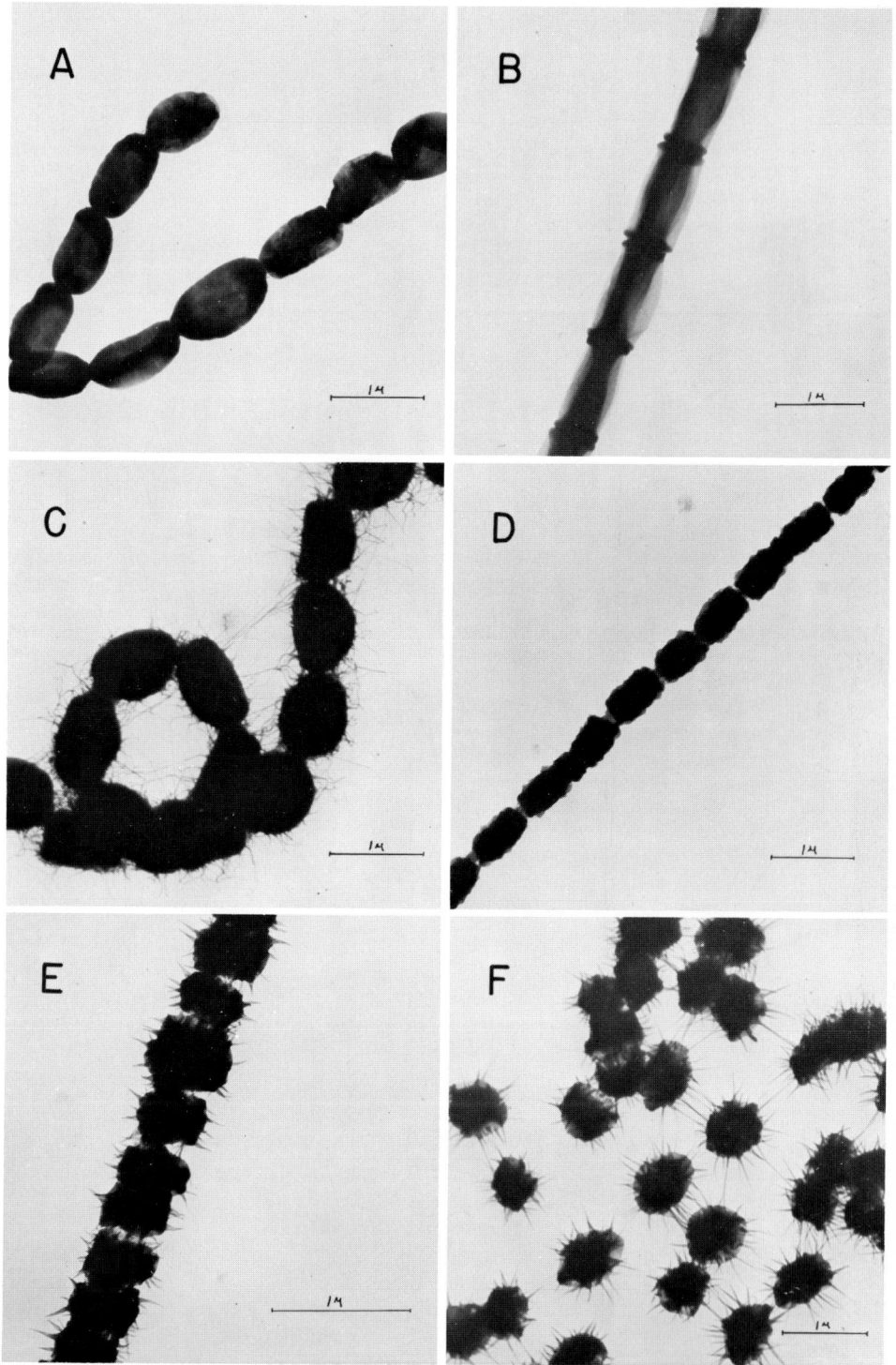

Plate 17.13. Spore surface morphology of streptomycetes

A, spore surfaces smooth. *B*, spore surfaces smooth; spores phalangiform. *C*, spore surfaces hairy. *D*, spore surfaces warty. *E*, spore surfaces with short stout spines. *F*, spore surfaces with long slender spines.

PART 18

THE RICKETTSIAS

ORDER I. **RICKETTSIALES** GIESZCZKIEWICZ 1939, 25

James W. Moulder

The majority are rod-shaped, coccoid and often pleomorphic microorganisms which have typical bacterial cell walls, no flagella, are Gram-negative and multiply only inside host cells. They may be cultivated in living tissues such as embryonated chicken eggs or vertebrate cell cultures. Except for binary fission, which is common to all members of this order, there are notable exceptions to any one of the characteristics listed above. For example, microorganisms are included that appear ring-shaped in stained preparations, or have a flagellum, or are Gram-positive or multiply on bacteriological media of moderate complexity. All are regarded as parasitic or mutualistic. The parasitic forms are associated with the reticuloendothelial and vascular endothelial cells or erythrocytes of vertebrates and often with various organs of arthropods which may act as vectors or primary hosts. May cause disease in man or in other vertebrate and invertebrate hosts. The mutualistic forms in insects are regarded as essential for development and reproduction of the host.

In the 7th edition of THE MANUAL, Class III *Microtatobiotes* included two orders, *Rickettsiales* and *Virales*. With the recognition that members of the order *Rickettsiales* are procaryotic microorganisms with no relationship to the non-organismic viruses, the class *Microtatobiotes* no longer serves any useful function. The order *Rickettsiales* is retained.

Three families are accepted in the order *Rickettsiales*, i.e. *Rickettsiaceae*, *Bartonellaceae* and *Anaplasmataceae*.

Ri.ckett.si.a′les. M.L. fem.n. *Rickettsia* type genus of order; -*ales* ending to denote order; M.L. fem.pl.n. *Rickettsiales* the *Rickettsia* order.

Key to the families of order **Rickettsiales**

I. Parasites, intracellular or intimately associated with tissue cells other than erythrocytes or with certain organs in arthropods; rarely extracellular in arthropods. Transmitted by arthropod vectors.

Family Rickettsiaceae, p. 883

II. Parasites, intracellular or facultatively extracellular; found characteristically in or on the erythrocytes of vertebrates, and in one genus in fixed tissue cells. Small, rod-shaped, bacteria-like cells. At least one species, when cultured, may show a single polar flagellum. Arthropod transmission established for some members of the family. Have been cultivated on media without living cells.

Family Bartonellaceae, p. 903

III. Very small, virus-like particles occurring in the erythrocytes of vertebrates. Transmitted by arthropods.

Family Anaplasmataceae, p. 906

FAMILY I. **RICKETTSIACEAE** PINKERTON 1936, 186

Ri.ckett.si.a′ce.ae. M.L. fem.n. *Rickettsia* type genus of family; *–aceae* ending to denote family; M.L. fem.pl.n. *Rickettsiaceae* the *Rickettsia* family.

Small, rod-shaped, coccoid and diplococcus-shaped, often pleomorphic **organisms which are often intimately associated with arthropod tissues, usually in an intracellular position.** Gram-negative. With one exception the species pathogenic for vertebrates **have not been cultivated to date in cell-free media.** May be parasitic in man and other animals causing disease (typhus and related ills) that may be transmitted by invertebrate vectors (chiefly lice, fleas, ticks and mites). Information is still inadequate for the systematic assignment of many of the species which inhabit arthropod hosts and which were originally described in this family.

The family *Rickettsiaceae* as accepted, includes three tribes, *Rickettsieae, Ehrlichieae* and *Wolbachieae.*

Key to the tribes of family **Rickettsiaceae**

I. Adapted to existence in arthropods but capable of infecting suitable vertebrate hosts including man; pathogenic for man who may be the primary host, but most frequently is an incidental host. Cells rod-shaped, coccoid and diplococcoid; occasionally filamentous.
Tribe I. *Rickettsieae,* p. 883

II. Only a few species adapted to invertebrate existence; pathogenic for certain mammals but not for man; cells spherical, occasionally pleomorphic.
Tribe II. *Ehrlichieae,* p. 893

III. Adapted to existence in arthropods as symbiotes but not in vertebrates as highly pathogenic parasites; cells pleomorphic, coccoid to short or long and curved rods.
Tribe III. *Wolbachieae,* p. 897

TRIBE I. RICKETTSIEAE PHILIP 1953, 486

Ri.ckett′.si.e.ae. M.L. fem.n. *Rickettsia* type genus of the tribe; *-eae* ending to designate a tribe; M.L. fem.pl.n. *Rickettsieae* the *Rickettsia* tribe.

Small, pleomorphic, usually intracellular organisms found in arthropods and pathogenic for man and certain other vertebrate hosts.

Key to the genera of tribe **Rickettsieae**

I. Rod-shaped to coccoid, sometimes filamentous, not cultivable in absence of host cells. Growth generally in cytoplasm, not in vacuoles, sometimes in nucleus of certain vertebrate and arthropod cells. Rapidly inactivated at 56 C, unstable when separated from host compounds.
Genus I. *Rickettsia*

II. Resembles *Rickettsia,* but usually in an extracellular environment in the arthropod host, and cultivable on certain bacteriological media.
Genus II. *Rochalimaea*

III. Growth preferentially in vacuoles of host cells, highly resistant to physical and chemical agents in an extracellular environment.
Genus III. *Coxiella*

Genus I. **Rickettsia** *da Rocha-Lima 1916, 567 Nom. gen. cons.* Opin. 19, Jud. Comm. 1958, 158

EMILIO WEISS AND JAMES W. MOULDER

Ri.ckett′si.a. M.L. fem.n. *Rickettsia* named for H. T. Ricketts, who first associated organisms of this description with spotted fever and typhus, and who died of typhus contracted in the course of his studies.

Short rods, 0.3–0.6 by 0.8–2.0 μm, some species up to 4 μm long prior to cell division. No flagella or capsules, but an outer layer of amorphous material is occasionally seen in electron micrographs of cells subjected to a minimum of laboratory manipulation. **Gram-negative.** Retain basic fuchsin when stained by the method of Giménez (1964), a modification of the procedure of Macchiavello (1937).

Have not been cultivated in the absence of host cells. Growth generally occurs in the cytoplasm, but not in vacuoles and, sometimes, in the nucleus of certain vertebrate and arthropod cells. The organisms grow to high titer in the entodermal cells lining the yolk sac of the developing chick embryo (Cox, 1941). Excellent growth has also been obtained *in vitro* in chick embryo cells and in several established mammalian cell lines. Plaques have been produced on chick embryo fibroblasts. Temperature optima for growth, 32–35 C. Multiplication by transverse binary fission only. Division time in chick embryos during phase of most rapid growth is about 8 hrs. Nutritional requirements for independent rickettsial cell growth not known.

Growth inhibited by *p*-aminobenzoic acid (approximate minimal inhibitory concentration for most species: 0.2–0.4 mg/egg); inhibition reversed by *p*-hydroxybenzoic acid. Sulfonamides do not affect growth. Growth is inhibited by several antibiotics. The following amounts (per egg) have a measurable effect on multiplication, i.e. delay the deaths of infected chick embryos by about 1½ days: chloramphenicol, 0.08 mg; chlortetracycline, 0.05 mg; oxytetracycline, 0.008 mg; erythromycin, 0.08 mg (in some species, 0.008 mg) (Ormsbee *et al.*, 1955). The organisms are affected to a lesser extent by penicillin and streptomycin in eggs, but these antibiotics may have a pronounced effect in cell cultures.

Rapidly inactivated at 56 C and generally unstable when separated from host components, except in diluents containing proteins such as skim milk or plasma albumin or a solution consisting of sucrose, potassium phosphate and glutamate (SPG) (Bovarnick *et al.*, 1950).

Human pathogens. Man is the reservoir of the type species, an incidental host of the other species. Small rodents and other vertebrates serve as reservoirs or disseminate the organisms. Arthropods play primary roles in the life cycles of all species, often as main reservoirs, and almost exclusively mediate natural transmission among vertebrates.

The guanine + cytosine (G + C) content of the DNA (in the species examined) ranges from **30–32.5 moles** %.

Type species: *Rickettsia prowazekii* da Rocha-Lima 1916, 567.

Further Comments

Despite the fact that members of the genus *Rickettsia* have not been grown in the absence of host cells, there is good basis for regarding them as bacteria. A cell wall has been demonstrated in all species examined. Three layers are usually described, but five layers have been detected in uninjured cells at high magnification (Anacker *et al.*, 1967). The chemical composition of rickettsial cell walls resembles that of other Gram-negative bacteria. Muramic acid, diaminopimelic acid, several sugars, amino sugars and at least 15 amino acids have been identified. The microorganisms also appear to have a functional cell membrane which is impermeable to sucrose and inorganic electrolytes (Myers *et al.*, 1967). Freezing and thawing or lengthy purification procedures alter the properties of the membrane, however, and the organisms become dependent for optimal biological activity on an extracellular concentration of K^+ of at least 0.13 M and on some osmotic protection. The internal structure consists of electron-dense polar zones and of a light central zone corresponding to the ribosome-filled cytoplasm and DNA strands demonstrated in other bacteria (Anderson *et al.*, 1965). A high DNA to RNA ratio has been reported in several instances, but this is possibly due to loss of RNA during purification procedures.

As first shown by Bovarnick and Snyder (1949), the species studied possess a moderately active energy-yielding metabolism that is independent of the host cell. Glutamate, the chief substrate, is transaminated with production of aspartate and oxidized via dicarboxylic and, possibly, tricarboxylic acids. Glucose is not utilized by intact cells; this substrate has not been tested with cell extracts. The electron transport system appears to be identical with that of other bacteria, except that nicotinamide adenine dinucleotide can be lost and, under certain conditions, regained by the cells. This effect is possibly a reflection of cell injury. Oxidative energy can be converted into high energy phosphate bonds by the formation of adenosine triphosphate (ATP). Limited protein and lipid syntheses have been demonstrated in a complex host cell-free medium which includes several cofactors. An unusual requirement for this activity is a dual source of ATP, one derived internally from glutamate and one which must be supplied with the medium. Nucleic acid synthesis has not been studied *in vitro*.

Rickettsiae actively penetrate host cells. It was shown that only viable rickettsiae were adsorbed onto tunica cells of the guinea pig or penetrated established cell lines. Hemolytic activity, demonstrable against rabbit, sheep or chicken erythrocytes and mouse toxicity also depend on rickettsial viability and are probably expressions of the same mechanism of penetration. Factors that stimulate or depress the metabolic activity of rickettsiae have the same effects on all of these interactions of rickettsiae with vertebrate cells. Acute mouse toxicity can be demonstrated with most rickettsial species, provided a sufficiently large inoculum can be injected intravenously. The only immediate lesion is a pronounced increase in vascular permeability and mice die of its consequences, usually within 1–8 hrs. In man, too, rickettsiae show a predilection for endothelial cells, and much of the pathogenesis is based on this phenomenon.

Strains comprising the genus *Rickettsia* fall into three distinct biotypes: typhus, spotted fever and scrub typhus (Table 18.1). The first two groups appear to be somewhat more closely related to each other than to the third group. Speciation within the first two biotypes is not based in every instance on multiple unrelated phenotypic differences. Because of the importance of these organisms as pathogens of man and because of their diverse ecology, painstaking work has produced a classification based on small variations in antigenic specificity and virulence for animals and man. Comparable differences have been encountered among the strains of the scrub typhus group, but separation into more than one species has not been proposed and is not recommended. A fourth biotype, the trench fever agent (*Rickettsia quintana* in the 7th edition) has been placed in the separate genus *Rochalimaea*. This separation is based on the extracellular position of the organism in the arthropod host and on its cultivation on host cell-free medium. The agent of Q fever, *Coxiella burnetii*, is maintained in a separate genus as in the 7th edition.

Several specific names have been proposed since the appearance of the 7th edition of THE MANUAL.

Rickettsia montana Lackman, Bell, Stoenner and Pickens 1965, 137 isolated by Bell *et al.* (1963) from *Dermacentor variabilis* and *D. andersoni*, is closely related to strain U, an avirulent variant of *R. rickettsii* (Price, 1953).

Rickettsia tamiyai Kawamura 1954, 49 is regarded as a strain of *R. tsutsugamushi* (Tamiya, 1962).

Rickettsia sennetsu Misao and Kobayashi, 1956, 456 does not resemble other rickettsiae in morphology and, unlike other rickettsiae, growth appears to take place in cellular vacuoles (Anderson *et al.*, 1965) and until its biological properties are better known is regarded as a *species incertae sedis*.

Description of the species of genus **Rickettsia**

1. **Rickettsia prowazekii** da Rocha-Lima 1916, 567. *Nom. cons.* Opin. 19, Jud. Comm. 1958, 158.

pro.wa.ze′ki.i. M.L. gen.n. *prowazekii* of Prowazek; named for S. von Prowazek, an early investigator of the etiology of typhus who died of typhus contracted in the course of his studies.

Rods 0.3–0.6 by 0.8–2.0 μm often up to 4 μm long, singly or in small chains.

Optimal growth in yolk sac of chick embryos incubated at 35 C. Highest viability just prior to embryo death, which occurs 6–13 days after inoculation, depending on the number of rickettsiae injected. Good growth also obtained in chick embryo cell cultures and in established lines such as mouse lymphoblasts, mouse L cells or monkey kidney cells but yields of viable rickettsiae are not as high as from yolk sac. Growth occurs throughout the cytoplasm (except the vacuoles) of the host cells.

Generally unstable, as other members of the genus, except that viability is retained for several months in louse feces if the temperature and humidity are low. Infected yolk sacs or crude yolk sac suspensions remain viable for several years at −70 C, but the material must be frozen and thawed very rapidly. Diluent SPG is very effective in stabilizing the activities of the organism under most conditions. Susceptibility to antibiotics is comparable to that of other members of the genus, except that the organism is inhibited by very small concentrations of erythromycin (inhibition demonstrable with 8 μg/egg). Strains unaffected by 2 mg/egg of erythromycin, 10 mg/egg of p-aminobenzoic acid or of increased resistance to chloramphenicol (minimal inhibitory concentration from 0.1–0.4 mg/egg) have been isolated in the laboratory (Weiss and Dressler, 1962).

Soluble antigens, released in the aqueous phase by ether treatment of infected triturated yolk sacs (Topping and Shear, 1945), react with antibodies elicited by this species and by *R. typhi*, but not by the other rickettsial species. When washed repeatedly, the organisms display a high degree of species specificity in complement fixation tests and can readily be differentiated from *R. typhi*.

The guinea pig is highly susceptible to infection and is commonly used for primary isolation. It develops a mild disease, manifested only by fever lasting approximately a week and, rarely, scrotal swelling. The organisms are harvested from the spleen or brain. The cotton rat, *Sigmodon hispidus*, is as susceptible to infection as the chick embryo, but signs of disease and death are produced only by doses in excess of 3 × 10⁵ egg infectious units. Inapparent infection elicits solid immunity to the homologous species and to *R. typhi*. True infection cannot usually be demonstrated in the mouse, but

TABLE 18.1

Differentiation of biotypes and species of the genus **Rickettsia**

Biotype	Typhus group	Spotted fever group	Scrub typhus group
Species[a]	1. R. prowazekii 2. R. typhi	4. R. rickettsii 5. R. sibirica 6. R. conorii 7. R. australis 8. R. akari	10. R. tsutsugamushi
Intracellular location			
Cytoplasm	+	+	+
Nucleus	−	+	−
Cultivation in chick embryos			
Opt. temp. C.	35	32–34	35
Peak titer			
Just prior to death	+	−	+
24–72 hrs after death	−	+	−
Plaque formation[b]			
Days required	8–10	5–8	11–17
Size, mm	1	2–3	1
Relative susceptibility to in-			
fection of			
Guinea pig[c]	+++	+++	+ to ++
Mouse	+(1)[d] ++(2)	+(4–6) ++(7) +++(8)	+++
Specificity of CF antigen			
Soluble, extracted by ether			
Group	+	+	−
Unsatisfactory	−	−	+
Washed cells			
Species	+	+	+
Type	−	−	+

  [a] *R. canada* and *R. parkeri* are not included, but are discussed in the text.

  [b] On monolayers of chick embryo monolayers (from McDade, Stakebake and Gerone, 1969; McDade and Gerone, 1970).

  [c] The following clinical manifestations of the disease in the guinea pig are often used as criteria for species differentiation; Fever only—*R. prowazekii;* Fever and scrotal swelling—*R. typhi, R. conorii, R. australis* and *R. akari;* Fever and scrotal necrosis—*R. rickettsii* and *R. sibirica.* These reactions are subject to considerable strain variation.

  [d] Species number in parentheses.

acute toxic death is produced by the intravenous injection of at least 10^6 egg infectious units.

Etiological agent of (epidemic) typhus fever and of its recrudescent form, Brill-Zinsser disease. Sera of typhus fever patients, but not of those with Brill-Zinsser disease, agglutinate *Proteus vulgaris* OX19 cells (*Weil-Felix reaction*). Strain variation in virulence for man or animals is relatively small, except that a variant (strain Madrid E) of low virulence for man was isolated following 255 serial passages in eggs. It has been used as a living vaccine on a moderate scale.

Man is the primary reservoir. Individuals who recover from typhus fever probably retain small numbers of organisms, presumably in their lymph nodes, for the rest of their lives. These organisms

may temporarily overcome host resistance and give rise to Brill-Zinsser disease.

Transmission is accomplished by the human louse *Pediculus humanus*, the body louse or, less frequently, the head louse. Both varieties are highly susceptible to the organisms and succumb to infection within 1–3 weeks. Heavily infected cells of the gut epithelium of the louse are discharged with the feces. They are the source of human infection when feces are driven into the skin in the process of scratching. Various species of fleas are also susceptible to infection, but they are not known to have a role in transmission. Aerosol infection may occur as a laboratory accident or under unusual circumstances.

The G + C content of the DNA is approximately

30 moles % (thermal denaturation and buoyant density, Tyeryar *et al.*, 1973).

2. Rickettsia typhi (Wolbach and Todd) Philip 1943, 304. (*Dermacentroxenus typhi* Wolbach and Todd 1920, 158; *Rickettsia mooseri* Monteiro 1931, 97; Not *Rickettsia typhi* do Amaral and Monteiro 1933, 806.)

ty'phi. Gr. n. *typhus* cloud, hence stupor arising from fever; M.L. n. *typhus* fever, typhus; M.L. gen.n. *typhi* of typhus.

Indistinguishable from prototype strains of *R. prowazekii*, except as noted below.

Although the soluble antigen is markedly similar to that of *R. prowazekii*, a specific antigen can be obtained from repeatedly washed organisms and the two species can be differentiated by complement fixation tests with appropriate antisera. Differentiation can also be accomplished by cross-challenge of vaccinated guinea pigs or by mouse neutralization tests. The results of the latter test must be interpreted with caution because some non-immune human and monkey sera neutralize *R. typhi* toxin (Bell *et. al.*, 1969).

More virulent for the guinea pig than *R. prowazekii*: scrotal swelling usually elicited. The organisms are harvested from spleen or testes. Mice surviving a sub-toxic dose die 3-8 days later as the result of infection. Unlike *R. prowazekii*, this species can be passed indefinitely in the white rat and may persist for months in rat brain.

Rats (*Rattus norvegicus* and *R. rattus*) and other rodents are the primary reservoirs of the organism. The rat louse, *Polyplax spinulosus*, and the rat flea, *Xenopsylla cheopis*, are believed to be chiefly responsible for the transmission of the organism from rat to rat. The rat flea is highly susceptible to infection, although not obviously harmed, and sheds engorged epithelial gut cells with its feces. Man is an incidental host and infection occurs through contact with the rat flea in the same manner as described for *R. prowazekii*. The human flea, *Pulex irritans*, and the human louse, *Pediculus humanus*, are highly susceptible to infection with *R. typhi* and may play roles in natural transmission to man.

Etiological agent of murine (endemic) typhus, a mild form of typhus fever.

The G + C content of the DNA is approximately 30 moles % (thermal denaturation and buoyant density, Tyeryar *et al.*, 1973).

3. Rickettsia canada McKiel, Bell and Lackman 1967, 509.

ca'na.da. M.L. n. Canada, the country where organism was first isolated.

It belongs to the typhus group biotype, although it has some properties in common with the spotted fever group.

Rods, 0.4 by 1.6 μm, average dimensions. Isolated from *Haemaphysalis leporispalustris*. Cultivated in *Dermacentor andersoni*, *D. variabilis* and *Amblyomma americanum* ticks and in chick embryos. Small plaques with indistinct perimeters are produced on chick embryo fibroblast monolayers after 11-14 days. Growth takes place in the cytoplasm and in the nucleus.

Elicits fever but no scrotal reaction in guinea pigs. It is toxic for mice when injected intravenously in large doses, but is not highly infectious. In complement fixation tests there are strong cross-reactions between *R. canada*, *R. prowazekii* and *R. typhi* antigens and corresponding immune sera obtained from guinea pigs, rabbits or hamsters. Antigenic differences can be demonstrated in complement fixation tests using immune mouse sera or in mouse serum neutralization tests using immune guinea pig sera. Cross-reactions with spotted fever antigens are not pronounced.

Human disease, indistinguishable from Rocky Mountain spotted fever, due to this species has been suspected on the basis of serological reactions.

The G + C content of the DNA is approximately 30 moles % (thermal denaturation and buoyant density, Tyeryar *et al.*, 1973)

4. Rickettsia rickettsii (Wolbach) Brumpt 1922, 757. (*Dermacentroxenus rickettsi* Wolbach 1919, 87.)

ri.ckett'si.i. M.L. gen.n. *rickettsii* of Ricketts; named for H. T. Ricketts for his classic studies of the etiology of Rocky Mountain spotted fever.

Rods somewhat shorter in average length (1.5 μm) and less variable in size (not exceeding 2.0 μm) than *R. prowazekii*, usually appearing singly.

Optimal conditions for growth in yolk sac of chick embryo defined as follows. Embryos, 4½ days old, are inoculated with sufficient organisms to kill most embryos within 104 hrs. Eggs are incubated at 33.5 C (lowest temperature compatible with survival of most embryos) and maintained in incubator at 32 C for 48 hrs after death of the embryos (Stoenner *et al.*, 1962). Yields of viable rickettsiae are not as high from cell cultures as from yolk sac. Growth occurs primarily in the cytoplasm but is often seen in the nucleus of vertebrate as well as arthropod host cells.

Generally unstable, as are other members of the genus. Diluent SPG is not effective as a stabilizing agent under all conditions. Viability and other activities are best maintained under microaerophilic conditions or in the presence of reduced glutathione or protein. Reversible changes in rickettsial activity occur in the tick: prolonged refrigeration reduces the virulence of the rickettsiae for the guinea pig, but virulence is restored when the ticks have a blood meal or are incubated at 37 C for 24 hrs. Similar physiological changes were reproduced with infected yolk sac suspensions. Loss of virulence was prevented or virulence was restored by nicotinamide adenine dinucleotide, coenzyme A or extracts of previously incubated normal tick extracts.

Soluble antigen, released in the aqueous phase by ether treatment of infected triturated yolk sacs, usually reacts with antibodies elicited by all species of the spotted fever group, but not by the other rickettsial species. When washed repeatedly, the organisms act as species-specific antigens in complement fixation tests. Mouse antibodies, although elicited with difficulty, are species-specific and can be used for species differentiation, even with soluble antigens. Species differentiation can also be made by toxin neutralization tests in mice and cross-challenge of immunized guinea pigs.

The guinea pig is highly susceptible to infection, but rickettsial isolates vary greatly in virulence for this animal. The more virulent strains elicit fever, scrotal swelling, hemorrhage and necrosis, and death. Virulence for man does not parallel virulence for the guinea pig. The mouse succumbs to toxic death when inoculated intravenously with heavy suspensions but is not generally susceptible to infection.

Ticks are the primary reservoirs of the organism and transmit it transovarially. Various rodents and rabbits play a role in dissemination. Man is an incidental host. R. rickettsii is confined to the Western Hemisphere. Serologically indistinguishable strains have been reported from most states of the U.S.A., from Canada, Mexico, Colombia and Brazil. The wood tick Dermacentor andersoni, the dog tick D. variabilis, the lone-star tick Amblyomma americanum and the rabbit ticks Haemaphysalis leporispalustris and D. parumapertus are the principal natural carriers in the U.S.A. and Canada. The dog ticks Rhipicephalus sanguineus and Amblyomma cajennense are among the ticks that have been most commonly implicated in Mexico and South America. Natural infection among vertebrates was demonstrated first in Brazil in some species of opossum (Didelphus), the wild rabbit (Sylvilagus minensis) and the Brazilian cavy (Cavia aperae) and more recently in the U.S.A. in various species of rodents and rabbits. Natural infection of dogs has been demonstrated or implied in Brazil and eastern U.S.A. Transmission to man generally occurs through the bite of an infected tick, which remains attached to the skin for a number of hours.

Etiological agent of Rocky Mountain spotted fever of North America and similar diseases of Central and South America. Strains vary greatly in virulence for man. They elicit an antibody response to Proteus OX19 and OX2 antigens (Weil-Felix reaction).

The G + C content of the DNA is approximately 32.5 moles % (thermal denaturation and buoyant density, Tyeryar et al., 1973).

5. Rickettsia sibirica Zdrodovskii 1948, 13.

si.bi′ri.ca. M.L. fem.adj. sibirica of Siberia, Siberian.

Resembles R. rickettsii more closely than any of the other species of the spotted fever group. Although serological cross-reactions are common, this species can be differentiated from R. rickettsii and R. conorii by mouse toxin neutralization tests using immune guinea pig sera, complement fixation tests using washed organisms and immune guinea pig sera or soluble antigen and mouse sera. Even in these tests cross-reactions do occur.

Resides in various species of ixodid ticks. Dermacentor nuttali, D. silvarum, D. marginatus, D. pictus, Haemaphysalis concinna and H. punctata have been found infected. Various domestic animals and rodents that are animal hosts of these ticks have been implicated in dissemination of the organism. Occurs mainly in Siberia, Armenia and the Central Asian republics of the U.S.S.R.

Etiological agent of North Asian tick typhus, a mild form of spotted fever. Elicits antibodies against Proteus OX19 antigen.

6. Rickettsia conorii Brumpt 1932, 1199.

co.no′ri.i. M.L. gen.n. conorii of Conor; named for A. Conor who in collaboration with A. Bruch, provided the first description of fièvre boutonneuse.

Indistinguishable from the two previously described species with the exceptions noted below.

Less virulent for the guinea pig than R. rickettsii or R. sibirica. Fever and scrotal swelling usually produced, but not scrotal necrosis and death. Convalescent guinea pigs are immune to challenge with the homologous species or R. rickettsii.

Can be differentiated from the other species of the spotted fever group by challenge of vaccinated guinea pigs, mouse toxin neutralization and complement fixation tests performed with antigens derived from washed cells. Strains of R. conorii collected in the Mediterranean region, Kenya, South Africa and India are antigenically indistinguishable.

It is primarily a parasite of the dog tick and the dog. Rhipicephalus sanguineus is the most important arthropod host in the Mediterranean region and in India, Haemaphysalis leachi and R. simus in Kenya, and, in addition to these ticks, Amblyomma hebraeum, R. appendiculatus, R. evertsi and Hyalomma aegypticum in South Africa. Several species of rodent have been found naturally infected.

Etiological agent of fièvre boutonneuse (Marseilles fever) of the Mediterranean region or tick bite fever or tick typhus of other regions. Elicits antibodies to *Proteus* OX19 and OX2 antigens.

The G + C content of the DNA is approximately 32.5 moles % (thermal denaturation and buoyant density, Tyeryar *et al.*, 1973).

7. Rickettsia australis Philip 1950, 786.

aus.tra'lis. L. fem.adj. *australis* southern.

Indistinguishable from *R. conorii*, with the exceptions noted below.

Acute mouse toxicity has not been demonstrated, but infection of the weaned mouse has been produced by the intraperitoneal injection of large numbers of organisms. Newborn mice are more highly susceptible to infection. Can be differentiated from the species described above by the immunological tests mentioned. Serological differentiation of *R. australis* and *R. akari* is best accomplished by complement fixation tests using immune mouse sera.

Epidemiological evidence implicates the Australian tick *Ixodes holocyclus* and serological evidence implicates several marsupials as the hosts of this organism.

Etiological agent of Queensland (Australia) tick typhus.

8. Rickettsia akari Huebner, Jellison and Pomerantz 1946, 1682. (*Dermacentroxenus murinus* Kulagin 1951, quoted by Zdrodovskii and Golinevich 1960, 15; *Gamasoxenus muris* Zhdanov 1953, 159.)

a'ka.ri. Gr. neut.n. *akari* a mite.

Most distantly related to *R. rickettsii* of all the recognized species of the spotted fever group.

Although acute toxicity has not been produced, the mouse is highly susceptible to infection with certain strains. Virulence for the guinea pig is comparable to that of *R. conorii*. Both animals are used for primary isolation.

See descriptions of other species of the spotted fever group for immunological tests of species differentiation.

Resides in the mite, *Allodermanyssus sanguineus*, in which transovarian passage occurs, and in its host, the house mouse, *Mus musculus*. The organism has been isolated from urban areas in the U.S.A. and U.S.S.R. The existence of a rural cycle is indicated by its recovery from a wild Korean rodent, *Microtus fortis pelliccus*. Man is an incidental host, becoming infected through the bite of larval or adult mites.

Etiological agent of rickettsialpox or vesicular (and varioliform) rickettsiosis. Antibodies to *Proteus* antigens not usually elicited.

The G + C content of the DNA is approximately 32.5 moles % (thermal denaturation and buoyant density, Tyeryar *et al.*, 1973).

9. Rickettsia parkeri Lackman, Bell, Stoenner and Pickens 1965, 137.

par'ke.ri. M.L. gen.n. *parkeri* of Parker; named for R. R. Parker, a founder of the Rocky Mountain Laboratory.

Member of the spotted fever biotype isolated from *Amblyomma maculatum* ticks by Parker *et al.* (1939). It can be differentiated from other species by appropriate mouse toxin neutralization tests and complement fixation tests with immune sera derived from mice.

10. Rickettsia tsutsugamushi (Hayashi) Ogata 1931, 252. (*Theileria tsutsugamushi* Hayashi 1920, 63; *Rickettsia orientalis* Nagayo *et al.* 1930, 317; *Rickettsia akamushi* Kawamura and Imagawa 1931, 258.)

tsu.tsu.ga.mu'shi. From two Japanese ideographs transliterated *tsutsuga* something small and dangerous, and *mushi* a creature, now known to be a mite.

Short rods with average length 1.2 μm, seldom exceeding 1.5 μm in length, but diplobacillary forms are frequently seen. Stain intensely blue-purple by the method of Giemsa. Other staining methods commonly used for rickettsiae, such as Giménez, not entirely satisfactory and require modification.

Established strains grow well in yolk sac of chick embryos. Excellent growth also obtained in several types of cell cultures, such as mouse lymphoblasts and fibroblasts, rat fibroblasts, monkey kidney cells and HeLa cells. Growth occurs preferentially in the perinuclear region of the host cells, where thick masses of organisms accumulate.

In an extracellular environment more heat-labile than other rickettsiae: at 37 C viability is greatly reduced in 2 or 3 hrs. Diluent SPG, glutamate, albumin and other proteins have a stabilizing effect. Unaffected by streptomycin, somewhat more resistant than other species to p-aminobenzoic acid, but of comparable susceptibility to the broad spectrum antibiotics.

Higher degree of internal phenotypic diversity with respect to antigenic composition than other rickettsial species. A species-specific antigen, analogous to the group-specific antigen of other rickettsiae, is demonstrated by fluorescent antibody techniques or isolation of a soluble antigen by density gradient. This antigen is relatively unstable and is lost during procedures such as ether extraction, commonly used for other rickettsial species. When purified organisms are used as antigens in complement fixation tests with guinea pig sera, three antigenic types, Karp, Gilliam and Kato, are usually demonstrated, but the occurrence of other types was shown by Elisberg *et al.* (1968). Infection elicits a weak antibody response of brief duration to the species-specific antigen and more pronounced response to the type-specific antigens.

The mouse is highly susceptible to infection by most routes and is the animal of choice in primary isolations. There is, however, considerable strain variation in virulence for the mouse. When injected with the more virulent strains, the mouse dies 7–14 days later with distended spleen and abdomen which contain numerous organisms. Acute toxic death was produced only with the Gilliam strain. Mice which recover following subcutaneous injection or chloramphenicol treatment are immune to all strains of *R. tsutsugamushi*. Vaccination with non-viable antigen, on the other hand, elicits only type-specific immunity. The guinea pig is susceptible to infection with only a

few of the strains, but it is used for the production of antisera. In serum or toxin neutralization tests immune sera are usually type-specific.

The organism is found in certain "ecological islands" which, because of climate and terrain, support a sufficiently large population of thrombiculid mites and their rodent hosts. These "ecological islands" occur in Japan, Southeast Asia, Northern Australia and extend as far as Pakistan, Siberia, Philippines and Korea. Both mites and rodents have been found infected, but it is not entirely clear which one is the more important reservoir of the rickettsiae. The mite can transmit the organisms transovarially and to the vertebrate host, which it infects during the larval stage. Man is an incidental host and acquires infection through the bite of the larva, or chigger. Established mite vectors are *Leptotrombidium akamushi*, *L. deliense*, *L. pallidum* and *L. scutellare*.

Etiological agent of scrub typhus. Elicits antibodies against *Proteus* OXK antigen.

Genus II. **Rochalimaea*** (*Macchiavello*) *Krieg 1961, 162*

EMILIO WEISS AND JAMES W. MOULDER

Ro.cha.li.mae′ a. M.L. fem.n. *Rochalimaea* named for H. da Rocha-Lima, one of the early investigators of the etiology of rickettsial diseases.

Closely resembling organisms of the genus *Rickettsia* in morphology, staining properties and in dependence on arthropod and vertebrate hosts for natural survival. **Usually reside in an extra-cellular environment in the arthropod host. Can be cultivated in host cell-free media.**

Type species: *Rochalimaea quintana* (Schmincke) Krieg 1961, 163.

Description of the species of genus **Rochalimaea**

1. **Rochalimaea quintana** (Schmincke) Krieg 1961, 163. (*Rickettsia quintana* Schmincke 1917, 961; *Rickettsia pediculi* Munk and da Rocha-Lima 1917, 1423; *Rickettsia wolhynica* Jungmann and Kuczynski 1918, 261; *Rickettsia weigli* Mosing 1936, 380; *Burnetia* (*Rocha-limae*) *wolhynica* Macchiavello 1947, 410; *Wolhynia quintanae* (*sic*) Zhdanov and Korenblit 1950, 42.)

quin.ta′na. M.L. fem.adj. *quintana* fifth; referring to 5-day fever, one of the vernacular names of the fever caused by the species.

Short rods, 0.2–0.5 by 1.0–1.6 μm. Trilaminar cell wall and plasma membrane, resembling those of Gram-negative bacteria (Ito and Vinson, 1965).

No flagella or capsules. Gram-negative. Retains basic fuchsin when stained by the method of Giménez (1964). Divides by transverse binary fission.

Cultivated on a medium consisting of blood-agar base, 6% inactivated horse serum and 4% lysed horse erythrocytes (Vinson and Fuller, 1961; Vinson, 1966). Crystalline hemoglobin or hemin can substitute for the lysed erythrocytes and bovine albumin or a colloidal "detoxifying agent" such as starch or charcoal can replace serum (Myers *et al.*, 1969). The organisms were also grown in a liquid medium in which fetal calf serum, but not calf serum, substituted for the hemoglobin

*** Editorial Note.** The name *Rocha-Limae* was proposed by Macchiavello 1943, 410 as that of a subgenus of the genus *Burnetia* with the species name *Burnetia* (*Rocha-Limae*) *wolhynica*. Some confusion has arisen because the subgeneric name from the patronymic Rocha-Lima is spelled *Rocha-Limae* and does not conform to the Recommendation 27d approved in the Moscow Code. This states that when the name of a genus or subgenus is taken from the name of a person whose name ends in a vowel or *y* the letter *a* is added, except when the name of the person ends in a when *ea* is added. The emendation to *Rochalimaea* was accepted by Philip in THE MANUAL 1957, 946 and by the editors of Index Bergeyana 1966, 915. Philip (1956, 265) proposed the transfer of subgenus *Rochalimaea* to the genus *Rickettsia*. Krieg (1961, 162) proposed its recognition as a generic name.

requirement (Mason, 1970). Aerobic conditions and increased CO_2 tension are required for growth.

Colonies, 65–200 μm in diameter, round, lenticular, translucent and mucoid appear on the surface of the agar following incubation at 37 C for 12–14 days in primary isolations, 3–5 days in subsequent passages. Can also be cultivated in liquid media of similar composition. Chick embryos and cell cultures are not very satisfactory for growth. The tetracyclines and chloramphenicol are highly effective inhibitors of growth. Penicillin is less effective and streptomycin is relatively ineffective.

Resting cells actively metabolize succinate and glutamine but not glucose. Glutamate-oxaloacetate transaminase activity was demonstrated (Huang, 1967). Experiments carried out prior to *in vitro* cultivation indicated that organisms are not completely inactivated by moist heat at 60 C for 30 min or by dry heat at 80 C for 20 min. Viability retained for several months in dry louse feces.

Guinea pigs and other common laboratory animals have not been infected. Experimental infection has only been produced in the rhesus monkey (*Macacus rhesus*) and in human volunteers. Convalescent sera react with the homologous antigen in fluorescent antibody and complement fixation tests, but not with antigens derived from species of the genera *Rickettsia* or *Proteus*.

Etiological agent of trench fever. Man is the primary host. Transmission is accomplished, as in the case of *Rickettsia prowazekii*, by the human louse, *Pediculus humanus*. The organisms grow in the lumen of the gut and are discharged with the feces, which are the source of human infection when they are driven into the skin in the process of scratching. There is no indication that the louse is harmed by the infection or that transovarian passage occurs.

The G + C content of the DNA is approximately 38.5 moles % (thermal denaturation and buoyant density, Tyeryar *et al.*, 1973).

Genus III. **Coxiella** *(Philip) Philip 1948, 58*

EMILIO WEISS AND JAMES W. MOULDER

(Subgenus *Coxiella* Philip 1943, 306; genus *Burnetia* Macchiavello 1947, 408.)

Co.xi.el'la. M.L. fem.dim. ending *-ella;* M.L. fem.dim.n. *Coxiella* named for H. R. Cox, who in collaboration with G. E. Davis first isolated this organism in the United States shortly after its discovery in Australia and who introduced the technique of yolk sac inoculation of the chick embryo which greatly facilitated the study of this and other genera.

Short rods, usually 0.2–0.4 by 0.4–1.0 μm, resembling organisms of the genus *Rickettsia* in staining properties, dependence on host cells for growth and close natural association with arthropod and vertebrate hosts.

Grows preferentially in the vacuoles of the host cell (rather than in cytoplasm or nucleus as

do the species of *Rickettsia*). **Grows well in the yolk sac of chick embryos. Resists drying and relatively elevated temperatures,** that generally destroy the viability of *Rickettsia*.

Type species: *Coxiella burnetii* (Derrick) Philip 1948, 58.

Description of the species of genus **Coxiella**

1. **Coxiella burnetii** (Derrick) Philip 1948, 58. (*Rickettsia burneti* (*sic*) Derrick 1939, 14; *Rickettsia diaporica* Cox 1939, 1826.)

bur.ne'ti.i. M.L. gen.n. *burnetii* of Burnet; named for F. M. Burnet, who first studied this organism.

Rods variable in size but smaller than those of *Rickettsia*, usually 0.2–0.4 by 0.4–1.0 μm, occasionally diplobacilli 1.0–1.6 μm long, or spheres 0.3–0.4 μm in diameter. No flagella or capsules. The fine structure is similar to that of *Rickettsia*. The cell wall contains muramic and diaminopimelic acids. Generally regarded as Gram-negative, under certain conditions of staining it can be shown to be Gram-positive (Giménez, 1965). Retains basic fuchsin when stained by the method

of Giménez (1964). Although there has been speculation that a developmental cycle takes place during early stages of development, which includes a filterable phase, the only clearly demonstrated mechanism of multiplication is transverse binary fission.

Grows well in yolk sac of chick embryos incubated at 35 C. Highest viability obtained at about the time of death of the embryos, but small variations in temperature or in time of harvest do not affect the yield as much as in the case of species of *Rickettsia*. Division time during phase of most rapid growth in chick embryos is about 12 hrs, but because of high stability, as many as 10^{11} viable particles can be obtained from a single egg. When the inoculum consists of one or just a few viable cells, a high yield of organisms is obtained only after two or three egg passages. Good growth is obtained in various types of cell cultures, such as chick embryo entodermal cells or fibroblasts, or mouse L cells but yields of viable organisms are not as high as from yolk sac. Growth in the egg is not affected by p-aminobenzoic acid or by moderate amounts of penicillin, streptomycin or erythromycin. The following amounts of antibiotics (per egg) have a measurable effect on growth, i.e. delay embryo death by about 1½ days; chloramphenicol, 0.25 mg; chlortetracycline, 0.06 mg; oxytetracycline, 0.006 mg (Ormsbee et al., 1955).

Cells separated from host components display a number of metabolic activities. Pyruvate is the substrate utilized most rapidly by intact cells (Ormsbee and Peacock, 1964). It also enhances growth of the organism in entodermal cell cultures. Glutamate and some of the intermediates of the citric acid cycle are catabolized at lower rates. Glucose is not metabolized by intact cells but is phosphorylated to glucose 6-phosphate and oxidized to 6-phosphogluconate by disrupted cells. The presence of folic acids and of ribosomes suggests that the cells have synthetic capabilities. Low to moderate levels of the following activities were demonstrated with disrupted preparations: synthesis of serine, citrulline and ureidosuccinate in the presence of appropriate precursors; incorporation of amino acids indicative of protein synthesis; polynucleotide formation indicative of polymerase activity (Paretsky, 1968).

More resistant to physical and chemical agents than the majority of non-sporogenic microorganisms. At 4 C viability is retained for 1 or more years in dried fomites, such as tick feces or wool, as well as in sterile skim milk or in unchlorinated water. Meats remain infected for at least 1 month. Organism is somewhat less stable when stored at 15–20 C. Complete inactivation is not always accomplished by exposure to 63 C for 30 min or 85–90 C for a few seconds. Viability of infected yolk sac suspensions is not entirely destroyed by 0.5% NaOH for 6 hrs (but is destroyed by similar exposure to HCl), 0.3% formaldehyde or 1% phenol for 24 hrs. It is inactivated by diethyl ether, but is considerably more resistant than Rickettsia to most of the commonly used antiseptics.

Phase variation (Stoker and Fiset, 1956), analogous to smooth to rough variation of other bacteria, detected primarily by immunological tests, is a prominent feature of this organism. All recent isolates are in phase I. Repeated passages in the yolk sac (usually 8–20) elicits the emergence of phase II. A single passage in the guinea pig is generally sufficient to re-establish phase I. Phase II cells, in contrast to those in phase I, readily agglutinate in salt solution or in the presence of normal serum and are phagocytized in the absence of specific antibodies. Phase I activity is attributed to a surface carbohydrate which is solubilized by treatment with dimethylsulfoxide or trichloroacetic acid and is destroyed by $NaIO_4$. Treatment with the latter two compounds leaves sediments which have the antigenic specificity of phase II cells. Phase II antigen has not been solubilized and its chemical composition has not been defined.

Phase I and II antigens differ not only in specificity, but also in the nature of their in vitro and in vivo reaction. Phase II antigen is by far the most satisfactory in complement fixation tests with sera of animals and man who have recovered from infection, especially when the infection is recent. The injection of phase II antigen elicits the production of phase II antibodies. Phase I, on the other hand, is by far the more powerful immunogen. In the guinea pig the dose can be reduced 100- to 300-fold when phase I replaces phase II antigen. Phase I organisms elicit antibodies against phase I and phase II antigens and, surprisingly, the antibody response to phase II is more rapid and longer lasting than that elicited by phase II organisms. These results suggest that phase I antigen contains masked phase II antigen and that the in vivo roles of phase I and II substances are, respectively, those of adjuvant and antigen (Fiset and Ormsbee, 1968).

Except for phase variation, the antigens from the various strains of C. burnetii differ slightly in specificity as demonstrated in complement fixation and cross-protection tests. There is no cross-reaction with antigens derived from Rickettsia or Proteus.

The mechanism of entry into vacuoles of host cells is not known. Multiplication takes place almost exclusively in vacuoles, which enlarge as the organisms become more numerous. An individual cell is usually surrounded by a membrane of host origin.

The guinea pig is the animal of choice in primary isolation. Although strains differ considerably in virulence for the guinea pig, death is usually caused only by relatively large inocula. Smaller doses cause fever and enlargement of the spleens from which the organisms can be isolated in large numbers. Rabbits, hamsters and mice are also highly susceptible. Acute toxic death has not been produced in mice.

Enzootic in domestic animals, such as cattle, sheep and goats, in most countries of the world. Widely disseminated among other animals, including birds. Numerous species of arthropod, especially tick ectoparasites of ungulates, rodents and marsupials, have been found infected and capable of transmitting the organism transovarially and to vertebrates. Organisms are shed in large numbers by infected animals in milk, excreta and particularly in placentae, which may contain as many as 10^9 infectious particles per g of tissue. Because of the high degree of stability of the organism in an extracellular environment, aerosol infection of cattle and of other animals is as important as vector transmission. In most animals, infection does not result in frank illness.

Etiological agent of Q fever in man, a moderately severe but rarely fatal pulmonary infection. Man is usually infected by the aerosol route. In some cases infection persists and in rare occasions gives rise to fatal subacute endocarditis. Virulence for man does not appear to vary widely among naturally occurring strains, but strains of limited virulence were produced in the laboratory by serial passage in eggs and animals (M-44 and 1/M-44 variants of the Grit strain, Genig, 1968).

The G + C content of the DNA is approximately 43 moles % (acid hydrolysis and chromatography, Smith and Stoker, 1951).

TRIBE II. EHRLICHIEAE PHILIP 1957, 948

CORNELIUS B. PHILIP

(Equivalent to family Ehrlichiaceae Moshkovski 1945, 18.)

Ehr.lich.'ie.ae. M.L. fem.n. *Ehrlichia* type genus of the tribe; *-eae* ending to denote a tribe; M.L. fem.pl.n. *Ehrlichieae* the *Ehrlichia* tribe.

Minute, **rickettsia-like organisms pathogenic for certain mammals,** but, unlike those of Tribe *Rickettsieae*, **not infectious for man.** Differential characteristics, such as cytotropisms, hosts and vectors, are given in Table 18.2. Cells grow in the cytoplasm, but not in the nucleus, usually as compact, single to multiple morula-like colonies. Less often seen as single cells (early termed initial bodies), or as cells scattered throughout the cytoplasm. Also adapted to existence in ticks or, in one case, in trematodes, but not in insects. **Cell walls present. Non-motile.** Gram-negative. No Weil-Felix (*Proteus* OX) antibodies elicited. Includes the genera *Ehrlichia*, *Cowdria* and *Neorickettsia*.

TABLE 18.2

Differential characteristics of genera of tribe **Ehrlichieae**

Genus	Mammalian hosts	Tissues parasitized	Invertebrate vectors	Maintenance in vectors	Distribution
Ehrlichia	Canidae; domestic ruminants	Circulating leukocytes	Ticks: *Rhipicephalus* spp., *Hyalomma* sp.	Transovarial and/or transstadial only	Various places in Old and New Worlds.
	?Equines	Ditto	?*Dermacentor* sp.	?	Calif.
Cowdria	Domestic and certain wild ruminants	Vascular endothelial cells	Ticks: *Amblyomma* spp.	Transstadial only	Africa and Madagascar; possibly Dalmatia
Neorickettsia	Canidae	Reticular cells of lymphoid tissues	Trematode: *Nanophyetus salmincola*	Transovarial and through all stages	Pacific Coast, U.S.A.

Genus IV. **Ehrlichia** *Moshkovski 1945, 18*

(*Rickettsia* subgen. *Ehrlichia* Moshkovski 1937, 382.)

Ehr.lich'ia. M.L. fem.n. *Ehrlichia* named for Paul Ehrlich, a German bacteriologist.

Small, often **pleomorphic, coccoid to ellipsoidal organisms occurring intracytoplasmi**cally, either singly or in compact colonies (morulae) **in certain circulating leukocytes** of susceptible mammalian hosts. Adapted to growth in certain species of ixodid ticks associated with such hosts. **Non-motile. Not cultivable in cell-free media or chicken embryos.** The etiological agents of tick-borne diseases chiefly of dogs, cattle, sheep, goats and probably horses.

Type species: *Ehrlichia canis* (Donatien and Lestoquard) Moshkovski 1945, 18.

Description of the species of genus **Ehrlichia**

1. **Ehrlichia canis** (Donatien and Lestoquard) Moshkovski 1945, 18. (*Rickettsia canis* Donatien and Lestoquard 1935, 419; *Ehrlichia* (*Rickettsia*) *canis* (*sic*) Moshkovski 1937, 382.)

ca′nis. M.L. n. *canis* the dog; L. gen.n. *canis* of the dog.

E. canis may be demonstrated by Romanowsky stain as cytoplasmic inclusions in circulating monocytes, lymphocytes and rarely neutrophils of the family *Canidae*, either as single "initial bodies" or more often morulae containing distinguishable groups of organisms; several such colonies may be found in one cell. More detailed morphological studies under the electron microscope not reported. Giemsa is the stain of choice; organisms are blue.

Transmitted by ticks of the *Rhipicephalus sanguineus* complex; transmission is both transstadial and transovarial. Since the same tick vectors commonly transmit *Babesia canis*, concurrent infections occur. Ewing (1969) found uncomplicated canine ehrlichiosis in Oklahoma, U.S.A., less fatal to adult dogs than usually reported. The organism has been labelled malignant in Africa, France, Syria and India. The pathology of a "highly fatal haemorrhagic disease of U.S. military dogs in Southeast Asia," apparently caused by *E. canis*, has been described by Huxsoll *et al.*, 1969. A similar outbreak probably occurred in Puerto Rico and the Virgin Islands. The organism persisted in a state of premunition in recovered animals for at least 13 months in dogs and 112 days in an African jackal. Adaptation to small laboratory animals and chicken embryos has not been accomplished. No specific antigen is available. Sera from dogs in various stages of convalescence showed no reaction to complement-fixing antigens of various *Rickettsia* spp. and *Coxiella*. *E. canis* is sensitive to sulfonamides and tetracycline antibiotics.

2. **Ehrlichia phagocytophila** (Foggie) Philip 1962, 42. (*Rickettsia phagocytophila ovis* Foggie 1949, vi, not validly published; *Rickettsia phagocytophila* Foggie 1951, 4; *Cytoecetes phagocytophila* (Foggie) Foggie 1962, 55, 59; *Cytoecetes bovis* Tuomi 1966, 419.)

pha.go.cy.to′phil.a. Gr. inf. *phagein* to eat, devour; Gr. *kytos* a vessel, enclosure; Gr. inf. *philein* to love; M.L. adj. *phagocytophila*, fond of devouring cells (in microbiology, attractive to phagocytes).

In Giemsa-stained blood and tissue smears the simplest forms, commonly referred to as 'initial bodies' by earlier workers appear in neutrophils, eosinophils, basophils and monocytes as bluish purple irregular-shaped bodies varying from 0.7–3.0 µm in size. These develop into large homogeneous masses 2.0–4.0 µm in diameter. Within these masses there appear darker stained particles in the form of morulae or clusters. The latter result from fragmentation of the homogeneous masses giving rise to a variable number of elementary bodies 0.5 µm in size.

Electron microscopy of ultrathin sections of infected sheep and calf leukocyte concentrates reveals large, intermediate and small particles in intracytoplasmic vesicles chiefly in cells of the granulocyte series. The larger initial bodies (Plate 18.1, Fig. 1) are considered the infective forms and have cell walls, plasma membranes, cytoplasm with ribosome-like granules, and nucleoid. Replication is by binary fission and release is by rupture of vesicles.

Not cultivable on cell-free media, in chicken embryos or in cell cultures, so that specific serology has not been developed except for demonstration of the relationship of Scottish and Finnish strains by immunofluorescence (Tuomi, 1966). Has remained infectious in citrated blood for 10 days at room temperature, 14 days at 4–8 C, and 18 months at −79 C in glycerol- or dimethylsulfoxide-treated infectious blood.

Can be passed serially in sheep and cattle, in which natural and experimental differences in strains and virulence have been demonstrated; cross-immunity between bovine and ovine strains is usually incomplete. From 6–50% of granulocytes may show parasites during the fastigium of the disease. Strains have been adapted to normal guinea pig passage after many transfers through splenectomized animals, and by blind passage; also serially passed in splenectomized albino mice but unstable in normal ones.

Transmission in Europe is by *Ixodes ricinus* in which the organisms are passed between active tick stages but not through ova.

Natural infection is reported chiefly for sheep

and cattle in Great Britain, the Netherlands and Finland, but isolates of *E. phagocytophila* have also been made from wild deer in Great Britain. Goats are susceptible.

Species incertae sedis

a. *Ehrlichia bovis* (Donatien and Lestoquard) Moshkovski 1945, 18. (*Rickettsia bovis* Donatien and Lestoquard 1936, 1061.)

Originally isolated from circulating monocytes of Moroccan cattle to which *Hyalomma* ticks were transferred from cattle imported from Iran. Further passage transstadially in ticks, as well as infection in cattle in other parts of Africa has been reported. Characterization of the organisms, most recently by Rioche (1967), has revealed their generic relationships, but other information, such as immunological and tick vector relationships, is still inadequate. Furthermore, Donatien and Lestoquard (1940) have reported infecting sheep which raises doubts of specific differentiation from *E. ovina*.

b. *Ehrlichia ovina* (Lestoquard and Donatien) Moshkovski 1945, 18. (*Rickettsia ovina* Lestoquard and Donatien 1936, 108.)

Observed originally in the peripheral monocytes of Moroccan sheep injected with ticks (*Rhipicephalus bursa*) from other sheep. Occurrence of the organisms in morular colonies resembles *E. bovis* but tick transmission by bite and adequate immunological relationships are unreported.

An unnamed, but obviously related, agent was recently reported by Gribble (1969) as causing a similar disease, "Equine Ehrlichiosis," in horses in northern California; organisms occurred in circulating granulocytes and were passed to horses, donkeys, dogs, sheep and goats, but calves proved refractory. Tick transmission is still under investigation.

c. *Ehrlichia kurlovi* Moshkovski 1937, 382.

Described from circulating monocytes of guinea pigs. The pathogenicity and rickettsial nature of so-called Kurlov's bodies has been disputed since their description by Moshkovski and later Zhdanov (1953), who compared their structure to *Cowdria ruminantium*. More recent isolates in yolk sacs of chicken embryos of what were possibly the same organisms, e.g. Bozeman *et al.* (1968), have focused new significance on these as distinct organisms cultivable in chicken embryos, without clarifying their systematic relationship. These, however, should not be confused with the non-organismal "intracytoplasmic inclusion called the Kurloff body" most recently discussed by Berendsen and Telford (1966) in light and electron micrographic studies which included supravital preparations.

d. *Cytoecetes microti* Tyzzer 1938, 254.

Described from the "granular leucocytes and rarely the lymphocytes" of a vole in Massachusetts. No subsequent isolates have been reported. Although originally differentiated from *Aegyptianella pullorum* Carpano, microscopic similarity to *Ehrlichia* is now obvious. Information on relationships and transmission is insufficient, however, to support recent assignment to *Cytoecetes* of *E. phagocytophila*.

e. *Cytoecetes ovis* var. *decani* Raghavarhari and Reddy 1959, 69.

A similar organism, described from the blood of sheep and goats in India, was reported also to cause "tick-borne fever" but otherwise insufficiently characterized for systematic assignment. Since a typical *C. ovis* has not been described, "var. *decani*" will remain illegitimate for any future use.

f. *Rickettsia* (*Donatienella*) *delpyi* Rousselot 1948, 112, is another organism on which there is no further information since its original isolation from the leucocytes of a splenectomized Iranian gerbille. However, the relationships are closer to *Cytoecetes* or *Ehrlichia* than to *Rickettsia sens. str.* Assignment of *E. canis*, *E. bovis* and *E. ovina* to *Donatienella* has even been suggested, although this is not tenable because of uncertainties of the systematic position of this subgenus.

g. *Rickettsia belgaumi* Manjrekar 1954, 219. Organisms found in monocytes which cause "rickettsiosis in sheep and goats in the State of Bombay," India. Differentiated "from tick-borne fever and *R. ovina*" by infection of guinea pigs and white rats. Although originally described in a mimeographed thesis, the name was validated by ample characterization in a published abstract.

Genus V. **Cowdria** *Moshkovski 1947, 62*

(*Ehrlichia* subgenus *Cowdria* Moshkovski 1945, 33; *Nicollea* Macchiavello 1947, 415; *Kurlovia* Zhdanov 1953, 166.)

Cow'dri.a. M.L. fem.n. *Cowdria* named for E. V. Cowdry, who first described the organism in heartwater diseases of sheep, goats and cattle.

Small, pleomorphic, coccoid or ellipsoidal, occasionally **rod-shaped, organisms** occurring **intracytoplasmically** but not intranuclearly, and characteristically localized in clusters inside vacuoles in the **cytoplasm of vascular endothelial cells** of ruminants (Table 18.2). **Not passed transovarially in tick vectors. Non-motile.**

Gram-negative. **Have not been cultivated in cell-free media.** The etiological agent of heartwater, a septicemic disease of domestic ruminants in Africa.

Type species: *Cowdria ruminantium* (Cowdry) Moshkovski 1947, 62.

Description of the species of genus **Cowdria**

1. **Cowdria ruminantium** (Cowdry) Moshkovski 1947, 62. (*Rickettsia ruminantium* Cowdry 1925, 231; *Kurlovia ruminantium* (Cowdry) Zhdanov 1953, 166; *Nicolœa ruminantium* (Cowdry) Macchiavello 1947, 415.)

ru.min.ant′i.um. M.L. neut.gen.pl.n. *ruminantium* of *Ruminantia*, formerly an ordinal name for cud-chewing mammals.

Differ morphologically from typical typhus-like rickettsiae, and show usually coccoid and ellipsoidal, occasionally short bacillary forms. Irregular pleomorphic forms occur, sometimes in densely packed masses. In the vascular endothelial cells of animals, cocci measure 0.2–0.5 μm in diameter and 0.2–0.3 μm in tick tissues; bacillary forms are 0.2–0.3 by 0.4–0.5 μm, and pairs are 0.2 by 0.8 μm. Stain dark blue with Giemsa; can also be stained by methylene blue and other basic aniline dyes. Sensitive to sulfonamides and tetracycline antibiotics.

Organisms persist in a state of subclinical premunition for periods up to 60 days in cattle, sheep and goats, during which re-exposure does not result in clinical relapse. After sterile immunity of 3 months to 5 years, *C. ruminantium* reintroduced by infected ticks, *Amblyomma* spp., results in renewed temporary subclinical infection. A significant characteristic of the agent is its **transstadial, but not transovarial, maintenance** in the **tick vectors.** At least five species of wild ruminants have been shown to be susceptible. The organisms persist up to 90 days as inapparent infections in laboratory mice and rats, but cannot be serially passed in them; this fact, however, facilitates isolation from infected animals in the field.

C. ruminantium is widespread in, but confined to, Africa and Madagascar; virulence varies, with fatality rates of 20–95%.

Genus VI. **Neorickettsia** *Philip, Hadlow and Hughes 1953, 257*

Ne.o.ri.ckett′si.a. Gr. pref. *neo-* new; M.L. fem.n. *Rickettsia* type genus of family *Rickettsiaceae*; M.L. fem.n. *Neorickettsia* the new *Rickettsia*.

Small, coccoid, often pleomorphic, intracytoplasmic (but not intranuclear) **organisms** which occur **primarily in reticular cells of lymphoid tissues** of *Canidae*. Also seen in certain tissues of mature **fluke vectors; all** other **fluke stages**, eggs, rediae-cercariae and metacercariae have been proven **infectious** by injection into susceptible vertebrate hosts, which confirms that the infectious cycle includes **transovarial trans-**

mission in the vector. Filterable forms not reported, but suspected. Gram-negative. Non-motile. **Not cultivable in cell-free media or in chicken embryos.** Sensitive to tetracycline antibiotics. The etiological agent of a helminth-borne disease of *Canidae* on the mid-West Coast of U.S.A.

Type species: *Neorickettsia helminthoeca* Philip, Hadlow and Hughes 1953, 257.

Description of the species of genus **Neorickettsia**

1. **Neorickettsia helminthoeca** Philip, Hadlow and Hughes 1953, 257. (*Neorickettsia helmintheca* (sic) Philip, Hadlow and Hughes 1953, 257.)

hel.minth′oe.ca. Gr. n. *helmins, helminthis* worm; Gr. n. *oikos* house; M.L. fem.adj. *helminthoeca* worm-dwelling.

Neorickettsia helminthoeca is characteristically found intracytoplasmatically in reticuloendothelial cells of the lymphoid tissues of infected *Canidae*. In Giemsa-stained preparations, the orga-

nisms are seen in single or multiple morula-like colonies (Plate 18.1, Fig. 2), or as scattered individuals throughout the cytoplasm. Coccoid forms of 0.3–0.4 μm are most common but pleomorphic forms include ellipsoids, short rods and even clubs and rings. There are occasional intracytoplasmatic homogeneously deep blue- or lilac-staining bodies of 1–2 μm in diameter or larger; these are suspected of equivalence to the so-called initial bodies reported for *Ehrlichia* spp., which are considered to fragment into infectious elementary bodies. No Weil-Felix (*Proteus* OX) antibodies elicited.

Not detected in circulating leukocytes (although blood is infectious by passage to susceptible hosts); in this respect, this agent differs from peripherally located *Ehrlichia* species. It is distinguished from both tick-borne *Ehrlichia* and *Cowdria* in its adaptation to the complicated life cycle of the trematode vector and its geographic isolation on the mid-Pacific Coast of U.S.A. *N. helminthoeca* also differs in not persisting premunitively after recovery of the canine host in which latent infection was not reactivated after splenectomy. Inflammatory changes in leptomeninges and brain have been likened to those caused by typhus-like rickettsiae.

The natural cycle of this pathogen was recently shown (Nyberg *et al.*, 1967) to include passage through trematode eggs, as previously postulated since other trematode stages in snails, salmonid fish and the dog had been shown to be infective. However, the evolutionary development of this rickettsia-like organism remains puzzling, since *Canidae* are the only known susceptible definitive hosts, and withdraw from aquatic environments when ill. Up to 90% mortality occurs in untreated animals, whereas other adult-trematode hosts, such as raccoons and bears are not susceptible. Dogs cannot account for the high infection rate among numerous metacercariae in salmonids in areas of abundance of the proper snail intermediate hosts.

Much remains to be elucidated in the biology of *N. helminthoeca* in spite of considerable information of its parasitic behavior.

"Elokomin fluke fever" has been applied to a second disease caused by a rickettsia-like agent which has recently been reported to cause concomitant disease in dogs and to be transmitted by the same trematodes (Farrell, 1966). Recovered animals do not show cross-immunity with *N. helminthoeca*, but etiological information is still insufficient to characterize this newly recognized fluke-borne pathogen adequately.

TRIBE III. WOLBACHIEAE PHILIP 1955, 271

EMILO WEISS

Wol.ba.chi′e.ae. M.L. fem.n. *Wolbachia* type genus of the tribe; -*eae* ending to denote a tribe; M.L. fem.pl.n. *Wolbachieae* the *Wolbachia* tribe.

A miscellaneous group of organisms associated with invertebrates, mostly arthropods, but not usually invading vertebrates. Most of these organisms have not been cultivated outside their hosts, and only in rare instances have they been grown on bacteriological media.

Four genera are accepted. They include many species which are rickettsia-like in morphology, staining properties and growth and which were formerly assigned to the genus *Rickettsia*. Characterization is usually not as adequate as for the preceding genera, which are pathogenic to vertebrates, and differentiation has been chiefly on the basis of presumed host specificity in arthropods, although differences in development and morphology are often noted. There is no evidence of an evolutionary relationship among the genera of this tribe.

Key to the genera of tribe **Wolbachieae**

I. Not pathogenic for larvae of insects or other invertebrate hosts.
  A. Associated with arthropods, seldom pathogenic for their hosts, seldom develop in mycetomata.
      Genus VII. *Wolbachia*
  B. Symbiotic; located in special organs (mycetomata) in the insect host, but found also in other tissues; highly pleomorphic.
      Genus VIII. *Symbiotes*
  C. Symbiotic in cockroaches; located in dispersed cells (mycetocytes) in the fat body and in ovaries; rod-shaped.
      Genus IX. *Blattabacterium*
II. Pathogenic for insect larvae and for other invertebrate hosts.
      Genus X. *Rickettsiella*

Genus VII. **Wolbachia** *Hertig 1936, 472*

Emilio Weiss

Wol.ba'chi.a. M.L. fem.n. *Wolbachia* named for S. B. Wolbach, who described the rickettsial agent of Rocky Mountain spotted fever.

Heterogeneous group of small coccoid forms, rods and spheres, which morphologically and tinctorially resemble either *Rickettsia* or *Chlamydia.* **Gram-negative.** Stain well with Giemsa's and Macchiavello's stains.

Only *Wolbachia melophagi* has been cultivated on a cell-free medium, but this work has not been confirmed since 1924. The other two species have not been cultivated and have been seen only in close association with host cells. Differential characteristics are summarized in Table 18.3.

Associated with arthropods, but unlike *Rickettsiella,* **seldom pathogenic for their**

hosts. Extracellular or intracellular but, unlike *Symbiotes,* seldom developing inside mycetomes. **Not associated with diseases of vertebrates.**

Type species: *Wolbachia pipientis* Hertig 1936, 472.

Further Comments

The amount of information available on this genus is small and future work may indicate reclassification is required. Only *Wolbachia persica* is available in the laboratory and can be identified by laboratory tests. The other two species can be recognized only by reference to the host of origin.

Description of the species of genus **Wolbachia**

1. **Wolbachia pipientis** Hertig 1936, 472.
pi.pi.en'tis. M.L. n. *pipiens* specific epithet of the host mosquito, *Culex pipiens;* M.L. gen.n. *pipientis* of *pipiens.*

Pleomorphic coccoid forms or rods often found in pairs or small packets. Electron microscopy reveals typical bacterial structure (Micks *et al.,* 1961).

Attempts to cultivate this organism in cell-free media have been unsuccessful. Cultivation in cell cultures or chick embryos has not been attempted.

The location in the host is primarily intracellular, but extracellular forms occur, possibly, as the result of disruption of the host cell. Degenerated host cells were seen in the gonad and in the gut epithelium, but there is no indication that the hosts are otherwise affected.

Found originally in the gonad cells of *Culex*

pipiens of North American and Chinese origin. Similar forms were also described in the epithelial lining of the stomach of *Culex fatigans* (Brumpt, 1938), in some other mosquitoes (Micks *et al.,* 1961) and in other Diptera (Krieg, 1961). The parasite of *Culex fatigans* was named *Rickettsia culicis* by Brumpt 1938, 154 (transferred to *Wolbachia* by Philip 1956, 267) and *Enterella culicis* by Krieg 1961, 166. Organisms closely resembling *W. pipientis* were also seen in a hymenopteran insect *Dahlbominus fuscipennis* (Zett.) (Byers and Wilkes, 1970). The basis for this separation in both cases was the organ parasitized, gonad in the case of *Culex pipiens,* gut epithelium in the other mosquitoes and an unsupported difference in degree of virulence for the host. Because of the limited information, separation of these microorganisms into more than one species is not justified.

TABLE 18.3
Characteristics differentiating the species of genus **Wolbachia**

	1. *W. pipientis*	2. *W. melophagi*	3. *W. persica*
Morphology	Coccoid forms to small rods, 0.25–0.5 by 0.5–1.3 μm	Small rods, 0.3 by 0.6 μm	Small spheres, 0.5 by 0.9 μm, often flattened
Cultivation	Not reported	On glucose-blood-bouillon-agar and in yolk sac of chick embryos	Profuse growth in yolk sac of chick embryos and cell culture
Host	Mosquito (gonads in *Culex pipiens,* gut epithelium in other species)	Sheep ked (*Melophagus ovinus*); possibly some species of lice	Tick (*Argas persicus* and *A. arboreus*)
Growth in host	Intracellular	Extracellular	Intracellular

The same applies to the designation *Wolbachia lynchiae* Krieg 1961, 170 from the fly *Lynchia maura* by Aschner (1931).

2. Wolbachia melophagi (Nöller) Philip 1956, 267. (*Rickettsia melophagi* Nöller 1917, 70.) me.lo'pha.gi. M.L. masc.n. *Melophagus* a genus of wingless flies commonly called sheep keds (sometimes incorrectly called "ticks"); M.L. gen.n. *melophagi* of *Melophagus*.

Rods fairly uniform in size and occurring characteristically in pairs, when seen in the intestinal epithelium of the host. Greater pleomorphism and greater average length was noted in smears from eggs of the host and from cultures.

Cultivated on glucose-blood-bouillon-agar by Nöller (1917), who obtained colonies as large as 0.4–0.6 mm after incubation for 35–40 days. Hertig and Wolbach (1924) detected minute colonies after 3–5 days. It has been grown repeatedly in the yolk of chick embryos by Steinhaus (1946). The entodermal cells of the yolk sac do not appear to be invaded. More recently Henneberg and Wolff (1963) claimed to have grown the organisms in chick embryo yolk sac *in vivo* and *in vitro* and in the guinea pig.

The location in the host appears to be strictly extracellular. Closely packed rows of small rods, arranged perpendicularly to the epithelial surface can be seen in the lumen of virtually every sheep ked, *Melophagus ovinus*.

There has been no report of injury by the organism to its insect host, *Melophagus ovinus*, and since the microorganism is almost universally present, injury is an unlikely occurrence. There is no clear evidence that the sheep, vertebrate host of the ectoparasite, becomes infected. It has not been possible to establish an infection in common laboratory animals (Steinhaus 1946).

Further Comments

Morphologically identical microorganisms were seen in a few instances in the gut lumina of other insects. Examples are the parasites of the biting louse (*Trichodectes pilosus*) and of the goat louse *Linognathus stenopsis*), named provisionally *Rickettsia trichodectae* Hindle 1921, 156 and *Rickettsia linognathi* Hindle 1921, 157 and transferred to *Wolbachia* by Philip (1956, 267). Other examples are listed by Krieg (1961, 164–166). Recognition of more than one species is not justified.

Krieg (1961), 165) placed this microorganism in a separate genus which he named *Rickettsoides*. This classification is not acceptable. His first criterion, lack of virulence for invertebrate and vertebrate animals, does not clearly separate *Wolbachia melophagi* from other species of *Wolbachia*. The second criterion, extracellular location,

although useful for the separation of *Wolbachia melophagi* from the other species of *Wolbachia*, by itself does not delineate a separate genus. Moreover, *Rickettsoides* is preoccupied as an orthographic variant of *Rickettsioides* da Rocha-Lima 1930, 1350.

3. Wolbachia persica Suitor and Weiss 1961, 105. per'si.ca. L. fem.adj. *persica* from the specific epithet of the reputed host tick, *Argas persicus*.

Spheres, relatively uniform in size when dried from the frozen state, somewhat flattened and variable in size when air-dried. The cell wall-membrane complex is about 12–13 nm thick and less rigid than the comparable structure of most other bacteria (Suitor, 1964). Although the cells are usually well separated, they appear typically in small clusters in the cells of the host, as shown by Jaschke (1933), or in smears of organisms cultivated in the yolk sac of chick embryos. The cells usually retain the basic fuchsin in Macchiavello's stains, but stain best by Giemsa's method following Carnoy's fixation.

No growth was obtained in any one of a large number of cell-free media. It grows profusely, however, in the yolk sac of chick embryos or explanted entodermal cells and in several lines of mammalian and insect cell cultures.

Relatively stable under ordinary conditions of storage or laboratory manipulation. It is unstable, however, when suspended in distilled water (Suitor, 1964) and metabolically inactive in hypotonic solutions. Under optimal conditions respiration is greatly stimulated by glucose, serine, glutamine and, to a lesser extent, by glycerol, pyruvate, acetate and glutamate. Synthesis of protein and lipid, but not of nucleic acid, was demonstrated with a number of substrates *in vitro*. Exogenous cofactors are not required for these activities (Weiss *et al.*, 1962, 1964; Neptune et al., 1964).

Growth in the chick embryo is inhibited by erythromycin (0.05 mg/egg), *p*-aminobenzoic acid (0.2 mg), chloramphenicol and chlortetracycline (0.5 mg). The inhibition by *p*-aminobenzoic acid is not reversed by *p*-hydroxybenzoic acid. Penicillin elicits the formation of aberrant forms.

Of a total of 15 isolations in chick embryos, 12 were obtained from organs of ticks parasitizing the buff-backed heron *Bubulcus ibis*, collected at the Nile Barrage Park, near Cairo, United Arab Republic. The ticks were regarded as *Argas persicus* Oken, but were later found to differ from prototype tick of this genus, placed in a new subgenus *Persicargas*, and were renamed *Argas (Persicargas) arboreus* by Kaiser, Hoogstraal and Kohls (1964, 62). The remaining three isolates were obtained from *Argas persicus* ticks, collected in a nearby location, parasitizing chickens but fed, in the laboratory, on herons. Isolation attempts from pools of tick organs in these experiments has been successful in about 60–70% of the cases. Also successful were isolations from the intestinal fluids of *Argas arboreus* rendered aposymbiotic and reinfected by feeding on chick embryos heavily infected with *Wolbachia persica*. Totally unsuccessful were scores of attempts at isolation from *Argas arboreus* fed on chicken, from *Argas persicus* obtained from various sources not fed on herons, from other species of *Argas* and from three species of *Ornithodoros*. Many of these ticks appeared to contain microorganisms similar to *Wolbachia persica* (Suitor, 1964).

The microorganisms were seen in considerable numbers in the Malpighian tubes of every *Argas arboreus* or *Argas persicus* examined and in lesser numbers in the reproductive organs. Isolates were obtained from both sources. There is no evidence that the microorganism harms or benefits its host. The role of heron blood, which makes isolation in chick embryos possible, or of chicken blood, which prevents it, is not known. There is no direct or serological evidence that the heron is infected.

Virulence for small laboratory animals is slight. Complement-fixing antibodies were produced in rabbits.

The G + C content of the DNA is approximately 30 moles % (T_m, buoyant density, Kingsbury and Weiss, 1968).

Species incertae sedis

a. *Wolbachia culicis* (Brumpt) Philip 1956, 267. (*Rickettsia culicis* Brumpt 1938, 153.)

Not sufficiently well differentiated from *W. pipientis*.

b. *Wolbachia trichodectae* (Hindle) Philip 1956, 267. (*Rickettsia trichodectae* Hindle 1921, 152).

c. *Wolbachia linognathi* (Hindle) Philip 1956, 267. (*Rickettsia linognathi* Hindle, 1921, 157.)

W. trichodectae and *W. linognathi* are not sufficiently well differentiated from *W. melophagi*.

d. *Wolbachia ctenocephali* (Sikora) Philip 1956, 267. (*Rickettsia ctenocephali* Sikora 1918, 445.)

e. *Wolbachia pulex* (Macchiavello) Philip 1956, 267. (*Cowdryia pulex* Macchiavello 1947, 418.)

W. ctenocephali and *W. pulex* have not been studied sufficiently to exclude the possibility that they are species of *Rickettsia*.

f. *Wolbachia sericea* (Giroud and Martin) Philip 1956, 267. (*Rickettsia sericea* Giroud and Martin 1946, 264.)

g. *Wolbachia dermacentrophilus* (Steinhaus) Philip 1956, 267. (*Rickettsia dermacentrophilus* Steinhaus 1942, 1376.)

Some think this is an avirulent strain of *Rickettsia rickettsii*.

Genus VIII. **Symbiotes** *Philip 1956, 267*

Marion A. Brooks

(*Cowdryia* Macchiavello 1947, 417.)

Sym.bi.o′tes. Gr. masc.n. *Symbiotes* one who lives with a companion, a partner.

Rickettsia-like, pleomorphic organisms of bedbugs (*Cimex*), living chiefly **intracellularly in** specialized paired organs called **mycetomes.** The intracellular forms coccoid or diplococcoid, 0.2 by 0.4–0.5 μm, in young hosts; extracellular bacillary, lanceolate or thread forms, 0.25–0.3 by 3.0–8.0 μm, migrate to other tissues in maturing hosts. Transmitted via ovaries to embryos.

Biochemical, nutritional and physiological characters unknown. Therefore, separation of the pleomorphic complex into pathogenic and symbiotic forms, as was done by Krieg (1961), is unwarranted.

Source and habitat identical. Members of the genus *Symbiotes* characteristically occurring as intracellular symbionts in mycetomes and much of the body tissue of species of *Cimex*.

Type species: *Symbiotes lectularius* (Arkwright et al.) Philip 1956, 267.

Description of the species of genus **Symbiotes**

1. **Symbiotes lectularius** (Arkwright *et al.*) Philip 1956, 267. (*Rickettsia lectularia* Arkwright, Atkin and Bacot 1921, 35; *Cowdryia lectularia* (Arkwright *et al.*) Macchiavello 1947, 417; *Wolbachia lectularia* (Arkwright *et al.*) Krieg 1961, 170.)

lec.tu.lar′i.us. M.L. n. *lectularius* the specific name of the host, the common bedbug, *Cimex lectularius*.

Description as for genus.

Genus IX. **Blattabacterium** *Hollande and Favre 1931, 754*

MARION A. BROOKS

Blat.ta.bac.te'ri.um. L. fem.n. *Blatta* generic name of cockroach; Gr. neut.n. *bakterion*, dim. of *baktron*, a stick or rod; M.L. neut.n. *Blattabacterium* a small rod found in cockroaches.

Straight or slightly curved rods, 0.8–0.9 by 1.5–8 μm, occurring singly, or occasionally in pairs in tandem **in the host insect.** The longer forms occur **in mycetocytes** in the abdominal fat body, the shorter forms in the **germ tissue and oocytes.** Microorganisms in mycetocytes are **encapsulated by a host-provided membrane** which is continuous with host-cell organelles. **Non-motile. No resting stages known. Gram-positive,** becoming indeterminate (speckled, barred or negative) in malnourished hosts. **Cell wall 5–10 nm composed of several alternating electron opaque and transparent layers.** Muramic acid and glucosamine present. **Complex fibrillar mesosomes** near site of formation of cross-walls.

Biochemical, nutritional and physiological characters unknown.

Growth in the insect host inhibited by penicillin, streptomycin, chloramphenicol, chlortetracycline, oxytetracycline, sulfathiazole and egg white lysozyme.

Source and habitat identical, members of the genus *Blattabacterium* characteristically occurring as **intracellular symbionts** in mycetocytes in abdominal fat body tissue and in ovaries and eggs, **of all species of cockroaches.**

Type species: *Blattabacterium cuenoti* (Mercier) Hollande and Favre 1931, 754.

Further Comments

The obligate intracellular symbionts, sometimes in the vernacular termed bacteroids, found in all species of cockroaches are considered at present to be members of the same genus (Brooks, 1970). Except for structural characteristics of those in a few species of cockroaches, precise determinative information is lacking. Culture of the cells has been unsuccessful, or unconfirmed.

The organisms are found in specialized cells, called mycetocytes, in the abdominal fat body, and in the germarium and oocytes, from which they normally are transmitted to the embryos. Certain factors in the diet of the female host insect may interfere with ovarian transmission. Embryos receive very few, if any, viable organisms if the diet of the mother is deficient in manganese, zinc or unsaturated fatty acids, while excessively high amounts of calcium or urea have the same effect. The antibiotics mentioned above also act primarily on the organisms in the ovary.

The cockroach host is dependent on the symbiotic microorganisms for growth factors as yet unidentified, normal cuticle color and egg viability.

The microorganisms are not known to be pathogenic or infective to any other insect, animal or man; they cannot orally infect natural hosts.

Description of the species of genus **Blattabacterium**

1. **Blattabacterium cuenoti** (Mercier) Hollande and Favre 1931, 754. (*Bacillus cuenoti* Mercier 1906, 684.)

cu.en.ot'i. L. gen.n. *cuenoti* of Cuenot; named for L. Cuenot who studied intracellular inclusions in orthopteran insects.

Description as for genus. Contain granules which reduce ammoniacal silver nitrate. Originally studied in *Blatta orientalis*, the oriental cockroach.

Genus X. **Rickettsiella** *Philip 1956, 267*

EMILIO WEISS

Ri.ckett.si.el'la. M.L. dim. ending *-ella*; M.L. fem.n. *Rickettsiella* a small rickettsia.

Gram-negative **rods,** usually smaller than those of the genus *Rickettsia.* **Pathogenic for their insect larval hosts and other invertebrate hosts,** but of very limited virulence for vertebrates.

Have not been cultivated in host cell-free media.

Type species: *Rickettsiella popilliae* (Dutky and Gooden) Philip 1956, 267.

Further Comments

At least nine specific names have been proposed in this genus, seven since the 7th edition of THE MANUAL, despite the fact that the information on these organisms is quite limited. The few comparative studies carried out thus far indicate that some of these organisms are similar to each other in morphology and in mode of growth in their respective host cells. They share an antigen, although they are not antigenically identical (Krieg, 1958). Host specificity has been studied in only a few instances: it appears to be of low level (Meynadier, 1964) and is not a promising criterion for species differentiation. The report that one strain grows in the host cell nucleus, while the other strains grow in the cytoplasm, is possibly in error (Dutky, 1959). Some strains, but not others, elicit the production of crystalline inclusions in their respective hosts. Most of the evidence indicates, however, that this is a host reaction and is not necessarily a valid criterion for differentiation. Observations made on individual strains reveal differences that are difficult to evaluate because they are not part of comparative studies.

One species, *Ricketsiella popilliae*, is recognized; most of the named organisms are considered strains of *R. popilliae*. *R. stethorae* may be different but is here considered as a *species incertae sedis* (q.v.).

Description of the species of genus **Rickettsiella**

1. **Rickettsiella popilliae** (Dutky and Gooden) Philip 1956, 267. (*Coxiella popilliae* Dutky and Gooden 1952, 749; *Rickettsia melolonthae* Krieg 1955, 34; *Rickettsiella tipulae* Müller-Kögler 1958, 250; *Rickettsiella grylli* Vago and Martoja 1963, 1047; *Rickettsiella chironomi* Weiser 1963, 122; *Rickettsiella blattae* Huger 1964, 22; *Rickettsiella tenebrionis* Krieg 1965, 145; *Rickettsiella schistocercae* Vago and Meynadier 1965, 309; *Rickettsiella cetonidarum* Meynadier and Monsarrat 1969, 405; *Rickettsiella armadillidii* Vago, Meynadier, Juchault, Legrand, Amargier and Duthoit 1970, 2063.)

pop.il'li.ae. M.L. gen.n. *popilliae* of *Popillia* generic name of the Japanese beetle.

Rods, 0.2 by 0.6 μm, or slightly larger, oval, somewhat curved or kidney-shaped with rounded edges. Some of the strains appear to be highly pleomorphic (Wille and Martignoni, 1952) and the average dimensions of the strain isolated from the cockroach are 0.4 by 1.2 μm (Huger, 1964). Since the equivalent of a pure culture has not been obtained with this organism, the significance of morphological variation is not known. Gram-negative; satisfactorily stained with Giemsa's or Macchiavello's stains. Not motile.

This organism has not been grown in cell-free media. The natural hosts of the microorganisms have provided the only consistent means of cultivation. The larvae have been infected by injection, by feeding or by holding them in soil inoculated with suspensions of the organism. Injection is the most effective method: less than six organisms constitute an LD₅₀ for a Japanese beetle larva. The time required for development of symptoms is dependent on dosage and on temperature of incubation of the larvae. The minimal doubling time has been estimated to be 6 hrs, but this growth rate is not sustained over a long period because at least 19 days intervene between injection of 10^4 organisms per larva and appearance of the first symptoms. Death of the larvae usually occurs 19–26 days after the appearance of symptoms. Yields of microorganisms have been as high as 3×10^{11} per larva. Growth takes place at temperatures ranging from 7–30 C with an optimum of about 27 C (Dutky and Gooden, 1952; Dutky, 1959). There is considerable variation in the time of appearance of symptoms and death of the larvae infected by the various strains. To what extent this variation reflects strain virulence or host susceptibility is not known.

Limited growth of two strains was obtained in chick embryo entodermal cell cultures and in a mammalian cell line incubated at 28 or 32 C. Typical organisms were first detected after 1 week and their numbers increased during the following 2 weeks. The organisms tended to be more elongated at the lower temperatures. During serial passage, infectivity was reduced progressively and no growth was demonstrated after four or five passages. Other cell lines and chick embryos failed to support the growth of the organisms even during the initial passage (Suitor, 1964).

Dihydrostreptomycin (20 μg/larva) or sulfadiazine (200 μg/larva) protected Japanese beetle larvae against infection with 7×10^3 organisms per larva. Penicillin and chlortetracycline did not protect the larvae. Organisms stored at 4 C remained viable for more than a year, but lost most of the viability after 3 years. They are inactivated by heat at 60 C for 10 min (Dutky, 1959).

The organisms are virulent for the mouse only when massive doses are inoculated intraperitoneally (Krieg, 1955) or intranasally. A limited number of serial passages was achieved by the latter route (Giroud *et al.*, 1958).

Natural infection was demonstrated or described on the basis of morphological evidence in many species of insects. Among them are the following: Coleoptera: the Japanese beetle (*Popillia japonica*), the June beetle (*Phyllophaga anxia* and *P. ephilida*), the cockchafer (*Melolontha vulgaris*, *M. melolontha* and *M. hippocastani*), four carabid beetles *Evarthrus alternans*, *Harpalus pennsylvanicus*, *H. erraticus* and *Amara carinata* (Sutter and Kirk, 1968), the grain beetle (*Tenebrio molitor*) and an unidentified species of the family Cetoniidae. Diptera: the crane fly (*Tipula paludosa*) and the midge (*Camptochironomus tentans*). Lepidoptera: the navel orangeworm (*Paramyelois transitella*) (Kellen and Lindegren, 1970). Orthoptera: the cricket (*Gryllus bimaculatus*, *G. capitatus* and *G. assimilis*), the desert locust (*Schistocerca gregaria*) and the common cockroach (*Blatta orientalis*). The infected Japanese and June beetles and the navel orangeworm were collected in the U.S.A., the cetonid in Madagascar, the desert locust in Jordan, and the rest in Europe.

More recently a crustacean, the isopode *Armadillidium vulgare* was found to be infected.

Infection is most commonly encountered among insect larvae, but is not confined to this stage. It usually starts in the cells of the fat body, but ultimately spreads to the blood and to the other organs. Multiplication occurs in the cytoplasm and leads to the formation of discrete foci of closely packed organisms arranged in an orderly manner. The foci eventually coalesce, involve the entire cytoplasm and disrupt the cell (Vago and Croissant, 1960). Crystalline bipyramidal bodies 0.8 by 1.2–2.2 by 3.8 μm in size are seen in the heavily infected cytoplasms of the beetles, the crane fly and the cockroach. Except for the absence of tyrosine, these bodies have approximately the same amino acid composition as the albuminoid spheres, found in greatest numbers in late larval and pupal stages in normal insects. It has been postulated that the crystalline bodies derive from the albuminoid reserve as the result of a disturbance of host cell metabolism (Huger, 1959; Krieg, 1959), but may also derive from the infecting microorganisms (Huger, 1962).

Etiological agent of blue disease of insect larvae. The name reflects the discoloration of the infected larvae, which is due to the increase in the turbidity of the blood (Dutky and Gooden, 1952). The designation "Lorscher Krankheit" (from Lorsch, Hesse, Germany) is also used (Wille and Martignoni, 1952). This organism is of particular interest since it might play a role in the control of certain agricultural pests.

Species incertae sedis

a. *Rickettsiella stethorae* Hall and Badgley 1957, 452. (*Enterella stethorae* (Hall and Badgley) Krieg 1961, 166.)

Slightly larger (0.4 by 1.0 μm) than *R. popilliae*, ovoid to elliptical in shape with tapered and rounded ends (instead of kidney-shaped with rounded ends) and grows preferentially in the gut epithelium (instead of the fat body). Isolated from several species of beetles of the genus *Stethorus*.

A somewhat similar organism, tentatively designated as a member of the genus *Enterella*, was isolated from two saturnid moths by Entwistle and Robertson (1968). This microorganism was shown to be antigenically related to a member of the *R. popilliae* group. Classification of these two organisms should be delayed until further information becomes available of their biological properties. Establishment of the genus *Enterella* does not seem to be warranted (cf. section on the genus *Wolbachia*).

FAMILY II. **BARTONELLACEAE** GIESZCZYKIEWICZ 1939, 25

DAVID WEINMAN

Bar.to.nel.la′ce.ae. M.L. fem.n. *Bartonella* type genus of the family; -*aceae* ending to denote a family; M.L. fem.pl.n. *Bartonellaceae* the *Bartonella* family.

Parasites of the erythrocytes in man and other vertebrates. **Cell wall present.** Rod-shaped, coccoid, ring- or disc-shaped, often beaded or filamentous and usually less than 3 μm in greatest diameter. Erythrocytic forms stain lightly with many aniline dyes but distinctly with Giemsa's stain after methyl alcohol fixation; and then do not exhibit a eucaryotic nucleus which distinguishes them from protozoa which parasitize red blood cells. Gram-negative. Not acid-alcohol-fast. At least one species bears unipolar flagella in culture. **Cultivated *in vitro* on non-living media.** Arthropod transmission has been established for one genus. Cause bartonellosis in man and grahamellosis in other vertebrates.

Key to the genera of family **Bartonellaceae**

I. Occur in or on erythrocytes and within fixed tissue cells. Often have flagella. Found in man and *Phlebotomus* spp.

Genus I. *Bartonella*

I. Intra-erythrocytic. Not known to multiply in fixed tissue cells. Flagella not described. Not found in man.

Genus II. *Grahamella*

Genus I. **Bartonella** *Strong, Tyzzer and Sellards 1915, 808*

(*Bartonia* Strong, Tyzzer, Brues, Sellards and Gastiaburú 1913, 1715.)
Bar.to.nel'la. M.L. dim. ending -*ella*; M.L. fem.dim.n. *Bartonella* named for Dr.
A. L. Barton, who described these organisms in 1909.

Microorganisms **found in fixed tissue cells** (Plate 18.2, Figs. 3 and 4) **and in or on erythrocytes** (Plate 18.2, Figs. 1 and 2). In stained blood films, the organisms appear as rounded or ellipsoidal forms or as slender, straight, curved or bent rods occurring either singly or in groups. **Characteristically occur in chains of several segmenting organisms,** sometimes swollen at one or both ends and frequently beaded (Strong *et al.*, 1913). In the tissues they are situated within the cytoplasm of endothelial cells as isolated elements or are grouped in rounded masses (Plate 18.2, Figs. 3, 6).

Has a cell wall, formation of which can be inhibited by penicillin. Reproduce by binary fission. In cultures possess a tuft of 1–10 unipolar flagella (Peters and Wigand, 1955) (Plate 18.2, Fig. 5); flagella have not been demonstrated in tissues.

Gram-negative. Not acid-fast. Stain poorly or not at all with many aniline dyes, but satisfactorily with Romanowsky's and Giemsa's stains.

May be cultivated by unlimited serial transfers on cell-free media.

Occur spontaneously **in man and in arthropod vectors** (*Phlebotomus* spp.) One species has been recognized, and it is known to be **established only on the South American continent** and perhaps in Central America. Human bartonellosis may be manifested clinically as a progressive anemia (Oroya fever) or as a cutaneous eruption (Verruga peruana), usually one after the other in succession, rarely simultaneously; asymptomatic infections also occur.

Type species: *Bartonella bacilliformis* (Strong *et al.*) Strong, Tyzzer and Sellards 1915, 808.

Description of the species of genus **Bartonella**

1. **Bartonella bacilliformis** (Strong *et al.*) Strong, Tyzzer and Sellards 1915, 808. (*Bartonia bacilliformis* Strong, Tyzzer, Brues, Sellards and Gastiaburú 1913, 1715.)

ba.cil.li.form'is. L. dim.n. *bacillus* a small staff, rodlet; L. n. *forma* shape, form; M.L. adj. *bacilliformis* rodlet-shaped.

Small polymorphic organisms: except that flagella have not been demonstrated on blood forms, maximum morphological range is seen in the blood of man where *B. bacilliformis* appears as a red-violet rod or coccoid form situated on or in the red cells when stained with Giemsa's stain. Bacilliform bodies are the most typical, measuring 0.25–0.5 by 1.0–3.0 μm. Often curved; may show polar enlargement and granules at one or both ends. Rounded organisms measure about 0.75 μm in diameter, and a ring-like variety is sometimes abundant (Plate 18.2, Fig. 2). Situated on the red cell as may be observed by light microscopy, or demonstrated by "stripping" in the pseudo-replica technique (Peters and Wigand, 1955); also

reported within erythrocytes in thin sections by electron microscopy (Cuadra and Takano, 1969).

In semi-solid media, a mixture of rods and granules appears; the organisms may occur singly or in large and small, irregular dense collections measuring up to 25 μm or more in length; punctiform, spindle-shaped and ellipsoidal forms occur which vary in size from 0.2–0.5 by 0.3–3.0 μm.

May be cultivated in semi-solid agar with fresh rabbit serum and rabbit hemoglobin or with the blood of man, horse or rabbit with or without the addition of fresh tissue and certain carbohydrates, in other culture media containing blood, serum or plasma, in Huntoon's hormone agar at 20%, in semi-solid gelatin media and in blood-glucose cystine agar. Also grows in certain tissue cultures and in the chorio-allantoic fluid and yolk sac of the chick embryo.

Gelatin not liquefied; hydrogen sulfide not detected with lead acetate.

No acid or gas from amygdalin, arabinose, dextrin, dulcitol, fructose, galactose, glucose, inulin, lactose, maltose, mannitol, mannose, raffinose, rhamnose, salicin, sucrose or xylose.

Obligately aerobic.

Grows at 28 and 37 C, with greater longevity at 28 C. Cultures viable after storage for 5 years at −70 C.

Natural immunity to infection has not been demonstrated in susceptible species. Acquired immunity is apparent both during and after the disease. Bartonellae from different sources appear to provoke similar responses; those from Oroya fever also protect against infection with organisms obtained from verruga cases.

Immune sera fix complement; when heterologous strains were employed, no significant titer differences were found in quantitative tests. Immune rabbit sera have not agglutinated *Proteus* OX19, OX2 or OXK at titers above 1:20. Agglutination of suspensions of *Bartonella* by sera from recovered cases has been reported.

Experimental Oroya fever has not been suc-cessfully produced in animals, except rarely in an atypical form in monkeys. Experimental Verruga peruana has been produced in man and in a number of species of monkeys.

Not sensitive *in vivo* to neosalvarsan nor in general to other arsenical compounds; sensitive to penicillin, streptomycin and chloramphenicol. Produces L-forms when grown with penicillin (Sharp, 1968). Inhibited *in vitro* by 0.1 μg of oxytetracycline per ml of semi-solid rabbit serum agar at pH 8.0.

Isolated from blood and endothelial cells of lymph nodes, spleen and liver of human cases of Oroya fever; also found in blood and in eruptive lesions in Verruga peruana; probably also found in sand flies (*Phlebotomus verrucarum*).

Not identical with rods observed on the red blood cells of anemic patients in Thailand; the latter have not been proved to be microorganisms (Whitaker *et al.*, 1966; Weinman, 1968).

Genus II. **Grahamella** Brumpt 1911, 517

(*Grahamia* Tartakowsky 1910, 243.)

Gra.ha.mel'la. M.L. dim. ending *ella*; M.L. fem.dim.n. *Grahamella* named for Dr. G. S. Graham Smith, who discovered these organisms in the blood of old world voles, in 1905.

Microorganisms observed **within the erythrocytes of mammals, excluding man.** Not known to multiply in fixed tissue cells. Morphologically these organisms bear a resemblance to but are less polymorphic than *Bartonella*. **Have a cell wall. Motility has not been demonstrated; flagella have not been seen.**

Gram-negative. Not acid-fast.

Both species have been cultivated on non-living media; growth is favored by the addition of hemoglobin. In cultures, rods and coccoid forms with indistinct contours commonly appear together in dense masses. Infective cultures were first grown by Tyzzer (1942); Wu Lien-Teh and Jettmar may have cultivated Grahamella in 1930 but their cultures failed to infect.

Aerobic.

Not affected by arsenicals.

Parasitic. The etiological agent of grahamellosis of rodents and other mammals. May possibly occur in other vertebrates and in insects. Not known in man. Wide geographical distribution.

Type species: *Grahamella talpae* Brumpt 1911, 517.

Description of the species of genus **Grahamella**

1. **Grahamella talpae** Brumpt 1911, 517. (*Grahamia talpae* Tartakowsky 1910, 243, Not Val. Pub.)

tal'pae. M.L. fem.n. *Talpa* a genus of moles; M.L. gen.n. *talpae* of Talpa.

Long or short rod-shaped organisms, many with a marked curve, usually near one end, lying within the red blood cells. One or both ends of the longer form is enlarged, giving a wedge- or club-shaped appearance. Some of the medium-sized forms are definitely dumbbell-shaped; small forms are nearly round. With Giemsa's stain, the protoplasm of the organism stains light blue with darker areas at the enlarged ends. Dark staining areas of the longer forms give the organism a banded appearance. Occasionally free in the plasma, but then usually occurring in groups. Most of the infected corpuscles contain between 6 and 20 organisms, but relatively few erythrocytes are infected (rarely more than 1% (Graham Smith, 1905)).

Infective for moles of the genus *Talpa* and found in these animals.

2. **Grahamella peromysci** Tyzzer 1942, 363.

per.ro.mys'ci. M.L. masc.n. *Peromyscus* a genus of mice; M.L. gen.n. *peromysci* of Peromyscus.

Uniform rods, spaced within red blood cells,

with no morphological features to distinguish it from other species.

Grows on non-living media containing blood, at temperatures varying from 20–28 C under aerobic conditions. Colonies rarely exceed 1.5 mm in diameter and are composed of rods as long as 1.5 μm, varying in thickness from 0.25–0.75 μm, and coccoids, 0.25–1.0 μm in diameter, occurring together in compact clumps. Older cultures may contain chains of rods and globoid bodies, the latter up to 12 μm in diameter. Organisms in cultures stain poorly with alkaline methylene blue solution (Loeffler's) but well with Giemsa's stain.

Blood not hemolyzed.

The natural host, the deer mouse (*Peromyscus leucopus novaboracensis*), may be infected by blood or cultures; the white Swiss mouse appears resistant. Monkeys (*Macaca mulatta*) are not infected by cultures.

Further Comments

Numerous species of *Grahamella* have been named according to their hosts (Kreier and Ristic, 1968), but there is no satisfactory evidence that they are different microorganisms. *Grahamella peromysci* has been described here to include information on the cultural characteristics of the genus *Grahamella*, and no opinion as to the validity of this species is expressed thereby.

Haemobartonella tyzzeri (Weinman and Pinkerton) Groot 1942, 279 (*Bartonella tyzzeri* Weinman and Pinkerton 1938, 217) requires restudy; more recent criteria suggest possible reclassification as a *Grahamella* (Weinman, 1957).

FAMILY III. ANAPLASMATACEAE PHILIP 1957, 980

MIODRAG RISTIC AND JULIUS P. KREIER

A.na.plas.ma.ta'ce.ae. M.L. neut.n. *Anaplasma* type genus of the family; -aceae ending to denote a family; M.L. fem.pl.n. *Anaplasmataceae* the *Anaplasma* family.

Obligately parasitic organisms found within or on erythrocytes or free in the plasma of various wild and domestic vertebrates. No demonstrable multiplication in other tissues.

In blood smears treated with Giemsa's stain, appear as rod-shaped, spherical, coccoid- or ring-shaped bodies staining reddish violet and measuring 0.2–0.4 μm in diameter. May occur in short chains or irregular groups in blood plasma or within erythrocytes. **Each organism enclosed in a membrane,** with internal structure resembling that of rickettsiae. Multiply by binary fission. Gram-negative; not acid-fast. **Have not been cultivated.** Organisms transmitted by arthropods. Blood or blood-containing tissue homogenates can cause infection by any parenteral route.

Infection may or may not cause disease but usually results in long term persistence of the agent, with concomitant resistance to clinically demonstrable reinfection. Anemia most prominent feature of clinical disease. Members of the family occur throughout the world.

Influenced by tetracycline compounds, but not by penicillin or streptomycin.

Further Comments

Decisions as to the taxonomic weight to be assigned to characteristics of microorganisms are subjective. This is particularly true for organisms for which sexuality has not been established and which have not been cultivated *in vitro*. The taxa here accepted for the family *Anaplasmataceae* are based on valid publication. While the characteristics on which some of the taxa are based are relatively minor, they are clearly defined. Thus, we feel that the groupings are justifiable.

When first described the parasitic bacteria were termed initial bodies and the surrounding structures which they produced were called inclusion bodies; both terms are still used.

Key to the genera of family **Anaplasmataceae**

I. Parasites form inclusions in erythrocytes.
   A. Inclusions in erythrocytes round, 0.3–1.0 μm in diameter. Several parasites may be found in each inclusion.
   1. No appendages to inclusions. Infect ruminants only.

   Genus I. *Anaplasma*, p. 907
   2. Appendages to inclusions. Infect cattle but not deer.

   Genus II. *Paranaplasma*, p. 908
   B. Inclusions in erythrocytes, 0.3–3.9 μm in diameter. Infect birds.

   Genus III. *Aegyptianella*, p. 909

II. Parasites within, or outside, erythrocytes.
  A. Parasites in and on erythrocytes. Stain intensely by Romanowsky methods. Ring structures rare or absent.

Genus IV. *Haemobartonella*, p. 910

  B. Parasites on erythrocytes and in plasma. Ring forms common in preparations stained by Romanowsky methods.

Genus V. *Eperythrozoon*, p. 912

Genus I. **Anaplasma** Theiler 1910, 7

Miodrag Ristic and Julius P. Kreier

A.na.plas′ma. Gr. pref. *an* without; Gr. n. *plasma* anything formed or molded; M.L. neut.n. *Anaplasma* a thing without form.

With Romanowsky-type stains, **this organism appears in the erythrocytes as dense, homogeneous,** bluish purple **round structures,** 0.3–1.0 μm in diameter. Electron microscopy reveals that these structures are inclusions separated from the cytoplasm of the erythrocyte by a limiting membrane. Each inclusion contains from one to eight subunits or initial bodies, which are the actual parasitic bacteria, each 0.3–0.4 μm in diameter that are dense aggregates of fine granular material embedded in an electron-lucid plasma and all enclosed in a double membrane (Ristic and Watrach, 1961).** The organism (initial body) enters erythrocyte by causing invagination of cytoplasmic membrane and subsequent formation of a vacuole. In the vacuole initial body multiplies by binary fusion and forms an inclusion. Thus, the **formation of the inclusion body,** which **is most frequently encountered during the acute and convalescent phases of infection,** represents only a phase in the developmental cycle of the initial body. **Spores or resistant stages not formed.** Apparently aerobic (Pilcher *et al.*, 1961), produce catalase (Wallace and Dimopoullos, 1965). Do not produce pigments. Incorporate amino acids from plasma *in vitro* (Mason and Ristic, 1966). The adenosine triphosphate and glutathione concentrations of erythrocytes remain essentially unchanged, regardless of the intensity of infection and only a small quantity of methemoglobin occurs in parasitized erythrocytes (Mann, 1967). **Histochemical analysis** of parasites **reveals DNA,** RNA, protein and organic iron (Moulton and Christensen, 1955).

Obligate parasites of vertebrates; transmitted by arthropod vectors. Ticks are probable biological vectors and tabanidae and other biting arthropods are mechanical vectors. **Host range limited to ruminants. Do not infect common laboratory animals.** A review of *Anaplasma* can be found in Ristic, 1968.

Type species: *Anaplasma marginale* Theiler 1910, 7.

Further Comments

Species differentiation in this genus is based entirely on host range and location of inclusions in the erythrocytes. Morphologically the initial bodies and individual erythrocytic inclusions of the species are indistinguishable by conventional microscopy and electron microscopy. Some antigens are common to all members of the genus; others are unique to one species (Kreier and Ristic, 1963; Schindler *et al.*, 1966; Kuttler, 1966, 1967). Antigenic analysis, if developed, could be used for species differentiation.

Anaplasma marginale is the most pathogenic of the three species and is the causative agent of severe bovine anaplasmosis. The disease is widespread in tropical and subtropical areas of the world. *Anaplasma centrale* causes a mild form of bovine anaplasmosis in Africa and *A. ovis* causes anaplasmosis of sheep and goats in various regions of the world. Common laboratory animals are all refractory: rabbits, guinea pigs, rats, mice, ferrets, dogs, cats and chickens.

Description of the species of genus **Anaplasma**

1. **Anaplasma marginale** Theiler 1910, 7. (*Anaplasma argentium* Ligniéres 1914, 153; *Anaplasma rossicum* Yakimoff and Bélawine 1927, 421; *Anaplasma theileri* Neitz 1957, 981: L'anaplasme Theileri Cardamatis 1911, 517.)

mar.gi.na′le. L. n. *margo, marginis*, edge, margin; M.L. neut.adj. *marginale* marginal.

Morphological characteristics as described for genus (Plate 18.3, Fig. 1).

Infectious for cattle, zebu, water buffalo (*Baba-*

TABLE 18.4

Differential characteristics of species of genus **Anaplasma**

Speics	Host	Location in erythrocyte	Disease
1a. *A. marginale* subsp. *marginale*	Cow Deer	Predominantly marginal	Severe anaplasmosis
1b. *A. marginale* subsp. *centrale*	Cow	Predominantly central	Generally mild anaplasmosis
2. *A. ovis*	Sheep Deer Goats	Predominantly marginal	Mild to severe disease

lus babalis), bison (*Bison bison*), African antelopes, gnu (*Connochaetes gnou*), blesbuck (*Damaliscus pygargus albifrons*) and duiker (*Sylvicapra grimmia*), American deer (southern black-tailed, Rocky Mountain mule deer, Virginia white-tailed deer (*Odocoileus* sp.), elk and camel (*Camelus bactrianus*). Sheep and goats may develop a submicroscopic infection. The African buffalo (*Syncerus caffer*) is refractory.

Infectivity of *A. marginale* can be destroyed by heating the organism at 60 C for at least 50 min, by exposure to sonic oscillation at 35 C for at least 90 min (Bedell and Dimopoullos, 1962, 1965) or to x-ray doses of 100,000 roentgens or higher (Wallace and Dimopoullos, 1965; Dommert *et al.*, 1965).

Anaplasma marginale in bovine blood can be preserved for many months by freezing infected blood to which glycerin has been added (Barnett, 1964).

Tetracycline compounds and dithiosemicarbazones inhibit multiplication and ameliorate the disease course (Miller, 1956; Barrett *et al.*, 1965). Penicillin, streptomycin, sulfonamides and arsenicals are inactive.

Parasite spread by ticks transstadially and transovarially and mechanically by horseflies.

The G + C content of the DNA has been reported as 51 moles % (buoyant density; Senitzer *et al.*, 1972).

Further Comments

In cattle infected with *A. marginale*, antibodies are found in IgG and IgM classes of immunoglobulins during the acute and convalescing phases of the infection (Murphy *et al.*, 1966). Erythrocytes from cows acutely infected with *A. marginale* yield two distinct antigens, a nonsedimentable

one (NS) and a sedimentable one (S). The NS antigen is soluble and lipoproteinaceous and is serologically active in the gel-precipitation system (Ristic *et al.*, 1963; Ristic and Mann, 1963). The particulate "S" antigen is a suspension of initial bodies (subunits of the marginal bodies) (Ristic, 1962). Friedhoff and Ristic (1966), using the fluorescent antibody technique, found that *A. marginale* organisms occur in the gut contents and in the malpighian tubules of engorged tick nymphs (*D. andersoni*).

1a. *Anaplasma marginale* subsp. *marginale*. Characteristics as for species.

1b. *Anaplasma marginale* subsp. *centrale* Theiler 1911, 19.

cen.tra'le. L. neut.adj. *centrale* central.

Characteristics as for the species except for the location of the bacteria in erythrocytes and in the nature of disease produced. In *A. marginale* subsp. *centrale* the cells are usually located centrally in the erythrocytes whereas in infections with subsp. *marginale* very few cells or cell groups are centrally located (Table 18.4).

The disease produced in cattle by *A. marginale* subsp. *centrale* is generally considered mild. In addition to common generic antigens, this subspecies possesses specific antigens (Schindler *et al.*, 1966).

2. Anaplasma ovis Lestoquard 1924, 784.

o'vis. L. n. *ovis* the sheep.

Characteristics as for *A. marginale* except for hosts infected. *A. ovis* produces disease only in sheep and goats and may produce inapparent infections in cattle.

Anaplasma ovis shares some antigens with *A. marginale*.

Genus II. **Paranaplasma** *Kreier and Ristic 1963, 701*

MIODRAG RISTIC AND JULIUS P. KREIER

Par.an'.a.plas.ma. Gr. pref. *para* near; M.L. n. *Anaplasma* a genus of *Anaplasmataceae*; M.L. neut.n. *Paranaplasma* resembling *Anaplasma*.

Inclusion bodies and initial bodies morphologically indistinguishable from species of *Anaplasma* in Giemsa-stained blood films studied by light microscopy. By techniques using wet preparations of blood or fluorescent antibody stain, and by electron microscopy, appendages to inclusion bodies are revealed which may resemble tails, loops or rings or connect two inclusion bodies in a "dumbbell" form. Fluorescent antibody and cross-immunity studies show organisms to be in part antigenically distinct from *A. marginale*. Infectious for cattle but not deer; may or may not grow in sheep (Kreier and Ristic, 1963).

Type species: *Paranaplasma caudatum* Kreier and Ristic 1963, 701.

Description of the species of genus **Paranaplasma**

1. **Paranaplasma caudatum** Kreier and Ristic 1963, 701. (*Paranaplasma caudata* (*sic*) Kreier and Ristic 1963, 701.)

cau'da.tum. L. n. *cauda* a tail; L. neut.adj. *caudatum* tailed, with a tail.

Inclusion bodies have appendages, usually in the form of a tapering tail, a loop or a ring and visible only through use of special techniques (Plate 18.3, Fig. 2). Infective for cattle, will not grow in deer or sheep. All isolations have been from mixed infections with *A. marginale* and *Paranaplasma discoides*.

Further Comments

In electron microscopic studies the tail of *P. caudatum* appears comet-shaped and may or may not seem to be attached to the head portion (Simpson *et al.*, 1965). The head portion of the parasite is a typical Anaplasma cell. The tail is of parasitic origin and may be stained with fluorescent-labeled antibody from cattle recovered from infection with the parasite (Kreier and Ristic, 1963).

2. **Paranaplasma discoides** Kreier and Ristic 1963, 701.

dis'oid'es. Gr. adj. *discoides* disc-shaped.

In Giemsa-stained blood films the inclusion bodies are indistinguishable from those of *A. marginale*. Examination of water-lysed erythrocytes by phase microscopy reveals an ovoid, disc-like structure with a dense mass at each pole (Kreier and Ristic, 1963).

Genus III. **Aegyptianella** Carpano 1929, 12

MIODRAG RISTIC AND JULIUS P. KREIER*

Ae.gyp'ti.an.el'la. Dim. ending *-ella*; M.L. fem.dim.n. *Aegyptianella* named for Egypt where the organism was described in 1928.

Infectious for cattle, grows in sheep but not in deer.

Initially found in Oregon cattle in a mixed infection with *A. marginale* and *P. caudatum*. Distinguished from *A. marginale* by means of phase-contrast microscopy, fluorescein-labeled antibody staining and cross-immunity studies (Kreier and Ristic, 1963).

Parasitic in species of wild and domestic birds and in some poikilothermal animals. With Romanowsky-type stains, purple staining intracytoplasmic inclusions containing bacteria occur in erythrocytes. Thin sections of inclusion bodies examined by electron microscopy revealed up to 26 bacteria or initial bodies enclosed in a double membrane (Gothe, 1967; Bird and Garnham, 1967) and with internal morphology strikingly similar to *Anaplasma*. Within the inclusion, initial bodies reproduce by binary fission. Has not been cultivated *in vitro*.

Organisms sensitive to dithiosemicarbazones and tetracyclines (Barrett *et al.*, 1965; Lämmler and Gothe, 1967).

Type species: *Aegyptianella pullorum* Carpano 1929, 12.

Description of the species of genus **Aegyptianella**

1. **Aegyptianella pullorum** Carpano 1929, 12. (*Spirochaeta granulosa penetrans* Balfour 1911, 114; *Aegytianella granulosa* Brumpt 1930, 1028; *Aegyptianella granulosa penetrans* Mesnil 1930, 125; *Babesia pullorum* Morcos 1935, 169; *Balfouria anserina* Dschunkowsky 1937, 315; *Balfouria gallinarium* Dschunkowsky 1937, 131.)

pul.lor'um. L. n. *pullus* a young fowl; L. gen. pl. *pullorum* of young fowl.

In blood smears stained with Giemsa cells appear in the host's erythrocytes in a great variety of forms: compact round and oval, ring- or horseshoe-shaped, polygonal or polymorphic, violet-reddish in color, with a diameter of 0.3–3.9 μm.

*Noteworthy assistance by Ranier Gothe.

In larger inclusions, clearly defined smaller round organisms (initial bodies) measuring up to 0.8 μm in diameter, can be distinguished.

Parasites may be found free in the plasma and in phagocytic cells. Parasites in phagocytic cells are probably ingested (Gothe, 1967).

In the erythrocytes the inclusions are separated from the cytoplasm by a single membrane. The internal structure of the initial bodies consists of dense aggregates of fine granular material embedded in an electron-lucid substance (Plate 18.4, Fig. 1).

Organismal RNA can easily be demonstrated histochemically; DNA-specific staining is possible only after treatment of the smears with RNase. The Fuelgen reaction is negative.

The infectivity of *A. pullorum* in chicken blood can be preserved up to 71 days by freezing with liquid nitrogen (Huchzermeyer, 1965). Multiplication of *A. pullorum* has not been observed in cell-free media or in tissue cultures. Continuous propagation of the organism in chicken embryos was not successful.

Causative agent of aegyptianellosis in birds. Chickens are naturally infected with *A. pullorum* by tick, *Argas* (*Persicargas*) *persicus*. Experimental infection can be achieved by subcutaneous, intramuscular, intravenous and intraperitoneal inoculation or by scarification with infected blood. Natural infections have also been described in geese, ducks, turkeys, guinea fowls, pigeons, quail and in the ostrich. Wild birds which have been experimentally infected are: *Turtur erythrophrys* and *Balearica pavonina* (Curasson and Andrjesky, 1929), *Turtur senegalensis*, *Milvus aegyptiacus* and *Vidua principalis* (Curasson, 1938).

Infection is transstadial in ticks (Gothe, 1967; Hadani and Dinur, 1968). Transovarial transmission has only occasionally been observed (Hadani and Dinur, 1968).

Species incertae sedis

The following species have been described but neither their true identity nor their relationship to *A. pullorum* has been established.

a. *Tunetella emydis* Brumpt and Lavier 1935, 548.

b. *Sogdianella moshkovskii* Shchurenkova 1938, 936; (*Babesia moshkovskii* (Shchurenkova) Laird and Lari 1957, 794.)

c. *Aegyptianella carpani* Batelli 1947, 212.

Other *Aegyptianella*- or *Anaplasma*-like parasites have been reported in a number of species of domestic and wild birds. The relation of these parasites to *A. pullorum* is not known.

Genus *IV*. Haemobartonella *Tyzzer and Weinman 1939, 143*

JULIUS P. KREIER AND MIODRAG RISTIC

Hae.mo.bar.to.nel'la. Gr. n. *haema* blood; M.L. fem.n. *Bartonella* a genus of the family *Bartonellaceae*; M.L. fem.dim.n. *Haemobartonella* the blood (-inhabiting) *Bartonella*.

Obligate parasites on or within erythrocytes of many vertebrate species. Organisms coccoid or rod-shaped as seen by light microscopy; rods seem to be chains of coccoids. Occur **singly, in pairs or in groups in shallow or deep indentations** on the erythrocyte surface, **sometimes in vacuoles within the erythrocytes, rarely in the plasma.** Stain well with Romanowsky-type stains; poorly with many other aniline dyes. **Possess single or a double limiting membrane, have neither cell wall nor distinct nuclear structures** (Tanaka, *et al.*, 1965); **Gram-negative;** not acid-fast. Have **not** been **cultivated** outside the host.

The experimental **host range** of any species is **restricted**; a species occurring naturally in a given host may be infective for closely related host species but not for the whole range of animals susceptible to other species of *Haemobartonella*. Morphology of the organism may vary in different hosts. **Growth inhibited by arsenicals and tetracyclines, not by penicillin or streptomycin.**

Most species are pathogenic but clinical disease, characterized by anemia, is usually not apparent unless the animal is splenectomized, except in the case of *Haemobartonella felis* in cats. The organisms are transmitted by arthropods and *H. felis* may also be transmitted by ingestion of blood (Flint *et al.*, 1958).

Type species: *Haemobartonella muris* (Mayer) Tyzzer and Weinman 1939, 143.

Description of the species of genus Haemobartonella

1. **Haemobartonella muris** (Mayer) Tyzzer and Weinman 1939, 143. (*Bartonella muris* Mayer 1921, 151; *Bartonella muris ratti* Regendanz and Kikuth 1928, 1578.)

mu'ris. L. n. *mus* the mouse; L. gen.n. *muris* of the mouse.

Infections in albino rat, albino mouse, some wild mice and hamsters. Appear under light microscopy of Romanowsky-stained blood films as slender rods with rounded ends. They frequently show granules or swellings at one or both ends, and as dumbbell, coccoid or diplococcoid forms. Occur singly, in pairs or in chains of three or four and, when abundant, in parallel groupings. Rods measure 0.1 by 0.3–0.7 µm; coccoids 0.1–0.2 µm diameter. Electron micrographs of 7000 magnification showed rods to be composed of coccoids 0.3–0.5 µm in diameter (Plate 18.4, Fig. 2A). No flagellae (Wigand and Peters, 1952) or cell wall present; organism non-motile.

With Giemsa's stain, appears intense red or blue with pink or purple shading. With Wright's stain appears bluish with red granules at ends. Stain bright red on blue-stained erythrocytes with Schilling's methylene blue-eosin stain. Stain faintly with Manson's stain, methyl green pyronin or fuchsin. Organisms stain with methyl blue (Giovannoni, 1946).

Non-filterable with Seitz or Berkefeld N. filter.

No convincing reports of cultivation *in vitro*. Enzyme treatment and histochemical staining indicate presence of both RNA and DNA, the former in greater quantity (Peters and Wigand, 1955).

Susceptible to organic arsenical compounds, chlortetracycline and oxytetracycline; these compounds will destroy the organisms in the host in both latent and clinical infections.

Haemobartonella muris is found worldwide in the blood of susceptible species. The rat louse (*Polyplax spinulosa*) is an important vector (Crystal, 1958).

Further Comments

Serum from *H. muris* infected rats does not yield a positive Weil-Felix reaction (Wigand, 1958). Complement fixation by *H. muris* infected rat serum does not occur with spotted fever, Q fever, psittacosis or vaccinia antigens. Weak complement binding in the presence of infected rat serum has been obtained with *Anaplasma marginale* and *Eperythrozoon coccoides* antigens.

Following exposure, either patent infection or a carrier state may be produced; the determining factors are not all known, although splenectomized or young animals seem more susceptible to patent infection. Latent infections may become patent following splenectomy or other processes which disturb the function of the reticuloendothelial and immunological systems of the host.

2. Haemobartonella felis (Clark) Flint and McKelvie 1956, 240. (*Eperythrozoon felis* Clark, 1942, 16.)

fe'lis. L. gen.n. *felis* of the cat.

Parasite of domestic cat. Occur as small round dots, short rods or coccoids, the last sometimes in chains. Coccoids 0.1–0.8 µm diameter; rods 0.2–0.5 by 0.9–1.5 µm (Plate 18.4, Fig. 2B). Do not have flagellae, cilia, nucleus or other intracellular organelles, no rigid cell wall; surrounded by two membranes. Frequently occur attached to and partly embedded in erythrocyte membrane, which may show erosion at point of contact (Small and Ristic, 1967). Stains deep purple with Giemsa stain. Fluorescent antibody or acridine orange stain may reveal organism when it cannot be demonstrated with Giemsa stain. Fluoresce bright orange with possibly an undertone of yellow green when stained with acridine orange, indicating high RNA content, and possible presence of DNA (Small and Ristic, 1967).

Organism has not been cultivated *in vitro*.

Chloramphenicol, tetracycline, oxytetracycline and neoarsphenamine are effective in suppressing the infection (Flint and McKelvie, 1956).

The infection can be transmitted by the intraperitoneal, intravenous or oral routes (Flint *et al.*, 1959), intrauterine infection may occur (Harbutt, 1963). It may be spread by biting during cat fights. Transmission by arthropod vectors has not yet been established. The organism is not infective for rats, mice, swine, cattle, sheep or dogs (Splitter *et al.*, 1956).

Further Comments

Eperythrozoon felis (Clark, 1942) was described as a predominantly ring- or ovoid-shaped organism, 0.5–1.0 µm in diameter, which with Giemsa stain stains a pale violet with non-staining centers, whereas *H. felis* stains a uniform deep purple. Simultaneous appearance of organisms of both types has been reported (Splitter *et al.*, 1956). Clark's description was based on a single blood film from a dead cat. The parasites have never been available for further study.

Haemobartonella felis is unique among species of this genus in its clinical importance. The organism produces patent parasitemia and severe, sometimes fatal, anemia in intact animals under field conditions. Splenectomy has relatively little effect on the course of the infection (Splitter *et al.*, 1956). The disease has been produced in susceptible cats by inoculations of pooled blood from clinically normal cats (Splitter *et al.*, 1956), an indication that the infection can produce a carrier state as well as patent disease.

3. Haemobartonella canis (Kikuth) Tyzzer and Weinman 1939, 151. (*Bartonella canis* Kikuth

1928, 1730; *Haemobartonella (Bartonella) canis* (Kikuth) Tyzzer and Weinman 1939, 151.)

can'is. L. gen.n. *canis* of the dog.

Has characteristics of the genus. A parasite usually seen on erythrocytes of dogs following splenectomy: does not produce disease.

Has been reported widely, but the classification is confused. Antigenic studies and additional investigation of morphology are needed. The reported infectivity of *H. canis* for cats (Lumb, 1961) suggests a close relationship between *H. canis* and *H. felis*.

Species incertae sedis

a. *Haemobartonella tyzzeri* (Weinman and Pinkerton) Groot 1942, 279. (*Bartonella tyzzeri* Weinman and Pinkerton 1938, 217.)

This organism has been grown in vitro and might more properly be placed in the genus *Grahamella* (Weinman, 1957; see p. 906).

A number of other species have been described, most of them named after the host (see Kreier and Ristic, 1968). The classification of many of these is in doubt.

Genus V. **Eperythrozoon** *Schilling 1928, 1854*

JULIUS P. KREIER AND MIODRAG RISTIC

Ep.e.ryth'ro.zo'on. Gr. pref. *epi* on. Gr. adj. *erythrus* red; Gr. n. *zoum* or *zoon* living thing, animal; M.L. neut.n. *Eperythrozoon* (presumably intended to mean) animals on red (blood cells).

Obligate parasites in the blood of various vertebrate species including some rodents, ruminants and pigs.

Stains bluish or pinkish violet with Giemsa's and Romanowsky-type stains. In such stained preparations cells show **no differentiation of nucleus and cytoplasm and appear as rings or coccoids, 0.4–1.5 µm in diameter, characteristically round with numerous annular or disc-shaped elements, rarely rods. Occur on erythrocytes and free in plasma with about equal frequency.**

Swarm-like clusters of ring-shaped eperythrozoa may occur on the surface of erythrocytes. Rod-shaped forms may occur partly or entirely circling an erythrocyte. In fresh preparations observed by **dark-field** and **phase microscopy, coccoids but no ring forms are seen** (Wigand, 1958; Kreier and Ristic, 1963). By electron microscopy organism appears pleomorphic, surrounded by a single limiting membrane, with no cell wall, nucleus or other organelles (Tanaka *et al.*, 1965). Has not been **cultivated in cell-free media.** Some species have been shown to be transmitted by arthropod

vectors. Experimental transmission by inoculation of blood occurs readily.

Splenectomy activates latent infection with most species.

Growth inhibited by arsenicals and tetracyclines but not by penicillin or streptomycin.

Type species: *Eperythrozoon coccoides* Schilling 1928, 1854.

Further Comments

Differentiation from *Haemobartonella* is in many cases difficult, and possibly arbitrary. Differentiation is based on the fact that haemobartonellae rarely occur as ring forms while eperythrozoa commonly do and on the fact that eperythrozoa occur with about equal frequency on the erythrocytes and free in the plasma while haemobartonallae rarely occur free in the plasma.

A number of species of *Eperythrozoon* have been described, most of them on the basis of observations of eperythrozoon-like bodies in stained blood films. The nature and classification of many of these is in doubt. The names and references to the original description are listed in Kreier and Ristic (1968).

Description of the species of genus **Eperythrozoon**

1. **Eperythrozoon coccoides** Shilling 1928, 1854. (*Gyromorpha musculi* Dinger 1928, 5905.)

coc.coi'des. Gr. n. *coccus* a berry; M.L. n. *coccus* a coccus Gr. n. *eidus* shape; M.L. adj. *coccoides* coccus-shaped.

A parasite of albino and wild mice, albino rats, rabbits and hamsters.

In stained blood films the cells appear as rings, coccoids and rods, the majority as rings of regular outline with clear centers. Organisms usually

measure 0.4–0.5 µm in greatest dimension. Electron microscopic studies of thin sections show only slight erosion of the erythrocyte surface at the point of contact with the parasite (Plate 18.5, Fig. 1A). In fresh blood stained by a vital acridine orange technique, organism is light yellow in color, coccoid and non-motile (Rafyi and Vercammen-Grandjean, 1964). High affinity for pyronine dye indicates RNA content; action of deoxyribonuclease on the organism suggests the presence of

DNA as well (Wigand and Peters, 1954). Propagation in embryonated hen's eggs has been reported (Seamer, 1959). Organism can be preserved at low temperatures.

Organism reported to pass collodion membranes of an average pore size of 0.36 μm (Niven et al., 1952), but, being non-rigid, may be drawn through pores which would retain rigid organisms of the same volume. Reproduction is probably by binary fission (Westphal, 1965).

Neoarsphenamine is a very effective therapeutic agent; chlortetracycline and oxytetracycline are active; sulfonamides, sulfones and penicillin show little to no activity.

Blood from E. coccoides infected animals has been reported infective by the oral route. The mouse louse (Polyplax serrata) is a natural vector. The organism has been reported from Europe, North and South America, and Africa.

Further Comments

Infection is usually maximal in young animals or in splenectomized adults. The organism is reported to cause fatal mouse hepatitis when associated with another etiological agent (a virus); otherwise, moderate to no anemic changes are reported. Virus titers are increased 100-fold in combined infections (Niven et al., 1952).

The immunological state is of the infection immunity (premunition) type. *Eperythrozoon coccoides* antigen has been reported non-reactive with serum from human beings with bartonellosis, spotted fever, Q fever, ornithosis and vaccinia, with serum from cattle infected with A. marginale and with serum from rats infected with *Haemobartonella muris*. Serum from mice infected with E. coccoides reacted with A. marginale and with H. muris antigens (Wigand, 1958).

2. **Eperythrozoon ovis** Neitz, Alexander and du Toit 1934, 267.

o'vis. L. n. ovis a sheep.

Parasite of domestic sheep, goats and deer. Stains pale purple with Giemsa's stain, appearing in stained blood films as delicate rings approximately 0.5–1.0 μm in diameter, or as discs of the same size range. Irregularly shaped forms may occur being ovoid-, comma-, rod-, dumbbell- or tennis racket-shaped. Rod forms when they occur are most commonly attached to the margin of the erythrocyte and may partly or even completely surround it (Plate 18.5, Figs. 1A, 1B). With acridine orange, fluoresce a bright orange color (Kreier, 1962). By phase-contrast microscopy appear as spheres, with neither rings nor rods apparent and no motility visible (Kreier and Ristic, 1963). In thin sections observed with the electron microscope, organisms are seen as round or oval structures 0.3–0.4 μm in diameter and par-

tially embedded in the erythrocyte. A peripheral dense region, 20–30 nm thick, is probably a surrounding membrane. No nucleus, endoplasmic reticulum, mitochondria or other organelles are seen (Kreier and Ristic, 1963). Organism has not been cultivated. Suggested modes of multiplication are budding and binary fission.

Eperythrozoon ovis has common antigens with E. wenyonii and possibly with A. marginale (Kreier and Ristic, 1963).

Infection can be transmitted by any parenteral route. Transmission by horsefly bite has been reported (Øverås, 1959), other arthropods may also be vectors. The organism has been reported from Africa, Australia, several European countries and is common throughout the U.S.A.

Further Comments

Eperythrozoon ovis can provoke mild symptoms in normal animals of receptive species without splenectomy. Mortality from infection is rare, but anemia and failure to gain weight have been observed in young lambs (Øverås, 1959; Foggie and Nisbit, 1964).

3. **Eperythrozoon suis** Splitter 1950, 513.

su'is. L. n. sus a pig; L. gen.n. suis of a pig.

Parasite of domestic pig. Occurs as rods, rings, coccoids and various budding forms in blood films treated with Giemsa's stain, a ring form averaging 0.8–1.0 μm being the most common. Larger ring and discoid types up to 2.5 μm also occur. This is the largest species in the genus. Organisms have been reported to pass "N" Berkefield and 8- and 14-pound Mandler filters (Splitter, 1952) indicating smaller size than revealed by microscopy. However, if non-rigid in form filtration characteristics would be affected. Has not been cultivated.

The parasite is the cause of "ictero-anemia" of swine, an economically important disease, which produces anemia, weakness, stunted growth and sometimes death in intact pigs under field conditions. Splenectomy causes relapse in carrier pigs, and unusually high parasitemia in pigs splenectomized before infection (Splitter, 1950). Infected animals produce a complement-fixing antibody (Splitter, 1958) possibly directed against erythrocyte stromal antigens. Neoarsphenamine controls the infection. The organism is common throughout the U.S.A.

Further Comments

Eperythrozoon suis can be differentiated from *Eperythrozoon parvum*, a non-pathogenic parasite of the pig, on the basis of size and morphology.

The parasite described in Taiwan as *Anaplasma taiwanensis* (Sugimoto, 1935) may be E. suis.

There is a single report of *E. suis* from the former Belgian Congo (Jansen, 1952).

4. Eperythrozoon parvum Splitter 1950, 513.

par'vum. L. neut.adj. *parvum* small.

Non-pathogenic parasite of domestic pigs. Occurs as disc or coccoid forms less than 0.5 μm in diameter or as ring forms 0.5–0.8 μm in diameter. Does not occur in cells or tissues other than blood. Will pass 8-, 12- and 14-pound Mandler filters. Passes gradacol membranes 0.57 and 0.41 μm but not 0.36 μm (Seamer, 1959).

Splenectomy causes relapse of carrier pigs.

Susceptible to arsenicals and tetracyclines but not to penicillin, streptomycin or sulfonamides.

May be readily transmitted by all parenteral routes and rarely by massive oral inoculations. The pig louse (*Haematopinus suis*) may transmit the disease (Jansen, 1952).

Infection has been reported from the United States, Europe and Africa.

5. Eperythrozoon wenyonii Adler and Ellenbogen 1934, 220. (*Haemobartonella weyonii* (Adler and Ellenbogen) Weinman 1944, 240.)

wen.yo'ni.i. M.L. gen.n. *wenyonii* of Wenyon; named for Dr. C. M. Wenyon, who studied these organisms.

Parasite of domestic cattle.

Morphologically similar to *Eperythrozoon coccoides*. In blood films treated with Giemsa's stain, appear as pinkish purple discs or ring-shaped structures, 0.3–0.5 μm in diameter, or, more rarely, as rod or ovoid forms. May be attached to erythrocytes or platelets, or occur free in the plasma (Kreier and Ristic, 1968). A number of ring forms may be attached to a single erythrocyte; rod or coccoid forms may completely encircle an erythrocyte.

Fluoresces bright orange with acridine orange stain, appearing primarily as coccoids 0.5 μm in diameter, and sometimes clustered like bunches of grapes. The rod-shaped structures seen in blood films stained by polychrome methods appear as chains of coccoid forms with acridine orange staining. Electron microscopic observations of dried, metal-shadowed blood preparations show flattened or collapsed spheres or short rods, indicating lack of a rigid cell wall. Masses in the cytoplasm, probably nucleoids, have been observed (Kreier and Ristic, 1963). Successful cultivation has not been reported.

The organism produces parasitemia and a mild anemia, but no other clinical signs. Latent infection is made manifest by splenectomy. Sheep, goats and deer (*Odocoileus virginiana*) are not susceptible.

E. wenyonii shares antigens with *E. ovis*, with *A. marginale* and *A. ovis* parasites.

Infection with *E. wenyonii* can be produced by parenteral injection of blood from a latently or patently infected animal. The natural means of transmission has not been determined. Distribution of the organism appears to be worldwide.

ORDER II. **CHLAMYDIALES** STORZ AND PAGE 1971, 334

L. A. PAGE

Chla.my.di.a'les. M.L. n. *Chlamydia* type genus of the order; *-ales* ending to denote an order; M.L. fem.pl.n. *Chlamydiales* the *Chlamydia* order.

Coccoid microorganisms whose obligately intracellular mode of multiplication is characterized by change of the small, rigid-walled infectious form of the organism (elementary body) into a larger, thin-walled, non-infectious form (initial body) that divides by fission. The developmental cycle is complete when daughter cells reorganize and condense to become elementary bodies which survive extracellularly to infect other host cells. Metabolically limited, Gram-negative parasites of vertebrates in which they may cause various diseases. Occasionally found in arthropods.

FAMILY I. **CHLAMYDIACEAE** RAKE 1957, 957

Chla.my.di.a'ce.ae. M.L. fem.n. *Chlamydia* type genus of the family; *-aceae* ending to denote a family; M.L. fem.pl.n. *Chlamydiaceae* the *Chlamydia* family.

Coccoid microorganisms, 0.2–1.5 μm in diameter, which multiply only within the cytoplasm of host cells by a developmental cycle characterized by change of a small elementary body into a larger initial body that divides by fission. The cycle is complete when daughter cells reorganize and condense to become elementary bodies which survive extracellularly to infect other host cells. Elementary bodies contain compactly arranged nuclear material and ribosomes and are bounded by a rigid, trilaminar cell wall that is chemically similar to that of Gram-negative bacteria. Initial bodies contain amorphous nuclear material less electron-dense than that of elementary bodies, have thin, fragile cell walls and apparently are non-infectious. Metabolically limited, Gram-negative parasites of vertebrates in which they may cause various diseases. Occasionally found in arthropods. Cultivatable in yolk sac of chicken embryos. Sensitive to tetracycline antibiotics. Stain with aniline dyes.

Genus I. Chlamydia *Jones, Rake and Stearns 1945, 55**

(*Miyagawanella* Brumpt 1938, 155 (not validly published); *Rickettsiaformis* Zhdanov and Korenblit 1950, 43 (not val. pub.); *Prowazekia* Coles 1953, 461 (illegitimate); *Bedsonia* Meyer 1953, 552 (illegitimate); *Rakeia* Levaditi, Roger and Destombes 1964, 658 (illegitimate).)

Chla.my′di.a. Gr. fem.n. *chlamys, chlamydis* a cloak; M.L. fem.dim.n. *Chlamydia* a cloak.

Non-motile, spheroidal organisms, 0.2–1.5 μm in diameter depending upon their stage of development in a unique, obligately intracellular growth cycle. The principal developmental stages are (1) the **elementary body** which is a small, electron-dense spherule, 0.2–0.4 μm in diameter containing a nucleus and numerous ribosomes surrounded by a multilaminated wall; (2) the **initial body** which is a large, thin walled, reticulated spheroid, 0.8–1.5 μm in diameter containing nuclear fibrils and ribosomal elements; (3) an **intermediate body** representative of a transitional stage between the initial body and elementary body (Plate 18.6). The **elementary body is the infectious form** of the organism. The **initial body is the vegetative form** which divides by fission intracellularly but is apparently non-infectious when separated from the host cell.

Growth, multiplication and maturation of the organisms occur over a period of 40 hrs or less as microcolonies within a cytoplasmic vesicle whose envelope originates from invaginated host cell membrane (Plate 18.6). The elementary body is phagocytized by the host cell and within several hours reorganizes and enlarges to become the initial body which divides by fission. Daughter cells continue to divide, then develop an electron-dense nuclear mass and laminated walls (intermediate form) and decrease in size to become the mature, infectious elementary bodies. When the wall of the vesicle and host cell are disrupted, organisms are released to repeat the cycle in other host cells. **Phagocytosis of more than one elementary body per host cell may occur** resulting in the presence of several developing chlamydial microcolonies in the cytoplasm of a host cell.

Gram-negative. Cell walls are similar in chemical composition to those of Gram-negative bacteria. May be stained by Giemsa's, Macchiavello's, Gimenez's or Castenada's methods, or may be seen readily in unstained wet mounts of infected cells with a phase-contrast optical system.

When separated from host cells, chlamydiae have **few detectable metabolic activities,** and these are dependent upon an exogenous source of organic and inorganic cofactors including high energy phosphates (e.g. ATP). Nevertheless, when extracellular chlamydiae are provided with cofactors, they can **catabolize glucose, pyruvic acid or glutamic acid and produce carbon dioxide.** They also can **synthesize certain lipids.** The growth of one species is inhibited by sodium sulfadiazine thereby reflecting the ability of the organisms to synthesize folates intracellularly. Since they are unable to synthesize their own high energy compounds, they have been described as "**energy parasites.**" To date, extracellular propagation of chlamydiae has not been achieved.

Multiplication is inhibited by tetracyclines, penicillin and 5-fluorouracil. An antigenically similar, 100 C-stable lipoglycoprotein is present in all organisms of the genus. A spectrum of specific antigens is present in the cell walls of various strains. Mosaics of five or six different antigens may be shared by strains isolated from different host species that have varying clinical syndromes.

After appropriate adaptation, all organisms of the genus **may be propagated in the yolk sac of chicken embryos or in cultures of vertebrate tissues.** Multiplication in chicken embryos

*An historical review of the classification and nomenclature of this genus has been given by Page (1966, 223).

occurs over a temperature range of 33–41 C but sensitivity to either end of the range or the optimum for growth varies with the species.

The organisms are **parasites of tissue cells of vertebrates.** However, they have been found in or on ectoparasitic arthropods. Many strains produce generalized infections in several host species, but some strains characteristically localize and cause pronounced inflammation in one or more tissues or organs of a certain host species. Groups of strains are found in **three major ecological niches: man,** where they cause oculo-urogenital and respiratory diseases; **birds,** where they cause respiratory disease and generalized infection; **mammals** (excluding primates) where

they cause respiratory, placental, arthritic and enteric diseases. Some crossing between niches by certain strains occurs, especially between birds and man.

RNA and DNA have been demonstrated in both initial and elementary bodies, but the ratio of RNA to DNA in the initial body is twice that in the elementary body. G + C content of the DNA ranges between 39 and 45 moles % (T_m). The DNA weight is approximately 6.6×10^8 daltons. The number of nucleotide pairs per chlamydial organism is less than 1×10^6.

Only two species are recognized.

Type species: *Chlamydia trachomatis* (Busacca) Rake 1957, 958.

Key to the species of genus **Chlamydia**

I. Forms compact microcolonies within cytoplasmic vesicles. Produces iodine-staining compounds within vesicle. Growth in the yolk sac of chicken embryos is inhibited by sodium sulfadiazine (1 mg/embryo).

1. *Chlamydia trachomatis*

II. Forms microcolonies within cytoplasmic vesicles which tend to rupture early in microcolony development, and the organisms in various stages of growth become distributed throughout the host cell cytoplasm. Does not produce iodine-staining compounds in vesicles. Growth in the yolk sac of chicken embryos is not inhibited by sodium sulfadiazine (1 mg/embryo).

2. *Chlamydia psittaci*

Description of the species of genus **Chlamydia**

1. Chlamydia trachomatis (Busacca) Rake 1957, 958. (*Rickettsia trachomae* (*sic*) Busacca 1935, 567; *Rickettsia trachomatis* (Busacca) Foley and Parrot 1937; 231; *Chlamydozoon trachomatis* (Busacca) Moshkovski 1945, 18.) Putative synonyms are listed by Page (1968, 58).

tra.cho′ma.tis. Gr. n. *trachoma* roughness; M.L. n. *trachoma* the disease trachoma; M.L. gen.n. *trachomatis* of trachoma.

Morphology and mode of multiplication same as the genus. Certain features of the microcolony are distinctive; continued multiplication of the organisms within a cytoplasmic vesicle results in the development of microcolonies that are densely packed with chlamydiae and that have enough rigidity to displace the host cell's nucleus to one side and distend the cell wall. Iodine-staining carbohydrate, probably glycogen, and lipid is formed by the organisms in the vesicle. Iodine-positive microcolonies are characteristic of this species and are readily detectable by staining with iodine-potassium iodide solution after fixation in methanol.

May be isolated and propagated in the yolk sac of chicken embryos incubated at 33–37 C with the optimum for most strains being 35 C. May also be propagated in vertebrate cell cultures, especially

those derived from murine tissues. Repeated blind passage of new isolates in chicken embryos may be necessary before embryo deaths are observed. Multiplication in chicken embryos is inhibited by sodium sulfadiazine (1 mg/embryo).

Many strains can be propagated in the lungs of mice and guinea pigs inoculated by the intranasal route, or in the skin of guinea pigs, or the eyes and conjunctiva of primates. Some strains can be propagated in mice inoculated intracerebrally.

The organisms may be preserved indefinitely in infected yolk sacs stored at −20 C or below. Viability of the organisms at 4 C may be preserved for several days by suspending them in a solution of 0.25 M sucrose buffered at pH 7.2 with 0.01 M potassium and sodium phosphate.

The G + C content of the DNA for six strains averaged 44.4 (range 43.6–45.0) moles % (T_m). Homology tests indicate that there is 10% or less binding between DNA strands of *C. trachomatis* and those of *C. psittaci* strains.

The organisms are parasites principally of man in which they cause a variety of oculo-urogenital diseases: trachoma, inclusion conjunctivitis, lymphogranuloma venereum, urethritis and proctitis. Occasionally they are associated with arthritis. One strain causes endemic pneumonitis in colonies of laboratory mice, probably transmitted via the respiratory tract.

In man, transmission occurs by contact contamination of the conjunctiva, oral or genital mucous membranes with infectious conjunctival or urogenital exudates during birth, during care of the young or during venereal contact.

Reference strains: organisms causing trachoma (three distinct serotypes) VR 571, VR 572, and VR 573; inclusion conjunctivitis VR 346, VR 575; lymphogranuloma venereum, VR 121; mouse pneumonitis, VR 123; all are ATCC numbers.

2. **Chlamydia psittaci** (Lillie) Page 1968, 60. (*Rickettsia psittaci* Lillie 1930, 778; *Ehrlichia psittaci* (Lillie) Moshkovski 1945, 18; *Rickettsiaformis psittacosis* Zhdanov and Korenblit 1950, 43; *Chlamydozoon psittaci* (Lillie) Ryzhkov 1950, 17.) Putative synonyms are listed by Page (1968, 60).

psit'ta.ci. Gr. n. *psittacus* a parrot; M.L. gen.n. *psittaci* of a parrot.

The morphology and mode of reproduction as for the genus. The intracellular colonial organization varies from that of *C. trachomatis* in that the vesicular membrane surrounding the *C. psittaci* microcolony ruptures earlier in the course of cellular infection releasing chlamydiae in various stages of development into the host cell's cytoplasm (Plate 18.6). Chlamydiae appear therefore to be developing throughout the cytoplasm of the host cell. Iodine-staining compounds are not formed in detectable amounts within microcolonies.

The organisms are readily isolated and propagated in the yolk sacs of chicken embryos incubated from 37–39 C with the optimum for many strains being 39 C. Multiplication in embryos is not inhibited by sodium sulfadiazine, 1 mg/embryo, or by phosphate-buffered saline containing 1 mg/ml of each of streptomycin sulfate, vancomycin and kanamycin sulfate. The latter solution is useful in eliminating non-chlamydial bacterial contaminants in infected tissues or cultures.

The organisms can also be propagated in cultures of veretebrate tissues, especially those of murine origin. All strains produce plaques of cell destruction in agar overlay cultures of murine L cells. Strains isolated from birds form plaques in primary cultures of chicken embryo cells, but most strains originally isolated from mammals do not.

The organisms can be propagated in the lungs, brains or peritoneal cavities of laboratory mice; however, strains from domestic herbivores often fail to grow in mice, but grow preferentially in guinea pigs.

The organisms may be preserved indefinitely as homogenates of infected yolk sac cultures suspended in beef heart infusion broth and stored at −20 C or below.

The organisms are parasites of tissue cells of vertebrates and have been detected in over 100 species of wild and domestic birds, most domesticated mammals and many wild mammals. The organisms generally produce systemic infection in a broad range of hosts although some strains appear to have a narrow host spectrum. The diseases caused by various strains of *C. psittaci* are: psittacosis, ornithosis, pneumonitis in cattle, sheep, goats, pigs, horses and cats; polyarthritis in sheep, cattle and pigs; placentitis leading to abortion in cattle and sheep; enteritis in cattle and hares; conjunctivitis in guinea pigs, cattle and sheep; widespread subclinical intestinal infections of cattle and sheep; encephalitis in opossums; encephalomyelitis in cattle.

Interspecies and intraspecies transmission of the organisms is accomplished by inhalation of fine dust containing dried cloacal or nasal excrement of infected birds; this method has been responsible for epidemics of psittacosis in man. Transmission may also occur by transovarian passage to the fetus in mammals, by venereal contact and possibly by the intravenous route following bites by sanguivorus arthropods. Chronic intestinal infections among cattle and sheep are maintained by oral-fecal transmission cycles. The organisms are perpetuated in nature probably by parent to offspring transmission cycles with a natural selection for resistant carriers. Stresses caused by unfavorable environments, reproductive activity or concurrent infection may lead to shedding of large numbers of chlamydiae into the environment by chronically infected reservoir hosts. Among avian species, members of the families *Psittacidae* and *Columbidae* are the most common reservoirs of *C. psittaci*.

The G + C content of the DNA for 10 strains averaged 41.2 with a range of 39.4–43.05 moles % (T_m).

Reference strains: bovine encephalomyelitis VR 189, VR 628; feline pneumonitis, VR 120; meningopneumonitis VR 122; pigeon ornithosis VR 574; turkey ornithosis VR 371; parakeet psittacosis VR 125; sheep polyarthritis VR 629; epizootic chlamydiosis of hares and muskrats VR 630; human psittacosis VR 601; all are ATCC numbers.

Genera and species incertae sedis

None of the following organisms has been isolated in pure culture and no cultures of any kind are available. The names are validly published and legitimate. If satisfactory confirmation of their existence can be obtained, description in future editions of THE MANUAL is not precluded.

a. *Colesiota conjunctivae* (Coles) Rake 1948, 1119. (*Rickettsia conjunctivae* Coles 1931, 179.)

b. *Colettsia pecoris* Rake 1957, 981.

c. *Ricolesia bovis* (Coles) Rake 1957, 960. (*Rickettsia conjunctivae bovis* Coles 1936, 223.)

d. *Ricolesia caprae* (Coles) Rake 1957, 960. (*Rickettsia conjunctivae caprae* Coles 1953, 460.)

e. *Ricolesia conjunctivae* (Coles) Rake 1957, 960. (*Rickettsia conjunctivae galli* Coles 1940, 474.)

f. *Ricolesia lestoquardii* (Donatien and Gayot) Rake 1957, 960. (*Rickettsia lestoquardi* Donatien and Gayot 1942, 325.)

Important Notes

for

Users of this Edition

1. Always read both generic and species descriptions because characters listed in the generic description are not usually listed in the species descriptions.

2. In tables, characters common to all taxa are not shown but may be listed in footnotes.

3. Generally in tables (exceptions are clearly indicated in footnotes, q.v.) the meanings of symbols are as follows:

+   more than 90 % strains positive

−   more than 90 % strains negative

d   11–89 % strains positive

( )   delayed reaction

w   weak reaction

D   Different reactions in different taxa (species of a genus or genera of a family)

v   strain instability (NOT differences between strains)

Plate 18.1. *Ehrlichia, Neorickettsia*

Fig. 1. *Ehrlichia phagocytophila* in cytoplasmic vesicle in eosinophil granulocyte of infected calf. The inclusion, ready to rupture, contains large particles of the organism. The structural units of the particles, cell wall, plasma membrane, nucleoid and cytoplasm may be seen. Ribosome-like granules occur in a ground substance of intermediate density. Almost complete division of particles is demonstrated in an oblique section (*arrow*). The vacuole is lined by a membrane. Cytoplasmic vesicles appear inside the vacuole and especially in the space between vacuole membrane and host cell membrane which are detached from each other. Budding of still smaller vesicles into the vacuole is also visible. × 45,000. (From Tuomi and von Bonsdorff (1965) J Bacteriol *92:* 1482.)

Fig. 2. *Neorickettsia helminthoeca* as seen in imprints of mesenteric lymph node of fatally infected dog. *A,* clumps and scattered organisms from ruptured cells. *B,* morula-like colonies in the cytoplasm of a disintegrating macrophage surrounded by lymphocytes. Giemsa stain. × 1500. (From Philip, Hadlow and Hughes (1954) Exp Parasitol *3:* 345.)

Plate 18.2. *Bartonella*

Fig. 1. *Bartonella bacilliformis,* in blood films from an Oroya fever patient. An intense infection involving the erythrocytes and, a rarer finding, the mononuclear leucocytes also.

Fig. 2. Human Oroya fever blood, showing rod and ring forms of *B. bacilliformis.*

Fig. 3. Section of an Oroya fever lymph node. Development of *B. bacilliformis* in distended endothelial cells lining the vein. Intracytoplasmic distribution in rounded clumps is distinct in some heavily infected cells (redrawn from C. Uribe).

Fig. 4. Section of a human verruga. *B. bacilliformis* is distinct and usually is clearly situated in the cytoplasm. It is not observed on or in erythrocytes. When stained by Giemsa stain the organism is bright red; for a color plate see Weinman (1968).

Fig. 5. *B. bacilliformis* from an 8-day culture, hemolyzed, palladium-shadowed (From Peters and Wigand (1952) Z Tropenmed Parasitol *3:* 313–326.)

Fig. 6. A 4-day culture of *B. bacilliformis* grown with an explant of guinea pig tunica vaginalis in Maitland medium. Intracytoplasmic clusters and discrete organisms. (From Pinkerton and Weinman (1937) Proc Soc Exp Biol Med *37:* 587–590.)

Plate 18.3. *Anaplasma, Paranaplasma*

Fig. 1. *Anaplasma marginale. A,* ultrathin section of an infected erythrocyte; an inclusion body containing five initial bodies is depicted. × 65,000. *B,* intraerythrocytic inclusion bodies stained by Giemsa. × 959. (From Ristic and Watrach (1961) Amer J Vet Res *22:* 109–116.)

Fig. 2. *Paranaplasma caudata. A,* ultrathin section of an infected erythrocyte. × 40,000. (From Simpson *et al.* (1965) J Cell Biol *27:* 225.) *B,* stained by fluorescent antibody technique. × 1800. (From Kreier and Ristic (1963) Amer J Vet Res *24:* 676–687.)

Plate 18.4. *Aegyptionella, Haemobartonella*

Fig. 1. *Aegyptionella pullorum. A,* ultrathin section of an infected erythrocyte, depicting an inclusion body with 12 initial bodies. × 100,000. *B,* intraerythrocytic inclusion bodies stained by Giemsa method. × 800. (From Gothe (1967) Z Parasitenk *29:* 119–129.)

Fig. 2. *Haemobartonella muris. A,* ultrathin section showing parasite on rat erythrocyte. There is some indentation of the erythrocyte membrane at point of attachment. × 63,920. (From Tanaka *et al.* (1965) J Bacteriol *90:* 1735–1749.) *B, Haemobartonella felis* on erythrocytes of an infected cat. Polychrome stain. × 950. (From Flint and McKelvie (1955) Proc 92 Ann Mtg Amer Vet Med Ass 240–242.)

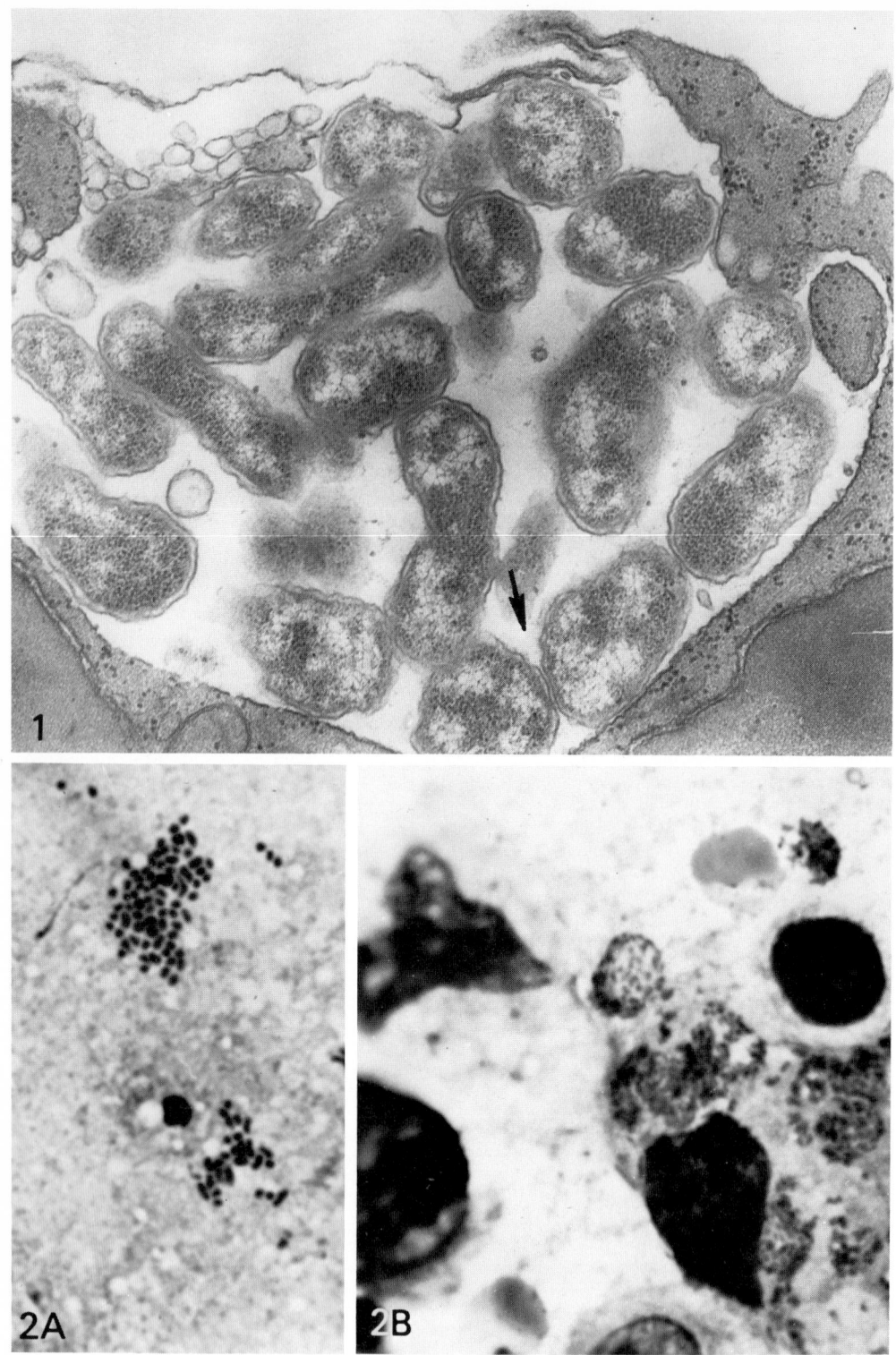

Plate 18.1.

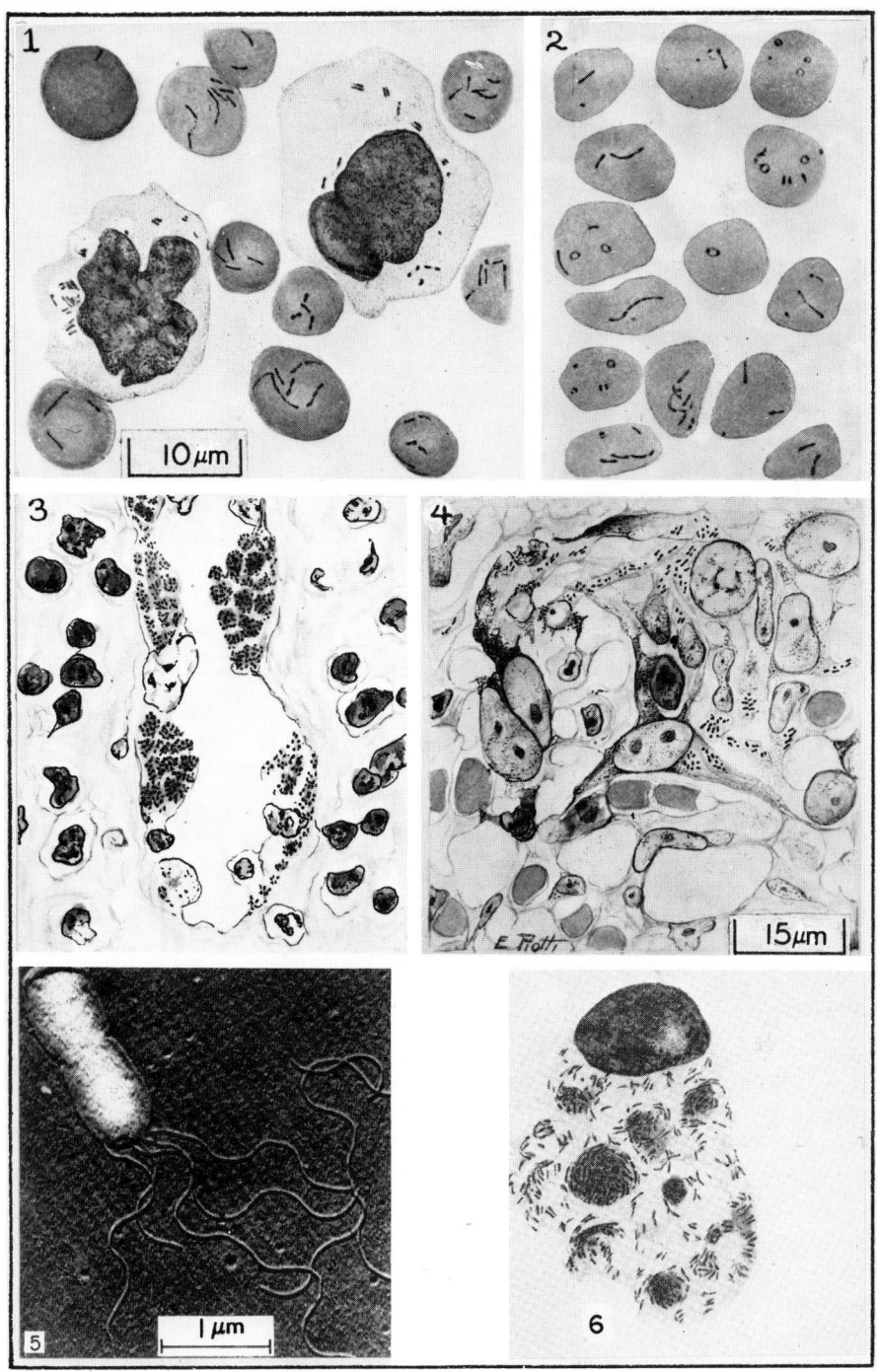

Plate 18.2.

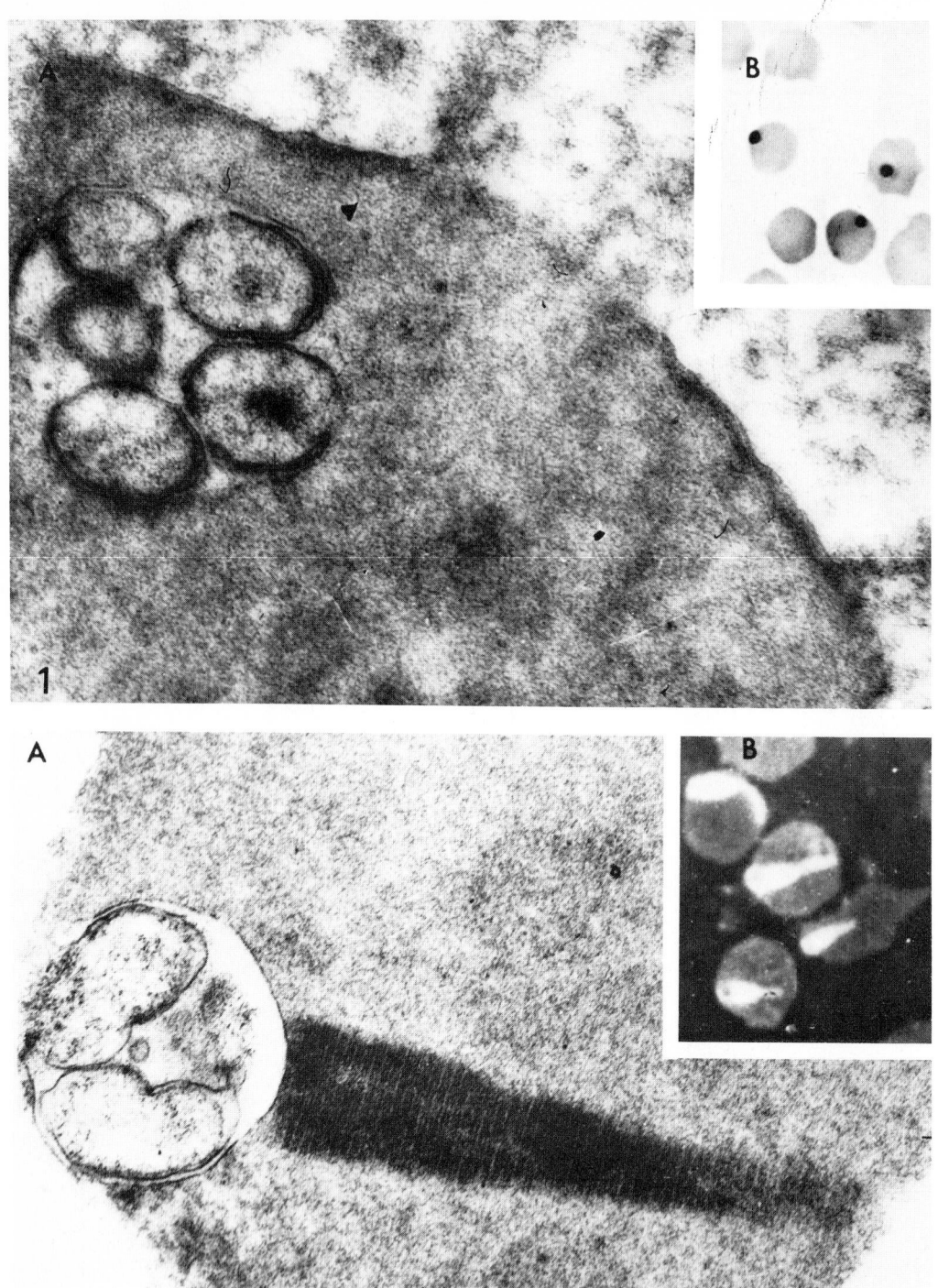

Plate 18.3.

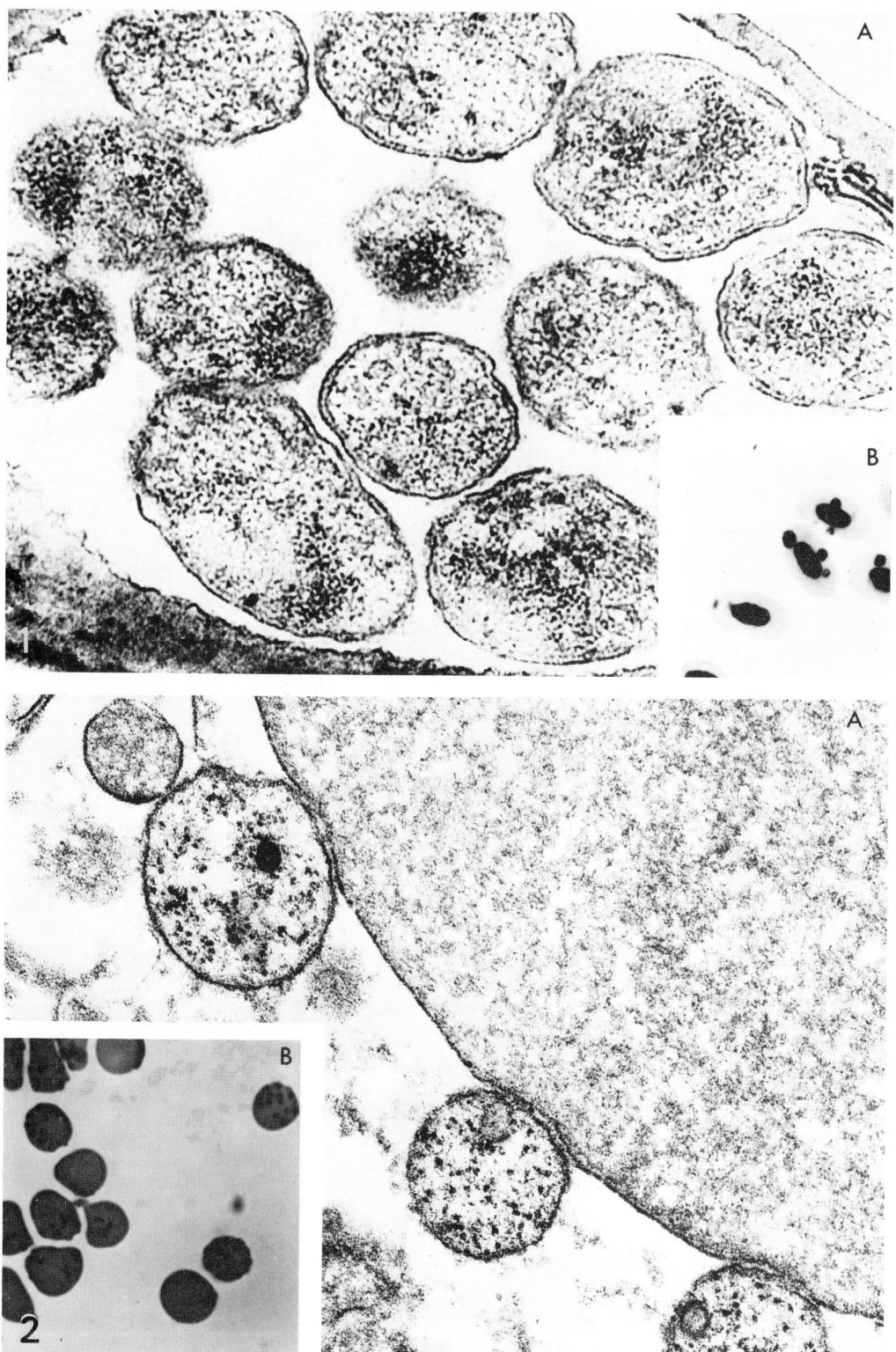

Plate 18.4.

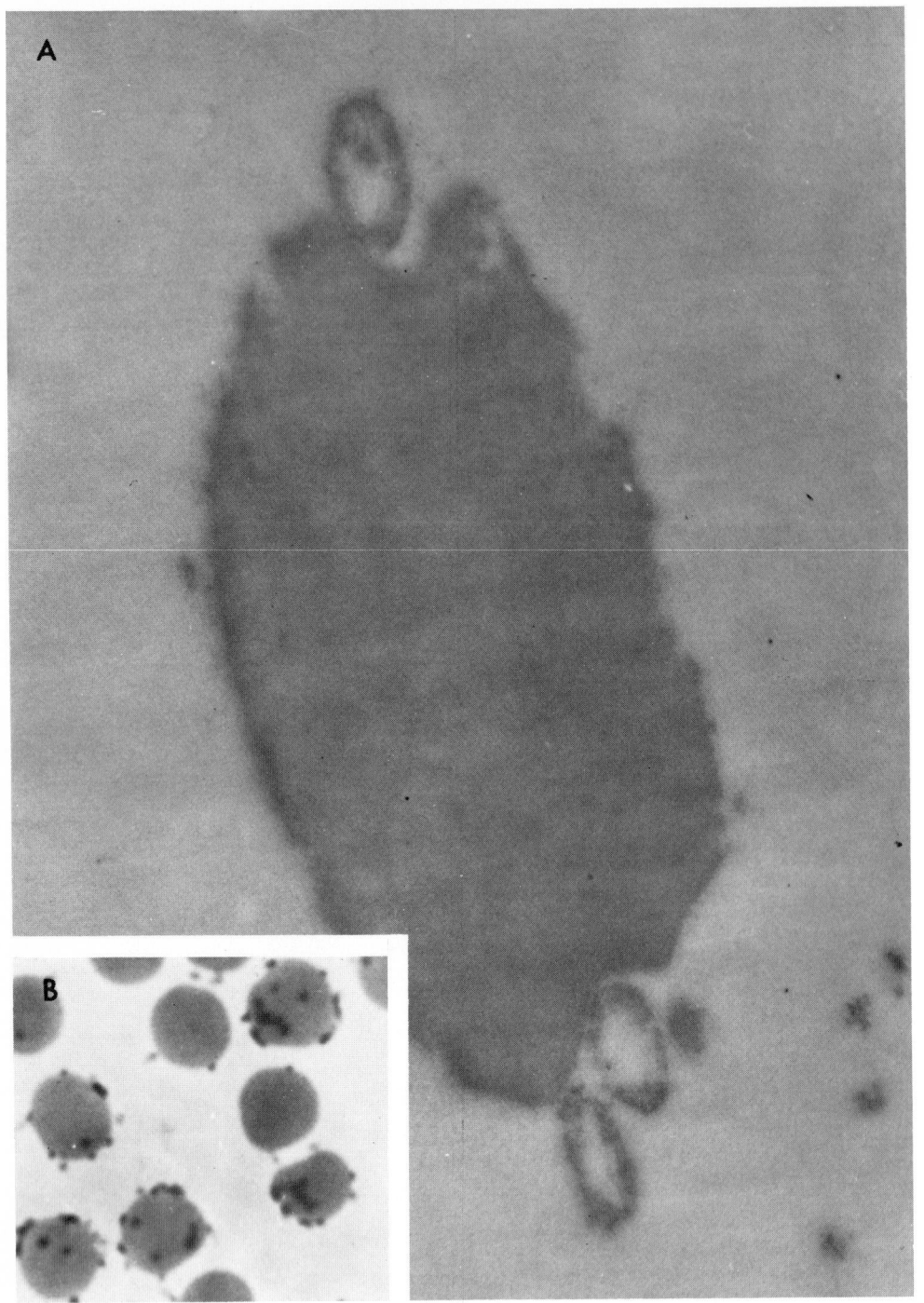

Plate 18.5. *Eperythrozoon*

Fig. 1. *Eperythrozoon ovis*. *A*, ultrathin section showing parasite on sheep erythrocytes; note erosion of the erythrocyte at point of attachment. × 21,500. *B*, Parasites on infected sheep blood. Giemsa stain. × 2400. (From Kreier and Ristic (1963) Amer J Vet Res *24*: 488–500.)

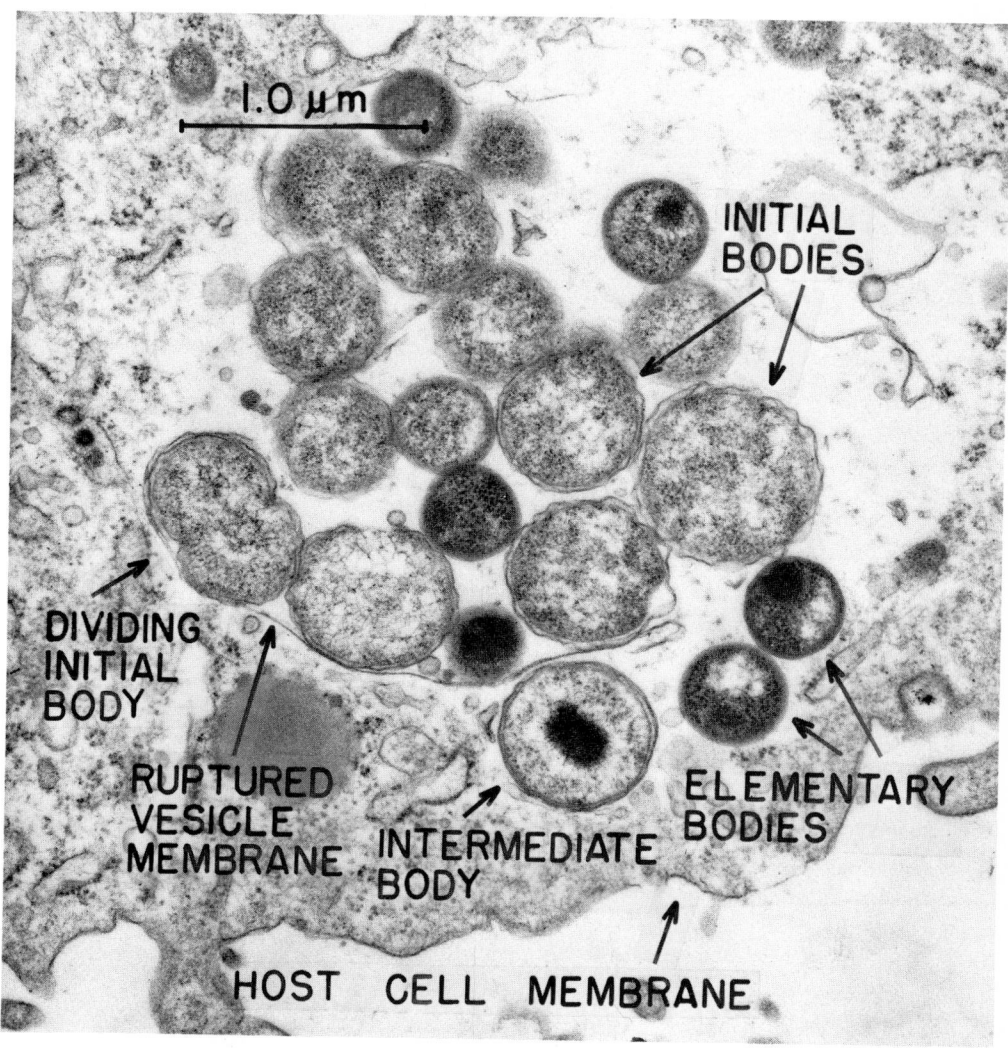

Plate 18.6. *Chlamydia*

Fig. 1. Electron micrograph of ultrathin section through a microcolony of *Chlamydia psittaci* in cytoplasm of a McCoy cell after 48-hrs incubation. The various developmental forms are labeled. The vesicular membrane has been ruptured and chlamydiae are being released into the cytoplasm. The multilaminated nature of the wall of the elementary bodies, the double unit membrane surrounding the intermediate and initial bodies are visible. × 28,700. (From Cutlip (1970) IAI *1*: 500.)

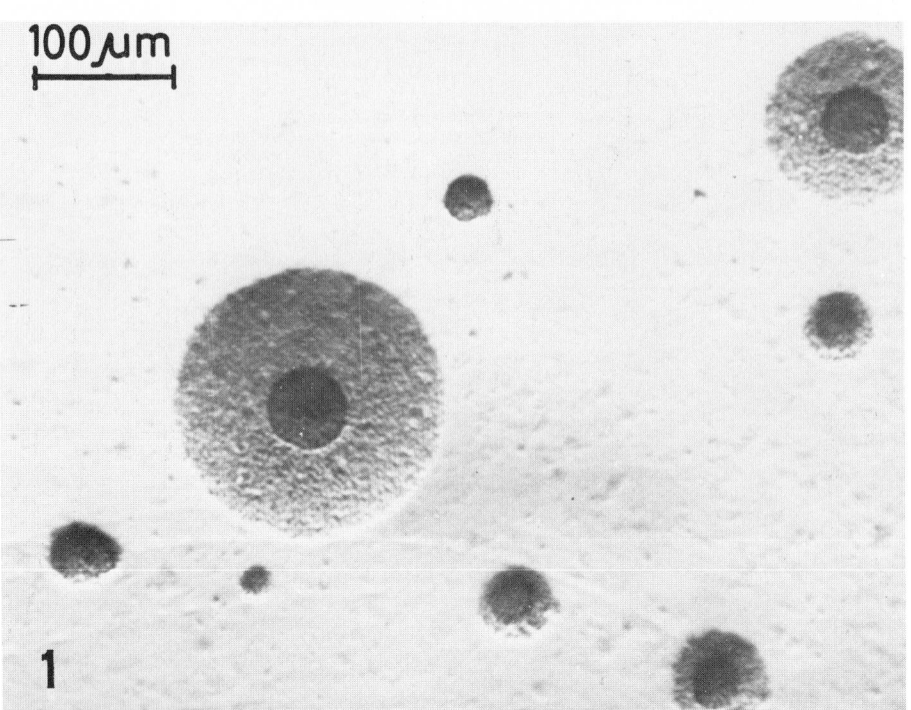

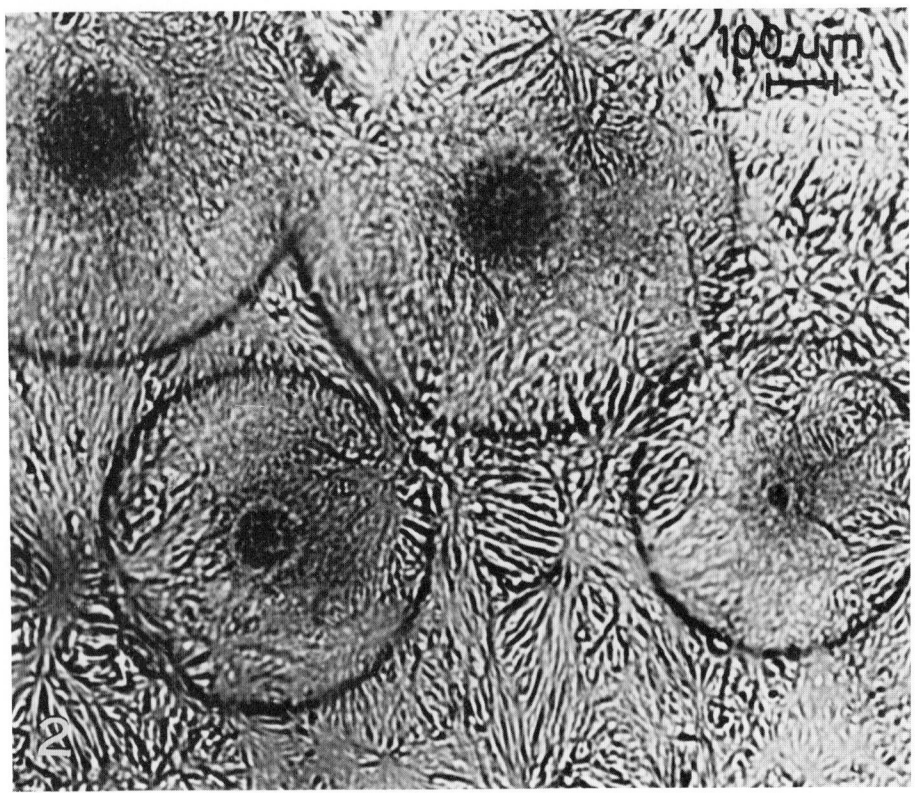

Plate 19.1.

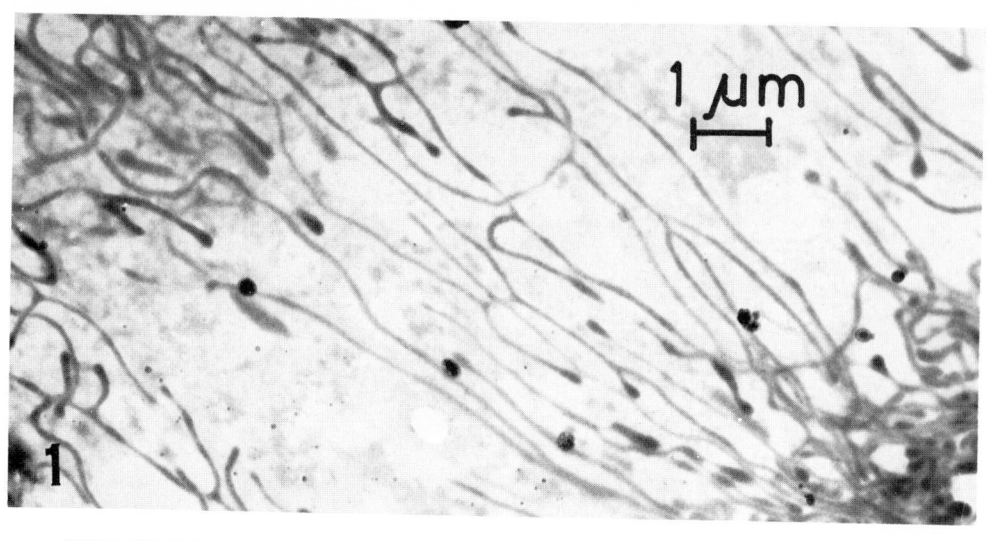

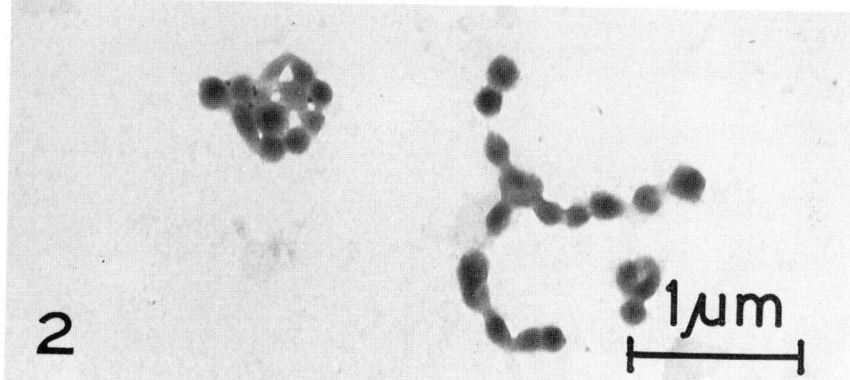

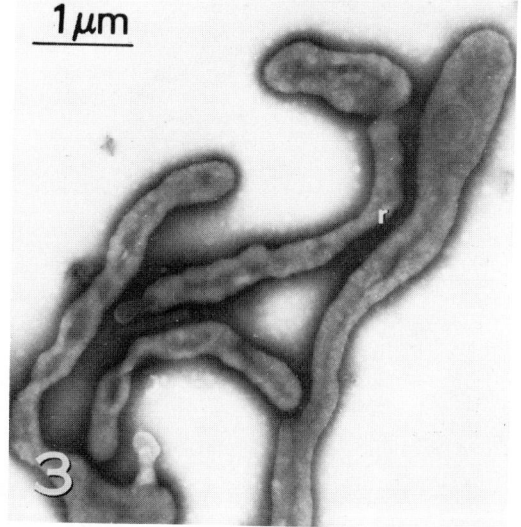

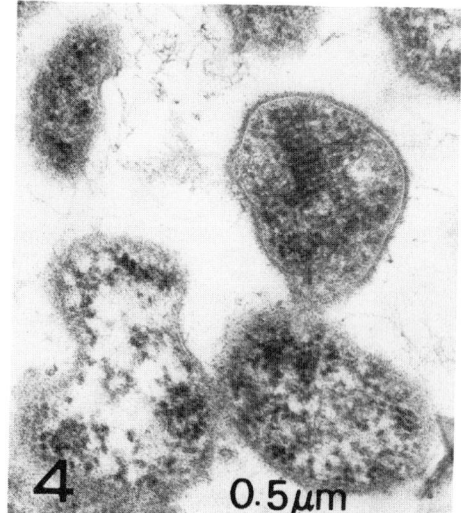

Plate 19.2.

Plate 19.1. *Mycoplasma*

Fig. 1. Biphasic colonies on horse serum agar showing typical "fried egg" appearance (*M. orale* type 1).

Fig. 2. "Film and spots phenomenon." Culture of *M. pulmonis* on horse serum agar showing biphasic colonies and pronounced development of a wrinkled, pearly film on the medium surface.

Figs. 1 and 2 from Freundt (1971) Medicinsk årbog XIV.

Plate 19.2. *Mycoplasma*

Fig. 1. Long branched filaments of *M. mycoides* subsp. *mycoides;* early exponential growth phase. Electron micrograph of cells on a collodion film floating upon serum broth.

Fig. 2. Chains of coccoid structures in liquid cultures of *M. mycoides* subsp. *mycoides;* late exponential growth phase.

Fig. 3. Short filaments of T-mycoplasma; exponential growth phase. Electron micrograph of a negatively stained specimen.

Fig. 4. Section of round to pleomorphic cells of T-mycoplasma showing triple-layered limiting membrane and arrangement of ribosomes in compact geometrical pattern; exponential growth phase.

Figs. 1 and 2 from Freundt (1958); Figs. 3 and 4 from Freundt (1971) Medicinsk årbog XIV.

PART 19

THE MYCOPLASMAS

E. A. FREUNDT*

CLASS **MOLLICUTES** EDWARD AND FREUNDT 1967, 267

(*Paramycetes* Sabin 1941, 58.)

Mol.li.cu′tes. L. adj. *mollis* soft, pliable; L. fem.n. *cutis* skin; M.L. fem.pl.n. *Mollicutes* class with pliable cell boundary.

Procaryotic organisms bounded by a single triple-layered (Plate 19.2, Fig. 4) membrane; they lack a true cell wall and are incapable of synthesizing cell wall precursors, such as muramic and diaminopimelic acids. Cells small, sometimes ultramicroscopic (about 200 nm), highly pleomorphic, coccoid to filamentous (Plate 19.2, Figs. 3 and 4), with a more or less pronounced tendency of the filaments to produce truly branched myceloid structures. The method of reproduction is controversial, but appears to take place by the development within the filaments of tiny coccoid structures ("elementary bodies") and their subsequent release by fragmentation and disintegration of the filaments, and/or by binary fission; also, reproduction through budding may occur. Usually non-motile. No resting stages known. Gram-negative.

The species recognized thus far can be grown on artificial cell-free media of diverse complexity. Colonies are minute; they have a marked tendency to grow down into solid media and usually have a characteristic "fried egg" appearance (Plate 19.1, Fig. 1). Most species completely resistant to penicillin and its analogues. Growth and metabolism specifically inhibited by antibody. Several of the above-mentioned characteristics of this Class, such as the morphological instability, the inability to retain the dye-iodine complex of the Gram stain, the tendency to penetrate into the depth of solid media, the insusceptibility to penicillin and the susceptibility to antibody, can be attributed to the lack of a cell wall. The mycoplasmas may be saprophytic, parasitic or pathogenic. The pathogens cause diseases of animals and possibly of plants.

Further Comments

It is still impossible to propose a clear-cut definition for *Mollicutes*, in terms of properties shared by all members of the Class, that would clearly distinguish them from cell wall-defective bacterial variants. Nevertheless, the existence of a number of fundamental differences with respect to morphology, mode of reproduction, biochemistry and physiology, collectively provides ample direct and indirect evidence in support of the view that the *Mollicutes* are phylogenetically and taxonomically unrelated to bacteria (Edward and Freundt, 1969).

In addition to the widespread occurrence of mycoplasmas as animal parasites and pathogens, there is increasingly strong evidence that related organisms play an important role as arthropod-borne plant pathogens. The evidence is based primarily on electron microscopic observations, together with the observation that plant diseases thought to be caused by mycoplasmas respond to treatment with tetracyclines and other broad spectrum antibiotics (see Addendum to Class Mollicutes, p. 953).

---

* Advisory Board: R. Chanock, D. G. ff. Edward, L. Hayflick and S. Razin.

ORDER I. **MYCOPLASMATALES** FREUNDT 1955, 71

(*Borrelomycetales* Turner 1935, 25; *Paramycetales* Sabin 1941, 58; *Pleuropneumoniales* Tulasne and Brisou 1955, 237; *Mollicutales* Edward 1955, 89.)

My.co.plas.ma.ta′les. M.L. neut.n. *Mycoplasma* type genus of the order; *-ales* ending to denote an order; M.L. fem.pl.n. *Mycoplasmatales* the *Mycoplasma* order.

Only one order, *Mycoplasmatales* is accepted in the class *Mollicutes;* the description of the order is therefore the same as for the class.

Two families are accepted in the order *Mycoplasmatales.*

Key to the families of order **Mycoplasmatales**

I. Require sterol for growth.

Family I. *Mycoplasmataceae*, p. 930

II. Do not require sterol for growth.

Family II. *Acholeplasmataceae*, p. 949

FAMILY I. **MYCOPLASMATACEAE** FREUNDT 1955, 71

(*Borrelomycetaceae* Turner 1935, 25; *Parasitaceae* Sabin 1941, 58; *Pleuropneumoniaceae* Tulasne and Brisou 1955, 237.)

My.co.plas.ma.ta′ce.ae. M.L. neut.n. *Mycoplasma* type genus of the family; *-aceae* ending to denote a family; M.L. fem.pl.n. *Mycoplasmataceae* the *Mycoplasma* family.

Sterol required for growth. Other characters as for the class and order.

Genus I. **Mycoplasma** *Nowak 1929, 1349 Nom. cons.* Jud. Comm. Opin. 22, 1958, 166

(*Asterococcus* Borrel, Dujardin-Beaumetz, Jeantet and Jouan 1910, 179; *Asteromyces* Wroblewski 1931, 105; *Borrelomyces* Turner 1935, 25; *Bovimyces* Sabin 1941, 57; *Pleuropneumonia* Tulasne and Brisou 1955, 237; the foregoing are all illegitimate.)

My.co.plas′ma. Gr. masc.n. *myces* a fungus; Gr. neut.n. *plasma* something formed or molded, a form; M.L. neut.n. *Mycoplasma* fungus form.

Cells highly pleomorphic, varying in shape **from spherical or slightly ovoid,** about 125–250 nm in diameter in early exponential growth phase, **to slender branched filaments** of uniform diameter ranging in length from a few to 150 μm (Plate 19.2, Fig. 1). Chains of coccoid bodies develop at a later growth phase (Plate 19.2, Fig. 2). **Possibly endowed with alternative modes of replication:** binary transverse fission of coccoid and filamentous cells, multiple release of elementary bodies produced within the filaments and budding. The filaments also branch, giving rise to a pseudo-mycelial structure; later dense corpuscles appear in the filaments which are transformed into chains of spherical bodies as a result of constrictions between the spheres but without formation of septa; the chains fragment liberating free elementary bodies.
Cells lack a true cell wall and are bounded by a **single triple-layered membrane,** about 7.5–10 nm thick. Usually non-motile but gliding motility has been described in some species. No resting stages known. **Gram-negative.**

Colonies on solid medium are hemispherical, slightly flattened and very small, their diameter generally ranging from 10–600 μm, although occasionally colonies are found up to 4 mm. **The typical colony is biphasic, with a "fried egg" appearance,** i.e. consisting of an opaque, granular central area which grows down into the medium and a flat, translucent peripheral zone (Plate 19.1, Fig. 1). On blood agar, colonies of most species produce zones of α,α′ or β hemolysis, under appropriate experimental conditions (Cole *et al.*, 1968; Aluotto *et al.*, 1970). Peroxide is the major hemolysin in most species. The colonies of some species are able to adsorb on their surfaces erythrocytes and/or other animal cells in suspension (Taylor-

TABLE 19.1
Characteristics useful in the differentiation of the species of genus **Mycoplasma**

All species require sterol for growth; none hydrolyzes urea or esculin. Do not grow at 45–62 C or at pH 1.0–3.0. All sensitive to 5% sodium-polyanethol-sulfonate except *M. anatis*, *M. gallinarum*, *M. iners* (strain PG30), *M. gateae* and *M. maculosum* which may be either totally resistant or show a slight partial inhibition only. The usual designation that + = 90% or more positive strains applies only to hydrolysis of arginine and glucose data. In tests marked *, generally only a few strains, including the type strain, of each species have been tested. d = variable reaction.

	Arginine hydrolyzed	Glucose acid	O-F test*	Mannose acid*	Phosphatase*	Film and spots on egg yolk*	Tetrazolium reduced* Aerobic/ anaerobic	G + C ratio*
1. *M. mycoides*	−	+	F	+	−	d	+/+	26.1–26.8
1b. *M. mycoides* subsp. *capri*	−	+		+	−	d	+/+	23.6–25.8
2. *M. bovirhinis*	−	+	F	−	−	−	+/+	24.5–25.5
3. *M. dispar*	−	+	O	+	−	−	+/+	28.5
4. *M. ovipneumoniae*	−	+					+/+	
5. *M. conjunctivae*	−	+	F	+	−	−	+/	
6. *M. gallisepticum*	−	+	F	+	−	−	+/+	32.0–35.5
7. *M. anatis*	−	+	F	+	+	+	−/+	
8. *M. synoviae*		+				+		34.0
9. *M. neurolyticum*	−	+	F	+	−	−	−/+	23.0–26.5
10. *M. pulmonis*	−	+	d	+	−	+	d/+	27.5–28.3
11. *M. felis*	−	+	F	−	+	+	−/+	25.0
12. *M. feliminutum*	−	−			−	+[1]	−/+	
13. *M. canis*	−	+	F	−	−	−	−/+	28.5–29.0
14. *M. edwardii*	−	+	F	−	−	+	−/+	29.2
15. *M. cynos*	−	+			+	+	∓/+	
16. *M. hyorhinis*	−	+	O	−	+	−	d/+	27.0–28.0
17A. *M. hyopneumoniae*	−	+						
18. *M. flocculare*	−	+			−		−/	
19. *M. pneumoniae*	−	+	O	+	−	+[2]	+/+	38.6–40.8
20. *M. bovigenitalium*	−	−		−	+	+	−/+	28.0–32.0
21. *M. agalactiae*	−	−		−	+	d	+/+	33.5–34.0
22. *M. alkalescens*	+	−		−	+		−/	25.9
23. *M. gallinarum*	+	−		−	−	+	+/+	27.0–28.0
24. *M. iners*	+	−		−	−	+	−/−	29.0–29.5
25. *M. meleagridis*	+	−		−	+	d	−/+	28.0–28.5
26. *M. arthritidis*	+	−		−	+	−	−/−	30.0–33.7
27. *M. gateae*	+	−		−	−	−	−/±	28.5
28. *M. maculosum*	+	−		−	+	+	−/+	26.5–29.5
29. *M. spumans*	+	−		−	+	−	−/−	28.5–29.0
30. *M. hyosynoviae*	+	−		−	−		−/	
31. *M. arginini*	+	−		−	−	−	−/+	28.6
32a. *M. orale* type 1	+	−		−	−	−	−/−	24.0–28.0
32b. type 2	+	−		−	+	−	−/+	26.1
32c. type 3	+	−		−	−			
33. *M. salivarium*	+	−		−	−	d	−/d[1]	27.0–31.5
34. *M. hominis*	+	−		−	−	−	−/−	27.3–29.2
35. *M. primatum*	+	−		−	+		−/	28.6
36. *M. fermentans*	+	+	F	−	−	+	−/+	27.5–28.5

[1] Weak reaction.
[2] Film but no spots.

Robinson and Manchee, 1967; Sobeslavsky et al., 1968; Manchee and Taylor-Robinson, 1969). Growth in fluid media is usually a barely visible opacity.

Chemoorganotrophs; metabolism is mainly fermentative, although a few species have an oxidative metabolism. **Most species utilize either glucose or arginine** as the major source of energy; rarely both are used, sometimes neither. Glucose-fermenting strains may or may not ferment mannose (Table 19.1); they usually also ferment fructose, galactose, maltose, glycogen, dextrin and starch. Reports on utilization of other carbohydrates are conflicting.

Generally not proteolytic; a few species liquefy gelatin, coagulated serum or casein (Edward, 1954; Freundt, 1958; Aluotto et al., 1970). Reduction of nitrate or production of indole has not been reported; hydrogen sulfide is produced by a few species only. **Urea not hydrolyzed.** Phosphatase activity a distinctive property of some species (Table 19.1).

2,3,5-Triphenyltetrazolium chloride, methylene blue and potassium tellurite are **frequently reduced under anaerobic conditions,** less frequently under aerobic.

All species require cholesterol, or certain other sterols, **for growth.** All species so far tested also require long chain fatty acids for growth and morphological and osmotic stability. Nutritional requirements are usually met by use of a complex medium containing serum or 1–2% PPLO Serum Fraction (Difco), although some species require yeast extract or yeast extract dialysate (see Aluotto et al., 1970; Hayflick and Chanock, 1965).

Most species are facultative anaerobes although growth is better aerobically. One or two species are aerobic, and several of primate origin prefer anaerobic conditions, especially on primary isolation. The addition of 5% CO_2 may or may not improve growth.

Temperature range 22–41 C; optimum about 36–37 C. Grows well at pH 7 or slightly acid or alkaline.

Relatively resistant to lysis by osmotic shock but very sensitive to lysis by surface-active agents and lipolytic agents. Lysed by taurocholate at concentrations ot 0.03% when suspended in M sucrose-0.05 M NaCl solution; by digitonin in concentrations of 10–15 µg/ml.

All species moderately to highly sensitive to 1.5% digitonin using a disc growth inhibition technique.

With a few notable exceptions (M. anatis, M. gallinarum, M. iners, (strain PG 30), M. gateae and M. maculosum) the growth of most species is inhibited by 5% sodium-polyanethol-sulfonate (Kunze, 1971; Andrews, Freundt, unpublished).

Resistant to 1000–10,000 µg/ml of penicillin G and to 1000–4000 µg/ml of ampicillin, chloraxillin, methicillin or nafcillin. Susceptibility to other antibiotics variable; susceptibility to erythromycin and certain other macrolides particularly variable and may be useful for differentiation (Taylor-Robinson and Path, 1967; Arai et al., 1967). Usually tolerates 1:2000–1:4000 thallium acetate.

Parasites and pathogens of a wide range of mammalian and avian hosts.

The G + C content of the DNA ranges from 23–40 moles % (T_m, buoyant density and chemical analysis). The genome size in the species examined is 4.4–4.8 × 10^8 daltons as determined by the renaturation method of Wetmur and Davidson (1968).

Type species: Mycoplasma mycoides (Borrel et al.) Freundt 1955, 73.

Further Comments

Glucose is catabolized by monolactic or heterolactic glycolytic pathways, the major end products being lactic acid with some pyruvic acid, acetic acid and acetylmethylcarbinol. In most species tested the respiratory pathway is flavin-terminated; heme compounds (cytochromes, catalase) are absent, although in M. arthriditis there is indirect evidence of the existence of cytochromes. Non-fermentative species usually contain the enzymes arginine deiminase, ornithine transcarboxylase and carbamate kinase, by which arginine is degraded to citrulline and ornithine to yield ATP, CO_2 and NH_3.

The lipid metabolism is complex and not yet completely understood. Cholesterol may be replaced by cholestanol, stigmasterol and certain other sterols, but esters such as cholesterol laurate are usually inactive. Cholesterol is incorporated into the cell membrane making up to 36% of the total membrane lipid.

The lipolytic ability of some strains is usually held responsible for the "film and spots phenomenon" (Plate 19.1, Fig. 2), a distinguishing characteristic; the film consists of a characteristic wrinkled, pearly film on the surface of the medium, containing cholesterol and phospholipids; the tiny black spots beneath and around the colonies are calcium and magnesium soaps. These were first noted on 20% horse serum agar (Edward, 1950; 1954) but appear more consistently on egg yolk emulsions (Edward, 1954; Fabricant and Freundt, 1967) or with Tween 80 and $CaCl_2$ (Razin and Rottem, 1963).

Susceptibility to optochin (ethylhydrocuprein hydrochloride) (Aluotto et al., 1970) has now been found to be of little value in differentiation (Black and Jørgensen, 1972).

Description of the species of genus **Mycoplasma**

1. **Mycoplasma mycoides** (Borrel, Dujardin-Beaumetz, Jeantet and Jouan) Freundt 1955, 73. (*Asterococcus mycoides* Borrel *et al.* 1910, 179; *Coccobacillus mycoides peripneumoniae* Martzinovski 1911, 917; *Micromyces peripneumoniae bovis contagiosae* Frosch 1923, 275; *Mycoplasma peripneumoniae* Nowak 1929, 1349; *Asteromyces peripneumoniae bovis* Wroblewski 1931, 105; *Borrelomyces peripneumoniae* (Nowak) Turner 1935, 25; *Bovimyces pleuropneumoniae* Sabin 1941, 57; *Pleuropneumonia bovis* Tulasne and Brisou 1955, 237: Le microbe de la péripneumonie Nocard and Roux 1898, 240.)

my.co.i'des. Gr. masc.n. *myces* a fungus; Gr. n. *eidus* shape; M.L. adj. *mycoides* fungus-like.

Most strains have a potential for producing repeatedly branching filaments of greatly varying length.

Does not produce film and spots on horse serum agar and only occasionally on egg yolk media; no clearing. Colonies on solid medium do not adsorb red cells from guinea pig or cattle.

Highly sensitive to erythromycin, minimal inhibitory concentrations ranging from 0.1–4.0 μg/ml.

Further Comment

Two subspecies are recognized; they are serologically distinct by agglutination, growth inhibition and metabolic inhibition tests; extensive to complete cross-reactivity is demonstrable by complement fixation and indirect hemagglutination tests. Common antigens also demonstrable by double immunodiffusion tests. The electrophoretic patterns of the cell proteins of the type strains resemble each other, but are not identical. The justification of recognizing the subspecies has been challenged (Turner, 1960; Hudson *et al.*, 1967).

1a. *Mycoplasma mycoides* subsp. *mycoides* Freundt 1955, 73.

Most strains may produce very long, repeatedly branching filaments up to 150 μm in length, average 40–50 μm (Plate 19.2, Fig. 1).

Slight liquefaction of coagulated serum.

The chemical nature of the hemolysin remains to be defined. On implantation of diffusion chambers into the peritoneal cavity of rabbits there is tissue necrosis suggesting production of a diffusible toxin.

The etiological agent of contagious pleuropneumonia of cattle. The natural disease can be reproduced by introducing the organism via the upper respiratory tract, preferably by exposure to infective aerosols. Subcutaneous inoculation causes inflammatory lesions varying in extent from a small necrotic nodule to a spreading edematous swelling accompanied by fever and other general symptoms. Mice, rats, guinea pigs, rabbits and hamsters are normally not susceptible, although infection may occur in mice and rabbits following inoculation of organisms suspended in an agar gel or mixed with mucin. The results of subcutaneous inoculation into rabbits are variable; the production of muscle lesions characterized by swelling, congestion, petechial and ecchymotic hemorrhages, edema and necrosis has been reported.

The G + C content of the DNA is about 26.1–26.8 moles % (T_m and buoyant density).

Proposed neotype strain: PG1; NCTC 10114; AMRC-C 01 (Edward and Freundt, 1973).

1b. *Mycoplasma mycoides* subsp. *capri* (Edward) Freundt 1953, 73. (*Borrelomyces peripneumoniae capri* Longley 1940, 195; *Borrelomyces peripneumoniae caprae* Longley 1951, 23; *Asteroccus mycoides* subsp. *capri* Edward 1953, 873; *Pleuropneumonia capri* Tulasne and Brisou 1955, 238; *Mycoplasma caprae* Turner 1960, 256; *Mycoplasma capri* Hudson, Cottew and Adler 1967, 288.)

Filaments of moderate length, 10–30 μm; filamentation is stimulated by addition of oleic acid (about 50 μg/ml) to growth medium.

Colonies on serum agar large, with a diameter of 1.5–2.5 mm after 3 days and may reach a maximum of 4 mm. Produces strong opalescence in fluid medium.

Liquefaction of coagulated horse serum is more rapid and extensive than with the primary variety. Also less exacting in its nutritional requirements, growing to a slight extent in media without serum. The hemolysin appears to contain an unidentified component in addition to peroxide.

The etiological agent of contagious pleuropneumonia of goats. Closely related strains have been identified as causal agents of highly fatal edema and cellulitis in goats and of outbreaks of a caprine disease characterized by septicemia and acute polyarthritis. However, further comparative studies are needed of the strains associated with various types of infectious diseases of goats.

Pneumonia can be reproduced experimentally in goats by intratracheal inoculation or exposure to nebulized cultures; subcutaneous inoculation produces extensive and often fatal cellulitis and spreading edematous lesions in both goats and sheep. Mice and rabbits are not normally susceptible, but resistance is markedly lowered when the organisms are injected together with mucin as an adjuvant.

The G + C content of the DNA ranges from 23.6–25.8 moles % (chemical analysis, T_m and buoyant density). The significance of the demonstration of a limited amount of heterologous nucleic acid hybridization (16% relatedness value) between this species and *Acholeplasma laidlawii* is difficult to assess at present. Similar reports on antigenic relationships between these organisms could not be confirmed, however, by other workers.

Type strain: PG 3, goat Turkey (monotype); NCTC 10137; AMRC-C 20.

2. Mycoplasma bovirhinis Leach 1967, 313.

bo.vi.rhi'nis. L. n. *bos, bovis* the ox; Gr. n. *rhis, rhinis* nose; M.L. gen.n. *bovirhinis* of the nose of the ox.

Morphology not described.

Does not hydrolyze gelatin. Reports on liquefaction of casein and coagulated horse serum are conflicting.

A common inhabitant of the upper respiratory tract of cattle. Although frequently isolated from pneumonic lesions there is no conclusive evidence regarding a possible pathogenic role in respiratory disease of cattle. Recovered on one occasion from a field case of mastitis; mastitis can be produced experimentally with at least some strains. Cytopathogenicity for bovine and canine embryo kidney tissue cultures has been reported.

The G + C content of the DNA is about 24.5–25.5 moles % (T_m and buoyant density).

Type strain: PG 43; Leach, 5M331; ATCC 19884; NCTC 10118; AMRC-C 03; NIH M-734-001-084.

3. Mycoplasma dispar Gourlay and Leach 1970, 111.

dis'par. L. adj. *dispar* dissimilar, different.

Morphology poorly defined. Typical mycoplasmal ultrastructure.

Colonies on solid medium have a granular, lacy or reticulated appearance with no or poorly defined central area; in early subcultures they attain a diameter of up to 1.5 mm after 1–3 days of incubation, while after several passages in artificial media they tend to be smaller with a more dense structure.

Although minimal nutritional requirements have not been precisely defined it is obviously more exacting than most other species; growth can be obtained on Hartley's broth agar containing 20% fetal calf serum, 0.5% lactalbumin hydrolysate and 0.002% DNA (calf thymus, B.D.H.).

Normal habitat not precisely defined at present. Although associated with calf pneumonia the etiological significance of this species remains unknown as yet.

The G + C content of the DNA is 28.5 moles % (T_m).

Type strain: 462/2; ATCC 27140; NCTC 10125; AMRC-C 05.

Further Comments

Serologically distinct by growth inhibition and metabolic inhibition tests from other recognized bovine mycoplasmas and from a wide range of *Mycoplasma* species from other sources.

4. Mycoplasma ovipneumoniae Carmichael, St. George, Sullivan and Horsfall 1972, 677.

o.vi.pneu.mo'ni.ae. L. fem.n. *ovis* a sheep; Gr. n. *pneumonia* inflammation of the lungs, pneumonia; M.L. gen.n. *ovipneumoniae* of sheep pneumonia.

Morphology not described.

Agar colonies slightly raised, granular, devoid of central nipple.

Does not produce film and spots on horse serum agar. Not tested for proteolytic activities. Agar colonies produce zones of β type hemolysis by the overlay technique using ox, sheep and horse red blood cells. Not tested for hemadsorption or hemagglutination.

Isolated in Australia from the nose, trachea, bronchi and lungs of lambs with proliferative interstitial pneumonia, and in one instance from the conjunctiva of a pneumonic lamb that also had keratoconjunctivitis. Isolated only occasionally from the nasal sinuses of healthy adult sheep. The pneumonia has been reproduced experimentally in gnotobiotic and normal lambs by intra-tracheal inoculation, aerosols and contact.

The G + C content of the DNA not determined.

Type strain: not formally designated.

Further Comment

Antigenically unrelated, by growth inhibition, metabolic inhibition and immunodiffusion tests, to other *Mycoplasma* and *Acholeplasma* species of ovine and caprine source.

5. Mycoplasma conjunctivae Barile, Del Giudice and Tully 1972, 74.

con.junc.tiv'ae. M.L. fem.n. *conjunctiva* the membrane "joining" the eye-ball to the lids; M.L. gen.n. *conjunctivae* of conjunctiva.

Pleomorphic, spherical, ring-shaped and coccobacillary forms. Appear coccoid with small clusters of cells joined together by short filaments, under phase-contrast microscopy.

Colonies on agar may have either a "fried egg" or granular appearance. Occasionally colonies have elevated central growth and greenish, brownish or olive color. Marked turbidity in broth cultures.

Colonies on solid media do not adsorb red cells from sheep or guinea pigs; α-hemolysis of both types of erythrocytes.

Urea not hydrolyzed; serum not digested.

Serum or cholesterol required. CO_2 (5%) in nitrogen preferred for primary isolation.

Differs from other species of *Mycoplasma* and *Acholeplasma* antigenically and in cell protein electrophoretic patterns.

Isolated from conjunctival tissues of goats and sheep with pink-eye in the United States, Canada and Australia. Pathogenicity not known.

Type strain: HRC 581; ATCC 25834; AMRC-C 23.

Further Comments

The strain (67R) which Surman (1968) named *M. conjunctivae* var. *ovis* but did not describe has been identified as *M. arginini* (Barile *et al.*, 1972).

6. **Mycoplasma gallisepticum** Edward and Kanarek 1960, 699. (Avian serotype S6, Adler *et al.* 1958, 444; avian serotype A, Kleckner 1960, 277.)

gal.li.sep'ti.cum. L. fem.n. *gallina* a hen; Gr. adj. *septicus* putrefactive, septic; M.L. neut.adj. *gallisepticum* hen-poisoning (-infecting).

Short branching filaments produced in media with an adequate and balanced supply of cholesterol and long chain fatty acids. The following ultrastructural details are distinctive for this species: a dense outer layer consisting of surface projections, 4–6 nm long, cellular protrusions ("blebs") of a highly differentiated internal structure and polysome-like arrangement of ribosomes ("corncobs").

No proteolytic activity. Most strains agglutinate guinea pig and turkey red blood cells; agar colonies adsorb red blood cells and tracheal epithelial cells from monkey, rat, guinea pig and chicken, as well as spermatozoa from man and bull and HeLa cells. Adsorption occurs at 37 C, less rapidly and less extensively at 22 C; it is specifically inhibited by antiserum. The red blood cell and epithelial cell receptors for *M. gallisepticum* are destroyed by neuraminidase; conversely, pretreatment of the mycoplasmas with neuraminic acid or neuraminelactose likewise prevents adsorption. Hemolysin identified as hydrogen peroxide.

Usually reported to be highly sensitive to erythromycin, the minimal inhibitory concentration ranging from 0.02–2.0 µg/ml.

More resistant to lysis by osmotic shock at 37 C than other species of this genus tested.

A common pathogen of poultry, being the primary etiological agent of chronic respiratory disease. In addition, some strains of this species (S6 and related strains) appear to be responsible for outbreaks of encephalitis in turkeys associated with polyarteritis of the cerebral arteries. The neurological disease and characteristic pathology can be reproduced experimentally in turkeys by intravenous inoculation of high doses of washed organisms of the S6 strain, but not by cell-free filtrates of the culture medium. The neurological

disease is believed to be caused by a toxic component of the mycoplasma cells.

Reports on the $G + C$ content of the DNA range from 32.0–35.5 moles % (T_m, buoyant density and chemical analysis). Nucleic acid homology tests have shown a very low level cross-reaction with *M. gallinarum* (1.3% relative homology).

Type strain: PG 31; X95; ATCC 19610; NCTC 10115; AMRC-C 72; NIH M-722-001-084 (Edward and Freundt, 1973).

Further Comments

Serologically distinct from other species of avian origin. The electrophoretic patterns of the cell proteins of three strains tested (strains PG 31, S6 and A5969) are essentially identical, while clearly different from those of other avian species, *M. gallinarum*, *M. iners*, *M. meleagridis* and *M. synoviae*.

7. **Mycoplasma anatis** Roberts 1964, 471.

a.na'tis. L. fem.n. *anas* a duck; L. gen.n. *anatis* of a duck.

Morphology poorly defined.

Film and spots produced on egg yolk media, and occasionally on horse serum agar; may or may not produce clearing. Not tested for proteolytic activities. Colonies on solid medium do not adsorb guinea pig or chicken red blood cells.

Growth occurs at 20 and 37 C, but not at 25 or 44 C.

Since the description of this species is based on a single isolate only, recovered from the sinuses and air sacs of a duck, the ecology remains to be determined.

Type strain: 1340 (monotype); ATCC 25524; AMRC-C 73; NIH M-716-001-084.

8. **Mycoplasma synoviae** Olson, Kerr and Campbell 1964, 209. (Avian serotype S, Dierks *et al.* 1967, 172.)

syn.ov'i.ae. M.L. n. *synovia* the joint fluid; M.L. gen.n. *synoviae* of joint fluid.

Morphology poorly defined.

Not tested for biochemical activities other than glucose fermentation. Agglutinates turkey red blood cells. Colonies on solid medium do not adsorb guinea pig red blood cells. Nicotinamide adenine dinucleotide (diphosphopyridine nucleotide; coenzyme I) in the reduced form has been defined as an essential nutritional requirement, the addition of 0.01% of this coenzyme to 10% swine serum medium ensuring satisfactory growth. Growth is further stimulated by the addition of 0.01% cysteine.

Although reported to occur in the respiratory tract of apparently healthy chickens, the actual incidence of healthy carriers remains undetermined as yet. Etiological agent of infectious synovitis in chickens and turkeys.

The G + C content of the DNA is about 34.0 moles % (buoyant density).

Type strain: WVU 1853; ATCC 25204; NCTC 10124; AMRC-C 75 (Edward and Freundt, 1973).

Further Comments

The electrophoretic pattern of the cell proteins is clearly distinct from those of other avian species including *M. gallinarum*, *M. gallisepticum*, *M. iners* and *M. meleagridis*.

9. **Mycoplasma neurolyticum** (Sabin) Freundt 1955, 73. (*Musculomyces neurolyticus* Sabin 1941, 57; *Pleuropneumonia cerebri-muris* Tulasne and Brisou 1955, 238; *Asterococcus neurolyticus* (Sabin) Prévot 1961, 721: L5, Findlay *et al.* 1938, 1511.)

neu.ro.ly'ti.cum. Gr. n. *neuron* nerve; Gr. adj. *lyticus* able to loosen, (dissolve); M.L. neut.adj. *neurolyticum* nerve-destroying.

Filaments vary in length from very short (2–5 μm) to extremely long structures (up to 160 μm reported).

Acid produced from glucose through fermentative degradation.

No proteolytic activity. No hemadsorption detected. A potent producer of hemolysin identified as hydrogen peroxide. Although minimal nutritional requirements are poorly defined, this species is apparently less exacting than most other *Mycoplasma* species as optimal growth is reported to be supported by as little as 1% whole serum or 2% agamma horse serum. Whole horse serum in the concentration of 1% exerts a transitory inhibitory effect on growth shortly after inoculation due to a heat-labile factor that can be replaced by complement.

The type strain of this species has been reported to be inhibited by moderate concentrations of penicillin G (40 units/ml and above) although it can be adapted to growth at high levels of penicillin on stepwise transfer to increasing concentrations. The effect of penicillin seems to be bacteriostatic rather than bactericidal.

A true exotoxin with neurotoxic effect on mice is produced by some but not all strains during the early growth phase. The toxin appears to be a protein with a molecular weight in excess of 200,000. It is thermolabile, being destroyed at 50 C in 10–30 min or at 45 C in 15–90 min. It is also inactivated by 0.025 mg/ml of trypsin within 10 min, or by binding *in vitro* to a sedimentable component of brain tissue, possibly a ganglioside. The toxin is antigenic and specifically neutralized *in vitro* by rabbit antiserum. The toxigenicity is often retained even after more than 100 subcultures although it may be lost on repeated subculturing.

Cell-free filtrates of growing cultures are effective only on intravenous, but not on intraperitoneal, subcutaneous or intracerebral inoculation of young mice, producing characteristic symptoms of rolling disease and neuropathological lesions as well as pulmonary hemorrhage. Less characteristic neurological symptoms are elicited in young rats, while hamsters, guinea pigs and chicken are insusceptible to the toxin.

A common inhabitant of healthy and diseased mice, apparently primarily localized to the mucous membranes of the upper respiratory tract. The organisms are apparently transmitted to young mice shortly after birth; these then become carriers of latent mycoplasmas. Frequently isolated in association with experimental infection of mice with a variety of other agents. For example, the first isolations reported were made from the brain of mice that had developed "rolling disease" during the course of intracerebral passage of *Toxoplasma gondii* (Sabin, 1938) and lymphocytic choriomeningitis and yellow fever viruses (Findlay *et al.*, 1938). Also repeatedly recovered from pneumonic lesions of mouse lungs after nasal instillation under anesthesia of various materials, and from the blood and tissues of mice under stress with malignant disease. Has been identified as the etiological agent of epidemic conjunctivitis occurring in mouse colonies. Except for this, the possible primary etiological role of this species in natural disease of mice remains to be determined. A cytopathic effect may occur on repeated passage in a variety of mammalian cell lines, provided large inocula are used. The production of neurotoxin in tissue cultures following several passages has been observed.

Reports on the G + C content of the DNA range from 23.0–26.5 moles % (T_m and buoyant density).

Type strain: Sabin type A; PG 39; ATCC 19988; AMRC-C 62; NIH M-723-001-084 (Edward and Freundt, 1973).

Further Comments

The electrophoretic pattern of the cell proteins of the toxigenic type strain of this species differed somewhat from that of a non-toxigenic strain.

10. **Mycoplasma pulmonis** (Sabin) Freundt 1955, 73. (*Murimyces pulmonis* Sabin 1941, 57; *Musculomyces histotropicus* (*sic*) Sabin 1941, 58; *Asterococcus pulmonis* (Sabin) Prévot 1961, 720; *Mycoplasma histotropicus* (*sic*) (Sabin) Tully 1965, 184; *Mycoplasma mergenhagen* Grace, Horoszewicz, Stim, Mirand and James 1965, 1369: L3, Klieneberger and Steabben 1937, 143.)

pul.mo'nis. L. masc.n. *pulmo* the lung; L. gen.n. *pulmonis* of the lung.

Filaments short (2–5 μm). Motility in the form of

gliding motion of bacilliform elements and constant rotation of coccoid structures reported. The ultrastructure is characterized by the occurrence of an outer layer of surface projections.

The central spot of the colonies is consistently less well defined than in most other Mycoplasma species; they have little tendency to grow into the agar and the surface is coarsely granulated or vacuolated.

Acid produced from glucose through fermentative or, in the case of at least one strain ("Negroni"), oxidative degradation.

Produces film and spots on horse serum and egg-yolk media, together with clearing of the latter. No proteolytic activity. Colonies on solid medium adsorb red blood cells, but not tracheal epithelial cells, from monkey, rat, guinea pig and chicken; they also adsorb spermatozoa from man and bull. The red blood cell receptor sites are not destroyed by neuraminidase. A potent producer of hemolysin identified as hydrogen peroxide.

Relatively resistant to erythromycin, minimum inhibitory concentration reported as 40 μg/ml.

A very common inhabitant of the respiratory tract of mice and of laboratory and wild rats. While newborn mice and rats usually appear to be free of M. pulmonis they quite regularly become infected during the early weeks of life. The organism may either lead a predominantly parasitic existence in the mucosa of the nasal passages or produce infectious catarrh that may or may not be complicated with otitis media and bronchopneumonia. Mice in particular have a pronounced tendency to develop generalized systemic infection. Rhinitis, otitis media and characteristic pneumonic lesions have been reproduced experimentally with M. pulmonis in gnotobiotic mice. On intraperitoneal inoculation into female mice the organism shows a predilection for the ovaries and oviducts and may produce inflammation of the reproductive system. Also associated with natural outbreaks of arthritis in mice and rats, particularly under conditions of stress. Intravenous inoculations of freshly isolated strains may also produce arthritis in mice. The natural host range of this species may not be restricted to small rodents; a similar organism has been recovered from the nares and oropharynx of rabbits.

The G + C content of the DNA is about 27.5–28.3 moles % (T_m).

Type strain: Ash; PG 34; ATCC 19612; AMRC-C 61; NIH M-717-001-084 (Edward and Freundt, 1973).

Further Comments

The murine strain C, originally assigned to a separate species *Musculomyces histrotopicus (sic)* Sabin 1941, 58, thought for a period of several years to be lost, was rediscovered in 1963 (Tully and Ruchman, 1964) and renamed *Mycoplasma histotropicus (sic)* Tully 1965, 184. Subsequent studies of the electrophoretic pattern of the cell proteins (Razin, 1968) and of the serology (Lemcke *et al.*, 1969) of Sabin's type C strain clearly identified it as *Mycoplasma pulmonis*.

Leach and Butler (1966) showed *Mycoplasma mergenhagen* Grace *et al.* 1965, 1369 to be a synonym of M. pulmonis.

Although a large majority of murine strains of this species appear to constitute a fairly homogeneous group serologically, minor differences were demonstrable, in gel diffusion tests, between the C strain and other murine strains. Also the demonstration of divergencies in the antigenic composition of certain strains of non-murine source (Deeb and Kenny, 1967) suggests the existence of serological subtypes of this species. The electrophoretic patterns of the cell proteins of three strains examined, including the type strain, were nearly identical.

11. **Mycoplasma felis** Cole, Golightly and Ward 1967, 1456.

fe′lis. L. fem.n. *felis* a cat; L. gen.n. *felis* of a cat.

Coccobacillary to short filamentous lobulated cells.

Produces film and spots on horse serum agar and egg yolk medium, together with clearing of the latter. No proteolytic activity. Hemolysin identified as hydrogen peroxide.

A relatively common inhabitant of the oral and nasal cavities, conjunctivae and the lower genital tract of cats. Not pathogenic as far as known, although occasionally associated with feline conjunctivitis.

The G + C content of the DNA is about 25 moles % (T_m).

Type strain: CO; ATCC 23391; AMRC-C 50 (Edward and Freundt, 1973).

Further Comments

Serologically this species is fairly closely related to M. canis, as revealed by complement fixation and double immunodiffusion tests, although clearly distinct by the disc growth inhibition test.

12. **Mycoplasma feliminutum** Heyward, Sabry and Dowdle 1969, 621.

fe.li.mi.nu′tum. L. fem.n. *felis* a cat; L. neut.-part. adj. *minutum* small; M. L. neut.adj. *feliminutum* apparently intended to designate a small-colony organism isolated from cats.

Morphology not defined.

Colonies relatively small and develop slowly compared to other mycoplasmas of feline origin. They are irregular in shape and lack a central core. On repeated subculturing these colonial characteristics are lost and larger colonies with a typical fried egg appearance develop.

Spots and a suggestion of film on egg yolk medium; no clearing. Not tested for proteolytic activities. Does not agglutinate sheep, guinea pig, rat, monkey or chicken red blood cells.

The description of this species is based on a single strain isolated from the oral cavity of a cat.

Genetic characters not determined.

Type strain: Ben (monotype); ATCC 25749; AMRC-C 52.

Further Comments

Serologically distinct by the growth inhibition test from 15 other species tested.

13. Mycoplasma canis Edward 1955, 90. (*Asterococcus canis* (Edward) Prévot 1961, 719: β strains from dogs, Edward and Fitzgerald 1951, 566.)

ca′nis. L. n. *canis* a dog; L. gen.n. *canis* of a dog.

Filaments short (2–5 μm) with only occasional branching and early disintegration into elementary bodies.

Although previously reported to be glucose negative, production of acid through fermentation of glucose was demonstrated by Aluotto *et al.* (1970) and subsequently confirmed in other laboratories.

May or may not hydrolyze gelatin. No liquefaction of casein or coagulated horse serum. Agar colonies adsorb guinea pig red blood cells.

A relatively common inhabitant of the upper respiratory tract, throat and genital tract of dogs. Not pathogenic as far as known.

The G + C content of the DNA is about 28.5–29.0 moles % (T_m and buoyant density).

Type strain: PG 14; C 55; ATCC 19525; AMRC-C 41; NIH M-725-001-084.

Further Comments

Serologically distinct from other species of canine provenance by agglutination, growth inhibition, indirect imunofluorescence and complement fixation tests. Common antigens shared to a wide extent with *M. felis* as shown by complement fixation and double immunodiffusion tests, while clearly distinct from this species by the disc growth inhibition test. A group of isolates found to have biological properties similar to *M. canis*, but seemingly only partially related to other strains of this species, was provisionally designated as a separate serotype (antigenic variant) of *M. canis* (Armstrong *et al.*, 1970). Evidence for the existence of some antigenic heterogeneity within this species was also provided by the original observations of Edward and Fitzgerald (1951).

14. Mycoplasma edwardii Tully, Barile, Del Giudice, Carski, Armstrong and Razin 1970, 346. (Strain No. 21 Edward and Fitzgerald 1951, 569.)

ed.ward′i.i. M.L. gen.n. *edwardii* of Edward; named for D.G. ff. Edward, who first isolated this species.

Filaments short.

Produces film and spots on egg yolk media on prolonged incubation (2–3 weeks), but not on horse serum agar; no clearing.

No proteolytic activities.

An inhabitant of the upper respiratory tract, throat and genital tract of dogs; although occasionally isolated from pneumonic lung lesions its etiological significance in respiratory disease of dogs remains undetermined as yet.

The G + C content of the DNA is 29.2 moles % (T_m).

Type strain: PG 24; C 21; ATCC 23462; NCTC 10132; AMR-C 43.

Further Comments

Serologically distinct by the growth inhibition test and by direct and indirect fluorescent antibody techniques from other species of canine provenance, as well as from a great variety of other *Mycoplasma* and *Acholeplasma* species tested.

15. Mycoplasma cynos Rosendal 1973, 53.

cy′nos. Gr.n. *cyon* a dog; Gr. gen.n. *cynos* of a dog.

Pleomorphic, coccoid to coccobacillary.

No liquefaction of coagulated horse serum. Agar colonies produce zones of β type hemolysis by the overlay technique using guinea pig red blood cells, and adsorb guinea pig red blood cells.

Isolated from the conjunctivae, upper respiratory tract and genital mucosae of dogs. Though isolated in one case from the pneumonic lung of a dog with "kennel cough" its pathogenicity remains to be determined.

The G + C content of the DNA not determined.

Type strain: H831; ATCC 27544; NCTC 10142; AMRC-C44.

Further Comment

Differs antigenically by growth inhibition, metabolic inhibition, indirect hemagglutination and immunofluorescent antibody tests from other canine and glucose fermenting non-canine *Mycoplasma* species.

16. Mycoplasma hyorhinis Switzer 1955, 544. (*Asterococcus hyorhinis* (Switzer) Prévot 1961, 719: Filterable agent from nasal cavity of swine with infectious, atrophic rhinitis Switzer 1953, 45.)

hy.o.rhi′nis. Gr. n. *hys, hyos* a swine; Gr. n. *rhis, rhinis* nose; M.L. gen.n. *hyorhinis* of a hog's nose.

Very short filaments with occasional branching.

Acid produced from glucose through oxidative degradation, although reported by some authors to be glucose negative. No proteolytic activity. Does not adsorb guinea pig red blood cells or HeLa cells. Satellite growth around staphylococci colonies observed with some strains; the growth factor supplied by the bacteria not defined.

Resistant to erythromycin in the concentration of at least 512 μg/ml.

A common inhabitant of the nasal cavity of swine, occurring in some herds with a frequency of 50–60%; not an etiological agent of atrophic rhinitis. Often recovered from lung lesions of pigs suffering from enzootic pneumonia. Although usually regarded mainly as a secondary invader in this condition recent reports of successful experimental production of pneumonia with certain strains provide suggestive evidence that it may, nevertheless, be etiologically incriminated in swine pneumonia. M. hyorhinis septicemia associated with arthritis-polyserositis has been reported in cases of enzootic pneumonia of young pigs under field conditions. Reported to produce irregular mortality and occasionally pericardial and peritoneal lesions in chick embryos. Most strains produce a well defined cytopathic effect in primary calf, swine and monkey kidney cell cultures as well as in fetal human lung diploid cell lines. Observations of apparently intracellular growth, particularly in monkey kidney cells, have been reported.

The G + C content of the DNA is about 27–28 moles % (T_m).

Type strain: BTS-7; ATCC 17981; NCTC 10130; AMRC-C 31; NIH M-718-001-084 (Edward and Freundt, 1973).

Further Comments

Data reported by several authors suggest the existence of some serological heterogeneity.

17A. Mycoplasma hyopneumoniae Maré and Switzer 1965, 841. (Possibly synonymous with *M. suipneumoniae* Goodwin *et al.* 1965, 1249; see species 17B.)

hy.o.pneu.mo'ni.ae. Gr. n. *hys, hyos* a hog; Gr. n. *pneumonia* inflammation of the lungs, pneumonia; M.L. gen.n. *hyopneumoniae* of hog pneumonia.

Coccoid to short filamentous.

Colonies very minute, with indistinct central elevation.

No production of film on swine serum agar. Does not hemolyze swine, horse, cattle or chicken red blood cells in agar.

Very fastidious. Although nutritional requirements are poorly defined, growth occurs in Hanks' balanced salt solution (Hanks and Wallace, 1949),

enriched with 20% heat-inactivated pig serum from pneumonia-free herds, 0.5% lactalbumin hydrolysate and 0.01% Difco yeast extract, and is further enhanced by 0.5% swine gastric mucin and by the satelliting effect of *Micrococcus* sp. colonies.

Apparently a strict aerobe.

Etiological agent of enzootic pneumonia of pigs; the natural disease can be reproduced experimentally.

Type strain: 11 (monotype); VPP 11; ATCC 25617; NCTC 10127.

Further Comments

Substantial evidence suggests the identity of this species with *M. suipneumoniae* Goodwin *et al.* 1965, 1249. The specific epithet *hyopneumoniae* has priority but its validity has been challenged. The type strain of *M. suipneumoniae* (strain J) was shown to be serologically indistinguishable from a culture of *M. hyopneumoniae*, but, since the latter was an uncloned broth culture, doubts have been raised as to whether it is the same as the original culture described by Maré and Switzer, or the same as the culture deposited with the ATCC (Minutes of Subcommittee on Taxonomy of Mycoplasmatales, 1971). The Subcommittee has arranged for comparative studies of the type strains and should they prove identical will request an Opinion on the valid name of the species from the Judicial Commission of the ICSB.

17B. Mycoplasma suipneumoniae Goodwin, Pomeroy and Whittlestone 1965, 1249.

su.i.pneu.mo'ni.ae. L. n. *sus, suis* the hog; Gr. n. *pneumonia* pneumonia; M.L. gen.n. *suipneumoniae* of hog pneumonia.

Coccoid to short filamentous.

Colonies on solid media attain a maximum diameter of about 400 μm at 7–10 days; they are convex, usually devoid of a central nipple, and show very little growth into the medium.

Biochemical characters other than glucose fermentation have not yet been determined.

The organism is very fastidious and difficult to grow. Growth requirements are similar to *M. hyopneumoniae*.

Apparently a strict aerobe.

Inhibition of growth by 0.01 mg/ml of penicillin G has been reported.

Etiological agent of enzootic pneumonia of pigs. Pneumonia of a similar type has been reproduced experimentally by intra-nasal inoculation of both conventional and gnotobiotic pigs.

Genetic characters not yet determined.

Type strain: J (monotype); ATCC 25934; NCTC 10110.

Further Comments

Serologically distinct by growth inhibition and metabolic inhibition tests from a great variety of mycoplasmas, including the porcine organisms *M. hyorhinis* and *Acholeplasma granularum*. On the other hand, it has been found serologically indistinguishable from *M. hyopneumoniae* (see "Further Comments" on *M. hyopneumoniae*).

18. **Mycoplasma flocculare** Friis 1972, 286.

floc′cu.lus. M.L. dim.n. *flocculus* a small floc or tuft of wool; apparently treated as the neuter *flocculare* of an adj. *floccularis* like a small floc of wool.

Pleomorphic, branched filaments.

Agar colonies coarsely granular, devoid of central nipple. Aggregates of cells produced during growth in broth, appearing as small floccular elements upon gentle shaking of the culture.

Does not produce film and spots on serum agar. Not tested for proteolytic activities, hemolysis, hemadsorption or hemagglutination.

Sensitive to 0.03 mg/ml of penicillin G but resistant to 1.0 mg/ml of methicillin.

Isolated from the lungs of pigs with enzootic or catarrhal pneumonia. Moderate histological lesions typical of enzootic pneumonia, but no overt disease, reproduced experimentally by aerosols in piglets.

The G + C contents of the DNA not determined.

Type strain: Ms42; ATCC 27399; NCTC 10143.

Further Comment

Antigenically distinct by growth inhibition, metabolic inhibition, immunofluorescent antibody and complement fixation tests from other porcine and glucose fermenting non-porcine *Mycoplasma* and *Acholeplasma* species. Shares an antigen with *M. suipneumoniae* as demonstrated by the double immunodiffusion technique.

19. **Mycoplasma pneumoniae** Somerson, Taylor-Robinson and Chanock 1963, 122. (*Schizoplasma pneumoniae* (Somerson *et al.*) Furness, Pipes and McMurtrey 1968, 12: filterable agent of primary atypical pneumonia, Eaton *et al.* 1944, 649.)

pneu.mo′ni.ae. Gr. n. *pneumonia* pneumonia; M.L. gen.n. *pneumoniae* of pneumonia.

Filaments short (2–5 μm). Motility through gliding motion of cells has been observed. The mode of reproduction, whether through release of elementary bodies from filaments, by binary fission, or possibly by both methods, is a point of controversy that has attracted particular attention in the case of this species.

Freshly isolated colonies on solid medium frequently lack the light peripheral zone, appearing as circular dome-shaped, granular structures measuring 30–100 μm in diameter. Grows more slowly than most other mycoplasmas, the colonies being demonstrable 5–10 days or more after inoculation. On adaptation to artificial media through repeated subcultivation colonies with the typical biphasic "fried egg" appearance predominate, and growth is more rapid.

No film or spots on horse serum agar; a film, but no spots or clearing, produced on egg yolk medium. No proteolytic activity. Agglutinates guinea pig erythrocytes. Colonies on solid medium adsorb red blood cells and tracheal epithelial cells from monkey, rat, guinea pig and chicken, as well as HeLa cells and cells from calf kidney and chick embryo tissue cultures and spermatozoa from man and bull. Adsorption occurs most readily at 37 C, less rapidly and less extensively at 22 C; it is specifically inhibited by antiserum. The erythrocyte and epithelial cell receptors for *M. pneumoniae* are destroyed by neuraminidase and pretreatment of the mycoplasmas with neuraminic acid or neuraminelactose prevents adsorption. A potent producer of hemolysin, identified as hydrogen peroxide. The affinity of *M. pneumoniae* for respiratory epithelium is believed to play a role in its virulence in the way that the firm binding of the organisms to the epithelial cell sheet may tend to counteract rapid destruction of the mycoplasma peroxide by peroxidase present in extracellular fluids. Fresh yeast extract is required for growth on primary isolation and in early subcultures. The replacement of crude horse serum by agamma horse serum in the growth medium has been reported to support more consistently the growth of this organism.

Highly sensitive to erythromycin, minimum inhibitory concentrations ranging from 0.001–1.0 μg/ml.

Temperature range about 30–39 C.

The etiological agent of cold-hemagglutinin associated primary atypical pneumonia of man. Infection may result in all degrees of respiratory involvement from inapparent infection to pneumonia, the incidence of clinically apparent lower respiratory tract involvement varying in different reports from 3–10% and up to about 80% of infections. Bullous myringitis, first observed in experimentally infected volunteers, occurs occasionally in naturally acquired infections. Involvement of the central and peripheral nervous system has been reported with increasing frequency during recent years. Produces distinct cytopathology, including loss of organized ciliary activity, in the respiratory epithelium of hamster trachea in organ culture, an effect that apparently distinguishes this species from other mycoplasmas of human provenance.

Reports on the G + C per cent of the DNA

range from 38.6–40.8 moles % (T_m and buoyant density). No cross-reaction between this and other species of human source by nucleic acid homology tests.

Type strain: FH; ATCC 15531; NCTC 10119; NIH M-710-001-084 (Edward and Freundt, 1973).

Further Comments

Antigenically quite distinct from other species of human source, no precipitation lines being demonstrable between this and other human mycoplasmas in double immunodiffusion tests.

On the basis of evidence obtained for reproduction by binary fission the suggestion has been made by Furness *et al.* (1968) that this species be recognized as the type species of a proposed new genus, *Schizoplasma*, of the order Mycoplasmatales. Since the mode of reproduction of the Mycoplasmatales is still controversial and since the capability of reproducing by binary fission may well be shared by all mycoplasmas there is no justification as yet for the establishment of a separate genus based on this criterion, and hence the proposal of Furness *et al.* has not received general approval; (for further discussion of this item, see Edward and Freundt, 1969).

20. Mycoplasma bovigenitalium Freundt 1955, 73. (*Pleuropneumonia bovigenitalis* Tulasne and Brisou 1955, 238; *Borrelomyces bovigenitalium* (Freundt) Freundt 1955, 74; *Asterococcus bovigenitalis* (Freundt) Prévot 1961, 718: P strains of the bovine genital tract, Edward 1950, 4.)

bo.vi.ge.ni.ta'lium. L. n. *bos*, *bovis* the ox; M.L. pl.n. *genitalia* the genitals; M.L. pl.gen.n. *bovigenitalium* of bovine genitalia.

Filaments short, 2–5 μm.

Apparently does not metabolize carbohydrates; the report of production of acid on repeated subcultures in glucose-containing medium, ascribed to adaptive enzymes, has not been confirmed. Agar colonies adsorb various species of red blood cells and bovine, but not human, spermatozoa. Hemolysin is identified as hydrogen peroxide. Growth on primary isolation markedly enhanced by DNA (0.002%), while laboratory strains grow equally well on conventional media.

Highly resistant to erythromycin, not being inhibited by 512 μg/ml. More sensitive to lysis when suspended in low tonicity media than other species tested of this genus.

A common inhabitant of the bovine lower genital tracts, both in females and males. Although frequently associated with vaginal disorders, including granular vulvovaginitis, that may or may not lower fertility, presently available evidence for the etiological significance of this species in genital tract disease of heifers is considered inconclusive.

On the other hand, it has occasionally been demonstrated as the cause of outbreaks of mastitis. The mastitis, which can be reproduced experimentally, is characterized in the acute stage by extensive accumulations of eosinophilic leucocytes in the interstitia and alveoli of the mammary gland. Frequently isolated from the semen and preputium of apparently healthy bulls. Eosinophilic acute seminal vesiculitis has been produced experimentally with a strain (K) isolated from a case of chronic vesiculitis. Generalized infection associated with low fever, mild diarrhoea and arthritic lesions characterized by eosinophilic granulomas of the joint capsules has been produced experimentally with the same strain on intravenous inoculation of calves. A marked cytopathogenic effect in calf, pig and monkey kidney cell cultures characterized by enlargement of the cells, the appearance of intracytoplasmic inclusions and partial destruction of the cell layer has been observed with at least one strain.

Reports on the G + C content of the DNA range from 28.0–32.0 moles % (T_m and buoyant density).

Type strain: PG 11; B2; ATCC 19852; NCTC 10122; AMRC-C 02; NIH M-731-001-084.

Further Comments

Antigenically this species appears to be relatively heterogeneous. In addition, partial serological crossings have been observed with other *Mycoplasma* species of bovine source, viz. *M. mycoides* subsp. *mycoides* and *M. bovirhinis*.

21. Mycoplasma agalactiae (Wroblewski) Freundt 1955, 73. (*Anulomyces agalaxiae* (*sic*) Wroblewski 1931, 111; *Borrelomyces agalactiae* (Wroblewski) Turner 1935, 25; *Capromyces agalactiae* (Wroblewski) Sabin 1941, 57; *Pleuropneumonia agalactiae* (Wroblewski) Tulasne and Brisou 1955, 238; *Asterococcus agalactiae* (Wroblewski) Prévot 1961, 718: Le microbe de l'agalaxie contagieuse, Bridré and Donatien 1923, 841.)

a.ga.lac'ti.ae. Gr. n. *agalactia* want of milk, agalactia: M.L. gen.n. *agalactiae* of agalactia.

Filaments very short to moderately long.

Colonies are relatively small with a well defined center.

No proteolytic activities. Agar colonies adsorb red blood cells from guinea pigs and cattle, but not HeLa cells or human spermatozoa. Highly resistant to erythromycin, the minimum inhibitory concentration being reported as 200–512 μg/ml or more. Two subspecies are recognized.

21a. Mycoplasma agalactiae subsp. *agalactiae* Freundt 1955, 73.

Short to moderately long filaments.

The etiological agent of contagious agalactia of sheep and goats. Goats are more susceptible than sheep to experimental infection by subcutaneous inoculation; the inflammatory lesions are localized in the udders of females and, in 10–20% of the cases, in the joints.

The G + C content of the DNA is about 33.5–34.0 moles % (T_m and buoyant density).

Proposed neotype strain: PG 2; NCTC 10123; AMRC-C 21 (Edward and Freundt, 1973).

21b. *Mycoplasma agalactiae* subsp. *bovis* Hale, Helmboldt, Plastridge and Stula 1962, 591. (*Mycoplasma bovimastitidis* Jain, Jasper and Dellinger 1967, 409: Bovine serotype 5, Leach 1967, 312.)

bo'vis. L. masc.n. *bos* the ox; L. gen.n. *bovis* of the ox.

Filaments very short.

Production of film and spots on horse serum and egg yolk media is variable; the type strain is negative in this respect. Not tested for proteolytic activities.

Growth optimum about 37 C; no growth at 28 C.

The etiological agent of a type of mastitis in cattle that is partially different, in its clinical and pathological characteristics, from the mastitis caused by *M. bovigenitalium*. The mastitis can be reproduced experimentally by inoculation into the udder; endometritis and salpingo-oophoritis, sometimes associated with impaired fertility, has been produced by inoculation of cultures or infectious semen into the uterus of heifers. The histopathology is characterized, i.a. by an eosinophilic cellular response as also found with *M. bovigenitalium*. Septic arthritis may occur as a complication of field cases of mastitis and of experimental infection. Some strains are cytopathogenic for bovine embryo tissue cultures.

The G + C content of the DNA is 33.0 moles % (T_m).

Type strain: Donetta; PG 45; ATCC 25523; NCTC 10131; AMRC-C 04 (Edward and Freundt, 1973).

Further Comments

Serologically the type strains of the two subspecies of this species are clearly distinct by the growth inhibition test, while cross-reactions to very low titers can be obtained with the metabolic inhibition test. More extensive crossings, although partly in the form of one-way crossings, are demonstrable by the complement fixation and indirect hemagglutination tests. Although a close similarity has been demonstrated in the electrophoretic pattern of cell proteins of the two subspecies, whether they should be regarded as varieties of one species, as originally proposed by Hale *et al.* (1962), or as two separate species as suggested by Jain *et al.* (1967), is still a question.

22. **Mycoplasma alkalescens** (*sic*) Leach 1973, 149. (Bovine serological group 8, Leach 1967, 312.)

al.ca.les'cens. M.L. v. *alcalesco* make alkaline; M.L. part.adj. *alcalescens* alkaline making, referring to reaction produced in arginine-containing media.

Pleomorphic, spherical, ring-shaped and coccobacillary forms.

No liquefaction of coagulated horse serum. No production of film and spots on horse serum agar. Agar colonies produce zones of α type hemolysis by the overlay technique, using ox, sheep and guinea pig red blood cells; do not adsorb ox or guinea pig red blood cells.

Isolated from the nose of cattle and from commercial bovine serum. Pathogenicity not known.

The G + C content of the DNA is 25.9 moles % (T_m).

Type strain: D12; PG 51; NCTC 10135.

Further Comment

Differs antigenically by the metabolic inhibition test from other *Mycoplasma* and *Acholeplasma* species.

23. **Mycoplasma gallinarum** Freundt 1955, 73. (*Borrelomyces gallinarum* Freundt 1955, 73; *Asterococcus gallinarum* (Freundt) Prévot 1961, 719: avian serotype B, Kleckner 1960, 277.)

gal.li.na'rum. L. fem.n. *gallina* a hen; L. fem.-gen.pl.n. *gallinarum* of hens.

Filaments very short, almost bacillary.

Produces film and spots on horse serum agar and egg yolk medium, together with clearing of the latter medium. No proteolytic activity. Does not adsorb or agglutinate guinea pig or chick red blood cells. Hemolysin identified as hydrogen peroxide.

Moderately to highly resistant to erythromycin, minimal inhibitory concentration ranging from 40–1000 μg/ml and above.

A common inhabitant of the upper respiratory tract of fowl. Not pathogenic.

The G + C content of the DNA is about 27.0–28.0 moles % (T_m and buoyant density). Nucleic acid homology tests have shown a very low level cross-reaction with *M. gallisepticum* (1.3% relative homology).

Type strain: PG 16; "Fowl" (monotype); ATCC 19708; NCTC 10120; AMRC-C 71; NIH M-721-001-084.

Further Comments

The electrophoretic pattern of the cell proteins is clearly distinct from those of other avian species including *M. gallisepticum*, *M. iners*, *M. meleagridis* and *M. synoviae*.

24. **Mycoplasma iners** Edward and Kanarek 1960, 699. (Avian serotype O, Adler *et al.* 1958, 444; avian serotype G, Kleckner 1960, 277.)

in'ers. L. adj. *iners* inactive, inert.

Morphology poorly defined.

Does not reduce tetrazolium, methylene blue or tellurite, aerobically or anaerobically. Film and spots produced on egg yolk medium, but not on 20% horse serum agar; clearing of egg yolk medium. No proteolytic activity. Does not adsorb guinea pig or chicken red blood cells.

An apparently parasitic inhabitant of the respiratory tract of fowl and turkeys. Although not known to be associated with natural disease, experimental production of joint lesions in chick embryos has been reported.

The G + C content of the DNA is about 29.0–29.5 moles % (T_m and buoyant density).

Type strain: PG 30; M; ATTC 19705; AMRC-C 71; NIH M-730-001-084 (Edward and Freundt, 1973).

Further Comments

The electrophoretic pattern of the cell proteins is clearly distinct from those of other avian species including *M. gallinarum*, *M. gallisepticum*, *M. meleagridis*, and *M. synoviae*.

25. **Mycoplasma meleagridis** Yamamoto, Bigland and Ortmayer 1965, 47. (Avian serotype N, Adler *et al.* 1958, 444; avian serotype H, Kleckner 1960, 277.)

me.le.a'gri.dis. L. fem.n. *meleagris* a turkey; L. gen.n. *meleagridis* of a turkey.

Morphology poorly defined.

Does not produce film or spots on horse serum and only occasionally on egg yolk media. Not tested for proteolytic activity. Most strains do not adsorb or agglutinate turkey, chicken or guinea pig red blood cells.

An inhabitant of the respiratory and urogenital tracts of turkeys. Produces air sacculitis in turkeys under field conditions and experimentally. Apparently not pathogenic to chicken.

The G + C content is about 28.0–28.5 moles % (buoyant density).

Type strain: 17529; ATCC 25294; AMRC-C 74; NIH M-736-001-084 (Edward and Freundt, 1973).

Further Comments

The electrophoretic pattern of the cell proteins is clearly distinct from those of other avian species including *M. gallinarum*, *M. gallisepticum*, *M. iners* and *M. synoviae*.

26. **Mycoplasma arthritidis** (Sabin) Freundt 1955, 73. (*Murimyces arthritidis* Sabin 1941, 57; *Pleuropneumonia arthritidis muris* Tulasne and Brisou 1955, 238; *Asterococcus arthritidis* (Sabin)

Prévot 1961, 720; *Schizoplasma arthritidis* (Sabin) Furness 1970, 156: L₄, Klieneberger 1938, 458.)

ar.thri'ti.dis. Gr. n. *arthritis* gout, arthritis; M.L. gen.n. *arthritidis* of arthritis.

Filaments vary from short (2–5 μm) to moderately long (10–30 μm).

Most strains cause liquefaction of gelatin, but do not digest casein or coagulated horse serum. No hemadsorption reported. Hemolysin identified as hydrogen peroxide.

A relatively common pathogen of rats causing purulent polyarthritis, sometimes occurring as localized outbreaks in stocks of laboratory rats. Repeatedly isolated also from submandibular abscesses, middle ear infections, ocular lesions, the nasal mucosa in purulent rhinitis, lung lesions and from a transplantable rat sarcoma. Recovery from normal rats not reported. Subcutaneous inoculation of the organisms suspended in agar emulsions produces localized abscesses and septicemia in rats. Localized arthritis may result from inoculation into the footpads, while widespread infection characterized by suppurative polyarthritis, conjunctivitis and urethritis can be produced on intravenous inoculation of particularly virulent strains. Flaccid paralysis related to inflammation of the interspinal articulations occurs in a high percentage of the infected animals. May or may not be pathogenic to mice on experimental inoculation. Not pathogenic for monkeys, rabbits and guinea pigs.

The G + C content of the DNA ranges from 30.0–33.7 moles % (T_m and buoyant density).

Proposed neotype: PG 6; Preston; ATCC 19611; AMRC-C 60; NIH M-715-011-084 (Edward and Freundt, 1973).

Further Comments

Although a fairly homogeneous group antigenically, variant strains differing in some ways serologically have been described (Lemcke, 1961; 1964). The Campo (PG 27) strain, and a small group of related strains, previously classified as *M. hominis* type 2 were later reclassified, on the basis of serological observations (Lemcke, 1964), as *M. arthritidis* (Edward and Freundt, 1965). The justification of this reclassification was subsequently confirmed by nucleic acid homology studies, by enzyme analysis and by an examination of the electrophoretic patterns of the cell proteins (Razin and Rottem, 1967) of the strains involved.

27. **Mycoplasma gateae** Cole, Golightly and Ward 1967, 1456.

ga'te.ae. Probably from Spanish gato = cat.

Morphology poorly defined.

Colonies are vacuolated and without a well defined central spot on primary isolation and in early subcultures. On repeated subculturing colonies achieve the typical fried egg appearance.

No proteolytic activity. Does not agglutinate sheep, guinea pig, rat, monkey or duck red blood cells; not tested for hemadsorption.

Apparently a common parasitic inhabitant of the upper respiratory tract, the conjunctivae and the genital mucosae of cats.

The G + C content of the DNA is 28.5 moles % (T_m).

Type strain: CS; ATCC 23392; AMRC-C 51 (Edward and Freundt, 1973).

28. **Mycoplasma maculosum** Edward 1955, 90. (*Asterococcus maculosus* (Edward) Prévot 1961, 721: γ strains of dogs, Edward and Fitzgerald 1951, 566.)

ma.cu.lo'sum. L. neut.adj. *maculosum* spotted.

Filaments short (2–5 μm) with occasional branching.

Produces film and spots on horse serum agar and egg yolk medium, together with clearing of the latter. No proteolytic activity. Does not adsorb guinea pig red blood cells.

Parasitic inhabitant of the genito-urinary and upper respiratory tracts and the throat of dogs. The significance of occasional isolations from lung tissues of dogs remains unknown.

The G + C content of the DNA ranges from about 26.5–29.5 moles % (T_m and buoyant density).

Type strain: PG 15; C27; ATCC 19327; AMRC-C 42; NIH M-727-001-084.

29. **Mycoplasma spumans** Edward 1955, 90. (*Asterococcus spumans* (Edward) Prévot 1961, 721: α strains of dogs, Edward and Fitzgerald 1951, 566.)

spu'mans. L. part.adj. *spumans* foaming.

Filaments short (2–5 μm). Gram-negative.

Colonies on horse serum agar are characterized, upon primary isolation and in early subcultures, by a coarsely reticulated and vacuolated appearance, the central spot tending to be hidden by the coarse markings. These colonial characteristics are lost on repeated subculture.

No proteolytic activity. Colonies on solid medium adsorb guinea pig red blood cells.

A parasitic inhabitant of the genito-urinary and upper respiratory tracts and the throat of dogs. Although occasionally isolated from lung tissues its possible etiological role in respiratory and other disease of dogs remains to be determined.

The G + C content of the DNA is about 28.5–29.0 moles % (T_m and buoyant density).

Type strain: PG 13; C48; ATCC 19526; AMRC-C 40; NIH M-726-001-084.

30. **Mycoplasma hyosynoviae** Ross and Karmon 1970, 710. (*Mycoplasma suidaniae* Friis 1970,

489: unreported serotype of porcine mycoplasma Roberts and Gois 1970, 214.)

hyo.syn'ov.i.ae. Gr. n. *hys*, *hyos* a swine; M.L. n. *synovia* fluid in joints; M.L. gen.n. *hyosynoviae* of joint fluid of swine.

Cells filamentous to coccoid.

Film and spots produced on turkey serum medium within 3–4 days and on horse serum medium within 14 days of incubation. Does not liquefy coagulated horse serum. Growth is significantly stimulated by the addition to the growth medium of 0.5% of mucin.

Isolated from synovial fluid, nasal secretions and tonsils of swine. Produces an acute synovitis and arthritis in swine.

Genetic characters not determined.

Type strain: S16; ATCC 25591; AMRC-C 104 (Edward and Freundt, 1973).

Further Comments

Serologically distinct by the disc growth inhibition test from a total of 14 other arginine-metabolizing *Mycoplasma* species tested, and from A. *laidlawii* and A. *granularum*. The serological non-relatedness to the porcine organisms, M. *hyorhinis* and A. *granularum* is further confirmed by metabolic inhibition and immunodiffusion tests. Also, the electrophoretic pattern of the cell proteins is dissimilar to those of the arginine-metabolizing *Mycoplasma* species and to A. *laidlawii* and A. *granularum*.

31. **Mycoplasma arginini** Barile, Del Giudice, Carski, Gibbs and Morris 1968, 489. (*Mycoplasma leonis* Heyward, Sabry and Dowdle 1969, 621.)

ar.gin.i'ni. Eng. n. *arginine* an amino acid; M.L. gen.n. *arginini* of arginine, referring to its hydrolysis.

Morphology poorly defined.

Not tested for proteolytic activities. Colonies on solid medium do not adsorb chicken red blood cells.

Mammalian parasite with an apparently wide host range having been isolated from a variety of different sources such as cattle, sheep, goat, chamois and mouse tissues, as well as from cell cultures derived from human, chimpanzee and dog tissues.

The G + C content of the DNA is 28.6 moles % (buoyant density).

Type strain: G 230; ATCC 23838; NCTC 10129; AMRC-C 22; NIH M-732-001-084 (Edward and Freundt, 1973).

Further Comments

Mycoplasma leonis Heyward *et al.* 1969, 615, the description of which was based on a single strain (LL) isolated from the brain and lung tissues of a lion, is a later synonym of this species as shown on the basis of biochemical properties, serology and electrophoretic pattern of cell proteins (Tully et al., 1972).

32. Mycoplasma orale Taylor-Robinson, Canchola, Fox and Chanock 1964, 141. (Mycoplasma pharyngis Clyde 1964, 579; Schizoplasma orale (Taylor-Robinson et al.) Furness 1968, 441.)

o.ra′le. L. neut.adj. orale pertaining to the mouth.

This species is presently subdivided into three groups of strains which differ serologically, and in some of their cultural and biochemical properties.

32a. Mycoplasma orale type 1 Taylor-Robinson et al. 1964, 135. (A new oral Mycoplasma, Herderscheé et al. 1963, 157).

Moderately long filaments (8–10 μm). Colonies on solid medium containing 20% of horse serum are usually characterized by a particularly well developed central dark area, surrounded by a narrow peripheral zone.

Does not reduce tetrazolium, methylene blue or tellurite, aerobically or anaerobically. No proteolytic activity. Colonies on solid medium adsorb chicken red blood cells, but not red blood cells from man, monkey, rat or guinea pig, and not tracheal epithelial cells from monkey, rat or chicken. Receptor sites on chicken red blood cells not destroyed by neuraminidase. Hemolysin identified as hydrogen peroxide. Yeast extract required for the growth of at least fresh isolates.

Facultative anaerobe, although growth is better anaerobically on primary isolation and in early subcultures.

Moderately to highly resistant to erythromycin, minimum inhibitory concentrations ranging from 25–512 μg/ml and above.

A common parasitic inhabitant of the human oropharynx and a rarer inhabitant of the oropharynx of non-human primates.

Reports on the G + C content of the DNA range from about 24–28 moles % (T_m and buoyant density). Nucleic acid homology tests show low level cross-reactions with M. salivarium (5%) and with M. hominis (2–3% relative homology).

Type strain: CH19299; ATCC 23714; NCTC 10112; AMRC-C 1006; NIH M-714-001-084 (Edward and Freundt, 1973).

Further Comments

This strain was found to be identical with Mycoplasma pharyngis Clyde 1964 (Kim et al., 1966). However this name was not validly published, and the Subcommittee on the Taxonomy of Mycoplasmatales recently recommended (Minutes, 1971) that the specific epithet orale be accepted as the first validly published name for this species.

32b. Mycoplasma orale type 2 Taylor-Robinson et al. 1965, 190.

Morphology poorly defined. Colonies on horse serum agar are distinguished from type 1 colonies by lacking a central area on primary isolation and possessing only a small central nipple in subcultures.

No proteolytic activity. Does not adsorb chicken red blood cells. Hemolysin identified as hydrogen peroxide. Yeast extract not required for growth.

Facultative anaerobe, although the growth of fresh isolates is better anaerobically. Resistance to erythromycin as for type 1.

An infrequent parasitic inhabitant of the human oropharynx but a common inhabitant of the oropharynx of non-human primates; occasionally recovered also from the vagina of the latter.

The G + C content of the DNA as for type 1. Nucleic acid homology tests have shown low level cross-reactions with M. orale type 1 (8%), M. hominis (8.4%), and M. salivarium (10% relative homology).

Reference strain: CH 20247 (cotype); ATCC 23636; AMRC-C 1007; NIH M-714-011-084.

32c. Mycoplasma orale type 3 Fox, Purcell and Chanock 1969, 42.

Morphology poorly defined. Gram-negative. Colonies on solid medium reported to grow more superficially and more loosely attached to the agar surface than other mycoplasmas.

Not tested for most biochemical activities. Resembles type 1 in the capability of adsorbing chicken red blood cells but not other cells tested, and in the growth requirement for yeast extract. Hemolysin identified as hydrogen peroxide. L-cysteine appears to stimulate growth significantly.

Facultative anaerobe, although preferring anaerobic conditions.

An infrequent member of the normal mycoplasmal flora of the human oropharynx.

The G + C content of the DNA of this type has not been determined.

Type strain: DC-333; ATCC 25293; AMRC-C 1008; NIH M-714-021-084.

Further Comments

The three types of Mycoplasma orale are serologically distinct by the growth inhibition tests; also, nucleic acid homology data indicate that types 1 and 2 at least are quite distinct from each other. With the complement fixation test a marked cross-reaction is demonstrable between types 1 and 2, and a one-way cross-relationship between types 1 and 3. Moreover, each of the three types of this species is serologically more or less closely related to *Mycoplasma salivarium*, while distinct from that species in nucleic acid homology tests. In fact, the authors who described the three "types" of *M. orale* regarded them as three distinct species rather than as subspecies of one species. The nomenclature proposed by them was guided, however, by the view that a numerical system of nomenclature is preferable to a binomial nomenclature. On the basis of the above data, the Subcommitte on the Taxonomy of Mycoplasmatales recently recommended (Minutes, 1971) that *M. orale* types 1 and 2 be recognized as distinct species and, in consequence, that a binomial name be proposed for *M. orale* type 2. The Subcommittee also agreed that a recommendation on the taxonomic status of *M. orale* type 3 should await the results of further comparative studies, based on nucleic acid homology and gel electrophoresis techniques.

33. Mycoplasma salivarium Edward 1955, 90.

(*Asterococcus salivarius* (Edward) Prévot 1961, 722; *Schizoplasma salivarium* (Edward) Furness 1970, 156: Human type 4 strains, Nicol and Edward 1953, 148.)

sa.li.va'ri.um. L. neut.adj. *salivarium* salivary, slimy; intended to mean of saliva.

Filaments very short, almost bacillary (0.6–1.0 μm) with only occasional branching.

Reduces tetrazolium weakly and inconsistently anaerobically, not aerobically. Production of film and spots on horse serum and egg yolk media, together with clearing of the latter, is variable. No proteolytic activity. Confluent colonies on solid medium adsorb HeLa cells and chick embryo tissue culture cells, but not red blood cells from man, monkey, rat, guinea pig, tracheal epithelial cells from monkey, rat or chicken and not spermatozoa from man or bull. The receptor sites on HeLa and chick embryo cells are not destroyed by neuraminidase. Hemolysin identified as hydrogen peroxide.

Facultative anaerobe, although at least on primary isolation and in early subcultures growth is definitely better under anaerobic conditions.

Highly resistant to erythromycin, not being inhibited by 512 μg/ml.

A common parasitic inhabitant of the oropharynx of man and non-human primates. Not pathogenic as far as known.

Reports on the G + C content of the DNA range from 27.0–31.5 moles % (T_m and buoyant density). Type strain: PG 20; H110; ATCC 23064; NCTC 10113; AMRC-C 1002; NIH M-712-001-084.

34. Mycoplasma hominis (Freundt) Edward 1955, 90.

(*Micromyces hominis* group I, Freundt 1953, 471; *Asterococcus hominis* (Freundt) Prévot 1961, 720; also see Freundt 1954, 143; *Schizoplasma*

hominis (Freundt) Furness 1970, 156: Human type 1 strains, Nicol and Edward 1953, 146.)

ho'mi.nis. L. masc.n. *homo* man; L. gen.n. *hominis* of man.

Filaments usually short (2–5 μm) although occasionally up to 30 μm in length.

Does not reduce tetrazolium, methylene blue or tellurite, aerobically or anaerobically. No proteolytic activity. Colonies on solid medium adsorb HeLa and chick embryo tissue culture cells, but not erythrocytes or tracheal epithelial cells from monkey, rat, guinea pig and chicken, and not spermatozoa from man or bull. The receptors on HeLa and chick cells are not destroyed by neuraminidase. Hemolysin defined as hydrogen peroxide.

Highly resistant to erythromycin, minimum inhibitory concentrations ranging from about 500–1000 μg/ml and above.

A very common parasitic inhabitant of the mucosae of the lower genito-urinary tract of man and non-human primates; more rarely encountered in the oropharynx. Potentially pathogenic as suggested by its recovery, with varying frequency, from the blood, uterine tube or ovary of patients with postpartum fever or pelvic inflammatory disease, and from the tissues of aborted fetuses. Particularly strong evidence for the etiological role of this species in inflammatory conditions of the female reproductive system was recently provided by its isolation in pure culture, in a high proportion of cases of acute salpingitis, from samples collected from the uterine tubes by laparoscopy. Serological studies further confirmed an etiological relationship. Further studies are needed to support available suggestive evidence for a possible etiological role of this species in acute exudative pharyngitis and respiratory disease of man.

Reports on the G + C content of the DNA range from 27.3–29.2 moles % (T_m and buoyant density). Determination of genetic relatedness by the nitrocellulose membrane filter DNA-RNA hybridization method of Nygaard and Hall (1963) showed a 39% or greater relatedness value among different strains of *M. hominis*, and 3% or less relatedness between this species and *M. fermentans*, *M. pneumoniae*, *M. orale* type 1 and *Acholeplasma laidlawii*.

Type strain: PG 21; H 50; ATCC 23114; NCTC 10111; AMRC-C 1003; NIH M-711-001-084 (Edward and Freundt, 1973).

Further Comments

Originally two serotypes, types 1 and 2, were recognized for this species. The Campo (PG 27) and other strains designated as "type 2" were later identified, on the basis of serology, nucleic acid homology, enzyme analysis and the electrophoretic patterns of cell proteins, as *M. arthritidis*. In consequence, *M. hominis* type 2 was withdrawn (Edward and Freundt, 1965; 1969).

The results of agglutination, growth inhibition and metabolic inhibition tests together with comparison of the electrophoretic patterns of cell proteins of a number of strains classified within this species suggest the existence of a fairly extensive antigenic heterogeneity sufficient enough to justify the recognition of different subspecies.

35. **Mycoplasma primatum** Del Giudice, Carski, Barile, Lemcke and Tully 1971, 442. (Navel strain, Ruiter and Wentholt 1955, 33.)

pri.mat'um. M.L. adj. *primas, primatis* chief, from which Primates, the highest order of mammals; M.L. pl.gen.n. *primatum* of primates.

Coccoid to short filamentous.

No digestion of coagulated horse serum. No hemolysis of sheep or guinea pig red blood cells by the agar overlay technique. Colonies on solid medium do not adsorb either sheep or guinea pig red blood cells.

A common parasitic inhabitant of the oral cavity and urogenital tract of monkeys (*Cercopithecus aethiops*). Although the first known strain (strain Navel) was isolated from an inflammatory skin lesion of a human umbilicus there are no other reports on the occurrence of this organism in man. Not pathogenic for man or non-human primates as far as known.

The G + C content of the DNA is 28.6 moles % (T_m).

Type strain: HRC292; ATCC 25948; AMRC-C 167.

Further Comments

Serologically distinct by growth inhibition, immunofluorescent antibody and complement fixation tests from other *Mycoplasma* and from *Acholeplasma* species.

36. **Mycoplasma fermentans** Edward 1955, 90. (*Asterococcus fermentans* (Edward) Prévot 1961, 722; *Schizoplasma fermentans* (Edward) Furness 1970, 156; *Micromyces hominis* group II, Freundt 1954, 143: G strains, Ruiter and Wentholt 1952, 332; Human type 3 strains, Nicol and Edward 1953, 147.)

fer.men'tans. L. part.adj. *fermentans* fermenting.

Short (1.5 µm) to moderately long (10–40 µm) filaments.

The type strain and most other strains tested are phosphatase negative. Produces film and spots on horse serum agar and egg yolk medium, clearing the latter. No proteolytic activity. No hemad-

sorption. Hemolysin identified as hydrogen peroxide.

While on primary isolation and in early subcultures growth is definitely enhanced by anaerobic conditions, it is equally good aerobically on repeated subculturing.

Moderately to highly resistant to erythromycin, minimum inhibitory concentrations ranging from 10–512 mg/ml and above.

A relatively rare parasitic inhabitant of the mucosa of the human male and female genital tracts and the oropharynx. Although on one occasion isolated in pure culture from the uterine tube in a case of subacute salpingitis, and on one occasion also from the tissues of an aborted fetus, the possible pathogenic significance of this species in gynecological disease requires further study. Evidence for the frequent etiological implication of the species in rheumatoid arthritis, based on isolation and immunochemical observations, was recently reported.

The G + C content of the DNA is about 27.5–28.5 moles % (T_m and buoyant density). No cross-reactivity with other species of human source by nucleic acid homology tests, except for a very low level reaction (1.7% relative homology), with *M. orale* type 1.

Type strain: PG 18; G (monotype); ATCC 19989; NCTC 10117; AMRC-C 1001; NIH M-713-001-084.

Species incertae sedis

A. Well defined groups which have not yet been assigned a valid name.

a. "T-mycoplasmas"

Among the more or less well defined groups of mycoplasmas that have not yet been formally assigned to a separate taxon the "T-mycoplasmas" ("T-form colonies of pleuropneumonia-like organisms," Shepard, 1956; also see Shepard, 1969) deserve particular mention. The "T" in the vernacular name stands for "tiny" and is intended to denote the small size of the colonies, the diameter of which is generally within the range of 15–25 µm, although on some media they may attain a size of 175–200 µm. While usually coarsely granular and without, or with only barely discernible central nipples, the colonies may, on appropriate media, become smoother and develop the typical "fried egg" appearance. The morphology and ultrastructure are essentially as for other mycoplasmas (Plate 19.2, Figs. 3 and 4).

The most distinctive metabolic property that is shared by all T-mycoplasmas, and in which they differ from all other mycoplasmas, **is the**

ability to hydrolyze urea, with the accumulation of ammonia. Apparently urea is an essential nutritional factor in addition to cholesterol. Carbohydrates and arginine are not metabolized. Tetrazolium is not reduced. Phosphatase positive. Colonies on solid medium do not adsorb human, bovine or guinea pig red blood cells.

Growth occurs throughout the pH range of 5.0–10.0, although with a marked optimum at 5.5–6.5. **Highly susceptible to erythromycin,** minimum inhibitory concentrations reported as low as 0.8–3.0 μg/ml. Generally more susceptible to thallium acetate than most other mycoplasmas, being inhibited by a concentration of about 1:2000, although the resistance may increase on adaptation.

While the electrophoretic patterns of the cell proteins of a wide range of strains tested show a remarkable similarity, at least seven more or less distinct serotypes of human T-mycoplasmas have been demonstrated by metabolic inhibition (Ford, 1967; Purcell *et al.*, 1969) and more recently by growth inhibition, indirect hemagglutination and indirect immunofluorescence tests (Black, 1970). The genome size of eight serotypes was within the range 4.1–4.8 $\times$ 10^8 daltons and the G + C content of the DNA 27.7–28.5 moles % (T_m, Black *et al.*, 1972).

T-Mycoplasmas are common parasitic inhabitants of the mucous membranes of the urogenital tracts of man, cattle, dogs and monkeys; occasionally recovered also from the oropharynx of man, and from the oropharynx and conjunctivae of cattle. The recognition of so far one binomially named species for the T-mycoplasmas, with strain T-960 as the type culture, was recently recommended by the Subcommittee on the Taxonomy of Mycoplasmatales (Minutes, 1971).

B. Organisms which have no validly published names or for which cultures are not available.

a. *Mycoplasma lipophilum* Del Giudice and Carski 1968, 67. (*Mycoplasma lipophiliae* (*sic*) Del Giudice and Carski 1968, 67.)

li.po'phi.lum. Gr. n. *lipus* animal fat; Gr. adj. *philus* loving; M. L. neut.adj. *lipophilum* fat-loving.

Morphology not described. Colonies growing on media containing Oil Red O are colored intensely red.

Biochemical properties poorly defined. No acid from glucose. Hydrolyzes arginine, although poorly. Does not reduce tetrazolium or methylene blue aerobically. No adsorption to colonies by monkey or chicken red blood cells and only slight hemolysis of guinea pig cells. Growth is slow and fresh yeast extract is required for growth on solid medium.

Since this organism has been isolated only twice from the oral cavity of man its ecology and normal habitat remains to be determined.

Serologically distinct from other species of human provenance.

Type strain: MaBy (monotype); AMRC-C 1009; NIH M-735-001-084; ATCC 27104.

b. *Mycoplasma hyoarthrinosa* Moore, Redmond and Livingston 1965, 20. (Also see Moore *et al.*, 1966.)

Reported to produce very small colonies with central elevations on solid media and a granular deposit together with an alkaline reaction in fluid cultures. Mucin required for growth. Some antigens shared with *M. hyorhinis*, as shown by direct plate agglutination and indirect hemagglutination tests, but not with *A. granularum*. Initially isolated from arthritic swine joints by inoculation into swine kidney cell cultures where it produced a cytopathic effect. Is further reported to produce arthritis in pigs on intravenous and intragastric inoculation resulting in lesions characterized by an excess of serosanguineous fluid in the joint cavity, hypertrophy of the synovial villi, and a perivascular cuffing with lymphocytes and plasma cells.

Thus far, other workers consistently failed to obtain any growth, except for pseudocolonies, from material provided by Moore *et al.*, and a type strain is not available. In consequence, the validity of this species must be questioned until the above observations have been confirmed and the organism adequately described.

c. *Mycoplasma hyogenitalium* Moore, Redmond and Livingston 1965, 21. (also see Moore *et al.*, 1965 and 1966.)

Morphological, cultural, biochemical and serological properties poorly defined, although claimed to be serologically distinct from *M. hyorhinis*, *M. hyoarthrinosa* and *A. granularum*. Reported to be isolated from the mammary gland and the mucosal surface of the uterus from swine suffering from mastitis and vaginal discharge; the clinical disease was reproduced experimentally.

Since other workers consistently failed to obtain growth of the organism, and since no type strain is available the recognition of this species must await further studies.

The two taxa listed above are *nomina dubia* and possibly *nomina nuda*.

C. Recently described species which require further assessment.

d. *Mycoplasma caviae* Hill 1971, 112.

Morphology not defined.

Acid produced from glucose. Does not hydrolyze arginine. Tetrazolium not reduced aerobically. Not tested for lipolytic or proteolytic activities.

Produces hemolysis of sheep and guinea pig red blood cells.

Isolated from the nasopharynx, genital tract and, on one occasion, the brain of guinea pigs. The pathogenicity remains to be defined.

Genetic characters not determined.

Type strain: G122; ATCC 27108; NCTC 10126; AMRC-C 224 (Edward and Freundt, 1973).

Further Comments

Although serologically relatively distinct, slight cross-reactions were found by growth inhibition, metabolic inhibition and gel-diffusion tests between the guinea pig isolates and some other species, in particular *M. neurolyticum*. Further studies are required, therefore, to determine more precisely the taxonomic position of this organism.

FAMILY II. **ACHOLEPLASMATACEAE** EDWARD AND FREUNDT 1970, 1

(*Saprophytaceae* (*sic*) Sabin 1941, 59; *Sapromycetaceae* Sabin 1941, 334.)

A.cho.le.plas.ma.ta′ce.ae. M.L. neut.n. *Acholeplasma* type genus of the family; -*aceae* ending to denote a family; M.L. fem.pl.n. *Acholeplasmataceae* the *Acholeplasma* family.

Sterol not required for growth. Other characters as for the class and order.

Genus I. **Acholeplasma** *Edward and Freundt 1970, 1*

(*Sapromyces* Sabin 1941, 59.)

A.cho.le.plas′ma. Gr. *a* not; Gr. *chole-* combining form denoting relationship to the bile; Gr. neut.n. *plasma* something formed or molded, a form; M.L. neut.n. *Acholeplasma* name intended to indicate cholesterol, a constituent of bile, is not required.

Cells spherical, with a minimum diameter of about 125–220 nm, **and filamentous,** usually about 2–5 μm in length. Modes of reproduction as for genus *Mycoplasma*. **Cells bounded by a triple-layered membrane,** approximately 7 nm thick. Non-motile. Gram-negative.

Colonies on solid medium usually fried egg appearance, relatively large, up to 3 mm in diameter.

Chemoorganotrophs; metabolism fermentative, carbohydrates being the fermentable substrates. **Serum or cholesterol not required for growth.** Minimal nutritional requirements only partially defined.

Arginine and urea not hydrolyzed. Phosphatase activity weak or negative. **Film or spots not produced** on horse serum agar or egg yolk medium; no clearing.

Facultative anaerobes.

Temperature range about 20–40 C.

More susceptible to lysis by osmotic shock at 37 C than genus *Mycoplasma*.

All species resistant or only very slightly sensitive to 1.5% digitonin. Highly resistant to 5% sodium-polyanethol-sulfonate (Kunze, 1971; Andrews, unpublished; Freundt, unpublished).

Absolutely resistant to penicillins, not being inhibited by penicillin G, ampicillin, cloxacillin or methicillin in concentrations of at least 4000 μg/ml.

Apparently free-living saprophytes as well as mammalian and avian parasites, and possibly pathogens, with a fairly wide host range.

The G + C content of the DNA is about 30.0–33.0 moles % (T_m and buoyant density). The genome size is 1.0×10^9 daltons as determined by the DNA renaturation method of Wetmur and Davidson.

Other characters as for genus *Mycoplasma*.

Type species: *Acholeplasma laidlawii* (Sabin) Edward and Freundt 1970, 1.

Further Comments

In one species of this genus (*A. laidlawii*, types A and B) the reduced nicotinamide adenine dinucleotide (NADH₂) oxidase activity has been found to be localized in the cell membrane, as it usually is in bacteria, while in nine strains, representing six different *Mycoplasma* species, the enzyme was associated with the soluble cell fraction. Although other *Acholeplasma* species have not yet been examined with respect to NADH₂ localization the above observations suggest another essential difference between the genera *Mycoplasma* and *Acholeplasma*.

Description of the species of genus **Acholeplasma**

1. Acholeplasma laidlawii (Sabin) Edward and Freundt 1970, 1. (*Sapromyces laidlawi* AB Sabin 1941, 59; *Mycoplasma laidlawii* (Sabin) Freundt 1955, 73; *Asterococcus laidlawii* (Sabin) Prévot 1961, 722; *Mycoplasma inocuum* Adler, Shifrine and Ortmayer 1961, 239: Types A and B, Laidlaw and Elford 1936, 292.)

laid.law′i.i. M.L. gen.n. *laidlawii* of Laidlaw; named for P. Laidlaw, one of the microbiologists who first isolated this species.

Filaments usually relatively short, 2–5 μm, although much longer branched filaments may develop in media with a proper ratio of saturated to unsaturated fatty acids. Colonies on solid medium up to 3 mm in diameter with a well developed central area of a yellowish or pale brown color. Relatively strong turbidity produced during growth in fluid medium.

Carotenoid pigments with absorption maxima at 414, 438 and 468 nm, synthesized as determined in ethanol extracts of organisms grown in medium supplemented with 0.05 M sodium acetate.

Acid produced from glucose, maltose, glycogen, dextrin, starch and cellobiose; no acid from mannose, sucrose, mannitol, lactose, dulcitol, xylose, salicin, glycerol or sorbitol. Most strains hydrolyze esculin (Williams and Wittler, 1971). Reduces tetrazolium anaerobically, and weakly aerobically.

Reports on liquefaction of gelatin are conflicting; no digestion of coagulated serum or casein. Agar colonies produce zones of β type hemolysis by the overlay technique, using guinea pig or sheep erythrocytes; hemolysin identified as hydrogen peroxide. No adsorption of guinea pig red blood cells.

Minimal nutritional requirements: potassium, magnesium and phosphate ions; glucose; 13 amino acids; nucleic acid precursors; nicotinic acid, riboflavin, folinic acid, pyridoxine, pyridoxol and thiamine; long chain fatty acids.

Facultative anaerobe, although freshly isolated strains grow better aerobically.

Temperature range about 20–41 C; optimum for organisms recently isolated from soil and sewage reported to be about 30 C, while laboratory adapted strains and strains from animal sources appear to grow as well or better at 37 C.

Very sensitive to osmotic lysis.

Growth and carotenoid pigment synthesis is inhibited by thallium acetate in concentrations of about 1:2000.

Although originally isolated from sewage, manure, humus and soil, the designation of *A. laidlawii* as a true saprophyte has been challenged by more recent findings of parasitic occurrence in a variety of mammals and birds, such as cattle (genital tract, nasal cavity), swine (nasal cavity), man (oral cavity) and chicken (sinus).

The G + C content of the DNA is about 32.0–32.5 moles % (T_m and buoyant density). Genome size 1.0×10^9 daltons. The degree of genetic homology between types A and B is about 70% as found by the agar column hybridization method of McCarthy and Bolton (1963), and about 90% as determined by the membrane filter method of Gillespie and Spiegelman (1965). A limited amount of heterologous hybridization (19% relatedness value) has been reported between this species and *Mycoplasma mycoides* subsp. *capri*.

Type strain: PG 8; Sewage A; ATCC 23206; NCTC 10116; AMRC-C 1020; NIH M-728-001-084 (Edward and Freundt, 1973).

Further Comments

Originally three types, designated A, B and C, were recognized on the basis of agglutination tests (Laidlaw and Elford, 1936). Strain C has been lost, but the relative antigenic distinctness of strains A and B was later confirmed, although some crossing is demonstrable even with highly specific serological tests such as the growth inhibition and indirect fluorescent antibody tests. Using a direct fluorescent antibody test on agar colonies they are indistinguishable. Also, the electrophoretic pattern of the cell proteins differs only slightly for strains A and B.

A subspecies of *A. laidlawii*, represented by a single isolate (strain H3-10) has been described under the name of *Mycoplasma laidlawii* subsp. *inocuum* Adler and Shifrine 1964, 1245. Although originally designated as a distinct species, *Mycoplasma inocuum* Adler *et al.* 1961, 239, it was later recognized to be serologically similar to *A. laidlawii*. However, since this strain was reported to differ from the type strain of *A. laidlawii* in its ability to metabolize glucose through a hexose monophosphate shunt it was regarded as a variety of this species. In a subsequent comparative study of a great number of *A. laidlawii* strains, strain H3-10 behaved similarly to the other strains in regard to carbohydrate fermentation, serology and electrophoretic pattern. Hence, its proposed status as a subspecies of *A. laidlawii* is questionable.

2. Acholeplasma granularum (Switzer) Edward and Freundt 1970, 2. (*Mycoplasma granularum* Switzer 1964, 504; *Sapromyces granularum* (Switzer) Edward and Freundt 1969, 393.)

gra.nu.la′rum. L. neut.n. *granulum* a small grain, a granule; L. gen.pl.n. *granularum* of small grains, made up of granules, granular.

Short filaments.

Growth rate somewhat slower than that of *A. laidlawii*. Synthesizes carotenoid pigments with adsorption maxima at 414, 418 and 468 nm.

Acid produced from glucose, but not mannose. Does not hydrolyze esculin. Reduces tetrazolium anaerobically, and weakly aerobically. Phosphatase negative. No proteolytic activity. Agar colonies produce zones of α type hemolysis by the overlay technique, using sheep red blood cells. No adsorption of guinea pig red blood cells. Minimal nutritional requirements not defined.

Temperature range and optimum not exactly defined, but growth occurs at 30–37 C.

Apparently a common inhabitant of the nasal cavity of swine. Although claimed to have been isolated from 85% of synovial fluids obtained from older pigs suffering from acute, non-febrile arthritis in certain geographical areas, recent reports by Ross and Karmon (1970) indicate that the agent has on some occasions been incorrectly identified, inasmuch as at least some of the isolates from arthritic pigs are *M. hyosynoviae* rather than *A. granularum*. Rarely encountered in arthritis in younger pigs, and rarely recovered from pneumonic lesions. No information at present to indicate the occurrence in soil or water.

The G + C content of the DNA is about 30.5–32.2 moles % (T_m and buoyant density). The per cent relative relatedness with *A. laidlawii* as determined by nucleic acid homology studies, is at the level of 15 to 20.

Type strain: BTS-39; ATTC 19168; NCTC 10128; AMRC-C 1022; NIH M-719-001-084 (Edward and Freundt, 1973).

Further Comments

Antigenically relatively closely related to *A. laidlawii*, a sharing of antigenic components being demonstrable by complement fixation, double immunodiffusion and indirect fluorescent-antibody tests and occasionally also to a slight extent even by the growth inhibition test, while on the other hand the two species are clearly distinguishable by direct fluorescent-antibody procedures performed on agar colonies. Also, although a comparison of the electrophoretic patterns of cell proteins reveals a basic similarity between *A. laidlawii* and *A. granularum*, they do differ with respect to some bands.

3. **Acholeplasma axanthum** Tully and Razin 1970, 751. (A new sterol-nonrequiring *Mycoplasma*, Tully and Razin 1969, 970.)

a.xan'thum. Gr. pref. *a* not, without; Gr. adj. *xanthus* yellow; M.L. neut.adj. *axanthum* without yellow (pigment).

Predominantly coccobacillary and coccoid with a few short myceloid elements usually 2–5 μm in length.

Large colonies with clearly marked centers on horse serum agar; colonies on serum-free media are smaller and usually lack the peripheral growth around their center. Altogether, growth in media devoid of serum or serum fraction is much poorer than that of either *A. laidlawii* or *A. granularum*.

Does not synthesize carotenoid pigments. Also differs from *A. laidlawii* in its ability to synthesize sphingolipids.

Acid produced from glucose, maltose, galactose, glycogen, starch, dextrin, salicin, glycerol and cellobiose; no acid from mannose, sucrose, mannitol, lactose, dulcitol, fructose, xylose or sorbitol. Hydrolyzes esculin. Reduces tetrazolium both aerobically and anaerobically. No phosphatase activity. No liquefaction of coagulated horse serum; not tested for other proteolytic activities. Agar colonies produce zones of β type hemolysis by the overlay technique; no adsorption or agglutination of guinea pig or human red blood cells.

Minimal nutritional requirements poorly defined; the marked stimulation of growth by "Tween 80" suggests a requirement for fatty acids. Although not dependent on cholesterol for growth, this species incorporates larger amounts of cholesterol from the growth medium than *A. laidlawii*, but smaller quantities than species of *Mycoplasma*.

Growth occurs at 30 and 37 C.

Very sensitive to osmotic lysis.

The ecology as yet undetermined; the description of this species is based on two strains, recovered from murine leukemia tissue culture cell lines.

The G + C content of the DNA is about 31 moles % (buoyant density). No DNA duplex formation between this species and either *A. laidlawii* or *A. granularum* in hybridization experiments.

Type strain: S-743; ATCC 25176; NCTC 10138; AMRC-C 1024.

Further Comments

Serologically distinct from *A. laidlawii* and *A. granularum* by complement fixation, growth inhibition and immunofluorescence tests. The acrylamide gel electrophoretic pattern of the cell proteins also shows major differences from those of *A. laidlawii* and *A. granularum*.

4. **Acholeplasma modicum** Leach 1973, 147. (Bovine serological group 6, Leach 1967, 312.)

mo.di'cum. L. neut.adj. *modicum* moderate, referring to moderate growth.

Pleomorphic, spherical, ring-shaped and coccobacillary forms.

Agar colonies distinctly smaller than those of

Acholeplasma laidlawii and *Acholeplasma granularum*.

Does not synthesize carotenoid pigments.

Acid produced from glucose, but not mannose. Does not hydrolyze esculin. Reduces tetrazolium aerobically and anaerobically. Phosphatase negative. No liquefaction of coagulated horse serum. Agar colonies produce zones of α to β type hemolysis by the overlay technique, using ox, sheep, and guinea pig red blood cells; do not adsorb ox or guinea pig red blood cells. Guinea pig red blood cells are agglutinated. Minimal nutritional requirements not defined.

Temperature range and optimum not exactly defined, but growth occurs at both 22 C and 37 C.

Isolated from the blood, bronchial lymph nodes and lungs of cattle with pneumoenteritis. Pathogenicity remains to be determined.

The G + C content of the DNA is 29.3 moles % (T_m).

Type strain: Squire; PG 49; NCTC 10134; AMRC-C 1025.

Further Comment

Differs from other *Acholeplasma* and *Mycoplasma* species antigenically, by the metabolic inhibition test, and in the electrophoretic patterns of its cell proteins.

5. Acholeplasma oculusi (*sic*) Al-Aubaidi, Dardiri, Muscoplatt and McCauley 1973, 126.

o.cu.lus'i. apparently meant as the genitive *oculi* of L. n. *oculus* the eye.

Morphology not described.

Carotenoid pigments synthesized.

Acid produced from glucose, galactose, xylose, cellobiose and fructose; no acid from mannose, sucrose, mannitol, glycerol, sorbitol or salicin. Esculin hydrolyzed. Not tested for liquefaction of coagulated horse serum or phosphatase activity. Reduces tetrazolium aerobically and anaerobically. Agar colonies produce zones of hemolysis by the overlay technique, using sheep red blood cells; not tested for hemadsorption or hemagglutination. Minimal nutritional requirements not defined.

Temperature range and optimum not exactly defined, but growth occurs at both 25 C and 37 C.

Resistant to 100 $\mu g/ml$ of kanamycin, sensitive to 10 $\mu g/ml$ of erythromycin. Sensitive to 1% bile salts.

Isolated from the conjunctivae of goats suffering from keratoconjunctivitis, but not from normal or recovered animals. Conjunctivitis, but not keratitis, has been reproduced experimentally by instillation into the conjunctival sac; conjunctivitis together with pneumonia and polyserositis can be produced on intravenous inoculation.

The G + C content of the DNA not determined.

Type strain: 19-L; ATCC 27350; AMRC-C 1026.

Further Comments

Differs antigenically by metabolic inhibition and double immunodiffusion tests from other *Acholeplasma* and *Mycoplasma* species.

ADDENDUM TO ORDER **MYCOPLASMATALES**
GENUS OF UNCERTAIN AFFILIATION

Genus **Thermoplasma** *Darland, Brock, Samsonoff and Conti 1970, 1418*

Ther.mo.plas'ma. Gr. n. *thermus* heat; Gr. neut.n. *plasma* something formed or molded, a form; M.L. neut.n. *Thermoplasma* heat (-loving) mycoplasma.

Cells pleomorphic, varying in shape from spherical (0.3–2 μm) to filamentous structures, believed to reproduce by budding. Cells lack a true cell wall and are surrounded by a single triple-layered membrane, approximately 10–12 nm thick. Non-motile as far as known. No resting stages known. Gram-variable.

Colonies on solid medium containing 1.2% Ion Agar attain a diameter of about 0.3 mm at pH 2; they are rather flat, coarsely granulated, dark brown in color and some of them exhibit a typical "fried egg" appearance with a translucent peripheral zone.

Biochemical and nutritional characters rela-

tively poorly defined. Growth occurs in fluid medium containing 0.02% $(NH_4)_2SO_4$, 0.05% $MgSO_4$, 0.025% $CaCl_2·2H_2O$, 0.3% KH_2PO_4, 0.1% yeast extract and 1.0% glucose. The cell yield is proportional to the concentration of yeast extract at concentrations less than 0.2%; no growth on the ether-extractable fraction of yeast. Growth is not stimulated by the addition of glucose, galactose, sucrose, ribose or glycerol. **Does not require sterol.**

Apparently strict aerobe.

Optimum temperature for growth about 59 C, no growth occurring at 37 or 65 C.

Grows well between pH 0.96 and 3.0; **the optimum pH is about 1–2.**

Cells rapidly lysed by sodium lauryl sulfate.

Insensitive to vancomycin, a specific inhibitor of cell wall synthesis, at concentrations of at least 5 mg/ml. Inhibited by novobiocin at a concentration of 0.1 μg/ml. Sensitivity to penicillin could not be determined because of the acid lability of this antibiotic.

Apparently a true saprophyte, having been isolated from a burning coal refuse pile and from acid hot springs.

Reports of the G + C content of the DNA are conflicting, ranging from 25 to 40 moles % (T_m and buoyant density).

Type species: *Thermoplasma acidophilum* Darland *et al.* 1970, 1418.

Further Comments

Although Darland *et al.* suggested that the genus belonged in the Order *Mycoplasmatales*, its affiliation with this order and its exact position must await further study.

Description of the species of genus **Thermoplasma**

1. **Thermoplasma acidophilum** Darland, Brock, Samsonoff and Conti 1970, 1418. (*Thermoplasma acidophila* Darland *et al.* 1970, 1418.)
a.ci.do'phil.um. M.L. n. *acidum* an acid; Gr. adj. *philus* loving; M.L. neut.adj. *acidophilum* acid-loving.

Description as for genus.

Type strain: 122-1B2; ATCC 25905; AMRC-C 165.

ADDENDUM TO CLASS **MOLLICUTES**
GENUS OF UNCERTAIN AFFILIATION

Genus **Spiroplasma** *Saglio, Lhospital, Laflèche, Dupont, Bové, Tully and Freundt 1973*, 201

Spi.ro.plas'ma. Gr. n. *spira* a coil, spiral; Gr. neut.n. *plasma* something formed or molded, a form; M.L. neut.n. *Spiroplasma* spiral form.

Cells pleomorphic, varying in shape **from spherical or slightly ovoid,** with a minimal diameter of about 100–250 nm, **to helical and branched non-helical filaments.** The helical forms, about 120 nm in diameter and 2–4 μm in length in the logarithmic and considerably longer in the postlogarithmic phase of growth, predominate in liquid cultures. Only non-helical and round cells have been demonstrated in agar cultures. Mode of replication not known. **Cells lack a true cell wall and are bounded by a single triple layered membrane,** about 7–8 nm thick. An **additional outer layer of short projections,** which often appear periodic, is present on the surface of the bounding membrane. Helical filaments are **motile,** showing two types of motility: a rapid rotary or "screw" motion, and a slow undulation. Flagella, axial filaments, or other organelles of locomotion not demonstrated. **Gram-positive.**

Colonies on solid medium attain a diameter of about 0.2 mm and exhibit a **typical biphasic "fried-egg" appearance.** Marked turbidity produced in both cultures.

Chemoorganotroph. **Acid produced from glucose and mannose.** Does not hydrolyze esculin, arginine or urea. Reduction of tetrazolium variable. Phosphatase positive. Spots, but no film, produced on horse serum agar. No liquefaction of coagulated horse serum. Agar colonies produce zones of α type hemolysis by the overlay technique, using guinea pig red blood cells. One of two strains tested adsorb guinea pig red blood cells. Does not synthesize pigmented carotenoids.

Cholesterol, or possibly other sterols, required for growth. Growth occurs in a conventional mycoplasma medium containing 10% yeast extract and 20% horse serum, modified by the addition of 7% sorbitol, 1% tryptone, 1% sucrose, 0.1% fructose, and 0.1% glucose. On adaptation, growth may occur in less complex media.

Facultative anaerobe, maximal growth occurring at 5% CO_2 in nitrogen.

Temperature range 20–37 C, growth being very sparse below 25 C and above 35 C; **sharp optimum at 32 C.**

Resistant to 10,000 i.u./ml of penicillin G; completely inhibited by 0.04–0.08 μg/ml of erythromycin, 0.16–0.32 μg/ml of tetracyclin, 20–40 μg/ml of neomycin, 15–25 μg/ml of amphotericin B, and 1,280–2,560 μg/ml of thallium acetate. Sensitive to 1.5% of digitonin, using a disc test; sensitivity to sodium-polyanethol-sulfonate varies with growth conditions.

Isolated from the leaves of Citrus plants affected by "Stubborn" disease, the sieve tubes of which contain morphologically identical structures. The disease has not been reproduced experimentally with the organism.

The G + C content of the DNA is 25.0–26.35 moles % (T_m and buoyant density); the **genome size is 10^9 daltons.**

Type species: *Spiroplasma citri* Saglio *et al.* 1973, 202.

Further Comments

Although several of the properties described for the genus would seem to allow its classification within the Class *Mollicutes*, Order *Mycoplas-* *matales*, other of its characteristics are apparently inconsistent with the definition of the *Mycoplasmatales*. While helical shape and motility need not rule out inclusion in the *Mycoplasmatales*, the demonstration of a surface layer of special structure, retention of the Gram stain complex, and infection by a tailed bacteriophage of classic type B morphology suggest the presence of a modified cell wall or a structure reminiscent of a cell wall. Hence, the affiliation of the genus with the Class *Mollicutes* and its exact position must await further ultrastructural and biochemical studies, including a search for cell wall precursors.

Description of the species of genus **Spiroplasma**

1. **Spiroplasma citri** Saglio, Lhospital, Lafléche, Dupont, Bové, Tully and Freundt 1973, 202.

cit′ri. M.L. n. *Citrus* generic name; M.L. gen.n. *citri* of citrus, to denote plant host.

Description as for genus. See also Cole *et al.*, 1973.

Type strain: Morocco (R8-A2); ATCC 27556.

Further Comment

Differs antigenically by growth inhibition and immunofluorescence tests from all *Mycoplasma* and *Acholeplasma* species. Serologically unrelated also to several species tested of the Order Spirochaetales.

MYCOPLASMA-LIKE BODIES IN PLANTS

Karl Maramorosch

For many years yellows-type diseases of plants, such as aster yellows, corn stunt, mulberry dwarf, rice yellow dwarf and stolbur, have been considered as virus diseases, although no virus-like particles were ever isolated or observed by electron microscopy techniques. Doi *et al.* (1967) and Ishiie *et al.* (1967) indicated the possible mycoplasma etiology of this group of plant diseases. Their suggestion was based on electron microscopy and chemotherapy studies. Thin sections of phloem tissues from plants with four yellows-type diseases revealed the presence of pleomorphic bodies, from 80–800 nm in diameter, bound by unit membranes, with ribosome-like granules and DNA-like strands, closely resembling in ultrastructure certain species of the Mycoplasmatales. When roots of yellows-diseased plants were immersed in a solution containing 100–1000 ppm of tetracycline hydrochloride or other tetracycline antibiotics, recovery from disease was observed. Furthermore, the unit membrane-bound bodies that were abundant in the phloem of untreated plants were no longer found in recovered plants. The recovery was not permanent and plants reverted to the diseased condition a few weeks after cessation of chemotherapy treatments. The investigators concluded that yellows-type diseases may be caused by mycoplasma agents.

During the following years a total of more than 40 different plant diseases were added to the list of those in which no virus particles were detected and in which bodies similar to those described by Doi *et al.* were observed in phloem tissue. Several reviews of this subject have been published (Maramorosch *et al.*, 1968, 1970; Casper, 1969; Shikata *et al.*, 1969; Bos, 1970; Whitcomb and Davis, 1970). It should be emphasized that while the evidence for a non-viral etiology of this group of diseases is very strong, the identification of the etiological agents as members of the Mycoplasmatales is incomplete. The presumptive mycoplasma etiology has been based primarily on electron microscopy of thin sections of tissues from diseased plants, and in some instances supported further by observations of thin sections of insect vectors. The bodies found in these alternate plant and arthropod hosts were indistinguishable in their fine structure and morphological characteristics from certain well defined species of animal mycoplasmas, provided that the fixation and electron microscopy were expertly carried out. The remission of signs of disease, reported in a few instances of yellows-type diseases after treatment with tetracycline antibiotics and the lack of remission following treatment with penicillin, although quoted by some plant pathologists as supporting a mycoplasma etiology, were not convincing since there is no typical antibiotic spectrum that would distinguish mycoplasmas from all other disease agents. The disappearance of the pleomorphic bodies from the phloem tissues following successful chemotherapy and their reappearance after cessation of treatment and recurrence of disease (Wolanski, unpublished observations) support the evidence that these bodies are the causative agents of the diseases, but the observations do not permit identification of the bodies as mycoplasmas.

The crucial tests concerning identification of the mycoplasma-like bodies require cultivation of these plant disease agents and comparison with known microorganisms. Laboratories in several countries have been engaged in such tests since 1967 without apparent success. Published reports from laboratories in the United States, Taiwan, France and India have described experiments in which retention of infectivity (from 48 hrs to 50 days) was obtained, but claims for cultivation are as yet unsubstantiated. Only recently have cultures of *Spiroplasma* sp. (Morrocco and California strains) been made available for comparison with established mycoplasma cultures. The negative-staining technique used in electron microscopy appears to be unreliable for the study of these mycoplasma-like bodies because artifacts formed in healthy and infected material may resemble negatively stained mycoplasmas (Wolanski and Maramorosch, 1970).

By the beginning of 1971 the agents of yellows-type diseases of plants could be classified safely as "mycoplasma-like," and the disease agents as "mycoplasma-like bodies" (MLB), but not as mycoplasmas. They may perhaps belong to a separate new group of the Class Mollicutes. Adequate media for their cultivation and improved isolation techniques are urgently needed.

Editorial Note. See "Addendum to Class *Mollicutes*—Genus *Spiroplasma*" and Cole *et al.*, 1973.

Important Notes

for

Users of this Edition

1. Always read both generic and species descriptions because characters listed in the generic description are not usually listed in the species descriptions.

2. In tables, characters common to all taxa are not shown but may be listed in footnotes.

3. Generally in tables (exceptions are clearly indicated in footnotes, q.v.) the meanings of symbols are as follows:

+   more than 90 % strains positive

−   more than 90 % strains negative

d   11–89 % strains positive

( )   delayed reaction

w   weak reaction

D   Different reactions in different taxa (species of a genus or genera of a family)

v   strain instability (NOT differences between strains)

Appendices

List of Culture Collections

Designations and addresses of culture collections mentioned in the descriptions. WFCC (World Federation of Culture Collections) numbers* are given in parentheses.

AMC—Walter Reed Army Medical Center, 6825 16th St., N.W., Washington, D.C. 20012 USA

AMRC-C (272)—FAO/WHO International Reference Centre for Animal Mycoplasmas, Institute for Medical Microbiology, University of Aarhus, DK-8000 Aarhus C, Denmark

ATCC (1)—American Type Culture Collection, 12301 Parklawn Drive, Rockville, Md. 20852 USA

BKM (342)—All-Union Collection of Microorganisms, Institute of Microbiology, USSR Academy of Sciences, Profsojuznaja 7, Moscow B133 USSR

BUCSAV—Biologický Ústav, Československá Akademie Věd, Prague, Czechoslovakia

CBS (133)—Centraalbureau voor Schimmelcultures, Baarn, The Netherlands

CCEB (86)—Culture Collection of Entomogenous Bacteria, Institute of Entomology CAS, Na cvičišti 2, Prague, 6 Czechoslovakia

CCM (65)—Czechoslovak Collection of Microorganisms, J. E. Purkyně University, tř. Obránců míru 10, Brno, Czechoslovakia

CDC—Center for Disease Control (Communicable Disease Center), 1600 Clifton Rd. NE, U. S. Dept. of Health, Education and Welfare, Atlanta, Ga. 30333 USA

CIP (245, 12)—Collection de l'Institut Pasteur, Paris, France

CMI (214)—Commonwealth Mycological Institute, Kew, England

CUB (124)—University of Bradford, Bradford, Yorkshire BD7 1DP England

D—Medical Research Council Microbial Systematics Unit Collection, Leicester University, Leicester, LE1 7RH England

DINR—División Immunología Vegetal, Sección Naturaleza de la Resistancia, Instituto de Fitotechnia M.A.G., Argentina

ETH (273)—Eidgenösische Technische Hochschule, Zürich, Switzerland

FAT—Faculty of Agriculture, Tokyo University, Tokyo, Japan

FDA—Food and Drug Administration, U. S. Dept. Health, Education and Welfare, Washington, D.C. 20204 USA

FI (177)—Farmitalia Research Laboratories, Milano, Italy

FUJI—Tokyo Laboratory, Fujisawa Pharm. Ind., Ltd. 957-3, Nukui-Kita-machi, Koganei-city, Tokyo, Japan

* *World Directory of Collections of Cultures of Microorganisms* (Martin and Skerman, editors) Wiley-Interscience, 1972.

HACC—Hindustani Antibiotics Ltd., Pimpri, Poona 18, India

HUT (195)—Hiroshima University, Faculty of Engineering, Hiroshima, Japan

IAM (190)—Institute of Applied Microbiology, University of Tokyo, 1-Chome, Yayoi, Bunkyo-ku, Tokyo, Japan

IAUR—Instituto de Antibióticos da Universidade de Recife, Brazil

IAW—Institute of Antibiotics, Warsaw, Poland

ICPB (74)—International Collection of Phytopathogenic Bacteria, University of California—Davis, Davis, California 95616 USA

IEM (130)—Institute of Epidemiology and Microbiology, Srobárova 48 Prague 10, Czechoslovakia

IFM—Institute of Food Microbiology, Chiba University, Chiba, Japan

IFO (191)—Institute for Fermentation, Jusonishinocho, Higashi-yodogawa-ku, Osaka, Japan

IHM—Instituto de Higiene Experimental, Montevideo, Uruguay

IMASP—Institute of Microbiology, Academy of Sciences, Peking, China

IMET (217)—Institutes für Mikrobiologie und Experimentelle Therapie, Deutsche Akademie der Wissenschaften zu Berlin, Beuthenbergstrasse 11, Jena 69, German Democratic Republic

IMRU (75)—Institute of Microbiology, Rutgers, The State University, New Brunswick, New Jersey 08903 USA

IMUR—Instituto de Micologia, Universidade de Recife, Brazil

INA—Institute for New Antibiotics, USSR Academy of Medical Sciences, Moscow, USSR

INMI (342)—Institute for Microbiology, USSR Academy of Sciences, Moscow, USSR

INTA—Instituto de Patologia Vegetal, Instituto Nacional de Microbiologia, Castelar, Argentina

IPCR—Institute for Physical and Chemical Research, 31, Komagome-Kamifujimae-cho, Bunkyo-ku, Tokyo, Japan

IPV—Istituto di Patologia Vegetale, Milan, Italy

IRNA—See INA

ISC—International Salmonella Center, Statens Seruminstitut, Amager Boulevard 80, 2300 Copenhagen S, Denmark

ISP—International Streptomyces Project (International Cooperative Project for Description and Deposition of Type Cultures of Streptomyces (Dr. Shirling).) (Sponsored by the Subcommittee on Actinomycetes of the Committee on Taxonomy, American Society for Microbiology and the corresponding committee of the ICSB).

ITCC—Indian Type Culture Collection, New Delhi, India

JFCC—Japanese Federation of Culture Collections of Micro-organisms, c/o Institute of Applied Microbiology, University of Tokyo, Bunkyo-ku, Tokyo, Japan

KIMG (274)—Kulturensammlung am Institut für Mikrobiologie, Gosslerstr. 16, 34 Göttingen, West Germany

KITA—Kitasato Institute for Infectious Diseases, Tokyo, Japan

LIA—Museum of Cultures, Leningrad Research Institute of Antibiotics, Leningrad, USSR

MTHU (MTU)—Dept. Bacteriology, Faculty of Medicine, Tohoku University, 4 Bancho, Kita, Sendaicity, Japan

NADL—National Animal Disease Laboratory, Ames, Iowa 50010 USA

NBL—Naval Biological Laboratory, Oakland, California 91627 USA

NCDC—See CDC

NCDO (118)—National Collection of Dairy Organisms, National Institute for Research in Dairying, Shinfield, Reading RG2 9AT, England

NCIB (239)—National Collection of Industrial Bacteria, Torry Research Station, P.O. Box 31, 135 Abbey Road, Aberdeen AB9 8DG, Scotland

NCIM (3)—National Collection of Industrial Microorganisms, National Chemical Laboratory, Poona, India

NCMB (238)—National Collection of Marine Bacteria, Torry Research Station, 135 Abbey Rd., Aberdeen AB9 8DG, Scotland

NCPPB (126)—National Collection of Plant Pathogenic Bacteria, Plant Pathology Laboratory, Ministry of Agriculture, Fisheries and Food, Hatching Green, Harpenden, Herts, England

NCTC (154)—National Collection of Type Cultures, Central Public Health Laboratory, Colindale Ave., London NW9 5HT, England

NIH—National Institutes of Health, Bethesda, Maryland 20014 USA

NIHJ—National Institute of Health, Tokyo, Japan

NIRD—See NCDO

NRC (242)—National Research Council, 100 Sussex Drive, Ottawa, Canada K1A 0R6

NRRL (97)—Northern Utilization Research and Development Division, U. S. Department of Agriculture, Peoria, Illinois 61604 USA

NTHC (40)—Culture Collection, Biochemistry Department, Technical University of Norway, Trondheim, Norway

OEU—Tennoji Branch, Osaka University of Liberal Arts and Education, Minami-Kawabori-cho, Tennoji-ku, Osaka, Japan

PD (D)—Parke, Davis Culture Bureau, Parke, Davis and Company, Joseph Campau Avenue, At the River, Detroit, Michigan 48232 USA

PD (W)—Plantenziektenkundige Dienst, Wageningen, The Netherlands

PSA—Progetto Sistematica Actinomiceti, Istituto "P. Stazzi," Milano, Italy

QM (129)—Quartermaster Research and Development Center, U. S. Army, Natick, Massachusetts 01760 USA

RIA (337)—All-Union Research Institute for Antibiotics, Moscow, USSR

RML—Rocky Mountain Laboratory, Hamilton, Montana 59840 USA

SMG (274)—Sammlung für Mikroorganismen Göttingen, Gosslerstr. 16, 34 Gottingen, West Germany

UC (K) (168)—Culture Collection, The Upjohn Company, 7171 Portage Rd., Kalamazoo, Michigan 49002 USA

UC (P) (254)—Instituto di Pathologia Vegetale, Universita Cattolica del S. Cuore, Piacenza, Italy

VNIIA—See RIA

VPI—Virginia Polytechnic Institute, Anaerobe Laboratory, P.O. Box 49, Blacksburg, Virginia 24060 USA

WRAIR—Walter Reed Army Institute of Research, Washington, D.C. 20012 USA

WRRL—Western Utilization Research and Development Division, U. S. Department of Agriculture, Albany, California 93306 USA

WVU—West Virginia University, Dept. of Bacteriology, Medical Center, Morgantown, W. Va. 26506 USA

Glossary

No attempt has been made to provide a dictionary of bacteriology. Words which in our opinion may not be too well known are given for convenience and to save the user time. Words found in the ordinary dictionary, e.g. *Webster's Seventh New Collegiate*, are not included.

Abbreviations (as used in etymologies).

Gr. = Greek, usually latinized Greek. The original word is transliterated into the Latin alphabet and, in most instances, is given in its Latin form.

L. = Latin; usually this indicates that the word is classic Latin and found in an unabridged Latin dictionary.

M.L. = Modern Latin; a word not necessarily classic Latin, but treated and used as a Latin word.

Med.L. = Medieval Latin; many words derived from languages other than Latin and Greek were latinized during the Middle Ages and used in fields such as alchemy, biology and pharmacy.

adj., adjective; dim., diminutive; n., noun; part., participle; part. adj., participial adjective; pl., plural (note: names of all taxa higher than genus are plural and feminine); sub., substantive or noun; fem., feminine gender; masc., masculine gender; neut., neuter gender; nom., nominative case; gen., genitive case.

Acrasin-like. Similar in action to acrasin, the diffusible substance which promotes cellular aggregation in a group of eukaryotic protists, the *Acrasiales.*

Acropetally. Produced successively in the direction of the apex, i.e. the apical member is the youngest.

Actinophage. A virus able to infect and eventually cause lysis of actinomycetes.

Aerial (of mycelium). Mycelial growth above the substrate.

Aleuriospore. Small, terminal spores whose position, color, form, structure and dimensions approach those of conidia. They differ from conidia in that they are not ephemeral and immediately separated from the mycelium by dehiscence but remain attached by a structure equal in diameter to the hyphal branch from which it was formed. They are normally solitary and terminal and present in great numbers, e.g. *Micromonospora, Thermomonospora.*

Alluvial Fan. An erosion deposit of a stream where it empties out from a gorge upon an open plain; delta-like.

Alpha-Reaction (or Alpha-hemolysis). The production of a green discoloration around a colony growing on a blood agar medium.

Andersen Sampler. A device for collecting, sizing and enumerating viable airborne particles (Andersen, 1958).

Apical Meristem. Growing point of a plant.

Arborescent. Branched in a treelike manner.

Arthrospore. An asexual spore (but not sporangiospore, chlamydospore or endospore) formed by the walling off a pre-existing hypha or branch to give a chain of spores, e.g. *Streptomyces.*

Articulate. Constrictions between cells in a chain so deep that cells appear connected by a thread.

Arylsulfatase. An enzyme characteristic of mycobacteria capable of splitting the bond between the sulfate group and the aromatic ring in compounds having the general formula $ROSO_3H$, wherein R represents the aromatic ring.

Asexual (of spores). Spores produced vegetatively, i.e. without involving the union of nuclei in a sexual process.

Auxotrophic (of mutants). Having a growth requirement for certain compound(s) or nutrients not required by the parental strain.

Axial Fibrils. Long, slender, threadlike processes of regular thickness which are intertwined with the protoplasmic cylinder of spirochetes. Each fibril arises from subterminal attachment discs at either end of the protoplasmic cylinder and overlaps fibrils originating at the opposite end for varying portions of their length. Probably responsible for motility.

959

Axial Filament. A term which has been used synonymously with axial fibril and axistyle.

Axistyle. The pair of axial fibrils observed in *Leptospira.*

Bacteroid. In *Rhizobium* a special deformed type of cell, probably associated with nitrogen fixation. In *Frankia* used for bacteria-like cells. In *Blattabacterium* bacteria-like intracellular microorganisms found in various species of insects.

Basionym. The name of the species whose specific epithet is adopted for a new combination, a senior synonym.

Basipetally. Development in the direction of the base, i.e. the apical part is oldest.

Battey Bacillus. Properly, a vernacular name for a strain of the species *Mycobacterium intracellulare,* but often used loosely for any strain of the group or related species, including *M. avium,* usually associated with human disease or resembling such bacteria.

Beta-Hemolytic Reaction (or Beta hemolysis). The elaboration of a hemolysin that lyses red blood cells with the resultant clearing around a colony growing on a blood agar medium.

Bifid. A rod with one or both ends forked. The ends may be short and clavate or short and rounded and of greater diameter than the rod.

Blastospore. Spores found as enlargements at the ends of aerial hyphae in acropetal succession.

Bloom. A surface texture characterized by the projection of very short aerial hyphae appearing minutely velvety or powdery.

Bulla. A bleb or bubble-like projection extending from the outer cell envelope of Spirochetes. May be observed with dark field or electron microscopy.

Camp Test. A blood agar plate is streak-inoculated across the center of the plate with a β-lysin producing strain of *Staphylococcus aureus* (*note:* β refers to the type of lysin produced and not the hemolytic reaction). At right angles to the *Staphylococcus* streak, but not touching it, a suspected *Streptococcus agalactiae* is streaked. After incubation (37 C for 24 hr), a synergistic clearing (complete hemolysis) in the proximity of the two organisms indicates a strong possibility that the unknown organism is *Streptococcus agalactiae* (usually used to determine the etiologic agent of bovine mastitis).

Cephalosporial. Like a rounded mass (head) of spores in a slimy matrix at tip of sporophore. Literally, "resembling the spore production in the fungal genus *Cephalosporium.*"

Chemolithotroph. An organism that uses carbon dioxide as its principal source of carbon for growth and obtains its energy by the oxidation of inorganic compounds.

Chemoorganotroph. An organism that obtains its energy by the oxidation of organic compounds. These are also its principal sources of carbon.

Chalmydospores. Terminal or intercalary, thick walled, resistant mycelial cells generally with a larger diameter than the parent hyphae. These cells differ from true conidia in their irregularity of size and shape and haphazard occurrence.

Chlorobium Vesicles. Photopigment-bearing elements of the green sulfur bacteria (Chlorobiaceae). The vesicles are oblong bodies, 30 to 40 nm wide and 100 to 150 nm long; they are completely bounded by a thin (3 nm) electron-dense nonunit membrane which lies adjacent to the cell membrane but is distinguishable from it (Plate 1.4).

Chromogenicity. Ability to produce color (pigment); often referring to the dark brown to blackish pigments produced *in situ* in certain tyrosine-containing media as a result of the formation of melanin or melanin-like compounds.

Coenobium. Several cells of an organism remain together after division to form an ordered (*Thiopedia*) or disordered clump.

Conidia (substrate). Asexual spores produced in the substrate medium.

Conidiospore. Any asexual spore (but not a sporangiospore or endospore). In general use for the spores formed by several genera of the Actinomycetales, e.g. *Streptomyces* and *Micromonospora,* but see definitions for aleuriospore and arthospore.

Cords. Term applied to mycobacteria compacted side by side and end to end in more or less curving ("serpentine") strands. Cord-forming strains of tubercle bacilli are more virulent than strains lacking cord formation.

Coremium (plural **coremia**). A fascicle of parallel aerial hyphae on the culture surface; usually the individual hyphae bear spores.

Cortex. Tissue of a plant located between the epidermis and the vascular system (from the Latin word for bark or rind).

Coxal Organs. Secretory organs of glandular nature located close to the attachment of the first pairs of legs of *Ornithodoros,* which excrete a filtrate of ingested blood through an outlet pore during or after feeding.

Crista. The large bundle of axial fibrils observed in *Cristispira,* forming a spiral around the cell, known as the "crest."

Cyst. In the Myxobacterales, the sporangium.

Dextrorse (of spore chains). Spiralling clockwise when viewed from above.

Diagnostic Titer for Fluorescent Antibody Work. One-half of the highest dilution of a conjugated antiserum which gives 4+ fluorescence with its homologous antigen.

Dictyospores. A muriform spore; a spore having both longitudinal and transverse septations, example: spores of the fungal genus *Alternaria.*

Diffusible (of pigments). Penetrating into the medium from their source. However, pigmentation in the medium does not prove the pigment to be diffusible. Generally, the melanoid pigments in a medium are developed *in situ*.

Diphtheroid. Cells of irregular morphology. Usually an irregularly staining rod of uneven diameter, frequently enlarged at one end, which may appear granular and may have lateral short branches. Frequently arranged as palisades or having a V, Y or T shape.

Endophyte. Microorganisms living in symbiosis with higher plants or lower plants (e.g. algae).

Endospore. A resting spore produced endogenously by bacteria with a characteristic fine structure and exhibiting degrees of resistance to heat, chemicals, and adverse environmental changes. Single endospores are normally formed within a vegetative bacterial cell (mother cell), e.g. *Bacillus, Clostridium*. Other species may form more than one spore per cell, e.g. *Metabacterium polyspora*, and certain genera of the mycelial Actinomycetales may produce many endospores within a thallus.

Epilimnion. The upper layer of lakes, subject to disturbance by winds, lying above the thermocline.

Etymology. The origin of a word. Those who name taxa often do not indicate the source of the name or do so in very general terms; the editors have indicated the possible derivation to the best of their ability. Pronunciation of names is not attempted; names are syllabilized and the usual place of emphasis indicated.

The letter k is rare in Latin and the Greek letter kappa has been transliterated as c.

According to Stearn (*Botanical Latin*, Thomas Nelson, London, 1966) the suffix-*cola* is a substantive and not an adjective; it is therefore not declined.

Eurythermal. Capable of good growth at 28 C as well as 50 C and above.

Extant. In existance; not destroyed; applied to cultures available and viable.

Filament. In this edition, any long threadlike form which may or may not be segmented and includes the forms sometimes referred to as trichomes (q.v.).

Floccose. Appearing like tufts of wool.

Fluorescent Antibody. Immune globulin conjugated to a fluorochrome.

Fragmentation. The division of a rod or filament into smaller rods by transverse fission.

Fragmentation Spores. Spores formed by breaking up of hyphae into separate cells (arthrospore).

Fruiting Body. In the Myxobacterales, the (usually) macroscopically visible structure containing the resting cells.

Gamma Reaction. Indicates no change in the appearance of red blood cells immediately surrounding a colony on a blood agar medium.

Gliding Motility. A continuous and regular movement across solid substrates performed by certain prokaryotes (e.g. myxobacteria, many blue-green algae), which do not possess detectable locomotor organelles.

Glossy Colony. Referring to a streptococcal colony that contains cells which are deficient in type-specific M capsular polysaccharide.

Granule. A term which has been used synonymously with spirochetal sphere.

Halophile (halophilic). Have a requirement for NaCl greater than marine forms. May be divided into (a) extreme halophiles which require from about 15% to saturated NaCl plus other ions (*Halobacterium, Halococcus*); (b) moderate halophiles which require from about 5 to 15% NaCl (*Paracoccus halodenitrificans, Vibrio costicola*). Marine strains usually require undiluted sea water or NaCl at concentrations of 1.5 to 5%, i.e. they have a requirement for added salt greater than that found in the usual laboratory media or in physiological saline (0.85%).

Heloplankton. Floating vegetation of a marsh.

Heptaene (heptaenic) (of antibiotics). With seven conjugated double bonds. Also see Polyene (antibiotics).

Heterokaryosis. A condition in which genetically different nuclei are associated in the same protoplast.

Hexaene (hexaenic). With six conjugated double bonds. Also see Polyene (antibiotics).

Hibitane (or chlorohexidine acetate). Bis-*p*-chlorophenyldiguanidohexane diacetate, a general murobicide.

Holdfast. A secreted mass of material which cements a cell or organism to the substrate or a surface.

Holocarpic. Condition in an organism in which the entire vegetative body (mycelium) is converted into a reproductive structure (sporangium).

Holotype Strain. The one strain designated by the author as the nomenclatural type of a species. A synonym of type.

Hormogonium (plural hormogonia). Groups of cells which detach from parent filaments, exhibit motility and develop into new filaments.

Hygroscopic (of aerial mycelium). Becoming moist and usually darkened.

Hypolimnion. The portion of certain lakes below thermocline which receives no heat from the sun and no aeration by circulation.

Hypothallus (plural hypothalli). A mass of hardened slime at the base of a fruiting body in species of the Myxobacterales.

Illegitimate Name. A name of a taxon that does *not* conform to the International Code of Nomenclature. Such names are to be rejected.

Immobilizin. An antibody which deprives an organism of its locomotory ability, i.e. immobilizes it.

Incurvation. The process of one end of a spirochete folding over and gliding closely around cell toward opposite end; may precede tranverse fission which then occurs at looped end of the intertwined organism.

Indicator Pigment. A pigment that changes color, depending upon the pH of the medium.

Indicator Strain. A strain used by several groups of investigators in lieu of a type of neotype strain.

Intercalary. A bud or spore or propagation cell formed *in* a filament or hypha, or chain of cells, but not terminally.

Intracytoplasmic Membrane Systems. The photopigment-bearing intracytoplasmic membrane system of purple nonsulfur and sulfur bacteria is an extension of the cytoplasmic unit membrane. 1. *Vesicular type:* The intracytoplasmic membrane system originates as vesicular invaginations at many points over the cytoplasmic membrane, and extends inward in the form of isolated and connected vesicles (Plate 1.2, Fig. 1). 2, *Lamellar type:* The intracytoplasmic membrane system originates in one or several points of the cytoplasmic membrane. Tubular intrusions extend to flattened discs which form stacks of lamellae, if several disc-shaped infoldings become appressed against each other (Plate 1.2, Figs. 2 and 3; Plate 1.3, Fig. 1). 3. *Tubular type:* The intracytoplasmic membrane system extends in the form of many tubular intrusions from the cytoplasmic membrane. There may either be many groups of parallel tubes lying in different directions within the cell (Plate 1.3, Fig. 2) or the tubes may form an irregular three-dimensional network. 4. *Thylakoids:* The term is used by some authors to designate photopigment-bearing intracytoplasmic membrane systems, preferably of the lamellar type. Also intracytoplasmic membrane systems of nitrifying bacteria have been called thylakoids.

Intrasporangial Hypha. A hyphal element within a sporangium.

Isoniazid. Isonicotinic acid hydrazide; an antituberculosis drug.

Johnin. A tuberculin-like agent for testing skin sensitivity to *Mycobacterium paratuberculosis* (*M. johnei*).

Lamellae. The flattened discs formed from infolding of the intracytoplasmic membrane systems (q.v.) of photosynthetic bacteria.

Lectotype. A specimen selected from the original material to serve as nomenclatural type when the holotype was not designated at the time of publication.

Macrolide (antibiotic). An antibiotic having a large ring lactone structure.

Meristem. Formative tissue adding new cells to the plant body.

Meromictic (lakes). Lakes partly mixed in which thermal turnover occurs only in top layers, bottom layers are stagnant and anaerobic in contrast to holomictic lakes in which mixing is complete.

Merosporangium. A tubular sporangium containing a single longitudinal row of sporangiospores, e.g. *Microellobosporia, Actinomycetes.*

Mesosomes. A membranous infolding derived from the cell membrane, found in the cells of certain bacteria. Synonym: plasmalemmosome (q.v.)

Metacyclic Forms. Small, irregular forms of *Borrelia*, often resembling granules, that appear after the "negative phase." The term is a survivor of the concept according to which *Borrelia* undergo a life cycle from fully developed forms through an invisible (by ordinary light microscope) stage.

Microarthrospores. Small spores resulting from the fragmentation of a hypha. According to Snell and Dick (1957) "an arthrospore produced in the substrate mycelium of actinomycetes."

Microconidium. A small conidium; a round, small, single celled spore produced either on a conidiophore or sessile.

Microcyst. In the Myxobacterales, a myxospore surrounded by a definite capsule of hardened slime. In the Spirochaetales, a term used synonomously with spirochetal sphere. In *Nocardia*, coccoid structures produced by shortening and condensing of bacillary elements; microcysts survive 80 C for an hour or more, vegetative cells are killed on exposure to 60 C for 30 min.

Mixotroph (mixotrophic). As used of phototrophic bacteria, the ability to grow in either the presence or the absence of light; also of organisms able to use various combinations of inorganic and organic substrates as energy and carbon sources; "a commingling of alternative modes of energy generation and/or carbon assimilation" (Rittenberg, 1969).

Monopodial. A main axis that continues its original line of growth giving off lateral branches (Jackson, 1928).

Monotype. A single strain, clearly designated by the author, upon which the species description was based.

Mother Cell. The part of the sporangium (vegetative bacterial cell) outside the developing spore (endospore) in *Bacillus* and *Clostridium*.

Mucoid Colony. A colony with a gummy consistency, resulting from the abundant synthesis of capsular material.

Muriform. Having both longitudinal and transverse septa; actually wall-like, resembling a stone or brick wall.

Mycetocytes. Specialized cells of some insects, usually in the fat body, containing symbiotic fungal or bacterial inclusions.

Mycetome. A specialized organ of some insects containing symbiotic microorganisms, an aggregation of mycetocytes.

Mycobactin. An agent, derived from some species of *Mycobacterium*, required for or stimulating *in vitro* growth of *M. paratuberculosis* and some other species; an iron-chelating growth factor.

Mycolic Acid. Long chain, branched fatty acids of high molecular weight, characteristic of species of *Mycobacterium;* similar acids of lower molecular weight occur in species of *Nocardia* and *Corynebacterium.*

Myxospore. The resting cell usually contained within the fruiting body of a myxobacter and substantially more resistant than a vegetative cell.

Necrohormones. A substance responsible for auto-antagonism of Actinomycetales.

Negative Phase. When *Borrelia* is taken up by an arthropod, it disappears from the stomach after a short time, and if found again in the celomic cavity after a few days. The interim is called the "negative phase." It may be due to the *Borrelia* becoming small and slender while penetrating the gut wall and their number so small that they are missed on microscopic examination. The arthropod is not considered infective during that period.

Neotype. A strain proposed in accordance with the Bacteriological Code to replace a type strain no longer extant or a type description of an organism which was not cultivated or grown in pure culture or was represented by a nonliving specimen. The present Code accepts a neotype as soon as published in the *International Journal of Systematic Bacteriology.* Some authors have included suggested neotypes—these had not been published in *International Journal of Systematic Bacteriology* when THE MANUAL went to press and until they are properly proposed have no nomenclatural significance and are synonymous with reference strain.

Nomen Dubium. A name representing a taxon that is not identifiable from the original diagnosis or from the type material.

Nomen Nudum. A name published without any definition, description or indication of the taxon to which it is intended to apply, and hence without status in nomenclature.

Nonphotochromogen. Not requiring light for pigment formation; usually applied to mycobacteria having (1) little or no pigmentation or (2) color developing late in colony development.

0/129. A vibriostatic compound, 2,4-diamino-6,7-diisopropyl pteridine.

Optical Antipode. The D and L isomers of an optically active organic compound.

Outer Envelope. A thin flexible structure surrounding the protoplasmic cylinder and axial fibrils of spirochetes.

Palisade. Rods lined-up side by side as pales in a picket fence.

Parenchyma. Fundamental or ground tissue of a plant; a tissue composed of living cells variable in their morphology and physiology, but generally having thin walls and a polyhedral shape.

Pedicel. A small stalk; in the Myxobacterales one of the ultimate divisions of a common stalk, bearing a single sporangium.

Pentaene (pentaenic) (antibiotics). With five conjugated double bonds. Also see Polyene (antibiotics).

Periplast. From Greek *plassein,* to form. Peripheral protoplasm of a cell; matrix of a part or organ; cell wall. (*Taber's Cyclopedic Medical Dictionary,* 12th edition). The use of this term by some authors originated probably at the time when *Borrelia* was classified with protozoa.

Phalangiform (of spores). Spores which, when joined in chains, give the appearance of phalanges (fingers).

Photochromogen. Organism (mycobacterium) which exhibits pigmentation (carotenoid) only if exposed to light during growth.

Photolithotrophic. Able to grow with light in a strictly inorganic medium.

Photoorganotrophic. Able to grow with light at the expense of organic compounds.

Pileus (plural **pilei**). The "cap" of a mushroom.

Plasmalemma. Synonym of cell (cytoplasmic) membrane (botanical usage).

Plasmalemmosome. Infolding of the cell membrane (see Mesosome).

Polyene (polyenic) (antibiotics). Characterized by having conjugated carbon to carbon double bonds, most of which give characteristic ultraviolet spectra.

PPD. A tuberculin ("purified protein derivative") derived from autoclaved culture filtrate by ammonium sulfate precipitation or comparable product from culture filtrates of mycobacteria other than tubercle bacilli.

Prostheca. A semirigid appendage, extending from a procaryotic cell, the diameter of which is always smaller than that of the mature cell and which is bounded by the cell wall. Includes the cellular stalks of caulobacters and the hyphae of hyphomicrobia but not the secreted stalks of *Gallionella* or *Nevskia.*

Protoplasmic Cylinder. The central structure of the spirochete which includes the nuclear apparatus, ribosomes and other intracellular organelles surrounded by a three layer unit membrane.

Pseudoplasmodium. In the Myxobacterales, a swarm or vegetative colony resembling the plasmodium of a slime mold (myxomycete) but composed of bacteria and slime. Cells not amoeboid.

Pseudostalk. In *Asticcacaulis,* an appendage similar in dimensions, fine structure and development to the stalk in *Caulobacter,* but which lacks adhesive material.

Purine (antibiotics). Antibiotics containing a purine, a nitrogen heterocyclic ring system.

Pyriform. Pearlike in form.

Rapid Grower (mycobacterium). One which exhibits complete growth to maturity within a week at 37 C and comparable growth at 25 C.

Rectus Flexibilis. A term to describe streptomycetes

having spore chains that are straight, flexuous or fascicled (see Plate 17.12).

Reference Strain. A strain which is used for reference in the production of a specific compound, antigen, or one which appears to meet the description of the type but has not been proposed as a neotype.

Retinaculum Apertum. A term to describe streptomycetes having spore chains that are hooks, open loops or greatly extended spirals (see Plate 17.12).

Rhizothamnion (plural **rhizothamnia**). Root nodules infected by two species of the genus *Frankia* become covered with negatively geotrophic rootlets growing from the apical meristem of each lobe of the nodule. A nodule covered with upward growing rootlets may be called a rhizothamnion.

Sclerotium (plural **sclerotia**). A hard resting body resistant to unfavorable conditions composed of a hardened mass of hyphae.

Scotochromogen. A mycobacterium exhibiting pigmentation in presence or absence of light.

Sedum-like. *Sedum*, a genus of fleshy green plants, a common example is stonecrop.

Segmentation Spores. Spores produced by the septation or successive contraction of the protoplasts of the tips of aerial filaments.

Sensitin. "... A non-antigenic substance, prepared from a microorganism (virus, bacterium or fungus), capable of revealing sensitivity of the delayed type evoked by the organism. Human and avian tuberculin, histoplasmin, and coccidioidin are examples of the 'sensitin'" (Magnusson, 1961, p. 57).

Serophilic. A bacterium that grows better when serum is added to the medium.

Sessile. Of cells, attached directly to the substrate or a surface without a stalk; of spores, budded directly from a hypha without an intervening sporophore.

Sinus, Apical. Crater-like concavity on top of colony of *Zoogloea ramigera*.

Sorption. The removal of antibodies from an antiserum by the addition of the antigen and then removal of the antigen-antibody complex to obtain a sorbed antiserum. Applicable to either adsorption or absorption.

Sorus (plural **sori**). In the Myxobacterales, a mass of sporangia enclosed in a common slime layer or envelope.

Spheroplast. In a general sense, used to designate a bacterial cell which has rounded up and become osmotically sensitive as a result of the partial or complete destruction of the peptidoglycan layer of the wall. In the genus *Actinomyces* it refers to the coccoid cells which occur in cultures of these normally filamentous organisms.

Spicule. A straight, pointed, rigid terminal extension of spirochetes; found on some cristispiras from fresh water bivalve mollusks.

Spider Colony. A colony composed almost entirely of radiating filaments which seem to originate from a central point.

Spira. Term to describe streptomycetes having spore chains that are either short, gnarled or compact spirals, or extended long and open spirals (see Plate 17.12).

Spirochetal Sphere. A form of spirochete in which the protoplasmic cylinder is either partially or completely coiled within a saclike outer envelope.

Sporangiole. A small sporangium generally having a small number of spores.

Sporangiospore. An asexual spore (but not an endospore) formed within a sporangium or spore vesicle, typically in the Actinoplanaceae.

Sporangium. A spore case or envelope, the sac in which spores are produced. (1) Spore vesicle containing one to many motile or nonmotile asexual spores (sporangiospores) in Actinoplanaceae, Actinomycetaceae. Differs from the sporangium formed in fungi, e.g. Mucorales; in the actinomycetes, hyphae grow inside the spore vesicle and eventually divide by annular ingrowth of the hyphal wall to give the individual sporangiospores; (2) the organ or part of the bacterium specially fashioned for the development of its spore (endospore) in *Bacillus, Thermoactinomyces* and *Actinobifida;* (3) in the Myxobacterales, a saclike structure delimited by a discrete slime layer or membrane and containing the myxospores.

Spore. A general term used for the reproductive or resting body produced in fungi and bacteria. See aleuriospore, arthrospore, chlamydospore, conidia, condiospore, endospore, microcyst, sporangiospore, zoospore.

Spore Vesicle. Term proposed by Cross (1970) for the envelope within which spores are produced by several of the Streptomycetaceae and which previously has been loosely referred to as a "sporangium."

Sporogenic. Producing spores.

Sporophore. Any structure which bears spores or spore chains.

Stalk. A filiform structure, either cellular or secreted, which anchors an organism to the substrate or a surface.

Streptotrichosis. A skin disease of sheep, cattle, horses and goats, especially in tropical Africa (also spelled streptothrichosis).

Subspecies. (variety). A subdivision of the species in which the name is a trinomial consisting of a generic name followed by a specific epithet and this, in turn, by a subspecific epithet. The epithets are preferably separated by the abbreviation "subsp.".

Substrate Mycelium. The portion of the mycelium of an actinomycete in contact with the substrate.

"Sulfur Granule." A small yellowish to white granule, 1 to 2 mm in diameter, usually present in pus from human or animal actinomycosis. Microscopically composed of branching filaments of *Actinomyces*; the ends of some of the filaments extend from the periphery and are encased in a protein-polysaccharide complex. The ashed granule contains CaO, P_2O_5 and acid phosphate probably occurring as poorly crystallized apatite.

Swarm (swarming). The spread of motile organisms over the surface of solid media due to the movement of individuals, groups or colonies; characteristic of the vegetative cells of myxobacteria and of a few flagellated bacteria, e.g. *Proteus, Bacillus, Clostridium*.

Sympodial. A dichotomy where at each forking one branch continues to develop and the other aborts (Jackson, 1928).

Synergistic (of antibiotics). The combined effect of two antibiotics when the total activity is greater than the sum of the two effects taken independently.

Synnema. A group of parallel hyphae joined together forming an elongated erect spore-bearing structure.

Tetraene (tetraenic) (antibiotics). With four conjugated double bonds. Also see Polyene (antibiotics).

Thallus. As used in *Dermatophilaceae*, a general term for the vegetative or assimilative phase of a non-vascular plant or organism.

Thermocline. The layer in a thermally stratified body of water within which the temperature decreases rapidly with increasing depth usually at a rate greater than 1 C per meter.

Thylakoids. See intracytoplasmic membrane systems.

Trichome. The meaning of this term in bacteriology seems to vary. It comes from the Greek n. *trichoma* a growth of hair or hair generally. In botany it usually refers to any hairlike outgrowth of the epidermis such as a hair or bristle. In the blue-green algae a trichome refers to a row of cells, i.e., a filament apart from its investing sheath. In the sheathed bacteria, trichome refers to a hairlike organism in which the septa are not necessarily a part of the fission mechanism. The septa may not be completely closed, allowing transfer of protoplasm from one compartment to the next. The term has also been used for cells, not necessarily filamentous or hairlike, in which at one time septation is found in various stages of completion. Because of this confusion the term is not used in this edition of The Manual

Tuberculin. Derivative of tubercle bacilli or their culture filtrate capable of eliciting skin reactions resulting from sensitization by tubercle bacilli.

Tweens. A group of non-ionic surface active agents. A trade name of Atlas Powder Co. Tween 20—polyoxyethylene sorbitan monolaurate; Tween 80—polyoxyethylene sorbitan monooleate.

Twitching. A movement on solid surfaces of both flagellated and nonflagellated bacteria but not due to action of flagella. Cell movement intermittent and jerky, does not follow the long axis of cell and is predominately of single cells.

Tycholimnetic. Fresh water forms which are at first fixed, but later break loose and float.

Umbel. The arrangement of three or more spore chains borne at the apex of each branch of a verticil in *Streptoverticillium*.

Umbellate (of spore chains). Branched spore chains in which the branches arise at about the same point from the spore-bearing hypha.

Vegetative Mycelium. See substrate mycelium.

Verticil. The three or more side branches which arise in a whorl at the same level and at more or less regular intervals along the main aerial hyphae in *Streptoverticillium*. Each side branch may subdivide to produce umbels (q.v.).

Verticillate (or spore chains). Whorled chains of spores arising radiately at nodes on a spore-bearing hypha like the spokes of a wheel.

V Factor. An accessory growth substance required by certain species of *Haemophilus* replaceable by niacinamide adenine dinucleotide or one of its riboside precursors. It is destroyed by autoclaving and deteriorates when stored, and especially rapidly above freezing temperatures.

Whole Cell Hydrolyzates (of actinomycetes). The residue from the acid hydrolysis of whole cells with 6 N HCl at 100 C for 18 hrs.

Xenodiagnosis. Diagnosis based on feeding of uninfected vectors on a patient suspected of harboring an organism and examining the vector for development in it of the organism in question.

X Factor. An accessory growth substance required by certain species of *Haemophilus* replaceable by hemin or other iron porphyrin compounds and stable to autoclaving.

Zoogloeal Floc. Growth form of *Zoogloea ramigera* in mature cultures in high C:N media, sewage or natural waters.

Zoospore. A spore motile by means of flagella. The term is used more appropriately in botany and zoology.

Cumulative List of References

Aaronson, S. 1955 Biotin assay with a coccus, *Micrococcus sodonensis*, nov. sp. J Bacteriol 69 (1): 67–69.

—. 1956 A biochemical-taxonomic study of a marine micrococcus, *Gaffkya homari*, and a terrestrial counterpart. J Gen Microbiol 15: 478–484.

Åarsleff, B., E. L. Biberstein, B. J. Shreeve and D. A. Thompson. 1970 A study of untypeable strains of *Pasteurella haemolytica*. J Comp Pathol 80: 493–498.

Aasen, A. J. and S. Liaaen-Jensen. 1967 Bacterial carotenoids. XXIII. The carotenoids of *Thiorhodaceae* 6. Total synthesis of okenone and related compounds. Acta Chem Scand 21: 970–982.

Abbott Laboratories. 1954 Improvements in or relating to antibiotics. British Patent 719,230, December 1.

Abd-el-Malek, Y. and T. Gibson. 1948 Studies in the bacteriology of milk. 1. The streptococci of milk. J Dairy Res 15: 233–248.

Abdou, M. A-F. 1969 Über eine neue Art eines laevanbildenden Bakteriums aus Zuckerruben, *Corynebacterium beticola*. Phytopathol Z 66: 147–167.

Abe, S., K. Takayama and S. Kinoshita. 1967 Taxonomical studies on glutamic acid-producing bacteria. J Gen Appl Microbiol 13(3): 279–301.

Abel, R. 1893 Bakteriologische Studien über Ozaena simplex. Zentrabl Bakteriol Parasitenk Infektionskr Hyg Abt I Orig 13: 161–173.

—. 1925 Genus XVIII *Klebsiella* Trevisan 1885, 263–266. *In* Bergey, Harrison, Breed, Hammer and Huntoon. *Bergey's Manual of Determinative Bacteriology*. Ed. 2, Williams & Wilkins Co., pp. 1–462.

Abhyankar, S. G., M. K. Patel and M. J. Thirumalachar. 1956 A new bacterial leafspot on *Cleome monophylla*. Curr Sci 25: 93.

Abo-Elnaga, I. G. and O. Kandler. 1965 Zur Taxonomie der Gattung *Lactobacillus* Beijerinck. I. Das Subgenus *Streptobacterium* Orla-Jensen. Zentrabl Bakteriol Parasitenk Infektionskr Hyg Abt II 119: 1–36.

— and —. 1965 Zur Taxonomie der Gattung *Lactobacillus* Beijerinck. II. Das Subgenus *Betabacterium* Orla-Jensen. Zentrabl Bakteriol Parasitenk Infektionskr Hyg Abt II 119: 117–129.

— and —. 1965 Zur Taxonomie der Gattung *Lactobacillus* Beijerinck. III. Das Vitaminbedurfnis. Zentrabl Bakteriol Parasitenk Infektionskr Hyg Abt II 119: 661–672.

Aburatani, I. 1959 On cycloserine produced by *Streptomyces nagasakiensis* n. sp. and on some by-products and decomposed products of cycloserine. (In Japanese) J Osaka City Med J 8(11): 175–189.

Achalme, P. and G. Rosenthal. 1906 Le *Bacillus gracilis ethylicus*, microbe anaérobe de l'estomac produit la fermentation alcoolique du lait. C R Soc Biol Paris 60: 1025–1027.

Adams, M. E. and J. R. Postgate. 1959 A new sulphate-reducing vibrio. J Gen Microbiol 20(2): 252–257.

Adamson, R. S. 1919 On the cultural characters of certain anaerobic bacteria isolated from war wounds. J Pathol Bacteriol 22: 345–400.

Adler, H. E., J. Fabricant, R. Yamamoto and J. Berg. 1958 Symposium on chronic respiratory diseases of poultry. I. Isolation and identification of pleuropneumonia-like organisms of avian origin. Amer J Vet Res 19: 440–447.

— and M. Shifrine. 1964 *Mycoplasma laidlawii* var. *inocuum* comb. nov. J Bacteriol 87(5): 1245.

—, — and H. Ortmayer. 1961 *Mycoplasma inocuum* n. sp., a saprophyte from chickens. J Bacteriol 82(2): 239–240.

Adler, S. and V. Ellenbogen. 1934 A note on two new blood parasites of cattle, *Eperythrozoon* and *Bartonella*. J Comp Pathol 47: 220–221.

Affanassieff, M. J. and M. Schulz. 1889 Ueber die Aetiologie der Actinomycosis. Zentrabl Bakteriol Parasitenk Infektionskr Hyg Abt I Orig 5: 683–684.

Agardh, C. A. 1827 Aufzählung einiger in Ostreichischen Ländern gefunden neuen Gattungen und Arten von Algen, nebst ihrer Diagnostik und beigefugten Bemerkungen; von Hrn Prof. Agardh in Lund. Flora Bot Ztg (Jena) 2: 625–640.

Agate, A. D. and W. V. Vishniac. 1973 Characterization of thiobacillus species by gas-liquid chromatography of cellular fatty acids. Arch Mikrobiol 89: 257–267.

Ahmad, K. and J. A. M. Bhuiyan. 1958 A new antifungal *Streptomyces* species, *Streptomyces corchorusii*. Pakistan J Biol Agr Sci 1(2): 137–143.

—, M. De and A. K. M. M. Rahman. 1955 A new antibiotic-producing *Streptomyces* sp. from a local soil sample. Ann Biochem Exp Med (Calcutta) 15: 175–180.

Ahrens, R. 1968 Taxonomische Untersuchungen an sternbildenden *Agrobacterium*-Arten aus der westlichen Ostsee. Kiel Meeresforsch 24: 147–173.

— and G. Moll. 1970 Ein neues Knospendes Bakterium aus der Ostsee. Arch Mikrobiol 70: 243–265.

— and G. Rheinheimer. 1967 Über einige sternbildende Bakterien aus der Ostsee. Kiel Meeresforsch 23: 127–136.

Ahvonen, P. and E. Jansson. 1968 Cross-reaction between Brucellae and a subtype of *Yersinia enterocolitica*. Scand J Clin Lab Invest Suppl 101 21: 57.

—, — and K. Aho. 1969 Marked cross-agglutination between Brucellae and a subtype of *Yersinia enterocolitica*. Acta Pathol Microbiol Scand 75: 291–295.

Aikimbaev, M. A. 1966 Taxonomy of the genus *Francisella*. Rep Acad Sci Kazakhstan SSR Biol Ser 5: 42–44.

Ainsworth, G. C. 1961 Ainsworth and Bisby's *Dictionary of the Fungi*. 5th ed. Commonwealth Mycological Institute, Kew.

Aiso, K., T. Arai, M. Suzuki and Y. Takamizawa. 1956 Gancidin, an antitumor substance derived from *Streptomyces* sp. I. J Antibiot (Tokyo) Ser A, 9: 97–101.

—, F. Yanagisawa and M. Nakajima. 1948–1949 Studies on the distribution of actinomycetes and their antagonistic strains in Japanese soil. J Antibiot (Tokyo) Ser A, 2(4): 240–248.

—, K. Miyake, F. Yanagisawa, T. Arai and M. Hayashi. 1950 "Flavomycin" an antibiotic produced by *Streptomyces* No. 320. J Antibiot (Tokyo) Ser A, 3(2): 87–92.

Akashi, A. 1960 Studies on the cellulose-decomposing bacteria in the rumen. J Agr Chem Soc Jap *34:* 895–900.

Akita, E., K. Maeda and H. Umezawa. 1963 Isolation and characterization of labilomycin, a new antibiotic. J Antibiot (Tokyo) Ser A, *16:* 147–151.

Akkada, A. R. A. and T. H. Blackburn. 1963 Some observations on the nitrogen metabolism of rumen proteolytic bacteria. J Gen Microbiol *31:* 461–469.

Alarie, A. M. and P. H. H. Gray. 1947 Aerobic bacteria that decompose cellulose, isolated from Quebec soils. I. Isolation and description of the species. Can J Res (Sect. C) *25:* 228–241.

Al-Aubaidi, J. M., A. H. Dardiri, C. C. Muscoplatt and E. H. McCauley. 1973 Identification and characterization of *Acholeplasma oculusi* spec. nov. from the eyes of goats with keratoconjunctivitis. Cornell Vet *63:* 117–129.

Albert, C. A. and V. M. Malaquias de Querioz. 1963 *Streptomyces tuirus* nov. sp., produtor do antibiotico tuoromicina. Rev Inst Antibiot Univ Recife *5 (1–2):* 43–51.

Albrecht, H. and A. Ghon. 1903 Zur Frage der morphologischen Charakterisierung des *Meningococcus intracellularis.* Zentrabl Bakteriol Parasitenk Infektionskr Hyg Abt I Orig *33:* 496–510.

Alcorn, S. M. 1961 Some hosts of *Erwinia carnegieana.* Plant Dis Rep *45:* 587–590.

Aldova, E., K. Laznickova, E. Stepankova and J. Lietava. 1968 Isolation of non-agglutinable vibrios from an enteritis outbreak in Czechoslovakia. J Infect Dis *118:* 25–31.

Ali-Cohen, C. H. 1889 Eigenbewegung bei Mikrokokken. Zentrabl Bakteriol Parasitenk Infektionskr Hyg Abt I *6:* 33–36.

Allen, E. K. and O. N. Allen. 1950 Biochemical and symbiotic properties of the rhizobia. Bacteriol Rev *14:* 273–330.

—, — and L. J. Klebesadel. 1964 An insight into symbiotic nitrogen-fixing plant associations in Alaska. Sci Alaska Proc Alaskan Sci Conf 54–63.

Allen, T. C. and A. J. Riker. 1932 A rot of apple fruit caused by *Phytomonas melophthora* n. sp. following invasion by the apple maggot. Phytopathology *22:* 557–571.

Allison, M. J., M. P. Bryant and R. N. Doetsch. 1962 Studies on the metabolic function of branched-chain volatile fatty acids, growth factors for ruminococci. I. Incorporation of isovalerate into leucine. J Bacteriol *83 (3):* 523–532.

—, —, I. Katz and M. Keeney. 1962 Metabolic function of branched-chain volatile fatty acids, growth factors for ruminococci. II. Biosynthesis of higher branched-chain fatty acids and aldehydes. J Bacteriol *83 (5):* 1084–1093.

Allsopp, A. 1969 Phylogenetic relationships of the Procaryota and the origin of the eucaryotic cell. New Phytol *68:* 591–612.

Alstatt, G. E. 1944 Tomato diseases in Texas. Plant Dis Rep *28:* 530.

Alton, G. G. and L. M. Jones. 1967 Laboratory techniques in Brucellosis. World Health Organ Monogr Ser No. 55.

Altson, R. A. 1936 Studies on *Azotobacter* in Malayan soils. J Agr Sci Camb *26:* 268–280.

Aluotto, B. B., R. G. Wittler, C. O. Williams and J. E. Faber. 1970 Standardized bacteriologic techniques for the characterization of

Mycoplasma species. Int J Syst Bacteriol *20 (1):* 35–58.

Alvarez, E. and E. Tavel. 1885 Recherches sur le bacille de Lustgarden. Arch Physiol Norm Pathol *6:* 303–321.

Ames, D. E., R. E. Bowman, J. F. Cavalla and D. D. Evans. 1955 Griseoviridin. Part I. J Chem Soc Jap *1955:* 4260–4263.

Amies, C. R. and M. Garabedian. 1963 The bacteriology of human vaginitis. Can J Public Health *54:* 50.

— and S. A. Jones. 1957 A description of *Haemophilus vaginalis* and its L-forms. Can J Microbiol *3:* 579–590.

Ammann, A., D. Gottlieb, T. D. Brock, H. E. Carter and G. B. Whitfield. 1955 Filipin, an antibiotic effective against fungi. Phytopathology *45:* 559–563.

Anacker, R. L., E. G. Pickens and D. B. Lackman. 1967 Details of the ultrastructure of *Rickettsia prowazekii* grown in the chick yolk sac. J Bacteriol *94 (1):* 260–262.

Andersen, A. A. 1958 New sampler for the collection, sizing and enumeration of viable airborne particles. J Bacteriol *76:* 471–484.

— and C. H. Werkman. 1940 Description of a dextro-lactic acid-forming organism of the genus *Bacillus.* Iowa State College J Sci *14 (2):* 187–194.

Anderson, D. R., H. E. Hopps, M. F. Barile and B. C. Bernheim. 1965 Comparison of the ultrastructure of several rickettsiae, ornithosis virus and mycoplasma in tissue culture. J. Bacteriol *90 (5):* 1387–1404.

Anderson, E. S. 1961 Slime-wall formation in the *Salmonellae.* Nature (London) *190:* 284–285.

—. 1964 The phage typing of *Salmonellae* other than *S. typhi. In* Van Oye *The world problem of salmonellosis.* Junk, The Hague, pp. 84–110.

Anderson, G. R. 1955 Nitrogen fixation by *Pseudomonas*-like soil bacteria. J Bacteriol *70 (2):* 129–133.

Anderson, H. 1954 The reddening of salted hides and fish. Appl Microbiol *21:* 64–69.

Anderson, H. W. and D. Gottlieb. 1952 Plant disease control with antibiotics. Econ Bot *6:* 294–308.

Anderson, K., W. DeMonbreun and E. Goodpasture. 1945 An etiologic consideration of *Donovania granulomatis* cultivated from granuloma inguinale. (Three cases) in embryonic yolk. J Exp Med *81:* 25–40.

Anderson, L. E., J. Ehrlich, S. H. Sun and P. R. Burkholder. 1956 Strains of *Streptomyces,* the sources of azaserine, elaiomycin, griseoviridin and viridogrisein. Antibiot Chemother *6 (2):* 100–115.

Anderson, R. L. and E. J. Ordal. 1961 *Cytophaga succinicans* sp. n., a facultatively anaerobic, aquatic myxobacterium. J Bacteriol *81 (1):* 130–138.

Andre. 1948 *Clostridium limosum* p. 277. *In* Prévot's Manual de Classification et de Determination des Bactéries Anaérobies. Masson and Co., Paris, pp. 1–290.

Andrewes, F. W., W. Bulloch, S. Douglas, G. Dreyer, A. Gardner, P. Fildes, J. Ledingham and C. Wolf. 1923 Diphtheria, its bacteriology, pathology and immunology. His Majesty's Stationery Office, London, pp. 7–544.

— and T. J. Horder. 1906 A study of the streptococci pathogenic for man. Lancet *2:* 708–713; 775–782; 852–855.

Andrews, J. and R. B. Gilliland. 1952 Superattenuation of beer: a study of three organisms capable of causing abnormal attenuations. J Inst Brew *58:* 189–196.

Angst, E. C. 1929 Some new agar-digesting bacteria. Publ Puget Sound Biol Sta *7:* 49–63.

Anthony, C. and L. J. Zatman. 1964 The microbial oxidation of methanol. 1. Isolation and properties of *Pseudomonas* sp. M27. Biochem J *92 (3):* 609–621.

Anzai, K. 1962 Studies on a new antibiotic, tuberin. II. Chemical structure. J Antibiot (Tokyo) Ser A *15 (3):* 117–122.

—, G. Nakamura and S. Suzuki. 1957 A new antibiotic, tubercidin. J Antibiot (Tokyo) Ser A *10:* 201–204.

—, K. Okuma, J. Nagatsu and S. Suzuki. 1962 Chemical structure of tuberin. J Antibiot (Tokyo) Ser A *15 (2):* 110–111.

Aoi, K. and J. Orikura. 1928 On the decomposition of agar, xylan, etc., and the sugars related to these hemicelluloses by *Vibrio andoi* n. sp. Zentrabl Bakteriol Parasitenk Infektionskr Hyg Abt II *74:* 31–333.

Aoyama, S. 1952 Studies on the thiamin decomposing bacterium. I. Bacteriological researches of a new thiamin decomposing bacillus, *Bacillus aneurinolyticus* Kimura et Aoyama. Acta Sch Med Univ Kioto *30:* 127–132.

Aragao, H. and G. Vianna. 1913 Resquizas sobre o *Granuloma venereo.* (Untersuchungen ueber das *Granuloma venereum*). Mem Inst Oswaldo Cruz Rio de Janeiro *5:* 211–238.

Arai, M., K. Karasawa, S. Nakamura, H. Yonehara and H. Umezawa. 1958 Studies on mikamycin. I. J Antibiot (Tokyo) Ser A *11:* 14–20.

—, S. Nakamura, Y. Sakagami, K. Fukuhara and H. Yonehara. 1956 A new antibiotic, mikamycin. J Antibiot (Tokyo) Ser A *9:* 193.

Arai, T. 1951 Studies of flavomycin: Taxonomic investigations on the strain, production of the antibiotic and application of cup method to the assay. J Antibiot (Tokyo) Ser A *4:* 215–221.

— and S. Kuroda. 1965 Transfer of *Nocardia narashinoensis* Aiso et Arai to the genus *Micromonospora* as *Micromonospora narshino* (Aiso et Arai) comb. nov. Annu Rep Inst Food Microbiol Chiba Univ *18:* 33–38.

—, —, S. Yamagishi and Y. Katoh. 1964 A new hydroxystreptomycin source, *Streptomyces subrutilus.* J. Antibiot (Tokyo) Ser A *17 (1):* 23–28.

—, T. Nakada and M. Suzuki. 1957 Production of viomycin-like substance by a *Streptomyces.* Antibiot Chemother *7:* 435–442.

Araki, C. and K. Arai. 1954 Studies on agar-digesting bacteria. The isolation of agar-digesting bacteria and their enzymatic activities. Mem Fac Indust Arts Kyoto Tech Univ *3B:* 7–23.

Araujo, W. D. de, E. Varah and S. E. Mergenhagen. 1963 Immunochemical analysis of human oral strains of *Fusobacterium* and *Leptotrichia.* J Bacteriol *86 (4):* 837–844.

Arbuthnott, J. P., T. Bauchop and E. A. Dawes. 1960 A clastic fission of pyruvic acid in *Zymosarcina ventriculi.* Biochem J *76:* 12.

Arcamone, F. 1960 Melanosporin and elaiophylin, new antibiotics produced by *Streptomyces*

melanosporus sp. n. *In* M. Herold and Z. Gabriel (editors) Proc Symp on Antibiotics Prague, Czechoslovakia, May 18–23, 1959. Statni Zdav. Nakladatelstvie, Prague, pp. 96–97.

Arcamone, F., C. Bertazzoli, G. Canevazzi, A. diMarco, M. Ghione and A. Grein. 1957 La etruscomicina nuovo antibiotico antifungino prodotto dallo *Streptomyes lucensis* n. sp. G Microbiol *4:* 119–128.

—, —, M. Ghione and T. Scotti. 1959 Melanosporin and elaiophylin, new antibiotics from *Streptomyces melanosporus* (sive *melanosporofaciens*) n. sp. (in Italian). G Microbiol *7 (3):* 207–216.

—, F. Bizioli, G. Canevazzi and A. Grien. 1959 Methods for production and preparation of antibiotics. German Patent 1039198.

—, T. Scotti, C. Bertazzoli and M. Ghione. 1961 F.I. 1600, a new antibiotic. British Patent 880,035, October 18.

Arishima, M. and J.-M. Sakamoto. 1961 Moldcidin-A, antibiotic substance (in Japanese). Japanese Patent 1,148, March 7.

— and —. 1961 Moldcidin-B (in Japanese). Japanese Patent 15,945, September 11.

—, — and T. Sato. 1956 Studies on an antibiotic *Streptomyces* No. 689 strain. Part I. Taxonomic studies (in Japanese). J Agr Chem Soc Jap *30:* 469–471.

—, Y. Sekizawa, T. Sato and K. Miwa. 1955 Studies on an antibiotic, *Streptomyces* No. 310 strain (in Japanese). J Agr Chem Soc Jap *29:* 810–817

Aristovskaya, T. V. 1961 Accumulation of iron by decomposing organo-mineral complexes of humic matter by microorganisms. Dokl Akad Nauk SSSR *136 (4):* 954–957.

—. 1963 Decomposition of organo-mineral compounds in podzolic soils. Pochvovedenie (Transl) *1:* 30–43.

— and O. M. Parinkina. 1963 A new soil microorganism *Seliberia stellata* n. gen. n. sp. (in Russian). Izv Akad Nauk SSSR Ser Biol *218 (1):* 49–56.

Ark, P. A. and J. T. Barrett. 1946 A new bacterial leaf-spot of greenhouse grown gardenias. Phytopathology *36:* 865–868.

Arkwright, J. A., E. E. Atkin and A. Bacot. 1921 An hereditary rickettsia-like parasite of the bedbug (*Cimex lectularius*). Parasitology *13:* 27–36.

Arloing, S., Cornevin and Thomas. 1887 Le charbon symptomatique du boeuf. 2nd ed. Asselin and Houzeau, Paris, pp. 1–281.

Armstrong, C. H. and J. B. Payne. 1969 A study of bacteria recovered from swine affected with cervical lymphadenitis (jowl abscess). Amer J Vet Res *30:* 1607–1612.

Armstrong, D., J. G. Tully, B. Yu, V. Morton, M. H. Friedman and L. Steger. 1970 Previously uncharacterized *Mycoplasma* isolates from an investigation of canine pneumonia. Infect Immun *1:* 1–7.

Arnaud, G. 1920 Une maladie bactérienne du lierre (*Hedera helix* L.). C R Hebd Seances Acad Sci Ser D (Paris) *171:* 121–122.

Arnaudi, C. 1939 Oxydation microbienne et enzymatique dans lasérie des hormones sexuelles par le "*Micrococcus dehydrogenans*" n. sp. Boll Sez Ital Soc Int Microbiol *11:* 208–211.

—. 1942 *Flavobacterium dehydrogenans (Micro-*

coccus dehydrogenans) und seine Fähigkeit zur Oxydation von Steroiden sowie Substanzen aus der Sexualhormonreihe. Zentrabl Bakteriol Parasitenk Infektionskr Hyg Abt II *105:* 352–366.

Aronson, J. D. 1926 Spontaneous tuberculosis in soft water fish. J Infect Dis *39:* 314–320.

—. 1929 Spontaneous tuberculosis in snakes, n. sp. *Mycobacterium thamnopheos.* J Infect Dis *44:* 215–223.

Arseculeratne, S. N. 1962 Actinobacillosis in joints of rabbits. J Comp Pathol *72:* 33–39.

Arthaud-Berthet and G. Bondar. 1915 Molestia bacteriana da mandioca. Bol Agr Sao Paulo *16:* 513–524.

Asahi, K., J. Nagatsu and S. Suzuki. 1966 Xanthocidin, a new antibiotic. J Antibiot (Tokyo) Ser A *19 (5):* 195–199.

Asai, R. and S. Haruda. 1943 Butanol-isopropanol fermentation. I. Production by *Clostridium butanologenum* nov.sp. and *Clostridium* no. K.M. J Agr Chem Soc Jap *91:* 872–878.

Asai, T. 1935 Taxonomic studies on acetic acid bacteria and allied oxidative bacteria isolated from fruits. A new classification of the oxidative bacteria. J Agr Chem Soc Jap *11:* 499–513, 610–620, 674–708.

—, H. Iizuka and K. Komagata. 1964 The flagellation and taxonomy of genera *Gluconobacter* and *Acetobacter* with reference to the existence of intermediate strains. J Gen Appl Microbiol *10:* 95–126.

— and K. Shōda. 1958 The taxonomy of *Acetobacter* and allied oxidative bacteria. J Gen Appl Microbiol *4:* 289–311.

Aschner, M. 1931 Die Bakterienflora der Pupiparen (*Diptera*). Eine Symbiosestudie am Blutausgenden Insekten. Z Morphol Oekol Tiere *20:* 368–442.

Ashby, S. F. 1929 Gumming disease of sugar cane. Trop Agr Trinidad *6:* 135–138.

Atkinson, G. F. 1892 The Genus *Frankia* in the United States. Bull Torrey Bot Club *19:* 171–177.

Audureau, A. 1940 Étude du genre *Moraxella.* Ann Inst Pasteur (Paris) *64:* 126–166.

Aue, B. J. and R. H. Deibel. 1967 Fumarate reductase activity of *Streptococcus faecalis.* J Bacteriol *93 (6):* 1770–1776.

Aujeszky, A. 1914 A Koeleria glauca bakteriozisarol. Bot Kozlem *13:* 87–93.

Auletta, A. E. and E. R. Kennedy. 1966 Deoxyribonucleic acid base composition of some members of the *Micrococcaceae.* J Bacteriol *92:* 28–34.

Austrian, R. 1953 Morphologic variation in *Pneumococcus.* I. An analysis of the bases for morphologic variation in *Pneumococcus* and description of a hitherto undefined morphologic variant. J Exp Med *98:* 21–34.

—. 1953 Morphologic variation in *Pneumococcus.* II. Control of pneumococcal morphology through transformation reactions. J Exp Med *98:* 35–40.

Austwick, P. K. C. 1958 Nomenclature of fungi pathogenic to man and animals. *In* Great Britain Medical Research Council Memorandum No. 23. 2nd ed. Her Majesty's Stationery Office, London, pp. 1–15.

—. 1958 Cutaneous streptothricosis, mycotic dermatitis and strawberry foot rot and the genus *Dermatophilus* Van Saceghem. Vet Rev Annot *4:* 33–48.

Axenfeld, T. 1897 Ueber die chronische Diplobacillen-Conjunctivitis. Zentrabl Bakteriol Parasitenk Infektionskr Hyg Abt I Orig *21:* 1–9.

Ayers, S. H. and W. T. Johnson. 1924 Studies on pasteurization. XII. Cause and significance of pin-point colonies from pasteurized milk. J Bacteriol *9:* 285–300.

—, — and C. S. Mudge. 1924 Streptococci of souring milk with special reference to *Streptococcus lactis.* J Infect Dis *34:* 29–53.

— and C. S. Mudge. 1922 The streptococci of the bovine udder. J Infect Dis *31:* 40–50.

Ayers, T. T., C. L. Lefebvre and H. W. Johnson. 1939 Bacterial wilt of lespedeza. U S Dept Agr Tech Bull *704:* 1–22.

Ayers, W. A. 1958 Nutrition and physiology of *Ruminococcus flavefaciens.* J Bacteriol *76 (5):* 504–509.

—. 1958 Phosphorylation of cellobiose and glucose by *Ruminococcus flavefaciens.* J Bacteriol *76 (5):* 515–517.

Ayoub, E. M. and J. J. Ferretti. 1966 Use of bisulfite in the streptococcal anti-nicotinamide adenine dinucleotidase test. Appl Microbiol *14:* 391–393.

Baalsrud, K. and K. Baalsrud. 1954 Studies on *Thiobacillus denitrificans.* Arch Mikrobiol *20:* 34–62.

Baars, J. K. 1930 Over sulfaatreductie door Bacteriën. W. D. Meinema, N.V. Delft, Holland, pp. 1–164.

Baba, T. 1943 Production of butanol, isopropanol and acetone by fermentation of sugary raw material. I. A new butanol-isopropanol bacterium. J Agr Chem Soc Jap *19:* 191–207.

Bachmann, B. J. 1955 Studies on *Cytophaga fermentans* n. sp., a facultatively anaerobic lower myxobacterium. J Gen Microbiol *13:* 541–551.

Bacigalupo, J. 1926 *Treponema uretritis* (nueva especie). Sem Med *33:* 1567–1569.

Backus, E. J., H. D. Tresner and T. H. Campbell. 1957 The nucleocidin and alazopeptin producing organisms: Two new species of *Streptomyces.* Antibiot Chemother *7 (10):* 532–541.

Bacon, M. F., W. G. Overend, P. H. Lloyd and A. R. Peacocke. 1967 The isolation, composition and physicochemical properties of deoxyribonucleic acid from *Bordetella pertussis.* Arch Biochem Biophys *118:* 352–361.

Bader, R. E. 1954 Über die Herstellung eines agglutinierenden Serums gegen die Rundform von *Shigella sonnei* mit einem Stamm der Gattung *Pseudomonas.* Z Hyg Infektionskr *140:* 450–456.

Badger, E. 1944 The structural specificity of choline for the growth of type III *Pneumococcus.* J Biol Chem *153:* 183–191.

Badian, J. 1930 Z cytologji miksobakterj (Zur Zytologie der Myxobakterien). Acta Soc Bot Pol *7:* 55–71.

Bahn, A. N., P. C. Y. Kung and J. A. Hayashi. 1960 Chemical composition and serological analysis of the cell wall of *Peptostreptococcus.* J Bacteriol *91 (5):* 1672–1676.

Bahr, L. 1919 Paratyfus hos honningbien samt nolge undersøgelser vedrørende forekomsten af bakterier-henhorende til Coli-tyfus-gruppe-i honningbiens tarm. Skand Vet Tidskr *9:* 25–40, 45–60.

Bailey, L. and D. C. Lee. 1962 *Bacillus larvae:* its cultivation *in vitro* and its growth *in vivo.* J Gen Microbiol *29 (4):* 711–717.

Bailie, W. E. 1969 Characterisation of *Haemophilus somnus* (n. sp.), a microorganism isolated from infectious thromboembolic meningoencephalomyelitis of cattle. Ph.D. thesis, Kansas State University.

Baillie, A., W. Hodgkiss and J. R. Norris. 1962 Flagellation of *Azotobacter* as demonstrated by electron microscopy. J Appl Bacteriol *25:* 116–119.

Baird-Parker, A. C. 1960 The classification of fusobacteria from the human mouth. J Gen Microbiol *22 (2):* 458–469.

—. 1963 A classification of micrococci and staphylococci based on physiological and biochemical tests. J Gen Microbiol *30 (3):* 409–427.

—. 1965 The classification of staphylococci and micrococci from world wide sources. J Gen Microbiol *39:* 363–387.

—. 1965 Staphylococci and their classification. Ann N Y Acad Sci *128:* 4–25.

Baker, E. E., H. Sommer, L. E. Foster, E. Meyer and K. F. Meyer. 1947 Antigenic structure of *Pasteurella pestis* and the isolation of a crystallin antigen. Proc Soc Exp Biol Med *64:* 139–141.

—, —, —, —, and —. 1952 Studies on immunization against plague. I. The isolation and characterization of the soluble antigen of *Pasteurella pestis.* J Immunol *68:* 131–145.

Baker, E. W. and G. W. Wharton. 1952 An Introduction to Acarology. The Macmillan Co., New York.

Baker, F., H. R. Papiska and L. L. Campbell. 1962 Choline fermentation by *Desulfovibrio desulfuricans.* J Bacteriol *84:* 973–978.

Baker, J. A. and W. A. Hagan. 1942 Tuberculosis of the Mexican platyfish (*Platypoecilus maculatus*). J Infect Dis *70:* 248–252.

Baker, J. L., F. E. Day and H. F. E. Hulton. 1912 Study of the organisms causing ropiness in beer and wort. J Inst Brew *18:* 651–672.

Bakoss, P. and B. Chorvath. 1965 Contribution à l'étude des propriétés immunologiques des lipases des Leptospires. Rapport préliminaire. Arch Inst Pasteur Tunis *42:* 171–178.

Balashova, V. V. 1967 Enrichment culture of *Gallionella filamenta* n. sp. Mikrobiologiya *36:* 646–650.

—. 1968 Taxonomy of genus *Gallionella.* Mikrobiologiya *37:* 715–723.

Balcke, J. 1884 Über faurigen Geruch des Bieres. Wochenschr Brau *1:* 257.

Baldacci, E. 1937 La conception d'espèce chez les actinomyces par rapport à leur classification et à leur determination. Boll Sez Ital Soc Int Microbiol *9:* 138–147.

—. 1939 Introduzione allo studio degli Attinomiceti. Mycopathologia *2:* 84–106.

—. 1944 Contributo alla sistematica degli attinomiceti: X-XVI—*Actinomyces madurae; Proactinomyces ruber; Proactinomyces pseudomadurae, Proactinomyces polychromogenus; Actinomyces violaceus; Actinomyces caeruleus;* con un elenco alfabetico delle specie e delle varieta finora studiate. Atti Ist Bot Univ Pavia Ser 5, *3:* 139–193.

—. 1958 Development in the classification of actinomycetes. G Microbiol *6:* 10–27.

—. 1961 The classification of the Actinomycetes in relation to their antibiotic activity. Advan Appl Microbiol *3:* 257–278.

—, G. Farina and R. Locci. 1966 Emendation of the genus *Streptoverticillium* Baldacci (1958) and revision of some species. G Microbiol *14:* 153–171.

—, —, — and G. Ragni. 1965 Description of a new species of *Streptomyces: Streptomyces capuensis* sp. nov. G Microbiol *13:* 45–62.

— and A. Grein. 1966 *Streptomyces avellaneus* and *Streptomyces libani:* two new species characterized by a hazel-nut brown (*avellaneus*) aerial mycelium. G Microbiol *14:* 185–198.

—, — and C. Spalla. 1955 Studio di una "Serie" de specie di attinomiceti: *A. diastaticus.* G Microbiol *1:* 127–143.

— and R. Locci. 1961 Osservazioni e ricerche su *Micromonospora melanosporea* comb. nov. e descrizione di una nuova sotto-specie: *M. melanosporea* subsp. *corymbica.* Ann Microbiol Enzimol *11:* 19–30.

— and —. 1966 A tentative arrangement of the genera in the actinomycetales. G Microbiol *14:* 131–139.

Baldwin, I. L. and E. B. Fred. 1929 Nomenclature of the root-nodule bacteria of the *Leguminosae.* J Bacteriol *17:* 141–150.

Balfour, A. 1911 The role of the infective granule in certain protozoal infections, as illustrated by the spirochaetosis of Sudanese fowls. J Trop Med Hyg *16:* 113–114.

Ballard, R. W., M. Doudoroff, R. Y. Stanier and M. Mandel. 1968 Taxonomy of the Aerobic Pseudomonads: *Pseudomonas diminuta* and *P. vesiculare.* J Gen Microbiol *53 (3):* 349–361.

—, N. J. Palleroni, M. Doudoroff, R. Y. Stanier and M Mandel. 1970 Taxonomy of the Aerobic Pseudomonads: *Pseudomonas cepacia, P. marginata, P. alliicola* and *P. caryophylli.* J Gen Microbiol *60 (2):* 199–214.

Baltazard, M. 1936 Individualité stricte des souches marocaines du spirochète Hispano-Africain. Bull Soc Pathol Exot *29:* 667–671.

—, M. Bahmanyar and C. Mofidi. 1950 *Ornithodorus erraticus* et fièvres récurrentes. Bull Soc Pathol Exot *43:* 595–601.

Bang, B. 1897 Die Aetiologie des seuchenhaften ("infectiosen") Verwerfens. Z Tiermed *1:* 241–278.

Banning, F. 1902 Zur Kenntnis der Oxalsäurebildung Durch Bakterien. Zentrabl Bakteriol Parasitenk Infektionskr Hyg Abt II, *8:* 395–398, 425–431, 453–456, 520–525, 556–567.

Barile, M. E., R. A. Del Giudice and J. G. Tully. 1972 Isolation and characterization of *Mycoplasma conjunctivae* sp. n. from sheep and goats with keratoconjunctivitis. Infect Immun *5 (1):* 70–76.

Barile, M. F., R. A. Del Giudice, T. R. Carski, C. F. Gibbs and J. A. Morris. 1968 Isolation and characterization of *Mycoplasma arginini:* spec. nov. Proc Soc Exp Biol Med *129:* 489–494.

—, T. C. Francis and E. A. Graykowski. 1968 *Streptococcus sanguis* in the pathogenesis of recurrent aphthous stomatitis. *In* Microbial protoplasts, spheroplasts and L-forms. Williams and Wilkins Co., Baltimore, pp. 444–456.

Barker, B. T. P. and V. F. Hillier. 1912 Cider sickness. J Agr Sci *5:* 67–85.

Barker, H. A. 1936 Studies upon the methane-producing bacteria. Arch Mikrobiol *7:* 420–438.

—. 1938 The fermentation of definite nitrogenous compounds by members of the genus *Clostridium*. J Bacteriol *36:* 322–323.

—. 1956 Bacterial fermentations. John Wiley and Sons, New York, pp. 1–95.

—. 1957 *Methanococcus* Kluyver and van Niel 1936, emend. Barker 1936. *In* Breed, Murray and Smith (Editors), Bergey's Manual of Determinative Bacteriology, 7th ed. The Williams and Wilkins Co., Baltimore, pp. 473–474.

— and V. Haas. 1944 *Butyribacterium*, a new genus of gram-positive, non-sporulating anaerobic bacteria of intestinal origin. J Bacteriol *47:* 301–305.

— and S. M. Taha. 1942 *Clostridium kluyverii*, an organism concerned in the formation of caproic acid from ethyl alcohol. J Bacteriol *43:* 347–363.

—, B. E. Volcani and B. P. Cardon. 1948 Tracer experiments on the mechanism of glycine fermentation by *Diplococcus glycinophilus*. J Biol Chem *173:* 803–804.

Barksdale, W. L., K. Li, C. S. Cummins and H. Harris. 1957 The mutation of *Corynebacterium pyogenes* to *Corynebacterium haemolyticum*. J. Gen Microbiol 16 (*3*): 749–758.

Barnes, E. M. 1964 Distribution and properties of serological types of *Streptococcus faecium*, *Streptococcus durans* and related strains. J Appl Bacteriol *27:* 461–470.

Barnes, I. J., H. W. Seeley and P. J. Van Demark. 1961 Nutrition of *Streptococcus bovis* in relation to dextran formation. J Bacteriol *82:* 85–93.

Barr, F. S. and P. E. Carman. 1956 *Streptomyces kentuckensis*, a new species, the producer of raisnomycin. Antibiot Chemother *6:* 286–289.

Barrett, P. A., E. Beveridge, P. L. Bradley, C. C. D. Brown, S. R. M. Bushby, M. L. Clarke, R. A. Neal, R. Smith and J. K. H. Wilde. 1965 Biological activites of some α-dithiosemicarbazones. Nature (London) *206:* 1340–1341.

Barnett, S. F. 1964 The preservation of *Babesia bigemina*, *Anaplasma centrale* and *Anaplasma marginale* by deep freezing. Vet Rec *76:* 4–7.

Bartholomew, J. W. and G. Paik. 1966 Isolation and identification of obligate thermophilic spore-forming bacilli from ocean basin cores. J. Bacteriol *92* (*3*): 635–638.

Bartnes, E., H. Zinnes, R. A. Moe and J. S. Kulesza. 1958 Studies on a new solubilized preparation of amphotericin B. Antibiot Ann 1957–1958, pp. 53–58.

Bartz, Q. R. 1948 Isolation and characterization of Chloromycetin. J Bacteriol *172* (*2*): 445–450.

—, C. C. Elder, R. P. Frohardt, S. A. Fusari, T. H. Haskell, O. W. Johannessen and A. Ryder. 1954 Isolation and characterization of azaserine. Nature (London) *173:* 72–73.

—, J. Ehrlich, J. D. Mold, M. A. Penner and R. M. Smith. 1951 Viomycin, a new tuberculostatic antibiotic. Amer Rev Tuberc *63* (*1*): 4–6.

—, J. Standiford, J. D. Mold, D. W. Johannessen, A. Ryder, A Maretzki and T. H. Haskell. 1955 Griseoviridin and viridogrisein: Isolation and characterization. Antibiot Ann 1954–1955, pp. 777–783.

Baruchello, L. 1887 Il virus dell' adenite equina; ricerche del Dott. Leopoldo Baruchello Tenente Veterinario al deposito alleva-mento cavalli di Palmanova. G Anat Fisiol Patalog Animal *19* (*5*): 241–270.

—. 1887 Il microbo dell adenite equina (*Bacillus adenitis equi*). Communicazione preventiva del socio dott. Leopoldo Baruchello, tenente veterina rio al deposito alevamento cavalli in Palmanova. Medico Veterin R Scuola Med Veter Torino *34:* 437–440.

Bascomb, S., S. P. Lapage, W. R. Willcox and M. A. Curtis. 1971 Numerical classification of the tribe *Klebsielleae*. J Gen Microbiol *66:* 279–295.

Baseman, J. B. and C. D. Cox. 1969 Terminal electron transport in *Leptospira*. J Bacteriol *97* (*3*): 1001–1004.

Bassalik, K. 1913 Uber die verarbeitung der oxalsaure durch *Bacillus extorquens* n. sp. Jahrb Wiss Bot *53:* 255–298.

Batchelor, M. D. 1919 Aerobic spore-bearing bacteria in the intestinal tract of children. J Bacteriol *4:* 23–34.

Bates, L. B. and J. H. St. John. 1922 Suggestion of *Spirochaeta neotropicalis* as name for spirochaete of relapsing fever found in Panama. J Amer Med Ass *79:* 575–576.

Batra, S. K. and B. S. Bajaj. 1965 *Streptomyces anandii*—a new species of *Streptomyces* isolated from soil. Indian J Exp Biol *3:* 240–242.

Battelli, C. 1947 Su di un piroplasma della *Naia nigricollis* (*Aegyptianella carpani* n. sp.). (A piroplasmid parasite of *N. nigricollis* (*A. carpani* n. sp.)). Riv Parassitol 8 (*4*): 205–212.

Batty, I. 1958 *Actinomyces odontolyticus*, a new species of actinomycete regularly isolated from deep carious dentine. J Pathol Bacteriol *75:* 455–459.

Bauchop, T. and R. W. Martucci. 1968 Ruminant-like digestion of the langur monkey. Science (Washington) *161:* 698–700.

Baudisch, O. 1936 Uber ein neues Schwefelbakterium aus den Thermen von Santa Rosalis, Mex. Svensk Kem Tidskr *47:* 191–204.

Bauer, V. L. 1962 Untersuchungen an *Sphaeromyxa xanthochlora*, n. sp., einer auf tropfkorpern vorkommenden myxobakterienart. Arch Hyg Bakteriol *146:* 392–400.

Baumann, P., L. Baumann and M. Mandel. 1971 Taxonomy of marine bacteria: the genus *Beneckea*. J Bacteriol *107*(*1*): 268–294.

—, —, — and R. D. Allen. 1971 Taxonomy of marine bacteria: *Beneckea nigrapulchrituda* sp. n. J Bacteriol *108* (*3*): 1380–1383.

—, M. Doudoroff and R. Y. Stanier. 1968 Study of the *Moraxella* group. I. Genus *Moraxella* and the *Neisseria catarrhalis* group. J Bacteriol *95* (*1*): 58–73.

—, — and —. 1968 A study of the *Moraxella* group. Oxidase-negative species (genus *Acinetobacter*). J Bacteriol *95* (*5*): 1520–1541.

Baumgarten, P. 1888 Jahresbericht über die Fortschritte in der Lehre von pathogenen Mikroorganismen umfassend Bacterien, Pilze und Protozöen. Dritter Jahrgang 1887, Jber Fortschr Pathol Microorg, pp. 1–517.

Baumgartner, J. G. 1937 The salt limits and thermal stability of new species of anaerobic halophile. Food Res *2:* 321–329.

Baur, E. 1905 Myxobakterien Studien. Arch Protistenk *5:* 92–121.

Bavendamm, W. 1924 Die farblosen und roten Schwefelbakterien des Süss-und Salzwassers. Pflanzenforsch edit. by Kolkwitz. Berlin-Dahlem, pp. 1–156.

Bayne-Jones, S. 1925 Club formation by *Actinomyces hominis* in glucose broth with a note on *B. actinomycetum-comitans*. J Bacteriol *10:* 569–575.

Baynes, I. D. and G. C. Simmons. 1960 Ovine epididymitis caused by *Actinobacillus seminis* n. sp. Aust Vet J *36:* 454–459.

Bazarewski, S. 1905 Ueber zwei neue farbstoffbildende Bakterien. Zentrabl Bakteriol Parasitenk Infektionskr Hyg Abt II *15:* 1–7.

Beale, G., H. A. Jurand and J. R. Preer. 1969 The classes of endosymbiont of *Paramecium aurelia*. J Cell Sci *5:* 65.

Beard, R. L. 1956 Two milky diseases of Australian Scarabaeidae. Can Entomol *88:* 640–647.

Beck, A. and J. L. Stanford. 1968 *Mycobacterium xenopei:* A study of sixteen strains. Tubercle *49:* 226–234.

Becker, B., M. P. Lechevalier, R. E. Gordon and H. A. Lechevalier. 1964 Rapid differentiation between *Nocardia* and *Streptomyces* by paper chromatography of whole-cell hydrolysates. Appl Microbiol *12:* 421–423.

—, — and H. A. Lechevalier. 1965 Chemical composition of cell wall preparations from strains of various form genera of aerobic actinomycetes. Appl Microbiol *13:* 236–242.

—, E. M. Wortzel and J. H. Nelson. 1967 Chemical composition of the cell wall of *Caryophanon latum*. Nature (London) *213:* 300.

Becking, J. H. 1961 Studies on nitrogen-fixing bacteria of the genus *Beijerinckia*. I. Geographical and ecological distribution in soils. Plant Soil *14:* 49–81.

—. 1961 Studies on nitrogen-fixing bacteria of the genus *Beijerinckia*. II. Mineral nutrition and resistance to high levels of certain elements in relation to soil type. Plant Soil *14:* 297–322.

—. 1961 A requirement of molybdenum for the symbiotic nitrogen fixation in alder (*Alnus glutinosa* Gaertn). Plant Soil *15:* 217–227.

—. 1961 Molybdenum and symbiotic nitrogen fixation by alder (*Alnus glutinosa* Gaertn). Nature (London) *192:* 1204–1205.

—. 1962 Species differences in molybdenum and vanadium requirements and combined nitrogen utilization by *Azotobacteriaceae*. Plant Soil *17:* 171–201.

—. 1965 *In vitro* cultivation of alder rootnodule tissue containing the endophyte. Nature (London) *207:* 885–887.

—. 1966 Interactions nutritionelles plantes-actinomycetes. Rapport général. Ann Inst Pasteur (Paris) Suppl *111:* 211–246.

—. 1968 Nitrogen fixation by nonleguminous plants. Dutch Nitrog Fert Rev *12:* 47–74.

—. 1970 Frankiaceae fam. nov. (*Actinomycetales*) with one new combination and six new species of the genus *Frankia* Brunschorst 1886, 174. Int J Syst Bacteriol *20 (2):* 201–220.

—. 1970 Plant-endophyte symbiosis in non-leguminous plants. Plant Soil *32:* 611–654.

—. 1971 Biological nitrogen fixation and its economic significance. *In* Nitrogen-15 in soil-plant studies, Proc Research co-ordination meeting Sofia 1969, Int Atomic Energy Agency (Vienna), pp. 189–222.

—, W. E. De Boer and A. L. Houwink. 1964 Electron microscopy of the endophyte of *Alnus glutinosa*. Antonie Van Leeuwenhoek. J Microbiol Serol *30:* 343–376.

Bedell, D. M. and G. T. Dimopoullos. 1962 Biologic properties and characteristics of *Anaplasma marginale*. I. Effects of temperature on infectivity of whole blood preparations. Amer J Vet Res *23:* 618–625.

— and —. 1965 Biologic properties and characteristics of *Anaplasma marginale:* Effect of sonic energy treatment of inoculums on the development of peak marginal body counts in splenectomized calves. Amer J Vet Res *26:* 889–891.

Beebe, J. M. 1941 The morphology and cytology of *Myxococcus xanthus* n. sp. J Bacteriol *42:* 193–223.

—. 1941 Studies on the myxobacteria. Iowa State J Sci *15:* 307–337.

Beerens, H. 1949 Contribution à l'étude de quelques bactéries anaérobies non sporulées. *Ristella insolita, Ristella pseudoinsolita (nov. sp.) Zuberella praeacuta, Vibrio crassus, Sphaerophorus varius, Corynebacterium anaerobium*. Ann Inst Pasteur Lille *2:* 1–11.

—, M. M. Castel and P. Fiévez. 1962 Classification des *Bacteroidaceae*. Int Congr Microbiol Abstr Montreal, p. 120.

— and R. Gaumont. 1953 Sur un nouveau bacille fusiforme: *Fusiformis hemolyticus* (nov. sp.) responsable d'une infection spontanée du lapin. Ann Inst Pasteur (Lille) *5:* 113–118.

— and M. Goudaert. 1952–1953 Etude de 30 souches d'une bactérie anaerobie nonsporulée gram positive: *Actinobacterium liquefaciens* (nov. comb.) Considerations taxonomiques. Ann Inst Pasteur (Lille) *5:* 119–132.

— and M. Tahon-Castel. 1965 Infections humaines à bactéries anaérobies non toxigènes. Presses Academiques Européennes, Brussels, pp. 1–194.

Beger, H. 1928 Über das Auftreten von Bakterien zoogloeen in Brunnenrohren. Kleine Mitt Ver Wasser-u Lufthyg Berlin *4:* 143–146.

—. 1935 *Leptothrix echinata*, ein neues, vorwiegend Mangan fällendes Eisenbakterium. Zentrabl Bakteriol Parasitenk Infektionskr Hyg Abt II *92:* 401–406.

—. 1941 *Naumanniella catenata* and *Sideronema globulifera*, zwei neue Eisenbakterien. Zentrabl Bakteriol Parasitenk Infektionskr Hyg Abt II *103:* 321–325.

—. 1949 Beitrage zur Systematik und geographischen Verbreitung der Eisenbakterien. Ber Deut Bot Ges *62 (1):* 7–13.

—. 1949 Uber *Zoogloea filipendula* = Formen und einige neue Eisenbakterien. Zentrabl Bakteriol Parasitenk Infektionskr Hyg Abt I Orig *154 (1):* 61–68.

—. 1953 *Toxothrix trichogenes* (Chol.) Beger nov. comb. p. 332. *In* Beger and Bringmann. Die Scheidenstruktur des Abwasserbakteriums *Sphaerotilus* und des Eisenbakteriums *Leptothrix* im elektronenmikroskopischen Bilde und ihre Bedeutung fur die Systematik dieser Gattungen. Zentrabl Bakteriol Parasitenk Infektionskr Hyg Abt *107:* 318–334.

—. 1957 Genus II *Gallionella* Ehrenberg 1838. *In* Breed, Murray and Smith (Editors), Bergey's Manual of Determinative Bacteriology, 7th ed. The Williams and Wilkins Co., Baltimore, pp. 214–216.

—. 1957 Family II. *Peloplocaceae* Beger Fam. Nov. *In* Breed, Murray and Smith (Editors),

Bergey's Manual of Determinative Bacteriology, 7th ed. The Williams and Wilkins Co., Baltimore, pp. 270–272.

— and G. Bringmann. 1953 Die Scheidenstruktur des Abwasserbakteriums *Sphaerotilus* und des Eisenbakteriums *Leptothrix* im elektronenmikroskopischen Bilde und ihre Bedeutung für die Systematik dieser Gattungen. Zentrabl Bakteriol Parasitenk Infektionskr Hyg Abt II *107:* 319–334.

Beijerinck, M. W. 1889 Le *Photobacterium luminosum*. Bactérie luminosum de la Mer Nord. Arch Neer Sci *23* (*2*): 401–427.

—. 1893 Ueber die Butylalkoholgärung und das Butylferment. Verh Kon Akad Wetensch (Sect 2) *1:* 1–51.

—. 1895 Über *Spirillum desulfuricans* als Ursache von Sulfat-reduktion. Zentrabl Bakteriol Parasitenk Infektionskr Hyg Abt I Orig *1:* 1–9; 49–59; 104–114.

—. 1898 Ueber die Arten der Essigbakterien. Zentrabl Bakteriol Parasitenk Infektionskr Hyg Abt II *4:* 209–216.

—. 1900 Der gegenwärtige Bestand der Kral'schen Sammlung von Mikroorganismen, Wien, pp. 1–34.

—. 1900 Schwefelwasserstoffbildung in den Stadtgräben und Aufstellung der Gattung *Aërobacter*. Zentrabl Bakteriol Parasitenk Infektionskr Hyg Abt II *6:* 193–206.

—. 1900 On different forms of hereditary variation in microbes. Proc Acad Sci Amst *3:* 352–365.

—. 1900 On indigofermentation. Proc Kon Akad Wetensch Amst *2:* 495–512.

—. 1901 Anhäufungsversuche mit Ureumbakterien. Ureumspaltung durch Urease und Katabolismus. Zentrabl Bakteriol Parasitenk Infektionskr Hyg Abt II *7:* 33–61.

—. 1901 Lichtbakteriën als reaktief bij het onderzoek der chlorophylfunctie. Proc Acad Sci Amst *4:* 45–49.

—. 1901 Sur les ferments lactiques de l'industrie. Arch Néer Sci (Sect 2) *6:* 212–243.

—. 1901 Ueber oligonitrophile Mikroben. Zentrabl Bakteriol Parasitenk Infektionskr Hyg Abt II *7:* 561–582.

—. 1903 Sur der Microbes oligonitrophiles. Arch Neer Sci (Sect 2) *8:* 190–217.

—. 1904 Phénomènes de reduction produits par les microbes. Arch Neer Sci (Sect 2) *9:* 131–157.

—. 1910 Bildung und Verbrauch von Stickoxydul durch Bakterien. Zentrabl Bakteriol Parasitenk Infektionskr Hyg Abt II *25:* 30–63.

—. 1911 Ueber pigmentbildung bei Essigbakterien. Zentrabl Bakteriol Parasitenk Infektionskr Hyg Abt II *29:* 169–176.

—. 1911 Pigments as products of oxidation by bacterial action. Proc Acad Sci Amst *13:* 1066–1077.

—. 1912 Mutation bei Mikroben. Folia Mikrobiol (Delft) *1:* 4–100.

—. 1912 Die durch Bakterien aus Rohrzucker erzeugten Schleimgen Wandstoffe. Folia Mikrobiol (Delft) *1:* 377–408.

—. 1916 Die Leuchtbakterien der Nordsee im August und September. Folia Mikrobiol (Delft) *4:* 15–40.

—. 1916 Formation of pyruvic acid from malic acid by microbes. Verslag gewone Vergad Akad Amst *18:* 1198–2000.

— and A. van Delden. 1902. Ueber die Assimilation des freien Stickstoffs durch Bakterien. Zentrabl Bakteriol Parasitenk Infektionskr Abt II *9:* 3–43.

— and —. 1903 On a colorless bacterium whose carbon food comes from the atmosphere. Proc Kon Akad Wetensch Amst *5:* 398–413.

Belaïch, J. P. and J. C. Senez. 1965 Influence of aeration and of pantothenate on growth yields of *Z. mobilis*. J Bacteriol *89* (*5*): 1195–1200.

Bell, E. J., G. M. Kohls, H. G. Stoenner and D. B. Lackman. 1963 Nonpathogenic rickettsiae related to the spotted fever group isolated from ticks, *Dermacentor variabilis* and *Dermacentor andersoni* from Eastern Montana. J Immunol *90:* 770–781.

—, D. B. Lackman, R. A. Ormsbee and M. Peacock. 1969 Neutralization of murine typhus toxin by serum of normal human beings and monkeys. Amer J Trop Med Hyg *18:* 559–567.

Bell, J. F., C. R. Owen and C. L. Larson. 1955 The virulence of *Bacterium tularense*. I. The virulence of *Bacterium tularense* in mice, guinea pigs and rabbits. J Infect Dis *97:* 162–166.

Bellenghi, M. 1956 Herstellung von tetracylin (in German). Deutches Patent Anmeldung L 21,111, March 8.

Belozersky, A. N. and A. B. Spirin. 1960 Chemistry of the nucleic acids of microorganisms. *In* Chargaff and Davidson (Editors), The nucleic acids, Vol. 3, Academic Press, New York, pp. 147–185.

Bendixen, H. C. and A. Jepsen. 1940 Fortgesetzte Untersuchungen über *Corynebacterium equi*, mit besonderer Berücksichtigung gewisser morphologischer und biologischer Verhältnisse sowie der Pathogenitäts-verhältnisse Schweinen gegenüber. Z Infektionskr Haustiere *57:* 9–36.

Benedict, R. G., W. Dvonch, O. L. Shotwell, T. G. Pridham and L. A Lindenfelser. 1952 Cinnamycin, an antibiotic from *Streptomyces cinnamoneus* nov. sp. Antibiot Chemother *2:* 591–594.

—, L. A. Lindenfelser, F. H. Stodola and D. H. Traufler. 1951 Studies on *Streptomyces griseocarneus* and the production of hydroxystreptomycin. J Bacteriol *62:* 487–497.

—, O. L. Shotwell, T. G. Pridham, L. A. Lindenfelser and W. C. Haynes. 1954 The production of the neomycin complex by *Streptomyces albogriseolus*, nov. sp. Antibiot Chemother *4:* 653–656.

—, F. H. Stodola, O. L. Shotwell, A. M. Borud and L. A. Lindenfelser. 1950 A new Streptomycin. Science (Washington) *112:* 77–78.

Benefey, B. 1950 Dissertation. University of Göttingen.

Bengtson, L. A. 1924 Studies of organisms concerned as causative factors in botulism. U S Public Health Serv Hyg Lab Bull *136:* 1–5.

Ben Gurion, R. and I. Hertman. 1958 Bacteriocin-like material produced by *Pasteurella pestis*. J Gen Microbiol *19* (*2*): 289–297.

Bennedsen, J. 1968 The specificity of circulating antibodies in experimental infection with *Mycobacterium avium* demonstrated by immunofluorescence. Acta Pathol Microbiol Scand *72:* 330–336.

Bennett, J. F. and E. Canale-Parola. 1965 The taxonomic status of *Lineola longa*. Arch Mikrobiol *52:* 197–205.

Berendsen, P. B. and I. R. Teford. 1966 A light

and microscopic study of Kurloff bodies in the blood and spleen of the guinea pig. Anat Rec *156:* 104–118.

Berestnev, N. M. 1897 Actinomycosis and its causes (in Russian). Moscow, pp. 1–206.

Bergan, T., K. Bøvre and B. Hovig. 1970 Present status of the species *Micrococcus freudenreichii* Guillebeau 1891. Int J Syst Bacteriol *20:* 249–254.

— and B. Hovig. 1968 A new species, *Sphaerophorus intermedius*, isolated from empyema. Acta Pathol Microbiol *74:* 421–430.

Bergeman, A. M. 1909 Die rote Beulenkrankheit des Aals. Ber Bayer Biol VersSta *2:* 10–54.

Berger, J., L. M. Jampolsky and M. W. Goldberg. 1949 Borrelidin, a new antibiotic with antiborrelia activity and penicillin enhancement properties. Arch Biochem *22:* 476–478.

Berger, U. 1960 *Neisseria animalis nov. spec.* Z Hyg Infektionskr *147:* 158–161.

—. 1960 *Neisseria haemolysans* (Thjøtta and Bøe 1938). Untersuchungen zur Stellung im System. Z Hyg Infektionskr Med Mikrobiol Immunol Virol *146:* 253–259.

—. 1961 A proposed new genus of gram-negative cocci: *Gemella.* Int Bull Bacteriol Nomencl Taxon *11 (1):* 17–19.

—. 1961 Untersuchungen an saprophytischen Neisserien. Z Hyg Infektionskr *147:* 257–268.

—. 1961 Untersuchungen über die Pigmentbildung durch Neisserien. Z Hyg Infektionskr Med Mikrobiol Immunol Virol *147:* 461–469.

—. 1961 Zur Variabilität der Zuckervergärungen durch Neisserien. Arch Hyg *145:* 296–301.

—. 1962 Untersuchungen über di Säurebildung durch saccharolytische anspruchslose Neisserien. Arch Hyg *146:* 55–60.

—. 1962 Über das Vorkommen von Neisserien bei einigen Tieren. Z Hyg Infektionskr *148:* 445–457.

—. 1963 Die anspruchslosen Neisserien. Ergeb Mikrobiol Immun Exp Ther *36:* 97–167.

—. 1963 Reinzuchtung von *Simonsiella* spp. Z Hyg Infektionskr Med Mikrobiol Immunol Virol *149:* 336–340.

— and H. Brunhoeber. 1961 *Neisseria flava* (Bergey *et al.* 1923). Art oder Varietät? Z Hyg Infektionskr *148:* 39–44.

— and M. Miersch. 1970 Zum normalen Vorkommen von *Neisseria mucosa* (Véron *et al.* 1959). Z Med Mikrobiol Immunol *155:* 186–191.

— and E. Paepcke. 1962 Untersuchungen über die asaccharolytischen Neisserien des menschlichen Nasopharanyx. Z Hyg Infektionskr Med Mikrobiol Immunol Virol *148:* 269–281.

—, S. Saeftel and K. Schlez. 1970 Untersuchungen über die Meningokokken-trägerquote im Hinblick auf Alter und Geschlecht. Z Med Mikrobiol Immunol *155:* 192–202.

— and K. Schlez. 1970 Untersuchungen auf Meningokokkenträger in Kindergärten und Kinderheimen. Z Kinderheilk *108:* 54–60.

— and B. Wulf. 1961 Untersuchungen an saprophytischen Neisserien. Z Hyg Infektionskr *147:* 257–268.

Bergey, D. H. and R. S. Breed. 1948 Genus III. *Flavobacterium.* Bergey et al. *In* Breed, Murray and Hitchens, (Editors), Bergey's Manual of Determinative Bacteriology, 6th ed. The Williams and Wilkins Co., Baltimore, pp. 427–442.

—, R. S. Breed, B. W. Hammer, F. M. Huntoon, E. G. D. Murray and F. C. Harrison. 1934

Bergey's Manual of Determinative Bacteriology, 4th ed. The Williams and Wilkins Co., Baltimore, pp. 1–664.

—, —, E. G. D. Murray and A. P. Hitchens. 1939 Bergey's Manual of Determinative Bacteriology, 5th ed. The Williams and Wilkins Co., Baltimore, pp. 1–1032.

—, F. C. Harrison, R. S. Breed, B. W. Hammer and F. M. Huntoon. 1923 Bergey's Manual of Determinative Bacteriology, 1st ed. The Williams and Wilkins Co., Baltimore, pp. 1–442.

—, —, —, — and —. 1925 Bergey's Manual of Determinative Bacteriology, 2nd ed. The Williams and Wilkins Co., Baltimore, pp. 1–462.

—, —, —, — and —. 1930 Bergey's Manual of Determinative Bacteriology, 3rd ed. The Williams and Wilkins Co., Baltimore, pp. 1–589.

—, W. C. Haynes and A. P. Hitchens. 1948 Genus I. *Vibrio. In* Breed, Murray and Hitchens (Editors), Bergey's Manual of Determinative Bacteriology, 6th ed. The Williams and Wilkins Co., Baltimore, pp. 1–196.

Bergonzini, C. 1879 I Bacteri. Annuar Soc Nat Modena, Ser 2, *13:* 19–100.

—. 1881 Sopra un nuovo bacterio colorato. Annuar Soc Nat Modena, Ser 2, *14:* 149–158.

Bergy, M. E., R. R. Herr and D. J. Mason. 1963 Antibiotic lincolnensin and method of production. United States Patent 3,086,912, April 23.

—, — and —. 1963 Lincomycin (in Belgian). Belgian Patent 619,645, January 2.

— and T. R. Pyke. 1967 Cremomycin and process for making. United States Patent 3,350,269, October 31.

Berkeley, M. J. 1857 Introduction to Cryptogamic Botany. H. Bailliere, London.

—. 1874 Notices of the North American Fungi. Grevillea *3 (25):* 1–17.

—. 1874 Notices of the North American Fungi. Grevillea *3 (26):* 49–64.

—. 1875 Notices of the North American Fungi. Grevillea *3 (27):* 97–112.

— and C. E. Broome. 1873 Enumeration of the fungi of Ceylon. J Linnean Soc London Bot *14:* 29–140.

Berliner, E. 1915 Über die Schlaffsucht der Mehlmottenraupe (*Ephestia kühniella* Zell) und ihren Erreger *Bacillus thuringiensis* n. sp. Z Angew Entomol *2:* 29–56.

Bernstein, A. and H. E. Morton. 1934 A new thermophilic *Actinomyces.* J Bacteriol *27:* 625–628.

Berridge, E. M. 1924 The influence of hydrogen ion concentration on the growth of certain bacterial plant parasites and saprophytes. Ann Appl Biol *11:* 73–85.

Berry, R. N. 1933 Some new heat resistant, acid tolerant organisms causing spoilage in tomato juice. J Bacteriol *25:* 72–73

Bersa, E. 1920 Ueber das Vorkommen von kohlensaurem Kalk in einer Gruppe von Schwefelbakterium. Sitzungsber Saechs Akad Wiss Leipzig Math-Naturwiss Kl *129:* 231–259.

—. 1926 Neue kalkführende Schwefelbakterien. Planta *2:* 373–379.

Bertrand, G. 1896. Préparation biochimique du Sorbose. Bull Soc Chim Biol *15:* 627–631.

Besdine, R. W. and L. Pine. 1968 Preparation

and description of high-molecular-weight soluble surface antigens from a group A streptococcus. J Bacteriol 96 (6): 1953–1960.

Bettelheim, K. A., J. F. Gordon and J. Taylor. 1968 The detection of a strain of Chromobacterium lividum in the tissues of certain leaf-nodulated plants by the immunofluorescence technique. J Gen Microbiol 54 (2): 177–184.

Bevan, L. E. W. 1930 Blood culture in undulant fever. Brit Med J 2: 267.

Beveridge, W. I. B. 1936 A study of Spirochaeta penortha (n. sp.) isolated from foot-rot in sheep. Aust J Exp Biol Med Sci 14: 307–318.

—. 1938 Foot-rot in sheep: A preliminary note on the probable causal agent. J Coun Sci Indust Res Aust 11 (1): 1–3.

—. 1941 Foot-rot in sheep: A transmissible disease due to infection with Fusiformis nodosus (n. sp.) Studies on its causes, epidemiology, and control. Counc Sci Indust Res Aust Bull 140: 1–56.

Bexon, J. and E. A. Dawes. 1970 The nutrition of Zymomonas anaerobia. J Gen Microbiol 60 (3): 421–425.

Bezjak, V. 1952 Sur deux espèces anaérobies nouvelles: Leptotrichia haemolytica n. sp. Terminosporus indologenes. Ann Inst Pasteur (Paris) 82: 98–101.

Bhandari, R. R. and T. K. Walker. 1953 Lactobacillus frigidus n. sp. isolated from brewery yeast. J Gen Microbiol 8: 330–332.

Bhat, J. F. and H. A. Barker. 1948 Studies on a new oxalate-decomposing bacterium, Vibrio oxaliticus. J Bacteriol 55: 359–368.

Bhatt, V. V., S. G. Abhyankar and M. K. Patel. 1956 A new bacterial leaf spot on Phaseolus trilobus. Curr Sci 25: 299.

— and M. K. Patel. 1954 Two new records of phytopathogenic bacteria from Bombay State. Curr Sci 23: 165.

— and —. 1955 Two new Xanthomonas species on legumes. Curr Sci 24: 94.

—, V. H. Pawar and R. S. Sukapure. 1960 A bacterial leaf spot disease of Teramnus labialis. Indian Phytopathol 13: 180–181.

Bhuyan, B. K. and A. Dietz. 1966 Fermentation, taxonomic and biological studies of nogalamycin. Antimicrob Agents Chemother 1965: 836–844.

—, — and C. G. Smith. 1962 Pactamycin, a new antitumor antibiotic. I. Discovery and biological properties. Antimicrob Agents Chemother 1961: 184–190.

—, R. B. Kelly and R. M. Smith. 1965 Antibiotic nogalamycin and method of producing. United States Patent 3,183,157, May 11.

—, S. P. Owen and A. Dietz. 1965 Rubradirin, a new antibiotic. I. Fermentation and biological properties. Antimicrob Agents Chemother 1964: 91–96.

Biberstein, E. L. and C. K. Francis. 1968 Nucleic acid homologies between the A and T types of Pasteurella haemolytica. J Med Microbiol 1: 105–108.

— and M. Gills. 1962 The relation of the antigenic types to the A and T types of Pasteurella haemolytica. J Comp Pathol 72: 316–320.

—, M. Gills and H. Knight. 1960 Serological types of Pasteurella haemolytica. Cornell Vet 50: 283–300.

—, P. D. Mini and M. G. Gills. 1963 Action of Haemophilus cultures on delta-aminolevulinic acid. J Bacteriol 86: 814–819.

— and D. A. Thompson. 1966 Epidemiological studies on Pasteurella haemolytica in sheep. J Comp Pathol 76: 83–94.

— and D. C. White. 1969 A proposal for the establishment of two new Haemophilus species. J Med Microbiol 2: 75–78.

Bidan, P. 1956 Sur quelques bactéries isolées de vin en fermentation malolactiques. Ann Technol Agr (Paris) 5: 597–617.

Biebl, H. and G. Drews. 1969 Das in-vivo-Spektrum als taxonomisches Merkmal bei Untersuchungen zur Verbreitung von Athiorhodaceae. Zentrabl Bakteriol Parasitenk Infektionskr Hyg Abt II 123: 425–452.

Bienstock, B. 1899 Recherches sur la putrefaction. Ann Inst Pasteur (Paris) 13: 854–864.

—. 1906 Bacillus putrificus. Ann Inst Pasteur (Paris) 20: 407–415.

Billing, E. and L. A. E. Baker. 1963 Characteristics of Erwinia-like organisms found in plant material. J Appl Bacteriol 26: 58–65.

Bird, R. G. and P. C. C. Garnham. 1967 Aegyptianella pullorum Carpano. Fine structure and taxonomy. J Protozool (Suppl.) 14: 42.

Bisset, K. A. and F. W. Moore. 1950 Jensenia, a new genus of Actinomycetales. J Gen Microbiol 4: 280.

Bitteeva, M. B. 1962 The taxonomic position of the actinomycete producing the antibiotic aurantin. Mikrobiologiya 31 (4): 601–607.

Bizio, B. 1823 Lettera di Bartolomeo Bizio al chiarissimo canonico Angelo Bellani sopra il fenomeno della polenta porporina. Bibl Ital G Lett, Sci Arti (Anno. VIII) 30: 275–295.

Black, F. T. 1970 Serological methods for classification of human T-mycoplasmas. Proc Vth Int Congr Infect Dis Vienna 1970, pp. 407–411.

—, C. Christiansen and G. Askaa. 1972 Genome size and base composition of deoxyribonucleic acid from eight human T-Mycoplasmas. Int J Syst Bacteriol 22: 241–242.

— and T. M. Jørgensen. 1972 Susceptibility of organisms of the order Mycoplasmatales to optochin. Int J Syst Bacteriol 22 (4): 237–240.

Blackburn, T. H. and P. N. Hobson. 1962 Further studies on the isolation of proteolytic bacteria from the sheep rumen. J Gen Microbiol 29 (1): 69–81.

Bladen, H. A., M. P. Bryant and R. N. Doetsch. 1961 A study of bacterial species of the rumen which produce ammonia from protein hydrolysate. Appl Microbiol 9: 175–180.

— and S. E. Mergenhagen. 1964 Ultrastructures of Veillonella and morphological correlation of an outer membrane with particles associated with endotoxic activity. J Bacteriol 88 (5): 1482–1492.

Blaizot, L. and P. Blaizot. 1928 Treponema podovis n. sp., agent pathogéne du piétin des moutons. C R Acad Sci Ser D (Paris) 187: 911–912.

Blakemore, F., S. D. Elliott and J. Hart-Mercer. 1941 Studies of suppurative polyarthritis (joint-ill) in lambs. J Pathol Bacteriol 52: 57–83.

Blanc, G. and A. Maurice. 1948 Contributions à l'étude du spirochète de Goulimine (Maroc méridional). Bull Soc Pathol Exot 41: 139–141.

Blanchard, R. 1896 Parasites végétaux a l'exclusion des bactéries. In C. Bouchard, Traité de Pathologie Genérale, Vol. 2, G. Masson, Paris, pp. 1–932.

—. 1906 Spirilles, spirochétes et autres microorganismes à corps spiralé. Sem Méd *26:* 1–5.

Bland, C. E. 1968 Structure and development in some genera of the Actinoplanaceae. Thesis, University of North Carolina.

Blank, C. H. and L. K. Georg. 1968 The use of fluorescent antibody methods for the detection and identification of *Actinomyces* species in clinical material. J Lab Clin Med *71:* 283–293.

Bloch, M. 1918 Beitrag sur Untersuchung über die *Zoogloea ramigera* (Itzigsohn) auf Grund von Reinkulturen. Zentrabl Bakteriol Parasitenk Infektionskr Hyg Abt II *48:* 44–62.

Boand, A. and M. Novak. 1949 Sensitivity changes of *Actinomyces bovis* to penicillin and streptomycin. J Bacteriol *57* (*5*)*:* 501–508.

Boatman, E. S. and H. C. Douglas. 1961 Fine structure of the photosynthetic bacterium *Rhodomicrobium vannielii*. J Biophys Biochem Cytol *11:* 469–483.

Bøe, J. 1941 *Fusobacterium*. Studies on its bacteriology, serology and pathogenicity. Skr Utgitt Norske Vidensk-Akad Oslo Mat-Naturvidensk Kl *9:* 1–191.

— and T. Thjøtta. 1944 The position of *Fusobacterium* and *Leptotrichia* in the bacteriological system. Acta Pathol Microbiol Scand *21:* 441–450.

Boháček, J. and O. Mráz. 1967 Basengehalt der Desoxyribonukleinsaüre bei den Arten *Pasteurella haemolytica*, *Actinobacillus lignieresii* und *Actinobacillus equuli*. Zentrabl Bakteriol Parasitenk Infektionskr Hyg Abt I Orig *202* (*4*)*:* 468–478.

Bohn, G. W. and J. C. Maloit. 1946 Bacterial spot of native golden current (*Ribes aureum*). J Agr Res *73:* 281–290.

Boissard, J. M. and P. J. Wormald. 1950 A new group of hemolytic steptococci for which the designation "group O" is proposed. J Pathol Bacteriol *62:* 37.

Boisvert, H. 1955 *Mycobacterium xenopei* (Marks et Schwabacher 1965), mycobactérie scotochromogene thermophile, dysgonique, eventuellment pathogène pour l'homme. Ann Inst Pasteur (Paris) *109:* 447–453.

Bojalil, L. F. 1959 Estudio comparativo entre *Mycobacterium marinum* y *Mycobacterium balnei*. Rev. Latinoamer Microbiol *2:* 169–174.

—, J. Cerbon and A. Trujillo. 1962 Adansonian classification of mycobacteria. J Gen Microbiol *28* (*2*)*:* 333–346.

Bokkenheuser, V. 1951 Étude d'une nouvelle espéce anaérobie du genre *Pasteurella: P. serophila* n. sp. Ann Inst Pasteur (Paris) *80:* 548–551.

Bollinger, O. 1870 Mycosis der Lunge beim Pferde. Virchows Arch Pathol Anat Physiol Klin Med *49:* 583–586.

Bolognesi, G. 1907 Die Anaërobiose des Fränkelschen Diplococcus in Beziehung zu einer seiner pathogener Eigenschaften. Zentrabl Bakteriol Parasitenk Infektionskr Hyg Abt I Orig *43:* 113–118.

Bond, G. 1951 The fixation of nitrogen associated with the root nodules of *Myrica gale L.*, with special reference to its pH relation and ecological significance. Ann Bot (London) *15:* 447–459.

—. 1955 An isotopic study of the fixation of

nitrogen associated with nodulated plants of *Alnus*, *Myrica*, and *Hippophaë*. J Exp Bot *6.* 303–311.

—. 1957 The development and significance of the root nodules of *Casuarina*. Ann Bot (London) *21:* 373–380.

—. 1957 Isotopic studies of nitrogen fixation in non-legume root nodules. Ann Bot (London) *21:* 513–521.

—. 1958 Symbiotic nitrogen fixation by non-legumes. *In* E. G. Hallsworth (Editor), Nutrition of the legumes, Proc. Univ. Nottingham, 5th Easter School Agr. Sci., Butterworth Scientific Publications, London, pp. 216–231.

—. 1959 Fixation of nitrogen in non-legume root-nodule plants. *In* H. K. Porter (Editor), Utilization of nitrogen and its compounds by plants, 13th Symp. Soc. Exp. Biol., Cambridge University Press, London, pp. 59–72.

—. 1961 The oxygen relation of nitrogen fixation in root nodules. Z Allg Mikrobiol *1:* 93–99.

—. 1962 Fixation of nitrogen in *Coriaria myrtifolia*. Nature (London) *193:* 1103–1104.

—. 1964 Isotopic investigations of nitrogen fixation in non-legume root nodules. Nature (London) *204:* 600–601.

—. 1967 Nitrogen fixation in some nonlegume root nodules. Phyton Rev Int Bot Exp *24:* 57–66.

—, J. F. Adams and E. H. Kennedy. 1965 Vitamin B$_{12}$ analogues in non-legume root nodules. Nature (London) *207:* 319–320.

— and E. J. Hewitt. 1961 Molybdenum and the fixation of nitrogen in *Myrica* root nodules. Nature (London) *190:* 1033–1034.

— and —. 1962 Cobalt and the fixation of nitrogen by root nodules of *Alnus* and *Casuarina*. Nature (London) *195:* 94–95.

—, J. T. MacConnell and A. H. McCallum. 1956 The nitrogen-nutrition of *Hippophaë rhamnoides* L. Ann Bot (London) *20:* 501–512.

— and P. Montserrat. 1958 Root nodules of *Coriaria*. Nature (London) *182:* 474–475.

Bönicke, R. 1961 Die Bedentung der Acylamidasen für die Identifizierung und Differenzierung der verschiedenen Arten der Gattung *Mycobacterium*. Jahresber Borstel *5:* 7–87.

—. 1962 Identification of mycobacteria by biochemical methods. Bull Inst Union Tuberc *32:* 13–68.

—. 1965 Beschreibung der neuen Species *Mycobacterium borstelense* n. sp. Zentrabl Bakteriol Parasitenk Infektionskr Hyg Abt I Orig *196:* 535–538.

— and S. E. Juhasz. 1964 Beschreibung der neuen Species *Mycobacterium vaccae* n. sp. Zentrabl Bakteriol Parasitenk Infektionskr Hyg Abt I Orig *192:* 133–135.

— and D. Stottmeier. 1965 Erkennung und Identifizierung von Stämmen der Species *Mycobacterium borstelense*. Beitr Klin Erforsch Tuberk Lungenkr *130:* 210–222.

Boone, C. J. and L. Pine. 1968 Rapid method for characterization of actinomycetes by cell wall composition. Appl Microbiol *16* (*2*)*:* 279–284.

Boguet, A. 1928 Sur le bacille de l'entérite paratuberculeuse des bovides. Ann Inst Pasteur (Paris) *42:* 495–528.

Bordet, J. and O. Gengou. 1906 Le microbe de la coqueluche. Ann Inst Pasteur (Paris) *20:* 731–741.

Borg, A. F. 1960 Studies on myxobacteria associated with diseases in salmonid fishes. Wildlife Dis *8:* 1–85.

Borg-Petersen, C. 1971 A thermo-labile antigen in the *Leptospira* strain Ictero No. 1. Trop Geogr Med *23:* 282–285.

Borman, E. K., C. A. Stuart and K. Wheeler. 1944 Taxonomy of the family *Enterobacteriaceae.* J Bacteriol *48 (3):* 351–367.

Bornside, G. H. and R. E. Kallio. 1956 Urea-hydrolyzing bacilli. II. Nutritional profiles. J Bacteriol *71 (6):* 655–660

Borowski, E., C. P. Schaffner, H. Lechevalier and B. S. Schwartz. 1961 Perimycin, a novel type of heptaene antifungal antibiotic. Antimicrob Agents Ann *1960:* 532–538.

Borrel, A., E. Dujardin-Baumetz, Jeantět and C. Jouan. 1910 Le Microbe de la péripneumonie. Ann Inst Pasteur (Paris) *24:* 168–179.

Bosanquet, W. C. 1911 Spirochaetes. A review of recent work with some original observations. Saunders, Philadelphia, pp. 1–152.

Boskamp, E. 1922 Ueber Bau, Lebensweise und systematische Stellung von *Selenomonas palpitans* (Simons). Zentrabl Bakteriol Parasitenk Infektionskr Hyg Abt I Orig *88:* 58–73.

Bouisset, L., J. Breuillard and G. Michel. 1963 Étude de l'ADN chez les *Actinomycetales* comparaison entre les valeurs du rapport A + T/C + G et les characteres bacteriologiques de *Corynebacterium.* Ann Inst Pasteur (Paris) *104:* 756–770.

—, —, — and G. Larrouy. 1968 Bases nucléiques des bactéries application au genne *Actinobacterium.* Ann Inst Pasteur (Paris) *115:* 1063–1081.

Bourne, B. A. 1970 Studies on the bacterial red stripe disease of sugarcane in Florida. Sugarcane Pathologists Newsletter No. 4, 27.

—. 1970 Supplemental notes on the causal bacterium of sugarcane red stripe disease in Florida. Sugarcane Pathologists Newsletter No. 5, 40.

Bousfield, E. G., G. G. H. Wright and T. K. Walker. 1947 Oxidation of glycerol by *Acetobacter* species. J Inst Brew *53:* 258–262.

Boutroux, M. L. 1881 Sur une fermentation nouvelle du glucose. Ann Sci Ec norm sup Paris *10:* 67–130.

Bovallius, A. and B. Zacharias. 1971 Variations in the metal content of some commercial media and their effect on microbial growth. Appl Microbiol *22 (3):* 260–262.

Bovarnick, M. R., J. C. Miller and J. C. Snyder. 1950 The influence of certain salts, amino acids, sugars and proteins on the stability of rickettsiae, J Bacteriol *59:* 509–522.

— and J. C. Snyder. 1949 Respiration of typhus rickettsiae. J Exp Med *89:* 561–565.

Bøvre, K. 1965 Studies on transformation in *Moraxella* and organisms assumed to be related to *Moraxella.* 4. Streptomycin resistance transformation between asaccharolytic *Neisseria* strains. Acta Pathol Microbiol Scand *64:* 229–242.

—. 1967 Transformation and DNA base composition in taxonomy, with special reference to recent studies in *Moraxella* and *Neisseria.* Acta Pathol Microbiol Scand *69:* 123–144.

—, M. Fiandt and W. Szybalski. 1969 DNA base composition of *Neisseria, Moraxella* and *Acinetobacter,* as determined by measurement of buoyant density in CsCl gradients. Can J Microbiol *15:* 335–338.

— and S. D. Henriksen. 1967 A new *Moraxella* species, *Moraxella osloensis,* and a revised description of *Moraxella nonliquefaciens.* Int J Syst Bacteriol *17 (2):* 127–135.

— and —. 1967 A revised description of *Moraxella polymorpha* Flamm 1957 with a proposal of a new name, *Moraxella phenylpyrouvica,* for this species. Int J Syst Bacteriol *17 (2):* 343–360.

— and E. Holten. 1970 *Neisseria elongata sp. nov.,* a rod-shaped member of the genus *Neisseria.* Re-evaluation of cell shape as a criterion in classification. J Gen Microbiol *60 (1):* 67–75.

Bowers, E. F. and L. R. Jeffries. 1955 Optochin in the identification of *Streptococcus pneumoniae.* J. Clin Pathol (London) *8:* 58–60.

Boyer, C. I., Jr., D. W. Bruner and J. A. Brown. 1958 A streptobacillus, the cause of tendon-sheath infection in turkeys. Avian Dis *2:* 418–427.

Boyle, A. M. 1949 Further studies of the bacterial necrosis of the giant cactus. Phytopathology *39:* 1029–1052.

Bozeman, F. M., J. W. Humphries and J. M. Campbell. 1968 A new group of rickettsialike agents recovered from guinea pigs. Acta Virol *12:* 87–93.

Braak, H. R. 1928 Onderzoekingen over Vergisting van Glycerine. Thesis, W. D. Meinema-Uitgeurer, Delft.

Bracco, R. M., M. R. Krauss, A. S. Rae and C. M. MacLeod. 1957 Transformation reactions between pneumococcus and three strains of streptococci. J Exp Med *106:* 247.

Bradbury, J. F. 1971 Nomenclature of the bacterial leaf streak pathogen of rice. Int J Syst Bacteriol *21:* 72.

Bradley, D. E. and J. G. Franklin. 1958 Electron microscope survey of the surface configuration of spores of the genus *Bacillus.* J Bacteriol *76 (6):* 618–630.

Branham, S. E. 1927 Anaerobic microorganisms in nasopharyngeal washings. Influenza studies XXXIII. J Infect Dis *41:* 203–207.

—. 1930 A new meningococcus-like organism (*Neisseria flavescens n. sp.*) from epidemic meningitis. U S Public Health Serv Publ Health Rep *45:* 845–849.

—. 1953 Serological relationships among meningococci. Bacteriol Rev *17:* 175–188.

—. 1958 Reference strains for the serologic groups of meningococcus (*Neisseria meningitidis*). Int Bull Bacteriol Nomencl Taxon *8 (1):* 1–15.

Braun, W. and A. Bonestell. 1947 Independent variation of characteristics in *Brucella abortus* variants and their detection. Amer J Vet Res *8:* 386–390.

Breed, R. S. 1939 Genus II. *Malleomyces* Pribram. *In* Bergey, Breed, Murray and Hitchens (Editors), Bergey's Manual of Determinative Bacteriology, 5th ed. The Williams & Wilkins Co., Baltimore, pp. 298–300.

—. 1948 Genus I. *Pseudomonas* Migula. *In* Breed, Murray and Hitchens (Editors), Bergey's Manual of Determinative Bacteriology, 6th ed. The Williams & Wilkins Co., Baltimore, pp. 82–150.

—. 1948 Appendix I. Genus II. *Xanthomonas* Dowson. *In* Breed, Murray and Hitchens (Editors), Bergey's Manual of Determinative

Bacteriology, 6th ed. The Williams & Wilkins Co., Baltimore, pp. 171–178.

—. 1948 Family I. *Chlamydobacteriaceae* Migula. *In* Breed, Murray and Hitchens (Editors), Bergey's Manual of Determinative Bacteriology, 6th ed. The Williams & Wilkins Co., Baltimore, pp. 981–986.

—. 1953 The *Brevibacteriaceae* fam. nov. of order Eubacteriales. Riass Communicazione, VI Congr Int Microbiol, Roma *1*: 13–14.

—. 1957 Family II. *Methanomonadaceae* Breed. *In* Breed, Murray and Smith (Editors), Bergey's Manual of Determinative Bacteriology, 7th ed. The Williams & Wilkins Co., Baltimore, pp. 74–78.

—. 1957 Genus III. *Chromobacterium* Bergonzini, 1881. *In* Breed, Murray and Smith (Editors), Bergey's Manual of Determinative Bacteriology, 7th ed. The Williams & Wilkins Co., Baltimore, pp. 292–296.

—. 1957 Genus IV. *Agarbacterium* Angst 1929. *In* Breed, Murray and Smith (Editors), Bergey's Manual of Determinative Bacteriology, 7th ed. The Williams & Wilkins Co., Baltimore, pp. 322–328.

— and E. F. Lessel. 1954 The classification of luminescent bacteria. Antonie Van Leeuwenhoek. J Microbiol Serol *20*: 58–64.

—, E. G. D. Murray and A. P. Hitchens. 1948 *Bergey's Manual of Determinative Bacteriology*, 6th ed. The Williams & Wilkins Co., Baltimore.

—, — and N. R. Smith. 1957 *Bergey's Manual of Determinative Bacteriology*, 7th ed. The Williams & Wilkins Co., Baltimore.

— and C. S. Pederson. 1938 The organisms causing rusty spot in cheddar cheese. J Bacteriol *36*: 667.

— and J. Smit. 1957 Genus IV. *Sarcina* Goodsir 1842. *In* Breed, Murray and Smith (Editors), Bergey's Manual of Determinative Bacteriology, 7th ed. The Williams & Wilkins Co., Baltimore, pp. 467–473.

Breinl, A. 1906 On the specific nature of the spirochaete of the African tick fever. Lancet *1*: 1690–1691.

Brenner, D. J., A. G. Steigerwalt, G. V. Miklos and G. R. Fanning. 1973 Deoxyribonucleic acid relatedness among Erwiniae and other Enterobacteriaceae: the soft-rot organisms (Genus *Pectobacterium* Waldee). Int J Syst Bacteriol *23*: 205–216.

Breznak, J. A. and E. Canale-Parola. 1969 *Spirochaeta aurantia*, a pigmented, facultatively anaerobic spirochete. J Bacteriol *97* (*1*): 386–395.

Bridré, J. and A. Donatien. 1923 Le Microbe de l'agalaxie contagieuse et sa culture in vitro. C R Acad Sci Paris *177*: 841–843.

Brindle, C. S. and S. T. Cowan. 1951 Flagellation and taxonomy of Whitmore's bacillus. J Pathol Bacteriol *63*: 571–575.

Brinley-Morgan, W. J. and S. G. M. Gower. 1966 Techniques in the identification and classification of *Brucella*. *In* Gibbs and Skinner (Editors), Identification methods for microbiologists. Academic Press, London, pp. 35–40.

—, D. Kay and D. E. Bradley. 1960 *Brucella* bacteriophage. Nature (London) *188*: 74–75.

Brinton, C. C. 1965 The structure, function, synthesis and genetic control of bacterial pili and a molecular model for DNA and RNA transport in gram-negative bacteria. Trans N Y Acad Sci *27*: 1003–1004.

Brion, A. 1942 L'actinomycose du chien et du chat. Rev Méd Vét *93*: 145–157.

— and H. Kayser. 1902 Ueber eine Erkrankung mit dem Befund eines typhus-ähn lichen Bakteriums (Paratyphus). Muenchen Med Wochenschr *49*: 611–615.

Brisou, J. 1953 Essai sur la systématique du genre *Achromobacter*. Ann Inst Pasteur (Paris) *84*: 812–814.

—. 1955 *Microbiologie du milieu marin*. E. Flammarion and Co., Paris, pp. 1–271.

—. 1957 Contribution à l'étude de la systématique des *Pseudomonadeaceae*. Ann Inst Pasteur (Paris) *93*: 397–404.

—. 1958 Étude de quelques *Pseudomonadaceae*. Classification. A Baillet, Bordeaux, pp. 1–214.

— and A. R. Prévot. 1954 Études de Systematique Bactérienne. X. Revision des espèces reuniés dans le genre *Achromobacter*. Ann Inst Pasteur (Paris) *86*: 722–728.

—, C. Tysset and A. Jacob. 1960 Étude d'un germe de la famille des *Pseudomonadaceae* (Tribu des *Chromobactereae*) *Empedobacter aquatile* isolé d'un produit frais de charcuterie. Arch Inst Pasteur Algerie *38*: 353–360.

—, —, — and L. Valette. 1960 Contribution à l'étude de deux germes saprophytes du genre *Empedobacter* isolés de yoghourt fabrique en Algére. Arch Inst Pasteur Algérie *38*: 487–499.

— and B. Vacher. 1959 Étude d'une souche d'Erwinia (*Erwinia salmonis* n. sp.) isolée d'une truite commune (*Salmo fario* L.). Ann Inst Pasteur (Paris) *97*: 241–244.

British Commonwealth Collections of Microorganisms. 1951 Directory of collections and list of species maintained in the United Kingdom and Crown colonies. His Majesty's Stationery Office, London, pp. 1–59.

Brocard, H. 1954 Sur les caractères microbiologiques de certains bacilles fusiformes pathogénes. C R Soc Biol Paris *148*: 83–84.

Brock, D. W. and L. K. Georg. 1969 Determination and analysis of *Actinomyces israelii* serotypes by fluorescent antibody procedures. J Bacteriol 97 (*2*): 581–588.

Brock, T. D. and H. Freeze. 1969 *Thermus aquaticus* gen. n. and sp. n. a non-sporulating extreme thermophile. J Bacteriol 98 (*1*): 289–297.

—, B. Reacher and D. Pierson. 1963 Survey of the bacteriocines of entercocci. J Bacteriol *86* (*4*): 702–707.

Brockmann, H. and K. Bauer. 1950 Rhodomycin, a red antibiotic from actinomycetes (in German). Naturwissenschaften *37* (*21*): 492–493.

—, — and I. Borchers. 1951 Rhodomycin, ein rotes Antibioticum (Antibiotica aus Actinomyceten VII Mitteil). Chem Ber *84* (*8*): 700–710.

—, H. Genth and R. Strufe. 1952 Antibiotica aus Actinomyceten. IX. Pikromycin II. Mitteil (in German). Chem Ber *85*: 426–433.

— and N. Grubhofer. 1949 Actinomycin C. Naturwissenschaften *36* (*12*): 376–377.

— and —. 1950 Zur Kenntnis des Actinomycins C. Naturwissenschaften *37* (*21*): 494–496.

—, —, H. Käss and H. Kalbe. 1951 Über das Actinomycin C (Antibiotica aus Actinomyceten, V. Mitteil) Chem Ber *84*: 260–284.

— and W. Henkel. 1950 Pikromycin, ein neues Antibiotikum aus Actinomyceten. Naturwissenschaften *37* (*6*): 138–139.

— and —. 1951 Pikromycin, ein bitter schmeck-endes Antibiotum aus Actinomyceten (Anti-biotica aus Actinomyceten VI. Mitteil). Chem Ber *84* (*3*): 284–288.

— and G. Schmidt-Kastner. 1951 Resistomycin, ein neues Antibioticum aus Actinomyceten. Naturwissenschaften *38:* 479–480.

Brooke, M. S. 1951 The biochemical properties of *Klebsiella* strains with new capsular anti-gens. Acta Pathol Microbiol Scand *28:* 328–337.

Brooke, W. F. 1941 The vole acid-fast *Bacillus.* I. Experimental studies on a new type of *Mycobacterium tuberculosis.* Amer Rev Tuberc *43:* 806–816.

Brooks, M. A. 1970 Comments on the classifica-tion of intracellular symbiotes of cockroaches and a description of the species. J Invertebr Pathol *16:* 249–258.

Brown, A. J. 1886 On an acetic ferment which forms cellulose. J Chem Soc (London) *49:* 432–439.

Brown, D. W. and W. E. C. Moore. 1960 Distri-bution of *Butyrivibrio fibrisolvens* in nature. J Dairy Sci *43:* 1570–1574.

Brown, J. H., W D. Frost and M. Shaw. 1926 Hemolytic streptococci of the beta type in certified milk. J Infect Dis *38:* 381–388.

Brown, L. R., R. J. Strawinski and C. S. McCles-key. 1964 The isolation and characteriza-tion of *Methanomonas methanooxidans* Brown and Strawinski. Can J Microbiol *10:* 791–799.

Brown, N. A. 1918 Some bacterial diseases of lettuce. J Agr Res *13:* 367–388.

—. 1923 Bacterial leafspot of geranium in the eastern United States. J Agr Res *23:* 361–372. 372.

—. 1934 A gall similar to crown gall produced on *Gypsophila* by a new bacterium. J Agr Res *48:* 1099–1112.

— and C. O. Jamieson. 1913 A bacterium caus-ing a disease of sugar-beet and nasturtium leaves. J Agr Res *1:* 189–210.

Brown, R. and E. L. Hazen. 1949 Activation of antifungal extracts of actinomycetes by ultrafiltration through Gradocol membranes. Proc Soc Exp Biol Med *71:* 454–457.

—, — and A. Mason. 1953 Effect of fungicidin (nystatin) in mice injected with lethal mix-tures of Aureomycin and *Candida albicans.* Science *117:* 609–610.

Brown, T. McP. and J. C. Nunemaker. 1942 Rat-bite fever. A review of the American cases with reevaluation of etiology; report of cases. Bull Johns Hopkins Hosp *70:* 201–236.

Brubaker, R. R. and M. J. Surgalla. 1962 Pesti-cins. II. Production of Pesticin I & II. J Bacteriol *84* (*3*): 539–545.

Bruce, D. 1887 Note on the discovery of a microorganism in Malta fever. Practitioner *39:* 161–170.

—. 1893 Sur une nouvelle forme de fièvre re-contreé sur les bords de la Mediterrannée. Ann Inst Pasteur (Paris) *7:* 289–304.

Brumpt, E. 1906 Les Mycétomes. Arch Para-sitol *10:* 489–527.

—. 1910 Précis de Parasitologie. 1st ed., Masson and Co., Paris.

—. 1911 Note sur le parasite des hematies de la taupe: *Grahamella talpae* n.g., n.sp. Bull Soc Pathol Exot *4:* 514–517.

—. 1921 Les parasites des invertébrés hemato-phages. *In* Lavier, Thèse, Paris, p. 207.

—. 1922 Les Spirochetoses. *In* Roger, Widal and Teissier, Nouveae Traité de Medicin. Fasc IV. Masson, Paris, pp. 491–531.

—. 1922 Précis de Parasitologie. 3rd ed., Mas-son and Co., Paris.

—. 1927 Précis de Parasitologie. 4th ed., Mas-son and Co., Paris.

—. 1930 Rechutes parasitaires intenses, dues à la splénectomie, au cours d'infections laten-tes à *Aegyptianella,* chez la poule. C R Acad Sci Paris *191:* 1028–1030.

—. 1932 Longévité de virus de la fièvre bou-tonneuse (*Rickettsia conori,* n.sp.) chez la Tique, *Rhipicephalus sanguineus.* C R Seances Soc Biol Filiales *110:* 1199–1202.

—. 1933 Étude du *Spirochaeta turicatae,* n. sp. agent de la fièvre récurrente sporadique des Etats-Unis transmis par *Ornithodorus turicata.* C R Soc Biol Paris *113:* 1369–1372.

—. 1937 Facteurs qui agissent sur la transmis-sion des infections par les arthropodes hémo-phages. Ann Parasitol Hum Comp *15:* 75–85.

—. 1938 Rickettsia intracellulaire stomacale (*Rickettsia culicis* n.sp.) of *Culex fatigans.* Ann Parasitol Hum Comp *16:* 153–158.

—. 1939 Un nouveau treponeme parasite de l'homme: *Treponema carateum,* agent des carates ou "mal del Pinto." C R Soc Biol Paris *130:* 942–945.

—. 1939 Une nouvelle fièvre récurrente hu-maine découverte dans la région de Babylone (Irak). C R Acad Sci Paris *208:* 2020–2031.

— and G. Lavier. 1935 Sur um piroplasmideé, Nouveau parasite de tortue, *Tunetella emydis* N.G., n.sp. Ann Parasitol Hum Comp *13* (*6*): 544–550.

Brunchorst, J. 1886 Uber einige Wurzelan-schwellugen, besonders diejenigen von *Alnus* und den *Elaegnaceen.* Bot Inst Tubingen *2:* 151–177.

Brundish, D. E. and J. Baddiley. 1968 Pneu-mococcal C-substance, a ribitol teichoic acid containing choline phosphate. Biochem J *110:* 573–582.

Bruner, D. W. and P. R. Edwards. 1941 Classi-fication of *Corynebacterium equi.* Ky Agr Exp Sta Bull *44:* 92–107.

Bryan, M. K. 1921 A bacterial budrot of cannas. J Agr Res *21:* 143–152.

—. 1926 Bacterial leaf-spot on hubbard squash. Science (Washington) *63:* 165.

— and F. P. McWhorter. 1930 Bacterial blight of poppy caused by *Bacterium papavericola* sp. nov. J Agr Res *40:* 1–9.

Bryant, M. P. 1952 The isolation and charac-teristics of a spirochete from the bovine ru-men. J Bacteriol *64:* 325–335.

—. 1956 The characteristics of strains of *Sele-nomonas* isolated from bovine rumen contents. J Bacteriol *72* (*2*): 162–167.

—. 1959 Bacterial species of the rumen. Bac-teriol Rev *23:* 125–153.

—. 1965 Rumen methanogenic bacteria. *In* Daugherty, Physiology of Digestion in the Ruminant. Butterworths, Washington, D.C., pp. 411–418.

—. 1969 Symbiotic associations of certain ethanol and lactate fermenting bacteria with methanogenic bacteria. Abstracts of papers, 158th National Meeting, Amer Chem Soc No.18.

— and R. N. Doetsch. 1954 A study of actively cellulolytic rod-shaped bacteria of the bovine rumen. J. Dairy Sci *37:* 1176–1183.

—, B. C. McBride and R. S. Wolfe. 1968 Hydrogen oxidizing methane bacteria. I. Cultivation and methanogenesis. J. Bacteriol 95 (3): 1118–1123.

— and I. M. Robinson. 1961 Some nutritional requirements of the genus *Ruminococcus*. Appl Microbiol 9: 91–95.

— and —. 1961 Studies on the nitrogen requirements of some ruminal cellulolytic cocci. Appl Microbiol 9: 96–103.

— and —. 1962 Some nutritional characteristics of predominant culturable ruminal bacteria. J Bacteriol 84 (4): 605–614.

— and —. 1963 Apparent incorporation of ammonia and amino acid carbon during growth of selected species of ruminal bacteria. J. Dairy Sci 46: 150–154.

—, — and H. Chu. 1959 Observations on the nutrition of *Bacteroides succinogenes*—a ruminal cellolytic bacterium. J Dairy Sci 42: 1831–1847.

—, — and I. L. Lindahl. 1961 A note on the flora and fauna in the rumen of steers fed a feedlot bloat-provoking ration and the effect of penicillin. Appl Microbiol 9: 511–515.

— and N. Small. 1956 The anaerobic monotrichous butyric acid-producing curved rod-shaped bacteria of the rumen. J Bacteriol 72: 16–21.

— and —. 1956 Characteristics of two new genera of anaerobic curved rods isolated from the rumen of cattle. J Bacteriol 72 (1): 22–26.

—, —, C. Bouma and H. Chu. 1958 *Bacteroides ruminicola n. sp.* and the new genus and species *Succinimonas amylolytica*. Species of succinic acid-producing anaerobic bacteria of the bovine rumen. J Bacteriol 76 (1): 15–23.

—, —, — and I. M. Robinson. 1958 Characteristics of ruminal anaerobic cellulolytic cocci and *Cillobacterium cellulosolvens* n. sp. J Bacteriol 76 (5): 529–537.

—, S. F. Tzeng, I. M. Robinson and A. E. Joyner, Jr. 1971 Nutrient requirements of methanogenic bacteria. *In* F. G. Pohland, Anaerobic Biological Treatment Processes. Advances in Chemistry, Series 105, American Chem Soc, Washington, D.C., pp. 23–40.

—, E. A. Wolin, M. J. Wolin and R. S. Wolfe. 1967 *Methanobacillus omelianskii*, a symbiotic association of two species of bacteria. Arch Mikrobiol 59: 20–31.

Brygoo, E. R. and N. Aladame. 1953 Étude d'une espèce nouvelle anaérobie stricte du genre *Eubacterium: E. crispatum* n. sp. Ann Inst Pasteur (Paris) 84: 640–651.

Bryner, J. H., A. H. Frank and P. A. O'Berry. 1962 Dissociation studies of *Vibrio* from the bovine genital tract. Amer J Vet Res 23: 32–41.

Brysk, M. M., W. A. Corpe and L. V. Hanks. 1969 β-Cyanoalamine formation by *Chromobacterium violaceum*. J Bacteriol 97 (1): 322–327.

Buchanan, B. B. and L. Pine 1962 Characterization of a propionic acid producing actinomycete, *Actinomyces propionicus*, sp. nov. J Gen Microbiol 28 (2): 305–323.

Buchanan, R. E. 1911 Veterinary Bacteriology. W. B. Saunders Co., Philadelphia.

—. 1918 Studies in the nomenclature and classification of the bacteria. V. Subgroups and genera of the *Bacteriaceae*. J Bacteriol 3 (1): 27–61.

—. 1918 Studies in the classification and nomenclature of the bacteria. VIII. The subgroups and genera of the *Actinomycetales*. J Bacteriol 3 (4): 403–406.

—. 1918 Studies in the classification and nomenclature of the bacteria. IX. The subgroups and genera of the *Thiobacteriales*. J Bacteriol 3 (5): 461–474.

—. 1925 General Systemic Bacteriology. The Williams and Wilkins Co., Baltimore.

—. 1926 What names should be used for the organisms producing nodules on the roots of leguminous plants? Proc Iowa Acad Sci 33: 81–90.

—. 1957 Family III. *Leucotrichaceae* Buchanan Fam. nov. *In* Breed, Murray and Smith (Editors), Bergey's Manual of Determinative Bacteriology, 7th ed. The Williams & Wilkins Co., Baltimore, pp. 850–851.

— and B. W. Hammer. 1915 Slimy and ropy milk. Bull Iowa State Coll Agr Mech Arts 22: 209–295.

—, J. G. Holt and E. F. Lessel. 1966 Index Bergeyana. The Williams and Wilkins Co., Baltimore.

Bucher, G. E. 1961 Artificial culture of *Clostridium brevifaciens* n. sp. and *C. malacosomae* n. sp., the causes of brachytosis of tent caterpillars. Can J Microbiol 7: 641–655.

Bucherer, H. 1942 Über den mikrobiellen Abbau von Giftstoffen. I. Mitteilung. Über den mikrobiellen Abbau von Nikotin. Zentrabl Bakteriol Parasitenk Infektionskr Hyg Abt I 105: 106–173.

Buchner, H. 1885 Über die Koch'schen und Finkler-Prior'schen "Kommabacillen". Sitzungsber Gesel Morphol Physiol München 1: 1–10.

Buchner, P. 1965 Endosymbiosis of animals with plant-like micro-organisms. John Wiley, New York.

Buck, J. D., S. P. Meyers and E. Leifson. 1963 *Pseudomonas (Flavobacterium) piscicida* Bein comb. nov. J. Bacteriol 86 (5): 1125–1126.

Buday, K. 1898 Zur Kenntnis der abnormen postmortalen Gasbildung. Zentrabl Bakteriol Parasitenk Infektionskr Hyg Abt I Orig 24: 369–375.

Buddle, M. B. 1956 Studies on *Brucella ovis* n.sp., a cause of genital disease of sheep in New Zealand and Australia. J. Hyg 54 (3): 351–364.

Buder, J. 1914 *Chloronium mirabile*. Ber Deut Bot Ges 31: 80–97.

—. 1915 Zur Kenntnis des *Thiospirillum jenense* und seiner Reaktionen auf Lichtreize. Jahrb Wiss Bot 56: 529–584.

Buhler, V. B. and A. Pollak. 1953 Human infection with atypical acid-fast organisms. Report of 2 cases with pathologic findings. Amer J Clin Pathol 23: 363–374.

Bull, L. B. 1929 Dermatomycosis of the sheep (lumpy or matted wool) due to *Actinomyces dermatonomus* (n.sp.) Aust J Exp Biol Med Sci 6: 301–314.

Bulla, L. A., G. St. Julian, R. A. Rhodes and C. W. Hesseltine. 1969 Scanning electron and phase-contrast microscopy of bacterial spores. Appl Microbiol 18 (3): 490–495.

Bulloch, W., W. E. Bullock, S. R. Douglas, H. Henry, J. McIntoch, R. A. O'Brien, M. Robertson and C. G. L. Wolf. 1919 Report on the anaerobic infections of wounds and the bacteriological and serological problems arising therefrom. Med Res Comm (Gt Brit) Spec Rep Ser *39:* 1–182.

Bullock, G. L. 1961 A schematic outline for the presumptive identification of bacterial diseases of fish. Prog Fish-Cult *23:* 147–151.

Bumm, E. 1885 Der Mikro-organismus der gonorrhoischen Schleimhault Erkrankungen "Gonococcus-Neisser," 1st ed., Bergman, Wiesbaden, pp. 1–146.

Bunker, H. J. 1936 Review of the physiology and biochemistry of the sulphur bacteria. Dept. Sci Ind Res Chem Res Spec Rept London *3:* 16–17.

Bunting, R. H. 1932 Actinomyces in cacao-beans. Ann Appl Biol *19:* 515–517.

Burgdorfer, W. 1951 Analyse des Infektionsverlaufes bei *O. moubata* (Murray) und der natürlichen Übertragung von *Sp. duttoni.* Acta Trop *8:* 193–262.

Burger, A., G. Drews and R. Ladwig. 1968 Wirtskreis und Infektionscyclus eines neu isolierten *Bdellovibrio bacteriovorus*-stammes. Arch Mikrobiol *61:* 261–279.

Burgvits, G. K. 1935 Phytopathogenic bacteria. Akad Nauk USSR, Moscow.

Burkholder, P. R., S. H. Sun, L. E. Anderson and J. Ehrlich. 1955 The identity of viomycin-producing cultures of *Streptomyces.* Bull Torrey Bot Club *82:* 108–117.

Burkholder, W. H. 1926 A new bacterial disease of the bean. Phytopathology *16:* 915–928.

—. 1930 The bacterial diseases of the bean. Cornell Agr Exp Sta Mem *127:* 1–88.

—. 1937 A bacterial leaf spot of geranium. Phytopathology *27:* 554–560.

—. 1939 Genus VI *Phytomonas* Bergey et al. *In* Bergey, Breed, Murray and Hitchens (Editors), Bergey's Manual of Determinative Bacteriology, 5th ed. The Williams & Wilkins Co., Baltimore, pp. 142–218.

—. 1941 The black rot of *Barbarea vulgaris.* Phytopathology *31:* 347–348.

—. 1942 Three bacterial plant pathogens, *Phytomonas caryophylli,* sp. n., *Phytomonas alliicola* sp. n., and *Phytomonas manihoti* (Arthaud-Berthet et Bondar) Viegas. Phytopathology *32:* 141–149.

—. 1944 *Xanthomonas vignicola* sp. nov. pathogenic on cowpeas and beans. Phytopathology *34:* 430–432.

—. 1948 Genus I. *Pseudomonas* Migula. Bacterial plant pathogens. *In* Breed, Murray and Hitchens (Editors), Bergey's Manual of Determinative Bacteriology, 6th ed. The Williams & Wilkins Co., Baltimore, pp. 82–150.

—. 1948 Genus II *Xanthomonas. In* Breed, Murray and Hitchens (Editors), Bergey's Manual of Determinative Bacteriology, 6th ed. The Williams & Wilkins Co., Baltimore, pp. 150–171.

—. 1948 Genus 1. *Corynebacterium* Lehmann & Neumann. *In* Breed *et al.* (Editors), Bergey's Manual of Determinative Bacteriology, 6th ed. The Williams & Wilkins Co., Baltimore, pp. 392–395; 398–400.

—. 1948 Genus I. *Erwinia* Winslow et al. *In* Breed, Murray and Hitchens (Editors), Bergey's Manual of Determinative Bacteriology, 6th ed. The Williams & Wilkins Co., Baltimore, pp. 463–478.

—. 1950 Sour skin, a bacterial rot of onion bulbs. Phytopathology *40:* 115–117.

—. 1957 Genus II *Xanthomonas. In* Breed, Murray and Smith (Editors), Bergey's Manual of Determinative Bacteriology, 7th ed. The Williams & Wilkins Co., Baltimore, pp. 152–180.

—. 1957 Genus VI. *Erwinia* Winslow et al. 1917. *In* Breed, Murray and Smith (Editors), Bergey's Manual of Determinative Bacteriology, 7th ed. The Williams & Wilkins Co., Baltimore pp. 349–359.

—. 1960 A bacterial grown rot of parsnip roots. Phytopathology *50:* 280–282.

—, L. A. McFadden and A. V. Dinock. 1953 A bacterial blight of chrysanthemums. Phytopathology *43:* 522–526.

— and P. P. Pirone. 1941 Bacterial leaf spot of gardenia. Phytopathology *31:* 192–194.

Burnham, J. C., T. Hashimoto and S. F. Conti. 1968 Electron microscopic observations on the penetration of *Bdellovibrio bacteriovorus* into Gram-negative bacterial hosts. J Bacteriol *96 (4):* 1366–1381.

—, — and —. 1970 Ultrastructure and cell division of a facultatively parasitic strain of *Bdellovibrio bacteriovorus.* J Bacteriol *101 (3):* 997–1004.

Burns, J. and D. F. Holtman. 1959 Tennecetin: A new antifungal antibiotic. Antibiot Chemother *9:* 398–405.

Burnside, C. E. and R. E. Foster. 1935 Studies on the bacteria associated with parafoulbrood. J. Econ Entomol *28:* 578–584.

Burr, S. 1928 Sprain or internal rust spot of potato. Ann Appl Biol *15:* 563–585.

Burri, R. and P. Ankersmit. 1906 *Bacterium clostridiiforme.* p. 115. *In* P. Ankersmit. Untersuchung über die Bakterien im Verdauungskanal des Rindes. Zentrabl Bakteriol Parasitenk Infektionskr Hyg Abt I Orig *40:* 100–118.

— and A. Stutzer. 1895 Ueber Nitrat Zerstörende Bakterien und den durch dieselben Bedington Stickstoffverlust. Zentrabl Bakteriol Parasitenk Infektionskr Hyg Abt II *1:* 257–265; 360–364; 392–398; 422–432.

Burrill, T. J. 1882 The Bacteria: an account of their nature and effects, together with a systematic description of the species. Illinois Indust Univ 11th Rep, pp. 93–157.

Burton, M. O. and A. G. Lochhead. 1953 Nutritional requirements of *Arthrobacter terregens.* Can J Bot *31:* 145–151.

—, F. J. Sowden and A. G Lochhead. 1954 Studies on the isolation and nature of the 'terregens factor.' Can J Biochem *32:* 400–406.

Burton, S. D., R. Y. Morita and W. Miller. 1966 Utilization of acetate by *Beggiatoa.* J Bacteriol *91 (3):* 1192–1200.

Busacca, A. 1935 Un germe aux caractères de rickettsies (*Rickettsia trachomae*) dans les tissus trachomateux. Arch Ophthalmol *52:* 567–572.

Büsing, K. H., W. Döll and K. Freytag. 1953 Die Bakterienflora der medizinischen Blutegel. Arch Mikrobiol *19:* 52–86.

— and K. Freytag. 1954 Die Bakterienflora der Blutegel-Harnblase. Zentrabl Bakteriol Parasitenk Infektionskr Hyg Abt I Orig *160:* 577–585.

Butkevich, V. S. 1928 Forming of ferro-manganic sea deposits and partaking microorganisms. Proc Sci Sea Inst *3 (3):* 5–62.

Butterfield, C. T. 1935 Studies of sewage purification. II. A zoogloea-forming bacterium isolated from activated sludge. U S Public Health Rep *50* (*20*): 671–684.

—, C. C. Ruchhoft and P. D. McNamee. 1937 Studies of sewage purification. VI. Biochemical oxidation by sludges developed by pure cultures of bacteria isolated from activated sludge. U S Public Health Rep *52*: 387–412.

Buttiaux, R., R. Osteux, R. Fresnoy and J. Moriamez. 1954 Les propriétés biochimiques charactéristiques du genre *Proteus*. Inclusion souhaitable des *Providencia* dans celui-ci. Ann Inst Pasteur (Paris) *87*: 375–386.

Buxton, J. B. 1929 A note on *Vibrio foetus ovis* in the ram. First Report of Director, Inst Anim Pathol, Univ. of Cambridge, 1929–1930, pp. 47–51.

Buyze, G. 1955 De koolhydraatstofwisseling van *Lactobacillus brevis*. Hexosemonophosphaat shunt en heterofermentatie bij *Lactobacteriaceae*. Thesis, Univ of Utrecht, pp. 1–109.

—, C. J. A. van den Hamer and P. G. de Haan. 1957 Correlation between hexose-monophosphate shunt, glycolytic system and fermentation-type in lactobacilli. Antonie Van Leeuwenhoek J Microbiol Serol *23*: 345–350.

Cabezas de Herrera, E. and R. Moreno. 1969 Amino acids and sugars of *Corynebacterium michiganensis* and its cell walls. Microbiol Espan *22*: 55–62.

Caldwell, D. R., M. Keeney and P. J. Van Voest. 1969 Effects of carbon dioxide on growth and maltose fermentation by *Bacteroides amylophilus*. J Bacteriol 98 (*2*): 668–676.

—, D. C. White, M. P. Bryant and R. N. Doetsch. 1965 Specificity of the heme requirement for growth of *Bacteroides ruminicola*. J Bacteriol *90*: 1645–1654.

Caldwell, M. E. and D. L. Ryerson. 1939 Salmonellosis in certain reptiles. J Infect Dis *65*: 242–245.

— and —. 1940 A new species of the genus *Pseudomonas* pathogenic for certain reptiles. J Bacteriol *39*: 323–335.

Calhoun, K. N. and L. E. Johnson. 1956 Taxonomic and microbiological studies of *Streptomyces chartreusis* n. sp. Antibiot Chemother *6* (*4*): 294–298.

Calot, L. and A. P. Cercos. 1963 *Streptomyces ornatus* nov. sp. et *Streptomyces erumpens* nov. sp. producteurs d'ornamicine et antibiotique 17732. Ann Inst Pasteur (Paris) *105*: 159–161.

— and —. 1963 *Streptomyces ornatus* nov. sp. y *Streptomyces erumpens* nov. sp. productores de los antibiotics 17044 y 17732. Rev Invest Agr *17* (*3*): 303–312.

Cameron, E. J., J. R. Esty and C. C. Williams. 1936 The cause of "black beets": an example of oligodynamic action as a contributory cause of spoilage. Food Res *1*: 73–85.

Camin, J. H. 1948 Mite transmission of hemorrhagic septicemia in snakes. J Parasitol *34*: 345–354.

Caminiti, R. 1907 Ueber eine neue Streptothrix species und die Streptothricheen im allgemeinen. Zentrabl Bakteriol Parasitenk Infektionskr Hyg Abt Orig *44*: 193–208.

Caminopetros, J. and E. Triantaphylopoulos. 1936 Existence en Grèce d'une fièvre récurrente dont le spirochète revet les caractères de "*Sp. hispanica*," agent de la fièvre récurrente hispano-africaine. Ann Parasitol Hum Comp *14*: 429–432.

Campbell, C. C., G. B. Hill and B. E. Brooks. 1956 Therapeutic activity of a new antibiotic, 1968, in mice with experimental histoplasmosis, sporotrichosis and moniliasis. Antibiot Ann *1955–1956*: 240–244.

Campbell, L. L. 1957 Genus V. *Beneckea* Campbell gen. nov. *In* Breed, Murray and Smith (Editors), Bergey's Manual of Determinative Bacteriology, 7th ed. The Williams & Wilkins Co., Baltimore, pp. 328–332.

—, H. A. Frank and E. R. Hall. 1957 Studies on thermophilic sulfate reducing bacteria. I. Identification of *Sporovibrio desulfuricans* as *Clostridium nigrificans*. J Bacteriol *73* (*4*): 516–521.

—, M. A. Kasprzycki and J. R. Postgate. 1966 *Desulfovibrio africanus* sp. n., a new dissimiliatory sulfate-reducing bacterium. J Bacteriol *92* (*4*): 1122–1127.

— and J. R. Postgate. 1965 Classification of the spore-forming sulfate-reducing bacteria. Bacteriol Rev *29* (*3*): 359–363.

— and —. 1969 Revision of the holotype strain of *Desulfotomaculum ruminis* (Coleman) Campbell and Postgate. Int J Syst Bacteriol *19* (*2*): 139–140.

— and O. B. Williams. 1951 A study of chitindecomposing microorganisms of marine origin. J Gen Microbiol *5*: 894–905.

Canale-Parola, E., R. Borasky and R. S. Wolfe. 1961 Studies on *Sarcina ventriculi*. III. Localization of cellulose. J Bacteriol *81* (*2*): 311–318.

—, S. C. Holt and Z. Udris. 1967 Isolation of free-living, anaerobic spirochetes. Arch Mikrobiol *59*: 41–48.

—, M. Mandel and D. G. Kupfer. 1967 The classification of sarcinae. Arch Mikrobiol *58*: 30–34.

—, Z. Udris and M. Mandel. 1968 The classification of free-living spirochetes. Arch Mikrobiol *63*: 385–397.

— and R. S. Wolfe. 1960 Studies on *Sarcina ventriculi*. I. Stock culture method. J Bacteriol *79* (*6*): 857–859.

— and —. 1960 Studies on *Sarcina ventriculi*. II. Nutrition. J Bacteriol *79* (*6*): 860–862.

Canevazzi, G. and T. Scotti. 1959 Descrizione di uno streptomicete (*Streptomyces chrestomyceticus*) sp. nova, produttore del nuovo antibiotico amminosidina. G Microbiol *7*: 242–250.

Cantacuzène, J. 1910 Sur un spirochète thermophile des eaux de Dax. C R Soc Biol (Paris) *68*: 75–77.

Carbone, D. and A. Tomboiato. 1917 Sulfa macerazione rustica della canapa. Sta Sperim Agr Ital *50*: 563–575.

Cardamatis, J. 1911 Des piroplasmiases et Leishmaniases. Zentrabl Bakteriol Parasitenk Infektionskr Hyg Abt I Orig *60*: 511–523.

Cardon, B. P. and H. A. Barker. 1946 Two new amino acid-fermenting bacteria, *Clostridium propionicum* and *Diplococcus glycinophilus*. J Bacteriol *52*: 629–634.

— and —. 1946 Amino acid fermentations by *Clostridium propionicum* and *Diplococcus glycinophilus*. Arch Biochem Biophys *12*: 165–180.

Carlson, A., A. Kellner, A. W. Bernheimer and E. B. Freeman. 1957 A streptococcal enzyme that acts especially on DPN; characterization on the enzyme and its separation from streptolysin O. J Exp Med *106:* 15–26.

Carlsson, J. 1965 Zoogloea-forming streptococci resembling *Streptococcus sanguis*, isolated from dental plaque in man. Odontol Revy *16:* 348–358.

—. 1967 Presence of various types of nonhaemolytic streptococci in dental plaque and in other sites of the oral cavity in man. Odontol Revy *18:* 55–74.

Carmichael, L. E. and D. W. Bruner. 1968 Characteristics of a newly recognized species of *Brucella* responsible for infectious canine abortions. Cornell Vet *58:* 579–592.

—, T. D. St. George, N. D. Sullivan and N. Horsfall. 1972 Isolation, propagation, and characterization studies of an ovine *Mycoplasma* responsible for proliferative interstitial pneumonia. Cornell Vet *62:* 654–679.

Carne, H. R. 1968 Action of bacteriophages obtained from *Corynebacterium diphtheriae* on *C. ulcerans* and *C. ovis.* Nature (London) *217:* 1066–1067.

—, N. Wickham and J. C. Kater. 1956 A toxic lipid from the surface of *Corynebacterium ovis.* Nature (London) *178:* 701–702.

Carpano, M. 1929 Su di un Piroplasma osservato nei Polli in Egitto (*Egyptianella pullorum*). Bull Minist Agr Egypt *86:* 1–12.

Carr, J. G. 1958 *Acetobacter estunense* nov. spec. An addition to Frateur's ten basic species. Antonie Van Leeuwenhoek J Microbiol Serol *24:* 157–160.

— and P. A. Davies. 1970 Homofermentative lactobacilli of ciders including *Lactobacillus mali* nov. spec. J Appl Bacteriol *33:* 768–774.

Carrier, E. B. 1963 A comparative study of members of the genus *Corynebacterium.* Ph.D. Thesis, Dept Bact Louisiana State Univ, pp. 1–216.

Carsner, E. 1918 Angular-leafspot of cucumber; dissemination, overwintering and control. J Agr Res *15 (3):* 201–220.

Carter, G. R. 1955 Studies on *Pasteurella multocida.* I. A haemagglutination test for the identification of serological types. Amer J Vet Res *16 (60):* 481–484.

—. 1957 Studies on *Pasteurella multocida.* II. Identification of antigenic characteristics and colonial variants. Amer J Vet Res *18 (66):* 210–213.

—. 1961 A new serological type of *Pasteurella multocida* from Central Africa. Vet Rec *73 (42):* 1052.

—. 1962 Further observations on typing *Pasteurella multocida* by the indirect haemagglutination test. Can J Comp Med *26:* 238–240.

—. 1963 Immunological differentiation of type B and E strains of *Pasteurella multocida.* Can Vet J *4:* 61–63.

— and R. V. S. Bain. 1960 Pasteurellosis (*Pasteurella multocida*). A review, stressing recent developments. Vet Rev Annot *6:* 105–128.

Carter, H. E., D. Gottlieb and H. W. Anderson. 1948 Chloromycetin and streptothricin. Science (Washington) *107:* 113.

Carter, H. V. 1888 Note on the occurrence of a minute blood-spirillum in an Indian rat. Sci Mem Med Officers Army India (Part 3), pp. 45–48.

Carter, J. C. 1945 Wetwood of elms. Ill. Natur Hist Surv Bull *23:* 407–448.

Caselitz, F. H. 1955 Ein neues Bacterium der Gattung: *Vibrio* Müller—*Vibrio jamaicensis.* Z Tropenmed Parasitol *6:* 52–63.

—. 1966 *Pseudomonas—Aeromonas* und ihre humanmedizinische Bedeutung. VEB Gustav Fischer Verlag, Jena, pp. 1–23.

Caspary, R. 1857 Bericht über die Verhandlungen der 33 Versammlung deutscher Naturforscher und Arzte gehalten in Bonn von 18 bis 24 September 1857. Bot Ztg *15:* 749–776; 784–792.

Cassinelli, G., A. Grein, P. Orezzi, P. Pennella and A. Sanfilippo. 1967 New antibiotics produced by *Streptoverticillium orinoci*, n. sp. Arch Mikrobiol *55:* 358–368.

Castellani, A. 1905 On the presence of spirochaetes in some cases of parangi (yaws, *Framboesia tropica*). Preliminary note. J Ceylon Br Brit Med Ass *2:* 54.

—. 1949 A new haemorrhagic ulcerative disease and its aetiology. J Trop Med Hyg *52:* 204–210.

—. 1951 Some little-known bacteria. Zentrabl Bakteriol Parasitenk Infektionskr Hyg Abt I Orig *157:* 74–80.

—. 1964 Macroulcus perstans (Persistent megaloulcer). Dermatologica *15:* 329–338.

— and A. J. Chalmers. 1910 Manual of Tropical Medicine, 1st ed. Baillière, Tindall and Cox, London.

— and —. 1913 Manual of Tropical Medicine, 2nd ed. Baillière, Tindall and Cox, London.

— and —. 1919 Manual of Tropical Medicine, 3rd ed. Williams Wood and Co., New York.

—, M. M. X. de Brito and M. R. Pinto. 1959 An actinomycete isolated from an autochthonous case of mycetoma in Portugal. J Trop Med Hyg *62:* 27–36.

Castelnuovo, G., A. Guadiano, M. Morellini, G. Penso and M. Polizzi-Sciarrone. 1959 La constituzione antigenica di alcuni micobatteri (The antigenic constitution of some mycobacteria). Ann Ist Forlanini *19:* 1–18.

—, —, — and C. Rossi. 1960 Gli antigeni dei micobatteri. Rend Ist Super Sanita *23:* 1222–1233.

—, —, — and M. Polizzi-Sciarrone. 1958 Differenziazione antigenica dei micobatteri studiata con le techniche di diffusione in agar, immunoelectroforesi deviazione del complemento. Riv Tuberc Mal App Resp *6:* 303–308.

— and M. Morellini. 1962 Gli antigen di alcuni dei cosidetti "Micobatteri atipici" O "anonimi". Ann Ist Forlanini *22:* 1–20.

Cataldi, M. S. 1939 Estudio fisiólogico y sistemático de algunas *Chlamydobacteriales.* Thesis Univ Buenos Aires, pp. 1–96.

—. 1940 Una nueva bacteria licuante del agar. Rev Inst Bacteriológico *9:* 366–377.

—, V. Lopez, O. L. Galmarini and J. Pahn. 1962 Curamycin and its production. United States Patent 3,015,607, January 2.

Catlin, B. W. 1967 Genetic studies of sulfadiazine-resistant and methionine-requiring *Neisseria* isolated from clinical material. J Bacteriol *94 (2):* 719–733.

—. 1970 Transfer of the organism named *Neisseria catarrhalis* to *Branhamella gen. nov.* Int J Syst Bacteriol *20 (2):* 155–159.

— and L. S. Cunningham. 1961 Transforming activities and base contents of deoxyribonucleate preparations from various *Neisseria*. J Gen Microbiol 26 (2): 303–312.

Cato, E., W. E. C. Moore and L. V. Holdeman. 1968 *Clostridium oroticum* comb. nov. Amended description. Int J Syst Bacteriol 18 (1): 9–13.

Cavara, F. 1905 Bacteriosi del fico. Atti Accad Gioenia Sci Natur Catania 18: Mem 14, 1–17.

Cayley, D. M. 1917 Bacterial disease of *Pisum sativum*. J Agr Sci 8: 461–479.

Cercos, A. P. 1953 Canfomicina, antibiotico producido por un "*Streptomyces*". Rev Argent Agron 20 (2): 53–62.

—. 1954 *Streptomyces rutgersensis* var. *castelarense* n. var. Nuevas propiedades de la canfomicina. Rev Invest Agr 8: 263–283.

—, B. L. Eilberg, J. G. Goyena, J. Souto, E. E. Vautier and I. Widuczynski. 1962 Misionina: antibiótico poliénico producido por *Streptomyces misionensis* n. sp. Rev Invest Agr 16 (1): 5–27.

— and A. Rosemblit. 1950 *Streptomyces* antibioticos aislados del aire y suelo de la Argentina. Rev Argent Agron 17 (2): 98–105.

Certes, A. 1882 Note sur les parasites et les commensaux de l'huître. Bull Soc Zool Fr 7: 347–353.

—. 1889 Note sur les microorganismes de la panse des ruminants. Bull Soc Zool Fr 14: 70–73.

Chalmers, A. J. and J. B. Christopherson. 1916 A Sudanese actinomycosis. Ann Trop Med Parasitol 10: 223–282.

Chamoiseau, G. and J. Asselineau. 1970 Examen des lipides de *Nocardia farcinica*: presence d'acides mycoloques. C R Acad Sci Paris 270 D: 2603–2604.

Chan, E. C. S. and I. L. Stevenson. 1962 On the biotin requirement of *Arthrobacter globiformis*. Can J Microbiol 8: 403–405.

Charles Pfizer & Company Inc. 1961 Antibiotic E73, cycloheximide and fungicidin. British Patent 866,600, April 26.

Charlton, D. B., M. E. Nelson and C. H. Werkman. 1934 Physiology of *Lactobacillus fructivorans* sp. nov. isolated from spoiled salad dressing. Iowa State J Sci 9: 1–11.

Charney, J., W. P. Fisher, C. Curran, R. A. Machlowitz and A. A. Tytell. 1953 Streptogramin, a new antibiotic. Antibiot Chemother 3: 1283–1286.

Chatterjee, K. R., B. Dasgupta, N. Mukherjee and H. N. Ray. 1956 Some cytochemical observations in *Mycobacterium leprae*. Bull Calcutta Sch Trop Med 4: 18.

Chatton, E. 1937 Titres et Travaux Scientifiques. Sète, Sottano.

— and C. Pérard. 1913 Schizophytes du caecum du cobaye. I. *Oscillospira guilliermondi* n.g., n.sp. C R Soc Biol Paris 74: 1159–1162.

— and —. 1913 Schizophytes du caecum du cobaye. II. *Metabacterium polyspora* n.g., n. sp. C R Soc Biol Paris 74: 1232–1234.

Chen, C. C. and R. C. Cleverdon. 1963 Some investigation on the growth of *Dialister pneumosintes*. J Appl Bacteriol 26: 107–111.

Chen, K. C. and A. W. Ravin. 1966 Heterospecific transformation of pneumococcus and streptococcus. J Mol Biol 22: 109.

Chen, P. L. 1964 The membrane system of *Streptomyces cinnamonensis*. Amer J Bot 51: 125–132.

Cherni, N. E., J. V. Solov'eva, V. D. Fedorova

and E. N. Kondrat'eva. 1969 The ultrastructure of cells of the species of purple sulfur bacteria. Mikrobiologiya 38 (3): 479–484.

Chesbro, W. R. and J. B. Evans. 1959 Factors affecting the growth of enterococci in highly alkaline media. J Bacteriol 78 (6): 858–862.

Chester, F. D. 1897 Report of the mycologist: Bacteriological work. Del Agr Exp Sta Bull 9: 38–145.

—. 1898 Report of the mycologist: Bacteriological work. Del Agr Exp Sta Bull 10: 47–137.

—. 1901 A Manual of Determinative Bacteriology. The Macmillan Co., New York, pp. 1–401.

—. 1938 A bacteriosis of dahlia, *Erwinia cytolytica*. Phytopathology 28: 427–432.

—. 1939 Genus IV. *Erwinia* Winslow et al. *In* Bergey, Breed, Murray and Hitchens (Editors), Bergey's Manual of Determinative Bacteriology, 5th ed. The Williams & Wilkins Co., Baltimore, pp. 404–420.

Chionglo, D. T. and J. A. Hayashi. 1969 Structural basis of group G streptococcal antigenicity. Arch Biochem Biophys 130: 39–47.

Cholodny, N. 1922 Uber Eisenbakterien und ihre Beziehungen zu den Algen. Ber Deut Bot Ges 40: 326–346.

—. 1924 Uber neue eisenbakterienarten aus der Gattung *Leptothrix* Kütz. Zentralbl Bakteriol Parasitenk Infektionskr Hyg Abt II 61: 292–298.

—. 1926 Die Eisenbakterien. Beiträge zu einer Monographie. Pflanzenforsch., H, 4. Hrsg. von Kolkwitz. G. Fischer, Jena, pp. 1–162.

—. 1953 Iron Bacteria, 2nd ed. Akad Nauk SSSR, Moscow

Choukevitch, J. 1911 Étude de la flore bactérienne du gros intestin du cheval. Ann Inst Pasteur (Paris) 25: 247–267; 345–368.

Christiansen, A. H. 1964 Studies on the antigenic structure of *T. pallidum*. Acta Pathol Microbiol Scand 60: 123–130.

Christiansen, M. 1917 En ejendommelig pyaemisk Lidelse hos Faar. Maandsskr Dyraeg 29: 449–458.

Christie, A. O. and J. W. Porteous. 1962 Growth of several strains of *Actinomyces israelii* in chemically defined media. Nature (London) 195: 408–409.

Christie, R., N. E. Atkins and E. Munch-Peterson. 1944 A note on a lytic phenomenon shown by group B streptococci. Aust J Exp Biol Med Sci 22: 197–200.

Chui, W.-F. and C.-A. Wu. 1963 The therapeutic and protectant effect of metabolites of some *Actinomyces* isolates on incidence of a mosaic virus disease of rape (*Brassica chinensis* L.). Acta Phytopathol Sinica 6 (2): 187–196.

CIBA Limited. 1951 Manufacture of an antibiotic agent. British Patent 651,269, March 14.

Ciferri, R. 1927 Studien über Kakao. Untersuchungen ueber den müffigen Geruch der Kakaobohnen. Zentralbl Bakteriol Parasitenk Infektionskr Hyg Abt II 71: 80–93.

Clara, F. M. 1930 A new bacterial leaf disease of tobacco in the Philippines. Phytopathology 20: 691–706.

—. 1932 A new bacterial disease of pears. Science (Washington) 75: 111.

—. 1934 A comparative study of the green-fluorescent bacterial plant pathogens. Cornell Agr Exp Sta Mem *159:* 1–36.

Clarenburg, A. 1925 *Bacillus pyosepticus* als Krankheitursache bei einem Schwein. Z Infektionskr Parassitar Hyg Haustiere *27:* 192–198.

Clark, F. E. 1951 The generic classification of certain cellulolytic bacteria. Proc Soil Sci Soc Amer *15:* 180–182.

—. 1952 The generic classification of the soil corynebacteria. Int Bull Bacteriol Nomencl Taxon *2* (*1*)*:* 45–56.

—. 1953 Criteria suitable for species differentiation in *Cellulomonas* and a revision of the genus. Int Bull Bacteriol Nomencl Taxon *3* (*4*)*:* 179–199.

—. 1955 The designation of *Corynebacterium ureafaciens* Krebs and Eggleston as *Arthrobacter ureafaciens* (Krebs and Eggleston) comb nov. Int Bull Bacteriol Nomencl Taxon *5* (*3*)*:* 111–113.

— and P. H. Carr. 1951 Motility and flagellation of the soil corynebacteria. J Bacteriol *62:* 1–6.

Clark, H. F. and C. C. Shepard. 1963 Effect of environmental temperatures on infection with *Mycobacterium marinum* (*balnei*) of mice and a number of poikilothermic species. J Bacteriol *86* (*5*)*:* 1057–1069.

Clark, R. 1942 *Eperythrozoon felis* (sp. nov.) in a cat. J Afr Vet Med Ass *13:* 15–16.

Clarke, J. K. 1924 On the bacterial factor in the aetiology of dental caries. Brit J Exp Pathol *5:* 141–147.

Claus, D. 1967 Taxonomy of some highly pleomorphic bacteria. Spisy Prirodoved Fak Univ J E Purkyne Brno *K40:* 254–257.

—, J. E. Bergendahl and M. Mandel. 1968 DNA base composition of *Microcyclus* species and organisms of similar morphology. Arch Mikrobiol *63:* 26–68.

Claus, K. D., J. H. Newhall and D. Mee. 1959 Isolation of *Pasteurella tularensis* from foals. J Bacteriol *78* (*2*)*:* 294–295.

Claussen, N. H. 1903 Études sur les bactéries dites sarcines et sur maladies quelles provoquent dans la bière. C R Trav Lab Carlsberg *6:* 64–83.

Claydon, T. J. and B. W. Hammer. 1939 A skunk-like odor of bacterial origin in butter. J Bacteriol *37:* 252–258.

Clements, F. E. 1909 The Genera of Fungi. H. W. Wilson Co., Minneapolis, pp. 1–227.

Clise, E. H. 1957 Genus V. *Cillobacterium* Prevot 1938. *In* Breed, Murray and Smith (Editors), Bergey's Manual of Determinative Bacteriology, 7th ed. The Williams & Wilkins Co., Baltimore, pp. 566–568.

Clyde, W. A. 1964 *Mycoplasma* species in the human pharynx. Fed Proc *23:* 579.

Cobb, N. A. 1893 Plant diseases and their remedy. Agr Gaz N S W *4:* 777–833.

Cobb, R. W. and J. K. Walley. 1962 *Corynebacterium bovis* as a probable cause of bovine mastitis. Vet Rec *74:* 101–102.

Cobet, A. B., C. Wirsen and G. E. Jones. 1970 The effect of nickel on a marine bacterium, *Arthrobacter marinus* sp. nov. J Gen Microbiol *62* (*2*)*:* 159–169.

Coerper, F. M. 1919 Bacterial blight of soybean. J Agr Res *18:* 179–194.

Coffey, G. L., L. E. Anderson, M. W. Fisher, M. M. Galbraith, A. B. Hillegas, D. L. Kohberger, P. E. Thompson, K. S. Weston and J. Ehrlich. 1959 Biological studies of paromomycin Antibiot Chemother *9* (*12*)*:* 730–738.

—, A. B. Hillegas, M. P. Knudsen, H. J. Koepsell, J. E. Oyaas and J. Ehrlich. 1954 Azaserine: Microbiological studies. Antibiot Chemother *4* (*7*)*:* 775–791.

Cohen, I. R. 1955 Ascosin and process of producing same. United States Patent 2,723,216, November 8.

Cohen, R. L., R. G. Wittler and J. E. Faber. 1968 Modified biochemical tests for characterization of L-phase variants of bacteria. Appl Microbiol *16:* 1655–1662.

Cohen-Bazire, G. 1963 Some observations on the organization of the photosynthetic apparatus in purple and green bacteria. *In* H. Gest, A. San Pietro, L. P. Vernon (Editors), Bacterial Photosynthesis. Antioch Press, Yellow Springs, Ohio, pp. 89–110.

— and R. Kunisawa. 1963 The fine structure of *Rhodospirillum rubrum*. J Cell Biol *16:* 401–420.

—, — and N. Pfennig. 1969 Comparative study of structure of gas vacuoles. J Bacteriol *100* (*2*)*:* 1049–1061.

—, N. Pfennig and R. Kunisawa. 1964 The fine structure of green bacteria. J Cell Biol *22:* 207–225.

— and W. R. Sistrom. 1966 The procaryotic photosynthetic apparatus. *In* L. P. Vernon and G. R. Seely (Editors), The Chlorophylls, Academic Press, New York, pp. 313–341.

Cohn, F. 1854 Untersuchungen über die Entwicklungsgeschichte der mikroskopischen Algen und Pilze. Nova Acta Leopold *24:* 101–256.

—. 1864 Über die Entstehung des Travertin in den Wasserfällen von Tivoli. Neues Jahrb Min Geol Palaeontol 580–610.

—. 1865 Zwei neue Beggiatoen. Hedwigia *4:* 81–84.

—. 1870 Über den Brunnenfaden (*Crenothrix polyspora*) mig Bemerkungen über die mikroskopische analyse des Brunnenwassers. Beitr Biol Pflan *1875 1* (*Heft 1*)*:* 108–131.

—. 1872 Untersuchungen über Bakterien. Beitr Biol Pflanz *1875 1* (*Heft 2*)*:* 127–224.

—. 1875 Untersuchungen über Bacterien. II. Beitr Biol Planz *1* (*Heft 3*)*:* 141–207.

—. 1878 Letter to J. Penn which describes *Micrococcus phosphoreum* Versameling van stukken betreffende het geneeskundig staats toerzicht, 126–130.

Cole, B. C., L. Golightly and J. R. Ward. 1967 Characterization of mycoplasma strains from cats. J Bacteriol *94* (*5*)*:* 1451–1458.

—, J. R. Ward and C. H. Martin. 1968 Haemolysin and peroxide activity of *Mycoplasma* species. J Bacteriol *95* (*6*)*:* 2022–2030.

Cole, R. M. 1968 Structure of the group A streptococcal cell and its L-form. pp 5–42. *In* R. Caravano, Current Research on Group A *Streptococcus*, Excerpta Medica Foundation, Amsterdam, N. Y., pp. 1–365.

—, J. G. Tully, T. J. Popkin and J. M. Bove. 1973 Morphology, Ultrastructure and Bacteriophage Infection of the Helical Mycoplasmalike Organism (*Spiroplasma citri gen. nov. sp. nov.*) Cultured from "Stubborn" Disease of Citrus. J Bacteriol *115:* 367–386.

Colebrook, L. 1920 The mycelial and other microorganisms associated with human actinomycosis. Brit J Exp Pathol *1:* 197–212.

Coleman, R. M. and L. K. Georg. 1969 Comparative pathogenicity of *Actinomyces naeslundii* and *Actinomyces israelii*. Appl Microbiol 18 (3): 427–432.

Coles, J. D. W. A. 1931 A rickettsia-like organism in the conjunctiva of sheep. Report Vet Serv Anim Ind Union S Afr, 17th, 175–186.

—. 1936 A rickettsia-like organism of the conjunctival epithelium of cattle. J S Afr Vet Med Ass 7: 221–225.

—. 1940 Conjunctivitis of the domestic fowl and as associated rickettsia-like organisms in the conjunctival epithelium. Onderstepoort J Vet Sci 14: 469–478.

—. 1953 Classification of rickettsiae pathogenic to vertebrates. Ann N Y Acad Sci 56: 457–483.

Collier, W. A. 1921 *Cristispira helgolandica* nov. spec. und ihre Fortpflanzung. Zentrabl Bakteriol Parasitenk Infektionskr Hyg Abt I Orig 86: 132–134.

Collin, B. 1913 Sur un ensemble de protistes parasites des batraciens (note préliminaire). Arch Zool Exp Gen Notes Rev 51: 59–76.

Collins, E. B., F. E. Nelson and C. E. Parmelee. 1950 Acetate and oleate requirements of the lactic group of streptococci. J Bacteriol 59 (1): 69–74.

Collins-Thompson, D. L., T. Sorhaug, L. D. Witter and Z. J. Ordal. 1972 Taxonomic consideration of *Microbacterium lacticum*, *Microbacterium flavum*, and *Microbacterium thermosphactum*. Int J Syst Bacteriol 22 (2): 65–72.

Colwell, R. R., R. V. Citarella and P. K. Chen. 1966. DNA base composition of *Cytophaga marinaflava* n. sp., determined by buoyant density measurements in cesium chloride. Can J Microbiol 12: 1099–1103.

— and N. E. Gibbons. 1972 Taxonomic studies of some halophilic bacteria. Antonie van Leewenhoek J Microbiol Serol 38: 27–46.

— and J. Liston. 1961 Taxonomic relationships among the pseudomonads. J Bacteriol 82 (1): 1–19.

Commercial Solvents Corp. 1957 A new antibiotic cycloserine and method of preparing the same. British Patent 768,007, February, 13.

Conn, H. J. 1917 Soil flora studies. V. Actinomycetes in soil. Bull N Y State Agr Exp Sta 60: 3–25.

—. 1928 A type of bacteria abundant in productive soils, but apparently lacking in certain soils of low productivity. N Y Agr Exp Sta Geneva Bull 138: 3–26.

—. 1938 Taxonomic relationships of certain non-sporeforming rods in soil. J Bacteriol 36: 320–321.

—. 1939 Genus III. *Alcaligenes* Castellani and Chalmers. *In* Bergey, Breed, Murray and Hitchens (Editors), Bergey's Manual of Determinative Bacteriology, 5th ed. The Williams & Wilkins Co., Baltimore, pp. 95–102.

—. 1942 Validity of the genus *Alcaligenes*. J Bacteriol 44: 353–360.

—. 1957 Staining methods. *In* M. J. Pelczar, Manual of Microbiological Methods, McGraw-Hill, New York, p. 10.

— and J. W. Bright. 1919 Ammonification of manure in soil. J Agr Res 16: 313–350.

— and I. Dimmick. 1947 Soil bacteria similar in morphology to *Mycobacterium* and *Corynebacterium*. J Bacteriol 54: 291–303.

Conn, H. W., W. M. Esten and W. A. Stocking.

1907 A classification of dairy bacteria. Storrs Agr Exp Sta 18th Ann Rep for year 1906, pp. 91–203.

Connell, W. T. 1898 Dairy Bacteriology. Discoloration of cheese. Rep Comm Agr Dairy, Ottawa, Canada for 1897. Part 16, 4–16.

Conti, S. F. and P. Hirsch. 1965 Biology of budding bacteria. III. Fine structure of *Rhodomicrobium* and *Hyphomicrobium* spp. J Bacteriol 89 (2): 503–512.

Copeland, H. F. 1938 The kingdoms of organisms. Quart Rev Biol 13: 384–420.

—. 1956 The classification of lower organisms. Pacific Book, Palo Alto, California.

Copeland, J. J. 1936. Yellowstone thermal *Myxophyceae*. Ann N Y Acad Sci 36: 1–232.

Corbaz, R., L. Ettlinger, E. Gäumann, W. Keller-Schierlein, F. Kradolfer, E. Kyburz, L. Neipp, V. Prelog, R. Reusser and H. Zähner. 1955 Stoffwechselprodukte von Actinomyceten. 1 Mitteil. Narbomycin. Helv Chim Acta 38: 935–942.

—, —, —, —, —, L. Neipp, V. Prelog, P. Reusser and H. Zähner. 1957 Stoffwechselprodukte von Actinomyceten. 7 Mitteil. Echinomycin. Helv Chim Acta 40: 199–204.

—, —, —, —, L. Neipp, V. Prelog, P. Reusser and H. Zähner. 1955 Stoffwechselprodukte von Actinomyceten. 2 Mitteil. Angolamycin. Helv Chim Acta 38: 1202–1209.

—, —, W. Keller-Schierlein and H. Zähner. 1957 Zur Systematik der Actinomyceten 1. Über streptomyceten mit rhodomycinartigen Pigmenten. Arch Mikrobiol 25: 325–332.

—, —, — and —. 1957 Zur Systematik der Actinomyceten. 2. Über Actinomycin bildende Streptomyceten. Arch Mikrobiol 26: 192–208.

Corda, A. C. J. 1839 Pracht-Flora Europaeischer Schimmelbildungen. Gerhard Fleischer, Leipzig, pp. 1–55.

Cordon, B. P. and H. A. Barker. 1946 Two new amino-acid-fermenting bacteria, *Clostridium propionicum* and *Diplococcus glycinophilus*. J Bacteriol 52: 629–634.

Corpe, W. A. 1951 A study of the wide spread distribution of *Chromobacterium* species in soil by a simple technique. J Bacteriol 62 (4): 515–517.

—. 1961 Accumulation of indole compounds in cultures of *Chromobacterium violaceum*. Nature (London) 190: 190–191.

—. 1963 Extracellular accumulation of pyrroles in bacterial cultures. Appl Microbiol 11 (2): 145–150.

—. 1964 Factors influencing growth and polysaccharide formation by strains of *Chromobacterium violaceum*. J Bacteriol 88 (5): 1433–1441.

Cosbie, A. J. C., J. Tošić and T. K. Walker. 1942 *Acetobacter turbidans*, a new species of acetic acid-producing bacterium. J Inst Brew 48: 82–86.

Cosenza, B. J. and J. D. Podgwaite. 1966 A new species of *Proteus* isolated from larvae of the gypsy moth *Porthetria dispar* (L). Antonie Van Leeuwenhoek J Microbiol Serol 32: 187–191.

Costilow, R. N. and T. W. Humphreys. 1955 Nitrate reduction by certain strains of *Lactobacillus plantarum*. Science (Washington) 121: 168.

Cotchin, E. 1943. *Corynebacterium equi* in the

submaxillary lymph nodes of swine. J Comp Pathol *53:* 298–309.

Cottet, J. 1900 Note sur un Microcoque stricte-ment anaérobie, trouvé dans les suppurations de l'appareil urinaire. C R Soc Biol Paris *52:* 421–423.

Cottew, G. S. and J. Francis. 1954 The isolation of *Shigella equuli* and *Salmonella newport* from normal horses. Aust Vet J *30:* 301–304.

Coty, V. F. 1967 Atmospheric nitrogen fixation by hydrocarbon-oxidizing bacteria. *In* Symposium on microbes and hydrocarbons. Biotechnol Bioeng 9 *(1):* 25–32.

Couch, J. N. 1950 *Actinoplanes.* A new genus of the *Actinomycetales.* J Elisha Mitchell Sci Soc *66:* 87–92.

—. 1955 A new genus and family of the Actinomycetales, with a revision of the genus *Actinoplanes.* J Elisha Mitchell Sci Soc *71:* 148–155.

—. 1955 *Actinosporangiaceae* should be *Actinoplanaceae.* J Elisha Mitchell Sci Soc *71:* 269.

—. 1955 A new genus and family of the *Actinomycetales,* with a revision of the genus *Actinoplanes.* J Elisha Mitchell Sci Soc *71:* 148–155.

—. 1963 Some new genera and species of the *Actinoplanaceae.* J Elisha Mitchell Sci Soc *79:* 53–70.

—. 1964 A proposal to replace the name *Ampullaria* Couch with *Ampullariella.* J Elisha Mitchell Sci Soc *80:* 29.

Coucke, P. 1969 Morphology and Morphogenesis of *Sorangium compositum.* J Appl Bacteriol *32:* 24–29.

Cowan, S. T., K. J. Steel, C. Shaw and J. P. Duguid. 1960 A classification of the *Klebsiella* group. J Gen Microbiol *23 (3):* 601–612.

Cowdry, E. V. 9125 Studies on the Etiology of heartwater. I. Observation of a Rickettsia, *Rickettsia ruminantium* (n. sp.) in the tissues of infected animals. J Exp Med *42:* 231–252.

Cox, H. R. 1939 Studies of a filter-passing infectious agent isolated from ticks. V. Further attempts to cultivate in cell-free media. Suggested classification. Pub Health Rep *54 (40):* 1822–1826.

—. 1941 Cultivation of rickettsia or Rocky Mountain spotted fever, typhus and Q-fever groups in the embryonic tissues of developing chicks. Science (Washington) *94:* 399–403.

Crabtree, K. T. 1965 Morphological and biochemical studies of *Zoogloea ramigera* species in pure culture. Ph. D. Thesis Univ Wisconsin.

Crabtree, K. and E. McCoy. 1967 *Zoogloea ramigera* Itzigsohn, identification and description. Int J Syst Bacteriol *17 (1):* 1–10.

—, —, W. C. Boyle and G. A. Rohlich. 1965 Isolation, identification, and metabolic role of the sudanophilic granules of *Zoogloea ramigera.* Appl Microbiol *13:* 218–226.

Craigie, J. and C. H. Yen. 1938. The demonstration of types of *B. typhosus* by means of preparation of type II Vi phage. Can J Public Health *29:* 448–463; 484–496.

Craveri, R., C. Coronelli, H. Ragani and P. Sensi. 1964 Termorubin, a new antibiotic from a termoactimomycetes. Nature (London) *195 (4843):* 832–833.

— and G. Giolitti. 1957 An antibiotic with

fungicidal and insecticidal activity produced by *Streptomyces.* Nature (London) *179:* 1307.

— and —. 1958. Antibiotique et procede pour le preparer (in French). Belgian Patent 560,930, March 18.

— and H. Pagani. 1962 Thermophilic microorganisms among Actinomycetes in the soil. Ann Microbiol Enzimol *12:* 115–130.

Crawford, J. P. 1954 A new fermentative pseudomonad, *Pseudomonas formicans* n. sp. J Bacteriol *68:* 734–738.

Crocker, R. L. and B. A. Dickson. 1957 Soil development on the recessional moraines of the Herbert and Mendenhall glaciers, South-Eastern Alaska. J Ecol *45:* 169–185.

— and J. Major. 1955 Soil development in relation to vegetation and surface age at Glacier Bay, Alaska. J Ecol *43:* 427–448.

Cron, M. J., D. F. Whitehead, I. R. Hooper, B. Heinemann and J. Lein. 1956 Bryamycin, a new antibiotic. Antibiot Chemother *6 (1):* 63–67.

Crookshank, E. M. 1887 Manual of Bacteriology. Revised 2nd ed. J. H. Vail and Co., New York.

Cross, T. 1962 *Streptomyces* species producing oxytetracycline. Nature (London) *195 (4843):* 832–833.

—. 1968 Thermophilic actinomycetes. J Appl Bacteriol *31:* 36–53.

—. 1970. The diversity of bacterial spores. J Appl Bacteriol *33:* 95–102.

— and J. Lacey. 1970 Studies on the genus *Thermomonospora. In* H. Prauser (Editor), The Taxonomy of the *Actinomycetales.* VEB Gustav Fischer, Verlag, Jena.

—, M. P. Lechevalier and H. A. Lechevalier. 1963. A new genus of the *Actinomycetales: Microellobosporia* gen. nov. J Gen Microbiol *31:* 421–429.

—, A. M. Maciver and J. Lacey. 1968 The thermophilic actinomycetes in mouldy hay; *Micropolyspora faeni* sp. nov. J Gen Microbiol *50:* 351–359.

—, P. D. Walker and G. W. Gould. 1968 Thermophilic actinomycetes producing resistant spores. Nature (London) *220:* 352–354.

Cruickshank, J. C. 1935 A study of the so-called *Bacterium typhi flavum.* J Hyg *35:* 354.

Crystal, M. M. 1958 The mechanism of transmission of *Haemobartonella muris* (Mayer) of rats by the spined rat louse, *Polyplax spinulosa* (Burmeister). J Parasitol *44:* 603–606.

Cuadra, M. and J. Takano. 1969 The relationship of *Bartonella bacilliformis* to the red blood cell as revealed by electron microscopy. Blood J Hematol *33:* 708–716.

Cullen, G. A. 1966 The ecology of *Streptococcus uberis.* Brit Vet J *122:* 333–339.

Cummins, C. S. 1962 Chemical composition and antigenic structure of cell walls of *Corynebacterium, Mycobacterium, Nocardia, Actinomyces* and *Arthrobacter.* J Gen Microbiol *28 (1):* 35–50.

—. 1971 Cell wall composition in *Corynebacterium bovis* and some other corynebacteria. J Bacteriol *105 (3):* 1227–1228.

—, G. Atfield, R. J. W. Rees and R. C. Valentine. Cell wall composition in *Mycobacterium lepraemurium.* J Gen Microbiol *49 (3):* 377–384.

— and H. Harris. 1956. The chemical composition of the cell wall in some Gram-positive bacteria and its possible value as a taxonomic character. J Gen Microbiol *14* (*3*): 583–600.

— and —. 1956 The relationships between certain members of the staphylococcus-micrococcus group as shown by their cell wall composition. Int Bull Bacteriol Nomencl Taxon *6* (*3*): 111–119.

— and —. 1958 Studies on the cell-wall composition and taxonomy of *Actinomycetales* and related groups. J Gen Microbiol *18* (*1*): 173–189.

— and —. 1959 Taxonomic position of *Arthrobacter*. Nature (London) *184*: 831–832.

Cunningham, A. and A. M. Smith. 1940 The microbiology of silage made by the addition of mineral acids to crops rich in protein. II. The microflora. J Dairy Res *11*: 243–265.

Curasson, G. 1938 Notes sur la piroplasmose aviare en A.O.F. Bull Serv Zootech Epiz A O F *1*: 33–35

— and P. Andrjesky. 1929 Sur les "corps de Balfour" du sang de la pule. Bull Soc Pathol Exot *22*: 316–317.

Curbelo y Hernandez, A. 1941 Las bacterias pathogenas del hombre. Elementos de Bacteriologica Medica, Havana, Cuba.

Curtis, S. N. and R. M. Krause. 1964 Antigenic relationships between groups B and G streptococci. J Exp Med *120*: 629–637.

Cushman, J. A. 1933 Some new foraminiferal genera. Contrib Cushman Found Foraminiferal Res *9*: 32–38.

Cutlip, R. C. 1970 Electron microscopy of cell cultures infected with a chlamydial agent causing polyarthritis of lambs. Infect Immun *1*: 499–502.

Cuttino, J. T. and A. M. McCabe. 1949 Pure granulomatous nocardiosis: A new fungus disease distinguished by intracellular parasitism. Amer J Clin Pathol *25*: 1–34.

Czurda, V. and E. Maresch. 1937 Beitrag zur Kenntnis der Athio-rhodobakterien-Gesellschaften. Arch Mikrobiol *8*: 99–124.

da Costa Cruz, J. C. 1938 *Mycobacterium fortuitum* um novo bacilo acido-resistance patogenico para o homen. Acta Med Rio de Janeiro *1*: 297–301.

Dain, J. A., A. K. Neal and H. W. Seeley. 1956 The effect of carbon dioxide on polysaccharide production by *Streptococcus bovis*. J Bacteriol *72* (*2*): 209–213.

Danga, F. and D. Gottlieb. 1959 English Translation of G. F. Gauze *et al.*, 1957. Problems in the classification of antagonistic actinomycetes. The American Institute of Biological Sciences, Washington, D. C.

Dangeard, P. A. 1926 Recherches sur les tubercles radicaux des Légumineuses. Botaniste (Paris) *16*: 1–275.

Darland, G., T. D. Brock, W. Samsonoff and S. F. Conti. 1970 A thermophilic, acidophilic mycoplasma isolated from a coal refuse pile. Science (Washington) *170* (*3965*): 1416–1418.

da Rocha-Lima, H. 1916 Zur Aetiologie des Fleckfeibers. Berlin Klin Wochenschr *53*: 567–569.

Dárzinš, E. 1950 Tuberculose das gias. (*Leptodactylus pentadactylus*). Arch Inst Bras Tuberc *9*: 29–37.

Da Silva, G. A. N. and J. G. Holt. 1965 Numerical taxonomy of certain coryneform bacteria. J Bacteriol *90* (*4*): 921–927.

David, H. 1927 Ueber eine durch cholera-ähnliche hervorgerufene Fischseuche. Zentrabl Bakteriol Parasitenk Infektionskr Hyg Abt I Orig *102*: 46–60.

Davidson, C. M. and E. F. Hartree. 1968 Cytochrome as a guide to classifying bacteria: taxonomy of *Microbacterium thermosphactum*. Nature (London) *220*: 502–504.

—, P. Mobbs and J. M. Stubbs. 1968 Some morphological and physiological properties of *Microbacterium thermosphactum*. J Appl Bacteriol *31*: 551–559.

Davis, B. R. and W. H. Ewing. 1964 Lipolytic, pectolytic and alginolytic activities of *Enterobacteriaceae*. J Bacteriol *88* (*1*): 16–19.

—, —, and R. W. Reavis. 1957 The biochemical reactions given by members of the *Serratia* group. Int Bull Bacteriol Nomencl Taxon *7* (*4*): 151–160.

— and J. M. Woodward. 1957 Some relationships of the somatic antigens of a group of *Serratia* cultures. Can J Microbiol *3*: 591–597.

Davis, D. H. 1969 *Alcaligenes paradoxus* sp. nov. *In* Davis, Doudoroff, Stanier and Mandel. Proposal to reject the genus *Hydrogenomonas*: taxonomic implications. Int J Syst Bacteriol *19* (*4*): 375–390.

—. 1969 *Paracoccus* Davis, gen. nov. *In* Davis, Douderoff, Stanier and Mandel. Proposal to reject the genus *Hydrogenomonas*: taxonomic implications. Int J Syst Bacteriol *19* (*4*): 375–390.

Davis, D. H., M. Doudoroff, R. Y. Stanier and M. Mandel. 1969 Proposal to reject the genus *Hydrogenomonas*: taxonomic implications. Int J Syst Bacteriol *19* (*4*): 375–390.

Davis, D. H., R. Y. Stanier, M. Doudoroff and M. Mandel. 1970 Taxonomic studies on some Gram-negative polarly flagellated "hydrogen bacteria" and related species. Arch Mikrobiol *70* (*1*): 1–13.

Davis, D. J. 1912 Relation of varieties of streptococci with especial reference to experimental arthritis. J Amer Med Ass *58* (*17*): 1283.

—. 1914 The growth and viability of streptococci of bovine and human origin in milk and milk products. J Infect Dis *15*: 378–388.

Davis, G. E. 1942 Species unity or plurality of the relapsing fever Spirochetes. Amer Ass Advan Sci Publ No. 18, pp. 41–47.

—. 1948 The spirochetes. Annu Rev Microbiol *2*: 305–334.

—. 1952 Observations on the biology of the Argasid tick, *Ornithodoros brasiliensis*. Aragão, 1923 with the recovery of a spirochete, *Borrelia brasiliensis* n. sp. J Parasitol *38*: 473–476.

—. 1956 A relapsing fever spirochete, *Borrelia mazzottii* (sp. nov.) from *Ornithodorus talaje* from Mexico. Amer J Hyg *63*: 13–17.

—. 1957 Order IX. Spirochaetales Buchanan 1918. *In* Breed, Murray and Smith (Editors), Bergey's Manual of Determinative Bacteriology, 7th ed. The Williams & Wilkins Co., Baltimore, pp. 892–907.

Davis, G. H. G. and J. H. Freer. 1960 Studies upon an oral aerobic actinomycete. J Gen Microbiol *23* (*1*): 163–178.

— and K. G. Newton. 1969 Numerical taxonomy of some named coryneform bacteria. J Gen Microbiol *56* (*2*): 195–214.

Davis, H. 1922 A new bacterial disease of fresh water fishes. U S Bur Fish Bull *38:* 261–280.

Davis, J. B., H. H. Chase and R. L. Raymond. 1956 *Mycobacterium paraffinicum*, n. sp. a bacterium isolated from soil. Appl Microbiol *4:* 310–315.

—, V. F. Coty and J. P. Stanley. 1964 Atmospheric nitrogen fixation by methane-oxidizing bacteria. J Bacteriol *88 (2):* 468–472.

Davis, J. G. 1937 A comparison on the growth activating effects on *Streptococcus* and *Lactobacillus* of various yeast preparations. J Pathol Bacteriol *45 (2):* 367–376.

—. 1939 *Chromobacterium iodinum (n. sp.)* Zentrabl Bakteriol Parasitenk Infektionskr Hyg Abt II *100:* 273–276.

— and A. T. R. Mattick. 1929 Rusty spot in cheddar and other cheeses. I. Description of the causative organism. J Dairy Res *1:* 50–57.

— and —. 1936 Proc Soc Agr Bacteriol *3* (cited by Breed and Pederson, 1938 but such reference in Proc Soc Agr Bacteriol *3* cannot be found).

Davisson, J. W. and A. C. Finlay. 1959 Catenulin and its production. United States Patent 2,895,876, July 21.

Davydov, N. N. 1961 Characteristics of the provoker of brucellosis in reindeer. Trudy resespyvzan Inst Exp Veter Moscow *27:* 24–31 (Vet Bull Weybridge 1962 *32:* 2189).

Dawes, E. A., D. W. Ribbons and P. J. Large. 1966 The route of ethanol formation in *Zymomonas mobilis*. Biochem J *98:* 795–803.

—, — and D. A. Rees. 1966 Sucrose utilization by *Z. mobilis*. Formation of a levan. Biochem J *98:* 804–812.

Day, W. R. 1924 The watermark disease of the cricket-bat willow. Oxford For Mem *3:* 1–30.

De, S. N. and D. N. Chatterjee. 1953 An experimental study of the mechanism of action of *Vibrio cholerae* on the intestinal mucous membrane. J Pathol Bacteriol *66:* 559–562.

de Almeida, F. 1940 Study of a black grain mycetoma due to *Actinomyces paraguayensis* Almeida, n. sp. Mycopathologia *2:* 201–203.

de Barjac, H. and A. Bonnefoi. 1967 Classification des souches de *Bacillus thuringiensis*. C R Acad Sci (Paris) *264:* 1811–1813.

— and —. 1968 A classification of strains of *Bacillus thuringiensis* Berliner with a key to their differentiation. J Invertebr Pathol *11:* 335–347.

de Bary, A. 1884 Vergleichende Morphologie und Biologie der Pilze, Mycetozoen und Bacterien. Wilhelm Engelmann, Leipzig.

—. 1887 Lectures on bacteria. (English Translation of Vorlesungen über Bacterien. 1885. W. Engelmann, Leipzig.), Clarendon Press, Oxford, pp. 1–193.

de Beaurepaire-Aragao, H. and G. Vianna. 1913 Sobre um novo treponema encontrado em ulceras, *Treponema minimum* nov. spec. Brasil-Med *27:* 61–62.

de Blieck, L. 1931 Een haemoglobinophile bacterie als oorzaak van Coryza infectiosa gallinarum. Tijdschr Diergeneesk *58:* 310–313.

— and T. van Heelsbergen. 1919 De *Bacillus pyosepticus equi* en de *B. abortus equi* als oorzaak van pyo-septicaemie z.g. lähme bij veulens in Nederland. Tijdschr Diergeneesk *46:* 492–496.

De Boer, C., E. L. Caron and J. W. Hinman. 1953 Amicetin, a new *Streptomyces* antibiotic. J Amer Chem Soc *75:* 499–500.

—, A. Dietz, T. E. Eble and C. N. Large. 1964 Streptolydigin and production thereof. United States Patent 3,160,560, December 8.

—, —, J. S. Evans and R. M. Michaels. 1960 Fervenulin, a new crystalline antibiotic. I. Discovery and biological activities. Antibiot Ann 220–226.

—, —, N. E. Lummis and G. M. Savage. 1961 Porfiromycin, a new antibiotic. I. Discovery and biological activites. Antimicrob Agents Ann 17–22.

—, —, W. S. Silver and G. M. Savage. 1956 Streptolydigin, a new antimicrobial antibiotic. I. Biologic studies of Streptolydigin. Antibiot Ann *1955–1956:* 886–892.

—, —, J. R. Wilkins, C. N. Lewis and G. M. Savage. 1955 Celesticetin—a new crystalline antibiotic. I. Biologic studies of celesticetin. Antibiot Ann *1954–1955:* 831–841.

de Boer, W. E. 1969 On Ultrastructures in *Rhodopseudomonas gelatinosa* and *Rhodospirillum tenue*. Antonie Van Leeuwenhoek J Microbiol Serol *35:* 241–242.

—, J. W. M. LaRivière and K. Schmidt. 1971 Some properties of *Achromatium oxaliferum*. Antonie Von Leeuwenhoek J Microbiol Serol *37:* 553–563.

Debono, M. 1912 On some anaerobical bacteria of the normal human intestine. Zentrabl Bakteriol Parasitenk Infektionskr Hyg Abt I Orig *62:* 229–234.

De Bord, G. G. 1942 Descriptions of *Mimeae* Trib. nov. with three genera and three species and two new species of *Neisseria* from conjunctivitis and vaginitis. Iowa State College J Sci *16:* 471–480.

de Buen, S. 1926 Note préliminaire sur l'épidémiologia de la fièvre récurrente espagnole. Ann Parasitol Hum Comp *4:* 185–192.

Dédié, K. 1949 Die säurelöslichen Antigene von *Erysipelothrix rhusiopathiae*. Monatsche Veterinaermed *4:* 7.

Deeb, B. J. and G. E. Kenny. 1967 Characterization of *Mycoplasma* variants isolated from rabbits. J Bacteriol *93 (4):* 1416–1424.

— and —. 1967 Characterization of *Mycoplasma pulmonis* variants isolated from rabbits. II. Basis for differentiation of antigenic serotypes. J Bacteriol *93 (4):* 1425–1429.

Dehnert, J. 1957 Untersuchung uber die Grampositive Stuhlflora des Brustmilchkindes. Zentrabl Bakteriol Parasitenk Infektionskr Hyg Abt I Orig *169:* 66–83.

—. 1960 Betrachtung zum Bifidumproblem. Zentrabl Bakteriol Parasitenk Infektionskr Hyg Abt I Orig *179:* 190–198.

—. 1965 Zur Systematik der "Bifidumgruppe." Ernaehrungsforschung *10:* 465–471.

Dehority, B. A. 1966 Characterization of several bovine rumen bacteria isolated with a xylan medium. J Bacteriol *91 (5):* 1724–1729.

—, H. W. Scott and P. Kowaluk. 1967 Volatile fatty acid requirements of cellulolytic rumen bacteria. J Bacteriol *94 (3):* 537–543.

Deibel, R. H. 1963 Hydrolysis of proteins and nucleic acids by Lancefield group A and other streptococci. J Bacteriol *86 (6):* 1270–1274.

—. 1964 The group D streptococci. Bacteriol Rev *28 (3):* 330–366.

—. 1964 Utilization of arginine as an energy

source for the growth of *Streptococcus faecalis.* J Bacteriol **87** *(5):* 988–992.

— and J. B. Evans. 1960 Modified benzidine test for the detection of cytochrome-containing respiratory systems in microorganisms. J Bacteriol *79:* 356–360.

— and M. J. Kvetkas. 1964 Fumarate reduction and its role in the diversion of glucose fermentation by *Streptococcus faecalis.* J Bacteriol **88** *(4):* 858–864.

—, D. E. Lake and C. F. Niven, Jr. 1963 Physiology of the enterococci as related to their taxonomy. J Bacteriol *86 (6):* 1275–1282.

— and C. F. Niven, Jr. 1955 Reciprocal replacement of oleic acid and CO_2 in the nutrition of the "minute streptococci" and *Lactobacillus leichmannii.* J Bacteriol *70 (2):* 134–140.

— and —. 1955 The "minute" streptococci: further studies on their nutritional requirements and growth characteristics on blood agar. J Bacteriol *70 (2):* 141–146.

— and —. 1960 Comparative study of *Gaffkya homari, Aerococcus viridans,* tetrad-forming cocci from meat curing brines and the genus *Pediococcus.* J Bacteriol *79 (2):* 175–180.

— and —. 1964 Pyruvate fermentation by *Streptococcus faecalis.* J Bacteriol *88 (1):* 4–10.

— and J. H. Silliker. 1963 Food-poisoning potential of the enterococci. J Bacteriol *85 (4):* 827–832.

—, J. Yao, N. J. Jacobs and C. F. Niven, Jr. 1964 Group E streptococci. I. Physiological characterization of strains isolated from swine cervical abscesses. J Infect Dis *114:* 327–332.

Delacroix, G. 1906 Sur quelques maladies bactériennes observées à la station de pathologie végétale. Ann Inst Nat Agron *5:* 354–368.

Delafield, F. P., M. Doudoroff, N. J. Palleroni, C. J. Lusty and C. R. Contopoulou. 1965 Decomposition of poly-β-hydroxybutyrate by pseudomonads. J Bacteriol *90 (5):* 1455–1466.

Delaplane, J. P., L. E. Erwin and H. O. Stuart. 1934 A hemophilic bacillus as the cause of an infectious rhinitis (coryza) of fowls. Rhode Island Agr Exp Sta Bull *244:* 1–12.

Delaporte, B. 1963 Un phénomène singulier: des "spores mobiles" chez des grandes bactéries. C R Acad Sci Paris *257:* 1414–1417.

—. 1964 Étude comparée de grands spirilles formant des spores: *Sporospirillum (Spirillum) praeclarum* (Collin) n.g., *Sporospirillum gyrini* n. sp. et *Sporospirillum bisporum* n. sp. Ann Inst Pasteur (Paris) *107:* 246–262.

—. 1964 Étude descriptive de bactéries de très grandes dimensions. Ann Inst Pasteur (Paris) *107:* 845–862.

—. 1967 Une bactérie nouvelle de l'océan Pacifique: *Bacillus pacificus* n. sp. C R Acad Sci Paris Ser D *264:* 3068–3071.

—. 1969 Description de *Bacillus medusa* n. sp. C R Acad Sci Paris Ser D *269:* 1129–1131.

— and P. Daste. 1956 Une bactérie du sol capable de décomposer la fraction fixe de certaines oléorésines *Flavobacterium resinovorum* n. sp. C R Acad Sci Paris *242:* 831–834.

—, M. Raymond and P. Daste. 1961 Une bactérie du sol capable d'utiliser, comme source de carbone, la fraction fixe de certaines oléorésines *Pseudomonas resinovorans* n. sp. C R Acad Sci Paris *252:* 1073–1075.

— and A. Sasson. 1967 Étude de bactéries des sols arides du Maroc: *Bacillus maroccanus* n. sp. C R Acad Sci Paris Ser D *264:* 2344–2346.

DeLey, J. 1961 Comparative carbohydrate metabolism and a proposal for a phylogenetic relationship of the acetic acid bacteria. J Gen Microbiol *24:* 31–50

—. 1962 Comparative biochemistry and enzymology in bacterial classification. *In* Microbial Classification. 12th Symp. Soc Gen Microbiol, Cambridge Univ Press, Cambridge, pp. 164–195.

—, M. Bernaerts, A. Rassel and J. Guimot. 1966 Approach to an improved taxonomy of the genus *Agrobacterium.* J Gen Microbiol *43 (1):* 7–17.

— and I. W. Park. 1966 Molecular biological taxonomy of some free-living nitrogen-fixing bacteria. Antonie Van Leeuwenhoek J Microbiol Serol *32:* 6–16.

—, —, R. Tijtgat and J van Ermengem. 1966 DNA homology and taxonomy of *Pseudomonas* and *Xanthomonas.* J Gen Microbiol *42 (1):* 43–56.

— and A. Rassel. 1965 DNA base composition, flagellation and taxonomy of the genus *Rhizobium.* J Gen Microbiol *41 (1):* 85–91.

Del Giudice, R. A. and T. R. Carski. 1968 Characterization of a new *Mycoplasma* species of human origin. Bacteriol Proc M10, 67.

—, —, M. F. Barile, R. M. Lemcke and J. Tully. 1971 Proposal for classifying human strain navel and related simian Mycoplasmas as *Mycoplasma primatum* sp. n. J Bacteriol *108 (1):* 439–445.

Delwiche, C. C., P. J. Zinke and C. M. Johnson. 1965 Nitrogen fixation by *Ceanothus.* Plant Physiol *40:* 1045–1047.

Delwiche, E. A. 1949 Vitamin requirements of the genus *Propionibacterium.* J Bacteriol *58:* 395–398.

—. 1957 Family XI. *Propionibacteriaceae* Delwiche. *In* Breed, Murray and Smith (Editors), Bergey's Manual of Determinative Bacteriology, 7th ed. The Williams & Wilkins Co., Baltimore, p. 569.

—. 1961. Catalase of *Pediococcus cerevisiae* J Bacteriol *81:* 416–417.

Demain, A. L. 1965 Contamination of commercial L-leucine preparations with methionine and cystine. J Bacteriol *89:* 1162.

de Man, J. C. 1956 De eigenschappen van enige stammen van bacterien behorende tot het geslacht *Thermobacterium* Orla-Jensen. Ned Melk Zuiveltijdschrift *10:* 190–198.

—, M. Rogosa and M. E. Sharpe. 1960 A medium for the cultivation of lactobacilli. J Appl Bacteriol *23:* 130–135.

Demaree, J. B. and N. R. Smith. 1952 *Nocardia vaccinii* n. sp. causing galls on blueberry plants. Phytopathology *42:* 249–252.

De Mello, F. 1921 Protozoaires parasites de *Pacheledra moesta* Reeve. C R Soc Biol Paris *84:* 241–242.

— and A. S. Ana Pais. 1918 Um caso de nocardiose pulmonar simulando a tisica. Arch Hyg Pathol Exot Lisboa *6:* 133–206.

DeMoor, C. E. 1963 Septicaemic infections in pigs, caused by haemolytic streptococci of new Lancefield groups designated R, S and T. Antonie Van Leeuwenhoek J Microbiol Serol *29:* 272–280.

DeMoss, R. D. 1967 Violacein. *In* Gottlieb and Shaw (Editors). Antibiotics, Vol. 2 Springer Verlag, New York, pp. 77–81.

den Dooren de Jong, L. E. 1926 Bijdrage tot de

kennis van het Mineralisatieproces. Thesis. Technische Hoogeschool, Delft, pp. 1–199.

—. 1927 Uber protaminophage Bakterien. Zentrabl Bakteriol Parasitenk Infektionskr Hyg Abt II 71: 193–232.

—. 1929 Über Bacillus fastidiosus. Zentrabl Bakteriol Parasitenk Infektionskr Hyg Abt II 79: 344–353.

deQuerioz, V. M. and C. A. Albert. 1962 Streptomyces iakyrus nov. sp., Produtor dos antibióticos, Iaquirina I, IIe, III. Rev Inst Antibiot Univ Recife 4 (1/2): 33–46.

Derby, H. A. and R. W. Hammer. 1931 Bacteriology of butter. IV. Bacteriological studies on surface taint butter. Iowa Agr Exp Sta Res Bull 145: 387–416.

de'Rossi, G. 1927 Microbiologia agraria e technica. Unione Tipographico-Editrice Torinese, Torino.

Derrick, E. H. 1939 Rickettsia burneti: The cause of Q-fever. Med J Aust 1: 14.

Derx, H. G. 1950 Further researches on Beijerinckia. Bogoriensis 1: 1–11.

—. 1950 Beijerinckia, a new genus of nitrogen-fixing bacteria occurring in tropical soils. Proc Kon Ned Akad Wetensch 53: 140–147.

—. 1951 Azotobacter insigne nov. spec. fixateur d'azote a flagellation polaire. Proc Kon Ned Akad Wetensch Ser C Biol Med Sci 54: 342–350.

Desai, A. J. and S. A. Dhala. 1967 Streptomyces thermonitrificans sp. n., a thermophilic streptomycete. Antonie Van Leeuwenhoek J Microbiol Serol 33: 137–144.

Desai, M. V. and H. M. Shah. 1959 A new bacterial leaf spot of Crotalaria juncea L. Curr Sci 28: 377–378.

— and —. 1960 Bacterial leaf spot disease of Desmodium rotundifolium DC. Curr Sci 29: 65–66.

— and —. 1963 Bacterial leaf blight of castor beans. Curr Sci 32: 474–475.

Desai, S. G., M. K. Patel, A. B. Gandhi and W. V. Kotasthane. 1967 Bacterial blight disease of Cynodon dactylon Pers. Curr Sci 36: 213–214.

—, M. J. Thirumalachar and M. K. Patel. 1965 Bacterial blight disease of Eleusine coracana Gaertn. Indian Phytopathol 18: 384–386.

de Soriano, A. M. 1935 Estudio sistemático de algunas bacterias esporuladas aerobias. Rev Inst Bacteriol Dep Nac Hig, Buenos Aires 6 (5): 507–642.

De Toni, J. B. and V. Trevisan. 1889 Schizomycetaceae Naeg. In P. A. Saccardo. Sylloge Fungorum, pp. 923–1073.

De Voe, S. E., N. E. Rigler, A. J. Shay, J. H. Martin, T. C. Boyd, E. J. Backus, J. H. Mowat and N. Bohonos. 1957 Alazopeptin: production, isolation and chemical characteristics. Antibiot Ann 1956–1957: 730–735.

De Vries, W. and A. H. Stouthamer. 1967 Pathway of glucose fermentation in relation to the taxonomy of bifidobacteria. J Bacteriol 93 (2): 574–576.

— and —. 1968 Fermentation of glucose, lactose, galactose, mannitol and xylose by bifidobacteria. J Bacteriol 96 (2): 472–478.

DeWeese, M. S., M. A. Gerencser and J. M. Slack. 1968 Quantitative analysis of Actinomyces cell walls. Appl Microbiol 16 (11): 1713–1718.

Dhala, S. A., F. M. Poonawalla and S. S. Bhatnagar. 1957 Antibiotic studies on Indian soil microorganisms: Part IV. Streptomyces caiusiae. A new antibiotic-producing species of Streptomyces. J Sci Ind Res (India) 16B (4): 76–80.

Dias, F. F. and J. V. Bhat. 1964 Microbiology of activated sludge. I. Dominant bacteria. Appl Microbiol 12: 412–417.

Diaz, R., L. M. Jones, D. Leong and J. B. Wilson. 1968 Surface antigens of smooth brucellae. J Bacteriol 96: 893–901.

—, — and J. B. Wilson. 1967 Antigenic relationship of Brucella ovis and Brucella melitensis. J Bacteriol 93: 1262–1268.

—, — and —. 1968 Antigenic relationship of the gram-negative organism causing canine abortion to smooth and rough brucellae. J Bacteriol 95: 618–624.

Diedrich, D. L., C. F. Denny, T. Hashimoto and S. F. Conti. 1970 Facultatively parasitic strain of Bdellovibrio bacteriovorus. J Bacteriol 101 (3): 989–996.

Dienes, L. 1946 Reproductive process in Proteus cultures. Proc Soc Exp Biol Med 63: 265–270.

— and G. Edsall. 1937 Observations on the L-organism of Kleineberger. Proc Soc Exp Biol Med 36: 740–744.

Dierks, R. E., J. A. Newman and B. S. Pomeroy. 1967 Characterization of avian Mycoplasma. Ann N Y Acad Sci 143 (1): 170–189.

Diernhofer, K. 1932 Asculinbouillon als Hilfsmittel fur die Differenzierung von Euter- und Milchstreptokokken bei Massenuntersuchungen. Milchw Forsch 13: 368–374.

—. 1949 Haemophile Bakterien im Geschlechtstrakt des Rindes. Wien Tierarztl Monatsschr 36: 582–588.

Di Marco, A., M. Gaetani, P. Orezzi and M. Soldati. 1963 Antitumor activity of a new antibiotic: daunomycin. Abstr Commun 3rd Int Cong Chemotherapy, Stuttgart, July, Abstract E-19.

— and C. Spalla. 1957 La produzione di cobalamine de fermentazione con una nuova specie di Nocardia: Nocardia rugosa. G Microbiol 4: 24–30.

Dimitroff, V. T. 1926 Spirochaetes in Baltimore market oysters. J Bacteriol 12: 135–177.

Dimock, W. W., P. R. Edwards and D. W. Bruner. 1947 Infections of fetuses and foals. Ky Agr Exp Sta Bull 509: 1–40.

Distaso, A. 1911 Sur les microbes protéolytiques de la flore intestinale de l'homme et des animaux. Zentrabl Bakteriol Parasitenk Infektionskr Hyg Abt I Orig 59: 97–103.

—. 1912 Contribution à l'étude sur l'intoxication intestinale. Zentrabl Bakteriol Parasitenk Infektionskr Hyg Abt I Orig 62: 433–468.

Ditky, S. R. and E. L. Gooden. 1952 Coxiella popilliae, n. sp., a rickettsia causing blue disease of Japanese beetle larvae. J Bacteriol 63: 743–750.

do Amaral, J. F., C. Teixeira and E. D. Pinheiro. 1956 O bacterio causador da mancha aureolada do caffeiro. Arq Inst Biol São Paulo 23: 151–155.

Dobell, C. C. 1910 On some parasitic protozoa from Ceylon. Spolia Zeylan Bull Nat Mus Ceylon 7: 65–87.

—. 1911 On Cristipira veneris nov. spec., and the affinities and classification of spirochaets. Quart J Microsep Sci 56: 507–541.

—. 1912 Researches on the spirochaets and

related organisms. Arch Protistenk *26:* 117–240.

Dobereiner, J. 1966 *Azotobacter paspali,* sp. n. una bacteria fixadora de nitragenio na rizosfera de *Paspalum.* Pesqui Agropecuar Brasil *1:* 357–365.

— and A. P. Ruschel. 1958 Uma nova especie de *Beijerinckia.* Rev Biol *1:* 261–272.

Dobrzanski, W. T., H. Osowiecki and M. A. Jagielski. 1968 Observations on intergeneric transformation between staphylococci and streptococci. J Gen Microbiol *53 (2):* 187–196.

Dock, G. M., L. R. Gragstedt, R. Johnson and N. B. McCullough. 1938 Comparison of *Bacterium necrophorum* from ulcerative colitis in man with strains from animals. J Infect Dis *62:* 169–180.

Dodd, R. L. 1949 Serologic relationship between *Streptococcus* group H and *Streptococcus sanguis.* Proc Soc Exp Biol Med *70:* 598–599.

Dodge, C. W. 1935 Medical mycology. Fungous diseases of men and other mammals. C. V. Mosby Co., St. Louis.

Doetsch, R. N., B. H. Howard, S. O. Mann and A. E. Oxford. 1957 Physiological factors in the production of an iodophilic polysaccharide from pentose by a sheep rumen bacterium. J Gen Microbiol *16:* 156–168.

Doidge, E. M. 1917 A bacterial blight of pear blossoms occurring in South Africa. Ann Appl Biol *4:* 50–74.

—. 1920 A tomato canker. J Dep Agr S Afr *1:* 718–721.

d'Oliveira, M. 1936 Una doenca bacteriana do *Ligustrum japonicum* (Thunb.). Rev Agron (Lisboa) *24:* 425–435.

Dolman, C. E., D. E. Kerr, H. Chang and A. R. Shearer. 1951 Two cases of rat-bite fever due to *Streptobacillus moniliformis.* Can J Public Health *42:* 228–241.

Domaradski, I. V., E. G. Grigoryan, V. I. Borzenkova and B. G. Val'kov. 1968 Multiplication of plague causative agents in sterile and nonsterile soil. J Microbiol Epidem Immunobiol *45 (8):* 104–108.

Dommert, A. R., W. R. Wallace, J. F. Finnerty, G. T. Schrader, T. F. Rogers, B. J. Gough, R. H. Cane and G. T. Dimopoullos. 1965 Comparative immunologic and pathologic studies in anaplasmosis. Progr Protozool 2nd Int Conf Protozool, London 1965. Int Congr Ser pp. 180–181. Excerpta Med Found, Amsterdam.

Domnas, A. 1968 Pigments of the *Actinoplanaceae.* I. Pigment production by *Spirillospora* No. 1655. J Elisha Mitchell Sci Soc *84:* 16–23.

Donatien, A. and G. Gayot. 1942 Conjonctivite rickettsienne du porc. Bull Soc Pathol Exot *35:* 325.

— and F. Lestoquard. 1935 Existence en Algerie d'une Rickettsia du chien. Bull Soc Pathol Exot *28:* 418–419.

— and —. 1936 *Rickettsia bovis* Nouvelle espèce pathogène pour le boeuf. Bull Soc Pathol Exot *29:* 1057–1061.

— and —. 1940 Rickettsiose bovine Algerienne à *R. bovis.* Bull Soc Pathol Exot *33:* 245–248.

Donk, P. J. 1920 A highly resistant thermophilic organism. J Bacteriol *5:* 373–374.

Donker, H. J. L. 1926 Bijdrage tot de Kennis

der Boterzuur-, Butylacoholen acctonigistingen. Diss., Delft. W. D. Meinema, Delft.

Donovick, R., J. F. Pagano and J. Vandeputte. 1957 Herstellung des antibiotikums. Thiostrepton und dessen Salzen. Deutsches Ausgelegeschrift 1,007,955, May 9.

Dorey, M. J. 1959 Some properties of a pectolytic soil *Flavobacterium.* J Gen Microbiol *20 (1):* 91–104.

Dorff, P. 1934 Die Eisenorganismen. Pflanzenforschung, Jena *16:* 1–62.

Dorofe'ev, K. A. 1947 Classification of the causative agent of tularemia. Symp Res Works Inst Epidemiol Mikrobiol Chita *1:* 170–180.

Dostálek, M. 1954 Propanové bakterie. Cesk Biol *3:* 162–169.

Doty, R. B., H. W. Dunne, J. F. Hokansen and J. J. Reid. 1964 A comparison of toxins produced by various isolates of *Corynebacterium pseudotuberculosis* and the development of a diagnostic skin test for cases of lymphadenitis of sheep and goats. Amer J Vet Res *25:* 1679–1685.

Doudoroff, M. 1940 The oxidative assimilation of sugars and related substances by *Pseudomonas saccharophila* with a contribution to the problem of the direct respiration of di- and polysaccharides. Enzymologia *9:* 59–72.

Douglas, H. C. 1951 Glycine fermentation by non-gas-forming anaerobic micrococci. J Bacteriol *62:* 517–518.

—. 1957 Genus VI. *Peptococcus* Kluyver and vanNiel 1936. *In* Breed, Murray and Smith (Editors), Bergey's Manual of Determinative Bacteriology, 7th ed. The Williams & Wilkins Co., Baltimore, pp. 474–480.

— and W. V. Cruess. 1936 A *Lactobacillus* from California wine: *Lactobacillus hilgardii.* Food Res *1:* 113–119.

Douglas, H. D. and S. E. Gunter. 1946 The taxonomic position of *Corynebacterium acnes.* J Bacteriol *52:* 15–23.

Douglas, S. R., A. Fleming and L. Colebrook. 1920 Studies in wound infections. Med Res Counc (Gr Brit) Spec Rep Ser *57:* 1–159.

Dowson, W. J. 1939 On the systematic position and generic names of the gram negative bacterial plant pathogens. Zentrabl Bakteriol Parasitenk Infektionskr Hyg Abt II *100:* 177–193.

—. 1942 On the generic name of the gram-positive bacterial plant pathogens. Trans Brit Mycol Soc *25:* 311–314.

—. 1943 On the generic names *Pseudomonas, Xanthomonas* and *Bacterium* for certain bacterial plant pathogens. Trans Brit Mycol Soc *26:* 1–14.

—. 1957 Plant Diseases due to Bacteria. 2nd ed., University Press, Cambridge.

— and A. C. Hayward. 1960 The bacterial mottle pathogen of Queensland sugar cane. Int Sugar J *62:* 275.

Doyle, L. P. 1948 The etiology of swine dysentery. Amer J Vet Res *9:* 50–51.

Draper, P. and P. D'Arcy Hart. 1968 The composition of normal and elongated *Mycobacterium lepraemurium.* J Gen Microbiol *52 (2):* 181–188.

Drechsler, C. 1919 Morphology of the Genus *Actinomyces* I and II. Bot Gaz *67:* 65–83; 147–168.

Dresel, E. G. and O. Strickl. 1928 Uber reversible Mutation-formen der Typusbazillen bein Menschen. Deut Med Wochenschr *54:* 517–519.

Drews, G. 1960 Untersuchungen zur Substruktur der "Chromatophoren" von *Rhodospirillum rubrum* und *Rhodospirillum molischianum.* Arch Mikrobiol *36:* 99–108.

— and P. Giesbrecht. 1963 Zur Morphogenese der Bakterien-"Chromatophoren" (-Thylakoide) und zur Synthese des Bakterien-Chlorophylls bei *Rhodopseudomonas spheroides* und *Rhodospirillum rubrum.* Zentrabl Bakteriol Parasitenk Infektionskr Hyg Abt I Orig *190:* 508–536.

— and —. 1966 *Rhodopseudomonas viridis* nov. spec., ein neu isoliertes, obligat phototrophes Bakterium. Arch Mikrobiol *53:* 255–262.

— and —. 1966 Die Bezeichnung des Stammes F von *Rhodopseudomonas viridis* als Typ-Stamm. Arch Mikrobiol *55:* 91–92.

Dschunkowsky, E. 1913 Das Rukfallfieber in Persien. Deut Med Wochenschr *39:* 419–420.

—. 1937 Balfoursche Granula als echte Geflügelparasiten, ihre Natur und Stellung in der Systematik: *Aegyptianella pullorum* Carpano, *Balfouria* n. gen. *Balfouria anserina* n. sp. and *Balfouria gallinarum* n. sp. Zentrabl Bakteriol Parasitenk Infektionskr Hyg Abt I Orig *140:* 131–136.

—. 1937 Pregled radova o Balfourovim granulama u vezi sa novim nazivima ovog parazita (*Aegyptianella* Carpano-*Balfouria* mihi). Jugosl Vet Glasn *17:* 315–321.

Dubos, R. and B. F. Miller. 1937 The production of bacterial enzymes capable of decomposing creatinine. J Biol Chem *121:* 429–445.

Duboscq, O. and P. Grassé. 1926 Les Schizophytes de *Devescovina hilli* n. sp. C R Soc Biol Paris *94:* 33–35.

— and —. 1927 Flagelles et Schizophytes de *Calotermes (Glyptotermes) irisdipennis,* Frogg. Arch Zool Exp Gen *66:* 451–496.

— and —. 1930 Protistologica xxi. *Coleomitus* n. g. au lieu de *Coleonema* pour le schizophyte. *C. pruvoti* Dub et Grassé, parasite d'un *Calotermes* des Iles Loyalty. Arch Zool Exp Gen Notes Rev *70:* 28.

— and C. Lebailly. 1912 Les spirochètes des poissons de mer. Arch Zool Exp Gén *50:* 331–369.

— and —. 1913 Sur les spirochètes des poissons. Arch Zool Exp Gen *52:* 9–24.

Dubost, M., P. Gautier, R. Moral, L. Ninet, S. Pinnert, J. Preud'homme and G. H. Werner. 1963 Un nouvel antibiotique à proprietés cytostatiques: la rubidomycine. C R Acad Sci Paris *257 (3):* 1813–1815.

Duché, J. 1934 Les Actinomyces du groupé albus. Encycl Mycol *6:* 1–375.

Duchow, E. and H. C. Douglas. 1949 *Rhodomicrobium vannielii,* a new photoheterotrophic bacterium. J Bacteriol *58 (4):* 409–416.

Ducrey, A. 1889 Il virus dell'ulcera venerea non e stato ancora coltivato. Rif Med *5:* 98.

—. 1889 Recherches expérimentales sur la nature intime du principe contagieux du chancre mou. Cong Int Derm Syph C R Paris *1890:* 229–250.

—. 1889 Ricerche sperimentali sulla natura intima del contagio dell'ulcera venerea sulla patogenesi del bubbone venereo. G Ital Mal Vener *30:* 377–425.

—. 1889 Experimentelle Untersuchungen uber den Ansteckungsstoff des weichen Schankers und uber die Bubonen. Monatsh Prakt Dermatol *9:* 387–405.

—. 1890 Recherches expérimentales sur la nature intime du principe contagieux du chancre mou. Ann Dermatol Syphiligr *3:* 56.

—. 1895 Noch einige Worte uber das Wesen des einfachen, kontagiosen Geschwures. Monatsch Prakt Dermatol *21:* 57–60.

Dugan, P. R. and D. G. Lundgren. 1960 Isolation of floc-forming organism, *Zoogloea ramigera,* and its culture in complex and synthetic media. Appl Microbiol *8 (6):* 357–361.

Duggar, B. M. 1909 Fungous diseases of plants. Ginn and Co., Boston.

—. 1948 Aureomycin: a product of the continuing search for new antibiotics. Ann N Y Acad Sci *51 (2):* 177–181.

—, E. J. Backus and T. H. Campbell. 1954 Types of variation in Actinomycetes. Ann N Y Acad Sci *60:* 71–85.

Duguid, J. P. 1959 Fimbriae and adhesive properties in *Klebsiella* strains. J Gen Microbiol *21 (1):* 271–286.

—, E. S. Anderson and I. Campbell. 1966 Fimbriae and adhesive properties in *Salmonellae.* J Pathol Bacteriol *92:* 107–138.

—, I. W. Smith, G. Dempster and P. N. Edmonds. 1955 Non-flagellar filamentous appendages ("fimbriae") and haemagglutinating activity in *Bacterium coli.* J Pathol Bacteriol *70:* 335–348.

Dujardin, F. 1841 Histoire naturelle des Zoophytes. Infusoires, comprenant la physiologie et la classification de ces animaux. De Roret, Paris.

Dukes, C. D. and H. L. Gardner. 1961 Identification of *Haemophilus vaginalis.* J Bacteriol *81 (2):* 277–283.

Dumbleton, L. J. 1945 Bacterial and nematode parasites of soil insects. N Z J Sci Technol Sect A *27:* 76–81.

Dunbar. 1893 Untersuchungen über choleraähnliche Wasserbacterien. Deut Med Wochenschr *19:* 799–800.

Dunican, L. K. and H. W. Seeley. 1962 Starch hydrolysis by *Streptococcus equinus.* J Bacteriol *83 (2):* 264–269.

Dunkelberg, W. E., Jr., R. Skaggs and D. S. Kellogg, Jr. 1970 A study and new description of *Corynebacterium vaginale (Haemophilus vaginalis).* Amer J Clin Pathol *53:* 370–377.

Dunkin, G. W. 1933 The preparation of Johnin from a synthetic medium without the addition of *B. phlei.* J Comp Pathol *46:* 159–164.

— and S. E. B. Balfour-Jones. 1935 Preliminary investigation of a disease of sheep possessing certain characteristics simulating Johne's disease. J Comp Pathol *48:* 236–240.

Dunn, C. G., G. J. Fuld, B. W. Kusmierek, P. G. Lim and D. I. C. Wang. 1964 Production of L-glutamic acid and α-ketoglutaric acids. (United States Patent 3,120,472). Offic Gaz U S Patent Off *799:* 147.

Dunn, L. H. 1927 Notes on two species of South American ticks, *Ornithodoros talaje* Guerin-Mene and *Ornithodoros venezuelensis* Brumpt. J Parasitol *13:* 177–182.

Dupaix, A. 1933 *B. caryocyaneus* Beijerinck-Dupaix 1930. Étude morphologique et biologique. Recherche sur le mécanisme du phénomène de Charrin et Roger (Agglutination serique des bactéries). Libraire Le Francois, Paris.

du Plessis, S. J. 1940 Bacterial blight of vines (Vlamiekte) in South Africa caused by *Erwinia vitivora* (Bacc.) Du P. S Afr Dep Tech Serv Sci Bull *214:* 1–105.

Dupouey, P. 1963 Étude immunologique de six espèces de treponemes anaérobies d'origine genitale: *Treponema phagedenes, refringens, calligyra, minutum,* Reiter, et *pallidum.* Ann Inst Pasteur (Paris) *105:* 725–736; 949–970.

Durlakowa, I., Z. Lachowicz and S. Ślopek. 1967 Biochemical properties of *Klebsiella* bacilli. Arch Immunol Ther Exp *15:* 490–496.

—, — and —. 1967 Serologic characterization of *Klebsiella* bacilli on the basis of properties of the capsular antigens. Arch Immunol Ther Exp *15:* 497–504.

Dutcher, J. D., F. E. Pansy and M. H. von Saltza. 1964 Septacidin and derivatives thereof. United States Patent 3,155,647, November 3.

—, M. H. von Saltza and F. E. Pansy. 1964 Septacidin, a new antitumor and antifungal antibiotic produced by *Streptomyces fimbriatus.* Antimicrob Agents Chemother *1963:* 83–88.

Dutky, S. R. 1940 Two new spore-forming bacteria causing milky diseases of Japanese beetle larvae. J Agr Res *61:* 57–68.

—. 1947 Preliminary observations on the growth requirements of *Bacillus popilliae* Dutky and *Bacillus lentimorbus* Dutky. J Bacteriol *54 (2):* 267.

—. 1959 Insect microbiology. Advan Appl Microbiol *1:* 175–200.

Dutton, J. E. and J. L. Todd. 1905 The nature of tick fever in the Eastern part of the Congo free state, with notes on the distribution and bionomics of the tick. Brit Med J *2 (2):* 1259–1260.

Dvonch, W O , O L Shotwell, R G Benedict, T. G. Pridham and L. A. Lindenfelser. 1954 Further studies on cynnamycin, a polypeptide antibiotic. Antibiot Chemother *4:* 1135–1142.

Dworkin, M. 1962 Nutritional requirements for vegetative growth of *Myxococcus xanthus.* J Bacteriol *84 (2):* 250–257.

—. 1963 Nutritional regulation of morphogenesis in *Myxococcus xanthus.* J Bacteriol *86 (1):* 67–72.

—. 1966 Biology of the Myxobacteria. Annu Rev Microbiol *20:* 75–106.

— and J. W. Foster. 1956 Studies on *Pseudomonas methanica* (Söhngen) nov. comb. J Bacteriol *72 (5):* 646–659.

— and S. M. Gibson. 1964 A system for studying microbial morphogenesis: Rapid formation of microcysts in *Myxococcus xanthus.* Science (Washington) *146:* 243–244.

Dye, D. W. 1962 The inadequacy of the usual determinative tests for the identification of *Xanthomonas* spp. N Z J Sci *5:* 393–416.

—. 1963 A bacterial disease of Pukatea (*Laurelia novae-zelandiae* A. Cunn.) caused by *Xanthomonas laureliae* n. sp. N Z J Sci *6:* 179–185.

—. 1963 The taxonomic position of *Xanthomonas stewartii* (Erw. Smith, 1914) Dowson 1939. N Z J Sci *6:* 495–506.

—. 1963 The taxonomic position of *Xanthomo-*

nas uredovorus Pon et al., 1954. N Z J Sci *6:* 146–149.

—. 1964 The taxonomic position of *Xanthomonas trifolii* (Huss, 1907) James, 1955. N Z J Sci *7:* 261–269.

—. 1966 A comparative study of some atypical "xanthomonads." N Z J Sci *9:* 843–854.

—. 1968 A taxonomic study of the genus *Erwinia.* I. The "amylovora" group. N Z J Sci *11:* 590–607.

—. 1969 A taxonomic study of the genus *Erwinia.* II. The "carotovora" group. N Z J Sci *12:* 81–97.

—. 1969 A taxonomic study of the genus *Erwinia.* III. The "herbicola" group. N Z J Sci *12:* 223–236.

—. 1969 A taxonomic study of the genus *Erwinia.* IV. "Atypical" erwinias. N Z J Sci *12:* 833–839.

Eaton, M. D., G. Meiklejohn and W. van Herick. 1944 Studies on the etiology of primary atypical pneumonia. A filterable agent transmissible to cotton rats, hamsters and chick embryos. J Exp Med *79:* 649–668.

Eberson, F. 1918 A bacteriologic study of the diphtheroid organisms with special reference to Hodgkin's disease. J Infect Dis *23:* 1–42.

Eble, T. E., H. Hoeksema, G. A. Boyack and G. M. Savage. 1959 Psicofuranine. I. Discovery, isolation and properties. Antibiot Chemother *9 (7):* 419–420.

Eckhardt, M. M., I. R. Baldwin and E. B. Fred. 1931 Studies on the root-nodule bacteria of *Lupinus.* J Bacteriol *21:* 273–285.

Eddy, B. P. 1960 Cephalotrichous, fermentative gram-negative bacteria: the genus *Aeromonas.* J Appl Bacteriol *23:* 216–249.

—. 1962 Further studies on *Aeromonas.* I. Additional strains and supplementary biochemical tests. J Appl Bacteriol *25:* 137–146.

— and K. P. Carpenter. 1964 Further studies on *Aeromonas.* II. Taxonomy of *Aeromonas* and C27 strains. J Appl Bacteriol *27:* 96–109.

Edmunds, P. N. 1960 *Haemophilus vaginalis:* morphology, cultural characters and viability. J Pathol Bacteriol *79:* 273–284.

—. 1960 The growth requirements of *Haemophilus vaginalis.* J Pathol Bacteriol *80:* 325–335.

—. 1962 The biochemical, serological and haemagglutinating reactions of "*Haemophilus vaginalis.*" J Pathol Bacteriol *83:* 411–422.

Edward, D. G. ff. 1950 An investigation of the biological properties of organisms of the pleuropneumonia group, with suggestions regarding the identification of strains. J Gen Microbiol *4:* 311–329.

—. 1953 Organisms of the pleuropneumonia group causing disease in goats. Vet Rec *65:* 873–874.

—. 1954 The pleuropneumonia group of organisms: a review, together with some new observations. J Gen Microbiol *10:* 27–64.

—. 1955 A suggested classification and nomenclature for organisms of the pleuropneumonia group. Int Bull Bacteriol Nomencl Taxon *5 (2):* 85–93.

— and W. A. Fitzgerald. 1951 The isolation of organisms of the pleuropneumonia group from dogs. J Gen Microbiol *5:* 566–575.

— and E. A. Freundt. 1956 The classification and nomenclature of organisms of the pleuropneumonia groups. J Gen Microbiol *14:* 197–207.

— and —. 1965 A note on the taxonomic status of strains like "Compo" hitherto classified as *Mycoplasma hominis*, type 2. J Gen Microbiol *41:* 263–265.

— and —. 1967 Proposal for *Mollicutes* as name of the class established for the order *Mycoplasmatales*. Int J Syst Bacteriol *17 (3):* 267–268.

— and —. 1969 Classification of the *Mycoplasmatales*. In Hayflick (editor). The *Mycoplasmatales* and the L-Phase of Bacteria. Appleton-Century-Crofts, New York, pp. 147–200.

— and —. 1969 Proposal for classifying organisms related to *Mycoplasma laidlawii* in a family *Sapromycetaceae*, Genus *Sapromyces*, within the *Mycoplasmatales*. J Gen Microbiol *57 (3):* 391–395.

— and —. 1970 Amended nomenclature for a classification of strains related to *Mycoplasma laidlawii*. J Gen Microbiol *62 (1):* 1–2.

— and —. 1973 Type strains of species of the order *Mycoplasmatales*, including designation of neotypes for *Mycoplasma mycoides* subsp. *mycoides*, *Mycoplasma agalactiae* subsp. *agalactiae* and *Mycoplasma arthritidis*. Int J Syst Bacteriol *23:* 55–61.

— and A. D. Kanarek. 1960 Organisms of the pleuropneumonia group of avian origin: their classification into species. Ann N Y Acad Sci *79:* 696–702.

Edwards, M. R. and R. W. Stevens. 1963 Fine structure of *Listeria monocytogenes*. J Bacteriol *86 (3):* 414–428.

Edwards, P. R. 1928 The relation of encapsulated bacilli found in metritis in mares to encapsulated bacilli from human sources. J Bacteriol *15:* 245–266.

—. 1929 Relationships of the encapsulated bacilli with special reference to *Bact. aerogenes* J Bacteriol *17:* 339–353.

—. 1931 Studies on *Shigella equirulis* (*Bact. viscosum equi*). Ky Agr Exp Sta Bull *320:* 291–330.

—. 1934 The differentiation of hemolytic streptococci of human and animal origin by group precipitin tests. J Bacteriol *27:* 527–534.

— and W. H. Ewing. 1955 Identification of *Enterobacteriaceae*. 1st ed., Burgess Publishing Co., Minneapolis, Minn.

— and —. 1962 Identification of *Enterobacteriaceae*. 2nd ed., Burgess Publishing Co., Minneapolis, Minn.

—, M. A. Fife and W. H. Ewing. 1965 Antigenic schema for the genus *Arizona*. Monograph Communicable Disease Center, Atlanta, Georgia.

—, — and C. H. Ramsey. 1959 Studies on the Arizona group *Enterobacteriaceae*. Bacteriol Rev *23 (4):* 155–174.

—, M. G. West and D. W. Bruner. 1947 *Arizona* group of paracolon bacteria. Ky Agr Exp Sta Bull *499:* 3–32.

—, — and —. 1948 Antigenic studies of a paracolon bacteria (Bethesda group). J Bacteriol *55(5):* 711–719.

Efthimion, M. H. and W. A. Corpe. 1969 Effect of cold temperatures on the viability of *Chromobacterium violaceum*. Appl Microbiol *17 (1):* 169–175.

Eggerth, A. H. 1935 The gram-positive nonspore-bearing anaerobic bacilli of human feces J Bacteriol *30:* 277–299.

— and B. H. Gagnon. 1933 The bacteroides of human feces. J Bacteriol *25:* 389–413.

Egorova, A. A. and Z. P. Deryurgina. 1963 On the sporulating thermophilic sulfur bacterium *Thiobacillus thermophilica imshenetskii* nov. sp. Mikrobiologiya *32:* 439–440.

Ehrenberg, C. G. 1835 Dritter Beitrag zur Erkenntniss grosser Organisation in der Richtung des kleinsten Raumes. Abh Preuss Akad Wiss Phys Kl Berlin aus den Jahre 1833–1835, pp. 143–336.

—. 1836 Vorlaufige Mitteilungen uber das wirkliche Vorkommen fossiler Infusorien und ihre grosse Verbreitung. Ann Physik Lpz (Poggendorfs Ann) *38:* 213–227.

—. 1838 Die Infusionsthierchen als volkommene Organismen. L. Voss, Leipzig.

—. 1840 Charakteristik von 274 neuen Arten von Infusorien. Ber Bekannt Verhandl Königl Preuss Akad Wiss Berlin *1840:* 197–219.

Ehrismann, O. 1933 Bakteriologische Differential-diagnose. Enzyklopedie der mikroskopischen Technik Dritte vermehrte und verbesserte Auflage. Verlag von Urban und Schwarzenberg, Berlin, pp. 137–412.

Ehrlich, J., L. E. Anderson, G. L. Coffey, A. B. Hillegas, M. P. Knudsen, H. J. Koepsell, D. L. Kohberger and J. E. Oyaas. 1954 Antibiotic studies of azaserine. Nature (London) *173:* 72.

—, Q. R. Bartz, R. M. Smith, D. A. Joslyn and P. R. Burkholder. 1947 Chloromycetin, a new antibiotic from a soil actinomycete. Science (Washington) *106:* 417.

—, G. L. Coffey, M. M. Galbraith, M. P. Knudsen, A. S. Schlingman and R. M. Smith. 1955 Griseoviridin and viridogrisein: biologic studies. Antibiot Ann *1954–1955:* 790–805.

—, D. Gottlieb, P. R. Burkholder, L. E. Anderson and T. G. Pridham. 1948 *Streptomyces venezuelae*, n. sp., the source of chloromycetin. J Bacteriol *56:* 467–477.

Eiken, M. 1958 Studies on an anaerobic, rod-shaped, gram-negative microorganism: *Bacteroides corrodens* n. sp. Acta Pathol Microbiol Scand *43:* 404–416.

Eimhjellen, K. E. 1965 Isolation of Extremely Halophilic Bacteria. Zentrabl Bakteriol, Parasitenk Infektionskr Hyg Abt I Suppl I 126–137.

—. 1970 *Thiocapsa pfennigii* sp. nov., a new species of phototrophic sulfur bacteria. Arch Mikrobiol *73:* 193–194.

—, O. Aasmundrud and A. Jensen. 1963 A new bacterial chlorophyll. Biochem Biophys Res Commun *10:* 232–236.

—, H. Steensland and J. Traetteberg. 1967 A *Thiococcus* sp. nov. gen., its pigments and internal membrane system. Arch Mikrobiol *59:* 82–92.

Eisenberg, G. M. 1942 Isolation of an aerobic sporulating lactose-fermenting organism from Philadelphia drinking water. J Amer Water Works Ass *34:* 365–366.

Eisenberg, J. 1891 Bakteriologische Diagnostik Hilfstabellen zum Gebrauche beim Praktischen Arbeiten. 3 Aufl. Leopold Voss, Hamburg.

Elazari-Volcani, B. 1940 Studies on the microflora of the Dead Sea. In Hebrew, English summary. Thesis, Hebrew University, Jerusalem.

—. 1957 Genus XII *Halobacterium. In* Breed, Murray and Smith. Bergey's Manual of Determinative Bacteriology, 7th ed. The Williams & Wilkins Co., Baltimore, pp. 207–212.

Eldering, G. and P. Kendrick. 1938 *Bacillus parapertussis:* A species resembling both *Bacillus pertussis* and *Bacillus bronchisepticus* but identical with neither. J Bacteriol *35:* 561–572.

Eli Lilly and Company. 1954 Improvements in or relating to Terramycin. British Patent 713,795, August 18.

Elisberg, B. L., J. M. Campbell and F. M. Bozeman. 1968 Antigenic diversity of *Rickettsia tsutsugamushi:* epidemiologic and ecologic significance. J Hyg Epidemiol Microbiol Immunol (Prague) *12:* 18–25.

Ellinghausen, H. C., Jr. and W. G. McCullough. 1965 Nutrition of *Leptospira pomona* and growth of 13 other serotypes: fractionation of oleic albumin complex and a medium of bovine albumin and polysorbate 80. Amer J Vet Res *26:* 45–51.

Ellinghausen, H. C. and M. J. Pelczar. 1955 Spectrophotometric characterization of *Neisseria* pigments. J Bacteriol *70 (4):* 448–453.

Elliott, C. 1920 Halo-blight of oats. J Agr Res *19:* 139–172.

—. 1923 A bacterial stripe disease of proso millet. J Agr Res *26:* 151–160.

—. 1930 Manual of bacterial plant pathogens. The Williams & Wilkins, Co., Baltimore, 1st ed.

—. 1930 Bacterial streak disease of sorghums. J Agr Res *40:* 963–976.

—. 1951 Manual of bacterial plant pathogens 2nd ed. Chronica Botanica Co., Waltham.

Elliott, S. D. 1945 A proteolytic enzyme produced by group A streptococci with special reference to its effect on the type-specific M antigen. J Exp Med *81:* 573–592.

—. 1960 Type and group polysaccharides of group D streptococci. J Exp Med *111:* 621–630.

—. 1962 Teichoic acid and the group antigen of group D streptococci. Nature (London) *193:* 1105–1106.

—. 1963 Teichoic acid and the group antigen of lactic streptococci (group N). Nature (London) *200:* 1184–1185.

—. 1966 Streptococcal infection in young pigs. I. An immunochemical study of the causative agent (P M streptococcus). J Hyg *64:* 205–212.

— and V. P. Dole. 1947 An inactive precursor of streptococcal proteinase. J Exp Med *85:* 305–320.

Ellis, D. 1907 Iron bacteria. Methuen, London.

—. 1907 A contribution to our knowledge of the thread—bacteria. 1. Leptothrix ochracea (Kützing). 2. Gallionella ferruginea (Ehrenberg). 3. Spirophyllum (Ellis). Zentrabl Bakteriol Parasitenk Infektionskr Hyg Abt II *19:* 502–518.

—. 1932 Sulphur bacteria. Longmans Green, London.

Elrod, R. P. and A. C. Braun. 1947 Serological studies of the genus *Xanthomonas.* I. Cross-agglutination relationships. J Bacteriol *53:* 509–518.

Elsden, S. R. and D. Lewis. 1953 The produc-

tion of fatty acids by a gram-negative coccus. Biochem J *55:* 183–189.

—, B. E. Volcani, F. M. C. Gilchrist and D. Lewis. 1956 Properties of a fatty acid forming organism isolated from the rumen of sheep. J Bacteriol *72 (5):* 681–689.

Elser, W. J. and F. M. Huntoon. 1909 Studies on Meningitis. J Med Res *20:* 371–541.

Emmerling, O. and E. Abderhalden. 1903 Ueber einen Chinasäure in Protokatechusäure überführenden Pilz. Zentrabl Bakteriol Parasitenk Infektionskr Hyg Abt II *10:* 337–339.

Emmons, C. W. 1938 The isolation of *Actinomyces bovis* from tonsillar granules. Publ Health Rep *53 (44):* 1967–1975.

Emoto, Y. 1928 Über eine neue schwefeloxydierende Bakterie. Bot Mag (Tokyo) *42:* 421–426.

—. 1929 Über drei neue Arten der Schwefeloxydierenden Bakterien. Proc Jap Acad *5:* 148–151.

Enderlein, G. 1917 Ein neues Bakteriensystem auf vergleichend morphologischen Grundlage. (Bakteriologische Studien IV). S B Ges Naturf Fr Berl Jahr, *1917:* 309–319.

—. 1925 Bakteriën-Cyclogenic Prolegomena zur untersuchungen ueber Bau, geschlechtliche und ungeschlechtliche Entwicklung der Bakterien. Walter de Gruyter and Co., Berlin.

Endo, T. 1956 Studies on the genus *Nocardia.* J Antibiot (Tokyo) Ser A *9:* 228.

Engbaek, H. C., A. Jespersen, D. Faber and D. W. Will. 1964 The pathology of joint disease in rabbits produced by atypical mycobacteria and *M. avium.* I. Macroscopical and bacteriological examination of organs, joints and tendon sheaths. Acta Tuberc Pneumol Scand *44:* 199–208.

—, B. Vergmann, I. Baess and M. W. Bentzon. 1968 *Mycobacterium avium:* A bacteriological and epidemiological study of *M. avium* isolated from animals and man in Denmark. Part I: Strains isolated from animals. Acta Pathol Microbiol Scand *72:* 277–294.

—, —, — and —. 1968 *Mycobacterium avium:* A bacteriological and epidemiological study of *M. avium* isolated from animals and man in Denmark. Part 2: Strains isolated from man. Acta Pathol Microbiol Scand *72:* 295–312.

—, —, — and D. W. Will. 1967 *M. xenopei*—a bacteriological study of *M. xenopei* including case reports of Danish patients. Acta Pathol Microbiol Scand *69:* 576–594.

Engelbrecht, H. 1969 Vaccination against strangles. J Amer Vet Med Ass *155:* 425–431.

Engler, A. 1883 Im Kieler Hafen in dem sogenannten "toten Grund" vorkommenden Pilzformen, Verh Bot Ver Brandenb *24:* 17–20.

—. 1883 Ueber die Pilzbegetation des weissen oder todten Grundes in der Kiel Bucht. Wiss Meeresuntersuch Abt I 187–193.

—. 1907 Syllabus der Pflanzenfamilien. 5th ed. Borntraeger Bros., Berlin.

Enlows, E. M. A. 1920 The generic names of Bacteria. U S Public Health Serv Hyg Lab Bull *121:* 1–115.

Ensign, J. C. and S. C. Rittenberg. 1963 A crystalline pigment produced from 2-hydroxypyridine by *Arthrobacter crystallopoietes* n. sp. Arch Mikrobiol *47:* 137–153.

— and R. S. Wolfe. 1964 Nutritional control of

morphogenesis in *Arthrobacter crystallopoietes*. J Bacteriol 87 (4): 924–932.

Ensminger, P. W. 1953 Pigment production by *Haemophilus parapertussis*. J Bacteriol 63 (5): 509–510.

Entwistle, P. F. and J. S. Robertson. 1968 The ultrastructure of a rickettsia pathogenic to a saturnid moth. J Gen Microbiol 54 (1): 97–104.

— and —. 1968 Rickettsiae pathogenic to two saturnid moths. J Invertebr Pathol 10: 345–354.

Eppinger, H. 1891 Über eine neue pathogene *Cladothrix* und eine durch sie nervorgerufene Pseudotuberculosis (Cladothrichica). Beitr Pathol Anat 9: 287–328.

Erikson, D. 1935 The pathogenic aerobic organisms of the actinomyces group. Med Res Counc (Gt Brit) Spec Rep Ser 203: 5–61.

—. 1940 Pathogenic anaerobic organisms of the *Actinomyces* group. Med Res Counc (Gt Brit) Spec Rep Ser 240: 1–63.

—. 1954 Factors promoting cell division in a "soft" mycelial type of Nocardia: *Nocardia turbata* n. sp. J Gen Microbiol 11: 198–208.

Errebo Larsen, H. and H. P. R. Seeliger. 1966 A mannitol fermenting *Listeria*, *Listeria grayi* sp. n. Proc. Third Intern. Sympos on Listeriosis July 13–16 Bilthoven, p. 35.

Escherich, T. 1885 Die Darmbacterien des Neugeborenen und Säuglings. Fortschr Med 3: 515–522; 547–554.

—. 1886 Die Darmbakterien des Säuglings und ihre Beziehungen zur Physiologie der Verdauung. F. Enke, Stuttgart.

Esmarch, E. 1887 Über die Reinkultur eines *Spirillum*. Zentrabl Bakteriol Parasitenk Infektionskr Hyg Abt I Orig 1: 225–230.

Etemadi, A. H. 1963 Isolement des acides *iso*-pentadecanoiques et *iso*-heptadecanoiques des lipids de *Corynebacterium parvum*. Bull Soc Chim Biol 45: 1423–1432.

— and E. Lederer. 1965 Sur la structure des acides α-mycoliques de la souche humaine test. Bull Soc Chim Fr Sept (9): 2640–2645.

Ettlinger, L., R. Corbaz and R. Hütter. 1958 Zur Systematik der Actinomyceten. 4. Eine arteinteilung der Gattung *Streptomyces* Waksman et Henrici. Arch Mikrobiol 31: 326–358.

—, E. Gäumann, R. Hütter, W. Keller-Schierlein, F. Kradolfer, L. Neipp, V. Prelog and H. Zähner. 1958 Stoffwechselprodukte von Actinomyceten. 12 Mitteilung. Über die Isolierung und Charakterisierung von Acetomycin. Helv Chim Acta 41: 216–219.

Evans, A. C. 1918 A study of the streptococci concerned with cheese ripening. J Agr Res 13: 235–252.

—. 1918 Further studies on *Bacterium abortus* and related bacteria. J Infect Dis 22: 580–593.

Evans, J. B., L. G. Buettner and C. F. Niven, Jr. 1952 Occurrence of streptococci that give a false-positive coagulase test. J Bacteriol 64 (3): 433–434.

— and C. F. Niven, Jr. 1951 Nutrition of the heterofermentative lactobacilli that cause greening of cured meat products. J Bacteriol 62 (5): 599–603.

Ewing, S. A. 1969 Canine ehrlichiosis. Adv Vet Sci Comp Med 13: 331–353.

Ewing, W. H. 1949 Shigella nomenclature. J Bacteriol 57 (6): 633–638.

—. 1958 The nomenclature and taxonomy of the *Proteus* and Providence groups. Int Bull Bacteriol Nomencl Taxon 8 (1): 17–22.

—. 1962 The tribe *Proteae*: its nomenclature and taxonomy. Int Bull Bacteriol Nomencl Taxon 12: 93–102.

—. 1963 An outline of nomenclature for the family *Enterobacteriaceae*. Int Bull Bacteriol Nomencl Taxon 13 (2): 95–110.

—. 1966 *Enterobacteriaceae* Taxonomy and Nomenclature. U S Dept Health Educ & Welfare, Public Health Serv, Washington.

— and B. R. Davis. 1961 The O antigen groups of *Escherichia coli* cultures from various sources. National Communicable Disease Center, Atlanta, Georgia.

—, —, M. A. Fife and E. F. Lessel. 1973 Biochemical characterization of *Serratia liquefaciens* (Grimes and Hennerty) Bascomb *et al.* (formerly *Enterobacter liquefaciens*) and *Serratia rubidaea* (Stapp) *comb. nov.* and designation of type and neotype strains. Int J Syst Bacteriol 23: 217–225.

—, — and J. G. Johnson. 1962. The genus *Serratia*: its taxonomy and nomenclature. Int Bull Bacteriol Nomencl Taxon 12 (2): 47–52.

—, — and W. J. Martin. 1967 The biochemical reactions of the genus *Escherichia*. National Communicable Disease Center, Atlanta, Georgia.

—, — and T. S. Montague. 1963. Studies on the occurrence of *Escherichia coli* serotypes associated with diarrhoeal disease. National Communicable Disease Center, Atlanta, Georgia.

—, — and R. W. Reavis. 1959 Studies on the *Serratia* group. CDC Laboratory Manual. Communicable Disease Center, Atlanta, Georgia.

— and M. Fife. 1968 *Enterobacter hafniae* (the Hafnia group). Int J Syst Bacteriol 18 (3): 263–271.

— and M. A. File. 1972 *Enterobacter agglomerans* (Beijernick) comb. nov. (the Herbicola-Lathyri Bacteria). Int J Syst Bacteriol 22 (1): 4–11.

—, R. Hugh and J. G. Johnson. 1961 Studies on the *Aeromonas* group. CDC Monograph U S Dept Health Educ & Welfare, Communicable Disease Center, Atlanta, Georgia.

— and A. C. McWhorter. 1965 Genus *Edwardsiella* and *E. tarda* p. 37. *In* Ewing, McWhorter, Escobar and Lubin. *Edwardsiella*, a new genus of *Enterobacteriaceae* based on a new species, *E. tarda*. Int Bull Bacteriol Nomencl Taxon 15 (1): 33–38.

—, K. E. Tanner and D. A. Dennard. 1954 The Providence group. J Infect Dis 94: 134–140.

Eymers, J. G. and K. L. van Schouwenburg. 1937 On the luminescence of bacteria. Enzymologia 3: 235–241.

Eyre, J. W. 1900 A clinical and bacteriological study of diplo-bacillary conjunctivitis. J Pathol Bacteriol 6: 1–13.

Fabricant, J. and E. A. Freundt. 1967 Importance of extension and standardization of laboratory tests for the identification and classification of *Mycoplasma*. Ann N. Y. Acad Sci 143 (1): 50–58.

Fairbrother, R. W. 1940 Coagulase production as a criterion for the classification of the staphylococci. J Pathol Bacteriol 50: 83–88.

Falcão de Morais, J. O. 1967 The genus *Elytrosporangium* and its relationship to *Microello-*

bosporia and *Streptomyces*. Hindustan Antibiot Bull *9* (*3*): 135–137.

—, A. Chaves Batista and D. M. G. Massa. 1966 *Elytrosporangium:* a new genus of the *Actinomycetales*. Mycopathol Mycol Appl *30* (*2*): 161–171.

— and M. H. Dália Maia. 1959 *S. erythrogriseus:* novo *Streptomyces* produtor de antibiotico. Rev Inst Antibiot Univ Recife *2* (*1-2*): 63–67.

— and —. 1961 Uma contribuição ao estudo taxonômico do gênero *Streptomyces*—Uma tentativa de simplificação. Rev Inst Antibiot Univ Recife *3* (*1*): 33–60.

—, — and M. E. Souto Maior Genn. 1958 Sôbre uma variedade de *Streptomyces* comum nos solos do Brasil: *Streptomyces venzuelae* var. *roseospori* nov. var. Rev. Inst Antibiot Univ Recife *1* (*2*): 99–106.

—, — and —. 1958 Um estudo taxonomico em torno do *Streptomyces lavendulae—S. lavendulae* var. *brasilicus* nov. var. Res Inst Antibiot Univ Recife *1*: 69–87.

—, Gonçalves de Lima and M. H. Dália Maia. 1957 Novo estudo sôbre *Nocardia recifei* Lima et al., e sua designacão como *Streptomyces recifensis*. An Soc Biol Pernambuco *15* (*1*): 239–253.

Falkow, S., I. R. Ryman and O. Washington. 1962 Deoxyribonucleic acid base composition of *Proteus* and Providence organisms. J Bacteriol *83* (*6*): 1318–1321.

Famintzin, A. 1892 Eine neue Bacterienform: *Nevskia ramosa*. Bull Acad Sci St Petersb New Ser 2 *34*: 481–486.

Fang, C. T., O. N. Allen, A. J. Riker and J. G. Dickson. 1950 The pathogenic physiological and serological reactions of the form species of *Xanthomonas translucens*. Phytopathology *40*: 44-64.

—, H. C. Ren, T. K. Chen, Y. K. Chu, H. C. Faan and S. C. Wu. 1957 A comparison of the rice bacterial leaf blight organism with the bacterial leaf streak organism of rice and *Leersia hexandra* Swartz. Acta Phytopathol Sinica *3*: 99–124.

Fantham, H. B. 1911 Some researches on the life-cycle of spirochaetes. Ann Trop Med Parasitol *5*: 479–496.

Farbenfabriken Bayer. 1952 An antibiotic picromycin and salts thereof. British Patent 682,045, November 5.

Farbwerke Hoechst A.-G. 1964 Antibiotic moenomycin. British Patent 977,327, December 9.

Farina, G. and R. Locci. 1966 Contribution to the study of *Streptoverticillium:* Description of a new species (*Streptoverticillium baldaccii* sp. nov.) and examination of previously illustrated species (In Italian). G Microbiol *14*: 33–52.

Farrell, R. K. 1966 Canine rickettsiosis, pp 285-288 *In* Kirk (editor), Current Veterinary Therapy 1966-1967, W. B. Saunders Co., Philadelphia.

Faull, J. 1916 *Chondromyces thaxteri*, a new myxobacterium. Bot Gaz *62*: 226–232.

Faust, L. and R. S. Wolfe. 1961 Enrichment and cultivation of *Beggiatoa alba*. J Bacteriol *81* (*1*): 99–106.

Fauve, R. M., C. H. Pierce-Chase and R. Dubos. 1964 Corynebacterial pseudotuberculosis in mice. II. Activation of natural and experimental latent infections. J Exp Med *120*: 283–304.

Feeley, J. C. and M. Pittman. 1963 Studies on the haemolytic activity of El Tor vibrios. Bull World Health Organ *28*: 347–356.

Feirer, W. A. 1927 Studies on some obligate thermophilic bacteria from soil. Soil Sci *23*: 47–56.

Feldman, W. H. 1938 Avian tuberculosis infections. The Williams & Wilkins Co., Baltimore.

— and R. E. Ritts. 1963 Pathogenicity studies of Group III (Battey) mycobacteria from pulmonary lesions of man. Dis Chest *43*: 26–33.

Felix, A. and B. R. Callow. 1943 Typing of paratyphoid B bacilli by means of Vi bacteriophage. Brit Med J *2*: 127.

Fellows, C. R. and R. W. Clough. 1925 Indol and skatol determination in bacterial cultures. J Bacteriol *10*: 105–133.

Felton, E. A., J. B. Evans and C. F. Niven, Jr. 1953 Production of catalase by the pediococci J Bacteriol *65* (*4*): 481–482.

— and C. F. Niven, Jr. 1953 The identity of *Leuconostoc citrovorum* strain 8081. J Bacteriol *65*: 482–483.

Fenner, F. 1950 The significance of the incubation period in infectious diseases. Med J Aust *2*: 813–818.

—. 1956 The pathogenic behavior of *Mycobacterium ulcerans* and *Mycobacterium balnei* in the mouse and the developing chick embryo. Amer Rev Tuberc Pulm Dis *73*: 650–673.

— and R. H. Leach. 1952 Studies on *Mycobacterium ulcerans*. I. Serological relationship to other mycobacteria. Aust J Exp Biol Med Sci *30*: 1–10.

— and —. 1952 Studies on *Mycobacterium ulcerans*. II. Cross reactivity in guinea pigs sensitized with *Mycobacterium ulcerans* and other mycobacterium. Aust J Exp Biol Med Sci *30*: 11–20.

Ferguson, W. W. and N. D. Henderson. 1947 Description of strain C 27: A motile organism with the major antigen of *Shigella sonnei* Phase I. J Bacteriol *54* (*2*): 179–181.

Ferrari, A. 1963 Nouve indagini sulla biologia di *Flavobacterium dehydrogenans*. Ann Microbiol Enzimol *13*: 1–12.

— and E. Zannini. 1958 Richerche sulle specie del genere *Flavobacterium*. Ann Microbiol Enzimol *8*: 138–204.

Ferraris, T. 1926 Trattato di patologia e terapia vegetable. 3rd ed Vol I, Heopli, Milan.

Ferry, N. S. 1911 Etiology of canine distemper. J Infect Dis *8*: 399–420.

—. 1912 Further studies on the *Bacillus bronchicanis*, the cause of canine distemper. Amer Vet Rev *41*: 77–79.

—. 1912 *Bacillus bronchisepticus (bronchicanis):* the cause of distemper in dogs and a similar disease in other animals. Vet J *68*: 376–391.

Fewson, C. A. 1967 The identity of the gram-negative bacterium NCIB 8250 ('Vibrio 01'). J Gen Microbiol *48* (*1*): 107–110.

Fitz-James, P. C. 1960 Participation of the cytoplasmic membrane in the growth and spore formation of Bacilli. J Biophys Biochem Cytol *8*: 507–528.

— and I. E. Young. 1958 Morphological and

chemical studies of the spores and parasporal bodies of *Bacillus laterosporus*. J Biophys Biochem Cytol *4:* 639–649.

Fiuczek, M. 1959 Wiazanie Azotu Atmosferycznego w czystuch kulturach *Streptomyces alni* (Fixation of atmospheric nitrogen in pure cultures of *Streptomyces alni*). Acta Microbiol Pol *8:* 283–287.

Flaig, E., E. Küster, H. Beutelspacher, I. Schlichting-Bauer, W. Politt-Runge and R. Kurz. 1955 Elektronmikroskopische Untersuchungen an Sporen verschiedener Streptomyceten. Zentrabl Bakteriol Parasitenk Infektionskr Hyg Abt II *108:* 376–382.

Fiedler, F., K. H. Schleifer, B. Cziharz, E. Interschick and O. Kandler. 1970 Murein types in Arthrobacter, Brevibacteria, Corynebacteria, and Microbacteria. Publ Fac Sciences Univ J. E. Purkyne, Brno *47:* 111–122. From Report on Taxonomy of Bacteria, Brno, Sept. 24–27, 1969

Field, H. I., D. Buntain and J. T. Done. 1954 Studies on piglet mortality. I. Streptococcal meningitis and arthritis. Vet Rec *66:* 453–455.

Fildes, P. 1924/25 The growth requirements of haemolytic influenza bacilli and the bearing of these upon the classification of related organisms. Brit J Exp Pathol *5:* 69–74.

—. 1938 The growth of *Proteus* on ammonium lactate plus nicotinic acid. Brit J Exp Pathol *19:* 239–244.

Finck, G. 1950 Biologische und Stoffwechselphysiologische Studien an Myxococcaceen. Arch Mikrobiol *15:* 358–388.

Finkelstein, R. A. and S. Mukerjee. 1963 Hemagglutination: A rapid method for differentiating *Vibrio cholerae* and El Tor vibrios. Proc Soc Exp Biol Med *112:* 355–359.

Finkler, D. and J. Prior. 1884 Über den Bacillus der Cholera nostras und seine Cultur. Deut Med Wochenschr *10:* 632–634.

Finlay, A. C., G. L. Hobby, S. Y. P'an, P. P. Regna, J. B. Routien, D. B. Seeley, G. M. Shull, B. A. Sobin, I. A. Solomons, J. W. Vinson and J. H. Kane. 1950 Terramycin, a new antibiotic. Science (Washington) *111:* 85.

—, F. A. Hochstein, B. A. Sobin and F. X. Murphy. 1951 Netropsin, a new antibiotic produced by a *Streptomyces*. J Amer Chem Soc *73:* 341–343.

— and B. A. Sobin. 1951 Verfahren zur Herstellung eines Antibiotikums. German Patent Application 352,176, July 19.

— and —. 1952 Netropsin and process for its production. United States Patent 2,586,762, February 19.

Findlay, G. M., E. Klieneberger, F. O. MacCallum and R. D. Mackenzie. 1938 Rolling disease. New syndrome in mice associated with a pleuro-pneumonia-like organism. Lancet *2:* 1511–1513.

Finnerty, W. R., E. Hawtrey and R. E. Kallio. 1962 Alkane oxidizing micrococci. Z Allg Mikrobiol *2:* 169–177.

Firehammer, B. D. 1965 The isolation of vibrios from ovine feces. Cornell Vet *55:* 482–494.

Fischer, B. 1888 Ueber einen neuen lichtentwickelnden *Bacillus*. Zentrabl Bakteriol Parasitenk Infektionskr Hyg Abt I Orig *3:* 105–108; 137–141.

Fiset, P. and R. A. Ormsbee. 1968 The anti-

body response to antigens of *Coxiella burneti*. Zentrabl Bakteriol Parasitenk Infektionskr Hyg Abt I Orig *206:* 321–329.

Fisher, C. A. and L. Barksdale. 1971 Elimination of the acid fastness but not the gram positivity of leprosy bacilli after extraction with pyridine. J Bacteriol *106 (2):* 707–708.

Fisher, P. J. 1963 The effect of freeze drying on the viability of *Chromobacterium lividum*. J Appl Bacteriol *26:* 502–503.

Flamm, H. 1956 *Moraxella saccharolytica* (sp. n.) aus dem Liquor eines Kindes mit Meningitis. Zentrabl Bakteriol Parasitenk Infektionskr Hyg Abt I Orig *166:* 498–502.

—. 1957 Eine weitere neue Species des Genus *Moraxella*, *M. polymorpha*, sp. n. Zentrabl Bakteriol Parasitenk Infektionskr Hyg Abt I Orig *168:* 261–267.

Fleming, A. 1922 On a remarkable bacteriolytic element found in tissues and secretions. Proc Roy Soc (Ser B) *93:* 306–317.

Fletcher, W. 1928 Recent work on leptospirosis, tsutsugamushi disease, and tropical typhus in the Federated Malay States. Trans Roy Soc Trop Med Hyg *21:* 265–282.

Flint, J. C. and D. K. McKelvie. 1956 Feline infectious anemia—diagnosis and treatment. Proc 92nd Ann Mtg Am Vet Med Assoc (1955), pp. 240–242.

—, M. H. Roepke and R. Jensen. 1958 Feline infectious anemia. I. Clinical aspects. Amer J Vet Res *19:* 164–168.

—, — and —. 1959 Feline infectious anemia. II. Experimental cases. Amer J Vet Res *20:* 33–40.

Florent, A. 1959 Les deux Vibrioses genitales de la bête bovine: La vibriose vénérienne, due à *Vibrio foetus venerialis*, et la vibriose d'origine intestinale due à *V. foetus intestinalis*. Proc 10th Int Vet Cong Madrid *2:* 953–957.

Florenzano, G., W. Balloni and R. Materassi. 1968 Nitrogen-fixing bacteria of the genus *Beijerinckia* in Venezuelan soils. *In* Transactions 9th Int Congr Soil Science, Adelaide, Australia, Vol 2 pp. 125–128.

Flossmann, K. D. and W. Erler. 1972 Serological, chemical and immunochemical studies on *Erysipelothrix rhusiopathiae* XI. Isolation and characterization of the DNA of *Erysipelothrix*. Arch Exp Veterinaermed *26:* 817–824.

Flügge, C. 1886 Die Mikroorganismen. F. C. W. Vogel, Leipzig.

Flynn, E. H., J. W. Hinman, E. L. Caron and D. O. Woolf, Jr. 1953 The chemistry of amicetin, a new antibiotic. J Amer Chem Soc *75 (23):* 5867–5871.

— and D. O. Woolf, Jr. 1953 The chemistry of amicetin, a new antibiotic. Abstr Papers, 123rd Meeting, Amer Chem Soc, Los Angeles, California, March 15–19, 4L-5L.

Foggie, A. 1949 Studies on tick-borne fever in sheep. J Gen Microbiol *3:* V–VI.

—. 1951 Studies on the infectious agent of tick-borne fever in sheep. J Pathol Bacteriol *63:* 1–15.

—. 1962 Studies on tick pyaemia and tick-borne fever. Symp Zool Soc London *6:* 51–58.

— and D. I. Nisbet. 1964 Studies on *Eperythrozoon* infection in sheep. J Comp Pathol *74:* 45–61.

Foley, H. and L. Parrot. 1937 Sur la Rickettsia du trachome. C R Soc Biol Paris *124:* 230–232.

Folkertsma, J. P., W. T. Sokolski and J. W. Snyder. 1958 Chromatographic strip eluate bioassay of streptovaricin. Antibiot Ann *1957–1958:* 114–118.

Ford, D. 1967 Relationships between mycoplasma and the etiology of non-gonococcal urethritis and Reiter's syndrome. Ann N Y Acad Sci *143 (1):* 501–504.

Ford, J. E. and M. Rogosa. 1961 The nutrition of a *Lactobacillus acidophilus* variant isolated from the duodenum of a chick. J Gen Microbiol *25 (2):* 249–252.

Ford, W. W. 1916 Studies on aerobic spore-bearing non-pathogenic bacteria. Part II. Miscellaneous cultures. J Bacteriol *1:* 518–526.

—. 1927 Text-book of Bacteriology, Saunders, Philadelphia.

Fornachon, J. C. M. 1943 Bacterial spoilage of fortified wines. Australian Wine Board, Adelaide.

Foster, J. W. 1944 Microbiological aspects of riboflavin. I. Introduction. II. Bacterial oxidation of riboflavin to lumichrome. J Bacteriol *47:* 27–41.

— and R. H. Davis. 1966 A methane-dependent coccus, with notes on classification and nomenclature of obligate, methane-utilizing bacteria. J Bacteriol *91 (5):* 1924–1931.

Fott, B. and J. Komarek. 1960 Das Phytoplankton der Teiche im Teschner Schlesien. Preslia (Praha) *32 (2):* 113–141.

Foubert, E. L., Jr. and H. C. Douglas. 1948 Studies on the anaerobic micrococci. I. Taxonomic considerations. J Bacteriol *56:* 25–34.

— and —. 1948 Studies on the anaerobic micrococci. II. The fermentation of lactate by *Micrococcus lactilyticus.* J Bacteriol 56 *(1):* 35–36.

Foulerton, A. G. R. 1905 New species of *Streptothrix* isolated from the air. Lancet *I:* 1199–1200

— and C. Price-Jones. 1901 *Streptothrix* infections in the lower animals. J Comp Pathol *14:* 45–59.

— and —. 1902 On the general characteristics and pathogenic action of the genus *Streptothrix.* Trans Pathol Soc London *53:* 56–127.

Fox, E. N. and M. K. Wittner. 1965 The multiple molecular structure of the M proteins of group A streptococci. Proc Nat Acad Sci U S A *54:* 1118–1125.

Fox, H., R. H. Purcell and R. M. Chanock. 1969 Characterization of a newly identified mycoplasma (*Mycoplasma orale* type 3) from the human oropharynx. J Bacteriol 98 *(1):* 36–43.

Francis, J. 1943 Infection of laboratory animals with *Mycobacterium johnei.* J Comp Pathol *53:* 140–150.

Frank, B. 1879 Ueber die Parasiten in den Wurzelanschwillungen der Papilionaceen. Ber Deut Bot Ges *37:* 376–387; 394–399.

—. 1889 Ueber die Pilzsymbiose der Leguminosen. Ber Deut Bot Ges *7:* 332–346.

Frankland, G. C. and P. F. Frankland. 1887 Studies on some new microorganisms obtained from air. Phil Trans Roy Soc London Ser B Biol Sci *178:* 257–287.

— and —. 1889 Ueber einige typische Mikroorganismen im Wasser und im Boden. Z Hyg Infektionskr Med Mikrobiol Immunol Virol *6:* 373–400.

— and —. 1894. Micro-organisms in water, their significance, identification and removal. Longmans, Green & Co., London, pp. 1–532.

Franklin, J. G. and M. E. Sharpe. 1964 Physiological characteristics and vitamin requirements of lactobacilli isolated from milk and cheese. J Gen Microbiol *34 (1):* 143–151.

Fraser, G. 1964 The effect on animal erythrocytes of combinations of diffusible substances produced by bacteria. J Pathol Bacteriol *88:* 43–53.

Frateur, J. 1950 Essai sur la systématique des acetobacters. Cellule *53:* 287–392.

— and P. Simonart. 1952 Etude de la flore bactérienne d'un acetificateur de vinaigre d'alcool. IX Congresso Inter Ind Agraric Roma, C. P. 15.

Frazier, P. D. and B. D. Fowler. 1967 X-ray diffraction and infrared study of the 'sulphur granules' of *Actinomyces bovis.* J Gen Microbiol *46 (3):* 445–450.

Fred, E. B., W. H. Peterson and J. A. Anderson. 1921 The characteristics of certain pentose destroying bacteria, especially as concerns their action on arabinose. J Biol Chem *48:* 385–412.

—, — and A. Davenport. 1919 Acid fermentation of xylose. J Biol Chem *39:* 347–384.

Fredericq, P. 1963 On the nature of colicinogenic factors: a review. J Theor Biol *4:* 159–165.

Frederiksen, W. 1964 A study of some *Yersinia pseudotuberculosis*-like bacteria (*Bacterium enterocoliticum* and *Pasteurella* X). Proc. XIV Scand Cong Path Microbiol Oslo, pp. 103–104.

Freeman, V. J. and G. H. Minzel. 1950 Serological studies of *C. diphtheriae.* II. Antigenic relationship between small colony types of *C. diphtheriae* and *C. diphtheriae*-like bacilli as determined by the method of slide agglutination. Amer J Hyg *51:* 305–309.

Freund, F. 1922 Ueber eine durch ein anaërobes Bakterium hervorgerufene Meningitis. Zentralbl Bakteriol Parasitenk Infektionskr Hyg Abt I Orig *88:* 9–24.

Freudenreich, E. von and S. Orla-Jensen. 1907 Ueber die im Emmentalerkäse stattfindende Propionsäuregarung. Zentralbl Bakteriol Parasitenk Infektionskr Hyg Abt II *17:* 529–546.

Freundt, E. A. 1952 Morphological studies of the peripneumonia organisms (*Micromyces peripneumoniae bovis*). Acta Pathol Microbiol Scand *31:* 508–529.

—. 1953 The occurrence of *Micromyces* (pleuropneumonia-like organisms) in the female genito-urinary tract. Acta Pathol Microbiol Scand *32:* 468–480.

—. 1954 Morphological and biochemical investigations of human pleuropneumonia-like organisms (*Micromyces*). Acta Pathol Microbiol Scand *34:* 127–144.

—. 1955 The classification of the pleuropneumonia group of organisms (*Borrelomycetales*). Int Bull Bacteriol Nomencl Taxon *5 (2):* 67–78.

—. 1956 *Streptobacillus moniliformis* infection in mice. Acta Pathol Microbiol Scand *38:* 231–245.

—. 1957 Genus V. *Streptobacillus* Levaditi et al. 1925. *In* Breed, Murray and Smith (Editors), Bergey's Manual of Determinative Bacteriology, 7th ed. The Williams & Wilkins Co., Baltimore, pp. 451–454.

—. 1958 The Mycoplasmataceae (The Pleuropneumonia group of Organisms) Morphology, Biology and Taxonomy. Munksgaard, Copenhagen.

—. 1971 Mykoplasmer. In Medicinsk årbog XIV. Munksgaard, Copenhagen.

Fried, J., D. Perlman, A. F. Langlykke and E. O. Titus. 1958 16-α-hydroxylation steroids. United States Patent 2,855,343, October 7.

Friedberger, E. 1902/03 Über ein neues zur Gruppe des Influenzabacillus gehoriges hämoglobinophiles Bakterium (Bacillus haemoglobinophilus-canis). Zentrabl Bakteriol Parasitenk Infektionskr Hyg Abt I Orig 33: 401–406.

Friedhoff, K. T. and M. Ristic. 1966 Anaplasmosis XIX. A preliminary study of Anaplasma marginale in Dermacentor andersoni (Stiles) by fluorescent antibody technique. Amer J Vet Res 27: 643–646.

Friedman, B. A., P. R. Dugan, R. M. Pfister and C. C. Remsen. 1968 Fine structure and composition of the zoogloeal matrix surrounding Zoogloea ramigera. J Bacteriol 96 (6): 2144–2153.

—, —, — and —. 1969 Structure of exocellular polymers and their relationship to bacterial flocculation. J Bacteriol 98 (3): 1328–1334.

Friis, N. F. 1970 A new porcine Mycoplasma species: Mycoplasma suidaniae. Acta Vet Scand 11: 487–490.

—. 1972 Isolation and characterization of a new porcine mycoplasma. Acta Vet Scand 13: 284–286.

Frobisher, M., L. Adams and W. J. Kuhns. 1945 Characteristics of diphtheria bacilli found in Baltimore since November, 1942. Proc Soc Exp Biol Med 58: 330–334.

Frobisher, M., Jr. and J. H. Brown. 1927 Transmissible toxicogenicity of streptococci. Bull Johns Hopkins Hosp 41: 167.

Frohardt, R. P., T. H. Haskell, J. Ehrlich and M. P. Knudsen. 1959 Antibiotic and methods for obtaining same. United States Patent 2,916,485, December 8.

Fromme, F. D. and T. J. Murray. 1919 Angular-leafspot of tobacco, an undescribed bacterial disease. J Agr Res 16: 219–228.

Frommer, W. 1959 Zur Systematik der Actinomycin bildenden Streptomyceten. Arch Mikrobiol 32: 187–206.

Frosch, P. 1923 Die Morphologie des Lungenseucheerregers (Eine mikrophotographische Studie). Arch Wiss Prakt Tierheilk 49: 35–48.

—. 1923 Zur Morphologie des Lungenseucheerregers. II. Mitteilung. Arch Wiss Prakt Tierheilk 49: 273–282.

— and W. Kolle. 1896 Die Mikrokokken. In C. Flügge. Die Mikroorganismen. 3rd ed. Verlag von Vogel, Leipzig, p. 154.

Frost, W. D. and M. A. Engelbrecht. 1936 A revision of the genus Streptococcus. Dept Agr Bacteriol, Univ Wisconsin, Madison, pp. 1–4.

— and —. 1940 The streptococci, their descriptions, classification and distribution, with special reference to those in milk. Willdof Book Co., Madison, Wisconsin, pp. 1–172.

Fuckel, L. 1869–1870 Symbolae Mycologicae. Julius Niedner, Wiesbaden.

Fügner, R., and G. Bradler. 1963 Studien an Polyenantibiotica bildenden Streptomyces-Stämmen. Z Allg Mikrobiol 3 (3): 173–194.

Fuhrmann, F. 1905 Morphologisch-biologische Untersuchungen ueber ein neues Essigsäure bildendes Bakterium. Beih Bot Zentrabl Abt 1 19: 1–33.

—. 1906 Zur Kenntnis der Bakterienflora des Flaschenbieres. I. Pseudomonas cerevisiae. Zentrabl Bakteriol Parasitenk Infektionskr Hyg Abt 2 16: 309–325.

—. 1907 Zur Kenntnis der Bakterienflora des Flaschenbieres. Zentrabl Bakteriol Parasitenk Infektionskr Hyg Abt 2 17: 453–467.

—. 1913 Vorlesungen über technische Mykologie. G. Fischer, Jena, pp. 1–454.

Fujii, S., H. Hitomi, M. Imanishi and K. Nakazawa. 1955. Studies on Streptomyces. Monilin, an antibiotic active against Candida albicans (in Japanese). Annu Rep Takeda Res Lab 14: 8–10.

Fujino, T., Y. Okuno, D. Nakada, A. Aoyama, K. Fukai, T. Mukai and T. Ueho. 1951 On the bacteriological examination of shirasu food poisoning (in Japanese). J Jap Ass Infect Dis 25: 11.

Fukumoto, J. 1943 Studies on the production of bacterial amylase. 1. Isolation of bacteria secreting potent amylases and their distribution (in Japanese). J Agr Chem Soc Jap 19: 487–503.

Fukunaga, K., T. Misato, I. Ishii and M. Asakawa. 1955 Blasticidin, a new anti-phytopathogenic fungal substance. Part I. Bull Agr Chem Soc Jap 19 (3): 181–188.

Fuller, R. 1966 Some morphologic characteristics of gram-negative anaerobic bacteria isolated from the alimentary tract of the pig. J Appl Bacteriol 29: 375–379.

— and L. G. M. Newland. 1963 The serological grouping of three strains of Streptococcus equinus. J Gen Microbiol 31 (3): 431–434.

Fuller, W. H. and A. G. Norman. 1943 Cellulose decomposition by aerobic mesophilic bacteria from soil. I. Isolation and description of organisms. J Bacteriol 46: 273–280.

— and —. 1943 Characteristics of some soil cytophagas. J Bacteriol 46: 565–572.

Fulton, MacD. 1943 The identity of Bacterium columbensis Castellani. J Bacteriol 46: 79–81.

Funaki, M. and F. Tsuchiya. 1958 On fradiomycin producing sp. Streptomyces rokugoensis (in Japanese). Jap J Pharm Chem 29: 1097–1098.

—, —, K. Maeda and T. Kamiya. 1958 Cyanomycin, a new antibiotic. J Antibiot (Tokyo) Ser A 11: 143–149.

Furness, G. 1968 Analysis of the growth cycle of Mycoplasma orale by synchronized division and by ultraviolet radiation. J Infect Dis 118: 436–442.

—. 1970 The growth and morphology of mycoplasmas replicating in synchrony. J Infect Dis 122: 146–158.

—, F. J. Pipes and M. J. McMurtrey. 1968 Analysis of the life-cycle of Mycoplasma pneumoniae by synchronized division and by ultraviolet and X irradiations. J Infect Dis 118: 7–13.

Furumai, T., H. Ogawa, and T. Okuda. 1968 Taxonomic study on Streptosporangium albidum Nov. sp. J Antibiot (Tokyo) 21: 179–181.

Furushiro, K., K. Shimizu, H. Sakai, M. Minogato and T. Fujisawa. 1958 Hygrostatin, a new antibiotic substance (in Japanese). Chem Abstr *54:* 10048.

Fusari, S. A., R. P. Frohardt, A. Ryder, T. H. Haskell, D. W. Johannessen, C. C. Elder and Q. R. Bartz. 1954 Azaserine, a new tumor-inhibitory substance. Isolation and characterization. J Amer Chem Soc *76:* 2878–2881.

—, T. H. Haskell, R. P. Frohardt and Q. R. Bartz. 1954 Azaserine, a new tumor-inhibitory substance. Structural studies. J Amer Chem Soc *76:* 2881–2883.

Gaertner, E. 1888 Ueber die Fleischuergiftung in Frankenhausen a. Kyffh. und der Erreger derselben. Korresp Allg Ärtzl Ver Thürigen *17:* 573–600.

Galarneault, T. P. and E. Leifson. 1956 Taxonomy of *Lophomonas* n. gen. Can J Microbiol *2:* 102–110.

— and —. 1964 *Pseudomonas vesiculare* (Busing et al.) nov comb. Int Bull Bacteriol Nomencl Taxon *14 (4):* 165–168.

Gallin, J. I. and E. R. Leadbetter. 1966 Morphogenesis of Sporocytophaga. Bacteriol Proc p. 75.

Gallo, P., N. Ivanov and H. Morris. 1961–62 Estudios Microbiologicos sobre una *Neisseria* (*Neisseria suis* n. sp.). Aislada de un caso de muerte cardiaca de suino en Venezuela. Rev Med Vet (Maracay) *9:* 55–57.

Galmarini, O. L. and V. Deulofeu. 1961 Curamycin. I. Isolation and characterization of some hydrolysis products. Tetrahedron *15:* 76–86.

Galton, M. M., R. W. Menges, E. B. Shotts, A. J. Mahmias and C. W. Heath. 1962 Leptospirosis: Epidemiology, clinical manifestations in man and animals, and methods in laboratory diagnosis. U S Public Health Serv Publ No. 951. U S Govt Printing Office, Washington, D. C.

Gamaléia, M. N. 1888 Sur l'étiologie de la pneumonie fibrineuse chez l'homme. Ann Inst Pasteur (Paris) *2:* 440–459.

—. 1888 *Vibrio metschnikovi* (n. sp.) et ses rapports avec le microbe du choléra asiatique. Ann Inst Pasteur (Paris) *2:* 482–488.

Gambaryan, M. E. 1962 A new species of purple sulfur bacteria, *Thiopedia servani* (sp. n.) Mikrobiologiya *31:* 282–283.

Gamova-Kayukova, N. I. 1945 Selection of the acetic acid bacteria, (in Russian). Mikrobiologiya *14:* 338–346.

Gan, K. H. and S. K. Tjia. 1963 A new method for the differentiation of *Vibrio comma* and *Vibrio eltor.* Amer J Hyg *77:* 184–186.

Gardner, A. D. and E. Chain. 1942 Proactinomycin: a "bacteriostatic" produced by a species of *Proactinomyces.* Brit J Exp Pathol *23:* 123–127.

Gardner, G. A. 1969 Physiological and morphological characteristics of *Kurthia zopfii* isolated from meat products. J Appl Bacteriol *32:* 371–380.

Gardner, H. L. and C. D. Dukes. 1955 *Haemophilus vaginalis;* a newly defined specific infection, previously classified as 'nonspecific' vaginitis. Amer J Obstet Gynecol *69:* 962–976.

Gardner, I. C. 1958 Nitrogen fixation in *Elaeagnus* root nodules. Nature (London) *181:* 717–718.

— and G. Bond. 1957 Observations on the root nodules of *Shepherdia.* Can J Bot *35:* 305–314.

— and —. 1966 Host plant-endophyte adaptation in *Myrica.* Naturwissenschaften *53:* 161.

Gardner, M. W. and J. B. Kendrick. 1923 Bacterial spot of cowpea. Science (Washington) *57:* 275.

Garnham, P. C. C. 1947 A new blood spirochaete in the grivet monkey, *Cercopithecus aethiops.* East Afr Med J *24:* 47–51.

Garnjobst, L. 1945 *Cytophaga columnaris* (Davis) in pure culture: A myxobacterium pathogenic to fish. J Bacteriol *49:* 113–128.

Garrod, L. P. 1952 Actinomycosis of the lung. Aetiology, diagnosis and chemotherapy. Tubercle *33:* 258–266.

Garvie, E. I. 1960 The genus *Leuconostoc* and its nomenclature. J Dairy Res *27:* 283–292.

—. 1967 *Leuconostoc oenos* sp. nov. J Gen Microbiol *48 (3):* 431–438.

—. 1967 The growth factor and amino acid requirements of species of the genus *Leuconostoc* including *Leuconostoc paramesenteroides* sp. nov. and *Leuconostoc oenos.* J Gen Microbiol *48 (3):* 439–447.

—. 1967 The production of L(+)- and D(−)-lactic acid in cultures of some lactic acid bacteria, with a special study of *Lactobacillus acidophilus* NCDO2. J Dairy Res *34:* 31–38.

—. 1969 Lactic dehydrogenases of strains of the genus *Leuconostoc.* J Gen Microbiol *58 (1):* 85–94.

—. 1969 Request for an opinion that the name *Leuconostoc citrovorum* be rejected and the name *Leuconostoc cremoris* be conserved. Int J Syst Bacteriol *19 (3):* 283–290.

— and L. A. Mabbitt. 1967 Stimulation of the growth of *Leuconostoc oenos* by tomato juice. Arch Mikrobiol *55:* 398–407.

Gasdorf, H. J., R. G. Benedict, M. C. Cadmus, R. F. Anderson and R. W. Jackson. 1965 Polymer-producing species of *Arthrobacter.* J Bacteriol *90 (1):* 147–150.

Gasperini, G. 1891 Sopra una nouva specie appartenente al gen. *Streptothrix* Cohn. P V Soc Tosc Sci Nat (Pisa) *7:* 267–277.

—. 1892 Ricerche morfologiche e biologiche sul genere *Actinomyces*-Harz come contributo allo studio delle relative Micosi. Ann Ist Igiene Sper Univ Roma *2:* 167–231.

—. 1894 Ulteriori ricerche sul genere *Actinomyces.* P V Soc Tosc Sci Nat (Pisa) *9:* 64–89.

—. 1894 Versuche über das Genus '*Actinomyces.*' Zentrabl Bakteriol Parasitenk Infektionskr Hyg Abt I Orig *15:* 684–686.

—. 1899 Sulla cosi detta *Crenothrix kuhniana o polyspora* in rapporto alla sorveglianza igienica delle acque potabili. Ann Ist Igiene Sper Univ Roma *9:* 1–102.

Gasser, F. 1970 Electrophoretic characterization of lactic dehydrogenases in the genus *Lactobacillus.* J Gen Microbiol *62 (2):* 223–239.

— and M. Mandel. 1968 Deoxyribonucleic acid base composition of the genus *Lactobacillus.* J Bacteriol *96:* 580–588.

—, — and M. Rogosa. 1970 *Lactobacillus jensenii* sp. nov., a new representative of the subgenus *Thermobacterium.* J Gen Microbiol *62 (2):* 219–222.

Gaston, L. W. and E. R. Stadtman. 1963 Fermentation of ethylene glycol by *Clostridium glycolicum.* J Bacteriol *85 (2):* 356–362.

Gaumann, E. 1921 Over een bacterieele

vaatbundelziek te der bananen in Nederlandsch-Indie. Meded Inst. PlZiekt, Buitenz.

—. 1923 Ueber zwei Bananenkrankheiten in Niederländisch Indien. Z Pflanzenkr Pflanzenpathol Pflanzenschutz *33:* 1–17.

Gauze, G., T. Makimova, L. Popova, M. Brazhnikova, T. Uspenskaya and O. Rossilimo. 1959 Mutomycin, A new antibiotic produced by *Actinomyces atroolivaceus*. Antibiotiki *4* (*3*): 20–23.

—, T. P. Preobrazhenskaya, V. K. Kovalenkova, N. P. Il'yicheva, M. G. Brazhinkova, N. N. Lomakina, I. N. Kovsharova, V. A. Shorin, I. A. Kunrat and S. P. Shapovalova. 1957 Crystallomycin, a new antibiotic. Antibiotiki *2* (*6*): 9–14.

—, —, E. S. Kudrina, N. O. Blinov, I. D. Ryabova and M. A. Sveshnikova. 1957 Problems in the classification of antagonistic Actinomycetes. State Publishing House of Medical Literature, Medzig, Moscow, USSR.

Gedoelst, L. 1902 Les champignons parasites de l'homme and des animaux domestiques. Joseph van In and Co., Lierre, Belgium.

Geilinger, H. 1921 Experimentalle Beitrage zur Mikrobiologie der Getreidamehle. I. Ueber koliartige Mehlbakterien. Mitt Lebensm Hyg, Bern *12:* 49–81; 105–119.

Geitler, L. 1924 Über *Polyangium parasiticum* n. sp., eine submerse parasitische *Myxobacteriaceae*. Arch Protistenk *50:* 67–88.

— and A. Pascher. 1925 *Cyanochloridinae-Chlorobacteriaceae*, 451–463. *In* Pascher, Die Süsswasser-Flora Deutschlands, Österreichs und der Schweiz, G. Fischer, Jena.

Gemmell, M. and W. Hodgkiss. 1964 The physiological characters and flagellar arrangement of homofermentative lactobacilli. J Gen Microbiol *35* (*3*): 519–526.

Genig, V. A. 1968 A live vaccine 1/M-44 against Q-fever for oral use. J Hyg Epidemiol Microbiol Immunol (Prague) *12:* 265–273.

Georg, L. K. and J. M. Brown. 1967 *Rothia*, gen. nov., an aerobic genus of the family *Actinomycetaceae*. Int J Syst Bacteriol *17* (*1*): 79–88.

— and M. R. Coleman. 1970 Comparative pathogenicity of various *Actinomyces* species. *In* H. Prauser (Editor), The *Actinomycetales*. The Jena International Symposium on Taxonomy, September 1968, pp. 35–45.

—, L. Pine and M. A. Gerencser. 1969 *Actinomyces viscosus* comb. nov. A catalase positive, facultative member of the genus *Actinomyces*. Int J Syst Bacteriol *19* (*3*): 291–293.

—, G. W. Robertstad and S. A. Brinkman. 1964 Identification of species of *Actinomyces*. J Bacteriol *88* (*2*): 477–490.

—, —, — and M. D. Hicklin. 1965 A new pathogenic anaerobic *Actinomyces* species. J Infect Dis *115:* 88–99.

Gerber, N. N. and M. P. Lechevalier. 1964 Phenazines and phenoxazinones from *Waksmania aerata* sp. nov. and *Pseudomonas iodina*. Biochemistry *3* (*3*): 598–602.

Gerber, P. 1910 Ueber Spirochäten in den oberen Luft und Verdauungswegen. Zentrabl Bakteriol Parasitenk Infektionskr Hyg Abt I Orig *56:* 508–521.

Gerencser, M. A. and J. M. Slack. 1967 Isolation and characterization of *Actinomyces propionicus*. J Bacteriol *94* (*1*): 109–115.

— and —. 1969 The identification of human strains of *Actinomyces viscosus*. Appl Microbiol *18* (*1*): 80–87.

Gerhardt, P. 1958 The nutrition of brucellae. Bacteriol Rev *22:* 81–98.

Getzel, D. 1941 Forme microbiche del sangue. Boll Zool Anno *12:* 171–181.

Ghon, A. and M. Sachs. 1905 Beiträge zur Kenntnis der anaeroben Bakterien des Menschen. III. Aetiologie der Peritonitis. Zentrabl Bacteriol Parasitenk Infektionskr Hyg Abt I Orig *38:* 1–136.

Ghuysen, J. M. 1968 Use of bacteriolytic enzymes in determination of wall structure and their role in cell metabolism. Bacteriol Rev *32* (*4*): 425–464.

Gibbons, N. E. 1957 Effect of salt concentration on the biochemical reactions of some halophilic bacteria. Can J Microbiol *3:* 249–255.

—. 1969 Isolation, growth and requirements of halophilic bacteria. *In* J. R. Norris and D. W. Ribbons (editors), Methods in Microbiology, Vol. 3B. Academic Press, New York, pp. 169–183.

Gibbs, S. P., W. R. Sistrom and P. B. Worden. 1965 The photosynthetic apparatus of *Rhodospirillum molischianum*. J Cell Biol *26:* 395–412.

Gibson, A. G. 1920 A new pathogenic form of *Streptothrix*. J Pathol Bacteriol *23:* 357–358.

Gibson, T. 1934 An investigation of the *Bacillus pasteuri* group. II. Special physiology of the organisms. J Bacteriol *28:* 313–322.

—. 1935 The urea-decomposing microflora of soils. I. Description and classification of the organisms. Zentrabl Bakteriol Parasitenk Infektionskr Hyg Abt II *92:* 364–380.

—. 1935 An investigation of the *Bacillus pasteuri* group. III. Systematic relationships of the group. J Bacteriol *29:* 491–502.

Gicklhorn, J. 1920 Über neue farblose Schwefelbakterien. Zentrabl Bakteriol Parasitenk Infektionskr Hyg Abt II *50:* 415–427.

—. 1921 Zur Morphologie und Mikrochemie einer neuen Gruppe der Purpurbakterien. Ber Deut Bot Ges *39:* 312–319.

Giesberger, G. 1947 Some observations on the culture, physiology and morphology of some brown-red *Rhodospirillum*-species. Antonie van Leeuwenhoek J Microbiol Serol *13:* 135–148.

Giesbrecht, P. and G. Drews. 1962 Elektromikroskopische Untersuchungen über die Entwicklung der "Chromatophoren" von *Rhodospirillum molischianum*. Arch Mikrobiol *43:* 152–161.

— and —. 1966 Über die Organisation und die makromolekulare Architektur der Thylakoide "lebender Bakterien." Arch Mikrobiol *54:* 297–330.

Gieszczykiewicz, M. 1939 Zagadniene systematihki w bakteriologii—Zur Frage der Bakterien-Systematic. Bull Acad Pol Sci Ser Sci Biol *1:* 9–27.

Gietzen, J. 1931 Untersuchungen über marine Thiorhodaceen. Zentrabl Bakteriol Parasitenk Infektionskr Hyg Abt II *83:* 183–218.

Gilardi, G. L., E. Bottone and M. Birnbaum. 1970 Unusual fermentative, gram-negative bacilli isolated from clinical specimens. I. Characterization of *Erwinia* strains of the "lathyri-herbicola group." Appl Microbiol *20* (*1*): 151–155.

Gilbert. 1904 Ueber *Actinomyces thermophilus* und andere Aktinomyceten. Z Hyg Infektionskr *47:* 383–406.

Gilbert, R. and F. C. Stewart. 1927 *Corynebacterium ulcerans;* a pathogenic microorganism resembling *C. diphtheriae.* J Lab Clin Med *12:* 756–761.

Gilbert, S. J. 1930 An unusual strain of *Brucella* causing abortion in cattle in Palestine. J Comp Pathol *43:* 118–124.

Gilchrist, T. C. 1901 A bacteriological and microscopical study of over three hundred vesicular and pustular lesions of the skin, with a research upon the etiology of *Acne vulgaris.* Johns Hopkins Hosp Rep *9:* 409–430.

Gill, J. W. and K. W. King. 1958 Nutritional characteristics of a butyrivibrio. J Bacteriol *75:* 666–673.

Gillespie, D. C. 1963 Cell wall carbohydrates of *Arthrobacter globiformis.* Can J Microbiol *9:* 509–514.

—. 1963 Composition of cell wall mucopeptide from *Arthrobacter globiformis.* Can J Microbiol *9:* 515–521.

Gillespie, D. and S. Spiegelman. 1965 A quantitative assay for DNA-RNA hybrids with DNA immobilized on a membrane. J Mol Biol *12:* 829–842.

Gilman, J. P. 1953 Studies on certain species of bacteria assigned to the genus *Chromobacterium.* J Bacteriol *65* (*1*): 48–52.

Gilmour, M. N., A. Howell, Jr. and B. G. Bibby. 1961 The classification of organisms termed *Leptotrichia* (*Leptothrix*) *buccalis.* I. Review of the literature and proposed separation into *Leptotrichia buccalis* Trevisan 1879 and *Bacterionema* gen nov., *B. matruchotii* (Mendel 1919) comb. nov. Bacteriol Rev. *25* (*2*): 131–141.

Giménez, D. F. 1964 Staining rickettsiae in yolk-sac cultures. Stain Technol *39:* 135–140.

—. 1965 Gram staining of *Coxiella burnetii.* J Bacteriol *90* (*3*): 834–835.

Gimesi, N. 1924 Hydrobiologiai talmanyok (Hydrobiolgische Studien). I. *Planktomyces Bekefii* Gim. nov. gen. et sp. Budapest, Kiadja a Magyar Ciszterci. Rend, pp. 1–8.

Gimpl, F. and M. Lanyi. 1965 Use of the gel precipitation method for determining the type of mycobacteria and the clinical diagnosis. Bull Inst Union Tuberc *36:* 22–25.

— and J. Vándor. 1967 Antigenic structure of *Mycobacterium kansasii* and its variants. Acta Microbiol Acad Sci Hung *14:* 271–276.

Gioelli, P. 1907 Di un particulare cocco anaerobico obbligato riscontrato in raccolta purulenta di pelvicellulite. Boll Accad Med di genova *22:* 159–169.

Giolitti, G. 1958 Studies on two strains of *Streptomyces* producers of flavensomycin, an antibiotic with fungicidal and insecticidal activity. Abstracts, VII Intern Congr Microbiol Abstract 22k, p. 38.

— and R. Craveri. 1958 On flavensomycin and *Streptomyces cavourensis* (in Flemish). Belgian Patent 560,930, March 18.

Giovannoni, M. 1946 Sobre a coloracao vital de *Bartonella muris* Mayer 1921. Arq Biol Tecnol Inst Biol Pesquisas Tecnol *1:* 181–194.

Giovannozzi-Sermanni, G. 1959 Una nuova species di *Arthrobacter* determinante la degradazione della nicotina: *Arthrobacter nicotinae:* Tabacco (Rome) *63:* 83–86.

Giroud, P., N. Dumas and B. Hurpin. 1958 Essais d'adaptation à la souris blanche de la rickettsie agent de la maladie bleue de *Melolontha melolontha* L.: voie pulmonaire et voie buccale. C R Acad Sci Paris *247:* 2499–2501.

— and R. Martin. 1946 Pseudo-rickettsies de la conjonctive du lapin. Bull Soc Pathol Exot *39:* 264–266.

Glage, F. 1903 Über den *Bazillus pyogenes suis* Grips, den *Bazillus pyogenes bovis* Kunneman und den bakteriologischen Befund bei den chronischen abszedierenden Euterenzundungen der Milchkuhe. Z Fleish-u Milchhyg *13:* 166–175.

Gledhill, W. E. and L. E. Casida. 1969 Predominant catalase-negative soil bacteria. II. Occurrence and characterization of *Actinomyces humiferus*, sp. n. Appl Microbiol *18* (*1*): 114–121.

Gochnauer, M. B. and D. J. Kushner. 1969 Growth and nutrition of extremely halophilic bacteria. Can J Microbiol *15:* 1157–1165.

Gochnauer, T. A. and J. B. Wilson. 1951 The production of hyaluronidase by Lancefield's group B streptococci. J Bacteriol *62* (*4*): 405–414.

Goddard, J. L. and J. R. Sokatch. 1964 2-Ketogluconate fermentation by *Streptococcus faecalis.* J Bacteriol *87* (*4*): 844–851.

Gold, W., H. A. Stout, J. F. Pagano and R. Donovick. 1956 Amphotericins A and B, antifungal antibiotics produced by a streptomycete. Antibiot Ann *1955-1956:* 579–585.

Goldberg, H. S., Ella M. Barnes and A. B. Charles. 1964 Unusual *Bacteroides*-like organism. J Bacteriol *87* (*3*): 737–742.

Gomes, J. M. 1923 Nocardiose de localiza cao Rava. Ann Paulist Med Cir *14:* 150–156.

Gonçalves de Lima, O., J. O. Falcão de Morais, and C. L. Carmona. 1955 Nova especia de genero *Acetobacter.* Ann Soc Biol Pernambuco *13:* 13–17.

—, F. D. Monache, I. L. d'Albuquerque and G. B. Marini Bettòlo. 1968 The identification of ciclacidine, an antibiotic from *Streptomyces capoamus* sp. nov. Tetrahedron Lett *4:* 471–473.

Gonçalves de Lima, V. O., C. A. Albert and O. Gonçalves de Lima. 1964 *Streptomyces capoamus* nov. sp., produtor da ciclamicina e das ciclacidinas A e B. An Acad Brasil Ciênc *36* (*3*): 317–322.

Gonder, R. 1908 Spirochäten aus dem Darmtraktus von Pinna: *Spirochaete pinnae* nov. spec. und *Spirochaete hartmanni* nov. spec. Zentrabl Bakteriol Parasitenk Infektionskr Hyg Abt I Orig *47:* 491–494.

—. 1912 Spirochaetenstudien. Zool Jahrb Suppl XV *1:* 485–514.

González, C. 1956 Estudios morfologico, fisiologico y bioquimico de una especie nueva de *Clostridium: Cl. albuminolyticum* nov sp. Microbiol Espan *9:* 327–332.

Goodfellow, M., A. Fleming and M. J. Sackin. 1972 Numerical classification of "*Mycobacterium*" *rhodochrous* and Runyon's Group IV Mycobacteria. Int J Syst Bacteriol *22:* 81–98.

Goodman, Y. E. 1972 Bacterial identification:

Clues from computer and chemistry. Can J Med Technol *34:* 2-16.

Goodsir, J. 1842 History of a case in which a fluid periodically ejected from the stomach contained vegetable organisms of an undescribed form. Edinburgh Med Surg J *57:* 430-443.

Goodwin, R. F. W., A. P. Pomeroy and P. Whittlestone. 1965 Production of enzootic pneumonia in pigs with a mycoplasma. Vet Rec *77:* 1247-1249.

Gordon, M. A. 1964 The genus *Dermatophilus.* J Bacteriol *88:* 509-522.

— and E. W. Lapa. 1966 Durhamycin, a pentaene antifungal antibiotic from *Streptomyces durhamensis* sp. n. Appl Microbiol *14* (*5*): 754-760.

Gordon, R. E. 1966 Some strains in search of a genus—*Corynebacterium, Mycobacterium, Nocardia,* or what? J Gen Microbiol *43* (*3*): 329-343.

—. 1968 The taxonomy of soil bacteria. *In* T. R. G. Gray and D. Parkinson (editors), The Ecology of Soil Bacteria. An International Symposium. University of Toronto Press, Toronto, Canada, pp. 293-321.

—, W. C. Haynes and C. H-N. Pang. 1973 The genus *Bacillus.* U S Dept Agr Handbook No. 427.

— and A. C. Horan. 1968 A piecemeal description of *Streptomyces griseus* (Krainsky) Waksman and Henrici. J Gen Microbiol *50:* 223-233.

— and J. M. Mihm. 1957 A comparative study of some strains received as *Nocardiae.* J Bacteriol *73* (*1*): 15-27.

— and —. 1959 A comparison of four species of Mycobacteria. J Gen Microbiol *21:* 736-748.

— and M. M. Smith. 1953 Rapidly growing acid fast bacteria. I. Species description of *Mycobacterium phlei* Lehmann and Neumann and *Mycobacterium smegmatis* (Trevisan) Lehmann and Neumann. J Bacteriol *66:* 41-48.

— and —. 1955 Rapidly growing, acid fast bacteria. II. Species description of *Mycobacterium fortuitum* Cruz. J Bacteriol *69:* 502-507.

Goresline, H. 1933 Studies on agar-digesting bacteria. J Bacteriol *26:* 435-457.

Goret, P. and L. Joubert. 1951 Sur une nouvelle espèce de *Streptomyces* (*Streptomyces galtieri* n. sp.). Isolée d'un cas d'actinomycose septicémique chez le chien. Ann Parasitol Hum Comp *26:* 118-127.

Gorini, C. 1894 Studi critico-sperimentali sulla sterilizazzione del latte. G Soc Ital Igiene *16:* 5-24.

Gorlenko, V. M. 1968 A new species of green sulfur bacteria. Dokl Akad Nauk SSSR *179:* 1229-1231.

—. 1969 Sporification in the budding photoheterotrophic bacteria. Mikrobiologiya *38:* 126-134.

—. 1970 A new phototrophic green sulfur bacterium, *Prosthecochloris aestuarii* nov gen., nov. spec. Z Allg Mikrobiol *10* (*2*): 147-149.

Goslings, W. R. O. and E. P. Snijders. 1936 Untersuchungen über das Scleroma respiratorium (Sklerom). IV. Mitteilung. Die antigene Struktur der Skleromstämme im Vergleich mit den anderen Kapselbakterien. Zentrabl Bak-

teriol Parasitenk Infektionskr Hyg Abt Orig I *13:* 161-173.

Gothe, R. 1967 Zur Entwicklung von *Aegyptianella pullorum* Carpano 1928, in der Lederzecke *Argas* (*Persicargas*) *persicus* (Oken 1818) und Übertragung. Z Parasitenk *29:* 103-118.

—. 1967 Ein Beitrag zur systematischen Stellung von *Aegyptianella pullorum* Carpano 1928. Z Parasitenk *29:* 119-129.

—. 1967 Untersuchungen über die Entwicklung und den infektions-verlauf von *Aegyptianella pullorum* Carpano 1928, in Huhn. Z Parasitenk *29:* 149-158.

Goto, M. and N. Okabe. 1958 Bacterial plant diseases in Japan, IX. 1. Bacterial stem rot of pea. 2. Halo blight of bean. 3. Bacterial spot of physalis plant. Rep Fac Sci Shizuoka Univ *8:* 33-49.

Goto, T., S. Nishio, H. Kojima, S. Hayakawa and H. Araki. 1967 Process for producing L-glutamic acid by using *Corynebacterium melassecola* (United States Patent 3,355,359). Offic Gaz U S Patent Off *844:* 1456.

Gotschlich, F. 1906 Ueber cholera- und choleraähnliche Vibrionen unter den aus Mekka züruckkehrenden Pilgern. Z Hyg Infektionskr *53:* 281-304.

Gottheil, O. 1901 Botanische Beschreibung einiger Bodenbakterien. Zentrabl Bakteriol Parasitenk Infektionskr Hyg Abt II *7:* 430-435; 449-465; 481-497; 529-544; 582-591; 627-637, 680-691; 717-730.

Gottlieb, D. (as chairman International Subcommittee on Taxonomy of the Actinomycetes). 1964 Recommendations for descriptions of some *Actinocetales* appearing in patent applications. Amer Soc Microbiol News *30* (*2*): 13-14.

—. 1968 Designation of type strains of 47 species of *Actinomyces* (*Streptomyces*). Int J Syst Bacteriol *18* (*1*): 19-20.

—, A. Ammann and H. E. Carter. 1955 A new antifungal agent, filipin. Plant Dis Rep *39* (*3*): 219.

—, P. K. Bhattacharyya, H. W. Anderson and H. E. Carter. 1948 Some properties of an antibiotic obtained from a species of *Streptomyces.* J Bacteriol *55* (*3*): 409-417.

—, —, H. E. Carter and H. W. Anderson. 1951 Endomycin, a new antibiotic. Phytopathology *41* (*5*): 393-400.

Gourevitch, A. and J. Lein. 1955 Production of tetracycline and substituted tetracyclines. United States Patent 2,712,517, July 5.

Gourlay, R. N. and R. H. Leach. 1970 A new *Mycoplasma* species isolated from pneumonic lungs of calves (*Mycoplasma dispar* sp. nov.). J Med Microbiol *3:* 111-123.

Grace, J. T., J. S. Horoszewicz, T. B. Stim, E. A. Mirand and C. James. 1965 Mycoplasmas (PPLO) and human leukemia and lymphoma. Cancer *18:* 1369-1376.

Gräf, H. and W. Wittneben. 1907 Zwei durch anaërobes Wachstum ausgezeichnete Streptokokken. Zentrabl Bakteriol Parasitenk Infektionskr Hyg Abt I Orig *44:* 97-110.

Gräf, W. 1961 Anaerobe Myxobakterien, neue Mikroben in der menschlichen Mundhöhle. Arch Hyg Bakteriol *145:* 405-459.

—. 1962 Über Wassermyxobakterien. Arch Hyg Bakteriol *146:* 114-125.

Graham, D. C. and W. J. Dowson. 1960 The

coliform bacteria associated with potato blackleg and other soft rots. I. Their pathogenicity in relation to temperature. Ann Appl Biol *48:* 51–57.

— and W. Hodgkiss. 1967 Identity of gram negative, yellow pigmented, fermentative bacteria isolated from plants and animals. J Appl Bacteriol *30:* 175–189.

— and E. C. Quinn. 1974 Identification of *Agrobacterium gypsophilae* (Brown) Starr and Weiss as *Erwinia herbicola* (Geilinger) Dye. Int J Syst Bacteriol *24:* 238–241.

Graham, P. H. and C. A. Parker. 1964 Diagnostic features in the characterisation of the root-nodule bacteria of legumes. Plant Soil *20:* 383–396.

Graham-Smith, G. S. 1905 A new form of parasite found in the red blood corpuscles of moles. J Hyg *5:* 453–459.

Gram, E., C. A. Jorgensen and S. Rostrup. 1929 Oversight over sygdomme hos landburgets og haveburgets kulturplanter i 1927. Zentrabl Bakteriol Parasitenk Infektionskr Hyg Abt II *78:* 463–465.

Grassé, P. P. 1924 Notes protistologiques. I. La sporulation des Oscillospiracées. II. Le genre *Alysiella* Langeron 1923. Arch Zool Exp Gen Notes Rev *62:* 25–34.

Grässer, R. 1957 Vergleichende Untersuchungen an Actinomyceten von Mensch, Rind und Schwein. Thesis. Leipzig.

—. 1962 Mikroaerophile Actinomyceten aus Gesäugeaktinomykosen des Schweines. Zentrabl Bakteriol Parasitenk Infektionskr Hyg Abt I Orig *184:* 478–492.

—. 1963 Untersuchungen über fermentative und serologische Eigenschaften mikroaerophiler Actinomyceten. Zentrabl Bakteriol Infektionskr Hyg Abt I Orig *188:* 251–263.

Gray, P. H. H. 1928 The formation of indigotin from indol by soil bacteria. Proc Roy Soc Ser B *102:* 263–280.

— and C. H. Chalmers. 1924 On the stimulating action of certain organic compounds on cellulose decomposition by means of a new aerobic microorganism that attacks both cellulose and agar. Ann Appl Biol *11:* 324–338.

— and H. G. Thornton. 1928 Soil bacteria that decompose certain aromatic compounds. Zentrabl Bakteriol Parasitenk Infektionskr Hyg Abt II *73:* 74–96.

Greco, N. V. 1916 Groupe des mycetomes (pied de madura) LXXI-LXXII. Origine Des Tumeurs, pp. 722–725.

Green, S. G., H. S. Goldberg and D. C. Blenden. 1967 Enzyme patterns in the study of *Leptospira*. Appl Microbiol *15* (5): 1104–1113.

Greer, F. E. 1928 The sanitary significance of lactose-fermenting bacteria not belonging to the *B. coli* group. I. Groups reported in the literature and isolated from water in Chicago. J Infect Dis *42:* 501–513.

Grein, A. and C. Spalla. 1962 Studio sui coremi formati in culture di *Streptomyces peucetius*. G Microbiol *10:* 175–184.

—, — and E. Cotta. 1965 *Streptomyces fradiae* var. *italicus:* a new microorganism which produces the antibiotic aminosidine. G Microbiol *13* (4): 299–305.

—, —, A. Di Marco and G. Canevazzi. 1963 Descrizione e classification di un attinomicete (*Streptomyces peucetius* sp. nova) produttore di una sostanza ad attività antitumorale: la daunomicina. G Microbiol *11:* 109–118.

Grenamen, J. and M. de Kam. 1970 *Erwinia solicis* as the cause of dieback in *Solix alba* in the Netherlands and its identity with *Pseudomonas saliciperda*. Neth J Plant Pathol *76:* 249–252.

Griffin, F. L. 1911 A bacterial gummosis of cherries. Science *34:* 615–616.

Griffin, P. J., S. F. Snieszko and S. B. Friddle. 1953 A more comprehensive description of *Bacterium salmonicida*. Trans Amer Fish Soc *82:* 129–138.

Griffith, F. 1934 The serological classification of *Streptococcus pyogenes*. J Hyg *34:* 542–584.

Griffith, J. W. 1853 On *Gallionella ferruginea* (Ehrenb). Natur Hist Mag *12* (*2*): 438.

Grigoroff, S. 1905 Étude sur un lait fermenté comestible. Le "Kisselomléko" de Bulgarie. Rev Méd Suisse Rom *25:* 714–721.

Grimes, M. 1927 An aerobic capsulated bacterium chromogenic on sugar media. Zentrabl Bakteriol Parasitenk Infektionskr Hyg Abt II *72:* 367–368.

—. 1961 Classification of the *Klebsiella-Aerobacter* group with special reference to the cold-tolerant mesophilic *Aerobacter* types. Int Bull Bacteriol Nomencl Taxon *11* (4): 111–129.

Grimme, A. 1902 Die wichtigsten Methoden der Bakterienfärbung in ihrer Wirkung auf die Membran, den Protoplasten und die Einschlüsse der Bakterienzelle. Zentrabl Bakteriol Parasitenk Infektionskr Abt I Orig *32:* 1–16; 81–90; 161–180; 241–255; 321–327.

Grimstone, A. V. 1963 A note on the fine structure of a spirochaete. Quart J Microscp Sci *104:* 145–153.

Grohmann, G. 1924 Zur Kenntnis Wasserstoff oxydierender Bakterien. Zentrabl Bakteriol Parasitenk Infektionskr Hyg Abt II *61:* 256–271.

Gromov, B. V. 1963 A new bacterium of the genus *Microcyclus*. Dokl Akad Nauk SSSR *152:* 733–734.

Groot, H. 1942 *Haemobartonella tyzzeri* in Colombia. Proc Soc Exp Biol Med *51:* 279.

Gross, J. 1910 *Cristispira* nov. gen. Ein Beitrag zur Spirachätenfrage. Mitt Zool Sta Neapel *20:* 41–93.

—. 1911 Über freilebende Spironemaceen. Mitt Zool Sta Neapel *20:* 188–203.

—. 1912 Ueber Systematik Struktur und Fortpflanzung der *Spironemaceae*. Zentrabl Bakteriol Parasitenk Infektionskr Hyg Abt I Orig *65:* 83–98.

—. 1912 Zur Nomenklatur der *Spirochaeta pallida* Schaud. u. Hoffm. Arch Protistenk *24:* 109–118.

Grover, A. A., L. H. Schmidt and J. Rehm. 1957 The pathogenicity of various atypical mycobacteria for the rhesus monkey. Abstr Annu Mtg Nat Tuberc Ass, p. 35.

Gruber, T. 1895 Die Arten der Gattung "Sarcina." Inaug. Diss. Univ. Basel. Emmendingen.

—. 1905 Ein weiterer Beitrag zur Aromabildung speziell zur Bildung des Erdbeergeruches in der Gruppe "Pseudomonas." *Pseudomonas fragariae* II. Zentrabl Bakteriol Parasitenk Infektionskr Hyg Abt II *14:* 122–123.

Grundy, W. E., A. L. Whitman, E. G. Rdzok, E. J.

Rdzok, M. E. Hanes and J. C. Sylvester. 1952 Actithiazic acid. I. Microbiological studies. Antibiot Chemother 2 (8): 399–408.

Grüter, W. 1933 Eine Pilzgeschwulst (Aktinomykose) im oberen Tränenröhrchen (Actinomyces discofoliatus). Z Augenheilk 79: 477–510.

Guélin, A., P. Lépine, D. Lamblin and J. Sisman. 1968 Isolement d'un parasit bactérien actif sur les germes Gram-positifs a partir d'echantillons d'eau polluée. C R Acad Sci Paris 266: 2508–2509.

Guignard, L. and C. Sauvageau. 1894 Sur un nouveau microbe chromogène, le Bacillus chloroaphis. C R Soc Biol Paris Ser 10 1: 841–843.

Guillebeau, A. 1890 Studien über Milchfehler und Euterentzundungen bei Rindern und Ziegen. I. Über Ursachen der Euterentzundung. Landwirt Jahrb Schweiz 4: 27–44.

Guillemot, L., J. Halli and E. Rist. 1904 Recherches bactériologiques et expérimentales sur les pleurésies putrides. Arch Med Exp 16: 571–640; 677–736.

Guirard, B. M. and E. E. Snell. 1964 Nutritional requirements of Lactobacillus 30a for growth and histidine decarboxylase production J Bacteriol 87: 370–376.

Guittonneau, G. 1925 Sur la transformation du soufre en sulfate par voie d'association microbienne. C R Acad Sci Paris 181: 261–262.

Gunnison, J. B., A. Larson and A. S. Lazarus. 1951 Rapid differentiation between Pasteurella pestis and Pasteurella pseudotuberculosis by action of bacteriophage. J Infect Dis 88: 254–255.

Gunsalus, J. C. and C. F. Niven, Jr. 1942 The effect of pH on the lactic acid fermentation. J Biol Chem 145: 131–136.

Gunther, H. L. and H. R. White. 1961 The cultural and physiological characters of the pediococci. J Gen Microbiol 26 (2): 185–197.

Gupta, K. C. and I. C. Chopra. 1963 Streptomyces katrae—a new species of Streptomyces isolated from soil. Indian J Microbiol 3 (1): 1–4.

— and —. 1963 A new whorl-forming species of Streptomyces. Hindustan Antibiot Bull 5: 110–112.

—, R. R. Sobti and I. C. Chopra. 1963 Actinomycin produced by a new species of Streptomyces. Hindustan Antibiot Bull 6 (1): 12–16.

Guss, M. L. and E. A. Delwiche. 1954 Streptococcus thermophilus. J Bacteriol 67 (6): 714–717.

Güssow, H. T. 1914 The systematic position of the common potato scab. Science (Washington) 39: 431–433.

Gustafson, K. L., A. V. Kraeger and E. M. K. Vaichulis. 1966 Chemical characteristics of Leptotrichia buccalis endotoxin. Nature (London) 212: 301–302.

Guthof, O. 1955 Über eine neue serologische Gruppe alphahämolytischen Streptokokken (seriologische Gruppe Q). Zentrabl Bakteriol Parasitenk Infektionskr Abt I Orig 164: 60–69.

Gutierrez, J., R. E. Davis, I. L. Lindahl and E. J. Warwick. 1959 Bacterial changes in the rumen during the onset of feed-lot bloat of cattle and characteristics of Peptostreptococcus elsdenii n. sp. Appl Microbiol 7: 16–22.

Haapala, D. K., M. Rogul, L. B. Evans and A. D. Alexander. 1969 Deoxyribonucleic acid base composition and homology studies of Leptospira. J Bacteriol 98 (2): 421–428.

Habs, H. and R. H. W. Schubert. 1962 Über die biochemische Merkmale und die taxonomische Stellung von Pseudomonas shigelloides (Bader). Zentrabl Bakteriol Parasitenk Infektionskr Hyg Abt I Orig 186: 316–327.

Hadani, A. and Y. Dinur. 1968 Studies on the translation of Aegyptianella pullorum. J Protozool Suppl 15: 45.

Hadley, P. 1918 The colon-typhoid intermediates as causative agents of disease in birds: 1. The paratyphoid bacteria. Bull R I Agr Exp Sta 174: 1–216.

Haeckel, E. H. 1894 Systematische Phylogenie der Protisten und Pflanzen I. G.Reimer, Berlin.

Hagborg, W. A. F. 1942 Classification revision in Xanthomonas translucens. Can J Res 20: 312–336.

Hagemann, G., L. Pénasse and J. Teillon. 1955 Sur un dérivé de la sérine, la O-carbamyl-D-sérine, produit par un Streptomyces. Biochim Biophys Acta 17: 240–243.

Hajek, J. P., M. J. Pelczar and J. E. Faber. 1950 Variations in the fermentative capacity of Neisseria. Amer J Clin Pathol 20: 630–636.

Hale, H. H., C. F. Helmboldt, W. N. Plastridge and E. F. Stula. 1962 Bovine mastitis caused by a Mycoplasma species. Cornell Vet 52: 582–591.

Hall, I. C. 1929 The occurrence of Bacillus sordellii in icterohemoglobinuria of cattle in Nevada. J Infect Dis 45: 156–162.

—. 1930 Micrococcus niger, a new pigment-forming anaerobic coccus recovered from urine in a case of general arteriosclerosis. J Bacteriol 20: 407–415.

—. 1948 Genus I. Micrococcus Cohn. In Breed, Murray and Hitchens (Editors), Bergey's Manual of Determinative Bacteriology, 6th ed. The Williams & Wilkins Co., Baltimore, pp. 246–248.

— and B. Howitt. 1925 Bacterial factors in pyorrhea-alveolaris. IV. Micrococcus gazogenes, a minute gram-negative, nonsporulating anaerobe prevalent in human saliva. J Infect Dis 37: 112–125.

— and E. O'Toole. 1935 Intestinal flora in newborn infants with a description of a new pathogenic anaerobe, Bacillus difficilis. Amer J Dis Child 49 (1): 390–402.

— and J. P. Scott. 1927 Bacillus sordellii, A cause of malignant edema in man. J Infect Dis 41: 329–335.

— and R. W. Whitehead. 1927 A pharmaco-bacteriologic study of African poisoned arrows. J Infect Dis 41: 51–69.

Hall, I. M. and M. E. Badgley. 1957 A rickettsial disease of larvae of species of Stethorus caused by Rickettsiella stethorae n. sp. J Bacteriol 74 (4): 452–455.

Hallé, J. 1898 Recherches sur la bactériologie du canal génital de la femme (état normal et pathologique). Thesis, Paris.

Hama, T. 1933 Studien über eine neue Rhodospirillum Art aus Yumoto bei Nikko. J Sci Hiroshima Univ Ser B Div 2 (Bot) 1: 135–155.

—. 1933 Nine species belonging to the order Thiobacteriales Buchanan, found in Hiroshima. J Sci Hiroshima Univ Ser B Div 2 (Bot) 1: 157–163.

Hamilton, R. D. and K. E. Austin. 1967 Physiological and cultural characteristics of *Chromobacterium marinum* sp. n. Antonie van Leeuwenhoek J Microbiol Serol *33:* 257–264.

— and S. A. Zahler. 1957 A study of *Leptotrichia buccalis.* J Bacteriol *73 (3):* 386–393.

Hamlin, L. J. and R. E. Hungate. 1956 Culture and physiology of a starch-digesting bacterium (*Bacteroides amylophilus* n. sp.) from the bovine rumen. J Bacteriol *72 (4):* 548–554.

Hamm, A. 1912 Die puerperale Wundinfektion. Julius Springer, Berlin.

Hammer, B. W. 1915 Bacteriological studies on the coagulation of evaporated milk. Iowa Agr Exp Sta Res Bull *19:* 119–131.

—. 1917 Fishiness in evaporated milk. Iowa Agr Exp Sta Res Bull *38:* 233–246.

—. 1920 Volatile acid production of *S. lacticus* and the organisms associated with it in starters. Iowa Agr Exp Sta Res Bull *63:* 60–96c.

Hammond, B. F. 1970 Isolation and serological characterization of a cell wall antigen of *Rothia dentocariosa.* J Bacteriol *103:* 634–640.

—. 1970 Deoxyribonucleic acid base composition of *Rothia dentocariosa* as determined by thermal denaturation. J Bacteriol *104:* 1024–1026.

Hampp, E. G. 1954 *Borrelia buccalis,* isolation, pure cultivation and morphologic characteristics. J Dent Res *33:* 660.

Hanada, K. 1954 Über die Blattknoten der *Ardisia*-Arten. Isolierung der Bakterien und ihre stickstoffbinde Kraft in Reinkultur. Jap J Bot *14:* 235–268.

Hanaoka, M., Y. Kato and T. Amano. 1969 Complementary examination of DNA's among *Vibrio* species. Biken J *12:* 181–185.

Hanert, H. 1968 Untersuchungen zur Isolierung, Stoffwechselphysiologie und Morphologie von *Gallionella ferruginea* Ehrenberg. Arch Mikrobiol *60:* 348–376.

Hanks, J. H. 1941 Behavior of leprosy bacilli in complex liquid media with highly available sources of nutrient and accessory substances. Int J Leprosy *9:* 275–298.

—. 1954 Relationship between the metabolic capacity and the infectiousness of *M. lepraemurium* refrigeration studies. Int J Leprosy *22:* 450–460.

Hansen, E. C. 1879 *Mycoderma aceti* (Kütz) Pasteur et *Myc. pasteurianum.* C R Trav Lab Carlsberg *1:* 96–100.

—. 1879 Bidrag til kundskab om hvilke organismer der kunne forekomme og leve i Øl og Ølurt. Medd Carlsberg Lab *1:* 185–234.

—. 1894 Undersøgelser over Eddikesyrerlaktenner. (Anden afhandling). Medd Carlsberg Lab *4:* 265–327.

Hansen, G. A. 1880 *Bacillus leprae.* Virchow's Arch *79:* 32–42.

Hansen, H. N. and R. E. Smith. 1937 A bacterial gall disease of Douglas fir *Pseudotsuga taxifolia.* Hilgardia *10:* 569–577.

Hansen, P. A. 1966 *Lactobacillus bulgaricus* and its relationships. Ernährungsforschung *10:* 485–488.

—. 1968 Type strains of *Lactobacillus* species. A report by the taxonomic subcommittee on lactobacilli and closely related organisms. American Type Culture Collections, Rockville, Md.

— and E. F. Lessel. 1971 *Lactobacillus casei*

(Orla Jensen) comb. nov. Int J Syst Bacteriol *21 (1):* 69–71.

— and G. Mocquot. 1970 *Lactobacillus acidophilus* (Moro) comb. nov. Int J Syst Bacteriol *20 (3):* 325–327.

Hansgirg, A. 1888 Beitrage zur Kenntniss der Kellerbacterien, nebst Bemerkungen zur Systematik der Spaltpilze (Bacteria). Österr Bot Z *38:* 227–230; 263–267.

—. 1890 Über neue Süsswasser- und Meeres-Algen und Bacterien, mit Bemerkungen zur Systematik dieser Phycophyten und über den Einfluss des Lichtes auf die Ortsbewegungen des *Bacillus pfeifferi* nob. Sitzungsber Böhm Ger Wiss Math-Naturwiss Kl *1:* 1–34.

—. 1892 Prodromus der Algenflora von Bohmen. Theil 2. Arch Naturw Land-Durchforsch Bohm Bot Abh *8 (4):* 1–268.

—. 1893 Neue Beitrage zur Kenntniss der Meeresalgen-Bacteriaceen-Flora der österreichisch ungarischen Kustenlander. Sitzungsber Böhm Ges Wiss Math-Naturwiss K1, 212–248.

Happold, F. C., K. I. Johnstone, H. J. Rogers and J. B. Yovatt. 1954 The isolation and characteristics of an organism oxidizing thiocyanate. J Gen Microbiol *10:* 261–266.

— and A. Keys. 1937 The bacterial purification of gas-works liquor. II. The biological oxidation of ammonium thiocyanate. Biochem J *31 (2):* 1323–1329.

Harada, T. and Y. Murooka. 1966 Formation of glycine from ethanolamine by a non-spore-forming soil bacterium. J Ferment Technol *44 (4):* 192–197.

—, — and Y. Izumi. 1967 *O*-Alkyl-L-homoserines: Three new amino acids formed from alcohols by *Corynebacterium ethanolaminophilum* sp. nov. Biochim Biophys Res Commun *28 (4):* 485–489.

—, K. Seto and Y. Murooka. 1968 Formation of glutamic acid from acetic acid by *Corynebacterium acetophilum* nov. sp. J Ferment Technol *46 (3):* 169–176.

Harada, Y., S. Kubo and S. Itagaki. 1956 Studies on carzinocidin, an antitumor substance, produced by *Streptomyces* sp. II. On antitumor effects of carzinocidin. J Antibiot (Tokyo) Ser A *9 (1):* 9–15.

—, T. Nara and F. Okamoto. 1956 Studies on carzinocidin, an antitumor substance produced by *Streptomyces* sp. I. On extraction, chemical and biological properties of carzinocidin. J Antibiot (Tokyo) Ser A *9 (1):* 6–8.

— and S. Tanaka. 1956 Studies on carzinocidin, antitumor substance produced by *Streptomyces* sp. III. On the taxonomic studies of the strain No. 48-B-3 identified as *S. kitazawaensis* nov. sp. J Antibiot (Tokyo) Ser A *9:* 113–117.

Harbutt, P. R. 1963 A clinical appraisal of feline infectious anemia and its transmission under natural conditions. Aust Vet J *39:* 401–404.

Hard, G. C. 1969 Electron microscopic examination of *C. ovis.* J Bacteriol *97 (3):* 1480–1485.

Hardman, J. K. and T. C. Stadtman. 1960 Metabolism of ω-amino acids. II. Fermentation of Δ-aminovaleric acid by *Clostridium aminovalericum* n. sp. J Bacteriol *79 (4):* 549–552.

Hardman, Y. and A. Henrici. 1939 Studies of fresh-water bacteria. V. The distribution of *Siderocapsa treubii* in some lakes and streams. J Bacteriol *37:* 97–104.

Hare, R. 1935 The classification of haemolytic streptococci from the nose and throat of normal human beings by means of precipitin and biochemical tests. J Pathol Bacteriol *41:* 499–512.

—. 1967 The anaerobic cocci. *In* H. P. Waterson (Editor), Recent Advances in Medical Microbiology. Little, Brown and Company, Boston, pp. 284–317.

—, P. Wildy, F. S. Billett and D. N. Twort. 1952 The anaerobic cocci: gas formation, fermentation reactions, sensitivity to antibiotics and sulphonamides; classification. J Hyg *50:* 295–319.

Hare, T. and R. M. Fry. 1938 Clinical observations of the beta hemolytic streptococcal infections of dogs. Vet Rec *50:* 1537–1548.

Harned, R. L. 1963 Process for the production of cycloserine. United States Patent 3,090,730, May 21.

—, P. H. Hidy and E. K. LaBaw. 1955 Cycloserine. I. A preliminary report. Antibiot Chemother *5 (4):* 204–205.

— and E. K. LaBaw. 1957 Herstellung des Antibiotikums Cycloserine. German Patent 958,242, February 14.

Harrington, A. A. and R. E. Kallio. 1960 Oxidation of methanol and formaldehyde by *Pseudomonas methanica.* Can J Microbiol *6:* 1–7.

Harris, D. A., M. Ruger, M. A. Reagan, F. J. Wolf, R. L. Peck, H. Wallick and H. B. Woodruff. 1955 Discovery, development and antimicrobial properties of D-4-amino-3-isoxazolidone (Oxamycin), a new antibiotic produced by *Streptomyces garyphalus* n. sp. Antibiot Chemother *5 (4):* 183–190.

— and H. B. Woodruff. 1953 A crystalline antiviral agent produced by a new species of *Nocardia.* I. Cultural characteristics, fermentation studies, and production of the antiviral agent. Antibiot Ann *1953-54:* 609–614.

Harris, G. P. and T. M. Morrison. 1958 Fixation of nitrogen-15 by excised nodules of *Coriaria arborea* Lindsay. Nature (London) *182:* 1812.

Harris, J. W. and J. H. Brown. 1927 Description of a new organism that may be a factor in the causation of puerperal infection. Bull Johns Hopkins Hosp *40:* 203–215.

Harris, N. M. 1901 *Bacillus mortiferus* (nov. spec.) J Exp Med *6:* 519–547.

Harrison, A. P., Jr. and P. A. Hansen. 1950 A motile *Lactobacillus* from the cecal feces of turkeys. J Bacteriol *59 (4):* 444–446.

— and —. 1950 Lactobacilli from turkeys. J Bacteriol *60 (5):* 543–555.

— and —. 1963 *Bacteroides hypermegas* nov. spec. Antonie van Leeuwenhoek J Microbiol Serol *29:* 22–28.

Harrison, F. C. 1904 A bacterial disease of cauliflower (*Brassica oleraceae*) and allied plants. Zentrabl Bakteriol Parasitenk Infektionskr Hyg Abt II *13:* 46–55.

—. 1905 The viscous fermentation of milk and beer. Rev Gen Lait *5:* 73–80; 97–103; 129–136; 145–152.

—. 1929 The discoloration of halibut. Can J Res *1:* 214–239.

— and M. E. Kennedy. 1922 The red discolouration of cured codfish. Proc Roy Soc Can Sect V *16:* 101–152.

Harry, E. G. 1969 *Pasteurella* (*Pfeifferella*) *anatipestifer* serotypes isolated from cases of anatipestifer septicaemia in ducks. Vet Rec *84 (20):* 673.

Hart, L. T., A. D. Larson and C. S. McClesky. 1965 Denitrification by *Corynebacterium nephridii.* J Bacteriol *89:* 1104–1108.

Hartsell, S. E. and L. F. Rettger. 1934 A taxonomic study of "*Clostridium putrificum*" and its establishment as a definite entity—*Clostridium lentoputrescens* nov. spec. J Bacteriol *27:* 497–514.

Harvey, J. M. 1952 Bacterial leaf spot of *Umbellularia californica.* Madroño S Francisco *11:* 195–198.

Harz, C. O. 1877/1878 *Actinomyces bovis* ein neuer Schimmel in den Geweben des Rindes. Deut Z Thiermed *5:* 125–140.

—. 1879 *Actinomyces bovis,* ein neuer Schimmel in den Geweben des Rindes. Jahresber K Cent Thierärzn Schule München (1877–1878), pp. 125–140.

Haskell, T. H. 1958 Amicetin, bamicetin and plicacetin. Chemical studies. J Amer Chem Soc *80:* 747–751.

—, A. Maretzki and Q. R. Bartz. 1955 Viridogrisein: chemical studies. Antibiot Ann *1954-1955:* 784–789.

—, A. Ryder, R. P. Frohardt, S. A. Fusari, Z. L. Jakubowski and Q. R. Bartz. 1958 The isolation and characterization of three crystalline antibiotics from *Streptomyces plicatus.* J Amer Chem Soc *80:* 743–747.

Hasse, C. H. 1915 *Pseudomonas citri,* the cause of citrus canker. J Agr Res *4:* 97–100.

Hässig, A., J. Karrer and F. Pusterla. 1949 Über Pseudotuberkulose beim Menschen. Schweiz Med Wochenschr *79 (41):* 971–973.

Hata, T., T. Higuchi, Y. Sano and K. Sawachika. 1949 Isolation of a new antibiotic substance "luteomycin" from a strain of *Streptomyces.* Kitasato Arch Exp Med *22 (4):* 229–242.

—, —, — and —. 1949–1950 Isolation of a new antibiotic substance "luteomycin." J Antibiot (Tokyo) Ser A *3 (5):* 313–325.

—, F. Koga, Y. Sano, K. Kanamori, A. Matsumae, R. Sugawara, T. Hoshi, T. Shima, S. Ito and S. Tomizawa. 1954 Carzinophilin, a new tumor inhibitory substance produced by *Streptomyces* I. J Antibiot (Tokyo) Ser A *7 (4):* 107–112.

—, N. Ohki and T. Higuchi. 1952 Studies on the antibiotic substance "luteomycin". On the strains and the cultural conditions. J Antibiot (Tokyo) Ser A *5 (10):* 529–534.

Hatt, H. D. and E. Zvirbulis. 1967 Status of names of bacterial taxa not evaluated in Index Bergeyana (1966). Int J Syst Bacteriol *17 (2):* 171–225.

Hauduroy, P. 1955 Derniers aspects du monde des mycobacteries. Masson et Cie, Paris.

—, G. Ehringer, G. Guillot, J. Magrou, A. R. Prévot, Rosset and A. Urbain. 1953 Dictionnaire des bactéries pathogènes. Masson and Co., Paris, 2nd ed.

—, —, A. Urbain, G. Guillot and J. Magrou.

1937 Dictionnaire des bactéries pathogènes. Masson and Co., Paris.

Haupt, H. 1932 Der gegenwärtige Stand der Systematik und Benennung der Bakterien und ihre Anwendung in der medizinischen Bakteriologie. Ergeb Hyg Bakteriol *13:* 641–685.

—. 1934 Zur Frage der Verwandschaft des *Actinobacillus Lignieresii* Brumpt 1910, des *Bacillus equuli* van Straaten 1918 und des *Bacillus mallei* Flügge 1886. Arch Wiss Prakt Tierheilk *67:* 513–524.

—. 1935 Zur Systematik der Bakterien. Die für Mensch und Tier pathogenen gram-negativen alkali-bildenden Stäbchenbakterien (*Aerobactereae* Pribram 1929 em.) Ergeb Hyg Bakteriol *17:* 175–230.

— and M. Zeki. 1933 Ist der erreger der knochenaktinomykose des rindes zur gattung *Actinomyces* order zur gattung *Corynebacterium* zu stellen. Zentralb Bakteriol Parasitenk Infektionskr Hyg Abt I Orig *130:* 91–102.

Hauser, G. 1885 Über Fäulnisbakterien und deren Beziehungen zur Septicämie. Ein Beitrag zur Morphologie der Spaltpilze. Vogel, Leipzig.

—. 1892 Über das Vorkommen von *Proteus vulgaris* bei einer jauchig-phlegmonösen Eiterung. Muenchen Med Wochenschr *39:* 104–105.

Hawiger, J. 1968 Frequency of staphylococcal lysozyme production tested by plate method. J Clin Pathol (London) *21:* 390–393.

—. 1968 Purification and properties of lysozyme produced by *Staphylococcus aureus*. J Bacteriol *95:* 376–384.

Hayashi, F. 1932 Mitsuda's skin reaction in leprosy. Int J Leprosy *1:* 31–38.

Hayashi, K., T. Kodaira, K. Baba and K. Kikuchi. 1966 Adansonian taxonomy and relationship of microorganisms based on the concept of similarity value and center species. III Studies on the tribe *Neisserieae* in the Family *Coccaceae* and isolation and certification of species in a new genus *Halococcus* induced theoretically. Jap J Bacteriol *21:* 633–639.

Hayashi, N. 1920 Etiology of tsutsugamushi disease. J Parasitol *7:* 53–68.

Hayashi, R. and H. Nakayama. 1953 Studies on the aerobic mesophilic bacteria with distinctly bulged sporangium. I. Special reference to *Bacillus thiaminolyticus*. Bull Yamaguchi Med Sch *1:* 57–63.

Hayflick, L. and R. M. Chanock. 1965 *Mycoplasma* species of man. Bacteriol Rev *29 (2):* 185–221.

Haynes, W. C. 1957 Genus I *Pseudomonas* Migula 1894. *In* Breed, Murray and Smith (Editors), Bergey's Manual of Determinative Bacteriology, 7th ed. The Williams & Wilkins Co., Baltimore, pp. 89–152.

— and L. J. Rhodes. 1963 A growth factor for *Bacillus lentimorbus* NRRL B-2522. Bacteriol Proc, 10.

— and —. 1966 Spore formation by *Bacillus popilliae* in liquid medium containing activated carbon. J Bacteriol *91 (6):* 2270–2274.

—, G. St. Julian, Jr., M. C. Shekleton, H. H. Hall and H. Tashiro. 1961 Preservation of infectious milky disease bacteria by lyophilization. J Insect Pathol *3:* 55–61.

Hayward, H. R. and T. C. Stadtman. 1959 Anaerobic degradation of choline. 1. Fermentation of choline by an anaerobic, cytochrome-producing bacterium, *Vibrio cholinicus*, n.sp. J Bacteriol *78 (4):* 557–561.

Hazen, E. L. and R. Brown. 1950 Two antifungal agents produced by a soil actinomycete. Science (Washington) *112:* 423.

— and —. 1951 Fungicidin, an antibiotic produced by a soil actinomycete. Proc Soc Exp Biol Med *76 (1):* 93–97.

— and G. N. Little. 1958 *Actinomyces bovis* and "anaerobic diphtheroids" pathogenicity for hamsters and some other differentiating characteristics. J Lab Clin Med *51:* 968–976.

Hazeu, W. and P. J. Steennis. 1970 Isolation and characterization of two vibrio-shaped methane-oxidizing bacteria. Antonie van Leeuwenhoek J Microbiol Serol *36 (1):* 67–72.

Heberlein, G. T., J. De Ley and R. Tijtgat. 1967 Deoxyribonucleic acid homology and taxonomy of *Agrobacterium*, *Rhizobium*, and *Chromobacterium*. J Bacteriol *94 (1):* 116–124.

Hedges, F. 1922 A bacterial wilt of the bean caused by *Bacterium flaccumfaciens* nov. sp. Science (Washington) *55:* 433–434.

—. 1922 Bacterial pustule of soy bean. Science (Washington) *56:* 111–112.

Hefferan, M. 1903–04 A comparative and experimental study of bacilli producing red pigment. Zentrabl Bakteriol Parasitenk Infektionskr Hyg Abt II *11:* 311–317; 397–404; 456–475; 520–540.

Hehre, E. J. and J. M. Neill. 1946 Formation of serologically reactive dextrans by streptococci from subacute bacterial endocarditis. J Exp Med *83:* 147–162.

Heiberg, B. 1934 Des réactions de fermentation chez les Vibrions. C R Soc Biol Paris *115:* 984–986.

Heidelberger, M. and S. Elliott. 1966 Cross-reactions of streptococcal group N teichoic acid in antipneumococcal horse sera of types VI, XIV, XVI, and XXVII. J Bacteriol *92 (1):* 281–283.

Heilman, F. R. 1941 A study of *Asterococcus muris* (*Streptobacillus moniliformis*). I. Morphologic aspects and nomenclature. J Infect Dis *69:* 32–44.

Heim, L. 1922 Lehrbuch der Bakteriologie 6 und 7 Aufl. Enke, Stuttgart.

Heim, R. 1951 Mémoire sur l'Antennopsis, ectoparasite du termite de Saintonge. IV. Étude du *Streptomyces termitum* n. sp., associé à l'antennopsis. Bull Soc Mycol Fr *67:* 359–364.

Heimpel, A. M. 1969 A critical review of *Bacillus thuringiensis* var. *thuringiensis* Berliner and other crystalliferous bacteria. Annu Rev Entomol *12:* 287–322.

— and T. A. Angus. 1958 The taxonomy of insect pathogens related to *Bacillus cereus* Frankland and Frankland. Can J Microbiol *4:* 531–541.

Heinemann, B., M. A. Kaplan, R. D. Muir and I. R. Hooper. 1953 Amphomycin, a new antibiotic. Antibiot Chemother *3:* 1239–1242.

Heinrich, S. and G. Pulverer. 1959 Zur Aetiologie und Mikrobiologie der Aktinomykose II. Definition und praktische Diagnostik des *Actinobacillus actinomycetem-comitans*. Zentrabl Bakteriol Parasitenk Infektionskr Hyg Abt I Orig *174:* 123–135.

Heisch, R. B. 1950 On *Spirochaeta dipodilli*

sp. nov., a parasite of pygmy gerbils (*Dipodillus* sp.). Ann Trop Med Parasitol *44:* 260–272.

—. 1953 On a spirochaete isolated from *Ornithodoros graingeri*. Parasitology *43:* 133–135.

— and A. E. C. Harvey. 1962 Experiments with *Sp. tillae* Brumpt. East Afr Med J *39:* 609–611.

Heller, H. H. 1922 Certain genera of the *Clostridiaceae*. Studies in pathogenic anaerobes. J Bacteriol *7:* 1–38.

Hellinger, E. 1944 Studies on a pink butyric acid *Clostridium*. Commemorative Vol. to Dr. Weizmann's 70th Birthday—Private print Nov. 1944, pp. 37–46.

Hellmann, G. 1913 Über die im Excretionsorgan der Ascidien der Gattung *Caesira* (*Molgula*) vorkommenden Spirochäten: *Spirochaeta caesirae septentrionalis* n. sp. und *Spirochaeta caesirae retortiformis* n. sp. Arch Protistenk *29:* 22–38.

Hellmers, E. 1958 Four wilt diseases of perpetual-flowering carnations in Denmark. Dan Bot Ark *18:* 1–200.

—. 1959 *Pectobacterium carotovorum* var. *atrosepticum* (van Hall) Dowson, the correct name of the potato blackleg pathogen: a historical and critical review. Eur Potato J *2:* 251–271.

— and W. J. Dowson. 1953 Further investigations of potato blackleg. Acta Agr Scand *3:* 103–112.

Hellmers, H. and J. M. Kelleher. 1959 *Ceanothus leucodermis* and soil nitrogen in Southern California Mountains. Forest Sci *5:* 275–278.

Hellmuth, H. 1956 Untersuchungen zur Bakterien-symbiose der Trypetiden (Diptera). Z Morphol Tiere *44:* 483–517.

Hemphill, H. E. and S. A. Zahler. 1968 Nutrition of *Myxococcus xanthus* FBa and some of its auxotrophic mutants. J Bacteriol *95 (3):* 1011–1017.

— and —. 1968 Nutritional induction and suppression of fruiting in *Myxococcus xanthus* FBa. J Bacteriol *95 (3):* 1018–1023.

Hendrickson, A. A., I. L. Baldwin and A. J. Riker. 1934 Studies on certain physiological characters of *Phytomonas tumefaciens*, *Phytomonas rhizogenes* and *Bacillus radiobacter*. Part II. J Bacteriol *28:* 597–618.

Hendrickson, J. M. and K. F. Hillbert. 1932 A new and serious septicemic disease of young ducks with a description of the causative organism, *Pfeifferella anatipestifer*. Cornell Vet *22 (3):* 239–252.

Hendrie, M. S., W. Hodgkiss and J. W. Shewan. 1970 The identification, taxonomy and classification of luminous bacteria. J Gen Microbiol *64 (2):* 151–169.

—, — and —. 1971 Proposal that the species *Vibrio anguillarum* Bergman 1909, *Vibrio piscium* David 1927 and *Vibrio ichthyodermis* (Wells and ZoBell) Shewan, Hobbs and Hodgkiss 1960 be combined as a single species, *Vibrio anguillarum*. Int J Syst Bacteriol *21 (1):* 64–68.

—, — and —. 1971 A proposal that *Vibrio marinus* (Russell) Ford 1927 be amalgamated with *Vibrio fischeri* (Beijerinck) Lehmann and Neumann 1896, and a request for an opinion. Int J Syst Bacteriol *21 (3):* 217–221.

—, A. J. Holding and J. M. Shewan. 1974 Emended descriptions of the genus *Alcaligenes*

and of *Alcaligenes faecalis* and proposal that the generic name *Achromobacter* be rejected. Status of the named species of *Alcaligenes* and *Achromobacter*. Int J Syst Bacteriol *24:* in press.

—, R. W. Horsley, A. R. MacKenzie, T. C. Mitchell, L. B. Perry and J. W. Shewan. 1966 Comments on Sneath, P. H. A. and Skerman, V. B. D. (1966) "A new list of type and reference strains of bacteria." Int J Syst Bacteriol *16 (1):* 1–133. Int J Syst Bacteriol *16:* 435–457.

—, J. M. Shewan and M. Véron. 1971 *Aeromonas shigelloides* (Bader) Ewing *et al.* 1961. A proposal that it be transferred to the genus *Vibrio*. Int J Syst Bacteriol *21 (1):* 25–27.

Henneberg, W. 1897 Beiträge zur Kenntnis der Essigbakteriën. Zentrabl Bakteriol Parasitenk Infektionskr Hyg Abt II *3:* 223–231.

—. 1898 Weitere Untersuchungen ueber Essigbakteriën. Zentrabl Bakteriol Parasitenk Infektionskr Hyg Abt II *4:* 14–20; 67–73; 138–147.

—. 1898 *Bacterium industrium* und *B. ascendens* und Ergänzungen zu den bisherigen untersuchungen ueber Essigbakteriën. Deut Essigindustrie *2:* 145–148; 153–155; 161–164; 169–172; 177–179.

—. 1901 Zur Kenntniss der Milchsäure-bakterien der Brennereimeische, der Milch, und des Bieres. (*Bacillus delbrücki* (Leichmann), *Bacillus delbrücki* var. α n. sp., *Pediococcus lactis acidi* (Lindner), *Bacillus lactis acidi* (Leichmann), *Bacterium lactis acidi* (Leichmann), *Saccharobacillus pastorianus* (v. Laer), *Saccharobacillus pastorianus* var. α n. sp., *Saccharobacillus pastorianus* var. *berolinensis* n. sp., *Bacillus lindneri* n. sp.) Wochenschr Brau *18:* 381–384.

—. 1903 Zur Kenntniss der Milchsäure-bakterien der Brenneriemaische, der Milch, des Bieres, der Presshefe, der Melasse, des Sauerkohls, der sauren Gurken und des Sauerteigs, sowie einige Bemerkungen über die Milchsäurebakterien des menschlichen Magens. Z Spiritusind *26 (22):* 226–227; *(23):* 243–244; *(24):* 255–257; *(25):* 270; *(26):* 277–279; *(27):* 288–291; *(28):* 302; *(29):* 315–318; *(30):* 329–332; *(31):* 341–344.

—. 1906 Zur Kenntnis der Schnellessig und Weinessigbakteriën. Deut Essigindustrie *10:* 89–93; 98–99; 106–108; 113–116; 121–124; 129–132; 137–140; 146–148.

—. 1922 Untersuchungen über die Darmflora des Menschen mit besonderer Berucksichtigung der jodophilen Bakterien im Menschen und Tierdarm sowie im Kimpostdünger. Zentrabl Bakteriol Parasitenk Infektionskr Hyg Abt II *55:* 242–281.

—. 1926 Handbuch der Garungsbakteriologie. Zweite, neubearbeitete und vermehrte Auflage. Zweiter Band. Spezielle Pilzkunde, Berlin.

Henrichsen, J. 1972 Bacterial Surface Translocation: a survey and a classification. Bacteriol Rev *36:* 478–503.

Henrici, A. T. 1939 The Biology of Bacteria, 2nd ed. D. C. Heath and Co., Chicago.

— and D. Johnson. 1935 Studies of Fresh water Bacteria. II. Stalked Bacteria, a new order of Schizomycetes. J Bacteriol *30 (1):* 61–92.

Henriksen, S. D. 1937 Studies on the bacterial flora of the respiratory tract. Skr Utgitt Norske Vidensk-Akad Oslo I Mat-Naturvidensk Kl 1936, No. 11, pp. 7–241.

—. 1954 Studies on the *Klebsiella* group (Kauffmann). II. Biochemical reactions. Acta Pathol Microbiol Scand *34:* 259–265.

—. 1962 Some *Pasteurella* strains from the human respiratory tract. A correction and supplement. Acta Pathol Microbiol Scand *55 (3):* 355–356.

—. 1969 Designation of the type strain of *Bacteroides corrodens* Eiken 1958. Int J Syst Bacteriol *19:* 165–166.

— and K. Bøvre. 1968 *Moraxella kingii* sp. nov., a haemolytic saccharolytic species of the genus *Moraxella*. J Gen Microbiol *51 (3):* 377–385.

— and —. 1968 The taxonomy of the genera *Moraxella* and *Neisseria*. J Gen Microbiol *51 (3):* 387–392.

— and K. Jyssum. 1961 A study of some *Pasteurella* strains from the human respiratory tract. Acta Pathol Microbiol Scand *51 (4):* 354–368.

Henry, B. S. 1933 Dissociation in the genus *Brucella*. J Infect Dis *52:* 374–402.

Henry, H. 1917 An investigation of the cultural reactions of certain anaerobes found in wounds. J Pathol Bacteriol *21:* 344–385.

Henry, J. N. 1952 Mycotic dermatitis, or lumpy wool. N S W Dep Agr Sci Bull *49:* 1–3.

Henssen, A. 1957 Beiträge zur Morphologie und Systematik der thermophilen Actinomyceten. Arch Mikrobiol *26:* 373–414.

— and E. Schneph. 1967 Zur Kenntnis thermophiler Actinomyceten. Arch Mikrobiol *57:* 214–231.

Herderscheê, A., C. Ruys and G. R. van Rhijn. 1963 A new oral *Mycoplasma* isolated from the tonsils of two patients suffering from scarlatina. Antonie van Leeuwenhoek J Microbiol Serol *29:* 157–162.

Hermann, S. 1928 Ueber die sogenannte "Kombucha." II. Biochem Z *192:* 188–199.

Herter, W. 1917 XVII. Schizomycetes 1910–1911. Mit einigen Nachträgen aus früheren Jahren. Just's Jber Abt 2 *39:* 506–805.

Hertig, M. 1936 The rickettsia, *Wolbachia pipientis* (gen. et sp. n.) and associated inclusions of the mosquito, *Culex pipiens*. Parasitology *28:* 453–486.

— and S. B. Wolbach. 1924 Studies on rickettsia-like microorganisms in insects. J Med Res *44:* 329–374.

Herzberg, K. 1928 Neue bakteriologische Befunde beim Scharlach. Zentrabl Bakteriol Parasitenk Infektionskr Hyg Abt I Ref *90:* 574–576.

—. 1929 Neue bakteriologische Befunde beim Scharlach. 1. Ueber sauerstaffscheue gramnegative Mikrokken des Rachens. Zentrabl Bakteriol Parasitenk Infektionskr Hyg Abt I Orig *111:* 373–384.

Hespell, R. B. and E. Canale-Parola. 1970 *Spirochaeta litoralis* sp. n., a strictly anaerobic marine spirochete. Arch Mikrobiol *74:* 1–18.

— and —. 1970 Carbohydrate metabolism in *Spirochaeta stenostrepta*. J Bacteriol *103 (1):* 216–226.

Hesseltine, C. W., R. G. Benedict and T. G. Pridham. 1954 Useful criteria for species differentiation in the genus *Streptomyces*. Ann N Y Acad Sci *60:* 136–151.

Heurlin, M. 1910 Bakteriologische Untersuchungen des Keimgehaltes im Genitalkanale der fiebernden Wochnerinnen mit Bericksichtung der Gesamtmorbidität im Laufe eines Jahres. Akademische Abhandlung, Helsigfors.

Heuschmann-Brunner, G. 1965 Ein Beitrag zur Erregerfrage der infektiosen Bauchwassersucht des Karpfens in: Der Fisch in Wissenschaft und Praxis. Festschrift, herausgegeben anlässlich des 50-jahrigen Bestehens der teichwirtschaftlichen Abteilung Wielenbach der Bayerischen Biologischen Versuchsanstalt (Demoll-Hofer-Institut), München, pp. 41–49.

Hewitt, E. J. and H. Bond. 1961 Molybdenum and the fixation of nitrogen in *Casuarina* and *Alnus* root nodules. Plant Soil *14:* 159–175.

— and —. 1966 The cobalt requirement of non-legume root nodule plants. J Exp Bot *17:* 480–491.

Hewitt, L. F. 1947 Serological typing of *C. diphtheriae*. Brit J Exp Pathol *28:* 338–346.

Hewitt, R. I., A. R. Gumble, L. H. Taylor and W. S. Wallace. 1957 The activity of a new antibiotic, nucleocidin, in experimental infections with *Trypanosoma equiperdum*. Antibiot Ann *1956–1957:* 722–729.

Heyward, J. T., M. Z. Sabry and W. R. Dowdle. 1969 Characterization of *Mycoplasma* species of feline origin. Amer J Vet Res *30:* 615–622.

Hickey, R. J., C. J. Corum, P. H. Hidy, I. R. Cohen, U. F. B. Nager and E. Kropp. 1952 Ascosin, an antifungal antibiotic produced by a streptomycete. Antibiot Chemother *2 (9):* 472–483.

Hickman, D. D. and A. W. Frenkel. 1959 The structure of *Rhodospirillum rubrum*. J Biophys Biochem Cytol *6:* 277–284.

Hidy, P. H., E. B. Hodge, V. V. Young, R. L. Harned, G. A. Brewer, W. F. Phillips, W. F. Runge, H. E. Stavely, A. Pohland, H. Boaz and H. R. Sullivan. 1955 Structure and reactions of cycloserine. J Amer Chem Soc *77 (8):* 2345–2346.

Higashide, E., T. Hasegawa, M. Shibata, K. Mizuno and H. Akaike. 1966 Studies on the streptomycetes. Streptomyces cuspidosporus nov. sp. and the antibiotics sparsomycin and tubercidin produced thereby (in Japanese and English). Ann Rep Takeda Res Lab *25:* 1–14.

Higgins, M. L., M. P. Lechevalier and H. A. Lechevalier. 1967 Flagellated actinomycetes. J Bacteriol *93 (4):* 1446–1451.

Hijmans, W. 1962 Absence of the group-specific and the cell-wall polysaccharide antigen in L-phase variants of group D streptococci. J Gen Microbiol *28 (1):* 177–179.

Hildebrand, D. C. and N. N. Schroth. 1967 A new species of *Erwinia* causing the drippy nut disease of live oaks. Phytopathology *57:* 250–253.

Hildebrand, E. M. 1940 Cane gall of brambles caused by *Phytomonas* n. sp. J Agr Res *61:* 685–696.

Hilger, F. 1965 Études sur la systématique du genre *Beijerinckia* Derx. Ann Inst Pasteur (Paris) *109:* 406–423.

Hill, A. 1971 *Mycoplasma caviae*, a new species. J Gen Microbiol *65 (1):* 109–113.

Hill, L. R. 1966 An index to deoxyribonucleic

acid base compositions of bacterial species. J Gen Microbiol *44* (*3*): 419–437.

—, J. J. S. Snell and S. P. Lapage. 1970 Identification and characterization of *Bacteroides corrodens*. J Med Microbiol *3:* 483–491.

Hindle, E. 1921 Notes on *Rickettsia*. Parasitology *13:* 152–159.

Hingorani, M. K. and N. J. Singh. 1959 *Xanthomonas punicae* sp. nov. on *Punica granatum* L. Indian J Agr Sci *29:* 45–48.

Hinman, J. W., E. L. Caron and C. De Boer. 1953 The isolation and purification of amicetin. J Amer Chem Soc *75* (*23*): 5864–5866.

—, H. Hoeksema and W. G. Jackson. 1958 Desalicetin and its salts, acids and degradation products. United States Patent 2,851,463, September 9.

Hino, S. and P. W. Wilson. 1958 Nitrogen fixation by a facultative bacillus. J Bacteriol *75* (*4*): 403–408.

Hinuma, Y. 1954 Zaomycin, a new antibiotic from a *Streptomyces* sp. Studies on the antibiotics substances from *Actinomyces*. III. J Antibiot (Tokyo) Ser A *7:* 134–136.

Hinze, G. 1903 *Thiophysa volutans*, ein neues Schwefelbakterium. Ber Deut Bot Ges *21:* 309–316.

—. 1913 Beitrage zur Kenntnis der farblosen Schwefelbakterien. Ber Deut Bot Ges *31:* 189–202.

Hirsch, A. 1951 Growth and nisin production of a strain of *Streptococcus lactis*. J Gen Microbiol *5:* 208–221.

Hirsch, P. 1958 Stoffwechselphysiologische Untersuchungen an *Nocardia petreophila* n. sp. Arch Mikrobiol *29:* 368–393.

—. 1960 Einige weitere von Luftverunreinigungen lebende Actinomyceten und ihre Klassifizierung. Arch Mikrobiol *35:* 391–414.

—. 1961 Wasserstoffactivierung und Chemoautotrophie bei Actinomyceten. Arch Microbiol *39:* 360–373.

—. 1972 Two identical genera of budding and stalked bacteria: *Planctomyces* Gimesi 1924 and *Blastocaulis* Henrici and Johnson 1935. Int J Syst Bacteriol *22* (*2*): 107–111.

—. 1972 Reevaluation of *Pasteuria ramosa* Metchnikoff 1888, a bacterium pathogenic for *Daphnia* species. Int J Syst Bacteriol *22* (*2*): 112–116.

— and H. Engel. 1956 Über oligocarbophile Actinomyceten. Ber Deut Bot Ges *69:* 441–454.

Hirschfeld, L. 1919 A new germ of paratyphoid. Lancet *196:* 296–297.

Hitchens, A. P. 1957 Genus III. *Dialister* Bergey et al. 1923. *In* Bergey, Harrison, Breed, Hammer and Huntoon (Editors), Bergey's Manual of Determinative Bacteriology, 1st ed. The Williams & Wilkins Co., Baltimore, pp. 440–441.

Hitchner, E. R. 1932 A cultural study of the propionic-acid bacteria. J Bacteriol *23:* 40–41.

—. 1934 Some physiological characteristics of the propionic acid bacteria. J Bacteriol *28:* 473–479.

— and S. F. Snieszko. 1947 A study of a microorganism causing a bacterial disease of lobsters. J Bacteriol *54:* 48.

Hobby, G. L., J. H. Hanks, M. A. Donikian and T. Backerman. 1954 An evaluation of chemotherapeutic agents in the control of experimental infections due to *Mycobac-*

terium leprae murium. Amer Rev Tuberc *69:* 173–191.

—, W. B. Redmond, E. H. Runyon, W. B. Schaefer, L. G. Wayne and R. H. Wichelhausen. 1967 A study on pulmonary disease associated with mycobacteria other than *Mycobacterium tuberculosis:* Identification and characterization of the mycobacteria. XVIII. A report of the Veterans Administration Armed Forces Cooperative Study. Amer Rev Resp Dis *95:* 954–971.

Hobson, P. N. and S. O. Mann. 1961 The isolation of glycerol-fermenting and lipolytic bacteria from the rumen of the sheep. J Gen Microbiol *25* (*2*): 227–240.

—, — and W. Smith. 1962 Serological tests of a relationship between rumen selenomonads *in vitro* and *in vivo*. J Gen Microbiol *29* (*2*): 265–270.

Hodge, E. B. 1969 Fermentative production of *O*-carbamyl-D-serine. United States Patent 3,428,525, February 18.

Hoeksema, H., J. L. Johnson and J. W. Hinman. 1955 Structural studies on streptonivicin, a new antibiotic. J Amer Chem Soc *77:* 6710–6711.

Hoelling, A. 1910 Die Kernverhältnisse von *Fusibacterium termitidis*. Arch Protistenk *19:* 239–245.

Hoeniger, J. F. M. 1965 Influence of pH on *Proteus* flagella. J Bacteriol *90* (*1*): 275–277.

Hof, T. 1935 An investigation of the microorganisms commonly present in salted beans. Rec Trav Bot Néerl *32:* 151–173.

Hofer, A. W. 1944 Flagellation of *Azotobacter*. J Bacteriol *48:* 697–701.

Hoffman, G. L. 1953 *Streptomyces leidnematis*, n. sp., growing on two species of nematodes of the cockroach. Trans Amer Microsc Soc *72* (*4*): 376–378.

Hoffman, H. 1957 Genus II. *Fusobacterium* Knorr. *In* Breed, Murray and Smith (Editors), Bergey's Manual of Determinative Bacteriology, 7th ed. The Williams & Wilkins Co., Baltimore, pp. 436–440.

Hoffmann, E. 1920 Ueber eine der weilschen Spirochäte ähnliche Zahnspirochäte des Menschen (*Spir. trimerodonta* und andere Mundspirochaten). Deut Med Wochenschr *46:* 257–259.

— and S. von Prowazek. 1906 Untersuchungen über die Balanitis- und Mundspirochäten. Zentrabl Bakteriol Parasitenk Infektionskr Hyg Abt I Orig *41:* 741–744.

Hofstad, T. 1967 An anaerobic oral filamentous organism possibly related to *Leptotrichia buccalis*. 1. Morphology, some physiological and serological properties. Acta Pathol Microbiol Scand *69:* 543–548.

—. 1967 An aerobic oral filamentous organism possibly related to *Leptotrichia buccalis*. II. Composition of cell walls. Acta Pathol Microbiol Scand *70:* 461–468.

—. 1970 *Leptotrichia buccalis*, a Gram-negative bacterium. Int J Syst Bacteriol *20* (*2*): 175–177.

— and K. A. Selvig. 1969 Ultrastructure of *Leptotrichia buccalis*. J Gen Microbiol *56* (*1*): 23–26.

Höhnl, G. 1955 Ernährung und Stoffwechselphysiologische Untersuchungen an *Sphaerotilus natans*. Arch Mikrobiol *23:* 207–250.

Holdeman, L. V., E. P. Cato and W. E. C. Moore.

1967 Amended description of *Ramibacterium alactolyticum* Prévot and Taffanel with proposal of a neotype strain. Int J Syst Bacteriol *17 (4):* 323–341.

—, — and —. 1971 *Clostridium ramosum* (Vuillemin) *comb. nov.:* emended description and proposed neotype strain. Int J Syst Bacteriol *21 (1):* 35–39.

—, — and —. 1971 *Eubacterium contortum* (Prévot) *comb. nov.:* Emendation of description and designation of the type strain. Int J Syst Bacteriol *21 (4):* 304–306.

— and W. E. C. Moore. 1970 *Bacteroides. In* E. P. Cato, C. S. Cummins, L. V. Holdeman, J. L. Johnson, W. E. C. Moore, R. M. Smibert and L. DS. Smith (Editors), Outline of clinical methods in anaerobic bacteriology, 2nd rev. Virginia Polytechnic Institute, Anaerobe Laboratory, Blacksburg, Va.

— and —. 1970 *Eubacterium* p. 23–30. *In* E. P. Cato, C. S. Cummins, L. V. Holdeman, J. L. Johnson, W. E. C. Moore, R. M. Smibert and L. DS. Smith (Editors), Outline of Clinical Methods in Anaerobic Bacteriology, 2nd rev. V.P.I. Anaerobe Laboratory, Blacksburg, Va.

— and —. 1972 *Bacteroides. In* L. V. Holdeman and W. E. C. Moore (Editors), Anaerobe Laboratory Manual. Virginia Polytechnic Institute Anaerobe Laboratory, Blacksburg, Va.

— and —. 1972 *Eubacterium. In* L. V. Holdeman and W. E. C. Moore (Editors), Anaerobe Laboratory Manual. Virginia Polytechnic Anaerobe Laboratory, Blacksburg, Va.

Holden, M. 1935 *Loefflerella;* glanders and melioidosis. *In* Gay, Agents of disease and host resistance. Charles C Thomas, Springfield, Illinois, p. 782.

Holland, D. F. 1920 V. Generic index of the commoner forms of bacteria (p. 215) *In* C.-E.A. Winslow, J. Broadhurst, R. E. Buchanan, C. Krumwiede, Jr., L. A. Rogers and G. H. Smith. The families and genera of the bacteria. J Bacteriol *5:* 191–229.

Hollande, A. C. 1921 Présence d'un spirochètoïde nouveau, *Cristispirella caviae* n. g., n. sp., à membrane ondulante très développée dans l'intestin du cobaye. C R Acad Sci Paris *172:* 1693–1696.

—. 1922 Les spirochetes des termites; processus de division: formation du schizoplaste. Arch Zool Exp Gén Notes Rev *61:* 23–28.

—. 1933 La structure cytologique des *Bacillus enterothrix camptospora* Collin et de *Bacillospira (Spirillum) praeclarum* Collin. C R Hebd Seances Acad Sci (Paris) *196:* 1830–1832.

—. 1934 Contribution à l'étude cytologique des microbes (*Coccus, Bacillus, Vibrio, Spirillum, Spirochaeta*). Arch Protistenk *83:* 465–468.

— and R. Favre. 1931 La structure cytologique de *Blattabacterium cuenoti* (Mercier) N. G., symbiote du tissu adipeux des Blattides. C R Soc Biol (Paris) *107:* 752–754.

Hollis, D. G., G. L. Wiggins and J. H. Schubert. 1968 Serological studies of ungroupable *Neisseria meningitidis.* J Bacteriol *95 (1):* 1–4.

—, — and R. E. Weaver. 1969 *Neisseria lactamicus* sp. n., a lactose-fermenting species resembling *Neisseria meningitidis.* Appl Microbiol *17:* 71–77.

Holloway, B. W. 1969 Genetics of *Pseudomonas.* Bacteriol Rev *33 (3):* 419–443.

Holm, P. 1948 Some investigations into the penicillin sensitivity of human pathogenic actinomycetes. Acta Pathol Microbiol Scand *25:* 376–404.

—. 1954 The influence of carbon-dioxide on the growth of *Actinobacillus actinomycetemcomitans (Bacterium actinomycetem comitans* [Klinger 1912]). Acta Pathol Microbiol Scand *34:* 235–248.

Holmes, R. K. and W. L. Barksdale. 1969 Genetic analysis of tox+ and tox− bacteriophage of *Corynebacterium diphtheriae.* J Virol *3:* 586–598.

Holt, J. G. and R. A. Lewin. 1968 *Herpetosiphon aurantiacus* gen. et sp. n., a new filamentous gliding organism. J Bacteriol *95 (6):* 2407–2408.

Holt, S. C. and E. Canale-Parola. 1967 Fine structure of *Sarcina maxima* and *Sarcina ventriculi.* J Bacteriol *93 (1):* 399–410.

— and —. 1968 Fine structure of *Spirochaeta stenostrepta,* a free-living anaerobic spirochete. J Bacteriol *96 (3):* 822–835.

—, S. F. Conti and R. C. Fuller. 1966 Photosynthetic apparatus in the green bacterium *Chloropseudomonas ethylicum.* J Bacteriol *91 (1):* 311–323.

—, H. G. Trüper and B. J. Takács. 1968 Fine structure of *Ectothiorhodospira mobilis* strain 8113 thylakoids: Chemical fixation and freeze-etching studies. Arch Mikrobiol *62:* 111–128.

Holtman, D. F. 1945 *Corynebacterium equi* in chronic pneumonia of the calf. J Bacteriol *49:* 159–162.

Holzapfel, W. and O. Kandler. 1969 Zur Taxonomie der Gattung *Lactobacillus* Beijerinck. VI. *Lactobacillus coprophilus* subsp. *confusus* nov. subsp., eine neue Unterart der Untergattung *Betabacterium.* Zentrabl Bakteriol Parasitenk Infektionskr Hyg Abt II *123:* 657–666.

Hopgood, M. F. and D. J. Walker. 1967 Succinic acid production by rumen bacteria. I. Isolation and metabolism of *Ruminococcus flavefaciens.* Aust J Biol Sci *20:* 165–182.

— and —. 1967 Succinic acid production by rumen bacteria. II. Radioisotope studies on succinate production by *Ruminococcus flavefaciens.* Aust J Biol Sci *20:* 183–192.

Hopkins, J. C. F. and W. J. Dowson. 1949 A bacterial leaf and flower disease in Southern Rhodesia. Trans Brit Mycol Soc *32:* 252–254.

Hori, S. 1911 A bacterial leaf-disease of tropical orchids. Zentrabl Bakteriol Parasitenk Infektionskr Hyg Abt II *31:* 85–92.

Horii, S., A. Miyake, K. Yamamoto and K. Nakazawa. 1959 On folimycin produced by *Streptomyces neyagawaensis.* Jap J Pharm Chem *31 (5):* 264–265.

Hormaeche, E. and P. R. Edwards. 1958 Observation on the genus *Aerobacter* with a description of two species. Int Bull Bacteriol Nomencl Taxon *8 (2):* 111–115.

— and —. 1960 A proposed genus *Enterobacter.* Int Bull Bacteriol Nomencl Taxon *10 (2):* 71–74.

Horowitz-Wlassowa, L. M. and N. W. Nowotelnow. 1932 Über eine sporogene Milchsäurebakterienart, *Lactobacillus sporogenes* n. sp. Zentrabl Bakteriol Parasitenk Infektionskr Hyg Abt II *87:* 331–333.

Hortobagyi, T. 1965 Neue *Planctomyces*-Arten. Botaniki Kozlem *52 (3):* 111–115.

Hoshina, T. 1957 Further observations on the causative bacteria of the epidemic disease like furunculosis of rainbow trout. J Tokyo Univ Fish *43:* 59–66.

Hosoya, S. 1955 On the new antitumor antibiotic carcinomycin: trade name-Gannmycin. Chemotherapy (Japan) *3:* 128–131.

—, N. Komatsu, M. Soeda and Y. Sonoda. 1952 Trichomycin, a new antibiotic produced by *Streptomyces hachijoensis* with trichomonadicidal and antifungal activity. Jap J Exp Med *22:* 505–509.

Howell, A. 1953 In vitro susceptibility of *Actinomyces* to terramycin. Antibiot Chemother *3:* 378–381.

—. 1963 A filamentous microorganism isolated from periodontal plaque in hamsters. I. Isolation, morphology and general cultural characteristics. Sabouraudia *3:* 81–92.

— and H. V. Jordan. 1963 A filamentous microorganism isolated from periodontal plaque in hamsters. II. Physiological and biochemical characteristics. Sabouraudia *3:* 93–105.

—, —, L. K. Georg and L. Pine. 1965 *Odontomyces viscosus* gen. nov. spec. nov. A filamentous microorganism isolated from periodontal plaque in hamsters. Sabouraudia *4:* 65–67.

—, W. C. Murphy, F. Paul and R. M. Stephan. 1959 Oral strains of *Actinomyces*. J Bacteriol *78 (1):* 82–95.

—, R. M. Stephan and F. Paul. 1962 Prevalence of *Actinomyces israelii. A. naeslundii, Bacterionema matruchotii* and *Candida albicans* in selected areas of the oral cavity. J Dent Res *41:* 1050–1059.

Hoyer, B. H. and N. B. McCullough. 1968 Polynucleotide homologies of *Brucella* deoxyribonucleic acids. J Bacteriol *95 (2):* 444–448.

— and —. 1968 Homologies of deoxyribonucleic acids from *Brucella ovis*, canine abortion organisms, and other *Brucella* species. J Bacteriol *96 (5):* 1783–1790.

Hoyer, D. P. 1898 Bijdrage tot de Kennis van de Azijnbacteriën. Dissertation, University of Leiden.

Hrubant, G. R. and R. A. Rhodes. 1968 Agglutinability of sporeforming insect pathogens with antiglobulins to milky disease bacteria. J Invertebr Pathol *11:* 371–376.

Huang, K. 1967 Metabolic activity of the trench fever rickettsia, *Rickettsia quintana*. J Bacteriol *93 (3):* 853–859.

Hübener, H. J. and H. Reiter. 1916 Beiträge zur Aetiologie der Weilschen Krankheit. Deut Med Wochenschr *42:* 1–2.

Huber, G., K. H. Wallhäusser, L. Fries, A. Steigler and H.-L. Weidenmüller. 1962 Niddamycin, ein neues Makrolid-Antibiotikum. Arzneimittel Forschung *12:* 1191–1195.

Huber-Pestalozzi, G. 1938 Das Phytoplankton des Süsswassers. Band XVI, Tel 1: Die Binnenge-wässer. Stuttgart 1938. E. Schweizerbart'sche Verlagsbuchhandlung, pp. 1–342.

Huchzermeyer, F. W. 1965 Das Tiefgefrieren von *Aegyptianella pullorum* in flüssigem Stickstoff mit einigen Bemerkungen über die künstliche infektion beim Huhn. Berlin Muenchen Tieraerztl Wochenschr *78:* 433–435.

Hucker, G. J. and C. S. Pederson. 1930 Studies on the *Coccaceae*. XVI. The genus *Leuconostoc*. N Y Agr Exp Sta Tech Bull *167:* 3–80.

Huddleson, I. F. 1929 The differentiation of the species of the genus *Brucella*. Mich State College Agr Exp Sta Tech Bull *100:* 1–16.

—. 1957 *In* Breed *et al.* (Editors), Bergey's Manual of Determinative Bacteriology, 7th ed. The Williams & Wilkins Co., Baltimore, p. 405.

Hudson, J. R. 1937 Cutaneous streptothricosis. Proc Roy Soc Med *30:* 1457–1460.

—, G. S. Cottew and H. E. Adler. 1967 Diseases of goats caused by *Mycoplasma*. A review of the subject with some new findings. Ann N Y Acad Sci *143 (1):* 287–297.

Huebner, R. J., W. L. Jellison and C. Pomerantz. 1946 Rickettsialpox, newly recognized rickettsial disease. IV. Isolation of a rickettsia apparently identical with the causative agent of Rickettsialpox from *Allodermanyssus sanguineus*, a rodent mite. Pub Health Rep *61:* 1677–1682.

Hueppe, F. 1886 Die Formen der Bakterien, und ihre Beziehungen zu den Gattungen und Arten. C. W. Kreidel's Verlag, Wiesbaden, pp. 1–152.

Huet, M. and N. Aladame. 1952 Recherches sur l'uréase des bactéries anaérobies. Ann Inst Pasteur (Paris) *82:* 766–767.

— and F. de Cadore. 1954 Technique d'isolement des bactéries anaérobies uréolytiques. Description d'une espèce nouvelle isolée par cette méthode. Ann Inst Pasteur (Paris) *86:* 241–243.

Huger, A. 1959 Histological observations on the development of crystalline inclusions of the rickettsial disease of *Tipula paludosa* Meigen. J Insect Pathol *1:* 60–66.

—. 1962 Zur Genese der Begleikristalle bei *Rickettsiella*-Infektionen von Insekten. Naturwissenschaften *49:* 358.

—. 1964 Eine Rickettsiose der Orientalischer Schabe, *Blatta orientalis* L. verursacht durch *Rickettsiella blattae* nov. spec. Naturwissenschaften *51:* 22.

Hugh, R. 1962 *Comamonas terrigena* comb. nov. with proposal of a neotype and request for an opinion. Int Bull Bact Nomencl Taxon *12 (2):* 33–35.

— and P. Ikari. 1964 The proposed neotype strain of *Pseudomonas alcaligenes* Monias (1928). Int Bull Bacteriol Nomencl Taxon *14 (3):* 103–107.

— and E. Leifson. 1953 The taxonomic significance of fermentative versus oxidative metabolism of carbohydrates by various gram negative bacteria. J Bacteriol *66 (1):* 24–26.

— and E. Ryschenkow. 1960 An alcaligenes-like *Pseudomonas* species. Bacteriol Proc. p. 78.

Hughes, D. E. and G. W. Pugh. 1970 Isolation and description of a *Moraxella* from horses with conjunctivitis. Amer J Vet Res *31:* 457–462.

Hughes, H. P. and E. L. Biberstein. 1959 Chronic equine abscesses associated with *Corynebacterium pseudo-tuberculosis*. J Amer Vet Med Ass *135:* 559–562.

Hughes, M. L. 1893 The natural history of certain fevers occurring in the Mediterranean. Mediterranean Nat *2:* 299–300; 325–327; 332–334.

Hughes, W. H. 1957 A reconsideration of the swarming of *Proteus vulgaris*. J Gen Microbiol *17 (1):* 49–58.

Humm, H. J. 1946 Marine agar-digesting bacteria of the South Atlantic Coast. Duke Univ Mar Sta Bull *3:* 45–75.

— and K. S. Shepard. 1946 Three new agar-digesting Actinomycetes. Duke Univ Mar Sta Bull *3:* 76–80.

Hungate, R. E. 1944 Cellulose-digesting bacteria in *Amitermes.* J Bacteriol *48:* 380–381.

—. 1944 Studies on cellulose fermentation. The culture and physiology of an anaerobic, cellulose-digesting bacterium. J Bacteriol *48:* 499–513.

—. 1946 Studies on cellulose fermentation. II. An anaerobic cellulose-decomposing actinomycete, *Micromonospora propionici* n. sp. J Bacteriol *51:* 51–56.

—. 1950 The anaerobic mesophilic cellulolytic bacteria. Bacteriol Rev *14:* 1–49.

—. 1957 Microorganisms in the rumen of cattle fed a constant ration. Can J Microbiol *3:* 289–311.

—. 1963 Polysaccharide storage and growth efficiency in *Ruminococcus albus.* J Bacteriol *86 (4):* 848–854.

—. 1966 The rumen and its microbes. Academic Press, New York.

Huntley, B. E., R. N. Philip and J. E. Maynard. 1963 Survey of brucellosis in Alaska. J Infect Dis *112:* 100–106.

Hurlbert, R. E. and J. Lascelles. 1963 Ribulose diphosphate carboxylase in *Thiorhodaceae.* J Gen Microbiol *33 (3):* 445–448.

Huss, H. 1907 Beitrag zur Kenntnis der Erdbeergeruch erzeugenden Bakterien. Zentrabl Bakteriol Parasitenk Infektionskr Hyg Abt II *19:* 661–674.

Hussaini, S. N. 1966 A taxonomic study of strains of *Pasteurella multocida* from a variety of animal and geographic sources. Ph.D. Thesis, University of London.

Hutchinson, C. M. 1917 A bacterial disease of wheat in the Punjab. Mem Dep Agr India Bact Ser *1:* 169–175.

Hutchinson, H. B. and J. Clayton. 1919 On the decomposition of cellulose by an aerobic organism (*Spirochaeta cytophaga,* n. sp.). J Agr Sci *9:* 143–173.

Hutchinson, P. B. 1949 A bacterial disease of *Dysoxylum spectabile* caused by the pathogen *Pseudomonas dysoxyli* n. sp. N Z J Sci Technol Sec B *30:* 274–286.

Hütter, R. 1961 Zur Systematik der Actinomyceten. 5. Die Art *Streptomyces albus* (Rossi-Doria emend. Krainsky) Waksman et Henrici 1943. Arch Mikrobiol *38:* 367–383.

—. 1962 Zur Systematik der Actinomyceten. 7. Streptomyceten mit blauen, blaugrünen und grünen Luftmycel. Arch Mikrobiol *43:* 23–49.

—. 1962 Zur Systematik der Actinomyceten. 8. Quirlbildende Streptomyceten. Arch Mikrobiol *43:* 365–391.

—. 1963 Zur Systematik der Actinomyceten. 10. Streptomyceten mit *griseus*-Luftmycel. G Microbiol *11:* 191–246.

—. 1964 Zur Systematik der Actinomyceten. 9. Streptomyceten mit *cinnamomeus* Luftmycel. Zentrabl Bakteriol Parasitenk Infektionskr Hyg Abt II *117:* 603–661.

—. 1967 Systematik der Streptomyceten unter besonderer Berucksichtigung der von ihnen gebildeten Antibiotica. Bibl Microbiol Fasc 6, 5. Karger AG, Basel, Switzerland.

—, W. Keller-Schierlein and H. Zähner. 1961 Zur Systematik der Actinomyceten. 6. Die Produzenten von Makrolid-Antibiotika. Arch Mikrobiol *39:* 158–194.

Hutyra, F. 1913 Septicaemia haemorrhagica. *In* Kolle and Wässerman (Editors), Handb Pathol Mikroorganismen, G. Fischer, Jena, pp. 64–95.

Huxsoll, D. L., P. K. Hildebrandt, R. M. Nims, J. A. Ferguson and J. S. Walker. 1969 *Ehrlichia canis*—the causative agent of a haemorrhagic disease of dogs? Vet Rec *85:* 587.

Hvid-Hansen, N. 1951 Sulfate-reducing and hydrocarbon-producing bacteria in groundwater. Acta Pathol Microbiol Scand *29:* 266–289.

Hylemon, P. B., J. S. Wells, J. H. Bowdre, T. O. MacAdoo and N. R. Krieg. 1973 Designation of *Spirillum volutans* Ehrenberg 1832 as type species of genus *Spirillum* Ehrenberg 1832 and designation of the neotype strain of *S. volutans.* Int J Syst Bacteriol *23:* 20–27.

—, — and N. R. Krieg. 1973 The genus *Spirillum:* A taxonomic study. Int J Syst Bacteriol *23 (4):* 340–380.

Iizuka, H. and K. Komagata. 1963 *Pseudomonas* isolated from rice, with special reference to the taxonomical studies of chromogenic group of genus *Pseudomonas.* J Agr Chem Soc Jap *37 (2):* 71–76.

— and —. 1964 Microbiological studies on petroleum and natural gas. I. Determination of hydrocarbon-utilizing bacteria. J Gen Appl Microbiol *10 (3):* 207–221.

—, I. Tanabe, R. Fukumura and K. Kato. 1967 Taxonomic study of the E-caprolactam-utilizing bacteria. J Gen Appl Microbiol *13 (2):* 125–137.

Ikata, S. and Y. Yamouchi. 1931 Bacterial streak of millet. J Plant Prot Tokyo *18:* 30–37.

Ikawa, M. and E. E. Snell. 1960 Cell wall composition of lactic acid bacteria. J Biol Chem *235:* 1376–1382.

Imamura, A., M. Hori, K. Nakazawa, M. Shibata, S. Tatsuoka and A. Miyake. 1956 A new species of *Streptomyces* producing dihydrostreptomycin. Proc Jap Acad Sci *32 (8):* 648–653.

Imshenetski, A. A. and L. Solntseva. 1936 On aerobic cellulose digesting bacteria. Izv Akad Nauk SSSR Ser Biol *6:* 1115–1172.

— and —. 1937 On cellulose-decomposing Myxobacteria (in Russian, English summary). Mikrobiologiya *6 (1):* 3–15.

— and —. 1945 The imperfect forms of myxobacteria. Mikrobiologiya *14:* 220–229.

Inada, R., Y. Ido, R. Hoki, R. Kaneko and H. Ito. 1916 The etiology, mode of infection, and specific therapy of Weil's disease (*Spirochaetosis icterohaemorrhagica*). J Exp Med *23:* 377–402.

International Code of Nomenclature of Bacteria. 1966 *In* Int J Syst Bacteriol *16 (4):* 459–490. Edited by The Editorial Board of the Judicial Commission of the International Committee on Nomenclature of Bacteria.

International Committee on Nomenclature of Bacteria. 1971 Minutes of Meeting of the Subcommittee on the Taxonomy of *Mycoplasmatales,* Mexico City, August 10, 1970. Int J Syst Bacteriol *21 (1):* 151–153.

International Enterobacteriaceae Subcommittee. 1968 Report and minutes of the meeting,

Moscow 1966. Int J Syst Bacteriol *18* (*3*): 191–196.

International Salmonella Subcommittee. 1934 The genus *Salmonella* Lignières 1900. J Hyg *34:* 333–350.

Irwin, J. and H. W. Seeley, Jr. 1958 Titration and partial characterization of a soluble hemolysin of a group D streptococcus. J Bacteriol *76* (*1*): 29–35.

Isachenko, B. L. 1914 Recherches sur les microbes de l'Océan Glacial Arctique. L'Expedition scientifique pour l'exploration des pêcheries de la côte Mourmáne. W. Kirschbaum, Petrograd I-VIII, pp. 1–297.

—. 1927 Microbiological studies of muddy lakes (in Russian, French summary). Mem Geolog Comm N.S., Issue *148:* 1–154.

—. 1929 Biological observations on sulfur bacteria. Collected works in honor of I. P. Borodin's anniversary (jubilee). Leningrad, pp. 1–15.

— and A. G. Saltimovskaya. 1928 On the morphology and physiology of the thionic acid bacteria. Bull Inst Hydrol, Leningrad *21:* 61–73.

Iseki, S., K. Furukawa and S. Yagiveroto. 1959 Blood group substance B-decomposing enzyme produced by an anaerobic bacterium. I. Serological action of the B-decomposing enzyme. Proc Jap Acad *35:* 507–512.

— and K. Kashiwagi. 1955 Induction of somatic 1 antigen by bacteriophage in *Salmonella* B group. Proc Jap Acad *31* (*8*): 558–563.

— and —. 1957 Lysogenic conversions and transduction of genetic characters by temperate phage iota in *Salmonella*. Proc Jap Acad *33* (*8*): 481–485.

— and T. Sakair. 1953 Artificial transformation of O antigens in *Salmonella* E group. Proc Jap Acad *29:* 121–126; 127–131.

Ishida, N., K. Kumagai, T. Niida, K. Hamamoto and T. Shomura. 1967 Nojirimycin, a new antibiotic. I. Taxonomy and fermentation. J Antibiot (Tokyo) Ser A *20:* 62–65.

Ishiyama, S. 1956 Bacterial leaf-blight of the rice plant. Proc Pan-Pacific Sci Congr Tokyo *2:* 2112.

Isono, K., K. Anzai and S. Suzuki. 1958 Tubermycins A and B, new antibiotics. I. J Antibiot (Tokyo) Ser A *11:* 264–267.

—, J. Nagatsu, Y. Kawashima and S. Suzuki. 1965 Studies on polyoxins, antifungal antibiotics. Part I. Isolation and characterization of polyoxins A and B. Agr Biol Chem *29* (*9*): 848–854.

— and S. Suzuki. 1960 9-β-D-Ribofuranosylpurine from a streptomycete. J Antibiot (Tokyo) Ser A *13* (*4*): 270–272.

—, S. Yamashita, Y. Tomiyama, S. Suzuki and H. Sakai. 1957 Studies on homomycin II. J Antibiot (Tokyo) Ser A *10:* 21–30.

Ito, S., K. Noguchi and F. Yasumura. 1963 Mekemycin, a new antibiotic substance. Preliminary report. Meij Yakkadaigaku Kenkyu Kiyo *2:* 1–8.

— and J. W. Vinson. 1965 Fine structure of *Rickettsia quintana* cultivated *in vitro* and in the louse. J Bacteriol *89* (*2*): 481–495.

Itzigsohn, H. 1868 Entwicklungsvorgänge von *Zoogloea, Oscillaria, Synedra, Staurastrum, Spirotaenia* und *Chroolepus*. S. B. Gesellschaft naturf Freunde, Berlin, 19 November 1867, pp. 30–31.

Ivanitskaya, L. P., G. D. Upiter, M. A. Sveshnikova and G. F. Gauze. 1966 Systematic position, variation and antibiotic properties of the producer of the antitumor antibiotic tavromycetin (in Russian) Antibiotiki *11* (*2*): 973–976.

Iwasaki, M. 1940 On the pentose-fermenting lactic acid bacteria. Bull Agr Chem Soc Jap *16:* 148–149.

Jablon, J. M., B. Brust and M. S. Saslaw. 1965 β-Hemolytic streptococci with Group A and Type II carbohydrate antigens. J Bacteriol *89* (*2*): 529–534.

Jackins, H. C. and H. A. Barker. 1951 Fermentative processes of the fusiform bacteria. J Bacteriol *61* (*2*): 101–114.

Jackson, B. D. 1928 A glossary of botanic terms, 4th ed. Hafner Pub. Co. Inc., New York (reprinted 1953).

Jackson, F. L. and Y. E. Goodman. 1972 Transfer of the facultatively anaerobic organism *Bacteroides corrodens* Eiken to a new genus *Eikenella*. Int J Syst Bacteriol *22* (*2*): 73–77.

Jackson, H. W. and M. E. Morgan. 1954 Identity and origin of the malty aroma substance from milk cultures of *Streptococcus lactis* var. *maltigenes*. J Dairy Sci *37:* 1316–1324.

Jackson, J. F., D. J. W. Moriarty and D. J. D. Nicholas. 1968 Deoxyribonucleic acid base composition and taxonomy of Thiobacilli and some nitrifying bacteria. J Gen Microbiol *53* (*1*): 53–60.

Jacob, A. 1961 Les Schizomycetes thermophiles. Physiologie, isolement, taxonomie, détermination. Bull Diplomes Microbiol Fac Pharm Nancy *83:* 31–44.

Jacob, F. M. and T. R. Helmbold. 1933 Bacteriologic studies on *Lichenplanus*. A preliminary report. Arch Dermatol Syph *27:* 472–480.

Jacobs, N. J. and P. J. Van Demark. 1960 Comparison of the mechanism of glycerol oxidation in aerobically and anaerobically grown *Streptococcus faecalis*. J Bacteriol *79* (*4*): 532–538.

Jacobsen, H. C. 1914 Die Oxydation von Schwefelwasserstoff durch Bakterien. Folia Microbiol *3:* 155–162.

Jacobsthal, E. 1920 Untersuchungen über eine syphilisahnliche Spontanerkrankung des Kaninchens (*Paralues cuniculi*). Derm Wochenschr *71:* 569–571.

The Jacques Loewe Research Foundation. 1955 Vegahren zur Erzeugung und Gewinnung eines Antibiotikums. German Patent 934,429, October 20.

—. 1956 New antibiotic-totomycin. British Patent 758,276, October 3.

—. 1957 On *Streptomyces crystallinus* and totomycin. Japanese Patent 5897, August 3.

Jagger, I. C. 1921 Bacterial leaf spot disease of celery. J Agr Res *21:* 185–188.

Jahn, E. 1906 Myxomycetenstudien. Ber Deut Bot Ges *24:* 538–541.

—. 1911 *Myxobacteriales*. Kryptogamenflora der Mark Brandenburg *5* (*1*): 187–206.

—. 1924 Beiträge zur botanischen Protistologie I. Die Polyangiden. Gebr. Borntraeger, Leipzig.

Jahn, T. L. and M. D. Landman. 1965 Locomotion of spirochetes. Trans Amer Microsc Soc *84:* 395–406.

Jain, N. C., D. E. Jasper and J. D. Dellinger. 1967 Cultural characteristics and serological relationships of some mycoplasma isolated from bovine sources. J Gen Microbiol *49 (3):* 401–410.

James, J. B. and C. F. Niven, Jr. 1951 Nutrition of the heterofermentative lactobacilli that cause greening of cured meat products. J Bacteriol *62 (5):* 599–603.

James, N. 1955 Yellow chromogenic bacteria on wheat. II. Determinative studies. Can J Microbiol *1:* 479–485.

Janke, A. 1924 Allgemeine Technische Mikrobiologie. I. Teil: Die Mikroorganismen. T. Steinkopf, Dresden und Leipzig.

—. 1950 *Acetobacter lafarianum* nov. nom. Arch Mikrobiol *15:* 116–118.

—. 1957 Zur Systematik der Essigbakterien. Zentrabl Bakteriol Parasitenk Infectionskr Hyg Abt II *110:* 728–739.

Janoschek, A. 1944 Zur Systematik der Propionsäurebakterien. Zentrabl Bakteriol Parasitenk Infektionskr Hyg Abt II *106:* 321–337.

Janot, M.-M., H. Pénau, G. Hagemann, H. Velu, J. Teillon and G. Bouet. 1954 Recherche de souches nouvelles de *Streptomyces* antibiotiques. Ann Pharm Fr. *12:* 440–447.

Jansen, B. C. 1952 The occurrence of *Eperythrozoon parvum* Splitter, 1950, in South African swine. Onderstepoort J Vet Res *24 (4):* 5–6.

Jarach, M. 1931 *Streptothrix rubescens* n. sp. Boll Sez Ital Soc Int Microbiol *3:* 43–45.

Jarvis, B. D. W. 1967 Antigenic relations of cellulolytic cocci in the sheep rumen. J Gen Microbiol *47 (2):* 309–319.

— and E. F. Annison. 1967 Isolation, classification and nutritional requirements of cellulolytic cocci in the sheep rumen. J Gen Microbiol *47 (2):* 295–307.

Jaschke, W. 1933 Beiträge zur Kenntnis der symbiotischen Einrichtungen bei Hirudineen und Ixodiden. Z Parasitenk *5:* 515–541.

Jawetz, E. 1950 A pneumotropic pasteurella of laboratory animals. I. Bacteriological and serological characteristics of the organism. J Infect Dis *86:* 172–183.

Jayne-Williams, D. J. and T. M. Skerman. 1966 Comparative studies on coryneform bacteria from milk and dairy sources. J Appl Bacteriol *29:* 72–92.

Jebb, W. H. H. 1948 Starch-fermenting gelatine-liquefying corynebacteria isolated from the human nose and throat. J Pathol Bacteriol *60:* 403–412.

Jeffers, E. E. 1964 Myxobacters of a freshwater lake and its environs. Int Bull Bacteriol Nomencl Taxon *14 (3):* 115–136.

— and J. G. Holt. 1961 The nomenclatural status of the taxa of the *Myxobacterales* (*Schizomycetes*). Int Bull Bacteriol Nomencl Taxon *11 (2):* 29–61.

Jennison, H. M. 1923 Potato blackleg with special reference to the etiological agent. Ann Rep Mo Bot Gard *10:* 1–72.

Jensen, A., O. Aasmundrud and K. E. Eimhjellen. 1964 Chlorophylls of photosynthetic bacteria. Biochim Biophys Acta *88:* 466–479.

Jensen, H. L. 1928 *Actinomyces acidophilus* n. sp., a group of acidophilic actinomycetes isolated from the soil. Soil Sci *25:* 225–233.

—. 1930 Actinomycetes in Danish soils. Soil Sci *30:* 59–77.

—. 1930 The genus *Micromonospora* Ørskov, a little known group of soil microorganisms. Proc Linnean Soc N S W *55:* 231–249.

—. 1931 Contributions to our knowledge of the Actinomycetales. II. The definition and subdivision of the genus *Actinomyces*, with a preliminary account of Australian soil Actinomycetes. Proc Linnean Soc N S W *56 (4):* 345–370.

—. 1932 Contributions to our knowledge of the Actinomycetales. III. Further observations on the genus *Micromonospora*. Proc Linnean Soc N S W *57:* 173–180.

—. 1932 Contributions to our knowledge of the *Actinomycetales* IV. The identity of certain species of *Mycobacterium* and *Proactinomyces* Proc Linnean Soc N S W *57:* 364–376.

—. 1934 Studies on saprophytic *Mycobacteria* and *Corynebacteria*. Proc Linnean Soc N S W *59:* 19–61.

—. 1953 The genus *Nocardia* (or *Proactinomyces*) and its separation from some other *Actinomycetales* with some reflection on the phylogeny of the *Actinomycetales*. Symp. Actinomycetales, Proc. VI. Int Congr Microbiol Rome Fond. Emanuele Paterno, pp. 69–88.

—. 1955 *Azotobacter macrocytogenes* n. sp. a nitrogen-fixing bacterium resistant to acid reaction. Acta Agr Scand *2-3:* 280–294.

—, P. K. De and R. Bhattacharya. 1959 A new nitrogen-fixing bacterium. Nature (London) *184:* 1743.

—, E. J. Petersen, P. K. De and R. Bhattacharya. 1960 A new nitrogen-fixing bacterium: *Derxia gummosa* nov. gen. nov. spec. Arch Mikrobiol *36:* 182–195.

Jensen, K. A., I. Klaer and L. Lundberg. 1966 Studies of the antigenic structure of mycobacteria. Acta Pathol Microbiol Scand *66:* 79–92.

—, — and —. 1968 Studies on the antigenic structure of mycobacteria. Report 5. Acta Pathol Microbiol Scand *73:* 450–458.

Jensen, R. G., K. L. Smith, J. E. Edmondson and C. P. Merilan. 1956 The characteristics of some rumen lactobacilli. J Bacteriol *72:* 253–258.

Jensen, V. 1960 *Arthrobacter ramosus* spec. nov. A new *Arthrobacter* species isolated from forest soils. Kgl Vet-Landbohojsk Arsskr, pp. 123–132.

Johne, A. and Frothingham. 1895 Ein Eigenthümlicher Fall von Tuberculose beim Rind. Deut Z Tiermed Verg Pathol *21:* 438–454.

Johns, A. T. 1952 The mechanism of propionic acid formation by *Clostridium propionicum*. J Gen Microbiol *6:* 123–127.

Johnson, A. H. 1966 Effect of amino acids and manganese on growth and metabolism of *Sphaerotilus discophorus*. Thesis, Washington State University.

Johnson, C. L. and W. Vishniac. 1970 Growth inhibition in *Thiobacillus neapolitanus* by histidine, methionine, phenylalanine, and threonine. J Bacteriol *104 (3):* 1145–1150.

Johnson, D. 1932 Some observations on chitin-destroying bacteria. J Bacteriol *24:* 335–340.

Johnson, F. H. and I. V. Shunk. 1936 An interesting new species of luminous bacteria. J Bacteriol *31:* 585–592.

Johnson, J. 1923 A bacterial leafspot of tobacco. J Agr Res *23:* 481–493.

Johnson, J. C. 1956 Pod twist: A previously unrecorded bacterial disease of French bean (*Phaseolus vulgaris* L.). Quart J Agr Sci *13:* 127–158.

Johnson, J. L., R. S. Anderson and E. J. Ordal. 1970 Nucleic acid homologies among oxidase-negative *Moraxella* species. J Bacteriol *101 (2):* 568–573.

— and C. S. Cummins. 1972 Cell wall composition and deoxyribonucleic acid similarities among the anaerobic coryneforms, classical propionibacteria, and strains of *Arachnia propionica*. J Bacteriol *109 (3):* 1047–1066.

— and E. J. Ordal. 1968 Deoxyribonucleic acid homology in bacterial taxonomy: Effect of incubation temperature on reaction specificity. J Bacteriol *95 (3):* 893–900.

Johnson, R. C. and V. G. Harris. 1967 Differentiation of pathogenic and saprophytic leptospires. I. Growth at low temperatures. J Bacteriol *94 (1):* 27–31.

— and —. 1968 Purine analogue sensitivity and lipase activity of leptospires. Appl Microbiol *16 (10):* 1584–1590.

—, — and J. K. Walby. 1969 Characterization of leptospires according to fatty acid requirements. J Gen Microbiol *55 (3):* 399–407.

Johnson, R. H. and K. L. Vosti. 1968 Purification of two fragments of M protein from a strain of group A, type 12 streptococcus. J Immunol *101:* 381–391.

Johnstone, D. B. and S. A. Waksman. 1947 Streptomycin II, an antibiotic substance produced by a new species of *Streptomyces*. Proc Soc Exp Biol Med *65:* 294–295.

Johnstone, K. I. and J. M. McLeod. 1949 Nomenclature of strains of *C. diphtheriae*. Pub Health Rep *64 (37):* 1181–1187.

Jolly, R. D. 1965 The pathogenic action of the exotoxin of *Corynebacterium ovis*. J Comp Pathol *75:* 417–431.

Jones, D., R. H. Deibel and C. F. Niven, Jr. 1963 Apparent pigment production by *Streptococcus faecalis* in the presence of metal ions. J Bacteriol *86 (1):* 171–172.

—, — and —. 1964 Catalase activity of two *Streptococcus faecalis* strains and its enhancement by aerobiosis and added cations. J Bacteriol *88 (3):* 602–610.

— and P. M. F. Shattock. 1960 The location of the group antigen of group D streptococcus. J Gen Microbiol *23 (2):* 335–343.

Jones, D. M. 1962 A pasteurella-like organism from the human respiratory tract. J Pathol Bacteriol *83:* 143–151.

Jones, F. S., M. Orcutt and R. B. Little. 1931 Vibrios (*Vibrio jejuni* n. sp.) associated with intestinal disorders of cows and calves. J Exp Med *53:* 853–864.

Jones, H., G. Rake and B. Stearns. 1945 Studies on lymphogranuloma venereum. III. The action of the sulfonamides on the agent of lymphogranuloma venereum. J Infect Dis *76:* 55–69.

Jones, H. E. and R. W. A. Park. 1967 The short and long form of *Proteus*. J Gen Microbiol *47 (3):* 359–367.

— and —. 1967 The influence of medium composition on the growth and swarming of *Proteus*. J Gen Microbiol *47 (3):* 369–378.

Jones, K. L. 1952 A new *Streptomyces* that produces vitamin B$_{12}$ actively. Pap Mich Acad Sci Arts Lett *37:* 47–48.

Jones, L. A. and S. G. Bradley. 1964 Phenetic classification of actinomycetes. Develop Ind Microbiol *5:* 267–272.

Jones, L. M. 1960 Comparison of phage typing with standard methods of species differentiation in brucellae. Bull World Health Organ *23:* 130–133.

—, M. Zanardi, D. Leong and J. B. Wilson. 1968 Taxonomic position in the genus *Brucella* of the causative agent of canine abortion. J Bacteriol *95:* 625–630.

Jones, L. R. 1901 *Bacillus carotovorus* n. sp., die Ursache einer weichen Faulnis der Mohre. Zentrabl Bakteriol Parasitenk Infektionskr Hyg Abt II *7:* 12–21.

—, A. G. Johnson and C. S. Reddy. 1917 Bacterial blight of barley. J Agr Res *11:* 625–644.

Jones, W. D., Jr., V. D. Abbott, A. L. Vestal and G. P. Kubica. 1966 A hitherto undescribed group of nonchromogenic mycobacteria. Amer Rev Resp Dis *94:* 790–795.

— and G. P. Kubica. 1963 The differential typing of certain rapidly growing mycobacteria based on their sensitivity to various dyes. Amer Rev Resp Dis *88:* 355–359.

— and —. 1965 Differential colonial characteristics of mycobacteria on oleic acid-albumin and modified corn meal agars. II. Investigation of rapidly growing mycobacteria. Zentrabl Bakteriol Parasitenk Infektionskr Hyg Abt I Orig *196:* 68–81.

— and —. 1968 Fluorescent antibody techniques with mycobacteria. III. Investigation of five serologically homogeneous groups of mycobacteria. Zentrabl Bakteriol Parasitenk Infektionskr Hyg Abt I Orig *207:* 58–62.

—, G. Saito and G. P. Kubica. 1965 Fluorescent antibody techniques with mycobacteria. Amer Rev Resp Dis *92:* 255–260.

Jordan, E. O. 1890 A report on certain species of bacteria observed in sewage. *In* Sedgewick, A report of the biological work of the Lawrence Experiment Station, including an account of methods employed and results obtained in the microscopical and bacteriological investigation of sewage and water. Report on water supply and sewerage (Part II). Rep Mass Bd Publ Hlth, pp. 821–844.

— and N. M. Harris. 1908 The cause of milk sickness or trembles. J Amer Med Ass *50:* 1665–1673.

Jordan, H. V., R. J. Fitzgerald and J. E. Faber, Jr. 1959 A survey of lactobacilli including pigmented strains from the oral cavity of the white rat. J Dent Res *38:* 611–617.

— and P. H. Keyes. 1964 Aerobic, gram-positive filamentous bacteria as etiologic agents of experimental periodontal disease in hamsters. Arch Oral Biol *9:* 401–414.

Jørgensen, B. V. 1960 Betragtninger over forskellige enterobacteriaceae gruppers biokemiske relationer og antigene struktur med saerligt henblik på differentiering ved isolation og identifikation af Salmonellabakterier. Nord Vet Med *12:* 197–229.

Joshi, J. G., W. R. Guild and P. Handler. 1963

The presence of two species of DNA in some halobacteria. J Mol Biol *6:* 34–38.

Joubert, L. 1958 Étude bactériologique et systématique de *Micrococcus abscedens ovis* (Morel 1911). Ann Inst Pasteur (Paris) *95:* 215–218.

Judicial Commission. 1952 Opinion 5. Conservation of the generic name *Pseudomonas* Migula 1894 and designation of *Pseudomonas aeruginosa* (Schroeter) Migula 1900 as type species. Int Bull Bacteriol Nomencl Taxon *2 (3):* 121–122.

—. 1954 Opinion 6. The bacterial generic name *Chlorobacterium* Lauterborn 1915 is conserved against the earlier homonym *Chlorobacterium* Guillebeau 1890. The generic name *Chlorobacterium* Guillebeau 1890 is placed in the list of *nomina generum rejicienda.* Int Bull Bacteriol Nomencl Taxon *4 (3–4):* 141–158.

—. 1954 Opinion 11. Nomenclature of species in the bacterial genus *Shigella.* Int Bull Bacteriol Nomencl Taxon *4 (3–4):* 148–149.

—. 1954 Opinion 12. Conservation of *Listeria* Pirie 1940 as a generic name in bacteriology. Int Bull Bacteriol Nomencl Taxon *4 (3–4):* 150–151.

—. 1954 Opinion 13. Conservation and rejection of names of genera of bacteria proposed by Trevisan 1842–1890. Int Bull Bacteriol Nomencl Taxon *4 (3–4):* 151–154.

—. 1954 Opinion 14. Names of bacterial genera to be rejected as later synonyms of names of genera of Protozoa. Int Bull Bacteriol Nomencl Taxon *4 (3–4):* 156–158.

—. 1958 Opinion 15. Conservation of the family name *Enterobacteriaceae,* of the name of the type genus and designation of the type species. Int Bull Bacteriol Nomencl Taxon *8 (1):* 73–74.

—. 1958 Opinion 16. Conservation of the generic name *Chromobacterium* Bergonzini 1880 and designation of the type species and the neotype culture of the type species. Int Bull Bacteriol Nomencl Taxon *8 (3–4):* 151–152.

—. 1958 Opinion 19. Conservation of the generic name *Rickettsia* da Rocha-Lima and of the species name *Rickettsia prowazekii* da Rocha-Lima. Int Bull Nomencl Taxon *8 (3–4):* 158–159.

—. 1958 Opinion 20. Status of new generic names of bacteria published without names of included species. Int Bull Bacteriol Nomencl Taxon *8 (3–4):* 160–162.

—. 1958 Opinion 21. Conservation of the generic name *Selenomonas* von Prowazek. Int Bull Bacteriol Nomencl Taxon *8 (3–4):* 163–165.

—. 1958 Opinion 22. Status of the generic name *Asterococcus* and conservation of the generic name *Mycoplasma.* Int Bull Bacteriol Nomencl Taxon *8 (3–4):* 166–168.

—. 1958 Opinion 23. Rejection of the generic names *Nitromonas* Winogradsky 1890 and *Nitromonas* Orla-Jensen 1909 conservation of the generic names *Nitrosomonas* Winogradsky 1892, *Nitrosococcus* Winogradsky 1892, and *Nitrobacter* Winogradsky 1892 and the designation of the type species of these genera. Int Bull Bacteriol Nomencl Taxon *8 (3–4):* 169–170.

—. 1962 Proposal for the recognition of the neotype strain of *Streptomyces albus* (Rossi-

Doria) Waksman and Henrici 1943, p. 339. Int Bull Bacteriol Nomencl Taxon *12:* 65.

—. 1963 Detailed minutes concerning actions taken on opinions. Minute 67. Int Bull Bacteriol Nomencl Taxon *13:* 23–30.

—. 1963 Opinion 26. Designation of neotype strains (cultures) of type species of the bacterial genera *Salmonella, Shigella Arizona, Escherichia, Citrobacter,* and *Proteus* of the family *Enterobacteriaceae.* Int Bull Bacteriol Nomencl Taxon *13 (1):* 35–36.

—. 1963 Opinion 28. The generic name *Cloaca* Castellani and Chalmers is rejected and replaced by the generic name *Enterobacter* Hormaeche and Edwards with the type species *Enterobacter cloacae* (Jordan) Hormaeche and Edwards; the basionym is *Bacillus cloacae* Jordan. Int Bull Bacteriol Nomencl Taxon *13 (1):* 38.

—. 1963 Opinion 29. Designation of strain ATCC 3004 (IMRU 3004) as the neotype strain of *Streptomyces albus* (Rossi Doria) Waksman and Henrici. Int Bull Bacteriol Nomencl Taxon *13 (2):* 123–124.

—. 1970 Opinion 32. Conservation of the specific epithet *rhusiopathiae* in the scientific name of the organism known as *Erysipelothrix rhusiopathiae.* (Migula) Buchanan 1918. Int J Syst Bacteriol *20 (1):* 9.

—. 1970 Opinion 33. Conservation of the generic name *Agrobacterium* Conn 1942. Int J Syst Bacteriol *20 (1):* 10.

—. 1970 Opinion 34. Conservation of the generic name *Rhizobium* Frank 1889. Int J Syst Bacteriol *20 (1):* 11–12.

—. 1970 Opinion 35. Conservation of the specific epithet *meningitidis* in the scientific name of the meningococcus. Int J Syst Bacteriol *20 (1):* 13–14.

—. 1971 Opinion 38. Conservation of the generic name *Lactobacillus* Beijerinck. Int J Syst Bacteriol *21 (1):* 104.

—. 1971 Opinion 39. Rejection of the generic name *Gaffkya* Trevisan. Int J Syst Bacteriol *21 (1):* 104–105.

—. 1971 Opinion 40. Rejection of the names *Mima* DeBord and *Herellea* DeBord and of the specific epithets *polymorpha* and *vaginicola* in *Mima polymorpha* DeBord and *Herellea vaginicola* DeBord, respectively. Int J Syst Bacteriol *21:* 105.

—. 1971 Opinion 41. Conservation of the generic name *Moraxella* Lwoff. Int J Syst Bacteriol *21 (1):* 106.

—. 1971 Opinion 42. Conservation of the specific epithet *"phenylpyruvica"* in the name of *Moraxella phenylpyruvica* Bøvre and Henriksen. Int J Syst Bacteriol *21 (1):* 107.

—. 1971 Opinion 45. Rejection of the name *Leuconostoc citrovorum* (Hammer) Hucker and Pederson. Int J Syst Bacteriol *21 (1):* 109–110.

Julianelle, L. A. 1926 A biological classification of encapsulatus pneumoniae (Friedländer's bacillus). J Exp Med *44:* 113–128.

Jungano, M. 1907 Sur un staphylocoque anaérobie. C R Soc Biol Paris *62:* 707–708.

—. 1908 *Bacillus parvus liquefaciens anaerobie.* C R Soc Biol Paris *65:* 618–620.

—. 1909 Sur la flore anaérobie du rat. C R Soc Biol Paris *66:* 112–114; 122–124.

Jungmann, P. and M. H. Luczynski. 1918 Zur Aetiologie und Pathogenese des Wolhynischer

Fiebers und Fleckfiebers. Z Klin Med *85:* 251–272.

Jyssum, K. 1959 Assimilation of nitrogen in meningococci grown with the ammonium ion as sole nitrogen source. Acta Pathol Microbiol Scand *46:* 320–332.

— and P. E. Joner. 1965 Regulation of the nitrogen assimilation from nitrate and nitrite in *Bacterium anitratum* (B5W). Acta Pathol Microbiol Scand *64:* 387–397.

— and S. Jyssum. 1968 Isolation of variants with increased mutability from *Neisseria meningitidis*. Acta Pathol Microbiol Scand *74:* 93–100.

Kaars Sijpesteijn, A. and G. Fahraeus. 1949 Adaptation of *Sporocytophaga myxococcoides* to sugars. J Gen Microbiol *3:* 224–234.

Kaczka, E. A., F. J. Wolf, F. P. Rathe and K. Folkers. 1955 Cathomycin. I. Isolation and characterization. J Amer Chem Soc *77 (23):* 6404–6405.

Kadota, H. 1951 Studies on the biochemical activities of marine bacteria. I. On the agar-decomposing bacteria in the sea. Mem Coll Agr Kyoto Univ *59:* 54–67.

—. 1953 The microbiological deterioration of fishing nets. Bull Jap Soc Sci Fish *19:* 476–489.

—. 1954 Microbiological studies on the weakening of fishing nets. V. A taxonomical study on marine cytophagas. Bull Jap Soc Sci Fish *20:* 125–129.

Kairies, A. 1935 Influenzaerkrankungen bei Frettchen und Beschreibung eines *Bacterium influenzae putoriorum multiforme*. Z Hyg Infektionskr *117:* 12–17.

— and K. Schwartzer. 1936 Studien zu einer bakteriellen Influenza der Mause und Beschreibung eines *"Bacterium influenzae murium"*. Zentrabl Bakteriol Parasitenk Infektionskr Hyg Abt I Orig *137:* 351–359.

Kaiser, M. N., H. Hoogstraal and G. M. Kohls. 1964 The subgenus *Persicargas*, new subgenus (*Ixodoidea, Argasidae, Argas*). I. *A. (P.) arboreus*, new species, an Egyptian *Persicus*-like parasite of wild birds, with a redefinition of the subgenus *Argas*. Ann Entomol Soc Amer *57:* 60–69.

Kalakutskii, L. V. 1964 A new species of the genus *Micropolyspora: Micropolyspora caesia* n. sp. Mikrobiologiya *33:* 765–768 (trans.).

—, N. S. Agre and N. A. Krasil'nikov. 1968 Comparative study on some oligosporic actinomycetes. Hindustan Antibiot Bull *10:* 254–268.

— and V. D. Kusnetsov. 1964 A new species of the Actinoplanes- *A. armeniacus* and some peculiarities of its mode of spore formation. Mikrobiologiya *33:* 613.

Kalbe, L., R. Keil and M. Thiele. 1965 Licht und elektronenmikroskopische Studien an Arten von *Leptothrix, Siderocapsa* und *Planctomyces*. Arch Protistenk *108:* 29–40.

Kalchbrenner, C. and M. C. Cooke. 1880 South African Fungi. Grevillea *9:* 17–34.

Kalinenko, V. O. 1933 Black necrosis on *Scarzonera tau-saghys* Lipsch et Bros. Mikrobiologiya *2:* 211–217.

—. 1934 Root macerations disease of tau-saghys. Mikrobiologiya *3:* 409–416.

Kalninš, A. 1930 Aerobic soil bacteria that decompose cellulose. Latv Ūniv. Rak Lauk-saimniecibas fakultātes serija I *11:* 221–312.

Kalonaros, J. V. and A. N. Bahn. 1965 Antigenic composition of the cell wall of *Streptococcus mitis*. Arch Oral Biol *10:* 625–633.

Kanazawa, K., K. Tsuchiya and T. Araki. 1960 A new antituberculous amino acid (α-hydroxy-γ-oxo-L-norvaline). Amer Rev Resp Dis *81:* 924.

Kanda, N. K., Asano and R. Shinobu. 1962 On *Streptomyces senoensis* nov. sp. Mem Osaka Univ Lib Arts Educ B Nat Sci *10:* 218–225.

Kandler, O. 1967 Taxonomie und technologische Bedeutung der Gattung *Lactobacillus* Beijerinck. Zentrabl Bakteriol Parasitenk Infektionskr Hyg Abt I Orig Suppl *2:* 139–164

—. 1970 Amino acid sequence of the murein and taxonomy of the genera *Lactobacillus, Bifidobacterium, Leuconostoc* and *Pediococcus*. Int J Syst Bacteriol *20 (4):* 491–507.

— and I. G. Abo-Elnaga. 1966 Zur Taxonomie der Gattung *Lactobacillus* Beijerinck. IV. *L. corynoides* ein Synonym von *L. viridescens*. Zentrabl Bakteriol Parasitenk Infektionskr Hyg Abt II *120:* 753–754.

— and —. 1966 Zur Taxonomie der Gattung *Lactobacillus* Beijerinck. V. *Lactobacillus coprophilus* nov. spec. eine neue Art der untergattung *Betabacterium*. Zentrabl Bakteriol Parasitenk Infektionskr Hyg Abt II *120:* 755–759.

—, D. Koch and K. H. Schliefer. 1968 Die Aminosaüresequenz eines glycinhaltigen Muriens einiger Stamme von *Lactobacillus bifidus*. Arch Mikrobiol *61:* 181–186.

—, R. Plapp and W. Holzapfel. 1967 Die aminosaüresequenz des serinhaltigen mureins von *Lactobacillus viridescens* und *Leuconostoc*. Biochim Biophys Acta *147:* 252.

—, K. H. Schleifer and R. Dandl. 1969 Differentiation of *Streptococcus faecalis* Andrewes and Horder and *Streptococcus faecium* Orla-Jensen based on the amino acid composition of their murein. J Bacteriol *96 (6):* 1935–1939.

—, —, E. Neibler, M. Nakel, H. Zahradnik and M. Reid. 1970 Murein types in micrococci and similar organisms. Publ Fac Sci Univ J. E. Purkyne Brno *47:* 143–156.

Kane, J. A. and W. W. Karakawa. 1969 Immunochemical analysis of *Streptococcus bovis* strain S19, cell walls. J Gen Microbiol *56 (2):* 157–164.

—, H. Lackland, W. W. Karakawa and R. M. Krause. 1969 Chemical studies on the structure of mucopeptide isolated from *Streptococcus bovis*. J Bacteriol *99 (1):* 175–179.

Kane, W. D. 1966 A new genus of the *Actinoplanaceae, Pilimelia*, with a description of two species, *Pilimelia terevasa* and *Pilimelia anulata*. J Elisha Mitchell Sci Soc *82:* 220–230.

Kaneda, T. and J. M. Roxburgh. 1959 Serine as an intermediate in the assimilation of methanol by *Pseudomonas*. Biochim Biophys Acta *33:* 106–110.

Kanegasaki, S. and H. Takahashi. 1967 Function of growth factors for rumen microorganisms. 1. Nutritional characteristics of *Selenomonas ruminantium*. J Bacteriol *93 (1):* 456–463.

Kaneko, T., K. Kitamura and Y. Yamamoto. 1969 *Arthrobacter luteus* nov. sp. isolated from brewery sewage. J Gen Appl Microbiol 15: 317–326.

Karassova, V. J. Weissfeiler and E. Krasznay. 1965 Occurrence of atypical mycobacteria in Macacus rhesus. Acta Microbiol Acad Sci Hung 12 (3): 275–282.

Karlson, A. G. and E. F. Lessel. 1970 Mycobacterium bovis nom. nov. Int J Syst Bacteriol 20 (3): 273–282.

—, H. E. Moses and W. H. Feldman. 1940 Corynebacterium equi (Magnusson 1923) in the submaxillary lymph nodes of swine. J Infect Dis 67: 243–251.

Kasai, G. J. 1961 A study of Leptotrichia buccalis. I. Morphology and preliminary observations. J Dent Res 40: 800–811.

—. 1965 A study of Leptotrichia buccalis. II. Biochemical and physiological observations. J Dent Res 44: 1015–1022.

Kåss, E., I. Lid and J. Molland. 1945 Investigations into the bacteria tribes Pseudomonadeae and Eschericheae with special attention to the alginic acid decomposition. Avh Norske Videnskaps-Akad Oslo Mat-Naturvidensk K1 11: 1–22.

Kassirsky, J. A. 1933 Diagnose und Klinik des mittel-asiatischen Zecken-Rückfallfiebers. Arch Schiffs-Trop Hyg 37: 380–387.

Katagiri, H., K. Kitahara and K. Fukami. 1934 The characteristics of the lactic acid bacteria isolated from moto, yeast mashes for saké manufacture. IV. Classification of the lactic acid bacteria. Bull Agr Chem Soc Jap 10: 156–157.

Katagiri, K. and J. Shoji. 1964 A new antibiotic with a chemoprophylactic effect against polio virus infection in mice. Proc III Intern Congr Chemother 1: 865–867.

Kataoka, T. 1930 On the significance of the root-nodules of Coriaria japonica, A. Gr. in the nitrogen nutrition of the plant. Jap J Bot 5: 209–218.

Kates, M., B. Palameta, C. N. Joo, D. J. Kushner and N. E. Gibbons. 1966 Aliphatic ether analogues of glyceride-derived lipids. The occurrence of di-o-dihydrophytolglycerol ether containing lipids in extremely halophilic bacteria. Biochemistry 5: 4092–4099.

Katô, H. 1949/1950 Studies on the antibiotic substances from Actinomyces. I. On the isolation of active strains and substances from strain No. 212. J Antibiot (Tokyo) 3: 579–581.

—. 1953 Studies on the antibiotic substances from Actinomyces. IV. On the taxonomic study and determination of the strain No. 212. J Antibiot (Tokyo) Ser B 6 (4): 205–209.

—. 1959 On Streptomyces sendaiensis n. sp. producing chloramphenicol. Ecol Rev 15: 79–82.

Kato, M. 1970 Action of a toxic glycolipid of Corynebacterium diphtheriae on mitochondrial structure and function. J Bacteriol 101 (3): 709–716.

Katznelson, H. 1950 Bacillus pulvifaciens, n. sp., an organism associated with powdery scale of honeybee larvae. J Bacteriol 59 (2): 153–155.

—. 1955 Bacillus apiarius n. sp., an aerobic spore-forming organism isolated from honeybee larvae. J Bacteriol 70 (6): 635–636.

— and C. A. Jamieson. 1952 Antibiotics and other chemotherapeutic agents in the control of bee diseases. Scientia Agr 32: 219–225.

— and A. G. Lochhead. 1948 Nutritional requirements of Bacillus larvae. J Bacteriol 55 (5): 763–764.

Kauffmann, F. 1933 Vergleichende Untersuchungen an Pseudotuberkulose-Paratyphus, Pasteurella and Pestbakterien. Z Hyg Infektionskr Med Mikrobiol Immunol Virol 114: 97–105.

—. 1941 Uber mehrere neue Salmonella types. Acta Pathol Microbiol Scand 18: 351–366.

—. 1941 Die Bakteriologie der Salmonella-Gruppe. E Munksgaard, Copenhagen.

—. 1949 On the serology of the Klebsiella group. Acta Pathol Microbiol Scand 26: 381–406.

—. 1954 Enterobacteriaceae. 2nd ed. E. Munksgaard, Copenhagen.

—. 1956 A new antigen of S. paratyphi B and S. typhi murium. Acta Pathol Microbiol Scand 39: 299–304.

—. 1960 Two biochemical subdivisions of the genus Salmonella. Acta Pathol Microbiol Scand 49: 393–396.

—. 1961 The species-definition in the Enterobacteriaceae. Int Bull Bacteriol Nomencl Taxon 11 (1): 5–6.

—. 1962 Supplement to the Kauffmann-White Scheme (V). Acta Pathol Microbiol Scand 55: 349–354.

—. 1963 Zur differential diagnose der Salmonella, sub-genera I, II und III. Acta Pathol Microbiol Scand 58: 109–113.

—. 1963 On the species definition. Int Bull Bacteriol Nomencl Taxon 13 (4): 181–186.

—. 1964 Vereinfachtes Antigen-Schema der Salmonella sub-genera II, III. Acta Pathol Microbiol Scand 62: 68–72.

—. 1965 Die Diagnose von Arizona-Kulturen nach dem originalen Kauffman-White Schema. Pathol Microbiol 28: 575–580.

—. 1966 Das Salmonella Sub-genus IV. Ann Immunol Hung 9: 77–80.

—. 1966 The bacteriology of Enterobacteriaceae. Munksgaard, Copenhagen.

— and P. R. Edwards. 1952 Classification and nomenclature of Enterobacteriaceae. Int Bull Bacteriol Nomencl Taxon 2 (1): 2–8.

— and F. Ørskov. 1956 Die Bakteriologie der Escherichia coli-Enteritis. In Säuglingsenteritis. Georg Thieme Verlag, Stuttgart.

— and R. Rohde. 1962 Eine Vereinfachung der serologischen Arizona, Diagnose. Acta Pathol Microbiol Scand 54: 473–478.

Kawaguchi, H., M. Okanishi and H. Tsukiura. 1963 Antibiotic substance. Belgian Patent 634,041, December 24.

—, H. Tsukiura, M. Okanishi, T. Miyaki, T. Ohmori, K. Fujisawa and H. Koshiyama. 1956 Studies on coumermycin, a new antibiotic. I. Production, isolation and characterization of coumermycin A₂. J. Antibiot (Tokyo) Ser A 18: 1–10.

Kawamura, A. 1954 Notes on the new strains of Rickettsia tamiyai n. sp. isolated from voles in Hokkaido, Japan. Jap J Exp Med 24: 47–50.

Kawamura, E. 1934 Bacterial blight of chestnut. Ann Phytopathol Soc Jap 3: 15–21.

Kawamura, R. and Y. Imagawa. 1931 Die Feststellung des Erregers bei der Tsutsugamushi-Krankheit. Zentrabl Bakteriol Parasitenk Infektionskr Hyg Abt Orig *122:* 253–261.

Kawato, N. and R. Shinobu. 1959 Concerning a new species of actinomycete. *Streptomyces viridofaciens.* Abstr., Lectures 24th Meeting, Japan Botan Soc Kawauchi Branch, Tohoku Univ, Sendai City, No. 53. Abstract B33.

— and —. 1959 On *Streptomyces herbaricolor* nov. sp. Supplement: A simple technique for the microscopical observation. Mem Osaka Univ Lib Arts Educ B Nat Sci *8:* 114–119.

— and —. 1960 On the utilization of N-sources of *Streptomyces.* I. On nitrate and nitrite when glycerine is used as C-source. Mem Osaka Univ Lib Arts Educ B Nat Sci *9:* 54–62.

— and —. 1962 On the utilization of N-sources of *Streptomyces.* II. On nitrate and nitrite when glucose is used as C-source. Mem Osaka Univ Lib Arts Educ B Nat Sci *10:* 211–217.

Kayser, H. 1902 Das Wachstum der zurischen *Bacterium typhi* und *coli* stehenden Spaltpilze auf dem v. Drigalski-Conradi'schen Agarboden. Zentrabl Bakteriol Parasitenk Infektionskr Hyg Abt I Orig *31:* 426–429.

Kazda, J. 1967 Mykobakterien im Trinkwasser als Ursache der Parallergie gegenuber Tuberkulinen bei Tieren. III. Mittilung: Taxonomische Studie einiger rasch wachsender Mykobakterien und Beschreibung einer neuen Art: *Mycobacterium brunense* n. sp. Zentrabl Bakteriol Parasitenk Infektionkr Hyg Abt I Orig *203 (2):* 199–211.

Keddie, R. M. 1959 The properties and classification of lactobacilli isolated from grass and silage. J Appl Bacteriol *22:* 403–416.

—, B. G. S. Leask and J. M. Grainger. 1966 A comparison of coryneform bacteria from soil and herbage: Cell wall composition and nutrition. J Appl Bacteriol *29:* 17–43.

Kele, R. A. and E. McCoy. 1970 Defined liquid medium for *Caryophanon latum.* Bacteriol Proc *1970:* 66.

Keléti, J., O. Lüderitz, D. Mlynarčík and J. Sedlák. 1971 Immunochemical studies on Citrobacter O-antigen. Eur J Biochem *20:* 237–244.

Kellen, W. R. and J. E. Lindegren. 1970 Previously unreported pathogens from the navel orangeworm *Paramyelois transitella* in California. J Invertebr Pathol *16:* 342–345.

Kellerman, K. F. 1915 Micrococci causing red deterioration of salted codfish. Zentrabl Bakteriol Parasitenk Infektionskr Hyg Abt II *42:* 398–402.

— and I. G. McBeth 1912 The fermentation of cellulose. Zentrabl Bakteriol Parasitenk Infektionskr Hyg Abt II *34:* 485–494.

—, —, F. M. Scales and N. R. Smith. 1913 Identification and classification of cellulose dissolving Bacteria. Zentrabl Bakteriol Parasitenk Infektionskr Hyg Abt II *39:* 502–522.

— and N. R. Smith. 1914 Bacterial precipitation of calcium carbonate. J Wash Acad Sci *4:* 400–402.

Kellogg, D. S., W. L. Peacock, W. E. Deacon, L. Brown, and C. I. Pirkle. 1963 *Neisseria gonorrhoeae.* I. Virulence genetically linked to clonal variation. J Bacteriol *85 (6):* 1274–1279.

Kelly, C. D. 1939 Genus I. *Acetobacter* Beijerinck 222–232. *In* Bergey, Breed, Murray and Hitchens (Editors), Bergey's Manual of Determinative Bacteriology, 5th ed. The Williams and Wilkins Co., Baltimore.

—. 1957 Genus I. *Bacteroides* Castellani and Chalmers 1919. *In* Breed, Murray and Smith (Editors), Bergey's Manual of Determinative Bacteriology, 7th ed. The Williams and Wilkins Co., Baltimore, pp. 424–436.

— and R. H. Vaughn. 1948 Genus IV. *Acetobacter.* Beijerinck 179–198. *In* Breed, Murray and Hitchens (Editors), Bergey's Manual of Determinative Bacteriology, 6th ed. The Williams and Wilkins Co., Baltimore.

Kelly, D. P. 1969 Regulation of chemoautotrophic metabolism. I. Toxicity of phenylalanine to thiobacilli. Arch Mikrobiol *69:* 330–342.

—. 1971 Autotrophy: Concepts of lithotrophic bacteria and their organic metabolism. Annu Rev Microbiol *25:* 177–210.

— and O. H. Tuovinen. 1972 Recommendation that the names *Ferrobacillus ferrooxidans* Leathen and Braley and *Ferrobacillus sulfooxidans* Kinsel be recognized as synonyms of *Thiobacillus ferrooxidans* Temple and Colmer. Int J Syst Bacteriol *22 (3):* 170–172.

Kelly, J., A. H. Kutscher and F. Tuoti. 1959 Thiostrepton, a new antibiotic: tube dilution sensitivity studies. Oral Surg Oral Med Pathol *12:* 1334–1339.

Kelly, M. and S. Liaaen-Jensen. 1967 Bacterial carotenoids. XXVI. C-50 carotenoids. 2. Bacterioruberin. Acta Chem Scand *21:* 2578–2580.

Kelterborn, E. 1967 *Salmonella*-Species. Hirzel Publ., Leipzig.

Kemenes, F. and B. Markói. 1959 Aktinobacillus Lignièresi okozta tályog kutya szájüregében. Magy Allatorv Lapja *14:* 31–32.

Kendrick, J. B. 1926 Holcus bacterial spot on species of *Holcus* and *Zea mays.* Phytopathology *16:* 236–237.

—. 1934 Bacterial blight of carrot. J Agr Res *49:* 483–510.

— and K. F. Baker. 1942 Bacterial blight of garden stocks and its control by hot-water seed treatment. Calif Agr Exp Sta Bull *665:* 1–23.

Kennedy, B. W. and T. H. King. 1962 Angular leaf spot of strawberry caused by *Xanthomonas frageriae* sp. nov. Phytopathology *52:* 873–875.

Kenny, C. P., F. E. Ashton, B. B. Deina and L. Greenberg. 1967 A chemically defined protein-free liquid medium for the cultivation of some species of *Neisseria.* Bull World Health Organ *37:* 569–573.

Kestle, D. G., V. D. Abbott and G. P. Kubica. 1967 Differential identification of mycobacteria. II. Subgroups of Groups II and III (Runyon) with different clinical significance. Amer Rev Resp Dis *95:* 1041–1052.

Keysselitz, G. *Spirochaeta anodontae* nov. spec. Arb GesundhAmt Berl *31:* 566–567.

Keyworth, W. G., J. Howell and W. J. Dowson. 1956 *Corynebacterium betae (sp. nov.)* The causal organism of silvering disease of red beet. Plant Pathol *5:* 88–90.

Khairat, O. 1940 Endocarditis due to new species of *Haemophilus.* J Pathol Bacteriol *50:* 497–505.

—. 1967 *Bacteroides corrodens* isolated from bacteriaemias. J Pathol Bacteriol *94:* 20–40.

Khambata, S. R. and J. V. Bhat. 1953 Studies on a new oxalate-decomposing bacterium, *Pseudomonas oxalaticus.* J. Bacteriol *66 (5):* 505–507.

Khokhlov, A. S. and T. S. Liberman. 1960 Isolation, purification and some characteristics of the new antibiotic albofungin. Proc Symp on Antibiotics, Prague, Czechoslovakia, May 18–23, 1959, pp. 81–83.

Kikuth, W. 1928 Über Einen neun anämieerreger, *Bartonella canis* nov. spec. Klin Wochenschr *1:* 1729–1730.

Kilburn, J. O., V. A. Silcox and G. P. Kubica. 1969 Differential identification of mycobacteria. V. The tellurite reduction test. Amer Rev Resp Dis *99:* 94–100.

Kim, K. S., W. A. Clyde and F. W. Denny. 1966 Physical properties of human *Mycoplasma* species. J Bacteriol *92:* 214–219.

Kimura, R. and T. H. Liao. 1953 A new thiamine decomposing anaerobic bacterium, *Clostridium thiaminolyticum* Kimura et Liao. Proc Jap Acad *29:* 132–133.

Kimura, T. 1969 A new subspecies of *Aeromonas salmonicida* as an etiological agent of furunculosis on "Sakuramasu" (*Oncorhynchus masou*) and Pink Salmon (*O. gorbuscha*) rearing for maturity. Part 1. On the morphological and physiological properties. Part 2. On the serological properties. Fish Pathol (Tokyo) *3:* 34–44; 45–52.

King, B. M. and D. L. Adler. 1964 A previously undescribed group of *Enterobacteriaceae.* Amer J Clin Pathol *41:* 230–232.

King, E. O. 1957 Human infections with *Vibrio fetus* and a closely related vibrio. J Infect Dis *101:* 119–128.

—. 1959 Studies on a previously unclassified bacteria associated with meningitis in infants. Amer J Clin Pathol *31:* 241–247.

—, M. K. Ward and D. E. Raney. 1954 Two simple media for the demonstration of pyocyanin and fluorescin. J Lab Clin Med *44:* 301.

King, S. and E. Meyer. 1957 Metabolic and serological differentiation of *Actinomyces bovis* and "anaerobic diphtheroids." J Bacteriol *74 (2):* 234–238.

— and —. 1963. Gel diffusion technique in antigen-antibody reactions of *Actinomyces* species and "anaerobic diphtheroids." J Bacteriol *85 (1):* 186–190.

King, W. E. and F. W. Baeslack. 1913 Studies on the virus of hog cholera. Preliminary report. J Infect Dis *12 (1):* 39–41.

—, — and G. L. Hoffman. 1913 Studies on the virus of hog cholera. J Infect Dis *12 (2):* 206–235.

— and R. H. Drake. 1915 Inoculation experiments with pure cultures of *Spirochaeta hyos.* Studies on hog cholera. J Infect Dis *16 (1):* 54–67.

Kingsbury, D. T. 1967 Deoxyribonucleic acid homologies among species of the genus *Neisseria.* J Bacteriol *94 (4):* 870–874.

— and J. F. Duncan. 1967 DNA homology of the genus *Neisseria.* Bacteriol Proc, p. 41.

—, G. R. Fanning, K. E. Johnson and D. J. Brenner. 1969 Thermal stability of interspecies *Neisseria* DNA duplexes. J Gen Microbiol *55 (2):* 201–208.

— and E. Weiss. 1968 Lack of deoxyribonucleic acid homology between species of the genus *Chlamydia.* J Bacteriol *96 (4):* 1421–1423.

Kingsley, V. V. 1968 Investigations into the Structure and Classification of Selenomonads. Ph.D. Thesis, U. Toronto, pp. 1–269.

Kinoshita, S., S. Nakayama and S. Akita. 1968 Taxomonical study of glutamic acid accumulating bacteria. *Micrococcus glutamicus,* nov. sp. Bull Agr Chem Soc Jap *22:* 176–185.

Kinsel, N. A. 1960 New sulfur oxidizing iron bacterium: *Ferrobacillus sulfooxidans* sp. n. J Bacteriol *80 (5):* 628–632.

Kirchner, O. 1878 Algen *In* Cohn, F. Kryptogamenflora von Schlesien *2:* 1–284.

—. 1896 Katalog der im Bodensee aufgefundenen Algen und Pilzen. *In* Schröter, C. and O. Kirchner (Editors). Die Vegetation des Bodensees. I. Schrift Ver Geschichte Bodensees. Suppl *25:* 53–122.

—. 1896 Die Wurzelknöllchen der Sojabohne. Beitr Biol Pflanz *7:* 213–224.

Kishitani, T. 1928 Drei Neue Arten von Leuchtbakterien. Proc Imp Acad Jap *4:* 69–72.

—. 1928 Ueber das Leuchtorgan von *Eupyrmna morsei* Verrill und die symbiotischen Leuchtbakterien. Proc Imp Acad Jap *4:* 306–309.

—. 1928 Preliminary report on the luminous symbiosis in *Sepiola birostrata* Sasaki. Proc Imp Acad Jap *4:* 393–396.

—. 1928 L'étude de l'organe photogène du *Loligo edulis* Hoyle. Proc Imp Acad Jap *4:* 609–612.

—. 1930 Studien über die Leuchtsymbiose in *Physiculus japonicus* Hilgendorf, mit bakterien. Sci Rep Tôhoku Univ (Ser IV Biol *5:* 801–823.

Kistner, A. 1953 On a bacterium oxidizing carbon monoxide. Proc Kon Ned Akad Wetensch Ser C Biol Med Sci *56 (4):* 443–350.

—. 1954 Conditions determining the oxidation of carbon monoxide and of hydrogen by *Hydrogenomonas carboxydovorans.* Proc Kon Ned Akad Wetensch Ser C Biol Med Sci *57 (2):* 186–195.

— and L. Gouws. 1964 Cellulolytic cocci occurring in the rumen of sheep conditioned to lucerne hay. J Gen Microbiol *34 (3):* 447–458.

Kitahara, K. 1938 Studies in lactic acid-forming bacteria in milk and milk products. J Agr Chem Soc Jap *14:* 1449–1465.

—. 1940 Studies on the lactic acid bacteria isolated from mashes of various kinds of cereals. J Agr Chem Soc Jap *16:* 697–714.

—. 1949 A new type lactic acid bacteria. Bull Res Inst Food Sci Kyoto Univ *2:* 23–36.

—, T. Kaneko and O. Goto. 1957 Taxonomic studies on the hiochi-bacteria, specific saprophytes of sake. I. Isolation and grouping of bacterial strains. J Gen Appl Microbiol *3:* 102–110.

—, — and —. 1957 Taxonomic studies on the hiochi-bacteria, specific saprophytes of sake. II. Identification and classification of hiochi-bacteria. J Gen Appl Microbiol *3:* 111–120.

— and C. L. Lai. 1967 On the spore formation of *Sporo-lactobacillus inulinus.* J Gen Appl Microbiol *13:* 197–203.

— and A. Nakagawa. 1958 *Pediococcus mevalovorus* nov. sp. isolated from beer. J Gen Appl Microbiol *4:* 21–30.

— and J. Suzuki. 1963 *Sporolactobacillus* nov. subgen. J Gen Appl Microbiol *9:* 59–71.

— and T. Toyota. 1972 Auto-spheroplastization and cell-permeation in *Sporolactobacillus inulinus.* J Gen Appl Microbiol *18:* 99–107.

Kitt, T. 1893 Bakterienkunde und pathologisch Mikroskopie für Thierärzte und Studirende der Thiermedicin. Moritz Perles, Wien *2:* 1–450.

—. 1899 Bacterienkunde und Pathologische Mikroskopie für Thierärzte und Studierende der Thiermedicin. Moritz Perles, Wein *1:* 1–525.

Klamann, Dr. 1887 Ueber einige Microorganismen im Secret bei Rhinitis chronica atrophicans (Ozäna) nebst Bemerkungen über die Therapie der Ozäna und über die Verschiedenheiten der Secrete. Allg Med Centralz *56:* 1346–1347.

Klarenbeek, A. 1921 Ueber das spontane Vorkommen der dem Syphilisparasiten ähnlichen spirochäte beim Kaninchen (*Treponema pallidum* var *cuniculi*). Zentrabl Bakteriol Parasitol Infektionskr Hyg Abt I Orig *87:* 203–216.

Klas, Z. 1936 Zwei neue Schwefelbakterien (*Thiothrix voukii* n. sp. et *Thiothrix longiarticulata* n. sp.). Arch Protistenk *88:* 121–126.

—. 1937 Ueber den Formenkreis von *Beggiatoa mirabilis.* Arch Mikrobiol *8:* 312–320.

Klebs, E. 1883 Ueber Diptherie. Verh Deut Ges Inn Med *2:* 139–154.

Kleckner, A. L. 1960 Serotypes of avian pleuropneumonia-like organisms. Amer J Vet Tes *21:* 274–280.

Kleczkowska, J., A. G. Norman and S. F. Śnieszko. 1940 Bacteriological studies on a new capsulated bacillus, *Bacillus krzemieniewski.* Soil Sci *49:* 185–190.

Klein, E. 1884 Micro-organisms and disease. Practitioner *32:* 170–186; 241–264; 321–352; 401–426.

—. 1884 Micro-organisms and diseases. Practitioner *33:* 21–40; 81–112; 160–180.

—. 1887 On the etiology of scarlatina. Loc Gov Brd Rep London *16:* 367–414.

—. 1889 Ueber eine epidemische Krankheit der Hühner, verursacht durch einen Bacillus - *Bacillus gallinarum.* Zentrabl Bakteriol Parasitenk Infektionskr Hyg Abt I Orig *5:* 689–693.

—. 1899 Ein Beitrag zur Bakteriologie der Leichenverwesung. Zentrabl Bakteriol Parasitenk Infektionskr Abt I Orig *25:* 278–284.

—. 1904 Ein neuer tierpathogener Mikrobe-*Bacillus carnis.* Zentrabl Bakteriol Parasitenk Infektionskr Hyg Abt I Orig *35:* 459–461.

—. 1908 On the nature and causes of taint in miscured hams (*Bacillus foedans*). Lancet *1:* 1832–1834.

Klein, R. M. and I. L. Tenebaum. 1955 A quantitative bioassay for crown-gall tumor formation. Amer J Bot *42:* 709–712.

Klein, S. M. and R. D. Sagers. 1962 Intermediary metabolism of *Diplococcus glycinophilus.* II. Enzymes of the acetate-generating system. J Bacteriol *83 (1):* 121–126.

Klement, Z. 1954 A new bacterial disease of rice caused by *Pseudomonas oryzicola* n. sp. Acta Microbiol Acad Sci Hung *2:* 265–274.

Klemme, J. H. 1968 Untersuchungen zur Photoautotrophie mit molekularem Wasserstoff bei neuisolierten schwefelfreien Purpurbakterien. Arch Mikrobiol *64:* 29–42.

Klieneberger, E. 1935 The natural occurrence of pleuropneumonia-like organisms in apparent symbiosis with *Streptobacillus moniliformis* and other bacteria. J Pathol Bacteriol *40:* 93–105.

—. 1938 Pleuropneumonia-like organisms of diverse provenance: some results of an enquiry into methods of differentiation. J Hyg *38:* 458–475.

—. 1942 Some new observations bearing on the nature of the pleuropneumonia-like organism known as L₁ associated with *Streptobacillus moniliformis.* J Hyg *42:* 485–497.

— and D. B. Steabben. 1937 On a pleuropneumonia-like organism in lung lesions of rats, with notes on the clinical and pathological features of the underlying condition. J Hyg *37:* 143–152.

Kliewer, M. and H. J. Evans. 1962 B₁₂ coenzyme content of the nodules from legumes, alder and of *Rhizobium meliloti.* Nature (London) *194:* 108–109.

Kline, L. and T. F. Sugihara. 1971 Microorganisms of the San Francisco Sour Dough Bread Process. II. Isolation and characterization of undescribed bacterial species responsible for the souring activity. Appl Microbiol *21:* 459–465.

Klinger, R. 1912 Untersuchungen über menschliche Atkinomykose. Zentrabl Bakteriol Parasitenk Infektionskr Hyg Abt I Orig *62:* 191–200.

Kluyver, A. J. 1942 De stofwisseling van de plantaardige cel. *In* V. J. Köningsberger (Editor), Leerbock der algemeene Plantkunde, Vol. II. Scheltema and Holkema, Amsterdam, p. 198.

—. 1956 *Pseudomonas aureofaciens* nov. spec. and its pigments. J Bacteriol *72 (3):* 406–411.

—. 1957 Genus VII. *Zymomonas* Kluyver and van Neil, 1926. *In* Breed Murry and Smith (Editors), Bergey's Manual of Determinative Bacteriology, 7th ed. The Williams & Wilkins Co., Baltimore, pp. 199–200.

— and F. J. G. de Leeuw. 1924 *Acetobacter suboxydans,* een merkwaardige azijnbacterie. Tijdschr Vergelikj Geneesk *10:* 170–281.

— and W. J. Hoppenbrouwers. 1931 Ein merkwürdiges Gärüngsbakterium: Lindner's *Termobacterium mobile.* Arch Mikrobiol *2:* 245–260.

— and A. Manten. 1942 Some observations on the metabolism of bacteria oxidizing molecular hydrogen. Antonie Van Leeuwenhoek J Microbiol Serol *8:* 71–85.

— and C. G. Schnellen. 1947 On the fermentation of carbon monoxide by pure cultures of methane bacteria. Arch Biochem *14:* 57–70.

— and C. B. van Niel. 1936 Prospects for a natural system of classification of bacteria. Zentrabl Bakteriol Parasitenk Infektionskr Hyg Abt II *94:* 369–403.

— and Van Reenen, W. J. 1933 Über *Azotobacter agilis* Beijerinck. Arch Mikrobiol *4:* 280–300.

Kmety, E. 1967 Faktorenanalyse von Leptospiren der Icterohaemorrhagiae und einiger verwandter Serogruppen. Slovak Academy of Sciences, Bratislava.

—, I. Plesko, P. Bakoss and B. Chorvath. 1966 Evaluation of methods for differentiating pathogenic and saprophytic leptospira strains. Ann Soc Belg Méd Trop *46:* 111–118.

Knapp, W. 1955 Die diagnostische Bedeutung der antigenen Beziehungen zwischen *Past. pseudotuberculosis* und der *Salmonella*-Gruppe. Zentrabl Bakteriol Parasitenk Infektionskr Hyg Abt I Orig *164:* 57–59.

—. 1968 Serologische kreuzreaktionen zwischen *Pasteurella pseudotuberculosis* (syn. *Yersinia pseudotuberculosis*), *Escherichia coli* und *Enterobacter cloacae*, International Symposium on *Pseudotuberculosis*. Symp Ser Immunobiol Standardization *9:* 179–186.

— and E. Thal. 1963 Untersuchungen über die kulturell-biochemischen, serologischen tier-experimentellen und immunologischen Eigenschaften einer Vorläufig "*Pasteurella* X" benannten Bakterienart. Zentrabl Bakteriol Parasitenk Infektionskr Hyg Abt I Orig *190 (4):* 472–484.

Knaysi, G. 1942 The demonstration of a nucleus in the cell of a *Staphylococcus*. J Bacteriol *43 (3):* 365–380.

Knight, B. C. J. G. and H. Proom. 1950 A comparative survey of the nutrition and physiology of mesophilic species in the genus *Bacillus*. J Gen Microbiol *4:* 508–538.

Knight, H. D. 1969 Corynebacterial infections in the horse: problems of prevention. J Amer Vet Med Ass *155:* 446–452.

Knipp, L. H. and J. R. Sokatch. 1969 The chemical composition of the cell envelope of *Streptobacillus moniliformis*. Can J Microbiol *15:* 665–669.

Knoll, H. and R. Horschak. 1964 Zur Ernährungsphysiologie der Gärungssarcinen. Monatsber Deut Akad Wiss Berlin *6:* 847–849.

Knorr, M. 1922 Über die fusospirillare Symbiose, die Gattung *Fusobacterium* (K. B. Lehmann) and *Spirillum sputigenum*. Die Gattung *Fusobacterium*. I. Mitteilung. Die Epidemiologie der fusospirillaren Symbiose, besonders der Plaut-Vincentschen Angina. Zentrabl Bakteriol Parasitenk Hyg Abt I Orig *87:* 536–545.

—. 1922 Ueber die fusospirillare Symbiose, die Gattung *Fusobacterium* (K. B. Lehmann) und *Spirillum sputigenum*. (Zugleich ein Beitrag zur Bakteriologie der Mundhöhle. II. Mitteilung. Die Gattung *Fusobacterium*. Zentrabl Bakteriol Parasitenk Infektionskr Hyg Abt I Orig *89:* 4–22.

Knösel, D. 1961 Eine an Kohl blattlfleckenerzeugende Varietas von *Xanthomonas campestris* (Pammel) Dowson. Z Pflanzenkr Pflanzenpathol Pflanzenschutz *68:* 1–6.

Kundsen, S. and A. Sörenson. 1929 Beiträge zur Bakteriologie der Säurewecker. Zentrabl Bakteriol Parasitenk Infektionskr Hyg Abt II *79:* 75–85.

Koch, A. 1888 Ueber Morphologieeund Entwickelungsgeschichte einiger endosporer Bacterienformen. Bot Ztg *46:* 277–287; 308–313; 347–349.

— and H. Hosaeus. 1894 Über einen neuen Froschlaich der Zuckerfabriken. Zentrabl Bakteriol Parasitenk Infektionskr Hyg Abt I Orig *16:* 225–228.

Koch, R. 1884 Conferenz zur Erörterung der Cholerafrage. Berlin Klin Wochenschr *21:* 477–483.

Kocur, M. and W. Hodgkiss. 1973 Taxonomic status of the genus *Halococcus* Schoop. Int J Syst Bacteriol *23:* 151–156.

— and T. Martinec. 1962 Taxonomická Studie Rodu *Micrococcus*. Folia Přirodov, Fak. J. E. Purkyne Brně Biol *3 (3):* 1–121.

— and —. 1965 Proposal for the rejection of the bacterial generic name *Gaffkya*. Int Bull Bacteriol Nomencl Taxon *15:* 177–179.

Kodama, R. 1956 Studies on the nutrition of lactic acid bacteria. Part I. *Lactobacillus fructosus* nov. sp., a new species of lactic acid bacteria. J Agr Chem Soc Jap *30:* 705–708.

Kofler, L. 1913 Die Myxobacterien der Umgebung von Wien. Sitzungsber Akad Wiss Math-Naturwiss Kl Abt I *122:* 845–876.

Kohler, W. 1955 *Corynebacterium citreum-mobilis*, ein neuer, farbstoffbildender saprophytarer Keim des Genus *Corynebacterium*. Zentrabl Bakteriol Parasitenk Infektionskr Hyg Abt I Orig *162 (445):* 275–280.

Kojima, M., S. Suda, S. Hotta and K. Hamada. 1968 Induction of pleomorphism in *Lactobacillus bifidus*. J Bacteriol *95 (2):* 710–711.

Kolk, L. A. 1938 A comparison of the filamentous iron organisms, *Clonothrix fusca* Roze and *Crenothrix polyspora* Cohn. Amer J Bot *25:* 11–17.

Kolkwitz, R. 1909 *Schizomycetes*. Kryptogamenflora der Mark Brandenburg *5:* 1–186.

Komagata, K., K. Yamada and H. Ogawa. 1969 Taxonomic studies on coryneform bacteria. I. Division of bacterial cells. J Gen Appl Microbiol *15:* 243–259.

Kondo, K. and M. Ameyama. 1958 Carbohydrate metabolism by *Acetobacter* Species. Part I. Oxidative activity for various carbohydrates. Bull Agr Chem Soc Jap *22:* 369–372.

Kondo, S.-I, T. Miyakawa, H. Yumoto, K. Sato, T. Niida and T. Hara. 1964 Bandamycin B, a novel antibiotic substance. Japanese Patent 10,244, June 11.

—, —, —, M. Sezaki, M. Shimura, K. Sato and T. Hara. 1962 Isolation and characterization of a new antibiotic, bandamycin B. J Antibiot (Tokyo) Ser A *15 (4):* 157–160.

—, T. Niida, J. Sakamoto, H. Yumoto and T. Hatakeyama. 1963 Bandamycin, a new antibiotic. Japanese Patent 26,948, December 28

—, J. M. J. Sakamoto and H. Yumoto. 1961 Bandamycin, a new antibiotic. J Antibiot (Tokyo) Ser A *14:* 365–366.

—, H. Yumoto, T. Miyakawa, K. Hamamoto, M. Sezaki, K. Sato and T. Niida. 1962 On streptomycin produced by *Streptomyces erythrochromogenes* var. *narutoensis*. Sci Rep Meiji Seika Kaisha *5:* 9–13.

Kondratieva, E. N. 1956 Assimilation of organic compounds by purple bacteria in the presence of light. Mikrobiologiya *25:* 393–400.

—. 1965 Photosynthetic Bacteria. Israel Program for Scientific Translations, Jerusalem, pp. 1–243.

—, V. D. Federov and K. P. Greshnyhk. 1958 Contribution to the study of the morphology of *Chlorobium thiosulfatophilum*. Dokl Akad Nauk SSSR *123:* 365.

—, E. N. Krasil'nikova and L. M. Novikova. 1968 Production of polysaccharides by green photosynthesizing bacteria. Mikrobiologiya *37:* 417–427.

— and I. V. Malofeeva. 1964 A contribution to

the carotenoids of purple sulfur bacteria. Mikrobiologiya *33:* 758–762.

Konev, I. E. 1964 Churn-staffed actinomycetes —the producents of pentaene antibiotics. Mikrobiologiya *33:* 622–630.

— and V. A. Tsyganov. 1962 A new species in the group of yellow actinomycetes, *Actinomyces xantholiticus* N. sp. Mikrobiologiya *31* (*6*): 1023–1028.

—, —, P. Minbayev and V. M. Morozov. 1965 Vydeleniye novogo poda aktinomitsetov *Microechinospora* gen. nov. Abstr Papers 4th Inst. Antibiot Conf., Leningrad, pp. 80–82.

—, —, — and —. 1967 New genus of actinomycetes, *Echinospora* gen. nov. Mikrobiologiya *36* (*2*): 308–317.

Koning, H. C. 1938 Bacterial canker of the poplar. Chron Bot *4:* 11–12.

Konova, I. V. 1962 Defense of Dissertations. Mikrobiologiya *31* (*1*): 188–189.

Koppe, F. 1924 Die Schlammflora der ostholsteinischen Seen und des Bodensees. Arch Hydrobiol *14:* 619–672.

—. 1924 Die Schlammflora der ostholsteinschen Seen und des Bodensees. Arch Hydrobiol *14:* 619–672.

Korenyako, A. I., A. I. Sokolova and N. I. Nikitina. 1960 Actinomycetes of the fluorescent group. Proc Symp on Antibiotics, Prague, Czechoslovakia May 18–23, 1959, pp. 59–60.

Korth, H., I. Ørskov and G. Pulverer. 1969 Farbstoffbildende *Klebsiella*-Stämme. Zentrabl Bakteriol Parasitenk Infektionskr Abt Orig I *211* (*1*): 105–107.

Korthof, G. 1932 Experimentelles Schlammfieber beim Menschen. Zentrabl Bakteriol Parasitenk Infektionskr Hyg Abt I Orig *125:* 429–434.

Koser, S. A. 1968 Vitamin requirements of Bacteria and Yeasts. Charles C Thomas, Springfield, Illinois.

Koshiyama, H., M. Okanishi, T. Ohmori, T. Miyaki, H. Tsukiura, M. Matsuzaki and H. Kawaguchi. 1963 Cirramycin, a new antibiotic. J. Antibiot (Tokyo) Ser A *16* (*1*): 59–66.

Kosmachev, A. E. 1958 Thesis. Mikrobiologiya *27* (*1*): 141–144.

—. 1962 A thermophilic *Micromonospora* and its production of antibiotic T-12 under conditions of surface and submerged fermentation at 50–60° C. Mikrobiologiya *31:* 52.

—. 1964 A new thermophilic actinomycete *Micropolispora thermovirida* n. sp. Mikrobiologiya *33:* 235–237 (trans.).

Kotte, W. 1930 Eine bakterielle Blattfäule der Winter-Endivie (*Cichorium endivia* L.). Phytopathol Z pp. 605–613.

Kovacs, N. 1956 Identification of *Pseudomonas pyocyanea* by the oxidase reaction. Nature (London) *178:* 703.

Kowallik, U. and E. G. Pringsheim. 1966 The oxidation of hydrogen sulfide by *Beggiatoa*. Amer J Bot *53:* 801–806.

Kozulis, J. A. and R. H. Parsons. 1958 *Acetobacter alcoholophilus* n. sp. - a new species isolated from storage beer. J Inst Brew *64:* 47–50.

Krainsky, A. 1914 Die Aktinomyceten und ihren Bedeutung in der Natur. Zentrabl Bakteriol Parasitenk Infektionskr Hyg Abt II *41:* 649–688.

Kramarenko, L. E. and I. I. Prisrenova. 1961 Denitrifying sulfur oxidizing bacteria in sulfide veins and methods of demonstrating them in prospecting. Proc All-Union Geol Res Inst (VSEGEI) *61:* 209–230.

Kramer, E. 1890 Die Bakteriologie in Ihren Beziehungen zur Landwirtschaft und den Landw.-technischen Gewerben. Erster Teil.

Kran, G., F. W. Schlote and H. G. Schlegel. 1963 Cytologische Untersuchungen an *Chromatium okenii* Perty. Naturwissenschaften *50:* 728–730.

Krasil'nikov, N. A. 1938 Proactinomyces. Bull Acad Sci USSR (Ser Biol) *1:* 139–172.

—. 1938 Ray Fungi and related organisms - *Actinomycetales*. Akad Nauk, SSSR Moscow, pp. 1–328.

—. 1941 Guide to the Bacteria and Actinomycetes (in Russian). Akad Nauk SSSR, Moscow, pp. 1–830.

—. 1941 Keys to *Actinomycetales* Inst. Microbiol., Acad Sci., USSR, Moscow-Leningrad. English translation by Israel Program for Scientific Translations for the United States Dept. of Agriculture and the National Science Foundation, Washington, D. C., 1966.

—. 1949 Diagnostik der Bakterien und Actinomyceten. Inst. Microbiol., Acad. Sci. USSR Moscow-Leningrad. German translation by R. Wittwer and R. Dickscheit. Veb. Gustav Fischer Verlag Jena Austria, 1959. English translation of section on Actinomycetes by Chas. Pfizer and Company, Inc. (J. B. Routien, ed.), 1959.

—. 1949 Guide to the bacteria and actinomycetes. Akad Nauk SSSR, Moscow, pp. 1–830.

—. 1958 The significance of antibiotics as specific characteristics of actinomycetes, and their determination by the method of experimental transformation. Folia Biol (Praha) *4* (*5*): 257–265.

—. 1960 Taxonomic principles in the actinomycetes. J Bacteriol *79* (*1*): 65–74.

—. 1960 Rules for the classification of antibiotic-producing actinomycetes. J Bacteriol *79* (*1*): 75–80.

—. 1962 A new genus of ray fungus—*Actinopycnidium* n. gen of family *Actinomycetaceae*. Mikrobiologiya *31* (*2*): 250–253; Engl transl 204–207.

—. 1964 Systematic position of ray fungi among the lower organisms. Hindustan Antibiot Bull *7:* 1–17.

—. (ed). 1965 Biology of selected groups of actinomycetes (in Russian). Institute of Microbiology, Academy of Science, Publishing Firm "Nauka," Moscow, USSR, pp. 1–372.

— and N. S. Agre. 1964 On two new species of *Thermopolyspora*. Hindustan Antibiot Bull *6:* 97–107.

— and —. 1964 A new genus of Actinomycetes *Actinobifida* n. gen. The yellow group— *Actinobifida dichotomica* n. sp. Mikrobiologiya *33:* 935–943.

— and —. 1965 The brown group of *Actinobifida chromogena* n. sp. Mikrobiologiya *34:* 284–291.

—, — and G. I. El-Registan. 1967 New thermophilic species of the genus *Micropolyspora*. Mikrobiologiya *37:* 1065–1072.

—, A. N. Belozerskii, Y. I. Rautenshtein, A. I. Korenyako, N. I. Nikitina, A. I. Sokolova and S. O. Uryson. 1956 Antibiotic grisine (grisemine) and its producers (in Russian) Dokl Akad Nauk SSSR *111:* 1117–1120.

—, —, —, —, —, —, and —. 1957 The antibiotic grizein (grizemin) and its producers. Mikrobiologiya *26* (*4*): 418–425,

— and Y. Chi-Shen. 1965 Composition of actinomycetes of the orange group (in Russian). *In* Biology of special Groups of Actinomycetes producers of antibiotics. Akad Nauk SSSR, Moscow, pp. 28–57.

— and V. I. Duda. 1963 Changes of nucleic structure with sporeformation in the anaerobic bacteria of the genus *Clostridium*. Dokl Akad Nauk SSSR *152* (*2*): 454–456.

—, — and A. A. Sokoloy. 1964 Protusions on the surface of spores of anaerobic bacteria of the genus *Clostridium*. Microbiologiya *33:* 454–458.

—, G. I. El-Registan and N. S. Agre. 1967 Pigments from *Actinobifida dichotomica*. Mikrobiologiya *36:* 602–607.

—, L. V. Kalakutskii and N. F. Kirillova. 1961 A new genus of *Actinomycetales, Promicromonospora*, gen. nov. Bull Acad Sci USSR (Ser Biol) *1:* 107–112.

—, A. I. Korenyako, M. M. Meksina, L. K. Valedinskaya and N. M. Veselov. 1957 On the culture of actinomycete No. 111, *Act. luridus* nov. sp. producing an antiviral antibiotic "luridin." Mikrobiologiya *26* (*5*): 558–564.

— and A. D. Koveshnikov. 1962 *Actinomyces tumemacerans* n. sp. A new species causing destruction of plant tumors. Mikrobiologiya *31* (*4*): 539–594 (Engl Transl 483–486).

—, A. G. Kutchayeva and G. K. Skryabin. 1959 Actinomycetes of the "Olivatus" group. Suppl. to Abstr Program Symp on Antibiotics, Prague, Czechoslovakia, May 18–23.

—, G. K. Skryabin and O. I. Artamonova. 1960 Actinomycetes synthesizing antivirus antibiotics. J Antibiot (Tokyo) Ser A *13* (*1*): 1–5.

— and Y. Tsi-Shen. 1961 *Actinosporangium*, A new genus of the *Actinoplanaceae* family. Bull Akad Nauk SSSR Ser Biol, pp. 113–116.

— and C.-S. Yuan. 1960 A new species of the *Actinomyces aurantiacus* group. Mikrobiologiya *29* (*4*): 482–489 (Engl Transl 354–358).

— and —. 1961 *Actinosporangium*, a new genus of the family *Actinoplanaceae*. Izv Akad Nauk SSSR Ser Biol 8 (*1*): 113–116.

— and T. Yuan. 1965 The species composition of orange-colored actinomycetes. *In* N. A. Krasil'nikov (Editor) Biology of Individual Groups of Actinomycetes (in Russian), pp. 28–57.

Krause, R. M. and M. McCarty. 1961 Studies on the chemical structure of the streptococcal cell wall. I. The identification of a mucopeptide in the cell walls of groups A and A-variant streptococci. J Exp Med *114:* 127–140.

— and —. 1962 Studies on the chemical structure of the streptococcal cell wall. II. The composition of Group C cell walls and chemical basis for serologic specificity of the carbohydrate moiety. J Exp Med *115:* 49–62.

Krebs, H. A. and L. V. Eggleston. 1939 Bac-

terial urea formation (Metabolism of *Corynebacterium ureafaciens*). Enzymologia *7:* 310–320.

Krehan, M. 1930 Beiträge zur Physiologie und Systematik der Essigbakterien. Arch Mikrobiol *1:* 493–536.

Kreier, J. P. 1962 A comparison of the antigenic and morphologic features of *Anaplasma marginale* with those of several other hemotropic parasites. Doctoral Dissertation, University of Illinois.

— and M. Ristic. 1963 Morphologic, antigenic and pathogenic characteristics of *Eperythrozoon ovis* and *Eperythrozoon wenyoni*. Amer J Vet Res *24:* 488–500.

— and —. 1963 Anaplasmosis. X. Morphologic characteristics of the parasites present in the blood of calves infected with the Oregon strain of *Anaplasma marginale*. Amer J Vet Res *24:* 676–687.

— and —. 1963 Anaplasmosis. X. Immunoserologic characteristics of the parasites present in the blood of calves infected with the Oregon strain of *Anaplasma marginale*. Amer J Vet Res *24:* 688–696.

— and —. 1963 Anaplasmosis. XII. The growth and survival in deer and sheep of the parasites present in the blood of calves infected with the Oregon Strain of *Anaplasma marginale*. Amer J Vet Res *24:* 697–702.

— and —. 1968 Haemobartonellosis, Eperythrozoonosis, Grahamellosis and Erlichiosis. *In* D. Weinman and M. Ristic (Editors), Infectious Blood Diseases of Man and Animals, Academic Press, Inc. N. Y., pp. 387–472.

Kremers, R. H. and R. W. Quinn. 1965 Growth media for typing group A streptococci. J Bacteriol 89 (*6*): 1619–1620.

Krieg, A. 1955 Licht und electronenmikroskopische Untersuchungen zur Pathologie der "Lorscher Erkrankung" von Engerlingen und zur Zytologie der *Rickettsia melolonthae* nov. spec. Z Naturforsch *10b:* 34–37.

—. 1955 Untersuchungen zur Wirbeltier-Pathogenität und sum serologischen Nachweis der *Rickettsia melolonthae* im Arthropod-Wirt. Naturwissenschaften *42:* 609–610.

—. 1959 Vergleichende taxonomische, morphologische und serologische Untersuchungen an insektenpathogenen Rickettsien. Z Naturforsch *13b:* 555–567.

—. 1959 On the problem of crystals associated with *Rickettsiella* infection. J Insect Pathol *1:* 95–98.

—. 1961 Grundlagen der Insektenpathologie. Dr. Dietrich Steinkopff. Verlag, Darmstadt.

—. 1965 Über eine neue Rickettsie aus Coleopteren, *Rickettsiella tenebrionis* nov. spec. Naturwissenschaften *52:* 144.

—. 1968 A taxonomic study of *Bacillus thuringiensis* Berliner. J Invertebr Pathol *12:* 366–378.

—. 1969 Transformations in the *Bacillus cereus, Bacillus thuringiensis* group. Description of a new subspecies: *Bacillus thuringiensis* var. *toumanoffii*. J Invertebr Pathol *14:* 279–281.

—, H. de Barjac and A. Bonnefoi. 1968 A new serotype of *Bacillus thuringiensis* isolated in Germany: *Bacillus thuringiensis* var. *darmstadiensis*. J Invertebr Pathol *10:* 428–430.

Kriss, A. E. 1938 Verticillate branching in

actinomycetes (*Act. verticillatus* n. sp.) (in Russian). Mikrobiologiya 7 (1): 105–111.

—. 1939 *Micromonospora*—an actinomycete-like organism (*Micromonospora globosa* n. sp.). Mikrobiologiya 8: 178–185.

Kristensen, M. 1931 Klassification dänischer und anderer Brucellestamme. Zentrabl Bakteriol Parasitenk Infektionskr Hyg Abt I Orig 120: 179–196.

Kritchewski, B. and P. Seguin. 1920 Spirochetoses buccales. Reproduction expérimentale et traitement. Rev Stomat Paris 22: 613–647.

— and —. 1921 Article on *Fusobacterium*. L'Odontologie 41: 720–742.

Krönig, Dr. 1895 Ueber die Natur der Scheidenkeime speciell über das Vorkommen anaërober Streptokokken im Schevdensekret Schwangerer. Zentrabl Gynaekol 19: 409–412.

Krüger, F. 1905 Untersuchungen über den Gürtelschorf der Zuckerrüben. Arb Biol Abt (Anst-Reichsanst.) Berl 4 (3): 254–318.

Krul, J. M., P. Hirsch and J. T. Staley. 1970 *Toxothrix trichogenes* (Chol.) Beger et Bringmann: the organism and its biology. Antonie van Leeuwenhoek J Microbiol Serol 36: 409–420.

Krulwich, T. A. and J. C. Ensign. 1968 Activity of an autolytic N-acetylmuraminidase during sphere-rod morphogenesis in *Arthrobacter crystallopoietes*. J Bacteriol 96: 857–859.

— and —. 1969 Alteration of glucose metabolism of *Arthrobacter crystallopoietes* by compounds which induce sphere to rod morphogenesis. J Bacteriol 97: 526–534.

—, —, D. J. Tipper and J. L. Strominger. 1967 Sphere-rod morphogenesis in *Arthrobacter crystallopoietes*. I. Cell wall composition and polysaccharides of the peptidoglycan. J Bacteriol 94: 734–740.

—, —, — and —. 1967 Sphere-rod morphogenesis in *Arthrobacter crystallopoietes*. II. Peptides of the cell wall peptidoglycan. J Bacteriol 94: 741–750.

Kruse, W. 1896 *Bacillus ulceris cancrosi* (Bacillus des weichen Schankers) in Flügge, C. Die Mikroorganismen, unter spezieller Berucksichtigung der Atiologie infektioser Krankheiten 3. Aufl 2: 456–458.

—. 1896 Systematik der Streptothricheen und Bakterien, pp. 48–96; 185–526. *In* C. Flügge, Die Mikroorganismen. 3rd Ed. Vol. 2, Vogel, Leipzig.

Krzemieniewska, H. 1933 Contribution à l'étude du genre *Cytophaga* (Winogradsky). Arch Mikrobiol 4: 394–408.

— and S. Krzemieniewski. 1926 Miksobakterje Polski (Die Myxobakterien von Polen). Acta Soc Bot Pol 4: 1–54.

— and —. 1927 Miksobacterje Polski Uzupelnienie. Acta Soc Bot Pol 5: 79–98.

— and —. 1928 Zur Morphologie der Myxobacterienzelle. Acta Soc Bot Pol 5: 46–90.

— and —. 1930 Miksobakterje Polski. (Czesc Trzecia) (Die myxobacterien von Polen. III Teil). Acta Soc Bot Pol 7: 250–273.

— and —. 1938 Die zelluloserzersetzenden Myxobacterien. Bull Int Acad Cracovie (Acad Pol Sci) Ser B Sci Nat I, pp. 11–31.

— and —. 1938 Über die Zersetzung der Zellulose durch Myxobacterien. Bull Int Acad Cracovie (Acad Pol Sci) Ser B Sci Nat I, pp. 33–59.

— and —. 1946 Myxobacteria of the species *Chondromyces* Berkeley and Curtis. Bull Int Acad Cracovie (Acad Pol Sci) Ser B Sci Nat I, pp. 31–48.

Kubica, G. P., I. Baess, R. E. Gordon, P. A. Jenkins, J. B. G. Kwapinski, C. McDurmont, S. R. Pattyn, H. Saito, V. Silcox, J. L. Stanford, K. Takeya and M. Tsukamura. 1972 A cooperative numerical analysis of the rapidly growing mycobacteria. J Gen Microbiol 73: 55–70.

—, T. H. Kim and F. B. Dunbar. 1972 Designation of strain H37Rv as the neotype of *Mycobacterium tuberculosis*. Int J Syst Bacteriol 22: 99–106.

— and G. L. Pool. 1960 Studies on catalase activity of acid-fast bacilli. Amer Rev Resp Dis 81: 387–391.

—, V. A. Silcox, J. O. Kilburn, R. W. Smithwick, R. E. Beam, W. D. Jones, Jr. and K. D. Stottmeier. 1970 Differential identification of mycobacteria. VI. *Mycobacterium triviale* sp. nov. Int J Syst Bacteriol 20 (2): 161–174.

Kubo, H., S. Suzuki and S. Tamura. 1964 Process for obtaining a new antibiotic piericidin. Japanese Patent 9443.

Kucera, S. and R. S. Wolfe. 1957 A selective enrichment method for *Gallionella ferruginea*. J Bacteriol 74: 344–349.

Kuchaeva, A. G. 1959 Use of antibiotics against insect pests of plants (in Russian). Vest Sel'Skokhoz Nauki (Moskva) 4 (7): 138–140.

—, N. A. Krasil'nikov, S. D. Taptykova and R. L. Gesheva. 1961 On systematics of actinomycetes of the *lavendulae* group (in Russian). Izv Microbiol Inst Bulg Acad Sci, Class Sci Biol Sofia 13: 103–124.

Kuchar, K. W. 1953 Der Formenkreis von *Pseudomonas punctata* (Zimm.) Chester. Zur Fassung der Gattung *Pseudomonas* in Bergey's Manual. Zentrabl Bakteriol Parasitenk Infektionskr Hyg Abt I Orig 160: 511–517.

—. 1954 *Pseudomonas jankei* n. sp. eine nichtverflüssigende Art mit Ringbildung in Peptonwasser. Zentrabl Bakteriol Parasitenk Infektionskr Hyg Abt I Orig 160: 517–521.

Kucsera, G. 1973 Proposal for standardization of the designations used for serotypes of *Erysipelothrix rhusiopathiae* (Migula) Buchanan. Int J Syst Bacteriol 23: 184–188.

Kuczynski, M. H. 1929 Der Erreger des Gelbfiebers, Wesen und Wirkung, Berlin.

Kudrina, E. S. and T. S. Maksimova. 1963 Some species of thermophilic actinomycetes from the soils of China and their antibiotic properties. Mikrobiologiya 32 (4): 623–631 (Engl Transl 532–538).

—, O. L. Ol'khovatova, L. I. Murav'eva and G. F. Gauze. 1966 Systematic position and variation of the organism producing bruneomycin, an antitumor antibiotic (in Russian). Antibiotiki 11 (5): 400–405.

Kuehl, F. A., Jr., F. J. Wolf, N. R. Trenner, R. L. Peck, P. E. Howe, R. P. Buhs, I. Putter, R. Ormond, J. E. Lyons and L. Chaiet. 1955

D-4-Amino-3-isoxazolidon, a new antibiotic. J Amer Chem Soc 77: 2344–2345.

Kufferath, H. 1911 Note sur les tropismes du *Bacterium zopfii* Kurth. Ann Inst Pasteur (Paris) 25: 601–618.

Kühlwein, H. 1950 Beitrage zur Biologie und Entwicklungsgeschichte der Myxobakterien Arch Mikrobiol 14: 678–704.

—. 1952 Untersuchungen über *Chondromyces apiculatus* Thaxter. Arch Mikrobiol 17: 403–408.

— and E. Gallwitz. 1958 *Polyangium violaceum* nov. spec., ein Beitrag zur Kenntnis Myxobakterien. Arch Mikrobiol 31: 139–145.

— and H. Reichenbach. 1964 Ein neuer Vertreter der Myxobakteriengattung *Archangium* Jahn. Arch Mikrobiol 48: 179–184.

— and —. 1966 Anreicherung und isolierung von Myxobakterien. Zentrabl Bakteriol Parasitenk Infektionskr Hyg Abt I (Supp I), pp. 57–80.

Kuhn, D. A. 1970 Proposal of *Cristispira pectinis* Gross 1910, 44 as the type species of the genus *Cristispira*. Int J Syst Bacteriol 20 (3): 301–303.

— and M. P. Starr. 1960 *Arthrobacter atrocyaneus* n. sp. and its blue pigment. Arch Mikrobiol 36: 175–181.

Kuhn, R., M. P. Starr, D. A. Kuhn, H. Bauer and H. Knackmuss. 1965 Indigoidine and other bacterial pigments related to 3,3'-bipyridyl. Arch Mikrobiol 51: 71–84.

Kulka, D., J. Singh, R. M. Nattrass, A. N. Hall and T. K. Walker. 1958 Studies of vinegar bacteria. J Sci Food Agr 9: 487–492.

Kulkarni, Y. S., M. K. Patel and S. G. Abhyankar. 1950 A new bacterial leaf spot and stem canker of pigeon pea. Curr Sci 19: 384.

—, — and G. W. Dhande. 1951 *Xanthomonas cassiae*, a new bacterial disease of *Cassia tora* L. Curr Sci 20: 47.

Kundrat, W. 1963 Zur Differenzierung aerober Sporebildner (Genus *Bacillus* Cohn). Zentrabl Veterinaermed Reihe B 10: 418–426.

Kung, H-F. and C. Wagner. 1970 Oxidation of C_1 compounds by *Pseudomonas* sp. MS. Biochem J 116: 357–365.

Kuno, Y. 1951 *Bacillus thiaminolyticus*, a new thiamin decomposing bacterium. Proc Jap Acad 27: 362–365.

Kupfer, D. G. and E. Canale-Parola. 1967 Pyruvate metabolism in *Sarcina maxima*. J Bacteriol 94 (4): 984–990.

— and —. 1968 Fermentation of glucose by *Sarcina maxima*. J Bacteriol 95 (1): 247–248.

Kurochkin, B. N. 1958 A new species of sporogenous bacteria isolated from the soil of several lakes (in Russian). Mikrobiologiya 27: 221–225.

Kurth, H. 1883 Ueber *Bacterium Zopfii*, eine neue Bacterienart. Ber Deut Bot Ges 1: 97–100.

—. 1883 *Bacterium Zopfii*. Ein Beitrag zur Kenntniss der Morphologie und Physiologie der Spaltpilze. Bot Ztg 41: 369–386; 393–405; 409–420; 425–440.

Kurylowicz, W. and F. Ulak. 1963 Oxytetracycline. Polish Patent 43,565, July 5.

— and W. Woźnicka. 1967 *Actinomyces (Streptomyces) varsoviensis*. I. Taxononic studies. Med Dōsw Mikrobiol 19 (1): 1–9.

Küster, E. 1953 Beitrag zur Genese und Mor-

phologie der Streptomyceten sporen. Atti VI, Intern. Congr. Microbiol. Rome, Italy 1: 114–116.

— and R. Locci. 1963 Taxonomic studies on thermophilic Actinomycetes isolated from peat. Arch Mikrobiol 45: 188–197.

Kutchayeva, A. G., N. A. Krasil'nikov and G. K. Skryabin. 1960 Actinomycetes of the *A. olivaceus* group. Proc Symp on Antibiotics, Prague, Czechoslovakia, May 18–28, 1959, pp. 57–58.

Kutscher, Dr. 1894 Ein Beitrag zur Kenntniss der bacillaren Pseudotuberculose der Nagethiere. Z Hyg Infektionskr 18: 327–342.

Kuttler, K. L. 1966 Clinical and hematologic comparison of *Anaplasma marginale* and *Anaplasma centrale* infections in cattle. Amer J Vet Res 27: 941–946.

—. 1967 Serological relationship of *Anaplasma marginale* and *Anaplasma centrale* as measured by the complement fixation and capillary tube agglutination tests. Res Vet Sci 8: 207–211.

Kützing, F. T. 1833 Beitrag zur Kenntnis über die Entstehung und Metamorphose der niedern vegetabilischen Organismen, nebst einer systematischen Zusammenstellung der hierher gehörigen niedern Algenformen. Linnaea 8: 335–387.

—. 1834 Algarum aquae dulcis germanicarum. Halis Saxonum, Decas 12: 1–2.

—. 1837 Microscopische Untersuchungen ueber die Hefe und Essigmutter, nebst mehreren andern dazu gehörigen vegetabilischen Gebilden. J Prakt Chem 11: 385–409.

—. 1843 Phycologia Generalis. Leipzig.

—. 1845 Phycologia germanica. Deutschlands Algen, pp. 1–240.

—. 1849 Species Algarum. Lipsiae, pp. 1–922.

Kutzner, H. J. and S. A. Waksman. 1959 *Streptomyces coelicolor* Müller and *Streptomyces violaceoruber* Waksman and Curtis, two distinctly different organisms. J Bacteriol 78 (4): 528–538.

Kuznetsov, V. D. 1962 A new species of genus *Chainia*. Mikrobiologiya 31 (3): 534–539.

—, N. M. Lyagina, N. B. Gracheva, E. V. Pivovarova and A. A. Sokolov. 1966 Systematic position of novobiocin producers. Mikrobiologiya 35 (5): 841–849.

LaCave, C., J. Asselineau and R. Toubiana. 1967 Sur quelques constituants lipidiques de *Corynebacterium ovis*. Eur J Biochem 2: 37–43.

Lacey, M. S. 1939 Studies in bacteriosis. XXIV. Studies on a bacterium associated with leafy galls, fasciations and cauliflower disease of various plants. Ann Appl Biol 26: 262–278.

—. 1955 The cytology and relationships of *Corynebacterium fascians*. Trans Brit Mycol Soc 38: 49–58.

—. 1961 The development of filter-passing organisms in *Corynebacterium fascians* cultures. Ann Appl Biol 49: 634–644.

Lachance, R. A. 1962 The amino acid requirements of *Corynebacterium sepedonicum* (Spieck. & Kott.) Skapt & Burkh. Can J Microbiol 8: 321–325.

Lachner-Sandoval, V. 1898 Über Strahlenpilze. Inaugural Dissertation Strassburg. Universitäts Buchdruckerei Von Carl Georgi, Bonn.

Lackey, J. B. and E. W. Lackey. 1961 The habitat and description of a new genus of sulphur bacterium. J Gen Microbiol 26 (1): 29–39.

Lackman, D. B., E. J. Bell, H. G. Stoenner and E. G. Pickens. 1965 The Rocky Mountain spotted fever group of rickettsias. Health Lab Sci 2 (3): 135–141.

Lacorte, J. G. 1932 Bacillus serositidis, nova especie. (Isolado em cultura pura, de um caso humano de inflamação primitava das serosas). Mem Inst Oswaldo Cruz Rio de Janeiro 26: 1–7.

Lafar, F. 1896 Die künstliche Säuerung des Hefegutes der Brennereien. Zentrabl Bakteriol Parasitenk Infektionskr Hyg Abt II 2: 194–196.

Lahelle, O. and T. Thjötta. 1945 A systematic study of Fusobacterium and Necrobacterium (i.e. Actinomyces necrophorus, Nekrosebacterium Bang) as to their biological relationships and proposal of a new and adequate name for the latter. Acta Pathol Microbiol Scand 22: 310–322.

Laidlaw, P. P. and W. J. Elford. 1936 A new group of filterable organisms. Proc Roy Soc Biol 20: 292–303.

Laird, M. and F. A. Lari. 1957 The avian blood parasite Babesia moshkovskii (Schurenkova 1938) with a record from Corvus splendens Vieillot in Pakistan. Can J Zool 35: 783–795.

La Macchia, E. H. and M. J. Pelczar. 1966 Analyses of deoxyribonucleic acid of Neisseria caviae and other Neisseria. J Bacteriol 91 (2): 514–516.

Lambert, F. W., J. M. Brown and L. K. Georg. 1967 Identification of Actinomyces israelii and Actinomyces naeslundii by fluorescent-antibody and agar-gel diffusion techniques. J Bacteriol 94 (5): 1287–1295.

Lämmler, G. and R. Gothe. 1967 Zur Chemotherapie der Aegyptianella pullorum infektion des Huhnes. Z Tropenmed Parasitol 18: 479–488.

Lancefield, R. C. 1933 A serological differentiation of human and other groups of hemolytic streptococci. J Exp Med 57: 571–595.

— and R. Hare. 1935 The serological differentiation of pathogenic and non-pathogenic strains of hemolytic streptococci from parturient women. J Exp Med 61: 335–349.

Lange, R. T. 1961 Nodule bacteria associated with the indigenous Leguminosae of South Western Australia. J Gen Microbiol 26 (2): 351–359.

Langenberg, K. F., M. P. Bryant and R. S. Wolfe. 1968 Hydrogen-oxidizing methane bacteria. II. Electron microscopy. J Bacteriol 95 (3): 1124–1129.

Lange-Posdeeva, I. P. 1930 On the oxidation of sulfur and thiosulfate by thionic acid bacteria. Arch Sci Biol USSR 30: 189–201.

Langeron, M. 1922. Les Mycetomes. Nouveau Traité de Medecin. Fasc 4.

—. 1923 Les oscillariées parasites du tube digestif de l'homme et des animaux. Ann Parasitol Hum Comp 1: 113–123.

Langford, G. C., Jr., J. E. Faber, Jr., and M. J. Pelczar, Jr. 1950 The occurrence of anaerobic gram-negative diplococci in the normal human mouth. J Bacteriol 59 (2): 349–356.

— and P. A. Hansen. 1953 Erysipelothrix insidiosa. Atti Del VI. Cong Int Microbiol Roma. Riassunti Communicazioni 1: 18.

Langston, C. W. and C. Bouma. 1960 A study of the microorganisms from grass silage. Appl Microbiol 8: 223–234.

— and P. P. Williams. 1962 Reduction of nitrate by streptococci. J Bacteriol 84 (3): 603.

Lankester, E. R. 1873 On a peach-coloured bacterium—Bacterium rubescens, n.s. Quart J Microscp Sci 13: 408–425.

Lankford, C. E. 1950 Chemically defined nutrient supplements for gonococcus culture media. Bacteriol Proc, pp. 40–41.

—. 1959 The Henry oblique light technique as an aid in bacteriological diagnosis of cholera. J Microbiol Soc Thailand 3: 10–13.

Lanzi, M. 1876 I Batteri parassiti di funghi. Nuovo G Bot Ital 8: 256–261.

Lapage, S. P. 1961 Haemophilus vaginalis and its role in vaginitis. Acta Pathol Microbiol Scand 52: 34–54.

Laplanche, J. and R. Saissac. 1948 Inflabilis lituseburense. In Prévot's Manual de Classification et de Detérmination des Bactéries Anaérobies. Masson and Co., Paris, p. 276.

Larkin, J. M. and J. L. Stokes. 1967 Taxonomy of psychrophilic strains of Bacillus. J Bacteriol 94 (4): 889–895.

Larsen, H. 1952 On the culture and general physiology of the green sulfur bacteria. J Bacteriol 64 (2): 187–196.

—. 1953 On the microbiology and biochemistry of the photosynthetic green sulfur bacteria. Kgl Norske Vidensk Selsk Skr 1: 1–199.

—. 1962 Halophilism. In I. C. Gunsalus and R. Y. Stanier (Editors), The Bacteria, Vol. 4. Academic Press, New York. pp. 275–286.

Larson, C. L., W. Wicht and W. L. Jellison. 1955 An organism resembling P. tularensis from water. Pub Health Rep 70: 253–258.

Lasseur, P. 1913 Contribution à l'étude de Bacillus lemonnieri, nov. spec. C R Soc Biol Paris 74: 47–48.

—, A. Dupaix-Lasseur and J. Melcion. 1944 Caractères antigèniques de Chromobacterium chocolatum Knutsen forme violette et forme orangée. Trav Lab Microbiol Fac Pharm Nancy 13: 293–312.

Laubach, C. A. 1916 Studies on aerobic spore-bearing non-pathogenic bacteria. Spore-bearing organisms in water. J Bacteriol 1: 505–512.

Laudien, L. 1923 Kotuntersuchung bei Pferden auf die Anwesenheit des Bakt. pyosepticum equi und von paratyphysbazillen. Inaug Diss Hannover, Germany.

Laughton, N. 1948 Canine beta haemolytic streptococci. J Pathol Bacteriol 60: 471–476.

Lauterborn, R. 1906 Zur Kenntnis der sapropelischen Flora. Allg Bot Z 12: 196–197.

—. 1907 Eine neue Gattung der Schwefelbakterien (Thioploca schmidlei nov. gen. nov. spec.). Ber Deut Bot Ges 25: 238–242.

—. 1913 Zur Kenntnis einiger sapropelischer Schizomyceten. Allg Bot Z 19: 97–100.

—. 1916 Die sapropelische Lebewelt. Ein Beitrag zur Biologie des Faulschlammes natürlicher Gewässer. Verh Naturh Mediz Ver Heidelb 13: 395–481.

Lautrop, H. 1956 Gelatin-liquefying Klebsiella strains (Bacterium oxytocum (Flügge)). Acta Pathol Microbiol Scand 39: 375–384.

—, K. Bøvre and W. Frederiksen. 1970 A Moraxella-like microorganism isolated from the genito-urinary tract of man. Acta Pathol

Microbiol Scand Sect B Microbiol Immunol *78:* 255–256.

Laveran, A. 1903 Sur la spirillose des bovides. C R Acad Sci Paris *136:* 939–941.

— and F. Mesnil. 1901 Sur la nature bactérienne du preétendu trypanosome des huîtres (*Tryp. balbianii* Certes). C R Séances Soc Biol Filiales *53:* 883–885.

Lawrence, D. B., R. E., Schoenike, A. Quispel and G. Bond. 1967 The role of *Dryas drummondii* in vegetation development following ice recession at Glacier Bay, Alaska, with special reference to its nitrogen fixation by root nodules. J Ecol *55:* 793–813.

Lawson, J. W. and H. Gooder. 1970 Growth and development of competence in group H streptococci. J Bacteriol *102 (3):* 820–825.

Laxa, O. 1900 Bakteriologische Studien über die Produkte des normalen Zuckerfabriksbetriebes. Zentrabl Bakteriol Parasitenk Infektionskr Hyg Abt II *6:* 286–295.

Lazar, I. 1968 Serological relationships of Corynebacteria. J Gen Microbiol *52 (1):* 77–88.

— and D. C. Graham. 1970 Comparative studies on *Corynebacterium poinsettiae*, *C. flaccumfaciens* and *C. flaccumfaciens* var. *aurantiacum*. Rev Roum Biol Ser Bot *15:* 287–293.

Leach, B. E., K. M. Calhoun, L. E. Johnson, C. M. Teeters and W. G. Jackson. 1953 Chartreusin, a new antibiotic produced by *Streptomyces chartreusis*, a new species. J Amer Chem Soc *75:* 4011–4012.

Leach, R. H. 1967 Comparative studies of mycoplasma of bovine origin. Ann N Y Acad Sci *143:* 305–316.

—. 1973 Further studies on classification of bovine strains of Mycoplasmatales, with proposals for new species, *Acholeplasma modicum* and *Mycoplasma alkalescens*. J Gen Microbiol *75:* 135–153.

Leadbetter, E. R. 1963 Growth and morphogenesis of *Sporocytophaga myxococcoides*. Bacteriol Proc, p. 42.

—. 1974 Substitution of *Methylomonas* gen. nov. and *Methylomonadaceae* fam. nov. for *Methanomonas* Orla Jensen and *Methanomonadaceae* Breed. Request for an opinion. Int J Syst Bacteriol *24:* in press.

— and J. A. Gottlieb. 1967 On methylamine assimilation in a bacterium. Arch Mikrobiol *59:* 211–217.

Leaf, G., I. C. Gardner and G. Bond. 1959 Observations on the composition and metabolism of the nitrogen-fixing root nodules of *Myrica*. Biochem J *72:* 662–667.

Leathen, W. W. and S. A. Braley. 1954 A new iron-oxidizing bacterium: *Ferrobacillus ferrooxidans*. Bacteriol Proc, p. 44.

Lebailly, C. 1913 Sur les spirochetès de l'intestin des oiseaux. C R Soc Biol Paris *75:* 389–391.

Leben, C., G. J. Stessel and G. W. Keitt. 1952 Helixin, an antibiotic active against certain fungi and bacteria. Mycologia *44 (2):* 159–169.

Lebert, F. 1949 Étude d'une bactérie anaérobie thermophile nouvelle des conserves de viande et legumes: *Plectridium causophilum* n. sp. Ann Inst Pasteur (Paris) *76:* 548–550.

Lebert, H. 1874 Rückfallstyphus und bilioses Typhoid. *In* Ziemssen's Handbuch der Speciellen Pathologie und Therapie, Ed. 2. F. C. W. Vogel, Leipzig, pp. 267–304.

LeBlaye, R. and H. Guggenheim. 1914 Manuel pratique de diagnostic bactériologique et de technique appliqué à la détermination des bactéries. Vigot Frères Edition, Paris, pp. 1–444.

Lechevalier, H. A. 1965 Priority of the generic name *Microbispora* over *Waksmania* and *Thermopolyspora*. Int Bull Bacteriol Nomencl Taxon *15:* 139–142.

— and M. P. Lechevalier. 1967 Biology of actinomycetes. Annu Rev Microbiol *21:* 71–100.

— and —. 1968 A critical evaluation of the genera of aerobic actinomycetes. *In* Prauser (editor), The Actinomycetales. G. Fischer, Jena, pp. 393–405.

— and —. 1969 Ultramicroscopic Structure of *Intrasporangium calvum* (Actinomycetales). J Bacteriol *100:* 522–525.

—, — and B. Becker. 1966 Comparison of chemical composition of cell walls of *Nocardia* with that of the aerobic actinomycetes. Int J Syst Bacteriol *16 (2):* 151–160.

—, — and N. N. Gerber. 1971 Chemical composition as a criterion in the classification of Actinomycetes. Advan Appl Microbiol *14:* 47–72.

—, M. Solotorovsky and C. I. McDurmont. 1961 A new genus of the *Actinomycetales: Micropolyspora* gen. nov. J Gen Microbiol *26:* 11–18

Lechevalier, M. P. 1968 Identification of aerobic actinomycetes of clinical importance. J Lab Clin Med *71:* 934–944.

—. 1972 Description of a new species, *Oerskovia xanthinolytica*, and emendation of *Oerskovia* Prauser et al. Int J Syst Bacteriol *22:* 260–264.

— and H. A. Lechevalier. 1957 A new genus of the *Actinomycetales: Waksmania* gen nov. J Gen Microbiol *17:* 104–111.

— and —. 1969 Composition of whole cell hydrolysates as a criterion in the classification of aerobic actinomycetes. *In* Prauser (Editor), The Taxonomy of the *Actinomycetales*. VEB Gustav Fischer Verlag, Jena.

— and —. 1970 Chemical composition as a criterion in the classification of aerobic actinomycetes. Int J Syst Bacteriol *20 (4):* 435–443.

—, — and P. E. Holbert. 1968 *Sporichthya*, un nouveau genre de *Streptomycetaceae*. Ann Inst Pasteur (Paris) *114:* 277–285.

Leclerc, H. 1962 Étude biochimique d'*Enterobacteriaceae* pigmentées. Ann Inst Pasteur (Paris) *102:* 726–741.

Lee, H. C. and W. E. C. Moore. 1959 Isolation and fermentation characteristics of strains of *Butyrivibrio* from ruminal ingesta. J Bacteriol *77:* 741–747.

Lee, K. Y., R. Wahl and E. Barbu. 1956 Conténu en bases puriques et pyrimidiques des acides déoxyribonucléiques des bactéries. Ann Inst Pasteur (Paris) *91:* 212–224.

Lee, M. and A. C. Chandler. 1941 A study of the nature, growth and control of bacteria in cutting compounds. J Bacteriol *41:* 373–386.

Lee, W. H. and R. C. Good. 1963 Amino acid synthesis (United States Patent 3,087,863). Abstr Offic Gaz U S Patent Off *789:* 1349.

Le Fevre, E. 1922 Pickle and sauerkraut experiments. Abstr Bacteriol *6:* 24–25.

Le Gall, J. 1963 A new species of *Desulfovibrio*. J Bacteriol *86 (5):* 1120

Leger, A. 1917 Spirochaete de la musaraigne

(*Crocidura stampfli* Tentink). Bull Soc Pathol Exot *10:* 280–281.

Lehmann, K. B. and R. Neumann. 1896 Atlas und Grundriss der Bakteriologie und Lehrbuch der speciellen bacteriologischen Diagnostik. 1st Ed. J. F. Lehmann, München.

— and —. 1899 Lehmann's Medizin, Haudetlanten X. Atlas und Grundriss der Bakteriologie und Lehrbuch der speciellen bakteriologischen Diagnsotik. 2 Aufl.

— and —. 1899 Lehmann's Medezin, Handatlanten. X. Atlas und Grundriss der Bakteriologie und Lehrbuch der speziellen Bakteriologischen Diagnostik. 3 Aufl.

— and —. 1907 Lehmann's Medizin, Handatlanten. X. Atlas und Grundriss der Bakteriologie und Lehrbuch der speciellen bakteriologischen Diagnostik. 4 Aufl.

— and —. 1912 Lehmann's Medizin. Handatlanten, X. Atlas und Grundriss der Bakteriologischen Diagnostik. 5 Aufl.

— and —. 1927 Bakteriologie insbesondere Bakteriologische Diagnostik. II. Allgemeine und spezielle Bakteriologie. 7 Aufl.

Leichmann, G. 1896 Ueber die im Brennereiprozess bei der Bereitung der Kunsthefe auftretende spontane Milchsäuregärung. Zentrabl Bakteriol Parasitenk Infektionskr Hyg Abt II *2:* 281–285.

—. 1896 Ueber die freiwillige Säurung der Milch. Zentrabl Bakteriol Parasitenk Infektionskr Hyg Abt II *2:* 777–780.

Leidy, J. 1850 On the existence of entophyta in healthy animals, as a natural condition. Proc Acad Natur Sci Philadelphia *4:* 225–229.

—. 1881 The parasites of the termites. J Acad Natur Sci Philadelphia (Ser 2) *8:* 425–447.

Leifson, E. 1954 The flagellation and taxonomy of species of *Acetobacter.* Antonie van Leeuwenhoek J Microbiol Serol *20:* 102–110.

—. 1956 Morphological and physiological characters of the genus *Chromobacterium.* J Bacteriol *71* (*4*): 393–400.

—. 1960 Atlas of bacterial flagellation. Academic Press, New York.

—. 1962 *Pseudomonas spinosa* n. sp. Int Bull Bacteriol Nomencl Taxon *12* (*3*): 89–92.

—. 1962 The bacterial flora of distilled and stored water. III. New species of the genera *Corynebacterium, Flavobacterium, Spirillum* and *Pseudomonas.* Int Bull Bacteriol Nomencl Taxon *12* (*4*): 161–170.

—. 1964 *Hyphomicrobium neptunium* sp. n. Antonie van Leeuwenhoek J Microbiol Serol *30:* 249–256.

— and R. Hugh. 1954 A new type of polar monotrichous flagellation. J Gen Microbiol *10:* 68–70.

Lelliott, R. A. 1956 Slow wilt of carnations caused by a species of *Erwinia.* Plant Pathol *5:* 19–23.

—. 1966 The plant pathogenic coryneform bacteria. J Appl Bacteriol *29:* 114–118.

—. 1968 The diagnosis of fireblight (*Erwinia amylovora*) and some diseases caused by *Pseudomonas syringae.* Rep Eur Medit Pl Prot Orgn Coaf Fireblight, 1967. *45:* 27–34.

—, E. Billing and A. C. Hayward. 1966 A determinative scheme for the fluorescent plant pathogenic pseudomonads. J Appl Bacteriol *29:* 470.

— and M. M. Wallace. 1955 A bacterial disease

of Shirley poppies in Tanganyika. Trans Brit Mycol Soc *38* (*1*): 88–91.

Lemcke, R. M. 1961 Association of PPLO infection and antibody response in rats and mice. J Hyg *59:* 401–412.

—. 1964 The serological differentiation of *Mycoplasma* strains (pleuropneumonia-like organisms) from various sources. J Hyg *62:* 199–219.

—, K. A. Forshaw and R. J. Fallon 1969 The serological identity of Sabin's murine type C *Mycoplasma* and *Mycoplasma pulmonis.* J J Gen Microbiol *58* (*1*): 95–98.

Le Minor, L. 1965 Conversions antigéniques chez les *Salmonella.* VI. Acquisitions des facteurs 6,14 par les sérotypes du groupe K (0:18) sous l'effet de la lysogénisation. Ann Inst Pasteur (Paris) *108:* 805–811.

—. 1968 Conversions antigéniques chez les *Salmonella.* Ann Inst. Pasteur (Paris) *109:* 505–515.

—. 1968 Lysogenie et classification des Salmonella. Int J Syst Bacteriol *18* (*3*): 197–201.

—, R. Rohde and J. Taylor. 1970 Nomenclature des *Salmonella.* Ann Inst Pasteur (Paris) *119* (*2*): 206–210.

Lemoigne, M., H. Girard and G. Jacobelli. 1952 Bactérie du sol utilisant facilement le 2-3 butanediol. Ann Inst Pasteur (Paris) *82:* 389–398.

Leon, L. A. 1940 El mal del pinto en el Ecuador. Rev Med Trop Habana *6:* 253–276

Leonard, R. C. 1968 A physiological study of *Pilimelia terevasa* (Actinoplanaceae). Thesis, University of North Carolina.

Leong, D., R. Diaz and J. B. Wilson. 1968 Identification of the toxic component of *Brucella abortus* endotoxin and its labeling with radioactive chromate. J Bacteriol *95:* 612–617.

Leon y Blanco, F. 1940 El *Treponema herrejoni.* Rev Med Trop Habana *6:* 5–12.

Leopold, S. 1953 Heretofore undescribed organism isolated from genitourinary system. U S Armed Forces Med J *4:* 263–266.

Lepetit, S. p. A. 1957 Production of tetracycline by fermentation. British Patent 755,139, May 22.

Lerche, M. and G. Reuter. 1961 Isolierung und Differenzierung anaerober *Lactobacilleae* aus dem Darm erwachsener Menschen (Beitrag zum *Lactobacillus bifidus* Problem). Zentrabl Bakteriol Parasitenk Infektionskr Hyg Abt I Orig *182:* 324–356.

Lerner, P. I. 1967 Susceptibility of *Actinomyces* to cephalosporins and lincomycin. Antimicrob Agents Chemother, pp. 730–735.

Lessel, E. F. and R. S. Breed. 1954 *Selenomonas* Boskamp, 1922. A genus that includes species showing an unusual type of flagellation. Bacteriol Rev *18:* 165–169.

— and M. Rogosa. 1971 Designation of the type strain of *Lactobacillus viridescens* Niven and Evans. Int J Syst Bacteriol *21* (*3*): 238–239.

Lestoquard, F. 1924 Deuxième note sur les Piraplasmoses du mouton en Algérie. L'anaplasmose: *Anaplasma ovis* nov. sp. Bull Soc Pathol Exot *17:* 784–787.

— and A. Donatien. 1936 Sur une nouvelle Rickettsia du mouton. Bull Soc Pathol Exot *29:* 105–108.

Levaditi, C., A. Marie and L. Isaicu. 1921

Recherches sur la spirochétose spontanée du lapin. C R Soc Biol Paris 85: 51.

—, S. Nicolau and P. Poincloux. 1925 Sur le rôle étiologique de *Streptobacillus moniliformis* (nov. spec.) dans l'érythème polymorph aigu septicémique. C R Hebd Séances Acad Sci (Paris) 180: 1188–1190.

Levaditi, J. C., F. Roger and P. Destombes. 1964 Tentative de classification des *Chlamydiaceae* (Rake 1955) tenante compte de leurs affinités tissulaires et de leur épidemiologie. Ann Inst Pasteur (Paris) 107: 656–662.

Levi, M. L. and E. Cotchin. 1950 *Bacillus actinoides:* its association with pneumonia in cattle and its relationship to *Streptobacillus moniliformis.* J Comp Pathol 60: 17–27.

Levin, R. A. 1971 Fatty acids of *Thiobacillus thiooxydans.* J Bacteriol 108: 992–995.

Levin, R. E. 1968 Detection and incidence of specific species of spoilage bacteria on fish. (I) Methodology. Appl Microbiol 16: 1734–1737.

Levine, M. 1920 Dysentery and allied bacilli. J Infect Dis 27: 31–39.

— and D. Q. Anderson. 1932 Two new species of bacteria causing mustiness in eggs. J Bacteriol 23: 337–347.

— and Soppeland. 1926 Bacteria in creamery wastes. Bull Iowa State Agr Coll 77: 1–72.

Levinthal, W. 1928 Ueber die anaërobe Flora der menschlichen Rachenschleimhaut. Zentrabl Bakteriol Parisitenk Infektionskr Hyg Abt I Orig 106: 195–200.

Lewandowsky, F. 1904 Die Pseudodiphtheriebacillen und ihre Beziehungen zu den Diphtheriebacillen. Zentrabl Bakteriol Parasitenk Infektionskr Hyg Abt I Orig 36: 336–351; 472–480.

Lewin, R. A. 1962 *Saprospira grandis* Gross; and suggestions for reclassifying helical, apochlorotic, gliding organisms. Can J Microbiol 8: 555–563.

—. 1965 Freshwater species of *Saprospira.* Can J Microbiol 11: 135–139.

—. 1969 A classification of flexibacteria. J Gen Microbiol 58 (2): 189–206.

—. 1970 *Flexithrix dorotheae* gen. et sp. nov. (*Flexibacterales*); and suggestions for reclassifying sheathed bacteria. Can J Microbiol 16: 511–515.

—. 1970 New *Herpetosiphon* species (*Flexibacterales*). Can J Microbiol 16: 517–520.

— and D. C. Lounsbery. 1969 Isolation, cultivation and characterization of flexibacteria. J Gen Microbiol 58 (2): 145–170.

— and M. Mandel. 1970 *Saprospira toviformis* nov. spec. (*Flexibacterales*) from a New Zealand seashore. Can J Microbiol 16: 507–510.

Lewis, C., H. R. Reames and L. E. Rhulend. 1959 Psicofuranine. II. Studies in experimental animal infections. Antibiot Chemother 9 (7): 421–426.

Lewis, I. M. 1914 A bacterial disease of *Erodium* and *Pelargonium.* Phytopathology 4 (4): 221–232.

— and E. Watson. 1927 A bacterial disease of *Bowlesia.* Phytopathology 17 (7): 507–512.

Lewis, P. A. and R. E. Shope. 1931 Swine influenza II. A hemophilic bacillus from the respiratory tract of infected swine. J Exp Med 54: 361–371.

Lewis, V. J., R. E. Weaver and D. G. Hollis. 1968 Fatty acid composition of *Neisseria* species as determined by gas chromatography. J Bacteriol 96 (1): 1–5.

Lewkowicz, X. 1901 Recherches sur la flore microbienne de la bouche des nourissons. Arch Med Exp 13: 633–660.

Li, Y. F. and L. K. Georg. 1968 Differentiation of *Actinomyces propionicus* from *Actinomyces israelii* and *Actinomyces naeslundii* by gas chromatography. Can J Microbiol 14: 749–753.

Liaaen-Jensen, S. 1963 Carotenoids of photosynthetic bacteria-distribution, structure and biosynthesis. In Gest, San Pietro and Vernon (Editors), Bacterial Photosynthesis. Antioch Press, Yellow Springs, Ohio, pp. 19–37.

—. 1965 Bacterial carotenoids. XVIII. Arylcarotenes from *Phaeobium.* Acta Chem Scand 19: 1025–1030.

—, G. Cohen-Bazire, T. O. Nakayama and R. Y. Stanier. 1958 The path of carotenoid synthesis in a photosynthetic bacterium. Biochim Biophys Acta 29: 477–498.

—, E. Hegge and L. M. Jackman. 1964 Bacterial carotenoids. XVII. The carotenoids of photosynthetic green bacteria. Acta Chem Scand 18: 1703–1718.

—, S. Hertzberg, O. B. Weeks and U. Schwieter. 1968 Bacterial carotenoids XXVI C_{50} carotenoids. 2. Structure determination of dehydrogenans P.-439. Acta Chem Scand 22: 1171–1186.

— and K. Schmidt. 1963 Die carotinoide der *Thiorhodaceae.* III. Die Carotinoide von *Chromatium warmingii* Migula. Arch Mikrobiol 46: 138–149.

Liebert, F. 1909 Het afbreken van urinezuur door bakterien. Versl Gewone Akad Amst 17: 990–1001.

Liégard, H. and M. Landrieu. 1911 Un cas de mycose conjunctivale. Ann Ocul 46: 418–426.

Lieske, R. 1911 Beitrage zur Kenntnis der Physiologie von *Spirophyllum ferrugineum* Ellis einem typischen Eisenbacterium. Jahrb Wiss Bot 49: 91–127.

—. 1919 Zur Ernährungsphysiologie der Eisenbakterien. Zentrabl Bakteriol Parasitenk Infektionskr Hyg Abt II 49: 413–425.

—. 1921 Morphologie und Biologie der Strahlenpilze (Actinomyceten). Borntraeger Bros., Leipzig.

—. 1928 Untersuchungen über die Krebskrankheit bei Pflanzen, Tieren und Menschen. Zentrabl Bakteriol Parasitenk Infektionskr Hyg Abt I Orig 108: 118–146.

Lignières, J. 1900 Maladies du porc. Bull Soc Centr Med Vet n s 18 (54): 389–431.

—. 1914 L'anaplasmose bovine en Argentine. Contribution à l'étude de cette maladie. Zentrabl Bakteriol Parasitenk Infektionskr Hyg Abt I Orig 74: 133–162.

—. 1924 Nouvelle contribution à l'études des champignons produisant les actinomycoses. Ann Parasitol Hum Comp 2: 1–25.

— and G. Spitz. 1902 L'actinobacillose. Bull Soc Centr. Méd Vét 20: 487–535; 546–565.

— and —. 1902 Contribución al estudio de las afecciones conocidas bajo el nombre de actinomicosis: Actinobacilosis. Bol Agr Ganad Buenos Aires 2: 169–230.

Lillie, R. D. 1930 Psittacosis-rickettsia-like

inclusions in man and in experimental animals. Pub Health Rep *45 (15):* 773–778.

Lin, F.-K. and L. L. Coriell. 1955 Streptonivicin: laboratory and clinical studies in the pediatric age group. Abstr 3rd Ann Symp on Antibiotics, November 2–4, Washington, D. C.

Lind, A. 1959 Serological studies of mycobacteria by means of the diffusion-in-gel technique. I. Preliminary investigations. Int Arch Allergy *14:* 264–278.

—. 1960 Serological studies of mycobacteria by means of the diffusion-in-gel techniques. IV. The precipitinogenic relationships between different species of mycobacteria with special reference to *M. tuberculosis, M. phlei, M. smegmatis* and *M. avium.* Int Arch Allergy *17:* 300–322.

— and M. Norlin. 1963 A comparative serological study of *M. avium, M. ulcerans, M. balnei* and *M. marinum* by means of double diffusion-in-gel methods, a preliminary investigation. Scand J Clin Lab Invest *15:* 152–163.

Lindau G. 1898 Schizomyceten. VI. Beziehungen der Bacterien zum Menschen und Thieren. A. Beiziehungen zum Menschen. (a) Coccen. Just's Bot Jahr *26 (1):* 1–139.

Lindeijer, E. J. 1932 De Bacterie-ziekte van den Wilg veroorzaakt door *Pseudomonas saliciperda* n. sp. Inaug Diss, Univ Amsterdam, pp. 1–82.

Lindenbein, W. 1952 Über einige chemisch interessante Aktinomyceten-stämme und ihre Klassifizierun. Arch Mikrobiol *17:* 361–383.

Lindenberg, A. 1909 Un Nouveau Mycetome. Arch Parasitol *13:* 265–282.

Lindner, F., G. Huber and K. H. Wallhäusser. 1960 Streptomycin F 3463 and its salts. German Patent 1,077,381, March 10.

—, R. Junk, G. Nesemann and J. Schmidt-Thomé. 1958 Gewinnung von 20 β-Hydroxysteroiden aus 17 α-21-Dihydroxy-20-Ketosteroiden durch mikrobiologische Hydrierung mit *Streptomyces hydrogenans.* Hoppe-Seyler's Z Physiol Chem *313:* 117–123.

—, J. Schmidt-Thomé, R. Junk and G. Nesemann. 1957 Vefahren zur Herstellung von 20-Oyxsteroiden. German Ausgelegeschrift 1,016,263, September 26.

—, K.-H Wallhäusser and G. Huber. 1961 Herstellung und Gewinnung des antibiotikums. Moenomycin. German Ausgelegeschrift 1,113,791, September 14.

—, — and —. 1963 Antibiotic and process of preparing it. Canadian Patent 672,917, October 22.

Lindner, P. 1887 Über ein neues in Malzmaischen vorkommendes, milchsäurebildendes Ferment. Wochenschr Brau *4:* 437–440.

—. 1888 Die Sarcina Organismen der Gärungsgewerben. Inaugural Dissertation, Friedrich-Wilhems Universitat, pp. 1–59.

—. 1895 Mikroskopische Betriebskontrolle in den Gärungsgewerben. Aufl. I. P. Parey, Berlin I-IX, pp. 1–278.

—. 1905 Mikroskopische Betriebskontrolle in den Garüngsgewerben mit einer Einführung in die technische Biologie, Hefenreinkultur und Infektionslehre. Vierte Aufl, Berlin, pp. 1–521.

—. 1930 Die Bakterienflora des Aguamiel.

Mikroskop Biol Betriebsk Garungsgew *6:* 584–593. P. Parey, Berlin.

Lindquist, K. 1960 A *Neisseria* species associated with infectious keratoconjunctivitis of sheep, *Neisseria ovis* nov. spec. J Infect Dis *106:* 162–165.

Linell, L. and A. Nordén. 1952 Hudinfektioner is imhall genom ny art av-*Mycobacterium.* Nord Med *47:* 888–891.

— and —. 1954 *Mycobacterium balnei:* a new acid-fast bacillus occurring in swimming pools and capable of producing skin lesions in humans. Acta Tuberc Pneumol Scand Suppl *33:* 1–84.

Link, H. F. 1809 Observationes in Ordines plantarum naturales. Dissertatio prima complectens Anandrarum ordines Epiphytas, Mucedines, Gastromycos et Fungos. Ges Nat Berlin *3:* 3–42.

Lipman, C. B. and E. McLee. 1940 A new species of sulfur-oxidizing bacteria from a coprolite. Soil Sci *50:* 429–432.

Lipman, J. G. 1903 Experiments on the transformation and fixation of nitrogen by bacteria. Rep N J Agr Exp Sta *24:* 217–285.

—. 1904 Soil bacteriological studies. Rep N J Agr Exp Sta *25:* 237–289.

Lippincott, J. A. and G. T. Heberlein. 1965 The quantitative determination of the infectivity of *Agrobacterium tumefaciens.* Amer J Bot *52:* 856–863.

— and B. B. Lippincott. 1969 Tumor initiating ability and nutrition in the genus *Agrobacterium.* J Gen Microbiol *59 (1):* 57–75.

Lister, J. 1873 A further contribution to the natural history of bacteria and the germ theory of fermentative changes. Quart J Microbiol Sci *13:* 380–408.

Liston, J., W. Wiebe and R. R. Colwell. 1963 Quantitative approach to the study of bacterial species. J Bacteriol *85:* 1061–1070.

Lo, T.-C., D.-W. Chen and J.-S. Huang. 1966 A new disease (bacterial wilt) of Taiwan giant bamboo. I. Studies on the causal organism (*Erwinia sinocalami* sp. nov.). Bot Bull Acad Sinica (Taipei) *7:* 14–22.

Locci, R., E. Baldacci and B. Petrolini. 1967 Contributions to the study of oligosporic actinomycetes. I. Description of a new species of *Actinobifida: Actinobifida alba* sp. nov. and revision of the genus. G Microbiol *15:* 79–91.

—, — and B. Petrolini Baldan. 1969 The genus *Streptoverticillium.* A taxonomic study. G Microbiol *17:* 1–60.

Lochhead, A. G. 1934 Bacteriological studies on the red discoloration of salted hides. Can J Res *10:* 275–286.

—. 1942 Growth factor requirements of *Bacillus larvae* White. J Bacteriol *44:* 185–189.

—. 1955 *Brevibacterium helvolum* (Zimmerman) comb. nov. Int Bull Bacteriol Nomencl Taxon *5 (3):* 115–119.

—. 1957 Genus VI. *Arthrobacter. In* Breed, Murray and Smith (Editors), Bergey's Manual of Determinative Bacteriology, 7th ed. The Williams & Wilkins Co., Baltimore, pp. 605–612.

—. 1958 Two new species of *Arthrobacter* requiring respectively vitamin B12 and the terregens factor. Arch Mikrobiol *31:* 163–170.

— and M. O. Burton 1953. An essential bacterial growth factor produced by microbial synthesis. Can J Bot *31:* 7–22.

— and —. 1955 Qualitative studies of soil microorganisms. XII. Characteristics of vitamin-B-12-requiring bacteria. Can J Microbiol *1* (*5*): 319–330.

Loeffler, F. 1884 Untersuchung über Bedeutung der Mikroorganismen für die Entstiehung der Diphtherie beim Manschen, bei der Taube und beim Kalbe. Arb GesundhAmt Berl *2:* 421–499.

—. 1886 Experimentelle Untersuchungen über Schweine-Rothlauf. Arb GesundhAmt Berl *1:* 46–57.

—. 1892 Ueber Epidemieen unter den im hygienischen Institut zu Greifswald gehaltenen Mäusen und über die Bekämpfung der Feldmausplage. Zentrabl Bakteriol Parasitenk Infektionskr Hyg Abt I Orig *11:* 129–141.

Loesche, W. J. and R. J. Gibbons. 1965 A practical scheme for identification of the most numerous oral gram negative anaerobic rods. Arch Oral Biol *10:* 723–725.

—, — and S. S. Socransky. 1965 Biochemical characteristics of *Vibrio sputorum* and relationship to *Vibrio bubulus* and *Vibrio fetus*. J Bacteriol *89* (*4*): 1109–1116.

—, S. S. Socransky and R. J. Gibbons. 1964 *Bacteroides oralis*, proposed new species isolated from the oral cavity of man. J Bacteriol *88* (*5*): 1329–1337.

Loewe, L., N. Plummer, C. F. Niven, Jr. and J. M. Sherman. 1946 *Streptococcus* S. B. E. in subacute bacterial endocarditis. J Amer Med Ass *130:* 257.

Löhnis, F. 1905 Beitrage zur Kenntnis der Stickstoffbacterien. Zentrabl Bakteriol Parasitenk Infektionskr Hyg Abt II *14:* 87–101; 582–604.

—. 1909 Die Benennung der Milchsaürebakterien. Zentrabl Bakteriol Parasitenk Infektionskr Hyg Abt II *22:* 553–555.

—. 1911 Landwirtschaftlichbakteriologisches Praktium. Gebrüder Borntraeger, Berlin, pp. 1–156.

— and J. Hanzawa. 1914 Die Stellung von *Azotobacter* im System. Zentrabl Bakteriol Parasitenk Infektionskr Hyg Abt II *42:* 1–8.

— and N. K. Pillai. 1907 Ueber stickstofffixierende Bakterien. II. Zentrabl Bakteriol Parasitenk Infektionskr Hyg Abt II *19:* 87–96.

— and T. Westermann. 1909 Ueber stickstofffixierende Bakterien. IV. Zentrabl Bakteriol Parasitenk Infektionskr Hyg Abt II *22:* 234–254.

Lombardo-Pellegrino, P. 1903 Di una streptothrix isolata dal sotto-suolo. Rif Med *19* (*39*): 1065–1068.

London, J. P. 1963 *Thiobacillus intermedius* nov. sp. A novel type of facultative autotroph. Arch Mikrobiol *46:* 329–337.

— and M. D. Appleman. 1962 Oxidative glucose and glycerol metabolism of two species of enterococci. J Bacteriol *84* (*3*): 597–598.

— and S. C. Rittenberg. 1964 Path of sulfur in sulfide and thiosulfate oxidation by thiobacilli. Proc Nat Acad Sci USA *52:* 1183–1190.

— and —. 1967 *Thiobacillus perometabolis* nov.

sp., a non-autotrophic thiobacillus. Arch Mikrobiol *59:* 218–225.

Long, H. F. and B. W. Hammer. 1941 Classification of the organisms important in dairy products. III. *Pseudomonas putrefaciens*. Res Bull Iowa Agr Exp Sta *285:* 176–195.

Long, P. H. and E. A. Bliss. 1934 Studies upon minute hemolytic streptococci. I. The isolation and cultural characteristics of minute hemolytic streptococci. J Exp Med *60:* 619–631.

Longley, E. O. 1940 Contagious pleuropneumonia of goats. Int J Vet Sci Delhi *10* (*2*): 127–197.

—. 1951 Contagious caprine pleuropneumonia. A study of the disease in Nigeria. Colon Res Pub (London) *7:* 23.

Lopez, R. and J. H. Becking. 1968 Polysaccharide production by *Beijerinckia* and *Azotobacter*. Microbiol Espan *21:* 53–75.

Lovell, R. 1937 Studies on *Corynebacterium pyogenes* with special reference to toxin production. J Pathol Bacteriol *45:* 339.

—. 1941 Studies on the toxin of *Corynebacterium pyogenes*. J Pathol Bacteriol *52:* 295.

—. 1944 Further studies on the toxin of *Corynebacterium pyogenes*. J Pathol Bacteriol *56:* 525.

—. 1946 Studies on *Corynebacterium renale*. I. A systematic study of a number of strains. J Comp Pathol *56:* 196–204.

— and M. M. Zaki. 1966 Studies on growth products of *Corynebacterium ovis*. I. The exotoxin and its lethal action on white mice. Res Vet Sci *7:* 302–306.

— and —. 1966 Studies on growth products of *Corynebacterium ovis*. II. Other activities and their relationship. Res Vet Sci *7:* 307–311.

Lovrekovich, O. and Z. Klement. 1960 Triphenyltetrazolium chloride tolerance of phytopathogenic bacteria. Phytopathol Z *39:* 129–133.

Lowe, J. 1937 Rat leprosy. A critical review of the literature. Int J Leprosy *5:* 311–328; 463–481.

Lowy, J. and J. Hanson. 1965 Electron microscope studies of bacterial flagella. J Mol Biol *11:* 293–313.

Lucet, A. 1893 Recherches bactériologiques sur la suppuration chez les animaux de l'espèce bovine. Ann Inst Pasteur (Paris) *7:* 325–330.

Luderitz, O., K. Jann and R. Wheat. 1968 Somatic and capsular antigens of gram-negative bacteria. Compr Biochem *26A:* 105–228.

—, A. M. Staub and O. Westphal. 1966 Immunochemistry of O and R antigens of *Salmonella* and related *Enterobacteriaceae*. Bacteriol Rev *30:* 192–255.

Ludwig, F. 1886 Ueber Alkoholgährung und Schleimflüss lebender Bäume und deren Urheber, *under* Protokoll der vierten General-Versammlung der Deutschen Botanischen Gesellschaft am 17 Sept. 1886 in Berlin. Ber Deut Bot Ges *4:* 17–27.

—. 1898 Review of Hoyer (1898). Bijdrage tot de Kennis van de Azijnbacteriën. Zentrabl Bakteriol Parasitenk Infektionskr Hyg Abt II *4:* 867–875.

—. 1899 Beobachtungen über Schleimflüsse der Bäume in Jahr 1898. Z Pflanzenkr Pflanzenpathol Pflanzenschutz *9:* 10–14.

Luedemann, G. M. 1968 *Geodermatophilus*, a

new genus of the *Dermatophilaceae* (Actinomycetales). J Bacteriol *96* (*5*): 1848–1858.

—. 1969 *Micromonospora* taxonomy. Advan Appl Microbiol *11:* 101–133.

—. 1971 *Micromonospora purpureochromogenes* (Waksman and Curtis 1916) comb. nov. (Subjective synonym: *Micromonospora fusca* Jensen 1932). Int J Syst Bacteriol *21* (*3*): 240–247.

—. 1971 Designation of neotype strains for *Micromonospora coerulea* Jensen 1932 and *Micromonospora chalcea* (Foulerton 1905) Ørskov 1923. Int J Syst Bacteriol *21* (*3*): 248–253.

— and B. C. Brodsky. 1964 Taxonomy of gentamicin-producing *Micromonospora*. Antimicrob Agents Chemother *1963:* 116–124.

— and —. 1965 *Micromonospora carbonacea* sp. n., an everninomicin-producing organism. Antimicrob Agents Chemother *1964:* 47–52.

Luerssen, A. and M. Kühn. 1907 Yoghurt, die bulgarische Sauermilch. Zentrabl Bakteriol Parasitenk Infektionskr Hyg Abt II *20:* 234–248.

Lumb, W. V. 1961 Canine haemobartonellosis and its feline counterpart. Calif Vet *14* (*5*): 24–25.

Lund, B. M. 1969 Properties of some pectolytic, yellow pigmented, gram negative bacteria isolated from fresh cauliflowers. J Appl Bacteriol *32* (*1*): 60–67.

Lundestad, J. 1928 Über einige an der norwegischen Küste isolierte Agar-spaltende Arten von Meerbakterien. Zentrabl Bakteriol Parasitenk Infektionsrk Hyg Abt II *75:* 321–344.

Luria, S. E. 1960 The bacterial protoplasm: composition and organization. *In* The Bacteria, Vol. 1 Academic Press, New York, pp. 1–34.

Lustig, A. 1890 Diagnostica dei batteri delle acque con una guida alle ricerche batteriologiche e microscopiche. Torino, pp. 1–121.

Lüthy, P. 1968 Untersuchungen an *Bacillus fribourgensis* Wille. Zentrabl Bakeriol Parasitenk Infektionskr Hyg Abt II *122:* 671–711.

Lütje, F. 1921 Fohlenkrankheiten. Deut Tieraertzl Wochenschr *29:* 463–471.

—. 1923 Ein weiterer Beitrag zum Vorkommen des Corynebakteriums pyogenes equi in Deutschland. Deut Tieraerztl Wochenschr *31:* 559–561.

Lutz, A. 1886 Zur Morphologie der Mikroorganismus der Lepra. Derm Stud Hamburg *1:* 77–100.

—, O. Grootten and T. Wurch. 1956 Étude des caractères culturaux et biochimiques de bacilles du type 'Haemophilus hemolyticus vaginalis'. Rev Immunol Ther Antimicrob *20:* 132–138.

—, T. Wurch and O. Grootten. 1956 Quelques données sur les 'petits bacilles gram-négatif' agents d'une leucorrhée individualisée. Gynecol Obstet *55:* 75–82.

L'vov, N. P. and V. I. L'vov. 1965 A study of physiology of a new nitrogen-fixing mycobacterium. *Mycobacterium azot-absorbtum* sp. n. Izv Akad Nauk SSSR Ser Biol *2:* 250–256.

Lwoff, A. 1939 Revision et démembrement des *Hemophilae*, le genre *Moraxella* nov. gen. Ann Inst Pasteur (Paris) *62:* 168–176.

—. 1964 Remarques sur les *Moraxella*. Ann Inst Pasteur (Paris) *106:* 483–484.

Lyons, A. J., Jr. and T. G. Pridham. 1962 Proposal to designate strain ATCC 3004 (IMRU 3004) as the neotype strain of *Streptomyces albus* (Rossi-Doria) Waksman and Henrici. J Bacteriol *83* (*2*): 370–380.

— and —. 1966 *Streptomyces griseus* (Krainsky) Waksman and Henrici. A taxonomic study of some strains. U S Dept Agr, Agr Res Serv Tech Bull *1360:* 1–31.

McBeth, I. G. 1916 Studies on the decomposition of cellulose in soils. Soil Sci *1:* 437–487.

— and F. M. Scales. 1913 The destruction of cellulose by bacteria and filamentous fungi. U S Bur Plant Ind *266:* 1–52.

McBryde, C. N. 1911 A bacteriological study of ham souring. U S Bur Anim Ind *132:* 1–55.

McCarthy, B. J. and E. T. Bolton. 1963 A general method for the isolation of RNA complementary to DNA. Proc Nat Acad Sci U S A *48:* 1390–1397.

McClung, L. S. 1935 Studies on anaerobic bacteria. IV. Taxonomy of cultures of thermophilic species causing "swells" of canned foods. J Bacteriol *29:* 189–203.

— and E. McCoy. 1957 Genus II. *Clostridium* Prazmowski 1880. *In* Breed, Murray and Smith (Editors), Bergey's Manual of Determinative Bacteriology, 7th ed. The William & Wilkins Co., Baltimore, pp. 634–693.

McClung, N. M. 1949 Morphological studies in the genus *Nocardia*. I. Developmental studies. Lloydia (Cincinnati) *12* (*3*): 137–177.

—. 1954 Morphological studies in the genus *Nocardia*. III. The morphology of young colonies. Ann N Y Acad Sci *60* (*1*): 168–181.

McCoy, E. E. B. Fred, W. H. Peterson and E. G. Hastings. 1926 A cultural study of the acetone butyl alcohol organisms. J Infect Dis *39:* 457–483.

McCoy, G. W. and C. W. Chapin. 1912 Further observations on a plague-like disease of rodents with a preliminary note on the causative agent, *Bacterium tularense*. J Infect Dis *10:* 61–72.

McCray, A. H. 1917 Spore-forming bacteria of the apiary. I. Description and comparison of species. J Agr Res *8:* 399–420.

McCulloch, L. 1911 A spot disease of cauliflower. Bull U S Bur Plant Ind *225:* 1–15.

—. 1918 A morphological and cultural note on the organism causing Stewart's disease of sweet corn. Phytopathology *8:* 440–441.

—. 1920 Basal glumerot of wheat. J Agr Res *18:* 543–552.

—. 1921 A bacterial disease of gladiolus. Science (Washington) *54:* 115–116.

—. 1924 Two bacterial diseases of gladiolus. Phytopathology *14:* 63–64.

—. 1925 *Aplanobacter insidiosum* n. sp. the cause of an alfalfa disease. Phytopathology *15:* 496–497.

—. 1929 A bacterial leaf-spot of horseradish caused by *Bacterium campestre* var. *armoraciae*, n. var. J Agr Res *38:* 269–298.

—. 1937 Bacterial leaf-spot of begonia. J Agr Res *54:* 583–590.

—. 1937 An iris leaf disease caused by *Bacterium tardicrescens* n. sp. Phytopathology *27:* 135.

— and J. B. Demaree. 1932 A bacterial disease of the tung-oil tree. J Agr Res *45:* 339–346.

— and P. P. Pirone. 1939 Bacterial leaf spot of dieffenbachia. Phytopathology *29:* 956–962.

McCullough, N. B. and L. A. Dick. 1943 Growth of *Brucella* in a simple chemically defined medium. Proc Soc Exp Biol Med *52:* 310–311.

McCurdy, H. D. 1963 A method for the isolation of myxobacteria in pure culture. Can J Microbiol *9:* 282–285.

—. 1964 Growth and fruiting body formation of *Chondromyces crocatus* in pure culture. Can J Microbiol *10:* 935–936.

—. 1968 Light and electron microscope studies on the fruiting bodies of *Chondromyces crocatus.* Arch Mikrobiol *65:* 380–390.

—. 1969 Studies on the taxonomy of the *Myxobacterales.* I. Record of Canadian isolates and survey of methods. Can J Microbiol *15:* 1453–1461.

—. 1970 Studies on the taxonomy of the *Myxobacterales.* II. *Polyangium* and the demise of the *Sporangiaceae.* Int J Syst Bacteriol *20 (3):* 283–296.

—. 1971 Studies on the taxonomy of the *Myxobacterales.* III. *Chondromyces* and *Stigmatella.* Int J Syst Bacteriol *21 (1):* 40–49.

—. 1971 Studies on the taxonomy of the *Myxobacterales.* IV. *Melittangium.* Int J Syst Bacteriol *21 (1):* 50–54.

— and B. T. Khouw. 1969 Studies on *Stigmatella brunnea.* Can J Microbiol *15:* 731–738.

— and S. Wolfe. 1967 Deoxyribonucleic acid base compositions of fruiting *Myxobacterales.* Can J Microbiol *13:* 1707–1708.

McDade, J. E. and P. J. Gerone. 1970 Plaque assay for Q fever and scrub typhus rickettsiae. Appl Microbiol *19:* 963–965.

—, J. R. Stakebake and P. J. Gerone. 1969 Plaque assay system for several species of *Rickettsia.* J Bacteriol *99 (3):* 910–912.

McDaniel, L. E., C. P. Schaffner and E. G. Bailey. 1965 Production of fungimycin. United States Patent 3,182,004, May 4.

McDonald, I. J. 1971 Relationship of *Micrococcus* sp. ATCC 407 to the status of *Micrococcus freudenreichii* Guillebeau. Int J Syst Bacteriol *21 (4):* 314–322.

McFadden, L. A. 1961 Bacterial stem and leaf rot of dieffenbachia in Florida. Phytopathology *51:* 663–667.

McGee, Z. A. and R. G. Wittler. 1969 The role of L-phase and other wall-defective microbial variants in disease. *In* Hayflick (Editor), The Mycoplasmatales and the L-Phase of Bacteria. Appleton-Century-Crofts, New York, pp. 697–720.

McGill, D. J., E. A. Dawes and D. W. Ribbons. 1965 Carbohydrate metabolism and growth yield coefficients of *Zymomonas anaerobia.* Biochem J *94:* 44P–45P.

McGuire, J. M. 1956 Oxytetracycline (Terramycin) and process of producing the same. Canadian Patent 520,836, January 17.

McIntyre, O. R., J. C. Feeley, W. B. Greenough, A. S. Benenson, S. I. Hassan and A. Saad. 1965 Diarrhea caused by non-cholera vibrios. Amer J Trop Med Hyg *14:* 412–418.

McKiel, J. A., E. J. Bell and D. B. Lackman. 1967 *Rickettsia canada:* A new member of the typhus group of rickettsiae isolated from *Haemaphysalis leporispalustris* ticks in Canada. Can J Microbiol *13:* 503–510.

McKray, G. A. and R. H. Vaughn. 1957 The fermentation of glucose by *Bacillus stearothermophilus.* Food Res *22:* 494–500.

McLean, R. A. and W. L. Sulzbacher. 1953 *Microbacterium thermosphactum,* spec. nov; a nonheat resistant bacterium from fresh pork sausage. J Bacteriol *65 (4):* 428–433.

—, — and S. Mudd. 1951 *Micrococcus cryophilus,* spec. nov., a large coccus especially suitable for cytologic study. J Bacteriol *62:* 723–728.

McLeod, J. W. 1943 The types Mitis, Intermedius and Gravis of *Corynebacterium diphtheriae.* Bacteriol Rev *7:* 1–41.

McNab, A. 1904 Über den *Diplobacillus liquefaciens* (Petit) und über sein Verhältnis zu dem Morax-Axenfeldschen Diplobazillus der Blepharonkonjunktivitis. Klin Monatsbl Augenheilk *42:* 54–63.

McVeigh, I. and C. R. Reyes. 1961 A new species of *Streptomyces* and its antibiotic activity. Antibiot Chemother *11:* 312–319.

Maassen, A. 1907 Ueber Gallertbildungen in den Säften Zuckerfabriken. Arb Biol Abt (Anst-Reichsaust) *5:* 1–30.

MacCallum, P. 1948 A new mycobacterial infection in man. I. Clinical aspects. J Pathol Bacteriol *60:* 93–122.

MacCallum, W. G. and T. W. Hastings. 1899 A case of acute endocarditis caused by *Micrococcus zymogenes* (nov. spec.) with a description of the microorganism. J Exp Med *4:* 521–534.

Macchiavello, A. 1947 Notes on the taxonomy of the Rickettsias and the classification of the Rickettsioses. Prim Reunion Interamer del Tifo, Mexico, pp. 405–426.

MacDonald, J. B. 1953 The Motile Non-Sporulating Anaerobic Rods of the Oral Cavity. Ph.D. Thesis, U of Toronto, pp. 1–95.

—, E. M. Madlener and S. S. Socransky. 1959 Observations on *Spirillum sputigenum* and its relationship to *Selenomonas* species with special reference to flagellation. J Bacteriol *77 (5):* 559–565.

MacDonald, R. E. and S. W. MacDonald. 1962 The physiology and natural relationships of the motile sporeforming sarcinae. Can J Microbiol *8:* 795–808.

Macé, E. 1889 Traité Pratique de Bactériologie, lst ed. J.-B. Ballière & Sons, Paris, pp. 1–711.

—. 1897 Traité Pratique de Bactériologie, 3rd. ed. Baillière, Paris, pp. 1–1144.

—. 1901 Traité pratique de Bactériologie, 4th ed. Baillière, Paris, pp. 1–1196.

—. 1913 Traité Pratique de Bactériologie, 6th ed. Ballèire, Paris, pp. 1–918.

Mach, F. and N. S. Agre. 1968 Die Feinstruktur der Sporen von *Actinobifida chromogena.* Acta Biol Med Ger *21:* 575–576.

MacLean, P. D., A. A. Liebow and A. A. Rosenberg. 1946 A haemolytic corynebacterium resembling *Corynebacterium ovis* and *C. pyogenes* in man. J Infect Dis *79:* 69–90.

Maeda, K., H. Kosaka, Y. Okami and H. Umezawa. 1953 A new antibiotic, pyridomycin. J Antibiot (Tokyo) Ser A *6:* 140

—, Y. Okami, H. Kosaka, O. Taya and H. Umezawa. 1952 On an anti-tubercular antibiotic produced by *Streptomyces cinnamonensis* n. sp. J Antibiot (Tokyo) *5:* 572–573.

—, T. Takeuchi, K. Nitta, K. Yagishita, R. Utahara, T. Osato, M. Ueda, S. Kondo, Y. Okami and H. Umezawa 1956 A new antitumor substance pluramycin. Studies on antitumor substances produced by actinomycetes. J Antibiot (Tokyo) Ser A *9:* 75–81.

Maggi, L. 1886 Essai d'une classification protistologique des ferments vivants. J Micrographie *10:* 80–85; 173–178; 327–333.

Magnuson, H. 1917 Om den infektiösa fölsjukans etiologi. Svensk VetTidskr, Op 81–99; 125–147.

—. 1923 Spezifische infektiose Pneumonie beim Fohlen. Ein neuer Eitreneger beim Pferde. Arch Wiss Prakt Tierheilk *50:* 22–38

—. 1938 Pyaemia in foals caused by *Corynebacterium equi*. Vet Rec *50:* 1459–1468.

Magnusson, M. 1961 Specificity of mycobacterial sensitins. I. Studies in guinea pigs with purified "Tuberculin" prepared from mammalian and avian tubercle bacilli, *Mycobacterium balnei*, and other acid-fast bacilli. Amer Rev Resp Dis *83:* 57–68.

—. 1962 Specificity of sensitins. III. Further studies in guinea pigs with sensitin of various species of *Mycobacterium* and *Nocardia*. Amer Rev Resp Dis *86:* 395–404.

—. 1967 Identification of species of Mycobacterium on the basis of specificity of the delayed type reaction in guinea pigs. Z Tuberk ErkranKungen Thoraxorgane *127:* 55–56.

Magrou, J. 1937 *Phytomonas gypsophilae. In* Hauduroy *et al.* (Editors), Dictionnaire des bactéries pathogènes. Masson and Co., Paris, p. 60.

—. 1937 *Phytomonas sepedonica* (Spieckermann) n. comb. *In* Hauduroy *et al.* (Editors), Dictionnaire des bactéries pathogènes. Masson & Co., Paris, p. 411.

— and A. R. Prévot. 1948 Études de systematique bactérienne. IX. Essai de classification des bactéries phytopathogènes et espèces voisines. Ann Inst Pasteur (Paris) *75:* 99–108

— and —. 1948 Étude taxonomique des bactéries phytopathogènes. C R Acad Sci Paris *226:* 1229–1230.

Maier, S. 1963 A cytological study of *Thioploca ingrica* Wislouch. PhD Dissertation, Ohio State University.

— and R. G. E. Murray. 1965 The fine structure of *Thioploca ingrica* and a comparison with *Beggiatoa*. Can J Microbiol *11:* 645–655.

Maire, R. and A. Tison. 1909 La cytologie des Plasmodiophoracées et la classe des Phytomyxinae. Ann Mycol *7:* 226–253.

Maksimova, T. S. and I. N. Kovsharova. 1964 Early identification of actinomycin antibiotics and the systematic position of their producers. (In Russian) Antibiotiki *9* (*2*): 110–115.

—, — and V. V. Proshylakova. 1965 Early identification of echinomycine antibiotics and systematic position of their producers (in Russian). Antibiotiki *10* (*4*): 298–304.

Málek, I. and Kazdová-Kožiškova. 1946 *Pseudomonas odorans* n. sp. novy mikrob z diagnostického materiálu. (A new microbe discovered from diagnostic material). Sb Lék *48* (*52*): 189–194.

—, M. Radochová and O. Lysenko. 1963 Taxonomy of the species *Pseudomonas odorans*. J Gen Microbiol *33:* 349–355.

Malkoff, K. 1906 Weitere Untersuchungen über die Bakterienkrankheit auf *Sesamum orientale*. Zentrabl Bakteriol Parasitenk Infektionskr Hyg Abt II *16:* 664–666.

Mamkaeva, K. A. 1966 Studies of lysis in culture of *Chlorella*. Microbiology *35:* 724–728.

Mamoli, L. 1939 Über Biochemische dehydrierungen in der Cortingruppe. Ber Deut Chem Ges *72:* 1863–1865.

— and A. Vercellone. 1939 Über die biochemische Dehydrierung von Kiemdrusenhormonen mit einem reinen Bakterienstamm Naturwissenschaften *27:* 319.

Manachini, P. L., A. Craveri and R. Craveri. 1966 *Thermonospora citrina*, una nuovo specie di Attinomicete Termofilo isolato da suolo. Ann Microbiol Enzimol *16:* 83–90.

Manchee, R. J. and D. Taylor-Robinson. 1969 Utilization of neuramic acid receptors by mycoplasmas. J Bacteriol *98* (*3*): 914–919.

Mancy-Courtillet, D. and S. Pinnert-Sindico. 1954 Une nouvelle espèce de *Streptomyces: Streptomyces armillatus*. Ann Inst Pasteur (Paris) *87* (*5*): 580–584.

Mandel, M. 1966 Deoxyribonucleic acid base composition in the genus *Pseudomonas*. J Gen Microbiol *43* (*2*): 273–293.

—, C. Bergendahl and N. Pfennig. 1965 Deoxyribonucleic acid base composition of isolates of the genus *Chlorobium*. J Bacteriol *89* (*3*): 917–918.

—, E. F. Guba and W. Litsky. 1961 The causal agent of bacterial blight of American Holly Bacteriol Proc *1961:* 61.

—, A. Johnson and J. L. Stokes. 1966 Desoxyribonucleic acid base composition of *Sphaerotilus natans* and *Sphaerotilus discophorus*. J Bacteriol *91* (*4*): 1657–1658.

— and E. R. Leadbetter. 1965 Deoxyribonucleic acid base composition of myxobacteria. J Bacteriol *90* (*6*): 1795–1796.

— and R. A. Lewin. 1969 Deoxyribonucleic acid base composition of flexibacteria. J Gen Microbiol *58* (*2*): 171–178.

Manjrekar, S. L. 1954 Rickettsia of domesticated animals. Indian J Vet Sci Anim Husb *24* (*4*): 217–222.

Mann, D. K. 1967 The effect of *Anaplasma marginale* infection on the metabolic and osmotic behavior of bovine erythrocytes. Ph.D. Thesis, University of Illinois, Urbana, Ill.

Mann, E. W. 1968 *Bacillus uniflagellatus:* sp. n. Its unusual characteristics. Southwest Natur *13:* 349–352.

Mann, S. O. and A. E. Oxford. 1954 Studies of some presumptive lactobacilli isolated from the rumens of young calves. J Gen Microbiol *11:* 83–90.

Mannheim, W. and W. Stenzel. 1962 Zur Systematik der obligat aeroben gram-negativen Diplobakterien des Menschen. Zentrabl Bakteriol Parasitenk Infektionskr Hyg Abt I Orig *185:* 55–83.

Manns, T. F. 1909 The blade blight of oats, a bacterial disease. Ohio Agr Exp Sta Bull *210:* 91–167.

Manson, P. 1907 Tropical diseases, 4th ed. Cassell London, pp. 1–876.

Manten, A. 1957 Antimicrobial susceptibility and some other properties of photochromogenic mycobacteria associated with pulmonary disease. Antonie Van Leeuwenhoek J Microbiol Serol 23: 357–363.

Marca, A. 1927 Contribution à l'étude de la flore bactérienne du lac de Genève. Thesis Univ. Genève, pp. 1–38.

Marchette, N. J. and P. S. Nicholes. 1961 Virulence and citrulline ureidase activity of Pasteurella tularensis. J Bacteriol 82: 26–32.

Marchoux, E. and F. Sorel. 1912 Recherches sur la lepre. Ann Inst Pasteur (Paris) 26: 675–700.

Marcus, P. I. and P. Talalay. 1956 Induction and purification of α- and β-hydroxysteroid dehydrogenases. J Biol Chem 218: 661.

Maré, C. J. and W. P. Switzer. 1965 New species: Mycoplasma hyopneumoniae a causative agent of virus pig pneumonia. Vet Med 60: 841–846.

Margalith, P. and G. Beretta. 1960 A new antibiotic producing Streptomyces: Str. bellus nov. sp. Mycopathol Mycol Appl 12 (Fasc 3): 189–195.

—, — and M. T. Timbal. 1959 Matamycin, a new antibiotic. I. Biological studies. Antibiot Chemother 9 (2): 71–75.

Margherita, S. S. and R. E. Hungate. 1963 Serological analysis of Butyrivibrio from the bovine rumen. J Bacteriol 86: 855–860.

Marini, F. and S. Merli. 1964 Ulteriori studi sul fattore F. T. (Fattore di crescita per Flavobacterium tirrenicum). G. Microbiol 12: 45–54.

— and C. Spalla. 1964 Un nuovo fattore di crescita per um batterio marino (Flavobacterium tirrenicum n. sp.) presente nella farina di pesci e prodotto da microorganismi. G Microbiol 12: 35–44.

Marks, J. 1964 Aspects of the epidemiology of infection of "anonymous" mycobacteria. Proc Roy Soc Med 57: 479–480.

— and M. Richards. 1962 Classification of the anonymous mycobacteria as a guide to their significance. Mon Bull Min Health Public Health Lab Ser 21: 200–208.

— and H. Schwabacher. 1965 Infection due to Mycobacterium xenopei. Brit Med J 1: 32–33.

Marmur, J. and P. Doty. 1962 Determination of the base composition of deoxyribonucleic acid from its thermal denaturation temperature. J Mol Biol 5: 109–118.

—, S. Falkow and M. Mandel. 1963 New approaches to bacterial taxonomy. Annu Rev Microbiol 17: 329–372.

Marsh, H. and B. D. Firehammer. 1953 Serologic relationships of strains of V. fetus. Amer J Vet Res 14: 396–398.

Marshall, B. J. and D. F. Ohye. 1966 Bacillus macquariensis n. sp., a psychrotrophic bacterium from sub-Antarctic soil. J Gen Microbiol 44 (1): 41–46.

Martin, J. E., W. L. Peacock, G. Reising, D. S. Kellogg, E. Ribi and J. D. Thayer. 1969 Preparation of cell walls and protoplasm of Neisseria with the Ribi Cell Fractionator. J Bacteriol 97 (3): 1009–1011.

Martin, S. M. and V. B. D. Sherman (Editors). 1972 World Directory of Collections of Cultures of Microorganisms. Wiley-Interscience, New York, pp. 1–560.

Martinec, T. and M. Kocur. 1960 The taxonomic status of S. plymuthica (Lehmann and Neumann) Bergey et al. and of S. indica (Eisenberg) Bergey et al. Int Bull Bacteriol Nomencl Taxon 10 (4): 247–254.

— and —. 1961 The taxonomic status of Serratia marcescens Bizio. Int Bull Bacteriol Nomenc Taxon 11 (1): 7–12.

— and —. 1961 Contribution to the taxonomic studies of Serratia kiliensis (Lehmann and Neumann) Bergey. Int Bull Bacteriol Nomencl Taxon 11 (3): 87–90.

— and —. 1963 Taxonomicka studie rodu Erwinia. Folia Biol (Praha) 4: 1–163.

Martinevski. I. L. 1969 Biologie et génétique du bacille de la peste et des germes voisins. Medicina, Moscow, pp, 1–296.

Marton, M. and I. Szabo. 1959 Nocardia uniformis a new species from solonetz soil. Acta Microbiol Acad Sci Hung 5: 131–134.

Martres, M., E. R. Brygoo and H. Thouvenot. 1952 Étude d'une espèce nouvelle du genre Zuberella: Z. constellata n. sp. Ann Inst Pasteur (Paris) 83: 139–141.

Martzinovski, E.-J. 1911 De l'etiologie de la péripneumonie. Ann Inst Pasteur (Paris) 25: 914–917.

Maruashvili, G. M. 1945 On the tick borne relapsing fever. Med Parazitol Parazit Bolez 14 (1): 24–27.

Mason, D. J., A. Dietz and C. DeBoer. 1963 Lincomycin, a new antibiotic. I. Discovery and biological properties. Antimicrob Agents Chemother 1962: 554–559.

—, — and L. J. Hanka. 1963 U-12898, a new antibiotic. I. Discovery, biological properties, and assay. Antimicrob Agents Chemother 1962: 607–613.

—, — and R. M. Smith. 1961 Actinospectacin, a new antibiotic. I. Discovery and biological properties. Antibiot Chemother 11 (2): 118–122.

—, W. L. Lummis and A. Dietz. 1965 U-22956, a new antibiotic. I. Discovery and biological activity. Antimicrob Agents Chemother, pp. 110–113.

Mason, R. A. and M. Ristic. 1966 In vitro incorporation of glycine by bovine erythrocytes infected with Anaplasma marginale. J Infect Dis 116: 335–342.

Massart, J. 1901 Recherches sur les Organismes Inférieurs. V. Sur le Protoplasme des Schizophytes. Rec Inst Bot "Leo Errera," Brux 5: 251–282.

Massini, R. 1914 Über anaerobe Bakterien. Z Gesamte Exp Med 2: 81–167.

Materassi, R., G. Florenzano, W. Balloni and F. Favilli. 1966 Su una nuova specie di Beijerinckia (Beijerinckia venezuelae nov. sp.) isolata da terreni venezuelani. Ann Microbiol 16: 201–215.

Mathey, W. J. and A. C. Rissberger. 1964 A turkey sinus vibrio (Vibrio maleagridis n. sp.) compared with the avian hepatitis vibrio (Vibrio hepaticus, n. sp.). Poultry Sci 43: 1339.

Matsumae, A. M. 1960 Cephalomycin, a new antiviral antibiotic. I. Studies on biological and chemical properties. J Antibiot (Tokyo) Ser A 13 (2): 143–154.

—, M. Ohtani, H. Takeshima and T. Hata. 1968 A new genus of *Actinomycetales: Kitasatoa* Gen. Nov. J Antibiot (Tokyo) *21:* 616–625.

Matsumoto, H. 1963 Studies on the Hafnia isolated from normal human. Jap J Microbiol *7:* 105–114.

—. 1964 Additional new antigens of Hafnia group. Jap J Microbiol *8:* 139–141.

Matsumoto, K. 1961 A vancomycin-related antibiotic from *Streptomyces* sp. K-288. J Antibiot (Tokyo) Ser A *14 (3):* 141–146.

Matteuzzi, D., F. Crociani, G. Zani and L. D. Trovatelli. 1971 *Bifidobacterium suis* n. sp.: a new species of the genus *Bifidobacterium* isolated from pig feces. Z Allg Mikrobiol *11:* 387–395.

Mattick, A. T. R. 1949 The lactic streptococci (including their antibiotic activity). Rep 4th Int Congr Microbiol Copenhagen, p. 519

Matuszewski, T., E. Pijanowski and J. Supinska. 1936 *Streptococcus diacetilactis* n. sp. and its application to butter-making. Prace Zakl Microbiol Prezem Roln Warsz *11:* 1–28.

Matzuschita, T. 1902 Bakteriologische Diagnostik. Gustav Fischer, Jena, pp. 1–690.

Maximescu, P. 1968 New host-strains for the lysogenic *Corynebacterium diphtheriae* Park-Williams No. 8 strain. J Gen Microbiol *53:* 125–133.

Maxted, W. R. and E. V. Potter. 1967 The presence of type 12 M-protein antigen in group G streptococci. J Gen Microbiol *49 (1):* 119–125.

Mayaki, T., H. Tsukiura, M. Wakae and H. Kawaguchi. 1962 Glebomycin, a new member of the streptomycin class. II. Isolation and physiochemical properties. J Antibiot (Tokyo) Ser A *15 (1):* 15–20.

Mayer, D. 1967 Ernährungsphysiologische Untersuchungen an *Archangium violaceum*. Arch Mikrobiol *58:* 186–200.

Mayer, G. 1926 *Corynebacterium parvum infectiosum*. Zentrabl Bakteriol Parasitenk Infektionskr Hyg Abt I Orig *98:* 370–371.

Mayer, H. D. 1938 Das "Tibi" Konsortium nebst einem Beitrag zur Kenntnis der Bakterien-Dissoziation. Thesis, Delft, The Netherlands, pp. 1–188.

Mayer, M. 1921 Über einige bakterienähnliche Parasiten der Erythrozyten bei Menschen und Tieren. Arch Schiffs-Trop Hyg *25:* 150–152.

Mayer, R. L., C. Crane, C. J. DeBoer, E. A. Konopka, J. S. Marsh and P. C. Eisman. 1951 Antibiotics from *Act. vinaceus* (nov. sp.). Microbiological Studies. XIIth Internat Cong Pure Appl Chem. Abstr Papers, pp. 283–284.

Mazé, P. 1915 Ferment forménique. Fermentation forménique de l'acétone. Procédé de culture simple du ferment forménique. C R Soc Biol Paris *78:* 398–405.

Mazurek, C. 1955 Étude d'une variété pigmentée de *Leptotrichia innominata*. Ann Inst Pasteur (Paris) *89 (2):* 208.

Mazzotti, L. 1949 Sobre una nueva espiroqueta de la fiebre recurrente, encontrada en Mexico. Rev Inst Salubr Inst Enferm Trop Méx *10:* 277–281.

Mechsner, K. 1957 Physiologische und morphologische Untersuchungen an Chlorobakterien. Arch Mikrobiol *26:* 32–51.

Medical Research Council Memorandum. 1951 List of species maintained in the National Collection of Type Cultures. His Majesty's Stationery Office, London, No. 21, 3rd ed. pp. 1–17.

Medrek, T. F. and E. M. Barnes. 1962 The physiological and serological properties of *Streptococcus bovis* and related organisms isolated from cattle and sheep. J App. Microbiol *25:* 169–179.

— and —. 1962 The influence of the growth medium on the demonstration of a group D antigen in fecal streptococci. J Gen Microbiol *28 (4):* 701–710.

Mees, R. H. 1934 Onderzoekingen over de Biersarcina. Thesis. Technical University, Delft, Holland, pp. 1–110.

Mehta, B. M., F. M. Sirotnak and D. J. Hutchison. 1967 Evidence of genetic transformation in *Streptococcus faecium* var. *durans* (SF/O). J Bacteriol *94 (4):* 1264–1265.

Mellon, R. R. 1917 A study of the diphtheroid group of organisms with special reference to their relation to the streptococci. Part I. Characteristics of a peculiar pleomorphic diphtheroid. J Bacteriol *2:* 81–107.

—. 1917 A study of the diphtheroid group of organisms with special reference to their relation to the streptococci. Part II. Classification of the diphtheroid group. J Bacteriol *2:* 269–307.

Mendel, J. 1919 *Cladothrix* et infection d' origin dentiare. C. R Séances Soc Biol Filiales *82:* 583–586.

Mercier, L. 1906 Les corps bacteroides de la blatte (*Periplanata orientalis: Bacillus cuenoti* (n. sp. L. Mercier)). C R Soc Bio. Paris *61 (58):* 682–684.

Merkel, J. R., E. D. Traganza, B. B. Mukherjee, T. B. Griffin and J. M. Prescott. 1964 Proteolytic activity and general characteristics of a marine bacterium, *Aeromonas proteolytica* sp. n. J Bacteriol *87 (4):* 1227–1233.

Mesnil, F. 1930 Piroplasmoses (Abstr) *In* M. Carpano, Su di un Piroplasma osservato nei polli in Egitto (*Aegyptianella pullorum*). Bull Inst Pasteur *28:* 125–138.

— and M. Caullery. 1916 Sur un organisme spirochétoide (*Cristispira polydorae* n. sp.) de l'intestin d'une annélide polychète. C R Seances Soc Biol Filiales *79:* 1118–1121.

Metcalf, G. 1940 *Bacterium rhaponticum* (Millard) Dowson, a cause of crown-rot disease of rhubarb. Ann Appl Biol *27:* 502–508.

— and M. Brown. 1957 Nitrogen fixation by new species of *Nocardia*. J Gen Microbiol *17:* 567–572.

Metchnikoff, E. 1888 Ueber die phagocytäre Rolle der Tuberkelriesenzellen. Virchows Arch *113:* 63–94.

—. 1888 *Pasteuria ramosa* un représentant des bactéries a division longitudinale. Ann Inst Pasteur (Paris) *2:* 165–170.

—. 1908 Études sur la flore intestinale. Ann Inst Pasteur (Paris) *22:* 929–955.

Meyen, F. J. F. 1827 *Actinomyce*, Strahlenpilz. Ein neue Pilz-Gattung. Linnaea *2:* 433–444.

Meyer, A. 1897 Studien über die Morphologie und Entwickelungsgeschichte der Bacterien, ausgeführt an *Astasia asterospora* A.M. und *Bacillus tumescens* Zopf. Flora Allg Bot Ztg (Jena) *84:* 185–248.

Meyer, E. and P. Verges. 1950 Mouse pathogenicity as a diagnostic aid in the identification of *Actinomyces bovis*. J Lab Clin Med *36:* 667–674.

Meyer, K. F. 1910 Experimental studies on a specific purulent nephritis. Transvaal Dept Agr Govt Rep for 1908–09, pp. 122–158.

—. 1953 Psittacosis Group. Ann N Y Acad Sci *56:* 545–556.

— and E. B. Shaw. 1920 A comparison of the morphologic, cultural and biochemical characteristics of *B. abortus* and *B. melitensis*. Studies on the genus *Brucella* nov. gen. J Infect Dis *27:* 173–184.

Meyer, M. E. 1961 Metabolic characterization of the genus *Brucella*. IV. Correlation of the oxidative metabolic patterns and susceptibility to *Brucella* bacteriophage, type abortus strain 3. J Bacteriol *82:* 950–953.

—. 1962 Metabolic and bacteriophage identification of *Brucella* strains described as *Brucella melitensis* from cattle. Bull World Health Organ *26:* 829–831.

—. 1964 The epizootiology of brucellosis and its relationship to the identification of *Brucella* organisms. Amer J Vet Res *25:* 553–557.

—. 1966 Identification and virulence studies of *Brucella* strains isolated from Eskimos and reindeer in Alaska, Canada and Russia. Amer J Vet Res *27:* 353–358.

Meyer, M. E. and H. S. Cameron. 1957 Species metabolism patterns in morphologically similar gram negative pathogens. J Bacteriol *73 (2):* 158–161.

— and —. 1961 Metabolic characterization of the genus *Brucella*. I. Statistical evaluation of the oxidative rates by which type 1 of each species can be identified. J Bacteriol *82:* 387–395.

— and —. 1961 Metabolic characterization of the genus *Brucella*. II. Oxidative metabolic patterns of the described biotypes. J Bacteriol *82:* 396–400.

Meyer, P. E. and E. F. Hunter. 1967 Antigenic relationships of 14 treponemes demonstrated by immunofluorescence. J Bacteriol *93 (3):* 784–789.

Meyer, W. 1967 A proposal for subdividing the species *Staphylococcus aureus*. Int J Syst Bacteriol *17:* 387–389.

—. 1967 *Staphylococcus aureus* strains of phage group IV. J Hyg *65:* 439–447.

Meynadier, G. 1964 Les Rickettsioses d'insectes Proc XIIth Int Congr Entomol, London, pp. 714–715.

— and P. Monsarrat. 1969 Une rickettsiose chez une Cetoine de Madagascar. Entomophaga *14:* 401–406.

Mez, C. 1898 Mikroskopische Wasseranalyse, Anleitung zur Untersuchung Wassers mit besonderer Berücksichtigung von Trink- und Abwasser. J. Springer, Berlin, pp. 1–69.

M'Fadyean, J. 1907 Johne's disease: a chronic bacterial enteritis of cattle. J Comp Pathol Therap *20:* 48–60.

— and A. L. Sheather. 1916 Johne's disease, the experimental transmission of the disease to cattle, sheep, goats, with notes regarding the occurrence of natural cases in sheep and goats. J Comp Pathol Therap *29:* 62–94.

Michel, M. F. and R. M. Krause. 1967 Immunochemical studies on the group F streptococci, and the identification of a group-like carbohydrate in a type II strain with an undesignated group antigen. J Exp Med *125:* 1075–1089.

— and J. M. Willers. 1964 Immunochemistry of group F streptococci; isolation of group specific oligosaccharides. J Gen Microbiol *37 (3):* 381–389.

Mickelson, M. N. 1964 Chemically defined medium for growth of *Streptococcus pyogenes*. J Bacteriol *88 (1):* 158–164.

Mickle, F. L. 1924 *Lactobacillus lycopersici* n. sp. the causative organism of the gaseous fermentation of tomato pulp and related products. J Bacteriol *8:* 403–404.

Micks, D. W., S. R. Julian, Jr., M. J. Ferguson and D. Duncan. 1961 Microorganisms associated with mosquitoes: II. Location and morphology of microorganisms in the mid gut of *Culex fatigans* Wiedemann and certain other species. J Insect Pathol *3:* 120–128.

Midgley, J. M., S. P. Lapage, B. A. G. Jenkins, G. I. Barrow, M. E. Roberts and A. G. Buck. 1970 *Cardiobacterium hominis* endocarditis. J Med Microbiol *3:* 91–98.

Miessner, H. 1921 Pyoseptikämie der Fohlen (sog. Fohlenlähme). Deut Tieraerztl Wochenschr *29:* 185–187.

— and R. Berge. 1922 Verfohlen und Fohlenkrankheiten. Das Verfohlen. A. Urachen des Verfohlens. Deut Tieraerztl Wochenschr *30:* 473–482.

— and —. Die Paratuberkulose des Rindes. In Kolle and Wassermann (Editors), Handbuck Pathogenen Mikroorganismen, Fischer, Jena. Vol. 6. pp. 779–798.

— and R. Wetzel. 1923 *Corynebacterium pyogenes (equi)* als Erreger einer infektiosen abszedierenden Pneumonie der Fohlen. Deut Tieraerztl Worchenschr *31:* 449–454.

Migula, W. 1894 Über den Zellinhalt von *Bacillus oxalaticus* Zopf. Arb Bakteriol Inst Karlsruhe *1:* 139–147.

—. 1894 Ueber ein neues System der Bakterien. Arb Bakteriol Inst Karlsruhe *1:* 235–238.

—. 1895 *Bacteriaceae* (Stabchenbacterien). In Engler and Prantl (Editors), Pfanzenfamilein, W. Engelmann, Leipzig, Teil I, Abt la. pp. 20–30.

—. 1895 *Chlamydobacteriaceae*. In A. Engler and K. Prantl (Editors), 1900 Die näturlichen Pflanzenfamilien, W. Engelmann, Leipzig, Vol. I. pp. 35–40.

—. 1895 *Schizomycetes* (Bacteria, Bacterien). In Engler and Prantl (Editors), Pflanzenfamilien, W. Engelmann, Leipzig, Tiel I, Abt. Ia. pp. 1–44.

—. 1895 *Spirillaceae* (Schraubenbacterien). In Engler and Prantl (Editors), Pflanzenfamilien, W. Engelmann, Leipzig, Teil 1, Abt. 1a. pp. 30–45.

—. 1900 System der Bakterien, Vol. 2. Gustav Fischer, Jena.

Miles, A. A. and E. G. Halnan. 1937 A new species of microorganism (*Proteus melanovogenes*) causing black rot in eggs. J Hyg *37:* 79–97.

Milhaud, G., J.-P. Aubert and C. B. van Niel. 1956 Étude de la glycolyse de *Zymosarcina ventriculi*. Ann Inst Pasteur (Paris) *91:* 363–368.

Millard, W. A. 1924 Crown rot of rhubarb. Yorks Council Agr Ed Bull *134*: 1–28.

— and F. Beeley. 1927 Mangel scab, its cause and histogeny. Ann Appl Biol *14*: 298–311.

— and S. Burr. 1926 A study of twenty-four strains of *Actinomyces* and their relation to types of common scab of potato. Ann Appl Biol *13*: 580–644.

Miller, A., W. E. Sandine and P. R. Elliker. 1970 DNA base composition of lactobacilli, determined by thermal denaturation. J Bacteriol *102*: 278–280.

Miller, J. D. A., J. E. Hughes, G. E. Saunders and L. L. Campbell. 1968 Physiological and biochemical characteristics of some strains of sulfate reducing bacteria. J Gen Microbiol *52* (*2*): 173–178.

— and A. M. Saleh. 1964 A sulphate-reducing bacterium containing cytochrome C₃ but lacking desulfoviridin. J Gen Microbiol *37* (*3*): 419–423.

Miller, J. G. 1956 The prevention and treatment of anaplasmosis. Ann N Y Acad Sci *64*: 49–55.

Miller, P. W., W. B. Bollen, J. E. Simmons, H. N. Gross and H. P. Barss. 1940 The pathogen of filbert bacteriosis compared with *Phytomonas juglandis*, the cause of walnut blight. Phytopathology *30*: 713–733.

Miller, T. L. and J. B. Evans. 1970 Nutritional requirements for growth of *Aerococcus viridans*. J Gen Microbiol *61*: 131–135.

Miller, W. D. 1886 Einige gasbildende Spaltpilze des Verdauungstractus, ihr Schicksal in Magen und ihr Reaktion auf verschiedene Spiesen. Deut Med Wochenschr *12*: 117–119.

—. 1889 Die Mikroorganismen der Mundhöhle. G. Thieme, Leipzig, pp. 1–305.

Millis, N. F. 1956 A study of the cider-sickness bacillus—a new variety of *Zymomonas anaerobia*. J Gen Microbiol *15*: 521–528.

Mills, C. K. and E. F. Lessel. 1973 Designation and description of the type strain of *Lactobacillus casei* subsp. *alactosus* Rogosa et al. Int J Syst Bacteriol *23*: 67–68.

Minett, F. C. 1932 Avian tuberculosis in cattle of Great Britain. J Comp Pathol Therap *45*: 317–330.

—. 1934 Streptococcus mastitis in cattle; bacteriology and preventive medicine. Twelfth Int Vet Cong New York *2*: 511–532.

Ming-Ching, C., H. Shiao-ching and L. Tung-yao. 1966 Description of a starch-fermenting strain of *Moraxella*. Acta Microbiol Sinica *12*: 11–14.

Miquel, P. 1889 Étude sur la fermentation ammoniacale et sur les ferments de l'urée. Ann Micrographie *1*: 414–425; 470–482; 506–519; 552–556; *2*: 13–33; 53–65; 122–133; 145–158.

—. 1890 Étude sur la fermentation ammoniacale et sur les ferments de l'urée. Ann Micrographie *2*: 367–376; 488–504.

— and R. Cambier. 1902 Traité de Bactériologie pure et appliquée à la Médecine et à l'Hygiène. C. Naud, Paris, pp. 1–1059.

Mirick, G. S., L. Thomas, E. C. Curnen and F. L. Horsfall. 1944 Studies on a non-hemolytic streptococcus isolated from the respiratory tract of human beings. I. Biological characteristics of *Streptococcus* MG. J Exp Med *80*: 391–406.

Misaghi, I. and R. G. Grogan. 1969 Nutritional and biochemical comparisons of plant pathogenic and saprophytic fluorescent pseudomonads. Phytopathology *59* (*10*): 1436–1450.

Misao, R. and Y. Kobayashi. 1956 Infectious mononucleosis (glandular fever). J Jap Ass Infect Dis *30*: 453–465.

Mishustin, E. N. 1938 Cellulose-decomposing myxobacteria (in Russian, English summary). Mikrobiologiya *7*: 427–444.

Mitchell, C. A. 1925 *Hemophilus ovis* (nov. sp.) as the cause of a specific disease in sheep. J Amer Vet Med Ass *68*: 8–18.

Mitchell, T. G., M. S. Hendrie and J. M. Shewan. 1969 The taxonomy, differentiation and identification of *Cytophaga* species. J Appl Bacteriol *32* (*1*): 40–50.

Mitsuoka, T. 1969 Vergleichende Untersuchungen über die Laktobazillen aus den Faeces von Menschen, Schweinen, und Huhnern. Zentrabl Bakteriol Parasitenk Infektionskr Hyg Abt I Orig *210*: 32–51.

—. 1969 Vergleichende Untersuchungen über die Bifidobacterien aus dem Verdauungstrakt von Menschen and Tieren. Zentrabl Bakteriol Parasitenk Infektionskr Hyg Abt I Orig *210*: 52–64.

Miyairi, N., M. Takashima, K. Shimizu and H. Sakai. 1966 Studies on new antibiotics, cineromycins A and B. J Antibiot (Tokyo) Ser A *19*: 56–62.

Miyamoto, Y., K. Nakamura and K. Takizawa. 1961 Pathogenic halophiles. Proposals of a new genus "*Oceanomonas*" and of the amended species names. Jap J Microbiol *5*: 477–486.

Miyoshi, M. 1897 Studien über die Schwefelrosenbildung und die Schwefelbacterien der Thermen von Yumoto bei Nikko. J Colloid Sci *10* (*2*): 143–173.

Moeller, A. 1898 Mikroorganismen, die den Tuberkelbacillen ähnlich sind und bei Thieren eine miliare Tuberkelkrankheit verursachten. Deut Med Wochenschr *24*: 376–379.

Moffett, M. L. and R. R. Colwell. 1968 Adansonian analysis of the Rhizobiaceae. J Gen Microbiol *51* (*2*): 245–266.

Mohanty, U. 1951 *Corynbacterium fascians* (Tilford) Dowson; its morphology, physiology, nutrition and taxonomic position. Trans Brit Mycol Soc *34*: 23–34.

Moir, R. J. and M. J. Masson. 1952 An illustrated scheme for the microscopic identification of the rumen micro-organisms of sheep. J Pathol Bacteriol *64*: 343–350.

Molisch, H. 1904 Die Leuchtbakterien im Hafen von Triest. Sitzungsber Saechs Akad Wiss Leipzig Math-Naturwiss Kl *63*: 513–527.

—. 1906 Zwei neue Purpurbakterien mit Schwebe körperchen. Bot Ztg Abt 1 *64*: 223–232.

—. 1907 Die purpurbakterien nach neuen Untersuchungen. G. Fischer, Jena I–VII, pp. 1–95.

—. 1910 *Siderocapsa treubii* Molisch, eine neue, weit verbreitete Eisenbakterie. Ann Jard Bot Buitenzorg Suppl *3* (*1*): 29–34.

—. 1910 Die Eisenbakterien. G. Fischer Verlag, Jena, pp. 1–83.

—. 1912 Leuchtende Pflanzen. Eine physiologische Studie. Zweite vermehrte Auflage, Jena, pp. 1–198.

—. 1912 Neue farblose Schwefelbakterien. Zentrabl Bakteriol Parasitenk Infektionskr Hyg Abt II *33*: 55–62.

—. 1925 Über Kalbakterien und andere Kalkfallende Pilze. Zentrabl Bakteriol Parasitenk Infektionskr Hyg Abt II *65:* 130–139.

Mollaret, H. H. 1963 Conservation expérimentale de la peste dans le sol. Bull Soc Pathol Exot *56 (6):* 1168–1182.

— and J. C. Guillon. 1965 Contribution à l'étude d'un nouveau groupe de germes (*Yersinia enterocolitica*) proches du bacille de Malassez et Vignal, Pouvoir pathogène expérimental. Ann Inst Pasteur (Paris) *109 (4):* 608–613.

— and A. Lucas. 1965 Sur les particularités biochimiques des souches de *Yersinia enterocolitica* isolées chez les lièvres. Ann Inst Pasteur (Paris) *108:* 121–125.

Moller, H. 1885 *Plasmodiophora alni.* Ber Deut Bot Ges *3:* 102–105.

—. 1890 Beitrag zur Kenntniss der *Frankia subtilis* Brunchorst. Ber Deut Bot Ges *8:* 215–224.

Møller, V. 1954 Distribution of amino acid decarboxylases in *Enterobacteriaceae.* Acta Pathol Microbiol Scand *35:* 259–277.

—. 1955 Simplified test for some amino acid decarboxylases and the arginine dihydrolase system. Acta Path Microbiol Scand *36:* 158–172.

Monias, B. L. 1928 Classification of *Bacterium alcaligenes pyocyaneum* and *fluorescens.* J Infect Dis *43:* 330–334.

Moniz, L. and M. K. Patel. 1958 Three new bacteriol diseases of plants from Bombay State. Curr Sci *27:* 494–495.

—, J. E. Sabley and W. D. More. 1964 A new bacterial canker of *Carissa congesta* in Maharashtra. Indian Phytopathol *17:* 256.

Montague, E. A. and K. W. Knox. 1968 Antigenic components of the cell wall of *Streptococcus salivarius.* J Gen Microbiol *54 (2):* 237–246.

Montecatini Società generale per l'industria mineraria e chimica. 1960 Flavensomycin. British Patent 850,325, October 5.

Monteiro, J. L. 1931 Estudos sobre o typho exanthematico de São Paulo. Mem Inst Butantan Sao Paulo *6:* 3–135.

Moor, C. E. de. 1957 Een nieuwe *Streptococcus haemolyticus* (Lancefield group R). Berichten Rijks Instituut Volksgezondheid, pp. 174–177.

—. 1963 A non-haemolytic El Tor vibrio as the cause of an outbreak of paracholera in West New Guinea. The El Tor problem and pandemic paracholera in the West Pacific. Trop Geogr Med *15:* 97–107.

—. 1963 Septicaemic infections in pigs, caused by hemolytic streptococci of new Lancefield groups designated R, S and T. Antonie van Leeuwenhoek J Microbiol Serol *29:* 272–280.

Moore, B. O. and J. T. Bryans. 1969 Antigenic classification of group C animal streptococci. J Amer Vet Med Ass *155:* 416–424.

Moore, J. A., J. R. Dice, E. D. Nicolaides, R. D. Westland and E. L. Wittle 1954 Azaserine, synthetic studies. I. J Amer Chem Soc *76 (11):* 2884–2887.

Moore, M. and J. B. Frerichs. 1953 An unusual acid-fast infection of the knee, with subcutaneous, abscess-like lesions of the gluteal region. J Invest Dermatol *20:* 133–169.

Moore, R. L. and B. J. McCarthy. 1969 Characterization of the deoxyribonucleic acid of various strains of halophilic bacteria. J Bacteriol *99 (1):* 248–254.

— and —. 1969 Base sequence homology and renaturation studies of the deoxyribonucleic acid of extremely halophilic bacteria. J Bacteriol *99 (1):* 255–262.

Moore, R. W., C. W. Livingston and H. E. Redmond. 1965 Cultural and serological techniques for the differential diagnosis of *Mycoplasma* in swine. 69th Animal Proc United States Livestock Sanitary Association, pp. 480–486.

—, H. E. Redmond and C. W. Livingston. 1965 Mycoplasmosis (PPLO) of swine. Southwest Vet *19:* 19–21.

—, — and —. 1966 Pathologic and serologic characteristics of a *Mycoplasma* causing arthritis in swine. J Vet Res *27:* 1649–1656.

Moore, W. E. C. 1966 Techniques for routine culture of fastidious anaerobes. Int J Syst Bacteriol *16:* 173–190.

— and E. P. Cato. 1965 Synonymy of *Eubacterium limosum* and *Butyribacterium rettgeri: Butyribacterium limosum* comb. nov. Int Bull Bacteriol Nomencl Taxon *15 (2):* 69–80.

—, — and L. V. Holdeman. 1971 *Eubacterium lentum* (Eggerth) Prevot 1938: Emendation of description and designation of the neotype strain. Int J Syst Bacteriol *21 (4):* 299–303.

—, — and —. 1971 *Eubacterium aerofaciens* Eggerth) Prevot 1938: Emendation of description and designation of the neotype strain. Int J Syst Bacteriol *21 (4):* 307–310.

— and L. V. Holdeman. 1969 Anaerobic diphtheroids. *In* Cato, Cummins, Holdeman, Johnson, Moore, Smibert and Smith (Editors), Outline of Clinical Methods in Anaerobic Bacteriology. V.P.I. Anaerobe Laboratory, Blacksburg, Va.

— and —. 1969 Anaerobic gram-negative nonsporeforming rods. *In* Cato, Cummins, Holdeman, Johnson, Moore, Smibert and Smith (Editors), Outline of clinical methods in anaerobic bacteriology, 1st rev. Virginia Polytechnic Institute Anaerobe Laboratory, Blacksburg, VA.

— and —. 1970 *Fusobacterium. In* Cato, Cummins, Holdeman, Johnson, Moore, Smibert and Smith (Editors), Outline of clinical methods in anaerobic bacteriology, 2nd rev. Virginia Polytechnic Institute Anaerobe Laboratory, Blacksburg, Va.

— and —. 1970 *Propionibacterium, Arachnia, Actinomyces, Lactobacillus* and *Bifidobacterium. In* Cato, Cummins, Holdeman, Johnson, Moore, Smibert and Smith (Editors), Outline of Clinical Methods in Anaerobic Bacteriology, 2nd rev. V.P.I. Anaerobe Laboratory, Blacksburg, Va., pp. 15–22.

— and —. 1972 *Fusobacterium. In* Holdeman and Moore (Editors), Anaerobe Laboratory Manual. Virginia Polytechnic Institute Anaerobe Laboratory, Blacksburg, Va.

— and —. 1972 *Lactobacillus. In* Holdeman and Moore (Editors). Anaerobe Laboratory Manual. Virginia Polytechnic Institute Anaerobe Laboratory, Blacksburg, VA.

— and —. 1973 New names and combinations in the genera *Bacteroides* Castellani and Chalmers, *Fusobacterium* Knorr, *Eubacterium* Prévot, *Propionibacterium* Delwiche, and *Lactobacillus* Orla-Jensen. Int J Syst Bacteriol *23:* 69–74.

Morax, V. 1896 Note sur un diplobacille pathogène pour la conjonctivite humain. Ann Inst Pasteur (Paris) *10:* 337–345.

Morcos, Z. 1935 Preliminary studies in fowl spirochaetosis in Egypt. Vet J *91:* 161–171.

Moreira-Jacob, M. 1956 The streptococci of Lancefield's group E; biochemical and serological identification of the hemolytic strains. J Gen Microbiol *14* (2): 268–280.

Morel, M. G. 1911 Contribution à l'étude de l'adenite caséeuse du mouton. J Med Vet Zool Technol *72:* 513–526.

Moreno-López, M. 1952 El genero *Bordetella.* Microbiol Espan *5:* 117–181.

Morgan, H. de R. 1906 Upon the bacteriology of the summer diarrhoea of infants. Brit Med J *1:* 908–912.

Morgan, K., W. J. Brinley and S. G. M. Gower. 1966 Techniques in the identification and classification of *Brucella. In* Gibbs and Skinner (Editors), Identification methods for microbiologists. Academic Press, London, pp. 35–40.

—, D. Kay and D. E. Bradley. 1960 *Brucella* bacteriophage. Nature (London) *188:* 74–75.

Mori, R. 1888 Ueber pathogene Bacterien im Canalwasser. Z Hyg Infektionskr Med Mikrobiol Immunol Virol *4:* 47–54.

Morita, R. Y. and P. W. Stave. 1963 Electron micrograph of an ultrathin section of *Beggiatoa.* J Bacteriol *85* (4): 940–942.

Moro, E. 1900 Ueber den *Bacillus acidophilus* n. sp. Jahrb Kinderheilk *52:* 38–55.

—. 1900 Ueber die nach Gram färbbaren Bacillen des Säuglingsstuhles. Wien Klin Wochenschr *13:* 114–115

Morquer, R. and L. Comby. 1943 Actinomycose du Cheval. Rev Med Vet Toulouse, Jan-Feb, pp. 5–47.

— and —. 1943 Affinités Systématiques du genre *Micromonospora* Ørskov. Bull Soc Hist Natur Toulouse *78:* 23–28.

Morris, J. G. 1960 Studies on the metabolism of *Arthrobacter globiformis.* J Gen Microbiol *22* (2): 564–582.

Morrison, N. E. 1965 Circumvention of the mycobactin requirement of *M. paratuberculosis.* J Bacteriol *89:* 762–767.

—, A. D. Antoine and E. E. Dewbrey. 1965 Synthetic metal chelators which replace the natural growth-factor requirements of *Arthrobacter terregens.* J Bacteriol *89:* 1630.

— and E. E. Dewbrey. 1966 Growth factor activity of Mycobactin for *Arthrobacter* species. J Bacteriol *92:* 1848.

Morrison, R. B. and A. Scott. 1966 Swarming of *Proteus*—a solution to an old problem. Nature (London) *211:* 255–256.

Morrison, T. M. 1961 Fixation of nitrogen-15 by excised nodules of *Discaria toumatou.* Nature (London) *189:* 945.

Morse, M. E. 1912 A study of the diphtheria group of organisms by the biometrical method. J Infect Dis *11:* 253–285.

Moshentseva, L. V. and E. N. Kondratieva. 1962 Chlorophyll production by purple and green bacteria in photoautotrophic and photoheterotrophic development. Mikrobiologiya *31:* 199–202.

Moshkovski, C. 1937 Sur l'existence chez le cobaye, d'une rickettsiose chronique déterminée par *Ehrlichia (Rickettsia) kurlovi* subg. nov. sp. nov. C R Soc Biol Paris *126:* 379–382.

Moshkovski, S. D. 1945 Cytotropic inducers of infection and the classification of the *Rickett-*

siae with *Chlamydozoa* (in Russian, English summary). Adv Mod Biol (Moscow) *19:* 1–44.

—. 1947 Comments by readers. Science (Washington *106:* 62.

Mosing, H. 1936 Une nouvelle infection à Rickettsia, *Rickettsia weigle* nov. sp. Arch Inst Pasteur Tunis *25:* 373–387.

Moss, C. W. and W. B. Cherry. 1968 Characterization of C_{15} branched-chain fatty acids by gas-chromatography. J Bacteriol *95 (1):* 241–242.

—, V. R. Dowell, Jr., D. Farshtchi, L. J. Raines and W. B. Cherry. 1969 Cultural characteristics and fatty acid composition of propionibacteria. J Bacteriol *97 (2):* 561–570.

— and W. E. Dunkelberg, Jr. 1969 Volatile and cellular fatty acids of *Haemophilus vaginalis.* J Bacteriol *100 (1):* 544–546.

Mosser, J. L. and A. Tomasz. 1970 Choline-containing teichoic acid as a structural component of pneumococcal cell wall and its role in sensitivity to lysis by an autolytic enzyme. J Biol Chem *245:* 287–298.

Moulton, J. E. and J. F. Christensen. 1955 The histochemical nature of *Anaplasma marginale.* Amer J Vet Res *16:* 377–380.

Mráz, O. 1963 Schizomycetes. *In* Mráz, Tesarcik and Varejka (Editors), Nomina und synonyma der pathogenen und saprophytären Mikroben, isoliert aus den wirtschaftlich oder epidemiologisch bedeutenden Wirbeltieren und Lebensmitteln tierischer Herkunft. VEB Gustav Fischer Verlag, Jena, pp. 53–334.

—. 1968 Reevaluation of original strains *Actinobacillus suis* and haemolytic strains *Actinobacillus lignieresii* ATCC isolated from organs of diseased pigs. Sb Vys Sk Zeměd Les Fak Brne (řada B) *37:* 277–290.

Mudd, S. and S. Warren. 1923 A readily cultivable vibrio, filterable through Berkefeld "V" candles, *Vibrio percolans* (new species). J Bacteriol *8:* 447–458.

Mukerjee, S. 1963 The bacteriophage susceptibility test in differentiating *Vibrio cholerae* and *Vibrio el tor.* Bull World Health Organ *28:* 333–336.

—. 1963 Bacteriophage typing of cholera. Bull World Health Organ *28:* 337–345.

Mulder, E. G. 1964 *Arthrobacter. In* Heukelekian and Dondero (Editors), Principles and Applications in Aquatic Microbiology. John Wiley and Sons, New York, pp. 254–279.

—. 1964 Iron bacteria, particularly those of the *Sphaerotilus-Leptothrix* group, and industrial problems, J Appl Bacteriol *27:* 151–173.

—, A. D. Adamse, J. Antheunisse, H. Deinema, J. W. Woldendorp and L. P. T. M. Zevenhuizen. 1966 The relationship between *Brevibacterium linens* and bacteria of the genus *Arthrobacter.* J Appl Bacteriol *29:* 44–71.

— and J. Antheunisse. 1963 Morphologie, physiologie et ecologie des *Arthrobacter.* Ann Inst Pasteur Paris *105:* 46–74.

— and W. L. van Veen. 1963 The *Sphaerotilus-Leptothrix* group. Antonie van Leeuwenhoek J Microbiol Serol *29:* 121–153.

— and —. 1965 Anreicherung von Organismen der *Sphaerotilus-Leptothrix* Gruppe. *In* Anreicherungskultur und Mutantenauslese; Symp. Göttingen, 1964. Zentrabl Bakteriol Infektionskr Hyg Abt I Suppl *1:* 28–46.

— and L. P. T. M. Zevenhuizen. 1967 Coryneform bacteria of the *Arthrobacter* type and their reserve material. Arch Mikrobiol *59:* 345–354.

Müller, E. 1950 Medizinische Mikrobiologie-Parasiten, Bakterien, Immunität, 4th ed. Urban and Schwartzenberg, Munich and Berlin.

Müller, G. 1961 Mikrobiologische Untersuchungen über die "Futterverpilzung durch Selbsterhitzung" III. Mitteilung: Ausfuhrliche Beschreibung neuer Bakerien-Species. Zentrabl Bakteriol Parasitenk Infektionskr Hyg Abt II *114:* 520–537.

Müller, L. 1923 Un nouveau milieu d'enrichment pour la recherche du bacille typhique et des paratyphiques. C R Soc Biol Paris *II:* 434.

Müller, O. F. 1773 Vermium Terrestrium et Fluviatilum, seu Animalium Infusoriorum, Helminthicorum et Testaceorum, Non Marionoram, Succincta Historia. *1 (1):* 1–135.

—. 1786 Animalcula infusoria, fluviatilia et marina. Hauniae, pp. 1–367.

Müller, R. 1908 Eine Diphtheridee und eine Streptothrix mit gleichem blauen Farbstoff sowie Untersuchungen über Streptothrixarten im all gemeinen. Zentrabl Bakteriol Parasiten Infektionskr Hyg Abt I Orig *46:* 195–222.

—. 1911 Zur Stellung der Krankheitserreger im Natursystem. Muenchen Med Wochenschr *58:* 2246–2247.

Müller-Kögler, E. 1958 Eine Rickettsiose von *Tipula paludosa* Meig. durch *Rickettsiella tipulae* nov spec. Naturwissenschaften *45 (10):* 248.

Müller - Thurgau, H. 1908 Bakterienblasen (Bacteriocysten). Zentrabl Bakteriol Parasitenk Infektionskr Hg Abt II *20:* 353–400.

—. and A. Osterwalder. 1913 Die Bakterien in Wein und Obstwein und die dadurch verursachten Veräuderungen. Zentrabl Bakteriol Parasitenk Infektionskr Hyg Abt II *36:* 129–338.

— and —. 1917 Weitere Beitrage zur Kentniss der Mannitbakterien im Wein. Zentrabl Bakteriol Parasitenk Infektionskr Hyg Abt II *48:* 1–35.

Mullohan, C. S. and M. S. Romer. 1961 Public health significance of swimming pool granuloma. Amer J Pub Health *51:* 883–891.

Munch-Petersen, E. 1954 A corynebacterial agent which protects ruminant erythrocytes against staphylococcal β-toxin. Aust J Exp Biol Med Sci *32:* 361–368.

Mundt, J. O., J. H. Coggin, Jr. and L. F. Johnson. 1962 Growth of *Streptococcus faecalis* var. *liquefaciens* on plants. Appl Microbiol *10:* 552–555.

—, W. F. Graham and I. E. McCarty. 1967 Spherical lactic acid producing bacteria of southern-grown raw and processed vegetables. Appl microbiol *15:* 1303–1308.

Mungelluzzi, G. 1966 *Actinomonospora lusitanica* Castellani, De Brito and Pinto (in Italian). Arch Ital Sci Med Trop Parassitol *47:* 33–38.

Munk, F. and H. da Rocha-Lima. 1917 Klink and Aetiologie des sogen "Wolhynischen Fiebers" (Werner-Hissche Krankheit). Muenchen Med Wochenschr *64 (1):* 1422–1426.

Muraschi, T. F., M. Friend and D. Bolles. 1965 *Erwinia*-like microorganisms isolated from animal and human hosts. Appl Microbiol *13:* 128–131.

Murase, M., T. Hikiji, K. Nitta, Y. Okami, T. Takeuchi and H. Umezawa. 1961 Peptimycin, a product of *Streptomyces* exhibiting apparent inhibition against Ehrlich carcinoma. J Antibiot (Tokyo) Ser A *14 (3):* 113–118.

Murat, A.-M., W. R. Stinebring, C. P. Schaffner and H. Lechevalier. 1959 Screening for antibiotics active against intracellular bacteria. Appl Microbiol *7:* 109–112.

Murphy, F. A., J. W. Osebold and O. Aalund. 1966 Kinetics of the antibody response to *Anaplasma marginale* infection. J Infect Dis *116:* 99–111.

Murray, E. G. D. 1929 The meningococcus. Med Res Conc (Gt Brit) Spec Rep Ser *124:* 7–142.

—. 1939 *Parvobacteriaceae. In* Bergey's Manual of Determinative Bacteriology, 5th ed. The Williams & Wilkins Co., Baltimore, p. 309.

—. 1939 Genus I *Neisseria* Trevisan. *In* Bergey, Breed, Murray and Hitchins (Editors), Bergey's Manual of Determinative Bacteriology, 5th ed. The Williams & Wilkins Co., Baltimore, pp. 278–288

—. 1939 Family VII. *Neisseriaceae* Prevot. *In* Bergey, Breed, Murray and Hitchens (Editors), Bergey's Manual of Determinative Bacteriology, 5th ed. The Williams & Wilkins Co., Baltimore, pp. 278–288.

—. 1948 Genus II *Moraxella* Lwoff. *In* Breed, Murray and Hitchens (Editors), Bergey's Manual of Determinative Bacteriology, 6th ed. The Williams & Wilkins Co., Baltimore, pp. 590–592.

—, A. A. Webb and M. B. R. Swann. 1926 A disease of rabbits characterized by a large mononuclear leucocytosis caused by a hitherto underscribed bacillus *Bacterium monocytogenes* n. sp. J Pathol Bacteriol *29:* 407–439.

Murray, I. G. and A. G. J. Procter. 1965 Paper chromatography as an aid to the identification of *Nocardia* species. J Gen Microbiol *41:* 163–167.

Murray, R. G. E. 1962 Fine structure and taxonomy of bacteria. *In* Ainsworth and Sneath (Editors), Microbial Classification. Cambridge University Press, p. 119.

—. 1968 Microbial structure as an aid to microbial classification and taxonomy. Spisy (Faculte des Sciences de l'Université J. E. Purkyne Brno) *43:* 249–252.

— and H. C. Douglas. 1950 The reproductive mechanism of *Rhodomicrobium vannielii* and the accompanying nuclear changes. J Bacteriol *59 (2):* 157–167.

Mushin, R. J. Naylor and N. Lahovary. 1959 Studies on plant pathogenic bacteria. II. Serology. Aust J Biol Sci *12:* 233–246.

Myers, W. F., L. D. Cutler and C. L. Wisseman, Jr. 1969 Role of erythrocytes and serum in the nutrition of *Rickettsia quintana* J. Bacteriol *97 (2):* 663–666.

—, P. J. Provost and C. L. Wisseman, Jr. 1967 Permeability properties of *Rickettsia mooseri.* J Bacteriol *93 (3):* 950–990.

Mylroie, R. L. and R. E. Hungate. 1954 Experiments on the methane bacteria in sludge. Can J Microbiol *1:* 55–64.

Nadson, G. A. 1906 The morphology of inferior Algae. III. *Chlorobium limicola* Nads., the green chlorophyll bearing microbe (in Russian). Bull Jard Bot St Peterb 6: 190.

—. 1913 On sulfuric microorganisms of the Hapsal Bay (in Russian). Bull Jard Bot St Petersb 13: 106–112.

— and N. A. Krasil'nikov. 1932 The structure and evolution of Pontothrix longissima Nads. and Krassiln. (Chlamydothrix longissima Molisch, a colorless alga of the group Schizophyceae). C R Acad Sci SSSR Leningrad 10: 243–247.

— and S. M. Visloukh. 1923 Formation and life of the gigantic bacterium Achromatium oxaliferum Schew. Bull Jard Bot URSS (Suppl 1) 22: 1–37.

Naegeli, C. 1857 In Caspary. Bericht ueber die Verhandlungen der 33, Versammlung deutscher Naturforsches und Aertzte, gehalten in Bonn von 18 bis 24 September 1857. Bot Ztg 15: 749–776; 784–792.

Naeslund, C. 1925 Studies of Actinomyces from the oral cavity. Acta Pathol Microbiol Scand 2: 110–140.

Nagatsu, J., K. Anzai, K. Ohkuma and S. Suzuki. 1963 Studies on a new antibiotic, tuberin. IV. Taxonomic studies on the tuberin producing organism, Streptomyces amakusaensis. J Antibiot (Tokyo) Ser A 16 (5): 207–210.

—, — and S. Suzuki. 1962 Pathocidin, a new antifungal antibiotic. II. Taxonomic studies on the pathocidin-producing organism Streptomyces albus var. pathocidicus. J Antibiot (Tokyo) Ser A 15 (2): 103–106.

— and S. Suzuki. 1963 Studies on an antitumor antibiotic, cervicarcin. III. Taxonomic studies on the cervicarcin producing organism, Streptomyces ogaensis nov. sp. J Antibiot (Tokyo) Ser A 16 (5): 203–206.

Nagayo, M., T. Tamiya, T. Mitamura and K. Sato. 1930 On the virus of Tsutsugamushi disease and its demonstration by a new method. Jap J Exp Med 8: 309–318.

Nakagawa, A. and K. Kitahara. 1959 Taxonomic studies on the genus Pediococcus. J Gen Appl Microbiol 5: 95–126.

— and —. 1962 Pleomorphism in bacterial cells. II. Giant cell formation in Pediococcus cerevisiae induced by hop-resins. J Gen Appl Microbiol 8: 142–148.

Nakahama, T. 1940 On the retting of vegetable fibre materials. Part XI. The useful anaerobes for the bacterial retting of flax. J Agr Chem Soc Jap 16: 39–42.

Nakamura, G. 1961 Studies on antibiotic actinomycetes. I. On Streptomyces producing a new antibiotic tubermycin. J Antibiot (Tokyo) Ser A 14 (2): 86–89.

—. 1961 Studies on antibiotic actinomycetes. II. On Streptomyces producing a new antibiotic tubercidin. J Antibiot (Tokyo) Ser A 14 (2): 90–93.

—. 1961 Studies on antibiotic actinomycetes. III. On Streptomyces producing 9-β-D-ribofuranosylpurine. J Antibiot (Tokyo) Ser A 14 (2): 94–97.

Nakano, K. 1919 Soybean leaf spot. J Plant Prot Tokyo 6: 217–221.

Nakayama, K., F. Okamoto and Y. Harada. 1956 Antimycin A: Isolation from a new Streptomyces and activity against rice plant blast fungi. J Antibiot (Tokyo) Ser A 9 (2): 63–66.

Nakayama, O. and M. Yanoshi. 1967 Spore-bearing lactic acid bacteria isolated from rhizosphere. I. Taxonomic studies on Bacillus laevolacticus nov. sp. and Bacillus racemilacticus nov. sp. J Gen Appl Microbiol 13: 139–153

— and —. 1967 Spore-bearing bacteria isolated from rhizosphere. II. Taxonomic studies on the catalase-negative strains. J Gen Appl Microbiol. 13: 155–165.

Nakazawa, K. 1955 Streptomyces albireticuli nov. sp. J Agr Chem Soc Jap 29: 647–649.

—. 1962 On the sexuality of a species of streptomycetes. Trans Mycol Soc Jap 4 (1): 9–10.

—. 1964 Studies on streptomycetes. Microbiological characteristics of Streptomyces No. 5866. Ann Rep Takeda Res Lab 23: 69–70.

—. 1964 Studies on the life history of Streptomyces No. 5866. (2) On a crystalline substance observed in the culture media. Ann Rep Takeda Res Lab 23: 71–75.

—. 1964 Studies on the life history of Streptomyces No. 5866. (1) Observation on the formation of the fruit body and gamete. Trans Mycol Soc Jap 4 (6): 147–151.

—. 1966 On the life history of Streptomyces No. 5866. (4). On the crystals of Streptomyces No. 5866. Trans Mycol Soc Jap 7 (2, 3): 241–244.

—. 1968 On the conversion of species of streptomycetes. Hindustan Antibiot Bull 10 (4): 278–279.

— and S. Fujii. 1957 Studies on streptomycetes. On Streptomyces sindenensis nov. sp. Ann Rep Takeda Res Lab 16: 109–110.

—, —, M. Inoue, H. Hitomi, A. Miyake and J. Kaneko. 1954 On a new anti-tuberculous substance produced by a Streptomyces. J Antibiot (Tokyo) Ser B 7: 168 (abstr).

—, M. Shibata, E. Higashide, T. Kanzaki, K. Yamamoto, A. Miyake, J. Ueyanagi and E. Iwasaki. 1961 Rufomycin, antibiotic substance. Japanese Patent 17,498, September 26.

—, —, K. Tanabe and H. Yamamoto. 1958 Studies on streptomycetes. Streptomyces humidus nov. sp., the source of dihydrostreptomycin. J Agr Chem Soc Jap 32 (4): 321–324.

—, K. Tanabe, M. Shibata, A. Miyake and T. Takewaka. 1956 Studies on streptomycetes. Cladomycin, a new antibiotic produced by Streptomyces lilacinus nov. sp. J Antibiot (Tokyo) 9 (B): 81.

Nakhimovskaya, M. I. 1948 Pseudomonas aurantiaca n. sp. Mikrobiologiya 17: 58–65.

Namioka, S. and M. Murata. 1961 Serological studies on Pasteurella multocida. I. A simplified method for capsule typing of the organism. Cornell Vet 51: 498–507.

— and R. Sakazaki. 1958 Étude sur les Rettgerella. Ann Inst Pasteur (Paris) 94: 485–499.

Namyslowski, B. 1912 Beitrag zur Kenntnis der Menschlichen Hornhautbakteriosen. Zentrabl Bakteriol Parasitenk Infektionskr Hyg Abt I Orig 62: 564–568.

Nannizzi, A. 1934 Repertorio sistematico dei miceti dell'umo e degli animali. In Pollacci, Tratt Micopat Umana. Poligrafica Meini, Siena, 4: 1–557.

Narasimha Rao, P. L. and B. N. Uma. 1958 Formation of antifungal antibiotics by a new Streptomyces species. Nature (London) 182 (4628): 115–116.

Natvig, H. 1905 Bakteriologische Verhältnisse in Weiblichen Genitalsekreten. Arch Gynaekol 76: 701–859.

Naumann, E. 1921 Untersuchungen über die Eisenorganismen Schwedens. I. Die Erscheinungen der Sideroplastie in den Gëwassern des Teichgebietes Aneboda. Kgl Svenska Vetenskapsakad Handl 62 (4): 1–68.

—. 1929 Die eisenspeichernden Bakterien. Kritische ubersicht der bisher bekannten Formen. Zentrabl Bakteriol Parasitenk Infektionskr Hyg Abt II 78. (24–26): 512–515.

Negroni, P. 1934 Microorganismes anaérobies dans les mycétoma humains. C R Séances Soc Biol Filiales 117: 1239–1240.

— and H. Bonfiglioli. 1937 Morphology and biology of Actinomyces israelii. J Trop Med Hyg 40: 226–232; 240–245.

— and I. Fischer. 1944 Estudio sobre el Lactobacillus bifidus (Tissier) Kulp y Rettger. Rev Soc Argent Biol 20: 315–327.

Neide, E. 1904 Botanische Beschreibung einiger sporenbildenden Bakterien. Zentrabl Bakteriol Parasitenk Infektionskr Hyg Abt II 12: 1–32; 161–176; 337–352; 539–554.

Neisser, A. 1879 Über eine der Gonorrhoe eigentümliche Micrococcus-form. Zentralbl Med Wiss 17 (28): 497–500.

Neitz, W. O. 1957 Genus I. Anaplasma Theiler 1910. In Breed, Murray and Smith (Editors), Bergey's Manual of Determinative Bacteriology, 7th ed. The Williams & Wilkins Co., Baltimore, pp. 981–986.

—, R. A. Alexander and P. J. du Toit. 1934 Eperythrozoon ovis (sp. nov.). Infection in sheep. Onderstepoort J Vet Sci 3: 263–274.

Nelson, D. H. 1931 Isolation and characterization Nitrosomonas and Nitrobacter. Zentrabl Bakteriol Parasitenk Infektionskr Hyg Abt I 83: 280–311.

Neptune, E. M., Jr., E. Weiss and J. A. Davis. 1964 Lipid metabolism of the rickettsialike microorganisms Wolbachia persica. II. Studies with labeled nonlipid substrates. J Infect Dis 114: 45–49.

—, —, and E. C. Suitor, Jr. 1964 Lipid metabolism of the rickettsialike microorganism Wolbachia persica. I. Incorporation of long chain fatty acids into phosphatides. J Infect Dis 114: 39–44.

Nesemann, G., H. J. Hübener, R. Junk and J. Schmidt-Thomé. 1960 20β-Hydroxysteroiddehydrogenase. I. Züchtung von Streptomyces hydrogenans und Induktion des Enzyms. Biochem Z 333: 88–94.

Neukirch, H. 1902 Über Strahlenpilze (Actinomyceten), 2nd ed. Ludolf, Beust Verlagsbuchhandlung, Strassburg.

Neveu-Lemaire, M. 1921 Précis de la parasitologie humaine, 5th ed. J Lemaire, Paris.

Nevin, T. A. 1954 The vitamin requirements of certain alpha-haemolytic streptococci isolated from the human mouth. J Bacteriol 67 (2): 217–219.

Newsome, I. E. and F. Cross. 1932 Some bipolar organisms found in pneumonia in sheep. J Amer Vet Med Ass 80: 711–719.

Ng, H. and R. W. Vaughn. 1963 Clostridium rubrum sp. n. and other pectinolytic clostridia from soil. J Bacteriol 85 (5): 1104–1113.

Nicol, C. S. and D. G. ff. Edward. 1953 Role of organisms of the pleuropneumonia group in human genital infections. Brit J Vener Dis 29: 141–150.

Nicolaeva, E. I. 1915 On the characterization of certain actinomycetes (in Russian). Arkh Biol Sci Nauk St Petersburg 18: 229.

Nicolle, C. and C. Anderson. 1927 Sur l'origine des fièvres recurrentes humaines. Bull Inst Pasteur 25: 657–665.

Niederhauser, J. S. 1943 A bacterial leaf spot and blight of the Russian dandelion. Phytopathology 33: 959–961.

Niemer, H. H. Bucherer, H. Zeitler and E. Stadler 1964 Über das "Nicotinblau". Hoppe-Seyler's Z Physiol Chem 337: 282–283.

Niida, T. and M. Ogasawara. 1960 Taxonomical study on a new Streptomyces producing taitomycin. Sci Rep Meiji Seika Kaisha 3: 23–26.

Nikitin, D. I. and L. V. Vasil'yeva. 1967 Rodshaped organisms with spheric inflations (in Russian). Izv Akad Nauk SSSR Ser Biol 2: 296–401.

— and —. 1968 A new soil microorganism Agrobacterium polyspheroidum n. sp. (in Russian, English summary). Izv Akad Nauk SSSR Ser Biol 3: 443–444.

Nikitina, N. N. 1958 Actinomycetes of the globisporus group. Thesis defended on July 12, 1957, at the Inst of Mikrobiology, Akad Nauk USSR. Abstracted by E. A. Kosmachev, Mikrobiologiya 27 (1): 141–144.

Niklewski, B. 1910 Über die wasserstoffoxydation durch Mikroorganismen. Jahrb Wiss Bot 48: 113–142.

Ninmich, W. 1968 Zur Isolierung und qualitativen Bausteinanalyse der K-Antigene von Klebsiellen. Z Med Mikrobiol Immunol 154: 117–131.

Niosi, F. 1911 Untersuchung eines streng anaeroben Bacillus ausschliesslichen Erregers einer eiterigen Pleuritis. Zentrabl Bakteriol Parasitenk Infektionskr Hyg Abt I Orig 58: 193–228.

Nishimura, H., T. Kimura and M. Kuroya. 1953 On a yellow crystalline antibiotic, identical with a aureothricin isolated from a new species of Streptomyces, 39a, and its taxonomic study. J Antibiot (Tokyo) Ser A 6 (2): 57–65.

—, —, K. Tawara, K. Sasaki, K. Nakajima, N. Shimaoka, S. Okamoto, M. Shimohara and J. Isono. 1957 Aburamycin, a new antibiotic. J Antibiot (Tokyo) Ser A 10 (5): 205–212.

—, M. Mayama, T. Kimura, A. Kimura, Y. Kawamura, K. Tawara, Y. Tanaka, S. Okamoto and H. Kyotani. 1964 Two antibiotics identical with nonactin and valinomycin obtained from a Streptomyces tsusimaensis, n. sp. J Antibiot (Tokyo) Ser A 17 (1): 11–22.

—, S. Okamoto, M. Mayama, H. Ohtsuka, K. Nakajima, K. Tawara, M. Shimohira and N. Shimaoka. 1961 Siomycin, a new thiostrepton-like antibiotic. J Antibiot (Tokyo) Ser A 14 (5): 255–263.

Niven, C. F., Jr. 1944 Nutrition of Streptococcus lactis. J Bacteriol 47: 343–350.

—, A. G. Castellani and V. Allanson. 1949 A study of the lactic acid bacteria that cause surface discoloration of sausages. J Bacteriol 58: 633–641.

— and J. B. Evans. 1957 Lactobacillus viridescens nov. spec. a heterofermentative species that produces a green discoloration of cured meat pigments. J Bacteriol 73 (6): 758–759.

—, Z. Kiziuta and J. C. White. 1946 Synthesis of polysaccharide from sucrose by Streptococcus S.B.E. J Bacteriol 51 (6): 711–716.

— and J. M. Sherman. 1944 Nutrition of the enterococci. J Bacteriol *47:* 335–342.

—, M. R. Washburn and J. M. Sherman. 1946 Folic acid requirements of the minute streptococci. J. Bacteriol *51 (1):* 128.

—, — and J. C. White. 1948 Nutrition of *Streptococcus bovis.* J Bacteriol *55 (5):* 601–606.

Niven, J. S. F., G. W. A. Dick, A. W. Gledhill and C. H. Andrewes. 1952 Further light on mouse hepatitis. Lancet *263:* 1061.

Nocard, E. and A. Mollereau. 1887 Sur une mammite contagieuse des vaches laitières. Ann Inst Pasteur (Paris) *1:* 109–126.

— and E. Roux. 1898 Le Microbe de la péripneumonie. Ann Inst Pasteur (Paris) *12:* 240–262.

Noguchi, H. 1912 Cultural studies on mouth spirochaetae (*Treponema microdentium* and *macrodentium*). J Esp Med *15:* 81–89.

—. 1912 *Treponema mucosum* (new species), a mucin-producing spirochaete from pyorrhea alveolaris, grown in pure culture. J Exp Med *16:* 194–198.

—. 1912 Pure cultivation of *Spirochaeta phagedenis* (new species), a spiral organism found in phagedenic lesions on human external genitalia. J Exp Med *16:* 261–268.

—. 1913 Cultivation of *Treponema calligyrum* (n. sp.) from condylomata of man. J Exp Med *17:* 89–98.

—. 1917 *Spirochaeta icterohaemorrhagiae* in American wild rats and its relation to the Japanese and European strains. J. Exp Med *25:* 755–763.

—. 1918 Morphological characteristics and nomenclature of *Leptospira (Spirochaeta) icterohemorrhagiae* (Inada and Ido). J Exp Med *27:* 575–592.

—. 1918 The spirochetal flora of the normal male genitalia. J Exp Med *27:* 667–678.

—. 1919 Etiology of yellow fever. II. Transmission experiments on yellow fever. J Exp Med *29:* 565–584.

—. 1921 A note on the venereal spirochetosis of rabbits. A new technique for staining *Treponema pallidum.* J Amer Med Ass *77:* 2052–2053.

—. 1923 Laboratory diagnosis of syphilis. Hoeber, New York.

—. 1927 *Cristispira* in North American shellfish. A note on a spirillum found in oysters. J Exp Med *34:* 295–316.

—. 1928 The Spirochetes. *In* Jordan and Falk (Editors), The newer knowledge of bacteriology and immunology. Univeristy of Chicago Press, Chicago, pp. 452–497.

Noguchi, T. T., R. Nachum and C. A. Lawrence. 1963 Acute purulent meningitis caused by chromogenic *Neisseria.* A case report and literature review. Med Arts Sci *17 (1):* 11–18.

Nöller, W. 1917 Blut- und Insektenflagellaten Züchtung auf Platten. Arch Schiffs-Trop Hyg *21:* 53–94.

Nolte, E. M. 1957 Untersuchungen über Ernährung und Fruchtkorperbildung von Myxobakterien. Arch Mikrobiol *28:* 191–246.

Nomenclature Committee of International Society for Microbiology. 1936 Opinion A. Rep Proc 2nd Int Congr Microbiol, London, pp. 28–29.

Nonomura, H. and Y. Ohara. 1957 Distribution of Actinomycetes in the soil. II. *Microbi-*

spora, a new genus of the *Streptomycetaceae.* J Ferment Technol *35:* 307–311.

— and —. 1960 Distribution of the actinomycetes in soil. IV. The isolation and classification of the genus *Microbispora.* J Ferment Technol *38:* 41; 401–409.

— and —. 1960 Distribution of the Actinomycetes in soil. (V). The isolation and classification of the genus *Streptosporangium.* J Ferment Technol *38:* 405–409.

— and —. 1969 Distribution of actinomycetes in soil. VII. A culture method effective for both of preferential isolation and enumeration of *Microbispora* and *Streptosporangium* strains in soil. J Ferment Technol *47 (11):* 701–709.

Norden, A. and F. Linell. 1951 A new type of pathogenic *Mycobacterium.* Nature (London) *168:* 826.

Norlin, M. 1965 Unclassified mycobacteria, a comparison between a serological and a biochemical classification method. Bull Int Union Tuberc *36:* 25–32.

Norman A. G. 1955 Inhibition of root growth by azaserine. Science (Washington) *121:* 213.

Norris, C., A. M. Pappenheimer and T. Flournoy. 1906 Study of a spirochaete obtained from a case of relapsing fever in man, with notes on morphology, animal reactions and attempts at cultivation. J Infect Dis *3:* 266–290.

Norris, J. R. 1964 The classification of *Bacillus thuringiensis.* J Appl Bacteriol *27:* 439–447.

Northrop. J. H., L. H. Ashe and J. K. Senior. 1919 Biochemistry of *Bacillus acetoethylicum* with reference to the formation of acetone. J Biol Chem *39:* 1–21.

Nottingham, P. M. and R. E. Hungate. 1968 Isolation of methanogenic bacteria from feces of man. J Bacteriol *96 (6):* 2178–2179.

Novacek, G. 1938 Cited by Fott and Komarek. 1960. Sb prir Kl Trebici *2:* 62–68.

Novogrudsky, D. M. 1950 Principles of series of antibiotics and its significance in finding new antimicrobial substances (in Russian). Izv Akad Nauk Kaz SSR Ser Microbiol *83:* 14–25.

Novotny, P. 1953 Plektridia. Cesk Epidemiol Mikrobiol Imunol *2:* 90–107.

Novy, F. G. and R. E. Knapp. 1906 Studies on *Spirillum obermeiri* and related organisms. J Infect Dis *3:* 291–393.

Nowak, J. 1929 Morphologie, nature et cycle évolutif du microbe de la péripneumonie des bovidés. Ann Inst Pasteur (Paris) *43:* 1330–1352.

Nowlan, S. S. and R. H. Deibel. 1967 Group Q streptococci. I. Ecology, serology, physiology and relationship to established enterococci. J. Bacteriol *94 (2):* 291–296.

— and —. 1967 Group Q streptococci. II Nutritional characteristics and growth relationship to thymine, folate and folinate. J Bacteriol *94 (2):* 297–299.

Nurmikko, V. and E. Karha. 1962 Nutritional requirements of lactic acid bacteria. II. Vitamin and amino acid requirements of *Streptococcus thermophilus* strains. Ann Acad Sci Fenn Ser A II Chem, pp. 1–21.

N. V. Koninklijke Nederlandsche Gist-en Spiritusfabriek. 1955 Process for the preparation of antibiotics. Australian Patent Application 3985/54, April 21.

—. 1956 Vengicide and processes for the preparation of the antibiotics oxytetracycline and vengicide. British Patent 764,198, December 19.

Nyberg, P. A., S. E. Knapp and R. E. Milleman. 1967 "Salmon poisoning" disease. IV. Transmission of the disease to dogs by *Nanophyetus salmincola* eggs. J Parasitol *53:* 694–699.

Nyfeldt, A. 1932 Klinische und experimentelle Untersuchungen über die Mononucleosis infectiosa. Foli Haematol (Leipzig) *47:* 1–144.

Nygaard, A. P. and B. D. Hall. 1963 A method for the detection of RNA-DNA complexes. Biochem Biophys Res Commun *12:* 98–104.

Oag, R. K. 1942 The biological properties of the Morax-Axenfeld *Bacillus* (*B. lacunata*) with particular reference to haemolysis. J Pathol Bacteriol *54:* 128–132.

Obst, M. M. 1919 A bacteriological study of sardines. J Bacteriol *24:* 158–169.

Oda, M. 1935 Microbiological studies on flora in washing water of saké brewery. (Report No. 6). I. Microbiological studies on Miyamizu (Holy Water) (8). (9). Micrococcus and Actinomyces isolated from Miyamizu (in Japanese). J Ferment Technol *13 (2):* 1202–1228.

Oersted, A. S. 1841 Beretning om en Excursion til Trindelen, en Alluvialdannelse i Odensefjord, i Esteraaret. Natuurwetensch Tijdschr *3:* 552–569.

—. 1844 De Regionibus Marinis. Elementa Topographiae Historiconaturalis Freti Oeresund. Inaug. Diss. J. C. Scharling, Copenhagen.

Oesterhelt, D. and W. Stoeckenius. 1971 Rhodopsin-like protein from the purple membrane of *Halobacterium halobium.* Nature New Biol *233:* 149–152.

Oetker, H. 1953 Untersuchungen über die Ernährung einiger Myxobacterien. Arch Mikrobiol *19:* 206–246.

O'Gara, P. J. 1916 A bacterial disease of western wheat grass. First account of the occurrence of a new type of bacterial disease in America. Science (Washington) *42:* 616–617.

Ogata, N. 1931 Aetiologie der Tsutsugamuchi-Krankheit: *Rickettsia tsutsugamushi.* Zentrabl Bakteriol Parasitenk Hyg Abt I Orig *122:* 249–253.

Ogawa, H., T. Itó, S. Inoue and M. Nishio. 1960 Chemical study on moldcidin B and its identification with pentamycin. J Antibiot (Tokyo) Ser A *13 (5):* 353–355.

Ogawa, T. 1937 Shoot drooping disease of *Acer trifidum* Hook et Arn. caused by *Pseudomonas acernea* n. sp. Ann Phytopathol Soc Jap *7:* 125–135.

Ogura, K. 1929 Ueber Druststreptococcus, mit besonderer Berucksichtigung seiner Spezifitat. J Jap Soc Vet Sci *8:* 174–203.

Ohad, I., D. Danon and S. Hestrin. 1962 Synthesis of cellulose by *Acetobacter xylinum.* V. Ultrastructure of polymer. J. Cell Biol *12:* 31–46.

Ohkuma, K., K. Anzai and S. Suzuki 1962 Studies on a new antibiotic, tuberin. I. Isolation and characterization. J Antibiot (Tokyo) Ser A *15 (3):* 115–116.

—, J. Nagatsu, C. Itakura, S. Suzuki and Y. Sumiki 1962 A new antitumor antibiotic, cervicarcin. J Antibiot (Tokyo) Ser A *15 (3):* 152–153.

Ohmori, T., M. Okanishi and H. Kawaguchi. 1962

Glebomycin, a new member of the streptomycin class. III. Taxonomic studies on Strain No. 12096, producer of glebomycin. J Antibiot (Tokyo) Ser A *15 (1):* 21–27.

Okabe, N. 1933 Bacterial diseases of plants occurring in Formosa. II. J Soc Trop Agr Formosa *5:* 26–36.

—. 1933 Bacterial diseases of plants occurring in Formosa. III. Bacterial leaf spot disease of jute plants. J Soc Trop Agr Taiwan *5:* 157–166.

—. 1934 Bacterial diseases of plants occurring in Formosa. IV. J Soc Trop Agr Formosa *6:* 54–63.

—. 1935 Bacterial disease of plants occurring in Taiwan (Formosa). V. J Soc Trop Agr Formosa *7:* 57–66.

Okami, Y. 1948 Studies on the characters of antibiotic *Streptomyces.* I. On the characters of a Chloromycetin-producing strain (0–163). Jap Med J *1 (6):* 499–503.

—. 1949 Studies on the characters of antibiotic *Streptomyces.* I. On the characters of a chloromycetin-producing strain (0–163). J Antibiot (Tokyo) *2 (9):* 593–595.

—. 1952 Classification of the antagonistic ray fungi of Japan of the family *Streptomycetaceae* (in Japanese). Doctoral Dissertation, Hokkaido University.

—, K. Maeda, H. Kondo, T. Tanaka and H. Umezawa. 1962 A *Streptomyces* producing *O*-carbamyl-D-serine. J Antibiot (Tokyo) Ser A *15 (3):* 147–151.

—, T. Okuda, T. Takeuchi, K. Nitta and H. Umezawa. 1953 Studies on anti-tumor substances produced by microorganisms. IV. Sarkomycin-producing *Streptomyces* and two other *Streptomyces* producing the antitumor substance, No- 289 and caryomycin. J Antibiot (Tokyo) Ser A *6 (4):* 153–157.

—, M. Suzuki and H. Umezawa. 1963 Taxonomical studies on a *Streptomyces* strain producing labilomycin. J Antibiot (Tokyo) Ser A *16 (4):* 152–154.

—, R. Utahara, S. Nakamura and H. Umezawa. 1954 Studies on antibiotic actinomycetes. IX. On *Streptomyces* producing a new antifungal substance of fungicidin-rimocidin-chromin group, eurocidin group and trichomycin-ascosin-candicidin group. J Antibiotics *7 (A):* 98–103.

—, —, H. Oyagi, S. Makamura, H. Umezawa, K. Yanagisawa and Y. Tsunematsu. 1955 The screening of anti-toxoplasmic substance produced by streptomycete and anti-toxoplasmic substance No. 534. J Antibiot (Tokyo) Ser A *8 (4):* 126–131.

Okamoto, H. 1962 Biochemical study of the streptolysin-S inducing effect of ribonucleic acid: a review. Ann Rep Res Inst Tuberc *19:* 165–197.

Okanishi, M., H. Koshiyama, T. Ohmori, M. Matsuzaki, S. Ohashi and H. Kawaguchi. 1962 Glebomycin, a new member of the streptomycin class. I. Biological studies. J Antibiot (Tokyo) Ser A *15 (1):* 7–14.

Okuda, T., T. Furumai, E. Watanabe, Y. Okugawa and S. Kimura. 1966 Taxonomic study on the sporaviridin-producing microorganism, *Streptosporangium viridogriseum* nov. sp. J Antibiot (Tokyo) *19:* 121–127.

Olin Mathieson Chemical Corporation. 1958

Thiostrepton and its salts. British Patent 795, 570, May 28.

Olitsky, P. K. and F. L. Gates. 1921 Experimental studies of the nasopharyngeal secretions from influenza patients. J Exp Med 33: 713–729.

Oliver, T. J., A. Goldstein, R. R. Bower, J. C. Holper and R. H. Otto. 1961 M-141, a new antibiotic. I. Antimicrobial properties, identity with actinospectacin, and production by Streptomyces flavopersicus, sp. n. Antimicrob Agents Chemother, pp. 495–502.

Oliver, W. W. and W. B. Wherry. 1921 Notes on some bacterial parasites of the human mucous membranes. J. Infect Dis 28: 341–344.

Olsen, E. 1944 En sporedannende maelkesyrebakterie Lactobacillus cereale (nov. sp.). Kem Maanedsbl Nord Handelsbl Kem Ind 25: 125–130.

Olson, N. O., K. M. Kerr and A. Campbell. 1964 Control of infectious synovitis. 13. The antigen study of three strains. Avian Dis 8: 209–214.

Olsufiev, N. G. 1968 Tularemia. Bull USSR Min Pub Hlth Cent Inst for Adv Med Studies, Moscow.

—, O. S. Emelyanova and T. N. Dunaeva. 1959 Comparative studies of strains of B. tularense in the Old and New World and their taxonomy. J Hyg Epidemiol Microbiol Immunol (Prague) 3: 138–149.

Omelianski, V. L. 1923 Aroma-producing microorganisms. J Bacteriol 8: 393–419.

Omelianski, W. 1905 Ueber eine neue Art farbloser Thiospirillen. Zentrabl Bakteriol Parasitenk Infektionskr Hyg Abt II 14: 769–772.

O'Neill, T. B., R. W. Drisko and H. Hochman. 1961 Pseudomonas creosotensis sp. n. a creosote-tolerant marine bacterium. Appl Microbiol 9 (6): 472–474.

Onishi, H. and M. Kamekura. 1972 Micrococcus halobius sp. n. Int J Syst Bacteriol 22: 233–236.

Onishi, M. 1949 Study on the actinomyces isolated from the deeper layer of carious dentine J. Dent Res 6: 273–282.

Ordal, E. J. and R. R. Rucker. 1944 Pathogenic myxobacteria. Proc Soc Exp Biol Med 65: 15–18.

Orian, G. 1972 A disease of Paspalum dilatum Poir. in Mauritius caused by a species of bacterium closely resembling Xanthomonas albilineans (Ashby) Dowson. Rev Agr Sucr Ile Maurice 41: 7–24.

Orla-Jensen, S. 1908 Hovedlinierne i det naturlige Bakteriesystem. Overs Danske Vidensk Selsk Forhandl 5: 267–330.

—. 1909 Die Hauptlinien des natürlichen Bakteriensystems. Zentrabl Bakteriol Parasitenk Infektionskr Hyg Abt II 22: 97–98; 305–346.

—. 1916 Maelkeri-Bakteriologi. Schønberske Forlag, Copenhagen.

—. 1919 The lactic acid bacteria. Høst, Copenhagen, pp. 1–196.

—. 1921 The main lines of the natural bacterial system. J Bacteriol 6: 263–273.

—. 1924 La Classification des bactéries lactiques. Lait 4: 468–474.

—. 1943 Die echten Milchsaürebakterien. Ejnar Munksgaard, Copenhagen.

—, A. D. Orla-Jensen and A. Kjaer. 1947 On the ensiling of lucerne by means of lactic acid

fermentation. Antonie van Leeuwenhoek J Microbiol Serol 12: 97–114.

—, — and O. Winther. 1936 Bacterium bifidum und Thermobacterium intestinale. Zentrabl Bakteriol Parasitenk Infektionskr Hyg Abt II 93: 321–343.

Ormsbee, R. A., H. Parker and E. G. Pickens. 1955 The comparative effectiveness of aureomycin, terramycin, chloramphenicol, erythromycin, and thiocymetin in suppressing experimental rickettsial infections in chick embryos. J Infect Dis 96: 162–167.

— and M. G. Peacock. 1964 Metabolic activity in Coxiella burnetii. J Bacteriol 88 (5): 1205–1210.

Ornstein, M. 1920 Zur Bakteriologie des Schmitz-bacillus. Z Hyg Infektionskr 91: 152–178.

Ørskov, F. and I. Ørskov. 1961 The fertility of Escherichia coli antigen test strains in crosses with K12. Acta Pathol Microbiol Scand 51: 280–290.

—, —, B. Jann, K. Jann, E. Müller-Seitz and O. Westphal. 1967 Immunochemistry of Escherichia coli O antigens. Acta Pathol Microbiol Scand 71: 339–358.

Ørskov, I. 1955 The biochemical properties of Klebsiella (Klebsiella aerogenes) strains. Acta Pathol Microbiol Scand 37: 353–368.

—. 1957 Biochemical types in the Klebsiella group. Acta Pathol Microbiol Scand 40: 155–162.

Ørskov, J. 1923 Investigations into the morphology of the ray fungi. Levin and Munksgaard, Copenhagen, Denmark.

—. 1928 Beschreibung eines neuen Microben, Microcyclus aquaticus, mit eigentümlicher Morphologie. Zentrabl Bakteriol Parasitenk Infektionskr Hyg Abt I Orig 107: 180–184.

Ortali, V. and L. Capocaccia. 1956 Una nuova specie di Corynebacterium: il Corynebacterium mycetoides (Castellani) Ortali e Capocaccia 1956. Rend Ist Super Sanita 19: 480–491.

Osnitskaya, L. K. 1954 A new species of purple bacteria from stratal water of oil fields (Rhodopseudomonas issatchenkoi). Tr Inst Mikrobiol Akad Nauk SSSR 3: 5–20.

Osono, T., Y. Oka, S. Watanabe, Y. Numazaki, K. Moriyama, H. Ishida, K. Suzaki, Y. Okami and H. Umezawa. 1967 A new antibiotic, josamycin. Isolation and physico-chemical characteristics. J Antibiot (Tokyo) Ser A 20 (3): 174–180.

Ostertag, H. 1952 Celluloseabbauende Mikroorganismen. Z Hyg Infektionskr Med Mikrobiol Immunol Virol 133: 489–509.

Osterwalder, A. 1909 Unbekannte Krankheiten an Kulturpflazen und deren Ursachen. Zentrabl Bakteriol Parasitenk Infektionskr Hyg Abt II 25: 260–270.

Ostrovskaya, N. N. and E. I. Kaitmasova. 1966 The Tb bacteriophage as a supplementary test for differentiation of Brucella species. Zh Mikrobiol Epidemiol Immunobiol 43: 75–79.

Oswald, E. J., R. J. Reedy and W. A. Randall. 1956 An antifungal agent, 1968, produced by a new Streptomyces species. Antibiot Ann 1955–1956: 236–239.

Otani, S. and T. Sugai. 1953 On the basic antibiotic substance produced by Streptomyces (in Japanese). J Antibiotic (Tokyo) Ser B 6 (5): 257.

Otani, Y. 1939 Microbiological studies on the "Nukamiso-pickles." Zentrabl Bakteriol Parasitenk Infektionskr Hyg Abt II *101:* 139–151.

Ottens, H. and K. C. Winkler. 1962 Indifferent and haemolytic streptococci possessing group F antigen. J. Gen Microbiol *28 (1):* 181–191.

Ouellette, C. A., R. H. Burris and P. W. Wilson. 1969 Deoxyribonucleic acid base composition of species of *Klebsiella, Azotobacter* and *Bacillus.* Antonie van Leeuwenhoek J Microbiol Serol *35:* 275–286.

Øverås, J. 1959 *Eperythrozoon ovis,* a new blood parasite in sheep in Norway. Nord Vet Med *11:* 791–800.

Owen, C. R., E. O. Buker, W. L. Jellison, D. B. Lackman and J. F. Bell. 1964 Comparative studies of *Francisella tularensis* and *Francisella novicida.* J Bacteriol *87 (3):* 676–683.

Owen, S. P., A. Dietz and G. W. Camiener. 1963 Sparsomycin, a new antitumor antibiotic. I. Discovery and Biological Properties. Antimicrob Agents Chemother *1962:* 772–779.

Owens, J. D. and R. M. Keddie. 1969 The nitrogen nutrition of soil and herbage coryneform bacteria. J Appl Bacteriol *32:* 338–347.

Oxford, A. E. 1944 Diplococcin, an anti-bacterial protein elaborated by certain milk streptococci. Biochem J *38:* 178–182.

Ozaki, Y. 1912 Zur Kenntnis der anaeroben Bakterien der Mundhöhle. Zentrabl Bakteriol Parasitenk Infektionskr Hyg Abt I Orig *62:* 76–88.

Pacini, F. 1854 Osservazione microscopiche e Deduzioni patologiche sul Cholera Asiatico. Gaz Med Ital Toscana Firenze *6 (2):* 405–412.

Packer, L. and W. Vishniac. 1955 Chemosynthetic fixation of carbon dioxide and characteristics of hydrogenase in resting cell suspensions of *Hydrogenomonas ruhlandii* nov. spec. J Bacteriol *70 (2):* 216–223.

Padhya, A. C. and M. K. Patel. 1962 A new bacterial leaf-spot on *Alangium lamarckii* Thw. Curr Sci *31:* 196–197.

— and —. 1963 A new bacterial leaf-spot on *Ionidium heterophyllum* Went. Indian Phytopathol *16:* 98–99.

— and —. 1963 A new bacterial leaf-spot on *Corchorus acutangulus* Lam. Curr Sci *32:* 326.

— and —. 1964 Bacterial leaf-spot on *Triumfetta pilosa* Roth. Curr Sci *33:* 342.

—, — and W. V. Kotasthane. 1965 A new bacterial leaf-spot disease of *Bauhinia racemosa* Lamk. Curr Sci *34:* 224–225.

—, — and —. 1965 A new bacterial leaf-spot on *Vitis trifolia.* Curr Sci *34:* 462–463.

Padula, J. F., R. R. Facklam and M. D. Moody. 1969 Effect of incubation temperature on T-agglutination typing of *Streptococcus pyogenes.* Appl Microbiol *17:* 878–880.

Pagano, J. F., M. J. Weinstein, H. A. Stout and R. Donovick. 1956 Thiostrepton, a new antibiotic. Antibiot Ann *1955–1956:* 554–559.

Page, L. A. 1961 Experimental ulcerative stomatitis in King snakes. Cornell Vet *51:* 258–266.

—. 1966 Revision of the family *Chlamydiaceae* Rake (*Rickettsiales*): unification of the psittacosis-lymphogranuloma venereum-trachoma group of organisms in the genus *Chlamydia* Jones, Rake and Stearns 1945. Int J Syst Bacteriol *16 (2):* 223–252.

—. 1968 Proposal for the recognition of two species in the genus *Chlamydia* Jones, Rake and Stearns 1945. Int J Syst Bacteriol *18 (1):* 51–66.

Paine, S. G. 1919 Studies in bacteriosis. II. A brown blotch disease of cultivated mushrooms. Ann Appl Biol *5:* 206–219.

—. 1923 Internal rust spot disease of the potato tuber. Rep Intern Conf Phytopath Econ Ent Holland, pp. 74–78.

— and H. Stansfield. 1919 Studies in Bacteriosis. III. A bacterial leaf spot disease of *Protea cynaroides,* exhibiting a host reaction of possibly bacteriolytic nature. Ann Appl Biol *6:* 27–39.

Pakula, R. 1963 Can transformation be used as a criterion in taxonomy? Recent Progr Microbiol *8:* 617–624.

—, E. Hulanicka and W. Walczak. 1958 Transformation reactions between streptococci, pneumococci and staphylococci. Bull Acad Pol Sci Ser Sci Biol *6:* 325.

Palleroni, N. J., M. Doudoroff, R. Y. Stanier, R. E. Solánes and M. Mandel. 1970 Taxonomy of the aerobic pseudomonads, the properties of the *Pseudomonas stutzeri* group. J Gen Microbiol *60 (2):* 215–231.

Pammel, L. H. 1895 Bacteriosis of rutabaga (*Bacillus campestris* n.sp.). Iowa State Coll Agr Exp Sta Bull *27:* 130–134.

Panagopoulos, C. G. 1969 The disease "Tsilik Marasi" of grapevine, its description and identification of the causal agent (*Xanthomonas ampelina* sp. nov.). Ann Inst Phytopathol Benaki *9:* 59–81.

Pande, P. G. and P. C. Sekariah. 1960 A preliminary note on the isolation of *Moraxella coprae* nov. sp. from an outbreak of infectious keratoconjunctivitis in goats. Curr Sci *29 (7):* 276–277.

Paquin, R. and R. A. Lachance. 1970 Sur la nutrition aminée de *Corynebacterium sepedonicum* (Spieck. et Kott.) Skapt. et Burkh. et la résistance de la pomme de terre au flétrissement bactérien. Can J Microbiol *16:* 719–726.

Pardo-Castello, V. 1940 El treponema del mal del pinto. Rev Med Trop Habana *6:* 117–118.

Paretsky, D. 1968 Biochemistry of rickettsiae and their infected hosts, with special reference to *Coxiella burneti.* Zentrabl Bakteriol Parasitenk Infektionskr Hyg Abt I Orig *206:* 284–291.

Park, W. E. and A. W. Williams. 1917 Pathogenic Microorganisms, 6th ed. Lea and Febiger, New York.

Parke, Davis and Company. 1954 Antibiotics and methods for obtaining same. British Patent 707,332, April 14.

—. 1954 Antibiotic and method of obtaining same. British Patent 712,547, July 28.

—. 1956 Antibiotique et procédé pour l'obtenir (in Flemish). Belgian Patent 547,976, June 15.

Parker, C. D. 1945 The corrosion of concrete. I. The isolation of a species of bacterium associated with the corrosion of concrete exposed to atmosphere containing hydrogen sulphide. Aust J Exp Biol Med Sci *23:* 81–90.

—. 1957 Genus V. *Thiobacillus* Beijerinck 1904. *In* Breed, Murray and Smith (Editors), Bergey's Manual of Determinative Bacteriology, 7th ed. The Williams & Wilkins Co., Baltimore, pp. 83–88.

— and J. Prisk. 1953 The oxidation of inorganic compounds of sulfur by various sulphur bacteria. J Gen Microbiol 8: 344–364.

Parker, F., Jr. and N. P. Hudson. 1926 The etiology of Haverhill fever (Erythema arthriticum epidemicum). Amer J Pathol 2: 357–379.

Parker, M. T. 1969 Streptococcal skin infection and acute glomerulonephritis. Brit J Dermatol 81: 37–46.

Parker, R. R., G. M. Kohls, C. W. Cox and G. E. Davis. 1939 Observations on an infectious agent from Amblyomma maculatum. Public Health Rep 54: 1482–1484.

Partansky, A. M. and B. S. Henry. 1935 Anaerobic bacteria capable of fermenting sulfite waste liquor. J Bacteriol 30: 559–571.

Partridge, S. M. and E. Klieneberger. 1941 Isolation of cholesterol from the oily droplets found in association with the L₁ organism separated from Streptobacillus moniliformis. J Pathol Bacteriol 42: 219–223.

Passet, J. 1885 Ueber Mikroorganismen der eiterigen Zellgewebsentzündung des Menschen. Fortschr Med 3 (2): 3–73.

Pasteur, L. 1864 Memoire sur la fermentation acétique. Ann Sci Éc norm sup Paris 1: 113–158.

—. 1876 Études sur la bière. Gauthier-Villars, Paris. pp. 1–383.

Pate, J. L. and E. J. Ordal. 1965 The fine structure of two unusual stalked bacteria. J Cell Biol 27: 133–150.

Patel, A. J. and M. K. Patel. 1958 A new bacterial blight of Cyamopsis tetragonoloba (L) Taub. Curr Sci 27: 258–259.

Patel, A. M., J. M. Chauhan, W. V. Kotasthane and M. V. Desai. 1969 A new bacterial disease of Biophytum sensitivum. Curr Sci 38: 274–275.

— and W. V. Kotasthane. 1969 Bacterial blight of Leea edgeworthii incited by Xanthomonas leeanum sp. nov. Curr Sci 38: 519–520.

— and —. 1969 Bacterial leaf spot disease of Corchorus fascicularis caused by Xanthomonas naktae var. fascicularis. Curr Sci 38: 596–597.

Patel, M. K. 1948 Xanthomonas uppallii sp. nov. pathogenic on Ipomoae muricata. Indian Phytopathol 1: 67–69.

—. 1949 Xanthomonas desmodii a new leaf-spot of Desmodium diffusum DC. Curr Sci 18: 213.

—, S. B. Allayyanavaramath and Y. S. Kulkarni. 1952 Bacterial shot-hole and fruit canker of Aegle marmelos Correa. Curr Sci 22: 216–217.

—, V. V. Bhatt and G. W. Dhande. 1954 A new variety of Xanthomonas alysicarpi. Indian Phytopathol 7: 182–183.

—, — and Y. S. Kulkarni. 1951 Three new bacterial diseases of plants from Bombay. Curr Sci 20: 326–327.

—, S. G. Desai and A. J. Patel. 1968 A new bacterial leaf-spot on Vernonia cinerea Less. Sci Cult 34: 220–221.

—, G. W. Dhande and Y. S. Kulkarni. 1953 Bacterial leafspot of Cyamopsis tetragonoloba (L) Taub. Curr Sci 22: 183.

— and Y. S. Kulkarni. 1951 Nomenclature of bacterial plant pathogens. Indian Phytopathol 4: 74–84.

—, — and G. W. Dhande. 1950 Xanthomonas badrii sp. nov. on Xanthium strumarium L. in India. Indian Phytopathol 3: 103–104.

—, — and —. 1951 Three bacterial diseases of plants. Curr Sci 20: 106.

—, —, and —. 1952 Two new bacterial diseases of plants. Curr Sci 21: 74–75.

—, — and —. 1952 Some new bacterial diseases of plants. Curr Sci 21: 345–346.

— and L. Moniz. 1948 Bacterial leafspot of Desmodium gangeticum DC. Indian Phytopathol 1: 137–141.

—, — and Y. S. Kulkarni. 1948 A new bacterial disease of Mangifera indica L. Curr Sci 17: 189–190.

—, M. J. Thirumalachar and V. V. Bhatt. 1955 A bacterial leaf-spot disease of Lochnera pusilla. Curr Sci 24: 172–175.

—, B. N. Wankar and Y. S. Kulkarni. 1952 Bacterial leaf-spot of Amaranthus viridis L. Curr Sci 21: 346–347.

Patel, P. N. and G. S. Shekhawat. 1971 Occurrence of Xanthomonas translucens f. sp. hordei in India and its pathogenicity to rice. Plant Dis Rep 55: 365–368.

Patelski, R. A. 1950 Terramycin and viomycin. Introductory remarks on their chemical, physical and antimicrobial properties. Ninth Streptomycin Conference. Trans. Veterans Administration, Central Office, Washington, D. C. 9: 186–188.

Patočka, F. and A. R. Prévot. 1947 Étude d'une nouvelle espèce anaérobie Capsularis stabilis. Ann Inst Pasteur (Paris) 73: 838–840.

— and V. Reynes. 1947 Étude d'une nouvelle espèce anaérobie: Leptotrichia vaginalis n. sp. Ann Inst Pasteur (Paris) 73: 599–600.

— and V. Sebek. 1951 Moving colonies in anaerobic microbe isolated from vaginal discharge. Bull Int Cl Sci Math Nat Med Acad Tchèque Sci 52 (1): 307–319.

Paton, A. M. 1959 An improved method for preparing pectate gels. Nature (London) 183: 1812–1813.

Pattyn, S. 1967 A study of Group III non-chromogenic mycobacteria; correlation of chicken virulence with other in vitro characters among 20 strains. Z Tuberk Erkrankungen Thoraxorgane 127: 41–46.

Pattyn S. R., M. T. Boveroulle, F. Gatti and J. Vandepitte. 1964–1966 Étude des souches de Mycobacterium ulcerans. Isolées au Congo (Leopoldville). Acad. Roy Sci. Outre-Mer, Bull des séances 1576–1599.

—, —, J. Mortelmans and J. Vercruysse. 1967 Mycobacteria in mammals and birds of the zoo of Antwerp. Acta Zool Pathol Antverpiensia 43: 125–134.

Payne, J. B. and C. H. Armstrong. 1970 Somatic antigens of Streptococcus group EII. Separation and a partial physico-chemical characterization. Appl Microbiol 19: 823–829.

Payne, W. J., R. G. Eagon and A. K. Williams. 1961 Some observations on the physiology of Pseudomonas natriegens nov. spec. Antonie van Leeuwenhoek J Microbiol Serol 27: 121–128.

Paynter, M. J. B. and R. E. Hungate. 1968 Characterization of Methanobacterium mobilis sp. n., isolated from the bovine rumen. J Bacteriol 95 (5): 1943–1951.

Peckham, M. C. 1966 An outbreak of streptococcosis (apoplectiform septicemia) in white rock chickens. Avian Dis 10 (4): 413–421.

Pederson, C. S. 1929 The types of organisms

found in spoiled tomato products. N Y Agr Exp Sta Tech Bull *150:* 1–46.

Peel, D. and J. R. Quayle. 1961 Microbial growth on C₁ compounds. I. Isolation and characterization of *Pseudomonas* AM 1. Biochem J *81:* 465–469.

Peklo, J. 1910 Die pflanzlichen Aktinomykosen. Zentrabl Bakteriol Parasitenk Infektionskr Hyg Abt II *27:* 451–579.

Pelczar, M. J. 1953 *Neisseria caviae* nov. spec. J Bacteriol *65 (6):* 744.

—. 1957 Genus II. *Veillonella* Prevot, 1933. *In* Breed, Murray and Smith (Editors), Bergey's Manual of Determinative Bacteriology, 7th ed. The Williams & Wilkins Co., Baltimore, pp. 485–490.

— and R. J. Porter. 1940 Pantothenic acid and nicotinic acid as essential growth substances for Morgan's bacillus (*Proteus morganii*). Proc Soc Exp Biol Med *43:* 151–154.

Pelsh, A. D. 1936 Hydrobiology of Karabugz Bay of the Kaspian Sea. Tr Vses Nauch-Issled Inst Galurgii Leningrad *5:* 49–126.

—. 1937 Photosynthetic sulfur bacteria of the Eastern reservoir of Lake Sakskoe. Mikrobiologiya *6:* 1090–1100.

Penso, G. 1947 Il rosso der baccalari-Etiologia, commestibilite bonificial e prevenzione. Rend Ist Super Sanita *10:* 563–605.

—. 1953 Criteri generali per determinare la posizione sistematica di un micobatterio. Symposium, Actinomycetales, Morphology, Biology and Systematics, pp. 89–101.

—. 1955 Sur quelques germes halophiles. Ann Inst Pasteur Lille *7:* 152–157.

—, G. Castelnuovo, A. Guadiano, M. Princivalle, L. Vella and A. Zampieri. 1952 Studi e recherche sui micobatteri. VIII. Un nuovo bacillo tubercolare: il *Mycobacterium minetti* n. sp., Studio microbiologica e patogenetico. Rend Ist Super Sanita *15:* 491–548.

—, R. Noel, M. Blanc and S. Marie-Suzanne. 1957 Études et recherches sur les mycobactéries. XV. Le *Mycobacterium marianum* (Penso 1953) Étude microbiologique, pathogénique et immunologique. Rend Accad Naz dei XL *8:* 1–75.

Perch, B., P. Kristjansen and K. Skadhauge. 1968 Group R streptococci pathogenic for man. Acta Pathol Microbiol Scand *74:* 69–76.

Pereira, A. L. G. 1969 Uma nova doenca bacteriana do Maracuja (*Passiflora edulis*, Sims) causada por *Xanthomonas passiflorae* n. sp. Arq Inst Biol Sao Paulo *36:* 163–174.

Perfil'ev, B. V. 1914 The chlorophyll-bearing microbe, *Pelodictyon clathratiforme*, of the green bacteria group (in Russian). J Mikrobiol Petrogr *1 (3–5):* 195–198.

—. 1914 On the theory of symbiosis of *Chlorochromatium aggregatum* Lauterb. (*Chloronium mirabile* Buder) and *Cylindrogloea bacterifera* nov. gen., nov. sp. (in Russian). J Mikrobiol Petrogr *1 (3–5):* 222–227.

—. 1921 On the knowledge of microorganisms of the Nevsky Bay. Bull Inst Hydrol Leningrad *1:* 84–96.

—. 1927 Die Rolle der Mikroben in der Erzbildung. Verh Int Ver Limnol *3:* 350–359.

—. 1952 Silting research and absolute geochronology. Izv Vses Geogr Obschest *84 (4):* 333–349.

— and D. R. Gabe. 1961 Capillary methods of investigating microorganisms. Russian Transl Oliver and Boyd 1969, Akad Nauk SSSR, pp. 1–534.

— and —. 1964 In The role of microorganisms in the formation of iron-manganese lake ore. Ed. by M. Gurevich. Akad. Nauk SSSR, Moskva, Leningrad, pp. 1–122.

—, —, A. M. Galperina, V. A. Rabinovich, A. A. Sapotnitskii, E. E. Sherman and E. P. Troshanov. 1964 Applied capillary microscopy. The role of microorganisms in the formation of iron-manganese deposits. Izdatelstvo Akad Nauk SSSR, Savarenskii Laboratory for Hydrogeological Problems. Moscow 1964 Russian (Transl. Consultants Bureau Enterprise Inc. 1965 New York).

Perkins, H. R. 1967 The use of photolysis of dinitrophenylpeptides in structural studies on the cell wall mucopeptide of *Corynebacterium poinsettiae*. Biochem J *102:* 29C.

—. 1970 Extraction procedures and cell wall composition, including some results with corynebacteria. Int J Syst Bacteriol *20 (4):* 379–382.

Perlman, D., J. Fried, E. O. Titus and A. F. Langlykke. 1955 16α-Hydroxyprogesterone, 16α-hydroxydihydroprogesterone and esters thereof. United States Patent 2,709,705, May 31.

Perrin, W. S. 1906 Researches upon the life-history of *Trypanosoma balbianii* (Certes). Arch Protistenk *7:* 131–156.

Perry, D. and H. D. Slade. 1964 Intraspecific and interspecific transformation in streptococci. J Bacteriol *88 (3):* 595–601.

Perry, K. D., L. G. M. Newland and C. A. E. Briggs. 1958 Group D rumen streptococci with type antigens of Group N. J Pathol Bacteriol *76:* 589–590.

Person, L. H. and W. J. Martin. 1940 Soil rot of sweet potatoes in Louisiana. Phytopathology *30:* 913–926.

Perty, M. 1852 Zur Kenntnis kleinster Lebensformen. Jent and Reinert, Bern I-VIII, pp. 1–228.

Peshkoff, M. A. 1939 Cytology, karyology and cycle of development of new microbes, *Caryophanon latum* and *Caryophanon tenue* (in Russian). Dokl Akad Nauk SSSR *25 (3):* 239–242.

—. 1948 Order *Caryophanales* Peshkoff. *In* Breed, Murray and Hitchens (Editors), Bergey's Manual of Determinative Bacteriology, 6th ed. The Williams & Wilkins Co., Baltimore, pp. 1002–1005.

Peters, D. and R. Wigand. 1955 *Bartonellaceae*. Bacteriol Rev *19 (3):* 150–155.

Petersen, E. J. 1959 Serological investigations on *Azotobacter* and *Beijerinckia*. Kgl Vet og Landbohøjskole Arsskr. Copenhagen, Denmark *1959:* 70–90.

Peterson, J. E. 1958 Two new fifty-year-old species of myxobacteria. Mycologia *50:* 628–633.

—. 1959 A monocystic genus of the *Myxobacterales* (*Schizomycetes*). Mycologia *51:* 1–8.

—. 1959 New species of myxobacter from the bark of living trees. Mycologia *51:* 163–172.

—. 1969 Isolation and maintenance of myxobacteria. *In* Norris and Ribbons (Editors), Methods in Microbiology. Vol. 3B, Academic Press, New York, pp. 185–210.

— and J. C. McDonald. 1966 The demise of the myxobacterial genus *Angiococcus*. Mycologia *58:* 962–965.

Pethybridge, C. H. and P. A. Murphy. 1911 A bacterial disease of the potato plant in Ireland and the organism causing it. Proc Roy Ir Acad *29:* 1–37.

Petrie, G. F. and D. McClean. 1934 Interrelations of *Corynebacterium ovis*, *Corynebacterium diphtheriae* and certain diphtheroid strains derived from the human nasopharynx. J Pathol Bacteriol *39:* 635.

Petruschky, J. 1896 *Bacillus faecalis alcaligenes* n. sp. Zentrabl Bakteriol Parasitenk Infektionskr Hyg Abt I *19:* 187–191.

Pette, J. W. and J. van Beynum. 1943 Boekelscheurbacterien. Rijkslandbauwproefstation te hoorn. Versl Landbouwk Onderz *49 (9) C:* 315–346.

Petter, H. F. M. 1931 On bacteria of salted fish. Proc Acad Sci Amsterdam *34:* 1417–1423.

Pettit, A. 1928 Contribution à l'étude des spirochetides. I. Morphologie, physiologie and chimiotherapie des spirochétides. II. Genres and espèces de spirochétides. III. *Spirocheta icterohemorragiae*. Chez 1' Auteur, Vanves (Seine), I: 1-119; II: 1–265; III: 1–207.

Peynaud, E. 1968 Études recentes sur les bactéries du Vin. Fermentation et vinification. 2nd Int Symp d'Oenologie Bordeaux, 1967, pp. 219-256.

— and S. Sapis-Domercq. 1970 Étude de deux cent-cinquante souches de bacilles héterolactiques isolés de vins. Arch Mikrobiol *70:* 348-360.

Pfeiffer, A. 1889 Über die bacilläre Pseudotuberkulose bei Nagethieren. Verlag von Georg Thieme, Leipzig, pp. 1–42.

Pfeiffer, R. 1892 Vorlaufige Mitteilungen über den Erreger der Influenza. Deut Med Wochenschr *18:* 28.

—. 1896 Die Spirillen. *In* Flugge (Editor), Die Mikroorganismen, 3 ed., 2 Thiel. F. C. W. Vogel, Leipzig, pp. 527–599.

Pfennig, N. 1967 Photosynthetic Bacteria. Annu Rev Microbiol *21:* 285–324.

—. 1968 *Chlorobium phaeobacteriodes* nov. spec. und *C. phaeovibrioides* nov. spec. Zwei neue Arten der grünen Schwefelbakterien. Arch Mikrobiol *63:* 224–226.

—. 1969 *Rhodopseudomonas acidophila* sp. n. a new species of the budding purple nonsulfur bacteria. J Bacteriol *99 (2):* 597–602.

—. 1969 *Rhodospirillum tenue* n. sp. a new species of the purple nonsulfur bacteria. J Bacteriol *99 (2):* 619–620.

— and G. Cohen-Bazire. 1967 Some properties of the green bacterium *Pelodictyon clathratiforme*. Arch Mikrobiol *59:* 226–236.

—, K. E. Eimhjellen and S. Liaaen-Jensen. 1965 A new isolate of the *Rhodospirillum fulvum* group and its photosynthetic pigments. Arch Mikrobiol *51:* 258–266.

— and K. D. Lippert. 1966 Über das Vitamin B₁₂-Bedürfnis phototropher Schwefelbakterien. Arch Mikrobiol *55:* 245–256.

—, M. C. Markham and S. Liaaen-Jensen. 1968 Carotenoids of *Thiorhodaceae*. 8. Isolation and characterization of a *Thiothece*, *Lamprocystis* and *Thiodictyon* strain and their carotenoid pigments. Arch Mikrobiol *62:* 178–191.

—. and H. G. Trüper. 1969 Proposal to declare *Rhodopseudomonas spheroides* and *Chloropseudomonas ethylica* as *nomina conservanda*. Int J Syst Bacteriol *19 (2):* 153–154.

— and —. 1970 Conservation of the family names *Thiorhodaceae* Molisch 1907, 27 and *Athiorhodaceae* Molisch 1907, 28 and designation of *Chromatium* Perty 1852, 174 as the type genus of the *Thiorhodaceae* and *Rhodospirillum* Molisch 1907, 24 as the type genus of the *Athiorhodaceae*. A recommendation to the Judicial Commission. Int J Syst Bacteriol *20 (1):* 31–33.

— and —. 1971 New nomenclatural combinations in the phototrophic sulfur bacteria. Int J Syst Bacteriol *21 (1):* 11–14.

— and —. 1971 Conservation of the family name *Chromatiaceae* Bavendamm 1924 with the type genus *Chromatium* Perty 1852. Int J Syst Bacteriol *21 (1):* 15–16.

— and —. 1971 Higher taxa of the phototrophic bacteria. Int J Syst Bacteriol *21 (1):* 17–18.

— and —. 1971 Type and neotype strains of the species of phototrophic bacteria maintained in pure culture. Int J Syst Bacteriol *21 (1):* 19–24.

Philip, C. B. 1943 Nomenclature of the pathogenic rickettsiae. Amer J Hyg *37:* 301–309.

—. 1948 Comments on the name of the Q-fever organism. Pub Health Rep *63:* 58.

—. 1950 Miscellaneous human rickettsioses. *In* R. L. Pullen, Communicable Diseases. Lea and Febiger Co., Philadelphia, pp. 781–788.

—. 1953 Nomenclature of the rickettsiaceae pathogenic to vertebrates. Ann N Y Acad Sci *56:* 484–494.

—. 1955 Changes in the Classification of order *Rickettsiales* p. 271. Symposium on Taxonomy. Bacteriol Rev *19 (4):* 270–274.

—. 1956 Comments on the classification of the order *Rickettsiales*. Can J Microbiol *2:* 261–270.

—. 1957 Class III. Microtatobiotes Philip 1956. *In* Breed, Murray and Smith (Editors), Bergey's Manual of Determinative Bacteriology, 7th ed. The Williams & Wilkins Co. Baltimore, p. 933.

—. 1957 Family IV. *Anaplasmataceae* Philip, fam. nov. *In* Breed, Murray and Smith (Editors), Bergey's Manual of Determinative Bacteriology, 7th ed. The Williams & Wilkins Co., Baltimore, pp. 980–984.

—. 1962 Appendix G. Summary of tickborne rickettsioses. Rept 2nd Meeting FAO/OIE Expert Panel Tick-borne Dis. Livestock, Cairo. United Nations, Rome 1962, pp. 41–43.

—, W. J. Hadlow and L. E. Hughes. 1953 *Neorickettsia helmintheca*, a new rickettsia-like disease agent in dogs in western United States transmitted by a helminth. Riass Comun VI Congr Int Microbiol, Roma *2:* 256–257.

—, — and —. 1954 Studies on salmon poisoning disease of canines. I. The rickettsial relationships and pathogenicity of *Neorickettsia helmintheca*. Exp Parasitol *3:* 336–350.

—, L. E. Hughes, B. Locker and W. J. Hadlow. 1954 Studies on salmon poisoning disease of canines. II. Further observations on etiologic agent. Proc Soc Exp Biol Med *87:* 397–400.

Phillips, H. C. 1953 Characterization of the soil globiforme bacteria. Iowa State J Sci *27:* 240–241.

Phillips, J. E. 1961 The commensal role of

Actinobacillus lignieresi. J Pathol Bacteriol *82:* 205–208.

—. 1964 Commensal actinobacilli from the bovine tongue. J Pathol Bacteriol *87:* 442–444.

—. 1967 Antigenic structure and serological typing of *Actinobacillus lignieresi.* J Pathol Bacteriol *93:* 463–475.

Pichinoty, F., C. Rigano, J. Bigliardi-Rouvier, L. Le Minor and M. Piéchaud. 1966 Recherche des nitrate-reductases A et B chez les *Enterobacteriaceae.* Ann Inst Pasteur (Paris) *110:* 126–130.

Pickett, M. J. and F. J. Cabelli. 1953 The precipitating antigens of Friedländer's bacillus. J Gen Microbiol *9:* 249–256.

Piéchaud, D., M. Piéchaud and L. Second. 1951 Étude de 26 souches de *Moraxella lwoffi.* Ann Inst Pasteur (Paris) *80:* 97–99.

—, — and —. 1956 Variétés protéolytiques de *Moraxella lwoffi* et de *Moraxella glucidolytica* (*Bact. anitratum*). Ann Inst Pasteur (Paris) *90:* 517–522.

Pierce, N. B. 1901 Walnut bacteriosis. Bot Gaz *31:* 272–273.

Pierce-Chase, C. H., R. M. Fauve and R. Dubos. 1964 Corynebacterial pseudotuberculosis in mice. I. Comparative susceptibility of mouse strains to experimental infections with *Corynebacterium kutscheri.* J Exp Med *120:* 267–281.

Pijper, A. and B. D. Pullinger. 1927 South African Nocardioses. J Trop Med Hyg *30:* 153–156.

Pike, J. E., P. F. Wiley and L. Slechta. 1965 Derivatives of 7-D-ribofuranosyl-pyrrolo-pyrimidines. United States Patent 3,167,540, January 26.

Pilcher, K. S., W. G. Wu and O. H. Muth. 1961 Studies on the morphology and respiration of *Anaplasma marginale.* Amer J Vet Res *22:* 298–307.

Pillai, S. C. 1938 A biochemical investigation of the tuberculation of water pipes. Proc Nat Inst Sci India *4:* 295–318.

Pillot, J. 1965 Contribution à l'étude du genre *Treponema.* Structures anatomique et antigènique. Lons-le-Saunier, Paris.

— and M. A. Ryter. 1965 Structure des spirochètes. I. Étude des Genres. *Treponema, Borrelia* et *Leptospira* au microscope électronique. Ann Inst Pasteur (Paris) *108:* 791–804.

Pilone, G. J., R. E. Kunkee and A. D. Webb. 1966 Chemical characterization of wines fermented with various malo-lactic bacteria. Appl Microbiol *14:* 608–615.

Pine, L. and C. J. Boone. 1967 Comparative cell wall analyses of morphological forms within the genus *Actinomyces.* J Bacteriol *94:* 875–883.

— and L. Georg. 1965 The classification and phylogenetic relationships of the *Actinomycetaceae.* Int Bull Bacteriol Nomencl Taxon *15 (3):* 143–163.

— and —. 1969 Reclassification of *Actinomyces propionicus.* Int J Syst Bacteriol *19 (3):* 267–272.

— and A. Howell, Jr. 1956 Comparison of physiological and biochemical characters of *Actinomyces* spp. with those of *Lactobacillus bifidus.* J Gen Microbiol *15 (3):* 428–445.

—, — and S. J. Watson. 1960 Studies of the morphological, physiological and biochemical

characters of *Actinomyces bovis.* J Gen Microbiol *23 (3):* 403–424.

— and J. R. Overman. 1966 Differentiation of capsules and hyphae in clubs of bovine sulphur granules. Sabouraudia *5:* 141–143.

Pinkerton, H. 1936 Criteria for the accurate classification of the rickettsial disease (Rickettsioses) and their etiological agents. Parasitology *28:* 172–189.

Pinnert-Sindico, S. 1954 Une nouvelle espèce de *Streptomyces* productrice d'antibiotiques; *Streptomyces ambofaciens* n. sp., caractères culturaux. Ann Inst Pasteur (Paris) *87:* 702–707.

Pinoy, E. 1913 Actinomycoses et Mycetomes. Bull Inst Pasteur *11:* 929–938; 977–984.

Pinto, M. A. 1945 Caractéristiques d'une souche de *Borrelia recurrentis* isolée au Portugal. Arq Inst Bacteriol Cam Pestana *9:* 224–227.

Pirie, J. H. H. 1927 A new disease of veld rodents; "Tiger River Disease". S Afr Inst Med Res Annu Rep *3:* 163–186.

—. 1940 The genus *Listerella* Pirie. Science (Washington) *91:* 383.

Pittman, K. A. and M. P. Bryant. 1964 Peptides and other nitrogen sources for growth of *Bacteroides ruminicola.* J Bacteriol *88 (2):* 401–410.

Pittman. M. 1953 A classification of the hemolytic bacteria of the genus *Haemophilus: Haemophilus haemolyticus* Bergey et al. and *Haemophilus parahaemolyticus* nov. spec. J Bacteriol *65:* 750–751.

—. 1970 *Bordetella pertussis*—bacterial and host factors in the pathogenesis and prevention of whooping cough. *In* Mudd (Editor), Infectious agents and host reactions. W. B. Saunders Co., Philadelphia, pp. 234–270.

— and D. J. Davis. 1950 Identification of the Koch-Weeks bacillus (*H. aegyptius*). J Bacteriol *59:* 413–426.

Pivnick, H. 1955 *Pseudomonas rubescens,* a new species from soluble oil emulsions. J Bacteriol *70 (1):* 1–6.

— and L. R. Sabina. 1957 Studies of *Aeromonas formicans* Crawford comb. nov. from soluble oil emulsions. J Bacteriol *73 (2):* 247–252.

Place, E. H., L. E. Sutton and O. Willner. 1926 Erythema arthriticum epidemicum; preliminary report. Boston Med Surg J *194:* 285–287.

Plagge and B. Proskauer. 1887 Bericht über die Untersuchung des Berliner Leitungswassers. Z Hyg Infektionskr *2:* 401–490.

Plastridge, W. N. and S. E. Hartsell. 1937 Biochemical and serological characteristics of streptococci of bovine origin. J Infect Dis *61:* 110–121.

Plotz, H. 1915 Bacteriological studies. *In* Plotz, Olitsky and Baehr, The etiology of typhus exanthematicus. J Infect Dis *17:* 1–68.

Plum, N. 1940 On corynebacterial infections in swine. Cornell Vet *30:* 14–20.

Pochon, J. 1942 Fermentation de la cellulose par un anaérobie thermophile. Ann Inst Pasteur (Paris) *68:* 353–354.

Pohl, F. 1892 Ueber Kultur und Eigenschaften einiger Sumpfwasser-Bacillen und über die Anwendung alkalischer Nährgelatine. Zentrabl Bakteriol Parasitenk Infektionskr Hyg Abt II *11:* 141–145.

Poindexter, J. S. 1964 Biological properties and

classification of the Caulobacter group. Bacteriol Rev *28:* 231–295.

— and R. F. Lewis. 1966 Recommendations for the revision of the taxonomic treatment of stalked bacteria. Int J Syst Bacteriol *16 (4):* 377–382.

Pollak, A. and V. B. Buhler. 1955 The cultural characteristics and animal pathogenicity of an atypical acid-fast organism which causes human disease. Amer Rev Tuberc *71:* 74–87.

Pon, D. S., C. E. Townsend, G. E. Wessman, C. G. Schmitt and C. H. Kingsolver. 1954 A Xanthomonas parasitic on uredia of cereal rusts. Phytopathology *44:* 707–710.

Pongratz, E. 1957 D'une bactèrie pediculée isolée d'un pus de sinus. Schweiz Z Allg Pathol Bakteriol *20:* 593–608.

Porter, J. N., R. I. Hewitt, C. W. Hesseltine, G. Krupka, J. A. Lowery, W. S. Wallace, N. Bohonos and J. H. Williams. 1952 Achromycin: A new antibiotic having trypanocidal properties. Antibiot Chemother *2 (8):* 409–410.

Porter, R., C. S. McCleskey and M. Levine. 1937 The facultative sporulating bacteria producing gas from lactose. J Bacteriol *33:* 163–183.

Postgate, J. R. 1963 A strain of *Desulfovibrio* able to use oxamate. Arch Mikrobiol *46:* 287–295.

—. 1967 Report of the Subcommittee on Sulfate-reducing bacteria (1962–1966) to the International Committee on Nomenclature of Bacteria. Int J Syst Bacteriol *17 (2):* 111–112.

— and L. L. Campbell. 1963 Identification of Coleman's sulfate-reducing bacterium as a mesophilic relative of *Clostridium nigrificans.* J Bacteriol *86 (2):* 274–279.

— and —. 1966 Classification of *Desulfovibrio* species, the nonsporulating sulfate-reducing bacteria. Bacteriol Rev *30:* 732–738.

Potekhina, L. I. 1965 *Streptosporangium rubum* n. sp. A new species of the *Streptosporangium* genus. Mikrobiologiya *34:* 292–299.

Potel, J. 1950 Die Morphologie, Kultur und Tierpathogenität des *Corynebacterium infantisepticum.* Zentrabl Bakteriol Parasitenk Infektionskr Hyg Abt I Orig *156:* 490–493.

Poulsen, V. A. 1879 Om nogle mikroscopiske Planteorganismer. Vidensk Medd Naturhist Forenkjobenhavn *1879–80:* 231–254.

Poyza, N. and E. D. Karn. 1956 The degradation of heparin by bacterial enzymes. I. Adaptation and lyophilized cells. J Biol Chem *223 (2):* 853–858.

Prabhakaran, K. W., W. F. Kirchheimer and E. B. Harris. 1968 Oxidation of phenolic compounds by *Mycobacterium leprae* and inhibition of phenolase by substrate analogues and copper chelators. J Bacteriol *95:* 2051–2053.

Pradip, I. S., A. D. Larson and C. S. McCleskey. 1966 Nutritional factors affecting growth and pigmentation of *Corynebacterium equi.* Bacteriol Proc, p. 20.

—, — and —. 1966 Glucose stimulation of *Corynebacterium equi* and *C. hoagii.* Bacteriol Proc, p. 25.

Prauser, H., M. P. Lechevalier and H. Lechevalier. 1970 Description of *Oerskovia* gen.n. to harbor Ørskov's motile *Nocardia.* Appl Microbiol *19:* 534.

— and S. Momirova. 1970 Phagensensibilität Zellwandzusammensetzung und Taxonomie

einiger thermophiler Actinomyceten. Z Allg Mikrobiol *10 (3):* 219–222.

—, L. Müller and R. Falta. 1967 On the taxonomic position of the genus *Microellobosporia* Cross, Lechevalier and Lechevalier 1963. Int J Syst Bacteriol *17 (4):* 361–366.

Prausnitz, C. 1922 Der *Bacillus mucosus anaerobius.* Zentrabl Bacteriol Parasitenk Infektionskr Hyg Abt I Orig *89:* 126–132.

Präve, P. 1957 Untersuchungen über die Stoffwechselphysiologie des Eisenbakteriums *Leptothrix ochracea* Kützing. Arch Mikrobiol *27:* 33–62.

Prazmowski, A. 1880 Untersuchung uber die Entwickelungsgeschichte und Fermentwirking einiger Bacterien-Arten. Inaug Diss Hugo Voigt, Leipzig, pp. 1–58.

Prebble, J. 1968 The carotenoids of *Corynebacterium fascians* Strain 2Y. J Gen Microbiol *52:* 15–24.

Preer, L. B., A. Jurand, J. R. Preer and B. M. Rudman. 1972 The classes of Kappa in *Paramecium aurelia.* J Cell Sci *11:* 581.

Preobrazhenskaya, T. P. 1966 Characteristics of actinomycetes-antagonists of azureus section (in Russian). Antibiotiki *11 (9):* 849–861.

—, E. S. Kudrina, T. S. Maksimova, M. A. Sveshnikova and R. V. Boyarskaya. 1960 Electron microscopy of spores of various actinomycete species (in Russian). Mikrobiologiya *29 (1):* 51–55.

—, T. S. Maksimova and N. O. Blinov. 1964 A study of green pigments from some actinomycetous species by the method of paper chromatography (in Russian). Antibiotiki *9 (11):* 963–970.

Prévot, A. R. 1924 Les Streptocoques Anaérobies. Thése. Amidie Legrand, Editeurs, Paris, pp. 1–144.

—. 1924 *Diplococcus constellatus* (n. sp.). C R Soc Biol Paris *91:* 426–428.

—. 1925 Les streptocoques anaérobies. Ann Inst Pasteur (Paris) *39:* 417–447.

—. 1933 Études de systématique bactérienne. I. Lois générales. II. Cocci anaérobius. Ann Sci Natur Zool Biol Anim *15:* 23–260.

—. 1938 Études de systématique bactérienne. Ann Inst Pasteur (Paris) *60:* 285–307.

—. 1938 Études de systématique bactérienne. IV. Critique de la conception actuelle du genre *Clostridium.* Ann Inst Pasteur (Paris) *61 (1):* 72–91.

—. 1940 Un anaérobie strict reduisant les nitrates en nitrites *Eubacterium nitritogenes* n. sp. C R Soc Biol Paris *134:* 353–355.

—. 1940 Manual de classification et de détermination des bactéries anaérobies. Masson and Co., Paris, pp. 1–223.

—. 1941 Recherches biochimiques sur les streptococques anaérobies gazogènes. Consequences taxonomiques. C R Soc Biol *135:* 103–105.

—. 1941 Sur une nouvelle espèce de streptocoque anaérobie gazogènes: *Streptococcus productus* nov. spec. C R Soc Biol *135:* 105–107.

—. 1947 Étude de quelques bactéries anaérobies nouvelles ou mal connués. Ann Inst Pasteur (Paris) *73 (5):* 409–418.

—. 1948 Étude des bactéries anaérobies d'Afrique occidentale française (Sénégal, Guinée,

Côte d'Ivoire). Ann Inst Pasteur (Paris) *74:* 157–170.

—. 1948 Manuel de classification et de détermination des bacteries anaérobies, 2nd ed. Masson and Co., Paris, pp. 1–290.

—. 1957 Manuel de Classification et de Determination des Bactéries Anaérobies, 3rd ed. Masson and Co., Paris, pp. 1–362.

—. 1961 Traité de Systématique Bactérienne, Vol. 2, Dunod, Paris, pp. 1–771.

—. 1966 Manual for the classification and determination of the anaerobic bacteria. 1st Amer. ed., transl. by V. Fredette. Lea and Febiger, Philadelphia, pp. 1–402.

—, M. Digeon and M. Enescu. 1946 Étude des caractères biochimiques des streptocoques anaérobies non gazogènes ni fétides. C R Soc Biol *140:* 5–6.

—, —, M. Peyre, J. Pantaleon and J. Senez. 1947 Étude de quelques bactéries anaérobies nouvelles ou mal connués. Ann Inst Pasteur (Paris) *73:* 409–418.

— and E. Kirchheiner. 1938 Recherches sur la nature de la fermentation du glucose par *Fusiformis biacutus.* C R Soc Biol Paris *128:* 963–964.

— and J. Laplanche. 1947 Étude d'une bactérie anaérobie nouvelle de Guinée Francaise *Cillobacterium combesi* n. sp. Ann Inst Pasteur (Paris) *73 (2):* 687–688.

— and R. Loth. 1941 Recherches sur les caractères biochimiques de deux diplocoques anaérobies: *D. magnus* et *D. plagarum belli.* Conséquences taxonomiques. C R Soc Biol *135:* 609–611.

—, M. Raynaud and M. Digeon. 1947 Sur une nouvelle espèce anaérobie: *Ramibacterium pleuriticum.* Ann Inst Pasteur (Paris) *73 (1):* 481–483.

— and J. Taffanel. 1942 Recherches sur une nouvelle espèce anaérobie *Ramibacterium alactolyticum* (nov. spec.). Ann Inst Pasteur (Paris) *68:* 259–262.

— and —. 1945 Recherches sur un nouveau coccus anaérobie *Staphylococcus activus* (n. sp.) Ann. Inst Pasteur (Paris) *71:* 152–154.

—, P. Tardieux, L. Joubert and F. de Cadore. 1956 Recherches sur *Fusiformis nucleatus* (Knorr) et son pouvoir pathogène pour l'homme et les animaux. Ann Inst Pasteur (Paris) *91:* 787–798.

—, H. Thouvenot and P. Kaiser. 1959 Étude de douze souches anaérobies pectinolytiques de l'intestin des poissons et des boues d'eau douce et salée. Ann Inst Pasteur (Paris) *93:* 429–434.

—, —, M. Patrigalla and R. Silloc. 1956 Une nouvelle espèce anaérobie des lacs de Ruwesori: *Inflabilis lacustris* n. sp. Ann Inst Pasteur (Paris) *91:* 929–932.

—, —, J. Pitre and M. Bressou. 1954 Étude d'une espèce thermophile anaérobie nouvelle: *Cillobacterium thermophilum* n. sp. Ann Inst Pasteur (Paris) *86:* 776–778.

—, A. Turpin and P. Kaiser. 1967 Les Bactéries Anaérobies. Dunod, Paris, pp. 1–2188.

— and Zimmes-Chaverou. 1947 Étude d'une nouvelle espèce anaérobie de Côte d'Ivoire: *Inflabilis mangenoti.* Ann Inst Pasteur (Paris) *73:* 602–604.

Pribram, E. 1919 Der gegenwärtige Bestand der vorm. Králschen Sammlung von Mikroorganismen, Wien, pp. 1–148.

—. 1929 A contribution to the classification of microorganisms. J Bacteriol *18:* 361–394.

—. 1933 Klassification der Schizomyceten. F. Deuticke, Leipzig, pp. 1–143.

Pridham, T. G. 1964 Taxonomic studies of *Streptomyces griseus* (Krainsky) Waksman et Henrici: A species comprising many subspecies. Antimicrob Agents Chemother *1963:* 104–115.

—. 1965 Color and streptomycetes. Report of an international workshop on determination of color of streptomycetes. Appl Microbiol *13 (1):* 43–61.

—. 1970 New names and new combinations in the order Actinomycetales Buchanan 1917. U S Dept Agr Tech Bull *1424:* 1–55.

— and D. Gottlieb. 1948 The utilization of carbon compounds by some *Actinomycetales* as an aid for species determination. J Bacteriol *56 (1):* 107–114.

—, C. W. Hesseltine and R. G. Benedict. 1958 A guide for the classification of streptomycetes according to selected groups: placement of strains in morphological sections. Appl Microbiol *6 (1):* 52–79.

— and A. J. Lyons, Jr. 1961 *Streptomyces albus* (Rossi-Doria) Waksman et Henrici: Taxonomic study of strains labelled *Streptomyces albus.* J Bacteriol *81:* 431–441.

— and —. 1962 Proposal to designate Strain ATCC 3004 (IMRU 3004) as the neotype strain of *Streptomyces albus* (Rossi-Doria) Waksman and Henrici. Int Bull Bacteriol Nomencl Taxon *12 (3):* 123–126.

— and —. 1965 Further taxonomic studies on straight to flexuous streptomycetes. J Bacteriol *89 (2):* 331–342.

— and —. 1969 Progress in clarification of the taxonomic and nomenclatural status of some problem actinomycetes. Develop Ind Microbiol *10:* 183–221.

—, — and H. L. Seckinger. 1965 Comparison of some dried holotype and neotype specimens of streptomycetes with their living counterparts. Int Bull Bacteriol Nomencl Taxon *15 (4):* 191–237.

—, G. St. Julian, Jr., G. L. Adams, H. H. Hall and R. W. Jackson. 1964 Infection of *Popillia japonica* Newman larvae with vegetative cells of *Bacillus popilliae* Dutky and *Bacillus lentimorbus* Dutky. J Insect Pathol *6:* 204–213.

—, O. L. Shotwell, F. H. Stodola, L. A. Lindenfelser, R. G. Benedict and R. V. Jackson. 1956 Antibiotics against plant disease. II. Effective agents produced by *Streptomyces cinnamomeus* forma *azacoluta* f. nov. Phytopathology *46:* 575–581.

Price, W. H. 1953 A quantitative analysis of the factors involved in the variations in virulence of rickettsiae. Science (Washington) *118:* 49–52.

Pringsheim, E. G. 1949 Iron bacteria. Biol Rev (Cambridge) *24:* 200–245.

—. 1949 The relationship between bacteria and *Myxophyceae.* Bacteriol Rev *13 (2):* 47–98.

—. 1949 The filamentous bacteria *Sphaerotilus, Leptothrix, Cladothrix* and their relation to iron and manganese. Trans Roy Soc (London) Ser B *233:* 453–482.

—. 1950 The bacterial genus *Lineola*. J Gen Microbiol *4:* 198–209.

—. 1951 The *Vitreoscillaceae:* a family of colourless, gliding, filamentous organisms. J Gen Microbiol *5:* 124–149.

—. 1953 Die Stellung der grünen Bakterien im System der Organismen. Arch Mikrobiol *19:* 353–364.

—. 1955 *Lampropedia hyalina* Schroeter 1886 and *Vannielia aggregata* n. g., n. sp., with remarks on natural and organized colonies in bacteria. J Gen Microbiol *13:* 285–291.

—. 1957 Observations on *Leucothrix mucor* and *Leucothrix cohaerens* nov. sp. Bacteriol Rev *21 (2):* 69–76.

—. 1964 Heterotrophism and species concepts in *Beggiatoa*. Amer J Bot *51:* 898–913.

—. 1967 Die mixotrophie von *Beggiatoa*. Arch Mikrobiol *59:* 247–254.

— and C. F. Robinow. 1947 Observations on two very large bacteria, *Caryophanon latum* Peshkoff and *Lineola longa (nomen provisiorum).* J Gen Microbiol *1:* 267–278.

Prissick, F. H. and A. M. Masson. 1956 Cervical lymphadenitis in children caused by chromogenic mycobacteria. Can Med Ass J *75:* 798–803.

— and —. 1957 Yellow-pigmented pathogenic mycobacteria from cervical lymphadenitis. Can J Microbiol *3:* 91–100.

Pritchett, I. W. and E. G. Stillman. 1919 The occurrence of bacillus influenzae in throats and saliva. J Exp Med *29:* 259–266.

Prokop, J. F. 1964 Method for the preparation of a composition of matter having antitumor and antifungal activity. Offic Gaz U S Patent Off Patent 3,117,916.

Proom, H. and B. C. J. G. Knight. 1950 *Bacillus pantothenticus* (n. sp.). J Gen Microbiol *4:* 539–541.

— and —. 1955 The minimal nutritional requirements of some species in the genus *Bacillus*. J Gen Microbiol *13 (3):* 474–480.

Protina, J., V. Schwarz and K. Syhara. 1964 Steroid derivatives. XXX. Transformation of steroids by microorganisms of the genus *Flavobacterium*. Folia Microbiol *9 (4):* 218–221.

Provost, P. J. and R. N. Doetsch. 1962 An appraisal of *Caryophanon latum*. J Gen Microbiol *28:* 547–557.

Pugh, E. L., M. K. Wassef and M. Kates. 1971 Inhibition of fatty acid synthetase in *Halobacterium cutirubrum* and *E. coli* by high salt concentrations. Can J Biochem *49:* 953–958.

Pugh, G. W., D. E. Hughes and T. J. McDonald. 1966 The isolation and characterisation of *Moraxella bovis*. Amer J Vet Res *27:* 957–962.

Puntoni, V. 1937 Sulle relazione fra il *B. bifido* e gl i attinomiceti anaerobi typo Wolff-Israel. Ann Igiene (Sperimentale) *47:* 157–168.

— and D. Leonardi. 1935 Sulla sistematica degli Attinomiceti *Asteroides* n. g. Boll Acad Med (Roma) *61:* 90–94.

Purcell, R. H., R. M. Chanock and D. Taylor-Robinson. 1969 Serology of the mycoplasmas of man. *In* Hayflick (Editors), The Mycoplasmatales and the L-Phase of Bacteria. Appleton-Century-Crofts, New York, pp. 221–264.

Purdom, M. R. 1963 Micromanipulation in the examination of rumen bacteria. Nature (London) *198:* 307–308.

Qadri, S. M. H. and D. S. Hoare. 1968 Formic hydrogen lyase and the photoassimilation of formate by a strain of *Rhodopseudomonas palustris*. J Bacteriol *95 (6):* 2344–2357.

Quadling, C. 1967 Evaluation of tests and groupings of cultures by a two-stage principle component method. Can J Microbiol *13:* 1379–1394.

— and S. M. Martin. 1968 Organization of information about micro-organisms. Progr Ind Microbiol *7:* 125–148.

Quehl, A. 1906 Untersuchungen über die Myxobacterien. Zentrabl Bakteriol Parasitenk Infektionskr Hyg Abt II *16:* 9–34.

Quick, C. R. 1944 Effects of snowbrush on the growth of Sierra gooseberry. J Forest *42:* 827–832.

Quincke, G. 1967 Untersuchungen über die O-Antigene der Plesiomonaden. Arch Hyg Berl *151:* 525–529.

Rabenhorst, L. 1854 Ein Notisblatt fur kryptogamische Studien. Hedwigia *1 (9):* 41–56.

—. 1863 Kryptogamen-Flora von Sachsen der Ober-Lausitz Thüringen und Nordbohmen mit Berücksichtung der benachbarten Länder Erste Abteilung, E. Kummer, Leipzig, p. 56.

—. 1865 Flora Europaea Algarum aquae dulcis et submarinae. E. Kummer, Leipzig, Sec. II, pp. 1–319.

—. 1868 Flora Europaea Algarum. Aquae dulcis et submarinae. Sectio III, E. Kummer, Leipzig, pp. 1–320.

Rafyi, A., O. Felsenfeld, J. Dupont and G. Maghami. 1965 Studies in *Borreliae*. Part I. A variant of *Borrelia parkeri*. Davis 1942 isolated in California and its tick vector. Ann Parasitol Hum Comp *40 (6):* 631–637.

— and P. H. Vercammen-Grandjean. 1964 Sur la morphologie et la transmission d'*Eperythrozoon coccoides*. Ann Inst Pasteur (Paris) *106:* 938–942.

Raghavarhari, K. and A. M. K. Reddy. 1959 *Cytoecetes ovis* var. *decani* (n. sp.) as the cause of tick-borne fever in sheep in India. Indian J Vet Sci *29:* 69–86.

Rahalkar, P. W. and M. J. Thirumalachar. 1968 Cultural characters and identity of some *Streptoverticillium* species producing polyene antibiotics. Hindustan Antibiot Bull *11 (2):* 90–96.

Rahn, O. 1937 New principles for the classification of bacteria. Zentrabl Bakteriol Parasitenk Infektionskr Hyg Abt II *96:* 273–286.

Raj, H. D. 1970 A new species-*Microcyclus flavus*. Int J Syst Bacteriol *20:* 61–81.

—, F. L. Duryee, A. M. Deeney, C. H. Wang, A. W. Anderson and P. R. Elliker. 1960 Utilization of carbohydrates and amino acids by *Micrococcus radiodurans*. Can J Microbiol *6 (3):* 289–298.

Rajagopalan, C. K. S. and G. Rangaswami. 1958 Bacterial leafspot of *Pennisetum typhoides*. Curr Sci *27:* 30–31.

Rake, G. 1948 Family III. *Chlamydozoaceae* Moshkovskiy. *In* Breed, Murray and Hitchens (Editors), Bergey's Manual of Determinative Bacteriology, 6th ed. The Williams & Wilkins Co., Baltimore, pp. 114–120.

—. 1957 Family II. *Chlamydiaceae* Fam. Nov. *In* Breed, Murray and Smith (Editors), Bergey's Manual of Determinative Bacteriology, 7th ed. The Williams & Wilkins Co., Baltimore pp. 957–968.

Ramaley, R. F. and J. Hixson. 1970 Isolation of a nonpigmented, thermophilic bacterium similar to *Thermus aquaticus.* J Bacteriol *103* (*2*): 527–528.

Ramalingam, M., H. D. Lewin, K. Sivapraksam and G. S. Krishnamurthy. 1965 Bacterial wilt of cotton (*Gossypium hirsutum* L. race *latifolium*) caused by *Xanthomonas celebensis* var. *gossypii.* Indian J Microbiol *5:* 51–54.

Ramamurthi, C. S. 1959 Comparative studies of some Gram-positive phytopathogenic bacteria and their relationship to the corynebacteria. Mem Cornell Univ Agr Exp Sta No 366.

Rancourt, M. and H. A. Lechevalier. 1963 Electron microscopic observation of the sporangial structure of an actinomycete, *Microellobosporia flavea.* J Gen Microbiol *31:* 495–498.

Rangaswami, G. 1962 Bacterial plant diseases in India. Asia Publishing House, New York, pp. i-xii; 1–163.

—. 1964 Occurrence of two bacterial plant diseases in South India. Curr Sci *33:* 286–287.

— and K. S. S. Eswaran. 1961 A bacterial leaf-spot disease of Jasmine. Curr Sci *30:* 352.

— and —. 1962 A bacterial leafspot disease of Bhendi or Okra. Andhra Agr J *9:* 1–2.

— and S. S. Gowda. 1963 On some bacterial diseases of ornamentals and vegetables in Madras State. Indian Phytopathol *16:* 74–85.

—, N. N. Prasad and K. S. S. Eswaran. 1961 A bacterial leaf blotch disease of cumbu (*Pennisetum typhoides* Stapf & Hubbard). Madras Agr J *48:* 180–181.

—, — and —. 1961 Bacterial leaf-spot diseases of *Eleusine coracana* and *Setaria italica* in Madras State. Indian Phytopathol *14:* 105–107.

—, — and —. 1961 A new bacterial leaf-spot disease of maize. Madras Agr J *48:* 392–393.

—, — and —. 1961 Two new bacterial diseases of sorghum. Andhra Agr J *8:* 269–272.

— and M. Ragarajan. 1965 A bacterial leaf spot disease of banana. Phytopathology *55:* 1035–1036.

Rao, K. V. 1960 E-73: An antitumor substance. Part II. Structure. J Amer Chem Soc *82* (*5*): 1129–1132.

—. 1962 E-73, an antitumor substance. III. Some derivatives. Antibiot Chemother *12* (*2*): 123–127.

—. 1963 3-2-(5-acetoxy-2-hydroxy-3,5-dimethylcyclohexy 1)-2-hydroxyethyl glutarimide and derivatives thereof. United States Patent 3,095,418, June 25.

— and W. P. Cullen. 1958 E-73, an antitumor substance. Isolation and characterization. Abstracts of papers, 134th meeting American Chemical Society, Sept. 7–12, Chicago, Illinois, *134:* 22-O-23-O. Abstract 36.

— and —. 1960 E-73: An antitumor substance. Part I. Isolation and characterization. J Amer Chem Soc *82* (*5*): 1127–1128.

—, W. S. Marsh and S. C. Brooks. 1964 Antibiotic narangomycin and method of production. United States Patent 3,155,583.

Rao, U. K. and P. L. Narasimha Rao. 1967 Actinomycetes Part II. Purification and pharmacological properties of champamycin-B from *Streptomyces champavatii.* Indian J Exp Biol *5* (*1*): 39–43.

Rao, Y. P. and S. K. Mohan. 1970 A new bacterial leaf stripe disease of arecanut (*Areca catechu*) in Mysore State. Indian Phytopathol *23:* 702–704.

Rauss, K. F. 1936 The systematic position of Morgan's bacillus. J Pathol Bacteriol *42:* 183–192.

—. 1962 A proposal for the nomenclature and classification of the *Proteus* and Providence groups. Int Bull Bacteriol Nomencl Taxon *12:* 53–64.

— and S. Vörös. 1959 The biochemical and serological properties of *Proteus morganii.* Acta Microbiol Acad Sci Hung. *6:* 233–248.

— and —. 1967 Five new serotypes of *Morganella morganii.* Acta Microbiol Acad Sci Hung *14:* 195–198.

Rautenshtein, Ya. I. 1960 Biology of special groups of actinomycetes. Producers of antibiotics. Tr Inst Microbiol Akad Nauk SSSR *8:* 1–344.

Ravenel, M. P. 1896 Notes on the bacteriological examination of the soil of Philadelphia. Mem Nat Acad Sci *8:* 3–41.

Ravin, A. W. and J. D. H. De Sa. 1964 Genetic linkage of mutational sites affecting similar characters in pneumococcus and streptococcus. J Bacteriol *87* (*1*): 86–96.

Raymond, J. C. and W. R. Sistrom. 1967 The isolation and preliminary characterization of a halophilic photosynthetic bacterium. Arch Mikrobiol *59:* 255–268.

— and —. 1969 *Ectothiorhodospira halophila,* a new species of the genus *Ectothiorhodospira.* Arch Mikrobiol *69* (*2*): 121–126.

Razin, S. 1968 Mycoplasma taxonomy studied by electrophoresis of cell proteins. J Bacteriol *96:* 687–694.

— and C. Boschwitz. 1968 The membrane of the *Streptobacillus moniliformis* L-phase. J Gen Microbiol *54:* 21–32.

— and S. Rottem. 1963 Fatty acid requirements of *Mycoplasma laidlawii.* J Gen Microbiol *33* (*3*): 459–470.

Razumov, A. S. 1949 *Gallionella Kljasmensis* sp. n. a component of the bacteriological plankton. Mikrobiologiya *18:* 442–446.

Reader, V. B. 1926 The identification of the so-called *B. mycoides corallinus* as a *Streptothrix.* J Pathol Bacteriol *29:* 1–4.

Reddish, G. and L. Rettger. 1922 *Clostridium putrificum* (*B. putrificus* Beinstock) a distinct species. Abstr Bacteriol *6:* 9.

Reddy, C. A., M. P. Bryant and M. J. Wolin. 1972 Characteristics of S organism isolated from *Methanobacillus omelianskii.* J Bacteriol *109* (*2*): 539–545.

Reddy, C. S., J. Godkin and A. G. Johnson. 1924 Bacterial blight of rye. J Agr Res *28:* 1039–1040.

Redfearn, M. S., N. J. Palleroni and R. Y. Stanier. 1966 A comparative study of *Pseudomonas pseudomallei* and *Bacillus mallei.* J Gen Microbiol *43* (*2*): 293–313.

Redinger, K. 1931 *Siderocapsa coronata* Redinger, eine neue Eisenbakterie aus dem Lunzer Obersee. Arch Hydrobiol *22:* 410–414.

Redmond, D. L. and E. Kotcher. 1963 Cultural and serological studies on *Haemophilus vaginalis.* J Gen Microbiol *33* (*1*): 77–87.

Reed, G. B. 1939 Genus III. *Proactinomyces* Jensen. *In* Bergey, Breed, Murray and Hitchens (Editors), Bergey's Manual of Determinative Bacteriology, 5th ed. The Williams & Wilkins Co., Baltimore, pp. 831–839.

—. 1957 Family I *Mycobacteriaceae* Chester 1897. *In* Breed, Murray and Smith (Editors), Bergey's Manual of Determinative Bacteriology, 7th ed. The Williams & Wilkins Co., Baltimore, pp. 695–713.

Rees, R. J. W., K. R. Chatterjee, J. Pepys and R. D. Tee. 1965 Some immunologic aspects of leprosy. Amer Rev Resp Dis *92:* 139–149.

—, R. C. Valentine and P. C. Wong. 1960 Application of quantitative electron microscopy to the study of *Mycobacterium lepraemurium* and *M. leprae.* J Gen Microbiol *22:* 443–457.

—, M. F. R. Waters, A. G. M. Wedell and E. Palmer. 1967 Experimental lepromatous leprosy. Nature (London) *215:* 599–602.

Regendanz, P. and W. Kikuth. 1928 Sur la *Bartonella muris ratti* (Mayer). C R Soc Biol (Paris) *98:* 1578–1579.

Reich, C. V. and J. H. Hanks. 1961 Use of *Arthrobacter terregens* for bioassay of Mycobactin. J Bacteriol *87:* 1317–1320.

Reichenbach, H. 1962 Über verschiedene Arten von Cysten mustern bei *Chondrococcus coralloides* (*Myxobacteriales*). Ber Deut Bot Ges *75:* 85–90.

—. 1966 *Myxococcus* spp. (*Myxobacteriales*). *In* Wolf (Editor) Encyclopedia Cinematographica. Inst Wiss Film, Göttingen, pp. 557–578.

—. 1970 *Nannocystis exedens*, gen. nov., spec. nov., a new myxobacterium of the family *Sorangiaceae.* Arch Mikrobiol *70:* 119–138.

— and M. Dworkin. 1969 Studies on *Stigmatella aurantiaca* (*Myxobacterales*). J Gen Microbiol *58 (1):* 3–14.

— and —. 1970 Induction of myxospore formation in *Stigmatella aurantiaca* (*Myxobacterales*) by monovalent cations. J Bacteriol *101 (1):* 325–326.

— and H. Voelz. 1969 Fine structure of fruiting bodies of *Stigmatella aurantiaca* (*Myxobacterales*). J Bacteriol *99 (3):* 856–866.

—, — and M. Dworkin. 1969 Structural changes in *Stigmatella aurantiaca* during myxospore induction. J Bacteriol *97 (2):* 905–911.

Reiderer-Henderson, M. A. and P. W. Wilson. 1970 Nitrogen-fixation by sulphate-reducing bacteria. J Gen Microbiol *61 (1):* 27–32.

Reinhold, L., E. M. Barnes and H. Beerens. 1967 Identification de *Ristella biacutus* (Prévot 1967) (ex *Fusiformis biacutus* Prévot) avec *Clostridium microsporum* (Spray 1947). Bull Off Int Epiz *67:* 7–8.

Reinke, J. and G. Berthold. 1879 Die zersetzung der Kartoffel durch Pilze. Untersuch Botan Lab Univ Gottingen *1:* 1–100.

Remsen, C. C., S. W. Watson, J. B. Waterbury and H. G. Trüper. 1968 Fine structure of *Ectothiorhodospira mobilis* Pelsh. J Bacteriol *95 (6):* 2374–2392.

Renco, P. 1942 Ricerche su un fermento lattico sporigeno (*Bac. thermoacidificans*). Ann Microbiol *2:* 109–114.

Renoux, G. 1952 Une nouvelle "espèce" de *Brucella: B. intermedia.* Ann Inst Pasteur (Paris) *83:* 814–815.

Repaci, G. 1910 Contribution à l'étude de la flore bactérienne anaérobic des gangrènes pulmonaires. Un streptobacille anaérobie. C R Soc Biol Paris *68:* 216–218; 292–293; 410–412.

Report of the Enterobacteriaceae Subcommittee of the Nomenclature Committee of the IAMS. 1958 Recommended biochemical methods for group differentiation within the *Enterobacteriaceae.* Int Bull Bacteriol Nomencl Taxon *8 (1):* 25–70.

Rettger, L. F. 1909 Further studies on fatal septicemia in young chickens, or "white diarrhea." J Med Res *21:* 115–123.

Reuter, G. 1963 Vergleichenden Untersuchung über die Bifidus-Flora im Sauglings und Erwachsenenstuhl. Zentrabl Bakteriol Parasitenk Infektionskr Hyg Abt I Orig *191:* 486–507.

—. 1971 Designation of type strains for *Bifidobacterium* species. Int J Syst Bacteriol *21 (4):* 273–275.

Reyn, A. 1970 Taxonomic position of *Neisseria haemolysans* (Thjøtta and Bøe 1938). Int J Syst Bacteriol *20 (1):* 19–22.

—, A. Birch-Andersen and U. Berger. 1970 Fine structure and taxonomic position of *Neisseria haemolysans* (Thjøtta and Bøe 1938) or *Gemella haemolysans* (Berger 1960). Acta Pathol Microbiol Scand Sec B Microbiol Immunol *78B (3):* 375–389.

—, — and S. P. Lapage. 1966 An electron microscope study of thin sections of *Haemophilus vaginalis* (Gardner and Dukes) and possibly related species. Can J Microbiol *12:* 1125–1136.

—, R. G. E. Murray and A. Birch-Andersen. 1969 Some electron microscopic features of *Cardiobacterium hominis.* J Gen Microbiol *57 (3):* xxiv.

Reynes, V. 1947 Étude d'une nouvelle espèce de *Neisseria* anaérobie isolée d'une vulvo-vaginite: *N. vulvo-vaginitis* n. sp. Ann Inst Pasteur (Paris) *73:* 601–602.

Reynolds, D. M. and S. A. Waksman. 1948 Grisein, an antibiotic produced by certain strains of *Streptomyces griseus.* J Bacteriol *55 (5):* 739–752.

Rhodes, R. A., M. S. Roth and G. R. Hrubant. 1965 Sporulation of *Bacillus popilliae* on solid media. Can J Microbiol *11:* 779–783.

Riccardo, S., M. Formisano and S. Cappala. 1966 Una nuova Micobatteriacea isolata dalle sponde del fiume Sarno: *Mycobacterium sarni* ed una nuova varieta: *Mycobacterium sarni rubrum.* Ann Fac Sci Agr Univ Stud Napoli Portici (Ser 4) *1:* 281–307.

Rich, L. G. 1955 Respiration studies on the organic nitrogen preferences of *Zoogloea ramigera.* Appl Microbiol *3:* 20–25.

Richmond, L. and M. M. Cummings. 1950 An evaluation of methods of testing the virulence of acid-fast bacilli. Amer Rev Tuberc *62:* 632–637.

Ridell, M. and M. Norlin. 1973 Serological study of *Nocardia* by using mycobacterial precipitation reference system. J Bacteriol *113:* 1–7.

Ridgway, R. 1912 Color Standards and Color Nomenclature. Washington, D. C.

Rifkind, D. and R. M. Cole. 1962 Non-beta-hemolytic group M-reacting streptococci of human origin. J Bacteriol *84 (1):* 163–168.

Riker, A. J., W. M. Banfield, W. H. Wright, G. W. Keitt and H. E. Sagen. 1930 Studies on infectious hairy-root of nursery apple trees. J Agr Res *41:* 507–540.

—, F. R. Jones and M. C. Davis. 1935 Bacterial leaf spot of alfalfa. J Agr Res *51:* 177–182.

Rioche, M. 1967 Lesions microscopiques de la rickettsiose générale bovine à *Rickettsia (Ehrlichia) bovis* (Donatien et Lestoquard 1936). Rev Elevage Med Vet des Pays Trop (Paris) *20:* 415–427.

Rippel, A. 1937 Über Eiweissbildung durch Bakterien. Arch Mikrobiol *8:* 41–65.

Ris, H. 1961 Ultrastructure and molecular organization of genetic systems. Can J Genet Cytol *3:* 95.

Rist, E. 1898 Études bactériologiques sur les infections d'origine otique. Thése George Carré et C. Navd, Paris, pp. 1–173.

—. 1902 Note sur sept cas de salpingite suppurée examines bactériologiquement. C R Soc Biol *54:* 305–306.

Ristic, M. 1968 Anaplasmosis. *In* Weinman and Ristic (Editors), Infectious blood diseases of man and animals, Vol. 2. Academic Press, New York, pp. 478–542.

— and D. K. Mann. 1963 Anaplasmosis IX. Immunoserologic properties of soluble *Anaplasma* antigens. Amer J Vet Res *24:* 478–482.

—, — and R. Kodras. 1963 Anaplasmosis VIII. Biochemical and biophysical characterization of soluble *Anaplasma* antigens. Amer J Vet Res *24:* 472–476.

— and A. Watrach. 1961 Studies in anaplasmosis: II. Electron microscopy of *Anaplasma marginale* in deer. Amer J Vet Res *22:* 109–116.

Rittenberg, S. C. 1969 The roles of exogenous organic matter in the physiology of chemolithotrophic bacteria. Adv Microbial Physiol *3:* 159–196.

Rivers, T. M. 1922 Influenza-like bacilli: Growth of influenza-like bacilli on media containing only an autoclave-labile substance as an accessory food factor. Bull Johns Hopkins Hosp *33:* 429–431.

Rivière, J. 1964 Isolement et purification des bactéries cellulolytiques aérobies du sol. II. Isolement et description d'une nouvelle espèce d'Arthrobacter associé avec *Sporocytophaga myxococcoides*. Ann Inst Pasteur (Paris) *101:* 793–800.

Rivolta, S. 1878 Sul cosi detto mal del rospo del Trutta e sull' *Actinomyces bovis* di Harz. Clin Vet Milano *1:* 201–208.

—. 1879 Sopra un micromicete del cavallo. Noto preventiva. G Guglielmo Salic Piacenza *1 (5):* 145–146.

Roach, A. W. and J. K. G. Silvey. 1958 The morphology and life cycle of fresh water actinomycetes. Trans Amer Microsc Soc *77 (1):* 36–47.

Roberg, M. 1934 Über den Erreger der Wurzelknollchen von *Alnus* und den Elaeagnaceen *Elaeagnus* und *Hippophae*. Jahrb Wiss Bot *79:* 472–492.

Roberts, D. H. 1964 The isolation of an influenza A virus and a *Mycoplasma* associated with duck sinusitis. Vet Rec *76:* 470–473.

Roberts, D. S. 1967 The pathogenic synergy of *Fusiformis necrophorus* and *Corynebacterium pyogenes*. I. Influence of leucocidal exotoxin of *F. necrophorus*. Brit J Exp Pathol *48:* 665–673.

—. 1967 The pathogenic synergy of *Fusiformis necrophorus* and *Corynebacterium pyogenes*. II. The response of *F. necrophorus* to a filterable product of *C. pyogenes*. Brit J Exp Pathol *48:* 674–679.

—, N. P. H. Graham, J. R. Egerton and I. M. Parsonson. 1968 Infective bulbar necrosis (heel abscess) of sheep, a mixed infection with *Fusiformis necrophorus* and *Corynebacterium pyogenes*. J Comp Pathol *78:* 1–8.

Roberts, J. L. 1935 A new species of the genus *Bacillus* exhibiting mobile colonies on the surface of nutrient agar. J Bacteriol *29:* 229–236.

Roberts, R. J. 1968 A numerical taxonomic study of 100 isolates of *Corynebacterium pyogenes*. J Gen Microbiol *53 (3):* 299–303.

Roberts, R. S. 1947 An immunological study of *Pasteurella septica*. J Comp Pathol *57:* 261–278.

Robin, C. 1853 Histoire naturelle des végétaux parasites qui croissent sur l'homme et sur les animaux vivants. J.-B. Baillière, Paris, pp. 1–702.

Robinson, J. and N. E. Gibbons. 1952 The effect of salts on the growth of *Micrococcus halodentrificans* n. sp. Can J Bot *30:* 147–154.

Robinson, J. A. and F. P. Meyer. 1966 Streptococcal fish pathogen. J Bacteriol *92 (2):* 512.

Robinson, K. 1966 Some observations on the taxonomy of the genus *Microbacterium*. I. Cultural and physiological reactions and heat resistance. J Appl Bacteriol *29:* 607–615.

—. 1966 Some observations on the taxonomy of the genus *Microbacterium*. II. Cell wall analysis, gel electrophoresis and serology. J Appl Bacteriol *29:* 616–624.

Rocchi, G. 1908 Lo stato attuale delle nostre cognizioni sui germi anaerobi. Bull Sci Méd *8:* 457–528.

Rocha-Lima, H. da. 1930 Rickettsien. *In* Kolle and Wasserman (Editors), Handbuch der pathogenen Mikroorganismen, Vol. 8. Gustav Fischer, Jena, pp. 1347–1386.

Rode, L. J., G. Oglesby and V. T. Schuhardt. 1950 The cultivation of brucellae on chemically defined media. J Bacteriol *60:* 661–668.

Rodenwaldt, E. and H. Jusatz. 1956 Welt-Seuchen Atlas. I. Map 58, II Map 47–48; III Map 86–87, Falk Verlag, Hamburg.

Rodriguez-Barrueco, C. 1966 Fixation of nitrogen in root nodules of *Alnus jorullensis* H.B. and K. Phyton Rev Int Bot Exp *23:* 103–110.

Roelofsen, P. A. 1934 On the metabolism of the purple sulfur bacteria. Proc Kon Ned Akad Wetensch *37:* 660–669.

—. 1941 De alkohol-bacterie in arensap. Natuurwetensch Tijdschr *101:* 374.

Roger, H. and M. Garnier. 1906 L'infection du sang dans l'occlusion intestinale. Bull Mem Soc Med Paris *23:* 870–874.

Rogosa, M. 1961 Experimental conditions for nitrate reduction by certain strains of the genus *Lactobacillus*. J Gen Microbiol *24 (3):* 401–408.

—. 1964 The genus *Veillonella*. I. General cultural, ecological, and biochemical considerations. J Bacteriol *87 (1):* 162–170.

—. 1965 The genus *Veillonella*. IV. Serological groupings, and genus and species emendations. J Bacteriol *90 (3):* 704–709.

—. 1969 *Acidaminococcus* gen. nov., *Acidaminococcus fermentans* sp. nov., anaerobic gramnegative diplococci using amino acids as the

sole energy source for growth. J Bacteriol 98 (2): 756–766.
—. 1971 Transfer of *Peptostreptococcus elsdenii* Gutierrez et al to a new genus, *Megasphaera* [*M. elsdenii* (Gutierrez et al) comb. nov.] Int J Syst Bacteriol 21 (2): 187–189.
—. 1971 Transfer of *Veillonella* Prévot and *Acidaminococcus* Rogosa from *Neisseriaceae* to *Veillonellaceae* fam. nov. and the inclusion of *Megasphaera* Rogosa in *Veillonellaceae*. Int J Syst Bacteriol 21 (3): 231–233.
—. 1971 *Peptococcaceae*, a new family to include the Gram-positive, anaerobic cocci of the genera *Peptococcus*, *Peptostreptococcus* and *Ruminococcus*. Int J Syst Bacteriol 21 (3): 234–237.
—. 1971 Transfer of *Sarcina* Goodsir from the family *Micrococcaceae* Pribram to the family *Peptococcaceae* Rogosa. Int J Syst Bacteriol 21 (4): 311–313.
—, J. G. Franklin and K. D. Perry. 1961 Correlation of the vitamin requirements with cultural and biochemical characteristics of *Lactobacillus* spp. J Gen Microbiol 25 (3): 473–482.
— and P. A. Hansen. 1971 Nomenclatural considerations of certain species of *Lactobacillus* Beijerinck. Int J Syst Bacteriol 21 (2): 177–186.
—, M. I. Krichevsky and R. R. Colwell. 1971 Method for coding date on microbial strains for computers. Int J Syst Bacteriol 21: A1–A185.
— and J. A. Mitchell. 1950 Induced colonial variation of a total population among certain lactobacilli. J Bacteriol 59 (2): 303–308.
— and M. E. Sharpe. 1959 An approach to the classification of the lactobacilli. J Appl Bacteriol 22: 329–340.
—, R. F. Wiseman, J. A. Mitchell and M. N. Disraely. 1953 Species differentiation of oral lactobacilli from man including descriptions of *Lactobacillus salivarius* nov. spec. and *Lactobacillus cellobiosus* nov. spec. J Bacteriol 65 (6): 681–699.
Roguinsky, M. 1969 Reactions de *Streptococcus uberis* avec les serums G et P. Ann Inst Pasteur (Paris) 117: 529–532.
Rohde, R. 1965 The identification, epidemiology and pathogenicity of the salmonellae of sub-genus II. J Appl Bacteriol 28: 368–372.
—. 1966 Neue serologische Befunde hinsichtlich der Subgenus Einteilung der Salmonellen. Zentrabl Bakteriol Parasitenk Infektionskr Hyg Abt I Orig 202: 484–503.
—. 1967 Zur serologischen Differentialdiagnose der *Salmonella* Subgenera I-IV. Zentrabl Bakteriol Parasitenk Infektionskr Hyg Abt I Orig 205: 404–424.
Roldan, E. F. 1931 A bacterial stem-rot of hybrid cane seedlings hitherto unreported. Philippine Agr 20: 247–260.
Rolly, F. 1911 Experimentelle bakteriologische Untersuchungen von rurschiedenen Streptokokkenstammen. Zentrabl Bakteriol Parasitenk Infektionskr Hyg Abt I Orig 61: 86–92.
Rosan, B. and N. B. Williams. 1966 Serology of strains of *Streptococcus faecalis* which produce hyaluronidase. Nature (London) 212: 1275–1276.
Rose, D. H. 1917 Blister spot of apples and its relation to a disease of apple bark. Phytopathology 7: 198–208.

Rosebury, T. 1944 The parasitic actinomycetes and other filamentous microorganisms of the mouth. Bacteriol Rev 8 (3): 189–223.
Rosen, H. R. 1922 A bacterial disease of foxtail (*Chaetochloa lutescens*). Ann Mo Bot Gard 9: 333–402.
—. 1922 The bacterial pathogen of corn stalk rot. Phytopathology 12: 497–499.
—. 1926 Bacterial stalk rot of corn. Phytopathology 16: 241–267.
Rosenbach, F. J. 1884 Micro-organismen bei den Wund-Infections-Krankheiten des Menschen. J. F. Bergmann, Wiesbaden, pp. 1–122.
—. 1909 Experimentelle, morphologische und klinische Studie über die krankheitserregenden Mikroorganismen der Schweinerotlaufs, der Erysipeloids und der Mäusesepsis. Z Hyg Infektionskr Med Mikrobiol Immunol Virol 63: 343–371.
Rosenbusch, C. T. and I. A. Merchant. 1939 A study of the haemorrhagic septicaemia pasteurellae. J Bacteriol 37: 69–89.
Rosendal, S. 1973 *Mycoplasma cynos*, a new canine *Mycoplasma* species. Int J Syst Bacteriol 23: 49–54.
Rosenfeld, G. S., V. M. Baikina, N. O. Blinov and A. S. Khokhlov. 1963 The fractionation of albofungin and antibiotic 660-15 (in Russian). Antibiotiki 8 (4): 320–326.
Rosenthal, S. A. and C. D. Cox. 1953 The somatic antigens of *Corynebacterium michiganense* and *Corynebacterium insidiosum*. J Bacteriol 65 (5): 532–537.
— and —. 1954 An antigenic analysis of some plant and soil corynebacteria. Phytopathology 44: 603–604.
Ross, R. F. and J. A. Karmon. 1970 Heterogeneity among strains of *Mycoplasma granularum* and identification of *Mycoplasma hyosynoviae* sp. n. J Bacteriol 103 (3): 707–713.
Rossi-Doria, T. 1891 Su di alcune specie di "Streptotrix" trovate nell'aria studiate in rapporto a quelle già note e specialment all' "Actinomyces." Ann Ist Igiene Sper Univ Roma 1: 399–438.
Rosypal, S. and A. Rosypalova. 1966 Genetic, phylogenetic and taxonomic relationships among bacteria as determined by their deoxyribonucleic acid base composition. Folia VII, Biologia 14, Spis 3, Publ Fac Sci Univ J E Purkyne, Brno, pp. 1–90.
Roth, A. W. 1797 Catalecta botanica quibus plantae novae et minus cognitae describuntur atque illustrantur. Lipsiae in Bibliopolio I. G. Mulleriano, Fasc. 1.
Roth, G. D. 1957 Proteolytic organisms of the carious lesion. Oral Surg Oral Med Oral Pathol 10: 1105–1117.
Rotmistrov, M. N. 1939 Isolation of pure cultures of thermophilic cellulose bacteria. Mikrobiologiya 8 (1): 56–68.
Rottgardt, A. 1926 Die Milch nach Tarozzi als Nahrboden und zur Differenxierung des Rauschbrand-bazillus und des *Vibrio septicus*. Deut Tieraerztl Wochenschr 34: 553–556.
Rouf, M. A. and J. L. Stokes. 1964 Morphology, nutrition and physiology of *Sphaerotilus discophorus*. Arch Mikrobiol 49: 132–149.
Roughgarden, J. W. 1965 Antimicrobial therapy of rat bite fever; a review. Arch Intern Med 116: 39–54.
Rousselot, R. 1948 *Rickettsia* (*Donatienella*)

delpyi n. sp. n. subgen. Bull Soc Pathol Exot *41:* 110–112.

Routien, J. B. and A. Hofmann. 1951 *Streptomyces californicus* productor de viomicina. Antibiot Chemother *1:* 387–389.

Rowatt, E. 1957 The growth of *Bordetella pertussis:* a review. J Gen Microbiol *17 (2):* 297–326.

Roy, A. B. 1958 A new species of *Azotobacter* producing heavy slime and acid. Nature (London) *182:* 120–121.

Roy, T. E. and C. D. Kelly. 1939 Genus VIII. *Bacteroides* Castellani and Chalmers. *In* Bergey, Breed, Murray and Hitchens (Editors), Bergey's Manual of Determinative Bacteriology, 5th ed. The Williams & Wilkins Co., Baltimore, pp. 556–558.

Roze, E. 1896 Le *Clonothrix*, un nouveau type générique de Cyanophycées. J Bot (Paris) *10:* 325–330.

Rubenchik, L. I. 1959 On systematics of bacteria of *Azotobacteriaceae* family. Mikrobiologiya *28 (3):* 328–335.

Rubentschik, L. 1928 Zur Frage der aeroben Zellulosezersetzung bei hohen Salzkonsentrationen (in German). Zentrabl Bakteriol Parasitenk Infektionskr Hyg Abt II *76:* 303–314.

Rucker, R. R. 1949 A streptomycete pathogenic to fish. J Bacteriol *58:* 659–664.

Ruhland, W. 1924 Beiträge zur Physiologie der Knallgasbakterien. Jahrb Wiss Bot *63:* 321–389.

Ruiter, M. and H. M. M. Wentholt. 1952 The occurrence of a pleuropneumonia-like organism in fusospirillary infections of the human genital mucosa. J Invest Dermatol *18:* 313–325.

— and —. 1955 Isolation of a pleuropneumonia-like organism from a skin lesion associated with a fusospirochetal flora. J Invest Dermatol *24:* 31–34.

Ruiz-Herrera, J. 1970 *Achromobacter starkey* n. sp. a methionine decomposing bacterium isolated from the soil. Antonie van Leeuwenhoek. J Microbiol Serol *36:* 329–333.

Rullmann, W. 1895 Chemische bakteriologische Untersuchungen von Zwischendecken-füllungen mit besondere Berücksichtung von *Cladotrix odorifera* (in German). Inaug Diss Akad Buchdruckerei von F. Strauv, Munich, pp. 1–47.

Runeberg, B. 1908 Studien über die bei peritonäaeln Infektionen appendikulären Ursprungs vorkommenden sauerstofftoleranten sowie obligat anaeroben Bakterienformen, met besonderer Berücksichtung ihrer Bedeutung für die Pathogenese derartiger Peritonitiden. Arb Pathol Inst Univ Helsingfors Finland *2:* 271–582.

Runge, W. F. 1957 Process of producing acetyl cycloserine. United States Patent 2,815,348, December 3.

Runyon, E. H. 1965 Pathogenic mycobacteria. Advan Tuberc Res *14:* 235–287.

—. 1967 *Mycobacterium intracellular.* Amer Rev Resp Dis *95:* 861–865.

—. 1968 Aerial hyphae of *Mycobacterium xenopei.* J Bacteriol *95:* 734–735.

—, R. Bonicke, R. E. Buchanan, J. H. Hanks, W. Kappler, A. G. Karlson, H. H. Kleeberg, G. P. Kubica, A. Lind, D. A. Mitchison, S. R.

Pattyn, W. B. Redmond, W. B. Schaefer, D. W. Smith, K. Takeys, R. L. Vollum, L. G. Wayne, E. Wolinsky and G. P. Youmans. 1967 *Mycobacterium tuberculosis, M. bovis* and *M. microti* species descriptions. Zentrabl Bakteriol Parasitenk Infektionskr Hyg Abt I Orig *204:* 405–413.

— and T. M. Dietz. 1971 Skin sensitivity in guinea pigs induced by Group II mycobacteria. Amer Rev Resp Dis *104:* 107–113.

Ruschmann, G. 1952 *Streptomyces mirabilis* und das das Miramycin. I. (in German). Pharmazie *7 (9):* 542–550.

Russ, V. R. 1905 Ueber ein Influenzabacillenahnliches anaerobes. Stabchen. Zentrabl Bakteriol Parasitenk Infektionskr Hyg Abt I Orig *39:* 357–359.

Russell, C. and T. K. Walker. 1953 *Lactobacillus malefermentans* n. sp. isolated from beer. J Gen Microbiol *8:* 160–162.

— and —. 1953 *Lactobacillus parvus* n. sp. isolated from beer. J Gen Microbiol *8:* 310–313.

Russell, H. L. 1892 Untersuchungen über im Golf von Neapel lebende Bacterien. Z Hyg Infektionskr *11:* 165–206.

Russell, S. A. and H. J. Evans. 1966 The nitrogen-fixing capacity of *Ceanothus velutinus.* Forest Sci *12:* 164–169.

Rustigian, R. and C. A. Stuart. 1943 Taxonomic relationships in the genus *Proteus.* Proc Soc Exp Biol Med *53:* 241–243.

Ryter, A. and J. Pillot. 1965 Structure des spirochetes. II. Étude du genre *Cristispira* au microscope optique et au microscope électronique. Ann Inst Pasteur (Paris) *108:* 552–562.

Ryu, E. 1963 A simple method for staining *Leptospira* and *Treponema.* Jap J Microbiol *7:* 81–85.

Ryzhkov, V. L. 1950 Study on systematics of viruses. Vop Med Virusol *3:* 9–19.

Saaltink, G. J. and H. P. Maas Geesteranus. 1969 A new disease in tulip caused by *Corynebacterium oortii* nov. sp. Neth J Plant Pathol *75:* 123–128.

Sabet, K. A. 1954 On the host range and systematic position of the bacteria responsible for the yellow shine disease of wheat (*Triticum vulgare* Vill.) and cocksfoot grass (*Dacytlis glomerata* L.). Ann Appl Bacteriol *41:* 606–611.

—. 1954 A new bacterial disease of maize in Egypt. Emp J Exp Agr *22:* 65–67.

—. 1957 Studies in the bacterial diseases of Sudan crops. I. Bacterial leaf spot of Jute (*Corchorus olitorius* L.). Ann Appl Biol *45:* 516–520.

—. 1959 Studies in the bacterial diseases of Sudan crops. III. On the occurrence, host range and taxonomy of the bacteria causing leaf blight diseases of certain leguminous plants. Ann Appl Biol *47:* 318–331.

—. 1959 Studies in the bacterial diseases of Sudan crops. IV. Bacterial leafspot and canker disease of Mahogany (*Khaya senegalensis* (Desr.) A. Juss and *K. grandifoliola* C. DC) Ann Appl Biol *47:* 658–665.

— and W. J. Dowson. 1960 Bacterial leaf spot of sesame (*Sesamum orientale* L.). Phytopathol Z *37:* 252–258.

—, F. Ishag and O. Khalil. 1969 Studies of the bacterial diseases of Sudan crops. VII. New records. Ann Appl Biol *63:* 357–369.

Sabin, A. B. 1938 Isolation of a filterable, transmissible agent with "neurolytic" properties from toxoplasma infected tissues. Science (Washington) 88: 189–191.

—. 1938 Identification of the filterable transmissible neurolytic agent isolated from toxoplasma-infected tissue as a new pleuropneumonia-like microbe. Science (Washington) 88: 575–576.

—. 1941 The filtrable microorganisms of the pleuropneumonia group. Bacteriol Rev 5: 1–66.

—. 1941 The filtrable microorganisms of the pleuropneumonia group (appendix on classification and nomenclature). Bacteriol Rev 5: 331–335.

Sabo, G. and T. P. Preobrazhenskaya. 1962 The characteristics of three strains of actinomycetes, producers of new antibiotics (in Russian). Antibiotiki 7 (4): 312–317.

Saccardo, P. A. 1889 Schizomycetaceae. Sylloge Fungorum 8: 923–1087.

Sachs, J. 1874 Lehrbuch der Botanik, 4 Aufl. W. Engelmann, Leipzig, pp. 1–928.

Sack, R. B. and C. C. J. Carpenter. 1969 Experimental canine cholera. I. Development of the model. J Infect Dis 119: 138–149.

Sackett, W. G. 1910 A bacterial disease of alfalfa caused by Pseudomonas medicaginis (Sackett) n. sp. Science (Washington) 31: 553.

—. 1916 A bacterial stem blight of field and garden peas. Colo Agr Exp Sta Bull 218: 1.

Sacks, L. E. 1954 Observations on the morphogenesis of Arthrobacter citreus, spec. nov. J Bacteriol 67: 342–345.

Sagers, R. D., M. Benziman and I. C. Gunsalus. 1961 Acetate formation in Clostridium acidiurici: acetokinase. J Bacteriol 82: 233–238.

— and I. C. Gunsalus. 1958 Glycine cleavage and one-carbon transfer reactions in Clostridium acidi-urici and Diplococcus glycinophilus. Bacteriol Proc, p. 119.

— and —. 1961 Intermediary metabolism of Diplococcus glycinophilus. I. Glycine cleavage and one-carbon interconversions. J Bacteriol 81: 541–549.

— and S. M. Klein. 1961 Conversion of glycine to acetate by Diplococcus glycinophilus. Bacteriol Proc, p. 187.

Saglio, P., M. Lhospital, D. Laflèche, G. Dupont, J. M. Bové, J. G. Tully and E. A. Freundt. 1973 Spiroplasma citri gen and sp. n.; a mycoplasma-like organism associated with "Stubborn" disease of Citrus. Int J Syst Bacteriol 23: 191–204.

St. John-Brooks, R. 1931 The aerobic actinomyces. In Bulloch, Colebrook, Griffith and St. John-Brooks, (Editors), A system of bacteriology in relation to medicine. Med. Res. Council, London 8: 72–78.

Saito, H. and G. P. Kubica. 1968 Serologic studies of avian Group III nonphotochromogen complex by agglutination test. Amer Rev Resp Dis 98: 47–59.

—, H. Tasaka and N. Takei. 1968 Studies on atypical acid-fast bacilli of the Group IV Mycobacterium—with special reference to the nonphotochromogenic, rapidly growing acid-fast organisms from natural sources Jap J Bacteriol 23: 758–766.

Saito, K. 1907 Mikrobiologische Studien über die Soyarbeiterung. Zentrabl Bakteriol Para-sitenk Infektionskr Hyg Abt II 17: 20–27; 101–109; 152–161.

Sakagami, Y., H. Sekine, S. Yamabayashi, Y. Kitaura, A. Ueda and Y. Kosaka. 1966 The studies on bramycin, a new antifungal antibiotic. I. Isolation, purification and properties of bramycin. J Antibiot (Tokyo) Ser A 19 (3): 99–103.

—, S. Yamabayashi and H. Sekine. 1966 The studies on bramycin, a new antifungal antibiotic. II. Taxonomic studies on bramycin-producing strain. J Antibiot (Tokyo) Ser A 19 (3): 104–109.

Sakaguchi, K. 1958 Taxonomic studies on Pediococcus soyae nov. sp. the soy sauce lactic acid bacteria. III. Studies on the activities of bacteria in soy sauce brewing. Bull Agr Chem Soc Jap 22: 353–362.

—. 1960 Betaine as a growth factor for Pediococcus soyae. VIII. Studies on the activities of bacteria in soy sauce brewing. Bull Agr Chem Soc Jap 24: 489–496.

—. 1960 Vitamins and amino acid requirements of Pediococcus soyae and Pediococcus acidilactici Kitahara's strain. IX. Studies on the activities of bacteria in soy sauce brewing. Bull Agr Chem Soc Jap 26: 638–643.

—. 1962 Carnitine as a growth factor for Pediococcis soyae. X. Studies on the activities of bacteria in soy sauce brewing. Bull Agr Chem Soc Jap 26: 72–74.

—, M. Iwasaki and S. Yamada. 1941 Studies on the propionic acid fermentation. J Agr Chem Soc Jap 17 (1): 127–138.

— and H. Mori. 1969 Comparative study on Pediococcus halophilus, P. soyae, P. homari, P. urinae-equi and related species. J Gen App Microbiol 15: 159–167.

Sakai, H., H. Yüntsen and F. Ishikawa. 1954 Studies on a new antibiotic, angustmycin. II. J Antibiot (Tokyo) Ser A 7 (4): 116–119.

Sakamoto, J. M. 1959 Étude sur antibiotique antifongique. III. La moldcidine A, un nouvel antibiotique produit par les streptomycete (in French). J Antibiot (Tokyo) Ser A 12 (4): 169–172.

—, S.-i Kondo, H. Yumoto and M. Arishima. 1962 Bundlins A and B, two antibiotics produced by Streptomyces griseofuscus nov. sp. J Antibiot (Tokyo) Ser A 15 (2): 98–102.

Sakazaki, R. 1961 Studies on the Hafnia group of Enterobacteriaceae. Jap J Med Sci Biol 14: 223–241.

—. 1965 A proposed group of the family Enterobacteriaceae, the Asakusa group. Int Bull Bacteriol Nomencl Taxon 15 (1): 45–48.

—. 1967 Studies on the Asakusa group of Enterobacteriaceae (Edwardsiella tarda). Jap J Med Sci Biol 20: 205–212.

—. 1968 Proposal of Vibrio alginolyticus for the biotype 2 of Vibrio parahaemolyticus. Jap J Med Sci Biol 21 (5): 359–362.

—, S. Iwanami and H. Fukumi. 1963 Studies on the enteropathogenic facultatively halophilic bacteria Vibrio parahaemolyticus. I. Morphological, cultural and biochemical properties and its taxonomical position. Jap J Med Sci Biol 16: 161–188.

— and Y. Murata. 1962 The new group of Enterobacteriaceae. The Asakusa group (in Japanese). Jap J Bacteriol 17: 616–617.

— and S. Namioka. 1957 Biochemical studies

on Voges-Proskauer positive enteric bacteria. Jap J Exp Med *27:* 273–282.

— and —. 1960 Serological studies on the Cloaca (Aerobacter) group of enteric bacteria. Jap J Med Sci Biol *13:* 1–12.

—, —, R. Nakaya and H. Fukumi. 1959 Studies on so-called Paracolon C 27 (Ferguson). Jap J Med Sci Biol *12:* 355–363.

Sakharoff, M. N. 1891 *Spirochaeta anserina* et la septicémie des oies. Ann Inst Pasteur (Paris) *5:* 564–566.

Saleh, A. M. 1964 Differences in the resistance of sulphate-reducing bacteria to inhibitors. J Gen Microbiol *37 (1):* 113–121.

Saltet, R. H. 1900 Ueber Reduktion von Sulfaten in Brackwasser durch Bakterien. Zentrabl Bakteriol Parasitenk Infektionskr Hyg Abt II *6:* 648–651; 695–703.

Salton, M. J. R. 1964 The Bacterial Cell Wall. Elsevier, New York.

Sambon, L. 1907 *Spiroschaudinnia. In* Manson (Editor), Tropical Diseases, 4th ed. Cassell, London, p. 833.

Sampietro, G. 1908 Sopra due casi di actinomicosi nell'uomo. Ann Igiene (Sperimentale) *18:* 391–416.

Sanaerelli, G. 1891 Ueber einen neuen Mikroorganismus des Wassers, welcher für Thiere mit veränderlicher und konstanter Temperatur pathogen ist. Zentrabl Bakteriol Parasitenk Infektionskr Hyg Abt I Orig *9:* 193–199; 222–228.

Sanchez-Marroquin, A. 1958 *Streptomyces lavendulae* and *Streptomyces venezuelae.* J Bacteriol *75 (4):* 383–389.

Sand, G. and C. O. Jensen. 1888 Die Aetiologie der Druise. Deut Z Tiermed Verg Pathol *13:* 437–464.

Sanderson, K. E. 1967 Revised linkage map of *S. typhimurium.* Bacteriol Rev *31 (4):* 354–372.

Sands, D. C., M. N. Schroth and D. C. Hildebrand. 1970 Taxonomy of phytopathogenic pseudomonads. J Bacteriol *101 (1):* 9–23.

Santer, M., J. Boyer and U. Santer. 1959 *Thiobacillus novellus* I. Growth on organic and inorganic media. J Bacteriol *78 (2):* 197–202.

Saperstein, S. and M. P. Starr. 1954 The ketonic carotenoid canthaxanthin isolated from a colour mutant of *Corynebacterium michiganense.* Biochem J *57:* 273–275.

Saragea, A. and P. Maximescu. 1966 Phage typing of *Corynebacterium diphtheriae.* Incidence of *C. diphtheriae* phage types in different countries. Bull World Health Organ *35:* 681–689.

—, E. Meitert and V. Bica-Popii. 1966 Lysogenisation et conversion toxigène chez *C. diphtheriae* par des phages provenant de Corynebacteries d'origine animal. Ann Inst Pasteur (Paris) *111:* 171–179.

Sartory, A. 1920 Champignons parasites de l'homme et des animaux. V. Arsant, Saint-Nicholas-du-Port, pp. 1–845.

Sarvas, M. 1967 Inheritance of Salmonella T₁ antigen. Ann Med Exp Biol Fenn *45:* 447–471.

Sasaki, Y. and T. Shiio. 1960 A study of amino acids fermenting bacteria. Amino Acids *2:* 30–36.

Sautet, J. 1937 Étiologie des fièvres recurrentes. Marseille Med *74:* 273–284.

Sauvageau, C.-F. and M. Radais. 1892 Sur les genres *Cladothrix, Streptothrix, Actinomyces* et description de deux *Streptothrix* nouveaux.

(Sur le genre *Oospora*) Ann Inst Pasteur (Paris) *6:* 242–273.

Săvulescu, T. 1947 Contribution à la classification des Bactériacées phytopathogénes. Anal Acad Romane Ser III *22 (4):* 1–26.

—. 1947 Contribution à la classification des Bactériacées phytopathogènes. Anal Acad Romane Ser III *22 (4):* 135–160.

Sawamura, S. 1906 On the micro-organisms of natto. Bull Coll Agr Tokyo Imp Univ *7:* 107–110.

Sawazaki, T., S. Suzuki, G. Nakamura, M. Kawasaki, S. Yamashita, K. Isono, K. Anzai, Y. Serizawa and Y. Sekiyama. 1955 Streptomycin production by a new strain *Streptomyces mashuensis.* J Antibiotics *8 (A):* 44–47.

Sawyer, S. J., J. B. Macdonald and R. J. Gibbons. 1962 Biochemical characteristics of *Bacteroides melaninogenicus.* A study of thirty-one strains. Arch Oral Biol *7:* 685–691.

Scammon, L., S. Froman and D. W. Will. 1964 Enhancement of virulence for chickens of Battey type mycobacteria by preincubation at 42°C. Amer Rev Resp Dis *90:* 804–805.

Scardovi, V. 1950 Un nuovo solfo-batterio fotosintezzante "*Rhodopseudomonas vannielii* n. sp." Ann Microbiol *4:* 77–102.

—. 1963 Studies in rumen microbiology 1. A succinic acid producing vibrio: main physiological characteristics and enzymology of its succinic acid forming system. Ann Microbiol Enzimol *13:* 171–187.

—. 1964 Studies in rumen bacteriology. IV. The formation of acetate through phosphorolytic cleavage of fructose-6-phosphate in a group of branched Gram-positive bacteria isolated from sheep rumen. Ann Microbiol Enzimol *14:* 189–198.

—, B. Sgorbati and G. Zani. 1971 Starch gel electrophoresis of fructose-6-phosphate phosphoketolase in the genus *Bifidobacterium.* J Bacteriol *106 (3):* 1036–1039.

— and L. D. Trovatelli. 1965 The fructose-6-phosphate shunt as peculiar pattern of hexose degradation in the genus *Bifidobacterium.* Ann Microbiol *15:* 19–29.

— and —. 1969 New species of bifid bacteria from *Apis mellifica* L. and *Apis indica* F. A contribution to the taxonomy and biochemistry of the genus *Bifidobacterium.* Zentrabl Bakteriol Parasitenk Infektionskr Hyg Abt II *123:* 64–88.

—, —, F. Crociani and B. Sgorbati. 1969 Bifid bacteria in bovine rumen. New species of the genus *Bifidobacterium: B. globosum* n. sp. and *B. ruminale* n. sp. Arch Mikrobiol *68:* 278–294.

—, —, G. Zani, F. Crociani and D. Matteuzzi. 1971 Deoxyribonucleic acid homology relationships among species of the genus *Bifidobacterium.* Int J Syst Bacteriol *21 (4):* 276–294.

—, G. Zani and L. D. Trovatelli. 1970 Deoxyribonucleic acid homology among the species of the genus *Bifidobacterium* isolated from animals. Arch Mikrobiol *72:* 318–325.

Scarlett. 1916 Infections cornéennes à diplobacilles. Note sur deux diplobacilles non encore décrits (*Bacillus duplex nonliquefaciens* et *Bacillus duplex* Josefi). Ann Ocul *153:* 100–110.

Schaaf, A. van der and M. Rosa. 1940 Brucellosis en onchocerciasis in verband met een chronisch gewrichtslijden bij runderen. Ned Ind Bl Diergeneesk *52:* 1–20.

Schachman, H. K., A. B. Pardee and R. Y. Stanier.

1952 Studies on the macromolecular organization of microbial cells. Arch Biochem Biophys *38:* 245–260.

Schaefer, W. B. 1965 Serologic identification and classification of the atypical mycobacteria by their agglutination. Amer Rev Resp Dis *92:* 85–93.

—. 1968 Incidence of the serotypes of *Mycobacterium avium* and atypical mycobacteria in human and animal diseases. Amer Rev Resp Dis *97:* 18–23.

— and C. L. Davis. 1961 A bacteriologic and histopathologic study of skin granuloma due to *Mycobacterium balnei.* Amer Rev Resp Dis *84:* 837–844.

Schafer, D. 1969 Eine neue *Streptosporangium* Art aus turkischer Steppenerde. Arch Mikrobiol *66:* 365–373.

Schäperclaus, W. 1930 *Pseudomonas punctata* als Krankheitserreger bei Fischen. Z Fisch Derem Hilfswiss *28:* 289–370.

Schardinger, F. 1905 *Bacillus macerans*, ein Aceton bildender Rottebacillus. Zentrabl Bakteriol Parasitenk Infektionskr Hyg Abt II *14:* 772–781.

Scharif, G. 1961 *Corynebacterium iranicum* sp. nov. on wheat (*Triticum vulgare* L.) in Iran, and a comparative study of it with *C. tritici* and *C. rathayi.* Entomol Phytopathol Appl Téhran *19:* 1–24.

Schatz, A. and C. R. Bovell. 1952 Growth and hydrogenase activity of a new bacterium, *Hydrogenomonas facilis.* J Bacteriol *63 (1):* 87–98.

Schaub, I. G. and F. D. Hauber. 1948 A biochemical and serological study of a group of identical unidentifiable gram negative bacilli from human sources. J Bacteriol *56:* 379–385.

Schaudinn, F. 1905 Korrespondenzen. Deut Med Wochenschr *31:* 1728.

— and E. Hoffmann. 1905 Vorläufiger Bericht über das Vorkommen for Spirochaeten in syphilitischen Krankheitsprodukten und bei Papillomen. Arb GesundhAmt Berl *22:* 528–534

Schefferle, H. E. 1957 An investigation of the microbiology of built-up poultry litter. Thesis Edinburgh University.

—. 1966 Coryneform bacteria in poultry deep litter. J Appl Bacteriol *29:* 147–160.

Schellack, C. 1908 Morphologische Beiträge zur Kenntnis der eurpäischen, amerikanischen und afrikanischen Rekurrensspirochaeten. Arb GesundhAmt Berl *27:* 364–387.

—. 1909 Studien zur Morphologie und Systematik der Spirochaeten aus Muscheln. Arb GesundhAmt Berl *30:* 379–428.

Scherago, M. 1936 An epizootic septicemia of young guinea pigs caused by *Pseudomonas caviae* n. sp. J Bacteriol *31:* 83.

Scherff, R. H., J. E. DeVay and T. W. Carroll. 1966 Ultrastructure of host parasitic relationships involving reproduction of *Bdellovibrio bacteriovorus* in host bacteria. Phytopathology *56:* 627–632.

Schewiakoff, W. 1893 Über einen neuen bacterienähnlichen organismus des Süsswassers., Habilitations-schrift, Universität Heidelberg C. Winter, pp. 1–36.

Schikora, F. 1899 Entwickelungs-Bedingungen einiger abwässerreinigender Pilze, insbes. *Sphaerotilus fluitans* nov. spec. und *Leptomitus lacteus* Ag. Z Fisch Deren Hilfswiss, *7:* 1–27.

Schildkraut, C. L., J. Marmur and P. Doty. 1962 Determination of the base composition of deoxyribonucleic acid from its buoyant density in CsCl. J Mol Biol *4:* 430–443.

Schilling, V. 1928 *Eperythrozoon coccoides*, eine neue durch splenektomie aktivierbare dauerinfektion der weissen maus. Klin Wochenschr *7:* 1854–1855.

Schindler, R., M. Ristic and R. Wokatsch. 1966 Vergleichende Untersuchungen mit *Anaplasma marginale* und *A. centrale.* Z Tropenmed Parasitol *17:* 337–360.

Schippers-Lammertse, A. F., A. D. Muijsers and K. B. Klatser-Oedekerk. 1963 *Arthrobacter polychromogenes*, its pigments and a bacteriophage of this species. Antonie van Leeuwenhoek J Microbiol Serol *29:* 1–15.

Schlegel, H. G. 1962 Die Speicherstoffe von *Chromatium okenii.* Arch Mikrobiol *42:* 110–116.

— and N. Pfennig. 1961 Die Anreicherungskultur einiger Schwefelpurpurbakterien. Arch Mikrobiol *38:* 1–39.

Schleifer, K. H. 1970 Die Mureintypen in der Gattung *Microbacterium.* Arch Mikrobiol *71:* 271–282.

— and O. Kandler. 1967 Zur chemischen Zusammensetzung der Zellwand der Streptokokken. I. Die Aminosaüresequenz des Mureins von *Str. thermophilus* und *Str. faecalis.* Arch. Mikrobiol *57:* 335–364.

— and —. 1967 Zur chemischen Zusammensetzung der Zellwand der Streptokokken. II. Die Aminosauresequenz des Mureins von *Str. lactis* und *cremoris.* Arch Mikrobiol *57:* 365–381.

— and —. 1972 Peptidoglycan types of bacterial cell walls and their taxonomic implications. Bacteriol Rev *36:* 407–477.

Schleifstein, J. and M. Coleman. 1943 *Bacterium enterocoliticum.* N Y State Dep Health Div Lab Res Annu Rep, pp. 56–57.

Schlirf, K. 1925 Zur Kentniss der "azidophilen" Bakterien. Zentrabl Bakteriol Parasitenk Infektionskr Hyg Abt I Orig *97:* 104–118.

Schmid, E. E., T. Ve. laudapillai and G. R. Niles. 1954 Study of Paracolon organisms with the major antigen of *Shigella sonnei*, Form I. J Bacteriol *68 (1):* 50–52.

Schmid, G. 1922 *Simonsiella filiformis* (G. Schmid 1922). *In* Simons, Saprophytische Oscillarien des Menschen und der Tiere. Zentrabl Bakteriol Parasitenk Infektionskr Hyg Abt I Orig *88:* 501–510.

Schmidle, W. 1901 Neue Algen aus dem Gibiete des Oberrheins. Beih Bot Zentrabl *10:* 179–180.

Schmidt, J. 1901 Familie *Bacteriaceae. In* Schmidt and Weis, Bakterienne. Naturhistorisk Grundlag for det Bakteriologiske Studium, Morten Porsild, København 1899–1901, pp. 248–296.

— and F. Weis. 1902 Die Bakterien. Naturhistorische Grundlage für das bakteriologische Studium. Verlag von Gustav Fischer, Jena.

Schmidt, K. 1963 Die Carotinoide der *Thiorhodaceae.* II. Carotinoids zusammensetzung von *Thiospirillum jenense* Winogradsky und *Chromatium vinosum* Winogradsky. Arch Mikrobiol *46:* 127–137.

—, S. Liaaen-Jensen and H. G. Schlegel. 1963 Die Carotinoide der *Thiorhodaceae.* I. Okenon als Hauptcarotinoid von *Chromatium okenii* Perty. Arch Mikrobiol *46:* 117–126.

—, N. Pfennig and S. Liaaen-Jensen. 1965 Carotenoids of *Thiorhodaceae*. IV. The carotenoid composition of 25 pure isolates. Arch Mikrobiol *52:* 132–146.

Schmincke, A. 1917 Histopathologischer Befund in Roseolen der Haut bei Wolhynischem Fieber. Muenchen Med Wochenschr *64:* 961.

Schmitz, H., S. B. Deak, K. E. Crook, Jr. and I. R. Hooper. 1964 Peliomycin, a new cytotoxic agent. I. Production, isolation and characterization. Antimicrob Agents Chemother *1963:* 89–94.

—, S. D. Jubinski, I. R. Hooper, K. E. Crook, Jr., K. E. Price and J. Lein. 1965 Ossamycin, a new cytotoxic agent. J Antibiot (Tokyo) Ser A *18 (2):* 82–88.

Schneider, P. 1894 Die Bedeutung der Bakterienfarbstoffe für die Unterscheidung der Arten. Arb Bakteriol Inst Karlsruhe *1:* 201–232.

Schnellen, C. G. T. P. 1947 Onderzoekingen over de methaangisting. Thesis, Delft, pp. 1–137.

Schoenike, R. E. 1957–58 Influence of mountain avens (*Dryas drummondii*) on growth of young cottonwoods (*Populus trichocarpa*) at Glacier Bay, Alaska. Proc Minn Acad Sci *25* and *26:* 55–58.

Scholl, H. 1891 Die Milch, ihre haüfigeren Zersetzungen und Verfälschungen mit spezieller Berücksichtung ihrer Beziehungen zur Hygiene. J. F. Bergmann, Wiesbaden pp. 1–133.

Schönfeld, F. and W. Rommel. 1902 Untersuchungen über Trübungen im Lagerbier verursachendes Stäbchen-Bakterium (*Bacillus fasciformis*). Wochenschr Brau *19:* 585–588.

Schönichen, W. 1925 Einfachsten Lebensformen des Tier- und Pflanzenteiches, 5th ed. Spaltpflanzen, Geissellinge, Algen H. Bermühler, Berlin, *1:* 1–519.

— and A. Kalberlah. 1900 IV. Fam. *Chlamydobacteriaceae*. *In* Eyferth's Lebensformen des Tier- und Pflanzenreiches. Benno Goeritz, Braunschweig, pp. 43–47.

Schoop, G. 1935 Obligat halophile Mikroben. Zentrabl Bakteriol Parasitenk Infektionskr Hyg Abt I Orig *134:* 14–26.

—. 1935 *Halococcus litoralis*, ein obligat halphiler Farbstoffbildner. Deut Tieraerztl Wochenschr *43:* 817–820.

Schopf, J. M., E. G. Ehlers, D. V. Stiles and J. D. Birle. 1965 Fossil iron bacteria preserved in pyrite. Proc Amer Phil Soc *109:* 288–308.

Schorler, B. 1904 Beiträge zur Kenntnis der Eisenbakterien. Zentrabl Bakteriol Parasitenk Infektionskr Hyg Abt II *12:* 681–695.

Schottmüller, H. 1910 Zur Bedeutung einiger Anaëroben in der Pathologie insbesandere die puerperalen Erkrankungen. Mitt Grenzegeb Med Chir *21:* 450–490.

—. 1912 Ein anaërober *Staphylococcus aërogenes* als Erreger von Puerperalfieber. Zentrabl Bakteriol Parasitenk Infektionskr Hyg Abt I Orig *64:* 270–284.

—. 1914 Zur Ätiologie und Klinik der Bisskrankheit (Ratten-, Katzen-, Eichhörnchen-Bisskrankheit). Derm Wochenschr *58* (Suppl): 77–103.

Schroeder, W. and H. Hoeksema. 1959 A new antibiotic, 6-amino-9-D-psico-furanosylpurine J Amer Chem Soc *81* (7): 1767–1768.

Schroeter, J. 1872 Ueber einige durch Bacterien gebildete Pigmente. pp. 109–126. *In* F. Cohn (1875). Beitrage Biologie der Pflanzen, J. U. Kern's Verlag, Breslau.

—. 1885–1889. *In* F. Cohn, Kryptogamenflora von Schlesien. Bd 3, Heft 3, Pilze J. U. Kern's Verlag, Breslau, pp. 1–814.

Schroth, M. N., J. P. Thompson and D. C. Hildebrand. 1965 Isolation of *Agrobacterium tumefaciens-A. radiobacter* group from soil. Phytopathology *55:* 645–647.

Schubert, R. H. W. 1960 Üntersuchungen über die Merkmale der Gattung *Aeromonas*. Zentrabl Bakteriol Parasitenk Infektionskr Hyg Abt I Orig *180:* 310–327.

—. 1964 Zur Taxonomie der anaerogenen Aeromonaden. Zentrabl Bakteriol Parasitenk Infektionskr Hyg Abt I Orig *193:* 343–352.

—. 1964 Zur Taxonomie der Voges-Proskauer-negativen "hydrophila–ähnlichen Aeromonaden. Zentrabl Bakteriol Parasitenk Infektionskr Hyg Abt I Orig *193:* 482–490.

—. 1967 The taxonomy and nomenclature of the genus *Aeromonas* Kluyver and van Niel 1936. Part I. Suggestions on the taxonomy and nomenclature of the aerogenic *Aeromonas* species. Int J Syst Bacteriol *17 (1):* 23–37.

—. 1967 The taxonomy and nomenclature of the genus *Aeromonas*. Part II. Suggestions on the taxonomy and nomenclature of the anaerogenic aeromonads. Int J Syst Bacteriol *17 (3):* 273–279.

—. 1968 The taxonomy and nomenclature of the genus *Aeromonas* Kluyver and van Niel 1936. Part III. Suggestions on the definition of the genus *Aeromonas* Kluyver and van Niel 1936. Int J Syst Bacteriol *18:* 1–7.

—. 1969 Über den Aminosäureverwendungstoffwechsel bei Angehörigen des Genus *Aeromonas*. Zentrabl Bakteriol Parasitenk Infektionskr Hyg Abt I Orig *211:* 403–405.

—. 1969 Infrasubspezifische Taxonomie von *Aeromonas hydrophila* (Chester 1901) Stanier 1943. Zentrabl Bakteriol Parasitenk Infektionskr Hyg Abt I Orig *211:* 406–409.

—. 1969 *Aeromonas hydrophila* subsp. *proteolytica* (Merkel et al. 1964) comb. nov. Zentrabl Bakteriol Parasitenk Infektionskr Hyg Abt I Orig *211:* 409–412.

—. 1971 Status of the names *Aeromonas* and *Aerobacter liquefaciens* Beijerinck and designation of a neotype strain for *Aeromonas hydrophila* Stanier. Int J Syst Bacteriol *21 (1):* 87–90.

Schuhardt, V. T. and M. Wilkerson. 1951 Relapse phenomenon in rats infected with single spirochetes (*B. recurrentis*). J Bacteriol *62:* 215–219.

Schultes, L. M. and J. B. Evans. 1971 Deoxyribonucleic acid homology of *Aerococcus viridans*. Int J Syst Bacteriol *21:* 207–209.

Schultz, E. W., M. C. Terry, A. T. Brice and L. P. Gebhardt. 1934 Bacteriological observations on a case of meningo-encephalitis. Proc Soc Exp Biol Med *3:* 1021–1023.

Schure, P. S. J. 1953 Attempts to control the kresek disease of rice by chemical treatment of the seedlings. Contrib Gen Agr Res Sta Bogor *136:* 1–17.

Schurmann, C. 1967 Growth of myxococci in suspension in liquid media. Appl Microbiol *15:* 971–974.

Schuster, M. L., A. K. Vidaver and M. Mandel. 1968 A purple-pigment-producing bean wilt bacterium *Corynebacterium flaccumfaciens* var. *violaceum* n. var. Can J Microbiol *14:* 423–427.

Schutze, H. 1908 Beitrage zur Kenntnis der thermopilen Aktinomyceten und ihrer Sporenbildung. Arch Hyg Berl *67:* 35–36.

—. 1932 Studies on *B. pestis* antigens. Brit J Exp Pathol *13:* 284–293.

Schuurmans, D. M., B. H. Olson and C. L. San Clemente. 1956 Production and isolation of Thermoviridin, an antibiotic produced by *Thermoactinomyces viridis* n. sp. Appl Microbiol *4:* 61–66.

Schwabacher, H. 1959 A strain of *Mycobacterium* isolated from skin lesions of a cold blooded animal, *Xenopus laevus*, and its relation to atypical acid-fast bacilli occurring in man. J Hyg *57:* 57–67.

—, D. R. Lucas and C. Rimington. 1947 *Bacterium melaninogenicum*, a misnomer. J Gen Microbiol *1:* 109–120.

Schwartz, P. H., Jr. and E. Sharpe. 1970 Infectivity of spores of *Bacillus popilliae* produced on a laboratory medium. J Invertebr Pathol *15:* 126–128.

Schwers, E. 1912 *Megalothrix discophora*, eine neue Eisenbakterie. Zentrabl Bakteriol Parasitenk Infektionskr Hyg Abt II *33:* 273–276.

Scopes, A. W. 1962 The infrared spectra of some acetic acid bacteria. J Gen Microbiol *28:* 69–79.

Scott, J. P. 1928 The etiology of blackleg and methods of determining *Clostridum chauvoei* from other anaerobic organisms found in cases of blackleg. Cornell Vet *18:* 259–271.

Scott, J., A. W. Turner and L. R. Vawter. 1935 Gas edema diseases. Twelfth Int Vet Congr *2:* 168–182.

Scotten, H. L. and J. L. Stokes. 1962 Isolation and properties of *Beggiatoa*. Arch Mikrobiol *42:* 353–368.

Seamer, J. 1959 The propagation and preservation of *Eperythrozoon coccoides*. J Gen Microbiol *21:* 344–351.

Sebald, M. 1962 Étude sur les bactéries anaérobies gram-négatives asporulées. Thèses de L'université Paris, Imprimerie Barnéoud S. A. Laval, France, pp. 1–171.

—, F. Gasser and H. Werner. 1965 Teneur GC% et classification. Application au groupe des bifidobactéries et à quelques genres voisins. Ann Inst Pasteur (Paris) *109:* 251–269.

— and A. R. Prévot. 1962 Étude d'une nouvelle espèce anáerobie stricte *Micromonospora acetoformici* n. sp. isolée de l'intestin posterieur de *Reticulitermes lucifugus* var. *saintonnensis*. Ann Inst Pasteur (Paris) *102:* 199–214.

— and M. Veron. 1963 Teneur en bases de l'ADN et classification des vibrions. Ann Inst Pasteur (Paris) *105:* 897–910.

Sebek, O. K. 1965 Microbiological method for the determination of L-tryptophan. J Bacteriol *90 (4):* 1026–1031.

— and H. Jager. 1962 Divergent pathways of indole metabolism in *Chromobacterium violaceum*. Nature (London) *196:* 793–795.

Sedlák, J., M. Puchmayerová, J. Keléti and O. Lüderitz. 1971 On the taxonomy, ecology and immunochemistry of genus *Citrobacter*. J Hyg Epidemiol Microbiol Immunol (Prague) *15:* 366–374.

— and M. Šlajsová. 1966 Antigenstruktur und Antigenbeziehungen der Gattung *Citrobacter*. Zentrabl Bakteriol Parasitenk Infektionskr Hyg Abt I Orig *200:* 369–374.

— and —. 1967 Taxonomie du "coliforme 1433". Ann Inst Pasteur (Paris) *112:* 119–121.

Seelemann, M. 1942 Uber die Bedeutung der hämolytischen streptokokken bei einigen Infektionen des Pferdes. Deut Tierarztl Wochenschr *50:* 8–12; 38–41.

Seeley, H. W. 1951 The physiology and nutrition of *Streptococcus uberis*. J Bacteriol *62 (1):* 107–115.

— and J. A. Dain. 1960 Starch hydrolyzing streptococci. J Bacteriol *79 (2):* 230–235.

— and P. J. Van Demark. 1951 An adaptive peroxidation by *Streptococcus faecalis*. J Bacteriol *61 (1):* 27–35.

Seeliger, H. P. R. 1961 Listeriosis. Karger, Basel.

—, R. H. W. Schubert and E. Schlieber. 1966 Zur Taxonomie und Benennung des *Bacterium anitratum* (Schaub und Hauber). Ann Immunol Hung *9:* 251–260.

— and H. Werner. 1963 Recherches qualitatives et quantitatives sur la flore intestinale de l'homme. Ann Inst Pasteur (Paris) *105:* 911–936.

Séguin, P. 1928 Culture du *Fusobacterium plauti* forme mobile du bacille fusiforme. C R Soc Biol (Paris) *99:* 439–442.

— and R. Vinzent. 1936 Étude systématique des spirochetes buccaux d'aprés les charactères de culture. C R Soc Biol Paris *121:* 408–411.

— and —. 1941 Les Spirochetes commensaux de l'homme. Ann Inst Pasteur (Paris) *67:* 37–86.

Seidler, R. J., M. Mandel and J. N. Baptist. 1972 Molecular heterogeneity of the bdellovibrios; evidence for two new species. J Bacteriol *109 (1):* 209–217.

— and M. P. Starr. 1969 Isolation and characterization of host-independent bdellovibrios J Bacteriol *100 (2):* 769–785.

—, —, and M. Mandel. 1969 Deoxyribonucleic acid characterization of bdellovibrios. J Bacteriol *100 (2):* 786–790.

Seidman, P. and E. C. S. Chan. 1970 Growth of *Arthrobacter citreus* in a chemically-defined medium and its requirement for chelating agents with schizokinen activity. J Gen Microbiol *80 (3):* 417–420.

Seitz, E. W., P. R. Elliker and W. E. Sandine. 1961 A pigment-producing spoilage bacterium responsible for violet discoloration of refrigerated market milk and cream. Appl Microbiol *9:* 287–290.

Seliskar, C. E. 1952 Wetwood organism in Aspen, Poplar is isolated. Colo Farm Home Res *2:* 6–20.

Sellers, W. 1964 Medium for differentiating the gram-negative nonfermenting bacilli of medical interest. J Bacteriol *87 (1):* 46–48.

Senn, M., T. Ioneda, J. Pudles and E. Lederer. 1967 Spectrométrie de masse de glycolipides. I. Structure du "cord-factor" de *Corynebacterium diphtheriae*. Eur J Biochem *1:* 353–356.

Sensi, P. and H. Pagani. 1963 Antibiotic L. A. 5937. British Patent 920,799, March 13.

— and M. T. Timbal. 1959 Isolation of two antibiotics of the grisein and albomycin group. Antibiot Chemother *9 (3):* 160–166.

Sermonti, G. and S. Casciano. 1963 Sexual

polarity in *Streptomyces coelicolor*. J Gen Microbiol *33* (*2*): 293–301.

Serrano, F. B. 1928 Bacterial fruitlet brown-rot of pineapple in the Philippines. Philippine J Sci *36:* 271–305.

—. 1934 Fruitlet black-rot of pineapple in the Philippines. Philippine J Sci *55:* 337–362.

Seubert, W. 1960 Degradation of isoprenoid compounds by microorganisms. I. Isolation and characterization of an isoprenoid-degrading bacterium, *Pseudomonas citronellolis* n. sp. J Bacteriol *79* (*3*): 426–434.

Severi, R. 1946 L'azione patogena delle Kurthie la loro systematica. Una nuova species: *Kurthia variabilis*. G Batteriol Immunol *34:* 107–114.

Sguros, P. L. 1954 Taxonomy and nutrition of a new species of nicotinophilic bacterium. Bacteriol Proc, pp. 21–22.

—. 1955 Microbial transformations of the tobacco alkaloids. I. Cultural and morphological characteristics of a nicotinophile. J Bacteriol *69:* 28–37.

—. 1957 New approach to the mode of formation of classical morphological configurations by certain coryneform bacteria. J Bacteriol *74:* 707–709.

Shafia, F. and R. F. Wilkinson, Jr. 1969 Growth of *Ferrobacillus ferrooxidans* on organic matter. J Bacteriol *97* (*1*): 256–260.

Shane, B. S., L. Gouws and A. Kistner. 1969 Cellulolytic bacteria occurring in the rumen of sheep conditioned to low-protein teff hay. J Gen Microbiol *55:* 445–457.

Shapiro, H. S. 1968 Handbook of Biochemistry. Chemical Rubber Co., Cleveland, Ohio, p. H-31.

Shaposhnikov, V. N., E. N. Kondratieva and V. D. Fedorov. 1958 A contribution to the study of green sulphur bacteria of the genus *Chlorobium*. Mikrobiologiya *27:* 521–527.

—, —, and —. 1960 A new species of green sulfur bacteria. Nature (London) *187:* 167–168.

-, —, E. N. Krasil'nikova and A. A. Ramenskaya. 1959 Green bacteria that assimilate organic compounds. Dokl Akad Nauk SSSR *129:* 1424–1426.

Sharma, M. P. 1968 Immunochemical studies of cellulases from several strains of *Ruminococcus flavefaciens*. Ph.D. Thesis, North Carolina State University, Raleigh, pp. 1–112.

Sharp, J. T. 1968 Isolation of L forms of *Bartonella bacilliformis*. Proc Soc Exp Biol Med *128:* 1072–1075.

Sharpe, E. S., G. St Julian and C. Crowell. 1970 Characteristics of a new strain of *Bacillus popilliae* sporogenic *in vitro*. Appl Microbiol *19:* 681–688.

Sharpe, M. E. 1955 A serological classification of lactobacilli. J Gen Microbiol *12:* 107–122.

—. 1962 Taxonomy of the lactobacilli. Dairy Sci Abstr *24:* 109–118.

—. 1964 Serological types of *Streptococcus faecalis* and its varieties and their cell wall type antigen. J Gen Microbiol *36* (*1*): 151–160.

—, E. I. Garvie and R. Tilbury. 1972 Some slime-forming heterofermentative species of the genus *Lactobacillus*. Appl Microbiol *23* (*2*): 389–397.

—, M. J. Latham, E. I. Garvie, J. Zirngibl and O. Kandler. 1973 Two new species of *Lactobacillus* isolated from the bovine rumen,

Lactobacillus ruminis sp. nov. and *Lactobacillus vitulinus* sp. nov. J Gen Microbiol *77:* 37–49.

— and D. M. Wheater. 1957 *Lactobacillus helveticus*. J Gen Microbiol *16* (*3*): 676–679.

Shattock, P. M. F. 1949 The streptococci of Group D; the serological grouping of *Streptococcus bovis* and observations on serologically refractory group D strains. J Gen Microbiol *3:* 80–92.

— and D. G. Smith. 1963 The location of the group D antigen in a strain of *Streptococcus faecalis* var. *liquefaciens*. J Gen Microbiol *31* (*1*): IV.

Shaw, C. and P. H. Clarke. 1955 Biochemical classification of *Proteus* and Providence cultures. J Gen Microbiol *13:* 155–161.

—, J. M. Stitt and S. T. Cowan. 1951 Staphylococci and their classification. J Gen Microbiol *5:* 1010–1023.

Shaw, N. and J. Baddiley. 1964 The teichoic acid from the walls of *Lactobacillus buchneri* N.C.I.B. 8007 Biochem J *93:* 317–321.

Shaw, W. V., L. Tsai and E. R. Stadtman. 1966 The enzymatic synthesis of N-methyl glutamic acid. J Biol Chem *241:* 935–945.

Shchepkina, T. A. 1940 Investigation and description of cotton fibre endoparasites (Trans.). Bull Acad Sci USSR (Ser Biol) *5:* 643–661.

Shchurenkova, A. 1938 *Sogdianella moshkovskii* gen. nov., sp. nov.—a parasite belonging to the *Piroplasmidea* in a raptatorial bird-*Gypäetus barbatus* L. Med Parasitol Moscow *7* (*6*): 932–937.

Shenberg, E. 1967 Growth of pathogenic leptospira in chemically defined media. J Bacteriol *93:* 1598–1606.

Shepard, C. C. 1960 The experimental disease that follows the injection of human leprosy bacilli into foot-pads of mice. J Exp Med *112:* 445–454.

— and Y. T. Chang. 1962 Effect of several antileprosy drugs on multiplication of human leprosy bacilli in foot-pads of mice. Proc Soc Exp Biol Med *109:* 636–638.

Shepard, M. C. 1956 T-form colonies of pleuropneumonia-like organisms. J Bacteriol *71:* 362–369.

—. 1969 Fundamental biology of the T-strains. *In* Hayflick (Editor), The *Mycoplasmatales* and the L-Phase of Bacteria. Appleton-Century-Crofts, New York, pp. 49–65.

Sherengovji, P. Z. 1968 Minor bacteriosis of raspberries. Zashch Rast, Moscow *13:* 39.

Sherman, J. M. 1921 The cause of eyes and characteristic flavor in Emmental or Swiss cheese. J Bacteriol *6:* 379–392.

—. 1937 The streptococci. Bacteriol Rev *1:* 3–97.

—, C. F. Niven, Jr. and K. L. Smiley. 1943 *Streptococcus salivarius* and other nonhemolytic streptococci of the human throat. J Bacteriol *45:* 249–263.

— and H. U. Wing. 1935 An unnoted hemolytic *Streptococcus* associated with milk products. J Dairy Sci *18* (*10*): 657–660.

— and —. 1937 *Streptococcus durans*. J Dairy Sci *20:* 165–167.

Shewan, J. M. 1971 The microbiology of fish and fishery products. A progress report. J Appl Bacteriol *34:* 299–315.

—. G. Hobbs and W. Hodgkiss. 1960 A determinative scheme for the identification of certain genera of Gram-negative bacteria, with special reference to the *Pseudomonadaceae*. J Appl Bacteriol *23:* 379–390.

—, W. Hodgkiss and J. Liston. 1954 A method for the rapid differentiation of certain nonpathogenic asporogenous bacilli. Nature (London) *173:* 208–209.

Shibata, M. 1959 On a new streptomycin-producing species, *Streptomyces rameus* n. sp. (in Japanese). J Antibiot (Tokyo) Ser B *12 (6):* 398–400.

—, E. Higashide, T. Kanzaki, H. Yamamoto and K. Nakazawa. 1961 Studies on streptomycetes. Part I. *Streptomyces pulveraceus* nov. sp., producing new antibiotics zygomycin A and B. Agr Biol Chem *25 (3):* 171–175.

—, —, H. Yamamoto and K. Nakazawa. 1962 Studies on streptomycetes. I. *Streptomyces atratus* nov. sp., producing new antituberculous antibiotics rufomycin A and B. Agr Biol Chem *26 (4):* 228–233.

—, M. Honjo, Y. Tokui and N. Nakazawa. 1954 On a new anti-fungal and anti-yeast substance, candimycin produced by a streptomyces. J Antibiotics 7 (B): 168.

—, K. Nakazawa, A. Miyake, M. Inoue and A. Okabori. 1957 Studies on streptomycetes. Croceomycin, a new antituberculous substance (in Japanese). Annu Rep Takeda Res Lab *16:* 32–37.

Shidara, I. 1955 Study on the antibiotic nocardorubin produced by *Nocardia*. J Chiba Med Soc *30:* 551–559.

Shieh, H. S. 1964 Aerobic degradation of choline. I. Fermentation of choline by a marine bacterium, *Achromobacter cholinophagum* n. sp. Can J Microbiol *10:* 837–842.

Shiga, K. 1898 Ueber den Dysenteriebacillus (*Bacillus dysenteriae*). Zentrabl Bakteriol Parasitenk Infektionskr Hyg Abt I Orig *24:* 817–828.

Shiio, I., K. Mitsugi, S. Otsuga and T. Tsunoda. 1964 Process for producing L-glutamic acid. (United States Patent 3,117,915). Offic Gaz U S Patent Off *798:* 433.

—, S. Otsuka, R. Ishii, N. Katsuya and H. Iizuka. Microbial production of amino acids from hydrocarbons. I. Preliminary screening of glutamic acid producing bacteria. J Gen Appl Microbiol 9 (1): 23–30.

Shilo, M. 1969 Morphological and physiological aspects of the interaction of *Bdellovibrio* with host bacteria. Curr Top Microbiol Immunol *50:* 174–204.

—. 1970 Lysis of blue-green algae by myxobacter. J Bacteriol *104* (1): 453–461.

Shimo, M., T. Shiga, T. Tomosugi and I. Kamoi. 1959 Studies on taitomycin, a new antibiotic produced by *Streptomyces* sp. No. 772 (*S. afghaniensis*). I. Studies on the strain and production of taitomycin. J Antibiot (Tokyo) Ser A *12* (1): 1–6.

Shimwell, J. L. 1936 Study of a new species of *Acetobacter* (*A. capsulatum*) producing ropiness in beer and beer-wort. J Inst Brew *42:* 585–595.

—. 1937 Study of a new type of beer disease bacterium (*Achromobacter anaerobium* sp. nov.) producing alcoholic fermentation of glucose. J InstBrew *43:* 507–509.

—. 1948 Brewing bacteriology. IV. The acetic acid bacteria (Family *Acetobacteriaceae*: Genus *Acetobacter*). Wallerstein Lab Commun *11:* 27–39.

—. 1949 A study of ropiness in beer. III. Ropiness due to lactobacilli. J Inst Brew *55:* 26–29.

—. 1950 *Saccharomonas*, a proposed new genus for bacteria producing a quantitative alcoholic fermentation of glucose. J Inst Brew *56:* 179–182.

—. 1957 A pattern of evolution in the genus *Acetobacter*. J Inst Brew *63:* 45–56.

—. 1959 A re-assessment of the genus *Acetobacter*. Antonie van Leeuwenhoek J Microbiol Serol *25:* 49–67.

—. 1963 *Obesumbacterium* gen. nov. Brew J *99:* 759–760.

— and J. G. Carr. 1959 The genus *Acetomonas*. Antonie van Leeuwenhoek J Microbiol Serol *25:* 353–368.

— and M. Grimes. 1936 The distinguishing characters of *Flavobacterium proteus* (sp. nov.) the common rod bacterium of brewer's yeast. J Inst Brew *42:* 348–350.

Shinobu, R. 1955 On *Streptomyces hiroshimensis* nov. sp. Seibutsugakkaishi *6:* 43–46.

—. 1956 Three new species of *Streptomyces* forming whirls. Mem Osaka Univ Lib Arts Educ *5B:* 84–93.

—. 1957 Two new species of *Streptomyces* (in Japanese). Mem Osaka Univ Lib Arts Educ B Nat Sci *6:* 63–73.

—. 1958 On *Streptomyces spiroverticillatus* nov. sp. Bot Mag (Tokyo) *71* (*837):* 87–93.

—. 1962 A new *Streptomyces* species producing fluorescent-yellow soluble pigment. Mem Osaka Univ Lib Arts Educ B Nat Sci *11:* 115–122.

—. 1965 Taxonomy of the whirl-forming *Streptomycetaceae*. Mem Osaka Univ Lib Arts Educ B Nat Sci *14:* 72–201.

— and M. Kawato. 1959 On *Streptomyces massasporeus* nov. sp. Bot Mag (Tokyo) *72* (*853-854):* 283–288.

— and —. 1960 On *Streptomyces indigoferus* nov. sp., producing blue to green soluble pigment on some synthetic media. Mem Osaka Univ Lib Arts Educ B Nat Sci *9:* 49–53.

— and —. 1960 On *Streptomyces aerocolonigenes* nov. sp. forming the secondary colonies on the aerial mycelia. Bot Mag (Tokyo) *73:* 212–216.

— and Y. Kayamura. 1964 On new whirl-forming species of *Streptomyces*. Bot Mag (Tokyo) *77:* 176–180.

— and Y. Shimada. 1962 On a new whirl-forming species of *Streptomyces*. Bot Mag (Tokyo) *75:* 170–175.

Shirling, E. B. 1968 Cooperative description of type cultures of *Streptomyces*. III. Additional species descriptions from first and second studies. Int J Syst Bacteriol *18* (*4):* 279–392.

— and D. Gottlieb. 1966 Methods for the characterization of *Streptomyces* species. Int J Syst Bacteriol *16* (*3):* 313–340.

— and —. 1968 Cooperative description of type cultures of *Streptomyces*. II. Species descriptions from first study. Int J Syst Bacteriol *18* (*2):* 69–189.

— and —. 1969 Cooperative description of type cultures of *Streptomyces*. IV. Species descriptions from the second, third and fourth studies Int J Syst Bacteriol 19 (4): 391–512.

— and —. 1972 Cooperative description of type strains of Streptomyces. V. Additonal descriptions. Int J Syst Bacteriol 22: 265–394.

Shreeve, B. J., I. N. Ivanov and D. A. Thompson. 1970 Biochemical reactions of different serotypes of Pasteurella haemolytica. J Med Microbiol 3: 356–358.

Shull, G. M. 1959 Preparation of Δ¹, ⁴-steroid compounds by micromonospora. United States Patent 2,890,153, June 9.

Sickles, G. M. and M. Shaw. 1934 A systematic study of microorganisms which decompose the specific carbohydrates of the pneumococcus. J Bacteriol 28: 415–431.

Sijderius, R. 1946 Heterotrophe bacterien, die thiosulfaat oxydeeren. Thesis, Univ. Amsterdam, pp. 1–146.

Sijpesteijn, A. K. 1948 Cellulose-decomposing bacteria from the rumen of cattle. Thesis, University of Leiden, The Netherlands.

—. 1949 Cellulose decomposing bacteria from the rumen of cattle. Antonie van Leeuwenhoek J Microbiol Serol 15: 49–52.

—. 1949 Bactéries cellulolytiques de la panse des ruminants. Antonie van Leeuwenhoek J Microbiol Serol 15: 49–52.

—. 1951 On Ruminococcus flavefaciens, a cellulose-decomposing bacterium from the rumen of sheep and cattle. J Gen Microbiol 5: 869–879.

Sikora, H. 1918 Beiträge zur Kenntnis der Rickettsien. Arch Schiffs-Trop Hyg 22: 442–446.

Siminoff, P., R. M. Smith, W. T. Sokolski and G. M. Savage. 1957 Streptovaricin. I. Discovery and biologic activity. Amer Rev Tuberc 75 (4): 576–583.

Simmons, J. S. 1927 An acidfast organism isolated from a mouse Mycobacterium muris n. sp. J Infect Dis 41: 13–15.

Simmons, R. T. and E. V. Keogh. 1940 Physiological characters and serological types of haemolytic streptococci of groups B, C, and G from human sources. Aust J Exp Biol Med Sci 18: 151–161.

Simonds, J., P. A. Hansen and S. Lakshmanan. 1971 Deoxyribonucleic acid hybridization among strains of lactobacilli. J Bacteriol 107 (1): 382–384.

Simons, H. 1920 Eine saprophytische Oscillarie im Darm des Meerschweinschens. Zentrabl Bakteriol Parasitenk Infektionskr Hyg Abt II 50: 356–368.

—. 1922 Saprophytische Oscillatorien des Menschen und der Tiere. Zentrabl Bakteriol Parasitenk Infektionskr Hyg Abt I Orig 88: 501–510.

Simpson, C. F., J. M. Kling and F. C. Neal. 1965 The nature of bands in parasitized bovine erythrocytes. J Cell Biol 27: 225–235.

Simpson, F. J. and J. Robinson. 1968 Some energy-producing systems in Bdellovibrio bacteriovorus strain 6-5-S Can J Biochem 46: 865–873.

Sims, W. 1964 A simple test for differentiating Streptococcus bovis from other streptococci. J Appl Bacteriol 27: 432–433.

Skadhauge, K. 1950 Studies of enterococci. Einar Munksgaards, Copenhagen, pp. 1–197.

— and B. Perch. 1959 Studies on the relation-

ship of some alpha-hemolytic streptococci of human origin to the Lancefield Group M. Acta Pathol Microbiol Scand 46: 239–250.

Skalinskii, E. A. and E. N. Kondratieva. 1961 A new species of green sulfur bacteria. Dokl Akad Nauk SSSR 138: 456–457.

Skaptasan, J. B. and W. H. Burkholder. 1942 Classification and nomenclature of the pathogen causing bacterial ring rot of potatoes. Phytopathology 32: 439–441.

Skerman, T. M. and D. J. Jayne-Williams. 1966 Nutrition of coryneform bacteria from milk and dairy sources. J Appl Bacteriol 29: 167–178.

Skoric, V. 1938 Jasenov rak i njegov uzrocnik. (The ash-canker disease and its causal organism.) Ann Exp For, Zagreb 6: 66–97.

Skripal, I. G. 1970 Bacterium nodoantrum nova sp—an agent of apple tuberculosis. Mikrobiol Zh (Kyyiv) 32 (1): 50–53.

Skuja, H. 1948 Taxonomie des Phytoplanktons einiger Seen in Uppland. Symb Bot Upsal 9 (3): 1–399.

—. 1956 Taxonomische und biologische Studien über das Phytoplankton schwedischer Binnenge-wasser. Nova Acta Reg Soc Sci Upsal Ser IV 16 (3): 1–404.

—. 1958 Die Pelonematacee Desmanthos thiokrenophilum, ein Vertreter der apochromatischen Blaualgen aus Schwefelquellen. Svensk Bot Tidskr 52: 437–444.

—. 1964 Grundzuge der Algenflora und Algenvegetation der Fjeldgegenden um Abisko in Schwedisch-Lappland. Nova Acta Reg Soc Sci Upsal Ser IV 18 (3): 1–465.

Skyring, G. W. and C. Quadling. 1968 Soil bacteria: comparisons of rhizosphere and nonrhizosphere populations. Can J Microbiol 15: 473–488.

— and —. 1969 Taxonomy of Arthrobacter-coryneform soil isolates in relation to named cultures. Bacteriol Proc, p. 19.

Slack, J. M. 1942 The source of infection in actinomycosis. J Bacteriol 43: 193–209.

—. 1968 Subgroup on taxonomy of microaerophilic actinomycetes. Report on organization, aims and procedures. Int J Syst Bacteriol 18 (3): 253–262.

— and M. A. Gerencser. 1966 Revision of serological grouping of Actinomyces. J Bacteriol 91 (5): 2107.

—, S. Landfried and M. A. Gerencser. 1969 Morphological biochemical and serological studies on 64 strains of Actinomyces israelii. J Bacteriol 97 (2): 873–884.

Slade, H. D. and G. D. Shockman. 1963 The protoplast membrane and the group D antigen of Streptococcus faecalis. Iowa State J Sci 38: 83–96.

— and W. C. Slamp. 1962 Cell wall composition and the grouping antigens of streptococci. J Bacteriol 84 (3): 345–351.

Slaterus, K. W. 1961 Serological typing of meningococci by means of microprecipitation. Antonie van Leeuwenhoek J Microbiol Serol 27: 305–315.

—, A. C. Ruys and I. G. Sieberg. 1963 Types of meningococci isolated from carriers and patients in a non-epidemic period in the Netherlands. Antonie van Leeuwenhoek J Microbiol Serol 29: 265–271.

Sloger, C. and W. S. Silver. 1965 Note on nitrogen fixation by excised root nodules and nodular homogenates of *Myrica cerifera* L. *In* San Pietro (Editor), Non-heme Iron Proteins: Role in Energy Conversion. Symp. Spons. by Charles F. Kettering Res. Lab. Antioch Press, Yellow Springs, Ohio, pp. 299–302.

Slopek, S. 1968 *Klebsiella*. *In* Sedlák and Rische, Enterobakteriaceae-Infektionen 2 Auflage. VEB Georg Thieme, Leipzig.

— and I. Durlakowa. 1967 Studies on the taxonomy of *Klebsiella* bacilli. Arch Immunol Ther Exp *15:* 481–487.

— and J. Maresz-Babczyszyn. 1967 A working scheme for typing *Klebsiella* bacilli by means of pneumocins. Arch Immunol Ther Exp *15:* 525–529.

—, A. Przonko-Hessek, A. Milch and S. Déak. 1967 A working scheme for bacteriophage typing of *Klebsiella* bacilli. Arch Immunol Ther Exp *15:* 589–599.

Slotnick, I. J. and M. Dougherty. 1964 Further characterization of an unclassified group of bacteria causing endocarditis in man: *Cardiobacterium hominis* gen. et sp. n. Antonie van Leeuwenhoek J Microbiol Serol *30:* 261–272.

— and —. 1965 Unusual toxicity of riboflavin and flavin mononucleotide for *Cardiobacterium hominis*. Antonie van Leeuwenhoek J Microbiol Serol *31:* 355–360.

—, J. A. Mertz and M. Dougherty. 1964 Fluorescent antibody detection of human occurrence of an unclassified bacterial group causing endocarditis. J Infect Dis *114:* 503–505.

Small, E. and M. Ristic. 1967 Morphologic features of *Haemobartonella felis*. Amer J Vet Res *28:* 845–851.

Smibert, R. M. 1965 *Vibrio fetus* var *intestinalis* isolated from fecal and intestinal contents of clinically normal sheep: Biochemical and cultural characteristics of microaerophilic vibrios isolated from the intestinal contents of sheep. Amer J Vet Res *26:* 320–327.

—. 1969 *Vibrio fetus* var *intestinalis* isolated from the intestinal content of birds. Amer J Vet Res *30:* 1437–1442.

Smiley, K. L., C. F. Niven, Jr. and J. M. Sherman. 1943 The nutrition of *Streptococcus salivarius*. J Bacteriol *45:* 445–454.

Smit, J. 1930 Die Garungssarcinen. Eine Monographie. Pflanzenforschung Heft 14, pp. 1–59.

Smith, C. G., A. Dietz, W. T. Sokolski and G. M. Savage. 1956 Streptonivicin, a new antibiotic. I. Discovery and biologic studies. Antibiot Chemother *6 (2):* 135–142.

Smith, C. O. 1913 Black pit of lemon. Phytopathology *3:* 69.

Smith, D. G. and P. M. F. Shattock. 1962 The serological grouping of *Streptococcus equinus*. J Gen Microbiol *29 (4):* 731–736.

— and —. 1964 The cellular location of antigens in streptococci of groups D, M and Q. J Gen Microbiol *34 (1):* 165–175.

Smith, D. T., D. S. Martin, N. F. Conant, J. W. Beard, G. Taylor, H. I. Kohn and M. A. Poston. 1948 Zinsser's Text Book of Bacteriology, 9th ed. Appleton-Century-Crofts, New York, pp. 1–992.

Smith, E. F. 1895 *Bacillus tracheiphilus* sp. nov. die Ursache des Verwelkens verschiedener Curcurbitaceen. Zentrabl Bakteriol Parasitenk Infektionskr Hyg Abt II *1:* 364–373.

—. 1896 A bacterial disease of the tomato, eggplant and Irish potato (*Bacillus solanacearum* n. sp.). U S Dept Agr Div Veg Phys Pathol Bull *12:* 1–28.

—. 1897 Description of *Bacillus phaseoli* n. sp. Bot Gaz *24:* 192.

—. 1897 *Pseudomonas campestris* (Pammel). The cause of a brown rot in cruciferous plants. Zentrabl Bakteriol Parasitenk Infektionskr Hyg Abt II *3:* 284–291; 408–415; 478–485.

—. 1898 Notes on Stewart's sweet corn germ, *Pseudomonas stewarti* n. sp. Proc Amer Ass Adv Sci *47:* 422–426.

—. 1901 The cultural characters of *Pseudomonas hyacinthi, Ps. campestris, Ps. phaseoli* and *Ps. stewarti*—four one-flagellate yellow bacteria parasitic on plants. U S Dept Agr Div Veg Phys Pathol Bull *28:* 1–153.

—. 1903 Observations on a hitherto unreported bacterial disease, the cause of which enters the plant through ordinary stomata. Science (Washington) *17:* 456–457.

—. 1904 Bacterial leaf spot diseases. Science (Washington) *19:* 417–418.

—. 1908 Recent studies of the olive-tubercle organism. Bull U S Bur Plant Ind *131:* 25–43.

—. 1910 A new tomato disease of economic importance. Science (Washington) *31:* 794–796.

—. 1911 Bacteria in relation to plant diseases. Carnegie Inst Wash Publ *2:* 1–368.

—. 1913 A new type of bacterial disease. Science (Washington) *38:* 926.

—. 1914 Bacteria in relation to plant diseases. Carnegie Inst Wash Publ *3:* 1–309.

—. 1920 An introduction to bacterial diseases of plants. W. B. Saunders, Philadelphia, pp. 1–688.

—, N. A. Brown and C. O. Townsend. 1911 Crown gall of plants: its cause and remedy. U S Dept Agr Bur Plant Ind Bull *213:* 1–215.

— and M. K. Bryan. 1915 Angular leaf-spot of cucumbers. J Agr Res *5:* 465–476.

—, L. R. Jones and C. S. Reddy. 1919 The black chaff of wheat. Science (Washington) *50:* 48.

— and C. O. Townsend. 1907 A plant-tumor of bacterial origin. Science (Washington) *25:* 671–673.

Smith, F. 1936 Anaerobic pneumococcus. Brit J Exp Pathol *17:* 329–334.

—. 1948 Genus I. *Salmonella* Ligniéres. Appendix II. *In* Breed, Murray and Hitchens (Editors), Bergey's Manual of Determinative Bacteriology, 6th ed. The Williams & Wilkins Co., Baltimore, p. 533.

Smith, F. B. 1938 An investigation of a taint in rib bones of bacon. The determination of halophilic vibrios. Proc Roy Soc Queensland *49 (3):* 29–52.

Smith, F. R. and J. M. Sherman. 1939 *Streptococcus acidominimus*. J Infect Dis *65:* 301–305.

Smith, G. R. 1961 The characteristics of two types of *Pasteurella haemolytica* associated with different pathological conditions in sheep. J Pathol Bacteriol *81 (2):* 431–440.

Smith, H. Williams and S. Halls. 1967 The transmissible nature of the genetic factor in *Escherichia coli* that controls haemolysin production. J Gen Microbiol *47:* 153–161.

— and —. 1968 The transmissible nature of

the genetic factor in *Escherichia coli* that controls enterotoxin production. J Gen Microbiol *52:* 319–324.

Smith, I. W. 1963 The classification of '*Bacterium salmonicida.*' J Gen Microbiol *33 (2):* 263–274.

Smith, J. B. and M. G. P. Stoker. 1951 The nucleic acids of *Rickettsia burneti.* Brit J Exp Pathol *32:* 433–441.

Smith, J. E. 1958 Studies on *Pasteurella septica.* II. Some cultural and biochemical properties of strains from different host species. J Comp Pathol *68:* 315–323.

— and E. Thal. 1965 A taxonomic study of the genus *Pasteurella* using a numeral technique. Acta Pathol Microbiol Scand *64:* 213–223.

Smith, J. L. and D. R. Pesetsky. 1967 The current status of *Treponema cuniculi.* Review of the literature. Brit J Vener Dis *43:* 117–127.

Smith, L. DS. 1970 *Clostridium oceanicum* n. sp., a spore-forming anaerobe isolated from marine sediments. J Bacteriol *103:* 811–813.

— and L. V. Holdeman. 1968 The Pathogenic Anaerobic Bacteria. Charles C Thomas, Springfield, Ill., pp. 1–423.

— and E. King. 1962 *Clostridium innocuum,* sp. n., a spore-forming anaerobe isolated from human infections. J Bacteriol *83 (4):* 938–939.

Smith, N. R., R. E. Gordon and F. E. Clark. 1946 Aerobic mesophilic sporeforming bacteria. U S Dep Agr Misc Publ *559:* 1–112.

—, —, and —. 1952 Aerobic sporeforming bacteria. Agr Monogr 16 U S Dep Agr, pp. 1–148.

Smith, P. H. 1966 The microbial ecology of sludge methanogenesis. Develop Ind Microbiol *7:* 156–161.

— and R. E. Hungate. 1958 Isolation and characterization of *Methanobacterium ruminantium* n. sp. J Bacteriol *75 (6):* 713–718.

Smith, R. F. 1969 Characterization of human cutaneous lipophilic diphtheroids. J Gen Microbiol *55 (3):* 433–443.

Smith, R. M., D. A. Joslyn, O. M. Gruhzit, I. W. McClean, Jr., M. A. Penner and J. Ehrlich. 1948 Chloromycetin: biological studies. J Bacteriol *55 (3):* 425–448.

Smith, T. 1894 The hog-cholera group of bacteria. U S Bur Anim Ind Bull *6:* 6–40.

—. 1896 Two varieties of the tubercle bacillus from mammals. Trans Ass Amer Physicians Philadelphia *11:* 75–95.

—. 1918 A pleomorphic bacillus from pneumonic lungs of calves simulating Actinomyces. J Exp Med *28:* 333–343.

— and M. S. Taylor. 1919 Some morphological and biochemical characters of the spirilla (*Vibrio fetus,* n. sp.) associated with disease of the fetal membranes in cattle. J Exp Med *30:* 299–311.

Sneath, P. H. A. 1956 Conservation of the generic name *Chromobacterium* and designation of type species and type strains. Request for an opinion. Int Bull Bacteriol Nomencl Taxon *6 (2):* 65–91; *6 (4):* 157–158.

—. 1956 Cultural and biochemical characteristics of the genus *Chromobacterium.* J Gen Microbiol *15 (1):* 70–98.

—. 1960 A study of the bacterial genus *Chromobacterium.* Iowa State J Sci *34 (3):* 243–500.

—. 1966 Identification methods applied to *Chromobacterium. In* Gibbs and Skinner (Editors), Identification Methods for Micro-biologists, Part A. Academic Press, New York, pp. 15–20.

— and V. B. D. Skerman. 1966 A list of type and reference strains of bacteria. Int J Syst Bacteriol *16 (1):* 1–134.

Snell, W. H. and E. A. Dick. 1957 A Glossary of Mycology. Harvard University Press, Cambridge, Mass.

Snieszko, S. F. 1957 Genus IV. *Aeromonas* Kluyver and van Niel 1936. *In* Breed, Murray and Smith (Editors), Bergey's Manual of Determinative Bacteriology, 7th ed. The Williams & Wilkins Co., Baltimore, pp. 189–193.

—, P. J. Griffin and S. B. Friddle. 1950 A new bacterium (*Hemophilus piscium* n. sp.) from ulcer disease of trout. J Bacteriol *59:* 699–710.

Snijders, A. S. and B. C. Jansen. 1955 A comparison of *Streptothrix bovis* and *Actinomyces dermatonomus.* Bull Epizoot Dis Afr *3:* 242–243.

Snijders, E. P. 1924 Cavia-scheefkopperij, een nocardiose. Geneesk. Tijdschr Ned Ind *64:* 85–87.

Snow, G. A. 1965 The structure of Mycobactin P a growth factor for *Mycobacterium johnei,* and the significance of its iron complex. Biochem J *94:* 160–165.

Snyder, E. M. 1925 *Eberthella viscosa (Bact viscosum equi):* Etiological factor in Joint-ill. J Amer Vet Met Ass *66:* 481–486.

Snyder, M. L. 1936 The serologic agglutination of the obligate anaerobes *Clostridium paraputricum* (Bienstock) and *Clostridium capitovalis* (Snyder and Hall). J Bacteriol *32:* 401–410.

—, M. S. Slawson, W. Bullock and R. B. Parker. 1967 Studies on oral filamentous bacteria. II. Serological relationships with the genera *Actinomyces, Nocardia, Bacterionema* and *Leptotrichia.* J Infect Dis *117:* 341–345.

Sobeslavsky, O., B. Prescott and R. M. Chanock. 1968 Adsorption of *Mycoplasma pneumoniae* to neuraminic acid receptors of various cells and possible role in virulence. J Bacteriol *96 (3):* 695–705.

Sobin, B. A., A. C. Finlay and J. H. Kane. 1950 Terramycin and its production. United States Patent 2,516,080, July 18.

Societá Farmaceutici Italia. 1962 Nuovo antibiotico, suoi derivati e procedimeto per la loro preparazione (in Italian). Italian Patent 22,651/62, November 16.

Socransky, S. S., M. Listgarten, C. Hubersak, J. Cotmore and A. Clark. 1969 Morphological and biochemical differentiation of three types of small oral spirochetes. J Bacteriol *98 (3):* 878–882.

Soeda, M. 1959 Studies on marinamycin, an antitumor antibiotic substance (in Japanese). J Antibiot (Tokyo) Ser B *12 (4):* 300–304.

— and H. Fujita. 1959 Studies on akitamycin, an antifungal antibiotic (in Japanese). J Antibiot (Tokyo) Ser B *12 (4):* 293–294.

— and —. 1959 Studies on akitamycin, an antifungal antibiotic. II. Mycological study of akitamycin-producing *Streptomyces* (in Japanese). J Antibiot (Tokyo) Ser B *12 (4):* 295–296.

Sofiev, M. S. 1941 *Spirochaeta latyschewi* n. sp. of relapsing fever type. Med Parasitol Moscow *10 (2):* 337–373.

Sohier, F., F. Benazet and M. Piéchaud. 1948 Sur un germe du genre *Listeria* apparement non pathogène. Ann Inst Pasteur (Paris) *74:* 54–57.

Söhngen, N. L. 1906 Het ontstaan en verdwijnen van waterstof en methaan onder den invloed van het organische leven. Thesis, Delft, pp. 1–138.

—. 1906 Ueber Bakterien, welche Methan als Kohlenstoffnahrung und Energiequelle gebrauchen. Zentrabl Parasitenk Infektionskr Hyg Abt II *15:* 513–517.

—. 1913 Benzin, Petroleum, Paraffinol und Paraffin als Kohlenstoff—und Energiequelle für Mikroben. Zentrabl Bakteriol Parasitenk Infektionskr Hyg Abt II *37:* 595–609.

Sojka, W. J. 1965 *Escherichia coli* in animals. Commonwealth Agricultural Bureaux, Farnham Royal, Bucks., England.

Sollied, P. R. 1903 Studien über ein Einfluss von Alkohol auf die verschiedenen Brauerei und Brennereimaterialen sich verfindenden Organismen, sowie Beschreibung einer gegen Alkohol sehr wiederstandfähigen neuen Pediokokkusarte (*Pediococcus hennebergi* n. sp.). Z Spiritusind *26:* 491–493.

Solntseva, L. I. 1941 Biology of the Myxobacteria. II. Genera *Melittangium* and *Chondromyces.* Mikrobiologiya *10:* 505–524.

Solov'eva, N. K. and S. M. Rudaya. 1959 Description of the producer of a new antifungal antibiotic albofungin (in Russian). Antibiotiki *4 (6):* 5–10.

Somerson, N. L., D. Taylor-Robinson and R. M. Chanock. 1963 Hemolysin production as an aid in the identification and quantitation of Eaton agent (*Mycoplasma pneumoniae*). Amer J Hyg *77:* 122–128.

Soong, P. and A. A. Au. 1962 Litmomycin, a new antibiotic. I. Taxonomic studies on *Streptomyces litmogenes*, litmomycin producing culture (in Chinese). Rep Taiwan Sugar Exp Sta *29:* 33–42.

Soppeland, L. 1924 *Flavobacterium suaveolens*, a new species of aromatic bacillus isolated from dairy water. J Agr Res *28:* 275–276.

Sordelli, A. 1923 Flore anaérobie de Buenos-Aires. C R Soc Biol Paris *89:* 53–55.

Soriano, S. 1945 Un nuevo orden de bacterias: *Flexibacteriales.* Cienc Invest (Buenos Aires) *1:* 92–93.

— and R. A. Lewin. 1965 Gliding microbes: Some taxonomic considerations. Antonie van Leeuwenhoek J Microbiol Serol *31:* 66–80.

— and A. Soriano. 1948 Nueva bacteria anaerobia productora de una alteracion en sordinas envasadas. Rev Asoc Argent Dietol *6:* 36–41.

Speckman, R. A. and E. B. Collins. 1968 Diacetyl biosynthesis in *Streptococcus diacetilactis* and *Leuconostoc citrovorum.* J Bacteriol *95 (1):* 174–180.

Spiekermann, A. and P. Kotthoff. 1914 Untersuchungen über die Kartoffelpflanze und ihre Krankheiten. Landhr Jb *46:* 659–732.

Splitter, E. J. 1950 *Eperythrozoon suis*, the etiologic agent of icteroanemia or an anaplasmosis-like disease in swine. Amer J Vet Res *11:* 324–329.

—. 1950 *Eperythrozoon suis* n. sp. and *Eperythrozoon parvum* n. sp. two new blood parasites of swine. Science (Washington) *111:* 513–514.

—. 1952 Eperythrozoonosis in swine—filtration studies. Amer J Vet Res *13:* 290–297.

—. 1958 The complement fixation test in diagnosis of eperythrozoonosis in swine. J Amer Vet Med Ass *132:* 47–49.

—, E. R. Castro and W. L. Kanawyer. 1956 Feline infectious anemia. Vet Med *51:* 17–22.

Spray, R. S. 1948 Genus II. *Clostridium* Prazmowski. *In* Breed, Murray and Hitchens (Editors), Bergey's Manual of Determinative Bacteriology, 6th ed. The Williams & Wilkins Co., Baltimore, pp. 763–827.

— and P. S. Laux. 1930 A peculiar lactose-fermenting organism from filtered and chlorinated water. J Amer Water Works Ass *22:* 235–241.

Sreenivasan, A. 1955 A new species of marine chitin digesting bacterium. Curr Sci *24:* 270–271.

Srinivasan, M. C. and M. K. Patel. 1956 Three undescribed species of *Xanthomonas.* Curr Sci *25:* 366–367.

— and —. 1957 Two new phytopathogenic bacteria on verbenaceous hosts. Curr Sci *26:* 90–91.

—, — and M. J. Thirumalachar. 1961 A new bacterial blight disease of *Argemone mexicana.* Proc Nat Inst Sci India Part B Biol Sci *27:* 104–107.

—, — and —. 1961 A bacterial blight disease of Coriander. Proc Indian Acad Sci Sect B *53:* 298–301.

—, — and —. 1962 Two bacterial leaf-spot diseases on *Physalis minima* and studies on their relationship to *Xanthomonas vesicatoria* (Doidge) Dowson. Proc Indian Acad Sci Sect B *56:* 93–96.

Stableforth, A. W. 1959 Diseases due to bacteria. *In* Stableforth and Galloway (Editors), Infectious Diseases of Animals, Vol. 1. Butterworths, London, p. 60.

—. 1959 Streptococcal diseases. *In* Stableforth and Galloway (Editors), Infectious Diseases of Animals, Vol. 2. Butterworths, London, pp. 589–650.

Stadtman, E. R., T. C. Stadtman, I. Pastan and L. Ds. Smith. 1972 *Clostridium barkeri* sp. n. J Bacteriol *110 (2):* 758–760.

Stadtman, T. C. and H. A. Barker. 1951 Studies on the methane fermentation. IX. The origin of methane in the acetate and methanol fermentation by methanosarcina. J Bacteriol *61 (1):* 81–86.

— and —. 1951 Studies on the methane fermentation. A new formate—decomposing bacterium. J Bacteriol *62 (3):* 269–280.

— and L. S. McClung. 1957 *Clostridium sticklandii* nov. spec. J Bacteriol *73 (2):* 218–219.

Stahel, G. 1933 The witchesbrooms of *Eugenia latifolia* Aubl. in Surinam caused by *Pseudomonas hypertrophicans* nov. spec. Phytopathol Z *6:* 441–452.

Staley, J. T. 1968 *Prosthecomicrobium* and *Ancalomicrobium*, new prosthecate freshwater bacteria. J Bacteriol *95 (5):* 1921–1942.

— and M. Mandel. 1973 Deoxyribonucleic acid base composition of *Prosthecomicrobium* and *Ancalomicrobium* strains. Int J Syst Bacteriol *23:* 271–273.

Standring, E. T. 1942 *In* Lightle, Standring and Brown. A bacterial necrosis of the giant cactus. Phytopathology *32:* 303–313.

Stanford, J. L., S. R. Pattyn, F. Portaels and W. J. Gunthorpe. 1972 Studies on *Mycobacterium chelonei*. J Med Microbiol *5:* 177–182.

Stanier, R. Y. 1940 Studies on the cytophagas. J Bacteriol *40:* 619–636.

—. 1941 Studies on marine agar-digesting bacteria. J Bacteriol *42:* 527–558.

—. 1942 The *Cytophaga* group: A contribution to the biology of Myxobacteria. Bacteriol Rev *6:* 143–196.

—. 1943 A note on the taxonomy of *Proteus hydrophilus*. J Bacteriol *46:* 213–214.

—. 1947 Studies on nonfruiting myxobacteria. I. *Cytophaga johnsonae*, n. sp., a chitin-decomposing myxobacterium. J Bacteriol *53:* 297–315.

—. 1957 Order VIII. *Myxobacterales* Jahn 1915. *In* Breed, Murray and Smith (Editors), Bergey's Manual of Determinative Bacteriology, 7th ed. The Williams & Wilkins Co., Baltimore, pp. 854–891.

—. 1961 La place des bactéries dans le monde vivant. Ann Inst Pasteur (Paris) *101:* 297.

—. 1970 Some aspects of the biology of cells and their possible evolutionary significance. *In* Charles and Knight (Editors), Organization and control in procaryotic and eucaryotic cells. Cambridge University Press.

—, R. Kunisawa, M. Mandel and G. Cohen-Bazire. 1971 Purification and properties of unicellular blue-green algae (order *Chroococcales*). Bacteriol Rev *35:* 171.

—, N. J. Palleroni and M. Doudoroff. 1966 The aerobic pseudomonads: a taxonomic study. J Gen Microbiol *43 (2):* 159–271.

— and C. B. van Niel. 1941 The main outlines of bacterial classification. J Bacteriol *42:* 437.

— and —. 1962 The concept of a bacterium. Arch Mikrobiol *42:* 17.

Stanton, A. T. and W. Fletcher. 1921 Melioidosis, a new disease of the tropics. Trans 4th Congr Far-East Ass Trop Med *2:* 196–198.

Stapley, E. O., J. M. Mata, I. M. Miller, T. C. Demny and H. B. Woodruff. 1964 Antibiotic MSD-235. I. Production by *Streptomyces avidinii* and *Streptomyces lavendulae*. Antimicrob Agents Chemother *1963:* 20–27.

Stapp, C. 1920 Botanische Untersuchung einiger neuer Bakterienspezies, welche mit reiner Harnsäure oder Hippursäure als alleinigem organischen Nährstoff auskommen. Zentrabl Bakteriol Parasitenk Infektionskr Hyg Abt II *51:* 1–71.

—. 1928 *Schizomycetes* (Spaltpilze oder Bakterien). *In* Sorauer (Editor), Handbuch Pflanzenkrankheiten 5th ed., Vol. 2, P. Parey, Berlin, pp. 1–5.

—. 1935 Contemporary understanding of bacterial plant diseases and their causal organisms. Bot Rev *1:* 405–418.

— and H. Bortels. 1934 Mikrobiologische Untersuchungen über die Zersetzung von Waldstreu. Zentrabl Bakteriol Parasitenk Infektionskr Hyg Abt II *90:* 28–66.

— and D. Knösel. 1954 Zur Genetik sternbildender Bakterien. Zentrabl Bakteriol Parasitenk Infektionskr Abt II *108:* 243–259.

— and G. Spicher. 1954 Untersuchungen über die Wirkung von 2.4-D im Boden. IV. Mitteilung. *Flavobacterium peregrinum* n. sp. und seine Fähigkeit zum abbau des Hormones. Zentrabl Bakteriol Parasitenk Infektionskr Hyg Abt II *108:* 113–126.

Stark, P. and J. M. Sherman. 1935 Concerning the habitat of *Streptococcus lactis*. J Bacteriol *30:* 639–646.

Starkey, R. L. 1934 Cultivation of organisms concerned in the oxidation of thiosulfate. J Bacteriol *28:* 365–386.

—. 1935 Isolation of some bacteria which oxidize thiosulfate. Soil Sci *39:* 197–219.

—. 1938 A study of spore formation and other morphological characteristics of *Vibrio desulfuricans*. Arch Mikrobiol *9:* 268–304.

—. 1948 Family I. *Nitrobacteriaceae* Buchanan. *In* Breed, Murray and Hitchens (Editors), Bergey's Manual of Determinative Bacteriology, 6th ed. The Williams & Wilkins Co., Baltimore, pp. 69–81.

— and P. K. De. 1939 A new species of *Azotobacter*. Soil Sci *47:* 329–343.

Starr, M. P. 1946 The nutrition of phytopathogenic bacteria. I. Minimal nutritive requirements of the genus *Xanthomonas*. J Bacteriol *51:* 131–143.

—. 1946 The nutrition of phytopathogenic bacteria. II. The genus *Agrobacterium*. J Bacteriol *52:* 187–194.

—. 1947 The causal agent of bacterial root and stem disease of guayule. Phytopathology *37:* 291–300.

—. 1949 The nutrition of phytopathogenic bacteria III. The Gram-positive phytopathogenic *Corynebacterium* species. J Bacteriol *57 (2):* 253–258.

— and N. L. Baigent. 1966 Parasitic interaction of *Bdellovibrio bacteriovorus* with other bacteria. J Bacteriol *91 (5):* 2006–2017.

—, W. Blau and G. Cosens. 1960 The blue pigment of *Pseudomonas lemonnieri*. Biochem Z *333 (4):* 328–334.

— and W. H. Burkholder. 1942 Lipolytic activity of phytopathogenic bacteria determined by means of spirit blue agar and its taxonomic significance. Phytopathology *32:* 598–604.

—, C. Cardona and D. Folsom. 1951 Bacterial fireblight of raspberry. Phytopathology *41:* 915–919.

— and C. Garces. 1950 El agente causante de la gomosis bacterial del pasto imperial en Colombia. Rev Fac Nac Agron Medellin Colombia *12:* 73–83.

— and D. A. Kuhn. 1962 On the origin of V-forms in *Arthrobacter atrocyaneus*. Arch Mikrobiol *42:* 289–298.

— and M. Mandel. 1969 DNA base composition and taxonomy of phytopathogenic and other Enterobacteria. J Gen Microbiol *56 (1):* 113–123.

— and P. P. Pirone. 1942 *Phytomonas poinsettiae* n. sp., the cause of a bacterial disease of Poinsettia. Phytopathology *32:* 1076–1081.

— and S. Saperstein. 1953 Thiamine and the carotenoid pigments of *Corynebacterium poinsettiae*. Arch Biochem *43:* 157–168.

— and V. B. D. Skerman. 1965 Bacterial diversity: the natural history of selected morphologically unusual bacteria. Annu Rev Microbiol *19:* 407–454.

— and J. E. Weiss. 1943 Growth of phytopathogenic bacteria in a synthetic asparagin medium. Phytopathology *33:* 314–318.

Staub, A. M. and O. Westphal. 1964 Étude chimique et biochimique de la spécificité immunologique des polysides bactériens. Bull Soc Chim Biol *46:* 1647–1684.

Steed, P. D. M. 1962 *Simonsiellaceae* fam. nov.

with characterization of *Simonsiella crassa* and *Alysiella filiformis*. J Gen Microbiol 29 (4): 615–624.

Steel, K. J. and S. T. Cowan. 1964 Le rattachement de *Bacterium anitratum*, *Moraxella lwoffi*, *Bacillus mallei* et *Haemophilus parapertussis* au genre *Acinetobacter* Brisou et Prévot. Ann Inst Pasteur (Paris) 106: 479–483.

— and J. Midgley. 1962 Decarboxylase and other reactions of some gram-negative rods. J Gen Microbiol 29 (1): 171–178.

Stefansky, W. K. 1903 Eine lepraahnliche Erkrankung der haut und der Lymphdrusen bei Wanderratten. Zentrabl Bakteriol Parasitenk Infektionskr Hyg Abt I Orig 33: 481–487.

Steinberg, B. A., W. P. Jambor, L. O. Suydam and A. Soriano. 1956 Thiostrepton, a new antibiotic. III. *In vivo* studies. Antibiot Ann 1955-1956: 562–565.

Steinberg, S. 1862 Studies of the white soft substance which accumulates on and between the teeth. Souremennaya Med 20: 377–380; 22: 417–423; 23: 433–439; 452–458.

Steinhaus, E. A. 1941 A study of the bacteria associated with thirty species of insects. J Bacteriol 42 (6): 757–790.

—. 1942 Rickettsia-like organism from normal *Dermacentor andersonii* Stiles. Pub Health Rep 57: 1375–1377.

—. 1946 Insect Microbiology. Comstock Pub Co., Ithaca, N. Y.

—, M. M. Batey and C. L. Boerke. 1956 Bacterial symbiotes from the caeca of certain *Heteroptera*. Hilgardia 24: 495–518.

Steinkraus, K. H. and H. Tashiro. 1955 Production of milky-disease spores (*Bacillus popilliae* Dutky and *Bacillus lentimorbus* Dutky) on artificial media. Science (Washington) 121: 873–874.

Stellmach-Helwig, R. 1961 Morphologische und physiologische Eigenschaften einiger aus Schafpansen isolierter Bakterienstämme. Arch Mikrobiol 38: 40–51.

Stephens, J. W. W. and S. R. Christophers. 1904 The practical study of malaria and other blood parasites. Williams and Northgate, London.

Sternberg, G. M. 1890 Report of etiology and prevention of yellow fever. United States Marine Hosp. Serv., Washington, pp. 1–271.

—. 1892 Manual of Bacteriology. W. Wood & Co., New York, pp. 1–874.

Stevens, C. L., P. Blumbergs, F. A. Daniher, R. W. Wheat, A. Kujomoto and E. L. Rollins. 1963 The identification and synthesis of the 4-amino sugar from *Chromobacterium violaceum*. J Amer Chem Soc 85: 3061.

Stevens, F. L. 1913 The fungi which cause plant disease. MacMillan, New York.

—. 1925 Plant disease fungi. MacMillan, New York.

Stevens, W. C. 1956 Taxonomic studies on the genus *Bacteroides* and similar forms. Thesis, Vanderbilt University.

Stevenson, I. L. 1961 Growth studies on *Arthrobacter globiformis*. Can J Microbiol 7: 569–575.

—. 1963 Some observations on the so-called 'cystites' of the genus *Arthrobacter*. Can J Microbiol 9: 467–472.

Stewart, F. S. and J. D. McKeever. 1963 Serological definition of Lancefield group K. J Pathol Bacteriol 85: 383–388.

— and M. McLaughlin. 1965 A red-cell sensitizing antigen of group D streptococci. Immunology 9: 319–326.

Stiles, C. W. and C. A. Pfender. 1905 The generic name *Spironema* Vuillemin 1905 (not Meek, 1864, Mullusk)—*Microspironema* Stiles and Pfender 1905 of the parasite of syphilis. Amer Med 10: 936.

Stillman, E. G. and M. Bourn. 1920 Biological study of the haemophilic bacilli. J Exp Med 32: 665–682.

Stimson, A. M. 1907 Note on an organism found in yellow-fever tissue. Pub Health Rep 22: 541.

Stirm, S., F. Ørskov and B. Mansa. 1967 Episome-carried surface antigen K88 of *Escherichia coli*. J Bacteriol 93 (2): 731–739.

Stitzenberger, E. 1860 Rabenhorst's Algen Sachsen's resp Mitteleuropa's. Decade I. C. Heinrich, Dresden, pp. 1–41.

Stock, C. C., H. C. Reilly, S. M. Buckley, D. A. Clarke and C. P. Rhoads. 1954 Azaserine, a new tumor-inhibitory substance. Nature (London) 173: 71–72.

Stocker, B. A. D. 1958 Lysogenic conversion by the A phages of *Salmonella typhimurium*. J Gen Microbiol 18 (1): IX.

—, R. G. Wilkinson and H. Mäkelä. 1966 Genetics aspects of biosynthesis and structure of *Salmonella* somatic polysaccharide. Ann N Y Acad Sci 133: 334–348.

Stocks, P. K. and C. S. McCleskey. 1964 Identity of the pink-pigmented methanol-oxidizing bacteria as *Vibrio extorquens*. J Bacteriol 88 (4): 1065–1070.

— and —. 1964 Morphology and physiology of *Methanomonas methanooxidans*. J Bacteriol 88 (4): 1071–1077.

Stoenner, H. G. and D. B. Lackman. 1957 A new species of *Brucella* isolated from the desert wood rat, *Neotoma lepida* Thomas. Amer J Vet Res 18: 947–951.

—, — and E. J. Bell. 1962 Factors affecting the growth of rickettsias of the spotted fever group in fertile hens' eggs. J Infect Dis 110: 121–128.

Stoker, M. G. P. and P. Fiset. 1956 Phase variation of Nine Mile and other strains of *Rickettsia burneti*. Can J Microbiol 2: 310–321.

Stokes, J. L. 1954 Studies on the filamentous sheathed iron bacterium *Sphaerotilus natans*. J Bacteriol 67: 278–291.

— and M. T. Powers. 1965 Formation of rough and smooth strains of *Sphaerotilus discophorus*. Antonie van Leeuwenhoek J Microbiol Serol 31: 157–164.

Stolp, H. J. and M. P. Starr. 1963 *Bdellovibrio bacteriovorus* gen. et sp. n., a predatory, ectoparasitic and bacteriolytic microorganism. Antonie van Leeuwenhoek J Microbiol Serol 29: 217–248.

Stonehill, E. H. and D. J. Hutchison. 1966 Chromosomal mapping by means of mutational induction in synchronous populations of *Streptococcus faecalis*. J Bacteriol 92 (1): 136–143.

Storz, J. and L. A. Page. 1971 Taxonomy of the *Chlamydiae*: Reasons for classifying organisms of the genus *Chlamydia*, family *Chlamydiaceae*, in a separate order, *Chlamydiales* ord. nov. Int J Syst Bacteriol 21 (4): 332–334.

Stow, I. 1954 A new disease of hop due to a bacterium. Proc Jap Acad 30: 226–251.

— and K. Ihara. 1955 On a new disease-germ of the hop plant. Proc Jap Acad *31:* 294–299.

Strauss, I. and N. Gamaléia. 1891 Recherches expérimentales sur la tuberculose, la tuberculose humaine, sa distinction de la tuberculose des oiseaux. Arch Med Exp *3:* 457–484.

Strong, R. P., E. E. Tyzzer, C. T. Brues, A. W. Sellards and J. C. Gastiaburú. 1913 Verruga peruviana, Oroya fever and uta. J Amer Med Ass *61:* 1713–1716.

—, — and A. W. Sellards. 1915 Oroya fever. Second report. J Amer Med Ass *64:* 806–808.

Struyk, A. P. and A. A. Stheeman. 1955 Process for the preparation of antibiotics. Canadian Patent 514,164, June 28.

— and —. 1957 Process for the preparation of antibiotics. Australian Patent 206,799, April 21

Strzeszewski, B. 1913 Beiträge zur Kenntnis der Schwefelflora in der Umgebung von Krakau. Bull Int Acad Sci Cracovie (Acad Pol Sci) Ser B Sci Nat I, pp. 309–334.

Stuart, C. A., S. Formal and V. McGann. 1949 Further studies on B5W, an anaerogenic group in the *Enterobacteriaceae*. J Infect Dis *84:* 235–239.

—, K. M. Wheeler, R. Rustigian and A. Zinnemann. 1943 Biochemical and antigenic relationships of the paracolon bacteria. J Bacteriol *45:* 101–119.

Stuart, R. D. 1946 The preparation and use of a simple culture medium for leptospira. J Pathol Bacteriol *58 (3):* 343–349.

Stuart, S. E. and H. J. Welshimer. 1973 Intrageneric relatedness of *Listeria* Pirie. Int J Syst Bacteriol *23:* 8–14.

— and —. 1974 A taxonomic reexamination of *Listeria* Pirie. Int J Syst Bacteriol *24:* in press.

Stubenrath, F. C. 1896 *Sarcina variabilis. In* Lehmann and Neumann (Editors), Atlas und Grundriss der Bakteriologie und Lehrbuch der speciellen bacteriologischen Diagnostik. Teil II. J. F. Lehmanns, München, p. 143.

Sturges, W. S. and E. T. Drake. 1927 A complete description of *Clostridium putrefaciens* (McBryde). J Bacteriol *14:* 175–179.

— and A. G. Heideman. 1924 Studies of halophilic organisms II. The flora of meat-curing solutions. Abstr Bacteriol *8:* 14–15.

Stutzer, A. and R. Hartleb. 1898 Untersuchungen über das im Alinit enthaltene Bakterium. Zentrabl Bakteriol Parasitenk Infektionskr Hyg Abt II *4:* 31–39.

Subcommittee on Actinomycetes of the Taxonomy Committee of the Amer. Soc. Microbiol. (Gottlieb, D., Chairman). 1964 Recommendations for description of some *Actinomycetales* appearing in patent applications. Amer Soc Microbiol News *30 (2):* 13–14.

Sugai, T. 1956 New antibiotics 229 and 229B of colorless, water-soluble and basic nature (in Japanese). J Antibiot (Tokyo) Ser B *9 (5):* 170–179.

Sugawara, R., A. Matsumae and T. Hata. 1957 Melanomycin, a new antitumor substance from streptomyces. I. J Antibiot (Tokyo) Ser A *10 (4):* 133–137.

— and M. Onuma. 1957 Melanomycin, a new antitumor substance from *Streptomyces*. II. Description of the strain. J Antibiot (Tokyo) Ser A *10 (4):* 138–142.

Sugimoto, M. 1935 Anaplasmosis-like disease in Formosan swine. J Soc Trop Agr Taiwan *7:* 240–244.

Sugiura, K. and C. C. Stock. 1955 IV. Effect of *O*-diazoacetyl-L-serine (azaserine) on growth of various mouse and rat tumors. Proc Soc Exp Biol Med *88 (1):* 127–129.

Suitor, E. C., Jr. 1964 Propagation of *Rickettsiella popilliae* (Dutky and Gooden) Philip and *Rickettsiella melolonthae* (Krieg) Philip in cell cultures. J Insect Pathol *6:* 31–40.

—. 1964 The relationship of *Wolbachia persica* Suitor and Weiss to its host. J Insect Pathol *6:* 111–124.

— and E. Weiss. 1961 Isolation of a rickettsialike microorganism (*Wolbachia persica* n. sp.) from *Argas persicus* (Oken). J Infect Dis *108:* 95–106.

Sukapure, R. S., M. P. Lechevalier, H. Reber, M. L. Higgins, H. A. Lechevalier and H. Prauser. 1970 Motile nocardial *Actinomycetales*. Appl Microbiol *19:* 527–533.

Sumiki, Y., G. Nakamura, M. Kawasaki, S. Yamashita, K. Anzai, K. Isono, Y. Serizawa, Y. Tomiyama and S. Suzuki. 1955 A new antibiotic, homomycin. J Antibiot (Tokyo) Ser A *8 (5):* 170.

—, K. Sakaguchi and T. Asai. 1957 Grasseriomycin, a new antibiotic for jaundice virus of *Bombyx mori* (in Japanese). Japanese Patent 6,296, August 15.

Sundman, V. 1958 Morphological comparison of some *Arthrobacter* species. Can J Microbiol *4:* 221–224.

—, K. Bjorksten and H. G. Gyllenberg. 1959 Morphology of bifid bacteria. J Gen Microbiol *21:* 371–384.

Surman, P. G. 1968 Cytology of "pink-eye" of sheep, including a reference to trachoma of man, by employing acridine orange and iodine stains, and isolation of *Mycoplasma* agents from infected sheep eyes. Aust J Biol Sci *21:* 447–467.

Suter, L. 1951 A new species of *Nocardia, N. fastidiosa* n. sp. isolated from a penile ulcer. Mycologia *4:* 658–676.

Suter, L. S. and B. F. Vaughan. 1955 The effect of antibacterial agents on the growth of *Actinomyces bovis*. Antibiot Chemother *5:* 557–560.

Sŭtić, D. and W. J. Dowson. 1959 An investigation of serious disease of hemp (*Cannabis sativa* L.) in Jugoslavia. Phytopathol Z *34:* 307–314.

Suto, R., K. Kurashima, Y. Namba, M. Matsuki and C. Furusaka. 1960 A new butanol producing anaerobic nitrogen fixer, *Clostridium aurantiacum*, nov. spec. J Agr Chem Soc Jap *34:* 117–121.

Suto, T. 1957 Some properties of an acid tolerant azotobacter, *Azotobacter indicum*. Tohoku J Agr Res *7:* 369–382.

Sutter G. R., and V. M. Kirk. 1968 Rickettsialike particles in fat-body cells of carabid beetles. J Invertebr Pathol *10:* 445–449.

Suzanne, M. and G. Penso. 1953 Sulla identita specifica del considdetto "ceppo Chauvire" *Mycobacterium marianum* n. sp. Riassunti della Comunicazioni, Vol. II. VI International Congress for Microbiology, Rome. pp. 382–383.

Suzuki, H., Y. Ozawa and O. Tanabe. 1964 Studies on the decomposition of raffinose by α-galactosidase of actinomycetes. Part II.

Identification of strains (in Japanese). J Agr Chem Soc Jap *38* (*7*): 334–336.

—, — and —. 1965 Studies on the decomposition of raffinose by α-galactosidase of actinomycetes. 2. Identification of strains (in Japanese). Rept No. 28, Ferm. Res. Inst. Agency Ind. Sci Technol (Japan), pp. 63–68.

— and O. Tanabe. 1963 Studies on the decomposition of raffinose by α-galactosidase of actinomycetes. Part I. Isolation and selection of strains (in Japanese). J Agr Chem Soc Jap *37* (*10*): 623–625.

Suzuki, J. and K. Kitahara. 1964 Base composition of deoxyribonucleic acid in *Sporolactobacillus inulinus* and other lactic acid bacteria. J Gen Appl Microbiol *10*: 305–311.

Suzuki, M. 1957 Studies on an antitumor substance, gancidin. Mycological study on the strain AAK-84 and production, purification (*sic*) of active fractions (in Japanese). J Chiba Med Soc *33* (*3*): 535–542.

Suzuki, S., K. Isono, J. Nagatsu, T. Mizutani, Y. Kawashima and T. Mizuno. 1965 A new antibiotic, polyoxin A. J Antibiot (Tokyo) Ser A *18* (*3*): 131.

— and S. Marumo. 1960 Chemical structure of tubercidin. J Antibiot (Tokyo) Ser A *13* (*5*): 360.

—, G. Nakamura, K. Okuma and Y. Tomiyama. 1958 Cellocidin, a new antibiotic. J Antibiot (Tokyo) Ser A *11* (*3*): 81–83.

Svenkerud, R. R., A. F. Rosted and K. Thorshaug. 1951 En lidelse hos sel som minner meget om grisens rodsyke. Nord Vet Med *3*: 147–149.

Swellengrebel, N. H. 1907 Sur la cytologie comparée des spirochètes et des spirilles. Ann Inst Pasteur (Paris) *21*: 448–466; 562–586.

Swift, H. F., A. T. Wilson and R. C. Lancefield. 1943 Typing group A hemolytic streptococci by M-precipitin reactions in capillary pipettes. J Exp Med *78*: 127–133.

Swingle, D. B. 1925 Center rot of "french endive" or wilt of chicory (*Cichorium intybos* L.). Phytopathology *15*: 730.

Switzer, W. P. 1953 Studies on infectious atrophic rhinitis of swine. I. Isolation of a filterable agent from the nasal cavity of swine with infectious atrophic rhinitis. J Amer Vet Med Ass *123*: 45–47.

—. 1955 Studies on infectious atrophic rhinitis. IV. Characterization of a pleuropneumonia-like organism isolated from the nasal cavities of swine. Amer J Vet Res *16*: 540–544.

—. 1964 Mycoplasmosis. *In* Dunne (Editor), Diseases of Swine. 2nd ed. Iowa State University Press, Ames, pp. 498–507.

Sylvester, C. J. and R. N. Costilow. 1964 Nutritional requirements of *Bacillus popilliae*. J Bacteriol *87* (*1*): 114–119.

Szabó, I. and M. Marton. 1958 A *Streptomyces vastus* és a *Streptomyces viridoniger* új sugárgomba fajokról (Adatok a szikestalajok mikrobiológiájához) (in Hungarian). Agrokem Talajtan 7 (*3*): 243–262.

— and —. 1964 Comments on the first results of the international cooperative work on criteria used in characterization of streptomycetes. Int Bull Bacteriol Nomencl Taxon *14* (*1*): 17–38.

—, I. Buti and G. Pártai. 1963 *Actinomyces finlayi* n. sp. Acta Microbiol Acad Sci Hung *10* (*3*): 207–214.

—, —, L. Ferenczy and I. Buti. 1967 Intestinal microflora of the larvae of St. Mark's fly. II. Computer analysis of intestinal actinomycetes from the larvae of a *Bibio*-population. Acta Microbiol Acad Sci Hung *14*: 239–249.

Szafer, W. 1911 Zur Kenntnis der Schwefelflora in der Umgebung von Lemberg. Bull Int Acad Sci Cracovie (Acad Pol Sci) Ser B Sci Nat I pp. 160–167.

Szaniszlo, P. J. 1967 Comparison of the cell-wall composition and intracellular pigmentation of some strains of Actinoplanaceae. Thesis, University of North Carolina, pp. 1–26.

Szilvinyi, A. 1936 *Pseudomonas nivalis* n. sp., ein Beitrag zur Kenntnis des Genus *Serratia* Bizio. Zentrabl Bakteriol Parasitenk Infektionskr Hyg Abt II *94*: 216–218.

Szulga, T., P. A. Jenkins and J. Marks. 1966 Thin-layer chromatography of mycobacterial lipids as an aid to classification; *Mycobacterium kansasii*, and *Mycobacterium marinum* (*balnei*). Tubercle *47*: 130–136.

Szulmajster, J. and P. Kaiser. 1960 Étude d'une nouvelle espèce anaérobie: *Eubacterium sarcosinogenum* nov. sp. Ann Inst Pasteur (Paris) *98*: 774–777.

Szybalski, W. 1950 A comparative study of bacteria causing mustiness in eggs. Nature (London) *165*: 733.

Tacquet, A., A. Andrejew and V. Macquet. 1958–59 Pouvoir pathogène pour la poule et le lapin des mycobactéries aviaires resistantes á de fortes concentrations d'isoniazide. Ann Inst Pasteur Lille *10*: 29–36.

—, A Collet, J. C. Martin, B. Devulder and C. Gernez-Rieux. 1967 Pulmonary dusting and infection of guinea pigs by *Mycobacterium kansasii*. Z Tuberk Erkrankungen Thoraxorgane *127*: 115–122.

—, F. Tison and B. Devulder. 1964 Quelques aspects actuels des infections bronchopulonaires provoquées par les mycobactéries dites "atypiques." Rev Tuberc *28*: 89–116.

Taguchi, H. and K. Yoshikawa. 1961 Antibiotic substances produced by microorganisms. VIII. A new antibiotic, ferromycin (in Japanese). J Ferment Technol *39* (*1*): 44–48.

Takahashi, T. 1906–1908 Studies on diseases of Saké. Bull Coll Agr Tokyo Imp Univ *7*: 531–563.

— and T. Asai. 1930 On the gluconic acid fermentation. Part I. On *Bacterium hoshigaki* var. *rosea*. Bull Agr Chem Soc Jap *6*: 83–99.

— and —. 1930 On gluconic acid fermentation. Part I. On *Bacterium hoshigaki* var. *rosea* nov. spec. Zentrabl Bakteriol Parasitenk Infektionskr Hyg Abt 2 *82*: 390–405.

Takebe, I. and K. Kitahara. 1963 Levels of nicotinamide nucleotide coenzymes in lactic acid bacteria. J Gen Appl Microbiol *9*: 31–40.

Takeda, Y. and T. Matsui. 1955 Mycological studies of acetone-butanol fermenting bacteria (KN-18) (*Clostridium kainatoi* n. sp.). J Agr Chem Soc Jap *29*: 78–82.

Takeuchi, T., T. Hikiji, K. Nitta and H. Umezawa. 1957 Effect of pluramycin A on Ehrlich carcinoma of mice. J Antibiot (Tokyo) Ser A *10* (*4*): 143–152.

—, —, —, S. Yamazaki, S. Abe, H. Takayama and H. Umezawa. 1957 Biological studies of kanamycin. J Antibiot (Tokyo) Ser A *10* (*3*): 107–114.

—, K. Nitta and H. Umezawa. 1956 Antitumor effect of pluramycin crude powder on Ehrlich carcinoma of mice. J Antibiot (Tokyo) Ser A *9* (*1*): 22–30.

Takikawa, I. 1958 Studies on pathogenic halophilic bacteria. Yokohama Med Bull *9:* 313–322.

Takimoto, S, 1920 On the bacterial leaf-spot of *Antirrhinum majus* L. Bot Mag (Tokyo) *34:* 253–257.

—. 1927 Bacterial black spot of burdock. J Plant Prot Tokyo *14:* 519–523.

—. 1933 The bacterial disease of New Zealand flax. J Plant Prot Tokyo *20:* 774–778.

—. 1934 Leaf-spot of begonia. J Plant Prot Tokyo *21:* 258–262.

Takita, T., K. Ohi, Y. Okami, K. Maeda and H. Umezawa 1962 New antibiotics, ilamycins. J Antibiot (Tokyo) Ser A *15* (*1*): 46–48.

Talbot, J. M. and P. H. A. Sneath. 1960 A taxonomic study of *Pasteurella septica* especially of strains from human sources. J Gen Microbiol *22* (*1*): 303–311.

Tamaguchi, T. 1965 Comparison of cell wall composition of morphologically distinct actinomycetes. J Bacteriol *89* (*2*): 444–453.

Tamiya, T. 1962 Recent advances in studies in tsutsugamushi disease in Japan. Med Culture Inc., Tokyo.

Tanaka, H., W. T. Hall, J. B. Sheffield and D. H. Moore. 1965 Fine structure of *Haemobartonella muris* as compared with *Eperythrozoon coccoides* and *Mycoplasma pulmonis*. J Bacteriol *90:* 1735–1749.

Tanaka, N., K. Karasawa, N. Miyairi, N. Shinjo, T. Nishimura and H. Umezawa. 1958 Raromycin, a new tumor-inhibitory antibiotic produced by a *Streptomyces*. II. Taxonomic studies of the raromycin-producing organism. J Gen Appl Microbiol *4:* 259–271.

—, H. Yamazaki, K. Okabe and H. Umezawa. 1957 Raromycin, a new tumor-inhibitory antibiotic produced by a *Streptomyces*. I. Studies with Ehrlich carcinoma and Croker (*sic*) sarcoma 180 of mice. J Antibiot (Tokyo) Ser A *10* (*5*): 189–194.

Tanner, F. W., J. W. Davisson, A. C. Finlay and J. H. Kane. 1954 Acetopyrrotine and preparation thereof. United States Patent 2,689, 854, September 21.

Tardieux, P. 1951 Étude de deux espèces nouvelles du genre *Spherophorus*. Ann Inst Pasteur (Paris) *80:* 275–280.

Tarr, H. A. 1958 Mechanical aids for the phage-typing of *Staphylococcus aureus*. Mon Bull Min Health Public Health Lab Serv *17:* 64–72.

Tartakowsky, J. G. 1910 Piroplasmose bei Fledermausen (*Vespertilio noctula*) und ihre Vermittler. Trav IXᵉ Cong Int Med Vet *4:* 242–244.

Tasman, A. and A. C. Brandwijk. 1938 Experiments on metabolism with diphtheria bacillus. II. J Infect Dis *63:* 10–20.

— and —. 1940 Experiments on metabolism with *C. diphtheriae*. III. J Infect Dis *67:* 282–291.

Tatsuoka, S., A. Miyake, H, Hitomi, J. Ueyanagi, H. Iwasaki, Y. Yamaguchi, K.-I. Kanazawa, T. Araki, K. Tsuchiya, F. Hiraiwa, K. Naka-

zawa and M. Shibata. 1961 HON, a new antibiotic produced by *Streptomyces akiyoshiensis* nov. sp. J Antibiot (Tokyo) Ser A *14* (*1*): 39–43.

Tauschel, H. D. and G. Drews 1968 Thylakoidmorphogenese bei *Rhodopseudomonas palustris*. Arch Mikrobiol *59:* 381–404.

Tavel, E. and O. Lanz. 1893 Ueber die Aetiologie der Peritonitis. Mitt Klin Medic Inst Schweiz, Basel and Leipzig, pp. 1–177.

Taylor, J., H. J. Bensted, J. S. K. Boyd, K. P. Carpenter, W. J. Dowson, R. Lovell, E. W. Taylor, H. G. Thornton, G. S. Wilson and C. Shaw. 1952 Classification of the *Bacteriaceae*. Int Bull Bacteriol Nomencl Taxon *2* (*4*): 137–140.

Taylor, M. M. 1972 *Eubacterium fissicatena* sp. nov. isolated from the alimentary tract of the goat. J Gen Microbiol *71:* 457–463.

— and R. Storck. 1964 Uniqueness of bacterial ribosomes. Proc Nat Acad Sci USA *52:* 958.

Taylor, R. M., M. Lisborne and G. Roman. 1932. Recherches sur l'identification des *Brucella* isolées en France par l'action bacteriostatique des matières colorantes et la production d'hydrogène sulfuré (Huddleson). Ann Inst Pasteur (Paris) *49:* 284–302.

Taylor-Robinson, D., J. Canchola, H. Fox and R. M. Chanock. 1964 A newly identified oral mycoplasma (*M. orale*) and its relationship to other human mycoplasmas. Amer J Hyg *80:* 135–148.

—, H. Fox and R. M. Chanock. 1965 Characterization of a newly identified mycoplasma from the human oropharynx. Amer J Epidemiol *81:* 180–191.

— and R. J. Manchee. 1967 Spermadsorption and spermagglutination by mycoplasmas. Nature (London) *215:* 484–487.

Tchan, Y. T. 1953 Studies of N-fixing bacteria. IV. Taxonomy of Genus *Azotobacter* (Beijerinck, 1901). Proc Linnean Soc N S W *78:* 85–89.

—. 1957 Studies of nitrogen-fixing bacteria. VI. A new species of nitrogen-fixing bacteria. Proc Linnean Soc N S W *82:* 314–316.

— and J. Pochon. 1950 Un espèce nouvelle de bactérie fixatrice d'azote moleculaire isolée du soil: *Endosporus azotophagus*. C R Acad Sci Paris *230:* 417–418.

Temple, K. L. and A. R. Colmer. 1951 The autotrophic oxidation of iron by a new bacterium, *Thiobacillus ferrooxidans*. J Bacteriol *62:* 605–611.

Ten, H.-M. 1969 A new manganese oxidizing soil microorganism. Dokl Akad Nauk SSSR *188:* (*3*) 697–699.

Tešić, Ž. 1966 IV. Determining keys for bacteria, actinomycetes and fungi (in Serbo-Croatian). *In* Tešić (Editor), Priručnik za Ispitivanje Zemljišta. Knjiga II. Mikrobiološke Metode Ispitivanja Zemljišta i Voda. Yugoslav Society of Soil Science, Beograd, pp. 85–122.

—. 1966 Determinative keys for bacteria, actinomycetes and fungi. *In* Manual for soil investigation. Microbiological methods for soil and water investigation. Knjiga II, Beograd, pp. 85–94.

— and M. Todorovic. 1952 Nouvelle bactérie

phytopathogène, *Bacterium sambuci*. Pl Prot Belgrade *11:* 3–12.

Tewfik, E. M. and S. G. Bradley. 1967 Characterization of deoxyribonucleic acids from streptomycetes and nocardiae. J Bacteriol *94:* (*6*)*:* 1994–2000.

Thal, E. 1954 Untersuchungen über *Pasteurella pseudotuberculosis*. Berlingska Boktryckeriet, Lund.

—. 1955 Immunisierung gegen *Pasteurella pestis* mit einem avirulenten Stämm der *Pasteurella pseudotuberculosis*. Nord Vet Med *7:* 151–153.

—. 1956 Relations immunologiques entre *Pasteurella pestis* et *Pasteurella pseudotuberculosis*. Ann Inst Pasteur (Paris) *91:* 68–73.

—. 1962 Oral immunization of guinea-pigs with avirulent "*Pasteurella pseudotuberculosis*." Nature (London) *194:* 490–491.

—. 1966 Immunobiologische Studien an *Yersinia pseudotuberculosis* (syn. *Pasteurella pseudotuberculosis*). Schweiz Arch Tierheilk *108* (*7*)*:* 372–388.

— and T. Chen. 1955 Two simple tests for the differentiation of plague and pseudotuberculosis bacilli. J Bacteriol. *69* (*1*)*:* 103–104.

— and L. O. Kallings. 1955 Zur Bestimuung des genus *Salmonella* mit Hilfe eines Bakteriophagen. Nord Vet Med *7:* 1063–1071.

— and W. Knapp. 1969 The revised antigenic scheme of *Yersinia pseudotuberculosis*. Symposia Series in Immunobiological Standardization, Bern. *15:* 219–222.

—, — and E. Hanko. 1967 Comparative immunization experiments with *Yersinia pseudotuberculosis* (syn. *Pasteurella pseudotuberculosis*). Zentrabl Bakteriol Parasitenk Infektionskr Hyg Abt I Orig *204:* 399–404.

Thaxter, R. 1890 The potato "scab." Presentation before Botanical Section at meeting of Assoc. of Agricultural Colleges and Experiment Stations, Champaign, Illinois. Nov. 12. Published *in* Thaxter, R. 1891 Report of the mycologist. Conn Agr Exp Sta Rept *1890:* 81–95.

—. 1892 Potato scab. Annu Rep Conn Agr Esp *1891:* 153–160.

—. 1892 On the *Myxobacteriaceae*, a new order of Schizomycetes. Bot Gaz *17:* 389–406.

—. 1893 A new order of *Schizomycetes*. Bot Gaz *18:* 29–30.

—. 1897 Further observations on the *Myxobacteriaceae*. Bot Gaz *23:* 395–411.

—. 1904 Notes on the *Myxobacteriaceae*. Bot Gaz *37:* 405–416.

Thaysen, A. C. and I. Thaysen. 1955 Comirin, a new fungistatic antibiotic. Atti VI Congr Intern Microbiol *1* (*2*)*:* 638.

Theilade, E. and M. N. Gilmour. 1961 An anaerobic oral filamentous microorganism. J Bacteriol *81* (*1*)*:* 661–666.

Theiler, A. 1910 *Anaplasma marginale* (gen. and spec. nov.). The marginal points in the blood of cattle suffering from a specific disease. Transvaal S Afr Rep Vet Bacteriol Dep Agr *1908-9* (*1910*)*:* 7–64.

—. 1911 Further investigations into anaplasmatosis of South Africa cattle. 1st Rept Dir Vet Res, August 1911, pp. 7–46.

Thiele, H. H. 1968 Die Verwertung einfacher organischer Substrate durch *Thiorhodaceae*. Arch Mikrobiol *60:* 124–138.

Thiemann, J. E. 1967 A new species of the genus *Amorphosporangium* isolated from Italian soil. Mycopathol Mycol Appl *33:* 233–240.

—. 1970 Study of some new genera and species of the *Actinoplanaceae*. *In* Prauser (Editor), The *Actinomycetales*, The Jena International Symposium on Taxonomy, Sept. 1968. VEB Gustav Fischer-Verlag Jena, 245–257.

— and G. Beretta. 1966 Alanosine, a new antiviral and antitumor antibiotic from *Streptomyces*. Description of the strain and antibiotic publication. J Antibiot (Tokyo) Ser A *19* (*4*)*:* 155–160.

— and —. 1968 A new genus of the *Actinoplanaceae: Planobispora* gen nov. Arch Mikrobiol *62:* 157–166.

—, C. Coronelli, H. Pagani, G. Beretta, G. Tamoni and V. Arioli. 1968 Antibiotic production by new form-genera of the *Actinomycetales*. I. Sporangiomycin, an antibacterial agent isolated from *Planomonospora parontospora* var. *antibiotica* var. nov. J Antibiot (Tokyo) Ser A *21* (*9*)*:* 525–531.

—, Y. K. S. Murthy and C. Coronelli. 1966 Antibiotic alanosine. South African Patent 65/3609, January 1.

—, H. Pagani and G. Beretta. 1967 A New Genus of the *Actinoplanaceae: Planomonospora*, gen. nov. G Microbiol *15:* 27–38.

—, — and —. 1967 A new Genus of the *Actinoplanaceae: Dactylosporangium*, gen. nov. Arch Mikrobiol *58:* 42–52.

—, — and —. 1968 A new genus of the *Actinomycetales: Microtetraspora* gen. nov. J Gen Microbiol *50:* 295–303.

Thiercelin, E. 1902 Procédés faciles pour isoler l'entérocoque des selles normales; filtration des selles; culture préalable en anaérobie. C R Soc Biol Paris *54:* 1082–1083.

— and L. Jouhaud. 1903 Reproduction de l'entérocoque; taches centrales; granulations périphériques et microblastes. C R Soc Biol Paris *55:* 686–688.

Thiroux, A. and J. Pelletier. 1912 Mycetome à grains rouges de la paroi thoracique. Isolement et culture d'une nouvelle oospora pathogène. Bull Soc Pathol Exot, pp. 585–589.

Thirumalachar, M. J. 1955 *Chainia*, a new genus of the *Actinomycetales*. Nature (London) *176:* 934–935.

—. 1966 Process of producing hamycin antibiotic and product produced. United States Patent 3,261,751, July 19.

— and V. V. Bhatt. 1960 Some *Streptomyces* species producing oxytetracycline. Hindustan Antibiot Bull *3* (*2*)*:* 61–63.

—, N. V. Bringi, P. V. Deshmukh and P. W. Rahalkar. 1964 Antiprotozoin, a new antifungal-antiprotozoal antibiotic. Hindustan Antibiot Bull *7* (*1*)*:* 25–29.

—, —, —, —, R. Indira and K. S. Gopalkrishnan. 1964 Streptorubin A and B. New antibiotics with cytostatic properties. Hindustan Antibiot Bull *7:* 18–24.

— and S. K. Menon. 1962 Dermostatin, a new antifungal antibiotic. I. Microbiological studies. Hindustan Antibiot Bull *4* (*3*)*:* 106–108.

—, — and V. V. Bhatt. 1961 Hamycin, a new antifungal antibiotic. I. Discovery and biological studies. Hindustan Antibiot Bull *3:* 136–138.

—, P. W. Rahalkar, P. V. Deshmukh and R. S. Sukapure. 1965 Production of aburamycin by *Chainia minutisclerotica*, a new species of actinomycete. Hindustan Antibiot Bull *8 (1)*: 6–9.

— and R. S. Sukapure. 1964 Studies on species of the genus *Chainia* from India. Hindustan Antibiot Bull *6 (4)*: 157–166.

—, —, P. W. Rahalkar and K. S. Gopalkrishnan. 1966 Studies on species of the genus *Chainia* from India II. Hindustan Antibiot Bull *9 (1)*: 10–14.

Thjøtta, T. and J. Bøe. 1938 *Neisseria hemolysans*. A hemolytic species of *Neisseria* Trevisan. Acta Pathol Microbiol Scand Suppl *37*: 527–531.

—, J. O. Hartman and J. Bøe. 1939 A study of the *Leptotrichia* Trevisan. Avh Norske Videnskaps-Akad Oslo Mat-Naturvidensk Kl *5*: 1–199.

— and S. Sydnes. 1951 *Actinobacillus actinomycetam comitans* as the sole infecting agent in a human being. Acta Pathol Microbiol Scand *28*: 27–35.

Thomas, C. G. A. and R. Hare. 1954 The classification of anaerobic cocci and their isolation in normal human beings and pathological processes. J Clin Pathol (London) *7*: 300–304.

Thomas, C. L. (Editor). 1973 Taber's Cyclopedic Medical Dictionary. F. A. Davis Co., Philadelphia.

Thomas, S. O., L. V. Singleton, J. A. Lowery, R. W. Sharpe, L. M. Pruess, J. N. Porter, J. H. Mowat and N. Bohonos. 1957 Nucleocidin, a new antibiotic with activity against trypanosomes. Antibiot Annu *1956–1957*: 716–721.

Thomas, W. D. and L. E. Dickens. 1953 A new bacterium causing leaf pimple of carnation. J Colo Wyo Acad Sci *4*: 22.

— and A. R. Weinhold. 1953 *Xanthomonas striaformans*, a new bacterial pathogen of onion. J Colo Wyo Acad Sci *4*: 23.

Thompson, L. 1933 The systematic relationships of *Actinobacillus*. J. Bacteriol *26*: 221–227.

—. 1950 Isolation and comparison of *Actinomyces* from human and bovine infections. Proc Staff Meetings Mayo Clinic *25*: 81–86.

— and S. A. Lovestedt. 1951 An actinomyceslike organism obtained from the human mouth. Proc Staff Meetings Mayo Clinic *26 (10)*: 169–175.

— and F. A. Willius. 1932 Actinobacillus bacteraemia. J Amer Med Ass *99*: 298–300.

Thompson, R. D. 1852 Über die Natur und die chemische Wirkungen der Essigmutter. Liebigs Ann Chem *83*: 89–93.

Thompson, R. E. M. and K. A. Bisset. 1957 *Polysepta*: A new genus and sub-order of bacteria. Nature (London) *179*: 590– 591.

Thompson, R. S. and E. R. Leadbetter. 1963 On the isolation of the dipicolinic acid from endospores of *Sarcina ureae*. Arch Mikrobiol *45*: 27–32.

Thomsen, A. 1929 Smitsom Kastningsenzooti (Bang Infektion) blandt sper i Midtjulland. Maandsskr Dyraeg *41*: 386–388.

Thomson, D. 1923 Discovery of a new type of germ isolated from cases of measles and scarlet fever (minute gram-negative anaerobic diplococcus). J Trop Med Hyg *26*: 227–228.

Thöni, J. and O. Allemann. 1910 Über das Vorkommen von gefärbten, makroskopischen Bakterienkolnien in Emmentalerkasen. Zentrabl Bakteriol Parasitenk Infektionskr Hyg Abt II *25*: 8–30.

Thornberry, H. H. and H. W. Anderson. 1937 Comparative studies of *Phytomonas lactucaescariolae* n. sp. and *Phytomonas pruni*. Phytopathology *27*: 109–110.

— and —. 1937 Some bacterial diseases of plants in Illinois. Phytopathology *27*: 946–949.

Thornley, M. J. 1967 A taxonomic study of *Acinetobacter* and related genera. J Gen Microbiol *49 (2)*: 211–257.

— and R. W. Horne. 1962 Electron microscopic observations on the structure of fimbriae with particular reference to *Klebsiella* strains by use of negative staining technique. J Gen Microbiol *28 (8)*: 51–56.

— and M. E. Sharpe. 1959 Microorganisms from chicken meat related to both lactobacilli and aerobic sporeformers. J Appl Bacteriol *22*: 368–376.

Thouvenot, H. and A. Florent. 1954 Étude d'un anaérobie du sperme du taureau et du vagin de la vache *Vibrio bubulus* Florent 1953. Ann Inst Pasteur (Paris) *86*: 237–240.

Thrum, H., K. Eckhardt, R. Fügner and G. Bradler. 1967 Pigment and antibiotic production of *Streptomyces californicus* (Waksman and Curtis 1916) Waksman and Henrici 1948 (in German). Z Allg Mikrobiol *7*: 121–127.

Tidel'skaya, I. L. 1939 A new anaerobic thermophilic species, *Cl. thermofermentans*. Mikrobiol Zh (Kyyiv) *6 (3)*: 175–191.

Tilford, P. E. 1936 Fasciation of sweet peas caused by *Phytomonas fascians* n. sp. J Agr Res *53*: 383–394.

Tillett, W. S. and T. Francis, Jr. 1930 Serological reactions in pneumonia with a non-protein somatic fraction of *Pneumococcus*. J Exp Med *52*: 561–571.

Tisdale, W. B. and M. M. Williamson. 1923 Bacterial spot of lima bean. J Agr Res *25*: 141–154.

Tissier, H. 1900 Recherches sur la flore intestinale des nourrissons. Thèses, Paris pp. 1–253.

—. 1908 Recherches sur la flore intestinale normale des enfants agés d'un an à cinq ans. Ann Inst Pasteur (Paris) *22*: 189–208.

—. 1918 Le bacille de barat. C R Seances Soc Biol Filiales *81*: 426–427.

—. 1926 Coccus anaérobie des selles de l'homme. C R Soc Biol Paris *94*: 447–448.

— and Martilly. 1902 Rescherches sur la putrifaction de la viande de boucherie. Ann Inst Pasteur (Paris) *16*: 865–903.

t'Mannetje, L. 1967 A re-examination of the taxonomy of the genus *Rhizobium* and related genera using numerical analysis. Antonie van Leeuwenhoek J Microbiol Serol *33*: 477–491.

Tochinai, Y. 1932 The black rot of rice grains caused by *Pseudomonas itoana* n. sp. Ann Phytopathol Soc Jap *2*: 253–257.

Tokhtamuratov, E. and A. B. Silaev. 1965 Recovery and purification of antibiotics produced by *Actinomyces tumemacerans* (in Russian). Antibiotiki *10*: 30–33.

— and S. M. Khodzhibaeva. 1964 Isolation of an antitumor substance from the culture fluid of *Actinomyces tumemacerans* P42 (in Russian). Antibiotiki *9*: 205–208.

Tomina, I. V. and V. D. Fedorov. 1967 Ultrafine structure of *Chlorobium thiosulfatophilum*, a green sulphur bacterium. Mikrobiologiya *36:* 663–666.

Tominaga, T. 1967 Bacterial blight of orchard grass caused by *Xanthomonas translucens* f. sp. *hordei* Hagborg. Jap J Bacteriol *22:* 628–633.

Topley, W. W. C. 1955 Topley and Wilson's Principles of Bacteriology and Immunity. I. Revised by G. S. Wilson and A. A. Miles. Edward Arnold and Co., London, pp. 1–642.

— and G. S. Wilson. 1929 The Principles of Bacteriology and Immunity. 1st ed. Edward Arnold and Co., London, pp. 1–587.

— and —. 1936 The Principles of Bacteriology and Immunity, 2nd ed. Edward Arnold and Co., London, pp. 1–1645.

Topping, L. E. 1937 The predominant microorganisms in soils. I. Description and classification of the organisms. Zentrabl Bakteriol Parasitenk Infektionskr Hyg Abt II *97:* 289–304.

Topping, N. H. and M. J. Shear. 1945 Studies of typhus fever vaccines. III. Studies of antigens in infected yolk sacs. Nat Inst Health Bull *183:* 13–17.

Tornabene, T. G., M. Kates, E. Gelpi and J. Oro. 1969 The occurrence of squalene, di- and tetrahydroxysqualenes and vitamin MK8 in *Halobacterium cutirubrum.* J Lipid Res *10:* 294–303.

Torry, J. C. and A. H. Rahe. 1915 A new member of the aciduric group of bacilli. J Infect Dis *17:* 437–441.

Tošić, J. and T. K. Walker. 1944 *Acetobacter* infection. I. *Acetobacter mobile* (sp. nov.) J Inst Brew *50:* 296–300.

— and —. 1950 *Acetobacter acidum-mucosum* Tošić and Walker, n. sp., an organism forming a starch-like polysaccharide. J Gen Microbiol *4:* 192–197.

Trapp, G. 1936 A bacillus isolated from diseased plants of *Aucuba japonica* (Thunb.). Phytopathology *26:* 257–265.

Traum, I. 1914 Immature and hairless pigs. Rep Dep Agr, Washington, p. 86.

Trejo, W. and R. E. Bennett. 1963 *Streptomyces* species comprising the blue-spore series. J Bacteriol 85 (*3*): 676–690.

Tresner, H. D. and E. J. Backus. 1956 A broadened concept of the characteristics of *Streptomyces hygroscopicus.* Appl Microbiol *4* (*5*): 243–250.

— and —. 1963 System of color wheels for streptomycete taxonomy. Appl Microbiol *11* (*4*): 335–338.

—, M. C. Davies and E. J. Backus. 1961 Electron microscopy of *Streptomyces* spore morphology and its role in species differentiation. J Bacteriol *81* (*1*): 70–80.

—, J. A. Hayes and E. J. Backus. 1966 *Streptomyces prasinosporus* sp. nov. A new green-spored species. Int J Syst Bacteriol *16* (*2*): 161–169.

—, — and —. 1968 Differential tolerance of streptomycetes to sodium chloride as a taxonomic aid. Appl Microbiol *16* (*8*): 1134–1136.

Trevisan, V. 1842 Prospetto della Flora Euganea. Coi Tipi Del Seminario, Padova, pp. 1–68.

—. 1845 Nomenclator Algarum. Impr du seminaire, Padone, pp. 58–59.

—. 1879 Prime linee d'introduzione allo die batterj italiani. R C Inst Lombardo (Ser 2) *12:* 133–151.

—. 1884 A proposito del bacillo del cholera. Koch o Pacini? Intorno al modo di agire del Bacille nel corpo umano. Gaz Med Ital Milano (Ser 8) *6:* 373–736.

—. 1885 Il fungo del cholera asiatico. Questioni risolte. Atti Accad Fis-Med-Stat Milano (Ser 4) *3:* 78–91.

—. 1885 Caratteri di alcuni nuovi generi di Batteriacee. Atti Accad Fis-Med-Stat Milano (Ser 4) *3:* 92–107.

—. 1887 Sul Micrococco della rabbia e sulla possiblità di riconoscere durante il periodo d'incubazione, dall'esame del sangue della persona moricata, se ha contratta l'infezione rabbica. Rend Ist Lombardo (Ser 2) *20:* 88–105.

—. 1889 I Generi e le Specie delle Battieriacee. Zanaboni and Gabuzzi, Milano.

Troili-Petersson, G. 1903 Studien über die Mikroorganismen des schwedischen güterkäses. Zentrabl Bakteriol Parasitentk Infektionskr Hyg Abt II *11:* 120–143.

Trotsenko, Y. A. 1966 Green photosynthetic bacteria from Sernoya Lake. Mikrobiologiya *35:* 1087–1093.

Trüper, H. G. 1968 *Ectothiorhodospira mobilis* Pelsh, a photosynthetic bacterium depositing sulfur outside the cells. J Bacteriol *95* (*5*): 1910–1920.

— and S. Genovese. 1968 Characterization of photosynthetic sulfur bacteria causing red water in Lake Faro (Messina, Sicily). Limnol Oceanogr *13:* 225–232.

— and H. W. Jannasch. 1968 *Chromatium buderi* nov. spec., eine neue Art der "grossen" *Thiorhodaceae.* Arch Mikrobiol *61:* 363–372.

— and N. Pfennig. 1966 Sulphur metabolism in *Thiorhodaceae.* III. Storage and turnover of thiosulphate sulphur in *Thiocapsa floridana* and *Chromatium* species. Antonie van Leeuwenhoek J Microbiol Serol *32:* 261–276.

— and —. 1969 Proposal to declare *Rhodopseudomonas spheroides* as *nomen conservandum.* Int J Syst Bacteriol *19* (*2*): 155–156.

— and —. 1971 Family of phototrophic green sulfur bacteria: *Chlorobiaceae* Copeland, the correct family name; rejection of *Chlorobacterium* Lauterborn; and the taxonomic situation of the consortium-forming species. Int J Syst Bacteriol *21* (*1*): 8–10.

Tsai, J.-s., T.-y. Su, C.-c Pao, S.-f. Liang, L.-c Wu, T.-c. Wu, M.-c. Liu, C.-c. Ch'u, P. Hsü and W. Kurylowicz. 1957 Actinomycin K. A new antitumorous substance (in Chinese). K'o Hsuen T'ung Pao K'un Ch'ung Hsuch Pao *1957:* 717–718.

Tschekan, L. 1929 Mikrobiologie der Busa. Zentrabl Bakteriol Parasitenk Infektionskr Hyg Abt II *78:* 74–93.

Tsenkovskii, L. 1878 Gel formation of sugar beet solutions (trans. title, orig. Russian). Proc Soc Sci Nat Imper Univ Kharkov *12:* 137–167.

Tsiklinsky, P. 1899 Sur les mucedinéés thermophiles. Ann Inst Pasteur (Paris) *13:* 500–505.

Tsilasani, G. A. 1966 Effect of irradiation on biological properties of causal organism of bacterial canker (*Xanthomonas gorlencovianum*) of tea bushes. Radiobiologiya *6:* 323–325.

Tsukamura, M. 1965 A group of mycobacteria

from soil sources resembling nonphotochromogens (Group 3). Med Biol 71: 110–113.

—. 1966 Mycobacterium parafortuitum: a new species. J Gen Microbiol 42 (1): 7–12.

—. 1966 Adansonian classification of mycobacteria. J Gen Microbiol 45 (2): 253–273.

—. 1966 Classification of mycobacteria I. Adansonian classification of slowly growing mycobacteria. Med Biol 72: 75–78.

—. 1966 Mycobacterium chitae, a new species. A preliminary report. Med Biol 73 (4): 203–205.

—. 1967 Two types of slowly growing, nonphotochromogenic mycobacteria obtained from soil by the mouse passage method: Mycobacterium terrae and Mycobacterium novum. Jap J Microbiol 11: 163–172.

—. 1971 Proposal of a new genus Gordona for slightly acid-fast organisms in sputum of patients with pulmonary disease and in soil. J Gen Microbiol 68: 15–26.

— and S. Mizuno. 1968 "Hypothetical Mean Organisms" of Mycobacteria. Jap J Microbiol 12 (4): 371–384.

—, — and S. Tsukamura. 1968 Classification of rapidly growing mycobacteria. Jap J Microbiol 12: 151–166.

Tsukiura, H., M. Okanishi, H. Koshiyama, T. Ohmori, T. Miyaki and H. Kawaguchi. 1964 Proceomycin, a new antibiotic. J Antibiot (Tokyo) Ser A 17 (6): 223–229.

—, —, T. Ohmor, H. Koshiyama, T. Miyaki, H. Kitazima and H. Kawaguchi. 1964 Danomycin, a new antibiotic. J Antibiot (Tokyo) Ser A 17 (2): 39–47.

Tsyganov, V. A., R A. Zhukova and K. A. Timofeeva. 1963 A new genus of actinomycetes. Macrospora gen. nov. (in Russian). Materials of the 3rd Scientific Symposium of the Institute of Antibiotics of Leningrad pp. 90–93.

—, — and —. 1964 Morphological and biochemical pecularities of a new series, actinomycetes 2732/3. Mikrobiologiya 33 (5): 863–869.

Tuboly, S. 1965 Studies on the antigenic structure of mycobacteria. I. Comparison of the antigenic structure of pathogenic and sprophytic mycobacteria. Acta Microbiol Acad Sci Hung 12: 233–240.

Tucker, D. N., I. J. Slotnick, E. O. King, B. Tynes, J. Nicholson and L. Crevasse. 1962 Endocarditis caused by a Pasteurella-like organism. Report of four cases. N Engl J Med 267: 913–916.

Tulasne, R. and J. Brisou. 1955 Les Pleuropneumoniales. Taxonomie des pleuropneumonia-like organisms et des formes L. Ann Inst Pasteur (Paris) 88: 237–239.

Tully, J. G. 1965 Biochemical, morphological and serological characterization of mycoplasma of murine origin. J Infect Dis 115: 171–185.

—, M. F. Barile, R. A. Del Giudice, T. R. Carski, D. Armstrong and S. Razin. 1970 Proposal for classifying strain PG-24 and related canine mycoplasmas as Mycoplasma edwardii sp. n. J Bacteriol 101 (2): 346–349.

—, R. A. Del Giudice and M. F. Barile. 1972 Synonymy of Mycoplasma arginini and Mycoplasma leonis. Int J Syst Bacteriol 22: 47–49.

— and S. Razin. 1969 Characteristics of a new sterol-nonrequiring Mycoplasma. J Bacteriol 98 (3): 970–978.

— and —. 1970 Acholeplasma axanthum, sp. n.: A new sterol-nonrequiring member of the Mycoplasmatales. J Bacteriol 103 (3): 751–754.

— and I. Ruchman. 1964 Recovery, identification and neurotoxicity of Sabin's type A and C mouse mycoplasma (PPLO) from lyophilized cultures. Proc Soc Exp Biol Med 115: 554–558.

Tunnicliff, E. A. 1941 A study of Actinobacillus lignieresi from sheep affected with actinobacillosis. J Infect Dis 69: 52–58.

Tunnicliff, R. 1915 Further observations on the bacteriology of rhinitis with special reference to an anaerobic organism, Bacillus rhinitis. J Infect Dis 16: 493–495.

—. 1917 The cultivation of a micrococcus from blood in pre-eruptive and eruptive stages of measles. J Amer Med Ass 68: 1028–1030.

—. 1933 Colony formation of Diplococcus rubeolae (measles). J Infect Dis 52: 39–53.

—. 1936 Opsonins for Diplococcus morbillorum and for Streptococcus scarlatinae in convalescent measles serum, convalescent scarlet fever and placental extract. J Infect Dis 58: 1–4.

— and L. Jackson. 1925 Bacillus gonidiaformans (n. sp.). An hitherto undescribed organism. J Infect Dis 36: 430–438.

Tuomi, J. 1966 Taxonomic position of pathogenic, tick-borne rickettsia-like organisms (in Finnish). Suomen Elainlaakarilehti 72: 415–422.

—. 1966 Studies on epidemiology of bovine tick-borne fever in Finland and a clinical description of field cases. Ann Med Exp Biol Fenn 44: Supp. 6, 1–62.

— and C. H. von Bonsdorff. 1966 Electron microscopy of tick-borne fever agent in bovine and ovine phagocytizing leukocytes. J Bacteriol 92: 1478–1492.

Turfitt, G. E. 1944 Microbiological agencies in the degradation of steroids. I. The cholesterol-decomposing organisms of soil. J Bacteriol 47: 487–493.

Turner, A. W. 1935 A study of the morphology and life cycles of the organism of Pleuropneumoniae contagiosa boum (Borrelomyces peripneumoniae nov. gen) by observation in the living state under dark-ground illumination. J Pathol Bacteriol 41: 1–32.

—. 1954 Bacterial oxidation of arsenite. I. Description of bacteria isolated from arsenical cattle-dipping fluids. Aust J Biol Sci 7: 452–478.

—. 1960 Letter. Int Bull Bacteriol Nomencl Taxon 10: 255–256.

Turner, L. H. 1966 Part of discussion. Ann Soc Belg Méd Trop 46: 83.

—. 1967 Leptospirosis I. Trans Roy Soc Trop Med Hyg 61: 842–855.

Twort, F. W. and G. L. Y. Ingram. 1913 A monograph on Johne's disease. Balliere, Tindall and Cox, London.

Tyeryar, F. J. and R. N. Doetsch. 1962 Protoplasts of the giant bacterium Caryophanon latum Peshkoff. Nature (London) 195: 1327–1328.

Tyeryar, F. J., Jr., E. Weiss, D. B. Millar, F. M. Bozeman and R. A. Ormsbee. 1973 DNA base composition of rickettsiae. Science (Washington) 180: 415–417.

Tyzzer, E. E. 1938 Cytoectes microti N. G., (n. sp.) a parasite developing in granulocytes and infective for small rodents. Parasitology *30:* 242–257.

—. 1942 A comparative study of *Grahamellae, Haemobartonellae* and *Eperythrozoa* in small mammals. Proc Amer Phil Soc *85:* 359–398.

— and D. Weinman. 1939 *Haemobartonella* n.g. (*Bartonella* olim pro parte), *H. microti* n. sp. of the field vole, *Microtus pennsylvanicus.* Amer J. Hyg *30:* 141–157.

Ucke, A. 1898 Ein Beitrag zur Kenntnis der Anäeroben. Zentrabl Bakteriol Parasitenk Infektionskr Hyg Abt I Orig *23:* 996–1001.

Ueda, K., S. Ishikawa, T. Itami and T. Asai. 1952 Studies on the aerobic mesophilic cellulose-decomposing bacteria. Part 5. I. Taxonomic study. J Agr Chem Soc Jap *25:* 543–549.

—, —, — and —. 1952 Studies on the aerobic mesophilic cellulose-decomposing bacteria. Taxonomic study of the Genus *Pseudomonas.* J Agr Chem Soc Jap *26:* 35–41.

Ueno, T. and S. Omata. 1961 On the anaerobic halotolerant and halophilic bacteria in soy-sauce mash (I) Taxonomic Studies. (Studies on the halophilic bacteria II). J Ferment Technol *39:* 360–370.

Uesaka, I. 1952 Studies on the antibiotic action of *Nocardia* III. Taxonomic studies of A 422 strain. J Antibiotics *5:* 75–79.

Uffen, R. L. and R. S. Wolfe. 1970 Anaerobic growth of purple nonsulfur bacteria under dark conditions. J Bacteriol *104 (1):* 426–472.

Uhlenhuth, P. and Fromme. 1916 Untersuchungen über die Aetiologie, Immunität and spezifische Behandlung der Weilschen Krankheit (*Icterus infectiosus*). Z Immunitaetsforsch Exp Ther *25:* 317–483.

Uma, B. N. 1961 Formation of champamycin-B by *Streptomyces champavatii.* Ph.D. Thesis, Indian Inst of Sci, Bangalore, India.

— and P. L. Narasimha Rao. 1959 Actinomycetes. I. Distribution of streptomycetes in Indian soils. Formation of antifungal antibiotics by *Streptomyces champavati* n. sp. Indian Inst Sci Golden Jubilee Res, Vol. 1909–1959. Bangalore, India, pp. 130–141.

Umezawa, H. 1967 Index of Antibiotics from Actinomycetes. University of Tokyo Press, Tokyo, Japan, and University Park Press, State College Pennsylvania.

—, S. Hayano, K. Maeda, Y. Ogata and Y. Okami. 1950 On a new antibiotic, griseolutein, produced by *Streptomyces.* Jap Med J *3:* 111–117.

—, — and Y. Ogata. 1948 On the differentiation of streptomycin and allied substances (streptothricin group) and rapid isolation of streptomycin-producing strain. Jap Med J *1 (4):* 339–346.

— and Y. Okami. 1957 Changing the name of pyridomycin producing streptomycete. J Antibiot (Tokyo) Ser A *10 (4):* 172.

—, T. Takeuchi, Y. Okami, K. Nitta and K. Maeda. 1959 Pluramycin, an antitumor substance (in Japanese). Japanese Patent 7598, August 29.

—, —, — and T. Tazaki. 1953 On screening of antiviral substances produced by *Streptomyces* and on an antiviral substance achromoviromycin. Jap J Med Sci Biol *6 (3):* 261–268.

—, T. Tazaki and S. Fukuyama. 1951 On antiviral substance, abikoviromycin, produced by *Streptomyces* species. Jap Med J *4:* 331–346.

—, —, H. Kanari, Y. Okami and S. Fukuyama.

1948 Isolation of crystalline antibiotic substance from a strain of *Streptomyces* and its identity with Chloromycetin. Jap Med J *1:* 358–363.

—, —, Y. Okami and S. Fukuyama. 1949/1950 On the new source of Chloromycetin, *Streptomyces omiyanensis* (in Japanese). J Antibiot (Tokyo) *3 (5):* 292–296.

—, M. Ueda, K. Maeda, K. Yagashita, S. Kondo, Y. Okami, R. Utahara, Y. Osato, K. Nitta and T. Takeuchi. 1957 Production and isolation of a new antibiotic, kanamycin. J Antibiot (Tokyo) Ser A *10 (5):* 181–188.

Umezawa, S., Y. Tanaka, M. Ooka and S. Shiotsu. 1958 A new antifungal antibiotic, pentamycin. J Antibiot (Tokyo) Ser A *11 (1):* 26–29.

Unz, R. F. 1971 Neotype strain of *Zoogloea ramigera* Itzigsohn. Int J Syst Bacteriol *21:* 91–99.

— and N. C. Dondero. 1967 The predominant bacteria in natural zoogloeal colonies. I. Isolation and identification. Can J Microbiol *13:* 1671–1682.

— and —. 1967 The predominant bacteria in natural zoogloeal colonies. II. Physiology and nutrition. Can J Microbiol *13:* 1683–1694.

Uphof, J. C. T. 1927 Zur Oekologie der Schwefelbakterien in den Schwefelquellen Mittelfloridas. Arch Hydrobiol *18:* 71–84.

The Upjohn Company. 1954 Antibiotic D-13 and its production. British Patent 708,686, May 5.

—. 1959 Antibiotic 101a. British Patent 811,757, April 8.

Urošefić, B. 1966 Canker of poplar caused by *Erwinia cancerogena* n. sp. Lesn Čas *12:* 493–505.

Uspenskaya, V. E. and E. N. Kondratieva. 1962 Relationship between photoautotrophic bacteria and vitamins and vitamin synthesis by these organisms. Mikrobiologiya *31:* 396–401.

Utermohl, H. 1924 Phaeobakterien. (Bakterien mit braunen Farbstoffen). Biol Zentrabl *43:* 605–610.

— and F. Koppe. 1924 Genus *Macromonas. In* Koppe, Die Schlammflora der ostholsteinischen Seen und des Bodensees. Arch Hydrobiol *14:* 619–672.

Uyeda, Y. 1907 Eine Bakterienkrankheit von *Zingiber officinale.* Zentrabl Bakteriol Parasitenk Infektionskr Hyg Abt II *17:* 383–384.

—. 1910 On the conjac leaf blight and some mannan-liquefying bacteria. Bot Mag (Tokyo) *24:* 177–182.

—. 1915 *In* Takimoto. J Plant Prot Tokyo *2:* 845.

Vago, C. and O. Croissant. 1960 Étude au microscope électronique de la pathogènese intracellulaire rickettsienne chez *Melolontha melolontha* L. Entomophaga *5:* 271–283.

— and R. Martoja. 1963 Une rickettsiose chez les *Gryllidae* (Orthoptera). C R Acad Sci Paris *256:* 1045–1047.

— and G. Meynadier. 1965 Une rickettsiose chez le criquet palerin (*Schistocerca gregaria* Forsk.). Entomophaga *10:* 307–310.

—, —, P. Juchault, J-J. Legrand, A. Amargier and J-L. Duthoit. 1970 Une maladie rickettsienne chez les Crustaces Isopodes. C R Acad Sci Paris *271:* 2061–2063.

Valentine, F. C. O. and T. M. Rivers. 1927 Further observations concerning growth requirements of hemophilic bacilli. J Exp Med *45:* 993–1002.

Vallée, A. and J.-A. Gaillard. 1953 Infection pyogène contagieuse de la souris déterminée par *Bacillus actinomycetem comitans.* Ann Inst Pasteur (Paris) *84:* 647–649.

—, P. Thibault and L. Second. 1963 Contribution à l'étude d'*A. lignierseii* et d'*A. equuli.* Ann Inst Pasteur (Paris) *104:* 108–114.

Vallee, H. 1903 Sur un nouveau *Streptothrix* (*Streptothrix polychromogene*). Ann Inst Pasteur (Paris) *17:* 288–292.

Valyi-Nagy, T., J. Uri and I. Szilágyi. 1956 Das Primycin, ein neues von Actinomyceten stammendes Antibioticum (in German). Pharmazie *11* (*5*): 304–312.

Van Beynum, J. and J. W. Pette. 1935 Zuckervergärend und Laktat vergärende Buttsäurebakterien. Zentrabl Bakteriol Parasitenk Infectionskr Hyg Abt II *93:* 198–212.

van Bijsterveld, O. P. 1970 New *Moraxella* strain isolated from angular conjunctivitis. Appl Microbiol *20* (*3*): 405–408.

van Boven, C. P. A., M. J. W. Kastelein and W. Hijams. 1967 A chemically defined medium for the L-phase of group A streptococci. Ann N Y Acad Sci *148:* 749–754.

Van Damme, P. A., A. G. Johannes, H. C. Cox and W. Berends. 1960 On toxoflavin, the yellow poison of *Pseudomonas cocovenenans.* Rec Trav Chim Pays-Bas *79:* 255–267.

van Delden, A. 1903 Beiträge zur Kenntnis der Sulfatreduktion durch Bakterien. Zentrabl Bakteriol Parasitenk Infektionskr Hyg Abt II *11:* 81–94.

Vandeputte, J., J. L. Wachtel and E. T. Stiller. 1956 Amphotericins A and B, antifungal antibiotics produced by a streptomycete. II. The isolation and properties of the crystalline amphotericins. Antibiot Annu *1955–1956:* 587–591.

van Dorssen, C. A. and F. H. J. Jaartsveld. 1962 *Actinobacillus suis* (novo species) een bij het varken voorkomende bacterie. Tijdschr Diergeneesk *87:* 448–450.

van Drimmelen, G. C. 1953 *Brucella melitensis* isolated from karakul sheep in South-West Africa. S Afr J Sci *49:* 299–302.

van Ermengem, E. 1896 Untersuchungen über Fälle von Fleischvergiftung mit Symptomen von Botulismus. Zentrabl Bakteriol Parasitenk Infektionskr Hyg Abt I Orig *19:* 442–444.

van Gylswyk, N. O. and J. P. L. Hoffman. 1970 Characteristics of cellulolytic cillobacteria from the rumens of sheep fed teff (*Eragrostis tef*) hay diets. J Gen Microbiol *60:* 381–386.

van Hall, C. J. J. 1902 Bijdragen tot de kennis der Bakterieele Plantenziekten. Inaug Diss., Amsterdam.

van Iterson, W. 1958 *Gallionella ferruginea* Ehrenberg in a different light. Verh Kon Ned Akad Wetensch Afd Naturk Tweeds Reeks 2, 52 nr 2, pp. 1–185.

van Laer, H. 1892 Contributions à l'histoire des ferments des hydrates de carbone. Acad Roy Belg Cl Sci Collect Octavo Mem *47:* 1–37.

Van Loghem, J. J. 1944 The classification of plague-bacillus. Antonie van Leeuwenhoek J Microbiol Serol *10:* 15–16.

van Niel, C. B. 1928 The Propionic Acid Bacteria. Uitgeverszaak and Boissevain and Co., Haarlem, Holland.

—. 1944 The culture, general physiology, morphology and classification of the nonsulfur purple and brown bacteria. Bacteriol Rev *8:* 1–118.

—. 1948 Suborder III. *Rhodobacteriineae* Breed, Murray and Hitchens. *In* Breed, Murray and Hitchens (Editors), Bergey's Manual of Determinative Bacteriology, 6th ed. The Williams & Wilkins Co., Baltimore, pp. 838–860.

—. 1948 Genus I. *Achromatium* Schewiakoff. *In* Breed, Murray and Hitchens (Editors), Bergey's Manual of Determinative Bacteriology, 6th ed. The Williams & Wilkins Co., Baltimore, pp. 997–1000.

—. 1957 Suborder I. *Rhodobacteriineae* Breed, Murray and Hitchens, 1944. *In* Breed, Murray and Smith, (Editors), Bergey's Manual of Determinative Bacteriology, 7th ed. The Williams & Wilkins Co., Baltimore, pp. 35–67.

— and M. B. Allen. 1952 A note on *Pseudomonas stutzeri.* J Bacteriol *64* (*3*): 413–422.

Van Oye, E. 1964 The world problem of salmonellosis. Junk Publ, The Hague.

Van Saceghem, R. 1915 Dermatose contagieuse (Impétigo contagieux). Bull Soc Pathol Exot *8:* 354–359.

—. 1915 Travaux du Laboratoire de Bactériologie vétérinaire de Zambi (Bas-Congo). III. Étude sur la Dermatose contagieuse (Impétigo contagieux). Bull Agr Congo *5:* 567–573 (1914).

—. 1934 La dermatose, dite contagieuse des Bovides. Impétigo tropical des Bovides. Bull Agr Congo *25:* 590–598.

van Steenberge, P. 1920 Les propriétés des microbes lactiques; leur classification. Ann Inst Pasteur (Paris) *34:* 803–870.

van Straaten, H. 1918 Bacteriologische bevindingen bij eenigo gevallen van pyo-septicaemie (Lähme) der veuluens. Verslag van den Werksaamheden der Rijksseruminrichting voor 1916–1917, Rotterdam. pp. 71–76.

van Tieghem, P. 1877 Sur le *Bacillus amylobacter* et son rôle dans la putrifaction des tissus végétaux. Bull Soc Bot Fr *24:* 128–135.

—. 1878 Sur la gomme du sucrerie (*Leuconostoc mesenteroides*). Ann Sci Natur Bot *7* (*6*): 180–203.

—. 1880 Observations sur des bactériacées vertes, sur des physochromacées blanches, et sur les affinités de ces deux familles. Bull Soc Bot Fr *27:* 174–179.

Vanyushin, B. F., A. N. Belozersky and N. A. Kokurina. 1966 Nucleic acids and plant evolution. Trans Moscow Soc Naturalists *24:* 7–25.

Vaucher, J. P. 1803 Histoire des conferves d'eau douce, contenant leurs différent modes de reproduction, et la description de leurs principales espèces. J. J. Paschoud, Geneva, pp. 1–285.

Vaughan, V. C. 1892 A bacteriological study of drinking water. Amer J Med Sci *104:* 167–198.

Vaughn, R. H. 1942 The acetic acid bacteria. Wallerstein Lab Commun *5:* 5–26.

—, H. C. Douglas and J. C. M. Fornachon. 1949 The taxonomy of *Lactobacillus hilgardii* and related heterofermentative lactobacilli. Hilgardia *19:* 133–139.

— and M. Levine. 1942 Differentiation of the

"intermediate" coli-like bacteria. J Bacteriol 44: 487–505.

Vavra, J. J. and M. E. Bergy. 1965 Antibiotic canarius and method of production. United States Patent 3,183,156, May 11.

— and —. 1965 Antibiotic canarius (in German). German Patent 1,187,767, February 25.

— and A. Dietz. 1965 U-13,714, a new antiviral agent. I. Discovery and biological properties. Antimicrob Agents Chemother 1964: 75–79.

—, —, B. W. Churchill, P. Siminoff and H. J. Koepsell. 1959 Psicofuranine. III. Production and biological studies. Antibiot Chemother 9 (7): 427–431.

Vawter, L. R. and E. Records. 1926 Recent studies on icters-hemoglobinuria of cattle. J Amer Vet Med Ass 68: 494–512.

Vedder, A. 1934 Bacillus alcalophilus n. sp.; benevens enkele ervaringen met sterk alcalische voedingsbodems. Antonie van Leeuwenhoek J Microbiol Serol 1: 141–147.

Vedros, N. A., J. Ng and G. Culver. 1968 A new serological group (E) of Neisseria meningitidis. J Bacteriol 95 (4): 1300–1304.

Veillon, A. and G. Repaci. 1912 Des infections secondaires dans la tuberculose ulcereuse du poumon. Ann Inst Pasteur (Paris) 26: 300–312.

— and A. Zuber. 1898 Recherches sur quelques microbes strictement anaérobies et leur role en pathologie. Arch Med Exp 10: 517–545.

Veillon, M. A. 1893 Sur un microcoque anaérobie trouvé dans de suppurations fétides. C R Soc Biol Paris 45: 807–809.

Veillon. R. 1922 Sur quelques microbes thermophiles strictement anaérobies. Ann Inst Pasteur (Paris) 36: 422–438.

Veldkamp, H. 1960 Isolation and characteristics of Treponema zuelzerae nov. spec., an anaerobic free-living spirochete. Antonie van Leeuwenhoek J Microbiol Serol 26: 103–125

—. 1961 A study of two marine agar-decomposing, facultatively anaerobic myxobacteria. J Gen Microbiol 26 (2): 331–342.

—, G. van den Berg and L. P. T. M. Zevenhuizen. 1963 Glutamic acid production by Arthrobacter globiformis. Antonie van Leeuwenhoek J Microbiol Serol 29: 35–51.

Venkataraman, R. and A. Sreenivasan. 1955 Utilization of various nitrogenous compounds by certain pseudomonas cultures from marine environments. Proc Indian Acad Sci 42: 31–38.

— and —. 1956 Red halophilic bacteria—the identity of some well-known species. Proc Indian Acad Sci 43: 264–270.

Verhoeven, W. 1952 Aerobic sporeforming nitrate reducing bacteria. Diss., Delft. Uitgeverij Waltman, Delft, pp. 1–160.

Vernon, T. R. 1955 Spore Formation in the Genus Streptomyces. Nature (London) 176: 935–936.

Véron, M. 1965 La position taxonomique des Vibrio et de certaines bactéries comparables. C R Acad Sci Paris 261: 5243–5246.

— and R. Chatelain. 1973 Taxonomic study of the genus Campylobacter Sebald and Véron and designation of the neotype strain for the type species, Campylobacter fetus (Smith and Taylor) Sebald and Véron. Int J Syst Bacteriol 23: 122–134.

—, P. Thibault and L. Second. 1959 Neisseria mucosa (Diplococcus mucosus Lingelsheim). I. Description bactériologique et étude du pouvoir pathogène. Ann Inst Pasteur (Paris) 97: 497–510.

—, — and —. 1961 Neisseria mucosa (Diplococcus mucosus Lingelsheim). II. Étude antigénique et classification. Ann Inst Pasteur (Paris) 100: 166–179.

Verona, O. 1934 Culture spontanee di cellulositici aerobi "Cytophaga winogradskii" n. sp. Rend Accad Lincei 19: 731–734.

Vetlugina, L. A. and M. Kh. Shigayeva. 1959 Vydelenie i nekotorye svoistva antibioticheskogo veshchestva, obrazuemogo Actinomyces antocyaneus (shtamm 1016) (in Russian). Trudy Inst Mikrobiol Virusol Akad Nauk Kaz SSR 3: 46–54.

Vickerstaff, J. M. and B. C. Cole. 1969 Characterization of Haemophilus vaginalis, Corynebacterium cervicis, and related bacteria. Can J Microbiol 15: 587–594.

Viehoever, A. 1913 Botanische Untersuchung harnstoffspaltender Bakterien mit besonderer Berücksichtigung der speziesdiagnostisch verwertbaren Merkmale und des Vermögens der Harnstoffspaltung. Zentrabl Bakteriol Parasitenk Infektionskr Hyg Abt II 39: 209–359.

Vieu, J. F., O. Croissant and C. Dauguet. 1965 Structure des bactériophages responsables des phénomènes de conversion chez les Salmonella. Ann Inst Pasteur (Paris) 109: 160–166.

Viljoen, J. A., E. B. Fred and W. H. Peterson. 1926 The fermentation of cellulose by thermophilic bacteria. J Agr Sci 16: 1–17.

Vinson, J. W. 1966 In vitro cultivation of the rickettsial agent of trench fever. Bull World Health Organ 35: 155–164.

— and H. S. Fuller. 1961 Studies on trench fever. I. Propagation of rickettsia-like microorganisms from a patient's blood. Pathol Microbiol Suppl 24: 152–166.

Vinzent, R. 1926 Isolement et culture de spirilles et de spirochètes des eaux. C R Soc Biol Paris 95: 1472–1474.

— and V. Reynes. 1947 Étude d'un nouvel anaérobie: Ramibacterium dentium. Ann Inst Pasteur (Paris) 73: 594–595.

— and P. Séguin. 1939 Contribution à l'étude des spirochètes intestinaux de l'homme. C R Soc Biol Paris 130: 12–13.

Virgilio, A. and C. Hengeller. 1960 Produzione di tetraciclina con Streptomyces psammoticus (in Italian). Farm Ed Sci 15 (3): 164–174.

Virtanen, A. I. and B. Bärlund. 1926 Die Oxydation des glycerins zu Dioxyaceton durch Bakterien. Biochem Z 169: 169–177.

Vishniac, W. V. 1952 The metabolism of Thiobacillus thioparus. I. The oxidation of thiosulfate. J Bacteriol 64 (3): 363–373.

— and M. Santer. 1957 The thiobacilli. Bacteriol Rev 21: 195–213.

Visloukh, S. M. 1911 A new sulfur-microorganism from the Neva, Thioploca ingrica Wisl. (trans. from Russian). Russkii Vrach 10: 2102–2104.

—. 1914 Spirillum kolkwitzii nov. sp. (in Russian). Zh Mikrobiol 1 (1–2): 42–51.

Visser't Hooft, F. 1925 Biochemische Onderzoekingen over het geslacht Acetobacter. Diss. Techn. Univ., Meinema, Delft, pp. 1–129.

Vlamis, J., A. M. Schultz and H. H. Biswell. 1964 Nitrogen fixation by root nodules of western mountain mahogany. J Range Manage 17: 73–74.

Voelz, H. 1966 The fate of the cell envelopes of *Myxococcus xanthus* during microcyst germination. Arch Mikrobiol *55:* 110–115.

— and M. Dworkin. 1962 Fine structure of *Myxococcus xanthus* during morphogenesis. J Bacteriol *84 (5):* 943–952.

Voets, J. P. and J. Debacker. 1956 *Pseudomonas azotogensis* nov. sp., a new free-living nitrogen-fixing bacterium. Naturwissenschaften *43:* 40–41.

Volk, W. A. and D. Pennington. 1950 The pigments of the photosynthetic bacterium *Rhodomicrobium vannielii*. J Bacteriol *59 (2):* 169–170.

von Freudenreich, E. 1889–1890 Recherches préliminaires sur le rôle des bactéries dans la maturation du fromage d'Emmenthal. Ann Microgr *2:* 257–283.

—. 1891 Bakteriologische Untersuchungen über den Reifungsprozess des Emmenthalerkäses. Landwirt Jahrb Schweiz *5:* 16–29.

—. 1895 Bakteriologische Untersuchungen über der Reifungsprozess des Emmenthalerkäses. Zentrabl Bakteriol Parasitenk Infektionskr Hyg Abt II *1:* 168–179.

—. 1897 Bakteriologische Untersuchungen über den Kefir. Zentrabl Bakteriol Parasitenk Infektionskr Hyg Abt II *3:* 47–54; 87–95; 135–141.

— and J. Thöni. 1904 Über die Wirkung verschiedener Milchsäurefermente auf Käsereifung. Landwirt Jahrb Schweiz *18:* 531–557.

von Holzhausen, G. F. 1927 Ein bisher unbekannter Erreger einer Mauserseptikamie. (*Corynebacterium murisepticum*. n. sp.) Zentrabl Bakteriol Parasitenk Infektionskr Hyg Abt I Orig *105:* 94–99.

von Lingelsheim, W. 1906 Die bakteriologischen Arbeiten der Kgl. hygienischen Station zu Beuthen O.-Schl. während der Genickstarreepidemie in Oberschlesien im Winter 1904/05. Klin Jahrb *15:* 373–489.

—. 1908 Beiträge zur Ätiologie der epidemischen Genickstarre nach Ergebnissen der letzten Jahre. Z Hyg Infektionskr *59:* 457–476.

von Prowazek, S. 1910 Parasitische Protozoen aus Japan, gesammelt von Herrn Dr. Mine in Fukuoka. Arch Schiffs-Trop Hyg *14:* 297–302.

—. 1913 Zur Parasitologie von Westafrika. Zentrabl Bakteriol Parasitenk Infektionskr Hyg Abt I Orig *70:* 32–36.

von Saltza, M. H., J. D. Dutcher and J. Reid. 1964 The aminoheptose moiety of septacidin. Abstr Papers 148th Meeting, Amer Chem Soc, Chicago, Illinois. Aug. 30–Sept. 4: 15Q–16Q. Abstract 32.

Von Tubeuf, K. 1895 Pflanzenkrankheiten durch Kryptogame Parasiten verursacht. Verlag J Springer, Berlin, pp. 1–599.

Von Wolzogen Kühr, C. A. H. 1932 Über eine Gärungsmikrobe in Fäkalien von Mückenlarven. Zentrabl Bakteriol Parasitenk Infektionskr Hyg Abt II *85:* 223–250.

Vouk, V. 1960 Ein neues Eisenbakterium aus der Gattung *Gallionella* in den Thermalquellen von Bad Gastein. Arch Mikrobiol *36:* 95–97.

Vuillax, I. 1963 *Streptomyces lusitanus* and the problem of classification of the various tetracycline-producing *Streptomyces*. Antimicrob Agents Chemother *1962:* 661–668.

Vuillemin, P. 1905 Sur la denomination de l'agent présume de la syphilis. C R Acad Sci Paris *140:* 1567–1568.

—. 1913 Genera Schizomycetum. Ann Mycolberl *11:* 512–527.

—. 1931 Les champignons parasites et les mycoses de l'homme. Encyclopédie Mycologique II. Paul LeChavalier and Sons, Paris, pp. 1–290.

Vzorov, V. I. 1930 Infektioeser Rost der Tabaks (*Nicotiana tabacum*). Bull N Caucas Pl Prot Sta *6-7:* 263–272.

Wachsman, J. T. and H. A. Barker. 1954 Characterization of an orotic acid fermenting bacterium, *Zymobacterium oroticum*, nov. gen., nov. spec. J Bacteriol *68 (4):* 400–404.

Waksman, S. A. 1919 Cultural studies of species of *Actinomyces*. Soil Sci *8:* 71–215.

—. 1923 Genus III. *Actinomyces* Harz *In* Bergey's Manual of Determinative Bacteriology, 1st ed. The Williams & Wilkins Co., Baltimore, pp. 339–371.

—. 1950 The actinomycetes: their nature, occurrence, activities, and importance. Ann Cryptogam Phytopathol *9:* 1–230.

—. 1957 Family *Actinomycetaceae* Buchanan and family *Streptomycetaceae* Waksman and Henrici. *In* Breed, Murray and Smith (Editors), Bergey's Manual of Determinative Bacteriology, 7th ed. The Williams & Wilkins Co., Baltimore, pp. 744–825.

—. 1959 Strain specificity and production of antibiotic substances. X. Characterization and classification of species within the *Streptomyces griseus* group. Proc Nat Acad Sci U S A *45 (7):* 1043–1047.

—. 1961 The Actinomycetes. Classification, identification and descriptions of genera and species Vol 2. The Williams & Wilkins Co., Baltimore, pp. 1–363.

— and C. T. Corke. 1953 *Thermoactinomyces Tsiklinsky*, a genus of thermophilic actinomycetes. J Bacteriol *66:* 377–378.

— and R. E. Curtis. 1916 The *Actinomyces* of the soil. Soil Sci *1 (2):* 99–134.

—, W. B. Geiger and E. Bugie. 1947 Micromonosporin, an antibiotic substance from a little-known group of microorganisms. J Bacteriol *53 (3):* 355–357.

— and F. J. Gregory. 1954 Actinomycin. II. Classification of organisms producing different forms of actinomycin. Antibiot Chemother *4 (10):* 1050–1056.

— and A. T. Henrici. 1943 The nomenclature and classification of the actinomycetes. J Bacteriol *46 (4):* 337–341.

— and —. 1948 Family II. *Actinomycetaceae* Buchanan. *In* Breed, Murray and Hitchens (Editors), Bergey's Manual of Determinative Bacteriology, 6th ed. The Williams & Wilkins Co., Baltimore, pp. 892–928.

— and —. 1948 Family III. *Streptomycetaceae* Waksman and Henrici. *In* Breed, Murray and Hitchens (Editors), Bergey's Manual of Determinative Bacteriology, 6th ed. The Williams & Wilkins Co., Baltimore, pp. 929–980.

—, E. S. Horning, M. Welsch and H. B. Woodruff. 1942 Distribution of antagonistic actinomycetes in nature. Soil Sci *54 (4):* 281–296.

— and J. S. Joffe. 1922 Microorganisms concerned in the oxidation of sulfur in the soil. II. The *Thiobacillus thiooxidans* a new sulfur oxidizing organism isolated from the soil. J Bacteriol 7: 239–256.

— and H. A. Lechevalier. 1953 Guide to the classification and identification of the actinomycetes and their antibiotics. The Williams and Wilkins Co., Baltimore, pp. 1–246.

— and —. 1962 The actinomycetes. Antibiotics of actinomycetes, Vol. III. The Williams and Wilkins Company, Baltimore, pp. 1–430.

—, W. W. Umbreit and T. C. Cordon. 1939 Thermophilic actinomycetes and fungi in soils and in composts. Soil Sci 47: 37–61.

— and H. B. Woodruff. 1940 The soil as a source of microorganisms antagonistic to disease-producing bacteria. J Bacteriol 40: 581–600.

— and —. 1941 Actinomyces antibioticus, a new soil organism antagonistic to pathogenic and non-pathogenic bacteria. J Bacteriol 42: 231–249.

Waldee, E. L. 1945 Comparative studies of some peritrichous phytopathogenic bacteria. Iowa State Coll J Sci 19: 435–484.

Wagle, R. F. and J. Vlamis. 1961 Nutrient deficiencies in two bitterbrush soils. Ecology 42: 745–752.

Wahrlich, W. 1890–1891 Bakteriologische Studien. I. Zur Frage über den Bau der Bakterienzelle. II. Bacillus nov. spec. Die Entwickelungsgeschichte und einige biologische Eigenthümlichkeiten desselben. S. -A from Scripta Botanica, St. Petersburg. (Abstr. by L. Klein. 1892 Zentrabl Bakteriol Parasitenk Infektionskr Hyg 11: 49–53).

Wakker, J. H. 1883 Vorläufige Mittheilungen über Hyacinthen Krankheiten. Bot Ztg 14: 315–317.

Walker, J. R. L. 1959 Pyruvate metabolism in Lactobacillus brevis. Biochem J 72: 188–192.

Wallace, A. L. and A. Harris. 1967 Reiter treponeme. A review of the literature. Bull World Health Organ 36: Suppl 2.

Wallace, W. R. and G. T. Dimopoullos. 1965 Biologic properties and characteristics of Anaplasma marginale: Effects of radiation on infectivity of partially purified marginal body preparations. J Amer Vet Res 26: 1356–1358.

Wallhäusser, K.-H. and G. Huber. 1963 Gewinnung des neuen Antibioticums Werramycin und von Nonactin sowie gegebenenfalls deren Trennung (in German). German Patent 1,142,989, January 31.

—, —, G. Nesemann, P. Präve, and K. Zepf. 1964 Die antibiotica FF 3582A und B und ihre Identität mit nonactin und seinen Homologen (in German.) Arzneimittel Forschung 14 (4): 356–360.

—, G. Nesemann, P. Prave and A. Steigler. 1966 Moenomycin, a new antibiotic. I. Fermentation and isolation. Antimicrob Agents Chemother 1965: 734–736.

Wallick, H., D. A. Harris, M. A. Reagan, M. Ruger and H. B. Woodruff. 1955 Discovery and antimicrobial properties of cathomycin, a new antibiotic produced by Streptomyces spheroides, n. sp. Abstr 3rd Ann Symp on Antibiotics, Washington, D. C., November 2–4. Abstract 90.

—, —, —, — and —. 1956 Discovery and antimicrobial properties of cathomycin, a new antibiotic produced by Streptomyces spheroides, n. sp. Antibiotics Annu 1955/56: 909–917.

Wallin, J. R. and C. S. Reddy. 1945 A bacterial streak disease of Phleum protense L. Phytopathology 35: 937–939.

Walsby, A. E. 1972 Structure and Function of gas vacuoles. Bacteriol Rev 36: 1–32.

Wang, E. L., M. Hamada, Y. Okami and H. Umezawa. 1966 A new antibiotic, spinamycin. J Antibiot (Tokyo) Ser A 19 (5): 216–221.

Wannamaker, L. W. 1962 Characterization of a fourth desoxyribonuclease of group A streptococci. Fed Proc 21: 231.

—, B. Hayes and W. Yasmineh. 1967 Streptococcal nucleases. II. Characterization of DNAse D. J Exp Med 126: 497–508.

Warke, G. M. and S. A. Dhala. 1968 Use of inhibitors for selective isolation and enumeration of cytophagas from natural substrates. J Gen Microbiol 51 (1): 43–48.

Warming, E. 1875 Om nogle ved Danmarks kyster levende bakterier. Vidensk Medd Naturhist Foren Kjobenhavn, pp. 306–420.

Warner, G. S., J. E. Faber and M. J. Pelczar. 1952 A serological study of certain members of the aerobic nonpathogenic Neisseria group. J Infect Dis 90: 97–103.

Warren, G. H. 1945 The antigenic structure and specificity of luminous bacteria. J Bacteriol 49: 547–561.

Warren, S. H. and W. M. Scott. 1930 A new serological type of Salmonella. J Hyg 29: 415–417.

Washburn, M. R., J. C. White and C. F. Niven, Jr. 1946 Streptococcus S.B.E.: immunological characteristics. J Bacteriol 51: 717–722.

Watanabe, K., T. Tanaka, K. Fukuhara, N. Miyairi, H. Yonehara and H. Umezawa. 1957 Blastomycin, a new antibiotic from Streptomyces sp. J Antibiot (Tokyo) 10 (A): 39–45.

Watson, E. A. 1935 Tuberculin, johnin and mallein derived from non-protein media. Can J Public Health 26: 268–275.

Watson, S. W. 1965 Characteristics of a marine nitrifying bacterium, Nitrosocystis oceanus sp. n. Limnol Oceanogr 10: 274–289.

—. 1971 Reisolation of Nitrosospira briensis. Arch Mikrobiol 75: 179–188.

—. 1971 Taxonomic considerations of the family Nitrobacteriaceae Buchanan. Requests for opinions. Int J Syst Bacteriol 21 (3): 254–270.

—, L. B. Graham, C. C. Remsen and F. W. Valois. 1971 A lobular ammonia-oxidizing bacterium Nitrosolobus multiformis Nov. sp. Arch Mikrobiol 76 (3): 83–203.

— and J. B. Waterbury. 1971 Characteristics of two marine nitrite oxidizing bacteria, Nitrospira gracilis nov. gen. nov. sp. and Nitrococcus mobilis nov. gen. nov. sp. Arch Mikrobiol 77: 203–230.

Wattie, E. 1943 Cultural characteristics of zoogloea-forming bacteria isolated from activated sludge and trickling filters. U S Public Health Rep 57: 1519–1533.

Wawrik, F. 1952 Planctomyces-Studien. Sydowia. Ann Mycol Ser II 6 (56): 443–452.

—. 1956 Siderocapsa arlbergensis nova species, eine Eisenbakterie aus den Hochgebirgs-Kleingewässern des Arlberggebietes. Osterr Bot Z 103: 19–23.

—. 1956 Neue Planktonorganismen aus den Waldvierteler Fischteichen. I. Oesterr Bot Z 103 (2–3): 291–299.

Wayne, L. G. 1959 Quantitative aspects of neutral red reactions of typical and "atypical" mycobacteria. Amer Rev Tuberc 79: 526–530.

—. 1962 Two varieties of *Mycobacterium kansasii* with different clinical significance. Amer Rev Resp Dis 86: 651–656.

—. 1966 Classification and identification of mycobacteria. III Species within Group III. Amer Rev Resp Dis 93: 919–928.

—. 1967 Selection of characters for an Adansonian analysis of mycobacterial taxonomy. J Bacteriol 93: 1382–1391.

—. 1971 Phenol soluble antigens from *M. kansasii*, *M. gastri* and *M. marinum*. Infect Immun 3: 36–40.

— and J. R. Doubek. 1965 Classification and identification of mycobacteria. II. Tests employing nitrate and nitrite as substrate. Amer Rev Resp Dis 91: 738–745.

— and —. 1968 Diagnostic key to mycobacteria encountered in clinical laboratories. Appl Microbiol 16: 925–931.

—, — and G. A. Diaz. 1967 Classification and identification of mycobacteria. IV Some important scotochromogens. Amer Rev Resp Dis 96: 88–95.

— and W. M. Gross. 1968 Base composition of deoxyribonucleic acid isolated from mycobacteria. J Bacteriol 96: 1915–1919.

—, I. Krasnow and M. Huppert. 1957 Characterization of atypical mycobacteria and of *Nocardia* species isolated from clinical specimens. I. Characterization of atypical mycobacteria by means of the microcolonial test. Amer Rev Tuberc 76: 451–467.

— and E. F. Lessel. 1969 On the synonymy of *Mycobacterium marianum* Penso 1953 and *Mycobacterium scrofulaceum* Prissick and Masson 1956 and the resolution of a nomenclatural problem. Int J Syst Bacteriol 19: 257–261.

Webster, S. R., C. T. Youngberg and A. G. Wollum, II. 1967 Fixation of nitrogen by bitterbrush (*Purshia tridentata* (Pursh) D.C.). Nature (London) 216: 392–393.

Weckesser, J., G. Drews and H. D. Tauschel. 1969 Zur Feinstruktur und Taxonomie von *Rhodopseudomonas gelatinosa*. Arch Mikrobiol 65: 346–358.

Weeks, O. B. 1955 *Flavobacterium aquatile* (Frankland and Frankland) Bergey *et al.*, type species of the genus *Flavobacterium* J Bacteriol 69 (6): 649–658.

—. 1969 Problems concerning the relationship of cytophagas and flavobacteria. J Appl Bacteriol 32 (1): 13–18.

—, A. G. Andrews, B. O. Brown and B. C. L. Weedon. 1969 Occurrence of C40 and C45 carotenoids in the C50 carotenoid system of *Flavobacterium dehydrogenans*. Nature (London) 224 (5222): 879–882.

— and S. M. Beck. 1960 Nutrition of *Flavobacterium aquatile* strain Taylor and a microbiological assay for thiamine. J Gen Microbiol 23 (2): 217–229.

— and L. M. Kelley. 1958 Observations on the growth of the bacterium *Caryophanon latum*. J Bacteriol 75: 326–330.

Wehmer, C. 1898 Untersuchungen über Kartoffelkrankheiten III. Die Bakterienfaule der Knollen (Nassfaule). Zentrablatt Bakteriol Parasitenk Infektionskr Hyg Abt II 4: 795–805.

—. 1903 Die Sauerkrautgärung. Zentrabl Bakteriol Parasitenk Infektionsk Hyg Abt II 10: 625–629.

Weichselbaum, A. 1886 Ueber die Aetiologie der akuten Lungen-und Rippenfellentzündungen. Med Jahr Wien 82 (8): 483–554.

—. 1887 Über die Atiologie der akuten Meningitis cerebrospinalis. Fortschr Med 5: 573–583.

Weigmann, H. 1890 Der Organismus der sogenannten langen Wei. Chem Zbl Berlin 61 (1): 431–432.

—. 1898 Ueber zwei an der Käsereifung beteiligte Bakterien. Zentrabl Bakteriol Parasitenk Infektionskr Hyg Abt II 4: 820–834.

Weinberg, M. and B. Ginsbourg. 1927 Données récéntes sur les microbes anaérobies et leur rôle en pathologie. Masson et Cie, Paris, pp. 1–291.

—, R. Nativelle and A. R. Prévot. 1937 Les Microbes Anaérobies. Masson and Co., Paris, pp. 1–1186.

— and A. R. Prévot. 1926 Recherches sur la flore microbienne de l'appendicite *Fusobacterium biacutum*. C R Soc Biol Paris 95: 519–522.

— and P. Séguin. 1915 Flore microbienne de la gangrène gazeuse. Le *B. fallax*. C R Seances Soc Biol Filiales 78: 686–689.

— and —. 1916 Contribution à l'étiologie de la gangrène gazeuse. C R Acad Sci Paris 163: 449–451.

— and —. 1918 La gangrène gazeuse-Bactériologie, Reproduction expérimentale, Séreothérapie. Masson and Co., Paris, pp. 1–444.

Weindling, R., H. D. Tresner and E. J. Backus. 1961 The host-range of a *Streptomyces aureofaciens* actinophage. Nature (London) 189 (4764): 603.

Weinfurtner, F., A. Uhl and R. Pöhlmann. 1955 The effect of oxidation-reduction potential, of oxygen and of carbon dioxide upon the propagation of *Pediococcus cerevisiae* (beer sarcina). Brauwissenschaft 8: 166–167; 192–199.

Weinman, D. 1944 Infectious anemias due to *Bartonella* and related red cell parasites. Trans Amer Phil Soc 33: 243–350.

—. 1957 *Bartonellaceae*. *In* Breed, Murray and Smith (Editors), Bergey's Manual of Determinative Bacteriology, 7th ed. The Williams and Wilkins Co., Baltimore, pp. 968–980.

—. 1968 Bartonellosis. *In* Infectious Blood Diseases of Man and Animals, Vol. 2. Academic Press, New York, pp. 3–24.

— and H. Pinkerton. 1938 A *Bartonella* of the guinea pig, *Bartonella tyzzeri* sp. nov. Ann Trop Med Parasitol 33: 215–224.

Weinstein, M. J., G. M. Luedemann, E. M. Oden and G. H. Wagman. 1968 Halomicin, a new *Micromonospora*-produced antibiotic. Antimicrob Agents Chemother 1967: 435–441.

Weiser, J. 1963 Diseases of insects of medical importance in Europe. Bull World Health Organ 28: 121–127.

Weisglass, H. and B. Gavrilović. 1963 *Photobacterium profundum* sp. n. Bull Sci Cons Acad RSF Yougoslavie Sec A Sci Natur Tech Med 8: 69–70.

— and Y. Skreb. 1963 *Vibrio noctiluca* n. sp.

Bull Sci Cons Acad RSF Yougoslavie Sect A Sci Natur Tech Med *8:* 9.

Weisrock, W. P. and R. M. Johnson. 1966 Marine species of *Hyphomicrobium*. Bacteriol Proc *G36:* 22.

Weiss, C. and D. G. Mercado. 1938 Studies of anaerobic streptococci from pulmonary abscesses. J Infect Dis *62:* 181–185.

Weiss, E. 1899 Über drei in gesäuerten Rübenschnitzeln gefundenen Milchsäurebakterien. Inaug Diss Vorber J Landw *47:* 141–161.

— and H. R. Dressler. 1962 Increased resistance to chloramphenicol in *Rickettsia prowazekii* with a note on failure to demonstrate genetic interaction among strains. J Bacteriol *83 (2):* 409–414.

—, W. F. Myers, E. C. Suitor, Jr. and E. M. Neptune, Jr. 1962 Respiration of a rickettsialike microorganism, *Wolbachia persica*. J Infect Dis *110:* 155–164.

—, E. M. Neptune, Jr. and J. A. Davies. 1964 Lipid metabolism of the rickettsialike microorganism *Wolbachia persica*. III. Comparison with other metabolic activities. J Infect Dis *114:* 50–54.

Weiss, J. E. and L. F. Rettger. 1934 *Lactobacillus bifidus*. J Bacteriol *28:* 501–524.

— and —. 1938 Taxonomic relationship of *Lactobacillus bifidus*, *Bacillus bifidus* (Tissier) and *Bacteroides bifidus* (Eggerth). J Bacteriol *35:* 17–18.

— and —. 1938 Taxonomic relationships of *Lactobacillus bifidus* (*B. bifidus* Tissier) and *Bacteroides bifidus*. J Infect Dis *62:* 115–120.

Weiss, N., R. Plapp and O. Kandler. 1967 Die Aminosäuresequenz des DAP-haltingen Mureins von *Lactobacillus plantarum* und *Lactobacillus inulinus*. Arch Mikrobiol *58:* 313–323.

Weissman, S. M., P. R. Reich, N. L. Somerson and R. M. Cole. 1966 Genetic differentiation by nucleic acid homology. IV. Relationships among Lancefield groups and serotypes of streptococci. J Bacteriol *92 (5):* 1372–1377.

Welby-Guisse, M., M. A. Laneelle and J. Asselineau. 1970 Structure des acides corynomycoliques de *C. hofmanii*, et leur implication biogénétique. Eur J Biochem *13:* 164–167.

Welch, H., L. E. Putnam and W. A. Randall. 1955 Antibacterial activity and blood and urine concentrations of cycloserine, a new antibiotic, following oral administration. Antibiot Med *1:* 72.

— and W. W. Wright. 1955 The common identity of cathomycin and streptonivicin. Antibiot Chemother *5 (12):* 670–673.

Weldin, J. C. 1927 The colon-typhoid group of bacteria and related forms. Relationships and classification. Iowa State J Sci *1:* 121–197.

— and M. Levine. 1923 An artificial key to the species and varieties of the colon—typhoid or intestinal group of bacilli. Abstr Bacteriol *7:* 13–16.

Welker, N. E. and L. L. Campbell. 1967 Unrelatedness of *Bacillus amyloliquefaciens* and *Bacillus subtilis*. J Bacteriol *94 (4):* 1124–1130.

— and —. 1967 Comparison of the α-amylase of *Bacillus subtilis* and *Bacillus amyloliquefaciens*. J Bacteriol *94 (4):* 1131–1135.

Wells, A. Q. 1937 Tuberculosis in wild voles. Lancet *232:* 1221.

—, E. Aquis and N. Smith. 1955 *Mycobacterium fortuitum*. Amer Rev Tuberc *72:* 53–63.

Wells, N. A. and C. E. ZoBell. 1934 *Achromobacter ichthyodermis* n. sp., the etiological agent of an infectious dermatitis of certain marine fishes. Proc Nat Acad Sci U S A *20:* 123–136.

Welsch, M. 1941 Bactericidal substances from sterile culture-media and bacterial cultures. With special reference to the bacteriolytic properties of actinomycetes. J Bacteriol *42 (6):* 801–814.

—. 1942 Bacteriostatic and bacteriolytic properties of actinomycetes. J Bacteriol *44 (5):* 571–588.

—. 1947 Phénomènes d'antibiose chez les actinomycètes. Université de Liège- Thése d'agrégation de l'enseignement supérieur. J Duculot, Gembloux, Belgium, pp. 1–315.

—. 1947 Actinomycetin. J Bacteriol *53 (1):* 101–102.

—. 1954 Activités staphylolytique et streptolytique des filtrats de culture de *Streptomyces* spp. C R Soc Biol Paris *148:* 604–608.

Welshimer, H. J. and A. Meredith. 1971 *Listeria murrayi* sp. n: nitrate-reducing mannitol fermenting *Listeria*. Int J Syst Bacteriol *21 (1):* 3–7.

Wenner, J. J. and L. F. Rettger. 1919 A systematic study of the *Proteus* group of bacteria. J Bacteriol *4:* 331–353.

Wenyon, C. M. 1926 Protozoology, Vol. 1. Bailliere, Tindall and Cox, London, pp. 1–778.

—. 1926 Spirochaetes. *In* Wenyon, Protozoology, Vol. 2. William Wood and Co., New York, pp. 1233–1288.

Werkman, C. H. and A. A. Anderson. 1938 D-lactic acid fermentation. Abstr Bacteriol *35:* 69–70.

— and R. W. Brown. 1933 The propionic acid bacteria. II. Classification. J Bacteriol *26:* 393–417.

— and G. F. Gillen. 1932 Bacteria producing trimethylene glycol. J Bacteriol *23:* 167–182.

— and S. E. Kendall. 1931 The propionic acid bacteria. I. Classification and Nomenclature. Iowa State J Sci *6:* 17–32.

— and H. J. Weaver. 1927 Studies in the bacteriology of sulphur stinker spoilage of canned sweet corn. Iowa State J Sci *2 (1):* 57–67.

— and P. W. Wilson. 1951 Bacterial Physiology. Academic Press Inc., New York, pp. 1–707.

Werner, H. 1966 The Gram-positive non-sporing anaerobic bacteria of the human intestine with particular reference to the corynebacteria and bifidobacteria. J Appl Bacteriol *29:* 138–146.

—. 1967 Untersuchungen über die Lipase- und Lecithinase-Aktivität von aeroben und anaeroben Corynebacterium- und von Propionibacterium-Arten. Zentralbl Bakteriol Parasitenk Infektionskr Hyg Abt I Orig *204:* 127–138.

—. 1970 Glutaminsäuredecarboxylaseaktivitat bei *Bacteroides* Arten. Zentralbl Bakteriol Parasitenk Infektionskr Hyg Abt I Orig *215:* 320–326.

—. 1970 Des Kulturell-biochemische Verhalten und die Antiobiotikaempfindlichkeit des *Bacteroides putredinis* (Weinberg et al., 1937) Kelly 1957. Zentralbl Bakteriol Parasitenk Infektionskr Hyg Abt I Orig *215:* 327–332.

—. 1972 A comparative study of 55 *Sphaerophorus* strains. Differentiation into 3 species: *Sphaerophorus necrophorus*, *Sph. varius* and

Sph. freundii. Med Microbiol Immunol *157:* 299–314.

—, G. Pulverer and C. Reichertz. 1971 Biochemical properties and antibiotic susceptibility of *Bacteroides melaninogenicus.* Med Microbiol Immunol *157:* 3–9.

— and H. P. R. Seeliger. 1964 Vergleichende Untersuchungen an Bifidus-Stammen verschiedener Herkunft. Pathol Mikrobiol *27:* 202–215.

Werner, W. 1933 Botanische Beschreibung häufiger am Buttersäureabbau beteiligter sporenbildender Bakterienspezies. Zentrabl Bakteriol Parasitenk Infektionskr Hyg Abt II *87:* 446–475.

Wertlake, P. T. and T. W. Williams. 1968 Septicemia caused by *Neisseria flavescens.* J Clin Pathol *21:* 437–439.

West, G. S. and B. M. Griffiths. 1909 *Hillhousia mirabilis,* a giant sulphur bacterium. Proc Roy Soc London B Biol Sci *81:* 398–405.

— and —. 1913 The lime-sulphur bacteria of the genus *Hillhousia.* Ann Bot London *27:* 83–91.

West, M. G. and P. R. Edwards. 1954 The Bethesda-Ballerup group of paracolon bacteria. Public Health Monograph No. 22, U.S. Dept Health, Education and Welfare.

West, S. E. H. and L. V. Holdeman. 1973 Placement of the name *Peptococcus anaerobius* (Hamm) Douglas on the list of *nomina rejicienda.* Request for an opinion. Int J Syst Bacteriol *23:* 283–289.

West, W. and G. S. West. 1898 Notes on fresh water algae. J Bot (London) *36:* 330–338.

Westerdijk, J. and C. Buisman. 1929 De Iepenziekte. Rapport over het onderzoek verricht op verzoek van de Nederlandsche Heidemaatschaapj. Arnhem, pp. 1–78.

Westphal, A. 1965 Nachweis und Entwicklung aktivierter *Eperythrozoon coccoides*-infektionen bei der Haltung von Protozoenstämmen in Mäusen. Z Tropenmed Parasit *16:* 53–65.

Westphal, O., F. Kauffmann, O. Luderitz and H. Stierlin. 1960 Zur Immunchemie der O-Antigene von *Enterobacteriaceae.* Zentrabl Bakteriol Parasitenk Infektionskr Hyg Abt I Orig *179:* 336–342.

Whitaker, J. A., E. Fort, D. Weinman, P. Tamasatit and K. Panas-Ampol. 1966 Acute febrile anaemia associated with *Bartonella*-like erythrocytic structures. Nature (London) *212:* 855–856.

White, A. H. 1940 A bacterial discoloration of print butter. Sci Agr *20:* 638–645.

White, A. J. and G. A. Snow. 1968 Methods for separation and identification of mycobactins from various species of mycobacteria. Biochem J *108:* 593–597.

White, B. 1926 Further studies of the *Salmonella* Group. Med Res Comm Spec Rep *103:* 3–160.

White, D., M. Dworkin and D. J. Tipper. 1968 Peptidoglycan of *Myxococcus xanthus:* Structure and relation to morphogenesis. J Bacteriol *95 (6):* 2186–2197.

White, D. C., M. P. Bryant and D. R. Caldwell. 1962 Cytochrome linked fermentation in *Bacteroides ruminicola.* J Bacteriol *84 (4):* 822–828.

—, G. Leidy, J. D. Jamieson and R. E. Shope. 1964 Porcine contagious pleuropneumonia. III. Interrelationship of *Hemophilus pleuropneu-*

moniae to other species of *Hemophilus:* Nutritional, metabolic, transformation and electron microscopy studies. J Exp Med *120:* 1–12.

White, G. F. 1906 The bacteria of the apiary, with special reference to bee diseases. Tech Ser Bur Entomol U S Dep Agr No. 14.

—. 1912 The cause of European foul brood. U S Dep Agr Bur Entomol Circ *157:* 1–15.

White, H. E. 1930 Bacterial spot of radish and turnip. Phytopathology *20:* 653–662.

White, H. L. 1936 Diseases of early vegetables. Rep Exp Res Sta Cheshunt *1935:* 42–43.

White, H. R. 1963 The effect of variation in pH on the heat resistance of cultures of *Streptococcus faecalis.* J Appl Bacteriol *26:* 91–99.

White, J. C. 1944 Streptococci from subacute bacterial endocarditis. Thesis, Cornell Univ., Ithaca, New York, pp. 1–47.

— and C. F. Niven, Jr. 1946 *Streptococcus* S.B.E.: a streptococcus associated with subacute bacterial endocarditis. J Bacteriol *51 (6):* 717–722.

White, J. N. and M. P. Starr. 1971 Glucose fermentation end products of *Erwinia* spp. and other Enterobacteria. J Appl Bacteriol *34:* 459–475.

White, P. B. 1921 The normal bacterial flora of the bee. J Pathol Bacteriol *24:* 64–78.

White, P. G. and J. B. Wilson. 1951 Differentiation of smooth and non-smooth colonies of brucellae. J Bacteriol *61:* 239–240.

Whiteley, H. R. 1952 The fermentation of purines by *Micrococcus aerogenes.* J Bacteriol *63:* 163–175.

—. 1957 Fermentation of amino acids by *Micrococcus aerogenes.* J Bacteriol *74 (3):* 324–330.

— and E. J. Ordal. 1957 Fermentation of alpha keto acids by *Micrococcus aerogenes* and *Micrococcus lactilyticus.* J Bacteriol *74:* 331–336.

Whitmore, A. 1913 An account of a glanderslike disease occurring in Rangoon. J Hyg *13:* 1–34.

Whittaker, R. H. 1959 On the broad classification of organisms. Quart Rev Biol *34:* 210.

Whittenberger, C. L. and R. Repaske. 1958 Studies on the electron transport system in *Hydrogenomonas eutropha.* Bacteriol Proc, p. 106.

Whittenbury, R. 1964 Hydrogen-peroxide formation and catalase activity in the lactic acid bacteria. J Gen Microbiol *35:* 13–26.

—. 1966 A study of the genus *Leuconostoc.* Arch Mikrobiol *53:* 317–327.

— and G. A. McLee. 1967 *Rhodopseudomonas palustris* and *R. viridis*—photosynthetic budding bacteria. Arch Mikrobiol *59:* 324–334.

—, K. C. Phillips and J. F. Wilkinson. 1970 Enrichment, isolation and some properties of methane-utilizing bacteria. J Gen Microbiol *61 (2):* 205–218.

Wiame, J. M., R. Harpigny and R. G. Dothey. 1959 A new type of *Acetobacter: Acetobacter acidophilum.* J Gen Microbiol *20:* 165–172.

Wicken, A. J. and J. Baddiley. 1963 Structure of intracellular teichoic acids from group D streptococci. Biochem J *87:* 54–62.

—, S. D. Elliott and J. Baddiley. 1963 The identity of streptococcal group D antigen with teichoic acid. J Gen Microbiol *31 (2):* 231–239.

Widra, A. 1963 Histochemical observations on *Actinomyces bovis* granules. Sabouraudia *2:* 264–267.

Wiehe, P. O. and W. J. Dowson. 1953 A bacterial disease of cassava (*Manihotis utilissima*) in Nyasaland. Emp J Exp Agr *21 (82):* 141–143.

Wieringa, K. T. 1935 Een bacterieziekte voorkomende bij begonias. Tijdschr Plantenziekten *41:* 309–313.

Wigand, R. 1958 Morphologische Biologische und Serologische Eigenschaften der Bartonellen. Georg Thieme, Stuttgart, pp. 1–95.

— and D. Peters. 1952 Neuere Untersuchungen über *Bartonella muris* Mayer. II. Mittellung. Z Tropenmed Parasitol *3:* 437–452.

— and —. 1954 Abbauversuche an *Haemobartonella muris* und *Eperythrozoon coccoides*. Z Tropenmed Parasitol *5:* 482–492.

Wildy, P. and R. Hare. 1953 The effect of fatty acids on the growth, metabolism and morphology of the anaerobic cocci. J Gen Microbiol *9:* 216–225.

Wiley, W. R. and J. L. Stokes. 1963 Effect of pH and ammonium ions on the permeability of *Bacillus pasteurii*. J Bacteriol *86:* 1152–1156.

Willcox, R. R. and T. Guthe. 1966 *Treponema pallidum.* A bibliographical review of the morphology, culture and survival of *T. pallidum* and associated organisms. Bull World Health Organ *35:* Suppl.

Wille, H. 1956 *Bacillus fribourgensis,* n. sp., Erreger einer "milky disease" im Engerling von *Melolontha melolontha* L. Mitt Schweiz Entomol Ges *29:* 271–282.

Willers, J. M. and G. H. Alderkamp. 1967 Loss of type antigen in a type III *Streptococcus* and identification of the determinant disaccharide of the remaining antigen. J Gen Microbiol *49:* 41–51.

—, P. A. Deddish and H. D. Slade. 1968 Transformation of type polysaccharide antigen synthesis and hemolysin synthesis in streptococci. J Bacteriol *96 (4):* 1225–1230.

— and M. F. Michel. 1966 Immunochemistry of the type antigen of *Streptococcus faecalis*. J Gen Microbiol *43 (3):* 375–382.

—, H. Ottens and M. F. Michel. 1964 Immunochemical relationship between *Streptococcus* MG, F II and *Streptococcus salivarius*. J Gen Microbiol *37 (3):* 425–431.

Willett, N. P., G. E. Morse and S. A. Carlisle. 1967 Requirements for growth of *Streptococcus agalactiae* in a chemically defined medium. J Bacteriol *94 (4):* 1247–1248.

Williams, A. S. 1964 Comparisons of Virginia isolates of *Corynebacterium* from orchard grass with *Corynebacterium rathayi* and *C. tritici*. Phytopathology *54:* 912.

Williams, C. O. and R. G. Wittler. 1971 Hydrolysis of aesculin and phosphatase production by members of the order *Mycoplasmatales* which do not require sterol. Int J Syst Bacteriol *21 (1):* 73–77.

—, — and C. Burris. 1969 Deoxyribonucleic acid base compositions of selected mycoplasmas and L-phase variants. J Bacteriol *99:* 341–343.

Williams, R. A. D. and S. A. Sadler. 1971 Electrophoresis of glucose-6-phosphate dehydrogenase, cell wall composition and the taxonomy of heterofermentative lactobacilli. J Gen Microbiol *65 (3):* 351–358.

Williams, R. E. O. 1956 *Streptococcus salivarius* (vel hominis) and its relation to Lancefield's group K. J Pathol Bacteriol *72:* 15–25.

—, A. Hirch and S. T. Cowan. 1953 *Aerococcus*, a new bacterial genus. J Gen Microbiol *8:* 475–480.

Williams, S. T. 1967 Sensitivity of streptomycetes to antibiotics as a taxonomic character. J Gen Microbiol *46:* 151–160.

Willis, A. T. 1964 Anaerobic bacteriology in clinical medicine. Butterworth & Co., London, pp. 1–234.

—. 1969 Clostridia of wound infection. Butterworth & Co., London.

Wilson, A. T. 1959 The relative importance of the capsule and the M-antigen in determining colony form of group A streptococci. J Exp Med *109:* 257–269.

Wilson, E. E., M. P. Starr and J. A. Berger. 1957 Bark canker, a bacterial disease of the Persian walnut tree. Phytopathology *47:* 669–673.

—, F. M. Zeitoun, and D. L. Fredrickson. 1967 Bacterial phloem canker, a new disease of Persian walnut trees. Phytopathology *57:* 618–621.

Wilson, G. S. 1933 The classification of the *Brucella* group: a systematic study. J Hyg *33:* 516–541.

— and A. A. Miles. 1932 The serological differentiation of smooth strains of the *Brucella* group. Brit J Exp Pathol *13:* 1–13.

— and —. 1946 Topley and Wilson's Principles of Bacteriology and Immunity. 3rd ed. Vol. 1. The Williams and Wilkins Co., Baltimore.

— and —. 1955 Topley and Wilson's Principles of Bacteriology and Immunity. 4th ed. The Williams and Wilkins Co., Baltimore.

— and M. M. Smith. 1928 Observations on the Gram-negative cocci of the nasopharynx, with a description of *Neisseria pharyngis*. J Pathol Bacteriol *31:* 597–608.

Wilson, M. M. 1955 A study of *Corynebacterium equi* infection in a stud of thoroughbred horses in Victoria. Aust Vet J *31:* 175–181.

Wilson, P. W. 1940 The Biochemistry of Symbiotic Nitrogen Fixation. University of Wisconsin Press, Madison, Wis., p. 69.

Wilson, S. N. 1953 Some carbohydrate fermenting organisms isolated from the rumen of sheep. J Gen Microbiol *9:* 1–11.

Wilson, S. P. 1928 An investigation on certain Gram-negative cocci met with in the nasopharynx, with special reference to their classification. J Pathol Bacteriol *31:* 477–492.

Winblad, S. 1968 Studies on O-antigen factors of *Yersinia enterocolitica*. International Symposium on Pseudotuberculosis, Paris 1967, Symposia series in Immunobiological Standardization. Vol. 9, pp. 337–342.

Winogradsky, H. 1935 Sur la flore nitrificatrice des boues activées de Paris. C R Acad Sci Paris *200:* 1880–1888.

—. 1937 Contribution à l'étude de la microflore nitrificatrice des boues activées de Paris. Ann Inst Pasteur (Paris) *58:* 326–341.

Winogradsky, S. 1887 Über Schwefelbacterien. Bot Ztg No. 31, Jahrg. *45:* 489–508; 513–523; 529–539; 545–559; 569–576; 585–594; 606–610.

—. 1888 Ueber Eisenbakterien. Bot Ztg *46:* 261–270.

—. 1888 Beiträge zur Morphologie und Physiologie der Bacterien. Heft I. Zur Morphologie und Physiologie der Schwefelbacterien. Arthur Felix, Leipzig, pp. 1–120.

—. 1890 Recherches sur les organismes de la nitrification. Ann Inst Pasteur (Paris) 4: 257–275.

—. 1892 Contributions à la morphologie des organismes de la nitrification. Arch Sci Biol (St Petersb) 1: 87–137.

—. 1902 Clostridum pastorianum seine Morphologie und seine Eigenschaften als Buttersäureferment. Zentrabl Bakteriol Parasitenk Infektionskr Hyg Abt II 9: 43–54; 107–112.

—. 1922 Eisenbakterien als Anorgoxydanten. Zentrabl Bakteriol Parasitenk Infektionskr Hyg Abt II 57: 1–21.

—. 1929 Études sur la microbiologie du sol—sur la dégradation de la cellulose dans le sol. Ann Inst Pasteur (Paris) 43: 549–633.

—. 1938 Études sur la microbiologie du sol et des eaux. Sur la morphologie et l'oecologie des Azotobacter. Ann Inst Pasteur (Paris) 60: 351–400.

— and H. Winogradsky. 1933 Études sur la microbiologie du sol. VII. Nouvelles recherches sur les organismes de la nitrification. Ann Inst Pasteur (Paris) 50: 350–434.

Winslow, C. E. A., J. Broadhurst, R. E. Buchanan, C. Krumwiede, Jr., L. A. Rogers and G. H. Smith. 1917 The families and genera of the bacteria. Preliminary report of the Committee of the Society of American Bacteriologists on Characterization and Classification of Bacterial Types. J Bacteriol 2: 505–566.

—, —, —, —, — and —. 1920 The families and genera of the bacteria. Final report of the Committee of the Society of American Bacteriologists on characterization and classification of bacterial types. J Bacteriol 5: 191–229.

—, I. J. Kligler and W. Rothberg. 1919 Studies on the classification of the colon-typhoid group of bacteria with special reference to their fermentative reactions. J Bacteriol 4: 429–503.

Winter, G. 1884 Die Pilze Deutschlands Oesterreichs und der Schueiz. I. Abt. Schizomyceten, Saccharomyceten und Basidiomyceten. Rabenhorst's Kryptogamen Flora 1: 1–924.

Winton, F. W. and N. S. Mair. 1969 Pasteurella pneumotropica isolated from a dog bite wound. Microbios 2: 155–162.

Witkin, S. S. and E. Rosenberg. 1970 Induction of morphogenesis by methionine starvation in Myxococcus xanthus: polyamine control. J Bacteriol 103 (3): 641–655.

Wittern, A. 1933 Beiträge zur Kenntnis der "Mikrobakterian" Orla-Jensen. Zentrabl Bakteriol Parasitenk Infektionskr Hyg Abt II 87: 412–446.

Wodehouse, R. P., E. J. Backus and C. Gussoni. 1967 Species and strain differentiation in Streptomyces by gel diffusion. Int Arch Allergy Appl Immunol 32: 378–395.

Wolbach, S. B. 1919 Studies on Rocky Mountain spotted fever. J Med Res 41: 1–197.

— and C. A. L. Binger. 1914 Notes on a filterable spirochaete from fresh water, Spirochaeta biflexa (new species). J Med Res 30: 23–26.

— and J. L. Todd. 1920 Note sur l'étiologie et l'anatomie pathologique du typhus exanthématique au Mexique. Ann Inst Pasteur (Paris) 34: 153–158.

Wolf, F. A. and A. C. Foster. 1917 Bacterial leaf spot of tobacco. Science (Washington) 46: 361–362.

Wolf, J. and A. N. Barker. 1968 The genus Bacillus: Aids to the identification of its species. In Gibbs and Shapton (Editors), Identification Methods for Microbiologists, Part B, No. 2. Academic Press, London, pp. 93–109.

Wolfe, R. S. 1960 Observations and studies of Crenothrix polyspora. J Amer Water Works Ass 52: 915–918.

—. 1971 Microbial formation of methane. Adv Microbial Physiol pp. 107–146.

Wolff, J. W. 1954 The Laboratory Diagnosis of Leptospirosis. Charles C Thomas, Springfield, Illinois.

— and J. C. Broom. 1954 The genus Leptospira Noguchi 1917. Problems of classification and a suggested system based on antigenic analysis. Doc Med Geogr Trop 6: 78–95.

Wolff, M. 1907 Pedioplana haeckeli n. g., n. sp. und Planosarcina schaudinni n. sp. zwei neue bewegliche Coccaceen. Zentrabl Bakteriol Parasitenk Infektionskr Hyg Abt II 18: 9–26.

Wolin, M. J., J. B. Evans and C. F. Niven, Jr. 1952 The oxidation of butyric acid by Streptococcus mitis. J Bacteriol 64 (4): 531–535.

—, G. B. Manning and W. O. Nelson. 1959 Ammonium salts as a sole source of nitrogen for the growth of Streptococcus bovis. J Bacteriol 78 (1): 147–149.

Wolinsky, E. and T. K. Rynearson. 1968 Mycobacteria in soil and their relation to disease-associated strains. Amer Rev Resp Dis 97: 1032–1037.

Wollenweber, H. W. 1920 Der Kartoffelschorf. Arb Forsch-Inst Kartoff., Verlagsbuchandlung Paul Parey, Berlin, No. 2, pp. 1–102.

Wood, E. J. 1950 The bacteriology of shark spoilage. Aust J Mar Freshwater Res 1: 129–138.

Wood, T. R. and D. Henlin. 1952 Process for production of vitamin B_{12}. United States Patent 2,595,499, May 6.

Woodcock, H. M. and G. Lapage. 1914 On a remarkable type of protisten parasite. Quart J Microscop Sci 59: 431–457.

Wooldridge, W. E. 1957 Fungicidal and antibiotic substances from Streptomyces aminophilus. German Patent 1,000,966, January 17.

—. 1960 Antifungal antibiotic. British Patent 828,792, February 24.

—. 1960 Antifungal antibiotic from S. coelicolor var. aminophilus. United States Patent 2,956,925, October 18.

World Health Organization. 1967 Current problems in leptospirosis research: Report of a WHO expert group. World Health Organ Tech Rep Ser No. 380 (Geneva), p. 32.

Wormald, H. 1914 A bacterial rot of celery. J Agr Sci 6: 203–219.

—. 1930 Bacterial diseases of stone fruit trees in Britain. II. Bacterial shoot wilt of plum trees. Ann Appl Biol 17: 725–744.

—. 1931 Bacterial diseases of stone fruit trees in Britain. III. The symptoms of bacterial canker in plum trees. J Hort Sci 9: 239–256.

Woronin, M. 1866 Über die bei der Schwarzerle (Alnus glutinosa) und der gewöhnlichen Garten-Lupine (Lupinus mutabilis) auftretenden Wurzelanschwellungen. Mem Acad Sci St Petersburg 10 (6): 1–10.

Worthington, R. W. and H. H. Kleeberg. 1964 Isolation of *Mycobacterium kansasii* from bovines. J S Afr Vet Med Ass *35:* 29–33.

— and —. 1967 Demonstration of species-specific fractions in mycobacterial purified protein derivative (PPD) sensitins. Tubercle *48:* 211–218.

Wright, J. H. 1895 Report on the results of an examination of the water supply of Phila-delphia. Mem Nat Acad Sci *7:* 422–484.

Wroblewski, W. 1931 Morphologie et cycle évolutif des microbes de la péripneumonie des bovides et de l'agalaxie contagieuse des chèvres et des moutons. Ann Inst Pasteur (Paris) *47:* 94–115.

Wu, Lien-Teh and H. J. Jettmar. 1929–1930 Bartellosen und Grahamellosen unter Ostasiatischen Wilden Nagetieren. Manchurian Plague Preventive Service Reports *7:* 24–26.

Yagishita, K. 1955 Studies on the pyridomycin production. II. X-ray irradiation on the pyridomycin-producing strain. J Antibiot (Tokyo) Ser A *8 (6):* 201–204.

Yakimoff, W. L. and W. S. Bélawine. 1927 L'anaplasmose des bovides en Russie (USSR). Zentrabl Bakteriol Parasitenk Infektionskr Hyg Abt I Orig *103:* 419–421.

Yale, M. W. 1939 Genus I. *Escherichia* Castellani and Chalmers. *In* Breed, Murray and Hitchens (Editors), Bergey's Manual of Determinative Bacteriology, 5th ed. The Williams & Wilkins Co., Baltimore, pp. 389–396.

—. 1939 Genus VI. *Proteus* Hauser. *In* Bergey, Breed, Murray and Hitchens (Editors), Bergey's Manual of Determinative Bacteriology, 5th ed. The Williams and Wilkins Co., Baltimore, pp. 430–436.

Yamada, K. and K. Komagata. 1968 Taxonomic studies on coryneform bacteria. International Conference of Culture Collections, Tokyo. Abstracts of Symposia, p. 23.

— and —. 1970 Taxonomic studies on coryne-bacteria. III. DNA base composition of coryneform bacteria. J Gen Appl Microbiol *16:* 215–224.

— and K. Yamagata. 1970 Taxonomic studies on corynebacteria. II. Principal amino acids in the cell wall and their taxonomic signifi-cance. J Gen Appl Microbiol *16:* 103–113.

Yamaguchi, H., Y. Nakayama, K. Takeda, K. Tawara, K. Maeda, T. Takeuchi and H. Umezawa. 1957 A new antibiotic, althio-mycin. J Antibiot (Tokyo) Ser A *10 (5):* 195–200.

Yamaguchi, T. and Y. Saburi. 1955 Studies on the anti-trichomonal actinomycetes and their classification. J Gen Appl Microbiol *1 (3):* 201–235.

Yamakawa, T. and N. Ueta. 1964 Gas chroma-tographic studies of microbial components. I. Carbohydrate and fatty acid constitution of *Neisseria.* Jap J Exp Med *34:* 361–374.

Yamamoto, H., I. Iwasa, M. Shibata, K. Mizuno and A. Miyake. 1965 *Streptomyces multi-spiralis* nov. sp. and a new antibiotic, neo-humidin. Agr Biol Chem *29 (4):* 360–358.

—, K. Nakazawa, S. Horii and A. Miyake. 1960 Studies on agricultural antibiotic. Folimycin, a new antifungal antibiotic produced by *Streptomyces neyagawaensis* nov. sp. J Agr Chem Soc Jap *34:* 268–272.

Yamamoto, M. 1951 Bacterial leaf spot of Jimson weed with special reference to the resistance of the causal organism to various chemicals. Forsch Pflkr, Tokyo *4:* 160–168.

Yamamoto, R., C. H. Bigland and H. B. Ortmayer. 1965 Characteristics of *Mycoplasma meleagri-dis* sp.n. isolated from turkeys. J Bacteriol *90 (1):* 47–49.

— and G. T. Clark. 1966 *Streptobacillus monili-formis* infection in turkeys. Vet Rec *79:* 95–100.

Yamamoto, S. 1934 Bacterial disease of aspara-gus lettuce (*Lactuca sativa* var *angustata*). J Plant Prot Tokyo *21:* 528–533.

Yamashita, K. 1955 Studies on pyridomycin production. II. X-ray irradiation on the pyridomycin-producing strain. J Antibiot (Tokyo) Ser A *8:* 201–204.

Yanagawa, R., H. Basri and K. Otsuki. 1967 Three types of *Corynebacterium renale*, classi-fied by precipitin reactions in gels. Jap J Vet Res *15:* 111–120.

— and K. Otsuki. 1970 Some properties of the pili of *Corynebacterium renale.* J Bacteriol *101 (3):* 1063–1069.

Yang, Hui-Fang. 1962 Morphological and phys-iological features of various strains of purple sulfur bacteria. Nauch Dokl Vysshei Shk Biol Nauki *3:* 163–170.

Yao, J., N. J. Jacobs, R. H. Deibel and C. F. Niven, Jr. 1964 Group E streptococci. II. Serological characterization of strains from swine cervical abscesses. J Infect Dis *114:* 333–340.

Yasaki, Y. and Y. Haneda. 1936 Ueber einen neuen Typus von Leuchtorgan im Fische. Proc Imp Acad Jap *12:* 55–57.

Yen, H.-C. 1956 Classification and determina-tion of *Actinomyces.* K'o Hsueh T'ung-Pao (Kexue Tungbao-Scientific Reporter) *1:* 75–78.

—. 1957 Description of *Actinomyces.* K'o Hsueh T'ung-Pao (Kexue Tungbao-Scientific Re-porter) *1:* 171–172.

—. 1957 Description of *Actinomyces.* II. K'o Hsueh T'ung-Pao (Kexue Tungbao-Scientific Reporter) *7:* 208–209.

—. 1957 Description of *Actinomyces.* IV. K'o Hsueh T'ung-Pao (Kexue Tungbao-Scientific Reporter) *15:* 474–475.

— and H.-C. Chou. 1964 Description of some new species and new varieties of *Actinomyces.* Acta Microbiol Sinica *10 (4):* 424–438.

— and Y.-Y. Lu. 1964 Studies on the classifica-tion of *Actinomyces.* V. Determination of *Actinomyces lavendulae* group. Acta Microbiol Sinica *10 (2):* 236–246.

— and G.-W. Zhang. 1964 Studies on the classi-fication of *Actinomyces.* VII. Determination of *Actinomyces chromogenes* group. Acta Micro-biol Sinica *10 (2):* 258–273.

Yen, Tun Ch'U. 1956 Nauchn. vestnik (China). *1:* 78 (ref. in Krasil'nikov, N. A. and Yuan Chi-Shen 1961. Mikrobiologiya *29:* 482–489.)

Yendo, Y. and T. Takase. 1932 On the root-nodule of *Elaeagnus* (in Japanese). Bull Seric Silk Indust, Uyeda (Japan) *4:* 114–134.

Yirgou, D. 1964 *Xanthomonas guizotiae* sp. nov. on *Guizotia abyssinica.* Phytopathology *54:* 1490–1491.

— and J. F. Bradbury. 1968 Bacterial wilt of enset (*Ensete ventricosum*) incited by *Xan-thomonas musacearum* sp. n. Phytopathology *58:* 111–112.

Yoshi, T. and S. Takimoto. 1928 Bacterial leafspot of hemp and its pathogen. J Plant Prot Tokyo *15:* 12–18.

Yuan-Chi-Shen. 1962 Biology of the group of orange-colored actinomycetes. Diss Inst Microbiol, Akad Nauk SSR (Abstr by Konova, I. V. 1962 Mikrobiologiya *31:* 188–189).

Yüntsen, H., K. Ohkuma, Y. Ishii and H. Yonehara. 1956 Studies on angustmycin. III. J Antibiot (Tokyo) Ser A *9 (6):* 195–201.

—, H. Yonehara and H. Ui. 1954 Studies on a new antibiotic, angustmycin. I. J Antibiot (Tokyo) Ser A *7 (4):* 113–115.

Zabriskie, J. B. 1964 The role of temperate bacteriophage in the production of erythrogenic toxin by group A streptococci. J Exp Med *119:* 761–779.

Zagallo, A. C. and C. H. Wang. 1962 Comparative carbohydrate catabolism in *Arthrobacter.* J Gen Microbiol *29 (3):* 389–401.

— and —. 1967 Comparative carbohydrate catabolism in Corynebacteria. J Gen Microbiol *47 (3):* 347–357.

Zagatto, A. G. and A. L. G. Pereira. 1963 Estudo do organismo causador da "murcha bacteriana" da Araruta (*Maranta arundinaceae* L.). Arq Inst Biol Sao Paulo *30:* 33–36.

Zarma, M. 1963 Contribution à l'étude des microorganismes marins anaérobies stricts. II. Bactéries ammonifiantes anaérobies isolées du fond de la Mer noire, au niveau de la Station zoologique marine "Prof. I. Borcea" d'Agigea (Dubroujda). Rappt. Proces Verbaux Reunions, Comm. Internat Exp Sci Mer Med. Monaco, *17:* 675–677.

Zaslavskii, A. S. 1952 On the halophilic thionic acid bacteria of salinas. Mikrobiologia *21:* 31–35.

Zavarzin, G. A. 1961 Symbiotic culture of new manganese oxidizing microorganism. Mikrobiologiya *30 (3):* 393–395.

—. 1961 Budding bacteria. Mikrobiologiya *30:* 774–791.

—. 1964 *Metallogenium symbioticum.* Z All Mikrobiol *4:* 390–395.

Zdrodovskii, P. R. and H. M. Golinevich. 1960 The Rickettsial Diseases (trans. fr. Russian). Pergamon Press, New York

Zeidler, A. 1896 Ueber eine Essigsäure bildende Termobakterie. Zentrabl Bakterien Parasitenk Infektionskr Hyg Abt II *2:* 729–739.

Zeikus, J. G. and R. S. Wolfe. 1972 *Methanobacterium thermoautotrophicus* sp. n., an anaerobic, autotrophic, extreme thermophile. J Bacteriol *109 (2):* 707–713.

Zettnow, E. 1915 Einigeneue Bakterien. Zentrabl Bakteriol Parasitenk Infektionskr Hyg Abt I Orig *77:* 209–234.

Zhavoronkova, I. P. 1932 Bacterial root-rot of red clover, alfalfa and lentil caused by *Bacterium radiciperda* n. sp. Bull Plant Prot Tokyo Ser II *5 (1):* 161–172.

Zhdanov, V. M. 1953 Opredelitel' Virusov Cheloveka i Zhivotnykh. Isdatel' stro Akademii Meditsinskikh Nauk USSR, Moscow.

— and S. Korenblit. 1950 Systematics and nomenclature of viruses. Zh Mikrobiol Epidemiol Immunobiol (transl) *9:* 40–44.

Zhukova, R. A. and V. A. Tsyganov. 1966 Comparative study of actinomycetes of the genera *Macrospora* and *Microellobosporia* (in Russian). Mikrobiologiya *35 (5):* 833–840.

—, — and V. M. Morozov. 1968 A new species of *Micropolyspora: Micropolyspora angiospora* sp. nov. Mikrobiologiya *37:* 724–728.

Ziegler, H. 1959–60 Die Rhizothamnien bei *Comptonia peregrina* (L.) Coult Mitt Deut Dendrol Ges *61:* 28–31.

Zierdt, C. H., C. Webster and W. S. Rude. 1968 Study of the anaerobic corynebacteria. Int J Syst Bacteriol *18:* 33–47.

Zillikin, F., P. N. Smith, C. S. Rose and P. Gyorgy. 1954 Enzymatic synthesis of a growth factor for *Lactobacillus bifidus* var *Penn.* J Biol Chem *208:* 299–305.

Zimmermann, O. E. R. 1890 Die Bakterien unserer Trink- und Nutzwässer, insbesondere des Wassers der Chemnitzer Wasserleitung. I Reihe. Elfter Ber Naturwiss Ges Chemnitz, pp. 54–154.

Zimmermann, T. 1964 Untersuchungen über die Actinobazillose des Schweines. Deut Tieraerztl Wochenschr *71:* 457–461.

Zinder, N. D. 1957 Lysogenic conversion in *S. typhimurium.* Science (Washington) *126 (3285):* 1237.

Zinnemann, K. 1960 *Haemophilus influenzae* and its pathogenicity. Ergeb Mikrobiol Immun Exp Ther *33:* 307–368.

—, K. B. Rogers, J. Frazer and J. M. H. Boyce. 1968 A new V-dependent *Haemophilus* species preferring increased CO_2 tension for growth and named *Haemophilus paraphrophilus.* J Pathol Bacteriol *96:* 413–419.

—, —, — and S. K. Devaraj. 1971 A haemolytic, V-dependent CO_2 preferring *Haemophilus* species (*Haemophilus paraphrohaemolyticus* nov. spec.). J Med Microbiol *4:* 139–143.

— and G. C. Turner. 1963 The taxonomic position of *Haemophilus vaginalis* (*Corynebacterium vaginale*). J Pathol Bacteriol *85:* 213–219.

Zinsser, H. and J. G. Hopkins. 1916 On a species of *Treponema* found in rabbits. J Bacteriol *1:* 489–492.

ZoBell, C. E. 1943 The effect of solid surfaces upon bacterial activity. J Bacteriol *46:* 39–56.

—. 1948 Genus II. *Desulfovibrio* Kluyver and Van Niel. *In* Breed, Murray and Hitchens (Editors), Bergey's Manual of Determinative Bacteriology, 6th ed. The Williams and Wilkins Co., Baltimore, pp. 207–209.

—. 1957 Genus II. *Desulfovibrio* Kluyver and van Niel, 1936. *In* Breed, Murray and Smith (Editors), Bergey's Manual of Determinative Bacteriology, 7th ed. The Williams and Wilkins Co., Baltimore, pp. 248–249.

— and E. C. Allen. 1935 The significance of marine bacteria in the fouling of submerged surfaces. J Bacteriol *29:* 239–251.

— and H. C. Upham. 1944 A list of marine bacteria including descriptions of sixty new species. Bull Scripps Inst Oceanogr Univ Calif (Tech Ser) *5:* 239–292.

Zopf, W. 1883 Die Spaltpilze, Edward Trewendt, Breslau, pp. 1–100.

—. 1884 Die Spaltpilze, 2nd ed. Edward Trewendt, Breslau, pp. 1–101.

—. 1885 Die Spaltpilze, 3rd ed. Edward Trewendt, Breslau, pp. 1–127.

Zuelzer, M. 1912 Über *Spirochaeta plicatilis* Ehrbg. und deren Verwandtschaftsbeziehungen. Arch Protistenk *24:* 1–59.

—. 1925 Die Spirochäten. *In* von Prowazek

(Editor), Handbuch der Pathogenen Protozoen, 3rd ed. Barth, Leipzig, pp. 1627–1798.

Zukal, H. 1896 *Myxobotrys variabilis* Zuk., als Reprasentant einer neuen Myxomyceten-Ordnung. Ber Deut Bot Ges *14:* 340–347.

—. 1897 Über Myxobacterien. Ber Deut Bot Ges *15:* 542–552.

Zumpt, F. and D. Organ. 1961 Strains of spirochaetes isolated from *Ornithodoros zumpti* Heisch and Guggisberg and from wild rats in the Cape province. A preliminary note. S Afr J Lab Clin Med *7:* 31–35.

Zvirbulis, E. and H. D. Hatt. 1967 Status of the generic name *Zoogloea* and its species. Int J Syst Bacteriol *17 (1):* 11–21.

— and —. 1969 Status of names not evaluated in Index Bergeyana (1966) Addendum II. *Acetobacter* to *Butyrivibrio*. Int J Syst Bacteriol *19 (1):* 57–115.

— and —. 1969 Status of names of bacterial taxa not evaluated in Index Bergeyana (1966) Addendum III. *Achromobacter* to *Lactobacterium*. Int J Syst Bacteriol *19 (3):* 309–370.

A Key for the
Determination of the
Generic Position of Organisms
Listed in the Manual

V. B. D. Skerman

The following key has been designed to enable the user to determine whether any isolated organism bears any resemblance to an organism described in THE MANUAL. Although the key terminates in most cases at a genus, it has been formulated on individual species descriptions. Every effort has been made to see that if the description of any species within THE MANUAL is applied to the key, the description will lead to the genus into which the species has been placed in THE MANUAL. Keys provided by the various contributors should be followed in deciding which species within the genus most closely agrees with the isolate. It is quite possible, owing to the limited description of many species, that the new isolate will be described more extensively and may fit into more than one species. It is unlikely, but not impossible, that it may fit into more than one genus.

No attempt has been made to fit the key to any system of classification. While it may undoubtedly act as a guide to the proper classification of an undescribed organism, it is designed solely for the purpose of identification of described species. The user must judge for himself whether an isolate is identical with a described species and, if not, determine its taxonomic position.

Characters were chosen solely for their suitability for purposes of differentiation. While ease of performance of tests is still an important consideration, still more important is the need for basic information which will provide not only for differentiation but also for assessing affinities. For this reason the information relating to such characters as cell wall composition, DNA base composition and metabolic byproducts have, of necessity, been used in the keys.

The author is fully aware of the developments in numerical taxonomy, of the use of computer methods to generate keys, of the need to consider a large series of characters and of the dangers associated with dichotomous keys based on single character differences. Although access to the necessary machinery and programs is readily available the use of the computer has severe limitations in the preparation of a general key partly due to the inadequacy of existing character coding systems but mainly to the fact that characters are either too few and/or not uniformly applied to all species descriptions. The forecast that

1098

future manuals will undoubtedly be compiled by computer-oriented methods may have some substance as far as general compilation is concerned. However, if computer methods are to be used to generate keys and provide better classifications—particularly across, rather than within, all the genera—there will be a need for a much higher degree of uniformity of approach to systematics than is presently the case.

The sequence of tests employed has largely eliminated the necessity to treat an organism as positive or negative with respect to a character for which the information was not given. In the few instances where this device has been employed, the species involved has been cited in the key, and the assumption made has been noted or the species has been traced through the key as far as the information permitted, a note having been made to this effect at the appropriate point.

In the present key, pathogenicity to animals has been used as a guide to the possible identity of the organism. However, it has been coupled with other characters in the separation of these genera. The plant pathogens and the rhizobia have been separated on the basis of pathogenicity and nodule formation, respectively. They have, in addition, been treated as organisms isolated from the soil, have been traced through the key as far as described characters would permit, and have been cited at the appropriate points. In the same way some pigmented organisms have been treated also as colorless organisms.

Some changes have been made in the approach used in previous keys. The initial differentiation on width of cells has been discontinued except to separate three genera with very large cells. The emphasis on lactose has been shifted onto glucose and on the oxidation or fermentation of this compound. Failure to provide data on the production of acid from glucose, particularly with the genus *Pseudomonas*, has provided some problems which may not have been entirely resolved.

DNA base ratios have been cited where they appeared to have good differential value and, in view of the decision to include illustrations in THE MANUAL, cross reference is made to these where considered desirable.

While every endeavor has been made to avoid inaccuracies in the key, some are inevitable. In order to avoid any repetition of these, users are requested to supply the author with a detailed statement of these as they are encountered. The author would also welcome reprints of any papers relating to subjects of taxonomic interest.

Definition and Use of Terms in the Key

Throughout the key the following terms have been used in the sense indicated.

Multicellular organism: A group of cells, arranged uniseriately, which are joined for the whole or major part of their width and result in an organism that has a lateral wall with little or no indentation and lack of articulation at the septa. The septa are clearly visible in unstained cells after the removal of inclusion substances, such as sulfur or fat.

In the key, small organisms that appear to be unicellular in unstained preparations and in preparations stained by Gram's method, but which are distinctly multicellular after being stained to show cell walls, are treated with the unicellular organisms.

Unicellular organism: An organism that shows no evidence of dividing septa, other than those involved in a normal division, in unstained preparations or in preparations stained to reveal cell walls.

Cell: A unicellular organism or a single component of a multicellular organism.

Filament: An elongated rod usually more than 10 μm long. It may be unicellular or multicellular. However, the term as used in THE MANUAL is a relative one, a 10 μm cell being 'filamentous' when the normal cell length is about 1–2 μm.

Chain of organisms: A group of unicellular or multicellular organisms, arranged uniseriately, which are completely separated or are attached for only a minor part of their width and are freely articulate at the points of attachment. They may or may not be ensheathed.

Trichome: This term is used rarely in the key to refer to very long multicellular filaments.

Sheath: A hollow structure surrounding a chain of cells or a filament. It may be close fitting but is not in intimate contact with the cells. Sheath-forming organisms usually produce a gum-like holdfast and a gum-like secretion resembling a capsule, but as the length of the chain or trichome increases, the gum-like secretion is replaced by a hollow structure, the sheath, which lacks intimate contact with the cells.

Capsule: A substance secreted by microorganisms that forms an envelope around the cell and remains in intimate contact with it. Its margin may be sharply defined or, owing to its relative solubility in water, merge imperceptibly into the surrounding fluids.

Diameter: In the assessment of diameter or width, measurement must be made of the cells themselves and not of any capsular structures or sheaths that may surround them.

USE OF THE KEY

First, determine the characters of the organism and then consult the key, always commencing from the beginning. The key poses a series of questions which can be answered in the affirmative or negative. **Boldface numbers** on the right-hand side of the key indicate the next number on the left to be consulted. The sequence should be followed until the right-hand number is replaced by a generic name. Keys to the particular genus in THE MANUAL should then be consulted for species identification.

The Key

1. Organisms green, yellowish green, brown or red; *eucaryotic;* contain plant chlorophyll "*a*" in well defined chloroplasts..............................**Algae**
   Not as above...**2**

2. Organisms blue-green to red; *procaryotic;* contain plant chlorophyll "*a*" and other pigments notably phycocyanin (blue) and phycoerythrin (red) in chromatophores but not in chloroplasts: if motile, show gliding motility......**Cyanobacteria**   *p. 22*
   (Blue-Green Algae)

   Not as above...**3**

3. Colorless organisms, which when stained with Giemsa, without preliminary acid hydrolysis, show a clearly differentiated nucleus: *eucaryotic;* amoeboid or flagellated (or ciliated): flagella clearly visible with a phase contrast microscope and reveal, in thin electron microscope sections, a pair of central fibrils surrounded by 9 pairs of peripheral fibrils...**Protozoa**
   Not as above...**4**

4. Submicroscopic bodies, devoid of a cell wall and other cytoplasmic components which characterize the bacteria: consisting of a genetic element containing *either* DNA or RNA; devoid of any metabolic systems; able to infect susceptible living plant, animal or bacterial cells in which they are replicated..............**Viruses**
   Not as above...**5**

5. Microscopic to submicroscopic organisms, with or without well defined cell walls, but with well developed cell membranes and differentiation into a procaryotic nucleus and cytoplasm; contain no mitochondria or other membrane-bound inclusions: may be photosynthetic but, if so, chlorophylls are of the bacterial type and are not contained in chloroplasts. The flagellum, if present cannot be resolved with the light microscope and lacks the complexity of the eucaryotic flagellum
   Bacteria   *p. 23*
   proceed to **Section A**

SECTION A

1. Parasites or symbiotes occurring intra- or extracellularly in the tissues or erythrocytes of animal hosts or arthopod vectors. With few exceptions (which are also included in other sections of the key) they cannot be cultivated on laboratory media. Some can be cultivated in the chick embryo or in animal tissue cell cultures. Stain with Gi-

emsa's, Macchiavello's, Gimenez's or Casteneda's stains: coccoid, ellipsoid, bacillary, triangular, ring-shaped, horse-shoe-shaped, filamentous and other pleomorphic forms may occur. Reproduction is by binary fission. Cell walls may be present or absent but when present are of the Gram-negative type. Differentiation of cellular contents within the cytoplasmic membrane varies considerably from one genus to another. Proliferation may occur on the surface, within vacuoles surrounded by the host cell membrane, within the cytoplasm or within the nucleus of the host cell

Section B   *p. 1102*

Not as above...2

2. Microorganisms which produce small spherical forms mostly within the range of 100–250 nm in diameter (one species 300–2000 nm) which lack a true cell wall and are bounded by triple layered membranes which range from 7–12 nm thick (mostly 7–8 nm); sensitive to lysis by surface active agents. In addition filaments are produced which range from 0.15–5.0 μm in length with different species—usually straight or undulate but spiral in one species: frequently branch. Reproduction appears to be by conversion of the filamentous forms to the spherical forms and the release of the latter by fragmentation: highly resistant to penicillin. The colony of most cultivated species has a dense granulated central area which penetrates into the medium and which is surrounded by a flat peripheral zone (the "fried egg" colony (Plate 19.1, Fig. 1)), translucent to dark brown; in some instances it consists of a pearly film containing cholesterol and phospholipids and tiny black spots in and around the colony composed of calcium and magnesium soaps (Plate 19.1, Fig. 2); vary in size from the very minute (15–25 μm for T-mycoplasmas) to 3 mm (*Acholeplasma*); in *Metallogenium* colonies are irregularly shaped brownish spots becoming black from the deposition of manganese...**Section C**   *p. 1104*

Note: (a) *L-phase colonies of some bacteria bear a strong resemblance to the colonies of* **Mycoplasma**. *They are generally more opaque and more heavily marked on the surface, tend to revert to the normal bacillary form in penicillin-free semisolid media, are more difficult to subculture, do not require cholesterol for growth, and ferment the same carbohydrates as the parent organism.*

(b) *Numerous reports have been made concerning the presence of mycoplasma-like bodies in the tissues of plants suffering from such diseases as aster yellows, corn stunt, mulberry dwarf, rice yellows, legume little leaf. They have the micromorphology of mycoplasmas and are sensitive to tetracycline drugs. However they have not been cultured or not confirmed as the causative agents of the diseases. See p. 954.*

Not as above...3

3. Organisms produce tubular extensions of the cell wall (prosthecae), which may act as stalks, *or* secrete substances which form stalks which are clearly distinguishable from the somatic portion of the cell.....................**Section D**   *p. 1105*

Not as above...4

4. Vegetative cells rod-shaped, cylindrical or tapered, not spirally twisted: frequently very flexible; Gram-negative; motile by a gliding action on solid surfaces, no flagella. At some stage during the development of the population, the cells either contract to produce spherical or oval refractile myxospores (microcysts) which are found at random among the vegetative forms *or* they aggregate to produce fruiting bodies (often macroscopically visible) in which the vegetative cells transform to myxospores. Myxospores may *either* resemble the vegetative cells in phase density or refractility and shape but are usually slightly shorter *or* be more dense or highly refractile under phase microscopy and differ morphologically from the vegetative cell. The myxospores and microcysts germinate to produce new vegetative cells

Section E   *p. 1107*

Not as above...5

5. Spiral cells..**Section F**   *p. 1109*

Not as above...6

6. Cells spherical to cylindrical, varying from spheres 5 μm in diameter to large cylindrical organisms 35–100 μm long; sulfur deposited internally when growing in the presence

of hydrogen sulfide. In one of the two recorded species, large crystals of calcium carbonate fill the cells; motile with a slow, jerky, rotating action when in contact with solid surfaces (see Plate 2.9, Figs. 1–4)................**Achromatium**   *p. 121*

Not as above...7

7. Cells spherical to ovoid, 5–25 μm in diameter, with the cytoplasm compressed in one end of the cell where a polar fibrillar organelle is visible in electron microscope sections; sulfur deposited in the cytoplasmic layer; exhibits an extremely rapid darting motion in free solution; peritrichate flagella; found in waters containing hydrogen sulfide, forming a tenacious web-like growth in a zone of critical hydrogen sulfide-oxygen concentration..**Thiovulum**   *p. 463*

Not as above...8

8. Large, cylindrical, pear-shaped or slightly curved rods 3–14 μm wide; motile by means of a single polar flagellum; normally contain large spherules of calcium carbonate and may also contain sulfur.............................**Macromonas**   *p. 462*

Not as above..9

9. Spherical cells which reproduce by binary fission or by budding..**Section G**   *p. 1112*

Not as above...10

10. Rod-shaped cells, each containing one or more sulfur globules. Cells embedded in a gelatinous mass which may be spherical, when free floating, or dendroid when attached to a solid substrate: non-motile.................**Thiobacterium**   *p. 462*

Not as above...11

11. Rod-shaped organisms occurring singly, in clusters, in chains, as unicellular filaments or as multicellular, branched or unbranched, filaments (or trichomes) which may be enclosed in a sheath.....................................**Section H**   *p. 1117*

SECTION B

1. Non-motile spheroidal organisms which exist as obligate intracellular parasites of tissue cells of vertebrates and are found also in or on ectoparasitic arthropods. An *elementary body* 0.2–0.4 μm in diameter, possessing a procaryotic nucleus and numerous ribosomes within a multilaminated wall, infects the host cell and develops to an *initial body* 0.8–1.5 μm in diameter within a vesicle originating from an invagination of the host cell membrane. This is followed by successive divisions by binary fission resulting in the formation of dense masses of *elementary bodies* which are released by disruption of the host cell and vesicle to act as new sources of infection. There is an *intermediate body* (Plate 18.6, Fig. 1) representing a transitional stage between the initial body and elementary body. Gram-negative; stain by Giemsa's, Macchiavello's, Giménez's or Casteneda's methods. Metabolically active in the extracellular state only in the presence of cofactors; inhibited by tetracyclines, penicillin and 5-fluorouracil; no growth in culture media but can be cultivated in the yolk sac of chick embryos or vertebrate tissue cultures; G + C content of the DNA ranges from 39–45% (T_m); cause diseases of the eye, the urogenital and respiratory systems of man, respiratory and generalized infections of birds and respiratory, placental, arthritic and enteric diseases of mammals (excluding primates)

                                                                 Chlamydia   *p. 915*

Not as above...2

2. Parasites of the erythrocytes of vertebrates.................................3

Not as above...9

3. Obligate parasites; not cultivable on laboratory media: transmitted by arthropods; cells 0.3–0.4 μm in diameter, lacking a cell wall; growth not affected by penicillin....4

Not obligate parasites; may be cultured on laboratory media containing blood or blood derivatives; cells are bounded by a cell wall, growth may be inhibited by penicillin. Rod-shaped, coccoid, ring- and disc-shaped cells, sometimes filamentous. Gram-negative...8

4. Endoparasites of erythrocytes; bacteria 0.3–0.4 μm in diameter which *enter* erythrocytes causing an invagination of the cytoplasmic membranes with the subsequent formation of membrane-bound vacuoles in which the organisms divide by binary fission to produce dense inclusion bodies; individual bacteria appear to have dense aggregates of fine granular material embedded in an electron transparent plasma and enclosed in a double membrane..5

Ectoparasites of erythrocytes—occasionally endoparasitic; individual bacteria have a limiting membrane without any apparent differentiation of nuclear and cytoplasmic bodies.......................................7

5. Inclusion bodies with appendages in the form of tails, loops or rings revealed in wet preparations of blood or by fluorescent antibody stain or by electron microscopy (Plate 18.3, Fig. 2). Infectious for cattle but not for deer..**Paranaplasma** *p. 908*

Inclusion bodies without appendages.......................................6

6. Organism infects ruminants only. Inclusion bodies in the erythrocytes are 0.3–1.0 μm in diameter (Plate 18.3, Fig. 1)............................**Anaplasma** *p. 907*

Organism infects birds and some cold-blood animals. Inclusion bodies in erythrocytes 0.3–3.9 μm in diameter (Plate 18.4, Fig. 1). Highly pleomorphic forms are common

Aegyptianella *p. 909*

7. Ectoparasitic only; may also be found in the blood plasma; appear as coccoid or ring forms 0.4–1.5 μm in diameter, the coccoid forms only being observed in fresh preparations examined by dark-field or phase contrast microscopy (Plate 18.5, Fig. 1). Cause infections in rodents, cattle, sheep or pigs..............**Eperythrozoon** *p. 912*

Predominantly ectoparasitic on the erythrocyte; when endoparasitic infections are confined to vacuoles surrounded by invaginated host cell membrane. Organisms coccoid or rod-shaped—the latter appearing under the electron microscope as a chain of coccoid bodies. Ring forms not present (Plate 18.4, Fig. 2); infections recorded in mice, cats and dogs...............................**Haemobartonella** *p. 910*

Note: *Attention is drawn in* THE MANUAL *to the very arbitrary differentiation of Eperythrozoon and Haemobartonella.*

8. Microorganisms found in fixed tissue cells and on or in erythrocytes (Plate 18.2, Figs. 1–6). Characteristically occur in chains of several segmenting organisms, sometimes swollen at one or both ends and frequently beaded in preparations stained by Giemsa, and other Romanowsky stains; flagellated with 1–10 polar flagella from cultures but not from tissues; occur in man and the arthropod *Phlebotomus* spp.; cause progressive anaemia or cutaneous eruptions in man...............**Bartonella** *p. 904*

Microorganisms occur only in the erythrocytes of mammals, excluding man. Similar to *Bartonella* but motility has not been detected. Infections have been recorded from moles and the deer mouse.............................**Grahamella** *p. 905*

9. Organisms occur only as pathogens or symbiotes in arthropods; do not infect man or other animals..10

Not as above.......................................11

10. Symbiotic in cockroaches; straight or slightly curved rods, 0.8–0.9 μm by 1.5–8 μm, occurring singly or in pairs, the longer forms occurring in mycetocytes and the shorter forms in germ tissue and oocytes...........**Blattabacterium** *p. 901*

Symbiotic in the bed bug, *Cimex lectularius*, where they occur mainly in coccoid and diplococcoid forms 0.2 μm by 0.4–0.5 μm in the paired host organs called *mycetomes;* also occur extracellularly as rods and filamentous forms 0.25–0.3 μm by 3.0–8 μm

Symbiotes *p. 900*

Pathogenic, causing "blue disease" in the larvae of Coleoptera, Lepidoptera, Diptera

Rickettsiella *p. 901*

Found in, but not necessarily pathogenic for the sheep ked, (*Melophagus ovinus*), mosquitoes (*Culex pipiens* and *Culex fatigans* and others) and some Diptera and the tick *Argas persicus*, where they occur extracellularly or intracellularly in various organs but are *not* usually associated with mycetomes........**Wolbachia** *p. 898*

11. Coccoid to ellipsoid microorganisms found intracytoplasmically (Plate 18.1, Fig. 1) either singly or in compact clusters in circulating monocytes and lymphocytes (in

dogs) and also in neutrophils, eosinophils and basophils (in domestic ruminants); also probably occur in horses; *not infectious for man;* cannot be cultured in cell-free media, chicken embryos or tissue cultures; non-motile; Gram-negative; transmitted by the ticks, *Rhipicephalus* spp. and *Hyalomma* sp............**Ehrlichia**   *p. 893*

Not as above...12

12. Coccoidal, ellipsoidal and occasionally rod-shaped pleomorphic organisms which occur intracytoplasmically but not intranuclearly in the vascular endothelial cells of ruminants; non-motile; Gram-negative; *not infectious for man;* transmitted by the ticks *Amblyomma* spp. in which maintenance is transstadial only

     Cowdria   *p. 895*

Not as above...13

13. Pleomorphic but predominantly coccoidal organisms 0.3–0.4 μm in diameter, found intracytoplasmically (Plate 18.1, Fig. 2) but not intranuclearly; mainly in the reticular cells of the lymphoid tissues of dogs and in tissues of the fluke vector, *Nanophyetus salmincola; not infectious for man;* Gram-negative; non-motile; not cultivated in laboratory media or in chicken embryos; does not give the Weil-Felix reaction

     Neorickettsia   *p. 896*

Not as above...14

14. Short rods, 0.2–0.4 μm by 0.4–1.0 μm or occasionally spheres 0.3–0.4 μm in diameter; non-motile; Gram-negative; grows preferentially in vacuoles of the host cell; not cultivated in laboratory media but grows well in the yolk sac of chick embryos; resists inactivation at 60 C for some minutes; infectious for guinea pigs, rabbits, hamsters and mice; enzootic in cattle, sheep and goats and other animals and birds; transmitted transovarially by numerous arthropods, especially ticks, and by aerosol sprays of infected material such as milk; causes Q fever in man. G + C content of the DNA is approximately 43 moles %........................**Coxiella**   *p. 891*

Not as above...15

15. Short rods, 0.2–0.5 μm by 1.0–1.6 μm; Gram-negative; non-motile; can be cultivated on blood-, or fetal serum-based media producing colonies 65–200 μm in diameter at 37 C in 12–14 days, reducing to 3–5 days on subsequent passages; poor or no growth in chick embryos; not infectious for guinea pigs and other common laboratory animals; *infect Rhesus monkeys and man* in whom it causes Trench fever; transmitted by the human louse, *Pediculus humanus* where it grows in the lumen of the gut and is discharged in feces. The G + C content of the DNA is approximately 38.5 moles %

     Rochalimaea   *p. 890*

Not as above...16

16. Short rods 0.3–0.6 by 0.8–2.0 μm; non-motile; Gram-negative; not cultivated in culture media; grow in entodermal cells of the yolk sac of the chick embryo and also in mammalian tissue cells where growth occurs mainly in the cytoplasm and sometimes in the nucleus; inactivated by heat at 56 C; G + C content of the DNA ranges from 30–32.5 moles %; pathogenic to man and other animals; transmitted by arthropods. The cause of epidemic typhus, murine typhus, Rocky Mountain spotted fever and similar infections...**Rickettsia**   *p. 883*

SECTION C

1. Microorganisms inhabiting the planktonic layers or bottom deposits of fresh water lakes and microzones in soils; cells coccoid 0.5–1.5 μm in diameter; germinate to produce filaments or to produce chains of similar cells; filaments 0.02–0.25 μm wide, highly refractile when growing in the presence of manganese; grow on ferric ammonium citrate agar and other media to produce irregularly shaped colonies after 4–6 weeks. In the presence of manganese and/or iron they become deep brown or black; also grow in media containing serum used for mycoplasmas (see Plate 4.4)

     Metallogenium   *p. 163*

Note: *Attention was drawn to the similarity between Metallogenium and the myco-*

plasmas by the author (Skerman, 1967) and later by Dubinina (1969) and Balashova (1969). The exact affinity has yet to be determined.

Not as above...2

2. Gram-positive; in liquid media only, filamentous forms are helical or branched and non-helical, the helical forms being motile; *sterols required for growth;* optimum temperature 32 C with growth best near neutrality; cell membrane 7–8 nm thick with an outer layer of short projections on the bounding membrane; isolated from the leaves of citrus affected by "stubborn disease" but pathogenicity not established experimentally; antigenically unrelated to *Mycoplasma* or *Acholeplasma*

Spiroplasma *p. 953*

Not as above..3

3. Spherical cell form 0.3–2.0 μm in diameter; optimum temperature of growth 59 C; no growth at 37 C or 65 C; non-motile; optimum pH of growth 1–2 (range to pH 3.0); growth in a mineral salts medium with yeast extract and glucose; *sterols not required for growth;* colony on ion-agar flat, coarsely granulated, and dark brown and may have the typical "fried-egg" appearance: found in acid hot springs and a burning coal refuse pile (*cf. Sulfolobus*)........................**Thermoplasma** *p. 952*

Not as above; optimum pH near 7.0...4

4. *Sterols required for growth;* most species with a moderate to high sensitivity to lysis by 1.5% digitonin; isolated principally from the upper respiratory tract of man and animals..**Mycoplasma** *p. 930*

Sterols not required for growth; most species resistant or showing only slight susceptibility to lysis by 1.5% digitonin; isolated from sewage, manure, humus soil and a variety of infections of domestic animals ....**Acholeplasma** *p. 949*

SECTION D

Note: *In this section the term prostheca is used to denote a tubular extension of the cell wall and cytoplasmic membrane. It may be short and conical or long and filamentous. It may attach to a substratum and thus function as a stalk.*

Stalk is used in a general sense to include both prosthecae which function as stalks and stalks which are formed as secretions from the cell.

Since some of the organisms included in this section have never been cultured, or examined in section under the electron microscope, the exact origin of a stalk is sometimes obscure.

1. Spherical, ovoid, rod-shaped, pear-shaped or bean-shaped microorganisms which produce one or more fine tubular prosthecae from the poles or other positions on the cells and upon the ends of which daughter cells (buds) are produced (Plate 4.1, Fig. 1). The buds may remain attached before producing prosthecae or may separate. In some forms, so far uncultured, the organisms appear as small spherical cells interconnected by fine threads...6

Not as above..2

2. Cells produce prosthecae, ranging from 0.1–3.0 μm long, *from two to several points on the mother cell* (Plate 4.2, Figs. 1–2); prosthecae may be bifurcated. The length of the prosthecae ranges from 0.2–3.0 times the width of the mother cell, never long and filamentous; reproduction is by budding or by binary fission of the mother cell, never by budding from the prosthecae...11

Not as above..3

3. Cells produce a single prostheca from one pole of the cell or from a position excentric to the pole or in a lateral position; the prostheca *may* be attached by a holdfast and function as a stalk. Reproduction is by asymmetric transverse binary fission *or* by budding; straight or curved rods or spherical to pear-shaped cells; no sulfur deposited from H_2S...13

Not as above..4

4. Cells produce a thin filament (prostheca ?) from one or both poles; cells spirally twisted with both ends tapered, 10–15 times longer than wide; rosettes may be formed by

terminal attachment of filaments; require H$_2$S for growth depositing sulfur externally
(?) (Plate 4.2, Figs. 3–4) . . . . **Thiodendron**   *p. 158*
Note: *NOT Thiodendron of Lackey and Lackey (J. Gen. Microbiol. 26: 29, 1961), a synonym
of Thiobacterium, p. 462.*
Not as above 5

5. Cells produce stalks by the secretion of substances which, in the process of accumulation, raise the cells above the substratum to which they may be attached; cells reproduce by binary fission; daughter cells continue to secrete with the result that the
initial stalk bifurcates; repetition of this process results in a multibranched dichotomous stalk; individual cells lie at a right angle to the long axis of the stalk 15
Note: *Organisms which simply produce a gum which fixes them to a surface in a sessile position are treated elsewhere in this key.*

6. Several prosthecae produced, arising independently from any point on the mother cell;
cells spherical, oval, rod-shaped, pear-shaped or bean-shaped 0.4–2.0 μm wide with
prosthecae 0.15–0.3 μm wide; grow in mineral salts media with organomineral complexes of fulvenic acid and sesquioxides or simple carbon sources; deposit iron and/or
manganese compounds mainly on the mother cell . . . . **Pedomicrobium**   *p. 151*
Prosthecae produced singly at one or both poles of the mother cells 7

7. Cell masses salmon pink to deep orange; free cells ovoid, 1.2 by 2.8 μm, motile with
peritrichous flagella; contain bacteriochlorophyll *a* and carotenoids: grow only under
anaerobic conditions exposed to light; prosthecae septate (see Plate 1.1, Fig. 3)
                                         Rhodomicrobium   *p. 33*
Organisms not photosynthetic; aerobic; prosthecae not septate; separated buds motile
or non-motile 8

8. Coccoid cells 0.5–1.5 μm in diameter interconnected by fine threads 0.1–0.2 μm in diameter; natural habitat lake waters and muds from which glass capillary cultures show
a deposition of oxides of manganese and/or iron 9
Not as above; cells rod-shaped, fusiform, oval, egg-shaped, pear-shaped or bean-shaped
with some coccal forms; metals apparently not deposited on the cells or
prosthecae 10

9. Coccoid cells 0.5 μm in diameter, connected by threads 0.1 μm wide. Multiply by budding; young cells motile; microcolonies 0.3 mm in width with a reticulate ribbed surface structure, irregularly shaped and round, almost black (common) or radiate-
lobate (rare) or trichospherical (very rare). The margins of colonies are often radially
fringed by dichotomously branched filaments . . . . **Caulococcus**   *p. 165*
Coccoid cells 0.5–1.5 μm in diameter, connected by filaments 0.1–0.2 μm wide: cells non-
motile, attached and multiply by budding . . . . **Kusnezovia**   *p. 166*
Note: *See also* **Metallogenium** *which is treated in these keys in* **Section C** *with* **Mycoplasma** *and similar organisms.*

10. Growth occurs in mineral salts media without added carbon or with simple carbon com-
pounds . . . . **Hyphomicrobium**   *p. 148*
Complex organic media required for growth . . . . **Hyphomonas**   *p. 150*
Note: *Two species of* **Hyphomicrobium** *(H. neptunium and H. indicum) which require
organic media for growth, Hirsch considers should probably be transferred to the genus*
Hyphomonas. *H. indicum does not show the typical production of prosthecae and buds. It
is described as being rod-shaped with occasional chains of spherical forms 1–2.5 μm wide
joined by threads 1–2.5 μm long, suggesting a budding process (e.g.* **Caulococcus** *and*
Kusnezovia*).*

11. Green photosynthetic organisms, 0.5–0.7 μm by 1.0–1.2 μm, each cell producing about
20 prosthecae 0.1–0.17 μm wide and 0.1–0.5 μm long; cells contain bacteriochlorophyll
c and carotenoids, both located in elongated, ovoid vesicles underlying and attached
to the cytoplasmic membrane . . . . **Prosthecochloris**   *p. 55*
Not photosynthetic 12

12. Length of prosthecae about three times the width of the mother cell (Plate 4.2, Fig. 1);
cells reproduced by budding from the mother cell, not from the prosthecae, the bud

separating by a transverse division when about the same size as the mother cell; non-motile..**Ancalomicrobium** *p. 156*

Length of prosthecae only 0.5–1.0 times the width of the mother cell; normally less than 2 μm long and conical in shape; mother cells divide by binary transverse division; daughter cells may be motile with a single polar to subpolar flagellum (Plate 4.2, Fig. 2)...**Prosthecomicrobium** *p. 157*

Note: *Siderococcus limoniticus* is described as cells having "*filamentous appendages but no capsules.*" *The statement is too vague to decide whether they represent prosthecae or not (see p. 468).*

13. Cells reproduce by budding which may be terminal or lateral; cells spherical to oblong or pear-shaped, 0.3–1.7 μm in diameter; prosthecae 0.3–0.9 μm wide and up to 11 μm long and function as stalks; often attach to a common holdfast to form rosettes; no sulfur deposited from H_2S...**Planctomyces** *p. 162*

Note: *The origin of the stalk has not been determined. It may be excreted. This genus was described under the name of* **Blastocaulis** *in the 7th edition of* THE MANUAL.

Cells reproduce by asymmetric binary fission...14

14. The prostheca (stalk) arises from the pole of the cell as an extension of the cell wall and cell membrane. The daughter cell, following separation, secretes adhesive material at the base of the flagellum, develops a stalk at this site and enters the immotile vegetative stage. The adhesive material allows the cell to attach to other microorganisms, to cells of similar type or to inanimate objects. Attachment of several cells by their holdfasts results in rosette formation. The cells are rod-shaped or vibrioid, tapered or cylindrical; colorless or contain yellow or orange or red pigments; the G + C content of the DNA is approximately 65 moles % (buoyant density); aerobic

Caulobacter *p. 153*

The stalk arises from a site on the cell that is not coincidental with the center of the pole of the cell. The stalk does not possess adhesive material. Multiplication occurs by division of the stalked cell, giving rise to a non-stalked sibling, which is smaller than the stalked sibling and is motile by means of a single flagellum which arises in an excentric position on the pole of the cell. This cell develops a stalk and enters the immotile vegetative phase. Adhesive material is secreted by cells in both phases at or near the pole of the cell at a site different from that at which the stalk develops. The adhesive material allows cells to attach to a variety of solid substrates, or to one another's holdfasts to form rosettes. The cells of known types are rod-shaped and colorless...**Asticcacaulis** *p. 155*

15. Cells rod-shaped, often bent, 0.7–2.0 μm by 2.4–12.0 μm; borne on the ends of lobose, dichotomously branched stalks composed of a gum which is soluble in 1% KOH. Cells lie with their long axes at right angles to the stalk (Figs. 4.B and 4.C). The gummy colonies may float on water or become attached: no metal is deposited in the stalks

Nevskia *p. 161*

Note: *The genus* **Siderophacus,** *included in the 7th edition of* THE MANUAL *as having horn-shaped stalks, round in cross-section, is here considered synonymous with* **Nevskia.** *Iron was deposited on the stalks.*

Stalks often flat and ribbon-like, secreted from the side of the kidney-shaped cells. Under the electron microscope they are seen to be composed of a mass of fibrils which are variously reported as being composed of, or later covered with, iron hydroxide. Division of the cells results in a bifurcation of the stalk as the two, often rounded, daughter cells continue to secrete stalks attached to the common origin; stalks usually disrupt completely in hydrochloric acid: liberated cells are motile with polar flagella...**Gallionella** *p. 160*

SECTION E

Note: *a. The arrangement of this key differs in some respects from those provided by Mc-Curdy and Zahler and McCurdy in* THE MANUAL. *Readers, in keying out the fruiting Myxo-*

bacterales, are advised to make use of both keys for the purpose of comparison. The McCurdy keys place primary emphasis on the morphology of the vegetative cells because of variability of the fruiting bodies. In these keys several characters are usually taken together, if this is possible, and the statements include the morphology of the vegetative cells.

b. Dimensions quoted in the following key cover the range of sizes encountered over all species described within a genus and do not apply to any single species. They are given to emphasize both approximate dimensions and the relative dimensions of vegetative cells and myxospores.

c. The term **myxospore** *is used uniformly in this key to refer to the resting cell stage as distinct from the vegetative stage.* **Sporangium** *is used in the same sense as* **cyst** *was used in the 7th edition of* THE MANUAL.

1. Myxospores produced at random among the vegetative cells; phase dense or highly refractile; no fruiting body formed; vegetative cells rod-shaped, tapered and flexible: G + C content of the DNA is 36 moles %...............**Sporocytophaga**   *p. 111*
   Myxospores are produced in fruiting bodies: G + C content of the DNA is 69–71 moles %...2
2. Myxospores spherical or ellipsoidal: phase dense or highly refractile (microcysts); vegetative cells slender rods with tapering or rounded ends, 0.4–0.7 by 2.0–10 μm. Fruiting bodies sessile or erect, the latter sometimes being constricted at the base, composed of a mass of slime which may remain moist or become quite firm and remain so or later deliquesce; colorless, yellow, buff, tan, flesh-colored to orange-red: not enclosed by a surrounding membrane: bacteriolytic, non-cellulolytic (Plate 2.1, Figs. 1–9)...**Myxococcus**   *p. 79*
   Not as above..3
3. Myxospores distinctly different from the vegetative cell; phase dense or refractile; vegetative cells do not erode or etch agar media: Congo red is absorbed............4
   Myxospores resemble the vegetative cells being the same width but possibly a little shorter; not phase dense or refractile; vegetative cells etch, erode and penetrate agar media; Congo red is not absorbed..7
4. Myxospores in fruiting bodies lie in parallel bundles in swollen, intertwined intestine-like masses *without a surrounding sporangial wall.* Fruiting bodies are sessile and depressed and usually disintegrate into clumps when compressed under a cover glass: vegetative cells slender and tapered, 0.4–0.7 μm by 4–15 μm; flexible. Myxospores short rods, ellipsoids or spheres 0.8–2.0 μm by 1.5–2.8 μm....**Archangium**   *p. 83*
   Note: *The fruiting bodies of some strains of* **Mellitangium** *lack the usual stalks and superficially resemble those of* **Archangium***. They differ, however, in possessing a sporangial wall.*
   Myxospores enclosed in sporangia (cysts) in the fruiting body. Sporangia may occur singly or in groups; sessile or borne on stalks....................................5
5. Sporangia are sessile, 20–150 μm in width; occur singly or in groups; rounded, elongated or coiled (*cf. Archangium*) and surrounded by a definite slime envelope. Vegetative cells slender, tapered, flexible rods, 0.6–0.8 μm by 3–20 μm. Myxospores are rod-shaped, fusiform or oval, 0.4–1.8 μm by 1.3–5.0 μm. Cellulose not digested (see Plate 2.2, Figs. 1–4)..**Cystobacter**   *p. 87*
   Note: *This genus includes some of the species of the genus* **Polyangium** *of the 7th edition of* THE MANUAL.
   Sporangia are borne on stalks...6
6. Sporangia borne singly on unbranched stalks; spherical, ellipsoidal, like the pileus of a mushroom; 15–50 μm minimum dimension; may be white, yellowish, brown, orange or orange-red. Vegetative cells tapered with rounded ends, 0.6–1.5 μm by 4.5–10 μm. Myxospores 0.7–1.0 μm by 1.5–5.0 μm (Plate 2.2, Figs. 6–10)
                                                               Melittangium   *p. 89*
   Note: *This genus replaces the genus* **Podangium** *of the 7th edition of* THE MANUAL. *Some sessile forms resemble* **Archangium.**
   Sporangia borne in clusters on branched or unbranched stalks; spherical, oval, or cylindrical, 16–30 μm minimum dimension; may be yellowish orange, bright orange-

red, reddish brown, chestnut to black at maturity. Vegetative cells are tapered 0.6–1.0 μm by 4.0–10.0 μm. Myxospores 0.8–1.5 μm by 1.5–4.0 μm (see Plate 2.2, Figs. 11–14).

Stigmatella   *p. 90*

Note: *1. The two species included in this genus were described in the 7th edition of* THE MANUAL *as species of* **Chondromyces** *and* **Podangium.**
2. The stalk of **Stigmatella** *withers rather quickly and mature fruiting bodies may appear sessile.*

7. Sporangia sessile 8

Sporangia borne on stalks; spindle-shaped, conical, spherical, apiculate, clavate, cylindrical or pyriform—occasionally with an apical tuft of fibrils; occur in spherical or umbel-shaped clusters or in chains; minimal dimension of single sporangia 10–40 μm; yellow, orange, white or light pink. Vegetative cells cylindrical, not tapered, with blunt rounded ends, 0.9–1.4 μm by 3–16 μm (mostly 3–6 μm). Myxospores 0.6–1.4 μm by 2.6–7.0 μm (see Plate 2.5, Figs. 1–8) . . . . **Chondromyces**   *p. 96*

Note: *This genus includes four species from* **Chondromyces** *and* **Synangium** *sessile from the 7th edition of* THE MANUAL.

8. Vegetative cells never coccoid; cylindrical with blunt ends, 0.7–1.2 μm by 3.0–10.0 μm. Myxospores 0.7–0.9 μm by 2.5–5.8 μm. Sporangia irregular, spherical, oval, cushion-shaped, elongated or polygonal—the latter resulting from close appression in the sori; minimum dimensions of the sporangia range from 5 μm (*P. sorediatum*) to 75 μm (*P. vitellinum*); occurring as solitary sporangia or in groups of two to several hundred in sori which are often covered with slime; yellow, reddish orange, brown, pink or orange-red (see Plate 2.3, Figs. 1–8) . . . . **Polyangium**   *p. 92*

Note: *This genus includes species from the genera* **Polyangium** *and* **Sorangium** *in the 7th edition of* THE MANUAL.

Vegetative cells short, blunt ended rods or cocci, 1.5 μm by 2.5–3.4 μm, becoming oval, cube-shaped or coccoid; myxospores similar to vegetative cells; sporangia solitary, oval to spherical, 3.5–6.0 by 40–110 μm. Vegetative colonies light orange to violet (see Plate 2.4, Figs. 2–5) . . . . **Nannocystis**   *p. 96*

SECTION F

1. Highly flexible helically coiled organisms; motile with a graceful undulating movement to a highly vigorous rotational movement, often rapidly reversing direction and flexing; no flagella. The electron microscope reveals that the cells are composed of a central helical protoplasmic body, to the subapical portions of which one or more axial fibrils are inserted which overlap toward the center of the cell, the whole structure being surrounded by a membranous envelope. With the light or phase contrast microscope the axial fibrils are not resolvable except in the larger morphological forms, where an "axial filament" *may* be observed wound round the protoplasmic body 2

Note: *Compare with Section F6.*

Not as above 6

2. Cells 5.0–300 μm or more long; 0.2–0.75 μm wide with one axial fibril inserted at each end of the cell; anaerobic or facultatively anaerobic; found in H₂S-containing fresh and marine waters, sewage and polluted water (see Plate 5.1, Fig. 1)

Spirochaeta   *p. 168*

Note: *4 of 5 species placed in this genus have been isolated and have the 2 axial fibrils indicated above. The type species has not been isolated.*

Not as above 3

3. Spiral cells, 0.5–3.0 μm wide and 30–150 μm long with a spiral amplitude of 6–8 μm; flexible cells characterized by a thin membrane or crista, which may be composed of over 100 axial fibrils, on one side of the body, extending the entire length of the cell; cross striations in stained cells are distinct; actively motile without flagella; re-

corded from the crystalline style sac and the alimentary canal of mollusks (see Plate 5.1, Fig. 2)...**Cristispira**   *p. 171*

Not as above...4

4. Aerobic; cells 0.1 μm thick and 6–20 μm long; wound in very fine coils 0.2–0.3 μm in overall diameter and with a pitch of 0.3–0.5 μm; bent or hooked at one or both ends; possibly only 2 axial filaments one inserted at each end; can be cultivated *in vitro* in semisolid rabbit plasma media; not readily stained; stain with Giemsa's stain or by silver impregnation methods; visible unstained by dark ground or phase contrast microscopy but rarely by ordinary light (see Plate 5.1, Fig. 5)....**Leptospira**   *p. 190*

Not as above...5

5. Anaerobic; 0.2–0.5 μm wide and 3–15 μm long; spirals frequently irregular, obtuse angled and of variable amplitude; stain readily, Gram-negative: The electron microscope reveals a foamy elastic envelope and 15–20 axial fibrils in each cell (see Plate 5.1, Fig. 4)...**Borrelia**   *p. 184*

Anaerobic: 0.09–0.5 μm wide, 5–15 μm long; tight regular or irregular coils with the ends of cells usually pointed; 1–3 axial fibrils inserted at each end (where specified); cells wider than 0.2 μm stain Gram-negative and are catalase and oxidase negative and fermentative (see Plate 5.1, Fig. 3)...........**Treponema**   *p. 175*

Note: *All but* **Treponema pallidum** *and three related species have been cultured and described in considerable detail.*

6. Tightly wound spirals of 3–20 turns; the spiral nature may not be resolved by the light field or phase contrast microscope—the shorter forms appearing as rods; cell width 0.3–0.4 μm; spiral amplitude 0.8–1.0 μm: if motile have 1–6 peritrichous flagella; obligate autotrophs oxidizing ammonia to nitrite; strictly aerobic; pH range 6.5–8.5; G + C content of the DNA of one strain is 54.1 moles % (buoyant density) (see Plate 12.2, Figs. 4–6)...**Nitrosospira**   *p. 454*

Not as above...7

7. Prosthecae or stalks produced.....................return to **Section D**   *p. 1105*

No prosthecae produced, holdfasts may be formed...............................8

8. Organisms contain bacteriochlorophylls with carotenoid pigments................9

Organisms do not contain photosynthetic pigments...........................12

9. Cells green; non-motile; contain bacteriochlorophyll *c* or *d* (long wave length absorption maxima 745–755 nm and 705–740 nm, respectively) and chlorobactene (carotenoid) pigments: anaerobic; deposit sulfur outside the cells (see Plate 1.1, Figs. 7–8)

                                                            Chlorobium   *p. 52*

Cells various shades of red, yellow, orange, brown or purple; contain bacteriochlorophyll *a* (absorption maxima 375, 590, 805, 830–890 nm) and carotenoids of the spirilloxanthin, spheroidenone and okenone series.................................10

10. Organisms will grow anaerobically when exposed to light or aerobically in the dark; when sulfide or thiosulfate is used as an electron donor elemental sulfur is not an oxidation product (see Plate 1.1, Fig. 1)...............**Rhodospirillum**   *p. 26*

Organisms grow only under anaerobic conditions exposed to light..................11

11. Sulfur deposited inside the cells following oxidation of hydrogen sulfide; pigments located in internal membranes of the vesicular type; G + C content of the DNA is 45.5 moles % (buoyant density)...........................**Thiospirillum**   *p. 41*

Sulfur deposited externally following oxidation of hydrogen sulfide: pigments located at lamellar membrane stacks continuous with the cytoplasmic membrane; G + C content of the DNA is 62.2–69.9 moles % (buoyant density)

                                                   Ectothiorhodospira   *p. 47*

12. Cells S-shaped, slender filaments, 0.23–0.25 by 9–30 μm, usually with a spiral twist; arranged side by side in flat sigmoid aggregates of 4 (Plate 6.1, Fig. 6) or multiples of 4; sometimes spread out in the shape of a fan; found in and on mud in fresh and brackish waters...**Pelosigma**   *p. 215*

Not as above..13

13. Motile with polar to subpolar flagella...14

Non-motile or motile with a gliding action: no flagella........................21

Motile with lateral flagella produced from the concave side of the cell curved rod
Selenomonas   *p. 424*

14. Aerobic to microaerophilic..15

  Anaerobic..19

15. Cells oxidize hydrogen sulfide depositing sulfur inside the cells: volutin granules toward the ends of the cells.............................**Thiospira**   *p. 464*

  Sulfur not deposited in the cells.....................................16

16. Rod-shaped, spirally twisted cells, 0.5–0.7 μm by 1–12 μm, which *bud* to produce star-shaped figures or rosettes (Plate 4.3, Figs. 1–4). Cells also divide by *transverse fission* producing motile cells each with a subpolar flagellum (Plate 4.3, Fig. 5): deposit iron hydroxide from media containing iron.................**Seliberia**   *p. 160*

  Not as above..17

17. Organisms parasitic on other bacteria. In the preinfection stage, cells are curved rods 0.25–0.4 μm by 0.8–1.2 μm; motile usually with a single polar sheathed flagellum (Plate 6.1, Figs. 1–2). In the host bacterium they grow in the periplasmic area as curved rods or coiled filaments (Plate 6.1, Figs. 3–4).....**Bdellovibrio**   *p. 212*

  Not as above...18

18. Cells 0.25–1.7 μm wide and 2–60 μm long; polytrichous polar flagella, usually at both poles; strictly aerobic or obligately microaerophilic; may oxidize but do not ferment carbohydrates; polyhydroxybutyrate granules usually present: G + C content of the DNA ranges from 38–65 moles % (T_m).................**Spirillum**   *p. 196*

  Note: *1. Hylemon et al. (Int J Syst Bacteriol October 1973) proposed a division of this genus into* **Spirillum, Aquaspirillum** *and* **Oceanospirillum**

  2. The 7th edition of THE MANUAL listed several species in the genus **Vibrio** *which showed a spiral morphology. Some of these have been removed to* **Campylobacter**. *See also* **V. fischeri** *which ferments carbohydrates.*

  Cells 0.2–0.8 μm wide and 0.5–5.0 μm long; single polar flagellum at one or both poles; polyhydroxybutyrate absent; microaerophilic to anaerobic; do not oxidize or ferment carbohydrates; G + C content of DNA ranges from 30–35 moles % (T_m)
Campylobacter   *p. 207*

19. No carbohydrates fermented or oxidized; cells 0.2–0.8 μm wide and 0.5–0.8 μm long; single polar flagellum at one or both poles; polyhydroxybutyrate absent; G + C content of the DNA ranges from 30–35 moles % (T_m)....**Campylobacter**   *p. 207*

  Not as above: glucose is fermented....................................20

20. Cells are helical, 0.3–0.7 μm by 1–7 μm, with pointed ends; an incomplete coil up to 3 coils per cell; Gram-negative; glucose is fermented with the production of succinic and acetic acid and also formic and sometimes lactic acid: no gas is produced: found in the rumen of cattle and sheep.......................**Succinivibrio**   *p. 422*

  Curved cells occasionally helically twisted: 0.4–0.6 μm by 2–4 μm with bluntly pointed ends; motile with lateral to subterminal flagella; weakly Gram-positive; glucose is fermented with the production of large amounts of ethanol, lactic, formic and acetic acid and CO_2 and a small amount of hydrogen; found in the rumen of cattle
Lachnospira   *p. 423*

21. Cells parasitic in the protozoan, *Paramecium*....................................22

  Not as above...23

22. Cells contain 1.5–2.5 spiral turns; tapered at the ends; parasitic within the micronucleus of *Paramecium aurelia*, causing marked enlargement of the micronucleus, which is filled with spirals...............................**Holospora**   *p. 383*
(*H. undulata*)

  Cells twisted in two spiral turns that are not abrupt; one end pointed and the other rounded; no flagella; movement helicoid; endospores are formed; parasitic in the cytoplasm of *Paramecium caudatum*...................**Drepanospira**   *p. 383*

23. Organisms produce multicellular filaments (trichomes) which *may* break up into shorter elements, particularly in aging cultures......................................24

  Not as above...27

24. Trichomes form regular spirals: *motile* with a gliding action on solid surfaces......25

Trichomes *non-motile:* do not form regular spirals, frequently being straight to undulating; free cells motile or non-motile..26

25. Sulfur deposited internally from the oxidation of H_2S; relatively slow rotational motility...**Thiospirillopsis**

Note: *This genus is not recognized in this edition of* THE MANUAL.

No sulfur deposited internally; spiral cells 0.8–1.2 μm wide and up to 500 μm long; width of helix 1.5–2.5 μm; pitch 4–17 μm; no axial filament or crista; Gram-negative; cell masses orange; orange-yellow or pink; easily cultured; found in oysters and also free living...**Saprospira** *p. 109*

Multicellular trichomes up to 250 μm long, with cells 0.3–0.5 μm ¹y 3–15 μm; not constricted at the cross walls; wound in loose spirals; trichomes fragment but not usually into arthrospore-like segments; active motility in fresh water, particularly the hypolimnion of lakes; not cultured as yet....................**Achroonema** *p. 123*

26. Trichomes non-motile; bound together in more or less spirally wound bundles; individual trichomes have a gelatinous sheath which may be vacuolated or ferruginous
Peloploca *p. 125*

Trichomes non-motile; ensheathed; impregnated with iron when growing in iron-bearing waters; spirally wound around themselves or algal filaments; free cells polar flagellated...**Leptothrix** *p. 129*

27. Spiral cells, occurring singly, capsulated, the capsule becoming impregnated with iron. The optical section of the edge of the cell appears a deeper color than the center giving the cells the appearance of a link in a chain; found in iron-bearing waters
Naumanniella *p. 467*
(*N. minor*)

Curved rods, 0.5–2.0 μm by 1.0 to more than 10 μm long which continue to grow in a circular, instead of a sigmoid, fashion as do other curved rods. Rings so formed are 1.5–10.0 μm in diameter. They may divide into two horse-shoe-shaped halves and then into smaller sections or continue to develop with the consequent formation of spirals..**Microcyclus** *p. 214*

Note: *A somewhat similar organism having chains of curved cells wound in a ball within a nearly spherical capsule was listed in the 7th edition of* THE MANUAL *as* **Myconostoc.**

SECTION G

1. Organisms contain bacteriochlorophylls and carotenoid pigments. Cell masses shades of yellow, green, brown, red and purple; grow anaerobically exposed to light; some also grow aerobically in the dark...........................**Section J** *p. 1134*

Spherical to cuboid cells 0.5–2.0 μm in diameter which undergo a succession of divisions in one to three planes to produce small chains of cocci to a muriformly divided compact mulberry-like mass, from which motile coccoid cells are produced. Occasionally produce pseudomycelium-like branching filaments which divide transversely and longitudinally to produce the coccoid forms (see Plate 17.10)
Geodermatophilus *p. 724*

Not as above...2

2. Single cells borne on the end of elongated stalks or united in chains by fine threads
return to **Section D** *p. 1105*

Not as above...3

3. Cells or their capsules impregnated with oxides of iron or manganese..............4

Note: *In the absence of further information, these organisms are identified on their iron-depositing characteristics. Most iron organisms studied in pure culture metabolize the organic compound that forms the iron chelate, and the liberated iron chelates with some cell component. Citrate-utilizing organisms, for example, will release iron from ferric ammonium citrate. Accumulation of the iron in or on the cell may depend only on the nature*

of the cell substance. Pure culture studies may place these organisms in more commonly recognized genera.
Not as above..6

4. Cells capsulated..5
   Cells not capsulated; may have filamentous appendages; 0.2–0.5 μm wide; occur singly or in small motile colonies; found in mud horizons in fresh waters with low oxygen concentrations and a neutral pH.......................**Siderococcus** *p. 468*

5. Cocci occur singly or in unordered groups in a common capsule; cells are colorless; *may* reproduce by budding; iron or manganese deposited on the capsule; common in fresh water, often attached to various substrates.........**Siderocapsa** *p. 465*
   Cells ellipsoidal to coccoid; occur singly in capsules which become impregnated with manganese (not iron), causing the cell to look like a ring in a chain. Multiplies by budding; buds motile and free of manganese...........**Naumanniella** *p. 467*
   (*N. polymorpha*)

6. Cells coccal only as a stage in a definite life cycle *or* as a pleomorphic phase of rod-shaped bacteria...7
   Not as above: cells coccal at all stages except during division..................13

7. Cells have a definite cyclic form of development; spherical cells germinate at one or more points to produce rod-shaped cells, which elongate and divide. At this point of division, growth of the cells continues at an angle to the original axis. When the side branch is equal in size to the parent cell, division occurs at the angle. This process is repeated during the growth of the colony. *In older colonies, the rods transform entirely into a mass of cocci.* Rods are most frequently Gram-negative with Gram-positive granules; cocci are frequently Gram-positive; soil inhabitants
   Arthrobacter *p. 618*
   Not as above..8

8. Gram-positive; organisms occur as cocci under anaerobic conditions in neutral media; in media becoming acid, they assume a diphtheroid form; extremely pleomorphic under aerobic conditions; *produce propionic acid from lactic acid*
   Propionibacterium *p. 633*
   Gram-negative..9

9. Anaerobic; non-motile; Gram-negative; principally spindle-shaped, rod-shaped cells exhibiting a coccoid phase; recorded from genital and alimentary tracts of man and other animals......................................**Fusobacterium** *p. 404*
   Aerobic..10

10. Obligate halophiles requiring 16 to 30% salt for growth; not luminescent; motile; colonies red, purple, yellow or orange...............**Halobacterium** *p. 270*
    Not obligate halophiles...11

11. Parasites attacking erythrocytes and endothelial cells of man; extremely pleomorphic within the host; straight and curved rods, ring forms, and cocci occur; grow in semi-solid rabbit serum agar mainly as rods and cocci; polar flagella
    Bartonella *p. 904*
    Note: *This genus is selected as an example of a large group of intracellular parasites that are pleomorphic and have coccal stages. Only an odd species has been cultivated. For other genera see Section B. p. 1102.*
    Animal parasites: produce tularemia or tularemia-like infections in rodents.
    a. Motile..**Yersinia** *p. 330*
    b. Non-motile; causes tularemia....................**Francisella** *p. 283*
    Organisms occurring as diplococci on primary isolation on solid media; give rise to a proportion of rod-shaped organisms in liquid, which increases as the culture ages. Rod forms may show bipolar staining and be capsulated
    Acinetobacter *p. 436*
    Not as above..12

12. Bioluminescent when grown on fish agar or meat infusion agar containing 3% salt
    Photobacterium *p. 349*

Organisms not fitting into any of the above groups are probably pleomorphic forms of rod-shaped cells...............................proceed to **Section I**   *p. 1120*

13. Obligate chemolithotrophs; will not grow on organic media; oxidize ammonia to nitrite, or nitrite to nitrate...14

Facultative chemolithotrophs; spherical cells which are distinctly lobulate on the surface, 0.8–1.0 μm in diameter; oxidize sulfur to sulfate lowering the pH to 2 or less; growth better with added yeast extract; optimum pH 2–3; optimum temperature 70–75 C, range of 55–85 C; cell wall devoid of peptidoglycan; G + C content of the DNA 60–68 moles % (buoyant density).....................**Sulfolobus**   *p. 461*

Heterotrophic..15

14. Oxidize ammonia to nitrite; motile or non-motile; large cells 1.5–2.2 μm in diameter; occur singly, in pairs or in tetrads: free or embedded in slime; G + C content of DNA of one strain 50.5–51.0 moles % (buoyant density) (Plate 12.2, Figs. 7–9)

Nitrosococcus   *p. 454*

Note: *It is possible that Nitrosolobus may appear spherical (see p. 455).*

Oxidize nitrite to nitrate: cells 1.5 μm or larger: motile with one or two flagella (Plate 12.1, Fig. 7): yellow to red due to high cytochrome content: G + C content of DNA of one strain is 61.2 moles % (buoyant density)............**Nitrococcus**   *p. 452*

15. Obligate anaerobes...16

Aerobes, facultative anaerobes, microaerophiles...............................24

16. Organisms arranged in cubical packets.....................................17

Not as above...18

17. Spheres 1.5–2.5 μm in diameter: non-motile: ferment acetate sometimes methanol, and possibly butyrate to methane and CO_2: may also reduce CO_2 with H_2 to produce methane......................................**Methanosarcina**   *p. 476*

Nearly spherical cells 1.8–3 μm in diameter non-motile: ferment glucose with the production of CO_2, H_2, acetic acid and either ethanol or butyric acid: G + C content of the DNA 28–31 moles % (buoyant density): isolated from mud, stomach contents and the surface of cereals...............................**Sarcina**   *p. 527*

18. Gram-negative...19

Gram-positive..20

19. Small cocci, 0.3–0.5 μm in diameter, occurring singly, as diplococci and in masses; non-motile; carbohydrates and polyols not fermented; produce acetate, propionate, CO_2 and H_2 from lactate; H_2S produced from organic sulfur compounds; G + C content of the DNA ranges from 40–44 moles % (buoyant density); parasitic in the mouth and the intestinal tract of man and animals.........**Veillonella**   *p. 446*

Cocci 0.6–1.0 μm in diameter: oval or kidney-shaped diplococci; non-motile: weak or no fermentation of glucose, cellobiose not fermented; lactate is not metabolized; amino acids can serve as sole energy source: acetic and butyric acid produced, with CO_2 from amino acids; no H_2S produced; G + C content of the DNA is 55.1–56.9 moles % (buoyant density): isolated from the intestinal tract of pigs

Acidaminococcus   *p. 447*

Large cocci; 2.4–2.6 μm; occur in pairs, sometimes arranged in chains; non-motile; produce lower fatty acids, CO_2 and H_2 from glucose and lactate; cellobiose not fermented; H_2S is produced; found in the rumen of cattle and sheep and cecum of pigs on starchy foods: G + C content of the DNA 53.1–54.1 moles % (buoyant density)

Megasphaera   *p. 448*

Not as above...20

20. Organisms arranged singly, in pairs and in chains...........................21

Organisms arranged singly, some pairs and in irregular masses: not in chains......23

21. Cocci or short lanceolate to oval rods, 0.7 μm by 0.8–1.8 μm; occur predominantly in pairs in chains; strongly Gram-positive; non-motile; reduces CO_2 with H_2 or formate to produce methane; isolated from the rumen and the gastrointestinal tract of man

Methanobacterium   *p. 473*

(*M. ruminantium*)   *p. 475*

Not as above...22

22. Cellulose digested; cellobiose fermented; spherical to elongated cocci; occur singly, in pairs or in chains; average 0.7–1.2 μm by 0.8–1.0 μm; non-motile; H_2S not produced; ammonia essential as the main nitrogen source, amino acids being very poorly utilized as a source of carbon and nitrogen; Gram-negative to Gram-positive—easily decolorized; cells have a Gram-positive cell structure; isolated from the rumen of cattle and sheep and the cecum of the guinea pig and rabbit; G + C content of the DNA is 39.8–45.4 moles % (T_m)..................**Ruminococcus** *p. 525*

Cellulose not digested; cellobiose may be fermented; amino acids actively metabolized often with the production of gas and a fetid odour; cells 0.7–1.0 μm mean diameter; non-motile; Gram-positive. Isolated from normal and pathological female genital tract and blood in puerperal fever, oral, respiratory and intestinal tracts and septic wounds; G + C content of the DNA 33.5 moles %....**Peptostreptococcus** *p. 522*

23. Organisms occur singly, in pairs or in irregular or regular cyst-like masses; in one species cells are 0.5–4.0 μm in diameter; ferment acetate and butyrate or ferment formate to produce methane and CO_2..................**Methanococcus** *p. 477*

Methane not produced; spherical cells 0.5–1.0 μm in diameter; occur singly, pairs, tetrads or irregular masses; non-motile; Gram-positive; can use peptones or amino acids as the sole energy source, sometimes with the production of CO_2 and H_2, lower fatty acids, and ammonia; glucose, if attacked, is not converted homofermentatively to lactic acid; G + C content of the DNA is 35.7–36.7 moles % (buoyant density)...**Peptococcus** *p. 518*

24. Gram-negative..**25**

Gram-positive..**30**

25. Colorless, spherical cells, 0.8–1.0 μm in diameter arranged in parallel rows in flat sheets on the surface of liquid manure and culture media. The sheets are derived from the synchronous or alternate division of cells in two planes, the cocci expanding to ovoid forms prior to division; sheets break characteristically into squares of 16 cells (Plate 10.3, Fig. 1). Single cells are rare; oxidative; the G + C content of the DNA *ca.* 61 moles % (buoyant density).................**Lampropedia** *p. 440*

Note: *Compare with* **Thiopedia** *(Section J p. 1134).*

Large coccal to ovoid cells 2 μm or more in diameter, occurring singly, in pairs and in irregular clumps; do not produce endospores or cysts; grow in mineral salts media with glucose or sucrose but no added nitrogen source.......**Azomonas** *p. 255*

Not as above...**26**

26. Cells kidney-shaped or hemispherical; 0.6–1.0 μm wide; occur singly or in pairs with concave or flat sides adjacent; divide in two planes at right angles to each other sometimes resulting in tetrad formation; non-motile; catalase and oxidase produced; sensitive to penicillin...**27**

Not as above...**28**

27. Acid from glucose (except *N. flavescens*); nitrate not reduced to nitrite (except *N. mucosa*); G + C content of the DNA 47.0–52.0 moles % parasites of mucous membranes of mammals..**Neisseria** *p. 428*

Note: *Compare with* **Gemella haemolysans** *which is morphologically similar and may stain Gram-negatively although it has the cell wall structure of a Gram-positive cell. However the G + C content of the DNA is 31.4–34.6 moles %.*

No acid from carbohydrates; nitrate reduced to nitrites and beyond without formation of N_2; G + C content of the DNA ranges from 40–45 moles %; parasites of mucous membranes of mammals.................................**Branhamella** *p. 432*

Note: *Species keying out to this genus should be carefully compared with those of the genus* **Moraxella** *which are phenotypically the same and have the same range for the G + C content of the DNA.* **Moraxella** *species are considered as plump rods with division only in one plane.*

28. Gram-negative cocci occurring in pairs; non-motile; capsulated; thermotolerant; temperature range 30–50 C; grow in a mineral salts base solely at the expense of methane or methanol...**Methylococcus** *p. 269*

Not as above...**29**

29. Cells occur in pairs, tetrads and clusters of tetrads; growth in not less than 2.4 M NaCl; colonies red due to the presence of carotenoids.............**Halococcus**   *p. 272*

Cells spherical; 0.5 μm (one species) to 1.0 μm (one species) in diameter; arranged singly, in pairs or aggregates; non-motile; accumulate polyhydroxybutyrate, especially in nitrogen-deficient media; oxidative requiring oxygen or nitrate as electron acceptors; nitrate is reduced to nitrous oxide and molecular nitrogen. G + C content of the DNA *ca.* 64–67 moles % (buoyant density). Isolated from soils and meat curing brines....................................**Paracoccus**   *p. 438*

30. Organisms arranged in cubical packets or tetrads; endospores produced; motile or non-motile.......................................**Sporosarcina**   *p. 573*

Note: *Some **Micrococcus** species produce cubical packets but do not produce endospores and are non-motile.*

Not as above..**31**

31. Organisms divide in one plane only producing pairs or chains of varying length. Catalase-negative..**32**

Organisms divide in several planes but are normally arranged in pairs with flat sides adjacent; Gram stain somewhat indeterminate; cell wall structure is of the Grampositive type; carbohydrates are fermented; no capsular polysaccharide produced in 5% sucrose; catalase negative; G + C content of the DNA is 33.5 moles % (buoyant density). Found in bronchial secretions and slime from the respiratory tract
Gemella   *p. 516*

Organisms divide in 2 or more planes to produce pairs, tetrads, and regular or irregular packets. Catalase positive or negative....................................**33**

32. Heterofermentative; glucose is fermented with the production of *levorotatory* lactic acid, CO_2, ethanol and/or acetic acid....................**Leuconostoc**   *p. 510*

Note: *Gas production is best detected in cultures in yeast extract-glucose-tryptone-phosphate broth dispensed in Eldredge tubes.*

Homofermentative; lactic acid produced is *dextrorotatory*..**Streptococcus**   *p. 490*

33. Spheres 1.0–2.0 μm in diameter with a strong tendency to form tetrads; colonies 0.5–1.0 mm and on blood agar produce a pronounced greening; will grow in the presence of 40% bile, 10% NaCl, 0.01% potassium tellurite and at pH 9.6; homofermentative yielding *dextrorotatory* lactic acid........................**Aerococcus**   *p. 515*

Note: *A. viridans is indistinguishable from the formerly recognized **Gaffkya homari**. The name **Gaffkya** which appeared in the 7th edition of THE MANUAL has been declared a **nomen rejiciendum**.*

Not as above..**34**

34. Organisms occur mainly in pairs and tetrads, particularly in acid media; homofermentative metabolism, the lactic acid produced being *optically inactive* (DL-lactic acid). Saprophytes found in fermenting plant material and spoiled beer
Pediococcus   *p. 513*

Not as above..**35**

35. Metabolism oxidative; organisms arranged singly, in pairs, triads and tetrads: motile with 1–4 flagella; catalase positive; benzidine test for porphyrins positive; colonies on agar 2–3 mm in diameter, yellowish orange; marine organism; G + C content of the DNA 48–52 moles % (T_m)........................**Planococcus**   *p. 489*

Not as above..**36**

36. Metabolism strictly oxidative; glucose oxidized to acetate or to CO_2 and H_2O; G + C content of DNA 66–75.5 moles % (T_m)...................**Micrococcus**   *p. 478*

Metabolism oxidative and fermentative; G + C content of the DNA 30–40 moles % (T_m).......................................**Staphylococcus**   *p. 483*

Note: *Under 'species incertae sedis' appended to the genus **Micrococcus**, the species **M. cryophilus** is described. It has a strictly respiratory metabolism but a G + C content of DNA characteristic of the genus **Staphylococcus**.*

SECTION H

1. Multicellular organisms (sheathed or not sheathed) *or* chains of unicellular organisms enclosed in a sheath..2

   Note: *This does not include chains of individually encapsulated cells or cells in zoogloea masses.*

   Not as above....................................proceed to **Section I**   *p. 1120*

2. Chains of rods, or multicellular filaments individually or collectively surrounded by a sheath...3

   Multicellular rods or filaments without sheaths.............................. **17**

3. Chains of rods within a sheath; single cells motile with polar or subpolar flagella....**16**

   Multicellular filaments or chains of cells; single cells, if motile, have no flagella.....**4**

4. Width of the sheath increasing from the base to the tip; cells within the sheath divide transversely and longitudinally toward the tip (Plate 3.4, Figs. 2–4 and Plate 3.3A) to produce large numbers of non-motile coccoid elements; sheaths attached by means of a holdfast..**5**

   Not as above..**6**

5. Cells within the basal portion of the sheath longer than wide (Plate 3.4, Fig. 1); when growing in iron- or manganese-bearing waters, become heavily impregnated with iron or manganese..**Crenothrix**   *p. 135*

   Cells within the basal portion of the sheath are wider than long, the diameter being 4–6 times the thickness (length) of the cell; no iron deposited

   Phragmidiothrix   *p. 134*

6. Sheaths become impregnated with iron or manganese when growing in iron- or manganese-bearing waters or media...**7**

   Not as above..**9**

7. Organisms in which the sheaths split longitudinally into fine, hair-like sections. The chain of cells remains attached to hairs at several points and, with continued growth, causes arching of the hairs and of the chain of cells within its new sheath, resulting in the formation of a helm-like mass (Plate 2.8, Fig. 7)..........**Toxothrix**   *p. 120*

   Note: *This entry is based on Cholodny's (1924) original concept of this organism. Krul et al. (1970) have studied the growth of **Toxothrix** in natural waters and provide a different concept for which provision is made elsewhere in the Key.*

   Cells within the base of the sheath 2 μm by 4.5–18.0 μm with rounded ends; divide transversely near the tip to produce spherical, non-motile cells, which are extruded either singly or in chains. The sheath is heavily impregnated with iron or manganese, becoming much wider at the base and tapering toward the tip; attached by a holdfast; false branching is common (Plate 3.3 B)....................**Clonothrix**   *p. 136*

   Not as above..**8**

8. Rod-shaped cells arranged in chains: usually two chains of zig-zagging cells are crisscrossed in the form of a series of XXXXX's, the whole embedded in a mass of slime which becomes encrusted with iron: the complex has a wriggling type of motility

   Lieskeella   *p. 134*

   Note: *In the author's opinion the surrounding slime does not constitute a sheath in the usual sense.*

   Multicellular organisms with gelatinous and often more or less ferruginous sheaths; characteristically in rigid, curved and twisted bundles; no gas vacuoles; not motile

   Peloploca   *p. 125*

   Note: *This entry applies to species 4 and 5.*

9. Multicellular filaments in which sulfur is deposited from the oxidation of sulfide....   **10**

   Not as above...**11**

10. Filaments occur in parallel or braided fascicles enclosed by a common sheath (Plate 2.6, Figs. A and B), often covered with detritus; independent filaments show independent gliding movement...............................**Thioploca**   *p. 115*

    Filaments do not aggregate in a common sheath: usually attached to a solid substrate;

motile gonidia are formed from the cells of the filament and may congregate to form rosettes; filaments are non-motile..........................**Thiothrix**   *p. 119*

     Note: *Thiothrix without sulfur is morphologically indistinguishable from* **Leucothrix** (*see under 13 of the key of this section*).

11. Multicellular filaments characteristically aggregated into bands; non-motile.......12

     Not as above..13

12. Bands of filaments attached by holdfasts to a solid substrate; basal portions of the bands are enclosed in a common hyaline gelatinous sheath, the filaments at the top being free and divergent; no gas vacuoles are formed......**Desmanthos**   *p. 127*

     Bands formed by lateral joining of filaments; no holdfast; individual filaments have a gelatinous sheath; cells without gas vacuoles; sheaths may be vacuolated

                                                          Peloploca   *p. 125*

13. Multicellular filaments or chains of rods; not attached to a substrate................14

     Multicellular filaments attached to a substrate by means of a holdfast; 3–5 μm wide with cross-walls clearly evident; colorless, unbranched; individual cells of the fila-ment may round up to produce spherical gonidia which exhibit a jerky gliding mo-tility and which may congregate and subsequently develop to produce rosettes; aerobic; require NaCl for growth with an optimum concentration of 1.5%; usually grow epiphytically on seaweeds: G + C content of the DNA ranges from 46–51 moles %...**Leucothrix**   *p. 118*

     Note: *1. The Manual states that no sheath is formed but that filaments "may become emptied of their contents giving the appearance of a sheath." This is not in accordance with the observations of Pringsheim and the author.*

     2. **Leucothrix** *is indistinguishable from sulfur-free forms of* **Thiothrix** (*see 10 of this section key*).

14. Thin rods, 0.35–0.45 μm by 3.2–4.6 μm, occurring in chains in a very delicate hyaline sheath 0.5–0.8 μm wide (Plate 3.2, Figs. 9–10). Motility never observed; free cells uncommon; branches, if they occur are very short compared with the main axis; Gram-negative; masses of cells are pink due to presence of carotenoid pigments; aerobic; oxidative; found in fresh water and activated sludge

                                                          Streptothrix   *p. 133*

     Not as above...15

15. Rods or filaments, 0.7–1.5 μm by 5–150 μm or more; sheath thin and hyaline; unsheathed segments motile; mass cell growth yellow or orange due to carotenoid pigments; G + C content of the DNA 48.1–53.1 moles %.........**Herpetosiphon**   *p. 107*

     In media with a low sugar content; rods 0.3–0.5 μm which form sheathed filaments 0.5 μm wide and up to 500 μm long; may show false branching; free cells, but not filaments, motile; aerobic; oxidative; cell masses yellow; G + C content of DNA 37.2 moles %

                                                          Flexithrix   *p. 109*

     Not as above; multicellular filaments ranging in width from 0.7–1.2 μm to 5.0–6.8 μm; cross-walls clearly visible; filaments may be constricted at the cross-walls; no gas vacuoles; cytoplasm may be distinctly granular and sometimes divided into a clearer ectoplasm and a denser endoplasm (as seen in many cyanobacteria); filaments motile (except species 9).....................................**Achroonema**   *p. 123*

     Note: *Most species do not reveal a sheath. The entry is made here as Table 2.7 cites the sheath as "being present in some species, not in others." A sheath is not mentioned in the generic or specific descriptions.*

16. Rods 0.7–2.4 μm by 3–10 μm; free cells motile by means of a bunch of subpolar flagella; aerobic; sheaths usually pale but may become encrusted with iron but *not* manganese oxides when growing in iron- and manganese-containing milieu. G + C content of the DNA is 69.5–70.5 moles %..........................**Sphaerotilus**   *p. 128*

     Rods 0.6–1.5 μm by 3–12 μm; motile as single cells, pairs or short chains; each cell has one polar flagellum (except *L. lopholea* which has a tuft of subpolar flagella); aerobic; sheaths have a pronounced tendency to become impregnated with or covered with hydrated ferric or manganic oxides; G + C content of the DNA of one species was

69.0–70.0 moles % (see Plate 3.1, Figs. 7–9; Plate 3.2, Figs. 1–8)

Leptothrix   *p. 129*

17. Microorganisms motile with lateral flagella..................................18

Microorganisms not flagellated; if motile, exhibit gliding motility..................19

18. Microorganisms 6–20 μm long and up to 3 μm wide; rigid, multicellular, bacillary forms, in which the stained cells are differentiated into a series of light and dark bands; end cells rounded; commonly form chains up to 200 μm long. Individual cells in a filament are less than 2 μm long and may separate as discoid elements which grow out into trichomes; no endospores produced; common in peat and cow dung

Caryophanon   *p. 598*

Microorganisms approximately 5 μm in width and straight to curved; individual cells disc-shaped and much wider than long; develop a large endospore apparently by fusion of several cells within the filament. The spore is normally centrally located. Division of the filament is preceded by formation of biconcave discs between certain cells somewhat similar to those produced by the blue-green alga *Oscillatoria;* found in large numbers in the cecum of the guinea pig.........**Oscillospira**   *p. 574*

19. Cells cylindrical, colorless, 0.5–0.75 μm by 3–6 μm, in filaments up to 400 μm long; a dense body is often located at each end of the cell (Plate 2.8, Fig. 1). Filaments bend in the shape of a U and rotate while slowly moving forward with the rounded part in the lead; a mucoid substance, excreted from several sites on the trailing ends, is deposited as a double track of twisted strings; each 0.2 μm wide (Plate 2.8, Figs. 8–10). Fan-shaped structures may be deposited laterally along the track as the arms of the U move from side to side, and between the tracks as a result of the middle section being lifted and then touched down again (Plate 2.8, Fig. 7). Oxidized iron may be deposited on the mucoid threads, rendering them yellowish brown and brittle and giving them a diameter of 2.5 μm......................**Toxothrix**   *p. 120*

Not as above..20

20. Multicellular filaments flat and ribbon-like; individual cells much wider than long and closely appressed; motility can only be effected when the broad surface is in contact with the medium; show pronounced curling of the filaments when on the narrow edge; isolated mainly from the mouth and upper respiratory tract of man, animals (including birds)..21

Multicellular filaments cylindrical; when motile often exhibit rotational as well as translational motion....................................22

21. Terminal cells of filaments rounded; filaments divide into hormogonia-like segments following the formation of biconcave discs between certain cells in the filament (Plate 2.7)..**Simonsiella**   *p. 116*

Terminal cells concave; cells occur typically in pairs in the filament which in aging cultures tend to disintegrate into groups of 4 cells (Plate 2.7)...**Alysiella**   *p. 117*

22. Whole filaments motile by a gliding action on solid surfaces......................23

Whole filaments non-motile; motility only in gonidia formed by the rounding and separation of individual cells from the ends of the filaments........................25

Filaments non-motile; no motile gonidia..26

23. Sulfur deposited in the cells from the oxidation of sulfide.......**Beggiatoa**   *p. 113*

Sulfide not oxidized; sulfur not deposited....................................24

24. Colorless filaments 1.2–2 μm by 3–70 μm; composed of cylindrical or barrel-shaped cells; reproduction is by binary fission of cells within the filament; filaments fragment readily; G + C content of the DNA 43.6 moles % for 2 strains (buoyant density)

Vitreoscilla   *p. 114*

Colorless filaments ranging in width from 0.3–0.5 μm to 5.0–6.8 μm; finely granular cytoplasm, sometimes clearly divided into an ecto- and endoplasm; length of individual cells in the filaments vary from 2–19 μm; frequently short and barrel-shaped

Achroonema   *p. 123*

Note: *The author finds it impossible from descriptions given in* THE MANUAL *to separate some species of* **Vitreoscilla** *and* **Achroonema.** *See also* **Pelonema tenue.**

25. Sulfide is oxidized and sulfur is deposited in the cells............**Thiothrix** *p. 119*
    Sulfide is not oxidized...................................**Leucothrix** *p. 118*
26. Filaments characteristically bound together laterally to form rigid bundles or flat
    ribbons: gas vacuoles may be present......................**Peloploca** *p. 125*
    Filaments remain free: each cell usually has a central gas vacuole
    Pelonema *p. 122*

SECTION I

1. Pear-shaped, rod-shaped, wedge-shaped or club-shaped; attached by their narrow
   poles to each other or to a substrate by an adhesive material; multiply by budding
   from the free pole or sometimes from the side of the cell.........................2
   Not as above...3
2. Cells 1–5 by 3–6 μm, non-motile; pear-shaped to nearly spherical; budding both polar
   and subpolar; isolated from *Daphnia pulex* and *Daphnia magna* and probably pathogenic (see Fig. 4A).....................................**Pasteuria** *p. 158*
   Cells rod-shaped, wedge-shaped or club-shaped often slightly curved; 0.7–1.0 μm by
   2.0–4.5 μm (greatest dimensions). Several cells attach by common poles to produce
   rosettes in the center of which there is usually a glistening corpuscle. Polar budding;
   described from a vessel containing iron water from north Russia and shredded filter
   paper: optimum pH 6.2.................................**Blastobacter** *p. 159*
3. Cells produce prosthecae or are borne on stalks.......return to **Section D** *p. 1105*
   Not as above...4
4. Cells contain bacteriochlorophylls and carotenoid pigments; mass growth and sometimes individual cells various shades of yellow, green, brown, red and purple; grow
   anaerobically exposed to light; some also grow aerobically in the dark
   Section J *p. 1134*
   Not as above...5
5. Organisms that store oxides of manganese or iron on the surface of the cell or the surrounding capsule or slime; found in water and mud................................6
   Note: *a. In the absence of further information, these organisms are identified on the basis
   of their iron-depositing characteristics. Most iron organisms studied in pure culture metabolize the organic compound that forms the iron chelate, and the liberated iron then chelates
   with some cell component. Citrate-utilizing organisms will, for example, release iron from
   ferric ammonium citrate. Accumulation of the iron in or on the cell may depend only on
   the nature of the cell substance. Pure culture studies may place these organisms in more
   commonly recognized genera. They should also be treated as non-iron-depositing cells and
   should be followed through the key.*
   b. In addition to the genus **Naumanniella** *and the genus* **Ochrobium** *listed in this
   edition of* THE MANUAL *others have been described and listed as "genera incertae sedis."
   For treatment of these see Skerman "Guide to the Identification of Genera of Bacteria,"
   2nd edition, 1967.*
   Not as above...8
6. Encapsulated cells occurring singly or in short chains, each capsule being completely
   surrounded by a ring (torus) heavily impregnated with iron or manganese, giving the
   general appearance of links in a chain..................**Naumanniella** *p. 467*
   Not as above...7
7. Cells surrounded by a ring (torus) which is open at one end; cells when motile have two
   unequal polar flagella...................................**Ochrobium** *p. 467*
   Note: *The unequal lengths of the flagella suggest that this may be an algal or a protozoan cell.*
   Cells arranged in two angulated and intertwined chains, embedded in mucus encrusted
   with iron; active wriggling motion amongst plant detritus....**Lieskeella** *p. 134*
8. Cells flexible: 5–100 μm or more long, not divided by any cross-walls; motile only by a
   gliding action on solid surfaces or along one another; cells may show a curious

twitching in liquids; motility can be observed at the glass-water interface on a slide or on the surface of agar plates, where cells move singly or in groups sometimes leaving a trail of slime behind. Cells are flexible, the extent of flexibility being determined by length and basic cell structure; unicellular; cell masses pink, red, yellow, orange. 9

Not as above... 10

9. Agar, cellulose or chitin attacked...........................**Cytophaga** *p. 101*

Agar, cellulose and chitin not attacked.....................**Flexibacter** *p. 105*

Note: 1. *Although the genus* **Herpetosiphon** *is included with the above genera in the* **Cytophagaceae** *it seems misplaced there. Although some short rods appear to be formed they are usually in sheathed or unsheathed chains. The genus* **Herpetosiphon** *is included in this key with the "sheathed or unsheathed multicellular and sheathed chain-forming unicellular organisms," in Section H.*

2. **Flexithrix,** *on rich media, resembles Cytophaga and Flexibacter.*

10. Gram-positive...**Section L** *p. 1138*

Gram-negative... 11

11. Obligate autotrophs; use CO_2 as the sole source of carbon and obtain energy from the oxidation of ammonia, nitrite, ferrous iron or inorganic sulfur compounds; will not grow on meat extract agar or other complex organic media...................... 12

Not as above.. 15

12. Organisms oxidize ammonia to nitrite.

Straight rods: 0.8–0.9 μm by 1.0–2.0 μm; when motile have 1–2 subpolar flagella; electron microscope sections reveal cytomembranes which form flattened lamellae in the peripheral regions of the cytoplasm (Plate 12.2, Fig. 3). The G + C content of the DNA ranges from 47.4–51.0 moles % (buoyant density)

Nitrosomonas *p. 453*

Cells pleomorphic and lobate (Plate 12.2, Fig. 10), 1.0–1.5 μm in diameter; divide by constriction; invagination of the cytoplasmic membrane and other segments of the cell envelope forms vesicular regions in the cytoplasm; no peripheral lamellae: cells yellowish to red; G + C content of the DNA ranges from 53.6–55.1 moles % (buoyant density)...**Nitrosolobus** *p. 455*

Not as above.. 13

13. Organisms oxidize nitrite to nitrate.

Cells short rods, often wedge- or pear-shaped; 0.6–0.8 μm by 1.0–2.0 μm (Plate 12.1, Fig. 1); rarely motile, but if so, have a single polar flagellum; occur singly or in clumps embedded in slime; electron microscope sections reveal a polar cap of cytomembranes forming flattened vesicles (Plate 12.1, Fig. 3); yellowish due to cytochrome pigments; G + C content of the DNA ranges from 60.7–61.7 moles %

Nitrobacter *p. 451*

Straight slender rods; 0.3–0.4 μm by 2.7–6.5 μm (Plate 12.1, Fig. 4); have no polar cap of cytomembranes; occur singly or in pairs; non-motile; dependent on sea water for growth; G + C content of the DNA 57.7 moles % (buoyant density)

Nitrospina *p. 452*

Not as above.. 14

14. Organisms oxidize reduced or partially reduced inorganic sulfur compounds and *may* also oxidize ferrous iron.............................**Thiobacillus** *p. 456*

Note: Ferrobacillus ferrooxidans, *originally described as unable to oxidize thiosulfate, has been shown to do so and has been transferred to the genus* **Thiobacillus.**

15. Aerobic to microaerophilic... 16

Anaerobic...**Section K** *p. 1136*

16. Pleomorphic organisms 1–2 μm long and usually encapsulated; isolated from cases of granuloma inguinale; grow only in the yolk sac of the developing chick embryo or in condensation water of a sloped medium prepared by adding 50% unheated embryonic yolk to melted and cooled nutrient agar from which they may be adapted to other media.......................................**Calymmatobacterium** *p. 381*

Pleomorphic organisms parasitic on or within erythrocytes of man and other vertebrates, causing infections of lower animals; stain by Giemsa's stain without visible

differentiation into nucleus and cytoplasm; grown with variable success in semisolid
agar containing whole blood.....................return to **Section B**   *p. 1102*
Organisms 1.0 μm in length occurring characteristically in pairs, extracellularly
lining the intestinal epithelium of the sheep ked, *Melophagus ovinus*: grow in glucose-
blood-bouillon agar producing colonies 0.4–0.6 mm in width 3–5 days or longer

                                                 Wolbachia   *p. 898*
                                                 (melophagi)

Rods 0.5–0.8 μm and 5–10 μm long which may form filaments of variable length; non-
motile; aerobic; grows at temperatures from 40–79 C with an optimum at 70–72 C;
colorless to yellow-orange...............................**Thermus**   *p. 285*
Not as above...**17**

17. Organisms isolated from nodules on the roots of leguminous plants and capable of producing nodules on the host plant. The G + C content of the DNA ranges from 59.1–
65.5 moles % (T_m)......................................**Rhizobium**   *p. 262*
Not as above...**18**

18. Endospores produced.............................**Bacillus**   *p. 529*
No endospores produced...**19**

19. Organisms capable of continued growth in a glucose, sucrose, or mannitol mineral salts
medium devoid of nitrogen compounds.....................................**20**
Not as above...**23**

20. Cells rod-shaped, within the ranges of 0.5–1.5 μm by 1.7–6.0 μm; distinctly rounded
ends; highly refractile sudanophilic bodies present in older cells (Plate 7.2, Fig. 5;
Plate 7.3, Fig. 4)..**21**
Cells large ovoid to cylindrical; 2 μm or more wide: catalase positive...............**22**

21. Cells 0.5–1.5 μm by 1.7–4.5 μm; when motile, have peritrichous flagella; very acid tolerant with growth from pH 3.0 to pH 9.5; catalase positive; G + C content of the DNA
ranges from 54.7–60.7 moles % (T_m) (see Plate 7.3, Figs. 2–5)

                                                 Beijerinckia   *p. 256*
Cells 1.0–1.2 μm by 3.0–6.0 μm occurring singly or in short chains; motile with a short
polar flagellum, particularly in a glucose liquid medium with combined nitrogen;
acid tolerant but no growth below pH 4.4; catalase negative; G + C content of the
DNA 70.4 ± 1.7 moles % (T_m) (see Plate 7.2, Figs. 4–7).........**Derxia**   *p. 260*

22. Thick-walled cysts produced in aging cultures; growth range pH 5.5–8.5; G + C content of the DNA 63–66 moles % (see Plate 7.2, Figs. 1–2)....**Azotobacter**   *p. 254*
No cysts produced; growth range pH 4.5–9.0; G + C content of the DNA 53–59 moles
% (see Plate 7.2, Fig. 3)................................**Azomonas**   *p. 255*

23. Organisms oxidize ethanol to acetic acid at pH 4.5. May oxidize acetate to CO_2 and
H_2O; growth very poor or absent in media containing no carbohydrates or alcohols;
may not utilize ethanol as the sole carbon source..............................**25**
Not as above...**24**

24. Organisms will grow in a mineral salts medium with methane or methanol as the sole
carbon and energy source; no growth on meat infusion agar; rods 0.5–1.0 μm by 1–4
μm, motile with a polar flagellum; catalase positive; oxidase positive

                                                 Methylomonas   *p. 268*
Not as above...**26**

25. Cells ellipsoidal to rod-shaped; 0.6–0.8 μm by 1.5–2.0 μm: solitary, in pairs or in chains;
pleomorphic in old cultures; motile with 3–8 polar flagella or non-motile; do not oxidize acetate to CO_2 and H_2O; do not oxidize lactate to carbonate; do not oxidize
amino acids; do not possess the Krebs cycle; highly ketogenic; G + C content of the
DNA 60–64 moles % (T_m).........................**Gluconobacter**   *p. 251*
Cells ellipsoidal to rod-shaped; straight or slightly curved; 0.6–0.8 μm by 1.0–3.0 μm;
solitary, in pairs or chains; often highly pleomorphic; motile with peritrichous
flagella or non-motile; oxidize acetate to CO_2 and H_2O; oxidize lactate to carbonate;
oxidize amino acids; possess the Krebs cycle enzymes; G + C content of the DNA
55–64 moles % (T_m)..................................**Acetobacter**   *p. 276*

26. Plump rod-shaped cells with rounded ends; motile with lophotrichous flagella; grow

poorly on peptone media in the absence of carbohydrates; ferment glucose and fructose with the production of ethanol and CO_2 with some lactate or acetaldehyde; G + C content of DNA is 47–48 moles %.................**Zymomonas** *p. 352*

    Not as above...27

27. Colonies with a purple pigment........................28

    Colonies with red pigments at 37 C or 25 C; sometimes a water-soluble red pigment produced..29

    Not as above...32

28. Purple pigment is violacein; soluble in ethanol but not water or chloroform; absorption maximum in ethanolic solution of 579 nm and a minimum at 430 nm. In 10% (v/v) H_2SO_4 in ethanol the solution is green with a maximum at 700 nm. Rods 0.6–1.2 μm by 1.5–6 μm; young cells from solid media have one polar flagellum and one to four lateral and antigenically distinct flagella; resistant to benzylpenicillin and O/129 (2,4-diamino-6,7-diisopropyl pteridine); G + C content of the DNA ranges from 63–72 moles % (T_m).............................**Chromobacterium** *p. 354*

    Not as above; flagella polar only; G + C content of the DNA 58–70 moles %
                        possibly **Pseudomonas** *p. 217*

    Note: *See p. 242.*

29. Red pigment is prodigiosin; spectroscopic examination of heavy suspensions reveals two absorption bands at 490–510 nm and 540–560 nm; the extracted pigment in acid ethanol has an absorption peak at 540 nm; produces deoxyribonuclease; fermentative, producing acid and sometimes gas from glucose; no acid or delayed fermentation from lactose; methyl red test negative; Voges-Proskauer test positive; H_2S not produced; urease negative; gelatin is hydrolyzed; citrate and acetate but not malonate are used; phenylalanine deaminase not produced: grows in the presence of KCN; G + C content of the DNA ranges from 53–59 moles %.........**Serratia** *p. 326*

    Note: *Colorless variants occur and are treated elsewhere in this key.*

    Not as above, red color is not due to prodigiosin.............................30

30. Obligate halophiles requiring 3.5 M NaCl (*ca.* 20%) for growth and maintenance of normal rod-shaped morphology; 0.6–1.0 μm by 1–6 μm; become spherical at 1.5 M NaCl through loss of the cell wall; very pleomorphic at intermediate concentrations; cell wall contains no diaminopimelic acid or muramic acid; red color due mainly to bacterioruberin. Motile with a tuft of polar flagella or non-motile; oxidase and catalase positive; H_2S produced; urease negative; indole positive; oxidative; G + C contents of the DNA's two components are 66–68 moles % and 57–60 moles % (buoyant density)...**Halobacterium** *p. 270*

    Not as above: not halophilic.................................31

31. Flagella polar, oxidative only........................**Pseudomonas** *p. 217*

    Note: *1. See p. 242 of* THE MANUAL.

    2. The genus **Protaminobacter** *is not accepted in* THE MANUAL.

    3. See also **Vibrio** *in the 7th edition of* THE MANUAL.

    Flagella polar, oxidative and fermentative....................**Vibrio** *p. 340*

    Note: *V. cholerae is stated to exhibit a green to red-bronze iridescence.*

    Flagella peritrichous.........................possibly **Flavobacterium** *p. 357*
                         (*F. tirrenicum*)
                  or   **Escherichia** *p. 293*
                  or   **Erwinia** *p. 332*

    Note: *The 7th edition of* THE MANUAL *lists reddish colonies for* **Escherichia aurescens** *and* **Erwinia rhapontici**. *The latter is listed in the 8th edition but the colonies are not described (see* E. rubrifaciens *and* E. rhapontici*)*

32. Pathogenic for plants.....................................33

    Not pathogenic or not known to be pathogenic for plants.....................36

    Note: *Organisms which cause plant diseases are inhabitants of soil or plant detritus. Isolated from these sources pathogenicity may not be suspected. For this reason plant pathogens are treated both as pathogens and non-pathogens in these keys.*

33. Organisms cause hyperplastic diseases of plants; enter plants through pre-existing

lesions or abrasions; rods 0.8 μm by 1.5–3.0 μm; motile with 1–4 peritrichous flagella; if only one, more usually lateral than polar; usually produce copious slime in carbo-hydrate-containing media; oxidative . . . .**Agrobacterium**   p. 264

Note: *See also* **Erwinia herbicola.**

Some strains of **Erwinia herbicola** *cause galls on* **Milletia japonica** *and on* **Gysophila paniculata.** *Strains causing the latter were classified as* **Agrobacterium gypsophilae** *in the 7th edition of* THE MANUAL.

Not as above 34

34. Flagella polar; oxidative 35

Flagella peritrichous (except *E. stewartii*); rods straight, 0.5–1.0 μm by 1.0–3.0 μm; fermentative; acid is produced from fructose, glucose, galactose, sucrose and β-methylglucoside and possibly from other carbohydrates; gas production weak or absent; oxidase negative; catalase positive; G + C content of the DNA ranges from 50–58 moles % (buoyant density and T_m); organisms cause mainly wilts and necrotic disease of plants . . . .**Erwinia**   p. 332

35. Water insoluble, yellow pigment produced; rods straight, 0.2–0.8 μm by 0.6–2.0 μm; oxidative; oxidase reaction negative or weak; catalase positive; nitrates not re-duced; Voges-Proskauer negative; indole negative; sodium hippurate is not hy-drolyzed; starch usually hydrolyzed; acid only produced from oxidizable carbohy-drates; growth on nutrient agar inhibited by 0.1% triphenyltetrazolium chloride; G + C content of the DNA of *X. campestris* ranges from 63.5–69.2 moles % (mostly T_m) . . . .**Xanthomonas**   p. 243

Note: *Starr and Stephens (J Bacteriol, 87: 293, 1964) state that species of* **Xanthomonas** *are characterized by the presence of a distinctive carotenoid alcohol with absorption peaks at 418, 437, and 463 nm in petroleum ether.*

Not as above . . . .**Pseudomonas**   p. 217

Note: *For species of* **Pseudomonas** *recorded as plant pathogens. None of the patho-genic species fully described in* THE MANUAL *has a non-diffusible yellow pigment but some may produce a yellow pigment which diffuses into the medium.*

36. Organisms digest cellulose in 0.5% peptone water; irregular rods 0.5 μm wide by 0.7–20 μm or more in length; may be straight, angular or slightly curved; occasionally club-shaped and beaded; sometimes arranged in palisades (Plate 17.1, Fig. 12); if motile have polar to subpolar flagella; Gram-positive cell wall structure; facultative anaerobes; catalase positive . . . .**Cellulomonas**   p. 629

Note: *These organisms are Gram-positive but decolorize very readily.*

Not as above 37

37. Cells with a distinctly curved axis 38

Not as above 47

Note: *A shunt is employed in the key at this point, some genera noted for specific arrange-ment of cells being selected, the remainder being returned for further treatment with straight (and other) axis organisms.*

38. Curved rods 0.5–2.0 μm wide by 1.0–10.0 μm or more long; elongation of the cells pro-duces a ring which later changes to two horse-shoe-shaped halves which divide into smaller curved rods; non-motile; oxidative; acid is produced from glucose, fructose, galactose and mannose by all species; found in ponds, lakes and wells

Microcyclus   p. 214

Note: *The G + C contents of the DNA of the three species are, respectively, 66.3–68.4, 51.0–51.5, and 39.5 moles %. This and other differences suggest that the genus is not homoge-neous.*

Not as above 39

39. Curved cells, 1.0 μm by 5–10 μm, which form chains that may twist around each other to form coiled, non-septate, non-motile, colorless bundles; enclosed in a spherical solid, gelatinous mass from 10–17 μm in diameter: found floating on water containing decomposing plant material . . . .**Myconostoc**

Note: *This genus was included in the 7th edition of* THE MANUAL *but has been excluded* *from the 8th. The resemblance of this description to that of* **Microcyclus flavus** *should be noted.*

Not as above **40**

40. Cells are S-shaped slender filaments, 0.23–0.35 μm wide and 9 μm or more long; aggregated into flat bands 8–10 μm wide and 20–25 μm long or one end of the aggregate is pointed and the other spread out like a fan with an apparently composite flagellum located at the pointed end; movement slow and trembling; found so far in mud of ponds in Germany and the USA . . . . **Pelosigma** *p. 215*

Not as above **41**

41. Cellulose is hydrolyzed, either in a mineral salts base or in 0.5% peptone water
possibly **Pseudomonas** *p. 217*

Note: Several species of **Pseudomonas,** *as defined in this edition of* The Manual, *attack cellulose (see p. 242). The genus* **Cellfalcicula** *and the genus* **Cellvibrio** *listed in the 7th edition of* The Manual *have been cited in Addendum IV (p. 241) as "genera apparently conforming to the present description of* **Pseudomonas."**

Not as above **42**

42. Predatory cells with the ability to attack, penetrate and develop within Gram-negative and Gram-positive bacteria. The preinfection stage consists of curved rods 0.25–0.4 μm by 0.8–1.2 μm (Plate 6.1, Fig. 1), with a single polar sheathed flagellum (sometimes two). Cells penetrate the host cell and develop in the periplasmic space as curved rods or coiled filaments without flagella. These divide into curved daughter cells which develop flagella . . . . **Bdellovibrio** *p. 212*

Note: Host independent strains develop in culture, leaving no simple means of identification. The G + C content of the DNA of **B. bacteriovorus** *is 49.5–51.3 moles % and that of the other two listed species 42–43.5 moles %.*

Not as above **43**

43. "Cells rod shaped and bent like a bow, 1 μm by 1.5–2.5 μm, colorless with several gas vesicles arranged in the center—occasionally with some minute sulfur granules; cells arranged in groups (coenobia) as a result of polar growth and median cross-division combined with tight attachment to surfaces by means of a mucoid substance. Coenobia of two (rings) and four (clover leaf appearance) or more cells quite common." Found in lakes and ponds in Lappland and the USA . . . . **Brachyarcus** *p. 216*

Note: The above description is cited verbatim from The Manual.

Not as above **44**

44. Curved rods, no action on carbohydrates; alkaline reaction in the Hugh and Leifson test; indole negative; cholera red negative; methyl red and Voges-Proskauer tests negative; gelatin not liquefied; citrate not utilized; no growth in the presence of KCN; phenylalanine deaminase and Moeller's lysine, arginine and ornithine tests negative; urease, catalase, and oxidase tests positive; no pigment or fluorescence under ultraviolet light; lophotrichous flagella; G + C content of the DNA *ca.* 64 moles % . . . . **Pseudomonas** *p. 217*

Note: This entry covers the genus **Comamonas** *(Gunther) Hugh listed in* The Manual *under Addendum IV as "apparently conforming to the present description of* **Pseudomonas."** *The genus* **Comamonas** *as circumscribed here included the following species of* **Vibrio** *in the 7th edition of* The Manual—V. **percolans, V. cyclosites, V. neocites** *and V.* **alcaligenes.**

Not as above **45**

45. Slender curved rods, 0.2–0.6 μm by 1.5–5.0 μm, comma, S-shaped, and gull-shaped—may form spirals in old cultures; ends of cells pointed; single polar flagellum at one or both ends; do not produce acid from sugars; reduce nitrate to nitrite; gelatin not hydrolyzed; urease negative; methyl red and Voges-Proskauer negative; oxidase positive; microaerophilic; metabolism of studied species oxidative. The G + C content of the DNA ranges from 30–35 moles % (T_m) . . . . **Campylobacter** *p. 207*

Note: This genus includes species listed as **Vibrio fetus,** *V.* **jejuni** *and V.* **bubulus** *from the 7th edition of* The Manual.

Not as above **46**

46. Fermentative and oxidative metabolism; rods curved, 0.5 μm by 1.5–3.0 μm; occurring singly or in chains; polar flagella; acid but no gas produced fermentatively from carbohydrates; oxidase positive; urease negative; usually sensitive to 2,4-diamino-6,7-diisopropyl pteridine (O/129); may be luminescent. G + C content of the DNA ranges from 40–50 moles %..............................**Vibrio**   *p. 340*

     Not as above...47

47. Acid and gas produced from glucose.....................................48

     Acid only or no acid produced from glucose..............................58

48. Motile..49

     Non-motile; acid from mucate and lactose; methyl red negative; Voges-Proskauer positive; citrate and malonate utilized; urease and lysine decarboxylase produced: growth with KCN....................................**Klebsiella**   *p. 321*

     Non-motile; no acid from mucate or lactose; methyl red positive; Voges-Proskauer negative; citrate and malonate not utilized; urease and lysine decarboxylase not produced; no growth with KCN...............**Shigella** (aerogenic)   *p. 318*

     Non-motile; phenylalanine deaminase negative; other than above....non-motile variants....proceed to...55

49. Flagella polar...50

     Flagella peritrichous..51

50. Not bioluminescent; cells rod-shaped with rounded ends to coccoid; both oxidative and fermentative metabolism; glucose, fructose, maltose and trehalose fermented; starch, dextrin and glycerol hydrolyzed; casein and gelatin hydrolyzed; deoxyribonuclease, arginine dehydrogenase and phosphatase produced: urease negative; nitrate reduced to nitrite; oxidase and catalase positive; not sensitive to vibriostatic agent O/129. G + C content of the DNA ranges from 57–63 moles %

                                          Aeromonas   *p. 345*

     Bioluminescent; coccobacilli and rods; metabolism oxidative and fermentative; nitrites produced from nitrates; indole negative; starch not hydrolyzed; sensitive to vibriostatic agent O/129................................**Photobacterium**   *p. 349*

     Note: *Lucibacterium harveyi—which normally has peritrichous flagella has sometimes in addition, or alone, a thick sheathed polar flagellum but it is indole positive; starch is hydrolyzed; insensitive to O/129.*

51. Pectate is hydrolyzed on Paton's medium in 3 days at 37 C*; acid and possibly gas produced from ribose, arabinose, rhamnose, xylose, fructose, galactose, sucrose, cellobiose, mannose, mannitol, sorbitol, salicin, and esculin; gelatin liquefied and acetoin produced (except species 12) urease negative: H$_2$S produced from cysteine: nitrate reduced to nitrite: deoxyribonuclease produced: oxidase negative: catalase positive..**Erwinia**   *p. 332*

     Note: * *This entry applies to species 10, 11 and 12 of the genus* **Erwinia**: *of these only* **E. cypripedii** *does not hydrolyze the pectate.*

     Not as above...52

52. *Phenylalanine deaminase produced;* growth with KCN; methyl red positive; no acid from arabinose, lactose, dulcitol or sorbitol; arginine dihydrolase not produced; glutamic acid decarboxylase produced......................**Proteus**   *p. 327*

     Phenylalanine deaminase not produced...................................53

53. *Growth with KCN: citrate utilized*....................................55

     No growth with KCN: citrate utilized....................**Salmonella**   *p. 298*

     No growth with KCN: citrate not utilized; indole produced; lysine decarboxylase produced; acid from maltose; methyl red positive; gelatin not hydrolyzed; urease negative...54

54. Hydrogen sulfide produced from triple sugar iron agar; no acid from arabinose, mannitol, trehalose or sorbitol; β-galactosidase not produced; mucate not utilized

                                       Edwardsiella   *p. 296*

     Hydrogen sulfide not produced from triple sugar iron agar; acid produced from arabinose, mannitol, trehalose and sorbitol; β-galactosidase produced; mucate utilized

                                           Escherichia   *p. 293*

55. Methyl red positive..**56**
    Methyl red negative...**57**
56. Lysine decarboxylase produced; urease negative..............**Salmonella** *p. 298*
    Lysine decarboxylase not produced; urease may or may not be hydrolyzed
    Citrobacter *p. 296*
57. Acid (and possibly gas) produced from arabinose, lactose and salicin; gelatin usually
    liquefied...**Enterobacter** *p. 324*
    Acid (and possibly gas) from arabinose but not from lactose or salicin; gelatin not
    liquefied...**Hafnia** *p. 325*
    Acid (and possibly gas) from salicin but not from arabinose or lactose; gelatin liquefied
    Serratia *p. 326*
58. Normal or pathogenic flora of man and animals..............................**59**
    Not as above..**84**
    Note: *The genera listed in the following section of the key to 1-83 are either pathogens, obligate parasites or commensals of man or animals and in practice would rarely be encountered from other sources unless a specific search for them was being made (e.g. for Brucella abortus on pastures or in milk). Parasitism, pathogenicity, and commensalism do not rate highly as taxonomic characters but there is no doubt that a knowledge of the habitat and disease symptoms can be extremely helpful in directing diagnosis.*

    Although much reliance is placed on host association in this section of the key every endeavor has been made to effect separation on more specific bases and where organisms are commonly encountered outside the host they are treated again in other sections of the key.
59. Motile at 37 C or 22 C with flagella.......................................**60**
    Non-motile or exhibit a twitching type of motility, no flagella...............**64**
60. Acid produced from carbohydrates..**61**
    No acid produced from carbohydrates......................................**63**
61. Flagella lateral or peritrichous...**62**
    Flagella polar, multitrichous; short rods 0.8 μm by 1.5 μm; accumulate polyhydroxy-butyrate granules, particularly in nitrogen-deficient media; colonies on agar small circular, raised, thick opaque and cream colored to bright orange; polyethylene sorbitan monooleate (Tween 80) hydrolyzed; gelatin and coagulated blood serum liquefied; some carbohydrate and a wide range of other carbon compounds oxidized: arginine dihydrolase produced: nitrate reduced to molecular nitrogen.
    Organisms cause a glanders-like infection (melioidosis) of man, rats, guinea pigs and rabbits
    Pseudomonas *(pseudomallei)* *p. 227*
62. Colonies mucoid; small, circular and translucent; litmus milk unchanged; gelatin not hydrolyzed; nitrates are not reduced and indole is not produced: *organisms cause trachoma in man and conjunctival folliculosis in rabbits and monkeys*......**Noguchia**
    Note: *This genus is not listed in this edition of THE MANUAL.*
    Cells coccoid, oval to rod-shaped; motile below 30 C only; grow well on meat infusion agar but dissociate readily on subculture; fermentative; acid from glucose, fructose, glycerol, maltose, mannitol, trehalose, arabinose and dextrin but not from lactose; methyl red positive; Voges-Proskauer negative; urease produced; β-galactosidase produced; citrate not utilized; G + C content of DNA ranges from 45.8–46.8 moles %.
    Organisms cause pseudotuberculous lesions, particularly of the mesenteric glands in rodents and man (Y. pseudotuberculosis) and entercolitis in a variety of animals, including man
    Yersinia *p. 330*
    Not as above..**86**
63. *Organisms cause human bartonellosis, which may be manifested clinically by one of the two syndromes constituting Carrion's disease (Oroya fever—a febrile anemia or Verruga peruana—a benign skin eruption) or by an asymptomatic infection.*
    Pleomorphic rods; do not grow on meat infusion agar; grow well in semisolid media containing horse, rabbit or human blood or other complex substances; rods and coccoid forms predominate in culture; flagella polar; multiply on erythrocytes and in fixed tissue cells; transmitted through the sandfly, *Phlebotomus verrucarum*
    Bartonella *p. 904*

Organisms cause conjunctival folliculosis in rabbits.

Rod-shaped organisms 0.2–0.3 μm by 0.5–1.2 μm; motile with lateral or peritrichous flagella; growth on blood and meat infusion agar; colonies small grayish, and translucent, may be mucoid; litmus milk unchanged; gelatin not liquefied; nitrates not reduced; indole not produced: optimum temperature 28–30 C..........**Noguchia**
Note: *This genus is not listed in this edition of* THE MANUAL.

Organisms cause bronchopneumia in rodents and dogs and possibly other animals: minute coccobacillary organisms 0.2–0.3 μm by 0.5–1.0 μm; motile with peritrichous flagella; oxidative; litmus milk alkaline; gelatin not liquefied; nitrate reduced to nitrite; urease produced; citrate utilized; grow on meat infusion agar; G + C content of the DNA 66 moles %...**Bordetella**   *p. 282*
Note: *This entry applies to* **B. bronchiseptica**. *It closely resembles* **Alcaligenes faecalis** *which has a G + C content of the DNA 62 moles %.*

64. *Organisms occur in the upper respiratory tract and lungs, in the eyes and in the urogenital system of man and animals, including birds and fish (H. piscium) where they may be normal commensals or assume a pathogenic state in which they have been held responsible for sore throats, acute pharyngitis, conjunctivitis, coryza (H. gallinarum), urethritis, and soft chancre (H. ducreyi); may also invade the blood stream and cerebrospinal canal causing endocarditis, meningitis, joint infections, pleuropneumonia and septicaemia.*

Organisms do not grow on nutrient agar; rods 0.2–0.5 μm by 0.5–2.5 μm but frequently produce long filaments; will grow on Levinthal agar base with the addition of X factor (hemin) or V factor (nicotinamide adenine dinucleotide) *or* of diphosphothiamine or adenosine triphosphate (*H. piscium*); colonies rarely more than 1 mm in diameter after 2 days of incubation; also grow well on Fildes' blood digest agar or chocolate agar but not always on blood agar; inhabitants of mucous membranes of vertebrates and responsible for a variety of infections (above); G + C content of the DNA ranges from 38–42 moles % (in species examined)...............**Haemophilus**   *p. 364*
Not as above..65

65. *Found in the human genital tract where it may cause non-specific vaginitis or urethritis: masses of cells cover the epithelial cells in discharges.*

Organisms 0.3–0.6 μm by 1–2 μm, arranged singly, in pairs end to end and also in palisades like corynebacteria; Gram-negative to Gram-variable but distinctly Gram-positive on inspissated serum; cell wall of Gram-positive type; no growth on meat infusion agar; X and V factors not required; poor or no growth on blood agar; growth on Casman's agar with colonies 0.1–0.2 mm in diameter in 24–48 hours, which may increase to 1 mm on further incubation; fermentative; acid but no gas from glucose and some other carbohydrates but not from lactose; catalase and cytochrome oxidase negative; nitrates not reduced; H₂S, indole and urease negative; does not liquefy gelatin or coagulated serum; does not decarboxylate lysine, ornithine or arginine, oxidize ethanol, digest cellulose or form starch from maltose; G + C content of the DNA is 41–43 moles % (Strain Dukes 594)

$$\text{Haemophilus } vaginalis \qquad p.\ 368$$

Not as above..66
66. Acid produced from carbohydrates...72
No acid produced from carbohydrates...67
67. *Organisms isolated from chronic pneumonia in calves and intratracheal injection causes lesions similar to those occurring naturally; also pathogenic for goats.*

Cells rod-shaped and arranged in groups in tissues; in cultures coccoid and bacillary forms occur; produce flakes of clubbed sheathed chains of rods in the condensation water on coagulated serum; poor or no growth on meat infusion agar without the addition of sterile guinea pig spleen and the provision of CO₂; optimum temperature 37 C; not pathogenic to laboratory animals

$$\textit{Actinobacillus actinoides} \qquad p.\ 376$$

Not as above..68
68. *Organisms cause a septicemic disease of ducks, ducklings, geese, turkeys and water fowl.*

Rods 0.5 μm by 1.0–2.0 μm; may become pleomorphic; poor growth in peptone media—
improved by serum; colonies small, transparent, glutinous; no growth on McCon-
key's agar; gelatin, *coagulated serum and egg liquefied;* no change in litmus milk;
nitrate not reduced; indole, citrate, malonate, phenylalanine, KCN and urease negative; catalase, oxidase, and arginine dihydrolase positive

Pasteurella antipestifer   p. 373

Not as above. 69

69. *Organisms cause contagious abortion in cattle, goats, sheep and pigs and possibly other*
animals, epididymitis in boars and rams and undulant fever in man.

Rods, 0.5–0.7 μm by 0.6–1.5 μm, occurring singly or rarely in chains; do not show bipolar staining; grow on meat infusion agar and tryptose or trypticase soya peptone
agars or liver extract agar, sometimes requiring increased CO_2 tension; litmus milk
unchanged; nitrates reduced to nitrite (except *B. ovis*); citrate not utilized; G + C
content of the DNA 56–58 moles % (buoyant density). . . . . .**Brucella**   p. 278

Not as above. 70

70. *Organisms cause whooping cough or whooping cough-like diseases in man.*

a. No growth on peptone agar; growth on Bordet-Gengou potato-blood-glycerol agar
in 3–4 days; litmus milk alkaline in 12–14 days; nitrates not reduced to nitrite;
oxidative; has a genus specific heat stable O antigen; sensitizes mice to histamine;
G + C content of the DNA 61 moles %. . . . .**Bordetella** *pertussis*   p. 282

b. Growth on meat infusion agar and peptone agar; produces browning of peptone
agar; citrate utilized; urease produced; litmus milk alkaline in 1–4 days; nitrates
not reduced to nitrite; oxidative; has a genus specific heat stable O antigen; does not
sensitize mice to histamine. . . . .**Bordetella** *parapertussis*   p. 283

Not as above. . . . . , ...71

71. Coagulated blood serum liquefied. 71a

Coagulated blood serum not liquefied. 71b

71a. *Organisms cause conjunctivitis in guinea pigs and man and keratoconjunctivitis (pink-*
eye) in cattle.

Cells vary from small lanceolate coccoids to plump rods 0.8–1.2 μm by 1.5–3.0 μm;
arranged predominantly in pairs and short chains; serum or oleic acid required for
growth; colonies on blood agar vary from 0.1–0.3 up to 3 mm both within and between species; pitting may occur under the colony; large dark zones around colonies
on chocolate agar; oxidative; no carbohydrates attacked; oxidase positive; indole,
acetoin and H_2S not produced; G + C content of the DNA ranges from 41–43 moles %

Moraxella   p. 433

(*M. lacunata* and *M. bovis*)

71b. *Organisms isolated from the upper respiratory tract and "different kinds of pathological*
specimens."

Description as for 71a except that serum or oleic acid is not required for one of the
three species which will grow in a mineral salts base containing acetate and ammonium salts; no hemolysis on blood agar; oxidative, using fatty acids and alcohols
but not carbohydrates as energy sources; G + C content of the DNA 40–44 moles %

Moraxella   p. 433

(*M. nonliquefaciens, M. phenylpyruvica, M. osloensis, "M. urethralis"*)

Note: *"M. urethralis" is similar to but smaller than the other species, rods being 0.6 μm*
by 1.0–1.5 μm and lack the plumpness of other species.

72. *Organisms inhabit the nasopharynx of wild and laboratory rats: cause streptobacillary-rat-*
bite (Haverhill) fever in man and streptobacillary arthritis in mice, rabbits and turkeys.

Rods 0.3–0.7 μm by 1–5 μm frequently in chains or filaments up to 150 μm long; often
highly pleomorphic; fastidious organisms which will not grow on meat infusion agar;
require the addition of blood or ascitic fluid; colonies on ascitic fluid agar 1–2.5 mm
in 3 days, circular, low convex, colorless; L-phase variants are common; fermentative; do not hydrolyze gelatin, casein or coagulated serum; indole and urease negative; H_2S produced; arginine hydrolyzed; phenylalanine not deaminated; catalase

and oxidase negative; phosphatase positive; G + C content of the DNA 24–25 moles
%..**Streptobacillus**   *p. 378*
Not as above...73

73. *Organism found in the human nose and throat: isolated from a case of endocarditis.*

Rods 0.5–0.75 μm by 1–3 μm arranged singly, in pairs, in short chains or in clusters;
pleomorphic; poor growth on meat infusion agar; colonies on horse blood agar 0.5
mm in 24–48 hours extending to 1–2 mm in 3–4 days, circular, low convex, smooth,
entire, shiny, opaque and butyrous; fermentative; acid but no gas from glucose,
fructose, mannose, sorbitol, and sucrose. Lactose and numerous other carbohydrates
not fermented; β-galactosidase not produced; no growth in citrate, KCN or Mac-
Conkey's agar; gelatin and coagulated serum not liquefied

Cardiobacterium   p. 377

Not as above..74

74. *Organisms cause tularemia and tularemia-like infections in man and/or other warm blooded
animals; transmitted by blood sucking arthropods, inhalation, ingestion and contact.
Buboes and areas of necrosis produced in organs and tissues; systemic infections follow
inhalation and anginal forms digestion: highly infectious for white mice and guinea pigs.
Enzootic in ground squirrels and many species of wild animals.*

Organisms 0.2 μm by 0.2–0.7 μm (*F. tularensis*) and 0.7 μm by 1.7 μm (*F. novicida* in
liquids); occur singly; very pleomorphic; poor or no growth on meat infusion agar;
growth on coagulated egg-yolk and blood-glucose-cystine agar gray and 1–4 mm
(*F. tularensis*) and up to 8 mm (*F. novicida*) in diameter in 2–5 days, easily emulsified;
fermentative; slight acid in glucose, mannose, and fructose and possibly other
carbohydrates and also in litmus milk; catalase negative; H_2S produced; cells
soluble in sodium lauryl sulfate and sodium ricinoleate....**Francisella**   *p. 283*
Not as above..75

75. *Organisms cause plague in man, rats, ground squirrels and other rodents; transmitted by
the rat flea; isolated from buboes, blood, sputum and lung exudate; infectious for rats
and guinea pigs.*

Coccoid to rod-shaped; pleomorphic; growth on meat infusion agar requires a heavy
inoculum; optimum temperature 27–28 C; pellicle and 'stalactite' growth in old broth
cultures; fermentative; acid from glucose, fructose, glycerol, maltose, mannitol,
mannose and trehalose; usually no acid from lactose but is β-galactosidase positive;
esculin hydrolyzed; methyl red positive; Voges-Proskauer negative; citrate and
malonate not utilized; gelatin not hydrolyzed; lysine and ornithine not decarboxylated: phenylalanine not deaminated: arginine not hydrolyzed: urease and indole not
produced...**Yersinia** *pestis*   *p. 331*
Not as above..76

76. *Organisms cause glanders and farcy in horses and donkeys; transmissible to man and other
animals.*

Organism 0.5 μm by 1.5–4.0 μm; occur singly, in pairs and in groups; store polyhydroxybutyrate in the cells, especially in nitrogen-deficient media; grow well on meat
infusion agar; colonies 0.5–1.0 mm, smooth, entire, butyrous and may increase in size
with further incubation; oxidative; utilize DL-arginine and betaine as sole carbon
sources; arginine dihydrolase produced; nitrate reduced to gaseous nitrogen; gelatin
is hydrolyzed.....................................**Pseudomonas** *mallei*   *p. 228*
Not as above. ﹐...77

77. Organisms produce sticky or cartilaginous colonies adherent to the medium; grow on
meat infusion agar *or* require blood *or* coagulated egg; nitrates reduced to nitrites;
urease positive; acid from lactose or β-galactosidase produced, or both..........78
Not as above..81

78. *Organisms cause granulomatous lesions of the upper alimentary tract of cattle—particularly
the tongue—and suppurative lesions of the skin and lungs of sheep (A. lignieresii);
and suppurative lesions of the kidneys and joints of foals and piglets (A. equuli); also
found in the mouth and intestinal tract; may cause disease in man and dogs; not pathogenic
for rabbits, mice or guinea pigs.*

Organisms 0.3–0.5 μm by 0.6–1.4 μm; rod-shaped with a tendency to form cocci which remain aligned with the main axis of the rod to give a semblance of morse code symbols; grow on meat infusion agar producing sticky colonies which are adherent to the medium; no well-defined capsules produced, cells are surrounded by slime; fermentative; acid but no gas produced from glucose, fructose and xylose and possibly other carbohydrates; β-galactosidase produced; indole not produced; grow on MacConkey's agar; G + C content of the DNA 40.6–42.0 moles % (T_m)

Actinobacillus *p. 373*

Not as above. 79

79. *Organisms isolated from lesions of the tarsal joints of rabbits.*

Rods, 0.6 μm by 1.2 μm, becoming filamentous and producing cocci; capsulated; no growth on meat infusion agar on primary isolation, later pinpoint colonies develop; very sticky colonies produced on sheep's blood agar; growth profuse on Dorset egg medium; fermentative; acid from glucose, lactose, maltose, sucrose, mannitol and galactose but not arabinose; gelatin not hydrolyzed; nitrates reduced to nitrites; urease produced; citrate not utilized; resistant to penicillin; sensitive to streptomycin, tetracycline and chloramphenicol. Pathogenic to rabbits and less so to guinea pigs. . . . .*Actinobacillus capsulatus* *p. 376*

Not as above. 80

80. *Organisms isolated from a variety of lesions in piglets.*

Rods 0.3–1.0 μm long occurring singly, in chains and in masses; colonies adherent to the agar but not markedly stringy; fermentative; acid from glucose, fructose, xylose, lactose, maltose, sucrose, arabinose, and salicin; methyl red positive; Voges-Proskauer negative; H_2S not produced; nitrates reduced to nitrites; indole not produced; catalase negative; urease positive; citrate not utilized; nonpathogenic for rabbits and guinea pigs. Pathogenic for mice by intraperitoneal injection.

Actinobacillus suis *p. 376*

(*van Dorssen* and *Jaartsveld*)

Note: *Another ill-described species of the same name is cited on p. 376*

Not as above. 81

81. *Organisms isolated from abscesses arising from a naturally occurring infection of laboratory mice and one recorded infection of the jaw of a woman.*

Rods 0.6–0.8 μm by 0.6–1.4 μm with marked tendency to produce coccoid forms particularly in broth and gelatin; colonies on agar 1 mm in diameter in 2–3 days, star-like, adherent and difficult to emulsify; growth improved by CO_2; granules produced in the base and up the sides of the tube in broth; fermentative; acid from glucose and some other carbohydrates; arabinose, inositol, glycerol, inulin, raffinose, salicin and xylose not fermented; H_2S not produced; indole and urease not produced; nitrate may be reduced: not pathogenic to laboratory animals

Actinobacillus actinomycetemcomitans *p. 375*

Not as above. 82

82. *Organisms produce hemorrhagic septicemia in ruminant animals, pneumonic conditions in cattle, sheep and pigs, wound infections and occasional respiratory and other infections of man, fowls, ducks and turkeys (***P. multocida***); respiratory diseases and abscess formation in laboratory animals (***P. pneumotropica***); enzootic pneumonia of sheep and septicemia of lambs (***P. haemolytica***) and occurs as a nasal commensal in man (***P. ureae***).*

Oval to rod-shaped organisms 0.3–0.5 μm by 1.0–1.8 μm; frequently in pairs and short chains; bipolar staining common in preparations from tissues; growth poor or absent on meat infusion agar, improved by addition of blood; fermentative; acid from glucose and other carbohydrates; lactose rarely fermented (except *P. pneumotropica*); catalase positive; nitrates reduced to nitrite; gelatin not hydrolyzed; G + C content of the DNA ranges from 36.5–43.0 moles %. . . .**Pasteurella** *p. 370*

Not as above. 83

83. *Organisms isolated from a variety of human pathological conditions.*

Short plump rods with square ends, occurring in pairs; require a complex medium for growth; colonies on blood agar 1.2–1.5 mm after 48 hours—sometimes mucoid; β-hemolytic; do not liquefy coagulated serum; oxidative; acid produced from glucose and maltose on ascites agar medium; urease and phenylalanine deaminase not produced; G + C content of DNA 44.5 moles %.......**Moraxella** (*kingii*)   *p. 435*

Not as above...96

Note: *Common enteric pathogens are treated with nonpathogens in the remainder of this section.*

84. Acid produced from glucose.................................85
    No acid produced from glucose..............................105
85. Motile..86
    Non-motile...96
86. Bioluminescent..87
    Not bioluminescent.......................................88
87. Flagella polar: sensitive to 2,4-diamino-6,7-diisopropyl pteridine (vibriostatic agent 0/129)...................................**Photobacterium**   *p. 349*

    Note: *Anaerogenic strains.*

    Flagella peritrichous; not sensitive to vibriostatic agent 0/129; fermentative; acid only from carbohydrates; starch hydrolyzed; nitrates reduced; indole produced; methyl red positive; Voges-Proskauer negative; oxidase and catalase positive; gelatin liquefied; trimethylamine oxide reduced to trimethylamine
                                  Lucibacterium   *p. 351*

88. Colonies yellow or yellow-orange, pigment not soluble in water...................89
    Not as above...91
89. Flagella polar...**Pseudomonas**   *p. 217*

    Note: *The following species of **Xanthomonas** may terminate here: X. campestris, X. fragariae, X. albilineans, X. axonopodis, and X. ampelina.*

    Flagella peritrichous or lateral...........................90
90. Oxidative metabolism, never fermentative; organisms motile by one or two flagella inserted in the lateral or subpolar position; G + C content of the DNA 68–70 moles %
                                  Alcaligenes (*paradoxus*)   *p. 275*

    Not as above; flagella peritrichous..............**Flavobacterium**   *p. 357*

    Note: *Erwinia species 7a, 7b and 9 terminate here.*

91. Flagella polar...92
    Flagella peritrichous and lateral..........................94
92. Metabolism oxidative only................................93
    Metabolism oxidative and fermentative; coccobacillary to rod-shaped organisms 0.8–1.0 μm by 3.0 μm, occurring singly, in pairs or short chains; motile by polar flagella; generally lophotrichous; acid from glucose, maltose, trehalose and glycerol; gluconate and malonate and citrate not utilized; no growth with KCN; H$_2$S not produced; nitrates reduced; gelatin not hydrolyzed; indole produced; phosphatase produced; phenylalanine deaminase not produced; G + C content of DNA is 51 moles %.......................................**Plesiomonas**   *p. 348*

    Note: *See also anaerogenic strains of **Aeromonas**.*

93. Organisms produce zoogleae; rod-shaped, 0.5–1.0 μm by 1.0–3.0 μm; young cells actively motile with a single polar flagellum. In natural waters the organisms produce cartilaginous dendritic masses frequently attached to submerged plants and other objects; in the laboratory culture flocs occur in media with a C/N ratio of >10:1, dispersed growth occurring in media with a C/N ratio of <5:1 (see Plate 7.1)
                                  Zoogloea   *p. 249*

    Not as above...**Pseudomonas**   *p. 217*

94. Oxidative; rods 0.8 μm by 1.5–3.0 μm: motile with 1–4 peritrichous or lateral flagella; growth on carbohydrate media usually accompanied by production of a copious extracellular slime...............................**Agrobacterium**   *p. 264*

    Fermentative...95

    Note: *Colorless forms of **Chromobacterium** terminate at this point in the key.*

95. No growth in KCN: nitrates not reduced: deoxyribonuclease not produced: no acid from maltose..**Erwinia** *p. 332*

Note: *Species 1 to 6 terminate here.*

Growth in KCN; nitrates reduced; deoxyribonuclease produced: acid from maltose, salicin and sucrose; no acid from arabinose; gelatin hydrolyzed; methyl red negative

Serratia *p. 326*

No growth in KCN, nitrate reduced; gelatin not hydrolyzed; methyl red positive; acid from maltose and arabinose but not from salicin or sucrose....**Salmonella** *p. 298*

96. Lactose fermented...97

Lactose not fermented...101

97. Colonies yellow or yellow-orange......................**Flavobacterium** *p. 357*

Colonies not as above...98

98. Note: *Some animal pathogens will key out to this point if you have by-passed the section between I-59 and I-84. Attention is drawn here to those which would have reached this point. These should be checked. See:*

Haemophilus I-64

Streptobacillus I-72

Actinobacillus I-78

Pasteurella I-82

Not as above...99

99. Short plump rods 1.0–1.5 μm by 1.5–2.5 μm becoming coccal in the stationary phase; occur mainly in pairs and short chains; may show a twitching movement on solid media; oxidative; oxidase negative; catalase positive; acetoin, H_2S and indole not produced; nitrates not reduced; resistant to penicillin; G + C content of the DNA 40–47 moles % (T_m).................................**Acinetobacter** *p. 436*

Not as above..100

100. Lactose fermentation delayed; growth with KCN; Voges-Proskauer positive; malonate and citrate utilized; urease produced.............**Klebsiella edwardsii** *p. 322*

Note: *This entry covers the Pa strain (Cowan et al (J Gen Microbiol 23: 601, 1960).*

Lactose fermentation delayed; no growth with KCN, Voges-Proskauer negative, malonate and citrate not utilized; urease not produced........**Shigella** *p. 318*

101. Colonies yellow..**Flavobacterium** *p. 357*

Note: *See also Aeromonas salmonicida (Smith, J Gen Microbiol 33: 263, 1963).*

Not as above..102

102. Note: *Some animal pathogens will key out to this point if you have by-passed the section from I-59 to I-84. Attention is drawn here to those which would have reached this point. These should be checked. See:*

Haemophilus I-64

Streptobacillus I-72

Cardiobacterium I-73

Franciscella I-74

Pseudomonas I-76

Actinobacillus I-78–79–80

Yersinia I-81

Pasteurella I-82

Not as above..103

103. Methyl red positive; Voges-Proskauer negative; no growth with KCN; no acid from sucrose, salicin, or adonitol; no growth in malonate............................104

Methyl red positive; Voges-Proskauer negative; growth with KCN; acid from sucrose, salicin, and adonitol; growth on malonate: urease negative....**Klebsiella**

(*K. rhinoscleromatis*) *p. 323*

Not as above..................................possibly **Acinetobacter** *p. 436*

104. Enteric pathogens of man; agglutinate with *Shigella* antisera; no growth on ammonium citrate; H_2S not produced.........................**Shigella** *p. 318*

Organisms pathogenic to birds; grow on ammonium citrate; H_2S produced; agglutinate with Salmonella group D "O" antiserum.................**Salmonella**

             (*S. gallinarum*) *p. 318*

105. Motile..106

  Non-motile...111

106. Colonies yellow or yellow-orange; flagella polar...........**Pseudomonas** *p. 217*

  Note: *Xanthomonas ampelina terminates here.*

  Not yellow; polar, lateral or peritrichous flagella.............................107

107. **Note:** *Some pathogens of man and animals will key out to this point if you have by-passed section I-59 to I-84. Attention is drawn here to those which would have reached this point. These should be checked. See:*

   Bordetella I-63

   Bartonella I-63

  Not as above..108

108. Organisms produce zoogleae; cells rod-shaped, 0.5–1.0 μm by 1.0–3.0 μm; young cell actively motile with single polar flagellum. In natural waters the organisms produce cartilaginous dendritic masses frequently attached to submerged detritus; in laboratory culture flocs occur in media with a C/N ratio of >10:1; dispersed growth occurring in media with a C/N ratio of <5:1 (see Plate 7.1).....**Zoogloea** *p. 249*

  Not as above..109

109. Flagella polar...110

  Flagella peritrichous or lateral............................**Alcaligenes** *p. 273*

110. Organisms obtain energy by the oxidation of inorganic sulfur compounds; some species will grow under strictly autotrophic conditions....**Thiobacillus** *p. 456*

  Not as above...**Pseudomonas** *p. 217*

  Note: *Alginomonas* and *Comamonas* *included here.*

111. Colonies yellow or yellow-orange..114

  Not as above...112

112. **Note:** *Some pathogens of man and animals will key out to this point if you have by-passed section I-59 to I-84. Attention is drawn here to those which would have reached this point. These should be checked. See:*

   Actinobacillus actinoides I-67

   Pasteurella anatipestifer I-68

   Brucella I-69

   Bordetella I-70

   Moraxella I-71

  Not as above..113

113. Organisms will oxidize inorganic sulfur compounds while using CO_2 as the sole source of carbon; thin short rods 0.4–1.0 μm by 0.6–4.0 μm; slow growth on nutrient agar or yeast extract agar....................................**Thiobacillus** *p. 456*

  Organisms plump rods usually occurring in pairs or short chains

           Acinetobacter *p. 436*

114. Metabolism oxidative.....................................**Pseudomonas** *p. 217*

  Metabolism fermentative.............................**Flavobacterium** *p. 357*

SECTION J

1. Cell masses green or brown; contain bacteriochlorophylls *c* or *d* with various carotenoids; photopigments are located in "chlorobium vesicles" which underlie and are attached to the cytoplasmic membrane; oxidize sulfide, depositing sulfur *outside* the cells; anaerobic..2

  Note: ***Rhodopseudomonas viridis*** *is green but has different pigments, see J-10.*

  Cell masses various shades of green, yellow, brown, red or purple; contain bacteriochlorophylls *a* or *b* with various carotenoids which usually dominate over the green pigments; photopigments arranged in internal membrane continuous with the cytoplasmic membrane...10

Note: *The spectrophotometric absorption peaks for the bacteriochlorophylls are shown in Table 1.1 (p. 24) and the various carotenoid pigments in Table 1.2 (p. 25). Of the various absorption peaks for bacteriochlorophylls a and b those occurring at 590 nm (bacteriochlorophyll a) and at 605 nm (bacteriochlorophyll b) are quite pronounced and usually readily observed in cell masses with a hand spectroscope. They are quite distinct from the absorption band for plant chloropyll a at 660–680 nm—also in the visible region of the spectrum. The peaks beyond 800 nm for bacteriochlorophylls a and b are not visible with the hand spectroscope.*

2. Organisms living in symbiotic association with other microorganisms 3
   Note: *The genera represented here are not recognized in this edition of* THE MANUAL *but are listed in the addenda on pages 58 and 59. Their taxonomic status is doubtful but they are included in the key for information purposes.*
   Not as above 6

3. Cells attached to the surface of protozoan cells . . . .**Chlorobacterium**   *p. 59*
   Cells attached to the surface of bacterial cells 4

4. Cell masses pink-brown; photosynthetic cells curved, 1 μm or less by 2 μm; 10–20 individual cells surrounding the central bacterium
   Pelochromatium roseum   *p. 58*
   Cell masses green 5

5. Aggregates barrel-shaped, 2.5–5.0 μm by 7–12 μm (one species) and 3–4 μm by 4–8 μm (other species); green cells 0.5–1.0 μm by 0.7–2.5 μm; 8–16 (one species) and 7–40 (other species) arranged around a central polarly flagellated cell
   Chlorochromatium   *p. 58*
   Aggregates long and cylindrical; non-motile; consist of green cells, 0.5–1.0 μm by 2.0–4.0 μm, lying on the surface of a slime capsule which covers the inner cylindrical cell. They are themselves covered by a layer of slime, the aggregate measuring 7.0–8.0 μm by up to 50 μm . . . .**Cylindrogloea**   *p. 59*

6. Cells spherical to ovoid: each producing about 20 prosthecae
   return to **Section D**   *p. 1105*
   Not as above 7

7. Cells contain vacuoles and are arranged into net-like or trellis-like aggregates or in a single layer on the surface of a hollow spherical or irregular colony; sulfur from oxidation of sulfide is deposited externally 8
   Cells do not contain vacuoles; not aggregated 9

8. Cells spherical; non-motile; arranged in chains in loose trellis-like aggregates
   Clathrochloris   *p. 57*
   Cells rod-shaped, straight or curved; aggregated into net-like structures (Plate 1.1, Fig. 9) or as a single layer on the surface of a hollow spherical or irregular colony
   Pelodictyon   *p. 56*

9. Rod-shaped, ovoid or curved; *non-motile;* contain bacteriochlorophyll c or d; cell suspensions green or brown . . . .**Chlorobium**   *p. 52*
   Rod-shaped; *motile* with polar flagella; contain mainly bacteriochlorophyll c; cell suspensions yellowish green . . . .**Chloropseudomonas**   *p. 55*

10. Rod-shaped, ovoid or spherical organisms (Plate 1.1, Fig. 2); reproduce by binary fission or by apical budding; motile with polar flagella; do not oxidize H₂S; capable of growth anaerobically when exposed to light in the presence of simple organic compounds; some species also grow aerobically in the dark
    Rhodopseudomonas   *p. 29*
    Note: *Vannielia aggregata, described by Pringsheim as forming rosettes in natural habitats, is considered as a species of* **Rhodopseudomonas.**
    Not as above:
    a. sulfide oxidized, sulfur deposited *outside* the cells; anaerobic
    Ectothiorhodospira   *p. 47*
    b. sulfide oxidized; sulfur deposited *inside* the cells, usually anaerobic 11

11. Cells arranged in cubical packets; spherical to ovoid; do not contain gas vacuoles; motile with polar flagella; contain bacteriochlorophyll a . . . .**Thiosarcina**   *p. 40*

Cells arranged in flat sheets in which cells occur in tetrads which form the structural unit (Plate 1.1, Fig. 6); cells spherical to ovoid; contain gas vacuoles in the center of each cell; non-motile; contain bacteriochlorophyll a . . . . . **Thiopedia**   *p. 45*

Not as above. 12

12. Cells do not contain gas vacuoles. 13

    Cells contain gas vacuoles. 15

13. Cells motile with polar flagella in actively growing cultures. 14

    Non-motile; cells spherical, typically diplococcus-shaped before division (Plate 1.1, Fig. 5); tetrads may be formed; may become embedded in slime under a variety of conditions; G + C content of the DNA 63.3–69.9 moles % (buoyant density)

                                                              Thiocapsa   *p. 42*

14. Cells ovoid to rod-shaped; usually solitary but may form aggregates embedded in slime, in which they are non-motile; internal sulfur globules very prominent when growing in the presence of sulfide (Plate 1.1, Fig. 4); swollen spindle-shaped and clubbed forms produced under unfavorable conditions

                                                             Chromatium   *p. 36*

    Note: *These pleomorphic forms were originally placed in the genera* **Rhabdomonas** *and* **Rhabdochromatium.**

    Cells spherical, becoming characteristically diplococcal before division may form aggregates embedded in slime in which cells are non-motile (*cf. Chromatium*)

                                                              Thiocystis   *p. 39*

15. Cells spherical: motile with polar flagella: typically diplococcal before division; cells may remain solitary or attached in tetrads which aggregate into areas of considerable size, the whole being embedded in slime; branched and net-like aggregates may form which later may break up into smaller clusters and spheroidal colonies which became motile. . . . . **Lamprocystis**   *p. 43*

    Cells non-motile. 16

16. Cells rod-shaped, with rounded ends, sometimes spindle-shaped; may form aggregates in which cells become arranged end to end in an irregular net-like structure which may form compact masses or break up into single cells. . . . **Thiodictyon**   *p. 44*

    Cells spherical; 1.5–3.0 μm in diameter; occur singly and in irregular aggregates which may be surrounded by a slime capsule. . . . . **Amoebobacter**   *p. 46*

SECTION K

1. Endospores produced. 2

    No endospores produced. 3

2. Straight to curved rods, motile with peritrichous flagella; spores oval to round, terminal to subterminal, slightly swelling the rods; oxidative; oxidize lactate and pyruvate while reducing sulfates, sulfites and reducible sulfur compounds to H_2S

                                             Desulfotomaculum   *p. 572*

    Not as above. . . . . **Clostridium**   *p. 551*

3. Motile. 4

    Non-motile. 13

4. Curved rods, motile with polar or lateral flagella. 5

    Straight rods. 10

5. Curved rods with a bunch of flagella inserted at or near the center of the concave side of the rod; fermentative; chiefly acetate, propionate, CO_2 and/or lactate produced from glucose; catalase negative; G + C content of the DNA 53–61 moles % (buoyant density); recorded from the alimentary canal of ruminants and guinea pigs and from the buccal cavity of man. . . . . **Selenomonas**   *p. 424*

    Not as above. 6

6. Curved rods, motile with polar flagella; reduce sulfates, sulfites, sulfur, thiosulfate, and hyposulfites to H_2S while oxidizing organic compounds, particularly lactate and

pyruvate, to acetate and CO_2; hydrogen may be oxidized; cells contain C_3 cytochromes and desulfoviridin (absorption band at 630 nm) which is responsible for a characteristic red fluorescence of cells when examined in light at 365 nm immediately after the addition of a few drops of 2.0 N NaOH.....**Desulfovibrio**   *p. 418*

Not as above..7

7. Glucose and other carbohydrates not fermented; curved rods 0.2–0.8 μm by 1.5–5.0 μm; single polar flagellum at one or both ends; oxidative; energy obtained from amino acids and tricarboxylic acids; gelatin and urea not hydrolyzed; methyl red and Voges-Proskauer negative; oxidase positive.......**Campylobacter**   *p. 207*
   Note: *Some strains only, most are microaerophilic.*
   Straight to curved rods occurring singly and in pairs—motile with a single polar flagellum; reduce CO_2 to methane while oxidizing hydrogen or formate; growth requires rumen fluid; glucose not fermented or oxidized.......**Methanobacterium**   *p. 473*
   (*M. mobile*)

   Not as above: glucose fermented..8

8. Curved rods 0.3–0.8 μm by 1–5 μm; polar flagella, monotrichous; ferment glucose, producing large quantities of *butyric acid*, a little formic and lactic acid, CO_2, and H_2; catalase negative....................................**Butyrivibrio**   *p. 420*
   Note: *Under some conditions of growth lactic acid may dominate and only a little butyric acid is formed (Gill and King, 1958).*

   Not as above..9

9. Curved rods 0.3–0.7 μm by 1–7 μm with pointed ends; single polar flagellum; glucose *fermented* with the production of *succinic* and *acetic* acid and also formic and sometimes lactic acid; no gas is produced; found in the rumen of cattle and sheep
   Succinivibrio   *p. 422*
   Note: *These organisms are described as spiral on primary isolation but may change to the appearance of curved rods on subculture.*
   Curved rods, 0.4–0.6 μm by 2–4 μm with bluntly pointed ends; motile with lateral to subterminal flagella; *ferment* glucose with the production of large amounts of ethanol, lactic, formic and acetic acids and CO_2 and a small amount of hydrogen; found in the rumen of cattle....................................**Lachnospira**   *p. 423*
   Note: *This organism is recorded as weakly Gram-positive, rapidly becoming Gram-negative.*

10. Straight rods 1.0–1.5 μm by 2.0–3.0 μm (one species) and 1.4–2.0 μm by 4.0–5.0 μm (other species); occur singly or in pairs; motile with lophotrichous flagella; ferment glucose and fructose and possibly sucrose with the production of equimolar amounts of ethanol and CO_2 (by the Entner-Doudoroff pathway); acetaldehyde—an intermediate product—may accumulate; catalase positive; H_2S is produced: gelatin not liquefied: indole not produced: nitrate not reduced: methyl red test negative; G + C content of the DNA 47–48 moles %....................**Zymomonas**   *p. 352*
   Straight to curved rods occurring singly and in pairs—motile with a single polar flagellum; reduce CO_2 to methane while oxidizing hydrogen or formate; growth requires rumen fluid; glucose not fermented or oxidized.......**Methanobacterium**   *p. 473*
   (*M. mobile*)

   Not as above..11

11. Straight rods with rounded ends, 1.0–1.5 μm by 1.0–3.0 μm, occurring singly in pairs and in clumps; single polar flagellum; *ferments* glucose with the production of a large amount of *succinic acid* and some acetate as major end products; no gas is formed; found in the rumen contents of cattle fed forage and grain
   Succinimonas   *p. 422*
   Not as above: organic acids produced in large quantities from either peptone or carbohydrate, or both...12

12. Rods 1.0 μm by 2–10 μm with rounded to pointed ends; often arranged in pairs and sometimes in chains; flagella peritrichous; *butyric acid* (without isobutyric acid or isovaleric acid) produced as the major product of fermentation
   Fusobacterium   *p. 404*
   Note: *F. aquatile, F. plauti and F. bullosum terminate here.*

Rods ranging in size from 2–20 μm in length depending on the species; major end product in peptone-yeast extract-glucose broth is acetic acid and butyric acid (with isobutyric acid and isovaleric acid (one species) or acetic and propionic acid (one species)) *or acetic acid with a small amount of ethanol* (two species)

Bacteroides *p. 385*

Note: *B. serpens, B. clostridiiformis, B. constellatus and B. praeacutus terminate here.*

13. Curved rods; ferment glucose producing large quantities of *butyric acid;* with some formic acid, lactic acid and CO_2; catalase negative............**Butyrivibrio** *p. 420*
Note: *This entry covers only some non-motile strains.*
Not as above...**14**

14. Straight to slightly curved rods, 1–1.5 μm by 5–15 μm, with one or both ends rounded or pointed; two or more cells arranged in septate filaments which may reach 200 μm long; no branching; *glucose is fermented with the production of about 90% lactic acid, 10% acetic acid.* No butyric acid is produced; H_2S and indole not produced; catalase, benzidine and Voges-Proskauer reactions negative; gelatin not liquefied; G + C content of the DNA of two strains ranges from 31.5–34 moles %

Leptotrichia *p. 416*

Straight to slightly curved rods occurring in long chains which are often in bundles; ferment acetate and butyrate with the production of carbon dioxide and methane

Methanobacterium *p. 473*
(*M. soehngenii*)

Not as above...**15**

15. Organisms produce (from peptone or glucose) *butyric acid* (without isobutyric acid and isovaleric acid) as the major end product; G + C content of DNA of most species examined ranges from 26–34 moles %.............**Fusobacterium** *p. 404*
Organisms produce (from peptone or glucose) mixtures of acids including succinic, acetic, formic, lactic and propionic. Butyric acid is not a major product but if produced, then both isobutyric acid and isovaleric acid are also produced

Bacteroides *p. 385*

SECTION L

Note: *Some acid-fast organisms, causing human and animal infections, have not been cultured. They key to this point and should be considered with Mycobacterium and Nocardia.*

1. Aerobic to microaerophilic...**2**
Anaerobic...**48**

2. Endospores produced...**3**
No endospores produced...**4**

3. Straight rods, 0.7–0.8 μm by 3–5 μm, occurring singly, in pairs or short chains; motile by means of a small number of long peritrichous flagella; homofermentative producing only lactic acid from glucose and other carbohydrates; limiting pH is 4.0; catalase negative and devoid of cytochrome pigments; microaerophilic; colonies on glucose-yeast extract-peptone agar 0.5 mm in diameter; gelatin not liquefied; nitrates not reduced; litmus milk unchanged; G + C content of the DNA 39.3 moles %...**Sporolactobacillus** *p. 550*
Not as above; organisms are either oxidative or fermentative or both; usually catalase positive...**Bacillus** *p. 529*
Note: *The conidia of some branching Gram-positive organisms have been shown to have all the characteristics of the bacterial endospore. These organisms are treated under the succeeding section on branching bacteria.*

4. Organisms do not branch or produce simple side branches or terminal bifurcations on isolated rods in the population...**14**
Organisms show distinct branching in young microcolonies.......................**5**
Note: *1. By 'microcolony' the author refers to the extremely early development of the colony which can only be examined by an oil immersion objective at a magnification of $\times$ 1000.*

Well dispersed cells or fragments of "mycelium" are inoculated onto the surface of an agar medium and the culture incubated and examined periodically until clear evidence of growth is discernible at a magnification of ✕ 100 with a phase contrast microscope. Such cultures are then examined periodically by impressing a coverglass over some of the microcolonies and examining these at ✕ 1000. A succession of such observations at appropriate intervals is needed to observe the sequential change referred to in some descriptions given in The Manual *and in these keys.*

These microcolonies are quite different to the "microcolonies" referred to in the text on Actinomyces *where the term refers to very small but relatively mature colonies requiring some magnification to be readily examined.*

2. In the section of the Key dealing with branching organisms repeated reference is made to the chemical composition of cell walls in terms of types (I-IV) and whole cell sugar patterns (A-D). Explanations of these symbols are given in Table 17.18 and 17.19, p. 658.

5. Growth occurs on or under the surface of the substrate (substrate mycelium); aerial mycelium is produced by some species but not others........................6
   Branching mycelium is of an aerial type only; no substrate mycelium produced; coccoid, rod-shaped or pisciform organisms, each with a collar-like scar at one pole from which pole 1–3 flagella are produced; on Czapek agar cells become sedentary and produce a holdfast which attaches them to the agar surface. Transverse division of both basal and intercalary cells, with occasional side branching, results in the production of a heavily segmented branched aerial mycelium which subsequently disintegrates to produce rods and cocci to complete the cycle; starch and casein hydrolyzed; gelatin not liquefied; cell wall contains L-diaminopimelic acid (Type 1)
   Sporichthya *p. 842*

Note: *On rich media the aerial growth may be replaced by a normal bacteria-like growth of rods and pisciform organisms on the surface of the agar. The aerial fungus-like growth is best observed on a restricted medium.*

6. Branches produced only as tubular out-growths from a subterminal position on rod-shaped elements produced by a process of fragmentation of previously formed mycelium *or*, in a similar manner, from the individual segments of unfragmented mycelium. Single rods, with the attached extended filaments, constitute the so-called 'whip-handles' (Plate 17.4, Fig. 1); rods, 1.5–2.5 μm by 3–10 μm produce filaments 20–200 μm long which later septate and fragment to produce the rods; non-motile; not acid-fast; catalase positive; can ferment glucose with the production of a small amount of CO_2 (soluble in the medium!) with propionic acid as the major and formic acetic and lactic acid as minor products; *cell walls contain diaminopimelic acid;* found in the oral cavity of man and primates particularly in calculus and plaque deposits on teeth.....................................**Bacterionema** *p. 676*
   Not as above; branches are produced at random or in a regular dichotomous manner throughout the developing mycelium..7

7. Organisms associated with the production of nodules in non-leguminous plants of the following genera—Alnus, Elaeagnus, Hippophaë, Shepherdia, Discaria, Ceanothus, Coriaria, Dryas, Purshia, Cercocarpus, Gale, Myrica, Comptonia and Casuarina; the organisms fix nitrogen in association with the host plant; "in active nitrogen fixing nodules, the center of the host cell is filled with a hyphal mass; near the periphery, spherical or club-shaped terminal swellings (Plate 17.6, Fig. 1) are formed on radially arranged hyphae close to the plant cell wall. In host cells which have not survived invasion by *Frankia* species, vesicles or club-shaped structures are not found; instead the mycelium fragments into small particles and bacteria-like cells (Plate 17.6, Fig. 2)." Not yet cultivated away from the plant host
   Frankia *p. 702*
   Not as above...8

8. Hyphae are produced by a process of budding similar in gross aspect to the production of pseudomycelium in yeasts. A constriction is produced just behind the tip of the developing hypha to produce a bud-like element which expands to produce the next segment (see Figs. 13 and 15, IJSB 1971, *21:* 32). Spores occur on both aerial and

substrate mycelium as 1) terminal or lateral chains or spores produced acropetally, 2) blastospores produced basipetally from each segment or 3) arthrospores resulting from fragmentation of the hyphae, usually restricted to the aerial mycelium; cell walls contain mesodiaminopimelic acid, arabinose and galactose (Type IV)

Pseudonocardia *p. 746*

Note: *See also Geodermatophilus.*

Growth of the mycelium is by extension of the hyphae with (septate) or without (non-septate) the progressive production of septa. Pseudomycelium not formed.......9

9. Branching filaments fragment into rods and cocci; no specialized reproductive cells produced (see "Note")...10

Branching filaments produce conidia, sporangiospores, zoospores or sometimes blasto-spores; mycelium does not fragment except following muriform division of the hyphae..18

Branching filaments produce single conidia or chains of conidia but also undergo frag-mentation within 24–48 hours to produce rods and cocci

Promicromonospora *p. 863*

Branching filaments which do not fragment: do not produce any conidia, sporangio-spores or blastospores..54

Note: *In microcolonies which are undisturbed and examined under phase contrast, frag-mentation is not evident until the rods resulting from the fragmentation begin to grow and the end-to-end pressure forces them out-of-line to produce a zig-zag arrangement, or the cells slip past each other to form small clusters—a process common to many mycobacteria. It is therefore a good practice, having first examined the undisturbed microcolony, to attempt to disrupt it by movement of the coverglass from side to side. Those filaments which have been progressively segmenting as septa are produced are quite readily disrupted.*

With organisms in which fragmentation is delayed for 5 days or more, it will be obvious that when fragmentation occurs there will no longer be a microcolony. In such cases low magnification (× 20–× 100) of colonies on agar will usually reveal a dense center, sur-rounded by a fringe of branching filaments. Preparations from the center reveal a mass of rods and cocci, some of the former with rudimentary branches.

10. No grossly visible aerial mycelium produced; fragmentation of substrate proceeds al-most simultaneously with filament formation; cells often slip past each other to form characteristic clusters; acid- and alcohol-fast in Ziehl-Neelsen stain: *cell walls con-tain diaminopimelic acid*.............................**Mycobacterium** *p. 682*

Note: *There is no clear way to separate these species of **Mycobacterium** from the acid-fast members of the genus **Nocardia** which have similar growth habits. The G + C con-tents of the DNAs are much the same and the cell wall types are the same (Type IV). It has been suggested that differentiation might be based on the chain length of the mycolic acids produced (see "Further Comments" to the description of the genus **Nocardia**).*

Not as above; a distinct development of a branching mycelium occurs before fragmen-tation..11

11. No aerial hyphae produced; catalase negative; not acid-fast; glucose is fermented to produce lactic, succinic, formic and acetic acid but no propionic acid; *cell wall con-tains no diaminopimelic acid*; the principal constituent being ornithine

Actinomyces *p. 660*

Note: *This entry covers **A. naeslundii** and **A. humiferus** (which grow well aerobically) and some microaerophilic organisms which are also treated as anaerobes in this section of the key.*

Not as above..12

12. No aerial mycelium; not acid-fast; catalase positive; nitrates reduced to nitrite; nitrite reduced; H_2S produced; esculin is hydrolyzed; no action on litmus milk; ferments glucose with the production of lactic acid, with some acetic, succinic and formic acid but no propionic; *cell walls contain no diaminopimelic acid or arabinose:* alanine, glutamic acid and lysine are present with large amounts of galactose; G + C content of the DNA is 65.4–69.7 moles % (T_m).......................... **Rothia** *p. 679*

Not as above..13

13. No aerial mycelium; catalase negative; indole negative; nitrates reduced to nitrites; urease negative; starch weakly hydrolyzed; glucose is fermented to yield propionic and acetic acid and carbon dioxide; *cell walls contain LL-diaminopimelic acid*, alanine, glutamic acid and glycine with galactose and a small amount of gluocse; G + C content of the DNA is 63–65 moles % (T_m): causes human actinomycosis and lacrimal canaliculitis: pathogenic to mice..........................**Arachnia** *p. 668*

Aerial mycelium may be absent or extensively produced; fragmentation of the mycelium may occur in any period from 8 hours up to 5 days, the time being reflected in the type colony which develops; obligate aerobes; majority of species are acid-fast or partially acid-fast on Dubos oleic acid medium; *cell wall contains meso-diaminopimelic acid, arabinose and galactose* (Type IV).................**Nocardia** *p. 726*

Note: *Three groups are recognized on the basis of time of fragmentation: 1) fragmentation commences within 12–14 hours (Plate 17.11, Fig. 1)—G + C content of DNA 61–63 moles %, 2) fragmentation commences after 20 hours (Plate 17.11, Fig. 2)—G + C content of the 66–68 moles %, and 3) fragmentation occurs after 5 days (Plate 17.11, Fig. 3)—G + C content of the DNA 68–72 moles %.*

14. Organisms produce rudimentary branches or appear to do so 15
Organisms do not branch..36

15. Acid- and acid alcohol-fast in the Ziehl-Neelsen stain: *cell walls contain meso-diaminopimelic acid*, alanine, glutamic acid, glucosamine, muramic acid, arabinose and galactose (Type IV)......................................**Mycobacterium** *p. 682*
Not acid-fast..16

16. Rudimentary branching may occur in cultures grown aerobically or in acid media anaerobically; chains of coccoid and rod-shaped organisms in neutral media under anaerobic conditions; pleomorphic; non-motile; microaerophilic; glucose fermented to produce CO_2 (usually soluble in the medium!), propionic and acetic acid as the major end products; lactic acid is converted to propionic acid by most strains; *cell walls contain either L-diaminopimelic acid or meso-diaminopimelic acid* (except for *P. lymphophilum* which has lysine): G + C content of DNA appears to lie within 2 ranges of 53–63 and 64–68 moles %.................**Propionibacterium** *p. 633*
Not as above...17

17. Non-motile; cells have a definite cyclic development (Plate 17.1, Figs. 1–11); spherical cells germinate at one or more points to produce rod-shaped cells, which elongate and divide. At the point of division, growth of the cells continues at an angle to the original axis. When side branches are equal in size to the parent cell, division occurs at the angle. This process is repeated during the growth of the colony. In undisturbed microcolonies the arrangement of the cells has a very deceptive appearance of branching—particularly at a magnification of × 100 (Plate 17.1, Fig. 8). *In older colonies the rods transform entirely into a mass of cocci.* Rods are most frequently Gram-positive but may decolorize readily to give Gram-negative cells with Gram-positive granules. The cocci are usually strongly Gram-positive; cellulose is not hydrolyzed; recorded mainly from soils..........................**Arthrobacter** *p. 618*
Motile; organisms disintegrate cellulose filter papers in 0.5% peptone water and produce clearing of precipitated cellulose agar plates.........**Cellulomonas** *p. 629*

18. Branching mycelium produced from coccoid bodies undergoes a primary subdivision by transverse septa laid down progressively 5–30 μm behind the tip of the hypha as growth proceeds. This is followed by progressive division of each cell by further transverse septa until individual cells are only 0.3–0.5 μm long. Then longitudinal and radial division of each cell occurs with progressive widening of the hypha sometimes 8- to 10-fold, producing sarcina or mulberry-like packets of coccoid cells. Individual cells may germinate *in situ* to produce hyphae or swell and are released as motile spores with a tuft of flagella (Plate 17.9, Figs. 1–4); *cell walls contain meso-diaminopimelic acid, without glycine or arabinose* (Type III).....................19
Not as above..20

19. Production of branching mycelium quite definite and of a true mycelial type (see Plate 17.9)..**Dermatophilus** *p. 723*

Production of mycelium rudimentary; resembles a pseudomycelium (see Plate 17.10)

Geodermatophilus *p. 724*

Note: *Quite often* **Geodermatophilus** *species produce no mycelium at all. These are included elsewhere in the key.*

20. Conidia are not produced in sporangia..21

Conidia are produced in sporangia..27

21. Conidia produced singly, sessile or on short or long conidiophores, or on the tips of hyphae...22

Conidia produced in longitudinal pairs, either sessile or on short conidiophores on aerial mycelium (Plate 17.8, Figs. 10–11)—rarely on the substrate mycelium; *cell walls contain meso-diaminopimelic acid without glycine or arabinose* (Type III)

Microbispora *p. 859*

Conidia produced in chains..25

22. Conidiophores dichotomously branched, borne on both aerial and substrate mycelium which also branches dichotomously; optimum temperature of growth lies between 45 C and 58 C. *Cell walls contain meso-diaminopimelic acid with* (Type II—one strain) *or without glycine* (Type III—most strains)..............**Actinobifida** *p. 856*

Note: *1. Conidia of two species examined have the properties and structure of bacterial endospores.*

2. **Thermomonospora** *spp. show occasional dichotomous branching of the conidiophores only on the aerial mycelium.*

Not as above...23

23. No aerial mycelium produced, or present irregularly as a sparse white to grayish bloom; branching of the sporophore is monopodial or less commonly sympodial; mycelium septate; no growth between 50 C and 65 C; spore masses dark brown to black, waxy or moist, never powdery; *cell walls contain meso-diaminopimelic acid and glycine* (Type II)..**Micromonospora** *p. 846*

Note: *THE MANUAL lists two species in the addendum to* **Micromonospora** *which could be consulted at this point. One,* **Promicromonospora citrea,** *produces aerial mycelium and single spores either sessile or on short conidiophores on the substrate mycelium. The spore masses are sarcina-like and yellow: the cell wall composition is not that of* **Micromonospora.** *The other,* **Actinomonospora lusitanica,** *produces conidia on the apices of the hyphae and also on conidiophores laterally on the hyphae. Cell walls contain meso-diaminopimelic acid without glycine or arabinose (i.e. Type III cf.* **Micromonospora** *Type II).*

Aerial mycelium is produced; sporulating colonies white, yellow, yellow-brown or green; optimum growth temperature lies between 45 C and 65 C...............24

24. Conidia produced on aerial mycelium only; *cell walls contain meso-diaminopimelic acid with* (Type IV) *or without* (Type III) *galactose and arabinose*

Thermomonospora *p. 858*

Conidia produced on both aerial and substrate mycelium: *cell wall may be Type III*

Thermoactinomyces *p. 855*

25. Conidia produced basipetally in chains from both substrate and aerial mycelium; *cell walls contain meso-diaminopimelic acid, arabinose and galactose* (Type IV)

Micropolyspora *p. 861*

Note: *This genus includes species of the genus* **Thermopolyspora** *Henssen.*

Conidia produced from aerial mycelium; the aerial hyphae are produced first and then undergo a process of internal reorganization with the simultaneous production of chains of conidia within the confines of the outer walls of the original hyphae which may then disintegrate to release the conidia, sometimes with portions of the old hyphal cells still attached; *cell walls contain LL-diaminopimelic acid and glycine* (Type I)...26

26. Conidia borne on verticillate conidiophores..........**Streptoverticillium** *p. 829*

Conidia not borne on verticillate conidiophores...........**Streptomyces** *p. 748*

27. Sporangia small, ranging from 1.5–3.6 μm wide and 2–9 μm long and containing only 1–5

sporangiospores; produced laterally and sometimes terminally from either aerial or substrate mycelium or both...28

Sporangia variable in size but usually large; produced terminally or laterally on special aerial branches produced from the substrate mycelium: contain many sporangiospores..31

28. Sporangia produced on both substrate and aerial mycelium.......................29

Sporangia produced on aerial mycelium only.................................30

Sporangia produced on substrate mycelium only; no or only sparse aerial mycelium formed. Sporangia, 1.0–1.2 μm by 4.0–6.0 μm, produced in clusters on the surface of the agar, and borne on short sporangiophores produced from the substrate mycelium (Plate 17.8, Figs. 5–9); each sporangium contains a single row of 3–5 sporangiospores; also large globose spores or "sporangioles" produced on the mycelium in the agar; spores motile with polar or peritrichous flagella; cell walls contain *meso-diaminopimelic acid*............................**Dactylosporangium** *p. 721*

29. Sporangia, 1.5–3.6 μm by 2–9 μm, are produced on short sporangiophores from both the substrate and aerial mycelium; 2–5 sporangiospores are produced simultaneously by division of an intrasporangial hypha; no arthrospores produced; cell wall contains *LL-diaminopimelic acid and glycine* (Type I)...........**Microellobosporia** *p. 843*

Sporangia club-like, 2–2.5 μm by 5 μm, produced *terminally* on aerial hyphae 1.2–1.5 μm in diameter and also on the substrate hyphae; in addition chains of conidiospores are produced on another type of aerial hyphae 1.0–1.2 μm in diameter; a single chain of diplococcal-like sporangiospores produced in each sporangium, motile with a single polar flagellum; cell wall composition not stated.........**Kitasatoa** *p. 722*

30. Sporangia 1.0–1.2 μm by 6–8 μm, formed singly along the aerial mycelium or in clusters on side branches, sessile or on sporangiophores 1–3 μm long; each sporangium contains two longitudinally arranged spores; (Plate 17.8, Fig. 4); spores motile with peritrichous flagella; cell walls contain *meso-diaminopimelic acid*

Planobispora *p. 720*

Sporangia, 1.0–1.5 μm by 4.5–5.5 μm, formed only on aerial mycelium; sporangiospores fusiform and produced singly in each sporangium; sporangia produced singly or in rows (Plate 17.8, Figs. 1–3) on characteristically arched hyphae; spores liberated by an operculum at the tip of the sporangium, motile with peritrichous flagella; cell walls contain *meso-diaminopimelic acid*...............**Planomonospora** *p. 719*

31. Sporangiospores arranged in parallel chains in the sporangia; sporangiospores motile..32

Sporangiospores are arranged in coiled chain(s) in the sporangia, sporangiospores motile or non-motile...33

32. Sporangia bottle-shaped, flask-shaped, bell-shaped, digitate or lobate; 5.0–20 μm by 8.0–30 μm; spores 0.5–1.0 μm, by 2.0–4.0 μm, motile with polytrichous flagella (Plate 17.7), produced in chains of 6–10 spores and released from the sporangium following rupture of the sporangium caused by swelling of an intersporal substance; sporangia persistent; good growth on Czapek agar and on peptone Czapek agar the latter supporting colonies 1.5–2.0 cm in diameter, orange, brown, green brown to black, purple-brown in color and of leathery consistency; absorption peaks of intrasporal pigments at 474 (maximum) 447, 504 and 418 nm: apparently no predilection for hair

Ampullariella *p. 716*

Sporangia globose to cylindrical, 2–24 μm by 2–35 μm; spores 0.3–0.7 μm by 0.8–1.5 μm, motile with a single polar flagellum, produced in chains of up to 30 spores and released by rupture of the sporangium which does not usually persist after spore release; do not grow on Czapek's agar but produce colonies 1–2 mm in diameter on peptone Czapek agar, which are yellow to yellow-gray in color and of a soft pasty texture; absorption peaks of intrasporal pigments at 452 (maximum) 425, 479 and 400 nm. Keratinophilic on primary isolation...............**Pilimelia** *p. 718*

Note: *Additional information and the base structure of this section of the key has been derived from a paper by Wilma D. Kane (J. Elisha Mitchell Sci Soc 82: 221 (Table 1), 1966).*

33. Sporangiospores motile...34

Sporangiospores non-motile (Plate 17.7); spherical, subspherical or short straight or bent rods, 1.8–2.0 μm in diameter; sporangia are spherical to ovoid 7–48 μm in diameter (usually between 7–20 μm) produced apically or at random laterally on the mycelium; vertical palisaded hyphae are not produced; colonies are white, pink to red, pink to olive-buff, yellow, orange, and yellow-brown

Streptosporangium   *p. 711*

34. Sporangiospores spherical to subspherical, 1.0–1.5 μm in diameter, motile with a tuft of polar flagella (Plate 17.7). Sporangia usually spherical to subspherical, 6–50 μm in diameter but may be irregular (*A. utahensis*) and only 5–18 μm at the widest part, produced at the tips of vertical hyphae arranged in palisades (except *A. missouriensis*) and arising, within the colony, from loosely arranged hyphae in the base of the somewhat leathery colony produced on Czapek's agar. Colonies orange to orange-red, yellow or purple..**Actinoplanes**   *p. 708*

Sporangiospores rod-shaped..35

35. Sporangiospores straight, curved, or spiral rods developed from one or more coiled hyphae within the sporangium; weakly but definitely motile by 1–3 subpolar flagella; sporangia spherical, subspherical to elongate, 5–24 μm in diameter (average 10 μm); and borne on branched or unbranched sporangiophores. The mycelium also forms compact regular and irregular coils without a surrounding walls, and such coils, when flooded, at times break up into motile spores; colonies white, pale yellow, light gray, gray or bright blue........................**Spirillospora**   *p. 711*

Sporangiospores straight rods, 0.5–0.7 μm by 1–1.5 μm, motile by two or three polar flagella (Plate 17.7). Sporangia very irregular in shape and much lobed, 6–25 μm wide by 8–15 μm high; good growth on Czapek's agar of brilliant apricot-orange or orange chrome colonies 10–13 mm wide in 6 weeks

Amorphosporangium   *p. 715*

36. Motile at 37 C or 20–25 C...37

Non-motile..41

37. Pathogenic to plants, causing wilts and/or leaf spots...........................38

Not pathogenic or not known to be pathogenic to plants.........................39

38. Yellow or pink colonies, polar to subpolar or lateral flagella

Corynebacterium   *p. 611*

(see *C. tritici, C. flaccumfaciens, C. oortii, C. betae, C. poinsettiae, C. ilicis*, and possibly *C. michiganense*)

Cream to colorless colonies, polar flagella............**Corynebacterium**   *p. 616*

(see *C. hypertrophicans, C. humuli*)

Cream to colorless, peritrichous flagella....possibly **Erwinia carnegiana**   *p. 338*

39. No acid produced from carbohydrates or polyols; regular unbranched rods with rounded ends, 0.8 μm by 3–4 μm (Plate 17.1; Figs. 13–14); motile with numerous peritrichous flagella; catalase positive; oxidase negative; nitrates not reduced to nitrites; oxidative, producing acid from ethanol but not glycerol; tests for hydrolysis of starch, gelatin and esculin and for the production of urease, lecithinase, H_2S and indole are all negative...**Kurthia**   *p. 631*

Not as above..40

40. Glucose oxidized and fermented; acid but no gas produced from glucose, fructose, maltose, mannose, cellobiose, trehalose, salicin, amygdalin, starch and dextrin; esculin is hydrolyzed and acid produced; rods small, not acid-fast, occurring singly or in short chains of 3–5 cells in young smooth cultures, filamentous in rough colonies and may show coryneform arrangements from old colonies; actively motile at 20–25 C with 3–5 peritrichous flagella; 3 of 4 species produce fewer flagella (often one or none) at 37 C; indole and urease not produced; gelatin and casein not hydrolyzed; catalase positive; G + C content of DNA 38 moles % (except *L. denitrificans* = 56 moles %). Cell wall of *L. monocytogenes* contains DL-diaminopimelic acid

Listeria   *p. 593*

Not as above.................................. 41

41. **Note:** *At this point the remainder of the genera with some motile species are merged with the non-motile ones for further examination.*

Acid-fast..**Mycobacterium**   *p. 682*

Not acid-fast...42

42. Cells have a definite cyclic development. In mature colonies the cells are mainly coccal. On transplantation to a new medium they germinate, often from more than one point, to produce rods which elongate and divide. At the point of division, growth of cells often continues at an angle to the original axis. When the angular extensions reach the size of the original rod division occurs. Repeated growth and division gives the appearance, in undisturbed microcolonies, of branching which does not occur. As the culture ages the rods become shorter and finally coccoid (Plate 17.1, Figs. 1–11): not acid-fast; cellulose is not hydrolyzed; rare strains are motile; catalase positive...**Arthrobacter**   *p. 618*

Not as above...43

43. Cellulose is hydrolyzed in 0.5% peptone; rods 0.5 μm by 0.7–2.0 μm; straight, angular or slightly curved, occasionally club-shaped and beaded and arranged in coryneform fashion (Plate 17.1, Fig. 12); flagella polar or subpolar or lateral, sometimes absent
Cellulomonas   *p. 629*

Cellulose is not hydrolyzed...44

44. Organisms 0.2–0.4 by 0.5–2.5 μm; long filaments common in rough colonies; occur singly and also in chains; pinpoint transparent colonies on agar in 24 hours at 37 C, extending on further incubation to 1.5 mm; oxidative, not fermentative; acid only produced from glucose and lactose and some other carbohydrates; esculin not hydrolyzed; final pH in glucose broth, approximately 6.0; hydrogen sulfide is produced; catalase negative; causes swine erysipelas, human erysipeloid, mouse septicemia, and infections in birds and fish..............................**Erysipelothrix**   *p. 597*

Not as above..45

45. Rods 0.4–0.7 by 1–3 μm; show granular staining with methylene blue; arranged in angular fashion similar to the corynebacteria; only two species were recorded in the 7th edition of THE MANUAL, both of which produced acid from glucose, fructose and mannose; catalase positive; resist heating to 72 C for 15 min; normally found in dairy products and equipment.............................**Microbacterium**   *p. 628*

Note: *This entry is retained here to indicate the basis upon which this genus was erected: See* MANUAL *for discussion on the status of the genus.*

Not as above..46

46. Straight or curved rods, occurring singly or in chains; rare strains motile; lactate is not metabolized; glucose is converted wholly to lactic acid (homofermentative species) or to a mixture of acids of which lactic acid forms at least 50%, the additional products being acetic, formic, and succinic acids, CO_2 or ethanol; catalase negative; cytochromes absent..................................**Lactobacillus**   *p. 576*

Not as above; organisms pleomorphic, often arranged in palisades and chinese letter forms...47

47. Organisms grow on agar under aerobic conditions only if a heavy inoculum is used; highly pleomorphic and often appearing in coryneform arrangement; propionic acid and acetic acids are the major products of glucose fermentation: lactate may be converted to propionate; catalase positive; non-motile; cell walls contain LL- *or meso-diaminopimelic acid* or *neither* (P. *lymphophilum*). The G + C content of the DNA ranges from 59–66 moles %.......................**Propionibacterium**   *p. 633*

Not as above...................................**Corynebacterium**   *p. 602*

Note: *Some species assigned to the genus* **Brevibacterium** *may terminate here; also strains of* **Cellulomonas** *which fail to hydrolyze cellulose.*

48. Endospores produced...................................**Clostridium**   *p. 551*

Endospores not produced...49

49. Organism produces a branching mycelium in young microcolonies.................50

Organisms do not branch or branching is limited to simple branches or bifurcations of isolated rods in a population...52

50. Branching mycelium disintegrates into rods and cocci: no conidia or sporangiospores produced...**51**

Branching mycelium is persistent; condidia are produced singly from short conidiospores on substrate mycelium: no aerial mycelium......**Micromonospora**   *p. 846*

51. Products of glucose fermentation are primarily propionic and acetic acid and CO_2; catalase negative; cell walls contain LL-*diaminopimelic acid, alanine glutamic acid and glycine*...**Arachnia**   *p. 668*

Products of glucose fermentation include acetic, formic, lactic and succinic acid but not propionic; cell walls *contain no diaminopimelic acid or arabinose*

Actinomyces   *p. 660*

52. Organisms rod-shaped, occurring in pairs, chains or filaments; obtain energy by the reduction of carbon dioxide to methane with hydrogen and possibly formate as the electron donor; glucose not fermented............**Methanobacterium**   *p. 473*

Not as above; glucose fermented...**53**

53. Propionic and acetic acids are the major products of glucose fermentation; cell walls contain LL- or *meso-diaminopimelic acid* (except *P. lymphophilum*)

Propionibacterium   *p. 633*

Acetic and lactic acids are produced in the molar ratio of *ca.* 3:2 as the major products of glucose fermentation; branched and bifurcated rods common; catalase negative; cell wall contains *no diaminopimelic acid or arabinose*....**Bifidobacterium**   *p. 669*

Lactic acid is the sole major product of glucose fermentation being produced solely or at least at a level of 50% of the glucose metabolized; catalase and cytochromes absent; organisms occur singly and in chains..............**Lactobacillus**   *p. 576*

Butyric, acetic and formic acids constitute the major products of glucose fermentation

Eubacterium   *p. 641*

Note: *The genus Eubacterium in this edition of* THE MANUAL *includes many of the species distributed amongst the genera Eubacterium, Ramibacterium, Cillobacterium, Catanebacterium and Butyribacterium in the 7th edition of* THE MANUAL. *The others have been reassigned to other genera, notably Lactobacillus or Clostridium, or discarded as unrecognizable.*

54. Some of the branching organisms do not fragment and produce no spores. They are impossible to identify. Some idea of their affinities may be obtained by cell wall analysis. Others form sclerotia and the genus *Chainia* (not recognized in this edition of THE MANUAL) was one of these.

INDEX OF SCIENTIFIC NAMES OF BACTERIA

Key to the fonts and symbols used in this index:

Nomenclature
CAPITALS:

BOLDFACE	accepted* names of taxa higher than genera
ROMAN	unaccepted names of taxa higher than genera

Lower Case:

Boldface	accepted names of genera, species and subspecies
Roman	unaccepted names of genera, species and subspecies, synonyms, *nomina ambigua* (n.a.), *nomina confusa* (n.c.), *nomina dubia* (n.d.), *nomina incertae sedis* (i.s.), and *species inquirendae* (*s.i.*)

Pagination

Boldface	pages on which accepted taxa are mentioned, e.g. type species, in Keys and Tables.
Roman	page reference to names of unaccepted taxa or incidental mention of accepted taxa
Italics	indicates page on which the organism is illustrated

Parentheses indicate the page on which the description of an accepted or unaccepted taxon begins.

Specific epithets, except as noted below, are listed individually and under the genus. Only accepted specific epithets and *species incertae sedis* are listed individually for the genera *Actinomyces*, *Bacillus*, *Pseudomonas*, *Streptomyces*, and *Xanthomonas*. The many synonyms in the unaccepted genus *Bacterium* and the serotypes of *Salmonella* appear only under the generic name.

Specific epithets which are orthographic variants, *lapsi calami*, incorrect endings, etc. are generally not listed.

---

* NOTE: The term "accepted" as used here implies: (1) that the taxon has been accepted for inclusion in THE MANUAL by the Advisory Committee for the group and by the author(s) and (2) that the name used is deemed to be the correct name by the author(s) and editors.

Salmonella—*Continued*
  brunei, 303
  budapest, 301
  bukavu, 311
  bukuru, 303
  bulawayo, 311
  bullbay, 306
  bunnik, 312
  burgas, 308
  bury, 301
  businga, 302
  butantan, 304
  butare, 315
  buzu, 307
  cairina, 305
  cairns, 313
  calabar, 306
  caledon, 301
  california, 301
  calvinia, 302
  camberene, 310
  cambridge, 305
  canada, 301
  canastel, 304
  cannonhill, 306
  cannstatt, 306
  canoga, 305
  cape, 302
  caracas, 307
  carletonville, 310
  carmel, 308
  carnac, 309
  carno, 306
  carrau, 298, 307
  casablanca, 313
  ceres, 309
  cerro, 298, *300*, 309
  ceyco, 304
  chagoua, 307
  chailey, 303
  chameleon, 308
  champaign, 311
  chandans, 306
  charity, 307
  chersina, 313
  chester, 301
  chicago, 309
  chincol, 303
  chingola, 306
  chinovum, 312
  chittagong, 306
  cholerae-suis, 298, 299, **(300),** 302
    var. **kuzendorf, (300)**
  christiansborg, 312
  chudleigh, 305
  clackamas, 301
  claibornei, 304
  clerkenwell, 305
  cleveland, 303
  clichy, 305
  clifton, 306
  clovelly, 313
  cocody, 303
  coeln, 301

  coleypark, 302
  coli, 297
  colindale, 302
  colobane, 306
  colombo, 311
  colorado, 302
  concord, 302
  congo, 307
  constantia, 308
  coquilhatville, 305
  corvallis, 303
  cotham, 309
  croft, 309
  cubana, 307
  curacáo, 303
  dahlem, 313
  dahomey, 313
  dakar, 298, 309
  dallgow, 306
  dan, 314
  dar-es-salaam, 318
  daressalamensis, 318
  daytona, 302
  degania, 311
  delplata, 318
  dembe, 310
  demerara, 307
  denver, 302
  derby, 301
  dessau, 306
  detroit, 312
  deversoir, 313
  diguel, 306
  diogoye, 303
  diourbel, 309
  djakarta, 298, 314
  djelfa, 303
  djermaia, 309
  djibouti, 308
  djugu, 302
  doncaster, 303
  donna, 310
  dougi, 314
  dou lassame, 309
  dresden, 309
  driffield, 311
  drypool, 305
  dublin, *298, 299*, 304
  dubrovnik, 311
  duesseldorf, 303
  dugbe, 313
  duisburg, 301
  duivenhoks, 304
  durban, 304
  durbanville, 302
  durham, 307
  duval, 311
  ealing, 310
  eastbourne, 304
  eberswalde, 309
  ebrie, 310
  echa, 310
  edinburg, 302
  edmonton, 303

Addendum and Corrigendum

EIGHTH EDITION, BERGEY'S MANUAL OF DETERMINATIVE BACTERIOLOGY

We regret that a number of errors and omissions were not detected until after publication. Some 5200 references were included but some 450 more were inadvertently omitted from the consolidated list. Those for the family *Micrococcaceae* were presumably lost in the mails between Ottawa and Ames and the loss was never detected. As soon as it became apparent that omissions were scattered throughout the volume, authors were asked to check their contributions and most have responded. References brought to our attention or detected by us until the first week of January, 1975, are included in an Addendum. A few others are known but we have been unable to obtain the exact citation in time for the second printing.

Errors that have been detected or brought to our attention are listed in a Corrigendum. A few authors have requested changes in Tables or text.

A number of accents have been left off names and words in foreign languages. We regret that only some can be corrected at this time.

Letters have been received from colleagues that work has been done on species not listed in the Manual and that further information was available on species listed. Such omissions, although unfortunate, are inevitable. It was evident before the Manual was ready for the press that a revision of many manuscripts could be made. Before the Manual appeared, one of our authors proposed that *Listeria grayi* and *L. murrayi* should be transferred to a new genus *Murraya* (*International Journal of Systematic Bacteriology 24*: 177–185) and his co-author now feels *L. denitrificans* belongs in yet another genus.

The editor is chastened by the extent of the Addendum and Corrigendum and apologizes for the inconvenience to readers. Your comments and criticisms are still welcomed.

THE EDITORIAL BOARD

Addendum

Aasen, J. and P. Oeding. 1971 Antigenic studies on *Staphylococcus epidermidis*. Acta Pathol Microbiol Scand *79B:* 827–834.

Abram, D. 1969 Basal structures and attachment of the polar flagella of *Spirillum serpens*. Bacteriol Proc G72.

Abramson, C. and H. Friedman. 1967 Enzymatic activity of primary isolates of staphylococci in relation to antibiotic resistance and phage type. J Infect Dis *117:* 242–248.

Adams, M. M., J. N. Adams and G. H. Brownell. 1970 The identification of *Jensenia canicruria* Bisset and More as a mating type of *Nocardia erythropolis* (Gray and Thornton) Waksman and Henrici. Int J Syst Bacteriol *20:* 133–147.

Ali-Cohen, C. H. 1889 Eigenbewegung bei Mikrokokken. Zentralbl Bakteriol Parasitenk Infectionskr Abt I *6:* 33–36.

Anderson, A. W., H. O. Nordan, R. F. Cain, G. Parrish and D. Duggan. 1956 Studies on radio-resistant Micrococcus 1. Isolation, morphology, cultural characteristics and resistance to gamma radiation. Food Technol *10:* 575–578.

Anderson, G. E. 1966 Identification of *Beijerinckia* from Pacific Northwest soils. J Bacteriol *91:* 2105–2106.

Andrewes, F. W. and M. H. Gordon. 1907 Report on the biological characters of the staphylococci pathogenic for man. Annu Rep Med Officer Loc Govt Bd 1905–1906: 543–560.

Arai, S., K. Y. Yuri, A. Kudo, M. Kikuchi, K. Kumagai and N. Ishida. 1967 Effect of antibiotics on the growth of various strains of *Mycoplasma*. J Antibiot (Tokyo) Ser A. *20:* 246–253.

Archibald, A. R., J. Baddiley and G. A. Shaukat. 1968 The glycerol teichoic acid from walls of *Staphylococcus epidermidis* I2. Biochem J *110:* 583–588.

Arvidson, S., T. Holme and T. Wadstrom. 1970 Formation of bacteriolytic enzymes in batch and continuous culture of *Staphylococcus aureus*. J Bacteriol *104:* 227–233.

Asselineau, J. 1966 The Bacterial Lipids. Herman, Paris.

Babenzien, H.-D. 1965 Über Vorkommen und Kultur von *Nevskia ramosa* in H.-G. Schlegel (Ed) Anreicherungskultur und Mutantenauslese. Zentralbl Bakteriol Parasitenk Infektionskr Abt. I, *Suppl. 1:* 111–116.

—. 1967 Zur Biologie von *Nevskia ramosa*. Z Allg Mikrobiol *7:* 89–96.

Baboolal, R. 1969 Cell wall analysis of oral filamentous bacteria. J Gen Microbiol *58:* 217–226.

Baer, E. F. 1968 Proposed revision of the method for isolating coagulase positive staphylococci from foods. J O A C *51:* 865–866.

Bailey, R. K., S. Wyles, M. Dingley, F. Hesse and G. W. Kent. 1970 The isolation of high catalase *Mycobacterium kansasii* from tap water. Amer Rev Resp Dis *101:* 430–431.

Baird-Parker, A. C. 1969 The use of Baird-Parker's medium for the isolation and enumeration of *Staphylococcus aureus*. In D. A. Shapton and G. W. Gould (Editors) Isolation Methods for Bacteriologists. Society for Applied Bacteriology Technical Series No. 3.

Academic Press, London and New York, pp. 1–8.

—. 1970 The relationship of cell wall composition to the current classification of staphylococci and micrococci. Int J Syst Bacteriol *20:* 483–490.

Balashova, V. V. 1969 The relationship of *Gallionella* to *Mycoplasma*. Dokl Akad Nauk SSSR *184:* 1429–1432.

Barkdale, W. L. 1970 *Corynebacterium diphtheriae* and its relatives. Bacteriol Rev *34:* 378–422.

Beaudoin, J. 1966 Contribution à l'étude de la genetique de *Nocardia restrictus*. Thesis. Fac de Med, Univ. Montreal, pp. 1–108.

Beerens, H., L. Fievez and P. Wattre. 1971 Observations concernant 7 souches appartenant aux espèces *Sphaerophorus necrophorus*, *Sphaerophorus funduliformis*, *Sphaerophorus pseudonecrophorus*. Ann Inst Pasteur (Paris) *121:* 37–41.

—, A. Gerard and J. Guillamme. 1957 Étude de 30 souches de *Bifidobacterium bifidum* (*Lactobacillus bifidus*). Caracterisation d'une variété buccale. Comparison avec les souches d'origine fecale. Ann Inst Pasteur Lille *9:* 77–85.

—, P. Wattre, T. Shinjo and C. Romond. 1971 Premiers résultats d'un essai de classification sérologique de 131 souches de *Bacteroides* du groupe *fragilis* (*Eggerthella*). Ann Inst Pasteur (Paris) *121:* 187–198.

Beeson, P. B. 1943 The problem of the etiology of rat-bite fever. J Amer Med Ass *123:* 332–334.

Beijerinck, M. W. 1904 Ueber die Bakterien welche sich im Dunkeln mit Kohlensäure als Kohlenstoffe quelle ernahren Können. Zentralbl Bakteriol Parasitenk Infektionskr Abt II, *11:* 593–599.

Bergdoll, M. S. 1972 The enterotoxins. In J. O. Cohen (Editor) The Staphylococci. J. Wiley and Sons Inc., New York, pp. 301–331.

Bergan, T., K. Bøvre and B. Hovig. 1970 Priority of *Micrococcus mucilaginosus* Migula 1900 over *Staphylococcus salivarius* Andrewes and Gordon 1907 with proposal of a neotype strain. Int J Syst Bacteriol *20:* 107–113.

Bergey, D. H. 1904 Source and nature of bacteria in milk. Bull Penn Agr Exp Sta *125:* 9–40.

— and R. S. Breed. 1948 Genus *Flavobacterium*. In Breed, Murray and Hitchens (Editors) Bergey's Manual of Determinative Bacteriology, 6th Ed. The Williams & Wilkins Co., Baltimore, pp. 427–442.

Bernaerts, M. J. and J. De Ley. 1963 A biochemical test for crown gall bacteria. Nature *197:* 406–407.

Bernheimer, A. W. 1968 Cytolytic toxins of bacterial origin. Science *159:* 847–851.

Bibby, B. G. 1935 A study of the filamentous bacteria of the mouth. Ph.D. thesis, University of Rochester, Rochester, N. Y.

— and G. P. Berry. 1939 A cultural study of filamentous bacteria obtained from the human mouth. J Bacteriol *38:* 263–274.

Blackstock, R., R. M. Hyde and F. C. Kelly. 1968 Inhibition of fibrinogen reaction by polysaccharide of encapsulated *Staphylococcus aureus*. J Bacteriol *96:* 799–803.

Blair, J. E. and M. T. Parker. 1967 Report of the subcommittee on phage-typing of staphylococci to the International Committee on Nomenclature of Bacteria, Moscow, July 1966. Int J Syst Bacteriol *17:* 113–125.

Blaurock, A. E. and W. Stoechenius. 1971 Structure of the purple membrane. Nature New Biol *233:* 152–154.

Blevins, W. T., J. J. Perry and J. B. Evans. 1969 Growth and macromolecular biosynthesis by *Micrococcus sodonensis* during the utilization of glucose and lactate. Can J Microbiol *15:* 383–388.

Boháček, J., M. Kocur and T. Martinec. 1967 DNA base composition and taxonomy of some micrococci. J Gen Microbiol *46:* 369–376.

—, —, and —. 1968 Deoxyribonucleic acid base composition of *Sporosarcina urea*. Arch Mikrobiol *364:* 23–27.

—, — and —. 1968 Deoxyribonucleic acid base composition of some marine and halophilic micrococci. J. Appl Bacteriol *31:* 215–219.

Bojalil, L. F. and J. Cerbón. 1959 Schema for the differentiation of *Nocardia asteroides* and *Nocardia braziliensis*. J Bacteriol *78:* 852–857.

Bönicke, R. and S. E. Juhasz. 1965 *Mycobacterium diernhoferi* n.sp., eine in der Umgebung des Rindes häufig vorkommende neue *Mycobacterium* Species. Zentralbl Bakteriol Parasitenk Infektionskr Abt Orig *197:* 292–294.

Bordet, C., A. H. Etemadi, G. Michel and E. Lederer. 1965 Structure des acides nocardiques de *Nocardia asteroides*. Bull Soc Chim Fr *1965:* 234–235.

Bos, L. 1970 Mycoplasma's, een nieuw hoofdstuk in de planteziektenkunde? Gewasbeschermig *1:* 45–54.

Boswell, P. A., G. P. Batstone and R. G. Mitchell. 1972 The oxidase reaction in the classification of the *Micrococcaceae*. J Med Microbiol *5:* 267–269.

Bovell, C. 1967 The effect of sodium nitrite on the growth of *Micrococcus denitrificans*. Arch Mikrobiol *59:* 13.

Bowden, G. H. 1969 The components of the cell walls and extracellular slime of four strains of *Staphylococcus salivarius* isolated from human dental plaque. Arch Oral Biol *14:* 685–697.

Braun, K. 1927 Bericht über das auftreten von Schädlingen und Krankheiten im Obstbau im Regierungsbezirk Stade wärend der Monate Juni, Juli, August 1927. Die Landwirtschaft 41, 42. 2 pp. (from Rev Appl Mycol *7:* 177. 1928).

Breed, R. S. 1952 The type species of the genus *Micrococcus*. Int Bull Bacteriol Nomencl Taxon *2:* 85–88.

Brock, T. D., K. M. Brock, R. T. Belly and R. L. Weiss. 1972 Sulfolobus: A new genus of sulfur-oxidizing bacteria living at low pH and high temperature. Arch Mikrobiol *84:* 54–68.

Brown, J. M., L. K. Georg and L. C. Waters. 1969 The laboratory identification of *Rothia dentocariosa* and its occurrence in human clinical materials. Appl Microbiol *17:* 150–156.

Brown, R. W., O. Sandvik, R. K. Scherer and D. L. Rose. 1967 Differentiation of strains of *Staphylococcus epidermidis* isolated from bovine udders. J Gen Microbiol *47:* 273–287.

Brownell, G. H., J. N. Adams and S. G. Bradley. 1967 Growth and characterization of nocardiophages for *Nocardia canicruria* and *No-*

cardia erythropolis mating types. J Gen Microbiol *47:* 247–256.

Brussoff, A. 1906 *Ferribacterium duplex*, eine stäbchenförmige Eisenbakterie. Zentralbl Bakteriol Parasitenk Infektionskr Abt II *45:* 547–554.

Buchanan, R. E. 1917 Studies on the nomenclature and classification of the bacteria. III. The families of the *Eubacteriales*. J Bacteriol *2:* 347–350.

Buddenhagen, I. W. 1965 The relation of plant pathogenic bacteria to the soil. *In* K. F. Baker and W. C. Snyder, (Editors) Ecology of Soil-Borne Pathogens: Prelude to Biological Control. Univ. California Press, Berkeley, California p. 269.

— and T. A. Elsasser. 1962 An insect spread bacterial wilt epiphytotic on Bluggoe banana. Nature *194:* 164.

— and A. Kelman. 1964 Biological and physiological aspects of bacterial wilt caused by *Pseudomonas solanacearum*. Ann Rev Phytopathol *2:* 203.

Bulleid, A. 1925 An experimental study of *Leptothrix buccalis*. Brit Dent J *46:* 289–300.

Burchard, G. 1898 Beiträge zur Morphologie und Entwickelungsgeschichte der Bakterien. Arb Bakt Inst Karlsruhe *2 (Heft 1):* 1–72.

Campbell, J. N., M. Leyh-Bouille and J. M. Ghuysen. 1969 Characterization of *Micrococcus lysodeikticus* type of peptidoglycan in walls of other *Micrococcaceae*. Biochemistry *8:* 193–200.

Canale-Parola, E. 1970 Biology of the sugar-fermenting sarcinae. Bacteriol Rev *34:* 82–97.

—, S. L. Rosenthal and D. G. Kupfer. 1966 Morphological and physiological characteristics of *Spirillum gracile* sp.n. Antonie Van Leeuwenhoek J Microbiol Serol *32:* 113–124.

Carter, G. R. 1952 The type specific antigen of *Pasteurella multocida*. Can J Med Sci *30:* 48.

Casman, E. P. 1967 Staphylococcal food poisoning. Health Lab Sci *4:* 199–206.

Casper, R. 1969 Mykoplasmen als Erreger von Pflanzenkrankheiten. Nachrichtenblatt Deutsch Pflanzenschutzdientes *21:* 177–182.

Castellani, A. 1927–1928 Some new micrococci: *M. levulosineritis, M. viscidus, M. enteroideus, M. afermentans*. Proc Soc Exp Biol Med *25:* 536–537.

—. 1955 Note preliminaire sur un nouveau micocoque isolé d'une dermatite axillaire superficielle tropicale. Ann Inst Pasteur *89:* 475–480.

Castelnuovo, G. and M. Morellini. 1962 The antigens of mycobacteria and their identification by immunoelectrophoretic analysis. Amer Rev Resp Dis *92 (#6, Pt 2):* 29–33.

Castets, M., N. Rist and H. Boisvert. 1969 La variété africaine du bacille tuberculeux humain. Medicine D'Afrique Noire No. 4. 321–322.

Cayton, H. R. and N. W. Preston. 1955 *Spirillum mancuniense* n. sp. J Gen Microbiol *12:* 519–525.

Centifanto, Y. M. and W. S. Silver. 1964 Leaf-nodule symbiosis. I. Endophyte of *Psychotria bacteriophila*. J Bacteriol *88:* 776–781.

Chapman, J. A. and M. R. J. Salton. 1962 The surface envelope of *Lampropedia hyalina*. *In* Proc. Fifth Int Congr Electron Microscopy, Academic Press, New York.

—, R. G. E. Murray and M. R. J. Salton. 1963

The surface anatomy of *Lampropedia hyalina*. Proc Roy Soc London B Biol Sci *158:* 498–513.

Chatelain, R. and L. Second. 1966 Taxonomie numérique de quelque Brevibacterium. Ann Inst Pasteur (Paris) *111:* 630–644.

Chesbro, W. R. and K. Auborn. 1967 Enzymatic detection of the growth of *Staphylococcus aureus* on foods. Appl Microbiol *15:* 1150–1159.

Cheshire, F. R. and W. W. Cheyne. 1885 The pathogenic history and history under cultivation of a new Bacillus (*B. alvei*), the cause of a disease of the hive bee hitherto known as foul brood. J Roy Microscop Soc, Ser II *5:* 581–601.

Clark, P. H. and S. T. Cowan. 1952 Biochemical methods for bacteriology. J Gen Microbiol *6:* 187–197.

Clark-Walker, G. D. 1969 Association of microcyst formation in *Spirillum itersonii* with the spontaneous induction of a defective bacteriophage. J Bacteriol *97:* 885–892.

Cocchi, P. and A. Ulivelli. 1968 Meningitis caused by *Neisseria catarrhalis*. Acta Paediat Scand *57:* 451–453.

Cock, D. J. and W. H. Bowen. 1967 Occurrence of *Bacterionema matruchotii* and *Bacteroides melaninogenicus* in gingival plaque from monkeys. J Periodontal Res *2:* 36–39.

Cole, M. 1959 Bacterial rotting of apple fruit. Ann Appl Biol *47:* 601.

Coleman, R. M., L. K. Georg and A. Z. Rozzell. 1969 *Actinomyces naeslundii* as an agent of human actinomycosis. Appl Microbiol *18:* 420–426.

Collins, F. M. and J. Lascelles. 1962 The effect of growth conditions on oxidative and dehydrogenase activity in *Staphylococcus aureus*. J Gen Microbiol *29:* 531–535.

Cooke, J. V. and H. R. Keith. 1927 A type of ureasplitting bacterium found in the human intestinal tract. J Bacteriol *13:* 315–319.

Cooney, J. J. and O. C. Thierry. 1966 A defined medium for growth and pigment synthesis of *Micrococcus roseus*. Can J Microbiol *12:* 83–89.

Copeland, W. R. 1899 Report upon bacteriological and microscopical investigations. Rep Filt Comm Pittsb Penn App *4:* 333–351.

Corbaz, R., P. H. Gregory and M. E. Lacey. 1963 Thermophilic and mesophilic actinomycetes in mouldy hay. J Gen Microbiol *32:* 449–455.

Cowan, S. T. and K. J. Steel. 1965 Manual for the Identification of Medical Bacteria, 1st ed. Cambridge University Press, Cambridge.

Craigie, J. and C. H. Yen. 1937 V. Bacteriophages for *B. typhosus*. Trans Roy Soc Canada, Ser V, *31:* 79–87.

Criswell, B. S., J. H. Marston, W. A. Stenback, S. H. Black and H. L. Gardner. 1971 *Haemophilus vaginalis* 594, a Gram-negative organism? Can J Microbiol *17:* 865–869.

Cross, T. 1969 The monosporic actinomycetes. Ph.D. Thesis, Univ Bradford.

Cruickshank, R., J. P. Duguid and R. H. A. Swain. 1968 Medical Microbiology, 11th ed. 819 pp. Edinburgh.

Cummins, C. S. 1954 Some observations on the nature of the antigen in the cell wall of *Corynebacterium diphtheriae*. Brit J Exp Pathol *35:* 166.

Dack, G. M., L. R. Dragstedt, R. Johnson and N. B. McCullough. 1938 Comparison of *Bacterium necrophorum* from ulcerative colitis in man with strains isolated from animals. J Infect Dis *62:* 169–180.

Darland, G. and T. D. Brock. 1971 *Bacillus acidocaldarius* sp. nov., an acidophilic thermophilic spore-forming bacterium. J Gen Microbiol *67:* 9–15.

Dave, C. V., S. V. Gogate and P. N. Kaul. 1962 Some polyene antibiotics as cardiac stimulants. Hindustan Antibiot Bull *5:* 28–30.

Davis, B. R. and W. H. Ewing. 1966 The biochemical reactions of *Citrobacter freundii*. Comm Dis Center Publ, Atlanta, Georgia.

Davis, G. H. G. and A. C. Baird-Parker. 1959 The classification of certain filamentous bacteria with respect to their chemical composition. J Gen Microbiol *21:* 612–621.

Davis, N. S., G. J. Silverman and E. B. Masurovsky. 1963 Radiation-resistant pigmented coccus isolated from haddock tissue. J Bacteriol *86:* 294–298.

Davison, A. L. and J. Baddiley. 1964 Glycerol teichoic acids in walls of *Staphylococcus epidermidis*. Nature (London) *202:* 874.

—, —, T. Hofstad, N. Losnegard and P. Oeding. 1964 Teichoic acids in walls of staphylococci. Serological investigations on teichoic acids from walls of staphylococci. Nature (London) *202:* 872–874.

Dawes, E. A. and W. J. Holmes. 1958 Metabolism of *Sarcina lutea* 1. Carbohydrate oxidation and terminal respiration. J Bacteriol *75:* 390–399.

Delaporte, B. and A. Sasson. 1967 Étude des bactéries des sols arides du Maroc: *Brevibacterium halotolerans* n. sp. et *Brevibacterium frigerotolerans*, n. sp. C R Acad Sci Paris Ser D *264:* 2257–2260.

de Vries, W., S. J. Gerbrandy and A. H. Stouthamer. 1967 Carbohydrate metabolism in *Bifidobacterium bifidum*. Biochim Biophys Acta *136:* 415–425.

Dimitroff, V. T. 1926 *Spirillum virginianum* nov. spec. J Bacteriol *12:* 19–48.

Doi, Y., M. Terenaka, K. Yora and H. Asuyama. 1967 Mycoplasma- or PLT group-like microorganisms found in the phloem elements of plants infected with mulberry dwarf, potato witches' broom, aster yellows, or Paulownia witches' broom. Ann Phytopathol Soc Jap *33:* 259–266.

Dorokhova, L. A., N. S. Agre, L. V. Kalakutski and N. A. Krasil'nikov. 1969 Fine structure of sporulating hyphae and spores in a thermophilic actinomycete, *Micropolyspora rectivergula*. J Microsc (Paris) *8:* 845–854.

—, —, — and —. 1970 A study on morphology of two cultures belonging to the genus *Micropolyspora*. Mikrobiologiya *39:* 95–100 (in Russian).

Drake, C. H. 1965 Occurrence of Siderocapsa treubii in certain waters of the Niderrhein. Gewässer und Abwässer *H39/40:* 41–61.

Dubinina, G. A. 1969 Inclusion of *Metallogenium* among the *Mycoplasmatales*. Dokl Akad Nauk SSSR *184:* 1433–1436.

—. 1970 Untersuchungen über die Morphologie von *Metallogenium* und die Beziehungen zu *Mycoplasma*. Z Allg Mikrobiol *10:* 309–320.

Dunkelberg, W. E., Jr. and I. McVeigh. 1969 Growth requirements of *Haemophilus vaginalis*. Antonie Van Leeuwenhoek *35:* 129–145.

Durr, F. E., D. W. Smith and D. P. Altman. 1959 A comparison of the virulence of various known and atypical mycobacteria for chickens, guinea pigs, hamsters and mice. Amer Rev Resp Dis *80:* 876–885.

Duryee, F. L., H. D. Raj, C. H. Wang, A. W. Anderson and P. R. Elliker. 1961 Carbohydrate metabolism in *Micrococcus radiodurans.* Can J Microbiol *7:* 700–805.

Eadie, J. M. 1962 The development of rumen microbial populations in lambs and calves under various conditions of management. J Gen Microbiol *29:* 563–578.

Eagon, R. G. and H. W. Cho. 1965 Major products of glucose dissimilation by *Pseudomonas natriegens.* J Bacteriol *84:* 1209.

Edward, D. G. ff. 1950 An investigation of pleuropneumonia-like organisms isolated from the bovine genital tract. J Gen Microbiol *4:* 4–15.

Ehrenberg, C. G. 1832 Beiträge zur Kenntnis der Organization der Infusorien und ihrer geographischen Verbreitung besonders in Sibirien. Abh Deut Akad Wiss Berlin *1830:* 1–88.

Eisenberg, R. C. and J. B. Evans. 1963 Energy and nitrogen requirements of *Micrococcus roseus.* Can J Microbiol *9:* 633–642.

Elek, S. D. 1959 *Staphylococcus pyogenes* and its relation to disease. E and S Livingstone, Edinburgh and London.

— and E. Levy. 1950 Distribution of haemolysins in pathogenic and non-pathogenic staphylococci. J Pathol Bacteriol *62:* 541–554.

Endo, T. 1955 Studies on the genus *Nocardia.* J Antibiot (Tokyo) Ser B *8:* 163–169.

Engbaek, H. C., E. H. Runyon and A. G. Karlson. 1971 *Mycobacterium avium* Chester. Designation of the neotype strain. Int J Syst Bacteriol *21:* 192–196.

Erickson, S. K. and G. L. Parker. 1969 The electron-transport system of *Micrococcus luteus (Sarcina lutea).* Biochim. Biophys Acta *180:* 59–62.

Ernst, W. 1906 Über Pyelonephritis diphtherica bovis und die Pyelonephritisbazillen. Zentralbl Bakteriol Parasitenk Infektionskr Abt I *39:* 349–558. Ibid *40:* 79–91.

Evans, A. C. 1916 The bacteria of milk freshly drawn from normal udders. J Infect Dis *18:* 437–476.

—. 1918 Further studies on *Bacterium abortus* and related bacteria.

1. The pathogenicity of *Bacterium lipolyticum* for guinea pigs. J Infect Dis *22:* 576–579.

2. A comparison of *Bacterium abortus* with *Bacterium bronchisepticus* and with the organism which causes Malta fever. J Infect Dis *22:* 580–593.

Evans, J. B. and W. E. Kloos. 1972 Use of shake cultures in a semisolid thioglycollate medium for differentiating staphylococci from micrococci. Appl Microbiol *23:* 326–331.

Ewing, W. H. 1971 *Citrobacter freundii* and *Citrobacter diversus.* Communicable Disease Center, Atlanta, Georgia.

— and B. R. Davis. 1971 Biochemical characterization of *Citrobacter freundii* and *Citrobacter diversus.* National Communicable Disease Center, Atlanta, Georgia.

Farlow, W. C. 1880 On the nature of the peculiar reddening of salted codfish during the summer season. U S Fish Comm Rept for 1878: 969–974.

Fildes, P. 1920 New medium for growth of *B. influenzae.* Brit J Exp Pathol *1:* 129–130.

—, G. M. Richardson, B. C. J. G. Knight and G. P. Gladstone. 1936 A nutrient mixture suitable for the growth of *Staphylococcus aureus.* Brit J Exp Pathol *17:* 481–484.

Finegold, S. M. and L. G. Miller. 1968 Susceptibility to antibiotics as an aid in classification of Gram-negative anaerobic bacilli. *In* V. Fredette (Editor) The Anaerobic Bacteria. Inst. of Microbiology and Hygiene, Montreal University, Montreal, Quebec, pp. 139–145.

Florman, A. L. 1968 Symposium on nonchemotherapeutic approaches to control of staphylococcal infection. Bull N Y Acad Med *44:* 1195–1227.

Fornachon, J. C. M., H. C. Douglas and R. H. Vaughn. 1949 *Lactobacillus trichodes* nov. spec., a bacterium causing spoilage in appetizer and dessert wines. Hilgardia *19:* 129–132.

Frederiksen, W. 1970 *Citrobacter koseri* (n. sp.), a new species within the genus *Citrobacter,* with a comment on the taxonomic position of *Citrobacter intermedium* (Werkman and Gillen). Publ Fac Sciences Univ J E Purkyne, Brno *47:* 89–94.

Frost, A. J. 1967 Phage typing of *Staphylococcus aureus* from daily cattle in Australia. J Hyg *65:* 311–319.

Garabedian, M. S. 1969 A study of *Haemophilus vaginalis* Gardner and Dukes. Can J Med Technol *31:* 144–150.

Gardner, J. F. and J. Lascelles. 1962 The requirement for acetate of a streptomycin-resistant strain of *Staphylococcus aureus.* J Gen Microbiol *29:* 157–164.

Garrity, F. L., B. Detrick and E. R. Kennedy. 1969 Deoxyribonucleic acid base composition in the taxonomy of staphylococcus. J Bacteriol *97:* 557–560.

Georg, L. K. 1970 Diagnostic procedures for the isolation and identification of the etiological agents of Actinomycosis. Proceedings, Int Symp on Mycoses. Science Publ Pan American Health Organization No. *205:* 71–81.

— and M. R. Coleman. 1968 Comparative pathogenicity of Actinomyces species. Int Symp on the Taxonomy of the Actinomycetales *68:* 35–45.

Ghon, A., V. Mucha and R. Müller. 1906 Beiträge zur Kenntnis der anaëroben Bakterien des Menschen, IV. Zur Aetiologie der akuten Meningitis. Zentralbl Bakteriol Parasitenk Infektionskr Abt I, Orig *41:* 145.

Ghuysen, J.-M. and J. L. Strominger. 1963 Structure of the cell wall of *Staphylococcus aureus* strain Copenhagen 2. Separation and structure of disaccharides. Biochemistry *2:* 1119–1125.

Gibson, T. 1967 The status of the genus *Micrococcus.* Int J Syst Bacteriol *17:* 231–233.

Giesberger, G. 1936 Beiträge zur Kenntnis der Gattung *Spirillum* Ehrb. Inaug Dissertation Utrecht 1–136.

Gilmour, M. N. 1961 The classification of organisms termed *Leptotrichia (Leptothrix) buccalis.* II. Reproduction of *Bacterionema matruchotii.* Bacteriol Rev *25:* 142–151.

—. 1962 Preliminary studies on the anaerobically isolated counterparts of *Bacterionema matruchotii*. J Dent Res *41:* 929.

— and P. H. Beck. 1961 The classification of organisms termed *Leptotrichia* (*Leptothrix*) *buccalis*. III. Growth and biochemical characteristics of *Bacterionema matruchotii*. Bacteriol Rev *25:* 152–161.

— and B. G. Bibby. 1966 A synthetic medium for *Bacterionema matruchotii*. J Dent Res *45:* 158.

—, A. Howell and B. G. Bibby. 1961 Proposal for designation of neotype strains of *Leptotrichia buccalis* and *Bacterionema matruchotii*. Int Bull Bacteriol Nomencl Taxon *11:* 161–164.

Girard, A. E. 1970 A comparative cytological and chemical study of some selected micrococci and neisseriae. Diss Abstr Int *30:* 3776-B.

Gochnauer, M. E., S. C. Kushwaha, M. Kates and D. J. Kushner. 1972 Nutritional control of pigment and isoprenoid compound formation in extremely halophilic bacteria. Arch Mikrobiol *84:* 339–349.

Gonzalez-Ochoa, A. and M. A. Sandoval. 1956 Revision determinativa de algunas especies de actinomicetes patogenos descritas como diferentes. Rev Inst Salubr Enferm Trop (Mex) *16:* 17–25.

Goodfellow, M. 1971 Numerical taxonomy of some nocardioform bacteria. J Gen Microbiol *69:* 33–80.

Gordon, D. F., Jr. 1967 Reisolation of *Staphylococcus salivarius* from the human oral cavity. J Bacteriol *94:* 1281–1286.

Gorlenko, M. V. 1961 Bacterial diseases of plants translated for the U. S. Dept of Agriculture and Natl Science Foundation, Washington, D. C., by the Israel Program for Scientific Translations, Jerusalem, 1965.

Grassé, P. P. 1925 *Anisomitus denisi* n. g., n. sp. schizophyte de l'intestin du canard domestique. Ann Parasitol Hum Comp *3:* 343–348.

Gremmen, J. and M. De Kam. 1970 *Erwinia salicis* as the cause of dieback in *Salix alba* in the Netherlands and its identity with *Pseudomonas saliciperda*. Neth J Plant Pathol *76:* 249–252.

Gretler, A. C., P. Muccolo, J. B. Evans and C. F. Niven, Jr. 1955 Vitamin nutrition of the staphylococci with special reference to their biotin requirements. J Bacteriol *70:* 44–49.

Griffith, A. S. 1957 The types of tubercle bacilli in lupus and scrofulodermia. J Hyg *55:* 1–26.

Grootten, O. 1929 Caractères d'un bacille anaérobie strict isolé du pus d'une rectite ulcéreuse. C R Soc Biol Paris *102:* 43.

Grov, A. and S. Rude. 1967 Immunochemical examination of phenylhydrazine-treated *Staphylococcus aureus* cell walls. Acta Pathol Microbiol Scand *71:* 417–423.

Grula, E. A. 1962 A comparative study of six cultures of *Micrococcus lysodeikticus*. Can J Microbiol *8:* 855–859.

Grün, von L. 1968 Zur bestimmung von standartvarianten der staphylokokken humaner und boviner herkunft. Milchwissenschaft *23:* 604–608.

Guillebeau, W. 1891 Beiträge zur Lehre von den Ursachen der fadenziehenden Milch. Landwirt Jahrb Schweiz *5:* 135–142.

Gupta, K. C. 1965 A new species of the genus *Streptosporangium* isolated from Indian soil. J Antibiotics Ser A *18:* 125–127.

Hájek, V. and E. Maršálek. 1971 The differentiation of pathogenic staphylococci and a suggestion for their taxonomic classification. Zentrabl Bakteriol Parasitenk Infectionskr Hyg Abt I Orig *217:* 176–182.

—, — and I. Cerná. 1968 Plazmakogulázosvá aktivita stafylokokú rózného púvodu. Cslká Epiderm Mikrobiol Immunol *17:* 39–46.

Hammond, R. K. and D. C. White. 1970 Carotenoid formation by *Staphylococcus aureus*. J Bacteriol *103:* 191–198

Hanks, J. H. and R. E. Wallace. 1949 Relation of oxygen and temperature in the preservation of tissues by refrigeration. Proc Soc Exp Biol Med *71:* 196–200.

Hansen, G. A. 1874 *Bacillus leprae*. Norsk Mag Laegevidensk *9:* 1.

Hanton, W. K. 1968 Amorphosporangium (Actinoplanaceae): Report of motility and additional characters. J Gen Microbiol *53:* 317–320.

Hartwig, E. C., R. Cacciatore and F. P. Dunbar. 1962 *M. fortuitum:* Its identification, incidence and significance in Florida. Amer Rev Resp Dis *85:* 84–91.

Harz, C. O. 1877 *In* O. Bollinger. Ueber eine neue Pilzkrankheit beim Rinde. Zentralbl Med Wissenschafter *15:* 481–485.

Hauduroy, P., G. Ehringer, G. Guillot, J. Magrou, A. R. Prévot and A. Urbain. 1953 Dictionnaire des bactéries pathogènes pour l'homme, les animaux et les plantes, 2nd ed. Masson et Cie, Paris.

Haukenes, G. 1967 Serological typing of *Staphylococcus aureus* 7. Technical aspects. Acta Pathol Microbiol Scand *70:* 590–600.

Hayward, A. C. 1964 Characteristics of *Pseudomonas solanacearum*. J Appl Bacteriol *27:* 265.

Hendricks, C. W. 1970 Formic hydrogenlyase induction as a basis for the Eijkman fecal coliform concept. Appl Microbiol *19:* 441–445.

Henriksen, S. D. and K. Jyssum. 1950. A new variety of *Pasteurella haemolytica* from the human respiratory tract. Acta Pathol Microbiol Scand *50:* 443.

Heumann, W. 1962 Die Methodik der Kreuzung Sternbildender Bakterien Biol Zentrabl *81:* 341.

— and R. Marx. 1964 Feinstruktur und Funktion der Fimbrien bei dem sternbildenden Bakterium *Pseudomonas echinoides*. Arch Mikrobiol *47:* 325.

Hill, L. R. 1959 The Adansonian classification of the staphylococci. J Gen Microbiol *20:* 277–283.

Hill, S. 1971 Influence of oxygen concentration on the colony type of *Derxia gummosa* grown on nitrogen-free media. J Gen Microbiol *67:* 77–83.

Hirsch, P. 1970 Budding nitrifying bacteria: The nomenclatural status of *Nitromicrobium germinans* Stutzer and Hartleb 1899 and *Nitrobacter winogradskyi* Winslow *et al.* 1917. Int J Syst Bacteriol *20:* 317–320.

Höflich. 1891 Die Pyelonephritis bacillosa des Rindes. Mh prakt Tierheilk *2:* 337–373.

Hofstad, T. 1970 *Leptotrichia buccalis* a Gram-negative bacterium. Int J Syst Bacteriol *20:* 175–177.

Holt, R. J. 1969 The classification of staphylococci from colonized ventriculo-atrial shunts. J Clin Pathol 22: 475–482.

—. 1972 The pathogenic role of coagulase-negative staphylococci. Brit J Dermatol 86 (Suppl. 8): 42–47.

Howell, A. and L. Pine. 1961 The classification of organisms termed Leptotrichia (Leptothrix) buccalis. IV. Physiological and biochemical characteristics of Bacterionema matruchotii. Bacteriol Rev 25: 162–171.

—, A. Rizzo and F. Paul. 1965 Cultivable bacteria in developing and mature human dental calculus. Arch Oral Biol 10: 307–313.

Hucker, G. J. 1924 Studies on the Coccaceae. IV. The classification of the genus Micrococcus. N Y Agr Exp Sta Tech Bull 102.

—. 1928 Studies on the Coccaceae. IX. Further studies on the classification of the micrococci. N Y State Agr Sta Tech Bull 135. 31 pp.

—. 1948 Genus I. Micrococcus. Cohn In Bergey's Manual of Determinative Bacteriology, 6th ed. The Williams & Wilkins Co., Baltimore, pp. 235–246.

Huet, M. and H. Thouvenot. 1964 Étude d'un bacteriophage actif sur une bactérie anaérobie: Sphaerophorus varius. Ann Inst Pasteur 106: 867.

Hugh, R. and M. A. Ellis. 1968 The neotype strain for Staphylococcus epidermidis (Winslow and Winslow 1908) Evans 1916. Int J Syst Bacteriol 18: 231–239.

Huisingh, D. and R. D. Durbin. 1967 Physical and physiological methods for differentiating among Agrobacterium rhizogenes, A. tumefaciens and A. radiobacter. Phytopathology 57: 922–923.

Hungate, R. E. 1960 Microbial ecology of the rumen. Bacteriol Rev 24: 353–364.

Hunter, D., J. N. Todd and M. Larkin. 1970 Exudative epidermitis of pigs. The serological identification and distribution of the associated Staphylococcus. Brit Vet J 126: 225–229.

Huss, H. 1908 Eine fettspaltende Bakterie (Bacteridium lipolyticum n. sp.). Zentralbl Bakteriol Parasitenk Infektionskr Abt II, 20: 474–484.

Hylemon, P. B. 1971 A taxonomic study of the genus Spirillum Ehrenberg. Doctoral Diss., Newman Library, Virginia Polytechnic Inst. and State Univ., Blacksburg, Va.

—, J. S. Wells, Jr., J. H. Bowdre, T. O. MacAdoo and N. R. Krieg. 1973 Designation of Spirillum volutans Ehrenberg 1932 as type species of the genus Spirillum Ehrenberg 1932 and designation of the neotype strain of S. volutans. Int J Syst Bacteriol 23: 20–27.

Iizuka, H. and K. Komagata. 1965 Microbiological studies on petroleum and natural gas III. Determination of Brevibacterium, Arthrobacter, Micrococcus, Sarcina, Alcaligenes and Achromobacter isolated from oil brines in Japan. J Gen Appl Microbiol 11: 1–14.

Inoue, M. and T. Niida. 1962 Taxonomical study for Brevibacterium chromogenes nov. sp. Transformer of adenine into adenosine. Sci Rep Meiji Seika Kaisha 5: 1–5.

Ishiie, T., Y. Doi, K. Yora and H. Asuyama. 1967 Suppressive effects of antibiotics of tetracycline group on symptom development of mulberry dwarf disease. Ann Phytopathol Soc Jap 33: 267–275.

Ivler, D. 1965 The staphylococci: Ecologic perspectives. Ann N Y Acad Sci 128: 1–456.

—, H. M. Preston and D. Portnoy. 1963 Direct hemagglutination by Haemophilus influenzae. Proc Soc Exp Biol Med 114: 232–234.

Jackson, F. L., Y. E. Goodman, F. R. Bel, P. C. Wong and R. L. S. Whitehouse. 1971 Taxonomic status of facultative and strictly anaerobic "corroding bacilli" that have been classified as Bacterioides corrodens. J Med Microbiol 4: 171–184.

—, R. L. S. Whitehouse, Y. Goodman and P. C. Wong. 1970 Comparison of certain agar-pitting organisms designated Bacteroides corrodens. Bacteriol Proc 75 (M3).

Jacobs, N. J., J. M. Jacobs and G. S. Sheng. 1969 Effect of oxygen on heme and porphyrin accumulation from δ-aminolevulinic acid by suspensions of anaerobically grown Staphylococcus epidermidis. J Bacteriol 99: 37–41.

Jeffries, L. 1969 Menaquinones in the classification of Micrococcaceae with observations on the application of lysozyme and novobiocin sensitivity tests. Int J Syst Bacteriol 19: 183–187.

—, M. A. Cawthorne, M. Harris, B. Cook and A. T. Diplock. 1968. Menaquinone determination in the taxonomy of Micrococcaceae. J Gen Microbiol 54: 365–380.

Jenkins, P. A., J. Marks and W. B. Schaefer. 1971 Lipid chromatography and seroagglutination in the classification of rapidly growing mycobacteria. Amer Rev Resp Dis 103: 179–187.

Jensen, C. O. 1897 Ueber die Kälberruhr und deren Aetiologie. Mh Prakt Tierheilk 4: 97–124.

Jensen, J. 1963 Apocatalase of catalase-negative staphylococci. Science 141: 45–46.

Jones, D., R. H. Deibel and C. F. Niven, Jr. 1963 Identity of Staphylococcus epidermidis. J Bacteriol 85: 62–67.

Judicial Commission. 1973 Opinion 47. Conservation of the specific epithet avium in the scientific name of the agent of avian tuberculosis. Int J Syst Bacteriol 23: 472.

—. 1973 Opinion 48. Rejection of the name Aerobacter liquefaciens Beijerinck and conservation of the name Aeromonas Stanier with Aeromonas hydrophila as the type species. Int J Syst Bacteriol 23: 273–274.

Julianelle, L. A. and C. W. Wieghard. 1935 The immunological specificity of staphylococci. 1. The occurrence of serological types. J Exp Med 62: 11–21.

Jungano, M. 1909 Sur la flore anaérobie du rat. C R Soc Biol Paris 66: 112–114, 122–124.

Kalakoutskii, L. V., I. P. Kirillova and N. A. Krasil'nikov. 1967 A new genus of the Actinomycetales, Intrasporangium gen. nov. J Gen Microbiol 48: 79–85.

Käppler, W. 1968 Zur Taxonomie der Gattung Mycobacterium. II. Klassifizierung langsam wachsender Mykobakterien. Z Tuberk Erkrankungen Thoraxorgane 129: 321–328.

Kates, M. 1972 Ether-linked lipids in extremely halophilic bacteria. pp. 351–398. In F. Snyder (Editor) Ether Lipids, Chemistry and Biology. Academic Press, New York and London. 433 pp.

—, M. K. Wassef and D. J. Kushner. 1968 Radio-isotope studies on the biosynthesis of glycerol diether lipids in *Halobacterium cutirubrum*. Can J Biochem *46:* 971–977.

Kauffmann, J. and P. Toussaint. 1951 Un nouveau germe fixateur de l'azote atmosphérique: *Azotobacter lacticogenes*. C R Acad Sci Paris *223:* 710–711.

Kersters, K. 1970 The application of simplified enzymatic methods in bacterial taxonomy. *In* Prauser (Editor) The Actinomycetales. G. Fischer Verlag, Jena, pp. 299–304.

Kinney, R. W. and C. H. Werkman. 1960 *Brevibacterium leucinophagum*. Int Bull Bacteriol Nomencl Taxon *10:* 213–217.

Klebahn, H. 1919 Die Schadlinge des Klippfisches. Ein Beitrag zur Kenntnis der Salzliebenden Organismen. Mitt Inst Allg Bot Ham *4:* 11–69.

Klebs, E. 1875 Beiträge zur Kenntniss der Pathogenen Schistomyceten. Arch Exp Path Pharmak *4:* 221.

Kleck, J. L. and J. A. Donahue. 1968 Production of thermostable haemolysin by cultures of *Staphylococcus epidermidis*. J Infect Dis *118:* 317–323.

Klesius, P. H. and V. T. Schuhardt. 1968 Use of lysostaphin in the isolation of highly polymerized deoxyribonucleic acid and in the taxonomy of aerobic *Micrococcaceae*. J Bacteriol *95:* 739–743.

Kloos, W. E. 1969 Transformation of *Micrococcus lysodeikticus* by various members of the family *Micrococcaceae*. J Gen Microbiol *59:* 247–255.

— and L. M. Schultes. 1969 Transformation in *Micrococcus lysodeikticus*. J Gen Microbiol *55:* 307–317.

Knöll, H. 1965 Zur Biologie der Gärungssarcinen. Monatsber Deut Akad Wiss Berlin *7:* 475–477.

Kocur, M., T. Bergan and N. Mortensen. 1971 DNA base composition of Gram-positive cocci. J Gen Microbiol *69:* 167–183.

— and T. Martinec. 1963 The taxonomic status of *Sporosarcina ureae* (Beijerinck) Orla Jensen. Int Bull Bacteriol Nomencl Taxon *13:* 201–209.

— and —. 1972 Taxonomic status of *Micrococcus varians* Migula 1900 and designation of the neotype strain. Int J Syst Bacteriol *22:* 228–232.

—, — and K. Mazanec. 1968 Fine structure of *Micrococcus denitrificans* and *M. halodenitrificans* in relation to their taxonomy. Antonie van Leeuwenhoek J Microbiol Serol *34:* 19.

— and Z. Páčová. 1970 The taxonomic status of *Micrococcus roseus* Flügge 1866. Int J Syst Bacteriol *20:* 233–240.

—, —, W. Hodgkiss and T. Martinec. 1970 The taxonomic status of the genus *Planococcus* Migula 1894. Int J Syst Bacteriol *20:* 241–248.

—, — and T. Martinec. 1972 Taxonomic status of *Micrococcus luteus* (Schroeter 1872) and designation of neotype strain. Int J Syst Bacteriol *22:* 218–223.

—, F. Přecechtěl and T. Martinec. 1966 Haemolysins in coagulase-negative staphylococci. J Pathol Bacteriol *92:* 331–336.

Kolenbrander, P. E. and J. C. Ensign. 1968 Isolation and chemical structure of the peptidoglycan of *Spirillum serpens* cell walls. J Bacteriol *95:* 201–210.

Komagata, K. and H. Iizuka. 1964 New species of *Brevibacterium* isolated from rice. (Studies

on the microorganisms of cereal grains, VII) J Agr Chem Soc Jap *38:* 496–502.

Kossaya, T. A. 1967 Composition of manganese oxides in cultures of *Metallogenium* and *Leptothrix*. Mikrobiologiya *36:* 1024–1029 (in Russian).

Krabbenhaft, K. L. 1966 Significance of pigments in radiation resistance of *Micrococcus radiodurans*. Diss Abstr *26:* 4969.

Krasil'nikov, N. A. 1938 A new genus of Actinomycetales—*Mycococcus* n. gen. Mikrobiologiya *7(1):* 335–352.

Krieg, N. R., J. P. Tomelty and J. S. Wells, Jr. 1967 Inhibition of flagellar coordination in *Spirillum volutans*. J Bacteriol *94:* 1431–1436.

Kroeger, A. V. and L. R. Sibal. 1961 Biochemical and serological reactions of an oral filamentous organism. J Bacteriol *81:* 581–585.

Kubica, G. P. and W. E. Dye. 1967 Laboratory Methods for Clinical and Public Health Mycobacteriology. Public Health Service Publication No. 1547. U S Govt Printing Office, Washington, D. C. 20402.

Kubín, M., E. Matušková and J. Kazda. 1969 *Mycobacterium brunense* n. sp. Identified as Serotype Davis of Group III (Runyon) Mycobacteria. Zentralbl Bakteriol Parasitenk Infektionskr Hyg Abt Orig *210:* 207–211.

Kuhn, D. A. and M. P. Starr. 1962 Developmental morphology of *Corynebacterium poinsettiae*. Bacteriol Proc *1962:* 46.

— and —. 1965 Clonol morphogenesis of *Lampropedia hyalina*. Arch Mikrobiol *52:* 360–375.

Künnemann, O. 1903 Ein Beitrag zur Kenntniss der Eitererreger des Rindes. Arch Wiss Prakt Tierheilk *29:* 128–157.

Kunze, M. 1971 Natrium-Polyanethol-sulfonat als diagnostiches Hilfmittel bei der Differenzierung von Mykoplasmen. Zentralbl Bakteriol Parasitenk Infektionskr Abt I Orig *216:* 501–505.

Kushwaha, S. C., M. B. Gochnauer, D. J. Kushner and M. Kates. 1974 Pigments and isoprenoid compounds in extremely and moderate halophilic bacteria. Can J Microbiol *20:* 241–245.

Lacey, J. 1971 *Thermoactinomyces sacchari* sp. nov. a thermophilic Actinomycete causing bagassosis. J Gen Microbiol *66:* 327–338.

Lachica, R. V. F., P. D. Hoeprich and C. Genigeorgis. 1971 Nuclease production and lysostaphin susceptibility of *Staphylococcus aureus* and other catalase-positive cocci. Appl Microbiol *21:* 823–826.

—, K. F. Weiss and R. H. Deibel. 1969 Relationships among coagulase, enterotoxin and heat-stable deoxyribonuclease production by *Staphylococcus aureus*. Appl Microbiol *18:* 126–127.

Lanéelle, M. A., J. Asselineau and G. Castelnuovo. 1965 Études sur les Mycobactéries et les Nocardiae. IV. Composition des lipides de *Mycobacterium rhodochrous*, *M. pellegrino* sp., et de quelques souches de Nocardiae. Ann Inst Pasteur (Paris) *108:* 69–81.

Langworthy, T. A., P. F. Smith and W. H. Mayberry. 1972 Lipids in *Thermoplasma acidophilum*. J Bacteriol *112:* 1193–1200.

La Rivière, J. W. M. 1963 Cultivation and properties of *Thiovulus majus* Hinze. *In* C. H. Oppenheimer (Editor) Symposium on marine microbiology. Charles C Thomas, Springfield, Ill. pp. 61–72.

Lautrop, H. and O. Jessen. 1964 On the distinction between polar monotrichous and lophotrichous flagellation in green fluorescent pseudomonads. Acta Pathol Microbiol Scand 60: 588.

Leach, R. H. and M. Butler. 1966 Comparison of mycoplasmas associated with human tumors, leukemia and tissue cultures. J Bacteriol 91: 934–940.

Lechevalier, M. P., A. C. Horan and H. Lechevalier. 1971 Lipid composition in the classification of nocardiae and mycobacteria. J Bacteriol 105: 313–318.

Leclerc, H. and R. Buttiaux. 1965 Les Citrobacters. Ann Inst Pasteur Lille 16: 67–74.

Leifson, E. 1963 Determination of carbohydrate metabolism of marine bacteria. J Bacteriol 85: 1183–1184.

—. 1964 Micrococcus eucinetus n. sp. Int Bull Bacteriol Nomencl Taxon 14: 41–44.

Levinthal, W. 1918 Bakteriologische und serologische Influenzastudien. Z Hyg Infektionskr 86: 1–24.

Lewis, N. F. 1971 Studies on a radio-resistant coccus isolated from Bombay Duck (Harpodon nehereus). J Gen Microbiol 66: 29–35.

Lind, A. and O. Obrant. 1962 A case of urogenital tuberculosis caused by low-virulent isoniazid sensitive tubercle bacteria. Acta Chir Scand 123: 484–489.

Lindner, P. 1928 Gärungsstudien über Pulque in Mexiko. Ber Westpreuss Bot Zool Vereins 50 (Jubiläum No.,): 253–255.

Live, I. 1972 Differentiation of Staphylococcus aureus of human and canine origins: Coagulation of human and of canine plasma, fibrinolysin activity and serologic reaction. Amer J Vet Res 33: 383–391.

Lombard, G. L. and V. R. Dowell, Jr. 1972 Preparation of fluorescent antibody reagents for identification of Bacteroides. Abstr Annu Mtg Amer Soc Microbiol, p. 95(M93).

Losnegard, N. and P. Oeding. 1963 Immunological studies on polysaccharides from Staphylococcus epidermidis 1. Isolation and chemical characterization. Acta Pathol Microbiol Scand 58: 482–492.

Louria, D. B. and R. E. Gordon. 1960 Pericarditis and pleuritis caused by a recently discovered organism, Waksmania rosea. Amer Rev Resp Dis 81: 83–88.

Ludwig, T. G. 1955 A study of some filamentous microorganisms isolated from the human mouth. Aust J Dentistry 59: 343–347.

Luedemann, G. M. 1971 Species concept and criteria in the genus Micromonospora. Trans N Y Acad Sci Ser II, 33 (2): 207–218.

Lukoyanova, M. A. and S. D. Taptykova. 1968 Cytochromes of Micrococcus lysodeikticus. Biochimia 33: 888–894.

Lysenko, O. 1959 The occurrence of species of the genus Brevibacterium in insects. J Insect Pathol 1: 34–42.

—. 1961 Pseudomonas—An attempt at a general classification. J Gen Microbiol 25: 379–408.

McBride, M. E., L. F. Montes and J. M. Knox. 1970 The characterization of fluorescent skin diphtheroids. Can J Microbiol 16: 941–946.

McClung, N. M. 1954 The utilization of carbon compounds by Nocardia species. J Bacteriol 68: 231–236.

McElroy, L. J. and N. R. Krieg. 1972 A serological method for the identification of Spirilla. Can J Microbiol 18: 57–64.

Macchiavello, A. 1937 Estudios sobre tifus exantematico. III. Un nuevo metodo para tenir Rickettsia. Revta Chil Hig 1: 101–106.

Magrou, J. 1957 in Hauduroy et al., Dictionnaire des bactéries pathogènes. Masson et Cie, Paris.

Mah, R. A., D. Y. C. Fung and S. A. Morse. 1967 Nutritional requirements of Staphylococcus aureus S-6. Appl Microbiol 15: 866–870.

Maheswaran, S. K. and R. K. Lindorfer. 1967 Staphylococcal β-haemolysin II. phospholipase C activity of purified β-haemolysis. J Bacteriol 94: 1313–1319.

Malveaux, F. J. and C. L. San Clemente. 1967 Elution of loosely bound acid phosphatase from Staphylococcus aureus. Appl Microbiol 15: 738–743.

Mandel, M., P. Hirsch and S. F. Conti. 1972 Deoxyribonucleic acid base composition of hyphomicrobia. Arch Mikrobiol 81: 289–294.

Maramorosch, K., R. R. Granados and H. Hirumi. 1970 Mycoplasma diseases of plants and insects. Adv Virus Res 16: 135–193.

—, E. Shikata and R. R. Granados. 1968 Structures resembling mycoplasma in diseased plants and in insect vectors. Trans N Y Acad Sci 30: 841–855.

Marandon, J. L. and P. Oesding. 1967 Investigations on animal Staphylococcus aureus strains. 2. Antigens. Acta Pathol Microbiol Scand 70: 300–304.

Marks, J., P. A. Jenkins and W. B. Schaefer. 1969 Identification and incidence of a third type of Mycobacterium avium. Tubercle, London 50: 394–395.

Marples, R. R. 1969 Violagabriellae variant of Staphylococcus epidermidis on normal human skin. J Bacteriol 100: 47–50.

Matsumoto, H. and T. Tazaki. 1970 Genetic recombination in Klebsiella pneumoniae. Jap J Microbiol 14: 129–141.

Mazanec, K., M. Kocur and T. Martinec. 1966 Electron microscopy of ultrathin sections of Micrococcus cryophilus. Can J Microbiol 12: 465–470.

Melville, T. H. 1965 A study of the overall similarity of certain actinomycetes mainly of oral origin. J Gen Microbiol 40: 309–315.

Meyer, M. E. and W. J. Brinley Morgan. 1973 Designation of neotype strains and of biotype strains for species of the genus Brucella Meyer and Shaw. Int J Syst Bacteriol 23: 135–141.

Meyerhof, O. and D. Burk. 1928 Über die Fixation des Luftstickstoffs durch Azotobacter. Z physiol Chem 139: 117–142.

Minor, T. E. and E. H. Marth. 1972 Staphylococcus aureus and staphylococcal food intoxications. A. Review II Entertoxins and epidemiology. J Food Milk Technol 35: 21–29.

Mitchell, R. G. 1964 Urinary tract infections due to coagulase negative staphylococci. J Clin Pathol 17: 105–106.

—. 1968 Classification of Staphylococcus albus strains isolated from the urinary tract. J Clin Pathol 21: 93–96.

— and A. C. Baird-Parker. 1967 Novobiocin resistance and the classification of staphylococci and micrococci. J Appl Bacteriol 30: 251–254.

Mitchison, D. A. 1970 Regional variation in the guinea pig virulence and other characteristics of tubercle bacilli. Pneumonology 142: 131–137.

—, J. B. Selkon and J. Lloyd. 1963 Virulence in the guinea pig, susceptibility to hydrogen peroxide and catalase activity of isoniazid-sensitive tubercle bacilli from south Indian and British patients. J Pathol Bacteriol 86: 277–386.

Molisch, H. 1925 Botanische Beobachtungen in Japan VIII Die Eisenorganismen in Japan. Sci Rep Tohoku Imp Univ Ser IV Biol 1: 135–168.

Moniz, L. 1963 Leaf spot of apple blossom. Curr Sci 32 (4): 177.

Montiel, F. and H. J. Blumenthal. 1965 Factors affecting the pathways of glucose catabolism and T. C. A. cycle in Staphylococcus aureus. Bacteriol Proc 1965: 77.

Moore, R. L. and P. Hirsch. 1972 Deoxyribonucleic acid base sequence homologies of some budding and prosthecate bacteria. J Bacteriol 110: 256–261.

Morris, E. O. 1954 The bacteriology of the oral cavity. V. Corynebacterium and Gram-positive filamentous organisms. Brit Dent J 97: 29–36.

Mortensen, N. 1969 Studies in urinary tract infections. III. Biochemical characteristics of coagulase negative staphylococci associated with urinary tract infections. Acta Med Scand 186: 47–51.

— and M. Kocur. 1967 Correlation of DNA base composition and acid formation from glucose of staphylococci and micrococci. Acta Pathol Microbiol Scand 69: 445–457.

Moseley, B. E. B. and A. H. Schein. 1964 Radiation-resistance and deoxyribonucleic acid base composition of Micrococcus radiodurans. Nature (London) 203: 1298–1299.

Moss, C. W., V. R. Dowell, V. J. Lewis and M. A. Schekter. 1967 Cultural characteristics and fatty acid composition of Corynebacterium acnes. J Bacteriol 94: 1300–1305.

Moss, F. J. and Y. T. Tchan. 1958 Studies on N-fixing bacteria. VII. Cytochromes of Azotobacteraceae. Proc Linnean Soc N S W 83: 161–164.

Murray, R. G. E. 1963 Role of superficial structures in the characteristic morphology of Lampropedia hyalina. Can J Microbiol 9: 593–606.

Nathanson, A. 1902 Uber eine neue Gruppe von Schwefelbakterien und ihre Stoffwechsel. Mitt Zool Sta Neapel 15: 655–680.

Nechaeva, N. B. 1949 Two methane oxidizing species of mycobacteria. Mikrobiologiya 18: 310–317.

Nimmich, W. 1971 Über die spezifischen polysaccharide (K-Antigene) der Klebsiella—Typen K73–K80. Acta Biol Med Ger 26: 297–403.

— and G. Korten. 1970 Die chemische Zusammensetzung der Klebsiella-Lipopolysaccharide (O-Antigene). Pathol et Microbiol (Basel) 36: 179–192.

Norlin, M., A. Lind and O. Ouchterlony. 1969 A serologically based taxonomic study of M. gastri. Z Immunitätsforsch Allergie Klin Immunol 137: 241–248.

Norton, C. F. and G. E. Jones. 1968 A marine isolate of Pseudomonas nigrifaciens. I. Classification and nutrition. Can J Microbiol 14: 1333–1340.

Nygrën, B., J. Haborn and P. Wåhlen. 1966 Phospholipase A production in Staphylococcus aureus. Acta Pathol Microbiol Scand 68: 429–433.

O'Connor, J. J., A. T. Willis and J. A. Smith. 1966 Pigmentation of Staphylococcus aureus. J Pathol Bacteriol 92: 585–588.

Oeding, P. 1957 Agglutinability of pyogenic staphylococci at various conditions. Acta Pathol Microbiol Scand 41: 310–324.

—. 1960 Antigenic properties of Staphylococcus aureus. Bacteriol Rev 24: 374–396.

—. 1965 Antigenic properties of staphylococci. Ann N Y Acad Sci 128: 183–190.

— and I. L. Hasselgren. 1972 Antigenic studies of genus Micrococcus 2. Double diffusion in agar gel with particular emphasis on techoic acids. Acta Pathol Microbiol Scand 80B: 265–269.

—, J. L. Maradon, V. Hájek and E. Maršálek. 1971 A comparison of phage patterns and antigenic structure with biochemical properties of Staphylococcus aureus strains isolated from cattle. Acta Pathol Microbiol Scand 79B: 357–364.

—, B. Myklestad and A. L. Davison. 1967 Serological investigations on teichoic acids from walls of Staphylococcus epidermidis and Micrococcus. Acta Pathol Microbiol Scand 69: 458–464.

O'Gara, P. J. 1916 A bacterial disease of western winter wheat Agropyron Smithii. Occurrence of a new type of bacterial disease in America. Phytopathology 6: 341–350.

Ogawa, C., M. Oide and Y. Midorikawa. 1959 Alanine-glutamic acid fermentation by Brevibacterium alanicum nov. sp. Part I. Identification of the strain and nutrition in seed culture (in Japanese). Amino Acids 1: 45–49.

Okabayashi, T. and E. Masuo. 1960 Occurrence of nucleotides in the culture fluid of microorganisms. I. Screening of purine autotrophs of Escherichia coli. Chem Pharm Bull (Tokyo) 8: 1084–1088.

Okumura, S., R. Tsugawa, T. Tsunoda, K. Kono, T. Matsui and N. Miyachi. 1962 Studies on the L-glutamic acid fermentation. 1. New bacteria of the genus Brevibacterium isolated from the nature to produce L-glutamic acid (in Japanese). J. Agr Chem Soc Jap 36: 141–159.

Omata, R. R. and R. C. Braunberg. 1960 Oral fusobacteria. J Bacteriol 80: 737–740.

Orskov, F. 1956 Escherichia coli. Nyt Nordisk Forlag, Copenhagen.

Ostwald, W. 1939 Die kleine Farbmesstafel Musterschmidt, Göttingen.

Ōta, S. and M. Tanaka. 1959 Studies on the species which accumulate glutamic acid. Part I. Research on classification of Brevibacterium aminogenes nov. sp. (in Japanese). Amino Acids 1: 50–55.

— and —. 1959 Study of L-glutamic acid fermentation (II) Bacteriological study of L-glutamic acid producing bacteria. (1) Taxonomic study of Brevibacterium aminogenes nov. sp. J Ferm Technol 37: 261–264.

Palleroni, N. J. and M. Doudoroff. 1971 Phenotypic characterization and deoxyribonucleic acid homologies of Pseudomonas solanacearum. J Bacteriol 107: 690–696.

Pangborn, J. and M. P. Starr. 1966 Ultrastructure of Lampropedia hyalina. J Bacteriol 91: 2025–2030.

Park, G. H., M. Fauber and C. B. Cook. 1968 Identification of Haemophilus vaginalis. Amer J Clin Pathol 49: 590–593.

Parker, M. T. 1962 Phage-typing and epidemiology of *Staphylococcus aureus* infection. J Appl Bacteriol *25:* 389–402.

Parker, R. B. 1967 Generation of electrical potentials by synthetic plaques of the indigenous oral microflora. Influence of species, temperature and load on voltage output. Arch Oral Biol *12:* 131–140.

Partridge, M. D., A. L. Davison and J. Baddiley. 1973 The distribution of teichoic acids and sugar-1-phosphates polymers in walls of micrococci. J Gen Microbiol *74:* 169–173.

Payza, A. N. and E. D. Korn. 1956 The degradation of heparin by bacterial cells. I. Adaptation of lyophilized cells. J Biol Chem *223:* 853–858.

Pederson C. S. and L. Ward. 1949 The effect of salt upon the bacteriological and chemical changes in fermenting cucumbers. N Y Agr Exp Sta Tech Bull *288:* 3–29.

Pennock, C. A. and R. B. Huddy. 1967 Phosphatase reaction of coagulase negative staphylococci and micrococci. J Pathol Bacteriol *93:* 685–688.

Penso, G., V. Ortali, A. Guadiano, M. Princivalle, L. Vella and A. Zampieri. 1951 Studi e ricerche sui micobatteri. VII. *Mycobacterium phlei* (Lehmann and Neumann 1899 proparte). Rend Ist Super Sanita *14:* 855–908.

Perkins, H. R. 1968 Cell wall mucopeptide of *Corynebacterium insidiosum* and *C. sepedonicum.* Biochem J *110:* 47P.

Perry, J. J. and J. B. Evans. 1966 Oxidation and assimilation of carbohydrates by *Micrococcus sodonensis.* J Bacteriol *91:* 33–38.

— and —. 1967 Glucose catabolism of *Micrococcus sodonensis.* J Bacteriol *93:* 1839–1846.

Perry, L. B. 1973 Gliding motility on some nonspreading *Flexibacteria.* J Appl Bacteriol *36:* 227–232.

Petit, J. F., E. Munoz and J. M. Ghuysen. 1966 Peptide cross-links in bacterial cell wall peptidoglycans studied with specific endopeptidases from *Streptomyces albus* G. Biochemistry *5:* 2764–2776.

Pike, E. B. 1965 A trial of statistical methods for selection of determinative characters from *Micrococcaceae* isolates. Publ Fac Sci Univ J. E. Purkyĕ Brno (Czechoslovakia) *K35:* 316–317.

Pine, L. 1963 Recent developments on the nature of the anaerobic actinomycetes. Ann Soc Belg Med Trop Parasitol Mycol Hum Anim *3:* 247–258.

— and H. Hardin. 1959 *Actinomyces israelii,* a cause of lacrimal canaliculitis in man. J Bacteriol *78:* 164–170.

Plum, N. 1926 Geflügeltuberkulose bei Säugetieren Ann Report Den Kongelige Veterinaerog Landbohjskole, Copenhagen, pp. 63–185.

Preer, J. R., Jr. 1971 Extrachromosomal inheritance: hereditary symbionts mitochondria, chloroplasts. Annu Rev Genet *5:* 361–406.

—, L. B. Preer and A. Jurand. 1974 Kappa and other endosymbionts in *Paramecium aurelia.* Bacteriol Rev *38:* 113–163.

Pretorius, W. A. 1963 A systematic study of genus *Spirillum* which occurs in oxidation ponds, with a description of a new species. J Gen Microbiol *32:* 403–408.

Prévot, A. R. 1940 Recherches sur la flore anaérobie de l'intestine humain: *Fusocillus girans* nov. sp. C R Soc Biol Paris *133:* 246–249.

—, P. Joubert, L. Tardieux and F. de Cadore. 1956 Recherches sur *Fusiformis nucleatus* (Knorr) et son pouvoir pathogène pour l'homme et les animaux. Ann Inst Pasteur *91:* 787–798.

— and M. Raynaud 1947 Étude d'une variété-de *Dialister pneumosintes* isolée d'une septicémie mortelle avec abcès du poumon et du cerveau. Ann Inst Pasteur *73:* 67–68.

Puttlitz, D. H. and H. W. Seeley, Jr. 1968 Physiology and nutrition of *Lampropedia hyalina.* J Bacteriol *96:* 931–938.

Quinn, E. L., F. Cox and E. H. Drake. 1966 Staphylococcal endocarditis. A disease of increasing importance. J Amer Med Ass *196:* 815–818.

Reid, J. D. and M. A. Joya. 1969 A study of the morphological and biochemical characteristics of certain anaerobic corynebacteria. Int J Syst Bacteriol *19:* 273–280.

Richardson, G. M. 1936 The nutrition of *Staphylococcus aureus.* Biochem J *30:* 2184.

Rittenberg, B. T. and S. C. Rittenberg. 1962 The growth of *Spirillum volutans* Ehrenberg in mixed and pure cultures. Arch Mikrobiol *42:* 138–153.

Roberts, A. P. 1967 Micrococcaceae from the urinary tract in pregnancy. J Clin Pathol *20:* 631–632.

Roberts, D. H. and M. Gois. 1970 A previously unreported serotype of porcine mycoplasma. Vet Rec *87:* 214–215.

Robertson, A. 1924 Observation of the causal organism of rat-bite fever in man. Ann Trop Med Parasitol *18:* 157–175.

Rosypal, S. and M. Kocur. 1963 The taxonomic significance of the oxidation of carbon compounds by different strains of *Micrococcus luteus.* Antonie van Leeuwenhoek J Microbiol Serol *29:* 313–318.

Rosypalová, A., J. Boháček and S. Rosypal. 1966 Deoxyribonucleic acid base composition of some micrococci and sarcinae. Antonie van Leeuwenhoek J Microbiol Serol *32:* 192–196.

Roth, G. D. and V. Flanagan. 1969 The pathogenicity of *Rothia dentocariosa* inoculated into mice. J Dent Res *49:* 957–958.

— and A. N. Thurn. 1962 Continued study of oral *Nocardia.* J Dent Res *41:* 1279–1292.

Roy, A. B. and S. Sen. 1962 A new species of *Derxia.* Nature (London) *194:* 604–605.

Runyon, E. H. 1972 Conservation of the specific epithet *fortuitum* in the name of the organisms known as *Mycobacterium fortuitum* da Costa Cruz. Request of an opinion. Int J Syst Bacteriol *22:* 50–51.

Salton, M. R. J. and J. G. Pavlik. 1960 Studies of the bacterial cell wall VI. Wall composition and sensitivity to lysozyme. Biochim Biophys Acta *39:* 398–407.

Saperstein, S., M. P. Starr and J. A. Filfus. 1954 Alterations in carotenoid synthesis accompanying mutation in *Corynebacterium michiganense.* J Gen Microbiol *10:* 85–92.

Sarkany, I., D. Taplan and H. Blank. 1961 Erythrasma—common bacterial infection of the skin. J Amer Med Ass *177:* 130–132.

—, — and —. 1962 Organisms causing erythrasma. Lancet *2:* 304–305.

Schad, G. A., R. Knowles and E. Meerovitch.

1964 The occurrence of *Lampropedia* in the intestines of some reptiles and nematodes. Can J Microbiol *10:* 801–803.

Schaefer, W. B. 1967 Type-specificity of atypical mycobacteria in agglutination and antibody absorption tests. Amer Rev Resp Dis *96:* 1165–1168.

Scharmann, W. and H. Blobel. 1968 Serological difference of staphylococcal nucleases. Z Naturforsch *23:* 1230–1235.

Schewiakoff, W. 1893 Über einen neuen bakterienähnlichen organismen des Süsswassers. Habilitation schrift Univ Heidelberg. C. Winter 1–36.

Schindler, C. A. and V. Schuhardt. 1964 Lysostaphin: A new bacteriolytic agent for the *Staphylococcus.* Proc Nat Acad Sci USA *51:* 414–421.

Schleifer, K. H. and O. Kandler. 1970 Amino acid sequence of the murein of *Planococcus* and other *Micrococcaceae.* J Bacteriol *103:* 387–392.

—, W. E. Kloos and A. Moore. 1972 Taxonomic status of *Micrococcus luteus* (Schroeter 1872) Cohn 1872: Correlation between peptidoglycan type and genetic compatibility. Int J Syst Bacteriol *22:* 224–227.

—, M. Reid and O. Kandler. 1968 The amino acid sequence of the murein of *Staphylococcus epidermidis.* Arch Mikrobiol *62:* 198–208.

Schmidt, J. M. and R. L. Richardson. 1962 Serology of *Bacterionema matruchotii.* J Bacteriol *83:* 584–589.

Schuster, M. L. and D. W. Christiansen. 1957 An orange coloured strain of *Corynebacterium flaccumfaciens* causing bean wilt. Phytopathology *47:* 51–53.

Scott, W. J. 1953 Water relations of *Micrococcus pyogenes* var. *aureus* at 30°C. Aust J Biol Sci *6:* 549–564.

Sedlák, J. 1969 Zur Systematik der Gattung *Citrobacter.* Zentralbl Bakteriol Parasitenk Infektionskr Abt I Orig *212:* 497–499.

—. 1973 Present knowledge and aspects of *Citrobacter.* Curr Top Microbiol Immunol *22:* 41–59.

Severini, G. 1913 Una bacteriosi dell' *Ixia maculata* e del *Gladiolus colvilli.* Ann Bot (Rome) *11:* 413.

Shikata, E., K. Maramorosch and K. C. Ling. 1969 Presumptive mycoplasma etiology of yellow diseases. FAO Plant Protection Bull *17:* 121–128.

Sibal, L. R., A. V. Kroeger, D. Kumarich and E. Meyer. 1962 Serological specificity of acid soluble antigens of *Bacterionema matruchotii.* J Bacteriol *83:* 811–818.

Slack, J. M. and M. A. Gerencser. 1970 Two new serological groups of *Actinomyces.* J Bacteriol *103:* 265–266.

Sleytr, U. and M. Kocur. 1971 Structure of *Micrococcus cryophilus* after freeze-etching. Arch Mikrobiol *78:* 353–357.

Smithies, W. R. and N. E. Gibbons. 1955 The deoxyribose nucleic acid slime layer of some halophilic bacteria. Can J Microbiol *1:* 614–621.

Socransky, S. S., C. Hubersak and D. Propos. 1970 Induction of peridontal destruction in gnotobiotic rats by a human oral strain of *Actinomyces naeslundii.* Arch Oral Biol *15:* 993.

Sokolova, G. A. and G. I. Karavaiko. 1964 Physiology and Geochemical activities of Thiobacteria. Izd Nauke (Russian). Translation Israel Program of Scientific Translation Jerusalem.

Sompolinsky, D. 1953 De l'impetigo contagiosa suis et du *Micrococcus hyicus* n. sp. Schweiz Arch Tierheilk *95:* 302–309.

Spaulding, E. H. and L. F. Rettger. 1937 *Fusobacterium* genus. I. Biochemical and serological classification. J Bacteriol *34:* 535–548.

Stanford, J. L. and A. Beck. 1969 Bacteriological and serological studies of fast growing mycobacteria identified as *Mycobacterium friedmannii.* J Gen Microbiol *58:* 99–106.

— and W. J. Gunthorpe. 1969 Serological and bacteriological investigation of *Mycobacterium ranae* (*fortuitum*). J Bacteriol *98:* 375–383.

Steenken, W., Jr., W. H. Oatway, Jr. and S. A. Petroff. 1934 Biological studies of tubercle bacillus. III. Dissociation and pathogenicity of the R and S variants of the human tubercle bacillus (H37). J Exp Med *60:* 515–540.

Stiles, C. W. 1905 The international code of zoological nomenclature as applied to medicine. Bull Hyg Laboratory No. 24, Washington, D.C.

Strasters, K. C. and K. C. Winkler. 1963 Carbohydrate metabolism of *Staphylococcus aureus.* J Gen Microbiol *33:* 213–229.

Stutzer, A and R. Hartleb. 1899 Untersuchungen über die bei der Bildung von Salpeter beobachteten Mikroorganisman. Mitt Landw Inst Univ Breslau *1:* 75–100, 197–232.

Su, Y. and K. Yamada. 1960 Studies on L-glutamic acid fermentation. Part I. Isolation of a L-glutamic acid producing strain and its taxonomical studies. Bull Agr Chem Soc Jap *24:* 69–74.

Subcommittee on Mycobacteria. American Society for Microbiology. 1962 J Bacteriol *83:* 931–932.

Subcommittee on the Taxonomy of Mycoplasmatales. 1971 Minutes of meeting, August 10, 1970. Int J Syst Bacteriol *21:* 151–153.

Subcommittee on Taxonomy of Staphylococci and Micrococci, International Committee on Nomenclature of Bacteria. 1965 Minutes of first meeting. Int Bull Bacteriol Nomencl Taxon *15:* 107–108.

—. 1967 Minutes of second meeting July 23, 1966. Int J Syst Bacteriol *17:* 381–382.

—. 1971 Minutes of meeting 2nd and 3rd April 1968. Int J Syst Bacteriol *21:* 161–163.

Suitor, E. C., Jr. 1964 Studies on the cell envelope of *Wolbachia persica.* J Infect Dis *114:* 125–134.

Sutić, D. and Z. Tešić. 1958 Jedna nova bakterioza bresta izazivač *Pseudomonas ulmi* n. sp. Zašt Bilja (Plant Protection) Beograd *45:* 13–25.

Suto, T. 1954 An acid fast *Azotobacter* in a volcanic ash soil. Sci Rep Res Inst Tohoku Univ *6:* 25–31.

Tager, M. and M. C. Drummond. 1965 Staphylocoagulase. Ann N Y Acad Sci *128:* 92–111.

Tai, P.-C. and H. Jackson. 1969 Mesophilic mutants of an obligate psychrophile *Micrococcus cryophilus.* Can J Microbiol *15:* 1145–1150.

Takayama, K., K. Udagawa and S. Abe. 1960 Studies on the lytic enzyme produced by *Brevibacterium.* J Agr Chem Soc Jap *34:* 652–656.

Takazoe, I. and J. Ennever. 1969 Ultrastructure of *Bacterionema matruchotii*. Bull Tokyo Dent Coll *10:* 45–59.

— and T. Nakamura. 1965 The relation between metachromatic granules and intracellular calcification of *Bacterionema matruchotii*. Bull Tokyo Dent Coll *6:* 29–42.

—, T. Takeuchi and T. Nakamura. 1963 A study of *Bacterionema matruchotii*. Bull Tokyo Dent Coll *4:* 42–49.

Taylor, B. F. and D. S. Hoare. 1969 New facultative *Thiobacillus* and a reevaluation of the heterotrophic potential of *Thiobacillus novellus*. J Bacteriol *100:* 487.

Taylor-Robinson, D. and M. C. Path. 1967 Mycoplasmas of various hosts and their antibiotic sensitivities. Postgrad Med J Suppl *43:* 100–104.

Terasaki, Y. 1961 On *Spirillum putridiconchylium* nov. sp. Bot Mag (Tokyo) *74:* 79–85.

—. 1961 On two new species of *Spirillum*. Bot Mag (Tokyo) *74:* 220–227.

Thirkell, D. 1969 Growth and pigmentation of *Micrococcus radiodurans*. J Gen Microbiol *55:* 337–340.

— and M. I. S. Hunter. 1969 Carotenoid-glycoprotein from *Sarcina flava* membrane. J Gen Microbiol *58:* 289–292.

— and —. 1969 The polar carotenoid fraction from *Sarcina flava*.

— and R. H. C. Strong. 1967 Analysis and comparison of the carotenoids of *Sarcina flava* and *Sarcina lutea*. J Gen Microbiol *49:* 53–57.

Thomas, E. T. 1964 A taxonomic study of the aerobic Micrococcaceae. Thesis: University of Texas.

Thompson, J. P. 1968 The occurrence of nitrogen-fixing bacteria of the genus *Beijerinckia* in Australia outside the tropical zone. Trans 9th Int Congr Soil Sci, Adelaide, Australia *2:* 129–139.

Tipper, D. J. 1969 Structure of the cell wall peptidoglycans of *Staphylococcus epidermidis* Texas 26 and *Staphylococcus aureus* Copenhagen 1. Structure of neutral and basic peptides from hydrolysis with Myxobacter AL-1 peptidase. Biochemistry *8:* 2192–2202.

— and M. F. Berman. 1969 Structure of the cell wall peptidoglycans of *Staphylococcus epidermidis* Texas 26 and *Staphylococcus aureus* Copenhagen 1. Chain lengths and average sequence of cross-bridge peptides. Biochemistry *8:* 2183–2191.

— and J. L. Strominger. 1966 Isolation of 4-O-β-N-acetylmuramyl-N-acetylglucosamine and 4-O-β-N, 6-O-diacetylmuramyl-N-acetylglucosamine and the structure of the cell wall polysaccharide of *Staphylococcus aureus*. Biochim Biophys Res Commun *22:* 48–56.

Tirunarayanan, M. O. 1968 Investigations on the enzymes and toxins of staphylococci. Study of phosphatase using *p*-nitrophenyl phosphate as substrate. Acta Pathol Microbiol Scand *74:* 573–590.

— and J. Lundbeck. 1967 Investigations on the enzymes and toxins of staphylococci. Identification of the substrate and products formed in the "egg yolk reaction." Acta Pathol Microbiol Scand *69:* 314–320.

Tissier, H. 1905 Répartition des microbes dans l'intestin du nourrisson. Ann Inst Pasteur *19:* 109–123.

Torres Pereira, A. 1961 Antigenic loss variation in *Staphylococcus aureus*. J Pathol Bacteriol *81:* 151–156.

—. 1962 Coagulase negative strains of staphylococcus possessing antigen 51 as agents of urinary infection. J Clin Pathol *15:* 252–253.

—. 1967 Aspects of virulence in *Staphylococcus aureus* in relation to antigenic structure. Int J Syst Bacteriol *17:* 395–402.

Torrey, J. C. 1916 Bacteria associated with certain types of abnormal lymph glands. J Med Res *34:* 65–80.

Tschäpe, H. and H. Rische. 1972 Phänotypische und genetische Analysen der "nutrient markers" von *Staphylococcus aureus* und ihre Taxonomische Bedeutung. Z Allg Mikrobiol *12:* 59–68.

Tsukamura, M. 1969 Numerical taxonomy of the genus *Nocardia*. J Gen Microbiol *56:* 265–287.

Tyler, P. A. and K. C. Marshall. 1967 Pleomorphy in stalked, budding bacteria. J Bacteriol *93:* 1132–1136.

Uesaka, I., K. Oiwa, K. Yasuhira, Y. Kobara and N. M. McClung. 1971 Studies on the pathogenicity of *Nocardia* isolates for mice. Jap J Exp Med *41:* 443–457.

Ungers, G. E. and J. J. Cooney. 1968 Isolation and characterization of carotenoid pigments of *Micrococcus roseus*. J Bacteriol *96:* 234–241.

van Wagtendonk, W. J., J. A. D. Clark and G. A. Godoy. 1963 The biological status of lambda and related particles in *Paramecium aurelia*. Proc Nat Acad Sci U S A *50:* 835–838.

Verhoef, J., C. P. H. Van Boven and K. C. Winkler. 1972 Phage typing of coagulase negative staphylococci. J Med Microbiol *5:* 9–19.

Visloukh, S. M. 1921 To the knowledge of the microorganisms of the Nevsky-bay. Gosud gidrolog Inst Izvestia Bull Inst Hydrolog Leningrad No I: 84–96.

von Freudenreich, E. and S. Orla-Jensen. 1907 Ueber die im Emmentalerkäse stattfindende Propiensäuregärung. Zentralbl Bakteriol Parasitenk Infektionskr Abt II *17:* 529–546.

Wadstrom, T. 1967 Studies on extracellular proteins from *Staphylococcus aureus* II. Separation of deoxyribonucleases by isoelectric focusing. Purification and properties of the enzymes. Biochim Biophys Acta *147:* 441–452.

Wallroth, F. G. 1833 Flora Cryptogamica Germaniae Algues et Champignona *2:* 182.

Walsh, F. and R. Mitchell. 1972 An acid-tolerant iron-oxidizing metallogenium. J Gen Microbiol *72:* 369–376.

Watanabe, N. 1959 On four new halophilic species of *Spirillum*. Bot Mag (Tokyo) *72:* 77–86

Weaver, R. E., P. S. Brachman and J. C. Feeley. 1970 Animal diseases transmissable to man. *In* Diagnostic Procedures for Bacterial, Mycotic, and Parasitic Infections, 5th ed. Amer Publ Health Ass, pp. 354–363.

Weiszfeiler, J., I. Jokay, E. Karczag, K. Almassy and P. Somos. 1968 Taxonomic studies on mycobacteria on the basis of their antigenic structure. Acta Microbiol Acad Sci Hung *15:* 69–76.

Welch, W. H. 1891 Conditions underlying the infection of wounds. Amer J Med Sci *102:* 439–465.

Wells, J. S., Jr. 1970 A morphological and

physiological classification of spirilla. Doctoral Diss. Newman Library, Virginia Polytechnic Inst and State Univ., Blacksburg, Va.

Welsch, M. and J. Thibaut. 1948 A study of corynebacteria from human sources, and especially their sensitivity to antibiotics. Leeuwenhoek ned Tijdschr *14:* 193–213

Wentworth, B. B. 1963 Bacteriophage typing of the staphylococci. Bacteriol Rev *27:* 253–272.

Wetmur, J. G. and N. Davidson. 1968 Kinetics of renaturation of DNA. J Mol Biol *31:* 349–370

Whitcomb, R. F. and R. E. Davis. 1970 Mycoplasma and phytarboviruses as plant pathogens persistently transmitted by insects. Annu Rev Entomol *15:* 405–464.

White D. C. 1963 Respiratory systems in the hemin-requiring *Haemophilus* species. J Bacteriol *85:* 84–96.

— and F. E. Frerman. 1967 Extraction, characterization and cellular localisation of the lipids of *Staphylococcus aureus.* J Bacteriol *94:* 1854–1867.

White, F., E. A. S. Rattray and D. J. G. Davidson. 1962 Serological typing of coagulase positive staphylococci isolated from the bovine udder. J Comp Pathol *72:* 19–28

Wilkins, T. D., L. V. Holdeman, I. J. Abramson and W. E. C. Moore. 1972 Standardized single-disc method for antibiotic susceptibility testing of anaerobic bacteria. Antimicrob Agents Chemother *1:* 451–459.

Williams, J. 1971 The growth *in vitro* of killer particles from *Paramecium aurelia* and the axenic culture of this protozoan. J Gen Microbiol *68:* 253–262.

Williams, M. A. and S. C. Rittenberg. 1956 Microcyst formation and germination in *Spirillum lunatum.* J Gen Microbiol *15:* 205–209.

— and —. 1957 A taxonomic study of the genus *Spirillum* Ehrenberg. Int Bull Bacteriol Nomencl Taxon *7:* 49–111.

Williams, R. E. O. 1969 *Staphylococcus aureus* on the skin. Brit J Dermatol *81 Suppl 1:* 33–36.

Willis, A. T., S. I. Jacobs and G. M. Goodburn. 1964 Pigment production, enzymatic activity and antibiotic sensitivity of staphylococci. Subdivision of the pathogenic group. J Pathol Bacteriol *87:* 157–167.

—, J. A. Smith and J. J. O'Connor. 1966 Properties of some epidemic strains of *Staphylococcus aureus.* J Pathol Bacteriol *92:* 345–358.

Winford, T. E. and S. Haberman. 1966 Isolation of aerobic Gram-positive filamentous rods from diseased gingivae. J Dent Res *45:* 1159–1167.

Winslow, C.-E. A. and A. F. Rogers. 1906 A statistical study of generic characteristics of coccaceae. J Infect Dis *3:* 485–546.

— and A. R. Winslow. 1908 The Systematic Relationships of the Coccaceae. John Wiley, New York.

Wolanski, B. and K. Maramorosch. 1970 Negatively stained mycoplasmas: fact or artifact? Virology *42:* 319–327.

Wolfe, A. 1910 Milchwirtschaftliche Bakteriologie. Zentralbl Bakteriol Parisitenk Infectionskr Abt II, *28:* 417–422.

Work, E. and H. Griffiths. 1968 Morphology and chemistry of cell walls of *Micrococcus radiodurans.* J Bacteriol *95:* 641–657.

Yamada, K. and Y. Hirose. 1960 Studies on the amino acid fermentation of pentose materials. I. Isolation of strains capable of producing amino acids from pentose and their taxonomical studies. Bull Agr Chem Soc Jap *24:* 621–632.

Yanagawa, R., K. Otsuki and T. Tokui. 1968 Electron microscopy of fine structure of *Corynebacterium renale* with special reference to pili. Jap J Vet Res *16:* 31–38.

Youssef, M., F. Mach, W. Muller-Beuthow and H. Haenel. 1970 Isolierung und Characterisierung von *Lactobacillus bifidus* phagen. Zentralbl Bakteriol Parisitenk Infectionskr Abt I Orig *213:* 495–509.

Zavarzin, G. A. 1961 Budding bacteria. Mikrobiologiya *30:* 952–975 (English Translation 774–791).

—. 1963 Structure of *Metallogenium.* Mikrobiologiya *32:* 1020–1023.

Zdrodovskii, P. F. 1948 Rickettsiae and rickettsioses. Akad Med Sci Moscow.

Corrigendum

Page	*Correction*
252, col. 2, l. 3 from bottom	1d. *Gluconobacter oxydans* subsp. *melanogenus* rather than *melanogenes* (Ed. note: Dr. de Ley and I have argued about this. I prefer the participle to agree with *oxydans*; he prefers the noun, the form of the original specific name.)
253, *Azotobacteraceae*, ll. 2–16	Add **Mesophilic**
256, col. 1, l. 7 from bottom col. 2, l. 7	*mobilis* instead of *mobile*
258, col. 1, *B. indica* l. 6 1.	For Peterson read Petersen
267 Footnote	should read ....during 1975. (delete probably in April)
268 *Genus I.* **Methylomomas** Leadbetter	should read *1975* not *1974*.
269 *Genus II.* **Methylococcus** Leadbetter	should read *1975* not *1974*
270, col. 2, l. 25	The Colwell and Gibbons paper listed here and on page 272 as 1972 has not been published. The information given was based on a preprint. Publication is expected in 1975.
272, col. 1, l. 18 *et seq.*	Bacteriorhodopsin has now been found in all strains of red-pigmented halophiles examined (Kushwaha *et al.*, 1974*)
274	Replace present footnote with: This paper has been published in the *International Journal of Systematic Bacteriology 24:* 534–550, 1974.
276, l. 3	*Termobacterium* Zeidler 1896, 739, not Lindner 1895, 243
Table 7.12	Under subsp. *liquefaciens* delete − sign for Cellulose
277, col. 1, ll. 11, 18	1. *Acetobacter aceti* (Pasteur) Beijerinck 1900, 503, not 1898, 215
col. 2, l. 18	Pasteur *in* Hansen 1879, 96, not 230
col. 2, l. 8 from bottom	(*Gluconoacetobacter liquifaciens* (sic) Asai 1935, 610; *Gluconobacter liquifaciens* (sic) Asai 1935, 679)
278, col. 2, l. 5	For *estunensis* read *estunense*.

Add to end of Description of species of genus *Acetobacter:*

Species incertae sedis

a. *Micrococcus oblongus* Boutroux 1881, 96.

b. *Bacterium oblongus* (Boutroux) Trevisan *in* de Toni and Trevisan 1889, 1021.

c. *Bacterium gluconicum* Boutroux *in* Miquel and Cambier 1902, 605. Possibly either *Acetobacter aceti* or *Gluconobacter oxydans*.

d. *Bacterium albuminosum* Lindner 1905, 501.

e. *Bacterium friabile* Lindner 1905, 501.

f. *Acetobacter plicatum* Fuhrmann 1906, 8 (spore-former, or contaminated with a spore-former).

---

* Reference given in Addendum.

Page	*Correction*

g. *Mycoderma acidificans* de 'Rossi 1927, 494.

h. *Acetobacter nikitinsky* Gamova-Kayukova 1945, 340.

i. *Acetobacter diversum* Humm 1946, 63.

j. *Acetobacter potens* Humm 1946, 63 (i and j are *Pseudomonas* according to Brisou 1955, 149).

k. *Acetobacter singulare* Humm 1946, 62 (*Protaminobacter* according to Brisou 1955, 197).

l. *Acetobacter aurantium* Kondo and Amayama 1958, 370 (possibly related to *A. aceti* var. *liquefaciens*).

m. *Acetobacter acidophilum* Wiame, Harpigny and Dothey 1959, 165 (possibly dwarf colonies of *A. pasteurianus* subsp. *acetosus*, *A. aceti* subsp. *orleanensis* or *A. aceti* subsp. *xylinum* according to Frateur and Simonart 1952).

286, Plate 7.2, caption — At end add (Photographs by J. H. Becking)

289, Plate 7.3, caption — At end add (Photographs by J. H. Becking)

297, col. 2, l. 17 — For Ewing and Davis (1971) read Ewing (1971)

col. 2, l. 21 — Replace paragraph beginning "H₂S is not produced" with:

Cultures fail to produce H_2S in TSI agar, lysine is not decarboxylated; about 20% of strains produce indole and 40% ferment sodium malonate. At least two biotypes can be distinguished (Table 8.4).

Reference strain: 5396/38 *S. coli* 1 (Kauffmann 1941; Sedlak 1969, 1973*).

Many aberrant cultures have been described by various authors. These decarboxylate lysine fail to grow in Simmons citrate medium and give a negative reaction in mucate and KCN (Leclerc and Buttiaux 1965*; Davis and Ewing, 1966*)

299, col. 1, l. 38 — Borman, Stuart and Wheeler (1944), not 1948

300, col. 1, l. 31 — Salmonella Subcommittee 1934 is listed in references (p. 1017) under International Salmonella Subcommittee 1934

335, Table 8.19, Footnote *1* — For Anon, 1958 read Report of Enterobacteriaceae Subcommittee, 1958

348, col. 1, l. 4 — Merkel *et al.* 1964, not 1965

357, col. 1, Flavobacterium l. 24 — For Fermentation behavior read Fermentative behavior

col. 2, l. 7 from bottom — Change to L. B. Perry, 1973*

363, col. 1, l. 23 — Jensen *et al.* (1968) is listed as Liaaen-Jensen *et al.*, p. 1034

368, col. 1, *vaginalis, Haemophilus* l. 4, and Syn. l. 4 — Redmond and Kotcher 1963, not 1961

389, Table 9.1, Footnote l. 5 — pH 5.4 to 5.7, not pG

Page	Correction
390 Table 9.2	Instead of symbols given read
	Under 1c. *B. fragilis* ss *distasonis* Gelatin digested $-^+$
	Under 2a. *B. ruminicola* ss *ruminicola*, Gas $-^2$
	Under 2b. *B. ruminicola* ss *brevis*, Gas $-^3$; and Growth enhanced by hrt
401, col. 1, ll. 4, 5, Further comments	For Henriksen read Hendricksen
419, col. 1, l. 11	For Hayward and Statman read Haywood and Stadtman 1959
l. 7 from bottom	ZoBell, 1957, not 1953
441, col. 2, l. 14	(Mandel, personal communication, 1970) reference not given
466, col. 1, l. 17	For Hartman read Hardman
485, col. 2, l. 5 from bottom	for Melveaux read Malveaux
486, col. 1, l. 1	For Files *et al.* read Fildes *et al.*, 1936*
col. 2, l. 4 from bottom	Bergdoll, 1972, not 1967
487, col. 2, l. 25	For High and Ellis read Hugh and Ellis
489, col. 2, l. 4	Gordon (1967),* not 1957
	Planococcus citreus, to description add:
	Suggested neotype strain: CCM 316; ATCC 14404 (Kocur *et al.*, 1970*)
552, Table 15.9	8. *C. pasteurianum*, reverse symbols under sorbitol and nitrate, i.e. sorbitol d, nitrate −
556, Table 15.12	14. *C. sordellii*, under Mannose change V to—
558, Table 15.13	Symbols for Cellobiose and Dulcitol under:
	21. *C. plagarum*, 22. *C. acetobutylicum*, 24. *C. aurantibutyricum*, and 30. *C. septicum* should be + and −, respectively; for 26. *C. perfringens* and 31. *C. difficile* d and −, respectively; 20. *C. botulinum* A, B, etc. should be − for Melibiose; the B, C, D, E, F strains d for Melibiose and Starch. 25. *C. novyi* Type B should be d for Mannose.
559, Table 15.14	25. *C. novyi* Type C should be d for blood hemolyzed
562, col. 1, l. 34	Reference strain-for NCIB 9747 read NCIB 11122
566, Table 15.16	Erythritol under 39. *C. thermosaccharolyticum* should be − and under 40. *C. pseudotetanicum* +
570, Table 15.18	53. *C. putrificum* Spore shape should read 0, not OS
572, col. 2, l. 3	n. *C. saturni-rubrum* Prévot 1946, not 1964
601, col. 2, l. 15	Lanéelle
602, col. 1, l. 1	For Dowell *et al.* read Moss *et al.*, 1967*
610, col. 2, l. 6 from bottom	For Collins-Thompson, Sporhaug (Sørhaug) and Witter, 1971 read Collins-Thompson *et al.*, 1972
617, col. 2, l. 30	13. *C. mediolanum* l. 3, reine dehydrierende Stamm. Mamoli, Koch and Teschen 1939, 319.)
619, col. 1, l. 1	Ensign and Wolfe, 1964, not 1963
622, col. 2, l. 12	change to read "compositions (Cummins and Harris, 1959; Yamada and Komagata, 1970) and G + C . . ."

Page	Correction
col. 2, ll. 22, 23	Ensign and Wolfe, 1964, not 1963
624, col. 1, l. 39	Delete 1969
col. 1, l. 46	Change to read "Type strain: ATCC 11624; NCIB 8915."
col. 2, l. 26	For NCIB 8908 read NCIB 8909
634, Key I.A.b.ii	For 2. *P. theoni* read 2. *P. thoeni*
637, Table 17.11	Under 7. *P. lymphophilum* nitrate reduced should read v, not d
649, Table 17.15	Instead of symbols given read:
	Under 11. *E. cylindroides*, Galactose $-$; Starch hydrolysis $-^+$.
	Under 14. *E. cellulosolvens*, Esculin $-$; Fructose $-$; Galactose $+^w$; Maltose $+^w$; Raffinose w$^-$; Salicin w$^-$
	Under 15. *E. combesii*, Glucose $+^-$
	Under 18. *E. contortum*, Glycerol $-^w$; Glycogen $-^+$; Sorbose $-^w$
650, Table 17.16	Instead of the symbols given read
	Under 14. *E. cellulosolvens*, Acetoin $-$
	Under 16. *E. tenue*, NH$_3$ $+^w$
	Under 17. *E. fissicatena*, Nitrate reduced $-^+$
656, col. 2, l. 8 from bottom	Bacillus of Ghon, Mucha and Müller 1906, 145*
677, col. 1, l. 5 from bottom	For Sartory 1920 read Sartory 1930, 813
684, col. 1, l. 9	For (Mitchison *et al.*, 1963) read (Mitchison *et al.*, 1963; Mitchison, 1970*)
col. 1, l. 10 from bottom	Castelnuovo and Morellini, 1962, not 1965
697, col. 2, l. 15 from bottom	For Runyon, 1971 read Runyon, 1972
702, col. 1, l. 1	for *Actinomyces* cell wall read actinomycete cell wall
705, col. 2, l. 27 and 12 from bottom	For sentence beginning "The presence of ..." read "Vesicles spherical to club-shaped; bacteria-like cells not yet known."
706, col. 1, l. 26	Change to read "*Myrica cerifera* and *Myrica cordifolia* have reduced compatibility"
715, **Amorhposporangium,** col. 1, l. 5	Insert after polar flagella (Hanton, 1968*)
727, col. 2, l. 5 from bottom	Lanéelle *et al.*, 1965,* not 1969
744, col. 1, l. 22	For Adams *et al.* (1967, 247) read Brownell *et al.* (1967)*
829, col. 2, l. 15	For Thirumalachar 1962 read Thirumalachar *in* Dave *et al.* 1962, 28*
847, col. 2, l. 4 from bottom	Luedemann, 1971,* not 1970
890	Editorial Note-for Macchiavello 1943, read Macchiavello 1947
897	Tribe Wolbachieae, author's name should be Emilio Weiss
909	The two opening paragraphs under Genus III **Aegyptianella,** col. 1 beginning "Infectious for cattle... cross-immunity studies (Kreier and Ristic, 1963)" belong at end of 2. **Paranaplasma** *discoides*, col. 2.
917, col. 2, l. 11	Should read psittacosis; ornithosis; pneumonitis in ...

References

Page	*Correction*
967	Allen and Klebesadel. 1964 Proc 14th Alaskan Sci Conf
969	Aujeszky, A. 1914 A *Koeleria glauca* bakteriozisarol
970	Balcke. Über fauligen Geruch
971	Baudische 1935, not 1936
972	Becking 1970, line 3 Brunchorst, not Brunschorst
973	Beijerinck 1900 On indigo fermentation, pages should read 495–520
974	Berger 1962 über die Säurebildung
976	Blöck 1918
	Boisvert 1965, not 1955
	Boquet 1928, not Boguet
979	Brown, N. A. 1924 delete second 372
	Brunchorst, J. 1886 Über einige Wurzelanschwellungen, besonders diejenigen von *Alnus* und den *Elaeagnaceae*. Unters Bot Inst Tübingen *2*: 151–177.
981	Burkholder, W. H. 1957 Genus II *Xanthomonas* Dowson 1939
	——, L. A. McFadden and A. V. Dimock, not Dinock
986	Collins-Thompson, D. L., T. Sørhaug, L. D. Witter and Z. J. Ordal. 1972
987	Delete second Couch. 1955 A new genus . . . .
	Cox 1941, rickettsia of Rocky, not or
	Cummins, Atfield, Rees and Valentine. 1967
990	De Ley, not DeLey
991	de 'Rossi 1927 for 'technica' read 'tecnica'. Add 1410 pp.
	de Toni, not De Toni
	Diernhofer. 1932 l. 2, für
	—, 1949 l. 2, Tierärztl
992	Döbereiner. 1966
993	Ducrey, 1889, col. 2, l. 1, 3, über
	—. 1895, über kontagiösen Geschwüres
994	du Plessis 1940 for "Vlamiekte" read "Vlamsiekte"
996	Ellis, D. 1919 Iron Bacteria, not 1907
997	Ewing, W. H. and M. A. Fife 1968, not M. Fife
	Ewing, W. H. and M. A. Fife 1972 *Enterobacter agglomerans* Beijerinck, not File, not Beijernick
998–999	References Fitz-James to Flaig should follow Fisher, P. J. on p. 999, col. 2
999	Findley *et al.* 1938 should follow Finck, G.
1000	Frateur and Simonart, acétificateur
1001	Friedberger l. 2, gehöriges
	Fuhrmann 1906, not 1905
1004	Giovannozzi-Sermanni 1959 l. 3, *Arthrobacter nicotianae*
1006	Gram *et al.* for "Oversight" read "Oversigt"
1007	Guillemot, Hallé and Rist 1904
1008	for — and S. A. Zahler read R. D. Hamilton and S. A. Zahler 1957; this R. D. Hamilton is not the R. D. Hamilton of Hamilton, R. D. and K. E. Austin 1967 on p. 1007
	Hansen, E. C. 1894, Eddikesyrerbakterier

ADDENDUM AND CORRIGENDUM

Page	Correction
	Hansgirg 1888 and 1893, Beiträge
	Happold *et al.* 1954, last name should be Youatt,
1010	Hazen and Steenis 1970
	Heimpel, A. M. 1967, not 1969
1011	Hendrie, Holding and Shewan 1974, for in press sub
1012	Hewitt, E. J. and G. Bond, not H. Bond
	Hildebrand, D. C. and M. N. Schroth, not N. N.
1016	For Hylemon, Wells and Krieg read Hylemon, We
	Jannasch. 1973
1017	Ishiyama for 1956 read 1926
	Iseki and Sakai, not Sakair
1020	Jungmann, P. and M. H. Kuczynski, not Luczynski
1021	Kairies and Schwartze l. 2, Mäuse
	Kandler and Ebo-Elnaga. 1966 l. 4, Untergattung
1023	Kennedy and King 1962 Phytopathology *52*, not *52*
	Keyworth *et al.* for J. Howell read J. S. Howell and f
1025	Klein, R. M. and I. L. Tenenbaum. 1955
	Kluyver and de Leeuw 1924 l. 3, Vergelijk
1026	Knöll
	Köhler
1027	Koppe, F. 1924 reference is repeated, delete second
1029	Kritchevsky and Séguin. 1921
	Kruse 1896 ll. 3, 4, Berücksichtigung der Ätiologie infek
	Krzemieniewska and Krzemieniewski, both references gi
	read 1937.
1032	Lechevalier, H. A. and M. P. Lechevalier. 1968 should be
	Lechevalier, M. P. and H. A. Lechevalier. 1969 should b
	The Prauser volume was published in 1970
1033	Lehmann and Neumann, 1899 Handatlanten, not Hau
	Lehmann and Neumann. 1899 Medizin, not Medezin
	Lelliott 1968 for Coaf read Conf
1035	Lipman, C. B. and E. McLees, not McLee
1036	Löhnis. 1911, for Praktium read Praktikum
	López, R. and J. H. Becking
	Lovrekovich's initial is L., not O.
1037	L'vov and Lyubimov 1965, not L'vov and L'vov; l. 3 *My*
	absorptum
1038	McCurdy, W. H. and S. Wolf, not Wolfe
	McGill, Dawes and Ribbons. 1965 Biochem J *97*, not *94*
	For MacCallum. 1948 read MacCallum, P., J. C. Tolhur
	H. A. Sissons. 1948
1039	For Mamoli and Vercellone read Mamoli, L., R. Koch
	1939
1041	Mercier, L. 1906 delete second parenthesis after Mercier
	Metcalf, G. should read Metcalfe, G.
1044	Möller, H. 1885
	Moniz and Patel 1958, for bacteriol read bacterial

ADDENDUM AND CORRIGENDUM

Correction

ıd da Rocha-Lima, for Klink and Aetiologie read Klinik und
ɔgie

, R., J. Naylor and . . . comma was omitted

et al. l., 3 for 950–990 read 950–960

i. 1857 l. 3, Naturforscher

o *et al.* l. 2, for Tsutsugamushi read tsutsugamushi

ederhauser 1943 read Niederhausser

nmich 1968 read Nimmich

a reference in Science is 1915, not 1916; see Addendum for 1916
rence

1933, 1934, 1935 The journal should read J Trop Agr Taiwan

G. 1962, not 1972

ır 1864 Mémoire

A. M. and Kotasthane 1969 for *naktae* read *nakatae*

, M. K., S. B. Allayyanavaramath and Y. S. Kulkarni, for 1952 read 1953

bridge, G. H., not C. H.

er 1892, Verläufige Mittheilungen

1896, Flügge

ick, H. 1954, not 1955

.am 1933, Klassifikation

iva, J., not Protina

ère, J. 1961, not 1964

erg 1934 l. 2, knöllchen, l. 3, Hippophaë

ıyon 1967, *Mycobacterium intracellulare*

et 1954, for *Dacytlis* read *Dactylis*

erstein and Starr. 1954, delete reference and replace with Saperstein,
Starr and Filfus 1954*

vulescu 1947, for phytopathogéne read phytopathogène

häfer, D. 1967

liskar 1952, for 6–20 read 9–11

ith, F. 1948, for Ligniéres read Lignières

app 1928, for pp. 1–5 read 1–295

silasani should read Tsilosani

rošefić should read Urošević

an Oye, not Van Oye

isser 't Hooft

on Wolzogen Kühr should read van Wolzogen Kühr

aksman and Waldee should follow Wakker

allin and Reddy 1945, for *protense* read *pratense*

elete Whittenberger and Repaske here and insert on p. 1094 as Wittenberger and Repaske

oronin 1866, not 1966 Mem Acad Sci St Petersburg, Ser 7 *10* (*6*): 1–10.

mada and Yamagata should read Yamada and Komagata (Actually
hould read — and — and second paper (II) should precede third (III)

atto and Pereira, 1963 should read 1964

arzin, G. A. 1961 Budding Bacteria. Mikrobiologiya *30:* 952–975
ranslation 774–791).

er 1959–60. *Comptonia peregrina* (L) Coult.